Elementary and Intermediate Algebra

Books in the Tussy and Gustafson Series

In paperback:

Basic Mathematics for College Students, Second Edition
Student edition: ISBN 0-534-37643-6
Instructor's edition: ISBN 0-534-38585-0

Developmental Mathematics for College Students
Student edition: ISBN 0-534-38031-X
Instructor's edition: ISBN 0-534-38584-2

Prealgebra, Second Edition
Student edition: ISBN 0-534-37642-8
Instructor's edition: ISBN 0-534-38309-2

Introductory Algebra, Second Edition
Student edition: ISBN 0-534-37641-X
Instructor's edition: ISBN 0-534-38571-0

Intermediate Algebra, Second Edition
Student edition: ISBN 0-534-37640-1
Instructor's edition: ISBN 0-534-38586-9

In hardcover:

Elementary Algebra, Second Edition
Student edition: ISBN 0-534-38629-6
Instructor's edition: ISBN 0-534-39119-2

Intermediate Algebra, Second Edition
Student edition: ISBN 0-534-38628-8
Instructor's edition: ISBN 0-534-39120-6

Elementary and Intermediate Algebra, Second Edition
Student edition: ISBN 0-534-38627-X
Instructor's edition: ISBN 0-534-39118-4

Elementary and Intermediate Algebra

Second Edition

Alan S. Tussy

Citrus College

R. David Gustafson

Rock Valley College

BROOKS/COLE

THOMSON LEARNING™

Australia • Canada • Mexico • Singapore • Spain • United Kingdom • United States

BROOKS/COLE

THOMSON LEARNING

Sponsoring Editor: Jennifer Huber/Robert W. Pirtle

Assistant Editor: Rachael Sturgeon

Marketing Team: Leah Thomson

Marketing Communications: Samantha Cabaluna

Marketing Assistant: Maria Salinas

Production Editors: Ellen Brownstein/Scott Brearton

Manuscript Editor: David Hoyt

Permissions Editor: Sue Ewing

Interior Design: Carolyn Deacy

Cover Design: Vernon T. Boes

Cover Illustration: George Abe

Interior Illustration: Lori Heckelman

Print Buyer: Kristine Waller

Typesetting: The Clarinda Company

Cover Printing: Phoenix Color Corp.

Printing and Binding: R.R. Donnelley–Willard

For more information about this or any other Brooks/Cole product, contact:
BROOKS/COLE
511 Forest Lodge Road
Pacific Grove, CA 93950 USA
www.brookscole.com
1-800-423-0563 (Thomson Learning Academic Resource Center)

Printed in the United States of America

10 9 8 7 6 5 4 3 2

Library of Congress Cataloging-in-Publication Data
Tussy, Alan S., [date]
  Elementary and intermediate algebra / Alan S. Tussy, R. David Gustafson.—2nd ed.
    p. cm.
  Includes index.
  ISBN 0-534-38627-X—ISBN 0-534-39118-4
  1. Algebra I. Gustafson, R. David (Roy David), [date] II. Title.

QA152.3.T86 2001
512.9—dc21   2001037426

To three good friends,

David,
Ellen,
and
Bob

Contents

Preface

▶

For the Instructor

An increasing number of schools are offering the traditional elementary algebra and intermediate algebra courses in combination. There are several advantages in doing this:

- Much of the redundancy encountered by teaching the elementary/intermediate sequence as two separate courses is eliminated. As a result, the students have more time to master the material.

- A combined approach promotes a smooth transition from the elementary algebra topics to the intermediate algebra topics.

- For many students, the purchase of a single textbook saves money.

However, there are several concerns inherent in offering a combination course:

- The textbook used in such a course must include enough elementary algebra to ensure that students who complete the first half of the book, and then transfer, will have the prerequisite skills to enroll in an intermediate algebra course at another college.

- The elementary algebra material should not get too difficult too fast.

- Students entering the second half of the combination course must get some review of the basic topics so that they can compete with students continuing from the first half of the course.

Elementary and Intermediate Algebra has been written to address these concerns. The first seven chapters of this book provide a complete course in elementary algebra. The standard beginning algebra topics are introduced at a reasonable pace that allows the student to develop a strong conceptual foundation on which the second half of the course can build. Chapter 8 serves as the transitional chapter. It quickly reviews the topics taught in the first part of the course and extends those topics to the intermediate algebra level. Chapters 9–14 provide a complete course in intermediate algebra.

The purpose of this textbook is to teach students how to read, write, speak, and think mathematically using the language of algebra. We have used a blend of the traditional and the reform instructional approaches to do this. In this book, you will find the vocabulary, practice, and well-defined pedagogy of a traditional approach. You will also find that we emphasize the reasoning, modeling, communicating, and technological skills that are such a big part of today's reform movement.

This textbook expands the students' mathematical reasoning abilities and gives them a set of mathematical survival skills that will help them succeed in a world that increasingly requires that every person become a better analytical thinker.

Features of the Text

Chapter 1   An Innovative Introduction to Algebra

The best way to learn a new language is to be immediately immersed in it. Therefore, Chapter 1 begins with an introduction of the fundamental algebraic concepts of variable, equation, function, and graphing. We show the students how to translate English phrases to mathematical symbols, and we introduce a problem-solving strategy that is used throughout the book. From the start, students see how algebra is a powerful tool that they can use to solve problems.

Interactivity

Most worked examples in the text are accompanied by Self Checks. This feature allows students to practice skills discussed in the example by working a similar problem. Because the Self Check problems follow the worked examples, students can easily refer to the solution and author's notes of the example as they solve the Self Check. Author's notes are used to explain the steps in the solutions of examples. The notes are extensive so as to increase the student's ability to read and write mathematics.

Example titles highlight the ▶
concept being discussed.

Author's notes explain the steps ▶
in the solution process.

Most examples have Self ▶
Checks. The answers are
provided.

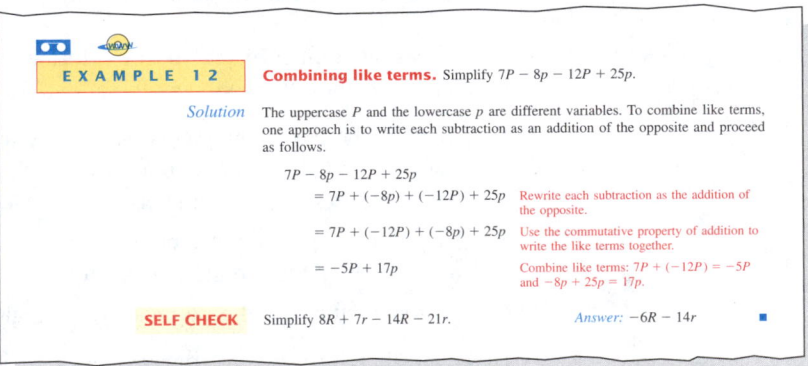

EXAMPLE 12   **Combining like terms.** Simplify $7P - 8p - 12P + 25p$.

Solution   The uppercase P and the lowercase p are different variables. To combine like terms, one approach is to write each subtraction as an addition of the opposite and proceed as follows.

$$7P - 8p - 12P + 25p$$
$$= 7P + (-8p) + (-12P) + 25p \quad \text{Rewrite each subtraction as the addition of the opposite.}$$
$$= 7P + (-12P) + (-8p) + 25p \quad \text{Use the commutative property of addition to write the like terms together.}$$
$$= -5P + 17p \quad \text{Combine like terms: } 7P + (-12P) = -5P \text{ and } -8p + 25p = 17p.$$

SELF CHECK   Simplify $8R + 7r - 14R - 21r$.   *Answer:* $-6R - 14r$ ■

Color is used to facilitate ▶
students' understanding.

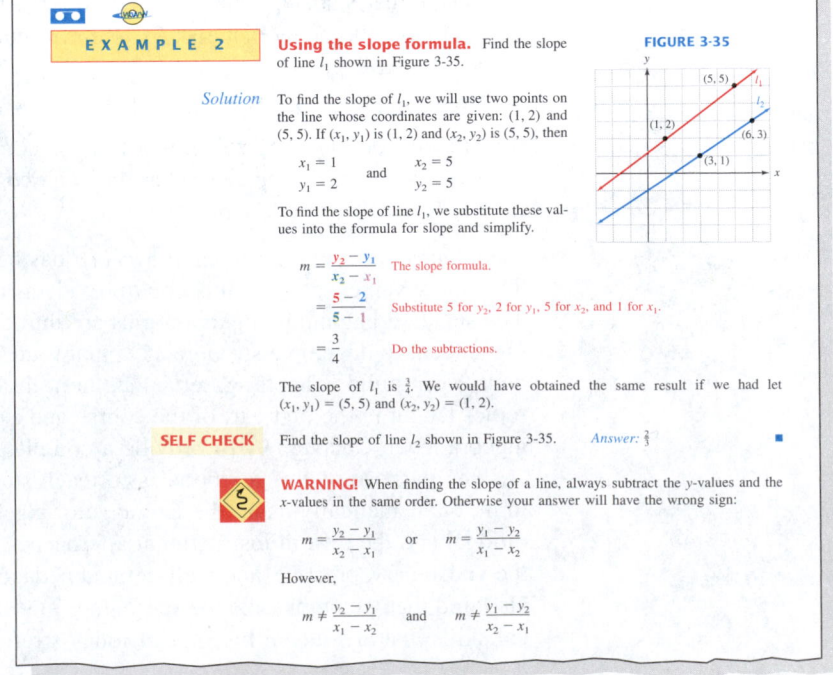

EXAMPLE 2   **Using the slope formula.** Find the slope of line l_1 shown in Figure 3-35.

FIGURE 3-35

Solution   To find the slope of l_1, we will use two points on the line whose coordinates are given: $(1, 2)$ and $(5, 5)$. If (x_1, y_1) is $(1, 2)$ and (x_2, y_2) is $(5, 5)$, then

$$x_1 = 1 \quad \text{and} \quad x_2 = 5$$
$$y_1 = 2 \qquad\qquad y_2 = 5$$

To find the slope of line l_1, we substitute these values into the formula for slope and simplify.

$$m = \frac{y_2 - y_1}{x_2 - x_1} \quad \text{The slope formula.}$$
$$= \frac{5 - 2}{5 - 1} \quad \text{Substitute 5 for } y_2, 2 \text{ for } y_1, 5 \text{ for } x_2, \text{ and 1 for } x_1.$$
$$= \frac{3}{4} \quad \text{Do the subtractions.}$$

The slope of l_1 is $\frac{3}{4}$. We would have obtained the same result if we had let $(x_1, y_1) = (5, 5)$ and $(x_2, y_2) = (1, 2)$.

SELF CHECK   Find the slope of line l_2 shown in Figure 3-35.   *Answer:* $\frac{2}{3}$ ■

WARNING! When finding the slope of a line, always subtract the y-values and the x-values in the same order. Otherwise your answer will have the wrong sign:

$$m = \frac{y_2 - y_1}{x_2 - x_1} \quad \text{or} \quad m = \frac{y_1 - y_2}{x_1 - x_2}$$

However,

$$m \neq \frac{y_2 - y_1}{x_1 - x_2} \quad \text{and} \quad m \neq \frac{y_1 - y_2}{x_2 - x_1}$$

In-Depth Coverage of Geometry

Perimeter, area, and volume, as well as many other geometric concepts, are used in a variety of contexts throughout the book. We have included many drawings to help students improve their ability to spot visual patterns in their everyday lives.

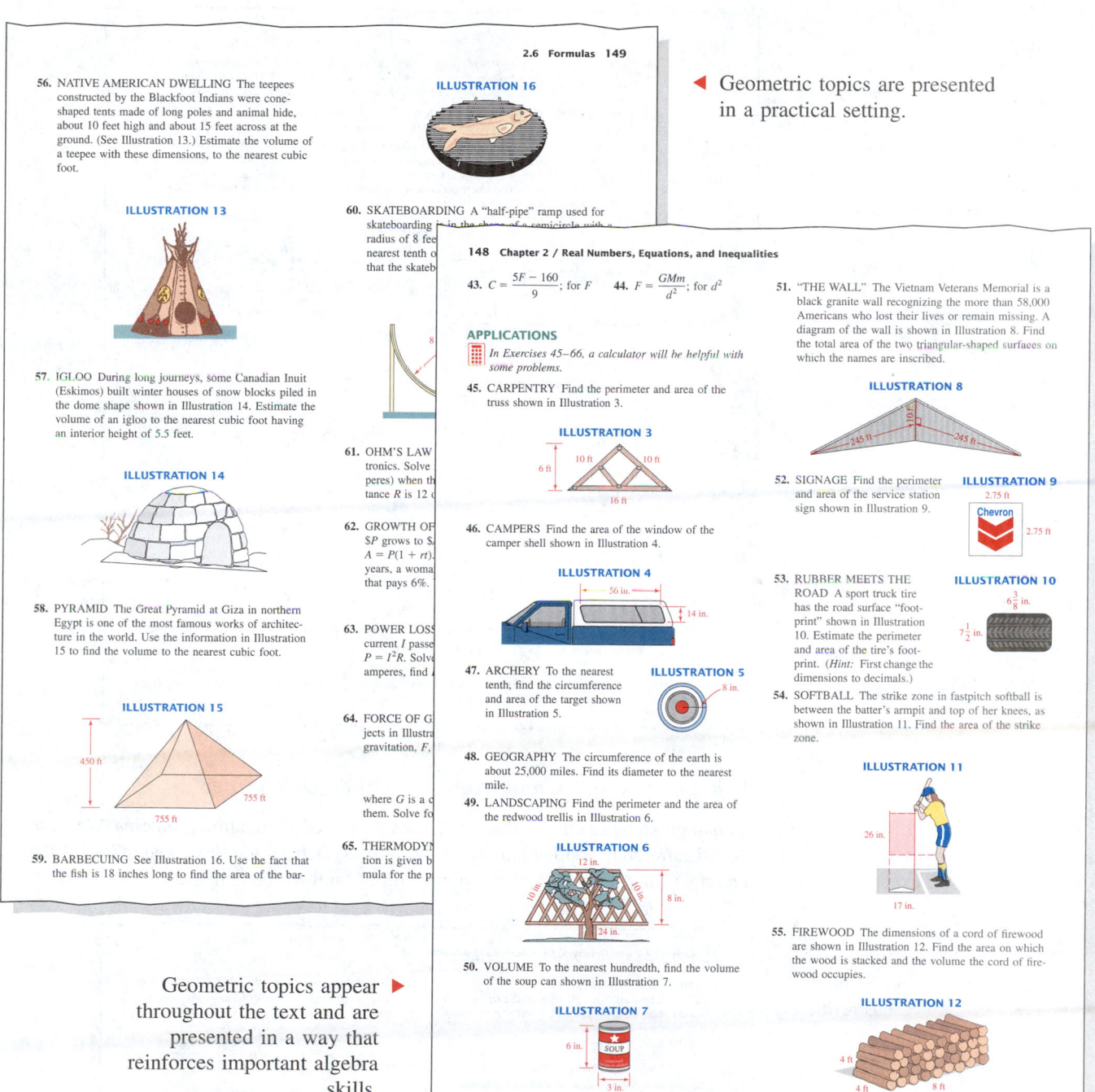

◀ Geometric topics are presented in a practical setting.

Geometric topics appear ▶ throughout the text and are presented in a way that reinforces important algebra skills.

The embedded textbook pages show:

2.6 Formulas 149

ILLUSTRATION 16

56. NATIVE AMERICAN DWELLING The teepees constructed by the Blackfoot Indians were cone-shaped tents made of long poles and animal hide, about 10 feet high and about 15 feet across at the ground. (See Illustration 13.) Estimate the volume of a teepee with these dimensions, to the nearest cubic foot.

ILLUSTRATION 13

57. IGLOO During long journeys, some Canadian Inuit (Eskimos) built winter houses of snow blocks piled in the dome shape shown in Illustration 14. Estimate the volume of an igloo to the nearest cubic foot having an interior height of 5.5 feet.

ILLUSTRATION 14

58. PYRAMID The Great Pyramid at Giza in northern Egypt is one of the most famous works of architecture in the world. Use the information in Illustration 15 to find the volume to the nearest cubic foot.

ILLUSTRATION 15

450 ft, 755 ft, 755 ft

59. BARBECUING See Illustration 16. Use the fact that the fish is 18 inches long to find the area of the bar-

60. SKATEBOARDING A "half-pipe" ramp used for skateboarding is in the shape of a semicircle with a radius of 8 fee... nearest tenth o... that the skateb...

61. OHM'S LAW ... tronics. Solve ... peres) when th... tance R is 12 ...

62. GROWTH OF ... $P grows to $... A = P(1 + rt)... years, a woma... that pays 6%.

63. POWER LOSS ... current I passe... P = I²R. Solve... amperes, find ...

64. FORCE OF G... jects in Illustr... gravitation, F,... where G is a ... them. Solve fo...

65. THERMODY... tion is given b... mula for the p...

148 Chapter 2 / Real Numbers, Equations, and Inequalities

43. $C = \dfrac{5F - 160}{9}$; for F **44.** $F = \dfrac{GMm}{d^2}$; for d^2

APPLICATIONS

In Exercises 45–66, a calculator will be helpful with some problems.

45. CARPENTRY Find the perimeter and area of the truss shown in Illustration 3.

ILLUSTRATION 3

6 ft, 10 ft, 10 ft, 16 ft

46. CAMPERS Find the area of the window of the camper shell shown in Illustration 4.

ILLUSTRATION 4

56 in., 14 in.

47. ARCHERY To the nearest tenth, find the circumference and area of the target shown in Illustration 5.

ILLUSTRATION 5

8 in.

48. GEOGRAPHY The circumference of the earth is about 25,000 miles. Find its diameter to the nearest mile.

49. LANDSCAPING Find the perimeter and the area of the redwood trellis in Illustration 6.

ILLUSTRATION 6

12 in., 10 in., 10 in., 8 in., 24 in.

50. VOLUME To the nearest hundredth, find the volume of the soup can shown in Illustration 7.

ILLUSTRATION 7

6 in., SOUP, 3 in.

51. "THE WALL" The Vietnam Veterans Memorial is a black granite wall recognizing the more than 58,000 Americans who lost their lives or remain missing. A diagram of the wall is shown in Illustration 8. Find the total area of the two triangular-shaped surfaces on which the names are inscribed.

ILLUSTRATION 8

245 ft, 245 ft

52. SIGNAGE Find the perimeter and area of the service station sign shown in Illustration 9.

ILLUSTRATION 9

Chevron, 2.75 ft, 2.75 ft

53. RUBBER MEETS THE ROAD A sport truck tire has the road surface "footprint" shown in Illustration 10. Estimate the perimeter and area of the tire's footprint. (*Hint:* First change the dimensions to decimals.)

ILLUSTRATION 10

$6\frac{3}{8}$ in., $7\frac{1}{2}$ in.

54. SOFTBALL The strike zone in fastpitch softball is between the batter's armpit and top of her knees, as shown in Illustration 11. Find the area of the strike zone.

ILLUSTRATION 11

26 in., 17 in.

55. FIREWOOD The dimensions of a cord of firewood are shown in Illustration 12. Find the area on which the wood is stacked and the volume the cord of firewood occupies.

ILLUSTRATION 12

4 ft, 4 ft, 8 ft

Coordinate Graphing Appears Early

The foundation for coordinate graphing is laid in Chapters 1 and 2, where the students graph many different types of real numbers on the number line. In Chapter 3, students learn how to graph lines. They quickly learn that the graph of an equation in two variables is not always a straight line.

Problem-Solving Strategy

One of the major objectives of this textbook is to make students better problem solvers. To this end, we use a five-step problem-solving strategy throughout the book. The five steps are: *Analyze the problem, Form an equation, Solve the equation, State the conclusion,* and *Check the result.*

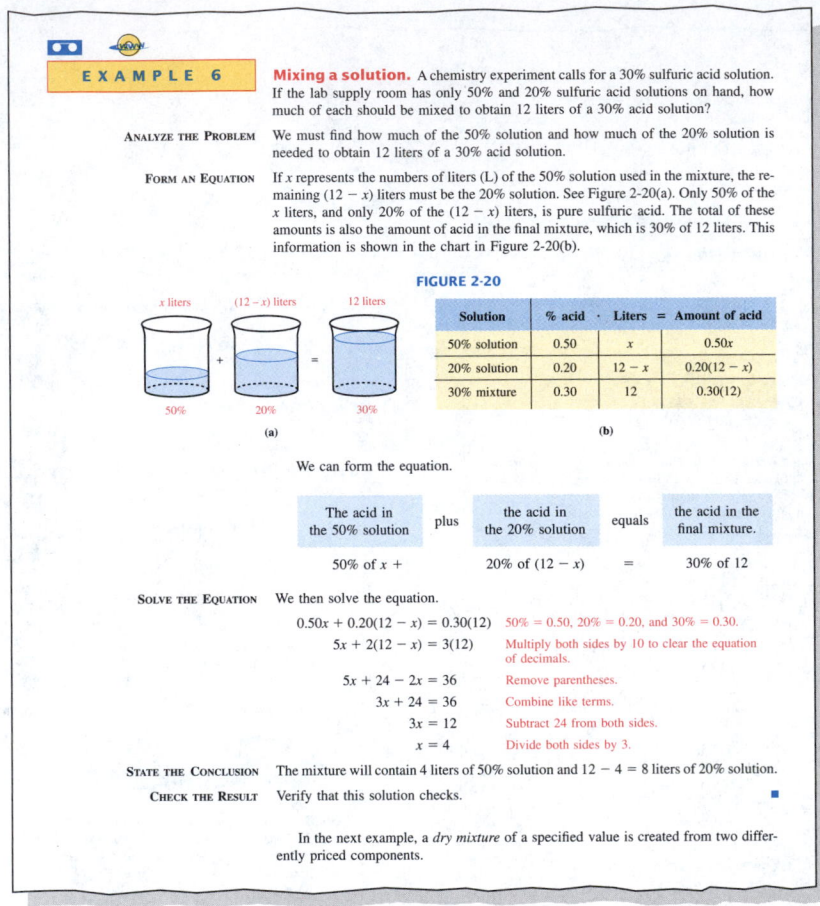

Applications and Connections to Other Disciplines

A distinguishing feature of this book is its wealth of application problems. We have included numerous applications from disciplines such as science, economics, business, manufacturing, history, and entertainment, as well as mathematics.

Every application problem ▶ has a title

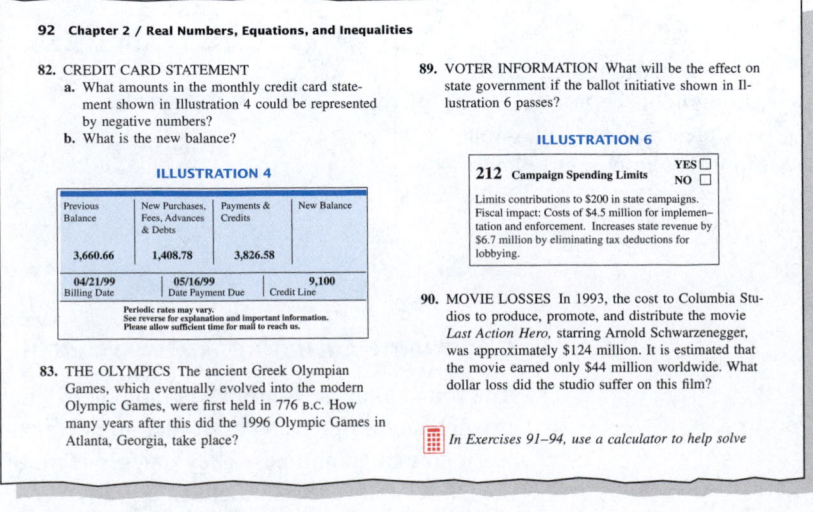

Study Sets—More Than Just Exercises

The problems at the end of each section are called Study Sets. Each Study Set includes Vocabulary, Notation, and Writing problems designed to help students improve their ability to read, write, and communicate mathematical ideas. The problems in the Concepts section of the Study Sets encourage students to engage in independent thinking and reinforce major ideas through exploration. In the Practice section of the Study Sets, students get the drill necessary to master the material. In the Applications section, students deal with real-life situations that involve the topics being studied. Each Study Set concludes with a Review section consisting of problems selected from previous sections.

◀ Each Study Set contains Vocabulary, Concepts, Notation, Practice, Applications, Writing, and Review sections.

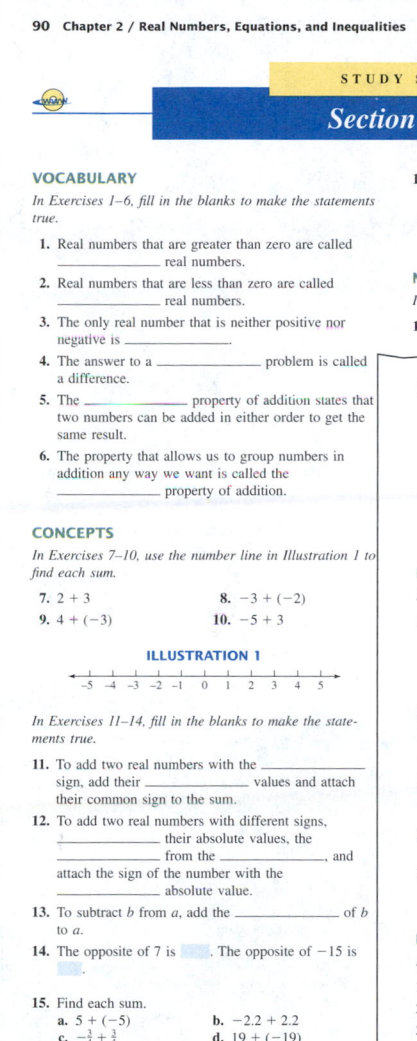

90 Chapter 2 / Real Numbers, Equations, and Inequalities

STUDY SET

Section 2.1

VOCABULARY

In Exercises 1–6, fill in the blanks to make the statements true.

1. Real numbers that are greater than zero are called _____ real numbers.

2. Real numbers that are less than zero are called _____ real numbers.

3. The only real number that is neither positive nor negative is _____.

4. The answer to a _____ problem is called a difference.

5. The _____ property of addition states that two numbers can be added in either order to get the same result.

6. The property that allows us to group numbers in addition any way we want is called the _____ property of addition.

CONCEPTS

In Exercises 7–10, use the number line in Illustration 1 to find each sum.

7. $2 + 3$ 8. $-3 + (-2)$
9. $4 + (-3)$ 10. $-5 + 3$

ILLUSTRATION 1

$$-5 \ -4 \ -3 \ -2 \ -1 \ 0 \ 1 \ 2 \ 3 \ 4 \ 5$$

In Exercises 11–14, fill in the blanks to make the statements true.

11. To add two real numbers with the _____ sign, add their _____ values and attach their common sign to the sum.

12. To add two real numbers with different signs, _____ their absolute values, the _____ from the _____, and attach the sign of the number with the _____ absolute value.

13. To subtract b from a, add the _____ of b to a.

14. The opposite of 7 is ____. The opposite of -15 is ____.

15. Find each sum.
 a. $5 + (-5)$ b. $-2.2 + 2.2$
 c. $-\frac{3}{4} + \frac{3}{4}$ d. $19 + (-19)$

16. a. Use the variables m and n to state the commutative property of addition.
 b. Use the variables r, s, and t to state the associative property of addition.

NOTATION

In Exercises 17–20, complete each solution.

17. Find $(-13 + 6) + 4$.
$$(-13 + 6) + 4 = \boxed{} + (6 + 4)$$

---

124 Chapter 2 / Real Numbers, Equations, and Inequalities

ILLUSTRATION 3   ILLUSTRATION 4

NOTATION

In Exercises 15–16, complete each solution.

15. $-7(a^2 + a - 5) = \boxed{} \cdot a^2 + \left(\boxed{}\right) \cdot a - \left(\boxed{}\right) \cdot 5$
$= -7a^2 + (-7a) - \left(\boxed{}\right)$
$= -7a^2 - \boxed{} - (-35)$
$= -7a^2 - 7a + 35$

16. $6(b - 5) + 12b + 7 = 6 \cdot \boxed{} - 6 \cdot \boxed{} + 12b + 7$
$= 6b - \boxed{} + 12b + 7$
$= 6b - \boxed{} + 12b + 7$
$= 6b + \boxed{} b - \boxed{} + 7$
$= 18b - 23$

17. a. Are $2K$ and $3k$ like terms?
 b. Are $-d$ and d like terms?

18. Fill in the blank to make the statement true.
$-(x + 10$

PRACTICE

In Exercises 19–2...

19. $9(7m)$
21. $5(-7q)$
23. $(-5p)(-4b)$
25. $-5(4r)(-2r)$

In Exercises 27–4...
remove parenthese...

27. $5(x + 3)$
29. $-2(b - 1)$
31. $(3t - 2)8$
33. $(2y - 1)6$
35. $0.4(x - 4)$

37. $-\frac{2}{3}(3w - 6)$
39. $r(r - 10)$
41. $-(x - 7)$
43. $17(2x - y + 2$

45. $-(-14 + 3p$

In Exercises 47–54, list the terms in each expression. Then identify the coefficient of each term.

47. $-5r + 4s$
48. $2m + n - 3m + 2n$
49. $-15r^2s$
50. $4b^2 - 5b + 6$
51. $50a + 2$
52. $a^2 - ab + b^2$
53. $x^3 - 125$
54. $-2.55x + 1.8$

In Exercises 55–58, identify the coefficient of each term.

55. $-b$ 56. $-9.9x^3$
57. $\frac{1}{4}x$ 58. $-\frac{2x}{3}$

In Exercises 59–62, tell whether the variable x is used as a factor or a term.

59. $24 - x$ 60. $24x$
61. $24 + 3x$ 62. $x - 12$

APPLICATIONS

91. THE AMERICAN RED CROSS In 1891, Clara Barton founded the Red Cross. Its symbol is a white flag bearing a red cross. If each side of the cross in Illustration 5 has length x, write an algebraic expression for the perimeter (the total distance around the outside) of the cross.

ILLUSTRATION 5

92. BILLIARDS Billiard tables vary in size, but all tables are twice as long as they are wide.
 a. If the billiard table in Illustration 6 is x feet wide, write an expression involving x that represents its length.
 b. Write an expression for the perimeter of the table.

ILLUSTRATION 6

x ft

93. PING-PONG Write an expression for the perimeter of the ping-pong table shown in Illustration 7.

2.5 Solving Equations 125

94. SEWING See Illustration 8. Write an expression for the length of the yellow trim needed to outline the pennant with the given side lengths.

ILLUSTRATION 8

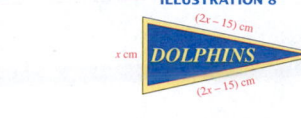
$(2x - 15)$ cm
x cm DOLPHINS
$(2x - 15)$ cm

WRITING

95. Explain why $3x^2y$ and $5x^2y$ are like terms.
96. Explain why $3x^2y$ and $5xy^2$ are not like terms.
97. Distinguish between a *factor* and a *term* of an algebraic expression. Give examples.
98. Tell how to combine like terms.

REVIEW

In Exercises 99–102, evaluate each expression given that $x = -3$, $y = -5$, and $z = 0$.

99. $x^2z(y^3 - z)$ 100. $z - y^3$
101. $\dfrac{x - y^2}{2y - 1 + x}$ 102. $\dfrac{2y + 1}{x} - x$

Group Work

A one-page feature called Accent on Teamwork appears near the end of each chapter. It gives the instructor a set of problems that can be assigned as group work or to individual students as outside-of-class projects.

Key Concepts

Fourteen key algebraic concepts are highlighted in one-page Key Concept features, appearing near the end of each chapter. Each Key Concept page summarizes a concept and gives students an opportunity to review the role it plays in the overall picture.

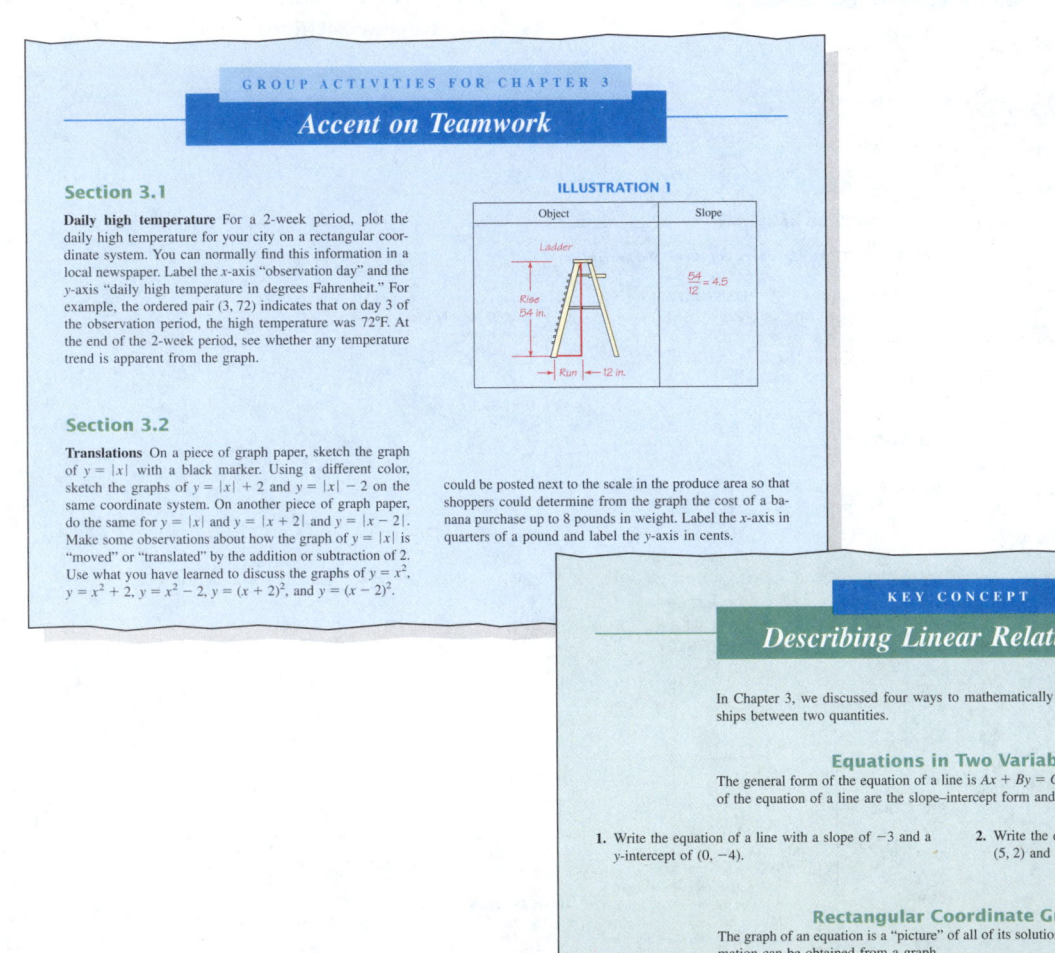

GROUP ACTIVITIES FOR CHAPTER 3

Accent on Teamwork

Section 3.1

Daily high temperature For a 2-week period, plot the daily high temperature for your city on a rectangular coordinate system. You can normally find this information in a local newspaper. Label the x-axis "observation day" and the y-axis "daily high temperature in degrees Fahrenheit." For example, the ordered pair (3, 72) indicates that on day 3 of the observation period, the high temperature was 72°F. At the end of the 2-week period, see whether any temperature trend is apparent from the graph.

ILLUSTRATION 1

Object	Slope
Ladder Rise 54 in. Run 12 in.	$\frac{54}{12} = 4.5$

Section 3.2

Translations On a piece of graph paper, sketch the graph of $y = |x|$ with a black marker. Using a different color, sketch the graphs of $y = |x| + 2$ and $y = |x| - 2$ on the same coordinate system. On another piece of graph paper, do the same for $y = |x|$ and $y = |x + 2|$ and $y = |x - 2|$. Make some observations about how the graph of $y = |x|$ is "moved" or "translated" by the addition or subtraction of 2. Use what you have learned to discuss the graphs of $y = x^2$, $y = x^2 + 2$, $y = x^2 - 2$, $y = (x + 2)^2$, and $y = (x - 2)^2$.

could be posted next to the scale in the produce area so that shoppers could determine from the graph the cost of a banana purchase up to 8 pounds in weight. Label the x-axis in quarters of a pound and label the y-axis in cents.

KEY CONCEPT

Describing Linear Relationships

In Chapter 3, we discussed four ways to mathematically describe linear relationships between two quantities.

Equations in Two Variables

The general form of the equation of a line is $Ax + By = C$. Two very useful forms of the equation of a line are the slope–intercept form and the point–slope form.

1. Write the equation of a line with a slope of -3 and a y-intercept of $(0, -4)$.

2. Write the equation of the line that passes through $(5, 2)$ and $(-5, 0)$. Answer in slope–intercept form.

Rectangular Coordinate Graphs

The graph of an equation is a "picture" of all of its solutions (x, y). Important information can be obtained from a graph.

3. Complete the table of solutions for $2x - 4y = 8$. Then graph the equation.

4. See Illustration 1.
   a. What information does the y-intercept of the graph give us?
   b. What is the slope of the line and what does it tell us?

$2x - 4y = 8$

x	y
0	
	0
-2	

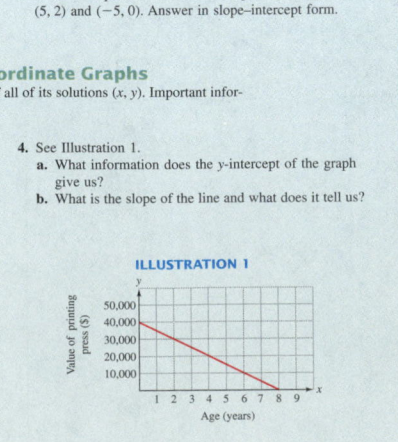

ILLUSTRATION 1

Value of printing press ($)

50,000
40,000
30,000
20,000
10,000

1 2 3 4 5 6 7 8 9
Age (years)

Functions and Modeling

The concept of function is introduced in Chapter 3 and is stressed throughout the text. Students learn to use function notation, graph functions, and write functions that mathematically model many interesting real-life situations. By the end of the course, students will recognize families of functions, their graphs, and areas of application.

Real data is integrated ▶
throughout the text.

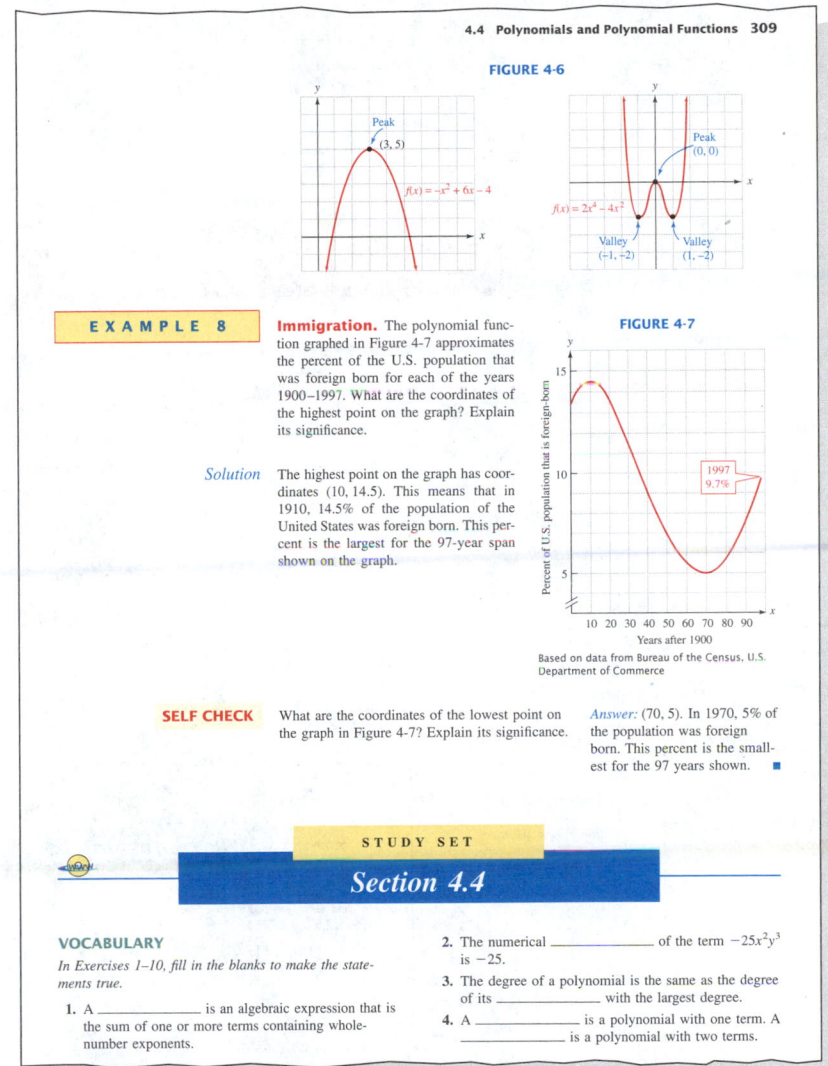

4.4 Polynomials and Polynomial Functions **309**

FIGURE 4-6

EXAMPLE 8

Immigration. The polynomial function graphed in Figure 4-7 approximates the percent of the U.S. population that was foreign born for each of the years 1900–1997. What are the coordinates of the highest point on the graph? Explain its significance.

Solution The highest point on the graph has coordinates (10, 14.5). This means that in 1910, 14.5% of the population of the United States was foreign born. This percent is the largest for the 97-year span shown on the graph.

FIGURE 4-7

Based on data from Bureau of the Census, U.S. Department of Commerce

SELF CHECK What are the coordinates of the lowest point on the graph in Figure 4-7? Explain its significance.

Answer: (70, 5). In 1970, 5% of the population was foreign born. This percent is the smallest for the 97 years shown. ■

STUDY SET

Section 4.4

VOCABULARY

In Exercises 1–10, fill in the blanks to make the statements true.

1. A _____ is an algebraic expression that is the sum of one or more terms containing whole-number exponents.

2. The numerical _____ of the term $-25x^2y^3$ is -25.

3. The degree of a polynomial is the same as the degree of its _____ with the largest degree.

4. A _____ is a polynomial with one term. A _____ is a polynomial with two terms.

Systematic Review

Each Study Set ends with a Review section that contains problems similar to those in previous sections. Each chapter ends with a Chapter Review and a Chapter Test. The chapter reviews have been designed to be "user friendly." In a unique format, the reviews list the important concepts of each section of the chapter in one column, with appropriate review problems running parallel in a second column. In addition, Cumulative Review Exercises appear after Chapters 2, 4, 6, 8, 10, 12, and 14.

CHAPTER REVIEW

Section 1.1

Describing Numerical Relationships

CONCEPTS

Tables, bar graphs, and *line graphs* are used to describe numerical relationships.

REVIEW EXERCISES

1. Illustration 1 lists the worldwide production of wide-screen TVs. Use the data to construct a bar graph. Describe the trend in the production in words.

ILLUSTRATION 1

Year	Production (millions of units)
'95	3
'96	5
'97	7
'98	11

Source: Electronic Industries Association of Japan

2. Consider the line graph in Illustra-

The result of an addition is called the *sum;* of a subtraction, the *difference;* of a multiplication, the *product;* and of a division, the *quotient.*

Variables are letters used to stand for numbers.

An *equation* is a mathematical sentence that contains an = sign. Variables and/or numbers can be combined with the operations of addition, subtraction, multiplication, and division to create *algebraic expressions.*

3.

4.

5.

6.

CHAPTER 2

Test

1. Add $(-6) + 8 + (-4)$.

2. Subtract $1.4 - (-0.8)$.

3. Multiply $(-2)(-3)(-5)$.

4. Divide $\dfrac{-22}{-11}$.

5. Evaluate $-7[(-5)^2 - 2(3 - 5^2)]$.

6. Evaluate $\dfrac{3(20 - 4^2)}{-2(6 - 2^2)}$.

In Problems 7–8, let $x = -2$, $y = 3$, and $z = 4$. Evaluate each expression.

7. $xy + z$

8. $\dfrac{z + 4y}{2x}$

9. What is the coefficient of the term $6x$?

10. How many terms are in the expression $3x^2 + 5x - 7$?

In Problems 11–14, simplify each expression.

11. $5(-4x)$

12. $-8(-7t)(4t)$

13. $3(x + 2) - 3(4 - x)$

14. $-1.1d^2 - 3.8d^2$

In Problems 15–22, solve each equa

15. $12x = -144$

16.

17. $\dfrac{c}{7} = -1$

18.

19. $2(x - 7) = -15$

20.

21. $23 - 5(x + 10) = -12$

22.

In Problems 23–24, solve each equa indicated.

23. $d = rt$; for t

24. $A = P + Prt$; for r

25. COMMERCIAL REAL ESTATE *Net absorption* is a term used to indicate how much office space in a city is being purchased. Use the information from the graph in Illustration 1 to determine the eight-quarter average net absorption figure for Long Beach, California.

ILLUSTRATION 1

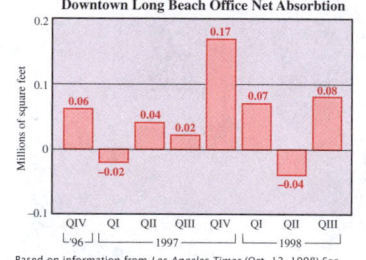

Downtown Long Beach Office Net Absorbtion

Based on information from *Los Angeles Times* (Oct. 13, 1998) Section C10.

CHAPTERS 1–2

Cumulative Review Exercises

In Exercises 1–4, tell whether the expression is an equation.

1. $m - 25$

2. $t = 25r$

3. $\dfrac{p + 5}{2}$

4. $x + 1 = 24$

In Exercises 5–6, classify each number as a natural number, a whole number, an integer, a rational number, an irrational number, and a real number. Each number may be in several classifications.

5. 3

6. -0.25

In Exercises 7–8, graph each set of numbers on the number line.

7. The natural numbers between 2 and 7

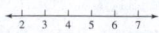

8. The real numbers between 2 and 7

In Exercises 9–12, evaluate each expression.

9. $|-12|$

10. $|15|$

11. $-|15|$

12. $-|-12|$

In Exercises 13–16, use a calculator to find each square root to the nearest hundredth.

13. $\sqrt{77}$

14. $\sqrt{0.26}$

15. $\sqrt{\pi}$

16. $\sqrt{\dfrac{7}{5}}$

In Exercises 17–20, write each product using exponents.

17. $3 \cdot 3 \cdot 3$

18. $8 \cdot \pi \cdot r \cdot r$

19. $4 \cdot x \cdot x \cdot y \cdot y$

20. $m \cdot m \cdot m \cdot n$

In Exercises 21–24, evaluate each expression.

21. $13 - 3 \cdot 4$

22. $-2 \cdot 7^2$

23. $6 + 3\left(\dfrac{5}{2}\right) - \dfrac{1}{2}$

24. $\dfrac{12^2 - 4^2 - 2}{2(7 - 4)}$

In Exercises 25–26, write each phrase as an algebraic expression.

25. The sum of the width w and 12

26. Four less than a number n

In Exercises 27–30, complete each table of values.

27.

t	$t^2 - 4$
0	
1	
3	

28.

t	$(t - 4)^2$
0	
1	
3	

Calculators

For instructors who wish to use calculators as part of the instruction in this course, the text includes an Accent on Technology feature that introduces keystrokes and shows how scientific calculators and graphing calculators can be used to solve problems. In the Study Sets, logos are used to denote problems that require a scientific calculator ▦ or a graphing calculator ▨ .

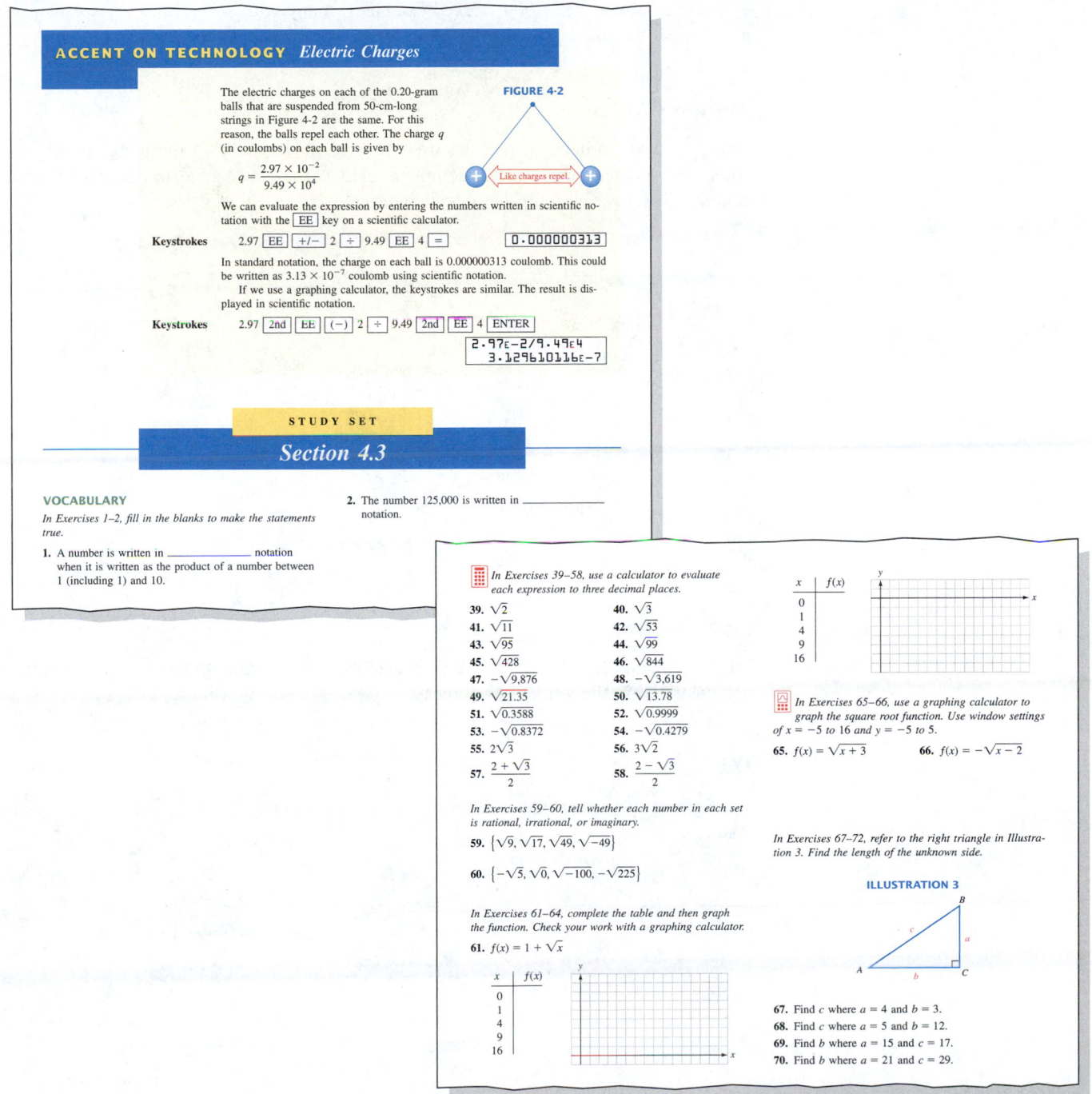

ACCENT ON TECHNOLOGY *Electric Charges*

The electric charges on each of the 0.20-gram balls that are suspended from 50-cm-long strings in Figure 4-2 are the same. For this reason, the balls repel each other. The charge q (in coulombs) on each ball is given by

$$q = \frac{2.97 \times 10^{-2}}{9.49 \times 10^{4}}$$

FIGURE 4-2

Like charges repel.

We can evaluate the expression by entering the numbers written in scientific notation with the ☐EE☐ key on a scientific calculator.

Keystrokes   2.97 ☐EE☐ ☐+/−☐ 2 ☐÷☐ 9.49 ☐EE☐ 4 ☐=☐   ☐0.000000313☐

In standard notation, the charge on each ball is 0.000000313 coulomb. This could be written as 3.13×10^{-7} coulomb using scientific notation.

If we use a graphing calculator, the keystrokes are similar. The result is displayed in scientific notation.

Keystrokes   2.97 ☐2nd☐ ☐EE☐ ☐(−)☐ 2 ☐÷☐ 9.49 ☐2nd☐ ☐EE☐ 4 ☐ENTER☐

☐2.97ε−2/9.49ε4☐
☐3.129610116ε−7☐

STUDY SET

Section 4.3

VOCABULARY

In Exercises 1–2, fill in the blanks to make the statements true.

1. A number is written in _____ notation when it is written as the product of a number between 1 (including 1) and 10.

2. The number 125,000 is written in _____ notation.

▦ *In Exercises 39–58, use a calculator to evaluate each expression to three decimal places.*

39. $\sqrt{2}$ **40.** $\sqrt{3}$

41. $\sqrt{11}$ **42.** $\sqrt{53}$

43. $\sqrt{95}$ **44.** $\sqrt{99}$

45. $\sqrt{428}$ **46.** $\sqrt{844}$

47. $-\sqrt{9,876}$ **48.** $-\sqrt{3,619}$

49. $\sqrt{21.35}$ **50.** $\sqrt{13.78}$

51. $\sqrt{0.3588}$ **52.** $\sqrt{0.9999}$

53. $-\sqrt{0.8372}$ **54.** $-\sqrt{0.4279}$

55. $2\sqrt{3}$ **56.** $3\sqrt{2}$

57. $\dfrac{2 + \sqrt{3}}{2}$ **58.** $\dfrac{2 - \sqrt{3}}{2}$

In Exercises 59–60, tell whether each number in each set is rational, irrational, or imaginary.

59. $\{\sqrt{9}, \sqrt{17}, \sqrt{49}, \sqrt{-49}\}$

60. $\{-\sqrt{5}, \sqrt{0}, \sqrt{-100}, -\sqrt{225}\}$

In Exercises 61–64, complete the table and then graph the function. Check your work with a graphing calculator.

61. $f(x) = 1 + \sqrt{x}$

x	$f(x)$
0	
1	
4	
9	
16	

x	$f(x)$
0	
1	
4	
9	
16	

▨ *In Exercises 65–66, use a graphing calculator to graph the square root function. Use window settings of $x = -5$ to 16 and $y = -5$ to 5.*

65. $f(x) = \sqrt{x + 3}$ **66.** $f(x) = -\sqrt{x - 2}$

In Exercises 67–72, refer to the right triangle in Illustration 3. Find the length of the unknown side.

ILLUSTRATION 3

67. Find c where $a = 4$ and $b = 3$.

68. Find c where $a = 5$ and $b = 12$.

69. Find b where $a = 15$ and $c = 17$.

70. Find b where $a = 21$ and $c = 29$.

Appendixes

A review of arithmetic fractions and decimal fractions is included in Appendix I. Appendix II covers synthetic division. For each, problem sets are included for student practice. Appendix III gives a table of roots and powers.

Student Support

We have included many features that make *Elementary and Intermediate Algebra* very accessible to students. (See the examples starting on page xii.)

Worked Examples

The text contains over 1,000 worked examples, many with several parts. Explanatory notes make the examples easy to follow.

Author's Notes

Author's notes, printed in red, are used to explain the steps in the solutions of examples. The notes are extensive; complete sentences are used so as to increase the students' ability to read and write mathematics.

A special logo shows which ▶
examples are included in the
videotape series.

Each step is explained using ▶
detailed author's notes.

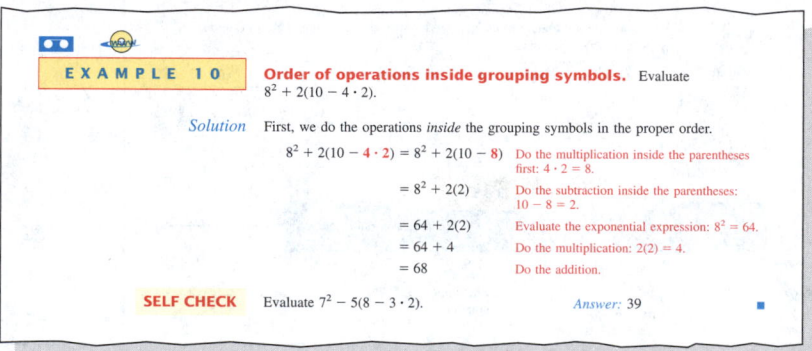

EXAMPLE 10   **Order of operations inside grouping symbols.**   Evaluate $8^2 + 2(10 - 4 \cdot 2)$.

Solution   First, we do the operations *inside* the grouping symbols in the proper order.

$$8^2 + 2(10 - 4 \cdot 2) = 8^2 + 2(10 - 8)$$ Do the multiplication inside the parentheses first: $4 \cdot 2 = 8$.

$$= 8^2 + 2(2)$$ Do the subtraction inside the parentheses: $10 - 8 = 2$.

$$= 64 + 2(2)$$ Evaluate the exponential expression: $8^2 = 64$.

$$= 64 + 4$$ Do the multiplication: $2(2) = 4$.

$$= 68$$ Do the addition.

SELF CHECK   Evaluate $7^2 - 5(8 - 3 \cdot 2)$.   *Answer:* 39

Self Checks

There are hundreds of Self Check problems that allow students to practice the skills demonstrated in the worked examples.

Warnings

Throughout the text, students are warned about common mistakes and how to avoid them.

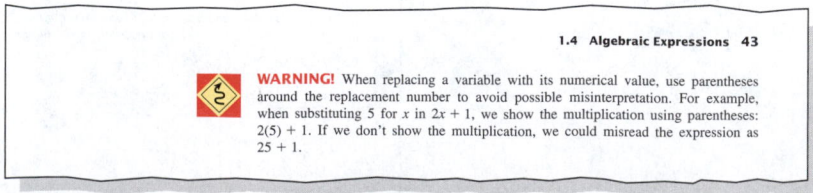

1.4 Algebraic Expressions 43

WARNING! When replacing a variable with its numerical value, use parentheses around the replacement number to avoid possible misinterpretation. For example, when substituting 5 for x in $2x + 1$, we show the multiplication using parentheses: $2(5) + 1$. If we don't show the multiplication, we could misread the expression as $25 + 1$.

Videotapes

The videotape series that accompanies this book uses eye-catching computer graphics to show students the steps in solving many examples in the text. A video logo 🔲 placed next to an example indicates that the example is taught on tape. In addition, the tapes present the solutions of two Study Set problems from each section.

Functional Use of Color

For easy reference, definition boxes (light blue with a green title), strategy boxes (light yellow with a red title), and rule or property boxes (bright yellow with royal blue title), are color-coded. In addition, the book uses color to highlight terms and expressions that you would point to in a classroom discussion.

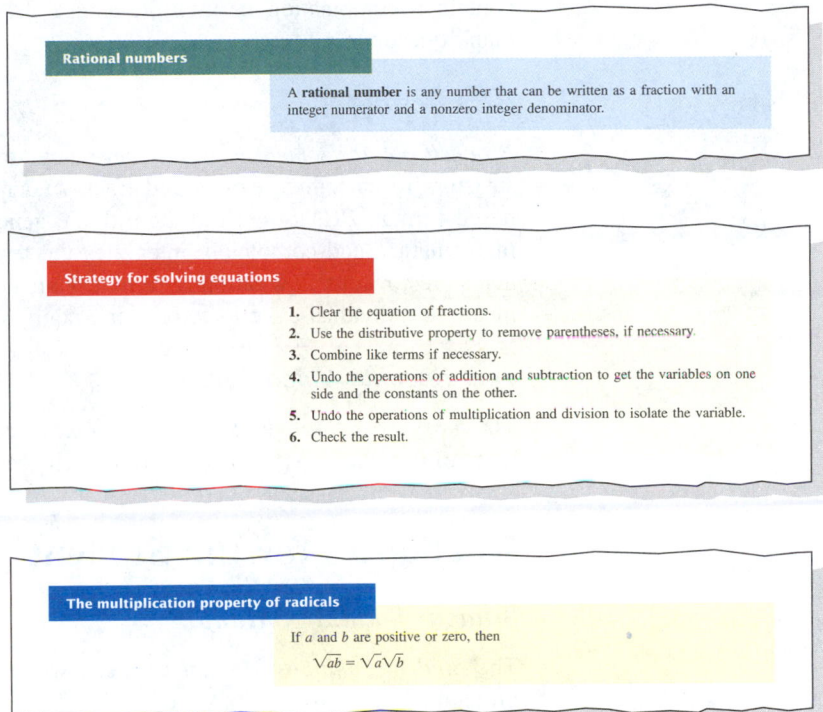

Rational numbers

A **rational number** is any number that can be written as a fraction with an integer numerator and a nonzero integer denominator.

Strategy for solving equations

1. Clear the equation of fractions.
2. Use the distributive property to remove parentheses, if necessary.
3. Combine like terms if necessary.
4. Undo the operations of addition and subtraction to get the variables on one side and the constants on the other.
5. Undo the operations of multiplication and division to isolate the variable.
6. Check the result.

The multiplication property of radicals

If a and b are positive or zero, then

$$\sqrt{ab} = \sqrt{a}\sqrt{b}$$

Problems and Answers

The book includes more than 10,000 carefully graded exercises. In the Student Edition, Appendix IV provides the answers to the odd-numbered exercises in the Study Sets, as well as all the answers to the Key Concept, Chapter Review, Chapter Test, and Cumulative Review problems.

Reading and Writing Mathematics

Also included (on pages xxvi–xxvii) are two features to help students improve their ability to read and write mathematics. "Reading Mathematics" helps students get the most out of the examples in this book by showing them how to read the solutions properly. "Writing Mathematics" highlights the characteristics of a well-written solution.

Study Skills and Math Anxiety

These two topics are discussed in detail in the section entitled "For the Student" at the end of this preface. In "Success in Algebra," students are asked to design a personal strategy for studying and learning the material. "Taking a Math Test" helps students prepare for a test and then gives them suggestions for improving their performance.

Ancillaries for the Instructor

Complete Solutions Manual

The *Complete Solutions Manual* provides worked-out solutions to all of the problems in the text.

Test Bank

The *Test Bank* includes three tests per chapter as well as three final exams. The tests contain a combination of multiple-choice, free-response, true/false, and fill-in-the-blank questions.

BCA Testing

Brooks/Cole Assessment is an internet-ready, text-specific testing suite that allows instructors to customize exams and track student progress in an accessible, browser-based format. *BCA* offers full algorithmic generation of problems and free-response mathematics, and completely integrates the testing and course management components. Test results flow automatically to your grade book and you can easily communicate to individuals, sections, or entire courses.

Text-Specific Video Series

The tapes work through examples from each section of the text, with additional solutions to two Study Set problems from each section.

Ancillaries for the Student

Student Solutions Manual

The *Student Solutions Manual* contains worked-out solutions to the odd-numbered problems in the text.

BCA Tutorial Student and Instructor Versions

This text-specific, interactive software is delivered via the Web (http://bca.brookscole.com) and is offered in both student and instructor versions. Like *BCA Testing,* it is browser based, making it an intuitive mathematical guide, even for students with little technological proficiency. *BCA Tutorial* allows students to work with real math notation in real time, and provides instant analysis and feedback. The tracking program built into the instructor version of the software enables instructors to monitor student progress carefully.

Interactive Video Skillbuilder CD

Packaged with the book, this single CD-ROM contains more than 8 hours of video instruction. The problems worked during each video lesson are listed next to the viewing screen, so students can work them ahead of time if they choose. In order to help students evaluate their progress, each section contains a 10-question Web quiz and each chapter contains a chapter test with answers provided.

Acknowledgments

We are grateful to the instructors who have reviewed the text at various stages of its development. Their comments and suggestions have proven invaluable in making this a better book. We sincerely thank all of them for lending their time and talent to this project.

Julia Brown *Atlantic Community College*	Elizabeth Morrison *Valencia Community College*
John Coburn *Saint Louis Community College– Florissant Valley*	Angelo Segalla *Orange Coast College*
Sally Copeland *Johnson County Community College*	June Strohm *Pennsylvania State Community College– DuBois*
Ben Cornelius *Oregon Institute of Technology*	Rita Sturgeon *San Bernardino Valley College*
James Edmondson *Santa Barbara Community College*	Jo Anne Temple *Texas Technical University*
Mary Lou Hammond *Spokane Community College*	Sharon Testone *Onondaga Community College*
Judith Jones *Valencia Community College*	Marilyn Treder *Rochester Community College*
Therese Jones *Amarillo College*	Betty Weissbecker *J. Sargeant Reynolds Community College*
Janice McFatter *Gulf Coast Community College*	Cathleen Zucco *SUNY–New Paltz*

We want to express our gratitude to Bob Billups, George Carlson, Robin Carter, Jim Cope, Terry Damron, Cathy Gong, Marion Hammond, Lin Humphrey, Karl Hunsicker, Doug Keebaugh, Arnold Kondo, John McKeown, Kent Miller, Donna Neff, Steve Odrich, Eric Rabitoy, Tanja Rinkel, Dave Ryba, Chris Scott, Liz Tussy, and the Citrus College library staff for their help with some of the application problems in the textbook.

We would also like to thank Bob Pirtle, Ellen Brownstein, David Hoyt, Lori Heckelman, Vernon Boes, Melissa Henderson, Erin Wickersham, Michelle Paolucci, and The Clarinda Company for their help in creating this book.

Alan S. Tussy
R. David Gustafson

▶

For the Student

Success in Algebra

To be successful in mathematics, you need to know how to study it. The following checklist will help you develop your own personal strategy to study and learn the material. The suggestions listed below require some time and self-discipline on your part, but it will be worth the effort. This will help you get the most out of this course.

As you read each of the following statements, place a check mark in the box if you can truthfully answer Yes. If you can't answer Yes, think of what you might do to make the suggestion part of your personal study plan. You should go over this checklist several times during the term to be sure you are following it.

Preparing for the Class

☐ I have made a commitment to myself to give this course my best effort.

☐ I have the proper materials: a pencil with an eraser, paper, a notebook, a ruler, a calculator, and a calendar or day planner.

☐ I am willing to spend a minimum of two hours doing homework for every hour of class.

☐ I will try to work on this subject every day.

☐ I have a copy of the class syllabus. I understand the requirements of the course and how I will be graded.

☐ I have scheduled a free hour after the class to give me time to review my notes and begin the homework assignment.

Class Participation

☐ I will regularly attend the class sessions and be on time.

☐ When I am absent, I will find out what the class studied, get a copy of any notes or handouts, and make up the work that was assigned when I was gone.

☐ I will sit where I can hear the instructor and see the chalkboard.

☐ I will pay attention in class and take careful notes.

☐ I will ask the instructor questions when I don't understand the material.

☐ When tests, quizzes, or homework papers are passed back and discussed in class, I will write down the correct solutions for the problems I missed so that I can learn from my mistakes.

Study Sessions

☐ I will find a comfortable and quiet place to study.

☐ I realize that reading a math book is different from reading a newspaper or a novel. Quite often, it will take more than one reading to understand the material.

☐ After studying an example in the textbook, I will work the accompanying Self Check.

☐ I will begin the homework assignment only after reading the assigned section.

☐ I will try to use the mathematical vocabulary mentioned in the book and used by my instructor when I am writing or talking about the topics studied in the course.

☐ I will look for opportunities to explain the material to others.

☐ I will check all of my answers to the problems with those provided in the back of the book (or with the *Student Solutions Manual*) and reconcile any differences.

☐ My homework will be organized and neat. My solutions will show all the necessary steps.

☐ I will work some review problems every day.

☐ After completing the homework assignment, I will read the next section to prepare for the coming class session.

☐ I will keep a notebook containing my class notes, homework papers, quizzes, tests, and any handouts—all in order by date.

Special Help

☐ I know my instructor's office hours and am willing to go in to ask for help.

☐ I have formed a study group with classmates that meets regularly to discuss the material and work on problems.

☐ When I need additional explanation of a topic, I view the video and check the web site.

☐ I take advantage of extra tutorial assistance that my school offers for mathematics courses.

☐ I have purchased the *Student Solutions Manual* that accompanies this text, and I use it.

To follow each of these suggestions will take time. It takes a lot of practice to learn mathematics, just as with any other skill.

No doubt, you will sometimes become frustrated along the way. This is natural. When it occurs, take a break and come back to the material after you have had time to clear your thoughts. Keep in mind that the skills and discipline you learn in this course will help make for a brighter future. Good luck!

Taking a Math Test

The best way to relieve anxiety about taking a mathematics test is to know that you are well-prepared for it and that you have a plan. Before any test, ask yourself three questions. When? What? How?

When Will I Study?

1. When is the test?

2. When will I begin to review for the test?

3. What are the dates and times that I will reserve for studying for the test?

What Will I Study?

1. What sections will the test cover?
2. Has the instructor indicated any types of problems that are guaranteed to be on the test?

How Will I Prepare for the Test?

Put a check mark by each method you will use to prepare for the test.

- ☐ Review the class notes.
- ☐ Outline the chapter(s) to see the big picture and to see how the topics relate to one another.
- ☐ Recite the important formulas, definitions, vocabulary, and rules into a tape recorder.
- ☐ Make flash cards for the important formulas, definitions, vocabulary, and rules.
- ☐ Rework problems from the homework assignments.
- ☐ Rework each of the Self Check problems in the text.
- ☐ Form a study group to discuss and practice the topics to be tested.
- ☐ Complete the appropriate Chapter Review(s) and the Chapter Test(s).
- ☐ Review the Warnings given in the text.
- ☐ Work on improving my speed in answering questions.
- ☐ Review the methods that can be used to check my answers.
- ☐ Write a sample test, trying to think of the questions the instructor will ask.
- ☐ Complete the appropriate Cumulative Review Exercises.
- ☐ Get organized the night before the test. Have materials ready to go so that the trip to school will not be hurried.
- ☐ Take some time to relax immediately before the test. Don't study right up to the last minute.

Taking the Test

Here are some tips that can help improve your performance on a mathematics test.

- ■ When you receive the test, scan it, looking for the types of problems you had expected to see. Do them first.
- ■ Write down any formulas or rules as soon as you receive the test.
- ■ Read the instructions carefully.
- ■ Don't spend too much time on any one problem until you have attempted all the problems.
- ■ If your instructor gives partial credit, at least try to begin a solution.

- Save the most difficult problems for last.
- Don't be afraid to skip a problem and come back to it later.
- If you finish early, go back over your work and look for mistakes.

Reading Mathematics

To get the most out of this book, you need to learn how to read it correctly. A mathematics textbook must be read differently than a novel or a newspaper. For one thing, you need to read it slowly and carefully. At times, you will have to reread a section to understand its content. You should also have pencil and paper with you when reading a mathematics book, so that you can work along with the text to understand the concepts presented.

Perhaps the most informative parts of a mathematics book are its examples. Each example in this textbook consists of a problem and its corresponding solution. One form of solution that is used many times in this book is shown in the diagram below. It is important that you follow the "flow" of its steps if you are to understand the mathematics involved. For this solution form, the basic idea is this:

- A property, rule, or procedure is applied to the original expression to obtain an equivalent expression. We show that the two expressions are equivalent by writing an equals sign between them. The property, rule, or procedure that was used is then listed next to the equivalent expression in the form of an author's note, printed in red.
- The process of writing equivalent expressions and explaining the reasons behind them continues, step by step, until the final result is obtained.

The solution in the following diagram consists of three steps, but solutions have varying lengths.

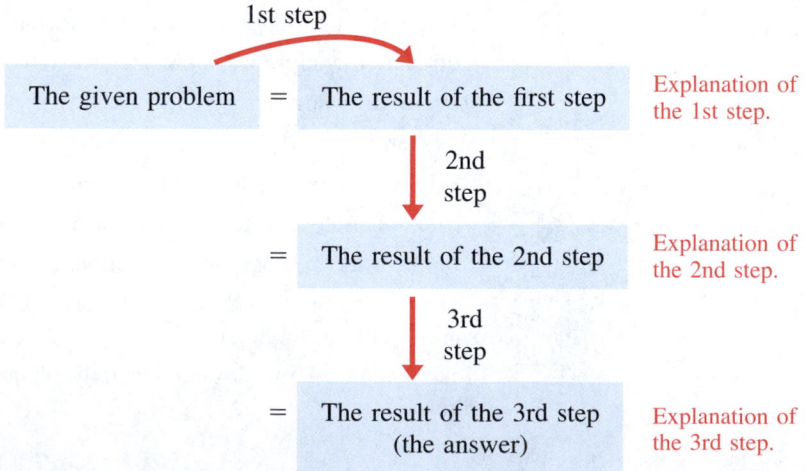

A solution (one of the basic forms)

1st step

| The given problem | = | The result of the first step | Explanation of the 1st step. |

2nd step

| | = | The result of the 2nd step | Explanation of the 2nd step. |

3rd step

| | = | The result of the 3rd step (the answer) | Explanation of the 3rd step. |

Writing Mathematics

One of the major objectives of this course is for you to learn how to write solutions to problems properly. A written solution to a problem should explain your thinking in a series of neat and organized mathematical steps. Think of a solution as a mathematical essay—one that your instructor and other students should be able to read and understand. Some solutions will be longer than others, but they must all be in the proper format and use the correct notation. To learn how to do this will take time and practice.

To give you an idea of what will be expected, let's look at two samples of student work. In the first, we have highlighted some important characteristics of a well-written solution. The second sample is poorly done and would not be acceptable.

$$\text{Evaluate: } 35 - 2^2 \cdot 3.$$

A well-written solution:

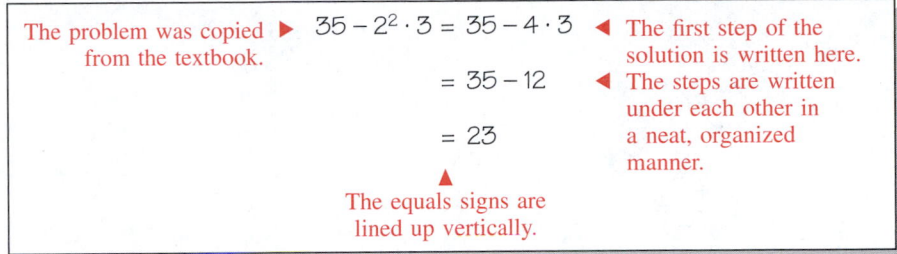

The problem was copied ▶ from the textbook.

$$35 - 2^2 \cdot 3 = 35 - 4 \cdot 3$$
$$= 35 - 12$$
$$= 23$$

◀ The first step of the solution is written here.

◀ The steps are written under each other in a neat, organized manner.

▲
The equals signs are lined up vertically.

A poorly written solution:

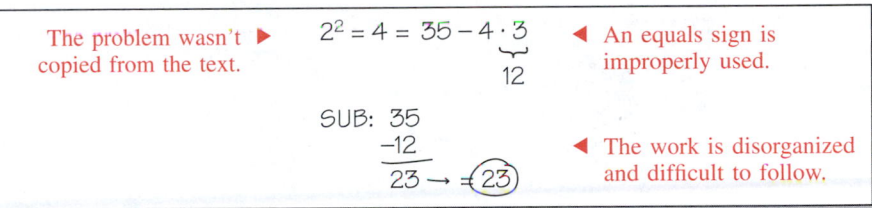

The problem wasn't ▶ copied from the text.

$$2^2 = 4 = 35 - 4 \cdot \underbrace{3}_{12}$$

SUB: 35
−12
‾‾‾‾
23 → 23

◀ An equals sign is improperly used.

◀ The work is disorganized and difficult to follow.

1

An Introduction to Algebra

CAMPUS CONNECTION

The Accounting Department

In accounting, students study some of the basic billing practices used in the health care industry. The students need to know how to work with percents, because many health care plans pay only a percentage of a patient's medical bill. For example, one popular coverage pays 80% of the costs and requires the patient to pay the remaining 20% (called the *copayment*). In this chapter, we discuss how the algebraic concepts of *variable* and *equation* can be used to solve percent problems from many different disciplines, including accounting.

Algebra is a mathematical language that can be used to solve many types of real-world problems.

▶ 1.1   Describing Numerical Relationships

In this section, you will learn about

   Tables ■ Bar graphs ■ Line graphs ■ Vocabulary ■ Symbols and notation ■ Variables, algebraic expressions, and equations ■ Constructing tables

Introduction   Using the vocabulary, symbols, and notation of algebra, we can mathematically describe (or **model**) the real world. From an algebraic model, we can make observations and predict outcomes. To solve problems using algebra, you will need to learn how to read it, write it, and speak it. In this section, we begin to explore the language of algebra by introducing some algebraic methods that are used to describe numerical relationships.

Tables

A production planner at a bicycle manufacturing plant must order parts for upcoming production runs. To order the correct number of tires, she uses the **table** in Figure 1-1. After locating the number of bicycles to be manufactured in the left-hand column, she scans across the table to the corresponding entry in the right-hand column to find the number of tires to order. For example, if 400 bikes are to be manufactured, the table shows that 800 tires should be ordered.

FIGURE 1-1

Bicycles to be manufactured	Tires to order
100	200
200	400
300	600
400	800

Bar Graphs

The information in the table in Figure 1-1 can also be presented in a **bar graph.** The bar graph in Figure 1-2 has a **horizontal axis** labeled "Number of bicycles to be manufactured" and has been scaled in units of 100 bicycles. The **vertical axis** of the graph, labeled "Number of tires to be ordered," is scaled in units of 100 tires. The bars directly over each of the production amounts (100, 200, 300, and 400 bicycles) extend to a height indicating the corresponding number of tires to order. For example, if 300 bikes are to be manufactured, the height of the bar indicates that 600 tires should be ordered.

FIGURE 1-2

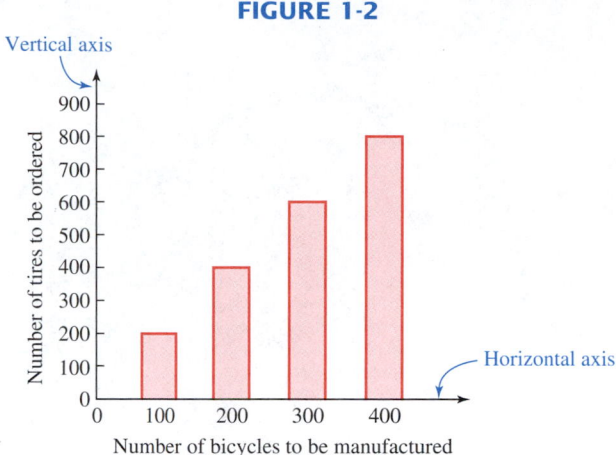

Number of bicycles to be manufactured

Line Graphs

A third way to present the information shown in the table in Figure 1-1 is with a **line graph.** Instead of using a bar to denote the number of tires to order for a given size of production run, we use a heavy dot drawn at the correct "height." See Figure 1-3(a). After drawing the four data points for 100, 200, 300, and 400 bicycles, we connect them with line segments to create the line graph. See Figure 1-3(b).

FIGURE 1-3

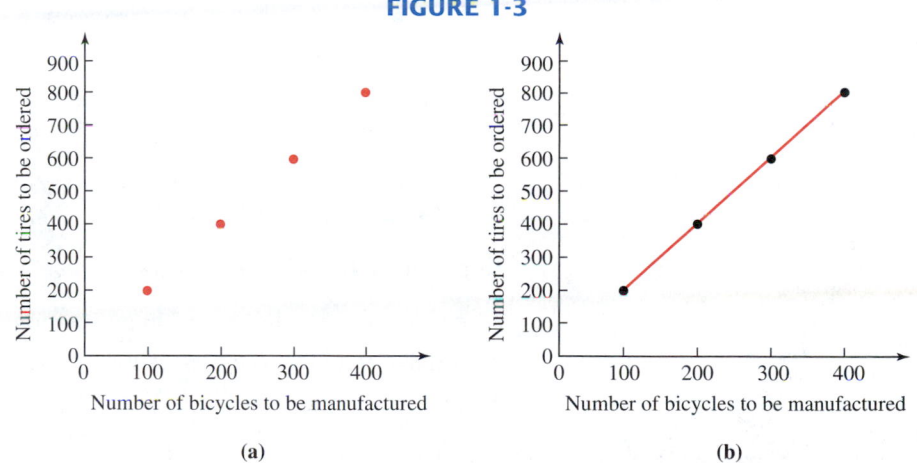

  (a)   (b)

The line graph not only presents all the information contained in the table and the bar graph, but it also provides additional information that they do not. We can use the line graph to find the number of tires to order for a production run of a size that is not shown in the table or the bar graph.

EXAMPLE 1

Reading a line graph. Use the line graph in Figure 1-3(b) to determine the number of tires needed if 250 bicycles are manufactured.

Solution First, locate the number 250 (between 200 and 300) on the horizontal axis. Then draw a line straight up to intersect the graph. (See Figure 1-4.) From the point of intersection, draw a horizontal line to the left that intersects the vertical axis. We see that the number of tires to order is 500.

FIGURE 1-4

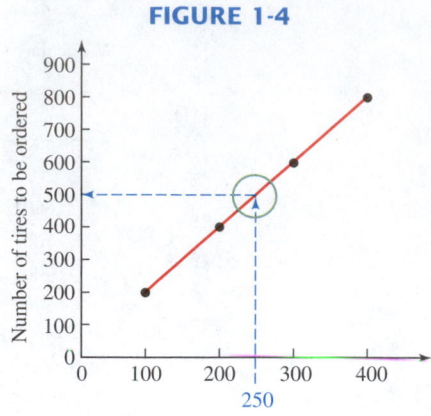

Number of bicycles to be manufactured

Use the line graph in Figure 1–4 to find the number of tires needed if 350 bicycles are manufactured.   *Answer:* 700   ■

Vocabulary

After working with a table, a bar graph, and a line graph, it is evident that there is a relationship between the number of tires to order and the number of bicycles to be manufactured. Using words, we can express this relationship as follows:

"The number of tires to order is two times the number of bicycles to be manufactured."

In algebra, the word **product** is used to indicate the answer to a multiplication. Therefore, we can restate the relationship this way:

"The number of tires to order is the product of two and the number of bicycles to be manufactured."

To indicate other arithmetic operations, we will use the following words:

■ **sum** to indicate the answer to an addition: The sum of 5 and 6 is 11.

■ **difference** to indicate the answer to a subtraction: The difference of 6 and 2 is 4.

■ **quotient** to indicate the answer to a division: The quotient of 6 and 3 is 2.

EXAMPLE 2

Vocabulary. Express each statement in words, using one of the words *sum, product, difference,* or *quotient:*   **a.** $22 \div 11 = 2$ and   **b.** $22 + 11 = 33$.

Solution   **a.** The quotient of 22 and 11 is 2.

b. The sum of 22 and 11 is 33.

Express each statement in words: **a.** $22 - 11 = 11$ and   **b.** $22 \times 11 = 242$.   *Answers:*   **a.** The difference of 22 and 11 is 11.   **b.** The product of 22 and 11 is 242.   ■

Symbols and Notation

In algebra, we will use many symbols and notations. Because the letter x is often used in algebra and could be confused with the multiplication sign $\times$, we usually write multiplication in another form.

Symbols that are used for multiplication

Symbol	Name	Example
$\times$	times sign	$6 \times 4 = 24$ or $\begin{array}{r} 451 \\ \times\ 53 \\ \hline 23{,}903 \end{array}$
$\cdot$	raised dot	$6 \cdot 4 = 24$ or $451 \cdot 53 = 23{,}903$
( )	parentheses	$(6)4 = 24$ or $6(4) = 24$ or $(6)(4) = 24$

There are several ways to indicate division. In algebra, the form most often used involves a fraction bar.

Symbols that are used for division

Symbol	Name	Example
$\div$	division sign	$16 \div 4 = 4$ or $3{,}400 \div 20 = 170$
$\overline{)}$	long division	$4\overline{)16}^{\,4}$ or $20\overline{)3{,}400}^{\,170}$
—	fraction bar	$\dfrac{16}{4} = 4$ or $\dfrac{3{,}400}{20} = 170$

Variables, Algebraic Expressions, and Equations

Another way to describe the relationship between the number of tires to order and the number of bicycles being manufactured uses *variables*. **Variables** are letters that stand for numbers. If we let the letter t stand for the number of tires to be ordered and b for the number of bicycles to be manufactured, we can translate the relationship from words to mathematical symbols.

The number of tires to order	is	two	times	the number of bicycles to be manufactured.
t	$=$	2	$\cdot$	b

The statement $t = 2 \cdot b$ is called an **equation.** Equations are mathematical sentences that contain an $=$ sign, which indicates that two quantities are equal. Some examples of equations are

$$3 + 5 = 8 \qquad x + 5 = 20 \qquad 17 - t = 14 - t$$

In the equation $t = 2 \cdot b$, the variable b is multiplied by 2. When we multiply a variable by another number or multiply a variable by another variable, we don't need to use a multiplication symbol.

$$2b \text{ means } 2 \cdot b \qquad xy \text{ means } x \cdot y \qquad abc \text{ means } a \cdot b \cdot c$$

Using this form, we can write the equation $t = 2 \cdot b$ as $t = 2b$. The notation $2b$ on the right-hand side of the equation is called an **algebraic expression** or, more simply, an **expression.** Algebraic expressions are the building blocks of equations.

Variables and/or numbers can be combined with the operations of addition, subtraction, multiplication, and division to create **algebraic expressions.**

Here are some examples of algebraic expressions.

$2a + 7$	This algebraic expression is a combination of the numbers 2 and 7, the variable a, and the operations of multiplication and addition.
$\dfrac{10 - y}{3}$	This algebraic expression is a combination of the numbers 10 and 3, the variable y, and the operations of subtraction and division.
$15mn(2m)$	This algebraic expression is a combination of the numbers 15 and 2, the variables m and n, and the operation of multiplication.

Using the equation $t = 2b$ to express the relationship between the number of tires to order and the number of bicycles to be manufactured has an advantage over the other methods we have discussed. It can be used to determine the exact number of tires to order for a production run of *any* size.

EXAMPLE 3

Using an equation. Find the number of tires needed for a production run of 178 bicycles.

Solution   To find the number of tires needed, we use the equation that describes this numerical relationship.

$t = 2\boldsymbol{b}$	The describing equation.
$t = 2(\boldsymbol{178})$	Replace b, which stands for the number of bicycles, with 178. Use parentheses to show the multiplication.
$t = 356$	Do the multiplication: $2(178) = 356$.

Therefore, 356 tires will be needed.

SELF CHECK   Find the number of tires needed if 604 bicycles are to be manufactured.   *Answer:* 1,208   ■

Constructing Tables

Equations such as $t = 2b$, which express a known relationship between two or more variables, are called **formulas.** Formulas are used in many fields, such as economics, biology, nursing, and construction. In the next example, we will see that the results found using the formula $t = 2b$ can be presented in table form.

EXAMPLE 4

Constructing a table. Find the number of tires to order for production runs of 233 and 852 bicycles. Present the results in a table.

Solution   We begin by constructing a table with appropriate column headings. The size of each production run (233 and 852) is entered in the left-hand column of the table.

Bicycles to be manufactured	Tires to order
233	
852	

Next, we use the equation $t = 2b$ to find the number of tires needed if 233 and 852 bikes are to be manufactured.

$t = 2b$ $t = 2b$
$t = 2(233)$ Replace b with 233. $t = 2(852)$ Replace b with 852.
$t = 466$ $t = 1,704$

Finally, we enter these results in the right-hand column of the table: 466 tires for 233 bicycles manufactured and 1,704 tires for 852 bicycles manufactured.

Bicycles to be manufactured	Tires to order
233	466
852	1,704

SELF CHECK   Find the number of tires to order for production runs of 87 and 487 bicycles. Present the results in a table.

Answer:

Bicycles to be manufactured	Tires to order
87	174
487	974

STUDY SET

Section 1.1

VOCABULARY

In Exercises 1–12, fill in the blanks to make the statements true.

1. The answer to an addition problem is called the _____.

2. The answer to a subtraction problem is called the _____.

3. The answer to a multiplication problem is called the _____.

4. The answer to a division problem is called the _____.

5. _____ are letters that stand for numbers.

6. The symbols (   ) are called _____.

7. Variables and/or numbers can be combined with the operations of addition, subtraction, multiplication, and division to create algebraic _____.

8. An _____ is a mathematical sentence that contains an = sign.

9. An equation such as $t = 2b$, which expresses a known relationship between two or more variables, is called a _____.

10. In Illustration 1, a _____ graph is shown.

11. In Illustration 1, the _____ axis of the graph has been scaled in units of ___ second.

12. In Illustration 1, the _____ axis of the graph has been scaled in units of 50 _____.

ILLUSTRATION 1

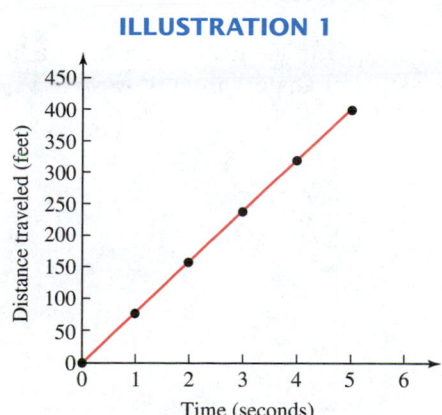

CONCEPTS

In Exercises 13–20, classify each item as either an algebraic expression or an equation.

13. $18 + m = 23$

14. $18 + m$

15. $y - 1$

16. $y - 1 = 2$

17. $30x$

18. $t = 16b$

19. $r = \dfrac{2}{3}$

20. $\dfrac{c - 7}{5}$

21. a. What operations does the expression $5x - 16$ contain?
   b. What variable(s) does it contain?

22. a. What operations does the expression $\frac{12 + t}{25}$ contain?
   b. What variable(s) does it contain?

23. a. What operations does the equation $4 + 1 = 20 - m$ contain?
   b. What variable(s) does it contain?

24. a. What operations does the equation $y + 14 = 5(6)$ contain?
   b. What variable(s) does it contain?

25. See Illustration 2. As the railroad crossing guard drops, the measure of angle 1 increases, while the measure of angle 2 decreases. At any instant, the sum of the measures of the two angles is 90°.
   a. Complete the table.

ILLUSTRATION 2

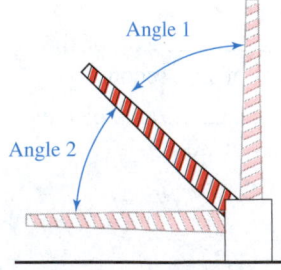

Angle 1	Angle 2
0°	
30°	
45°	
60°	
90°	

b. Use the data in the table to construct a line graph for values of angle 1 from 0° to 90°.

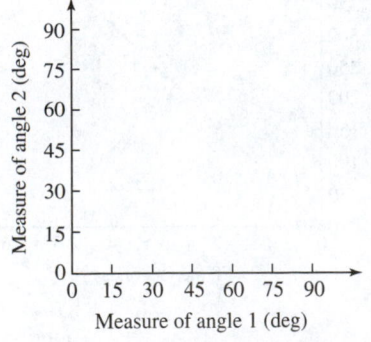

26. See Illustration 3. As the legs on the keyboard stand are widened, the measure of angle 1 will increase, and in turn, the measure of angle 2 will decrease. For any position, the sum of the measures of the angles is 180°.
   a. Complete the table.

ILLUSTRATION 3

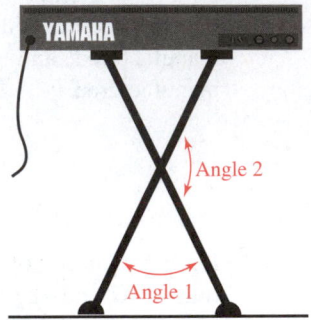

Angle 1	Angle 2
50°	
60°	
70°	
80°	
90°	

b. Use the data in the table to construct a line graph for values of angle 1 from 50° to 90°.

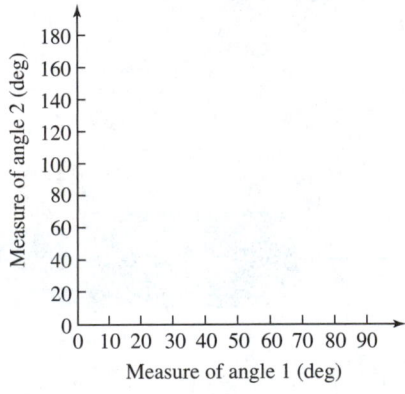

27. a. Explain what the dotted lines help us find in the graph in Illustration 4.

   b. According to the graph, as the machinery ages, what happens to its value?

ILLUSTRATION 4

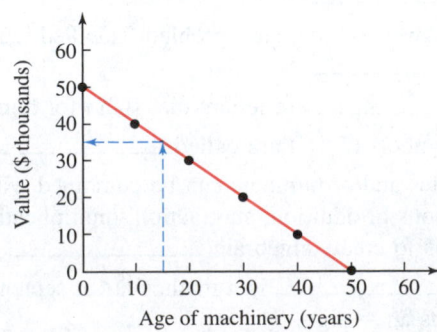

28. a. Use the line graph in Illustration 5 to find the income received from 30, 50, and 70 customers.

b. According to the graph, as the number of customers increases, what happens to the income?

ILLUSTRATION 5

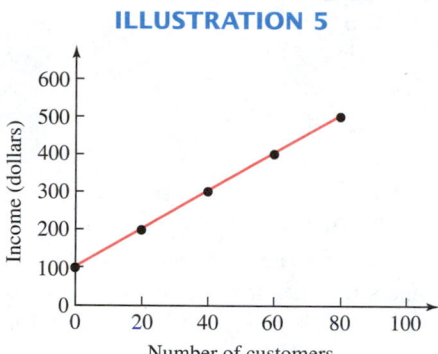

In Exercises 53–56, translate the word model into an equation. (Hint: You will need to use variables.)

53.

| The sale price | is | $100 | minus | the discount. |

54.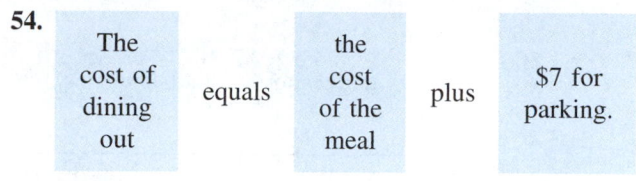

| The cost of dining out | equals | the cost of the meal | plus | $7 for parking. |

55.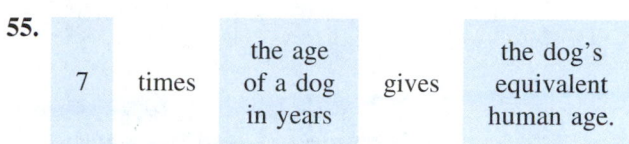

| 7 | times | the age of a dog in years | gives | the dog's equivalent human age. |

56.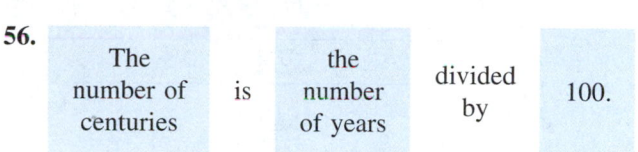

| The number of centuries | is | the number of years | divided by | 100. |

NOTATION

In Exercises 29–32, write each multiplication using a raised dot ⋅ and then using parentheses ( ).

29. 5×6 **30.** 4×7

31. 34×75 **32.** 90×12

In Exercises 33–40, write each multiplication without using a multiplication symbol.

33. $4 \cdot x$ **34.** $5 \cdot y$

35. $3 \cdot r \cdot t$ **36.** $22 \cdot q \cdot s$

37. $l \cdot w$ **38.** $b \cdot h$

39. $P \cdot r \cdot t$ **40.** $l \cdot w \cdot h$

In Exercises 41–44, write each division using a fraction bar.

41. $32 \div x$ **42.** $y \div 15$

43. $30\overline{)90}$ **44.** $20\overline{)80}$

PRACTICE

In Exercises 45–52, express each statement using one of the words sum, difference, product, or quotient.

45. $18(24)$ **46.** $45 \cdot 12$

47. $11 - 9$ **48.** $65 + 89$

49. $2x$ **50.** $16t$

51. $\dfrac{66}{11}$ **52.** $12 \div 3$

In Exercises 57–64, translate the word model into an equation. (Hint: You will need to use variables.)

57. The amount of sand that should be used is the product of 3 and the amount of cement used.

58. The number of waiters needed is the quotient of the number of customers and 10.

59. The weight of the truck is the sum of the weight of the engine and 1,200.

60. The number of classes that are still open is the difference of 150 and the number of classes that are closed.

61. The profit is the difference of the revenue and 600.

62. The distance is the product of the rate and 3.

63. The quotient of the number of laps run and 4 is the number of miles run.

64. The sum of the tax and 35 is the total cost.

In Exercises 65–68, use the given equation (formula) to complete each table.

65. $d = 360 + l$

Lunch time (minutes)	School day (minutes)
30	
40	
45	

66. $b = 1,024k$

Kilobytes	Bytes
1	
5	
10	

72.

Benefit package ($)	Total compensation ($)
4,000	39,000
5,000	40,000
6,000	41,000

67. $t = 1,500 - d$

Deductions	Take-home pay
200	
300	
400	

APPLICATIONS

73. CHAIR PRODUCTION Use the diagram shown in Illustration 6 to write six formulas that could be used by planners to order the necessary number of legs, arms, seats, backs, arm pads, and screws for a production run of c chairs.

68. $w = \dfrac{s}{12}$

Inches of snow	Inches of water
12	
24	
72	

ILLUSTRATION 6

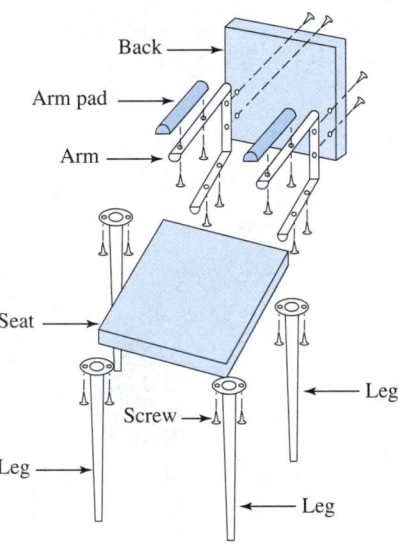

In Exercises 69–72, use the data to find an equation that describes the relationship between the two quantities. Then state the relationship in words.

69.

Eggs	Dozen
24	2
36	3
48	4

70.

Input voltage	Output voltage
60	50
110	100
220	210

74. STAIRCASE PRODUCTION Write four formulas that could be used by the job superintendent to order the necessary number of staircase parts for a tract of h homes, each of which will have a staircase as shown in Illustration 7.

ILLUSTRATION 7

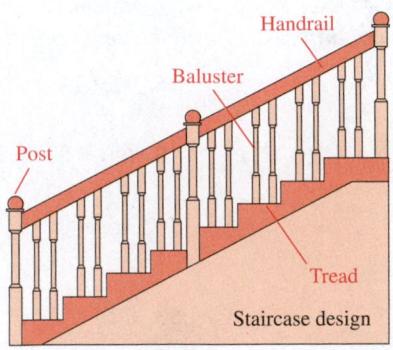

Staircase design

71.

Couples	Individuals
20	40
100	200
200	400

75. SPARE PARTS When an automobile company has parts manufactured for the assembly-line production of a car, it orders 500 more than necessary, to stock dealership service departments. Write an equation to find the number of left-side front doors that should be manufactured for a production run of x cars.

76. NO-SHOWS Park rangers have noticed that each weekend in the summer, on average, 15 people who had made campground reservations fail to show up. Write an equation the rangers could use to predict the weekend occupancy of the campground if r reservations have been made.

77. RELIGIOUS BOOKS Illustration 8 shows the annual sales of books on religion, sprituality, and inspiration for the years 1991–1997. Graph the data using a bar graph. Then describe any trend that is apparent. On the horizontal axis, use the label 1 to represent the year 1991, 2 to represent 1992, and so on.

ILLUSTRATION 8

Year	Book sales
1991	36,651,000
1992	50,104,000
1993	60,449,000
1994	70,541,000
1995	74,794,000
1996	78,022,000
1997	91,627,000

Based on data from the
Book Industry Study Group

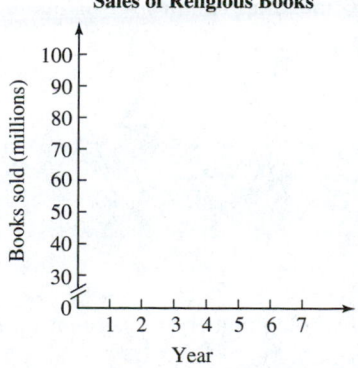

Sales of Religious Books

78. U.S. CRIME STATISTICS Property crimes include burglary, theft, and motor vehicle theft. Graph the property crime rates listed in Illustration 9 using a bar graph. Is an overall trend apparent?

ILLUSTRATION 9

Year	Victimizations per 1,000 households
1991	354
1992	325
1993	319
1994	310
1995	291
1996	266

Based on data from the Bureau of Justice Statistics

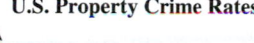
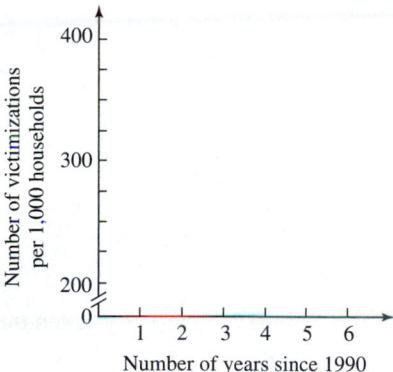

U.S. Property Crime Rates

WRITING

79. Many people misuse the word *equation* when discussing mathematics. What is an equation? Give an example.

80. Explain the difference between an algebraic expression and an equation. Give an example of each.

81. Which do you think is more informative, a bar graph or a line graph? Explain your reasoning.

82. Create a bar graph that shows, on average, how many hours of television you watch each day of the week. Let Sunday be day 1, Monday be day 2, and so on.

▶ 1.2   The Real Numbers

In this section, you will learn about

Sets of numbers ▪ Order on the number line ▪ Rational numbers (fractions and mixed numbers) ▪ Decimals ▪ Irrational numbers ▪ The real numbers ▪ Opposites ▪ Absolute value

Introduction   In this course, we will work with many types of numbers. We will solve problems involving temperatures that are negative, see units of time expressed as fractions, and express amounts of money using decimals. Some of the geometric figures we will encounter will have dimensions that are expressed as square roots. In this section, we define each type of number we will use in the course and show that they are part of a larger collection of numbers called **real numbers.**

Sets of Numbers

Table 1-1 shows the low temperatures for Rockford, IL during the first week of January. In the left column, we have used the numbers 1, 2, 3, 4, 5, 6, and 7 to denote the calendar days of the month. This collection of numbers is called a **set,** and the members (or **elements**) of the set can be listed within **braces** { }.

{1, 2, 3, 4, 5, 6, 7}

TABLE 1-1

Day of the month	Low temperature (°F)
1	4
2	−5
3	−7
4	0
5	3
6	6
7	6

Each of the numbers 1, 2, 3, 4, 5, 6, and 7 is a member of a basic set of numbers called the **natural numbers.** The natural numbers are the numbers that we count with.

Natural numbers

The set of **natural numbers** is {1, 2, 3, 4, 5, 6, 7, 8, 9, 10, . . .}.

The three dots used in this definition indicate that the list of natural numbers continues forever.

The natural numbers together with 0 make up another set of numbers called the **whole numbers.**

Whole numbers

The set of **whole numbers** is {0, 1, 2, 3, 4, 5, 6, 7, 8, 9, 10, . . .}.

Since every natural number is also a whole number, we say that the set of natural numbers is a **subset** of the whole numbers.

Table 1-1 contains positive and negative temperatures. For example, on the 2nd day of the month, the low temperature was −5° (5° below zero). On the 5th day, the low temperature was 3° (3° above zero). The numbers used to represent the temperatures listed in the table are members of a set of numbers called the **integers.**

Integers

The set of **integers** is {. . . , −4, −3, −2, −1, 0, 1, 2, 3, 4, . . .}.

Since every whole number is also an integer, we say that the set of whole numbers is a subset of the set of integers. The natural numbers are also a subset of the integers.

Order on the Number Line

We can illustrate sets of numbers with a **number line.** Like a ruler, a number line is straight and has uniform markings, as in Figure 1-5. The arrowheads indicate that the number line continues forever to the left and to the right. Numbers to the left of 0 have values that are less than 0; they are called **negative numbers.** Numbers to the right of 0 have values that are greater than 0; they are called **positive numbers.** The number 0 is neither positive nor negative.

FIGURE 1-5

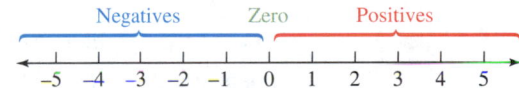

Negative numbers can be used to describe amounts that are less than 0, such as a checking account that is $75 overdrawn (−$75), an elevation of 200 feet below sea level (−200 ft), and 5 seconds before liftoff (−5 sec).

Using a process known as **graphing,** a single number or a set of numbers can be represented on a number line. The **graph of a number** is the point on the number line that corresponds to that number. *To graph a number* means to locate its position on the number line and then to highlight it by using a heavy dot.

EXAMPLE 1

Graphing on the number line. Graph the integers between −4 and 5.

Solution   To graph these integers, we locate their positions on the number line and highlight each position by drawing a dot.

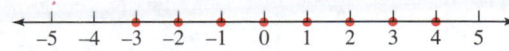

SELF CHECK   Graph the integers between −2 and 2.   *Answer:*

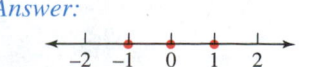

As we move to the right on a number line, the values of the numbers increase. As we move to the left, the values of the numbers decrease. In Figure 1-6, we know that 5 is greater than −3, because the graph of 5 lies to the right of the graph of −3. We also know that −3 is less than 5, because its graph lies to the left of the graph of 5.

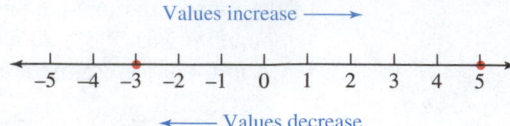

FIGURE 1-6

The **inequality symbol** > ("is greater than") can be used to show that 5 is greater than −3, and the inequality symbol < ("is less than") can be used to show that −3 is less than 5.

$5 > -3$ Read as "5 is greater than −3."

$-3 < 5$ Read as "−3 is less than 5."

To distinguish between these two inequality symbols, remember that each one points to the smaller of the two numbers involved.

$5 > -3$ $-3 < 5$

Points to the smaller number.

EXAMPLE 2

Inequality symbols. Use one of the symbols > or < to make each statement true: **a.** −4 ⬚ 4   and   **b.** −2 ⬚ −3.

Solution   **a.** Since −4 is to the left of 4 on the number line, $-4 < 4$.

b. Since −2 is to the right of −3 on the number line, $-2 > -3$.

SELF CHECK   Use one of the symbols > or < to make each statement true: **a.** 1 ⬚ −1   and

b. −5 ⬚ −4.   *Answers:* **a.** >,   **b.** <   ■

By extending the number line to include negative numbers, we can represent more situations graphically. In Figure 1-7, the line graph illustrates the low temperatures listed in Table 1-1. The vertical axis is scaled in units of degrees Fahrenheit, and temperatures below zero (negative temperatures) are graphed. For example, for the second day of the month, the low was −5°F.

FIGURE 1-7

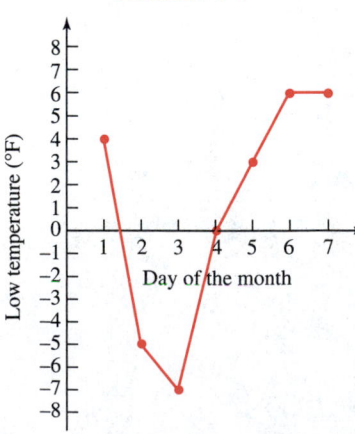

Rational Numbers (Fractions and Mixed Numbers)

In this course, we will work with positive and negative fractions. For example, the time it takes for a motorist to complete the commute home might be $\frac{3}{4}$ of an hour, or a surveyor might indicate that a building's foundation has fallen below finished grade by expressing its elevation as $-\frac{7}{8}$ of an inch. In a fraction, the number above the fraction bar is called the **numerator.** The number below is called the **denominator.** If the numerator is less than the denominator, the fraction is called a **proper fraction.** If the numerator is greater than or equal to the denominator, the fraction is called an **improper fraction.**

$$\text{Proper fraction} \quad \frac{3}{4} \; {\color{red}\longleftarrow \text{ Numerator } \longrightarrow} \; \frac{25}{12} \quad \text{Improper fraction}$$
$$\quad\quad\quad\quad {\color{red}\longleftarrow \text{ Denominator } \longrightarrow}$$

We will also work with **mixed numbers**—numbers that are the sum of a whole number and a proper fraction. For example, a piece of fabric could be $5\frac{1}{2}$ yards long, or a job could take $3\frac{2}{5}$ days to complete. Fractions and mixed numbers are part of the set of numbers called the **rational numbers.**

Rational numbers

A **rational number** is any number that can be written as a fraction with an integer numerator and a nonzero integer denominator.

Fractions such as $\frac{3}{4}$ and $\frac{25}{12}$ are rational numbers, because they have an integer numerator and a nonzero integer denominator. The fraction $-\frac{7}{8}$ is a rational number, because it can be written in the form $\frac{-7}{8}$. Mixed numbers such as $5\frac{1}{2}$ and $3\frac{2}{5}$ are also rational numbers, because they can be written as fractions with an integer numerator and a nonzero integer denominator:

$$5\frac{1}{2} = \frac{11}{2} \quad \text{and} \quad 3\frac{2}{5} = \frac{17}{5}$$

All integers are rational numbers, because they can be written as fractions with a denominator of 1. For example, $-4 = \frac{-4}{1}$ and $0 = \frac{0}{1}$. Therefore, the set of integers is a subset of the rational numbers.

EXAMPLE 3

Graphing fractions and mixed numbers. Graph $-4\frac{1}{4}, 3\frac{7}{8}, -\frac{5}{3}$, and $\frac{1}{10}$.

Solution It is easier to locate the graph of $-\frac{5}{3}$ if we express it as $-1\frac{2}{3}$.

Since the graph of $-\frac{5}{3}$ lies to the left of the graph of $\frac{1}{10}$, we can write $-\frac{5}{3} < \frac{1}{10}$.

SELF CHECK Graph $-\frac{2}{3}, -\frac{7}{4}$, and $2\frac{1}{8}$.   *Answer:*

Decimals

We will also work with decimals. Here are some examples of decimals.

- The price of apples is $0.89 a pound.
- A dragster was clocked at 203.156 miles per hour.
- The first-quarter loss for a business was $-$2.7 million.

EXAMPLE 4

Graphing decimals. Graph -3.75, 2.5, -0.9, and 1.25.

Solution   Recall that $0.75 = \frac{3}{4}$, so $-3.75 = -3\frac{3}{4}$, and $0.25 = \frac{1}{4}$, so $1.25 = 1\frac{1}{4}$.

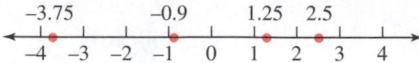

Notice that the graph of -0.9 lies to the right of the graph of -3.75. Therefore, we can write $-0.9 > -3.75$.

SELF CHECK   Graph -1.25, -2.5, and 1.75.   *Answer:*

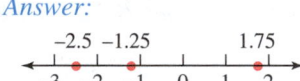

Decimals such as 0.89, $203.15\overset{3}{6}$, and -2.7 are called **terminating decimals.** Since they can be written as fractions with integer numerators and nonzero integer denominators, they are rational numbers.

$$0.89 = \frac{89}{100} \qquad 203.156 = 203\frac{156}{1,000} = \frac{203,156}{1,000} \qquad -2.7 = -2\frac{7}{10} = \frac{-27}{10}$$

We will also work with **repeating decimals** such as $0.33333\ldots$ and $2.161616\ldots$. (The three dots indicate that the digits continue in the pattern shown forever.) Any repeating decimal can be expressed as a fraction with an integer numerator and a nonzero integer denominator. For example, $0.333\ldots = \frac{1}{3}$, and $2.161616\ldots = 2\frac{16}{99} = \frac{214}{99}$. Since every repeating decimal can be written as a fraction, repeating decimals are also rational numbers.

Irrational Numbers

Some decimals cannot be written as fractions. In Figure 1-8, the length of each wire used to anchor the volleyball net is $\sqrt{2}$ (read as "the **square root** of 2") yards. Expressed in decimal form,

$$\sqrt{2} = 1.414213562\ldots$$

FIGURE 1-8

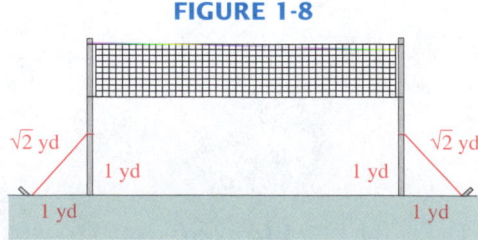

This **nonterminating, nonrepeating decimal** cannot be written as a fraction with an integer for its numerator and a nonzero integer for its denominator. Therefore, $\sqrt{2}$ is not a rational number. It is called an **irrational number.** When working with irrational numbers, it is often beneficial to approximate them. We can use a scientific calculator to approximate square roots.

ACCENT ON TECHNOLOGY *Approximating Irrational Numbers*

We can use the $\boxed{\sqrt{}}$ key (the square root key) on a calculator to find square roots. For example, to find $\sqrt{2}$ using a scientific calculator, we enter 2 and press the $\boxed{\sqrt{}}$ key.

Keystrokes $\quad 2\,\boxed{\sqrt{}}$ $\qquad\qquad\qquad\qquad$ `1.414213562`

We see that $\sqrt{2} \approx 1.414213562$. The symbol $\approx$ means "is approximately equal to." If the problem calls for an approximate answer, we can round this decimal. For example, to the nearest hundredth, $\sqrt{2} \approx 1.41$.

To approximate $\sqrt{2}$ using a graphing calculator, we use the following keystrokes:

Keystrokes $\quad \boxed{\text{2nd}}\ \boxed{\sqrt{}}\ \boxed{2}\ \boxed{)}\ \boxed{\text{ENTER}}$ $\qquad$ `√(2)`
$\qquad\qquad\qquad\qquad\qquad\qquad\qquad\qquad\qquad\qquad\qquad$ `1.414213562`

An irrational number often used in geometry is π (read as "pi"). To find the approximate value of π using a scientific calculator, we simply press the $\boxed{\pi}$ key.

Keystrokes $\quad \boxed{\pi}$ (you may have to use a $\boxed{\text{2nd}}$ or $\boxed{\text{Shift}}$ key first)
$\qquad\qquad\qquad\qquad\qquad\qquad\qquad\qquad\qquad\qquad\qquad$ `3.141592654`

Rounded to the nearest thousandth, $\pi \approx 3.142$.

To approximate π with a graphing calculator, we press the following keys:

Keystrokes $\quad \boxed{\text{2nd}}\ \boxed{\pi}\ \boxed{\text{ENTER}}$ $\qquad\qquad\qquad\qquad$ π
$\qquad\qquad\qquad\qquad\qquad\qquad\qquad\qquad\qquad\qquad\qquad$ `3.141592654`

Some other examples of irrational numbers are 2π (this means $2 \cdot \pi$), $\sqrt{3}$, $3\sqrt{2}$ (this means $3 \cdot \sqrt{2}$), $-\sqrt{7}$, and $\sqrt{27}$.

EXAMPLE 5

Graphing irrational numbers. Graph $2\sqrt{2}$, $-\sqrt{7}$, and π.

Solution To locate these numbers on a number line, we can use a calculator to approximate them. To the nearest tenth, $2\sqrt{2} = 2 \cdot \sqrt{2} \approx 2.8$, $-\sqrt{7} \approx -2.6$, and $\pi \approx 3.1$.

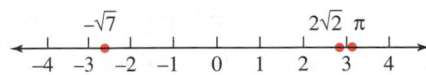

SELF CHECK Graph $-\sqrt{3}$ and $\frac{\pi}{2}$. $\qquad\qquad\qquad$ *Answer:*

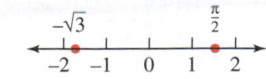

The Real Numbers

The set of rational numbers together with the set of irrational numbers form the set of **real numbers.** This means that every real number can be written as either a terminating; a repeating; or a nonterminating, nonrepeating decimal. Thus, the set of real numbers is the set of all decimals. All of the points on a number line represent the set of real numbers.

The real numbers

A **real number** is any number that is either a rational or an irrational number.

Figure 1-9 shows how the sets of numbers discussed in this section are related; it also gives some specific examples of each type of number.

FIGURE 1-9

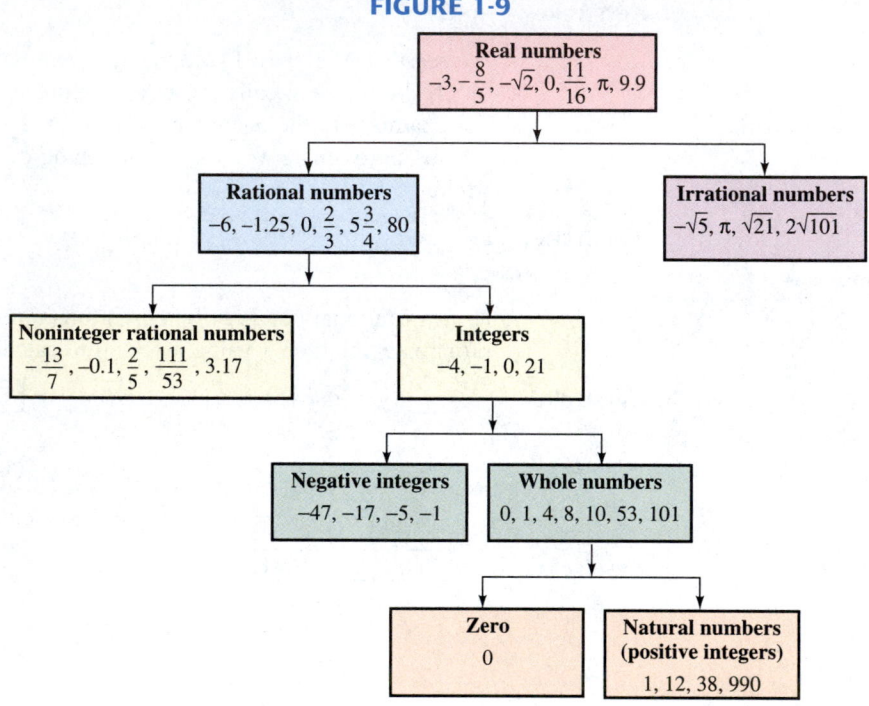

EXAMPLE 6

Classifying real numbers. Tell which numbers in the following set are natural numbers, whole numbers, integers, rational numbers, irrational numbers, and real numbers: $\left\{-3.4, \frac{12}{5}, 0, -6, 1\frac{3}{4}, -\pi, 16\right\}$.

Solution

Natural numbers: 16

Whole numbers: 0, 16

Integers: 0, −6, 16

Rational numbers: $-3.4, \frac{12}{5}, 0, -6, 1\frac{3}{4}, 16$

Irrational numbers: $-\pi$

Real numbers: $-3.4, \frac{12}{5}, 0, -6, 1\frac{3}{4}, -\pi, 16$

SELF CHECK

Use the instructions for Example 6 with the set $\left\{0.4, -\sqrt{2}, -\frac{2}{7}, 45, -2, \frac{13}{4}\right\}$

Answers: natural numbers: 45; whole numbers: 45; integers: 45, −2; rational numbers: 0.4, $-\frac{2}{7}$, 45, −2, $\frac{13}{4}$; irrational numbers: $-\sqrt{2}$; real numbers: 0.4, $-\sqrt{2}$, $-\frac{2}{7}$, 45, −2, $\frac{13}{4}$ ■

Opposites

In Figure 1-10, we see that −4 and 4 are both a distance of 4 away from 0. Because of this, we say that −4 and 4 are **opposites** or **additive inverses.**

FIGURE 1-10

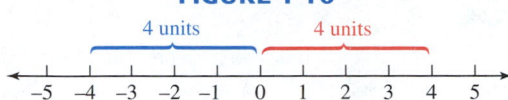

Opposites

Two numbers represented by points on a number line that are the same distance away from the origin, but on opposite sides of it, are called **opposites** or **additive inverses**.

To write the opposite of a number, a − symbol is used. For example, the opposite, or additive inverse, of 6 can be written as −6. The opposite of 0 is 0, so −0 = 0. Since the opposite of −6 is 6, we have −(−6) = 6. In general, if a represents any real number; then

$$-(-a) = a$$

Absolute Value

The **absolute value** of a number gives the distance between the number and 0 on a number line. To indicate absolute value, a number is inserted between two vertical bars. For the example shown in Figure 1-10, we would write $|-4| = 4$. This notation is read as "The absolute value of negative 4 is 4," and it tells us that the distance between −4 and 0 is 4 units. In Figure 1-10, we also see that $|4| = 4$.

Absolute value

The **absolute value** of a number is the distance on a number line between the number and 0.

 WARNING! Absolute value expresses distance. The absolute value of a number is always positive or zero—never negative.

EXAMPLE 7

Evaluating absolute values. Evaluate each absolute value: **a.** $|-18|$, **b.** $\left|-\frac{7}{8}\right|$, **c.** $-|-\pi|$.

Solution **a.** Since −18 is a distance of 18 from 0 on the number line,

$$|-18| = 18$$

b. Since $-\frac{7}{8}$ is a distance of $\frac{7}{8}$ from 0 on the number line,

$$\left|-\frac{7}{8}\right| = \frac{7}{8}$$

c. Since $-\pi$ is a distance of π from 0 on the number line, $|-\pi| = \pi$. Therefore, $-|-\pi| = -(\pi) = -\pi$.

SELF CHECK Evaluate each absolute value: **a.** $|100|$, **b.** $\left|-\frac{1}{2}\right|$, and **c.** $-|-\sqrt{2}|$.

Answers: **a.** 100, **b.** $\frac{1}{2}$, **c.** $-\sqrt{2}$

EXAMPLE 8

Comparing real numbers. Insert one of the symbols $>$, $<$, or $=$ in the blank to make a true statement: **a.** $-(-3.9)$ ▨ 3, **b.** $-\left|-\frac{4}{5}\right|$ ▨ $\left|\sqrt{5}\right|$.

Solution
a. $-(-3.9) > 3$, because $-(-3.9) = 3.9$ and $3.9 > 3$.
b. $-\left|-\frac{4}{5}\right| < \left|\sqrt{5}\right|$, because $-\left|-\frac{4}{5}\right| = -\frac{4}{5}$, $\left|\sqrt{5}\right| = \sqrt{5} \approx 2.2$, and $-\frac{4}{5} < 2.2$.

SELF CHECK

Insert one of the symbols $>$, $<$, or $=$ in the blank to make a true statement:
a. $-(-7)$ ▨ 12 and **b.** $3\frac{3}{4}$ ▨ $\left|-\frac{5}{4}\right|$. *Answers:* **a.** $<$, **b.** $>$ ■

STUDY SET

Section 1.2

VOCABULARY

In Exercises 1–12, fill in the blanks to make the statements true.

1. The set of _____ numbers is $\{0, 1, 2, 3, 4, 5, \ldots\}$.

2. The set of _____ numbers is $\{1, 2, 3, 4, 5, \ldots\}$.

3. Numbers less than zero are _____, and numbers greater than zero are _____.

4. The set of _____ numbers is the set of all decimals.

5. A _____ number can be written as a fraction with an _____ numerator and a nonzero integer denominator.

6. A decimal such as 0.25 is called a _____ decimal, while 0.333. . . is called a _____ decimal.

7. The set of _____ is $\{\ldots, -2, -1, 0, 1, 2, \ldots\}$.

8. The symbols $<$ and $>$ are _____ symbols.

9. An _____ number cannot be written as a fraction with an integer for its numerator and denominator.

10. If the numerator of a fraction is less than the denominator, it is called a _____ fraction.

11. The _____ of a number is the distance on a number line between the number and 0.

12. Two numbers represented by points on a number line that are the same distance away from the origin, but on opposite sides of it, are called _____ or _____.

CONCEPTS

13. Show that each of the following numbers is a rational number by expressing it as a fraction with an integer in its numerator and a nonzero integer in its denominator: 6, -9, $-\frac{7}{8}$, $3\frac{1}{2}$, -0.3, 2.83.

14. Represent each situation using a signed number.
    a. A trade deficit of \$15 million
    b. A rainfall total 0.75 inch below average
    c. A score $12\frac{1}{2}$ points under the standard
    d. A building foundation $\frac{5}{16}$ inch above grade

15. What two numbers are a distance of 8 away from 5 on the number line?

16. Suppose the variable m stands for a negative number. Use an inequality to state this fact.

17. The variables a and b represent real numbers. Use an inequality symbol, $<$ or $>$, to make each statement true.

    a.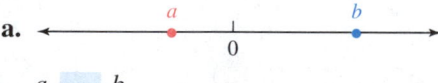
    a ▨ b

    b.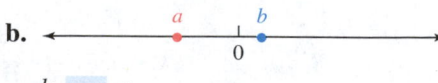
    b ▨ a

    c.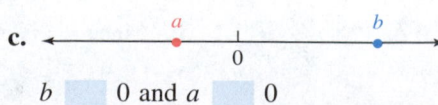
    b ▨ 0 and a ▨ 0

18. Each year from 1990–1997, the United States imported more goods and services from Japan than it exported to Japan. This caused trade *deficits*, which can be represented by negative numbers. See Illustration 1.

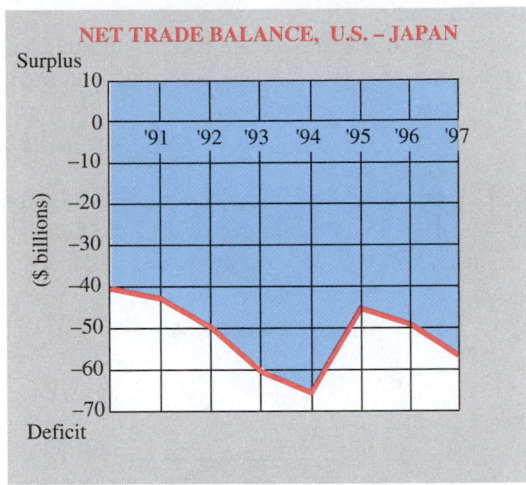

ILLUSTRATION 1

NET TRADE BALANCE, U.S. – JAPAN

Based on data from the Department of Commerce and the Bureau of Economic Analysis

a. In what year was the deficit the worst? Estimate the deficit then.

b. In what year was the deficit the smallest? Estimate the deficit then.

19. The diagram in Illustration 2 can be used to show how the natural numbers, whole numbers, integers, rational numbers, and irrational numbers make up the set of real numbers. If the natural numbers are represented as shown, label each of the other sets.

ILLUSTRATION 2

Real Numbers

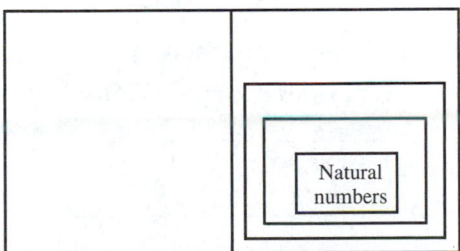

20. Which number graphed below has the largest absolute value?

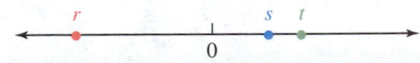

NOTATION

21. $\sqrt{5}$ is read "the _____ of 5."

22. $|-15|$ is read "the _____ of -15."

23. The symbol $\approx$ means _____.

24. The symbols { }, called _____, are used when writing a set.

25. In $\frac{3}{4}$, 3 is the _____, and 4 is the _____ of the fraction.

26. The grouping symbols ( ) are called _____.

27. Explain what 4π means, and then use a calculator to approximate it to the nearest tenth.

28. Explain what $2\sqrt{3}$ means, and then use a calculator to approximate it to the nearest tenth.

In Exercises 29–44, insert one of the symbols $>$, $<$, or $=$ in the blank to make each statement true.

29. -2 ___ -3

30. 0 ___ 32

31. $|3.4|$ ___ $\sqrt{10}$

32. 0.08 ___ 0.079

33. $-|-1.1|$ ___ -1

34. $-(-5.5)$ ___ $-\left(-5\frac{1}{2}\right)$

35. $-\left(-\frac{5}{8}\right)$ ___ $-\left(-\frac{3}{8}\right)$

36. $-19\frac{2}{3}$ ___ $-19\frac{1}{3}$

37. $\left|-\frac{15}{2}\right|$ ___ 7.5

38. $\sqrt{39}$ ___ 3π

39. $\frac{99}{100}$ ___ 0.99

40. $|2|$ ___ $-|-2|$

41. $0.333\ldots$ ___ 0.3

42. $\left|-2\frac{2}{3}\right|$ ___ $-\left(-\frac{3}{2}\right)$

43. $-(-1)$ ___ $\left|-\frac{15}{16}\right|$

44. $-0.666\ldots$ ___ 0

PRACTICE

In Exercises 45–46, tell which numbers in the given set are natural numbers, whole numbers, integers, rational numbers, irrational numbers, and real numbers.

45. $\left\{-\dfrac{5}{6},\ 35.99,\ 0,\ 4\dfrac{3}{8},\ \sqrt{2},\ -50,\ \dfrac{17}{5}\right\}$

46. $\left\{-0.001,\ 10\dfrac{1}{2},\ 6,\ 3\pi,\ \sqrt{7},\ -23,\ -5.6\right\}$

In Exercises 47–48, tell whether each statement is true or false.

47. a. Every whole number is an integer.
b. Every integer is a natural number.
c. Every integer is a whole number.
d. Irrational numbers are nonterminating, nonrepeating decimals.

48. a. Irrational numbers are real numbers.
b. Every whole number is a rational number.
c. Every rational number can be written as a fraction.
d. Every rational number is a whole number.

49. a. Write the statement $-6 < -5$ using an inequality symbol that points in the other direction.

b. Write the statement $16 > -25$ using an inequality symbol that points in the other direction.

50. If we begin with the number -4 and find its opposite, and then find the opposite of that result, what number do we obtain?

In Exercises 51–52, graph each set of numbers on the number line.

51. $\left\{-\pi, 4.25, -1\frac{1}{2}, -0.333. . . , \sqrt{2}, -\frac{35}{8}\right\}$

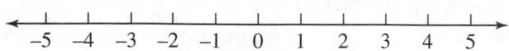

52. $\left\{\pi, -2\frac{1}{8}, 2.75, -\sqrt{17}, \frac{17}{4}, -0.666. . .\right\}$

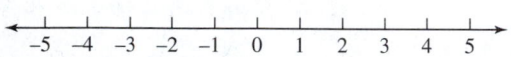

In Exercises 53–60, use a calculator to approximate each irrational number to the nearest thousandth.

53. $\sqrt{5}$ **54.** $\sqrt{19}$

55. $\sqrt{99}$ **56.** $\sqrt{42}$

57. $2\sqrt{5}$ **58.** $\dfrac{\sqrt{3}}{2}$

59. 5π **60.** $\dfrac{\pi}{4}$

In Exercises 61–68, write each expression in simpler form.

61. The opposite of 5   **62.** The opposite of -9

63. The opposite of $-\dfrac{7}{8}$ **64.** The opposite of 6.56

65. $-(-10)$ **66.** $-(-1)$

67. $-(-2.3)$ **68.** $-\left(-\frac{3}{4}\right)$

APPLICATIONS

69. BANKING Later in this course, we will use a table such as the one in Illustration 3 to solve banking problems. Which numbers shown here are natural numbers, whole numbers, integers, rational numbers, irrational numbers, and real numbers?

ILLUSTRATION 3

Type of account	Principal	Rate	Time (years)	Interest
Checking	$135.75	0.0275	$\frac{31}{365}$	$0.32
Savings	$5,000	0.06	$2\frac{1}{2}$	$750

70. DRAFTING The drawing in Illustration 4 shows the dimensions of an aluminum bracket.
 a. Which numbers shown are natural numbers, whole numbers, integers, rational numbers, irrational numbers, and real numbers?

 b. 🔲 Use a calculator to approximate all the irrational numbers in the drawing to the nearest thousandth.

ILLUSTRATION 4

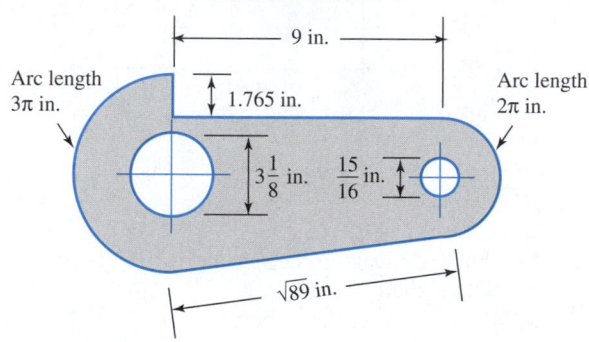

71. AUTOMOBILE INDUSTRY See Illustration 5.
 a. Estimate the net income ($ billions) for the Chrysler Corporation for each of the years from 1990 to 1997.

 b. What does a negative net income indicate?

ILLUSTRATION 5
Chrysler Corporation

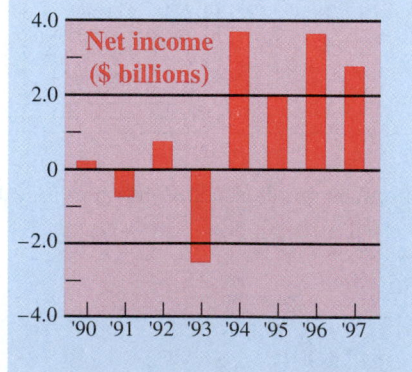

Based on data from *Business Week* and the American Automobile Manufacturers Association

72. GOVERNMENT DEBT A budget *deficit* indicates that the government's outlays (expenditures) were more than the receipts (revenue) it took in that year. See Illustration 6.
 a. For the years 1980–1998, when was the federal budget deficit the worst? Estimate the size of the deficit?

b. The 1998 budget *surplus* was the first in nearly three decades. Estimate it. Explain what it means to have a budget surplus.

ILLUSTRATION 6
Federal Budget Deficit/Surplus

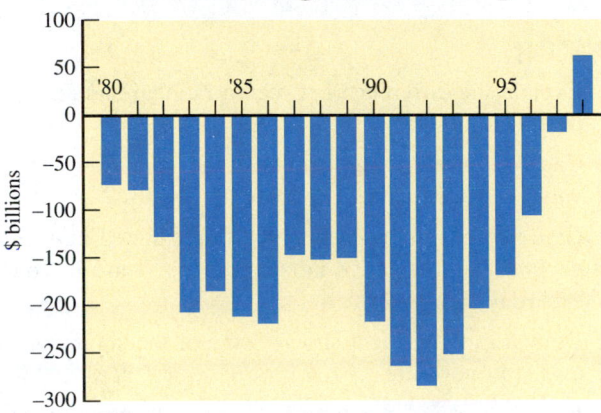

Based on data from U.S. Census Bureau and the *Los Angeles Times* (Oct. 1, 1998)

73. TIRES The distance a tire rolls in one revolution can be found by computing the circumference of the circular tire using the formula $C = \pi d$, where d is the diameter of the tire. How far will the tire shown in Illustration 7 roll in one revolution? Answer to the nearest tenth of an inch.

ILLUSTRATION 7

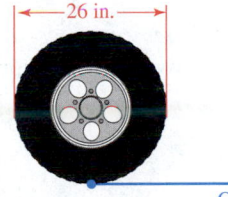

26 in.

One revolution

74. HULA HOOP The length of plastic pipe needed to form a hula hoop can be found by computing the circumference of the circular hula hoop using the formula $C = \pi d$, where d is its diameter. Find the length of pipe needed to form the hula hoop shown in Illustration 8. Answer to the nearest tenth of an inch.

ILLUSTRATION 8

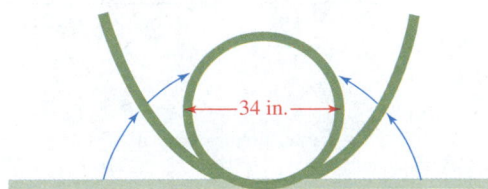

34 in.

75. FROZEN FOODS In freezing some baked goods for cold storage, the temperature of the product is

lowered, as shown in Illustration 9. Use the data to draw a line graph.

ILLUSTRATION 9

Time since removed from oven (min)	Temperature of product (° Celsius)
5	50
10	10
15	0
20	−5
25	−10

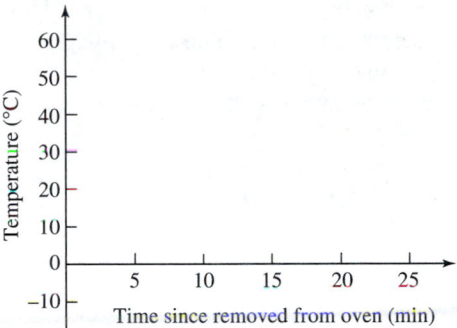

76. APPLE COMPUTER Use the data given in Illustration 10 to draw a bar graph of the company's net income for each quarter of 1997 and the first three quarters of 1998.

ILLUSTRATION 10

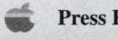 **Press Release—July 15, 1998**

Cupertino, California— Apple Computer, Inc. today announced a profit of $101 million for the fiscal 1998 third quarter that ended June 26. With a 1998 first-quarter profit of $47 million and a second-quarter profit of $55 million, Apple has now rebounded from a difficult 1997 in which the company experienced four consecutive quarters with losses of $120 million, $708 million, $56 million, and $161 million, respectively.

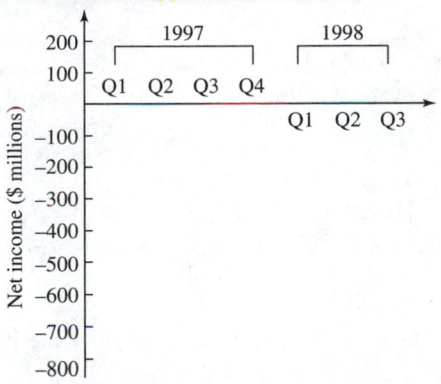

77. VELOCITY AND SPEED In science, a positive velocity normally indicates movement to the right, and a negative velocity indicates movement to the left. The speed of an object is the absolute value of its velocity. Find the speed of the car and the motorcycle in Illustration 11.

ILLUSTRATION 11

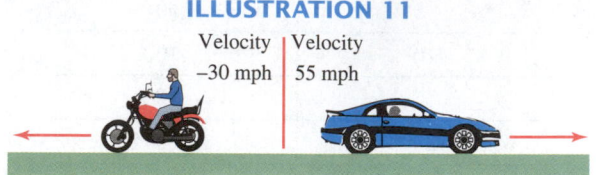

Velocity | Velocity
−30 mph | 55 mph

78. TARGET PRACTICE In Illustration 12, which artillery shell landed farther from the target? How can the concept of absolute value be applied to answer this question?

ILLUSTRATION 12

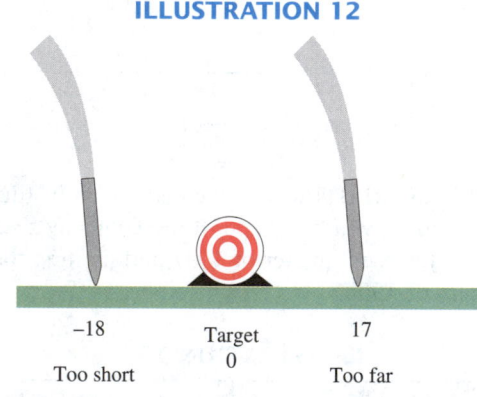

−18   17
Too short   Too far
      Target
        0

In Exercises 79–82, refer to the historical time line in Illustration 13.

ILLUSTRATION 13
MAYA CIVILIZATION

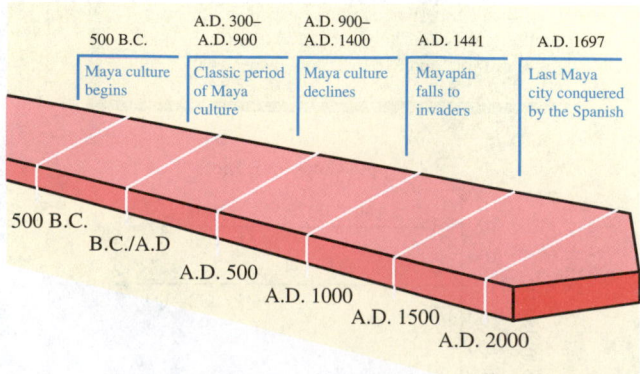

Based on data from *People in Time and Place, Western Hemisphere* (Silver Burdett & Ginn Inc., 1991), p. 129

79. What basic unit was used to scale the time line?

80. On the time line, what symbolism was used to represent zero?

81. On the time line, what could be thought of as positive and what could be thought of as negative numbers?

82. Express the dates for the Maya civilization using positive and negative numbers.

WRITING

83. Explain the difference between a rational and an irrational number.

84. Can two different numbers have the same absolute value? Explain.

85. Give two examples each of fractions, mixed numbers, decimals, and negative numbers that you use in your everyday life.

86. Draw a time line that shows some of the significant events in your life.

87. In some parts of the country, there is negative unemployment in the computer information systems field. Explain what is meant by a −5% unemployment rate.

88. In writing courses, students are warned not to use double negatives in their compositions. Identify the double negative in the following sentence. Then rewrite the sentence so that it conveys the same idea without using a double negative. "No one didn't turn in the homework."

REVIEW

In Exercises 89–92, express each statement in words, using one of the words sum, difference, product, *and* quotient.

89. $7 - 5 = 2$

90. $5(6) = 30$

91. $30 \div 15 = 2$

92. $12 + 12 = 24$

In Exercises 93–94, use the given equation to complete the table.

93. $T = 15g$

Number of gears	Number of teeth
10	
12	
15	

94. $p = r - 200$

Revenue	Profit
1,000	
5,000	
10,500	

▶ 1.3

Exponents and Order of Operations

In this section, you will learn about

> Exponents ■ Order of operations ■ Grouping symbols
> ■ Applications

Introduction In this course, we will perform six operations with real numbers: addition, subtraction, multiplication, division, raising to a power, and finding a root. Quite often, we will have to **evaluate** (find the value of) expressions containing more than one operation. In that case, we need to know the order in which the operations are to be performed. That is the focus of this section.

Exponents

The multiplication statement $3 \cdot 5 = 15$ has two parts: the two numbers that are being multiplied, and the answer. The answer (15) is called the **product,** and the numbers that are being multiplied (3 and 5) are called **factors.**

In the expression $3 \cdot 3 \cdot 3 \cdot 3 \cdot 3$, the number 3 is used as a factor five times. We call 3 a *repeated factor.* To express a repeated factor, we can use an **exponent.**

Exponent and base

> An **exponent** is used to indicate repeated multiplication. It tells how many times the **base** is used as a factor.

The exponent is 5.
$$\underbrace{3 \cdot 3 \cdot 3 \cdot 3 \cdot 3}_{\text{Five repeated factors of 3.}} = 3^5$$
The base is 3.

In the **exponential expression** a^n, a is the base, and n is the exponent. The expression a^n is called a **power of a.** Some examples of powers are

5^2 Read as "5 to the second power" or "5 squared." Here, $a = 5$ and $n = 2$.

9^3 Read as "9 to the third power" or "9 cubed." Here, $a = 9$ and $n = 3$.

$(-2)^5$ Read as "−2 to the fifth power." Here, $a = -2$ and $n = 5$.

EXAMPLE 1

Repeated factors. Write each expression using exponents.

a. $4 \cdot 4 \cdot 4$

b. $8 \cdot 8 \cdot 15 \cdot 15 \cdot 15 \cdot 15$

c. Sixteen cubed

Solution

a. In $4 \cdot 4 \cdot 4$, the 4 is repeated as a factor 3 times. So $4 \cdot 4 \cdot 4 = 4^3$.

b. $8 \cdot 8 \cdot 15 \cdot 15 \cdot 15 \cdot 15 = 8^2 \cdot 15^4$

c. 16^3

SELF CHECK

SELF CHECK   Write each expression using exponents:
a. (12)(12)(12)(12)(12)(12),   **b.** $2 \cdot 9 \cdot 9 \cdot 9$,   *Answers:*   **a.** 12^6,   **b.** $2 \cdot 9^3$,
and   **c.** fifty squared.   **c.** 50^2 ∎

In the next example, we use exponents to rewrite expressions involving repeated variable factors.

E X A M P L E   2   **Using exponents with variables.** Write each product using exponents.

a. $(a)(a)(a)(a)(a)(a)$

b. $4 \cdot \pi \cdot r \cdot r$

Solution   **a.** $(a)(a)(a)(a)(a)(a) = a^6$ *a* is repeated as a factor 6 times.

b. $4 \cdot \pi \cdot r \cdot r = 4\pi r^2$ *r* is repeated as a factor 2 times.

SELF CHECK   Write each product using exponents:
a. $y \cdot y \cdot y \cdot y$ and   **b.** $12 \cdot b \cdot b \cdot b \cdot c$.   *Answers:*   **a.** y^4,   **b.** $12b^3 c$ ∎

E X A M P L E   3   **Evaluating exponential expressions.** Find the value of each of the following: **a.** 3^2,   **b.** 5^3,   **c.** 10^1,   **d.** 2^5.

Solution   We write the base as a factor the number of times indicated by the exponent and then do the multiplication.

a. $3^2 = 3 \cdot 3 = 9$ The base is 3, the exponent is 2.

b. $5^3 = 5 \cdot 5 \cdot 5 = 125$ The base is 5, the exponent is 3.

c. $10^1 = 10$ The base is 10, the exponent is 1.

d. $2^5 = 2 \cdot 2 \cdot 2 \cdot 2 \cdot 2 = 32$ The base is 2, the exponent is 5.

SELF CHECK   Which of the numbers 3^4, 4^3, and 5^2 is the largest?   *Answer:* $3^4 = 81$ ∎

ACCENT ON TECHNOLOGY *The Squaring Key*

A plastic tarp, used to cover the infield of a baseball field when it rains, is in the shape of a square with sides 125 feet long. To find the area covered by the tarp, we can use the formula for the area of a square.

$A = s^2$ The formula for the area of a square with side length *s*.

$A = \mathbf{125}^2$ Substitute 125 for *s*.

We can use the squaring key $\boxed{x^2}$ on a scientific calculator to find the square of a number. To find 125^2, we enter these numbers and press these keys:

Keystrokes   125 $\boxed{x^2}$ $\boxed{ 15625}$

Using a graphing calculator, we can find 125^2 with the following keystrokes:

Keystrokes   125 $\boxed{x^2}$ $\boxed{\text{ENTER}}$ $\boxed{\begin{array}{r} 125^2 \\ 15625 \end{array}}$

The plastic tarp covers an area of 15,625 square feet (ft^2).

ACCENT ON TECHNOLOGY   *The Exponential Key*

A store owner sent two friends a letter advertising her store's low prices. The ad closed with the following request: "Please send a copy of this letter to two of your friends." If all those receiving letters respond, how many letters will be circulated in the 10th level of the mailing?

Table 1-2 shows how a pattern develops. On the first level, 2 letters are mailed by the owner to her two friends. On the second level, the two friends mail out 2 letters each, for a total of 4 (or 2^2) letters. On the third level, those four people mail two letters each, for a total of 8 (or 2^3) letters. Therefore, the tenth level will mail out 2^{10} letters. We can use the exponential key $\boxed{y^x}$ (on some calculators, it is labeled x^y) to raise a number to a power. To find 2^{10} using a scientific calculator, we enter these numbers and press these keys:

TABLE 1-2

Level	Number of letters circulated
1st	$2 = 2^1$
2nd	$4 = 2^2$
3rd	$8 = 2^3$
10th	$? = 2^{10}$

Keystrokes $2 \boxed{y^x} 10 \boxed{=}$ $\boxed{ 1024}$

To evaluate 2^{10} using a graphing calculator, we press the following keys:

Keystrokes $2 \boxed{\wedge} 10 \boxed{\text{ENTER}}$ $\boxed{\begin{array}{l} 2\text{^}10 \\ 1024 \end{array}}$

On the 10th level of this chain letter mailing, $2^{10} = 1{,}024$ letters will be circulated.

Order of Operations

Suppose you have been asked to contact a friend if you see a certain type of oriental rug for sale while you are traveling in Turkey. While in Turkey, you spot the rug and send the following E-mail message.

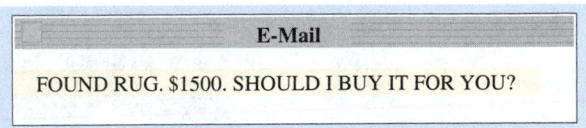

E-Mail
FOUND RUG. $1500. SHOULD I BUY IT FOR YOU?

The next day, you get this response from your friend.

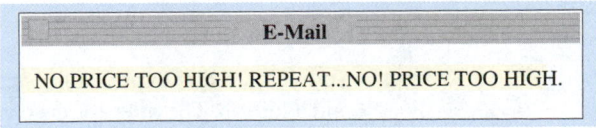

E-Mail
NO PRICE TOO HIGH! REPEAT...NO! PRICE TOO HIGH.

Something is wrong. One part of the response says to buy the rug at any price. The other part of the response says not to buy it, because it's too expensive. The placement of the exclamation point makes us read the two parts of the response differently, resulting in different interpretations. When reading a mathematical statement, the same kind of confusion is possible. For example, consider the expression

$$2 + 3 \cdot 6$$

This expression contains two operations: addition and multiplication. We can do the calculations in two ways. We can do the addition first and then do the multiplication. Or we can do the multiplication first and then do the addition. However, we get different results.

Method 1: Add first		*Method 2: Multiply first*	
$2 + 3 \cdot 6 = 5 \cdot 6$	Add 2 and 3 first.	$2 + 3 \cdot 6 = 2 + 18$	Multiply 3 and 6 first.
$= 30$	Multiply 5 and 6.	$= 20$	Add 2 and 18.

If we don't establish a uniform order of operations, the expression $2 + 3 \cdot 6$ has two different answers. To avoid this possibility, we always use the following set of priority rules.

Order of operations

1. Evaluate all exponential expressions.
2. Do all multiplications and divisions, working from left to right.
3. Do all additions and subtractions, working from left to right.

It may not be necessary to apply all of these steps in every problem. For example, the expression $2 + 3 \cdot 6$ does not contain any exponential expressions. So we next look for multiplications and divisions to perform. To correctly evaluate $2 + 3 \cdot 6$, we apply the rules for the order of operations:

$$2 + 3 \cdot 6 = 2 + 18 \quad \text{Do the multiplication first: } 3 \cdot 6 = 18.$$
$$= 20 \quad \text{Do the addition.}$$

Therefore, the correct result when evaluating $2 + 3 \cdot 6$ is 20.

EXAMPLE 4

Order of operations. Evaluate $3 \cdot 2^3 - 4$.

Solution   To find the value of this expression, we must do the operations of multiplication, raising to a power, and subtraction. The rules for the order of operations tell us to begin by evaluating the exponential expression.

$$3 \cdot 2^3 - 4 = 3 \cdot 8 - 4 \quad \text{Evaluate the exponential expression: } 2^3 = 8.$$
$$= 24 - 4 \quad \text{Do the multiplication: } 3 \cdot 8 = 24.$$
$$= 20 \quad \text{Do the subtraction.}$$

SELF CHECK   Evaluate $2 \cdot 3^2 + 17$.   *Answer:* 35   ■

EXAMPLE 5

Order of operations. Evaluate $30 - 4 \cdot 5 + 9$.

Solution   To evaluate this expression, we must do the operations of subtraction, multiplication, and addition. The rules for the order of operations tell us to begin with the multiplication.

$$30 - 4 \cdot 5 + 9 = 30 - 20 + 9 \quad \text{Do the multiplication: } 4 \cdot 5 = 20.$$
$$= 10 + 9 \quad \text{Working from left to right, do the subtraction:}$$
$$30 - 20 = 10.$$
$$= 19 \quad \text{Do the addition.}$$

SELF CHECK   Evaluate $40 - 9 \cdot 4 + 10$.   *Answer:* 14   ■

EXAMPLE 6   **Order of operations.** Evaluate $\dfrac{160}{4} - 6(2)3$.

Solution   This expression contains the operations of division, subtraction, and multiplication. Working from left to right, we begin with the division.

$$\frac{\mathbf{160}}{\mathbf{4}} - 6(2)3 = \mathbf{40} - 6(2)3 \qquad \text{Do the division: } \tfrac{160}{4} = 40.$$

$$= 40 - (12)3 \qquad \text{Do the multiplication: } 6(2) = 12.$$

$$= 40 - 36 \qquad \text{Do the multiplication: } (12)3 = 36.$$

$$= 4 \qquad \text{Do the subtraction.}$$

SELF CHECK   Evaluate $\dfrac{240}{8} - 3(2)4$.   *Answer:* 6   ■

EXAMPLE 7   **Order of operations.** Evaluate $5^5 - 3 \cdot 2^4$.

Solution   This expression contains the operations of raising to a power, subtraction, and multiplication. We begin by evaluating the exponential expressions.

$$5^5 - 3 \cdot \mathbf{2^4} = \mathbf{3{,}125} - 3 \cdot \mathbf{16} \qquad \text{Use a calculator to evaluate the exponential}$$
$$\text{expressions: } 5^5 = 3{,}125 \text{ and } 2^4 = 16.$$

$$= 3{,}125 - 48 \qquad \text{Do the multiplication: } 3 \cdot 16 = 48.$$

$$= 3{,}077 \qquad \text{Do the subtraction.}$$

SELF CHECK   Evaluate $6^4 - 4 \cdot 3^4$.   *Answer:* 972   ■

Grouping Symbols

Grouping symbols serve as mathematical punctuation marks. They help determine the order in which an expression is to be evaluated. Examples of grouping symbols are parentheses ( ), brackets [ ], braces { }, and the fraction bar —.

Order of operations when grouping symbols are present

If the expression contains grouping symbols, do all calculations within each pair of grouping symbols, working from the innermost pair to the outermost pair, in the following order:

1. Evalute all exponential expressions.

2. Do all multiplications and divisions, working from left to right.

3. Do all additions and subtractions, working from left to right.

If the expression *does not* contain grouping symbols, begin with step 1.

In a fraction, first simplify the numerator and denominator separately. Then simplify the fraction, whenever possible.

In the next example, there are two similar-looking expressions. However, because of the parentheses, we evaluate them in a different order.

EXAMPLE 8

Working with grouping symbols. Evaluate each expression:
a. $15 - 6 + 4$ and **b.** $15 - (6 + 4)$.

Solution   **a.** This expression does not contain any grouping symbols.

$$15 - 6 + 4 = 9 + 4 \quad \text{Working from left to right, we first do the subtraction:}$$
$$ 15 - 6 = 9.$$
$$ = 13 \quad \text{Do the addition.}$$

b. Since the expression contains grouping symbols, we must do the operation inside the parentheses first.

$$15 - (6 + 4) = 15 - 10 \quad \text{Do the addition inside the parentheses: } 6 + 4 = 10.$$
$$ = 5 \quad \text{Do the subtraction.}$$

SELF CHECK   Evaluate each expression: **a.** $30 - 9 + 3$ and
b. $30 - (9 + 3)$.   *Answers:* **a.** 24, **b.** 18   ■

EXAMPLE 9

Evaluating expressions containing grouping symbols. Evaluate $(6 - 3)^2$.

Solution   This expression contains parentheses. By the rules for the order of operations, we must do the operation within the parentheses first.

$$(6 - 3)^2 = 3^2 \quad \text{Do the subtraction inside the parentheses: } 6 - 3 = 3.$$
$$ = 9 \quad \text{Evaluate the exponential expression.}$$

SELF CHECK   Evaluate $(12 - 6)^3$.   *Answer:* 216   ■

EXAMPLE 10

Order of operations inside grouping symbols. Evaluate $8^2 + 2(10 - 4 \cdot 2)$.

Solution   First, we do the operations *inside* the grouping symbols in the proper order.

$$8^2 + 2(10 - 4 \cdot 2) = 8^2 + 2(10 - 8) \quad \text{Do the multiplication inside the parentheses}$$
$$ \text{first: } 4 \cdot 2 = 8.$$
$$ = 8^2 + 2(2) \quad \text{Do the subtraction inside the parentheses:}$$
$$ 10 - 8 = 2.$$
$$ = 64 + 2(2) \quad \text{Evaluate the exponential expression: } 8^2 = 64.$$
$$ = 64 + 4 \quad \text{Do the multiplication: } 2(2) = 4.$$
$$ = 68 \quad \text{Do the addition.}$$

SELF CHECK   Evaluate $7^2 - 5(8 - 3 \cdot 2)$.   *Answer:* 39   ■

EXAMPLE 11

Working with a fraction bar. Evaluate $\dfrac{3(15) - 12}{2 + 3^2}$.

Solution   We evaluate the numerator and denominator separately. Then we do the division indicated by the fraction bar.

$$\frac{3(15) - 12}{2 + 3^2} = \frac{45 - 12}{2 + 9} \quad \begin{array}{l}\text{In the numerator, do the multiplication: } 3(15) = 45.\\ \text{In the denominator, evaluate } 3^2 = 9.\end{array}$$

$$\phantom{\frac{3(15) - 12}{2 + 3^2}} = \frac{33}{11} \quad \begin{array}{l}\text{In the numerator, do the subtraction.}\\ \text{In the denominator, do the addition.}\end{array}$$

$$\phantom{\frac{3(15) - 12}{2 + 3^2}} = 3 \quad \text{Do the division.}$$

SELF CHECK Evaluate $\dfrac{5(10) + 15}{4^2 - 3}$.   *Answer:* 5 ∎

ACCENT ON TECHNOLOGY *Order of Operations and Parentheses*

Calculators have the rules for the order of operations built in. A left parenthesis key $($ and a right parenthesis key $)$ should be used when grouping symbols are needed. To evaluate $\frac{320}{20 - 16}$ with a scientific calculator, we enter these numbers and press these keys:

Keystrokes   320 $\div$ $($ 20 $-$ 16 $)$ $=$ | 80 |

To evaluate $\frac{320}{20 - 16}$ with a graphing calculator, we enter these numbers and press these keys:

Keystrokes   320 $\div$ $($ 20 $-$ 16 $)$ ENTER | 320/(20-16)
                                                             80 |

The answer is 80.

If an expression contains more than one pair of grouping symbols, we begin by working inside the innermost pair and then work to the outermost pair.

Innermost parentheses
↓   ↓
$25 + 2[13 - 3(4 - 2)]$
↑   ↑
Outermost brackets

EXAMPLE 12   **Grouping symbols inside grouping symbols.**   Evaluate $25 + 2[13 - 3(4 - 2)]$.

Solution   We begin by working inside the innermost grouping symbols, the parentheses.

$25 + 2[13 - 3(\mathbf{4 - 2})] = 25 + 2[13 - 3(\mathbf{2})]$ Do the subtraction inside the parentheses: $4 - 2 = 2$.

$= 25 + 2[13 - 6]$ Do the multiplication inside the brackets: $3(2) = 6$.

$= 25 + 2(7)$ Do the subtraction inside the brackets: $13 - 6 = 7$.

$= 25 + 14$ Do the multiplication.

$= 39$ Do the addition.

SELF CHECK   Evaluate $9 + 4[16 - 2(9 - 2)]$.   *Answer:* 17 ∎

Applications

EXAMPLE 13   **Travelers' checks.** The following table shows the number of each denomination of travelers' check contained in a booklet of 25 checks. Find the total value of the booklet.

Denomination	$20	$50	$100	$500
Number	15	5	3	2

Solution   We can find the total value by adding the values of each of the denominations. First, we describe this in words; then we translate to mathematical symbols.

Total value	=	value of all the 20's	+	value of all the 50's	+	value of all the 100's	+	value of all the 500's.
Total value	=	15(20)	+	5(50)	+	3(100)	+	2(500)

Multiply each denomination by the number of checks.

Total value = 300 + 250 + 300 + 1,000   *Do the multiplications.*

Total value = 1,850   *Do the additions.*

The total value of the booklet is $1,850.   ∎

The **arithmetic mean** (or **average**) of a set of numbers is a value around which the values of the numbers are grouped. When finding the mean, we usually need to apply the rules for the order of operations.

Finding an arithmetic mean

> To find the **mean** of a set of values, divide the sum of the values by the number of values.

E X A M P L E   1 4

Customer service. To measure its effectiveness in serving customers, a store had the telephone company electronically record the number of times the telephone rang before an employee answered it. The results of the week-long survey are shown in Table 1-3. Find the average number of times the phone rang before an employee answered it that week.

TABLE 1-3

Number of rings	Occurrences
1	11
2	46
3	45
4	28
5	20

Solution   To find the total number of rings, we multiply each of the *number of rings* (1, 2, 3, 4, and 5 rings) by the respective number of occurrences and add those subtotals.

Total number of rings = 11(1) + 46(2) + 45(3) + 28(4) + 20(5)

To find the total number of calls received, we add the number of occurrences in the right-hand column of the table.

Total number of calls received = 11 + 46 + 45 + 28 + 20

To find the average, we divide the total number of rings by the total number of calls and apply the rules for the order of operations to evaluate the expression.

$$\text{Average} = \frac{11(1) + 46(2) + 45(3) + 28(4) + 20(5)}{11 + 46 + 45 + 28 + 20}$$

$$\text{Average} = \frac{11 + 92 + 135 + 112 + 100}{150}$$ *In the numerator, do the multiplication.*
In the denominator, do the addition.

$$\text{Average} = \frac{450}{150}$$ *Do the addition.*

Average = 3   *Do the division.*

The average number of times the phone rang before it was answered was 3.   ∎

STUDY SET

Section 1.3

VOCABULARY

In Exercises 1–6, fill in the blanks to make the statements true.

1. In the multiplication statement $6 \cdot 7 = 42$, the numbers being multiplied are called _____.

2. In the exponential expression x^2, x is the _____, and 2 is the _____.

3. 10^2 can be read as ten _____, and 10^3 can be read as ten _____.

4. 7^5 is the fifth _____ of seven.

5. The arithmetic _____ or _____ of a set of numbers is a value around which the values of the numbers are grouped.

6. An _____ is used to represent repeated multiplication.

CONCEPTS

7. Given: $4 + 5 \cdot 6$.
   a. What operations does this expression contain?

   b. Evaluate the expression in two different ways and state the two possible results.
   c. Which result from part b is correct, and why?

8. a. What repeated multiplication does 5^3 represent?

   b. Write a multiplication statement where the factor x is repeated 4 times. Then write the expression in simpler form using an exponent.
   c. How can we represent the repeated *addition* $3 + 3 + 3 + 3 + 3$ in a simpler form?

9. a. How is the mean (or average) of a set of scores found?

   b. Find the average of 75, 81, 47, and 53.

10. In the expression $8 + 2[15 - (6 + 1)]$, which grouping symbols are *innermost* and which are *outermost*?

11. a. What operations does the expression $12 + 5^2 \cdot 3$ contain?
    b. In what order should they be performed?

12. a. What operations does the expression $20 - (2)^2 + 3(1)$ contain?

    b. In what order should they be performed?

13. Consider the expression $\frac{36 - 4(7)}{2(10 - 8)}$. In the numerator, what operation should be done first? In the denominator, what operation should be done first?

14. Explain the differences in evaluating $4 \cdot 2^2$ and $(4 \cdot 2)^2$.

15. To evaluate each expression, what operation should be performed first?
    a. $80 - 3 + 5 - 2^2$
    b. $80 - (3 + 5) - 2^2$
    c. $80 - 3 + (5 - 2)^2$

16. To evaluate each expression, what operation should be performed first?
    a. $(65 - 3)^3$
    b. $65 - 3^3$
    c. $6(5) - (3)^3$

NOTATION

17. Write an exponential expression with a base of 12 and an exponent of 6.

18. Tell the name of each grouping symbol: ( ), [ ], —.

In Exercises 19–22, complete each solution.

19. $50 + 6 \cdot 3^2 = 50 + 6\left(\boxed{}\right)$
    $= 50 + \boxed{}$
    $= 104$

20. $100 - (25 - 8 \cdot 2) = 100 - \left(25 - \boxed{}\right)$
    $= 100 - \boxed{}$
    $= 91$

21. $19 - 2[(1 + 2) \cdot 3] = 19 - 2\left[\left(\boxed{}\right) \cdot 3\right]$
    $= 19 - 2\left(\boxed{}\right)$
    $= 19 - \boxed{}$
    $= 1$

22. $\dfrac{46 - 2^3}{3(5) + 4} = \dfrac{46 - \boxed{}}{\boxed{} + 4}$
    $= \dfrac{\boxed{}}{\boxed{}}$
    $= 2$

PRACTICE

In Exercises 23–24, translate each word model to an equation. (Hint: You will need to use variables.)

23.

| The distance fallen by a skydiver | is | 16 | times | the square of the time he has been falling. |

24.

| The loudness of a stereo speaker | is | 2,000 | divided by | the square of the distance of the listener from the speaker. |

25. See Illustration 1. The formula for the volume of a cube is $V = s^3$. Complete the table.

ILLUSTRATION 1

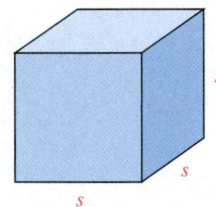

Side length of cube (in.)	Volume of cube (in.³)
1	
2	
3	
4	

26. See Illustration 2. Examine the data in the table in the right column and then write an equation that mathematically describes the relationship.

ILLUSTRATION 2

Length of a side of a painter's tarp (ft)	Area covered by tarp (ft²)
5	25
6	36
7	49
8	64

In Exercises 27–34, write each product using exponents.

27. $b \cdot b \cdot b$　　　　　　**28.** $m \cdot m \cdot m \cdot m \cdot m$

29. $10 \cdot 10 \cdot k \cdot k \cdot k$　　**30.** $5(5)(5)(i)(i)$

31. $8 \cdot \pi \cdot r \cdot r \cdot r$　　　**32.** $8 \cdot \pi \cdot r \cdot r \cdot r$

33. $6 \cdot x \cdot x \cdot y \cdot y \cdot y$　　**34.** $76 \cdot s \cdot s \cdot s \cdot s \cdot t$

In Exercises 35–84, evaluate each expression.

35. $12 - 2 \cdot 3$　　　　　**36.** $9 + 5 \cdot 3$

37. $15 + (30 - 4)$　　　**38.** $(15 + 30) - 4$

39. $100 - 8(10) + 60$　**40.** $50 - 2(5) - 7$

41. $22 - (15 - 3)$　　　**42.** $(33 - 8) - 10$

43. $2(9) - 2(5)$　　　　**44.** $75 - 7^2$

45. $5^2 + 13^2$　　　　　**46.** $3^3 - 2^3$

47. $3 \cdot 8^2$　　　　　　**48.** $(3 \cdot 4)^2$

49. $8 \cdot 5 - 4 \div 2$　　　**50.** $9 \cdot 5 - 6 \div 3$

51. $14 + 3(7 - 5)$　　　**52.** $5(10 + 2) - 1$

53. $4 + 2[26 - 5(3)]$　**54.** $64 - 6[15 - (3)3]$

55. $(10 - 3)^2$　　　　**56.** $(12 - 2)^3$

57. $10 - 3^2$　　　　　**58.** $12 - 2^3$

59. $19 - (45 - 41)^2$　**60.** $200 - (6 - 5)^3$

61. $2 + 3 \cdot 2^2 \cdot 4$　　**62.** $5 \cdot 2^2 \cdot 4 - 30$

63. $3(4)(5)(6)$　　　　**64.** $1(2)(3)(4)$

65. $5[9(2) - 2(8)]$　　**66.** $[6(5) - 5(5)]4$

67. $75 - 3 \cdot 1^2$　　　**68.** $175 - 2 \cdot 3^4$

69. $5(150 - 3^3)$　　　**70.** $6(130 - 4^3)$

71. $(4 + 2 \cdot 3)^4$　　　**72.** $(17 - 5 \cdot 2)^3$

73. $3(2)^5(2)^2$　　　　**74.** $5(2)^3(3)^2$

75. $6\left(\dfrac{25}{5}\right) - \dfrac{36}{9} + 1$　**76.** $2\left(\dfrac{15}{5}\right) - \dfrac{6}{2} + 9$

77. $\dfrac{5(68 - 32)}{9}$　　　**78.** $\dfrac{5 \cdot 50 - 160}{9}$

79. $\dfrac{(6 - 5)^4 + 21}{27 - 4^2}$　　**80.** $\dfrac{(4^3 - 10) - 4}{5^2 - 4(5)}$

81. $\dfrac{13^2 - 5^2}{3(9 - 5)}$　　　**82.** $\dfrac{72 - (2 - 2 \cdot 1)}{10^2 - (90 + 2^2)}$

83. $\dfrac{8^2 - 10}{2(3)(4) - 5(3)}$　　**84.** $\dfrac{40 - 1^3 - 2^4}{3(2 + 5) + 2}$

In Exercises 85–88, evaluate each expression using a calculator.

85. $5^7 - (45 \cdot 489)$

86. $\dfrac{3(3,246 - 1,111)}{561 - 546}$

87. $54^3 - 16^4 + 19(3)$

88. $\dfrac{36^2 - 2(48)}{25^2 - 325}$

APPLICATIONS

Write a numerical expression describing each situation and evaluate it using the rules for the order of operations. A calculator may be helpful for some problems.

89. CASH AWARDS A contest is to be part of a promotional kickoff for a new children's cereal. The prizes to be awarded are shown in Illustration 3.
 a. How much money will be awarded in the promotion?
 b. What is the average cash prize?

ILLUSTRATION 3

Coloring Contest

Grand prize: Disney World vacation plus $2,500
Four 1st place prizes of $500
Thirty-five 2nd place prizes of $150
Eighty-five 3rd place prizes of $25

90. DISCOUNT COUPONS An entertainment book contains 50 coupons good for discounts on dining, lodging, and attractions. (See Illustration 4.)
 a. Overall, how much money can a person save if all of the coupons are used?
 b. What is the average savings per coupon?

ILLUSTRATION 4

Type of coupon	Amount of discount	Number of coupons
Dining	$10	32
Lodging	$25	11
Attractions	$15	7

91. STAKES RACES One weekend, a thoroughbred race track held eight stakes races, with cash awards to the race winners of $300,000, $200,000, $175,000, $300,000, $200,000, $175,000, $150,000, and $150,000. What was the average stakes award that weekend?

92. AUTO INSURANCE See the premium comparison in Illustration 5. What is the average six-month insurance premium?

ILLUSTRATION 5

Allstate	$2,672	Mercury	$1,370
Auto Club	$1,680	State Farm	$2,737
Farmers	$2,485	20th Century	$1,692

Criteria: Six-month premium. Husband, 45, drives a 1995 Explorer, 12,000 annual miles. Wife, 43, drives a 1996 Dodge Caravan, 12,000 annual miles. Son, 17, is an occasional operator. All have clean driving records.

93. SPREADSHEETS The spreadsheet in Illustration 6 contains data collected by a chemist. For each row, the sum of the values in columns A and B is to be subtracted from the product of 6 and the value in column C. That result is then to be divided by 12 and entered in column D. Use this information to complete the spreadsheet.

ILLUSTRATION 6

	A	B	C	D
1	20	4	8	
2	9	3	16	
3	1	5	11	

94. DOG SHOWS The final score for each dog competing in a "toy breeds" competition is computed by dividing the sum of the judges' marks, after the highest and lowest have been dropped, by 6. (See Illustration 7.)
 a. What was their order of finish?
 b. Did any judge rate all the dogs the same?

ILLUSTRATION 7

Judge	1	2	3	4	5	6	7	8
Terrier	14	11	11	10	12	12	13	13
Pekinese	10	9	8	11	11	12	9	10
Pomeranian	15	14	13	11	14	12	10	14

95. CABLE TELEVISION The number of new households that subscribed each month to a newly installed cable television system are shown in the line graph in Illustration 8.
 a. Express the number of new subscribers each month as a power of 5.
 b. If the trend continues, predict the number of new subscribers in the 5th month.

ILLUSTRATION 8

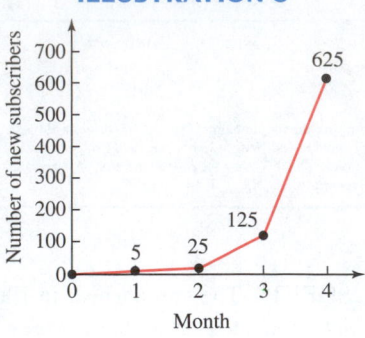

96. TURNTABLES The number of record turntables sold by a large music store declined over a six-year period, as shown in Illustration 9. Express the number of turntables sold in each of the years as a power.

ILLUSTRATION 9

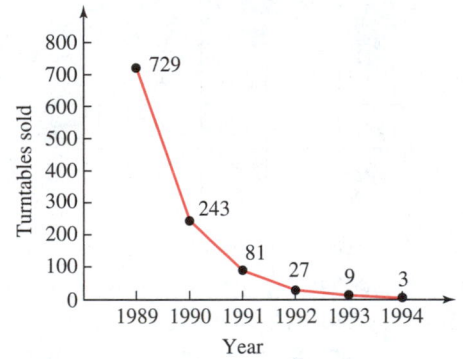

WRITING

97. Explain the difference between 2^3 and 3^2.

98. Explain why rules for the order of operations are necessary.

99. Consider the process of evaluating $25 - 3 \cdot 5$. Explain how the insertion of grouping symbols, $(25 - 3) \cdot 5$, causes the expression to be evaluated in a different order.

100. In what settings do you encounter or use the concept of arithmetic mean (average) in your everyday life?

REVIEW

101. Find $|-5|$.

102. List the set of integers.

103. True or false: Every real number can be expressed as a decimal.

104. True or false: Irrational numbers are nonterminating, nonrepeating decimals.

105. What two numbers are a distance of 6 away from -3 on the number line?

106. Graph $\left\{ -2.5, \sqrt{5}, \frac{11}{3}, -0.333\ldots, 0.75 \right\}$ on the number line.

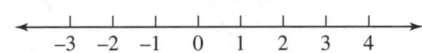

▶ **1.4**   # Algebraic Expressions

In this section, you will learn about

   Translating from English to mathematical symbols ■ Representing two unknown quantities with a variable ■ Looking for hidden operations ■ Algebraic expressions involving more than one operation ■ Number and value ■ Evaluating algebraic expressions

Introduction   When solving problems, we must often write an equation that mathematically describes the situation in question. In this section, you will see how the facts of a problem can be *translated* to mathematical symbols, which are then assembled to form algebraic expressions—the building blocks of equations.

Translating from English to Mathematical Symbols

To describe numerical relationships, we can translate the words of a problem into mathematical symbols. Below, we list some key words and phrases that are used to represent the operations of addition, subtraction, multiplication, and division and show how they can be translated to form algebraic expressions.

Addition

The phrase	translates to the algebraic expression
the sum of a and 8	$a + 8$
4 plus c	$4 + c$
16 added to m	$m + 16$
4 more than t	$t + 4$
20 greater than F	$F + 20$
T increased by r	$T + r$
Exceeds y by 35	$y + 35$

Subtraction

The phrase	translates to the algebraic expression
the difference of 23 and P	$23 - P$
550 minus h	$550 - h$
w less than 108	$108 - w$
7 decreased by j	$7 - j$
M reduced by x	$M - x$
12 subtracted from L	$L - 12$
5 less f	$5 - f$

Multiplication

The phrase	translates to the algebraic expression
the product of 4 and x	$4x$
20 times B	$20B$
twice r	$2r$
$\frac{3}{4}$ of m	$\frac{3}{4}m$

Division

The phrase	translates to the algebraic expression
the quotient of R and 19	$\dfrac{R}{19}$
s divided by d	$\dfrac{s}{d}$
the ratio of c to d	$\dfrac{c}{d}$
k split into 4 equal parts	$\dfrac{k}{4}$

EXAMPLE 1

Writing algebraic expressions. Write each phrase as an algebraic expression.

   a. The sum of the length l and the width 20

   b. b less than the capacity c

   c. The product of the weight w and 2,000

Solution **a.** **Key word:** *sum*   **Translation:** add
      The phrase translates to $l + 20$.

     b. **Key phrase:** *less than*   **Translation:** subtract
      The capacity c is to be made less, so we subtract b from it: $c - b$.

     c. **Key word:** *product*   **Translation:** multiply
      The weight w is to be multiplied by 2,000: $2,000w$.

SELF CHECK

Write each phrase as an algebraic expression:
a. The distance d divided by r,   **b.** 80 cents less than t cents,   and   **c.** $\frac{2}{3}$ of the time T.   *Answers:* **a.** $\frac{d}{r}$, **b.** $t - 80$, **c.** $\frac{2}{3}T$ ∎

Representing Two Unknown Quantities with a Variable

In the next two examples, we will use a variable to describe two unknown quantities.

EXAMPLE 2

Writing an algebraic expression. A butcher trims 4 ounces of fat from a roast that originally weighed x ounces. Write an algebraic expression that represents the weight of the roast after it is trimmed.

Solution We let x = the original weight of the roast (in ounces).

     Key word: *trimmed*   **Translation:** subtract

After 4 ounces of fat have been trimmed, the weight of the roast is $(x - 4)$ ounces.

SELF CHECK

When a secretary rides the bus to work, it takes her m minutes. If she drives her own car, her travel time exceeds this by 15 minutes. How can we represent the time it takes her to get to work by car?   *Answer:* $(m + 15)$ minutes ∎

EXAMPLE 3

Writing an algebraic expression. The swimming pool in Figure 1-11 is x feet wide. If it is to be sectioned into 8 equally wide swimming lanes, write an algebraic expression that represents the width of each lane.

FIGURE 1-11

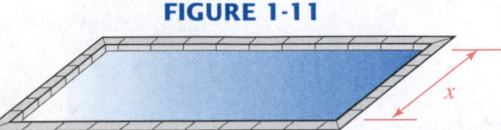

Solution Let x = the width of the swimming pool.

     Key phrase: *sectioned into 8 equally wide lanes*   **Translation:** divide

The width of each lane is $\dfrac{x}{8}$ feet.

SELF CHECK

A handyman estimates that it will take the same amount of time to sand as to paint some kitchen cabinets. If the entire job takes x hours, how can we express the time it will take him to do the painting?

Answer: $\dfrac{x}{2}$ hours ■

When solving problems, the variable to be used is rarely specified. You must decide what the unknown quantities are and how they will be represented using variables. The next example illustrates how to approach these situations.

EXAMPLE 4

Naming two unknown quantities. The value of a collectible doll is three times that of an antique toy truck. Express the value of each, using a variable.

Solution

There are two unknown quantities. Since the doll's value is related to the truck's value, we will let $x =$ the value of the toy truck.

Key phrase: *3 times*   **Translation:** multiply by 3

The value of the doll is $3x$.

SELF CHECK

The McDonald's Chicken Deluxe sandwich has 5 less grams of fat than the Quarter-Pounder hamburger. Express the grams of fat in each sandwich using a variable.

Answers: $x =$ the number of grams of fat in the hamburger, $x - 5 =$ the number of grams of fat in the chicken sandwich ■

WARNING! A variable is used to represent an unknown number. Therefore, in Example 4, it would be incorrect to write, "Let $x =$ toy truck," because the truck is not a number. We need to write, "Let $x =$ the *value* of the toy truck."

Looking for Hidden Operations

When analyzing problems, we aren't always given key words or key phrases to help establish what mathematical operation to use. Sometimes a careful reading of the problem is needed to determine the hidden operations.

EXAMPLE 5

Hidden operations. Disneyland, located in Anaheim, California, was in operation 16 years before the opening of Walt Disney World, in Orlando, Florida. Euro Disney, in Paris, France, was constructed 21 years after Disney World. Use algebraic expressions to express the ages (in years) of each of these Disney attractions.

Solution

The ages of Disneyland and Euro Disney are both related to the age of Walt Disney World. Therefore, we will let $x =$ the age of Walt Disney World.

In carefully reading the problem, we find that Disneyland was built 16 years *before* Disney World, so its age is more than that of Disney World.

Key phrase: *more than*   **Translation:** add

In years, the age of Disneyland is $x + 16$. Euro Disney was built 21 years *after* Disney World, so its age is less than that of Disney World.

Key phrase: *less than*   **Translation:** subtract

In years, the age of Euro Disney is $x - 21$. The results are summarized in Table 1-4.

TABLE 1-4

Attraction	Age
Disneyland	$x + 16$
Disney World	x
Euro Disney	$x - 21$

■

<table>
<tr><td>

EXAMPLE 6

</td><td>

Hidden operations.   **a.** How many months are in x years?   **b.** How many yards are in i inches?

</td></tr>
<tr><td>

Solution

</td><td>

a. Since there are no key words, we must decide what operation is needed to find the number of months. Each of the x years contains 12 months; this indicates multiplication. Therefore, the number of months is $12 \cdot x$, or $12x$.

b. Since there are no key words, we must decide what operation is called for. 36 inches make up one yard. We need to see how many groups of 36 inches are in i inches. This indicates division; therefore, the number of yards is $\frac{i}{36}$.

</td></tr>
<tr><td>

SELF CHECK

</td><td>

a. h hours is how many days?   **b.** How many dimes equal the value of x dollars?   *Answers:*   **a.** $\dfrac{h}{24}$,   **b.** $10x$ ■

</td></tr>
</table>

Algebraic Expressions Involving More Than One Operation

In all of the preceding examples, each algebraic expression contained only one operation. We now examine expressions involving two operations.

<table>
<tr><td>

EXAMPLE 7

</td><td>

An expression involving two operations. In the second semester, student enrollment in a retraining program at a college was 32 more than twice that of the first semester. Use a variable to express the student enrollment in the program each semester.

</td></tr>
<tr><td>

Solution

</td><td>

Since the second-semester enrollment is expressed in terms of the first-semester enrollment, we let x = the enrollment in the first semester.

> **Key phrase:** *more than*   **Translation:** add
>
> **Key word:** *twice*   **Translation:** multiply by 2

The enrollment for the second semester is $2x + 32$.

</td></tr>
<tr><td>

SELF CHECK

</td><td>

The number of votes received by the incumbent in a congressional election was 55 less than three times the challenger's vote. Use a variable to express the number of votes received by each candidate.

</td><td>

Answers: x = the number of votes received by the challenger, $3x - 55$ = the number of votes received by the incumbent   ■

</td></tr>
</table>

Number and Value

Some problems deal with quantities that have value. In these problems, we must distinguish between *the number of* and *the value of* the unknown quantity. For example, to find the value of 3 quarters, we multiply the number of quarters by the value (in cents) of one quarter. Therefore, the value of 3 quarters is $3 \cdot 25¢ = 75¢$.

The same distinction must be made if the number is unknown. For example, the value of n nickels is not $n¢$. The value of n nickels is $n \cdot 5¢ = (5n)¢$. For problems of this type, we will use the relationship

> Number $\cdot$ value = total value

<table>
<tr><td>

EXAMPLE 8

</td><td>

Number–value problems. Suppose a roll of paper towels sells for 79¢. Find the cost of   **a.** five rolls of paper towels,   **b.** x rolls of paper towels,   and   **c.** $x + 1$ rolls of paper towels.

</td></tr>
</table>

Solution In each case, we will multiply the *number* of rolls of paper towels by the *value* of one roll (79¢) to find the total cost.

a. The cost of 5 rolls of paper towels is $5 \cdot 79¢ = 395¢$, or \$3.95.

b. The cost of x rolls of paper towels is $x \cdot 79¢ = (79 \cdot x)¢ = (79x)¢$.

c. The cost of $x + 1$ rolls of paper towels is
$(x + 1) \cdot 79¢ = 79 \cdot (x + 1)¢ = 79(x + 1)¢$.

SELF CHECK Find the value of a. six \$50 savings bonds, b. t \$100 savings bonds, and c. $(x - 4)$ \$1,000 savings bonds.

Answers: a. \$300, b. \$100t, c. \$1,000$(x - 4)$ ∎

Evaluating Algebraic Expressions

To **evaluate** an algebraic expression, we replace its variable or variables with specific numbers and then apply the rules for the order of operations. Consider the algebraic expression $x^2 - 2x$. If we are told that $x = 3$, we can substitute 3 for x in $x^2 - 2x$ and evaluate the resulting expression.

$$x^2 - 2x = (3)^2 - 2(3) \quad \text{Replace } x \text{ with 3.}$$
$$= 9 - 2(3) \quad \text{Find the power: } (3)^2 = 9.$$
$$= 9 - 6 \quad \text{Do the multiplication: } 2(3) = 6.$$
$$= 3 \quad \text{Do the subtraction.}$$

It is often necessary to evaluate an algebraic expression for *several* values of its variable. When doing that, we can show the results in a **table of values.** Suppose we want to evaluate the expression $x^2 - 2x$ for $x = 2$, $x = 4$, and $x = 5$. (See Table 1-5.)

In the column headed "x," we list each value of the variable to be used in the evaluations. In the column headed "$x^2 - 2x$," we write the result of each evaluation.

TABLE 1-5

x	$x^2 - 2x$
2	0
4	8
5	15

Evaluate for $x = 2$:
$$x^2 - 2x = (2)^2 - 2(2)$$
$$= 4 - 2(2)$$
$$= 4 - 4$$
$$= 0$$

Evaluate for $x = 4$:
$$x^2 - 2x = (4)^2 - 2(4)$$
$$= 16 - 2(4)$$
$$= 16 - 8$$
$$= 8$$

Evaluate for $x = 5$:
$$x^2 - 2x = (5)^2 - 2(5)$$
$$= 25 - 2(5)$$
$$= 25 - 10$$
$$= 15$$

The two columns of a table of values are sometimes headed with the terms **input** and **output,** as shown in Table 1-6. The x-values are the "inputs" into the expression $x^2 - 2x$, and the resulting values are thought of as the "outputs."

TABLE 1-6

Input x	Output $x^2 - 2x$
2	0
4	8
5	15

EXAMPLE 9

Ballistics. If a toy rocket is shot into the air with an initial velocity of 80 feet per second, its height (in feet) after t seconds in flight is given by the algebraic expression

$$80t - 16t^2$$

How many seconds after the launch will it hit the ground?

Solution   We can substitute positive values for t, the time in flight, until we find the one that gives a height of 0. At that time, the rocket will be on the ground. We will begin by finding the height after the rocket has been in flight for 1 second ($t = 1$) and record the result in a table.

t	$80t - 16t^2$
1	64

Evaluate for $t = 1$:
$$80t - 16t^2 = 80(1) - 16(1)^2$$
$$= 64$$

After 1 second in flight, the height of the rocket is 64 feet. We continue to pick more values of t until we find out when the height is 0.

t	$80t - 16t^2$
2	96
3	96
4	64
5	0

Evaluate for $t = 2$:
$$80t - 16t^2 = 80(2) - 16(2)^2$$
$$= 96$$

Evaluate for $t = 3$:
$$80t - 16t^2 = 80(3) - 16(3)^2$$
$$= 96$$

Evaluate for $t = 4$:
$$80t - 16t^2 = 80(4) - 16(4)^2$$
$$= 64$$

Evaluate for $t = 5$:
$$80t - 16t^2 = 80(5) - 16(5)^2$$
$$= 0$$

We see that for $t = 5$, the height of the rocket is 0. Therefore, the rocket will hit the ground in 5 seconds.

SELF CHECK   In Example 9, suppose the height of the rocket was given by $112t - 16t^2$. Complete the table to find out how many seconds after launch it would hit the ground.

t	$112t - 16t^2$
1	
3	
5	
7	

Answer: 7 (the heights are 96, 192, 160, and 0)

WARNING! When replacing a variable with its numerical value, use parentheses around the replacement number to avoid possible misinterpretation. For example, when substituting 5 for x in $2x + 1$, we show the multiplication using parentheses: $2(5) + 1$. If we don't show the multiplication, we could misread the expression as $25 + 1$.

EXAMPLE 10

Evaluating algebraic expressions. If $a = 3$, $b = 3$, $c = 2$, and $d = 9$, evaluate $\dfrac{6b - a}{d - c^2}$.

Solution

The given expression contains four variables. We substitute the values for a, b, c, and d into the expression and apply the rules for the order of operations.

$$\frac{6b - a}{d - c^2} = \frac{6(3) - 3}{9 - (2)^2} \qquad \text{In the numerator, replace } b \text{ with 3 and } a \text{ with 3.}$$

In the denominator, replace d with 9 and c with 2.

$$= \frac{18 - 3}{9 - 4} \qquad \text{In the numerator, do the multiplication: } 6(3) = 18.$$

In the denominator, find the power: $(2)^2 = 4$.

$$= \frac{15}{5} \qquad \text{In the numerator, do the subtraction.}$$

In the denominator, do the subtraction.

$$= 3 \qquad \text{Do the division.}$$

SELF CHECK

If $e = 4$, $f = 10$, $g = 6$, and $h = 2$, evaluate $\dfrac{f + 4e}{h^3 - g}$.

Answer: 13

ACCENT ON TECHNOLOGY *Evaluating Algebraic Expressions*

The rotating drum of a clothes dryer is a cylinder. (See Figure 1-12.) To find the capacity of the dryer, we can find its volume by evaluating the algebraic expression $\pi r^2 h$, where r represents the radius and h represents the height (although the cylinder is lying on its side) of the drum. If we substitute 13.5 for r and 20 for h, we obtain $\pi(13.5)^2(20)$. Using a scientific calculator, we can evaluate the expression by entering these numbers and pressing these keys:

FIGURE 1-12

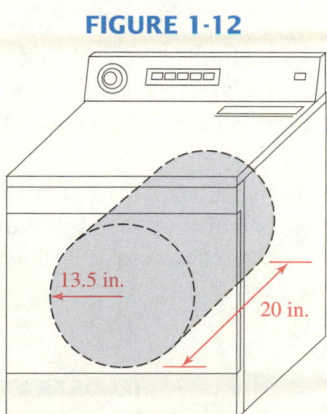

13.5 in.

20 in.

Keystrokes 13.5 $\boxed{x^2}$ $\boxed{\times}$ $\boxed{\pi}$ $\boxed{\times}$ 20 $\boxed{=}$ $\qquad$ $\boxed{\text{11451.10522}}$

Using a graphing calculator, we can evaluate the expression by entering these numbers and pressing these keys:

Keystrokes 13.5 $\boxed{x^2}$ $\boxed{\times}$ $\boxed{\text{2nd}}$ $\boxed{\pi}$ $\boxed{\times}$ 20 $\boxed{\text{ENTER}}$

$$\boxed{\begin{array}{l} 13.5^2 * \pi * 20 \\ \qquad\qquad 11451.10522 \end{array}}$$

To the nearest cubic inch, the capacity of the dryer is 11,451 in.^{3}.

VOCABULARY

In Exercises 1–4, fill in the blanks to make the statements true.

1. To _____ an algebraic expression, we substitute the values for the variables and then apply the rules for the order of operations.

2. Variables and/or numbers can be combined with the operation symbols of addition, subtraction, multiplication, and division to create algebraic _____.

3. When translated to mathematical symbols, words such as *decreased* and *reduced* indicate the operation of _____.

4. A variable is a letter that stands for a _____.

CONCEPTS

5. Write two algebraic expressions that contain the variable x and the numbers 6 and 20.

6. **a.** How many days are in w weeks?
   b. D days is how many weeks?

7. When evaluating $3x - 6$ for $x = 4$, what misunderstanding can occur if we don't write parentheses around 4 when it is substituted for the variable?

8. To find the amount of money that will accumulate in a savings account over time, we can use the algebraic expression $P + Prt$. How many variables does it contain?

9. **a.** In Illustration 1, the weight of the van is 500 pounds less than twice the weight of the car. Express the weight of the van and the car using the variable x.

ILLUSTRATION 1

   b. Suppose you learn that the actual weight of the car is 2,000 pounds. What is the weight of the van?

10. See Illustration 2.
   a. If we let b represent the length of the beam, write an algebraic expression for the length of the pipe.

   b. If we let p represent the length of the pipe, write an algebraic expression for the length of the beam.

ILLUSTRATION 2

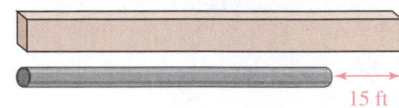

15 ft

11. **a.** In Illustration 3, what algebraic expression was evaluated?
    b. For what values of x was it evaluated?
    c. What was the value of the expression when it was evaluated for $x = 4$?

ILLUSTRATION 3

x	$8x - x^2$
3	15
4	16
5	15

12. Complete the table in Illustration 4.

ILLUSTRATION 4

Type of coin	Number	Value in cents	Total value in cents
Penny	12		
Nickel	n		
Dime	d		
Quarter	5		
Half dollar	$x + 5$		

NOTATION

In Exercises 13–14, complete each solution.

13. Evaluate the expression $9a - a^2$ for $a = 5$.
$$9a - a^2 = 9(\boxed{}) - (\boxed{})^2$$
$$= 9(5) - \boxed{}$$
$$= \boxed{} - 25$$
$$= 20$$

14. Evaluate the expression $b^2 - 4ac$ for $a = 1$, $b = 6$, and $c = 5$.
$$b^2 - 4ac = (\boxed{})^2 - 4(\boxed{})(5)$$
$$= \boxed{} - 4(1)(\boxed{})$$
$$= 36 - \boxed{}$$
$$= 16$$

PRACTICE

In Exercises 15–40, write each phrase as an algebraic expression. If no variable is given, use x as the variable.

15. The sum of the length l and 15

16. The difference of a number and 10

17. The product of a number and 50

18. Three-fourths of the population p

19. The ratio of the amount won w and lost l

20. The tax t added to c

21. P increased by p

22. 21 less than the total height h

23. The square of k minus 2,005

24. s subtracted from S

25. J reduced by 500

26. Twice the attendance a

27. 1,000 split n equal ways

28. Exceeds the cost c by 25,000

29. 90 more than the current price p

30. 64 divided by the cube of y

31. The total of 35, h, and 300

32. x decreased by 17

33. 680 fewer than the entire population p

34. Triple the number of expected participants

35. The product of d and 4, decreased by 15

36. Forty-five more than the quotient of y and 6

37. Twice the sum of 200 and t

38. The square of the quantity 14 less than x

39. The absolute value of the difference of a and 2

40. The absolute value of a, decreased by 2

In Exercises 41–44, if n represents a number, write a word description of each algebraic expression. (Answers may vary.)

41. $n - 7$ **42.** $n^2 + 7$

43. $7n + 4$ **44.** $3(n + 1)$

45. Express how many minutes there are in   **a.** 5 hours and   **b.** h hours.

46. A woman watches television x hours a day. Express the number of hours she watches TV   **a.** in a week and   **b.** in a year.

47. a. Express how many feet are in y yards.
    b. Express how many yards are in f feet.

48. If a car rental agency charges 29¢ a mile, express the rental fee if a car is driven x miles.

49. A model's skirt is x inches long. The designer then lets the hem down 2 inches. How can we express the length (in inches) of the altered skirt?

50. A soft drink manufacturer produced c cans of cola during the morning shift. Write an expression for how many six-packs of cola can be assembled from the morning shift's production.

51. The tag on a new pair of 36-inch-long jeans warns that after washing, they will shrink x inches in length. Express the length (in inches) of the jeans after they are washed.

52. A caravan of b cars, each carrying 5 people, traveled to the state capital for a political rally. Express how many people were in the car caravan.

53. A sales clerk earns $\$x$ an hour. Express how much he will earn in   **a.** an 8-hour day and   **b.** a 40-hour week.

54. A caterer always prepares food for 10 more people than the order specifies. If p people are to attend a reception, write an expression for the number of people she should prepare for.

55. Tickets to a circus cost $5 each. Express how much tickets will cost for a family of x people if they also pay for two of their neighbors.

56. If each egg is worth e¢, express the value (in cents) of a dozen eggs.

In Exercises 57–60, evaluate each algebraic expression for the given value of the variable.

57. $6x$ for $x = 7$ **58.** $\dfrac{p - 15}{30}$ for $p = 75$

59. $3(t - 6)$ for $t = 8$ **60.** $y^3 - 2y^2 - 2$ for $y = 3$

In Exercises 61–68, complete each table of values.

61.

g	$g^2 - 7g + 1$
0	
7	
10	

62.

f	$5(16 - f)^2$
7	
8	
9	

63.

s	$\dfrac{5s + 36}{s}$
1	
6	
12	

64.

a	$2,500a + a^3$
2	
4	
5	

65.

Input x	Output $2x - \frac{x}{2}$
100	
300	

66.

Input x	Output $\frac{x}{3} + \frac{x}{4}$
12	
36	

67.

Input a	Output $3a^2 + 1$
4	
8	

68.

Input b	Output $50 - 2b^2$
3	
5	

In Exercises 69–74, evaluate each algebraic expression.

69. $a^2 + b^2$ for $a = 5$ and $b = 12$

70. $\dfrac{x + y}{2}$ for $x = 8$ and $y = 10$

71. $\dfrac{s + t}{s - t}$ for $s = 23$ and $t = 21$

72. $3r^2 h$ for $r = 4$ and $h = 10$

73. $\dfrac{h(b + c)}{2}$ for $h = 5$, $b = 7$, and $c = 9$

74. $b^2 - 4ac$ for $a = 2$, $b = 8$, and $c = 4$

In Exercises 75–76, use a calculator to evaluate each expression. Round to the nearest tenth.

75. Find the volume of a basketball having a radius r of 4.5 inches by evaluating

$$\frac{4\pi r^3}{3}$$

76. Find the volume of a cone with a radius r of 3 inches and a height h of 8 inches by evaluating

$$\frac{\pi r^2 h}{3}$$

APPLICATIONS

In Exercises 77–80, use a variable to represent one unknown. Then write an algebraic expression to describe the second unknown.

77. TRAINING PROGRAM Of the original group that entered a lifeguard training program, six candidates dropped out. After graduation, those remaining were divided into eight equal-size squads for beach patrol duty. Write an expression for the number of lifeguards in each patrol squad.

78. INVESTMENTS The value of a gold coin is $350 more than twice that of a silver coin. Write an expression for the value of the gold coin.

79. OCCUPANCY RATE An apartment owner lowered the monthly rent, and soon the number of apartments rented doubled. Then water damage forced three units to be vacated. Write an expression for the number of apartments that are now occupied.

80. SCHEDULING The manager of a restaurant schedules 4 times as many workers per weekend shift as per weekday shift. If a holiday falls on a weekend, 6 additional workers are scheduled. Write an expression for the number of employees needed for a holiday weekend shift.

In Exercises 81–86, solve each problem.

81. ROCKETRY The algebraic expression $64t - 16t^2$ gives the height of a toy rocket (in feet) t seconds after being launched. Find the height of the rocket for each of the times shown in Illustration 5. Present your results in an input/output table.

ILLUSTRATION 5

t	h
0	
0.5	
1	
1.5	
2	
2.5	
3	
3.5	
4	

82. FREE FALL A tennis ball is dropped from the 1,024-foot-tall Chrysler Building in New York. The algebraic expression $1{,}024 - 16t^2$ gives the height of the ball (in feet) t seconds after it is dropped. Complete the input/output table in Illustration 6; then approximate *when* the ball would have fallen half of the height of the building.

ILLUSTRATION 6

t	$1{,}024 - 16t^2$
2	
3	
4	
5	
6	
7	
8	

83. PACKAGING The lengths, widths, and heights (in inches) of three cardboard shipping boxes are recorded in columns B, C, and D (respectively) of the spreadsheet in Illustration 7. To find the *surface area* for the box described in row 1, the computer uses the formula

SUM(2*B1*C1, 2*B1*D1, 2*C1*D1)

where 2*B1*C1 means 2 *times* the number in cell B1 *times* the number in cell C1. The result of the computation is recorded in cell E1.

a. Use this information to complete the spreadsheet.

b. What are the units of your answers?

ILLUSTRATION 7

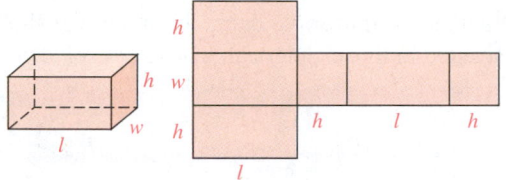

	B	C	D	E
1	12	6	6	
2	18	12	12	
3	18	24	18	

84. GRAPHIC ARTS An artist's design of a logo for a company used words to describe the relative sizes of each of the elements. See Illustration 8.

ILLUSTRATION 8

Base is 2 inches longer than side.

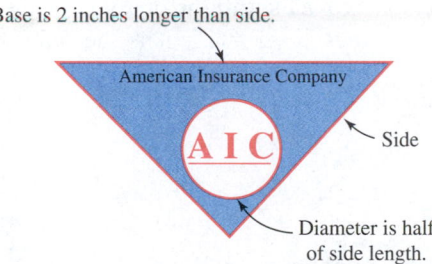

American Insurance Company

AIC

Side

Diameter is half of side length.

a. Describe the lengths of the base and sides, and the diameter, using a variable.

b. If the company decides to have the sides of the logo be 8 inches long, what would be the dimensions of the other elements of the design?

85. ENERGY CONSERVATION A fiberglass blanket wrapped around a water heater helps prevent heat loss. See Illustration 9. Find the number of square feet of heater surface the blanket covers by evaluating the algebraic expression $2\pi rh$. Round to the nearest square foot.

ILLUSTRATION 9

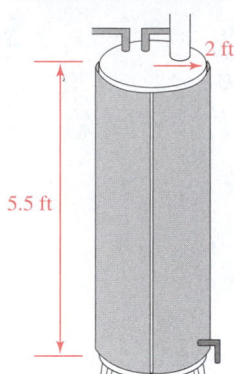

2 ft

5.5 ft

86. LANDSCAPING A grass strip is to be planted around a tree, as shown in Illustration 10. Find the number of square feet of sod to order by evaluating the expression $\pi(R^2 - r^2)$. Round to the nearest square foot.

ILLUSTRATION 10

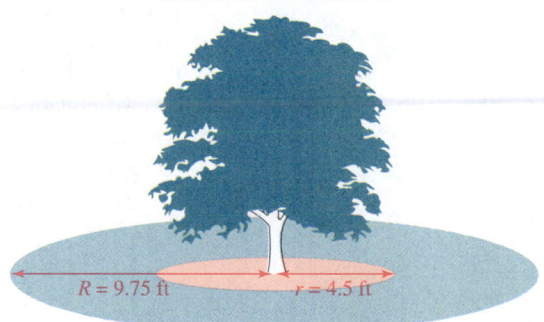

$R = 9.75$ ft $r = 4.5$ ft

WRITING

87. What is an algebraic expression? Give some examples.

88. What is a variable? How are variables used in this section?

REVIEW

89. Simplify -0.

90. Is the statement $-5 > -4$ true or false?

91. Evaluate $\left|-\frac{2}{3}\right|$.

92. Evaluate $2^3 \cdot 3^2$.

93. Write $c \cdot c \cdot c \cdot c$ in exponential form.

94. Evaluate $15 + 2[15 - (12 - 10)]$.

95. Find the mean (average) of the three test scores 84, 93, and 72.

96. Fill in the blanks to make the statement true: In the multiplication statement $5 \cdot x = 5x$, 5 and x are called _____, and $5x$ is called the _____.

▶ **1.5**

Solving Equations

In this section, you will learn about

> Equations ■ Checking solutions ■ The subtraction property of equality ■ The addition property of equality ■ The division property of equality ■ The multiplication property of equality ■ Using algebra to solve percent problems

Introduction   The idea of an equation is one of the most useful concepts in all of algebra. Equations are mathematical sentences that can be used to describe real-life situations. In this section, we will introduce some basic types of equations and discuss four fundamental properties that are used to solve them. Then we will show how some of the algebraic concepts discussed in this chapter can be used to solve problems involving percent.

Equations

Recall that an **equation** is a statement indicating that two expressions are equal. In the equation $x + 5 = 15$, the expression $x + 5$ is called the **left-hand side,** and 15 is called the **right-hand side.**

An equation can be true or false. For example, $10 + 5 = 15$ is a true equation, whereas $11 + 5 = 15$ is a false equation. An equation containing a variable can be true or false, depending upon the value of the variable. If $x = 10$, the equation $x + 5 = 15$ is true, because

$$10 + 5 = 15 \quad \text{Substitute 10 for } x.$$

However, this equation is false for all other values of x.

Any number that makes an equation true when substituted for its variable is said to **satisfy** the equation. Such numbers are called **solutions** or **roots** of the equation. Because 10 is the only number that satisfies $x + 5 = 15$, it is the only solution of the equation.

Checking Solutions

EXAMPLE 1

Checking a solution. Is 9 a solution of $3y - 1 = 2y + 7$?

Solution   We substitute 9 for y in the equation and simplify each side. If 9 is a solution, we will obtain a true statement.

$$3y - 1 = 2y + 7 \quad \text{The original equation.}$$
$$3(9) - 1 \stackrel{?}{=} 2(9) + 7 \quad \text{Substitute 9 for } y.$$
$$27 - 1 \stackrel{?}{=} 18 + 7 \quad \text{Do the multiplication.}$$
$$26 = 25 \quad \text{Do the subtraction and the addition.}$$

Since the resulting equation is not true, 9 is *not* a solution.

SELF CHECK   Is 25 a solution of $2(46 - x) = 41$?   *Answer:* no   ■

EXAMPLE 2

Checking a solution. Verify that 6 is a solution of the equation $x^2 - 5x - 6 = 0$.

Solution   We substitute 6 for x in the equation and simplify.

$$x^2 - 5x - 6 = 0 \quad \text{The original equation.}$$
$$(6)^2 - 5(6) - 6 \overset{?}{=} 0 \quad \text{Substitute 6 for } x.$$
$$36 - 30 - 6 \overset{?}{=} 0 \quad \text{Evaluate the power: } (6)^2 = 36. \text{ Do the multiplication: } 5(6) = 30.$$
$$0 = 0 \quad \text{Do the subtractions on the left-hand side.}$$

Since the resulting equation is true, 6 is a solution.

SELF CHECK   Is 8 a solution of the equation $\frac{m-4}{4} = \frac{m+4}{12}$?   *Answer:* yes   ■

The Subtraction Property of Equality

In practice, we will not be told the solutions of an equation. We will need to find the solutions (that is, solve the equation) ourselves. To develop an understanding of the procedures used to solve an equation, we will first examine $x + 2 = 5$ and make some observations as we solve it.

We can think of the scales shown in Figure 1-13(a) as representing the equation $x + 2 = 5$. The weight (in grams) on the left-hand side of the scale is $x + 2$, and the weight (in grams) on the right-hand side is 5. Because these weights are equal, the scale is in balance. To find x, we need to isolate it. That can be accomplished by removing 2 grams from the left-hand side of the scale. Common sense tells us that we must also remove 2 grams from the right-hand side if the scales are to remain in balance. In Figure 1-13(b), we can see that x grams will be balanced by 3 grams. We say that we have *solved* the equation and that the *solution* is 3.

FIGURE 1-13

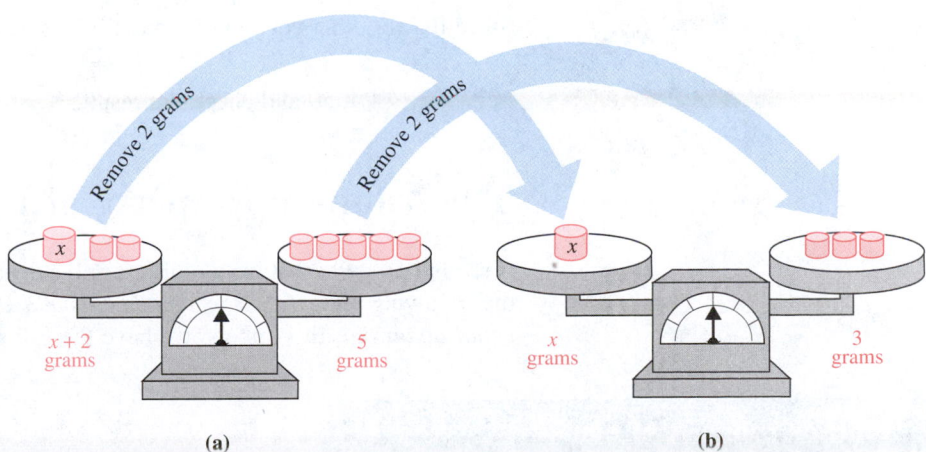

(a)   (b)

These observations suggest a property of equality: *If the same quantity is subtracted from equal quantities, the results will be equal quantities.* We can express this property, called the **subtraction property of equality**, in symbols.

Subtraction property of equality

Let a, b, and c be any real numbers.

If $a = b$, then $a - c = b - c$.

When we use this property, the resulting equation will be equivalent to the original equation.

Equivalent equations

Two equations are **equivalent** when they have the same solutions.

In the previous example, we found that $x + 2 = 5$ is equivalent to $x = 3$. This is true because both of these equations have a solution of $x = 3$.

We now show how to solve the equation using an algebraic approach.

E X A M P L E 3

Solving an equation. Solve $x + 2 = 5$ and check the result.

Solution   To isolate x on the left-hand side of the equation, we use the subtraction property of equality. We can undo the addition of 2 by subtracting 2 from both sides.

$$x + 2 = 5$$
$$x + 2 - 2 = 5 - 2 \quad \text{Subtract 2 from both sides.}$$
$$x = 3 \quad \text{Do the subtractions: } 2 - 2 = 0 \text{ and } 5 - 2 = 3.$$

We check by substituting 3 for x in the original equation and simplifying. If 3 is the solution, we will obtain a true statement.

$$x + 2 = 5 \quad \text{The original equation.}$$
$$3 + 2 \overset{?}{=} 5 \quad \text{Substitute 3 for } x.$$
$$5 = 5 \quad \text{Do the addition: } 3 + 2 = 5.$$

Since the resulting equation is true, 3 is a solution.

SELF CHECK   Solve $x + 24 = 50$ and check the result.   *Answer:* 26   ■

The Addition Property of Equality

A second property that we will use to solve equations involves addition. It is based on the following idea: *If the same quantity is added to equal quantities, the results will be equal quantities.* In symbols, we have the following property.

Addition property of equality

Let a, b, and c be any real numbers.

If $a = b$, then $a + c = b + c$.

To illustrate the addition property of equality, we can think of the scales shown in Figure 1-14(a) as representing the equation $x - 2 = 3$. To find x, we need to add 2 grams of weight to each side. The scales will remain in balance. From the scales in Figure 1-14(b), we can see that x grams will be balanced by 5 grams. Thus, $x = 5$.

We now show how to solve a similar equation using an algebraic approach.

FIGURE 1-14

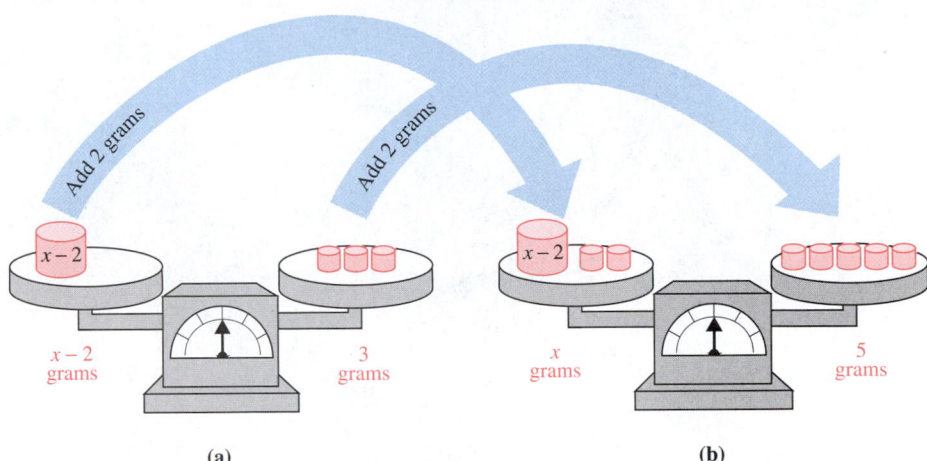

(a)   (b)

EXAMPLE 4

Solving an equation. Solve $19 = y - 7$ and check the result.

Solution   To isolate the variable y on the right-hand side, we use the addition property of equality. We can undo the subtraction of 7 by adding 7 to both sides.

$$19 = y - 7$$
$$19 + 7 = y + 7 - 7 \quad \text{Add 7 to both sides.}$$
$$26 = y \quad \text{Do the operations: } 19 + 7 = 26 \text{ and } 7 - 7 = 0.$$
$$y = 26 \quad \text{If } 26 = y, \text{ then } y = 26.$$

We check by substituting 26 for y in the original equation and simplifying.

$$19 = y - 7 \quad \text{The original equation.}$$
$$19 \overset{?}{=} 26 - 7 \quad \text{Substitute 26 for } y.$$
$$19 = 19 \quad \text{Do the subtraction: } 26 - 7 = 19.$$

This is a true statement, so 26 is a solution.

SELF CHECK   Solve $75 = b - 38$ and check the result.   *Answer:* 113   ■

The Division Property of Equality

We can think of the scales in Figure 1-15(a) as representing the equation $2x = 8$. Since $2x$ means $2 \cdot x$, the equation can be written as $2 \cdot x = 8$. The weight (in grams) on the left-hand side of the scale is $2 \cdot x$, and the weight (in grams) on the right-hand side is 8. Because these weights are equal, the scale is in balance.

To find x, we need to isolate it on the left-hand side of the scale. That can be accomplished by removing half of the weight on the left-hand side. Common sense tells us that if the scale is to remain in balance, we need to remove half of the weight from the right-hand side. We can think of the process of removing half of the weight as dividing the weight by 2. In Figure 1-15(b), we can see that x grams will be balanced by 4 grams. Thus, $x = 4$.

These observations suggest a property of equality: *If equal quantities are divided by the same nonzero quantity, the results will be equal quantities.* We can express this property, called the **division property of equality,** in symbols.

FIGURE 1-15

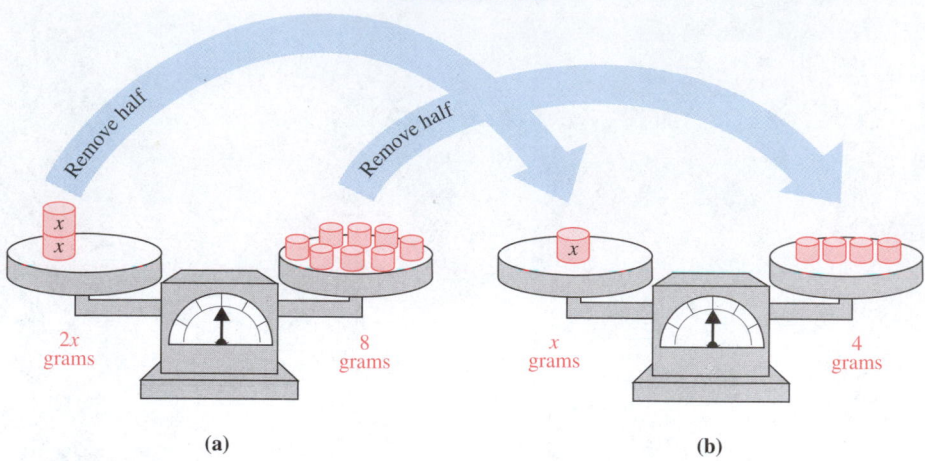

(a)   (b)

Division property of equality

Let a, b, and c be any real numbers.

If $a = b$, then $\dfrac{a}{c} = \dfrac{b}{c}$ $(c \neq 0)$

EXAMPLE 5   **Solving an equation.** Solve the equation $2x = 8$ and check the result.

Solution   Recall that $2x = 8$ means $2 \cdot x = 8$. To isolate x on the left-hand side of the equation, we use the division property of equality: We undo the multiplication by 2 by dividing both sides of the equation by 2.

$2x = 8$

$\dfrac{2x}{2} = \dfrac{8}{2}$ To undo the multiplication by 2, divide both sides by 2.

$x = 4$ Do the divisions: $\frac{2}{2} = 1$ and $\frac{8}{2} = 4$.

The solution is 4. Check it as follows:

$2x = 8$ The original equation.

$2 \cdot 4 \overset{?}{=} 8$ Substitute 4 for x.

$8 = 8$ Do the multiplication: $2 \cdot 4 = 8$.

SELF CHECK   Solve the equation $16x = 176$ and check the result.   *Answer:* 11   ∎

The Multiplication Property of Equality

Sometimes we need to multiply both sides of an equation by the same nonzero number to solve it. This concept is summed up in the multiplication property of equality. It is based on the following idea: *If equal quantities are multiplied by the same nonzero quantity, the results will be equal quantities.* In symbols, we have the following property.

Multiplication property of equality

Let a, b, and c be any real numbers.

If $a = b$, then $ca = cb$ $\quad (c \neq 0)$

To illustrate the multiplication property of equality, we can think of the scales shown in Figure 1-16(a) as representing the equation $\frac{x}{3} = 25$. The weight on the left-hand side of the scale is $\frac{x}{3}$ grams, and the weight on the right-hand side is 25 grams. Because these weights are equal, the scale is in balance. To find x, we triple (or multiply by 3) the weight on each side. The scales will remain in balance. From the scales shown in Figure 1-16(b), we can see that x grams will be balanced by 75 grams. Thus, $x = 75$.

FIGURE 1-16

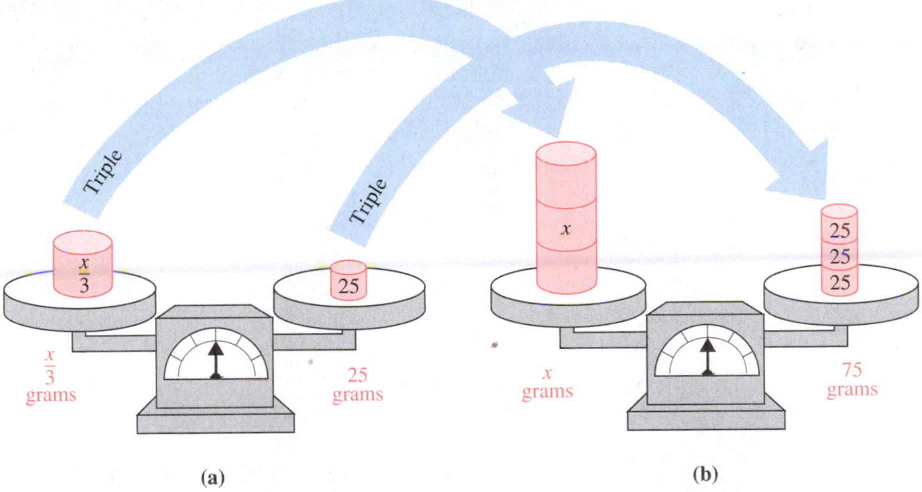

(a)   (b)

We now show how to solve a similar equation using an algebraic approach.

EXAMPLE 6

Solving an equation. Solve the equation $\frac{s}{5} = 15$ and check the result.

Solution   To isolate s on the left-hand side, we use the multiplication property of equality. We can undo the division of the variable by 5 by multiplying both sides by 5.

$$\frac{s}{5} = 15$$

$$5 \cdot \frac{s}{5} = 5 \cdot 15 \quad \text{Multiply both sides by 5.}$$

$$s = 75 \quad \text{Do the multiplication: } 5 \cdot 15 = 75.$$

Check: $\dfrac{s}{5} = 15$ $\quad$ The original equation.

$$\frac{75}{5} \overset{?}{=} 15 \quad \text{Substitute 75 for } s.$$

$$15 = 15 \quad \text{Do the division: } \tfrac{75}{5} = 15.$$

SELF CHECK   Solve the equation $\dfrac{t}{24} = 3$ and check the result. $\qquad$ *Answer:* 72

Using Algebra to Solve Percent Problems

Percents are often used to present numeric information. Stores use them to advertise discounts; manufacturers use them to describe the content of their products; and banks use them to list interest rates for loans and savings accounts. Percent problems occur in three types. Examples of these are shown below.

- What is 28% of 270?
- 14 is what percent of 52?
- 80 is 20% of what number?

Using the concept of a variable, along with the translating skills studied in Section 1.4 and the equation-solving skills of this section, we will now solve these problems.

EXAMPLE 7

Hours of sleep. Figure 1-17 shows the average number of hours United States residents sleep each night. Using these results, how many of the approximately 270 million residents would we expect to get 8 hours of sleep each night?

FIGURE 1-17

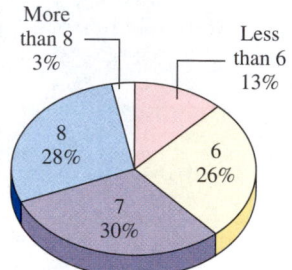

Based on data from *The Macmillan Visual Almanac* (Blackbirch Press, 1996), p. 121

Solution The circle graph tells us that 28% of U.S. residents get 8 hours of sleep each night. Since there are approximately 270 million residents, we need to find:

What number is 28% of 270 million?

To do this, we let a variable represent this unknown number.

Let x = the number of residents who get 8 hours of sleep.

Then we translate the words of the problem into an equation. In this case, the word *of* indicates multiplication, and the word *is* means equals.

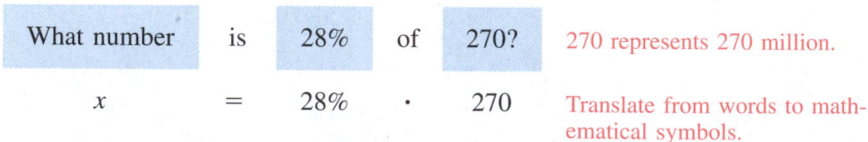

| What number | is | 28% | of | 270? | 270 represents 270 million. |
| x | $=$ | 28% | $\cdot$ | 270 | Translate from words to mathematical symbols. |

To find the number of people who get 8 hours of sleep, we need to find x.

$x = 28\% \cdot 270$ The original equation.

$x = 0.28 \cdot 270$ Write 28% as a decimal: 28% = 0.28.

$x = 75.6$ Use a calculator to do the multiplication on the right-hand side.

According to the survey, 75.6 million U.S. residents would get, on average, 8 hours of sleep each night.

We can check this result using estimation: 28% is approximately 25% $\left(\text{or } \frac{1}{4}\right)$, and 270 million is approximately 280 million. If we find $\frac{1}{4}$ of 280 $\left(\frac{1}{4} \cdot 280\right)$, the result is 70. This is close to 75.6, so our answer seems reasonable. ∎

In the statement "28% of 270 is 75.6," the number 75.6 is called the **amount,** 28% is the **percent,** and 270 is the **base.** The relationship between the amount, the percent, and the base is shown in the **percent formula.**

The percent formula

$$\text{Amount} = \text{percent} \cdot \text{base}$$

EXAMPLE 8

Best-selling song. In 1993, Whitney Houston's "I Will Always Love You" was at the top of Billboard's music charts for 14 weeks. What percent of the year did she have the #1 song? (Round to the nearest percent.)

Solution   First, we write the problem in a form that can be translated to an equation. We are asked to find the percent, knowing that for 14 out of the 52 weeks in a year, she had the #1 song. Therefore, we let $x =$ the unknown percent and translate the words of the problem into an equation.

14	is	what percent	of	52?
14	=	x	$\cdot$	52

14 is the amount, x is the percent, and 52 is the base.

To find x, we solve the equation.

$14 = x \cdot 52$ The original equation.

$14 = 52x$ Write $x \cdot 52$ as $52x$.

$\dfrac{14}{52} = \dfrac{52x}{52}$ To isolate x, undo the multiplication by 52 by dividing both sides by 52.

$0.2692307 \approx x$ Use a calculator to do the division.

$26.92307\% \approx x$ To write the decimal as a percent, multiply by 100 and write a % sign.

$x \approx 27\%$ Round to the nearest percent.

To the nearest percent, Whitney Houston had the best-selling song 27% of the year. We can check this result using estimation. 14 weeks out of 52 weeks is approximately $\frac{14}{50}$ or $\frac{28}{100}$, which is 28%. The answer of 27% seems reasonable.

SELF CHECK   In 1956, Elvis Presley's hit "Hound Dog" was the best-selling song for 11 weeks. To the nearest percent, what percent of the year was this?   *Answer:* 21%   ■

EXAMPLE 9

Aging population. By the year 2050, the U.S. Census Bureau estimates that about 20%, or 80 million, of the U.S. population will be over 65 years of age. If this is true, what is the population of the country predicted to be at that time?

Solution   First, we write the problem in words. Then we let $x =$ the predicted population in 2050 and translate to form an equation.

80	is	20%	of	what number?
80	=	20%	$\cdot$	x

80 is the amount, 20 is the percent, and x is the base.

Now we solve the equation to find x.

$80 = 20\% \cdot x$ The original equation.

$80 = 0.20 \cdot x$ Write 20% as a decimal by dividing by 100 and dropping the % sign.

$80 = 0.20x$ Write $0.20 \cdot x$ as $0.20x$.

$\dfrac{80}{0.20} = \dfrac{0.20x}{0.20}$ To undo the multiplication by 0.20, divide both sides by 0.20.

$400 = x$ Use a calculator to do the division.

The Census Bureau is predicting a population of 400 million, of which 80 million are expected to be over the age of 65. We can check the reasonableness of this result using estimation. 80 million out of a population of 400 million could be expressed as $\frac{80}{400}$ or $\frac{40}{200}$ or $\frac{20}{100}$, which is 20%. The answer checks.

SELF CHECK

In 1997, Census Bureau records indicated that about 1.49%, or 4 million, of the U.S. population were over 85 years old. To the nearest million, what was the bureau using as the country's estimated population?   *Answer:* 268 million   ■

Percents are often used to describe how a quantity has changed. For example, a health care provider might increase the cost of medical insurance by 3%, or a police department might decrease the number of officers assigned to street patrols by 10%. To describe such changes, we use **percent increase** or **percent decrease.**

EXAMPLE 10

Minimum wage. In August of 1996, Congress passed, and President Clinton signed, a bill that raised the federal hourly minimum wage to $5.15 on September 1, 1997. See Table 1-7. Find the percent increase in the minimum wage that this legislation provided.

TABLE 1-7

The Hourly Federal Minimum Wage	
1996	**September 1, 1997**
$4.75	$5.15

Source: U.S. Bureau of Labor Statistics

Solution   The minimum wage increased from 1996 to 1997. To find the percent increase, we first find the *amount* of increase by subtracting the smaller number from the larger.

$\$5.15 - \$4.75 = \$0.40$ Subtract the 1996 minimum wage from the 1997 minimum wage.

Next, we find what percent of the *original* wage ($4.75) the $0.40 increase is. We let x = the unknown percent and translate the words to an equation.

What percent	of	$4.75	is	$0.40?
x	$\cdot$	4.75	$=$	0.40

x is the percent, 4.75 is the base, and 0.40 is the amount.

$x \cdot 4.75 = 0.40$ The original equation.

$4.75x = 0.40$ Rewrite $x \cdot 4.75$ as $4.75x$.

$\dfrac{4.75x}{4.75} = \dfrac{0.40}{4.75}$ To isolate x, divide both sides by 4.75.

$x \approx 0.0842105263$ Use a calculator to do the division.

$x \approx 8.4\%$ To write the decimal as a percent, multiply by 100 and insert a % sign. Round to the nearest tenth of a percent.

The hourly federal minimum wage increased 8.4% from 1996 to 1997.   ■

VOCABULARY

In Exercises 1–8, fill in the blanks to make the statements true.

1. An _____ is a statement indicating that two expressions are equal.

2. Any number that makes an equation true when substituted for its variable is said to _____ the equation. Such numbers are called _____.

3. To _____ the solution of an equation, we substitute the value for the variable in the original equation and see whether the result is a _____ statement.

4. In the statement "10 is 50% of 20," 10 is called the _____, 50% is the _____, and 20 is the _____.

5. Two equations are _____ when they have the same solutions.

6. In mathematics, the word *of* often indicates _____, and _____ means equals.

7. To solve an equation, we _____ the variable on one side of the equals sign.

8. When solving an equation, the objective is to find all the values for the _____ that will make the equation true.

CONCEPTS

9. For each equation, tell what operation is performed on the variable. Then tell how to undo that operation to isolate the variable.
   a. $x - 8 = 24$
   b. $x + 8 = 24$
   c. $\dfrac{x}{8} = 24$
   d. $8x = 24$

10.
ILLUSTRATION 1

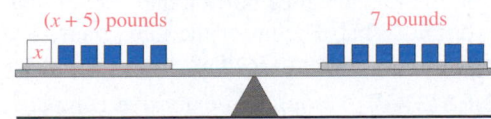

(x + 5) pounds   7 pounds

x

   a. What equation is represented by the scale in Illustration 1?
   b. How will the scale react if 5 pounds is removed from the left side?
   c. How will the scale react if 5 pounds is then removed from the right side?

d. What algebraic property do the steps listed in parts b and c illustrate?

e. After parts b and c, what equation will the scale represent, and what is its significance?

11. Given $x + 6 = 12$,
   a. What forms the left-hand side of the equation?

   b. Is this equation true or false?
   c. Is $x = 5$ a solution?
   d. Does $x = 6$ satisfy the equation?

12. Complete the following properties, and then give their names.
   a. If $x = y$ and c is any number, then $x + c =$ _____.

   b. If $x = y$ and c is any nonzero number, then $cx =$ _____.

NOTATION

In Exercises 13–14, complete the solution of each equation.

13. Solve $x + 15 = 45$.
$$x + 15 = 45$$
$$x + 15 - \boxed{} = 45 - \boxed{}$$
$$x = 30$$

14. Solve $8x = 40$.
$$8x = 40$$
$$\frac{8x}{\boxed{}} = \frac{40}{\boxed{}}$$
$$x = 5$$

In Exercises 15–16, translate each sentence into an equation.

15. 12 is 40% of what number?

16. 99 is what percent of 200?

17. When computing with percents, the percent must be changed to a decimal or a fraction. Change each percent to a decimal.
   a. 35%   b. 3.5%
   c. 350%   d. $\frac{1}{2}$%

18. Change each decimal to a percent.
   a. 0.9   b. 0.09
   c. 9   d. 0.999

PRACTICE

In Exercises 19–36, tell whether the given number is a solution of the equation. (Exercises 35 and 36 require a calculator.)

19. $x + 12 = 18; x = 6$
20. $x - 50 = 60; x = 110$
21. $2b + 3 = 15; b = 5$
22. $5t - 4 = 16; t = 4$
23. $0.5x = 2.9; x = 5$
24. $1.2 + x = 4.7; x = 3.5$
25. $33 - \dfrac{x}{2} = 30; x = 6$
26. $\dfrac{x}{4} + 98 = 100; x = 8$
27. $|c - 8| = 10; c = 20$
28. $|30 - r| = 15; r = 20$
29. $3x - 2 = 4x - 5; x = 12$
30. $5y + 8 = 3y - 2; y = 5$
31. $x^2 - x - 6 = 0; x = 3$
32. $y^2 + 5y - 3 = 0; y = 2$
33. $\dfrac{2}{a + 1} + 5 = \dfrac{12}{a + 1}; a = 1$
34. $\dfrac{2t}{t - 2} - \dfrac{4}{t - 2} = 1; t = 4$
35. $\sqrt{x - 5} + 1 = 15; x = 201$
36. $\sqrt{15 + y} - 3 = 20; y = 514$

In Exercises 37–64, use a property of equality to solve each equation. Check all solutions.

37. $x + 7 = 10$
38. $15 + y = 24$
39. $a - 5 = 66$
40. $x - 34 = 19$
41. $0 = n - 9$
42. $3 = m - 20$
43. $9 + p = 90$
44. $16 + k = 71$
45. $9 + p = 9$
46. $88 + j = 88$
47. $203 + f = 442$
48. $y - 34 = 601$
49. $4x = 16$
50. $5y = 45$
51. $369 = 9c$
52. $840 = 105t$
53. $4f = 0$
54. $0 = 60k$
55. $23b = 23$
56. $16 = 16h$
57. $\dfrac{x}{15} = 3$
58. $\dfrac{y}{7} = 12$
59. $\dfrac{l}{24} = 2$
60. $\dfrac{k}{17} = 8$
61. $35 = \dfrac{y}{4}$
62. $550 = \dfrac{w}{3}$
63. $0 = \dfrac{v}{11}$
64. $\dfrac{d}{49} = 0$

In Exercises 65–76, translate each problem from words to an equation, and then solve the equation.

65. What number is 48% of 650?
66. What number is 60% of 200?
67. What percent of 300 is 78?
68. What percent of 325 is 143?
69. 75 is 25% of what number?
70. 78 is 6% of what number?
71. What number is 92.4% of 50?
72. What number is 2.8% of 220?
73. What percent of 16.8 is 0.42?
74. What percent of 2,352 is 199.92?
75. 128.1 is 8.75% of what number?
76. 1.12 is 140% of what number?

APPLICATIONS

77. LAND OF THE RISING SUN The flag of Japan is a red disc (representing sincerity and passion) on a white background (representing honesty and purity).
    a. What is the area of the rectangular-shaped flag in Illustration 2?
    b. To the nearest tenth of a square foot, what is the area of the red disc? (*Hint: $A = \pi r^2$*)
    c. Use the results from parts a and b to find what percent of the area of the Japanese flag is occupied by the red disc.

ILLUSTRATION 2

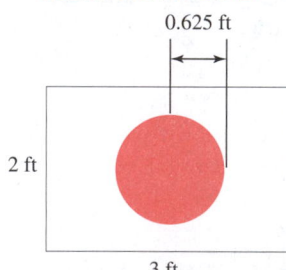

0.625 ft

2 ft

3 ft

78. COLLEGE ENTRANCE EXAMS On the Scholastic Aptitude Test, or SAT, a high school senior scored 550 on the mathematics portion and 700 on the verbal portion. What percent of the maximum 1,600 points did this student receive?

79. GENEALOGY Through an extensive computer search, a genealogist determined that worldwide, 180 out of every 10 million people had his last name. What percent is this?

80. AREA The total area of the 50 states and the District of Columbia is 3,618,770 square miles. If Alaska covers 591,004 square miles, what percent is this of the U.S. total (to the nearest percent)?

81. FEDERAL OUTLAYS Illustration 3 shows the breakdown of U.S. federal budget for fiscal year 1997. If total spending was approximately $1,601 billion, how much was paid as interest on the national debt (to the nearest billion dollars)?

ILLUSTRATION 3

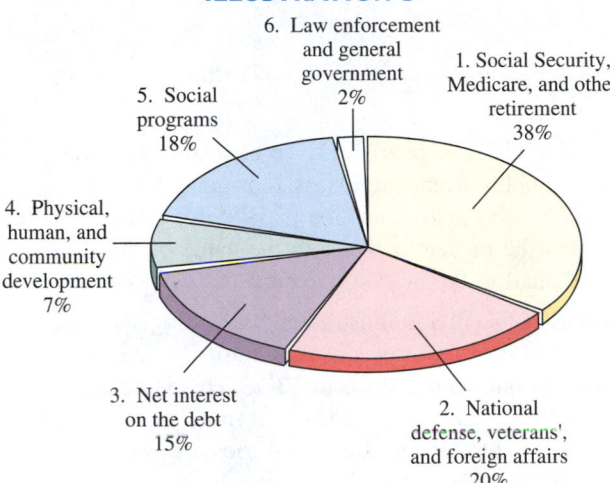

Based on 1998 Federal Income Tax Form 1040

82. DENTAL RECORDS The dental chart for an adult patient is shown in Illustration 4. The dentist marks each tooth that has had a filling. To the nearest percent, what percent of this patient's teeth have fillings?

ILLUSTRATION 4

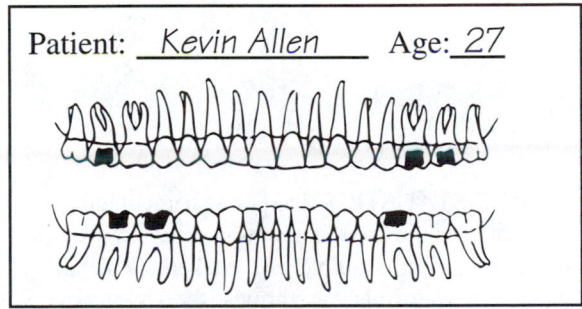

Patient: _Kevin Allen_   Age: _27_

83. TIPPING When paying with a Visa Card, the user must fill in the amount of the gratuity (tip) and then compute the total. Complete the sales draft in Illustration 5 if a 15% tip, rounded up to the nearest dollar, is to be left for the waiter.

ILLUSTRATION 5

STEAK STAMPEDE
Bloomington, MN
Server #12\ AT

VISA	67463777288
NAME	DALTON/ LIZ
AMOUNT	$75.18
GRATUITY $	_____
TOTAL $	_____

84. INCOME TAX Use the Schedule X Tax Table shown in Illustration 6 to compute the amount of federal income tax if the amount of taxable income entered on Form 1040, line 39, is $39,909.

ILLUSTRATION 6

If the amount on Form 1040, line 39, is Over—	But not over—	Enter on Form 1040, line 40		of the amount over—
$0	$25,350	....	15%	$0
25,350	61,400	$3,802.50 +	28%	25,350
61,400	128,100	13,896.50 +	31%	61,400
128,100	278,450	34,573.50 +	36%	128,100
278,450	....	88,699.50 +	39.6%	278,450

85. INSURANCE COST A college student's good grades earned her a student discount on her car insurance premium. What was the percent decrease, to the nearest percent, if her annual premium was lowered from $1,050 to $925?

86. POPULATION The magazine *American Demographics* estimated the population of the United States as 270,539,805 on August 31, 1998. For the month of September, 1998, it estimated that the number of births was 427,810, the number of deaths was 219,341, and net immigration was 75,279.
   a. What was the magazine's estimate of the U.S. population as of September 30, 1998?

   b. What was the percent increase in the U.S. population for the month of September, 1998, according to this magazine?

87. CHARITABLE GIVING Nonprofit organizations receive contributions from individuals, corporations, foundations, and bequests. In 1997, individuals contributed $109 billion to nonprofit organizations. If this was 76% of all charitable giving for that year, what was the total amount given to nonprofit organizations in 1997 (to the nearest billion dollars)?

88. NUTRITION The Nutrition Facts label from a can of New England Clam Chowder is shown in Illustration 7.
   a. Use the information on the label to determine the number of grams of fat and the number of grams of saturated fat that should be consumed daily. Round to the nearest gram.

   b. To the nearest percent, what percent of the calories in a serving of clam chowder come from fat?

89. COPAYMENT "80/20 health care coverage" means that the plan pays 80% of the medical bill, and the patient pays the remaining 20% (called the *copayment*). Under this plan, what was the cost of an office visit to a dermatologist if the patient's copayment was $11.30?

ILLUSTRATION 7

Nutrition Facts

Serving Size 1 cup (240mL)
Servings Per Container about 2

Amount per serving	
Calories 240 Calories from Fat 140	
	% Daily Value*
Total Fat 15 g	23%
Saturated Fat 5 g	25%
Cholesterol 10 mg	3%
Sodium 980 mg	41%
Total Carbohydrate 21 g	7%
Dietary Fiber 2 g	8%
Sugars 1 g	
Protein 7 g	

90. EARTHQUAKE INSURANCE A homeowner received a settlement check of $21,568.50 from her insurance company to cover damages caused by an earthquake. If her policy required her to pay a 10% deductible, what was the total dollar amount of the damages to the home?

91. EXPORTS The bar graph in Illustration 8 shows United States exports to Mexico for the years 1992 to 1997.

   a. Between what two years was there a decline in U.S. exports? Find the percent decrease, to the nearest percent.

   b. Between what two years was there the most dramatic increase in exports? Find the percent increase, to the nearest percent.

ILLUSTRATION 8

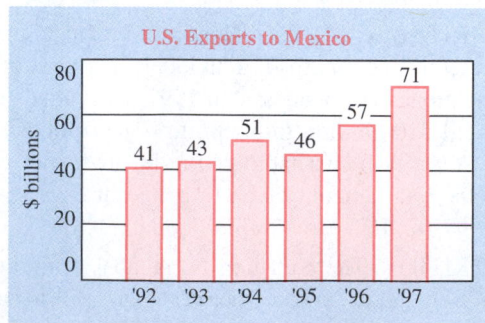

U.S. Exports to Mexico

Based on data from *Los Angeles Times* (May 5, 1997), Business Section

92. PROJECTIONS A market research company was hired to project the demand for professional house-cleaning services in the future. A survey was taken, and projections were made. Complete the last two columns in Illustration 9, rounding to the nearest tenth. Would you classify the projected growth in this industry as small, moderate, or large?

ILLUSTRATION 9

Households using a maid, housekeeper, or professional cleaning service				
Year	Year	Year	% change	% change
1996	2000	2006	1996–2000	2000–2006
9,436,000	9,999,000	10,740,000		

Based on data from *Demographics Magazine* (Nov. 1996), p. 4

93. AUCTION A pearl necklace of former First Lady Jacqueline Kennedy Onassis, originally valued at $700, was sold at auction in 1996 for $211,500. What was the percent increase in the value of the necklace? (Round to the nearest percent.)

94. GEOLOGY See Illustration 10. Geologists have found that a pile of loose sand always takes the form of a cone with a slope of $33\frac{1}{3}$%. This is called the natural **angle of repose.** When more sand is added to the top of the pile, some will trickle down the sides until the $33\frac{1}{3}$% slope is restored. Find the slope of this cone of sand by finding what percent of the horizontal length of 45 feet the 15-foot vertical rise represents.

ILLUSTRATION 10

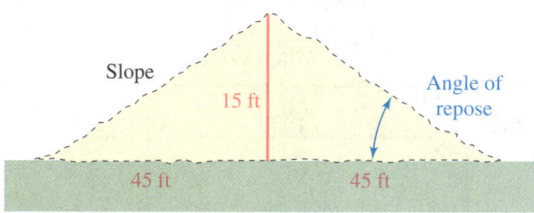

Slope

15 ft

Angle of repose

45 ft   45 ft

95. SPORTS STATISTICS In sports, percentages are most often expressed as three-place decimals instead of percents. For example, if a basketball player makes 75.8% of his free throws, the sports page will list this as .758. Use this format to complete Illustration 11.

ILLUSTRATION 11

All-Time Best Regular-Season Winning Percentages		
Team	**Won–lost record**	**Winning percentage**
1996 Chicago Bulls Basketball	72–10	
1972 Miami Dolphins Football	14–0	
1906 Chicago Cubs Baseball	116–36	

96. TAXES AND PENALTIES
   a. What is the hotel room tax rate if a $6.84 tax is charged on a room costing $72 a night?

   b. What is the late penalty, expressed as a percent, if a charge of $70.75 is applied to a car registration fee of $283 that was not paid to the Motor Vehicle Department on time?

WRITING

97. What does it mean to solve an equation?

98. Write a real-life situation that could be described by "45 is what percent of 50?"

99. After solving an equation, how do we check the solution?

100. Explain what it would mean if the attendance at an amusement park reached 105% of capacity.

REVIEW

101. What is the output of the expression $9 - 3x$ if 3 is the input?

102. Write a formula that would give the number of eggs in d dozen.

103. Translate to symbols: the difference of 45 and x.

104. Evaluate $\dfrac{2^3 + 3(5 - 3)}{15 - 4 \cdot 2}$.

105. Approximate 3π to the nearest tenth.

106. True or false? $-23 > -24$

▶ 1.6   Problem Solving

In this section, you will learn about

   Writing equations ■ A problem-solving strategy ■ Drawing diagrams ■ Constructing tables

Introduction   One of the objectives of this course is for you to become a better problem solver. The key to problem solving is to understand the problem and then to devise a plan for solving it. In this section, we will introduce a five-step problem-solving strategy. We will also show how drawing a diagram or constructing a table is often helpful in visualizing the facts of a given problem.

Writing Equations

To solve a problem, it is often necessary to write an equation that describes the given situation. To write an equation, we must analyze the facts of the problem, looking for two different ways to describe the same quantity. As an introduction to this procedure, let's consider the following statement.

   If a charity fundraising drive can raise $13 million more, the goal of $87 million will be reached.

   Since we are not told the amount already raised, we will let the variable m represent this unknown amount. See Figure 1-18. The key word *more* tells us that if we add $13 million to the m dollars already raised, the goal will be reached. Therefore, the goal can be represented by the algebraic expression $m + 13$. We now have two ways to represent the fundraising goal: $m + 13$ and 87. We can use an equation to state this.

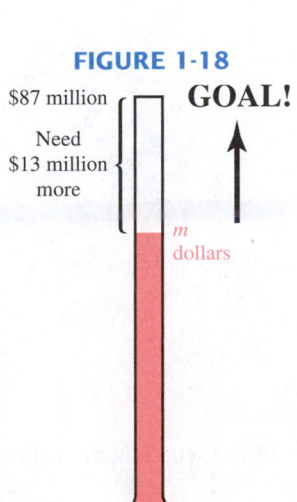

FIGURE 1-18

$87 million

Need $13 million more

m dollars

GOAL!

$$m + 13 = 87$$

This describes the fundraising goal in one way.

This describes the fundraising goal in another way.

EXAMPLE 1

Writing equations. Using the facts in the following statement, write an equation that describes the same quantity in two different ways.

> If Nina had twice as much money in the bank as she has now, she would have enough to pay her tuition bill of $1,500.

Solution
We don't know how much money Nina has in the bank, so we let $x =$ the amount currently in her account. The key word *twice* tells us to *multiply* the amount currently in her account by 2 to find an expression for the cost of tuition. Therefore, $2 \cdot x$, or $2x$, is the cost of tuition in dollars. We now have two ways to represent the tuition: $2x$ and 1,500. Thus,

$$2x = 1,500$$

SELF CHECK
Based on the following statement, write an equation that describes the same quantity in two different ways. "After having 5 feet trimmed off the top, the height of a pine tree is 46 feet."
(*Hint:* Let $x =$ the original height of the tree.)   *Answer:* $x - 5 = 46$ ■

EXAMPLE 2

Writing equations. Using the facts in the following situation, write an equation that describes the same quantity in two different ways.

> Three friends pooled their money to purchase some Lotto tickets. When one of the tickets won, they split the cash prize evenly. Each of them received $95.

Solution
We don't know how much money the winning Lotto ticket was worth, so we let $x =$ the amount of money won. The key phrase *split evenly* tells us to *divide* the amount of money won by 3 to find an expression for each person's share. Therefore, each person receives $\frac{x}{3}$ dollars. We are told that each person won $95, so we now have two ways to represent the amount won by each person: $\frac{x}{3}$ and 95. Thus,

$$\frac{x}{3} = 95$$

SELF CHECK
Based on the following facts, write an equation that describes the same amount in two different ways. "The attendance at a political rally was 250 people more than the organizers had anticipated. 800 people were present." (*Hint:* Let $x =$ the number of people that were anticipated.)   *Answer:* $x + 250 = 800$ ■

A Problem-Solving Strategy

To become a good problem solver, you need a plan to follow, such as the following five-step problem-solving strategy.

Strategy for problem solving

1. **Analyze the problem** by reading it carefully to understand the given facts. What information is given? What are you asked to find? What vocabulary is given? Often, a diagram or table will help you visualize the facts of the problem.

2. **Form an equation** by picking a variable to represent the quantity to be found. Then express all other unknown quantities as expressions involving that variable. Key words or phrases can be helpful. Finally, write an equation expressing a quantity in two different ways.

3. **Solve the equation.**

4. **State the conclusion.**

5. **Check the result** in the words of the problem.

E X A M P L E   3

Systems analysis. An engineer found that a company's telephone use would have to increase by 350 calls per hour before the system would reach the maximum capacity of 1,500 calls per hour. Currently, how many calls are being made each hour on the system?

ANALYZE THE PROBLEM We are asked to find the number of calls currently being made each hour. We are given two facts:

- The maximum capacity of the system is 1,500 calls per hour.
- If the number of calls increases by 350, the system will reach capacity.

FORM AN EQUATION Let n = the number of calls currently being made each hour. To form an equation involving n, we look for a key word or phrase in the problem.

Key phrase: *increase by 350*   **Translation:** addition

The key phrase tells us to add 350 to the current number of calls to obtain an expression for the maximum capacity of the system. Therefore, we can write the maximum capacity of the system in two ways.

The current number of calls per hour	increased by	350	is	the maximum capacity of the system.
n	$+$	350	$=$	1,500

SOLVE THE EQUATION

$$n + 350 = 1,500$$
$$n + 350 \mathbf{- 350} = 1,500 \mathbf{- 350} \qquad \text{To undo the addition of 350, subtract 350 from both sides.}$$
$$n = 1,150 \qquad \text{Do the subtractions.}$$

STATE THE CONCLUSION Currently, 1,150 calls per hour are being made.

CHECK THE RESULT If 1,150 calls are currently being made each hour and an increase of 350 calls per hour occurs, then $1,150 + 350 = 1,500$ calls will be made each hour. This is the capacity of the system. The solution checks.

EXAMPLE 4

Marine recruitment. The annual number of Marine recruits from a certain county tripled after an intense recruiting program was conducted at the area's high schools. If 384 students from the county decided to join the Marines, what was the previous year's county recruitment total?

ANALYZE THE PROBLEM   We are asked to find the county recruitment total for last year. We are given two facts about the situation:

- 384 recruits were signed this year.
- The number of recruits this year is triple that of last year.

FORM AN EQUATION   Let r = the number of recruits last year. To form an equation involving r, we look for a key word or phrase in the problem.

Key word: *tripled*   **Translation:** multiplication by 3

The key word tells us that we can multiply last year's number of recruits by 3 to obtain an expression for the number of recruits this year. Therefore, we can write this year's number of recruits in two ways.

3	times	the number of recruits last year	is	384.
3	·	r	=	384

SOLVE THE EQUATION

$$3r = 384$$
$$\frac{3r}{3} = \frac{384}{3} \qquad \text{To undo the multiplication by 3, divide both sides by 3.}$$
$$r = 128 \qquad \text{Do the divisions.}$$

STATE THE CONCLUSION   The number of recruits last year was 128.

CHECK THE RESULT   If we multiply last year's number of recruits (128) by 3, we get $3 \cdot 128 = 384$. This is the number of recruits for this year. The solution checks.   ■

Drawing Diagrams

When solving problems, diagrams are often helpful; they allow us to visualize the facts of the problem.

EXAMPLE 5

Airline travel. On a book tour that took her from New York City to Chicago to Los Angeles and back to New York City, an author flew a total of 4,910 miles. The flight from New York to Chicago was 714 miles, and the flight from Chicago to L.A. was 1,745 miles. How long was the direct flight back to New York City?

ANALYZE THE PROBLEM   We are asked to find the length of the flight from L.A. to New York City. We are given the following three facts.

- The total miles flown on the tour was 4,910.
- The flight from New York to Chicago was 714 miles.
- The flight from Chicago to L.A. was 1,745 miles.

In the diagram in Figure 1-19, we see that the three parts of the tour form a triangle. We know the lengths of two of the sides of the triangle (714 and 1,745) and the perimeter of the triangle (4,910).

FIGURE 1-19

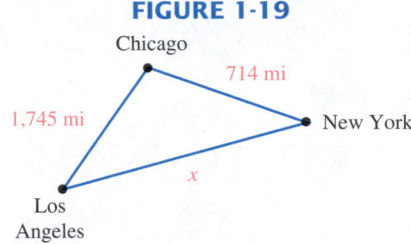

FORM AN EQUATION   We will let x = the length (in miles) of the flight from L.A. to New York and label the appropriate side of the triangle in Figure 1-19. There are two ways to describe the total number of miles traveled by the author on the book tour.

The miles from New York to Chicago	plus	the miles from Chicago to L.A.	plus	the miles from L.A. to New York	is	4,910.
714	+	1,745	+	x	=	4,910

SOLVE THE EQUATION

$$714 + 1,745 + x = 4,910$$
$$2,459 + x = 4,910 \qquad \text{Simplify the left-hand side of the equation: } 714 + 1,745 = 2,459.$$
$$2,459 - \mathbf{2{,}459} + x = 4,910 - \mathbf{2{,}459} \qquad \text{Subtract 2,459 from both sides to isolate } x.$$
$$x = 2,451 \qquad \text{Do the subtractions.}$$

STATE THE CONCLUSION   The flight from L.A. to New York was 2,451 miles.

CHECK THE RESULT   If we add the three flight lengths, we get $714 + 1{,}745 + 2{,}451 = 4{,}910$. This was the total number of miles flown on the book tour. The solution checks. ∎

EXAMPLE 6

Eye surgery. A surgical technique called **radial keratotomy** is sometimes used to correct nearsightedness. This procedure involves equally spaced incisions in the cornea, as shown in Figure 1-20. Find the angle between each incision.

FIGURE 1-20

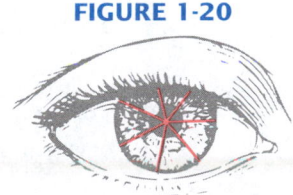

Solution

ANALYZE THE PROBLEM   From the diagram in Figure 1-21, we see that there are 7 angles of equal measure. We also know that 1 complete revolution is 360°.

FIGURE 1-21

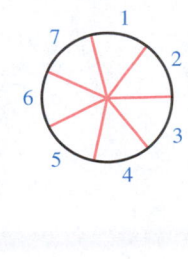

FORM AN EQUATION   We will let x = the measure of one of the angles. We then have:

7	times	the measure of one of the angles	is	360°.
7	·	x	=	360

SOLVE THE EQUATION

$$7x = 360$$
$$\frac{7x}{7} = \frac{360}{7} \qquad \text{Divide both sides by 7.}$$
$$x \approx 51.42857143 \qquad \text{Use a calculator to do the division.}$$
$$x \approx 51.4° \qquad \text{Round to the nearest tenth of a degree.}$$

STATE THE CONCLUSION   The incisions are approximately 51.4° apart.

CHECK THE RESULT   If we multiply 51 by 7, we obtain 357. This is close to 360, so the result of 51.4° seems reasonable.

Constructing Tables

Sometimes it is helpful to organize the given facts of the problem in a table.

<table>
<tr><td>EXAMPLE 7</td><td>**Labor statistics.** The number of women in the U.S. labor force has grown steadily over the past 100 years. From 1900 to 1940, the number grew by 8 million. By 1980, it had increased by an additional 32 million. By 1997, the number rose another 15 million; by the end of that year, 60 million women were in the labor force. How many women were in the labor force in 1900?</td></tr>
</table>

ANALYZE THE PROBLEM
We are to find the number of women in the U.S. labor force in 1900. We know the following:

- The number grew by 8 million, increased by 32 million, and rose another 15 million.

- The number of women in the labor force in 1997 was 60 million.

We will let x represent the number of women (in millions) in the labor force in 1900. We can write algebraic expressions to represent the number of women in the work force in 1940, 1980, and 1997 by translating key words. See Table 1-8.

TABLE 1-8

Year	Women in the labor force (millions)	
1900	x	
1940	$x + 8$	**Key word:** *grew* **Translation:** addition
1980	$x + 8 + 32$	**Key word:** *increased* **Translation:** addition
1997	$x + 8 + 32 + 15$	**Key word:** *rose* **Translation:** addition

FORM AN EQUATION
There are two ways to represent the number of women (in millions) in the 1997 labor force: $x + 8 + 32 + 15$ and 60. Therefore,

$$x + 8 + 32 + 15 = 60$$

SOLVE THE EQUATION

$$x + 8 + 32 + 15 = 60$$
$$x + 55 = 60 \qquad \text{Simplify: } 8 + 32 + 15 = 55.$$
$$x + 55 - 55 = 60 - 55 \qquad \text{To undo the addition of 55, subtract 55 from both sides.}$$
$$x = 5 \qquad \text{Do the subtractions.}$$

STATE THE CONCLUSION
There were 5 million women in the U.S. labor force in 1900.

CHECK THE RESULT
Adding the number of women (in millions) in the labor force in 1900 and the increases, we get $5 + 8 + 32 + 15 = 60$. In 1997, there were 60 million, so the solution checks.

■

STUDY SET

Section 1.6

VOCABULARY

In Exercises 1–4, fill in the blanks to make the statements true.

1. A letter that is used to represent a number is called a _____.

2. To _____ an equation means to find all the values of the variable that make the equation true.

3. An _____ is a mathematical statement that two quantities are equal.

4. Phrases such as *increased by* and *more than* indicate the operation of _____.

CONCEPTS

5. Put the steps of the five-step problem-solving strategy listed below in the proper order.

State the result. Solve the equation. Check the result. Analyze the problem. Form an equation.

6. Fill in the blanks to make the statements true. When solving a real-world problem, we let a _____ represent the unknown quantity. Then we write an _____ that describes the same quantity in two different ways. Finally, we _____ the equation for the variable to find the unknown.

In Exercises 7–8, a diagram is given. Tell what the variable represents, and then write an equation that describes the same quantity in two ways.

7.

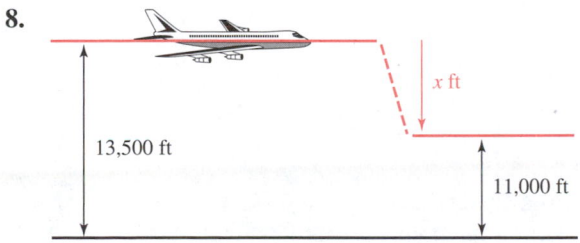

Triathlon–16 mi

Swimmimg 1 mi   Biking 10 mi   Running x mi

8.

13,500 ft

x ft

11,000 ft

In Exercises 9–12, a statement about a situation and a table containing facts about the situation are given. Tell what the variable represents, and then write an equation that describes the same quantity in two ways.

9. The sections of a 430-page book are being assembled.

Section	Number of pages
Table of Contents	4
Preface	x
Text	400
Index	12

10. A hamburger chain sold a total of 31 million hamburgers in its first four years of business.

Years in business	Cumulative number of hamburgers sold (millions)
1	x
2	$x + 5$
3	$x + 5 + 8$
4	$x + 5 + 8 + 16$

11. Two accounts earned a total of $678 in interest for the year.

Account	Interest earned ($)
Savings	x
Checking	73

12. A motorist traveled from Toledo, OH to Columbus, OH at 55 mph and from Columbus to Cincinnati, OH at 50 mph. The entire trip covered 253 miles.

	Speed (mph)	Distance (mi)
Toledo to Columbus	55	145
Columbus to Cincinnati	50	x

PRACTICE

In Exercises 13–26, an occupation is listed along with a statement that someone with that occupation might make. Identify the key word or phrase in the sentence and the operation it indicates.

13. *Financial planner:* The profits from the sale were disbursed equally among the investors.

14. *Auditor:* The cost of the project skyrocketed by a factor of 10.

15. *Park ranger:* Two feet of the snow pack had melted before February 1.

16. *Lawyer:* The city annexed 15 uninhabited houses that were under county jurisdiction.

17. *Ecologist:* After the flood, 6 acres of the marshy land were reclaimed.

18. *Race car driver:* The new tires helped shave off 2.5 seconds from each lap.

19. *Landscaper:* Four feet of the embankment had eroded.

20. *Developer:* Plans were drawn up to bisect the lot with a one-way street.

21. *Archaeologist:* The area to be excavated was sectioned off uniformly.

22. *Carpenter:* The rough-cut lumber needed to be sanded down a sixteenth of an inch.

23. *School administrator:* Two of the cheerleaders had to be cut from the squad for disciplinary reasons.

24. *Producer:* Because of overflow crowds, the play's run was extended two weeks.

25. *Optometrist:* The patient's pupils dilated to twice their size.

26. *Nurse:* The amount of required paperwork has quadrupled over the last few years.

In Exercises 27–34, write an equation that describes the same quantity in two ways.

27. An existing 1,000-foot water line had to be extended to a length of 1,525 feet to reach a new restroom facility. Let x = the length of the extension.

28. Because of overgrazing, state agriculture officials determined that the 4,500 head of cattle currently on the ranch had to be reduced to 2,750. Let x = the number of head of cattle to be removed.

29. Because 4 people didn't show up for their appointments, the dentist saw only 8 patients on Wednesday. Let x = the number of scheduled appointments.

30. After the management of a preschool decided to open up an additional 15 spaces, the total number of children attending the school reached 260. Let x = the number of children who attended before the enrollment was increased.

31. A length of gold chain, cut into 12-inch-long pieces, makes five bracelets. Let x = the length of the chain.

32. The 24 ounces of walnuts used to make a fruitcake was twice what was called for in the recipe. Let x = the number of ounces of walnuts called for in the recipe.

33. When the value of 15 postage stamps was calculated, the total was found to be 495 cents. Let x = the value of one stamp.

34. The overtime hours available for 8 employees were distributed equally, so that each of them got to work an additional 11 hours. Let x = the total number of overtime hours.

In Exercises 35–36, use the outline of the five-step problem-solving strategy to help answer each question.

35. MAJOR REQUIREMENTS The business department of a college reduced by 6 the number of units of course work needed to obtain a degree. The department now requires the completion of 28 units. What was the old unit requirement?

ANALYZE THE PROBLEM
What are you asked to find? _____

We know that

- The unit requirement was reduced by ▆ .
- The new unit requirement is ▆ .

FORM AN EQUATION
Let x = _____

   Key word: *reduced* **Translation:** _____

We can express the new requirement in two ways.

The old unit requirement	reduced by	6	is	the new unit requirement.
▆	−	6	=	▆

SOLVE THE EQUATION

$$\boxed{} - 6 = 28$$
$$x + \boxed{} - 6 = 28 + \boxed{}$$
$$x = 34$$

STATE THE CONCLUSION
The old unit requirement was ▆ units.

CHECK THE RESULT
If we reduce the old unit requirement of ▆ by 6, we have $34 - 6 = 28$. The answer checks.

36. BUSINESS LOSSES After the membership fee to a health spa was raised, the number of new members joining each week was half of what it used to be. If, on average, 27 people per week are now joining, how many used to join each week?

ANALYZE THE PROBLEM
What are you asked to find? _____

We know that

- ▆ people are now joining each week.
- The number of people now joining is _____ of what it used to be.

FORM AN EQUATION
Let $x =$ _____

   Key phrase: *half of* **Translation:** _____

We can express the number now joining in two ways.

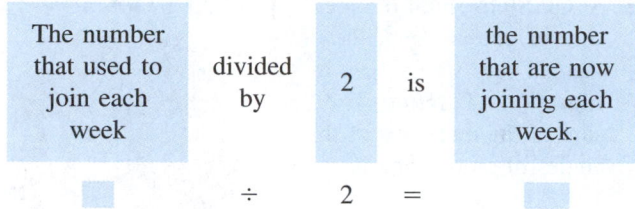

| The number that used to join each week | divided by | 2 | is | the number that are now joining each week. |

| ▮ | ÷ | 2 | = | ▮ |

SOLVE THE EQUATION

$$\frac{x}{2} = ▮$$

$$▮\left(\frac{x}{2}\right) = ▮ \quad (27)$$

$$x = 54$$

STATE THE CONCLUSION
The number of people that used to join each week was ▮.

CHECK THE RESULT
If we divide the number of people that used to join each week by ▮, we have $\frac{54}{2} = 27$. The answer checks.

APPLICATIONS

You can probably solve Exercises 37–46 without using algebra. Nevertheless, you should use the methods discussed in this section to solve the problems, so that you can gain experience in applying these concepts and procedures. This will prepare you for more challenging problems later in the course.

37. ICE CREAM People in the United States consume more ice cream per capita than any other people in the world—47 pints per year. If this is 20 more pints than Canadians eat, what is the yearly per-capita consumption of ice cream in Canada?

38. ENTERTAINMENT According to *Forbes* magazine, Oprah Winfrey made an estimated $125 million in 1998. This was $67 million more than Harrison Ford's estimated earnings for that year. How much did Harrison Ford make in 1998?

39. TENNIS Billie Jean King won 40 Grand Slam tennis titles in her career. This is 14 less than the all-time leader, Martina Navratilova. How many Grand Slam titles did Navratilova win?

40. MONARCHY George III reigned as king of Great Britain for 59 years. This is four years less than the longest-reigning British monarch, Queen Victoria. For how many years did Queen Victoria rule?

41. COST OVERRUN At completion, the construction cost of a subway system totaled $564 million. This was larger than the original estimate by a factor of 3. What was the original estimate for the subway system?

42. FLOODING Torrential rains caused the width of a river to swell to 84 feet. If this was twice its normal size, how wide was the river before the flooding?

43. ATM RECEIPT Use the information on the automatic-teller receipt in Illustration 1 to find the balance in the account before the withdrawal.

ILLUSTRATION 1

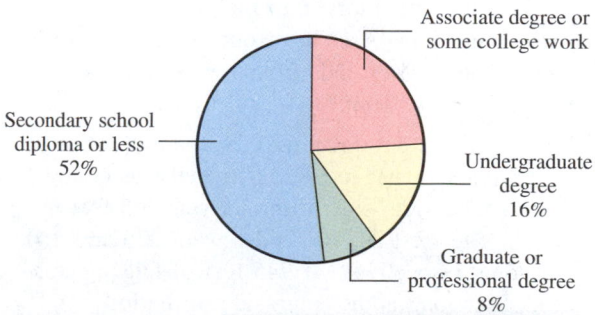

HOME SAVINGS OF AMERICA

Thank you for letting us
serve all your financial needs.

TRAN.	DATE	TIME	TERM
0286.	1/16/99	11:46 AM	HSOA822

CARD NO.   6125 8
WITHDRAWAL OF $35.00
FROM CHECKING ACCT.   3325256-612

CHECKING BAL. $287.00

44. EDUCATION IN THE UNITED STATES One category in the circle graph shown in Illustration 2 is not labeled with a percent. What percent should be written there?

ILLUSTRATION 2

**Highest level of education attained
by persons 25 years and older**

Associate degree or
some college work

Secondary school
diploma or less
52%

Undergraduate
degree
16%

Graduate or
professional degree
8%

Based on data from *Digest of Education Statistics*

45. TV NEWS An interview with a world leader was edited into equally long segments and broadcast in parts over a three-day period on a TV news program. If each daily segment of the interview lasted 9 minutes, how long was the original interview?

46. MISSING CHILD Six search and rescue squads were called to a mountain wilderness park to look for a lost child. If each squad was able to cover 16 acres, over how large an area was the search conducted?

In Exercises 47–50, use a table to help organize the facts of the problem; then find the solution.

47. STATEHOOD From 1800 to 1850, 15 states joined the Union. From 1851 to 1900, an additional 14 states entered. Three states joined from 1901 to 1950. Since then, Alaska and Hawaii are the only others to enter the Union. How many states were part of the Union prior to 1800?

48. STUDIO TOUR Over a four-year span, improvements in a Hollywood movie studio tour caused it to take longer. The first year, 10 minutes were added to the tour length. The second, third, and fourth years, 5 minutes were added. If the tour now lasts 135 minutes, how long was it originally?

49. ANATOMY A premed student has to know the names of all 206 bones that make up the human skeleton. So far, she has memorized the names of the 60 bones in the feet and legs, the 31 bones in the torso, and the 55 bones in the neck and head. How many more names does she have to memorize?

50. ORCHESTRA A 98-member orchestra is made up of a woodwind section with 19 musicians, a brass section with 23 players, a two-person percussion section, and a large string section. How many musicians make up the string section of the orchestra?

In Exercises 51–54, draw a diagram to help organize the facts of the problem, and then find the solution.

51. BERMUDA TRIANGLE The Bermuda Triangle is a triangular region in the Atlantic Ocean where many ships and airplanes have disappeared. The perimeter of the triangle is about 3,075 miles. It is formed by three imaginary lines. The first, 1,100 miles long, is from Melbourne, Florida, to Puerto Rico. The second, 1,000 miles long, stretches from Puerto Rico to Bermuda. The third extends from Bermuda back to Florida. Find its length.

52. FENCING To cut down on vandalism, a lot on which a house was to be constructed was completely fenced. The north side of the lot was 205 feet in length. The west and east sides were 275 and 210 feet long, respectively. If 945 feet of fencing was used, how long is the south side of the lot?

53. SPACE TRAVEL The 364-foot-tall Saturn V rocket carried the first astronauts to the moon. Its first, second, and third stages were 138, 98, and 46 feet tall, respectively. Atop the third stage was the Lunar Module, and from it extended a 28-foot escape tower. How tall was the Lunar Module?

54. PLANETS Mercury, Venus, and the earth have approximately circular orbits about the sun. Earth is the farthest from the sun at 93 million miles, and Mercury is the closest, at 36 million miles. The orbit of Venus is about 31 million miles from that of Mercury. How far is the earth's orbit from that of Venus?

In Exercises 55–56, solve each problem.

55. STOP SIGN Find the measure of one angle of the octagonal stop sign shown in Illustration 3. (*Hint:* The sum of the measures of the angles of an octagon is 1080°.)

ILLUSTRATION 3

56. FERRIS WHEEL What is the measure of the angle between each of the "spokes" of the Ferris wheel shown in Illustration 4?

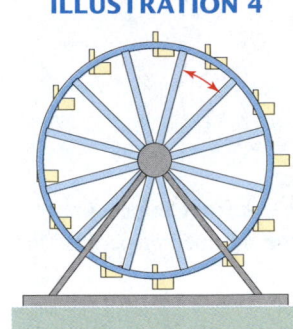

ILLUSTRATION 4

WRITING

57. How can a variable and an equation be used to find an unknown quantity?

58. How can a diagram or a table be helpful when solving problems?

59. From what you have learned in this section, what is the key to problem solving?

60. Briefly explain what should occur in each step of the five-step problem-solving strategy.

REVIEW

61. If an 85-seat restaurant added an additional 25 seats, what is the percent increase in seating (to the nearest percent)?

62. What two numbers are a distance of 8 away from 4 on the number line?

63. What is 35% of 7,800?

64. Solve: $\dfrac{t}{25} = 6$.

65. Is $x = 34$ a solution of $x - 12 = 20$?

66. Write 0.898 as a percent.

67. Evaluate $2x^2 - 3x$ for $x = 4$.

68. Evaluate $2 + 3[24 - 2(5 - 2)]$.

Variables

One of the major objectives of this course is for you to become comfortable working with **variables**. In Chapter 1, we have used the concept of variable in four ways.

Stating Mathematical Properties

Variables have been used to state properties of mathematics in a concise, "shorthand" notation.

1. Complete the statement of the subtraction property of equality: Let a, b, and c be any three numbers. If $a = b$, then $a - c = $ _____.

2. Complete the statement of the multiplication property of equality: Let a, b, and c be any three numbers $(c \neq 0)$. If $a = b$, then $ca = $ _____.

Stating Relationships between Quantities

Variables are letters that stand for numbers. We have used variables to express known relationships between two or more quantities. These written relationships are called **formulas.**

3. Translate the word model to an equation (formula) that mathematically describes the situation.

The total cost is the sum of the purchase price of the item and the sales tax.

4. Use the data in the table to state the relationship between the quantities using a formula.

Picnic tables	Benches needed
2	4
3	6
4	8

Writing Algebraic Expressions

Variables and numbers (called constants) have been combined with the operations of addition, subtraction, multiplication, and division to create **algebraic expressions.**

5. One year, a cruise company did x million dollars worth of business. After a television celebrity was signed as a spokeswoman for the company, its business increased by $4 million the next year. Use an algebraic expression to indicate the amount of business the cruise company had in the year the celebrity was the spokeswoman.

6. Evaluate the algebraic expression for the given values of the variable, and enter the results in the table.

x	$3x^2 - 2x + 1$
0	
4	
6	

Writing Equations to Solve Problems

To solve problems, we have let a variable represent an unknown quantity, written an equation containing the variable, and then solved the equation to find the unknown. In Exercises 7 and 8, complete the statement: Let $x = $ _____.

7. AREA The total area covered by the United States is 3,618,770 square miles. If Maine covers 33,265 square miles, what percent of the U.S. total is this?

8. CANVASSING Eight volunteers were hired to conduct a survey of the residents of a city. If each volunteer was able to canvass 4 city blocks, how many city blocks were canvassed by the volunteers?

Accent on Teamwork

Section 1.1

Production planning In the Study Set for Section 1.1, Exercise 73 on page 10 asks you to write a series of equations that would assist a production planner in ordering the correct number of parts for a production run of c chairs. See Illustration 1.

a. Decide on a product for which you will be the production planner. Make a detailed drawing of it, like the one in Illustration 1.

b. For one component of your product, create a table giving the number that should be ordered for production runs of 50, 100, 150, and 200 units. Do the same for a second component of your product using a bar graph, and for a third component using a line graph.

c. For each of the three components in part b, write an equation describing the number of components that need to be ordered for a production run of u units.

ILLUSTRATION 1

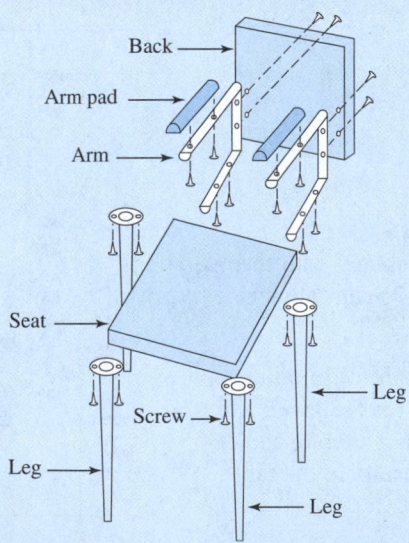

Section 1.2

Real numbers Give some examples of situations in your everyday life where you encounter the types of numbers listed below.

- Whole numbers
- Zero
- Negative numbers
- Fractions
- Decimals

Section 1.3

Order of operations To make a cake from a mix, the instructions must be followed carefully. Otherwise, the results can be disastrous. Think of two other multistep processes and explain why the steps must be performed in the proper order, or the outcome is adversely affected. Think of two processes where the order in which the steps are performed does not affect the outcome.

Section 1.4

Evaluating algebraic expressions Find five examples of cylinders. Measure and record the diameter d of their bases and their heights h. Express the measurements as decimals. Find the radius r of each base by dividing the diameter by 2. Then find each volume by evaluating the expression $\pi r^2 h$. Round to the nearest tenth of a cubic unit. See the Accent on Technology on page 43 for an example. Present your results in a table of the form shown in Illustration 2.

ILLUSTRATION 2

Cylinder	d	r	h	Volume
Container of salt	$3\frac{1}{4}$ in. (3.25 in.)	$1\frac{5}{8}$ in. (1.625 in.)	$5\frac{3}{8}$ in. (5.375 in.)	44.6 in.^{3}

Section 1.5

Subtraction property of equality Check out a scale and some weights from your school's science department. Use them as part of a class presentation to explain how the subtraction property of equality is used to solve the equation $x + 2 = 5$. See the discussion and Figure 1-13 on page 49 for some suggestions on how to do this.

Section 1.6

Translation Study Exercises 13–26 of the Study Set for Section 1.6 on page 67. These problems contain key words and phrases used in occupations. When translated to mathematical symbols, the words and phrases suggest addition, subtraction, multiplication, and division. Give two new examples for each operation.

Section 1.1

Describing Numerical Relationships

CONCEPTS

Tables, bar graphs, and *line graphs* are used to describe numerical relationships.

REVIEW EXERCISES

1. Illustration 1 lists the worldwide production of wide-screen TVs. Use the data to construct a bar graph. Describe the trend in the production in words.

ILLUSTRATION 1

Year	Production (millions of units)
'95	3
'96	5
'97	7
'98	11

Source: Electronic Industries Association of Japan

2. Consider the line graph in Illustration 2 that shows the number of cars parked in a mall parking structure from 6 P.M. to 12 midnight on a Saturday.

  a. What units are used to scale the horizontal and vertical axes?

  b. How many cars were in the parking structure at 11 P.M.?

  c. At what time did the parking structure have 500 cars in it?

ILLUSTRATION 2

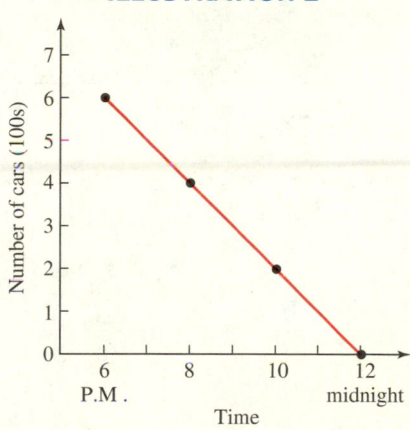

The result of an addition is called the *sum;* of a subtraction, the *difference;* of a multiplication, the *product;* and of a division, the *quotient.*

3. Express each statement in words.
  a. $15 - 3 = 12$
  b. $15 + 3 = 18$
  c. $15 \div 3 = 5$
  d. $15 \cdot 3 = 45$

4. a. Write the multiplication 4×9 in two ways: first with a raised dot $\cdot$, and then using parentheses.
  b. Write the division $9 \div 3$ without using the symbols $\div$ or $\overline{)}$.

Variables are letters used to stand for numbers.

5. Write each multiplication without a multiplication symbol.
  a. $8 \cdot b$
  b. $x \cdot y$
  c. $2 \cdot l \cdot w$
  d. $P \cdot r \cdot t$

An *equation* is a mathematical sentence that contains an $=$ sign. Variables and/or numbers can be combined with the operations of addition, subtraction, multiplication, and division to create *algebraic expressions.*

6. Classify each item as either an algebraic expression or an equation.
  a. $5 = 2x + 3$
  b. $2x + 3$
  c. $\dfrac{t + 6}{12}$
  d. $P = 2l + 2w$

Equations that express a known relationship between two or more variables are called *formulas.*

7. Translate the word model to an equation that mathematically describes this situation: The number of tellers needed by a bank is the quotient of the number of bank accounts and 350.

8. Use the equation (formula) $n = b + 5$ to complete the table in Illustration 3.

9. Use the data in Illustration 4 to write an equation (formula) that mathematically describes the relationship between the two quantities; then state the relationship in words.

ILLUSTRATION 3	
Number of brackets	Number of nails
5	
10	
20	

ILLUSTRATION 4	
Number of children	Total fees (dollars)
1	50
2	100
4	200

Section 1.2

The Real Numbers

The *natural numbers:*
{1, 2, 3, 4, 5, 6, . . .}
The *whole numbers:*
{0, 1, 2, 3, 4, 5, 6, . . .}
The *integers:*
{. . . , −3, −2, −1, 0, 1, 2, 3, . . .}

10. Which number is a whole number but not a natural number?

11. Graph each member of the set {−3, 5, 0, −1} on the number line.

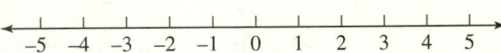

12. Represent each of these situations with a signed number.
    a. A budget deficit of $65 billion   **b.** 206 feet below sea level

Two *inequality symbols* are
  > "is greater than"
  < "is less than"

13. Use one of the symbols > or < to make each statement true.
    a. 0 ☐ 5   **b.** 6 ☐ 4
    c. −12 ☐ −13   **d.** −3 ☐ 2

A *rational number* is any number that can be written as a fraction with an integer in its numerator and a nonzero integer in its denominator.

14. Show that each of the following numbers is a rational number by expressing it as a fraction.
    a. 5   **b.** −12
    c. 0.7   **d.** $4\frac{2}{3}$

Rational numbers are either *terminating* or *repeating* decimals.

An *irrational number* is a nonterminating, nonrepeating decimal.

15. Graph each member of the set $\left\{-\pi, 0.333. . . , 3.75, -\frac{17}{4}, \frac{7}{8}\right\}$ on the number line.

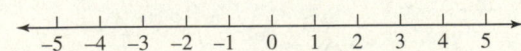

16. Use a calculator to approximate 3π and $2\sqrt{2}$ to the nearest hundredth.

A *real number* is any number that is either a rational or an irrational number.

The natural numbers are a *subset* of the whole numbers. The whole numbers are a subset of the integers. The integers are a subset of the rational numbers.

17. Tell whether each statement is true or false.
  a. All integers are whole numbers.
  b. π is an irrational number.
  c. The real numbers are the set of all decimals.
  d. A real number is either rational or irrational.

18. Tell which numbers in the given set are natural numbers, whole numbers, integers, rational numbers, irrational numbers, and real numbers.
$$\left\{-\tfrac{4}{5},\ 99.99,\ 0,\ \sqrt{2},\ -12,\ 4\tfrac{1}{2},\ 0.666.\ .\ .\ ,\ 8\right\}$$

Two numbers represented by points on a number line that are the same distance away from the origin, but on opposite sides of it, are called *opposites* or *additive inverses*.

19. Write the expression in simpler form.
  a. The opposite of 10
  b. The opposite of -3
  c. $-\left(-\tfrac{9}{16}\right)$
  d. -0

The *absolute value* of a number is the distance on the number line between the number and 0.

20. Insert one of the symbols $>$, $<$, or $=$ in the blank to make each statement true.
  a. $|-6|\ \square\ |5|$
  b. $|-39|\ \square\ 39$
  c. $-9\ \square\ -|-10|$
  d. $\left|-\tfrac{1}{4}\right|\ \square\ 0.333.\ .\ .$

Section 1.3   Exponents and Order of Operations

An *exponent* is used to represent repeated multiplication. In the *exponential expression* a^n, a is the base, and n is the exponent.

21. Write each expression using exponents.
  a. $8 \cdot 8 \cdot 8 \cdot 8 \cdot 8$
  b. $(2)(2)(2)$
  c. $5 \cdot 5 \cdot 5 \cdot 9 \cdot 9$
  d. $a \cdot a \cdot a \cdot a$
  e. $9 \cdot \pi \cdot r \cdot r$
  f. $x \cdot x \cdot x \cdot y \cdot y \cdot y \cdot y$
  g. one hundred squared
  h. the sixth power of one

22. Evaluate each expression.
  a. 9^2
  b. 2 cubed
  c. 2^5
  d. 15^1

23. CHECKERBOARD Use the formula $A = s^2$ to find the area of a checkerboard that has sides of length 15 inches.

24. STORAGE At a preschool, each child's jacket, sack lunch, and papers are kept in individual cubicles called "cubbyholes" by the kids. The cubbyholes are plywood cubes with an open front and have sides 1.75 feet long. Draw a picture of a cubicle and then find its volume using the formula $V = s^3$. Round the result to the nearest tenth of a cubic foot.

Order of operations
If an expression contains grouping symbols, do all calculations working from the innermost pair to the outermost pair, in the following order:
1. Evaluate all exponential expressions.

25. How many operations does the expression $5 \cdot 4 - 3^2 + 1$ contain, and in what order should they be performed?

26. Evaluate each expression.
  a. $24 - 3 \cdot 6$
  b. $3^2 + 10^3$
  c. $6 \cdot 5 - 18 \div 9$
  d. $2(3)(4)(5)$
  e. $800 - 3 \cdot 4^4$
  f. $8\left(\tfrac{12}{3}\right) - \tfrac{30}{6} + 2$
  g. $3(5)^2\left(\tfrac{24}{6}\right)$
  h. $60 \div 20 + 10$

2. Do all multiplications and divisions, working from left to right.

3. Do all additions and subtractions, working from left to right.

If the expression does not contain grouping symbols, begin with step 1.

In a fraction, simplify the numerator and the denominator separately. Then simplify the fraction, whenever possible.

The *arithmetic mean* (or *average*) is a value around which number values are grouped.

$$\text{Mean} = \frac{\text{sum of values}}{\text{number of values}}$$

27. Tell the name of each type of grouping symbol: ( ), [ ], —.

28. Evaluate each expression.

a. $(3 + 4 \cdot 2) - 3^2$

b. $2 + 3[2 \cdot 8 - (3^2 + 2 \cdot 3)]$

c. $(200 - 3^3 + 1) - 64$

d. $2(5 - 4)^2$

e. $\dfrac{8^2 - 10}{2(3)(4) - 2 \cdot 3^2}$

f. $\dfrac{5 + (80 - 8 + 4)}{3 + (6 - 2 \cdot 3)}$

29. COMPARISON SHOPPING Using the prices listed in Illustration 5, find the average (mean) cost of a new Ford Explorer.

30. WALK-A-THON Use the data in Illustration 6 to find the average (mean) donation to a charity walk-a-thon.

ILLUSTRATION 5

Dealer	Price of Explorer
Inland Motors	$19,981
Ford City	$21,456
Central Ford	$20,599
C.J. Smith Ford	$22,500

ILLUSTRATION 6

Donation	Number received
$5	20
$10	65
$20	25
$50	5
$100	10

Section 1.4

In order to describe numerical relationships, we need to translate the words of a problem into mathematical symbols.

Key words and *key phrases* are used to represent the operations of addition, subtraction, multiplication, and division.

Algebraic Expressions

31. Write each phrase as an algebraic expression.

a. 25 more than the height h

b. 15 less than the cutoff score s

c. $\frac{1}{2}$ of the time t

d. the product of 6 and x

32. An advertisement stated that the regular price p of a living room set was slashed $500 during a spring clearance sale. Express the sale price of the furniture, in dollars, using an algebraic expression.

33. The time a man spent standing in line to pay his fees was three times longer than it took to complete his driving test. Express each length of time using a variable.

34. See Illustration 7.

    a. If we let n represent the length of the nail, write an algebraic expression for the length of the bolt (in cm).

    b. If we let b represent the length of the bolt, write an algebraic expression for the length of the nail (in cm).

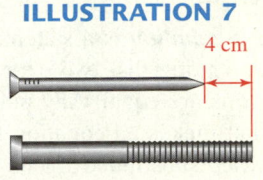

ILLUSTRATION 7

4 cm

Sometimes you must rely on common sense and insight to find hidden operations.

35. **a.** How many years are in d decades?

    b. If you have x donuts, how many dozen donuts do you have?

    c. Five years after a house was constructed, a patio was added. How old, in years, is the patio if the house is x years old?

Number · value = total value

36. Complete the table in Illustration 8.

ILLUSTRATION 8

Type of coin	Number	Value in cents	Total value in cents
Nickel	6		
Dime	d		

When we replace the variable, or variables, in an algebraic expression with specific numbers and then apply the rules for the order of operations, we are evaluating the algebraic expression.

37. Complete the table of values in Illustration 9.

ILLUSTRATION 9

x	$20x - x^3$
0	
1	
4	

ILLUSTRATION 10

input	output
0	
	15
	60

38. Complete the input/output table in Illustration 10 for the algebraic expression $\dfrac{x}{3} - \dfrac{x}{5}$.

39. Evaluate each algebraic expression for the given value(s) of the variable(s).

    a. $6x$ for $x = 6$

    b. $7x^2 - \frac{x}{2}$ for $x = 4$

    c. $b^2 - 4ac$ for $b = 10$, $a = 3$, and $c = 5$

    d. $2(24 - 2c)^3$ for $c = 9$

    e. $3r^2h + 1$ for $r = 7$ and $h = 5$

    f. $\dfrac{x + y}{x - y}$ for $x = 19$ and $y = 17$

40. Use a calculator to find the volume, to the nearest tenth of a cubic inch, of the ice cream waffle cone in Illustration 11 by evaluating the algebraic expression $\frac{\pi r^2 h}{3}$.

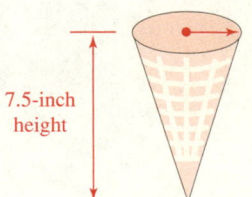

ILLUSTRATION 11

1.5-inch radius

7.5-inch height

Section 1.5

An *equation* is a statement indicating that two expressions are equal. Any number that makes an equation true when substituted for its variable is said to *satisfy* the equation. Such numbers are called *solutions* or *roots*.

To *solve an equation,* isolate the variable on one side of the equation by undoing the operations performed on it.

If the same number is added to, or subtracted from, both sides of an equation, an equivalent equation results.
If $a = b$, then $a + c = b + c$.
If $a = b$, then $a - c = b - c$.

If both sides of an equation are multiplied, or divided, by the same nonzero number, an *equivalent* equation results.
If $a = b$, then $ca = cb$ $(c \neq 0)$.
If $a = b$, then $\dfrac{a}{c} = \dfrac{b}{c}$ $(c \neq 0)$.

We can translate a percent problem from words into an equation. A variable is used to stand for the unknown number; *is* can be translated to an $=$ sign; and *of* means multiply.

The percent formula:
   Amount = percent · base

Solving Equations

41. Tell whether the given number is a solution of the equation.

a. $x - 34 = 50$; $x = 84$ **b.** $5y + 2 = 12$; $y = 3$

c. $\frac{x}{5} = 6$; $x = 30$ **d.** $a^2 - a - 1 = 0$; $a = 2$

e. $5b - 2 = 3b + 3$; $b = 3$ **f.** $\dfrac{2}{y + 1} = \dfrac{12}{y + 1} - 5$; $y = 1$

42. Fill in the blanks to make the statement true: When solving the equation $x + 8 = 10$, we are to find all the values of the _____ that make the equation a _____ statement.

43. Solve each equation. Check all solutions.

a. $x - 9 = 12$ **b.** $y + 15 = 32$

c. $4 = v - 1$ **d.** $100 = 7 + x$

e. $2x = 40$ **f.** $120 = 15c$

g. $x - 11 = 0$ **h.** $p + 3 = 3$

i. $\dfrac{t}{8} = 12$ **j.** $3 = \dfrac{q}{26}$

k. $6b = 0$ **l.** $\dfrac{x}{14} = 0$

44. Translate "16 is 5% of an unknown number" into an equation.

45. COST OF LIVING A retired trucker receives a monthly Social Security check of $764. If she is to receive a 3.5% cost-of-living increase soon, how much larger will her check be?

46. 4.81 is 2.5% of what number?

47. FAMILY BUDGET It is recommended that a family pay no more than 30% of its monthly income (after taxes) on housing. If a family has an after-tax income of $1,890 per month and pays $625 in housing costs each month, are they within the recommended range?

48. AUTOMOBILE SALES In 1997, 8,272,074 passenger cars were sold in the United States. Use the circle graph in Illustration 12 to determine the number sold by General Motors.

ILLUSTRATION 12

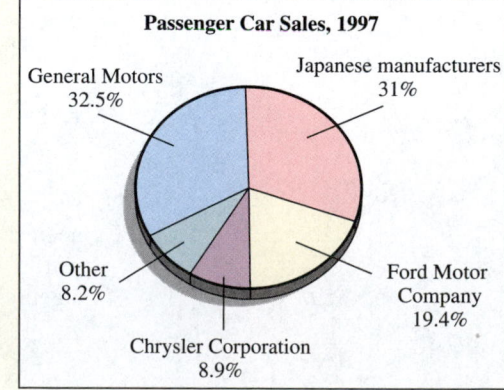

Passenger Car Sales, 1997

General Motors 32.5%
Japanese manufacturers 31%
Other 8.2%
Ford Motor Company 19.4%
Chrysler Corporation 8.9%

Based on data from the American Automobile Manufacturers Association

49. COLLECTIBLES A collector of football trading cards paid $6 for a 1984 Dan Marino rookie card several years ago. If the card is now worth $100, what is the percent of increase in the card's value? (Round to the nearest percent.)

Section 1.6

An *equation* is a mathematical statement that two quantities are equal.

To solve a problem, follow these steps:
1. Analyze the problem.
2. Form an equation.
3. Solve the equation.
4. State the conclusion.
5. Check the result.

Drawing a diagram or creating a table is often helpful in problem solving.

Problem Solving

50. Write an equation that describes the same quantity in two ways: At the end of the holiday season, a bakery had sold 165 cherry cheesecakes. This was 28 cheesecakes less than the projected sales total. Let x = the number of cheesecakes the bakery had hoped to sell.

51. SOCIAL WORK A human services program assigns each of its social workers a caseload of 75 clients. How many clients are served by this program if it employs 20 social workers?

52. HISTORIC TOUR A driving tour of three historic cities is an 858-mile round trip. Beginning in Boston, the drive to Philadelphia is 296 miles. From Philadelphia to Washington, DC is another 133 miles. How long would the return trip to Boston be?

53. CARDS A standard deck of 54 playing cards contains 2 jokers, 4 aces, and 12 face cards; the remainder are numbered cards. How many numbered cards are there in a standard deck?

54. CHROME WHEELS Find the measure of the angle between each of the spokes on the wheel shown in Illustration 13.

ILLUSTRATION 13

CHAPTER 1

Test

The graph in Illustration 1 shows the cost to hire a security guard. Use the graph to answer Problems 1 and 2.

ILLUSTRATION 1

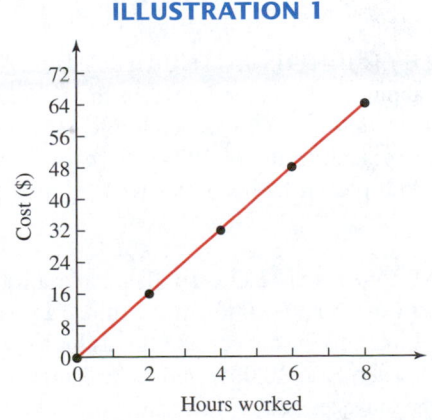

1. What will it cost to hire a security guard for 3 hours?

2. If a school was billed $40 for hiring a security guard for a dance, for how long did the guard work?

3. Use the formula $f = \frac{a}{5}$ to complete the table.

Area in square miles	Number of fire stations
15	
100	
350	

4. Graph each member of the set on the number line.
$$\left\{-1\tfrac{1}{4}, \sqrt{2}, -3.75, \tfrac{7}{2}, 0.5\right\}$$

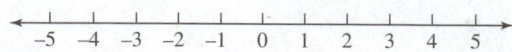

5. Tell whether each statement is true or false.
   a. Every integer is a rational number.
   b. Every rational number is an integer.
   c. π is an irrational number.
   d. 0 is a whole number.

6. Insert the proper symbol, $>$ or $<$, in the blanks to make each statement true.
   a. -2 ___ -3
   b. $-|-7|$ ___ 8
   c. $|-4|$ ___ $-(-5)$
   d. $\left|-\tfrac{7}{8}\right|$ ___ 0.5

7. Find the volume of a styrofoam ice chest that is a cube with sides of length 10 inches.

8. Rewrite each product using exponents:
   a. $9 \cdot 9 \cdot 9 \cdot 9 \cdot 9$
   b. $3 \cdot x \cdot x \cdot z \cdot z \cdot z$.

9. Evaluate 5^6.

10. Evaluate $8 + 2 \cdot 3^4$.

11. Evaluate $9^2 - 3[45 - 3(6 + 4)]$.

12. Evaluate $\dfrac{4^3 - 2 \cdot 13}{4\left(\tfrac{25}{5}\right) - 1^4}$.

13. What is a real number?

14. Explain the difference between an expression and an equation.

15. SICK DAYS Use the data in Illustration 2 to find the average (mean) number of sick days used by this group of employees this year.

ILLUSTRATION 2

Name	Sick days	Name	Sick days
Chung	4	Ryba	0
Cruz	8	Nguyen	5
Damron	3	Tomaka	4
Hammond	2	Young	6

16. Complete the table in Illustration 3.

ILLUSTRATION 3

x	$2x - \tfrac{30}{x}$
5	
10	
30	

17. Evaluate the expression $2lw + w^2$ for $l = 4$ and $w = 8$.

18. A rock band recorded x songs for a CD. Technicians had to delete two songs from the album, because of poor sound quality. Express the number of songs on the CD using an algebraic expression.

19. What is the value of q quarters in cents?

20. Is $x = 3$ a solution of the equation $2x + 3 = 4x - 6$?

21. Solve $x + 11 = 24$.
22. Solve $22t = 110$.

23. Solve $5 = x - 25$.
24. Solve $\tfrac{c}{10} = 55$.

25. DOWN PAYMENT To buy a house, a woman was required to make a down payment of \$11,400. What did the house sell for if this was 15% of the purchase price?

26. VIDEO SALES Illustration 4 shows the number of videos sold nationwide during the fourth quarter of 1997 and the first quarter of 1998. Find the percent of decrease in the number of videos sold (to the nearest percent). What do you think is the reason for such a dramatic decrease?

ILLUSTRATION 4

Quarterly Video Sales (in millions of units)	
4th qtr, 1997	1st qtr, 1998
97	61

27. MULTIPLE BIRTHS IN THE UNITED STATES In 1994, about 4,500 women gave birth to three or more babies at one time. This is quadruple (4 times) the number of such births in 1974, 20 years earlier. How many multiple births occurred in the United States in 1974?

28. GREAT LAKES The Great Lakes have a total surface area of about 94,000 square miles. In square miles, Lake Huron covers 23,000, Lake Michigan 22,000, Lake Erie 10,000, and Lake Ontario 7,000. Find the surface area of Lake Superior.

2

Real Numbers, Equations, and Inequalities

CAMPUS CONNECTION

The Automotive Technology Department

In automotive courses, instructors stress that mechanics need a solid understanding of mathematics to service today's technologically advanced cars. During their training, automotive students will use different types of *formulas* dealing with such things as carburetors, brakes, and electronic ignition systems. One basic formula they will use is $d = rt$, where d is the distance traveled by a car, t is the time traveled, and r is the rate or speed. This chapter will help you build the mathematical foundation necessary to be able to work formulas from many different disciplines, including automotive technology.

In this chapter, we will learn how to add, subtract, multiply, and divide real numbers, and we will use these skills to solve equations and inequalities.

▶ **2.1**

Adding and Subtracting Real Numbers

In this section, you will learn about

Adding two real numbers with the same sign ▪ Adding two real numbers with different signs ▪ Properties of addition ▪ Subtracting real numbers ▪ Solving equations

Introduction   Recall that all of the points on a number line represent the set of real numbers. Real numbers that are greater than zero are *positive real numbers.* (See Figure 2-1.) Positive numbers can be written with or without a + sign. For example, $2 = +2$ and $4.75 = +4.75$. Real numbers that are less than zero are *negative real numbers.* They are always written with a − sign. For example, negative $2 = -2$ and negative $4.75 = -4.75$.

FIGURE 2-1

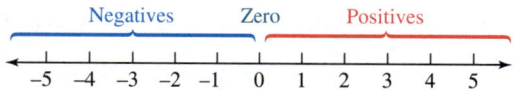

 WARNING! Zero is neither positive nor negative.

We use real numbers to describe many situations. Words such as *gain, above, up, to the right,* and *in the future* indicate positive numbers. Words such as *loss, below, down, to the left,* and *in the past* indicate negative numbers.

In words	*In symbols*	*Meaning*
16 degrees above 0	$+16°$	positive sixteen degrees
5 degrees below 0	$-5°$	negative five degrees
a balance of $590.80	$590.80	positive five hundred ninety dollars and eighty cents
$104.93 overdrawn	$-\$104.93$	negative one hundred four dollars and ninety-three cents
750 feet above sea level	750	positive seven hundred fifty
$25\frac{1}{2}$ feet below sea level	$-25\frac{1}{2}$	negative twenty-five and one-half

The bar graph in Figure 2-2 shows the 1997 quarterly profits and losses of Greyhound Bus Lines. In the figure, the first-quarter loss of $17 million and the second-quarter loss of $29 million are represented by the negative numbers −17 and −29. The third-quarter profit of $5 million and a fourth-quarter profit of $25 million are represented by 5 and 25. To find Greyhound's 1997 annual net income, we must add these positive and negative numbers:

$$\text{Annual income} = -17 + (-29) + 5 + 25$$

In this section, we will discuss how to add and subtract positive and negative real numbers and find the annual profit (or loss) of Greyhound.

FIGURE 2-2

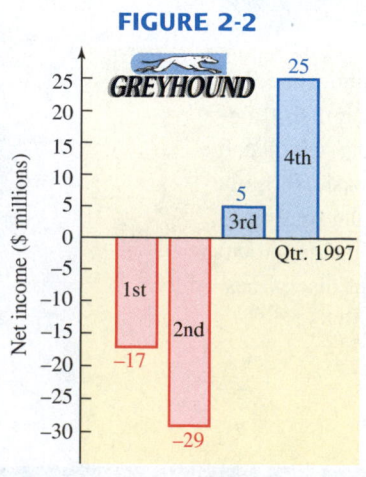

Adding Two Real Numbers with the Same Sign

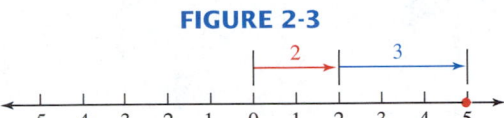

To explain the addition of signed numbers, we can use a number line, as shown in Figure 2-3. To compute $2 + 3$, we begin at the **origin** (the zero point) and draw an arrow 2 units long, pointing to the right. This represents 2. From that point, we draw an arrow 3 units long, also pointing to the right. This represents 3. We end up at 5; therefore, $2 + 3 = 5$.

FIGURE 2-3

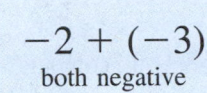

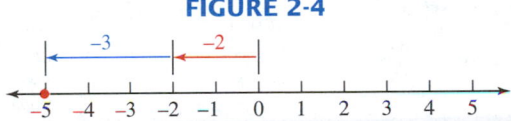

To compute $-2 + (-3)$ on a number line, we begin at the origin and draw an arrow 2 units long, pointing to the left. This represents -2. From there, we draw an arrow 3 units long, also pointing to the left. This represents -3. We end up at -5, as shown in Figure 2-4; therefore, $-2 + (-3) = -5$.

FIGURE 2-4

As a check, think of this problem in terms of money. If you had a debt of \$2 (-2) and incurred another debt of \$3 (-3), you would have a debt of \$5 (-5).

From the first two examples, we observe that both arrows point in the same direction and build on each other. The result has the same sign as the two real numbers that are being added.

$$2 \quad + \quad 3 \quad = \quad 5 \qquad \text{and} \qquad -2 \quad + \quad (-3) \quad = \quad -5$$

positive + positive = positive   negative + negative = negative

These observations suggest the following rule.

Adding two real numbers with the same sign

To add two real numbers with the same sign, add their absolute values and attach their common sign to the sum.

If both real numbers are positive, the sum is positive. If both real numbers are negative, the sum is negative.

E X A M P L E 1   **Adding real numbers with the same sign.** Find the sum: $-25 + (-18)$.

Solution   Since both real numbers are negative, the answer will be negative.

$$-25 + (-18) = -43 \quad \text{Add their absolute values, 25 and 18, to get 43. Use their common sign.}$$

SELF CHECK   Find the sum: $-45 + (-12)$.   *Answer:* -57 ■

Adding Two Real Numbers with Different Signs

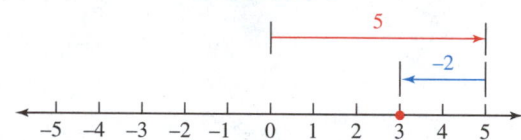

$$5 + (-2)$$

one positive, one negative

To compute $5 + (-2)$ on a number line, we start at the origin and draw an arrow 5 units long, pointing to the right; this represents 5. From there, we draw an arrow 2 units long, pointing to the left; this represents -2. We end up at 3, as shown in Figure 2-5. Therefore, $5 + (-2) = 3$. In terms of money, if you had \$5 (+5) and lost \$2 (-2), you would have \$3 (+3) left.

FIGURE 2-5

To compute $-5 + 3$ on a number line, we start at the origin and draw an arrow 5 units long, pointing to the left; this represents -5. From there, we draw an arrow 3 units long, pointing to the right; this represents 3. We end up at -2, as shown in Figure 2-6.

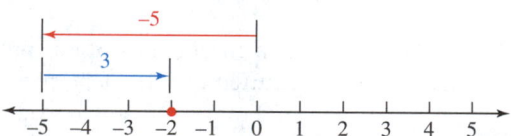

$$-5 + 3$$

one negative, one positive

Therefore, $-5 + 3 = -2$. In terms of money, if you owed a friend \$5 (-5) and paid back \$3 (+3), you would still owe your friend \$2 (-2).

FIGURE 2-6

From the previous examples, we observe that the arrows point in opposite directions and that the longer arrow determines the sign of the result.

$$5 \;+\; (-2) \;=\; 3 \qquad \text{and} \qquad -5 \;+\; 3 \;=\; -2$$

positive + negative = positive   negative + positive = negative

These observations suggest the following rule.

> **Adding two real numbers with different signs**
>
> To add two real numbers with different signs, subtract their absolute values (the smaller from the larger). To this result, attach the sign of the number with the larger absolute value.

EXAMPLE 2   **Adding real numbers with different signs.** Find each sum:
a. $-17 + 32$,   **b.** $5.4 + (-7.7)$.

Solution   **a.** Since 32 has the larger absolute value, the answer will be positive.

$$-17 + 32 = 15 \qquad \text{Subtract their absolute values, 17 from 32, to get 15.}$$

b. Since -7.7 has the larger absolute value, the answer will be negative.

$$5.4 + (-7.7) = -2.3 \qquad \text{Subtract their absolute values, 5.4 from 7.7, to get 2.3. Use a } - \text{ sign.}$$

SELF CHECK   Find each sum: **a.** $63 + (-87)$ and
b. $-\frac{1}{5} + \frac{3}{5}$.   *Answers:*   **a.** -24,   **b.** $\frac{2}{5}$ ■

EXAMPLE 3   **Adding several real numbers.**   Find the 1997 annual net income of Greyhound Bus Lines from the data given in the graph in Figure 2-2.

Solution   To find the annual net income, add the 1997 quarterly profits and losses. We use the rules for order of operations and do the additions as they occur from left to right.

$$-17 + (-29) + 5 + 25 = -46 + 5 + 25 \quad \text{Add: } -17 + (-29) = -46.$$
$$= -41 + 25 \quad \text{Add: } -46 + 5 = -41.$$
$$= -16$$

In 1997, Greyhound lost $16 million.

SELF CHECK   Add $-7 + 13 + (-5) + 10$.   *Answer:* 11   ■

ACCENT ON TECHNOLOGY *The Sign Change Key*

A scientific calculator can add positive and negative numbers.

■ You don't have to do anything special to enter positive numbers. When you press 5, for example, a positive 5 is entered.

■ To enter a negative 17 on a scientific calculator, you must press the $\boxed{+/-}$ key after entering 17. This key is called the *opposite* or *sign change* key.

To find the annual net income of Greyhound Bus Lines, we must find the sum $-17 + (-29) + 5 + 25$. To do so, we enter these numbers and press these keys:

Keystrokes   17 $\boxed{+/-}$ $\boxed{+}$ 29 $\boxed{+/-}$ $\boxed{+}$ 5 $\boxed{+}$ 25 $\boxed{=}$ $\boxed{\qquad -16}$

Using a graphing calculator, we enter a negative value by first pressing the negation key $\boxed{(-)}$. To find the sum, we press these keys:

Keystrokes $\boxed{(-)}$ 17 $\boxed{+}$ $\boxed{(-)}$ 29 $\boxed{+}$ 5 $\boxed{+}$ 25 $\boxed{\text{ENTER}}$

$$\boxed{\begin{array}{r} ^-17 + ^-29 + 5 + 25 \\ ^-16 \end{array}}$$

The sum is -16. Greyhound's 1997 net loss was $16 million.

Properties of Addition

The operation of addition has some special properties. The first property, called the **commutative property,** states that two real numbers can be added in either order to get the same result. For example, when adding the numbers 10 and -25, we see that

$$10 + (-25) = -15 \quad \text{and} \quad -25 + 10 = -15$$

To state the **commutative property of addition** concisely, we use variables.

The commutative property of addition

If a and b represent any real numbers, then

$$a + b = b + a$$

To find the sum of three numbers, we first add two of them and then add the third to that result. For example, we can add $-3 + 7 + 5$ in two ways.

Method 1: Group -3 and 7

$(-3 + 7) + 5 = 4 + 5$ Because of the parentheses, add -3 and 7 first to get 4.

$= 9$ Then add 4 and 5.

Method 2: Group 7 and 5

$-3 + (7 + 5) = -3 + 12$ Because of the parentheses, add 7 and 5 first to get 12.

$= 9$ Then add -3 and 12.

Either way, the sum is 9, which suggests that it doesn't matter how we group or "associate" numbers in addition. This property is called the **associative property of addition.**

The associative property of addition

If a, b, and c represent any real numbers, then

$$(a + b) + c = a + (b + c)$$

EXAMPLE 4

"Jeopardy." A contestant on the game show "Jeopardy" answered the first question correctly to win $100, missed the second question to lose $200, answered the third question correctly to win $300, and answered the fourth question incorrectly to lose $400. Find her net gain or loss after four questions.

Solution   "To win $100" can be represented by 100. "To lose $200" can be represented by -200. "To win $300" can be represented by 300, and "to lose $400" can be represented by -400. Her net gain or loss is the sum of these four numbers. We can find the sum by doing the additions from left to right. An alternate method, which uses the commutative and associative properties of addition, is to add the positives, then add the negatives, and finally add those results.

$100 + (-200) + 300 + (-400) = 100 + 300 + (-200) + (-400)$ Reorder the terms.

$= (100 + 300) + [(-200) + (-400)]$ Group the positives together. Group the negatives together.

$= 400 + (-600)$ Add the positives. Add the negatives.

$= -200$

After four questions, she had a net loss of $200. ∎

Whenever we add zero to a number, the number remains the same. For example,

$$0 + 8 = 8, \qquad 2.3 + 0 = 2.3, \qquad \text{and} \qquad -16 + 0 = -16$$

These examples suggest the **addition property of zero.**

Addition property of zero

If a represents any real number, then

$$a + 0 = a \qquad \text{and} \qquad 0 + a = a$$

Two numbers that are the same distance away from the origin, but on opposite sides of it, are called **opposites** or **additive inverses**. For example, 10 is the additive inverse of -10, and -10 is the additive inverse of 10. Whenever we add opposites or additive inverses, the result is 0.

$$10 + (-10) = 0, \qquad -\tfrac{4}{5} + \tfrac{4}{5} = 0 \qquad 56.8 + (-56.8) = 0$$

Adding opposites (additive inverses)

If a represents any number, then

$$a + (-a) = 0$$

Subtracting Real Numbers

The subtraction $5 - 2$ can be thought of as taking 2 away from 5. We can use the number line shown in Figure 2-7 to illustrate this. Beginning at the origin, we draw an arrow of length 5 units pointing to the right. From that point, we move back 2 units to the left. The result, 3, is called the **difference.**

FIGURE 2-7

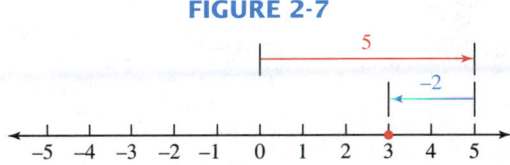

Figure 2-7 looks like the illustration for the addition problem $5 + (-2)$ shown in Figure 2-5. In the problem $5 - 2$, we subtracted 2 from 5. In the problem $5 + (-2)$, we added -2 (which is the opposite of 2) to 5. In each case, the result is 3.

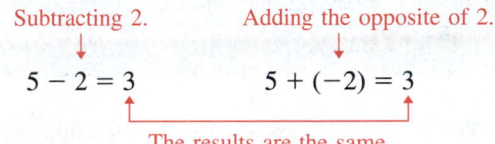

This observation suggests the following rule.

Subtracting real numbers

If a and b represent any real numbers, then

$$a - b = a + (-b)$$

This rule indicates that *subtraction is the same as adding the opposite of the number to be subtracted.* We won't need this rule for every subtraction problem. For example, $5 - 2$ is obviously 3. However, for more complicated problems such as $-8 - (-3)$, where the result is not obvious, the subtraction rule will be helpful.

$$-8 - (-3) = -8 + 3 \qquad \text{To subtract } -3, \text{ add the opposite of } -3, \text{ which is 3.}$$
$$= -5 \qquad \text{Do the addition.}$$

EXAMPLE 5

Adding the opposite. Find   **a.** $-13 - 18$,   **b.** $-45 - (-27)$,   and
c. $\frac{1}{4} - \left(-\frac{1}{8}\right)$.

Solution   **a.** To subtract 18 from -13, we use the subtraction rule.

$$-13 - 18 = -13 + (-18)$$ To subtract 18, add the opposite of 18, which is -18.

$$= -31$$ Add their absolute values, 13 and 18, to get 31. Keep their common sign.

b. To subtract -27 from -45, we use the subtraction rule.

$$-45 - (-27) = -45 + 27$$ To subtract -27, add the opposite of -27, which is 27.

$$= -18$$ Subtract their absolute values, 27 from 45, to get 18. Use the sign of the number with the greater absolute value, which is -45.

c. The lowest common denominator (LCD) for the fractions is 8.

$$\frac{1}{4} - \left(-\frac{1}{8}\right) = \frac{2}{8} - \left(-\frac{1}{8}\right)$$ Express $\frac{1}{4}$ in terms of eighths: $\frac{1}{4} = \frac{2}{8}$.

$$= \frac{2}{8} + \frac{1}{8}$$ Add the opposite of $-\frac{1}{8}$, which is $\frac{1}{8}$.

$$= \frac{3}{8}$$ Add the numerators: $2 + 1 = 3$. Write the sum over the common denominator 8.

SELF CHECK   Find: **a.** $-32 - 25$,   **b.** $1.7 - (-1.2)$,   and   *Answers:*   **a.** -57,   **b.** 2.9,
c. $-\frac{1}{2} - \frac{1}{8}$.   **c.** $-\frac{5}{8}$

ACCENT ON TECHNOLOGY *U.S. Temperature Extremes*

The record high temperature in the United States was 134°F in Death Valley, California, on July 10, 1913. The record low was -80°F at Prospect Creek, Alaska, on January 23, 1971. See Figure 2-8. To find the difference between these two temperatures, we subtract:

$$134 - (-80)$$

We can subtract positive and negative real numbers using a scientific calculator. To find $134 - (-80)$, we enter these numbers and press these keys:

Keystrokes   134 $\boxed{-}$ 80 $\boxed{+/-}$ $\boxed{=}$

If we use a graphing calculator, we enter these numbers and press these keys:

Keystrokes   134 $\boxed{-}$ $\boxed{(-)}$ 80 $\boxed{\text{ENTER}}$

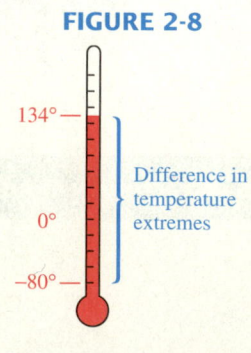

FIGURE 2-8

134° —

Difference in temperature extremes

0°

−80° —

$\boxed{\qquad\qquad 214}$

$\boxed{\begin{array}{l} 134 - {}^-80 \\ \qquad\qquad 214 \end{array}}$

The difference in the record high and low temperatures is 214°F.

EXAMPLE 6

Constructing a table of values. Evaluate the expression $5 - x$ for $x = -10$ and $x = 7$. Show the results in an input/output table of values.

Solution

To evaluate $5 - x$ for $x = -10$ and $x = 7$, we substitute these numbers for x and simplify.

x	$5 - x$
-10	15
7	-2

Evaluate for $x = -10$:
$$5 - x = 5 - (-10)$$
$$= 15$$

Evaluate for $x = 7$:
$$5 - x = 5 - 7$$
$$= -2$$

SELF CHECK

Complete the following input/output table of values.

x	$7 + x$
-8	
-4	

Answers: $-1, 3$ ■

Solving Equations

The following examples will show that solutions of equations can be negative numbers.

EXAMPLE 7

Solving equations. Solve $x + 5 = -13$ and check the result.

Solution

To isolate x on the left-hand side of the equation, we use the subtraction property of equality. To undo the addition of 5, we subtract 5 from both sides.

$$x + 5 = -13$$
$$x + 5 - 5 = -13 - 5 \quad \text{Subtract 5 from both sides.}$$
$$x = -18 \quad \text{Do the subtractions: } 5 - 5 = 0 \text{ and } -13 - 5 = -18.$$

We check by substituting -18 for x in the original equation.

$$x + 5 = -13 \quad \text{The original equation.}$$
$$-18 + 5 \overset{?}{=} -13 \quad \text{Substitute } -18 \text{ for } x.$$
$$-13 = -13 \quad \text{Do the addition: } -18 + 5 = -13.$$

Since the result is true, -18 is a solution.

SELF CHECK

Solve $-22 = 7 + y$ and check the result. *Answer:* -29 ■

EXAMPLE 8

Solving equations. Solve $-23 = y - 14$.

Solution

To isolate y on the right-hand side of the equation, we use the addition property of equality. To undo the subtraction of 14, we add 14 to both sides.

$$-23 = y - 14$$
$$-23 + 14 = y - 14 + 14 \quad \text{Add 14 to both sides.}$$
$$-9 = y \quad \text{Do the additions: } -23 + 14 = -9 \text{ and } -14 + 14 = 0.$$
$$y = -9 \quad \text{If } -9 = y, \text{ then } y = -9.$$

Check the result.

SELF CHECK

Solve $-43 = -7 + p$ and check the result. *Answer:* -36 ■

STUDY SET

Section 2.1

VOCABULARY

In Exercises 1–6, fill in the blanks to make the statements true.

1. Real numbers that are greater than zero are called _____ real numbers.

2. Real numbers that are less than zero are called _____ real numbers.

3. The only real number that is neither positive nor negative is _____.

4. The answer to a _____ problem is called a difference.

5. The _____ property of addition states that two numbers can be added in either order to get the same result.

6. The property that allows us to group numbers in addition any way we want is called the _____ property of addition.

CONCEPTS

In Exercises 7–10, use the number line in Illustration 1 to find each sum.

7. $2 + 3$ 8. $-3 + (-2)$

9. $4 + (-3)$ 10. $-5 + 3$

ILLUSTRATION 1

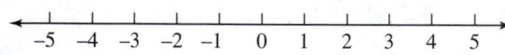

In Exercises 11–14, fill in the blanks to make the statements true.

11. To add two real numbers with the _____ sign, add their _____ values and attach their common sign to the sum.

12. To add two real numbers with different signs, _____ their absolute values, the _____ from the _____, and attach the sign of the number with the _____ absolute value.

13. To subtract b from a, add the _____ of b to a.

14. The opposite of 7 is ⬜. The opposite of -15 is ⬜.

15. Find each sum.
   a. $5 + (-5)$ b. $-2.2 + 2.2$
   c. $-\frac{3}{4} + \frac{3}{4}$ d. $19 + (-19)$

16. a. Use the variables m and n to state the commutative property of addition.
   b. Use the variables r, s, and t to state the associative property of addition.

NOTATION

In Exercises 17–20, complete each solution.

17. Find $(-13 + 6) + 4$.
$$(-13 + 6) + 4 = \boxed{} + (6 + 4)$$
$$= -13 + \boxed{}$$
$$= -3$$

18. Find $-9 + (9 + 43)$.
$$-9 + (9 + 43) = \left(\boxed{} + 9\right) + 43$$
$$= \boxed{} + 43$$
$$= 43$$

19. Solve $x + 23 = -12$.
$$x + 23 = -12$$
$$x + 23 - \boxed{} = -12 - \boxed{}$$
$$x = -35$$

20. Solve $-17 = b - 9$.
$$-17 = b - 9$$
$$-17 + \boxed{} = b - 9 + \boxed{}$$
$$-8 = b$$
$$b = -8$$

PRACTICE

In Exercises 21–38, find each sum.

21. $6 + (-8)$ 22. $4 + (-3)$

23. $-6 + 8$ 24. $-21 + (-12)$

25. $-65 + (-12)$ 26. $75 + (-13)$

27. $-10.5 + 2.3$ 28. $-2.1 + 0.4$

29. $-\dfrac{9}{16} + \dfrac{7}{16}$ 30. $-\dfrac{3}{4} + \dfrac{1}{4}$

31. $-\dfrac{1}{4} + \dfrac{2}{3}$ 32. $\dfrac{3}{16} + \left(-\dfrac{1}{2}\right)$

33. $8 + (-5) + 13$ 34. $17 + (-12) + (-23)$

35. $21 + (-27) + (-9)$ 36. $-32 + 12 + 17$

37. $-27 + (-3) + (-13) + 22$

38. $53 + (-27) + (-32) + (-7)$

In Exercises 39–42, use a calculator to find each sum.

39. $3,718 + (-5,237)$

40. $-5,235 + (-17,235)$

41. $-237.37 + (-315.07) + (-27.4)$

42. $-587.77 + (-1,732.13) + 687.39$

In Exercises 43–58, find each difference.

43. $8 - (-3)$

44. $17 - (-21)$

45. $-12 - 9$

46. $-25 - 17$

47. $-19 - (-17)$

48. $-30 - (-11)$

49. $-1.5 - 0.8$

50. $-1.5 - (-0.8)$

51. $-25 - (-25)$

52. $13 - (-13)$

53. $0 - 4$

54. $0 - (-3)$

55. $-\dfrac{1}{8} - \dfrac{3}{8}$

56. $-\dfrac{3}{4} - \dfrac{1}{4}$

57. $-\dfrac{9}{16} - \left(-\dfrac{1}{4}\right)$

58. $-\dfrac{1}{2} - \left(-\dfrac{1}{4}\right)$

In Exercises 59–62, use a calculator to find each difference.

59. $8,713 - (-3,753)$

60. $-2,727 - 1,208$

61. $-27,357.875 - 17,213.376$

62. $-45,307.039 - (-27,592.47)$

In Exercises 63–66, complete each input/output table.

63.

x	$x + 4$
-8	
-4	
20	

64.

x	$x - 5$
4	
0	
-8	

65.

x	$8 + x$
-8	
2	
-13	

66.

x	$-5 - x$
3	
-5	
-15	

In Exercises 67–70, tell whether the given number is a solution of the equation.

67. $57 + x = 12$; -45

68. $p + 37 = 65$; -26

69. $-23 + t = -12$; -11

70. $-51 = y + 6$; -57

In Exercises 71–78, solve each equation.

71. $x + 7 = -12$

72. $5 + x = -11$

73. $20 = -31 + x$

74. $-5 = -12 + x$

75. $x - 9 = -23$

76. $-9 = y - 5$

77. $a - 7 = -3$

78. $-5 = b - 12$

APPLICATIONS

In Exercises 79–90, solve each problem.

79. **MILITARY SCIENCE** During a battle, an army retreated 1,500 meters, regrouped, and advanced 2,400 meters. The next day, it advanced another 1,250 meters. Find the army's net gain.

80. **MEDICAL QUESTIONNAIRE** Determine the risk of contracting heart disease for the woman whose responses are shown in Illustration 2.

ILLUSTRATION 2

Age		Total Cholesterol	
Age	Points	Reading	Points
35	-4	280	3

Cholesterol		Blood Pressure	
HDL	Points	Systolic/Diastolic	Points
62	-3	124/100	3

Diabetic		Smoker	
	Points		Points
Yes	4	Yes	2

10-Year Heart Disease Risk

Total Points	Risk	Total Points	Risk
-2 or less	1%	5	4%
-1 to 1	2%	6	6%
2 to 3	3%	7	6%
4	4%	8	7%

Source: National Heart, Lung, and Blood Institute

81. **GOLF** Illustration 3 shows the top four finishers from the 1997 Masters Golf Tournament. The scores for each round are related to *par*, the standard number of strokes deemed necessary to complete the course. A score of -2, for example, indicates that the golfer used 2 strokes less than par to complete the course. A score of $+5$ indicates the golfer used five strokes more than par.

  a. Determine the tournament total for each golfer.

  b. Tiger Woods won by the largest margin in the history of the Masters. What was the margin?

ILLUSTRATION 3

Leaderboard

	Round				
	1	2	3	4	Total
Tiger Woods	-2	-6	-7	-3	
Tom Kite	$+5$	-3	-6	-2	
Tommy Tolles	0	0	0	-5	
Tom Watson	$+3$	-4	-3	0	

82. CREDIT CARD STATEMENT
  a. What amounts in the monthly credit card statement shown in Illustration 4 could be represented by negative numbers?
  b. What is the new balance?

ILLUSTRATION 4

Previous Balance	New Purchases, Fees, Advances & Debts	Payments & Credits	New Balance
3,660.66	1,408.78	3,826.58	

04/21/99 Billing Date	05/16/99 Date Payment Due	9,100 Credit Line

Periodic rates may vary.
See reverse for explanation and important information.
Please allow sufficient time for mail to reach us.

83. THE OLYMPICS The ancient Greek Olympian Games, which eventually evolved into the modern Olympic Games, were first held in 776 B.C. How many years after this did the 1996 Olympic Games in Atlanta, Georgia, take place?

84. SUBMARINE A submarine was cruising at a depth of 1,250 feet. The captain gave the order to climb 550 feet. Relative to sea level, find the new depth of the sub.

85. TEMPERATURE RECORDS Find the difference between the record high temperature of 108°F set in 1926 and the record low of −52°F set in 1979 for New York State.

86. LIE DETECTOR TEST A burglar scored −18 on a lie detector test, a score that indicates deception. However, on a second test, he scored +3, a score that is inconclusive. Find the difference in the scores.

87. LAND ELEVATIONS The elevation of Death Valley, California, is 282 feet below sea level. The elevation of the Dead Sea in Israel is 1,312 feet below sea level. Find the difference in their elevations.

88. STOCK EXCHANGE Many newspapers publish daily summaries of the stock market's activity. (See Illustration 5.) The last entry on the line for October 5 indicates that one share of Walt Disney Co. stock lost $\$\frac{3}{16}$ in value that day. How much did the value of a share of Disney stock rise or fall over the five-day period shown?

ILLUSTRATION 5

Oct. 5	42³/₄	23⁷/₈	Disney	.21	0.8	26	42172	25¹/₁₆	−³/₁₆
Oct. 6	42³/₄	23⁷/₈	Disney	.21	0.8	26	46600	25³/₈	+⁵/₁₆
Oct. 7	42³/₄	23⁷/₈	Disney	.21	0.8	26	46404	24¹⁵/₁₆	−³/₈
Oct. 8	42³/₄	23⁷/₈	Disney	.21	0.9	24	97098	23¹/₂	−1⁷/₁₆
Oct 9	42³/₄	23⁷/₈	Disney	.21			61333	23¹¹/₁₆	+³/₁₆

Based on data from *Los Angeles Times*

89. VOTER INFORMATION What will be the effect on state government if the ballot initiative shown in Illustration 6 passes?

ILLUSTRATION 6

212 **Campaign Spending Limits**	YES ☐ NO ☐

Limits contributions to $200 in state campaigns. Fiscal impact: Costs of $4.5 million for implementation and enforcement. Increases state revenue by $6.7 million by eliminating tax deductions for lobbying.

90. MOVIE LOSSES In 1993, the cost to Columbia Studios to produce, promote, and distribute the movie *Last Action Hero,* starring Arnold Schwarzenegger, was approximately $124 million. It is estimated that the movie earned only $44 million worldwide. What dollar loss did the studio suffer on this film?

 In Exercises 91–94, use a calculator to help solve each problem.

91. SAHARA DESERT From 1980 to 1990, a satellite was used to trace the expansion and contraction of the southern boundary of the Sahara Desert in Africa (see Illustration 7). If movement southward is represented with a negative number and movement northward with a positive number, use the data in the table to determine the net movement of the Sahara Desert boundary over the 10-year period.

ILLUSTRATION 7

Years	Distance/Direction
1980–1984	240 km/South
1984–1985	110 km/North
1985–1986	30 km/North
1986–1987	55 km/South
1987–1988	100 km/North
1988–1990	77 km/South

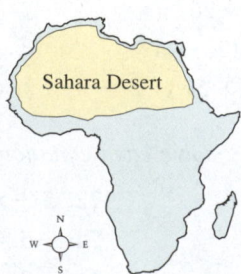

Based on data from A. Dolgoff, *Physical Geology* (D. C. Heath, 1996), p. 496

92. BANKING On February 1, Marta had $1,704.29 in a checking account. During the month, she made deposits of $713.87 and $1,245.57, wrote checks for $813.45, $937.49, and $1,532.79, and had a total of $500 in ATM withdrawals. Find her checking account balance at the end of the month.

93. CARD GAME In the second hand of a card game, Gonzalo was the winner and earned 50 points. Matt and Hydecki had to deduct the value of each of the cards left in their hands from their running point total. Use the information in Illustration 8 to update the score sheet. (Face cards are counted as 10 points and aces as 1 point.)

ILLUSTRATION 8

Matt   Hydecki

Running point total	Hand 1	Hand 2
Matt	+50	
Gonzalo	−15	
Hydecki	−2	

94. PROFITS AND LOSSES Odwalla®, a juice maker, reported a large first-quarter loss in 1997 because of a juice recall. Approximate the net loss during the nine quarters shown in Illustration 9.

ILLUSTRATION 9

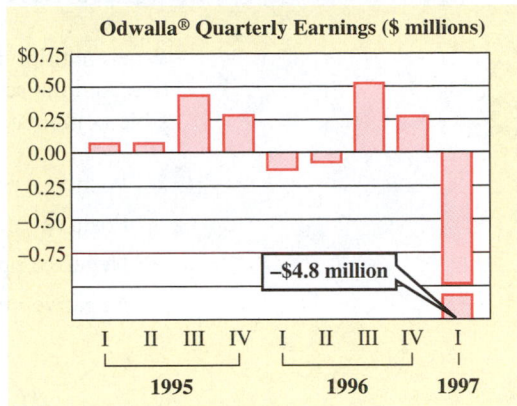

Odwalla® Quarterly Earnings ($ millions)

−$4.8 million

1995   1996   1997

Based on company reports

WRITING

95. Explain why the sum of two positive numbers is always positive and why the sum of two negative numbers is always negative.

96. Is subtracting 2 from 10 the same as subtracting 10 from 2? Explain.

97. Explain why we need to subtract when we add two real numbers with different signs.

98. Explain why we can subtract by adding the opposite.

REVIEW

In Exercises 99–102, write each expression using exponents.

99. $c \cdot c \cdot c \cdot c$ **100.** $3 \cdot d \cdot d \cdot d$

101. $a \cdot a \cdot b \cdot b \cdot b$ **102.** $3 \cdot x \cdot x \cdot x \cdot y \cdot y$

▶ 2.2   Multiplying and Dividing Real Numbers

In this section, you will learn about

   Multiplication of real numbers ■ Properties of multiplication ■ Division of real numbers ■ Properties of division ■ Division and zero ■ Solving equations

Introduction   In this course, we will often need to multiply or divide positive and negative numbers. For example,

   If the temperature drops 4° per hour for 5 hours, we can find the total drop in temperature by doing the multiplication $5(-4)$.

   If the temperature uniformly drops 30° over a 5-hour period, we can find the number of degrees it drops each hour by doing the division $\frac{-30}{5}$.

In this section, we will show how to do such multiplications and divisions.

Multiplication of Real Numbers

When multiplying two nonzero real numbers, the first factor can be positive or negative, and the second factor can be positive or negative. This means there are four possible combinations to consider.

Positive · positive

Positive · negative

Negative · positive

Negative · negative

$4(3)$
like signs, both positive

We begin by considering the product $4(3)$. Since both factors are positive, they have *like signs*. Because multiplication represents repeated addition, $4(3)$ equals the sum of four 3's.

$$4(3) = 3 + 3 + 3 + 3$$ Multiplication is repeated addition. Write 3 four times.

$$= 12$$ The result is $+12$.

This example suggests that *the product of two positive numbers is positive.*

As a check, let's think of this problem in terms of money. If someone gave you $3 four times, you would have $12.

Multiplying two positive real numbers

To multiply two positive real numbers, multiply their absolute values. The product is positive.

$4(-3)$
unlike signs
one positive, one negative

Next, we consider $4(-3)$. The signs of these factors are *unlike*. According to the definition of multiplication, $4(-3)$ means that we are to add -3 four times.

$$4(-3) = (-3) + (-3) + (-3) + (-3)$$ Multiplication is repeated addition. Write -3 four times.

$$= (-6) + (-3) + (-3)$$ Add: $-3 + (-3) = -6$.

$$= (-9) + (-3)$$ Add: $-6 + (-3) = -9$.

$$= -12$$ Add: $-9 + (-3) = -12$. The result is negative.

This example suggests that *the product of a positive number and a negative number is negative.*

In terms of money, if you lost $3 four times in the lottery, you would lose a total of $12, which is denoted as $-\$12$.

$-3(4)$
unlike signs
one negative, one positive

Next, consider $-3(4)$. The signs of these factors are *unlike*. Because changing the order when multiplying does not change the result, $-3(4) = 4(-3)$. Since $4(-3) = -12$, we know that $-3(4) = -12$. This suggests that *the product of a negative number and a positive number is negative.*

Multiplying two real numbers with unlike signs

To multiply two real numbers with unlike signs, multiply their absolute values. Then make the product negative.

E X A M P L E 1

Multiplying two numbers with unlike signs. Multiply: **a.** $8(-12)$ and **b.** $(-15)(25)$.

Solution　**a.**　$8(-12) = -96$ 　 Multiply the absolute values, 8 and 12, to get 96. Since the numbers have unlike signs, make the answer negative.

b.　$(-15)(25) = -375$ 　 Multiply the absolute values, 15 and 25, to get 375. Make the answer negative.

SELF CHECK　Multiply　**a.** $20(-30)$　and　**b.** $(-0.4)(2)$ 　　*Answers:* **a.** -600, **b.** -0.8

$$-4(-3)$$
like signs, both negative

Finally, consider the product $(-4)(-3)$. To develop a rule for multiplying two negative numbers, we examine the following pattern, in which we multiply -4 and a series of factors that decrease by 1. After finding the first four products, we graph them on a number line, as shown in Figure 2-9.

This factor decreases
by 1 as you read down
the column.

Look for a
pattern here.

$$
\begin{aligned}
-4(\mathbf{3}) &= -12 \\
-4(\mathbf{2}) &= -8 \\
-4(\mathbf{1}) &= -4 \\
-4(\mathbf{0}) &= 0 \\
-4(\mathbf{-1}) &= ? \\
-4(\mathbf{-2}) &= ? \\
-4(\mathbf{-3}) &= ?
\end{aligned}
$$

FIGURE 2-9

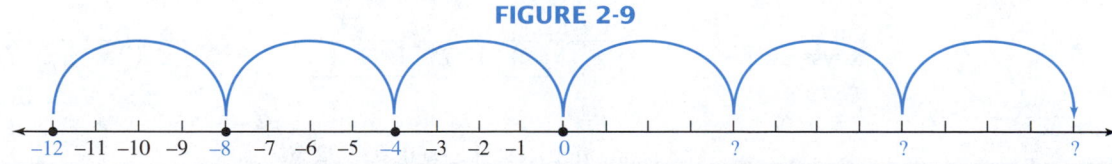

From the pattern, we see that the product increases by 4 each time. Thus,

$$-4(\mathbf{-1}) = 4, \qquad -4(\mathbf{-2}) = 8, \qquad \text{and} \qquad -4(\mathbf{-3}) = 12$$

These results suggest that *the product of two negative numbers is positive.*

Multiplying two negative real numbers

To multiply two negative real numbers, multiply their absolute values. The product is positive.

EXAMPLE 2   **Multiplying two negative numbers.** Multiply   **a.** $(-5)(-6)$,   and
b. $\left(-\frac{1}{2}\right)\left(-\frac{5}{8}\right)$.

Solution   **a.** $(-5)(-6) = 30$ Multiply the absolute values, 5 and 6, to get 30. Since both numbers are negative, the answer is positive.

b. $\left(-\frac{1}{2}\right)\left(-\frac{5}{8}\right) = \frac{5}{16}$ Multiply the absolute values, $\frac{1}{2}$ and $\frac{5}{8}$, to get $\frac{5}{16}$. The product is positive.

SELF CHECK   Multiply: **a.** $(-15)(-8)$ and   **b.** $-\frac{1}{4}\left(-\frac{1}{3}\right)$ *Answers:* **a.** 120,   **b.** $\frac{1}{12}$ ■

We now summarize the rules for multiplying two real numbers.

Multiplying two real numbers

To multiply two real numbers, multiply their absolute values.

1. The product of two numbers with *like signs* is positive.
2. The product of two numbers with *unlike signs* is negative.

ACCENT ON TECHNOLOGY   *Bank Promotion*

To attract business, a bank gave a clock radio to each customer who opened a checking account. The radios cost the bank $12.75 each, and 230 new accounts were opened. Each of the 230 radios was given away at a cost of $12.75, which can be expressed as -12.75. To find how much money the promotion cost the bank, we need to find the product of 230 and -12.75.

We can multiply positive and negative numbers with a scientific calculator. To find the product $(230)(-12.75)$, we enter these numbers and press these keys:

Keystrokes   230 $\boxed{\times}$ 12.75 $\boxed{+/-}$ $\boxed{=}$ $\boxed{-2932.5}$

Using a graphing calculator, we enter the following sequence:

Keystrokes   230 $\boxed{\times}$ $\boxed{(-)}$ 12.75 $\boxed{\text{ENTER}}$

```
230*-12.75
           -2932.5
```

The promotion cost the bank $2,932.50.

Properties of Multiplication

A special property of multiplication is that two real numbers can be multiplied in either order to get the same result. For example, when multiplying -6 and 5, we see that

$$-6(5) = -30 \quad \text{and} \quad 5(-6) = -30$$

This property is called the **commutative property of multiplication.**

The commutative property of multiplication

If a and b represent any real numbers, then

$$ab = ba$$

To find the product of three numbers, we multiply two of them and then multiply the third number by that result. For example, we can multiply $-3 \cdot 7 \cdot 5$ in two ways.

Method 1: Group −3 and 7

$(-3 \cdot 7)5 = (-21)5$ Because of the parentheses, multiply -3 and 7 first.

$\qquad\qquad = -105$ Then multiply -21 and 5.

Method 2: Group 7 and 5

$-3(7 \cdot 5) = -3(35)$ Because of the parentheses, multiply 7 and 5 first.

$\qquad\qquad = -105$ Then multiply -3 and 35.

Either way, the product is -105, which suggests that it doesn't matter how we group or "associate" numbers in multiplication. This property is called the **associative property of multiplication.**

The associative property of multiplication

If a, b, and c represent any real numbers, then

$$(ab)c = a(bc)$$

EXAMPLE 3

Multiplying more than two numbers. Multiply   **a.** $-5(-37)(2)$ and
b. $2(-3)(-2)(-3)$

Solution Using the commutative and associative properties of multiplication, we can reorder and regroup the factors to simplify the computations.

a. $-5(-37)(2) = -10(-37)$ Think of the problem as $-5(2)(-37)$ and then multiply -5 and 2.

$\qquad\qquad\qquad = 370$

b. $2(-3)(-2)(-3) = -6(6)$ Multiply the first two factors and multiply the last two factors.

$\qquad\qquad\qquad\quad = -36$

SELF CHECK Multiply: **a.** $-25(-3)(-4)$ and
b. $-1(-2)(-3)(-3)$.   *Answers:* **a.** -300, **b.** 18 ■

Whenever we multiply a number and 0, the product is 0. For example,

$$0 \cdot 8 = 0, \qquad 6.5(0) = 0, \qquad \text{and} \qquad 0(-12) = 0$$

We also see that whenever we multiply a number by 1, the number remains the same. For example,

$$6 \cdot 1 = 6, \qquad 4.53(1) = 4.53, \qquad \text{and} \qquad 1(-9) = -9$$

These examples suggest the **multiplication properties of 0 and 1.**

Multiplication properties of 0 and 1

If a represents any real number, then

$$a \cdot 0 = 0 \qquad \text{and} \qquad 0 \cdot a = 0$$
$$a \cdot 1 = a \qquad \text{and} \qquad 1 \cdot a = a$$

Division of Real Numbers

Every division fact containing three numbers can be written as an equivalent multiplication fact containing the same three numbers. For example,

$$\frac{15}{5} = 3 \qquad \text{because} \qquad 5(3) = 15$$

We will use this relationship between multiplication and division to develop the rules for dividing signed numbers. There are four cases to consider.

$$\frac{15}{5}$$

like signs, both positive

From the previous example, $\frac{15}{5} = 3$, we see that *the quotient of two positive numbers is positive.*

$$\frac{-15}{-5}$$

like signs, both negative

To determine the quotient of two negative numbers, we consider the division $\frac{-15}{-5} = ?$. We can do the division by examining its related multiplication fact: $-5(?) = -15$. To find the integer that should replace the question mark, we use the rules for multiplying signed numbers discussed earlier in this section.

Multiplication fact	*Division fact*
$-5(?) = -15$	$\dfrac{-15}{-5} = 3$
This must be *positive* 3 if the product is to be *negative* 15.	So the quotient is *positive* 3.

From this example, we see that *the quotient of two negative numbers is positive.*

$$\frac{15}{-5}$$

unlike signs
one positive, one negative

To determine the quotient of a positive number and a negative number, we consider $\frac{15}{-5} = ?$ and its equivalent multiplication fact $-5(?) = 15$.

Multiplication fact	*Division fact*
$-5(?) = 15$	$\dfrac{15}{-5} = -3$
This must be *negative* 3 if the product is to be *positive* 15.	So the quotient is *negative* 3.

From this example, we see that *the quotient of a positive number and a negative number is negative.*

$$\frac{-15}{5}$$

unlike signs
one negative, one positive

To determine the quotient of a negative number and a positive number, we consider $\frac{-15}{5} = ?$ and its equivalent multiplication fact $5(?) = -15$.

Multiplication fact

$$5(?) = -15$$

↑

This must be
negative 3 if the
product is to be
negative 15.

Division fact

$$\frac{-15}{5} = -3$$

↑

So the quotient
is *negative* 3.

From this example, we see that *the quotient of a negative number and a positive number is negative.*

We can now summarize the results from the previous discussion. Note that the rules for division are similar to those for multiplication.

Dividing two real numbers

To divide two real numbers, divide their absolute values.

1. The quotient of two numbers with *like signs* is positive.

2. The quotient of two numbers with *unlike signs* is negative.

EXAMPLE 4

Dividing two real numbers. Find each quotient: **a.** $\frac{66}{11}$,   **b.** $\frac{-81}{-9}$,

c. $\frac{-45}{9}$,   and   **d.** $\frac{28}{-7}$.

Solution   To divide numbers with like signs, we find the quotient of their absolute values and make the quotient positive.

a. $\frac{66}{11} = 6$ Dividing the absolute values, 66 by 11, we get 6. The answer is positive.

b. $\frac{-81}{-9} = 9$ Dividing the absolute values, 81 by 9, we get 9. The answer is positive.

To divide numbers with unlike signs, we find the quotient of their absolute values and make the quotient negative.

c. $\frac{-45}{9} = -5$ Dividing the absolute values, 45 by 9, we get 5. The answer is negative.

d. $\frac{28}{-7} = -4$ Dividing the absolute values, 28 by 7, we get 4. The answer is negative.

SELF CHECK   Find each quotient: **a.** $\frac{48}{12}$,   **b.** $\frac{-63}{-9}$,

c. $\frac{40}{-8}$,   and   **d.** $\frac{-49}{7}$.

Answers: **a.** 4, **b.** 7,
c. −5, **d.** −7

Properties of Division

The examples

$$\frac{12}{1} = 12, \qquad \frac{-80}{1} = -80 \qquad \text{and} \qquad \frac{7.75}{1} = 7.75$$

illustrate that any number divided by 1 *is the number itself.* The examples

$$\frac{35}{35} = 1, \qquad \frac{-4}{-4} = 1, \qquad \text{and} \qquad \frac{0.9}{0.9} = 1$$

illustrate that *any number (except 0) divided by itself is 1.*

Division properties

If a represents any real number, then

$$\frac{a}{1} = a \qquad \text{and} \qquad \frac{a}{a} = 1 \quad (a \neq 0)$$

Division and Zero

We will now consider two types of division that involve zero. In the first case, we will examine division *of* zero; in the second case, division *by* zero.

To help explain the concept of division of zero, we consider the division $\frac{0}{5} = ?$ and its equivalent multiplication fact $5(?) = 0$.

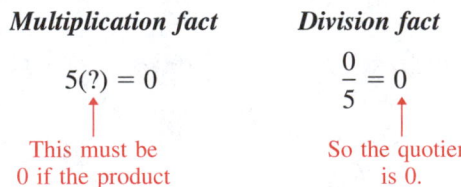

Multiplication fact   *Division fact*

$$5(?) = 0 \qquad\qquad \frac{0}{5} = 0$$

This must be   So the quotient
0 if the product   is 0.
is to be 0.

This example suggests that the *quotient of zero divided by any nonzero number is zero.*

To illustrate that division by zero is not permitted, we consider the division $\frac{5}{0} = ?$ and its equivalent multiplication fact $0(?) = 5$.

Multiplication fact   *Division fact*

$$0(?) = 5 \qquad\qquad \frac{5}{0} = \text{undefined}$$

There is no number   There is no
that gives 5 when   quotient.
multiplied by 0.

This example illustrates that the *quotient of any nonzero number divided by zero is undefined.*

EXAMPLE 5   **Division involving zero.** Find each quotient, if possible: **a.** $\dfrac{0}{13}$ and

b. $\dfrac{-13}{0}$.

Solution **a.** $\dfrac{0}{13} = 0$ Because $13(0) = 0$.

b. Since $\dfrac{-13}{0}$ involves division by zero, the division is undefined.

SELF CHECK   Find each quotient, if possible: **a.** $\dfrac{4}{0}$ and   *Answers:* **a.** undefined, **b.** 0
b. $\dfrac{0}{17}$.   ■

ACCENT ON TECHNOLOGY *Depreciation of a House*

During a period of 17.5 years, the value of a $124,930 house fell at a uniform rate to $97,105. To find how much the house depreciated per year, we must first find the change in its value by subtracting $124,930 from $97,105. To calculate this difference, we enter these numbers and press these keys on a scientific calculator:

Keystrokes   97105 $\boxed{-}$ 124930 $\boxed{=}$ $\boxed{\text{-27825}}$

-27825 represents a drop in value of $27,825. Since this depreciation occurred in 17.5 years, we divide $-27,825$ by 17.5 to find the amount of depreciation per year. With $-27,825$ already displayed, we only need to enter these numbers and press these keys:

Keystrokes $\boxed{\div}$ 17.5 $\boxed{=}$ $\boxed{\text{-1590}}$

If we use a graphing calculator to compute the amount of depreciation per year, we enter these numbers and press these keys:

Keystrokes   97105 $\boxed{-}$ 124930 $\boxed{\text{ENTER}}$ $\boxed{\div}$ 17.5 $\boxed{\text{ENTER}}$

```
97105-124930
               -27825
Ans/17.5
               -1590
```

The amount of depreciation per year was $1,590.

Solving Equations

In the following examples, we will use the division and multiplication properties of equality to solve equations involving negative numbers.

EXAMPLE 6   **The division property of equality.** Solve $-4x = 48$ and check the result.

Solution   Recall that $-4x$ indicates multiplication: $-4 \cdot x$. To undo the multiplication of x by -4, we divide both sides by -4.

$$-4x = 48$$

$$\frac{-4x}{-4} = \frac{48}{-4} \qquad \text{Divide both sides by } -4$$

$$x = -12$$

Check: $-4x = 48$

$$-4(-12) \stackrel{?}{=} 48 \qquad \text{Substitute } -12 \text{ for } x.$$

$$48 = 48 \qquad \text{Do the multiplication: } -4(-12) = 48.$$

SELF CHECK   Solve $-45 = 9x$ and check the result.   *Answer:* -5 ∎

EXAMPLE 7   **The multiplication property of equality.** Solve $\dfrac{b}{-15} = 3$ and check the result.

Solution   Recall that $\frac{b}{-15}$ indicates that b is to be divided by -15. To undo the division of b by -15, we multiply by -15.

$$\frac{b}{-15} = 3$$

$$-15\left(\frac{b}{-15}\right) = -15(3) \quad \text{Multiply both sides by } -15.$$

$$b = -45 \quad \text{Simplify each side.}$$

Check: $\dfrac{b}{-15} = 3$

$$\frac{-45}{-15} \stackrel{?}{=} 3 \quad \text{Substitute } -45 \text{ for } b.$$

$$3 = 3 \quad \text{Do the division.}$$

SELF CHECK   Solve $\frac{y}{23} = -12$ and check the result.   *Answer:* -276 ∎

STUDY SET

Section 2.2

VOCABULARY

In Exercises 1–8, fill in the blanks to make the statements true.

1. The answer to a multiplication problem is called a _____.

2. The answer to a division problem is called a _____.

3. The numbers -4 and -6 are said to have _____ signs.

4. The numbers -10 and $+12$ are said to have _____ signs.

5. The _____ property of multiplication states that two numbers can be multiplied in either order to get the same result.

6. _____ numbers are greater than zero, and _____ numbers are less than zero.

7. Division by zero is _____.

8. The statement $(ab)c = a(bc)$ expresses the _____ property of _____.

CONCEPTS

In Exercises 9–14, fill in the blanks to make the statements true.

9. The division fact $\frac{25}{-5} = -5$ is related to the multiplication fact _____.

10. The expression $-5 + (-5) + (-5) + (-5)$ can be represented by the multiplication statement _____.

11. The quotient of two numbers with _____ signs is negative.

12. The product of two negative numbers is _____.

13. The product of zero and any number is _____.

14. The product of _____ and any number is that number.

15. Draw a number line from -6 to 6. Graph each of these products on the number line. What is the distance between each product?

$$-3(2), \quad -3(1), \quad -3(0), \quad -3(-1), \quad -3(-2)$$

16. a. Find $-1(8)$. In general, what is the result when a number is multiplied by -1?

   b. Find $\frac{8}{-1}$. In general, what is the result when a number is divided by -1?

In Exercises 17–18, POS stands for a positive number and NEG stands for a negative number. Determine the sign of each result, if possible.

17. a. POS · NEG   **b.** POS + NEG

   c. POS − NEG   **d.** $\dfrac{\text{POS}}{\text{NEG}}$

18. a. NEG · NEG   **b.** NEG + NEG

   c. NEG − NEG   **d.** $\dfrac{\text{NEG}}{\text{NEG}}$

19. Is -6 a solution of $\dfrac{x}{2} = -3$?

20. Is -6 a solution of $-2x = -12$?

NOTATION

In Exercises 21–24, complete the solution.

21. Find $(-37 \cdot 5)2$.

$$(-37 \cdot 5)2 = -37(\boxed{} \cdot 2)$$
$$= -37(\boxed{})$$
$$= -370$$

22. Find $-20[5(-79)]$.

$$-20[5(-79)] = (-20 \cdot 5)(\boxed{})$$
$$= \boxed{}(-79)$$
$$= 7{,}900$$

23. Solve $-3x = 36$.

$$-3x = 36$$
$$\frac{-3x}{\boxed{}} = \frac{36}{\boxed{}}$$
$$x = \boxed{}$$

24. Solve $\dfrac{x}{-7} = -5$.

$$\frac{x}{-7} = -5$$
$$\boxed{}\left(\frac{x}{-7}\right) = \boxed{}(-5)$$
$$x = \boxed{}$$

PRACTICE

In Exercises 25–70, find each product or quotient, if possible.

25. $(-6)(-9)$ **26.** $(-8)(-7)$

27. $12(-5)$ **28.** $(-9)(11)$

29. $-6 \cdot 4$ **30.** $-8 \cdot 9$

31. $-20(40)$ **32.** $-10(10)$

33. $-0.6(-4)$ **34.** $-0.7(-8)$

35. $1.2(-0.4)$ **36.** $0(-0.2)$

37. $\dfrac{1}{2}\left(-\dfrac{3}{4}\right)$ **38.** $\dfrac{1}{3}\left(-\dfrac{5}{16}\right)$

39. $-1\dfrac{1}{4}\left(-\dfrac{3}{4}\right)$ **40.** $-1\dfrac{1}{8}\left(-\dfrac{3}{8}\right)$

41. $0(-22)$ **42.** $-8 \cdot 0$

43. $-3(-4)(0)$ **44.** $15(0)(-22)$

45. $3(-4)(-5)$ **46.** $(-2)(-4)(-5)$

47. $(-4)(3)(-7)$ **48.** $5(-3)(-4)$

49. $(-2)(-3)(-4)(-5)$ **50.** $(-3)(-4)(5)(-6)$

51. $\dfrac{-6}{-2}$ **52.** $\dfrac{-36}{9}$

53. $\dfrac{4}{-2}$ **54.** $\dfrac{-9}{3}$

55. $\dfrac{80}{-20}$ **56.** $\dfrac{-66}{33}$

57. $\dfrac{-110}{-110}$ **58.** $\dfrac{-200}{-200}$

59. $\dfrac{-160}{40}$ **60.** $\dfrac{-250}{-50}$

61. $\dfrac{320}{-16}$ **62.** $\dfrac{-180}{36}$

63. $\dfrac{0}{150}$ **64.** $\dfrac{225}{0}$

65. $\dfrac{-17}{0}$ **66.** $\dfrac{0}{-12}$

67. $-\dfrac{1}{3} \div \dfrac{4}{5}$ **68.** $-\dfrac{1}{8} \div \dfrac{2}{3}$

69. $-\dfrac{3}{16} \div \left(-\dfrac{2}{3}\right)$ **70.** $-\dfrac{3}{25} \div \left(-\dfrac{2}{3}\right)$

In Exercises 71–76, use a calculator to do each operation.

71. $(-23.5)(47.2)$

72. $(-435.7)(-37.8)$

73. $(-6.37)(-7.2)(-9.1)$

74. $(5.2)(-8.2)(7.75)$

75. $\dfrac{204.6}{-37.2}$

76. $\dfrac{-30.56625}{-4.875}$

In Exercises 77–80, complete each input/output table.

77.

x	$3x$
-1	
-5	
-10	

78.

x	$-3x$
4	
0	
-4	

79.

x	$\dfrac{x}{2}$
-2	
-6	
-8	

80.

x	$\dfrac{x}{4}$
8	
0	
-12	

In Exercises 81–88, solve each equation. Check each solution.

81. $3x = -3$

82. $-4x = 36$

83. $-54 = -18z$

84. $-57 = -19x$

85. $\dfrac{b}{3} = -5$

86. $-3 = \dfrac{s}{11}$

87. $-6 = \dfrac{t}{-7}$

88. $\dfrac{v}{12} = -7$

APPLICATIONS

In Exercises 89–100, use signed numbers to solve each problem.

89. TEMPERATURE CHANGE In a lab, the temperature of a fluid was decreased 6° per hour for 12 hours. What signed number indicates the change in temperature?

90. BACTERIAL GROWTH To slowly warm a bacterial culture, biologists programmed a heating pad under the culture to increase the temperature 4° every hour for 6 hours. What signed number indicates the change in the temperature of the pad?

91. GAMBLING A gambler places a $40 bet and loses. He then decides to go "double or nothing," and loses again. Feeling that his luck has to change, he goes "double or nothing" once more and, for the third time, loses. What signed number indicates his gambling losses?

92. REAL ESTATE A house has depreciated $1,250 each year for 8 years. What signed number indicates its change in value over that time period?

93. PLANETS The temperature on Pluto gets as low as $-386°$ F. This is twice as low as the lowest temperature reached on Jupiter. What is the lowest temperature on Jupiter?

94. CAR RADIATOR The instructions on the back of a container of antifreeze state, "A 50/50 mixture of antifreeze and water protects against freeze-ups down to $-34°$ F, while a 60/40 mix protects against freeze-ups down to one and one-half times that temperature." To what temperature does the 60/40 mixture protect?

95. AIRLINE INCOME For the fourth quarter of 1997, Trans World Airlines' total net income was $-$31$ million. The company's losses for the first quarter of 1998 were even worse, by a factor of about 1.8. What signed number indicates the company's total net income that quarter?

96. ACCOUNTING Illustration 1 shows the income statement for Converse Inc., the sports shoe company. The numbers in parentheses indicate losses. What signed number describes the company's average net income per quarter for the four quarters shown?

ILLUSTRATION 1

Converse® Inc.	INCOME STATEMENT			
All dollar amounts in millions	2nd Qtr **Jun 98**	1st Qtr **Mar 98**	4th Qtr **Dec 97**	3rd Qtr **Sep 97**
Total Net Income	(1.5)	(1.2)	(17.9)	0.2

Based on information from Hoover's Online

97. QUEEN MARY The ocean liner Queen Mary was commissioned in 1936 and cost $22,500,000 to build. In 1967, the ship was purchased by the city of Long Beach, California for $3,450,000 and now serves as a hotel and convention center. What signed number indicates the annual average depreciation of the Queen Mary over the 31-year period from 1936 to 1967? Round to the nearest dollar.

98. COMPUTER SPREADSHEET The "formula" = SUM(A1:C1)/3 in cell D1 of the spreadsheet shown in Illustration 2 instructs the computer to add the values in cells A1, B1, and C1, then to divide that sum by 3, and finally to print the result *in place of the formula* in cell D1. What values will the computer print in the cells D1, D2, and D3?

ILLUSTRATION 2

	Microsoft Excel-Book 1			▲ ▼
	File Edit View Insert Format Tools			▲ ▼
	A	**B**	**C**	**D**
1	4	−5	−17	= SUM(A1:C1)/3
2	22	−30	14	= SUM(A2:C2)/3
3	−60	−20	−34	= SUM(A3:C3)/3
4				
◀▶	Sheet 1 / Sheet 2 / Sheet 3 / Sheet 4 / Sheet 5 /			➡

99. PHYSICS An oscilloscope is an instrument that displays electrical signals, which appear as wavy lines on a fluorescent screen. (See Illustration 3.) By switching the magnification setting (MAGNIFN.) to × 2, for example, the "height" of the crest and the "depth" of the trough of a graph will be doubled. Use signed numbers to indicate the crest height and the trough depth for each setting of the magnification dial.

   a. normal   **b.** × 0.5
   c. × 1.5   **d.** × 2

ILLUSTRATION 4

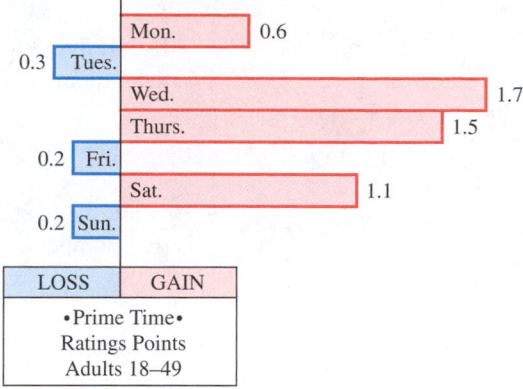

ILLUSTRATION 3

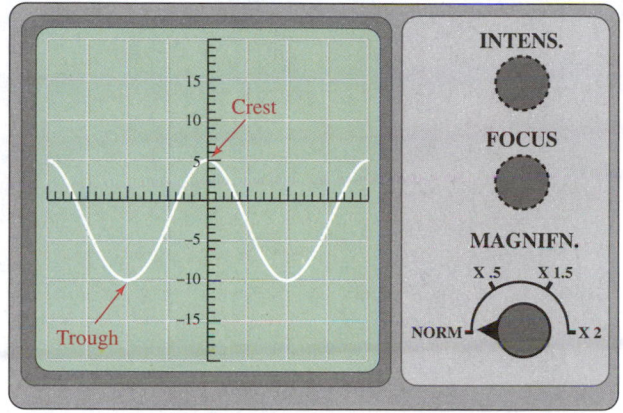

100. SWEEPS WEEK During "sweeps week," television networks make a special effort to gain viewers by showing unusually flashy programming. Use the information in Illustration 4 to determine the average daily gain (or loss) of ratings points by a network for the 7-day "sweeps period."

WRITING

101. Explain how you would decide whether the product of several numbers was positive or negative.

102. If the product of five numbers is negative, how many of them could be negative? Explain.

REVIEW

103. Use the formula $V = s^3$ to find the volume of a cube with a side of length 4 inches.

104. Write the subtraction statement $-3 - (-5)$ as addition of the opposite.

105. What is a real number?

106. Describe the balance in a checking account that is overdrawn $65 using a signed number.

107. Find 34% of 612.

108. 18 is what percent of 90?

▶ 2.3   Order of Operations and Evaluating Algebraic Expressions

In this section, you will learn about

    Powers of real numbers ■ Order of operations ■ Evaluating algebraic expressions ■ Making tables

Introduction In Section 1.3, we saw that the order in which we do arithmetic operations is important. To guarantee that a simplification of a numerical expression has only one answer, we need to perform arithmetic operations in an agreed-upon order. In this section, we will review the rules for the order of operations as we evaluate expressions involving real numbers.

Powers of Real Numbers

Recall that 3 to the fourth power (or 3^4) is a shorthand way of writing $3 \cdot 3 \cdot 3 \cdot 3$. In the exponential expression 3^4, the base is 3 and the exponent is 4.

Base $\longrightarrow 3^4 \longleftarrow$ Exponent

In the next example, we evaluate exponential expressions with bases that are negative numbers.

| **EXAMPLE 1** | **Powers of integers.** Find each power: **a.** $(-3)^4$, and **b.** $(-3)^5$. |

Solution

a. $(-3)^4 = (-3)(-3)(-3)(-3)$ Write -3 as a factor four times.

$= 9(-3)(-3)$ Work from left to right: $(-3)(-3) = 9$.

$= -27(-3)$ Work from left to right: $9(-3) = -27$.

$= 81$ Do the multiplication.

b. $(-3)^5 = (-3)(-3)(-3)(-3)(-3)$ Write -3 as a factor five times.

$= 9(-3)(-3)(-3)$ Work from left to right: $(-3)(-3) = 9$.

$= -27(-3)(-3)$ Work from left to right: $(9)(-3) = -27$.

$= 81(-3)$ Work from left to right: $(-27)(-3) = 81$.

$= -243$ Do the multiplication.

SELF CHECK   Find each power: **a.** $(-6)^2$ and **b.** $(-5)^3$.   *Answers:* **a.** 36, **b.** -125 ∎

In part a of Example 1, -3 was raised to an even power, and the result was positive. In part b, -3 was raised to an odd power, and the result was negative. This suggests a general rule.

Even and odd powers of a negative number

When a negative number is raised to an even power, the result is positive.

When a negative number is raised to an odd power, the result is negative.

| **EXAMPLE 2** | **Powers of fractions and decimals.** Find each power: **a.** $\left(-\dfrac{2}{3}\right)^3$ and **b.** $(0.6)^2$. |

Solution   **a.** In the expression $\left(-\frac{2}{3}\right)^3$, $-\frac{2}{3}$ is the base and 3 is the exponent.

$$\left(-\frac{2}{3}\right)^3 = \left(-\frac{2}{3}\right)\left(-\frac{2}{3}\right)\left(-\frac{2}{3}\right)$$ Write $-\frac{2}{3}$ as a factor three times.

$$= \frac{4}{9}\left(-\frac{2}{3}\right)$$ Multiply: $\left(-\frac{2}{3}\right)\left(-\frac{2}{3}\right) = \frac{4}{9}$.

$$= -\frac{8}{27}$$ Do the multiplication.

b. In the expression $(0.6)^2$, 0.6 is the base and 2 is the exponent.

$(0.6)^2 = (0.6)(0.6)$ Write 0.6 as a factor two times.

$= 0.36$ Do the multiplication.

SELF CHECK   Find each power: **a.** $\left(-\frac{3}{4}\right)^3$ and   **b.** $(-0.3)^2$ *Answers:*   **a.** $-\frac{27}{64}$,   **b.** 0.09 ■

 WARNING! Although the expressions -4^2 and $(-4)^2$ look somewhat alike, they are not. In -4^2, the base is 4 and the exponent is 2. The $-$ sign in front of 4^2 means the opposite of 4^2. In $(-4)^2$, the base is -4 and the exponent is 2. When we find the value of each expression, it becomes clear that they are not equivalent.

$-4^2 = -(4 \cdot 4)$ Write 4 as a factor two times.

$= -16$ Multiply inside the parentheses.

$(-4)^2 = (-4)(-4)$ Write -4 as a factor two times.

$= 16$ The product of two negative numbers is positive.

Different results

ACCENT ON TECHNOLOGY *Raising a Negative Number to a Power*

We can use a scientific calculator to raise negative numbers to powers. We need to use the change-sign key $\boxed{+/-}$ and the power key $\boxed{y^x}$. For example, to evaluate $(-7.36)^5$, we enter these numbers and press these keys:

Keystrokes   7.36 $\boxed{+/-}$ $\boxed{y^x}$ 5 $\boxed{=}$ $\boxed{-21596.78335}$

When using a graphing calculator to raise a negative number to a power, we must enter the parentheses that contain the base.

Keystrokes $\boxed{(}$ $\boxed{(-)}$ 7.36 $\boxed{)}$ $\boxed{\wedge}$ 5 $\boxed{ENTER}$ $\boxed{\begin{array}{l}(-7.36)^5 \\ -21596.78335\end{array}}$

Thus, $(-7.36)^5 = -21,596.78335$.

Order of Operations

To illustrate that the rules for the order of operations are necessary when working with negative numbers, let's evaluate $-2 + 3(-4)$. If we multiply first, we obtain -14. However, if we add first, we obtain -4.

Method 1: Multiply first

$-2 + 3(-4) = -2 + (-12)$ Multiply 3 and -4 first: $3(-4) = -12$.

$= -14$ Add.

Method 2: Add first

$-2 + 3(-4) = 1(-4)$ Add -2 and 3 first: $-2 + 3 = 1$.

$= -4$ Multiply.

Different results

To make sure that problems involving exponents, multiplication, division, addition, and subtraction have only one answer, we must apply the rules for the order of operations.

Order of operations

If the expression contains grouping symbols, do all calculations within each pair of grouping symbols, working from the innermost pair to the outermost pair, in the following order:

1. Evaluate all exponential expressions.

2. Do all multiplications and divisions, working from left to right.

3. Do all additions and subtractions, working from left to right.

If the expression does not contain grouping symbols, begin with step 1.

In a fraction, first simplify the numerator and denominator separately. Then simplify the fraction, whenever possible.

Because we do multiplications before additions, the correct calculation of $-2 + 3(-4)$ is

$$-2 + \mathbf{3(-4)} = -2 + (\mathbf{-12}) \quad \text{Multiply first: } 3(-4) = -12.$$
$$= -14 \quad \text{Do the addition.}$$

EXAMPLE 3

Order of operations. Evaluate $-5 + 4(-3)^2$.

Solution
To evaluate (find the value of) this expression, we must perform three operations: addition, multiplication, and raising a number to a power. By the rules for the order of operations, we find the power first.

$$-5 + 4(\mathbf{-3})^2 = -5 + 4(\mathbf{9}) \quad \text{Evaluate the exponential expression: } (-3)^2 = 9.$$
$$= -5 + 36 \quad \text{Do the multiplication: } 4(9) = 36.$$
$$= 31 \quad \text{Do the addition.}$$

SELF CHECK
Evaluate $-9 + 2(-4)^2$. 　　　　　　　　　　　　　*Answer:* 23 ■

EXAMPLE 4

An expression containing grouping symbols. Evaluate $5^3 + 2(-8 - 3 \cdot 2)$.

Solution
We do the work within the parentheses first.

$$5^3 + 2(-8 - \mathbf{3 \cdot 2}) = 5^3 + 2(-8 - \mathbf{6}) \quad \text{Do the multiplication within the parentheses: } 3 \cdot 2 = 6.$$
$$= 5^3 + 2(-14) \quad \text{Do the subtraction within the parentheses: } -8 - 6 = -14.$$
$$= 125 + 2(-14) \quad \text{Evaluate the exponential expression: } 5^3 = 125.$$
$$= 125 + (-28) \quad \text{Do the multiplication: } 2(-14) = -28.$$
$$= 97 \quad \text{Do the addition.}$$

SELF CHECK
Evaluate $-3[5 + 3(-2)] + 3^2$. 　　　　　　　　*Answer:* 12 ■

EXAMPLE 5

An expression containing two pairs of grouping symbols. Evaluate $-4[-2 - 3(4 - 8^2)] - 2$.

Solution We do the work within the innermost grouping symbols (the parentheses) first.

$$-4[-2 - 3(4 - \mathbf{8^2})] - 2$$

$= -4[-2 - 3(4 - \mathbf{64})] - 2$	Evaluate the exponential expression within the parentheses: $8^2 = 64$.
$= -4[-2 - 3(-60)] - 2$	Do the subtraction within the parentheses: $4 - 64 = -60$.
$= -4[-2 - (-180)] - 2$	Do the multiplication within the brackets: $3(-60) = -180$.
$= -4(178) - 2$	Do the subtraction within the brackets by adding the opposite of -180, which is 180: $-2 + 180 = 178$.
$= -712 - 2$	Do the multiplication: $-4(178) = -712$.
$= -714$	Do the subtraction by adding the opposite of 2, which is -2: $-712 + (-2) = -714$.

SELF CHECK Evaluate $-5[2(5^2 - 15) + 4] - 10$. *Answer:* -130 ∎

EXAMPLE 6

Simplifying a fractional expression. Evaluate $\dfrac{-3(3 + 2) + 5}{17 - 3(-4)}$.

Solution We simplify the numerator and the denominator separately.

$\dfrac{-3(\mathbf{3 + 2}) + 5}{17 - \mathbf{3(-4)}} = \dfrac{-3(\mathbf{5}) + 5}{17 - (-12)}$	In the numerator, do the addition within the parentheses. In the denominator, do the multiplication.
$= \dfrac{-15 + 5}{17 + 12}$	In the numerator, do the multiplication. In the denominator, write the subtraction as addition of the opposite of -12, which is 12.
$= \dfrac{-10}{29}$	Do the additions.
$= -\dfrac{10}{29}$	$\dfrac{-10}{29} = -\dfrac{10}{29}$.

SELF CHECK Evaluate $\dfrac{-4(-2 + 8) + 6}{8 - 5(-2)}$. *Answer:* -1 ∎

Evaluating Algebraic Expressions

We will often need to evaluate algebraic expressions that contain variables. To do so, we substitute the values for the variables and simplify the resulting expression.

EXAMPLE 7

Surface area of a swim fin. Divers use swim fins because they provide a much larger surface area to push against the water than do bare feet. Consequently, the diver can swim faster wearing them. In Figure 2-10, we see that the fin is in the shape of a

FIGURE 2-10

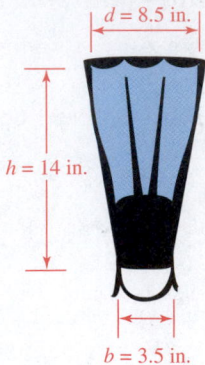

$d = 8.5$ in.

$h = 14$ in.

$b = 3.5$ in.

trapezoid. The algebraic expression $\frac{1}{2}h(b + d)$ gives the area of a trapezoid, where h is the height and b and d are the lengths of the lower and upper bases, respectively. To find the area of the top portion of the fin shown here, we evaluate the algebraic expression for $h = 14$, $b = 3.5$, and $d = 8.5$.

$$\frac{1}{2}h(b + d) = \frac{1}{2}(14)(3.5 + 8.5)$$ Substitute 14 for h, 3.5 for b, and 8.5 for d.

$$= \frac{1}{2}(14)(12)$$ Do the addition within the parentheses.

$$= 7(12)$$ Work from left to right: $\frac{1}{2}(14) = 7$.

$$= 84$$ Do the multiplication.

The fin has an area of 84 square inches.   ∎

EXAMPLE 8

Evaluating algebraic expressions. Evaluate   **a.** $-y$ and   **b.** $-3(y + x^2)$ when $x = 3$ and $y = -4$.

Solution   **a.** $-y = -(-4)$ Substitute -4 for y.

$\quad\quad = 4$ The opposite of -4 is 4.

b. $-3(y + x^2) = -3(-4 + 3^2)$ Substitute 3 for x and -4 for y.

$\quad\quad\quad = -3(-4 + 9)$ Do the work within the parentheses first. Evaluate the exponential expression.

$\quad\quad\quad = -3(5)$ Do the addition within the parentheses.

$\quad\quad\quad = -15$ Do the multiplication.

SELF CHECK   Evaluate   **a.** $-x$ and   **b.** $5(x - y)$ when $x = -2$ and $y = 3$.   *Answers:*   **a.** 2,   **b.** -25 ∎

EXAMPLE 9

Evaluating an algebraic expression containing an absolute value. Evaluate $\dfrac{3a^2 + 2b}{-2|a + b|}$ when $a = -4$ and $b = -3$.

Solution   In the denominator, the absolute value bars serve as grouping symbols.

$$\frac{3a^2 + 2b}{-2|a + b|} = \frac{3(-4)^2 + 2(-3)}{-2|-4 + (-3)|}$$ Substitute -4 for a and -3 for b.

$$= \frac{3(16) + 2(-3)}{-2|-7|}$$ Find the value of $(-4)^2$ in the numerator. In the denominator, do the addition inside the absolute value bars.

$$= \frac{48 + (-6)}{-2(7)}$$ Do the multiplications in the numerator. Find the absolute value in the denominator.

$$= \frac{42}{-14}$$ In the numerator, do the addition. Do the multiplication in the denominator.

$$= -3$$ Do the division.

SELF CHECK   Evaluate $\dfrac{4s^2 + 3t - 1}{-3|s - t|}$ when $s = -5$ and $t = 2$.   *Answer:* -5 ∎

Making Tables

When we evaluate an algebraic expression in one variable for several values of the variable, we can keep track of the results in an input/output table.

EXAMPLE 10

Temperature conversion. The equation

$$F = \frac{9C + 160}{5}$$

is used to change temperatures given in degrees Celsius to temperatures in degrees Fahrenheit. Change each temperature to degrees Fahrenheit and give the results in a table.

a. The coldest temperature on the moon: $-170°C$.

b. The coldest recorded temperature on earth; Vostok, Antarctica, July 21, 1983: $-89.2°C$.

Solution

C	$F = \frac{9C + 160}{5}$
-170	-274
-89.2	-128.56

Evaluate for C = −170:

$$\frac{9C + 160}{5} = \frac{9(-170) + 160}{5}$$
$$= \frac{-1{,}530 + 160}{5}$$
$$= \frac{-1{,}370}{5}$$
$$= -274$$

Evaluate for C = −89.2:

$$\frac{9C + 160}{5} = \frac{9(-89.2) + 160}{5}$$
$$= \frac{-802.8 + 160}{5}$$
$$= \frac{-642.8}{5}$$
$$= -128.56$$

The corresponding temperatures in degrees Fahrenheit are $-274°F$ and approximately $-128.6°F$.

SELF CHECK

On January 22, 1943, the temperature in Spearfish, South Dakota, changed from $-20°C$ to $7.2°C$ in 2 minutes. Change each temperature to degrees Fahrenheit.

C	$F = \frac{9C + 160}{5}$
-20	
7.2	

Answers: $-4°F$, approximately $45°F$

STUDY SET

Section 2.3

VOCABULARY

In Exercises 1–8, fill in the blanks to make the statements true.

1. The rules for the _____ of operations guarantee that a simplification of a numerical expression results in a single answer.

2. $2x + 5$ is an example of an algebraic _____, whereas $2x + 5 = 7$ is an example of an _____.

3. To _____ an algebraic expression means to substitute the values for the variables and then apply the rules for the order of operations.

4. When we evaluate an algebraic expression in one variable for several values of the variable, we can keep track of the results in an input/output _____.

5. In the expression $(-9)^3$, -9 is the _____ and ▨ is the exponent.

6. In the expression -8^4, ▨ is the base and 4 is the _____.

7. Some examples of grouping symbols are ( ) _____, [ ] _____, and | | _____ bars.

8. The expression $(-4.5)^6$ is called a _____ of -4.5.

CONCEPTS

In Exercises 9–10, fill in the blanks to make the statements true.

9. To simplify an expression with no grouping symbols, evaluate all _____ expressions before doing any multiplications.

10. When simplifying an expression containing grouping symbols, do all calculations within each pair of grouping symbols, working from the _____ pair to the _____ pair.

11. Complete each input/output table and make an observation about the signs of the outputs.

a.

x	x^2
-3	
-4	
-5	

b.

x	x^3
-3	
-4	
-5	

12. Complete the table and make an observation about the outputs.

x	x^2	x^3	x^4	x^5	x^6	x^7
-1						

13. a. How many operations need to be performed to evaluate the expression $-3[5^2 - 6(3 - 2)]$?
    b. List the operations in the order in which they should be performed.

14. If $x = -9$, find the value of
    a. $-x$ b. $-(-x)$
    c. $-x^2$ d. $(-x)^2$

NOTATION

In Exercises 15–16, complete each solution.

15. Evaluate $(-6)^2 - 2(5 - 4 \cdot 2)$.

$$(-6)^2 - 2(5 - 4 \cdot 2) = (-6)^2 - 2\left(5 - \boxed{}\right)$$
$$= (-6)^2 - 2\left(\boxed{}\right)$$
$$= \boxed{} - 2(-3)$$
$$= 36 - \left(\boxed{}\right)$$
$$= 42$$

16. Evaluate $\dfrac{4x^2 - 3y}{9(x - y)}$ when $x = 4$ and $y = -3$.

$$\frac{4x^2 - 3y}{9(x - y)} = \frac{4(4)^2 - 3(-3)}{9[4 - (-3)]}$$
$$= \frac{4\left(\boxed{}\right) - 3\left(\boxed{}\right)}{9\left[\boxed{}\right]}$$
$$= \frac{\boxed{} - \left(\boxed{}\right)}{\boxed{}}$$
$$= \frac{73}{63}$$

PRACTICE

In Exercises 17–28, evaluate each expression.

17. $(-6)^2$ 18. -6^2

19. -4^4 20. $(-4)^4$

21. $(-5)^3$ 22. -5^3

23. $-(-6)^4$ 24. $-(-7)^2$

25. $(-0.4)^2$ **26.** $(-0.5)^2$

27. $\left(-\dfrac{2}{5}\right)^3$ **28.** $\left(-\dfrac{1}{4}\right)^3$

In Exercises 29–58, evaluate each expression.

29. $3 - 5 \cdot 4$ **30.** $-4 \cdot 6 + 5$

31. $-3(5 - 4)$ **32.** $-4(6 + 5)$

33. $3 + (-5)^2$ **34.** $4^2 - (-2)^2$

35. $(-3 - 5)^2$ **36.** $(-5 - 2)^2$

37. $2 + 3\left(-\dfrac{25}{5}\right) - (-4)$ **38.** $12 + 2\left(-\dfrac{9}{3}\right) - (-2)$

39. $(-2)^3\left(\dfrac{-6}{-2}\right)(-1)$ **40.** $(-3)^3\left(\dfrac{-4}{-2}\right)(-1)$

41. $\dfrac{-7 - 3^2}{2 \cdot 4}$ **42.** $\dfrac{-5 - 3^3}{2^3}$

43. $\dfrac{1}{2}\left(\dfrac{1}{8}\right) + \left(-\dfrac{1}{4}\right)^2$ **44.** $-\dfrac{1}{9}\left(\dfrac{1}{4}\right) + \left(-\dfrac{1}{6}\right)^2$

45. $-2|4 - 8|$ **46.** $-5|1 - 8|$

47. $|7 - 8(4 - 7)|$ **48.** $|9 - 5(1 - 8)|$

49. $3 + 2[-1 - 4(5)]$ **50.** $4 + 2[-7 - 3(9)]$

51. $-3[5^2 - (7 - 3)^2]$ **52.** $3 - [3^3 + (3 - 1)^3]$

53. $-(2 \cdot 3 - 4)^3$ **54.** $-(3 \cdot 5 - 2 \cdot 6)^2$

55. $\dfrac{(3 + 5)^2 + |-2|}{-2(5 - 8)}$ **56.** $\dfrac{|-25| - 8(-5)}{2^4 - 29}$

57. $\dfrac{2[-4 - 2(3 - 1)]}{3[(3)(2)]}$ **58.** $\dfrac{3[-9 + 2(7 - 3)]}{(8 - 5)(9 - 7)}$

In Exercises 59–70, evaluate each expression given that $x = 3$, $y = -2$, and $z = -4$.

59. $2x - y$ **60.** $2z - y$

61. $-y + yz$ **62.** $-z + x - 2y$

63. $(3 + x)y$ **64.** $(4 + z)y$

65. $(x + y)^2(x - y)$ **66.** $[(z - 1)(z + 1)]^2$

67. $(4x)^2 + 3y^2$ **68.** $4x^2 + (3y)^2$

69. $\dfrac{2x + y^3}{y + 2z}$ **70.** $\dfrac{2z^2 - y}{2x - y^2}$

In Exercises 71–76, evaluate each expression for the given values of the variables.

71. $b^2 - 4ac$; $a = -1$, $b = 5$, and $c = -2$

72. $(x - a)^2 + (y - b)^2$; $x = -2$, $y = 1$, $a = 5$, and $b = -3$

73. $a^2 + 2ab + b^2$; $a = -5$ and $b = -1$

74. $\dfrac{x - a}{y - b}$; $x = -2$, $y = 1$, $a = 5$, and $b = 2$

75. $\dfrac{n}{2}[2a + (n - 1)d]$; $n = 10$, $a = -4$, and $d = 6$

76. $\dfrac{a(1 - r^n)}{1 - r}$; $a = -5$, $r = 2$, and $n = 3$

In Exercises 77–82, use a calculator to find the value of each expression.

77. $(-5.6)^4$ **78.** $(-8.3)^4$

79. $(-1.12)^5$ **80.** $(-4.07)^5$

81. $(23.1)^2 - (14.7)(-61.9)$ **82.** $12 - 7\left(-\dfrac{85.684}{34.55}\right)^3$

APPLICATIONS

In Exercises 83–88, a calculator may be helpful.

83. TRANSLATION When an American company exports its Advanced Formula Antifreeze to China, the product description on the back of the container, seen in Illustration 1, must be translated into Chinese. Since China uses the metric system of measurement, the temperatures in the description must be converted to degrees Celsius. Use the formula

$$C = \frac{5(F - 32)}{9}$$

to make the conversions to the nearest degree Celsius.

ILLUSTRATION 1

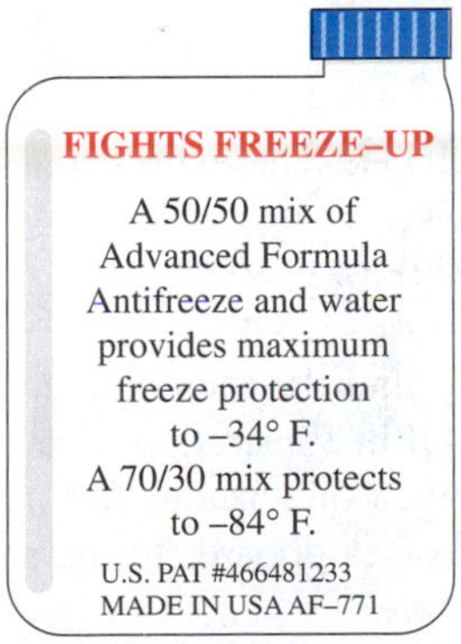

FIGHTS FREEZE–UP

A 50/50 mix of Advanced Formula Antifreeze and water provides maximum freeze protection to –34° F. A 70/30 mix protects to –84° F.

U.S. PAT #466481233
MADE IN USA AF–771

84. TEMPERATURE ON MARS On Mars, maximum summer temperatures can reach 20°C. However, daily temperatures average −33°C. Convert each of these temperatures to degrees Fahrenheit. See Example 10. Round to the nearest degree.

85. TIDES After reaching a high tide mark of 80 centimeters (cm) on the graduated pole shown in Illustration 2, the water level fell for 5 hours at an average rate of 21 centimeters per hour to reach the low tide mark. Write an expression that gives the low tide reading on the pole, then evaluate it.

ILLUSTRATION 2

Graduated pole for
determining tidal range

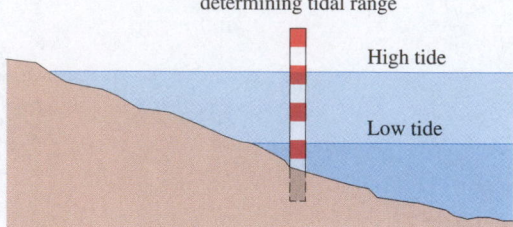

86. GROWING SOD To determine the number of square feet of sod *remaining* in a field after filling an order (see Illustration 3), the manager of a sod farm uses the expression $20,000 - 3s$ (where s is the number of 1-foot-by-3-foot strips the customer has ordered). To sod a soccer field, a city orders 7,000 strips of sod. Evaluate the expression for this value of s and explain the result.

ILLUSTRATION 3

1-ft-by-3-ft strips of sod, cut and ready to be loaded on a truck for delivery

87. TRUMPET MUTE The expression

$$\pi[b^2 + d^2 + (b + d)s]$$

can be used to find the total surface area of the trumpet mute shown in Illustration 4. Evaluate the expression for the given dimensions to find the number of square inches of cardboard (to the nearest tenth) used to make the mute.

ILLUSTRATION 4

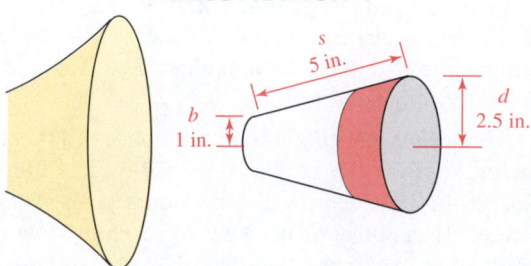

88. PET SUPPLIES The volume of the dog's water bowl shown in Illustration 5 can be found using the expression

$$\frac{\pi h(3a^2 + 3b^2 + h^2)}{6}$$

Evaluate the expression for the given dimensions to find the capacity of the bowl in cubic inches. Round to the nearest tenth.

ILLUSTRATION 5

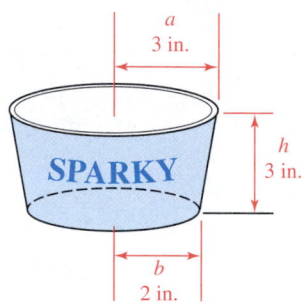

WRITING

89. Explain why $3 + 5 \cdot 4$ can have two answers if we don't use the rules for the order of operations.

90. Explain how to evaluate $x^2 - y$ when $x = 2$ and $y = -3$.

91. Does your calculator have the rules for the order of operations "built in"? Explain how you can tell.

92. Explain the difference in how the two expressions -3^2 and $(-3)^2$ are evaluated.

REVIEW

93. On the number line, graph the integers from -4 to 2.

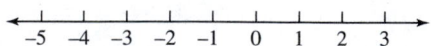

94. Write the inequality $7 < 12$ as an inequality that uses the symbol $>$.

95. Write the expression $a \cdot a \cdot b \cdot b \cdot b$ using exponents.

96. Is 7 a solution of the equation $3x + 4 = 25$?

97. Solve $12 + x = 17$.

98. Solve $x - 14 = -25$.

▶ 2.4   Simplifying Algebraic Expressions

In this section, you will learn about

> Simplifying algebraic expressions involving multiplication ■ The distributive property ■ Extending the distributive property ■ Terms of an algebraic expression ■ Coefficients of a term ■ Terms and factors ■ Like terms ■ Combining like terms

Introduction   In this section, we will simplify algebraic expressions by replacing them with equivalent expressions that are less complicated. This will often require that we use the distributive property to combine like terms.

Simplifying Algebraic Expressions Involving Multiplication

To **simplify algebraic expressions,** we use properties of algebra to write the expressions in a less complicated form. Two properties used to simplify algebraic expressions are the associative and commutative properties of multiplication. Recall that the associative property of multiplication enables us to change the grouping of factors involved in a multiplication. The commutative property of multiplication enables us to change the order of the factors.

As an example, let's consider the expression $8(4x)$ and simplify it as follows:

$8(4x) = 8 \cdot (4 \cdot x)$ $4x = 4 \cdot x$.

$\qquad = (8 \cdot 4) \cdot x$ Apply the associative property of multiplication to group 4 with 8, instead of with x.

$\qquad = 32x$ Do the multiplication inside the parentheses: $8 \cdot 4 = 32$.

Since $8(4x) = 32x$, we say that $8(4x)$ simplifies to $32x$.

EXAMPLE 1   **Simplifying algebraic expressions involving multiplication.** Simplify each expression: **a.** $5(8t)$,   **b.** $15a(-7)$.

Solution   **a.** $5(8t) = (5 \cdot 8)t$ Use the associative property of multiplication to regroup the factors.

$\qquad = 40t$ Do the multiplication inside the parentheses: $5 \cdot 8 = 40$.

b. $15a(-7) = 15(-7)a$ Use the commutative property of multiplication to change the order of the factors.

$\qquad = [15(-7)]a$ Use the associative property of multiplication to group the numbers together.

$\qquad = -105a$ Do the multiplication within the brackets: $15(-7) = -105$.

SELF CHECK   Simplify each expression: **a.** $9 \cdot 6s$ and
b. $-5b(13)$.   *Answers:*   **a.** $54s$,   **b.** $-65b$ ■

In the next example, we will work with expressions involving two variables.

EXAMPLE 2

Simplifying algebraic expressions involving multiplication. Simplify each expression: **a.** $-5r(-6s)$ and   **b.** $3(7p)(-5p)$.

Solution   **a.** $-5r(-6s) = [-5(-6)][r \cdot s]$ Use the commutative and associative properties to group the numbers and to group the variables.

$$= 30rs$$ Do the multiplications within the brackets: $-5(-6) = 30$ and $r \cdot s = rs$.

b. $3(7p)(-5p) = [3(7)(-5)][p \cdot p]$ Use the commutative and associative properties to change the order and to regroup the factors.

$$= (-105)(p^2)$$ Do the multiplications within the brackets: $3(7)(-5) = -105$ and $p \cdot p = p^2$.

$$= -105p^2$$ Write the multiplication without parentheses.

SELF CHECK   Simplify each expression: **a.** $7p(-3q)$ and   *Answers:*   **a.** $-21pq$,
b. $-4(6m)(-2m)$.   **b.** $48m^2$

The Distributive Property

To introduce the **distributive property,** we will examine the expression $4(5 + 3)$, which can be evaluated in two ways.

Method 1: Rules for the order of operations: In this method, we compute the sum inside the parentheses first.

$$4(\mathbf{5 + 3}) = 4(\mathbf{8})$$ Do the addition inside the parentheses first: $5 + 3 = 8$.

$$= 32$$ Do the multiplication.

Method 2: The distributive property: In this method, we distribute the 4 across 5 and 3, find each product separately, and add the results.

Distribute the 4.

$$4(5 + 3) = 4(5 + 3)$$ To apply the distributive property, we multiply each term inside the parentheses by the factor outside the parentheses.

First product   Second product

$$= \quad \mathbf{4}(5) \quad + \quad \mathbf{4}(3)$$

$$= \quad 20 \quad + \quad 12$$ Do the multiplications first: $4(5) = 20$ and $4(3) = 12$.

$$= \quad 32$$ Do the addition.

Notice that each method gives a result of 32.

We can interpret the distributive property geometrically. Figure 2-11 shows three rectangles that are divided into squares. Since the area of the rectangle on the left-hand side of the equals sign can be found by multiplying its width by its length, its area is $4(5 + 3)$ square units. We can evaluate this expression or we can count squares; either way, we see that the area is 32 square units.

FIGURE 2-11

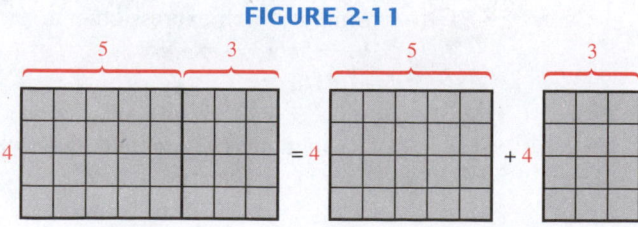

The area shown on the right-hand side is the sum of the areas of two rectangles: $4(5) + 4(3)$. Either by evaluating this expression or by counting squares, we see that this area is also 32 square units. Therefore,

$$4(5 + 3) = 4(5) + 4(3)$$

Figure 2-12 shows the general case where the width is a and the length is $b + c$.

FIGURE 2-12

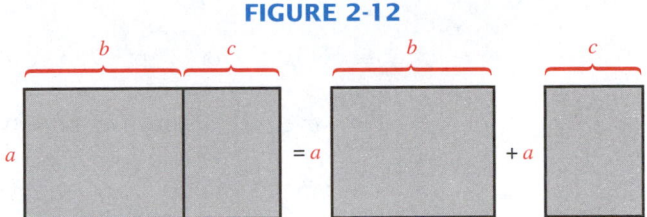

Using Figure 2-12 as a basis, we can state the distributive property in symbols.

The distributive property

If a, b, and c are real numbers, then

$$a(b + c) = ab + ac$$

Since subtraction is the same as adding the opposite, the distributive property also holds for subtraction.

The distributive property

If a, b, and c are real numbers, then

$$a(b - c) = ab - ac$$

We can use the distributive property to *remove the parentheses* when an expression is multiplied by a quantity. For example, to remove the parentheses in the expression $-5(x + 2)$, we proceed as follows:

$$\begin{aligned} -5(x + 2) &= -5(x) + (-5)(2) & \text{Distribute the multiplication by } -5. \\ &= -5x + (-10) & \text{Do the multiplications.} \\ &= -5x - 10 & \text{Write the addition of } -10 \text{ as subtraction of } 10. \end{aligned}$$

EXAMPLE 3

Applying the distributive property. Use the distributive property to remove parentheses: **a.** $3(x + 8)$, **b.** $-6(3y - 2)$, and **c.** $x(x + 2)$.

Solution
a. $\begin{aligned}[t] 3(x + 8) &= 3 \cdot x + 3 \cdot 8 & \text{Distribute the 3.} \\ &= 3x + 24 & \text{Do the multiplications.} \end{aligned}$

b. $\begin{aligned}[t] -6(3y - 2) &= -6(3y) - (-6)(2) & \text{Distribute the } -6. \\ &= -18y - (-12) & \text{Do the multiplications.} \\ &= -18y + 12 & \text{Add the opposite of } -12, \text{ which is } 12. \end{aligned}$

c. $\begin{aligned}[t] x(x + 2) &= x \cdot x + x \cdot 2 & \text{Distribute the } x. \\ &= x^2 + 2x & \text{Do the multiplications. Recall that } x \cdot x = x^2. \end{aligned}$

SELF CHECK   Remove parentheses: **a.** $5(p + 2)$,   *Answers:*   **a.** $5p + 10$,
b. $-8(2x - 4)$,   and   **c.** $p(p - 5)$.   **b.** $-16x + 32$,   **c.** $p^2 - 5p$ ■

WARNING! If an expression contains parentheses, it does not necessarily mean that the distributive property can be applied. For example, the distributive property does not apply to the expressions

$$6(5x) \qquad \text{or} \qquad 6(-7 \cdot y) \qquad \text{\color{red}{Here a product is multiplied by 6.}}$$

However, the distributive property does apply to the expressions

$$6(5 + x) \qquad \text{or} \qquad 6(-7 - y) \quad \text{\color{red}{Here a sum or difference is multiplied by 6.}}$$

Extending the Distributive Property

The distributive property can be extended to situations where there are more than two terms within parentheses.

The extended distributive property

If a, b, c, and d are real numbers, then

$$a(b + c + d) = ab + ac + ad \qquad \text{and} \qquad a(b - c - d) = ab - ac - ad$$

EXAMPLE 4   **Applying the extended distributive property.** Remove parentheses:
a. $5(x + y - z)$ and   **b.** $-0.3(3x - 4y + 7z)$.

Solution   **a.** $5(x + y - z) = \mathbf{5} \cdot x + \mathbf{5} \cdot y - \mathbf{5} \cdot z$ Use the extended distributive
                                                                              property.

$$= 5x + 5y - 5z \qquad \text{Write the multiplication without the}$$
$$\text{multiplication dots.}$$

b. $-\mathbf{0.3}(3x - 4y + 7z)$

$$= -\mathbf{0.3}(3x) - (-\mathbf{0.3})(4y) + (-\mathbf{0.3})(7z) \quad \text{Distribute the } -0.3.$$
$$= -0.9x - (-1.2y) + (-2.1z) \qquad \text{Do the multiplications.}$$
$$= -0.9x + 1.2y + (-2.1z) \qquad \text{Do the subtraction of } -1.2y \text{ by add-}$$
$$\text{ing its opposite, which is } 1.2y.$$
$$= -0.9x + 1.2y - 2.1z \qquad \text{Write the addition of } -2.1z \text{ as a sub-}$$
$$\text{traction.}$$

SELF CHECK   Remove parentheses in the expression
$-7(2r + 5s - 8t)$.   *Answer:* $-14r - 35s + 56t$ ■

Since multiplication is commutative, we can write the distributive property in the following forms.

$$(b + c)a = ba + ca, \qquad (b - c)a = ba - ca, \qquad (b + c + d)a = ba + ca + da$$

EXAMPLE 5	**Applying the distributive property.** Multiply **a.** $(6 + 4p)5$ and **b.** $(-12 - 6x + 4y)\dfrac{1}{2}$.

Solution **a.** $(6 + 4p)\mathbf{5} = 6 \cdot \mathbf{5} + 4p \cdot \mathbf{5}$ Use the distributive property.

$\qquad\qquad\quad = 30 + 20p$ Do the multiplications.

b. $(-12 - 6x + 4y)\dfrac{\mathbf{1}}{\mathbf{2}}$

$\qquad = -12 \cdot \dfrac{\mathbf{1}}{\mathbf{2}} - 6x \cdot \dfrac{\mathbf{1}}{\mathbf{2}} + 4y \cdot \dfrac{\mathbf{1}}{\mathbf{2}}$ Distribute the $\frac{1}{2}$.

$\qquad = -6 - 3x + 2y$ Do the multiplications.

SELF CHECK   Multiply $(-5x - 4y)8$.   *Answer:* $-40x - 32y$ ■

To use the distributive property to simplify $-(x + 10)$, we note that the negative sign in front of the parentheses represents -1.

The $-$ sign represents -1.

$-(x + 10) = \mathbf{-1}(x + 10)$

$\qquad\quad\; = \mathbf{-1}(x) + (\mathbf{-1})(10)$ Use the distributive property to distribute -1.

$\qquad\quad\; = -x + (-10)$ Do the multiplications.

$\qquad\quad\; = -x - 10$ Write the addition of -10 as a subtraction.

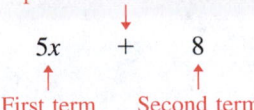

EXAMPLE 6	**Distributing a negative sign.** Simplify $-(-12 - 3p)$.

Solution $\qquad -(-12 - 3p)$

$\qquad\quad = \mathbf{-1}(-12 - 3p)$ Change the $-$ sign in front of the parentheses to -1.

$\qquad\quad = \mathbf{-1}(-12) - (\mathbf{-1})(3p)$ Use the distributive property to distribute -1.

$\qquad\quad = 12 - (-3p)$ Do the multiplications.

$\qquad\quad = 12 + 3p$ To subtract $-3p$, add the opposite of $-3p$, which is $3p$.

SELF CHECK   Simplify $-(-5x + 18)$.   *Answer:* $5x - 18$ ■

Terms of an Algebraic Expression

Addition signs separate algebraic expressions into parts called **terms.** The expression $5x + 8$ contains two terms, $5x$ and 8.

The $+$ signs separates the expression into two terms.

$\qquad 5x \quad + \quad 8$

$\qquad$ First term $\quad$ Second term

A term may be

- a number. Examples are 8, 98.6, and -45.
- a variable, or a product of variables (which may be raised to powers). Examples are x, s^3, rt, and a^2bc^4.
- a product of a number and one or more variables (which may be raised to powers). Examples are $-35x$, $\frac{1}{2}bh$, and $\pi r^2 h$.

Since subtraction can be expressed as addition of the opposite, the expression $6x - 5$ can be written in the equivalent form $6x + (-5)$. We can then see that $6x - 5$ contains two terms, $6x$ and -5.

EXAMPLE 7

Identifying terms. List the terms in each expression: **a.** $-4p + 7 + 5p$, **b.** $-12r^2st$, and **c.** $y^3 + 8y^2 - 3y + 24$

Solution
a. $-4p + 7 + 5p$ has three terms: $-4p$, 7, and $5p$.

b. The expression $-12r^2st$ has one term: $-12r^2st$.

c. $y^3 + 8y^2 - 3y + 24$ can be written as $y^3 + 8y^2 + (-3y) + 24$. It contains four terms: y^3, $8y^2$, $-3y$, and -24.

SELF CHECK
List the terms in each expression: **a.** $\frac{1}{3}Bh$, **b.** $3q + 5q - 1.2$, and **c.** $b^2 - 4ac$.

Answers: **a.** $\frac{1}{3}Bh$, **b.** $3q$, $5q$, -1.2, **c.** b^2, $-4ac$ ∎

Coefficients of a Term

In a term that is the product of a number and one or more variables, the number factor is called the **numerical coefficient**, or simply the **coefficient**. In the expression $5x$, 5 is the coefficient and x is the variable part. Other examples are shown in Table 2-1.

TABLE 2-1

Term	Coefficient	Variable part
$8y^2$	8	y^2
$-0.9pq$	-0.9	pq
$\frac{3}{4}b$	$\frac{3}{4}$	b
$-\frac{x}{6}$	$-\frac{1}{6}$	x
x	1	x
$-t$	-1	t
15	15	none

Notice that when there is no number in front of a variable, the coefficient is 1. For example, $x = 1x$. When there is only a negative sign in front of the variable, the coefficient is -1. For example, $-t = -1t$.

EXAMPLE 8

Identifying coefficients of terms. Identify the coefficient of each term in the expression $-7x^2 + 3x - 6$.

Solution

Term	Coefficient
$-7x^2$	-7
$3x$	3
-6	-6

SELF CHECK
Identify the coefficient of each term in the expression $p^3 - 12p^2 + 3p - 4$.

Answers: 1, -12, 3, -4 ∎

Terms and Factors

It is important to distinguish between a *term* of an expression and a *factor* of a term. Terms are separated by an addition sign +. Factors are numbers or variables that are multiplied together.

EXAMPLE 9

Determine whether the variable x is a factor or a term: **a.** $18 + x$,   **b.** $18x$,   and **c.** $18 - 3x$.

Solution

a. x is a *term* of $18 + x$, since x and 18 are separated by a + sign.

b. x is a *factor* of $18x$, since x and 18 are multiplied.

c. The expression $18 - 3x$ can be written $18 + (-3x)$. The + sign separates the expression into two terms, 18 and $-3x$. The variable x is a *factor* of the second term.

SELF CHECK

Determine whether y is a factor or a term: *Answers:*   **a.** factor,   **b.** term, **a.** $32y$,   **b.** $-45 + y$,   and   **c.** $-45y - 25$.   **c.** factor   ■

Like Terms

The expression $5p + 7q - 3p + 12$, which can be written $5p + 7q + (-3p) + 12$, contains four terms, $5p$, $7q$, $-3p$, and 12. Since the variable of $5p$ and $-3p$ are the same, we say that these terms are **like** or **similar terms.**

Like terms (similar terms)

Like terms (or **similar terms**) are terms with exactly the same variables raised to exactly the same powers. Any numbers (called **constants**) in an expression are considered to be like terms.

Like terms

$4x$ and $7x$

↑   ↑

Same variable

$-10p^2$, $25p^2$, and $150p^2$

Same variable to the same power

Unlike terms

$4x$ and $3y$

↑   ↑

Different variables

$15p$ and $23p^2$

Different exponents on the variable p

 WARNING! When looking for like terms, don't look at the coefficients of the terms. Consider only their variable parts.

EXAMPLE 10

Identifying like terms. List like terms: **a.** $7r + 5 + 3r$, **b.** $x^4 - 6x^2 - 5$,   and   **c.** $-7m + 7 - 2 + m$.

Solution

a. $7r + 5 + 3r$ contains the like terms $7r$ and $3r$.

b. $x^4 - 6x^2 - 5$ contains no like terms.

c. $-7m + 7 - 2 + m$ contains two pairs of like terms: $-7m$ and m are like terms, and the constants, 7 and -2, are like terms.

SELF CHECK

List like terms: **a.** $5x - 2y + 7y$ and   *Answers:*   **a.** $-2y$ and $7y$, **b.** $-5pq + 17p - 12q - 2pq$.   **b.** $-5pq$ and $-2pq$ ■

Combining Like Terms

If we are to add (or subtract) objects, they must have the same units. For example, we can add dollars to dollars and inches to inches, but we cannot add dollars to inches. The same is true when we work with terms of an algebraic expression. They can be added or subtracted only when they are like terms.

This expression can be simplified, because it contains like terms.	This expression cannot be simplified, because its terms are not like terms.
$3x + 4x$	$3x + 4y$
↑ ↑	↑ ↑
Like terms	Unlike terms
The variable parts are identical.	The variable parts are not identical.

To simplify an expression containing like terms, we use the distributive property. For example, we can simplify $3x + 4x$ as follows:

$$3x + 4x = (\mathbf{3 + 4})x \quad \text{Apply the distributive property.}$$
$$= 7x \qquad \text{Do the addition in the parentheses: } 3 + 4 = 7.$$

We have simplified the expression $3x + 4x$ by **combining like terms**. The result is the equivalent expression $7x$. This example suggests the following general rule.

Combining like terms

To add or subtract like terms, combine their coefficients and keep the same variables with the same exponents.

EXAMPLE 11

Simplifying algebraic expressions. Simplify by combining like terms:
a. $-8p + (-12p)$, and **b.** $0.5s^2 - 0.3s^2$.

Solution **a.** $-8p + (-12p) = -20p$ Add the coefficients of the like terms: $-8 + (-12) = -20$. Keep the variable p.

b. $0.5s^2 - 0.3s^2 = 0.2s^2$ Subtract: $0.5 - 0.3 = 0.2$. Keep the variable part s^2.

SELF CHECK Simplify by combining like terms:
a. $5n + (-8n)$ and **b.** $-1.2a^3 + (1.4a^3)$. *Answers:* **a.** $-3n$, **b.** $0.2a^3$ ∎

EXAMPLE 12

Combining like terms. Simplify $7P - 8p - 12P + 25p$.

Solution The uppercase P and the lowercase p are different variables. To combine like terms, one approach is to write each subtraction as an addition of the opposite and proceed as follows.

$$7P - 8p - 12P + 25p$$
$$= 7P + (-8p) + (-12P) + 25p \quad \text{Rewrite each subtraction as the addition of the opposite.}$$
$$= 7P + (-12P) + (-8p) + 25p \quad \text{Use the commutative property of addition to write the like terms together.}$$
$$= -5P + 17p \qquad \text{Combine like terms: } 7P + (-12P) = -5P \text{ and } -8p + 25p = 17p.$$

SELF CHECK Simplify $8R + 7r - 14R - 21r$. *Answer:* $-6R - 14r$ ∎

The expression in Example 12 contained two sets of like terms, and we rearranged the terms so that like terms were next to each other. With practice, you will be able to combine like terms without having to write them next to each other and without having to write each subtraction as addition of the opposite.

EXAMPLE 13 **Combining like terms without rearranging terms.** Simplify $4(x + 5) - 3(2x - 4)$.

Solution

$$4(x + 5) - 3(2x - 4)$$
$$= 4x + 20 - 6x + 12 \quad \text{Use the distributive property twice.}$$
$$= -2x + 32 \quad \text{Combine like terms: } 4x - 6x = -2x \text{ and } 20 + 12 = 32.$$

SELF CHECK Simplify $-5(y - 4) + 2(4y + 6)$. *Answer:* $3y + 32$ ■

STUDY SET

Section 2.4

VOCABULARY

In Exercises 1–6, fill in the blanks to make the statements true.

1. To _____ an algebraic expression, we use properties of algebra to write the expression in a less complicated form.

2. A _____ is a number or a product of a number and one or more variables.

3. In the term $3x^2$, the number factor 3 is called the _____ and x^2 is called the _____ part.

4. Two terms with exactly the same variables and exponents are called _____ terms.

5. We can use the distributive property to _____ the parentheses when an expression is multiplied by a quantity.

6. The _____ property of multiplication enables us to change the order of the factors involved in a multiplication.

CONCEPTS

7. What property does the statement $a(b + c) = ab + ac$ illustrate?

8. Complete this statement:
   $a(b + c + d) =$ _____

9. What application of the distributive property is demonstrated in Illustration 1?

ILLUSTRATION 1

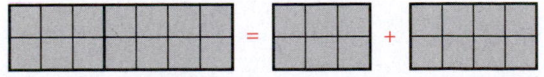

10. Complete this statement: To add or subtract like terms, combine their _____ and keep the same variables and _____.

11. A board was cut into two pieces, as shown in Illustration 2. Add the lengths of the two pieces. How long was the original board?

ILLUSTRATION 2

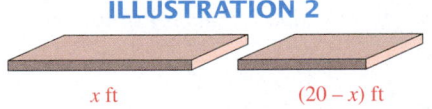

x ft $(20 - x)$ ft

12. Let x equal the number of miles driven the first day of a 2-day driving trip. Translate the verbal model to mathematical symbols and simplify by combining like terms.

the miles driven day 1	plus	100 miles more than the miles driven day 1

13. Two angles are **complementary angles** if the sum of their measures is 90°. If one of the angles has a measure of $a°$ (see Illustration 3), use an algebraic expression to represent the measure of its complement.

14. Two angles are **supplementary angles** if the sum of their measures is 180°. If one of the angles has a measure of $a°$ (see Illustration 4), use an algebraic expression to represent the measure of its supplement.

ILLUSTRATION 3   **ILLUSTRATION 4**

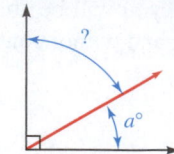   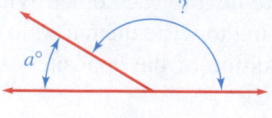

NOTATION

In Exercises 15–16, complete each solution.

15. $-7(a^2 + a - 5) = \boxed{} \cdot a^2 + \left(\boxed{}\right) \cdot a - \left(\boxed{}\right) \cdot 5$

$$= -7a^2 + (-7a) - \left(\boxed{}\right)$$
$$= -7a^2 - \boxed{} - (-35)$$
$$= -7a^2 - 7a + 35$$

16. $6(b - 5) + 12b + 7 = 6 \cdot \boxed{} - 6 \cdot \boxed{} + 12b + 7$

$$= 6b - \boxed{} + 12b + 7$$
$$= 6b + \boxed{} b - \boxed{} + 7$$
$$= 18b - 23$$

17. a. Are $2K$ and $3k$ like terms?
 b. Are $-d$ and d like terms?

18. Fill in the blank to make the statement true.

$$-(x + 10) = - \boxed{} (x + 10)$$

PRACTICE

In Exercises 19–26, simplify each expression.

19. $9(7m)$ **20.** $12n(8)$

21. $5(-7q)$ **22.** $-7(5t)$

23. $(-5p)(-4b)$ **24.** $(-7d)(-7c)$

25. $-5(4r)(-2r)$ **26.** $7t(-4t)(-2)$

In Exercises 27–46, use the distributive property to remove parentheses.

27. $5(x + 3)$ **28.** $4(x + 2)$

29. $-2(b - 1)$ **30.** $-7(p - 5)$

31. $(3t - 2)8$ **32.** $(2q + 1)9$

33. $(2y - 1)6$ **34.** $(3w - 5)5$

35. $0.4(x - 4)$ **36.** $-2.2(2q + 1)$

37. $-\dfrac{2}{3}(3w - 6)$ **38.** $\dfrac{1}{2}(2y - 8)$

39. $r(r - 10)$ **40.** $h(h + 4)$

41. $-(x - 7)$ **42.** $-(y + 1)$

43. $17(2x - y + 2)$ **44.** $-12(3a + 2b - 1)$

45. $-(-14 + 3p - t)$ **46.** $-(-x - y + 5)$

In Exercises 47–54, list the terms in each expression. Then identify the coefficient of each term.

47. $-5r + 4s$

48. $2m + n - 3m + 2n$

49. $-15r^2 s$

50. $4b^2 - 5b + 6$

51. $50a + 2$

52. $a^2 - ab + b^2$

53. $x^3 - 125$

54. $-2.55x + 1.8$

In Exercises 55–58, identify the coefficient of each term.

55. $-b$ **56.** $-9.9x^3$

57. $\dfrac{1}{4}x$ **58.** $-\dfrac{2x}{3}$

In Exercises 59–62, tell whether the variable x is used as a factor or a term.

59. $24 - x$ **60.** $24x$

61. $24 + 3x$ **62.** $x - 12$

In Exercises 63–90, simplify each expression by combining like terms.

63. $3x + 17x$ **64.** $12y - 15y$

65. $8x^2 - 5x^2$ **66.** $17x^2 + 3x^2$

67. $-4x + 4x$ **68.** $-16y + 16y$

69. $-7b^2 + 7b^2$ **70.** $-2c^3 + 2c^3$

71. $3x + 5x - 7x$ **72.** $-y + 3y + 2y$

73. $13x^2 + 2x^2 - 5x^2$ **74.** $8x^3 - x^3 + 2x^3$

75. $1.8h - 0.7h$ **76.** $-5.7m + 4.3m$

77. $\dfrac{3}{5}t + \dfrac{1}{5}t$ **78.** $\dfrac{3}{16}x - \dfrac{5}{16}x$

79. $4(y + 9) - 6y$ **80.** $-3(3 + z) + 2z$

81. $2z + 5(z - 3)$ **82.** $12(m + 11) - 11$

83. $8(c + 7) - 2(c - 3)$ **84.** $9(z + 2) + 5(3 - z)$

85. $2x + 4(X - x) + 3X$ **86.** $3p - 6(p + z) + p$

87. $(a + 2) - (a - b)$ **88.** $3z + 2(Z - z) + Z$

89. $x(x + 3) - 3x^2$ **90.** $2x + x(x - 3)$

APPLICATIONS

91. THE AMERICAN RED CROSS In 1891, Clara Barton founded the Red Cross. Its symbol is a white flag bearing a red cross. If each side of the cross in Illustration 5 has length x, write an algebraic expression for the perimeter (the total distance around the outside) of the cross.

ILLUSTRATION 5

92. BILLIARDS Billiard tables vary in size, but all tables are twice as long as they are wide.
   a. If the billiard table in Illustration 6 is x feet wide, write an expression involving x that represents its length.
   b. Write an expression for the perimeter of the table.

ILLUSTRATION 6

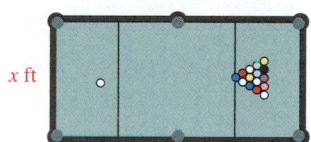

93. PING-PONG Write an expression for the perimeter of the ping-pong table shown in Illustration 7.

ILLUSTRATION 7

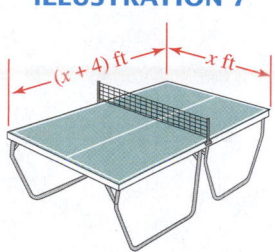

94. SEWING See Illustration 8. Write an expression for the length of the yellow trim needed to outline the pennant with the given side lengths.

ILLUSTRATION 8

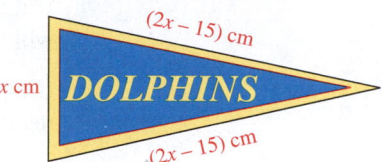

WRITING

95. Explain why $3x^2y$ and $5x^2y$ are like terms.

96. Explain why $3x^2y$ and $5xy^2$ are not like terms.

97. Distinguish between a *factor* and a *term* of an algebraic expression. Give examples.

98. Tell how to combine like terms.

REVIEW

In Exercises 99–102, evaluate each expression given that $x = -3$, $y = -5$, and $z = 0$.

99. $x^2z(y^3 - z)$

100. $z - y^3$

101. $\dfrac{x - y^2}{2y - 1 + x}$

102. $\dfrac{2y + 1}{x} - x$

▶ 2.5

Solving Equations

In this section, you will learn about

Solving equations by using more than one property of equality ■ Simplifying expressions to solve equations ■ Identities and impossible equations ■ Problem solving

Introduction In Chapter 1, we solved some simple equations using the properties of equality. In this section, we will solve more complicated equations. Our objective is to develop a general strategy that can be used to solve any kind of linear equation.

Solving Equations by Using More Than One Property of Equality

Recall the following properties of equality:

■ If the same quantity is added to (or subtracted from) equal quantities, the results will be equal quantities.

■ If equal quantities are multiplied (or divided) by the same nonzero quantity, the results will be equal quantities.

We have already solved many simple equations using one of the properties listed above. For example, to solve $x + 6 = 10$, we isolate x by subtracting 6 from both sides.

$$x + 6 = 10$$
$$x + 6 - 6 = 10 - 6 \qquad \text{To undo the addition of 6, subtract 6 from both sides.}$$
$$x = 4 \qquad \text{Do the subtractions.}$$

To solve $2x = 10$, we isolate x by dividing both sides by 2.

$$2x = 10$$
$$\frac{2x}{2} = \frac{10}{2} \qquad \text{To undo the multiplication by 2, divide both sides by 2.}$$
$$x = 5 \qquad \text{Do the divisions: } \tfrac{2}{2} = 1 \text{ and } \tfrac{10}{2} = 5.$$

Sometimes several properties of equality must be applied in succession to solve an equation. For example, on the left-hand side of $2x + 6 = 10$, x is first multiplied by 2 and then 6 is added to that product. To isolate x, we use the rules for the order of operations in reverse. First, we undo the addition of 6, and then we undo the multiplication by 2.

$$2x + 6 = 10$$
$$2x + 6 - 6 = 10 - 6 \qquad \text{To undo the addition of 6, subtract 6 from both sides.}$$
$$2x = 4 \qquad \text{Do the subtractions.}$$
$$\frac{2x}{2} = \frac{4}{2} \qquad \text{To undo the multiplication by 2, divide both sides by 2.}$$
$$x = 2 \qquad \text{Do the divisions: } \tfrac{2}{2} = 1 \text{ and } \tfrac{4}{2} = 2.$$

EXAMPLE 1

Using two properties of equality. Solve and check: $-12x + 5 = 17$.

Solution The left-hand side indicates that we must multiply x by -12 and then add 5. To isolate x, we undo these operations in the opposite order.

■ To undo the addition of 5, we subtract 5 from both sides.

■ To undo the multiplication by -12, we divide both sides by -12.

$$-12x + 5 = 17$$
$$-12x + 5 - 5 = 17 - 5 \qquad \text{Subtract 5 from both sides.}$$
$$-12x = 12 \qquad \text{Do the subtractions: } 5 - 5 = 0 \text{ and } 17 - 5 = 12.$$
$$\frac{-12x}{-12} = \frac{12}{-12} \qquad \text{Divide both sides by } -12.$$
$$x = -1 \qquad \text{Do the divisions: } \tfrac{-12}{-12} = 1 \text{ and } \tfrac{12}{-12} = -1.$$

Check: $-12x + 5 = 17$ The original equation.

$-12(\mathbf{-1}) + 5 \overset{?}{=} 17$ Substitute -1 for x.

$12 + 5 \overset{?}{=} 17$ Do the multiplication: $-12(-1) = 12$.

$17 = 17$ Do the addition.

Since the statement $17 = 17$ is true, -1 satisfies the equation.

SELF CHECK   Solve and check: $8x - 13 = 43$.   *Answer:* 7   ■

EXAMPLE 2   **Using two properties of equality.** Solve and check: $\dfrac{2x}{3} = -6$.

Solution   The left-hand side indicates that we must multiply x by 2 and divide that product by 3. To solve this equation, we must undo these operations in the opposite order.

■ To undo the division of 3, we multiply both sides by 3.
■ To undo the multiplication by 2, we divide both sides by 2.

$$\frac{2x}{3} = -6$$

$$3\left(\frac{2x}{3}\right) = 3(-6)$$ Multiply both sides by 3.

$$2x = -18$$ Simplify: $\frac{3}{3} = 1$ and $3(-6) = -18$.

$$\frac{2x}{2} = \frac{-18}{2}$$ Divide both sides by 2.

$$x = -9$$ Do the divisions: $\frac{2}{2} = 1$ and $\frac{-18}{2} = -9$.

Check: $\dfrac{2x}{3} = -6$ The original equation.

$$\frac{2(\mathbf{-9})}{3} \overset{?}{=} -6$$ Substitute -9 for x.

$$\frac{-18}{3} \overset{?}{=} -6$$ Do the multiplication: $2(-9) = -18$.

$$-6 = -6$$ Do the division.

Since we obtain a true statement, -9 is a solution.

SELF CHECK   Solve and check: $\dfrac{7h}{16} = -14$.

Answer: -32 ■

   Another approach can be used to solve the equation from Example 2, $\frac{2x}{3} = -6$. To isolate the variable, we will use the fact that the product of a number and its **reciprocal,** or **multiplicative inverse,** is 1. Since $\frac{2x}{3} = \frac{2}{3}x$, the equation can be rewritten as

$$\frac{2}{3}x = -6$$

To isolate x, we multiply both sides by $\frac{3}{2}$, the reciprocal (multiplicative inverse) of $\frac{2}{3}$.

$$\mathbf{\frac{3}{2}}\left(\frac{2}{3}x\right) = \mathbf{\frac{3}{2}}(-6)$$ The coefficient of x is $\frac{2}{3}$. Multiply both sides by the reciprocal of $\frac{2}{3}$, which is $\frac{3}{2}$.

$$\left(\frac{3}{2} \cdot \frac{2}{3}\right)x = \frac{3}{2}(-6)$$ Apply the associative property of multiplication to regroup factors.

$$1x = -9$$ Do the multiplications: $\frac{3}{2} \cdot \frac{2}{3} = 1$ and $\frac{3}{2}(-6) = -9$.

$$x = -9$$ $1x = x$.

EXAMPLE 3

Isolating the variable. Solve $-0.2 = -0.8 - y$.

Solution   To solve the equation, we begin by eliminating -0.8 from the right-hand side. We can do this by adding 0.8 to both sides.

$$-0.2 = -0.8 - y$$

$$-0.2 \mathbf{+ 0.8} = -0.8 - y \mathbf{+ 0.8} \quad \text{Add 0.8 to both sides.}$$

$$0.6 = -y \qquad\qquad \text{Do the additions: } -0.2 + 0.8 = 0.6 \text{ and} \\ -0.8 + 0.8 = 0.$$

Since $-y$ means $-1y$, the equation can be rewritten as $0.6 = -1y$. To isolate y, either multiply both sides or divide both sides by -1.

$$0.6 = -1y \quad \text{Write } -y \text{ as } -1y.$$

$$\frac{0.6}{\mathbf{-1}} = \frac{-1y}{\mathbf{-1}} \quad \text{Divide both sides by } -1.$$

$$-0.6 = y \qquad \text{Do the divisions: } \tfrac{0.6}{-1} = -0.6 \text{ and } \tfrac{-1}{-1} = 1.$$

$$y = -0.6$$

Verify that -0.6 satisfies the equation.

SELF CHECK

Solve $-6.6 - m = -2.7$. *Answer:* -3.9 ∎

EXAMPLE 4

Using three properties of equality.   Solve $\dfrac{3x}{4} + 2 = -7$.

Solution   The left-hand side indicates that we must multiply x by 3, divide that product by 4, and then add 2. To solve the equation, we must undo these operations in the opposite order.

- ■ To undo the addition of 2, we subtract 2 from both sides.
- ■ To undo the division by 4, we multiply both sides by 4.
- ■ To undo the multiplication by 3, we divide both sides by 3.

$$\frac{3x}{4} + 2 = -7$$

$$\frac{3x}{4} + 2 \mathbf{- 2} = -7 \mathbf{- 2} \quad \text{Subtract 2 from both sides.}$$

$$\frac{3x}{4} = -9 \qquad \text{Do the subtractions: } 2 - 2 = 0 \text{ and } -7 - 2 = -9.$$

$$\mathbf{4}\left(\frac{3x}{4}\right) = 4(-9) \qquad \text{Multiply both sides by 4.}$$

$$3x = -36 \qquad \text{Simplify: } \tfrac{4}{4} = 1 \text{ and } 4(-9) = -36.$$

$$\frac{3x}{\mathbf{3}} = \frac{-36}{\mathbf{3}} \qquad \text{Divide both sides by 3.}$$

$$x = -12 \qquad \text{Do the divisions: } \tfrac{3}{3} = 1 \text{ and } \tfrac{-36}{3} = -12.$$

Verify that -12 satisfies the equation.

SELF CHECK

Solve $\dfrac{2}{3}b - 3 = -15$. *Answer:* -18 ∎

Simplifying Expressions to Solve Equations

In the next example, we must use the distributive property to remove parentheses.

EXAMPLE 5

Combining like terms. Solve and check: $3(k + 1) - 5k = 0$.

Solution

$$3(k + 1) - 5k = 0$$

$$3k + 3(1) - 5k = 0 \qquad \text{Use the distributive property to remove parentheses.}$$

$$3k - 5k + 3 = 0 \qquad \text{Do the multiplication and rearrange terms.}$$

$$-2k + 3 = 0 \qquad \text{Simplify the left-hand side by combining like terms.}$$

$$-2k + 3 - 3 = 0 - 3 \qquad \text{To undo the addition of 3, subtract 3 from both sides.}$$

$$-2k = -3 \qquad \text{Do the subtractions: } 3 - 3 = 0 \text{ and } 0 - 3 = -3.$$

$$\frac{-2k}{-2} = \frac{-3}{-2} \qquad \text{To undo the multiplication by } -2, \text{ divide both sides by } -2.$$

$$k = \frac{3}{2} \qquad \text{Simplify: } \frac{-2}{-2} = 1 \text{ and } \frac{-3}{-2} = \frac{3}{2}.$$

Check: $3(k + 1) - 5k = 0$ The original equation.

$$3\left(\frac{3}{2} + 1\right) - 5\left(\frac{3}{2}\right) \overset{?}{=} 0 \qquad \text{Substitute } \tfrac{3}{2} \text{ for } k.$$

$$3\left(\frac{3}{2} + \frac{2}{2}\right) - 5\left(\frac{3}{2}\right) \overset{?}{=} 0 \qquad 1 = \tfrac{2}{2}.$$

$$3\left(\frac{5}{2}\right) - 5\left(\frac{3}{2}\right) \overset{?}{=} 0 \qquad \text{Do the addition inside the parentheses.}$$

$$\frac{15}{2} - \frac{15}{2} \overset{?}{=} 0 \qquad \text{Do the multiplications.}$$

$$0 = 0 \qquad \text{Do the subtraction.}$$

SELF CHECK   Solve and check: $-5(x - 3) + 3x = 11$.   *Answer:* 2   ■

EXAMPLE 6

Variables on both sides of the equation. Solve and check: $3x - 15 = 4x + 36$.

Solution   To solve for x, all the terms containing x must be on the same side of the equation. We can eliminate $3x$ from the left-hand side by subtracting $3x$ from both sides.

$$3x - 15 = 4x + 36$$

$$3x - 15 - 3x = 4x + 36 - 3x \qquad \text{Subtract } 3x \text{ from both sides.}$$

$$-15 = x + 36 \qquad \text{Combine like terms: } 3x - 3x = 0 \text{ and } 4x - 3x = x.$$

$$-15 - 36 = x + 36 - 36 \qquad \text{To undo the addition of 36, subtract 36 from both sides.}$$

$$-51 = x \qquad \text{Do the subtractions: } -15 - 36 = -51 \text{ and } 36 - 36 = 0.$$

$$x = -51$$

Check: $3x - 15 = 4x + 36$ The original equation.

$3(-51) - 15 \stackrel{?}{=} 4(-51) + 36$ Substitute -51 for x.

$-153 - 15 \stackrel{?}{=} -204 + 36$ Do the multiplications.

$-168 = -168$ Simplify each side.

SELF CHECK   Solve and check: $3n + 48 = -4n - 8$.   *Answer:* -8 ∎

EXAMPLE 7

Clearing an equation of fractions. Solve $\dfrac{x}{6} - \dfrac{5}{2} = -\dfrac{1}{3}$.

Solution   Since integers are easier to work with, we will clear the equation of the fractions by multiplying both sides by the least common denominator (LCD), which in this case is 6.

$$\frac{x}{6} - \frac{5}{2} = -\frac{1}{3}$$

$$6\left(\frac{x}{6} - \frac{5}{2}\right) = 6\left(-\frac{1}{3}\right)$$ 6 is the smallest number that each denominator will divide exactly. Multiply both sides by 6 to clear the fractions.

$$6\left(\frac{x}{6}\right) - 6\left(\frac{5}{2}\right) = 6\left(-\frac{1}{3}\right)$$ On the left-hand side, remove parentheses.

$$x - 15 = -2$$ Do the multiplications.

$$x - 15 + 15 = -2 + 15$$ To undo the subtraction of 15, add 15 to both sides.

$$x = 13$$ Do the additions: $-15 + 15 = 0$ and $-2 + 15 = 13$.

Verify that 13 satisfies the equation.

SELF CHECK   Solve $\dfrac{x}{4} + \dfrac{1}{2} = -\dfrac{1}{8}$.   *Answer:* $-\dfrac{5}{2}$ ∎

The preceding examples suggest the following strategy for solving equations.

Strategy for solving equations

1. Clear the equation of fractions.
2. Use the distributive property to remove parentheses, if necessary.
3. Combine like terms if necessary.
4. Undo the operations of addition and subtraction to get the variables on one side and the constants on the other.
5. Undo the operations of multiplication and division to isolate the variable.
6. Check the result.

EXAMPLE 8

Applying the equation-solving strategy. Solve and check:
$$\frac{3x + 11}{5} = x + 3.$$

Solution

$$\frac{3x + 11}{5} = x + 3$$

$$5\left(\frac{3x + 11}{5}\right) = 5(x + 3)$$ Clear the equation of the fraction by multiplying both sides by 5.

$$3x + 11 = 5x + 15$$ On the left-hand side, simplify: $\frac{\cancel{5}}{1}\left(\dfrac{3x + 11}{\cancel{5}}\right)$. On the right-hand side, remove parentheses.

$$3x + 11 - 11 = 5x + 15 - 11$$ Subtract 11 from both sides.

$$3x = 5x + 4$$ Do the subtractions.

$$3x - 5x = 5x + 4 - 5x$$ To eliminate $5x$ from the right-hand side, subtract $5x$ from both sides.

$$-2x = 4$$ Combine like terms: $3x - 5x = -2x$ and $5x - 5x = 0$.

$$\frac{-2x}{-2} = \frac{4}{-2}$$ To undo the multiplication by -2, divide both sides by -2.

$$x = -2$$ Do the divisions.

Verify that -2 satisfies the equation.

SELF CHECK Solve and check: $\dfrac{3x + 23}{7} = x + 5$.

Answer: -3 ■

 WARNING! Remember that when you multiply one side of an equation by a nonzero number, you must multiply the other side of the equation by the same number.

Identities and Impossible Equations

Equations in which some numbers satisfy the equation and others don't are called **conditional equations.** The equations in Examples 1–8 are conditional equations.

An equation that is true for all values of its variable is called an **identity.**

$$x + x = 2x$$ This is an identity, because it is true for all values of x.

An equation that is not true for any values of its variable is called an **impossible equation** or a **contradiction.** Such equations are said to have no solution.

$$x = x + 1$$ Because no number is 1 greater than itself, this is an impossible equation.

EXAMPLE 9 **Identities.** Solve $3(x + 8) + 5x = 2(12 + 4x)$.

Solution

$$3(x + 8) + 5x = 2(12 + 4x)$$

$$3x + 24 + 5x = 24 + 8x$$ Remove parentheses.

$$8x + 24 = 24 + 8x$$ Combine like terms.

$$8x + 24 - 8x = 24 + 8x - 8x$$ Subtract $8x$ from both sides.

$$24 = 24$$ Combine like terms: $8x - 8x = 0$.

Since the result $24 = 24$ is true for every number x, all values of x satisfy the original equation. This equation is an identity.

SELF CHECK Solve $3(x + 5) - 4(x + 4) = -x - 1$.

Answer: all values of x; this equation is an identity ■

EXAMPLE 10

Impossible equations. Solve $3(d + 7) - d = 2(d + 10)$.

Solution

$3(d + 7) - d = 2(d + 10)$	
$3d + 21 - d = 2d + 20$	Remove parentheses.
$2d + 21 = 2d + 20$	Combine like terms.
$2d + 21 - 2d = 2d + 20 - 2d$	Subtract $2d$ from both sides.
$21 = 20$	Combine like terms.

Since the result $21 = 20$ is false, the original equation has no solution. It is an impossible equation.

SELF CHECK

Solve $-4(c - 3) + 2c = 2(10 - c)$.

Answer: No solution. This equation is an impossible equation. ∎

Problem Solving

We can use equations to solve many types of problems. To set up and solve the equations, we will follow these steps:

Strategy for problem solving

1. **Analyze the problem** by reading it carefully to understand the given facts. What information is given? What vocabulary is given? What are you asked to find? Often, a diagram will help you visualize the facts of the problem.

2. **Form an equation** by picking a variable to represent the quantity to be found. Then express all other unknown quantities as expressions involving that variable. Finally, write an equation expressing a quantity in two different ways.

3. **Solve the equation.**

4. **State the conclusion.**

5. **Check the result** in the words of the problem.

EXAMPLE 11

California coastline. The first part of California's magnificent 17-Mile Drive scenic tour, shown in Figure 2-13, begins at the Pacific Grove entrance and travels to Seal Rock. It is 1 mile longer than the second part of the drive, which extends from Seal Rock to the Lone Cypress. The final part of the tour winds through the hills of the Monterey Peninsula, eventually returning to the entrance. This part of the drive is 1 mile longer than four times the length of the second part. How long is each of the three parts of 17-Mile Drive?

ANALYZE THE PROBLEM

In Figure 2-14, we "straighten out" the winding 17-Mile Drive so that it can be modeled with a line segment. The drive is composed of three parts. We need to find the length of each part.

FIGURE 2-13

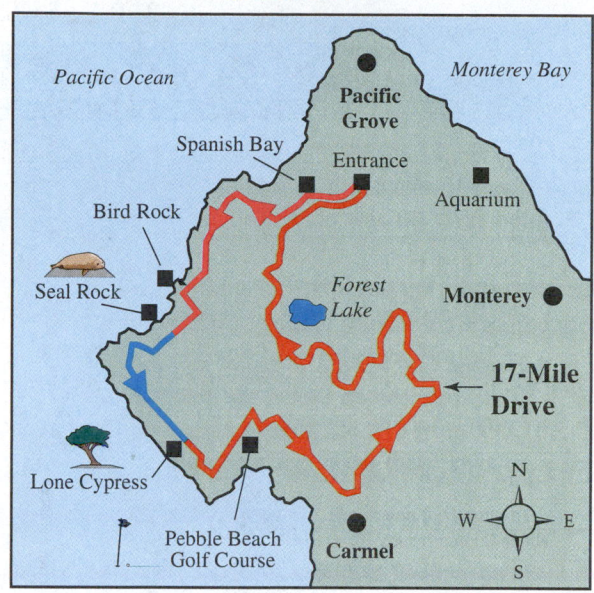

FORM AN EQUATION Since the lengths of the first part and of the third part of the scenic drive are related to the length of the second part, we will let x represent the length of that part. We then express the other lengths in terms of that variable.

$x + 1$ represents the length of the first part of the drive.

$4x + 1$ represents the length of the third part of the drive.

FIGURE 2-14

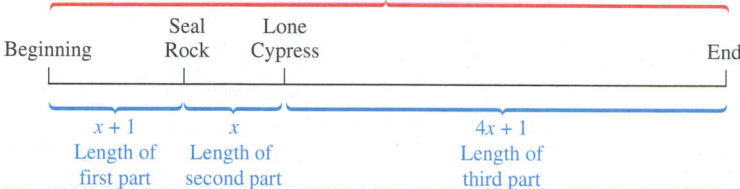

The sum of the lengths of the three parts of the drive must equal the total length of 17-Mile Drive.

The length of part 1	plus	the length of part 2	plus	the length of part 3	equals	the total length.
$x + 1$	$+$	x	$+$	$4x + 1$	$=$	17

SOLVE THE EQUATION

$x + 1 + x + 4x + 1 = 17$	The equation to solve.
$6x + 2 = 17$	Combine like terms: $x + x + 4x = 6x$.
$6x = 15$	To undo the addition of 2, subtract 2 from both sides.
$\dfrac{6x}{6} = \dfrac{15}{6}$	To undo the multiplication by 6, divide both sides by 6.
$x = 2.5$	Do the divisions.

STATE THE CONCLUSION The second part of the scenic drive is 2.5 miles long. Because the first part of the drive is 1 mile longer than the second, it is 3.5 miles long. The length of the third part of the drive is 1 mile more than four times the second part, so it is $4(2.5) + 1$, or 11 miles long.

CHECK THE RESULT Because the sum of 3.5 miles, 2.5 miles, and 11 miles is 17 miles, the solution checks.

Section 2.5

VOCABULARY

In Exercises 1–6, fill in the blanks to make the statements true.

1. An _____ is a statement that two quantities are equal.

2. If a number is a solution of an equation, the number is said to _____ the equation.

3. In an equation that is an identity, all _____ satisfy the equation.

4. If no numbers satisfy an equation, the equation is called an _____ equation.

5. In $2(x - 7)$, "to remove parentheses" means to apply the _____ property.

6. To solve an equation, we must _____ the variable on one side of the equation.

CONCEPTS

In Exercises 7–10, fill in the blanks to make the statements true.

7. To solve the equation $2x - 7 = 21$, we first undo the _____ of 7 by adding 7 to both sides. We then undo the _____ by 2 by dividing both sides by 2.

8. To solve the equation $\frac{x}{2} + 3 = 5$, we first undo the _____ of 3 by subtracting 3 from both sides. We then undo the _____ by 2 by multiplying both sides by 2.

9. To solve $3b - 8 = 2b + 1$, all the terms containing b should be on the same side. To do this, we can eliminate $2b$ on the right side by _____ $2b$ from both sides.

10. To solve $\frac{s}{3} + \frac{1}{4} = -\frac{1}{2}$, we can clear the equation of the fractions by _____ both sides of the equation by 12.

11. **a.** Simplify $3x + 5 - x$.
    b. Solve $3x + 5 - x = 9$.
    c. Evaluate $3x + 5 - x$ for $x = 9$.

12. **a.** Simplify $3(x - 4) - 4x$.
    b. Solve $3(x - 4) - 4x = 0$.
    c. Evaluate $3(x - 4) - 4x$ for $x = 0$.

13. The amount of seating in an auditorium can be described using algebraic expressions, as in Illustration 1.

ILLUSTRATION 1

Seating area	Number of seats
Main floor	$x + 200$
Box seats	$x - 50$
Balcony	x

a. The number of seats in the table is based on which seating area?

b. How many more seats are there on the main floor than in the balcony?

c. What is the total seating in the auditorium?

14. The circulation for a large newspaper can be described using algebraic expressions, as in Illustration 2.

ILLUSTRATION 2

Day	Circulation
Weekday	x
Saturday	$x - 75,000$
Sunday	$2x$

a. On which day are the circulation numbers based?

b. On which day is the circulation the smallest?

c. What is the total weekly circulation of the paper?

15. What is the LCD for the fractions in the equation $\frac{x}{3} - \frac{4}{5} = \frac{1}{2}$?

16. One method of solving $-\frac{4}{5}x = 8$ is to multiply both sides of the equation by the reciprocal of $-\frac{4}{5}$. What is the reciprocal of $-\frac{4}{5}$?

NOTATION

In Exercises 17–18, complete each solution.

17. Solve the equation $2x - 7 = 21$.

$$2x - 7 = 21$$
$$2x - 7 + \boxed{} = 21 + \boxed{}$$
$$2x = \boxed{}$$
$$\frac{2x}{\boxed{}} = \frac{28}{\boxed{}}$$
$$x = 14$$

18. Solve the equation $\dfrac{x}{2} + 3 = 5$.

$$\frac{x}{2} + 3 = 5$$

$$\frac{x}{2} + 3 - \boxed{} = 5 - \boxed{}$$

$$\frac{x}{2} = \boxed{}$$

$$\boxed{}\left(\frac{x}{2}\right) = \boxed{} (2)$$

$$x = 4$$

19. Fill in the blanks to make the statements true.

a. $-x = \boxed{}\ x.$ **b.** $\dfrac{3x}{5} = \boxed{}\ x.$

20. When checking a solution of an equation, the symbol $\stackrel{?}{=}$ is used. What does it mean?

PRACTICE

In Exercises 21–68, solve each equation. Check the result.

21. $2x + 5 = 17$ **22.** $3x - 5 = 13$

23. $-5q - 2 = 1$ **24.** $4p + 3 = 2$

25. $0.6 = 4.1 - x$ **26.** $1.2 - x = -1.7$

27. $-g = -4$ **28.** $-u = -20$

29. $-8 - 3c = 0$ **30.** $-5 - 2d = 0$

31. $-\dfrac{5}{6}k = 10$ **32.** $\dfrac{2c}{5} = 2$

33. $-\dfrac{t}{3} + 2 = 6$ **34.** $\dfrac{x}{5} - 5 = -12$

35. $\dfrac{2x}{3} - 2 = 4$ **36.** $\dfrac{2}{5}y + 3 = 9$

37. $\dfrac{x + 5}{3} = 11$ **38.** $\dfrac{x + 2}{13} = 3$

39. $\dfrac{y - 2}{7} = -3$ **40.** $\dfrac{x - 7}{3} = -1$

41. $2(-3) + 4y = 14$ **42.** $4(-1) + 3y = 8$

43. $-2x - 4(1) = -6$ **44.** $-5x - 3(5) = 0$

45. $3(x + 2) - x = 12$ **46.** $2(x - 4) + x = 7$

47. $-3(2y - 2) - y = 5$ **48.** $-(3a + 1) + a = 2$

49. $3x + 2.5 = 2x$ **50.** $5x + 7.2 = 4x$

51. $9y - 3 = 6y$ **52.** $8y + 4 = 4y$

53. $\dfrac{1}{2} + \dfrac{x}{5} = \dfrac{3}{4}$ **54.** $\dfrac{1}{3} + \dfrac{c}{5} = -\dfrac{3}{2}$

55. $\dfrac{2}{3} = -\dfrac{2x}{3} + \dfrac{3}{4}$ **56.** $-\dfrac{2}{9} = \dfrac{5x}{6} - \dfrac{1}{3}$

57. $3(a + 2) = 2(a - 7)$ **58.** $9(t - 1) = 6(t + 2) - t$

59. $9(x + 11) + 5(13 - x) = 0$

60. $3(x + 15) + 4(11 - x) = 0$

61. $\dfrac{3t - 21}{2} = t - 6$ **62.** $\dfrac{2t + 18}{3} = t - 8$

63. $\dfrac{10 - 5s}{3} = s + 6$ **64.** $\dfrac{40 - 8s}{5} = -2s$

65. $2 - 3(x - 5) = 4(x - 1)$

66. $2 - (4x + 7) = 3 + 2(x + 2)$

67. $15x = x$ **68.** $-7y = -8y$

In Exercises 69–76, solve each equation. If it is an identity or an impossible equation, so indicate.

69. $8x + 3(2 - x) = 5(x + 2) - 4$

70. $5(x + 2) = 5x - 2$

71. $-3(s + 2) = -2(s + 4) - s$

72. $21(b - 1) + 3 = 3(7b - 6)$

73. $2(3z + 4) = 2(3z - 2) + 13$

74. $x + 7 = \dfrac{2x + 6}{2} + 4$

75. $4(y - 3) - y = 3(y - 4)$

76. $5(x + 3) - 3x = 2(x + 8)$

APPLICATIONS

77. CARPENTRY The 12-foot board in Illustration 3 has been cut into two sections, one twice as long as the other. How long is each section?

ILLUSTRATION 3

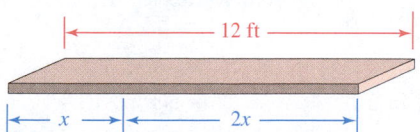

78. PLUMBING A 20-foot pipe has been cut into two sections, one 3 times as long as the other. How long is each section?

79. ROBOTICS The robotic arm shown in Illustration 4 will extend a total distance of 18 feet. Find the length of each section.

ILLUSTRATION 4

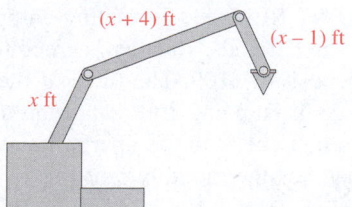

80. WINDOW DESIGN A window in the tank of an aquarium is shown in Illustration 5. If the total distance around the window is 36 feet, how long is each side?

ILLUSTRATION 5

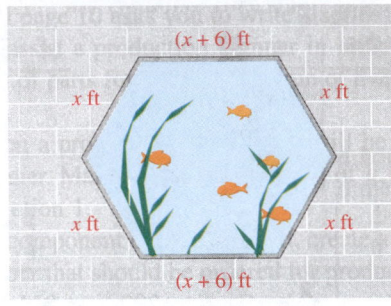

81. TOURING An American rock group plans to travel for a total of 38 weeks, making three major concert tours. They will be in Japan for 4 more weeks than they will be in Australia. Their stay in Sweden will be 2 weeks less than that in Australia. How many weeks will they be in each country?

82. HOURLY PAY After an evaluation, an employee who was making $5 an hour was promised a raise. This week, the employee worked 54 hours. If he is paid time and a half for any hours over 40 and his paycheck was $332.45, did he receive the raise? If so, what is his new hourly rate?

83. COUNTING CALORIES A slice of pie with a scoop of ice cream has 850 calories. The calories in the pie alone are 100 more than twice the calories in the ice cream alone. How many calories are in the ice cream?

84. PUBLISHER'S INVENTORY A novel can be purchased in a hardcover edition for $15.95 or in paperback for $4.95. The publisher printed 11 times as many paperbacks as hardcover books, a total of 114,000 copies for both books. How many hardcover books were printed?

85. ATTORNEY'S FEES An attorney and her client each took half of the cash award that she negotiated in an out-of-court settlement. After paying her assistant $1,000, the attorney ended up making $12,000 from the case. What was the amount of the out-of-court settlement?

86. APARTMENT RENTAL In renting an apartment with two other friends, Jonathan agreed to pay the security deposit of $100. The three of them also agreed to contribute equally toward the monthly rent. Jonathan's first check to the apartment owner was for $225. What was the monthly rent for the apartment?

87. SOLAR HEATING One solar panel in Illustration 6 is 3.4 feet wider than the other. Find the width of each panel.

ILLUSTRATION 6

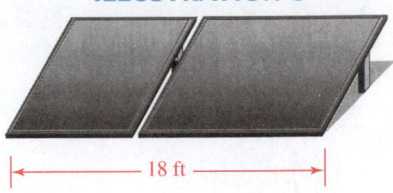

88. WASTE DISPOSAL Two tanks hold a total of 45 gallons of a toxic solvent. One tank holds 6 gallons more than twice the amount in the other. Can the smaller tank be emptied into a 10-gallon waste disposal canister?

89. SHOPPING If you buy one bottle of vitamins, you can buy a second one for half price. If two bottles cost $2.25, find the regular price for one bottle.

90. BOTTLED WATER DELIVERY A driver left the plant with 300 bottles of drinking water on his truck. His route consisted of office buildings, each of which received 3 bottles of water. The driver returned to the plant at the end of the day with 117 bottles on the truck. To how many office buildings did he deliver?

91. CORPORATE DOWNSIZING In an effort to cut costs, a corporation has decided to lay off 5 employees every month until the number of employees totals 465. If 510 people are currently employed, how many months will it take to reach the employment goal?

92. NET INCOME From the information given in Illustration 7, determine the net income of Sears, Roebuck and Co. for each quarter of 1997.

ILLUSTRATION 7

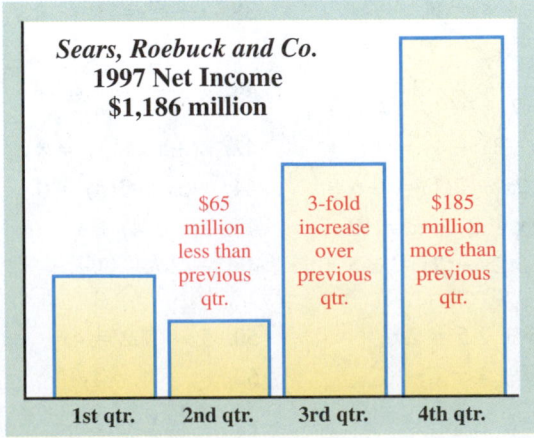

Based on data from Hoover's Online

93. Explain the difference between *simplifying* an expression and *solving* an equation. Give some examples.

94. Briefly explain what should be accomplished in each of the five steps of the problem-solving strategy studied in this section.

 In Exercises 95–98, use a calculator to help solve each problem.

95. Find 82% of 168.

96. 29.05 is what percent of 415?

97. What percent of 200 is 30?

98. A woman bought a coat for $98.95 and some gloves for $7.95. If the sales tax was 6%, how much did the purchase cost her?

▶ 2.6   Formulas

In this section, you will learn about

Formulas from business ■ Formulas from science ■ Formulas from geometry ■ Solving formulas

Introduction   A **formula** is an equation that is used to state a known relationship between two or more variables. Formulas are used in many fields: economics, physical education, anthropology, biology, automotive repair, and nursing, to name a few. In this section, we will consider formulas from business, science, and geometry.

Formulas from Business

A formula to find the sale price: To find the sale price of an item that has been discounted, we subtract the discount from the original price.

$$\boxed{\text{Sale price}} \;=\; \boxed{\text{original price}} \;-\; \boxed{\text{discount}}$$

Using variables to represent the sale price s, the original price p, and the discount d, this formula can be written as

$$\boxed{s = p - d}$$

EXAMPLE 1

Clearance sale. At a year-end clearance sale, a Jeep Wrangler sold for $13,450. If it was discounted $2,300, what was its original price?

Solution   Since the sale price (s) was $13,450 and the discount (d) was $2,300, we substitute 13,450 for s and 2,300 for d in the formula $s = p - d$. Then we solve for p.

$$s = p - d$$
$$13{,}450 = p - 2{,}300 \qquad \text{Substitute 13,450 for } s \text{ and 2,300 for } d.$$
$$13{,}450 + 2{,}300 = p - 2{,}300 + 2{,}300 \qquad \text{To undo the subtraction of 2,300, add 2,300 to both sides.}$$
$$15{,}750 = p \qquad \text{Do the additions.}$$

The original price of the Jeep Wrangler was $15,750.

Because she maintained a 3.0 GPA, a high school senior received a "good student discount" on her auto insurance premium. The $124 discount lowered the six-month premium to $468. What would the premium be if she did not receive the discount? *Answer:* $592 ∎

A formula to find the retail price: To make a profit, a merchant must sell a product for more than he or she paid for it. The price at which the merchant sells the product, called the **retail price,** is the sum of what the item cost the merchant plus the **markup**.

| Retail price | = | cost | + | markup |

Using r to represent the retail price, c the wholesale cost, and m the markup, we can write this formula as

$$r = c + m$$

As an example, suppose a jeweler purchases a gold ring at a wholesale jewelry mart for $612.50. Then she sets the price of the ring at $837.95 for sale in her store. We can find the markup on the ring as follows:

$$r = c + m$$
$$837.95 = 612.50 + m \qquad \text{Substitute 837.95 for } r \text{ and 612.50 for } c.$$
$$837.95 - 612.50 = 612.50 + m - 612.50 \qquad \text{To undo the addition of 612.50, subtract 612.50 from both sides.}$$
$$225.45 = m \qquad \text{Do the subtractions.}$$

The markup on the ring is $225.45.

A formula for profit: The profit a business makes is the difference between the revenue (the money it takes in) and the costs.

| Profit | = | revenue | − | costs |

Using p to represent the profit, r the revenue, and c the costs, we can write this formula as

$$p = r - c$$

EXAMPLE 2 **Charitable giving.** In 1997, the Salvation Army collected $1.7 billion in donations. Of that amount, $1.4 billion went directly to the support of its programs. What were the 1997 administrative costs of the organization?

Solution   The charity collected $1.7 billion in revenue. We can think of the $1.4 billion that was spent on programs as profit. We need to find the administrative costs, c.

$$p = r - c \qquad \text{The formula for profit.}$$
$$1.4 = 1.7 - c \qquad \text{Substitute 1.4 for } p \text{ and 1.7 for } r.$$
$$1.4 - 1.7 = 1.7 - c - 1.7 \qquad \text{To eliminate 1.7, subtract 1.7 from both sides.}$$
$$-0.3 = -c \qquad \text{Combine like terms on both sides.}$$
$$\frac{-0.3}{-1} = \frac{-c}{-1} \qquad \text{Divide both sides by } -1$$
$$0.3 = c \qquad \text{Do the divisions.}$$

In 1997, the Salvation Army had administrative costs of $0.3 billion.

SELF CHECK   A PTA spaghetti dinner made a profit of $275.50. If the cost to host the dinner was $1,235, how much revenue did it generate?   *Answer:* $1,510.50   ■

A formula for simple interest: When money is borrowed, the lender expects to be paid back the amount of the loan plus an additional charge for the use of the money. The additional charge is called **interest**. When money is deposited in a bank, the depositor is paid for the use of the money. The money the deposit earns is also called interest. In general, interest is the money that is paid for the use of money.

Interest is calculated in two ways: either as **simple interest** or as **compound interest.** To find simple interest, we use the formula

$$\text{Interest} \;=\; \text{principal} \;\cdot\; \text{rate} \;\cdot\; \text{time}$$

Using I to represent the simple interest, P the principal (the amount of money that is invested, deposited, or borrowed), r the annual interest rate, and t the length of time in years, we can write the formula as

$$\boxed{I = Prt}$$

EXAMPLE 3   **Retirement income.** One year after investing $15,000 in a mini-mall development, a retired couple received a check for $1,125 in interest. What interest rate did their money earn that year?

Solution   The couple invested $15,000 (the principal) for 1 year (the time) and made $1,125 (the interest). We need to find the annual interest rate.

$$I = Prt$$
$$1,125 = 15,000r(1) \qquad \text{Substitute 1,125 for } I, \text{ 15,000 for } P, \text{ and 1 for } t.$$
$$1,125 = 15,000r \qquad \text{Simplify.}$$
$$\frac{1,125}{15,000} = \frac{15,000r}{15,000} \qquad \text{To solve for } r, \text{ undo the multiplication by 15,000 by dividing both sides by 15,000.}$$
$$0.075 = r \qquad \text{Do the divisions.}$$
$$7.5\% = r \qquad \text{To write 0.075 as a percent, multiply 0.075 by 100 and insert a \% sign.}$$

The couple received an annual rate of 7.5% that year.

A father loaned his daughter and son-in-law $12,200 at a 2% annual simple interest rate for a down payment on a house. If the interest on the loan amounted to $610, for how long was the loan?

Answer: 2.5 years ▪

Formulas from Science

A formula for distance traveled: If we know the average rate (speed) at which we will be traveling and the time we will be traveling at that rate, we can find the distance traveled by using the formula

$$\boxed{\text{Distance}} = \boxed{\text{rate}} \cdot \boxed{\text{time}}$$

Using d to represent the distance, r the average rate (speed), and t the time, we can write this formula as

$$\boxed{d = rt}$$

 WARNING! When using this formula, the units must be the same. For example, if the rate is given in miles per hour, the time must be expressed in hours.

EXAMPLE 4

Finding the rate. As they migrate from the Bering Sea to Baja California, gray whales swim for about 20 hours each day, covering a distance of approximately 70 miles. Estimate their average swimming rate in miles per hour (mph).

Solution

Since the distance d is 70 miles and the time t is 20 hours, we substitute 70 for d and 20 for t in the formula $d = rt$, and then solve for r.

$$d = rt$$
$$70 = r(20) \qquad \text{Substitute 70 for } d \text{ and 20 for } t.$$
$$\frac{70}{20} = \frac{20r}{20} \qquad \text{To undo the multiplication by 20, divide both sides by 20.}$$
$$3.5 = r \qquad \text{Do the divisions.}$$

The whales' average swimming rate is 3.5 mph.

An elevator in a building travels at an average rate of 288 feet per minute. How long will it take it to climb 30 stories, a distance of 360 feet?

Answer: 1.25 minutes ▪

A formula for converting degrees Fahrenheit to degrees Celsius: Many marquees, like the one shown in Figure 2-15, flash two temperature readings, one in degrees Fahrenheit and one in degrees Celsius. The Fahrenheit scale is used in the American system of measurement. The Celsius scale is used in the metric system. As we noted in the Study Set in Section 2.3, there is a formula that shows how a Fahrenheit temperature F relates to a Celsius temperature C:

$$\boxed{C = \frac{5(F - 32)}{9}}$$

FIGURE 2-15

▪CITY SAVINGS

TEMP 30°C

EXAMPLE 5

Changing degrees Celsius to degrees Fahrenheit. Change the temperature reading on the sign in Figure 2-15 to degrees Fahrenheit.

Solution   Since the temperature C in degrees Celsius is 30°, we substitute 30 for C in the formula and solve for F.

$$C = \frac{5(F - 32)}{9}$$

$$30 = \frac{5(F - 32)}{9} \qquad \text{Substitute 30 for } C.$$

$$9(30) = 9\left[\frac{5(F - 32)}{9}\right] \qquad \text{To undo the division by 9, multiply both sides by 9.}$$

$$270 = 5(F - 32) \qquad \text{Simplify: } 9(30) = 270 \text{ and } \frac{9}{9} = 1.$$

$$270 = 5F - 5(32) \qquad \text{Distribute the 5.}$$

$$270 = 5F - 160 \qquad \text{Do the multiplication: } 5(32) = 160.$$

$$270 + 160 = 5F - 160 + 160 \qquad \text{To undo the subtraction of 160, add 160 to both sides.}$$

$$430 = 5F \qquad \text{Do the additions.}$$

$$\frac{430}{5} = \frac{5F}{5} \qquad \text{To undo the multiplication by 5, divide both sides by 5.}$$

$$86 = F \qquad \text{Do the divisions.}$$

Thus, 30°C is equivalent to 86°F.

SELF CHECK   Change −175°C, the temperature on Saturn, to degrees Fahrenheit.

Answer: −283°F

Formulas from Geometry

The **perimeter** of a geometric figure is the distance around it. The **area** of the figure is the amount of surface that it encloses. Table 2-2 shows the formulas for the perimeter (P) and area (A) of several geometric figures.

TABLE 2-2

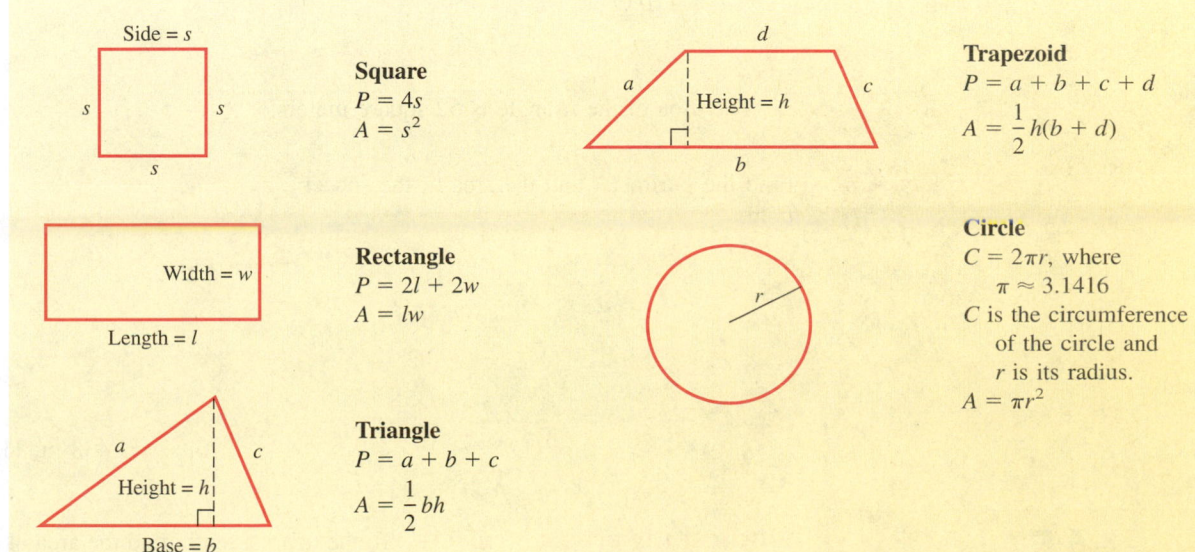

Square
$P = 4s$
$A = s^2$

Trapezoid
$P = a + b + c + d$
$A = \frac{1}{2}h(b + d)$

Rectangle
$P = 2l + 2w$
$A = lw$

Circle
$C = 2\pi r$, where
$\pi \approx 3.1416$
C is the circumference of the circle and r is its radius.
$A = \pi r^2$

Triangle
$P = a + b + c$
$A = \frac{1}{2}bh$

FIGURE 2-16

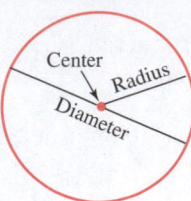

A **circle** (see Figure 2-16) is the set of all points in a plane that are a fixed distance from a point called its **center**. A segment drawn from the center of a circle to a point on the circle is called a **radius**. Since a **diameter** of a circle is a segment passing through the center that joins two points on the circle, the diameter D of a circle is twice as long as its radius r.

$$D = 2r$$

The perimeter of a circle is called its **circumference**. The formula for the circumference of a circle is

$$C = 2\pi r$$

EXAMPLE 6

Finding perimeters and areas. Find **a.** the perimeter of a square with sides 6 inches long and **b.** the area of a triangle with base 8 meters and height 13 meters.

Solution   **a.** The perimeter of a square is given by the formula $P = 4s$, where P is the perimeter and s is the length of one side. Since the sides of the square are 6 inches long, we substitute 6 for s and simplify.

$$P = 4s$$
$$P = 4(6) \quad \text{Substitute 6 for } s.$$
$$ = 24 \quad \text{Do the multiplication.}$$

The perimeter of the square is 24 inches.

b. The area of a triangle is given by the formula $A = \frac{1}{2}bh$. Since the base of the triangle is 8 meters and the height is 13 meters, we substitute 8 for b and 13 for h and simplify.

$$A = \frac{1}{2}bh$$
$$A = \frac{1}{2}(8)(13) \quad \text{Substitute 8 for } b \text{ and 13 for } h.$$
$$ = 4(13) \quad \tfrac{1}{2}(8) = \tfrac{8}{2} = 4.$$
$$ = 52 \quad \text{Do the multiplication.}$$

The area of the triangle is 52 square meters.

SELF CHECK   Find the perimeter and the area of the soccer field.

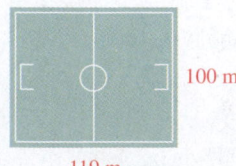

100 m

119 m

Answers: 438 m, 11,900 m² ■

EXAMPLE 7

Finding the area of a circle. To the nearest tenth, find the area of a circle with a diameter of 14 feet.

Solution   Since the radius of a circle is one-half its diameter, the radius of this circle is 7 feet. We can then substitute 7 for r in the formula for the area of a circle and simplify.

$$A = \pi r^2$$
$$A = \pi (7)^2$$
$$\quad = 49\pi \qquad \text{First, evaluate the exponential expression: } 7^2 = 49.$$
$$\quad \approx 153.93804 \quad \text{Use a calculator to do the multiplication. Enter these numbers and press these keys on a scientific calculator: } 49 \;\boxed{\times}\; \boxed{\pi}\; \boxed{=}.$$

To the nearest tenth, the area is 153.9 square feet.

SELF CHECK   To the nearest hundredth, find the circumference of the circle of Example 7.   *Answer:* 43.98 ft   ■

The **volume** of a three-dimensional geometric solid is the amount of space it encloses. Table 2-3 shows the formula for the volume (V) of several solids.

TABLE 2-3

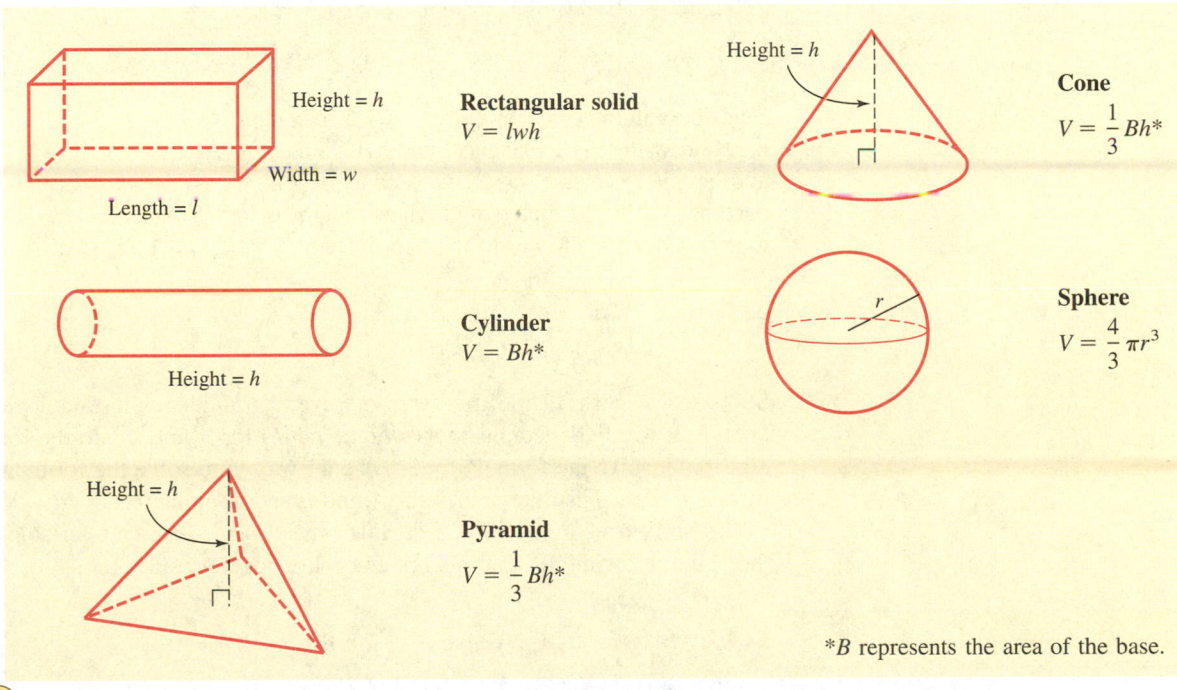

Rectangular solid
$$V = lwh$$

Cone
$$V = \frac{1}{3}Bh^*$$

Cylinder
$$V = Bh^*$$

Sphere
$$V = \frac{4}{3}\pi r^3$$

Pyramid
$$V = \frac{1}{3}Bh^*$$

*B represents the area of the base.

EXAMPLE 8   **Finding volumes.** To the nearest tenth, find the volume of each figure.

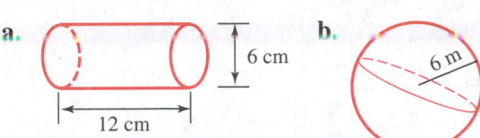

a.   6 cm   12 cm   b.   6 m

Solution   **a.** The formula for the volume of a cylinder is $V = Bh$, where B is the area of the circular base and h is the height. We first find the area of the circular base.

$$A = \pi r^2 \qquad \text{The formula for the area of a circle.}$$
$$A = \pi (3)^2 \qquad \text{The radius is one-half the diameter: } \tfrac{1}{2}(6) = 3.$$
$$\quad = 9\pi$$

We then substitute 9π for B and 12 for h in the formula for the volume of a cylinder.

$V = \boldsymbol{Bh}$	The formula for the volume of a cylinder.
$V = (\boldsymbol{9\pi})(\boldsymbol{12})$	Substitute 9π for B and 12 for h.
$= 108\pi$	Multiply: $9(12) = 108$.
≈ 339.2920066	Use a calculator.

To the nearest tenth, the volume is 339.3 cubic centimeters.

b. To find the volume of the sphere, we substitute 6 for r into the formula for the volume of a sphere.

$V = \dfrac{4}{3}\pi r^3$	The formula for the volume of a sphere.
$V = \dfrac{4}{3}\pi(\boldsymbol{6})^3$	Substitute 6 for r.
$= \dfrac{4}{3}\pi(216)$	$6^3 = 6 \cdot 6 \cdot 6 = 216$.
$= 288\pi$	$\frac{4}{3}(216) = \frac{4(216)}{3} = \frac{864}{3} = 288$.
≈ 904.7786842	Use a calculator.

To the nearest tenth, the volume is 904.8 cubic meters.

SELF CHECK Find the volume of each figure: **a.** a rectangular solid with length 7 inches, width 12 inches, and height 15 inches and **b.** a cone whose base has radius 12 meters and whose height is 9 meters. Give the answer to the nearest tenth.

Answers: **a.** 1,260 in.^{3}, **b.** 1,357.2 m^3 ∎

Solving Formulas

Suppose we wish to find the bases of several triangles whose areas and heights are known. It would be tedious to substitute values for A and h into the formula and then repeatedly solve the formula for b. A better way is to solve the formula $A = \frac{1}{2}bh$ for b first, and then substitute values for A and h and compute b directly.

To **solve an equation for a variable** means to isolate that variable on one side of the equation, with all other quantities on the opposite side.

EXAMPLE 9 **Solving formulas.** Solve $A = \frac{1}{2}bh$ for b.

Solution To solve for b, we must isolate b on one side of the equation.

$A = \dfrac{1}{2}bh$	
$\boldsymbol{2}A = \boldsymbol{2} \cdot \dfrac{1}{2}bh$	To clear the equation of the fraction, multiply both sides by 2.
$2A = bh$	Simplify: $2 \cdot \frac{1}{2} = \frac{2}{2} = 1$.
$\dfrac{2A}{\boldsymbol{h}} = \dfrac{bh}{\boldsymbol{h}}$	To undo the multiplication by h, divide both sides by h.
$\dfrac{2A}{h} = b$	Simplify: $\frac{h}{h} = 1$.
$b = \dfrac{2A}{h}$	

SELF CHECK Solve $A = \frac{1}{2}bh$ for h.

Answer: $h = \frac{2A}{b}$ ∎

EXAMPLE 10

Solving formulas. Solve $r = c + m$ for m.

Solution   To isolate m on the left-hand side, we undo the addition of c by subtracting c from both sides.

$$r = c + m$$
$$r - c = c + m - c \quad \text{Subtract } c \text{ from both sides.}$$
$$r - c = m \qquad\qquad \text{Combine like terms on the right-hand side: } c - c = 0.$$
$$m = r - c$$

SELF CHECK   Solve $p = r - c$ for r.   *Answer:* $r = p + c$ ■

EXAMPLE 11

Solving formulas. Solve $P = 2l + 2w$ for l.

Solution   To solve for l, we must isolate l on one side of the equation.

$$P = 2l + 2w$$
$$P - 2w = 2l + 2w - 2w \quad \text{To undo the addition of } 2w, \text{ subtract } 2w \text{ from both sides.}$$
$$P - 2w = 2l \qquad\qquad\qquad \text{Combine like terms.}$$
$$\frac{P - 2w}{2} = \frac{2l}{2} \qquad\qquad \text{To undo the multiplication by 2, divide both sides by 2.}$$
$$\frac{P - 2w}{2} = l \qquad\qquad \text{Simplify the right-hand side.}$$

If we solve the formula for l, we obtain $l = \dfrac{P - 2w}{2}$.

SELF CHECK   Solve $P = 2l + 2w$ for w.   *Answer:* $w = \frac{P - 2l}{2}$ ■

EXAMPLE 12

Solving formulas. Solve $C = \frac{5}{9}(F - 32)$ for F.

Solution   To solve for F, we must isolate F on one side of the equation.

$$C = \frac{5}{9}(F - 32)$$
$$9(C) = 9\left[\frac{5}{9}(F - 32)\right] \quad \text{Clear the fraction by multiplying both sides by 9.}$$
$$9C = 5(F - 32) \qquad\qquad \text{Simplify: } \tfrac{9}{9} = 1.$$
$$\frac{9C}{5} = \frac{5(F - 32)}{5} \qquad\qquad \text{To undo the multiplication by 5, divide both sides by 5.}$$
$$\frac{9}{5}C = F - 32 \qquad\qquad \text{Do the division: } \tfrac{5}{5} = 1.$$
$$\frac{9}{5}C + 32 = F - 32 + 32 \qquad \text{To undo the subtraction of 32, add 32 to both sides.}$$
$$\frac{9}{5}C + 32 = F \qquad\qquad \text{Do the addition.}$$
$$F = \frac{9}{5}C + 32$$

SELF CHECK   Solve $F = \frac{9C + 160}{5}$ for C.   *Answer:* $C = \frac{5F - 160}{9}$ ■

STUDY SET

Section 2.6

VOCABULARY

In Exercises 1–8, fill in the blanks to make the statements true.

1. A _____ is an equation that is used to state a known relationship between two or more variables.

2. The _____ of a three-dimensional geometric solid is the amount of space it encloses.

3. The distance around a geometric figure is called its _____.

4. A _____ is the set of all points in a plane that are a fixed distance from a point called its center.

5. A segment drawn from the center of a circle to a point on the circle is called a _____.

6. The amount of surface that is enclosed by a geometric figure is called its _____.

7. The perimeter of a circle is called its _____.

8. A segment passing through the center of a circle and connecting two points on the circle is called a _____.

CONCEPTS

9. Tell which geometric concept—perimeter, circumference, area, or volume—should be used to find the following:
   a. The amount of storage in a freezer
   b. How far a bicycle tire rolls in one revolution
   c. The amount of land making up the Sahara Desert
   d. The distance around a Monopoly game board

10. Tell which unit of measurement—ft, ft², or ft³— would be appropriate when finding the following:
    a. The amount of storage inside a safe
    b. The ground covered by a sleeping bag lying on the floor
    c. The distance the tip of an airplane propeller travels in one revolution
    d. The size of the trunk of a car

11. Write an expression for the area of the figure shown in Illustration 1.

ILLUSTRATION 1

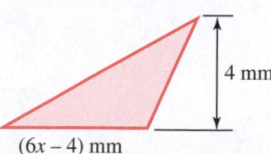

4 mm

(6x − 4) mm

12. Write an expression for the area of the figure shown in Illustration 2.

ILLUSTRATION 2

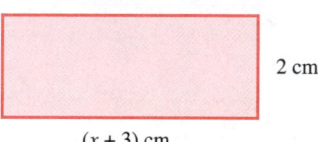

2 cm

(x + 3) cm

13. a. Explain what it means to solve the equation $P = 2l + 2w$ for w.

    b. Why can't we say that this equation is solved for a?
       $a = 2b - a$

14. What type of figure is the base of each of these three-dimensional solids, and, for each base, what is the formula used to find its area B?
    a.   b.

 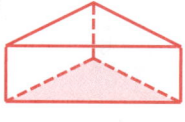

15. a. If the radius r of a circle is known, how can its diameter be found?
    b. If the diameter D of a circle is known, how can its radius r be found?

16. Use variables to write the formula relating the following:
    a. Time, distance, rate
    b. Sale price, discount, original price
    c. Markup, retail price, cost
    d. Costs, revenue, profit
    e. Interest rate, time, interest, principal

17. Complete the table. (mi/sec means miles per second.)

	Rate	· time	= distance
Light	186,282 mi/sec	60 sec	
Sound	1,088 ft/sec	60 sec	

18. Complete the table.

Principal ·	rate ·	time =	interest
$2,500	5%	2 yr	
$15,000	4.8%	1 yr	

NOTATION

In Exercises 19–20, complete each solution.

19. Solve $V = \frac{1}{3}Bh$ for B.

$$V = \frac{1}{3}Bh$$

$$\boxed{}(V) = \boxed{}\left(\frac{1}{3}\right)Bh$$

$$3V = \boxed{}$$

$$\frac{3V}{\boxed{}} = \frac{Bh}{\boxed{}}$$

$$\frac{3V}{h} = \boxed{}$$

$$B = \frac{3V}{h}$$

20. Solve $A = \frac{1}{2}h(B + b)$ for B.

$$A = \frac{1}{2}h(B + b)$$

$$\boxed{}(A) = \boxed{}\left[\frac{1}{2}h(B + b)\right]$$

$$2A = \boxed{}$$

$$\frac{2A}{\boxed{}} = \frac{h(B + b)}{\boxed{}}$$

$$\frac{2A}{h} = \boxed{}$$

$$\frac{2A}{h} - \boxed{} = B + b - \boxed{}$$

$$\frac{2A}{h} - b = B$$

PRACTICE

In Exercises 21–32, use a formula discussed in this section to solve each problem. A calculator will be helpful with some problems.

21. SENIOR CITIZEN DISCOUNT Anyone 65 years of age or older receives $1.75 off the regular price of admission at a certain zoo. If a senior pays $5.75, what is the regular price of admission?

22. REBATE A breadmaking machine that regularly sells for $119.99 costs only $99.99 if the customer takes advantage of a mail-in rebate. How much is the rebate offer worth?

23. VALENTINE'S DAY Find the markup on a dozen roses if a florist buys them wholesale for $12.95 and sells them for $37.50.

24. STICKER PRICE The factory invoice for a minivan shows that the dealer paid $16,264.55 for the vehicle. If the sticker price of the van is $18,202, how much over factory invoice is the sticker price?

25. HOLLYWOOD Figures for the summer of 1998 showed that the movie *Saving Private Ryan* had U.S. box-office receipts of $190 million. What were the production costs to make the movie, if, at that time, the studio had made a $125 million profit?

26. SERVICE CLUB After expenses of $55.15 were paid, a Rotary Club donated $875.85 in proceeds from a pancake breakfast to a local health clinic. How much did the pancake breakfast gross?

27. ENTREPRENEURS To start a mobile dog-grooming service, a woman borrowed $2,500. If the loan was for 2 years and the amount of interest was $175, what simple interest rate was she charged?

28. BANKING Three years after opening an account that paid 6.45% annually, a depositor withdrew the $3,483 in interest earned. How much money was left in the account?

29. SWIMMING In 1930, a man swam down the Mississippi River from Minneapolis to New Orleans, a total of 1,826 miles. He was in the water for 742 hours. To the nearest tenth, what was his average rate swimming?

30. ROSE PARADE Rose Parade floats travel down the 5.5-mile-long parade route at a rate of 2.5 mph. How long will it take a float to complete the parade if there are no delays?

31. METALLURGY Change 2,212°C, the temperature at which silver boils, to degrees Fahrenheit. Round to the nearest degree.

32. LOW TEMPERATURES Cryobiologists freeze living matter to preserve it for future use. They can work with temperatures as low as −270°C. Change this to degrees Fahrenheit.

In Exercises 33–44, solve each formula for the given variable.

33. $E = IR$; for R

34. $I = Prt$; for r

35. $V = lwh$; for w

36. $d = rt$; for t

37. $y = mx + b$; for x

38. $P = 2l + 2w$; for l

39. $A = P + Prt$; for t

40. $V = \pi r^2 h$; for h

41. $V = \frac{1}{3}\pi r^2 h$; for h

42. $K = \frac{wv^2}{2g}$; for w

43. $C = \dfrac{5F - 160}{9}$; for F **44.** $F = \dfrac{GMm}{d^2}$; for d^2

APPLICATIONS

In Exercises 45–66, a calculator will be helpful with some problems.

45. CARPENTRY Find the perimeter and area of the truss shown in Illustration 3.

ILLUSTRATION 3

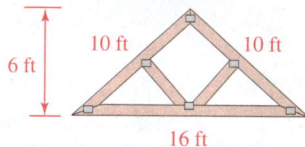

46. CAMPERS Find the area of the window of the camper shell shown in Illustration 4.

ILLUSTRATION 4

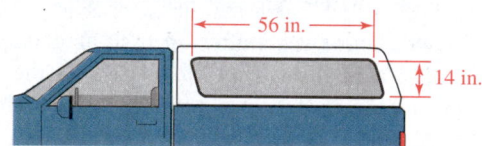

47. ARCHERY To the nearest tenth, find the circumference and area of the target shown in Illustration 5.

ILLUSTRATION 5

48. GEOGRAPHY The circumference of the earth is about 25,000 miles. Find its diameter to the nearest mile.

49. LANDSCAPING Find the perimeter and the area of the redwood trellis in Illustration 6.

ILLUSTRATION 6

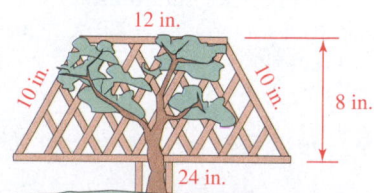

50. VOLUME To the nearest hundredth, find the volume of the soup can shown in Illustration 7.

ILLUSTRATION 7

51. "THE WALL" The Vietnam Veterans Memorial is a black granite wall recognizing the more than 58,000 Americans who lost their lives or remain missing. A diagram of the wall is shown in Illustration 8. Find the total area of the two triangular-shaped surfaces on which the names are inscribed.

ILLUSTRATION 8

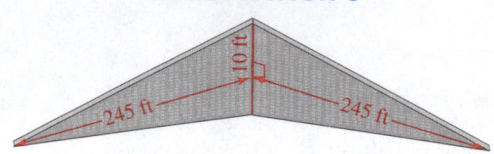

52. SIGNAGE Find the perimeter and area of the service station sign shown in Illustration 9.

ILLUSTRATION 9

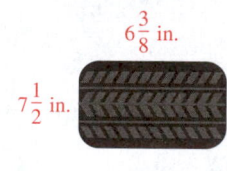

53. RUBBER MEETS THE ROAD A sport truck tire has the road surface "footprint" shown in Illustration 10. Estimate the perimeter and area of the tire's footprint. (*Hint:* First change the dimensions to decimals.)

ILLUSTRATION 10

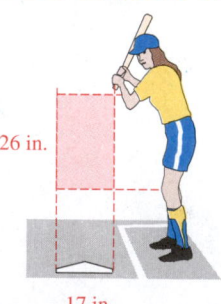

54. SOFTBALL The strike zone in fastpitch softball is between the batter's armpit and top of her knees, as shown in Illustration 11. Find the area of the strike zone.

ILLUSTRATION 11

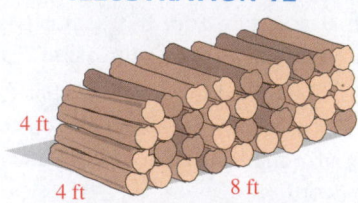

55. FIREWOOD The dimensions of a cord of firewood are shown in Illustration 12. Find the area on which the wood is stacked and the volume the cord of firewood occupies.

ILLUSTRATION 12

56. NATIVE AMERICAN DWELLING The teepees constructed by the Blackfoot Indians were cone-shaped tents made of long poles and animal hide, about 10 feet high and about 15 feet across at the ground. (See Illustration 13.) Estimate the volume of a teepee with these dimensions, to the nearest cubic foot.

ILLUSTRATION 13

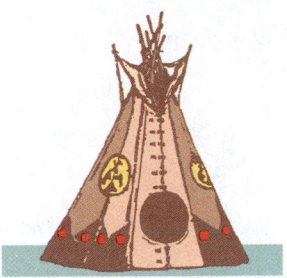

57. IGLOO During long journeys, some Canadian Inuit (Eskimos) built winter houses of snow blocks piled in the dome shape shown in Illustration 14. Estimate the volume of an igloo to the nearest cubic foot having an interior height of 5.5 feet.

ILLUSTRATION 14

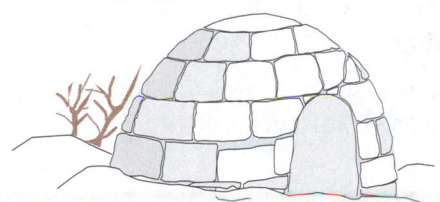

58. PYRAMID The Great Pyramid at Giza in northern Egypt is one of the most famous works of architecture in the world. Use the information in Illustration 15 to find the volume to the nearest cubic foot.

ILLUSTRATION 15

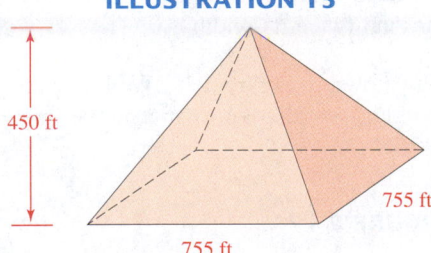

450 ft

755 ft

755 ft

59. BARBECUING See Illustration 16. Use the fact that the fish is 18 inches long to find the area of the barbecue grill to the nearest square inch.

ILLUSTRATION 16

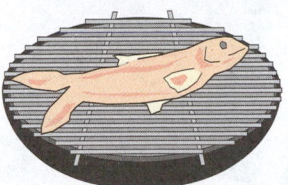

60. SKATEBOARDING A "half-pipe" ramp used for skateboarding is in the shape of a semicircle with a radius of 8 feet, as shown in Illustration 17. To the nearest tenth of a foot, what is the length of the arc that the skateboarder travels on the ramp?

ILLUSTRATION 17

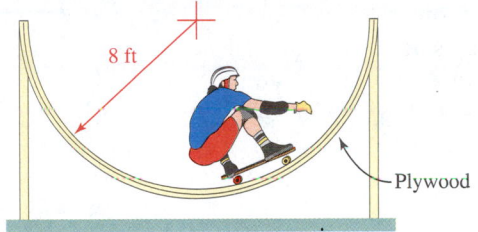

8 ft

Plywood

61. OHM'S LAW The formula $E = IR$ is used in electronics. Solve it for I and find the current I (in amperes) when the voltage E is 48 volts and the resistance R is 12 ohms.

62. GROWTH OF MONEY At a simple interest rate r, \$$P$ grows to \$$A$ in t years according to the formula $A = P(1 + rt)$. Solve the formula for P. After $t = 3$ years, a woman has \$4,357 on deposit in an account that pays 6%. What amount P did she start with?

63. POWER LOSS The power P lost when an electrical current I passes through a resistance R is given by $P = I^2R$. Solve for R. If P is 2,700 watts and I is 14 amperes, find R to the nearest hundredth of an ohm.

64. FORCE OF GRAVITY The masses of the two objects in Illustration 18 are m and M. The force of gravitation, F, between the masses is given by

$$F = \frac{GmM}{d^2}$$

where G is a constant and d is the distance between them. Solve for m.

65. THERMODYNAMICS The Gibbs free-energy function is given by $G = U - TS + pV$. Solve this formula for the pressure p.

ILLUSTRATION 18

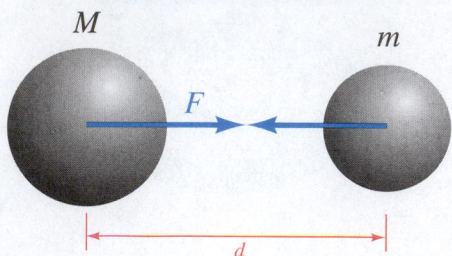

66. **GEOMETRY** The measure a of an interior angle of a regular polygon with n sides is given by the formula

$$a = 180° \left(1 - \frac{2}{n} \right)$$

See Illustration 19. Solve the formula for n. How many sides does a regular polygon have if an interior angle is 108°? (*Hint:* Distribute first.)

ILLUSTRATION 19

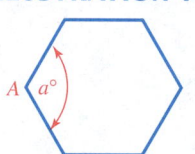

67. The formula $P = 2l + 2w$ is also an equation, but an equation such as $2x + 3 = 5$ is not a formula. What equations do you think should be called formulas?

68. To solve the equation $s - A(s - 5) = r$ for the variable s, one student simply added $A(s - 5)$ to both sides to get $s = r + A(s - 5)$. Explain why this is not correct.

REVIEW

In Exercises 69–70, complete each input/output table.

69.

x	$x^2 - 3$
-2	
0	
3	

70.

x	$\frac{x}{3} + 2$
-6	
0	
12	

In Exercises 71–74, classify each item as either an equation or an expression.

71. $18 + y = 7$

72. $2y + 7$

73. $\dfrac{3x - 2}{7}$

74. $\dfrac{3x}{5} = 12$

▶ 2.7   Problem Solving

In this section, you will learn about

- Solving geometric problems ■ Solving number–value problems
- ■ Solving investment problems ■ Solving uniform motion problems
- ■ Solving mixture problems

Introduction   In this section, we will solve several different types of problems using the five-step problem-solving strategy.

Solving Geometric Problems

EXAMPLE 1

Dimensions of a garden. A gardener wants to use 62 feet of fencing bought at a garage sale to enclose a rectangular-shaped garden. Find the dimensions of the garden if its length is to be 4 feet longer than twice its width.

ANALYZE THE PROBLEM   We can make a sketch of the garden, as shown in Figure 2-17. We know that its length is to be 4 feet longer than twice its width. We also know that its perimeter is to be 62 feet.

FIGURE 2-17

$2w + 4$

FORM AN EQUATION   If we let w represent the width of the garden, then $2w + 4$ represents its length. Since the formula for the perimeter of a rectangle is $P = 2l + 2w$, the perimeter of the garden is $2(2w + 4) + 2w$, which is also 62. This fact enables us to form the equation.

2 times	the length	plus	2 times	the width	is	the perimeter.
$2 \cdot$	$(2w + 4)$	$+$	$2 \cdot$	w	$=$	62

SOLVE THE EQUATION   We then solve the equation.

$2(2w + 4) + 2w = 62$ The equation to solve.

$4w + 8 + 2w = 62$ Use the distributive property to remove parentheses.

$6w + 8 = 62$ Combine like terms.

$6w = 54$ To undo the addition of 8, subtract 8 from both sides.

$w = 9$ To undo the multiplication by 6, divide both sides by 6.

STATE THE CONCLUSION   The width of the garden is 9 feet. Since $2w + 4 = 2 \cdot 9 + 4 = 22$, the length is 22 feet.

CHECK THE RESULT   If the garden has a width of 9 feet and a length of 22 feet, its length is 4 feet longer than twice the width ($2 \cdot 9 + 4 = 22$). Since its perimeter is $(2 \cdot 22 + 2 \cdot 9)$ feet $= 62$ feet, the solution checks.   ■

EXAMPLE 2

Isosceles triangles. If the vertex angle of an isosceles triangle is 56°, find the measure of each base angle.

ANALYZE THE PROBLEM   An **isosceles triangle** has two sides of equal length, which meet to form the **vertex angle.** In this case, the measurement of the vertex angle is 56°. We can sketch the triangle as shown in Figure 2-18. The **base angles** opposite the equal sides are also equal. We need to find their measure.

FIGURE 2-18

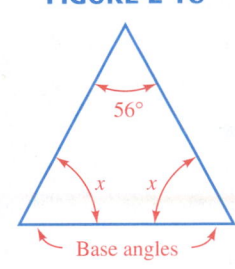

FORM AN EQUATION   If we let x represent the measure of one base angle, the measure of the other base angle is also x. Since the sum of the angles of any triangle is 180°, the sum of the base angles and the vertex angle is 180°. Therefore, we can form the equation.

One base angle	plus	the other base angle	plus	the vertex angle	is	180°.
x	$+$	x	$+$	56	$=$	180

SOLVE THE EQUATION   We then solve the equation.

$x + x + 56 = 180$ The equation to solve.

$2x + 56 = 180$ Combine like terms.

$2x = 124$ To undo the addition of 56, subtract 56 from both sides.

$x = 62$ To undo the multiplication by 2, divide both sides by 2.

STATE THE CONCLUSION   The measure of each base angle is 62°.

CHECK THE RESULT   The measure of each base angle is 62°, and the vertex angle measures 56°. Since $62° + 62° + 56° = 180°$, the solution checks.   ■

Solving Number–Value Problems

Some problems deal with quantities that have a monetary value. In these problems, we must distinguish between the *number of* and the *value of* the unknown quantity. For problems of this type, we will use the relationship

Number · value = total value

EXAMPLE 3

Dining area improvements. A restaurant owner needs to purchase some new tables, chairs, and dinner plates for the dining area of her establishment. She plans to buy four chairs and four plates for each new table. She also needs 20 additional plates to keep in case of breakage. If a table costs $100, a chair $50, and a plate $5, how many of each can she buy if she takes out a small business loan for $6,500 to pay for the new items?

ANALYZE THE PROBLEM We know the *value* of each item: Tables cost $100, chairs cost $50, and plates cost $5 each. We need to find the *number* of tables, chairs, and plates she can purchase for $6,500.

FORM AN EQUATION The number of chairs and plates she needs depends on the number of tables she buys. So we let t be the number of tables to be purchased. Since every table requires four chairs and four plates, she needs to order $4t$ chairs. Because an additional 20 plates are needed, she should order $4t + 20$ plates. The total value of each purchase is the *product* of the number of items bought and the price, or value, of each item.

Item	Number purchased ·	Price per item =	Total value
Table	t	$100	$100t$
Chair	$4t$	$50	$50(4t)$
Plate	$4t + 20$	$5	$5(4t + 20)$

The total purchase can be expressed in two ways:

The value of the tables	+	the value of the chairs	+	the value of the plates	is	the total value of the purchase.
$100t$	+	$50(4t)$	+	$5(4t + 20)$	=	$6,500$

SOLVE THE EQUATION We then solve the equation.

$$100t + 50(4t) + 5(4t + 20) = 6,500 \quad \text{The equation to solve.}$$
$$100t + 200t + 20t + 100 = 6,500 \quad \text{Do the multiplications.}$$
$$320t + 100 = 6,500 \quad \text{Combine like terms.}$$
$$320t = 6,400 \quad \text{Subtract 100 from both sides.}$$
$$t = 20 \quad \text{Divide both sides by 320.}$$

STATE THE CONCLUSION The purchases are summarized as follows:

Item	Number purchased	Price per item	Total value
Table	$t = 20$	$100	$2,000
Chair	$4t = 80$	$50	$4,000
Plate	$4t + 20 = 100$	$5	$500
Total			$6,500

CHECK THE RESULT Because the total purchase is $6,500, the solution checks.

■

Solving Investment Problems

To find the amount of interest I an investment earns, we use the formula

$$I = Prt$$

where P is the principal, r is the annual rate, and t is the time in years. When $t = 1$, the formula simplifies to $I = Pr$.

EXAMPLE 4

Paying tuition. A college student invested the $12,000 inheritance he received and decided to use the annual interest earned to pay his yearly tuition costs of $945. The highest rate offered by a savings and loan at that time was 6% annual simple interest. At this rate, he could not earn the needed $945, so he invested some of the money in a riskier, but more lucrative, investment offering a 9% return. How much did he invest at each rate?

ANALYZE THE PROBLEM We know $12,000 was invested for 1 year at two rates: 6% and 9%. We are asked to find the amount invested at each rate so that the total return would be $945.

FORM AN EQUATION Let x represent the amount invested at 6%. Then $12,000 - x$ represents the amount invested at 9%.

If $\$x$ (the principal P) is invested at 6% (the rate r), the interest earned would be Pr or $\$0.06x$. At 9%, the rest of the inheritance money, $\$(12,000 - x)$, would earn $\$0.09(12,000 - x)$ interest. These facts are summarized in the following table.

	P	$\cdot$ r	$=$ I
Savings and loan	x	0.06	$0.06x$
Riskier investment	$12,000 - x$	0.09	$0.09(12,000 - x)$

The total interest earned can be expressed in two ways.

The interest earned at 6%	plus	the interest earned at 9%	is	the total interest.
$0.06x$	$+$	$0.09(12,000 - x)$	$=$	945

SOLVE THE EQUATION We then solve the equation.

$$0.06x + 0.09(12,000 - x) = 945 \qquad \text{The equation to solve.}$$

$$\mathbf{100}[0.06x + 0.09(12,000 - x)] = \mathbf{100}(945) \qquad \text{Multiply both sides by 100 to clear the equation of decimals.}$$

$$100(0.06x) + 100(0.09)(12,000 - x) = 100(945) \qquad \text{Distribute the 100.}$$

$$6x + 9(12,000 - x) = 94,500 \qquad \text{Do the multiplications by 100.}$$

$$6x + 108,000 - 9x = 94,500 \qquad \text{Remove parentheses.}$$

$$-3x + 108,000 = 94,500 \qquad \text{Combine like terms.}$$

$$-3x = -13,500 \qquad \text{Subtract 108,000 from both sides.}$$

$$x = 4,500 \qquad \text{Divide both sides by } -3.$$

STATE THE CONCLUSION The student invested $4,500 at 6% and $12,000 - \$4,500 = \$7,500$ at 9%.

CHECK THE RESULT The first investment earned 6% of $4,500, or $270. The second earned 9% of $7,500, or $675. The total return was $270 + $675 = $945. The solution checks. ∎

Solving Uniform Motion Problems

If we know the rate r at which we will be traveling and the time t we will be traveling at that rate, we can find the distance d traveled by using the formula

$$d = rt$$

EXAMPLE 5

Coast Guard rescue. A cargo ship, heading into port, radios the Coast Guard that it is experiencing engine trouble and that its speed has dropped to 3 knots. Immediately, a Coast Guard cutter leaves the port and speeds at a rate of 25 knots directly toward the disabled craft, which is 21 nautical miles away. How long will it take the Coast Guard cutter to reach the cargo ship?

ANALYZE THE PROBLEM The diagram in Figure 2-19(a) shows the situation.

FIGURE 2-19

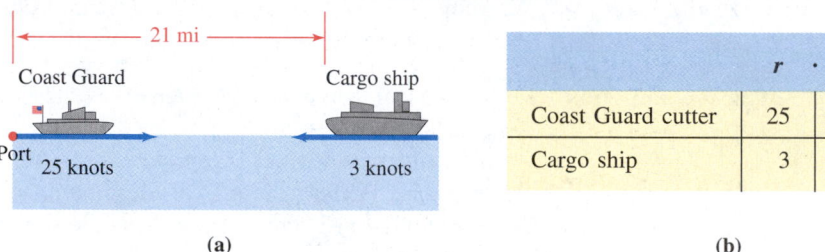

	r	$\cdot$ t	$=$ d
Coast Guard cutter	25	t	$25t$
Cargo ship	3	t	$3t$

(a)   (b)

We know the *rate* of each ship (25 knots and 3 knots), and we know that they must close a *distance* of 21 nautical miles between them. We don't know the *time* it will take them to do this.

FORM AN EQUATION Let t represent the time it takes for the ships to meet. Using $d = rt$, we find that $25t$ represents the distance traveled by the Coast Guard cutter and $3t$ represents the distance traveled by the cargo ship. This information is recorded in the table in Figure 2-19(b). We can form the equation.

The distance the Coast Guard cutter travels	plus	the distance the cargo ship travels	is	the initial distance between the two ships.
$25t$	$+$	$3t$	$=$	21

SOLVE THE EQUATION We then solve the equation.

$25t + 3t = 21$ The equation to solve.

$28t = 21$ Combine like terms.

$t = \dfrac{21}{28}$ Divide both sides by 28.

$t = \dfrac{3}{4}$ Simplify the fraction: $\dfrac{21}{28} = \dfrac{\overset{1}{\cancel{7}} \cdot 3}{\underset{1}{\cancel{7}} \cdot 4}$.

STATE THE CONCLUSION The ships will meet in three-quarters of an hour, or 45 minutes.

CHECK THE RESULT In three-quarters of an hour, the Coast Guard cutter travels $25 \cdot \frac{3}{4} = \frac{75}{4}$ nautical miles and the cargo ship travels $3 \cdot \frac{3}{4} = \frac{9}{4}$ nautical miles. Together, they travel $\frac{75}{4} + \frac{9}{4} = \frac{84}{4} = 21$ nautical miles. Since this is the initial distance between the ships, the solution checks.

Solving Mixture Problems

We now discuss how to solve two types of mixture problems. In the first type, a *liquid mixture* of a desired strength is made from two solutions with different concentrations.

EXAMPLE 6

Mixing a solution. A chemistry experiment calls for a 30% sulfuric acid solution. If the lab supply room has only 50% and 20% sulfuric acid solutions on hand, how much of each should be mixed to obtain 12 liters of a 30% acid solution?

ANALYZE THE PROBLEM We must find how much of the 50% solution and how much of the 20% solution is needed to obtain 12 liters of a 30% acid solution.

FORM AN EQUATION If x represents the numbers of liters (L) of the 50% solution used in the mixture, the remaining $(12 - x)$ liters must be the 20% solution. See Figure 2-20(a). Only 50% of the x liters, and only 20% of the $(12 - x)$ liters, is pure sulfuric acid. The total of these amounts is also the amount of acid in the final mixture, which is 30% of 12 liters. This information is shown in the chart in Figure 2-20(b).

FIGURE 2-20

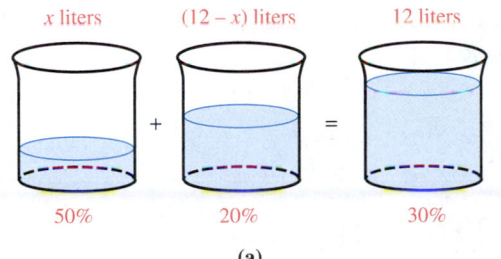

Solution	% acid ·	Liters =	Amount of acid
50% solution	0.50	x	$0.50x$
20% solution	0.20	$12 - x$	$0.20(12 - x)$
30% mixture	0.30	12	$0.30(12)$

(a) (b)

We can form the equation.

The acid in the 50% solution	plus	the acid in the 20% solution	equals	the acid in the final mixture.
50% of x	+	20% of $(12 - x)$	=	30% of 12

SOLVE THE EQUATION We then solve the equation.

$0.50x + 0.20(12 - x) = 0.30(12)$ 50% = 0.50, 20% = 0.20, and 30% = 0.30.

$5x + 2(12 - x) = 3(12)$ Multiply both sides by 10 to clear the equation of decimals.

$5x + 24 - 2x = 36$ Remove parentheses.

$3x + 24 = 36$ Combine like terms.

$3x = 12$ Subtract 24 from both sides.

$x = 4$ Divide both sides by 3.

STATE THE CONCLUSION The mixture will contain 4 liters of 50% solution and $12 - 4 = 8$ liters of 20% solution.

CHECK THE RESULT Verify that this solution checks. ■

In the next example, a *dry mixture* of a specified value is created from two differently priced components.

EXAMPLE 7

Snack food. Because fancy cashews priced at $9 per pound were not selling, a market produce clerk decided to combine them with less expensive filberts and sell the mixture for $7 per pound. How many pounds of filberts, selling at $6 per pound, should be mixed with 50 pounds of cashews to obtain such a mixture?

ANALYZE THE PROBLEM We know the value of the cashews ($9 per pound) and the filberts ($6 per pound). We also know that 50 pounds of cashews are to be mixed with an unknown number of pounds of filberts to obtain a mixture worth $7 per pound.

FORM THE EQUATION To solve this problem, we use the formula $v = pn$, where v is value, p is the price per pound, and n is the number of pounds.

Suppose that x pounds of filberts are used in the mixture. At $6 per pound, they are worth $6x$. At $9 per pound, the 50 pounds of cashews are worth $9 \cdot 50 = \$450$. Their combined value will be $\$(6x + 450)$. We also know that the mixture weighs $(50 + x)$ pounds. At $7 per pound, that mixture will be worth $\$7(50 + x)$. This information is recorded in the table in Figure 2-21.

FIGURE 2-21

	p $\cdot$	n $=$	v
Filberts	6	x	$6x$
Cashews	9	50	450
Mixture	7	$50 + x$	$7(50 + x)$

We can form the equation.

The value of the filberts	plus	the value of the cashews	equals	the value of the mixture.
$6x$	$+$	450	$=$	$7(50 + x)$

SOLVE THE EQUATION We then solve the equation.

$6x + 450 = 7(50 + x)$ The equation to solve.
$6x + 450 = 350 + 7x$ Remove parentheses.
$100 = x$ Subtract $6x$ and 350 from both sides.

STATE THE CONCLUSION Thus, 100 pounds of filberts should be used in the mixture.

CHECK THE RESULT

The value of 100 pounds of filberts at $6 per pound is $600
The value of 50 pounds of cashews at $9 per pound is $450
The value of the mixture is $1,050

The value of 150 pounds of the mixture at $7 per pound is also $1,050. The solution checks. ■

STUDY SET

Section 2.7

VOCABULARY

In Exercises 1–4, fill in the blanks to make the statements true.

1. The _____ of a triangle or a rectangle is the distance around it.

2. An _____ triangle is a triangle with two sides of the same length.

3. The equal sides of an isosceles triangle meet to form the _____ angle.

4. The angles opposite the equal sides in an isosceles triangle are called _____ angles, and they have _____ measures.

CONCEPTS

5. What is the sum of the measures of the angles of any triangle?

6. Use a ruler to draw an isosceles triangle with sides 3 inches long and a base that is 2 inches long. Label the vertex and the base angles.

7. a. Complete Illustration 1, which shows the inventory of nylon brushes that a paint store carries.

ILLUSTRATION 1

Paint brush	Number ·	Value =	Total value
1 inch	$\frac{x}{2}$	$4	
2 inch	x	$5	
3 inch	$x + 10$	$7	

b. Which type of brush does the store have the largest number of?
c. What is the least expensive brush?
d. What is the total value of the inventory of nylon brushes?

8. In the advertisement in Illustration 2, what are the principal, the rate, and the time for the investment opportunity shown?

ILLUSTRATION 2

> **Invest in Mini Malls!**
> Builder seeks daring people who want to earn big $$$$$. In just 1 year, you will earn a gigantic 14% on an investment of only $30,000! Call now.

9. a. Complete Illustration 3, which gives the details about two investments that were made by a retired couple.

ILLUSTRATION 3

	P ·	r =	I
Certificate of deposit	x	0.04	
Brother-in-law's business	$2x$	0.06	

b. How much more money was invested in the brother-in-law's business than in the certificate of deposit?
c. What is the total amount of interest the couple will make from these investments?

10. a. Complete Illustration 4, which gives the details of each morning's commute by a husband and wife who travel in opposite directions.

ILLUSTRATION 4

	r ·	t =	d
Husband	35 mph	t hr	
Wife	45 mph	t hr	

b. Who is able to drive to work at the faster rate?

c. Which person is on the road longer?

d. Write an algebraic expression that represents the distance between their workplaces.

11. Suppose the contents of the two barrels shown in Illustration 5 are poured into an empty third barrel.

ILLUSTRATION 5

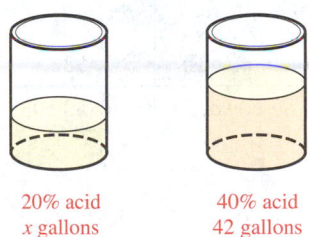

20% acid
x gallons

40% acid
42 gallons

a. How many gallons of liquid will the third barrel contain?
b. What would be a *reasonable* estimate of the concentration of the solution in the third barrel—19%, 32%, or 43% acid?

12. Each bottle of dressing shown in Illustration 6 contains a mixture of oil and vinegar. After sitting overnight, the liquids separate completely, with the oil rising to the top. On each bottle, draw the line estimating where the separation would occur and shade the vinegar.

ILLUSTRATION 6

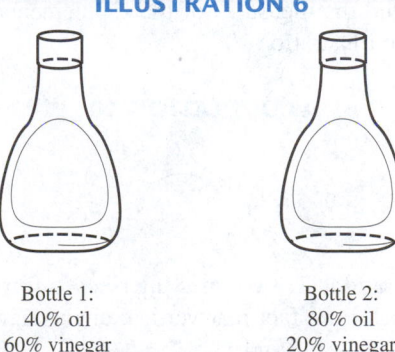

Bottle 1:
40% oil
60% vinegar

Bottle 2:
80% oil
20% vinegar

13. a. Complete Illustration 7, which gives the details about the ingredients of a box of breakfast cereal.

ILLUSTRATION 7

	Price ($/oz)	Amount (oz)	Value
Blueberries	$0.38	x	
Bran Flakes	$0.08	14	
Blueberries & Bran Flakes Cereal	$0.21	$14 + x$	

b. What does the cereal cost per ounce?

14. The ingredients of a weight-gain drink powder and their prices are shown in Illustration 8. Why can't the weight-gain powder's value be less than $3.50 a pound or more than $8.25 a pound?

ILLUSTRATION 8

Ingredients	Price (per lb)
Protein powder	$8.25
Carob powder	$3.50

PRACTICE

In Exercises 15–16, solve the equation by first clearing it of decimals.

15. $0.08x + 0.07(15{,}000 - x) = 1{,}110$

16. $0.108x + 0.07(16{,}000 - x) = 1{,}500$

17. Two angles are called **complementary angles** when the sum of their measures is 90°. Find the measures of the complementary angles shown in Illustration 9.

ILLUSTRATION 9

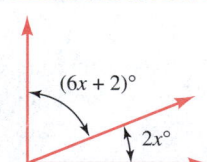

18. Two angles are called **supplementary angles** when the sum of their measures is 180°. Find the measures of the supplementary angles shown in Illustration 10.

ILLUSTRATION 10

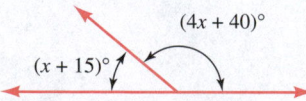

19. In Illustration 11, two lines intersect to form **vertical angles.** Use the fact that vertical angles have the same measure to find x.

ILLUSTRATION 11

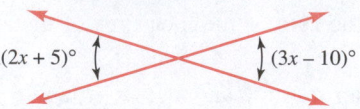

20. Find the measures of the vertical angles shown in Illustration 12. (See Exercise 19.)

ILLUSTRATION 12

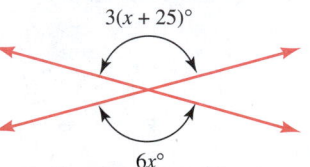

APPLICATIONS

In Exercises 21–54, solve each problem; a diagram or table may be helpful in organizing the facts of the problem.

21. TRIANGULAR BRACING The outside perimeter of the triangular brace in Illustration 13 is 57 feet. If all three sides are equal, find the length of each side.

ILLUSTRATION 13

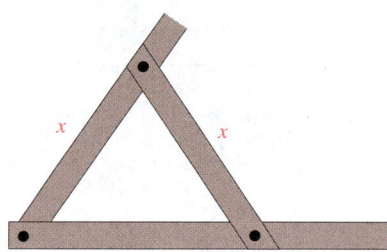

22. CIRCUIT BOARD The perimeter of the circuit board in Illustration 14 is 90 centimeters. Find its dimensions.

ILLUSTRATION 14

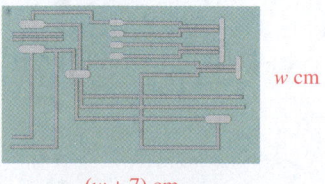

23. TRUSS The truss in Illustration 15 is in the form of an isosceles triangle. Each of the two equal sides is 4 feet less than the third side. If the perimeter is 25 feet, find the lengths of the sides.

ILLUSTRATION 15

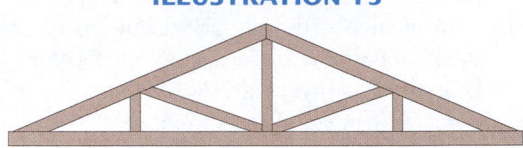

24. FIRST AID The sling shown in Illustration 16 is in the shape of an isosceles triangle with a perimeter of 144 inches. The longest side of the sling is 18 inches longer than either of the other two sides. Find the lengths of each side.

ILLUSTRATION 16

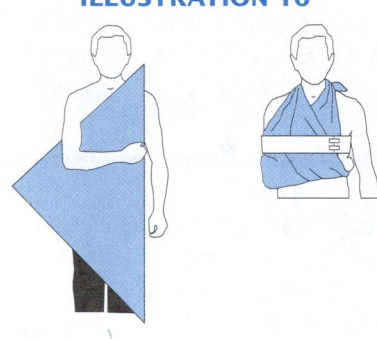

25. SWIMMING POOL The seawater Orthlieb Pool in Casablanca, Morocco is the largest swimming pool in the world. With a perimeter of 1,110 meters, this rectangular-shaped pool has a length that is 30 meters more than 6 times its width. Find its dimensions.

26. ART The *Mona Lisa,* shown in Illustration 17, was completed by Leonardo da Vinci in 1506. The length of the picture is 11.75 inches less than twice the width. If the perimeter of the picture is 102.5 inches, find its dimensions.

ILLUSTRATION 17

27. GUY WIRES The two guy wires shown in Illustration 18 form an isosceles triangle. One of the two equal angles of the triangle is 4 times the third angle (the vertex angle). Find the measure of the vertex angle.

ILLUSTRATION 18

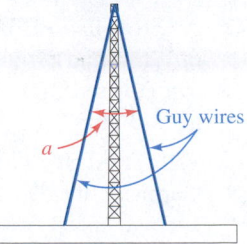

a → Guy wires

28. MOUNTAIN BICYCLE For the bicycle frame in Illustration 19, the angle that the horizontal crossbar makes with the seat support is 15° less than twice the angle at the steering column. The angle at the pedal gear is 25° more than the angle at the steering column. Find these three angle measures.

ILLUSTRATION 19

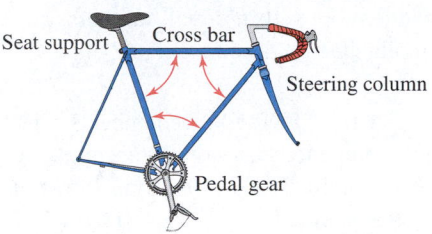

Seat support • Cross bar • Steering column • Pedal gear

29. WAREHOUSING COSTS A store warehouses 40 more portables than big-screen TV sets, and 25 fewer consoles than portables. Storage costs for the different TV sets are shown in Illustration 20. If storage costs $276 per month, how many big-screen sets are in stock?

ILLUSTRATION 20

Type of TV	Monthly cost
Portable	$1.50
Console	$4.00
Big screen	$7.50

30. APARTMENT RENTAL The owners of an apartment building rent 1-, 2-, and 3-bedroom units. They rent equal numbers of each, with the monthly rents given in Illustration 21. If the total monthly income is $36,550, how many of each type of unit are there?

ILLUSTRATION 21

Unit	Rent
One-bedroom	$550
Two-bedroom	$700
Three-bedroom	$900

31. SOFTWARE SALES Three software applications are priced as shown in Illustration 22. Spreadsheet and database programs sold in equal numbers, but 15 more word processing applications were sold than the other two combined. If the three applications generated sales of $72,000, how many spreadsheets were sold?

ILLUSTRATION 22

Software	Price
Spreadsheet	$150
Database	$195
Word processing	$210

32. INVENTORY With summer approaching, the number of air conditioners sold is expected to be double that of stoves and refrigerators combined. Stoves sell for $350, refrigerators for $450, and air conditioners for $500, and sales of $56,000 are expected. If stoves and refrigerators sell in equal numbers, how many of each appliance should be stocked?

33. INTEREST INCOME On December 31, 1997, Terrell Washington opened two savings accounts. At the end of 1998, his bank mailed him the form shown in Illustration 23, for income tax purposes. If a total of $12,000 was initially deposited and if no further deposits or withdrawals were made, how much money was originally deposited in account number 721-94?

ILLUSTRATION 23

USA HOME SAVINGS	Copy B For Recipient Interest Income
This is important tax information and is being furnished to the Internal Revenue Service.	OMB No. 1545-0112 **1998** Form 1099–iNT
RECIPIENT'S name **TERRELL WASHINGTON**	

Acct. Number	Annual Percent Yield	Early Withdrawal Penalty
822–06	6%	.00
721–94	4.5%	.00
	Total Interest Income 637.50	

34. MAKING A PRESENTATION A financial planner recommends a plan for a client who has $65,000 to invest. (See Illustration 24.) At the end of the presentation, the client asks, "How much will be invested at each rate?" Answer this question using the given information.

ILLUSTRATION 24

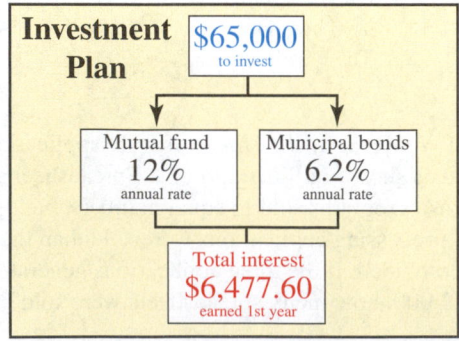

35. INVESTMENTS Equal amounts are invested in each of three accounts paying 7%, 8%, and 10.5%. If one year's combined interest income is $1,249.50, how much is invested in each account?

36. RETIREMENT A professor wants to supplement her retirement income with investment interest. If she invests $15,000 at 6% interest, how much more would she have to invest at 7% to achieve a goal of $1,250 per year in supplemental income?

37. FINANCIAL PLANNING A plumber has a choice of two investment plans:

- An insured fund that pays 11% interest
- A risky investment that pays a 13% return

If the same amount invested at the higher rate would generate an extra $150 per year, how much does the plumber have to invest?

38. INVESTMENTS The amount of annual interest earned by $8,000 invested at a certain rate is $200 less than $12,000 would earn at a rate 1% lower. At what rate is the $8,000 invested?

39. TORNADO During a storm, two teams of scientists leave a university at the same time in specially designed vans to search for tornadoes. The first team travels east at 20 mph and the second travels west at 25 mph, as shown in Illustration 25. If their radios have a range of up to 90 miles, how long will it be before they lose radio contact?

ILLUSTRATION 25

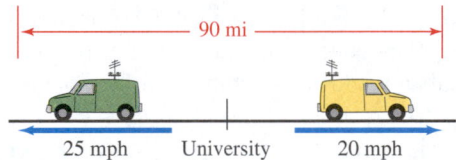

40. SEARCH AND RESCUE Two search-and-rescue teams leave base at the same time looking for a lost boy. The first team, on foot, heads north at 2 mph and the other, on horseback, south at 4 mph. How long will it take them to search a distance of 21 miles between them?

41. SPEED OF TRAINS Two trains are 330 miles apart, and their speeds differ by 20 mph. Find the speed of each train if they are traveling toward each other and will meet in 3 hours.

42. AVERAGE SPEED A car averaged 40 mph for part of a trip and 50 mph for the remainder. If the 5-hour trip covered 210 miles, for how long did the car average 40 mph?

43. AIR TRAFFIC CONTROL An airliner leaves Berlin, Germany, headed for Montreal, Canada, flying at an average speed of 450 mph. At the same time, an airliner leaves Montreal headed for Berlin, averaging 500 mph. If the airports are 3,800 miles apart, when will the air traffic controllers have to make the pilots aware that the planes are passing each other?

44. ROAD TRIP A bus, carrying the members of a marching band, and a truck, carrying their instruments, leave a high school at the same time. The bus travels at 65 mph and the truck at 55 mph. In how many hours will they be 75 miles apart?

45. SALT SOLUTION How many gallons of a 3% salt solution must be mixed with 50 gallons of a 7% solution to obtain a 5% solution?

46. MAKING CHEESE To make low-fat cottage cheese, milk containing 4% butterfat is mixed with 10 gallons of milk containing 1% butterfat to obtain a mixture containing 2% butterfat. How many gallons of the richer milk must be used?

47. ANTISEPTIC SOLUTION A nurse wants to add water to 30 ounces of a 10% solution of benzalkonium chloride to dilute it to an 8% solution. How much water must she add?

48. PHOTOGRAPHIC CHEMICALS A photographer wishes to mix 2 liters of a 5% acetic acid solution with a 10% solution to get a 7% solution. How many liters of 10% solution must be added?

49. MIXING FUELS How many gallons of fuel costing $1.15 per gallon must be mixed with 20 gallons of a fuel costing $0.85 per gallon to obtain a mixture costing $1 per gallon? See Illustration 26.

ILLUSTRATION 26

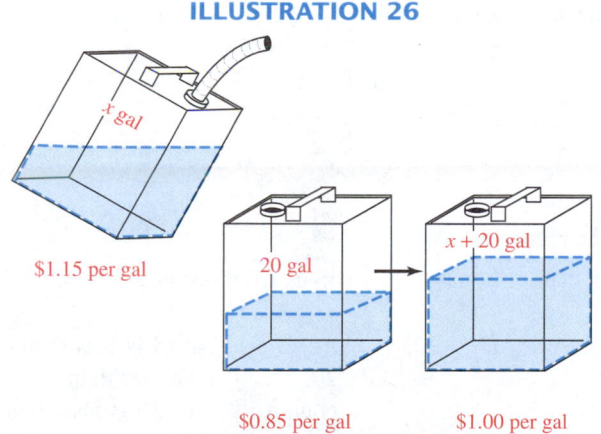

$1.15 per gal   20 gal   x + 20 gal

$0.85 per gal $1.00 per gal

50. MIXING PAINT Paint costing $19 per gallon is to be mixed with 5 gallons of a $3-per-gallon thinner to make a paint that can be sold for $14 per gallon. How much paint will be produced?

51. MIXING CANDY Lemon drops worth $1.90 per pound are to be mixed with jelly beans that cost $1.20 per pound to make 100 pounds of a mixture worth $1.48 per pound. How many pounds of each candy should be used?

52. MIXING CANDY See Illustration 27. Twenty pounds of lemon drops are to be mixed with cherry chews to make a mixture that will sell for $1.80 per pound. How much of the more expensive candy should be used?

ILLUSTRATION 27

Candy	Price per pound
Peppermint patties	$1.35
Lemon drops	$1.70
Licorice lumps	$1.95
Cherry chews	$2.00

53. BLENDING COFFEE A store sells regular coffee for $4 a pound and gourmet coffee for $7 a pound. To get rid of 40 pounds of the gourmet coffee, a shopkeeper makes a blend to put on sale for $5 a pound. How many pounds of regular coffee should he use?

54. BLENDING LAWN SEED A store sells bluegrass seed for $6 per pound and ryegrass seed for $3 per pound. How much ryegrass must be mixed with 100 pounds of bluegrass to obtain a blend that will sell for $5 per pound?

WRITING

55. Create a mixture problem of your own, and solve it.

56. Use an example to explain the difference between the quantity and the value of the materials being combined in a mixture problem.

57. A car travels at 60 mph for 15 minutes. Why can't we multiply the rate, 60, and the time, 15, to find the distance traveled by the car?

58. Create a geometry problem that could be answered by solving the equation $2w + 2(w + 5) = 26$.

REVIEW

In Exercises 59–60, write each expression without using parentheses.

59. $-25(2x - 5)$

60. $-12(3a + 4b - 32)$

In Exercises 61–62, combine like terms.

61. $8p - 9q + 11p + 20q$

62. $-5(t - 120) - 7(t + 5)$

▶2.8   Inequalities

In this section, you will learn about

Inequality symbols ■ Graphing inequalities ■ Interval notation
■ Solving inequalities ■ Graphing compound inequalities ■ Solving
compound inequalities ■ An application

Introduction   **Inequalities** are expressions indicating that two quantities are not necessarily equal. They appear in many situations:

■ An airplane is rated to fly at altitudes that are less than 36,000 feet.

■ To thaw ice, the temperature must be greater than 32°F.

■ To earn a B, I need a final exam score of at least 80%.

Inequality Symbols

We can use **inequality symbols** to show that two expressions are not equal.

Inequality symbols

$\neq$	means	"is not equal to"
$<$	means	"is less than"
$>$	means	"is greater than"
$\leq$	means	"is less than or equal to"
$\geq$	means	"is greater than or equal to"

EXAMPLE 1

Reading inequalities.

a. $6 \neq 9$ is read as "6 is not equal to 9."

b. $x > 5$ is read as "x is greater than 5."

c. $5 \leq 5$ is read as "5 is less than or equal to 5." This is true, because $5 = 5$.

SELF CHECK   Write each inequality in words: **a.** $15 < 20$,   *Answers:*   **a.** 15 is less than
b. $y \geq 9$,   and   **c.** $30 \leq 30$.   20.   **b.** y is greater than or
equal to 9.   **c.** 30 is less than
or equal to 30.   ■

If two numbers are graphed on a number line, the one to the right is the greater. For example, from Figure 2-22, we see that $-1 > -4$, because -1 lies to the right of -4.

FIGURE 2-22

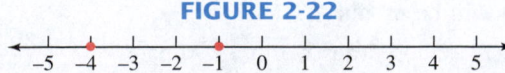

Inequalities can be written so that the inequality symbol points in the opposite direction. For example, the following statements both indicate that 27 is a smaller number than 32.

$27 < 32$ (27 is less than 32)   and $32 > 27$ (32 is greater than 27)

The following statements both indicate that 9 is greater than or equal to 6.

$9 \geq 6$ (9 is greater than or equal to 6)   and $6 \leq 9$ (6 is less than or equal to 9)

Variables can be used with inequality symbols to show mathematical relationships. For example, consider the statement, "You must be taller than 54 inches to ride the roller coaster." If we let h represent a person's height in inches, then to ride the roller coaster, $h > 54$ inches.

EXAMPLE 2

Writing inequalities. Express the following situation using an inequality symbol: "The occupancy of the dining room cannot exceed 200 people."

Solution   If p represents the number of people that can occupy the room, then p cannot be greater than (exceed) 200. Another way to state this is that p must be *less than or equal to* 200:

$$p \leq 200$$

SELF CHECK   Express the following statement using an inequality symbol: "The thermostat on the pool heater is set so that the water temperature t is at least 72°."

Answer: $t \geq 72$ ■

Graphing Inequalities

Graphs of inequalities involving real numbers are **intervals** on the number line. For example, two versions of the graph of all real numbers x such that $x > -3$ are shown in Figure 2-23.

FIGURE 2-23

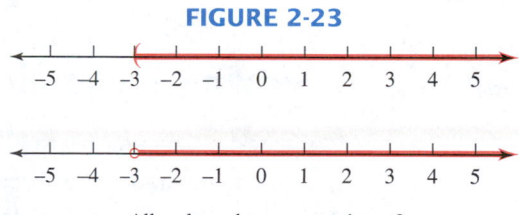

All real numbers greater than –3

The thick arrow pointing to the right shows that all numbers to the right of -3 are in the graph. The **parenthesis** (or open circle) at -3 indicates that -3 is not in the graph.

Interval Notation

The interval shown in Figure 2-23 can be expressed in **interval notation** as $(-3, \infty)$. Again, the first parenthesis indicates that -3 is not included in the interval. The infinity symbol ∞ does not represent a number. It indicates that the interval continues on forever to the right.

Figure 2-24 shows two versions of the graph of $x \leq 2$. The thick arrow pointing to the left shows that all numbers to the left of 2 are in the graph. The **bracket** (or closed circle) at 2 indicates that 2 is included in the graph. We can express this interval as $(-\infty, 2]$. Here the bracket indicates that 2 is included in the interval.

From now on, we will use parentheses or brackets when graphing intervals, because they are consistent with interval notation.

FIGURE 2-24

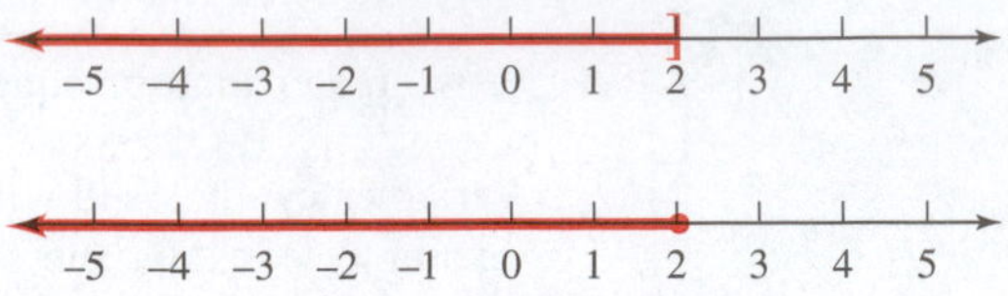

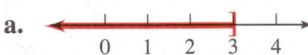

All real numbers less than or equal to 2

EXAMPLE 3

Writing an inequality from its graph. What inequality is represented by each graph?

a.

b.

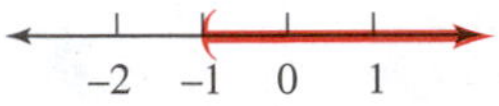

Solution
a. This is the interval $(-\infty, 3]$, which consists of all real numbers less than or equal to 3. The inequality is $x \leq 3$.

b. This is the interval $(-1, \infty)$, consisting of all real numbers greater than -1. The inequality is $x > -1$.

SELF CHECK

What inequality is represented by each graph?

a.

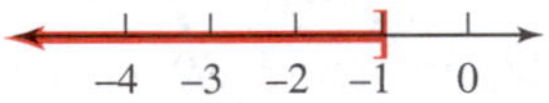

b.

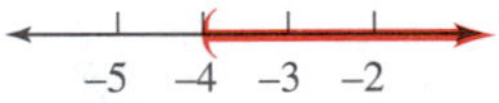

Answers:
a. $x \leq -1$, $(-\infty, -1]$
b. $x > -4$, $(-4, \infty)$ ∎

Solving Inequalities

A **solution of an inequality** is any number that makes the inequality true. For example, 2 is a solution of $x \leq 3$, because $2 \leq 3$.

To solve more complicated inequalities, we will use the addition, subtraction, multiplication, and division properties of inequality. When we use one of these properties, the resulting inequality will always be equivalent to the original one.

Addition and subtraction properties of inequality

For real numbers a, b, and c,

If $a < b$, then $a + c < b + c$.
If $a < b$, then $a - c < b - c$.

Similar statements can be made for the symbols $>$, $\leq$, and $\geq$.

The **addition property of inequality** can be stated this way: *If a quantity is added to both sides of an inequality, the resulting inequality will have the same direction as the original one.*

The **subtraction property of inequality** can be stated this way: *If a quantity is subtracted from both sides of an inequality, the resulting inequality will have the same direction as the original one.*

EXAMPLE 4

Solving inequalities. Solve $x + 3 > 2$ and graph its solution.

Solution

To isolate the x on the left-hand side of the $>$ sign, we proceed as we would when solving equations.

$$x + 3 > 2$$
$$x + 3 - 3 > 2 - 3 \quad \text{To undo the addition of 3, subtract 3 from both sides.}$$
$$x > -1 \quad \text{Do the subtractions: } 3 - 3 = 0 \text{ and } 2 - 3 = 2 + (-3) = -1.$$

The graph (see Figure 2-25) includes all points to the right of -1 but does not include -1. This represents all real numbers greater than -1. Expressed as an interval, we have $(-1, \infty)$.

FIGURE 2-25

To check, we pick several numbers in the graph, such as 1 and 3, substitute each number for x in the inequality, and see whether it satisfies the inequality.

$$x + 3 > 2 \qquad\qquad x + 3 > 2$$
$$1 + 3 \overset{?}{>} 2 \quad \text{Substitute 1 for } x. \qquad 3 + 3 \overset{?}{>} 2 \quad \text{Substitute 3 for } x.$$
$$4 > 2 \quad \text{Do the addition.} \qquad\quad 6 > 2 \quad \text{Do the addition.}$$

Since $4 > 2$, 1 satisfies the inequality. Since $6 > 2$, 2 satisfies the inequality. The solution appears to be correct.

SELF CHECK

Solve $x - 3 \leq -2$ and graph its solution. Then use interval notation to describe the solution.

Answer: $x \leq 1$, $(-\infty, 1]$

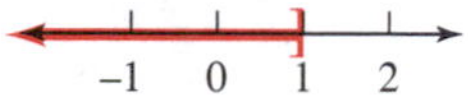

If both sides of the inequality $2 < 5$ are multiplied by a *positive* number, such as 3, another true inequality results.

$$2 < 5$$
$$3 \cdot 2 < 3 \cdot 5 \quad \text{Multiply both sides by 3.}$$
$$6 < 15 \quad \text{Do the multiplications: } 3 \cdot 2 = 6 \text{ and } 3 \cdot 5 = 15.$$

However, if we multiply both sides of $2 < 5$ by a *negative* number, such as -3, the direction of the inequality symbol must be reversed to produce another true inequality.

$$2 < 5$$
$$-3 \cdot 2 > -3 \cdot 5 \quad \text{Multiply both sides by the } negative \text{ number } -3 \text{ and reverse the direction of the inequality.}$$
$$-6 > -15 \quad \text{Do the multiplications: } -3 \cdot 2 = -6 \text{ and } -3 \cdot 5 = -15.$$

The inequality $-6 > -15$ is true, because -6 is to the right of -15 on the number line.

Multiplication and division properties of inequalities

For real numbers a, b, and c,

If $a < b$ and $c > 0$,   then $ac < bc$.

If $a < b$ and $c < 0$,   then $ac > bc$.

If $a < b$ and $c > 0$,   then $\frac{a}{c} < \frac{b}{c}$.

If $a < b$ and $c < 0$,   then $\frac{a}{c} > \frac{b}{c}$.

Similar statements can be made for the symbols $>$, $\leq$, and $\geq$.

The **multiplication property of inequality** can be stated this way:

If both sides of an inequality are multiplied by the same positive number, the resulting inequality will have the same direction as the original one.

If both sides of an inequality are multiplied by the same negative number, the resulting inequality will have the opposite direction from the original one.

The **division property of inequality** can be stated this way:

If both sides of an inequality are divided by the same positive number, the resulting inequality will have the same direction as the original one.

If both sides of an inequality are divided by the same negative number, the resulting inequality will have the opposite direction from the original one.

EXAMPLE 5

Solving inequalities. Solve $3x + 7 \leq -5$ and graph the solution.

Solution

$3x + 7 \leq -5$

$3x + 7 \mathbf{- 7} \leq -5 \mathbf{- 7}$ To undo the addition of 7, subtract 7 from both sides.

$3x \leq -12$ Do the subtractions: $7 - 7 = 0$ and $-5 - 7 = -5 + (-7) = -12$.

$\dfrac{3x}{\mathbf{3}} \leq \dfrac{-12}{\mathbf{3}}$ To undo the multiplication by 3, divide both sides by 3.

$x \leq -4$ Do the divisions.

The graph (shown in Figure 2-26) consists of all real numbers less than or equal to -4. Using interval notation, we have $(-\infty, -4]$.

FIGURE 2-26

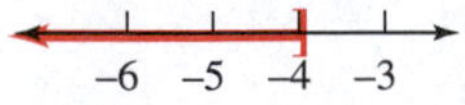

$\xleftarrow{\hspace{2cm}} \underset{-6 \quad -5 \quad -4 \quad -3}{\rule{3cm}{0.4pt}}$

To check, we can pick a number in the graph, such as -6, and see whether it satisfies the inequality.

$3x + 7 \leq -5$

$3(-6) + 7 \overset{?}{\leq} -5$ Substitute -6 for x.

$-18 + 7 \overset{?}{\leq} -5$ Do the multiplication.

$-11 \leq -5$ Do the addition.

Since $-11 \leq -5$, -6 satisfies the inequality. The solution appears to be correct.

Solve $2x - 7 > -13$ and graph the solution. Then use interval notation to describe the solution.

Answer: $x > -3$, $(-3, \infty)$

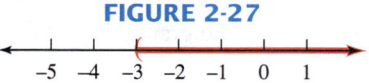

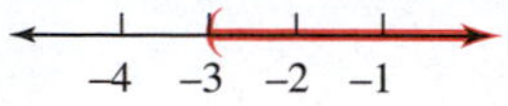

E X A M P L E   6

Reversing the inequality symbol. Solve $5 - 3x < 14$ and graph the solution.

Solution

$$5 - 3x < 14$$

$$5 - 3x - 5 < 14 - 5$$ To isolate $-3x$ on the left-hand side, subtract 5 from both sides.

$$-3x < 9$$ Do the subtractions: $5 - 5 = 0$ and $14 - 5 = 9$.

$$\frac{-3x}{-3} > \frac{9}{-3}$$ To undo the multiplication by -3, divide both sides by -3. Since we are dividing by -3, we reverse the direction of the $<$ symbol.

$$x > -3$$

The graph is shown in Figure 2-27. This is the interval $(-3, \infty)$, which consists of all real numbers greater than -3.

FIGURE 2-27

Check the result.

Solve $-2x - 5 \geq -7$ and graph the solution. Then use interval notation to describe the solution.

Answer: $x \leq 1$, $(-\infty, 1]$

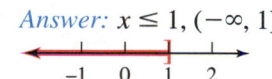

E X A M P L E   7

Solving inequalities. Solve $5(x + 1) \leq 2(x - 3)$ and graph the solution.

Solution

$$5(x + 1) \leq 2(x - 3)$$

$$5x + 5 \leq 2x - 6$$ Remove the parentheses on both sides.

$$5x + 5 - 2x \leq 2x - 6 - 2x$$ To eliminate $2x$ from the right side, subtract $2x$ from both sides.

$$3x + 5 \leq -6$$ Combine like terms on both sides.

$$3x + 5 - 5 \leq -6 - 5$$ To undo the addition of 5, subtract 5 from both sides.

$$3x \leq -11$$ Do the subtractions.

$$\frac{3x}{3} \leq \frac{-11}{3}$$ To undo the multiplication by 3, divide both sides by 3.

$$x \leq -\frac{11}{3}$$

The graph is shown in Figure 2-28. This is the interval $\left(-\infty, -\frac{11}{3}\right]$, which consists of all real numbers less than or equal to $-\frac{11}{3}$. We note that $-\frac{11}{3} = -3\frac{2}{3}$.

FIGURE 2-28

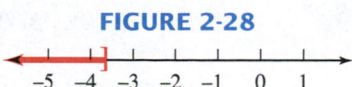

Check the result.

SELF CHECK   Solve $3(x - 2) > -(x + 1)$ and graph the solution. Then use interval notation to describe the solution.

Answer: $x > \dfrac{5}{4}$, $\left(\dfrac{5}{4}, \infty\right)$

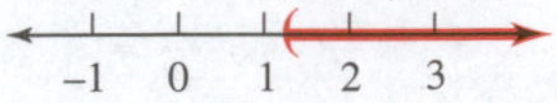

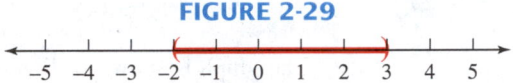

■

Graphing Compound Inequalities

Two inequalities can be combined into a **compound inequality** to indicate that numbers lie *between* two fixed values. For example, $-2 < x < 3$, is a combination of

$$-2 < x \quad \text{and} \quad x < 3$$

It indicates that x is greater than -2 and that x is also less than 3. The solution of $-2 < x < 3$ consists of all numbers that lie *between* -2 and 3. The graph of this interval appears in Figure 2-29. We can express this interval as $(-2, 3)$.

FIGURE 2-29

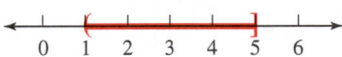

E X A M P L E   8   **Writing an inequality from its graph.** What inequality is represented by the graph below?

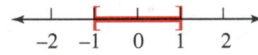

Solution $1 < x \leq 5$. This is the interval $(1, 5]$.

SELF CHECK   What inequality is represented by the graph below?

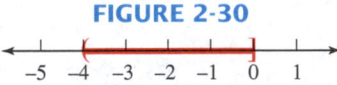

Answer: $-1 \leq x \leq 1$. This is the interval $[-1, 1]$.

■

E X A M P L E   9   **Graphing compound inequalities.** Graph the interval $-4 < x \leq 0$.

Solution   The interval $-4 < x \leq 0$ consists of all real numbers between -4 and 0, including 0. The graph appears in Figure 2-30. This is the interval $(-4, 0]$.

FIGURE 2-30

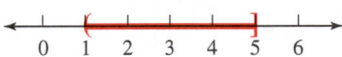

To check, we pick a number, such as -2, in the graph and see whether it satisfies the inequality. Since $-4 < -2 \leq 0$, the solution appears to be correct.

SELF CHECK   Graph the interval $-2 \leq x < 1$. Then use interval notation to describe the solution.

Answer: $[-2, 1)$

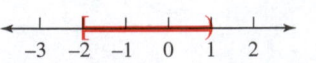

■

Solving Compound Inequalities

To solve compound inequalities, we use the same methods we used for solving equations. However, instead of applying the properties of equality to both sides of an equation, we will apply the properties of inequality to all three parts of the inequality.

EXAMPLE 10

Solving compound inequalities. Solve $-4 < 2(x - 1) \leq 4$ and graph the solution.

Solution

$$-4 < 2(x - 1) \leq 4$$

$-4 < 2x - 2 \leq 4$ Use the distributive property to remove parentheses.

$-2 < 2x \leq 6$ To undo the subtraction of 2, add 2 to all three parts.

$-1 < x \leq 3$ To undo the multiplication by 2, divide all three parts by 2.

The graph of the solution appears in Figure 2-31. This is the interval $(-1, 3]$.

FIGURE 2-31

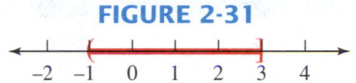

Check the solution.

SELF CHECK

Solve $-6 \leq 3(x + 2) \leq 6$ and graph the solution. Then use interval notation to describe the solution.   *Answer:* $-4 \leq x \leq 0$, $[-4, 0]$

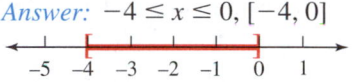

An Application

When solving problems, phrases such as "not more than," "at least," or "should exceed" suggest that an *inequality* should be written instead of an *equation*.

EXAMPLE 11

A student has scores of 72%, 74%, and 78% on three exams. What percent score does he need on the last exam to earn no less than a grade of B (80%)?

ANALYZE THE PROBLEM   We know three of the student's scores. We are to find what he must score on the last exam to earn at least a B grade.

FORM AN INEQUALITY   We can let x represent the score on the fourth (and last) exam. To find the average grade, we add the four scores and divide by 4. To earn no less than a grade of B, the student's average must be greater than or equal to 80%.

The average of the four grades	must be greater than or equal to	80.
$\dfrac{72 + 74 + 78 + x}{4}$	$\geq$	80

SOLVE THE INEQUALITY   We can solve this inequality for x.

$\dfrac{224 + x}{4} \geq 80$ $72 + 74 + 78 = 224$.

$224 + x \geq 320$ Multiply both sides by 4.

$x \geq 96$ Subtract 224 from both sides.

STATE THE CONCLUSION   To earn a B, the student must score 96% or better on the last exam. Of course, the student cannot score higher than 100%. The graph appears in Figure 2-32 on the next page. This is the interval $[96, 100]$.

FIGURE 2-32

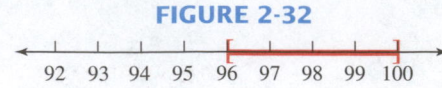

CHECK THE RESULT   Pick some numbers in the interval and verify that the average of the four scores will be 80% or greater. ∎

STUDY SET

Section 2.8

VOCABULARY

In Exercises 1–4, fill in the blanks to make the statements true.

1. An expression containing one of the symbols $>$, $<$, $\geq$, $\leq$, or $\neq$ is called an _____.

2. Graphs of inequalities involving real numbers are called _____ on the number line.

3. A _____ of an inequality is any real number that makes the inequality true.

4. The inequality $-4 < x \leq 12$ is an example of a _____ inequality.

CONCEPTS

In Exercises 5–8, fill in the blanks to make the statements true.

5. If a quantity is added to or subtracted from both sides of an inequality, the resulting inequality will have the _____ direction as the original one.

6. If both sides of an inequality are multiplied or divided by a positive number, the resulting inequality will have the _____ direction as the original one.

7. If both sides of an inequality are multiplied or divided by a negative number, the resulting inequality will have the _____ direction from the original one.

8. To solve compound inequalities, the properties of inequalities are applied to all _____ parts of the inequality.

9. The solution of an inequality is graphed below.

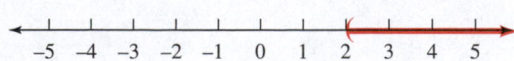

   a. If 3 is substituted for the variable in the inequality, what type of statement will result?

   b. If -3 is substituted for the variable in the inequality, what type of statement will result?

10. The solution of an inequality is graphed below.

   a. If 3 is substituted for the variable in the inequality, what type of statement will result?

   b. If -3 is substituted for the variable in the inequality, what type of statement will result?

11. Solve the inequality $2x - 4 > 12$ and give the solution:
   a. in words
   b. using a graph

   c. using interval notation

12. Solve the compound inequality $-4 < 2x < 12$ and give the solution:
   a. in words
   b. using a graph

   c. using interval notation

NOTATION

In Exercises 13–18, fill in the blanks to make the statements true.

13. The symbol $<$ means "_____."

14. the symbol $>$ means "_____."

15. The symbol $\geq$ means "_____ or equal to."

16. The symbol $\leq$ means "is less than _____."

17. The symbol $\neq$ means "_____."

18. In the interval [4, 8), the endpoint 4 is _____, but the endpoint 8 is not included.

19. Suppose you solve an inequality and obtain $-2 < x$. Rewrite this inequality so that x is on the left-hand side.

20. Explain what is wrong with the compound inequality $8 < x < -1$.

In Exercises 21–22, write each inequality so that the inequality symbol points in the opposite direction.

21. $17 \geq -2$

22. $-32 < -10$

In Exercises 23–24, complete each solution.

23. Solve $4x - 5 \geq 7$.

$$4x - 5 \geq 7$$
$$4x - 5 + \boxed{} \geq 7 + \boxed{}$$
$$4x \geq \boxed{}$$
$$\frac{4x}{\boxed{}} \geq \frac{12}{\boxed{}}$$
$$x \geq 3$$

24. Solve $\dfrac{-x}{2} + 4 < 5$.

$$\frac{-x}{2} + 4 < 5$$
$$\frac{-x}{2} + 4 - \boxed{} < 5 - \boxed{}$$
$$\frac{-x}{2} < \boxed{}$$
$$\boxed{}\left(\frac{-x}{2}\right) < \boxed{}(1)$$
$$\boxed{} < 2$$
$$\frac{-x}{\boxed{}} \boxed{} \frac{2}{-1}$$
$$x > -2$$

PRACTICE

In Exercises 25–28, graph each inequality. Then describe the graph using interval notation.

25. $x < 5$

26. $x \geq -2$

27. $-3 < x \leq 1$

28. $-1 \leq x \leq 3$

In Exercises 29–32, write the inequality that is represented by each graph. Then describe the graph using interval notation.

29.

30.

31.

32.

In Exercises 33–62, solve each inequality, graph the solution, and then use interval notation to describe the solution.

33. $x + 2 > 5$

34. $x + 5 \geq 2$

35. $-x - 3 \leq 7$

36. $-x - 9 > 3$

37. $3 + x < 2$

38. $5 + x \geq 3$

39. $2x - 0.3 \leq 0.5$

40. $-3x - 0.5 < 0.4$

41. $-3x - 7 > -1$

42. $-5x + 7 \leq 12$

43. $-4x + 6 > 17$

44. $7x - 1 > 5$

45. $\dfrac{2}{3}x \geq 2$

46. $\dfrac{3}{4}x < 3$

47. $-\dfrac{7}{8}x \le 21$

48. $-\dfrac{3}{16}x \ge -9$

49. $2x + 9 \le x + 8$

50. $3x + 7 \le 4x - 2$

51. $9x + 13 \ge 8x$

52. $7x - 16 < 6x$

53. $8x + 4 > 3x + 4$

54. $7x + 6 \ge 4x + 6$

55. $5x + 7 < 2x + 1$

56. $7x + 2 \ge 4x - 1$

57. $7 - x \le 3x - 2$

58. $9 - 3x \ge 6 + x$

59. $3(x - 8) < 5x + 6$

60. $9(x - 11) > 13 + 7x$

61. $8(5 - x) \le 10(8 - x)$

62. $17(3 - x) \ge 3 - 13x$

In Exercises 63–76, solve each inequality, graph the solution, and use interval notation to describe the solution.

63. $2 < x - 5 < 5$

64. $3 < x - 2 < 7$

65. $-5 < x + 4 \le 7$

66. $-9 \le x + 8 < 1$

67. $0 \le x + 10 \le 10$

68. $-8 < x - 8 < 8$

69. $4 < -2x < 10$

70. $-4 \le -4x < 12$

71. $-3 \le \dfrac{x}{2} \le 5$

72. $-12 < \dfrac{x}{3} < 0$

73. $3 \le 2x - 1 < 5$

74. $4 < 3x - 5 \le 7$

75. $0 < 10 - 5x \le 15$

76. $1 \le -7x + 8 \le 15$

APPLICATIONS

In Exercises 77–88, express each solution as an inequality.

77. CALCULATING GRADES A student has test scores of 68%, 75%, and 79% in a government class. What must she score on the last exam to earn a B (80% or better) in the course?

78. OCCUPATIONAL TESTING Before taking on a client, an employment agency requires the applicant to average at least 70% on a battery of four job skills tests. If an applicant scored 70%, 74%, and 84% on the first three exams, what must he score on the fourth test to maintain a 70% or better average?

79. FLEET AVERAGES A car manufacturer produces three models in equal quantities. One model has an economy rating of 17 miles per gallon, and the second model is rated for 19 mpg. If governmental regulations require the manufacturer to have a fleet average of at least 21 mpg, what economy rating is required for the third model?

80. SERVICE CHARGES When the average daily balance of a customer's checking account falls below $500 in any week, the bank assesses a $5 service charge. Illustration 1 shows the daily balances of one customer. What must Friday's balance be to avoid the service charge?

ILLUSTRATION 1

Day	Balance
Monday	$540.00
Tuesday	$435.50
Wednesday	$345.30
Thursday	$310.00

81. DOING HOMEWORK A Spanish teacher requires that students devote no less than 1 hour a day to their homework assignments. Write an inequality that describes the number of minutes a student should spend each week on Spanish homework.

82. CHILD LABOR A child labor law reads, "The number of hours a full-time student under 16 years of age can work on a weekday shall not exceed 4 hours." Write an inequality that describes the number of hours such a student can work Monday through Friday.

83. SAFETY CODE Illustration 2 shows the acceptable and preferred angles of "pitch" or slope for ladders, stairs, and ramps. Use a compound inequality to describe each safe-angle range:
a. Ramps or inclines
b. Stairs
c. Preferred range for stairs
d. Ladders with cleats

ILLUSTRATION 2

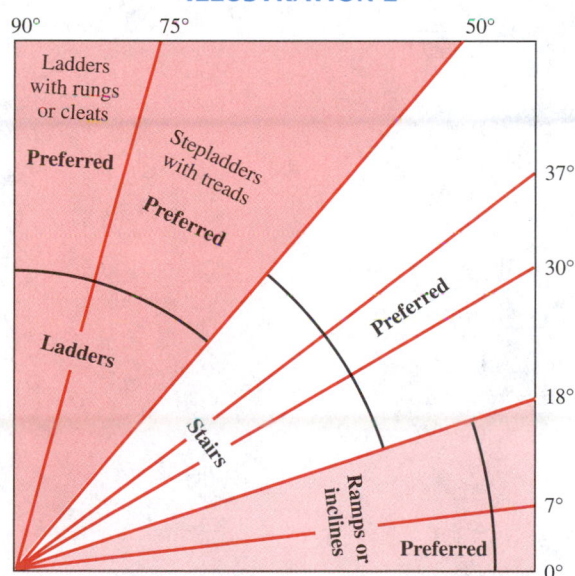

84. WEIGHT CHART Illustration 3 is used to classify the weight of a baby boy from birth to 1 year. Estimate the weight range w for boys in the following classifications, using a compound inequality:
a. 10 months old, "heavy"
b. 5 months old, "light"
c. 8 months old, "average"
d. 3 months old, "moderately light"

85. LAND ELEVATIONS The land elevations in Nevada range from the 13,143-foot height of Boundary Peak to the Colorado River at 470 feet. Use a compound inequality to express the range of these elevations:
a. in feet
b. in miles (round to the nearest tenth) (*Hint:* 1 mile is 5,280 feet.)

ILLUSTRATION 3

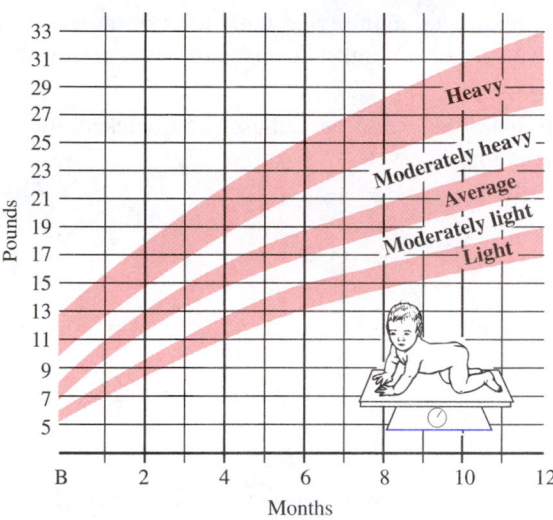

Based on data from *Better Homes and Gardens Baby Book* (Meredith Corp., 1969)

86. COMPARING TEMPERATURES To hold the temperature of a room between 19° C and 22°C, what Fahrenheit temperatures must be maintained? ($Hint:$ Fahrenheit temperature F and Celsius temperature C are related by the formula $F = \frac{9C + 160}{5}$.)

87. DRAFTING In Illustration 4, the ± (read "plus or minus") symbol means that the width of a plug a manufacturer produces can range from $1.497 - 0.001$ inches to $1.497 + 0.001$ inches. Write the range of acceptable widths for the plug and the opening it fits into using compound inequalities.

ILLUSTRATION 4

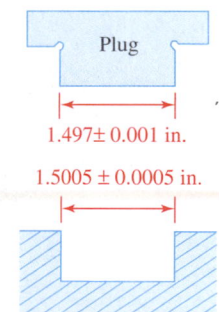

Plug

1.497± 0.001 in.

1.5005 ± 0.0005 in.

88. COUNTER SPACE In a large discount store, a rectangular counter is being built for the customer service department. If designers have determined that the outside perimeter of the counter (shown in red) needs to be at least 150 feet, use the plan in Illustration 5 to determine the acceptable values for x.

ILLUSTRATION 5

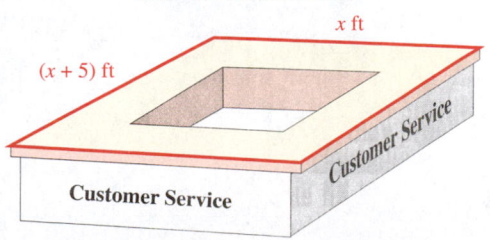

x ft

$(x + 5)$ ft

Customer Service

Customer Service

WRITING

89. Explain why multiplying both sides of an inequality by a negative number reverses the direction of the inequality.

90. Explain the use of parentheses and brackets for graphing intervals.

REVIEW

In Exercises 91–94, find each power.

91. -5^3

92. $(-3)^4$

93. -4^4

94. $(-4)^4$

Simplify and Solve

Two of the most often used instructions in this book are **simplify** and **solve.** In algebra, we *simplify expressions* and we *solve equations and inequalities.*

To simplify an expression, we write it in a less complicated form. To do so, we apply the rules of arithmetic, as well as algebraic concepts such as combining like terms, the distributive property, and the properties of 0 and 1.

To solve an equation or an inequality means to find the numbers that make the equation or inequality true when substituted for its variable. We use the addition, subtraction, multiplication, and division properties of equality or inequality to solve equations and inequalities. Quite often, we must simplify expressions on the left- or right-hand sides of an equation or inequality when solving it.

In Exercises 1–4, use the procedures and the properties that we have studied to simplify the expression in part a and to solve the equation or inequality in part b.

Simplify

1. a. $-3x + 2 + 5x - 10$

2. a. $4(y + 2) - 3(y + 1)$

3. a. $\frac{1}{3}a + \frac{1}{3}a$

4. a. $-(2x + 10)$

Solve

b. $-3x + 2 + 5x - 10 = 4$

b. $4(y + 2) = 3(y + 1)$

b. $\frac{1}{3}a + \frac{1}{3} = \frac{1}{2}$

b. $-2x \geq -10$

5. In the student's work on the right, where was the mistake made? Explain what the student did wrong.

Simplify $2(x + 3) - x - 12$.

$2(x + 3) - x - 12 = 2x + 6 - x - 12$
$= x - 6$
$0 = x - 6$
$0 + 6 = x - 6 + 6$
$\boxed{6 = x}$

Accent on Teamwork

Section 2.1

Temperature extremes Check out a current world almanac from a library. Look up the record high and low temperatures for each state and calculate the difference in temperature extremes. Rank the states from largest to smallest differences in their record highs and lows.

Section 2.2

Operations with integers Prepare a presentation for the class in which you explain why the *sum* of -3 and -2 is negative and why the *product* of -3 and -2 is positive.

Section 2.3

Exponents Use a scientific calculator and the exponential key $\boxed{y^x}$ (on some calculators, it is labeled $\boxed{x^y}$) to decide whether each statement is true or false.

1. $7^5 = 5^7$

2. $2^3 + 7^3 = (2 + 7)^3$

3. $(-4)^4 = -4^4$

4. $\dfrac{10^3}{5^3} = 2^3$

5. $8^4 \cdot 9^4 = (8 \cdot 9)^4$

6. $2^3 \cdot 3^3 = 6^3$

7. $\dfrac{3^{10}}{3^2} = 3^5$

8. $[(1.2)^3]^2 = [(1.2)^2]^3$

Section 2.4

The distributive property Use colored paper to make models like those in Figures 2-11 and 2-12 on pages 116–117 to help you explain why $a(b + c) = ab + ac$ and $a(b - c) = ab - bc$.

Section 2.5

Solving equations Make a presentation to the class explaining how we "undo" operations to isolate the variable when solving an equation. As a visual aid, bring in a box, tied shut with string, that contains a toy wrapped in tissue paper. Compare the three-step process a person would use to get to the toy inside the box to the three-step process we could use to solve the equation $\frac{2x}{3} - 4 = 2$.

Section 2.6

Geometry gourmet Find snack foods that have the shapes of the geometric figures in Table 2-2 and Table 2-3 on pages 141 and 143. For example, tortilla chips can be triangular in shape, and malted milk balls are spheres. If you are unable to find a particular shape already available, decide on a way to make a snack in that shape. Make up several trays of the snacks to bring to class. Split the class up into groups, with each group having its own tray of snacks. Call out a geometric shape and have the students pick out the snack in that shape from the tray and eat it.

Section 2.7

Mixtures Get several cans of the same brand of orange juice concentrate and make up four pitchers of concentrate and water mixtures that are 10%, 30%, 50%, and 70% orange juice concentrate. For example, a 30% solution would consist of three small paper cups of concentrate and seven small paper cups of water. Pour small amounts of each mixture into clean cups. Have students not in your group taste each solution and see whether they can put the mixtures in order from least concentrated to most concentrated.

Section 2.8

Inequalities In most states, a person must be at least 16 years of age to have a driver's license. We can mathematically describe this situation with the inequality $a \geq 16$, where a represents a person's age in years. Think of other situations that can be described using an inequality. Try to come up with some examples that require compound inequalities.

Section 2.1

Adding and Subtracting Real Numbers

CONCEPTS

To add two real numbers with *like signs,* add their absolute values and attach their common sign to the sum.

To add two real numbers with *unlike signs,* subtract their absolute values, the smaller from the larger. To that result, attach the sign of the number with the larger absolute value.

Properties of the real numbers—the *commutative* and *associative* properties of addition:

$a + b = b + a$
$(a + b) + c = a + (b + c)$

To *subtract* real numbers, add the opposite:

$a - b = a + (-b)$

Solutions of equations can be negative numbers.

REVIEW EXERCISES

1. Add the numbers.
 a. $12 + 33$
 b. $-45 + (-37)$
 c. $-15 + 37$
 d. $25 + (-13)$
 e. $12 + (-8) + (-15)$
 f. $-25 + (-14) + 35$
 g. $-9.9 + (-2.4)$
 h. $\dfrac{5}{16} + \left(-\dfrac{1}{2}\right)$
 i. $35 + (-13) + (-17) + 6$
 j. $-21 + (-11) + 32 + (-45)$
 k. $0 + (-7)$
 l. $-7 + 7$

2. Tell what property of addition guarantees that the quantities are equal.
 a. $-2 + 5 = 5 + (-2)$
 b. $(-2 + 5) + 1 = -2 + (5 + 1)$

3. Subtract the numbers.
 a. $45 - 64$
 b. $-17 - 32$
 c. $-27 - (-12)$
 d. $3.6 - (-2.1)$

4. Complete the following input/output tables.

 a.

x	$x + 5$
-2	
0	
-9	

 b.

x	$7 - x$
8	
0	
-2	

5. Solve each equation.
 a. $x + 12 = -17$
 b. $-1.7 = y - 1.3$
 c. $17 + p = -8$
 d. $-8 + q = -5$

6. ASTRONOMY *Magnitude* is a term used in astronomy to designate the brightness of celestial objects as viewed from the earth. Smaller magnitudes are associated with brighter objects, and larger magnitudes refer to fainter objects. See Illustration 1. For each of the following pairs of objects, by how many magnitudes do their brightnesses differ?
 a. A full moon and the sun
 b. The star Beta Crucis and a full moon

ILLUSTRATION 1

Object	Magnitude
Sun	-26.5
Full moon	-12.5
Beta Crucis	1.28

Based on data from *Exploration of the Universe* (Abell, Morrison, and Wolf; Saunders College Publishing, 1987)

7. GEOGRAPHY The tallest peak on earth is Mt. Everest at 29,028 feet. The greatest ocean depth is the Mariana Trench at $-36,205$ feet. Find the difference in the two elevations.

Section 2.2

When multiplying two real numbers:
1. The product of two real numbers with *like signs* is positive.
2. The product of two real numbers with *unlike signs* is negative.

Multiplying and Dividing Real Numbers

8. Multiply the numbers.
    a. $-8 \cdot 7$ **b.** $(-9)(-6)$
    c. $2(-3)(-2)$ **d.** $(-3)(4)(2)$
    e. $(-3)(-4)(-2)$ **f.** $(-4)(-1)(-3)(-3)$
    g. $-1.2(-5.3)$ **h.** $0.002(-1,000)$
    i. $-\dfrac{2}{3}\left(\dfrac{1}{5}\right)$ **j.** $2\dfrac{1}{4}\left(-\dfrac{1}{3}\right)$
    k. $-6 \cdot 0$ **l.** $(-3)(1)$

Properties of the real numbers—the *commutative* and *associative* properties of multiplication:

$ab = ba$
$(ab)c = a(bc)$

9. Tell what property of multiplication guarantees that the quantities are equal.
    a. $(2 \cdot 3)5 = 2(3 \cdot 5)$ **b.** $(-5)(-6) = (-6)(-5)$

When dividing two real numbers:
1. The quotient of two real numbers with *like signs* is positive.
2. The quotient of two real numbers with *unlike signs* is negative.

Division *of zero* by a non-zero number is zero. Division *by zero* is undefined.

10. Do each division.
    a. $\dfrac{88}{44}$ **b.** $\dfrac{-100}{25}$
    c. $\dfrac{-81}{-27}$ **d.** $\dfrac{0}{37}$
    e. $-\dfrac{3}{5} \div \dfrac{1}{2}$ **f.** $\dfrac{-60}{0}$
    g. $\dfrac{-4.5}{1}$ **h.** $\dfrac{-5}{-5}$

11. Solve each equation.
    a. $-12x = 24$ **b.** $36a = -108$
    c. $\dfrac{b}{-5} = -4$ **d.** $\dfrac{f}{17} = -3$

Section 2.3

An *exponent* is used to indicate repeated multiplication.

Order of Operations and Evaluating Algebraic Expressions

12. Find each power.
    a. 2^5 **b.** $(-2)^5$
    c. $(-3)^4$ **d.** $(-5)^3$
    e. $(-0.8)^2$ **f.** $\left(-\dfrac{2}{3}\right)^3$

Order of operations:
Work from the innermost pair of grouping symbols to the outermost pair in the following order:

1. Evaluate all exponential expressions.
2. Do all multiplications and divisions, working from left to right.
3. Do all additions and subtractions, working from left to right.

If the expression does not contain grouping symbols, begin with step 1. In a fraction, simplify the numerator and denominator separately. Then simplify the fraction, if possible.

13. Evaluate each expression.

a. $4^3 + 2(-6 - 2 \cdot 2)$
b. $-5[-3 - 2(5 - 7^2)] - 5$
c. $\dfrac{-4(4 + 2) - 4}{18 - 4(-5)}$
d. $(-3)^3\left(\dfrac{-8}{2}\right) + 5$
e. $\dfrac{|-35| - 2(-7)}{2^4 - 23}$
f. $-9^2 + (-9)^2$

14. Evaluate $3(x - y) - 5(x + y)$ when

a. $x = 2$ and $y = -5$
b. $x = -3$ and $y = 3$

15. Find the volume of a sphere with a radius r of 2 inches by evaluating the expression $\dfrac{4\pi r^3}{3}$. Round to the nearest tenth.

16. Complete each input/output table.

a.

t	$50t - 16t^2$
1	
-2	
3	

b.

x	$x^3 - \frac{x+2}{2}$
-2	
-10	
0	

Section 2.4

Simplifying Algebraic Expressions

"To *simplify* an algebraic expression" means to write it in less complicated form.

The *distributive property:*
$a(b + c) = ab + ac$
$a(b - c) = ab - ac$

A *term* is a number or a product of a number and one or more variables. Addition signs separate algebraic expressions into terms.

In a term, the numerical factor is called the *coefficient.*

Like terms are terms with exactly the same variables raised to exactly the same powers.

17. Simplify each expression.

a. $-4(7w)$
b. $-3r(-5r)$
c. $3(-2x)(-4y)$
d. $0.4(5.2f)$

18. Write each expression without parentheses.

a. $5(x + 3)$
b. $-2(2x + 3 - y)$
c. $-(a - 4)$
d. $\dfrac{3}{4}(4c - 8)$

19. How many terms are in each expression?

a. $3x^2 + 2x - 5$
b. $-12xyz$

20. Identify the coefficient of each term.

a. $2x - 5$
b. $16x^2 - 5x + 25$
c. $\frac{1}{2}x + y$
d. $9.6t^2 - t$

21. Simplify each expression by combining like terms.

a. $8p + 5p - 4p$
b. $-5m + 2n - 2m - 2n$
c. $6a + 2b - 8a - 12b$
d. $5(p - 2q) - 2(3p + 4q)$
e. $x^2 - x(x - 1)$
f. $8a^3 + 4a^3 - 20a^3$

22. Write an algebraic expression in simplified form for the perimeter of the triangle in Illustration 2.

ILLUSTRATION 2

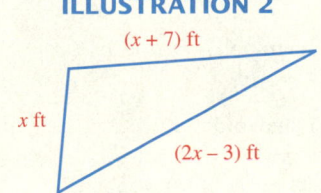

$(x + 7)$ ft

x ft

$(2x - 3)$ ft

Section 2.5

To solve an equation means to find all the values of the variable that, when substituted for the variable, make a true statement.

An equation that is true for all values of its variable is called an *identity*.

An equation that is not true for any values of its variable is called an *impossible equation*.

Solving Equations

23. Solve each equation.

a. $5x + 4 = 14$

b. $-12y + 8 = 20$

c. $\dfrac{n}{5} - 2 = 4$

d. $\dfrac{b - 5}{4} = -6$

e. $5(2x - 4) - 5x = 0$

f. $-2(x - 5) = 5(-3x + 4) + 3$

g. $\dfrac{3}{4} = \dfrac{1}{2} + \dfrac{d}{5}$

h. $-\dfrac{2}{3}f = 4$

i. $3(a + 8) = 6(a + 4) - 3a$

j. $2(y + 10) + y = 3(y + 8)$

24. SOUND SYSTEM A 45-foot-long speaker wire is to be cut into three pieces. One piece is to be 15 feet long. Of the remaining pieces, one must be 2 feet less than 3 times the length of the other. Find the length of the shorter piece of wire.

Section 2.6

A *formula* is an equation that is used to state a known relationship between two or more variables.

Sale price: $s = p - d$

Retail price: $r = c + m$

Profit: $p = r - c$

Distance: $d = rt$

Temperature: $C = \dfrac{5(F - 32)}{9}$

Formulas from geometry:

Square: $P = 4s$, $A = s^2$

Rectangle: $P = 2l + 2w$, $A = lw$

Formulas

25. A boat that is on sale for $13,998 has been discounted by $2,100. Find its original price.

26. Find the markup on a CD player whose wholesale cost is $219 and whose retail price is $395.

27. One month, a restaurant had sales of $13,500 and made a profit of $1,700. Find the expenses for the month.

28. INDY 500 In 1996, the winner of the Indianapolis 500-mile automobile race averaged 147.956 mph. To the nearest hundredth of an hour, how long did it take him to complete the race?

29. JEWELRY MAKING Gold melts at about 1,065°C. Change this to degrees Fahrenheit.

30. CAMPING Find the perimeter of the air mattress in Illustration 3.

31. CAMPING Find the amount of sleeping area on the top surface of the air mattress in Illustration 3.

ILLUSTRATION 3

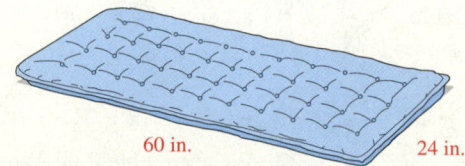

60 in.   24 in.

Triangle: $P = a + b + c$
$A = \frac{1}{2}bh$

Trapezoid:
$P = a + b + c + d$
$A = \frac{1}{2}h(b + d)$

32. Find the area of a triangle with a base 17 meters long and a height of 9 meters.

33. Find the area of a trapezoid with bases 11 inches and 13 inches long and a height of 12 inches.

Circle: $D = 2r$
$C = 2\pi r$
$A = \pi r^2$

Rectangular solid: $V = lwh$

Cylinder: $V = Bh$

Pyramid: $V = \frac{1}{3}Bh$
Cone: $V = \frac{1}{3}Bh$

Sphere: $V = \frac{4}{3}\pi r^3$

34. To the nearest hundredth, find the circumference of a circle with a radius of 8 centimeters.

35. To the nearest hundredth, find the area of the circle in Exercise 34.

36. CAMPING Find the approximate volume of the air mattress in Illustration 3 if it is 3 inches thick.

37. Find the volume of a 12-foot cylinder whose circular base has a radius of 0.5 feet. Give the result to the nearest tenth.

38. Find the volume of a pyramid that has a square base, measuring 6 feet on a side, and a height of 10 feet.

39. HALLOWEEN After being cleaned out, a spherical-shaped pumpkin has an inside diameter of 9 inches. To the nearest hundredth, what is its volume?

40. Solve each formula for the required variable.
    a. $A = 2\pi rh$ for h **b.** $P = 2l + 2w$ for l

Section 2.7

Problem Solving

To solve problems, use the five-step problem-solving strategy.
1. Analyze the problem.
2. Form an equation.
3. Solve the equation.
4. State the conclusion.
5. Check the result.

41. UTILITY BILLS The electric company charges $17.50 per month, plus 18 cents for every kilowatt hour of energy used. One resident's bill was $43.96. How many kilowatt hours were used that month?

42. ART HISTORY *American Gothic,* shown in Illustration 4, was painted in 1930 by American artist Grant Wood. The length of the rectangular painting is 5 inches more than the width. Find the dimensions of the painting if it has a perimeter of $109\frac{1}{2}$ inches.

ILLUSTRATION 4

43. Find the missing angle measures of the triangle in Illustration 5.

ILLUSTRATION 5

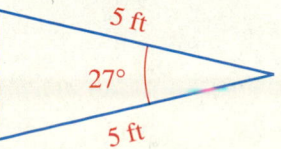

5 ft

27°

5 ft

Total value = number · value

Interest = principal · rate · time
$I = Prt$

44. What is the value of x video games each costing $45?

45. INVESTMENT INCOME A woman has $27,000. Part is invested for one year in a certificate of deposit paying 7% interest, and the remaining amount in a cash management fund paying 9%. After 1 year, the total interest on the two investments is $2,110. How much is invested at each rate?

Distance = rate · time
$$d = rt$$

46. WALKING AND BICYCLING A bicycle path is 5 miles long. A man walks from one end at the rate of 3 mph. At the same time, a friend bicycles from the other end, traveling at 12 mph. In how many minutes will they meet?

The value v of a commodity is its price per pound p times the number of pounds n: $v = pn$.

47. MIXTURE A store manager mixes candy worth 90¢ per pound with gumdrops worth $1.50 per pound to make 20 pounds of a mixture worth $1.20 per pound. How many pounds of each kind of candy does he use?

48. SOLUTION How much acetic acid is in x gallons of a solution that is 12% acetic acid?

Section 2.8

Inequalities

An *inequality* is a mathematical expression that contains a $>$, $<$, $\geq$, $\leq$, or $\neq$ symbol.

A *solution of an inequality* is any number that makes the inequality true.

A *parenthesis* indicates that a number is not on the graph. A *bracket* indicates that a number is included in the graph.

Interval notation can be used to describe a set of real numbers.

49. Solve each inequality, graph the solution, and use interval notation to describe the solution.

a. $3x + 2 < 5$

b. $-5x - 8 > 7$

c. $5x - 3 \geq 2x + 9$

d. $7x + 1 \leq 8x - 5$

e. $5(3 - x) \leq 3(x - 3)$

f. $-\dfrac{3}{4}x \geq -9$

g. $8 < x + 2 < 13$

h. $0 \leq 2 - 2x < 6$

50. Graph the interval represented by $[-13, \infty)$.

51. SPORTS EQUIPMENT The acceptable weight of ping-pong balls used in competition can range between 2.40 and 2.53 grams. Express this range using a compound inequality.

CHAPTER 2

Test

1. Add $(-6) + 8 + (-4)$.

2. Subtract $1.4 - (-0.8)$.

3. Multiply $(-2)(-3)(-5)$.

4. Divide $\dfrac{-22}{-11}$.

5. Evaluate $-7[(-5)^2 - 2(3 - 5^2)]$.

6. Evaluate $\dfrac{3(20 - 4^2)}{-2(6 - 2^2)}$.

In Problems 7–8, let $x = -2$, $y = 3$, and $z = 4$. Evaluate each expression.

7. $xy + z$

8. $\dfrac{z + 4y}{2x}$

9. What is the coefficient of the term $6x$?

10. How many terms are in the expression $3x^2 + 5x - 7$?

In Problems 11–14, simplify each expression.

11. $5(-4x)$

12. $-8(-7t)(4t)$

13. $3(x + 2) - 3(4 - x)$

14. $-1.1d^2 - 3.8d^2$

In Problems 15–22, solve each equation.

15. $12x = -144$

16. $\dfrac{4}{5}t = -4$

17. $\dfrac{c}{7} = -1$

18. $3x = 5 - 2x$

19. $2(x - 7) = -15$

20. $\dfrac{m}{2} - \dfrac{1}{3} = \dfrac{1}{4}$

21. $23 - 5(x + 10) = -12$

22. $5t + 7.2 = 12.7$

In Problems 23–24, solve each equation for the variable indicated.

23. $d = rt$; for t

24. $A = P + Prt$; for r

25. COMMERCIAL REAL ESTATE *Net absorption* is a term used to indicate how much office space in a city is being purchased. Use the information from the graph in Illustration 1 to determine the eight-quarter average net absorption figure for Long Beach, California.

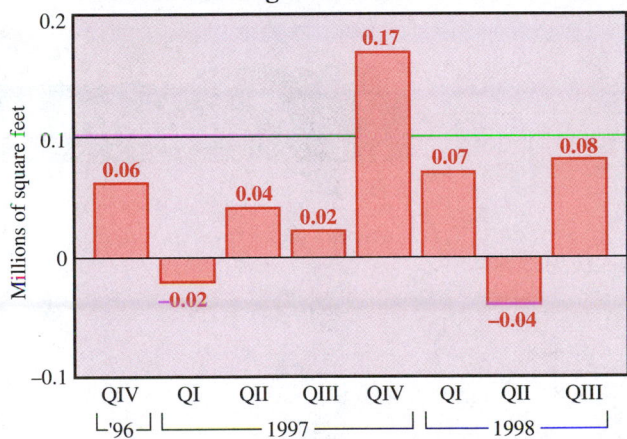

ILLUSTRATION 1

Downtown Long Beach Office Net Absorbtion

Based on information from *Los Angeles Times* (Oct. 13, 1998) Section C10.

26. PETS The spherical fishbowl shown in Illustration 2 is three-quarters full of water. To the nearest cubic inch, what is the volume of water in the bowl?

ILLUSTRATION 2

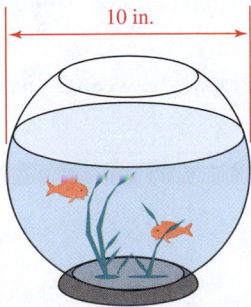

10 in.

27. INVESTMENT PROBLEM Part of $13,750 is invested at 9% annual interest, and the rest is invested at 8%. After one year, the accounts paid $1,185 in interest. How much was invested at the lower rate?

28. TRAVEL TIMES A car leaves Rockford, Illinois at the rate of 65 mph, bound for Madison, Wisconsin. At the same time, a truck leaves Madison at the rate of 55 mph, bound for Rockford. If the cities are 72 miles apart, how long will it take for the car and the truck to meet?

29. MIXTURE PROBLEM How many liters of a 2% brine solution must be added to 30 liters of a 10% brine solution to dilute it to an 8% solution?

30. GEOMETRY If the vertex angle of an isosceles triangle is 44°, find the measure of each base angle.

In Problems 31–32, solve each inequality, graph its solution, and use interval notation to describe the solution.

31. $-8x - 20 \leq 4$

32. $-4 \leq 2(x + 1) < 10$

33. Explain the difference between an equation and an inequality.

34. Explain this statement: "Subtraction is the same as addition of the opposite."

CHAPTERS 1–2

Cumulative Review Exercises

In Exercises 1–4, tell whether the expression is an equation.

1. $m - 25$ **2.** $t = 25r$

3. $\dfrac{p + 5}{2}$ **4.** $x + 1 = 24$

In Exercises 5–6, classify each number as a natural number, a whole number, an integer, a rational number, an irrational number, and a real number. Each number may be in several classifications.

5. 3

6. -0.25

In Exercises 7–8, graph each set of numbers on the number line.

7. The natural numbers between 2 and 7

8. The real numbers between 2 and 7

In Exercises 9–12, evaluate each expression.

9. $|-12|$ **10.** $|15|$

11. $-|15|$ **12.** $-|-12|$

In Exercises 13–16, use a calculator to find each square root to the nearest hundredth.

13. $\sqrt{77}$ **14.** $\sqrt{0.26}$

15. $\sqrt{\pi}$ **16.** $\sqrt{\dfrac{7}{5}}$

In Exercises 17–20, write each product using exponents.

17. $3 \cdot 3 \cdot 3$ **18.** $8 \cdot \pi \cdot r \cdot r$

19. $4 \cdot x \cdot x \cdot y \cdot y$ **20.** $m \cdot m \cdot m \cdot n$

In Exercises 21–24, evaluate each expression.

21. $13 - 3 \cdot 4$ **22.** $-2 \cdot 7^2$

23. $6 + 3\left(\dfrac{5}{2}\right) - \dfrac{1}{2}$ **24.** $\dfrac{12^2 - 4^2 - 2}{2(7 - 4)}$

In Exercises 25–26, write each phrase as an algebraic expression.

25. The sum of the width w and 12

26. Four less than a number n

In Exercises 27–30, complete each table of values.

27.

t	$t^2 - 4$
0	
1	
3	

28.

t	$(t - 4)^2$
0	
1	
3	

29.

a	$\frac{a}{3} + 2$
0	
3	
6	

30.

x	$\frac{x}{4} + \frac{x}{3}$
0	
12	
24	

In Exercises 31–34, a = 2 and b = 5. Evaluate each expression.

31. $a^2 + b$

32. $b^3 - 12a^2$

33. $\dfrac{a + b}{b + 2}$

34. $\dfrac{3(b^2 - 1)}{2a}$

In Exercises 35–38, solve each equation.

35. $x + 9 = 13$

36. $5x = 25$

37. $\dfrac{y}{4} = 5$

38. $t - 5 = 17$

In Exercises 39–42, solve each problem.

39. Find 45% of 640.

40. What percent of 200 is 30?

41. 45 is 15% of what number?

42. 20% of what number is 240?

In Exercises 43–46, let x = −5, y = 3, and z = 0. Evaluate each expression.

43. $(3x - 2y)z$

44. $\dfrac{x - 3y + |z|}{2 - x}$

45. $x^2 - y^2 + z^2$

46. $\dfrac{x}{y} + \dfrac{y + 2}{3 - z}$

In Exercises 47–50, find each power.

47. $(-6)^3$

48. -6^2

49. $-(-5)^2$

50. $\left(-\dfrac{2}{3}\right)^3$

In Exercises 51–54, evaluate each expression.

51. $2(5 - 3)^2$

52. $5^2 - (8 - 4)^2$

53. $\dfrac{2 + (2 + 6)^2}{2(9 - 6)}$

54. $\dfrac{2[4 + 2(5 - 3)]}{3[3(2 \cdot 4 - 6)]}$

In Exercises 55–60, simplify each expression.

55. $-8(4d)$

56. $5(2x - 3y + 1)$

57. $2x + 3x$

58. $3a + 6a - 17a$

59. $q(q - 5) + 7q^2$

60. $5(t - 4) + 3t$

61. What is the length of the longest side of the triangle in Illustration 1?

62. Write an algebraic expression in simplest form for the perimeter of the triangle in Illustration 1.

ILLUSTRATION 1

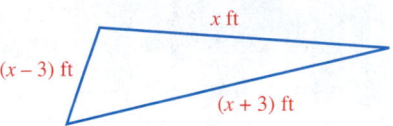

In Exercises 63–68, solve each equation.

63. $3x - 4 = 23$

64. $\dfrac{x}{5} + 3 = 7$

65. $-5p + 0.7 = 3.7$

66. $\dfrac{y - 4}{5} = 3$

67. $-\dfrac{4}{5}x = 16$

68. $-9(n + 2) - 2(n - 3) = 10$

 *In Exercises 69–70, find the area of each figure.*

69. A rectangle with sides of 5 meters and 13 meters

70. A circle with a radius of 5 centimeters (Give the answer to the nearest hundredth.)

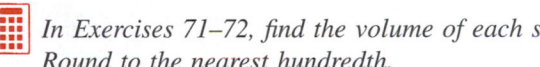 *In Exercises 71–72, find the volume of each solid. Round to the nearest hundredth.*

71. A 12-foot-long cylinder with a circular base with a radius of 0.5 feet

72. A cone that is 10 centimeters tall and has a circular base whose diameter is 12 centimeters

In Exercises 73–74, solve each formula for the given variable.

73. $P = 2l + 2w$ (for w)

74. $A = P + Prt$ (for t)

75. WORK Physicists say that *work* is done when an object is moved a distance d by a force F. To find the work done, we can use the formula

$W = Fd$

Find the work done in lifting the bundle of newspapers shown in Illustration 2 onto the workbench. (*Hint:* The force that must be applied to lift the papers is the weight of the newspapers.)

ILLUSTRATION 2

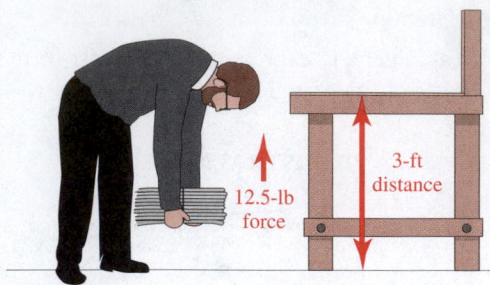

76. WORK See Exercise 75. Find the weight of a 1-gallon can of paint if the amount of work done to lift it onto the workbench is 28.35 foot-pounds.

In Exercises 77–78, find each unknown angle measure represented by a variable.

77.

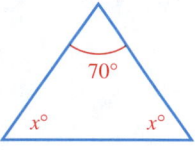

78.

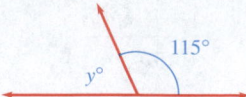

In Exercises 79–82, solve each inequality, graph the solution, and use interval notation to describe the solution.

79. $x - 4 > -6$

80. $-6x \geq -12$

81. $8x + 4 \geq 5x + 1$

82. $-1 \leq 2x + 1 < 5$

3

Graphs, Linear Equations, and Functions

CAMPUS CONNECTION

The Drafting Department

In drafting, students learn how to use AutoCAD, a computer-aided drafting program. As part of their coursework, they will draw the floor plan of a house on a grid, or *coordinate system*, displayed on a monitor. In this chapter, we will discuss the rectangular coordinate system. It allows us to represent mathematical relationships visually by graphing them. Learning how to use the rectangular coordinate system will give you the insight you'll need when you encounter coordinate systems in other disciplines such as drafting.

Relationships between two quantities can be described by a table, a graph, or an equation.

▶ 3.1

Graphing Using the Rectangular Coordinate System

In this section, you will learn about

The rectangular coordinate system ■ Graphing mathematical relationships ■ Reading graphs ■ Step graphs

Introduction   It is often said, "A picture is worth a thousand words." In this section, we will show how numerical relationships can be described using mathematical pictures called **graphs.** We will also show how graphs are constructed and how we can obtain important information by reading graphs.

The Rectangular Coordinate System

When designing the Gateway Arch in St. Louis, shown in Figure 3-1(a), architects created a mathematical model called a **rectangular coordinate graph.** This graph, shown in Figure 3-1(b), is drawn on a grid called a **rectangular coordinate system.** This coordinate system is sometimes called a **Cartesian coordinate system,** after the 17th-century French mathematician René Descartes.

FIGURE 3-1

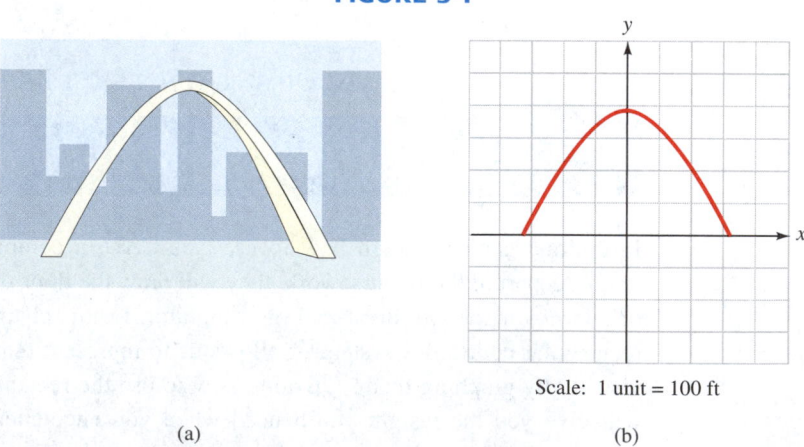

Scale: 1 unit = 100 ft

(a)   (b)

A rectangular coordinate system (see Figure 3-2) is formed by two perpendicular number lines. The horizontal number line is called the **x-axis,** and the vertical number line is called the **y-axis.** The positive direction on the x-axis is to the right, and the positive direction on the y-axis is upward. The scale on each axis should fit the data. For example, the axes of the graph of the arch shown in Figure 3-1(b) are scaled in units of 100 feet. If no scale is indicated on the axes, we assume that the axes are scaled in units of 1.

The point where the axes cross is called the **origin.** This is the zero point on each axis. The axes form a **coordinate plane** and divide it into four regions called **quadrants,** which are numbered as shown in Figure 3-2.

FIGURE 3-2

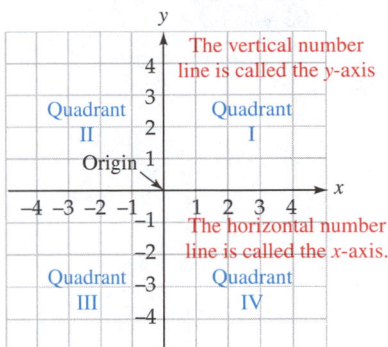

The vertical number line is called the y-axis.

The horizontal number line is called the x-axis.

Each point in a coordinate plane can be identified by a pair of real numbers x and y written in the form (x, y). The first number x in the pair is called the **x-coordinate,** and the second number y is called the **y-coordinate.** The numbers in the pair are called the **coordinates** of the point. Some examples of such pairs are $(3, -4)$, $\left(-1, -\frac{3}{2}\right)$, and $(0, 2.5)$.

$$(3, -4)$$
$$\uparrow \quad \uparrow$$

The x-coordinate   The y-coordinate
is listed first.   is listed second.

The process of locating a point in the coordinate plane is called **graphing** or **plotting** the point. In Figure 3-3(a), we show how to graph the point with coordinates of $(3, -4)$. Since the **x-coordinate,** 3, is positive, we start at the origin and move 3 units to the *right* along the x-axis. Since the **y-coordinate,** -4, is negative, we then move *down* 4 units to locate point A. Point A is the **graph** of $(3, -4)$ and lies in quadrant IV.

To plot the point $(-4, 3)$, we start at the origin, move 4 units to the left along the x-axis, and then move up 3 units to locate point B. Point B lies in quadrant II.

FIGURE 3-3

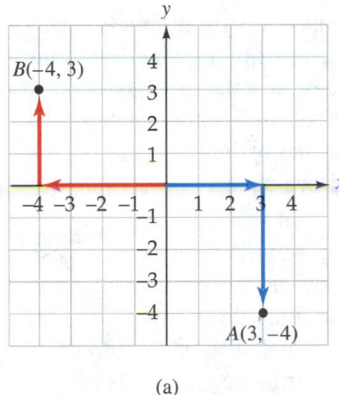

(a)

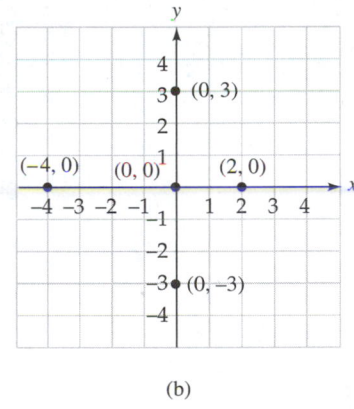

(b)

WARNING! Note that point A with coordinates $(3, -4)$ is not the same as point B with coordinates $(-4, 3)$. Since the order of the coordinates of a point is important, we call the pairs **ordered pairs.**

In Figure 3-3(b), we see that the points $(-4, 0)$, $(0, 0)$, and $(2, 0)$ lie on the x-axis. In fact, all points with a y-coordinate of zero will lie on the x-axis. We also see that the points $(0, -3)$, $(0, 0)$, and $(0, 3)$ lie on the y-axis. All points with an x-coordinate of zero lie on the y-axis. We can also see that the coordinates of the origin are $(0, 0)$.

EXAMPLE 1

Graphing points. Plot the points: **a.** $A(-2, 3)$,   **b.** $B\left(-1, -\frac{3}{2}\right)$,   **c.** $C(0, 2.5)$, and   **d.** $D(4, 2)$.

Solution   See Figure 3-4. (Note: If no scale is indicated on the axes, we assume that the axes are scaled in units of 1.)

a. To plot point A with coordinates $(-2, 3)$, we start at the origin, move 2 units to the *left* on the x-axis, and move 3 units *up*. Point A lies in quadrant II.

b. To plot point B with coordinates $\left(-1, -\frac{3}{2}\right)$, we start at the origin and move 1 unit to the *left* and $\frac{3}{2}$ $\left(\text{or } 1\frac{1}{2}\right)$ units *down*. Point B lies in quadrant III.

FIGURE 3-4

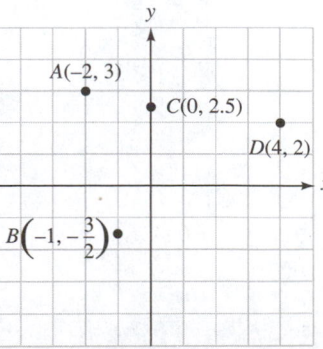

c. To graph point *C* with coordinates (0, 2.5), we start at the origin and move 0 units on the *x*-axis and 2.5 units *up*. Point *C* lies on the *y*-axis.

d. To graph point *D* with coordinates (4, 2), we start at the origin and move 4 units to the *right* and 2 units *up*. Point *D* lies in quadrant I.

SELF CHECK

Plot the points: **a.** $E(2, -2)$, **b.** $F(-4, 0)$, **c.** $G\left(1.5, \frac{5}{2}\right)$, and **d.** $H(0, 5)$.

Answers:

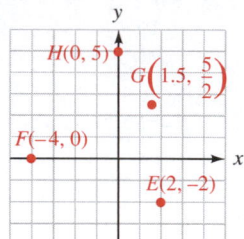

EXAMPLE 2

Orbit of the earth. The circle shown in Figure 3-5 is an approximate **graph** of the orbit of the earth. The graph is made up of infinitely many points, each with its own *x*- and *y*-coordinates. Use the graph to find the coordinates of the earth's position during the months of February, May, August, and December.

FIGURE 3-5

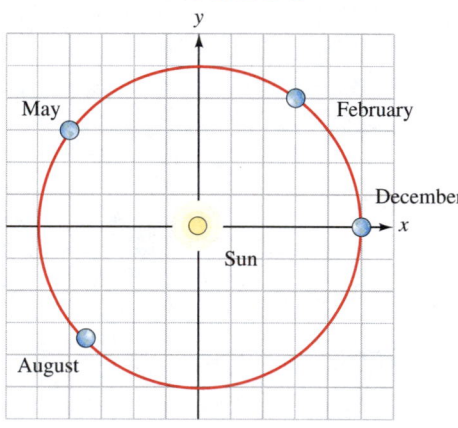

Scale: 1 unit = 18,600,000 mi

Solution

To find the coordinates of each position, we start at the origin and move left or right along the *x*-axis to find the *x*-coordinate and then up or down to find the *y*-coordinate.

Month	Position of earth on graph	Coordinates
February	3 units to the *right*, then 4 units *up*	(3, 4)
May	4 units to the *left*, then 3 units *up*	(−4, 3)
August	3.5 units to the *left*, then 3.5 units *down*.	(−3.5, −3.5)
December	5 units *right*, no units *up* or *down*	(5, 0)

Graphing Mathematical Relationships

Every day, we deal with quantities that are related:

■ The distance that we travel depends on how fast we are going.

■ Our weight depends on how much we eat.

■ The amount of water in a tub depends on how long the water has been running.

We often use graphs to visualize relationships between two quantities. For example, suppose we know the number of gallons of water that are in a tub at several time intervals after the water has been turned on. We can list that information in a **table of values** (see Figure 3-6).

The information in the table can be used to construct a graph that shows the relationship between the amount of water in the tub and the time the water has been run-

FIGURE 3-6

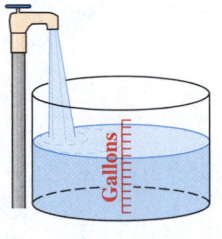

Time (min)	Water in tub (gal)	
0	0	→ (0, 0)
1	8	→ (1, 8)
3	24	→ (3, 24)
4	32	→ (4, 32)

At various times, the amount of water in the tub was measured and recorded in the table of values.

↑ *x*-coordinate   ↑ *y*-coordinate   ↑ The data in the table can be expressed as ordered pairs (*x*, *y*).

(a)   **(b)**

ning. Since the amount of water in the tub depends on the time, we will associate *time* with the *x*-axis and *amount of water* with the *y*-axis.

To construct the graph in Figure 3-7, we plot the four ordered pairs and draw a line through the resulting data points. The *y*-axis is scaled in larger units (4 gallons) because the data range from 0 to 32 gallons.

FIGURE 3-7

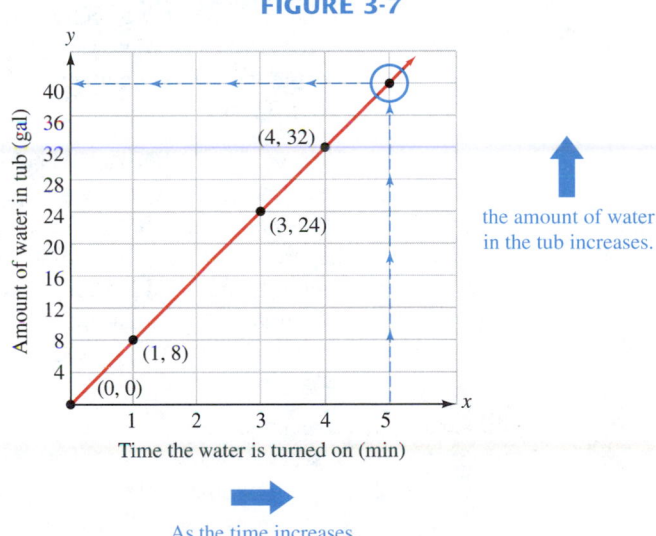

the amount of water in the tub increases.

As the time increases,

From the graph, we can see that the amount of water in the tub steadily increases as the water is allowed to run. We can also use the graph to make observations about the amount of water in the tub at other times. For example, the dashed line on the graph shows that in 5 minutes, the tub will contain 40 gallons of water.

Reading Graphs

Valuable information can be obtained from a graph, as can be seen in the next example.

EXAMPLE 3

Reading a graph. The graph in Figure 3-8 shows the number of people in an audience before, during, and after the taping of a television show. On the *x*-axis, zero represents the time when taping began. Use the graph to answer the following questions, and record each result in a table of values.

a. How many people were in the audience when taping began?

b. What was the size of the audience 10 minutes before taping began?

c. At what times were there exactly 100 people in the audience?

FIGURE 3-8

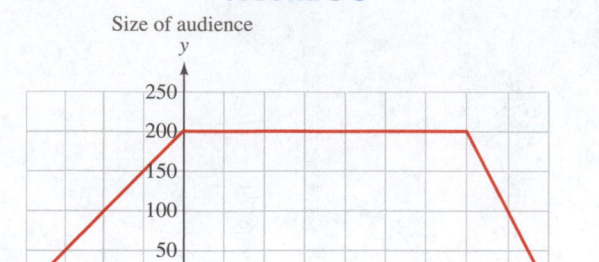

Size of audience

Taping begins

Taping ends

Time (min)

Solution

Time	Audience
0	200
−10	150
−20	100
80	100

a. The time when taping began is represented by zero on the *x*-axis. Since the point on the graph directly above zero has a *y*-coordinate of 200, the point (0, 200) is on the graph. The *y*-coordinate of this point indicates that 200 people were in the audience when the taping began. We enter this result in the table at the left.

b. Ten minutes before taping began is represented by −10 on the *x*-axis. Since the point on the graph directly above −10 has a *y*-coordinate of 150, the point (−10, 150) is on the graph. The *y*-coordinate of this point indicates that 150 people were in the audience 10 minutes before the taping began. We enter this result in the table.

c. We can draw a horizontal line passing through 100 on the *y*-axis. This line intersects the graph twice, at (−20, 100) and (80, 100). So there are two times when 100 people were in the audience. The first time was 20 minutes before taping began (−20), and the second time was 80 minutes after taping began (80). The *y*-coordinates of these points indicate that there were 100 people in the audience 20 minutes before and 80 minutes after taping began. We enter these results in the table.

SELF CHECK

Use the graph in Figure 3-8 to answer the following questions.   **a.** At what times were there exactly 50 people in the audience?   **b.** What was the size of the audience that watched the taping?   **c.** How long did it take for the audience to leave the studio after the taping ended?

Answers: **a.** 30 min before and 85 min after taping began,   **b.** 200,   **c.** 20 min   ■

Step Graphs

The graph in Figure 3-9 shows the cost of renting a trailer for different periods of time. For example, the cost of renting the trailer for 4 days is $60, which is the *y*-coordinate of the point (4, 60). The cost of renting the trailer for a period lasting over 4 and up to 5 days jumps to $70. Since the jumps in cost form steps in the graph, we call this graph a **step graph**.

EXAMPLE 4

Use the information in Figure 3-9 to answer the following questions. Write the results in a table of values.

a. Find the cost of renting the trailer for 2 days.

b. Find the cost of renting the trailer for $5\frac{1}{2}$ days.

c. How long can you rent the trailer if you have $50?

d. Is the rental cost per day the same?

FIGURE 3-9

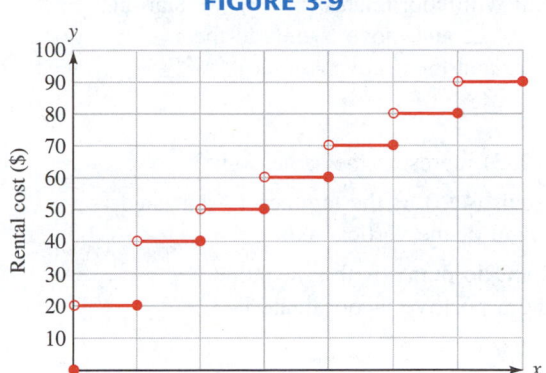

Length of rental (days)

Solution   **a.** The solid dot at the end of each step indicates the rental cost for 1, 2, 3, 4, 5, 6, or 7 days. Just as when we graphed inequalities, an open circle indicates that that point is not on the graph. We locate 2 days on the *x*-axis and move up to locate the point on the graph directly above the 2. Since the point has coordinates (2, 40), a 2-day rental would cost $40. We enter this ordered pair in the table at the left.

Length of rental (days)	Cost (dollars)
2	40
$5\frac{1}{2}$	80
3	50

b. We locate $5\frac{1}{2}$ days on the *x*-axis and move straight up to locate the point with coordinates $\left(5\frac{1}{2}, 80\right)$, which indicates that a $5\frac{1}{2}$-day rental would cost $80. We then enter this ordered pair in the table.

c. We draw a horizontal line through the point labeled 50 on the *y*-axis. Since this line intersects one step in the graph, we can look down to the *x*-axis to find the *x*-values that correspond to a *y*-value of 50. From the graph, we see that the trailer can be rented for more than 2 and up to 3 days for $50. We write (3, 50) in the table.

d. No. If we look at the *y*-coordinates, we see that for the first day, the rental fee is $20. The second day, the cost jumps another $20. The third day, and all subsequent days, the cost jumps only $10.   ∎

STUDY SET

Section 3.1

VOCABULARY

In Exercises 1–6, fill in the blanks to make the statements true.

1. The pair of numbers (−1, −5) is called an _____ pair.

2. In the ordered pair $\left(-\frac{3}{2}, -5\right)$, the −5 is called the _____.

3. The point with coordinates (0, 0) is called the _____.

4. The *x*- and *y*-axes divide the coordinate plane into four regions called _____.

5. The point with coordinates (4, 2) can be graphed on a _____ coordinate system.

6. The process of locating the position of a point on a coordinate plane is called _____ the point.

CONCEPTS

In Exercises 7–8, fill in the blanks to make the statements true.

7. To plot the point with coordinates (−5, 4.5), we start at the _____ and move 5 units to the _____ and then move 4.5 units ___.

8. To plot the point with coordinates $\left(6, -\frac{3}{2}\right)$, we start at the _____ and move 6 units to the _____ and then move $\frac{3}{2}$ units _____.

9. Do (3, 2) and (2, 3) represent the same point?

10. In the ordered pair (4, 5), is the number 4 associated with the horizontal or the vertical axis?

11. In which quadrant do points with a negative *x*-coordinate and a positive *y*-coordinate lie?

12. In which quadrant do points with a positive *x*-coordinate and a negative *y*-coordinate lie?

13. Use the graph in Illustration 1 to complete the table.

ILLUSTRATION 1

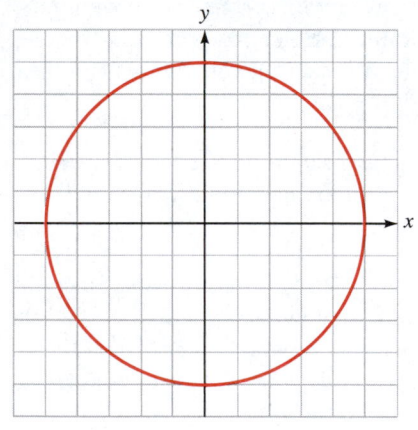

x	y
4	
4	
0	
0	
−3	
−3	
−4	
−4	
5	

14. Use the graph in Illustration 2 to complete the table.

ILLUSTRATION 2

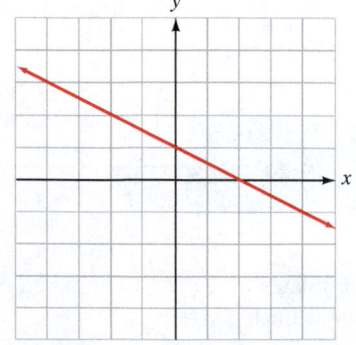

x	y
	0
	2
	−1
−4	
	1

The graph in Illustration 3 gives the heart rate of a woman before, during, and after an aerobic workout. In Exercises 15–22, use the graph to answer the questions.

15. What information does the point (−10, 60) give us?

16. After beginning her workout, how long did it take the woman to reach her training-zone heart rate?

17. What was the woman's heart rate half an hour after beginning the workout?

ILLUSTRATION 3

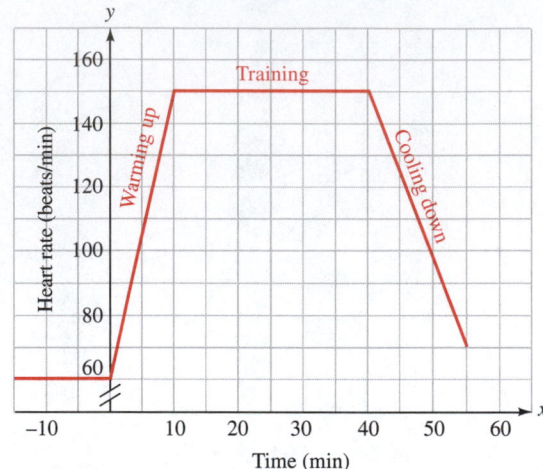

18. For how long did the woman work out at her training zone?

19. At what time was her heart rate 100 beats per minute?

20. How long was her cool-down period?

21. What was the difference in the woman's heart rate before the workout and after the cool-down period?

22. What was her approximate heart rate 8 minutes after beginning?

NOTATION

23. Explain the difference between (3, 5), 3(5), and 5(3 + 5).

24. In the table, which column contains values associated with the vertical axis of a graph?

x	y
2	0
5	−2
−1	$-\frac{1}{2}$

25. Do these ordered pairs name the same point?
$\left(2.5, -\frac{7}{2}\right)$, $\left(2\frac{1}{2}, -3.5\right)$, $\left(2.5, -3\frac{1}{2}\right)$

26. Do these ordered pairs name the same point?
$(-1.25, 4)$, $\left(-1\frac{1}{4}, 4.0\right)$, $\left(-\frac{5}{4}, 4\right)$

PRACTICE

In Exercises 27–28, graph each point on the coordinate grid provided.

27. $A(-3, 4)$, $B(4, 3.5)$, $C\left(-2, -\frac{5}{2}\right)$, $D(0, -4)$, $E\left(\frac{3}{2}, 0\right)$, $F(3, -4)$

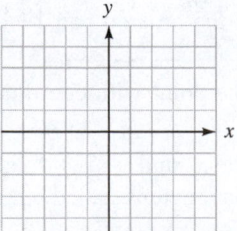

28. $G(4, 4)$, $H(0.5, -3)$, $I(-4, -4)$, $J(0, -1)$, $K(0, 0)$, $L(0, 3)$, $M(-2, 0)$

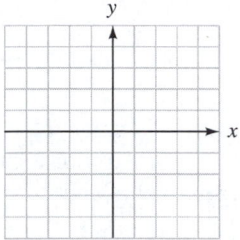

APPLICATIONS

29. CONSTRUCTION The graph in Illustration 4 shows a side view of a bridge design. Make a table with three columns; label them *rivets, welds,* and *anchors.* List the coordinates of the points at which each category is located.

ILLUSTRATION 4

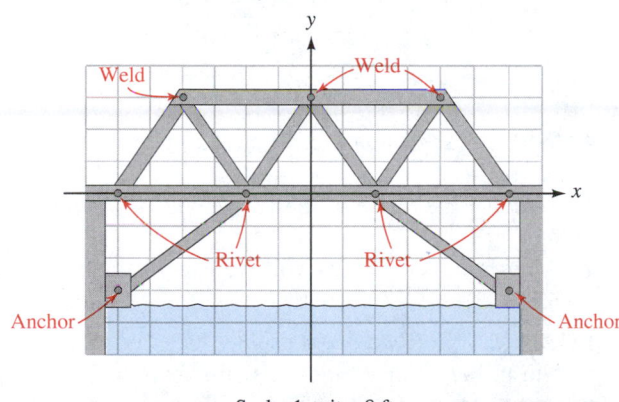

Scale: 1 unit = 8 ft

30. WATER PRESSURE The graph in Illustration 5 shows how the path of a stream of water changes when the hose is held at two different angles.
  a. At which angle does the stream of water shoot up higher? How much higher?
  b. At which angle does the stream of water shoot out farther? How much farther?

ILLUSTRATION 5

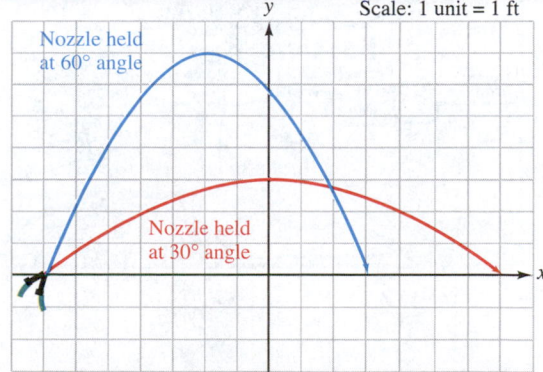

Scale: 1 unit = 1 ft

31. GOLF SWING To correct her swing, a golfer is videotaped and then has her image displayed on a computer monitor so that it can be analyzed by a golf pro. (See Illustration 6.) Give the coordinates of the points that are highlighted on the arc of her swing.

ILLUSTRATION 6

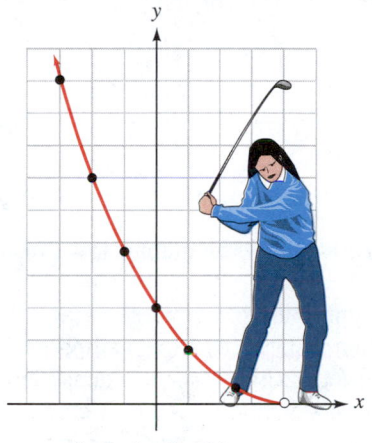

Scale: 1 unit = 6 in.

32. MEDICINE Scoliosis is a lateral curvature of the spine that can be more easily detected when a grid is superimposed over an X ray. In Illustration 7, find the coordinates of the "center points" of the indicated vertebrae. Note that T3 means the third thoracic vertebra, L4 means the fourth lumbar vertebra, and so on.

ILLUSTRATION 7

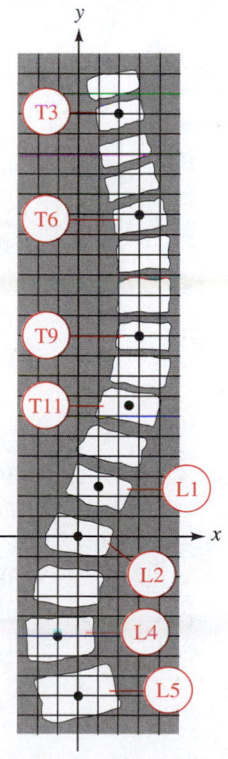

Scale: 1 unit = 0.5 in.

33. VIDEO RENTAL The charges for renting a video are shown in the graph in Illustration 8.
  a. Find the charge for a 1-day rental.
  b. Find the charge for a 2-day rental.
  c. What is the charge if a tape is kept for 5 days?
  d. What is the charge if a tape is kept for a week?

ILLUSTRATION 8

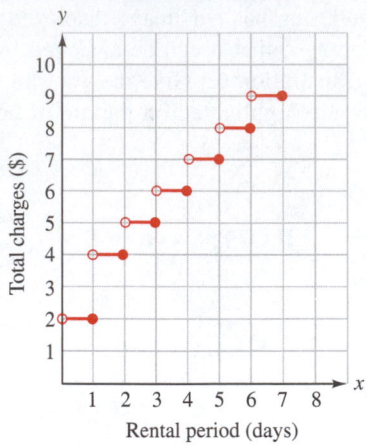

Rental period (days)

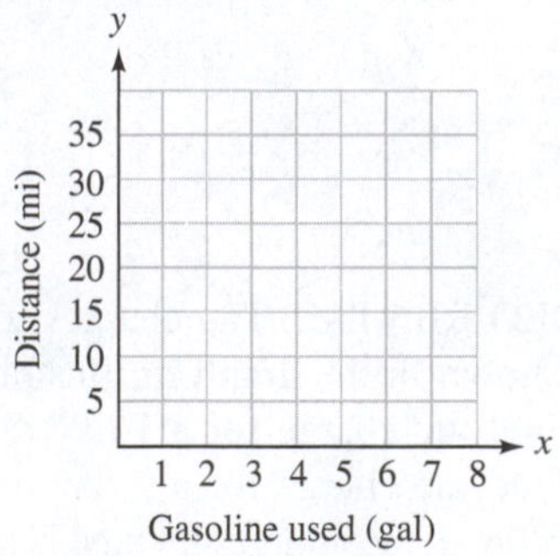

a. Estimate how far the truck can go on 7 gallons of gasoline.
b. How many gallons of gas are needed to travel a distance of 20 miles?
c. How far can the truck go on 6.5 gallons of gasoline?

36. VALUE OF A CAR The table in Illustration 11 shows the value y (in thousands of dollars) of a car that is x years old. Plot the ordered pairs and draw a line connecting the points.

34. POSTAGE RATES The graph shown in Illustration 9 gives the first-class postage rates in 1999 for mailing items weighing up to 5 ounces.
a. Find the postage costs for mailing each of the following letters first class: a 1-ounce letter, a 4-ounce letter, and a $2\frac{1}{2}$-ounce letter.
b. Find the difference in postage for a 3.75-ounce letter and a 4.75-ounce letter.
c. What is the heaviest letter that could be mailed for 55¢ first class?

ILLUSTRATION 11

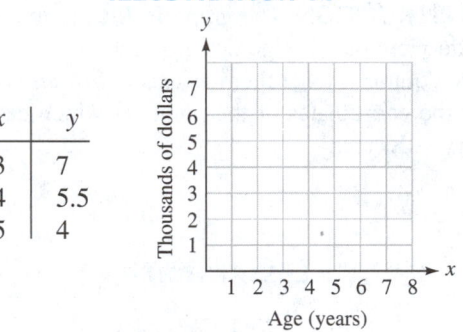

x	y
3	7
4	5.5
5	4

Age (years)

a. What does the point (3, 7) on the graph tell you?

b. Estimate the value of the car when it is 7 years old.

c. After how many years will the car be worth $2,500?

37. ROAD MAPS Road maps usually have a coordinate system to help locate cities. Use the map in Illustration 12 to locate Rockford, Mount Carroll, Harvard, and the intersection of state Highway 251 and U.S. Highway 30. Express each answer in the form (number, letter).

ILLUSTRATION 9

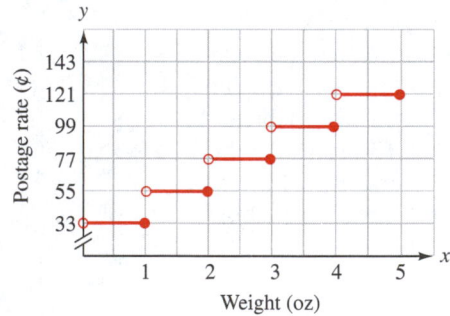

Weight (oz)

35. GAS MILEAGE The table in Illustration 10 gives the number of miles (y) that a truck can be driven on x gallons of gasoline. Plot the ordered pairs and draw a line connecting the points.

ILLUSTRATION 12

ILLUSTRATION 10

x	y
2	10
3	15
5	25

Gasoline used (gal)

38. BATTLESHIP In the game Battleship, the player uses coordinates to drop depth charges from a battleship to hit a hidden submarine. What coordinates should be used to make three hits on the exposed submarine shown in Illustration 13? Express each answer in the form (letter, number).

ILLUSTRATION 13

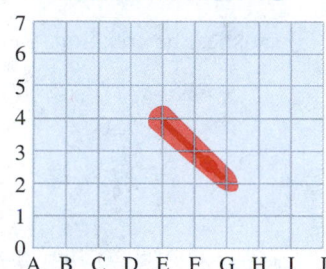

WRITING

39. Explain why the point $(-3, 3)$ is not the same as the point $(3, -3)$.

40. Explain what is meant when we say that the rectangular coordinate graph of the St. Louis Gateway Arch is made up of *infinitely many* points.

41. Explain how to plot the point $(-2, 5)$.

42. Explain why the coordinates of the origin are $(0, 0)$.

REVIEW

43. Evaluate $-3 - 3(-5)$.

44. Evaluate $(-5)^2 + (-5)$.

45. What is the opposite of -8?

46. Simplify $|-1 - 9|$.

47. Solve $-4x + 7 = -21$.

48. Solve $P = 2l + 2w$ for w.

49. Evaluate $(x + 1)(x + y)^2$ for $x = -2$ and $y = -5$.

50. Simplify $-6(x - 3) - 2(1 - x)$.

▶ **3.2**

Equations Containing Two Variables

In this section, you will learn about

> Solving equations with two variables ■ Constructing tables of values ■ Graphing equations ■ Using different variables

Introduction   In this section, we will discuss equations that contain two variables. Such equations are often used to describe relationships between two quantities. To see a mathematical picture of these relationships, we will construct graphs of their equations.

Solving Equations with Two Variables

We have previously solved equations containing one variable. For example, we can show that the solution of each of the following equations is $x = 3$.

$$2x + 3 = 9, \quad -5x + 1 = 4 - 6x, \quad \text{and} \quad -3(x + 1) = 2x - 18$$

If we graph the solution $x = 3$ on a number line, we get the graph shown in Figure 3-10.

FIGURE 3-10

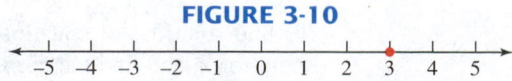

To describe relationships between two quantities mathematically, we use equations with two variables. Some examples of equations in two variables are

$$y = x - 1, \quad y = x^2, \quad y = |x|, \quad \text{and} \quad y = x^3$$

Solutions of equations in two variables are ordered pairs. For example, one solution of $y = x - 1$ is the ordered pair $(5, 4)$, because the equation is true when $x = 5$ and $y = 4$.

$$y = x - 1 \quad \text{The original equation.}$$
$$4 \overset{?}{=} 5 - 1 \quad \text{Substitute 5 for } x \text{ and 4 for } y.$$
$$4 = 4 \quad \text{Do the subtraction on the right-hand side: } 5 - 1 = 4.$$

Since $4 = 4$ is a true statement, the ordered pair $(5, 4)$ is a solution, and we say that $(5, 4)$ **satisfies** the equation.

To see whether the ordered pair $(-1, -3)$ satisfies the equation, we substitute -1 for x and -3 for y.

$$y = x - 1 \quad \text{The original equation.}$$
$$-3 \overset{?}{=} -1 - 1 \quad \text{Substitute } -1 \text{ for } x \text{ and } -3 \text{ for } y.$$
$$-3 = -2 \quad \text{Do the subtraction on the right-hand side: } -1 - 1 = -2.$$

Since $-3 = -2$ is a false statement, $(-1, -3)$ does not satisfy the equation.

EXAMPLE 1

Verifying a solution. Is the ordered pair $(-2, 4)$ a solution of $y = 3x + 9$?

Solution   We substitute -2 for x and 4 for y and see if a true statement results.

$$y = 3x + 9 \quad \text{The original equation.}$$
$$4 \overset{?}{=} 3(-2) + 9 \quad \text{Substitute } -2 \text{ for } x \text{ and 4 for } y.$$
$$4 \overset{?}{=} -6 + 9 \quad \text{Do the multiplication: } 3(-2) = -6.$$
$$4 = 3 \quad \text{Do the addition: } -6 + 9 = 3.$$

Since the equation $4 = 3$ is false, $x = -2$ and $y = 4$ is not a solution.

SELF CHECK   Is $(-1, -5)$ a solution of $y = 5x$?   *Answer:* yes   ■

EXAMPLE 2

Verifying a solution. Is the ordered pair $(3, -4)$ a solution of $y = 5 - x^2$?

Solution   We substitute 3 for x and -4 for y and see whether the resulting equation is a true statement.

$$y = 5 - x^2 \quad \text{The original equation.}$$
$$-4 \overset{?}{=} 5 - (3)^2 \quad \text{Substitute 3 for } x \text{ and } -4 \text{ for } y.$$
$$-4 \overset{?}{=} 5 - 9 \quad \text{Find the power: } (3)^2 = 9.$$
$$-4 = -4 \quad \text{Do the subtraction: } 5 - 9 = -4.$$

Since the equation $-4 = -4$ is true, $x = 3$ and $y = -4$ is a solution.

SELF CHECK   Is the ordered pair $(-2, 0)$ a solution of $y = x^2 + 4$?   *Answer:* no   ■

Constructing Tables of Values

To find solutions of equations in x and y, we can pick numbers at random, substitute them for x, and find the corresponding values of y. For example, to find some ordered pairs that satisfy the equation $y = x - 1$, we can let $x = -4$ (called the **input value**), substitute -4 for x, and solve for y (called the **output value**).

$y = x - 1$

x	y	(x, y)
-4	-5	$(-4, -5)$

$$y = x - 1 \quad \text{The original equation.}$$
$$y = -4 - 1 \quad \text{Substitute the input } -4 \text{ for } x.$$
$$y = -5 \quad \text{The output is } -5.$$

The ordered pair $(-4, -5)$ is a solution. We list this ordered pair in the **table of values** (or **table of solutions**), shown on the previous page.

To find another ordered pair that satisfies $y = x - 1$, we let $x = -2$.

$y = x - 1$ The original equation.
$y = -2 - 1$ Substitute the input -2 for x.
$y = -3$ The output is -3.

A second solution is $(-2, -3)$, and we list it in the table of values.

If we let $x = 0$, we can find a third ordered pair that satisfies $y = x - 1$.

$y = x - 1$ The original equation.
$y = 0 - 1$ Substitute the input 0 for x.
$y = -1$ The output is -1.

A third solution is $(0, -1)$, which we also add to our table of values.

If we let $x = 2$, we can find a fourth solution.

$y = x - 1$ The original equation.
$y = 2 - 1$ Substitute the input 2 for x.
$y = 1$ The output is 1.

A fourth solution is $(2, 1)$, and we add it to our table of values.

If we let $x = 4$, we have

$y = x - 1$ The original equation.
$y = 4 - 1$ Substitute the input 4 for x.
$y = 3$ The output is 3.

A fifth solution is $(4, 3)$.

Since we can choose any real number for x, and since any choice of x will give a corresponding value of y, it is apparent that the equation $y = x - 1$ has *infinitely many solutions*. We have found five of them: $(-4, -5)$, $(-2, -3)$, $(0, -1)$, $(2, 1)$, and $(4, 3)$.

$y = x - 1$

x	y	(x, y)
-4	-5	$(-4, -5)$
-2	-3	$(-2, -3)$

$y = x - 1$

x	y	(x, y)
-4	-5	$(-4, -5)$
-2	-3	$(-2, -3)$
0	-1	$(0, -1)$

$y = x - 1$

x	y	(x, y)
-4	-5	$(-4, -5)$
-2	-3	$(-2, -3)$
0	-1	$(0, -1)$
2	1	$(2, 1)$

$y = x - 1$

x	y	(x, y)
-4	-5	$(-4, -5)$
-2	-3	$(-2, -3)$
0	-1	$(0, -1)$
2	1	$(2, 1)$
4	3	$(4, 3)$

ACCENT ON TECHNOLOGY *Generating a Table of Values with a Graphing Calculator*

Constructing a table of values can be a tedious job. Instead of doing the work by hand (as we did above), we can use a graphing calculator to quickly generate solutions of an equation in two variables. Several brands of graphing calculators are available, and each has its own sequence of keystrokes to make a table of solutions. The instructions in this discussion are for a TI-83 graphing calculator, shown in Figure 3-11. For specific details about your calculator, please consult your owner's manual.

To construct a table of solutions for $y = x - 1$, we begin by entering the x-values that are to appear in the table. To do this, we press the keys 2nd TBLSET and enter the first value for x on the line labeled TblStart =.

FIGURE 3-11

Courtesy of Texas Instruments

(continued)

In Figure 3-12(a), the number -4 has been entered on that line. The other values for x that are to appear in the table are determined by entering an **increment** value on the line labeled $\Delta\text{Tbl} =$. Figure 3-12(a) shows that an increment of 2 was entered. This means that each x-value in the table will be 2 larger than the previous x-value.

To enter the equation $y = x - 1$, we press $\boxed{y =}$ and enter $x - 1$, as shown in Figure 3-12(b). (Ignore the subscript 1 on y_1; it is not relevant at this time.)

The final step is to the press the keys $\boxed{\text{2nd}}$ $\boxed{\text{TABLE}}$. This will display a table of solutions, as shown in Figure 3-12(c). The table contains all of the entries that we obtained by hand, as well as two additional solutions: $(6, 5)$ and $(8, 7)$.

FIGURE 3-12

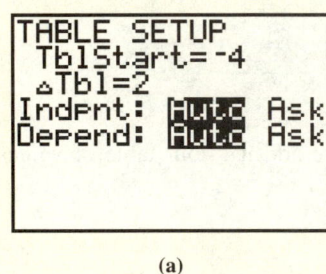
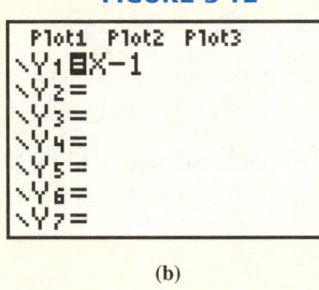
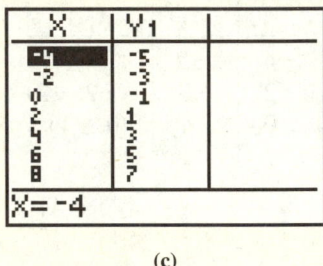

(a)   (b)   (c)

Graphing Equations

To graph the equation $y = x - 1$, we plot the ordered pairs listed in the table on a rectangular coordinate system, as shown in Figure 3-13(a). From the figure, we can see that the five points lie on a line.

In Figure 3-13(b), we draw a line through the points, because the graph of any solution of $y = x - 1$ will lie on this line. The arrowheads show that the line continues forever in both directions. The line is a picture of all the solutions of the equation $y = x - 1$. This line is called the **graph** of the equation.

FIGURE 3-13

$y = x - 1$

x	y	(x, y)
-4	-5	$(-4, -5)$
-2	-3	$(-2, -3)$
0	-1	$(0, -1)$
2	1	$(2, 1)$
4	3	$(4, 3)$

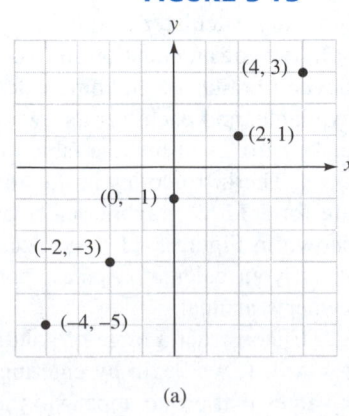
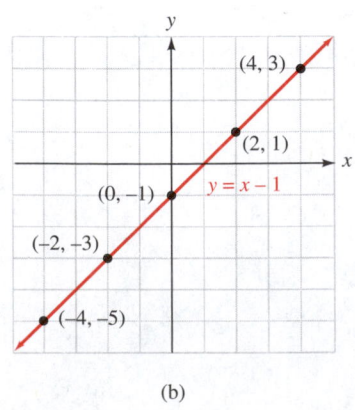

(a)   (b)

To graph an equation in x and y, we follow these steps.

Graphing an equation in x and y

1. Make a table of values containing several ordered pairs of numbers (x, y) that satisfy the equation. Do this by picking values for x and finding the corresponding values for y.
2. Plot each ordered pair on a rectangular coordinate system.
3. Carefully draw a line or smooth curve through the points.

Since we will usually choose a number for x and then find the corresponding value of y, the value of y depends on x. For this reason, we call y the **dependent variable** and x the **independent variable.** The value of the independent variable is the input value, and the value of the dependent variable is the output value.

EXAMPLE 3

Graphing equations. Graph $y = x^2$.

Solution To make a table of values, we will choose numbers for x and find the corresponding values of y. If $x = -3$, we have

$$y = x^2 \qquad \text{The original equation.}$$
$$y = (-3)^2 \quad \text{Substitute the input } -3 \text{ for } x.$$
$$y = 9 \qquad \text{The output is 9.}$$

Thus, $x = -3$ and $y = 9$ is a solution. In a similar manner, we find the corresponding y-values for x-values of -2, -1, 0, 1, 2, and 3. If we plot the ordered pairs listed in the table in Figure 3-14 and join the points with a smooth curve, we get the graph shown in the figure, which is called a **parabola.**

FIGURE 3-14

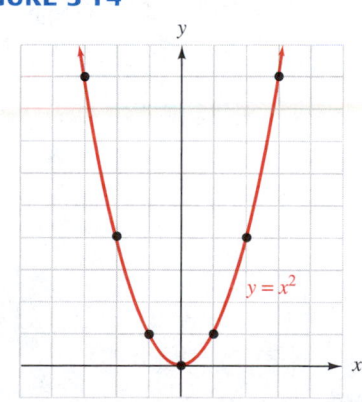

$y = x^2$

x	y	(x, y)
-3	9	$(-3, 9)$
-2	4	$(-2, 4)$
-1	1	$(-1, 1)$
0	0	$(0, 0)$
1	1	$(1, 1)$
2	4	$(2, 4)$
3	9	$(3, 9)$

SELF CHECK Graph $y = x^2 - 2$ and compare the result to the graph of $y = x^2$. What do you notice?

Answer: The graph has the same shape, but is 2 units lower.

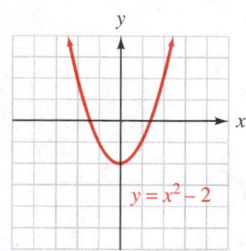

EXAMPLE 4

Graphing equations. Graph $y = |x|$.

Solution To make a table of values, we will choose numbers for x and find the corresponding values of y. If $x = -5$, we have

$y = |x|$ The original equation.

$y = |-5|$ Substitute the input -5 for x.

$y = 5$ The output is 5.

The ordered pair $(-5, 5)$ satisfies the equation. This pair and several others that satisfy the equation are listed in the table of values in Figure 3-15. If we plot the ordered pairs in the table, we see that they lie in a "V" shape. We join the points to complete the graph shown in the figure.

FIGURE 3-15

$y = |x|$

x	y	(x, y)
-5	5	$(-5, 5)$
-4	4	$(-4, 4)$
-3	3	$(-3, 3)$
-2	2	$(-2, 2)$
-1	1	$(-1, 1)$
0	0	$(0, 0)$
1	1	$(1, 1)$
2	2	$(2, 2)$
3	3	$(3, 3)$

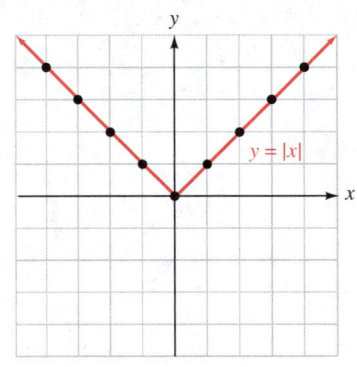

SELF CHECK

Graph $y = |x| + 2$ and compare the result to the graph of $y = |x|$. What do you notice?

Answer: The graph has the same shape, but is 2 units higher.

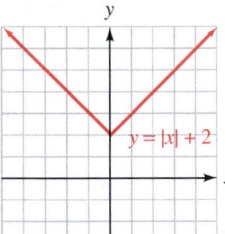

EXAMPLE 5

Graphing equations. Graph $y = x^3$.

Solution If we let $x = -2$, we have

$y = x^3$ The original equation.

$y = (-2)^3$ Substitute the input -2 for x.

$y = -8$ The output is -8.

The ordered pair $(-2, -8)$ satisfies the equation. This ordered pair and several others that satisfy the equation are listed in the table of values in Figure 3-16. Plotting the ordered pairs and joining them with a smooth curve gives us the graph shown in the figure.

FIGURE 3-16

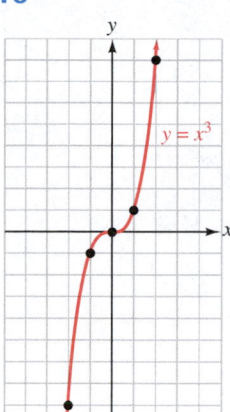

$y = x^3$

x	y	(x, y)
-2	-8	$(-2, -8)$
-1	-1	$(-1, -1)$
0	0	$(0, 0)$
1	1	$(1, 1)$
2	8	$(2, 8)$

SELF CHECK   Graph $y = (x - 2)^3$ and compare the result to the graph of $y = x^3$. What do you notice?

Answer: The graph has the same shape but is 2 units to the right.

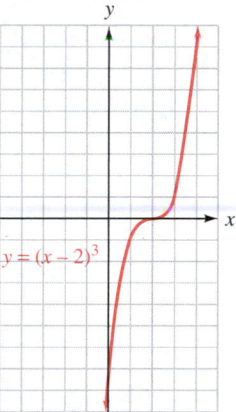

ACCENT ON TECHNOLOGY   *Using a Graphing Calculator to Graph an Equation*

So far, we have graphed equations by making tables of values and plotting points. The task of graphing is made much easier when we use a graphing calculator. The instructions in this discussion will be general in nature. For specific details about your calculator, please consult your owner's manual.

The Viewing Window   All graphing calculators have a viewing **window,** used to display graphs. The **standard window** has settings of

$$\text{Xmin} = -10, \quad \text{Xmax} = 10, \quad \text{Ymin} = -10, \quad \text{and} \quad \text{Ymax} = 10$$

which indicate that the minimum x- and y-coordinates used in the graph will be -10, and that the maximum x- and y-coordinates will be 10.

Graphing an Equation   To graph the equation $y = x - 1$ using a graphing calculator, we press the $\boxed{Y=}$ key and enter the right-hand side of the equation after the symbol Y_1. The display will show the equation

$$Y_1 = x - 1$$

Then we press the $\boxed{\text{GRAPH}}$ key to produce the graph shown in Figure 3-17.

(continued)

Next, we will graph the equation $y = |x - 4|$. Since absolute values are always nonnegative, the minimum y-value is zero. To obtain a reasonable viewing window, we set the Ymin value slightly lower, at Ymin = -3. We set Ymax to be 10 units greater than Ymin, at Ymax = 7. The minimum value of y occurs when $x = 4$. To center the graph in the viewing window, we set the Xmin and Xmax values 5 units to the left and right of 4. Therefore, Xmin = -1 and Xmax = 9.

After entering the right-hand side of the equation, we obtain the graph shown in Figure 3-18. Consult your owner's manual to learn how to enter an absolute value.

FIGURE 3-17   **FIGURE 3-18**

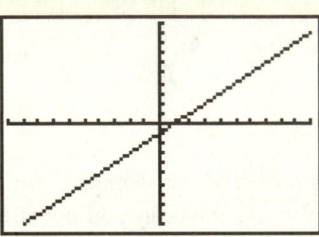   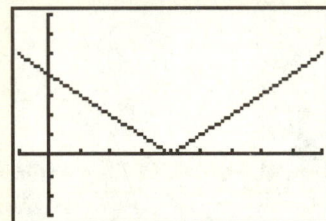

Changing the Viewing Window

The choice of viewing windows is extremely important when graphing equations. To show this, let's graph $y = x^2 - 25$ with x-values from -1 to 6 and y-values from -5 to 5.

To graph this equation, we set the x and y window values and enter the right-hand side of the equation. The display will show

$$Y_1 = x^2 - 25$$

Then we press the $\boxed{\text{GRAPH}}$ key to produce the graph shown in Figure 3-19(a). Although the graph appears to be a straight line, it is not. Actually, we are only seeing part of a parabola. If we pick a viewing window with x-values of -6 to 6 and y-values of -30 to 2, as in Figure 3-19(b), we can see that the graph is a parabola.

FIGURE 3-19

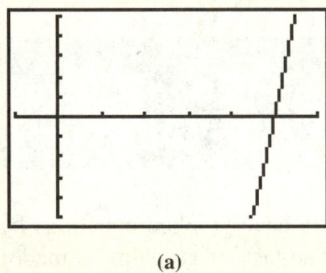   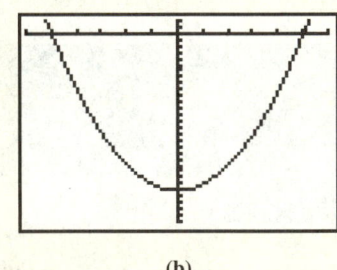

(a)   (b)

Using Different Variables

We will often encounter equations with variables other than x and y. When we make tables of values and graph these equations, we must know which is the independent variable (the input values) and which is the dependent variable (the output values). The independent variable is usually associated with the horizontal axis of the coordinate system, and the dependent variable is usually associated with the vertical axis.

EXAMPLE 6

Advertising. The profit a certain company can make depends on the amount of money it spends on advertising. This relationship is described by the equation $P = 6m - m^2$, where P represents the profit (in millions of dollars) and m represents the amount spent on advertising (in tens of thousands of dollars). Graph the equation and interpret the results.

Solution   Since P depends on m in the equation $P = 6m - m^2$, m is the independent variable (the input) and P is the dependent variable (the output). Therefore, we will choose values for m and find the corresponding values of P. Since m represents the amount of money spent on advertising, we will not choose negative values for m.

If $m = 0$, we have

$$P = 6\boldsymbol{m} - \boldsymbol{m}^2$$

$P = 6(\mathbf{0}) - (\mathbf{0})^2$ Substitute the input 0 for x.

$\quad = 6(0) - 0$ Find the power: $(0)^2 = 0$.

$\quad = 0 - 0$ Do the multiplication: $6(0) = 0$.

$\quad = 0$ The output is 0.

The pair $m = 0$ and $P = 0$, or $(0, 0)$, is a solution. This ordered pair and others that satisfy the equation are listed in the table of values shown in Figure 3-20. If we plot the ordered pairs as in Figure 3-20(a) and join them with a smooth curve, we will obtain the graph shown in Figure 3-20(b). Here the graph is a parabola opening downward.

From the graph, we see that the profit increases to a point and then decreases. Since the graph peaks at the point $(3, 9)$, we know that \$30,000 spent on advertising will generate a maximum profit of \$9 million.

FIGURE 3-20

Plot the profit on this axis in millions of dollars.

The profit is a maximum at this "peak" in the graph, which occurs at $(3, 9)$.

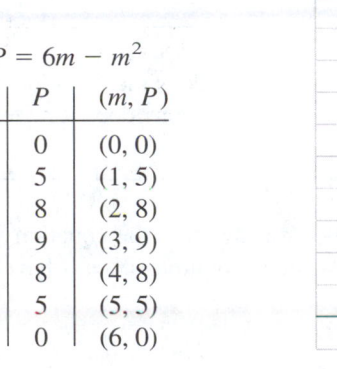

$P = 6m - m^2$

m	P	(m, P)
0	0	$(0, 0)$
1	5	$(1, 5)$
2	8	$(2, 8)$
3	9	$(3, 9)$
4	8	$(4, 8)$
5	5	$(5, 5)$
6	0	$(6, 0)$

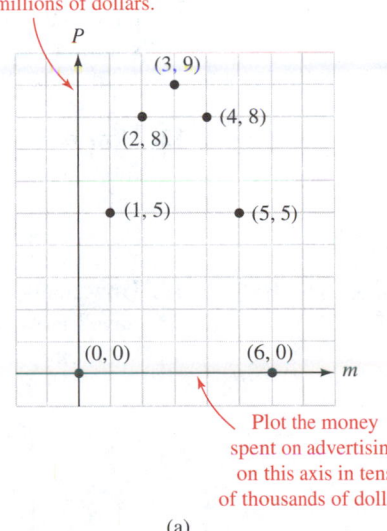

Plot the money spent on advertising on this axis in tens of thousands of dollars

(a)

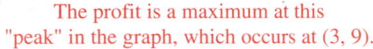
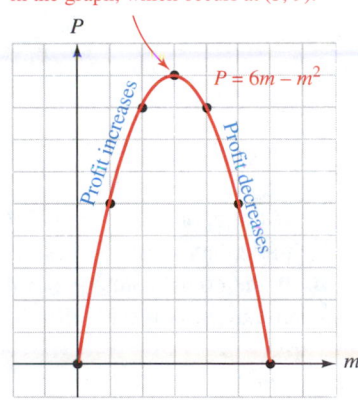

(b)

STUDY SET

Section 3.2

VOCABULARY

In Exercises 1–4, fill in the blanks to make the statements true.

1. The equation $y = x + 1$ is an equation in ____ variables.

2. An ordered pair is a _____ of an equation if the numbers in the ordered pair satisfy the equation.

3. In equations containing the variables x and y, x is called the _____ variable and y is called the _____ variable.

4. When constructing a _____ of values, the values of x are the _____ values and the values of y are the _____ values.

CONCEPTS

5. Consider the equation $y = -2x + 6$.
  a. How many variables does the equation contain?
  b. Does the ordered pair $(4, -2)$ satisfy the equation?

  c. Is $x = -3$ and $y = 12$ a solution?
  d. How many solutions does this equation have?

6. Consider the equation $8 = -2x + 6$.
  a. How many variables does the equation contain?
  b. Is $x = -1$ a solution of the equation?
  c. How many solutions does this equation have?

7. Consider the graph of an equation shown in Illustration 1.

ILLUSTRATION 1

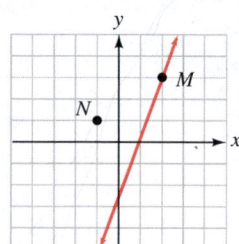

  a. If the coordinates of point M are substituted into the equation, is the result a true statement?
  b. If the coordinates of point N are substituted into the equation, is the result a true statement?

8. Complete each table of values.
  a. $y = x^3$

x (inputs)	y (outputs)
0	
-1	
-2	
1	
2	

  b. $y = x^4$

x (inputs)	y (outputs)
0	
-1	
-2	
1	
2	

9. To graph $y = x^2 - 4$, a table of values is constructed and a graph is drawn, as shown in Illustration 2. Explain the error made here.

ILLUSTRATION 2

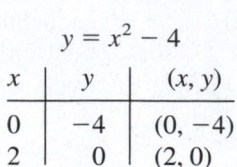

$y = x^2 - 4$

x	y	(x, y)
0	-4	$(0, -4)$
2	0	$(2, 0)$

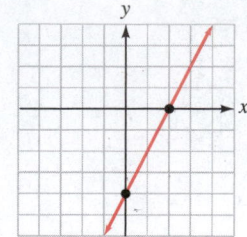

10. Several solutions of an equation are listed in the table of values. When graphing them, how should the horizontal and vertical axes of the graph be labeled?

t	s	(t, s)
0	4	$(0, 4)$
1	5	$(1, 5)$
2	10	$(2, 10)$

11. How many variables does the equation $x + 2 = 6$ have? How many solutions does it have? Graph the solution(s).

$$\xleftarrow{\qquad} \begin{array}{cccccccc} | & | & | & | & | & | & | & | \\ -2 & -1 & 0 & 1 & 2 & 3 & 4 & 5 \end{array} \xrightarrow{\qquad}$$

12. How many variables does the equation $x + 3 = 1$ have? How many solutions does it have? Graph the solution(s).

$$\xleftarrow{\qquad} \begin{array}{cccccccc} | & | & | & | & | & | & | & | \\ -5 & -4 & -3 & -2 & -1 & 0 & 1 & 2 \end{array} \xrightarrow{\qquad}$$

13. How many variables does the equation $y = x^2 + 2$ have? How many solutions does it have? Graph the solution(s).

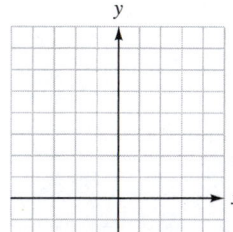

14. How many variables does the equation $y = |x + 3|$ have? How many solutions does it have? Graph the solution(s).

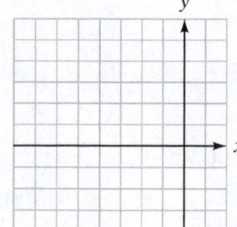

NOTATION

In Exercises 15–16, complete each solution.

15. Verify that $(-2, 6)$ satisfies $y = -x + 4$.

$$y = -x + 4$$
$$\boxed{} \overset{?}{=} -\left(\boxed{}\right) + 4$$
$$6 \overset{?}{=} \boxed{} + 4$$
$$6 = 6$$

16. For the equation $y = |x - 2|$, if $x = -3$, find y.

$$y = |x - 2|$$
$$y = |\boxed{} - 2|$$
$$y = |\boxed{}|$$
$$y = 5$$

PRACTICE

In Exercises 17–20, tell whether the ordered pair satisfies the equation.

17. $x - 2y = -4$; $(4, 4)$

18. $y = 2x - x^2$; $(8, 48)$

19. $y = |5 - 2x|$; $(4, -3)$

20. $y = 3 - x^3$; $(-2, 11)$

In Exercises 21–24, complete each table of values.

21. $y = x - 3$

x	y
0	
1	
-2	

22. $y = |x - 3|$

x	$|x - 3|$
0	
-1	
3	

23. $y = x^2 - 3$

Input	Output
0	
2	
-2	

24. $y = x + 1$

Input	Output
0	
2	
-1	

In Exercises 25–28, construct a table of values, then graph each equation.

25. $y = 2x - 3$

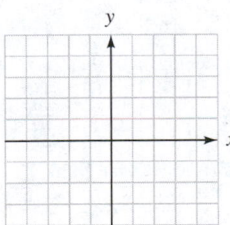

26. $y = 3x + 1$

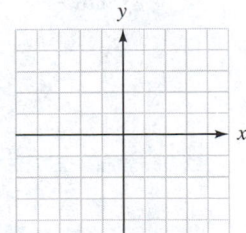

27. $y = -2x + 1$

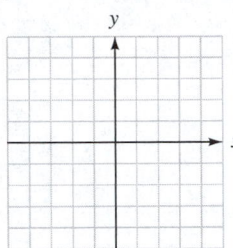

28. $y = -3x + 2$

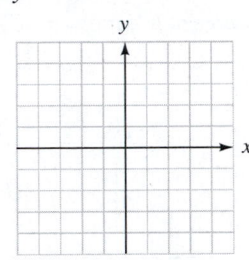

In Exercises 29–32, construct a table of values, then graph each equation. Compare it to the graph of $y = x^2$.

29. $y = x^2 + 1$

30. $y = -x^2$

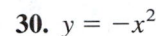
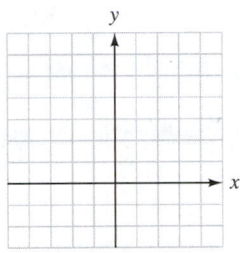
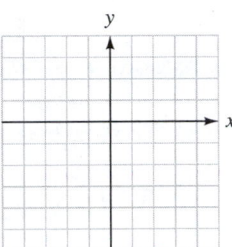

31. $y = (x - 2)^2$

32. $y = (x + 2)^2$

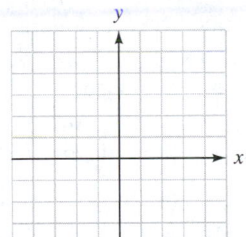
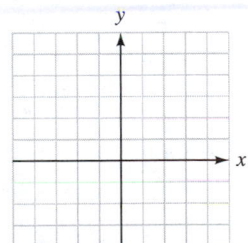

In Exercises 33–36, construct a table of values, then graph each equation. Compare it to the graph of $y = |x|$.

33. $y = -|x|$

34. $y = |x| - 2$

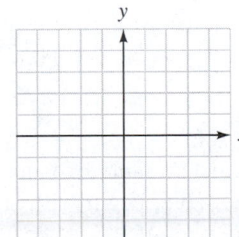
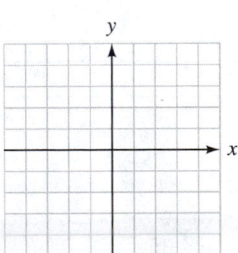

35. $y = |x + 2|$

36. $y = |x - 2|$

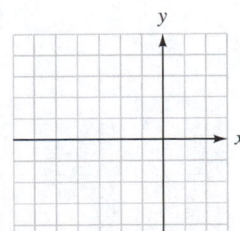
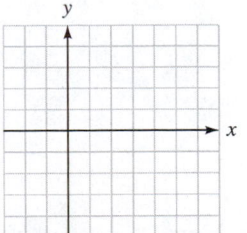

In Exercises 37–40, construct a table of values, then graph each equation. Compare it to the graph of $y = x^3$.

37. $y = -x^3$

38. $y = x^3 + 2$

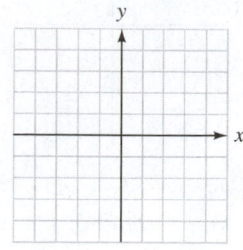

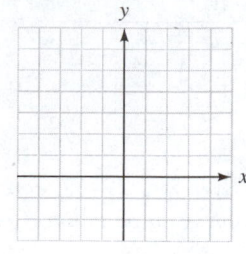

39. $y = x^3 - 2$

40. $y = (x + 2)^3$

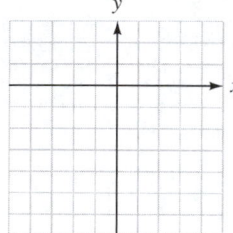

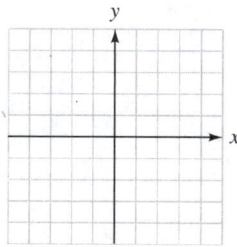

41. The table of values for $y = 0.5x - 1.75$ shown in Illustration 3 was generated by a graphing calculator.

  a. What increment was used to generate the x-values?

  b. Graph the equation by plotting the ordered pairs represented in the table.

ILLUSTRATION 3

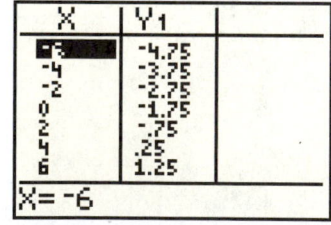

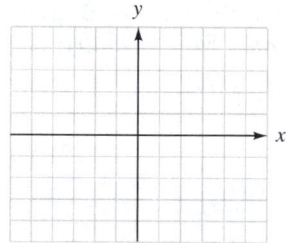

42. Use a graphing calculator to create a table of values for $y = -0.25x^2 + 0.75$. In the table, begin with an x-value of -2 and use an increment of 0.5. Then graph the equation by plotting the ordered pairs represented in the table.

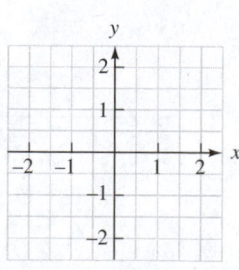

In Exercises 43–46, use a graphing calculator to graph each equation. Use a viewing window of $x = -5$ to 5 and $y = -5$ to 5.

43. $y = 2.1x - 1.1$

44. $y = 1.12x^2 - 1$

45. $y = |x + 0.7|$

46. $y = 0.1x^3 + 1$

In Exercises 47–50, graph each equation in a viewing window of $x = -4$ to 4 and $y = -4$ to 4. This graph is not what it appears to be. Pick a better viewing window and find a better representation of the true graph.

47. $y = -x^3 - 8.2$

48. $y = -|x - 4.01|$

49. $y = x^2 + 5.9$

50. $y = -x + 7.95$

APPLICATIONS

51. SUSPENSION BRIDGES The suspension cables of a bridge hang in the shape of a parabola, as shown in Illustration 4. Use the information in the illustration to complete the table of values.

ILLUSTRATION 4

x	0	2	4	-2	-4
y					

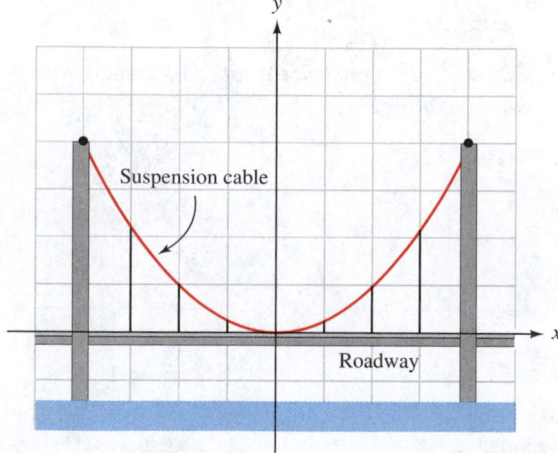

52. FIRE BOAT A stream of water from a high-pressure hose on a fire boat travels in the shape of a parabola, as shown in Illustration 5. Use the information in the graph to complete the table of values.

ILLUSTRATION 5

x	1	2	3	4
y				

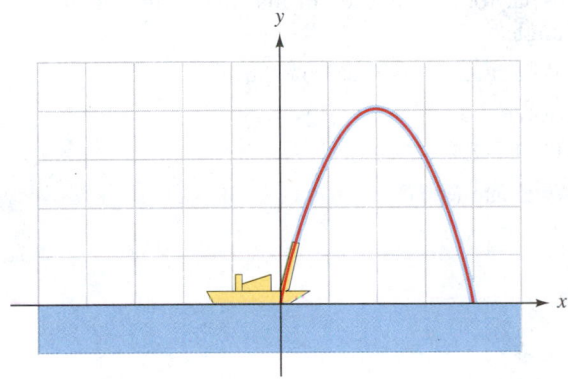

53. MANUFACTURING The graph in Illustration 6 shows the relationship between the length l (in inches) of a machine bolt and the cost C (in cents) to manufacture it.
   a. What information does the point (2, 8) on the graph give us?
   b. How much does it cost to make a 7-inch bolt?
   c. What length bolt is the least expensive to make?

   d. Describe how the cost changes as the length of the bolt increases.

ILLUSTRATION 6

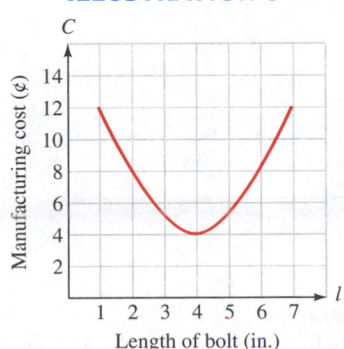

Length of bolt (in.)

54. SOFTBALL The graph in Illustration 7 shows the relationship between the distance d (in feet) traveled by a batted softball and the height h (in feet) it attains.
   a. What information does the point (40, 40) on the graph give us?

   b. At what distance from home plate does the ball reach its maximum height?
   c. Where will the ball land?

ILLUSTRATION 7

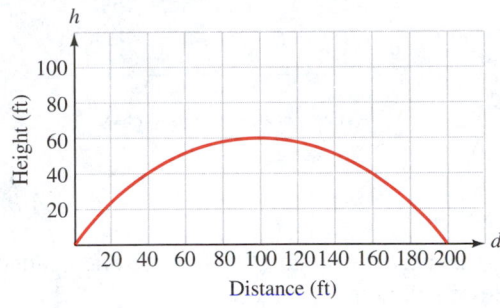

Distance (ft)

55. MARKET VALUE OF A HOUSE The graph in Illustration 8 shows the relationship between the market value v of a house and the time t since it was purchased.

ILLUSTRATION 8

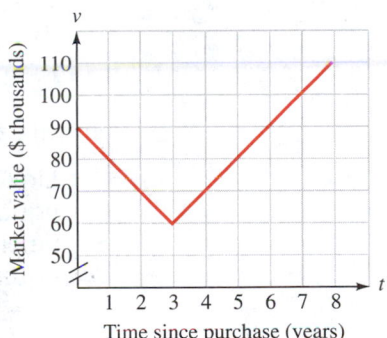

Time since purchase (years)

   a. What was the purchase price of the house?
   b. When did the value of the house reach its lowest point?
   c. When did the value of the house begin to surpass the purchase price?
   d. Describe how the market value of the house changed over the 8-year period.

56. POLITICAL SURVEY The graph in Illustration 9 shows the relationship between the percent P of those surveyed who rated their senator's job performance as satisfactory or better and the time t she had been in office.
   a. When did her job performance rating reach a maximum?
   b. When was her job performance rating at or above the 60% mark?
   c. Describe how her job performance rating changed over the 12-month period.

ILLUSTRATION 9

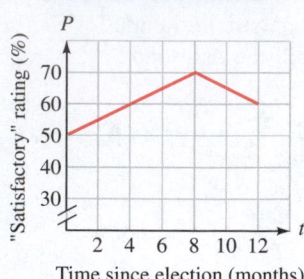

Time since election (months)

WRITING

57. What is a table of values? Why is it often called a table of solutions?

58. To graph an equation in two variables, how many solutions of the equation must be found?

59. Give an example of an equation in one variable and an equation in two variables. How do their solutions differ?

60. What does it mean when we say that an equation in two variables has infinitely many solutions?

REVIEW

61. Solve $\dfrac{x}{8} = -12$.

62. Combine like terms: $3t - 4T + 5T - 6t$.

63. Is $\dfrac{x + 5}{6}$ an expression or an equation?

64. What formula is used to find the perimeter of a rectangle?

65. What number is 0.5% of 250?

66. Solve $-3x + 5 > -7$.

67. Find $-2.5 - (-2.6)$.

68. Evaluate $(-5)^3$.

3.3   Graphing Linear Equations

In this section, you will learn about

> Linear equations ■ Solutions of linear equations ■ Graphing linear equations ■ The intercept method ■ Graphing horizontal and vertical lines ■ An application of linear equations ■ Solving equations graphically

Introduction   In Section 3.2, we graphed the equations shown in Figure 3-21. Because the graph of the equation $y = x - 1$ is a line, we call it a *linear equation*. Since the graphs of $y = x^2$, $y = |x|$, and $y = x^3$ are *not* lines, they are *nonlinear equations*.

FIGURE 3-21

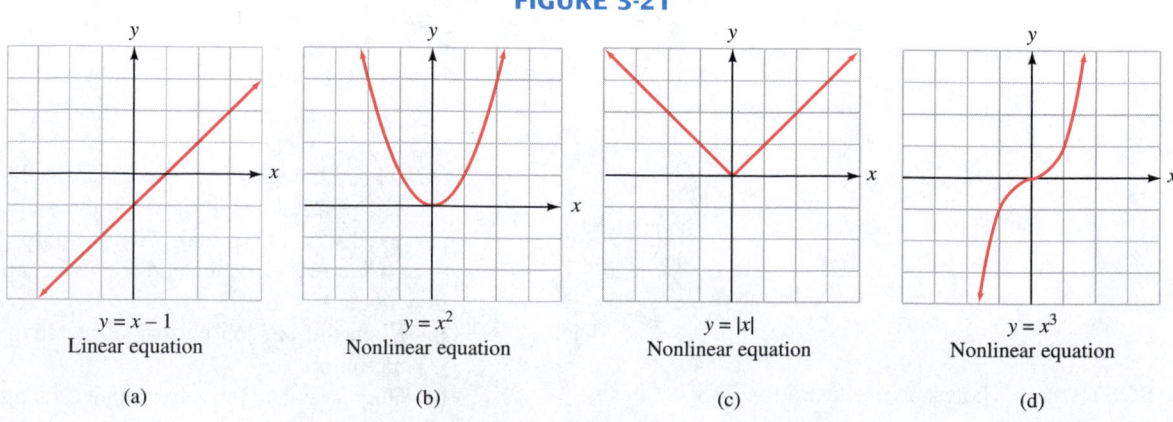

$y = x - 1$	$y = x^2$	$y = |x|$	$y = x^3$
Linear equation	Nonlinear equation	Nonlinear equation	Nonlinear equation
(a)	(b)	(c)	(d)

In this section, we will discuss how to graph linear equations and show how to use their graphs to solve problems.

Linear Equations

Any equation, such as $y = x - 1$, whose graph is a line is called a **linear equation in x and y.** Some other examples of linear equations are

$$y = \frac{1}{2}x + 2, \qquad 2x + 3y = 6, \qquad 5y - x + 2 = 0, \qquad \text{and} \qquad y = -2$$

A linear equation in x and y is any equation that can be written in a special form, called **general** (or **standard**) form.

General form of a linear equation

If A, B, and C are real numbers, the equation

$$Ax + By = C \qquad (A \text{ and } B \text{ are not both zero})$$

is called the **general form** (or **standard form**) of the equation of a line.

Whenever possible, we will write the general form $Ax + By = C$ so that A, B, and C are integers and $A \geq 0$. Note that in a linear equation in x and y, the exponents on x and y are 1.

EXAMPLE 1

Identifying linear equations. Which of the following equations are linear equations?   **a.** $3x = 1 - 2y$ **b.** $y = x^3 + 1$ **c.** $y = -\frac{1}{2}x$

Solution   **a.** Since the equation $3x = 1 - 2y$ can be written in $Ax + By = C$ form, it is a linear equation.

$$
\begin{aligned}
3x &= 1 - 2y & &\text{The original equation.} \\
3x + 2y &= 1 - 2y + 2y & &\text{Add } 2y \text{ to both sides.} \\
3x + 2y &= 1 & &\text{Simplify the right-hand side: } -2y + 2y = 0.
\end{aligned}
$$

Here $A = 3$, $B = 2$, and $C = 1$.

b. Since the exponent on x in $y = x^3 + 1$ is 3, the equation is a nonlinear equation.

c. Since the equation $y = -\frac{1}{2}x$ can be written in $Ax + By = C$ form, it is a linear equation.

$$
\begin{aligned}
y &= -\frac{1}{2}x & &\text{The original equation.} \\
-2(y) &= -2\left(-\frac{1}{2}x\right) & &\text{Multiply both sides by } -2 \text{ so that the coefficient of } x \text{ will be 1.} \\
-2y &= x & &\text{Simplify the right-hand side: } -2\left(-\frac{1}{2}\right) = 1. \\
0 &= x + 2y & &\text{Add } 2y \text{ to both sides.} \\
x + 2y &= 0 & &\text{Write the equation in general form.}
\end{aligned}
$$

Here $A = 1$, $B = 2$, and $C = 0$.

SELF CHECK   Tell which of the following are linear equations and which are nonlinear: **a.** $y = |x|$,   **b.** $-x = 6 - y$,   and   **c.** $y = x$.

Answers: **a.** nonlinear,   **b.** linear,   **c.** linear   ∎

Solutions of Linear Equations

To find solutions of linear equations, we substitute arbitrary values for one variable and solve for the other.

EXAMPLE 2

Finding solutions of linear equations. Complete the table of values for $3x + 2y = 5$.

x	y	(x, y)
7		(7,)
	4	(, 4)

Solution   In the first row, we are given the x-value of 7. To find the corresponding y-value, we substitute 7 for x and solve for y.

$3x + 2y = 5$	The original equation.
$3(7) + 2y = 5$	Substitute 7 for x.
$21 + 2y = 5$	Do the multiplication: $3(7) = 21$.
$2y = -16$	Subtract 21 from both sides: $5 - 21 = -16$.
$y = -8$	Divide both sides by 2.

A solution of $3x + 2y = 5$ is $(7, -8)$.

In the second row, we are given a y-value of 4. To find the corresponding x-value, we substitute 4 for y and solve for x.

$3x + 2y = 5$	The original equation.
$3x + 2(4) = 5$	Substitute 4 for y.
$3x + 8 = 5$	Do the multiplication: $2(4) = 8$.
$3x = -3$	Subtract 8 from both sides: $5 - 8 = -3$.
$x = -1$	Divide both sides by 3.

Another solution is $(-1, 4)$. The completed table is as follows:

x	y	(x, y)
7	-8	$(7, -8)$
-1	4	$(-1, 4)$

SELF CHECK

Complete the table of values for $3x + 2y = 5$:

x	y	(x, y)
	-2	(, -2)
5		(5,)

Answer:

x	y	(x, y)
3	-2	$(3, -2)$
5	-5	$(5, -5)$

■

Graphing Linear Equations

Since two points determine a line, only two points are needed to graph a linear equation. However, we will often plot a third point as a check. If the three points do not lie on a line, then at least one of them is in error.

Graphing linear equations

1. Find three pairs (x, y) that satisfy the equation by picking arbitrary numbers for x and finding the corresponding values of y.

2. Plot each resulting pair (x, y) on a rectangular coordinate system. If the three points do not lie on a line, check your computations.

3. Draw the line passing through the points.

<table>
<tr><td>**EXAMPLE 3**</td><td>**Graphing linear equations.** Graph $y = -3x$.</td></tr>
</table>

Solution   To find three ordered pairs that satisfy the equation, we begin by choosing three x-values: -2, 0, and 2.

If x = −2	*If x = 0*	*If x = 2*
$y = -3x$	$y = -3x$	$y = -3x$
$y = -3(-2)$	$y = -3(0)$	$y = -3(2)$
$y = 6$	$y = 0$	$y = -6$

We enter the results in a table of values, plot the points, and draw a line through the points. The graph appears in Figure 3-22. Check this work with a graphing calculator.

FIGURE 3-22

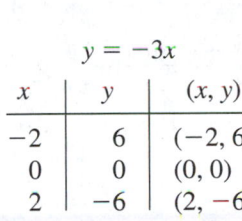

$y = -3x$

x	y	(x, y)
-2	6	$(-2, 6)$
0	0	$(0, 0)$
2	-6	$(2, -6)$

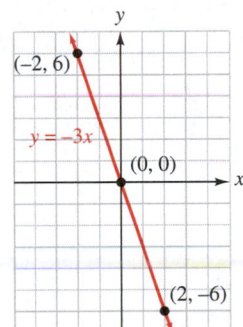

<table>
<tr><td>**SELF CHECK**</td><td>Graph $y = -3x + 2$ and compare the result to the graph of $y = -3x$. What do you notice?</td><td>*Answer:* It is a line 2 units above the graph of $y = -3x$.</td></tr>
</table>

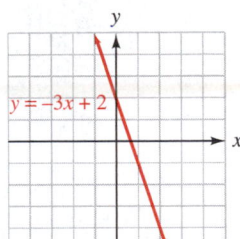

When graphing linear equations, it is often easier to find solutions of the equation if it is first solved for y.

<table>
<tr><td>**EXAMPLE 4**</td><td>**Solving for y.** Graph $2y = 4 - x$.</td></tr>
</table>

Solution   We first solve the equation for y.

$$2y = 4 - x$$

$$\frac{2y}{2} = \frac{4}{2} - \frac{x}{2}$$ To isolate y, divide both sides by 2.

$$y = 2 - \frac{x}{2}$$ Simplify: $\frac{2y}{2} = y$ and $\frac{4}{2} = 2$.

Since each value of x will be divided by 2, we will choose values of x that are divisible by 2: -4, 0, and 4. If $x = -4$, we have

$$y = 2 - \frac{x}{2}$$

$$y = 2 - \frac{-4}{2} \qquad \text{Substitute } -4 \text{ for } x.$$

$$y = 2 - (-2) \quad \text{Divide: } \tfrac{-4}{2} = -2.$$

$$y = 4 \qquad\qquad \text{Simplify.}$$

A solution is $(-4, 4)$. This pair and two others satisfying the equation are shown in the table of values in Figure 3-23. If we plot the points and draw a line through them, we will obtain the graph shown in the figure. Check this work with a graphing calculator.

FIGURE 3-23

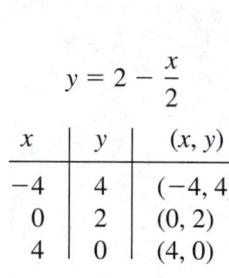

$$y = 2 - \frac{x}{2}$$

x	y	(x, y)
-4	4	$(-4, 4)$
0	2	$(0, 2)$
4	0	$(4, 0)$

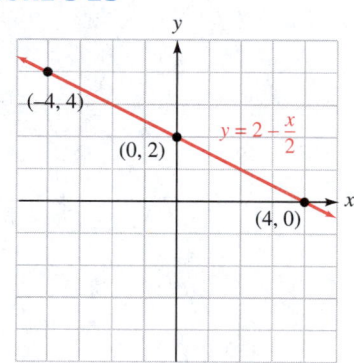

SELF CHECK Solve $3y = 3 + x$ for y, then graph the equation. *Answer:* $y = 1 + \frac{x}{3}$

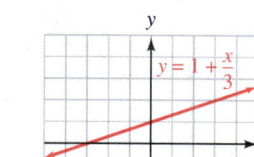

The Intercept Method

In Figure 3-24(a), the graph of $3x + 4y = 12$ intersects the y-axis at the point $(0, 3)$; we call this point the **y-intercept** of the graph. Since the graph intersects the x-axis at

FIGURE 3-24

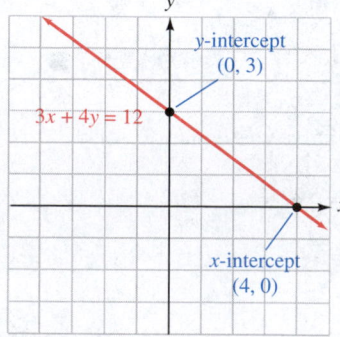

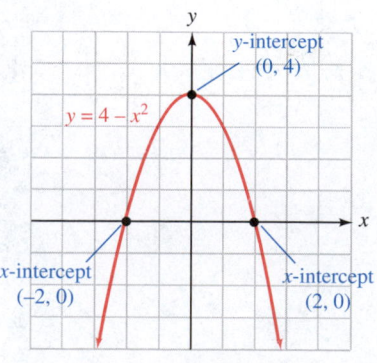

(a)   (b)

(4, 0), the point (4, 0) is the **x-intercept.** In Figure 3-24(b), we see that the graph of $y = 4 - x^2$ has two x-intercepts and one y-intercept.

In general, we have the following definitions.

y- and x-intercepts

The **y-intercept** of a line is the point $(0, b)$ where the line intersects the y-axis. To find b, substitute 0 for x in the equation of the line and solve for y.

The **x-intercept** of a line is the point $(a, 0)$ where the line intersects the x-axis. To find a, substitute 0 for y in the equation of the line and solve for x.

Plotting the x- and y-intercepts of a graph and drawing a line through them is called the **intercept method of graphing a line.** This method is useful when graphing equations written in general form.

EXAMPLE 5

The intercept method. Graph $3x - 2y = 8$.

Solution To find the x-intercept, we let $y = 0$ and solve for x.

$$3x - 2y = 8$$
$$3x - 2(0) = 8 \qquad \text{Substitute 0 for } y.$$
$$3x = 8 \qquad \text{Simplify the left-hand side.}$$
$$x = \frac{8}{3} \qquad \text{Divide both sides by 3.}$$
$$x = 2\frac{2}{3} \qquad \text{Write } \tfrac{8}{3} \text{ as a mixed number.}$$

The x-intercept is $\left(2\frac{2}{3}, 0\right)$. This ordered pair is entered in the table in Figure 3-25. To find the y-intercept, we let $x = 0$ and solve for y.

$$3x - 2y = 8$$
$$3(0) - 2y = 8 \qquad \text{Substitute 0 for } x.$$
$$-2y = 8 \qquad \text{Simplify the left-hand side.}$$
$$y = -4 \qquad \text{Divide both sides by } -2.$$

The y-intercept is $(0, -4)$. It is entered in the table below. As a check, we find one more point on the line. If $x = 4$, then $y = 2$. We plot these three points and draw a line through them. The graph of $3x - 2y = 8$ is shown in Figure 3-25.

FIGURE 3-25

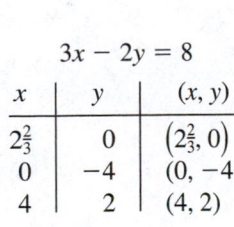

$3x - 2y = 8$

x	y	(x, y)
$2\frac{2}{3}$	0	$\left(2\frac{2}{3}, 0\right)$
0	-4	$(0, -4)$
4	2	$(4, 2)$

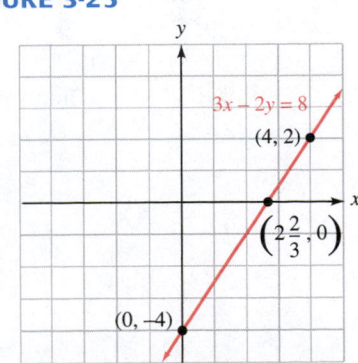

SELF CHECK   Graph $4x + 3y = 6$ using the intercept method.   *Answer:*

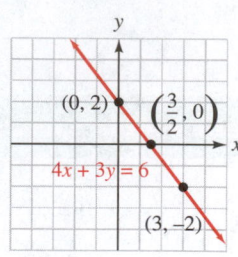

Graphing Horizontal and Vertical Lines

Equations such as $y = 4$ and $x = -3$ are linear equations, because they can be written in the general form $Ax + By = C$.

| $y = 4$ | is equivalent to | $0x + 1y = 4$ |
| $x = -3$ | is equivalent to | $1x + 0y = -3$ |

We now discuss how to graph these types of linear equations.

EXAMPLE 6

Graphing horizontal lines. Graph $y = 4$.

Solution   We can write the equation in general form as $0x + y = 4$. Since the coefficient of x is zero, the numbers chosen for x have no effect on y. The value of y is always 4. For example, if $x = 2$, we have

$0x + y = 4$	The original equation written in general form.
$0(2) + y = 4$	Substitute 2 for x.
$y = 4$	Simplify the left-hand side.

The table of values shown in Figure 3-26 contains three ordered pairs that satisfy the equation $y = 4$. If we plot the points and draw a line through them, the result is a horizontal line. The y-intercept is $(0, 4)$, and there is no x-intercept.

FIGURE 3-26

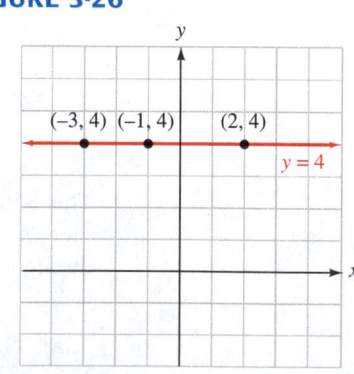

SELF CHECK   Graph $y = -2$.   *Answer:*

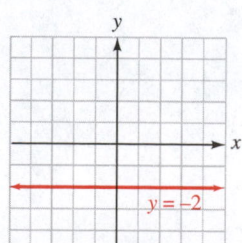

EXAMPLE 7

Graphing vertical lines. Graph $x = -3$.

Solution We can write the equation in general form as $x + 0y = -3$. Since the coefficient of y is zero, the numbers chosen for y have no effect on x. The value of x is always -3. For example, if $y = -2$, we have

$$x + 0y = -3 \quad \text{The original equation written in general form.}$$
$$x + 0(-2) = -3 \quad \text{Substitute } -2 \text{ for } y.$$
$$x = -3 \quad \text{Simplify the left-hand side.}$$

The table of values shown in Figure 3-27 contains three ordered pairs that satisfy the equation $x = -3$. If we plot the points and draw a line through them, the result is a vertical line. The x-intercept is $(-3, 0)$, and there is no y-intercept.

FIGURE 3-27

$x = -3$

x	y	(x, y)
-3	-2	$(-3, -2)$
-3	0	$(-3, 0)$
-3	3	$(-3, 3)$

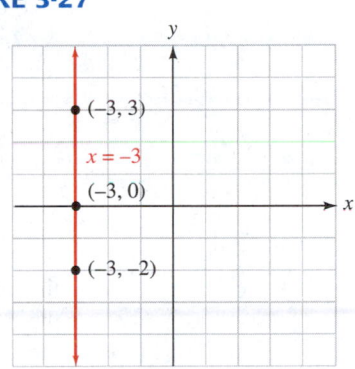

SELF CHECK Graph $x = 4$.

Answer:

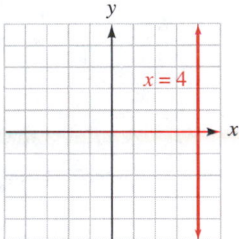

From the results of Examples 6 and 7, we have the following facts.

Equation of horizontal and vertical lines

The equation $y = b$ represents the horizontal line that intersects the y-axis at $(0, b)$. If $b = 0$, the line is the x-axis.

The equation $x = a$ represents the vertical line that intersects the x-axis at $(a, 0)$. If $a = 0$, the line is the y-axis.

An Application of Linear Equations

EXAMPLE 8

Birthday parties. A restaurant offers a party package that includes food, drinks, cake, and party favors for a cost of $25 plus $3 per child. Write a linear equation that will give the cost for a party of any size, and then graph the equation.

Solution   We can let c represent the cost of the party. The cost c is the sum of the basic charge of \$25 and the cost per child times the number of children attending. If the number of children attending is n, at \$3 per child, the total cost for the children is \$3n.

The cost	is	the basic \$25 charge	plus	\$3	times	the number of children.
c	$=$	25	$+$	3	$\cdot$	n

For the equation $c = 25 + 3n$, the independent variable (input) is n, the number of children. The dependent variable (output) is c, the cost of the party. We will find three points on the graph of the equation by choosing n-values of 0, 5, and 10 and finding the corresponding c-values. The results are recorded in the table.

If $n = 0$
$c = 25 + 3(0)$
$c = 25$

If $n = 5$
$c = 25 + 3(5)$
$c = 25 + 15$
$c = 40$

If $n = 10$
$c = 25 + 3(10)$
$c = 25 + 30$
$c = 55$

$c = 25 + 3n$

n	c
0	25
5	40
10	55

Next, we graph the points and draw a line through them (Figure 3-28). Note that the c-axis is scaled in units of \$5 to accommodate costs ranging from \$0 to \$65. We don't draw an arrowhead on the left, because it doesn't make sense to have a negative number of children attend a party. We can use the graph to determine the cost of a party of any size. For example, to find the cost of a party with 8 children, we locate 8 on the horizontal axis and then move up to find a point on the graph directly above the 8. Since the coordinates of that point are (8, 49), the cost for 8 children would be \$49.

FIGURE 3-28

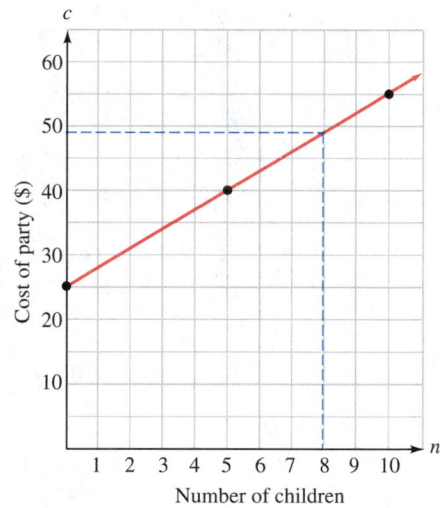

Solving Equations Graphically

Some of the graphing concepts discussed in this chapter can be used to solve equations. For example, the solution of $-2x - 4 = 0$ is the number x that will make y equal to 0 in the equation $y = -2x - 4$. To find this number, we inspect the graph of $y = -2x - 4$ and locate the point on the graph that has a y-coordinate of 0. In Figure 3-29, we see that the point is $(-2, 0)$, which is the x-intercept of the graph. We can conclude that the x-coordinate of the x-intercept, $x = -2$, is the solution of $-2x - 4 = 0$.

FIGURE 3-29

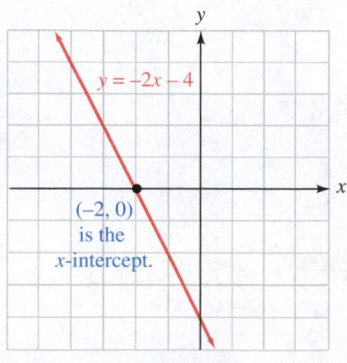

ACCENT ON TECHNOLOGY *Solving Equations Graphically*

To solve the equation $2(x - 3) - 8 = -2x$ using a graphing calculator, we first add $2x$ to both sides so that the right-hand side is 0.

$$2(x - 3) - 8 = -2x$$
$$2(x - 3) - 8 + 2x = -2x + 2x$$
$$2(x - 3) - 8 + 2x = 0$$

Next, we enter the left-hand side of the equation into the calculator in the form $y = 2(x - 3) - 8 + 2x$. We do not need to simplify the left-hand side to graph it. See Figure 3-30(a).

If we use window settings where x is between -5 and 5 and y is between -5 and 5 and press $\boxed{\text{GRAPH}}$, we will obtain the graph in Figure 3-30(b).

To find the x-intercept, we trace as shown in Figure 3-30(c). After repeated zooms and traces, we will see that the x-coordinate of the x-intercept is 3.5, so $x = 3.5$ is the solution of $2(x - 3) - 8 = -2x$.

FIGURE 3-30

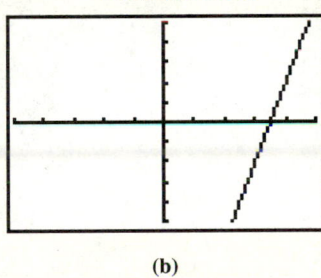

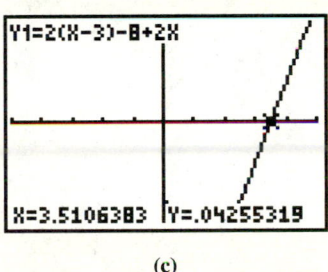

(a)   (b)   (c)

An easier way to find the solution of $2(x - 3) - 8 + 2x = -2x$ is using the zero feature, found under the CALC menu. With this option, the cursor automatically locates the x-intercept of the graph of $y = 2(x - 3) - 8 + 2x$ and displays its coordinates. See Figure 3-30(d). The x-coordinate of the x-intercept of the graph is called a *zero* of $y = 2(x - 3) - 8 + 2x$, and the zero (in this case, 3.5) is the solution of the given equation. Consult your owner's manual for specific instructions on how to use this feature.

FIGURE 3-30(d)

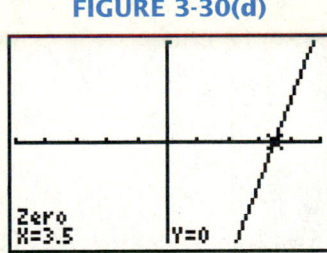

STUDY SET

Section 3.3

VOCABULARY

In Exercises 1–6, fill in the blanks to make the statements true.

1. An equation whose graph is a line and whose variables are to the first power is called a _____ equation.

2. The equation $Ax + By = C$ is the _____ form of the equation of a line.

3. The _____ of a line is the point $(0, b)$ where the line intersects the y-axis.

4. The _____ of a line is the point $(a, 0)$ where the line intersects the x-axis.

5. Lines _____ to the *y*-axis are vertical lines.

6. Lines parallel to the *x*-axis are _____ lines.

CONCEPTS

7. Classify each equation as linear or nonlinear.
   a. $y = x^3$
   b. $2x + 3y = 6$
   c. $y = |x + 2|$
   d. $x = -2$
   e. $y = 2x - x^2$

8. What information can be obtained by finding the *x*- and *y*-intercepts of the graph shown in Illustration 1?

ILLUSTRATION 1

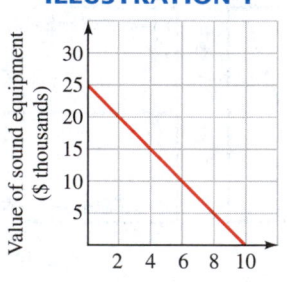

Age of equipment (years)

In Exercises 9–12, complete each table of values.

9. $5y = 2x + 10$

x	y
10	
	0
5	

10. $2x + 4y = 24$

x	y
4	
	7
−4	

11. $x - 2y = 4$

x	y
0	
	0
1	

12. $5x - y = 3$

x	y
0	
	0
1	

In Exercises 13–14, consider the graph of a linear equation shown in Illustration 2.

13. Why will the coordinates of point *A*, when substituted into the equation, yield a true statement?

14. Why will the coordinates of point *B*, when substituted into the equation, yield a false statement?

ILLUSTRATION 2

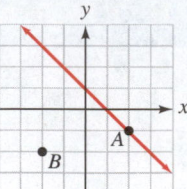

15. To what power is each variable in the equation $y = 2x - 6$?

16. To what power is each variable in the equation $y = x^2 - 6$?

17. Give the *x*- and *y*-intercepts of the graph in Illustration 3.

ILLUSTRATION 3

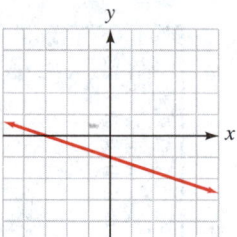

18. Give the *x*- and *y*-intercepts of the graph in Illustration 4.

ILLUSTRATION 4

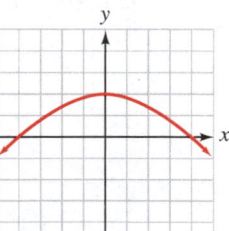

19. What is the general form of an equation of a horizontal line?

20. What is the general form of an equation of a vertical line?

21. A student found three solutions of a linear equation and plotted them as shown in Illustration 5. What conclusion can be made?

ILLUSTRATION 5

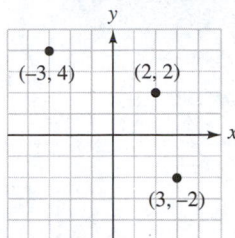

(−3, 4)   (2, 2)

(3, −2)

22. On the same coordinate system:
   a. Draw the graph of a line with no *x*-intercept.
   b. Draw the graph of a line with no *y*-intercept.
   c. Draw a line with an *x*-intercept of (2, 0).
   d. Draw a line with a *y*-intercept of $\left(0, -\frac{5}{2}\right)$.

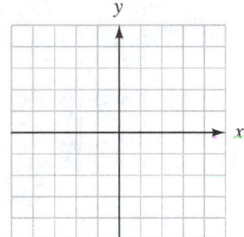

In Exercises 23–24, fill in the blanks to make the statements true.

23. To find the *y*-intercept of the graph of a linear equation, we let ▢ = 0 and solve for ▢.

24. To find the *x*-intercept of the graph of a linear equation, we let ▢ = 0 and solve for ▢.

25. Consider the linear equation $y = 6x$.
   a. Find the *x*-intercept of the graph.
   b. Find the *y*-intercept of the graph.
   c. Explain why your answers to parts a and b are not enough information to graph the line.

26. **a.** What is another name for the line $x = 0$?

   b. What is another name for the line $y = 0$?

27. How can the solution of the equation $-3x - 3 = 0$ be determined from the graphing calculator display of the graph of $y = -3x - 3$? What is the solution?

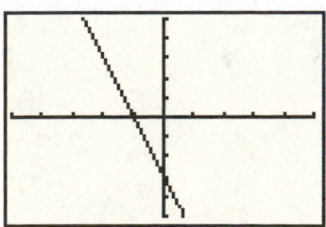

28. What is the solution of the equation $5x - 3(x + 1) = x$ if the calculator display shows the graph of $y = 5x - 3(x + 1) - x$?

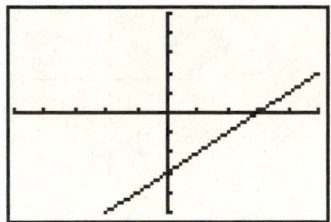

NOTATION

29. Write each equation in general form.
   a. $-4x = -y - 6$
   b. $y = \dfrac{1}{2}x$
   c. $3 = \dfrac{x}{3} + y$
   d. $x = 12$

30. Solve each equation for *y*.
   a. $x + y = 8$
   b. $2x - y = 8$
   c. $3x + \dfrac{y}{2} = 4$
   d. $y - 2 = 0$

PRACTICE

In Exercises 31–34, find three solutions of the equation, then graph it.

31. $y = -x + 2$

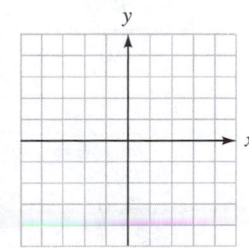

32. $y = -x - 1$

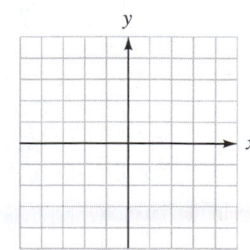

33. $y = 2x + 1$

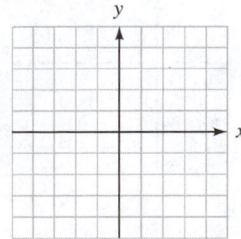

34. $y = 3x - 2$

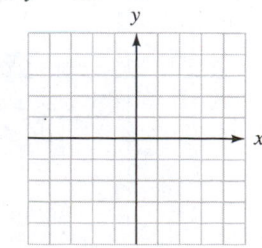

In Exercises 35–38, solve each equation for y, find three solutions of the equation, and then graph it.

35. $2y = 4x - 6$

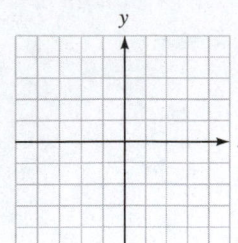

36. $3y = 6x - 3$

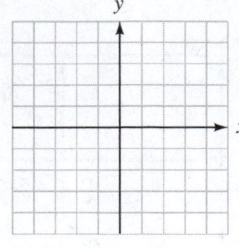

37. $2y = x - 4$

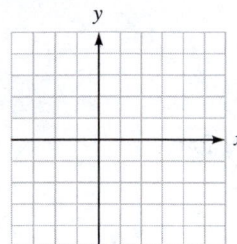

38. $4y = x + 16$

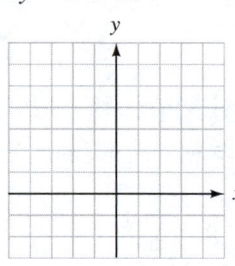

In Exercises 39–46, graph each equation using the intercept method.

39. $2y - 2x = 6$

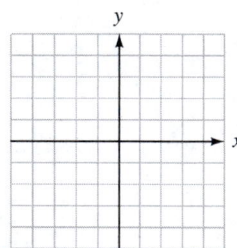

40. $3x - 3y = 9$

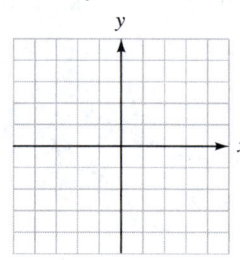

41. $15y + 5x = -15$

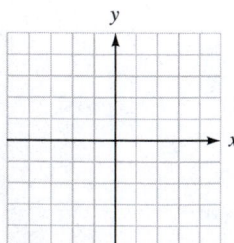

42. $8x + 4y = -24$

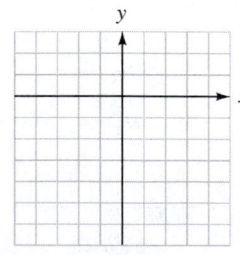

43. $3x + 4y = 8$

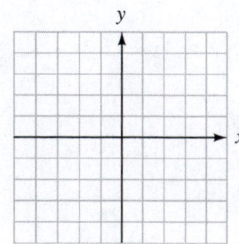

44. $2x + 3y = 9$

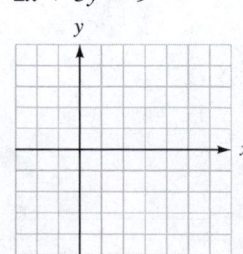

45. $-4y + 9x = -9$

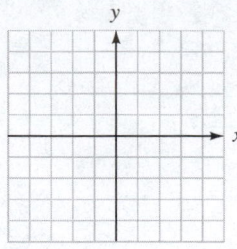

46. $-4y + 5x = -15$

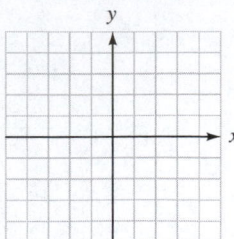

In Exercises 47–58, graph each equation.

47. $y = 4$

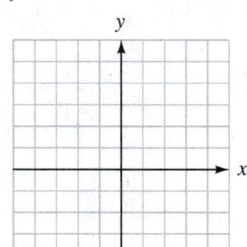

48. $y = -3$

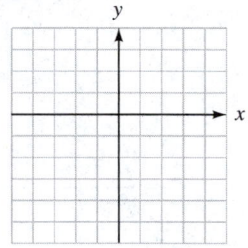

49. $x = -2$

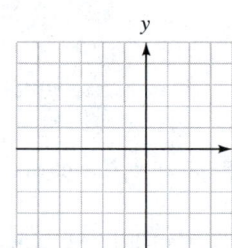

50. $x = 5$

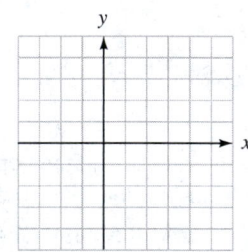

51. $y = -\dfrac{1}{2}$

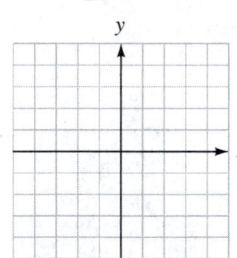

52. $y = \dfrac{5}{2}$

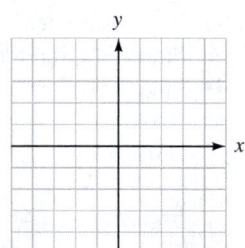

53. $x = \dfrac{4}{3}$

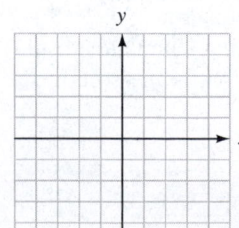

54. $x = -\dfrac{5}{3}$

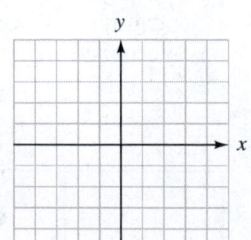

55. $y = 2x$

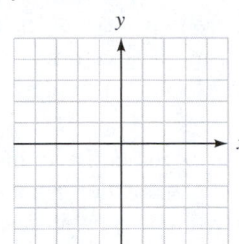

56. $y = 3x$

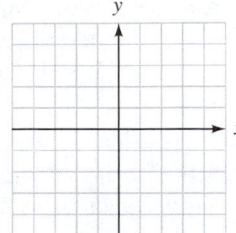

57. $y = -2x$

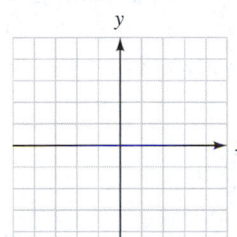

58. $y = -3x$

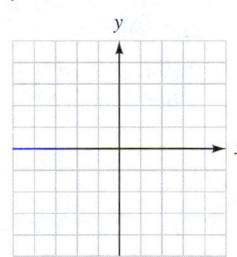

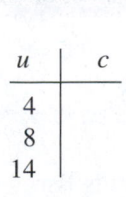 *In Exercises 59–62, solve each equation graphically.*

59. $3(x + 2) - x = 12$

60. $-(3x + 1) + x = 2$

61. $10 - 5x = 3x + 18$

62. $0.5x + 0.2(12 - x) = 0.3(12)$

APPLICATIONS

63. EDUCATION COSTS Each semester, a college charges a services fee of $50 plus $25 for each unit taken by a student.

 a. Write a linear equation that gives the total enrollment cost c for a student taking u units.

 b. Complete the table of values and graph the equation. (See Illustration 6.)

 c. What does the y-intercept of the line tell you?

 d. Use the graph to find the total cost for a student taking 18 units the first semester and 12 units the second semester.

ILLUSTRATION 6

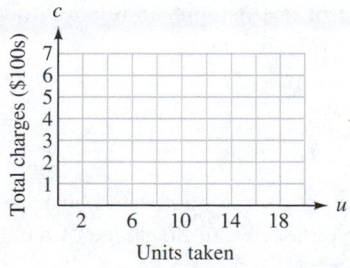

u	c
4	
8	
14	

Units taken

64. GROUP RATES To promote the sale of tickets for a cruise to Alaska, a travel agency reduces the regular ticket price of $3,000 by $5 for each individual traveling in the group.

 a. Write a linear equation that would find the ticket price t for the cruise if a group of p people travel together.

 b. Complete the table of values and graph the equation. (See Illustration 7.)

 c. As the size of the group increases, what happens to the ticket price?

 d. Use the graph to determine the cost of an individual ticket if a group of 25 will be traveling together.

ILLUSTRATION 7

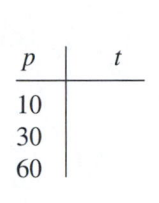

p	t
10	
30	
60	

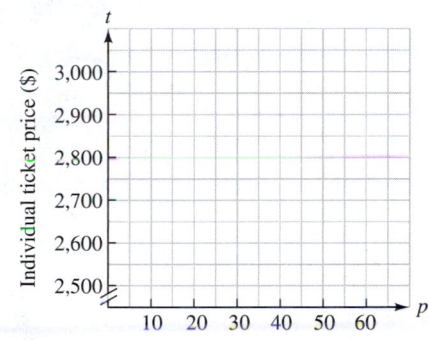

Number of persons in the group

65. PHYSIOLOGY Physiologists have found that a woman's height h in inches can be approximated using the linear equation $h = 3.9r + 28.9$, where r represents the length of her radius bone in inches.

 a. Complete the table of values in Illustration 8. Round to the nearest tenth and then graph the equation.

 b. Complete this sentence: From the graph, we see that the longer the radius bone, the . . . .

 c. From the graph, estimate the height of a woman whose radius bone is 7.5 inches long.

ILLUSTRATION 8

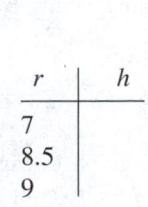

r	h
7	
8.5	
9	

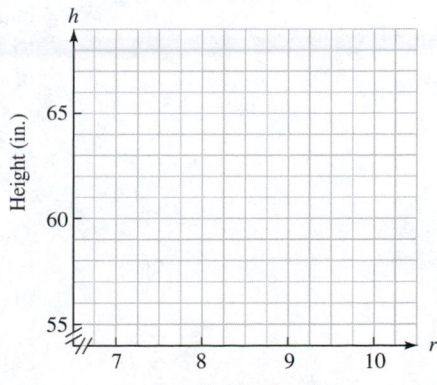

Length of radius (in.)

66. RESEARCH EXPERIMENT A psychology major found that the time t (in seconds) that it took a white rat to complete a maze was related to the number of trials n the rat had been given. The resulting equation was $t = 25 - 0.25n$.

a. Complete the table of values in Illustration 9 and then graph the equation.

b. Complete this sentence: From the graph, we see that the more trials the rat had, the . . . .

c. From the graph, estimate the time it will take the rat to complete the maze on its 32nd trial.

ILLUSTRATION 9

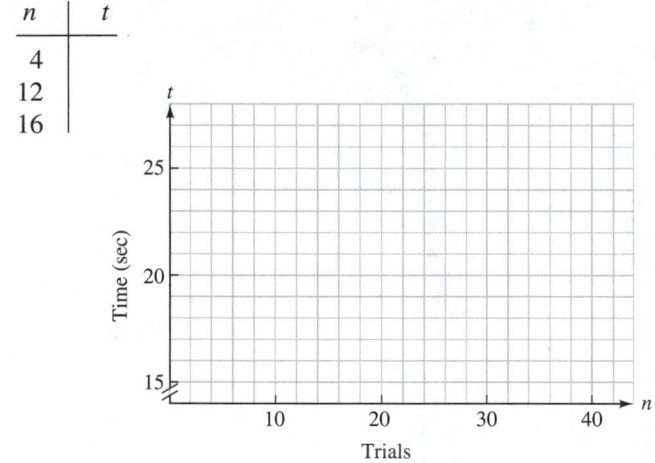

n	t
4	
12	
16	

WRITING

67. A linear equation and a graph are two ways of mathematically describing a relationship between two quantities. Which do you think is more informative and why?

68. From geometry, we know that two points determine a line. Explain why it is a good practice when graphing linear equations to find and plot three points instead of just two.

69. How can we tell by looking at an equation if its graph will be a straight line?

70. Can the x-intercept and the y-intercept of a line be the same point? Explain.

REVIEW

71. Simplify $-(-5 - 4c)$.

72. List the integers.

73. Solve $\dfrac{x + 6}{2} = 1$.

74. Evaluate $-2^2 + 2^2$.

75. Write a formula that relates profit, revenue, and costs.

76. Find the volume, to the nearest tenth, of a sphere with radius 6 feet.

77. Evaluate $1 + 2[-3 - 4(2 - 8^2)]$.

78. Evaluate $\dfrac{x + y}{x - y}$ if $x = -2$ and $y = -4$.

▶ **3.4**   # Rate of Change and the Slope of a Line

In this section, you will learn about

> Rates of change ■ Slope of a line ■ The slope formula ■ Positive and negative slope ■ Slopes of horizontal and vertical lines ■ Using slope to graph a line

Introduction   Since our world is one of constant change, we must be able to describe change so that we can plan effectively for the future. In this section, we will show how to describe the amount of change of one quantity in relation to the amount of change of another quantity by finding a *rate of change*.

Rates of Change

The line graph in Figure 3-31(a) shows the number of business permits issued each month by a city over a 12-month period. From the shape of the graph, we can see that the number of permits issued *increased* each month.

For situations such as the one graphed in Figure 3-31(a), it is often useful to calculate a rate of increase (called a **rate of change**). We do so by finding the **ratio** of the change in the number of business permits issued each month to the number of months over which that change took place.

FIGURE 3-31

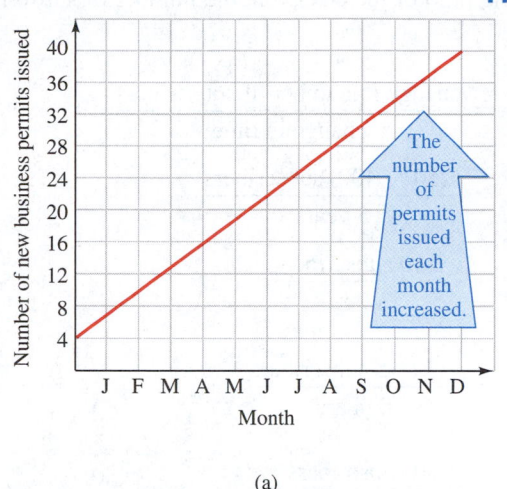

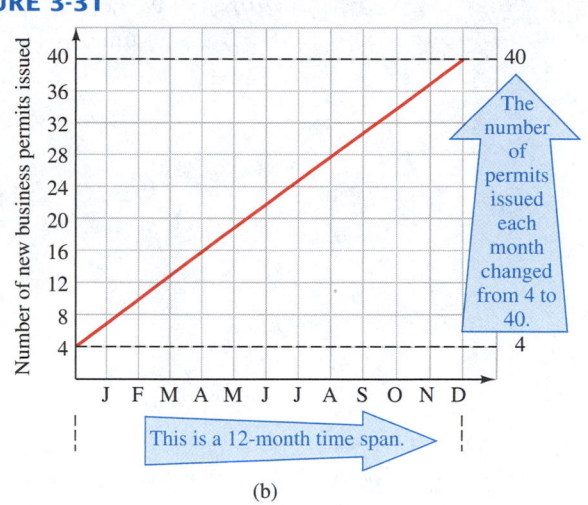

(a)

(b)

Ratios and rates

A **ratio** is a comparison of two numbers by their indicated quotient. In symbols, if a and b are two numbers, the ratio of a to b is $\frac{a}{b}$. Ratios that are used to compare quantities with different units are called **rates.**

In Figure 3-31(b), we see that the number of permits issued prior to the month of January was 4. By the end of the year, the number of permits issued during the month of December was 40. This is a change of $40 - 4$, or 36, over a 12-month period. So we have

$$\text{Rate of change} = \frac{\text{change in number of permits issued each month}}{\text{change in time}} \qquad \text{The rate of change is a ratio.}$$

$$= \frac{36 \text{ permits}}{12 \text{ months}}$$

$$= \frac{\overset{1}{\cancel{12}} \cdot 3 \text{ permits}}{\underset{1}{\cancel{12}} \text{ months}} \qquad \text{Factor 36 as } 12 \cdot 3 \text{ and divide out the common factor of 12.}$$

$$= \frac{3 \text{ permits}}{1 \text{ month}}$$

The number of business permits being issued increased at a rate of 3 per month, denoted as 3 permits/month.

FIGURE 3-32

E X A M P L E   1

Finding rate of change.
The graph in Figure 3-32 shows the number of subscribers to a newspaper. Find the rate of change in the number of subscribers over the first 5-year period. Write the rate in simplest form.

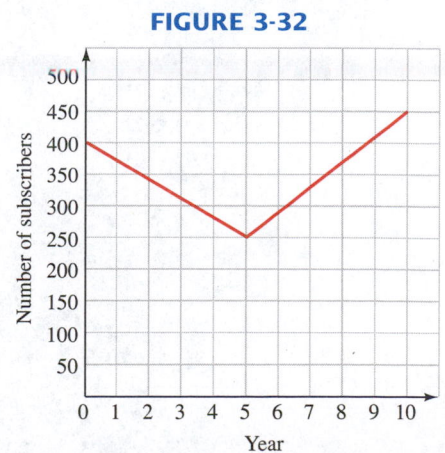

Solution   We need to write the ratio of the change in the number of subscribers over the change in time.

$$\text{Rate of change} = \frac{\text{change in number of subscribers}}{\text{change in time}} \qquad \text{Set up the ratio.}$$

$$= \frac{(250 - 400) \text{ subscribers}}{5 \text{ years}} \qquad \begin{array}{l}\text{Subtract the earlier number of}\\ \text{subscribers from the later}\\ \text{number of subscribers.}\end{array}$$

$$= \frac{-150 \text{ subscribers}}{5 \text{ years}} \qquad 250 - 400 = -150$$

$$= \frac{-30 \cdot \overset{1}{\cancel{5}} \text{ subscribers}}{\underset{1}{\cancel{5}} \text{ years}} \qquad \begin{array}{l}\text{Factor } -150 \text{ as } -30 \cdot 5 \text{ and}\\ \text{divide out the common factor}\\ \text{of } 5.\end{array}$$

$$= \frac{-30 \text{ subscribers}}{1 \text{ year}}$$

The number of subscribers for the first 5 years *decreased* by 30 per year, as indicated by the negative sign in the result. We can write this as -30 subscribers/year.

SELF CHECK   Find the rate of change in the number of subscribers over the second 5-year period. Write the
rate in simplest form.   *Answer:* 40 subscribers/year ■

Slope of a Line

The **slope** of a nonvertical line is a number that measures the line's steepness. We can calculate the slope by picking two points on the line and writing the ratio of the vertical change (called the **rise**) to the corresponding horizontal change (called the **run**) as we move from one point to the other. As an example, we will find the slope of the line that was used to describe the number of building permits issued and show that it gives the rate of change.

In Figure 3-33 (a modified version of Figure 3-31(a)), the line passes through points $P(0, 4)$ and $Q(12, 40)$. Moving along the line from point P to point Q causes the value of y to change from $y = 4$ to $y = 40$, an increase of $40 - 4 = 36$ units. We say that the *rise* is 36.

FIGURE 3-33

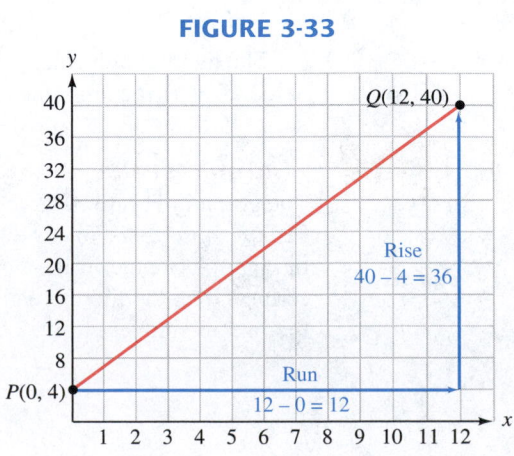

Moving from point P to point Q, the value of x increases from $x = 0$ to $x = 12$, an increase of $12 - 0 = 12$ units. We say that the *run* is 12. The slope of a line, usually denoted with the letter m, is defined to be the ratio of the change in y to the change in x.

$$m = \frac{\text{change in } y\text{-values}}{\text{change in } x\text{-values}}$$ Slope is a ratio.

$$= \frac{40 - 4}{12 - 0}$$ To find the change in y (the rise), subtract the y-values.
To find the change in x (the run), subtract the x-values.

$$= \frac{36}{12}$$ Do the subtractions.

$$= 3$$ Do the division.

This is the same value we obtained when we found the rate of change of the number of business permits issued over the 12-month period. Therefore, by finding the slope of the line, we found a rate of change.

The Slope Formula

The slope of a line can be described in several ways.

$$\text{Slope} = m = \frac{\text{vertical change}}{\text{horizontal change}} = \frac{\text{rise}}{\text{run}} = \frac{\text{change in } y}{\text{change in } x}$$

To distinguish between the coordinates of two points, say points P and Q (see Figure 3-34), we often use **subscript notation.**

■ Point P is denoted as $P(x_1, y_1)$.   Read as "point P with coordinates of x sub 1 and y sub 1."

■ Point Q is denoted as $Q(x_2, y_2)$.   Read as "point Q with coordinates of x sub 2 and y sub 2."

FIGURE 3-34

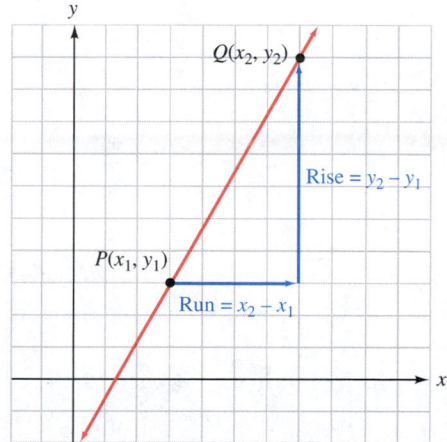

As a point on the line in Figure 3-34 moves from P to Q, its y-coordinate changes by the amount $y_2 - y_1$ (the rise), while its x-coordinate changes by $x_2 - x_1$ (the run). Since the slope is the ratio $\frac{\text{rise}}{\text{run}}$, we have the following formula for calculating slope.

Slope of a nonvertical line

If $P(x_1, y_1)$ and $Q(x_2, y_2)$ are two points on a nonvertical line, the slope m of line PQ is given by the formula

$$m = \frac{y_2 - y_1}{x_2 - x_1}$$

EXAMPLE 2

Using the slope formula. Find the slope of line l_1 shown in Figure 3-35.

FIGURE 3-35

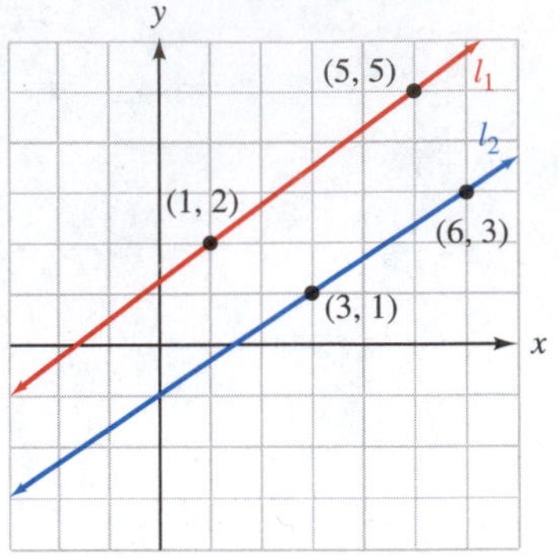

Solution To find the slope of l_1, we will use two points on the line whose coordinates are given: $(1, 2)$ and $(5, 5)$. If (x_1, y_1) is $(1, 2)$ and (x_2, y_2) is $(5, 5)$, then

$$x_1 = 1 \quad \text{and} \quad x_2 = 5$$
$$y_1 = 2 \qquad\qquad y_2 = 5$$

To find the slope of line l_1, we substitute these values into the formula for slope and simplify.

$$m = \frac{y_2 - y_1}{x_2 - x_1} \qquad \text{The slope formula.}$$

$$= \frac{5 - 2}{5 - 1} \qquad \text{Substitute 5 for } y_2\text{, 2 for } y_1\text{, 5 for } x_2\text{, and 1 for } x_1.$$

$$= \frac{3}{4} \qquad \text{Do the subtractions.}$$

The slope of l_1 is $\frac{3}{4}$. We would have obtained the same result if we had let $(x_1, y_1) = (5, 5)$ and $(x_2, y_2) = (1, 2)$.

SELF CHECK Find the slope of line l_2 shown in Figure 3-35. *Answer:* $\frac{2}{3}$ ■

WARNING! When finding the slope of a line, always subtract the y-values and the x-values in the same order. Otherwise your answer will have the wrong sign:

$$m = \frac{y_2 - y_1}{x_2 - x_1} \qquad \text{or} \qquad m = \frac{y_1 - y_2}{x_1 - x_2}$$

However,

$$m \neq \frac{y_2 - y_1}{x_1 - x_2} \qquad \text{and} \qquad m \neq \frac{y_1 - y_2}{x_2 - x_1}$$

EXAMPLE 3

Using the slope formula. Find the slope of the line that passes through $(-2, 4)$ and $(5, -6)$ and draw its graph.

Solution Since we know the coordinates of two points on the line, we can find its slope. If (x_1, y_1) is $(-2, 4)$ and (x_2, y_2) is $(5, -6)$, then

$$x_1 = -2 \quad \text{and} \quad x_2 = 5$$
$$y_1 = 4 \qquad\qquad y_2 = -6$$

FIGURE 3-36

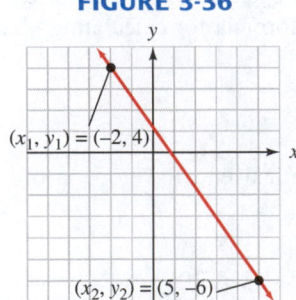

$$m = \frac{y_2 - y_1}{x_2 - x_1} \qquad \text{The slope formula.}$$

$$m = \frac{-6 - 4}{5 - (-2)} \qquad \text{Substitute } -6 \text{ for } y_2\text{, 4 for } y_1\text{, 5 for } x_2\text{, and } -2 \text{ for } x_1.$$

$$m = -\frac{10}{7} \qquad \begin{array}{l}\text{Simplify the numerator: } -6 - 4 = -10.\\ \text{Simplify the denominator: } 5 - (-2) = 7.\end{array}$$

The slope of the line is $-\frac{10}{7}$. Figure 3-36 shows the graph of the line. Note that the line "falls" from left to right—a fact that is indicated by its negative slope.

SELF CHECK   Find the slope of the line that passes through $(-1, -2)$ and $(1, -7)$.   *Answer:* $-\frac{5}{2}$ ■

Positive and Negative Slope

In Example 2, the slope of line l_1 was positive $\left(\frac{3}{4}\right)$. In Example 3, the slope of the line was negative $\left(-\frac{10}{7}\right)$. In general, lines that rise from left to right have a positive slope, and lines that fall from left to right have a negative slope, as shown in Figure 3-37.

FIGURE 3-37

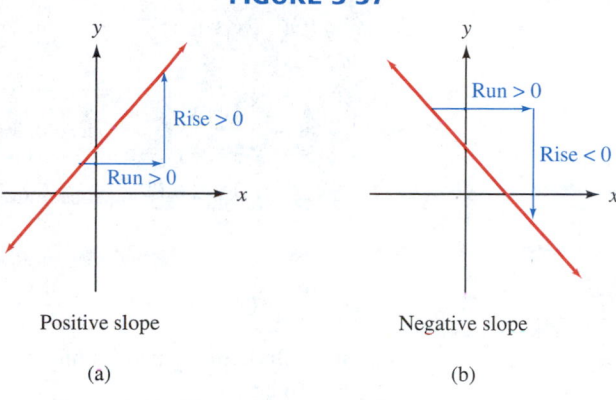

Positive slope

(a)

Negative slope

(b)

Slopes of Horizontal and Vertical Lines

In the next two examples, we will calculate the slope of a horizontal line and show that a vertical line has no defined slope.

EXAMPLE 4   **Slope of a horizontal line.**   Find the slope of the line $y = 3$.

Solution   To find the slope of the line $y = 3$, we need to know two points on the line. In Figure 3-38, we graph the horizontal line $y = 3$ and label two points on the line: $(-2, 3)$ and $(3, 3)$.

If (x_1, y_1) is $(-2, 3)$ and (x_2, y_2) is $(3, 3)$, we have

FIGURE 3-38

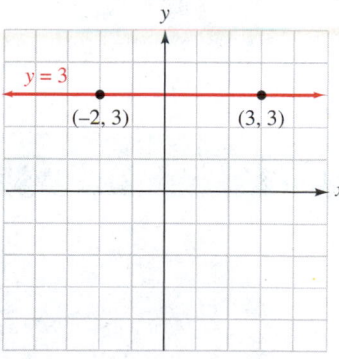

$$m = \frac{y_2 - y_1}{x_2 - x_1} \quad \text{The slope formula.}$$

$$m = \frac{3 - 3}{3 - (-2)} \quad \begin{array}{l}\text{Substitute 3 for } y_2, \text{ 3 for } y_1, \\ \text{3 for } x_2, \text{ and } -2 \text{ for } x_1.\end{array}$$

$$m = \frac{0}{5} \quad \begin{array}{l}\text{Simplify the numerator and} \\ \text{the denominator.}\end{array}$$

$$m = 0$$

The slope of the line $y = 3$ is 0.

SELF CHECK   Find the slope of the line $y = -5$.   *Answer:* $m = 0$ ■

The y-values of any two points on any horizontal line will be the same, and the x-values will be different. Thus, the numerator of

$$\frac{y_2 - y_1}{x_2 - x_1}$$

will always be zero, and the denominator will always be nonzero. Therefore, the slope of a horizontal line is zero.

EXAMPLE 5

Slope of a vertical line. If possible, find the slope of the line $x = -2$.

FIGURE 3-39

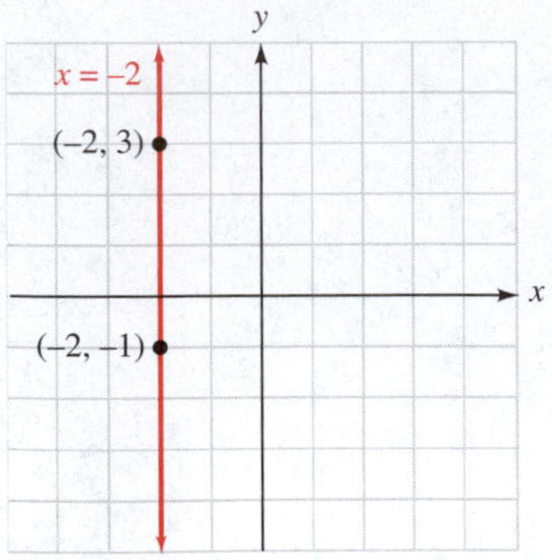

Solution To find the slope of the line $x = -2$, we need to know two points on the line. In Figure 3-39, we graph the vertical line $x = -2$ and label two points on the line: $(-2, -1)$ and $(-2, 3)$.

If (x_1, y_1) is $(-2, -1)$ and (x_2, y_2) is $(-2, 3)$, we have

$$m = \frac{y_2 - y_1}{x_2 - x_1} \qquad \text{The slope formula.}$$

$$m = \frac{3 - (-1)}{-2 - (-2)} \qquad \text{Substitute 3 for } y_2, -1 \text{ for } y_1, \\ -2 \text{ for } x_2, \text{ and } -2 \text{ for } x_1.$$

$$m = \frac{4}{0} \qquad \text{Simplify the numerator and the denominator.}$$

Since division by zero is undefined, $\frac{4}{0}$ has no meaning. The slope of the line $x = -2$ is undefined.

SELF CHECK If possible, find the slope of the line $x = 1$. *Answer:* undefined slope ■

The y-values of any two points on a vertical line will be different, and the x-values will be the same. Thus, the numerator of

$$\frac{y_2 - y_1}{x_2 - x_1}$$

will always be nonzero, and the denominator will always be zero. Therefore, the slope of a vertical line is undefined.

We now summarize the results from Examples 4 and 5.

Slopes of horizontal and vertical lines

Horizontal lines (lines with equations of the form $y = b$) have a slope of zero.

Vertical lines (lines with equations of the form $x = a$) have undefined slope.

FIGURE 3-40

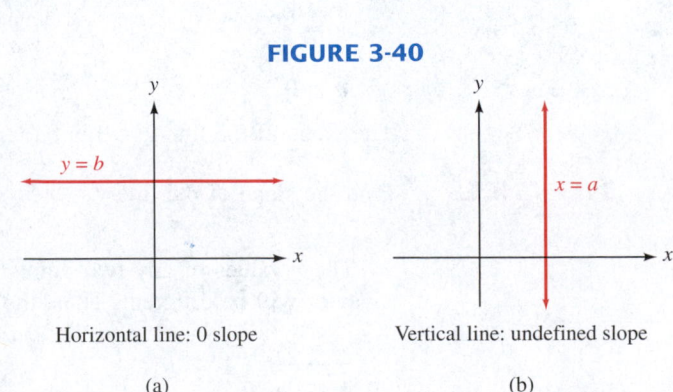

Horizontal line: 0 slope

Vertical line: undefined slope

(a)

(b)

FIGURE 3-41

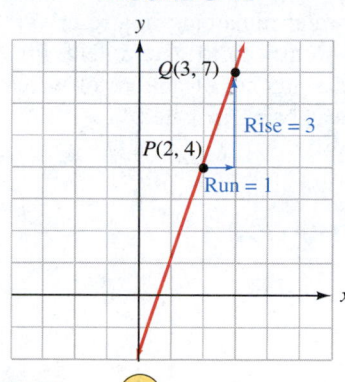

Using Slope to Graph a Line

We can graph a line whenever we know the coordinates of one point on the line and the slope of the line. For example, to graph the line that passes through $P(2, 4)$ and has a slope of 3, we first plot $P(2, 4)$, as in Figure 3-41. We can express the slope of 3 as a fraction: $3 = \frac{3}{1}$. Therefore, the line *rises* 3 units for every 1 unit it *runs* to the right. We can find a second point on the line by starting at $P(2, 4)$ and moving 1 unit to the right (run) and then 3 units up (rise). This brings us to a point that we will call Q with coordinates $(2 + \mathbf{1}, 4 + \mathbf{3})$ or $(3, 7)$. The required line must pass through points P and Q.

EXAMPLE 6

Solution

Using slope to graph a line. Graph the line that passes through the point $(-3, 4)$ with slope $-\frac{2}{5}$.

We plot the point $(-3, 4)$ as shown in Figure 3-42. Then, after writing slope $-\frac{2}{5}$ as $\frac{-2}{5}$, we see that the *rise* is -2 and the *run* is 5. From the point $(-3, 4)$, we can find a second point on the line by moving 5 units to the right (run) and then 2 units down (a rise of -2 means to move down 2 units). This brings us to the point with coordinates of $(-3 + 5, 4 - 2) = (2, 2)$. We then draw a line that passes through the two points.

FIGURE 3-42

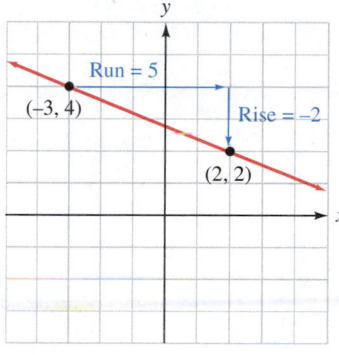

SELF CHECK

Graph the line that passes through the point $(-4, 2)$ with slope -4.

Answer:

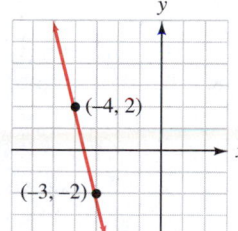

STUDY SET

Section 3.4

VOCABULARY

In Exercises 1–6, fill in the blanks to make the statements true.

1. A _____ is a comparison of two numbers by their indicated quotient.

2. Ratios used to compare quantities with different units are called _____.

3. The _____ of a line is defined to be the ratio of the change in y to the change in x.

4. $m = \dfrac{\boxed{} \text{ change}}{\text{horizontal change}} = \dfrac{\text{rise}}{\boxed{}} = \dfrac{\text{change in } \boxed{}}{\text{change in } \boxed{}}$

5. The rate of _____ of a linear relationship can be found by finding the slope of the graph of the line.

6. _____ lines have a slope of zero. Vertical lines have _____ slope.

CONCEPTS

7. Which line graphed in Illustration 1 has
   a. a positive slope?
   b. a negative slope?
   c. zero slope?
   d. undefined slope?

ILLUSTRATION 1

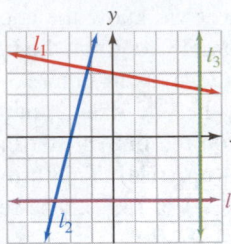

8. For the line graphed in Illustration 2:
   a. Find its slope using points A and B.
   b. Find its slope using points B and C.
   c. Find its slope using points A and C.
   d. What observation is suggested by your answers to parts a, b, and c?

ILLUSTRATION 2

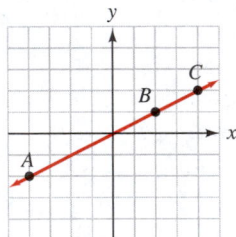

9. Use the information in the table of values for a linear equation to determine what the slope of the line would be if it was graphed.

x	y
-4	2
5	-7

10. Fill in the blanks to make the statements true.
   a. A line with positive slope _____ from left to right.
   b. A line with negative slope _____ from left to right.

11. GROWTH RATE Use the graph in Illustration 3 to find the rate of change of a boy's height over the time period shown.

ILLUSTRATION 3

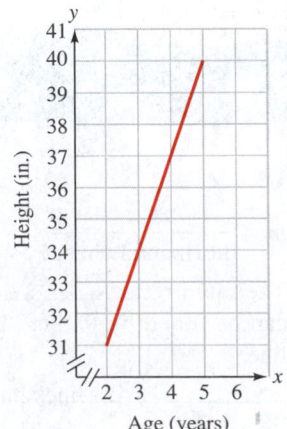

12. IRRIGATION The graph in Illustration 4 shows the number of gallons of water remaining in a reservoir as water is discharged from it to irrigate a field. Find the rate of change in the number of gallons of water for the time the field was being irrigated.

ILLUSTRATION 4

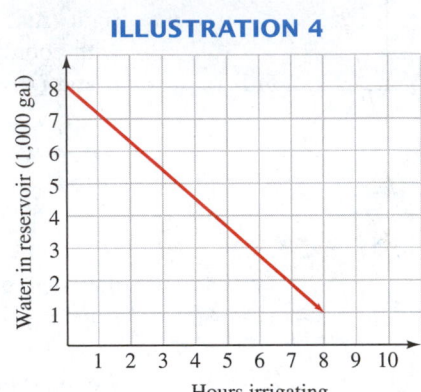

13. RECORD-SETTING FILMS Prior to the release of the movie *Titanic*, in 1997, the three movies that earned $100 million at the box office the fastest are shown in Illustration 5.
   a. Which film reached the $100-million mark fastest? Explain how you can tell.

   b. Find the rate of earnings of each film by finding the slope of each line. Round to the nearest tenth.

ILLUSTRATION 5

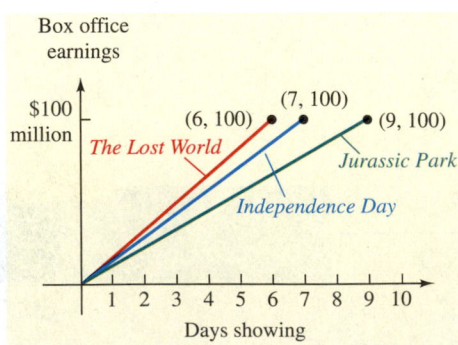

14. WAL-MART On the graph in Illustration 6, draw a straight line through the points (1991, 34) and (1998, 118). This line approximates Wal-Mart's annual net sales for the years 1991–1998. Find the rate of increase in sales over this period by finding the slope of the line.

ILLUSTRATION 6

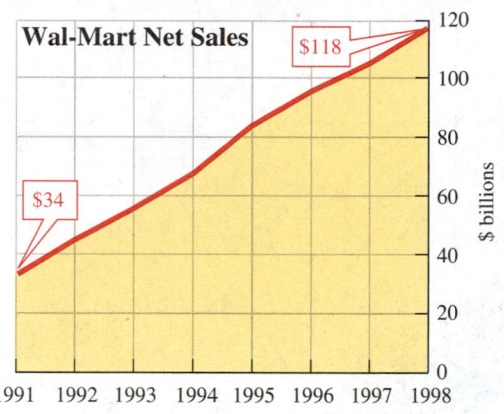

Based on data from Wal-Mart and *USA TODAY* (November 6, 1998)

ILLUSTRATION 8

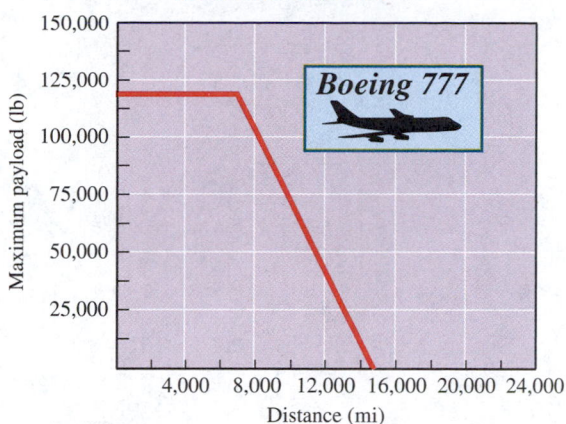

Based on data from Lawrence Livermore National Laboratory and *Los Angeles Times* (October 22, 1998)

15. THE UNCOLA On the graph in Illustration 7, draw a straight line through the points (90, 200) and (97, 200). This line approximates the number of cases of 7-Up sold annually for the years 1990–1997. Find the slope of the line. What important information does the slope give about 7-Up sales?

ILLUSTRATION 7

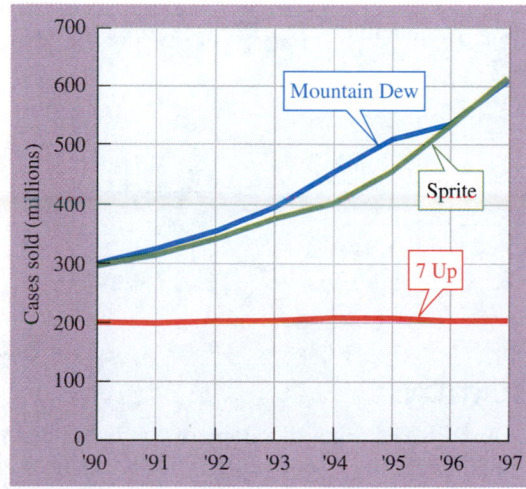

Based on data from *Beverage Digest* and *Los Angeles Times* (November 5, 1998)

16. COMMERCIAL JETS Examine the graph in Illustration 8 and consider trips of more than 7,000 miles by a Boeing 777. Use a rate of change to estimate how the maximum payload decreases as the distance traveled increases. Explain your result in words.

NOTATION

17. What is the formula used to find the slope of the line passing through (x_1, y_1) and (x_2, y_2)?

18. Explain the difference between y^2 and y_2.

PRACTICE

In Exercises 19–36, find the slope of the line passing through the given points, when possible.

19. (2, 4) and (1, 3)

20. (1, 3) and (2, 5)

21. (3, 4) and (2, 7)

22. (3, 6) and (5, 2)

23. (0, 0) and (4, 5)

24. (4, 3) and (7, 8)

25. (−3, 5) and (−5, 6)

26. (6, −2) and (−3, 2)

27. (−2, −2) and (−12, −8)

28. (−1, −2) and (−10, −5)

29. (5, 7) and (−4, 7)

30. (−1, −12) and (6, −12)

31. (8, −4) and (8, −3)

32. (−2, 8) and (−2, 15)

33. (−6, 0) and (0, −4)

34. (0, −9) and (−6, 0)

35. (−2.5, 1.75) and (−0.5, −7.75)

36. (6.4, −7.2) and (−8.8, 4.2)

In Exercises 37–40, find the slope of each line.

37.

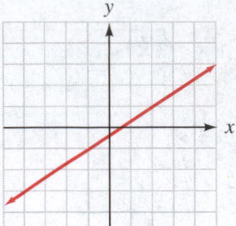

38.

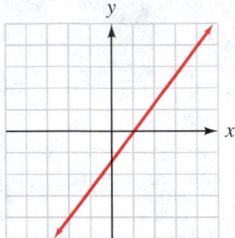

39.

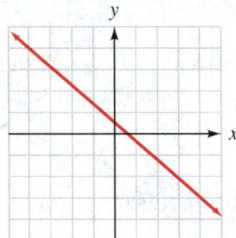

40.

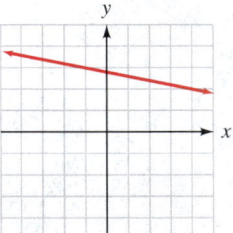

In Exercises 41–52, graph the line that passes through the given point and has the given slope.

41. $(0, 1)$, $m = 2$

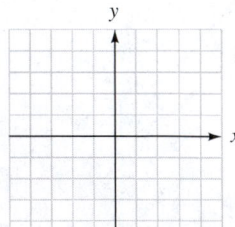

42. $(-4, 1)$, $m = -3$

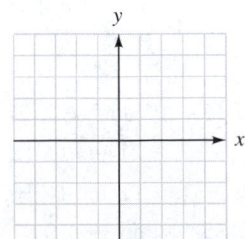

43. $(-3, -3)$, $m = -\dfrac{3}{2}$

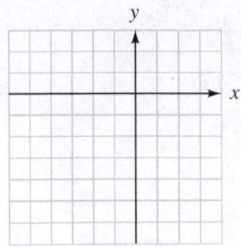

44. $(-2, -1)$, $m = \dfrac{4}{3}$

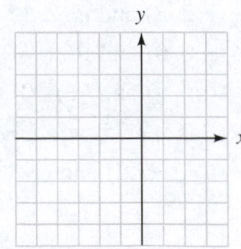

45. $(5, -3)$, $m = \dfrac{3}{4}$

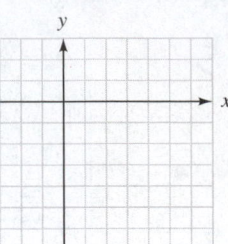

46. $(2, -4)$, $m = \dfrac{2}{3}$

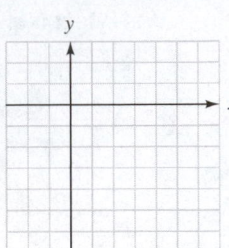

47. $(0, 0)$, $m = -4$

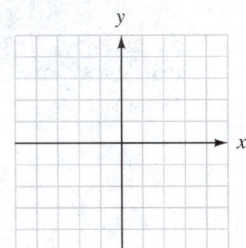

48. $(0, 0)$, $m = 5$

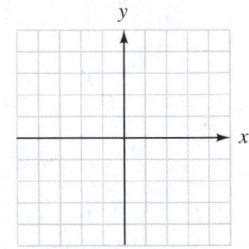

49. $(-5, 1)$, $m = 0$

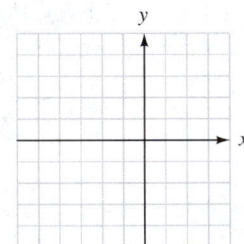

50. $(0, 3)$, undefined slope

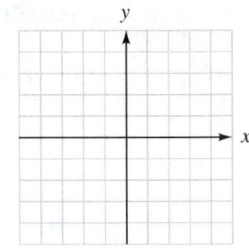

51. $(-1, -4)$, undefined slope

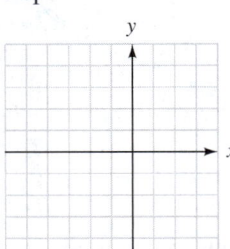

52. $(-3, -2)$, $m = 0$

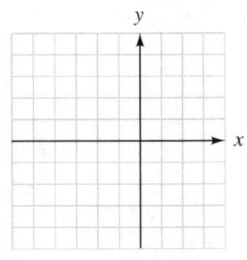

APPLICATIONS

53. POOL DESIGN Find the slope of the bottom of the swimming pool as it drops off from the shallow end to the deep end, as shown in Illustration 9.

ILLUSTRATION 9

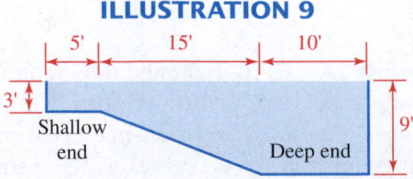

54. DRAINAGE To measure the amount of fall (slope) of a concrete patio slab in Illustration 10, a 10-foot-long 2-by-4, a 1-foot ruler, and a level were used. Find the amount of fall in the slab. Explain what it means.

ILLUSTRATION 10

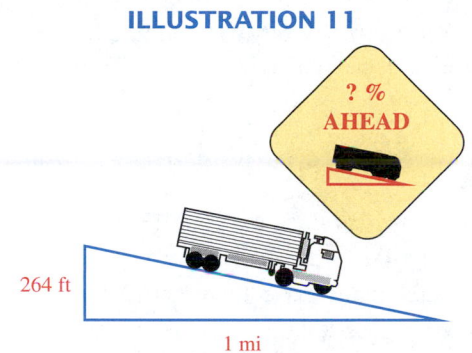

Patio slab

55. GRADE OF A ROAD The vertical fall of the road shown in Illustration 11 is 264 feet for a horizontal run of 1 mile. Find the slope of the decline and use that fact to complete the roadside warning sign for truckers. (*Hint:* 1 mile = 5,280 feet.)

ILLUSTRATION 11

? %
AHEAD

264 ft

1 mi

56. TREADMILL For each height setting listed in the table, find the resulting slope of the jogging surface of the treadmill shown in Illustration 12. Express each incline as a percent.

ILLUSTRATION 12

Height setting	% incline
2 inches	
4 inches	
6 inches	

Height setting

50 in.

57. ACCESSIBILITY Illustration 13 shows two designs to make the upper level of a stadium wheelchair-accessible.
   a. Find the slope of the ramp in design 1.
   b. Find the slopes of the ramps in design 2.
   c. Give one advantage and one drawback of each design.

ILLUSTRATION 13

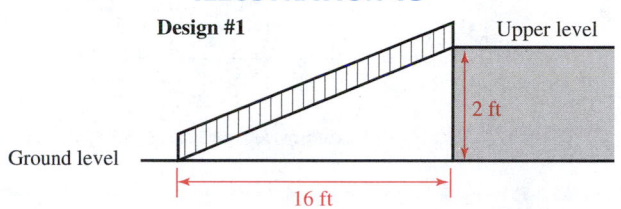

Design #1   Upper level

Ground level

2 ft

16 ft

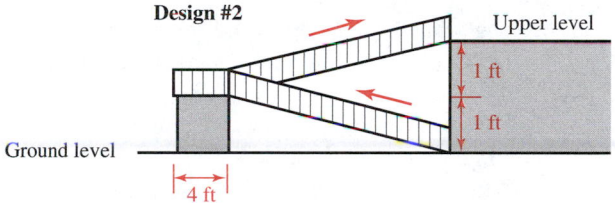

Design #2   Upper level

Ground level

1 ft

1 ft

4 ft

58. ARCHITECTURE Since the slope of the roof of the house shown in Illustration 14 is to be $\frac{2}{5}$, there will be a 2-foot rise for every 5-foot run. Draw the roof line if it is to pass through the given black points. Find the coordinates of the peak of the roof.

ILLUSTRATION 14

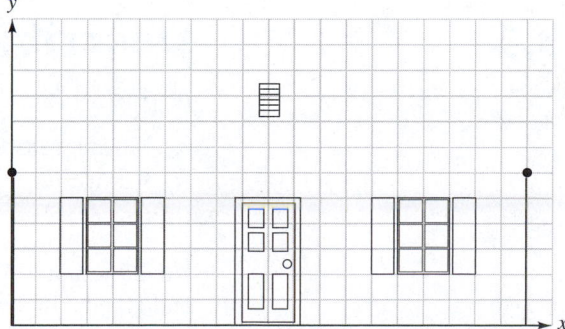

59. ENGINE OUTPUT Use the graph in Illustration 15 to find the rate of change in the horsepower (hp) produced by an automobile engine for engine speeds in the range of 2,400–4,800 revolutions per minute (rpm).

ILLUSTRATION 15

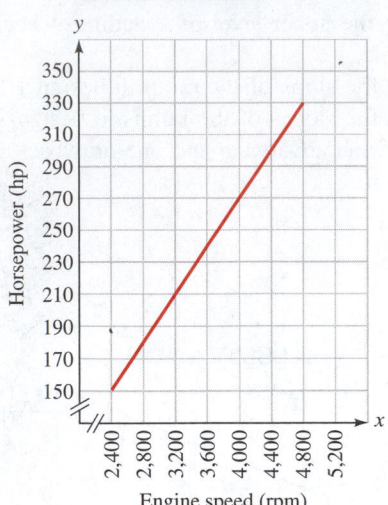

Engine speed (rpm)

ILLUSTRATION 16

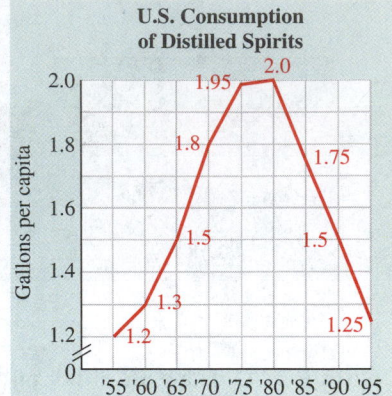

Based on data from *Business Week* (Nov. 25, 1996), p. 46

60. 📱 LIQUOR CONSUMPTION Refer to Illustration 16.
   a. During what 5-year period did the per capita consumption of distilled spirits increase the most? Find the rate of change.
   b. In what year did per capita consumption reach a maximum? What was it?
   c. Find the rate of change in per capita consumption for 1980–1995. What does it mean?

WRITING

61. Explain why the slope of a vertical line is undefined.

62. How do we distinguish between a line with positive slope and a line with negative slope?

63. Give an example of a rate of change that government officials might be interested in knowing so they can plan for the future needs of our country.

64. Explain the difference between a rate of change that is positive and one that is negative. Give an example of each.

REVIEW

65. In what quadrant does the point $(-3, 6)$ lie?

66. What is the name given the point $(0, 0)$?

67. Is $(-1, -2)$ a solution of $y = x^2 + 1$?

68. What basic shape does the graph of the equation $y = |x - 2|$ have?

69. Is the equation $y = 2x + 2$ linear or nonlinear?

70. Solve $-3x \le 15$.

▶ **3.5**

Describing Linear Relationships

In this section, you will learn about

▪ Linear relationships ▪ Slope–intercept form of the equation of a line
▪ Parallel lines ▪ Perpendicular lines

Introduction   We have seen that numerical relationships are often presented in tables and graphs. In this section, we will begin a discussion of a special type of relationship between two quantities whose graph is a straight line. Our objective is to learn how to write an equation that describes these *linear relationships* using a given table or graph.

Linear Relationships

A company manufactures concrete drain pipe. Various lengths of pipe and their corresponding weights are listed in the table in Figure 3-43. When this information is plotted as ordered pairs of the form (length, weight), we see that the points lie in a line. We say that the relationship between length and weight is *linear.*

FIGURE 3-43

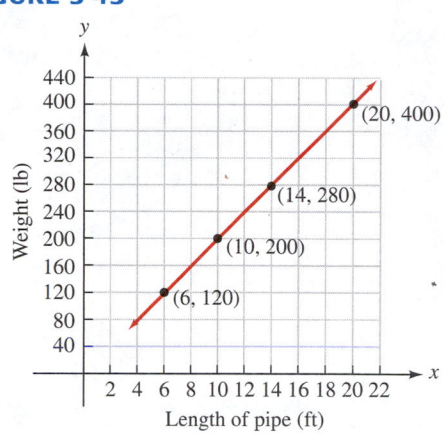

Length (ft)	Weight (lb)
6	120
10	200
14	280
20	400

Figure 3-44 shows a graph of the *time* a cup of coffee has been sitting on a kitchen counter and its *temperature.* Since the graph is not a straight line, the relationship between time and temperature in this example is not linear.

FIGURE 3-44

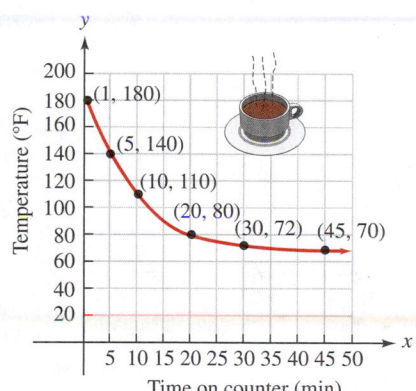

Time on counter (min)	Temperature of coffee (°F)
1	180
5	140
10	110
20	80
30	72
45	70

Slope–Intercept Form of the Equation of a Line

The graph of $2x + 3y = 12$ shown in Figure 3-45 enables us to see that the slope of the line is $-\frac{2}{3}$ and that the y-intercept is $(0, 4)$.

FIGURE 3-45

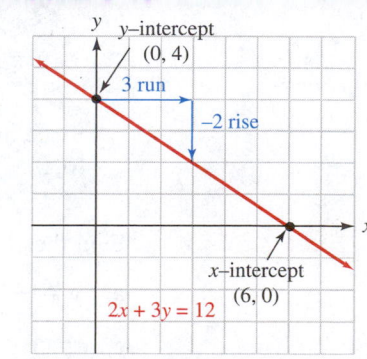

$$2x + 3y = 12$$

x	y	(x, y)
6	0	(6, 0)
0	4	(0, 4)

If we solve the equation for y, we will observe some interesting results.

$$2x + 3y = 12$$

$3y = -2x + 12$ Subtract $2x$ from both sides.

$$\frac{3y}{3} = \frac{-2x}{3} + \frac{12}{3}$$ Divide both sides by 3.

$y = -\frac{2}{3}x + 4$ Do the divisions. Rewrite $\frac{-2x}{3}$ as $-\frac{2}{3}x$.

In the equation $y = -\frac{2}{3}x + 4$, the *slope* of the graph $\left(-\frac{2}{3}\right)$ is the coefficient of x, and the constant (4) is the y-coordinate of the *y-intercept* of the graph.

$$y = -\frac{2}{3}x + 4$$

    ↑   ↑

The slope   The y-intercept
of the line.   is $(0, 4)$.

These observations suggest the following form of an equation of a line.

Slope–intercept form of the equation of a line

If a linear equation is written in the form

$$y = mx + b$$

where m and b are constants, the graph of the equation is a line with slope m and y-intercept $(0, b)$.

E X A M P L E   1

Slope–intercept form. Find the slope and the y-intercept of the graph of each equation: **a.** $y = 6x - 2$ and   **b.** $y = -\frac{5}{4}x$.

Solution   **a.** If we write the subtraction as the addition of the opposite, the equation will be in $y = mx + b$ form:

$$y = 6x + (-2)$$

Since $m = 6$ and $b = -2$, the slope of the line is 6 and the y-intercept is $(0, -2)$.

b. Writing $y = -\frac{5}{4}x$ in slope–intercept form, we have

$$y = -\frac{5}{4}x + 0$$

Since $m = -\frac{5}{4}$ and $b = 0$, the slope of the line is $-\frac{5}{4}$ and the y-intercept is $(0, 0)$.

SELF CHECK   Find the slope and the y-intercept:

a. $y = -5x - 1$,   **b.** $y = \frac{7}{8}x$,   and

c. $y = 5 - 2x$.

Answers: **a.** $m = -5$, $(0, -1)$;   **b.** $m = \frac{7}{8}$, $(0, 0)$;   **c.** $m = -2$, $(0, 5)$ ∎

EXAMPLE 2

Slope–intercept form. Find the slope and the y-intercept of the line determined by $6x - 3y = 9$. Then graph it.

Solution

To find the slope and the y-intercept of the line, we need to write the equation in slope–intercept form. We do this by solving for y.

$$6x - 3y = 9$$

$$-3y = -6x + 9 \qquad \text{Subtract } 6x \text{ from both sides.}$$

$$\frac{-3y}{-3} = \frac{-6x}{-3} + \frac{9}{-3} \qquad \text{Divide both sides by } -3.$$

$$y = 2x - 3 \qquad \text{Do the divisions.}$$
Note: $m = 2$ and $b = -3$.

FIGURE 3-46

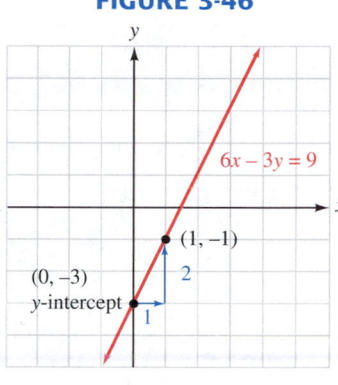

From the equation, we see that the slope is 2 and the y-intercept is $(0, -3)$.

To graph $y = 2x - 3$, we plot the y-intercept $(0, -3)$, as shown in Figure 3-46. Since the slope is $\frac{\text{rise}}{\text{run}} = 2 = \frac{2}{1}$, the line rises 2 units for every unit it moves to the right. If we begin at $(0, -3)$ and move 1 unit to the right (run) and then 2 units up (rise), we locate the point $(1, -1)$, which is a second point on the line. We then draw a line through $(0, -3)$ and $(1, -1)$.

SELF CHECK

Find the slope and the y-intercept of the line determined by $8x - 2y = -2$. Then graph it.

Answer: $m = 4$, $(0, 1)$

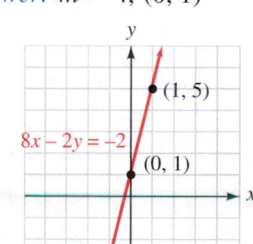

If we are given the slope and y-intercept of a line, we can write its equation, as in the next example.

EXAMPLE 3

Limo service. On weekends, a limousine service charges a fee of $100, plus 50¢ per mile, for the rental of a stretch limo. Write a linear equation that describes the relationship between the rental cost and the number of miles driven. Graph the result.

Solution

To write an equation describing this relationship, we will let x represent the number of miles driven and y represent the cost (in dollars). We can make two observations:

■ The cost increases by $0.50 (50¢) for each mile driven. This is the *rate of change* of the rental cost to miles driven, and it will be the *slope* of the graph of the equation. Thus, $m = 0.50$.

■ The basic fee is $100. Before driving any miles (that is, when $x = 0$), the cost y is 100. The ordered pair $(0, 100)$ will be the y-intercept of the graph of the equation. So we know that $b = 100$.

We substitute 0.50 for m and 100 for b in the slope–intercept form to get

$$y = 0.50x + 100$$

Here the cost y depends on x (the number of miles driven).

$m = 0.50 \qquad b = 100$

To graph $y = 0.50x + 100$, we plot its y-intercept, $(0, 100)$, as shown in Figure 3-47. Since the slope is $0.50 = \frac{50}{100} = \frac{5}{10}$, we can start at $(0, 100)$ and locate a second point on the line by moving 10 units to the right (run) and then 5 units up (rise). This point will have coordinates $(0 + 10, 100 + 5)$ or $(10, 105)$. We draw a line through these two points to get a graph that illustrates the relationship between the rental cost and the number of miles driven.

FIGURE 3-47

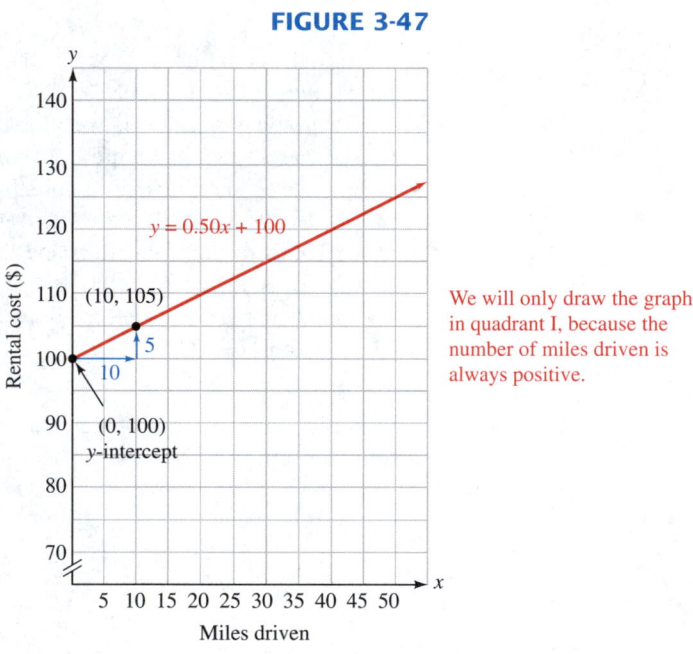

We will only draw the graph in quadrant I, because the number of miles driven is always positive.

EXAMPLE 4

Videotapes. A VHS videocassette contains 800 feet of tape. In the long play (LP) mode, it plays 10 feet of tape every 3 minutes. Write a linear equation that relates the number of feet of tape yet to be played and the number of minutes the tape has been playing. Graph the equation.

Solution   The number of feet yet to be played depends on the time the tape has been playing. To write an equation describing this relationship, we let x represent the number of minutes the tape has been playing and y represent the number of feet of tape yet to be played. We can make two observations:

- Since the VCR plays 10 feet of tape every 3 minutes, the number of feet remaining is constantly *decreasing*. This rate of change $\left(-\frac{10}{3}\right.$ feet per minute$\left.\right)$ will be the slope of the graph of the equation. Thus, $m = -\frac{10}{3}$.

- The cassette tape is 800 feet long. Before any of the tape is played (that is, when $x = 0$), the amount of tape yet to be played is $y = 800$. Written as an ordered pair, we have $(0, 800)$. Thus, $b = 800$.

Writing the equation in slope–intercept form, we have $y = -\frac{10}{3}x + 800$. Its graph is shown in Figure 3-48.

FIGURE 3-48

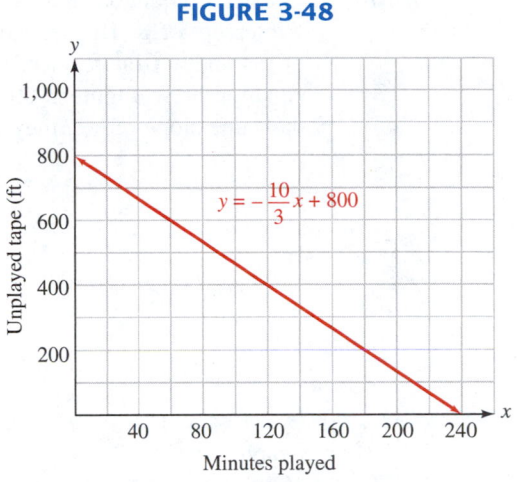

<table>
<tr><td>SELF CHECK</td></tr>
</table>

Answer the problem in Example 4, where the VCR is in super long play (SLP) mode, which plays 11 feet every 5 minutes. Graph the equation on the graph in Figure 3-48 and make an observation.

Answer: $y = -\frac{11}{5}x + 800$; the graphs have the same y-intercept but different slopes. ■

Parallel Lines

Suppose it costs $75, plus 50¢ per mile, to rent the limo discussed in Example 3 on a weekday. If we substitute 0.50 for m and 75 for b in the slope–intercept form of a line, we have

$$y = 0.50x + 75$$

The graph of this equation and the graph of the equation

$$y = 0.50x + 100$$

appear in Figure 3-49.

From the figure, we see that the lines, each with slope 0.50, are parallel (do not intersect). This observation suggests the following fact.

FIGURE 3-49

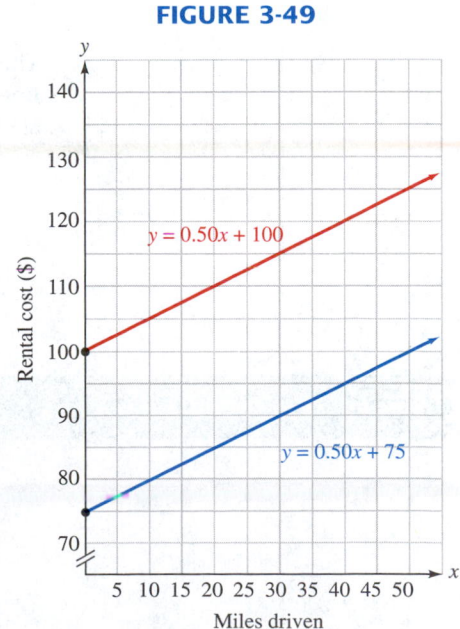

Slopes of parallel lines

Two lines with the same slope are parallel.

EXAMPLE 5

Parallel lines. Graph $y = -\frac{2}{3}x$ and $y = -\frac{2}{3}x + 3$ on the same set of axes.

Solution　The first equation has a slope of $-\frac{2}{3}$ and a y-intercept of 0. The second equation has a slope of $-\frac{2}{3}$ and a y-intercept of $(0, 3)$. We graph each equation as in Figure 3-50. Since the lines have the same slope of $-\frac{2}{3}$, they are parallel.

FIGURE 3-50

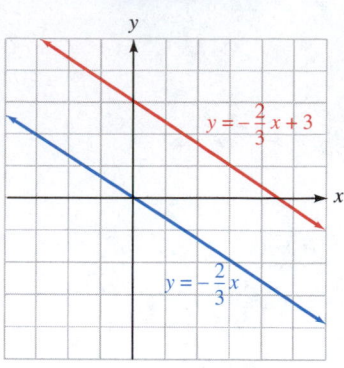

SELF CHECK

Graph $y = \frac{5}{2}x - 2$ and $y = \frac{5}{2}x$ on the same set of axes.

Answer:

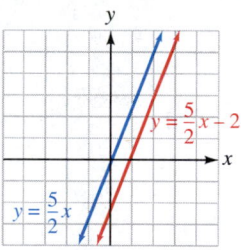

FIGURE 3-51

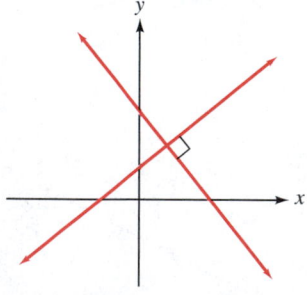

Perpendicular Lines

The two lines shown in Figure 3-51 meet at right angles and are called **perpendicular lines.** In the figure, the symbol ⌐ is used to denote a right angle. Each of the four angles that are formed has a measure of 90°.

The slopes of two (nonvertical) perpendicular lines are related by the following fact.

Slopes of perpendicular lines

The product of the slopes of perpendicular lines is -1.

If the product of two numbers is -1, they are called **negative reciprocals.** For example, 3 and $-\frac{1}{3}$ are negative reciprocals, because their product is -1.

$$3\left(-\frac{1}{3}\right) = -\frac{3}{3} = -1$$

Perpendicular lines have slopes that are negative reciprocals.

ACCENT ON TECHNOLOGY *Exploring Graphs with a Graphing Calculator*

Graphing calculators can be used to explore some important features of the graph of a linear equation, such as its *x*- and *y*-intercepts and its slope. To graph a linear equation such as $2x - 3y = 12$ using a graphing calculator, we must enter the equation into the calculator using the $\boxed{Y=}$ key. This requires that the equation be solved for *y*.

$$2x - 3y = 12$$

$$-3y = 12 - 2x \qquad \text{To eliminate } 2x \text{ from the left-hand side, subtract } 2x \text{ from both sides.}$$

$$\frac{-3y}{-3} = \frac{12}{-3} - \frac{2x}{-3} \qquad \text{To undo the multiplication by } -3, \text{ divide both sides by } -3.$$

$$y = \frac{2}{3}x - 4 \qquad \text{Simplify and write the equation in slope-intercept form: } y = mx + b.$$

Using the standard viewing window, we press the $\boxed{Y=}$ key and enter the equation as $(2/3)x - 4$. The display will show

$$Y_1 = (2/3)\, x - 4$$

Then we press the $\boxed{GRAPH}$ key to obtain the graph shown in Figure 3-52.

FIGURE 3-52

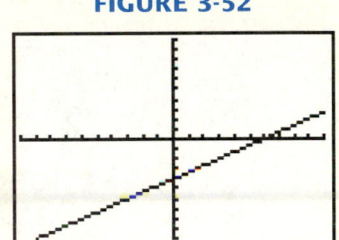

Determining the Intercepts of a Graph

To determine the *x*-intercept of the graph of $y = \frac{2}{3}x - 4$, we can use the zero feature, found under the CALC menu. After we guess left and right bounds, the cursor automatically moves to the *x*-intercept of the graph, and the coordinates of that point are displayed at the bottom of the screen. See Figure 3-53(a).

To determine the *y*-intercept of the graph, we can use the value feature, which is also found under the CALC menu. With this option, we first enter an *x*-value of 0, as shown in Figure 3-53(b). Then the cursor highlights the *y*-intercept and the coordinates of the *y*-intercept are given. See Figure 3-53(c).

FIGURE 3-53

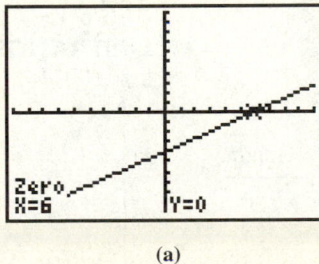

(a)

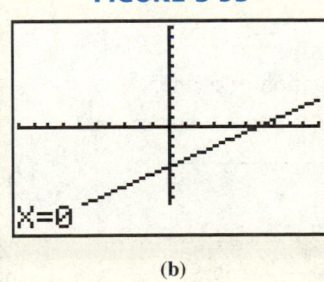

(b)

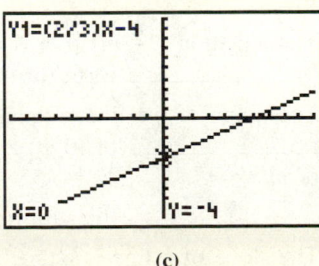

(c)

Inputting Several Equations

Graphing calculators can graph several equations on the same set of axes. For example, we can graph the equations $y = -2x$, $y = -2x - 4$, and $y = -2x + 5$ by pressing the $\boxed{Y=}$ key and entering the equations, one at a time, on the first three lines.

$$Y_1 = -2x$$

$$Y_2 = -2x - 4$$

$$Y_3 = -2x + 5$$

(continued)

The resulting three graphs are shown in Figure 3-54. The lines are parallel, which is a fact we can confirm by examining their equations. In each case, $m = -2$, which implies that each line has a slope of -2.

FIGURE 3-54

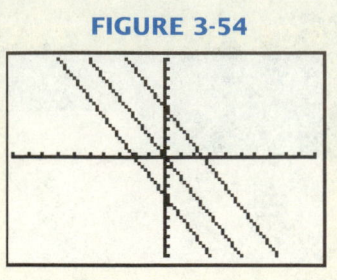

EXAMPLE 6

Parallel and perpendicular lines. Determine whether the graphs of $y = -5x + 6$ and $y = \frac{x}{5} - 2$ are parallel, perpendicular, or neither.

Solution The slope of the line $y = -5x + 6$ is -5. The slope of the line $y = \frac{x}{5} - 2$ is $\frac{1}{5}$. (Recall that $\frac{x}{5} = \frac{1}{5}x$.) Since the slopes are not equal, the lines are not parallel. If we find the product of their slopes, we have

$$-5\left(\frac{1}{5}\right) = -\frac{5}{5} = -1$$

Since the product of their slopes is -1, the lines are perpendicular.

SELF CHECK Determine whether the graphs of $y = 4x + 4$ and $y = \frac{1}{4}x$ are parallel, perpendicular, or neither.   *Answer:* neither ■

STUDY SET

Section 3.5

VOCABULARY

In Exercises 1–6, fill in the blanks to make the statements true.

1. The equation $y = mx + b$ is called the _____ form for the equation of a line.

2. The graph of the linear equation $y = mx + b$ has a _____ of $(0, b)$ and a _____ of m.

3. _____ lines do not intersect.

4. The slope of a line is a _____ of change.

5. The numbers $\frac{5}{6}$ and $-\frac{6}{5}$ are called negative _____. Their product is -1.

6. The product of the slopes of _____ lines is -1.

CONCEPTS

7. TREE GROWTH Graph the values shown in Illustration 1 and connect the points with a smooth curve.

Does the graph indicate a linear relationship between the age of the tree and its height? Explain your answer.

ILLUSTRATION 1

Age	Height
0	0
5	8
10	15
15	28
20	45
25	62
30	85
35	100
40	112
45	118

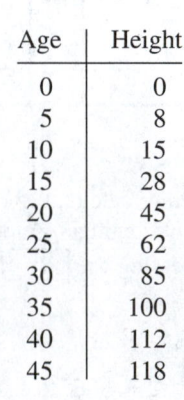

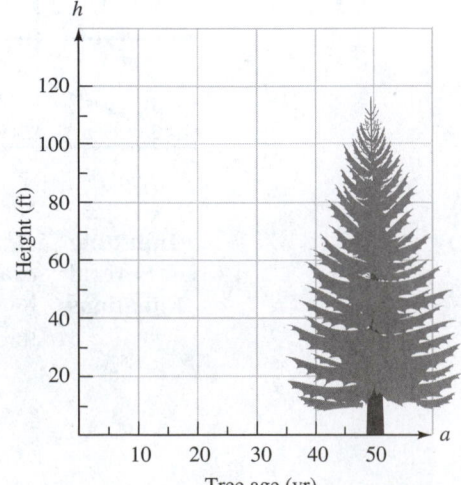

Tree age (yr)

8. See Illustration 2.
   a. What is the slope of the line?
   b. What is the *y*-intercept of the line?
   c. Write the equation of the line.

ILLUSTRATION 2

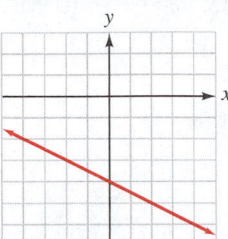

9. NAVIGATION The performance graph in Illustration 3 shows the recommended speed at which a ship should proceed into head waves of various heights.
   a. What information does the *y*-intercept of the graph give?

   b. What is the rate of change in the recommended speed of the ship as the wave height increases?

   c. Write the equation of the graph.

ILLUSTRATION 3

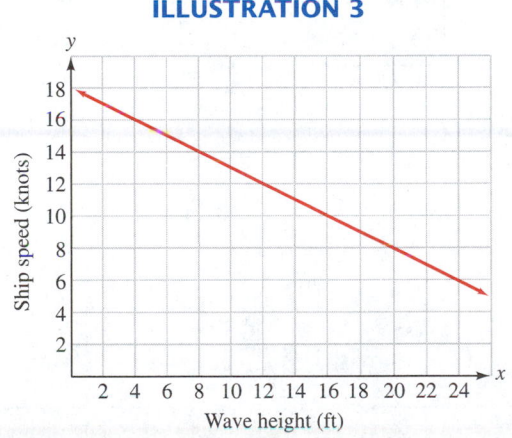

10. In Illustration 4, the slope of line l_1 is 2.
   a. What is the slope of line l_2?
   b. What is the slope of line l_3?
   c. What is the slope of line l_4?
   d. Which lines have the same *y*-intercept?

ILLUSTRATION 4

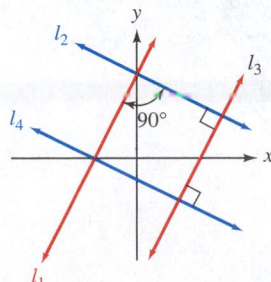

11. a. What is the *y*-intercept of line l_1 graphed in Illustration 5?
   b. What do lines l_1 and l_2 have in common? How are they different?

ILLUSTRATION 5

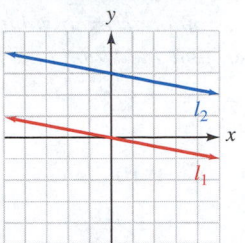

12. Use the graph in Illustration 6 to find *m* and *b*; then write the equation of the line in slope–intercept form.

ILLUSTRATION 6

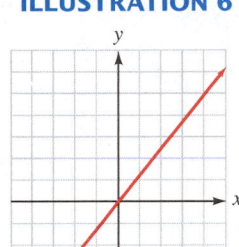

13. What is the slope of the line defined by each equation?
   a. $y = \dfrac{-2x}{3} - 2$ **b.** $y = \frac{x}{4} + 1$
   c. $y = 2 - 8x$ **d.** $y = 3x$
   e. $y = x$ **f.** $y = -x$

14. Without graphing, tell whether the graphs of each pair of lines are parallel, perpendicular, or neither.
   a. $y = 0.5x - 3$; $y = \dfrac{1}{2}x + 3$

   b. $y = 0.75x$; $y = -\dfrac{4}{3}x + 2$

   c. $y = -x$; $y = x$

In Exercises 15–16, a graphing calculator was used to graph $y = -2.5x - 1.25$, and the TRACE *key was pressed. What important feature of the graph does the TRACE cursor show?*

15.

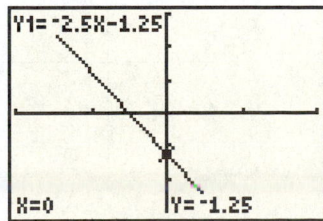

16.

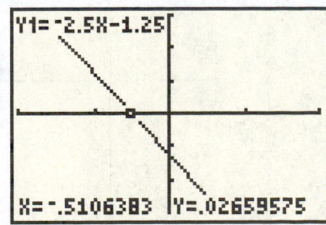

NOTATION

In Exercises 17–18, complete each solution by solving the equation for y. Then find the slope and the y-intercept of its graph.

17. $6x - 2y = 10$

$6x -$ [ ] $- 2y = -6x + 10$

$-2y =$ [ ] $+ 10$

$y =$ [ ] $- 5$

The slope is [ ] and the *y*-intercept is [ ].

18. $2x + 5y = 15$

$2x + 5y -$ [ ] $= -$ [ ] $+ 15$

[ ] $= -2x + 15$

$y = -\dfrac{2}{5}x + 3$

The slope is [ ] and the *y*-intercept is [ ].

PRACTICE

In Exercises 19–20, find the slope and the y-intercept of the graph of each equation.

19. a. $y = 4x + 2$ **b.** $y = -4x - 2$

   c. $4y = x - 2$ **d.** $4x - 2 = y$

20. a. $y = \dfrac{1}{2}x + 6$ **b.** $y = 6 - x$

   c. $6y = x - 6$ **d.** $6x - 1 = y$

In Exercises 21–26, write the equation of the line with the given slope and y-intercept. Then graph it.

21. $m = 5$, $(0, -3)$ **22.** $m = -2$, $(0, 1)$

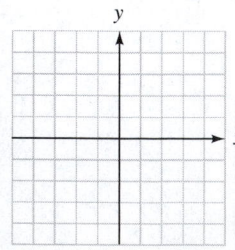   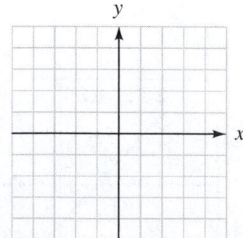

23. $m = \dfrac{1}{4}$, $(0, -2)$ **24.** $m = \dfrac{1}{3}$, $(0, -5)$

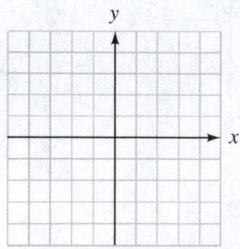   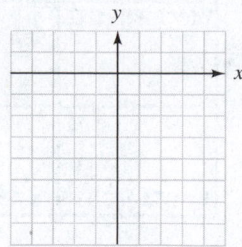

25. $m = -\dfrac{8}{3}$, $(0, 5)$ **26.** $m = -\dfrac{7}{6}$, $(0, 2)$

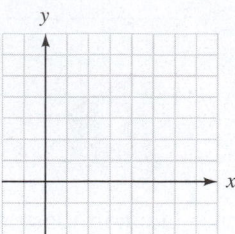   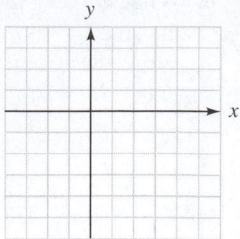

In Exercises 27–30, find the slope and the y-intercept of the graph of each equation. Then graph it.

27. $3x + 4y = 16$ **28.** $2x + 3y = 9$

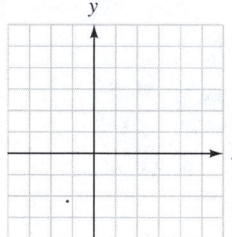   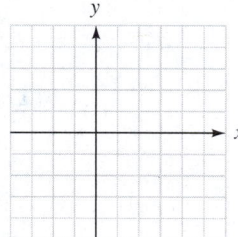

29. $10x - 5y = 5$ **30.** $4x - 2y = 6$

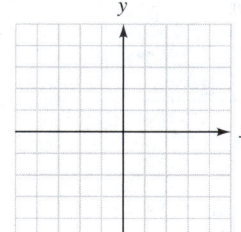   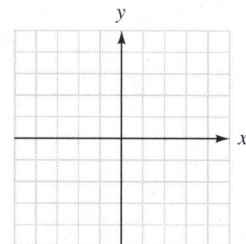

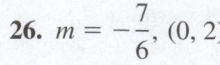 *In Exercises 31–32, solve each equation for y. Then use a graphing calculator to graph it. Use a viewing window of x = −5 to 5 and y = −5 to 5.*

31. $2y + 5x = 4$ **32.** $1.2x - 3.2y + 4.7 = 0$

 In Exercises 33–34, estimate the x-intercept of the graph of each equation to the nearest hundredth.

33. $y = 2.3x + 3.76$ **34.** $y = \dfrac{7}{8}x + 8$

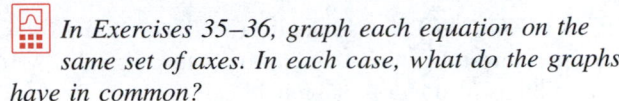

In Exercises 35–36, graph each equation on the same set of axes. In each case, what do the graphs have in common?

35. $y = 4x - 2.1$

$y = 4x$

$y = 4x + 3.75$

36. $y = -x + 3$

$y = 5.34x + 3$

$y = \dfrac{7}{2}x + 3$

APPLICATIONS

37. PRODUCTION COSTS A television production company charges a basic fee of $5,000 and then $2,000 an hour when filming a commercial.

  a. Write a linear equation that describes the relationship between the total production costs y and the hours of filming x.

  b. Use your answer to part a to find the production costs if a commercial required 8 hours of filming.

38. COLLEGE FEES Each semester, students enrolling at a community college must pay tuition costs of $20 per unit as well as a $40 student services fee.

  a. Write a linear equation that gives the total fees y to be paid by a student enrolling at the college and taking x units.

  b. Use your answer to part a to find the enrollment cost for a student taking 12 units.

39. CHEMISTRY EXPERIMENT A portion of a student's chemistry lab manual is shown in Illustration 7. Use the information to write a linear equation relating the temperature y (in degrees Fahrenheit) of the compound to the time x (in minutes) elapsed during the lab procedure.

ILLUSTRATION 7

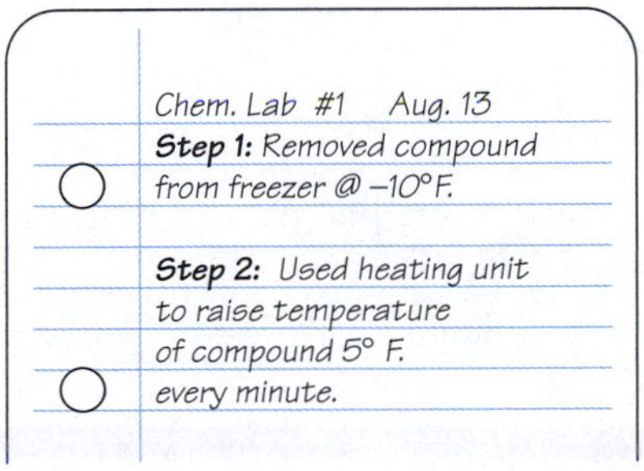

> Chem. Lab #1   Aug. 13
> **Step 1:** Removed compound
> from freezer @ –10° F.
>
> **Step 2:** Used heating unit
> to raise temperature
> of compound 5° F.
> every minute.

40. INCOME PROPERTY Use the information in the newspaper advertisement in Illustration 8 to write a linear equation that gives the amount of income y (in dollars) the apartment owner will receive when the unit is rented for x months.

ILLUSTRATION 8

> **APARTMENT FOR RENT**
> 1 bedroom/1 bath,
> with garage
> $500 per month +
> $250 nonrefundable
> security fee.

41. SALAD BAR For lunch, a delicatessen offers a "Salad and Soda" special where customers serve themselves at a well-stocked salad bar. The cost is $1.00 for the drink and 20¢ an ounce for the salad.

  a. Write a linear equation that will find the cost y of a "Salad and Soda" lunch when a salad weighing x ounces is purchased.

  b. Graph the equation (see Illustration 9).

  c. How would the graph from part b change if the delicatessen began charging $2.00 for the drink?

  d. How would the graph from part b change if the cost of the salad changed to 30¢ an ounce?

ILLUSTRATION 9

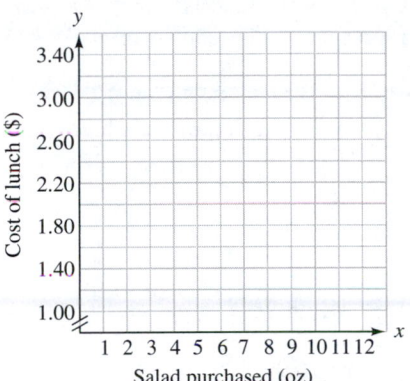

Salad purchased (oz)

42. SEWING COSTS A tailor charges a basic fee of $20.00 plus $2.50 per letter to sew an athlete's name on the back of a jacket.

  a. Write a linear equation that will find the cost y to have a name containing x letters sewn on the back of a jacket.

  b. Graph the equation (see Illustration 10).

  c. Suppose the tailor raises the basic fee to $30. On your graph from part b, draw the new graph showing the increased cost.

ILLUSTRATION 10

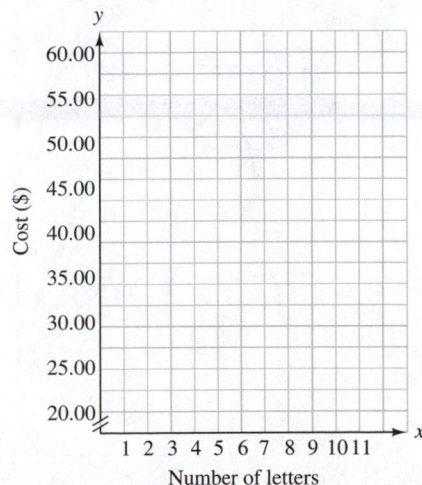

Number of letters

43. EMPLOYMENT SERVICE The policy statement of LIZCO, Inc., is shown in Illustration 11. Suppose a secretary had to pay an employment service $500 to get placed in a new job at LIZCO. Write a linear equation that tells the secretary the actual cost y of the employment service to her x months after being hired.

ILLUSTRATION 11

> **Policy no. 23452**– A new hire will be reimbursed by LIZCO for any employment service fees paid by the employee at the rate of $20 per month.

44. POPULATION Use the data in Illustration 12 to write a linear equation that approximates South Korea's population y (in millions) for the years 1980–2000. On the horizontal axis, $x = 0$ represents 1980, $x = 5$ represents 1985, and so on.

ILLUSTRATION 12

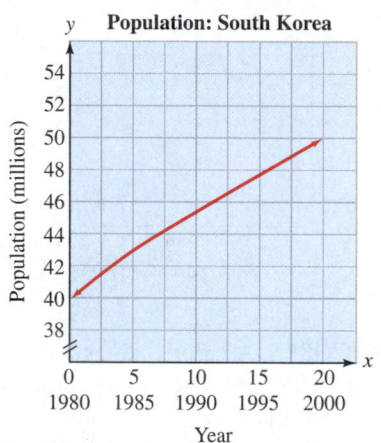

Population: South Korea

Based on data from *Los Angeles Times* (Apr. 9, 1997), p. A5

45. 📟 **COMPUTER DRAFTING** Illustration 13 shows a computer-generated drawing of an automobile engine mount. When the designer clicks the mouse on a line of the drawing, the computer finds the equation of the line. Determine whether the two lines selected in the drawing are perpendicular.

ILLUSTRATION 13

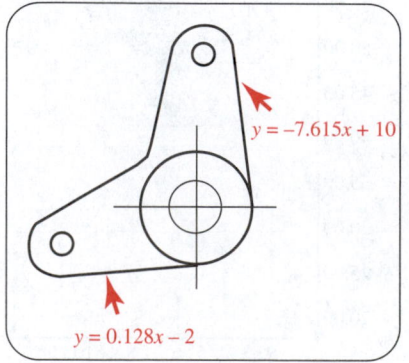

46. CALLING CARDS Use the data in Illustration 14 to write a linear equation that approximates the sales y (in billions of dollars) of prepaid calling cards for the years 1995–2000. On the horizontal axis, $x = 0$ represents 1995, $x = 1$ represents 1996, and so on.

ILLUSTRATION 14

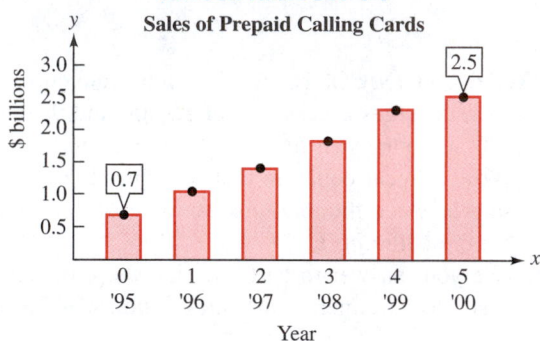

Sales of Prepaid Calling Cards

Based on data from *Los Angeles Times* (Apr. 17, 1997), p. D1

WRITING

47. Explain the advantages of writing the equation of a line in slope–intercept form ($y = mx + b$) as opposed to general form ($Ax + By = C$).

48. To describe a linear relationship between two quantities, we can use a graph or an equation. Which method do you think is better? Explain why.

49. What is the minimum number of points needed to draw the graph of a line? Explain why.

50. List some examples of parallel and perpendicular lines that you see in your daily life.

REVIEW

51. Find the slope of the line passing through the points $(6, -2)$ and $(-6, 1)$.

52. Is $(3, -7)$ a solution of $y = 3x - 2$?

53. Evaluate $-4 - (-4)$.

54. Solve $2(x - 3) = 3x$.

55. To evaluate $[-2(4 - 8) + 4^2]$, which operation should be performed first?

56. Translate to mathematical symbols: four less than twice the price p.

57. What percent of 6 is 1.5?

58. Does $x = -6.75$ make $x + 1 > -9$ true?

▶ **3.6**

Writing Linear Equations

In this section, you will learn about

Point–slope form of the equation of a line ■ Writing the equation of a line through two points ■ Horizontal and vertical lines

Introduction If we know the slope of a line and its y-intercept, we can use the slope–intercept form to write the equation of the line. The question that now arises is, can *any* point on the line be used in combination with its slope to write its equation? In this section, we will answer this question.

Point–Slope Form of the Equation of a Line

FIGURE 3-55

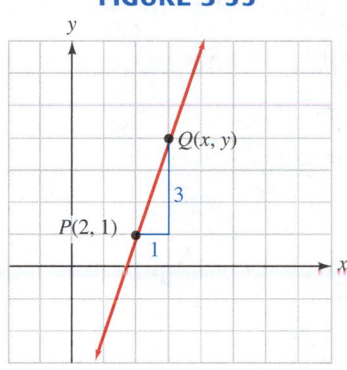

For the line shown in Figure 3-55, suppose we know that it has a slope of 3 and that it passes through the point $P(2, 1)$. If we pick another point on the line and call it $Q(x, y)$, we can find the slope of the line by using the coordinates of points P and Q. Using the slope formula, we have

$$\frac{y_2 - y_1}{x_2 - x_1} = m \qquad \text{The slope formula.}$$

$$\frac{y - 1}{x - 2} = m \qquad \text{Substitute } y \text{ for } y_2, 1 \text{ for } y_1, x \text{ for } x_2, \text{ and } 2 \text{ for } x_1.$$

Since the slope of the line is given to be 3, we can substitute 3 for m in the previous equation.

$$\frac{y - 1}{x - 2} = m$$

$$\frac{y - 1}{x - 2} = 3$$

We then multiply both sides by $(x - 2)$ to get

$$\frac{y - 1}{x - 2}(x - 2) = 3(x - 2) \qquad \text{Clear the equation of the fraction.}$$

$$y - 1 = 3(x - 2) \qquad \text{Simplify the left-hand side.}$$

The resulting equation displays the slope of the line and the coordinates of one point on the line:

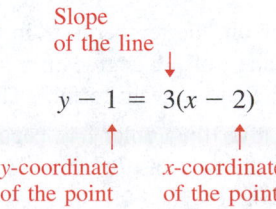

In general, suppose we know that the slope of a line is m and that the line passes through the point (x_1, y_1). Then if (x, y) is any other point on the line, we can use the definition of slope to write

$$\frac{y - y_1}{x - x_1} = m$$

If we multiply both sides by $x - x_1$, we have

$$y - y_1 = m(x - x_1)$$

This form of a linear equation is called the **point–slope form.** It can be used to write the equation of a line when the slope and one point on the line are known.

> **Point–slope form of the equation of a line**
>
> If a line with slope m passes through the point (x_1, y_1), the equation of the line is
>
> $$y - y_1 = m(x - x_1)$$

EXAMPLE 1

Point–slope form. Write the equation of a line that has a slope of -3 and passes through $(-1, 5)$. Express the result in slope–intercept form.

Solution Since we are given the slope and a point on the line, we will use the point–slope form.

$$y - y_1 = m(x - x_1) \qquad \text{The point–slope form.}$$
$$y - 5 = -3[x - (-1)] \qquad \text{Substitute } -3 \text{ for } m, -1 \text{ for } x_1, \text{ and } 5 \text{ for } y_1.$$
$$y - 5 = -3(x + 1) \qquad \text{Simplify inside the brackets.}$$

We can write this result in slope–intercept form, as follows:

$$y - 5 = -3x - 3 \qquad \text{Distribute the } -3.$$
$$y = -3x + 2 \qquad \text{Add 5 to both sides: } -3 + 5 = 2.$$

In slope–intercept form, the equation is $y = -3x + 2$.

SELF CHECK

Write the equation of a line that has a slope of -2 and passes through $(4, -3)$. Write the result in slope–intercept form.   *Answer:* $y = -2x + 5$ ∎

EXAMPLE 2

Temperature drop. A refrigeration unit can lower the temperature in a railroad car by 6°F every 5 minutes. One day, the temperature in a car was 76°F after the cooler had run for 10 minutes. Find a linear equation that describes the relationship between the time the cooler has been running and the temperature in the car.

Graph the equation and use it to find the temperature in the car before the cooler was turned on and the temperature in the car after the cooler had run for 25 minutes.

Solution We will let x represent the time, in minutes, that the cooler was running, and y will represent the air temperature in the car. We can make two observations:

- With the cooler on, the temperature in the railroad car drops 6° every 5 minutes. The rate of change of $-\frac{6}{5}$ degrees per minute is the slope of the graph of the linear equation that we want to find. Thus, $m = -\frac{6}{5}$.

- We know that after the cooler had been running for 10 minutes ($x = 10$), the temperature in the car was 76° ($y = 76$). We can express these facts with the ordered pair $(10, 76)$.

To write the linear equation, we substitute $-\frac{6}{5}$ for m, 10 for x_1, and 76 for y_1, into the point–slope form of the equation of a line.

$$y - y_1 = m(x - x_1) \qquad \text{The point–slope form.}$$
$$y - 76 = -\frac{6}{5}(x - 10) \qquad \text{Substitute: } m = -\frac{6}{5}, x_1 = 10, \text{ and } y_1 = 76.$$
$$5(y - 76) = 5\left(-\frac{6}{5}\right)(x - 10) \qquad \text{Multiply both sides by 5 to eliminate the fraction.}$$

$$5y - 380 = -6(x - 10)$$ — Distribute the 5 on the left-hand side. On the right-hand side, $5\left(-\frac{6}{5}\right) = -6$.

$$5y - 380 = -6x + 60$$ — Distribute the -6.

$$5y = -6x + 440$$ — Add 380 to both sides: $60 + 380 = 440$.

$$\frac{5y}{5} = \frac{-6x}{5} + \frac{440}{5}$$ — Divide both sides by 5.

$$y = -\frac{6}{5}x + 88$$ — Do the divisions. Write $\frac{-6x}{5}$ as $-\frac{6x}{5}$.

The graph of $y = -\frac{6}{5}x + 88$ is shown in Figure 3-56. From the graph, we see that the temperature in the railroad car before the cooler was turned on was 88°F. This is given by the y-intercept of the graph, $(0, 88)$. If we locate 25 on the x-axis and move straight up to intersect the graph, we will see that the temperature in the car was 58°F. This shows that after the cooler ran for 25 minutes, the temperature was 58°F.

FIGURE 3-56

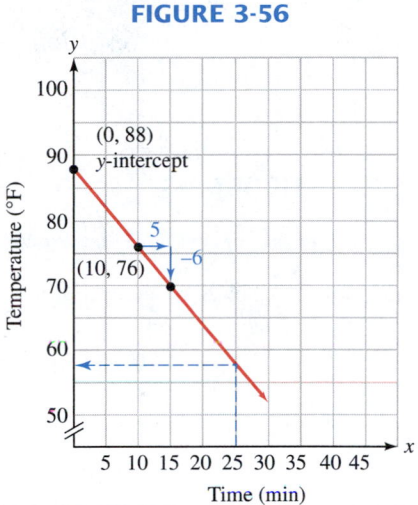

Writing the Equation of a Line Through Two Points

In the next example, we will show that it is possible to write the equation of a line when we know the coordinates of two points on the line.

EXAMPLE 3

Given two points on a line. Write the equation of the line passing through $P(4, 0)$ and $Q(6, -8)$.

Solution First we find the slope of the line.

$$m = \frac{y_2 - y_1}{x_2 - x_1}$$ — The slope formula.

$$= \frac{-8 - 0}{6 - 4}$$ — Substitute -8 for y_2, 0 for y_1, 6 for x_2, and 4 for x_1.

$$= \frac{-8}{2}$$ — Simplify.

$$= -4$$

Since the line passes through both P and Q, we can choose either point and substitute its coordinates into the point–slope form. If we choose $P(4, 0)$, we substitute 4 for x_1, 0 for y_1, and -4 for m and proceed as follows.

$$y - y_1 = m(x - x_1)$$ — Point–slope form.

$$y - 0 = -4(x - 4)$$ — Substitute -4 for m, 4 for x_1, and 0 for y_1.

$$y = -4x + 16$$ — Remove parentheses: distribute -4.

The equation of the line is $y = -4x + 16$.

SELF CHECK

Write the equation of the line passing through $R(0, -3)$ and $S(2, 1)$.
Answer: $y = 2x - 3$ ∎

EXAMPLE 4

Market research. A company that makes a breakfast cereal has found that the number of discount coupons redeemed for its product is linearly related to the coupon's value. In one advertising campaign, 10,000 "10¢ off" coupons were redeemed. In another campaign, 45,000 "50¢ off" coupons were redeemed. How many coupons can the company expect to be redeemed if it issues a "35¢ off" coupon?

Solution

If we let x represent the value of a coupon and y represent the number of coupons that will be redeemed, ordered pairs will have the form

(coupon value, number redeemed)

FIGURE 3-57

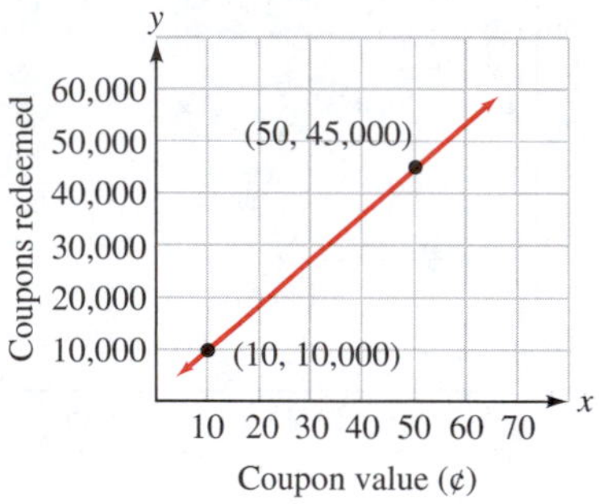

Two points on the graph of the equation are (10, 10,000) and (50, 45,000). These points are plotted on the graph shown in Figure 3-57. To write the equation of the line passing through the points, we first find the slope of the line.

$$m = \frac{y_2 - y_1}{x_2 - x_1} \qquad \text{The slope formula.}$$

$$= \frac{45,000 - 10,000}{50 - 10} \qquad \begin{array}{l}\text{Substitute 45,000 for } y_2\text{, 10,000 for } y_1, \\ \text{50 for } x_2\text{, and 10 for } x_1.\end{array}$$

$$= \frac{35,000}{40}$$

$$= 875$$

We then substitute 875 for m and the coordinates of one known point—say, (10, 10,000)—into the point–slope form of the equation of a line and proceed as follows:

$$y - y_1 = m(x - x_1) \qquad \text{The point–slope form.}$$
$$y - 10,000 = 875(x - 10) \qquad \text{Substitute for } m, x_1, \text{ and } y_1.$$
$$y - 10,000 = 875x - 8,750 \qquad \text{Distribute the 875.}$$
$$y = 875x + 1,250 \qquad \text{Add 10,000 to both sides.}$$

To find the expected number of coupons that will be redeemed, we substitute the value of the coupon, 35¢, into the equation $y = 875x + 1,250$ and find y.

$$y = 875x + 1,250$$
$$y = 875(35) + 1,250 \qquad \text{Substitute 35 for } x.$$
$$y = 30,625 + 1,250 \qquad \text{Do the multiplication.}$$
$$y = 31,875$$

The company can expect 31,875 of the 35¢ coupons to be redeemed.   ∎

Horizontal and Vertical Lines

We have graphed horizontal and vertical lines. We will now discuss how to write their equations.

EXAMPLE 5

Equations of horizontal and vertical lines. Write the equation of each line and then graph it: **a.** A horizontal line passing through $(-2, -4)$ and   **b.** A vertical line passing through $(1, 3)$.

Solution **a.** The equation of a horizontal line can be written in the form $y = b$. Since the y-coordinate of $(-2, -4)$ is -4, the equation of the line is $y = -4$. The graph is shown in Figure 3-58.

b. The equation of a vertical line can be written in the form $x = a$. Since the x-coordinate of $(1, 3)$ is 1, the equation of the line is $x = 1$. The graph is shown in Figure 3-58.

FIGURE 3-58

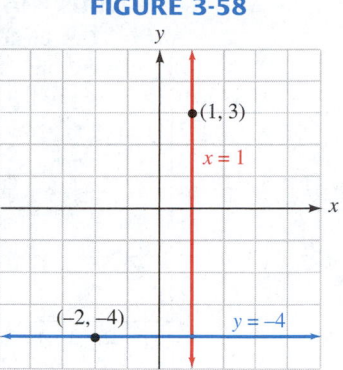

SELF CHECK Write the equation of each line and then graph it:
a. a horizontal line passing through $(3, 2)$ and
b. a vertical line passing through $(-1, -3)$.

Answers: **a.** $y = 2$,
b. $x = -1$

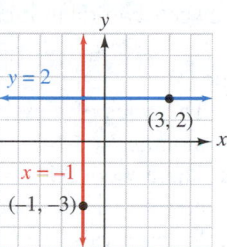

STUDY SET

Section 3.6

VOCABULARY

In Exercises 1–4, fill in the blanks to make the statements true.

1. $y - y_1 = m(x - x_1)$ is called the _____ form of the equation of a line.

2. The line in Illustration 1 _____ through point P.

3. In Illustration 1, point P has an _____ of 2 and a _____ of -1.

4. The _____ of a line gives a rate of change.

ILLUSTRATION 1

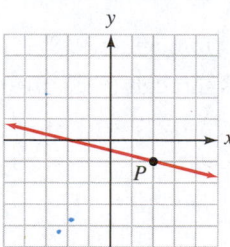

CONCEPTS

5. a. The linear equation $y = 2x - 3$ is written in *slope–intercept* form. What are the slope and the y-intercept of the graph of this line?

b. The linear equation $y - 4 = 6(x - 5)$ is written in *point–slope* form. What point does the graph of this equation pass through, and what is the line's slope?

6. Is the following statement true or false? The equations
$$y - 1 = 2(x - 2)$$
$$y = 2x - 3$$
$$2x - y = 3$$
all describe the same line.

7. See Illustration 2.
a. What information can be obtained from the fact that the line passes through the point $(8, 108)$?

b. What is the slope of the line and what does it tell us?

ILLUSTRATION 2

U.S. Birth Rates, 1997

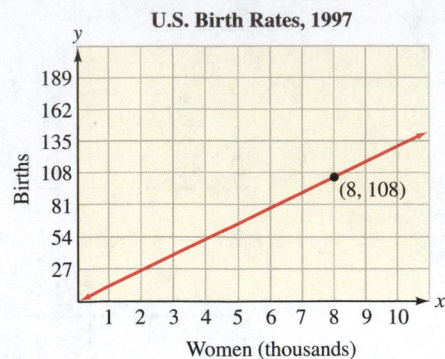

Women (thousands)

Based on data from the National Center for Health Statistics

8. In each of the following cases, a linear relationship between two quantities is described. If the relationship were graphed, what would be the slope of the line?
   a. The sales of new cars increased by 15 every 2 months.
   b. There were 35 fewer robberies for each dozen police officers added to the force.
   c. Withdrawals were occurring at the rate of $700 every 45 minutes.
   d. One acre of forest is being destroyed every 30 seconds.

9. In each of the following cases, is the given information sufficient to write the equation of the line?
   a. It passes through $(2, -7)$.
   b. Its slope is $-\frac{3}{4}$.
   c. It has the following table of values:

x	y
2	3
-3	-6

10. In each of the following cases, is the given information sufficient to write the equation of the line?
    a. It is horizontal.
    b. It is vertical and passes through $(-1, 1)$.
    c. It has the following table of values:

x	y
4	5

NOTATION

11. Fill in the blank. In $y - y_1 = m(x - x_1)$, we read x_1 as "x _____ one."

12. Write the equation of a horizontal line passing through $(0, b)$.

In Exercises 13–14, write the equation in slope–intercept form.

13. $y - 2 = -3(x - 4)$

   $y - \boxed{} = \boxed{} + 12$

   $y = -3x + 14$

14. $y + 2 = \frac{1}{2}(x + 2)$

   $\boxed{}(y + 2) = \boxed{}\left(\frac{1}{2}\right)(x + 2)$

   $2y + \boxed{} = \boxed{} + 2$

   $2y = x - \boxed{}$

   $y = \frac{1}{2}x - 1$

In Exercises 15–16, complete each solution.

15. Write the equation of the line with slope -2 that passes through the point $(-1, 5)$.

   $y - y_1 = m(x - x_1)$

   $y - \boxed{} = -2\left[x - \left(\boxed{}\right)\right]$

   $y - 5 = \boxed{} - 2$

   $y = -2x + 3$

16. Write the equation of the line with slope 4 that passes through the point $(0, 3)$.

   $y - y_1 = m(x - x_1)$

   $y - \boxed{} = 4\left(x - \boxed{}\right)$

   $y - 3 = \boxed{}$

   $y = 4x + 3$

PRACTICE

In Exercises 17–20, use the point–slope form to write the equation of the line with the given slope and point.

17. $m = 3$, passes through $(2, 1)$

18. $m = 2$, passes through $(4, 3)$

19. $m = -\dfrac{4}{5}$, passes through $(-5, -1)$

20. $m = -\dfrac{7}{8}$, passes through $(-2, -9)$

In Exercises 21–32, use the point–slope form to first write the equation of the line with the given slope and point. Then write your result in slope–intercept form.

21. $m = \dfrac{1}{5}$, passes through $(10, 1)$

22. $m = \dfrac{1}{4}$, passes through $(8, 1)$

23. $m = -5$, passes through $(-9, 8)$

24. $m = -4$, passes through $(-2, 10)$

25. $m = -\dfrac{4}{3}$,

x	y
6	-4

26. $m = -\dfrac{3}{2}$,

x	y
-2	1

27. $m = -\dfrac{2}{3}$, passes through $(3, 0)$

28. $m = -\dfrac{2}{5}$, passes through $(15, 0)$

29. $m = 8$, passes through $(0, 4)$

30. $m = 6$, passes through $(0, -4)$

31. $m = -3$, passes through the origin

32. $m = -1$, passes through the origin

In Exercises 33–38, write the equation of the line that passes through the two given points. Write your result in slope–intercept form.

33. Passes through $(1, 7)$ and $(-2, 1)$

34. Passes through $(-2, 2)$ and $(2, -8)$

35.

x	y
-4	3
2	0

36.

x	y
-1	-4
1	-2

37. Passes through $(5, 5)$ and $(7, 5)$

38. Passes through $(-2, 1)$ and $(-2, 15)$

In Exercises 39–42, write the equation of the line with the given characteristics.

39. vertical, passes through $(4, 5)$

40. vertical, passes through $(-2, -5)$

41. horizontal, passes through $(4, 5)$

42. horizontal, passes through $(-2, -5)$

APPLICATIONS

43. POLE VAULT See Illustration 3.
   a. For each of the four positions of the vault shown, give two points that the pole passes through.

ILLUSTRATION 3

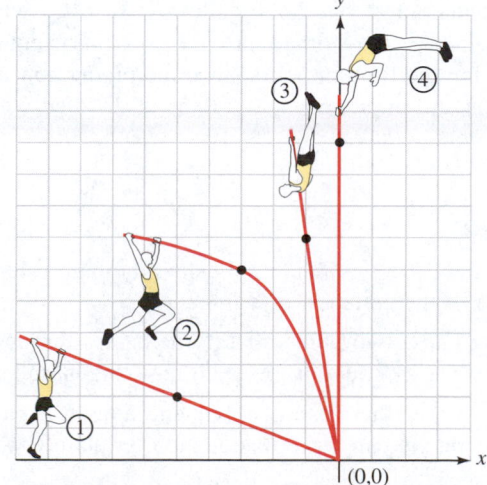

b. Write the equations of the lines that describe the position of the pole for parts 1, 3, and 4 of the jump.
   c. Why can't we write a linear equation describing the position of the pole for part 2?

44. FREEWAY DESIGN The graph in Illustration 4 shows the route of a proposed freeway.
   a. Give the coordinates of the points where the proposed freeway will join Interstate 25 and Highway 40.
   b. Write the equation of the line that mathematically describes the route of the proposed freeway. Answer in slope–intercept form.

ILLUSTRATION 4

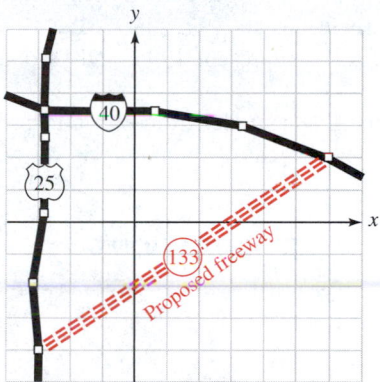

45. TOXIC CLEANUP Three months after cleanup began at a dump site, 800 cubic yards of toxic waste had yet to be removed. Two months later, that number had been lowered to 720 cubic yards.
   a. Write an equation that mathematically describes the linear relationship between the length of time x (in months) the cleanup crew has been working and the number of cubic yards y of toxic waste remaining.
   b. Use your answer to part a to predict the number of cubic yards of waste that will still be on the site one year after the cleanup project began.

46. DEPRECIATION To lower its corporate income tax, accountants of a large company depreciated a word processing system over several years using a linear model, as shown in the worksheet in Illustration 5.

ILLUSTRATION 5

Tax Worksheet

Method of depreciation: *Linear*

Property	Value	Years after purchase
Word processing system	$60,000	2
"	$30,000	4

a. Use the information in Illustration 5 to write a linear equation relating the years since the system was purchased x and its value y, in dollars.

b. Find the purchase price of the system by substituting $x = 0$ into your answer from part a.

47. COUNSELING In the first year of her practice, a family counselor saw 75 clients. In her second year, the number of clients grew to 105. If a linear trend continues, write an equation that gives the number of clients c the counselor will have t years after beginning her practice.

48. HEALTH CARE SPENDING The graph in Illustration 6 can be approximated by a straight line.
a. Use the given data to write a linear equation that approximates the per-person health care expenditures for the years 1992–1997. Let $x = 0$ represent 1992, $x = 2$ represent 1994, and so on.

b. Use the linear model from part a to predict the per person health care expenditure in the year 2020.

ILLUSTRATION 6

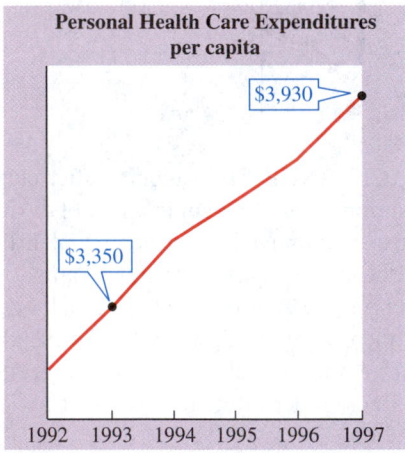

Personal Health Care Expenditures per capita

$3,930

$3,350

1992   1993   1994   1995   1996   1997

Based on data from the Health Care Financing Administration

49. CONVERTING TEMPERATURES The relationship between Fahrenheit temperature, F, and Celsius temperature, C, is linear.
a. Use the data in Illustration 7 to write two ordered pairs of the form (C, F).

ILLUSTRATION 7

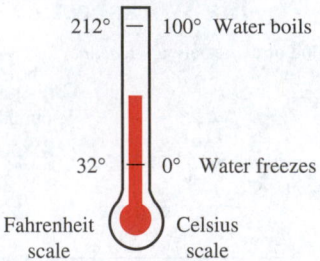

212° — 100° Water boils

32° — 0° Water freezes

Fahrenheit scale   Celsius scale

b. Use your answer to part a to write a linear equation relating the Fahrenheit and Celsius scales.

50. TRAMPOLINE The relationship between the circumference of a circle and its radius is linear. For instance, the length of the protective pad that wraps around a trampoline is related to the radius of the trampoline. Use the data in Illustration 8 to write a linear equation that approximates the length of pad needed for any trampoline radius.

ILLUSTRATION 8

Radius (ft)	Approximate length of padding (ft)
3	19
7	44

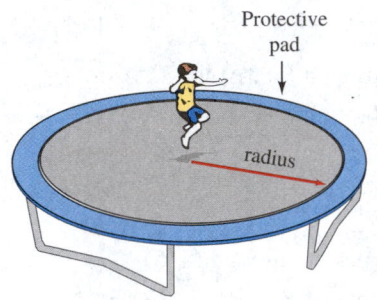

Protective pad

radius

51. AIR CONDITIONING An air-conditioning unit can lower the air temperature in a classroom 4° every 15 minutes. After the air conditioner had been running for half an hour, the air temperature in the room was 75°F. Write a linear equation relating the time in minutes the unit had been on and the temperature of the classroom. (*Hint:* How many minutes are there in half an hour?)

52. AUTOMATION An automated production line uses distilled water at a rate of 300 gallons every 2 hours to make shampoo. After the line had run for 7 hours, planners noted that 2,500 gallons of distilled water remained in the storage tank. Write a linear equation relating the time in hours since the production line began and the number of gallons of distilled water in the storage tank.

WRITING

53. Why is $y - y_1 = m(x - x_1)$ called the point–slope form of the equation of a line?

54. If we know two points that a line passes through, we can write its equation. Explain how this is done.

55. If we know the slope of a line and a point it passes through, we can write its equation. Explain how this is done.

56. Think of several points on the graph of the horizontal line $y = 4$. What do the points have in common? How do they differ?

REVIEW

57. Find the slope of the line passing through the points $(2, 4)$ and $(-6, 8)$.

58. Is the graph of $y = x^2$ a line?

59. Find the area of a circle with a diameter of 12 feet. Round to the nearest tenth.

60. If a 15-foot board is cut into two pieces and we let x represent the length of one piece (in feet), how long is the other piece?

61. Evaluate $(-1)^5$.

62. Solve $\dfrac{x - 3}{4} = -4$.

63. What is the coefficient of the second term of $-4x^2 + 6x - 13$?

64. Simplify $(-2p)(-5)(4x)$.

▶ 3.7 Functions

In this section, you will learn about

Functions ■ Domain and range of a function ■ Function notation ■ Graphs of functions ■ The vertical line test ■ Determining the domain and range

---

Introduction   In everyday life, we see a wide variety of situations where one quantity depends on another:

■ The distance traveled by a car depends on its speed.

■ The cost of renting a video depends on the number of days it is rented.

■ A state's number of representatives in Congress depends on the state's population.

In this section, we will discuss many situations where one quantity depends on another according to a specific rule, called a *function*. For example, the equation $y = 2x - 3$ sets up a rule where each value of y depends on the choice of some number x. The rule is: *To find y, double the value of x and subtract 3*. In this case, y (the *dependent variable*) depends on x (the *independent variable*).

Functions

We have previously described relationships between two quantities in different ways:

The number of tires to order	is	two	times	the number of bicycles to be manufactured.

Here words are used to state that the number of bicycle tires to order depends on the number of bicycles to be manufactured.

This rectangular coordinate graph shows many ordered pairs (x, y) that satisfy the equation $y = x^2$, where the value of the y-coordinate depends on the value of the x-coordinate.

Acres	Schools
400	4
800	8
1,000	10
2,000	20

This table shows that the number of schools needed depends on the size of the housing development.

$$t = 1{,}500 - d$$

This equation describes how the amount of take-home pay t depends on the amount of deductions d.

Two observations can be made about these examples:

■ Each one establishes a relationship between two sets of values. For example, the number of bicycle tires that must be ordered depends on the number of bicycles to be manufactured.

■ In these relationships, each value in one set is assigned a *single* value of a second set. For example, for each number of bicycles to be manufactured, there is exactly one number of tires to order.

Relationships between two quantities that exhibit both of these characteristics are called **functions.**

Functions

A **function** is a rule that assigns to each number x (the input) a single value y (the output). In this case, we say y is a *function of x.*

We can restate the previous definition as follows: For y to be a function of x, each value of x must determine exactly one value of y.

EXAMPLE 1

Identifying functions. **a.** Does $y = 4x + 1$ define a function?   **b.** Is age a function of body weight?

Solution   **a.** For each number x, we apply the rule: *Multiply x by 4 and add 1.* Since this arithmetic gives a single value of y, the equation defines a function.

b. To answer this question, we ask ourselves: *Does each body weight determine exactly one age?* Since a person weighing 130 pounds could be almost any age, the answer is no. This statement does not define a function.

SELF CHECK   **a.** Does $y = 2 - x^2$ define a function?   **b.** Is the temperature in a city a function of the time of day?   *Answers:* **a.** yes, **b.** yes   ■

Domain and Range of a Function

We have seen that functions can be represented by equations in two variables. Most often, x and y are used, but any letters are acceptable. Some examples of functions are

$$y = 2x - 10, \qquad y = x^2 + 2x - 3, \qquad \text{and} \qquad s = 5 - 16t$$

For a function, the set of all possible values of the independent variable x (the inputs) is called the **domain of the function.** The set of all possible values of the dependent variable y (the outputs) is called the **range of the function.**

EXAMPLE 2

Finding the domain and range of a function. Find the domain and range of $y = |x|$.

Solution   To find the domain of $y = |x|$, we determine which real numbers are allowable inputs for x. Since we can find the absolute value of any real number, the domain is the set of all real numbers. Since the absolute value of any real number x is greater than or equal to zero, the range of $y = |x|$ is the set of all real numbers greater than or equal to zero.

SELF CHECK Find the domain and range of the function *y* = −*x*. *Answer:* domain: all real numbers; range: all real numbers ∎

Function Notation

There is a special notation that we use to denote functions.

Function notation

The notation $y = f(x)$ denotes that y is a function of x.

The notation $y = f(x)$ is read as "*y* equals *f* of *x*." Note that y and $f(x)$ are two different notations for the same quantity. Thus, the equations $y = 4x + 1$ and $f(x) = 4x + 1$ represent the same relationship.

 WARNING! The symbol $f(x)$ denotes a function. It does not mean "*f* times *x*."

The notation $y = f(x)$ provides a way of denoting the value of y that corresponds to some number x. For example, if $f(x) = 4x + 1$, the value of y that is determined when $x = 2$ is denoted by $f(2)$.

$$f(x) = 4x + 1 \qquad \text{The function.}$$
$$f(2) = 4(2) + 1 \qquad \text{Replace } x \text{ with 2.}$$
$$= 8 + 1$$
$$= 9$$

Thus, $f(2) = 9$.

The letter f used in the notation $y = f(x)$ represents the word *function*. However, other letters can be used to represent functions. For example, $y = g(x)$ and $y = h(x)$ also denote functions involving the variable x.

EXAMPLE 3   **Evaluating functions.** For $g(x) = 3 - 2x$ and $h(x) = x^3 - 1$, find   **a.** $g(3)$ and   **b.** $h(-2)$.

Solution   **a.** To find $g(3)$, we use the function rule $g(x) = 3 - 2x$ and replace x with 3.

$$g(x) = 3 - 2x$$
$$g(3) = 3 - 2(3)$$
$$= 3 - 6$$
$$= -3$$

b. To find $h(-2)$, we use the function rule $h(x) = x^3 - 1$ and replace x with -2.

$$h(x) = x^3 - 1$$
$$h(-2) = (-2)^3 - 1$$
$$= -8 - 1$$
$$= -9$$

SELF CHECK Find $g(0)$ and $h(4)$ using the functions in Example 3.   *Answers:*   **a.** 3,   **b.** 63 ∎

We can think of a function as a machine that takes some input x and turns it into some output $f(x)$, as shown in Figure 3-59(a). The machine in Figure 3-59(b) turns the input value of -2 into the output value of -9, and we can write $f(-2) = -9$.

FIGURE 3-59

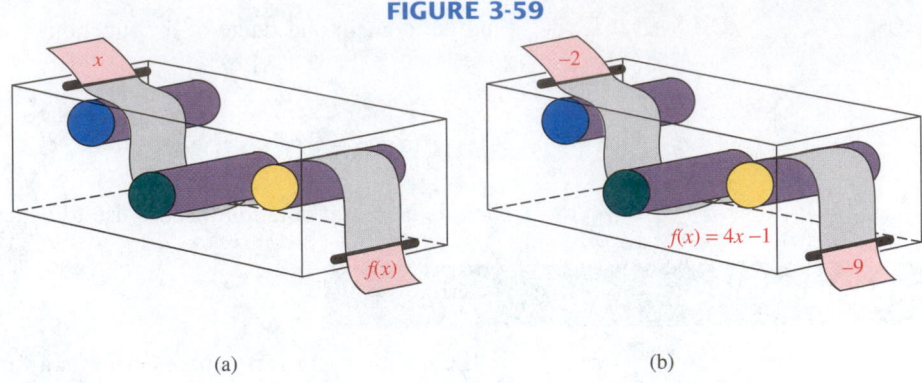

(a)   (b)

ACCENT ON TECHNOLOGY *Predicting Business Profits*

Accountants have found that the function $f(x) = -0.000065x^2 + 12x - 278,000$ estimates the profit a bowling alley will make when x games are bowled per year. Suppose that management predicts that 90,000 games will be bowled in the upcoming year. The expected profit for that year can be found by evaluating $f(90,000)$.

$$f(\mathbf{90,000}) = -0.000065(\mathbf{90,000})^2 + 12(\mathbf{90,000}) - 278,000$$

On a scientific calculator, we enter these numbers and press these keys:

Keystrokes   .000065 $\boxed{+/-}$ $\boxed{\times}$ 90000 $\boxed{x^2}$ $\boxed{+}$ 12 $\boxed{\times}$ 90000 $\boxed{-}$ 278000 $\boxed{=}$

$$\boxed{275500}$$

The expected profit is $275,500.

Using the Table mode on a graphing calculator, we can quickly evaluate a function for several values of the independent variable. For example, to predict the profit that the bowling alley would make if 93,000, 94,000 or 95,000 games are bowled, we first press the $\boxed{Y=}$ key and enter the function, as shown in Figure 3-60(a).

Next, we press $\boxed{2nd}$ $\boxed{TBLSET}$ and use the $\boxed{\blacktriangledown}$ key and the $\boxed{\blacktriangleright}$ key, in combination, to highlight the Ask option on the line labeled Indpnt and press ENTER. See Figure 3-60(b).

Finally, we press $\boxed{2nd}$ $\boxed{TABLE}$, enter the first value for x (93,000), and press $\boxed{ENTER}$. The predicted profit will appear under the column headed Y_1. See Figure 3-60(c). The dark cursor will drop down a line, and the calculator will "ask" you for another value of the independent variable. Enter 94,000, and then 95,000. In each case, the corresponding profit (275,660 and 275,375) will be displayed to the right of the input value. This process can be used to evaluate the function for any value of the independent variable.

FIGURE 3-60

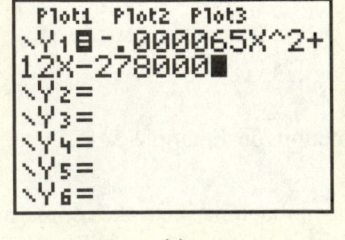

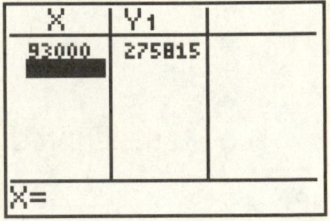

(a)   (b)   (c)

Graphs of Functions

The graph of the function $f(x) = 4x + 1$ is the same as the graph of the equation $y = 4x + 1$. So we can graph the function by making a table of values, plotting the points, and drawing the graph. A table of values and the graph of $f(x) = 4x + 1$ are shown in Figure 3-61.

FIGURE 3-61

$$y = 4x + 1$$

or

$$f(x) = 4x + 1$$

x	$f(x)$	$(x, f(x))$
-1	-3	$(-1, -3)$
0	1	$(0, 1)$
2	9	$(2, 9)$

Pick input values from the domain.   Find each corresponding output value.   Form ordered pairs.

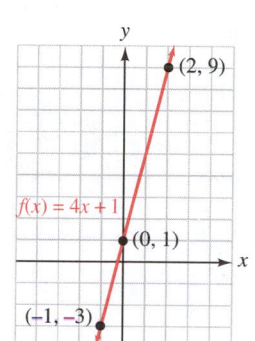

Any linear equation, except those of the form $x = a$, can be written using function notation by writing it in slope–intercept form ($y = mx + b$) and then replacing y with $f(x)$. We call this type of function a **linear function.**

Figure 3-62 shows the graphs of four basic functions.

FIGURE 3-62

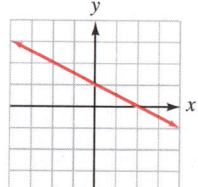

Linear function
$f(x) = mx + b$

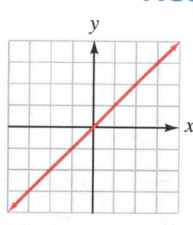

Identity function
$f(x) = x$

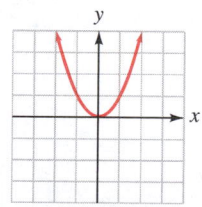

Squaring function
$f(x) = x^2$

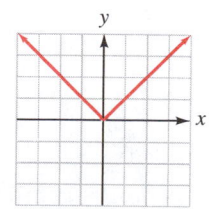

Absolute value function
$f(x) = |x|$

The Vertical Line Test

We can use the **vertical line test** to determine whether a given graph is the graph of a function. If any vertical line intersects a graph more than once, the graph cannot represent a function, because to one value of x, there corresponds more than one value of y. The graph in Figure 3-63(a), shown in red, is not the graph of a function, because the x-value -1 is assigned to three different y-values: 3, -1, and -4.

The graph shown in Figure 3-63(b) does represent a function, because every vertical line intersects the graph exactly once.

FIGURE 3-63

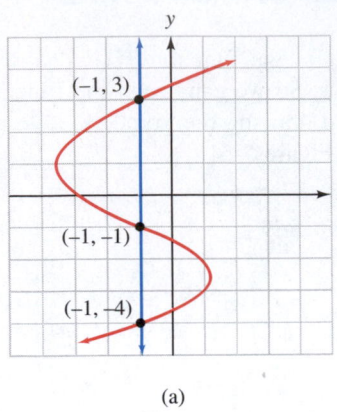

(a)

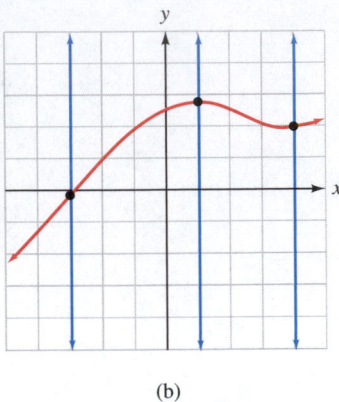

(b)

E X A M P L E   4

The vertical line test. Which of the following graphs are graphs of functions?

a.

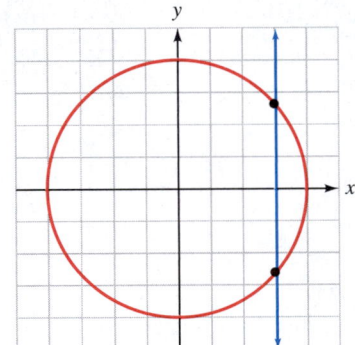

b.

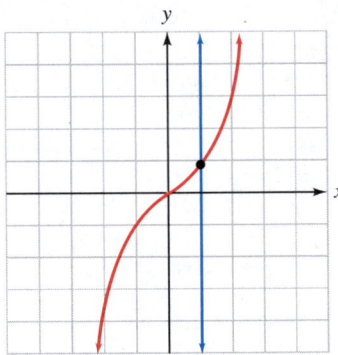

Solution **a.** This graph is not the graph of a function, because the vertical line intersects the graph at more than one point.

b. This graph is the graph of a function, because no vertical line will intersect the graph at more than one point.

SELF CHECK   Which of the following graphs are graphs of functions?

a.

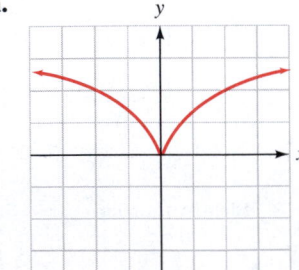

b.

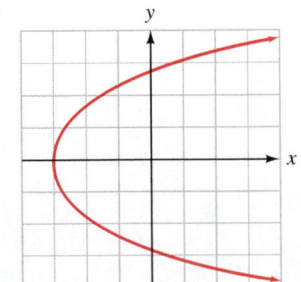

Answers:
a. function,
b. not a function ■

FIGURE 3-64

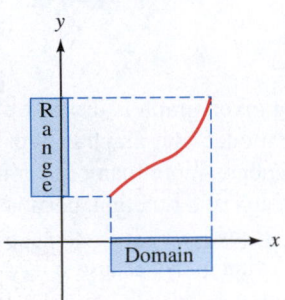

Determining the Domain and Range

In a function of the form $y = f(x)$, the numbers that we can substitute for x make up the domain of the function. The resulting values of y make up the range.

In a table of values, the numbers in the x-column are a partial listing of the numbers in the domain of a function. The numbers in the y-column are a partial listing of the numbers in the range of the function.

We can also determine the domain and range by looking at the graph of a function. For the function graphed in Figure 3-64, the domain is highlighted on the x-axis, and the range is highlighted on the y-axis.

EXAMPLE 5

Finding the domain and range. Find the domain and range of the function $f(x) = 4 - x^2$.

Solution The graph of $f(x) = 4 - x^2$ is shown in Figure 3-65. From the graph, we see that all numbers x on the x-axis are used. Thus, the domain is the set of all real numbers.

Since the values for y are always less than or equal to 4, the range is the set of real numbers less than or equal to 4.

FIGURE 3-65

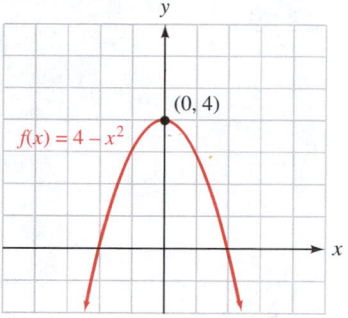

SELF CHECK Graph $f(x) = x^2 + 2$. Find the domain and range of the function.

Answer: domain: all real numbers; range: all real numbers greater than or equal to 2 ■

STUDY SET

Section 3.7

VOCABULARY

In Exercises 1–4, fill in the blanks to make the statements true.

1. A _____ is a rule that assigns to each value of the input set a single value of the output set.

2. The set of all possible input values for a function is called the _____, and the set of all possible output values is called the _____.

3. For $y = 2x + 8$, x is called the _____ variable, and y is called the _____ variable.

4. $f(x) = 6 - 5x$ is an example of _____ notation.

CONCEPTS

5. Consider the function $f(x) = x^2$.
   a. If positive real numbers are substituted for x, what type of numbers result?
   b. If negative real numbers are substituted for x, what type of numbers result?
   c. If zero is substituted for x, what number results?
   d. What are the domain and range of the function?

6. Consider the function $g(x) = x^4$.
   a. What type of numbers can be input in this function? What is the special name for this set?
   b. What type of numbers will be output by this function? What is the special name for this set?

7. See Illustration 1.
   a. Give the coordinates of the points where the given vertical line intersects the graph.
   b. Is this the graph of a function? Explain your answer.

ILLUSTRATION 1

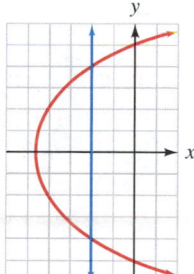

8. A function can be thought of as a machine that converts inputs into outputs. Use the terms *domain, range, input,* and *output* to label the diagram of a function machine in Illustration 2. Then find $f(2)$.

ILLUSTRATION 2

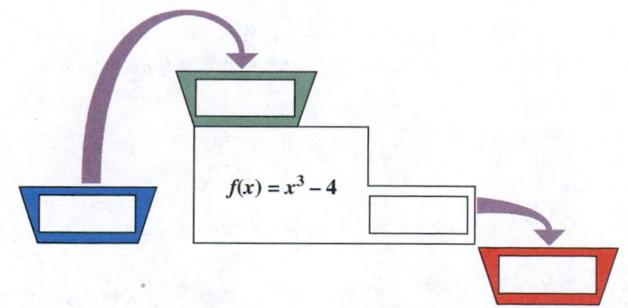

$f(x) = x^3 - 4$

9. Use the graph in Illustration 3 to find:
   a. $f(2)$ **b.** $f(0)$
   c. $f(-4)$

ILLUSTRATION 3

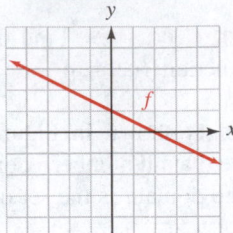

10. Use the graph in Illustration 4 to find:
   a. $g(-2)$ **b.** $g(1)$
   c. $g(-3)$

ILLUSTRATION 4

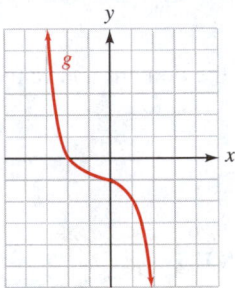

25.

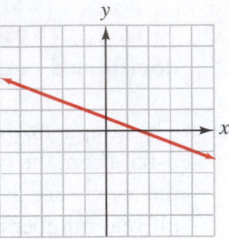

26.

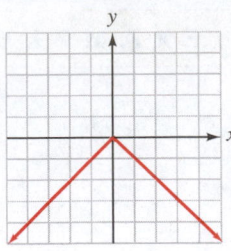

27.

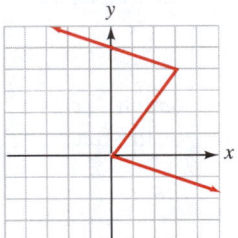

28.

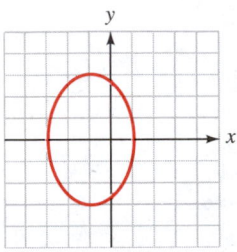

29. Is your age a function of your shoe size?

30. Is the number of phone calls you receive during the day a function of the time you wake up?

NOTATION

11. In the function $f(x) = x^2 + 1$, what is the input (independent) variable?

12. In the function $v(t) = -32t + 1,000$, what is the input (independent) variable?

13. Fill in the blanks to make the statements true. The function notation $f(4) = -5$ states that when 4 is substituted for ▢ in function f, the result is ▢. This fact can be illustrated graphically by plotting the point (▢ , ▢).

14. Fill in the blank: $f(x) = 6 - 5x$ is read as "f ___ x is $6 - 5x$."

15. Fill in the blanks: If $f(x) = 6 - 5x$, then $f(0) = 6$ is read as "f ___ zero ___ 6."

16. Tell whether this statement is true or false: The equations $y = 3x + 5$ and $f(x) = 3x + 5$ are the same.

PRACTICE

In Exercises 17–30, tell whether a function is defined. If it is not, indicate an input for which there is more than one output.

17. $y = 2x + 10$ **18.** $y = x - 15$

19. $y = x^2$ **20.** $y = |x|$

21. $y^2 = x$ **22.** $|y| = x$

23. $y = x^3$ **24.** $y = -x$

In Exercises 31–34, find the domain and range of the function.

31. $y = x + 1$ **32.** $y = 3x - 2$

33. $y = x^2$ **34.** $y = -|x|$

In Exercises 35–42, find each value.

35. $f(x) = 4x - 1$
   a. $f(1)$ **b.** $f(-2)$
   c. $f\left(\dfrac{1}{4}\right)$ **d.** $f(50)$

36. $g(x) = 1 - 5x$
   a. $g(0)$ **b.** $g(-75)$
   c. $g(0.2)$ **d.** $g\left(-\dfrac{4}{5}\right)$

37. $h(t) = 2t^2$
   a. $h(0.4)$ **b.** $h(-3)$
   c. $h(1,000)$ **d.** $h\left(\dfrac{1}{8}\right)$

38. $v(t) = 6 - t^2$
   a. $v(30)$ **b.** $v(6)$
   c. $v(-1)$ **d.** $v(0.5)$

39. $s(x) = |x - 7|$
   a. $s(0)$ **b.** $s(-7)$
   c. $s(7)$ **d.** $s(8)$

40. $f(x) = |2 + x|$
   a. $f(0)$ **b.** $f(2)$
   c. $f(-2)$ **d.** $f(-99)$

41. $f(x) = x^3 - x$
   a. $f(1)$ **b.** $f(10)$
   c. $f(-3)$ **d.** $f(6)$

42. $g(x) = x^4 + x$
   a. $g(1)$ **b.** $g(-2)$
   c. $g(0)$ **d.** $g(10)$

In Exercises 43–46, complete the table and graph the function. Then give the domain and range of the function.

43. $f(x) = -2 - 3x$

x	$f(x)$
0	
1	
−1	
−2	

44. $h(x) = |1 - x|$

x	$h(x)$
0	
1	
2	
3	
−1	
−2	

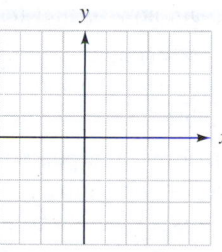

45. $s(x) = 2 - x^2$

x	$s(x)$
0	
1	
2	
−1	
−2	

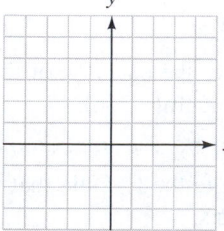

46. $g(x) = 1 + x^3$

x	$g(x)$
0	
1	
2	
−1	
−2	

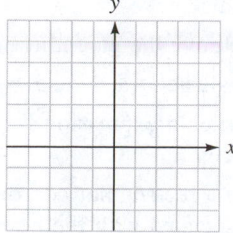

APPLICATIONS

47. REFLECTIONS When a beam of light hits a mirror, it is reflected off the mirror at the same angle that the incoming beam strikes the mirror, as shown in Illustration 5. What type of function could serve as a mathematical model for the path of the light beam shown here?

ILLUSTRATION 5

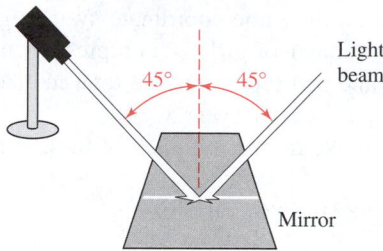

48. MATHEMATICAL MODELS Illustration 6 shows the path of a basketball shot taken by a player. What type of function could be used to mathematically model the path of the basketball?

ILLUSTRATION 6

49. TIDES Illustration 7 shows the graph of a function f, which gives the height of the tide for a 24-hour period in Seattle, Washington. (Note that military time is used on the x-axis: 3 A.M. = 3, noon = 12, 3 P.M. = 15, 9 P.M. = 21, and so on.)
   a. Find the domain of the function.
   b. Find $f(3)$.

ILLUSTRATION 7

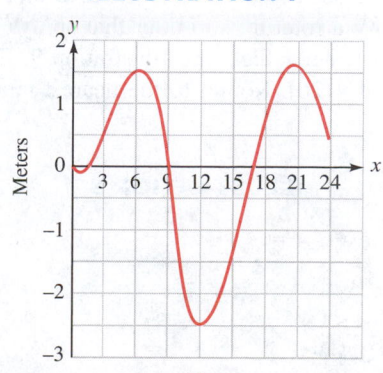

c. Find $f(6)$.

d. Find $f(15)$.

e. What information does $f(12)$ give?

f. Estimate $f(21)$.

50. SOCCER Illustration 8 shows the graphs of three functions on the same coordinate system: $g(x)$ represents the number of girls, $f(x)$ represents the number of boys, and $t(x)$ represents the total number playing high school soccer in year x.

a. What is the domain of each of these functions?

b. Find $g(87)$, $f(86)$, and $t(93)$.

c. Estimate $g(95)$, $f(95)$, and $t(95)$.

d. For what year x was $g(x) = 75,000$?

e. For what year x was $f(x) = 225,000$?

f. For what year x was $t(x)$ first greater than 350,000?

ILLUSTRATION 8

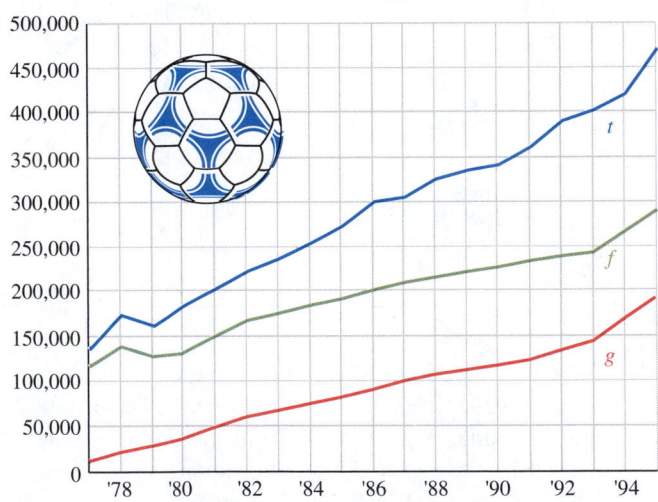

U.S. High School Soccer Participation 1977-1995

Based on data from *Los Angeles Times* (Dec. 12, 1996), p. A5

51. LAWN SPRINKLERS The function $A(r) = \pi r^2$ can be used to determine the area that will be watered by a rotating sprinkler that sprays out a stream of water r feet. See Illustration 9. Find $A(5)$, $A(10)$, and $A(20)$. Round to the nearest tenth.

ILLUSTRATION 9

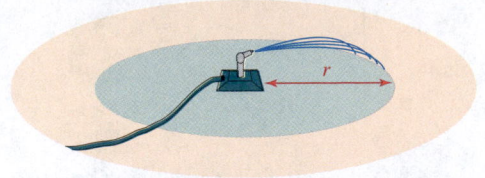

52. PARTS LIST The function

$$f(r) = 2.30 + 3.25(r + 0.40)$$

approximates the length (in feet) of the belt that joins the two pulleys shown in Illustration 10. r is the radius (in feet) of the smaller pulley. Find the belt length needed for each pulley in the parts list.

ILLUSTRATION 10

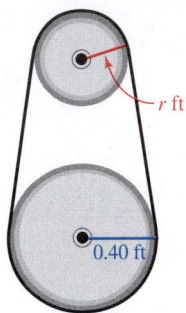

PARTS LIST		
Pulley	r	**Belt length**
P-45M	0.32	
P-08D	0.24	
P-00A	0.18	
P-57X	0.38	

WRITING

53. What is a function? Give an example.

54. Explain how to find the domain of a function.

55. In the function $y = -5x + 2$, why do you think x is called the *independent* variable and y the *dependent* variable?

56. Explain what a politician meant when she said, "The speed at which the downtown area will be redeveloped is a function of the number of low-interest loans made available to the property owners."

REVIEW

57. Give the equation of the horizontal line passing through $(-3, 6)$.

58. Is $t = -3$ a solution of $t^2 - t + 1 = 13$?

59. Write the formula that relates profit, revenue, and costs.

60. What is the word used to represent the perimeter of a circle?

61. Remove the parentheses in $-3(2x - 4)$.

62. Evaluate $r^2 - r$ for $r = -0.5$.

63. Write an expression for how many eggs there are in d dozen.

64. On a rectangular coordinate graph, what variable is associated with the horizontal axis?

Describing Linear Relationships

In Chapter 3, we discussed four ways to mathematically describe linear relationships between two quantities.

Equations in Two Variables

The general form of the equation of a line is $Ax + By = C$. Two very useful forms of the equation of a line are the slope–intercept form and the point–slope form.

1. Write the equation of a line with a slope of -3 and a y-intercept of $(0, -4)$.

2. Write the equation of the line that passes through $(5, 2)$ and $(-5, 0)$. Answer in slope–intercept form.

Rectangular Coordinate Graphs

The graph of an equation is a "picture" of all of its solutions (x, y). Important information can be obtained from a graph.

3. Complete the table of solutions for $2x - 4y = 8$. Then graph the equation.

4. See Illustration 1.
   a. What information does the y-intercept of the graph give us?
   b. What is the slope of the line and what does it tell us?

$2x - 4y = 8$

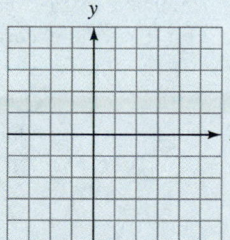

x	y
0	
	0
-2	

ILLUSTRATION 1

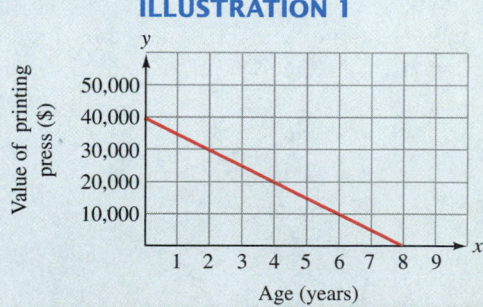

Linear Functions

We can use the notation $f(x) = mx + b$ to describe linear functions.

5. The function $f(x) = 25 + 35x$ gives the cost (in dollars) to rent a cement mixer for x days. Find $f(3)$. What does it represent?

6. The function $T(c) = \frac{1}{4}c + 40$ predicts the outdoor temperature T in degrees Fahrenheit using the number of cricket chirps c per minute. Find $T(160)$.

Accent on Teamwork

Section 3.1

Daily high temperature For a 2-week period, plot the daily high temperature for your city on a rectangular coordinate system. You can normally find this information in a local newspaper. Label the *x*-axis "observation day" and the *y*-axis "daily high temperature in degrees Fahrenheit." For example, the ordered pair (3, 72) indicates that on day 3 of the observation period, the high temperature was 72°F. At the end of the 2-week period, see whether any temperature trend is apparent from the graph.

ILLUSTRATION 1

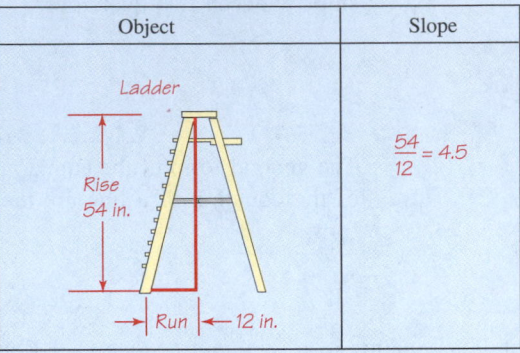

Section 3.2

Translations On a piece of graph paper, sketch the graph of $y = |x|$ with a black marker. Using a different color, sketch the graphs of $y = |x| + 2$ and $y = |x| - 2$ on the same coordinate system. On another piece of graph paper, do the same for $y = |x|$ and $y = |x + 2|$ and $y = |x - 2|$. Make some observations about how the graph of $y = |x|$ is "moved" or "translated" by the addition or subtraction of 2. Use what you have learned to discuss the graphs of $y = x^2$, $y = x^2 + 2$, $y = x^2 - 2$, $y = (x + 2)^2$, and $y = (x - 2)^2$.

could be posted next to the scale in the produce area so that shoppers could determine from the graph the cost of a banana purchase up to 8 pounds in weight. Label the *x*-axis in quarters of a pound and label the *y*-axis in cents.

Section 3.6

Matching game Have a student in your group write 10 linear equations on 3×5 note cards, one equation per card. Then have him or her graph each equation on a separate set of 10 cards. Shuffle each set of cards. Then put all the equation cards on one side of a table and all the cards with graphs on the other side. Work together to match each equation with its proper graph.

Section 3.3

Computer graphing programs If your school has a mathematics computer lab, ask the lab supervisor whether there is a graphing program on the system. If so, familiarize yourself with the operation of the program and then graph each of the equations from Figure 3-21 on page 210 and from Examples 3–6 in Section 3.3. Print out each graph and compare with those in the textbook.

Section 3.4

Measuring slope Use a tape measure (and a level if necessary) to find the slopes of five objects by finding $\frac{\text{rise}}{\text{run}}$. See the applications in Study Set 3.4 for some ideas about what you can measure. Record your results in a chart like the one shown in Illustration 1. List the examples in increasing order of magnitude, starting with the smallest slope.

Section 3.7

Table of values Consider the table of values shown at the right. After carefully inspecting each *x*-input and its corresponding output, we see that the values in the table are generated by the function $f(x) = \frac{x}{2}$.

x	$f(x)$
-4	-2
-2	-1
0	0
2	1
4	2

Have each member of your group construct a table of values for a function, write it on 3 × 5 cards, and distribute the cards to the other members of the group. (The function should be such that only one operation is performed on the input to obtain the output.) For each card you receive, determine a function that will generate the values in the table.

Section 3.5

Shopping Visit a local grocery store and find the price per pound of bananas. Make a rectangular coordinate graph that

Section 3.1 — Graphing Using the Rectangular Coordinate System

CONCEPTS

A *rectangular coordinate system* is composed of a horizontal number line called the *x*-axis and a vertical number line called the *y*-axis.

The coordinates of the *origin* are $(0, 0)$.

To *graph* ordered pairs means to locate their position on a coordinate system.

The two axes divide the coordinate plane into four distinct regions called *quadrants*.

REVIEW EXERCISES

1. a. Graph the points with coordinates $(-1, 3)$, $(0, 1.5)$, $(-4, -4)$, $\left(2, \frac{7}{2}\right)$, and $(4, 0)$.

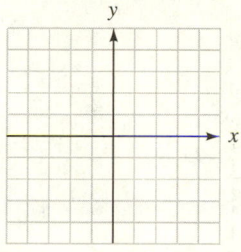

b. Use the graph in Illustration 1 to complete the table.

x	y
3	
	0
-3	

ILLUSTRATION 1

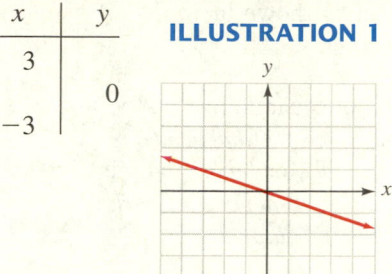

2. In what quadrant does the point $(-3, -4)$ lie?

3. SNOWFALL The amount of snow on the ground at a mountain resort was measured once each day over a 7-day period. (See Illustration 2.)
   a. On the first day, how much snow was on the ground?
   b. What was the difference in the amount of snow on the ground when the measurements were taken the second and third day?
   c. How much snow was on the ground on the sixth day?

ILLUSTRATION 2

ILLUSTRATION 3

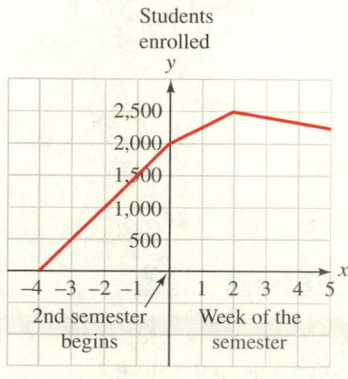

4. COLLEGE ENROLLMENT The graph in Illustration 3 gives the number of students enrolled at a college for the period from 4 weeks before to 5 weeks after the semester began.
   a. What was the maximum enrollment and when did it occur?
   b. How many students had enrolled 2 weeks before the semester began?
   c. When was enrollment 2,250?

Section 3.2

Equations Containing Two Variables

An ordered pair is a *solution* if, after substituting the values of the ordered pair for the variables in the equation, the result is a true statement.

Solutions of an equation can be shown in a *table of values.*

In an equation in x and y, x is called the *independent variable,* or *input,* and y is called the *dependent variable,* or *output.*

To graph an equation in two variables:
1. Make a table of values that contains several solutions written as ordered pairs.
2. Plot each ordered pair.
3. Draw a line or smooth curve through the points.

In many application problems, we encounter equations that contain variables other than x and y.

5. Check to see whether $(-3, 5)$ is a solution of $y = |2 + x|$.

6. a. Complete the table of values and graph the equation $y = -x^3$.

$$y = -x^3$$

x	y	(x, y)
-2		
-1		
0		
1		
2		

b. How would the graph of $y = -x^3 + 2$ compare to the graph of the equation given in part a?

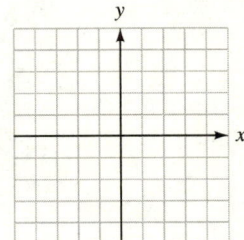

7. The graph in Illustration 4 shows the relationship between the number of oranges O an acre of land will yield if t orange trees are planted on it.
  a. If $t = 70$, what is O?
  b. What importance does the point $(40, 18)$ on the graph have?

ILLUSTRATION 4

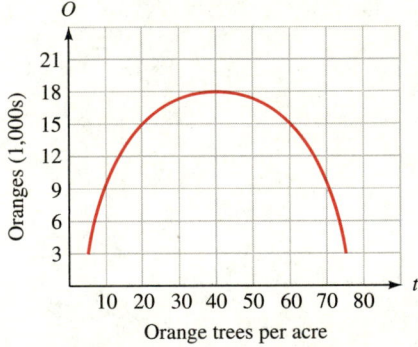

Orange trees per acre

Section 3.3

Graphing Linear Equations

An equation whose graph is a straight line and whose variables are raised to the first power is called a *linear equation.*

The *general* or *standard form* of a linear equation is $Ax + By = C$ where A, B, and C are real numbers and A and B are not both zero.

8. Classify each equation as either linear or nonlinear.
  a. $y = |x + 2|$
  b. $3x + 4y = 12$
  c. $y = 2x - 3$
  d. $y = x^2 - x$

9. The equation $5x + 2y = 10$ is in general form; what are A, B, and C?

10. Complete the table of solutions for the equation $3x + 2y = -18$.

x	y	(x, y)
-2		$(-2, \quad)$
	3	$(\quad, 3)$

To graph a linear equation:
1. Find three (x, y) pairs that satisfy the equation by picking three arbitrary x-values and finding their corresponding y-values.
2. Plot each ordered pair.
3. Draw a line through the points.

11. Solve the equation $x + 2y = 6$ for y, find three solutions, and then graph it.

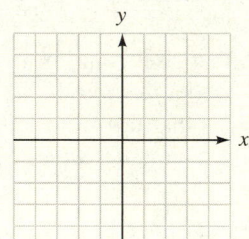

To find the *y-intercept* of a linear equation, substitute zero for x in the equation of the line and solve for y. To find the *x-intercept* of a linear equation, substitute zero for y in the equation of the line and solve for x.

12. Graph $-4x + 2y = 8$ by finding its x- and y-intercepts.

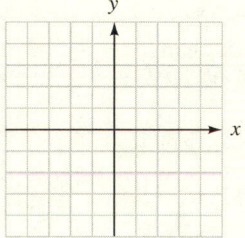

The equation $y = b$ represents the horizontal line that intersects the y-axis at $(0, b)$. The equation $x = a$ represents the vertical line that intersects the x-axis at $(a, 0)$.

13. Graph each equation.
    a. $y = 4$
    b. $x = -1$

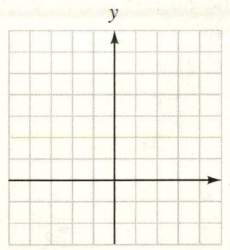
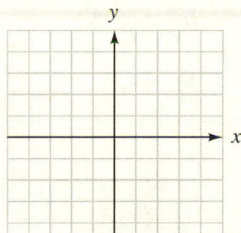

We can solve equations graphically by determining the x-coordinate of the x-intercept of the associated graph.

14. What is the solution of the equation $-3(x - 1) - 4x = 17$ if the graphing calculator display shows the graph of $y = -3(x - 1) - 4x - 17$?

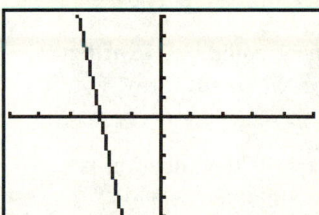

Section 3.4 Rate of Change and the Slope of a Line

The *slope m* of a nonvertical line is a number that measures "steepness" by finding the ratio $\frac{\text{rise}}{\text{run}}$.

$$m = \frac{\text{change in the } y\text{-values}}{\text{change in the } x\text{-values}}$$

15. In each case, find the slope of the line.

a.

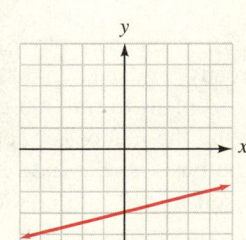

b. The line with the table of values shown here.

x	y	(x, y)
2	-3	$(2, -3)$
4	-17	$(4, -17)$

If $P(x_1, y_1)$ and $Q(x_2, y_2)$ are two points on a nonvertical line, the slope m of line PQ is

$$m = \frac{y_2 - y_1}{x_2 - x_1}$$

Lines that rise from left to right have a *positive slope*, and lines that fall from left to right have a *negative slope*.

Horizontal lines have a slope of zero. Vertical lines have *undefined* slope.

The slope of a line gives a rate of change.

c. The line passing through the points $(2, -5)$ and $(5, -5)$.

d. The line passing through the points $(1, -4)$ and $(3, -7)$.

16. Graph the line that passes through $(-2, 4)$ and has slope $m = -\frac{4}{5}$.

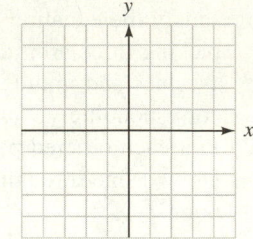

17. SALMON FISHING The graph in Illustration 5 shows the annual British Columbia salmon catch, in millions of fish.

a. When did the largest decline in the number of salmon caught occur? What was the rate of change over this time period?

b. When did the largest increase in the number of salmon caught occur? What was the rate of change over this time period?

ILLUSTRATION 5

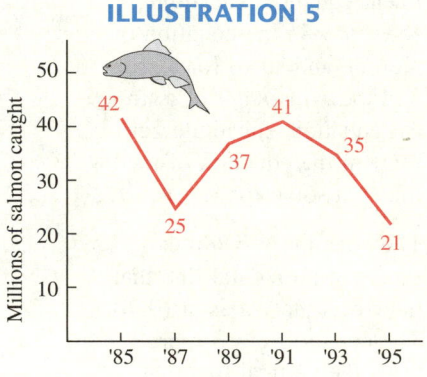

Based on data from *Los Angeles Times* (June 13, 1996)

Section 3.5

If a linear equation is written in *slope–intercept* form,

$$y = mx + b$$

the graph of the equation is a line with slope m and y-intercept $(0, b)$.

Describing Linear Relationships

18. Find the slope and the y-intercept of each line.

a. $y = \frac{3}{4}x - 2$

b. $y = -4x$

19. Find the slope and the y-intercept of the line determined by $9x - 3y = 15$. Then graph it.

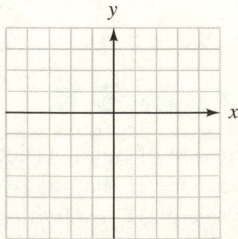

The *rate of change* is the slope of the graph of a linear equation.

20. COPIER USAGE A business buys a used copy machine that, when purchased, has already produced 75,000 copies.

a. If the business plans to run 300 copies a week, write a linear equation that would find the number of copies c the machine has made in its lifetime after the business has used it for w weeks.

b. Use your result to part a to predict the total number of copies that will have been made on the machine 1 year, or 52 weeks, after being purchased by the business.

Two lines with the same slope are *parallel*.

The product of the slopes of *perpendicular* lines is -1.

21. Without graphing, tell whether graphs of the given pair of lines would be parallel, perpendicular, or neither.

a. $y = -\dfrac{2}{3}x + 6$

$y = -\dfrac{2}{3}x - 6$

b. $x + 5y = -10$

$y = 5x$

Section 3.6

Writing Linear Equations

If a line with slope m passes through the point (x_1, y_1), the equation of the line in *point–slope* form is

$y - y_1 = m(x - x_1)$

22. Write the equation of a line with the given slope that passes through the given point. Express the result in slope–intercept form and graph the equation.

a. $m = 3$, $(1, 5)$

b. $m = -\dfrac{1}{2}$, $(-4, -1)$

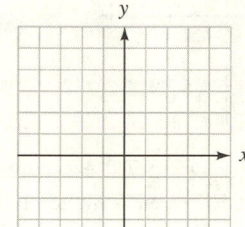 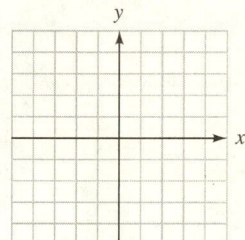

23. Write the equation of the line with the following characteristics. Express the result in slope–intercept form.

a. passing through $(3, 7)$ and $(-6, 1)$

b. horizontal, passing through $(6, -8)$

24. CAR REGISTRATION When it was 2 years old, the annual registration fee for a Dodge Caravan was $380. When it was 4 years old, the registration fee dropped to $310. If the relationship is linear, write an equation that gives the registration fee f in dollars for the van when it is x years old.

Section 3.7

Functions

A *function* is a rule that assigns to each input value a single output value.

For a function, the set of all possible values of the independent variable x (the inputs) is called the *domain,* and the set of all possible values of the dependent variable y (the outputs) is called the *range.*

25. In each case, tell whether a function is defined.

a. $y = 3x - 2$

b. Is your age a function of your height?

26. Find the domain and range of each function.

a.

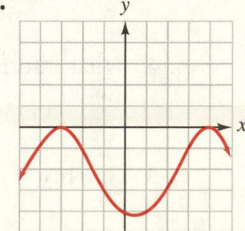

b. $y = x^2$

The notation $y = f(x)$ denotes that y is a function of x.

27. For the function $g(x) = 1 - 6x$, find each value.

a. $g(1)$

b. $g(-6)$

c. $g(0.5)$

d. $g\left(\dfrac{3}{2}\right)$

Four basic functions are

Linear: $f(x) = mx + b$
Identity: $f(x) = x$
Squaring: $f(x) = x^2$
Absolute value: $f(x) = |x|$

28. Complete the table and graph the function.

$$h(x) = 1 - |x|$$

x	$h(x)$
0	
1	
2	
−1	
−2	
−3	

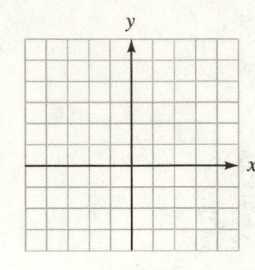

We can use the *vertical line test* to determine whether a graph is the graph of a function.

29. Tell whether each graph is the graph of a function.

a.

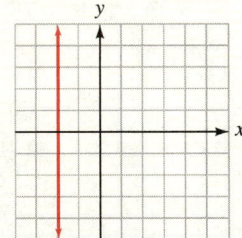

b.

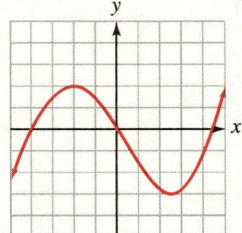

30. The function $f(r) = 15.7r^2$ estimates the volume in cubic inches of a can 5 inches tall with a radius of r inches. Find the volume of the can in Illustration 6.

ILLUSTRATION 6

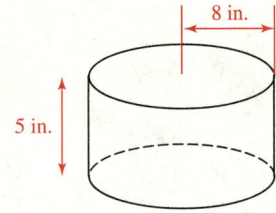

CHAPTER 3

Test

The graph in Illustration 1 shows the number of dogs being boarded in a kennel over a 3-day holiday weekend. Use the graph to answer Problems 1–4.

ILLUSTRATION 1

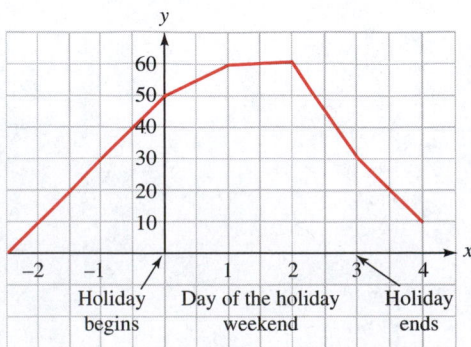

1. How many dogs were in the kennel 2 days before the holiday?

2. What is the maximum number of dogs that were boarded on the holiday weekend?

3. When were there 30 dogs in the kennel?

4. What information does the y-intercept of the graph give?

5. Graph $y = x^2 - 4$.

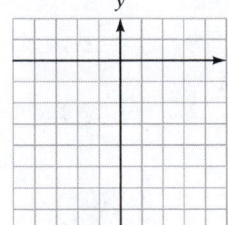

6. Graph $8x + 4y = -24$.

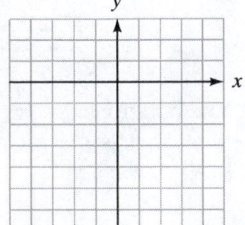

7. Is $(-3, -4)$ a solution of $3x - 4y = 7$?

8. Is $y = x^3$ a linear equation?

9. What are the x- and y-intercepts of the graph of $2x - 3y = 6$?

10. Find the slope and the y-intercept of $x + 2y = 8$.

11. Graph $x = -4$.

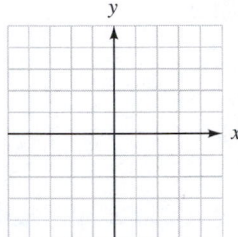

12. Graph the line passing through $(-2, -4)$ having a slope of $\frac{2}{3}$.

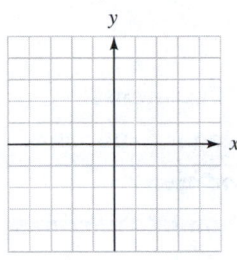

13. What is the slope of the line passing through $(-1, 3)$ and $(3, -1)$?

14. What is the slope of a vertical line?

15. What is the slope of a line that is perpendicular to a line with slope $-\frac{7}{8}$?

16. When graphed, are the lines $y = 2x + 6$ and $6x - 3y = 0$ parallel, perpendicular, or neither?

19. DEPRECIATION After it is purchased, a $15,000 computer loses $1,500 in resale value every year. Write an equation that gives the resale value v of the computer x years after being purchased.

20. Write the equation of the line passing through $(-2, 5)$ and $(-3, -2)$. Answer in slope–intercept form.

21. Is this the graph of a function?

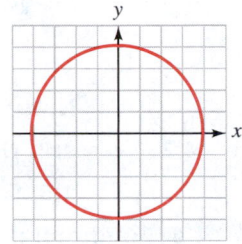

22. Find the domain and range of the function.

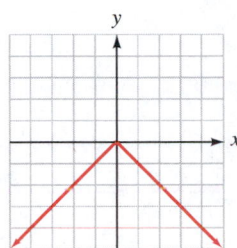

23. Does the equation $y = 2x - 8$ define a function?

24. If $f(x) = 2x - 7$, find $f(-3)$.

25. If $g(s) = 3.5s^3$, find $g(6)$.

26. Explain what is meant by the statement slope $= \frac{\text{rise}}{\text{run}}$.

In Problems 17–18, refer to the graph in Illustration 2, which shows the elevation changes in a 26-mile marathon course. Give the rate of change of the part of the course that has . . .

17. the steepest incline   **18.** the steepest decline

ILLUSTRATION 2

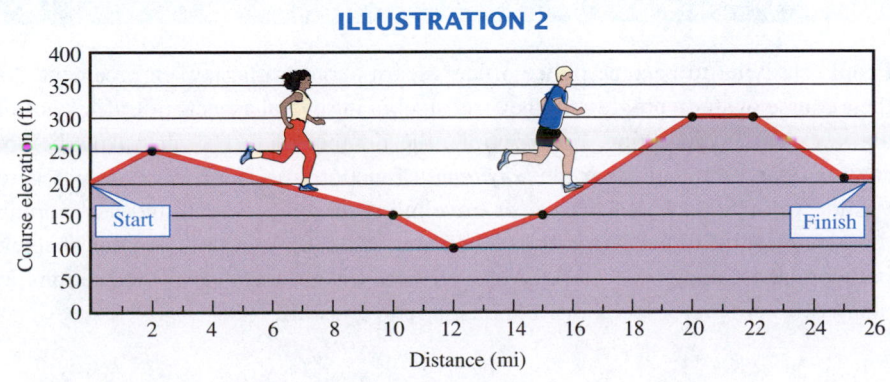

4

Exponents and Polynomials

CAMPUS CONNECTION

The Administration of Justice Department

People studying to become police officers learn about traffic law enforcement. As their course of study progresses, they use algebra in several aspects of traffic control and accident investigation. For example, the number of feet a car travels before stopping can be found using the *polynomial* function $f(v) = 0.04v^2 + 0.9v$, where v is the velocity of the car. For a car traveling at 30 mph, we can find its stopping distance by substituting 30 for v and evaluating $0.04(30)^2 + 0.9(30)$. In this chapter, polynomial functions are introduced. You will see that they have applications in many different areas, including administration of justice.

> *In this chapter, we introduce the rules for exponents and use them to add, subtract, multiply, and divide polynomials.*

▶ 4.1 Natural-Number Exponents

In this section, you will learn about

> Powers of expressions ■ The product rule for exponents ■ The power rules for exponents ■ The quotient rule for exponents

Introduction   We have used natural-number exponents to indicate repeated multiplication. For example,

$$2^5 = 2 \cdot 2 \cdot 2 \cdot 2 \cdot 2 = 32 \qquad (-7)^3 = (-7)(-7)(-7) = -343$$
$$x^4 = x \cdot x \cdot x \cdot x \qquad\qquad -y^5 = -y \cdot y \cdot y \cdot y \cdot y$$

These examples suggest a definition for x^n, where n is a natural number.

Natural-number exponents

If n is a natural number, then

$$x^n = \overbrace{x \cdot x \cdot x \cdot \cdots \cdot x}^{n \text{ factors of } x}$$

In the exponential expression x^n, x is called the **base** and n is called the **exponent.** The entire expression is called a **power of x.**

Base $\rightarrow x^n \leftarrow$ Exponent

Powers of Expressions

If an exponent is a natural number, it tells how many times its base is to be used as a factor. An exponent of 1 indicates that its base is to be used one time as a factor, an exponent of 2 indicates that its base is to be used two times as a factor, and so on.

$$3^1 = 3, \qquad (-y)^1 = -y, \qquad (-4z)^2 = (-4z)(-4z), \qquad \text{and} \qquad (t^2)^3 = t^2 \cdot t^2 \cdot t^2$$

EXAMPLE 1   **Powers of integers.** Show that -2^4 and $(-2)^4$ have different values.

Solution   We find each power and show that the results are different. In the first expression, the base is 2. In the second expression, the base is -2.

$$-2^4 = -(2^4) \qquad\qquad (-2)^4 = (-2)(-2)(-2)(-2)$$
$$= -(2 \cdot 2 \cdot 2 \cdot 2) \qquad\qquad\quad = 16$$
$$= -16$$

Since $-16 \neq 16$, it follows that $-2^4 \neq (-2)^4$.

SELF CHECK   Show that $(-4)^3$ and -4^3 have the same value.   *Answer:* $(-4)^3 = -4^3 = -64$ ■

EXAMPLE 2

Powers of algebraic expressions. Write each expression without using exponents: **a.** r^3, **b.** $(-2s)^4$, and **c.** $\left(\frac{1}{3}ab\right)^5$.

Solution

a. $r^3 = r \cdot r \cdot r$ Write the base r as a factor three times.

b. $(-2s)^4 = (-2s)(-2s)(-2s)(-2s)$ Write the base $-2s$ as a factor four times.

c. $\left(\frac{1}{3}ab\right)^5 = \left(\frac{1}{3}ab\right)\left(\frac{1}{3}ab\right)\left(\frac{1}{3}ab\right)\left(\frac{1}{3}ab\right)\left(\frac{1}{3}ab\right)$ Write the base $\frac{1}{3}ab$ as a factor five times.

SELF CHECK

Write each expression without using exponents: *Answers:* **a.** $x \cdot x \cdot x \cdot x$,
a. x^4 and **b.** $\left(-\frac{1}{2}xy\right)^3$.   **b.** $\left(-\frac{1}{2}xy\right)\left(-\frac{1}{2}xy\right)\left(-\frac{1}{2}xy\right)$ ∎

The Product Rule for Exponents

To develop a rule for multiplying exponential expressions with the same base, we consider the product $x^2 \cdot x^3$. Since the expression x^2 means that x is to be used as a factor two times and the expression x^3 means that x is to be used as a factor three times, we have

$$x^2 \cdot x^3 = \overbrace{x \cdot x}^{2 \text{ factors of } x} \cdot \overbrace{x \cdot x \cdot x}^{3 \text{ factors of } x}$$

$$= \overbrace{x \cdot x \cdot x \cdot x \cdot x}^{5 \text{ factors of } x}$$

$$= x^5$$

In general,

$$x^m \cdot x^n = \overbrace{x \cdot x \cdot x \cdots \cdots x}^{m \text{ factors of } x} \overbrace{x \cdot x \cdot x \cdot x \cdots \cdots x}^{n \text{ factors of } x}$$

$$= \overbrace{x \cdot x \cdot x \cdot x \cdot x \cdot x \cdots \cdots x \cdot x \cdot x}^{m + n \text{ factors of } x}$$

$$= x^{m+n}$$

This discussion suggests the following rule: *To multiply two exponential expressions with the same base, keep the common base and add the exponents.*

Product rule for exponents

If m and n are natural numbers, then

$$x^m x^n = x^{m+n}$$

EXAMPLE 3

The product rule for exponents. Simplify by writing each expression using one base and one exponent: **a.** $x^3 x^4$ and **b.** $y^2 y^4 y$.

a. $x^3 x^4 = x^{3+4}$ Keep the base and add the exponents.

     $= x^7$ Do the addition: $3 + 4 = 7$.

b. $y^2 y^4 y = (y^2 y^4)y$ Use the associative property to group y^2 and y^4.

     $= (y^{2+4})y$ Keep the base and add the exponents.

     $= y^6 y$ Do the addition: $2 + 4 = 6$.

     $= y^{6+1}$ Keep the base and add the exponents. Recall that $y = y^1$.

     $= y^7$ Do the addition: $6 + 1 = 7$.

Simplify **a.** zz^3 and **b.** $x^2x^3x^6$. *Answers:* **a.** z^4, **b.** x^{11} ■

WARNING! We cannot simplify $x^3 + x^4$ or $x^3 - x^4$ because x^3 and x^4 are not like terms—they don't have exactly the same variables *and* exponents. However, we can simplify $x^3 \cdot x^4$ using the product rule, because the exponential expressions have the same base.

E X A M P L E   4   **The product rule for exponents.** Simplify $(2y^3)(3y^2)$.

Solution $(2y^3)(3y^2) = 2(3)y^3y^2$ Use the commutative and associative properties to group the coefficients together and the variables together.

$= 6y^{3+2}$ Multiply 2 and 3. Keep the base y and add the exponents.

$= 6y^5$ Do the addition: $3 + 2 = 5$.

Simplify $(4x)(-3x^2)$.   *Answer:* $-12x^3$ ■

WARNING! The product rule for exponents applies to exponential expressions with the same base. An expression such as x^2y^3 cannot be simplified, because x^2 and y^3 have different bases.

The Power Rules for Exponents

To find another rule for exponents, we consider the expression $(x^3)^4$, which can be written as $x^3 \cdot x^3 \cdot x^3 \cdot x^3$. Because each of the four factors of x^3 contains three factors of x, there are $4 \cdot 3$ (or 12) factors of x. This product can be written as x^{12}.

$$(x^3)^4 = x^3 \cdot x^3 \cdot x^3 \cdot x^3$$

$$\overbrace{= \underbrace{x \cdot x \cdot x}_{x^3} \cdot \underbrace{x \cdot x \cdot x}_{x^3} \cdot \underbrace{x \cdot x \cdot x}_{x^3} \cdot \underbrace{x \cdot x \cdot x}_{x^3}}^{12 \text{ factors of } x}$$

$$= x^{12}$$

In general,

$$(x^m)^n = \overbrace{x^m \cdot x^m \cdot x^m \cdot \cdots \cdot x^m}^{n \text{ factors of } x^m}$$

$$= \overbrace{x \cdot x \cdot x \cdot x \cdot x \cdot x \cdot x \cdot \cdots \cdot x}^{m \cdot n \text{ factors of } x}$$

$$= x^{m \cdot n}$$

This discussion suggests the following rule: *To raise an exponential expression to a power, keep the base and multiply the exponents.*

The first power rule for exponents

If m and n are natural numbers, then

$$(x^m)^n = x^{m \cdot n}$$

EXAMPLE 5

The first power rule for exponents. Simplify by writing each expression using one base and one exponent.

a. $(2^3)^7 = 2^{3 \cdot 7}$ Keep the base and multiply the exponents.

$= 2^{21}$ Do the multiplication: $3 \cdot 7 = 21$.

b. $(z^7)^7 = z^{7 \cdot 7}$ Keep the base and multiply the exponents.

$= z^{49}$ Do the multiplication: $7 \cdot 7 = 49$.

SELF CHECK

Write each expression using one exponent:
a. $(y^5)^2$ and   **b.** $(u^x)^y$.   *Answers:* **a.** y^{10},   **b.** u^{xy} ■

EXAMPLE 6

Applying two rules for exponents. Use the product and power rules of exponents to write each expression using one base and one exponent.

a. $(x^2 x^5)^2 = (x^7)^2$ **b.** $(y^6 y^2)^3 = (y^8)^3$

$= x^{14}$ $= y^{24}$

c. $(z^2)^4(z^3)^3 = z^8 z^9$ **d.** $(x^3)^2(x^5 x^2)^3 = x^6(x^7)^3$

$= z^{17}$ $= x^6 x^{21}$

$= x^{27}$

SELF CHECK

Write each expression using one exponent:
a. $(a^4 a^3)^3$ and   **b.** $(a^3)^3(a^4)^2$.   *Answers:* **a.** a^{21},   **b.** a^{17} ■

To find two more rules for exponents, we consider the expressions $(2x)^3$ and $\left(\frac{2}{x}\right)^3$.

$(2x)^3 = (2x)(2x)(2x)$ $\left(\frac{2}{x}\right)^3 = \left(\frac{2}{x}\right)\left(\frac{2}{x}\right)\left(\frac{2}{x}\right)$ $(x \neq 0)$

$= (2 \cdot 2 \cdot 2)(x \cdot x \cdot x)$ $= \frac{2 \cdot 2 \cdot 2}{x \cdot x \cdot x}$ Multiply the numerators.
Multiply the denominators.

$= 2^3 x^3$ $= \frac{2^3}{x^3}$

$= 8x^3$ $= \frac{8}{x^3}$

These examples suggest the following rules: *To raise a product to a power, we raise each factor of the product to that power,* and *to raise a fraction to a power, we raise both the numerator and the denominator to that power.*

More power rules for exponents

If n is a natural number, then

$(xy)^n = x^n y^n$ and if $y \neq 0$, then $\left(\frac{x}{y}\right)^n = \frac{x^n}{y^n}$

EXAMPLE 7

Power rules for exponents. Simplify by writing each expression without using parentheses.

a. $(ab)^4 = a^4 b^4$ **b.** $(3c)^3 = 3^3 c^3$

$= 27c^3$

c. $(x^2 y^3)^5 = (x^2)^5(y^3)^5$ **d.** $(-2x^3 y)^2 = (-2)^2(x^3)^2 y^2$

$= x^{10} y^{15}$ $= 4x^6 y^2$

e. $\left(\dfrac{4}{k}\right)^3 = \dfrac{4^3}{k^3}$

$= \dfrac{64}{k^3}$

f. $\left(\dfrac{3x^2}{2y^3}\right)^5 = \dfrac{3^5(x^2)^5}{2^5(y^3)^5}$

$= \dfrac{243x^{10}}{32y^{15}}$

SELF CHECK Write each expression without using parentheses:

a. $(3x^2y)^2$ and **b.** $\left(\dfrac{2x^3}{3y^2}\right)^4$

Answers: **a.** $9x^4y^2$, **b.** $\dfrac{16x^{12}}{81y^8}$

The Quotient Rule for Exponents

We now consider the fraction

$$\dfrac{4^5}{4^2}$$

where the exponent in the numerator is greater than the exponent in the denominator. We can simplify this fraction as follows:

$$\dfrac{4^5}{4^2} = \dfrac{4 \cdot 4 \cdot 4 \cdot 4 \cdot 4}{4 \cdot 4}$$

$$= \dfrac{\overset{1}{\cancel{4}} \cdot \overset{1}{\cancel{4}} \cdot 4 \cdot 4 \cdot 4}{\underset{1}{\cancel{4}} \cdot \underset{1}{\cancel{4}}} \qquad \text{\color{red}Divide out the common factors of 4.}$$

$$= 4^3$$

The result of 4^3 has a base of 4 and an exponent of $5 - 2$ (or 3). This suggests that *to divide exponential expressions with the same base, we keep the common base and subtract the exponents.*

Quotient rule for exponents

If m and n are natural numbers, $m > n$ and $x \neq 0$, then

$$\dfrac{x^m}{x^n} = x^{m-n}$$

E X A M P L E 8 **Quotient rule for exponents.** Simplify each expression. Assume that there are no divisions by zero.

a. $\dfrac{x^4}{x^3} = x^{4-3}$

$= x^1$

$= x$

b. $\dfrac{8y^2y^6}{4y^3} = \dfrac{8y^8}{4y^3}$

$= \dfrac{8}{4}y^{8-3}$

$= 2y^5$

c. $\dfrac{a^3a^5a^7}{a^4a} = \dfrac{a^{15}}{a^5}$

$= a^{15-5}$

$= a^{10}$

d. $\dfrac{(a^3b^4)^2}{ab^5} = \dfrac{a^6b^8}{ab^5}$

$= a^{6-1}b^{8-5}$

$= a^5b^3$

SELF CHECK   Simplify each expression: **a.** $\dfrac{a^5}{a^3}$,

b. $\dfrac{6b^2b^3}{2b^4}$,   and   **c.** $\dfrac{(x^2y^3)^2}{x^3y^4}$.

Answers: **a.** a^2,   **b.** $3b$,
c. xy^2

The rules for natural-number exponents are summarized as follows.

Rules for exponents

If n is a natural number, then

$$\overbrace{x^n = x \cdot x \cdot x \cdot \cdots \cdot x}^{n \text{ factors of } x}$$

If m and n are natural numbers and there are no divisions by zero, then

$$x^m x^n = x^{m+n} \qquad (x^m)^n = x^{m \cdot n} \qquad (xy)^n = x^n y^n$$

$$\left(\dfrac{x}{y}\right)^n = \dfrac{x^n}{y^n} \qquad \dfrac{x^m}{x^n} = x^{m-n} \quad (\text{provided } m > n)$$

STUDY SET

Section 4.1

VOCABULARY

In Exercises 1–4, fill in the blanks to make the statements true.

1. The _____ of the exponential expression $(-5)^3$ is -5. The _____ is 3.

2. The _____ expression x^4 represents a repeated multiplication where x is to be written as a _____ four times.

3. x^n is called a _____ of x.

4. $\{1, 2, 3, 4, 5, \ldots\}$ is the set of _____ numbers.

CONCEPTS

In Exercises 5–14, fill in the blanks to make the statements true.

5. $(3x)^4$ means $\boxed{} \cdot \boxed{} \cdot \boxed{} \cdot \boxed{}$

6. Using an exponent, $(-5y)(-5y)(-5y)$ can be written as $\boxed{}$.

7. $x^m x^n = \boxed{}$

8. $(xy)^n = \boxed{}$

9. $\left(\dfrac{a}{b}\right)^n = \boxed{}$

10. $(a^b)^c = \boxed{}$

11. $\dfrac{x^m}{x^n} = \boxed{}$

12. $x = x^{\boxed{}}$

13. $(xy)^{\boxed{}} = (xy)$

14. $(t^3)^2 = \boxed{} \cdot \boxed{}$

In Exercises 15–18, simplify each expression, if possible.

15. **a.** $x^2 + x^2$ **b.** $x^2 - x^2$
    c. $x^2 \cdot x^2$ **d.** $\dfrac{x^2}{x^2}$

16. **a.** $x^2 + x$ **b.** $x^2 - x$
    c. $x^2 \cdot x$ **d.** $\dfrac{x^2}{x}$

17. **a.** $6x^3 + 2x^2$ **b.** $6x^3 - 2x^2$
    c. $6x^3 \cdot 2x^2$ **d.** $\dfrac{6x^3}{2x^2}$

18. **a.** $-8x^4 + 4x^4$ **b.** $-8x^4 - 4x^4$
    c. $-8x^4(4x^4)$ **d.** $\dfrac{-8x^4}{4x^4}$

In Exercises 19–22, find the area or volume of each figure, whichever is appropriate. You may leave π in your answer.

19.

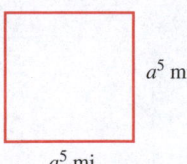

a^5 mi

a^5 mi

20.

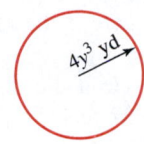

21.

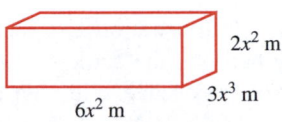

22.

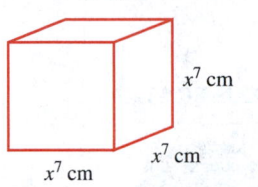

In Exercises 23–26, complete each table.

23.

x	3^x
1	
2	
3	
4	

24.

x	6^x
1	
2	
3	
4	

25.

x	$(-4)^x$
1	
2	
3	
4	

26.

x	$(-5)^x$
1	
2	
3	
4	

In Exercises 27–28, use the graph to determine the missing y-coordinates in the table.

27.

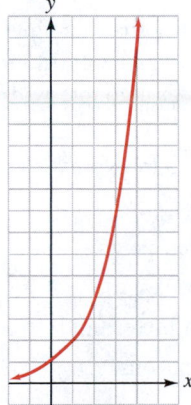

x	y
1	
2	
3	
4	

28.

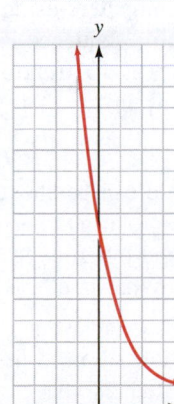

x	y
−1	
0	
1	
2	

NOTATION

In Exercises 29–30, complete each solution.

29. $(x^4x^2)^3 = \left(\right)^3$
$$= x^{18}$$

30. $\dfrac{12a^3a^4}{2a^2} = \dfrac{12}{2a^2}$
$$= 6a^{-2}$$
$$= 6a^5$$

PRACTICE

In Exercises 31–38, identify the base and the exponent in each expression.

31. 4^3

32. $(-8)^2$

33. x^5

34. $(2y)^3$

35. $(-3x)^2$

36. $-x^4$

37. $-\dfrac{1}{3}y^6$

38. $3.14r^4$

In Exercises 39–46, write each expression without using exponents.

39. 5^3

40. -4^5

41. x^6

42. $3x^3$

43. $-\dfrac{3}{4}x^5$

44. $(2.1y)^4$

45. $\left(\dfrac{1}{3}t\right)^3$

46. a^3b^2

In Exercises 47–50, write each expression using exponents.

47. $4t(4t)(4t)(4t)$

48. $-5u(-5u)$

49. $-4 \cdot t \cdot t \cdot t$

50. $-5 \cdot u \cdot u$

In Exercises 51–54, evaluate each expression using the rules for the order of operations.

51. $2(5^4 - 4^3)$

52. $2(4^3 + 3^2)$

53. $-5^2(3^4 + 4^3)$

54. $-5^2(4^3 - 2^6)$

In Exercises 55–74, write each expression as an expression involving only one base and one exponent.

55. x^4x^3

56. y^5y^2

57. a^3aa^5

58. b^2b^3b

59. $y^3(y^2y^4)$

60. $(y^4y)y^6$

61. $4x^2(3x^5)$

62. $-2y(y^3)$

63. $(-y^2)(4y^3)$

64. $(-4x^3)(-5x)$

65. $(3^2)^4$

66. $(4^3)^3$

67. $(y^5)^3$

68. $(b^3)^6$

69. $(x^2x^3)^5$

70. $(y^3y^4)^4$

71. $(3zz^2z^3)^5$

72. $(4t^3t^6t^2)^2$

73. $(x^5)^2(x^7)^3$

74. $(y^3y)^2(y^2)^2$

In Exercises 75–90, simplify by writing each expression without using parentheses.

75. $(xy)^3$

76. $(uv)^4$

77. $(r^3s^2)^2$

78. $(a^3b^2)^3$

79. $(4ab^2)^2$

80. $(3x^2y)^3$

81. $(-2r^2s^3)^3$

82. $(-3x^2y^4)^2$

83. $\left(\dfrac{a}{b}\right)^3$

84. $\left(\dfrac{r}{s}\right)^4$

85. $\left(\dfrac{x^2}{y^3}\right)^5$

86. $\left(\dfrac{u^4}{v^2}\right)^6$

87. $\left(\dfrac{-2a}{b}\right)^5$

88. $\left(\dfrac{-2t}{3}\right)^4$

89. $\left(\dfrac{b^2}{3a}\right)^3$

90. $\left(\dfrac{a^3b}{c^4}\right)^5$

In Exercises 91–102, simplify each expression.

91. $\dfrac{x^5}{x^3}$

92. $\dfrac{a^6}{a^3}$

93. $\dfrac{y^3y^4}{yy^2}$

94. $\dfrac{b^4b^5}{b^2b^3}$

95. $\dfrac{12a^2a^3a^4}{4(a^4)^2}$

96. $\dfrac{16(aa^2)^3}{2a^2a^3}$

97. $\dfrac{(ab^2)^3}{(ab)^2}$

98. $\dfrac{(m^3n^4)^3}{(mn^2)^3}$

99. $\dfrac{20(r^4s^3)^4}{6(rs^3)^3}$

100. $\dfrac{15(x^2y^5)^5}{21(x^3y)^2}$

101. $\left(\dfrac{y^3y}{2yy^2}\right)^3$

102. $\left(\dfrac{3t^3t^4t^5}{4t^2t^6}\right)^3$

APPLICATIONS

103. ART HISTORY Leonardo da Vinci's drawing relating a human figure to a square and a circle is shown in Illustration 1.

ILLUSTRATION 1

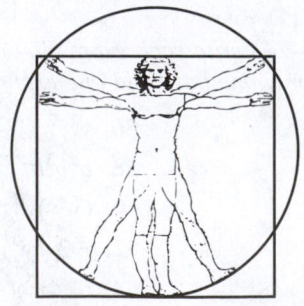

a. Find the area of the square if the man's height is $5x$ feet.

b. Find the area of the circle if the distance from his waist to his feet is $3x$ feet. You may leave π in your answer.

104. PACKAGING Use Illustration 2 to find the volume of the bowling ball and the cardboard box it is packaged in. You may leave π in your answer.

ILLUSTRATION 2

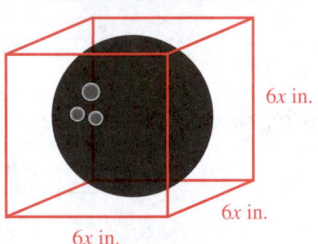

$6x$ in.

$6x$ in.

$6x$ in.

105. BOUNCING BALL A ball is dropped from a height of 32 feet. Each rebound is one-half of its previous height.

a. Draw a diagram of the path of the ball, showing four bounces.

b. Explain why the expressions $32\left(\frac{1}{2}\right)$, $32\left(\frac{1}{2}\right)^2$, $32\left(\frac{1}{2}\right)^3$, and $32\left(\frac{1}{2}\right)^4$ represent the height of the ball on the first, second, third, and fourth bounces, respectively. Find the heights of the first four bounces.

106. HAVING BABIES The probability that a couple will have n baby boys in a row is given by the formula $\left(\frac{1}{2}\right)^n$. Find the probability that a couple will have four baby boys in a row.

107. PACKAGING Use a power of 12 to express the number of pencils in a case if

■ each package contains a dozen pencils,

■ each box contains a dozen packages,

■ each carton contains a dozen boxes, and

■ each case contains a dozen cartons.

108. COMPUTERS Text is stored by computers using a sequence of eight 0's and 1's. Such a sequence is called a **byte**. An example of a byte is 10101110.

a. Write four other bytes, all ending in 1.

b. Each of the eight digits of a byte can be chosen in *two* ways (either 0 or 1). The total number of different bytes can be represented by an exponential expression with base 2. What is it?

109. INVESTMENT On average, a certain investment doubles every 7 years. Use this information to find the value of a $1,000 investment after 28 years.

110.  INVESTING Guess the answer to the following problem. Then use a calculator to find the correct answer. Were you close?

If the value of 1¢ is to double every day, what will the penny be worth after 31 days?

WRITING

111. Describe how you would multiply two exponential expressions with like bases.

112. Are the expressions $2x^3$ and $(2x)^3$ equivalent? Explain.

113. Is the operation of raising to a power commutative? That is, is $a^b = b^a$? Explain.

114. When a number is raised to a power, is the result always larger than the original number? Support your answer with some examples.

REVIEW

In Exercises 115–118, match each equation with its graph in the right column.

115. $y = 2x - 1$

116. $y = 3x - 1$

117. $y = 3$

118. $x = 3$

a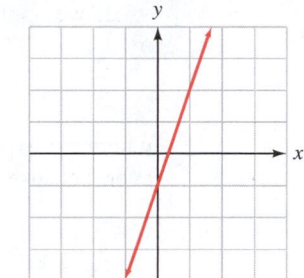

b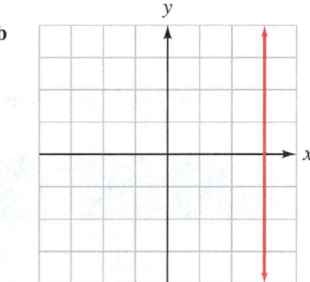

c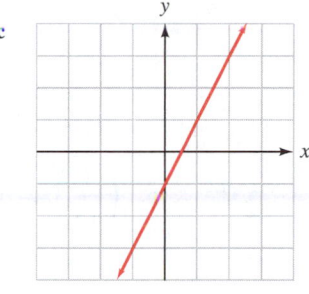

d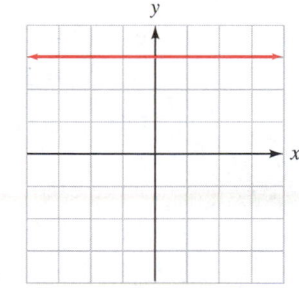

▶ **4.2**

Zero and Negative Integer Exponents

In this section, you will learn about

Zero exponents ■ Negative integer exponents ■ Variable exponents

Introduction In the previous section, we discussed natural-number exponents. We now extend the discussion to include exponents that are zero and exponents that are negative integers.

Zero Exponents

When we discussed the quotient rule for exponents in the previous section, the exponent in the numerator was always greater than the exponent in the denominator. We now consider what happens when the exponents are equal.

If we apply the quotient rule to the fraction

$$\frac{5^3}{5^3}$$

where the exponents in the numerator and denominator are equal, we obtain 5^0. However, because any nonzero number divided by itself equals 1, we also obtain 1.

$$\frac{5^3}{5^3} = 5^{3-3} = \mathbf{5^0} \qquad \frac{5^3}{5^3} = \frac{\overset{1}{\cancel{5}} \cdot \overset{1}{\cancel{5}} \cdot \overset{1}{\cancel{5}}}{\underset{1}{\cancel{5}} \cdot \underset{1}{\cancel{5}} \cdot \underset{1}{\cancel{5}}} = \mathbf{1}$$

These are equal.

For this reason, we will define 5^0 to be equal to 1. This example suggests the following.

Zero exponents

If x is any nonzero real number, then

$$x^0 = 1$$

EXAMPLE 1

Zero exponents. Write each expression without using exponents.

a. $\left(\dfrac{1}{13}\right)^0 = 1$ **b.** $\dfrac{x^5}{x^5} = x^{5-5}$ $(x \neq 0)$

$$= x^0$$
$$= 1$$

c. $3x^0 = 3(\mathbf{1})$ **d.** $(3x)^0 = 1$
$$= 3$$

Parts c and d point out that $3x^0 \neq (3x)^0$.

SELF CHECK

Write each expression without using exponents:
a. $(-0.115)^0$ and **b.** $-5a^0 b$.   *Answers:* **a.** 1, **b.** $-5b$ ∎

Negative Integer Exponents

If we apply the quotient rule to

$$\frac{6^2}{6^5}$$

where the exponent in the numerator is less than the exponent in the denominator, we obtain 6^{-3}. However, by dividing out two factors of 6, we also obtain $\frac{1}{6^3}$.

$$\frac{6^2}{6^5} = 6^{2-5} = \mathbf{6^{-3}} \qquad \frac{6^2}{6^5} = \frac{\overset{1}{\cancel{6}} \cdot \overset{1}{\cancel{6}}}{\underset{1}{\cancel{6}} \cdot \underset{1}{\cancel{6}} \cdot 6 \cdot 6 \cdot 6} = \frac{\mathbf{1}}{\mathbf{6^3}}$$

These are equal.

For these reasons, we define 6^{-3} to be equal to $\dfrac{1}{6^3}$.

$$6^{-3} = \frac{1}{6^3}$$

Negative exponents

If x is any nonzero number and n is a natural number, then

$$x^{-n} = \frac{1}{x^n}$$

E X A M P L E   2

Negative exponents. Simplify by using the definition of negative exponents.

a. $3^{-5} = \dfrac{1}{3^5}$

$\phantom{3^{-5}} = \dfrac{1}{243}$

b. $(-2)^{-3} = \dfrac{1}{(-2)^3}$

$\phantom{(-2)^{-3}} = -\dfrac{1}{8}$

c. $\dfrac{1}{5^{-2}} = \dfrac{1}{\dfrac{1}{5^2}}$

$\phantom{\dfrac{1}{5^{-2}}} = 1 \div \dfrac{1}{5^2}$

$\phantom{\dfrac{1}{5^{-2}}} = 1 \cdot \dfrac{5^2}{1}$

$\phantom{\dfrac{1}{5^{-2}}} = 5^2$

$\phantom{\dfrac{1}{5^{-2}}} = 25$

d. $\dfrac{2^{-3}}{3^{-4}} = \dfrac{\dfrac{1}{2^3}}{\dfrac{1}{3^4}}$

$\phantom{\dfrac{2^{-3}}{3^{-4}}} = \dfrac{1}{2^3} \cdot \dfrac{3^4}{1}$

$\phantom{\dfrac{2^{-3}}{3^{-4}}} = \dfrac{3^4}{2^3}$

$\phantom{\dfrac{2^{-3}}{3^{-4}}} = \dfrac{81}{8}$

SELF CHECK   Simplify by using the definition of negative exponents: **a.** 4^{-4},   **b.** $(-5)^{-3}$,   and   **c.** $\dfrac{8^{-2}}{7^{-1}}$.   *Answers:*   **a.** $\frac{1}{256}$,   **b.** $-\frac{1}{125}$,   **c.** $\frac{7}{64}$

The results from parts c and d of Example 2 suggest that in a fraction, we can move factors that have negative exponents between the numerator and denominator if we change the sign of their exponents. For example,

$$\frac{3^{-3}}{b^{-1}} = \frac{b^1}{3^3} = \frac{b}{27}$$ This property applies only if there is one term in the numerator and one term in the denominator.

E X A M P L E   3

Negative exponents. Simplify by using the definition of negative exponents. Assume that no denominators are zero.

a. $x^{-4} = \dfrac{1}{x^4}$

b. $\dfrac{x^{-3}}{y^{-7}} = \dfrac{y^7}{x^3}$

c. $(2x)^{-2} = \dfrac{1}{(2x)^2}$

$\phantom{(2x)^{-2}} = \dfrac{1}{4x^2}$

d. $2x^{-2} = 2\left(\dfrac{1}{x^2}\right)$

$\phantom{2x^{-2}} = \dfrac{2}{x^2}$

e. $(-3a)^{-4} = \dfrac{1}{(-3a)^4}$

$\phantom{(-3a)^{-4}} = \dfrac{1}{81a^4}$

f. $(x^3x^2)^{-3} = (x^5)^{-3}$

$\phantom{(x^3x^2)^{-3}} = \dfrac{1}{(x^5)^3}$

$\phantom{(x^3x^2)^{-3}} = \dfrac{1}{x^{15}}$

Simplify by using the definition of negative exponents: **a.** a^{-5}, **b.** $3y^{-3}$, and **c.** $(a^4a^3)^{-2}$.

Answers: **a.** $\dfrac{1}{a^5}$, **b.** $\dfrac{3}{y^3}$, **c.** $\dfrac{1}{a^{14}}$ ∎

The rules for exponents discussed in Section 4.1 (the product, power, and quotient rules) are also true for zero and negative exponents.

Rules for exponents

If m and n are integers and there are no divisions by zero, then

$$x^m x^n = x^{m+n} \qquad (x^m)^n = x^{m \cdot n} \qquad (xy)^n = x^n y^n \qquad \left(\dfrac{x}{y}\right)^n = \dfrac{x^n}{y^n}$$

$$x^0 = 1 \quad (x \neq 0) \qquad x^{-n} = \dfrac{1}{x^n} \qquad \dfrac{x^m}{x^n} = x^{m-n}$$

EXAMPLE 4 **Applying rules for exponents.** Write $\left(\frac{5}{16}\right)^{-1}$ without using exponents.

Solution

$$\left(\frac{5}{16}\right)^{-1} = \frac{5^{-1}}{16^{-1}}$$

$$= \frac{16^1}{5^1}$$

$$= \frac{16}{5}$$

Write $\left(\frac{3}{7}\right)^{-2}$ without using exponents.

Answer: $\frac{49}{9}$ ∎

EXAMPLE 5 **Applying rules for exponents.** Simplify and write the result without using negative exponents. Assume that no denominators are zero.

a. $(x^{-3})^2 = x^{-6}$

$\qquad\qquad = \dfrac{1}{x^6}$

b. $\dfrac{x^3}{x^7} = x^{3-7}$

$\qquad\quad = x^{-4}$

$\qquad\quad = \dfrac{1}{x^4}$

c. $\dfrac{y^{-4}y^{-3}}{y^{-20}} = \dfrac{y^{-7}}{y^{-20}}$

$\qquad\qquad = y^{-7-(-20)}$

$\qquad\qquad = y^{-7+20}$

$\qquad\qquad = y^{13}$

d. $\dfrac{12a^3b^4}{4a^5b^2} = 3a^{3-5}b^{4-2}$

$\qquad\qquad = 3a^{-2}b^2$

$\qquad\qquad = \dfrac{3b^2}{a^2}$

e. $\left(-\dfrac{x^3y^2}{xy^{-3}}\right)^{-2} = (-x^{3-1}y^{2-(-3)})^{-2}$

$\qquad\qquad\qquad = (-x^2y^5)^{-2}$

$\qquad\qquad\qquad = \dfrac{1}{(-x^2y^5)^2}$

$\qquad\qquad\qquad = \dfrac{1}{x^4y^{10}}$

SELF CHECK   Simplify and write the result without using negative exponents: **a.** $(x^4)^{-3}$,   **b.** $\dfrac{a^4}{a^8}$,   **c.** $\dfrac{a^{-4}a^{-5}}{a^{-3}}$,   and   **d.** $\dfrac{20x^5y^3}{5x^3y^6}$.

Answers: **a.** $\dfrac{1}{x^{12}}$,   **b.** $\dfrac{1}{a^4}$,   **c.** $\dfrac{1}{a^6}$,   **d.** $\dfrac{4x^2}{y^3}$

Variable Exponents

We can apply the rules for exponents to simplify expressions involving variable exponents.

EXAMPLE 6   **Variable exponents.** Simplify each expression.

a. $\dfrac{6^n}{6^n} = 6^{n-n}$

$\qquad = 6^0$

$\qquad = 1$

b. $\dfrac{y^m}{y^m} = y^{m-m} \quad (y \neq 0)$

$\qquad = y^0$

$\qquad = 1$

c. $x^{2m}x^{3m} = x^{2m+3m}$

$\qquad = x^{5m}$

d. $\dfrac{y^{2m}}{y^{4m}} = y^{2m-4m} \quad (y \neq 0)$

$\qquad = y^{-2m}$

$\qquad = \dfrac{1}{y^{2m}}$

e. $a^{2m-1}a^{2m} = a^{2m-1+2m}$

$\qquad = a^{4m-1}$

f. $(b^{m+1})^{2m} = b^{(m+1)2m}$

$\qquad = b^{2m^2+2m}$

SELF CHECK   Simplify each expression: **a.** $\dfrac{x^m}{x^m}$,   **b.** $z^{3n}z^{2n}$,   **c.** $\dfrac{z^{3n}}{z^{5n}}$,   and   **d.** $(x^{m+2})^{3m}$

Answers: **a.** 1,   **b.** z^{5n},   **c.** $\dfrac{1}{z^{2n}}$,   **d.** x^{3m^2+6m}

ACCENT ON TECHNOLOGY   *Finding Present Value*

As a gift for their newborn grandson, the grandparents want to deposit enough money in the bank now so that when he turns 18, the young man will have a college fund of $20,000 waiting for him. How much should they deposit now if the money will earn 6% annually?

To find how much money P must be invested at an annual rate i (expressed as a decimal) to have $$A$ in n years, we use the formula $P = A(1 + i)^{-n}$. If we substitute 20,000 for A, 0.06 (6%) for i, and 18 for n, we have

$P = A(1 + i)^{-n}$ *P is called the present value.*

$P = 20,000(1 + 0.06)^{-18}$

To find P with a scientific calculator, we enter these numbers and press these keys:

Keystrokes $(\ 1\ +\ .06\ )\ \boxed{y^x}\ 18\ \boxed{+/-}\ \boxed{\times}\ 20000\ \boxed{=}$ $\boxed{7006.875823}$

(continued)

To evaluate the expression with a graphing calculator, we use the following keystrokes:

Keystrokes   20000 $\times$ $($ 1 $+$ .06 $)$ $\wedge$ $(-)$ 18 $\boxed{\text{ENTER}}$

```
20000*(1+.06)^-1
8
        7006.875823
```

They must invest approximately $7,006.88 to have $20,000 in 18 years.

STUDY SET

Section 4.2

VOCABULARY

In Exercises 1–4, fill in the blanks to make the statements true.

1. In the exponential expression 8^{-3}, 8 is the _____ and -3 is the _____.

2. The set $\{\ldots, -3, -2, -1, 0, 1, 2, 3, \ldots\}$ is the set of _____.

3. In the exponential expression 5^{-1}, the exponent is a _____ integer.

4. In the exponential expression z^m, the exponent is a _____.

CONCEPTS

5. In parts a and b, fill in the blanks as you simplify the fraction in two different ways. Then complete the sentence in part c.

   a. $\dfrac{6^4}{6^4} = 6^{\boxed{}}$

   $\quad = 6^{\boxed{}}$

   b. $\dfrac{6^4}{6^4} = \dfrac{\boxed{} \cdot \boxed{} \cdot \boxed{} \cdot \boxed{}}{6 \cdot 6 \cdot 6 \cdot 6}$

   $\quad = \boxed{}$

   c. So we define 6^0 to be $\boxed{}$, and in general, if x is any nonzero real number, then $x^0 = \boxed{}$.

6. In parts a and b, fill in the blanks as you simplify the fraction in two different ways. Then complete the sentence in part c.

   a. $\dfrac{8^3}{8^5} = 8^{\boxed{}}$

   $\quad = 8^{\boxed{}}$

   b. $\dfrac{8^3}{8^5} = \dfrac{\boxed{} \cdot \boxed{} \cdot \boxed{}}{8 \cdot 8 \cdot 8 \cdot 8 \cdot 8}$

   $\quad = \dfrac{1}{8^{\boxed{}}}$

c. So we define 8^{-2} to be $\boxed{}$, and in general, if x is any nonzero real number, then $x^{-n} = \boxed{}$.

In Exercises 7–10, complete each table.

7.
x	3^x
2	
1	
0	
-1	
-2	

8.
x	4^x
2	
1	
0	
-1	
-2	

9.
x	$(-9)^x$
2	
1	
0	
-1	
-2	

10.
x	$(-5)^x$
2	
1	
0	
-1	
-2	

In Exercises 11–12, first use the graph to determine the missing y-coordinates in the table. Then express each y-coordinate as a power of 2.

11.

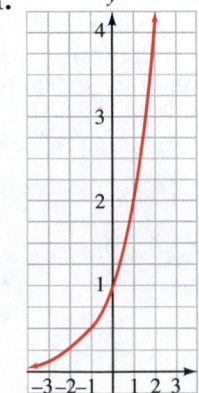

x	y	Power
2		
1		
0		
-1		
-2		

12.

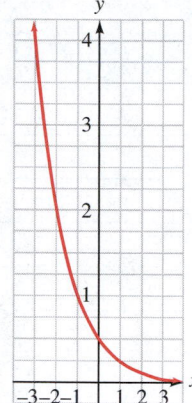

x	y	Power
1		
0		
−1		
−2		
−3		

NOTATION

In Exercises 13–14, complete each solution.

13. $(y^5 y^3)^{-5} = ()^{-5}$

$$= y^{}$$

$$= \frac{1}{y^{40}}$$

14. $\left(\dfrac{a^2 b^3}{a^{-3} b}\right)^{-3} = \left(a^{2-(-3)} b^{-1}\right)^{-3}$

$$= \left(a^{} b^{}\right)^{-3}$$

$$= \frac{1}{(a^5 b^2)^{}}$$

$$= \frac{1}{a^{15} b^6}$$

15. In the expression $3x^{-2}$, what is the base and what is the exponent?

16. In the expression $-3x^{-2}$, what is the base and what is the exponent?

17. First tell the base and the exponent, then evaluate each expression.
   a. -4^2
   b. 4^{-2}
   c. -4^{-2}

18. First tell the base and the exponent, then evaluate each expression.
   a. $(-7)^2$
   b. -7^{-2}
   c. $(-7)^{-2}$

PRACTICE

In Exercises 19–84, simplify each expression. Write each answer without using parentheses or negative exponents.

19. 7^0

20. 9^0

21. $\left(\dfrac{1}{4}\right)^0$

22. $\left(\dfrac{3}{8}\right)^0$

23. $2x^0$

24. $(2x)^0$

25. $(-x)^0$

26. $-x^0$

27. $\left(\dfrac{a^2 b^3}{ab^4}\right)^0$

28. $\dfrac{2}{3}\left(\dfrac{xyz}{x^2 y}\right)^0$

29. $\dfrac{5}{2x^0}$

30. $\dfrac{4}{3a^0}$

31. 12^{-2}

32. 11^{-2}

33. $(-4)^{-1}$

34. $(-8)^{-1}$

35. $\dfrac{1}{5^{-3}}$

36. $\dfrac{1}{3^{-3}}$

37. $\dfrac{2^{-4}}{3^{-1}}$

38. $\dfrac{7^{-2}}{2^{-3}}$

39. -4^{-3}

40. -6^{-3}

41. $-(-4)^{-3}$

42. $-(-4)^{-2}$

43. x^{-2}

44. y^{-3}

45. $-b^{-5}$

46. $-c^{-4}$

47. $(2y)^{-4}$

48. $(-3x)^{-1}$

49. $(ab^2)^{-3}$

50. $(m^2 n^3)^{-2}$

51. $2^5 \cdot 2^{-2}$

52. $10^2 \cdot 10^{-4}$

53. $4^{-3} \cdot 4^{-2} \cdot 4^5$

54. $3^{-4} \cdot 3^5 \cdot 3^{-3}$

55. $\left(\dfrac{7}{8}\right)^{-1}$

56. $\left(\dfrac{16}{5}\right)^{-1}$

57. $\dfrac{3^5 \cdot 3^{-2}}{3^3}$

58. $\dfrac{6^2 \cdot 6^{-3}}{6^{-2}}$

59. $\dfrac{y^4}{y^5}$

60. $\dfrac{t^7}{t^{10}}$

61. $\dfrac{(r^2)^3}{(r^3)^4}$

62. $\dfrac{(b^3)^4}{(b^5)^4}$

63. $\dfrac{y^4 y^3}{y^4 y^{-2}}$

64. $\dfrac{x^{12} x^{-7}}{x^3 x^4}$

65. $\dfrac{10a^4 a^{-2}}{5a^2 a^0}$

66. $\dfrac{9b^0 b^3}{3b^{-3} b^4}$

67. $(ab^2)^{-2}$

68. $(c^2 d^3)^{-2}$

69. $(x^2 y)^{-3}$

70. $(-xy^2)^{-4}$

71. $(x^{-4} x^3)^3$

72. $(y^{-2} y)^3$

73. $(a^{-2}b^3)^{-4}$

74. $(y^{-3}z^5)^{-6}$

75. $(-2x^3y^{-2})^{-5}$

76. $(-3u^{-2}v^3)^{-3}$

77. $\left(\dfrac{a^3}{a^{-4}}\right)^2$

78. $\left(\dfrac{a^4}{a^{-3}}\right)^3$

79. $\left(\dfrac{b^5}{b^{-2}}\right)^{-2}$

80. $\left(\dfrac{b^{-2}}{b^3}\right)^3$

81. $\left(\dfrac{4x^2}{3x^{-5}}\right)^4$

82. $\left(\dfrac{-3r^4r^{-3}}{r^{-3}r^7}\right)^3$

83. $\left(\dfrac{12y^3z^{-2}}{3y^{-4}z^3}\right)^2$

84. $\left(\dfrac{6xy^3}{3x^{-1}y}\right)^3$

In Exercises 85–96, write each expression with a single exponent.

85. $x^{2m}x^m$

86. $y^{3m}y^{2m}$

87. $u^{2m}u^{-3m}$

88. $r^{5m}r^{-6m}$

89. $y^{3m+2}y^{-m}$

90. $x^{m+1}x^m$

91. $\dfrac{y^{3m}}{y^{2m}}$

92. $\dfrac{z^{4m}}{z^{2m}}$

93. $\dfrac{x^{3n}}{x^{6n}}$

94. $\dfrac{x^m}{x^{5m}}$

95. $(x^{m+1})^2$

96. $(y^2)^{m+1}$

APPLICATIONS

97. THE DECIMAL NUMERATION SYSTEM Decimal numbers are written by putting digits into place-value columns that are separated by a decimal point. Express the value of each of the columns shown in Illustration 1 using a power of 10.

ILLUSTRATION 1

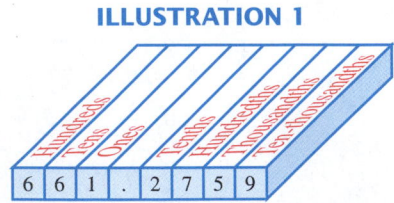

98. UNIT COMPARISON Consider the relative sizes of the items listed in the table in Illustration 2. In the column titled "measurement," write the most appropriate number from the following list. Each number is used only once.

10^0 meter

10^{-1} meter

10^{-2} meter

10^{-3} meter

10^{-4} meter

10^{-5} meter

ILLUSTRATION 2

Item	Measurement (m)
Thickness of a dime	
Height of a bathroom sink	
Length of a pencil eraser	
Thickness of soap bubble film	
Width of a video cassette	
Thickness of a piece of paper	

99. RETIREMENT YEARS How much money should a young married couple invest now at an 8% annual rate if they want to have $100,000 in the bank when they reach retirement age in 40 years? (See the Accent on Technology in this section for the formula.)

100. BIOLOGY During bacterial reproduction, the time required for a population to double is called the **generation time.** If b bacteria are introduced into a medium, then after the generation time has elapsed, there will be $2b$ bacteria. After n generations, there will be $b \cdot 2^n$ bacteria. Explain what this expression represents when $n = 0$.

WRITING

101. Explain how you would help a friend understand that 2^{-3} is not equal to -8.

102. Describe how you would verify on a calculator that

$$2^{-3} = \frac{1}{2^3}$$

REVIEW

103. IQ TEST An IQ (intelligence quotient) is a score derived from the formula

$$IQ = \frac{\text{mental age}}{\text{chronological age}} \cdot 100$$

Find the mental age of a 10-year-old girl if she has an IQ of 135.

104. DIVING When under water, the pressure you feel in your ears is given by the formula

$$\text{Pressure} = \text{depth} \cdot \text{density of water}$$

Find the density of water (in lb/ft^3) if, at a depth of 9 feet, the pressure on your eardrum is 561.6 lb/ft^2.

105. Write the equation of the line having slope $\frac{3}{4}$ and y-intercept -5.

106. Find $f(-6)$ if $f(x) = x^2 - 3x + 1$.

▶ 4.3 Scientific Notation

In this section, you will learn about

Scientific notation ■ Writing numbers in scientific notation
■ Changing from scientific notation to standard notation ■ Using
scientific notation to simplify computations

Introduction Scientists often deal with extremely large and extremely small numbers. Two examples are shown in Figure 4-1.

FIGURE 4-1

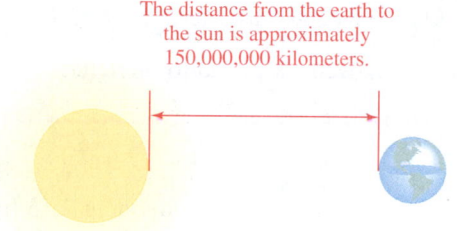

The distance from the earth to
the sun is approximately
150,000,000 kilometers.

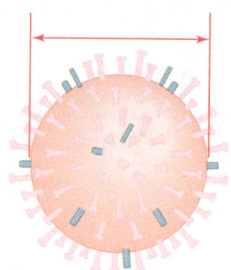

The influenza virus, which causes
"flu" symptoms of cough, sore throat,
headache, and congestion, has a
diameter of 0.00000256 inch.

The large number of zeros in 150,000,000 and 0.00000256 makes them difficult to read and hard to remember. In this section, we will discuss a notation that will make these numbers easier to work with.

Scientific Notation

Scientific notation provides a compact way of writing large and small numbers.

Scientific notation

A number is written in **scientific notation** if it is written as the product of a number between 1 (including 1) and 10 and an integer power of 10.

These numbers are written in scientific notation:

$$3.67 \times 10^6, \quad 2.24 \times 10^{-4}, \quad \text{and} \quad 9.875 \times 10^{22}$$

Every number written in scientific notation has the following form:

An integer exponent

$$\boxed{} \cdot \boxed{} \times 10^{\boxed{}}$$

A decimal between 1 and 10

Writing Numbers in Scientific Notation

EXAMPLE 1

Writing numbers in scientific notation. Change 150,000,000 to scientific notation.

Solution   We note that 1.5 lies between 1 and 10. To obtain 150,000,000, the decimal point in 1.5 must be moved eight places to the right.

$$1.\underbrace{50000000}_{\text{8 places to the right}}$$

Because multiplying a number by 10 moves the decimal point one place to the right, we can accomplish this by multiplying 1.5 by 10 eight times. We can show the multiplication of 1.5 by 10 eight times using the notation 10^8. Thus, 150,000,000 written in scientific notation is 1.5×10^8.

SELF CHECK   The distance from the earth to the sun is approximately 93,000,000 miles. Write this number in scientific notation.   *Answer:* 9.3×10^7 ■

EXAMPLE 2   **Writing numbers in scientific notation.** Change 0.00000256 to scientific notation.

Solution   We note that 2.56 is between 1 and 10. To obtain 0.00000256, the decimal point in 2.56 must be moved six places to the left.

$$\underbrace{000002}_{\text{6 places to the left}}.56$$

We can accomplish this by dividing 2.56 by 10^6, which is equivalent to multiplying 2.56 by $\frac{1}{10^6}$ (or by 10^{-6}). Thus, 0.00000256 written in scientific notation is 2.56×10^{-6}.

SELF CHECK   The *Salmonella* bacterium, which causes food poisoning, is 0.00009055 inch long. Write this number in scientific notation.   *Answer:* 9.055×10^{-5} ■

EXAMPLE 3   **Writing numbers in scientific notation.** Write   **a.** 235,000   and   **b.** 0.00000235 in scientific notation.

Solution   **a.** $235,000 = 2.35 \times 10^5$ Because $2.35 \times 10^5 = 235,000$ and 2.35 is between 1 and 10.

b. $0.00000235 = 2.35 \times 10^{-6}$ Because $2.35 \times 10^{-6} = 0.00000235$ and 2.35 is between 1 and 10.

SELF CHECK   Write   **a.** 17,500   and   **b.** 0.657 in scientific notation.   *Answers:*   **a.** 1.75×10^4,   **b.** 6.57×10^{-1} ■

From Examples 1, 2, and 3, we see that in scientific notation, a positive exponent is used when writing a number that is greater than 1. A negative exponent is used when writing a number that is between 0 and 1.

EXAMPLE 4   **Writing numbers in scientific notation.** Write 432.0×10^5 in scientific notation.

Solution   The number 432.0×10^5 is not written in scientific notation, because 432.0 is not a number between 1 and 10. To write this number in scientific notation, we proceed as follows:

$$432.0 \times 10^5 = 4.32 \times 10^2 \times 10^5 \quad \text{Write 432.0 in scientific notation.}$$
$$= 4.32 \times 10^7 \quad 10^2 \times 10^5 = 10^{2+5} = 10^7.$$

SELF CHECK   Write 85×10^{-3} in scientific notation.   *Answer:* 8.5×10^{-2} ■

ACCENT ON TECHNOLOGY *Calculators and Scientific Notation*

When displaying a very large or a very small number as an answer, most scientific calculators express it in scientific notation. To show this, we will find the values of $(453.46)^5$ and $(0.0005)^{12}$. To find these powers of decimals, we enter these numbers and press these keys:

Keystrokes   453.46 $\boxed{y^x}$ 5 $\boxed{=}$ $\boxed{1.917321395 \quad 13}$

.0005 $\boxed{y^x}$ 12 $\boxed{=}$ $\boxed{2.44140625 \quad -40}$

Since these numbers in standard notation require more space than the calculator display has, the calculator gives each result in scientific notation. The first display represents $1.917321395 \times 10^{13}$, and the second display represents $2.44140625 \times 10^{-40}$.

If we evaluate the same two expressions using a graphing calculator, we see that the letter E is used when displaying a number in scientific notation.

Keystrokes   453.46 $\boxed{\wedge}$ 5 $\boxed{\text{ENTER}}$ $\boxed{\begin{array}{l} 453.46^5 \\ \quad 1.917321395 \text{E} 13 \end{array}}$

.0005 $\boxed{\wedge}$ 12 $\boxed{\text{ENTER}}$ $\boxed{\begin{array}{l} .0005^{12} \\ \quad 2.44140625 \text{E} -40 \end{array}}$

Changing from Scientific Notation to Standard Notation

We can change a number written in scientific notation to **standard notation.** For example, to write 9.3×10^7 in standard notation, we multiply 9.3 by 10^7.

$9.3 \times 10^7 = 9.3 \times 10,000,000$ 10^7 is equal to 1 followed by 7 zeros.

$= 93,000,000$

EXAMPLE 5   **Writing numbers in standard notation.** Write **a.** 3.4×10^5 and **b.** 2.1×10^{-4} in standard notation.

Solution   **a.** $3.4 \times 10^5 = 3.4 \times 100,000$ **b.** $2.1 \times 10^{-4} = 2.1 \times \dfrac{1}{10^4}$

$= 340,000$

$= 2.1 \times \dfrac{1}{10,000}$

$= 2.1 \times 0.0001$

$= 0.00021$

SELF CHECK   Write **a.** 4.76×10^5 and **b.** 9.8×10^{-3} in standard notation.   *Answers:* **a.** 476,000, **b.** 0.0098 ■

The following numbers are written in both scientific and standard notation. In each case, the exponent gives the number of places that the decimal point moves, and the sign of the exponent indicates the direction that it moves.

$$5.32 \times 10^5 = 5\,3\,2\,0\,0\,0.\qquad \text{5 places to the right.}$$
$$2.37 \times 10^6 = 2\,3\,7\,0\,0\,0\,0.\qquad \text{6 places to the right.}$$
$$8.95 \times 10^{-4} = 0.0\,0\,0\,8\,9\,5\qquad \text{4 places to the left.}$$
$$8.375 \times 10^{-3} = 0.0\,0\,8\,3\,7\,5\qquad \text{3 places to the left.}$$
$$9.77 \times 10^0 = 9.77\qquad \text{No movement of the decimal point.}$$

Using Scientific Notation to Simplify Computations

Another advantage of scientific notation becomes apparent when we evaluate products or quotients that contain very large or very small numbers.

EXAMPLE 6

Stars. Except for the sun, the nearest star visible to the naked eye from most parts of the United States is Sirius. Light from Sirius reaches the earth in about 70,000 hours. If light travels at approximately 670,000,000 miles per hour, how far from the earth is Sirius?

Solution   We are given the *rate* at which light travels (670,000,000 mi/hr) and the *time* it takes the light to travel from Sirius to earth (70,000 hr). We can find the *distance* the light travels using the formula $d = rt$.

$$d = rt$$
$$d = 670{,}000{,}000 \cdot 70{,}000 \qquad \text{Substitute 670,000,000 for } r \text{ and 70,000 for } t.$$
$$= 6.7 \times 10^8 \cdot 7.0 \times 10^4 \qquad \text{Write each number in scientific notation.}$$
$$= (6.7 \cdot 7.0) \times (10^8 \cdot 10^4) \qquad \text{Apply the commutative and associative properties of multiplication to group the numbers together and the powers of 10 together.}$$
$$= (6.7 \cdot 7.0) \times 10^{8+4} \qquad \text{For the powers of 10, keep the base and add the exponents.}$$
$$= 46.9 \times 10^{12} \qquad \text{Do the multiplication. Do the addition.}$$

We note that 46.9 is not between 0 and 1, so 46.9×10^{12} is not written in scientific notation. To answer in scientific notation, we proceed as follows.

$$= 4.69 \times 10^1 \times 10^{12} \qquad \text{Write 46.9 in scientific notation as } 4.69 \times 10^1.$$
$$= 4.69 \times 10^{13} \qquad \text{Keep the base of 10 and add the exponents.}$$

Sirius is approximately 4.69×10^{13} or 46,900,000,000,000 miles from the earth   ■

EXAMPLE 7

Atoms. Scientific notation is used in chemistry. As an example, the approximate weight (in grams) of one atom of the heaviest naturally occurring element, uranium, is given by

$$\frac{2.4 \times 10^2}{6.0 \times 10^{23}}$$

Evaluate the expression.

Solution

$$\frac{2.4 \times 10^2}{6.0 \times 10^{23}} = \frac{2.4}{6.0} \times \frac{10^2}{10^{23}}$$

Divide the numbers and the powers of 10 separately.

$$= \frac{2.4}{6.0} \times 10^{2-23}$$

For the powers of 10, keep the base and subtract the exponents.

$$= 0.4 \times 10^{-21}$$

Do the division. Subtract the exponents: $2 - 23 = -21$.

$$= 4.0 \times 10^{-1} \times 10^{-21}$$

Write 0.4 in scientific notation as 4.0×10^{-1}.

$$= 4.0 \times 10^{-22}$$

For the powers of 10, keep the base and add the exponents: $-1 + (-21) = -22$.

One atom of uranium weighs 4.0×10^{-22} gram. Written in standard notation, this is 0.00000000000000000000004 g.

SELF CHECK

Find the approximate weight (in grams) of one atom of gold by evaluating $\dfrac{1.98 \times 10^2}{6.0 \times 10^{23}}$.

Answer: 3.3×10^{-22} g ∎

ACCENT ON TECHNOLOGY *Electric Charges*

The electric charges on each of the 0.20-gram balls that are suspended from 50-cm-long strings in Figure 4-2 are the same. For this reason, the balls repel each other. The charge q (in coulombs) on each ball is given by

$$q = \frac{2.97 \times 10^{-2}}{9.49 \times 10^4}$$

FIGURE 4-2

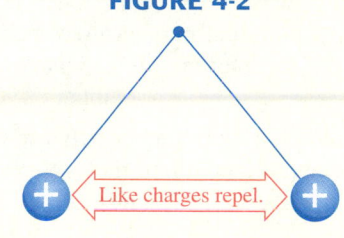

Like charges repel.

We can evaluate the expression by entering the numbers written in scientific notation with the $\boxed{\text{EE}}$ key on a scientific calculator.

Keystrokes   2.97 $\boxed{\text{EE}}$ $\boxed{+/-}$ 2 $\boxed{\div}$ 9.49 $\boxed{\text{EE}}$ 4 $\boxed{=}$ $\boxed{\text{0.000000313}}$

In standard notation, the charge on each ball is 0.000000313 coulomb. This could be written as 3.13×10^{-7} coulomb using scientific notation.

If we use a graphing calculator, the keystrokes are similar. The result is displayed in scientific notation.

Keystrokes   2.97 $\boxed{\text{2nd}}$ $\boxed{\text{EE}}$ $\boxed{(-)}$ 2 $\boxed{\div}$ 9.49 $\boxed{\text{2nd}}$ $\boxed{\text{EE}}$ 4 $\boxed{\text{ENTER}}$

$$\boxed{\begin{array}{l} 2.97\text{E}{-}2/9.49\text{E}4 \\ 3.129610116\text{E}{-}7 \end{array}}$$

STUDY SET

Section 4.3

VOCABULARY

In Exercises 1–2, fill in the blanks to make the statements true.

1. A number is written in _____ notation when it is written as the product of a number between 1 (including 1) and 10.

2. The number 125,000 is written in _____ notation.

CONCEPTS

In Exercises 3–16, fill in the blanks to make the statements true.

3. $2.5 \times 10^2 = $ ▮

4. $2.5 \times 10^{-2} = $ ▮

5. $2.5 \times 10^{-5} = $ ▮

6. $2.5 \times 10^5 = $ ▮

7. $387{,}000 = 3.87 \times$ ▮

8. $38.7 = 3.87 \times$ ▮

9. $0.00387 = 3.87 \times$ ▮

10. $0.000387 = 3.87 \times$ ▮

11. When we multiply a decimal by 10^5, the decimal point moves ▮ places to the _____.

12. When we multiply a decimal by 10^{-7}, the decimal point moves ▮ places to the _____.

13. Dividing a decimal by 10^4 is equivalent to multiplying it by ▮.

14. Multiplying a decimal by 10^0 does not move the decimal point, because $10^0 = $ ▮.

15. When a real number greater than 1 is written in scientific notation, the exponent on 10 is a _____ number.

16. When a real number between 0 and 1 is written in scientific notation, the exponent on 10 is a _____ number.

NOTATION

In Exercises 17–18, complete each solution.

17. Write 63.7×10^5 in scientific notation.

$$63.7 \times 10^5 = \boxed{} \times 10^5$$
$$= 6.37 \times 10^{\boxed{}+5}$$
$$= 6.37 \times 10^6$$

18. Simplify $\dfrac{64{,}000}{0.00004}$.

$$\frac{64{,}000}{0.00004} = \frac{6.4 \times \boxed{}}{4 \times \boxed{}}$$
$$= \frac{\boxed{}}{\boxed{}} \times \frac{10^4}{10^{-5}}$$
$$= 1.6 \times 10^{\boxed{}-(-5)}$$
$$= 1.6 \times 10^9$$

PRACTICE

In Exercises 19–30, write each number in scientific notation.

19. $23{,}000$

20. $4{,}750$

21. $1{,}700{,}000$

22. $290{,}000$

23. 0.062

24. 0.00073

25. 0.0000051

26. 0.04

27. 42.5×10^2

28. 0.3×10^3

29. 0.25×10^{-2}

30. 25.2×10^{-3}

In Exercises 31–42, write each number in standard notation.

31. 2.3×10^2

32. 3.75×10^4

33. 8.12×10^5

34. 1.2×10^3

35. 1.15×10^{-3}

36. 4.9×10^{-2}

37. 9.76×10^{-4}

38. 7.63×10^{-5}

39. 25×10^6

40. 0.07×10^3

41. 0.51×10^{-3}

42. 617×10^{-2}

43. ASTRONOMY The distance from earth to Alpha Centauri (the nearest star outside our solar system) is about 25,700,000,000,000 miles. Express this number in scientific notation.

44. SPEED OF SOUND The speed of sound in air is 33,100 centimeters per second. Express this number in scientific notation.

45. GEOGRAPHY The largest ocean in the world is the Pacific Ocean, which covers 6.38×10^7 square miles. Express this number in standard notation.

46. ATOMS The number of atoms in 1 gram of iron is approximately 1.08×10^{22}. Express this number in standard notation.

47. LENGTH OF A METER One meter is approximately 0.00622 mile. Use scientific notation to express this number.

48. ANGSTROM One angstrom is 1.0×10^{-7} millimeter. Express this number in standard notation.

In Exercises 49–54, use scientific notation and the rules for exponents to simplify each expression. Give all answers in standard notation. Use a calculator to check each result.

49. $(3.4 \times 10^2)(2.1 \times 10^3)$

50. $(4.1 \times 10^{-3})(3.4 \times 10^4)$

51. $\dfrac{9.3 \times 10^2}{3.1 \times 10^{-2}}$

52. $\dfrac{7.2 \times 10^6}{1.2 \times 10^8}$

53. $\dfrac{96{,}000}{(12{,}000)(0.00004)}$

54. $\dfrac{(0.48)(14{,}400{,}000)}{96{,}000{,}000}$

In Exercises 55–60, use a calculator to evaluate each expression.

55. $(456.4)^6$

56. $(0.053)^8$

57. $(0.009)^{-6}$

58. 225^{-5}

59. $\dfrac{(3.12 \times 10^{16})(4.50 \times 10^{-6})}{2.40 \times 10^{-5}}$

60. $(7.35 \times 10^5)(3.84 \times 10^{-7}) \cdot \dfrac{1}{2.10 \times 10^{12}}$

APPLICATIONS

61. WAVELENGTH Transmitters, vacuum tubes, and lights emit energy that can be modeled as a wave, as shown in Illustration 1. Examples of the most common types of electromagnetic waves are given in the table. List the wavelengths in order from shortest to longest.

ILLUSTRATION 1

This distance between the two crests of the wave is called the wavelength.

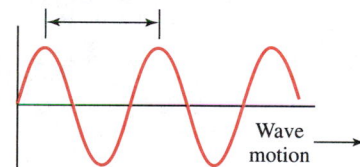

Wave motion →

Type	Use	Wavelength (m)
visible light	lighting	9.3×10^{-6}
infrared	photography	3.7×10^{-5}
x-ray	medical	2.3×10^{-11}
radio wave	communication	3.0×10^{2}
gamma ray	treating cancer	8.9×10^{-14}
microwave	cooking	1.1×10^{-2}
ultraviolet	sun lamp	6.1×10^{-8}

62. EXPLORATION On July 4, 1997, the Pathfinder, carrying the rover vehicle called Sojourner, landed on Mars to perform a scientific investigation of the planet. The distance from Mars to the earth is approximately 3.5×10^7 miles. Use scientific notation to express this distance in feet. (*Hint:* 5,280 feet = 1 mile.)

63. PROTON The mass of one proton is approximately 1.7×10^{-24} gram. Use scientific notation to express the mass of 1 million protons.

64. SPEED OF SOUND The speed of sound in air is approximately 3.3×10^4 centimeters per second. Use scientific notation to express this speed in kilometers per second. (*Hint:* 100 centimeters = 1 meter and 1,000 meters = 1 kilometer.)

65. LIGHT YEAR One light year is about 5.87×10^{12} miles. Use scientific notation to express this distance in feet. (*Hint:* 5,280 feet = 1 mile.)

66. OIL RESERVES Saudi Arabia is believed to have crude oil reserves of about 2.61×10^{11} barrels. A barrel contains 42 gallons of oil. Use scientific notation to express its oil reserves in gallons.

67. INTEREST EARNED As of December 31, 1997, the Federal Deposit Insurance Corporation (FDIC) reported that the total insured deposits in U.S. banks and savings and loans was approximately 3.42×10^{12} dollars. If this money was invested at a rate of 4% simple annual interest, how much would it earn in one year? (Use scientific notation to express the answer.)

68. CURRENCY As of March 31, 1998, the U.S. Treasury reported that the number of $20 bills in circulation was approximately 4.136×10^9. What was the total value of the currency? (Use scientific notation to express the answer.)

69. SIZE OF THE MILITARY The graph in Illustration 2 shows the number of U.S. troops for 1979–1997. Estimate each of the following and express your answers in scientific and standard notation.
  a. The number of troops in 1993

  b. The smallest and largest numbers of troops during these years

ILLUSTRATION 2

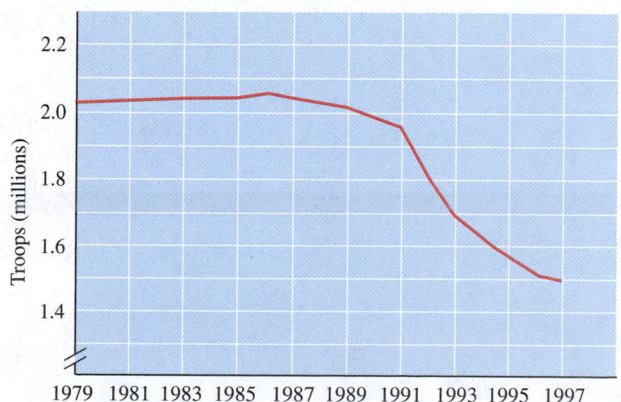

Based on data from the U.S. Department of Defense

70. THE NATIONAL DEBT The graph in Illustration 3 shows the growth of the national debt for the fiscal years 1992–1997.

ILLUSTRATION 3

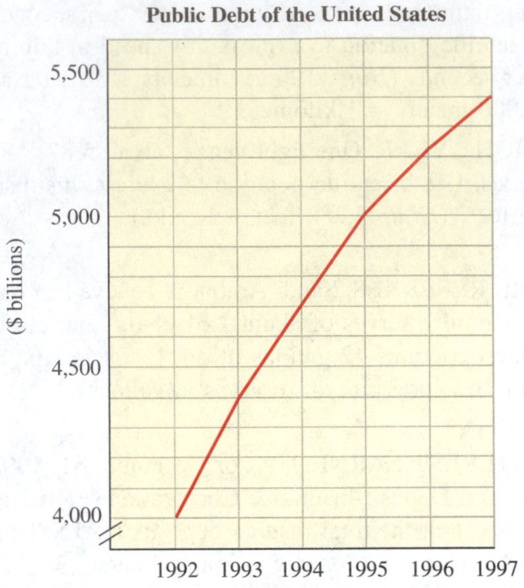

Public Debt of the United States

Based on data from the U.S. Department of the Treasury

a. Use scientific notation to express the debt as of 1994, 1995, and 1997.

b. In 1997, the population of the United States was about 2.68×10^8. To the nearest dollar, what was the share of the debt for each man, woman, and child in the United States? Answer in standard notation.

WRITING

71. In what situations would scientific notation be more convenient than standard notation?

72. To multiply a number by a power of 10, we move the decimal point. Which way, and how far? Explain.

73. 2.3×10^{-3} contains a negative sign but represents a positive number. Explain.

74. Is this a true statement? $2.0 \times 10^3 = 2 \times 10^3$. Explain.

REVIEW

75. If $y = -1$, find the value of $-5y^{55}$.

76. What is the y-intercept of the graph of $y = -3x - 5$?

In Exercises 77–78, tell which property of real numbers justifies each statement.

77. $5 + z = z + 5$

78. $7(u + 3) = 7u + 7 \cdot 3$

In Exercises 79–80, solve each equation.

79. $3(x - 4) - 6 = 0$

80. $8(3x - 5) - 4(2x + 3) = 12$

▶ **4.4**

Polynomials and Polynomial Functions

In this section, you will learn about

> Polynomials ■ Monomials, binomials, and trinomials ■ Degree of a polynomial ■ Evaluating polynomial functions ■ Making tables ■ Graphing polynomial functions ■ Graphs of polynomial functions

Introduction   In arithmetic, we learned how to add, subtract, multiply, divide, and find powers of numbers. In algebra, we will learn how to perform these operations on *polynomials*. In this section, we will introduce polynomials, classify them into groups, define their degrees, and show how to evaluate them at specific values of their variables. Finally, we will show how to graph polynomial functions.

Polynomials

Recall that a **term** is a number or a product of a number and one or more variables, which may be raised to powers. Examples of terms are

$$3x, \qquad -4y^2, \qquad \frac{1}{2}a^2b^3, \qquad t, \qquad \text{and} \qquad 25$$

The **numerical coefficients,** or simply **coefficients,** of the first four of these terms are $3, -4, \frac{1}{2}$, and 1, respectively. Because $25 = 25x^0$, 25 is considered to be the numerical coefficient of the term 25.

Polynomials

A **polynomial** is a term or a sum of terms in which all variables have whole-number exponents.

Here are some examples of polynomials:

$$3x + 2, \qquad -4y^2 - 2y - 3, \qquad 8xy^2, \qquad \text{and} \qquad 3a - 4b - 4c + 8d$$

The polynomial $3x + 2$ has two terms, and we say that it is a **polynomial in** x. Since $-4y^2 - 2y - 3$ can be written as $-4y^2 + (-2y) + (-3)$, it is the sum of three terms, $-4y^2, -2y, -3$. It is written in **decreasing powers** of y, because the powers on y decrease from left to right. $8xy^2$ is a single term, and it is in two variables, x and y.

 WARNING! The expression $2x^3 - 3x^{-2} + 5$ is not a polynomial, because the second term contains a variable with an exponent that is not a whole number. Similarly, $y^2 - \frac{7}{y}$ is not a polynomial, because $\frac{7}{y}$ can be written $7y^{-1}$.

EXAMPLE 1

Identifying polynomials. Tell whether each expression is a polynomial.

a. $x^2 + 2x + 1$ Yes.

b. $3a^{-1} - 2a - 3$ No. In the first term, the exponent on the variable is not a whole number.

c. $\dfrac{1}{2}x^3 - 2.3x$ Yes, since it can be written as the sum $\frac{1}{2}x^3 + (-2.3x)$.

d. $\dfrac{p + 3}{p - 1}$ No. Variables cannot be in the denominator of a fraction.

SELF CHECK   Tell whether each expression is a polynomial:
a. $3x^{-4} + 2x^2 - 3$ and   **b.** $7.5p^3 - 4p^2 - 3p + 4$.   *Answers:*   **a.** no,   **b.** yes ∎

Monomials, Binomials, and Trinomials

We classify some polynomials by the number of terms they contain. A polynomial with one term is called a **monomial.** A polynomial with two terms is called a **binomial.** A polynomial with three terms is called a **trinomial.** Here are some examples. There is no special name for a polynomial with four or more terms.

Monomials	Binomials	Trinomials
$5x^2y$	$3u^3 - 4u^2$	$-5t^2 + 4t + 3$
$-6x$	$18a^2b + 4ab$	$27x^3 - 6x - 2$
29	$-29z^{17} - 1$	$-32r^6 + 7y^3 - z$

EXAMPLE 2

Classifying polynomials. Classify each polynomial as a monomial, binomial, or trinomial.

a. $5.2x^4 + 3.1x$ Since the polynomial has two terms, $5.2x^4$ and $3.1x$, it is a binomial.

b. $7g^4 - 5g^3 - 2$ Since the polynomial has three terms, $7g^4$, $-5g^3$, and -2, it is a trinomial.

c. $-5x^2y^3$ Since the polynomial has one term, it is a monomial.

SELF CHECK

Classify each polynomial as a monomial, binomial, or trinomial: **a.** $5x$,   **b.** $-5x^2 + 2x - 0.5$, and   **c.** $16x^2 - 9y^2$.

Answers: **a.** monomial,   **b.** trinomial,   **c.** binomial

Degree of a Polynomial

The monomial $7x^6$ is called a **monomial of sixth degree** or a **monomial of degree 6,** because the variable x occurs as a factor six times. The monomial $3x^3y^4$ is a monomial of seventh degree, because the variables x and y occur as factors a total of seven times. Here are some more examples:

$2.7a$ is a monomial of degree 1.

$-2x^3$ is a monomial of degree 3.

$47x^2y^3$ is a monomial of degree 5.

8 is a monomial of degree 0, because $8 = 8x^0$.

These examples illustrate the following definition.

Degree of a monomial

If a is a nonzero constant, the **degree of the monomial** ax^n is n.

The **degree of a monomial** in several variables is the sum of the exponents on those variables.

 WARNING! Note that the degree of ax^n is not defined when $a = 0$. Since $ax^n = 0$ when $a = 0$, the constant 0 has no defined degree.

Because each term of a polynomial is a monomial, we define the degree of a polynomial by considering the degrees of each of its terms.

Degree of a polynomial

The **degree of a polynomial** is determined by the term with the largest degree.

Here are some examples:

$x^2 + 2x$ is a binomial of degree 2, because the degree of its first term is 2 and the degree of its second term is less than 2.

$d^3 - 3d^2 + 1$ is a trinomial of degree 3, because the degree of its first term is 3 and the degree of each of its other terms is less than 3.

$25y^{13} - 15y^8z^{10} - 32y^{10}z^8 + 4$ is a polynomial of degree 18, because its second and third terms are of degree 18. Its other terms have degree less than 18.

EXAMPLE 3

Degree of a polynomial. Find the degree of each polynomial:
a. $-4x^3 - 5x^2 + 3x$, **b.** $1.6w - 1.6$, and **c.** $-17a^2b^3 + 12ab^6$.

Solution **a.** The trinomial $-4x^3 - 5x^2 + 3x$ has terms of degree 3, 2, and 1. Therefore, its degree is 3.

b. The first term of $1.6w - 1.6$ has degree 1 and the second term has degree 0, so the binomial has degree 1.

c. The degree of the first term of $-17a^2b^3 + 12ab^6$ is 5 and the degree of the second term is 7, so the binomial has degree 7.

SELF CHECK

Find the degree of each polynomial:
a. $15p^3 - 25p^2 - 3p + 4$ and **b.** $-14st^4 + 12s^3t$. *Answers:* **a.** 3, **b.** 5 ■

If written in descending powers of the variable, the **leading term** of a polynomial is the term of highest degree. For example, the leading term of $-4x^3 - 5x^2 + 3x$ is $-4x^3$. The coefficient of the leading term (in this case, -4) is called the **leading coefficient.**

Evaluating Polynomial Functions

Each of the equations below defines a function, because each input x-value gives exactly one output value. Since the right-hand side of each equation is a polynomial, these functions are called **polynomial functions.**

$$f(x) = 6x + 4 \qquad g(x) = 3x^2 + 4x - 5 \qquad h(x) = -x^3 + x^2 - 2x + 3$$

This polynomial has two terms. Its degree is 1. | This polynomial has three terms. Its degree is 2. | This polynomial has four terms. Its degree is 3.

To evaluate a polynomial function for a specific value, we replace the variable in the defining equation with the value, called the **input**. Then we simplify the resulting expression to find the **output**. For example, suppose we wish to evaluate the polynomial function $f(x) = 6x + 4$ for $x = 1$. Then $f(1)$ (read as "f of 1") represents the value of $f(x) = 6x + 4$ when $x = 1$. We find $f(1)$ as follows.

$f(x) = 6x + 4$ The given function.

$f(1) = 6(1) + 4$ Substitute 1 for x. The number 1 is the input.

$\quad\;\; = 6 + 4$ Do the multiplication.

$\quad\;\; = 10$ Do the addition. 10 is the output.

Thus, $f(1) = 10$.

EXAMPLE 4

Evaluating polynomial functions. Consider the function $g(x) = 3x^2 + 4x - 5$. Find **a.** $g(0)$ and **b.** $g(-2)$.

Solution **a.** $g(x) = 3x^2 + 4x - 5$ The given function.

$g(0) = 3(0)^2 + 4(0) - 5$ To find $g(0)$, substitute 0 for x.

$= 3(0) + 4(0) - 5$ Evaluate the power.

$= 0 + 0 - 5$ Do the multiplications.

$g(0) = -5$

b. $g(x) = 3x^2 + 4x - 5$ The given function.

$g(-2) = 3(-2)^2 + 4(-2) - 5$ To find $g(-2)$, substitute -2 for x.

$= 3(4) + 4(-2) - 5$ Evaluate the power.

$= 12 + (-8) - 5$ Do the multiplications.

$g(-2) = -1$

SELF CHECK

Consider the function $h(x) = -x^3 + x^2 - 2x + 3$.
Find **a.** $h(0)$ and **b.** $h(-3)$.   *Answers:* **a.** 3, **b.** 45 ∎

EXAMPLE 5

Supermarket display. The polynomial function

$$f(c) = \frac{1}{3}c^3 + \frac{1}{2}c^2 + \frac{1}{6}c$$

gives the number of cans used in a display shaped like a square pyramid, having a square base formed by c cans per side. Find the number of cans of soup used in the display shown in Figure 4-3.

FIGURE 4-3

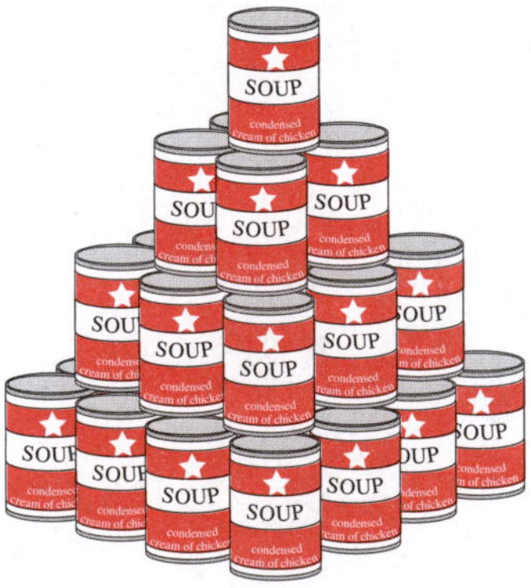

Solution Since each side of the square base of the display is formed by 4 cans, $c = 4$. We can find the number of cans used in the display by finding $f(4)$.

$f(c) = \frac{1}{3}c^3 + \frac{1}{2}c^2 + \frac{1}{6}c$ The given function.

$f(4) = \frac{1}{3}(4)^3 + \frac{1}{2}(4)^2 + \frac{1}{6}(4)$ Substitute 4 for c.

$= \frac{1}{3}(64) + \frac{1}{2}(16) + \frac{1}{6}(4)$ Find the powers.

$= \frac{64}{3} + 8 + \frac{2}{3}$ Do the multiplication, then simplify: $\frac{4}{6} = \frac{2}{3}$.

$= \frac{66}{3} + 8$ Add the fractions.

$= 22 + 8$

$= 30$

30 cans of soup were used in the display.

∎

ACCENT ON TECHNOLOGY *100-Meter Sprint*

In the 1996 Olympics, Donovan Bailey of Canada set the world record in the 100 meters with a time of 9.84 seconds. Suppose the polynomial function $f(t) = -0.2t^2 + 11.12t$ gives the distance in meters run in t seconds by another sprinter in the 100 meters. To predict how far behind Bailey this sprinter would have finished in the Olympic competition, we can substitute Bailey's time of 9.84 seconds for t and evaluate $f(9.84)$.

$$f(t) = -0.2t^2 + 11.12t$$
$$f(\mathbf{9.84}) = -0.2(\mathbf{9.84})^2 + 11.12\,(\mathbf{9.84})$$

To evaluate $f(9.84)$ using a scientific calculator, we enter these numbers and press these keys:

Keystrokes .2 $\boxed{+/-}$ $\boxed{\times}$ 9.84 $\boxed{x^2}$ $\boxed{+}$ 11.12 $\boxed{\times}$ 9.84 $\boxed{=}$ $\boxed{\text{90.05568}}$

In 9.84 seconds, the sprinter can run about 90 meters and would finish approximately 10 meters behind Bailey.

Making Tables

When we evaluate a polynomial function for several values of its variable, we can write the results in a table.

EXAMPLE 6

Constructing a table. Find $f(-2), f(-1)$, and $f(0)$, where $f(x) = x^3 - 3x^2 + 4$. Write the results in a table.

Solution

In the first column of the table, we write each of the input values for x. (These are the numbers inside the parentheses of the function notation: $-2, -1, 0$.) We then find each corresponding output value by substituting an input value for x in $x^3 - 3x^2 + 4$ and evaluating the expression. To review how to generate a table using a graphing calculator, see page 199.

x	$f(x)$
-2	-16
-1	0
0	4

$f(-2) = (-2)^3 - 3\,(-2)^2 + 4$
$\quad = -8 - 3(4) + 4$
$\quad = -8 - 12 + 4$
$\quad = -16$

$f(-1) = (-1)^3 - 3(-1)^2 + 4$
$\quad = -1 - 3(1) + 4$
$\quad = -1 - 3 + 4$
$\quad = 0$

$f(0) = (0)^3 - 3(0)^2 + 4$
$\quad = 0 - 3(0) + 4$
$\quad = 0 - 0 + 4$
$\quad = 4$

SELF CHECK Find $f(1), f(2)$, and $f(3)$, where $f(x) = x^3 - 3x^2 + 4$. Write the results in a table.   *Answers:*

x	$f(x)$
1	2
2	0
3	4

Graphing Polynomial Functions

We can graph polynomial functions by making a table of values, plotting points, and drawing a smooth curve that passes through those points.

EXAMPLE 7

Graphing a polynomial function. Graph $f(x) = x^3 - 3x^2 + 4$.

Solution   We substitute numbers for x, compute the corresponding values of $f(x)$, and list the results in a table as ordered pairs. (Note that this work was done in Example 6 and the Self Check.) We then plot the pairs (x, y) and draw a smooth curve through the points, as shown in Figure 4-4.

FIGURE 4-4

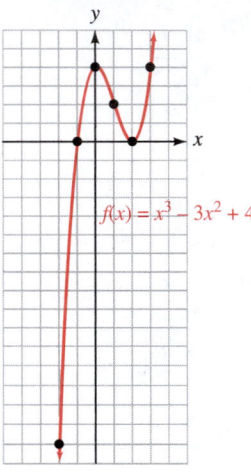

$$f(x) = x^3 - 3x^2 + 4$$

— This column can also be labeled $(x, f(x))$.

x	$f(x)$	(x, y)
-2	-16	$(-2, -16)$
-1	0	$(-1, 0)$
0	4	$(0, 4)$
1	2	$(1, 2)$
2	0	$(2, 0)$
3	4	$(3, 4)$

The value of $f(x)$ is the y-coordinate of the point.

ACCENT ON TECHNOLOGY *Graphing Polynomial Functions*

It is possible to use graphing calculators to generate tables and graphs for polynomial functions. For example, Figure 4-5 (a) shows how to enter the function from Example 7, $f(x) = x^3 - 3x^2 + 4$. Figures 4-5(b) and (c) show the calculator display of a table and the graph.

FIGURE 4-5

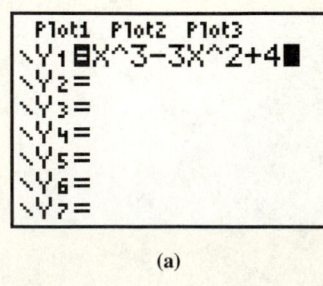

(a)

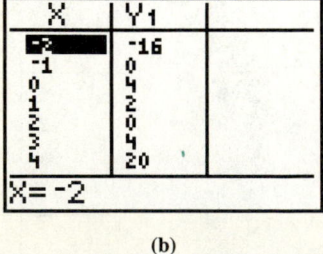

(b)

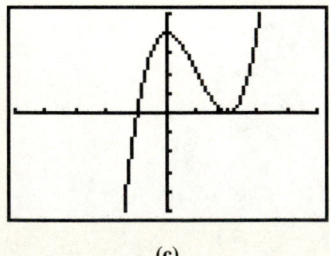

(c)

Graphs of Polynomial Functions

The graphs of two polynomial functions are shown in Figure 4-6. These graphs have characteristic "peaks" and "valleys." When such graphs describe real-life situations, locating the highest and lowest points on the graph can give valuable information.

FIGURE 4-6

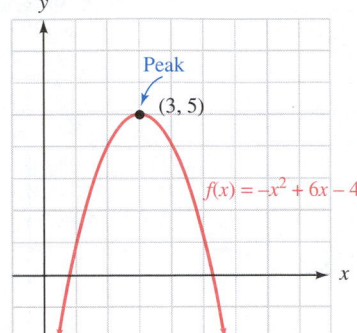

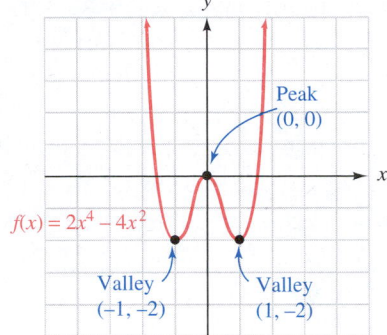

EXAMPLE 8

Immigration. The polynomial function graphed in Figure 4-7 approximates the percent of the U.S. population that was foreign born for each of the years 1900–1997. What are the coordinates of the highest point on the graph? Explain its significance.

FIGURE 4-7

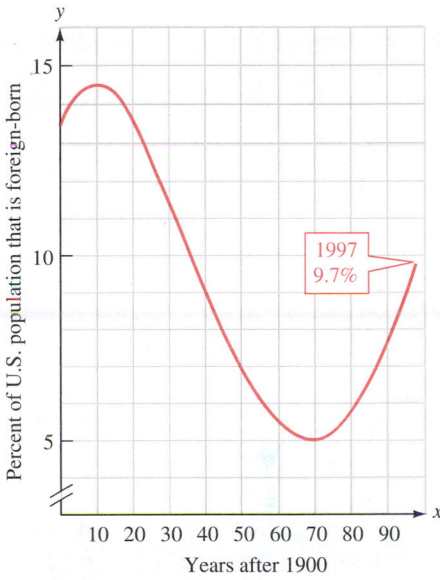

Based on data from Bureau of the Census, U.S. Department of Commerce

Solution

The highest point on the graph has coordinates (10, 14.5). This means that in 1910, 14.5% of the population of the United States was foreign born. This percent is the largest for the 97-year span shown on the graph.

SELF CHECK

What are the coordinates of the lowest point on the graph in Figure 4-7? Explain its significance.

Answer: (70, 5). In 1970, 5% of the population was foreign born. This percent is the smallest for the 97 years shown. ∎

STUDY SET

Section 4.4

VOCABULARY

In Exercises 1–10, fill in the blanks to make the statements true.

1. A _____ is an algebraic expression that is the sum of one or more terms containing whole-number exponents.

2. The numerical _____ of the term $-25x^2y^3$ is -25.

3. The degree of a polynomial is the same as the degree of its _____ with the largest degree.

4. A _____ is a polynomial with one term. A _____ is a polynomial with two terms.

5. The _____ of the monomial $3x^7$ is 7.

6. For the polynomial $6x^2 + 3x - 1$, the _____ term is $6x^2$, and the _____ coefficient is 6.

7. $-x^3 - 6x^2 + 9x - 2$ is a polynomial ___ x and is written in _____ powers of x.

8. A _____ is a polynomial with three terms.

9. The notation $f(x)$ is read as f ___ x.

10. $f(2)$ represents the _____ of a function when $x = 2$.

CONCEPTS

In Exercises 11–14, tell whether each expression is a polynomial.

11. $x^3 - 5x^2 - 2$ 12. $x^{-4} - 5x$

13. $\dfrac{1}{2x} + 3$ 14. $x^3 - 1$

In Exercises 15–26, classify each polynomial as a monomial, binomial, or trinomial, if possible.

15. $3x + 7$ 16. $3y - 5$

17. $y^2 + 4y + 3$ 18. $3xy$

19. $3z^2$ 20. $3x^4 - 2x^3 + 3x - 1$

21. $5t - 32$ 22. $9x^2y^3z^4$

23. $s^2 - 23s + 31$ 24. $2x^3 - 5x^2 + 6x - 3$

25. $3x^5 - x^4 - 3x^3 + 7$ 26. x^3

In Exercises 27–38, find the degree of each polynomial.

27. $3x^4$ 28. $3x^5$

29. $-2x^2 + 3x + 1$ 30. $-5x^4 + 3x^2 - 3x$

31. $3x - 5$ 32. $y^3 + 4y^2$

33. $-5r^2s^2 - r^3s + 3$ 34. $4r^2s^3 - 5r^2s^8$

35. $x^{12} + 3x^2y^3$ 36. $17ab^5 - 12a^3b$

37. 38 38. -25

39. Give the coordinates of the "valley" of the graph of the polynomial function shown in Illustration 1.

ILLUSTRATION 1

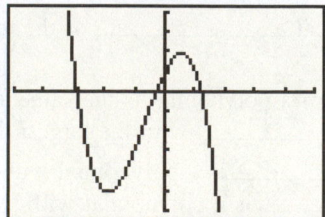

40. Give the coordinates of the "peaks" of the graph of the polynomial function shown in Illustration 2.

ILLUSTRATION 2

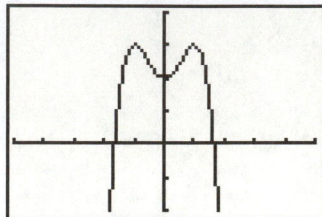

NOTATION

In Exercises 41–42, complete each solution.

41. If $f(x) = -2x^2 + 3x - 1$, find $f(2)$.
$$f(2) = -2(\;\;)^2 + 3(\;\;) - 1$$
$$= -2(\;\;) + \;\; - 1$$
$$= -8 + 6 - \;\;$$
$$= \;\; - 1$$
$$= -3$$

42. If $f(x) = -2x^2 + 3x - 1$, find $f(-2)$.
$$f(-2) = -2(\;\;)^2 + 3(\;\;) - 1$$
$$= -2(\;\;) + (\;\;) - 1$$
$$= \;\; + (-6) - 1$$
$$= \;\; - 1$$
$$= -15$$

43. Explain why $f(x) = x^3 + 2x^2 - 3$ is called a polynomial function.

44. Give another way to label the last column of the table shown below.

x	$f(x)$	$(x, f(x))$

PRACTICE

In Exercises 45–52, let $f(x) = 5x - 3$. Find each value.

45. $f(2)$ 46. $f(0)$

47. $f(-1)$ 48. $f(-2)$

49. $f\left(\dfrac{1}{5}\right)$ 50. $f\left(\dfrac{4}{5}\right)$

51. $f(-0.9)$ 52. $f(-1.2)$

In Exercises 53–60, let $g(x) = -x^2 - 4$. Find each value.

53. $g(0)$ 54. $g(1)$

55. $g(-1)$ 56. $g(-2)$

57. $g(1.3)$ **58.** $g(2.4)$

59. $g(-13.6)$ **60.** $g(-25.3)$

 In Exercises 61–68, let $h(x) = x^3 - 2x + 3$. Find each value.

61. $h(0)$ **62.** $h(3)$

63. $h(-2)$ **64.** $h(-1)$

65. $h(0.9)$ **66.** $h(0.4)$

67. $h(-8.1)$ **68.** $h(-7.7)$

In Exercises 69–72, complete each table and then graph the polynomial function.

69. $f(x) = x^2 + 2x$

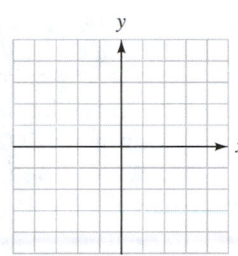

x	$f(x)$
-3	
-2	
-1	
0	
1	

70. $f(x) = -x^2 + 2x + 3$

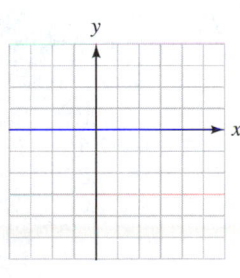

x	$f(x)$
-2	
-1	
0	
1	
2	
3	
4	

71. $f(x) = x^3 + 3x^2$

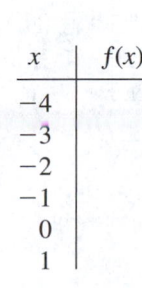
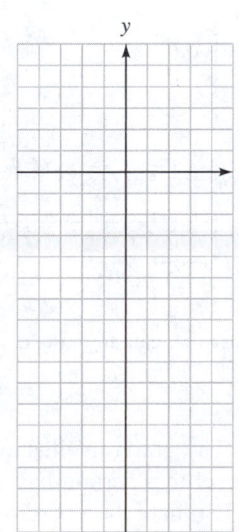

x	$f(x)$
-4	
-3	
-2	
-1	
0	
1	

72. $f(x) = -x^3 + 3x^2 - 4$

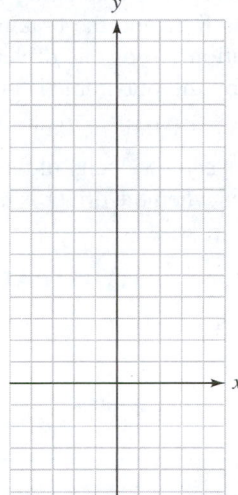

x	$f(x)$
-2	
-1	
0	
1	
2	
3	

 In Exercises 73–74, use a graphing calculator to graph each polynomial function. Use window settings of $x = -5$ to 5 and $y = -10$ to 10.

73. $f(x) = -0.5x^2 + 1.2x - 2.5$

74. $f(x) = x^3 + x^2 - 6x$

APPLICATIONS

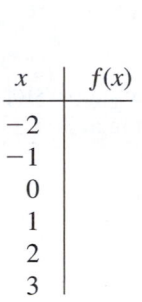 *In Exercises 75–82, use a calculator to help solve each problem.*

75. TRACK RECORDS By how many meters would the slower sprinter mentioned in the Accent on Technology feature in this section finish behind Florence Griffith Joyner if she ran her world record time of 10.49 seconds in the 100 meters?

76. MAXIMIZING REVENUE The revenue (in dollars) that a manufacturer of office desks receives is given by the polynomial function

$$f(d) = -0.08d^2 + 100d$$

where d is the number of desks manufactured.
a. Find the total revenue if 625 desks are manufactured.
b. Does increasing the number of desks being manufactured to 650 increase the revenue?

77. WATER BALLOONS Some college students launched water balloons from the balcony of their dormitory on unsuspecting sunbathers on the college quad. The height in feet of the balloons at a time t seconds after being launched is given by the polynomial function

$$f(t) = -16t^2 + 12t + 20$$

What was the height of the balloons 0.5 second and 1.5 seconds after being launched?

78. STOPPING DISTANCE The number of feet that a car travels before stopping depends on the driver's reaction time and the braking distance, as shown in Illustration 3. For one driver, the stopping distance is given by the polynomial function

$$f(v) = 0.04v^2 + 0.9v$$

where v is the velocity of the car. Find the stopping distance when the driver is traveling at 30 mph.

ILLUSTRATION 3

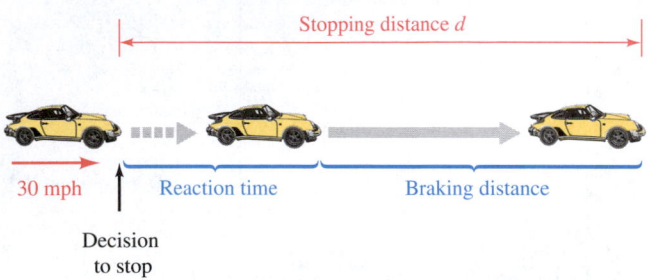

Stopping distance d

30 mph — Decision to stop — Reaction time — Braking distance

79. SUSPENSION BRIDGE See Illustration 4. The function

$$f(s) = 400 + 0.0066667s^2 - 0.0000001s^4$$

approximates the length of the cable between the two vertical towers of a suspension bridge, where s is the sag in the cable. Estimate the length of the cable if the sag is 24.6 feet.

ILLUSTRATION 4

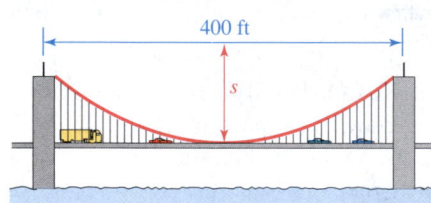

400 ft

s

80. PRODUCE DEPARTMENT Suppose a grocer is going to set up a pyramid-shaped display of cantaloupes like that shown in Figure 4–3 in Example 5. If each side of the square base of the display is made of 6 cantaloupes, how many will be used in the display?

81. DOLPHINS At a marine park, three trained dolphins jump in unison over an arching stream of water whose path can be described by the polynomial function $f(x) = -0.05x^2 + 2x$. See Illustration 5. Given the takeoff points for each dolphin, how high must each dolphin jump to clear the stream of water?

82. TUNNEL The arch at the entrance to a tunnel is described by the polynomial function $f(x) = -0.25x^2 + 23$. See Illustration 6. What is the height of the arch at the edge of the pavement?

ILLUSTRATION 5

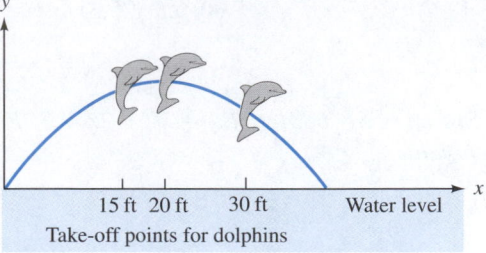

15 ft   20 ft   30 ft   Water level

Take-off points for dolphins

ILLUSTRATION 6

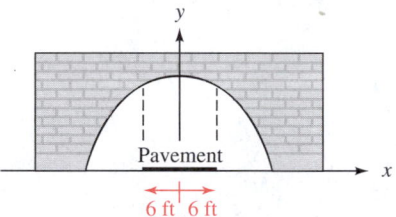

Pavement

6 ft   6 ft

83. IMPORTED STEEL Plot the data for each quarter on the graph in Illustration 7. Connect the points with a smooth curve. When were the imports the lowest? When were they the highest?

U.S. Steel Imports (100,000 tons)			
1995	**1996**	**1997**	**1998**
Q1 5.0	Q1 2.0	Q1 8.0	Q1 10.0
Q2 3.5	Q2 2.5	Q2 9.0	Q2 18.0
Q3 2.0	Q3 4.0	Q3 8.0	Q3 23.5
Q4 1.5	Q4 7.0	Q4 7.5	

ILLUSTRATION 7

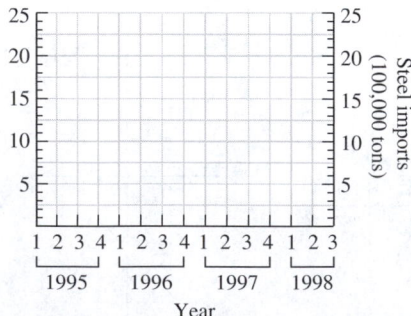

Year

Based on data from the U.S. Department of Commerce

84. TIDES The graph in Illustration 8 shows how the water level in a certain bay changed over a 24-hour period. Estimate the coordinates of each of the following: the lower high water mark and the higher low water mark.

ILLUSTRATION 8

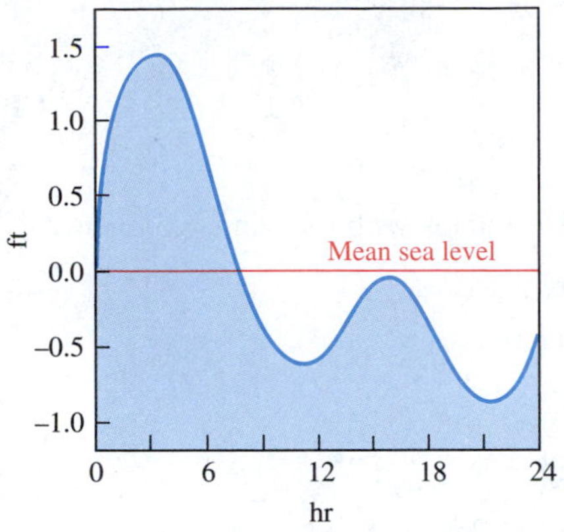

ft

Mean sea level

hr

In Exercises 87–88, solve each inequality and graph the solution set.

87. $-4(3y + 2) \le 28$

88. $-5 < 3t + 4 \le 13$

In Exercises 89–92, write each expression without using parentheses or negative exponents.

89. $(x^2 x^4)^3$

90. $(a^2)^3 (a^3)^2$

91. $\left(\dfrac{y^2 y^5}{y^4} \right)^3$

92. $\left(\dfrac{2t^3}{t} \right)^{-4}$

WRITING

85. Describe how to determine the degree of a polynomial.

86. List some words that contain the prefixes *mono*, *bi*, or *tri*.

▶ 4.5   Adding and Subtracting Polynomials

In this section, you will learn about

> Adding monomials ■ Subtracting monomials ■ Adding polynomials ■ Subtracting polynomials ■ Adding and subtracting multiples of polynomials ■ An application of adding polynomials

Introduction   In Figure 4-8(a), the heights of the Seattle Space Needle and the Eiffel Tower in Paris are given. Using rules from arithmetic, we can find the difference in the heights of the towers by subtracting two numbers.

FIGURE 4-8

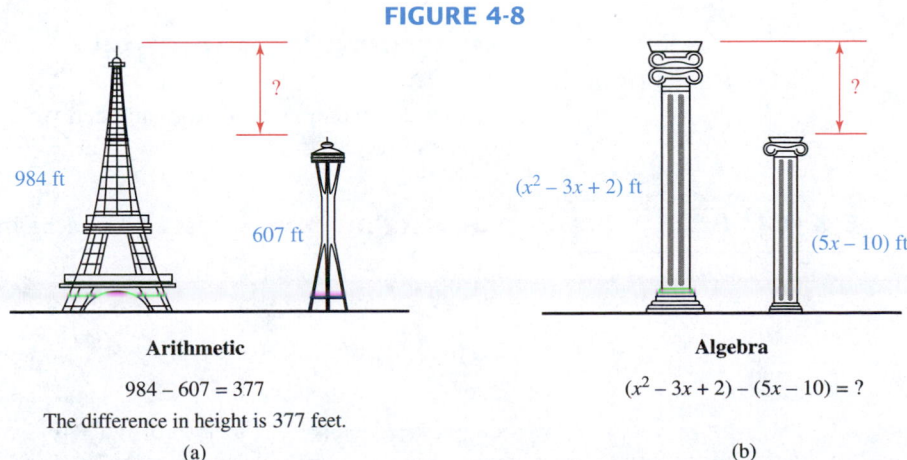

984 ft

607 ft

$(x^2 - 3x + 2)$ ft

$(5x - 10)$ ft

Arithmetic

$984 - 607 = 377$

The difference in height is 377 feet.

(a)

Algebra

$(x^2 - 3x + 2) - (5x - 10) = ?$

(b)

In Figure 4-8(b), the heights of two types of classical Greek columns are expressed using *polynomials*. To find the difference in their heights, we must subtract the polynomials. In this section, we will discuss the algebraic rules that are used to do this. Since any subtraction can be written in terms of addition, we will consider the procedures used to add polynomials first. We begin with monomials, which are polynomials with just one term.

Adding Monomials

Recall that like terms have the same variables with the same exponents:

Like terms	*Unlike terms*
$-7x$ and $15x$	$-7x$ and $15a$
$4y^3$ and $16y^3$	$4y^3$ and $16y^2$
$\dfrac{1}{2}xy^2$ and $-\dfrac{1}{3}xy^2$	$\dfrac{1}{2}xy^2$ and $-\dfrac{1}{3}x^2y$

Also recall that to combine like terms, we combine their coefficients and keep the same variables with the same exponents. For example,

$$4y + 5y = (4 + 5)y \qquad \text{and} \qquad 8x^2 - x^2 = (8 - 1)x^2$$
$$= 9y \qquad\qquad\qquad\qquad = 7x^2$$

Likewise,

$$3a + 4b - 6a + 3b = -3a + 7b \qquad \text{and} \qquad -4cd^3 + 9cd^3 = 5cd^3$$

These examples suggest that to add like monomials, we simply combine like terms.

EXAMPLE 1   **Adding monomials.** Do the following additions.

a. $4x^4 + 81x^4 = 85x^4$

b. $-8x^2y^2 + 6x^2y^2 + x^2y^2 = -2x^2y^2 + x^2y^2$ Work from left to right.
$$= -x^2y^2 \qquad\qquad\qquad \text{Combine like terms.}$$

c. $32c^2 + 10c + 4c^2 = 32c^2 + 4c^2 + 10c$ Group like terms together.
$$= 36c^2 + 10c \qquad\qquad \text{Combine like terms. The terms of the answer are written in descending order.}$$

SELF CHECK   Do the following additions: **a.** $27x^6 + 8x^6$,   *Answers:* **a.** $35x^6$,
b. $-12pq^2 + 5pq^2 + 8pq^2$, and **c.** $6a^3 + 15a + a^3$. **b.** pq^2, **c.** $7a^3 + 15a$ ■

Subtracting Monomials

To subtract one monomial from another, we add the opposite of the monomial that is to be subtracted. In symbols, $x - y = x + (-y)$.

EXAMPLE 2   **Subtracting monomials.** Find each difference.

a. $8x^2 - 3x^2 = 8x^2 + (-3x^2)$ Add the opposite of $3x^2$, which is $-3x^2$.
$$= 5x^2 \qquad\qquad\qquad \text{Combine like terms.}$$

b. $6xy - 9xy = 6xy + (-9xy)$
$$= -3xy$$

c. $-3r - 5 - 4r = -3r + (-5) + (-4r)$ Add the opposite of 5 and $4r$.
$$= -3r + (-4r) + (-5) \quad \text{Group like terms together.}$$
$$= -7r - 5 \qquad\qquad\quad \text{Combine like terms. Write the addition of } -5 \text{ as a subtraction of 5.}$$

SELF CHECK   Find each difference: **a.** $12m^3 - 7m^3$ and   *Answers:* **a.** $5m^3$,
b. $-4pq - 27p - 8pq$.   **b.** $-12pq - 27p$ ■

Adding Polynomials

Because of the distributive property, we can remove parentheses enclosing several terms when the sign preceding the parentheses is a + sign. We simply drop the parentheses.

$$+(3x^2 + 3x - 2) = +1(3x^2 + 3x - 2)$$
$$= 1(3x^2) + 1(3x) + 1(-2) \quad \text{Use the distributive property.}$$
$$= 3x^2 + 3x + (-2)$$
$$= 3x^2 + 3x - 2$$

We can add polynomials by removing parentheses, if necessary, and then combining any like terms that are contained within the polynomials.

EXAMPLE 3

Adding polynomials. Add $(3x^2 - 3x + 2) + (2x^2 + 7x - 4)$.

Solution
$$(3x^2 - 3x + 2) + (2x^2 + 7x - 4)$$
$$= 3x^2 - 3x + 2 + 2x^2 + 7x - 4 \quad \text{Drop the parentheses.}$$
$$= 3x^2 + 2x^2 - 3x + 7x + 2 - 4 \quad \text{Write the like terms together.}$$
$$= 5x^2 + 4x - 2 \quad \text{Combine like terms.}$$

SELF CHECK Add $(2a^2 - a + 4) + (5a^2 + 6a - 5)$. *Answer:* $7a^2 + 5a - 1$ ∎

Problems such as Example 3 are often written with like terms aligned vertically. We can then add column by column.

$$\begin{array}{r} 3x^2 - 3x + 2 \\ + \ 2x^2 + 7x - 4 \\ \hline 5x^2 + 4x - 2 \end{array}$$

EXAMPLE 4

Adding polynomials vertically. Add $(4x^2 - 3)$ and $(3x^2 - 8x + 8)$.

Solution Since the first polynomial does not have an x-term, we leave a space so that the constant terms can be aligned.

$$\begin{array}{r} 4x^2 \quad\ \ - 3 \\ + \ 3x^2 - 8x + 8 \\ \hline 7x^2 - 8x + 5 \end{array}$$

SELF CHECK Add $(4q^2 - 7)$ and $(2q^2 - 8q + 9)$ vertically. *Answer:* $6q^2 - 8q + 2$ ∎

Subtracting Polynomials

Because of the distributive property, we can remove parentheses enclosing several terms when the sign preceding the parentheses is a − sign. We simply drop the minus sign and the parentheses, and *change the sign of every term within the parentheses.*

$$-(3x^2 + 3x - 2) = -1(3x^2 + 3x - 2)$$
$$= -1(3x^2) + (-1)(3x) + (-1)(-2)$$
$$= -3x^2 + (-3x) + 2$$
$$-(3x^2 + 3x - 2) = -3x^2 - 3x + 2$$

This suggests that the way to subtract polynomials is to remove parentheses and combine like terms.

EXAMPLE 5

Subtracting polynomials. Find each difference.

a. $(3x - 4) - (5x + 7) = 3x - 4 - 5x - 7$ Change the sign of each term inside $(5x + 7)$.

$$= -2x - 11 \qquad \text{Combine like terms.}$$

b. $(3x^2 - 4x - 6) - (2x^2 - 6x) = 3x^2 - 4x - 6 - 2x^2 + 6x$

$$= x^2 + 2x - 6$$

c. $(-t^3 - 2t^2 - 1) - (-t^3 - 2t^2) = -t^3 - 2t^2 - 1 + t^3 + 2t^2$

$$= -1$$

SELF CHECK

Find the difference: $(-2a^2 + 5) - (-5a^2 - 7)$.   *Answer:* $3a^2 + 12$ ■

To subtract polynomials in vertical form, we add the opposite of the **subtrahend** (the bottom polynomial) to the **minuend** (the top polynomial).

EXAMPLE 6

Subtracting polynomials vertically. Subtract $3x^2 - 2x$ from $2x^2 + 4x$.

Solution

Since $3x^2 - 2x$ is to be subtracted from $2x^2 + 4x$, we write $3x^2 - 2x$ below $2x^2 + 4x$ in vertical form. Then we change the signs of the terms of $3x^2 - 2x$ and add:

$$
\begin{array}{r}
2x^2 + 4x \\
- \quad 3x^2 - 2x \\
\hline
\end{array}
\quad \longrightarrow \quad
\begin{array}{r}
2x^2 + 4x \\
+ \quad -3x^2 + 2x \\
\hline
-x^2 + 6x
\end{array}
$$

SELF CHECK

Subtract $2p^2 + 2p - 8$ from $5p^2 - 6p + 7$.   *Answer:* $3p^2 - 8p + 15$ ■

EXAMPLE 7

Combining polynomials. Subtract $(12a - 7)$ from the sum of $(6a + 5)$ and $(4a - 10)$.

Solution

We will use brackets to show that $(12a - 7)$ is to be subtracted from the *sum* of $(6a + 5)$ and $(4a - 10)$.

$$[(6a + 5) + (4a - 10)] - (12a - 7)$$

Next, we remove the grouping symbols to obtain

$$= 6a + 5 + 4a - 10 - 12a + 7$$
$$= -2a + 2 \qquad \text{Combine like terms.}$$

SELF CHECK

Subtract $(-2q^2 - 2q)$ from the sum of $(q^2 - 6q)$ and $(3q^2 + q)$.   *Answer:* $6q^2 - 3q$ ■

Adding and Subtracting Multiples of Polynomials

Because of the distributive property, we can remove parentheses enclosing several terms when a monomial precedes the parentheses. We simply multiply every term within the parentheses by that monomial. For example, to add $3(2x + 5)$ and $2(4x - 3)$, we proceed as follows:

$$3(2x + 5) + 2(4x - 3) = 6x + 15 + 8x - 6 \qquad \text{Use the distributive property to remove parentheses.}$$

$$= 6x + 8x + 15 - 6 \qquad 15 + 8x = 8x + 15.$$

$$= 14x + 9 \qquad \text{Combine like terms.}$$

EXAMPLE 8

Adding and subtracting multiples of polynomials. Remove parentheses and simplify.

a. $3(x^2 + 4x) + 2(x^2 - 4) = 3x^2 + 12x + 2x^2 - 8$
$$= 5x^2 + 12x - 8$$

b. $-8(y^2 - 2y + 3) - 4(2y^2 + y - 6) = -8y^2 + 16y - 24 - 8y^2 - 4y + 24$
$$= -16y^2 + 12y$$

c. $-4x(x^2 - x + 3) - x(x^2 - 2) + 3(x^2 + 2x)$
$$= -4x^3 + 4x^2 - 12x - x^3 + 2x + 3x^2 + 6x$$
$$= -5x^3 + 7x^2 - 4x$$

SELF CHECK

Remove parentheses and simplify:
a. $2(a^2 - 3a) + 5(a^2 + 2a)$ and
b. $5x(x^2 + 2x + 1) - x(x - 3)$.

Answers: **a.** $7a^2 + 4a$,
b. $5x^3 + 9x^2 + 8x$ ■

An Application of Adding Polynomials

EXAMPLE 9

Property values. A house purchased for $95,000 is expected to appreciate according to the polynomial function $y = 2{,}500x + 95{,}000$, where y is the value of the house after x years. A second house purchased for $125,000 is expected to appreciate according to the equation $y = 4{,}500x + 125{,}000$. Find one polynomial function that will give the total value of both properties after x years.

Solution

The value of the first house after x years is given by the polynomial $2{,}500x + 95{,}000$. The value of the second house after x years is given by the polynomial $4{,}500x + 125{,}000$. The value of both houses will be the sum of these two polynomials.
$$2{,}500x + 95{,}000 + 4{,}500x + 125{,}000 = 7{,}000x + 220{,}000$$

The total value y of the properties is given by the polynomial function $y = 7{,}000x + 220{,}000$. ■

STUDY SET

Section 4.5

VOCABULARY

In Exercises 1–4, fill in the blanks to make the statements true.

1. The expression $(x^2 - 3x + 2) + (x^2 - 4x)$ is the sum of two _____.

2. _____ terms have the same variables and the same exponents.

3. "To add or subtract like terms" means to combine their _____ and keep the same variables with the same exponents.

4. If two polynomials are subtracted in vertical form, the bottom polynomial is called the _____, and the top polynomial is called the _____.

CONCEPTS

In Exercises 5–12, fill in the blanks to make the statements true.

5. To add like monomials, combine like _____.

6. $a - b = a +$

7. To add two polynomials, combine any _____ terms contained in the polynomials.

8. To subtract two polynomials, remove parentheses and combine _____ terms.

9. When the sign preceding parentheses is a $-$ sign, we can remove the parentheses by dropping the sign and the parentheses, and _____ the sign of every term within the parentheses.

10. When a monomial precedes parentheses, we can remove the parentheses by _____ every term within the parentheses by that monomial.

11. $-(-2x^2 - 3x + 4) = $

12. $-3(-2x^2 - 3x + 4) = $

13. JETS Find the polynomial representing the length of the passenger jet in Illustration 1.

ILLUSTRATION 1

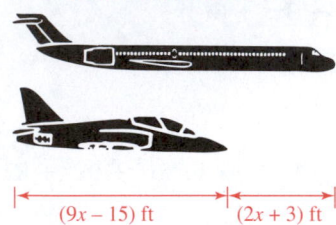

$(9x - 15)$ ft $(2x + 3)$ ft

14. WATER SKIING Find the polynomial representing the distance of the water skier from the boat in Illustration 2.

ILLUSTRATION 2

$(15y - 3)$ m

$(6y + 1)$ m

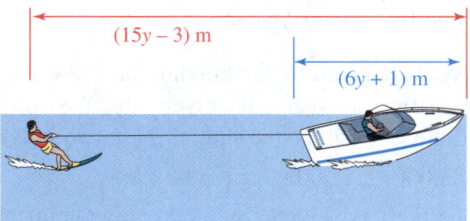

NOTATION

In Exercises 15–16, complete each solution.

15. $(5x^2 + 3x) - (7x^2 - 2x)$

$= 5x^2 + \boxed{} - 7x^2 + \boxed{}$

$= 5x^2 - \boxed{} + 3x + 2x$

$= -2x^2 + 5x$

16. $4(3x^2 - 2x) - x(2x + 4)$

$= 12x^2 - \boxed{} - \boxed{} - 4x$

$= 12x^2 - 2x^2 - \boxed{} - 4x$

$= 10x^2 - 12x$

PRACTICE

In Exercises 17–32, simplify each expression, if possible.

17. $4y + 5y$

18. $-2x + 3x$

19. $8t^2 + 4t^2$

20. $15x^2 + 10x^2$

21. $-32u^3 - 16u^3$

22. $-25x^3 - 7x^3$

23. $1.8x - 1.9x$

24. $1.7y - 2.2y$

25. $\frac{1}{2}s + \frac{3}{2}s$

26. $\frac{2}{5}a + \frac{1}{5}a$

27. $3r - 4r + 7r$

28. $-2b + 7b - 3b$

29. $-4ab + 4ab - ab$

30. $xy - 4xy - 2xy$

31. $(3x)^2 - 4x^2 + 10x^2$

32. $(2x)^4 - (3x^2)^2$

In Exercises 33–46, do the operations.

33. $(3x + 7) + (4x - 3)$

34. $(2y - 3) + (4y + 7)$

35. $(4a + 3) - (2a - 4)$

36. $(5b - 7) - (3b - 5)$

37. $(2x + 3y) + (5x - 10y)$

38. $(5x - 8y) - (-2x + 5y)$

39. $(-8x - 3y) - (-11x + y)$

40. $(-4a + b) + (5a - b)$

41. $(3x^2 - 3x - 2) + (3x^2 + 4x - 3)$

42. $(3a^2 - 2a + 4) - (a^2 - 3a + 7)$

43. $(2b^2 + 3b - 5) - (2b^2 - 4b - 9)$

44. $(4c^2 + 3c - 2) + (3c^2 + 4c + 2)$

45. $(2x^2 - 3x + 1) - (4x^2 - 3x + 2) + (2x^2 + 3x + 2)$

46. $(-3z^2 - 4z + 7) + (2z^2 + 2z - 1) - (2z^2 - 3z + 7)$

In Exercises 47–52, add the polynomials.

47. $\begin{array}{r} 3x^2 + 4x + 5 \\ +\ 2x^2 - 3x + 6 \\ \hline \end{array}$

48. $\begin{array}{r} 2x^3 + 2x^2 - 3x + 5 \\ +\ 3x^3 - 4x^2 -\ x - 7 \\ \hline \end{array}$

49. $\begin{array}{r} 2x^3 - 3x^2 + 4x - 7 \\ +\ -9x^3 - 4x^2 - 5x + 6 \\ \hline \end{array}$

50. $\begin{array}{r} -3x^3 + 4x^2 - 4x + 9 \\ +\ 2x^3 \qquad + 9x - 3 \\ \hline \end{array}$

51. $\begin{array}{r} -3x^2 + 4x + 25 \\ +\ 5x^2 \qquad - 12 \\ \hline \end{array}$

52. $\begin{array}{r} -6x^3 - 4x^2 + 7 \\ +\ -7x^3 + 9x^2 \\ \hline \end{array}$

In Exercises 53–58, find each difference.

53. $\begin{array}{r} 3x^2 + 4x - 5 \\ -\ -2x^2 - 2x + 3 \\ \hline \end{array}$

54. $\begin{array}{r} 3y^2 - 4y +\ 7 \\ -\ 6y^2 - 6y - 13 \\ \hline \end{array}$

55. $\quad -\dfrac{\begin{array}{r} 4x^3 + 4x^2 - 3x + 10 \\ 5x^3 - 2x^2 - 4x - 4 \end{array}}{}$

56. $\quad -\dfrac{\begin{array}{r} 3x^3 + 4x^2 + 7x + 12 \\ -4x^3 + 6x^2 + 9x - 3 \end{array}}{}$

57. $\quad -\dfrac{\begin{array}{r} -2x^2y^2 \qquad + 12y^2 \\ 10x^2y^2 + 9xy - 24y^2 \end{array}}{}$

58. $\quad -\dfrac{\begin{array}{r} 25x^3 \qquad + 31xz^2 \\ 12x^3 + 27x^2z - 17xz^2 \end{array}}{}$

59. Find the difference when $t^3 - 2t^2 + 2$ is subtracted from the sum of $3t^3 + t^2$ and $-t^3 + 6t - 3$.

60. Find the difference when $-3z^3 - 4z + 7$ is subtracted from the sum of $2z^2 + 3z - 7$ and $-4z^3 - 2z - 3$.

61. Find the sum when $3x^2 + 4x - 7$ is added to the sum of $-2x^2 - 7x + 1$ and $-4x^2 + 8x - 1$.

62. Find the difference when $32x^2 - 17x + 45$ is subtracted from the sum of $23x^2 - 12x - 7$ and $-11x^2 + 12x + 7$.

In Exercises 63–70, simplify each expression.

63. $2(x + 3) + 4(x - 2)$

64. $3(y - 4) - 5(y + 3)$

65. $-2(x^2 + 7x - 1) - 3(x^2 - 2x + 2)$

66. $-5(y^2 - 2y - 6) + 6(2y^2 + 2y - 5)$

67. $2(2y^2 - 2y + 2) - 4(3y^2 - 4y - 1) + 4y(y^2 - y - 1)$

68. $-4(z^2 - 5z) - 5(4z^2 - 1) + 6(2z - 3)$

69. $2a(ab^2 - b) - 3b(a + 2ab) + b(b - a + a^2b)$

70. $3y(xy + y) - 2y^2(x - 4 + y) + 2(y^3 + y^2)$

In Exercises 71–72, find the polynomial that represents the perimeter of the figure.

71.

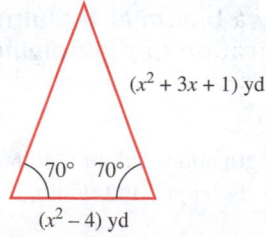

$(x^2 + 3x + 1)$ yd

$70°$ $70°$

$(x^2 - 4)$ yd

72.

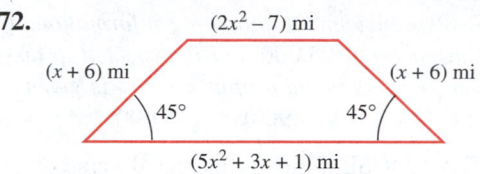

$(2x^2 - 7)$ mi

$(x + 6)$ mi $\qquad$ $(x + 6)$ mi

$45°$ $\qquad$ $45°$

$(5x^2 + 3x + 1)$ mi

APPLICATIONS

73. GREEK ARCHITECTURE Find the difference in the heights of the columns shown in Figure 4-8 at the beginning of this section.

74. CLASSICAL GREEK COLUMNS If the columns shown in Figure 4-8 at the beginning of this section were stacked one atop the other, to what height would they reach?

75. AUTO MECHANICS Find the polynomial representing the length of the fan belt shown in Illustration 3. The dimensions are in inches. Your answer will involve π.

ILLUSTRATION 3

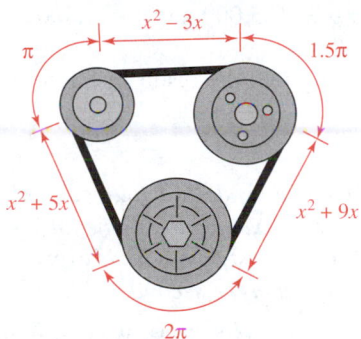

$x^2 - 3x$

π $\qquad$ 1.5π

$x^2 + 5x$ $\qquad$ $x^2 + 9x$

2π

76. READING BLUEPRINTS
   a. What is the difference in the length and width of the one-bedroom apartment shown in Illustration 4?
   b. Find the perimeter of the apartment.

ILLUSTRATION 4

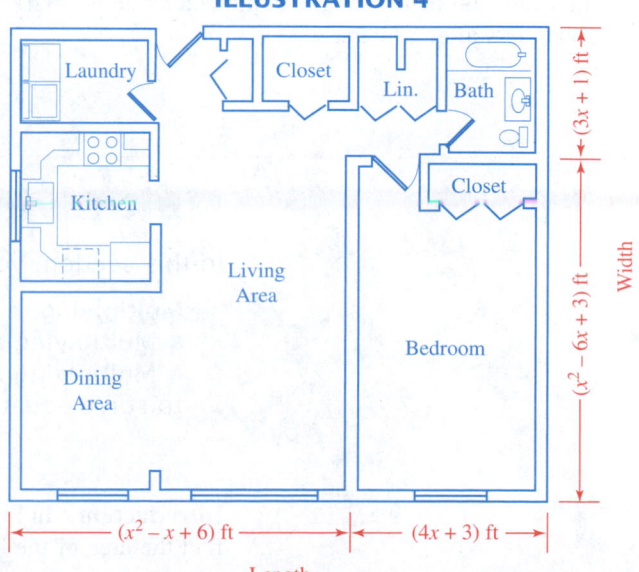

Laundry $\qquad$ Closet $\qquad$ Lin. $\qquad$ Bath $\qquad$ $(3x + 1)$ ft

Closet

Kitchen

Living Area $\qquad$ $(x^2 - 6x + 3)$ ft $\quad$ Width

Bedroom

Dining Area

$(x^2 - x + 6)$ ft $\qquad$ $(4x + 3)$ ft

Length

In Exercises 77–80, consider the following information: If a house is purchased for $105,000 and is expected to appreciate $900 per year, its value y after x years is given by the polynomial function y = 900x + 105,000.

77. VALUE OF A HOUSE Find the expected value of the house in 10 years.

78. VALUE OF A HOUSE A second house is purchased for $120,000 and is expected to appreciate $1,000 per year.
   a. Find a polynomial function that will give the value y of the house in x years.

   b. Find the value of this second house after 12 years.

79. VALUE OF TWO HOUSES Find one polynomial function that will give the combined value y of both houses after x years.

80. VALUE OF TWO HOUSES Find the value of the two houses after 20 years by
   a. substituting 20 into the polynomial functions $y = 900x + 105,000$ and $y = 1,000x + 120,000$ and adding.
   b. substituting into the result of Exercise 79.

In Exercises 81–84, consider the following information: A business purchases two computers, one for $6,600 and the other for $9,200. The first computer is expected to depreciate $1,100 per year and the second $1,700 per year.

81. VALUE OF A COMPUTER Write a polynomial function that gives the value of the first computer after x years.

82. VALUE OF A COMPUTER Write a polynomial function that gives the value of the second computer after x years.

83. VALUE OF TWO COMPUTERS Find one polynomial function that gives the combined value of both computers after x years.

84. VALUE OF TWO COMPUTERS In two ways, find the combined value of the two computers after 3 years.

WRITING

85. How do you recognize like terms?

86. How do you add like terms?

87. Explain the concept that is illustrated by the statement
$$-(x^2 + 3x - 1) = -1(x^2 + 3x - 1)$$

88. Explain the mistake made in the solution.
   Simplify: $(12x - 4) - (3x - 1)$.

   $(12x - 4) - (3x - 1) = 12x - 4 - 3x - 1$
   $= 9x - 5$

REVIEW

89. What is the sum of the measures of the angles of a triangle?

90. What is the sum of the measures of two complementary angles?

91. Solve the inequality $-4(3x - 3) \geq -12$ and graph the solution.

92. CURLING IRON A curling iron is plugged into a 110-volt electrical outlet and used for $\frac{1}{4}$ hour. If its resistance is 10 ohms, find the electrical power (in kilowatt hours, kwh) used by the curling iron by applying the formula
$$kwh = \frac{(volts)^2}{1,000 \cdot ohms} \cdot hours$$

▶ **4.6**   # Multiplying Polynomials

In this section, you will learn about

   Multiplying monomials ■ Multiplying a polynomial by a monomial
   ■ Multiplying a binomial by a binomial ■ The FOIL method
   ■ Multiplying a polynomial by a binomial ■ Multiplying binomials
   to solve equations ■ An application of multiplying polynomials

---

Introduction   In Figure 4-9(a), the length and width of a dollar bill are given. We can find the area of the bill by multiplying its length and width.

FIGURE 4-9

6.5 cm

15.6 cm

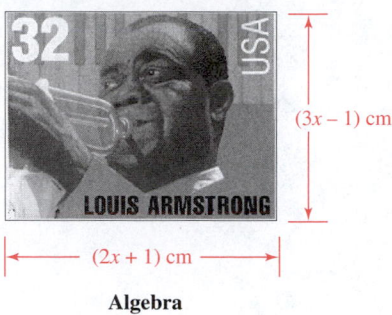
$(3x - 1)$ cm

$(2x + 1)$ cm

Arithmetic	Algebra
$15.6 \cdot 6.5 = 101.4$	$(2x + 1)(3x - 1) = ?$
The area is 101.4 cm².	
(a)	(b)

In Figure 4-9(b), the length and the width of a postage stamp are represented by two-term polynomials called *binomials*. To find the area of the stamp, we must multiply the binomials. In this section, we will discuss the algebraic rules that are used to do this. We begin the discussion of multiplication of polynomials with the simplest case, the product of two monomials.

Multiplying Monomials

In Section 4.1, we multiplied monomials by other monomials. For example, to multiply $8x^2$ by $-3x^4$, we use the commutative and associative properties of multiplication to group the numerical factors and the variable factors. Then we multiply the numerical factors and multiply the variable factors.

$$8x^2(-3x^4) = 8(-3)x^2x^4$$
$$= -24x^6$$

This example suggests the following rule.

Multiplying monomials

To multiply two monomials, multiply the numerical factors and then multiply the variable factors.

EXAMPLE 1

Multiplying monomials. Multiply **a.** $3x^4(2x^5)$, **b.** $-2a^2b^3(5ab^2)$, and **c.** $-4y^5z^2(2y^3z^3)(3yz)$.

Solution **a.** $3x^4(2x^5) = 3(2)x^4x^5$ Multiply the numerical factors, 3 and 2. Multiply the variable factors, x^4 and x^5. Use the product rule for exponents: $x^4x^5 = x^{4+5}$.

$$= 6x^9$$

b. $-2a^2b^3(5ab^2) = -2(5)a^2ab^3b^2$
$$= -10a^3b^5$$

c. $-4y^5z^2(2y^3z^3)(3yz) = -4(2)(3)y^5y^3yz^2z^3z$
$$= -24y^9z^6$$

SELF CHECK Multiply: **a.** $(5a^2b^3)(6a^3b^4)$ and **b.** $(-15p^3q^2)(5p^3q^2)$.

Answers: **a.** $30a^5b^7$, **b.** $-75p^6q^4$

Multiplying a Polynomial by a Monomial

To find the product of a polynomial (with more than one term) and a monomial, we use the distributive property. To multiply $2x + 4$ by $5x$, for example, we proceed as follows:

$$5x(2x + 4) = 5x \cdot 2x + 5x \cdot 4 \quad \text{Use the distributive property.}$$
$$= 10x^2 + 20x \quad \text{Multiply the monomials:}$$
$$5x \cdot 2x = 10x^2 \text{ and}$$
$$5x \cdot 4 = 20x.$$

This example suggests the following rule.

Multiplying polynomials by monomials

To multiply a polynomial with more than one term by a monomial, use the distributive property to remove parentheses and simplify.

EXAMPLE 2

Multiplying a polynomial by a monomial. Multiply **a.** $3a^2(3a^2 - 5a)$ and **b.** $-2xz^2(2x - 3z + 2z^2)$.

Solution   **a.** $3a^2(3a^2 - 5a) = 3a^2 \cdot 3a^2 - 3a^2 \cdot 5a$ Use the distributive property.
$$= 9a^4 - 15a^3 \quad \text{Multiply: } 3a^2 \cdot 3a^2 = 9a^4 \text{ and } 3a^2 \cdot 5a = 15a^3.$$

b. $-2xz^2(2x - 3z + 2z^2)$
$$= -2xz^2 \cdot 2x - (-2xz^2) \cdot 3z + (-2xz^2) \cdot 2z^2 \quad \text{Use the distributive property.}$$
$$= -4x^2z^2 - (-6xz^3) + (-4xz^4) \quad \text{Multiply: } -2xz^2 \cdot 2x = -4x^2z^2,$$
$$-2xz^2 \cdot 3z = -6xz^3, \text{ and}$$
$$-2xz^2 \cdot 2z^2 = -4xz^4.$$

$$= -4x^2z^2 + 6xz^3 - 4xz^4$$

SELF CHECK   Multiply: **a.** $2p^3(3p^2 - 5p)$ and   *Answers:* **a.** $6p^5 - 10p^4$,
b. $-5a^2b(3a + 2b - 4ab)$.   **b.** $-15a^3b - 10a^2b^2 + 20a^3b^2$

Multiplying a Binomial by a Binomial

To multiply two binomials, we must use the distributive property more than once. For example, to multiply $2a - 4$ by $3a + 5$, we proceed as follows.

$$(2a - 4)(3a + 5) = (2a - 4) \cdot 3a + (2a - 4) \cdot 5 \quad \text{Distribute the binomial } (2a - 4).$$

$$= 3a(2a - 4) + 5(2a - 4) \quad \text{Use the commutative property of multiplication.}$$

$$= 3a \cdot 2a - 3a \cdot 4 + 5 \cdot 2a - 5 \cdot 4 \quad \text{Use the distributive property twice.}$$

$$= 6a^2 - 12a + 10a - 20 \quad \text{Do the multiplications.}$$

$$= 6a^2 - 2a - 20 \quad \text{Combine like terms: } -12a + 10a = -2a$$

This example suggests the following rule.

Multiplying two binomials

To multiply two binomials, multiply each term of one binomial by each term of the other binomial and combine like terms.

The FOIL Method

We can use a shortcut method, called the **FOIL** method, to multiply binomials. FOIL is an acronym for **F**irst terms, **O**uter terms, **I**nner terms, and **L**ast terms. To use the FOIL method to multiply $2a - 4$ by $3a + 5$, we

1. multiply the **F**irst terms $2a$ and $3a$ to obtain $6a^2$,
2. multiply the **O**uter terms $2a$ and 5 to obtain $10a$,
3. multiply the **I**nner terms -4 and $3a$ to obtain $-12a$, and
4. multiply the **L**ast terms -4 and 5 to obtain -20.

Then we simplify the resulting polynomial, if possible.

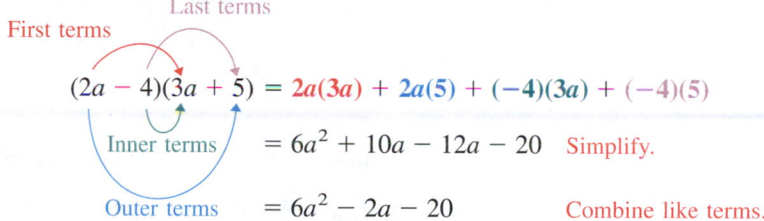

$$(2a - 4)(3a + 5) = 2a(3a) + 2a(5) + (-4)(3a) + (-4)(5)$$

$$= 6a^2 + 10a - 12a - 20 \quad \text{Simplify.}$$

$$= 6a^2 - 2a - 20 \quad\quad \text{Combine like terms.}$$

EXAMPLE 3

Using the FOIL method. Find each product.

a. $(3x + 4)(2x - 3) = 3x(2x) + 3x(-3) + 4(2x) + 4(-3)$

$$= 6x^2 - 9x + 8x - 12$$

$$= 6x^2 - x - 12$$

b. $(2y - 7)(5y - 4) = 2y(5y) + 2y(-4) + (-7)(5y) + (-7)(-4)$

$$= 10y^2 - 8y - 35y + 28$$

$$= 10y^2 - 43y + 28$$

c. $(2r - 3s)(2r + t) = 2r(2r) + 2r(t) - 3s(2r) - 3s(t)$

$$= 4r^2 + 2rt - 6rs - 3st$$

SELF CHECK Find each product: **a.** $(2a - 1)(3a + 2)$ and *Answers:* **a.** $6a^2 + a - 2$,
b. $(5y - 2z)(2y + 3z)$. **b.** $10y^2 + 11yz - 6z^2$

EXAMPLE 4

Simplifying expressions. Simplify each expression.

a. $3(2x - 3)(x + 1) = 3(2x^2 + 2x - 3x - 3)$ Use FOIL to multiply the binomials.

$\qquad\qquad\qquad\quad = 3(2x^2 - x - 3)$ Combine like terms.

$\qquad\qquad\qquad\quad = 6x^2 - 3x - 9$ Use the distributive property to remove parentheses.

b. $(x + 1)(x - 2) - 3x(x + 3) = x^2 - 2x + x - 2 - 3x^2 - 9x$ Use FOIL to find $(x + 1)(x + 2)$.

$\qquad\qquad\qquad\qquad\qquad = -2x^2 - 10x - 2$ Combine like terms.

SELF CHECK   Simplify $(x + 3)(2x - 1) + 2x(x - 1)$.   *Answer:* $4x^2 + 3x - 3$ ∎

The products discussed in Example 5 are called **special products.**

EXAMPLE 5

Special products. Find each product.

a. The square of the sum of two quantities has three terms:

$$(x + y)^2 = (x + y)(x + y)$$
$$= x^2 + xy + xy + y^2$$
$$= \quad x^2 \quad + \quad 2xy \quad + \quad y^2$$

The square of the first quantity   Twice the product of the quantities   The square of the second quantity

b. The square of the difference of two quantities has three terms:

$$(x - y)^2 = (x - y)(x - y)$$
$$= x^2 - xy - xy + y^2$$
$$= \quad x^2 \quad - \quad 2xy \quad + \quad y^2$$

The square of the first quantity   Twice the product of the quantities   The square of the second quantity

c. The product of a sum and a difference of two quantities is a binomial.

$$(x + y)(x - y) = x^2 - xy + xy - y^2$$
$$= \quad x^2 \quad - \quad y^2$$

The product of the first quantities   The product of the second quantities

SELF CHECK   Find each product: **a.** $(p + 2)^2$,   **b.** $(p - 2)^2$, and   **c.** $(p + 2q)(p - 2q)$.   *Answers:*   **a.** $p^2 + 4p + 4$, **b.** $p^2 - 4p + 4$,   **c.** $p^2 - 4q^2$ ∎

Because the products discussed in Example 5 occur so often, it is wise to learn their forms.

Special products

$$(x + y)^2 = x^2 + 2xy + y^2$$
$$(x - y)^2 = x^2 - 2xy + y^2$$
$$(x + y)(x - y) = x^2 - y^2$$

EXAMPLE 6

Road signs. The area of a regular octagon is approximately $3.314r^2$, where r is as shown in Figure 4-10(a). Find the polynomial that approximates the area of the stop sign shown in Figure 4-10(b).

FIGURE 4-10

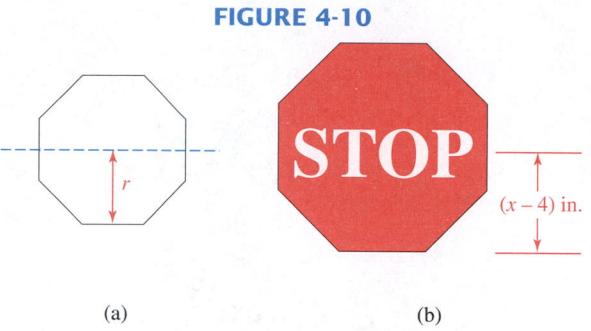

(a)   (b)

Solution   To find the approximate area of the stop sign, we substitute $(x - 4)$ for r in the formula.

$$A \approx 3.314r^2$$
$$A \approx 3.314(x - 4)^2$$

We note that $(x - 4)^2$ is a special product. The result is a trinomial consisting of the first term (x) squared, minus twice the product of x and 4, plus the square of 4.

$$A \approx 3.314[x^2 - 2(x)(4) + (4)^2]$$
$$\approx 3.314[x^2 - 8x + 16] \qquad \text{Simplify inside the brackets.}$$
$$\approx 3.314x^2 - 26.512x + 53.024 \quad \text{Distribute 3.314.}$$

The approximate area of the stop sign is $(3.314x^2 - 26.512x + 53.024)$ in.^{2}

SELF CHECK

See Example 6. Find the approximate area of a stop sign where $r = (y + 6)$ inches.

Answer:
$(3.314y^2 + 39.768y + 119.304)$ in.^{2} ∎

WARNING! A common error when squaring a binomial is to forget the middle term of the product. For example, $(x + 2)^2 \neq x^2 + 4$ and $(x - 2)^2 \neq x^2 + 4$. Applying the special product formulas, we have $(x + 2)^2 = x^2 + 4x + 4$ and $(x - 2)^2 = x^2 - 4x + 4$.

Multiplying a Polynomial by a Binomial

We must use the distributive property more than once to multiply a polynomial by a binomial. For example, to multiply $3x^2 + 3x - 5$ by $2x + 3$, we proceed as follows:

$$(2x + 3)(3x^2 + 3x - 5) = (2x + 3)3x^2 + (2x + 3)3x - (2x + 3)5$$
$$= 3x^2(2x + 3) + 3x(2x + 3) - 5(2x + 3)$$
$$= 6x^3 + 9x^2 + 6x^2 + 9x - 10x - 15$$
$$= 6x^3 + 15x^2 - x - 15$$

This example suggests the following rule.

Multiplying polynomials

To multiply one polynomial by another, multiply each term of one polynomial by each term of the other polynomial and combine like terms.

It is often convenient to organize the work vertically.

E X A M P L E   7

Multiplying polynomials using vertical form.

a. Multiply:

$$
\begin{array}{r}
3a^2 - 4a + 7 \\
2a + 5 \\
\hline
\end{array}
$$

$2a(3a^2 - 4a + 7) \rightarrow \quad 6a^3 - 8a^2 + 14a$

$5(3a^2 - 4a + 7) \rightarrow \quad \underline{+ 15a^2 - 20a + 35}$

$6a^3 + 7a^2 - 6a + 35$

b. Multiply:

$$
\begin{array}{r}
3y^2 - 5y + 4 \\
-4y^2 - 3 \\
\hline
\end{array}
$$

$-4y^2(3y^2 - 5y + 4) \rightarrow \quad -12y^4 + 20y^3 - 16y^2$

$-3(3y^2 - 5y + 4) \rightarrow \quad \underline{- 9y^2 + 15y - 12}$

$-12y^4 + 20y^3 - 25y^2 + 15y - 12$

SELF CHECK   Multiply: **a.** $(3x + 2)(2x^2 - 4x + 5)$ and
b. $(-2x^2 + 3)(2x^2 - 4x - 1)$.

Answers:
a. $6x^3 - 8x^2 + 7x + 10$,
b. $-4x^4 + 8x^3 + 8x^2 - 12x - 3$

Multiplying Binomials to Solve Equations

To solve an equation such as $(x + 2)(x + 3) = x(x + 7)$, we can first use the FOIL method to remove the parentheses on the left-hand side, then use the distributive property to remove parentheses on the right-hand side, and proceed as follows:

$$(x + 2)(x + 3) = x(x + 7)$$

$$x^2 + 3x + 2x + 6 = x^2 + 7x$$

$x^2 + 3x + 2x + 6 - x^2 = x^2 + 7x - x^2$ Subtract x^2 from both sides.

$5x + 6 = 7x$ Combine like terms: $x^2 - x^2 = 0$ and $3x + 2x = 5x$.

$6 = 2x$ Subtract $5x$ from both sides.

$3 = x$ Divide both sides by 2.

Check: $(x + 2)(x + 3) = x(x + 7)$

$(3 + 2)(3 + 3) \stackrel{?}{=} 3(3 + 7)$ Replace x with 3.

$5(6) \stackrel{?}{=} 3(10)$ Do the additions within parentheses.

$30 = 30$

An Application of Multiplying Polynomials

E X A M P L E   8

A square painting is surrounded by a border 2 inches wide. If the area of the border is 96 square inches, find the dimensions of the painting.

ANALYZE THE PROBLEM
Refer to Figure 4-11, which shows a square painting surrounded by a border 2 inches wide. We know that the area of this border is 96 square inches, and we are to find the dimensions of the painting.

FORM AN EQUATION
Let x represent the length of each side of the square painting. Since the border is 2 inches wide, the length and the width of the outer rectangle are both $(x + 2 + 2)$ inches. Then the outer rectangle is also a square, and its dimensions are $(x + 4)$ by $(x + 4)$ inches. Since the area of a square is the product of its length and width, the area of the larger square is $(x + 4)(x + 4)$, and the area of the painting is $x \cdot x$. If we subtract the area of the painting from the area of the larger square, the difference is 96.

FIGURE 4-11

The area of the large square	minus	the area of the square painting	is	the area of the border.
$(x + 4)(x + 4)$	$-$	$x \cdot x$	$=$	96

SOLVE THE EQUATION

$(x + 4)(x + 4) - x^2 = 96$ $x \cdot x = x^2$.

$x^2 + 8x + 16 - x^2 = 96$ $(x + 4)(x + 4) = x^2 + 8x + 16$.

$8x + 16 = 96$ Combine like terms: $x^2 - x^2 = 0$.

$8x = 80$ Subtract 16 from both sides.

$x = 10$ Divide both sides by 8.

STATE THE CONCLUSION
The dimensions of the painting are 10 inches by 10 inches.

CHECK THE RESULT
Verify that the 2-inch-wide border of a 10-inch-square painting would have an area of 96 square inches. ∎

STUDY SET

Section 4.6

VOCABULARY

In Exercises 1–4, fill in the blanks to make the statements true.

1. The expression $(2a - 4)(3a + 5)$ is the product of two _____.

2. The expression $(2a - 4)(3a^2 + 5a - 1)$ is the product of a _____ and a _____.

3. When multiplying a monomial and a polynomial, the _____ property is used to remove parentheses.

4. In the acronym FOIL, F stands for _____ terms, O for _____ terms, I for _____ terms, and L for _____ terms.

CONCEPTS

In Exercises 5–8, consider the product $(2x + 5)(3x - 4)$.

5. The product of the first terms is [   ].

6. The product of the outer terms is [   ].

7. The product of the inner terms is [   ].

8. The product of the last terms is [   ].

9. STAMPS Find the area of the stamp shown in Figure 4-9 at the beginning of this section.

10. LUGGAGE Find the volume of the garment bag shown in Illustration 1.

ILLUSTRATION 1

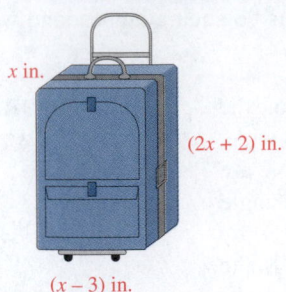

x in.

$(2x + 2)$ in.

$(x - 3)$ in.

NOTATION

In Exercises 11–12, complete each solution.

11. $7x(3x^2 - 2x + 5) = \boxed{} \cdot 3x^2 - \boxed{} \cdot 2x + \boxed{} \cdot 5$

$\qquad = 21x^3 - 14x^2 + 35x$

12. $(2x + 5)(3x - 2) = 2x \cdot 3x - \boxed{} \cdot 2 + \boxed{} \cdot 3x - \boxed{} \cdot 2$

$\qquad = 6x^2 - \boxed{} + \boxed{} - 10$

$\qquad = 6x^2 + 11x - 10$

PRACTICE

In Exercises 13–20, find each product.

13. $(3x^2)(4x^3)$ **14.** $(-2a^3)(3a^2)$

15. $(3b^2)(-2b)(4b^3)$ **16.** $(3y)(2y^2)(-y^4)$

17. $(2x^2y^3)(3x^3y^2)$ **18.** $(-5x^3y^6)(x^2y^2)$

19. $(x^2y^5)(x^2z^5)(-3z^3)$ **20.** $(-r^4st^2)(2r^2st)(rst)$

In Exercises 21–34, find each product.

21. $3(x + 4)$ **22.** $-3(a - 2)$

23. $-4(t + 7)$ **24.** $6(s^2 - 3)$

25. $3x(x - 2)$ **26.** $4y(y + 5)$

27. $-2x^2(3x^2 - x)$ **28.** $4b^3(2b^2 - 2b)$

29. $3xy(x + y)$ **30.** $-4x^2z(3x^2 - z)$

31. $2x^2(3x^2 + 4x - 7)$ **32.** $3y^3(2y^2 - 7y - 8)$

33. $(3x)(-2x^2)(x + 4)$ **34.** $(-2a^2)(-3a^3)(3a - 2)$

In Exercises 35–50, find each product.

35. $(a + 4)(a + 5)$ **36.** $(y - 3)(y + 5)$

37. $(3x - 2)(x + 4)$ **38.** $(t + 4)(2t - 3)$

39. $(2a + 4)(3a - 5)$ **40.** $(2b - 1)(3b + 4)$

41. $(3x - 5)(2x + 1)$ **42.** $(2y - 5)(3y + 7)$

43. $(x + 3)(2x - 3)$ **44.** $(2x + 3)(2x - 5)$

45. $(2t + 3s)(3t - s)$ **46.** $(3a - 2b)(4a + b)$

47. $(x + y)(x + z)$ **48.** $(a - b)(x + y)$

49. $(4t - u)(-3t + u)$ **50.** $(-3t + 2s)(2t - 3s)$

In Exercises 51–58, simplify each expression.

51. $4(2x + 1)(x - 2)$

52. $-5(3a - 2)(2a + 3)$

53. $3a(a + b)(a - b)$

54. $-2r(r + s)(r + s)$

55. $2t(t + 2) + 3t(t - 5)$

56. $3y(y + 2) + (y + 1)(y - 1)$

57. $(x + y)(x - y) + x(x + y)$

58. $(3x + 4)(2x - 2) - (2x + 1)(x + 3)$

In Exercises 59–76, find each special product.

59. $(x + 4)(x + 4)$ **60.** $(a + 3)(a + 3)$

61. $(t - 3)(t - 3)$ **62.** $(z - 5)(z - 5)$

63. $(r + 4)(r - 4)$ **64.** $(b + 2)(b - 2)$

65. $(4x + 5)(4x - 5)$ **66.** $(5z + 1)(5z - 1)$

67. $(2s + 1)(2s + 1)$ **68.** $(3t - 2)(3t - 2)$

69. $(x + 5)^2$ **70.** $(y - 6)^2$

71. $(x - 2y)^2$ **72.** $(3a + 2b)^2$

73. $(2a - 3b)^2$ **74.** $(2x + 5y)^2$

75. $(4x + 5y)^2$ **76.** $(6p - 5q)^2$

In Exercises 77–80, find the area of each figure. You may leave π in your answer.

77.

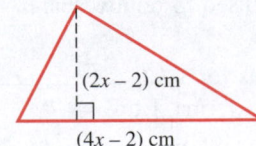

$(2x - 2)$ cm

$(4x - 2)$ cm

78.

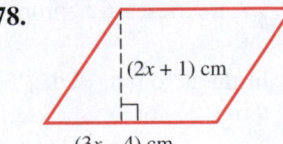

$(2x + 1)$ cm

$(3x - 4)$ cm

79.

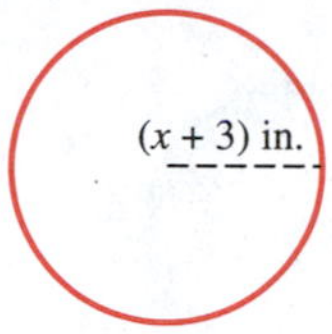

$(x + 3)$ in.

80.

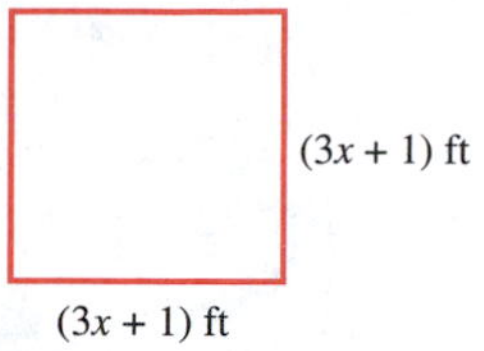

$(3x + 1)$ ft

$(3x + 1)$ ft

In Exercises 81–86, find each product.

81. $(x + 2)(x^2 - 2x + 3)$

82. $(x - 5)(x^2 + 2x - 3)$

83. $(4t + 3)(t^2 + 2t + 3)$

84. $(3x + 1)(2x^2 - 3x + 1)$

85. $(-3x + y)(x^2 - 8xy + 16y^2)$

86. $(3x - y)(x^2 + 3xy - y^2)$

In Exercises 87–90, find each product.

87. $x^2 - 2x + 1$
$\underline{x + 2}$

88. $5r^2 + r + 6$
$\underline{2r - 1}$

89. $4x^2 + 3x - 4$
$\underline{3x + 2}$

90. $x^2 - x + 1$
$\underline{x + 1}$

In Exercises 91–98, solve each equation.

91. $(s - 4)(s + 1) = s^2 + 5$

92. $(y - 5)(y - 2) = y^2 - 4$

93. $z(z + 2) = (z + 4)(z - 4)$

94. $(z + 3)(z - 3) = z(z - 3)$

95. $(x + 4)(x - 4) = (x - 2)(x + 6)$

96. $(y - 1)(y + 6) = (y - 3)(y - 2) + 8$

97. $(a - 3)^2 = (a + 3)^2$

98. $(b + 2)^2 = (b - 1)^2$

APPLICATIONS

99. TOYS Find the perimeter and the area of the screen of the Etch A Sketch® shown in Illustration 2.

ILLUSTRATION 2

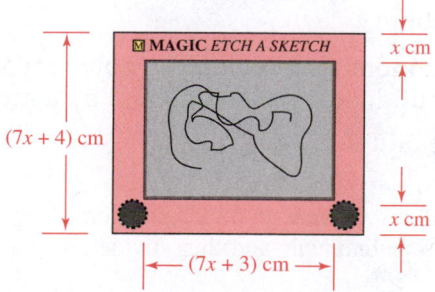

x cm

$(7x + 4)$ cm

x cm

$(7x + 3)$ cm

100. SUNGLASSES An ellipse is an oval-shaped closed curve. The area of an ellipse is approximately $3.14ab$, where a is its length and b is its width. Find the polynomial that approximates the total area of the elliptical-shaped lenses of the sunglasses shown in Illustration 3.

ILLUSTRATION 3

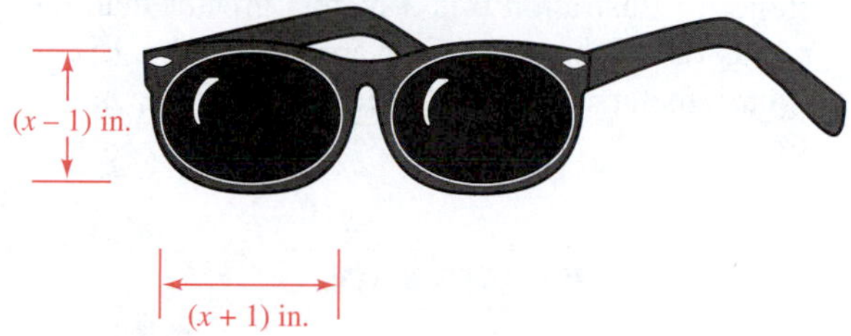

$(x - 1)$ in.

$(x + 1)$ in.

101. GARDENING See Illustration 4.
  a. What is the area of the region planted with corn? tomatoes? beans? carrots? Use your answers to find the total area of the garden.

  b. What is the length of the garden? What is its width? Use your answers to find its area.

  c. How do the answers from parts a and b for the area of the garden compare?

ILLUSTRATION 4

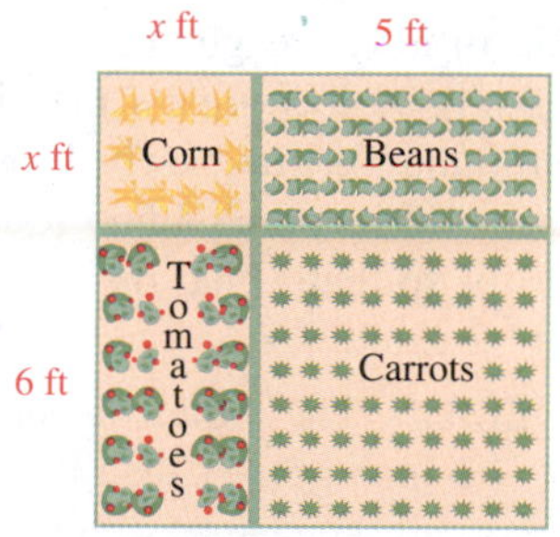

x ft   5 ft

x ft   Corn   Beans

6 ft   Tomatoes   Carrots

102. PAINTING See Illustration 5. To purchase the correct amount of enamel to paint these two garage doors, a painter must find their areas. Find a polynomial that gives the number of square feet to be painted. All dimensions are in feet, and the windows are squares with sides of x feet.

ILLUSTRATION 5

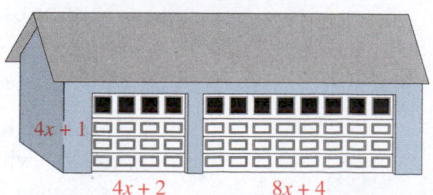

$4x + 1$

$4x + 2$ $8x + 4$

103. **INTEGER PROBLEM** The difference between the squares of two consecutive integers is 11. Find the integers.

104. **INTEGER PROBLEM** If 3 less than a certain integer is multiplied by 4 more than the integer, the product is 6 less than the square of the integer. Find the integer.

105. **STONE-GROUND FLOUR** The radius of one millstone in Illustration 6 is 3 meters greater than the radius of another, and their areas differ by 15π square meters. Find the radius of the larger millstone.

ILLUSTRATION 6

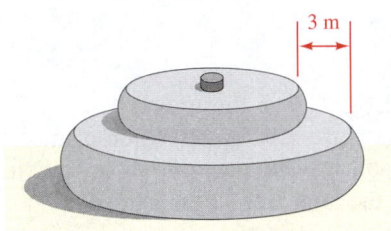

106. **BOOKBINDING** Two square sheets of cardboard used for making book covers differ in area by 44 square inches. An edge of the larger square is 2 inches greater than an edge of the smaller square. Find the length of an edge of the smaller square.

107. **BASEBALL** In major league baseball, the distance between bases is 30 feet greater than it is in softball. The bases in major league baseball mark the corners of a square that has an area 4,500 square feet greater than for softball. Find the distance between the bases in baseball.

108. **PULLEY DESIGN** The radius of one pulley in Illustration 7 is 1 inch greater than the radius of the second pulley, and their areas differ by 4π square inches. Find the radius of the smaller pulley.

ILLUSTRATION 7

WRITING

109. Describe the steps involved in finding the product of $x + 2$ and $x - 2$.

110. Writing $(x + y)^2$ as $x^2 + y^2$ illustrates a common error. Explain.

REVIEW

In Exercises 111–116, refer to Illustration 8.

111. What is the slope of line AB?

112. What is the slope of line BC?

113. What is the slope of line CD?

114. What is the slope of the x-axis?

115. What is the y-intercept of line AB?

116. What is the x-intercept of line AB?

ILLUSTRATION 8

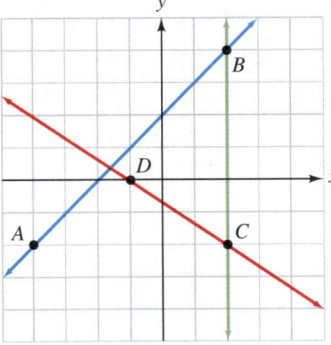

▶ 4.7   Dividing Polynomials by Monomials

In this section, you will learn about

> Dividing a monomial by a monomial ■ Dividing a polynomial by a monomial ■ An application of dividing a polynomial by a monomial

Introduction In this section, we will discuss how to divide polynomials by monomials. We will first divide monomials by monomials and then divide polynomials with more than one term by monomials.

Dividing a Monomial by a Monomial

Recall that to simplify a fraction, we write both its numerator and denominator as the product of several factors and then divide out all common factors:

$$\frac{4}{6} = \frac{2 \cdot 2}{2 \cdot 3} \quad \text{Factor: } 4 = 2 \cdot 2 \text{ and } 6 = 2 \cdot 3. \qquad \frac{20}{25} = \frac{4 \cdot 5}{5 \cdot 5} \quad \text{Factor: } 20 = 4 \cdot 5 \text{ and } 25 = 5 \cdot 5.$$

$$= \frac{\overset{1}{\cancel{2}} \cdot 2}{\underset{1}{\cancel{2}} \cdot 3} \quad \text{Divide out the common factor of 2.} \qquad = \frac{4 \cdot \overset{1}{\cancel{5}}}{\underset{1}{\cancel{5}} \cdot 5} \quad \text{Divide out the common factor of 5.}$$

$$= \frac{2}{3} \quad \tfrac{2}{2} = 1. \qquad = \frac{4}{5} \quad \tfrac{5}{5} = 1.$$

We can use the same method to simplify algebraic fractions that contain variables.

$$\frac{3p^2}{6p} = \frac{3 \cdot p \cdot p}{2 \cdot 3 \cdot p} \quad \text{Factor: } p^2 = p \cdot p \text{ and } 6 = 2 \cdot 3.$$

$$= \frac{\overset{1}{\cancel{3}} \cdot \overset{1}{\cancel{p}} \cdot p}{2 \cdot \underset{1}{\cancel{3}} \cdot \underset{1}{\cancel{p}}} \quad \text{Divide out the common factors of 3 and } p.$$

$$= \frac{p}{2} \quad \tfrac{3}{3} = 1 \text{ and } \tfrac{p}{p} = 1.$$

To divide monomials, we can use either the preceding method for simplifying arithmetic fractions or the rules for exponents.

EXAMPLE 1

Dividing monomials. Simplify: **a.** $\dfrac{x^2y}{xy^2}$ and **b.** $\dfrac{-8a^3b^2}{4ab^3}$.

Solution

By simplifying fractions

a. $\dfrac{x^2y}{xy^2} = \dfrac{x \cdot x \cdot y}{x \cdot y \cdot y}$

$$= \frac{\overset{1}{\cancel{x}} \cdot x \cdot \overset{1}{\cancel{y}}}{\underset{1}{\cancel{x}} \cdot y \cdot \underset{1}{\cancel{y}}}$$

$$= \frac{x}{y}$$

b. $\dfrac{-8a^3b^2}{4ab^3} = \dfrac{-2 \cdot 4 \cdot a \cdot a \cdot a \cdot b \cdot b}{4 \cdot a \cdot b \cdot b \cdot b}$

$$= \frac{-2 \cdot \overset{1}{\cancel{4}} \cdot \overset{1}{\cancel{a}} \cdot a \cdot a \cdot \overset{1}{\cancel{b}} \cdot \overset{1}{\cancel{b}}}{\underset{1}{\cancel{4}} \cdot \underset{1}{\cancel{a}} \cdot \underset{1}{\cancel{b}} \cdot \underset{1}{\cancel{b}} \cdot b}$$

$$= -\frac{2a^2}{b}$$

Using the rules of exponents

a. $\dfrac{x^2y}{xy^2} = x^{2-1}y^{1-2}$

$$= x^1y^{-1}$$

$$= \frac{x}{y}$$

b. $\dfrac{-8a^3b^2}{4ab^3} = \dfrac{-2^3a^3b^2}{2^2ab^3}$

$$= -2^{3-2}a^{3-1}b^{2-3}$$

$$= -2^1a^2b^{-1}$$

$$= -\frac{2a^2}{b}$$

SELF CHECK

Simplify $\dfrac{-5p^2q^3}{10pq^4}$.

Answer: $-\dfrac{p}{2q}$

EXAMPLE 2

Dividing monomials. Simplify $\dfrac{25(s^2t^3)^2}{15(st^3)^3}$. Write the result using positive exponents only.

Solution To divide these monomials, we will use the method for simplifying fractions and several rules for exponents.

$$\frac{25(s^2t^3)^2}{15(st^3)^3} = \frac{25s^4t^6}{15s^3t^9}$$
Use the power rules for exponents: $(xy)^n = x^ny^n$ and $(x^m)^n = x^{m \cdot n}$.

$$= \frac{5 \cdot 5 \cdot s^{4-3}t^{6-9}}{5 \cdot 3}$$
Factor 25 and 15. Use the quotient rule for exponents: $\dfrac{x^m}{x^n} = x^{m-n}$.

$$= \frac{5 \cdot \overset{1}{\cancel{5}} \cdot s^1t^{-3}}{\underset{1}{\cancel{5}} \cdot 3}$$
Divide out the common factors of 5. Do the subtractions.

$$= \frac{5s}{3t^3}$$
Use the negative integer exponent rule: $x^{-n} = \dfrac{1}{x^n}$.

SELF CHECK Simplify $\dfrac{-24(h^3p)^5}{20(h^2p^2)^3}$.

Answer: $-\dfrac{6h^9}{5p}$ ∎

Dividing a Polynomial by a Monomial

Dividing by a number is equivalent to multiplying by its reciprocal. For example, dividing the number 8 by 2 gives the same answer as multiplying 8 by $\frac{1}{2}$.

$$\frac{8}{2} = 4 \qquad \text{and} \qquad \frac{1}{2} \cdot 8 = 4$$

In general, the following is true.

Division

$$\frac{a}{b} = \frac{1}{b} \cdot a \qquad (b \neq 0)$$

To divide a polynomial with more than one term by a monomial, we write the division as a product of the numerator times the reciprocal of the denominator, use the distributive property to remove parentheses, and then simplify each resulting fraction.

EXAMPLE 3

Dividing a binomial by a monomial. Simplify $\dfrac{9x + 6}{3}$.

Solution

$$\frac{9x + 6}{3} = \frac{1}{3}(9x + 6)$$
Division by 3 is equivalent to multiplication by $\frac{1}{3}$.

$$= \frac{9x}{3} + \frac{6}{3}$$
Remove parentheses.

$$= 3x + 2$$
Simplify each fraction.

SELF CHECK Simplify $\dfrac{4 - 8b}{4}$.

Answer: $1 - 2b$ ∎

EXAMPLE 4

Dividing a trinomial by a monomial. Simplify $\dfrac{6x^2y^2 + 4x^2y - 2xy}{2xy}$.

Solution

$$\dfrac{6x^2y^2 + 4x^2y - 2xy}{2xy}$$

$$= \dfrac{1}{2xy}(6x^2y^2 + 4x^2y - 2xy) \quad \text{Multiply by the reciprocal of } 2xy, \text{ which is } \tfrac{1}{2xy}.$$

$$= \dfrac{6x^2y^2}{2xy} + \dfrac{4x^2y}{2xy} - \dfrac{2xy}{2xy} \quad \text{Remove parentheses.}$$

$$= 3xy + 2x - 1 \quad \text{Simplify each fraction.}$$

SELF CHECK Simplify $\dfrac{9a^2b - 6ab^2 + 3ab}{3ab}$.

Answer: $3a - 2b + 1$ ■

EXAMPLE 5

Dividing a trinomial by a monomial. Simplify $\dfrac{12a^3b^2 - 4a^2b + a}{6a^2b^2}$.

Solution

$$\dfrac{12a^3b^2 - 4a^2b + a}{6a^2b^2}$$

$$= \dfrac{1}{6a^2b^2}(12a^3b^2 - 4a^2b + a) \quad \text{Multiply by the reciprocal of } 6a^2b^2.$$

$$= \dfrac{12a^3b^2}{6a^2b^2} - \dfrac{4a^2b}{6a^2b^2} + \dfrac{a}{6a^2b^2} \quad \text{Remove parentheses.}$$

$$= 2a - \dfrac{2}{3b} + \dfrac{1}{6ab^2} \quad \text{Simplify each fraction.}$$

SELF CHECK Simplify $\dfrac{14p^3q + pq^2 - p}{7p^2q}$.

Answer: $2p + \dfrac{q}{7p} - \dfrac{1}{7pq}$ ■

EXAMPLE 6

Dividing by a monomial. Simplify $\dfrac{(x - y)^2 - (x + y)^2}{xy}$.

Solution

$$\dfrac{(x - y)^2 - (x + y)^2}{xy}$$

$$= \dfrac{x^2 - 2xy + y^2 - (x^2 + 2xy + y^2)}{xy} \quad \text{Square the binomials in the numerator.}$$

$$= \dfrac{x^2 - 2xy + y^2 - x^2 - 2xy - y^2}{xy} \quad \text{Remove parentheses.}$$

$$= \dfrac{-4xy}{xy} \quad \text{Combine like terms.}$$

$$= -4 \quad \text{Divide out } xy.$$

SELF CHECK Simplify $\dfrac{(x + y)^2 - (x - y)^2}{xy}$.

Answer: 4 ■

An Application of Dividing a Polynomial by a Monomial

FIGURE 4-12

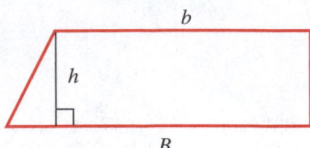

The area of the trapezoid shown in Figure 4-12 is given by the formula $A = \frac{1}{2}h(B + b)$, where B and b are its bases and h is its height. To solve the formula for b, we proceed as follows.

$$A = \frac{1}{2}h(B + b)$$

$$\mathbf{2} \cdot A = \mathbf{2} \cdot \frac{1}{2}h(B + b)$$ Multiply both sides by 2 to clear the equation of the fraction.

$$2A = h(B + b)$$ Simplify: $2 \cdot \frac{1}{2} = \frac{2}{2} = 1$.

$$2A = hB + hb$$ Use the distributive property to remove parentheses.

$$2A - hB = hB + hb - hB$$ Subtract hB from both sides.

$$2A - hB = hb$$ Combine like terms: $hB - hB = 0$.

$$\frac{2A - hB}{h} = \frac{hb}{h}$$ Divide both sides by h.

$$\frac{2A - hB}{h} = b$$ $\frac{h}{h} = 1$.

EXAMPLE 7

Confirming answers. Another student worked the previous problem in a different way and got a result of $b = \frac{2A}{h} - B$. Is this result correct?

Solution To determine whether this result is correct, we must show that

$$\frac{2A - hB}{h} = \frac{2A}{h} - B$$

We can do this by dividing $2A - hB$ by h.

$$\frac{2A - hB}{h} = \frac{1}{h}(2A - hB)$$

$$= \frac{2A}{h} - \frac{hB}{h}$$ Use the distributive property to remove parentheses.

$$= \frac{2A}{h} - B$$ Simplify: $\dfrac{\overset{1}{\cancel{h}}B}{\underset{1}{\cancel{h}}} = B$.

The results are the same.

SELF CHECK Suppose another student got $b = 2A - \frac{hB}{h}$. Is this result correct? *Answer:* no

STUDY SET

Section 4.7

VOCABULARY

In Exercises 1–6, fill in the blanks to make the statements true.

1. A _____ is an algebraic expression that is the sum of one or more terms containing whole-number exponents.

2. A _____ is a polynomial with one term.

3. A binomial is a polynomial with _____ terms.

4. A trinomial is a polynomial with _____ terms.

5. Dividing by a number is equivalent to multiplying by its _____.

6. To _____ a fraction, we divide out common factors of the numerator and denominator.

CONCEPTS

In Exercises 7–8, fill in the blanks to make the statements true.

7. $\dfrac{1}{b} \cdot a = \boxed{}$

8. $\dfrac{15x - 6y}{6xy} = \boxed{}\,(15x - 6y)$

9. What do the slashes and the small 1's mean?

$$\frac{4}{6} = \frac{\overset{1}{\cancel{2}} \cdot 2}{\underset{1}{\cancel{2}} \cdot 3}$$

10. Complete each rule of exponents.

a. $\dfrac{x^m}{x^n} = \boxed{}$

b. $x^{-n} = \boxed{}$

11. a. Solve the formula $d = rt$ for t.

b. Use your answer from part a to complete the table.

	r	$\cdot$	t	$=$	d
Motorcycle	$2x$				$6x^3$

12. a. Solve the formula $I = Prt$ for r.

b. Use your answer from part a to complete the table.

	P	$\cdot$	r	$\cdot$	t	$=$	I
Savings account	$8x^3$				$2x$		$24x^6$

13. How many nickels would have a value of $(10x + 35)$ cents?

14. How many twenty-dollar bills would have a value of $\$(60x - 100)$?

NOTATION

In Exercises 15–16, complete each solution.

15. $\dfrac{a^2 b^3}{a^3 b^2} = \dfrac{a \cdot a \cdot \boxed{} \cdot \boxed{} \cdot \boxed{}}{\boxed{} \cdot \boxed{} \cdot \boxed{} \cdot b \cdot b}$

$$= \frac{\overset{1}{\cancel{a}} \cdot \overset{1}{\cancel{a}} \cdot \overset{1}{\cancel{b}} \cdot \overset{1}{\cancel{b}} \cdot \boxed{}}{\underset{1}{\cancel{a}} \cdot \underset{1}{\cancel{a}} \cdot \boxed{} \cdot \underset{1}{\cancel{b}} \cdot \underset{1}{\cancel{b}}}$$

$$= \frac{b}{a}$$

16. $\dfrac{6pq^2 - 9p^2q^2 + pq}{3p^2q} = \boxed{}\,(6pq^2 - 9p^2q^2 + pq)$

$$= \frac{6pq^2}{\boxed{}} - \frac{9p^2q^2}{\boxed{}} + \frac{pq}{\boxed{}}$$

$$= \frac{6 \cdot p \cdot q \cdot q}{3 \cdot p \cdot p \cdot q} - \frac{\boxed{}}{3 \cdot p \cdot p \cdot q} + \frac{p \cdot q}{3 \cdot p \cdot p \cdot q}$$

$$= \frac{2q}{p} - 3q + \frac{1}{3p}$$

PRACTICE

In Exercises 17–24, simplify each fraction.

17. $\dfrac{5}{15}$

18. $\dfrac{64}{128}$

19. $\dfrac{-125}{75}$

20. $\dfrac{-98}{21}$

21. $\dfrac{120}{160}$

22. $\dfrac{70}{420}$

23. $\dfrac{-3{,}612}{-3{,}612}$

24. $\dfrac{-288}{-112}$

In Exercises 25–48, do each division by simplifying the fraction.

25. $\dfrac{x^5}{x^2}$

26. $\dfrac{a^{12}}{a^8}$

27. $\dfrac{r^3 s^2}{rs^3}$

28. $\dfrac{y^4 z^3}{y^2 z^2}$

29. $\dfrac{8x^3 y^2}{4xy^3}$

30. $\dfrac{-3y^3 z}{6yz^2}$

31. $\dfrac{12u^5 v}{-4u^2 v^3}$

32. $\dfrac{16rst^2}{-8rst^3}$

33. $\dfrac{-16r^3 y^2}{-4r^2 y^4}$

34. $\dfrac{35xyz^2}{-7x^2 yz}$

35. $\dfrac{-65rs^2 t}{15r^2 s^3 t}$

36. $\dfrac{112u^3 z^6}{-42u^3 z^6}$

37. $\dfrac{x^2 x^3}{xy^6}$

38. $\dfrac{x^2 y^2}{x^2 y^3}$

39. $\dfrac{(a^3 b^4)^3}{ab^4}$

40. $\dfrac{(a^2 b^3)^3}{a^6 b^6}$

41. $\dfrac{15(r^2 s^3)^2}{-5(rs^5)^3}$

42. $\dfrac{-5(a^2 b)^3}{10(ab^2)^3}$

43. $\dfrac{-32(x^3 y)^3}{128(x^2 y^2)^3}$

44. $\dfrac{68(a^6 b^7)^2}{-96(abc^2)^3}$

45. $\dfrac{-(4x^3 y^3)^2}{(x^2 y^4)^3}$

46. $\dfrac{(2r^3 s^2)^2}{-(4r^2 s^2)^2}$

47. $\dfrac{(a^2 a^3)^4}{(a^4)^3}$

48. $\dfrac{(b^3 b^4)^5}{(bb^2)^2}$

In Exercises 49–62, do each division.

49. $\dfrac{6x + 9}{3}$

50. $\dfrac{8x + 12y}{4}$

51. $\dfrac{5x - 10y}{25xy}$

52. $\dfrac{2x - 32}{16x}$

53. $\dfrac{3x^2 + 6y^3}{3x^2y^2}$

54. $\dfrac{4a^2 - 9b^2}{12ab}$

55. $\dfrac{15a^3b^2 - 10a^2b^3}{5a^2b^2}$

56. $\dfrac{9a^4b^3 - 16a^3b^4}{12a^2b}$

57. $\dfrac{4x - 2y + 8z}{4xy}$

58. $\dfrac{5a^2 + 10b^2 - 15ab}{5ab}$

59. $\dfrac{12x^3y^2 - 8x^2y - 4x}{4xy}$

60. $\dfrac{12a^2b^2 - 8a^2b - 4ab}{4ab}$

61. $\dfrac{-25x^2y + 30xy^2 - 5xy}{-5xy}$

62. $\dfrac{-30a^2b^2 - 15a^2b - 10ab^2}{-10ab}$

In Exercises 63–72, simplify each numerator and do the division.

63. $\dfrac{5x(4x - 2y)}{2y}$

64. $\dfrac{9y^2(x^2 - 3xy)}{3x^2}$

65. $\dfrac{(-2x)^3 + (3x^2)^2}{6x^2}$

66. $\dfrac{(-3x^2y)^3 + (3xy^2)^3}{27x^3y^4}$

67. $\dfrac{4x^2y^2 - 2(x^2y^2 + xy)}{2xy}$

68. $\dfrac{-5a^3b - 5a(ab^2 - a^2b)}{10a^2b^2}$

69. $\dfrac{(3x - y)(2x - 3y)}{6xy}$

70. $\dfrac{(2m - n)(3m - 2n)}{-3m^2n^2}$

71. $\dfrac{(a + b)^2 - (a - b)^2}{2ab}$

72. $\dfrac{(x - y)^2 + (x + y)^2}{2x^2y^2}$

APPLICATIONS

73. POOL The rack shown in Illustration 1 is used to set up the balls when beginning a game of pool. If the perimeter of the rack, in inches, is given by the polynomial $6x^2 - 3x + 9$, what is the length of one side?

ILLUSTRATION 1

74. CHECKERBOARD If the perimeter (in inches) of the checkerboard, shown in Illustration 2 is $12x^2 - 8x + 32$, what is the length of one side?

ILLUSTRATION 2

75. AIR CONDITIONING If the volume occupied by the air conditioning unit shown in Illustration 3 is $(36x^3 - 24x^2)$ cubic feet, find its height.

ILLUSTRATION 3

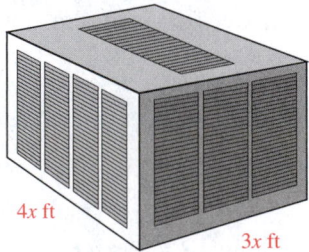

4x ft

3x ft

76. MINI BLINDS The area covered by the mini blinds shown in Illustration 4 is $(3x^3 - 6x)$ square feet. How long are the blinds?

ILLUSTRATION 4

3x ft

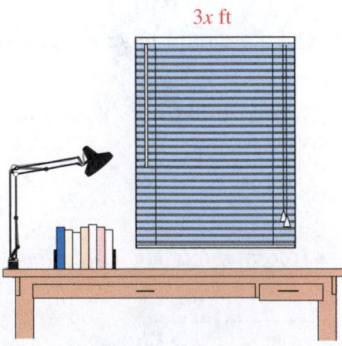

77. CONFIRMING FORMULAS Are these formulas the same?

$$l = \frac{P - 2w}{2} \quad \text{and} \quad l = \frac{P}{2} - w$$

78. CONFIRMING FORMULAS Are these formulas the same?

$$r = \frac{G + 2b}{2b} \quad \text{and} \quad r = \frac{G}{2b} + b$$

79. ELECTRIC BILLS On an electric bill, the following two formulas are used to compute the average cost of x kwh of electricity. Are the formulas equivalent?

$$\frac{0.08x + 5}{x} \quad \text{and} \quad 0.08x + \frac{5}{x}$$

80. PHONE BILLS On a phone bill, the following two formulas are used to compute the average cost per minute of x minutes of phone usage. Are the formulas equivalent?

$$\frac{0.15x + 12}{x} \quad \text{and} \quad 0.15 + \frac{12}{x}$$

WRITING

81. What would you say to a student to explain the error in the following solution?

Simplify $\dfrac{3x + 5}{5}$.

$$\frac{3x + 5}{5} = \frac{3x + \overset{1}{\cancel{5}}}{\underset{1}{\cancel{5}}}$$
$$= 3x$$

82. How do you simplify this fraction?

$$\frac{4x^2y + 8xy^2}{4xy}$$

REVIEW

In Exercises 83–86, identify each polynomial as a monomial, binomial, trinomial, or none of the above.

83. $5a^2b + 2ab^2$

84. $-3x^3y$

85. $-2x^3 + 3x^2 - 4x + 12$

86. $17t^2 - 15t + 27$

87. What is the degree of the trinomial $3x^2 - 2x + 4$?

88. What is the numerical coefficient of the the second term of the trinomial $-7t^2 + 5t + 17$?

▶ 4.8   Dividing Polynomials by Polynomials

In this section, you will learn about

> Dividing polynomials by polynomials ▪ Writing powers in descending order ▪ Missing terms

Introduction In this section, we will conclude our discussion on operations with polynomials by discussing how to divide one polynomial by another.

Dividing Polynomials by Polynomials

To divide one polynomial by another, we use a method similar to long division in arithmetic. We illustrate the method with several examples.

EXAMPLE 1   **Dividing polynomials.** Divide $x^2 + 5x + 6$ by $x + 2$.

Solution   Here the divisor is $x + 2$, and the dividend is $x^2 + 5x + 6$.

Step 1: $x + 2 \overline{)x^2 + 5x + 6}$ with x above the division symbol

How many times does x divide x^2? $x^2 \div x = x$. Place the x above the division symbol.

Step 2: $x + 2 \overline{)x^2 + 5x + 6}$ with x above and $x^2 + 2x$ below

Multiply each term in the divisor by x. Place the product under $x^2 + 5x$ and draw a line.

Step 3: $x + 2\overline{\smash{)}x^2 + 5x + 6}$

$\phantom{x + 2\overline{)}}\underline{x^2 + 2x}$

$\phantom{x + 2\overline{)}x^2 + }3x + 6$

Subtract $x^2 + 2x$ from $x^2 + 5x$. Work vertically, column by column: $x^2 - x^2 = 0$ and $5x - 2x = 3x$.

Bring down the 6.

Step 4: $x + 2\overline{\smash{)}x^2 + 5x + 6}$

$\phantom{x + 2\overline{)}}\underline{x^2 + 2x}$

$\phantom{x + 2\overline{)}x^2 + }3x + 6$

How many times does x divide $3x$? $3x \div x = +3$. Place the $+3$ above the division symbol.

Step 5: $x + 2\overline{\smash{)}x^2 + 5x + 6}$

$\phantom{x + 2\overline{)}}\underline{x^2 + 2x}$

$\phantom{x + 2\overline{)}x^2 + }3x + 6$

$\phantom{x + 2\overline{)}x^2 + }\underline{3x + 6}$

Multiply each term in the divisor by 3. Place the product under $3x + 6$ and draw a line.

Step 6: $x + 2\overline{\smash{)}x^2 + 5x + 6}$

$\phantom{x + 2\overline{)}}\underline{x^2 + 2x}$

$\phantom{x + 2\overline{)}x^2 + }3x + 6$

$\phantom{x + 2\overline{)}x^2 + }\underline{3x + 6}$

$\phantom{x + 2\overline{)}x^2 + 3x + }0$

Subtract $3x + 6$ from $3x + 6$. Work vertically: $3x - 3x = 0$ and $6 - 6 = 0$.

The quotient is $x + 3$ and the remainder is 0.

Step 7: Check the work by verifying that $x + 2$ times $x + 3$ is $x^2 + 5x + 6$.

$$(x + 2)(x + 3) = x^2 + 3x + 2x + 6$$
$$= x^2 + 5x + 6$$

The answer checks.

SELF CHECK Divide $x^2 + 7x + 12$ by $x + 3$. *Answer:* $x + 4$ ∎

EXAMPLE 2 **Dividing polynomials.** Divide $\dfrac{6x^2 - 7x - 2}{2x - 1}$.

Solution Here the divisor is $2x - 1$ and the dividend is $6x^2 - 7x - 2$.

Step 1: $2x - 1\overline{\smash{)}6x^2 - 7x - 2}$

How many times does $2x$ divide $6x^2$? $6x^2 \div 2x = 3x$. Place the $3x$ above the division symbol.

Step 2: $2x - 1\overline{\smash{)}6x^2 - 7x - 2}$

$\phantom{2x - 1\overline{)}}\underline{6x^2 - 3x}$

Multiply each term in the divisor by $3x$. Place the product under $6x^2 - 7x$ and draw a line.

Step 3: $2x - 1\overline{\smash{)}6x^2 - 7x - 2}$

$\phantom{2x - 1\overline{)}}\underline{6x^2 - 3x}$

$\phantom{2x - 1\overline{)}6x^2 - }-4x - 2$

Subtract $6x^2 - 3x$ from $6x^2 - 7x$. Work vertically: $6x^2 - 6x^2 = 0$ and $-7x - (-3x) = -7x + 3x = -4x$.

Bring down the -2.

Step 4: $2x - 1\overline{\smash{)}6x^2 - 7x - 2}$

$\phantom{2x - 1\overline{)}}\underline{6x^2 - 3x}$

$\phantom{2x - 1\overline{)}6x^2 - }-4x - 2$

How many times does $2x$ divide $-4x$? $-4x \div 2x = -2$. Place the -2 above the division symbol.

Step 5:
$$
\begin{array}{r}
3x \;-\; 2 \\
2x - 1 \overline{) 6x^2 - 7x - 2} \\
\underline{6x^2 - 3x} \\
-4x - 2 \\
\underline{-4x + 2}
\end{array}
$$
Multiply each term in the divisor by -2. Place the product under $-4x - 2$ and draw a line.

Step 6:
$$
\begin{array}{r}
3x \;-\; 2 \\
2x - 1 \overline{) 6x^2 - 7x - 2} \\
\underline{6x^2 - 3x} \\
-4x - 2 \\
\underline{-4x + 2} \\
-4
\end{array}
$$
Subtract $-4x + 2$ from $-4x - 2$. Work vertically: $-4x - (-4x) = -4x + 4x = 0$ and $-2 - 2 = -4$.

Here the quotient is $3x - 2$ and the remainder is -4. It is common to write the answer in this form:

$$3x - 2 + \dfrac{-4}{2x - 1} \qquad \text{Quotient} + \dfrac{\text{remainder}}{\text{divisor}}.$$

Step 7: To check the answer, we multiply

$$3x - 2 + \dfrac{-4}{2x - 1} \qquad \text{by} \qquad 2x - 1$$

The product should be the dividend.

$$
\begin{aligned}
(2x - 1)\left(3x - 2 + \dfrac{-4}{2x - 1}\right) &= (2x - 1)(3x - 2) + (2x - 1)\left(\dfrac{-4}{2x - 1}\right) \\
&= (2x - 1)(3x - 2) - 4 \\
&= 6x^2 - 4x - 3x + 2 - 4 \\
&= 6x^2 - 7x - 2
\end{aligned}
$$

Because the result is the dividend, the answer checks.

SELF CHECK   Divide $\dfrac{8x^2 + 6x - 3}{2x + 3}$.   *Answer:* $4x - 3 + \dfrac{6}{2x + 3}$ ∎

Writing Powers in Descending Order

The division method works best when the exponents of the terms in the divisor and the dividend are written in descending order. This means that the term involving the highest power of x appears first, the term involving the second-highest power of x appears second, and so on. For example, the terms in

$$3x^3 + 2x^2 - 7x + 5$$

have their exponents written in descending order.

If the powers in the dividend or divisor are not in descending order, we use the commutative property of addition to write them that way.

EXAMPLE 3   **Dividing polynomials.** Divide $4x^2 + 2x^3 + 12 - 2x$ by $x + 3$.

Solution   We write the dividend so that the exponents are in descending order.

$$
\begin{array}{r}
2x^2 - 2x \;+\; 4 \\
x + 3 \overline{) 2x^3 + 4x^2 - 2x + 12} \\
\underline{2x^3 + 6x^2} \\
-2x^2 - 2x \\
\underline{-2x^2 - 6x} \\
4x + 12 \\
\underline{4x + 12}
\end{array}
$$

Check: $(x + 3)(2x^2 - 2x + 4) = 2x^3 - 2x^2 + 4x + 6x^2 - 6x + 12$
$$= 2x^3 + 4x^2 - 2x + 12$$

SELF CHECK Divide $x^2 - 10x + 6x^3 + 4$ by $2x - 1$. *Answer:* $3x^2 + 2x - 4$ ∎

Missing Terms

When we write the terms of a dividend in descending powers of x, we must determine whether some powers of x are missing. For example, in the dividend of

$$x + 1 \overline{) 3x^4 - 7x^2 - 3x + 15}$$

the term involving x^3 is missing. When this happens, we should either write the term with a coefficient of 0 or leave a blank space for it. In this case, we would write the dividend as

$$3x^4 + 0x^3 - 7x^2 - 3x + 15 \qquad \text{or} \qquad 3x^4 \qquad - 7x^2 - 3x + 15$$

EXAMPLE 4

Dividing polynomials. Divide $\dfrac{x^2 - 4}{x + 2}$.

Solution Since $x^2 - 4$ does not have a term involving x, we must either include the term $0x$ or leave a space for it.

$$
\begin{array}{r}
x - 2 \\
x + 2 \overline{) x^2 + 0x - 4} \\
\underline{x^2 + 2x} \\
-2x - 4 \\
\underline{-2x - 4}
\end{array}
$$

Check: $(x + 2)(x - 2) = x^2 - 2x + 2x - 4$
$$= x^2 - 4$$

SELF CHECK Divide $\dfrac{x^2 - 9}{x - 3}$ *Answer:* $x + 3$ ∎

STUDY SET

Section 4.8

VOCABULARY

In Exercises 1–4, fill in the blanks to make the statements true.

1. In the division $x + 1 \overline{) x^2 + 2x + 1}$, $x + 1$ is called the _____ and $x^2 + 2x + 1$ is called the _____.

2. The answer to a division problem is called the _____.

3. If a division does not come out even, the leftover part is called a _____.

4. The exponents in $2x^4 + 3x^3 + 4x^2 - 7x - 2$ are said to be written in _____ order.

CONCEPTS

In Exercises 5–8, write each polynomial with the powers in descending order.

5. $4x^3 + 7x - 2x^2 + 6$

6. $5x^2 + 7x^3 - 3x - 9$

7. $9x + 2x^2 - x^3 + 6x^4$

8. $7x^5 + x^3 - x^2 + 2x^4$

In Exercises 9–10, identify the missing terms in each polynomial.

9. $5x^4 + 2x^2 - 1$

10. $-3x^5 - 2x^3 + 4x - 6$

In Exercises 11–12, without doing the division, determine which of the three possible quotients seems reasonable.

11. $\dfrac{x^4 - 81}{x - 3}$

$x^2 + 3x + 9$
$x^3 + 3x^2 + 9x + 27$
$x^4 + 3x^3 + 9x^2 + 27x + 1$

12. $\dfrac{8x^3 - 27}{2x - 3}$

$4x^2 + 6x + 9$
$4x^3 - 6x^2 - 9$
$4x^4 - 6x^3 - 9x^2 + 1$

13. a. Solve $d = rt$ for r.
  b. Use your answer to part a and the long division method to complete the chart.

	r	$\cdot$	t	$=$	d
Subway			$x + 4$		$x^2 + x - 12$

14. a. Solve $I = Prt$ for P.
  b. Use your answer to part a and the long division method to complete the chart.

	P	$\cdot$	r	$\cdot$	$t =$	I
Bonds			$x + 4$		1	$x^2 + 7x + 12$

15. Using long division, a student found that

$$\frac{3x^2 + 8x + 4}{3x + 2} = x + 2$$

Use multiplication to see whether the result is correct.

16. Using long division, a student found that

$$\frac{x^2 + 4x - 21}{x - 3} = x - 7$$

Use multiplication to see whether the result is correct.

NOTATION

In Exercises 17–18, complete each division.

17.
$$
\begin{array}{r}
\ \blacksquare + 2 \\
x + 2 \overline{)\,x^2 + 4x + 4} \\
\underline{x^2 + \blacksquare} \\
\blacksquare + 4 \\
\underline{2x + 4} \\
0
\end{array}
$$

18.
$$
\begin{array}{r}
\blacksquare + x - 2 + \frac{7}{2x+1} \\
2x + 1 \overline{)\,2x^3 + 3x^2 - 3x + 5} \\
\underline{+ x^2} \\
2x^2 - 3x \\
\underline{2x^2 + \blacksquare} \\
\blacksquare + 5 \\
\underline{-4x - \blacksquare} \\
7
\end{array}
$$

PRACTICE

In Exercises 19–26, do each division.

19. Divide $x^2 + 4x - 12$ by $x - 2$.

20. Divide $x^2 - 5x + 6$ by $x - 2$.

21. Divide $y^2 + 13y + 12$ by $y + 1$.

22. Divide $z^2 - 7z + 12$ by $z - 3$.

23. $\dfrac{6a^2 + 5a - 6}{2a + 3}$

24. $\dfrac{8a^2 + 2a - 3}{2a - 1}$

25. $\dfrac{3b^2 + 11b + 6}{3b + 2}$

26. $\dfrac{3b^2 - 5b + 2}{3b - 2}$

In Exercises 27–34, write the terms so that the powers of x are in descending order. Then do each division.

27. $5x + 3 \overline{)\,11x + 10x^2 + 3}$

28. $2x - 7 \overline{)\,-x - 21 + 2x^2}$

29. $4 + 2x \overline{)\,-10x - 28 + 2x^2}$

30. $1 + 3x \overline{)\,9x^2 + 1 + 6x}$

31. $2x - 1 \overline{)\,x - 2 + 6x^2}$

32. $2 + x \overline{)\,3x + 2x^2 - 2}$

33. $3 + x \overline{)\,2x^2 - 3 + 5x}$

34. $x - 3 \overline{)\,2x^2 - 3 - 5x}$

In Exercises 35–40, do each division.

35. $2x + 3 \overline{)\,2x^3 + 7x^2 + 4x - 3}$

36. $2x - 1 \overline{)\,2x^3 - 3x^2 + 5x - 2}$

37. $3x + 2 \overline{)\,6x^3 + 10x^2 + 7x + 2}$

38. $4x + 3 \overline{)\,4x^3 - 5x^2 - 2x + 3}$

39. $2x + 1 \overline{)\,2x^3 + 3x^2 + 3x + 1}$

40. $3x - 2 \overline{)\,6x^3 - x^2 + 4x - 4}$

In Exercises 41–50, do each division. If there is a remainder, write the answer in quotient $+ \dfrac{remainder}{divisor}$ form.

41. $\dfrac{2x^2 + 5x + 2}{2x + 3}$

42. $\dfrac{3x^2 - 8x + 3}{3x - 2}$

43. $\dfrac{4x^2 + 6x - 1}{2x + 1}$

44. $\dfrac{6x^2 - 11x + 2}{3x - 1}$

45. $\dfrac{x^3 + 3x^2 + 3x + 1}{x + 1}$

46. $\dfrac{x^3 + 6x^2 + 12x + 8}{x + 2}$

47. $\dfrac{2x^3 + 7x^2 + 4x + 3}{2x + 3}$

48. $\dfrac{6x^3 + x^2 + 2x + 1}{3x - 1}$

49. $\dfrac{2x^3 + 4x^2 - 2x + 3}{x - 2}$

50. $\dfrac{3y^3 - 4y^2 + 2y + 3}{y + 3}$

In Exercises 51–60, do each division.

51. $\dfrac{x^2 - 1}{x - 1}$

52. $\dfrac{x^2 - 9}{x + 3}$

53. $\dfrac{4x^2 - 9}{2x + 3}$

54. $\dfrac{25x^2 - 16}{5x - 4}$

55. $\dfrac{x^3 + 1}{x + 1}$

56. $\dfrac{x^3 - 8}{x - 2}$

57. $\dfrac{a^3 + a}{a + 3}$

58. $\dfrac{y^3 - 50}{y - 5}$

59. $3x - 4 \overline{)15x^3 - 23x^2 + 16x}$

60. $2y + 3 \overline{)21y^2 + 6y^3 - 20}$

APPLICATIONS

61. FURNACE FILTER The area of the furnace filter shown in Illustration 1 is $(x^2 - 2x - 24)$ square inches.
a. Find its length.
b. Find its perimeter.

ILLUSTRATION 1

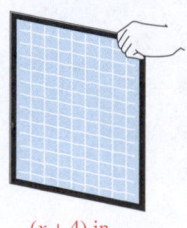

$(x + 4)$ in.

62. SHELF SPACE The formula $V = Bh$ gives the volume of a cylinder where B is the area of the base and h is the height. Find the amount of shelf space that the container of potato chips shown in Illustration 2 occupies if its volume is $(2x^3 - 4x - 2)$ cubic inches.

ILLUSTRATION 2

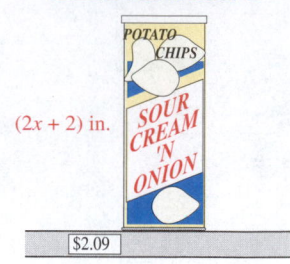

$(2x + 2)$ in.

$2.09

63. COMMUNICATION See Illustration 3. Telephone poles were installed every $(2x - 3)$ feet along a stretch of railroad track $(8x^3 - 6x^2 + 5x - 21)$ feet long. How many poles were used?

ILLUSTRATION 3

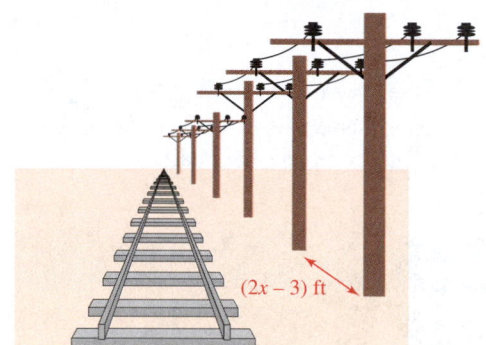

$(2x - 3)$ ft

64. CONSTRUCTION COSTS Find the price per square foot to remodel each of the three rooms listed in the chart.

Room	Remodeling cost	Area (ft^2)	Cost (per ft^2)
Kitchen	$(3x^3 - 9x - 6)$	$3x + 3$	
Bathroom	$(2x^2 + x - 6)$	$2x - 3$	
Bedroom	$(x^2 + 9x + 20)$	$x + 4$	

WRITING

65. Distinguish between *dividend, divisor, quotient,* and *remainder.*

66. How would you check the results of a division?

REVIEW

67. Simplify $(x^5 x^6)^2$.

68. Simplify $(a^2)^3(a^3)^4$.

In Exercises 69–70, simplify each expression.

69. $3(2x^2 - 4x + 5) + 2(x^2 + 3x - 7)$

70. $-2(y^3 + 2y^2 - y) - 3(3y^3 + y)$

71. What can be said about the slopes of two parallel lines?

72. What is the slope of a line perpendicular to a line with a slope of $\frac{3}{4}$?

Polynomials

A **polynomial** is an algebraic expression that is the sum of one or more terms containing whole-number exponents. Some examples are

$$-16a^2b, \qquad y + 2, \qquad x^2 - 3x + 9, \qquad \text{and} \qquad 3st - 6r + 5st - 8r$$

Operations with Polynomials

Polynomials are the numbers of algebra. Just like numbers in arithmetic, they can be added, subtracted, multiplied, divided, and raised to powers. In Chapter 4, we have discussed some rules to be used when performing operations with polynomials.

In Exercises 1–4, fill in the blanks to make the statements true. They are rules for working with polynomials containing more than one term.

1. To add polynomials, remove the parentheses and then _____ any like terms.

2. To subtract polynomials, drop the minus sign and the parentheses, and _____ the sign of every term within the parentheses. Then combine like terms.

3. To multiply polynomials, multiply _____ term of one polynomial by _____ term of the other polynomial and then combine like terms.

4. To divide polynomials, use the _____ division method.

In Exercises 5–12, do the operations.

5. $(2x + 3) + (x - 8)$

6. $(2x + 3) - (x - 8)$

7. $(2x + 3)(x - 8)$

8. $(2x^2 + 3)^2$

9. $(y^2 + y - 6) + (y + 3)$

10. $(y^2 + y - 6) - (y + 3)$

11. $(y^2 + y - 6)(y + 3)$

12. $(y^2 + y - 6) \div (y + 3)$

Polynomial Functions

Polynomial functions can be used to mathematically describe such situations as the stopping distance of a car, the appreciation of a house, and the area of a geometric figure.

13. Complete the table and then graph the polynomial function $f(x) = -x^3 - 1$.

14. YOUTH DRUG USE Find the coordinates of the lowest point on the graph in Illustration 1 and explain its significance.

x	$f(x)$	(x, y)
-2		
-1		
0		
1		
2		

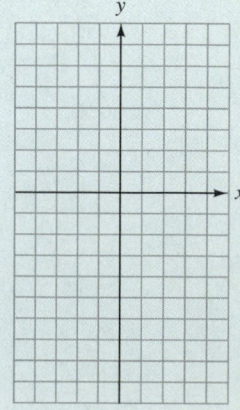

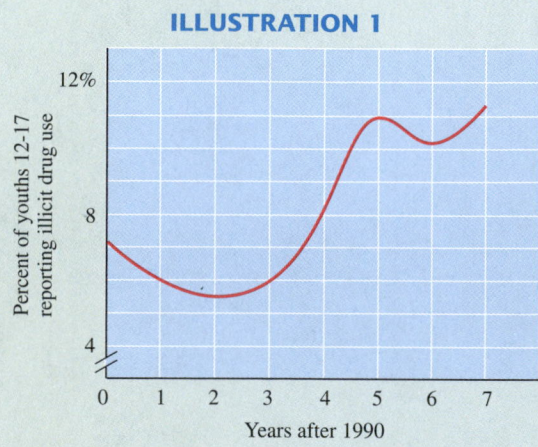

ILLUSTRATION 1

Percent of youths 12-17 reporting illicit drug use

Years after 1990

Accent on Teamwork

Section 4.1

Rules for exponents Have a student in your group write each of the five rules for exponents listed on page 282 on separate 3×5 cards. On a second set of cards, write an explanation of each rule using words. On a third set of cards, write a separate example of the use of each rule for exponents. Shuffle the cards and work together to match the symbolic description, the word description, and the example for each of the five rules for exponents.

Section 4.2

Graphing Complete Table 1, plot the ordered pairs on a rectangular coordinate system, and then draw a smooth curve through the points. Do the same for Table 2, using the same coordinate system. Compare the graphs. How are they similar and how do they differ?

TABLE 1		**TABLE 2**	
x	2^x	x	2^{-x}
-2		-2	
-1		-1	
0		0	
1		1	
2		2	
3		3	

Section 4.3

Scientific notation Go to the library and find five examples of extremely large and five examples of extremely small numbers. Encyclopedias, government statistics books, and science books are good places to look. Write each number in scientific notation on a separate piece of paper. Include a brief explanation of what the number represents. Present the ten examples in numerical order, beginning with the smallest number first.

Section 4.4

Polynomial functions The height (in feet) of a rock from the floor of the Grand Canyon t seconds after being thrown downward from the rim with an initial velocity of 6 feet per second is given by the polynomial

$$f(t) = -16t^2 - 6t + 5{,}292$$

a. Find $f(0)$ and $f(18)$ and explain their significance.

b. Find $f(3)$, $f(6)$, $f(9)$, $f(12)$, and $f(15)$. Use this information to show the position of the rock for these times on the scale shown in Illustration 1.

c. Are the distances the rock fell during each 3-second time interval the same?

ILLUSTRATION 1

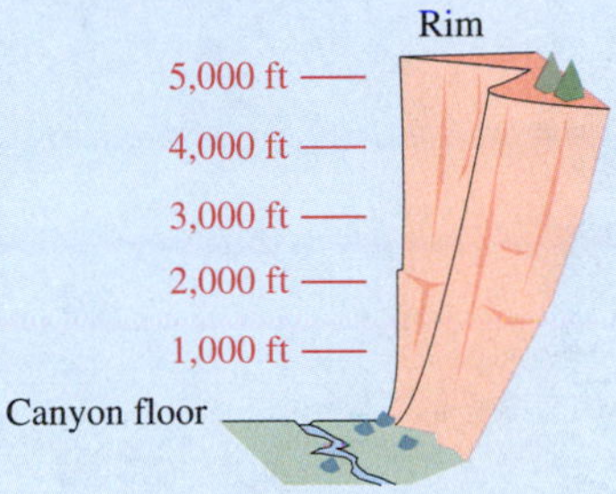

Rim

5,000 ft ——

4,000 ft ——

3,000 ft ——

2,000 ft ——

1,000 ft ——

Canyon floor

Section 4.5

Adding polynomials An old adage is that "You can't add apples and oranges." Give an example of how this concept applies when adding two polynomials.

Section 4.6

Multiplying binomials Use colored construction paper to make a model that can be used in a presentation explaining what it means to multiply the binomials $x + 3$ and $x + 4$. See Exercise 101 in Study Set 4.6 for an example.

Section 4.7

Working with monomials For the monomials $15a^3$ and $5a^2$, show, if possible, how they are added, subtracted, multiplied, and divided. If an operation cannot be done, explain why this is so.

Section 4.8

Working with polynomials Pick a certain binomial that divides a certain trinomial evenly. Then show how they are added, subtracted, multiplied, and divided.

Section 4.1

Natural-Number Exponents

CONCEPTS

If n is a natural number, then

$$x^n = \overbrace{x \cdot x \cdot x \cdot \dots \cdot x}^{n \text{ factors of } x}$$

where x is called the *base* and n is called the *exponent*.

Rules for exponents:
If m and n are integers, then

$$x^m x^n = x^{m+n}$$

$$(x^m)^n = x^{m \cdot n}$$

$$(xy)^n = x^n y^n$$

$$\left(\frac{x}{y}\right)^n = \frac{x^n}{y^n} \quad (y \neq 0)$$

$$\frac{x^m}{x^n} = x^{m-n} \quad (x \neq 0)$$

REVIEW EXERCISES

1. Write each expression without using exponents.

    a. $-3x^4$ **b.** $\left(\frac{1}{2}pq\right)^3$

2. Evaluate each expression.

    a. 5^3 **b.** $(-8)^2$

    c. -8^2 **d.** $(5-3)^2$

3. Simplify each expression.

    a. $x^3 x^2$ **b.** $-3y(y^5)$

    c. $(y^7)^3$ **d.** $(3x)^4$

    e. $b^3 b^4 b^5$ **f.** $-z^2(z^3 y^2)$

    g. $(-16s)^2 s$ **h.** $(2x^2 y)^2$

    i. $(x^2 x^3)^3$ **j.** $\left(\frac{x^2 y}{xy^2}\right)^2$

    k. $\dfrac{x^7}{x^3}$ **l.** $\dfrac{(5y^2 z^3)^3}{25(yz)^5}$

4. Find the area or the volume of each figure, whichever is appropriate.

    a.   **b.**

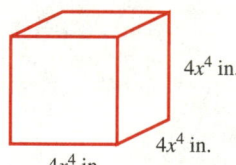

$4x^4$ in.
$4x^4$ in.
$4x^4$ in.

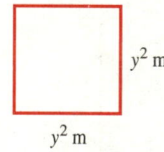

y^2 m
y^2 m

Section 4.2

Zero and Negative Integer Exponents

Zero exponents:
$$x^0 = 1 \quad (x \neq 0)$$

Negative integer exponents:
$$x^{-n} = \frac{1}{x^n} \quad (x \neq 0)$$

5. Write each expression without using negative exponents or parentheses.

    a. x^0 **b.** $(3x^2 y^2)^0$

    c. $(3x^0)^2$ **d.** 10^{-3}

    e. $\left(\dfrac{3}{4}\right)^{-1}$ **f.** -5^{-2}

    g. x^{-5} **h.** $-6y^4 y^{-5}$

    i. $\dfrac{x^{-3}}{x^7}$ **j.** $(x^{-3} x^{-4})^{-2}$

    k. $\left(\dfrac{x^2}{x}\right)^{-5}$ **l.** $\left(\dfrac{15z^4}{5z^3}\right)^{-2}$

6. Write each expression with a single exponent.

    a. $y^{3n} y^{4n}$ **b.** $\dfrac{z^{8c}}{z^{10c}}$

Section 4.3

A number is written in *scientific notation* if it is written as the product of a number between 1 (including 1) and 10 and an integer power of 10.

Scientific notation provides an easier way to do some computations.

Scientific Notation

7. Write each number in scientific notation.
    a. 728   **b.** 9,370,000
    c. 0.0136   **d.** 0.00942
    e. 0.018×10^{-2} **f.** 753×10^{3}

8. Write each number in standard notation.
    a. 7.26×10^{5} **b.** 3.91×10^{-4}
    c. 2.68×10^{0} **d.** 5.76×10^{1}

9. Simplify each fraction by first writing each number in scientific notation, then do the arithmetic. Express the result in standard notation.
    a. $\dfrac{(0.00012)(0.00004)}{0.00000016}$ **b.** $\dfrac{(4,800)(20,000)}{600,000}$

10. WORLD POPULATION As of 1998, the world's population was estimated to be 5.927 billion. Write this number in standard notation and in scientific notation.

11. ATOMS Illustration 1 shows a cross section of an atom. How many nuclei, placed end-to-end, would it take to stretch across the atom?

ILLUSTRATION 1

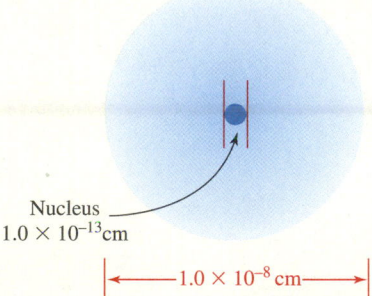

Nucleus
1.0×10^{-13} cm

$\longleftarrow 1.0 \times 10^{-8}$ cm $\longrightarrow$

Section 4.4

A *polynomial* is a term or a sum of terms in which all variables have whole-number exponents.

The *degree of a monomial* ax^n is n. The *degree of a monomial* in several variables is the sum of the exponents on those variables. The *degree of a polynomial* is the same as the degree of its term with the largest degree.

If $f(x)$ is a polynomial funtion in x, then $f(3)$ is the value of the function when $x = 3$.

Polynomials and Polynomial Functions

12. Tell whether each expression is a polynomial.
    a. $x^3 - x^2 - x - 1$ **b.** $x^{-2} - x^{-1} - 1$
    c. $\dfrac{11}{y} + 4y$ **d.** $-16x^2y + 5xy^2$

13. Find the degree of each polynomial and classify it as a monomial, binomial, trinomial, or none of these.
    a. $13x^7$ **b.** $-16a^2b$
    c. $5^3x + x^2$ **d.** $-3x^5 + x - 1$
    e. $9xy^2 + 21x^3y^3$ **f.** $4s^4 - 3s^2 + 5s + 4$

14. Let $f(x) = 3x^2 + 2x + 1$. Find each value.
    a. $f(3)$ **b.** $f(0)$
    c. $f(-2)$ **d.** $f(-0.2)$

15. Complete the table and then graph the polynomial function $f(x) = x^3 - 3x + 2$.

x	$f(x)$	(x, y)
-3		
-2		
-1		
0		
1		
2		
3		

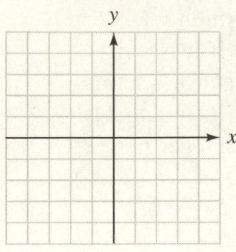

16. DIVING See Illustration 2. The number of inches that the woman deflects the diving board is given by the function

$$f(x) = 0.1875x^2 - 0.0078125x^3$$

where x is the number of feet that she stands from the front anchor point of the board. Find the amount of deflection if she stands on the end of the diving board, 8 feet from the anchor point.

ILLUSTRATION 2

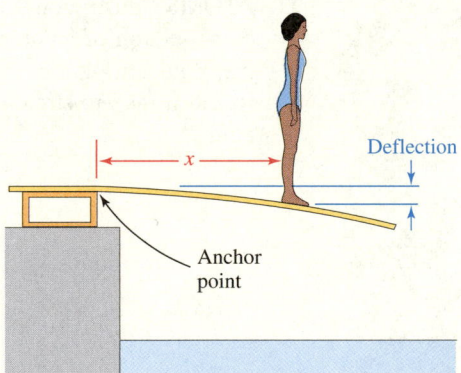

Section 4.5

When *adding* or *subtracting polynomials,* add or subtract like terms by combining the numerical coefficients and using the same variables and the same exponents.

Adding and Subtracting Polynomials

17. Simplify each expression.
  a. $3x^6 + 5x^5 - x^6$ **b.** $x^2y^2 - 3x^2y^2$
  c. $(3x^2 + 2x) + (5x^2 - 8x)$
  d. $3(9x^2 + 3x + 7) - 2(11x^2 - 5x + 9)$

Polynomials can be added or subtracted *vertically.*

18. Do the operations.

  a. $\begin{array}{r} 3x^2 + 5x + 2 \\ + \quad x^2 - 3x + 6 \\ \hline \end{array}$

  b. $\begin{array}{r} 20x^3 \qquad\quad + 12x \\ - \quad 12x^3 + 7x^2 - \ 7x \\ \hline \end{array}$

Section 4.6

To multiply two monomials, first multiply the numerical factors and then multiply the variable factors.

Multiplying Polynomials

19. Find each product.
  a. $(2x^2)(5x)$ **b.** $(-6x^4z^3)(x^6z^2)$
  c. $(2rst)(-3r^2s^3t^4)$ **d.** $5b^3 \cdot 6b^2 \cdot 4b^6$

To multiply a polynomial with more than one term by a monomial, multiply each term of the polynomial by the monomial and simplify.

20. Find each product.
a. $5(x + 3)$
b. $x^2(3x^2 - 5)$
c. $x^2y(y^2 - xy)$
d. $-2y^2(y^2 - 5y)$
e. $2x(3x^4)(x + 2)$
f. $-3x(x^2 - x + 2)$

To multiply two binomials, use the *FOIL method:*
F: First
O: Outer
I: Inner
L: Last

21. Find each product.
a. $(x + 3)(x + 2)$
b. $(2x + 1)(x - 1)$
c. $(3a - 3)(2a + 2)$
d. $6(a - 1)(a + 1)$
e. $(a - b)(2a + b)$
f. $(-3x - y)(2x + y)$

Special products:
$(x + y)^2 = x^2 + 2xy + y^2$
$(x - y)^2 = x^2 - 2xy + y^2$
$(x + y)(x - y) = x^2 - y^2$

22. Find each product.
a. $(x + 3)(x + 3)$
b. $(x + 5)(x - 5)$
c. $(a - 3)^2$
d. $(x + 4)^2$
e. $(-2y + 1)^2$
f. $(y^2 + 1)(y^2 - 1)$

To multiply one polynomial by another, multiply each term of one polynomial by each term of the other polynomial, and simplify.

23. Find each product.
a. $(3x + 1)(x^2 + 2x + 1)$
b. $(2a - 3)(4a^2 + 6a + 9)$

24. Solve each equation.
a. $x^2 + 3 = x(x + 3)$
b. $x^2 + x = (x + 1)(x + 2)$
c. $(x + 2)(x - 5) = (x - 4)(x - 1)$
d. $(x + 5)(3x + 1) = x^2 + (2x - 1)(x - 5)$

25. APPLIANCES Find the perimeter of the base, the area of the base, and the volume occupied by the dishwasher shown in Illustration 3.

ILLUSTRATION 3

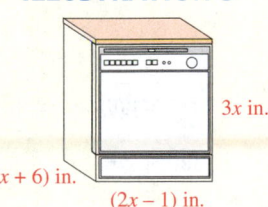

$3x$ in.
$(x + 6)$ in.
$(2x - 1)$ in.

Section 4.7

Dividing Polynomials by Monomials

To divide monomials, use the method for simplifying fractions or use the rules for exponents.

26. Simplify each expression ($x > 0$, $y > 0$).
a. $\dfrac{-14x^2y}{21xy^3}$
b. $\dfrac{(x^2)^2}{xx^4}$

To divide a polynomial by a monomial, write the division as a product, use the distributive property to remove parentheses, and simplify each resulting fraction.

27. Do each division. All the variables represent positive numbers.
a. $\dfrac{8x + 6}{2}$
b. $\dfrac{14xy - 21x}{7xy}$
c. $\dfrac{15a^2b + 20ab^2 - 25ab}{5ab}$
d. $\dfrac{(x + y)^2 + (x - y)^2}{-2xy}$

28. SAVINGS BONDS How many $50 savings bonds would have a total value of $(50x + 250)$?

Section 4.8   Dividing Polynomials by Polynomials

Long division is used to divide one polynomial by another.

When a division has a remainder, write the answer in the form

$$\text{Quotient} + \frac{\text{remainder}}{\text{divisor}}$$

The division method works best when the exponents of the terms of the divisor and the dividend are written in descending order.

When the dividend is missing a term, write it with a coefficient of zero or leave a blank space.

29. Do each division.

a. $x + 2 \overline{)x^2 + 3x + 5}$

b. $x - 1 \overline{)x^2 - 6x + 5}$

c. $\dfrac{2x^2 + 3 + 7x}{x + 3}$

d. $\dfrac{3x^2 + 14x - 2}{3x - 1}$

e. $2x - 1 \overline{)6x^3 + x^2 + 1}$

f. $3x + 1 \overline{)-13x - 4 + 9x^3}$

30. Use multiplication to show that the answer when dividing $3y^2 + 11y + 6$ by $y + 3$ is $3y + 2$.

31. ZOOLOGY The distance in inches traveled by a certain type of snail in $(2x - 1)$ minutes is given by the polynomial $8x^2 + 2x - 3$. At what rate did the snail travel?

CHAPTER 4

Test

1. Use exponents to rewrite $2xxxyyyy$.

2. Evaluate $(3 + 5)^2$.

In Problems 3–4, write each expression as an expression containing only one exponent.

3. $y^2(yy^3)$

4. $(2x^3)^5(x^2)^3$

In Problems 5–8, simplify each expression. Write answers without using parentheses or negative exponents.

5. $3x^0$

6. $2y^{-5}y^2$

7. $\dfrac{y^2}{yy^{-2}}$

8. $\left(\dfrac{a^2b^{-1}}{4a^3b^{-2}}\right)^{-3}$

9. What is the volume of a cube that has sides of length $10y^4$ inches?

10. Rewrite 4^{-2} using a positive exponent and then evaluate the result.

11. ELECTRICITY One ampere (amp) corresponds to the flow of 6,250,000,000,000,000,000 electrons per second past any point in a direct current (DC) circuit. Write this number in scientific notation.

12. Write 9.3×10^{-5} in standard notation.

13. Identify $3x^2 + 2$ as a monomial, binomial, or trinomial.

14. Find the degree of the polynomial $3x^2y^3 + 2x^3y - 5x^2y$.

15. If $f(x) = x^2 + x - 2$, find $f(-2)$.

16. Simplify $(xy)^2 + 5x^2y^2 - (3x)^2y^2$.

17. Simplify $-6(x - y) + 2(x + y) - 3(x + 2y)$.

18. Subtract: $\begin{array}{r} 2x^2 - 7x + 3 \\ 3x^2 - 2x - 1 \end{array}$

In Problems 19–24, find each product.

19. $(-2x^3)(2x^2y)$

20. $3y^2(y^2 - 2y + 3)$

21. $(x - 9)(x + 9)$

22. $(3y - 4)^2$

23. $(2x - 5)(3x + 4)$

24. $(2x - 3)(x^2 - 2x + 4)$

25. Solve the equation $(a + 2)^2 = (a - 3)^2$.

26. Simplify $\dfrac{8x^2y^3z^4}{16x^3y^2z^4}$

27. Simplify $\dfrac{6a^2 - 12b^2}{24ab}$.

28. Divide $2x + 3\overline{)2x^2 - x - 6}$.

29. In your own words, explain this rule for exponents:

$$x^{-n} = \frac{1}{x^n}$$

30. A rectangle has an area of $(x^2 - 6x + 5)$ ft^2 and a length of $(x - 1)$ feet. Show how division can be used to find the width of the rectangle. Explain your steps.

31. Complete the table and then graph the polynomial function $f(x) = -x^3 + 3x^2 - 5$.

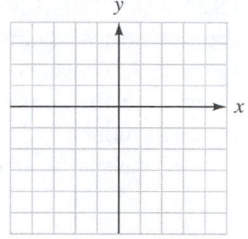

x	$f(x)$	(x, y)
-2		
-1		
0		
1		
2		
3		

32. ROCK SALT Illustration 1 shows U.S. sales of rock salt, which is used for road de-icing. Find the coordinates of the lowest and highest points on the graph and explain their significance.

ILLUSTRATION 1

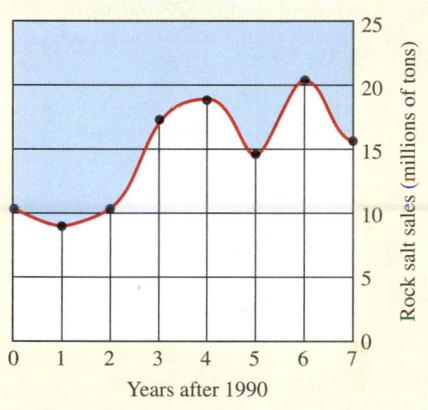

CHAPTERS 1–4

Cumulative Review Exercises

1. PERSONAL SAVINGS RATE The graph in Illustration 1 shows a situation occurring in September of 1998 that hadn't occurred since the Great Depression. Explain what was unusual.

ILLUSTRATION 1

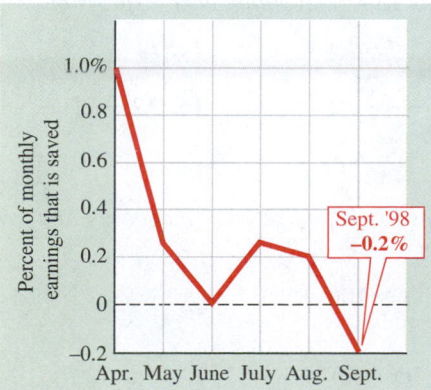

Based on data from the U.S. Department of Commerce

2. CLINICAL TRIALS In a clinical test of Aricept, a drug to treat Alzheimer's disease, one group of patients took a placebo (a sugar pill) while another group took the actual medication. See Illustration 2. Find the number of patients in each group who experienced nausea. Round to the nearest whole number.

ILLUSTRATION 2

Comparison of rates of adverse events in patients

Adverse event	Group 1—Placebo (number = 315)	Group 2—Aricept (number = 311)
Nausea	6%	5%

In Exercises 3–4, consider the algebraic expression $3x^3 + 5x^2y + 37y$.

3. Find the coefficient of the second term.

4. What is the third term?

In Exercises 5–8, simplify each expression.

5. $3x - 5x + 2y$

6. $3(x - 7) + 2(8 - x)$

7. $2x^2y^3 - xy(xy^2)$

8. $x^2(3 - y) + x(xy + x)$

In Exercises 9–10, solve each equation.

9. $3(x - 5) + 2 = 2x$ **10.** $\dfrac{x - 5}{3} - 5 = 7$

In Exercises 11–12, solve each formula for the variable indicated.

11. $A = \dfrac{1}{2}h(b + B)$; for h

12. $y = mx + b$; for x

In Exercises 13–16, evaluate each expression.

13. $4^2 - 5^2$ **14.** $(4 - 5)^2$

15. $\dfrac{-3 - (-7)}{2^2 - 3}$ **16.** $12 - 2[1 - (-8 + 2)]$

In Exercises 17–18, solve each inequality and graph the solution set.

17. $8(4 + x) > 10(6 + x)$

18. $-9 < 3(x + 2) \le 3$

In Exercises 19–22, graph each equation.

19. $y = x^2$ **20.** $y = |x|$

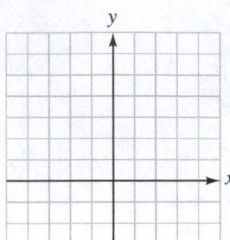   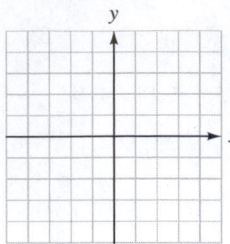

21. $4x - 3y = 12$ **22.** $3x = 12$

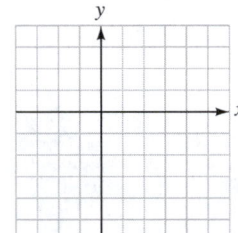   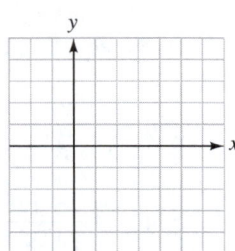

In Exercises 23–26, find the slope of the line with the given properties.

23. Passing through $(-2, 4)$ and $(6, 8)$

24. A line that is horizontal

25. An equation of $y = -4x + 3$

26. An equation of $2x - 3y = 12$

In Exercises 27–30, write the equation of the line with the following properties.

27. Slope $= \dfrac{2}{3}$, y-intercept $= (0, 5)$

28. Passing through $(-2, 4)$ and $(6, 10)$

29. A horizontal line passing through $(2, 4)$

30. A vertical line passing through $(2, 4)$

In Exercises 31–32, are the graphs of the lines parallel or perpendicular?

31. $\begin{cases} y = -\dfrac{3}{4}x + \dfrac{15}{4} \\ 4x - 3y = 25 \end{cases}$

32. $\begin{cases} y = -\dfrac{3}{4}x + \dfrac{15}{4} \\ 6x = 15 - 8y \end{cases}$

In Exercises 33–34, tell whether each equation defines a function.

33. $y = x^3 - 4$ **34.** $x = |y|$

In Exercises 35–38, $f(x) = 2x^2 - 3$. Find each value.

35. $f(0)$ **36.** $f(3)$

37. $f(-2)$ **38.** $f(0.5)$

39. Find the domain and range of the function graphed in Illustration 3.

ILLUSTRATION 3

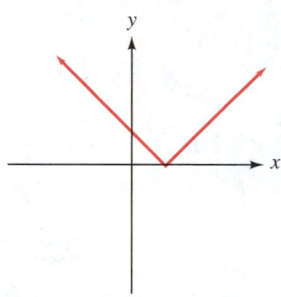

40. Tell whether the graph is the graph of a function.

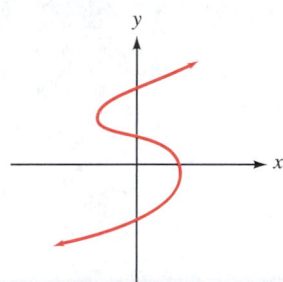

41. Give an example of a positive rate of change and a negative rate of change.

42. Fill in the blank: $f(x) = 2 - 10x$ is read as "f ___ x is $2 - 10x$."

In Exercises 43–50, write each expression using one positive exponent.

43. $(y^3y^5)y^6$

44. $(x^3x^4)^2$

45. $\dfrac{x^3x^4}{x^2x^3}$

46. $\dfrac{a^4b^0}{a^{-3}}$

47. x^{-5}

48. $(-2y)^{-4}$

49. $(x^{-4})^2$

50. $\left(-\dfrac{x^3}{x^{-2}}\right)^3$

In Exercises 51–52, write each number in scientific notation.

51. 615,000

52. 0.0000013

In Exercises 53–54, write each number in standard notation.

53. 5.25×10^{-4}

54. 2.77×10^3

In Exercises 55–56, give the degree of each polynomial.

55. $3x^2 + 2x - 5$

56. $-3x^3y^2 + 3x^2y^2 - xy$

57. MUSICAL INSTRUMENTS The gong shown in Illustration 4 is a percussion instrument used throughout Southeast Asia. The amount of deflection of the horizontal support (in inches) is given by the polynomial function

$$f(x) = 0.01875x^4 - 0.15x^3 + 1.2x$$

where x is the distance (in feet) that the gong is hung from one end of the support. Find the deflection if the gong is hung in the middle of the support.

ILLUSTRATION 4

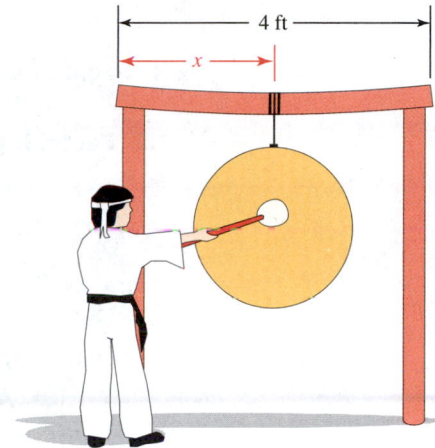

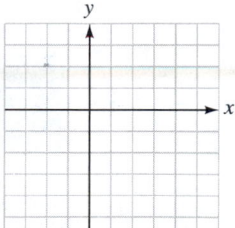

58. Complete the table and then graph the function $f(x) = 4x^2 - 8x - 1$.

x	$f(x)$	(x, y)
-1		
0		
1		
2		
3		

In Exercises 59–68, do the operations.

59. $(3x^2 + 2x - 7) - (2x^2 - 2x + 7)$

60. $(2x^2 - 3x + 4) + (2x^2 + 2x - 5)$

61. $-5x^2(7x^3 - 2x^2 - 2)$

62. $(3x^3y^2)(-4x^2y^3)$

63. $(3x - 7)(2x + 8)$

64. $(5x - 4y)(3x + 2y)$

65. $(3x + 1)^2$

66. $(x - 2)(x^2 + 2x + 4)$

67. $\dfrac{6x^2 - 8x}{2x}$

68. $x - 3\overline{)2x^2 - 5x - 3}$

5

Factoring and Quadratic Equations

CAMPUS CONNECTION

The Business Department

The relationship between production and revenue is one that business students study in great detail. For example, a company that manufactures radios has found that as the number of radios produced increases, the amount of revenue increases to reach a maximum, and then it steadily decreases. These observations come from a graph of a *quadratic function,* which mathematically describes the relationship between the number of radios manufactured and the revenue the company receives. In this chapter, we introduce quadratic functions, and we discuss a procedure used to determine maximum values for such situations.

The solutions of many problems require the use of quadratic equations—equations of the form $ax^2 + bx + c = 0$, where a, b, and c are real numbers and $a \neq 0$.

▶ 5.1   Factoring Out the Greatest Common Factor and Factoring by Grouping

In this section, you will learn about

Prime factorization ■ The greatest common factor (GCF) ■ Finding the GCF of several monomials ■ Factoring out the greatest common factor from a polynomial ■ Factoring out a negative factor ■ Factoring by grouping

Introduction   Recall that the distributive property provides a way to multiply the monomial $4y$ and the binomial $3y + 5$.

$$4y(3y + 5) = 4y \cdot 3y + 4y \cdot 5$$
$$= 12y^2 + 20y$$

In this section, we will reverse the operation of multiplication. Given a polynomial such as $12y^2 + 20y$, we will ask ourselves, "What factors were multiplied to obtain $12y^2 + 20y$?" The process of finding the individual factors of a known product is called **factoring**.

The multiplication process

Given the factors . . . find the product

$$4y(3y + 5) \quad = \quad ?$$

The factoring process

Given the product . . . find the factors

$$12y^2 + 20y \quad = \quad ?(\quad ? \quad)$$

Factoring can be used to solve certain types of equations and to simplify certain kinds of algebraic expressions. To begin the discussion of factoring, we consider two procedures that are used to factor natural numbers.

Prime Factorization

Because 4 divides 12 exactly, 4 is called a **factor** of 12. The numbers 1, 2, 3, 4, 6, and 12 are the natural-number factors of 12, because each divides 12 exactly.

Prime numbers

A **prime number** is a natural number greater than 1 whose only factors are 1 and itself.

For example, 17 is a prime number, because

1. 17 is a natural number greater than 1, and
2. the only two natural-number factors of 17 are 1 and 17.

The prime numbers less than 50 are

2, 3, 5, 7, 11, 13, 17, 19, 23, 29, 31, 37, 41, 43, and 47

A natural number is said to be in **prime-factored form** if it is written as the product of factors that are prime numbers.

To find the prime-factored form of a natural number, we can use a **factoring tree.** The following examples show two ways to proceed to find the prime-factored form of 90 using factoring trees. The factoring process stops when a row of the tree contains only prime-number factors.

1. Start with 90.	**1.** Start with 90.	
2. Factor 90 as $9 \cdot 10$.	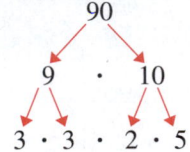	**2.** Factor 90 as $6 \cdot 15$.	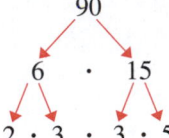
3. Factor 9 and 10.		**3.** Factor 6 and 15.	

Since the prime factors in either case are $2 \cdot 3 \cdot 3 \cdot 5$, the prime-factored form, or the **prime factorization**, of 90 is $2 \cdot 3^2 \cdot 5$. This example illustrates the **fundamental theorem of arithmetic,** which states that there is only one prime factorization for every natural number greater than 1.

We can also find the prime factorization of a natural number using the **division method.** For example, to find the prime factorization of 42, we begin by choosing the *smallest* prime number that will divide the given number exactly. We continue this process until the result of the division is a prime number.

Step 1: 2 divides 42 exactly. The result is 21, which is not prime. We continue the process.

$$2\,\overline{\smash{\big)}\,42} \\ 21$$

Step 2: We choose the smallest prime number that divides 21. The prime number 2 does not divide 21 exactly, but 3 does. The result is 7, which is prime. We are done.

$$2\,\overline{\smash{\big)}\,42} \\ 3\,\overline{\smash{\big)}\,21} \\ 7$$

The prime factorization of 42 is $2 \cdot 3 \cdot 7$.

EXAMPLE 1

Prime factorizations. Find the prime factorization of 150.

Solution   We can use a factoring tree or the division method to find the prime factorization of 150.

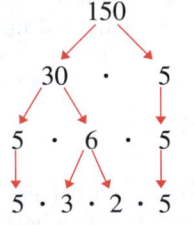

$$2\,\overline{\smash{\big)}\,150} \\ 3\,\overline{\smash{\big)}\,75} \\ 5\,\overline{\smash{\big)}\,25} \\ 5$$

The prime factorization of 150 is $2 \cdot 3 \cdot 5^2$.

SELF CHECK   Find the prime factorization of 225.   *Answer:* $3^2 \cdot 5^2$ ■

The Greatest Common Factor (GCF)

The right-hand sides of the equations

$$90 = \mathbf{2} \cdot \mathbf{3} \cdot 3 \cdot 5$$

$$42 = \mathbf{2} \cdot \mathbf{3} \cdot 7$$

show the prime-factored forms of 90 and 42. The color highlighting indicates that 90 and 42 have one prime factor of 2 and one prime factor of 3 in common. We can conclude that $2 \cdot 3 = 6$ is the largest natural number that divides 90 and 42 exactly, and we say that 6 is their **greatest common factor (GCF).**

$$\frac{90}{6} = 15 \quad \text{and} \quad \frac{42}{6} = 7$$

EXAMPLE 2

Finding the GCF of three numbers. Find the greatest common factor of 24, 60, and 96.

Solution We write each prime factorization and highlight the prime factors the three numbers have in common.

$$24 = 2 \cdot 2 \cdot 2 \cdot 3$$
$$60 = 2 \cdot 2 \cdot 3 \cdot 5$$
$$96 = 2 \cdot 2 \cdot 2 \cdot 2 \cdot 2 \cdot 3$$

Since 24, 60, and 96 each have two factors of 2 and one factor of 3, their greatest common factor is $2 \cdot 2 \cdot 3 = 12$.

SELF CHECK Find the GCF for 45, 60, 75. *Answer:* $3 \cdot 5 = 15$ ■

Finding the GCF of Several Monomials

The right-hand sides of the equations

$$12y^2 = 2 \cdot 2 \cdot 3 \cdot y \cdot y$$
$$20y = 2 \cdot 2 \cdot 5 \cdot y$$

show the prime factorizations of $12y^2$ and $20y$. Since the monomials have two factors of 2 and one factor of y in common, their GCF is

$$2 \cdot 2 \cdot y \quad \text{or} \quad 4y$$

To find the GCF of several monomials, we follow these steps.

Strategy for finding the greatest common factor (GCF)

1. Find the prime factorization of each monomial.
2. List each common factor the least number of times it appears in any one monomial.
3. Find the product of the factors in the list to obtain the GCF.

EXAMPLE 3

Finding the GCF of three monomials. Find the GCF of $10x^3y^2$, $60x^2y$, and $30xy^2$.

Solution **Step 1:** Find the prime factorization of each monomial.

$$10x^3y^2 = 2 \cdot 5 \cdot x \cdot x \cdot x \cdot y \cdot y$$
$$60x^2y = 2 \cdot 2 \cdot 3 \cdot 5 \cdot x \cdot x \cdot y$$
$$30xy^2 = 2 \cdot 3 \cdot 5 \cdot x \cdot y \cdot y$$

Step 2: List each common factor the least number of times it appears in any one monomial: 2, 5, x, and y.

Step 3: Find the product of the factors in the list:

$$2 \cdot 5 \cdot x \cdot y = 10xy \qquad \text{The GCF is } 10xy.$$

SELF CHECK   Find the GCF of $20a^2b^3$, $12ab^4$, and $8a^3b^2$.   *Answer:* $4ab^2$ ■

Factoring Out the Greatest Common Factor from a Polynomial

To factor $12y^2 + 20y$, we find the GCF of $12y^2$ and $20y$ (which we earlier determined to be $4y$) and use the distributive property.

$$12y^2 + 20y = \mathbf{4y} \cdot 3y + \mathbf{4y} \cdot 5 \qquad \text{Write each term of the polynomial as the product of the GCF, } 4y, \text{ and one other factor.}$$

$$= \mathbf{4y}(3y + 5) \qquad\qquad 4y \text{ is a common factor of both terms, and we can use the distributive property to factor it out.}$$

This process is called **factoring out the greatest common factor.**

EXAMPLE 4   **Factoring out the greatest common factor.** Factor $25 - 5m$.

Solution   To find the GCF of 25 and $5m$, we find their prime factorizations.

$$\left.\begin{array}{l} 25 = \mathbf{5} \cdot 5 \\ 5m = \mathbf{5} \cdot m \end{array}\right\} \quad \text{GCF} = 5$$

We can use the distributive property to factor out the GCF.

$$25 - 5m = \mathbf{5} \cdot 5 - \mathbf{5} \cdot m \qquad \text{Factor each monomial using 5 and one other factor.}$$

$$= \mathbf{5}(5 - m) \qquad\qquad \text{Factor out the common factor of 5.}$$

We check by verifying that $5(5 - m) = 25 - 5m$.

SELF CHECK   Factor $18x - 24$.   *Answer:* $6(3x - 4)$ ■

EXAMPLE 5   **Factoring out the GCF.** Factor $35a^3b^2 - 14a^2b^3$.

Solution   To find the GCF, we find the prime factorizations of $35a^3b^2$ and $14a^2b^3$.

$$\left.\begin{array}{l} 35a^3b^2 = 5 \cdot \mathbf{7} \cdot \mathbf{a} \cdot \mathbf{a} \cdot a \cdot \mathbf{b} \cdot \mathbf{b} \\ 14a^2b^3 = 2 \cdot \mathbf{7} \cdot \mathbf{a} \cdot \mathbf{a} \cdot \mathbf{b} \cdot \mathbf{b} \cdot b \end{array}\right\} \quad \text{GCF} = 7 \cdot a \cdot a \cdot b \cdot b = 7a^2b^2$$

We factor out the GCF of $7a^2b^2$.

$$35a^3b^2 - 14a^2b^3 = \mathbf{7a^2b^2} \cdot 5a - \mathbf{7a^2b^2} \cdot 2b$$

$$= \mathbf{7a^2b^2}(5a - 2b)$$

We check by verifying that $7a^2b^2(5a - 2b) = 35a^3b^2 - 14a^2b^3$.

SELF CHECK   Factor $40x^2y^3 + 15x^3y^2$.   *Answer:* $5x^2y^2(8y + 3x)$ ■

EXAMPLE 6   **An implied coefficient of 1.** Factor $4x^3y^2z - 2x^2yz + xz$.

Solution   The expression has three terms. We factor out the GCF, which is xz.

$$4x^3y^2z - 2x^2yz + xz = \mathbf{xz} \cdot 4x^2y^2 - \mathbf{xz} \cdot 2xy + \mathbf{xz} \cdot 1$$

$$= \mathbf{xz}(4x^2y^2 - 2xy + 1)$$

The last term of $4x^3y^2z - 2x^2yz + xz$ has an implied coefficient of 1. That is, $xz = 1xz = xz \cdot 1$. When xz is factored out, we must write this coefficient of 1. We check by verifying that $xz(4x^2y^2 - 2xy + 1) = 4x^3y^2z - 2x^2yz + xz$.

SELF CHECK Factor $2ab^2c + 4a^2bc - ab$.   *Answer:* $ab(2bc + 4ac - 1)$ ∎

E X A M P L E   7

Lava lamp. Figure 5-1 shows a lava lamp. The formula that gives the volume of the glass dome of the lamp is given below. Factor the expression on the right-hand side of the equation. (Read r_1^2 as "r sub 1, squared.")

$$V = \frac{1}{3}\pi r_1^2 h + \frac{1}{3}\pi r_2^2 h + \frac{1}{3}\pi r_1 r_2 h$$

FIGURE 5-1

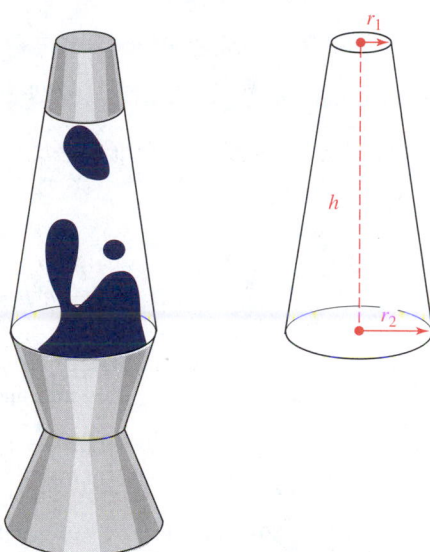

Solution The expression on the right-hand side of the formula contains three terms. By inspection, we see that each term contains a factor of $\frac{1}{3}$, π, and h. Note that neither r_1 nor r_2 is common to all three terms.

$$V = \frac{1}{3}\pi r_1^2 h + \frac{1}{3}\pi r_2^2 h + \frac{1}{3}\pi r_1 r_2 h \quad \text{The given formula.}$$

$$= \frac{1}{3}\pi h(r_1^2 + r_2^2 + r_1 r_2) \quad \begin{array}{l}\text{Factor out the}\\\text{GCF, } \frac{1}{3}\pi h.\end{array}$$

∎

E X A M P L E   8

Factoring out a common binomial factor. Factor $x(x + 4) + 3(x + 4)$.

Solution The given polynomial has two terms:

$$\underbrace{x(x + 4)}_{\substack{\text{The first}\\\text{term}}} + \underbrace{3(x + 4)}_{\substack{\text{The second}\\\text{term}}}$$

The GCF of the terms is the binomial $x + 4$. We can factor it out.

$$x(x + 4) + 3(x + 4) = (x + 4)(x + 3)$$

SELF CHECK   Factor $2y(y-1) - 7(y-1)$.   *Answer:* $(y-1)(2y-7)$ ■

Factoring Out a Negative Factor

It is often useful to factor out a common factor having a negative coefficient.

EXAMPLE 9   **Factoring out −1.** Factor -1 out of $-a^3 + 2a^2 - 4$.

Solution   First, we write each term of the polynomial as the product of -1 and another factor: $-a^3 = (-1)a^3$, $2a^2 = (-1)(-2a^2)$, and $-4 = (-1)4$. Then we factor out the common factor of -1.

$$-a^3 + 2a^2 - 4 = (\mathbf{-1})a^3 + (\mathbf{-1})(-2a^2) + (\mathbf{-1})4$$
$$= \mathbf{-1}(a^3 - 2a^2 + 4) \quad \text{Factor out } -1.$$
$$= -(a^3 - 2a^2 + 4) \quad \text{The coefficient of 1 need not be written.}$$

We check by verifying that $-(a^3 - 2a^2 + 4) = -a^3 + 2a^2 - 4$.

SELF CHECK   Factor -1 out of $-b^4 - 3b^2 + 2$.   *Answer:* $-(b^4 + 3b^2 - 2)$ ■

EXAMPLE 10   **Factoring out the negative of the GCF.** Factor out the negative (opposite) of the GCF in $-18a^2b + 6ab^2 - 12a^2b^2$.

Solution   The GCF is $6ab$. To factor out its negative, we write each term of the polynomial as the product of $-6ab$ and another factor. Then we factor out $-6ab$.

$$-18a^2b + 6ab^2 - 12a^2b^2 = (\mathbf{-6ab})3a - (\mathbf{-6ab})b + (\mathbf{-6ab})2ab$$
$$= \mathbf{-6ab}(3a - b + 2ab)$$

We check by verifying that $-6ab(3a - b + 2ab) = -18a^2b + 6ab^2 - 12a^2b^2$.

SELF CHECK   Factor out the negative (opposite) of the GCF in $-27xy^2 - 18x^2y + 36x^2y^2$.   *Answer:* $-9xy(3y + 2x - 4xy)$ ■

Factoring by Grouping

Suppose we wish to factor

$$ax + ay + cx + cy$$

Although no factor is common to all four terms, there is a common factor of a in $ax + ay$ and a common factor of c in $cx + cy$. We can factor out the a and c to obtain

$$ax + ay + cx + cy = a(\mathbf{x + y}) + c(\mathbf{x + y})$$
$$= (\mathbf{x + y})(a + c) \quad \text{Factor out } x + y.$$

We can check the result by multiplication.

$$(x + y)(a + c) = ax + cx + ay + cy$$
$$= ax + ay + cx + cy \quad \text{Rearrange the terms.}$$

Thus, $ax + ay + cx + cy$ factors as $(x + y)(a + c)$. This type of factoring is called **factoring by grouping.** Polynomials having four terms can be factored by grouping if the polynomials can be split into two groups of terms and both groups share a common factor.

Factoring by grouping

1. Group the terms of the polynomial so that the first two terms have a common factor and the last two terms have a common factor.

2. Factor out the common factor from each group.

3. Factor out the resulting common binomial factor. If there is no common binomial factor, regroup the terms of the polynomial and repeat steps 2 and 3.

EXAMPLE 11

Factoring by grouping. Factor $2c - 2d + cd - d^2$.

Solution   No factor is common to all four term, but 2 is common to the first two terms and d is common to the last two terms.

$$2c - 2d + cd - d^2 = 2(c - d) + d(c - d) \quad \text{Factor out 2 from } 2c - 2d \text{ and } d \text{ from } cd - d^2.$$

$$= (c - d)(2 + d) \quad \text{Factor out } c - d.$$

We check by verifying that

$$(c - d)(2 + d) = 2c + cd - 2d - d^2$$
$$= 2c - 2d + cd - d^2 \quad \text{Rearrange the terms.}$$

SELF CHECK   Factor $7x - 7y + xy - y^2$.   *Answer:* $(x - y)(7 + y)$ ■

EXAMPLE 12

Factoring by grouping. Factor $x^2y - ax - xy + a$.

Solution   No factor is common to all four terms, but x is common to $x^2y - ax$.

$$x^2y - ax - xy + a = x(xy - a) - xy + a \quad \text{Factor out } x \text{ from } x^2y - ax.$$

If we factor -1 from $-xy + a$, a common binomial factor $xy - a$ appears.

$$x^2y - ax - xy + a = x(xy - a) - 1(xy - a)$$
$$= (xy - a)(x - 1) \quad \text{Factor out } xy - a.$$

Check by multiplication.

SELF CHECK   Factor $7b + 3c - 7bt - 3ct$.   *Answer:* $(7b + 3c)(1 - t)$ ■

 WARNING! When factoring expressions such as those in the previous two examples, don't think that $2(c - d) + d(c - d)$ or $x(xy - a) - 1(xy - a)$ are in factored form. To be in factored form, the result must be a product.

The next example illustrates that when factoring a polynomial, we should always look for a common factor first.

EXAMPLE 13

Factoring out the GCF first. Factor $10k + 10m - 2km - 2m^2$.

Solution   The four terms have a common factor of 2. We factor it out first. Then we use factoring by grouping to factor the polynomial in the parentheses. The first two terms have a common factor of 5. The last two terms have a common factor of $-m$.

$$10k + 10m - 2km - 2m^2 = 2(5k + 5m - km - m^2) \quad \text{Factor out the GCF, 2.}$$
$$= 2[5(k + m) - m(k + m)]$$
$$= 2[(k + m)(5 - m)] \quad \text{Factor out } k + m.$$
$$= 2(k + m)(5 - m)$$

Use multiplication to check the result.

SELF CHECK   Factor $-4t - 4s - 4tz - 4sz$.   *Answer:* $-4(t + s)(1 + z)$ ∎

STUDY SET

Section 5.1

VOCABULARY

In Exercises 1–6, fill in the blanks to make the statements true.

1. A natural number greater than 1 whose only factors are 1 and itself is called a _____ number.

2. When we write 24 as $2^3 \cdot 3$, we say that 24 has been written in _____ form.

3. The GCF of several natural numbers is the _____ number that divides each of the numbers exactly.

4. When we write $15x^2 - 25x$ as $5x(3x - 5)$, we say that we have _____ the greatest common factor.

5. The process of finding the individual factors of a known product is called _____.

6. The numbers 1, 2, 3, 4, 6, and 12 are the natural-number _____ of 12.

CONCEPTS

*In Exercises 7–10, explain what is **wrong** with each solution.*

7. Factor $6a + 9b + 3$.

$$6a + 9b + 3 = 3(2a + 3b + 0)$$
$$= 3(2a + 3b)$$

8. Prime factor 100.

$$10 \underline{| 100}$$
$$5 \underline{| 10}$$
$$2$$
$$100 = 2 \cdot 5 \cdot 10$$

9. Factor out the GCF: $30a^3 - 12a^2$.

$$30a^3 - 12a^2 = 6a(5a^2 - 2a)$$

10. Factor $ab + b + a + 1$.

$$ab + b + a + 1 = b(a + 1) + (a + 1)$$
$$= (a + 1)b$$

---

11. **a.** What property is illustrated here?

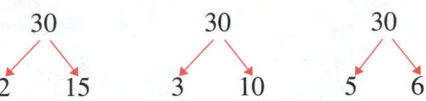

$$2x(x - 3) = 2x^2 - 6x$$

   b. Explain how we use the distributive property in reverse to factor $2x^2 - 6x$.

12. **a.** Complete each tree diagram to prime-factor 30.

     30 → 2, 15   30 → 3, 10   30 → 5, 6

   b. Complete the statement: The fundamental theorem of arithmetic states that every natural number greater than 1 has _____ prime factorization.

13. The prime factorizations of three monomials are shown here. Find their GCF.

$$3 \cdot 3 \cdot 5 \cdot x \cdot x$$
$$2 \cdot 3 \cdot 5 \cdot x \cdot y$$
$$2 \cdot 2 \cdot 3 \cdot x \cdot y \cdot y$$

14. Consider the polynomial $2k - 8 + hk - 4h$.
   a. How many terms does the polynomial have?
   b. Is there a common factor of all the terms?
   c. What is the common factor of the first two terms?

   d. What is the common factor of the last two terms?

15. How can we check the answer of the factoring problem shown below?

     Factor $3j^3 + 6j^2 + 2j + 4$.

$$3j^3 + 6j^2 + 2j + 4 = 3j^2(j + 2) + 2(j + 2)$$
$$= (j + 2)(3j^2 + 2)$$

16. List the first 12 prime numbers.

NOTATION

In Exercises 17–18, complete each factorization.

17. Factor $b^3 - 6b^2 + 2b - 12$.
$$b^3 - 6b^2 + 2b - 12 = \boxed{}(b - 6) + 2\boxed{}$$
$$= (b - 6)\boxed{}$$

18. Factor $12b^3 - 6b^2 + 2b - 12$.

$$12b^3 - 6b^2 + 2b - 12 = \boxed{}(6b^3 - 3b^2 + b - 6)$$

19. In the expression $4x^2y + xy$, what is the implied coefficient of the last term?

20. True or false?

$$-(x^2 - 3x + 1) = -1(x^2 - 3x + 1)$$

PRACTICE

In Exercises 21–32, find the prime factorization of each number.

21. 12	**22.** 24
23. 15	**24.** 20
25. 40	**26.** 62
27. 98	**28.** 112
29. 225	**30.** 144
31. 288	**32.** 968

In Exercises 33–36, complete each factorization.

33. $4a + 12 = \boxed{}(a + 3)$

34. $r^4 + r^2 = r^2\left(\boxed{} + 1\right)$

35. $4y^2 + 8y - 2xy = 2y\left(2y + \boxed{} - \boxed{}\right)$

36. $3x^2 - 6xy + 9xy^2 = \boxed{}\left(\boxed{} - 2y + 3y^2\right)$

In Exercises 37–60, factor out the GCF.

37. $3x + 6$	**38.** $2y - 10$
39. $2\pi R - 2\pi r$	
40. $\frac{1}{3}\pi R^2 h - \frac{4}{3}\pi R^3$	
41. $t^3 + 2t^2$	**42.** $b^3 - 3b^2$
43. $a^3 - a^2$	**44.** $r^3 + r^2$
45. $24x^2y^3 + 8xy^2$	
46. $3x^2y^3 - 9x^4y^3$	
47. $12uvw^3 - 18uv^2w^2$	
48. $14xyz - 16x^2y^2z$	
49. $3x + 3y - 6z$	
50. $2x - 4y + 8z$	
51. $ab + ac - ad$	
52. $rs - rt + ru$	
53. $12r^2 - 3rs + 9r^2s^2$	
54. $6a^2 - 12a^3b + 36ab$	
55. $\pi R^2 - \pi ab$	
56. $\frac{1}{3}\pi R^2 h - \frac{1}{3}\pi rh$	
57. $3(x + 2) - x(x + 2)$	
58. $t(5 - s) + 4(5 - s)$	
59. $h^2(14 + r) + 14 + r$	
60. $k^2(14 + v) - 7(14 + v)$	

In Exercises 61–68, factor out -1 from each polynomial.

61. $-a - b$	**62.** $-x - 2y$
63. $-2x + 5y$	**64.** $-3x + 8z$
65. $-3m - 4n + 1$	
66. $-3r + 2s - 3$	
67. $-3ab - 5ac + 9bc$	
68. $-6yz + 12xz - 5xy$	

In Exercises 69–74, factor each polynomial by factoring out the negative of the GCF.

69. $-3x^2 - 6x$

70. $-4a^2 + 6a$

71. $-4a^2b^3 + 12a^3b^2$

72. $-25x^4y^3 + 30x^2y^3$

73. $-4a^2b^2c^2 + 14a^2b^2c - 10ab^2c^2$

74. $-10x^4y^3z^2 + 8x^3y^2z - 20x^2y$

In Exercises 75–92, factor by grouping.

75. $2x + 2y + ax + ay$

76. $bx + bz + 5x + 5z$

77. $7r + 7s - kr - ks$

78. $9p - 9q + mp - mq$

79. $xr + xs + yr + ys$

80. $pm - pn + qm - qn$

81. $2ax + 2bx + 3a + 3b$

82. $3xy + 3xz - 5y - 5z$

83. $2ab + 2ac + 3b + 3c$

84. $3ac + a + 3bc + b$

85. $6x^2 - 2x - 15x + 5$

86. $6x^2 + 2x + 9x + 3$

87. $9mp + 3mq - 3np - nq$

88. $ax + bx - a - b$

89. $2xy + y^2 - 2x - y$

90. $2xy - 3y^2 + 2x - 3y$

91. $8z^5 + 12z^2 - 10z^3 - 15$

92. $2a^4 + 2a^3 - 4a - 4$

In Exercises 93–98, factor by grouping. Factor out the GCF first.

93. $ax^3 + bx^3 + 2ax^2y + 2bx^2y$

94. $x^3y^2 - 2x^2y^2 + 3xy^2 - 6y^2$

95. $4a^2b + 12a^2 - 8ab - 24a$

96. $-4abc - 4ac^2 + 2bc + 2c^2$

97. $x^3y - x^2y - xy^2 + y^2$

98. $2x^3z - 4x^2z + 32xz - 64z$

APPLICATIONS

99. PICTURE FRAMING The dimensions of a family portrait and the frame in which it is mounted are given in Illustration 1. Write an algebraic expression that describes
a. the area of the picture frame.
b. the area of the portrait.
c. the area of the mat used in the framing. Express the result in factored form.

ILLUSTRATION 1

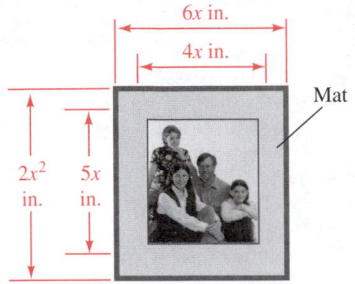

100. REARVIEW MIRRORS The dimensions of the three rearview mirrors on an automobile are given in Illustration 2. Write an algebraic expression that gives
a. the area of the rearview mirror mounted on the windshield.
b. the total area of the two side mirrors.
c. the total area of all three mirrors. Express the result in factored form.

ILLUSTRATION 2

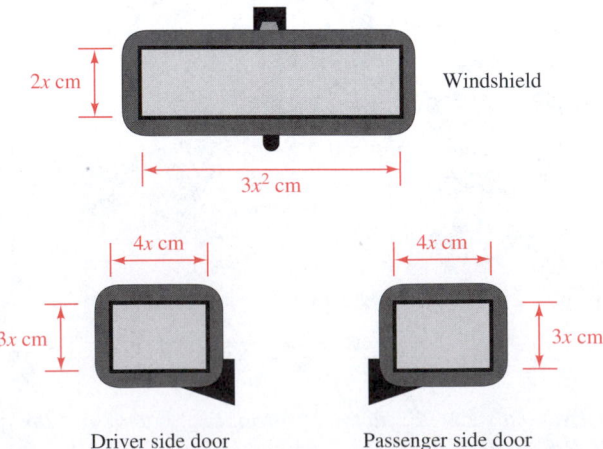

101. COOKING See Illustration 3.
a. What is the length of a side of the square griddle, in terms of r? What is the area of the cooking surface of the griddle, in terms of r?

b. How many square inches of the cooking surface do the pancakes cover, in terms of r?
c. Find the amount of cooking surface that is not covered by the pancakes. Express the result in factored form.

ILLUSTRATION 3

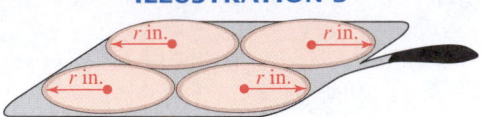

102. U.S. NAVY Illustration 4 shows the deck of the aircraft carrier *Enterprise*. The rectangular-shaped landing area of $(x^3 + 4x^2 + 5x + 20)$ ft^2 is shaded. What is the width of the landing area? (*Hint:* Factor the expression that represents the area.)

ILLUSTRATION 4

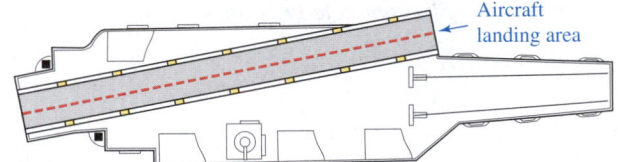

WRITING

103. To add $5x$ and $7x$, we combine like terms: $5x + 7x = 12x$. Explain how this is related to factoring out a common factor.

104. One student commented, "Factoring undoes the distributive property." What do you think she meant? Give an example.

105. If asked to write $ax + ay - bx - by$ in factored form, explain why $a(x + y) - b(x + y)$ is not an acceptable answer.

106. When asked to factor $rx - sy + ry - sx$, a student wrote the expression as $rx + ry - sx - sy$. Then she factored it by grouping. Can the terms be rearranged in this manner? Explain your answer.

REVIEW

107. Find the slope of the line whose equation is $y = 13x - 2$.

108. Find the slope of the line passing through the points $(3, 5)$ and $(-2, -7)$.

109. Does the point $(3, 5)$ lie on the graph of the line $4x - y = 7$?

110. Simplify $(a^{-2}a^5)^3$.

▶ **5.2**

Factoring the Difference of Two Squares

In this section, you will learn about

Factoring the difference of two squares ▪ Multistep factoring

Introduction   When discussing multiplication of polynomials in Section 4.6, we considered the following special product formula:

$$(x - y)(x + y) = x^2 - y^2$$

In this section, we will reverse the multiplication process and factor binomials that are in the form $x^2 - y^2$.

Factoring the Difference of Two Squares

Whenever we multiply a binomial of the form $x + y$ by a binomial of the form $x - y$, we obtain a binomial of the form $x^2 - y^2$.

$$(x + y)(x - y) = x^2 - xy + xy - y^2 \qquad \text{Use the FOIL method.}$$
$$= x^2 - y^2 \qquad \text{Combine like terms: } -xy + xy = 0.$$

The binomial $x^2 - y^2$ is called the **difference of two squares,** because x^2 is the square of x and y^2 is the square of y. The difference of the squares of two quantities always factors into the sum of those two quantities multiplied by the difference of those two quantities.

> **Factoring the difference of two squares**
>
> $$x^2 - y^2 = (x + y)(x - y)$$

If we think of the difference of two squares as the square of a **F**irst quantity minus the square of a **L**ast quantity, we have the formula

$$F^2 - L^2 = (F + L)(F - L)$$

and we say: *To factor the square of a First quantity minus the square of a Last quantity, we multiply the First plus the Last by the First minus the Last.*

To factor $x^2 - 9$, we note that it can be written in the form $x^2 - 3^2$ and use the formula for factoring the difference of two squares:

$$F^2 - L^2 = (F + L)(F - L)$$
$$x^2 - 3^2 = (x + 3)(x - 3) \qquad \text{Substitute } x \text{ for F and 3 for L.}$$

We can check by verifying that $(x + 3)(x - 3) = x^2 - 9$. Because of the commutative property of multiplication, we can also write this factorization as $(x - 3)(x + 3)$.

To factor the difference of two squares, it is helpful to know the integers that are perfect squares. The number 400, for example, is a perfect square, because $20^2 = 400$. The perfect integer squares less than 400 are

1, 4, 9, 16, 25, 36, 49, 64, 81, 100, 121, 144, 169, 196, 225, 256, 289, 324, 361

Expressions containing variables such as $25x^2$ are also perfect squares, because they can be written as the square of a quantity:

$$25x^2 = (5x)^2$$

EXAMPLE 1

Factoring the difference of two squares. Factor $25x^2 - 49$.

Solution We can write $25x^2 - 49$ in the form $(5x)^2 - 7^2$ and use the formula for factoring the difference of two squares:

$$\mathbf{F}^2 - \mathbf{L}^2 = (\mathbf{F} + \mathbf{L})(\mathbf{F} - \mathbf{L})$$
$$(\mathbf{5x})^2 - \mathbf{7}^2 = (\mathbf{5x} + \mathbf{7})(\mathbf{5x} - \mathbf{7}) \quad \text{Substitute } 5x \text{ for F and 7 for L.}$$

We can check by multiplying $5x + 7$ and $5x - 7$.

$$(5x + 7)(5x - 7) = 25x^2 - 35x + 35x - 49$$
$$= 25x^2 - 49$$

SELF CHECK Factor $16a^2 - 81$. *Answer:* $(4a + 9)(4a - 9)$ ∎

EXAMPLE 2

Factoring the difference of two squares. Factor $4y^4 - 121z^2$.

Solution We can write $4y^4 - 121z^2$ in the form $(2y^2)^2 - (11z)^2$ and use the formula for factoring the difference of two squares:

$$\mathbf{F}^2 - \mathbf{L}^2 = (\mathbf{F} + \mathbf{L})(\mathbf{F} - \mathbf{L})$$
$$(\mathbf{2y^2})^2 - (\mathbf{11z})^2 = (\mathbf{2y^2} + \mathbf{11z})(\mathbf{2y^2} - \mathbf{11z})$$

Check by multiplication.

SELF CHECK Factor $9m^2 - 64n^4$. *Answer:* $(3m + 8n^2)(3m - 8n^2)$ ∎

Multistep Factoring

We can often factor out a greatest common factor before factoring the difference of two squares. To factor $8x^2 - 32$, for example, we factor out the GCF of 8 and then factor the resulting difference of two squares.

$$8x^2 - 32 = 8(x^2 - 4) \qquad \text{Factor out 8.}$$
$$= 8(x^2 - 2^2) \qquad \text{Write 4 as } 2^2.$$
$$= 8(x + 2)(x - 2) \quad \text{Factor the difference of two squares.}$$

We can check by multiplication:

$$8(x + 2)(x - 2) = 8(x^2 - 4) \quad \text{Multiply the binomials first.}$$
$$= 8x^2 - 32 \qquad \text{Distribute the 8.}$$

EXAMPLE 3

Factoring out the GCF first. Factor $5a^2x^3y - 20b^2xy$.

Solution We factor out the GCF of $5xy$ and then factor the resulting difference of two squares.

$$5a^2x^3y - 20b^2xy$$
$$= \mathbf{5xy} \cdot a^2x^2 - \mathbf{5xy} \cdot 4b^2 \qquad \text{The GCF is } 5xy.$$
$$= \mathbf{5xy}(a^2x^2 - 4b^2) \qquad \text{Factor out } 5xy.$$
$$= 5xy[(\mathbf{ax})^2 - (\mathbf{2b})^2] \qquad \text{Write } a^2x^2 \text{ as } (ax)^2 \text{ and } 4b^2 \text{ as } (2b)^2.$$
$$= 5xy(\mathbf{ax} + \mathbf{2b})(\mathbf{ax} - \mathbf{2b}) \quad \text{Factor the difference of two squares.}$$

We check by multiplication.

SELF CHECK Factor $6p^2q^2s^2 - 54r^2s^2$. *Answer:* $6s^2(pq + 3r)(pq - 3r)$ ■

Sometimes we must factor a difference of two squares more than once to completely factor a polynomial. For example, the binomial $625a^4 - 81b^4$ can be written in the form $(25a^2)^2 - (9b^2)^2$, which factors as

$$625a^4 - 81b^4 = (25a^2)^2 - (9b^2)^2$$
$$= (25a^2 + 9b^2)(25a^2 - 9b^2)$$

Since the factor $25a^2 - 9b^2$ can be written in the form $(5a)^2 - (3b)^2$, it is the difference of two squares and can be factored as $(5a + 3b)(5a - 3b)$. Thus,

$$625a^4 - 81b^4 = (25a^2 + 9b^2)(5a + 3b)(5a - 3b)$$

 WARNING! The binomial $25a^2 + 9b^2$ is the **sum of two squares,** because it can be written in the form $(5a)^2 + (3b)^2$. If we are limited to integer coefficients and there are no common factors, binomials that are the sum of two squares cannot be factored. If a polynomial cannot be factored using only integers, it is called a **prime polynomial.**

E X A M P L E 4 **Multistep factoring.** Factor $2x^4y - 32y$.

Solution

$$2x^4y - 32y = 2y \cdot x^4 - 2y \cdot 16 \qquad \text{The GCF is } 2y.$$
$$= 2y(x^4 - 16) \qquad \text{Factor out the GCF of } 2y.$$
$$= 2y(x^2 + 4)(x^2 - 4) \qquad \text{Factor } x^4 - 16.$$
$$= 2y(x^2 + 4)(x + 2)(x - 2) \qquad \text{Factor } x^2 - 4. \text{ Note that } x^2 + 4 \text{ does not factor.}$$

SELF CHECK Factor $48a^5 - 3ab^4$. *Answer:* $3a(4a^2 + b^2)(2a + b)(2a - b)$ ■

Example 5 requires the techniques of factoring out a common factor, factoring by grouping, and factoring the difference of two squares.

E X A M P L E 5 **Multistep factoring.** Factor $2x^3 - 8x + 2yx^2 - 8y$.

Solution

$$2x^3 - 8x + 2yx^2 - 8y = 2(x^3 - 4x + yx^2 - 4y) \qquad \text{Factor out 2.}$$
$$= 2[x(x^2 - 4) + y(x^2 - 4)] \qquad \text{Factor out } x \text{ from } x^3 - 4x \text{ and } y \text{ from } yx^2 - 4y.$$
$$= 2[(x^2 - 4)(x + y)] \qquad \text{Factor out } x^2 - 4.$$
$$= 2(x + 2)(x - 2)(x + y) \qquad \text{Factor } x^2 - 4.$$

SELF CHECK Factor $3a^3 - 12a + 3a^2b - 12b$. *Answer:* $3(a + 2)(a - 2)(a + b)$ ■

S T U D Y S E T

Section 5.2

VOCABULARY

Fill in the blanks to make the statements true.

1. A binomial of the form $a^2 - b^2$ is called the _____ of two squares.

2. A binomial of the form $a^2 + b^2$ is called the _____ of two squares.

CONCEPTS

3. Complete the statement:

$$F^2 - L^2 = (F + L)(\qquad)$$

4. a. Multiply: $(x + 5)(x - 5)$.
   b. Multiply: $(x - 5)(x + 5)$.

5. Find the missing part of each statement:
   a. $(?)^2 = 36x^2$
   b. $(?)^2 = 4y^4$
   c. $(?)^2 = 49a^2b^4$

6. List the first 12 integers that are perfect squares.

7. Why is $x^2 - 25$ called a difference of two squares?

8. Is the following statement true or false?
   $(-x^2 + y^2) = -(x^2 - y^2)$

NOTATION

In Exercises 9–10, complete each factorization.

9. $5x^2 - 20y^2 = 5\left(x^2 - \boxed{}\right)$

$$= \boxed{}\,[x^2 - (2y)^2]$$

$$= 5\left(x + \boxed{}\right)\!\left((x - \boxed{}\right)$$

10. $16a^4 - 625b^4 = \left(\boxed{}\right)^2 - \left(\boxed{}\right)^2$

$$= \left(4a^2 + \boxed{}\right)\!\left(\boxed{} - 25b^2\right)$$

$$= \left(\boxed{}\right)[(2a)^2 - (5b)^2]$$

$$= (4a^2 + 25b^2)(2a + 5b)(2a - 5b)$$

PRACTICE

In Exercises 11–16, complete each factorization.

11. $y^2 - 49 = \left(y + \boxed{}\right)\!\left(y - \boxed{}\right)$

12. $y^2 - 81 = \left(\boxed{} + 9\right)\!\left(\boxed{} - 9\right)$

13. $t^2 - w^2 = \left(\boxed{} + \boxed{}\right)(t - w)$

14. $p^4 - q^2 = (p^2 + q)\!\left(\boxed{} - \boxed{}\right)$

15. $25t^2 - 36u^2 = \left(5t + \boxed{}\right)\!\left(\boxed{} - 6u\right)$

16. $49u^2 - 64v^2 = \left(\boxed{} + 8v\right)\!\left(7u - \boxed{}\right)$

In Exercises 17–28, factor each expression, if possible.

17. $x^2 - 16$

18. $x^2 - 25$

19. $4y^2 - 1$

20. $9z^2 - 1$

21. $9x^2 - y^2$

22. $4x^2 - z^2$

23. $16a^2 - 25b^2$

24. $36a^2 - 121b^2$

25. $a^2 + b^2$

26. $121a^2 + 144b^2$

27. $a^4 - 4b^2$

28. $9y^4 - 16z^2$

In Exercises 29–38, factor each expression.

29. $8x^2 - 32y^2$

30. $2a^2 - 200b^2$

31. $2a^2 - 2$

32. $32x^2 - 8$

33. $3r^2 - 12s^2$

34. $45u^2 - 20v^2$

35. $x^3 - xy^2$

36. $a^2b - b^3$

37. $4a^2x - 9b^2x$

38. $4b^2y - 16c^2y$

In Exercises 39–46, factor each expression.

39. $x^4 - 81$

40. $y^4 - 625$

41. $a^4 - 16$

42. $b^4 - 256$

43. $81r^4 - 256s^4$

44. $16y^8 - 81z^4$

45. $2x^4 - 2y^4$

46. $a^5 - ab^4$

In Exercises 47–56, factor each expression.

47. $a^3 - 9a + 3a^2 - 27$

48. $b^3 - 25b - 2b^2 + 50$

49. $y^3 - 16y - 3y^2 + 48$

50. $a^3 - 49a + 2a^2 - 98$

51. $3x^3 - 12x + 3x^2 - 12$

52. $2x^3 - 18x - 6x^2 + 54$

53. $3m^3 - 3mn^2 + 3am^2 - 3an^2$

54. $ax^3 - axy^2 - bx^3 + bxy^2$

55. $2m^3n^2 - 32mn^2 + 8m^2 - 128$

56. $2x^3y + 4x^2y - 98xy - 196y$

APPLICATIONS

57. FALLING OBJECTS The height h of a ball above the ground t seconds after it falls off a 256-foot-tall building is given by the formula

$$h = 256 - 16t^2$$

Factor the right-hand side of the equation.

58. PHYSICS Illustration 1 shows a time-sequence picture of a falling apple. Factor the expression, which gives the difference in the distance the apple falls during the time interval from t_1 to t_2 seconds.

ILLUSTRATION 1

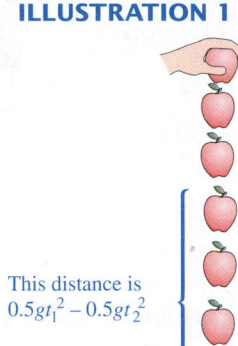

This distance is $0.5gt_1^2 - 0.5gt_2^2$

59. DARTS A circular dartboard has a series of rings around a solid center, called the bullseye. (See Illustration 2.) To find the area of the outer white ring, we can use the formula

$$A = \pi R^2 - \pi r^2$$

Factor the expression on the right-hand side of the equation.

ILLUSTRATION 2

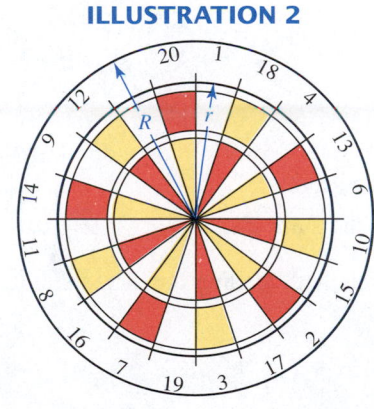

60. SPACE TRAVEL The first Soviet manned spacecraft, Vostok, is shown in Illustration 3. The surface area of the spherical part of the craft is given by $(36\pi r^2 - 48\pi r + 16\pi)$ m^2. Factor the expression.

ILLUSTRATION 3

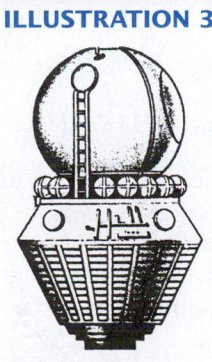

WRITING

61. When asked to factor $x^2 - 25$, one student wrote $(x + 5)(x - 5)$, and another student wrote $(x - 5)(x + 5)$. Are both answers correct? Explain.

62. Write a comment to the student whose work is shown below, explaining the **error** that was made.

Factor $4x^2 - 16y^2$.

$$(2x + 4y)(2x - 4y)$$

REVIEW

In Exercises 63–68, solve each equation.

63. $\dfrac{a}{3} - 6 = -2$

64. $\dfrac{y - 7}{3} = 9$

65. $\dfrac{3t}{4} + 2 = -7$

66. $3(p - 5) = 4p + 36$

67. $\dfrac{2x - 5}{4} = x - 2$

68. $0.2x + 0.4(50 - x) = 19$

In Exercises 69–70, solve each inequality. Give each answer in interval notation.

69. $3(y - 8) < 5y + 6$

70. $\dfrac{3(y - 1)}{4} \geq y + 1$

In Exercises 71–72, solve each formula for the given variable.

71. $C = \dfrac{5}{9}(F - 32)$ for F

72. $P = 2l + 2w$ for l

▶ **5.3**

Factoring Trinomials

In this section, you will learn about

> Factoring trinomials of the form $x^2 + bx + c$ ■ Prime trinomials ■ Factoring trinomials of the form $ax^2 + bx + c$ ■ Factoring trinomials by the grouping method ■ Factoring perfect square trinomials ■ Summary of factoring techniques

Introduction   The product of two binomials is often a trinomial. For example,

$$(x + 3)(x + 3) = x^2 + 6x + 9$$
$$(3x - 4)(2x + 3) = 6x^2 + x - 12$$

For this reason, we should not be surprised that many trinomials factor into the product of two binomials. In this section, we will discuss how to factor trinomials into the product of two binomials. First we will factor trinomials of the form $x^2 + bx + c$, where the coefficient of x^2 is 1. Then we will factor trinomials of the form $ax^2 + bx^2 + c$, where the coefficient of x^2 is not necessarily 1.

Factoring Trinomials of the Form $x^2 + bx + c$

To develop a method for factoring trinomials, we multiply $x + a$ and $x + b$.

$$
\begin{aligned}
(x + a)(x + b) &= x \cdot x + bx + ax + ab && \text{Use the FOIL method.} \\
&= x^2 + ax + bx + ab && \text{Write } x \cdot x \text{ as } x^2. \text{ Write } bx + ax \text{ as } ax + bx. \\
&= x^2 + (a + b)x + ab && \text{Factor } x \text{ out of } ax + bx. \text{ The result has three terms.}
\end{aligned}
$$

First term   Middle term   Last term

From the result, we can see that

■ the first term is the product of x and x,

■ the last term is the product of a and b, and

■ the coefficient of the middle term is the sum of a and b.

We can use these facts to factor trinomials with leading coefficients of 1.

EXAMPLE 1

A positive last term. Factor $x^2 + 5x + 6$.

Solution   To factor $x^2 + 5x + 6$, we will write it as the product of two binomials. Since the first term of the trinomial is x^2, the first term of each of its binomial factors must be x.

$$x^2 + 5x + 6 = \left(x + \right)\left(x + \right)$$

To fill in the blanks, we must find two integers such that

■ their *product* is 6 (because the last term of $x^2 + 5x + 6$ is 6), and

■ their *sum* is 5 (because the coefficient of the middle term of $x^2 + 5x + 6$ is 5).

To determine the integers, we list the two-integer factorizations of 6 in a table. Since the integers must have a positive sum, we need not list $(-1)(-6)$ and $(-2)(-3)$, because their sums are -7 and -5, respectively.

Product of the factors of 6	Sum of the factors of 6
1(6)	$1 + 6 = 7$
2(3)	$2 + 3 = 5$

The last row contains the integers 2 and 3, whose product is 6 and whose sum is 5. So we can fill in the blanks with 2 and 3.

$$x^2 + 5x + 6 = (x + \boxed{2})(x + \boxed{3})$$

To check the result, we find the product of $x + 2$ and $x + 3$ and verify that it is $x^2 + 5x + 6$.

$$(x + 2)(x + 3) = x^2 + 3x + 2x + 6$$
$$= x^2 + 5x + 6$$

SELF CHECK   Factor $y^2 + 7y + 6$.   *Answer:* $(y + 1)(y + 6)$ ■

In Example 1, the factors can be written in either order. An equivalent factorization is

$$x^2 + 5x + 6 = (x + 3)(x + 2)$$

EXAMPLE 2

A positive last term. Factor $y^2 - 7y + 12$.

Solution   Since the first term is y^2, the first term of each factor must be y. The last term of the trinomial is 12 and the coefficient of the middle term is -7. To fill in the blanks, we must find two integers whose product is 12 and whose sum is -7.

$$y^2 - 7y + 12 = (y + \boxed{})(y + \boxed{})$$

The two-integer factorizations of 12 and the sums of the factors are shown in the following table. Since the integers must have a negative sum, we need not list 1(12), 2(6), and 3(4), because their sums are positive 13, positive 8, and positive 7, respectively.

Product of the factors of 12	Sum of the factors of 12
$-1(-12)$	$-1 + (-12) = -13$
$-2(-6)$	$-2 + (-6) = -8$
$-3(-4)$	$-3 + (-4) = -7$

The last row contains the integers -3 and -4, whose product is 12 and whose sum is -7. So we can fill in the blanks with -3 and -4 and simplify.

$$y^2 - 7y + 12 = (y + \boxed{(-3)})(y + \boxed{(-4)})$$
$$= (y - 3)(y - 4)$$

To check the result, we multiply $y - 3$ by $y - 4$ and verify that the product is $y^2 - 7y + 12$.

$$(y - 3)(y - 4) = y^2 - 4y - 3y + 12$$
$$= y^2 - 7y + 12$$

SELF CHECK   Factor $p^2 - 5p + 6$.   *Answer:* $(p - 3)(p - 2)$ ■

EXAMPLE 3

A negative last term. Factor $a^2 + 2a - 15$.

Solution   Since the first term is a^2, the first term of each factor must be a. To fill in the blanks, we must find two integers whose product is -15 and whose sum is 2.

$$a^2 + 2a - 15 = (a + \boxed{})(a + \boxed{})$$

The possible factorizations of -15 and the sum of the factors are shown in the following table.

Product of the factors of −15	Sum of the factors of −15
1(−15)	1 + (−15) = −14
3(−5)	3 + (−5) = −2
5(−3)	5 + (−3) = 2
15(−1)	15 + (−1) = 14

The third row contains the integers 5 and -3, whose product is -15 and whose sum is 2. So we can fill in the blanks with 5 and -3 and simplify.

$$a^2 + 2a - 15 = (a + \boxed{5})(a + \boxed{(-3)})$$
$$= (a + 5)(a - 3)$$

We can check by multiplying $a + 5$ and $a - 3$.

$$(a + 5)(a - 3) = a^2 - 3a + 5a - 15$$
$$= a^2 + 2a - 15$$

SELF CHECK   Factor $p^2 + 3p - 18$.   *Answer:* $(p + 6)(p - 3)$ ■

EXAMPLE 4

A negative last term. Factor $z^2 - 4z - 21$.

Solution   Since the first term is z^2, the first term of each factor must be z. To fill in the blanks, we must find two integers whose product is -21 and whose sum is -4.

$$z^2 - 4y - 21 = (z + \boxed{})(z + \boxed{})$$

The factorizations of -21 and the sums of the factors are shown in the following table.

Product of the factors of −21	Sum of the factors of −21
1(−21)	1 + (−21) = −20
3(−7)	3 + (−7) = −4
7(−3)	7 + (−3) = 4
21(−1)	21 + (−1) = 20

The second row contains the integers 3 and -7, whose product is -21 and whose sum is -4. So we can fill in the blanks with 3 and -7 and simplify.

$$z^2 - 4z - 21 = (z + \boxed{3})(z + \boxed{(-7)})$$
$$= (z + 3)(z - 7)$$

To check, we multiply $z + 3$ and $z - 7$.

$$(z + 3)(z - 7) = z^2 - 7z + 3z - 21$$
$$= z^2 - 4z - 21$$

SELF CHECK   Factor $q^2 - 2q - 24$.   *Answer:* $(q + 4)(q - 6)$ ∎

When the coefficient of the first term is -1, we begin by factoring out -1.

EXAMPLE 5   **Factoring out -1 first.** Factor $-h^2 + 2h + 15$.

Solution   We factor out -1 and then factor $h^2 - 2h - 15$.

$$-h^2 + 2h + 15 = \mathbf{-1}(h^2 - 2h - 15) \quad \text{Factor out } -1.$$
$$= -(h^2 - 2h - 15)$$
$$= -(h - 5)(h + 3) \quad \text{Use the integers } -5 \text{ and } 3, \text{ because their}$$
$$\text{product is } -15 \text{ and their sum is } -2.$$

We check by multiplying.

$$-(h - 5)(h + 3) = -(h^2 + 3h - 5h - 15) \quad \text{Use the FOIL method first.}$$
$$= -(h^2 - 2h - 15)$$
$$= -h^2 + 2h + 15$$

SELF CHECK   Factor $-x^2 + 11x - 18$.   *Answer:* $-(x - 9)(x - 2)$ ∎

Prime Trinomials

If a trinomial cannot be factored using only integers, it is called a **prime trinomial**.

EXAMPLE 6   **Trinomials that do not factor.** Factor $x^2 + 2x + 3$, if possible.

Solution   To factor the trinomial, we must find two integers whose product is 3 and whose sum is 2. The possible factorizations of 3 and the sums of the factors are shown in the following table.

Product of the factors of 3	Sum of the factors of 3
$1(3)$	$1 + 3 = 4$
$-1(-3)$	$-1 + (-3) = -4$

Since two integers whose product is 3 and whose sum is 2 do not exist, $x^2 + 2x + 3$ cannot be factored. It is a prime trinomial.

SELF CHECK   Factor $x^2 - 4x + 6$, if possible.   *Answer:* not possible; prime trinomial ∎

Factoring Trinomials of the Form $ax^2 + bx + c$

We must consider more combinations of factors when we factor trinomials with leading coefficients other than 1.

EXAMPLE 7

A leading coefficient of 2. Factor $2x^2 + 5x + 3$.

Solution
Since the first term is $2x^2$, the first term of the binomial factors must be $2x$ and x. To fill in the blanks, we must find two factors of 3 that will give a middle term of $5x$.

$$(2x + \boxed{})(x + \boxed{})$$

Because each term of the trinomial is positive, we need only consider positive factors of the last term (3). Since the positive factors of 3 are 1 and 3, there are two possible factorizations:

$$(2x + 1)(x + 3) \qquad \text{or} \qquad (2x + 3)(x + 1)$$

The first possibility is incorrect, because it gives a middle term of $7x$. The second possibility is correct, because it gives a middle term of $5x$. Thus,

$$2x^2 + 5x + 3 = (2x + 3)(x + 1)$$

Check by multiplication.

SELF CHECK
Factor $3x^2 + 7x + 2$. *Answer:* $(3x + 1)(x + 2)$ ■

EXAMPLE 8

A leading coefficient of 6. Factor $6x^2 - 17x + 5$.

Solution
Since the first term is $6x^2$, the first terms of the binomial factors must be $6x$ and x or $3x$ and $2x$. To fill in the blanks, we must find two factors of 5 that will give a middle term of $-17x$.

$$(6x + \boxed{})(x + \boxed{}) \qquad \text{or} \qquad (3x + \boxed{})(2x + \boxed{})$$

Because the sign of the third term is positive and the sign of the middle term is negative, we need only consider negative factors of the last term (5). Since the negative factors of 5 are -1 and -5, there are four possible factorizations:

$$(6x - 1)(x - 5) \qquad\qquad (6x - 5)(x - 1)$$
$$\textcolor{red}{(3x - 1)(2x - 5)} \qquad\qquad (3x - 5)(2x - 1)$$

Only the possibility printed in color gives the correct middle term of $-17x$. Thus,

$$6x^2 - 17x + 5 = (3x - 1)(2x - 5)$$

Check by multiplication.

SELF CHECK
Factor $6x^2 - 7x + 2$. *Answer:* $(3x - 2)(2x - 1)$ ■

EXAMPLE 9

A leading coefficient of 3. Factor $3y^2 - 4y - 4$.

Solution
Since the first term is $3y^2$, the first terms of the binomial factors must be $3y$ and y. To fill in the blanks, we must find two factors of -4 that will give a middle term of $-4y$.

$$(3y + \boxed{})(y + \boxed{})$$

Since the sign of the third term is negative, the signs inside the binomial factors will be different. Because the factors of the last term (4) are 1, 2, and 4, there are six possibilities to consider:

$$(3y + 1)(y - 4) \qquad\qquad (3y + 4)(y - 1)$$
$$(3y - 1)(y + 4) \qquad\qquad (3y - 4)(y + 1)$$
$$(3y - 2)(y + 2) \qquad\qquad \textcolor{red}{(3y + 2)(y - 2)}$$

Only the possibility printed in color gives the correct middle term of $-4y$. Thus,

$$3y^2 - 4y - 4 = (3y + 2)(y - 2)$$

Check by multiplication.

SELF CHECK   Factor $5a^2 - 7a - 6$. *Answer:* $(5a + 3)(a - 2)$ ∎

EXAMPLE 10   **A leading coefficient of 6.** Factor $6b^2 + 7b - 20$.

Solution   Since the first term is $6b^2$, the first terms of the binomial factors must be $6b$ and b or $3b$ and $2b$. To fill in the blanks, we must find two factors of -20 that will give a middle term of $7b$.

$$\left(6b + \right)\left(b + \right) \quad \text{or} \quad \left(3b + \right)\left(2b + \right)$$

Since the factors of the last term (20) are 1, 2, 5, 10, and 20, there are many possible combinations for the last terms of the binomial factors. We must try to find one that will (in combination with our choice of first terms) give a last term of -20 and a sum of the products of the outer terms and inner terms of $7b$.

If we pick factors of $6b$ and b for the first terms and -5 and 4 for the last terms, we have

$$(6b - 5)(b + 4)$$

$$\begin{array}{r} -5b \\ 24b \\ \hline 19b \end{array}$$

which gives a middle term of $19b$, so it is incorrect.

If we pick factors of $3b$ and $2b$ for the first terms and 4 and -5 for the last terms, we have

$$(3b + 4)(2b - 5)$$

$$\begin{array}{r} 8b \\ -15b \\ \hline -7b \end{array}$$

which gives a middle term of $-7b$, so it is incorrect.

If we pick factors of $3b$ and $2b$ for the first terms and -4 and 5 for the last terms, we have

$$(3b - 4)(2b + 5)$$

$$\begin{array}{r} -8b \\ 15b \\ \hline 7b \end{array}$$

which gives the correct middle term of $7b$ and a last term of -20, so it is correct.

$$6b^2 + 7b - 20 = (3b - 4)(2b + 5)$$

Check by multiplication.

SELF CHECK   Factor $4x^2 + 4x - 3$. *Answer:* $(2x + 3)(2x - 1)$ ∎

Because some guesswork is often necessary, it is difficult to give specific rules for factoring trinomials. However, the following hints are often helpful.

1. Write the trinomial in descending powers of the variable.

2. Factor out any GCF (including -1 if that is necessary to make the coefficient of the first term positive).

3. If the sign of the third term is positive, the signs between the terms of the binomial factors are the same as the sign of the middle term. If the sign of the third term is negative, the signs between the terms of the binomial factors are opposite.

4. Try combinations of first terms and last terms until you find one that works, or until you exhaust all the possibilities. If no combination works, the trinomial is prime.

5. Check the factorization by multiplication.

E X A M P L E   1 1

Multistep factoring. Factor $2x^2 - 8x^3 + 3x$.

Solution **Step 1:** Write the trinomial in descending powers of x.

$$-8x^3 + 2x^2 + 3x$$

Step 2: Factor out the negative (opposite) of the GCF, which is $-x$.

$$-8x^3 + 2x^2 + 3x = -x(8x^2 - 2x - 3)$$

Step 3: Because the sign of the third term of the trinomial factor is negative, the signs within its binomial factors will be different.

Step 4: Find the binomial factors of the trinomial.

$$-8x^3 + 2x^2 + 3x = -x(8x^2 - 2x - 3)$$
$$= -x(2x + 1)(4x - 3)$$

Step 5: Check by multiplication.

$$-x(2x + 1)(4x - 3) = -x(8x^2 - 6x + 4x - 3)$$
$$= -x(8x^2 - 2x - 3)$$
$$= -8x^3 + 2x^2 + 3x$$
$$= 2x^2 - 8x^3 + 3x$$

SELF CHECK Factor $12y - 2y^3 - 2y^2$. *Answer:* $-2y(y + 3)(y - 2)$ ■

Factoring Trinomials by the Grouping Method

We can use the method of factoring by grouping to factor trinomials of the form $ax^2 + bx + c$. For example, to use this method to factor $6x^2 + 7x - 3$, we first write the trinomial in the form

$$6x^2 + 7x - 3 = 6x^2 + \boxed{} x + \boxed{} x - 3$$

First Middle Last
term terms term

To find the numbers to write in the blanks, we proceed as follows:

1. Find the product ac: $6(-3) = -18$. This product is called the **key number**.
2. Find two factors of the key number -18 whose sum is $b = 7$:

$$9(-2) = -18 \quad \text{and} \quad 9 + (-2) = 7$$

3. Use the factors 9 and -2 to fill in the blanks:

$$6x^2 + 7x - 3 = 6x^2 + \boxed{9}\, x + \boxed{(-2)}\, x - 3$$
$$= 6x^2 + 9x - 2x - 3$$

4. Factor by grouping:

$$6x^2 + 9x - 2x - 3 = 3x(2x + 3) - 1(2x + 3)$$
$$= (2x + 3)(3x - 1)$$

A check will show that the factorization is correct.

EXAMPLE 12

Factoring by grouping. Factor $6b^2 - 17b + 5$.

Solution First, we write the trinomial in the form

$$6b^2 - 17b + 5 = 6b^2 + \boxed{}\, b + \boxed{}\, b + 5$$

and find the key number: $ac = 6(5) = 30$. We then find two factors of 30 whose sum is -17.

$$-2(-15) = 30 \quad \text{and} \quad -2 + (-15) = -17$$

We use the factors -2 and -15 to fill in the blanks:

$$6b^2 - 17b + 5 = 6b^2 + \boxed{(-2)}\, b + \boxed{(-15)}\, b + 5$$
$$= 6b^2 - 2b - 15b + 5$$

We can now factor by grouping.

$$6b^2 - 2b - 15b + 5 = 2b(3b - 1) - 5(3b - 1)$$
$$= (3b - 1)(2b - 5)$$

SELF CHECK Factor $21t^2 - 13t + 2$.   *Answer:* $(7t - 2)(3t - 1)$ ■

Factoring Perfect Square Trinomials

In Section 4.6, we discussed the following special product formulas used to square binomials.

Special product formulas

$$(x + y)^2 = x^2 + 2xy + y^2$$
$$(x - y)^2 = x^2 - 2xy + y^2$$

In words, the first formula states that *the square of the sum of two quantities is the square of the first quantity, plus twice the product of the quantities, plus the square of the second quantity.*

In words, the second formula states that *the square of the difference of two quantities is the square of the first quantity, minus twice the product of the quantities, plus the square of the second quantity.*

These formulas can be used in reverse order to factor perfect square trinomials.

$$x^2 + 2xy + y^2 = (x + y)^2$$
$$x^2 - 2xy + y^2 = (x - y)^2$$

In words, the first formula states that *if a trinomial is the square of one quantity, plus the square of a second quantity, plus twice the product of the quantities, it factors into the square of the sum of the quantities.*

The second formula states that *if a trinomial is the square of one quantity, plus the square of a second quantity, minus twice the product of the quantities, it factors into the square of the difference of the quantities.*

The trinomials on the left-hand sides of the previous equations are called **perfect square trinomials,** because they are the results of squaring a binomial. Although we can factor perfect square trinomials by using the techniques discussed earlier in this section, we can usually factor them by inspecting their terms. For example, $9x^2 + 24x + 16$ is a perfect square trinomial, because

- The first term $9x^2$ is the square of $3x$: $(3x)^2 = 9x^2$.
- The last term 16 is the square of 4: $4^2 = 16$.
- The middle term $24x$ is twice the product of $3x$ and 4.

Thus,

$$9x^2 + 24x + 16 = (3x)^2 + 2(3x)(4) + 4^2$$
$$= (3x + 4)^2$$

EXAMPLE 13

Factoring perfect square trinomials. Factor $4x^2 - 20x + 25$.

Solution $4x^2 - 20x + 25$ is a perfect square trinomial, because

- The first term $4x^2$ is the square of $2x$: $(2x)^2 = 4x^2$.
- The last term 25 is the square of 5: $5^2 = 25$.
- The middle term $-20x$ is the negative of twice the product of $2x$ and 5.

Thus,

$$4x^2 - 20x + 25 = (2x)^2 - 2(2x)(5) + 5^2$$
$$= (2x - 5)^2$$

SELF CHECK   Factor $x^2 + 10x + 25$.   *Answer:* $(x + 5)^2$ ■

Summary of Factoring Techniques

To identify a random factoring problem, we follow these steps.

Steps for factoring a polynomial

1. Factor out all common factors.
2. If a polynomial has two terms, check to see if it is the difference of two squares:

   $$x^2 - y^2 = (x + y)(x - y)$$

3. If a polynomial has three terms, check to see if it is a perfect square trinomial:

   $$x^2 + 2xy + y^2 = (x + y)^2$$
   $$x^2 - 2xy + y^2 = (x - y)^2$$

   If the trinomial is not a perfect square, attempt to factor it as a general trinomial.

STUDY SET

Section 5.3

VOCABULARY

In Exercises 1–6, fill in the blanks to make the statements true.

1. A polynomial, such as $x^2 - x - 6$, that has exactly three terms is called a _____.

2. A polynomial, such as $x - 3$, that has exactly two terms is called a _____.

3. The statement $x^2 - x - 12 = (x - 4)(x + 3)$ shows that $x^2 - x - 12$ _____ into the product of two binomials.

4. Since $10 = (-5)(-2)$, we say -5 and -2 are _____ of 10.

5. A _____ polynomial cannot be factored by using only integers.

6. If a trinomial is the square of a binomial, it is called a _____ square trinomial.

CONCEPTS

In Exercises 7–12, fill in the blanks to make the statements true.

7. Two factorizations of 4 that involve only positive numbers are _____ and _____.

8. Two factorizations of 4 that involve only negative numbers are _____ and _____.

9. $x^2 + 2xy + y^2 =$ _____.

10. $x^2 - 2xy + y^2 =$ _____.

11. Before attempting to factor a trinomial, be sure that the exponents are written in _____ order.

12. Before attempting to factor a trinomial into two binomials, always factor out any common _____ first.

13. Complete the table.

Product of the factors of 8	Sum of the factors of 8
1(8)	
2(4)	
−1(−8)	
−2(−4)	

14. If we use the FOIL method to do the multiplication $(x + 5)(x + 4)$, we obtain $x^2 + 9x + 20$.
    a. What step of the FOIL process produced 20?

    b. What steps of the FOIL process produced $9x$?

15. Given $x^2 - 2x - 15$:
    a. What is the coefficient of the x^2-term?
    b. What is the last term? The last term is the product of what two integers?

    c. What is the coefficient of the middle term? It is the sum of what two integers?

16. Given $x^2 + 8x + 15$:
    a. What is the coefficient of the x^2-term?
    b. What is the last term? The last term is the product of what two integers?

    c. What is the coefficient of the middle term? It is the sum of what two integers?

NOTATION

In Exercises 17–18, complete each factorization.

17. $3x^2 + 15x + 18 = (x^2 + 5x + 6)$
    $$= 3(x +)(x +)$$

18. $2y^3 - y^2 - 6y = (2y^2 - y - 6)$
    $$= y(+ 3)(- 2)$$

PRACTICE

In Exercises 19–24, complete each factorization.

19. $x^2 + 3x + 2 = (x +)(x +)$
20. $y^2 + 4y + 3 = (y 3)(y +)$
21. $t^2 - 9t + 14 = (- 7)(t -)$
22. $c^2 - 9c + 8 = (- 8)(c -)$
23. $a^2 + 6a - 16 = (a 8)(a 2)$
24. $x^2 - 3x - 40 = (x 8)(x 5)$

In Exercises 25–42, factor each trinomial. If a trinomial can't be factored, write "prime."

25. $z^2 + 12z + 11$
26. $x^2 + 7x + 10$
27. $m^2 - 5m + 6$
28. $n^2 - 7n + 10$

29. $a^2 - 4a - 5$

30. $b^2 + 6b - 7$

31. $x^2 + 5x - 24$

32. $t^2 - 5t - 50$

33. $a^2 - 10a - 39$

34. $r^2 - 9r - 12$

35. $u^2 + 10u + 15$

36. $v^2 + 9v + 15$

37. $s^2 + 11s - 26$

38. $y^2 + 8y + 12$

39. $r^2 - 2r + 4$

40. $m^2 + 3m - 10$

41. $m^2 - m - 12$

42. $u^2 + u - 42$

In Exercises 43–50, factor each trinomial.

43. $x^2 + 6x + 9$

44. $x^2 + 10x + 25$

45. $y^2 - 8y + 16$

46. $z^2 - 2z + 1$

47. $t^2 + 20t + 100$

48. $r^2 + 24r + 144$

49. $u^2 - 18u + 81$

50. $v^2 - 14v + 49$

In Exercises 51–56, factor each trinomial. Factor out -1 first.

51. $-x^2 - 7x - 10$

52. $-x^2 + 9x - 20$

53. $-t^2 - 15t + 34$

54. $-t^2 - t + 30$

55. $-r^2 + 14r - 40$

56. $-r^2 + 14r - 45$

In Exercises 57–62, complete each factorization.

57. $3a^2 + 13a + 4 = (3a +)(a + 4)$

58. $2b^2 + 7b + 6 = (2b + 3)(b +)$

59. $4z^2 - 13z + 3 = (z -)(- 1)$

60. $4t^2 - 4t + 1 = (- 1)(- 1)$

61. $2m^2 + 5m - 12 = (2m 3)(m 4)$

62. $10u^2 - 13u - 3 = (2u 3)(5u 1)$

In Exercises 63-78, factor each trinomial.

63. $2x^2 - 3x + 1$

64. $2y^2 - 7y + 3$

65. $6y^2 + 7y + 2$

66. $4x^2 + 8x + 3$

67. $6x^2 - 7x + 2$

68. $4z^2 - 9z + 2$

69. $2x^2 - 3x - 2$

70. $12y^2 - y - 1$

71. $10y^2 - 3y - 1$

72. $6m^2 + 19m + 3$

73. $12y^2 - 5y - 2$

74. $10x^2 + 21x - 10$

75. $5t^2 + 13t + 6$

76. $16y^2 + 10y + 1$

77. $16m^2 - 14m + 3$

78. $16x^2 + 16x + 3$

In Exercises 79–84, write each trinomial in descending powers of one variable and factor.

79. $4 - 5x + x^2$

80. $y^2 + 5 + 6y$

81. $10y + 9 + y^2$

82. $x^2 - 13 - 12x$

83. $-r^2 + 2 + r$

84. $u^2 - 3 + 2u$

In Exercises 85–92, completely factor each trinomial. Factor out any common monomials first (including -1 if necessary).

85. $2x^2 + 10x + 12$

86. $3y^2 - 21y + 18$

87. $3y^3 + 6y^2 + 3y$

88. $4x^4 + 16x^3 + 16x^2$

89. $-5a^2 + 25a - 30$

90. $-2b^2 + 20b - 18$

91. $3z^2 - 15z + 12$

92. $5m^2 + 45m - 50$

WRITING

93. Explain how you would write a trinomial with its powers in descending order.

94. Explain how to use the FOIL method to check the factoring of a trinomial.

95. Two students factor $2x^2 + 20x + 42$ and get two different answers:

$$(2x + 6)(x + 7) \quad \text{and} \quad (x + 3)(2x + 14)$$

Do both answers check? Why don't they agree? Is either answer completely correct? Explain.

96. Find and explain the error:

$x = y$	Given this fact.
$x^2 = xy$	Multiply both sides by x.
$x^2 - y^2 = xy - y^2$	Subtract y^2 from both sides.
$(x + y)(x - y) = y(x - y)$	Factor.
$x + y = y$	Divide both sides by $(x - y)$.
$y + y = y$	Substitute y for its equal, x.
$2y = y$	Combine like terms.
$2 = 1$	Divide both sides by y.

REVIEW

In Exercises 97–100, graph the solution of each inequality on a number line.

97. $x - 3 > 5$

98. $x + 4 \leq 3$

99. $-3x - 5 \geq 4$

100. $2x - 3 < 7$

▶ **5.4**

Quadratic Equations

In this section, you will learn about

  Quadratic equations ■ Solving quadratic equations by factoring
  ■ Applications

Introduction   Equations that involve first-degree polynomials, such as $9x - 6 = 0$, are called *linear equations*. Equations that involve second-degree polynomials, such as $9x^2 - 6x = 0$, are called **quadratic equations.** In this section, we will define quadratic equations and learn how to solve many of them by factoring.

Quadratic Equations

If a polynomial contains one variable with an exponent to the second (but no higher) power, it is called a **second-degree polynomial.** Equations in which a second-degree polynomial is equal to zero are called **quadratic equations.** Some examples are

$$9x^2 - 6x = 0, \qquad x^2 - 2x - 63 = 0, \qquad \text{and} \qquad 2x^2 + 3x - 2 = 0$$

Quadratic equations

A **quadratic equation** is an equation that can be written in the form

$$ax^2 + bx + c = 0 \quad (a \neq 0) \quad \text{This form is called \textbf{quadratic form.}}$$

where a, b, and c are real numbers.

Quadratic equations don't always appear in quadratic form. To write a quadratic equation such as $21x = 10 - 10x^2$ in $ax^2 + bx + c = 0$ form, we use the addition and subtraction properties of equality to get 0 on the right-hand side.

$21x = 10 - 10x^2$	This is a quadratic equation: It contains one variable with an exponent of 2, but none larger.
$10x^2 + 21x = 10 - 10x^2 + 10x^2$	Add $10x^2$ to both sides.
$10x^2 + 21x = 10$	On the right-hand side, combine like terms: $-10x^2 + 10x^2 = 0$.
$10x^2 + 21x - 10 = 0$	Subtract 10 from both sides.

When it is written in quadratic form, we see that the equation $21x = 10 - 10x^2$ is $10x^2 + 21x - 10 = 0$, where $a = 10$, $b = 21$, and $c = -10$.

The techniques we have used to solve linear equations cannot be used to solve a quadratic equation, because those techniques cannot isolate x on one side of the equation. However, we can often solve quadratic equations using factoring and the following property of real numbers.

The zero-factor property of real numbers

Suppose a and b represent two real numbers. Then

If $ab = 0$, then $a = 0$ or $b = 0$.

The zero-factor property states that when the product of two numbers is zero, at least one of them must be zero. For example, the equation $(x - 4)(x + 5) = 0$ indicates that a product is equal to zero. By the zero-factor property, one of the factors must be zero:

$$x - 4 = 0 \quad \text{or} \quad x + 5 = 0$$

We can solve each of these linear equations to get

$$x = 4 \quad \text{or} \quad x = -5$$

The equation $(x - 4)(x + 5) = 0$ has two solutions: 4 and -5.

Solving Quadratic Equations by Factoring

We can use the following steps to solve a quadratic equation by factoring.

Factoring method

1. Write the equation in $ax^2 + bx + c = 0$ form (called **quadratic form**).
2. Factor the left-hand side of the equation.
3. Use the zero-factor property to set each factor equal to zero.
4. Solve each resulting linear equation.
5. Check the solutions in the original equation.

EXAMPLE 1   **Solving quadratic equations by factoring.** Solve $9x^2 - 6x = 0$.

Solution   We begin by factoring the left-hand side of the equation.

$$9x^2 - 6x = 0$$
$$3x(3x - 2) = 0 \quad \text{Factor out the GCF of } 3x.$$

By the zero-factor property, we have

$$3x = 0 \quad \text{or} \quad 3x - 2 = 0$$

We can solve each of the linear equations to get

$$x = 0 \quad \text{or} \quad x = \frac{2}{3}$$

To check, we substitute the results for x in the original equation and simplify.

For $x = 0$

$$9x^2 - 6x = 0$$

$$9(0)^2 - 6(0) \stackrel{?}{=} 0$$

$$0 - 0 \stackrel{?}{=} 0$$

$$0 = 0$$

For $x = \frac{2}{3}$

$$9x^2 - 6x = 0$$

$$9\left(\frac{2}{3}\right)^2 - 6\left(\frac{2}{3}\right) \stackrel{?}{=} 0$$

$$9\left(\frac{4}{9}\right) - 6\left(\frac{2}{3}\right) \stackrel{?}{=} 0$$

$$4 - 4 \stackrel{?}{=} 0$$

$$0 = 0$$

Both solutions check.

SELF CHECK   Solve $5x^2 + 10x = 0$.   *Answer:* $0, -2$ ■

Sometimes we must factor a difference of two squares to solve a quadratic equation.

EXAMPLE 2   **Solving quadratic equations by factoring.** Solve $4x^2 = 36$.

Solution   Before we can use the zero-factor property, we must subtract 36 from both sides to make the right-hand side zero.

$$4x^2 = 36$$

$$4x^2 - 36 = 0 \qquad \text{Subtract 36 from both sides.}$$

$$x^2 - 9 = 0 \qquad \text{Divide both sides by 4.}$$

$$(x + 3)(x - 3) = 0 \qquad \text{Factor the difference of two squares.}$$

$$x + 3 = 0 \quad \text{or} \quad x - 3 = 0 \qquad \text{Set each factor equal to zero.}$$

$$x = -3 \qquad\qquad x = 3 \qquad \text{Solve each linear equation.}$$

Check each possible solution by substituting it into the original equation.

For $x = -3$

$$4x^2 = 36$$

$$4(-3)^2 \stackrel{?}{=} 36$$

$$4(9) \stackrel{?}{=} 36$$

$$36 = 36$$

For $x = 3$

$$4x^2 = 36$$

$$4(3)^2 \stackrel{?}{=} 36$$

$$4(9) \stackrel{?}{=} 36$$

$$36 = 36$$

Both solutions check.

SELF CHECK   Solve $9x^2 - 81 = 0$.   *Answer:* $3, -3$ ■

EXAMPLE 3   **Solving quadratic equations by factoring.** Solve $x^2 - 2x - 63 = 0$.

Solution   In this case, we must factor a trinomial to solve the equation.

$$x^2 - 2x - 63 = 0$$

$$(x + 7)(x - 9) = 0 \qquad \text{Factor the trinomial } x^2 - 2x - 63.$$

$$x + 7 = 0 \quad \text{or} \quad x - 9 = 0 \qquad \text{Set each factor equal to zero.}$$

$$x = -7 \qquad\qquad x = 9 \qquad \text{Solve each linear equation.}$$

The solutions are -7 and 9. Check each one.

SELF CHECK   Solve $x^2 + 5x + 6 = 0$.   *Answer:* $-2, -3$ ■

EXAMPLE 4

Writing an equation in quadratic form. Solve $2x^2 + 3x = 2$.

Solution We write the equation in the form $ax^2 + bx + c = 0$ and then solve for x.

$$2x^2 + 3x = 2$$

$$2x^2 + 3x - 2 = 0$$ Subtract 2 from both sides so that the right-hand side is zero.

$$(2x - 1)(x + 2) = 0$$ Factor $2x^2 + 3x - 2$.

$$2x - 1 = 0 \quad \text{or} \quad x + 2 = 0$$ Set each factor equal to zero.

$$2x = 1 \qquad\qquad x = -2$$ Solve each linear equation.

$$x = \frac{1}{2}$$

Check each solution.

SELF CHECK

Solve $3x^2 - 6 = -7x$. *Answer:* $\frac{2}{3}, -3$ ∎

EXAMPLE 5

A repeated solution. Solve $-2 = \frac{1}{2}x(9x - 12)$.

Solution First, we need to write the equation in the form $ax^2 + bx + c = 0$.

$$-2 = \frac{1}{2}x(9x - 12)$$

$$-4 = x(9x - 12)$$ Multiply both sides of the equation by 2 to clear the equation of the fraction.

$$-4 = 9x^2 - 12x$$ Remove parentheses.

$$0 = 9x^2 - 12x + 4$$ Add 4 to both sides to make the left-hand side zero.

$$0 = (3x - 2)(3x - 2)$$ Factor the trinomial.

$$3x - 2 = 0 \quad \text{or} \quad 3x - 2 = 0$$ Set each factor equal to zero.

$$3x = 2 \qquad\qquad 3x = 2$$ Add 2 to both sides.

$$x = \frac{2}{3} \qquad\qquad x = \frac{2}{3}$$ Divide both sides by 3.

The equation has two solutions that are the same. We call $\frac{2}{3}$ a *repeated solution*. Check by substituting it into the original equation.

SELF CHECK

Solve $\frac{1}{3}x(4x + 12) = -3$. *Answer:* $-\frac{3}{2}, -\frac{3}{2}$ ∎

EXAMPLE 6

An equation with three solutions. Solve $6x^3 + 12x = 17x^2$.

Solution This is not a quadratic equation, because it contains the term x^3. However, we can solve it using factoring and an extension of the zero-factor property.

$$6x^3 + 12x = 17x^2$$

$$6x^3 - 17x^2 + 12x = 0$$ Add $-17x^2$ to both sides to get 0 on the right-hand side.

$$x(6x^2 - 17x + 12) = 0$$ Factor out the GCF of x.

$$x(2x - 3)(3x - 4) = 0$$ Factor $6x^2 - 17x + 12$.

$$x = 0 \quad \text{or} \quad 2x - 3 = 0 \qquad \text{or} \qquad 3x - 4 = 0$$ Set each factor equal to zero.

$$2x = 3 \qquad\qquad 3x = 4$$ Solve the linear equations.

$$x = \frac{3}{2} \qquad\qquad x = \frac{4}{3}$$

This equation has three solutions. Check all three.

SELF CHECK Solve $10x^3 + x^2 - 2x = 0$. *Answer:* $0, \frac{2}{5}, -\frac{1}{2}$ ■

Applications

The solutions of many problems involve the use of quadratic equations.

E X A M P L E 7

Softball. A softball pitcher can throw a "fastball" underhand at about 55 mph (80 feet per second). If she throws a ball up into the air with that velocity, as in Figure 5-2, its height h in feet, t seconds after being released, is given by the formula

$$h = 80t - 16t^2$$

After the ball is thrown, in how many seconds will it hit the ground?

FIGURE 5-2

Solution When the ball hits the ground, its height will be zero. Thus, we set h equal to zero and solve for t.

$$h = 80t - 16t^2$$
$$0 = 80t - 16t^2$$
$$0 = 16t(5 - t) \quad \text{Factor out the GCF of } 16t.$$

$16t = 0 \quad \text{or} \quad 5 - t = 0 \quad$ Set each factor equal to zero.

$t = 0 \qquad\qquad t = 5 \quad$ Solve each linear equation.

When $t = 0$, the ball's height above the ground is 0 feet. When $t = 5$, the height is again 0 feet, and the ball has hit the ground. The solution is 5 seconds. ■

E X A M P L E 8

Perimeter of a rectangle. Assume that the rectangle in Figure 5-3 has an area of 52 square centimeters and that its length is 1 centimeter more than 3 times its width. Find the perimeter of the rectangle.

FIGURE 5-3

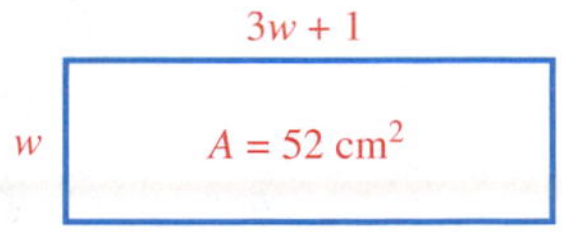

$3w + 1$

w | $A = 52 \text{ cm}^2$

ANALYZE THE PROBLEM The area of the rectangle is 52 square centimeters. Recall that the formula that gives the area of a rectangle is $A = lw$. To find the perimeter of the rectangle, we need to know its length and width. We are told that its length is related to its width; the length is 1 centimeter more than 3 times the width.

FORM AN EQUATION Let w represent the width of the rectangle. Then $3w + 1$ represents its length. Because the area is 52 square centimeters, we substitute 52 for A and $3w + 1$ for l in the formula $A = lw$.

$$A = lw$$
$$52 = (3w + 1)w$$

SOLVE THE EQUATION Now we solve the equation for w.

$$52 = (3w + 1)w \qquad \text{The equation to solve.}$$
$$52 = 3w^2 + w \qquad \text{Remove parentheses.}$$
$$0 = 3w^2 + w - 52 \qquad \text{Subtract 52 from both sides to make the left-hand side zero.}$$
$$0 = (3w + 13)(w - 4) \qquad \text{Factor the trinomial.}$$

$3w + 13 = 0 \quad \text{or} \quad w - 4 = 0 \quad$ Set each factor equal to zero.

$3w = -13 \qquad\qquad w = 4 \quad$ Solve each linear equation.

$$w = -\frac{13}{3}$$

STATE THE CONCLUSION   Since the width cannot be negative, we discard the result $w = -\frac{13}{3}$. Thus, the width of the rectangle is 4, and the length is given by

$$3w + 1 = 3(4) + 1 \quad \text{Substitute 4 for } w.$$
$$= 12 + 1$$
$$= 13$$

The dimensions of the rectangle are 4 centimeters by 13 centimeters. We find the perimeter by substituting 13 for l and 4 for w in the formula for the perimeter of a rectangle.

$$P = 2l + 2w$$
$$= 2(13) + 2(4)$$
$$= 26 + 8$$
$$= 34$$

The perimeter of the rectangle is 34 centimeters.

CHECK THE RESULT   A rectangle with dimensions of 13 centimeters by 4 centimeters does have an area of 52 square centimeters, and the length is 1 centimeter more than 3 times the width. A rectangle with these dimensions has a perimeter of 34 centimeters.   ■

STUDY SET

Section 5.4

VOCABULARY

In Exercises 1–2, fill in the blanks to make the statements true.

1. Any equation that can be written in the form $ax^2 + bx + c = 0$ is called a _____ equation.

2. To _____ a binomial or trinomial means to write it as a product.

CONCEPTS

In Exercises 3–6, fill in the blanks to make the statements true.

3. When the product of two numbers is zero, at least one of them is _____. Symbolically, we can state this: If $ab = 0$, then $a = $ ▢ or $b = $ ▢.

4. The techniques used to solve linear equations cannot be used to solve quadratic equations, because those techniques cannot _____ the variable on one side of the equation.

5. To write a quadratic equation in *quadratic form* means that one side of the equation must be _____ and the other side must be in the form $ax^2 + bx + c$.

6. In the quadratic equation $ax^2 + bx + c = 0$, $a \neq$ ▢.

7. Classify each equation as quadratic or linear.
   a. $3x^2 + 4x + 2 = 0$ **b.** $3x + 7 = 0$
   c. $2 = -16 - 4x$ **d.** $-6x + 2 = x^2$

8. Check to see whether the given number is a solution of the given quadratic equation.
   a. $x^2 - 4x = 0$; $x = 4$
   b. $x^2 + 2x - 4 = 0$; $x = -2$
   c. $4x^2 - x + 3 = 0$; $x = 1$

9. **a.** Evaluate $x^2 + 6x - 16$ for $x = 0$.
   b. Factor $x^2 + 6x - 16$.
   c. Solve $x^2 + 6x - 16 = 0$.

10. The equation $3x^2 - 4x + 5 = 0$ is written in $ax^2 + bx + c = 0$ form. What are a, b, and c?

11. What is the first step that should be performed to solve each equation?
    a. $x^2 + 7x = -6$
    b. $\frac{1}{2}x(x + 7) = -3$

12. **a.** How many solutions does the linear equation $2a + 3 = 2$ have?
    b. How many solutions does the quadratic equation $2a^2 + 3a = 2$ have?

NOTATION

In Exercises 13–14, complete each solution.

13. $7y^2 + 14y = 0$

 $\boxed{}(y + 2) = 0$

 $7y = 0$ or $\boxed{} = 0$

 $y = 0$ $y = -2$

14. $12p^2 - p - 6 = 0$

 $\left(\boxed{} - 3\right)\left(3p + \boxed{}\right) = 0$

 $\boxed{} = 0$ or $3p + 2 = \boxed{}$

 $4p = \boxed{}$ $3p = \boxed{}$

 $p = \dfrac{3}{4}$ $p = -\dfrac{2}{3}$

PRACTICE

In Exercises 15–72, solve each equation.

15. $(x - 2)(x + 3) = 0$

16. $(x - 3)(x - 2) = 0$

17. $(2s - 5)(s + 6) = 0$

18. $(3h - 4)(h + 1) = 0$

19. $(x - 1)(x + 2)(x - 3) = 0$

20. $(x + 2)(x + 3)(x - 4) = 0$

21. $x(x - 3) = 0$ 22. $x(x + 5) = 0$

23. $x(2x - 5) = 0$ 24. $x(5x + 7) = 0$

25. $w^2 - 7w = 0$ 26. $p^2 + 5p = 0$

27. $3x^2 + 8x = 0$ 28. $5x^2 - x = 0$

29. $8s^2 - 16s = 0$ 30. $15s^2 - 20s = 0$

31. $x^2 - 25 = 0$ 32. $x^2 - 36 = 0$

33. $4x^2 - 1 = 0$ 34. $9y^2 - 1 = 0$

35. $9y^2 - 4 = 0$ 36. $16z^2 - 25 = 0$

37. $x^2 = 100$ 38. $z^2 = 25$

39. $4x^2 = 81$ 40. $9y^2 = 64$

41. $x^2 - 13x + 12 = 0$

42. $x^2 + 7x + 6 = 0$

43. $x^2 - 4x - 21 = 0$

44. $x^2 + 2x - 15 = 0$

45. $x^2 - 9x + 8 = 0$

46. $x^2 - 14x + 45 = 0$

47. $a^2 + 8a = -15$

48. $a^2 - a = 56$

49. $2y - 8 = -y^2$

50. $-3y + 18 = y^2$

51. $x^3 + 3x^2 + 2x = 0$

52. $x^3 - 7x^2 + 10x = 0$

53. $k^3 - 27k - 6k^2 = 0$

54. $j^3 - 22j - 9j^2 = 0$

55. $(x - 1)(x^2 + 5x + 6) = 0$

56. $(x - 2)(x^2 - 8x + 7) = 0$

57. $2x^2 - 5x + 2 = 0$

58. $2x^2 + x - 3 = 0$

59. $5x^2 - 6x + 1 = 0$

60. $6x^2 - 5x + 1 = 0$

61. $4r^2 + 4r = -1$

62. $9m^2 + 6m = -1$

63. $-15x^2 + 2 = -7x$

64. $-8x^2 - 10x = -3$

65. $\dfrac{1}{2}x(2x - 3) = 10$

66. $\dfrac{1}{2}x(2x - 3) = 7$

67. $(d + 1)(8d + 1) = 18d$

68. $4h(3h + 2) = h + 12$

69. $2x(3x^2 + 10x) = -6x$

70. $2x^3 = 2x(x + 2)$

71. $x^3 + 7x^2 = x^2 - 9x$

72. $x^2(x + 10) = 2x(x - 8)$

APPLICATIONS

In Exercises 73–74, an object has been thrown straight up into the air. The formula $h = vt - 16t^2$ gives the height h of the object above the ground after t seconds, when it is thrown upward with an initial velocity v.

73. TIME OF FLIGHT After how many seconds will the object hit the ground if it is thrown with a velocity of 144 feet per second?

74. TIME OF FLIGHT After how many seconds will the object hit the ground if it is thrown with a velocity of 160 feet per second?

75. OFFICIATING Before a football game, a coin toss is used to determine which team will kick off. See Illustration 1. The height *h* (in feet) of a coin above the ground *t* seconds after being flipped up into the air is given by

ILLUSTRATION 1

 $h = -16t^2 + 22t + 3$

How long does a team captain have to call heads or tails if it must be done while the coin is in the air?

76. DOLPHINS See Illustration 2. The height h in feet reached by a dolphin t seconds after breaking the surface of the water is given by

$$h = -16t^2 + 32t$$

How long will it take the dolphin to jump out of the water and touch the trainer's hand?

ILLUSTRATION 2

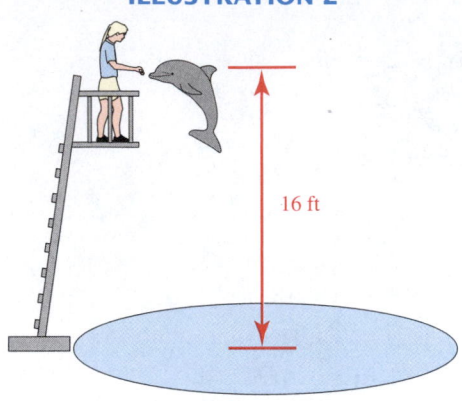

16 ft

77. EXHIBITION DIVING In Acapulco, Mexico, men diving from a cliff to the water 64 feet below are quite a tourist attraction. A diver's height, h, above the water t seconds after diving is given by $h = -16t^2 + 64$. How long does a dive last?

78. FORENSIC MEDICINE The kinetic energy E of a moving object is given by $E = \frac{1}{2}mv^2$, where m is the mass of the object (in kilograms) and v is the object's velocity (in meters per second). Kinetic energy is measured in joules. Examining the damage done to a victim, a police pathologist determines that the energy of a 3-kilogram mass at impact was 54 joules. Find the velocity at impact.

79. CHOREOGRAPHY For the finale of a musical, 36 dancers are to assemble in a triangular-shaped series of rows, where each successive row has one more dancer than the previous row. Illustration 3 shows the beginning of such a formation. The relationship between the number of rows r and the number of dancers d is given by

$$d = \frac{1}{2}r(r + 1)$$

Determine the number of rows in the formation.

ILLUSTRATION 3

80. CRAFTS Illustration 4 shows how a geometric wall hanging can be created by stretching yarn from peg to peg across a wooden ring. The relationship between the number of pegs p placed evenly around the ring and the number of yarn segments s that criss-cross the ring is given by the formula

$$s = \frac{p(p - 3)}{2}$$

How many pegs are needed if the designer wants 27 segments to criss-cross the ring?

ILLUSTRATION 4

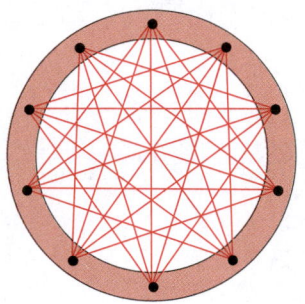

81. INSULATION The area of the rectangular slab of foam insulation in Illustration 5 is 36 square meters. Find the dimensions of the slab.

ILLUSTRATION 5

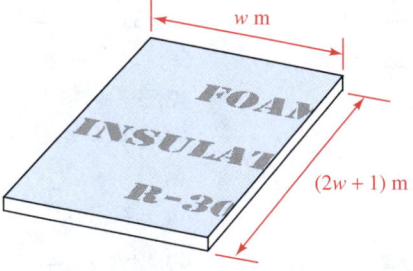

w m

$(2w + 1)$ m

82. SHIPPING PALLETS The length of a rectangular shipping pallet is 2 feet less than 3 times its width. Its area is 21 square feet. Find the dimensions of the pallet.

83. COOKING The electric griddle shown in Illustration 6 has a cooking surface of 160 square inches. Find the length and the width of the griddle.

ILLUSTRATION 6

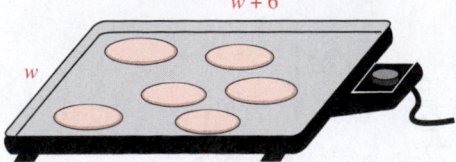

$w + 6$

w

84. INTEGER PROBLEM The product of two consecutive even integers is 288. Find the integers. (*Hint:* Let x = the first even integer. Then represent the second even integer in terms of x.)

85. WINTER RECREATION The length of the rectangular ice-skating rink in Illustration 7 is 20 meters greater than twice its width. Find the width.

ILLUSTRATION 7

Area = 6,000 m² w

86. TRAFFIC ACCIDENTS Investigators at a traffic accident used the function $d(v) = 0.04v^2 + 0.8v$, where v is the velocity of the car (in mph) and $d(v)$ is the stopping distance of the car (in feet), to reconstruct the events leading up to a collision. From physical evidence, it was concluded that it took one car 32 feet to stop. At what velocity was the car traveling prior to the accident?

87. DESIGNING A TENT The length of the base of the triangular sheet of canvas above the door of the tent in Illustration 8 is 2 feet more than twice its height. The area is 30 square feet. Find the height and the length of the base of the triangle.

ILLUSTRATION 8

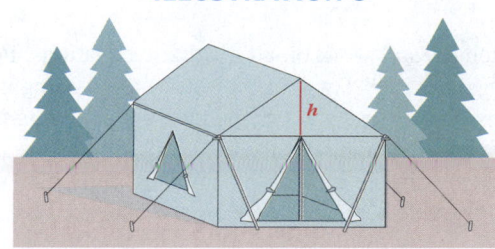

88. DIMENSIONS OF A TRIANGLE The height of a triangle is 2 inches less than 5 times the length of its base. The area is 36 square inches. Find the length of the base and the height of the triangle.

89. TUBING A piece of cardboard in the shape of a parallelogram is twisted to form the tube for a roll of paper towels. (See Illustration 9.) The parallelogram has an area of 60 square inches. If its height h is 7 inches more than the length of the base b, what is the circumference of the tube? (*Hint:* The formula for the area of a paralellogram is $A = bh$.)

ILLUSTRATION 9

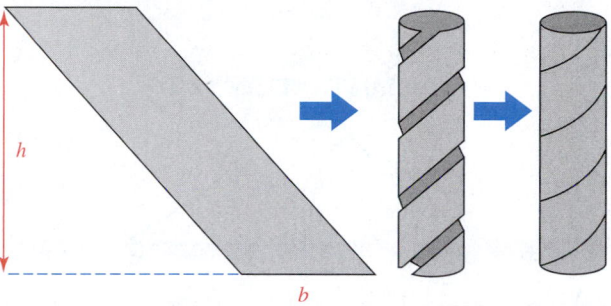

h b

90. SWIMMING POOL BORDER The owners of the rectangular swimming pool in Illustration 10 want to surround the pool with a crushed-stone border of uniform width. They have enough stone to cover 74 square meters. How wide should they make the border? (*Hint:* The area of the larger rectangle minus the area of the smaller is the area of the border.)

ILLUSTRATION 10

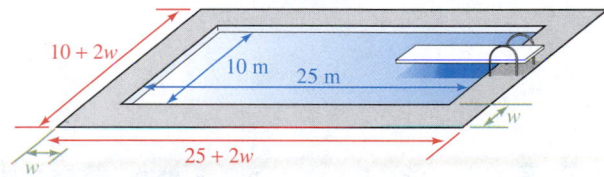

$10 + 2w$ 10 m 25 m w $25 + 2w$ w

91. HOUSE CONSTRUCTION The formula for the area of a trapezoid is

$$A = \frac{h(B + b)}{2}$$

The area of the trapezoidal truss in Illustration 11 is 24 square meters. Find the height of the trapezoid if one base is 8 meters and the other base is the same as the height.

ILLUSTRATION 11

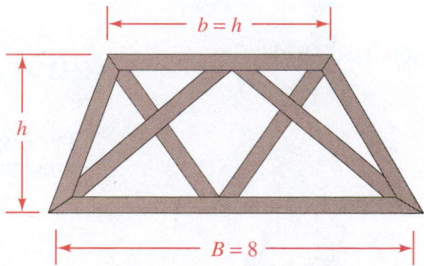

$b = h$ h $B = 8$

92. VOLUME OF A PYRAMID The volume of a pyramid is given by the formula

$$V = \frac{Bh}{3}$$

where B is the area of its base and h is its height. The volume of the pyramid in Illustration 12 is 192 cubic centimeters. Find the dimensions of its rectangular base if one edge of the base is 2 centimeters longer than the other and the height of the pyramid is 12 centimeters.

ILLUSTRATION 12

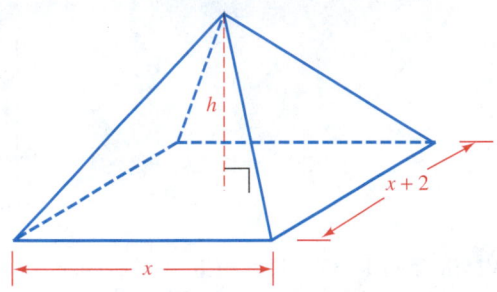

WRITING

93. What is **wrong** with the logic used by a student to "solve" $x^2 + x = 6$?

$$x(x + 1) = 6$$
$$x = 6 \quad \text{or} \quad x + 1 = 6$$
$$x = 5$$
So the solutions are
6 or 5.

94. Suppose that to find the length of the base of a triangle, you write a quadratic equation and solve it to find $b = 6$ or $b = -8$. Explain why one solution should be discarded.

REVIEW

95. A doctor advises one patient to exercise at least 15 minutes but less than 30 minutes per day. Use a compound inequality to express the range of these times in minutes.

96. A bag of peanuts is worth $0.30 less than a bag of cashews. Equal amounts of peanuts and cashews are used to make 40 bags of a mixture that is worth $1.05 per bag. How much is a bag of cashews worth?

97. A rectangle is 3 times as long as it is wide, and its perimeter is 120 centimeters. Find its area.

98. A woman invests $15,000, part at 7% annual interest and part at 8% annual interest. If she receives $1,100 interest per year, how much did she invest at 7%?

▶ 5.5   The Quadratic Formula

In this section, you will learn about

- Square roots ■ The square root method ■ The quadratic formula
- ■ Applications

Introduction The factoring method doesn't work on all quadratic equations. For example, the trinomial in the equation $x^2 + 5x + 1 = 0$ cannot be factored using integer coefficients. To solve such equations, we can use another method, called *the quadratic formula*. To solve quadratic equations by the quadratic formula, we will use the *square root method,* a method used for solving simple quadratic equations. Since the square root method involves square roots, we will begin by reviewing square roots.

Square Roots

FIGURE 5-4

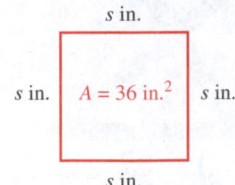

Suppose we know that the area of the square shown in Figure 5-4 is 36 square inches. To find the length of each side, we substitute 36 for A in the formula $A = s^2$ and solve for s.

$$A = s^2 \quad \text{The formula for the area of a square.}$$
$$36 = s^2 \quad \text{Substitute 36 for } A.$$

To solve for s, we must find a positive number whose square is 36. Since 6 is such a number, the sides of the square are 6 inches long. The number 6 is called a *square root* of 36 . because $6^2 = 36$.

Here are more examples of square roots.

- 3 is a square root of 9, because $3^2 = 9$.
- -3 is a square root of 9, because $(-3)^2 = 9$.
- 12.5 is a square root of 156.25, because $(12.5)^2 = 156.25$.
- -12.5 is a square root of 156.25, because $(-12.5)^2 = 156.25$.
- 0 is a square root of 0, because $0^2 = 0$.

In general, we have the following.

Square root

The number b is a **square root of a** if $b^2 = a$.

All positive numbers have two square roots. The two square roots of 9 are 3 and -3. The two square roots of 156.25 are 12.5 and -12.5. Zero is the only number with a single square root, which is 0.

The symbol $\sqrt{156.25}$ represents the positive square root of 156.25, which is 12.5. The symbol $-\sqrt{12.5}$ represents the negative square root of 156.25, which is -12.5. The symbol $\sqrt{}$ is called a **radical sign**, and the number under a radical sign is called a **radicand**.

$$\sqrt{156.25} = 12.5 \qquad \text{and} \qquad -\sqrt{156.25} = -12.5$$

Recall that to find square roots, we can use a calculator.

ACCENT ON TECHNOLOGY *Finding Square Roots*

We can use the square root key on a calculator to find positive square roots of numbers. For example, to find $\sqrt{315.725}$ with a scientific calculator, we enter 315.725 and press the $\boxed{\sqrt{}}$ key.

Keystrokes   315.725 $\boxed{\sqrt{}}$ $\boxed{\text{17.76865217}}$

To use a graphing calculator, we press the $\boxed{\sqrt{}}$ key, enter 315.725, and press $\boxed{\text{ENTER}}$.

Keystrokes $\boxed{\text{2nd}}$ $\boxed{\sqrt{}}$ 315.725 $\boxed{\text{ENTER}}$ $\boxed{\begin{array}{l}\sqrt{(315.725}\\ 17.76865217\end{array}}$

To the nearest thousandth, $\sqrt{315.725} = 17.769$.

EXAMPLE 1   **Finding square roots.** Use a calculator.

a. $\sqrt{0} = 0$ b. $\sqrt{25} = 5$

c. $\sqrt{2.25} = 1.5$ d. $-\sqrt{1.5376} = -1.24$

e. $-\sqrt{610.09} = -24.7$ f. $\sqrt{2,533.45} \approx 50.33338852$

Find each square root: **a.** $\sqrt{121}$, **b.** $\sqrt{56.25}$, *Answers:* **a.** 11, **b.** 7.5, **c.** $-\sqrt{0.64}$, and **d.** $\sqrt{1,075.32}$. **c.** -0.8, **d.** approximately 32.79207221 ∎

The Square Root Method

If $x^2 = 9$, x is a number whose square is 9. Since $3^2 = 9$ and $(-3)^2 = 9$, the equation $x^2 = 9$ has two solutions, $x = \sqrt{9} = 3$ and $x = -\sqrt{9} = -3$. In general, any equation of the form $x^2 = c$ ($c > 0$) has two solutions.

The square root method

> If $c > 0$, the equation $x^2 = c$ has two solutions:
>
> $$x = \sqrt{c} \quad \text{or} \quad x = -\sqrt{c}$$

We can write this result with double-sign notation. The equation $x = \pm\sqrt{c}$ (read as "x equals plus or minus $\sqrt{c}$") means that $x = \sqrt{c}$ or $x = -\sqrt{c}$.

EXAMPLE 2 **Solving equations using the square root method.** Solve $x^2 = 16$.

Solution We use the square root method to find that the equation $x^2 = 16$ has two solutions:

$$x = \sqrt{16} \quad \text{or} \quad x = -\sqrt{16}$$
$$x = 4 \qquad\qquad x = -4$$

Using double-sign notation, we have $x = \pm 4$.

Check: For $x = 4$ **For $x = -4$**
$$x^2 = 16 \qquad\qquad x^2 = 16$$
$$4^2 \stackrel{?}{=} 16 \qquad\qquad (-4)^2 \stackrel{?}{=} 16$$
$$16 = 16 \qquad\qquad 16 = 16$$

Solve $x^2 = 25$. *Answer:* ± 5 ∎

The equation in Example 2 can also be solved by factoring.

$$x^2 = 16$$
$$x^2 - 16 = 0 \qquad \text{Subtract 16 from both sides.}$$
$$(x + 4)(x - 4) = 0 \qquad \text{Factor the difference of two squares.}$$
$$x + 4 = 0 \quad \text{or} \quad x - 4 = 0$$
$$x = -4 \qquad\qquad x = 4$$

EXAMPLE 3 **Solving equations using the square root method.** Solve $3x^2 - 12 = 0$.

Solution We first isolate x^2.

$$3x^2 - 12 = 0$$
$$3x^2 = 12 \quad \text{Add 12 to both sides to isolate } 3x^2.$$
$$x^2 = 4 \quad \text{Divide both sides by 3 to isolate } x.$$

We can find the two solutions of this equation by the square root method.

$$x = \sqrt{4} \quad \text{or} \quad x = -\sqrt{4}$$

$$x = 2 \qquad\qquad x = -2 \quad \textcolor{red}{\sqrt{4} = 2.}$$

Check: *For x = 2*   *For x = −2*

$$3x^2 - 12 = 0 \qquad\qquad 3x^2 - 12 = 0$$

$$3(2)^2 - 12 \overset{?}{=} 0 \qquad 3(-2)^2 - 12 \overset{?}{=} 0$$

$$3(4) - 12 \overset{?}{=} 0 \qquad 3(4) - 12 \overset{?}{=} 0$$

$$12 - 12 \overset{?}{=} 0 \qquad 12 - 12 \overset{?}{=} 0$$

$$0 = 0 \qquad\qquad 0 = 0$$

SELF CHECK Solve $3x^2 - 300 = 0$.   *Answer:* ± 10 ■

EXAMPLE 4 **Solving equations using the square root method.** Solve $x^2 + 2x + 1 = 9$.

Solution We begin by factoring the trinomial on the left-hand side to get

$$(x + 1)^2 = 9$$

We can find the two solutions of this equation by the square root method.

$$x + 1 = \sqrt{9} \quad \text{or} \quad x + 1 = -\sqrt{9}$$

$$x + 1 = 3 \qquad\qquad x + 1 = -3 \quad \textcolor{red}{\sqrt{9} = 3.}$$

$$x = 2 \qquad\qquad x = -4 \quad \textcolor{red}{\text{Subtract 1 from both sides.}}$$

Check each solution.

SELF CHECK Solve $x^2 + 4x + 4 = 4$.   *Answer:* $0, -4$ ■

EXAMPLE 5 **Solving equations using the square root method.** Solve $(x - 2)^2 - 18 = 0$. Give each result to the nearest thousandth.

Solution
$$(x - 2)^2 - 18 = 0$$

$$(x - 2)^2 = 18 \quad \textcolor{red}{\text{Add 18 to both sides to isolate } (x - 2)^2.}$$

The two solutions are

$$x - 2 = \sqrt{18} \qquad\qquad \text{or} \qquad x - 2 = -\sqrt{18}$$

$$x = 2 + \sqrt{18} \qquad\qquad x = 2 - \sqrt{18} \quad \textcolor{red}{\text{Add 2 to both sides.}}$$

$$x \approx 2 + 4.242640687 \qquad x \approx 2 - 4.242640687 \quad \textcolor{red}{\sqrt{18} \approx 4.242640687.}$$

$$x \approx 6.242640687 \qquad\qquad x \approx -2.242640687 \quad \textcolor{red}{\text{Do the operations.}}$$

$$x \approx 6.243 \qquad\qquad\qquad x \approx -2.243 \quad \textcolor{red}{\text{Round to the nearest thousandth.}}$$

Check: Because the results that will be substituted for x are approximations, the left-hand side of each part of the check might not be exactly zero. Usually, it will be an approximation of zero.

For x ≈ 6.243   *For x ≈ −2.243*

$$(x - 2)^2 - 18 = 0 \qquad\qquad (x - 2)^2 - 18 = 0$$

$$(\textbf{6.243} - 2)^2 - 18 \overset{?}{=} 0 \qquad (\textbf{-2.243} - 2)^2 - 18 \overset{?}{=} 0$$

$$(4.243)^2 - 18 \overset{?}{=} 0 \qquad (-4.243)^2 - 18 \overset{?}{=} 0$$

$$18.003049 - 18 \overset{?}{=} 0 \qquad 18.003049 - 18 \overset{?}{=} 0$$

$$0.003049 \approx 0 \qquad\qquad 0.003049 \approx 0$$

Since 0.003049 is approximately equal to zero, the check suggests that the answers are correct.

Solve $(x + 3)^2 - 12 = 0$. Give each result to the nearest thousandth. *Answer: 0.464, −6.464* ■

EXAMPLE 6

Hurricane. In 1998, Hurricane Mitch dumped heavy rains on Central America, causing extensive flooding in Honduras, Belize, and Guatemala. Figure 5-5 shows the position of the storm on October 26, at which time the weather service estimated that it covered an area of about 71,000 square miles. What was the diameter of the storm?

Solution We can use the formula for the area of a circle to find the *radius* of the circular-shaped storm.

$$A = \pi r^2$$

$$\mathbf{71{,}000} = \pi r^2 \qquad \text{Substitute 71,000 for the area } A.$$

$$\frac{71{,}000}{\pi} = r^2 \qquad \text{Divide both sides by } \pi \text{ to isolate } r^2.$$

FIGURE 5-5

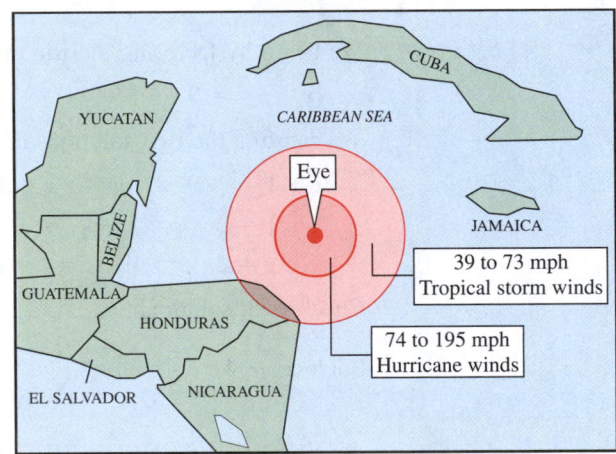

Now we use the square root method to solve for r.

$$r = \sqrt{\frac{71{,}000}{\pi}} \qquad \text{or} \qquad r = -\sqrt{\frac{71{,}000}{\pi}}$$

Using a calculator to approximate the square root, we have

$$r \approx 150.3329702 \qquad \text{The second solution is discarded, because the radius cannot be negative.}$$

If we multiply the radius by 2, we find that the diameter of the storm was about 300 miles.

■

The Quadratic Formula

To develop the quadratic formula, we can use factoring and the square root method to solve the **general quadratic equation** $ax^2 + bx + c = 0 \; (a \neq 0)$.

$$ax^2 + bx + c = 0$$

$$ax^2 + bx = -c \qquad \text{Subtract } c \text{ from both sides.}$$

$$\mathbf{4a}(ax^2) + \mathbf{4a}(bx) = (\mathbf{4a})(-c) \qquad \text{Multiply both sides by } 4a.$$

$$4a^2x^2 + 4abx + \mathbf{b^2} = \mathbf{b^2} - 4ac \qquad \text{Simplify and add } b^2 \text{ to both sides.}$$

$$(2ax + b)(2ax + b) = b^2 - 4ac \qquad \text{Factor the left-hand side.}$$

$$(2ax + b)^2 = b^2 - 4ac \qquad \text{Write } (2ax + b)(2ax + b) \text{ as } (2ax + b)^2.$$

$$2ax + b = \pm\sqrt{b^2 - 4ac} \qquad \text{Use the square root method.}$$

$$2ax = -b \pm \sqrt{b^2 - 4ac} \qquad \text{Subtract } b \text{ from both sides.}$$

$$x = \frac{-b \pm \sqrt{b^2 - 4ac}}{2a} \qquad \text{Since } a \neq 0, \text{ we can divide both sides by } 2a.$$

This solution is called the **quadratic formula.**

Quadratic formula

The solutions of the quadratic equation $ax^2 + bx + c = 0$ are

$$x = \frac{-b \pm \sqrt{b^2 - 4ac}}{2a} \qquad (a \neq 0)$$

 WARNING! When you write the quadratic formula, be careful to draw the fraction bar so that it includes the complete numerator. Do not write

$$x = -b \pm \frac{\sqrt{b^2 - 4ac}}{2a}$$

EXAMPLE 7

Solving equations using the quadratic formula. Solve $x^2 + 5x + 6 = 0$.

Solution   The equation is written in $ax^2 + bx + c = 0$ form with $a = 1$, $b = 5$, and $c = 6$. We substitute these values into the quadratic formula and simplify.

$$x = \frac{-b \pm \sqrt{b^2 - 4ac}}{2a} \qquad \text{The quadratic formula.}$$

$$= \frac{-5 \pm \sqrt{5^2 - 4(1)(6)}}{2(1)} \qquad \text{Substitute 1 for } a, \text{ 5 for } b, \text{ and 6 for } c.$$

$$= \frac{-5 \pm \sqrt{25 - 24}}{2} \qquad \text{Evaluate the power and do the multiplication within the radical.}$$

$$= \frac{-5 \pm \sqrt{1}}{2} \qquad \text{Do the subtraction within the radical.}$$

$$x = \frac{-5 \pm 1}{2} \qquad \text{Simplify: } \sqrt{1} = 1.$$

This notation represents two solutions. We simplify them separately, first using the $+$ sign and then using the $-$ sign.

$$x = \frac{-5 + 1}{2} \qquad \text{or} \qquad x = \frac{-5 - 1}{2}$$

$$x = \frac{-4}{2} \qquad\qquad\qquad x = \frac{-6}{2}$$

$$x = -2 \qquad\qquad\qquad x = -3$$

Check both solutions.

SELF CHECK   Solve $x^2 - 4x - 12 = 0$.   *Answer:* 6, -2 ∎

 WARNING! Be sure to write a quadratic equation in quadratic form $(ax^2 + bx + c = 0)$ before identifying the values of a, b, and c.

EXAMPLE 8

Writing equations in quadratic form. Solve $2x^2 = 5x + 3$.

Solution   We begin by writing the equation in quadratic form.

$$2x^2 = 5x + 3$$

$$2x^2 - 5x - 3 = 0 \qquad \text{Subtract } 5x \text{ and 3 from both sides.}$$

In this equation, $a = 2$, $b = -5$, and $c = -3$. We substitute these values into the quadratic formula and simplify.

$$x = \frac{-b \pm \sqrt{b^2 - 4ac}}{2a}$$ The quadratic formula.

$$= \frac{-(-5) \pm \sqrt{(-5)^2 - 4(2)(-3)}}{2(2)}$$ Substitute 2 for a, -5 for b, and -3 for c.

$$= \frac{5 \pm \sqrt{25 - (-24)}}{4}$$ $-(-5) = 5$. Evaluate the power and do the multiplication within the radical.

$$= \frac{5 \pm \sqrt{49}}{4}$$ Do the subtraction within the radical; $25 - (-24) = 25 + 24 = 49$.

$$= \frac{5 \pm 7}{4}$$ Simplify: $\sqrt{49} = 7$.

Thus,

$$x = \frac{5 + 7}{4} \qquad \text{or} \qquad x = \frac{5 - 7}{4}$$ Evaluate each solution separately.

$$x = \frac{12}{4} \qquad\qquad\qquad x = \frac{-2}{4}$$

$$x = 3 \qquad\qquad\qquad x = -\frac{1}{2}$$

Check both solutions.

SELF CHECK   Solve $4x^2 + 4x = 3$.   *Answer:* $\frac{1}{2}$, $-\frac{3}{2}$ ■

E X A M P L E   9   **Approximating solutions.** Solve $5x^2 = -7x - 1$. Round each solution to the nearest hundredth.

Solution   To determine a, b, and c, we must write the equation in $ax^2 + bx + c = 0$ form.

$$5x^2 = -7x - 1$$

$$5x^2 + 7x = -7x - 1 + 7x$$ To eliminate $-7x$ from the right-hand side, add $7x$ to both sides.

$$5x^2 + 7x = -1$$ Combine like terms: $-7x + 7x = 0$.

$$5x^2 + 7x + 1 = -1 + 1$$ To get 0 on the right-hand side, add 1 to both sides.

$$5x^2 + 7x + 1 = 0$$ Do the addition: $-1 + 1 = 0$.

In this equation, $a = 5$, $b = 7$, and $c = 1$. We substitute these values into the quadratic formula and simplify.

$$x = \frac{-b \pm \sqrt{b^2 - 4ac}}{2a}$$ The quadratic formula.

$$= \frac{-7 \pm \sqrt{7^2 - 4(5)(1)}}{2(5)}$$ Substitute 5 for a, 7 for b, and 1 for c.

$$= \frac{-7 \pm \sqrt{49 - 20}}{10}$$ Simplify within the radical. In the denominator, do the multiplication.

$$= \frac{-7 \pm \sqrt{29}}{10}$$ Do the subtraction within the radical.

Thus, the two solutions are

$$x = \frac{-7 + \sqrt{29}}{10} \quad \text{or} \quad x = \frac{-7 - \sqrt{29}}{10}$$

An alternate form for expressing the solutions is

$$x = -\frac{7}{10} + \frac{\sqrt{29}}{10} \quad \text{or} \quad x = -\frac{7}{10} - \frac{\sqrt{29}}{10}$$

We can use a calculator to approximate each of the solutions. To the nearest hundredth,

$$x = \frac{-7 + \sqrt{29}}{10} \approx -0.16 \quad \text{or} \quad x = \frac{-7 - \sqrt{29}}{10} \approx -1.24$$

SELF CHECK   Solve $3x^2 = -2 - 9x$. Round each solution to the nearest hundredth.

Answers: $\dfrac{-9 + \sqrt{57}}{6} \approx -0.24$, $\dfrac{-9 - \sqrt{57}}{6} \approx -2.76$ ∎

Applications

Quadratic equations are used to solve many types of problems. In the following examples, when we get to the "solve the equation" step in the solution, we will need to determine the most efficient method to use to solve the equation. To do this, we can use the following strategy.

Strategy for solving quadratic equations

1. First, see whether the equation is in a form such that the **square root method** is easily applied.

2. If the square root method can't be used, write the equation in $ax^2 + bx + c = 0$ form.

3. Then see whether the equation can be solved using the **factoring method.**

4. If you can't factor the quadratic, solve the equation by the **quadratic formula.**

EXAMPLE 10   **Oriental rug.** The oriental rug shown in Figure 5-6 is 3 feet longer than it is wide. Find the dimensions of the rug if its area is 180 square feet.

FIGURE 5-6

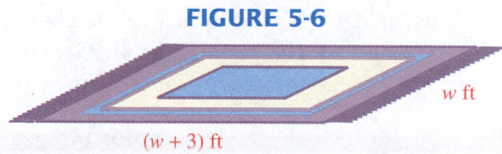

w ft

$(w + 3)$ ft

ANALYZE THE PROBLEM   We are given the area of the rectangular-shaped rug and asked to find its dimensions (width and length).

FORM AN EQUATION   We can let w represent the width of the rug. Then $w + 3$ will represent its length. Since the area of a rectangle is given by the formula $A = lw$, this gives the equation

The length of the rug	times	the width of the rug	is	the area of the rug.
$(w + 3)$	$\cdot$	w	$=$	180

SOLVE THE EQUATION   We must solve the equation $(w + 3)w = 180$. It is not in a form such that the square root method can be used to solve it, so we write it in quadratic form.

$$(w + 3)w = 180$$
$$w^2 + 3w = 180 \quad \text{Use the distributive property to remove parentheses.}$$
$$w^2 + 3w - 180 = 0 \quad \text{Subtract 180 from both sides. The equation is now in quadratic form.}$$

By inspection, we see that -180 has factors -12 and 15 and that their sum is 3. Therefore, we can use the factoring method to solve the equation.

$$(w - 12)(w + 15) = 0 \quad \text{Factor } w^2 + 3w - 180.$$
$$w - 12 = 0 \quad \text{or} \quad w + 15 = 0$$
$$w = 12 \qquad\qquad w = -15$$

STATE THE CONCLUSION   Because the rug cannot have a negative width, we discard the solution $w = -15$. When $w = 12$, the length, $w + 3$, is 15. So the width is 12 feet, and the length is 15 feet.

CHECK THE RESULT   With a length of 15 feet and a width of 12 feet, the rug is 3 feet longer than it is wide. Its area is $15 \cdot 12 = 180$ square feet. The solution checks. ■

E X A M P L E   1 1

Action movies. As part of an action scene in a movie, a stuntman is to fall from the top of a 95-foot-tall building into a large airbag directly below him on the ground, as shown in Figure 5-7. If an object falls s feet in t seconds, where $s = 16t^2$, and if the bag is inflated to a height of 10 feet, how long will the stuntman fall before making contact with the airbag?

FIGURE 5-7

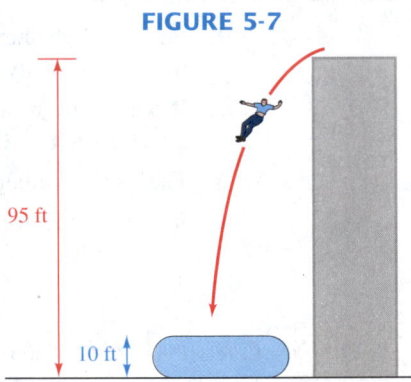

95 ft

10 ft

Solution   If we subtract the height of the airbag from the height of the building, we find that the stuntman will fall $95 - 10 = 85$ feet. We substitute 85 for s in the formula and find that the equation is in a form that allows us to use the square root method.

$$s = 16t^2 \quad \text{The given formula.}$$
$$85 = 16t^2 \quad \text{Substitute 85 for } s.$$
$$\frac{85}{16} = t^2 \quad \text{Divide both sides by 16.}$$
$$\pm\sqrt{\frac{85}{16}} = t \quad \text{Use the square root method to solve the equation.}$$
$$\pm\frac{\sqrt{85}}{\sqrt{16}} = t \quad \text{The square root of a quotient is the quotient of the square roots.}$$
$$\pm\frac{\sqrt{85}}{4} = t \quad \sqrt{16} = 4.$$

The stuntman will fall for $\frac{\sqrt{85}}{4}$ seconds before making contact with the airbag. To the nearest tenth, this is 2.3 seconds. We discard the other solution, $-\frac{\sqrt{85}}{4}$, because a negative time does not make sense in this context. ∎

EXAMPLE 12

Finance. If $\$P$ is invested at an annual rate r, it will grow to an amount of $\$A$ in n years according to the formula $A = P(1 + r)^n$. What interest rate is needed so that a \$5,000 investment will grow to \$5,618 after 2 years?

Solution   We can substitute 5,000 for P, 5,618 for A, and 2 for n in the formula and solve for r.

$$A = P(1 + r)^n$$
$$5{,}618 = 5{,}000(1 + r)^2$$
$$5{,}618 = 5{,}000(1 + 2r + r^2) \qquad \text{Use FOIL: } (1 + r)^2 = 1 + 2r + r^2.$$
$$5{,}618 = 5{,}000 + 10{,}000r + 5{,}000r^2 \quad \text{Remove parentheses.}$$
$$0 = 5{,}000r^2 + 10{,}000r - 618 \quad \text{Subtract 5,618 from both sides.}$$

We can use a calculator and solve this equation with the quadratic formula.

$$r = \frac{-b \pm \sqrt{b^2 - 4ac}}{2a}$$
$$= \frac{-10{,}000 \pm \sqrt{10{,}000^2 - 4(5{,}000)(-618)}}{2(5{,}000)}$$
$$= \frac{-10{,}000 \pm \sqrt{100{,}000{,}000 + 12{,}360{,}000}}{10{,}000}$$
$$= \frac{-10{,}000 \pm \sqrt{112{,}360{,}000}}{10{,}000}$$
$$= \frac{-10{,}000 \pm 10{,}600}{10{,}000}$$

$$r = \frac{-10{,}000 + 10{,}600}{10{,}000} \quad \text{or} \quad r = \frac{-10{,}000 - 10{,}600}{10{,}000}$$
$$= \frac{600}{10{,}000} \qquad\qquad\qquad = \frac{-20{,}600}{10{,}000}$$
$$= 0.06 \qquad\qquad\qquad\qquad = -2.06$$
$$r = 6\% \qquad\qquad\qquad\qquad r = -206\%$$

The required rate is 6%. The rate of -206% has no meaning in this problem. ∎

STUDY SET

Section 5.5

VOCABULARY

In Exercises 1–6, fill in the blanks to make the statements true.

1. If $b^2 = a$, then b is called a _____ of a.

2. In the symbol $\sqrt{10}$, 10 is called the _____.

3. If the polynomial in the equation $ax^2 + bx + c = 0$ doesn't factor, we can solve the equation by using the _____ formula.

4. The general _____ equation is $ax^2 + bx + c = 0$.

5. The formula
$$x = \frac{-b \pm \sqrt{b^2 - 4ac}}{2a}$$
is called the _____ formula.

6. To _____ a quadratic equation means to find all the values of the variable that make the equation true.

CONCEPTS

In Exercises 7–16, fill in the blanks to make the statements true, or answer the question.

7. The equation $x^2 = c$ $(c > 0)$ has _____ solutions.

8. The solutions of $x^2 = c$ $(c > 0)$ are [   ] and [   ].

9. Solve $x^2 = 81$ by
  a. the square root method.
  b. the factoring method.

10. $n^2 - 4n + 6 = 0$ is written in $ax^2 + bx + c = 0$ form. What is b?

11. To solve $x^2 - 2x - 1 = 0$, we must use the quadratic formula. Why can't we use the factoring method?

12. In the quadratic equation $ax^2 + bx + c = 0$, a cannot equal [   ].

13. Before we can determine a, b, and c for $x = 3x^2 - 1$, we must write the equation in _____ form.

14. In the quadratic equation $3x^2 - 5 = 0$, $a =$ [   ], $b =$ [   ], and $c =$ [   ].

15. In the quadratic equation $-4x^2 + 8x = 0$, $a =$ [   ], $b =$ [   ], and $c =$ [   ].

16. In evaluating the numerator of

$$\frac{-5 \pm \sqrt{5^2 - 4(2)(1)}}{2(2)}$$

what operation should be performed first?

17. Approximate each solution to the nearest hundredth.

$$x = \frac{3 + \sqrt{2}}{3} \qquad x = \frac{3 - \sqrt{2}}{3}$$

18. A student used the quadratic formula to solve an equation and obtained

$$x = \frac{-3 \pm \sqrt{15}}{2}$$

  a. How many solutions does the equation have?
  b. What are they *exactly?*

  c. Approximate them to the nearest hundredth.

NOTATION

In Exercises 19–20, complete each solution.

19. Solve $(y - 1)^2 = 9$.

$$(y - 1)^2 = 9$$

$$y - 1 = [\quad] \quad \text{or} \quad y - [\ ] = -\sqrt{9}$$

$$[\quad] = 3 \qquad y - 1 = [\quad]$$

$$y = 4 \qquad\qquad y = -2$$

20. Solve $x^2 - 5x - 6 = 0$.

$$x = \frac{-b \pm \sqrt{b^2 - 4ac}}{2a}$$

$$= \frac{-([\ \ ]) \pm \sqrt{(-5)^2 - 4(1)(-6)}}{2(1)}$$

$$= \frac{[\ \ ] \pm \sqrt{25 + [\ \ ]}}{2}$$

$$= \frac{5 \pm \sqrt{[\ \ ]}}{2}$$

$$x = \frac{[\ \ ] \pm 7}{2}$$

$$x = \frac{5 \;[\ ]\; 7}{2} = 6 \quad \text{or} \quad x = \frac{5 \;[\ ]\; 7}{2} = -1$$

21. When solving a quadratic equation, a student obtains $x = \pm\sqrt{10}$. How many solutions are represented by this notation? Find each one to the nearest tenth.

22. In solving a quadratic equation, a student obtains $x = 8 \pm \sqrt{3}$. Find each solution, rounded to the nearest hundredth.

23. What is **wrong** with this student's work?
Solve $x^2 + 4x - 5 = 0$.

$$x = -4 \pm \frac{\sqrt{16 - 4(1)(-5)}}{2}$$

24. In reading

$$\frac{-b \pm \sqrt{b^2 - 4ac}}{2a}$$

we say, "The _____ of b, plus or _____ the _____ root of b _____ minus 4 _____ a times c, all _____ $2a$."

PRACTICE

In Exercises 25–28, use a calculator to find each positive square root to the nearest hundredth.

25. $\sqrt{65}$ **26.** $\sqrt{3{,}067}$

27. $\sqrt{0.00897}$ **28.** $\sqrt{721.36982}$

In Exercises 29–40, use the square root method to solve each equation. If an answer is not exact, round to the nearest tenth.

29. $x^2 = 1$ **30.** $r^2 = 4$

31. $x^2 = 27$ **32.** $x^2 = 32$

33. $t^2 = 20$ **34.** $x^2 = 0$

35. $3m^2 = 27$ **36.** $4x^2 = 64$

37. $x^2 = 45.82$ **38.** $x^2 = 6.05$

39. $(x + 2)^2 = 90.04$

40. $(x - 5)^2 = 33.31$

In Exercises 41–46, use the square root method to solve each equation for x. If an answer is not exact, round to the nearest hundredth.

41. $(x + 1)^2 = 25$

42. $(x - 1)^2 = 49$

43. $(x + 2)^2 = 81$

44. $(x + 3)^2 = 16$

45. $(x - 2)^2 = 8$

46. $(x + 2)^2 = 50$

In Exercises 47–50, factor the trinomial square and use the square root method to solve each equation.

47. $y^2 + 4y + 4 = 4$

48. $y^2 - 6y + 9 = 9$

49. $9x^2 - 12x + 4 = 16$

50. $4x^2 - 20x + 25 = 36$

In Exercises 51–58, change each equation into quadratic form, if necessary, and find the values of a, b, and c. **Do not solve the equation.**

51. $x^2 + 4x + 3 = 0$

52. $x^2 - x - 4 = 0$

53. $3x^2 - 2x + 7 = 0$

54. $4x^2 + 7x - 3 = 0$

55. $4y^2 = 2y - 1$

56. $2x = 3x^2 + 4$

57. $x(3x - 5) = 2$

58. $y(5y + 10) = 8$

In Exercises 59–74, use the quadratic formula to find all real solutions of each equation. If an answer is not exact, round to the nearest thousandth.

59. $x^2 - 5x + 6 = 0$

60. $x^2 + 5x + 4 = 0$

61. $x^2 + 7x + 12 = 0$

62. $x^2 - x - 12 = 0$

63. $2x^2 - x - 1 = 0$

64. $2x^2 + 3x - 2 = 0$

65. $3x^2 + 5x + 2 = 0$

66. $3x^2 - 4x + 1 = 0$

67. $4x^2 + 4x - 3 = 0$

68. $4x^2 + 3x - 1 = 0$

69. $x^2 - 10x = -18$

70. $x^2 + 1 = -8x$

71. $x^2 = 1 - 2x$

72. $x^2 = 4 + 2x$

73. $x^2 + 3x + 1 = 0$

74. $x^2 + 3x - 2 = 0$

In Exercises 75–86, use the most convenient method to find all real solutions of each equation. Use a calculator when necessary. If an answer is not exact, round to the nearest hundredth.

75. $(2y - 1)^2 = 25$

76. $m^2 + 14m + 49 = 0$

77. $b^2 = 18$

78. $t^2 - 1 = 0$

79. $x^2 - 2x - 1 = 0$

80. $25x - 50x^2 = 0$

81. $2x^2 + x = 5$

82. $x^2 + 5x + 3 = 0$

83. $x^2 - 2x - 35 = 0$

84. $3x^2 - x = 1$

85. $2.4x^2 - 9.5x + 6.2 = 0$

86. $-1.7x^2 + 0.5x + 0.9 = 0$

APPLICATIONS

Use a calculator when necessary.

87. BICYCLE SAFETY A bicycle training program for children uses a figure-eight course to help them improve their balance and steering. The course is laid out over a paved area covering 800 square feet, as shown in Illustration 1. Find its dimensions.

ILLUSTRATION 1

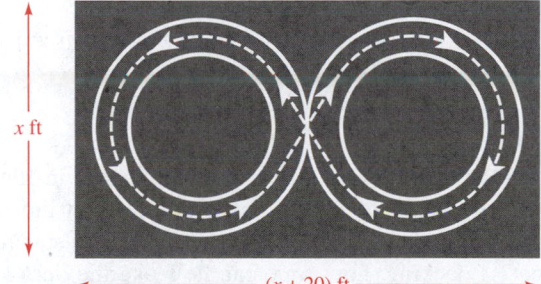

x ft

(x + 20) ft

88. BADMINTON The badminton court shown in Illustration 2 occupies 880 square feet of the floor space of a gymnasium. If its length is 4 feet more than twice its width, find its dimensions.

ILLUSTRATION 2

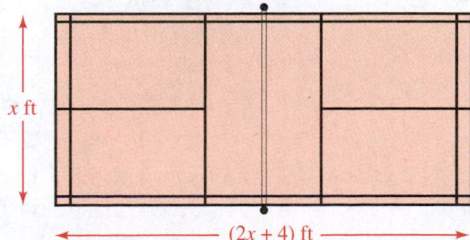

x ft

(2x + 4) ft

89. GEOMETRY PROBLEM A rectangular mural is 4 feet longer than it is wide. Find its dimensions if its area is 32 square feet.

90. GEOMETRY PROBLEM The length of a 220-square-foot rectangular garden is 2 feet more than twice its width. Find its perimeter.

91. HEIGHT OF A TRIANGLE The triangle shown in Illustration 3 has an area of 30 square inches. Find its height.

ILLUSTRATION 3

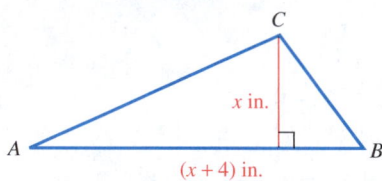

92. BASE OF A TRIANGLE Find the length of the base of the triangle shown in Illustration 3.

93. AREA OF A GARDEN The rectangular garden shown in Illustration 4 is surrounded by a walk of uniform width. Find the dimensions of the garden if its area is 180 square feet.

ILLUSTRATION 4

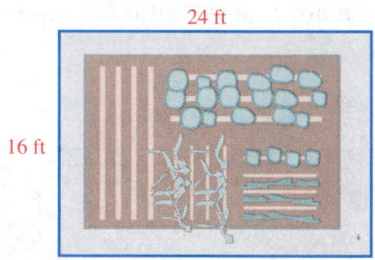

94. AREA OF A POOL The owner of the pool in Illustration 5 wants to surround it with a deck of uniform width (shown in gray). If he can afford 368 square feet of decking, how wide can he make the deck?

ILLUSTRATION 5

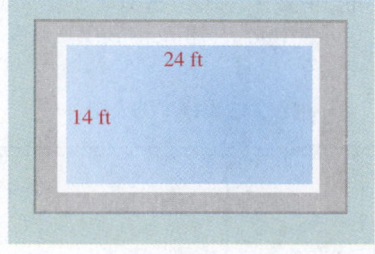

95. FALLING OBJECTS An object will fall s feet in t seconds, where $s = 16t^2$. If a workman 1,454 feet

above the ground at the top of the Sears Tower drops a hammer, how long will it take for the hammer to hit the ground?

96. FALLING OBJECTS A tourist drops a penny from the observation deck of the World Trade Center, 1,377 feet above the ground. How long will it take for the penny to hit the ground? (See Exercise 95.)

97. INVESTING We can use the formula $A = P(1 + r)^2$ to find the amount $\$A$ that $\$P$ will become when invested at an annual rate r for 2 years. What interest rate is needed for $5,000 to grow to $5,724.50 in 2 years?

98. INVESTING What interest rate is needed for $7,000 to grow to $8,470 in 2 years? (See Exercise 97.)

99. MANUFACTURING An electronics firm has found that its revenue for manufacturing and selling x television sets is given by the formula $R = -\frac{1}{6}x^2 + 450x$. How many TVs must be sold to generate $210,000 in revenue?

100. RETAILING When a wholesaler sells n CD players, her revenue R is given by the formula $R = 150n - \frac{1}{2}n^2$. How many players would she have to sell to receive $11,250?

101. METAL FABRICATION A piece of tin, 12 inches on a side, is to have four equal squares cut from its corners, as shown in Illustration 6. If the edges are then to be folded up to make a box with a floor area of 64 square inches, find the depth of the box.

ILLUSTRATION 6

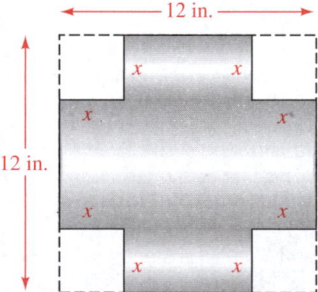

102. MAKING GUTTERS A piece of sheet metal, 18 inches wide, is bent to form the gutter shown in Illustration 7. If the cross-sectional area is 36 square inches, find the depth of the gutter.

ILLUSTRATION 7

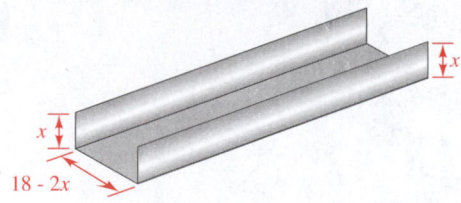

103. Explain how to solve $x^2 = 81$ using the square root method.

104. Rounded to the nearest hundredth, one solution of $x^2 + 4x + 1 = 0$ is -0.27. Use a calculator to check it. How could it be a solution if it doesn't make the left-hand side zero? Explain.

105. Why is the quadratic formula useful in solving some types of quadratic equations?

106. Explain the meaning of the $\pm$ symbol.

107. Use the quadratic formula to solve $x^2 - 2x - 4 = 0$. What is an exact solution, and what is an approximate solution of this equation? Explain the difference.

Solve each equation for the indicated variable.

108. $A = p + prt$; for r

109. $F = \dfrac{GMm}{d^2}$, for M

Write the equation of the line with the given properties in general form.

110. Slope of $\frac{3}{5}$ and passing through $(0, 12)$

111. Passes through $(6, 8)$ and the origin

▶ 5.6

Graphing Quadratic Functions

In this section, you will learn about

Quadratic functions ■ Finding the vertex and the intercepts of a parabola ■ A strategy for graphing quadratic functions ■ Finding a maximum value ■ Solving quadratic equations graphically

---

Introduction In Chapter 4, we discussed polynomial functions. We now consider a special type of polynomial function called a *quadratic function.* When graphing polynomial functions, we constructed a table of values and plotted points. In this section, we will develop a more general strategy for graphing quadratic functions by analyzing the given function and determining the important characteristics of its graph.

Quadratic Functions

FIGURE 5-8

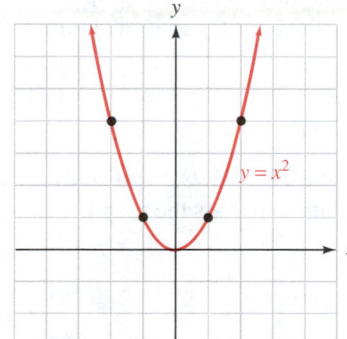

Quadratic functions are defined by equations of the form $y = ax^2 + bx + c$ $(a \neq 0)$ where the right-hand side is a second-degree polynomial in the variable x. Three examples of quadratic functions are

$$y = x^2 - 3 \qquad y = x^2 - 2x - 3 \qquad y = -2x^2 - 4x + 2$$

We can replace y with the function notation $f(x)$ to express the defining equation in the form $f(x) = ax^2 + bx + c$. For the functions above, we can write

$$f(x) = x^2 - 3 \qquad f(x) = x^2 - 2x - 3 \qquad f(x) = -2x^2 - 4x + 2$$

In Section 3.2, we constructed the graph of $y = x^2$ by plotting points. The result was the **parabola** shown in Figure 5-8.

EXAMPLE 1

Graphing quadratic functions by plotting points. Graph $y = x^2 - 3$. Compare the graph to that of $y = x^2$.

Solution   The function is written in $y = ax^2 + bx + c$ form, where $a = 1$, $b = 0$, and $c = -3$. To find ordered pairs (x, y) that satisfy the equation, we pick several numbers x and find the corresponding values of y. If we let $x = 3$, we have

$$y = x^2 - 3$$
$$= 3^2 - 3 \quad \text{Substitute 3 for } x.$$
$$= 6$$

The ordered pair (3, 6) and six others satisfying the equation appear in the table shown in Figure 5-9. To graph the equation, we plot each point and draw a smooth curve passing through them. The resulting parabola is the graph of $y = x^2 - 3$. The parabola opens upward, and the lowest point on the graph, called the **vertex of the parabola,** is the point $(0, -3)$.

Note that the graph of $y = x^2 - 3$ looks just like the graph of $y = x^2$, except that it is 3 units lower.

FIGURE 5-9

$y = x^2 - 3$

x	y	(x, y)
3	6	(3, 6)
2	1	(2, 1)
1	-2	(1, -2)
0	-3	(0, -3)
-1	-2	($-1, -2$)
-2	1	($-2, 1$)
-3	6	($-3, 6$)

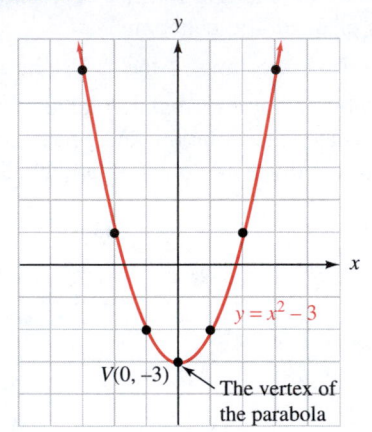

SELF CHECK   Graph $y = x^2 + 2$. Compare the graph to that of $y = x^2$.

Answer: The graph has the same shape as the graph of $y = x^2$, but it is 2 units higher.

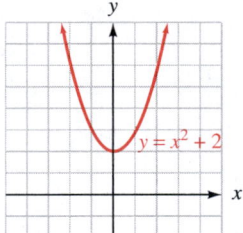

If we draw a vertical line through the vertex of a parabola and fold the graph on this line, the two sides of the graph will match. We call the vertical line the **axis of symmetry.**

EXAMPLE 2   **Graphing quadratic functions.** Graph $f(x) = -2x^2 - 4x + 2$, find its vertex, and draw its axis of symmetry.

Solution   The function is written in $f(x) = ax^2 + bx + c$ form, where $a = -2, b = -4,$ and $c = 2$. We construct the table shown in Figure 5-10, plot the points, and draw the graph.

FIGURE 5-10

$f(x) = -2x^2 - 4x + 2$

x	$f(x)$	$(x, f(x))$
-3	-4	($-3, -4$)
-2	2	($-2, 2$)
-1	4	($-1, 4$)
0	2	(0, 2)
1	-4	(1, -4)

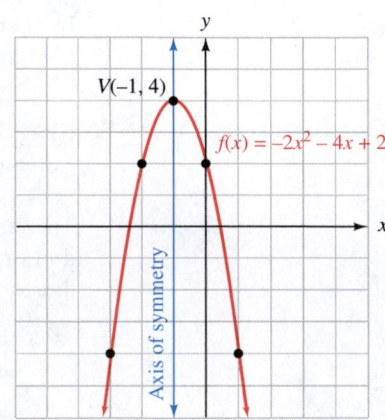

The parabola opens downward, so its vertex is its highest point, the point $(-1, 4)$.

SELF CHECK   Graph $f(x) = -x^2 - 4x - 4$, find its vertex, and draw its axis of symmetry.   *Answer:* The vertex is at $(-2, 0)$.

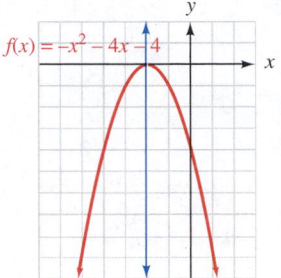

In Example 1, the coefficient of the x^2 term in $y = x^2 - 3$ is positive $(a = 1)$. In Example 2, the coefficient of the x^2 term in $f(x) = -2x^2 - 4x + 2$ is negative $(a = -2)$. The results of these first two examples suggest the following fact.

Graphs of quadratic functions

The graph of the function $y = ax^2 + bx + c$ or $f(x) = ax^2 + bx + c$ $(a \neq 0)$ is a parabola. It opens upward when $a > 0$ and downward when $a < 0$.

The cup-like shape of a parabola can be seen in a wide variety of real-world settings. Some examples are shown in Figure 5-11.

FIGURE 5-11

The path of a thrown object

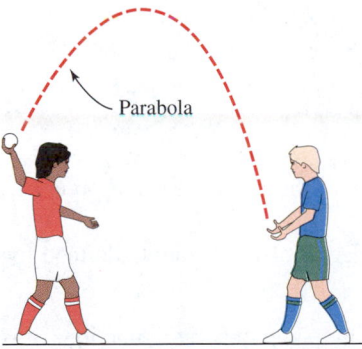

The pursuit path of a shark seeking its prey

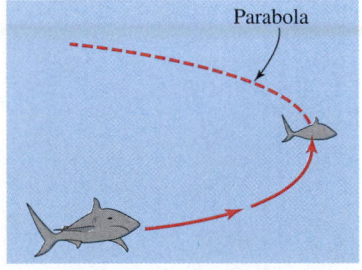

The shape of a satellite antenna dish

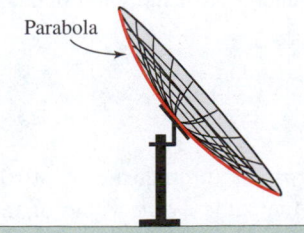

The path of a stream of water

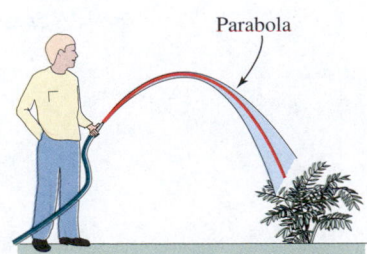

Finding the Vertex and the Intercepts of a Parabola

It is easier to graph a quadratic function when we know the coordinates of the vertex of its parabolic graph. For a parabola defined by $y = ax^2 + bx + c$ or $f(x) = ax^2 + bx + c$, it can be shown that the x-coordinate of the vertex is given by $-\frac{b}{2a}$. This fact enables us to find the coordinates of its vertex.

Finding the vertex of a parabola

> The graph of the quadratic function $y = ax^2 + bx + c$ or $f(x) = ax^2 + bx + c$ is a parabola whose vertex has an x-coordinate of $-\frac{b}{2a}$. To find the y-coordinate of the vertex, substitute $-\frac{b}{2a}$ into the defining equation and find y.

EXAMPLE 3

Finding the vertex of a parabola. Find the vertex of the parabola defined by $y = x^2 - 2x - 3$.

Solution   For $y = x^2 - 2x - 3$, we have $a = 1$, $b = -2$, and $c = -3$. To find the x-coordinate of the vertex, we substitute the values for a and b into the formula $x = -\frac{b}{2a}$.

$$x = -\frac{b}{2a}$$

$$x = -\frac{-2}{2(1)}$$

$$= 1$$

The x-coordinate of the vertex is $x = 1$. To find the y-coordinate, we substitute 1 for x:

$$y = x^2 - 2x - 3$$
$$y = 1^2 - 2(1) - 3$$
$$= 1 - 2 - 3$$
$$= -4$$

The vertex of the parabola is the point $(1, -4)$. See Figure 5-12(a) on the next page.

SELF CHECK   Find the vertex of the parabola defined by $y = -x^2 + 6x - 8$.   *Answer:* $(3, 1)$ ■

When graphing quadratic functions, it is often helpful to find the x- and y-intercepts of the parabola.

EXAMPLE 4

Finding the intercepts of a parabola. Find the x- and y-intercepts of the parabola defined by $y = x^2 - 2x - 3$.

Solution   To find the y-intercept of the parabola, we let $x = 0$ and solve for y.

$$y = x^2 - 2x - 3$$
$$y = 0^2 - 2(0) - 3$$
$$y = -3$$

The parabola passes through the point $(0, -3)$. We note that the y-coordinate of the y-intercept is the same as the value of the constant term c on the right-hand side of $y = x^2 - 2x - 3$.

To find the x-intercepts of the graph, we set y equal to 0 and solve the resulting quadratic equation.

$$y = x^2 - 2x - 3$$
$$0 = x^2 - 2x - 3 \qquad \text{Substitute 0 for } y.$$
$$0 = (x - 3)(x + 1) \qquad \text{Factor the trinomial.}$$
$$x - 3 = 0 \quad \text{or} \quad x + 1 = 0 \qquad \text{Set each factor equal to 0.}$$
$$x = 3 \qquad\qquad x = -1$$

Since there are two solutions, the graph has two x-intercepts: $(3, 0)$ and $(-1, 0)$. See Figure 5-12(a).

SELF CHECK Find the x- and y-intercepts of the parabola defined by $y = -x^2 + 6x - 8$.

Answers: y-intercept: $(0, -8)$; x-intercepts: $(2, 0)$, $(4, 0)$ ∎

A Strategy for Graphing Quadratic Functions

There is another way to graph quadratic functions. Rather than plotting random points, we can determine the important characteristics of a parabola. For example, to graph $y = x^2 - 2x - 3$, we note that the coefficient of the x^2 term is positive ($a = 1$). Therefore, the parabola defined by this function opens upward. In Examples 3 and 4, we found that the vertex of the graph of $y = x^2 - 2x - 3$ is at $(1, -4)$ and that it has a y-intercept of $(0, -3)$ and the x-intercepts of $(3, 0)$ and $(-1, 0)$. See Figure 5-12(a).

We can locate other points on the parabola by noting that the graph has the axis of symmetry shown in Figure 5-12(a). If the point $(0, -3)$, which is 1 unit to the left of the axis of symmetry, is on the graph, the point $(2, -3)$, which is 1 unit to the right of the axis of symmetry, is also on the graph.

FIGURE 5-12

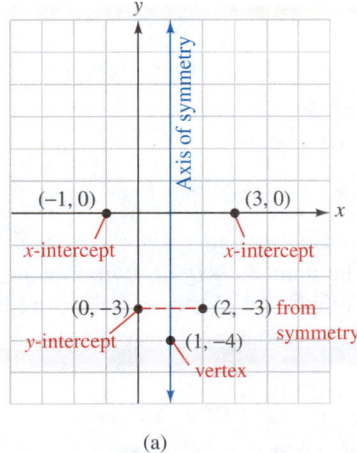

(a)

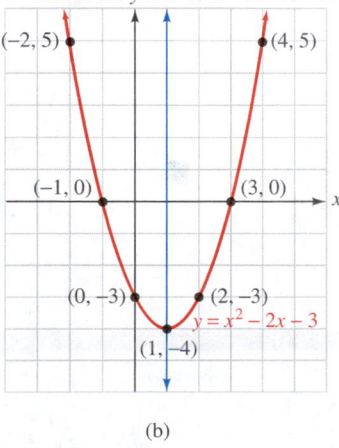

(b)

$$y = x^2 - 2x - 3$$

x	y	(x, y)
-2	5	$(-2, 5)$

We can complete the graph by plotting two more points. If $x = -2$, then $y = 5$, and the parabola passes through $(-2, 5)$. Again using symmetry, the parabola must also pass through $(4, 5)$. The completed graph of $y = x^2 - 2x - 3$ is shown in Figure 5-12(b).

Much can be determined about the graph of $y = ax^2 + bx + c$ from the coefficients a, b, and c. This information is summarized below.

Graphing a quadratic function
$y = ax^2 + bx + c \ (a \neq 0)$

If $a > 0$, the parabola opens upward. If $a < 0$, the parabola opens downward.

The x-coordinate of the vertex of the parabola is $x = -\frac{b}{2a}$.

To find the y-coordinate of the vertex, substitute $-\frac{b}{2a}$ for x into the equation and find y.

The axis of symmetry is the vertical line passing throught the vertex.

The y-intercept is determined by the value of y attained when $x = 0$: the y-intercept is $(0, c)$.

The x-intercepts (if any) are determined by the numbers x that make $y = 0$. To find them, solve the quadratic equation $ax^2 + bx + c = 0$.

EXAMPLE 5

Graphing quadratic functions. Graph $f(x) = -2x^2 - 8x - 8$.

Solution   **Step 1:** *Determine whether the parabola opens upward or downward.* The equation is in the form $f(x) = ax^2 + bx + c$, with $a = -2$, $b = -8$, and $c = -8$. Since $a < 0$, the parabola opens downward.

Step 2: *Find the vertex and draw the axis of symmetry.* To find the x-coordinate of the vertex, we substitute the values for a and b into the formula $x = -\frac{b}{2a}$.

$$x = -\frac{b}{2a}$$

$$x = -\frac{-8}{2(-2)}$$

$$= -2$$

The x-coordinate of the vertex is -2. To find the y-coordinate, we substitute -2 for x in the equation and find $f(x)$.

$$f(x) = -2x^2 - 8x - 8$$
$$f(x) = -2(-2)^2 - 8(-2) - 8$$
$$= -8 + 16 - 8$$
$$= 0 \qquad\qquad \text{If } f(x) = 0 \text{ for } x = -2, \text{ then } y = 0 \text{ for } x = -2.$$

The vertex of the parabola is the point $(-2, 0)$. This point is the blue dot in Figure 5-13.

Step 3: *Find the x- and y-intercepts.* Since $c = -8$, the y-intercept of the parabola is $(0, -8)$. The point $(-4, -8)$, two units to the left of the axis of symmetry, must also be on the graph. We plot both points in black in Figure 5-13.

To find the x-intercepts, we set $f(x)$ equal to 0 and solve the resulting quadratic equation.

$$f(x) = -2x^2 - 8x - 8$$
$$0 = -2x^2 - 8x - 8 \quad \text{Set } f(x) = 0.$$
$$0 = x^2 + 4x + 4 \quad \text{Divide both sides by } -2.$$
$$= (x + 2)(x + 2) \quad \text{Factor the trinomial.}$$
$$x + 2 = 0 \quad \text{or} \quad x + 2 = 0 \quad \text{Set each factor equal to 0.}$$
$$x = -2 \qquad\qquad x = -2$$

Since the solutions are the same, the graph has only one x-intercept: $(-2, 0)$. This point is the vertex of the parabola and has already been plotted.

Step 4: Plot another point. Finally, we find another point on the parabola. If $x = -3$, then $y = -2$. We plot $(-3, -2)$ in Figure 5-13 and use symmetry to determine that $(-1, -2)$ is also on the graph. Both points are in green.

FIGURE 5-13

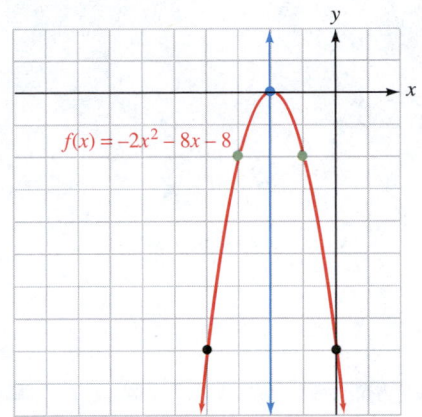

$$f(x) = -2x^2 - 8x - 8$$

x	$f(x)$	$(x, f(x))$
-3	-2	$(-3, -2)$

SELF CHECK   Graph the function $y = -x^2 + 6x - 8$.   *Answer:*

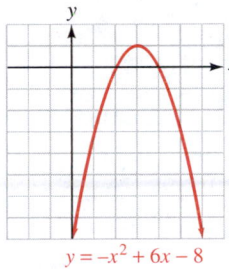

$y = -x^2 + 6x - 8$ ∎

Finding a Maximum Value

EXAMPLE 6

Finding maximum revenue. An electronics firm manufactures radios. Over the past 10 years, the firm has learned that it can sell x radios at a price of $\left(200 - \frac{1}{5}x\right)$ dollars. How many radios should the firm manufacture and sell to maximize its revenue? Find the maximum revenue.

Solution   The revenue obtained is the product of the number of radios sold (x) and the price of each radio $\left(200 - \frac{1}{5}x\right)$. Thus, the revenue R is given by the function

$$R = x\left(200 - \frac{1}{5}x\right) \quad \text{or} \quad R = -\frac{1}{5}x^2 + 200x$$

Since the graph of this function is a parabola that opens downward, the *maximum* value of R will be the value of R determined by the vertex of the parabola. Because the x-coordinate of the vertex is at $x = -\frac{b}{2a}$, we have

$$x = -\frac{b}{2a}$$

$$= -\frac{200}{2\left(-\frac{1}{5}\right)} \qquad \text{Substitute 200 for } b \text{ and } -\frac{1}{5} \text{ for } a.$$

$$= -\frac{200}{-\frac{2}{5}} \qquad \text{Do the multiplication in the denominator.}$$

$$= (-200)\left(-\frac{5}{2}\right) \quad \text{Division by } -\frac{2}{5} \text{ is the same as multiplication by its reciprocal, which is } -\frac{5}{2}.$$

$$= 500$$

If the firm manufactures 500 radios, the maximum revenue will be

$$R = -\frac{1}{5}x^2 + 200x \qquad \text{The revenue formula.}$$

$$= -\frac{1}{5}(\textbf{500})^2 + 200(\textbf{500}) \quad \text{Substitute 500 for } x, \text{ the number of radios.}$$

$$= 50{,}000$$

The firm should manufacture 500 radios to get a maximum revenue of $50,000. This fact is verified by examining the graph of $R = -\frac{1}{5}x^2 + 200x$, which appears in Figure 5-14.

FIGURE 5-14

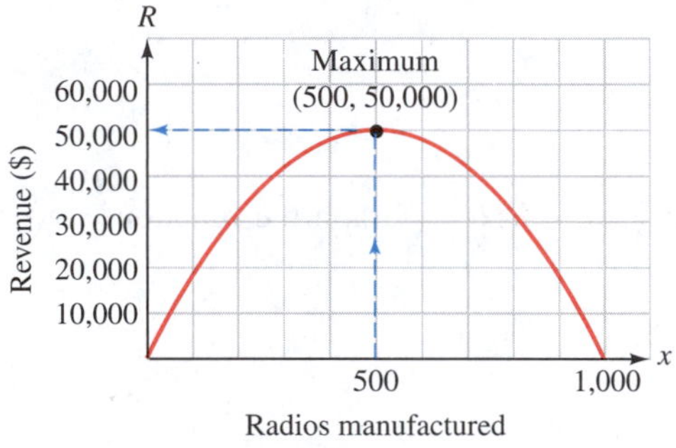

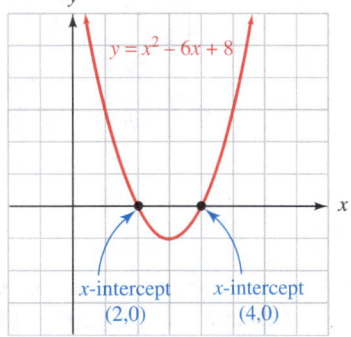

FIGURE 5-15

Solving Quadratic Equations Graphically

We can also solve quadratic equations graphically. For example, the solutions of $x^2 - 6x + 8 = 0$ are the numbers x that will make y equal to 0 in the quadratic function $y = x^2 - 6x + 8$. To find these numbers, we inspect the graph of $y = x^2 - 6x + 8$ and locate the points on the graph with a y-coordinate of 0. In Figure 5-15, these points are $(2, 0)$ and $(4, 0)$, the x-intercepts of the graph. We can conclude that the x-coordinates of the x-intercepts, $x = 2$ and $x = 4$, are the solutions of $x^2 - 6x + 8 = 0$.

ACCENT ON TECHNOLOGY *Solving Quadratic Equations Graphically*

The solutions of $-2x^2 + 5x + 3 = 0$ are the numbers x that will make y equal to 0 in the quadratic function $y = -2x^2 + 5x + 3$. To find these numbers, we can graph the function and read the x-intercepts of the graph.

Using the standard window settings, we press the $\boxed{Y=}$ key and enter $-2x^2 + 5x + 3$. The screen should read

$$Y_1 = -2x^2 + 5x + 3$$

The resulting graph is shown in Figure 5-16(a). We then trace and move the cursor to identify each x-intercept. From the calculator, we can then find the x-coordinate of the x-intercept. See Figure 5-16(b) and 5-16(c), where the approximate solutions of the quadratic equation $-2x^2 + 5x + 3 = 0$ are the x-values shown on the screen.

FIGURE 5-16

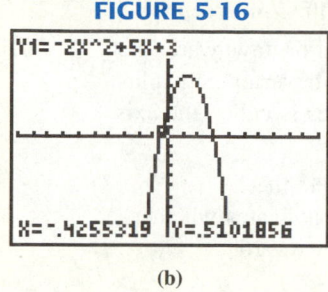

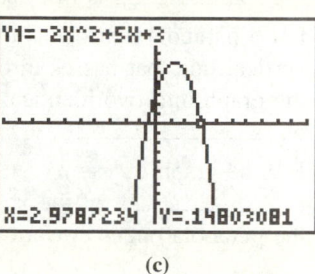

(a)   (b)   (c)

For better results, we can use the ZOOM feature. We would then conclude that the two solutions of $-2x^2 + 5x + 3 = 0$ are

$$x = -0.5 \qquad \text{and} \qquad x = 3$$

Another way to find the solutions of a quadratic equation is by using the ZERO feature of the calculator. With this feature, the x-coordinate of the x-intercepts of the graph of $y = -2x^2 + 5x + 3$ can be found. The *zeros*, in this case -0.5 and 3, are solutions of $-2x^2 + 5x + 3 = 0$. Consult your owner's manual for the specific instructions to use this feature.

When solving quadratic equations graphically, there are three possibilities to consider. If the graph of the associated quadratic function has two x-intercepts, the quadratic equation has two real-number solutions. Figure 5-17(a) shows an example of this. If the graph has one x-intercept, as shown in Figure 5-17(b), the equation has one real-number solution. Finally, if the graph does not have an x-intercept, as shown in Figure 5-17(c), the equation does not have any real-number solutions.

FIGURE 5-17

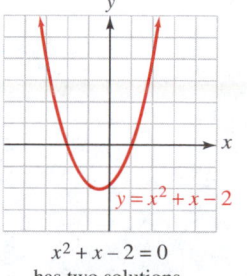

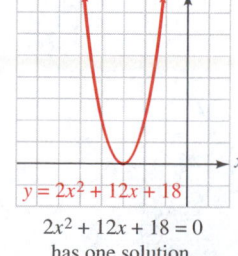

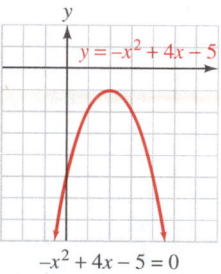

$x^2 + x - 2 = 0$
has two solutions,
$x = -2$ and $x = 1$.

$2x^2 + 12x + 18 = 0$
has one solution,
$x = -3$.

$-x^2 + 4x - 5 = 0$
has no real-number
solutions.

(a)   (b)   (c)

STUDY SET

Section 5.6

VOCABULARY

In Exercises 1–6, fill in the blanks to make the statements true.

1. A function defined by the equation $y = ax^2 + bx + c$ ($a \neq 0$) is called a _____ function.

2. The lowest (or highest) point on a parabola is called the _____ of the parabola.

3. The point where a parabola intersects the y-axis is called the _____.

4. The point (or points) where a parabola intersects the _____ is (are) called the x-intercept(s).

5. For a parabola that opens upward or downward, the vertical line that passes through its vertex and splits the graph into two identical pieces is called the axis of _____.

6. For the graph of $y = ax^2 + bx + c$, the _____ of the x^2 term indicates whether the parabola opens upward or downward.

CONCEPTS

In Exercises 7–10, fill in the blanks to make the statements true.

7. The graph of $y = ax^2 + bx + c$ $(a \neq 0)$ opens upward when _____ .

8. The graph of $f(x) = ax^2 + bx + c$ $(a \neq 0)$ opens downward when _____ .

9. The y-intercept of the graph of $f(x) = ax^2 + bx + c$ is the point _____ .

10. The x-coordinate of the vertex of the parabola that results when we graph $y = ax^2 + bx + c$ is $x =$ _____ .

11. Refer to the graph in Illustration 1.
    a. What do we call the curve shown there?
    b. What are the x-intercepts of the graph?
    c. What is the y-intercept of the graph?
    d. What is the vertex?

ILLUSTRATION 1

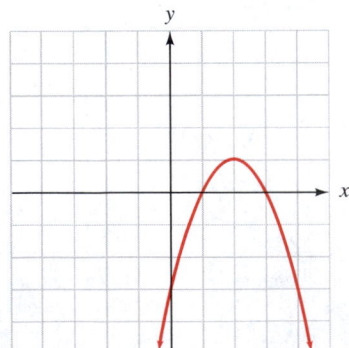

12. The vertex of a parabola is at $(1, -3)$, its y-intercept is $(0, -2)$, and it passes through the point $(3, 1)$, as shown in Illustration 2. Draw the axis of symmetry and use it to help determine two other points on the parabola.

ILLUSTRATION 2

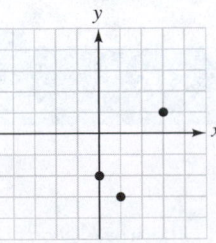

13. Sketch the graphs of parabolas with zero, one, and two x-intercepts.

14. Sketch the graph of a parabola that doesn't have a y-intercept, if possible.

15. HEALTH DEPARTMENT The number of cases of flu seen by doctors at a county health clinic each week during a 10-week period is described by the quadratic function graphed in Illustration 3. Write a brief summary report about the flu outbreak. What important piece of information does the vertex give?

ILLUSTRATION 3

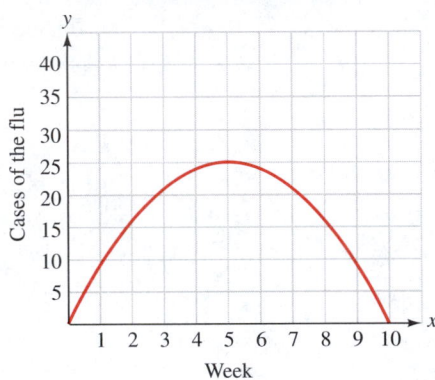

16. COST ANALYSIS A company has found that when it assembles x carburetors in a production run, the manufacturing cost $\$y$ per carburetor is given by the quadratic function graphed in Illustration 4. What important piece of information does the vertex give?

ILLUSTRATION 4

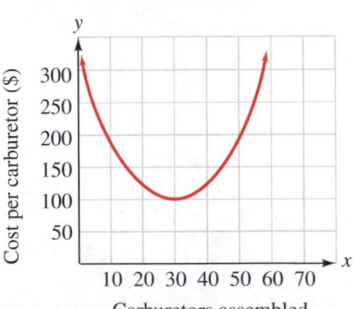

17. The graph of $f(x) = -x^2 + 4x - 3$ is shown in Illustration 5. What are the solutions of $-x^2 + 4x - 3 = 0$?

ILLUSTRATION 5

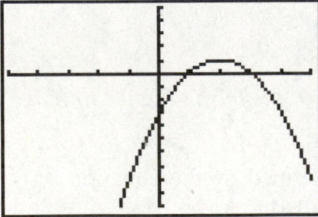

18. The graph of $y = 2x^2 - 4x - 6$ is shown in Illustration 6. What are the solutions of $2x^2 - 4x - 6 = 0$?

ILLUSTRATION 6

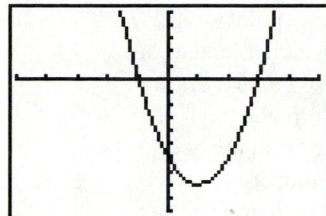

NOTATION

19. Tell whether this statement is true or false: The equations $y = 2x^2 - x - 2$ and $f(x) = 2x^2 - x - 2$ are the same.

20. The function $y = -x^2 + 3x - 5$ is written in $y = ax^2 + bx + c$ form. What are a, b, and c?

PRACTICE

In Exercises 21–24, graph each quadratic function and compare the graph to the graph of $y = x^2$.

21. $y = x^2 + 1$

22. $y = x^2 - 4$

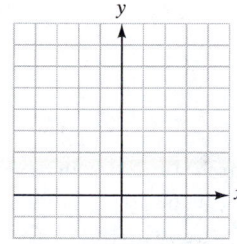 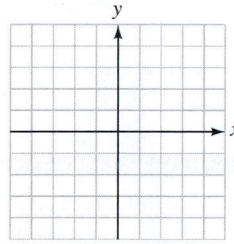

23. $f(x) = -x^2$

24. $f(x) = (x - 1)^2$

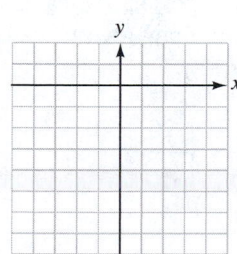 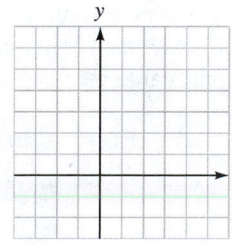

In Exercises 25–28, find the vertex of the graph of each quadratic function.

25. $y = -x^2 + 6x - 8$

26. $y = -x^2 - 2x - 1$

27. $f(x) = 2x^2 - 4x + 1$

28. $f(x) = 2x^2 + 8x - 4$

In Exercises 29–32, find the x- and y-intercepts of the graph of the quadratic function.

29. $f(x) = x^2 - 2x + 1$

30. $f(x) = 2x^2 - 4x$

31. $y = -x^2 - 10x - 21$

32. $y = 3x^2 + 6x - 9$

In Exercises 33–50, graph each quadratic function. Use the method discussed in Example 5.

33. $y = x^2 - 2x$

34. $f(x) = -x^2 - 4x$

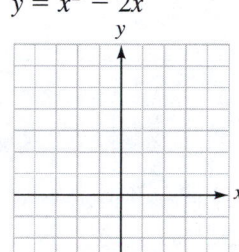 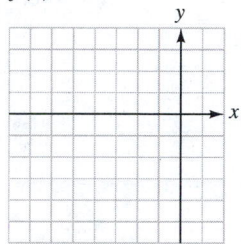

35. $f(x) = -x^2 + 2x$

36. $y = x^2 + x$

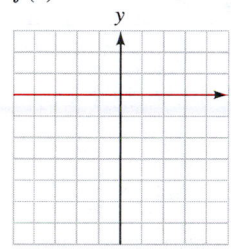 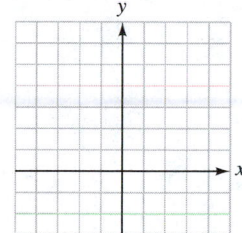

37. $f(x) = x^2 + 4x + 4$

38. $f(x) = x^2 - 6x + 9$

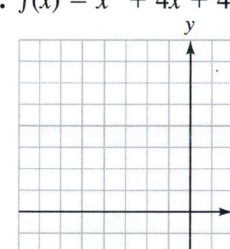 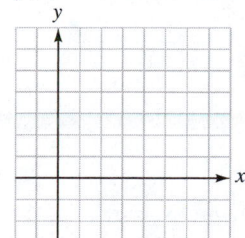

39. $y = -x^2 - 2x - 1$

40. $y = -x^2 + 2x - 1$

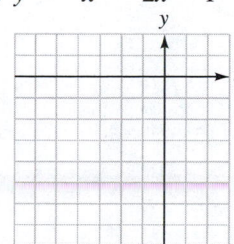 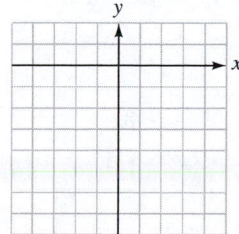

41. $y = x^2 + 2x - 3$

42. $y = x^2 + 6x + 5$

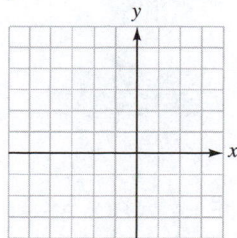 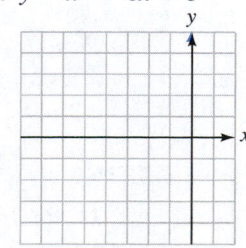

43. $f(x) = 2x^2 + 8x + 6$

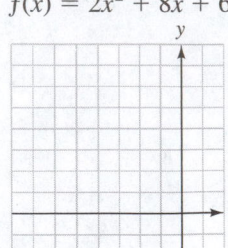

44. $f(x) = 3x^2 - 12x + 9$

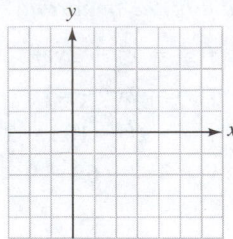

45. $y = x^2 - 2x - 8$

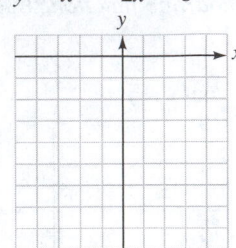

46. $y = -x^2 + 2x + 3$

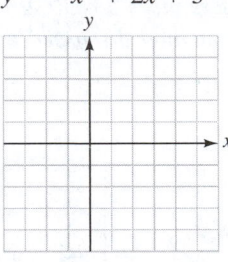

47. $y = x^2 - x - 2$

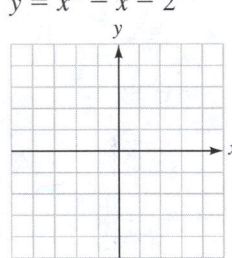

48. $y = -x^2 + 5x - 4$

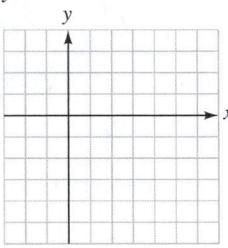

49. $f(x) = 2x^2 + 3x - 2$

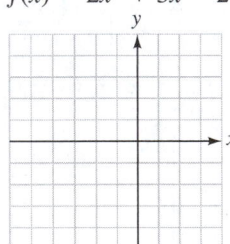

50. $f(x) = 3x^2 - 7x + 2$

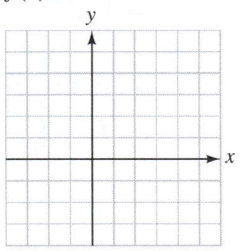

In Exercises 51–60, solve each quadratic equation using a graphing calculator.

51. $x^2 - 3x + 2 = 0$

52. $x^2 + x - 6 = 0$

53. $-4x^2 - 8x - 3 = 0$

54. $16x^2 + 8x - 3 = 0$

55. $-6x^2 - 11x + 10 = 0$

56. $5x^2 + 11x = 0$

57. $0.5x^2 - 0.7x - 3 = 0$

58. $x^2 - 2x + 1 = 0$

59. $x^2 - x + 0.25 = 0$

60. $-16x^2 + 56x - 49 = 0$

APPLICATIONS

61. REFLECTIVE PROPERTY OF PARABOLAS In Illustration 7, plot (0, 0), (1, 0.2), (2, 0.8), (3, 1.8), (4, 3.2), (4.5, 4.1), (−1, 0.2), (−2, 0.8), (−3, 1.8), (−4, 3.2), and (−4.5, 4.1) and connect the points with a smooth curve. This is a side view of a mirrored reflector of a flashlight. Draw the parabola's axis of symmetry. Plot the point (0, 1.25), which locates the filament of the lightbulb. Finally, draw a line representing a "ray" of light coming from the bulb, striking the mirrored surface at (2, 0.8) and then reflecting outward, parallel to the axis of symmetry.

ILLUSTRATION 7

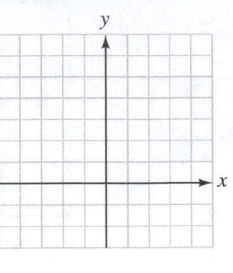

62. PARABOLIC DISH In Illustration 8, plot (0, 0), (0.1, 1), (0.4, 2), (0.9, 3), (1.6, 4), (2.5, 5), (0.1, −1), (0.4, −2), (0.9, −3), (1.6, −4), and (2.5, −5) and connect the points with a smooth curve. This is a side view of a parabolic dish used to pick up and amplify sound waves. In this case, the parabola opens to the right. Draw the axis of symmetry. Plot the point (2.5, 0), which locates the microphone that picks up the reflected sound waves. Finally, draw a line representing a single wave of sound, coming *into* the dish, parallel to the axis of symmetry. Show it striking the dish at (0.9, −3) and reflecting into the microphone.

ILLUSTRATION 8

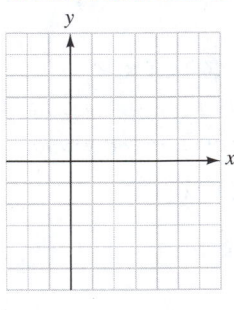

63. BRIDGES The shapes of the suspension cables in certain types of bridges are parabolic. The suspension cable for the bridge shown in Illustration 9 is described by $y = 0.005x^2$. Finish the mathematical model of the bridge by completing the table of values using a calculator, plotting the points, and drawing a smooth curve through them to represent the cable. Finally, from each plotted point, draw a vertical support cable attached to the roadway.

x	−80	−60	−40	−20	0	20	40	60	80
y									

ILLUSTRATION 9

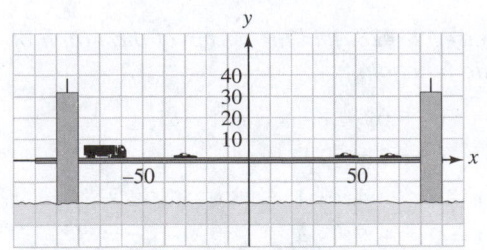

64. PROJECTILES If we disregard air resistance and other outside factors, the path of a projectile, such as a kicked soccer ball, is parabolic. Suppose the path of the soccer ball after it is kicked is given by the quadratic function $y = -0.5x^2 + 2x$. Use a calculator to complete the table of values in Illustration 10 and then plot the points and draw a smooth curve through them to depict the ball's path.

x	0	0.5	1	1.5	2	2.5	3	3.5	4
y									

ILLUSTRATION 10

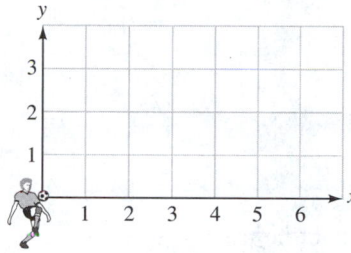

65. SELLING TVS A company has found that it can sell x TVs at a price of $\$(450 - \frac{1}{6}x)$.
  a. How many TVs must the company sell to maximize its revenue?
  b. Find the maximum revenue.

66. SELLING CD PLAYERS A wholesaler sells CD players for $150 each. However, she gives volume discounts on purchases of 500 to 1,000 units according to the formula $\left(150 - \frac{1}{10}n\right)$, where n represents the number of units purchased.
  a. How many units would a retailer have to buy for the wholesaler to obtain maximum revenue?

  b. Find the maximum revenue.

WRITING

67. Explain why the y-intercept of the graph of $y = ax^2 + bx + c$ $(a \neq 0)$ is $(0, c)$.

68. Use the example of a stream of water from a drinking fountain to explain the concept of the vertex of a parabola.

69. Explain why parabolas that open left or right aren't the graphs of functions.

70. Is it possible for the graph of a parabola not to have a y-intercept? Explain.

REVIEW

In Exercises 71–74, do each multiplication.

71. $3x(x - 2)$

72. $(x - 3)(x + 2)$

73. $(2x + 1)(3x - 2)$

74. $(2x + 3)(x^2 + 4)$

Quadratic Equations

In this chapter, we have studied several ways to solve quadratic equations. We have also graphed quadratic functions and seen that their graphs are parabolas.

What Is a Quadratic Equation?

A quadratic equation can be written in the form $ax^2 + bx + c = 0$ where $a \neq 0$. In Exercises 1–12, which are quadratic equations?

1. $y = 3x + 7$

2. $4(x + 5) = 2x$

3. $2x^2 - 3x + 4 = 0$

4. $y(y - 6) = 0$

5. $a^2 + 7a - 1 > 0$

6. $3y^2 - y + 4$

7. $5 = y - y^2$

8. $|x - 8|$

9. $\sqrt{x + 7} = 4$

10. $x^2 = 16$

11. $\dfrac{m}{2} - \dfrac{1}{3} = \dfrac{1}{4}$

12. $C = \dfrac{5}{9}(F - 32)$

Solving Quadratic Equations

The techniques we have used to solve linear equations cannot be used to solve a quadratic equation, because those techniques cannot isolate the variable on one side of the equation. Exercises 13 and 14 show examples of student work to solve quadratic equations. In each case, what did the student do **wrong** and why is it incorrect?

13. Solve $x^2 = 6$.

$$\frac{x^2}{2} = \frac{6}{2}$$

$$x = 3$$

14. Solve $x^2 - x = 10$.

$$x(x - 1) = 10$$
$$x = 0 \quad \text{or} \quad x - 1 = 10$$
$$x = 11$$

In Exercises 15–20, solve each quadratic equation using the method listed. If an answer is not exact, round to the nearest hundredth.

15. $4x^2 - x = 0$; factoring method

16. $x^2 + 3x + 1 = 0$; quadratic formula

17. $x^2 = 36$; square root method

18. $a^2 - a - 56 = 0$; factoring method

19. $x^2 + 4x + 1 = 0$; quadratic formula

20. $(x + 3)^2 = 16$; square root method

Quadratic Functions

The graph of the quadratic function $y = ax^2 + bx + c$ $(a \neq 0)$ is a parabola. It opens upward when $a > 0$ and downward when $a < 0$.

21. The formula $R = 4x - x^2$ gives the revenue R (in tens of thousands of dollars) that a business obtains from the manufacture and sale of x patio chairs (in hundreds). Graph $R = 4x - x^2$ in Illustration 1.

22. Find the vertex of the parabola. What is its significance concerning the revenue the business brings in?

ILLUSTRATION 1

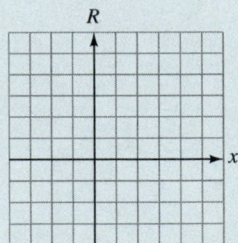

Accent on Teamwork

Section 5.1

Prime numbers We can use a procedure called the *sieve of Eratosthenes* to find all the prime numbers in the set of the first 100 whole numbers. Give each member in your group a copy of the table shown in Illustration 1. Cross out 1, since it is not a prime number by definition. Cross out any numbers divisible by 2, 3, 5, 7, or 11, because they have a factor of 2, 3, 5, 7, or 11 and thus would not be prime. Don't cross out 2, 3, 5, 7, or 11, because they are prime numbers. At the end of this process, you should end up with the first 25 prime numbers.

ILLUSTRATION 1

1	2	3	4	5	6	7	8	9	10
11	12	13	14	15	16	17	18	19	20
21	22	23	24	25	26	27	28	29	30
31	32	33	34	35	36	37	38	39	40
41	42	43	44	45	46	47	48	49	50
51	52	53	54	55	56	57	58	59	60
61	62	63	64	65	66	67	68	69	70
71	72	73	74	75	76	77	78	79	80
81	82	83	84	85	86	87	88	89	90
91	92	93	94	95	96	97	98	99	100

Section 5.2

Factoring the difference of two squares Use Illustration 2 to show that

$$x^2 - y^2 = (x + y)(x - y)$$

ILLUSTRATION 2

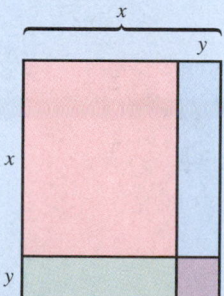

Section 5.3

Factoring trinomials Use colored paper to create the model shown in Illustration 3. Note that the total area of the figure is $(x + 1)(x + 1)$ square units. Disassemble the model and show that it is composed of four pieces with areas of x^2, x, x, and 1 square units. Label each piece with its area. What is the relationship between $x^2 + 2x + 1$ and $(x + 1)(x + 1)$? Next, create a model with an area of $(x + 2)(x + 2)$ square units. What observations do you have about this example?

ILLUSTRATION 3

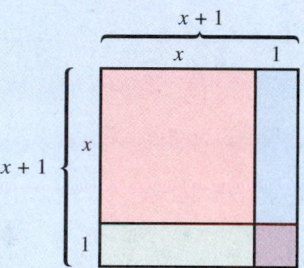

Section 5.4

Solving quadratic equations The techniques used to solve linear equations cannot be used to solve a quadratic equation. Write a brief note to the student whose work in "solving" $x^2 + 3x - 1 = 0$ is shown here and explain the **misunderstanding.**

$$x^2 + 3x - 1 = 0$$
$$x^2 + 3x = 1 \qquad \text{Add 1.}$$
$$3x = 1 - x^2 \qquad \text{Subtract } x^2.$$
$$\frac{3x}{3} = \frac{1 - x^2}{3} \qquad \text{Divide by 3.}$$
$$\text{So } x = \frac{1 - x^2}{3}$$

Section 5.5

Solving quadratic equations Solve the quadratic equation $2x^2 - x - 1 = 0$ using two methods: first by factoring, and then by the quadratic formula. Write each solution on a separate piece of poster board. Under each solution, in two columns, list the advantages and the drawbacks of each method.

Section 5.6

Parabolas Use a home video camera to make a "documentary" showing examples of parabolic shapes you see in everyday life. Write a script for your video and have a narrator explain the setting, point out the vertex, and tell whether the parabola opens upward or downward in each case.

Section 5.1

Factoring Out the Greatest Common Factor and Factoring by Grouping

CONCEPTS

A *prime number* is a natural number greater than 1 whose only factors are 1 and itself. A natural number is in *prime-factored form* when it is written as the product of prime numbers.

To find the *greatest common factor* (GCF) of several monomials:
1. Prime-factor each monomial.
2. List each common factor the least number of times it appears in any one monomial.
3. Find the product of the factors in the list to obtain the GCF.

To *factor by grouping*, arrange the polynomial so that the first two terms have a common factor and the last two terms have a common factor. Factor out the common factor from both groups. Then factor out the resulting common binomial factor.

REVIEW EXERCISES

1. Find the prime factorization of each number.
  a. 35   **b.** 45
  c. 96   **d.** 99
  e. 2,050   **f.** 4,096

2. Factor out the GCF.
  a. $3x + 9y$ **b.** $5ax^2 + 15a$
  c. $7s^2 + 14s$ **d.** $\pi ab - \pi ac$
  e. $2x^3 + 4x^2 - 8x$ **f.** $x^2yz + xy^2z + xyz$
  g. $-5ab^2 + 10a^2b - 15ab$
  h. $4(x - 2) - x(x - 2)$

3. Factor out -1 from each polynomial.
  a. $-a - 7$ **b.** $-4t^2 + 3t - 1$

4. Factor by grouping:
  a. $2c + 2d + ac + ad$ **b.** $3xy + 9x - 2y - 6$
  c. $2a^3 - a + 2a^2 - 1$ **d.** $4m^2n + 12m^2 - 8mn - 24m$

Section 5.2

Factoring the Difference of Two Squares

To factor the *difference of two squares,* use the formula

$$\mathbf{F^2 - L^2 = (F + L)(F - L)}$$

When factoring a polynomial, always factor out any common monomials first.

5. Factor each polynomial completely, if possible.
  a. $x^2 - 9$ **b.** $49t^2 - 25y^2$
  c. $x^2y^2 - 400$ **d.** $8at^2 - 32a$
  e. $c^4 - 64$ **f.** $h^2 + 36$

6. Factor each expression completely.
  a. $6x^2y - 24y^3$
  b. $2x^4 - 162$
  c. $-m^2 + 100$

Section 5.3

To *factor a trinomial* of the form $x^2 + bx + c$ means to write it as the product of two binomials.

To factor $x^2 + bx + c$, find two integers whose product is c and whose sum is b.

$$(x + \boxed{})(x + \boxed{})$$

Write the trinomial in descending powers of the variable and factor out -1 when applicable.

If a trinomial cannot be factored using only integers, it is called a *prime polynomial*.

The *GCF* should always be factored out first. A trinomial is *factored completely* when it is expressed as a product of prime polynomials.

To factor $ax^2 + bx + c$ using the *trial-and-check* factoring method, we must determine four integers. Use the FOIL method to check your work.

Factors
of a

$$\left(\boxed{}x + \boxed{}\right)\left(\boxed{}x + \boxed{}\right)$$

Factors
of c

To factor $ax^2 + bx + c$ using the *grouping method*, we write it as

$$ax^2 + \boxed{}x + \boxed{}x + c$$

Factoring Trinomials

7. Complete the table.

Product of the factors of 6	Sum of the factors of 6
1(6)	
2($\boxed{}$)	
$\boxed{}$(−6)	
−2(−3)	

8. Factor each trinomial, if possible.
   a. $x^2 + 2x - 24$ **b.** $x^2 - 4x - 12$
   c. $n^2 - 7n + 10$ **d.** $t^2 + 10t + 15$
   e. $-y^2 + 9y - 20$ **f.** $10y + 9 + y^2$

   g. $c^2 + 3cd - 10d^2$ **h.** $-3mn + m^2 + 2n^2$

9. Explain how we can check to see if $(x - 4)(x + 5)$ is the factorization of $x^2 + x - 20$.

10. Completely factor each trinomial.
   a. $5a^2 + 45a - 50$ **b.** $-4x^2y - 4x^3 + 24xy^2$

11. Factor each trinomial completely, if possible.
   a. $2x^2 - 5x - 3$ **b.** $10y^2 + 21y - 10$
   c. $-3x^2 + 14x + 5$ **d.** $8a^2 + 16a + 6$

   e. $-9p^2 - 6p + 6p^3$ **f.** $4b^2 + 15bc - 4c^2$

   g. $3y^2 + 7y - 11$ **h.** $7r^4 + 31r^2 + 12$

12. ENTERTAINING The rectangular-shaped area occupied by a table setting shown in Illustration 1 is $(12x^2 - x - 1)$ square inches. Factor the expression to find the binomials that represent the length and width of the table setting.

ILLUSTRATION 1

Section 5.4

A *quadratic equation* is an equation of the form $ax^2 + bx + c = 0$ $(a \neq 0)$, where a, b, and c are real numbers.

To use the *factoring method* to solve a quadratic equation:

1. Write the equation in $ax^2 + bx + c = 0$ form.
2. Factor the left-hand side.
3. Use the *zero-factor property* (if $ab = 0$, then $a = 0$ or $b = 0$) and set each factor equal to zero.
4. Solve each resulting linear equation.

Quadratic Equations

13. Solve each quadratic equation by factoring.

  a. $x^2 + 2x = 0$ **b.** $2x^2 - 6x = 0$

  c. $x^2 - 9 = 0$ **d.** $36p^2 = 25$

  e. $a^2 - 7a + 12 = 0$ **f.** $t^2 + 4t + 4 = 0$

  g. $2x - x^2 + 24 = 0$ **h.** $(t + 1)(8t + 1) = 18t$

  i. $x^2 - 11x = 12$ **j.** $2p^3 = 2p(p + 2)$

14. CONSTRUCTION The face of the triangular preformed concrete panel shown in Illustration 2 has an area of 45 square meters, and its base is 3 meters longer than twice its height. How long is its base?

ILLUSTRATION 2

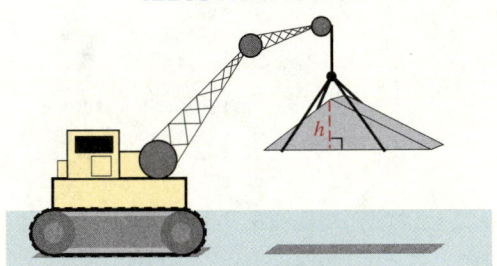

15. GARDENING A rectangular flower bed occupies 27 square feet and is 3 feet longer than twice its width. Find its dimensions.

Section 5.5

For the *general quadratic equation* $ax^2 + bx + c = 0$, where $a \neq 0$,

$$x = \frac{-b \pm \sqrt{b^2 - 4ac}}{2a}$$

This expression is called the *quadratic formula*.

The Quadratic Formula

16. Use the quadratic formula to solve each quadratic equation.

  a. $x^2 - 2x - 15 = 0$ **b.** $x^2 - 6x - 7 = 0$

  c. $6x^2 - 7x - 3 = 0$ **d.** $x^2 - 6x + 7 = 0$

17. Use the quadratic formula to solve $3x^2 + 2x - 2 = 0$. Give the solutions in exact form and then rounded to the nearest hundredth.

18. Use the quadratic formula to solve $10x^2 + 2x + 1 = 0$.

19. SECURITY GATE The length of the frame for the iron gate in Illustration 3 is 14 feet longer than the width. The area of the gate is 240 square feet. Find the width and length of the gate frame.

ILLUSTRATION 3

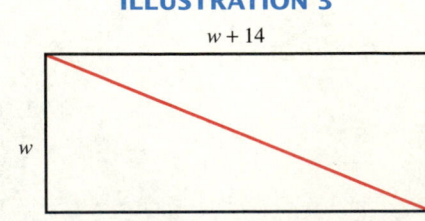

20. MILITARY A pilot releases a bomb from an altitude of 3,000 feet. The bomb's height h above the target t seconds after its release is given by the formula

$$h = 3,000 + 40t - 16t^2$$

How long will it be until the bomb hits its target?

Strategy for solving quadratic equations:
1. Try the square root method.
2. If it doesn't apply, write the equation in $ax^2 + bx + c = 0$ form.
3. Try the factoring method.
4. If it doesn't work, use the quadratic formula.

21. Use the most convenient method to find all real solutions of each equation.

a. $x^2 + 5x + 2 = 0$

b. $(y + 3)^2 = 16$

c. $x^2 + 5x = 0$

d. $2x^2 + x = 5$

e. $g^2 - 21 = 0$

f. $a^2 = 4a - 4$

Section 5.6

Graphing Quadratic Functions

The *vertex* of a parabola is the lowest (or highest) point on the parabola.

22. See the graph in Illustration 4.

   a. What are the *x*-intercepts of the parabola?

   b. What is the *y*-intercept of the parabola?

   c. What is the vertex of the parabola?

   d. Draw the axis of symmetry of the parabola on the graph.

A vertical line through the vertex of a parabola that opens upward or downward is its *axis of symmetry*.

23. What important information can be obtained from the vertex of the parabola in Illustration 5?

ILLUSTRATION 4

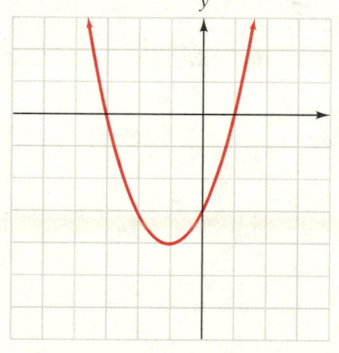

ILLUSTRATION 5

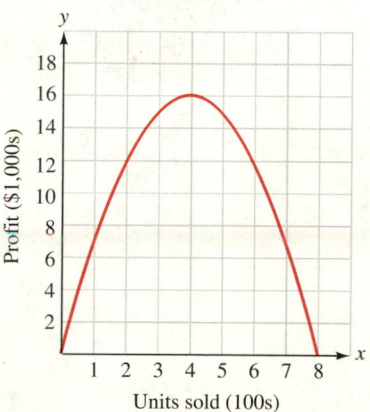

Units sold (100s)

The graph of the quadratic function $y = ax^2 + bx + c$ is a parabola. It opens upward when $a > 0$ and downward when $a < 0$.

24. Find the vertex of the graph of each quadratic function and tell which direction the parabola opens. **Do not draw the graph.**

   a. $y = 2x^2 - 4x + 7$ **b.** $y = -3x^2 + 18x - 11$

The *x*-coordinate of the vertex of the parabola $y = ax^2 + bx + c$ is $x = -\frac{b}{2a}$. To find the *y*-coordinate of the vertex, substitute $-\frac{b}{2a}$ for x in the equation of the parabola and find y.

25. Find the *x*- and *y*-intercepts of the graph of $y = x^2 + 6x + 5$.

The x-intercepts of a parabola are determined by solving $ax^2 + bx + c = 0$. The y-intercept is $(0, c)$.

26. Graph each quadratic function by finding the vertex, x- and y-intercepts, and axis of symmetry of its graph.

a. $y = x^2 + 2x - 3$

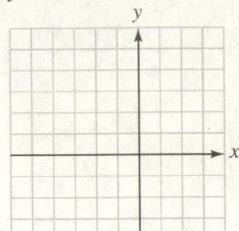

b. $f(x) = -2x^2 + 4x - 2$

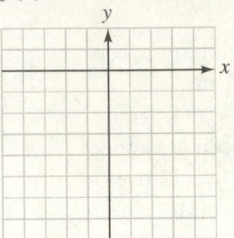

27. Use the graph in Illustration 6 to solve the quadratic equation $x^2 + 2x - 3 = 0$.

ILLUSTRATION 6

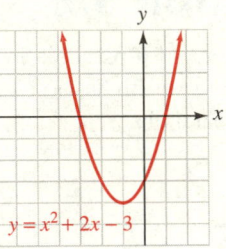

$y = x^2 + 2x - 3$

CHAPTER 5

Test

In Problems 1–2, find the prime factorization of each number.

1. 196

2. 111

In Problems 3–12, factor each expression, if possible.

3. $4x + 16$

4. $30a^2b^3 - 20a^3b^2 + 5abc$

5. $q^2 - 81$

6. $x^2 + 9$

7. $16x^4 - 81$

8. $x^2 + 4x + 3$

9. $-x^2 + 9x + 22$

10. $3x^2 + 4 + 13x$

11. $2a^2 + 5a - 12$

12. $12p^2 + 6p - 36$

In Problems 13–16, solve each equation by factoring.

13. $6x^2 - x = 0$

14. $x^2 + 6x + 9 = 0$

15. $6x^2 + x - 1 = 0$

16. $10x^2 + 43x = 9$

In Problems 17–18, solve each equation by the square root method.

17. $x^2 = 16$

18. $(x - 2)^2 = 3$

In Problems 19–21, use the quadratic formula to solve each equation.

19. $x^2 + 3x - 10 = 0$

20. $2x^2 - 5x = 12$

21. $x^2 = 5x + 2$

22. The base of a triangle with an area of 40 square meters is 2 meters longer than the triangle is high. Find the length of the base.

23. ADVERTISING When a business runs x advertisements per week on television, the number y of air conditioners it sells is given by the quadratic function graphed in Illustration 1. What important information can be obtained from the vertex?

ILLUSTRATION 1

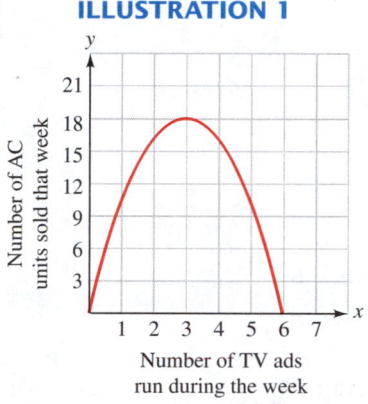

Number of TV ads
run during the week

24. Graph the function $y = x^2 + x - 2$ by finding the vertex, x- and y-intercepts, and axis of symmetry.

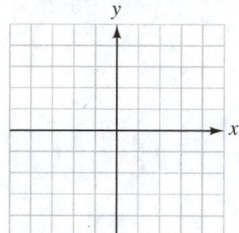

25. Find the vertex of the parabola determined by the equation $y = -4x^2 - 40x - 104$.

26. Explain how the FOIL method can be used to check the factorization of $x^2 - 3x - 54$.

6

Proportion and Rational Expressions

CAMPUS CONNECTION

The Public Works Department

In Public Works classes, students learn about water systems and water management. As their instruction progresses, they see that algebra is used by people who work in the field as well as those in the office. For instance, if a manager knows the time it takes each of two inlet pipes to fill a water storage tank working alone, he can use algebra to find the time it would take the two pipes working together to fill the empty tank. This is done by solving an *equation that contains algebraic fractions*, known as rational expressions. In this chapter, we will use such equations to solve problems concerning finances, travel, and public works projects.

Ratios can be used to solve problems involving pricing, travel, mixtures, and manufacturing. They are examples of a broader set of algebraic expressions called rational expressions.

▶ 6.1 Ratios

In this section, you will learn about

Ratios ■ Unit costs ■ Rates

Introduction The concept of *ratio* is often applicable in real-life situations. For example,

■ To prepare fuel for a Lawnboy lawnmower, gasoline must be mixed with oil in the ratio of 50 to 1.

■ To make 14-karat jewelry, gold is mixed with other metals in the ratio of 14 to 10.

■ In the stock market, winning stocks might outnumber losing stocks in the ratio of 7 to 4.

■ At Citrus College, the ratio of students to faculty is 21 to 1.

In this section, we will discuss ratios and use them to solve problems. Since ratios are fractions, you may want to read Appendix I, which reviews the basic properties of fractions.

Ratios

Ratios give us a way to compare numerical quantities.

Ratios

A **ratio** is the quotient of two numbers or two quantities.

There are three common ways to write a ratio: as a fraction, with the word *to*, or with a colon.

Some examples of ratios are

$$\frac{7}{9}, \qquad 21 \text{ to } 27, \qquad \text{and} \qquad 2{,}290\!:\!1{,}317$$

■ The fraction $\frac{7}{9}$ can be read as "the ratio of 7 to 9."

■ 21 to 27 can be read as "the ratio of 21 to 27" and can be written as $\frac{21}{27}$.

■ 2,290:1,317 can be read as "the ratio of 2,290 to 1,317" and can be written as $\frac{2{,}290}{1{,}137}$.

Because the fractions $\frac{7}{9}$ and $\frac{21}{27}$ represent equal numbers, they are **equal ratios.**

 WARNING! Unlike the fraction $\frac{a}{b}$, b can be zero in the ratio $\frac{a}{b}$. For example, the ratio of women to men on a women's softball team could be 25 to 0. However, these applications are rare.

EXAMPLE 1

Writing ratios. Express each phrase as a ratio in lowest terms: **a.** the ratio of 15 to 12   and   **b.** the ratio of 0.3 to 1.2.

Solution   **a.** The ratio of 15 to 12 can be written as the fraction $\frac{15}{12}$. After simplifying, the ratio is $\frac{5}{4}$.

$$\frac{15}{12} = \frac{\overset{1}{\cancel{3}} \cdot 5}{\underset{1}{\cancel{3}} \cdot 4} \qquad \text{Factor 15 and 12. Then divide out the common factor of 3.}$$

$$= \frac{5}{4}$$

b. The ratio of 0.3 to 1.2 can be written as the fraction $\frac{0.3}{1.2}$. We can simplify this fraction as follows:

$$\frac{0.3}{1.2} = \frac{0.3 \cdot \mathbf{10}}{1.2 \cdot \mathbf{10}} \qquad \text{To clear the fraction of decimals, multiply both the numerator and the denominator by 10.}$$

$$= \frac{3}{12} \qquad \text{Multiply: } 0.3 \cdot 10 = 3 \text{ and } 1.2 \cdot 10 = 12.$$

$$= \frac{1}{4} \qquad \text{Simplify the fraction: } \frac{3}{12} = \frac{\overset{1}{\cancel{3}} \cdot 1}{\underset{1}{\cancel{3}} \cdot 4} = \frac{1}{4}.$$

SELF CHECK   Express each ratio in lowest terms: **a.**   The ratio of 8 to 12   and   **b.** The ratio of 3.2 to 16.   *Answers:*   **a.** $\frac{2}{3}$,   **b.** $\frac{1}{5}$ ■

EXAMPLE 2

Writing ratios. Express each phrase as a ratio in lowest terms: **a.** the ratio of 3 meters to 8 meters   and   **b.** the ratio of 4 ounces to 1 pound.

Solution   **a.** The ratio of 3 meters to 8 meters can be written as the fraction $\frac{3 \text{ meters}}{8 \text{ meters}}$, or simply $\frac{3}{8}$.

b. When possible, we should express ratios in the same units. Since there are 16 ounces in 1 pound, the proper ratio is $\frac{4 \text{ ounces}}{16 \text{ ounces}}$, which simplifies to $\frac{1}{4}$.

SELF CHECK   Express each ratio in lowest terms: **a.** the ratio of 8 ounces to 2 pounds   and   **b.** the ratio of 1 foot to 2 yards.   (*Hint:* 3 feet = 1 yard.)   *Answers:*   **a.** $\frac{1}{4}$,   **b.** $\frac{1}{6}$ ■

EXAMPLE 3

Student-to-faculty ratios. At a college, there are 2,772 students and 154 faculty members. Write a fraction in simplified form to express the ratio of students per faculty member.

Solution   The ratio of students to faculty is 2,772 to 154. We can write this ratio as the fraction $\frac{2,772}{154}$ and simplify it.

$$\frac{2,772}{154} = \frac{2 \cdot 2 \cdot 3 \cdot 3 \cdot \overset{1}{\cancel{7}} \cdot \overset{1}{\cancel{11}}}{\underset{1}{\cancel{2}} \cdot \underset{1}{\cancel{7}} \cdot \underset{1}{\cancel{11}}} \qquad \text{Prime factor 2,772 and 154. Divide out common factors.}$$

$$= \frac{18}{1}$$

The ratio of students to faculty is 18 to 1.

SELF CHECK

In a college graduating class, 224 students out of 632 went on to graduate school. Write a fraction in simplified form to express the ratio of the number of students continuing their education to the number in the graduating class.

Answer: $\frac{28}{79}$ ∎

Unit Costs

The *unit cost* of an item is the ratio of its cost to its quantity. For example, the unit cost (the cost per pound) of 5 pounds of coffee priced at $20.75 is given by the ratio

$$\frac{\$20.75}{5 \text{ pounds}} = \frac{\$2,075}{500 \text{ pounds}}$$ To eliminate the decimal, multiply the numerator and the denominator by 100.

$$= \$4.15 \text{ per pound}$$ Do the division.

The unit cost is $4.15 per pound.

EXAMPLE 4

Comparison shopping. Olives come packaged in a 12-ounce jar, which sells for $3.09, or in a 6-ounce jar, which sells for $1.53. Which is the better buy?

Solution

To find the better buy, we must compare the unit costs. The unit cost of the 12-ounce jar is

$$\frac{\$3.09}{12 \text{ ounces}} = \frac{309¢}{12 \text{ ounces}}$$ Change $3.09 to 309¢.

$$= 25.75¢ \text{ per ounce.}$$ Simplify by dividing.

The unit cost of the 6-ounce jar is

$$\frac{\$1.53}{6 \text{ ounces}} = \frac{153¢}{6 \text{ ounces}}$$ Change $1.53 to 153¢.

$$= 25.5¢ \text{ per ounce.}$$ Do the division.

Since the unit cost is less when olives are packaged in 6-ounce jars, that is the better buy.

SELF CHECK

A fast-food restaurant sells a 12-ounce Coke for 79¢ and a 16-ounce Coke for 99¢. Which is the better buy? *Answer:* the 16-oz Coke ∎

Rates

When ratios are used to compare quantities with different units, they are called *rates*. For example, if the 495-mile drive from New Orleans to Dallas takes 9 hours, the average rate of speed is the ratio of the miles driven to the length of time of the trip.

$$\text{Average rate of speed} = \frac{495 \text{ miles}}{9 \text{ hours}} = \frac{55 \text{ miles}}{1 \text{ hour}} \qquad \frac{495}{9} = \frac{\overset{1}{\cancel{9}} \cdot 55}{\underset{1}{\cancel{9}} \cdot 1} = \frac{55}{1}.$$

The ratio $\frac{55 \text{ miles}}{1 \text{ hour}}$ can be expressed in any of the following forms:

$$55 \frac{\text{miles}}{\text{hour}}, \qquad 55 \text{ miles per hour}, \qquad 55 \text{ miles/hour}, \qquad \text{or} \qquad 55 \text{ mph}$$

EXAMPLE 5

Finding hourly rates of pay. Find the hourly rate of pay for a student who earns $370 for working 40 hours.

Solution   We can write the rate of pay as the ratio

$$\text{Rate of pay} = \frac{\$370}{40 \text{ hours}}$$

and simplify by dividing 370 by 40.

$$\text{Rate of pay} = 9.25 \frac{\$}{\text{hour}}$$

The rate is $9.25 per hour.

SELF CHECK

In 1997, the average weekly pay of a member of a union working full-time was $640. Based on a 40-hour work week, what was the average rate of pay per hour?   *Answer:* $16 per hour   ■

EXAMPLE 6

Electric bill. The amount of electricity a household uses is measured in kilowatt hours by a meter like that shown in Figure 6-1. Determine the rate of energy consumption in kilowatt hours per day from the readings listed in the table.

FIGURE 6-1

Meter Reading

Previous From 4/1/98	Present To 4/30/98
4589	5384

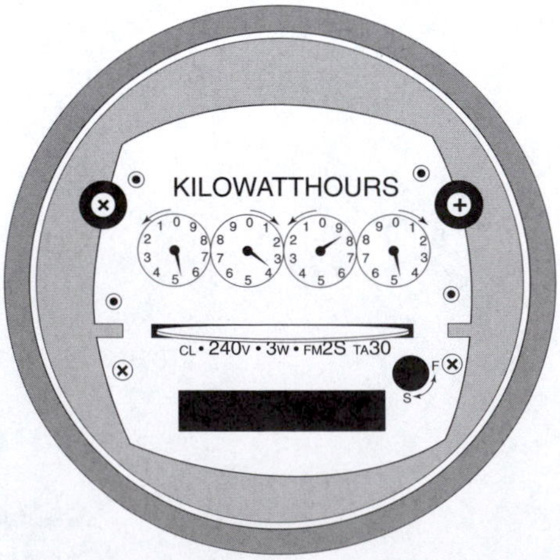

Solution   To find the number of kilowatt hours of electricity used, we subtract the meter reading on the first day of the month from the reading on the last day of the month.

$$5,384 - 4,589 = 795$$

We can write the rate of energy consumption as the ratio

$$\text{Rate of energy consumption} = \frac{795 \text{ kilowatt hours}}{30 \text{ days}}$$

From 4/1/98 to 4/30/98 is 30 days.

and simplify by dividing 795 by 30.

$$\text{Rate of energy consumption} = 26.5 \frac{\text{kilowatt hours}}{\text{day}}$$

The rate of consumption in April was 26.5 kilowatt hours per day. ∎

EXAMPLE 7

Tax rates. A book cost $49.22, including $3.22 sales tax. Find the sales tax rate.

Solution   Since the tax was $3.22, the cost of the book alone was

$$\$49.22 - \$3.22 = \$46.00$$

We can write the sales tax rate as the ratio

$$\text{Sales tax rate} = \frac{\text{amount of sales tax}}{\text{cost of the book, without tax}}$$

$$= \frac{\$3.22}{\$46}$$

and simplify by dividing 3.22 by 46.

$$\text{Sales tax rate} = 0.07$$

The tax rate is 0.07, or 7%.

SELF CHECK   A married couple, filing jointly, had a taxable income of $22,420. They paid $3,363 in federal income tax. What tax bracket were they in? That is, what income tax rate did they have to pay?   *Answer:* 15% ∎

ACCENT ON TECHNOLOGY *Computing Gas Mileage*

A man drove a total of 775 miles. Along the way, he stopped for gas three times, pumping 10.5, 11.3, and 8.75 gallons of gas. He started with the tank half full and ended with the tank half full. To find how many miles he got per gallon, we need to divide the total distance by the total number of gallons of gas consumed.

$$\frac{775}{10.5 + 11.3 + 8.75} \quad \begin{array}{l} \leftarrow \text{Total distance} \\ \leftarrow \text{Total number of gallons consumed} \end{array}$$

Using a scientific calculator, we enter these numbers and press these keys:

Keystrokes   775 ÷ (10.5 + 11.3 + 8.75) = | 25.36824877 |

Using a graphing calculator, we enter these numbers and press these keys:

Keystrokes   775 ÷ (10.5 + 11.3 + 8.75) ENTER

```
775/(10.5+11.3+8
.75)
         25.36824877
```

To the nearest hundredth, he got 25.37 mpg.

STUDY SET

Section 6.1

VOCABULARY

In Exercises 1–4, fill in the blanks to make the statements true.

1. A ratio is a _____ of two numbers by their indicated _____.

2. The _____ cost of an item is the ratio of its cost to its quantity.

3. The ratios $\frac{2}{3}$ and $\frac{4}{6}$ are _____ ratios.

4. The ratio $\frac{500 \text{ miles}}{15 \text{ hours}}$ is called a _____.

CONCEPTS

5. Give three examples of ratios that you have encountered this past week.

6. Suppose that a basketball player made 8 free throws out of 12 tries. The ratio of $\frac{8}{12}$ can be simplified as $\frac{2}{3}$. Interpret this result.

NOTATION

7. **a.** Write the ratio 3 to 11 as a fraction.
   b. Write the ratio $27:32$ as a fraction.

8. Express the ratio $\frac{468 \text{ miles}}{9 \text{ hours}}$ in two different ways.

PRACTICE

In Exercises 9–24, express each phrase as a ratio in lowest terms.

9. 5 to 7
10. 3 to 5
11. $17:34$
12. $19:38$
13. 22 to 33
14. 14 to 21
15. 7 to 24.5
16. 0.65 to 0.15
17. 4 ounces to 12 ounces
18. 3 inches to 15 inches
19. 12 minutes to 1 hour
20. 8 ounces to 1 pound
21. 3 days to 1 week
22. 4 inches to 2 yards
23. 18 months to 2 years
24. 8 feet to 4 yards

In Exercises 25–28, refer to the monthly family budget shown in Illustration 1. Give each ratio in lowest terms.

25. Find the total amount of the budget.

26. Find the ratio of the amount budgeted for rent to the total budget.

27. Find the ratio of the amount budgeted for entertainment to the total budget.

28. Find the ratio of the amount budgeted for phone to the amount budgeted for entertainment. Then complete this statement: For every $25 spent on the phone, $ ____ is spent on entertainment.

ILLUSTRATION 1

Item	Amount
Rent	$750
Food	$652
Gas and Electric	$188
Phone	$125
Entertainment	$110

In Exercises 29–32, refer to the tax deductions listed in Illustration 2. Give each ratio in lowest terms.

29. Find the total amount of deductions.

30. Find the ratio of the real estate tax deduction to the total deductions.

31. Find the ratio of the contributions to the total deductions.

32. Find the ratio of the mortgage deduction to the union dues deduction.

ILLUSTRATION 2

Item	Amount
Medical	$ 995
Real estate tax	$1,245
Contributions	$1,680
Mortgage	$4,580
Union dues	$ 225

APPLICATIONS

In Exercises 33–52, find each ratio and express it in lowest terms.

33. FACULTY-TO-STUDENT RATIO At a college, there are 125 faculty members and 2,000 students. Find the faculty-to-student ratio.

34. U.S. SENATE In the 105th Congress, which convened January 7, 1997, there were 9 women in the 100-member Senate. Find the ratio of women to men senators.

35. BLOOD PRESSURE A person's blood pressure is expressed as the ratio of the systolic pressure (when the heart contracts to empty its blood) to the diastolic pressure (when the heart relaxes to fill with blood). This ratio, written as a fraction, is not simplified. Complete the table in Illustration 3, which lists the "normal" readings for three age groups. Do you know your blood pressure ratio?

ILLUSTRATION 3

	Systolic	Diastolic	Ratio
Infant	80	45	
Age 30	120	80	
Age 40	140	85	

36. BICYCLING Illustration 4 shows a "high wheeler" bicycle, which was popular in the 1870s. Find the ratio of the circumference of the front wheel to the circumference of the rear wheel.

ILLUSTRATION 4

48 in.

15 in.

37. UNIT COST OF GASOLINE A driver pumped 17 gallons of gasoline into her tank at a cost of $21.59. Write a ratio of dollars to gallons, and give the unit cost of gasoline.

38. UNIT COST OF GRASS SEED A 50-pound bag of grass seed costs $222.50. Write a ratio of dollars to pounds, and give the unit cost of grass seed.

39. ENTERTAINMENT Disneyland has 63 rides and attractions. If an adult admission ticket costs $36, what is the unit cost of each ride and attraction if a person rides each of them once? Round to the nearest cent.

40. UNIT COST OF BEANS A 24-ounce package of green beans sells for $1.29. Give the unit cost in cents per ounce.

41. COMPARISON SHOPPING A 6-ounce can of orange juice sells for 89¢, and an 8-ounce can sells for $1.19. Which is the better buy?

42. COMPARING SPEEDS A car travels 345 miles in 6 hours, and a truck travels 376 miles in 6.2 hours. Which vehicle is going faster?

43. COMPARING READING SPEEDS One seventh-grader read a 54-page book in 40 minutes, and another read an 80-page book in 62 minutes. If the books were equally difficult, which student reads faster?

44. COMPARISON SHOPPING A 30-pound bag of fertilizer costs $12.25, and an 80-pound bag costs $30.25. Which is the better buy?

45. EMPTYING A TANK An 11,880-gallon tank can be emptied in 27 minutes. Write a ratio of gallons to minutes, and give the rate of flow in gallons per minute.

46. PRESIDENTIAL PAY The president of the United States receives a salary of $200,000 per year. Write a ratio of dollars to days, and give the daily rate of presidential pay, rounded to the nearest cent.

47. SALES TAX A sweater cost $36.75 after sales tax had been added. Find the tax rate as a percent if the sweater retailed for $35.

48. REAL ESTATE TAXES The real estate taxes on a summer home assessed at $75,000 were $1,500. Find the tax rate as a percent.

49. RATE OF SPEED A car travels 325 miles in 5 hours. Find its rate of speed in miles per hour.

50. RATE OF SPEED An airplane travels from Chicago to San Francisco, a distance of 1,883 miles, in 3.5 hours. Find the rate of speed of the plane.

51. COMPARING GAS MILEAGE One car went 1,235 miles on 51.3 gallons of gasoline, and another went 1,456 miles on 55.78 gallons. Which car has the better mpg rating?

52. COMPARING ELECTRIC RATES In one community, a bill for 575 kilowatt hours (kwh) of electricity was $38.81. In a second community, a bill for 831 kwh was $58.10. In which community is electricity cheaper?

53. DIVORCE RATE See the graph in Illustration 5.
   a. Express the 1920 U.S. marriage/divorce ratio as a fraction. Explain what it means.

   b. According to the graph, in 1996 for every ___ marriages, there was (were) ___ divorce(s).
   c. How has the marriage/divorce ratio changed from 1920 to 1996?

ILLUSTRATION 5

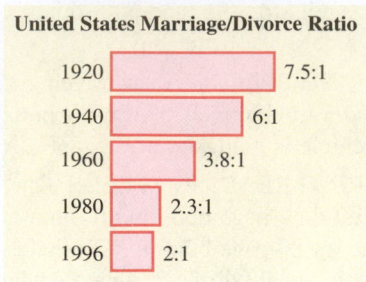

United States Marriage/Divorce Ratio

Year		Ratio
1920		7.5:1
1940		6:1
1960		3.8:1
1980		2.3:1
1996		2:1

54. SPREADING TECHNOLOGY For each of the regions listed in Illustration 6, write the ratio of the number of homes with TV to the number of homes without TV.

ILLUSTRATION 6

	United States	Western Europe	Asia-Pacific	Latin America
Homes (millions)	97	160	587	96
Homes with TV (millions)	95	142	390	83

Based on data from *Los Angeles Times* (June 6, 1997), page D4

55. SCALE DRAWINGS The map of Tokyo in Illustration 7 provides a reduced representation of a very large area. To express the relationship between a length on the map and the actual length of the geographic area it portrays, a scale is used. The scale for this map is written as a ratio. Explain its meaning. The units used are inches.

ILLUSTRATION 7

Scale 1 : 1,000,000

56. HAWAII'S AGRICULTURE The graph in Illustration 8 shows sales of sugar and pineapple and all other agricultural products of Hawaii.
   a. What was the ratio of sugar and pineapple sales to the sales of all other agricultural products in 1986 and in 1995?

b. Approximately when was the ratio 1:1?

c. Explain what we can learn about Hawaii's agriculture from the change in the ratios.

ILLUSTRATION 8

Hawaiian Agriculture

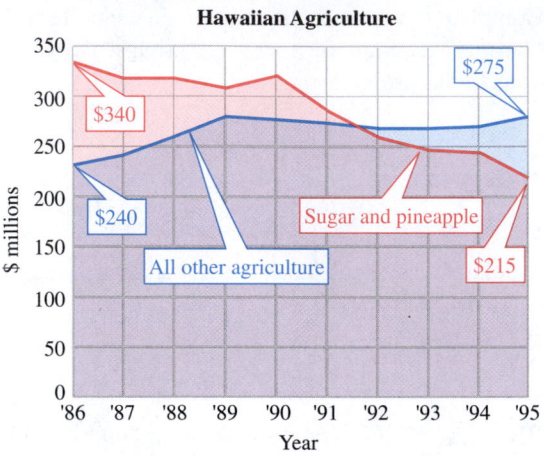

Based on data from the Hawaii Agricultural Statistics Service

WRITING

57. Some people think that the word *ratio* comes from the words *rational number.* Explain why this may be true.

58. In the fraction $\frac{a}{b}$, b cannot be zero. Explain why. In the ratio $\frac{a}{b}$, b can be zero. Explain why.

REVIEW

In Exercises 59–62, solve each equation.

59. $0.2x + 4 = 3.8$

60. $\frac{x}{2} - 4 = 38$

61. $3(x + 2) = 24$

62. $\frac{x - 6}{3} = 20$

In Exercises 63–64, find each square root.

63. $\sqrt{16}$

64. $\sqrt{25}$

▶6.2 Proportions and Similar Triangles

In this section, you will learn about

Proportions ■ Solving proportions ■ Problem solving ■ Similar triangles

Introduction   Consider the following table, in which we are given the costs of various numbers of gallons of gasoline.

Gallons of gas	Cost
2	$ 2.72
5	$ 6.80
8	$10.88
12	$16.32
20	$27.20

If we find the ratios of the costs to the number of gallons purchased, we will see that they are equal. In this example, each ratio represents the cost of 1 gallon of gasoline, which is $1.36 per gallon.

$$\frac{\$2.72}{2} = \$1.36, \qquad \frac{\$6.80}{5} = \$1.36, \qquad \frac{\$10.88}{8} = \$1.36,$$

$$\frac{\$16.32}{12} = \$1.36, \qquad \text{and} \qquad \frac{\$27.20}{20} = \$1.36$$

When two ratios such as $\frac{\$2.72}{2}$ and $\frac{\$6.80}{5}$ are equal, they form a *proportion*. In this section, we will discuss proportions and use them to solve problems.

Proportions

Proportions

A **proportion** is a statement that two ratios are equal.

Some examples of proportions are

$$\frac{1}{2} = \frac{3}{6}, \qquad \frac{7}{3} = \frac{21}{9}, \qquad \frac{8x}{1} = \frac{40x}{5}, \qquad \text{and} \qquad \frac{a}{b} = \frac{c}{d}$$

- The proportion $\frac{1}{2} = \frac{3}{6}$ can be read as "1 is to 2 as 3 is to 6."
- The proportion $\frac{7}{3} = \frac{21}{9}$ can be read as "7 is to 3 as 21 is to 9."
- The proportion $\frac{8x}{1} = \frac{40x}{5}$ can be read as "8x is to 1 as 40x is to 5."
- The proportion $\frac{a}{b} = \frac{c}{d}$ can be read as "a is to b as c is to d."

The terms of the proportion $\frac{a}{b} = \frac{c}{d}$ are numbered as follows:

$$\text{First term} \longrightarrow \frac{a}{b} = \frac{c}{d} \longleftarrow \text{Third term}$$
$$\text{Second term} \longrightarrow \phantom{\frac{a}{b} = \frac{c}{d}} \longleftarrow \text{Fourth term}$$

In the proportion $\frac{1}{2} = \frac{3}{6}$, the numbers 1 and 6 are called the **extremes,** and the numbers 2 and 3 are called the **means.**

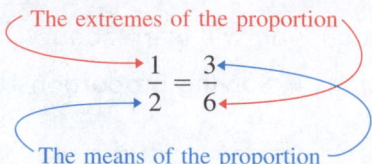

The extremes of the proportion

$$\frac{1}{2} = \frac{3}{6}$$

The means of the proportion

In this proportion, the product of the extremes is equal to the product of the means.

$$1 \cdot 6 = 6 \qquad \text{and} \qquad 2 \cdot 3 = 6$$

This illustrates a fundamental property of proportions.

Fundamental property of proportions

In any proportion, the product of the extremes is equal to the product of the means.

In the proportion $\frac{a}{b} = \frac{c}{d}$, a and d are the extremes, and b and c are the means. We can show that the product of the extremes (ad) is equal to the product of the means (bc) by multiplying both sides of the proportion by bd and observing that $ad = bc$.

$$\frac{a}{b} = \frac{c}{d}$$

$$\frac{bd}{1} \cdot \frac{a}{b} = \frac{bd}{1} \cdot \frac{c}{d} \qquad \text{To eliminate the fractions, multiply both sides by } \frac{bd}{1}.$$

$$\frac{abd}{b} = \frac{bcd}{d} \qquad \text{Multiply the numerators and multiply the denominators.}$$

$$ad = bc \qquad \text{Divide out the common factors: } \frac{b}{b} = 1 \text{ and } \frac{d}{d} = 1.$$

Since $ad = bc$, the product of the extremes equals the product of the means.

To determine whether an equation is a proportion, we can check to see whether the product of the extremes is equal to the product of the means.

EXAMPLE 1

Determining whether an equation is a proportion. Determine whether each equation is a proportion: **a.** $\frac{3}{7} = \frac{9}{21}$ and   **b.** $\frac{8}{3} = \frac{13}{5}$.

Solution   In each case, we check to see whether the product of the extremes is equal to the product of the means.

a. The product of the extremes is $3 \cdot 21 = 63$. The product of the means is $7 \cdot 9 = 63$. Since the products are equal, the equation is a proportion: $\frac{3}{7} = \frac{9}{21}$.

b. The product of the extremes is $8 \cdot 5 = 40$. The product of the means is $3 \cdot 13 = 39$. Since the products are not equal, the equation is not a proportion: $\frac{8}{3} \neq \frac{13}{5}$.

SELF CHECK   Determine whether this equation is a proportion: $\frac{6}{13} = \frac{24}{53}$.   *Answer:* no   ■

When two pairs of numbers such as 2, 3 and 8, 12 form a proportion, we say that they are **proportional.** To show that 2, 3 and 8, 12 are proportional, we check to see whether the equation

$$\frac{2}{3} = \frac{8}{12}$$

is a proportion. To do so, we find the product of the extremes and the product of the means:

$$2 \cdot 12 = 24 \qquad 3 \cdot 8 = 24$$

Since the products are equal, the equation is a proportion, and the numbers are proportional.

EXAMPLE 2	**Determining whether numbers are proportional.** Determine whether 3, 7 and 36, 91 are proportional.

Solution   We check to see whether $\frac{3}{7} = \frac{36}{91}$ is a proportion by finding two products:

$$3 \cdot 91 = 273 \quad \text{The product of the extremes.}$$
$$7 \cdot 36 = 252 \quad \text{The product of the means.}$$

Since the products are not equal, the numbers are not proportional.

SELF CHECK   Determine whether 6, 11 and 54, 99 are proportional.   *Answer:* yes   ■

Solving Proportions

Suppose that we know three terms in the proportion

$$\frac{x}{5} = \frac{24}{20}$$

To find the unknown term, we multiply the extremes and multiply the means, set them equal, and solve for x:

$$\frac{x}{5} = \frac{24}{20}$$

$$20 \cdot x = 5 \cdot 24 \quad \text{In a proportion, the product of the extremes is equal to the product of the means.}$$

$$20x = 120 \quad \text{Multiply: } 5 \cdot 24 = 120.$$

$$\frac{20x}{20} = \frac{120}{20} \quad \text{To undo the multiplication by 20, divide both sides by 20.}$$

$$x = 6 \quad \text{Simplify: } \frac{120}{20} = 6.$$

The first term is 6.

EXAMPLE 3	**Solving proportions.** Solve $\dfrac{12}{18} = \dfrac{3}{x}$.

Solution

$$\frac{12}{18} = \frac{3}{x}$$

$$12 \cdot x = 18 \cdot 3 \quad \text{In a proportion, the product of the extremes equals the product of the means.}$$

$$12x = 54 \quad \text{Multiply: } 18 \cdot 3 = 54.$$

$$\frac{12x}{12} = \frac{54}{12} \quad \text{To undo the multiplication by 12, divide both sides by 12.}$$

$$x = \frac{9}{2} \quad \text{Simplify: } \frac{54}{12} = \frac{9}{2}.$$

Thus, $x = \frac{9}{2}$.

SELF CHECK   Solve $\frac{15}{x} = \frac{25}{40}$ *Answer:* 24   ■

EXAMPLE 4

Solving proportions. Find the third term of the proportion $\dfrac{3.5}{7.2} = \dfrac{x}{15.84}$.

Solution

$$\frac{3.5}{7.2} = \frac{x}{15.84}$$

$3.5(15.84) = 7.2x$ In a proportion, the product of the extremes equals the product of the means.

$55.44 = 7.2x$ Multiply: $3.5(15.84) = 55.44$.

$$\frac{55.44}{\mathbf{7.2}} = \frac{7.2x}{\mathbf{7.2}}$$ To undo the multiplication by 7.2, divide both sides by 7.2.

$7.7 = x$ Simplify: $\frac{55.44}{7.2} = 7.7$.

The third term is 7.7

SELF CHECK

Find the second term of the proportion $\frac{6.7}{x} = \frac{33.5}{38}$ *Answer:* 7.6   ■

ACCENT ON TECHNOLOGY   *Solving Proportions with a Calculator*

To solve the proportion in Example 4 with a calculator, we can proceed as follows.

$$\frac{3.5}{7.2} = \frac{x}{15.84}$$

$$\frac{3.5(15.84)}{7.2} = x$$ Multiply both sides by 15.84.

We can find x by entering these numbers into a scientific calculator:

Keystrokes   3.5 $\boxed{\times}$ 15.84 $\boxed{\div}$ 7.2 $\boxed{=}$ $\boxed{\qquad\qquad 7.7}$

Using a graphing calculator, we enter these numbers and press these keys:

Keystrokes   3.5 $\boxed{\times}$ 15.84 $\boxed{\div}$ 7.2 $\boxed{\text{ENTER}}$ $\boxed{\begin{array}{l} 3.5*15.84/7.2 \\ \qquad\qquad\qquad 7.7 \end{array}}$

Thus, $x = 7.7$.

EXAMPLE 5

Solving proportions. Solve $\dfrac{2x+1}{4} = \dfrac{10}{8}$.

Solution

$$\frac{2x+1}{4} = \frac{10}{8}$$

$8(2x+1) = 40$ In a proportion, the product of the extremes equals the product of the means.

$16x + 8 = 40$ Use the distributive property to remove parentheses.

$16x + 8 \mathbf{-8} = 40 \mathbf{-8}$ To undo the addition of 8, subtract 8 from both sides.

$16x = 32$ Combine like terms.

$$\frac{16x}{\mathbf{16}} = \frac{32}{\mathbf{16}}$$ To undo the multiplication by 16, divide both sides by 16.

$x = 2$ Simplify: $\frac{32}{16} = 2$.

Thus, $x = 2$.

SELF CHECK   Solve $\frac{3x-1}{2} = \frac{12.5}{5}$.   *Answer:* 2 ∎

Problem Solving

We can use proportions to solve problems.

EXAMPLE 6   **Grocery shopping.** If 6 apples cost $1.38, how much will 16 apples cost?

Solution

ANALYZE THE PROBLEM   We know the cost of 6 apples; we are to find the cost of 16 apples.

FORM AN EQUATION   Let c represent the cost of 16 apples. The ratios of the numbers of apples to their costs are equal.

6 apples is to $1.38 as 16 apples is to $c.

$$\begin{array}{l} \text{6 apples} \longrightarrow \\ \text{Cost of 6 apples} \longrightarrow \end{array} \frac{6}{1.38} = \frac{16}{c} \begin{array}{l} \longleftarrow \text{16 apples} \\ \longleftarrow \text{Cost of 16 apples} \end{array}$$

SOLVE THE EQUATION $6 \cdot c = 1.38(16)$ In a proportion, the product of the extremes is equal to the product of the means.

$6c = 22.08$ Do the multiplication: $1.38(16) = 22.08$.

$\dfrac{6c}{6} = \dfrac{22.08}{6}$ To undo the multiplication by 6, divide both sides by 6.

$c = 3.68$ Simplify: $\frac{22.08}{6} = 3.68$.

STATE THE CONCLUSION   Sixteen apples will cost $3.68.

CHECK THE RESULT   If 16 apples are bought, this is *about* 3 times as many as a purchase of 6 apples, which cost $1.38. If we multiply $1.38 by 3, we get an estimate of the cost of 16 apples. $1.38 \cdot 3 = \$4.14$. The result of $3.68 seems reasonable.

SELF CHECK   If 9 tickets to a concert cost $112.50, how much will 15 tickets cost?   *Answer:* $187.50 ∎

EXAMPLE 7   **Mixing solutions.** A solution contains 2 quarts of antifreeze and 5 quarts of water. How many quarts of antifreeze must be mixed with 18 quarts of water to have the same concentration?

Solution

ANALYZE THE PROBLEM   We want to maintain the same concentration of antifreeze to water. This problem can be solved using a proportion.

FORM AN EQUATION   Let q represent the number of quarts of antifreeze to be mixed with the water. The ratios of the quarts of antifreeze to the quarts of water are equal.

2 quarts antifreeze is to 5 quarts water as q quarts antifreeze is to 18 quarts water.

$$\begin{array}{l} \text{2 quarts antifreeze} \longrightarrow \\ \text{5 quarts water} \longrightarrow \end{array} \frac{2}{5} = \frac{q}{18} \begin{array}{l} \longleftarrow \text{quarts of antifreeze} \\ \longleftarrow \text{18 quarts of water} \end{array}$$

SOLVE THE EQUATION $2 \cdot 18 = 5q$ In a proportion, the product of the extremes is equal to the product of the means.

$36 = 5q$ Do the multiplication: $2 \cdot 18 = 36$.

$\dfrac{36}{5} = \dfrac{5q}{5}$ To undo the multiplication by 5, divide both sides by 5.

$\dfrac{36}{5} = q$

STATE THE CONCLUSION   The mixture should contain $\frac{36}{5}$ or 7.2 quarts of antifreeze.

CHECK THE RESULT   See if the result seems reasonable.

SELF CHECK   A solution should contain 2 ounces of alcohol for every 7 ounces of water. How much alcohol should be added to 20 ounces of water to get the proper concentration?   *Answer:* $\frac{40}{7}$ oz   ∎

E X A M P L E   8   **Baking.** A recipe for rhubarb cake calls for $1\frac{1}{4}$ cups of sugar for every $2\frac{1}{2}$ cups of flour. How many cups of flour are needed if the baker intends to use 3 cups of sugar?

Solution

ANALYZE THE PROBLEM   The baker needs to maintain the same ratio between the sugar and flour as is called for in the original recipe.

FORM AN EQUATION   Let f represent the number of cups of flour to be mixed with the 3 cups of sugar. The ratios of the cups of sugar to the cups of flour are equal.

$1\frac{1}{4}$ cups sugar is to $2\frac{1}{2}$ cups flour as 3 cups sugar is to f cups flour.

$1\frac{1}{4}$ cups sugar ⟶ $\dfrac{1\frac{1}{4}}{2\frac{1}{2}} = \dfrac{3}{f}$ ⟵ 3 cups sugar
$2\frac{1}{2}$ cups flour ⟶ ⟵ f cups flour

SOLVE THE EQUATION

$\dfrac{1.25}{2.5} = \dfrac{3}{f}$ Change the fractions to decimals.

$1.25f = 2.5 \cdot 3$ In a proportion, the product of the extremes is equal to the product of the means.

$1.25f = 7.5$ Do the multiplication: $2.5 \cdot 3 = 7.5$.

$\dfrac{1.25f}{\mathbf{1.25}} = \dfrac{7.5}{\mathbf{1.25}}$ To undo the multiplication by 1.25, divide both sides by 1.25.

$f = 6$ Divide: $\frac{7.5}{1.25} = 6$.

STATE THE CONCLUSION   The baker should use 6 cups of flour.

CHECK THE RESULT   The recipe calls for about 2 cups of flour for about 1 cup of sugar. If 3 cups of sugar are used, 6 cups of flour seems reasonable.

SELF CHECK   How many cups of sugar will be needed to make several cakes that will require a total of 25 cups of flour?   *Answer:* $12\frac{1}{2}$ ∎

Similar Triangles

If two angles of one triangle have the same measures as two angles of a second triangle, the triangles will have the same shape. Triangles with the same shape are called **similar triangles.** In Figure 6-2, $\triangle ABC \sim \triangle DEF$. (Read the symbol $\sim$ as "is similar to.")

FIGURE 6-2

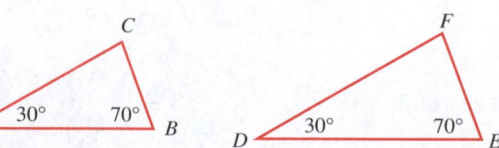

Property of similar triangles

If two triangles are **similar,** all pairs of corresponding sides are in proportion.

In the similar triangles shown in Figure 6-2, the following proportions are true.

$$\frac{AB}{DE} = \frac{BC}{EF}, \qquad \frac{BC}{EF} = \frac{CA}{FD}, \qquad \text{and} \qquad \frac{CA}{FD} = \frac{AB}{DE}$$

Read AB as "the length of segment AB."

EXAMPLE 9

Finding the height of a tree. A tree casts a shadow 18 feet long at the same time as a woman 5 feet tall casts a shadow 1.5 feet long. Find the height of the tree.

Solution

ANALYZE THE PROBLEM Figure 6-3 shows the triangles determined by the tree and its shadow and the woman and her shadow. Since the triangles have the same shape, they are similar, and the lengths of their corresponding sides are in proportion.

FIGURE 6-3

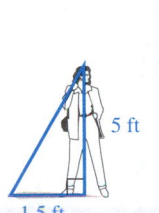 

5 ft

h

1.5 ft

18 ft

FORM AN EQUATION If we let h represent the height of the tree, we can find h by solving the following proportion.

$$\frac{h}{5} = \frac{18}{1.5} \qquad \frac{\text{Height of the tree}}{\text{Height of the woman}} = \frac{\text{Length of shadow of the tree}}{\text{Length of shadow of the woman}}$$

SOLVE THE EQUATION

$1.5h = 5(18)$ In a proportion, the product of the extremes is equal to the product of the means.

$1.5h = 90$ Do the multiplication.

$h = 60$ To undo the multiplication by 1.5, divide both sides by 1.5 and simplify.

STATE THE CONCLUSION The tree is 60 feet tall.

CHECK THE RESULT $\frac{18}{1.5} = 12$ and $\frac{60}{5} = 12$. The ratios are the same. The solution checks.

SELF CHECK Find the height of the tree in Example 9 if the woman is 5 feet, 6 inches tall. *Answer:* 66 ft ■

STUDY SET

Section 6.2

VOCABULARY

In Exercises 1–6, fill in the blanks to make the statements true.

1. A _____ is a statement that two ratios are equal.

2. In the proportion $\frac{a}{b} = \frac{c}{d}$, a and d are called the _____ of the proportion.

3. The second and third terms of a proportion are called the _____ of the proportion.

4. When two pairs of numbers form a proportion, we say that the numbers are _____.

5. If two triangles have the same _____, they are said to be similar.

6. If two triangles are similar, their corresponding sides are in _____.

CONCEPTS

In Exercises 7–8, fill in the blanks to make the statements true.

7. The equation $\frac{a}{b} = \frac{c}{d}$ is a proportion if the product ▢ is equal to the product ▢.

8. If $3 \cdot 10 = x \cdot 17$, then ▢ is a proportion.

9. DEFENSE SPENDING See Illustration 1.
  a. Find the ratio of defense spending to the gross domestic product for the United States and then for China.
  b. Set the two ratios from part a equal. Is this equation a proportion? Explain what this means in terms of defense spending by the two countries.

ILLUSTRATION 1

Country	Defense spending (billions)	Gross domestic product (billions)
United States	$278	$7,316
China	$ 32	$ 561

Based on data from the International Institute for Strategic Studies

10. IMPORT/EXPORT RATIO Examine the graph in Illustration 2 to determine the year when the ratio of imports to exports was the closest to $1:1$. Explain in words what this $1:1$ ratio means.

ILLUSTRATION 2

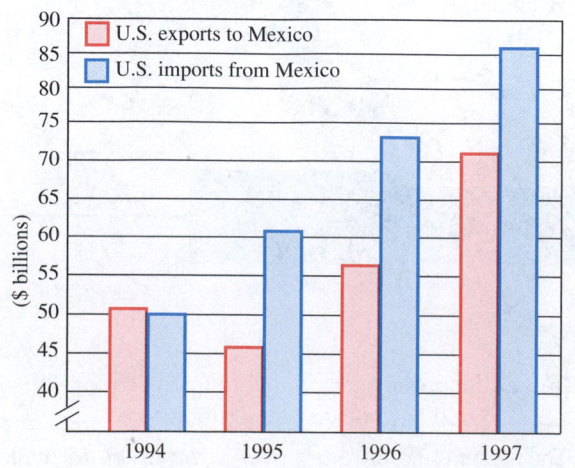

Based on data from the Census Bureau, Foreign Trade Division

NOTATION

In Exercises 11–12, complete each solution.

11. Solve for x: $\dfrac{12}{18} = \dfrac{x}{24}$.

$$12 \cdot 24 = 18 \cdot \boxed{}$$
$$\boxed{} = 18x$$
$$\dfrac{288}{\boxed{}} = \dfrac{18x}{\boxed{}}$$
$$16 = x$$

12. Solve for x: $\dfrac{14}{x} = \dfrac{49}{17.5}$.

$$14 \cdot \boxed{} = 49x$$
$$\boxed{} = 49x$$
$$\dfrac{245}{\boxed{}} = \dfrac{49x}{\boxed{}}$$
$$5 = x$$

13. Read $\triangle ABC$ as _____ ABC.

14. The symbol $\sim$ is read as _____.

PRACTICE

In Exercises 15–22, tell whether each statement is a proportion.

15. $\dfrac{9}{7} = \dfrac{81}{70}$

16. $\dfrac{5}{2} = \dfrac{20}{8}$

17. $\dfrac{7}{3} = \dfrac{14}{6}$

18. $\dfrac{13}{19} = \dfrac{65}{95}$

19. $\dfrac{9}{19} = \dfrac{38}{80}$

20. $\dfrac{40}{29} = \dfrac{29}{22}$

21. $\dfrac{10.4}{3.6} = \dfrac{41.6}{14.4}$

22. $\dfrac{13.23}{3.45} = \dfrac{39.96}{11.35}$

In Exercises 23–38, solve for the variable in each proportion.

23. $\dfrac{2}{3} = \dfrac{x}{6}$

24. $\dfrac{3}{6} = \dfrac{x}{8}$

25. $\dfrac{5}{10} = \dfrac{3}{c}$

26. $\dfrac{7}{14} = \dfrac{2}{x}$

27. $\dfrac{6}{x} = \dfrac{8}{4}$

28. $\dfrac{4}{x} = \dfrac{2}{8}$

29. $\dfrac{x}{3} = \dfrac{9}{3}$

30. $\dfrac{x}{2} = \dfrac{18}{6}$

31. $\dfrac{x+1}{5} = \dfrac{3}{15}$

32. $\dfrac{x-1}{7} = \dfrac{2}{21}$

33. $\dfrac{x+3}{12} = \dfrac{-7}{6}$

34. $\dfrac{x+7}{-4} = \dfrac{1}{4}$

35. $\dfrac{4 - x}{13} = \dfrac{11}{26}$ **36.** $\dfrac{5 - x}{17} = \dfrac{13}{34}$

37. $\dfrac{2x + 1}{18} = \dfrac{14}{3}$ **38.** $\dfrac{2x - 1}{18} = \dfrac{9}{54}$

APPLICATIONS

In Exercises 39–58, set up and solve a proportion. Use a calculator when it is helpful.

39. GROCERY SHOPPING If 3 pints of yogurt cost $1, how much will 51 pints cost?

40. SHOPPING FOR CLOTHES If shirts are on sale at two for $25, how much will five shirts cost?

41. ADVERTISING In 1997, a 30-second TV ad during the Super Bowl telecast cost $1.2 million. At this rate, what was the cost of a 45-second ad?

42. COOKING A recipe for spaghetti sauce requires four 16-ounce bottles of ketchup to make two gallons of sauce. How many bottles of ketchup are needed to make 10 gallons of sauce?

43. MIXING PERFUME A perfume is to be mixed in the ratio of 3 drops of pure essence to 7 drops of alcohol. How many drops of pure essence should be mixed with 56 drops of alcohol?

44. CPR A first aid handbook states that when performing cardiopulmonary resuscitation on an adult, the ratio of chest compressions to breaths should be 5:2. If 210 compressions were administered to an adult patient, how many breaths should have been given?

45. COOKING A recipe for wild rice soup is shown in Illustration 3. Find the amount of each ingredient needed to make 15 servings.

ILLUSTRATION 3

Wild Rice Soup
A sumptuous side dish with a nutty flavor

3 cups chicken broth	1 cup light cream
$\frac{2}{3}$ cup uncooked rice	2 tablespoons flour
$\frac{1}{4}$ cup sliced onions	$\frac{1}{8}$ teaspoon pepper
$\frac{1}{2}$ cup shredded carrots	Serves: 6

46. PHOTO ENLARGEMENTS In Illustration 4, the 3-by-5 photo is to be blown up to the larger size. Find *x*.

ILLUSTRATION 4

5 in. $6\frac{1}{4}$ in.

3 in.   *x* in.

47. QUALITY CONTROL In a manufacturing process, 95% of the parts made are to be within specifications. How many defective parts would be expected in a run of 940 pieces?

48. QUALITY CONTROL Out of a sample of 500 men's shirts, 17 were rejected because of crooked collars. How many crooked collars would you expect to find in a run of 15,000 shirts?

49. GAS CONSUMPTION If a car can travel 42 miles on 1 gallon of gas, how much gas is needed to travel 315 miles?

50. RAPPERS According to the *Guinness Book of World Records 1998*, Rebel X.D. of Chicago, IL rapped 674 syllables in 54.9 seconds. At this rate, how many syllables could he rap in 1 minute? Round to the nearest syllable.

51. BANKRUPTCY After filing for bankruptcy, a company was only able to pay its creditors 15 cents on the dollar. If the company owed a lumberyard $9,712, how much could the lumberyard expect to be paid?

52. COMPUTING PAYCHECKS Billie earns $412 for a 40-hour week. If she missed 10 hours of work last week, how much did she get paid?

53. MODEL RAILROADING A model railroad engine is 9 inches long. If the scale is 87 feet to 1 foot, how long is a real engine?

54. MODEL RAILROADING A model railroad caboose is 3.5 inches long. If the scale is 169 feet to 1 foot, how long is a real caboose?

55. NUTRITION Illustration 5 shows the nutritional facts about a 10-oz chocolate milkshake sold by a fast-food restaurant. Use the information to complete the table for the 16-oz shake. Round to the nearest unit when an answer is not exact.

ILLUSTRATION 5

	Calories	Fat (gm)	Protein (gm)
10-oz chocolate milkshake	355	8	9
16-oz chocolate milkshake			

56. DRIVER'S LICENSES Of the fifty states, Oregon has the largest ratio of licensed drivers per 1,000 residents. If the ratio is 824 to 1,000 and Oregon's population is 3,140,000, how many Oregonians have a driver's license?

57. MIXING FUEL The instructions on a can of oil intended to be added to lawnmower gasoline read as follows:

Recommended	Gasoline	Oil
50 to 1	6 gal	16 oz

Are these instructions correct? (*Hint:* There are 128 ounces in 1 gallon.)

58. MIXING FUEL In Exercise 57, how much oil should be mixed with 28 gallons of gas?

In Exercises 59–64, use similar triangles to solve each problem.

59. HEIGHT OF A TREE A tree casts a shadow of 26 feet at the same time as a 6-foot man casts a shadow of 4 feet. (See Illustration 6.) Find the height of the tree.

ILLUSTRATION 6

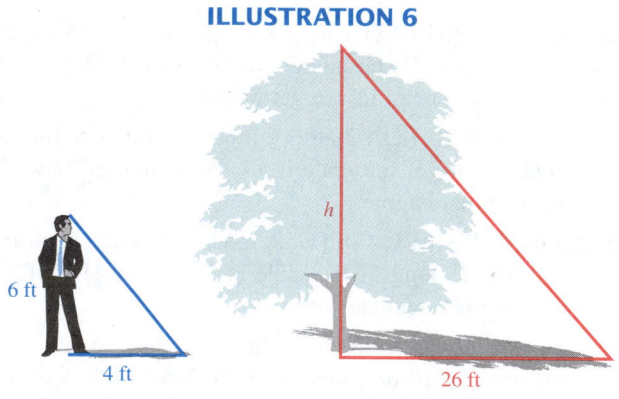

6 ft
4 ft
26 ft
h

60. HEIGHT OF A BUILDING A man places a mirror on the ground and sees the reflection of the top of a building, as shown in Illustration 7. The two triangles in the illustration are similar. Find the height, *h*, of the building.

ILLUSTRATION 7

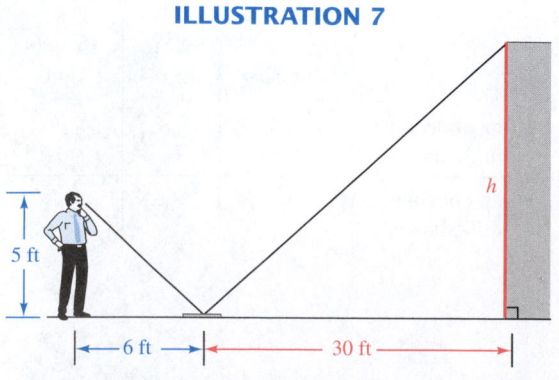

5 ft
6 ft
30 ft
h

61. WIDTH OF A RIVER Use the dimensions in Illustration 8 to find *w*, the width of the river. The two triangles in the illustration are similar.

ILLUSTRATION 8

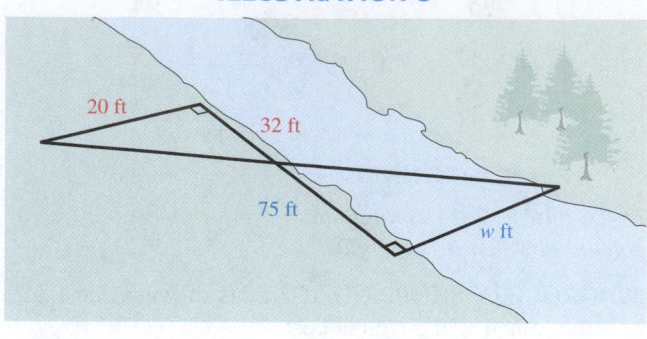

20 ft
32 ft
75 ft
w ft

62. FLIGHT PATH An airplane ascends 100 feet as it flies a horizontal distance of 1,000 feet. How much altitude will it gain as it flies a horizontal distance of 1 mile? See Illustration 9. (*Hint:* 5,280 feet = 1 mile.)

ILLUSTRATION 9

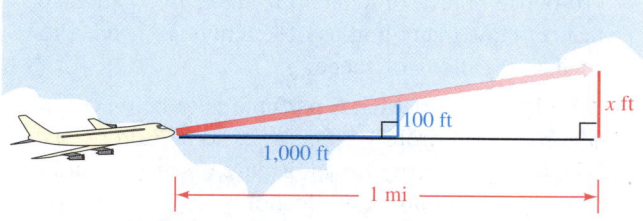

100 ft
1,000 ft
x ft
1 mi

63. FLIGHT PATH An airplane descends 1,350 feet as it flies a horizontal distance of 1 mile. How much altitude is lost as it flies a horizontal distance of 5 miles?

64. SKI RUNS A ski course falls 100 feet in every 300 feet of horizontal run. If the total horizontal run is $\frac{1}{2}$ mile, find the height of the hill.

WRITING

65. Explain the difference between a ratio and a proportion.

66. Explain how to tell whether the equation $\frac{3.2}{3.7} = \frac{5.44}{6.29}$ is a proportion.

REVIEW

67. Change $\frac{9}{10}$ to a percent.

68. Change $\frac{7}{8}$ to a percent.

69. Change $33\frac{1}{3}\%$ to a fraction.

70. Change 75% to a fraction.

71. Find 30% of 1,600.

72. Find $\frac{1}{2}\%$ of 520.

73. SHOPPING Maria bought a dress for 25% off the original price of $98. How much did the dress cost?

74. SHOPPING Liz purchased a shirt on sale for $17.50. Find the original cost of the shirt if it was marked down 30%.

▶ 6.3 Rational Expressions and Rational Functions

In this section, you will learn about

Simplifying rational expressions ■ Division by 1 ■ Dividing polynomials that are negatives ■ Rational functions ■ Graphing rational functions

Introduction   Fractions such as $\frac{1}{2}$ and $\frac{3}{4}$ that are the quotient of two integers are *rational numbers*. Fractions such as

$$\frac{3}{2y}, \qquad \frac{x}{x+2}, \qquad \text{and} \qquad \frac{5a^2+b^2}{3a-b}$$

where the numerators and denominators are polynomials, are called **algebraic fractions** or **rational expressions.**

Simplifying Rational Expressions

A fraction can be simplified by dividing out common factors shared by its numerator and denominator. For example,

$$\frac{18}{30} = \frac{3 \cdot 6}{5 \cdot 6} = \frac{3 \cdot \overset{1}{\cancel{6}}}{5 \cdot \underset{1}{\cancel{6}}} = \frac{3}{5} \qquad \text{and} \qquad -\frac{6}{15} = -\frac{3 \cdot 2}{3 \cdot 5} = -\frac{\overset{1}{\cancel{3}} \cdot 2}{\underset{1}{\cancel{3}} \cdot 5} = -\frac{2}{5}$$

When all common factors have been divided out, we say that the fraction has been **expressed in lowest terms**. The generalization of this idea is called the *fundamental property of fractions.*

Fundamental property of fractions

If a is a real number and b and c are nonzero real numbers, then

$$\frac{ac}{bc} = \frac{a}{b}$$

We can use the same procedure to simplify rational expressions.

EXAMPLE 1

Simplifying rational expressions. Simplify $\dfrac{21x^2y}{14xy^2}$.

Solution   We look for common factors in the numerator and denominator that can be divided out.

$$\frac{21x^2y}{14xy^2} = \frac{3 \cdot 7 \cdot x \cdot x \cdot y}{2 \cdot 7 \cdot x \cdot y \cdot y} \qquad \text{Factor the numerator and denominator.}$$

$$= \frac{3 \cdot \overset{1}{\cancel{7}} \cdot \overset{1}{\cancel{x}} \cdot x \cdot \overset{1}{\cancel{y}}}{2 \cdot \underset{1}{\cancel{7}} \cdot \underset{1}{\cancel{x}} \cdot y \cdot \underset{1}{\cancel{y}}} \qquad \text{Divide out the common factors of 7, } x, \text{ and } y.$$

$$= \frac{3x}{2y} \qquad \text{Do the multiplications in the numerator and in the denominator.}$$

We can also simplify by using the rules of exponents:

$$\frac{21x^2y}{14xy^2} = \frac{3 \cdot 7}{2 \cdot 7}x^{2-1}y^{1-2} \qquad \frac{x^2}{x} = x^{2-1} \text{ and } \frac{y}{y^2} = y^{1-2}.$$

$$= \frac{3}{2}xy^{-1} \qquad\qquad 2 - 1 = 1 \text{ and } 1 - 2 = -1.$$

$$= \frac{3}{2} \cdot \frac{x}{y} \qquad\qquad y^{-1} = \frac{1}{y}.$$

$$= \frac{3x}{2y} \qquad\qquad \text{Multiply.}$$

SELF CHECK   Simplify $\dfrac{32a^3b^2}{24ab^4}$.   *Answer:* $\dfrac{4a^2}{3b^2}$ ■

E X A M P L E   2   **Factoring to simplify rational expressions.** Write $\dfrac{x^2 + 3x}{3x + 9}$ in lowest terms.

Solution   We note that the terms of the numerator have a common factor of x and the terms of the denominator have a common factor of 3.

$$\frac{x^2 + 3x}{3x + 9} = \frac{x(x + 3)}{3(x + 3)} \qquad \text{Factor the numerator and the denominator.}$$

$$= \frac{x\cancel{(x + 3)}^{\,1}}{3\cancel{(x + 3)}_{\,1}} \qquad \text{Divide out the common factor of } x + 3.$$

$$= \frac{x}{3} \qquad\qquad \text{Simplify the numerator and the denominator.}$$

SELF CHECK   Simplify $\dfrac{x^2 - 5x}{5x - 25}$.   *Answer:* $\dfrac{x}{5}$ ■

E X A M P L E   3   **Factoring to simplify rational expressions.** Simplify $\dfrac{x^2 + 13x + 12}{x^2 - 144}$.

Solution   The numerator is a trinomial, and the denominator is a difference of two squares.

$$\frac{x^2 + 13x + 12}{x^2 - 144} = \frac{(x + 1)(x + 12)}{(x + 12)(x - 12)} \qquad \text{Factor the numerator and the denominator.}$$

$$= \frac{(x + 1)\cancel{(x + 12)}^{\,1}}{\cancel{(x + 12)}_{\,1}(x - 12)} \qquad \text{Divide out the common factor of } x + 12.$$

$$= \frac{x + 1}{x - 12}$$

SELF CHECK   Simplify $\dfrac{3x^2 - 8x - 3}{x^2 - 9}$.   *Answer:* $\dfrac{3x + 1}{x + 3}$ ■

Divison by 1

Any number or algebraic expression divided by 1 remains unchanged. For example,

$$\frac{37}{1} = 37, \qquad \frac{5x}{1} = 5x, \qquad \text{and} \qquad \frac{3x + y}{1} = 3x + y$$

In general, we have the following.

Division by 1

For any real number a, $\dfrac{a}{1} = a$

E X A M P L E 4 **Simplifying rational expressions.** Simplify $\dfrac{x^3 + x^2}{x + 1}$.

Solution
$$\frac{x^3 + x^2}{x + 1} = \frac{x^2(x + 1)}{x + 1} \qquad \text{Factor the numerator.}$$

$$= \frac{x^2\overset{1}{\cancel{(x + 1)}}}{\underset{1}{\cancel{x + 1}}} \qquad \text{Divide out the common factor of } x + 1.$$

$$= \frac{x^2}{1} \qquad \text{Simplify the numerator.}$$

$$= x^2 \qquad \text{Denominators of 1 need not be written.}$$

SELF CHECK Simplify $\dfrac{a^2 + a - 2}{a - 1}$.   *Answer: $a + 2$*   ∎

 WARNING! Remember that only *factors* that are common to the *entire numerator* and the *entire denominator* can be divided out. For example, consider the correct simplification

$$\frac{5 + 8}{5} = \frac{13}{5}$$

It would be incorrect to divide out the common *term* of 5 in this simplification. Doing so gives an incorrect answer:

$$\frac{5 + 8}{5} = \frac{\overset{1}{\cancel{5}} + 8}{\underset{1}{\cancel{5}}} = \frac{1 + 8}{1} = 9$$

When simplifying algebraic fractions, it is also incorrect to divide out terms common to both the numerator and denominator.

$$\frac{\overset{1}{\cancel{x}} + 5}{\underset{1}{\cancel{x}} + 6} \qquad\qquad \frac{a^2 - 3\overset{1}{\cancel{a}} + \overset{1}{\cancel{2}}}{\underset{1}{\cancel{a}} + \underset{1}{\cancel{2}}} \qquad\qquad \frac{\overset{1}{\cancel{y^2}} - 36}{\underset{1}{\cancel{y^2}} - y - 7}$$

EXAMPLE 5

Dividing out common factors. Write $\dfrac{5(x+3)-5}{7(x+3)-7}$ in lowest terms.

Solution   We cannot divide out $x + 3$, because it is not a factor of the entire numerator, nor is it a factor of the entire denominator. Instead, we simplify the numerator and denominator, factor them, and then divide out any common factors.

$$\frac{5(x+3)-5}{7(x+3)-7} = \frac{5x+15-5}{7x+21-7} \qquad \text{Remove parentheses.}$$

$$= \frac{5x+10}{7x+14} \qquad \text{Combine like terms.}$$

$$= \frac{5(x+2)}{7(x+2)} \qquad \text{Factor the numerator and the denominator.}$$

$$= \frac{5\cancel{(x+2)}^{\,1}}{7\cancel{(x+2)}_{\,1}} \qquad \text{Divide out the common factor of } x+2.$$

$$= \frac{5}{7}$$

SELF CHECK   Simplify $\dfrac{4(x-2)+4}{3(x-2)+3}$.   *Answer:* $\frac{4}{3}$ ■

EXAMPLE 6

Combining like terms. Simplify $\dfrac{x(x+3)-3(x-1)}{x^2+3}$.

Solution
$$\frac{x(x+3)-3(x-1)}{x^2+3} = \frac{x^2+3x-3x+3}{x^2+3} \qquad \text{Remove parentheses in the numerator.}$$

$$= \frac{x^2+3}{x^2+3} \qquad \begin{array}{l}\text{Combine like terms in the numerator:}\\ 3x - 3x = 0.\end{array}$$

$$= \frac{\cancel{x^2+3}^{\,1}}{\cancel{x^2+3}_{\,1}} \qquad \text{Divide out the common factor of } x^2+3.$$

$$= 1$$

SELF CHECK   Simplify $\dfrac{a(a+2)-2(a-1)}{a^2+2}$.   *Answer:* 1   ■

Sometimes a fraction does not simplify. For example, to attempt to simplify

$$\frac{x^2+x-2}{x^2+x}$$

we factor the numerator and the denominator.

$$\frac{x^2+x-2}{x^2+x} = \frac{(x+2)(x-1)}{x(x+1)}$$

Because there are no factors common to the numerator and denominator, this fraction is already in lowest terms.

Dividing Polynomials That Are Negatives

If the terms of two polynomials are the same, except for sign, the polynomials are called **negatives** (**opposites**) of each other. For example, the following pairs are negatives of each other:

$$x - y \quad \text{and} \quad -x + y$$
$$2a - 1 \quad \text{and} \quad -2a + 1$$
$$-3x^2 - 2x + 5 \quad \text{and} \quad 3x^2 + 2x - 5$$

Example 7 shows why the quotient of two binomials that are negatives is always -1.

EXAMPLE 7 **Quotients of negatives.** Simplify **a.** $\dfrac{x - y}{y - x}$ and **b.** $\dfrac{2a - 1}{1 - 2a}$.

Solution We can rearrange terms in each numerator, factor out -1, and proceed as follows:

a.
$$\frac{x - y}{y - x} = \frac{-y + x}{y - x}$$
$$= \frac{-(y - x)}{y - x}$$
$$= \frac{-\overset{1}{\cancel{(y - x)}}}{\underset{1}{\cancel{y - x}}}$$
$$= -1$$

b.
$$\frac{2a - 1}{1 - 2a} = \frac{-1 + 2a}{1 - 2a}$$
$$= \frac{-(1 - 2a)}{1 - 2a}$$
$$= \frac{-\overset{1}{\cancel{(1 - 2a)}}}{\underset{1}{\cancel{1 - 2a}}}$$
$$= -1$$

SELF CHECK Simplify $\dfrac{3p - 2q}{2q - 3p}$. *Answer:* -1 ∎

In general, we have this important fact.

Division of negatives

The quotient of any nonzero expression and its negative is -1.

Rational Functions

Rational expressions can be used to model many situations. In Figure 6-4, for example, the time t (in minutes) that it takes the cardiac rehabilitation patient to complete his $\frac{1}{4}$-mile treadmill workout is given by

$$t = \frac{15}{r}$$

FIGURE 6-4

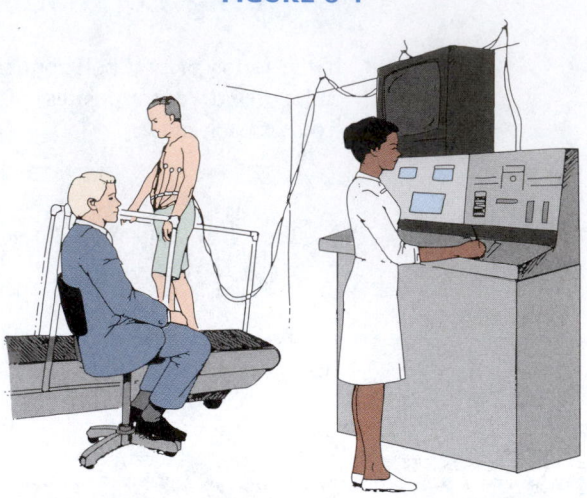

where r is the rate that he walks (in miles per hour). The expression on the right-hand side of this equation is a rational expression.

The *rational function* that gives the time it takes for the patient to complete the workout, when walking at a rate of r mph, can be written

$$f(r) = \frac{15}{r}$$ Assume that the patient walks at a constant rate r and that $r > 0$.

Rational functions

A **rational function** is a function whose equation is defined by a rational expression in one variable. The value of the polynomial in the denominator of the expression cannot be zero.

EXAMPLE 8

Evaluating rational functions. Use the function $f(r) = \frac{15}{r}$ to find the time it will take the patient to complete the treadmill workout if he walks at a rate of 3 mph.

Solution   To find the workout time if the patient walks at a rate of 3 mph, we find $f(3)$:

$$f(\mathbf{3}) = \frac{15}{\mathbf{3}}$$ Input 3 for r.

$$= 5$$ Do the division. The units of the output are minutes.

It will take the patient 5 minutes to complete the workout.

SELF CHECK   Find the time it will take the patient to complete the workout if he walks at a rate of 4 mph.   *Answer:* $3\frac{3}{4}$ min   ■

Graphing Rational Functions

To graph the rational function $f(r) = \frac{15}{r}$, we substitute values for r (the inputs) in the equation, compute the corresponding values of $f(r)$ (the outputs), and express the results as ordered pairs. From the evaluation in Example 8 and its self check, we know two ordered pairs that satisfy the equation: $(3, 5)$ and $\left(4, 3\frac{3}{4}\right)$. These pairs and others are listed in the table shown in Figure 6-5. (To show the entire graph of the rational function, we have chosen rates of 6, 10, and 15 mph, although they are unrealistic rates for a rehabilitation patient.) We plot each point and draw a smooth curve through them to get the graph.

FIGURE 6-5

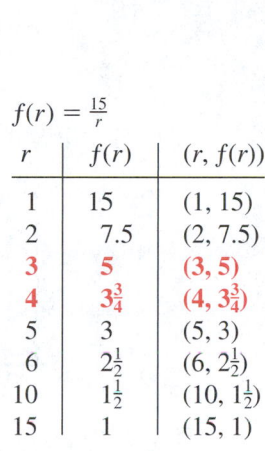

$$f(r) = \frac{15}{r}$$

r	$f(r)$	$(r, f(r))$
1	15	$(1, 15)$
2	7.5	$(2, 7.5)$
3	**5**	**(3, 5)**
4	**$3\frac{3}{4}$**	**$\left(4, 3\frac{3}{4}\right)$**
5	3	$(5, 3)$
6	$2\frac{1}{2}$	$\left(6, 2\frac{1}{2}\right)$
10	$1\frac{1}{2}$	$\left(10, 1\frac{1}{2}\right)$
15	1	$(15, 1)$

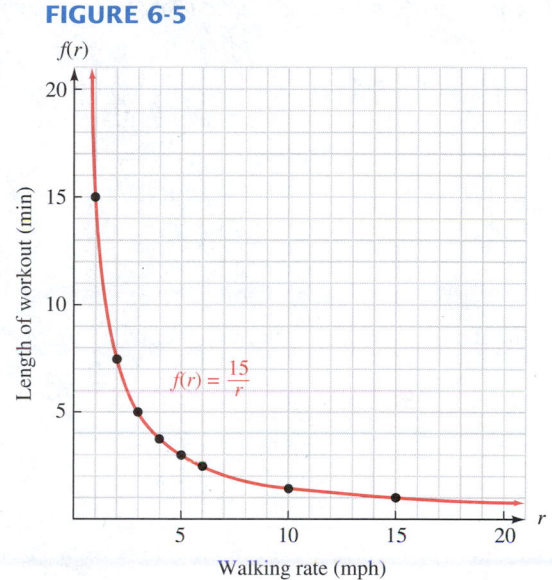

From the graph, we can see that the time to complete the treadmill workout decreases as the patient's rate of walking increases. We note that the graph approaches the x-axis as r increases without bound. When a graph approaches a line, we call the line an **asymptote**. The x-axis is a **horizontal asymptote** of the graph.

As r gets smaller and approaches 0, the graph approaches the y-axis. The y-axis is a **vertical asymptote** of the graph.

ACCENT ON TECHNOLOGY *Graphing Rational Functions*

We can use a graphing calculator to generate tables and graphs for rational functions. For example, Figure 6-6(a) shows how to enter the function $f(r) = \frac{15}{r}$, using x as the independent variable instead of r. Figures 6-6(b) and (c) show the calculator display of a table and the graph of the function.

FIGURE 6-6

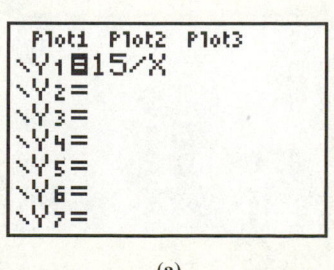

(a)

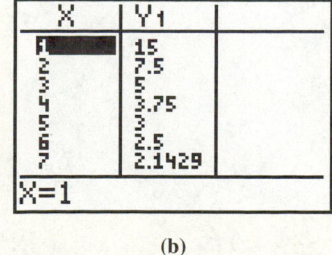

(b)

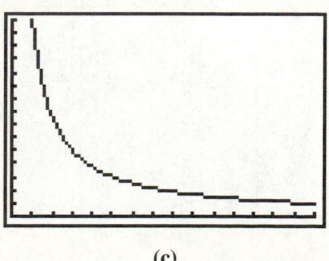

(c)

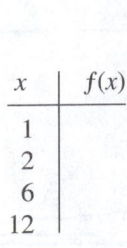

STUDY SET

Section 6.3

VOCABULARY

In Exercises 1–6, fill in the blanks to make the statements true.

1. In a fraction, the part above the fraction bar is called the _____. In a fraction, the part below the fraction bar is called the _____.

2. A fraction that has polynomials in its numerator and denominator, such as $\dfrac{x+2}{x-3}$, is called a _____ expression.

3. The denominator of a fraction cannot be ▨.

4. $x - 2$ and $2 - x$ are called _____ of each other.

5. "To simplify a fraction" means to write it in _____ terms.

6. $f(x) = \dfrac{3}{x}$ is a _____ function, because its equation is defined by a rational expression in one variable.

CONCEPTS

In Exercises 7–10, fill in the blanks to make the statements true.

7. The fundamental property of fractions states that $\frac{ac}{bc} = ▨$.

8. Any number x divided by 1 is ▨.

9. To simplify a rational expression, we _____ the numerator and denominator and divide out _____ factors.

10. A rational expression cannot be simplified when it is written in _____ terms.

11. Use the graph of a rational function shown below to complete the table.

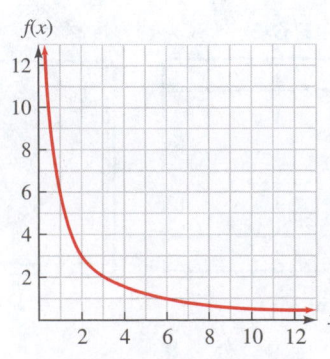

x	$f(x)$
1	
2	
6	
12	

12. Illustration 1 shows a table of values for the rational function $f(x) = \frac{5}{x}$. Explain why the word ERROR appears in the second column.

ILLUSTRATION 1

X	Y₁
0	ERROR
1	5
2	2.5
3	1.6667
4	1.25
5	1
6	.83333

X=0

NOTATION

In Exercises 13–14, complete each solution.

13. $\dfrac{x^2 + 5x - 6}{x^2 - 1} = \dfrac{(x + ▨)(x - 1)}{(x + 1)(x - ▨)}$

    $= \dfrac{x + 6}{x + 1}$

14. $\dfrac{5(x + 2) - 5}{4(x + 2) - 4} = \dfrac{5x + ▨ - 5}{4x + ▨ - 4}$

    $= \dfrac{5x + ▨}{4x + ▨}$

    $= \dfrac{5▨}{4(x + 1)}$

    $= \dfrac{5}{4}$

PRACTICE

In Exercises 15–82 simplify each rational expression. If it is already in lowest terms, so indicate. Assume that no denominators are zero.

15. $\dfrac{8}{10}$

16. $\dfrac{16}{28}$

17. $\dfrac{28}{35}$

18. $\dfrac{14}{20}$

19. $\dfrac{8}{52}$

20. $\dfrac{15}{21}$

21. $\dfrac{10}{45}$

22. $\dfrac{21}{35}$

23. $\dfrac{-18}{54}$

24. $\dfrac{-16}{40}$

25. $\dfrac{4x}{2}$

26. $\dfrac{2x}{4}$

27. $-\dfrac{6x}{18}$

28. $-\dfrac{25y}{5}$

29. $\dfrac{45}{9a}$

30. $\dfrac{48}{16y}$

31. $\dfrac{5+5}{5z}$

32. $\dfrac{(3-18)k}{25}$

33. $\dfrac{(3+4)a}{24-3}$

34. $\dfrac{x+x}{2}$

35. $\dfrac{2x}{3x}$

36. $\dfrac{5y}{7y}$

37. $\dfrac{6x^2}{4x^2}$

38. $\dfrac{9xy}{6xy}$

39. $\dfrac{2x^2}{3y}$

40. $\dfrac{5y^2}{2y^2}$

41. $\dfrac{15x^2y}{5xy^2}$

42. $\dfrac{12xz}{4xz^2}$

43. $\dfrac{28x}{32y}$

44. $\dfrac{14xz^2}{7x^2z^2}$

45. $\dfrac{x+3}{3(x+3)}$

46. $\dfrac{2(x+7)}{x+7}$

47. $\dfrac{5x+35}{x+7}$

48. $\dfrac{x-9}{3x-27}$

49. $\dfrac{x^2+3x}{2x+6}$

50. $\dfrac{xz-2x}{yz-2y}$

51. $\dfrac{15x-3x^2}{25y-5xy}$

52. $\dfrac{3y+xy}{3x+xy}$

53. $\dfrac{6a-6b+6c}{9a-9b+9c}$

54. $\dfrac{3a-3b-6}{2a-2b-4}$

55. $\dfrac{x-7}{7-x}$

56. $\dfrac{d-c}{c-d}$

57. $\dfrac{6x-3y}{3y-6x}$

58. $\dfrac{3c-4d}{4c-3d}$

59. $\dfrac{a+b-c}{c-a-b}$

60. $\dfrac{x-y-z}{z+y-x}$

61. $\dfrac{x^2+3x+2}{x^2+x-2}$

62. $\dfrac{x^2+x-6}{x^2-x-2}$

63. $\dfrac{x^2-8x+15}{x^2-x-6}$

64. $\dfrac{x^2-6x-7}{x^2+8x+7}$

65. $\dfrac{2x^2-8x}{x^2-6x+8}$

66. $\dfrac{3y^2-15y}{y^2-3y-10}$

67. $\dfrac{xy+2x^2}{2xy+y^2}$

68. $\dfrac{3x+3y}{x^2+xy}$

69. $\dfrac{x^2+3x+2}{x^3+x^2}$

70. $\dfrac{6x^2-13x+6}{3x^2+x-2}$

71. $\dfrac{x^2-8x+16}{x^2-16}$

72. $\dfrac{3x+15}{x^2-25}$

73. $\dfrac{2x^2-8}{x^2-3x+2}$

74. $\dfrac{3x^2-27}{x^2+3x-18}$

75. $\dfrac{x^2-2x-15}{x^2+2x-15}$

76. $\dfrac{x^2+4x-77}{x^2-4x-21}$

77. $\dfrac{x^2-3(2x-3)}{9-x^2}$

78. $\dfrac{x(x-8)+16}{16-x^2}$

79. $\dfrac{4(x+3)+4}{3(x+2)+6}$

80. $\dfrac{4+2(x-5)}{3x-5(x-2)}$

81. $\dfrac{x^2-9}{(2x+3)-(x+6)}$

82. $\dfrac{x^2+5x+4}{2(x+3)-(x+2)}$

In Exercises 83–84, complete the table of values for each rational function (round to the nearest tenth when applicable). Then graph it. Each function is defined for $x > 0$.

83. $f(x) = \dfrac{4}{x}$

x	$f(x)$
1	
2	
3	
4	
5	
6	
8	

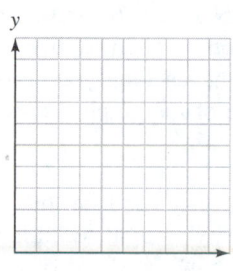

84. $f(x) = \dfrac{5}{2x}$

x	$f(x)$
0.5	
1	
2	
3	
4	
5	

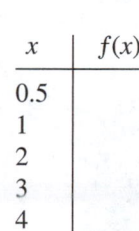
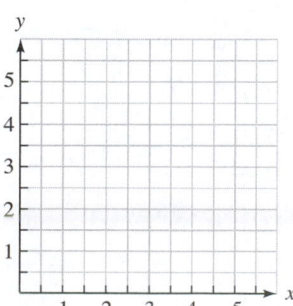

APPLICATIONS

85. ROOFING The *pitch* of a roof is a measure of how steep or how flat the roof is. If pitch $= \frac{\text{rise}}{\text{run}}$, find the pitch of the roof of the cabin shown in Illustration 2. Express the result in lowest terms.

ILLUSTRATION 2

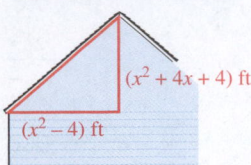

$(x^2 + 4x + 4)$ ft

$(x^2 - 4)$ ft

86. GRAPHIC DESIGN A chart of the basic food groups, in the shape of an equilateral triangle, is to be enlarged and distributed to schools for display in their health classes. (See Illustration 3.) What is the ratio of the length of a side of the original design to a length of a side of the enlargement? Express the result in lowest terms.

ILLUSTRATION 3

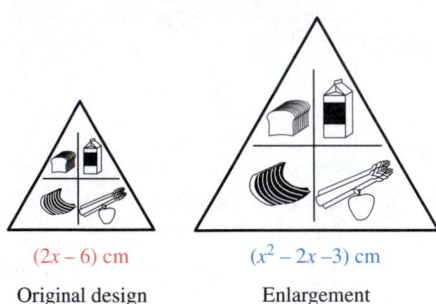

$(2x - 6)$ cm $(x^2 - 2x - 3)$ cm

Original design   Enlargement

87. LIGHTING See Illustration 4.

a. As you move away from a light bulb, the intensity of light reaching you decreases. Explain how the shape of the graph shows this.

b. When you stand far away from a light bulb, the intensity of light reaching you is almost zero. Explain how the shape of the graph shows this.

ILLUSTRATION 4

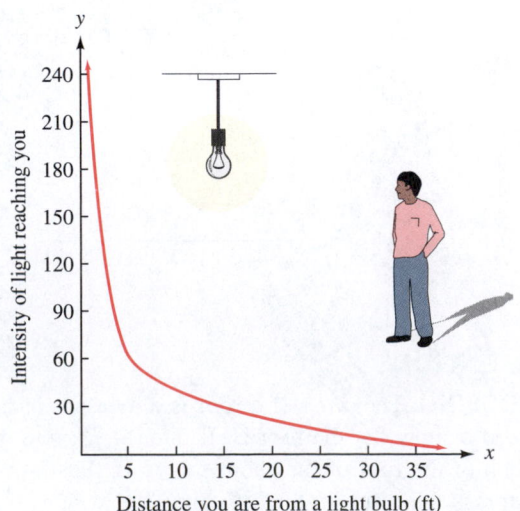

Distance you are from a light bulb (ft)

88. WORD PROCESSOR For the word processor shown in Illustration 5, the number of words that can be typed on a piece of paper is given by

$$f(x) = \frac{8,000}{x}$$

where x is the font size used. Find the number of words that can be typed on a page for each font size choice shown.

ILLUSTRATION 5

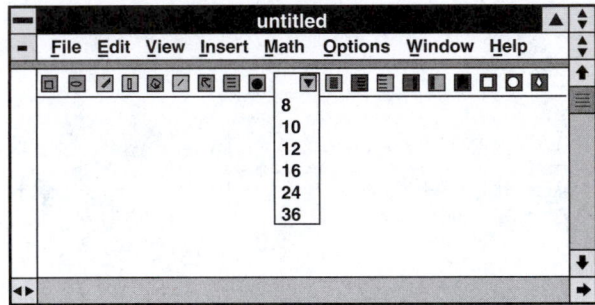

WRITING

89. Explain why $\dfrac{x - 7}{7 - x} = -1$.

90. Explain the difference between a factor and a term. Give several examples.

91. Explain the **error.**

$$\frac{\overset{1}{\cancel{x}} + 6}{\underset{1}{\cancel{x}} + 1} = \frac{7}{2}$$

92. Explain the **error.**

$$\frac{3(\cancel{x + 1}) - x}{\underset{1}{\cancel{x + 1}}} = 3 - x$$

REVIEW

93. State the associative property of addition using the variables a, b, and c.

94. State the distributive property using the variables x, y, and z.

95. If $ab = 0$, what must be true about a or b?

96. What is the product of a number and 1?

97. What is the opposite of $-\dfrac{5}{3}$?

98. What is the cube of 2 squared?

99. What is the sum of a number and zero?

100. What is the quotient of a nonzero number and itself?

▶ 6.4 Multiplying and Dividing Rational Expressions

In this section, you will learn about

> Multiplying rational expressions ■ Multiplying a rational expression by a polynomial ■ Dividing rational expressions ■ Dividing a rational expression by a polynomial ■ Combined operations

Introduction   In this section, we will extend the rules for multiplying and dividing numerical fractions to problems involving multiplication and division of rational expressions.

Multiplying Rational Expressions

To multiply fractions, we multiply their numerators and multiply their denominators. For example,

$$\frac{4}{7} \cdot \frac{3}{5} = \frac{4 \cdot 3}{7 \cdot 5} \qquad \text{Multiply the numerators and multiply the denominators.}$$

$$= \frac{12}{35} \qquad \begin{array}{l}\text{Do the multiplication in the numerator: } 4 \cdot 3 = 12. \\ \text{Do the multiplication in the denominator: } 7 \cdot 5 = 35.\end{array}$$

In general, we have the following rule.

Rule for multiplying fractions

If a, b, c, and d are real numbers and $b \neq 0$ and $d \neq 0$,

$$\frac{a}{b} \cdot \frac{c}{d} = \frac{ac}{bd}$$

We use the same procedure to multiply rational expressions.

EXAMPLE 1

Multiplying rational expressions. Multiply   **a.** $\dfrac{x}{3} \cdot \dfrac{2}{5}$, **b.** $\dfrac{7}{9} \cdot \dfrac{-5}{3x}$, **c.** $\dfrac{x^2}{2} \cdot \dfrac{3}{y^2}$, and **d.** $\dfrac{t+1}{t} \cdot \dfrac{t-1}{t-2}$.

Solution

a. $\dfrac{x}{3} \cdot \dfrac{2}{5} = \dfrac{x \cdot 2}{3 \cdot 5}$

$= \dfrac{2x}{15}$

b. $\dfrac{7}{9} \cdot \dfrac{-5}{3x} = \dfrac{7(-5)}{9 \cdot 3x}$

$= \dfrac{-35}{27x}$

$= -\dfrac{35}{27x}$

c. $\dfrac{x^2}{2} \cdot \dfrac{3}{y^2} = \dfrac{x^2 \cdot 3}{2 \cdot y^2}$

$= \dfrac{3x^2}{2y^2}$

d. $\dfrac{t+1}{t} \cdot \dfrac{t-1}{t-2} = \dfrac{(t+1)(t-1)}{t(t-2)}$

SELF CHECK   Multiply $\dfrac{3x}{4} \cdot \dfrac{x-3}{5}$.

Answer: $\dfrac{3x(x-3)}{20}$ ∎

EXAMPLE 2

Multiplying rational expressions. Multiply $\dfrac{35x^2y}{7y^2z} \cdot \dfrac{z}{5xy}$.

Solution

$$\dfrac{35x^2y}{7y^2z} \cdot \dfrac{z}{5xy} = \dfrac{35x^2y \cdot z}{7y^2z \cdot 5xy}$$ Multiply the numerators and multiply the denominators.

$$= \dfrac{5 \cdot 7 \cdot x \cdot x \cdot y \cdot z}{7 \cdot y \cdot y \cdot z \cdot 5 \cdot x \cdot y}$$ Factor: $35x^2 = 5 \cdot 7 \cdot x \cdot x$.
Factor: $y^2 = y \cdot y$.

$$= \dfrac{\overset{1}{\cancel{5}} \cdot \overset{1}{\cancel{7}} \cdot \overset{1}{\cancel{x}} \cdot x \cdot \overset{1}{\cancel{y}} \cdot \overset{1}{\cancel{z}}}{\underset{1}{\cancel{7}} \cdot y \cdot y \cdot \underset{1}{\cancel{z}} \cdot \underset{1}{\cancel{5}} \cdot \underset{1}{\cancel{x}} \cdot y}$$ Divide out the common factors of 5, 7, x, y, and z.

$$= \dfrac{x}{y^2}$$ Do the multiplications in the numerator and the denominator.

SELF CHECK   Multiply $\dfrac{a^2b^2}{2a} \cdot \dfrac{9a^3}{3b^3}$.

Answer: $\dfrac{3a^4}{2b}$ ∎

EXAMPLE 3

Factoring to simplify a product. Multiply $\dfrac{x^2-x}{2x+4} \cdot \dfrac{x+2}{x}$.

Solution

$$\dfrac{x^2-x}{2x+4} \cdot \dfrac{x+2}{x} = \dfrac{(x^2-x)(x+2)}{(2x+4)(x)}$$ Multiply the numerators and multiply the denominators.

We now factor the numerator and denominator to see if this product can be simplified.

$$\dfrac{x^2-x}{2x+y} \cdot \dfrac{x+2}{x} = \dfrac{x(x-1)(x+2)}{2(x+2)x}$$ Factor the numerator: $(x^2-x) = x(x-1)$.
Factor the denominator: $(2x+4) = 2(x+2)$.

$$= \dfrac{x(x-1)\overset{1}{\cancel{(x+2)}}}{2\underset{1}{\cancel{(x+2)}}x}$$ Divide out common factors.

$$= \dfrac{x-1}{2}$$

SELF CHECK   Multiply $\dfrac{x^2+x}{3x+6} \cdot \dfrac{x+2}{x+1}$.

Answer: $\dfrac{x}{3}$ ∎

EXAMPLE 4

Factoring to simplify a product. Multiply $\dfrac{x^2-3x}{x^2-x-6} \cdot \dfrac{x^2+x-2}{x^2-x}$.

Solution

$$\dfrac{x^2-3x}{x^2-x-6} \cdot \dfrac{x^2+x-2}{x^2-x}$$

$$= \dfrac{(x^2-3x)(x^2+x-2)}{(x^2-x-6)(x^2-x)}$$ Multiply the numerators and multiply the denominators.

$$= \dfrac{x(x-3)(x+2)(x-1)}{(x+2)(x-3)x(x-1)}$$ Factor the numerator and denominator to see if the result can be simplified.

$$= \dfrac{\overset{1}{\cancel{x}}\overset{1}{\cancel{(x-3)}}\overset{1}{\cancel{(x+2)}}\overset{1}{\cancel{(x-1)}}}{\underset{1}{\cancel{(x+2)}}\underset{1}{\cancel{(x-3)}}\underset{1}{\cancel{x}}\underset{1}{\cancel{(x-1)}}}$$ Divide out common factors.

$$= 1$$

SELF CHECK   Multiply $\dfrac{a^2 + a}{a^2 - 4} \cdot \dfrac{a^2 - a - 2}{a^2 + 2a + 1}$.   *Answer:* $\dfrac{a}{a + 2}$ ∎

Multiplying a Rational Expression by a Polynomial

Since any number divided by 1 remains unchanged, we can write any polynomial as a fraction by inserting a denominator of 1.

E X A M P L E   5   **Multiplying a rational expression by a binomial.** Multiply

$$\dfrac{x^2 + x}{x^2 + 8x + 7} \cdot (x + 7).$$

Solution

$$\dfrac{x^2 + x}{x^2 + 8x + 7} \cdot (x + 7)$$

$$= \dfrac{x^2 + x}{x^2 + 8x + 7} \cdot \dfrac{x + 7}{1}$$ Write $x + 7$ as a fraction with a denominator of 1.

$$= \dfrac{x(x + 1)(x + 7)}{(x + 1)(x + 7)1}$$ Multiply the numerators and multiply the denominators. Factor where possible.

$$= \dfrac{x\cancel{(x + 1)}\cancel{(x + 7)}}{\cancel{(x + 1)}\cancel{(x + 7)}1}$$ Divide out common factors.

$$= x$$

SELF CHECK   Multiply $(a - 7) \cdot \dfrac{a^2 - a}{a^2 - 8a + 7}$.   *Answer:* a ∎

Dividing Rational Expressions

Division by a nonzero number is equivalent to multiplying by its reciprocal. Thus, to divide two fractions, we can invert the divisor (the fraction following the ÷ sign) and multiply. For example,

$$\dfrac{4}{7} \div \dfrac{3}{5} = \dfrac{4}{7} \cdot \dfrac{5}{3}$$ Invert $\frac{3}{5}$ and change the division to a multiplication.

$$= \dfrac{20}{21}$$ Multiply the numerators and multiply the denominators.

In general, we have the following rule.

Division of fractions

If a is a real number and b, c, and d are nonzero real numbers, then

$$\dfrac{a}{b} \div \dfrac{c}{d} = \dfrac{a}{b} \cdot \dfrac{d}{c}$$

We use the same procedures to divide rational expressions.

EXAMPLE 6

Dividing rational expressions. Divide **a.** $\dfrac{a}{13} \div \dfrac{17}{26}$ and
b. $\dfrac{-9x}{35y} \div \dfrac{15x^2}{14}$.

Solution **a.** $\dfrac{a}{13} \div \dfrac{17}{26} = \dfrac{a}{13} \cdot \dfrac{26}{17}$ Invert the divisor, which is $\frac{17}{26}$, and change the division to a multiplication.

$= \dfrac{a \cdot 2 \cdot 13}{13 \cdot 17}$ Multiply. Then factor where possible.

$= \dfrac{a \cdot 2 \cdot \overset{1}{\cancel{13}}}{\underset{1}{\cancel{13}} \cdot 17}$ Divide out common factors.

$= \dfrac{2a}{17}$

b. $\dfrac{-9x}{35y} \div \dfrac{15x^2}{14} = \dfrac{-9x}{35y} \cdot \dfrac{14}{15x^2}$ Multiply by the reciprocal of $\dfrac{15x^2}{14}$.

$= \dfrac{-3 \cdot 3 \cdot x \cdot 2 \cdot 7}{5 \cdot 7 \cdot y \cdot 3 \cdot 5 \cdot x \cdot x}$ Multiply. Then factor where possible.

$= \dfrac{-3 \cdot \overset{1}{\cancel{3}} \cdot x \cdot 2 \cdot \overset{1}{\cancel{7}}}{5 \cdot \underset{1}{\cancel{7}} \cdot y \cdot \underset{1}{\cancel{3}} \cdot 5 \cdot x \cdot x}$ Divide out common factors.

$= -\dfrac{6}{25xy}$ Multiply the remaining factors.

SELF CHECK Divide $\dfrac{-8a}{3b} \div \dfrac{16a^2}{9b^2}$. *Answer:* $-\dfrac{3b}{2a}$ ■

EXAMPLE 7

Dividing rational expressions. Divide $\dfrac{x^2 + x}{3x - 15} \div \dfrac{x^2 + 2x + 1}{6x - 30}$.

Solution $\dfrac{x^2 + x}{3x - 15} \div \dfrac{x^2 + 2x + 1}{6x - 30}$

$= \dfrac{x^2 + x}{3x - 15} \cdot \dfrac{6x - 30}{x^2 + 2x + 1}$ Invert the divisor and multiply.

$= \dfrac{x(x + 1) \cdot 2 \cdot 3(x - 5)}{3(x - 5)(x + 1)(x + 1)}$ Multiply. Then factor.

$= \dfrac{x\overset{1}{\cancel{(x + 1)}} \cdot 2 \cdot \overset{1}{\cancel{3}}\overset{1}{\cancel{(x - 5)}}}{\underset{1}{\cancel{3}}\underset{1}{\cancel{(x - 5)}}\underset{1}{\cancel{(x + 1)}}(x + 1)}$ Divide out common factors.

$= \dfrac{2x}{x + 1}$

SELF CHECK Divide $\dfrac{z^2 - 1}{z^2 + 4z + 3} \div \dfrac{z - 1}{z^2 + 2z - 3}$. *Answer:* $z - 1$ ■

Dividing a Rational Expression by a Polynomial

To divide a rational expression by a polynomial, we write the polynomial as a fraction by inserting a denominator of 1, and then we divide the fractions.

EXAMPLE 8 **Dividing by a polynomial.** Divide $\dfrac{2x^2 - 3x - 2}{2x + 1} \div (4 - x^2)$.

Solution
$$\frac{2x^2 - 3x - 2}{2x + 1} \div (4 - x^2)$$

$$= \frac{2x^2 - 3x - 2}{2x + 1} \div \frac{4 - x^2}{1} \qquad \text{Write } 4 - x^2 \text{ as a fraction with a denominator of 1.}$$

$$= \frac{2x^2 - 3x - 2}{2x + 1} \cdot \frac{1}{4 - x^2} \qquad \text{Invert the divisor and multiply.}$$

$$= \frac{(2x + 1)(x - 2) \cdot 1}{(2x + 1)(2 + x)(2 - x)} \qquad \text{Multiply. Then factor where possible.}$$

$$= \frac{\overset{1}{\cancel{(2x + 1)}}\overset{-1}{\cancel{(x - 2)}} \cdot 1}{\underset{1}{\cancel{(2x + 1)}}(2 + x)\underset{1}{\cancel{(2 - x)}}} \qquad \begin{array}{l}\text{Divide out common factors. The bino-}\\ \text{mials } x - 2 \text{ and } 2 - x \text{ are negatives:}\\ \frac{x - 2}{2 - x} = -1.\end{array}$$

$$= \frac{-1}{2 + x}$$

$$= -\frac{1}{2 + x}$$

SELF CHECK Divide $(b - a) \div \dfrac{a^2 - b^2}{a^2 + ab}$.

Answer: −a ∎

Combined Operations

Unless parentheses indicate otherwise, we do multiplications and divisions in order from left to right.

EXAMPLE 9 **Multiplying and dividing rational expressions.** Simplify $\dfrac{x^2 - x - 6}{x - 2} \div \dfrac{x^2 - 4x}{x^2 - x - 2} \cdot \dfrac{x - 4}{x^2 + x}$.

Solution Since there are no parentheses to indicate otherwise, we do the division first.

$$\frac{x^2 - x - 6}{x - 2} \div \frac{x^2 - 4x}{x^2 - x - 2} \cdot \frac{x - 4}{x^2 + x}$$

$$= \frac{x^2 - x - 6}{x - 2} \cdot \frac{x^2 - x - 2}{x^2 - 4x} \cdot \frac{x - 4}{x^2 + x} \qquad \begin{array}{l}\text{Invert the divisor, which is}\\ \frac{x^2 - 4x}{x^2 - x - 2}, \text{ and change the division}\\ \text{to a multiplication.}\end{array}$$

$$= \frac{(x + 2)(x - 3)(x + 1)(x - 2)(x - 4)}{(x - 2)x(x - 4)x(x + 1)} \qquad \text{Multiply. Then factor.}$$

$$= \frac{(x + 2)(x - 3)\overset{1}{\cancel{(x + 1)}}\overset{1}{\cancel{(x - 2)}}\overset{1}{\cancel{(x - 4)}}}{\cancel{(x - 2)}x\cancel{(x - 4)}x\cancel{(x + 1)}} \qquad \text{Divide out common factors.}$$

$$= \frac{(x + 2)(x - 3)}{x^2}$$

SELF CHECK Simplify $\dfrac{a^2 + ab}{ab - b^2} \cdot \dfrac{a^2 - b^2}{a^2 + ab} \div \dfrac{a + b}{b}$.

Answer: 1 ∎

EXAMPLE 10

Multiplying and dividing rational expressions. Simplify
$$\frac{x^2 + 6x + 9}{x^2 - 2x}\left(\frac{x^2 - 4}{x^2 + 3x} \div \frac{x + 2}{x}\right).$$

Solution   We do the division within the parentheses first.

$$\frac{x^2 + 6x + 9}{x^2 - 2x}\left(\frac{x^2 - 4}{x^2 + 3x} \div \frac{x + 2}{x}\right)$$

$$= \frac{x^2 + 6x + 9}{x^2 - 2x}\left(\frac{x^2 - 4}{x^2 + 3x} \cdot \frac{x}{x + 2}\right)$$ Invert the divisor and change the division to a multiplication.

$$= \frac{(x + 3)(x + 3)(x - 2)(x + 2)x}{x(x - 2)x(x + 3)(x + 2)}$$ Multiply and factor where possible.

$$= \frac{\overset{1}{\cancel{(x + 3)}}(x + 3)\overset{1}{\cancel{(x - 2)}}\overset{1}{\cancel{(x + 2)}}x}{x\underset{1}{\cancel{(x - 2)}}x\underset{1}{\cancel{(x + 3)}}\underset{1}{\cancel{(x + 2)}}}$$ Divide out common factors.

$$= \frac{x + 3}{x}$$

SELF CHECK   Simplify $\dfrac{x^2 - 2x}{x^2 + 6x + 9} \div \left(\dfrac{x^2 - 4}{x^2 + 3x} \cdot \dfrac{x}{x + 2}\right)$ *Answer:* $\dfrac{x}{x + 3}$ ■

STUDY SET

Section 6.4

VOCABULARY

In Exercises 1–2, fill in the blanks to make the statements true.

1. In a fraction, the part above the fraction bar is called the _____.

2. In a fraction, the part below the fraction bar is called the _____.

CONCEPTS

In Exercises 3–8, fill in the blanks to make the statements true.

3. To multiply fractions, we multiply their _____ and multiply their _____.

4. $\dfrac{a}{b} \cdot \dfrac{c}{d} = $

5. To write a polynomial in fractional form, we insert a denominator of ▮.

6. $\dfrac{a}{b} \div \dfrac{c}{d} = \dfrac{a}{b} \cdot $

7. To divide fractions, we invert the _____ and _____.

8. The _____ of $\dfrac{x}{x + 2}$ is $\dfrac{x + 2}{x}$.

NOTATION

In Exercises 9–10, complete each solution.

9. $\dfrac{x^2 + x}{3x - 6} \cdot \dfrac{x - 2}{x + 1} = \dfrac{(x^2 + x)}{(x + 1)}$

$$= \dfrac{(x - 2)}{(x + 1)}$$

$$= \dfrac{x}{3}$$

10. $\dfrac{x^2 - x}{4x + 12} \div \dfrac{x - 1}{x + 3} = \dfrac{x^2 - x}{4x + 12} \cdot \rule{1cm}{0.6cm}$

$= \dfrac{\rule{1cm}{0.6cm}\,(x + 3)}{(4x + 12)\,\rule{1cm}{0.6cm}}$

$= \dfrac{\rule{1cm}{0.6cm}\,(x + 3)}{\rule{1cm}{0.6cm}\,(x - 1)}$

$= \dfrac{x}{4}$

PRACTICE

In Exercises 11–54, do the multiplications. Simplify answers if possible.

11. $\dfrac{5}{7} \cdot \dfrac{9}{13}$
12. $\dfrac{2}{7} \cdot \dfrac{5}{11}$

13. $\dfrac{25}{35} \cdot \dfrac{-21}{55}$
14. $-\dfrac{27}{24} \cdot \left(-\dfrac{56}{35}\right)$

15. $\dfrac{2}{3} \cdot \dfrac{15}{2} \cdot \dfrac{1}{7}$
16. $\dfrac{2}{5} \cdot \dfrac{10}{9} \cdot \dfrac{3}{2}$

17. $\dfrac{3x}{y} \cdot \dfrac{y}{2}$
18. $\dfrac{2y}{z} \cdot \dfrac{z}{3}$

19. $\dfrac{5y}{7} \cdot \dfrac{7x}{5z}$
20. $\dfrac{4x}{3y} \cdot \dfrac{3y}{7x}$

21. $\dfrac{7z}{9z} \cdot \dfrac{4z}{2z}$
22. $\dfrac{8z}{2x} \cdot \dfrac{16x}{3x}$

23. $\dfrac{2x^2 y}{3xy} \cdot \dfrac{3xy^2}{2}$
24. $\dfrac{2x^2 z}{z} \cdot \dfrac{5x}{z}$

25. $\dfrac{8x^2 y^2}{4x^2} \cdot \dfrac{2xy}{2y}$
26. $\dfrac{9x^2 y}{3x} \cdot \dfrac{3xy}{3y}$

27. $\dfrac{-2xy}{x^2} \cdot \dfrac{3xy}{2}$
28. $\dfrac{-3x}{x^2} \cdot \dfrac{2xz}{3}$

29. $\dfrac{ab^2}{a^2 b} \cdot \dfrac{b^2 c^2}{abc} \cdot \dfrac{abc^2}{a^3 c^2}$
30. $\dfrac{x^3 y}{z} \cdot \dfrac{xz^3}{x^2 y^2} \cdot \dfrac{yz}{xyz}$

31. $\dfrac{10r^2 st^3}{6rs^2} \cdot \dfrac{3r^3 t}{2rst} \cdot \dfrac{2s^3 t^4}{5s^2 t^3}$

32. $\dfrac{3a^3 b}{25cd^3} \cdot \dfrac{-5cd^2}{6ab} \cdot \dfrac{10abc^2}{2bc^2 d}$

33. $\dfrac{z + 7}{7} \cdot \dfrac{z + 2}{z}$

34. $\dfrac{a - 3}{a} \cdot \dfrac{a + 3}{5}$

35. $\dfrac{x - 2}{2} \cdot \dfrac{2x}{x - 2}$
36. $\dfrac{y + 3}{y} \cdot \dfrac{3y}{y + 3}$

37. $\dfrac{x + 5}{5} \cdot \dfrac{x}{x + 5}$
38. $\dfrac{y - 9}{y + 9} \cdot \dfrac{y}{9}$

39. $\dfrac{(x + 1)^2}{x + 1} \cdot \dfrac{x + 2}{x + 1}$
40. $\dfrac{(y - 3)^2}{y - 3} \cdot \dfrac{y - 3}{y - 3}$

41. $\dfrac{2x + 6}{x + 3} \cdot \dfrac{3}{4x}$
42. $\dfrac{3y - 9}{y - 3} \cdot \dfrac{y}{3y^2}$

43. $\dfrac{x^2 - x}{x} \cdot \dfrac{3x - 6}{3x - 3}$
44. $\dfrac{5z - 10}{z + 2} \cdot \dfrac{3}{3z - 6}$

45. $\dfrac{7y - 14}{y - 2} \cdot \dfrac{x^2}{7x}$
46. $\dfrac{y^2 + 3y}{9} \cdot \dfrac{3x}{y + 3}$

47. $\dfrac{x^2 + x - 6}{5x} \cdot \dfrac{5x - 10}{x + 3}$
48. $\dfrac{z^2 + 4z - 5}{5z - 5} \cdot \dfrac{5z}{z + 5}$

49. $\dfrac{m^2 - 2m - 3}{2m + 4} \cdot \dfrac{m^2 - 4}{m^2 + 3m + 2}$

50. $\dfrac{p^2 - p - 6}{3p - 9} \cdot \dfrac{p^2 - 9}{p^2 + 6p + 9}$

51. $\dfrac{abc^2}{a + 1} \cdot \dfrac{c}{a^2 b^2} \cdot \dfrac{a^2 + a}{ac}$

52. $\dfrac{x^3 yz^2}{4x + 8} \cdot \dfrac{x^2 - 4}{2x^2 y^2 z^2} \cdot \dfrac{8yz}{x - 2}$

53. $\dfrac{3x^2 + 5x + 2}{x^2 - 9} \cdot \dfrac{x - 3}{x^2 - 4} \cdot \dfrac{x^2 + 5x + 6}{6x + 4}$

54. $\dfrac{x^2 - 25}{3x + 6} \cdot \dfrac{x^2 + x - 2}{2x + 10} \cdot \dfrac{6x}{3x^2 - 18x + 15}$

In Exercises 55–84, do each division. Simplify answers when possible.

55. $\dfrac{1}{3} \div \dfrac{1}{2}$
56. $\dfrac{3}{4} \div \dfrac{1}{3}$

57. $\dfrac{21}{14} \div \dfrac{5}{2}$
58. $\dfrac{14}{3} \div \dfrac{10}{3}$

59. $\dfrac{2}{y} \div \dfrac{4}{3}$
60. $\dfrac{3}{a} \div \dfrac{a}{9}$

61. $\dfrac{3x}{2} \div \dfrac{x}{2}$
62. $\dfrac{y}{6} \div \dfrac{2}{3y}$

63. $\dfrac{3x}{y} \div \dfrac{2x}{4}$
64. $\dfrac{3y}{8} \div \dfrac{2y}{4y}$

65. $\dfrac{4x}{3x} \div \dfrac{2y}{9y}$
66. $\dfrac{14}{7y} \div \dfrac{10}{5z}$

67. $\dfrac{x^2}{3} \div \dfrac{2x}{4}$
68. $\dfrac{z^2}{z} \div \dfrac{z}{3z}$

69. $\dfrac{x^2 y}{3xy} \div \dfrac{xy^2}{6y}$
70. $\dfrac{2xz}{z} \div \dfrac{4x^2}{z^2}$

71. $\dfrac{x + 2}{3x} \div \dfrac{x + 2}{2}$
72. $\dfrac{z - 3}{3z} \div \dfrac{z + 3}{z}$

73. $\dfrac{(z-2)^2}{3z^2} \div \dfrac{z-2}{6z}$

74. $\dfrac{(x+7)^2}{x+7} \div \dfrac{(x-3)^2}{x+7}$

75. $\dfrac{(z-7)^2}{z+2} \div \dfrac{z(z-7)}{5z^2}$

76. $\dfrac{y(y+2)}{y^2(y-3)} \div \dfrac{y^2(y+2)}{(y-3)^2}$

77. $\dfrac{x^2-4}{3x+6} \div \dfrac{x-2}{x+2}$

78. $\dfrac{x^2-9}{5x+15} \div \dfrac{x-3}{x+3}$

79. $\dfrac{x^2-1}{3x-3} \div \dfrac{x+1}{3}$

80. $\dfrac{x^2-16}{x-4} \div \dfrac{3x+12}{x}$

81. $\dfrac{x^2-2x-35}{3x^2+27x} \div \dfrac{x^2+7x+10}{6x^2+12x}$

82. $\dfrac{x^2-x-6}{2x^2+9x+10} \div \dfrac{x^2-25}{2x^2+15x+25}$

83. $\dfrac{2d^2+8d-42}{d-3} \div \dfrac{2d^2+14d}{d^2+5d}$

84. $\dfrac{5x^2+13x-6}{x+3} \div \dfrac{5x^2-17x+6}{x-2}$

In Exercises 85–96, do the operations.

85. $\dfrac{x}{3} \cdot \dfrac{9}{4} \div \dfrac{x^2}{6}$

86. $\dfrac{y^2}{2} \div \dfrac{4}{y} \cdot \dfrac{y^2}{8}$

87. $\dfrac{x^2}{18} \div \dfrac{x^3}{6} \div \dfrac{12}{x^2}$

88. $\dfrac{y^3}{3y} \cdot \dfrac{3y^2}{4} \div \dfrac{15}{20}$

89. $\dfrac{z^2-4}{2z+6} \div \dfrac{z+2}{4} \cdot \dfrac{z+3}{z-2}$

90. $\dfrac{2}{3x-3} \div \dfrac{2x+2}{x-1} \cdot \dfrac{5}{x+1}$

91. $\dfrac{x-x^2}{x^2-4}\left(\dfrac{2x+4}{x+2} \div \dfrac{5}{x+2}\right)$

92. $\dfrac{2}{3x-3} \div \left(\dfrac{2x+2}{x-1} \cdot \dfrac{5}{x+1}\right)$

93. $\dfrac{y^2}{x+1} \cdot \dfrac{x^2+2x+1}{x^2-1} \div \dfrac{3y}{xy-y}$

94. $\dfrac{x^2-y^2}{x^4-x^3} \div \dfrac{x-y}{x^2} \div \dfrac{x^2+2xy+y^2}{x+y}$

95. $\dfrac{x^2+x-6}{x^2-4} \cdot \dfrac{x^2+2x}{x-2} \div \dfrac{x^2+3x}{x+2}$

96. $\dfrac{x^2-x-6}{x^2+6x-7} \cdot \dfrac{x^2+x-2}{x^2+2x} \div \dfrac{x^2+7x}{x^2-3x}$

APPLICATIONS

97. INTERNATIONAL ALPHABET The symbols representing the letters A, B, C, D, E, and F of an international code used at sea are printed six to a sheet and then cut into separate cards. If each card is a square, find the area of the large printed sheet shown in Illustration 1.

ILLUSTRATION 1

$\left.\dfrac{2x+1}{2}\right\}$ in.

98. PHYSICS EXPERIMENT The table in Illustration 2 contains algebraic expressions for the rate an object travels, and the time traveled at that rate, in terms of a constant k. Complete the table.

ILLUSTRATION 2

Rate (mph)	Time (hr)	Distance (mi)
$\dfrac{k^2+k-6}{k-3}$	$\dfrac{k^2-9}{k^2-4}$	

WRITING

99. Explain how to multiply two fractions and how to simplify the result.

100. Explain why any mathematical expression can be written as a fraction.

101. To divide fractions, you must first know how to multiply fractions. Explain.

102. Explain how to do the division $\dfrac{a}{b} \div \dfrac{c}{d} \div \dfrac{e}{f}$.

REVIEW

In Exercises 103–106, simplify each expression. Write all answers without using negative exponents.

103. $2x^3y^2(-3x^2y^4)$

104. $\dfrac{8x^4y^5}{-2x^3y^2}$

105. $(3y)^{-4}$

106. $\dfrac{x^{3m}}{x^{4m}}$

In Exercises 107–108, do the operations and simplify.

107. $-4(y^3-4y^2+3y-2)-4(-2y^3-y)$

108. $y-5\overline{)5y^3-3y^2+4y-1}$

▶ 6.5 Adding and Subtracting Rational Expressions

In this section, you will learn about

Adding rational expressions with like denominators ■ Subtracting rational expressions with like denominators ■ Combined operations ■ The LCD ■ Adding rational expressions with unlike denominators ■ Subtracting rational expressions with unlike denominators ■ Combined operations

Introduction   In this section, we will extend the rules for adding and subtracting numerical fractions to problems involving addition and subtraction of rational expressions.

Adding Rational Expressions with Like Denominators

To add fractions with a common denominator, we add their numerators and keep the common denominator. For example,

$$\frac{2}{7} + \frac{3}{7} = \frac{2+3}{7} \qquad \text{Add the numerators and keep the common denominator.}$$

$$= \frac{5}{7}$$

In general, we have the following rule.

Adding fractions with like denominators

If a, b, and d represent real numbers, then

$$\frac{a}{d} + \frac{b}{d} = \frac{a+b}{d} \qquad (d \neq 0)$$

We use the same procedure to add rational expressions with like denominators.

EXAMPLE 1

Adding rational expressions.   Do each addition.

a. $\dfrac{x}{8} + \dfrac{3x}{8} = \dfrac{x + 3x}{8}$ Add the numerators and keep the common denominator.

$$= \frac{4x}{8} \qquad \text{Combine like terms: } x + 3x = 4x.$$

$$= \frac{\overset{1}{4} \cdot x}{\underset{1}{4} \cdot 2} \qquad \text{Factor and divide out the common factor of 4.}$$

$$= \frac{x}{2} \qquad \text{Simplify.}$$

b. $\dfrac{3x + y}{5x} + \dfrac{x + y}{5x} = \dfrac{3x + y + x + y}{5x}$ Add the numerators and keep the common denominator.

$$= \dfrac{4x + 2y}{5x}$$ Combine like terms.

SELF CHECK Add: **a.** $\dfrac{x}{7} + \dfrac{4x}{7}$ and **b.** $\dfrac{3x}{7y} + \dfrac{4x}{7y}$. *Answers:* **a.** $\frac{5x}{7}$, **b.** $\frac{x}{y}$ ∎

EXAMPLE 2 **Adding rational expressions.** Add $\dfrac{3x + 21}{5x + 10} + \dfrac{8x + 1}{5x + 10}$.

Solution Because the fractions have the same denominator, we add their numerators and keep the common denominator.

$$\dfrac{3x + 21}{5x + 10} + \dfrac{8x + 1}{5x + 10} = \dfrac{3x + 21 + 8x + 1}{5x + 10}$$ Add.

$$= \dfrac{11x + 22}{5x + 10}$$ Combine like terms.

$$= \dfrac{\overset{1}{\cancel{11(x + 2)}}}{\underset{1}{\cancel{5(x + 2)}}}$$ Simplify the result by factoring the numerator and denominator. Divide out the common factor of $x + 2$.

$$= \dfrac{11}{5}$$

SELF CHECK Add $\frac{x+4}{6x-12} + \frac{x-8}{6x-12}$. *Answer:* $\frac{1}{3}$ ∎

Subtracting Rational Expressions with Like Denominators

To subtract fractions with a common denominator, we subtract their numerators and keep the common denominator.

Subtracting fractions with like denominators

If a, b, and d represent real numbers, then

$$\dfrac{a}{d} - \dfrac{b}{d} = \dfrac{a - b}{d} \quad (d \neq 0)$$

We use the same procedure to subtract rational expressions.

EXAMPLE 3 **Subtracting rational expressions.** Subtract **a.** $\dfrac{5x}{3} - \dfrac{2x}{3}$ and

b. $\dfrac{5x + 1}{x - 3} - \dfrac{4x - 2}{x - 3}$.

Solution   In each part, the fractions have the same denominator. To subtract them, we subtract their numerators and keep the common denominator.

a. $\dfrac{5x}{3} - \dfrac{2x}{3} = \dfrac{5x - 2x}{3}$ Subtract.

$= \dfrac{3x}{3}$ Combine like terms: $5x - 2x = 3x$.

$= \dfrac{x}{1}$ Divide out the common factor of 3.

$= x$ Denominators of 1 need not be written.

b. $\dfrac{5x + 1}{x - 3} - \dfrac{4x - 2}{x - 3} = \dfrac{(5x + 1) - (4x - 2)}{x - 3}$ Subtract. Write each numerator in parentheses.

$= \dfrac{5x + 1 - 4x + 2}{x - 3}$ Remove parentheses: $-(4x - 2) = -4x + 2$.

$= \dfrac{x + 3}{x - 3}$ Combine like terms.

SELF CHECK   Subtract $\dfrac{2y + 1}{y + 5} - \dfrac{y - 4}{y + 5}$.   *Answer:* 1   ■

Combined Operations

To add and/or subtract three or more rational expressions, we follow the rules for the order of operations.

EXAMPLE 4   **Combined operations.** Simplify $\dfrac{3x + 1}{x^2 + x + 1} - \dfrac{5x + 2}{x^2 + x + 1} + \dfrac{2x + 1}{x^2 + x + 1}$.

Solution   This example combines addition and subtraction. Unless parentheses indicate otherwise, we do additions and subtractions from left to right.

$\dfrac{3x + 1}{x^2 + x + 1} - \dfrac{5x + 2}{x^2 + x + 1} + \dfrac{2x + 1}{x^2 + x + 1}$

$= \dfrac{(3x + 1) - (5x + 2) + (2x + 1)}{x^2 + x + 1}$ Combine the numerators and keep the common denominator.

$= \dfrac{3x + 1 - 5x - 2 + 2x + 1}{x^2 + x + 1}$ Remove parentheses: $-(5x + 2) = -5x - 2$.

$= \dfrac{0}{x^2 + x + 1}$ Combine like terms.

$= 0$ If the numerator of a fraction is zero and the denominator is not zero, the fraction's value is zero.

SELF CHECK   Simplify $\dfrac{2a^2 - 3}{a - 5} + \dfrac{3a^2 + 2}{a - 5} - \dfrac{5a^2}{a - 5}$.   *Answer:* $-\dfrac{1}{a - 5}$ ■

The LCD

Since the denominators of the fractions in the addition $\frac{4}{7} + \frac{3}{5}$ are different, we cannot add the fractions in their present form.

four-sevenths   +   three-fifths

└── Different denominators ──┘

To add these fractions, we need to find a common denominator. The smallest common denominator (called the **least** or **lowest common denominator**) is usually the easiest one to work with.

Least common denominator

The **least common denominator (LCD)** for a set of fractions is the smallest number that each denominator will divide exactly.

In the addition $\frac{4}{7} + \frac{3}{5}$, the denominators are 7 and 5. The smallest number that 7 and 5 will divide exactly is 35. This is the LCD. We now **build** each fraction into a fraction with a denominator of 35. To do so, we use the fundamental property of fractions to multiply both the numerator and the denominator of each fraction by some appropriate number.

$$\frac{4}{7} + \frac{3}{5} = \frac{4 \cdot 5}{7 \cdot 5} + \frac{3 \cdot 7}{5 \cdot 7}$$ Multiply numerator and denominator of $\frac{4}{7}$ by 5, and multiply numerator and denominator of $\frac{3}{5}$ by 7.

$$= \frac{20}{35} + \frac{21}{35}$$ Do the multiplications.

Now that the fractions have a common denominator, we can add them.

$$\frac{20}{35} + \frac{21}{35} = \frac{20 + 21}{35} = \frac{41}{35}$$

EXAMPLE 5

Building fractions. Change each fraction into one with a denominator of $30y$:

a. $\dfrac{1}{2y}$,   **b.** $\dfrac{3y}{5}$,   and   **c.** $\dfrac{7+x}{10y}$.

Solution   To build each fraction, we multiply the numerator and denominator by the factor that makes the denominator $30y$.

a. $\dfrac{1}{2y} = \dfrac{1 \cdot \mathbf{15}}{2y \cdot \mathbf{15}} = \dfrac{15}{30y}$ Multiply numerator and denominator by 15, because $2y \cdot 15 = 30y$.

b. $\dfrac{3y}{5} = \dfrac{3y \cdot \mathbf{6y}}{5 \cdot \mathbf{6y}} = \dfrac{18y^2}{30y}$ Multiply numerator and denominator by $6y$, because $5 \cdot 6y = 30y$.

c. $\dfrac{7+x}{10y} = \dfrac{(7+x)\mathbf{3}}{(10y)\mathbf{3}} = \dfrac{21 + 3x}{30y}$ Multiply numerator and denominator by 3, because $10y \cdot 3 = 30y$.

SELF CHECK   Change $\frac{5}{6b}$ into a fraction with a denominator of $30ab$.   *Answer:* $\frac{25a}{30ab}$ ■

There is a process that we can use to find the least common denominator of several fractions.

Finding the least common denominator (LCD)

1. List the different denominators that appear in the fraction.
2. Completely factor each denominator.
3. Form a product using each different factor obtained in step 2. Use each different factor the *greatest* number of times it appears in any one factorization. The product formed by multiplying these factors is the LCD.

EXAMPLE 6

Finding the LCD. Find the LCD of $\dfrac{5}{24b}$ and $\dfrac{11}{18b}$.

Solution We list and factor each denominator into the product of prime numbers.

$$24b = 2 \cdot 2 \cdot 2 \cdot 3 \cdot b$$
$$18b = 2 \cdot 3 \cdot 3 \cdot b$$

To find the LCD, we use each of these factors the greatest number of times it appears in any one factorization. We use 2 three times, because it appears three times as a factor of 24. We use 3 twice, because it occurs twice as a factor of 18. We use b once.

$$LCD = 2 \cdot 2 \cdot 2 \cdot 3 \cdot 3 \cdot b$$
$$= 8 \cdot 9 \cdot b$$
$$= 72b$$

SELF CHECK Find the LCD of $\frac{3}{28z}$ and $\frac{5}{21z}$.   *Answer:* $84z$ ∎

Adding Rational Expressions with Unlike Denominators

The following steps summarize how to add fractions that have unlike denominators.

Adding fractions with unlike denominators

To add fractions with different denominators,

1. Find the LCD.
2. Write each fraction as a fraction whose denominator is the LCD.
3. Add the resulting fractions and simplify the result, if possible.

EXAMPLE 7

Adding rational expressions. Add $\dfrac{4x}{7} + \dfrac{3x}{5}$.

Solution The LCD is 35. We build each fraction so that it has a denominator of 35 and then add the resulting fractions.

$$\frac{4x}{7} + \frac{3x}{5} = \frac{4x \cdot \mathbf{5}}{7 \cdot \mathbf{5}} + \frac{3x \cdot \mathbf{7}}{5 \cdot \mathbf{7}}$$ Multiply the numerator and the denominator of $\frac{4x}{7}$ by 5 and the numerator and denominator of $\frac{3x}{5}$ by 7.

$$= \frac{20x}{35} + \frac{21x}{35}$$ Do the multiplication.

$$= \frac{41x}{35}$$ Add the numerators and keep the common denominator.

SELF CHECK Add $\frac{y}{2} + \frac{6y}{7}$.   *Answer:* $\frac{19y}{14}$ ∎

EXAMPLE 8

Adding rational expressions. Add $\dfrac{5}{24b} + \dfrac{11}{18b}$.

Solution In Example 6, we saw that the LCD of these fractions is $2 \cdot 2 \cdot 2 \cdot 3 \cdot 3 \cdot b = 72b$. To add them, we first factor each denominator:

$$\frac{5}{24b} + \frac{11}{18b} = \frac{5}{2 \cdot 2 \cdot 2 \cdot 3 \cdot b} + \frac{11}{2 \cdot 3 \cdot 3 \cdot b}$$

In each resulting fraction, we multiply the numerator and the denominator by whatever it takes to build the denominator to the LCD of $2 \cdot 2 \cdot 2 \cdot 3 \cdot 3 \cdot b$.

$$= \frac{5 \cdot 3}{2 \cdot 2 \cdot 2 \cdot 3 \cdot b \cdot 3} + \frac{11 \cdot 2 \cdot 2}{2 \cdot 3 \cdot 3 \cdot b \cdot 2 \cdot 2}$$

$$= \frac{15}{72b} + \frac{44}{72b} \qquad \text{Do the multiplications.}$$

$$= \frac{59}{72b} \qquad \text{Add the numerators and keep the common denominator.}$$

SELF CHECK Add $\frac{3}{28z} + \frac{5}{21z}$. *Answer:* $\frac{29}{84z}$ ∎

EXAMPLE 9   **Adding rational expressions.** Add $\dfrac{x+4}{x^2} + \dfrac{x-5}{4x}$.

Solution   First we find the LCD.

$$\left.\begin{array}{l} x^2 = x \cdot x \\ 4x = 2 \cdot 2 \cdot x \end{array}\right\} \qquad \text{LCD} = x \cdot x \cdot 2 \cdot 2 = 4x^2$$

$$\frac{x+4}{x^2} + \frac{x-5}{4x} = \frac{(x+4)4}{(x^2)4} + \frac{(x-5)x}{(4x)x} \qquad \begin{array}{l}\text{Build the fractions to get the common}\\ \text{denominator of } 4x^2.\end{array}$$

$$= \frac{4x+16}{4x^2} + \frac{x^2-5x}{4x^2} \qquad \text{Do the multiplications.}$$

$$= \frac{4x+16+x^2-5x}{4x^2} \qquad \begin{array}{l}\text{Add the numerators and keep the common}\\ \text{denominator.}\end{array}$$

$$= \frac{x^2-x+16}{4x^2} \qquad \text{Combine like terms.}$$

SELF CHECK Add $\dfrac{a-1}{9a} + \dfrac{2-a}{a^2}$. *Answer:* $\dfrac{a^2-10a+18}{9a^2}$ ∎

Subtracting Rational Expressions with Unlike Denominators

To subtract fractions with unlike denominators, we first change them into fractions with the same denominators. We use the same procedure to subtract rational expressions.

EXAMPLE 10   **Subtracting rational expressions.** Subtract $\dfrac{x}{x+1} - \dfrac{3}{x}$.

Solution   By inspection, the least common denominator is $(x+1)x$.

$$\frac{x}{x+1} - \frac{3}{x} = \frac{x(x)}{(x+1)x} - \frac{3(x+1)}{x(x+1)} \qquad \begin{array}{l}\text{Build the fractions to get the common denom-}\\ \text{inator.}\end{array}$$

$$= \frac{x(x)-3(x+1)}{x(x+1)} \qquad \begin{array}{l}\text{Subtract the numerators and keep the common}\\ \text{denominator.}\end{array}$$

$$= \frac{x^2-3x-3}{x(x+1)} \qquad \text{Do the multiplications in the numerator.}$$

SELF CHECK Subtract $\dfrac{a}{a-1} - \dfrac{5}{a}$. *Answer:* $\dfrac{a^2-5a+5}{a(a-1)}$ ∎

EXAMPLE 11

Simplifying the result. Subtract $\dfrac{a}{a-1} - \dfrac{2}{a^2-1}$.

Solution   We factor $a^2 - 1$ to see that the LCD $= (a+1)(a-1)$.

$$\frac{a}{a-1} - \frac{2}{a^2-1}$$

$$= \frac{a(a+1)}{(a-1)(a+1)} - \frac{2}{(a+1)(a-1)} \qquad \text{Build the first fraction to get the LCD.}$$

$$= \frac{a(a+1)-2}{(a-1)(a+1)} \qquad \text{Subtract the numerators and keep the common denominator.}$$

$$= \frac{a^2+a-2}{(a-1)(a+1)} \qquad \text{Remove parentheses.}$$

$$= \frac{(a+2)\overset{1}{\cancel{(a-1)}}}{\underset{1}{\cancel{(a-1)}}(a+1)} \qquad \text{Simplify the result by factoring } a^2+a-2. \text{ Divide out the common factor of } a-1.$$

$$= \frac{a+2}{a+1}$$

SELF CHECK   Subtract $\dfrac{b}{b-2} - \dfrac{8}{b^2-4}$.   *Answer:* $\dfrac{b+4}{b+2}$ ∎

EXAMPLE 12

Factoring to find the LCD. Subtract $\dfrac{2a}{a^2+4a+4} - \dfrac{1}{2a+4}$.

Solution   Find the least common denominator by factoring each denominator.

$$\left.\begin{array}{l} a^2+4a+4 = (a+2)(a+2) \\ 2a+4 = 2(a+2) \end{array}\right\} \qquad \text{LCD} = (a+2)(a+2)2$$

We build each fraction into a new fraction with a denominator of $2(a+2)(a+2)$.

$$\frac{2a}{a^2+4a+4} - \frac{1}{2a+4}$$

$$= \frac{2a}{(a+2)(a+2)} - \frac{1}{2(a+2)} \qquad \text{Write the denominators in factored form.}$$

$$= \frac{2a \cdot 2}{(a+2)(a+2)2} - \frac{1(a+2)}{2(a+2)(a+2)} \qquad \text{Build each fraction to get a common denominator.}$$

$$= \frac{4a - 1(a+2)}{2(a+2)^2} \qquad \text{Subtract the numerators and keep the common denominator. Write } (a+2)(a+2) \text{ as } (a+2)^2.$$

$$= \frac{4a-a-2}{2(a+2)^2} \qquad \text{Remove parentheses.}$$

$$= \frac{3a-2}{2(a+2)^2} \qquad \text{Combine like terms.}$$

SELF CHECK   Subtract $\dfrac{a}{a^2-2a+1} - \dfrac{1}{6a-6}$.   *Answer:* $\dfrac{5a+1}{6(a-1)^2}$. ∎

EXAMPLE 13

Denominators that are opposites. Subtract $\dfrac{3}{x-y} - \dfrac{x}{y-x}$.

Solution   We note that the second denominator is the negative of the first. So we can multiply the numerator and denominator of the second fraction by -1 to get

$$\dfrac{3}{x-y} - \dfrac{x}{y-x} = \dfrac{3}{x-y} - \dfrac{-1x}{-1(y-x)} \qquad \text{Multiply numerator and denominator by } -1.$$

$$= \dfrac{3}{x-y} - \dfrac{-x}{-y+x} \qquad \text{Remove parentheses: } -1(y-x) = -y+x.$$

$$= \dfrac{3}{x-y} - \dfrac{-x}{x-y} \qquad \begin{array}{l}-y+x = x-y.\text{ The fractions have a} \\ \text{common denominator of } x-y.\end{array}$$

$$= \dfrac{3-(-x)}{x-y} \qquad \begin{array}{l}\text{Subtract the numerators and keep the com-} \\ \text{mon denominator.}\end{array}$$

$$= \dfrac{3+x}{x-y} \qquad -(-x)=x.$$

SELF CHECK   Subtract $\dfrac{5}{a-b} - \dfrac{2}{b-a}$.   *Answer:* $\dfrac{7}{a-b}$ ■

Combined Operations

To add and/or subtract three or more rational expressions, we follow the rules for the order of operations.

EXAMPLE 14

Addition and subtraction. Do the operations: $\dfrac{3}{x^2y} + \dfrac{2}{xy} - \dfrac{1}{xy^2}$.

Solution   Find the least common denominator.

$$\left.\begin{array}{l} x^2y = x \cdot x \cdot y \\ xy = x \cdot y \\ xy^2 = x \cdot y \cdot y \end{array}\right\} \quad \text{Factor each denominator.}$$

In any one of these denominators, the factor x occurs at most twice, and the factor y occurs at most twice. Thus,

$$\begin{aligned} \text{LCD} &= x \cdot x \cdot y \cdot y \\ &= x^2y^2 \end{aligned}$$

We build each fraction into one with a denominator of x^2y^2.

$$\dfrac{3}{x^2y} + \dfrac{2}{xy} - \dfrac{1}{xy^2}$$

$$= \dfrac{3 \cdot y}{x \cdot x \cdot y \cdot y} + \dfrac{2 \cdot x \cdot y}{x \cdot y \cdot x \cdot y} - \dfrac{1 \cdot x}{x \cdot y \cdot y \cdot x} \qquad \begin{array}{l}\text{Factor each denominator and build} \\ \text{each fraction.}\end{array}$$

$$= \dfrac{3y + 2xy - x}{x^2y^2} \qquad \begin{array}{l}\text{Do the multiplications and combine} \\ \text{the numerators.}\end{array}$$

SELF CHECK   Combine $\dfrac{5}{ab^2} - \dfrac{b}{a} + \dfrac{a}{b}$.   *Answer:* $\dfrac{5 - b^3 + a^2b}{ab^2}$ ■

Section 6.5

VOCABULARY

In Exercises 1–2, fill in the blanks to make the statements true.

1. The _____ for a set of fractions is the smallest number that each denominator divides exactly.

2. When we multiply the numerator and denominator of a fraction by some number to get a common denominator, we say that we are _____ the fraction.

CONCEPTS

In Exercises 3–4, fill in the blanks to make the statements true.

3. To add two fractions with like denominators, we add their _____ and keep the _____.

4. To subtract two fractions with _____ denominators, we need to find a common denominator.

NOTATION

In Exercises 5–6, complete each solution.

5. $\dfrac{6a - 1}{4a + 1} + \dfrac{2a + 3}{4a + 1} = \dfrac{6a - 1 + \boxed{}}{4a + 1}$

$= \dfrac{8a + \boxed{}}{4a + 1}$

$= \dfrac{2\boxed{}}{4a + 1}$

$= 2$

6. $\dfrac{x}{2x + 1} - \dfrac{1}{3x} = \dfrac{x\boxed{}}{(2x + 1)(3x)} - \dfrac{1(2x + 1)}{3x\boxed{}}$

$= \dfrac{x(3x) - 1\boxed{}}{3x(2x + 1)}$

$= \dfrac{3x^2 - \boxed{} - \boxed{}}{3x(2x + 1)}$

$= \dfrac{(3x + 1)(x - 1)}{3x(2x + 1)}$

PRACTICE

In Exercises 7–18, do each addition. Simplify answers, if possible.

7. $\dfrac{x}{9} + \dfrac{2x}{9}$

8. $\dfrac{5x}{7} + \dfrac{9x}{7}$

9. $\dfrac{2x}{y} + \dfrac{2x}{y}$

10. $\dfrac{4y}{3x} + \dfrac{2y}{3x}$

11. $\dfrac{4}{7y} + \dfrac{10}{7y}$

12. $\dfrac{x^2}{4y} + \dfrac{x^2}{4y}$

13. $\dfrac{y + 2}{10z} + \dfrac{y + 4}{10z}$

14. $\dfrac{x + 3}{2x^2} + \dfrac{x + 5}{2x^2}$

15. $\dfrac{3x - 5}{x - 2} + \dfrac{6x - 13}{x - 2}$

16. $\dfrac{8x - 7}{x + 3} + \dfrac{2x + 37}{x + 3}$

17. $\dfrac{a}{a^2 + 5a + 6} + \dfrac{3}{a^2 + 5a + 6}$

18. $\dfrac{b}{b^2 - 4} + \dfrac{2}{b^2 - 4}$

In Exercises 19–30, do each subtraction. Simplify answers, if possible.

19. $\dfrac{35y}{72} - \dfrac{44y}{72}$

20. $\dfrac{13t}{99} - \dfrac{35t}{99}$

21. $\dfrac{2x}{y} - \dfrac{x}{y}$

22. $\dfrac{7y}{5} - \dfrac{4y}{5}$

23. $\dfrac{9y}{3x} - \dfrac{6y}{3x}$

24. $\dfrac{5r^2}{2r} - \dfrac{r^2}{2r}$

25. $\dfrac{6x - 5}{3xy} - \dfrac{3x - 5}{3xy}$

26. $\dfrac{7x + 7}{5y} - \dfrac{2x + 7}{5y}$

27. $\dfrac{3y - 2}{2y + 6} - \dfrac{2y - 5}{2y + 6}$

28. $\dfrac{5x + 8}{3x + 15} - \dfrac{3x - 2}{3x + 15}$

29. $\dfrac{2c}{c^2 - d^2} - \dfrac{2d}{c^2 - d^2}$

30. $\dfrac{3t}{t^2 - 8t + 7} - \dfrac{3}{t^2 - 8t + 7}$

In Exercises 31–38, do the operations. Simplify answers if possible.

31. $\dfrac{13x}{15} + \dfrac{12x}{15} - \dfrac{5x}{15}$

32. $\dfrac{13y}{32} + \dfrac{13y}{32} - \dfrac{10y}{32}$

33. $-\dfrac{x}{y} + \dfrac{2x}{y} - \dfrac{x}{y}$

34. $\dfrac{5y}{8x} + \dfrac{4y}{8x} - \dfrac{9y}{8x}$

35. $\dfrac{3x}{y + 2} - \dfrac{3y}{y + 2} + \dfrac{x + y}{y + 2}$

36. $\dfrac{3y}{x-5} + \dfrac{x}{x-5} - \dfrac{y-x}{x-5}$

37. $\dfrac{x+1}{x-2} - \dfrac{2(x-3)}{x-2} + \dfrac{3(x+1)}{x-2}$

38. $\dfrac{3xy}{x-y} - \dfrac{x(3y-x)}{x-y} - \dfrac{x(x-y)}{x-y}$

In Exercises 39–50, build each fraction into an equivalent fraction with the indicated denominator.

39. $\dfrac{25}{4}$; $20x$

40. $\dfrac{5}{y}$; y^2

41. $\dfrac{8}{x}$; x^2y

42. $\dfrac{7}{y}$; xy^2

43. $\dfrac{3x}{x+1}$; $(x+1)^2$

44. $\dfrac{5y}{y-2}$; $(y-2)^2$

45. $\dfrac{2y}{x}$; $x^2 + x$

46. $\dfrac{3x}{y}$; $y^2 - y$

47. $\dfrac{z}{z-1}$; $z^2 - 1$

48. $\dfrac{y}{y+2}$; $y^2 - 4$

49. $\dfrac{2}{x+1}$; $x^2 + 3x + 2$

50. $\dfrac{3}{x-1}$; $x^2 + x - 2$

In Exercises 51–60, several denominators are given. Find the LCD.

51. $2x$, $6x$

52. $3y$, $9y$

53. $6y$, $9xy^2$

54. $6y$, $3x^2y$

55. $x^2 - 1$, $x + 1$

56. $y^2 - 9$, $y - 3$

57. $x^2 + 6x$, $x + 6$, x

58. $xy^2 - xy$, xy, $y - 1$

59. $x^2 - 4x - 5$, $x^2 - 25$

60. $x^2 - x - 6$, $x^2 - 9$

In Exercises 61–96, do the operations. Simplify answers, if possible.

61. $\dfrac{2y}{9} + \dfrac{y}{3}$

62. $\dfrac{8a}{15} - \dfrac{5a}{12}$

63. $\dfrac{21x}{14} - \dfrac{5x}{21}$

64. $\dfrac{7y}{6} + \dfrac{10y}{9}$

65. $\dfrac{4x}{3} + \dfrac{2x}{y}$

66. $\dfrac{2y}{5x} - \dfrac{y}{2}$

67. $\dfrac{2}{x} - 3x$ $\left(Hint:\ 3x = \tfrac{3x}{1}\right)$

68. $14 + \dfrac{10}{y^2}$ $\left(Hint:\ 14 = \tfrac{14}{1}\right)$

69. $\dfrac{y+2}{5y^2} + \dfrac{y+4}{15y}$

70. $\dfrac{x+3}{x^2} + \dfrac{x+5}{2x}$

71. $\dfrac{x+5}{xy} - \dfrac{x-1}{x^2y}$

72. $\dfrac{y-7}{y^2} - \dfrac{y+7}{2y}$

73. $\dfrac{x}{x+1} + \dfrac{x-1}{x}$

74. $\dfrac{3x}{xy} + \dfrac{x+1}{y-1}$

75. $\dfrac{x-1}{x} + \dfrac{y+1}{y}$

76. $\dfrac{a+2}{b} + \dfrac{b-2}{a}$

77. $\dfrac{x}{x-2} + \dfrac{4+2x}{x^2-4}$

78. $\dfrac{y}{y+3} - \dfrac{2y-6}{y^2-9}$

79. $\dfrac{x+1}{x-1} + \dfrac{x-1}{x+1}$

80. $\dfrac{2x}{x+2} + \dfrac{x+1}{x-3}$

81. $\dfrac{5}{a-4} + \dfrac{7}{4-a}$

82. $\dfrac{4}{b-6} - \dfrac{b}{6-b}$

83. $\dfrac{t+1}{t-7} - \dfrac{t+1}{7-t}$

84. $\dfrac{r+2}{r^2-4} + \dfrac{4}{4-r^2}$

85. $\dfrac{2x+2}{x-2} - \dfrac{2x}{2-x}$

86. $\dfrac{y+3}{y-1} - \dfrac{y+4}{1-y}$

87. $\dfrac{b}{b+1} - \dfrac{b+1}{2b+2}$

88. $\dfrac{4x+1}{8x-12} + \dfrac{x-3}{2x-3}$

89. $\dfrac{2}{a^2+4a+3} + \dfrac{1}{a+3}$

90. $\dfrac{1}{c+6} - \dfrac{-4}{c^2 + 8a + 12}$

91. $\dfrac{x+1}{2x+4} - \dfrac{x^2}{2x^2 - 8}$

92. $\dfrac{x+1}{x+2} - \dfrac{x^2+1}{x^2 - x - 6}$

93. $\dfrac{2x}{x^2 - 3x + 2} + \dfrac{2x}{x-1} - \dfrac{x}{x-2}$

94. $\dfrac{4a}{a-2} - \dfrac{3a}{a-3} + \dfrac{4a}{a^2 - 5a + 6}$

95. $\dfrac{2x}{x-1} + \dfrac{3x}{x+1} - \dfrac{x+3}{x^2 - 1}$

96. $\dfrac{a}{a-1} - \dfrac{2}{a+2} + \dfrac{3(a-2)}{a^2 + a - 2}$

APPLICATIONS

In Exercises 97–98, refer to Illustration 1.

97. Find the total height of the funnel.

98. What is the difference between the diameter of the opening at the top of the funnel and the diameter of its spout?

ILLUSTRATION 1

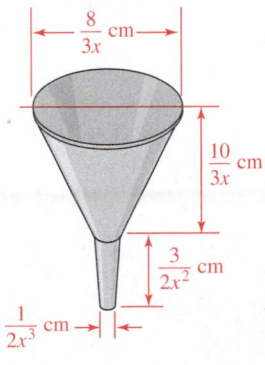

$\frac{8}{3x}$ cm

$\frac{10}{3x}$ cm

$\frac{3}{2x^2}$ cm

$\frac{1}{2x^3}$ cm

WRITING

99. Explain how to add fractions with the same denominator.

100. Explain how to find a lowest common denominator.

101. Explain what is **wrong** with the following solution:

$$\frac{2x+3}{x+5} - \frac{x+2}{x+5} = \frac{2x+3-x+2}{x+5}$$

$$= \frac{x+5}{x+5}$$

$$= 1$$

102. Explain what is **wrong** with the following solution:

$$\frac{5x-4}{y} + \frac{x}{y} = \frac{5x-4+x}{y+y}$$

$$= \frac{6x-4}{2y}$$

$$= \frac{2(x-2)}{2y}$$

$$= \frac{3x-2}{y}$$

REVIEW

In Exercises 103–106, write each number in prime-factored form.

103. 49   **104.** 64

105. 136   **106.** 315

▶ **6.6**

Complex Fractions

In this section, you will learn about

Simplifying complex fractions ■ Simplifying fractions with terms containing negative exponents

---

Introduction   Rational expressions such as

$$\frac{\frac{5x}{3}}{\frac{2y}{9}}, \qquad \frac{x + \frac{1}{2}}{3 - x}, \qquad \text{and} \qquad \frac{\frac{x+1}{2}}{x + \frac{1}{x}}$$

which contain fractions in their numerators or denominators, are called **complex fractions.** In this section, we will show how to use the properties of fractions to simplify complex fractions.

Simplifying Complex Fractions

Complex fractions can often be simplified.

$$\cfrac{\dfrac{5x}{3}}{\dfrac{2y}{9}} \quad \leftarrow \text{The main fraction bar indicates division.}$$

We can simplify the complex fraction by doing the division:

$$\cfrac{\dfrac{5x}{3}}{\dfrac{2y}{9}} = \frac{5x}{3} \div \frac{2y}{9} = \frac{5x}{3} \cdot \frac{9}{2y} = \frac{5x \cdot 3 \cdot \overset{1}{\cancel{3}}}{\underset{1}{\cancel{3}} \cdot 2y} = \frac{15x}{2y}$$

There are two ways to simplify complex fractions.

> **Methods for simplifying complex fractions**

Method 1: Write the numerator and denominator of the complex fraction as single fractions. Then divide the fractions and simplify.

Method 2: Multiply the numerator and denominator of the complex fraction by the LCD of the fractions in its numerator and denominator. Then simplify the results, if possible.

To simplify the complex fraction

$$\cfrac{\dfrac{3x}{5} + 1}{2 - \dfrac{x}{5}}$$

using method 1, we proceed as follows:

$$\cfrac{\dfrac{3x}{5} + \mathbf{1}}{\mathbf{2} - \dfrac{x}{5}} = \cfrac{\dfrac{3x}{5} + \dfrac{\mathbf{5}}{\mathbf{5}}}{\dfrac{\mathbf{10}}{\mathbf{5}} - \dfrac{x}{5}} \qquad \text{Change 1 to } \tfrac{5}{5} \text{ and 2 to } \tfrac{10}{5} \text{ so that we can write the numerator and denominator as single fractions.}$$

$$= \cfrac{\dfrac{3x + 5}{5}}{\dfrac{10 - x}{5}} \qquad \text{Add the fractions in the numerator and subtract the fractions in the denominator.}$$

$$= \frac{3x + 5}{5} \div \frac{10 - x}{5} \qquad \text{Write the complex fraction as an equivalent division problem.}$$

$$= \frac{3x + 5}{5} \cdot \frac{5}{10 - x} \qquad \text{Invert the divisor and multiply.}$$

$$= \frac{(3x + 5)5}{5(10 - x)} \qquad \text{Multiply the fractions.}$$

$$= \frac{3x + 5}{10 - x} \qquad \text{Divide out the common factor of 5.}$$

To use method 2, we proceed as follows:

$$\dfrac{\dfrac{3x}{5}+1}{2-\dfrac{x}{5}}=\dfrac{5\left(\dfrac{3x}{5}+1\right)}{5\left(2-\dfrac{x}{5}\right)}$$ Multiply both the numerator and denominator of the complex fraction by 5, the LCD of $\frac{3x}{5}$ and $\frac{x}{5}$.

$$=\dfrac{5\cdot\dfrac{3x}{5}+5\cdot 1}{5\cdot 2-5\cdot\dfrac{x}{5}}$$ Remove parentheses.

$$=\dfrac{3x+5}{10-x}$$ Do the multiplications.

In this example, method 2 is easier than method 1. Either method can be used to simplify complex fractions. With practice, you will be able to see which method is best in a given situation.

EXAMPLE 1 **Simplifying complex fractions.** Simplify $\dfrac{\dfrac{x}{3}}{\dfrac{y}{3}}$.

Solution

Method 1

$$\dfrac{\dfrac{x}{3}}{\dfrac{y}{3}}=\dfrac{x}{3}\div\dfrac{y}{3}$$

$$=\dfrac{x}{3}\cdot\dfrac{3}{y}$$

$$=\dfrac{3x}{3y}$$

$$=\dfrac{x}{y}$$

Method 2

$$\dfrac{\dfrac{x}{3}}{\dfrac{y}{3}}=\dfrac{3\left(\dfrac{x}{3}\right)}{3\left(\dfrac{y}{3}\right)}$$

$$=\dfrac{x}{y}$$

SELF CHECK Simplify $\dfrac{\dfrac{a}{4}}{\dfrac{5}{b}}$.

Answer: $\frac{ab}{20}$ ■

EXAMPLE 2 **Simplifying complex fractions.** Simplify $\dfrac{\dfrac{x}{x+1}}{\dfrac{y}{x}}$.

Solution

Method 1

$$\dfrac{\dfrac{x}{x+1}}{\dfrac{y}{x}}=\dfrac{x}{x+1}\div\dfrac{y}{x}$$

$$=\dfrac{x}{x+1}\cdot\dfrac{x}{y}$$

$$=\dfrac{x^2}{y(x+1)}$$

Method 2

$$\dfrac{\dfrac{x}{x+1}}{\dfrac{y}{x}}=\dfrac{x(x+1)\left(\dfrac{x}{x+1}\right)}{x(x+1)\left(\dfrac{y}{x}\right)}$$

$$=\dfrac{x^2}{y(x+1)}$$

SELF CHECK Simplify $\dfrac{\dfrac{x}{y}}{\dfrac{x}{y+1}}$.

Answer: $\dfrac{y+1}{y}$

EXAMPLE 3 **Simplifying complex fractions.** Simplify $\dfrac{1+\dfrac{1}{x}}{1-\dfrac{1}{x}}$.

Solution

Method 1

$$\frac{1+\dfrac{1}{x}}{1-\dfrac{1}{x}} = \frac{\dfrac{x}{x}+\dfrac{1}{x}}{\dfrac{x}{x}-\dfrac{1}{x}}$$

$$= \frac{\dfrac{x+1}{x}}{\dfrac{x-1}{x}}$$

$$= \frac{x+1}{x} \div \frac{x-1}{x}$$

$$= \frac{x+1}{x} \cdot \frac{x}{x-1}$$

$$= \frac{(x+1)\cancel{x}}{\cancel{x}(x-1)}$$

$$= \frac{x+1}{x-1}$$

Method 2

$$\frac{1+\dfrac{1}{x}}{1-\dfrac{1}{x}} = \frac{x\left(1+\dfrac{1}{x}\right)}{x\left(1-\dfrac{1}{x}\right)}$$

$$= \frac{x+1}{x-1}$$

SELF CHECK Simplify $\dfrac{\dfrac{1}{x}+1}{\dfrac{1}{x}-1}$.

Answer: $\dfrac{1+x}{1-x}$

EXAMPLE 4 **Simplifying complex fractions.** Simplify $\dfrac{1}{1+\dfrac{1}{x+1}}$.

Solution We use method 2.

$$\frac{1}{1+\dfrac{1}{x+1}} = \frac{(x+1) \cdot 1}{(x+1)\left(1+\dfrac{1}{x+1}\right)}$$ Multiply the numerator and the denominator of the complex fraction by $x+1$.

$$= \frac{x+1}{(x+1)1+1}$$ In the denominator, distribute $x+1$.

$$= \frac{x+1}{x+2}$$ Simplify.

SELF CHECK Simplify $\dfrac{2}{\dfrac{1}{x+2}-2}$.

Answer: $\dfrac{2(x+2)}{-2x-3}$

Simplifying Fractions with Terms Containing Negative Exponents

Many fractions with terms containing negative exponents are complex fractions in disguise.

EXAMPLE 5   **Simplifying complex fractions.** Simplify $\dfrac{x^{-1} + y^{-2}}{x^{-2} - y^{-1}}$.

Solution   Write the fraction in complex fraction form and simplify using method 2.

$$\frac{x^{-1} + y^{-2}}{x^{-2} - y^{-1}} = \frac{\dfrac{1}{x} + \dfrac{1}{y^2}}{\dfrac{1}{x^2} - \dfrac{1}{y}}$$

$$= \frac{x^2 y^2 \left(\dfrac{1}{x} + \dfrac{1}{y^2} \right)}{x^2 y^2 \left(\dfrac{1}{x^2} - \dfrac{1}{y} \right)}$$

Multiply the numerator and denominator by $x^2 y^2$, which is the LCD of the fractions in the numerator and the denominator.

$$= \frac{xy^2 + x^2}{y^2 - x^2 y}$$

Remove parentheses.

$$= \frac{x(y^2 + x)}{y(y - x^2)}$$

Attempt to simplify the fraction by factoring the numerator and the denominator. The result cannot be simplified.

SELF CHECK   Simplify $\dfrac{x^{-2} - y^{-1}}{x^{-1} + y^{-2}}$.   *Answer:* $\dfrac{y(y - x^2)}{x(y^2 + x)}$ ■

STUDY SET

Section 6.6

VOCABULARY

In Exercises 1–2, fill in the blanks to make the statements true.

1. If a fraction has a fraction in its numerator or denominator, it is called a _____.

2. The denominator of the complex fraction $\dfrac{\frac{3}{x} + \frac{x}{y}}{\frac{1}{x} + 2}$ is _____.

CONCEPTS

In Exercises 3–4, fill in the blanks to make the statements true.

3. In method 1, we write the numerator and denominator of a complex fraction as _____ fractions and then _____.

4. In method 2, we multiply the numerator and denominator of the complex fraction by the _____ of the fractions in its numerator and denominator.

NOTATION

In Exercises 5–6, complete each solution.

5.
$$\frac{\dfrac{2}{a} - \dfrac{1}{b}}{\dfrac{1}{a} + \dfrac{2}{b}} = \frac{\dfrac{}{ab}}{\dfrac{}{ab}}$$

$$= \frac{\dfrac{2b - a}{ab}}{\dfrac{b + 2a}{ab}}$$

$$= \frac{2b - a}{ab} \cdot \frac{}{b + 2a}$$

$$= \frac{(2b - a)}{ab}$$

$$= \frac{2b - a}{b + 2a}$$

6. $\dfrac{\dfrac{2}{a} - \dfrac{1}{b}}{\dfrac{1}{a} + \dfrac{2}{b}} = \dfrac{\left(\dfrac{2}{a} - \dfrac{1}{b}\right)}{\left(\dfrac{1}{a} + \dfrac{2}{b}\right)}$

$= \dfrac{2b - a}{b + 2a}$

PRACTICE

In Exercises 7–40, simplify each complex fraction.

7. $\dfrac{\dfrac{2}{3}}{\dfrac{3}{4}}$

8. $\dfrac{\dfrac{3}{5}}{\dfrac{2}{7}}$

9. $\dfrac{\dfrac{4}{5}}{\dfrac{32}{15}}$

10. $\dfrac{\dfrac{7}{8}}{\dfrac{49}{4}}$

11. $\dfrac{\dfrac{2}{3} + 1}{\dfrac{1}{3} + 1}$

12. $\dfrac{\dfrac{3}{5} - 2}{\dfrac{2}{5} - 2}$

13. $\dfrac{\dfrac{1}{2} + \dfrac{3}{4}}{\dfrac{3}{2} + \dfrac{1}{4}}$

14. $\dfrac{\dfrac{2}{3} - \dfrac{5}{2}}{\dfrac{2}{3} - \dfrac{3}{2}}$

15. $\dfrac{\dfrac{x}{y}}{\dfrac{1}{x}}$

16. $\dfrac{\dfrac{y}{x}}{\dfrac{x}{xy}}$

17. $\dfrac{\dfrac{5t^2}{9x^2}}{\dfrac{3t}{x^2 t}}$

18. $\dfrac{\dfrac{5w^2}{4tz}}{\dfrac{15wt}{z^2}}$

19. $\dfrac{\dfrac{1}{x} - 3}{\dfrac{5}{x} + 2}$

20. $\dfrac{\dfrac{1}{y} + 3}{\dfrac{3}{y} - 2}$

21. $\dfrac{\dfrac{2}{x} + 2}{\dfrac{4}{x} + 2}$

22. $\dfrac{\dfrac{3}{x} - 3}{\dfrac{9}{x} - 3}$

23. $\dfrac{\dfrac{3y}{x} - y}{y - \dfrac{y}{x}}$

24. $\dfrac{\dfrac{y}{x} + 3y}{y + \dfrac{2y}{x}}$

25. $\dfrac{\dfrac{1}{x + 1}}{1 + \dfrac{1}{x + 1}}$

26. $\dfrac{\dfrac{1}{x - 1}}{1 - \dfrac{1}{x - 1}}$

27. $\dfrac{\dfrac{x}{x + 2}}{\dfrac{x}{x + 2} + x}$

28. $\dfrac{\dfrac{2}{x - 2}}{\dfrac{2}{x - 2} - 1}$

29. $\dfrac{1}{\dfrac{1}{x} + \dfrac{1}{y}}$

30. $\dfrac{1}{\dfrac{b}{a} - \dfrac{a}{b}}$

31. $\dfrac{\dfrac{2}{x}}{\dfrac{2}{y} - \dfrac{4}{x}}$

32. $\dfrac{\dfrac{2y}{3}}{\dfrac{2y}{3} - \dfrac{8}{y}}$

33. $\dfrac{3 + \dfrac{3}{x - 1}}{3 - \dfrac{3}{x}}$

34. $\dfrac{2 - \dfrac{2}{x + 1}}{2 + \dfrac{2}{x}}$

35. $\dfrac{\dfrac{3}{x} + \dfrac{4}{x + 1}}{\dfrac{2}{x + 1} - \dfrac{3}{x}}$

36. $\dfrac{\dfrac{5}{y - 3} - \dfrac{2}{y}}{\dfrac{1}{y} + \dfrac{2}{y - 3}}$

37. $\dfrac{\dfrac{2}{x} - \dfrac{3}{x + 1}}{\dfrac{2}{x + 1} - \dfrac{3}{x}}$

38. $\dfrac{\dfrac{5}{y} + \dfrac{4}{y + 1}}{\dfrac{4}{y} - \dfrac{5}{y + 1}}$

39. $\dfrac{\dfrac{1}{y^2 + y} - \dfrac{1}{xy + x}}{\dfrac{1}{xy + x} - \dfrac{1}{y^2 + y}}$

40. $\dfrac{\dfrac{2}{b^2 - 1} - \dfrac{3}{ab - a}}{\dfrac{3}{ab - a} - \dfrac{2}{b^2 - 1}}$

In Exercises 41–50, simplify each complex fraction.

41. $\dfrac{x^{-2}}{y^{-1}}$

42. $\dfrac{a^{-4}}{b^{-2}}$

43. $\dfrac{1 + x^{-1}}{x^{-1} - 1}$

44. $\dfrac{y^{-2} + 1}{y^{-2} - 1}$

45. $\dfrac{a^{-2} + a}{a}$

46. $\dfrac{t - t^{-2}}{t^{-1}}$

47. $\dfrac{2x^{-1} + 4x^{-2}}{2x^{-2} + x^{-1}}$

48. $\dfrac{x^{-2} - 3x^{-3}}{3x^{-2} - 9x^{-3}}$

49. $\dfrac{1 - 25y^{-2}}{1 + 10y^{-1} + 25y^{-2}}$

50. $\dfrac{1 - 9x^{-2}}{1 - 6x^{-1} + 9x^{-2}}$

APPLICATIONS

51. GARDENING TOOLS In Illustration 1, what is the ratio of the opening of the cutting blades to the opening of the handles for the lopping shears? Express the result in simplest form.

ILLUSTRATION 1

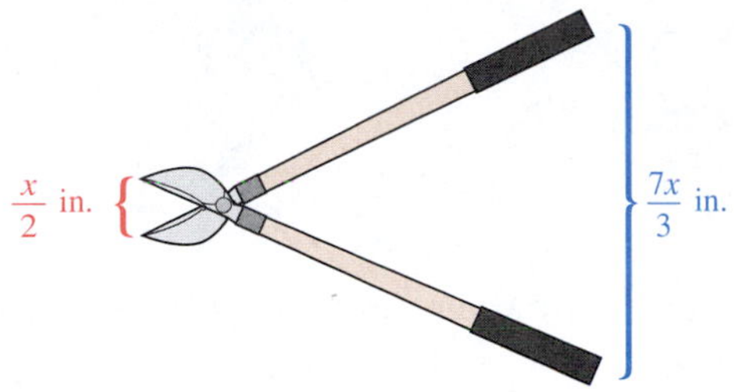

$\frac{x}{2}$ in. $\frac{7x}{3}$ in.

52. EARNED RUN AVERAGE The earned run average (ERA) is a statistic that gives the average number of earned runs a pitcher allows. For a softball pitcher, this is based on a six-inning game. The formula for ERA is

$$\text{ERA} = \frac{\dfrac{\text{earned runs}}{\text{innings pitched}}}{6}$$

Simplify the complex fraction on the right-hand side of the equation.

53. ELECTRONICS In electronic circuits, resistors oppose the flow of an electric current. To find the total resistance of a parallel combination of two resistors (see Illustration 2), we can use the formula

$$\text{Total resistance} = \frac{1}{\dfrac{1}{R_1} + \dfrac{1}{R_2}}$$

ILLUSTRATION 2

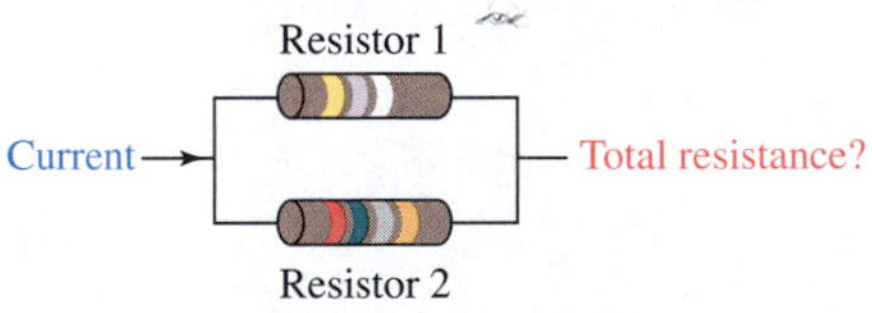

where R_1 is the resistance of the first resistor and R_2 is the resistance of the second. Simplify the complex fraction on the right-hand side of the formula.

54. DATA ANALYSIS Use the data in Illustration 3 to find the average measurement for the three-trial experiment.

ILLUSTRATION 3

	Trial 1	Trial 2	Trial 3
Measurement	$\dfrac{k}{2}$	$\dfrac{k}{3}$	$\dfrac{k}{2}$

WRITING

55. Explain how to use method 1 to simplify

$$\frac{1 + \dfrac{1}{x}}{3 - \dfrac{1}{x}}$$

56. Explain how to use method 2 to simplify the expression in Exercise 55.

REVIEW

In Exercises 57–60, write each expression as an expression involving only one exponent.

57. $t^3 t^4 t^2$

58. $(a^0 a^2)^3$

59. $-2r(r^3)^2$

60. $(s^3)^2 (s^4)^0$

In Exercises 61–64, write each expression without using parentheses or negative exponents.

61. $\left(\dfrac{3r}{4r^3}\right)^4$

62. $\left(\dfrac{12y^{-3}}{3y^2}\right)^{-2}$

63. $\left(\dfrac{6r^{-2}}{2r^3}\right)^{-2}$

64. $\left(\dfrac{4x^3}{5x^{-3}}\right)^{-2}$

▶ **6.7**

Solving Equations That Contain Rational Expressions

In this section, you will learn about

Solving equations that contain rational expressions ■ Extraneous solutions ■ Literal equations

Introduction In this section, we will discuss how to solve equations containing rational expressions. We will make use of the properties of equality and several concepts from this chapter, one of which is LCD.

Solving Equations That Contain Rational Expressions

To solve equations containing fractions, it's usually best to eliminate those fractions. To do so, we multiply both sides of the equation by the LCD. For example, to solve $\frac{x}{3} + 1 = \frac{x}{6}$, we multiply both sides of the equation by 6:

$$\frac{x}{3} + 1 = \frac{x}{6}$$

$$6\left(\frac{x}{3} + 1\right) = 6\left(\frac{x}{6}\right)$$

We then use the distributive property to remove parentheses, simplify, and solve the resulting equation for x.

$$6 \cdot \frac{x}{3} + 6 \cdot 1 = 6 \cdot \frac{x}{6}$$

$$2x + 6 = x$$

$$x + 6 = 0 \qquad \text{Subtract } x \text{ from both sides.}$$

$$x = -6 \qquad \text{Subtract 6 from both sides.}$$

Check:

$$\frac{x}{3} + 1 = \frac{x}{6}$$

$$\frac{-6}{3} + 1 \stackrel{?}{=} \frac{-6}{6} \qquad \text{Substitute } -6 \text{ for } x.$$

$$-2 + 1 \stackrel{?}{=} -1 \qquad \text{Simplify.}$$

$$-1 = -1$$

EXAMPLE 1

Solving equations containing rational expressions. Solve $\frac{4}{x} + 1 = \frac{6}{x}$.

Solution To clear the equation of fractions, we multiply both sides by the LCD of $\frac{4}{x}$ and $\frac{6}{x}$, which is x.

$$\frac{4}{x} + 1 = \frac{6}{x}$$

$$x\left(\frac{4}{x} + 1\right) = x\left(\frac{6}{x}\right) \qquad \text{Multiply both sides by } x.$$

$$x \cdot \frac{4}{x} + x \cdot 1 = x \cdot \frac{6}{x} \qquad \text{Remove parentheses.}$$

$$4 + x = 6 \qquad \text{Simplify.}$$

$$x = 2 \qquad \text{Subtract 4 from both sides.}$$

Check:

$$\frac{4}{x} + 1 = \frac{6}{x}$$

$$\frac{4}{2} + 1 \stackrel{?}{=} \frac{6}{2} \qquad \text{Substitute 2 for } x.$$

$$2 + 1 \stackrel{?}{=} 3 \qquad \text{Simplify.}$$

$$3 = 3$$

SELF CHECK Solve $\frac{6}{x} - 1 = \frac{3}{x}$.   *Answer:* 3   ∎

Extraneous Solutions

If we multiply both sides of an equation by an expression that involves a variable, as we did in Example 1, we must check the apparent solutions. The next example shows why.

EXAMPLE 2

Checking apparent solutions. Solve $\dfrac{x+3}{x-1} = \dfrac{4}{x-1}$.

Solution   To clear the equation of fractions, we multiply both sides by the LCD, which is $x - 1$.

$$\frac{x+3}{x-1} = \frac{4}{x-1}$$

$$(x-1)\frac{x+3}{x-1} = (x-1)\frac{4}{x-1} \qquad \text{Multiply both sides by } x-1.$$

$$x + 3 = 4 \qquad \text{Simplify each side of the equation.}$$

$$x = 1 \qquad \text{Subtract 3 from both sides.}$$

Because both sides were multiplied by an expression containing a variable, we must check the apparent solution.

$$\frac{x+3}{x-1} = \frac{4}{x-1}$$

$$\frac{1+3}{1-1} \stackrel{?}{=} \frac{4}{1-1} \qquad \text{Substitute 1 for } x.$$

$$\frac{4}{0} \stackrel{?}{=} \frac{4}{0} \qquad \text{Simplify.}$$

Since zeros appear in the denominators, the fractions are undefined. Thus, 1 is a false solution, and the equation has no solutions. Such false solutions are often called **extraneous solutions.**

SELF CHECK   Solve $\frac{x+5}{x-2} = \frac{7}{x-2}$.   *Answer:* 2 is extraneous.   ■

EXAMPLE 3

Equations containing rational expressions. Solve $\dfrac{3x+1}{x+1} - 2 = \dfrac{3(x-3)}{x+1}$.

Solution   To clear the equation of fractions, we multiply both sides by the LCD, which is $x + 1$.

$$\frac{3x+1}{x+1} - 2 = \frac{3(x-3)}{x+1}$$

$$(x+1)\left(\frac{3x+1}{x+1} - 2\right) = (x+1)\left[\frac{3(x-3)}{x+1}\right]$$

$$3x + 1 - 2(x+1) = 3(x-3) \qquad \text{Use the distributive property to remove parentheses.}$$

$$3x + 1 - 2x - 2 = 3x - 9 \qquad \text{Remove parentheses.}$$

$$x - 1 = 3x - 9 \qquad \text{Combine like terms.}$$

$$-2x = -8 \qquad \text{Subtract } 3x \text{ and add 1 to both sides.}$$

$$x = 4 \qquad \text{Divide both sides by } -2.$$

Check:
$$\frac{3x+1}{x+1} - 2 = \frac{3(x-3)}{x+1}$$

$$\frac{3(4)+1}{4+1} - 2 \stackrel{?}{=} \frac{3(4-3)}{4+1} \qquad \text{Substitute 4 for } x.$$

$$\frac{13}{5} - \frac{10}{5} \stackrel{?}{=} \frac{3(1)}{5}$$

$$\frac{3}{5} = \frac{3}{5}$$

SELF CHECK Solve $\frac{12}{x+1} - 5 = \frac{2}{x+1}$. *Answer:* 1 ∎

Many times, we will have to factor a denominator to find the LCD.

EXAMPLE 4 **Factoring to find the LCD.** Solve $\dfrac{x+2}{x+3} + \dfrac{1}{x^2+2x-3} = 1$.

Solution To find the LCD, we must factor the second denominator.

$$\frac{x+2}{x+3} + \frac{1}{x^2+2x-3} = 1$$

$$\frac{x+2}{x+3} + \frac{1}{(x+3)(x-1)} = 1 \quad \text{Factor } x^2+2x-3.$$

To clear the equation of fractions, we multiply both sides by the LCD, which is $(x+3)(x-1)$.

$$(x+3)(x-1)\left[\frac{x+2}{x+3} + \frac{1}{(x+3)(x-1)}\right] = (x+3)(x-1)1 \quad \text{Multiply both sides by } (x+3)(x-1).$$

$$(x+3)(x-1)\frac{x+2}{x+3} + (x+3)(x-1)\frac{1}{(x+3)(x-1)} = (x+3)(x-1)1 \quad \text{Remove brackets.}$$

$$(x-1)(x+2) + 1 = (x+3)(x-1) \quad \text{Simplify.}$$

$$x^2+x-2+1 = x^2+2x-3 \quad \text{Use the FOIL method to remove parentheses.}$$

$$x-2+1 = 2x-3 \quad \text{Subtract } x^2 \text{ from both sides.}$$

$$x-1 = 2x-3 \quad \text{Combine like terms.}$$

$$-x-1 = -3 \quad \text{Add } -2x \text{ to both sides.}$$

$$-x = -2 \quad \text{Add 1 to both sides.}$$

$$x = 2 \quad \text{Divide both sides by } -1.$$

Verify that 2 is a solution of the given equation. ∎

EXAMPLE 5 **Factoring to find the LCD.** Solve $\dfrac{4}{5} + y = \dfrac{4y-50}{5y-25}$.

Solution

$$\frac{4}{5} + y = \frac{4y-50}{5y-25} \quad \text{To clear the equation of fractions, we must find the LCD.}$$

$$\frac{4}{5} + y = \frac{4y-50}{5(y-5)} \quad \text{Factor } 5y-25 \text{ to help determine the LCD.}$$

$$5(y-5)\left[\frac{4}{5} + y\right] = 5(y-5)\left[\frac{4y-50}{5(y-5)}\right] \quad \text{Multiply both sides by the LCD, which is } 5(y-5).$$

$$4(y-5) + 5y(y-5) = 4y-50 \quad \text{Remove brackets and simplify.}$$

$$4y-20+5y^2-25y = 4y-50 \quad \text{Remove parentheses.}$$

$$5y^2-25y-20 = -50 \quad \text{Subtract } 4y \text{ from both sides and rearrange terms.}$$

$$5y^2-25y+30 = 0 \quad \text{Add 50 to both sides.}$$

$$y^2-5y+6 = 0 \quad \text{Divide both sides by 5.}$$

$$(y-3)(y-2) = 0 \quad \text{Factor } y^2-5y+6.$$

$$y-3 = 0 \quad \text{or} \quad y-2 = 0 \quad \text{Set each factor equal to zero.}$$

$$y = 3 \qquad y = 2 \quad \text{Solve each equation.}$$

Verify that 3 and 2 satisfy the original equation.

SELF CHECK Solve $\frac{x-6}{3x-9} - \frac{1}{3} = \frac{x}{2}$. *Answer:* 1, 2 ∎

Literal Equations

Many formulas are equations that contain rational expressions.

EXAMPLE 6

Solving formulas. The formula

$$\frac{1}{r} = \frac{1}{r_1} + \frac{1}{r_2}$$

is used in electronics to calculate parallel resistances. Solve the equation for r.

Solution Clear the equation of fractions by multiplying both sides by the LCD, which is rr_1r_2.

$$\frac{1}{r} = \frac{1}{r_1} + \frac{1}{r_2}$$

$$rr_1r_2\left(\frac{1}{r}\right) = rr_1r_2\left(\frac{1}{r_1} + \frac{1}{r_2}\right) \qquad \text{Multiply both sides by } rr_1r_2.$$

$$\frac{rr_1r_2}{r} = \frac{rr_1r_2}{r_1} + \frac{rr_1r_2}{r_2} \qquad \text{Remove parentheses.}$$

$$r_1r_2 = rr_2 + rr_1 \qquad \text{Simplify each fraction.}$$

$$r_1r_2 = r(r_2 + r_1) \qquad \text{Factor out } r.$$

$$\frac{r_1r_2}{r_2 + r_1} = r \qquad \text{To isolate } r, \text{ divide both sides by } r_2 + r_1.$$

or

$$r = \frac{r_1r_2}{r_2 + r_1}$$

SELF CHECK Solve the formula in Example 6 for r_1. *Answer:* $r_1 = \dfrac{rr_2}{r_2 - r}$ ∎

STUDY SET

Section 6.7

VOCABULARY

In Exercises 1–2, fill in the blanks to make the statements true.

1. False solutions that result from multiplying both sides of an equation by a variable are called _____ solutions.

2. In science and other fields, an equation that states a known relationship between two or more variables is called a _____.

CONCEPTS

In Exercises 3–6, fill in the blanks to make the statements true.

3. To clear an equation of fractions, we multiply both sides by the _____ of the fractions of the equation.

4. If you multiply both sides of an equation by an expression that involves a variable, you must _____ the solution.

5. To clear the equation

$$\frac{1}{x} + \frac{2}{x} = 5$$

of fractions, we multiply both sides by ▨.

6. To clear the equation

$$\frac{x}{x-2} - \frac{x}{x-1} = 5$$

of fractions, we multiply both sides by _____.

7. a. Simplify the expressions.

$$\frac{3}{2} - \frac{3}{r} + \frac{1}{3}$$

b. Solve the equation.

$$\frac{3}{2} - \frac{3}{r} = \frac{1}{3}$$

8. Is $x = 5$ a solution of the following equations?

a. $\dfrac{1}{x-1} = 1 - \dfrac{3}{x-1}$

b. $\dfrac{x}{x-5} = 3 + \dfrac{5}{x-5}$

NOTATION

In Exercises 9–10, complete each solution.

9.
$$\frac{2}{a} + \frac{1}{2} = \frac{7}{2a}$$

$$\boxed{}\left(\frac{2}{a} + \frac{1}{2}\right) = \boxed{}\left(\frac{7}{2a}\right)$$

$$\frac{4a}{a} + \frac{\boxed{}}{2} = \frac{14a}{2a}$$

$$\boxed{} + a = 7$$

$$4 + a - \boxed{} = 7 - \boxed{}$$

$$a = 3$$

10.
$$\frac{3}{5} + \frac{7}{a+2} = 2$$

$$\boxed{}\left(\frac{3}{5} + \frac{7}{a+2}\right) = \boxed{}2$$

$$\frac{5(a+2)3}{5} + \frac{\boxed{}7}{a+2} = \boxed{}(a+2)$$

$$3(a+2) + \boxed{} = 10(a+2)$$

$$3a + \boxed{} + 35 = 10a + \boxed{}$$

$$3a + \boxed{} = 10a + 20$$

$$-7a = \boxed{}$$

$$a = 3$$

PRACTICE

In Exercises 11–62, solve each equation and check the solution. If an equation has no solution, so indicate.

11. $\dfrac{x}{2} + 4 = \dfrac{3x}{2}$

12. $\dfrac{y}{3} + 6 = \dfrac{4y}{3}$

13. $\dfrac{2y}{5} - 8 = \dfrac{4y}{5}$

14. $\dfrac{3x}{4} - 6 = \dfrac{x}{4}$

15. $\dfrac{3a}{2} + \dfrac{a}{3} + 22 = 0$

16. $\dfrac{x}{2} + x - \dfrac{9}{2} = 0$

17. $\dfrac{5(x+1)}{8} = x + 1$

18. $\dfrac{3(x-1)}{2} + 2 = x$

19. $\dfrac{x+1}{3} + \dfrac{x-1}{5} = \dfrac{2}{15}$

20. $\dfrac{y-5}{7} + \dfrac{y-7}{5} = \dfrac{-2}{5}$

21. $\dfrac{3x-1}{6} - \dfrac{x+3}{2} = \dfrac{3x+4}{3}$

22. $\dfrac{2x+3}{3} + \dfrac{3x-4}{6} = \dfrac{x-2}{2}$

23. $\dfrac{3}{x} + 2 = 3$

24. $\dfrac{2}{x} + 9 = 11$

25. $\dfrac{5}{a} - \dfrac{4}{a} = 8 + \dfrac{1}{a}$

26. $\dfrac{11}{b} + \dfrac{13}{b} = 12$

27. $\dfrac{3}{4h} + \dfrac{2}{h} = 5 - 4$

28. $\dfrac{5}{3k} + \dfrac{1}{k} = 2 - 4$

29. $\dfrac{a}{4} - \dfrac{4}{a} = 0$

30. $0 = \dfrac{t}{3} - \dfrac{12}{t}$

31. $\dfrac{2}{y+1} + 5 = \dfrac{12}{y+1}$

32. $\dfrac{1}{t-3} = \dfrac{-2}{t-3} + 1$

33. $\dfrac{1}{x-1} + \dfrac{3}{x-1} = 1$

34. $\dfrac{3}{p+6} - 2 = \dfrac{7}{p+6}$

35. $\dfrac{a^2}{a+2} - \dfrac{4}{a+2} = a$

36. $\dfrac{z^2}{z+1} + 2 = \dfrac{1}{z+1}$

37. $\dfrac{x}{x-5} - \dfrac{5}{x-5} = 3$

38. $\dfrac{3}{y-2} + 1 = \dfrac{3}{y-2}$

39. $\dfrac{3r}{2} - \dfrac{3}{r} = \dfrac{3r}{2} + 3$

40. $\dfrac{2p}{3} - \dfrac{1}{p} = \dfrac{2p-1}{3}$

41. $\dfrac{1}{3} + \dfrac{2}{x-3} = 1$

42. $\dfrac{3}{5} + \dfrac{7}{x+2} = 2$

43. $\dfrac{u}{u-1} + \dfrac{1}{u} = \dfrac{u^2+1}{u^2-u}$

44. $\dfrac{v}{v+2} + \dfrac{1}{v-1} = 1$

45. $\dfrac{3}{x-2} + \dfrac{1}{x} = \dfrac{2(3x+2)}{x^2-2x}$

46. $\dfrac{5}{x} + \dfrac{3}{x+2} = \dfrac{-6}{x(x+2)}$

47. $\dfrac{7}{q^2-q-2} + \dfrac{1}{q+1} = \dfrac{3}{q-2}$

48. $\dfrac{-5}{s^2 + s - 2} + \dfrac{3}{s + 2} = \dfrac{1}{s - 1}$

49. $\dfrac{3y}{3y - 6} + \dfrac{8}{y^2 - 4} = \dfrac{2y}{2y + 4}$

50. $\dfrac{x - 3}{4x - 4} + \dfrac{1}{9} = \dfrac{x - 5}{6x - 6}$

51. $y + \dfrac{2}{3} = \dfrac{2y - 12}{3y - 9}$ **52.** $y + \dfrac{3}{4} = \dfrac{3y - 50}{4y - 24}$

53. $\dfrac{5}{4y + 12} - \dfrac{3}{4} = \dfrac{5}{4y + 12} - \dfrac{y}{4}$

54. $\dfrac{3}{5x - 20} + \dfrac{4}{5} = \dfrac{3}{5x - 20} - \dfrac{x}{5}$

55. $\dfrac{x}{x - 1} - \dfrac{12}{x^2 - x} = \dfrac{-1}{x - 1}$

56. $1 - \dfrac{3}{b} = \dfrac{-8b}{b^2 + 3b}$

57. $\dfrac{z - 4}{z - 3} = \dfrac{z + 2}{z + 1}$

58. $\dfrac{a + 2}{a + 8} = \dfrac{a - 3}{a - 2}$

59. $\dfrac{n}{n^2 - 9} + \dfrac{n + 8}{n + 3} = \dfrac{n - 8}{n - 3}$

60. $\dfrac{x - 3}{x - 2} - \dfrac{1}{x} = \dfrac{x - 3}{x}$

61. $\dfrac{b + 2}{b + 3} + 1 = \dfrac{-7}{b - 5}$

62. $\dfrac{x - 4}{x - 3} + \dfrac{x - 2}{x - 3} = x - 3$

63. Solve the formula $\dfrac{1}{a} + \dfrac{1}{b} = 1$ for a.

64. Solve the formula $\dfrac{1}{a} - \dfrac{1}{b} = 1$ for b.

APPLICATIONS

65. OPTICS The focal length f of a lens is given by the formula

$$\dfrac{1}{f} = \dfrac{1}{d_1} + \dfrac{1}{d_2}$$

where d_1 is the distance from the object to the lens and d_2 is the distance from the lens to the image.

Solve the formula for f.

66. OPTICS Solve the formula in Exercise 65 for d_1.

67. MEDICINE Radioactive tracers are used for diagnostic work in nuclear medicine. The **effective half-life,** H, of a radioactive material in a biological organism is given by the formula

$$H = \dfrac{RB}{R + B}$$

where R is the radioactive half-life and B is the biological half-life of the tracer. Solve the formula for R.

68. CHEMISTRY Charles's Law describes the relationship between the volume and the temperature of a gas that is kept at a constant pressure. It states that as the temperature of the gas increases, the volume of the gas will increase:

$$\dfrac{V_1}{V_2} = \dfrac{T_1}{T_2}$$

Solve the equation for V_2.

WRITING

69. Explain how you would decide what to do first to solve an equation that involves fractions.

70. Explain why it is important to check your solutions to an equation that contains fractions with variables in the denominator.

REVIEW

In Exercises 71–76, factor each expression.

71. $x^2 + 4x$

72. $x^2 - 16y^2$

73. $2x^2 + x - 3$

74. $6a^2 - 5a - 6$

75. $x^4 - 16$

76. $4x^2 + 10x - 6$

▶ **6.8**

Applications of Equations That Contain Rational Expressions

In this section, you will learn about

> Solving number problems ■ Solving shared-work problems
> ■ Solving uniform motion problems ■ Solving investment problems

Introduction   Many problems involve equations that contain rational expressions. In this section, we will consider several of these applications.

Solving Number Problems

E X A M P L E   1

Number problem. If the same number is added to both the numerator and the denominator of the fraction $\frac{3}{5}$, the result is $\frac{4}{5}$. Find the number.

ANALYZE THE PROBLEM   We are asked to find a number. If we add it to both the numerator and the denominator of a fraction, we will get $\frac{4}{5}$.

FORM AN EQUATION   Let n represent the unknown number and add n to both the numerator and the denominator of $\frac{3}{5}$. Then set the result equal to $\frac{4}{5}$ to get the equation

$$\frac{3+n}{5+n} = \frac{4}{5}$$

SOLVE THE EQUATION   To solve the equation, we proceed as follows:

$$\frac{3+n}{5+n} = \frac{4}{5}$$

$$\mathbf{5(5+n)}\frac{3+n}{5+n} = \mathbf{5(5+n)}\frac{4}{5} \qquad \text{Multiply both sides by } 5(5+n), \text{ which is the LCD of the fractions appearing in the equation.}$$

$$5(3+n) = (5+n)4 \qquad \text{Simplify.}$$

$$15 + 5n = 20 + 4n \qquad \text{Use the distributive property to remove parentheses.}$$

$$15 + n = 20 \qquad \text{Subtract } 4n \text{ from both sides.}$$

$$n = 5 \qquad \text{Subtract } 15 \text{ from both sides.}$$

STATE THE CONCLUSION   The number is 5.

CHECK THE RESULT   When we add 5 to both the numerator and denominator of $\frac{3}{5}$, we get

$$\frac{3+5}{5+5} = \frac{8}{10} = \frac{4}{5}$$

The result checks.   ■

Solving Shared-Work Problems

E X A M P L E   2

Draining an oil tank. An inlet pipe can fill an oil tank in 7 days, and a second inlet pipe can fill the same tank in 9 days. If both pipes are used, how long will it take to fill the tank?

ANALYZE THE PROBLEM   The key is to note what each pipe can do in 1 day. If we add what the first pipe can do in 1 day to what the second pipe can do in 1 day, the sum is what they can do together in 1 day. Since the first pipe can fill the tank in 7 days, it can do $\frac{1}{7}$ of the job in 1 day.

Since the second pipe can fill the tank in 9 days, it can do $\frac{1}{9}$ of the job in 1 day. If it takes x days for both pipes to fill the tank, together they can do $\frac{1}{x}$ of the job in 1 day.

FORM AN EQUATION Let x represent the number of days it will take to fill the tank if both inlet pipes are used. Then form the equation.

What the first inlet pipe can do in 1 day	plus	what the second inlet pipe can do in 1 day	equals	what they can do together in 1 day.
$\frac{1}{7}$	$+$	$\frac{1}{9}$	$=$	$\frac{1}{x}$

SOLVE THE EQUATION To solve the equation, we proceed as follows:

$$\frac{1}{7} + \frac{1}{9} = \frac{1}{x}$$

$$63x\left(\frac{1}{7} + \frac{1}{9}\right) = 63x\left(\frac{1}{x}\right) \quad \text{Multiply both sides by } 63x \text{ to clear the equation of fractions.}$$

$$9x + 7x = 63 \quad \text{Use the distributive property to remove parentheses and simplify.}$$

$$16x = 63 \quad \text{Combine like terms.}$$

$$x = \frac{63}{16} \quad \text{Divide both sides by 16.}$$

STATE THE CONCLUSION It will take $\frac{63}{16}$ or $3\frac{15}{16}$ days for both inlet pipes to fill the tank.

CHECK THE RESULT In $\frac{63}{16}$ days, the first pipe fills $\frac{1}{7}\left(\frac{63}{16}\right)$ of the tank and the second pipe fills $\frac{1}{9}\left(\frac{63}{16}\right)$ of the tank. The sum of these efforts, $\frac{9}{16} + \frac{7}{16}$, is equal to one full tank. ∎

Solving Uniform Motion Problems

EXAMPLE 3

Track and field. A coach can run 10 miles in the same amount of time that his best student-athlete can run 12 miles. If the student can run 1 mile per hour faster than the coach, how fast can the student run?

ANALYZE THE PROBLEM This is a uniform motion problem. We use the formula $d = rt$, where d is the distance traveled, r is the rate, and t is the time. If we solve this formula for t, we obtain

$$t = \frac{d}{r}$$

FORM AN EQUATION It will take $\frac{10}{r}$ hours for the coach to run 10 miles at some unknown rate of r mph. It will take $\frac{12}{r+1}$ hours for the student to run 12 miles at some unknown rate of $(r + 1)$ mph. We can organize the information of the problem in a table, as shown in Figure 6-7.

FIGURE 6-7

	r	$\cdot$	t	$=$	d
Student	$r + 1$		$\frac{12}{r+1}$		12
Coach	r		$\frac{10}{r}$		10

The time it takes the student to run 12 miles	equals	the time it takes the coach to run 10 miles.

$$\frac{12}{r+1} = \frac{10}{r}$$

SOLVE THE EQUATION   We can solve the equation as follows:

$$\frac{12}{r+1} = \frac{10}{r}$$

$$r(r+1)\frac{12}{r+1} = r(r+1)\frac{10}{r} \qquad \text{Multiply both sides by } r(r+1).$$

$$12r = 10(r+1) \qquad \text{Simplify.}$$

$$12r = 10r + 10 \qquad \text{Use the distributive property to remove parentheses.}$$

$$2r = 10 \qquad \text{Subtract } 10r \text{ from both sides.}$$

$$r = 5 \qquad \text{Divide both sides by 2.}$$

STATE THE CONCLUSION   The coach can run 5 mph. The student, running 1 mph faster, can run 6 mph.

CHECK THE RESULT   Verify that these results check.   ■

Solving Investment Problems

EXAMPLE 4

Comparing investments. At one bank, a sum of money invested for one year will earn $96 interest. If invested in bonds, that money would earn $108, because the interest rate paid by the bonds is 1% greater than that paid by the bank. Find the bank's rate.

ANALYZE THE PROBLEM   This interest problem is based on the formula $I = Pr$, where I is the interest earned in 1 year, P is the principal (the amount invested), and r is the annual rate of interest. If we solve this formula for P, we obtain

$$P = \frac{I}{r}$$

FORM AN EQUATION   If we let r represent the bank's rate of interest, then $r + 0.01$ represents the rate paid by the bonds. If a person earns $96 interest at a bank at some unknown rate r, the principal invested was $\frac{96}{r}$. If a person earns $108 interest in bonds at some unknown rate $(r + 0.01)$, the principal invested was $\frac{108}{r+0.01}$. We can organize the information of the problem in a table, as shown in Figure 6-8.

FIGURE 6-8

	Principal	·	Rate	=	Interest
Bank	$\frac{96}{r}$		r		96
Bonds	$\frac{108}{r+0.01}$		$r + 0.01$		108

Because the same principal would be invested in either account, we can set up the following equation:

$$\frac{96}{r} = \frac{108}{r+0.01}$$

SOLVE THE EQUATION We can solve the equation as follows:

$$\frac{96}{r} = \frac{108}{r + 0.01}$$

$$r(r + 0.01) \cdot \frac{96}{r} = r(r + 0.01) \cdot \frac{108}{r + 0.01} \qquad \text{Multiply both sides by } r(r + 0.01).$$

$$96(r + 0.01) = 108r$$

$$96r + 0.96 = 108r \qquad \text{Remove parentheses.}$$

$$0.96 = 12r \qquad \text{Subtract } 96r \text{ from both sides.}$$

$$0.08 = r \qquad \text{Divide both sides by 12.}$$

STATE THE CONCLUSION The bank's interest rate is 0.08, or 8%. The bonds pay 9% interest, a rate 1% greater than that paid by the bank.

CHECK THE RESULT Verify that these rates check. ■

STUDY SET

Section 6.8

VOCABULARY

In Exercises 1–4, fill in the blanks to make the statements true.

1. In the formula $I = Pr$, I stands for the amount of _____ earned in one year, P stands for the _____, and r stands for the interest _____.

2. In the formula $d = rt$, d stands for the _____ traveled, r is the _____, and t is the _____.

3. We can clear an equation of fractions by multiplying both sides by the _____ of the fractions of the equation.

4. The distributive property is used to _____ parentheses.

CONCEPTS

5. List the five steps used in problem solving.

6. Write 6% as a decimal.

7. Solve $d = rt$
   a. for r **b.** for t

8. Solve $I = Pr$
   a. for r **b.** for P

9. Illustration 1 shows the length of time it takes each of two hardware store employees to assemble a metal storage shed working alone.

a. Complete the table.

ILLUSTRATION 1

	Time to assemble the shed (hr)	Amount of the shed assembled in 1 hr
Marvin	6	
Kyla	5	

b. If we assume that working together would not change their individual rates, how much of the shed could they assemble in one hour if they worked together?

10. **a.** Complete the table in Illustration 2.

ILLUSTRATION 2

	r	$\cdot$	t	$=$	d
Snowmobile	r				4
4×4 truck	$r - 5$				3

b. Complete the table in Illustration 3.

ILLUSTRATION 3

	P	$\cdot$	r	$=$	I
City Savings			r		50
Credit Union			$r - 0.02$		75

11. When two ice machines are both running, they can fill a supermarket's order in x hours. At this rate, how much of the order do they fill in 1 hour?

12. If the exits at the front of a theater are opened, a full theater can be emptied of all occupants in 6 minutes. How much of the theater is emptied in 1 minute?

PRACTICE

In Exercises 13–18, use the statement to find the number or numbers.

13. If the denominator of $\frac{3}{4}$ is increased by a number and the numerator of the fraction is doubled, the result is 1.

14. If a number is added to the numerator of $\frac{7}{8}$ and the same number is subtracted from the denominator, the result is 2.

15. If a number is added to the numerator of $\frac{3}{4}$ and twice as much is added to the denominator, the result is $\frac{4}{7}$.

16. If a number is added to the numerator of $\frac{5}{7}$ and twice as much is subtracted from the denominator, the result is 8.

17. The sum of a number and its reciprocal is $\frac{13}{6}$.

18. The sum of the reciprocals of two consecutive even integers is $\frac{7}{24}$.

APPLICATIONS

19. FILLING A POOL An inlet pipe can fill an empty swimming pool in 5 hours, and another inlet pipe can fill the pool in 4 hours. How long will it take both pipes to fill the pool?

20. FILLING A POOL One inlet pipe can fill an empty pool in 4 hours, and a drain can empty the pool in 8 hours. How long will it take the pipe to fill the pool if the drain is left open?

21. ROOFING A HOUSE A homeowner estimates that it will take her 7 days to roof her house. A professional roofer estimates that he could roof the house in 4 days. How long will it take if the homeowner helps the roofer?

22. SEWAGE TREATMENT A sludge pool is filled by two inlet pipes. One pipe can fill the pool in 15 days, and the other can fill it in 21 days. However, if no sewage is added, continuous waste removal will empty the pool in 36 days. How long will it take the two inlet pipes to fill an empty sludge pool?

23. TOURING A woman can bicycle 28 miles in the same time as it takes her to walk 8 miles. If she can ride 10 mph faster than she can walk, how much time should she allow to walk a 30-mile trail? See Illustration 4. (*Hint:* How fast can she walk?)

ILLUSTRATION 4

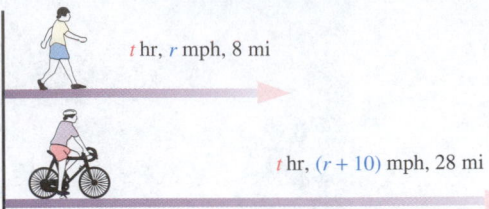

t hr, r mph, 8 mi

t hr, $(r + 10)$ mph, 28 mi

24. COMPARING TRAVEL A plane can fly 300 miles in the same time as it takes a car to go 120 miles. If the car travels 90 mph slower than the plane, find the speed of the plane.

25. BOATING A boat that travels 18 mph in still water can travel 22 miles downstream in the same time as it takes to travel 14 miles upstream. Find the speed of the current in the river. (See Illustration 5.)

ILLUSTRATION 5

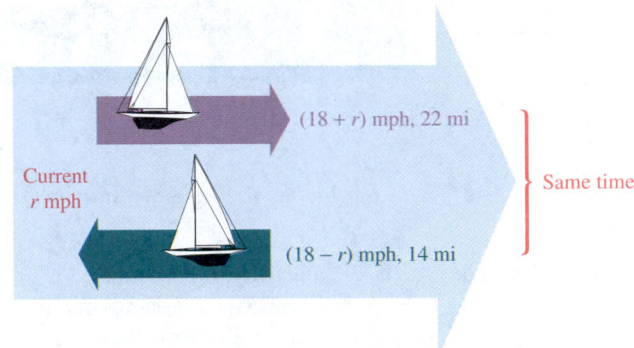

$(18 + r)$ mph, 22 mi

Current r mph

$(18 - r)$ mph, 14 mi

Same time

26. WIND SPEED A plane can fly 300 miles downwind in the same time as it can travel 210 miles upwind. Find the velocity of the wind if the plane can fly 255 mph in still air.

27. COMPARING INVESTMENTS Two certificates of deposit (CDs) pay interest at rates that differ by 1%. Money invested for one year in the first CD earns $175 interest. The same principal invested in the second CD earns $200. Find the two rates of interest.

28. COMPARING INTEREST RATES Two bond funds pay interest at rates that differ by 2%. Money invested for one year in the first fund earns $315 interest. The same amount invested in the second fund earns $385. Find the lower rate of interest.

29. SHARING COSTS Several office workers bought a $35 gift for their boss. If there had been two more employees to contribute, everyone's cost would have been $2 less. How many workers contributed to the gift?

30. SALES A dealer bought some radios for a total of $1,200. She gave away 6 radios as gifts, sold the rest for $10 more than she paid for each radio, and broke even. How many radios did she buy?

31. SALES A bookstore can purchase several calculators for a total cost of $120. If each calculator cost $1 less, the bookstore could purchase 10 additional calculators at the same total cost. How many calculators can be purchased at the regular price?

32. FURNACE REPAIR A repairman purchased several furnace-blower motors for a total cost of $210. If his cost per motor had been $5 less, he could have purchased one additional motor. How many motors did he buy at the regular rate?

33. RIVER TOURS A river boat tour begins by going 60 miles upstream against a 5-mph current. There, the boat turns around and returns with the current. What still-water speed should the captain use to complete the tour in 5 hours?

34. TRAVEL TIME A company president flew 680 miles one way in the corporate jet but returned in a smaller plane that could fly only half as fast. If the total travel time was 6 hours, find the speeds of the planes.

WRITING

35. The key to solving shared-work problems is to ask, "How much of the job can be done in 1 hour?" Explain.

36. It's difficult to check the solution of a shared-work problem. Explain how you could decide whether an answer is at least reasonable.

REVIEW

In Exercises 37–44, solve each equation.

37. $x^2 - 5x - 6 = 0$

38. $x^2 - 25 = 0$

39. $(t + 2)(t^2 + 7t + 12) = 0$

40. $2(y - 4) = -y^2$

41. $y^3 - y^2 = 0$

42. $5a^3 - 125a = 0$

43. $(x^2 - 1)(x^2 - 4) = 0$

44. $6t^3 + 35t^2 = 6t$

The Language of Algebra

Algebra is a language in its own right. One of the keys to becoming a good algebra student is to know the vocabulary of algebra.

In Exercises 1–19, match each instruction in column I with the most appropriate problem in column II. Each letter in column II is used only once.

Column I

1. Use the FOIL method.

2. Graph the function.

3. Simplify the expression.

4. Factor completely.

5. Evaluate the expression for $a = -1$ and $b = -6$.

6. Express in lowest terms.

7. Solve for t.

8. Combine like terms.

9. Remove parentheses.

10. Solve the equation by clearing it of fractions.

11. Prime factor.

12. Solve using the quadratic formula.

13. Identify the base and the exponent.

14. Write without a radical sign.

15. Write the equation of the line.

16. Solve the inequality.

17. Find the slope of the line passing through the given points.

18. Translate into mathematical symbols.

19. Name the property shown.

Column II

a. $R = cd + 2t$

b. 5^3

c. 144

d. $2(3x - 4)$

e. $4x - 7 > -3$

f. $\dfrac{x}{3} + \dfrac{1}{2} = \dfrac{1}{6}$

g. $(x - 1)(x + 8)$

h. $a + (b + c) = (a + b) + c$

i. $\dfrac{1}{2x^2} + \dfrac{5}{4x}$

j. $\sqrt{4}$

k. the sum of four cubed and y

l. $2x - 8 + 6y - 14$

m. $(3, -2)$ and $(0, -5)$

n. $f(x) = x^3 + 1$

o. $\dfrac{4x^2}{16x}$

p. $x^2 - 3x - 4 = 0$

q. $2a^2 - 3b$

r. with $m = \frac{2}{3}$ and $b = 2$

s. $y^4 - 81$

Accent on Teamwork

Section 6.1

Pi The Greek letter pi (π) represents the ratio of the circumference C of any circle to its diameter d. That is, $\pi = \frac{C}{d}$. Use a tape measure to find the circumference and the diameter of various objects that are circular in shape. You can measure anything round: for example, a swimming pool spa, the top of a can, or a ring. Enter your results in a table like that in Illustration 1. Convert each measurement to a decimal and use a calculator to compute the ratio of C to d. Make some observations about your results.

ILLUSTRATION 1

Object	Circumference	Diameter	$\dfrac{C}{d}$
Quarter-dollar	$2\frac{15}{16}$ in. 2.9375 in.	$\frac{15}{16}$ in. 0.9375 in.	3.13333. . .

Section 6.2

Cooking Find a simple recipe for a treat that you can make for your class. Use a proportion to determine the amount of each ingredient needed to make enough for the exact number of people in your class. See Exercise 45 in Study Set 6.2 for an example. Write the old recipe and the new recipe on separate pieces of poster board. Did the recipe serve the correct number of people? Share with the class how you made the calculations, as well as any difficulties you encountered.

Section 6.3

Warning In Section 6.3, the following two-part warning was given: "Remember that only *factors* that are common to the entire numerator and the entire denominator can be divided out. *Terms* that are common to both the numerator and denominator cannot be divided out." Explain how each part of the warning applies to the rational expression $\frac{2x+8}{8}$.

Section 6.4

The rule of four The trend in mathematics education today is to present a topic in four ways:

- symbolically (using variables)
- geometrically (using a diagram)
- numerically (using numbers)
- verbally (using words)

Use the rule of four as a guide in developing a presentation explaining the concept of multiplying fractions, as shown in the beginning of Section 6.4.

Section 6.5

First, add

$$\frac{1}{2x^2} + \frac{1}{8x}$$

by expressing each fraction in terms of a common denominator $16x^3$. (This is the *product* of their denominators.) Then add the fractions again by expressing each of them in terms of their lowest common denominator. What is one advantage and one drawback of each method?

Section 6.6

Unit analysis Simplify each complex fraction. The units can be divided out just as in the case of common factors.

$\dfrac{\dfrac{36 \text{ inches}}{3 \text{ feet}}}{\dfrac{1 \text{ yard}}{3 \text{ feet}}}$	$\dfrac{\dfrac{60 \text{ minutes}}{1 \text{ hour}}}{\dfrac{3{,}600 \text{ seconds}}{1 \text{ hour}}}$	$\dfrac{\dfrac{4 \text{ quarts}}{1 \text{ gallon}}}{\dfrac{8 \text{ pints}}{1 \text{ gallon}}}$

Section 6.7

Simplify and solve The two problems below look similar. Explain why their one-word instructions can't be switched. Write a solution for each problem and then identify the major similarity and the major difference in the solution methods.

Simplify	*Solve*
$\dfrac{2}{4x-4} + \dfrac{3}{x-1}$	$\dfrac{2}{4x-4} + \dfrac{3}{x-1} = \dfrac{7}{4}$

Section 6.8

Problem solving Problems such as the filling of a water tank (Section 6.8) are often called *shared-work problems*. For each of the following equations, write a shared-work problem that could be solved using it.

$$\frac{1}{3} + \frac{1}{8} = \frac{1}{x}$$

$$\frac{1}{3} - \frac{1}{8} = \frac{1}{x}$$

$$\frac{1}{3} + \frac{1}{8} - \frac{1}{16} = \frac{1}{x}$$

Section 6.1

Ratios

CONCEPTS

A *ratio* is the comparison of two numbers by their indicated quotient. The ratio of a to b is written $\frac{a}{b}$ or $a:b$.

Where possible, express ratios using the same units.

REVIEW EXERCISES

1. Write each ratio as a fraction in lowest terms.
   a. 3 to 6
   b. $12x$ to $15x$
   c. 2 feet to 1 yard
   d. 5 pints to 3 quarts

2. BICYCLE GEARS See Illustration 1. Find the ratio of the number of teeth on the front sprocket of the bicycle to the number of teeth on the smallest rear sprocket. Express the ratio in lowest terms using $a:b$ form. Then write ratios for the following sprockets: the front to the middle rear and the front to the large rear.

ILLUSTRATION 1

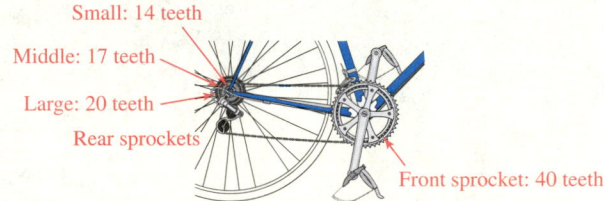

Small: 14 teeth
Middle: 17 teeth
Large: 20 teeth
Rear sprockets
Front sprocket: 40 teeth

The *unit cost* of an item is the ratio of its cost to its quantity.

3. If three pounds of coffee cost $8.79, find its unit cost (the cost per pound).

When ratios are used to compare quantities with different units, they are called *rates*.

4. If a factory used 2,275 kilowatt hours of electricity in February, what was the rate of energy consumption in kilowatt hours per week?

5. SUPREME COURT The annual salary of a U.S. Supreme Court Justice is $164,100. What is their weekly rate of pay? Round to the nearest dollar.

Section 6.2

Proportions and Similar Triangles

A *proportion* is a statement that two ratios are equal.

In the proportion $\frac{a}{b} = \frac{c}{d}$, a and d are the *extremes*, and b and c are the *means*.

6. Determine whether the following equations are proportions.
   a. $\frac{4}{7} = \frac{20}{34}$
   b. $\frac{5}{7} = \frac{30}{42}$

In any proportion, the product of the extremes is equal to the product of the means.

7. Solve each proportion.
   a. $\frac{3}{x} = \frac{6}{9}$
   b. $\frac{x}{3} = \frac{x}{5}$
   c. $\frac{x-2}{5} = \frac{x}{7}$
   d. $\frac{4x-1}{18} = \frac{x}{6}$

The measures of corresponding sides of *similar triangles* are in proportion.

8. A telephone pole casts a shadow 12 feet long at the same time that a man 6 feet tall casts a shadow of 3.6 feet. How tall is the pole?

9. DENTISTRY The diagram in Illustration 2 was displayed in a dentist's office. According to the diagram, if the dentist has 340 adult patients, how many will develop gum disese?

ILLUSTRATION 2

3 out of 4 adults will develop gum disease.

Section 6.3

Rational Expressions and Rational Functions

A *rational expression* is a ratio of two polynomials.

The fundamental property of fractions:
If b and c are not zero, then

$$\frac{ac}{bc} = \frac{a}{b}$$

When all common factors have been divided out, a fraction is in *lowest terms*.

$$\frac{a}{1} = a$$

$$\frac{a}{0} \text{ is not defined}$$

If the terms of two polynomials are the same, except for sign, the polynomials are called *negatives* of each other.

The quotient of any nonzero expression and its negative is -1.

10. Write each fraction in lowest terms. If it is already in lowest terms, so indicate.

a. $\dfrac{10}{25}$

b. $-\dfrac{12}{18}$

c. $\dfrac{3x^2}{6x^3}$

d. $\dfrac{5xy^2}{2x^2y^2}$

e. $\dfrac{x^2}{x^2 + x}$

f. $\dfrac{x + 2}{x^2 - 4}$

g. $\dfrac{3p - 2}{2 - 3p}$

h. $\dfrac{8 - x}{x^2 - 5x - 24}$

i. $\dfrac{2x^2 - 16x}{2x^2 - 18x + 16}$

j. $\dfrac{x^2 + x - 2}{x^2 - x - 2}$

11. Explain why it would be incorrect to divide out the common x's in $\frac{x+1}{x}$.

12. Complete the table and graph the rational function for $x > 0$. Round to the nearest tenth when necessary.

$$f(x) = \frac{8}{x}$$

x	$f(x)$
1	
2	
3	
4	
5	
6	
7	
8	

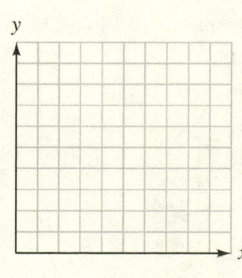

Section 6.4

Multiplying and Dividing Rational Expressions

Rule for multiplying fractions:

$$\frac{a}{b} \cdot \frac{c}{d} = \frac{ac}{bd} \quad (b, d \neq 0)$$

13. Do each multiplication and simplify.

a. $\dfrac{3xy}{2x} \cdot \dfrac{4x}{2y^2}$

b. $\dfrac{3x}{x^2 - x} \cdot \dfrac{2x - 2}{x^2}$

c. $\dfrac{x^2 - 1}{x^2 + 2x} \cdot \dfrac{x}{x + 1}$

d. $\dfrac{x^2 + x}{3x - 15} \cdot \dfrac{6x - 30}{x^2 + 2x + 1}$

Rule for dividing fractions:

$$\frac{a}{b} \div \frac{c}{d} = \frac{a}{b} \cdot \frac{d}{c} \quad (b, c, d \neq 0)$$

To write the *reciprocal* of a fraction, we invert the fraction.

14. Do each division and simplify.

a. $\dfrac{3x^2}{5x^2 y} \div \dfrac{6x}{15xy^2}$

b. $\dfrac{x^2 + 5x}{x^2 + 4x - 5} \div \dfrac{x^2}{x - 1}$

c. $\dfrac{x^2 - x - 6}{2x - 1} \div \dfrac{x^2 - 2x - 3}{2x^2 + x - 1}$

d. $\dfrac{x^2 - 3x}{x^2 - x - 6} \div \dfrac{x^2 - x}{4 - x^2}$

e. $\dfrac{b^2 + 4b + 4}{b^2 + b - 6} \left(\dfrac{b - 2}{b - 1} \div \dfrac{b + 2}{b^2 + 2b - 3} \right)$

Section 6.5

Adding and Subtracting Rational Expressions

Adding and subtracting fractions with like denominators:

$$\frac{a}{d} + \frac{b}{d} = \frac{a + b}{d} \quad (d \neq 0)$$

$$\frac{a}{d} - \frac{b}{d} = \frac{a - b}{d} \quad (d \neq 0)$$

15. Do each operation. Simplify all answers.

a. $\dfrac{x}{x + y} + \dfrac{y}{x + y}$

b. $\dfrac{3x}{x - 7} - \dfrac{x - 2}{x - 7}$

c. $\dfrac{a}{a^2 - 2a - 8} + \dfrac{2}{a^2 - 2a - 8}$

To find the *LCD*, completely factor each denominator. Form a product using each different factor the greatest number of times it appears in any one factorization.

16. Several denominators are given. Find the lowest common denominator (LCD).

a. $2x^2, 4x$

b. $3y^2, 9x, 6x$

c. $x + 1, x + 2$

d. $y^2 - 25, y - 5$

To add or subtract fractions with unlike denominators, first find the LCD of the fractions. Then express each fraction in equivalent form with a common denominator. Finally, add or subtract the fractions.

17. Do each operation. Simplify all answers.

a. $\dfrac{x}{x - 1} + \dfrac{1}{x}$

b. $\dfrac{1}{7} - \dfrac{1}{c}$

c. $\dfrac{x + 2}{2x} - \dfrac{2 - x}{x^2}$

d. $\dfrac{2t + 2}{t^2 + 2t + 1} - \dfrac{1}{t + 1}$

e. $\dfrac{x}{x + 2} + \dfrac{3}{x} - \dfrac{4}{x^2 + 2x}$

f. $\dfrac{6}{b - 1} - \dfrac{b}{1 - b}$

Section 6.6

Complex fractions contain fractions in their numerators or denominators.

To simplify a complex fraction, use either of these methods:

1. Write the numerator and denominator of the complex fraction as single fractions, do the division of the fractions, and simplify.

2. Multiply both the numerator and the denominator of the complex fraction by the LCD of the fractions that appear in the numerator and denominator, then simplify.

Complex Fractions

18. Simplify each complex fraction.

a. $\dfrac{\dfrac{3}{2}}{\dfrac{2}{3}}$

b. $\dfrac{\dfrac{3}{2} + 1}{\dfrac{2}{3} + 1}$

c. $\dfrac{\dfrac{1}{y} + 1}{\dfrac{1}{y} - 1}$

d. $\dfrac{1 + \dfrac{3}{x}}{2 - \dfrac{1}{x^2}}$

e. $\dfrac{\dfrac{2}{x-1} + \dfrac{x-1}{x+1}}{\dfrac{1}{x^2 - 1}}$

f. $\dfrac{x^{-2} + 1}{x^{-2} - 1}$

Section 6.7

To solve an equation that contains fractions, change it to another equation without fractions. Do so by multiplying both sides by the LCD of the fractions. Check all solutions.

An apparent solution that does not satisfy the original equation is called an *extraneous* solution.

Solving Equations That Contain Rational Expressions

19. Solve each equation and check all answers.

a. $\dfrac{3}{x} = \dfrac{2}{x-1}$

b. $\dfrac{5}{r+4} = \dfrac{3}{r+2}$

c. $\dfrac{2}{3t} + \dfrac{1}{t} = \dfrac{5}{9}$

d. $\dfrac{2x}{x+4} = \dfrac{3}{x-1}$

e. $a = \dfrac{3a - 50}{4a - 24} - \dfrac{3}{4}$

f. $\dfrac{4}{x+2} - \dfrac{3}{x+3} = \dfrac{6}{x^2 + 5x + 6}$

20. The efficiency E of a Carnot engine is given by the formula

$$E = 1 - \dfrac{T_2}{T_1}$$

Solve the formula for T_1.

21. Solve for r_1: $\dfrac{1}{r} = \dfrac{1}{r_1} + \dfrac{1}{r_2}$.

Section 6.8

To solve a problem, follow these steps:

1. Analyze the problem.
2. Form an equation.
3. Solve the equation.
4. State the conclusion.
5. Check the result.

Interest = principal · rate · time

Distance = rate · time

Applications of Equations That Contain Rational Expressions

22. **NUMBER PROBLEM** If a number is subtracted from the denominator of $\frac{4}{5}$ and twice as much is added to the numerator, the result is 5. Find the number.

23. If a maid can clean a house in 4 hours, how much of the house does she clean in 1 hour?

24. **PAINTING HOUSES** If a homeowner can paint a house in 14 days and a professional painter can paint it in 10 days, how long will it take if they work together?

25. **INVESTMENTS** In one year, a student earned $100 interest on money she deposited at a savings and loan. She later learned that the money would have earned $120 if she had deposited it at a credit union, because the credit union paid 1% more interest at the time. Find the rate she received from the savings and loan.

26. **EXERCISE** A jogger can bicycle 30 miles in the same time that it takes her to jog 10 miles. If she can ride 10 mph faster than she can jog, how fast can she jog?

27. **WIND SPEED** A plane flies 400 miles downwind in the same amount of time as it takes to travel 320 miles upwind. If the plane can fly at 360 mph in still air, find the velocity of the wind.

CHAPTER 6

Test

1. Express as a ratio in lowest terms: 6 feet to 3 yards.

2. Is the equation $\dfrac{3}{5} = \dfrac{6xt}{10xt}$ a proportion?

3. Solve the proportion for y: $\dfrac{y}{y-1} = \dfrac{y-2}{y}$.

4. **HEALTH RISKS** A medical newsletter states that a "healthy" waist-to-hip ratio for men is $19:20$ or less. Does the patient shown in Illustration 1 fall within the "healthy" range?

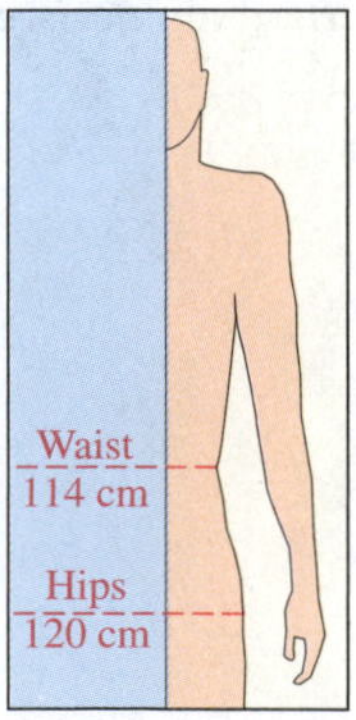

5. Simplify $\dfrac{48x^2y}{54xy^2}$.

6. Simplify $\dfrac{2x^2 - x - 3}{4x^2 - 9}$.

ILLUSTRATION 1

Waist
114 cm

Hips
120 cm

7. Simplify $\dfrac{3(x+2) - 3}{2x - 4 - (x - 5)}$.

8. Multiply and simplify $-\dfrac{12x^2y}{15xy} \cdot \dfrac{25y^2}{16x}$.

9. Multiply and simplify $\dfrac{x^2 + 3x + 2}{3x + 9} \cdot \dfrac{x + 3}{x^2 - 4}$.

10. Divide and simplify $\dfrac{8x^2}{25x} \div \dfrac{16x^2}{30x}$.

11. Divide and simplify $\dfrac{x - x^2}{3x^2 + 6x} \div \dfrac{3x - 3}{3x^3 + 6x^2}$.

12. Simplify $\dfrac{x^2 + x}{x - 1} \cdot \dfrac{x^2 - 1}{x^2 - 2x} \div \dfrac{x^2 + 2x + 1}{x^2 - 4}$.

13. Add $\dfrac{5x - 4}{x - 1} + \dfrac{5x + 3}{x - 1}$.

14. Subtract $\dfrac{3y + 7}{2y + 3} - \dfrac{3(y - 2)}{2y + 3}$.

15. Add $\dfrac{x + 1}{x} + \dfrac{x - 1}{x + 1}$.

16. Subtract $\dfrac{a + 3}{a - 1} - \dfrac{a + 4}{1 - a}$.

17. Subtract $\dfrac{2n}{5m} - \dfrac{n}{2}$.

18. Simplify $\dfrac{1 + \dfrac{y}{x}}{\dfrac{y}{x} - 1}$.

19. Solve for c: $\dfrac{2}{3} = \dfrac{2c - 12}{3c - 9} - c$.

20. Solve for x: $3x - \dfrac{2(x + 3)}{3} = 16 - \dfrac{x + 2}{2}$.

21. Solve for x: $\dfrac{7}{x + 4} - \dfrac{1}{2} = \dfrac{3}{x + 4}$.

22. Solve for B: $H = \dfrac{RB}{R + B}$.

23. CLEANING HIGHWAYS One highway worker can pick up all the trash on a strip of highway in 7 hours, and his helper can pick up the trash in 9 hours. How long will it take them if they work together?

24. BOATING A boat can motor 28 miles downstream in the same amount of time as it can motor 18 miles upstream. Find the speed of the current if the boat can motor at 23 mph in still water.

25. FLIGHT PATH A plane drops 575 feet as it flies a horizontal distance of $\frac{1}{2}$ mile, as shown in Illustration 2. How much altitude will it lose as it flies a horizontal distance of 7 miles?

ILLUSTRATION 2

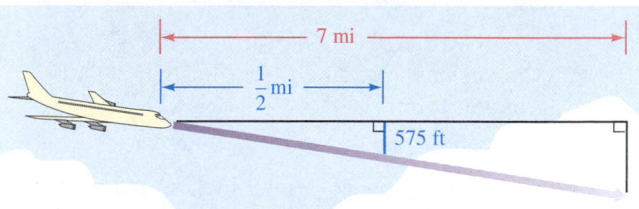

26. Explain why we can divide out the 5's in $\frac{5x}{5}$ and why we can't divide them out in $\frac{5 + x}{5}$.

27. Complete the table and graph the rational function for $x > 0$. Round to the nearest tenth when necessary.

$$f(x) = \frac{2}{x}$$

x	$f(x)$
$\frac{1}{2}$	
1	
2	
3	
4	
5	
6	

CHAPTERS 1–6

Cumulative Review Exercises

In Exercises 1–4, graph each equation or inequality.

1. $3x - 4y = 12$

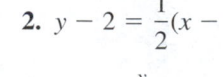

2. $y - 2 = \dfrac{1}{2}(x - 4)$

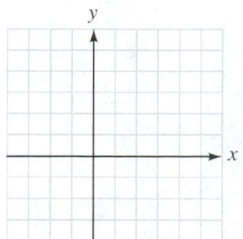

3. $y = x^2 - 4$

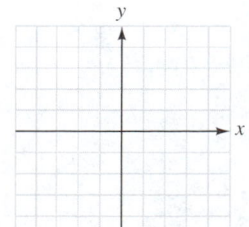

4. $3x + 4y \leq 12$

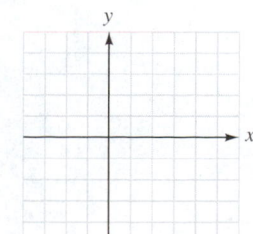

5. Find the slope of the line passing through points $(-2, 5)$ and $(4, 8)$.

6. Write the equation of the line passing through $(-2, 5)$ and $(4, 8)$.

7. ASTRONOMY The **parsec,** a unit of distance used in astronomy, is 3×10^{16} meters. The distance to Betelgeuse, a star in the constellation Orion, is 1.6×10^2 parsecs. Use scientific notation to express this distance in meters.

8. SURFACE AREA The total surface area A of a box with dimensions l, w, and d (see Illustration 1) is given by the formula

$$A = 2lw + 2wd + 2ld$$

If $A = 202$ square inches, $l = 9$ inches, and $w = 5$ inches, find d.

ILLUSTRATION 1

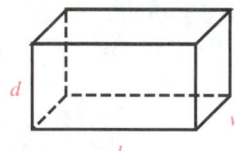

9. SHOPPING SURGE On the graph in Illustration 2, draw a line through the points (1991, 724) and (1996, 974). The line approximates the total annual sales at U.S. shopping centers for the years 1991–1996. Find the rate of increase in sales over this period by finding the slope of the line.

ILLUSTRATION 2

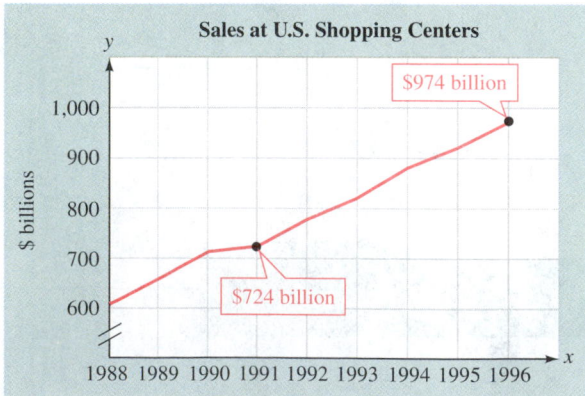

Based on data from International Council of Shopping Centers

10. Simplify $\sqrt{49}$.

In Exercises 11–14, solve each inequality and graph the solution set.

11. $5x - 3 > 7$

12. $7x - 9 < 5$

13. $-2 < -x + 3 \le 5$

14. $0 \le \dfrac{4 - x}{3} \le 2$

In Exercises 15–22, factor each expression.

15. $3x^2y - 6xy^2$

16. $x^2 - 64$

17. $25p^4 - 16q^2$

18. $3x^3 - 243x$

19. $x^2 - 11x - 12$

20. $x^2 - x - 6$

21. $6a^2 - 7a - 20$

22. $16m^2 - 20m - 6$

In Exercises 23–28, solve each equation.

23. $\dfrac{4}{5}x + 6 = 18$

24. $5 - \dfrac{x + 2}{3} = 7 - x$

25. $6x^2 - x - 2 = 0$

26. $5x^2 = 10x$

27. $x^2 + 3x + 2 = 0$

28. $2y^2 + 5y - 12 = 0$

In Exercises 29–30, use the quadratic formula to solve each equation.

29. $x^2 + 2x - 8 = 0$

30. $-6x = x^2 + 4$

In Exercises 31–32, graph each equation.

31. $y = -(x - 2)^2$

32. $y = x^2 + x - 6$

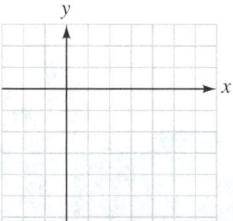
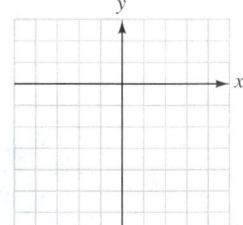

In Exercises 33–34, solve each proportion.

33. $\dfrac{4 - a}{13} = \dfrac{11}{26}$

34. $\dfrac{3a - 2}{7} = \dfrac{a}{28}$

In Exercises 35–38, simplify each fraction.

35. $\dfrac{x^2 + 2x + 1}{x^2 - 1}$

36. $\dfrac{-15a^2}{25a^3}$

37. $\dfrac{x^2 + 2x + 1}{x^2 + 4x + 3}$

38. $\dfrac{9y^2(y - z)}{21y(z - y)}$

In Exercises 39–48, do the operation(s) and simplify when possible.

39. $\dfrac{x^2 + x - 6}{5x - 5} \cdot \dfrac{5x - 10}{x + 3}$

40. $\dfrac{p^2 - p - 6}{3p - 9} \div \dfrac{p^2 + 6p + 9}{p^2 - 9}$

41. $\dfrac{x^2 y^2}{cd} \cdot \dfrac{d^2}{c^2 x}$

42. $\dfrac{x^2 - 4}{x - 1} \div \dfrac{x - 2}{x - 1}$

43. $\dfrac{x + y}{3} + \dfrac{x - y}{7}$

44. $\dfrac{x + 2}{x + 5} - \dfrac{x - 3}{x + 7}$

45. $\dfrac{3x}{x + 2} + \dfrac{5x}{x + 2} - \dfrac{7x - 2}{x + 2}$

46. $\dfrac{3a}{2b} - \dfrac{2b}{3a}$

47. $\dfrac{a + 1}{2a + 4} - \dfrac{a^2}{2a^2 - 8}$

48. $\dfrac{\dfrac{1}{x} + \dfrac{1}{y}}{\dfrac{1}{x} - \dfrac{1}{y}}$

In Exercises 49–50, solve each equation.

49. $\dfrac{4}{a} = \dfrac{6}{a} - 1$

50. $\dfrac{a + 2}{a + 3} - 1 = \dfrac{-1}{a^2 + 2a - 3}$

7

Solving Systems of Equations and Inequalities

CAMPUS CONNECTION

The Theater Arts Department

Theater arts students often use mathematics to help them plan and produce stage plays. Using algebra, the ticket sales department can determine the number of regular admissions and the number of senior citizen tickets sold for a performance if the total attendance and the total dollar receipts are known. This is done by solving a *system of equations* in which each equation contains two variables. In this chapter, we will introduce systems of equations. You will see that they can be used to solve a wide variety of problems, including some from theater arts.

To solve many problems, we must use two variables. This requires that we solve a system of equations.

▶ 7.1

Solving Systems of Equations by Graphing

In this section, you will learn about

> Systems of equations ■ The graphing method ■ Inconsistent systems ■ Dependent equations

Introduction   The lines graphed in Figure 7-1 approximate the per-person consumption of chicken and beef in the United States for the years 1990–1997. We can see that consumption of chicken increased, while that of beef decreased.

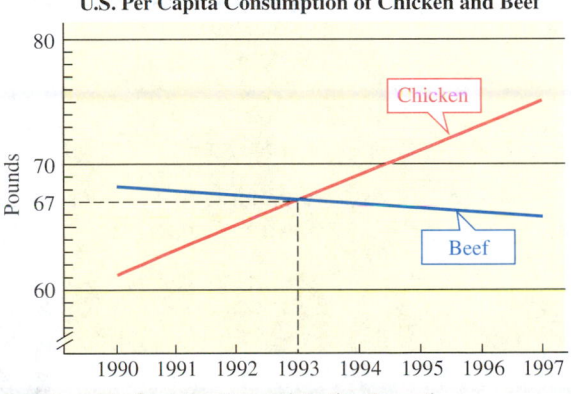

FIGURE 7-1

U.S. Per Capita Consumption of Chicken and Beef

Based on data from the National Broiler Council

By graphing this *pair* of lines on the same coordinate system, it is apparent that Americans consumed equal amounts of chicken and beef in 1993—about 67 pounds of each. In this section, we will work with pairs of linear equations. We call such a pair of equations a *system of equations*.

Systems of Equations

We have considered equations that contain two variables, such as $x + y = 3$. Because there are infinitely many pairs of numbers whose sum is 3, there are infinitely many pairs (x, y) that satisfy this equation. Some of these pairs are

$$x + y = 3$$

x	y
0	3
1	2
2	1
3	0

Likewise, there are infinitely many pairs (x, y) that satisfy the equation $3x - y = 1$. Some of these pairs are

$$3x - y = 1$$

x	y
0	-1
1	2
2	5
3	8

Although there are infinitely many pairs that satisfy each of these equations, only the pair $(1, 2)$ satisfies both equations at the same time. The pair of equations

$$\begin{cases} x + y = 3 \\ 3x - y = 1 \end{cases}$$

is called a **system of equations.** Because the ordered pair $(1, 2)$ satisfies both equations simultaneously, it is called a **simultaneous solution,** or a **solution of the system of equations.** In this chapter, we will discuss three methods for finding the simultaneous solution of a system of two equations that each have two variables. First, we consider the graphing method.

The Graphing Method

To use the graphing method to solve the system

$$\begin{cases} x + y = 3 \\ 3x - y = 1 \end{cases}$$

we graph both equations on one set of coordinate axes using the intercept method, as shown in Figure 7-2.

FIGURE 7-2

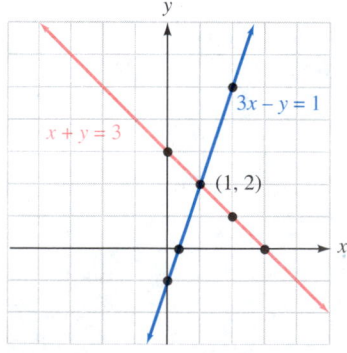

$x + y = 3$			$3x - y = 1$		
x	y	(x, y)	x	y	(x, y)
0	3	$(0, 3)$	0	-1	$(0, -1)$
3	0	$(3, 0)$	$\frac{1}{3}$	0	$\left(\frac{1}{3}, 0\right)$
2	1	$(2, 1)$	2	5	$(2, 5)$

Although there are infinitely many pairs (x, y) that satisfy $x + y = 3$, and infinitely many pairs (x, y) that satisfy $3x - y = 1$, only the coordinates of the point where their graphs intersect satisfy both equations simultaneously. Thus, the solution of the system is $x = 1$ and $y = 2$, or $(1, 2)$.

To check this solution, we substitute 1 for x and 2 for y in each equation and verify that the pair $(1, 2)$ satisfies each equation.

First equation	*Second equation*
$x + y = 3$	$3x - y = 1$
$1 + 2 \overset{?}{=} 3$	$3(1) - 2 \overset{?}{=} 1$
$3 = 3$	$3 - 2 \overset{?}{=} 1$
	$1 = 1$

When the graphs of two equations in a system are different lines, the equations are called **independent equations.** When a system of equations has a solution, the system is called a **consistent system.**

To solve a system of equations in two variables by graphing, we follow these steps.

The graphing method

1. Carefully graph each equation.

2. When possible, find the coordinates of the point where the graphs intersect.

3. Check the solution in the equations of the original system.

EXAMPLE 1

Solving systems by graphing. Use graphing to solve $\begin{cases} 2x + 3y = 2 \\ 3x = 2y + 16 \end{cases}$.

Solution   Using the intercept method, we graph both equations on one set of coordinate axes, as shown in Figure 7-3.

FIGURE 7-3

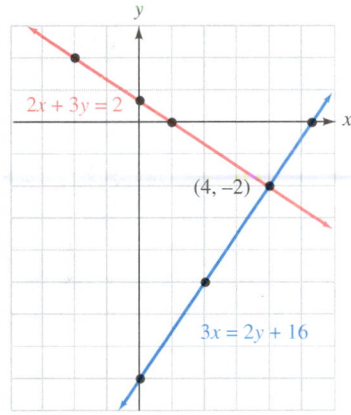

2x + 3y = 2

x	y	(x, y)
0	$\frac{2}{3}$	$\left(0, \frac{2}{3}\right)$
1	0	$(1, 0)$
-2	2	$(-2, 2)$

3x = 2y + 16

x	y	(x, y)
0	-8	$(0, -8)$
$\frac{16}{3}$	0	$\left(\frac{16}{3}, 0\right)$
2	-5	$(2, -5)$

Although there are infinitely many pairs (x, y) that satisfy $2x + 3y = 2$, and infinitely many pairs (x, y) that satisfy $3x = 2y + 16$, only the coordinates of the point where the graphs intersect satisfy both equations at the same time. The solution is $x = 4$ and $y = -2$, or $(4, -2)$.

To check, we substitute 4 for x and -2 for y in each equation and verify that the pair $(4, -2)$ satisfies each equation.

$$2x + 3y = 2 \qquad\qquad 3x = 2y + 16$$
$$2(4) + 3(-2) \stackrel{?}{=} 2 \qquad 3(4) \stackrel{?}{=} 2(-2) + 16$$
$$8 - 6 \stackrel{?}{=} 2 \qquad\qquad 12 \stackrel{?}{=} -4 + 16$$
$$2 = 2 \qquad\qquad\qquad 12 = 12$$

The equations in this system are independent equations, and the system is a consistent system of equations.

SELF CHECK   Use graphing to solve $\begin{cases} 2x = y - 5 \\ x + y = -1 \end{cases}$.   *Answer:* $(-2, 1)$

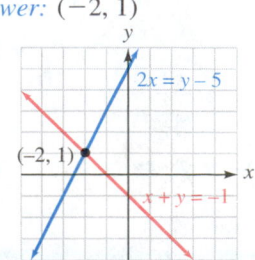

Inconsistent Systems

Sometimes a system of equations has no solution. Such systems are called **inconsistent systems.**

EXAMPLE 2

A system having no solution. Solve $\begin{cases} y = -2x - 6 \\ 4x + 2y = 8 \end{cases}$.

Solution   Since $y = -2x - 6$ is written in slope-intercept form, we can graph it by plotting the y-intercept $(0, -6)$ and then drawing a slope of -2. (The run is 1, and the rise is -2.) We graph $4x + 2y = 8$ using the intercept method.

$$y = -2x - 6 \qquad\qquad 4x + 2y = 8$$

so

$$m = -2 = -\frac{2}{1}$$

and

$$b = -6$$

x	y	(x, y)
0	4	$(0, 4)$
2	0	$(2, 0)$
1	2	$(1, 2)$

The system is graphed in Figure 7-4. Since the lines in the figure are parallel, they have the same slope. We can verify this by writing the second equation in slope–intercept form and observing that the coefficients of x in each equation are equal.

$$y = -2x - 6 \qquad 4x + 2y = 8$$
$$2y = -4x + 8$$
$$y = -2x + 4$$

Because parallel lines do not intersect, this system has no solution and is inconsistent. Since the graphs are different lines, the equations of the system are independent.

FIGURE 7-4

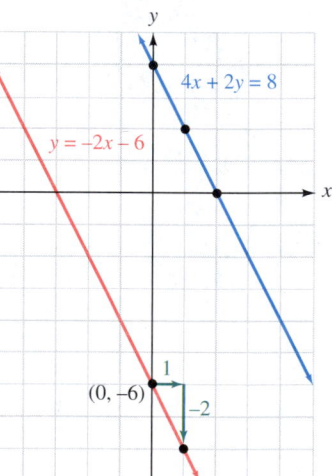

SELF CHECK   Solve $\begin{cases} y = \frac{3}{2}x \\ 3x - 2y = 6 \end{cases}$.

Answer: The lines are parallel; therefore, the system has no solution.

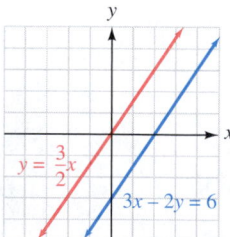

Dependent Equations

Sometimes a system has an infinite number of solutions. In this case, we say that the equations of the system are **dependent equations.**

EXAMPLE 3

Infinitely many solutions. Solve $\begin{cases} y - 4 = 2x \\ 4x + 8 = 2y \end{cases}$.

Solution   We graph both equations on one set of axes, as shown in Figure 7-5.

FIGURE 7-5

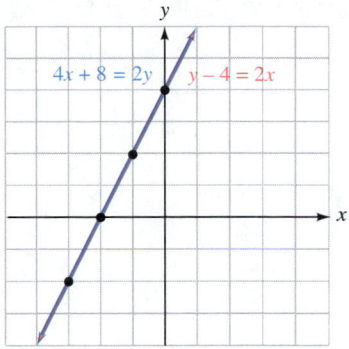

$y - 4 = 2x$				$4x + 8 = 2y$		
x	y	(x, y)		x	y	(x, y)
0	4	$(0, 4)$		0	4	$(0, 4)$
-2	0	$(-2, 0)$		-2	0	$(-2, 0)$
-1	2	$(-1, 2)$		-3	-2	$(-3, -2)$

The lines in Figure 7-5 coincide (they are the same line). Because the lines intersect at infinitely many points, there is an infinite number of solutions. Any pair (x, y) that satisfies one of the equations also satisfies the other.

From the graph, we can see that some possible solutions are $(0, 4)$, $(-1, 2)$, and $(-3, -2)$, since each of these points lies on the one line that is the graph of both equations.

SELF CHECK Solve $\begin{cases} 6x - 2y = 4 \\ y + 2 = 3x \end{cases}$.

Answer: The graphs are the same line. There is an infinite number of solutions.

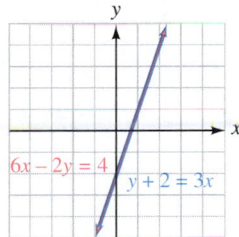

The possibilities that can occur when graphing two linear equations, each with two variables, are summarized as follows.

Possible graph	If the	then
	lines are different and intersect,	the equations are independent and the system is consistent. One solution exists.
	lines are different and parallel,	the equations are independent and the system is inconsistent. No solutions exist.
	lines coincide,	the equations are dependent and the system is consistent. Infinitely many solutions exist.

EXAMPLE 4

Solving an equivalent system. Solve $\begin{cases} -\frac{x}{2} - 1 = \frac{y}{2} \\ \frac{1}{3}x - \frac{1}{2}y = -4 \end{cases}$.

Solution   We can multiply both sides of the first equation by 2 to clear it of fractions.

$$-\frac{x}{2} - 1 = \frac{y}{2}$$

$$2\left(-\frac{x}{2} - 1\right) = 2\left(\frac{y}{2}\right)$$

1. $-x - 2 = y$ We will call this Equation 1.

We then multiply both sides of the second equation by 6 to clear it of fractions.

$$\frac{1}{3}x - \frac{1}{2}y = -4$$

$$6\left(\frac{1}{3}x - \frac{1}{2}y\right) = 6(-4)$$

2. $2x - 3y = -24$ We will call this Equation 2.

Equations 1 and 2 form the following **equivalent system**, which has the same solutions as the original system:

$$\begin{cases} -x - 2 = y \\ 2x - 3y = -24 \end{cases}$$

In Figure 7-6, we graph $-x - 2 = y$ by plotting the y-intercept $(0, -2)$ and then drawing a slope of -1. We graph $2x - 3y = -24$ using the intercept method. We find that $(-6, 4)$ is the point of intersection. The solution is $x = -6$ and $y = 4$, or $(-6, 4)$.

$y = -x - 2$

so

$$m = -1 = \frac{-1}{1}$$

and

$$b = -2$$

$2x - 3y = -24$

x	y	(x, y)
0	8	$(0, 8)$
-12	0	$(-12, 0)$
-3	6	$(-3, 6)$

FIGURE 7-6

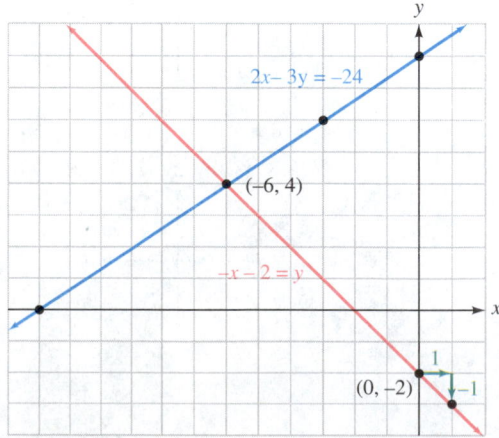

A check will show that when the coordinates of $(-6, 4)$ are substituted into the two original equations, true statements result. Therefore, the equations are independent and the system is consistent.

SELF CHECK Solve $\begin{cases} -\frac{x}{2} = \frac{y}{4} \\ \frac{1}{4}x - \frac{3}{8}y = -2 \end{cases}$.

Answer: $(-2, 4)$

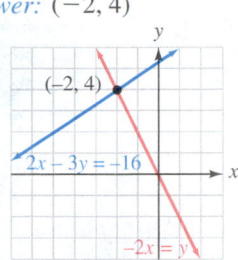

ACCENT ON TECHNOLOGY *Solving Systems with a Graphing Calculator*

We can use a graphing calculator to solve the system

$$\begin{cases} 2x + y = 12 \\ 2x - y = -2 \end{cases}$$

However, before we can enter the equations into the calculator, we must solve them for y.

$$2x + y = 12 \qquad\qquad 2x - y = -2$$
$$y = -2x + 12 \qquad\qquad -y = -2x - 2$$
$$\qquad\qquad\qquad\qquad y = 2x + 2$$

We enter the resulting equations and graph them on the same coordinate axes. If we use the standard window settings, their graphs will look like Figure 7-7(a). We can use the $\boxed{\text{TRACE}}$ key to find that the coordinates of the intersection point are

$$x = 2.5531915 \qquad \text{and} \qquad y = 6.893617$$

See Figure 7-7(b). For better results, we can zoom in on the intersection point, use the $\boxed{\text{TRACE}}$ key again, and find that

$$x = 2.5 \qquad \text{and} \qquad y = 7$$

See Figure 7-7(c). Check each solution.

FIGURE 7-7

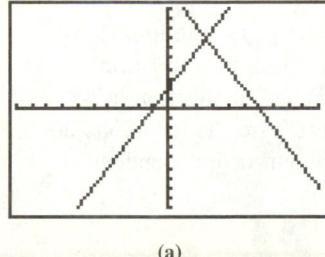   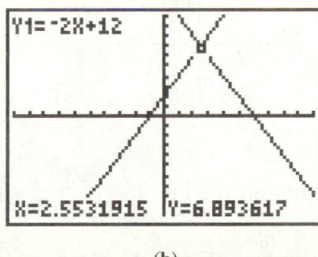   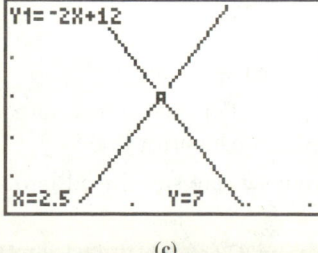

(a)   (b)   (c)

An easier way to find the coordinates of the point of intersection of two graphs is using the INTERSECT feature that is found on most graphing calculators. With this option, the cursor automatically moves to the point of intersection shown on the screen. Then the coordinates of the point are displayed. See Figure 7-8. Consult your owner's manual for the specific keystrokes to use INTERSECT.

FIGURE 7-8

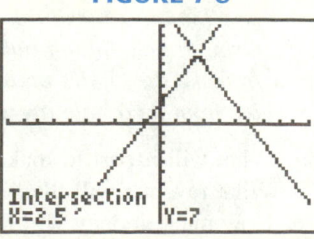

STUDY SET

Section 7.1

VOCABULARY

In Exercises 1–6, fill in the blanks to make the statements true.

1. The pair of equations $\begin{cases} x - y = -1 \\ 2x - y = 1 \end{cases}$ is called a _____ of equations.

2. Because the ordered pair (2, 3) satisfies both equations in Exercise 1, it is called a _____ of the system of equations.

3. When the graphs of two equations in a system are different lines, the equations are called _____ equations.

4. When a system of equations has a solution, the system is called a _____.

5. Systems of equations that have no solution are called _____ systems.

6. When a system has infinitely many solutions, the equations of the system are said to be _____ equations.

CONCEPTS

In Exercises 7–12, refer to Illustration 1. Tell whether a true or false statement would be obtained when the coordinates of

7. point A are substituted into the equation for line l_1.

8. point B are substituted into the equation for line l_2.

9. point A are substituted into the equation for line l_2.

10. point B are substituted into the equation for line l_1.

ILLUSTRATION 1

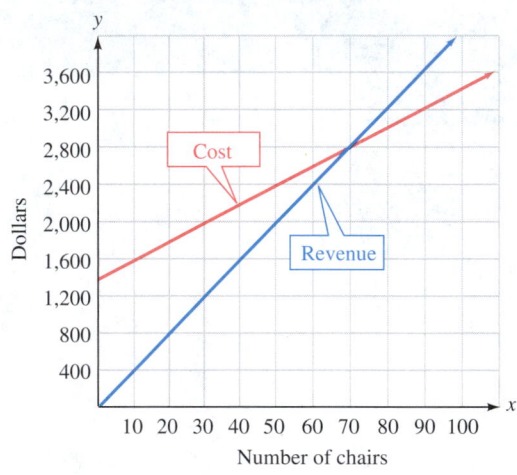

11. point C are substituted into the equation for line l_1.

12. point C are substituted into the equation for line l_2.

In Exercises 13–14, a furniture company is considering manufacturing a new line of oak chairs. Graphs showing the cost to make the chairs and the revenue the company will receive from their sale are given in Illustration 2.

13. **a.** What will it cost to make 30 chairs?
    b. What revenue will their sale bring?
    c. How much money will the company make or lose in this case?

14. **a.** How many chairs must be built and sold so that the costs and the revenue are the same?
    b. Why do you think (70, 2,800) is called the "break-even point"?

ILLUSTRATION 2

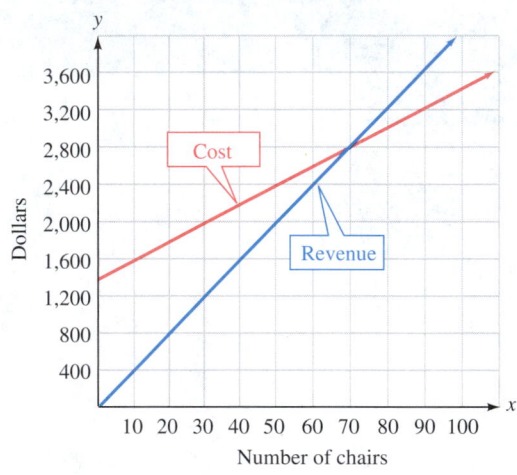

15. How many solutions does the system of equations graphed in Illustration 3 have? Is the system consistent or inconsistent?

ILLUSTRATION 3

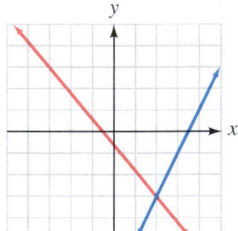

16. How many solutions does the system of equations graphed in Illustration 4 have? Are the equations dependent or independent?

ILLUSTRATION 4

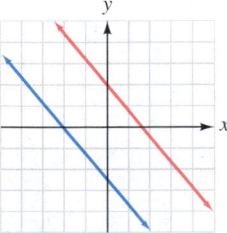

17. The solution of the system of equations graphed in Illustration 5 is $\left(\frac{2}{5}, -\frac{1}{3}\right)$. Knowing this, can you see any disadvantages to the graphing method?

ILLUSTRATION 5

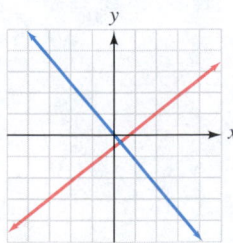

18. Draw the graphs of two linear equations so that the system has

    a. one solution $(-3, -2)$.

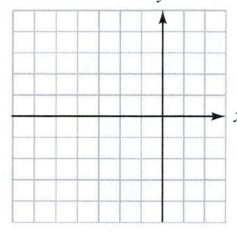

    b. infinitely many solutions, three of which are $(-2, 0)$, $(1, 2)$, and $(4, 4)$.

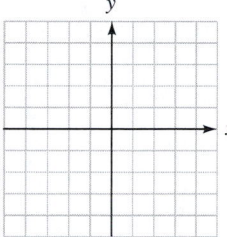

NOTATION

In Exercises 19–20, clear the equation of the fractions.

19.
$$\frac{1}{6}x - \frac{1}{3}y = \frac{11}{2}$$

$$\boxed{}\left(\frac{1}{6}x - \frac{1}{3}y\right) = \boxed{}\left(\frac{11}{2}\right)$$

$$\boxed{}\left(\frac{1}{6}x\right) - 6\boxed{} = 6\left(\frac{11}{2}\right)$$

$$x - 2y = 33$$

20.
$$\frac{3x}{5} - \frac{4y}{5} = -1$$

$$\boxed{}\left(\frac{3x}{5} - \frac{4y}{5}\right) = \boxed{}(-1)$$

$$\boxed{}\left(\frac{3x}{5}\right) - 5\left(\boxed{}\right) = 5(-1)$$

$$3x - 4y = -5$$

PRACTICE

In Exercises 21–32, tell whether the ordered pair is a solution of the given system.

21. $(1, 1)$, $\begin{cases} x + y = 2 \\ 2x - y = 1 \end{cases}$

22. $(1, 3)$, $\begin{cases} 2x + y = 5 \\ 3x - y = 0 \end{cases}$

23. $(3, -2)$, $\begin{cases} 2x + y = 4 \\ y = 1 - x \end{cases}$

24. $(-2, 4)$, $\begin{cases} 2x + 2y = 4 \\ 3y = 10 - x \end{cases}$

25. $(-2, -4)$, $\begin{cases} 4x + 5y = -23 \\ -3x + 2y = 0 \end{cases}$

26. $(-5, 2)$, $\begin{cases} -2x + 7y = 17 \\ 3x - 4y = -19 \end{cases}$

27. $\left(\frac{1}{2}, 3\right)$, $\begin{cases} 2x + y = 4 \\ 4x - 11 = 3y \end{cases}$

28. $\left(2, \frac{1}{3}\right)$, $\begin{cases} x - 3y = 1 \\ -2x + 6 = -6y \end{cases}$

29. $\left(-\frac{2}{5}, \frac{1}{4}\right)$, $\begin{cases} x - 4y = -6 \\ 8y = 10x + 12 \end{cases}$

30. $\left(-\frac{1}{3}, \frac{3}{4}\right)$, $\begin{cases} 3x + 4y = 2 \\ 12y = 3(2 - 3x) \end{cases}$

31. $(0.2, 0.3)$, $\begin{cases} 20x + 10y = 7 \\ 20y = 15x + 3 \end{cases}$

32. $(2.5, 3.5)$, $\begin{cases} 4x - 3 = 2y \\ 4y + 1 = 6x \end{cases}$

In Exercises 33–52, solve each system by the graphing method. If the equations of a system are dependent or if a system is inconsistent, so indicate.

33. $\begin{cases} x + y = 2 \\ x - y = 0 \end{cases}$ **34.** $\begin{cases} x + y = 4 \\ x - y = 0 \end{cases}$

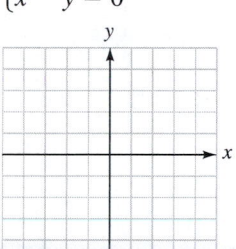
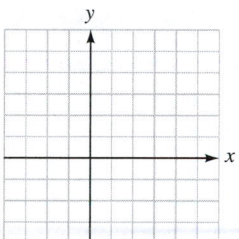

35. $\begin{cases} x + y = 2 \\ y = x - 4 \end{cases}$ **36.** $\begin{cases} x + y = 1 \\ y = x + 5 \end{cases}$

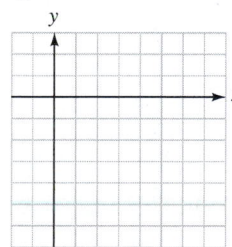
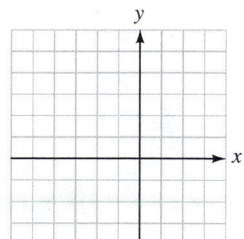

37. $\begin{cases} 3x + 2y = -8 \\ 2x - 3y = -1 \end{cases}$ **38.** $\begin{cases} x + 4y = -2 \\ y = -x - 5 \end{cases}$

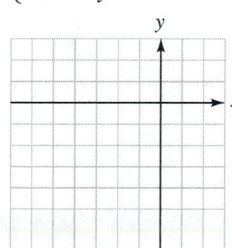
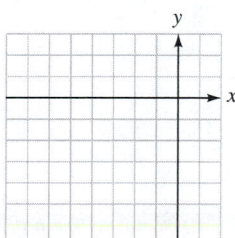

39. $\begin{cases} 4x - 2y = 8 \\ y = 2x - 4 \end{cases}$ **40.** $\begin{cases} 3x - 6y = 18 \\ x = 2y + 3 \end{cases}$

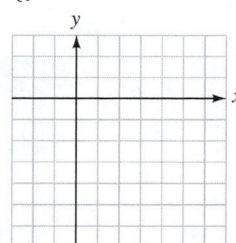
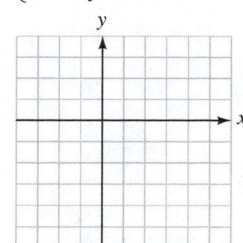

41. $\begin{cases} 2x - 3y = -18 \\ 3x + 2y = -1 \end{cases}$

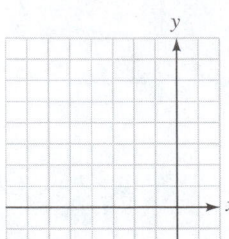

42. $\begin{cases} -x + 3y = -11 \\ 3x - y = 17 \end{cases}$

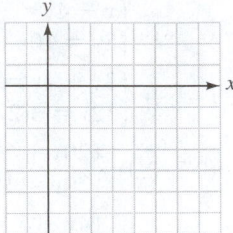

43. $\begin{cases} x = 4 \\ 2y = 12 - 4x \end{cases}$

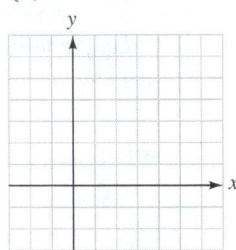

44. $\begin{cases} x = 3 \\ 3y = 6 - 2x \end{cases}$

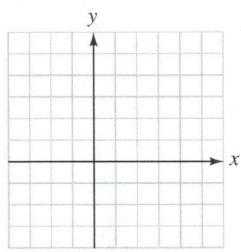

45. $\begin{cases} x + 2y = -4 \\ x - \frac{1}{2}y = 6 \end{cases}$

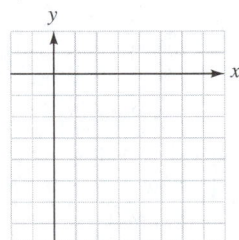

46. $\begin{cases} \frac{2}{3}x - y = -3 \\ 3x + y = 3 \end{cases}$

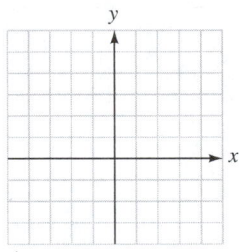

47. $\begin{cases} -\frac{3}{4}x + y = 3 \\ \frac{1}{4}x + y = -1 \end{cases}$

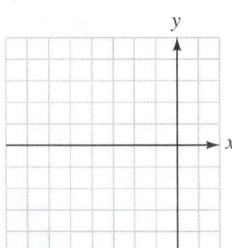

48. $\begin{cases} \frac{1}{3}x + y = 7 \\ \frac{2x}{3} - y = -4 \end{cases}$

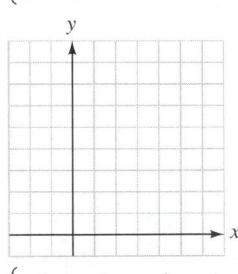

49. $\begin{cases} 2y = 3x + 2 \\ \frac{3}{2}x - y = 3 \end{cases}$

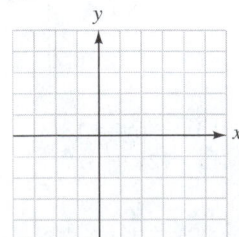

50. $\begin{cases} -\frac{3}{5}x - \frac{1}{5}y = \frac{6}{5} \\ x + \frac{y}{3} = -2 \end{cases}$

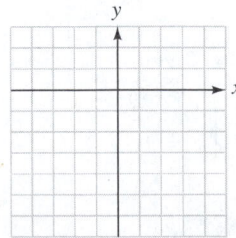

51. $\begin{cases} \frac{1}{3}x - \frac{1}{2}y = \frac{1}{6} \\ \frac{2x}{5} + \frac{y}{2} = \frac{13}{10} \end{cases}$

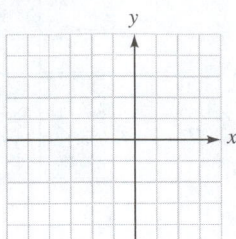

52. $\begin{cases} \frac{3x}{4} + \frac{2y}{3} = -\frac{19}{6} \\ 3y = -x \end{cases}$

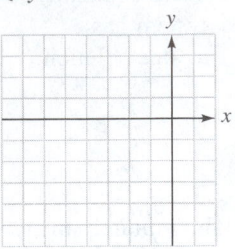

In Exercises 53–56, use a graphing calculator to solve each psystem, if possible.

53. $\begin{cases} y = 4 - x \\ y = 2 + x \end{cases}$

54. $\begin{cases} 3x - 6y = 4 \\ 2x + y = 1 \end{cases}$

55. $\begin{cases} 6x - 2y = 5 \\ 3x = y + 10 \end{cases}$

56. $\begin{cases} x - 3y = -2 \\ 5x + y = 10 \end{cases}$

APPLICATIONS

57. TRANSPLANTS See Illustration 6.
   a. What was the relationship between the number of donors and those awaiting a transplant in 1989?

   b. In what year were the number of donors and the number waiting for a transplant the same? Estimate the number.
   c. Explain the most recent trend.

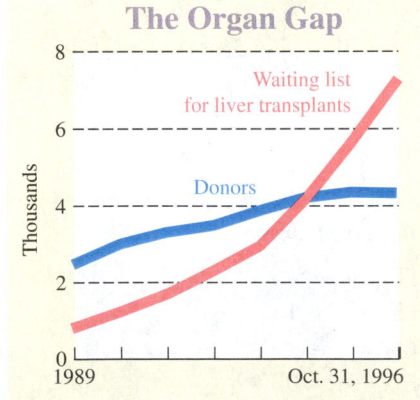
58. DAILY TRACKING POLL See Illustration 7.
   a. Which political candidate was ahead on October 28 and by how much?
   b. On what day did the challenger pull even with the incumbent?
   c. If the election was held November 4, who did the poll predict would win, and by how many percentage points?

ILLUSTRATION 7

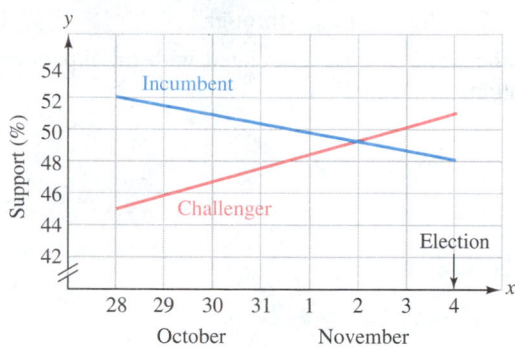

61. AIR TRAFFIC CONTROL The equations describing the paths of two airplanes are $y = -\frac{1}{2}x + 3$ and $3y = 2x + 2$. Graph each equation on the radar screen shown in Illustration 10. Is there a possibility of a mid-air collision? If so, where?

ILLUSTRATION 10

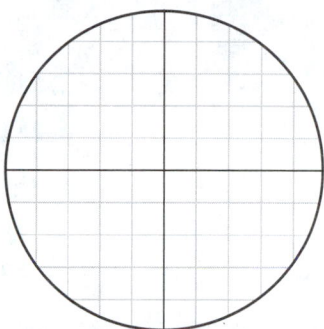

59. LATITUDE AND LONGITUDE See Illustration 8.
   a. Name three American cities that lie on a latitude line of 30° north.

   b. Name three American cities that lie on a longitude line of 90° west.
   c. What city lies on both lines?

ILLUSTRATION 8

62. TV COVERAGE A television camera is located at $(-2, 0)$ and will follow the launch of a space shuttle, as shown in Illustration 11. (Each unit in the illustration is 1 mile.) As the shuttle rises vertically on a path described by $x = 2$, the farthest the camera can tilt back is a line of sight given by $y = \frac{5}{2}x + 5$. For how many miles of the shuttle's flight will it be in view of the camera?

ILLUSTRATION 11

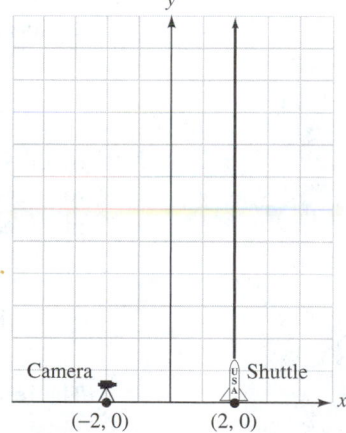

60. ECONOMICS The graph in Illustration 9 illustrates the law of supply and demand.
   a. Complete this sentence: As the price of an item increases, the *supply* of the item _____.
   b. Complete this sentence: As the price of an item increases, the *demand* for the item _____.

   c. For what price will the supply equal the demand? How many items will be supplied for this price?

WRITING

63. Look up the word *simultaneous* in a dictionary and give its definition. In mathematics, what is meant by a simultaneous solution of a system of equations?

64. Suppose the solution of a system is $\left(\frac{1}{3}, -\frac{3}{5}\right)$. Do you think you would be able to find the solution using the graphing method? Explain.

REVIEW

65. What is the slope and the y-intercept of the graph of the line $y = -3x + 4$?

ILLUSTRATION 9

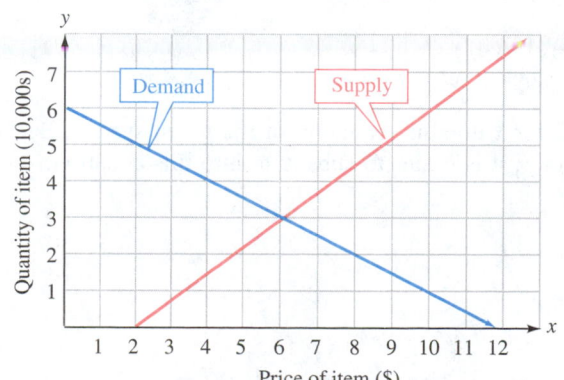

66. Are the graphs of the lines $y = 5x$ and $y = -\frac{1}{5}x$ parallel, perpendicular, or neither?

67. If $f(x) = -4x - x^2$, find $f(3)$.

68. In what quadrant does $(-12, 15)$ lie?

69. Write the equation for the y-axis.

70. Does $(1, 2)$ lie on the line $2x + y = 4$?

71. What point does the line with equation $y - 2 = 7(x - 5)$ pass through?

72. Is the word *domain* associated with the inputs or the outputs of a function?

▶ **7.2**

Solving Systems of Equations by Substitution

In this section, you will learn about

The substitution method ■ Inconsistent systems ■ Dependent equations

---

Introduction   When solving a system of equations by the graphing method, it is often difficult to determine the exact coordinates of the point of intersection. For example, it would be virtually impossible to distinguish that two lines intersect at the point $\left(\frac{1}{16}, -\frac{3}{5}\right)$. In this section, we introduce an algebraic method that finds *exact* solutions. It is called the **substitution method.** This method is based on the **substitution principle,** which states: If $a = b$, then a may replace b or b may replace a in any statement.

The Substitution Method

To solve the system

$$\begin{cases} y = 3x - 2 \\ 2x + y = 8 \end{cases}$$

by the substitution method, we note that the first equation, $y = 3x - 2$, is *solved for y* (or *y is expressed in terms of x*). Because $y = 3x - 2$, we can substitute $3x - 2$ for y in the equation $2x + y = 8$ to get

$2x + y = 8$ The second equation.

$2x + (3x - 2) = 8$ Substitute $3x - 2$ for y. Write parentheses around the expression that was substituted for y.

The resulting equation has only one variable and can be solved for x.

$2x + (3x - 2) = 8$

$2x + 3x - 2 = 8$ Remove the parentheses.

$5x - 2 = 8$ Combine like terms: $2x + 3x = 5x$.

$5x = 10$ Add 2 to both sides.

$x = 2$ Divide both sides by 5.

We can find y by substituting 2 for x in either equation of the given system. Because $y = 3x - 2$ is already solved for y, it is easier to substitute into this equation.

$y = 3x - 2$ The first equation.

$= 3(2) - 2$ Substitute 2 for x.

$= 6 - 2$

$y = 4$

The solution to the given system is $x = 2$ and $y = 4$, or $(2, 4)$.

Check: **First equation**   **Second equation**

$$y = 3x - 2 \qquad\qquad 2x + y = 8$$
$$4 \stackrel{?}{=} 3(2) - 2 \qquad 2(2) + 4 \stackrel{?}{=} 8$$
$$4 \stackrel{?}{=} 6 - 2 \qquad\qquad 4 + 4 \stackrel{?}{=} 8$$
$$4 = 4 \qquad\qquad\qquad 8 = 8$$

If we graphed the lines represented by the equations of the given system, they would intersect at the point (2, 4). The equations of this system are independent, and the system is consistent.

To solve a system of equations in x and y by the substitution method, we follow these steps.

The substitution method

1. Solve one of the equations for either x or y. (This step will not be necessary if an equation is already solved for x or y.)

2. Substitute the resulting expression for the variable obtained in step 1 into the remaining equation and solve that equation.

3. Find the value of the other variable by substituting the solution found in step 2 into any equation containing both variables.

4. Check the solution in the equations of the original system.

EXAMPLE 1

Solving systems by substitution. Solve $\begin{cases} 2x + y = -10 \\ x = -3y \end{cases}$.

Solution   The second equation, $x = -3y$, tells us that x and $-3y$ have the same value. Therefore, we may substitute $-3y$ for x in the first equation.

$$2x + y = -10 \qquad \text{The first equation.}$$
$$2(-3y) + y = -10 \qquad \text{Replace } x \text{ with } -3y.$$
$$-6y + y = -10 \qquad \text{Do the multiplication.}$$
$$-5y = -10 \qquad \text{Combine like terms.}$$
$$y = 2 \qquad \text{Divide both sides by } -5.$$

We can find x by substituting 2 for y in the equation $x = -3y$.

$$x = -3y \qquad \text{The second equation.}$$
$$= -3(2) \qquad \text{Substitute 2 for } y.$$
$$= -6$$

The solution is $x = -6$ and $y = 2$, or $(-6, 2)$.

Check: **First equation**   **Second equation**

$$2x + y = -10 \qquad\qquad x = -3y$$
$$2(-6) + 2 \stackrel{?}{=} -10 \qquad -6 \stackrel{?}{=} -3(2)$$
$$-12 + 2 \stackrel{?}{=} -10 \qquad -6 = -6$$
$$-10 = -10$$

SELF CHECK   Solve $\begin{cases} y = -2x \\ 3x - 2y = -7 \end{cases}$.   *Answer:* $(-1, 2)$ ∎

Solving for a variable first. Solve $\begin{cases} 2x + y = -5 \\ 3x + 5y = -4 \end{cases}$.

Solution We solve one of the equations for one of the variables. Since the term y in the first equation has a coefficient of 1, we solve the first equation for y.

$$2x + y = -5 \qquad \text{The first equation.}$$
$$y = -5 - 2x \qquad \text{Subtract } 2x \text{ from both sides to isolate } y.$$

We then substitute $-5 - 2x$ for y in the second equation and solve for x.

$$3x + 5y = -4 \qquad \text{The second equation.}$$
$$3x + 5(\mathbf{-5 - 2x}) = -4 \qquad \text{Substitute } -5 - 2x \text{ for } y.$$
$$3x - 25 - 10x = -4 \qquad \text{Remove parentheses.}$$
$$-7x - 25 = -4 \qquad \text{Combine like terms: } 3x - 10x = -7x.$$
$$-7x = 21 \qquad \text{Add 25 to both sides.}$$
$$x = -3 \qquad \text{Divide both sides by } -7.$$

We can find y by substituting -3 for x in the equation $y = -5 - 2x$.

$$y = -5 - 2\mathbf{x}$$
$$= -5 - 2(\mathbf{-3}) \qquad \text{Substitute } -3 \text{ for } x.$$
$$= -5 + 6$$
$$= 1$$

The solution is $(-3, 1)$. Check it in the original equations.

SELF CHECK Solve $\begin{cases} 2x - 3y = 13 \\ 3x + y = 3 \end{cases}$.   *Answer:* $(2, -3)$ ■

Systems of equations are sometimes written in variables other than x and y. For example, the system

$$\begin{cases} 3a - 3b = 5 \\ 3 - a = -2b \end{cases}$$

is written in a and b. Regardless of the variables used, the procedures used to solve the system remain the same. The solution should be expressed in the form (a, b).

Solving for a variable first. Solve $\begin{cases} 3a - 3b = 5 \\ 3 - a = -2b \end{cases}$.

Solution Since the coefficient of a in the second equation is -1, we will solve that equation for a.

$$3 - a = -2b \qquad \text{The second equation.}$$
$$-a = -2b - 3 \qquad \text{Subtract 3 from both sides.}$$

To obtain a on the left-hand side, we can multiply (or divide) both sides of the equation by -1.

$$\mathbf{-1}(-a) = \mathbf{-1}(-2b - 3) \qquad \text{Multiply both sides by } -1.$$
$$a = 2b + 3 \qquad \text{Do the multiplications.}$$

We then substitute $2b + 3$ for a in the first equation and proceed as follows:

$$3a - 3b = 5$$
$$3(2b + 3) - 3b = 5 \qquad \text{Substitute.}$$
$$6b + 9 - 3b = 5 \qquad \text{Remove parentheses.}$$
$$3b + 9 = 5 \qquad \text{Combine like terms.}$$
$$3b = -4 \qquad \text{Subtract 9 from both sides: } 5 - 9 = -4.$$
$$b = -\frac{4}{3} \qquad \text{Divide both sides by 3.}$$

To find a, we substitute $-\frac{4}{3}$ for b in $a = 2b + 3$ and simplify.

$$a = 2b + 3$$
$$= 2\left(-\frac{4}{3}\right) + 3 \qquad \text{Substitute.}$$
$$= -\frac{8}{3} + \frac{9}{3} \qquad \text{Do the multiplication: } 2\left(-\frac{4}{3}\right) = -\frac{8}{3}. \text{ Write 3 as } \frac{9}{3}.$$
$$= \frac{1}{3} \qquad \text{Add the numerators and keep the common denominator.}$$

The solution is $\left(\frac{1}{3}, -\frac{4}{3}\right)$. Check it in the original equations.

SELF CHECK Solve $\begin{cases} 2s - t = 4 \\ 3s - 5t = 2 \end{cases}$. *Answer:* $\left(\frac{18}{7}, \frac{8}{7}\right)$ ∎

EXAMPLE 4

Solving an equivalent system. Solve $\begin{cases} \frac{x}{2} + \frac{y}{4} = -\frac{1}{4} \\ 2x - y = 2 + y - x \end{cases}$.

Solution It is helpful to rewrite each equation in simpler form before performing a substitution. We begin by clearing the first equation of fractions.

$$\frac{x}{2} + \frac{y}{4} = -\frac{1}{4}$$
$$4\left(\frac{x}{2} + \frac{y}{4}\right) = 4\left(-\frac{1}{4}\right) \qquad \text{Multiply both sides by the LCD, 4.}$$
$$2x + y = -1$$

We can write the second equation in general form ($Ax + By = C$) by adding x and subtracting y from both sides.

$$2x - y = 2 + y - x$$
$$2x - y + x - y = 2 + y - x + x - y$$
$$3x - 2y = 2 \qquad \text{Combine like terms.}$$

The two results form the following equivalent system, which has the same solution as the original one.

1. $\begin{cases} 2x + y = -1 \\ \textbf{2.} \quad 3x - 2y = 2 \end{cases}$

To solve this system, we solve Equation 1 for y.

$$2x + y = -1$$
$$2x + y - 2x = -1 - 2x \qquad \text{Subtract } 2x \text{ from both sides.}$$
$$\textbf{3.} \qquad y = -1 - 2x \qquad \text{Combine like terms.}$$

To find x, we substitute $-1 - 2x$ for y in Equation 2 and proceed as follows:

$$3x - 2y = 2$$
$$3x - 2(\mathbf{-1 - 2x}) = 2 \quad \text{Substitute.}$$
$$3x + 2 + 4x = 2 \quad \text{Remove parentheses.}$$
$$7x + 2 = 2 \quad \text{Combine like terms.}$$
$$7x = 0 \quad \text{Subtract 2 from both sides.}$$
$$x = 0 \quad \text{Divide both sides by 7.}$$

To find y, we substitute 0 for x in Equation 3.

$$y = -1 - 2\mathbf{x}$$
$$y = -1 - 2(\mathbf{0})$$
$$y = -1$$

The solution is $(0, -1)$. Check it in the original equations.

SELF CHECK Solve $\begin{cases} \frac{1}{3}x - \frac{1}{6}y = -\frac{1}{3} \\ x + y = -3 - 2x - y \end{cases}$.   *Answer:* $(-1, 0)$ ■

Inconsistent Systems

EXAMPLE 5

A system with no solution. Solve $\begin{cases} 0.01x = 0.12 - 0.04y \\ 2x = 4(3 - 2y) \end{cases}$.

Solution   The first equation contains decimal coefficients. We can clear the equation of decimals by multiplying both sides by 100.

$$\begin{cases} x = 12 - 4y \\ 2x = 4(3 - 2y) \end{cases}$$

FIGURE 7-9

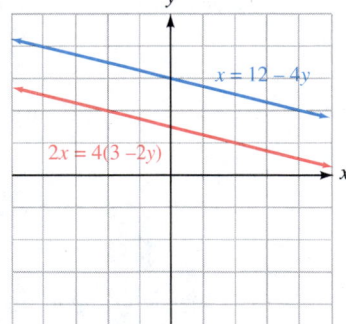

Since $x = 12 - 4y$, we can substitute $12 - 4y$ for x in the second equation and solve for y.

$$2\mathbf{x} = 4(3 - 2y) \quad \text{The second equation.}$$
$$2(\mathbf{12 - 4y}) = 4(3 - 2y) \quad \text{Substitute.}$$
$$24 - 8y = 12 - 8y \quad \text{Remove parentheses.}$$
$$24 \neq 12 \quad \text{Add } 8y \text{ to both sides.}$$

This result indicates that the equations are independent and also that the system is inconsistent. As we see in Figure 7-9, when the equations are graphed, the graphs are parallel lines. This system has no solution.

SELF CHECK Solve $\begin{cases} 0.1x - 0.4 = 0.1y \\ -2y = 2(2 - x) \end{cases}$.   *Answer:* no solution   ■

Dependent Equations

EXAMPLE 6

Infinitely many solutions. Solve $\begin{cases} x = -3y + 6 \\ 2x + 6y = 12 \end{cases}$.

Solution   We can substitute $-3y + 6$ for x in the second equation and proceed as follows:

$$2x + 6y = 12 \quad \text{The second equation.}$$
$$2(\mathbf{-3y + 6}) + 6y = 12 \quad \text{Substitute.}$$
$$-6y + 12 + 6y = 12 \quad \text{Remove parentheses.}$$
$$12 = 12 \quad \text{Combine like terms.}$$

Although $12 = 12$ is true, we did not find y. This indicates that the equations are dependent. As we see in Figure 7-10, when these equations are graphed, their graphs are identical.

Because any ordered pair that satisfies one equation of the system also satisfies the other, the system has infinitely many solutions. To find some, we substitute 0, 3, and 6 for x in either equation and solve for y. The pairs $(0, 2)$, $(3, 1)$, and $(6, 0)$ are some of the solutions.

FIGURE 7-10

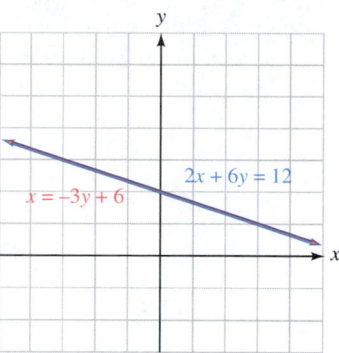

SELF CHECK   Solve $\begin{cases} y = 2 - x \\ 3x + 3y = 6 \end{cases}$.

Answer: infinitely many solutions ∎

STUDY SET

Section 7.2

VOCABULARY

In Exercises 1–6, fill in the blanks to make the statements true.

1. We say that the equation $y = 2x + 4$ is solved for ___ or that y is expressed in _____ of x.

2. "To _____ a solution of a system" means to see whether the coordinates of the ordered pair satisfy both equations.

3. Consider $2(x - 6) = 2x - 12$. The distributive property was applied to _____ parentheses.

4. In mathematics, "to _____" means to replace an expression with one that is equivalent to it.

5. A dependent system has _____ many solutions.

6. In the term y, the _____ is understood to be 1.

CONCEPTS

7. Consider the system $\begin{cases} 2x + 3y = 12 \\ y = 2x + 4 \end{cases}$.

   a. How many variables does each equation of the system contain?

   b. Substitute $2x + 4$ for y in the first equation. How many variables does the resulting equation contain?

8. For each equation, solve for y.

   a. $y + 2 = x$

   b. $2 - y = x$

   c. $2 + x + y = 0$

9. Given the equation $x - 2y = -10$,

   a. solve it for x.

   b. solve it for y.

   c. which variable was easier to solve for, x or y? Explain.

10. Which variable in which equation should be solved for in step 1 of the substitution method?

   a. $\begin{cases} x - 2y = 2 \\ 2x + 3y = 11 \end{cases}$

   b. $\begin{cases} 2x - 3y = 2 \\ 2x - y = 11 \end{cases}$

   c. $\begin{cases} 7x - 3y = 2 \\ 2x - 8y = 0 \end{cases}$

11. **a.** Find the **error** in the following work when $x - 4$ is substituted for y.

$$x + 2y = 5 \quad \text{The first equation of the system.}$$
$$x + 2x - 4 = 5 \quad \text{Substitute for } y: y = x - 4.$$
$$3x - 4 = 5 \quad \text{Combine like terms.}$$
$$3x = 9 \quad \text{Add 4 to both sides.}$$
$$x = 3 \quad \text{Do the divisions.}$$

   b. Rework the problem to find the correct value of x.

12. A student uses the substitution method to solve the system $\begin{cases} 4a + 5b = 2 \\ b = 3a - 11 \end{cases}$. She finds that $a = 3$. What is the easiest way for her to determine the value of b?

13. Consider the system $\begin{cases} x - 2y = 0 \\ 6x + 3y = 5 \end{cases}$.

    a. Graph the equations on the same coordinate system. Why is it difficult to determine the solution of the system?

    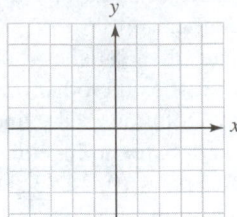

    b. Solve the system by the substitution method.

14. The equation $-2 = 1$ is the result when a system is solved by the substitution method. Which graph in Illustration 1 is a possible graph of the system?

ILLUSTRATION 1

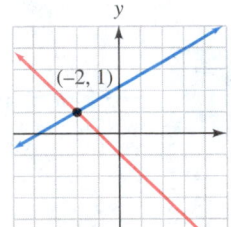 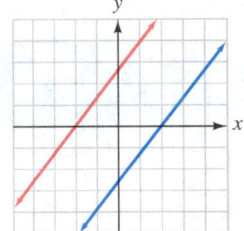

NOTATION

In Exercises 15–16, complete the solution of each system.

15. Solve $\begin{cases} y = 3x \\ x - y = 4 \end{cases}$.

$$x - y = 4$$
$$x - (\boxed{}) = 4$$
$$-2x = \boxed{}$$
$$x = -2$$
$$y = 3x$$
$$y = 3(\boxed{})$$
$$y = -6$$

    The solution of the system is $\boxed{}$.

16. Solve $\begin{cases} 2x + y = -5 \\ 2 - 2y = x \end{cases}$.

$$2x + y = -5$$
$$2(\boxed{}) + y = -5$$
$$4 - \boxed{} + y = -5$$
$$\boxed{} - 3y = -5$$
$$-3y = \boxed{}$$
$$y = 3$$
$$2 - 2y = x$$
$$2 - 2(\boxed{}) = x$$
$$2 - 6 = x$$
$$-4 = x$$

    The solution of the system is $\boxed{}$.

PRACTICE

In Exercises 17–52, use the substitution method to solve each system. If the equations of a system are dependent or if a system is inconsistent, so indicate.

17. $\begin{cases} y = 2x \\ x + y = 6 \end{cases}$

18. $\begin{cases} y = 3x \\ x + y = 4 \end{cases}$

19. $\begin{cases} y = 2x - 6 \\ 2x + y = 6 \end{cases}$

20. $\begin{cases} y = 2x - 9 \\ x + 3y = 8 \end{cases}$

21. $\begin{cases} y = 2x + 5 \\ x + 2y = -5 \end{cases}$

22. $\begin{cases} y = -2x \\ 3x + 2y = -1 \end{cases}$

23. $\begin{cases} 2a + 4b = -24 \\ a = 20 - 2b \end{cases}$

24. $\begin{cases} 3a + 6b = -15 \\ a = -2b - 5 \end{cases}$

25. $\begin{cases} 2a = 3b - 13 \\ -b = -2a - 7 \end{cases}$

26. $\begin{cases} a = 3b - 1 \\ -b = -2a - 2 \end{cases}$

27. $\begin{cases} r + 3s = 9 \\ 3r + 2s = 13 \end{cases}$

28. $\begin{cases} x - 2y = 2 \\ 2x + 3y = 11 \end{cases}$

29. $\begin{cases} 0.4x + 0.5y = 0.2 \\ 3x - y = 11 \end{cases}$

30. $\begin{cases} 0.5u + 0.3v = 0.5 \\ 4u - v = 4 \end{cases}$

31. $\begin{cases} 6x - 3y = 5 \\ 2y + x = 0 \end{cases}$

32. $\begin{cases} 5s + 10t = 3 \\ 2s + t = 0 \end{cases}$

33. $\begin{cases} 3x + 4y = -7 \\ 2y - x = -1 \end{cases}$

34. $\begin{cases} 4x + 5y = -2 \\ x + 2y = -2 \end{cases}$

35. $\begin{cases} 9x = 3y + 12 \\ 4 = 3x - y \end{cases}$

36. $\begin{cases} 8y = 15 - 4x \\ x + 2y = 4 \end{cases}$

37. $\begin{cases} 0.02x + 0.05y = -0.02 \\ -\frac{x}{2} = y \end{cases}$

38. $\begin{cases} y = -\frac{x}{2} \\ 0.02x - 0.03y = -0.07 \end{cases}$

39. $\begin{cases} b = \frac{2}{3}a \\ 8a - 3b = 3 \end{cases}$

40. $\begin{cases} a = \frac{2}{3}b \\ 9a + 4b = 5 \end{cases}$

41. $\begin{cases} y - x = 3x \\ 2x + 2y = 14 - y \end{cases}$

42. $\begin{cases} y + x = 2x + 2 \\ 6x - 4y = 21 - y \end{cases}$

43. $\begin{cases} 2x - y = x + y \\ -2x + 4y = 6 \end{cases}$

44. $\begin{cases} x = -3y + 6 \\ 2x + 4y = 6 + x + y \end{cases}$

45. $\begin{cases} 3(x - 1) + 3 = 8 + 2y \\ 2(x + 1) = 8 + y \end{cases}$

46. $\begin{cases} 4(x - 2) = 19 - 5y \\ 3(x - 2) - 2y = -y \end{cases}$

47. $\begin{cases} \frac{1}{2}x + \frac{1}{2}y = -1 \\ \frac{1}{3}x - \frac{1}{2}y = -4 \end{cases}$

48. $\begin{cases} \frac{2}{3}y + \frac{1}{5}z = 1 \\ \frac{1}{3}y - \frac{2}{5}z = 3 \end{cases}$

49. $\begin{cases} 5x = \frac{1}{2}y - 1 \\ \frac{1}{4}y = 10x - 1 \end{cases}$

50. $\begin{cases} \frac{2}{3}x = 1 - 2y \\ 2(5y - x) + 11 = 0 \end{cases}$

51. $\begin{cases} \dfrac{6x-1}{3} - \dfrac{5}{3} = \dfrac{3y+1}{2} \\ \dfrac{1+5y}{4} + \dfrac{x+3}{4} = \dfrac{17}{2} \end{cases}$ **52.** $\begin{cases} \dfrac{5x-2}{4} + \dfrac{1}{2} = \dfrac{3y+2}{2} \\ \dfrac{7y+3}{3} = \dfrac{x}{2} + \dfrac{7}{3} \end{cases}$

ILLUSTRATION 3

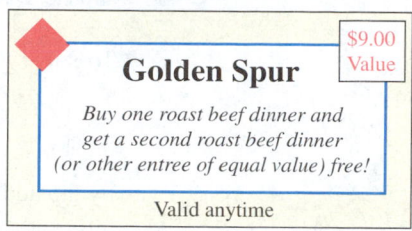

Golden Spur $9.00 Value

Buy one roast beef dinner and get a second roast beef dinner (or other entree of equal value) free!

Valid anytime

APPLICATIONS

53. DINING See the breakfast menu in Illustration 2. What substitution from the a la carte menu will the restaurant owner allow customers to make if they don't want hash browns with their country breakfast? Why?

ILLUSTRATION 2

Village Vault Restaurant		
Country Breakfast $5.95		
Includes 2 eggs, 3 pancakes, sausage, bacon hash browns, and coffee		
A la Carte Menu–Single Servings		
Strawberries $1.25	Melon	$0.95
Croissant $1.70	Orange juice	$1.65
Hash browns $0.95	Oatmeal	$1.95
Muffin $1.30	Ham	$1.80

54. DISCOUNT COUPON In mathematics, the substitution property states:

If a = b, then a may replace b or b may replace a in any statement.

Where on the coupon in Illustration 3 is there an application of the substitution property? Explain.

WRITING

55. Explain how to use substitution to solve a system of equations.

56. If the equations of a system are written in general form, why is it to your advantage to solve for a variable whose coefficient is 1 when using the substitution method?

57. When solving a system, what advantages and disadvantages are there with the graphing method? With the substitution method?

58. In this section, the substitution method for solving a system of two equations was discussed. List some other uses of the word *substitution*, or *substitute*, that you encounter in everyday life.

REVIEW

59. What is the slope of the line $y = -\frac{5}{8}x - 12$?

60. If $g(x) = -3x + 9$, find $g(-3)$.

61. Find the y-intercept of $2x - 3y = 18$.

62. Write the equation of the line passing through $(-1, 5)$ with a slope of -3.

63. Can a circle represent the graph of a function?

64. On what axis does $(0, -2)$ lie?

▶ 7.3

Solving Systems of Equations by Addition

In this section, you will learn about

The addition method ■ Inconsistent systems ■ Dependent equations

Introduction In step 1 of the substitution method for solving a system of equations, we solve one equation for one of the variables. At times, this can be difficult— especially if neither variable has a coefficient of 1 or -1. In cases such as these, we can use another algebraic method called the **addition** or **elimination method** to find the exact solution of the system. This method is based on the addition property of equality: *When equal quantities are added to both sides of an equation, the results are equal.*

The Addition Method

To solve the system

$$\begin{cases} x + y = 8 \\ x - y = -2 \end{cases}$$

by the addition method, we see that the coefficients of y are opposites and then add the left- and right-hand sides of the equations to eliminate the variable y.

$$\begin{array}{l} x + y = 8 \\ \underline{x - y = -2} \end{array}$$ Equal quantities, $x - y$ and -2, are added to both sides of the equation $x + y = 8$. By the addition property of equality, the results will be equal.

Now, column by column, we add like terms. The terms y and $-y$ are eliminated.

$$\begin{array}{l} \downarrow\downarrow \downarrow \text{ Combine like terms: } x + x = 2x, \ y + (-y) = 0, \text{ and } 8 + (-2) = 6. \\ x + y = 8 \\ \underline{x - y = -2} \\ 2x = 6 \ \leftarrow \text{ Write each result here.} \end{array}$$

We can then solve the resulting equation for x.

$$2x = 6$$
$$x = 3 \quad \text{Divide both sides by 2.}$$

To find y, we substitute 3 for x in either equation and solve it for y.

$$x + y = 8 \quad \text{The first equation of the system.}$$
$$3 + y = 8 \quad \text{Substitute 3 for } x.$$
$$y = 5 \quad \text{Subtract 3 from both sides.}$$

We check the solution by verifying that $(3, 5)$ satisfies each equation of the system. To solve an equation in x and y by the addition method, we follow these steps.

The addition method

1. Write both equations in general form: $Ax + By = C$.
2. If necessary, multiply one or both of the equations by nonzero quantities to make the coefficients of x (or the coefficients of y) opposites.
3. Add the equations to eliminate the term involving x (or y).
4. Solve the equation resulting from step 3.
5. Find the value of the other variable by substituting the solution found in step 4 into any equation containing both variables.
6. Check the solution in the equations of the original system.

E X A M P L E 1

Solving systems by addition. Solve $\begin{cases} 5x + y = -4 \\ -5x + 2y = 7 \end{cases}.$

Solution When the equations are added, the terms $5x$ and $-5x$ drop out. We can then solve the resulting equation for y.

$$\begin{array}{l} 5x + y = -4 \quad \text{Combine like terms: } 5x + (-5x) = 0, \ y + 2y = 3y, \text{ and } -4 + 7 = 3. \\ \underline{-5x + 2y = 7} \\ 3y = 3 \\ y = 1 \quad \text{Divide both sides by 3.} \end{array}$$

To find x, we substitute 1 for y in either equation. If we use $5x + y = -4$, we have

$$5x + y = -4 \qquad \text{The first equation of the system.}$$
$$5x + (1) = -4 \qquad \text{Substitute 1 for } y.$$
$$5x = -5 \qquad \text{Subtract 1 from both sides.}$$
$$x = -1 \qquad \text{Divide both sides by 5.}$$

Verify that $(-1, 1)$ satisfies each original equation.

SELF CHECK   Solve $\begin{cases} x + 3y = 7 \\ 2x - 3y = -22 \end{cases}.$

Answer: $(-5, 4)$ ∎

EXAMPLE 2

Solving systems by addition. Solve $\begin{cases} 3x + y = 7 \\ x + 2y = 4 \end{cases}.$

Solution   If we add the equations as they are, neither variable will be eliminated. We must write the equations so that the coefficients of one of the variables are opposites. To eliminate x, we can multiply both sides of the second equation by -3 to get

$$\begin{cases} 3x + y = 7 \\ -3(x + 2y) = -3(4) \end{cases} \longrightarrow \begin{cases} 3x + y = 7 \\ -3x - 6y = -12 \end{cases}$$

The coefficients of the terms $3x$ and $-3x$ are now opposites. When the equations are added, x is eliminated.

$$\begin{array}{r} 3x + y = 7 \\ -3x - 6y = -12 \\ \hline -5y = -5 \\ y = 1 \end{array} \qquad \text{Divide both sides by } -5.$$

To find x, we substitute 1 for y in the equation $x + 2y = 4$.

$$x + 2y = 4 \qquad \text{The second equation of the original system.}$$
$$x + 2(1) = 4 \qquad \text{Substitute 1 for } y.$$
$$x + 2 = 4 \qquad \text{Do the multiplication.}$$
$$x = 2 \qquad \text{Subtract 2 from both sides.}$$

Check the solution $(2, 1)$ in the original system of equations.

SELF CHECK   Solve $\begin{cases} 3x + 4y = 25 \\ 2x + y = 10 \end{cases}.$

Answer: $(3, 4)$ ∎

EXAMPLE 3

Solving systems by addition. Solve $\begin{cases} 2a - 5b = 10 \\ 3a - 2b = -7 \end{cases}.$

Solution   The equations in the system must be written so that one of the variables will be eliminated when the equations are added.

To eliminate a, we can multiply the first equation by 3 and the second equation by -2 to get

$$\begin{cases} 3(2a - 5b) = 3(10) \\ -2(3a - 2b) = -2(-7) \end{cases} \longrightarrow \begin{cases} 6a - 15b = 30 \\ -6a + 4b = 14 \end{cases}$$

When these equations are added, the terms $6a$ and $-6a$ are eliminated.

$$\begin{aligned} 6a - 15b &= 30 \\ -6a + 4b &= 14 \\ \hline -11b &= 44 \\ b &= -4 \quad \text{\color{red}Divide both sides by } -11. \end{aligned}$$

To find a, we substitute -4 for b in the equation $2a - 5b = 10$.

$$\begin{aligned} 2a - 5b &= 10 && \text{\color{red}The first equation of the original system.} \\ 2a - 5(-4) &= 10 && \text{\color{red}Substitute } -4 \text{ for } b. \\ 2a + 20 &= 10 && \text{\color{red}Simplify.} \\ 2a &= -10 && \text{\color{red}Subtract 20 from both sides.} \\ a &= -5 && \text{\color{red}Divide both sides by 2.} \end{aligned}$$

Check the solution $(-5, -4)$ in the original equations.

SELF CHECK   Solve $\begin{cases} 2a + 3b = 7 \\ 5a + 2b = 1 \end{cases}$.   *Answer:* $(-1, 3)$ ■

EXAMPLE 4   **Equations containing fractions.** Solve $\begin{cases} \frac{5}{6}x + \frac{2}{3}y = \frac{7}{6} \\ \frac{10}{7}x - \frac{4}{9}y = \frac{17}{21} \end{cases}$.

Solution   To clear the equations of fractions, we multiply both sides of the first equation by 6 and both sides of the second equation by 63. This gives the equivalent system

1. $\begin{cases} 5x + 4y = 7 \\ **2.** \quad 90x - 28y = 51 \end{cases}$

We can solve for x by eliminating the terms involving y. To do so, we multiply Equation 1 by 7 and add the result to Equation 2.

$$\begin{aligned} 35x + 28y &= 49 \\ 90x - 28y &= 51 \\ \hline 125x &= 100 \end{aligned}$$

$$x = \frac{100}{125} \quad \text{\color{red}Divide both sides by 125.}$$

$$x = \frac{4}{5} \quad \text{\color{red}Simplify. Divide out the common factor of 25.}$$

To solve for y, we substitute $\frac{4}{5}$ for x in Equation 1 and simplify.

$$\begin{aligned} 5x + 4y &= 7 \\ 5\left(\frac{4}{5}\right) + 4y &= 7 \\ 4 + 4y &= 7 && \text{\color{red}Simplify.} \\ 4y &= 3 && \text{\color{red}Subtract 4 from both sides.} \\ y &= \frac{3}{4} && \text{\color{red}Divide both sides by 4.} \end{aligned}$$

Check the solution of $\left(\frac{4}{5}, \frac{3}{4}\right)$ in the original equations.

SELF CHECK   Solve $\begin{cases} \frac{1}{3}x + \frac{1}{6}y = 1 \\ \frac{1}{2}x - \frac{1}{4}y = 0 \end{cases}$.   *Answer:* $\left(\frac{3}{2}, 3\right)$ ■

EXAMPLE 5

Writing equations in general form. Solve $\begin{cases} 2(2x + y) = 13 \\ 8x = 2y - 16 \end{cases}$.

Solution   We begin by writing each equation in $Ax + By = C$ form. For the first equation, we need only remove the parentheses. To write the second equation in general form, we subtract $2y$ from both sides.

$$\begin{array}{ll} \mathbf{2}(2x + y) = 13 & 8x = 2y - 16 \\ 4x + 2y = 13 & 8x - \mathbf{2y} = 2y - 16 - \mathbf{2y} \\ & 8x - 2y = -16 \end{array}$$

The two resulting equations form the following system.

1. $\begin{cases} 4x + 2y = 13 \\ 8x - 2y = -16 \end{cases}$
2.

When the equations are added, the terms involving y are eliminated.

$$\begin{array}{r} 4x + 2y = 13 \\ \underline{8x - 2y = -16} \\ 12x = -3 \end{array}$$

$$x = -\frac{1}{4} \quad \text{Divide both sides by 12 and simplify the fraction: } -\frac{3}{12} = -\frac{1}{4}.$$

We can use Equation 1 to find y.

$$\begin{array}{ll} 4x + 2y = 13 & \\ 4\left(-\frac{1}{4}\right) + 2y = 13 & \text{Substitute } -\frac{1}{4} \text{ for } x. \\ -1 + 2y = 13 & \text{Do the multiplication.} \\ 2y = 14 & \text{Add 1 to both sides.} \\ y = 7 & \text{Divide both sides by 2.} \end{array}$$

Verify that $\left(-\frac{1}{4}, 7\right)$ satisfies each original equation.

SELF CHECK   Solve $\begin{cases} -3y = -5 - x \\ 3(x - y) = -11 \end{cases}$.
    Answer: $\left(-3, \frac{2}{3}\right)$ ■

Inconsistent Systems

EXAMPLE 6

A system with no solutions. Solve $\begin{cases} 3x - 2y = 8 \\ -3x + 2y = -12 \end{cases}$.

Solution   We can add the equations to eliminate the term involving x.

$$\begin{array}{r} 3x - 2y = 8 \\ \underline{-3x + 2y = -12} \\ 0 = -4 \end{array}$$

Here the terms involving both x and y drop out, and a false result of $0 = -4$ is obtained. This shows that the equations of the system are independent and that the system is inconsistent. This system has no solution.

SELF CHECK   Solve $\begin{cases} 2t - 7v = 5 \\ -2t + 7v = 3 \end{cases}$.
    Answer: no solution   ■

Dependent Equations

EXAMPLE 7

Infinitely many solutions. Solve $\begin{cases} \frac{2x - 5y}{2} = \frac{19}{2} \\ -0.2x + 0.5y = -1.9 \end{cases}$.

Solution   We can multiply both sides of the first equation by **2** to clear it of the fractions and both sides of the second equation by **10** to clear it of the decimals.

$$\begin{cases} \mathbf{2}\left(\dfrac{2x - 5y}{2}\right) = \mathbf{2}\left(\dfrac{19}{2}\right) \\ \mathbf{10}(-0.2x + 0.5y) = \mathbf{10}(-1.9) \end{cases} \longrightarrow \begin{cases} 2x - 5y = 19 \\ -2x + 5y = -19 \end{cases}$$

We add the resulting equations to get

$$\begin{array}{r} 2x - 5y = 19 \\ -2x + 5y = -19 \\ \hline 0 = 0 \end{array}$$

As in Example 6, both x and y drop out. However, this time a true result is obtained, $0 = 0$. This shows that the equations are dependent and that the system has infinitely many solutions.

Any ordered pair that satisfies one equation also satisfies the other equation. Some solutions are $(2, -3)$, $(12, 1)$, and $\left(0, -\frac{19}{5}\right)$.

SELF CHECK

Solve $\begin{cases} \frac{3x + y}{6} = \frac{1}{3} \\ -0.3x - 0.1y = -0.2 \end{cases}$.

Answer: infinitely many solutions ■

STUDY SET

Section 7.3

VOCABULARY

In Exercises 1–4, fill in the blanks to make the statements true.

1. The _____ of the term $-3x$ is -3.

2. The _____ of 4 is -4.

3. $Ax + By = C$ is the _____ form of the equation of a line.

4. When adding the equations
$$\begin{array}{r} 5x - 6y = 10 \\ -3x + 6y = 24 \\ \hline \end{array}$$
the variable y will be _____.

CONCEPTS

5. If the addition method is to be used to solve this system, what is wrong with the form in which it is written?
$$\begin{cases} 2x - 5y = -3 \\ -2y + 3x = 10 \end{cases}$$

6. Can the system
$$\begin{cases} 2x + 5y = -13 \\ -2x - 3y = -5 \end{cases}$$
be solved more easily using the addition method or the substitution method? Explain.

7. What algebraic step should be performed to clear this equation of the fractions?
$$\frac{2}{3}x + 4y = -\frac{4}{5}$$

8. If the addition method is used to solve
$$\begin{cases} 3x + 12y = 4 \\ 6x - 4y = 8 \end{cases}$$
   a. By what would we multiply the first equation to eliminate x?
   b. By what would we multiply the second equation to eliminate y?

9. Solve $\begin{cases} 4x + 2y = 2 \\ 3x - 2y = 12 \end{cases}$.

   a. Use the graphing method.

   b. Use the substitution method.

   c. Use the addition method.

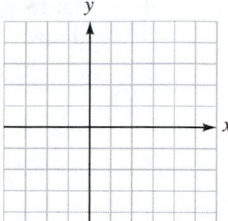

10. The addition method was used to solve three different systems. The results after x was eliminated in each case are listed here. Match each result with a possible graph of the system.

   Result when solving

   a. System 1:   b. System 2:   c. System 3:
     $-1 = -1$ $y = -1$ $-1 = -2$

Possible graph of the system

   i   ii   iii

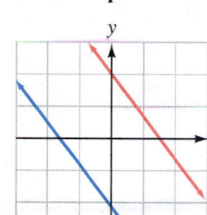 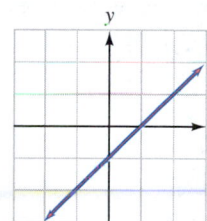 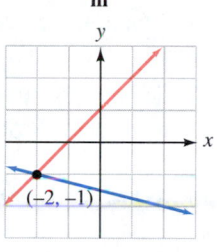

NOTATION

In Exercises 11–12, complete the solution of each system.

11. Solve $\begin{cases} x + y = 5 \\ x - y = -3 \end{cases}$.

$$\begin{array}{r} x + y = 5 \\ x - y = -3 \\ \hline \rule{0pt}{1.2em}\boxed{} = 2 \\ x = \boxed{} \end{array}$$

$$x + y = 5$$
$$\left(\boxed{}\right) + y = 5$$
$$y = 4$$

The solution of the system is $\boxed{}$.

12. Solve $\begin{cases} x - 2y = 8 \\ -x + 5y = -17 \end{cases}$.

$$\begin{array}{r} x - 2y = 8 \\ -x + 5y = -17 \\ \hline \rule{0pt}{1.2em}\boxed{} = -9 \\ y = \boxed{} \end{array}$$

$$x - 2y = 8$$
$$x - 2\left(\boxed{}\right) = 8$$
$$x + 6 = 8$$
$$x = 2$$

The solution of the system is $\boxed{}$.

PRACTICE

In Exercises 13–20, use the addition method to solve each system.

13. $\begin{cases} x - y = -5 \\ x + y = 1 \end{cases}$ 14. $\begin{cases} x + y = 1 \\ x - y = 5 \end{cases}$

15. $\begin{cases} 2r + s = -1 \\ -2r + s = 3 \end{cases}$ 16. $\begin{cases} 3m + n = -6 \\ m - n = -2 \end{cases}$

17. $\begin{cases} 2x + y = -2 \\ -2x - 3y = -6 \end{cases}$ 18. $\begin{cases} 3x + 4y = 8 \\ 5x - 4y = 24 \end{cases}$

19. $\begin{cases} 4x + 3y = 24 \\ 4x - 3y = -24 \end{cases}$ 20. $\begin{cases} 5x - 4y = 8 \\ -5x - 4y = 8 \end{cases}$

In Exercises 21–48, use the addition method to solve each system of equations. If the equations of a system are dependent or if a system is inconsistent, so indicate.

21. $\begin{cases} x + y = 5 \\ x + 2y = 8 \end{cases}$ 22. $\begin{cases} x + 2y = 0 \\ x - y = -3 \end{cases}$

23. $\begin{cases} 2x + y = 4 \\ 2x + 3y = 0 \end{cases}$ 24. $\begin{cases} 2x + 5y = -13 \\ 2x - 3y = -5 \end{cases}$

25. $\begin{cases} 3x - 5y = -29 \\ 3x + 4y = 34 \end{cases}$ 26. $\begin{cases} 3x - 5y = 16 \\ 4x + 5y = 33 \end{cases}$

27. $\begin{cases} 2a - 3b = -6 \\ 2a - 3b = 8 \end{cases}$ 28. $\begin{cases} 3a - 4b = 6 \\ 2(2b + 3) = 3a \end{cases}$

29. $\begin{cases} 8x - 4y = 18 \\ 3x - 2y = 8 \end{cases}$ 30. $\begin{cases} 4x + 6y = 5 \\ 8x - 9y = 3 \end{cases}$

31. $\begin{cases} 2x + y = 10 \\ 0.1x + 0.2y = 1.0 \end{cases}$ 32. $\begin{cases} 0.3x + 0.2y = 0 \\ 2x - 3y = -13 \end{cases}$

33. $\begin{cases} 2x - y = 16 \\ 0.03x + 0.02y = 0.03 \end{cases}$

34. $\begin{cases} -5y + 2x = 4 \\ -0.02y + 0.03x = 0.04 \end{cases}$

35. $\begin{cases} 6x + 3y = 0 \\ 5y = 2x + 12 \end{cases}$

36. $\begin{cases} 0 = 4x - 3y \\ 5x = 4y - 2 \end{cases}$

37. $\begin{cases} -2(x + 1) = 3y - 6 \\ 3(y + 2) = 10 - 2x \end{cases}$

38. $\begin{cases} 3x + 2y + 1 = 5 \\ 3(x - 1) = -2y - 4 \end{cases}$

39. $\begin{cases} 4(x + 1) = 17 - 3(y - 1) \\ 2(x + 2) + 3(y - 1) = 9 \end{cases}$

40. $\begin{cases} 5(x - 1) = 8 - 3(y + 2) \\ 4(x + 2) - 7 = 3(2 - y) \end{cases}$

41. $\begin{cases} \frac{3}{5}s + \frac{4}{5}t = 1 \\ -\frac{1}{4}s + \frac{3}{8}t = 1 \end{cases}$

42. $\begin{cases} \frac{1}{2}s - \frac{1}{4}t = 1 \\ \frac{1}{3}s + t = 3 \end{cases}$

43. $\begin{cases} \frac{3}{5}x + y = 1 \\ \frac{4}{5}x - y = -1 \end{cases}$

44. $\begin{cases} \frac{1}{2}x + \frac{4}{7}y = -1 \\ 5x - \frac{4}{5}y = -10 \end{cases}$

45. $\begin{cases} \frac{x}{2} - \frac{y}{3} = -2 \\ \frac{2x-3}{2} + \frac{6y+1}{3} = \frac{17}{6} \end{cases}$

46. $\begin{cases} \frac{x+2}{4} + \frac{y-1}{3} = \frac{1}{12} \\ \frac{x+4}{5} - \frac{y-2}{2} = \frac{5}{2} \end{cases}$

47. $\begin{cases} \frac{x-3}{2} + \frac{y+5}{3} = \frac{11}{6} \\ \frac{x+3}{3} - \frac{5}{12} = \frac{y+3}{4} \end{cases}$

48. $\begin{cases} \frac{x+2}{3} = \frac{3-y}{2} \\ \frac{x+3}{2} = \frac{2-y}{3} \end{cases}$

WRITING

49. Why is it usually to your advantage to write the equations of a system in general form before using the addition method to solve it?

50. How would you decide whether to use substitution or addition to solve a system of equations?

51. In this section, we discussed the addition method for solving a system of two equations. Some instructors call it the *elimination method*. Why do you think it would be known by this name?

52. Write a note to the student whose work is shown below, explaining his **error.**

Solve $\begin{cases} x + y = 1 \\ x - y = 5 \end{cases}$.

$$\begin{array}{r} x + y = 1 \\ + \quad x - y = 5 \\ \hline 2x \quad\quad = 6 \end{array}$$

$$\frac{2x}{2} = \frac{6}{2}$$

$\boxed{x = 3}$ —Done—

REVIEW

53. Solve $8(3x - 5) - 12 = 4(2x + 3)$.

54. Solve $3y + \dfrac{y + 2}{2} = \dfrac{2(y + 3)}{3} + 16$.

55. Simplify $x - x$.

56. Simplify $3.2m - 4.4 + 2.1m + 16$.

57. Find the area of a triangular-shaped sign with a base of 4 feet and a height of 3.75 feet.

58. Translate to mathematical symbols: *the product of the sum of x and y and the difference of x and y.*

59. What is 10 less than x?

60. Factor $6x^2 + 7x - 20$.

▶ 7.4   Applications of Systems of Equations

In this section, you will learn about

     Solving problems using two variables

Introduction   We have previously formed equations involving one variable to solve problems. In this section, we consider ways to solve problems by using two variables.

Solving Problems Using Two Variables

The following steps are helpful when solving problems involving two unknown quantities.

Problem-solving strategy

1. Read the problem several times and *analyze* the facts. Occasionally, a sketch, table, or diagram will help you visualize the facts of the problem.

2. Pick different variables to represent two unknown quantities. *Form* two equations involving each of the two variables. This will give a system of two equations in two variables.

3. *Solve* the system of equations using the most convenient method: graphing, substitution, or addition.

4. *State* the conclusion.

5. *Check* the result in the words of the problem.

EXAMPLE 1

Farming. A farmer raises wheat and soybeans on 215 acres. If he wants to plant 31 more acres in wheat than in soybeans, how many acres of each should he plant?

ANALYZE THE PROBLEM

A farmer plants two fields, one in wheat and one in soybeans. We know that the number of acres of wheat planted plus the number of acres of soybeans planted will equal a total of 215 acres.

FORM TWO EQUATIONS

If w represents the number of acres of wheat and s the number of acres of soybeans to be planted, we can form the two equations.

The number of acres planted in wheat	plus	the number of acres planted in soybeans	is	215 acres.
w	$+$	s	$=$	215

Since the farmer wants to plant 31 more acres in wheat than in soybeans, we have

The number of acres planted in wheat	less	the number of acres planted in soybeans	is	31 acres.
w	$-$	s	$=$	31

SOLVE THE SYSTEM

We can now solve the system

$$\begin{cases} \textbf{1.} & w + s = 215 \\ \textbf{2.} & w - s = 31 \end{cases}$$

using the addition method.

$$\begin{aligned} w + s &= 215 \\ \underline{w - s} &= \underline{31} \\ 2w &= 246 \\ w &= 123 \quad \text{Divide both sides by 2.} \end{aligned}$$

To find s, we substitute 123 for w in Equation 1.

$$\begin{aligned} w + s &= 215 \\ 123 + s &= 215 \quad \text{Substitute.} \\ s &= 92 \quad \text{Subtract 123 from both sides.} \end{aligned}$$

STATE THE CONCLUSION

The farmer should plant 123 acres of wheat and 92 acres of soybeans.

CHECK THE RESULT

The total acreage planted is $123 + 92$, or 215 acres. The area planted in wheat is 31 acres greater than that planted in soybeans, because $123 - 92 = 31$. The answers check. ■

EXAMPLE 2

Lawn care. An installer of underground irrigation systems wants to cut a 20-foot length of plastic tubing into two pieces. The longer piece is to be 2 feet longer than twice the shorter piece. Find the length of each piece.

FIGURE 7-11

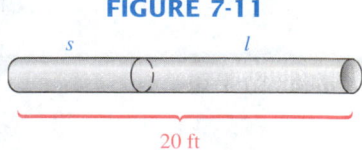

ANALYZE THE PROBLEM

Refer to Figure 7-11, which shows the pipe.

FORM TWO EQUATIONS

We can let s represent the length of the shorter piece and l the length of the longer piece. Then we can form the two equations.

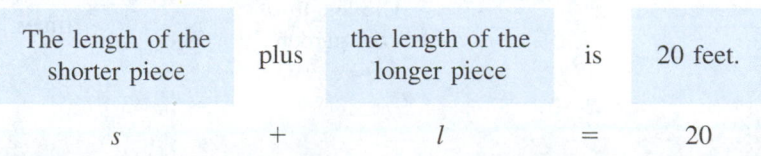

The length of the shorter piece	plus	the length of the longer piece	is	20 feet.
s	$+$	l	$=$	20

Since the longer piece is 2 feet longer than twice the shorter piece, we have

The length of the longer piece	is	2	times	the length of the shorter piece	plus	2 feet.
l	$=$	2	$\cdot$	s	$+$	2

SOLVE THE SYSTEM   We can use the substitution method to solve the system.

$$\textbf{1.} \quad \begin{cases} s + l = 20 \\ l = 2s + 2 \end{cases}$$
$$\textbf{2.}$$

$$s + (2s + 2) = 20 \quad \text{Substitute } 2s + 2 \text{ for } l \text{ in Equation 1.}$$
$$3s + 2 = 20 \quad \text{Combine like terms.}$$
$$3s = 18 \quad \text{Subtract 2 from both sides.}$$
$$s = 6 \quad \text{Divide both sides by 3.}$$

The shorter piece should be 6 feet long. To find the length of the longer piece, we substitute 6 for s in Equation 2 and find l.

$$l = 2s + 2$$
$$= 2(6) + 2 \quad \text{Substitute.}$$
$$= 12 + 2 \quad \text{Simplify.}$$
$$l = 14$$

STATE THE CONCLUSION   The longer piece should be 14 feet long, and the shorter piece 6 feet long.

CHECK THE RESULT   The sum of 6 and 14 is 20, and 14 is 2 more than twice 6. The answers check.   ∎

EXAMPLE 3   **Gardening.** Tom has 150 feet of fencing to enclose a rectangular garden. If the garden's length is to be 5 feet less than 3 times its width, find the area of the garden.

FIGURE 7-12

ANALYZE THE PROBLEM   To find the area of a rectangle, we need to know its length and width.

FORM TWO EQUATIONS   We can let l represent the length of the garden and w its width, as shown in Figure 7-12. Since the perimeter of a rectangle is two lengths plus two widths, we can form the two equations.

2	times	the length of the garden	plus	2	times	the width of the garden	is	150 feet.
2	$\cdot$	l	$+$	2	$\cdot$	w	$=$	150

Since the length is 5 feet less than 3 times the width,

The length of the garden	is	3	times	the width of the garden	minus	5 feet.
l	$=$	3	$\cdot$	w	$-$	5

SOLVE THE SYSTEM   We can use the substitution method to solve this system.

1. $\begin{cases} 2l + 2w = 150 \\ l = 3w - 5 \end{cases}$
2.

$2(3w - 5) + 2w = 150$ Substitute $3w - 5$ for l in Equation 1.

$6w - 10 + 2w = 150$ Remove parentheses.

$8w - 10 = 150$ Combine like terms.

$8w = 160$ Add 10 to both sides.

$w = 20$ Divide both sides by 8.

The width of the garden is 20 feet. To find the length, we substitute 20 for w in Equation 2 and simplify.

$l = 3w - 5$

$\quad = 3(20) - 5$ Substitute.

$\quad = 60 - 5$

$l = 55$

Now we find the area of the rectangle with dimensions 55 feet by 20 feet.

$A = l \cdot w$ The formula for the area of a rectangle.

$\quad = 55 \cdot 20$ Substitute 55 for l and 20 for w.

$A = 1{,}100$

STATE THE CONCLUSION   The garden covers an area of 1,100 square feet.

CHECK THE RESULT   Because the dimensions of the garden are 55 feet by 20 feet, the perimeter is

$P = 2l + 2w$

$\quad = 2(55) + 2(20)$ Substitute for l and w.

$\quad = 110 + 40$

$P = 150$

It is also true that 55 feet is 5 feet less than 3 times 20 feet. The answers check. ∎

EXAMPLE 4   **Manufacturing.**   The setup cost of a machine that mills brass plates is $750. After setup, it costs $0.25 to mill each plate. Management is considering the purchase of a larger machine that can produce the same plate at a cost of $0.20 per plate. If the setup cost of the larger machine is $1,200, how many plates would the company have to produce to make the purchase worthwhile?

ANALYZE THE PROBLEM   We need to find the number of plates (called the **break point**) that will cost equal amounts to produce on either machine.

FORM TWO EQUATIONS   We can let c represent the cost of milling p plates. If we call the machine currently being used machine 1, and the new, larger one machine 2, we can form the two equations.

The cost of making p plates on machine 1	is	the setup cost of machine 1	plus	the cost per plate on machine 1	times	the number of plates p to be made.
c	$=$	750	$+$	0.25	$\cdot$	p

The cost of making p plates on machine 2	is	the setup cost of machine 2	plus	the cost per plate on machine 2	times	the number of plates p to be made.
c	$=$	1,200	$+$	0.20	$\cdot$	p

SOLVE THE SYSTEM   Since the costs are equal, we can use the substitution method to solve the system

1. $\begin{cases} c = \mathbf{750 + 0.25p} \\ c = 1{,}200 + 0.20p \end{cases}$
2.

$750 + 0.25p = 1{,}200 + 0.20p$ Substitute $750 + 0.25p$ for c in the second equation.

$\qquad 0.25p = 450 + 0.20p$ Subtract 750 from both sides.

$\qquad 0.05p = 450$ Subtract $0.20p$ from both sides.

$\qquad\qquad p = 9{,}000$ Divide both sides by 0.05.

STATE THE CONCLUSION   If 9,000 plates are milled, the cost will be the same on either machine. If more than 9,000 plates are milled, the cost will be cheaper on the larger machine, because it mills the plates less expensively than the smaller machine.

CHECK THE RESULT   We check the solution by substituting 9,000 for p in Equations 1 and 2 and verifying that 3,000 is the value of c in both cases.

If we graph the two equations, we can illustrate the break point. (See Figure 7-13.)

FIGURE 7-13

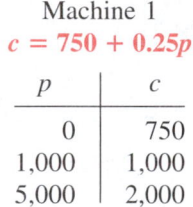

Machine 1
$c = \mathbf{750 + 0.25p}$

p	c
0	750
1,000	1,000
5,000	2,000

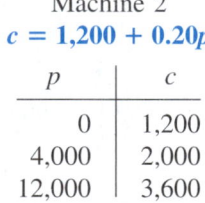

Machine 2
$c = \mathbf{1{,}200 + 0.20p}$

p	c
0	1,200
4,000	2,000
12,000	3,600

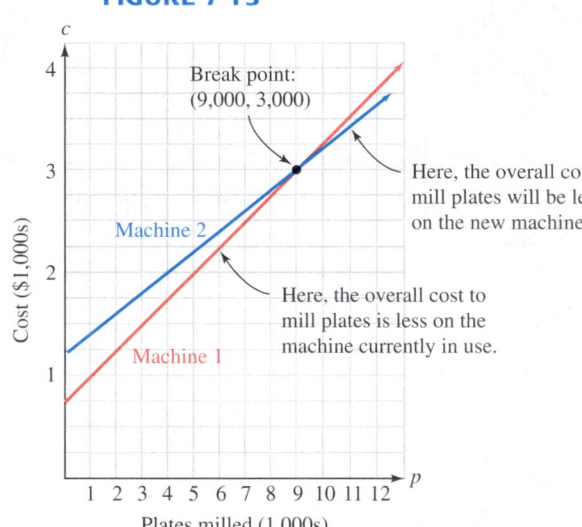

EXAMPLE 5

White-collar crime. Federal investigators discovered that a company secretly moved $150,000 out of the country to avoid paying corporate income tax on it. Some of the money was invested in a Swiss bank account that paid 8% interest annually. The remainder was deposited in a Cayman Islands account, paying 7% annual interest. The investigation also revealed that the combined interest earned the first year was $11,500. How much money was invested in each account?

ANALYZE THE PROBLEM   We are told that an unknown part of the $150,000 was invested at an annual rate of 8% and the rest at 7%. Together, the accounts earned $11,500 in interest.

FORM TWO EQUATIONS   We can let x represent the amount invested in the Swiss bank account and y represent the amount invested in the Cayman Islands account. Because the total investment was $150,000, we have

The amount invested in the Swiss account	+	the amount invested in the Cayman Is. account	is	$150,000.
x	+	y	=	150,000

Since the annual income on x dollars invested at 8% is $0.08x$, the income on y dollars invested at 7% is $0.07y$, and the combined income is $11,500, we have

The income on the 8% investment	+	the income on the 7% investment	is	$11,500.
$0.08x$	+	$0.07y$	=	11,500

The resulting system is

1. $\begin{cases} x + y = 150{,}000 \\ 0.08x + 0.07y = 11{,}500 \end{cases}$
2.

SOLVE THE SYSTEM To solve the system, we use the addition method to eliminate x.

$$\begin{aligned} -8x - 8y &= -1{,}200{,}000 \quad \text{Multiply both sides of Equation 1 by } -8. \\ 8x + 7y &= 1{,}150{,}000 \quad \text{Multiply both sides of Equation 2 by } 100. \\ \hline -y &= -50{,}000 \\ y &= 50{,}000 \quad \text{Multiply (or divide) both sides by } -1. \end{aligned}$$

To find x, we substitute 50,000 for y in Equation 1 and simplify.

$$\begin{aligned} x + y &= 150{,}000 \\ x + \mathbf{50{,}000} &= 150{,}000 \quad \text{Substitute.} \\ x &= 100{,}000 \quad \text{Subtract 50,000 from both sides.} \end{aligned}$$

STATE THE CONCLUSION $100,000 was invested in the Swiss bank account, and $50,000 was invested in the Cayman Islands account.

CHECK THE RESULT

$$\begin{aligned} \$100{,}000 + \$50{,}000 &= \$150{,}000 \quad \text{The two investments total } \$150{,}000. \\ 0.08(\$100{,}000) &= \$8{,}000 \quad \text{The Swiss bank account earned } \$8{,}000. \\ 0.07(\$50{,}000) &= \$3{,}500 \quad \text{The Cayman Islands account earned } \$3{,}500. \end{aligned}$$

The combined interest is $8,000 + $3,500 = $11,500. The answers check. ∎

EXAMPLE 6

Boating. A boat traveled 30 kilometers downstream in 3 hours and made the return trip in 5 hours. Find the speed of the boat in still water.

ANALYZE THE PROBLEM Traveling downstream, the speed of the boat will be faster than it would be in still water. Traveling upstream, the speed of the boat will be less than it would be in still water.

FORM TWO EQUATIONS We can let s represent the speed of the boat in still water and c the speed of the current. Then the rate of the boat going downstream is $s + c$, and its rate going upstream is $s - c$. We can organize the information as shown in Figure 7-14.

FIGURE 7-14

	Rate	·	Time	=	Distance
Downstream	$s + c$		3		30
Upstream	$s - c$		5		30

Because $d = r \cdot t$, the information in the table gives two equations in two variables.

$$\begin{cases} 3(s + c) = 30 \\ 5(s - c) = 30 \end{cases}$$

After removing parentheses, we have

1. $\begin{cases} 3s + 3c = 30 \end{cases}$
2. $\begin{cases} 5s - 5c = 30 \end{cases}$

SOLVE THE SYSTEM   To solve this system by addition, we multiply Equation 1 by 5, multiply Equation 2 by 3, add the equations, and solve for s.

$$\begin{array}{r} 15s + 15c = 150 \\ 15s - 15c = 90 \\ \hline 30s = 240 \end{array}$$

$$s = 8 \qquad \text{Divide both sides by 30.}$$

STATE THE CONCLUSION   The speed of the boat in still water is 8 kilometers per hour.

CHECK THE RESULT   We leave the check to the reader. ■

EXAMPLE 7

Medical technology. A laboratory technician has one batch of antiseptic that is 40% alcohol and a second batch that is 60% alcohol. She would like to make 8 liters of solution that is 55% alcohol. How many liters of each batch should she use?

ANALYZE THE PROBLEM   Some 60% alcohol solution must be added to some 40% alcohol solution to make a 55% alcohol solution.

FORM TWO EQUATIONS   We can let x represent the number of liters to be used from batch 1 and y the number of liters to be used from batch 2. We then organize the information as shown in Figure 7-15.

FIGURE 7-15

	Fractional part that is alcohol	Number of liters of solution	Number of liters of alcohol
Batch 1	0.40	x	0.40x
Batch 2	0.60	y	0.60y
Mixture	0.55	8	0.55(8)

One equation comes from information in this column.

Another equation comes from information in this column.

The information in Figure 7-15 provides two equations.

1. $\begin{cases} x + y = 8 & \text{The number of liters of batch 1 plus the number of liters of batch 2 equals the total number of liters in the mixture.} \end{cases}$
2. $\begin{cases} 0.40x + 0.60y = 0.55(8) & \text{The amount of alcohol in batch 1 plus the amount of alcohol in batch 2 equals the amount of alcohol in the mixture.} \end{cases}$

SOLVE THE SYSTEM   We can use addition to solve this system.

$$\begin{array}{rl} -40x - 40y = -320 & \text{Multiply both sides of Equation 1 by } -40. \\ 40x + 60y = 440 & \text{Multiply both sides of Equation 2 by 100.} \\ \hline 20y = 120 \end{array}$$

$$y = 6 \qquad \text{Divide both sides by 20.}$$

To find x, we substitute 6 for y in Equation 1 and simplify.

$$x + y = 8$$
$$x + 6 = 8 \qquad \text{Substitute.}$$
$$x = 2 \qquad \text{Subtract 6 from both sides.}$$

STATE THE CONCLUSION   The technician should use 2 liters of the 40% solution and 6 liters of the 60% solution.

CHECK THE RESULT   The check is left to the reader. ■

VOCABULARY

In Exercises 1–4, fill in the blanks to make the statements true.

1. A _____ is a letter that stands for a number.

2. An _____ is a statement indicating that two quantities are equal.

3. $\begin{cases} a + b = 20 \\ a = 2b + 4 \end{cases}$ is a _____ of linear equations.

4. A _____ of a system of linear equations satisfies both equations simultaneously.

CONCEPTS

5. For each case in Illustration 1, write an algebraic expression that represents the speed of the canoe in miles per hour if its speed in still water is x miles per hour.

ILLUSTRATION 1

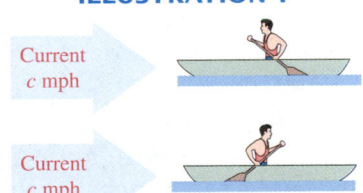

Current
c mph

Current
c mph

6. See Illustration 2.

a. If the contents of the two test tubes are poured into a third test tube, how much solution will the third test tube contain?

b. Which is the best estimate of the concentration of the solution in the third test tube— 25%, 35%, or 45% acid solution?

ILLUSTRATION 2

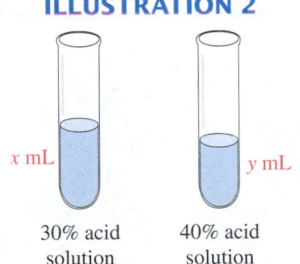

x mL y mL

30% acid   40% acid
solution   solution

7. Use the information in the table to answer the questions about two investments.

	Principal ·	Rate ·	Time =	Interest
City Bank	x	5%	1 yr	
USA Savings	y	11%	1 yr	

a. How much money was deposited in the USA Savings account?

b. What interest rate did the City Bank account earn?

c. Complete the table.

8. Use the information in the table to answer the questions about a plane flying in windy conditions.

	Rate ·	Time =	Distance
With	$x + y$	3 hr	450 mi
Against	$x - y$	5 hr	450 mi

a. For how long did the plane fly against the wind?

b. At what rate did the plane travel when flying with the wind?

c. Write two equations that could be used to solve for x and y.

9. a. If a problem contains two unknowns, and if two variables are used to represent them, how many equations must be written to find the unknowns?

b. Name three methods that can be used to solve a system of linear equations.

10. Put the steps of the five-step problem-solving strategy listed below in the correct order.

State the conclusion   Form two equations
Analyze the problem   Check the result
Solve the system

NOTATION

In Exercises 11–14, write a formula that relates the given quantities.

11. length, width, area of a rectangle

12. length, width, perimeter of a rectangle

13. rate, time, distance traveled

14. principal, rate, time, interest earned

In Exercises 15–16, translate each verbal model into mathematical symbols. Use variables to represent any unknowns.

15. $2 \cdot \boxed{\text{length of pool}} + 2 \cdot \boxed{\text{width of pool}}$ is $\boxed{90}$ yards.

16. $6 · [number of adults] + $2 · [number of children] is $26.

PRACTICE

In Exercises 17–20, use two equations in two variables to find the integers.

17. One integer is twice another. Their sum is 96.

18. The sum of two integers is 38. Their difference is 12.

19. Three times one integer plus another integer is 29. The first integer plus twice the second is 18.

20. Twice one integer plus another integer is 21. The first integer plus 3 times the second is 33.

APPLICATIONS

In Exercises 21–46, use two equations in two variables to solve each problem.

21. TREE TRIMMING When fully extended, the arm on the tree service truck shown in Illustration 3 is 51 feet long. If the upper part of the arm is 7 feet shorter than the lower part, how long is each part of the arm?

ILLUSTRATION 3

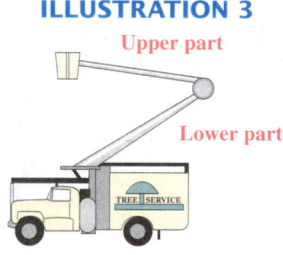

Upper part

Lower part

22. TV PROGRAMMING The producer of a 30-minute TV documentary about World War I divided it into two parts. Four times as much program time was devoted to the causes of the war as to the outcome. How long is each part of the documentary?

23. EXECUTIVE BRANCH The salaries of the president and vice president of the United States total $371,500 a year. If the president makes $28,500 more than the vice president, find each of their salaries.

24. CAUSES OF DEATH In 1993, the number of Americans dying from cancer was 6 times the number who died from accidents. If the number of deaths from these two causes totaled 630,000, how many Americans died from each cause?

25. BUYING PAINTING SUPPLIES Two partial receipts for paint supplies are shown in Illustration 4. How much does each gallon of paint and each brush cost?

ILLUSTRATION 4

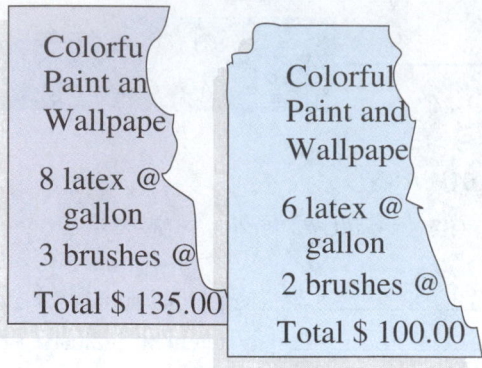

Colorfu Paint an Wallpape
8 latex @ gallon
3 brushes @
Total $ 135.00

Colorful Paint and Wallpape
6 latex @ gallon
2 brushes @
Total $ 100.00

26. WEDDING PICTURES A photographer sells the two wedding picture packages shown in Illustration 5.

ILLUSTRATION 5

Package #1	Package #2
1 10 x 14	1 10 x 14
10 8 x 10	5 8 x 10
color photos	color photos
Cost $239.50	Cost $134.50

How much does a 10×14 photo cost? An 8×10 photo?

27. BUYING TICKETS If receipts for the movie advertised in Illustration 6 were $720 for an audience of 190 people, how many senior citizens attended?

ILLUSTRATION 6

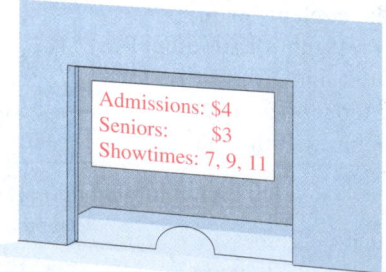

Admissions: $4
Seniors: $3
Showtimes: 7, 9, 11

28. SELLING ICE CREAM At a store, ice cream cones cost $0.90 and sundaes cost $1.65. One day the receipts for a total of 148 cones and sundaes were $180.45. How many cones were sold?

29. MARINE CORPS The Marine Corps War Memorial in Arlington, Virginia, portrays the raising of the U.S. flag on Iwo Jima during World War II. Find the two angles shown in Illustration 7 if the measure of one of the angles is 15° less than twice the other.

ILLUSTRATION 7

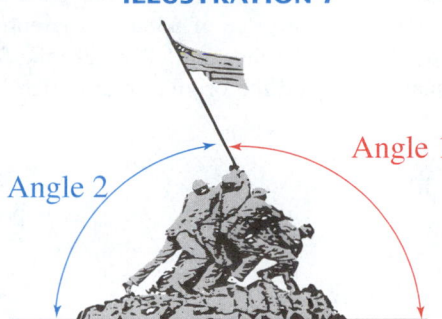

Angle 1

Angle 2

30. PHYSICAL THERAPY To rehabilitate her knee, an athlete does leg extensions. Her goal is to regain a full 90° range of motion in this exercise. Use the information in Illustration 8 to determine her current range of motion in degrees.

ILLUSTRATION 8

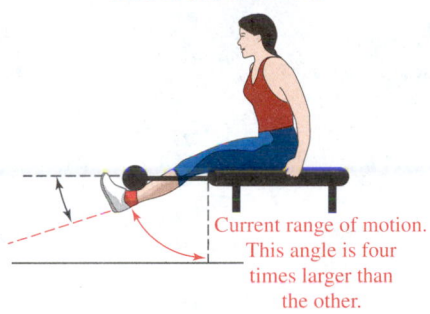

Current range of motion. This angle is four times larger than the other.

31. THEATER SCREEN At an IMAX theater, the giant rectangular movie screen has a width 26 feet less than its length. If its perimeter is 332 feet, find the area of the screen.

32. GEOMETRY A 50-meter path surrounds the rectangular garden shown in Illustration 9. The width of the garden is two-thirds its length. Find its area.

ILLUSTRATION 9

33. MAKING TIRES A company has two molds to form tires. One mold has a setup cost of $1,000, and the other has a setup cost of $3,000. The cost to make each tire with the first mold is $15, and the cost to make each tire with the second mold is $10.
a. Find the break point.
b. Check your result by graphing both equations on the coordinate system in Illustration 10.

ILLUSTRATION 10

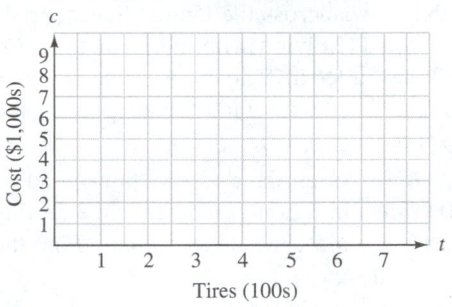

Cost ($1,000s)

Tires (100s)

c. If a production run of 500 tires is planned, determine which mold should be used.

34. CHOOSING A FURNACE A high-efficiency 90+ furnace can be purchased for $2,250 and costs an average of $412 per year to operate in Rockford, Illinois. An 80+ furnace can be purchased for only $1,715, but it costs $466 per year to operate.
a. Find the break point.
b. If you intended to live in a Rockford house for 7 years, which furnace would you choose?

35. STUDENT LOANS A college used a $5,000 gift from an alumnus to make two student loans. The first was at 5% annual interest to a nursing student. The second was at 7% to a business major. If the college collected $310 in interest the first year, how much was loaned to each student?

36. FINANCIAL PLANNING In investing $6,000 of a couple's money, a financial planner put some of it into a savings account paying 6% annual interest. The rest was invested in a riskier mini-mall development plan paying 12% annually. The combined interest earned for the first year was $540. How much money was invested at each rate?

37. GULF STREAM The Gulf Stream is a warm ocean current of the North Atlantic Ocean that flows northward, as shown in Illustration 11. Heading north with the Gulf Stream, a cruise ship traveled 300 miles in 10 hours. Against the current, it took 15 hours to make the return trip. Find the speed of the current.

ILLUSTRATION 11

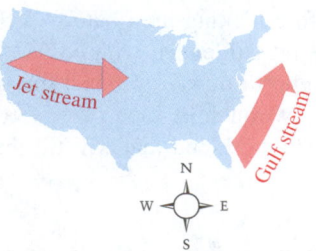

Jet stream

Gulf stream

N

W   E

S

38. JET STREAM The jet stream is a strong wind current that flows across the United States, as shown in Illustration 11. Flying with the jet stream, a plane flew 3,000 miles in 5 hours. Against the same wind, the trip took 6 hours. Find the airspeed of the plane (the speed in still air).

39. AVIATION An airplane can fly downwind a distance of 600 miles in 2 hours. However, the return trip against the same wind takes 3 hours. Find the speed of the wind.

40. BOATING A boat can travel 24 miles downstream in 2 hours and can make the return trip in 3 hours. Find the speed of the boat in still water.

41. MARINE BIOLOGY A marine biologist wants to set up an aquarium containing 3% salt water. He has two tanks on hand that contain 6% and 2% salt water. How much water from each tank must he use to fill a 16-liter aquarium with a 3% saltwater mixture?

42. COMMEMORATIVE COINS A foundry has been commissioned to make souvenir coins. The coins are to be made from an alloy that is 40% silver. The foundry has on hand two alloys, one with 50% silver content and one with a 25% silver content. How many kilograms of each alloy should be used to make 20 kilograms of the 40% silver alloy?

43. MIXING NUTS A merchant wants to mix peanuts with cashews, as shown in Illustration 12, to get 48 pounds of mixed nuts that will be sold at $4 per pound. How many pounds of each should the merchant use?

ILLUSTRATION 12

44. COFFEE SALES A coffee supply store waits until the orders for its special coffee blend reach 100 pounds before making up a batch. Coffee selling for $8.75 a pound is blended with coffee selling for $3.75 a pound to make a product that sells for $6.35 a pound. How much of each type of coffee should be used to make the blend that will fill the orders?

45. MARKDOWN A set of golf clubs has been marked down 40% to a sale price of $384. Let r represent the retail price and d the discount. Then use the following equations to find the original retail price.

| Retail price | $-$ | discount | $=$ | sale price |

| Discount | $=$ | discount rate | $\cdot$ | retail price |

46. MARKUP A stereo system retailing at $565.50 has been marked up 45% from wholesale. Let w represent the wholesale cost and m the markup. Then use the following equations to find the wholesale cost.

| Wholesale cost | $+$ | markup | $=$ | retail price |

| Markup | $=$ | markup rate | $\cdot$ | wholesale cost |

WRITING

47. When solving a problem using two variables, why isn't one equation sufficient to find the two unknown quantities?

48. Describe an everyday situation in which you might need to make a mixture.

REVIEW

In Exercises 49–52, graph each inequality.

49. $x < 4$　　　　　　**50.** $x \geq -3$

51. $-1 < x \leq 2$　　　　**52.** $-2 \leq x \leq 0$

In Exercises 53–56, solve each equation. If an answer is not exact, round to the nearest hundredth.

53. $x^2 - 4 = 0$　　　　**54.** $x^2 - 4x = 0$

55. $x^2 - 4x + 4 = 0$　　**56.** $x^2 - 4x - 2 = 0$

▶ 7.5   Graphing Linear Inequalities

In this section, you will learn about

- Solutions of linear inequalities ■ Graphing linear inequalities
- An application of linear inequalities

Introduction   We have seen that the solutions of a linear *equation* in x and y can be expressed as ordered pairs (x, y) and that when graphed, the ordered pairs form a line. In this section, we consider linear *inequalities*. Solutions of linear inequalities can also be expressed as ordered pairs and graphed.

Solutions of Linear Inequalities

A linear equation in x and y is an equation that can be written in the form $Ax + By = C$. A **linear inequality** in x and y is an inequality that can be written in one of four forms:

$$Ax + By > C, \qquad Ax + By < C, \qquad Ax + By \geq C, \qquad \text{or} \qquad Ax + By \leq C$$

where A, B, and C are real numbers and A and B are not both zero. Some examples of linear inequalities are

$$2x - y > -3, \qquad y < 3, \qquad x + 4y \geq 6, \qquad \text{and} \qquad x \leq -2$$

As with linear equations, an ordered pair (x, y) is a solution of an inequality in x and y if a true statement results when the variables in the inequality are replaced by the coordinates of the ordered pair.

EXAMPLE 1

Verifying a solution. Determine whether each ordered pair is a solution of $x - y \leq 5$. Then graph each solution: **a.** $(4, 2)$, **b.** $(0, -6)$, and **c.** $(1, -4)$.

Solution   In each case, we substitute the x-coordinate for x and the y-coordinate for y in the inequality $x - y \leq 5$. If the ordered pair is a solution, a true statement will be obtained.

a. For $(4, 2)$:

$$x - y \leq 5 \qquad \text{The original inequality.}$$
$$4 - 2 \leq 5 \qquad \text{Replace } x \text{ with 4 and } y \text{ with 2.}$$
$$2 \leq 5 \qquad \text{True.}$$

Because $2 \leq 5$ is true, $(4, 2)$ is a solution of the inequality, and we graph it in Figure 7-16.

b. For $(0, -6)$:

$$x - y \leq 5 \qquad \text{The original inequality.}$$
$$0 - (-6) \leq 5 \qquad \text{Replace } x \text{ with 0 and } y \text{ with } -6.$$
$$6 \leq 5 \qquad \text{False.}$$

Because $6 \leq 5$ is false, $(0, -6)$ is not a solution.

c. For $(1, -4)$:

$$x - y \leq 5 \qquad \text{The original inequality.}$$
$$1 - (-4) \leq 5 \qquad \text{Replace } x \text{ with 1 and } y \text{ with } -4.$$
$$5 \leq 5 \qquad \text{True.}$$

Because $5 \leq 5$ is true, $(1, -4)$ is a solution, and we graph it in Figure 7-16.

FIGURE 7-16

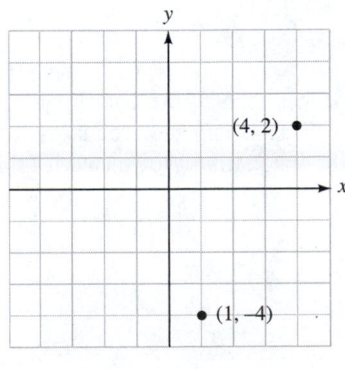

Using the inequality in Example 1, determine whether each ordered pair is a solution. If it is, graph that solution on the coordinate system in Figure 7-16.   **a.** $(8, 2)$,   **b.** $(4, -1)$,   **c.** $(-2, 4)$,   and   **d.** $(-3, -5)$.

Answers: **a.** not a solution,   **b.** solution,   **c.** solution,   **d.** solution   ■

The graph in Figure 7-16 contains several solutions of the inequality $x - y \leq 5$. Intuition tells us that there are many more ordered pairs (x, y) such that $x - y$ is less than or equal to 5. How then do we get a complete graph of the solutions of $x - y \leq 5$? We address this question in the following discussion.

Graphing Linear Inequalities

The graph of $x - y = 5$ is a line consisting of the points whose coordinates satisfy the equation. The graph of the inequality $x - y \leq 5$ is not a line, but an area bounded by a line, called a **half-plane.** The half-plane consists of the points whose coordinates satisfy the inequality.

EXAMPLE 2

Graphing a linear inequality. Graph $x - y \leq 5$.

Solution   Since the inequality symbol $\leq$ includes an equals sign, the graph of $x - y \leq 5$ includes the graph of $x - y = 5$. So we begin by graphing the equation $x - y = 5$, as shown in Figure 7-17(a).

FIGURE 7-17

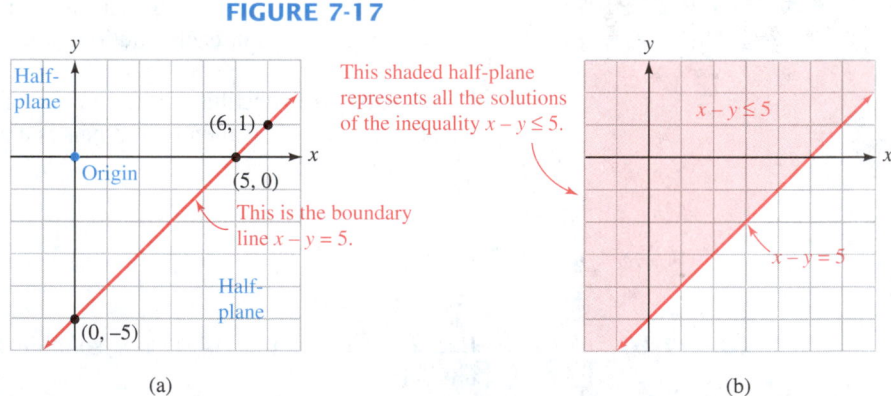

$x - y = 5$

x	y	(x, y)
0	−5	$(0, -5)$
5	0	$(5, 0)$
6	1	$(6, 1)$

(a)

(b)

Since the inequality $x - y \leq 5$ allows $x - y$ to be less than 5, the coordinates of points other than those shown on the line in Figure 7-17(a) satisfy the inequality. For example, the coordinates of the origin $(0, 0)$ satisfy the inequality. We can verify this by letting x and y be zero in the given inequality:

$x - y \leq 5$

$0 - 0 \leq 5$ Substitute 0 for x and 0 for y.

$0 \leq 5$ True.

Because $0 \leq 5$, the coordinates of the origin satisfy the original inequality. In fact, the coordinates of every point on the *same side* of the line as the origin satisfy the inequality. The graph of $x - y \leq 5$ is the half-plane that is shaded in Figure 7-17(b). Since the **boundary line** $x - y = 5$ is included, we draw it with a solid line.   ■

EXAMPLE 3

Graphing a linear inequality. Graph $2(x - 3) - (x - y) \geq -3$.

Solution   We begin by simplifying the inequality as follows:

$$2(x - 3) - (x - y) \geq -3$$

$$2x - 6 - x + y \geq -3 \quad \text{Remove the parentheses.}$$

$$x - 6 + y \geq -3 \quad \text{Combine like terms.}$$

$$x + y \geq 3 \quad \text{Add 6 to both sides.}$$

To graph the inequality $x + y \geq 3$, we graph the boundary line whose equation is $x + y = 3$. Since the graph of $x + y \geq 3$ includes the line $x + y = 3$, we draw the boundary with a solid line. Note that it divides the coordinate plane into two half-planes. See Figure 7-18(a).

To decide which half-plane to shade, we substitute the coordinates of some point that lies on one side of the boundary line into the inequality. If we use the origin $(0, 0)$ for the **test point,** we have

$$x + y \geq 3$$

$$\mathbf{0} + \mathbf{0} \geq 3 \quad \text{Substitute 0 for } x \text{ and 0 for } y.$$

$$0 \geq 3 \quad \text{False.}$$

Since $0 \geq 3$ is a false statement, the origin is not in the graph. In fact, the coordinates of *every* point on the origin's side of the boundary line will not satisfy the inequality. However, every point on the other side of the boundary line will satisfy the inequality. We shade that half-plane. The graph of $x + y \geq 3$ is the half-plane that appears in color in Figure 7-18(b).

FIGURE 7-18

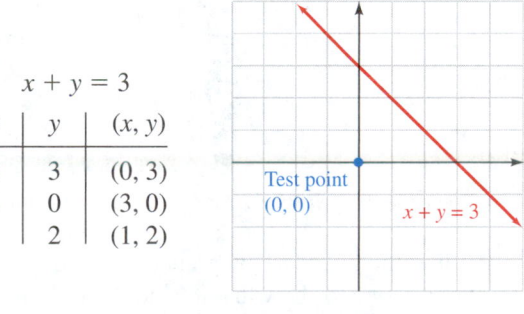

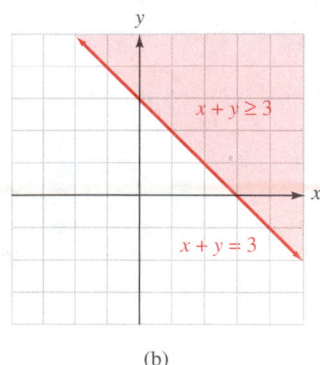

$x + y = 3$

x	y	(x, y)
0	3	$(0, 3)$
3	0	$(3, 0)$
1	2	$(1, 2)$

(a)   (b)

SELF CHECK   Graph $3(x - 1) - (2x + y) \leq -5$.   *Answer:*

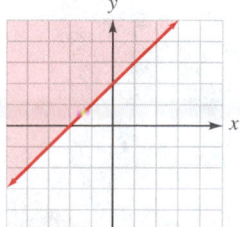

EXAMPLE 4

Graphing a linear inequality. Graph $y > 2x$.

Solution   To find the boundary line, we graph $y = 2x$. Since the symbol $>$ does not include an equals sign, the points on the graph of $y = 2x$ are not part of the graph of $y > 2x$. We draw the boundary line as a broken line to show this, as in Figure 7-19(a).

To determine which half-plane to shade, we substitute the coordinates of some point that lies on one side of the boundary line into $y > 2x$. Since the origin is on the boundary, we cannot use it as a test point. Point $T(2, 0)$, for example, is below the boundary line. See Figure 7-19(a). To see whether point $T(2, 0)$ satisfies $y > 2x$, we substitute 2 for x and 0 for y in the inequality.

$$y > 2x$$

$$\mathbf{0} > 2(\mathbf{2}) \quad \text{Substitute 2 for } x \text{ and 0 for } y.$$

$$0 > 4 \quad \text{False.}$$

Since $0 > 4$ is a false statement, the coordinates of point T do not satisfy the inequality, and point T is not on the side of the line we wish to shade. Instead, we shade the other side of the boundary line. The graph of the solution set of $y > 2x$ is shown in Figure 7-19(b).

FIGURE 7-19

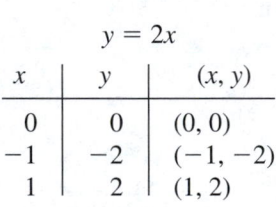

$y = 2x$

x	y	(x, y)
0	0	$(0, 0)$
-1	-2	$(-1, -2)$
1	2	$(1, 2)$

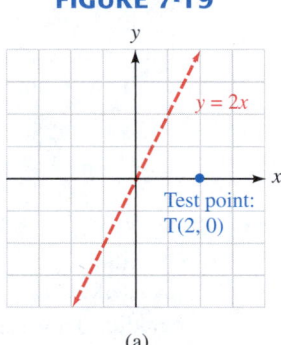

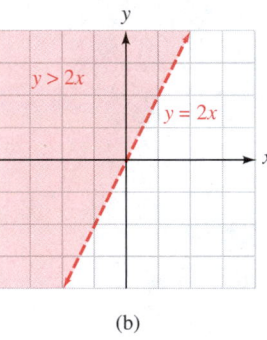

(a)   (b)

SELF CHECK   Graph $y < 3x$.   *Answer:*

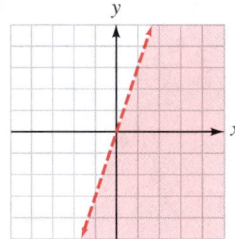

E X A M P L E   5   **Graphing a linear inequality.** Graph $x + 2y < 6$.

Solution   We find the boundary by graphing the equation $x + 2y = 6$. We draw the boundary as a broken line to show that it is not part of the solution. We then choose a test point not on the boundary and see whether its coordinates satisfy $x + 2y < 6$. The origin is a convenient choice.

$$x + 2y < 6$$

$$\mathbf{0} + 2(\mathbf{0}) < 6 \quad \text{Substitute 0 for } x \text{ and 0 for } y.$$

$$0 < 6 \quad \text{True.}$$

Since $0 < 6$ is a true statement, we shade the side of the line that includes the origin. The graph is shown in Figure 7-20.

SELF CHECK   Graph $2x - y < 4$.   *Answer:*

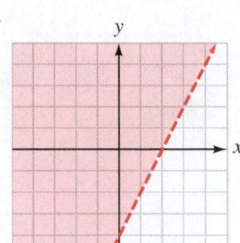

FIGURE 7-20

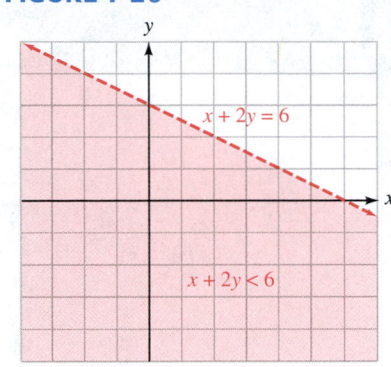

$x + 2y = 6$

x	y	(x, y)
0	3	$(0, 3)$
6	0	$(6, 0)$
4	1	$(4, 1)$

EXAMPLE 6

Graphing a linear inequality. Graph $y \geq 0$.

Solution   We find the boundary by graphing the equation $y = 0$. We draw the boundary as a solid line to show that it is part of the solution. We then choose a test point not on the boundary and see whether its coordinates satisfy $y \geq 0$. The point $T(0, 1)$ is a convenient choice.

$y \geq 0$

$1 \geq 0$ Substitute 1 for y.

Since $1 \geq 0$ is a true statement, we shade the side of the line that includes point T. The graph is shown in Figure 7-21.

FIGURE 7-21

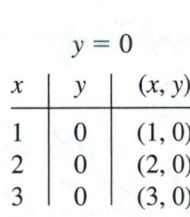

$y = 0$

x	y	(x, y)
1	0	$(1, 0)$
2	0	$(2, 0)$
3	0	$(3, 0)$

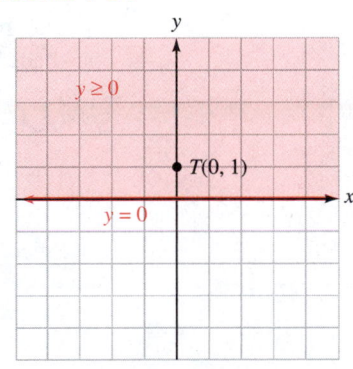

SELF CHECK   Graph $x \geq 2$.   *Answer:*

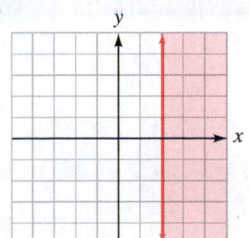

The following is a summary of the procedure for graphing linear inequalities.

Graphing linear inequalities in two variables

1. Graph the boundary line of the region. If the inequality allows the possibility of equality (the symbol is either ≤ or ≥), draw the boundary line as a solid line. If equality is not allowed (< or >), draw the boundary line as a broken line.

2. Pick a test point that is on one side of the boundary line. (Use the origin if possible.) Replace x and y in the inequality with the coordinates of that point. If the inequality is satisfied, shade the side that contains that point. If the inequality is not satisfied, shade the other side of the boundary.

An Application of Linear Inequalities

EXAMPLE 7

Earning money. Carlos has two part-time jobs, one paying $5 per hour and another paying $6 per hour. He must earn at least $120 per week to pay his expenses while attending college. Write an inequality that shows the various ways he can schedule his time to achieve his goal.

Solution   If we let x represent the number of hours Carlos works on the first job and y the number of hours he works on the second job, we have

The hourly rate on the first job	times	the hours worked on the first job	plus	the hourly rate on the second job	times	the hours worked on the second job	is at least	$120.
$5	·	x	+	6	·	y	≥	$120

The graph of the inequality $5x + 6y \geq 120$ is shown in Figure 7-22. Any point in the shaded region indicates a possible way Carlos can schedule his time and earn $120 or more per week. For example, if he works 20 hours on the first job and 10 hours on the second job, he will earn

$$\$5(20) + \$6(10) = \$100 + \$60$$
$$= \$160$$

Since Carlos cannot work a negative number of hours, a graph showing negative values of x or y would have no meaning. ■

FIGURE 7-22

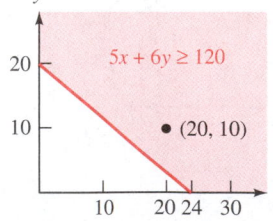

STUDY SET

Section 7.5

VOCABULARY

In Exercises 1–4, fill in the blanks to make the statements true.

1. $2x - y \leq 4$ is a linear _____ in x and y.

2. The symbol ≤ means _____ or _____.

3. In the graph in Illustration 1, the line $2x - y = 4$ is the _____.

4. In Illustration 1, the line $2x - y = 4$ divides the rectangular coordinate system into two _____.

ILLUSTRATION 1

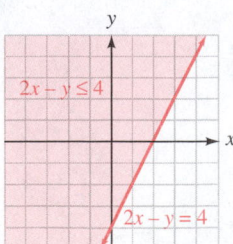

CONCEPTS

5. Tell whether each ordered pair is a solution of $5x - 3y \geq 0$.
   a. $(1, 1)$ **b.** $(-2, -3)$
   c. $(0, 0)$ **d.** $\left(\dfrac{1}{5}, \dfrac{4}{3}\right)$

6. Tell whether each ordered pair is a solution of $x + 4y < -1$.
   a. $(3, 1)$ **b.** $(-2, 0)$
   c. $(-0.5, 0.2)$ **d.** $\left(-2, \dfrac{1}{4}\right)$

7. Tell whether the graph of each linear inequality includes the boundary line.
   a. $y > -x$ **b.** $5x - 3y \leq -2$

8. If a false statement results when the coordinates of a test point are substituted into a linear inequality, which half-plane should be shaded to represent the solution of the inequality?

9. A linear inequality has been graphed in Illustration 2. Tell whether each point satisfies the inequality.

ILLUSTRATION 2

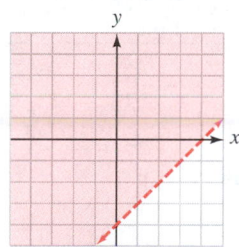

   a. $(1, -3)$
   b. $(-2, -1)$
   c. $(2, 3)$
   d. $(3, -4)$

10. A linear inequality has been graphed in Illustration 3. Tell whether each point satisfies the inequality.

ILLUSTRATION 3

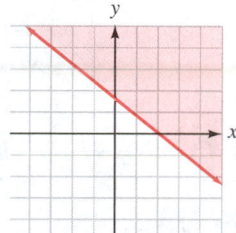

   a. $(2, 1)$
   b. $(-2, -4)$
   c. $(4, -2)$
   d. $(-3, 4)$

11. The boundary for the graph of a linear inequality is shown in Illustration 4. Why can't the origin be used as a test point to decide which side to shade?

ILLUSTRATION 4

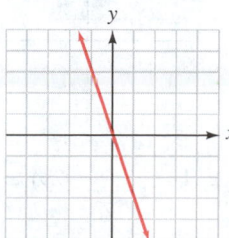

12. To decide how many pallets (x) and barrels (y) a delivery truck can hold, a dispatcher refers to the loading sheet in Illustration 5. Can a truck make a delivery of 4 pallets and 10 barrels in one trip?

ILLUSTRATION 5

Truck Loading Sheet
(acceptable load combinations)

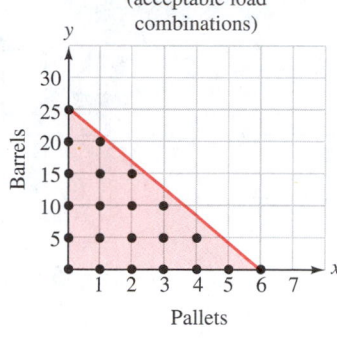

PRACTICE

In Exercises 13–20, complete the graph by shading the correct side of the boundary.

13. $y \leq x + 2$

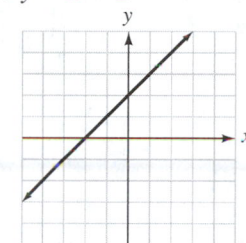

14. $y > x - 3$

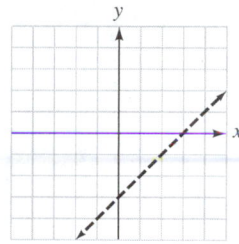

15. $y > 2x - 4$

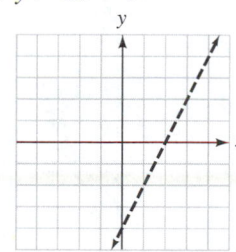

16. $y \leq -x + 1$

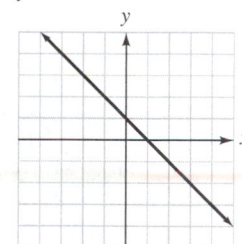

17. $x - 2y \geq 4$

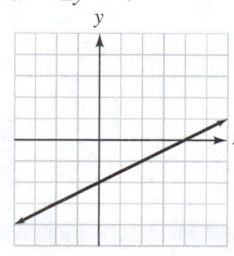

18. $3x + 2y > 12$

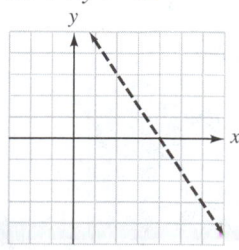

19. $y \leq 4x$

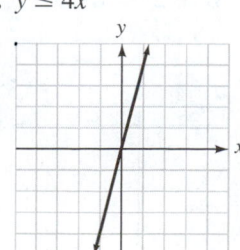

20. $y + 2x < 0$

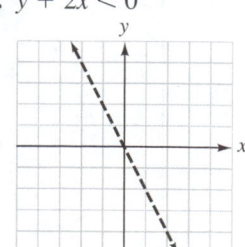

In Exercises 21–40, graph each inequality.

21. $y \geq 3 - x$

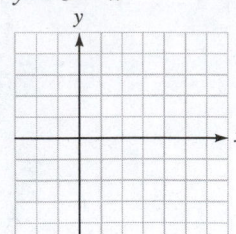

22. $y < 2 - x$

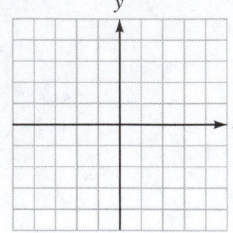

23. $y < 2 - 3x$

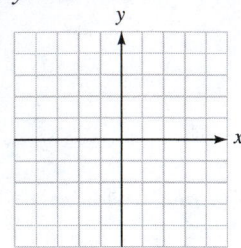

24. $y \geq 5 - 2x$

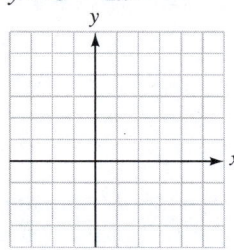

25. $y \geq 2x$

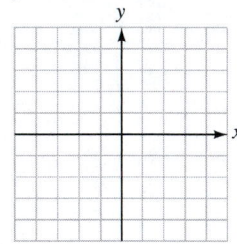

26. $y < 3x$

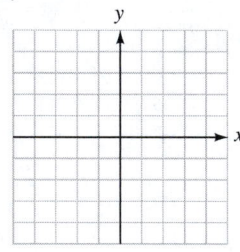

27. $2y - x < 8$

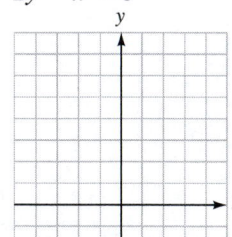

28. $y + 9x \geq 3$

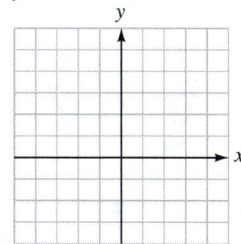

29. $y - x \geq 0$

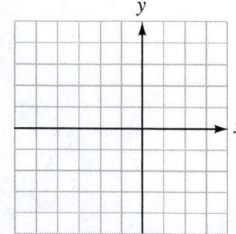

30. $y + x < 0$

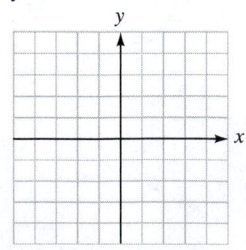

31. $2x + y > 2$

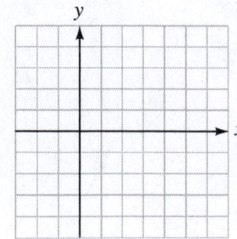

32. $3x - 2y > 6$

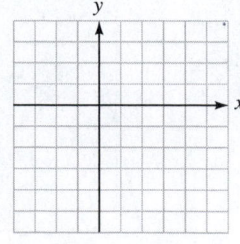

33. $3x - 4y > 12$

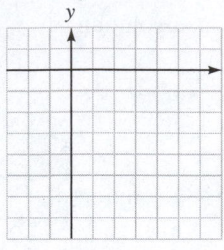

34. $4x + 3y \leq 12$

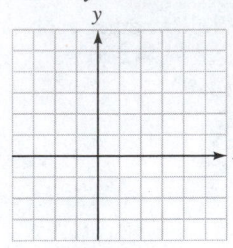

35. $5x + 4y \geq 20$

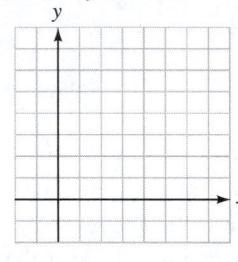

36. $7x - 2y < 21$

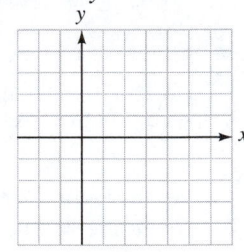

37. $x < 2$

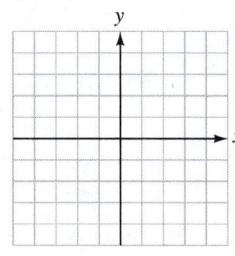

38. $y > -3$

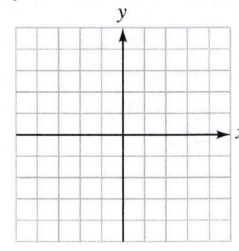

39. $y \leq 1$

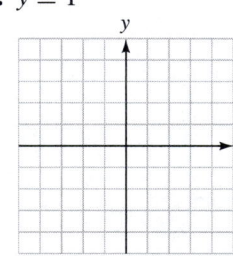

40. $x \geq -4$

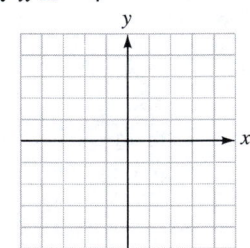

In Exercises 41–44, simplify each inequality and then graph it.

41. $3(x + y) + x < 6$

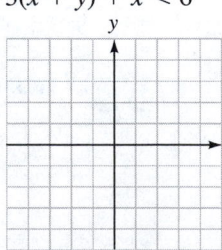

42. $2(x - y) - y \geq 4$

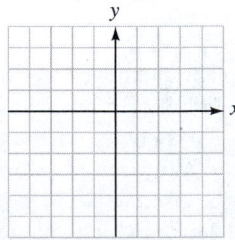

43. $4x - 3(x + 2y) \geq -6y$

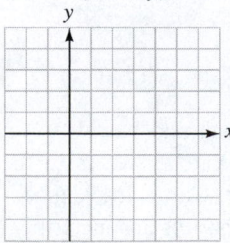

44. $3y + 2(x + y) < 5y$

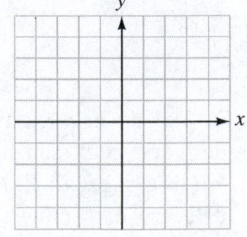

APPLICATIONS

45. NATO In March of 1999, NATO aircraft and cruise missiles targeted Serbian military forces that were south of the 44th parallel of north latitude in Yugoslavia, Montenegro, and Kosovo. See Illustration 6. Shade the geographic area that NATO was trying to rid of Serbian forces.

ILLUSTRATION 6

Based on data from *Los Angeles Times* (March 24, 1999)

46. U.S. HISTORY When he ran for president in 1844, the campaign slogan of James K. Polk was "54-40 or fight!" It meant that Polk was willing to fight Great Britain for the possession of the Oregon Territory north to the 54°40' parallel, as shown in Illustration 7. In 1846, Polk accepted a compromise to establish the 49th parallel as the permanent boundary of the United States. Shade the area of land that Polk conceded to the British.

ILLUSTRATION 7

In Exercises 47–52, write an inequality and graph it for nonnegative values of x and y. Then give three ordered pairs that satisfy the inequality.

47. PRODUCTION PLANNING It costs a bakery $3 to make a cake and $4 to make a pie. Production costs cannot exceed $120 per day. Use Illustration 8 to graph an inequality that shows the possible combinations of cakes (x) and pies (y) that can be made.

ILLUSTRATION 8

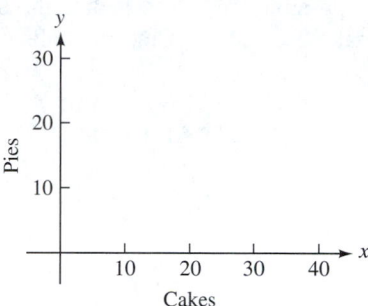

48. HIRING BABYSITTERS Mary has a choice of two babysitters. Sitter 1 charges $6 per hour, and sitter 2 charges $7 per hour. If Mary can afford no more than $42 per week for sitters, use Illustration 9 to graph an inequality that shows the possible ways that she can hire sitter 1 (x) and sitter 2 (y).

ILLUSTRATION 9

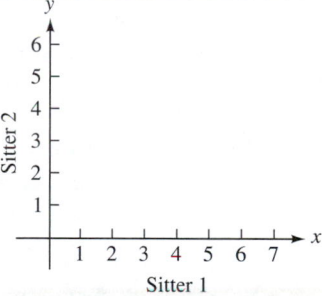

49. INVENTORY A clothing store advertises that it maintains an inventory of at least $4,400 worth of men's jackets. If a leather jacket costs $100 and a nylon jacket costs $88, use Illustration 10 to graph an inequality that shows the possible ways that leather jackets (x) and nylon jackets (y) can be stocked.

ILLUSTRATION 10

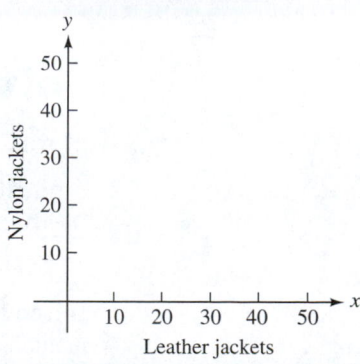

50. MAKING SPORTING GOODS To keep up with demand, a sporting goods manufacturer allocates at least 2,400 units of production time per day to make baseballs and footballs. If it takes 20 units of time to make a baseball and 30 units of time to make a football, use Illustration 11 to graph an inequality that shows the possible ways to schedule the production time to make baseballs (x) and footballs (y).

ILLUSTRATION 11

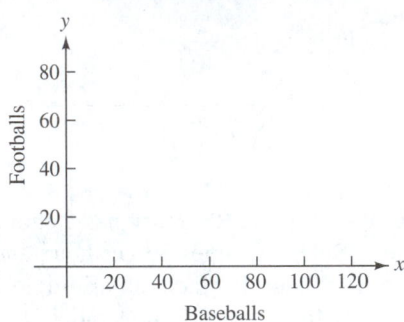

51. INVESTING IN STOCKS Robert has up to $8,000 to invest in two companies. If stock in Robotronics sells for $40 per share and stock in Macrocorp sells for $50 per share, use Illustration 12 to graph an inequality that shows the possible ways that he can buy shares of Robotronics (x) and Macrocorp (y).

ILLUSTRATION 12

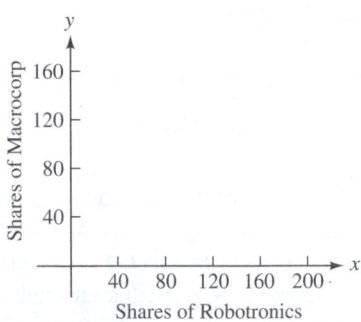

52. BUYING BASEBALL TICKETS Tickets to the Rockford Rox baseball games cost $6 for reserved

seats and $4 for general admission. If nightly receipts must average at least $10,200 to meet expenses, use Illustration 13 to graph an inequality that shows the possible ways that the Rox can sell reserved seats (x) and general admission tickets (y).

ILLUSTRATION 13

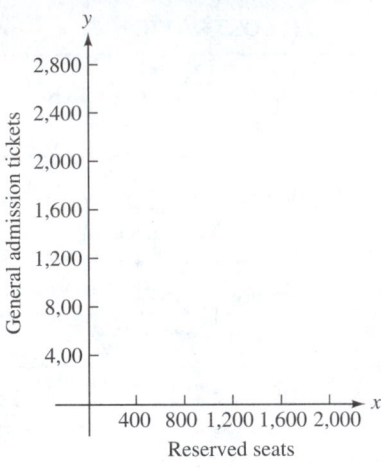

WRITING

53. Explain how to find the boundary for the graph of an inequality.

54. Explain how to decide which side of the boundary line to shade when graphing an inequality.

REVIEW

55. Let $g(x) = 3x^2 - 4x + 3$. Find $g(2)$.

56. Solve $2(x - 4) \le -12$.

57. Multiply $(x + 3)(x^2 - 2x + 4)$.

58. Factor $9p - 9q + mp - mq$.

59. Write a formula relating distance, rate, and time.

60. What is the slope of the line $2x - 3y = 2$?

61. Solve $A = P + Prt$ for t.

62. What is the sum of the measures of the three angles of any triangle?

▶ **7.6**　　　　　**Solving Systems of Linear Inequalities**

In this section, you will learn about

　　Systems of linear inequalities ■ An application of systems of linear inequalities

---

Introduction　We have previously solved systems of linear *equations* by the graphing method. The solution of such a system is the point of intersection of the straight lines.

We now consider how to solve systems of linear *inequalities* graphically. When the solution of a linear inequality is graphed, the result is a half-plane. Therefore, we would expect to find the graphical solution of a system of inequalities by looking for the intersection, or "overlap," of shaded half-planes.

Systems of Linear Inequalities

To solve the system

$$\begin{cases} x + y \geq 1 \\ x - y \geq 1 \end{cases}$$

we first graph each inequality. For instructional purposes, we will initially graph each inequality on a separate set of axes, although in practice we will draw them on the same axes.

The graph of $x + y \geq 1$ includes the graph of the equation $x + y = 1$ and all points above it. Because the boundary line is included, we draw it with a solid line, as shown in Figure 7-23(a).

The graph of $x - y \geq 1$ includes the graph of the equation $x - y = 1$ and all points below it. Because the boundary line is included here also, it is drawn with a solid line, as shown in Figure 7-23(b).

FIGURE 7-23

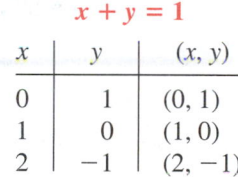

$x + y = 1$		
x	y	(x, y)
0	1	$(0, 1)$
1	0	$(1, 0)$
2	-1	$(2, -1)$

$x - y = 1$		
x	y	(x, y)
0	-1	$(0, -1)$
1	0	$(1, 0)$
2	1	$(2, 1)$

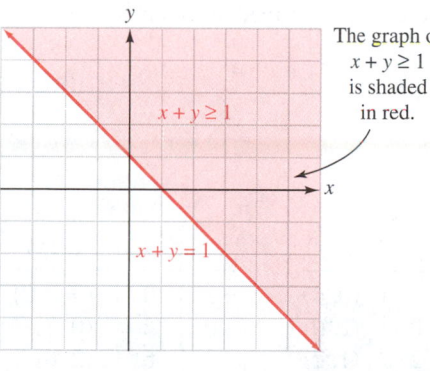

The graph of $x + y \geq 1$ is shaded in red.

(a)

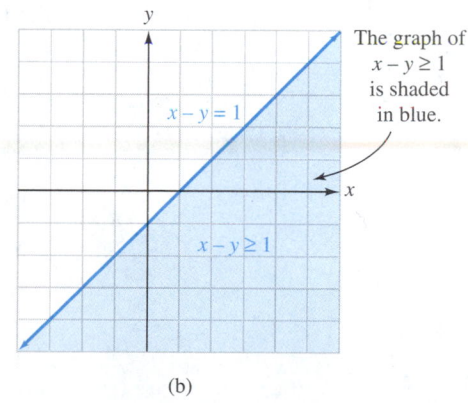

The graph of $x - y \geq 1$ is shaded in blue.

(b)

In Figure 7-24 on the next page, we show the result when the inequalities $x + y \geq 1$ and $x - y \geq 1$ are graphed one at a time on the same coordinate axes. The area that is shaded twice represents the set of simultaneous solutions of the given system of inequalities. Any point in the doubly shaded region has coordinates that satisfy both inequalities of the system.

To see whether this is true, we can pick a point, such as point A, that lies in the doubly shaded region and show that its coordinates satisfy both inequalities. Because point A has coordinates $(4, 1)$, we have

$$\begin{aligned} x + y &\geq 1 & \text{and} && x - y &\geq 1 \\ 4 + 1 &\geq 1 & && 4 - 1 &\geq 1 \\ 5 &\geq 1 & && 3 &\geq 1 \end{aligned}$$

FIGURE 7-24

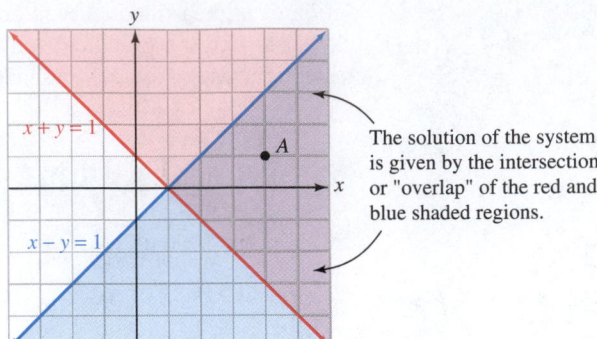

The solution of the system is given by the intersection or "overlap" of the red and blue shaded regions.

Since the coordinates of point A satisfy each equation, point A is a solution of the system. If we pick a point that is not in the doubly shaded region, its coordinates will fail to satisfy at least one of the inequalities.

In general, to solve systems of linear inequalities, we will follow these steps.

Solving systems of inequalities

1. Graph each inequality in the system on the same coordinate axes.
2. Find the region that is common to every graph.
3. Pick a test point from the region to verify the solution.

EXAMPLE 1

Solving systems of inequalities. Graph the solution of $\begin{cases} 2x + y < 4 \\ -2x + y > 2 \end{cases}$.

Solution   First, we graph each inequality on one set of axes, as shown in Figure 7-25.

FIGURE 7-25

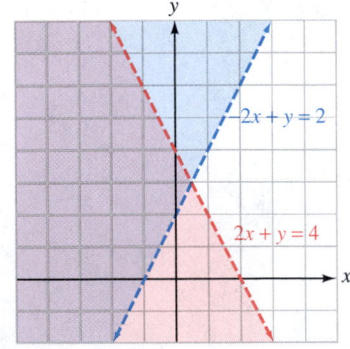

$2x + y = 4$				$-2x + y = 2$		
x	y	(x, y)		x	y	(x, y)
0	4	$(0, 4)$		-1	0	$(-1, 0)$
2	0	$(2, 0)$		0	2	$(0, 2)$
1	2	$(1, 2)$		2	6	$(2, 6)$

We note that

■ The graph of $2x + y < 4$ includes all points below the line $2x + y = 4$. Since the boundary is not included, we draw it as a broken line.

■ The graph of $-2x + y > 2$ includes all points above the line $-2x + y = 2$. Since the boundary is not included, we also draw it as a broken line.

The area that is shaded twice (the region in purple) is the solution of the given system of inequalities. Any point in the doubly shaded region has coordinates that will satisfy both inequalities of the system.

Pick a point in the doubly shaded region and show that it satisfies both inequalities.

SELF CHECK Graph the solution of $\begin{cases} x + 3y < 3 \\ -x + 3y > 3 \end{cases}$.

Answer:

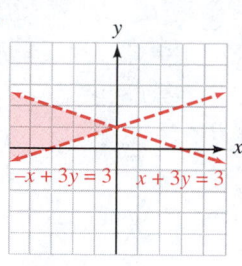

EXAMPLE 2

Solving systems of inequalities. Graph the solution of $\begin{cases} x \leq 2 \\ y > 3 \end{cases}$.

Solution We graph each inequality on one set of axes, as shown in Figure 7-26.

FIGURE 7-26

$x = 2$		
x	y	(x, y)
2	0	(2, 0)
2	2	(2, 2)
2	4	(2, 4)

$y = 3$		
x	y	(x, y)
0	3	(0, 3)
1	3	(1, 3)
4	3	(4, 3)

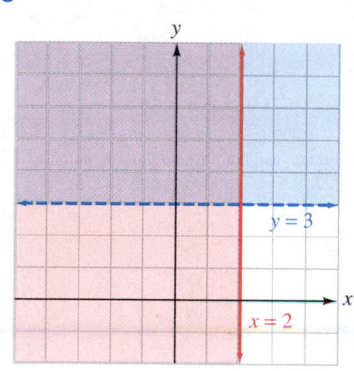

We note that

■ The graph of $x \leq 2$ includes all points to the left of the line $x = 2$. Since the boundary line is included, we draw it as a solid line.

■ The graph $y > 3$ includes all points above the line $y = 3$. Since the boundary is not included, we draw it as a broken line.

The area that is shaded twice is the solution of the given system of inequalities. Any point in the doubly shaded region (purple) has coordinates that will satisfy both inequalities of the system. Pick a point in the doubly shaded region and show that this is true.

SELF CHECK Graph the solution of $\begin{cases} y \leq 1 \\ x > 2 \end{cases}$.

Answer:

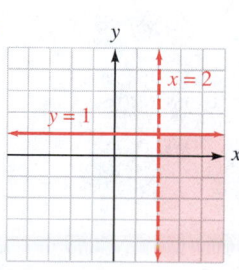

EXAMPLE 3

Solving systems of inequalities. Graph the solution of $\begin{cases} y < 3x - 1 \\ y \geq 3x + 1 \end{cases}$.

Solution We graph each inequality as shown in Figure 7-27 and make the following observations:

■ The graph of $y < 3x - 1$ includes all points below the broken line $y = 3x - 1$.

■ The graph of $y \geq 3x + 1$ includes all points on and above the solid line $y = 3x + 1$.

FIGURE 7-27

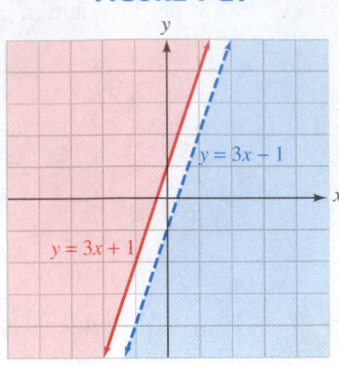

Because the graphs of these inequalities do not intersect, the solution set is empty. There are no solutions.

SELF CHECK   Graph the solution of $\begin{cases} y \ge -\frac{1}{2}x + 1 \\ y < -\frac{1}{2}x - 1 \end{cases}$.   *Answer:* no solutions

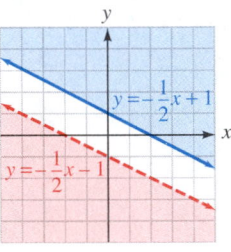

EXAMPLE 4   **Solving systems of inequalities.** Graph the solution of $\begin{cases} x \ge 0 \\ y \ge 0 \\ x + 2y \le 6 \end{cases}$.

Solution   We graph each inequality as shown in Figure 7-28 and make the following observations:

- The graph of $x \ge 0$ includes all points on the y-axis and to the right.
- The graph of $y \ge 0$ includes all points on the x-axis and above.
- The graph of $x + 2y \le 6$ includes all points on the line $x + 2y = 6$ and below.

The solution is the region that is shaded three times. This includes triangle OPQ and the triangular region it encloses.

FIGURE 7-28

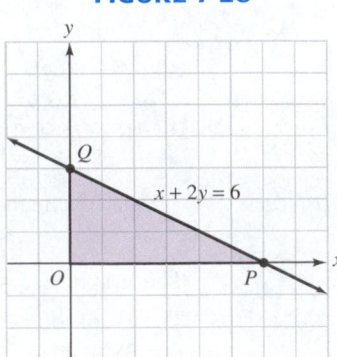

SELF CHECK Graph the solution of $\begin{cases} x \le 1 \\ y \le 2 \\ 2x - y \le 4 \end{cases}$. *Answer:*

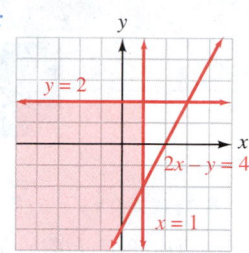

An Application of Systems of Linear Inequalities

E X A M P L E   5

Landscaping. A homeowner budgets from $300 to $600 for trees and bushes to landscape his yard. After shopping around, he finds that good trees cost $150 and mature bushes cost $75. What combinations of trees and bushes can he afford to buy?

ANALYZE THE PROBLEM The homeowner wants to spend *at least* $300 but *not more than* $600 for trees and bushes.

FORM TWO INEQUALITIES We can let x represent the number of trees purchased and y the number of bushes purchased. We then form the following system of inequalities:

The cost of a tree	times	the number of trees purchased	plus	the cost of a bush	times	the number of bushes purchased	should at least be	$300.
$150	·	x	+	$75	·	y	$\ge$	$300

The cost of a tree	times	the number of trees purchased	plus	the cost of a bush	times	the number of bushes purchased	should not be more than	$600.
$150	·	x	+	$75	·	y	$\le$	$600

SOLVE THE SYSTEM We graph the system

$$\begin{cases} 150x + 75y \ge 300 \\ 150x + 75y \le 600 \end{cases}$$

FIGURE 7-29

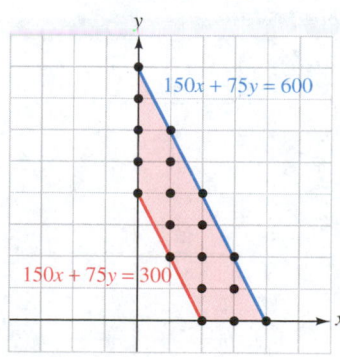

as shown in Figure 7-29. The coordinates of each point shown in the graph give a possible combination of the number of trees (x) and the number of bushes (y) that can be purchased. These possibilities are

$(0, 4), (0, 5), (0, 6), (0, 7), (0, 8)$

$(1, 2), (1, 3), (1, 4), (1, 5), (1, 6)$

$(2, 0), (2, 1), (2, 2), (2, 3), (2, 4)$

$(3, 0), (3, 1), (3, 2), (4, 0)$

Only these points can be used, because the homeowner cannot buy a portion of a tree or a bush.

STUDY SET

Section 7.6

VOCABULARY

In Exercises 1–4, fill in the blanks to make the statements true.

1. $\begin{cases} x + y > 2 \\ x + y < 4 \end{cases}$ This is a system of linear _____.

2. The _____ of a system of linear inequalities is all the ordered pairs that make all inequalities of the system true at the same time.

3. Any point in the _____ region of the graph of the solution of a system of two linear inequalities has coordinates that satisfy both inequalities of the system.

4. To graph a linear inequality such as $x + y > 2$, first graph the boundary. Then pick a test _____ to determine which half-plane to shade.

CONCEPTS

5. In Illustration 1, the solution of linear inequality 1 is shaded in red, and the solution of linear inequality 2 is shaded in blue. Tell whether a true or a false statement results when the coordinates of the given point are substituted into the given inequality.
   ILLUSTRATION 1
   a. A, inequality 1
   b. A, inequality 2
   c. B, inequality 1
   d. B, inequality 2
   e. C, inequality 1
   f. C, inequality 2

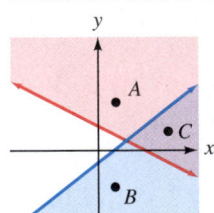

6. Match each equation, inequality, or system with the graph of its solution.
   a. $x + y = 2$
   b. $x + y \geq 2$
   c. $\begin{cases} x + y = 2 \\ x - y = 2 \end{cases}$
   d. $\begin{cases} x + y \geq 2 \\ x - y \leq 2 \end{cases}$

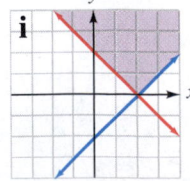
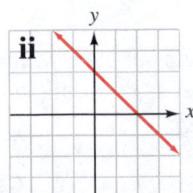
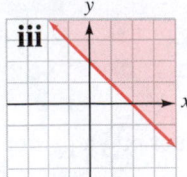
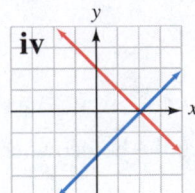

7. The graph of the solution of a system of linear inequalities is shown in Illustration 2. Tell whether each point is a part of the solution set.
   a. $(4, -2)$
   b. $(1, 3)$
   c. the origin

   ILLUSTRATION 2
   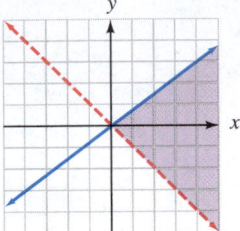

8. Use a system of inequalities to describe the shaded region in Illustration 3.

   ILLUSTRATION 3
   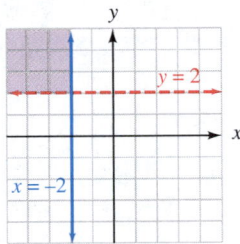

NOTATION

9. Fill in the blank to make the statement true: The graph of the solution of a system of linear inequalities shown in Illustration 4 can be described as the triangle _____ and the triangular region it encloses.

   ILLUSTRATION 4
   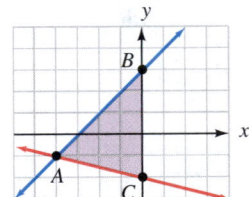

10. Represent each phrase using either $>$, $<$, $\geq$, or $\leq$.
    a. is not more than
    b. must be at least
    c. should not surpass
    d. cannot go below

PRACTICE

In Exercises 11–34, graph the solution set of each system of inequalities, when possible.

11. $\begin{cases} x + 2y \leq 3 \\ 2x - y \geq 1 \end{cases}$

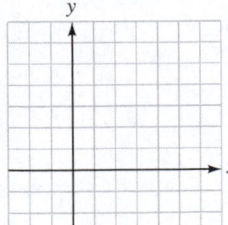

12. $\begin{cases} 2x + y \geq 3 \\ x - 2y \leq -1 \end{cases}$

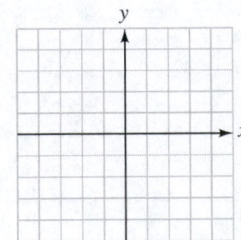

13. $\begin{cases} x + y < -1 \\ x - y > -1 \end{cases}$

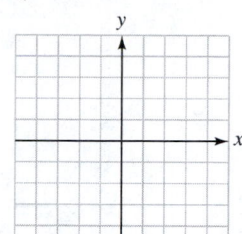

14. $\begin{cases} x + y > 2 \\ x - y < -2 \end{cases}$

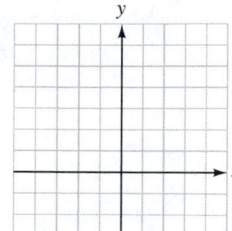

23. $\begin{cases} 3x + 4y \geq -7 \\ 2x - 3y \geq 1 \end{cases}$

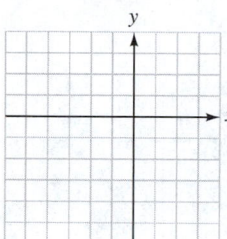

24. $\begin{cases} 3x + y \leq 1 \\ 4x - y \geq -8 \end{cases}$

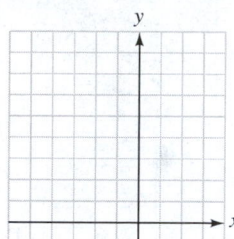

15. $\begin{cases} x \geq 2 \\ y \leq 3 \end{cases}$

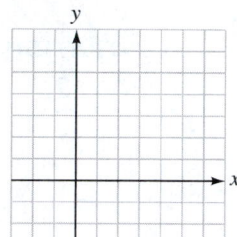

16. $\begin{cases} x \geq -1 \\ y > -2 \end{cases}$

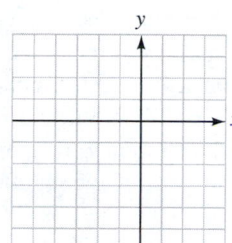

25. $\begin{cases} 2x + y < 7 \\ y > 2(1 - x) \end{cases}$

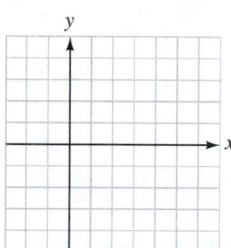

26. $\begin{cases} 2x + y \geq 6 \\ y \leq 2(2x - 3) \end{cases}$

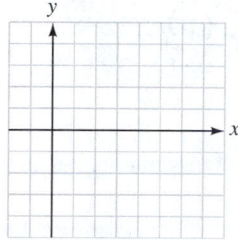

17. $\begin{cases} 2x - 3y \leq 0 \\ y \geq x - 1 \end{cases}$

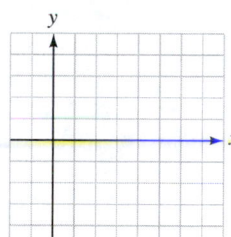

18. $\begin{cases} y > 2x - 4 \\ y \geq -x - 1 \end{cases}$

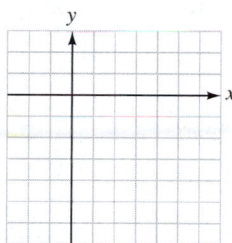

27. $\begin{cases} 2x - 4y > -6 \\ 3x + y \geq 5 \end{cases}$

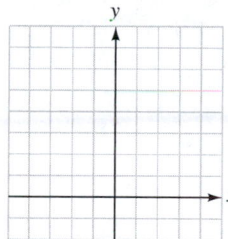

28. $\begin{cases} 2x - 3y < 0 \\ 2x + 3y \geq 12 \end{cases}$

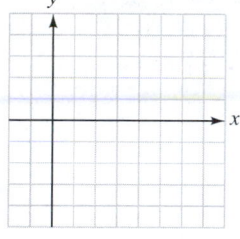

19. $\begin{cases} y < -x + 1 \\ y > -x + 3 \end{cases}$

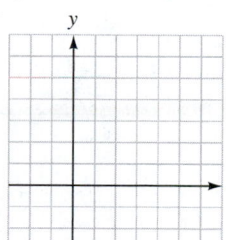

20. $\begin{cases} y > -x + 2 \\ y < -x + 4 \end{cases}$

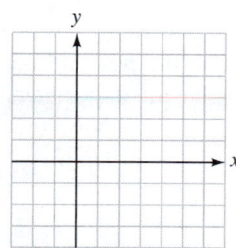

29. $\begin{cases} 3x - y \leq -4 \\ 3y > -2(x + 5) \end{cases}$

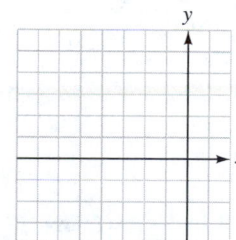

30. $\begin{cases} 3x + y < -2 \\ y > 3(1 - x) \end{cases}$

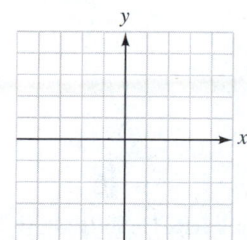

21. $\begin{cases} x > 0 \\ y > 0 \end{cases}$

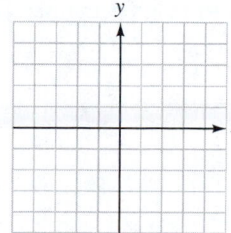

22. $\begin{cases} x \leq 0 \\ y < 0 \end{cases}$

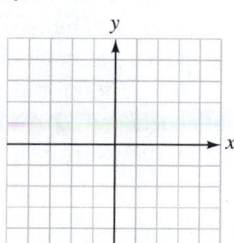

31. $\begin{cases} \dfrac{x}{2} + \dfrac{y}{3} \geq 2 \\ \dfrac{x}{2} - \dfrac{y}{2} < -1 \end{cases}$

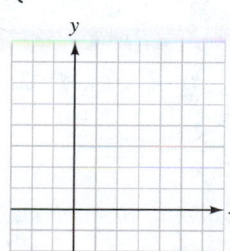

32. $\begin{cases} \dfrac{x}{3} - \dfrac{y}{2} < -3 \\ \dfrac{x}{3} + \dfrac{y}{2} > -1 \end{cases}$

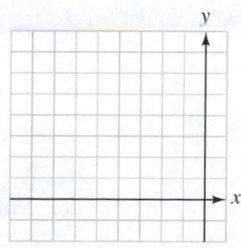

33. $\begin{cases} x \geq 0 \\ y \geq 0 \\ x + y \leq 3 \end{cases}$

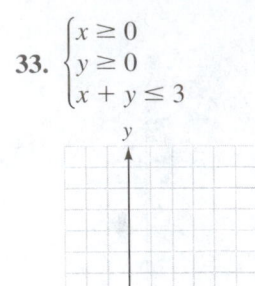

34. $\begin{cases} x - y \leq 6 \\ x + 2y \leq 6 \\ x \geq 0 \end{cases}$

APPLICATIONS

35. BIRDS OF PREY Parts a and b of Illustration 5 show the individual fields of vision for each eye of an owl. In part c, shade the area where the fields of vision overlap—that is, the area that is seen by both eyes.

ILLUSTRATION 5

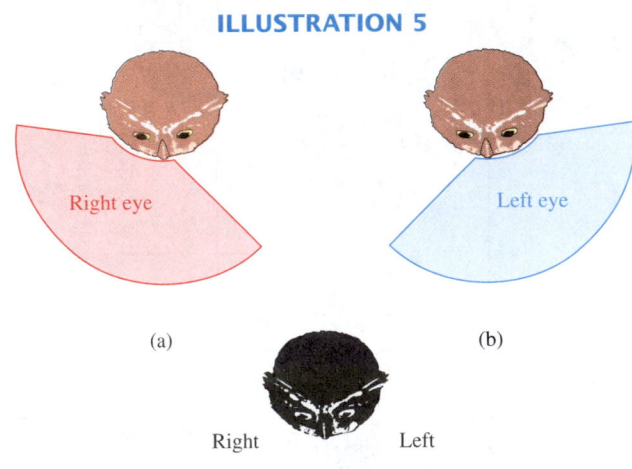

(a) (b)

(c)

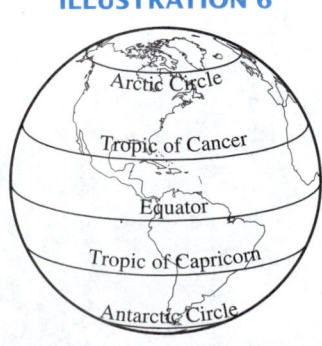

Right Left

36. EARTH SCIENCE In Illustration 6, shade the area of the earth's surface that is north of the Tropic of Capricorn and south of the Tropic of Cancer.

ILLUSTRATION 6

In Exercises 37–40, graph each system of inequalities and give two possible solutions.

37. BUYING COMPACT DISCS Melodic Music has compact discs on sale for either $10 or $15. If a customer wants to spend at least $30 but no more than

$60 on CDs, use Illustration 7 to graph a system of inequalities showing the possible combinations of $10 CDs ($x$) and $15 CDs ($y$) that the customer can buy.

ILLUSTRATION 7

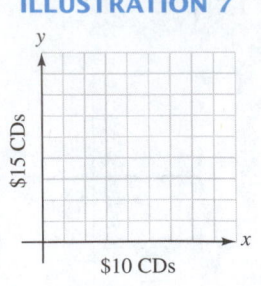

38. BUYING BOATS Dry Boatworks wholesales aluminum boats for $800 and fiberglass boats for $600. Northland Marina wants to make a purchase totaling at least $2,400, but no more than $4,800. Use Illustration 8 to graph a system of inequalities showing the possible combinations of aluminum boats (x) and fiberglass boats (y) that can be ordered.

ILLUSTRATION 8

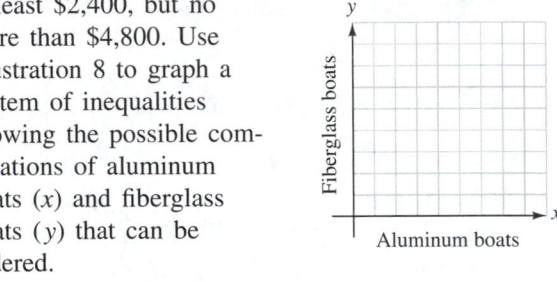

39. BUYING FURNITURE A distributor wholesales desk chairs for $150 and side chairs for $100. Best Furniture wants its order to total no more than $900; Best also wants to order more side chairs than desk chairs. Use Illustration 9 to graph a system of inequalities showing the possible combinations of desk chairs (x) and side chairs (y) that can be ordered.

ILLUSTRATION 9

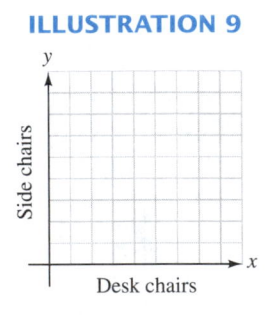

40. ORDERING FURNACE EQUIPMENT J. Bolden Heating Company wants to order no more than $2,000 worth of electronic air cleaners and humidifiers from a wholesaler that charges $500 for air cleaners and $200 for humidifiers. If Bolden wants more humidifiers than air cleaners, use Illustration 10 to graph a system of inequalities showing the possible combinations of air cleaners (x) and humidifiers (y) that can be ordered.

ILLUSTRATION 10

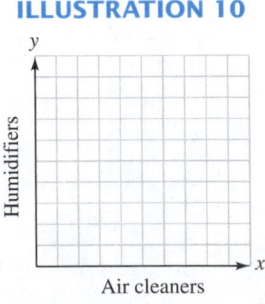

41. PESTICIDE To eradicate a fruit fly infestation, helicopters sprayed an area of a city that can be described by $y \geq -2x + 1$ (within the city limits). In another two weeks, more spraying was ordered over the area described by $y \geq \frac{1}{4}x - 4$ (within the city limits). In Illustration 11, show the part of the city that was sprayed twice.

ILLUSTRATION 11

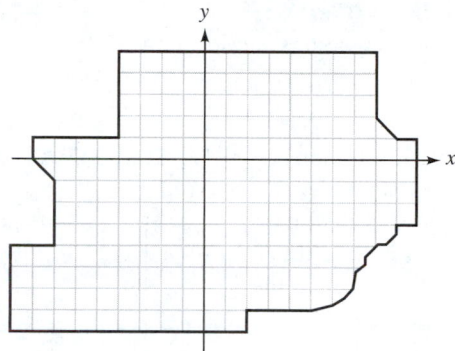

42. REDEVELOPMENT A government agency has declared an area of a city east of First Street, north of Second Avenue, south of Sixth Avenue, and west of Fifth Street as eligible for federal redevelopment funds. See Illustration 12. Describe this area of the city mathematically using a system of four inequalities, if the corner of Central Avenue and Main Street is considered the origin.

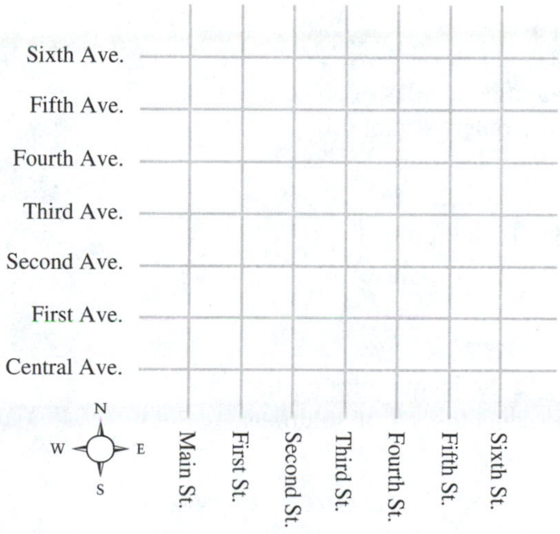

WRITING

43. Explain how to use graphing to solve a system of inequalities.

44. Explain when a system of inequalities will have no solutions.

45. Describe how the graphs of the solutions of these systems are similar and how they differ.

$$\begin{cases} x + y = 4 \\ x - y = 4 \end{cases} \quad \text{and} \quad \begin{cases} x + y \geq 4 \\ x - y \geq 4 \end{cases}$$

46. When a solution of a system of linear inequalities is graphed, what does the shading represent?

REVIEW

In Exercises 47–50, complete each table of values.

47. $y = 2x^2$

x	y
8	
-2	

48. $t = -|s + 2|$

s	t
-3	
-10	

49. $f(x) = 4 + x^3$

Input	Output
0	
-3	

50. $g(x) = 2x - x^2$

x	g(x)
5	
-5	

ILLUSTRATION 12

Sixth Ave.

Fifth Ave.

Fourth Ave.

Third Ave.

Second Ave.

First Ave.

Central Ave.

N W E S

Main St. First St. Second St. Third St. Fourth St. Fifth St. Sixth St.

Systems of Equations and Inequalities

In Chapter 7, we have solved problems that required the use of two variables to represent two unknown quantities. To find the unknowns, we write a pair of equations (or inequalities) called a **system.**

Solving Systems of Equations by Graphing

A system of linear equations can be solved by graphing both equations and locating the point of intersection of the two lines.

1. FOOD SERVICE The two equations in the table give the fees two different catering companies charge a Hollywood studio for on-location meal service.

Caterer	Setup fee	Cost per meal	Equation
Sunshine	$1,000	$4	$y = 4x + 1,000$
Lucy's	$500	$5	$y = 5x + 500$

Complete Illustration 1, using the graphing method to find the break point. That is, find the number of meals and the corresponding fee for which the two caterers will charge the studio the same amount.

ILLUSTRATION 1

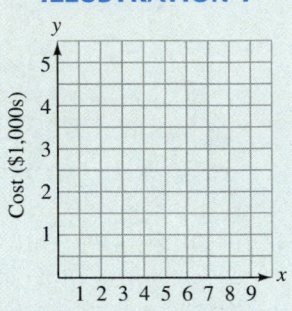

Solving Systems of Equations by Substitution

The substitution method for solving a system of equations works well when a variable in either equation has a coefficient of 1 or -1.

2. Solve by substitution: $\begin{cases} y = 2x - 9 \\ x + 3y = 8 \end{cases}$.

3. Solve by substitution: $\begin{cases} 3x + 4y = -7 \\ 2y - x = -1 \end{cases}$.

Solving Systems of Equations by Addition

With the addition method, equal quantities are added to both sides of an equation to eliminate one of the variables. Then we solve for the other variable.

4. Solve by addition: $\begin{cases} x + y = 1 \\ x - y = 5 \end{cases}$.

5. Solve by addition: $\begin{cases} 2x - 3y = -18 \\ 3x + 2y = -1 \end{cases}$.

Solving Systems of Inequalities

To solve a system of two linear inequalities, we graph the inequalities on the same coordinate axes. The area that is shaded twice represents the set of solutions.

6. This system of inequalities describes the number of $20 shirts ($x$) and $40 pants ($y$) a person can buy if he or she plans to spend not less than $80 but not more than $120. Using Illustration 2, graph the system. Then give three solutions.

$$\begin{cases} 20x + 40y \geq 80 \\ 20x + 40y \leq 120 \end{cases}$$

ILLUSTRATION 2

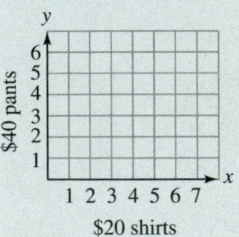

Accent on Teamwork

Section 7.1

TV ratings Illustration 1 shows the Nielsen ratings of the four major television networks from 1995 to 1997. The ratings indicate the number of viewers a network had. Write a report that describes the ratings of each network over this period. Use descriptive phrases such as *sharp decrease* and *moderate gain* to tell how the ratings changed. Explain the significance of the points of intersection of the graphs, and estimate the dates when they occurred. As of 1997, which network do you think had reason to be the most optimistic about its future?

ILLUSTRATION 1

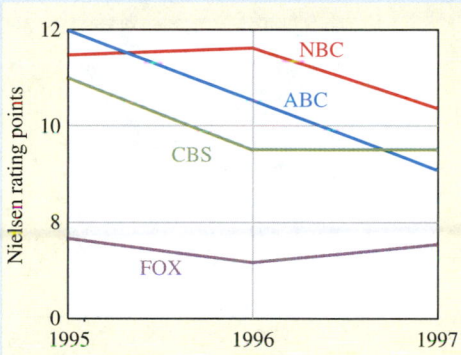

Based on data from Nielsen Media Research, Trade-Line

Section 7.2

Substitutions The word *substitution* is used in several ways. Explain how it is used in the context of a sporting event such as a basketball game. Explain how it is sometimes used when ordering food at a restaurant. Explain what is meant by a *substitute* teacher. Finally, explain how the substitution method is used to solve a system such as

$$\begin{cases} y = -2x - 5 \\ 3x + 5y = -4 \end{cases}$$

What is the difference in the mathematical meaning of the word *substitution* as opposed to the everyday usage of the word?

Section 7.3

Solving systems Solve the system

$$\begin{cases} 2x + y = 4 \\ 2x + 3y = 0 \end{cases}$$

using the graphing method, the substitution method, and the addition method. Which method do you think is the best to use in this case? Why?

Section 7.4

Two unknowns Consider the following problem: A man paid $89 for two white shirts and four pairs of black socks. Find the cost of a white shirt.

If we let x represent the cost of a white shirt and y represent the cost of a pair of black socks, an equation describing the situation is $2x + 4y = 89$. Explain why there is not enough information to solve the problem.

Section 7.5

Matching game Have a student in your group write ten linear inequalities on 3×5 note cards, one inequality per card. Then have him or her graph each of the inequalities on separate cards. Mix up the cards and put all the inequality cards on one side of a table and all the cards with graphs on the other side. Work together to match each inequality with its proper graph.

Section 7.6

Systems of inequalities The points A, B, C, D, E, F, and G, are labeled in the graph of the solution of the system of inequalities in Illustration 2. Tell whether the coordinates of each point make the first inequality (whose solution is shown in red) and the second inequality (whose solution is shown in blue) true or false. Use a table of the following form to keep track of your results.

Point	Coordinates	1st inequality	2nd inequality
A			

ILLUSTRATION 2

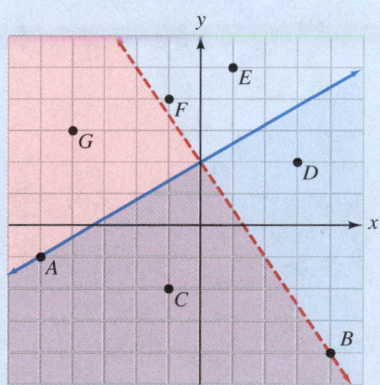

Section 7.1

Solving Systems of Equations by Graphing

CONCEPTS

An ordered pair that satisfies both equations simultaneously is a *solution* of the system.

REVIEW EXERCISES

1. Tell whether the ordered pair is a solution of the system.

   a. $(2, -3)$, $\begin{cases} 3x - 2y = 12 \\ 2x + 3y = -5 \end{cases}$

   b. $\left(\dfrac{7}{2}, -\dfrac{2}{3}\right)$, $\begin{cases} 4x - 6y = 18 \\ \dfrac{x}{3} + \dfrac{y}{2} = \dfrac{5}{6} \end{cases}$

2. INJURY COMPARISON Illustration 1 shows the number of skiing and snowboarding injuries nationally for the years 1993–1997. If the number of injuries continued to occur at the 1996–1997 rates, estimate when they would be the same for skiing and snowboarding. About how many injuries would that be?

ILLUSTRATION 1

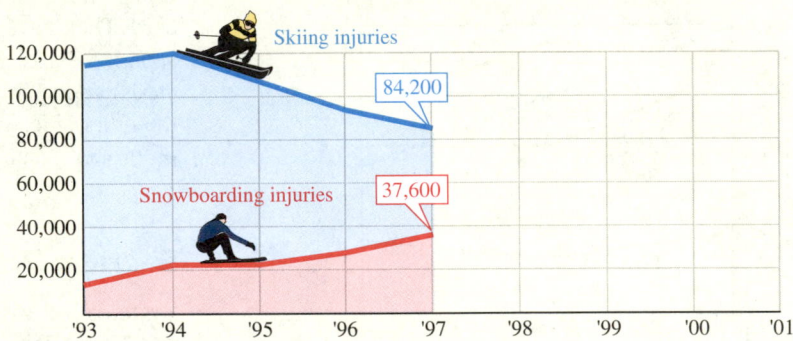

Based on data from *Los Angeles Times* (March 23, 1999)

To *solve a system graphically:*
1. Carefully graph each equation.
2. If the lines intersect, the coordinates of the point of intersection give the solution of the system.
3. Check the solution in the original equations.

When a system of equations has a solution, it is a *consistent* system. Systems with no solutions are *inconsistent*.

When the graphs of two equations in a system are different lines, the equations are *independent* equations.

The equations of a system with infinitely many solutions are *dependent*.

3. Use the graphing method to solve each system.

   a. $\begin{cases} x + y = 7 \\ 2x - y = 5 \end{cases}$

   b. $\begin{cases} y = -\dfrac{x}{3} \\ 2x + y = 5 \end{cases}$

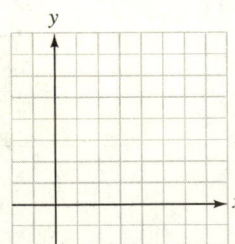

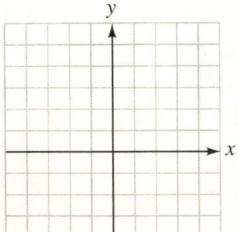

   c. $\begin{cases} 3x + 6y = 6 \\ x + 2y = 2 \end{cases}$

   d. $\begin{cases} 6x + 3y = 12 \\ 2x + y = 2 \end{cases}$

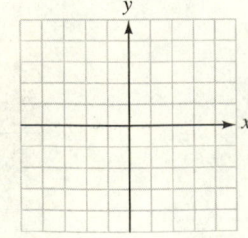

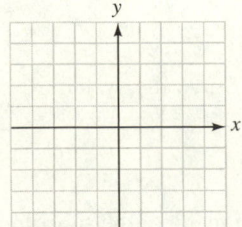

Section 7.2

Solving Systems of Equations by Substitution

To solve a system of equations in x and y by the *substitution method*:

1. Solve one of the equations for either x or y.
2. Substitute the resulting expression for the variable found in step 1 into the other equation, and solve the equation.
3. Find the value of the other variable by substituting the solution found in step 2 in any equation containing x and y.
4. Check the solution in the original equations.

4. Use the substitution method to solve each system.

a. $\begin{cases} x = y \\ 5x - 4y = 3 \end{cases}$

b. $\begin{cases} y = 15 - 3x \\ 7y + 3x = 15 \end{cases}$

c. $\begin{cases} 0.2x + 0.2y = 0.6 \\ 3x = 2 - y \end{cases}$

d. $\begin{cases} 6(r + 2) = s - 1 \\ r - 5s = -7 \end{cases}$

e. $\begin{cases} 9x + 3y = 5 \\ 3x + y = \dfrac{5}{3} \end{cases}$

f. $\begin{cases} \dfrac{x}{6} + \dfrac{y}{10} = 3 \\ \dfrac{5x}{16} - \dfrac{3y}{16} = \dfrac{15}{8} \end{cases}$

5. In solving a system using the substitution method, suppose you obtain the result of $8 = 9$.

a. How many solutions does the system have?

b. Describe the graph of the system.

c. What term is used to describe the system?

Section 7.3

Solving Systems of Equations by Addition

To solve a system of equations using the *addition method*:

1. Write each equation in $Ax + By = C$ form.
2. Multiply one or both equations by nonzero quantities to make the coefficients of x (or y) opposites.
3. Add the equations to eliminate the term involving x (or y).
4. Solve the equation resulting from step 3.
5. Find the value of the other variable by substituting the value of the variable found in step 4 into any equation containing both variables.
6. Check the solution in the original equations.

6. Solve each system using the addition method.

a. $\begin{cases} 2x + y = 1 \\ 5x - y = 20 \end{cases}$

b. $\begin{cases} x + 8y = 7 \\ x - 4y = 1 \end{cases}$

c. $\begin{cases} 5a + b = 2 \\ 3a + 2b = 11 \end{cases}$

d. $\begin{cases} 11x + 3y = 27 \\ 8x + 4y = 36 \end{cases}$

e. $\begin{cases} 9x + 3y = 15 \\ 3x = 5 - y \end{cases}$

f. $\begin{cases} \dfrac{x}{3} + \dfrac{y + 2}{2} = 1 \\ \dfrac{x + 8}{8} + \dfrac{y - 3}{3} = 0 \end{cases}$

g. $\begin{cases} 0.02x + 0.05y = 0 \\ 0.3x - 0.2y = -1.9 \end{cases}$

h. $\begin{cases} -\dfrac{1}{4}x = 1 - \dfrac{2}{3}y \\ 6(x - 3y) + 2y = 5 \end{cases}$

7. For each system, tell which method, substitution or addition, would be easier to use to solve the system and why.

a. $\begin{cases} 6x + 2y = 5 \\ 3x - 3y = -4 \end{cases}$

b. $\begin{cases} x = 5 - 7y \\ 3x - 3y = -4 \end{cases}$

Section 7.4

In this section, we considered ways to solve problems by using *two* variables.

To solve problems involving two unknown quantities:

1. *Analyze* the facts of the problem. Make a table or diagram if necessary.
2. Pick different variables to represent two unknown quantities. *Form* two equations involving the variables.
3. *Solve* the system of equations.
4. *State* the conclusion.
5. *Check* the results.

The *break point* of a linear system is the point of intersection of the graph.

Applications of Systems of Equations

In Exercises 8–16, use two equations in two variables to solve each problem.

8. CAUSES OF DEATH In 1993, the number of Americans dying from heart disease was 5 times more than the number dying from a stroke. If the total number of deaths from these causes was 900,000, how many deaths were attributed to each?

9. PAINTING EQUIPMENT When fully extended, the ladder shown in Illustration 2 is 35 feet in length. If the extension is 7 feet shorter than the base, how long is each part of the ladder?

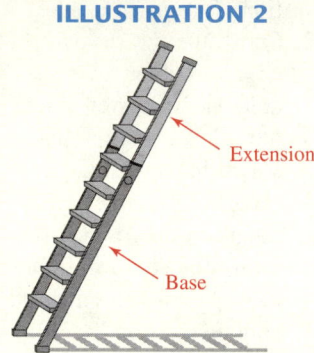

ILLUSTRATION 2

Extension

Base

10. CRASH INVESTIGATION In an effort to protect evidence, investigators used 420 yards of yellow "Police Line—Do Not Cross" tape to seal off a large rectangular-shaped area around an airplane crash site. How much area will the investigators have to search if the width of the rectangle is three-fourths of the length?

11. CELEBRITY ENDORSEMENT A company selling a home juicing machine is contemplating hiring either an athlete or an actor to serve as a spokesperson for a product. The terms of each contract would be as follows:

Celebrity	Base pay	Commission per item sold
Athlete	$30,000	$5
Actor	$20,000	$10

a. For each celebrity, write an equation giving the money (y) the celebrity would earn if x juicers were sold.

b. For what number of juicers would the athlete and the actor earn the same amount?

c. Using Illustration 3, graph the equations from part a. The company expects to sell over 3,000 juicers. Which celebrity would cost the company the least money to serve as a spokesperson?

12. CANDY OUTLET STORE A merchant wants to mix gummy worms worth $3 per pound and gummy bears worth $1.50 per pound to make 30 pounds of a mixture worth $2.10 per pound. How many pounds of each type of candy should he use?

13. BOATING It takes a motorboat 4 hours to travel 56 miles down a river, and 3 hours longer to make the return trip. Find the speed of the current.

14. SHOPPING Packages containing two bottles of contact lens cleaner and three bottles of soaking solution cost $29.40, and packages containing three bottles of cleaner and two bottles of soaking solution cost $28.60. Find the cost of a bottle of cleaner and a bottle of soaking solution.

ILLUSTRATION 3

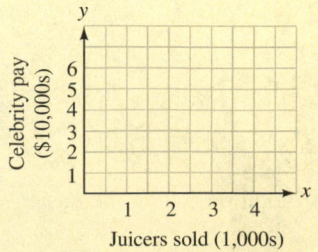

Celebrity pay ($10,000s)

Juicers sold (1,000s)

15. INVESTING Carlos invested part of $3,000 in a 10% certificate account and the rest in a 6% passbook account. The total annual interest from both accounts is $270. How much did he invest at 6%?

16. ANTIFREEZE How much of a 40% antifreeze solution must a mechanic mix with a 70% antifreeze solution if he needs 20 gallons of a 50% antifreeze solution?

Section 7.5

An ordered pair (x, y) is a *solution* of an inequality in x and y if a true statement results when the variables are replaced by the coordinates of the ordered pair.

To graph a linear inequality:
1. Graph the *boundary line*. Draw a solid line if the inequality contains $\leq$ or $\geq$ and a broken line if it contains $<$ or $>$.

2. Pick a *test point* on one side of the boundary. Use the origin if possible. Replace x and y with the coordinates of that point. If the inequality is satisfied, shade the side that contains the point. If the inequality is not satisfied, shade the other side.

Graphing Linear Inequalities

17. Determine whether each ordered pair is a solution of $2x - y \leq -4$.
    a. $(0, 5)$ **b.** $(2, 8)$

    c. $(-3, -2)$ **d.** $\left(\frac{1}{2}, -5\right)$

18. Graph each inequality.
    a. $x - y < 5$

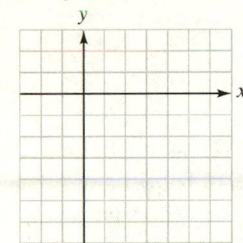

    b. $2x - 3y \geq 6$

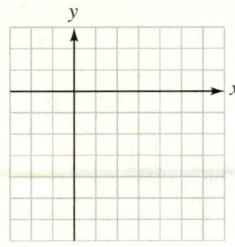

    c. $y \leq -2x$

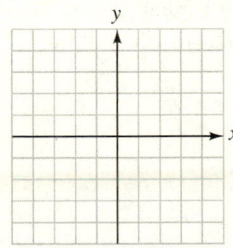

    d. $y < -4$

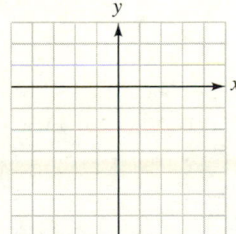

19. In Illustration 4, the graph of a linear inequality is shown. Would a true or a false statement result if the coordinates of
    a. point A were substituted into the inequality?

    b. point B were substituted into the inequality?

    c. point C were substituted into the inequality?

ILLUSTRATION 4

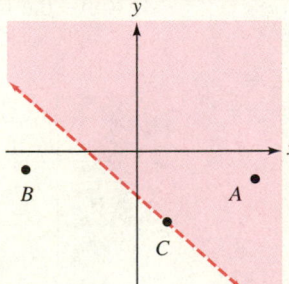

ILLUSTRATION 5

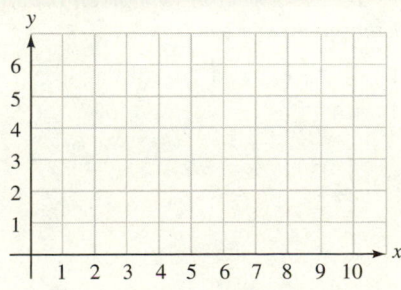

20. WORK SCHEDULE A student told her employer that during the school year, she would be available for up to 30 hours a week, working either 3- or 5-hour shifts. Find an inequality that shows the possible ways to schedule the number of 3-hour (x) and 5-hour shifts (y) she can work, and graph it in Illustration 5 on the previous page. Give three ordered pairs that satisfy the inequality.

Section 7.6

To graph a system of linear inequalities:
1. Graph the individual inequalities of the system on the same coordinate axes.
2. The final solution, if one exists, is that region where all individual graphs intersect.

Systems of linear inequalities can be used to solve application problems.

Solving Systems of Linear Inequalities

21. Solve each system of inequalities.

a. $\begin{cases} 5x + 3y < 15 \\ 3x - y > 3 \end{cases}$

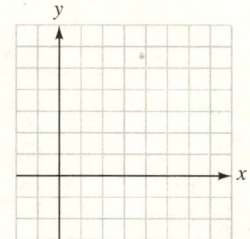

b. $\begin{cases} x \geq 3y \\ y < 3x \end{cases}$

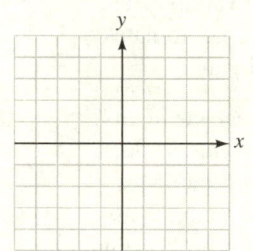

22. GIFT SHOPPING A grandmother wants to spend at least $40 but no more than $60 on school clothes for her grandson. If T-shirts sell for $10 and pants sell for $20, write a system of inequalities that describes the possible combinations of T-shirts (x) and pants (y) she can buy. Graph the system in Illustration 6. Give two possible solutions.

ILLUSTRATION 6

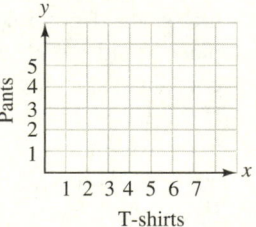

CHAPTER 7

Test

In Problems 1–2, tell whether the given ordered pair is a solution of the given system.

1. (5, 3), $\begin{cases} 3x + 2y = 21 \\ x + y = 8 \end{cases}$

2. (−2, −1), $\begin{cases} 4x + y = -9 \\ 2x - 3y = -7 \end{cases}$

3. Solve the system by graphing: $\begin{cases} 3x + y = 7 \\ x - 2y = 0 \end{cases}$

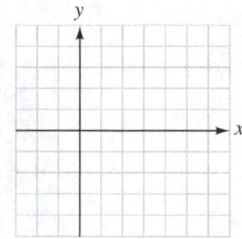

4. To solve a system of two linear equations in x and y, a student used a graphing calculator. From the calculator display in Illustration 1, determine whether the system has a solution. Explain your answer.

ILLUSTRATION 1

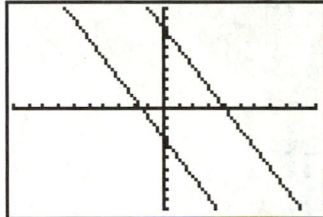

In Problems 5–6, solve each system by substitution.

5. $\begin{cases} y = x - 1 \\ 2x + y = -7 \end{cases}$

6. $\begin{cases} 3a + 4b = -7 \\ 2b - a = -1 \end{cases}$

In Problems 7–8, solve each system by addition.

7. $\begin{cases} 3x - y = 2 \\ 2x + y = 8 \end{cases}$

8. $\begin{cases} 4x + 3y = -3 \\ -3x = -4y + 21 \end{cases}$

In Problems 9–10, classify each system as consistent or inconsistent.

9. $\begin{cases} x + y = 4 \\ x + y = 6 \end{cases}$

10. $\begin{cases} \dfrac{x}{3} + y = 4 \\ x + 3y = 12 \end{cases}$

11. Which method would be most efficient to solve the following system?

$$\begin{cases} 5x - 3y = 5 \\ 3x + 3y = 3 \end{cases}$$

Explain your answer. (You do not need to solve the system.)

12. FINANCIAL PLANNING A woman invested some money at 8% and some at 9%. The interest for 1 year on the combined investment of $10,000 was $840. How much was invested at 9%? Use a system of equations in two variables to solve this problem.

In Problems 13–14, tell whether the given ordered pair is a solution of $2x - 4y > 8$.

13. (7, 1)

14. (0, −2)

15. Graph the inequality $x - y > -2$.

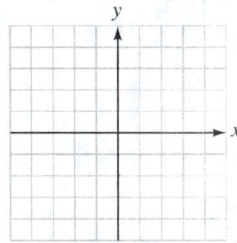

16. Solve the system by graphing.

$$\begin{cases} 2x + 3y \le 6 \\ x \ge 2 \end{cases}$$

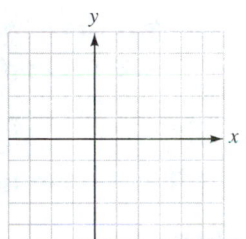

For Problems 17–18, see the graph in Illustration 2, which shows two different ways in which a salesperson can be paid according to the number of items he or she sells.

17. What is the point of intersection of the graphs? Explain its significance.

18. Which plan do you think is better for the salesperson? Explain why.

ILLUSTRATION 2

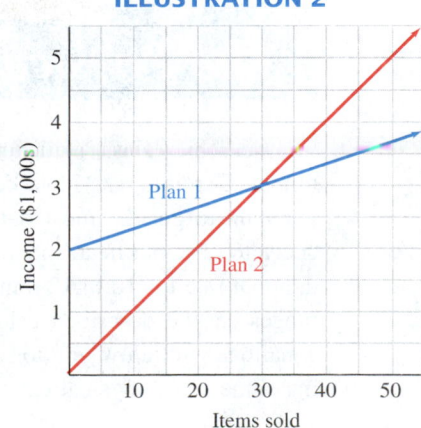

8

More Equations, Inequalities, and Factoring

CAMPUS CONNECTION

The Physical Education Department

In an aerobics class, students perform exercises that improve their cardiovascular fitness. Cardiologists have determined what heart rate range participants need to maintain to get the most out of the training. For example, a 30-year-old woman needs to raise her heart rate to between 137 and 165 beats per minute and then sustain that rate for 10 to 12 minutes. In this chapter, you will see that these suggested ranges can be described using *inequalities*. Learning how to write and solve inequalities will allow you to solve many new types of application problems, including some from physical education.

In this chapter, we will review many of the ideas covered in the first seven chapters and then extend them to the intermediate algebra level.

▶ 8.1 Review of Equations and Inequalities

In this section, you will learn about

Solving equations ■ Identities and impossible equations ■ Solving formulas ■ Problem solving ■ Inequalities ■ Compound inequalities ■ Problem solving

Introduction In this section, we will review how to solve equations and formulas. We will then extend our review to include inequalities.

Solving Equations

Recall that an **equation** is a statement indicating that two mathematical expressions are equal. The set of numbers that satisfy an equation is called its **solution set,** and the elements in the solution set are called **solutions** or **roots** of the equation. Finding the solution set of an equation is called **solving the equation.**

To solve an equation, we will use the following properties of equality to replace the equation with simpler equivalent equations that have the same solution set. We continue this process until we have isolated the variable on one side of the = sign.

1. If any quantity is added to (or subtracted from) both sides of an equation, a new equation is formed that is equivalent to the original equation.

2. If both sides of an equation are multiplied (or divided) by the same nonzero constant, a new equation is formed that is equivalent to the original equation.

EXAMPLE 1

Solving an equation. Solve $3(2x - 1) = 2x + 9$.

Solution We use the distributive property to remove parentheses and then isolate x on the left-hand side of the equation.

$$3(2x - 1) = 2x + 9$$

$$6x - 3 = 2x + 9 \qquad \text{To remove parentheses, use the distributive property.}$$

$$6x - 3 + 3 = 2x + 9 + 3 \qquad \text{To undo the subtraction by 3, add 3 to both sides.}$$

$$6x = 2x + 12 \qquad \text{Combine like terms.}$$

$$6x - 2x = 2x + 12 - 2x \qquad \text{To eliminate } 2x \text{ from the right-hand side, subtract } 2x \text{ from both sides.}$$

$$4x = 12 \qquad \text{Combine like terms.}$$

$$x = 3 \qquad \text{To undo the multiplication by 4, divide both sides by 4.}$$

Check: We substitute 3 for x in the original equation to see whether it satisfies the equation.

$$3(2x - 1) = 2x + 9$$

$$3(2 \cdot 3 - 1) \overset{?}{=} 2 \cdot 3 + 9$$

$$3(5) \overset{?}{=} 6 + 9 \qquad \text{On the left-hand side, do the work in parentheses first: } 2 \cdot 3 - 1 = 5.$$

$$15 = 15$$

Since 3 satisfies the original equation, it is a solution. The solution set is {3}.

SELF CHECK
Solve $2(3x - 2) = 3x - 13$.   *Answer:* -3 ■

To solve more complicated linear equations, we will follow these steps.

Solving equations

1. If an equation contains fractions, multiply both sides of the equation by their least common denominator (LCD) to eliminate the denominators.

2. Use the distributive property to remove all grouping symbols and combine like terms.

3. Use the addition and subtraction properties to get all variables on one side of the equation and all numbers on the other side. Combine like terms, if necessary.

4. Use the multiplication and division properties to make the coefficient of the variable equal to 1.

5. Check the result by replacing the variable with the possible solution and verifying that the number satisfies the equation.

EXAMPLE 2

Solving an equation containing fractions. Solve $\frac{5}{3}(x - 3) = \frac{3}{2}(x - 2) + 2$.

Solution

Step 1: Since 6 is the smallest number that can be divided by both 2 and 3, we multiply both sides of the equation by 6 to eliminate the fractions:

$$\frac{5}{3}(x - 3) = \frac{3}{2}(x - 2) + 2$$

$$6\left[\frac{5}{3}(x - 3)\right] = 6\left[\frac{3}{2}(x - 2) + 2\right] \quad \text{Multiply both sides by the LCD, which is 6.}$$

$$10(x - 3) = 9(x - 2) + 12 \quad 6 \cdot \frac{5}{3} = 10, \ 6 \cdot \frac{3}{2} = 9, \ 6 \cdot 2 = 12.$$

Step 2: We use the distributive property to remove parentheses and then combine like terms.

$$10x - 30 = 9x - 18 + 12$$
$$10x - 30 = 9x - 6$$

Step 3: We use the addition and subtraction properties by adding 30 to both sides and subtracting $9x$ from both sides.

$$10x - 30 - 9x + 30 = 9x - 6 - 9x + 30$$
$$x = 24 \quad \text{Combine like terms.}$$

Since the coefficient of x in the above equation is 1, step 4 is unnecessary.

Step 5: We check by substituting 24 for x in the original equation and simplifying:

$$\frac{5}{3}(x - 3) = \frac{3}{2}(x - 2) + 2$$

$$\frac{5}{3}(24 - 3) \stackrel{?}{=} \frac{3}{2}(24 - 2) + 2$$

$$\frac{5}{3}(21) \stackrel{?}{=} \frac{3}{2}(22) + 2$$

$$5(7) \stackrel{?}{=} 33 + 2$$

$$35 = 35$$

Since 24 satisfies the equation, the solution set is {24}.

SELF CHECK Solve $\frac{2}{3}(x - 2) = \frac{5}{2}(x - 1) + 3$. *Answer:* -1 ■

Identities and Impossible Equations

The equations discussed so far have been **conditional equations.** For these equations, some numbers x are solutions and others are not. An **identity** is an equation that is satisfied by every number x for which both sides of the equation are defined.

EXAMPLE 3 **Solving an equation that is an identity.** Solve $-2(x - 1) - 4 = -4(1 + x) + (2x + 2)$.

Solution

$$-2(x - 1) - 4 = -4(1 + x) + (2x + 2)$$

$-2x + 2 - 4 = -4 - 4x + 2x + 2$ Use the distributive property to remove parentheses.

$-2x - 2 = -2x - 2$ Combine like terms.

$-2 = -2$ Subtract $2x$ from both sides.

Since $-2 = -2$, the equation is true for every number x. Since every number x satisfies this equation, it is an identity.

SELF CHECK Solve $3(a + 1) - (20 + a) = 5(a - 1) - 3(a + 4)$. *Answer:* an identity ■

An **impossible equation** or a **contradiction** is an equation that has no solution.

EXAMPLE 4 **An impossible equation.** Solve $\frac{x - 1}{3} + 4x = \frac{3}{2} + \frac{13x - 2}{3}$.

Solution

$$\frac{x - 1}{3} + 4x = \frac{3}{2} + \frac{13x - 2}{3}$$

$6\left(\frac{x - 1}{3} + 4x\right) = 6\left(\frac{3}{2} + \frac{13x - 2}{3}\right)$ To eliminate the fractions, multiply both sides by 6.

$2(x - 1) + 6(4x) = 9 + 2(13x - 2)$ Use the distributive property to remove parentheses.

$2x - 2 + 24x = 9 + 26x - 4$ Remove parentheses.

$26x - 2 = 26x + 5$ Combine like terms.

$-2 = 5$ Subtract $26x$ from both sides.

Since $-2 = 5$ is false, no number x satisfies the equation. The solution set is $\emptyset$, called the **empty set.**

SELF CHECK Solve $\frac{x - 2}{3} - 3 = \frac{1}{5} + \frac{x + 1}{3}$. *Answer:* no solution ■

Solving Formulas

To solve a formula for a variable means to isolate that variable on one side of the equation and isolate all other quantities on the other side.

EXAMPLE 5 **Selling merchandise.** A sales clerk earns $200 per week plus a 5% commission on the value of the merchandise she sells. What dollar volume must she sell each week to earn $250, $300, and $350 in three successive weeks?

Solution   The weekly earnings e are computed using the formula

1. $e = 200 + 0.05v$

where v represents the value of the merchandise sold. To find v for the three values of e, we first solve Equation 1 for v.

$$e = 200 + 0.05v$$

$$e - 200 = 0.05v \qquad \text{Subtract 200 from both sides.}$$

$$\frac{e - 200}{0.05} = v \qquad \text{Divide both sides by 0.05.}$$

We can now substitute \$250, \$300, and \$350 for e and compute v.

$$v = \frac{e - 200}{0.05} \qquad\qquad v = \frac{e - 200}{0.05} \qquad\qquad v = \frac{e - 200}{0.05}$$

$$v = \frac{250 - 200}{0.05} \qquad\qquad v = \frac{300 - 200}{0.05} \qquad\qquad v = \frac{350 - 200}{0.05}$$

$$v = 1{,}000 \qquad\qquad\qquad v = 2{,}000 \qquad\qquad\qquad v = 3{,}000$$

She must sell \$1,000 worth of merchandise the first week, \$2,000 worth in the second week, and \$3,000 worth in the third week.

SELF CHECK   In Example 5, what dollar volume must the clerk sell to earn \$500?   *Answer:* \$6,000   ■

Problem Solving

EXAMPLE 6

Building a dog run. A man has 28 meters of fencing to make a rectangular dog run. If he wants the dog run to be 6 feet longer than it is wide, find its dimensions.

ANALYZE THE PROBLEM   The perimeter P of a rectangle is the distance around it. If w is chosen to represent the width of the dog run, then $w + 6$ represents its length (See Figure 8-1.) The perimeter can be expressed either as $2w + 2(w + 6)$ or as 28.

FIGURE 8-1

FORM AN EQUATION   We let w represent the width of the dog run. Then $w + 6$ represents its length.

Two widths	plus	two lengths	is	the perimeter.
$2 \cdot w$	$+$	$2 \cdot (w + 6)$	$=$	28

SOLVE THE EQUATION   We can solve this equation as follows:

$$2w + 2(w + 6) = 28$$

$$2w + 2w + 12 = 28 \qquad \text{Use the distributive property to remove parentheses.}$$

$$4w + 12 = 28 \qquad \text{Combine like terms.}$$

$$4w = 16 \qquad \text{Subtract 12 from both sides.}$$

$$w = 4 \qquad \text{Divide both sides by 4.}$$

$$w + 6 = 10$$

STATE THE CONCLUSION The dimensions of the dog run are 4 meters by 10 meters.

CHECK THE RESULT If a dog run has a width of 4 meters and a length of 10 meters, its length is 6 meters longer than its width, and the perimeter is $2(4) + 2(10) = 28$. ∎

For more examples of using equations to solve problems, see Chapter 2, Section 2.5.

Inequalities

Traffic signs like the one shown in Figure 8-2 often appear near schools. From the sign, a motorist knows that

FIGURE 8-2

- a speed that *is greater than* 25 mph breaks the law and could result in a speeding ticket.
- a speed that *is less than or equal to* 25 mph is within the posted speed limit.

Recall that statements such as these can be expressed mathematically by using inequality symbols.

Inequality symbols

$a \neq b$ means "a is not equal to b."

$a < b$ means "a is less than b."

$a > b$ means "a is greater than b."

$a \leq b$ means "a is less than or equal to b."

$a \geq b$ means "a is greater than or equal to b."

We have seen that many inequalities can be graphed as regions on the number line, called **intervals.** For example, the graph of the inequality $-4 < x < 2$ is shown in Figure 8-3(a). Since neither endpoint is included, we say that the graph is an **open interval.** In **interval notation,** this interval is denoted as $(-4, 2)$, where the parentheses indicate that endpoints are not included.

The graph of the inequality $-2 \leq x \leq 5$ is shown in Figure 8-3(b). Since both endpoints are included, we say that the graph is a **closed interval.** This interval is denoted as $[-2, 5]$, where the brackets indicate that the endpoints are included.

Since one endpoint is included and one is not in the interval shown in Figure 8-3(c), we call the interval an **half-open interval.** This interval is denoted as $[-10, 10)$. Since the interval shown in Figure 8-3(d) extends forever in one direction, it is called an **unbounded interval.** This interval is denoted as $[-6, \infty)$, where the symbol ∞ is read as "infinity."

FIGURE 8-3

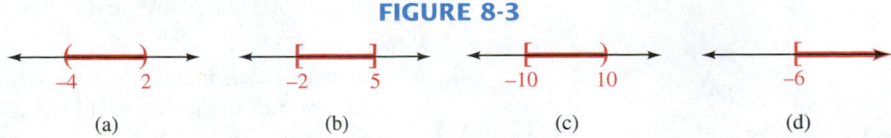

| (a) | (b) | (c) | (d) |

If a and b are real numbers, Table 8-1 shows the types of intervals that can occur.

TABLE 8-1

Kind of interval	Inequality	Graph	Interval
Open interval	$a < x < b$		(a, b)
Half-open intervals	$a \leq x < b$		$[a, b)$
	$a < x \leq b$		$(a, b]$
Closed interval	$a \leq x \leq b$		$[a, b]$
Unbounded intervals	$x > a$		(a, ∞)
	$x \geq a$		$[a, \infty)$
	$x < a$		$(-\infty, a)$
	$x \leq a$		$(-\infty, a]$
	$-\infty < x < \infty$		$(-\infty, \infty)$

Inequalities that are true for all numbers x, such as $x + 1 > x$, are called **absolute inequalities.** Inequalities that are true for some numbers x but not all numbers x, such as $3x + 2 < 8$, are called **conditional inequalities.**

If a and b are two real numbers, then $a < b$, $a = b$, or $a > b$. This property, called the **trichotomy property,** indicates that one and only one of three statements is true about any two real numbers. Either

- the first number is less than the second,
- the first numbers is equal to the second, or
- the first number is greater than the second.

If a, b, and c are real numbers with $a < b$ and $b < c$, then $a < c$. This property, called the **transitive property,** indicates that if we have three numbers and

- the first number is less than the second and
- the second number is less than the third, then
- the first number is less than the third.

To solve an inequality, we use the following properties of inequality.

Properties of inequalities

1. Any real number can be added to (or subtracted from) both sides of an inequality to produce another inequality with the same direction.
2. If both sides of an inequality are multiplied (or divided) by a positive number, another inequality results with the same direction as the original inequality.
3. If both sides of an inequality are multiplied (or divided) by a negative number, another inequality results, but with the opposite direction from the original inequality.

Property 1 indicates that any number can be added to (or subtracted from) both sides of a true inequality to get another true inequality with the same direction. For example, if 4 is added to (or subtracted from) both sides of the inequality $3 < 12$, we get

$$3 + 4 < 12 + 4 \qquad 3 - 4 < 12 - 4$$
$$7 < 16 \qquad\qquad -1 < 8$$

and the $<$ symbol remains an $<$ symbol. These operations do not change the direction (sometimes called the **order**) of the inequality.

Property 2 indicates that both sides of a true inequality can be multiplied (or divided) by any positive number to get another true inequality with the same direction. For example, if both sides of the true inequality $-4 < 6$ are multiplied (or divided) by 2, we get

$$2(-4) < 2(6) \qquad \frac{-4}{2} < \frac{6}{2}$$
$$-8 < 12 \qquad\qquad -2 < 3$$

and the $<$ symbol remains an $<$ symbol. These operations do not change the direction of the inequality.

Property 3 indicates that if both sides of a true inequality are multiplied (or divided) by any negative number, another true inequality results, but with the opposite direction. For example, if both sides of the true inequality $-4 < 6$ are multiplied (or divided) by -2, we get

$$-4 < 6 \qquad\qquad -4 < 6$$
$$-2(-4) > -2(6) \qquad \frac{-4}{-2} > \frac{6}{-2}$$
$$8 > -12 \qquad\qquad 2 > -3$$

and the $<$ symbol becomes an $>$ symbol. These operations reverse the direction of the inequality.

WARNING! We must remember to reverse the inequality symbol every time we multiply or divide both sides by a negative number.

A **linear inequality** is any inequality that can be expressed in the form

$$ax + b < c, \qquad ax + b > c, \qquad ax + b \le c, \qquad \text{or} \qquad ax + b \ge c \quad (a \ne 0)$$

We can solve linear inequalities by using the same steps that we use for solving linear equations, with one exception. If we multiply or divide both sides by a *negative* number, we must reverse the direction of the inequality.

EXAMPLE 7   **Solving inequalities.** Solve **a.** $3(2x - 9) < 9$ and **b.** $-4(3x + 2) \le 16$.

Solution   **a.** We solve the inequality as if it were an equation:

$$3(2x - 9) < 9$$
$$6x - 27 < 9 \qquad \text{Use the distributive property to remove parentheses.}$$
$$6x < 36 \qquad \text{Add 27 to both sides.}$$
$$x < 6 \qquad \text{Divide both sides by 6.}$$

The solution set is the interval $(-\infty, 6)$. The graph of the solution set is shown in Figure 8-4(a) on the next page.

b. We solve the inequality as if it were an equation:

$$-4(3x + 2) \le 16$$
$$-12x - 8 \le 16 \qquad \text{Use the distributive property to remove parentheses.}$$
$$-12x \le 24 \qquad \text{Add 8 to both sides.}$$
$$x \ge -2 \qquad \text{Divide both sides by } -12 \text{ and reverse the } \le \text{ symbol.}$$

The solution set is the interval $[-2, \infty)$. The graph of the solution set is shown in Figure 8-4(b).

FIGURE 8-4

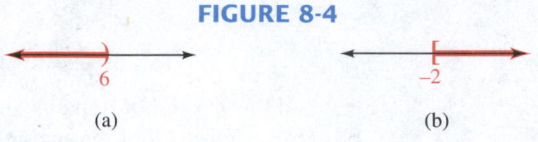

(a)   (b)

SELF CHECK   Solve $-3(2x + 1) > 9$.   *Answer:*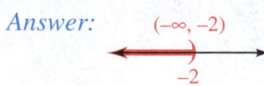

E X A M P L E   8   **Solving inequalities.** Solve $\frac{2}{3}(x + 2) > \frac{4}{5}(x - 3)$.

Solution

$$\frac{2}{3}(x + 2) > \frac{4}{5}(x - 3)$$

$$15 \cdot \frac{2}{3}(x + 2) > 15 \cdot \frac{4}{5}(x - 3) \quad \text{To eliminate the fractions, multiply both sides by 15.}$$

$$10(x + 2) > 12(x - 3) \quad \quad 15 \cdot \frac{2}{3} = 10 \text{ and } 15 \cdot \frac{4}{5} = 12.$$

$$10x + 20 > 12x - 36 \quad \quad \text{Use the distributive property to remove parentheses.}$$

$$-2x + 20 > -36 \quad \quad \quad \text{Add } -12x \text{ to both sides.}$$

$$-2x > -56 \quad \quad \quad \quad \text{Subtract 20 from both sides.}$$

$$x < 28 \quad \quad \quad \quad \quad \text{Divide both sides by } -2 \text{ and reverse the } > \text{ symbol.}$$

FIGURE 8-5

The solution set is the interval $(-\infty, 28)$, whose graph is shown in Figure 8-5.

SELF CHECK   Solve $\frac{1}{2}(x - 1) \leq \frac{2}{3}(x + 1)$.   *Answer:*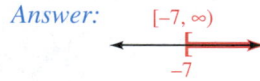

Compound Inequalities

To say that x is between -3 and 8, we write a double inequality:

$-3 < x < 8$ Read as "-3 is less than x and x is less than 8."

This double inequality contains two different linear inequalities:

$-3 < x$ and $x < 8$

These two inequalities mean that $x > -3$ and $x < 8$. The word *and* indicates that these two inequalities are true at the same time.

Double inequalities

The double inequality $c < x < d$ is equivalent to $c < x$ and $x < d$.

   WARNING! The inequality $c < x < d$ cannot be expressed as

$c < x$ or $x < d$

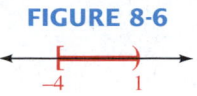

EXAMPLE 9

Solving double inequalities. Solve $-3 \le 2x + 5 < 7$.

Solution This inequality means that $2x + 5$ is between -3 and 7. We can solve it by isolating x between the inequality symbols:

FIGURE 8-6

$$-3 \le 2x + 5 < 7$$
$$-8 \le 2x < 2 \qquad \text{Subtract 5 from all three parts.}$$
$$-4 \le x < 1 \qquad \text{Divide all three parts by 2.}$$

The solution set is the interval $[-4, 1)$. Its graph is shown in Figure 8-6.

SELF CHECK Solve $-3 < 2x - 5 \le 9$.

Answer: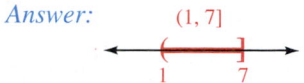

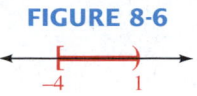

EXAMPLE 10

Solving double inequalities. Solve $x + 3 < 2x - 1 < 4x - 3$.

Solution Since it is impossible to isolate x between the inequality symbols, we solve each of the linear inequalities separately.

$$x + 3 < 2x - 1 \qquad \text{and} \qquad 2x - 1 < 4x - 3$$
$$4 < x \qquad\qquad\qquad\qquad 2 < 2x$$
$$\qquad\qquad\qquad\qquad\qquad 1 < x$$

FIGURE 8-7

Only those numbers x where $x > 4$ and $x > 1$ are in the solution set. Since all numbers greater than 4 are also greater than 1, the solutions are the numbers x where $x > 4$. The solution set is the interval $(4, \infty)$. The graph is shown in Figure 8-7.

SELF CHECK Solve $x - 5 \le 3x - 1 \le 5x + 5$.

Answer: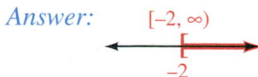

The solution set of a compound inequality containing the word *or* contains all of the numbers that make *one or the other or both* inequalities true. When solving these inequalities, the solution set is the **union** of the solution sets of the two inequalities. We can denote the union of two sets by using the symbol $\cup$, which is read as "union." For the compound inequality $x \le -3$ or $x \ge 8$, we can write the solution set in interval notation as

$$(-\infty, -3] \cup [8, \infty)$$

We can express the solution set of $x \le -3$ or $x \ge 8$ in two other ways:

1. As a graph:

2. In words: *all real numbers less than or equal to -3 or greater than or equal to 8.*

WARNING! In the statement $x \le -3$ or $x \ge 8$, it is incorrect to string the equalities together as $8 \le x \le -3$, because that would imply that $8 \le -3$, which is false.

EXAMPLE 11

Solving compound inequalities. Solve the compound inequality $\frac{x}{3} > \frac{2}{3}$ or $-(x - 2) > 3$.

Solution

We can solve each inequality separately.

$$\frac{x}{3} > \frac{2}{3} \qquad \text{or} \qquad -(x - 2) > 3$$
$$x > 2 \qquad\qquad -x + 2 > 3$$
$$-x > 1$$
$$x < -1$$

FIGURE 8-8

The solution set is the interval $(-\infty, -1) \cup (2, \infty)$. The graph is shown in Figure 8-8.

SELF CHECK

Solve $\frac{x}{2} > 2$ or $-3(x - 2) > 0$. *Answer:* $(-\infty, 2) \cup (4, \infty)$

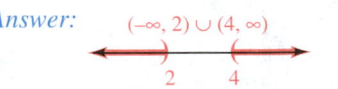

Problem Solving

EXAMPLE 12

Political contributions. Some volunteers are making long-distance telephone calls to solicit contributions for their candidate. The calls are billed at the rate of 42¢ for the first thee minutes and 11¢ for each additional minute or part thereof. If the campaign chairperson has ordered that the cost of each call is not to exceed $2, for how many minutes can a volunteer talk to a prospective donor?

ANALYZE THE PROBLEM

We are given the rate at which a call is billed. Since the cost of a call is not to exceed $2, the cost must be *less than or equal to* $2. This phrase indicates that we should write an *inequality* to find how long a volunteer can talk to a prospective donor.

FORM AN INEQUALITY

We can let x represent the total number of minutes that a call can last. Then the cost of a call will be 42¢ for the first three minutes plus 11¢ times the number of additional minutes, where the number of *additional* minutes is $x - 3$ (the total number of minutes minus the first three minutes). With this information, we can form an inequality.

The cost of the first three minutes	plus	the cost of the additional minutes	is not to exceed	$2.
0.42	+	0.11(x − 3)	≤	2

SOLVE THE INEQUALITY

To simplify the computations, we first clear the inequality of decimals.

$$0.42 + 0.11(x - 3) \le 2$$
$$42 + 11(x - 3) \le 200 \qquad \text{To eliminate the decimals, multiply both sides by 100.}$$
$$42 + 11x - 33 \le 200 \qquad \text{Use the distributive property to remove parentheses.}$$
$$11x + 9 \le 200 \qquad \text{Combine like terms.}$$
$$11x \le 191 \qquad \text{Subtract 9 from both sides.}$$
$$x \le 17.\overline{36} \qquad \text{Divide both sides by 11.}$$

STATE THE CONCLUSION

Since the phone company doesn't bill for part of a minute, the longest time a call can last is 17 minutes. If a call lasts for $17.\overline{36}$ minutes, it will be charged as an 18-minute call, and the cost will be $0.42 + $0.11(15) = $2.07.

CHECK THE RESULT

If the call last 17 minutes, the cost will be $0.42 + $0.11(14) = $1.96. Since this is less than $2, the result checks.

Section 8.1

VOCABULARY

Fill in the blanks to make the statements true.

1. An _____ is a statement indicating that two mathematical expressions are equal.

2. An _____ is an equation that is true for all values of its variable.

3. An _____ equation is an equation that is true for no values of its variable.

4. A _____ equation is true for some but not all values of the variable.

5. $(-\infty, 5)$ is an example of an unbounded _____.

6. An _____ is a statement indicating that two quantities are not equal.

NOTATION

In Exercises 15–18, fill in the blanks to make the statements true.

15. The symbol $<$ is read as _____.

16. The symbol $\leq$ is read as _____ or equal to.

17. ∞ is a symbol representing positive _____.

18. The symbol for _____ is $\geq$.

19. Express the inequality determined by the sentence, "Raise her heart rate to between 137 and 165 beats per minute."

20. Express the inequality determined by "Sustain that rate for 10 to 12 minutes."

CONCEPTS

In Exercises 7–13, fill in the blanks to make the statements true.

7. If any quantity is _____ to (or _____ from) both sides of an equation, a new equation is formed that is equivalent to the original one.

8. If both sides of an equation are _____ (or _____) by the same nonzero number, a new equation is formed that is equivalent to the original one.

9. If both sides of an inequality are multiplied by a _____ number, a new inequality is formed that has the same direction as original one.

10. If both sides of an inequality are multiplied by a _____ number, a new inequality is formed that has the opposite direction from the original one.

11. An open interval has no _____.

12. A _____ interval has one endpoint.

13. If $a < b$ and $b < c$, then _____.

14. Match each interval with its graph.

   a. $(-\infty, -1]$ i
   b. $(-\infty, 1)$ ii
   c. $[-1, \infty)$ iii

PRACTICE

In Exercises 21–40, solve each equation.

21. $2x + 1 = 13$

22. $2x - 4 = 16$

23. $3(x + 1) = 15$

24. $-2(x + 5) = 30$

25. $2r - 5 = 1 - r$

26. $5s - 13 = s - 1$

27. $3(2y - 4) - 6 = 3y$

28. $2x + (2x - 3) = 5$

29. $5(5 - a) = 37 - 2a$

30. $4a + 17 = 7(a + 2)$

31. $4(y + 1) = -2(4 - y)$

32. $5(r + 4) = -2(r - 3)$

33. $2(a - 5) - (3a + 1) = 0$

34. $8(3a - 5) - 4(2a + 3) = 12$

35. $\dfrac{x}{2} - \dfrac{x}{3} = 4$

36. $\dfrac{x}{2} + \dfrac{x}{3} = 10$

37. $\dfrac{x}{6} + 1 = \dfrac{x}{3}$

38. $\dfrac{3}{2}(y + 4) = \dfrac{20 - y}{2}$

39. $\dfrac{a + 1}{3} + \dfrac{a - 1}{5} = \dfrac{2}{15}$

40. $\dfrac{2z + 3}{3} + \dfrac{3z - 4}{6} = \dfrac{z - 2}{2}$

In Exercises 41–48, solve each equation. If the equation is an identity or an impossible equation, so indicate.

41. $4(2 - 3t) + 6t = -6t + 8$

42. $2x - 6 = -2x + 4(x - 2)$

43. $\dfrac{a+1}{4} + \dfrac{2a-3}{4} = \dfrac{a}{2} - 2$

44. $\dfrac{y-8}{5} + 2 = \dfrac{2}{5} - \dfrac{y}{3}$

45. $3(x-4) + 6 = -2(x+4) + 5x$

46. $2(x-3) = \dfrac{3}{2}(x-4) + \dfrac{x}{2}$

47. $y(y+2) = (y+1)^2 - 1$

48. $x(x-3) = (x-1)^2 - (5+x)$

In Exercises 49–58, solve each formula for the indicated variable.

49. $V = \dfrac{1}{3} Bh$ for B

50. $A = \dfrac{1}{2} bh$ for b

51. $p = 2l + 2w$ for w

52. $p = 2l + 2w$ for l

53. $z = \dfrac{x-\mu}{\sigma}$ for x

54. $z = \dfrac{x-\mu}{\sigma}$ for μ

55. $y = mx + b$ for x

56. $y = mx + b$ for m

57. $P = L + \dfrac{s}{f} i$ for s

58. $P = L + \dfrac{s}{f} i$ for f

In Exercises 59–84, solve each inequality. Give the result in interval notation and graph the solution set.

59. $5x - 3 > 7$

60. $7x - 9 < 5$

61. $-3x - 1 \le 5$

62. $-2x + 6 \ge 16$

63. $8 - 9y \ge -y$

64. $4 - 3x \le x$

65. $-3(a+2) > 2(a+1)$

66. $-4(y-1) < y + 8$

67. $\dfrac{1}{2} y + 2 \ge \dfrac{1}{3} y - 4$

68. $\dfrac{1}{4} x - \dfrac{1}{3} \le x + 2$

69. $-2 < -b + 3 < 5$

70. $4 < -t - 2 < 9$

71. $15 > 2x - 7 > 9$

72. $25 > 3x - 2 > 7$

73. $-6 < -3(x - 4) \le 24$

74. $-4 \le -2(x + 8) < 8$

75. $0 \ge \dfrac{1}{2} x - 4 > 6$

76. $-6 \le \dfrac{1}{3} a + 1 \le 0$

77. $0 \le \dfrac{4 - x}{3} \le 2$

78. $-2 \le \dfrac{5 - 3x}{2} \le 2$

79. $3x + 2 < 8$ or $2x - 3 > 11$

80. $3x + 4 < -2$ or $3x + 4 > 10$

81. $-4(x + 2) \ge 12$ or $3x + 8 < 11$

82. $5(x - 2) \ge 0$ and $-3x < 9$

83. $x < -3$ and $x > 3$

84. $x < 3$ or $x > -3$

APPLICATIONS

85. CUTTING A BOARD The carpenter in Illustration 1 saws a board into two pieces. He wants one piece to be 1 foot longer than twice the length of the shorter piece. Find the length of each piece.

ILLUSTRATION 1

22 ft

86. CUTTING A BEAM A 30-foot steel beam is to be cut into two pieces. The longer piece is to be 2 feet more than 3 times as long as the shorter piece. Find the length of each piece.

87. FINDING DIMENSIONS The rectangular garden shown in Illustration 2 is twice as long as it is wide. Find its dimensions.

ILLUSTRATION 2

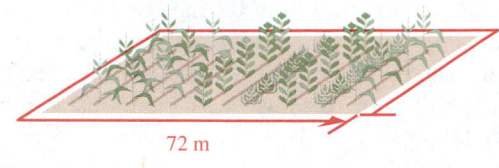

72 m

88. FENCING A PASTURE A farmer has 624 feet of fencing to enclose the pasture shown in Illustration 3. Because a river runs along one side, fencing will be needed on only three sides. Find the dimensions of the pasture if its length is parallel to the river and is double its width.

ILLUSTRATION 3

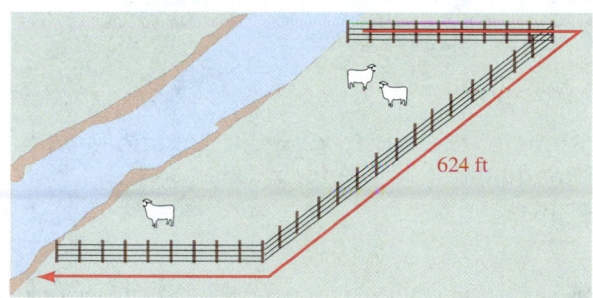

624 ft

89. FENCING A PEN A man has 150 feet of fencing to build the pen shown in Illustration 4. If one end is a square, find the outside dimensions of the entire pen.

ILLUSTRATION 4

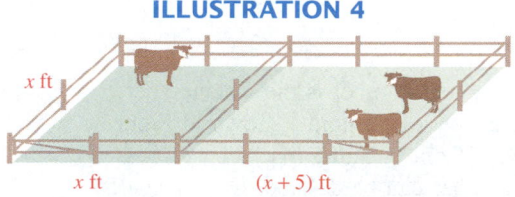

x ft

x ft $(x + 5)$ ft

90. ENCLOSING A SWIMMING POOL A woman wants to enclose the swimming pool shown in Illustration 5 and have a walkway of uniform width all the way around. How wide will the walkway be if the woman uses 180 feet of fencing?

ILLUSTRATION 5

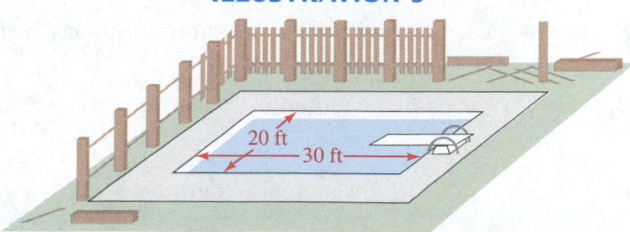

20 ft   30 ft

91. FINDING PROFIT The wholesale cost of a clock is $27. A store owner knows that in order to sell, the clock must be priced under $42. If p is the profit, express the possible profit as an inequality.

92. INVESTING MONEY If a woman invests $10,000 at 8% annual interest, how much more must she invest at 9% so that her annual income will exceed $1,250?

93. BUYING COMPACT DISCS A student can afford to spend up to $330 on a stereo system and some compact discs. If the stereo costs $175 and the discs cost $8.50 each, find the greatest number of discs the student can buy.

94. GRADE AVERAGE A student has scores of 70, 77, and 85 on three exams. What score is needed on a fourth exam to make the average 80 or better?

95. BABY FURNITURE See Illustration 6. A company makes various playpens having perimeters between 128 and 192 inches, inclusive.
   a. Complete the double inequality that describes the range of the perimeters of the playpens.
   $$? \le 4s \le ?$$
   b. Solve the double inequality to find the range of the side lengths of the playpens.

ILLUSTRATION 6

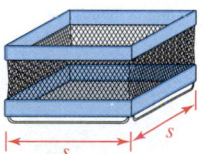

s s

96. THERMOSTAT The *Temp range* control on the thermostat shown in Illustration 7 directs the heater to come on when the room temperature gets 5° below the *Temp setting;* it directs the air conditioner to come on when the room temperature gets 5° above the *Temp setting.* Use interval notation to describe
   a. the temperature range for the room when neither the heater nor the air conditioner will be on.

ILLUSTRATION 7

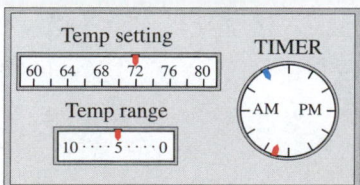

   b. the temperature range for the room when either the heater or the air conditioner will be on. (*Note:* The lowest theoretical temperature possible is −460° F, called **absolute zero.**)

WRITING

97. Explain the difference between a conditional equation, an identity, and an impossible equation.

98. The techniques for solving linear equations and linear inequalities are similar, yet different. Explain.

99. Find the **error.**

$$4(x + 3) = 16$$
$$4x + 3 = 16$$
$$4x = 13$$
$$x = \frac{13}{4}$$

100. Which of these relations is transitive?
a. $=$ **b.** $\leq$ **c.** $\not\leq$ **d.** $\neq$

REVIEW

Simplify each expression.

101. $\left(\dfrac{t^3 t^5 t^{-6}}{t^2 t^{-4}} \right)^{-3}$ **102.** $\left(\dfrac{a^{-2} b^3 a^5 b^{-2}}{a^6 b^{-5}} \right)^{-4}$

103. A man invests \$1,200 in baking equipment to make pies. Each pie requires \$3.40 in ingredients. If the man can sell all the pies he can make for \$5.95 each, how many pies will he have to make to earn a profit?

104. A woman invested \$15,000, part at 7% annual interest and the rest at 8%. If she earns \$2,200 in income over a two-year period, how much did she invest at 7%?

▶ 8.2   Solving Absolute Value Equations and Inequalities

In this section, you will learn about

Absolute value ■ Equations of the form $|x| = k$ ■ Equations with two absolute values ■ Inequalities of the form $|x| < k$
■ Inequalities of the form $|x| > k$

Introduction   Many quantities studied in mathematics, science, and engineering are expressed as positive numbers. To guarantee that a quantity is positive, we often use the concept of absolute value. In this section, we will work with equations and inequalities containing expressions involving absolute value. Using the definition of absolute value, we will develop procedures to solve absolute value equations and absolute value inequalities.

Absolute Value

FIGURE 8-9

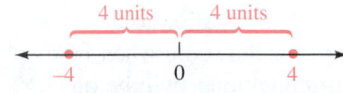

Recall that the absolute value of any real number is the distance between the number and zero on the number line. For example, the points shown in Figure 8-9 with coordinates of 4 and -4 both lie 4 units from 0. Thus, $|4| = |-4| = 4$.

The absolute value of a real number can be defined more formally.

Absolute value

If $x \geq 0$, then $|x| = x$.

If $x < 0$, then $|x| = -x$.

This definition gives a way to associate a nonnegative real number with any real number.

■ If $x \geq 0$, then x (which is positive or 0) is its own absolute value.

■ If $x < 0$, then $-x$, (which is positive) is the absolute value.

Either way, $|x|$ is positive or 0. That is, $|x| \geq 0$, for all real numbers x.

EXAMPLE 1

Finding absolute values. Find **a.** $|9|$, **b.** $|-5.68|$, and **c.** $|0|$.

Solution
a. Since $9 \geq 0$, the number 9 is its own absolute value: $|9| = 9$.

b. Since $-5.68 < 0$, the negative of -5.68 is the absolute value:
$$|-5.68| = -(-5.68) = 5.68$$

c. Since $0 \geq 0$, 0 is its own absolute value: $|0| = 0$.

SELF CHECK
Find **a.** $|-3|$, **b.** $|100.99|$, and **c.** $|-2\pi|$.

Answers: **a.** 3, **b.** 100.99, **c.** 2π ∎

WARNING! The placement of a $-$ sign in an expression containing an absolute value symbol is important. For example, $|-19| = 19$, but $-|19| = -19$.

Equations of the Form $|x| = k$

FIGURE 8-10

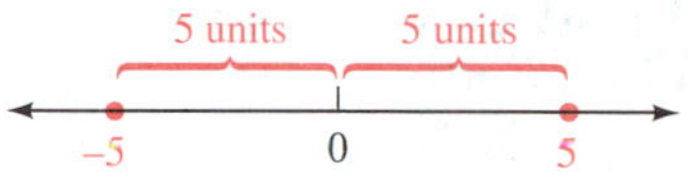

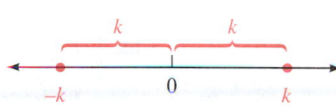

The absolute value of a real number represents the distance on the number line from a point to the origin. To solve the **absolute value equation** $|x| = 5$, we must find the coordinates of all points on the number line that are exactly 5 units from zero. See Figure 8-10. The only two points that satisfy this condition have coordinates 5 and -5. That is, $x = 5$ or $x = -5$.

FIGURE 8-11

In general, the solution set of the absolute value equation $|x| = k$, where $k \geq 0$, includes the coordinates of the points on the number line that are k units from the origin. (See Figure 8-11.)

Absolute value equations

If $k \geq 0$, then

$$|x| = k \quad \text{is equivalent to} \quad x = k \text{ or } x = -k$$

EXAMPLE 2

Solving an absolute value equation. Solve **a.** $|x| = 8$ and **b.** $|s| = 0.003$.

Solution
a. If $|x| = 8$, then $x = 8$ or $x = -8$.

b. If $|s| = 0.003$, then $s = 0.003$ or $s = -0.003$.

SELF CHECK
Solve **a.** $|y| = 24$ and **b.** $|x| = \frac{1}{2}$.

Answers: **a.** 24, -24 **b.** $\frac{1}{2}$, $-\frac{1}{2}$ ∎

The equation $|x - 3| = 7$ indicates that a point on the number line with a coordinate of $x - 3$ is 7 units from the origin. Thus, $x - 3$ can be either 7 or -7.

$$x - 3 = 7 \quad \text{or} \quad x - 3 = -7$$
$$x = 10 \qquad\qquad x = -4$$

The solutions of the absolute value equation are 10 and -4. We can graph them on a number line, as shown in Figure 8-12. If either of these numbers is substituted for x in $|x - 3| = 7$, the equation is satisfied.

FIGURE 8-12

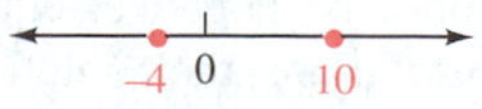

$$|x - 3| = 7 \qquad\qquad |x - 3| = 7$$
$$|10 - 3| \overset{?}{=} 7 \qquad |-4 - 3| \overset{?}{=} 7$$
$$|7| \overset{?}{=} 7 \qquad\qquad |-7| \overset{?}{=} 7$$
$$7 = 7 \qquad\qquad\quad 7 = 7$$

EXAMPLE 3

Solving an absolute value equation. Solve $|3x - 2| = 5$.

Solution We can write $|3x - 2| = 5$ as

$$3x - 2 = 5 \quad \text{or} \quad 3x - 2 = -5$$

and solve each equation for x:

$$3x - 2 = 5 \quad \text{or} \quad 3x - 2 = -5$$
$$3x = 7 \qquad\qquad\quad 3x = -3$$
$$x = \frac{7}{3} \qquad\qquad\quad x = -1$$

Verify that both solutions check.

SELF CHECK Solve $|2x - 3| = 7$. *Answer:* 5, -2 ■

When solving an absolute value equation, we want the absolute value isolated on one side. If this is not the case in a given equation, we use the equation-solving procedures studied earlier to isolate the absolute value first.

EXAMPLE 4

Isolating the absolute value. Solve $\left|\frac{2}{3}x + 3\right| + 4 = 10$.

Solution We can isolate $\left|\frac{2}{3}x + 3\right|$ on the left-hand side of the equation by subtracting 4 from both sides.

$$\left|\frac{2}{3}x + 3\right| + 4 = 10$$

1. $\qquad \left|\frac{2}{3}x + 3\right| = 6 \qquad$ Subtract 4 from both sides.

Now that the absolute value is isolated, we can write Equation 1 as

$$\frac{2}{3}x + 3 = 6 \quad \text{or} \quad \frac{2}{3}x + 3 = -6$$

and solve each equation for x:

$$\frac{2}{3}x + 3 = 6 \quad \text{or} \quad \frac{2}{3}x + 3 = -6$$
$$\frac{2}{3}x = 3 \qquad\qquad\quad \frac{2}{3}x = -9$$
$$2x = 9 \qquad\qquad\quad 2x = -27$$
$$x = \frac{9}{2} \qquad\qquad\quad x = -\frac{27}{2}$$

Verify that both solutions check.

SELF CHECK Solve $|0.4x - 2| - 0.6 = 0.4$. *Answer:* 7.5, 2.5 ■

 WARNING! Since the absolute value of a quantity cannot be negative, equations such as $|7x + \frac{1}{2}| = -4$ have no solution. Since there are no solutions, their solution sets are empty.

EXAMPLE 5

An absolute value equal to 0. Solve $3\left|\frac{1}{2}x - 5\right| - 4 = -4$.

Solution   We first isolate $\left|\frac{1}{2}x - 5\right|$ on the left-hand side.

$$3\left|\frac{1}{2}x - 5\right| - 4 = -4$$

$$3\left|\frac{1}{2}x - 5\right| = 0 \qquad \text{Add 4 to both sides.}$$

$$\left|\frac{1}{2}x - 5\right| = 0 \qquad \text{Divide both sides by 3.}$$

Since 0 is the only number whose absolute value is 0, $\frac{1}{2}x - 5$ must be 0, and we have

$$\frac{1}{2}x - 5 = 0$$

$$\frac{1}{2}x = 5 \qquad \text{Add 5 to both sides.}$$

$$x = 10 \qquad \text{Multiply both sides by 2.}$$

Verify that 10 satisfies the original equation.

SELF CHECK   Solve $-5\left|\frac{2x}{3} + 4\right| + 1 = 1$.   *Answer:* -6 ∎

Equations with Two Absolute Values

The equation $|a| = |b|$ is true when $a = b$ or when $a = -b$. For example,

$$|3| = |3| \qquad \text{or} \qquad |3| = |-3|$$

↑   ↑   ↑   ↑

The same number.   These numbers are opposites.

In general, the following statement is true.

Equations with two absolute values

If a and b represent algebraic expressions, the equation $|a| = |b|$ is equivalent to the pair of equations

$$a = b \qquad \text{or} \qquad a = -b$$

EXAMPLE 6   **Solving equations with two absolute values.** Solve $|5x + 3| = |3x + 25|$.

Solution   This equation is true when $5x + 3 = 3x + 25$, or when $5x + 3 = -(3x + 25)$. We solve each equation for x.

$$
\begin{array}{lll}
5x + 3 = 3x + 25 & \text{or} & 5x + 3 = -(3x + 25) \\
2x = 22 & & 5x + 3 = -3x - 25 \\
x = 11 & & 8x = -28 \\
& & x = -\dfrac{28}{8} \\
& & x = -\dfrac{7}{2}
\end{array}
$$

Verify that both solutions check.

SELF CHECK   Solve $|2x - 3| = |4x + 9|$.   *Answer:* $-1, -6$ ∎

Inequalities of the Form $|x| < k$

FIGURE 8-13

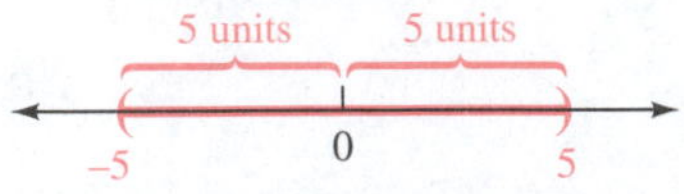

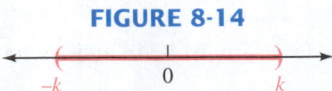

To solve the **absolute value inequality** $|x| < 5$, we must find the coordinates of all points on a number line that are less than 5 units from the origin. See Figure 8-13. Thus, x is between -5 and 5, and

$$|x| < 5 \quad \text{is equivalent to} \quad -5 < x < 5$$

FIGURE 8-14

In general, the solution set of the absolute value inequality $|x| < k$ $(k > 0)$ includes the coordinates of the points on the number line that are less than k units from the origin. See Figure 8-14.

Solving $|x| < k$ and $|x| \le k$

$$|x| < k \quad \text{is equivalent to} \quad -k < x < k \quad (k > 0)$$
$$|x| \le k \quad \text{is equivalent to} \quad -k \le x \le k \quad (k \ge 0)$$

EXAMPLE 7   **Solving an absolute value inequality.** Solve $|2x - 3| < 9$ and graph the solution set.

Solution   We write the absolute value inequality as a double inequality and solve for x.

$$|2x - 3| < 9 \quad \text{is equivalent to} \quad -9 < 2x - 3 < 9$$

$$-9 < 2x - 3 < 9$$
$$-6 < 2x < 12 \qquad \text{Add 3 to all three parts.}$$
$$-3 < x < 6 \qquad \text{Divide all parts by 2.}$$

FIGURE 8-15

Any number between -3 and 6 is in the solution set. This is the interval $(-3, 6)$, whose graph is shown in Figure 8-15.

SELF CHECK   Solve $|3x + 2| < 4$ and graph the solution set.   *Answer:* $\left(-2, \frac{2}{3}\right)$

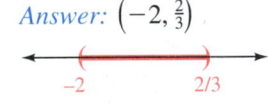

EXAMPLE 8   **Tolerances.** When manufactured parts are inspected by a quality control engineer, they are classified as acceptable if each dimension falls within a given *tolerance range* of the dimensions listed on the blueprint. For the bracket shown in Figure 8-16, the distance between the two drilled holes is given as 2.900 inches. Because the tolerance is ± 0.015 inch, this distance can be as much as 0.015 inch longer or 0.015 inch shorter, and the part will be considered acceptable. The acceptable distance d between holes can be represented by the absolute value inequality $|d - 2.900| \le 0.015$. Solve the inequality and explain the result.

FIGURE 8-16

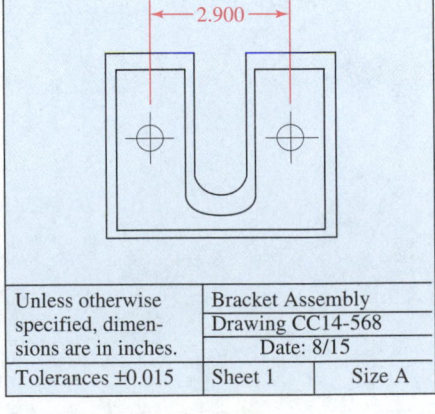

Unless otherwise specified, dimensions are in inches.	Bracket Assembly	
	Drawing CC14-568	
	Date: 8/15	
Tolerances ±0.015	Sheet 1	Size A

Solution   We can write the absolute value inequality as a double inequality and solve for d:

$$|d - 2.900| \leq 0.015 \quad \text{is equivalent to} \quad -0.015 \leq d - 2.900 \leq 0.015$$

$$-0.015 \leq d - 2.900 \leq 0.015$$

$$2.885 \leq d \leq 2.915 \qquad \text{Add 2.900 to all three parts.}$$

The solution set is the interval [2.885, 2.915]. This means that the distance between the two holes should be between 2.885 and 2.915 inches, inclusive. If the distance is less than 2.885 inches or more than 2.915 inches, the part should be rejected.   ∎

Inequalities of the Form $|x| > k$

To solve the *absolute value inequality* $|x| > 5$, we must find the coordinates of all points on a number line that are more than 5 units from the origin. See Figure 8-17.

FIGURE 8-17

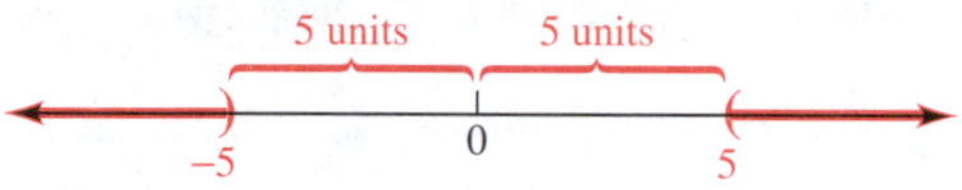

Thus, $x < -5$ or $x > 5$.

In general, the solution set of $|x| > k$ includes the coordinates of the points on the number line that are more than k units from the origin. See Figure 8-18. Thus,

$$|x| > k \quad \text{is equivalent to} \quad x < -k \text{ or } x > k$$

The *or* indicates an either/or situation. It is only necessary that x satisfy one of the two conditions to be in the solution set.

FIGURE 8-18

Solving $|x| > k$ and $|x| \geq k$

If $k \geq 0$, then

$$|x| > k \quad \text{is equivalent to} \quad x < -k \text{ or } x > k$$
$$|x| \geq k \quad \text{is equivalent to} \quad x \leq -k \text{ or } x \geq k$$

EXAMPLE 9

Solving an absolute value inequality. Solve  $\left| \dfrac{3 - x}{5} \right| \geq 6$ and graph the solution set.

Solution   We write the absolute value inequality as two separate inequalities connected with the word "or."

$$\left| \frac{3 - x}{5} \right| \geq 6 \quad \text{is equivalent to} \quad \frac{3 - x}{5} \leq -6 \text{ or } \frac{3 - x}{5} \geq 6$$

Then we solve each inequality for x:

$\dfrac{3 - x}{5} \leq -6$	or	$\dfrac{3 - x}{5} \geq 6$	
$3 - x \leq -30$		$3 - x \geq 30$	Multiply both sides by 5.
$-x \leq -33$		$-x \geq 27$	Subtract 3 from both sides.
$x \geq 33$		$x \leq -27$	Divide both sides by -1 and reverse the direction of the inequality symbol.

The solution set is the interval $(-\infty, -27] \cup [33, \infty)$, whose graph appears in Figure 8-19.

FIGURE 8-19

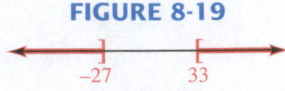

SELF CHECK Solve $\left| \dfrac{2 - x}{4} \right| \geq 1$ and graph the solution set.   *Answer:* $(-\infty, -2] \cup [6, \infty)$

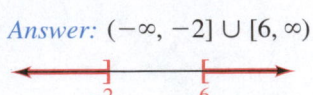

EXAMPLE 10   **Solving an absolute value inequality.** Solve $\left| \dfrac{2}{3}x - 2 \right| - 3 > 6$ and graph the solution set.

Solution   We begin by adding 3 to both sides to isolate the absolute value on the left-hand side.

$$\left| \frac{2}{3}x - 2 \right| - 3 > 6$$

$$\left| \frac{2}{3}x - 2 \right| > 9 \quad \text{Add 3 to both sides to isolate the absolute value.}$$

We then proceed as follows:

$$\frac{2}{3}x - 2 < -9 \qquad \text{or} \qquad \frac{2}{3}x - 2 > 9$$

$$\frac{2}{3}x < -7 \qquad\qquad \frac{2}{3}x > 11 \quad \text{Add 2 to both sides.}$$

$$2x < -21 \qquad\qquad 2x > 33 \quad \text{Multiply both sides by 3.}$$

$$x < -\frac{21}{2} \qquad\qquad x > \frac{33}{2} \quad \text{Divide both sides by 2.}$$

FIGURE 8-20

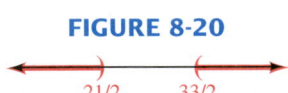

The solution set is $\left(-\infty, -\frac{21}{2}\right) \cup \left(\frac{33}{2}, \infty\right)$. Its graph appears in Figure 8-20.

SELF CHECK Solve $\left| \dfrac{3}{4}x + 2 \right| - 1 > 3$ and graph the solution set.   *Answer:* $(-\infty, -8) \cup \left(\frac{8}{3}, \infty\right)$

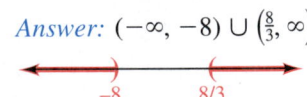

ACCENT ON TECHNOLOGY *Solving Absolute Value Inequalities*

We can also solve absolute value inequalities using a graphing calculator. For example, to solve $|2x - 3| < 9$, we graph the equations $y = |2x - 3|$ and $y = 9$ on the same coordinate system. If we use settings of $[-5, 15]$ for x and $[-5, 15]$ for y, we will get the graph shown in Figure 8-21.

FIGURE 8-21

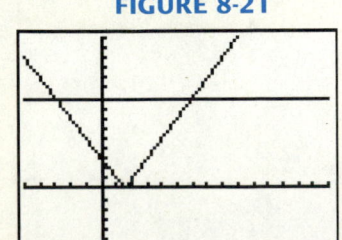

The inequality $|2x - 3| < 9$ will be true for all x-coordinates of points that lie on the graph of $y = |2x - 3|$ and below the graph of $y = 9$. Using the TRACE feature, we can see that these values of x are in the interval $(-3, 6)$.

> STUDY SET

Section 8.2

VOCABULARY

In Exercises 1–4, fill in the blanks to make the statements true.

1. $|2x - 1| = 10$ is an absolute value _____.
2. $|2x - 1| > 10$ is an absolute value _____.
3. To _____ the absolute value in $|3 - x| - 4 = 5$, we add 4 to both sides.
4. $|x| = 2$ is _____ to $x = 2$ or $x = -2$.

CONCEPTS

In Exercises 5–10, fill in the blanks to make the statements true.

5. $|x| \geq$ ___ for all real numbers x.
6. If $x < 0$, $|x| =$ _____.
7. To solve $|x| > 5$, we must find the coordinates of all points on a number line that are _____ 5 units from the origin.
8. To solve $|x| < 5$, we must find the coordinates of all points on a number line that are _____ 5 units from the origin.
9. To solve $|x| = 5$, we must find the coordinates of all points on a number line that are ___ units from the origin.
10. The equation $|a| = |b|$ is true when _____ or when _____.

11. Tell whether $x = -3$ is a solution of the given equation or inequality.
    a. $|x - 1| = 4$ b. $|x - 1| > 4$
    c. $|x - 1| \leq 4$ d. $|5 - x| = |x + 12|$
12. Write each absolute value equation or inequality in its equivalent form.
    a. $|x| = 8$
    b. $|x| \geq 8$
    c. $|x| \leq 8$
    d. $|5x - 1| = |x + 3|$

NOTATION

13. Match each equation or inequality with its graph.
    a. $|x| = 1$ i
    b. $|x| > 1$ ii
    c. $|x| < 1$ iii

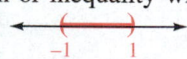

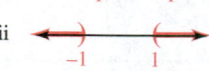

14. Match each graph with its corresponding equation or inequality.
    a.   i $|x| \geq 2$
    b.   ii $|x| \leq 2$
    c.   iii $|x| = 2$

In Exercises 15–18, write each compound inequality as an inequality containing an absolute value.

15. $-4 < x < 4$
16. $x < -4$ or $x > 4$
17. $x + 3 < -6$ or $x + 3 > 6$
18. $-5 \leq x - 3 \leq 5$

PRACTICE

In Exercises 19–26, find the value of each expression.

19. $|8|$ 20. $|-18|$
21. $-|0.02|$ 22. $-|-3.14|$
23. $-\left|-\dfrac{31}{16}\right|$ 24. $-\left|\dfrac{25}{4}\right|$
25. $|\pi|$ 26. $\left|-\dfrac{\pi}{2}\right|$

In Exercises 27–42, solve each equation, if possible.

27. $|x| = 23$ 28. $|x| = 90$
29. $|x - 3.1| = 6$ 30. $|x + 4.3| = 8.9$
31. $|3x + 2| = 16$ 32. $|5x - 3| = 22$
33. $\left|\dfrac{7}{2}x + 3\right| = -5$ 34. $\left|\dfrac{2x}{3} + 10\right| = 0$
35. $|3 - 4x| = 5$ 36. $|8 - 5x| = 18$
37. $2|3x + 24| = 0$ 38. $5|x - 21| = -8$
39. $\left|\dfrac{3x + 48}{3}\right| = 12$ 40. $\left|\dfrac{4x - 64}{4}\right| = 32$
41. $|x + 3| + 7 = 10$ 42. $|2 - x| + 3 = 5$

In Exercises 43–50, solve each equation, if possible.

43. $|2x + 1| = |3x + 3|$ **44.** $|5x - 7| = |4x + 1|$

45. $|2 - x| = |3x + 2|$ **46.** $|4x + 3| = |9 - 2x|$

47. $\left|\dfrac{x}{2} + 2\right| = \left|\dfrac{x}{2} - 2\right|$ **48.** $|7x + 12| = |x - 6|$

49. $\left|x + \dfrac{1}{3}\right| = |x - 3|$ **50.** $\left|x - \dfrac{1}{4}\right| = |x + 4|$

In Exercises 51–76, solve each inequality. Write the solution set in interval notation and graph it.

51. $|x| < 4$

52. $|x| < 9$

53. $|x + 9| \leq 12$

54. $|x - 8| \leq 12$

55. $|3x - 2| \leq 10$

56. $|4 - 3x| \leq 13$

57. $|3x + 2| \leq -3$

58. $|5x - 12| < -5$

59. $|x| > 3$

60. $|x| > 7$

61. $|x - 12| > 24$

62. $|x + 5| \geq 7$

63. $|3x + 2| > 14$

64. $|2x - 5| > 25$

65. $|4x + 3| > -5$

66. $|7x + 2| > -8$

67. $|2 - 3x| \geq 8$

68. $|-1 - 2x| > 5$

69. $-|2x - 3| < -7$

70. $-|3x + 1| < -8$

71. $\left|\dfrac{x - 2}{3}\right| \leq 4$

72. $\left|\dfrac{x - 2}{3}\right| > 4$

73. $|3x + 1| + 2 < 6$

74. $1 + \left|\dfrac{1}{7}x + 1\right| \leq 1$

75. $\left|\dfrac{1}{3}x + 7\right| + 5 > 6$

76. $-2|3x - 4| < 16$

APPLICATIONS

77. TEMPERATURE RANGE The temperatures on a sunny summer day satisfied the inequality $|t - 78°| \leq 8°$, where t is a temperature in degrees Fahrenheit. Solve this inequality and express the range of temperatures as a double inequality.

78. OPERATING TEMPERATURE A car CD player has an operating temperature of $|t - 40°| < 80°$, where t is a temperature in degrees Fahrenheit. Solve the inequality and express this range of temperatures as an interval.

79. AUTO MECHANICS On most cars, the bottoms of the front wheels are closer together than the tops, creating a *camber angle*. This lessens road shock to the steering system. (See Illustration 1.) The specifications for a certain car state that the camber angle c of its wheels should be $0.6° \pm 0.5°$.
   a. Express the range with an inequality containing absolute value symbols.
   b. Solve the inequality and express this range of camber angles as an interval.

ILLUSTRATION 1

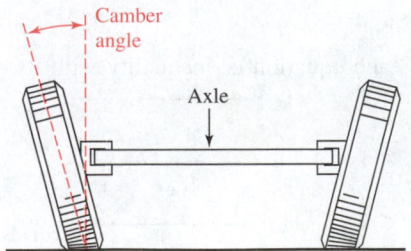

80. STEEL PRODUCTION A sheet of steel is to be 0.250 inch thick with a tolerance of 0.025 inch.
a. Express this specification with an inequality containing absolute value symbols, where x is the thickness of a sheet of steel.
b. Solve the inequality and express the range of thicknesses as an interval.

81. ERROR ANALYSIS In a lab, students measured the percent of copper p in a sample of copper sulfate. The students know that copper sulfate is actually 25.46% copper by mass. They are to compare their results to the actual value and find the amount of *experimental error.*
a. Which measurements shown in Illustration 2 satisfy the absolute value inequality $|p - 25.46| \leq 1.00$?
b. What can be said about the amount of error for each of the trials listed in part a?

ILLUSTRATION 2

Lab 4	Section A

Title:
"Percent copper (Cu) in copper sulfate ($CuSO_4 \cdot 5H_2O$)"

Results

	% Copper
Trial #1:	22.91%
Trial #2:	26.45%
Trial #3:	26.49%
Trial #4:	24.76%

82. ERROR ANALYSIS See Exercise 81.
a. Which measurements satisfy the absolute value inequality $|p - 25.46| > 1.00$?

b. What can be said about the amount of error for each of the trials listed in part a?

WRITING

83. Explain how to find the absolute value of a given number.

84. Explain why the equation $|x - 4| = -5$ has no solutions.

85. Explain the use of parentheses and brackets when graphing inequalities.

86. Explain the differences between the solution sets of $|x| < 8$ and $|x| > 8$.

REVIEW

87. RAILROAD CROSSING The warning sign in Illustration 3 is to be painted on the street in front of a railroad crossing. If y is 30° more than twice x, find x and y.

ILLUSTRATION 3

88. GEOMETRY Refer to Illustration 3. What is $2x + 2y$?

▶ **8.3**

Review of Factoring

In this section, you will learn about

Factoring out the greatest common factor ■ Factoring by grouping ■ Formulas ■ Factoring the difference of two squares ■ Factoring trinomials ■ Test for factorability ■ Using grouping to factor trinomials

Introduction In Chapter 5, we discussed how to factor many polynomials. In this section, we will review that material and introduce a test for factorability. We will also extend the ideas to include variable exponents.

Factoring Out the Greatest Common Factor

Factoring out a common monomial factor is based on the distributive property.

EXAMPLE 1

Factoring out a common monomial. Factor $3xy^2z^3 + 6xz^2 - 9xyz^4$.

Solution   We begin by factoring each monomial:

$$3xy^2z^3 = 3 \cdot x \cdot y \cdot y \cdot z \cdot z \cdot z$$
$$6xz^2 = 3 \cdot 2 \cdot x \cdot z \cdot z$$
$$9xyz^4 = 3 \cdot 3 \cdot x \cdot y \cdot z \cdot z \cdot z \cdot z$$

Since each term has one factor of 3, one factor of x, and two factors of z, and there are no other common factors, $3xz^2$ is the greatest common factor of the three terms. We can use the distributive property to factor out $3xz^2$.

$$3xy^2z^3 + 6xz^2 - 9xyz^4 = 3xz^2 \cdot y^2z + 3xz^2 \cdot 2 - 3xz^2 \cdot 3yz^2$$
$$= 3xz^2(y^2z + 2 - 3yz^2)$$

SELF CHECK   Factor $-6a^2b^2 + 4ab^3$.   *Answer:* $-2ab^2(3a - 2b)$ ∎

EXAMPLE 2

Factoring out a negative factor. Factor the negative of the greatest common factor from $-6u^2v^3 + 8u^3v^2$.

Solution   Because the greatest common factor of the two terms is $2u^2v^2$, the negative of the greatest common factor is $-2u^2v^2$. To factor out $-2u^2v^2$, we proceed as follows:

$$-6u^2v^3 + 8u^3v^2 = -2u^2v^2 \cdot 3v + 2u^2v^2 \cdot 4u$$
$$= -2u^2v^2 \cdot 3v - (-2u^2v^2)4u$$
$$= -2u^2v^2(3v - 4u)$$

SELF CHECK   Factor the negative of the greatest common factor from $-3p^3q + 6p^2q^2$.   *Answer:* $-3p^2q(p - 2q)$ ∎

We can also factor out a common factor with a variable exponent.

EXAMPLE 3

Factoring out an expression with variable factors. Factor x^{2n} from $x^{4n} + x^{3n} + x^{2n}$.

Solution   We can write the trinomial in the form

$$x^{2n} \cdot x^{2n} + x^{2n} \cdot x^n + x^{2n} \cdot 1$$

and factor out x^{2n}.

$$x^{4n} + x^{3n} + x^{2n} = x^{2n} \cdot x^{2n} + x^{2n} \cdot x^n + x^{2n} \cdot 1$$
$$= x^{2n}(x^{2n} + x^n + 1)$$

SELF CHECK   Factor $2a^n$ from $6a^{2n} - 4a^{n+1}$.   *Answer:* $2a^n(3a^n - 2a)$ ∎

Factoring by Grouping

Suppose we wish to factor

$$ac + ad + bc + bd$$

Although there is no factor common to all four terms, there is a factor of a in the first two terms and a factor of b in the last two terms. We can factor out these common factors to get

$$ac + ad + bc + bd = a(c + d) + b(c + d)$$

We can now factor out the common factor of $c + d$:

$$ac + ad + bc + bd = (c + d)(a + b)$$

The grouping in this type of problem is not always unique. For example, if we write the expression $ac + ad + bc + bd$ in the form

$$ac + bc + ad + bd$$

and factor c from the first two terms and d from the last two terms, we obtain

$$ac + bc + ad + bd = c(a + b) + d(a + b)$$
$$= (a + b)(c + d)$$

EXAMPLE 4

Factoring out a common factor. Factor $3ax^2 + 3bx^2 + a + 5bx + 5ax + b$.

Solution Although there is no factor common to all six terms, $3x^2$ can be factored out of the first two terms, and $5x$ can be factored out of the fourth and fifth terms to get

$$3ax^2 + 3bx^2 + a + 5bx + 5ax + b = 3x^2(a + b) + a + 5x(b + a) + b$$

This result can be written in the form

$$3ax^2 + 3bx^2 + a + 5bx + 5ax + b = 3x^2(a + b) + 5x(a + b) + (a + b)$$

Since $a + b$ is common to all three terms, it can be factored out to get

$$3ax^2 + 3bx^2 + a + 5bx + 5ax + b = (a + b)(3x^2 + 5x + 1)$$

SELF CHECK Factor $2mp - np + 2mq - nq$.   *Answer:* $(p + q)(2m - n)$ ∎

Formulas

Factoring is often required to solve a formula for one of its variables.

EXAMPLE 5

Solving formulas. The formula $r_1r_2 = rr_2 + rr_1$ is used in electronics to relate the combined resistance r of two resistors wired in parallel. The variable r_1 represents the resistance of the first resistor, and the variable r_2 represents the resistance of the second. Solve the formula for r_2.

Solution To isolate r_2 on one side of the equation, we get all terms involving r_2 on the left-hand side and all terms not involving r_2 on the right-hand side. We then proceed as follows:

$$r_1r_2 = rr_2 + rr_1$$
$$r_1r_2 - rr_2 = rr_1 \qquad \text{Subtract } rr_2 \text{ from both sides.}$$
$$r_2(r_1 - r) = rr_1 \qquad \text{Factor out } r_2 \text{ on the left-hand side.}$$
$$r_2 = \frac{rr_1}{r_1 - r} \qquad \text{Divide both sides by } r_1 - r.$$

SELF CHECK Solve $f_1f_2 = ff_1 + ff_2$ for f_1.   *Answer:* $f_1 = \dfrac{ff_2}{f_2 - f}$ ∎

Factoring the Difference of Two Squares

Recall the formula for factoring the difference of two squares.

Factoring the difference of two squares

$$x^2 - y^2 = (x + y)(x - y)$$

If we think of the difference of two squares as the square of a **First** quantity minus the square of a **Last** quantity, we have the formula

$$F^2 - L^2 = (F + L)(F - L)$$

In words, we say

*To factor the square of a **First** quantity minus the square of a **Last** quantity, we multiply the **First** plus the **Last** by the **First** minus the **Last**.*

To factor $49x^2 - 16$, for example, we write $49x^2 - 16$ in the form $(7x)^2 - (4)^2$ and use the formula for factoring the difference of two squares:

$$49x^2 - 16 = (7x)^2 - (4)^2$$
$$= (7x + 4)(7x - 4)$$

We can verify this result by multiplying $7x + 4$ and $7x - 4$.

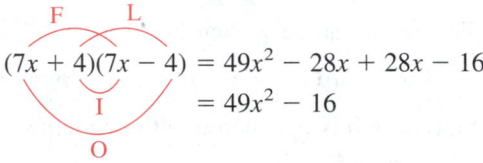

$$(7x + 4)(7x - 4) = 49x^2 - 28x + 28x - 16$$
$$= 49x^2 - 16$$

We note that if $49x^2 - 16$ is divided by $7x + 4$, the quotient is $7x - 4$, and that if $49x^2 - 16$ is divided by $7x - 4$, the quotient is $7x + 4$.

 WARNING! Expressions that represent the sum of two squares, such as $(7x)^2 + (4)^2$, cannot be factored in the real number system. The binomial $49x^2 + 16$ is a prime binomial.

EXAMPLE 6

Factoring the difference of two squares. Factor $(x + y)^4 - z^4$.

Solution

This expression is the difference of two squares and can be factored:

$$(x + y)^4 - z^4 = [(x + y)^2]^2 - (z^2)^2$$
$$= [(x + y)^2 + z^2][(x + y)^2 - z^2]$$

The factor $(x + y)^2 + z^2$ is the sum of two squares and is prime. However, the factor $(x + y)^2 - z^2$ is the difference of two squares and it can then be factored as $(x + y + z)(x + y - z)$. Thus,

$$(x + y)^4 - z^4 = [(x + y)^2 + z^2][(x + y)^2 - z^2]$$
$$= [(x + y)^2 + z^2](x + y + z)(x + y - z)$$

SELF CHECK

Factor $a^4 - (b + c)^4$.　　　　*Answer:* $[a^2 + (b + c)^2](a + b + c)(a - b - c)$ ∎

When possible, we will always factor out a common factor before factoring the difference of two squares. The factoring process is easier when all common factors are factored out first.

EXAMPLE 7

Mixed factoring. Factor $2x^4y - 32y$.

Solution

$$\begin{array}{lll} 2x^4y - 32y & = 2y(x^4 - 16) & \text{Factor out } 2y. \\ & = 2y(x^2 + 4)(x^2 - 4) & \text{Factor } x^4 - 16. \\ & = 2y(x^2 + 4)(x + 2)(x - 2) & \text{Factor } x^2 - 4. \end{array}$$

SELF CHECK

Factor $3ap^4 - 243a$.　　　　*Answer:* $3a(p^2 + 9)(p + 3)(p - 3)$ ∎

Factoring Trinomials

Recall that to factor trinomials with leading coefficients of 1, we follow these steps.

> **Factoring trinomials**
>
> 1. Write the trinomial in descending powers of one variable.
> 2. List the factorizations of the third term of the trinomial.
> 3. Pick the factorization where the sum of the factors is the coefficient of the middle term.

EXAMPLE 8

Factoring trinomials. Factor $x^2 - 6x + 8$.

Solution Since this trinomial is already written in descending powers of x, we can move to step 2 and list the possible factorizations of the third term, which is 8.

The one to choose

$$8(1) \qquad 4(2) \qquad -8(-1) \qquad \mathbf{-4(-2)}$$

In this trinomial, the coefficient of the middle term is -6. The only factorization where the sum of the factors is -6 is $-4(-2)$. Thus,

$$x^2 - 6x + 8 = (x - 4)(x - 2)$$

The factorization of $x^2 - 6x + 8$ is $(x - 4)(x - 2)$. Because of the commutative property of multiplication, the order of the factors is not important. We can verify this result by multiplication:

$$(x - 4)(x - 2) = x^2 - 2x - 4x + 8$$
$$= x^2 - 6x + 8$$

SELF CHECK Factor $x^2 + 5x + 6$. *Answer:* $(x + 3)(x + 2)$ ∎

EXAMPLE 9

Factoring trinomials. Factor $30x - 4xy - 2xy^2$.

Solution We begin by writing the trinomial in descending powers of y:

$$30x - 4xy - 2xy^2 = -2xy^2 - 4xy + 30x$$

Each term in this trinomial has a common factor of $-2x$, which can be factored out.

$$30x - 4xy - 2xy^2 = -2x(y^2 + 2y - 15)$$

To factor $y^2 + 2y - 15$, we list the factors of -15 and find the pair whose sum is 2.

The one to choose

$$15(-1) \qquad \mathbf{5(-3)} \qquad 1(-15) \qquad 3(-5)$$

The only factorization where the sum of the factors is 2 (the coefficient of the middle term of $y^2 + 2y - 15$) is $5(-3)$. Thus,

$$30x - 4xy - 2xy^2 = -2x(y^2 + 2y - 15)$$
$$= -2x(y + 5)(y - 3)$$

Verify this result by multiplication.

Factor $16a - 2ap^2 - 4ap$.   *Answer:* $-2a(p + 4)(p - 2)$ ∎

 WARNING! In Example 9, be sure to include all factors in the final answer. It is a common error to forget to write the $-2x$.

There are more combinations of factors to consider when factoring trinomials with leading coefficients other than 1. To factor $5x^2 + 7x + 2$, for example, we must find two binomials of the form $ax + b$ and $cx + d$ such that

$$5x^2 + 7x + 2 = (ax + b)(cx + d)$$

Since the first term of the trinomial $5x^2 + 7x + 2$ is $5x^2$, the first terms of the binomial factors must be $5x$ and x.

$$5x^2 + 7x + 2 = (5x + b)(x + d)$$

Since the product of the last terms must be 2, and the sum of the products of the outer and inner terms must be $7x$, we must find two numbers whose product is 2 that will give a middle term of $7x$.

$$5x^2 + 7x + 2 = (5x + b)(x + d)$$

$$O + I = 7x$$

Because both $2(1)$ and $(-2)(-1)$ give a product of 2, there are four possible combinations to consider:

$$(5x + 2)(x + 1) \qquad (5x - 2)(x - 1)$$
$$(5x + 1)(x + 2) \qquad (5x - 1)(x - 2)$$

Of these possibilities, only the first one gives the correct middle term of $7x$.

1. $5x^2 + 7x + 2 = (5x + 2)(x + 1)$

We can verify this result by multiplication:

$$(5x + 2)(x + 1) = 5x^2 + 5x + 2x + 2$$
$$= 5x^2 + 7x + 2$$

Test for Factorability

If a trinomial has the form $ax^2 + bx + c$, with integer coefficients and $a \neq 0$, we can test to see whether it is factorable. If the value of $b^2 - 4ac$ is a perfect square, the trinomial can be factored using only integers. If the value is not a perfect square, the trinomial is prime and cannot be factored using only integers.

For example, $5x^2 + 7x + 2$ is a trinomial in the form $ax^2 + bx + c$ with

$$a = 5, \qquad b = 7, \qquad \text{and} \qquad c = 2$$

For this trinomial, the value of $b^2 - 4ac$ is

$$b^2 - 4ac = 7^2 - 4(5)(2)$$
$$= 49 - 40$$
$$= 9$$

Since 9 is a perfect square, the trinomial is factorable. Its factorization is shown in Equation 1 on the previous page.

Test for factorability

A trinomial of the form $ax^2 + bx + c$, with integer coefficients and $a \neq 0$, will factor into two binomials with integer coefficients if the value of

$$b^2 - 4ac$$

is a perfect square. If $b^2 - 4ac = 0$, the factors will be the same. If $b^2 - 4ac$ is not a perfect square, the trinomial is prime.

EXAMPLE 10   **Factoring trinomials.** Factor $3p^2 - 4p - 4$.

Solution   In this trinomial, $a = 3$, $b = -4$, and $c = -4$. To see whether it factors, we evaluate $b^2 - 4ac$.

$$b^2 - 4ac = (-4)^2 - 4(3)(-4)$$
$$= 16 + 48$$
$$= 64$$

Since 64 is a perfect square, the trinomial is factorable.

To factor the trinomial, we note that the first terms of the binomial factors must be $3p$ and p to give the first term of $3p^2$.

$$3p^2 - 4p - 4 = (3p + ?)(p + ?)$$

The product of the last terms must be -4, and the sum of the products of the outer terms and the inner terms must be $-4p$.

$$3p^2 - 4p - 4 = (3p + ?)(p + ?)$$

$$O + I = -4p$$

Because $1(-4)$, $-1(4)$, and $-2(2)$ all give a product of -4, there are six possible combinations to consider:

$(3p + 1)(p - 4)$ $(3p - 4)(p + 1)$
$(3p - 1)(p + 4)$ $(3p + 4)(p - 1)$
$(3p - 2)(p + 2)$ $(3p + 2)(p - 2)$

Of these possibilities, only the last gives the correct middle term of $-4p$. Thus,

$$3p^2 - 4p - 4 = (3p + 2)(p - 2)$$

Verify this result by multiplying.

SELF CHECK   Factor $2m^2 - 3m - 9$, if possible.   *Answer:* $(2m + 3)(m - 3)$ ■

Recall the following hints for factoring trinomials.

Factoring a general trinomial

1. Write the trinomial in descending powers of one variable.

2. Test the trinomial for factorability.

3. Factor out any greatest common factor (including -1 if that is necessary to make the coefficient of the first term positive).

4. When the sign of the first term of a trinomial is $+$ and the sign of the third term is $+$, the signs between the terms of each binomial factor are the same as the sign of the middle term of the trinomial.

   When the sign of the first term is $+$ and the sign of the third term is $-$, the signs between the terms of the binomial are opposite.

5. Try various combinations of first terms and last terms until you find one that works.

6. Check the factorization by multiplication.

EXAMPLE 11

Mixed factoring. Factor $24y + 10xy - 6x^2y$.

Solution We write the trinomial in descending powers of x and factor out the common factor of $-2y$:

$$24y + 10xy - 6x^2y = -6x^2y + 10xy + 24y$$
$$= -2y(3x^2 - 5x - 12)$$

In the trinomial $3x^2 - 5x - 12$, $a = 3$, $b = -5$, and $c = -12$.

$$b^2 - 4ac = (-5)^2 - 4(3)(-12) \quad \text{Substitute 3 for } a, -5 \text{ for } b, \text{ and } -12 \text{ for } c.$$
$$= 25 + 144$$
$$= 169$$

Since 169 is a perfect square, the trinomial will factor.

Because the sign of the third term of $3x^2 - 5x - 12$ is $-$, the signs between the binomial factors will be opposite. Because the first term is $3x^2$, the first terms of the binomial factors must be $3x$ and x.

$$\overset{3x^2}{24y + 10xy - 6x^2y = -2y(3x \qquad)(x \qquad)}$$

The product of the last terms must be -12, and the sum of the outer terms and the inner terms must be $-5x$.

$$\overset{-12}{24y + 10xy - 6x^2y = -2y(3x \quad ?)(x \quad ?)}$$
$$O + I = -5x$$

Because $1(-12)$, $2(-6)$, $3(-4)$, $12(-1)$, $6(-2)$, and $4(-3)$ all give a product of -12, there are 12 possible combinations to consider.

$(3x + 1)(x - 12)$ **$(3x - 12)(x + 1)$**

$(3x + 2)(x - 6)$ **$(3x - 6)(x + 2)$**

$(3x + 3)(x - 4)$ $(3x - 4)(x + 3)$

$(3x + 12)(x - 1)$ $(3x - 1)(x + 12)$

$(3x + 6)(x - 2)$ $(3x - 2)(x + 6)$

The one to choose → **$(3x + 4)(x - 3)$**   **$(3x - 3)(x + 4)$**

The six combinations marked in blue cannot work, because one of the factors has a common factor. This implies that $3x^2 - 5x - 12$ would have a common factor, which it doesn't.

After mentally trying the remaining combinations, we find that only $(3x + 4)(x - 3)$ gives the correct middle term of $-5x$.

$$24y + 10xy - 6x^2y = -2y(3x^2 - 5x - 12)$$
$$= -2y(3x + 4)(x - 3)$$

Verify this result by multiplication.

SELF CHECK Factor $18a - 6ap^2 + 3ap$. *Answer:* $-3a(2p + 3)(p - 2)$ ∎

EXAMPLE 12 **Factoring trinomials with variable exponents.** Factor $x^{2n} + x^n - 2$.

Solution Since the first term is x^{2n}, the first terms of the binomial factors must be x^n and x^n.

$$x^{2n} + x^n - 2 = (x^n \qquad)(x^n \qquad)$$

overbrace: x^{2n}

Since the third term is -2, the last terms of the binomial factors must have opposite signs, have a product of -2, and lead to a middle term of x^n. The only combination that works is

$$x^{2n} + x^n - 2 = (x^n + 2)(x^n - 1)$$

Verify this result by multiplication.

SELF CHECK Factor $a^{2n} + 2a^n - 3$. *Answer:* $(a^n + 3)(a^n - 1)$ ∎

EXAMPLE 13 **Mixed factoring.** Factor $x^2 + 6x + 9 - z^2$.

Solution We group the first three terms together and factor the trinomial to get

$$x^2 + 6x + 9 - z^2 = (x + 3)(x + 3) - z^2$$
$$= (x + 3)^2 - z^2$$

We can now factor the difference of two squares to get

$$x^2 + 6x + 9 - z^2 = (x + 3 + z)(x + 3 - z)$$

SELF CHECK Factor $y^2 + 4y + 4 - t^2$. *Answer:* $(y + 2 + t)(y + 2 - t)$ ∎

Using Grouping to Factor Trinomials

The method of factoring by grouping can be used to help factor trinomials of the form $ax^2 + bx + c$. For example, to factor $6x^2 + 5x - 4$, we proceed as follows:

1. First determine the product *ac:* $6(-4) = -24$. This number is called the **key number.**

2. Find two factors of the key number -24 whose sum is $b = 5$:

$$8(-3) = -24 \quad \text{and} \quad 8 + (-3) = 5$$

3. Use the factors 8 and -3 as coefficients of terms to be placed between $6x^2$ and -4:

$$6x^2 + 5x - 4 = 6x^2 + 8x - 3x - 4$$

4. Factor by grouping:

$$6x^2 + 8x - 3x - 4 = 2x(3x + 4) - (3x + 4)$$
$$= (3x + 4)(2x - 1) \qquad \text{Factor out } 3x + 4.$$

We can verify this factorization by multiplication.

STUDY SET

Section 8.3

VOCABULARY

Fill in the blanks to make the statements true.

1. When we write $2x + 4$ as $2(x + 2)$, we say that we have _____ $2x + 4$.

2. If a polynomial cannot be factored, it is called a _____ or irreducible polynomial.

3. When the polynomial $4x^2 - 25$ is written as $(2x)^2 - (5)^2$, we see that it is the difference of two _____.

4. A polynomial with two terms is called a _____.

5. A polynomial with three terms is called a _____.

6. For $a^2 - a - 6$, the _____ coefficient (the coefficient of the a^2 term) is 1.

CONCEPTS

In Exercises 7–10, fill in the blanks to make the statements true.

7. Factoring out a common monomial is based on the distributive property, which is $a(b + c) =$ _____.

8. Factoring the difference of two squares is based on the formula $F^2 - L^2 =$ _____.

9. A trinomial of the form $ax^2 + bx + c$, with integer coefficients and $a \neq 0$, will factor if the value of $b^2 - 4ac$ is a _____ square.

10. If $b^2 - 4ac$ in Exercise 9 is not a perfect square, the trinomial is _____.

11. Explain why each factorization of $30t^2 - 20t^3$ is not complete.
    a. $5t^2(6 - 4t)$

    b. $10(3t - 2t^2)$

12. Explain why each factorization is not complete.
    a. $4g^2 - 16 = (2g + 4)(2g - 4)$

    b. $1 - z^8 = (1 + z^4)(1 - z^4)$

13. Consider $3x^2 - x + 16$. What is the sign of
    a. the first term?
    b. the middle term?
    c. the last term?

14. If $b^2 - 4ac$ is a perfect square, the trinomial $ax^2 + bx + c$ can be factored by using what type of coefficients?

NOTATION

Complete each factorization.

15. $3a - 12 = 3\left(a - \boxed{}\right)$

16. $x^3 - x^2 + 2x - 2 = \boxed{}(x - 1) + \boxed{}(x - 1)$
    $= (x - 1)(x^2 + 2)$

17. $p^2 - q^2 = (p + q)\left(\boxed{}\right)$

18. $2y^2 + 10y + 12 = 2(y^2 + 5y + 6)$
    $= 2\left(y + \boxed{}\right)\left(\boxed{} + 2\right)$

PRACTICE

In Exercises 19–28, factor each expression.

19. $2x + 8$
20. $3y - 9$
21. $2x^2 - 6x$
22. $3y^3 + 3y^2$
23. $15x^2y - 10x^2y^2$
24. $63x^3y^2 + 81x^2y^4$
25. $27z^3 + 12z^2 + 3z$
26. $25t^6 - 10t^3 + 5t^2$
27. $24s^3 - 12s^2t + 6st^2$
28. $18y^2z^2 + 12y^2z^3 - 24y^4z^3$

In Exercises 29–34, factor out the negative of the greatest common factor.

29. $-3a - 6$
30. $-6b + 12$
31. $-6x^2 - 3xy$
32. $-15y^3 + 25y^2$
33. $-63u^3v^6 + 28u^2v^7 - 21u^3v^3$
34. $-56x^4y^3z^2 - 72x^3y^4z^5 + 80xy^2z^3$

In Exercises 35–38, factor out the designated common factor.

35. x^2 from $x^{n+2} + x^{n+3}$
36. y^3 from $y^{n+3} + y^{n+5}$
37. y^n from $2y^{n+2} - 3y^{n+3}$
38. x^n from $4x^{n+3} - 5x^{n+5}$

In Exercises 39–44, factor by grouping.

39. $ax + bx + ay + by$
40. $ar - br + as - bs$
41. $x^2 + yx + 2x + 2y$
42. $2c + 2d - cd - d^2$
43. $3c - cd + 3d - c^2$
44. $x^2 + 4y - xy - 4x$

In Exercises 45–48, solve for the indicated variable.

45. $r_1r_2 = rr_2 + rr_1$ for r_1

46. $r_1r_2 = rr_2 + rr_1$ for r

47. $S(1 - r) = a - lr$ for r

48. $Sn = (n - 2)180°$ for n

In Exercises 49–64, factor each expression. When possible, factor out any common factors first.

49. $x^2 - 4$ **50.** $y^2 - 9$

51. $9y^2 - 64$ **52.** $16x^4 - 81y^2$

53. $81a^4 - 49b^2$ **54.** $64r^6 - 121s^2$

55. $(x + y)^2 - z^2$ **56.** $a^2 - (b - c)^2$

57. $x^4 - y^4$

58. $16a^4 - 81b^4$

59. $2x^2 - 288$ **60.** $8x^2 - 72$

61. $2x^3 - 32x$ **62.** $3x^3 - 243x$

63. $x^{2m} - y^{4n}$

64. $a^{4m} - b^{8n}$

In Exercises 65–68, factor each expression by grouping.

65. $a^2 - b^2 + a + b$ **66.** $x^2 - y^2 - x - y$

67. $2x + y + 4x^2 - y^2$ **68.** $m - 2n + m^2 - 4n^2$

In Exercises 69–74, test each trinomial for factorability and factor it, if possible.

69. $x^2 + 5x + 6$ **70.** $y^2 + 7y + 6$

71. $x^2 - 7x + 10$ **72.** $c^2 - 7c + 12$

73. $a^2 + 5a - 52$ **74.** $b^2 + 9b - 38$

In Exercises 75–82, factor each trinomial. If the coefficient of the first term is negative, begin by factoring out −1.

75. $3x^2 + 12x - 63$ **76.** $2y^2 + 4y - 48$

77. $a^2b^2 - 13ab^2 + 22b^2$

78. $a^2b^2x^2 - 18a^2b^2x + 81a^2b^2$

79. $-a^2 + 4a + 32$ **80.** $-x^2 - 2x + 15$

81. $-3x^2 + 15x - 18$ **82.** $-2y^2 - 16y + 40$

In Exercises 83–100, factor each trinomial. Factor out all common monomials first (including −1 if the first term is negative). If a trinomial is prime, so indicate.

83. $6y^2 + 7y + 2$ **84.** $6x^2 - 11x + 3$

85. $8a^2 + 6a - 9$ **86.** $15b^2 + 4b - 4$

87. $5x^2 + 4x + 1$ **88.** $6z^2 + 17z + 12$

89. $8x^2 - 10x + 3$ **90.** $4a^2 + 20a + 3$

91. $a^2 - 3ab - 4b^2$ **92.** $b^2 + 2bc - 80c^2$

93. $2y^2 + ty - 6t^2$ **94.** $3x^2 - 10xy - 8y^2$

95. $-3a^2 + ab + 2b^2$ **96.** $-2x^2 + 3xy + 5y^2$

97. $3x^3 - 10x^2 + 3x$ **98.** $6y^3 + 7y^2 + 2y$

99. $-4x^3 - 9x + 12x^2$ **100.** $6x^2 + 4x + 9x^3$

In Exercises 101–106, factor each trinomial.

101. $x^4 + 8x^2 + 15$ **102.** $x^4 + 11x^2 + 24$

103. $y^4 - 13y^2 + 30$ **104.** $y^4 - 13y^2 + 42$

105. $a^4 - 13a^2 + 36$

106. $b^4 - 17b^2 + 16$

In Exercises 107–114, factor each expression. Assume that n is a natural number.

107. $x^{2n} + 2x^n + 1$ **108.** $x^{4n} - 2x^{2n} + 1$

109. $2a^{6n} - 3a^{3n} - 2$ **110.** $b^{2n} - b^n - 6$

111. $x^{4n} + 2x^{2n}y^{2n} + y^{4n}$ **112.** $y^{6n} + 2y^{3n}z + z^2$

113. $6x^{2n} + 7x^n - 3$ **114.** $12y^{4n} + 10y^{2n} + 2$

In Exercises 115–120, factor each expression.

115. $x^2 + 4x + 4 - y^2$ **116.** $x^2 - 6x + 9 - 4y^2$

117. $x^2 + 2x + 1 - 9z^2$ **118.** $x^2 + 10x + 25 - 16z^2$

119. $c^2 - 4a^2 + 4ab - b^2$

120. $4c^2 - a^2 - 6ab - 9b^2$

In Exercises 121–128, use factoring by grouping to help factor each trinomial.

121. $a^2 - 17a + 16$ **122.** $b^2 - 4b - 21$

123. $2u^2 + 5u + 3$ **124.** $6y^2 + 5y - 6$

125. $20r^2 - 7rs - 6s^2$ **126.** $6s^2 + st - 12t^2$

127. $20u^2 + 19uv + 3v^2$ **128.** $12m^2 + mn - 6n^2$

APPLICATIONS

129. GEOMETRIC FORMULAS
  a. Write an expression that gives the area of the portion of Illustration 1 that is shaded in red.
  b. Do the same for the portion of the illustration that is shaded in blue.
  c. Add the results from parts a and b and then factor that expression. What formula from geometry do you obtain?

ILLUSTRATION 1

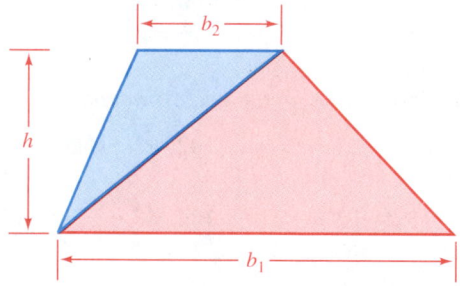

130. PACKAGING The amount of cardboard needed to make the cereal box shown in Illustration 2 can be found by finding the area A, which is given by the formula

$$A = 2wh + 4wl + 2lh$$

where w is the width, h the height, and l the length. Solve the equation for the width.

ILLUSTRATION 2

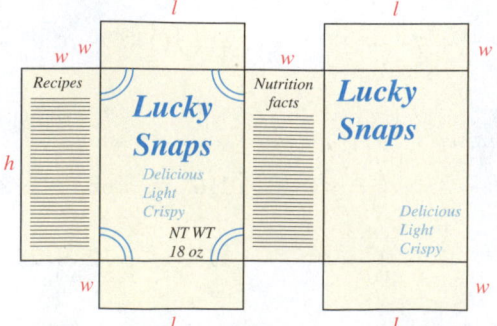

131. LANDSCAPING See Illustration 3. The combined area of the portions of the square lot that the sprinkler doesn't reach is given by $4r^2 - \pi r^2$, where r is the radius of the circular spray. Factor this expression.

ILLUSTRATION 3

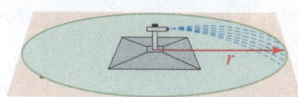

132. CRAYONS The amount of colored wax used to make the crayon shown in Illustration 4 can be found by computing its volume using the formula

$$V = \pi r^2 h_1 + \frac{1}{3}\pi r^2 h_2$$

Factor the expression on the right-hand side of this equation.

ILLUSTRATION 4

133. CANDY To find the amount of chocolate used in the outer coating of the malted-milk ball shown in Illustration 5, we can find the volume V of the chocolate shell using the formula

$$V = \frac{4}{3}\pi r_1^{\,3} - \frac{4}{3}\pi r_2^{\,3}$$

Factor the expression on the right-hand side of the formula.

ILLUSTRATION 5

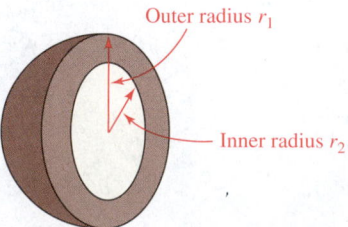

134. MOVIE STUNTS See Illustration 6. The function that gives a stuntwoman's distance above the ground t seconds after falling over the side of a 144-foot-tall building is

$$h(t) = 144 - 16t^2$$

Factor the right-hand side of the equation.

ILLUSTRATION 6

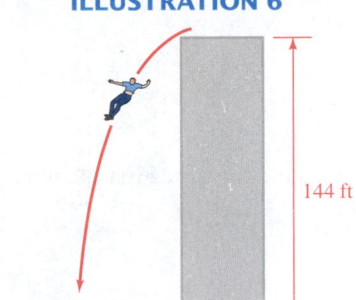

144 ft

135. ICE The surface area of the ice cube shown in Illustration 7 is $6x^2 + 36x + 54$. Find the length of an edge of the cube.

ILLUSTRATION 7

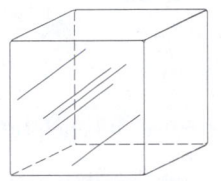

136. CHECKERS The area of the square checkerboard in Illustration 8 is $25x^2 - 40x + 16$. Find the length of each side.

ILLUSTRATION 8

WRITING

137. Explain how you would factor -1 from a trinomial.

138. Explain how you would test the polynomial $ax^2 + bx + c$ for factorability.

139. Because it is the difference of two squares, $x^2 - q^2$ always factors. Does the test for factorability predict this?

140. The polynomial $ax^2 + ax + a$ factors; a is a common factor. Does the test for factorability predict this? Is there something wrong with the test? Explain.

REVIEW

Do each multiplication.

141. $(x + 1)(x^2 - x + 1)$

142. $(m + 3)(m^2 - 3m + 9)$

143. $(r - 2)(r^2 + 2r + 4)$

144. $(a - 5)(a^2 + 5a + 25)$

Solve each equation.

145. $\dfrac{2}{3}(5t - 3) = 38$

146. $2q^2 - 9 = q(q + 3) + q^2$

▶ 8.4 Factoring the Sum and Difference of Two Cubes

In this section, you will learn about

> Factoring the sum of two cubes ■ Factoring the difference of two cubes ■ Multistep factoring

Introduction Recall that the difference of the squares of two quantities factors into the product of two binomials. One binomial is the sum of the quantities, and the other is the difference of the quantities.

$$x^2 - y^2 = (x + y)(x - y) \qquad \text{or} \qquad F^2 - L^2 = (F + L)(F - L)$$

There are similar formulas for factoring the sum of two cubes and the difference of two cubes.

Factoring the Sum of Two Cubes

To find the formula for factoring the sum of two cubes, we need to find the following product:

$$(x + y)(x^2 - xy + y^2) = (x + y)x^2 - (x + y)xy + (x + y)y^2 \quad \text{Use the distributive property.}$$

$$= x^3 + x^2y - x^2y - xy^2 + xy^2 + y^3$$
$$= x^3 + y^3$$

This result justifies the formula for factoring the **sum of two cubes.**

Factoring the sum of two cubes

$$x^3 + y^3 = (x + y)(x^2 - xy + y^2)$$

If we think of the sum of two cubes as the cube of a **First** quantity plus the cube of a **Last** quantity, we have the formula

$$F^3 + L^3 = (F + L)(F^2 - FL + L^2)$$

In words, we say, *To factor the cube of a* **First** *quantity plus the cube of a* **Last** *quantity, we multiply the* **First** *plus the* **Last** *by*

- *the* **First** *squared*
- *minus the* **First** *times the* **Last**
- plus the **Last** *squared.*

To factor the sum of two cubes, it is helpful to know the cubes of the numbers from 1 to 10:

1, 8, 27, 64, 125, 216, 343, 512, 729, 1,000

Expressions containing variables, such as x^6y^3, are also perfect cubes, because they can be written as the cube of a quantity:

$$x^6y^3 = (x^2y)^3$$

EXAMPLE 1

Factoring the sum of two cubes. Factor $x^3 + 8$.

Solution Since $x^3 + 8$ can be written as $x^3 + 2^3$, we have the sum of two cubes, which factors as follows:

$$F^3 + L^3 = (F + L)(F^2 - F \; L + L^2)$$
$$x^3 + 2^3 = (x + 2)(x^2 - x \cdot 2 + 2^2)$$
$$= (x + 2)(x^2 - 2x + 4)$$

We can check by multiplication.

$$(x + 2)(x^2 - 2x + 4) = (x + 2)x^2 - (x + 2)2x + (x + 2)4$$
$$= x^3 + 2x^2 - 2x^2 - 4x + 4x + 8$$
$$= x^3 + 8$$

SELF CHECK

Factor $p^3 + 64$.   *Answer:* $(p + 4)(p^2 - 4p + 16)$ ∎

EXAMPLE 2

Factoring the sum of two cubes. Factor $8b^3 + 27c^3$.

Solution Since $8b^3 + 27c^3$ can be written as $(2b)^3 + (3c)^3$, we have the sum of two cubes, which can be factored as

$$F^3 + L^3 = (F + L)(F^2 - F \; L + L^2)$$
$$(2b)^3 + (3c)^3 = (2b + 3c)[(2b)^2 - (2b)(3c) + (3c)^2]$$
$$= (2b + 3c)(4b^2 - 6bc + 9c^2)$$

We can check by multiplication.

$$(2b + 3c)(4b^2 - 6bc + 9c^2)$$
$$= (2b + 3c)4b^2 - (2b + 3c)6bc + (2b + 3c)9c^2$$
$$= 8b^3 + 12b^2c - 12b^2c - 18bc^2 + 18bc^2 + 27c^3$$
$$= 8b^3 + 27c^3$$

SELF CHECK   Factor $1{,}000p^3 + 125q^3$.   *Answer:* $(10p + 5q)(100p^2 - 50pq + 25q^2)$ ∎

Factoring the Difference of Two Cubes

To find the formula for factoring the difference of two cubes, we need to find the following product:

$$(x - y)(x^2 + xy + y^2) = (x - y)x^2 + (x - y)xy + (x - y)y^2 \quad \text{Use the distributive}$$
$$\text{property.}$$
$$= x^3 - x^2y + x^2y - xy^2 + xy^2 - y^3$$
$$= x^3 - y^3$$

This result justifies the formula for factoring the **difference of two cubes.**

Factoring the difference of two cubes

$$x^3 - y^3 = (x - y)(x^2 + xy + y^2)$$

If we think of the difference of two cubes as the cube of a **First** quantity minus the cube of a **Last** quantity, we have the formula

$$F^3 - L^3 = (F - L)(F^2 + FL + L^2)$$

In words, we say, *To factor the cube of a **First** quantity minus the cube of a **Last** quantity, we multiply the **First** minus the **Last** by*

- *the **First** squared*
- *plus the **First** times the **Last***
- *plus the **Last** squared.*

E X A M P L E   3   **Factoring the difference of two cubes.** Factor $a^3 - 64b^3$.

Solution   Since $a^3 - 64b^3$ can be written as $a^3 - (4b)^3$, we have difference of two cubes, which can be factored as

$$F^3 - L^3 = (F - L)(F^2 + F\ L + L^2)$$
$$a^3 - (4b)^3 = (a - 4b)[a^2 + a(4b) + (4b)^2]$$
$$= (a - 4b)(a^2 + 4ab + 16b^2)$$

We can check by multiplication.

$$(a - 4b)(a^2 + 4ab + 16b^2)$$
$$= (a - 4b)a^2 + (a - 4b)4ab + (a - 4b)16b^2$$
$$= a^3 - 4a^2b + 4a^2b - 16ab^2 + 16ab^2 - 64b^3$$
$$= a^3 - 64b^3$$

SELF CHECK Factor $27p^3 - 8$.

Answer: $(3p - 2)(9p^2 + 6p + 4)$ ■

Multistep Factoring

Sometimes we must factor out a greatest common factor before factoring a sum or difference of two cubes.

EXAMPLE 4

Multistep factoring. Factor $-2t^5 + 128t^2$.

Solution
$$-2t^5 + 128t^2 = -2t^2(t^3 - 64) \qquad \text{Factor out } -2t^2.$$
$$= -2t^2(t - 4)(t^2 + 4t + 16) \quad \text{Factor } t^3 - 64.$$

Verify this factorization by multiplication.

SELF CHECK Factor $-3p^4 + 81p$.

Answer: $-3p(p - 3)(p^2 + 3p + 9)$ ■

EXAMPLE 5

Multistep factoring. Factor $x^6 - 64$.

Solution The binomial $x^6 - 64$ is both the difference of two squares and the difference of two cubes. Since it is easier to factor the difference of two squares first, the expression factors into the product of a sum and a difference.

$$x^6 - 64 = (x^3)^2 - 8^2$$
$$= (x^3 + 8)(x^3 - 8)$$

Because $x^3 + 8$ is the sum of two cubes and $x^3 - 8$ is the difference of two cubes, each of these binomials can be factored.

$$x^6 - 64 = (x^3 + 8)(x^3 - 8)$$
$$= (x + 2)(x^2 - 2x + 4)(x - 2)(x^2 + 2x + 4)$$

Verify this factorization by multiplication.

SELF CHECK Factor $a^6 - 1$.

Answer: $(a + 1)(a^2 - a + 1)(a - 1)(a^2 + a + 1)$ ■

STUDY SET

Section 8.4

VOCABULARY

Fill in the blanks to make the statements true.

1. The binomial $x^3 - y^3$ is called the _____ of two cubes.

2. The binomial $x^3 + y^3$ is called the ___ of two cubes.

CONCEPTS

Complete each formula.

3. $x^3 + y^3 = (x + y)$_____

4. $x^3 - y^3 = (x - y)$_____

NOTATION

Complete each factorization.

5. $a^3 + b^3 = ($ _____ $)(a^2 - ab + b^2)$

6. $4p^3 + 32q^3 =$ ____ $(p^3 + 8q^3)$
   $= 4(p + 2q)($ _____ $)$

PRACTICE

In Exercises 7–26, factor each expression.

7. $y^3 + 1$

8. $x^3 - 8$

9. $a^3 - 27$

10. $b^3 + 125$

11. $8 + x^3$

12. $27 - y^3$

13. $s^3 - t^3$

14. $8u^3 + w^3$

15. $27x^3 + y^3$

16. $x^3 - 27y^3$

17. $a^3 + 8b^3$

18. $27a^3 - b^3$

19. $64x^3 - 27$

20. $27x^3 + 125$

21. $27x^3 - 125y^3$

22. $64x^3 + 27y^3$

23. $a^6 - b^3$

24. $a^3 + b^6$

25. $x^9 + y^6$

26. $x^3 - y^9$

In Exercises 27–42, factor each expression. Factor out any greatest common factors first.

27. $2x^3 + 54$

28. $2x^3 - 2$

29. $-x^3 + 216$

30. $-x^3 - 125$

31. $64m^3x - 8n^3x$

32. $16r^4 + 128rs^3$

33. $x^4y + 216xy^4$

34. $16a^5 - 54a^2b^3$

35. $81r^4s^2 - 24rs^5$

36. $4m^5n + 500m^2n^4$

37. $125a^6b^2 + 64a^3b^5$

38. $216a^4b^4 - 1{,}000ab^7$

39. $y^7z - yz^4$

40. $x^{10}y^2 - xy^5$

41. $2mp^4 + 16mpq^3$

42. $24m^5n - 3m^2n^4$

In Exercises 43–46, factor each expression completely. Factor a difference of two squares first.

43. $x^6 - 1$

44. $x^6 - y^6$

45. $x^{12} - y^6$

46. $a^{12} - 64$

In Exercises 47–54, factor each expression completely.

47. $3(x^3 + y^3) - z(x^3 + y^3)$

48. $x(8a^3 - b^3) + 4(8a^3 - b^3)$

49. $(m^3 + 8n^3) + (m^3x + 8n^3x)$

50. $(a^3x + b^3x) - (a^3y + b^3y)$

51. $(a^4 + 27a) - (a^3b + 27b)$

52. $(x^4 + xy^3) - (x^3y + y^4)$

53. $y^3(y^2 - 1) - 27(y^2 - 1)$

54. $z^3(y^2 - 4) + 8(y^2 - 4)$

APPLICATIONS

55. The area of the rectangle in Illustration 1 is $8a^3 + 27b^3$. Find its length.

ILLUSTRATION 1

$4a^2 - 6ab + 9b^2$

56. The area of the triangle in Illustration 2 is $x^3 + 8y^3$. Find its height.

ILLUSTRATION 2

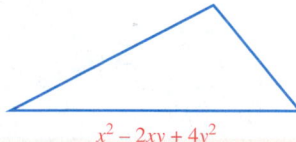

$x^2 - 2xy + 4y^2$

WRITING

57. Explain how to factor $a^3 + b^3$.

58. Explain the difference between $x^3 - y^3$ and $(x - y)^3$.

59. Use a calculator to verify that
$$a^3 - b^3 = (a - b)(a^2 + ab + b^2)$$
when $a = 11$ and $b = 7$.

60. What difficulty do you encounter when you solve $x^3 - 8 = 0$ by factoring?

REVIEW

61. A length of one Fermi is 1×10^{-13} centimeter, approximately the radius of a proton. Express this number in standard notation.

62. In the 14th century, the Black Plague killed about 25,000,000 people, which was 25% of the population of Europe. Find the population at that time, expressed in scientific notation.

▶**8.5**

Review of Rational Expressions

In this section, you will learn about

> Simplifying rational expressions ■ Multiplying and dividing rational expressions ■ Adding and subtracting rational expressions ■ Complex fractions

Introduction In this section, we will review the arithmetic of rational expressions (fractions), first studied in Chapter 6.

Simplifying Rational Expressions

Recall that rational expressions are algebraic fractions with polynomial numerators and polynomial denominators. To manipulate rational expressions, we use the same rules as we use to simplify, multiply, divide, add, and subtract arithmetic fractions.

EXAMPLE 1 **Simplifying fractions.** Simplify $\dfrac{-8y^3z^5}{6y^4z^3}$.

Solution We factor the numerator and denominator and divide out all common factors:

$$\frac{-8y^3z^5}{6y^4z^3} = \frac{-2 \cdot 4 \cdot y \cdot y \cdot y \cdot z \cdot z \cdot z \cdot z \cdot z}{2 \cdot 3 \cdot y \cdot y \cdot y \cdot y \cdot z \cdot z \cdot z}$$

$$= \frac{\overset{1}{-\cancel{2}} \cdot 4 \cdot \overset{1}{\cancel{y}} \cdot \overset{1}{\cancel{y}} \cdot \overset{1}{\cancel{y}} \cdot \overset{1}{\cancel{z}} \cdot \overset{1}{\cancel{z}} \cdot \overset{1}{\cancel{z}} \cdot z \cdot z}{\underset{1}{\cancel{2}} \cdot 3 \cdot \underset{1}{\cancel{y}} \cdot \underset{1}{\cancel{y}} \cdot \underset{1}{\cancel{y}} \cdot y \cdot \underset{1}{\cancel{z}} \cdot \underset{1}{\cancel{z}} \cdot \underset{1}{\cancel{z}}}$$

$$= -\frac{4z^2}{3y}$$

SELF CHECK Simplify $\dfrac{10k}{25k^2}$.

Answer: $\frac{2}{5k}$ ■

The fractions in Example 1 and the Self Check can be simplified by using the rules of exponents:

$$\frac{-8y^3z^5}{6y^4z^3} = \frac{-2 \cdot 4 \cdot y^{3-4} \cdot z^{5-3}}{2 \cdot 3}$$
$$= \frac{-4 \cdot y^{-1} \cdot z^2}{3}$$
$$= -\frac{4z^2}{3y}$$

$$\frac{10k}{25k^2} = \frac{5 \cdot 2 \cdot k^{1-2}}{5 \cdot 5}$$
$$= \frac{2 \cdot k^{-1}}{5}$$
$$= \frac{2}{5k}$$

EXAMPLE 2 **Simplifying fractions.** Simplify $\dfrac{2x^2 + 11x + 12}{3x^2 + 11x - 4}$.

Solution　We factor the numerator and denominator and divide out all common factors:

$$\frac{2x^2 + 11x + 12}{3x^2 + 11x - 4} = \frac{(2x + 3)\overset{1}{\cancel{(x + 4)}}}{(3x - 1)\underset{1}{\cancel{(x + 4)}}}$$

$$= \frac{2x + 3}{3x - 1} \qquad \frac{x+4}{x+4} = 1.$$

 WARNING! Do not divide out the x's in the fraction $\frac{2x + 3}{3x - 1}$. The x in the numerator is a factor of the first term only. It is not a factor of the entire numerator. Likewise, the x in the denominator is not a factor of the entire denominator.

SELF CHECK　Simplify $\dfrac{2x^2 + 5x + 2}{3x^2 + 5x - 2}$. 　　　　　　*Answer:* $\dfrac{2x + 1}{3x - 1}$ ∎

EXAMPLE 3　**Simplifying fractions.** Simplify $\dfrac{3x^2 - 10xy - 8y^2}{4y^2 - xy}$.

Solution　We factor the numerator and denominator and proceed as follows:

$$\frac{3x^2 - 10xy - 8y^2}{4y^2 - xy} = \frac{(3x + 2y)\overset{-1}{\cancel{(x - 4y)}}}{y\underset{1}{\cancel{(4y - x)}}} \qquad \begin{array}{l}\text{Because } x - 4y \text{ and } 4y - x \text{ are} \\ \text{negatives, their quotient is } -1.\end{array}$$

$$= \frac{-(3x + 2y)}{y}$$

$$= \frac{-3x - 2y}{y}$$

SELF CHECK　Simplify $\dfrac{-2a^2 - ab + 3b^2}{a^2 - ab}$. 　　　　　*Answer:* $\dfrac{-2a - 3b}{a}$ ∎

Multiplying and Dividing Rational Expressions

To multiply two fractions, we multiply the numerators and multiply the denominators.

EXAMPLE 4　**Multiplying fractions.** Multiply $\dfrac{x^2 - 6x + 9}{x} \cdot \dfrac{x^2}{x - 3}$.

Solution　We multiply the numerators and multiply the denominators. Then we simplify the resulting fraction.

$$\frac{x^2 - 6x + 9}{x} \cdot \frac{x^2}{x - 3} = \frac{(x^2 - 6x + 9)(x^2)}{x(x - 3)} \qquad \begin{array}{l}\text{Multiply the numerators and} \\ \text{multiply the denominators.}\end{array}$$

$$= \frac{(x - 3)(x - 3)xx}{x(x - 3)} \qquad \begin{array}{l}\text{Factor the numerator and} \\ \text{the denominator.}\end{array}$$

$$= \frac{\overset{1}{\cancel{(x - 3)}}(x - 3)x\overset{1}{\cancel{x}}}{\underset{1}{\cancel{x}}\underset{1}{\cancel{(x - 3)}}} \qquad \text{Divide out common factors.}$$

$$= x(x - 3)$$

SELF CHECK　Multiply $\dfrac{a^2 - 2a + 1}{a} \cdot \dfrac{a^3}{a - 1}$. 　　　　*Answer:* $a^2(a - 1)$ ∎

EXAMPLE 5

Multiplying fractions. Multiply $\dfrac{6x^2 + 5x - 4}{2x^2 + 5x + 3} \cdot \dfrac{8x^2 + 6x - 9}{12x^2 + 7x - 12}$.

Solution   We multiply the fractions, factor each polynomial, and simplify.

$$\dfrac{6x^2 + 5x - 4}{2x^2 + 5x + 3} \cdot \dfrac{8x^2 + 6x - 9}{12x^2 + 7x - 12}$$

$$= \dfrac{(6x^2 + 5x - 4)(8x^2 + 6x - 9)}{(2x^2 + 5x + 3)(12x^2 + 7x - 12)}$$ Multiply the numerators and multiply the denominators.

$$= \dfrac{(3x + 4)(2x - 1)(4x - 3)(2x + 3)}{(2x + 3)(x + 1)(3x + 4)(4x - 3)}$$ Factor the polynomials.

$$= \dfrac{\overset{1}{\cancel{(3x + 4)}}(2x - 1)\overset{1}{\cancel{(4x - 3)}}\overset{1}{\cancel{(2x + 3)}}}{\underset{1}{\cancel{(2x + 3)}}(x + 1)\underset{1}{\cancel{(3x + 4)}}\underset{1}{\cancel{(4x - 3)}}}$$ Divide out the common factors.

$$= \dfrac{2x - 1}{x + 1}$$

SELF CHECK   Multiply $\dfrac{2x^2 + 5x + 3}{3x^2 + 5x + 2} \cdot \dfrac{2x^2 - 5x + 3}{4x^2 - 9}$.

Answer: $\frac{x - 1}{3x + 2}$ ∎

In Examples 4 and 5, we would have obtained the same answers if we had factored first and divided out the common factors before we multiplied.

To divide two fractions, we invert the divisor and multiply.

EXAMPLE 6

Dividing fractions. Divide $\dfrac{x^3 + 8}{x + 1} \div \dfrac{x^2 - 2x + 4}{2x^2 - 2}$.

Solution   Using the rule for division of fractions, we invert the divisor and multiply.

$$\dfrac{x^3 + 8}{x + 1} \div \dfrac{x^2 - 2x + 4}{2x^2 - 2}$$

$$= \dfrac{x^3 + 8}{x + 1} \cdot \dfrac{2x^2 - 2}{x^2 - 2x + 4}$$

$$= \dfrac{(x^3 + 8)(2x^2 - 2)}{(x + 1)(x^2 - 2x + 4)}$$

$$= \dfrac{(x + 2)\overset{1}{\cancel{(x^2 - 2x + 4)}}2\overset{1}{\cancel{(x + 1)}}(x - 1)}{\underset{1}{\cancel{(x + 1)}}\underset{1}{\cancel{(x^2 - 2x + 4)}}}$$ $2x^2 - 2 = 2(x^2 - 1) = 2(x + 1)(x - 1)$.

$$= 2(x + 2)(x - 1)$$

SELF CHECK   Divide $\dfrac{x^3 + 27}{x^2 - 4} \div \dfrac{x^2 - 3x + 9}{x + 2}$.

Answer: $\frac{x + 3}{x - 2}$ ∎

EXAMPLE 7

Simplify $\dfrac{x^2 + 2x - 3}{6x^2 + 5x + 1} \div \dfrac{2x^2 - 2}{2x^2 - 5x - 3} \cdot \dfrac{6x^2 + 4x - 2}{x^2 - 2x - 3}$.

Solution   We change the division to a multiplication. Since multiplications and divisions are done from left to right, only the middle fraction should be inverted. Finally, we multiply the fractions, factor each polynomial, and divide out the common factors.

$$\frac{x^2 + 2x - 3}{6x^2 + 5x + 1} \div \frac{2x^2 - 2}{2x^2 - 5x - 3} \cdot \frac{6x^2 + 4x - 2}{x^2 - 2x - 3}$$

$$= \frac{x^2 + 2x - 3}{6x^2 + 5x + 1} \cdot \frac{2x^2 - 5x - 3}{2x^2 - 2} \cdot \frac{6x^2 + 4x - 2}{x^2 - 2x - 3}$$

$$= \frac{(x^2 + 2x - 3)(2x^2 - 5x - 3)(6x^2 + 4x - 2)}{(6x^2 + 5x + 1)(2x^2 - 2)(x^2 - 2x - 3)}$$

$$= \frac{(x + 3)\overset{1}{\cancel{(x - 1)}}\overset{1}{\cancel{(2x + 1)}}\overset{1}{\cancel{(x - 3)}}\overset{1}{2}(3x - 1)\overset{1}{\cancel{(x + 1)}}}{(3x + 1)\underset{1}{\cancel{(2x + 1)}}\underset{1}{2}(x + 1)\underset{1}{\cancel{(x - 1)}}\underset{1}{\cancel{(x - 3)}}\underset{1}{\cancel{(x + 1)}}}$$

$$= \frac{(x + 3)(3x - 1)}{(3x + 1)(x + 1)}$$ ∎

Adding and Subtracting Rational Expressions

To add or subtract fractions with like denominators, we add or subtract the numerators and keep the same denominator. Whenever possible, we should simplify the result.

EXAMPLE 8

Adding fractions with like denominators. Simplify $\dfrac{4x}{x + 2} + \dfrac{7x}{x + 2}$.

Solution

$$\frac{4x}{x + 2} + \frac{7x}{x + 2} = \frac{4x + 7x}{x + 2}$$

$$= \frac{11x}{x + 2}$$

SELF CHECK   Simplify $\dfrac{4a}{a + 3} + \dfrac{2a}{a + 3}$.

Answer: $\frac{6a}{a + 3}$ ∎

To add or subtract fractions with unlike denominators, we must convert them to fractions with the same denominator.

EXAMPLE 9

Adding fractions with unlike denominators. Simplify $\dfrac{4x}{x + 2} - \dfrac{7x}{x - 2}$.

Solution

$$\frac{4x}{x + 2} - \frac{7x}{x - 2}$$

$$= \frac{4x(x - 2)}{(x + 2)(x - 2)} - \frac{(x + 2)7x}{(x + 2)(x - 2)} \qquad \frac{x-2}{x-2} = 1; \frac{x+2}{x+2} = 1.$$

$$= \frac{(4x^2 - 8x) - (7x^2 + 14x)}{(x + 2)(x - 2)} \qquad \text{Subtract the numerators and keep the common denominator.}$$

$$= \frac{4x^2 - 8x - 7x^2 - 14x}{(x + 2)(x - 2)} \qquad \text{To remove parentheses, use the distributive property.}$$

$$= \frac{-3x^2 - 22x}{(x + 2)(x - 2)} \qquad \text{Combine like terms.}$$

$$= \frac{-x(3x + 22)}{(x + 2)(x - 2)} \qquad \text{Factor the numerator to see if the result simplifies.}$$

 WARNING! The − sign between the fractions in step 2 applies to both terms of $7x^2 + 14x$.

SELF CHECK   Simplify $\dfrac{3a}{a+3} - \dfrac{2a}{a-3}$.   *Answer:* $\dfrac{a^2 - 15a}{(a+3)(a-3)}$ ∎

EXAMPLE 10   **Adding fractions with unlike denominators.** Add

$$\dfrac{x}{x^2 - 2x + 1} + \dfrac{3}{x^2 - 1}.$$

Solution   We factor each denominator and find the LCD:

$$x^2 - 2x + 1 = (x - 1)(x - 1) = (x - 1)^2$$
$$x^2 - 1 = (x + 1)(x - 1)$$

The LCD is $(x - 1)^2(x + 1)$.

We now write each fraction with its denominator in factored form and convert the fractions to fractions with an LCD of $(x - 1)^2(x + 1)$. Finally, we add the fractions.

$$\dfrac{x}{x^2 - 2x + 1} + \dfrac{3}{x^2 - 1} = \dfrac{x}{(x-1)(x-1)} + \dfrac{3}{(x+1)(x-1)}$$

$$= \dfrac{x(x+1)}{(x-1)(x-1)(x+1)} + \dfrac{3(x-1)}{(x+1)(x-1)(x-1)}$$

$$= \dfrac{x^2 + x + 3x - 3}{(x-1)(x-1)(x+1)}$$

$$= \dfrac{x^2 + 4x - 3}{(x-1)^2(x+1)} \quad \text{This result does not simiplify.}$$

SELF CHECK   Add $\dfrac{3}{a^2 + a} + \dfrac{2}{a^2 - 1}$.   *Answer:* $\dfrac{5a - 3}{a(a+1)(a-1)}$ ∎

Complex Fractions

Recall that a *complex fraction* is a fraction with a fraction in its numerator or its denominator or both. Some examples of complex fractions are

$$\dfrac{\frac{3}{5}}{\frac{6}{7}}, \qquad \dfrac{\frac{x+2}{3}}{x - 4}, \qquad \text{and} \qquad \dfrac{\frac{3x^2 - 2}{2x}}{3x - \frac{2}{y}}$$

EXAMPLE 11   **Simplifying complex fractions.** Simplify $\dfrac{\frac{3a}{b}}{\frac{6ac}{b^2}}$.

Solution   **Method 1:** We write the complex fraction as a division and proceed as follows:

$$\dfrac{\frac{3a}{b}}{\frac{6ac}{b^2}} = \dfrac{3a}{b} \div \dfrac{6ac}{b^2}$$

$$= \dfrac{3a}{b} \cdot \dfrac{b^2}{6ac} \qquad \text{Invert the divisor and multiply.}$$

$$= \dfrac{b}{2c} \qquad \text{Multiply the fractions and simplify.}$$

Method 2: We multiply the numerator and denominator by b^2, the LCD of $\frac{3a}{b}$ and $\frac{6ac}{b^2}$, and simplify:

$$\frac{\dfrac{3a}{b}}{\dfrac{6ac}{b^2}} = \frac{\dfrac{3a}{b} \cdot \boldsymbol{b^2}}{\dfrac{6ac}{b^2} \cdot \boldsymbol{b^2}} \qquad \frac{b^2}{b^2} = 1.$$

$$= \frac{\dfrac{3ab^2}{b}}{\dfrac{6ab^2c}{b^2}}$$

$$= \frac{3ab}{6ac} \qquad \text{Simplify the fractions in the numerator and denominator.}$$

$$= \frac{b}{2c} \qquad \text{Divide out the common factor of } 3a.$$

SELF CHECK   Simplify $\dfrac{\dfrac{2x}{y^2}}{\dfrac{6xz}{y}}$.

Answer: $\frac{1}{3yz}$ ∎

E X A M P L E   1 2   **Simplifying complex fractions.** Simplify $\dfrac{\dfrac{1}{x} + \dfrac{1}{y}}{\dfrac{1}{x} - \dfrac{1}{y}}$.

Solution   ***Method 1:*** We add the fractions in the numerator and in the denominator and proceed as follows:

$$\frac{\dfrac{1}{x} + \dfrac{1}{y}}{\dfrac{1}{x} - \dfrac{1}{y}} = \frac{\dfrac{1y}{xy} + \dfrac{x1}{xy}}{\dfrac{1y}{xy} - \dfrac{x1}{xy}}$$

$$= \frac{\dfrac{y + x}{xy}}{\dfrac{y - x}{xy}}$$

$$= \frac{y + x}{xy} \div \frac{y - x}{xy}$$

$$= \frac{y + x}{xy} \cdot \frac{xy}{y - x}$$

$$= \frac{y + x}{y - x} \qquad \text{Multiply and then divide out the factor of } xy.$$

Method 2: We multiply the numerator and denominator by xy (the LCD of the fractions appearing in the complex fraction) and simplify.

$$\frac{\dfrac{1}{x} + \dfrac{1}{y}}{\dfrac{1}{x} - \dfrac{1}{y}} = \frac{xy\left(\dfrac{1}{x} + \dfrac{1}{y}\right)}{xy\left(\dfrac{1}{x} - \dfrac{1}{y}\right)} \qquad \frac{xy}{xy} = 1.$$

$$= \dfrac{\dfrac{xy}{x} + \dfrac{xy}{y}}{\dfrac{xy}{x} - \dfrac{xy}{y}}$$

$$= \dfrac{y + x}{y - x} \qquad \text{Simplify the fraction.}$$

SELF CHECK Simplify $\dfrac{\dfrac{1}{x} - \dfrac{1}{y}}{\dfrac{1}{x} + \dfrac{1}{y}}$.

Answer: $\frac{y-x}{y+x}$ ∎

E X A M P L E   1 3 **Simplifying complex fractions.** Simplify $\dfrac{x^{-1} + y^{-1}}{x^{-2} - y^{-2}}$.

Solution **Method 1:** We proceed as follows:

$$\dfrac{x^{-1} + y^{-1}}{x^{-2} - y^{-2}} = \dfrac{\dfrac{1}{x} + \dfrac{1}{y}}{\dfrac{1}{x^2} - \dfrac{1}{y^2}} \qquad \begin{array}{l}\text{Write the fraction without using negative}\\ \text{exponents.}\end{array}$$

$$= \dfrac{\dfrac{y}{xy} + \dfrac{x}{xy}}{\dfrac{y^2}{x^2y^2} - \dfrac{x^2}{x^2y^2}} \qquad \begin{array}{l}\text{Get a common denominator in the numerator}\\ \text{and denominator.}\end{array}$$

$$= \dfrac{\dfrac{y + x}{xy}}{\dfrac{y^2 - x^2}{x^2y^2}} \qquad \begin{array}{l}\text{Add the fractions in the numerator and}\\ \text{denominator.}\end{array}$$

$$= \dfrac{y + x}{xy} \div \dfrac{y^2 - x^2}{x^2y^2} \qquad \text{Write the fraction as a division.}$$

$$= \dfrac{y + x}{xy} \cdot \dfrac{xxyy}{(y - x)(y + x)} \qquad \text{Invert and multiply.}$$

$$= \dfrac{(y + x)xxyy}{xy(y - x)(y + x)} \qquad \text{Multiply the numerators and the denominators.}$$

$$= \dfrac{xy}{y - x} \qquad \begin{array}{l}\text{Divide out the common factors of } x, y, \text{ and}\\ y + x \text{ in the numerator and denominator.}\end{array}$$

Method 2: We multiply both numerator and denominator by x^2y^2, the LCD of the fractions in the problem, and proceed as follows:

$$\dfrac{x^{-1} + y^{-1}}{x^{-2} - y^{-2}} = \dfrac{\dfrac{1}{x} + \dfrac{1}{y}}{\dfrac{1}{x^2} - \dfrac{1}{y^2}} \qquad \text{Write the fraction without negative exponents.}$$

$$= \dfrac{x^2y^2\left(\dfrac{1}{x} + \dfrac{1}{y}\right)}{x^2y^2\left(\dfrac{1}{x^2} - \dfrac{1}{y^2}\right)} \qquad \frac{x^2y^2}{x^2y^2} = 1.$$

$$= \frac{xy^2 + x^2y}{y^2 - x^2}$$ Use the distributive property to remove parentheses and simplify.

$$= \frac{xy(y + x)}{(y + x)(y - x)}$$ Factor the numerator and denominator.

$$= \frac{xy}{y - x}$$ Divide out $y + x$.

SELF CHECK   Simplify $\dfrac{x^{-1} - y^{-1}}{x^{-2}}$.   *Answer:* $\dfrac{xy - x^2}{y}$ ∎

 WARNING! $x^{-1} + y^{-1}$ means $\frac{1}{x} + \frac{1}{y}$ and $(x + y)^{-1}$ means $\frac{1}{x+y}$. Thus,

$$x^{-1} + y^{-1} \neq \frac{1}{x + y} \qquad \text{and} \qquad (x + y)^{-1} \neq x^{-1} + y^{-1}$$

EXAMPLE 14   **Simplifying complex fractions.** Simplify $\dfrac{\dfrac{2x}{1 - \dfrac{1}{x}} + 3}{3 - \dfrac{2}{x}}$.

Solution   We begin by multiplying the numerator and denominator of the fraction

$$\frac{2x}{1 - \dfrac{1}{x}}$$

by x. This will eliminate the complex fraction in the numerator of the given fraction.

$$\frac{\dfrac{2x}{1 - \dfrac{1}{x}} + 3}{3 - \dfrac{2}{x}} = \frac{\dfrac{x2x}{x\left(1 - \dfrac{1}{x}\right)} + 3}{3 - \dfrac{2}{x}} \qquad \tfrac{x}{x} = 1.$$

$$= \frac{\dfrac{2x^2}{x - 1} + 3}{3 - \dfrac{2}{x}}$$

We then multiply the numerator and denominator of the previous fraction by $x(x - 1)$, the LCD of $\frac{2x^2}{x-1}$, 3, and $\frac{2}{x}$, and simplify:

$$\frac{\dfrac{2x}{1 - \dfrac{1}{x}} + 3}{3 - \dfrac{2}{x}} = \frac{x(x - 1)\left(\dfrac{2x^2}{x - 1} + 3\right)}{x(x - 1)\left(3 - \dfrac{2}{x}\right)} \qquad \tfrac{x(x-1)}{x(x-1)} = 1.$$

$$= \frac{2x^3 + 3x(x - 1)}{3x(x - 1) - 2(x - 1)}$$

$$= \frac{2x^3 + 3x^2 - 3x}{3x^2 - 5x + 2}$$

This result does not simplify.

SELF CHECK Simplify $\dfrac{\dfrac{3}{1 - \frac{2}{x}} + 1}{2 - \frac{1}{x}}$.

Answer: $\dfrac{4x^2 - 2x}{x^2 - 5x + 2}$

STUDY SET

Section 8.5

VOCABULARY

Fill in the blanks to make the statements true.

1. The expression above a fraction bar is called the ______________ of the fraction.
2. The expression below a fraction bar is called the ______________ of the fraction.
3. A ______________ fraction has a fraction in its numerator or denominator, or both.
4. The denominator of $\dfrac{\frac{2x}{3} + 2}{\frac{3x}{2} - 1}$ is ______.

CONCEPTS

Fill in the blanks to make the statements true.

5. The denominator of a fraction can never be ___.
6. $x^{-2} =$ ______
7. $\dfrac{ax}{bx} =$ ______ $(b \neq 0, x \neq 0)$
8. $\dfrac{a}{b} \cdot \dfrac{c}{d} =$ ______ $(b \neq 0, d \neq 0)$
9. $\dfrac{a}{b} \div \dfrac{c}{d} =$ ______ $(b \neq 0, c \neq 0, d \neq 0)$
10. $\dfrac{a}{b} + \dfrac{c}{b} =$ ______ $(b \neq 0)$

NOTATION

Complete each solution.

11. $\dfrac{x^2 + 3x}{5x - 25} \cdot \dfrac{x - 5}{x + 3} = \dfrac{(x^2 + 3x)\ ____}{____\ (x + 3)}$
$= \dfrac{____\ (x - 5)}{____\ (x + 3)}$
$= \dfrac{x}{5}$

12. $\dfrac{6x - 1}{3x - 1} + \dfrac{3x - 2}{3x - 1} = \dfrac{6x - 1 + ____}{3x - 1}$
$= \dfrac{9x - ____}{3x - 1}$
$= \dfrac{3(____)}{3x - 1}$
$= 3$

PRACTICE

In Exercises 13–36, simplify each fraction.

13. $\dfrac{12x^3}{3x}$
14. $-\dfrac{15a^2}{25a^3}$
15. $\dfrac{-24x^3y^4}{18x^4y^3}$
16. $\dfrac{15a^5b^4}{21b^3c^2}$
17. $\dfrac{9y^2(y - z)}{21y(y - z)^2}$
18. $\dfrac{-3ab^2(a - b)}{9ab(b - a)}$
19. $\dfrac{(a - b)(b - c)(c - d)}{(c - d)(b - c)(a - b)}$
20. $\dfrac{(p + q)(p - r)(r + s)}{(r - p)(r + s)(p + q)}$
21. $\dfrac{x + y}{x^2 - y^2}$
22. $\dfrac{x - y}{x^2 - y^2}$
23. $\dfrac{12 - 3x^2}{x^2 - x - 2}$
24. $\dfrac{x^2 + 2x - 15}{x^2 - 25}$
25. $\dfrac{x^3 + 8}{x^2 - 2x + 4}$
26. $\dfrac{x^2 + 3x + 9}{x^3 - 27}$

27. $\dfrac{x^2 + 2x + 1}{x^2 + 4x + 3}$

28. $\dfrac{6x^2 + x - 2}{8x^2 + 2x - 3}$

29. $\dfrac{3m - 6n}{3n - 6m}$

30. $\dfrac{ax + by + ay + bx}{a^2 - b^2}$

31. $\dfrac{4x^2 + 24x + 32}{16x^2 + 8x - 48}$

32. $\dfrac{a^2 - 4}{a^3 - 8}$

33. $\dfrac{3x^2 - 3y^2}{x^2 + 2y + 2x + yx}$

34. $\dfrac{x^2 + 2xy}{x + 2y + x^2 - 4y^2}$

35. $\dfrac{x - y}{x^3 - y^3 - x + y}$

36. $\dfrac{2x^2 + 2x - 12}{x^3 + 3x^2 - 4x - 12}$

In Exercises 37–106, do the operations and simplify.

37. $\dfrac{x^2 y^2}{cd} \cdot \dfrac{c^{-2} d^2}{x}$

38. $\dfrac{a^{-2} b^2}{x^{-1} y} \cdot \dfrac{a^4 b^4}{x^2 y^3}$

39. $\dfrac{-x^2 y^{-2}}{x^{-1} y^{-3}} \div \dfrac{x^{-3} y^2}{x^4 y^{-1}}$

40. $\dfrac{(a^3)^2}{b^{-1}} \div \dfrac{(a^3)^{-2}}{b^{-1}}$

41. $\dfrac{x^2 + 2x + 1}{x} \cdot \dfrac{x^2 - x}{x^2 - 1}$

42. $\dfrac{a + 6}{a^2 - 16} \cdot \dfrac{3a - 12}{3a + 18}$

43. $\dfrac{2x^2 - x - 3}{x^2 - 1} \cdot \dfrac{x^2 + x - 2}{2x^2 + x - 6}$

44. $\dfrac{9x^2 + 3x - 20}{3x^2 - 7x + 4} \cdot \dfrac{3x^2 - 5x + 2}{9x^2 + 18x + 5}$

45. $\dfrac{x^2 - 16}{x^2 - 25} \div \dfrac{x + 4}{x - 5}$

46. $\dfrac{a^2 - 9}{a^2 - 49} \div \dfrac{a + 3}{a + 7}$

47. $\dfrac{a^2 + 2a - 35}{12x} \div \dfrac{ax - 3x}{a^2 + 4a - 21}$

48. $\dfrac{x^2 - 4}{2b - bx} \div \dfrac{x^2 + 4x + 4}{2b + bx}$

49. $\dfrac{3t^2 - t - 2}{6t^2 - 5t - 6} \cdot \dfrac{4t^2 - 9}{2t^2 + 5t + 3}$

50. $\dfrac{2p^2 - 5p - 3}{p^2 - 9} \cdot \dfrac{2p^2 + 5p - 3}{2p^2 + 5p + 2}$

51. $\dfrac{3n^2 + 5n - 2}{12n^2 - 13n + 3} \div \dfrac{n^2 + 3n + 2}{4n^2 + 5n - 6}$

52. $\dfrac{8y^2 - 14y - 15}{6y^2 - 11y - 10} \div \dfrac{4y^2 - 9y - 9}{3y^2 - 7y - 6}$

53. $(2x^2 - 15x + 25) \div \dfrac{2x^2 - 3x - 5}{x + 1}$

54. $(x^2 - 6x + 9) \div \dfrac{x^2 - 9}{x + 3}$

55. $\dfrac{x^3 + y^3}{x^3 - y^3} \div \dfrac{x^2 - xy + y^2}{x^2 + xy + y^2}$

56. $\dfrac{x^2 - 6x + 9}{4 - x^2} \div \dfrac{x^2 - 9}{x^2 - 8x + 12}$

57. $\dfrac{m^2 - n^2}{2x^2 + 3x - 2} \cdot \dfrac{2x^2 + 5x - 3}{n^2 - m^2}$

58. $\dfrac{x^2 - y^2}{2x^2 + 2xy + x + y} \cdot \dfrac{2x^2 - 5x - 3}{yx - 3y - x^2 + 3x}$

59. $\dfrac{ax + ay + bx + by}{x^3 - 27} \cdot \dfrac{x^2 + 3x + 9}{xc + xd + yc + yd}$

60. $\dfrac{x^2 + 3x + yx + 3y}{x^2 - 9} \cdot \dfrac{x - 3}{x + 3}$

61. $\dfrac{x^2 - x - 6}{x^2 - 4} \cdot \dfrac{x^2 - x - 2}{9 - x^2}$

62. $\dfrac{2x^2 - 7x - 4}{20 - x - x^2} \div \dfrac{2x^2 - 9x - 5}{x^2 - 25}$

63. $\dfrac{2x^2 + 3xy + y^2}{y^2 - x^2} \div \dfrac{6x^2 + 5xy + y^2}{2x^2 - xy - y^2}$

64. $\dfrac{p^3 - q^3}{q^2 - p^2} \cdot \dfrac{q^2 + pq}{p^3 + p^2 q + pq^2}$

65. $\dfrac{3x^2 y^2}{6x^3 y} \cdot \dfrac{-4x^7 y^{-2}}{18x^{-2} y} \div \dfrac{36x}{18y^{-2}}$

66. $\dfrac{9ab^3}{7xy} \cdot \dfrac{14xy^2}{27z^3} \div \dfrac{18a^2 b^2 x}{3z^2}$

67. $(4x + 12) \cdot \dfrac{x^2}{2x - 6} \div \dfrac{2}{x - 3}$

68. $(4x^2 - 9) \div \dfrac{2x^2 + 5x + 3}{x + 2} \div (2x - 3)$

69. $\dfrac{2x^2 - 2x - 4}{x^2 + 2x - 8} \cdot \dfrac{3x^2 + 15x}{x + 1} \div \dfrac{4x^2 - 100}{x^2 - x - 20}$

70. $\dfrac{6a^2 - 7a - 3}{a^2 - 1} \div \dfrac{4a^2 - 12a + 9}{a^2 - 1} \cdot \dfrac{2a^2 - a - 3}{3a^2 - 2a - 1}$

71. $\dfrac{2x^2 + 5x + 3}{x^2 + 2x - 3} \div \left(\dfrac{x^2 + 2x - 35}{x^2 - 6x + 5} \div \dfrac{x^2 - 9x + 14}{2x^2 - 5x + 2} \right)$

72. $\dfrac{x^2 - 4}{x^2 - x - 6} \div \left(\dfrac{x^2 - x - 2}{x^2 - 8x + 15} \cdot \dfrac{x^2 - 3x - 10}{x^2 + 3x + 2} \right)$

73. $\dfrac{x^2 - x - 12}{x^2 + x - 2} \div \dfrac{x^2 - 6x + 8}{x^2 - 3x - 10} \cdot \dfrac{x^2 - 3x + 2}{x^2 - 2x - 15}$

74. $\dfrac{4x^2 - 10x + 6}{x^4 - 3x^3} \div \dfrac{2x - 3}{2x^3} \cdot \dfrac{x - 3}{2x - 2}$

75. $\dfrac{3}{a + b} - \dfrac{a}{a + b}$

76. $\dfrac{x}{x + 4} + \dfrac{5}{x + 4}$

77. $\dfrac{3x}{2x + 2} + \dfrac{x + 4}{2x + 2}$

78. $\dfrac{4y}{y - 4} - \dfrac{16}{y - 4}$

79. $\dfrac{5x}{x + 1} + \dfrac{3}{x + 1} - \dfrac{2x}{x + 1}$

80. $\dfrac{4}{a + 4} - \dfrac{2a}{a + 4} + \dfrac{3a}{a + 4}$

81. $\dfrac{3(x^2 + x)}{x^2 - 5x + 6} + \dfrac{-3(x^2 - x)}{x^2 - 5x + 6}$

82. $\dfrac{2x + 4}{x^2 + 13x + 12} - \dfrac{x + 3}{x^2 + 13x + 12}$

83. $\dfrac{a}{2} + \dfrac{2a}{5}$

84. $\dfrac{b}{6} + \dfrac{3a}{4}$

85. $\dfrac{3}{4x} + \dfrac{2}{3x}$

86. $\dfrac{2}{5a} + \dfrac{3}{2b}$

87. $\dfrac{a + b}{3} + \dfrac{a - b}{7}$

88. $\dfrac{x - y}{2} + \dfrac{x + y}{3}$

89. $\dfrac{3}{x + 2} + \dfrac{5}{x - 4}$

90. $\dfrac{2}{a + 4} - \dfrac{6}{a + 3}$

91. $\dfrac{x + 2}{x + 5} - \dfrac{x - 3}{x + 7}$

92. $\dfrac{7}{x + 3} + \dfrac{4x}{x + 6}$

93. $x + \dfrac{1}{x}$

94. $2 - \dfrac{1}{x + 1}$

95. $\dfrac{x + 8}{x - 3} - \dfrac{x - 14}{3 - x}$

96. $\dfrac{3 - x}{2 - x} + \dfrac{x - 1}{x - 2}$

97. $\dfrac{x}{x^2 + 5x + 6} + \dfrac{x}{x^2 - 4}$

98. $\dfrac{x}{3x^2 - 2x - 1} + \dfrac{4}{3x^2 + 10x + 3}$

99. $\dfrac{8}{x^2 - 9} + \dfrac{2}{x - 3} - \dfrac{6}{x}$

100. $\dfrac{x}{x^2 - 4} - \dfrac{x}{x + 2} + \dfrac{2}{x}$

101. $1 + x - \dfrac{x}{x - 5}$

102. $2 - x + \dfrac{3}{x - 9}$

103. $\dfrac{3}{x + 1} - \dfrac{2}{x - 1} + \dfrac{x + 3}{x^2 - 1}$

104. $\dfrac{2}{x - 2} + \dfrac{3}{x + 2} - \dfrac{x - 1}{x^2 - 4}$

105. $\dfrac{x - 2}{x^2 - 3x} + \dfrac{2x - 1}{x^2 + 3x} - \dfrac{2}{x^2 - 9}$

106. $\dfrac{2}{x - 1} - \dfrac{2x}{x^2 - 1} - \dfrac{x}{x^2 + 2x + 1}$

In Exercises 107–130, simplify each complex fraction.

107. $\dfrac{\dfrac{4x}{y}}{\dfrac{6xz}{y^2}}$

108. $\dfrac{\dfrac{5t^4}{9x}}{\dfrac{2t}{18x}}$

109. $\dfrac{\dfrac{x - y}{xy}}{\dfrac{y - x}{x}}$

110. $\dfrac{\dfrac{x^2 + 5x + 6}{3xy}}{\dfrac{x^2 - 9}{6xy}}$

111. $\dfrac{\dfrac{1}{a} + \dfrac{1}{b}}{\dfrac{1}{a}}$

112. $\dfrac{\dfrac{1}{b}}{\dfrac{1}{a} - \dfrac{1}{b}}$

113. $\dfrac{\dfrac{y}{x} - \dfrac{x}{y}}{\dfrac{1}{x} + \dfrac{1}{y}}$

114. $\dfrac{\dfrac{y}{x} - \dfrac{x}{y}}{\dfrac{1}{y} - \dfrac{1}{x}}$

115. $\dfrac{\dfrac{1}{a} - \dfrac{1}{b}}{\dfrac{a}{b} - \dfrac{b}{a}}$

116. $\dfrac{\dfrac{1}{a} + \dfrac{1}{b}}{\dfrac{a}{b} - \dfrac{b}{a}}$

117. $\dfrac{1 + \dfrac{6}{x} + \dfrac{8}{x^2}}{1 + \dfrac{1}{x} - \dfrac{12}{x^2}}$

118. $\dfrac{1 - x - \dfrac{2}{x}}{\dfrac{6}{x^2} + \dfrac{1}{x} - 1}$

119. $\dfrac{\dfrac{1}{a + 1} + 1}{\dfrac{3}{a - 1} + 1}$

120. $\dfrac{2 + \dfrac{3}{x + 1}}{\dfrac{1}{x} + x + x^2}$

121. $\dfrac{x^{-1} + y^{-1}}{x^{-1} - y^{-1}}$

122. $\dfrac{(x + y)^{-1}}{x^{-1} + y^{-1}}$

123. $\dfrac{x + y}{x^{-1} + y^{-1}}$

124. $\dfrac{x - y}{x^{-1} - y^{-1}}$

125. $\dfrac{x - y^{-2}}{y - x^{-2}}$

126. $\dfrac{x^{-2} - y^{-2}}{x^{-1} - y^{-1}}$

127. $\dfrac{1 + \dfrac{a}{b}}{1 - \dfrac{a}{1 - \dfrac{a}{b}}}$

128. $\dfrac{1 + \dfrac{2}{1 + \dfrac{a}{b}}}{1 - \dfrac{a}{b}}$

129. $a + \dfrac{a}{1 + \dfrac{a}{a + 1}}$

130. $b + \dfrac{b}{1 - \dfrac{b + 1}{b}}$

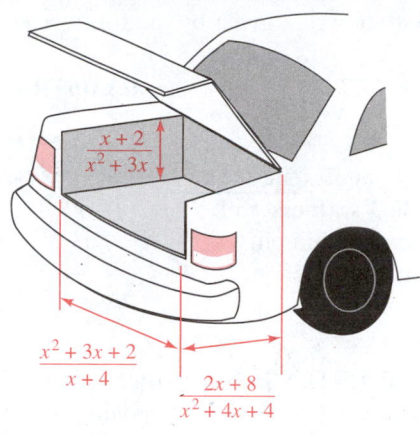

APPLICATIONS

131. PHYSICS EXPERIMENT Illustration 1 contains data from a physics experiment. k_1 and k_2 are constants. Complete the table.

ILLUSTRATION 1

Trial	Rate (m/sec)	Time (sec)	Distance (m)
1	$\dfrac{k_1^2 + 3k_1 + 2}{k_1 - 3}$	$\dfrac{k_1^2 - 3k_1}{k_1 + 1}$	
2	$\dfrac{k_2^2 + 6k_2 + 5}{k_2 + 1}$		$k_2^2 + 11k_2 + 30$

132. TRUNK CAPACITY The shape of the storage space in the trunk of the car shown in Illustration 2 is approximately a rectangular solid. Write a simplified rational expression that gives the number of cubic units of storage space in the trunk.

133. DRAFTING Among the tools used in drafting are the 45°–45°–90° and the 30°–60°–90° triangles shown in Illustration 3. Find the perimeter of each triangle and express each result as a rational expression.

134. THE AMAZON The Amazon River flows in a general eastern direction to the Atlantic Ocean. In Brazil, when the river is at low stage, the rate of flow is about 5 mph. Suppose that a river guide can canoe in still water at a rate of r mph.

a. Complete the table in Illustration 4 to find rational expressions that represent the time it would take the guide to canoe 3 miles downriver and to canoe 3 miles upriver.

ILLUSTRATION 4

	Rate (mph)	Time (hr)	Distance (mi)
Downriver	$r + 5$		3
Upriver	$r - 5$		3

b. Find the difference in the times for the trips downriver and upriver. Express the result as a rational expression.

ILLUSTRATION 2

$\dfrac{x+2}{x^2+3x}$
$\dfrac{x^2+3x+2}{x+4}$
$\dfrac{2x+8}{x^2+4x+4}$

ILLUSTRATION 3

45°
10
$\dfrac{10}{r}$
90°
45°
30°
$\dfrac{6}{t}$
3
90°
60°
For a 45°-45°-90° triangle, these two sides are the same length.
For a 30°-60°-90° triangle, this side is half as long as the hypotenuse.

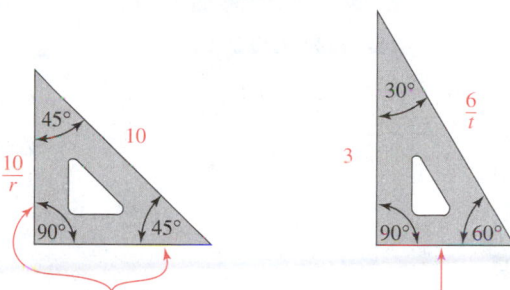

135. ENGINEERING The stiffness k of the shaft shown in Illustration 5 is given by the formula

$$k = \dfrac{1}{\dfrac{1}{k_1} + \dfrac{1}{k_2}}$$

ILLUSTRATION 5

Section 1   Section 2

where k_1 and k_2 are the individual stiffnesses of each section. Simplify the complex fraction.

136. KITCHEN UTENSILS See Illustration 6. What is the ratio of the width of the opening of the ice tongs to the width of the opening of the handles? Express the results in simplest form.

ILLUSTRATION 6

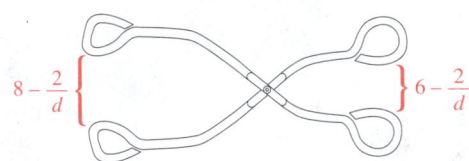

$8 - \dfrac{2}{d}$ $6 - \dfrac{2}{d}$

WRITING

137. A student compared his answer, $\frac{a - 3b}{2b - a}$, with the answer, $\frac{3b - a}{a - 2b}$, in the back of the text. Is the student's work correct? Explain.

138. Another student shows this work:

$$\dfrac{3x^2 + 6}{3y} = \dfrac{\cancel{3}x^2 + \overset{2}{\cancel{6}}}{\cancel{3}y} = \dfrac{x^2 + 2}{y}$$

Is the student's work correct? Explain.

139. In which parts can you divide out the 4's? Explain

 a. $\dfrac{4x}{4y}$ b. $\dfrac{4x}{x + 4}$

 c. $\dfrac{4 + x}{4 + y}$ d. $\dfrac{4x}{4 + 4y}$

140. In which parts can you divide out the 3's? Explain.

 a. $\dfrac{3x + 3y}{3z}$ b. $\dfrac{3(x + y)}{3x + y}$

 c. $\dfrac{x + 3}{3y}$ d. $\dfrac{3x + 3y}{3a - 3b}$

REVIEW

Graph each interval.

141. $(-\infty, -4) \cup [5, \infty)$

142. $(4, 8]$

Solve each formula for the indicated letter.

143. $P = 2l + 2w$; for w

144. $S = \dfrac{a - lr}{1 - r}$; for a

Solve each equation.

145. $a^4 - 13a^2 + 36 = 0$

146. $|2x - 1| = 9$

▶8.6   Review of Linear Functions

In this section, you will review

 ■ The rectangular coordinate system ■ The midpoint formula ■ Slope of a nonvertical line ■ Horizontal and vertical lines ■ Slopes of parallel and perpendicular lines ■ Writing equations of lines ■ Functions and function notation ■ Linear functions ■ An application

Introduction In Section 3.7, we saw that if x and y are real numbers, an equation in x and y determines a correspondence between the values of x and y. When a correspondence between x and y is set up by an equation, a table, or a graph, in which only one y-value corresponds to each x-value, we call the correspondence a **function**. In this section, we will review the concepts of graphing and functions.

The Rectangular Coordinate System

Recall that the rectangular coordinate system is composed of two perpendicular number lines, one horizontal and one vertical, that divide the plane into four quadrants, numbered as in Figure 8-22. The horizontal number line is the **x-axis,** and the vertical number line is the **y-axis.** The point where the axes intersect is called the **origin.**

FIGURE 8-22

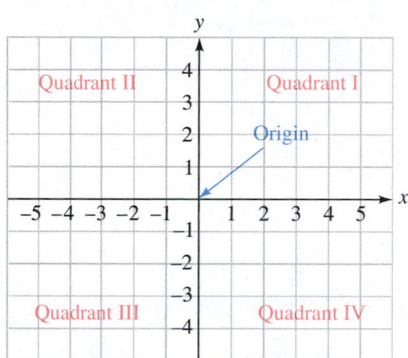

To plot the point associated with the ordered pair of real numbers (2, 3), we start at the origin and count 2 units to the right and then 3 units up, as in Figure 8-23. The point P, which lies in the first quadrant, is the **graph** of the ordered pair (2, 3). The ordered pair (2, 3) gives the **coordinates** of point P.

FIGURE 8-23

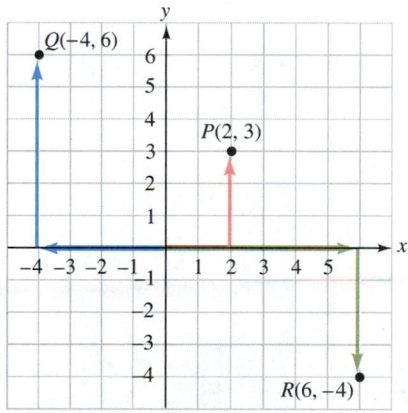

To plot point Q with coordinates (−4, 6), we start at the origin and count 4 units to the left and then 6 units up. Point Q lines in the second quadrant. The point R with coordinates (6, −4) lies in the fourth quadrant.

The **graph of an equation** in the variables x and y is the set of all points with coordinates (x, y) that satisfy the equation.

EXAMPLE 1   **Graphing equations.** Graph $3x + 2y = 12$.

Solution   We pick values for either x or y, substitute them in the equation, and solve for the other variable. For example, if $x = 2$, then

$$3x + 2y = 12$$
$$3(2) + 2y = 12 \quad \text{Substitute 2 for } x.$$
$$6 + 2y = 12 \quad \text{Simplify.}$$
$$2y = 6 \quad \text{Subtract 6 from both sides.}$$
$$y = 3 \quad \text{Divide both sides by 2.}$$

One ordered pair that satisfies the equation is $(2, 3)$. If $y = 6$, we will find that $x = 0$, and a second ordered pair that satisfies the equation is $(0, 6)$.

The ordered pairs $(2, 3)$ and $(0, 6)$ and others that satisfy the equation are shown in the table in Figure 8-24. We plot each of these pairs and join them to get the line shown in the figure. This line is the graph of the equation.

FIGURE 8-24

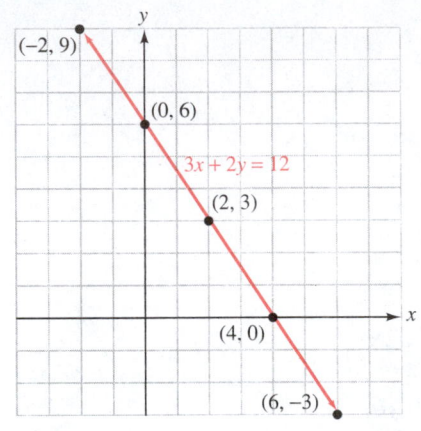

$$3x + 2y = 12$$

x	y	(x, y)
2	3	$(2, 3)$
0	6	$(0, 6)$
4	0	$(4, 0)$
6	-3	$(6, -3)$
-2	9	$(-2, 9)$

SELF CHECK   Graph $2x + 3y = -6$.

Answer:

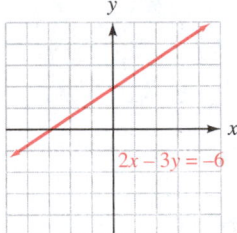

In Example 1, the **y-intercept** of the line is the point where the line intersects the y-axis, the point with coordinates of $(0, 6)$. The **x-intercept** is the point where the line intersects the x-axis, the point with coordinates of $(4, 0)$.

The Midpoint Formula

FIGURE 8-25

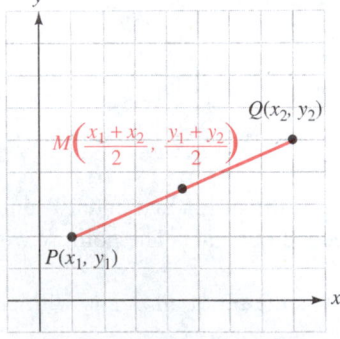

To distinguish between the coordinates of two points on a line, we often use subscript notation. Point $P(x_1, y_1)$ is read as "point P with coordinates of x sub 1 and y sub 1." Point $Q(x_2, y_2)$ is read as "point Q with coordinates of x sub 2 and y sub 2."

If point M in Figure 8-25 lies midway between points $P(x_1, y_1)$ and $Q(x_2, y_2)$, point M is called the **midpoint** of segment PQ. To find the coordinates of M, we average the x-coordinates and average the y-coordinates of P and Q.

The midpoint formula

The **midpoint** of the line segment with endpoints at $P(x_1, y_1)$ and $Q(x_2, y_2)$ is the point M with coordinates of

$$\left(\frac{x_1 + x_2}{2}, \frac{y_1 + y_2}{2} \right)$$

EXAMPLE 2

Finding midpoints. Find the midpoint of the line segment joining $P(-2, 3)$ and $Q(3, -5)$.

Solution

To find the midpoint, we average the x-coordinates and average the y-coordinates to get

$$\frac{x_1 + x_2}{2} = \frac{-2 + 3}{2} \quad \text{and} \quad \frac{y_1 + y_2}{2} = \frac{3 + (-5)}{2}$$

$$= \frac{1}{2} \qquad\qquad\qquad = -1$$

The midpoint of segment PQ is the point $M\left(\frac{1}{2}, -1\right)$.

SELF CHECK

Find the midpoint of the segment joining $P(5, -4)$ and $Q(-3, 5)$. *Answer:* $\left(1, \frac{1}{2}\right)$ ∎

Slope of a Nonvertical Line

Recall that the **slope of a nonvertical line** is the ratio of the change in y divided by the change in x between any two points on a line. For any line, this ratio is always constant.

Slope of a nonvertical line

> The **slope of the nonvertical line** passing through points $P(x_1, y_1)$ and $Q(x_2, y_2)$ is
>
> $$m = \frac{\text{change in } y}{\text{change in } x} = \frac{y_2 - y_1}{x_2 - x_1} \quad (x_2 \neq x_1)$$

 WARNING! When calculating slope, always subtract the y-values and the x-values in the same order.

The change in y (often denoted as Δy) is the **rise** of the line between points P and Q. The change in x (often denoted as Δx) is the **run**. Using this terminology, we can define slope to be the ratio of the rise to the run:

$$m = \frac{\Delta y}{\Delta x} = \frac{\text{rise}}{\text{run}} \quad (\Delta x \neq 0)$$

EXAMPLE 3

Finding slopes. Find the slope of the line determined by $3x - 4y = 12$.

Solution

First, we find the coordinates of two points on the line.

- If $x = 0$, then $y = -3$. The point $(0, -3)$ is on the line.
- If $y = 0$, then $x = 4$. The point $(4, 0)$ is on the line.

We then refer to Figure 8-26 and find the slope of the line between $P(0, -3)$ and $Q(4, 0)$ by substituting 0 for y_2, -3 for y_1, 4 for x_2, and 0 for x_1 in the formula for slope.

$$m = \frac{\text{change in } y}{\text{change in } x}$$

$$= \frac{y_2 - y_1}{x_2 - x_1}$$

$$= \frac{0 - (-3)}{4 - 0}$$

$$= \frac{3}{4}$$

FIGURE 8-26

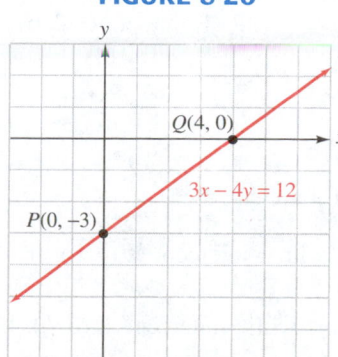

The slope of the line is $\frac{3}{4}$.

SELF CHECK   Find the slope of the line determined by $2x + 5y = 12$.   *Answer:* $-\frac{2}{5}$ ∎

If a line rises as we follow it from left to right, as in Figure 8-27(a), its slope is positive. If a line drops as we follow it from left to right, as in Figure 8-27(b), its slope is negative. If a line is horizontal, as in Figure 8-27(c), its slope is 0. If a line is vertical, as in Figure 8-27(d), it has no defined slope.

FIGURE 8-27

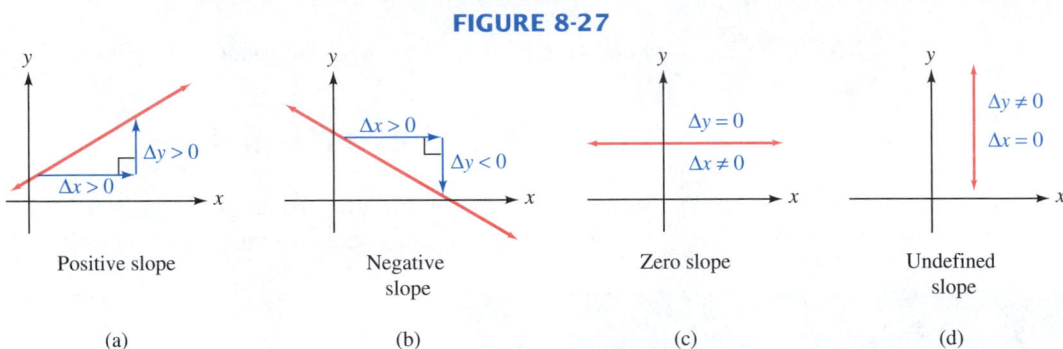

Positive slope	Negative slope	Zero slope	Undefined slope
(a)	(b)	(c)	(d)

Horizontal and Vertical Lines

To graph $y = 3$, we note that the equation $y = 3$ does not contain an x. Thus, the numbers chosen for x have no effect on y. The value of y is always 3. After plotting the pairs (x, y) shown in Figure 8-28, we see that the graph is a horizontal line, parallel to the x-axis, with a y-intercept of $(0, 3)$. The line has no x-intercept. Its slope is 0.

To graph $x = -2$, we note that the equation does not contain a y. Thus, the value of y can be any number. After plotting the pairs (x, y) shown in Figure 8-28, we see that the graph is a vertical line, parallel to the y-axis, with an x-intercept of $(-2, 0)$. The line has no y-intercept and no defined slope.

FIGURE 8-28

$y = 3$

x	y	(x, y)
-3	3	$(-3, 3)$
0	3	$(0, 3)$
2	3	$(2, 3)$
4	3	$(4, 3)$

$x = -2$

x	y	(x, y)
-2	-2	$(-2, -2)$
-2	0	$(-2, 0)$
-2	2	$(-2, 2)$
-2	6	$(-2, 6)$

The previous results show that if a and b are real numbers, then

■ The graph of the equation $x = a$ is a vertical line with x-intercept at $(a, 0)$. If $a = 0$, the line is the y-axis.

■ The graph of the equation $y = b$ is a horizontal line with y-intercept at $(0, b)$. If $b = 0$, the line is the x-axis.

ACCENT ON TECHNOLOGY *Graphing Lines*

To graph $3x + 2y = 12$ with a graphing calculator, we must first solve the equation for y.

$$3x + 2y = 12$$

$$2y = -3x + 12 \quad \text{Subtract } 3x \text{ from both sides.}$$

$$y = -\frac{3}{2}x + 6 \quad \text{Divide both sides by 2.}$$

We enter the right-hand side of the equation and press the $\boxed{\text{GRAPH}}$ key to get the graph shown in Figure 8-29.

FIGURE 8-29

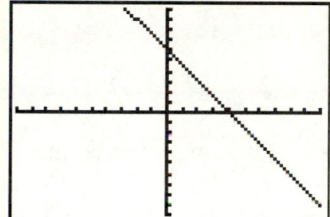

Slopes of Parallel and Perpendicular Lines

Recall the following facts about parallel and perpendicular lines.

1. If two nonvertical lines are parallel, they have the same slope.
2. If two lines have the same slope, they are parallel.
3. If two nonvertical lines are perpendicular, their slopes are negative reciprocals.
4. If the slopes of two lines are negative reciprocals, the lines are perpendicular.

EXAMPLE 4

Determining whether two lines are perpendicular. See Figure 8-30 on the next page. Are the lines shown in the figure perpendicular?

Solution We find the slopes of the lines and see if they are negative reciprocals.

$$\text{Slope of } OP = \frac{\Delta y}{\Delta x} \qquad \text{and} \qquad \text{Slope of } PQ = \frac{\Delta y}{\Delta x}$$

$$= \frac{y_2 - y_1}{x_2 - x_1} \qquad\qquad\qquad = \frac{y_2 - y_1}{x_2 - x_1}$$

$$= \frac{-4 - 0}{3 - 0} \qquad\qquad\qquad = \frac{4 - (-4)}{0 - 3}$$

$$= -\frac{4}{3} \qquad\qquad\qquad\qquad = \frac{8}{6}$$

$$\qquad\qquad\qquad\qquad\qquad = \frac{4}{3}$$

Since their slopes are not negative reciprocals, the lines are not perpendicular.

FIGURE 8-30

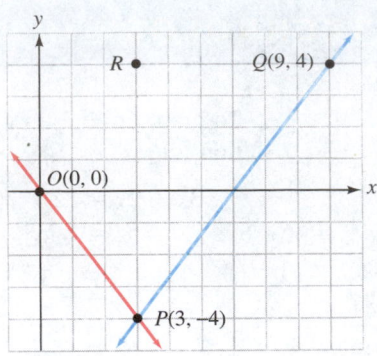

In Figure 8-30, is OR·parallel to PQ? (*Hint:* Find the slopes and see whether they are equal.)   *Answer:* yes

Writing Equations of Lines

Suppose that line l shown in Figure 8-31 has a slope of m and passes through the point $P(x_1, y_1)$. If $Q(x, y)$ is a second point on line l, we have

FIGURE 8-31

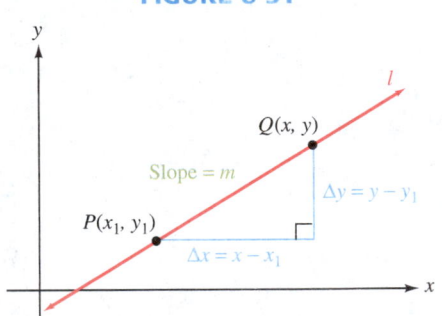

$$m = \frac{y - y_1}{x - x_1}$$

or if we multiply both sides by $x - x_1$, we have

1. $y - y_1 = m(x - x_1)$

Because Equation 1 displays the coordinates of the point (x_1, y_1) on the line and the slope m of the line, it is called the **point–slope form** of the equation of a line.

Point–slope form

The equation of the line passing through $P(x_1, y_1)$ and with slope m is

$$y - y_1 = m(x - x_1)$$

E X A M P L E 5   **Writing equations of lines.** Write the equation of the line passing through $P(-5, 4)$ and $Q(8, -6)$.

Solution   First, we find the slope of the line.

$$m = \frac{y_2 - y_1}{x_2 - x_1}$$

$$= \frac{-6 - 4}{8 - (-5)} \qquad \text{Substitute } -6 \text{ for } y_2, 4 \text{ for } y_1, 8 \text{ for } x_2, \text{ and } -5 \text{ for } x_1.$$

$$= -\frac{10}{13}$$

Because the line passes through both P and Q, we can choose either point and substitute its coordinates into the point–slope form. If we choose $P(-5, 4)$, we substitute -5 for x_1, 4 for y_1, and $-\frac{10}{13}$ for m and proceed as follows.

$$y - y_1 = m(x - x_1)$$

$$y - 4 = -\frac{10}{13}[x - (-5)] \quad \text{Substitute } -\tfrac{10}{13} \text{ for } m, -5 \text{ for } x_1, \text{ and } 4 \text{ for } y_1.$$

$$y - 4 = -\frac{10}{13}(x + 5) \quad -(-5) = 5.$$

$$y - 4 = -\frac{10}{13}x - \frac{50}{13} \quad \text{Remove parentheses.}$$

$$y = -\frac{10}{13}x + \frac{2}{13} \quad \text{Add 4 to both sides and simplify.}$$

The equation of the line is $y = -\frac{10}{13}x + \frac{2}{13}$.

SELF CHECK   Write the equation of the line passing through $(-2, 5)$ and $(4, -3)$.

Answer: $y = -\frac{4}{3}x + \frac{7}{3}$ ∎

Since the y-intercept of the line shown in Figure 8-32 is the point $(0, b)$, we can write the equation of the line by substituting 0 for x_1 and b for y_1 in the point–slope form and simplifying.

$$y - y_1 = m(x - x_1)$$
$$y - b = m(x - 0)$$
$$y - b = mx$$
$$\textbf{2.} \qquad y = mx + b$$

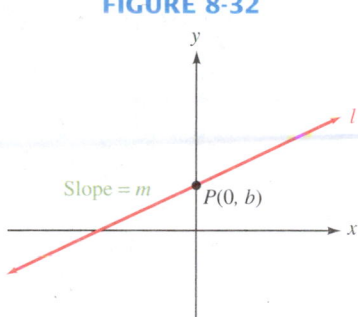

FIGURE 8-32

Because Equation 2 displays the slope m and the y-coordinate b of the y-intercept, it is called the **slope–intercept form** of the equation of a line.

Slope–intercept form

The equation of the line with slope m and y-intercept $(0, b)$ is

$$y = mx + b$$

EXAMPLE 6   **Writing equations of lines** Write the equation of the line that has a slope of 4 and passes through the point $(5, 9)$.

Solution   Since we are given that $m = 4$ and that the pair $(5, 9)$ satisfies the equation, we can substitute 5 for x, 9 for y, and 4 for m in the equation $y = mx + b$ and solve for b.

$$y = mx + b$$
$$9 = 4(5) + b \quad \text{Substitute 9 for } y, 4 \text{ for } m, \text{ and 5 for } x.$$
$$9 = 20 + b \quad \text{Simplify.}$$
$$-11 = b \quad \text{Subtract 20 from both sides.}$$

Because $m = 4$ and $b = -11$, the equation is $y = 4x - 11$.

Write the equation of the line that has slope -2 and passes through the point $(-2, 8)$. *Answer:* $y = -2x + 4$ ■

EXAMPLE 7

Writing equations of lines. Write the equation of the line that passes through $(-2, 5)$ and is parallel to the line $y = 8x - 3$.

Solution   Since the slope of the line given by $y = 8x - 3$ is the coefficient of x, the slope is 8. Since the desired equation is to have a graph that is parallel to the graph of $y = 8x - 3$, its slope must also be 8.

We substitute -2 for x, 5 for y, and 8 for m in the point–slope form and simplify.

$$y - y_1 = m(x - x_1)$$
$$y - 5 = 8[x - (-2)] \quad \text{Substitute 5 for } y_1, \text{ 8 for } m, \text{ and } -2 \text{ for } x_1.$$
$$y - 5 = 8(x + 2) \quad -(-2) = 2.$$
$$y - 5 = 8x + 16 \quad \text{Use the distributive property to remove parentheses.}$$
$$y = 8x + 21 \quad \text{Add 5 to both sides.}$$

The equation is $y = 8x + 21$.

SELF CHECK   Write the equation of the line that passes through $(-2, 5)$ and is perpendicular to the line $y = 8x - 3$. (*Hint:* Remember that the slopes of perpendicular lines are negative reciprocals.)   *Answer:* $y = -\frac{1}{8}x + \frac{19}{4}$ ■

Any linear equation that is written in the form $Ax + By = C$, where A, B, and C are constants, is said to be written in **general form.** When writing equations in general form, we usually clear the equation of fractions and make A positive. We also make A, B, and C as small as possible. For example, the equation $6x + 12y = 24$ can be changed to $x + 2y = 4$ by dividing both sides by 6.

Finding the slope and *y*-intercept from the general form

If A, B, and C are real numbers, the graph of the equation

$$Ax + By = C \quad (B \neq 0)$$

is a nonvertical line with slope of $-\dfrac{A}{B}$ and y-intercept of $\left(0, \dfrac{C}{B}\right)$.

You will be asked to justify the previous results in the exercises. You will also be asked to show that if $B = 0$, the equation $Ax + By = C$ represents a vertical line with x-intercept of $\left(\frac{C}{A}, 0\right)$.

EXAMPLE 8

Determining whether two lines are perpendicular. Are the lines represented by $4x + 3y = 7$ and $3x - 4y = 12$ perpendicular?

Solution   To show that the lines are perpendicular, we will show that their slopes are negative reciprocals. The first equation, $4x + 3y = 7$, is written in general form, with $A = 4$, $B = 3$, and $C = 7$. By the previous formula, the slope of the line is

$$m_1 = -\frac{A}{B} = -\frac{4}{3}$$

The second equation, $3x - 4y = 12$, is also written in general form, with $A = 3$, $B = -4$, and $C = 12$. The slope of this line is

$$m_2 = -\frac{A}{B} = -\frac{3}{-4} = \frac{3}{4}$$

Since the slopes are negative reciprocals, the lines are perpendicular.

SELF CHECK   Are the lines $4x + 3y = 7$ and $y = -\frac{4}{3}x + 2$ parallel?   *Answer:* yes   ■

Functions and Function Notation

Functions

A **function** is a correspondence between a set of input values x (called the **domain**) and a set of output values y (called the **range**), where exactly one y-value in the range is assigned to each number x in the domain.

The equation $y = 2x - 3$ defines y to be a function of x, because to every number x, there corresponds one number y. Since the input x can be any real number, the domain of the function is the set of real numbers, denoted by the interval $(-\infty, \infty)$. Since the output y can be any real number, the range is the set of real numbers, denoted as $(-\infty, \infty)$. A table of values and the graph appear in Figure 8-33.

FIGURE 8-33

$y = 2x - 3$

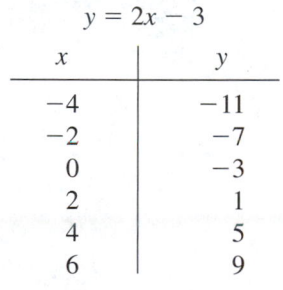

x	y
-4	-11
-2	-7
0	-3
2	1
4	5
6	9

The inputs can be any real number.

The outputs can be any real number.

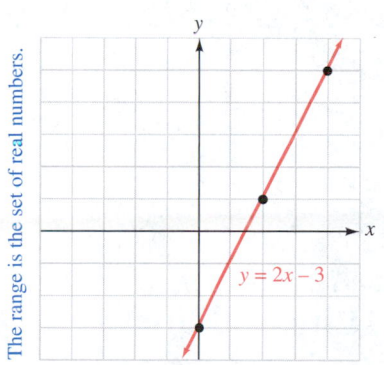

The domain is the set of real numbers.

Recall the vertical line test: Since every vertical line that intersects the graph of $y = 2x - 3$ does so only once, the graph represents a function.

Recall that there is a special notation to denote functions.

Function notation

The notation $y = f(x)$ denotes that the variable y is a function of x.

The notation $y = f(x)$ is read as "y equals f of x." Note that y and $f(x)$ are two notations for the same quantity. Thus, the equations $y = 4x + 3$ and $f(x) = 4x + 3$ are equivalent.

The notation $y = f(x)$ provides a way of denoting the value of y (the **dependent variable**) that corresponds to some number x (the **independent variable**). For example, if $y = f(x)$, the value of y that is determined by $x = 3$ is denoted by $f(3)$.

EXAMPLE 9

Function notation. Let $f(x) = 4x + 3$. Find **a.** $f(3)$ and **b.** $f(r + 2)$.

Solution

a. We replace x with 3:

$$f(x) = 4x + 3$$
$$f(3) = 4(3) + 3$$
$$= 12 + 3$$
$$= 15$$

b. We replace x with $r + 2$:

$$f(x) = 4x + 3$$
$$f(r + 2) = 4(r + 2) + 3$$
$$= 4r + 8 + 3$$
$$= 4r + 11$$

SELF CHECK

If $f(x) = -2x - 1$, find **a.** $f(-2)$ and **b.** $f(t - 3)$.

Answers: **a.** 3, **b.** $-2t + 5$ ∎

We can think of a function as a machine that takes some input x and turns it into some output $f(x)$, as shown in Figure 8-34(a). The machine shown in Figure 8-34(b) turns the input number 2 into the output value -3 and turns the input number 6 into the output value -11. The set of numbers that we can put into the machine is the domain of the function, and the set of numbers that comes out is the range.

FIGURE 8-34

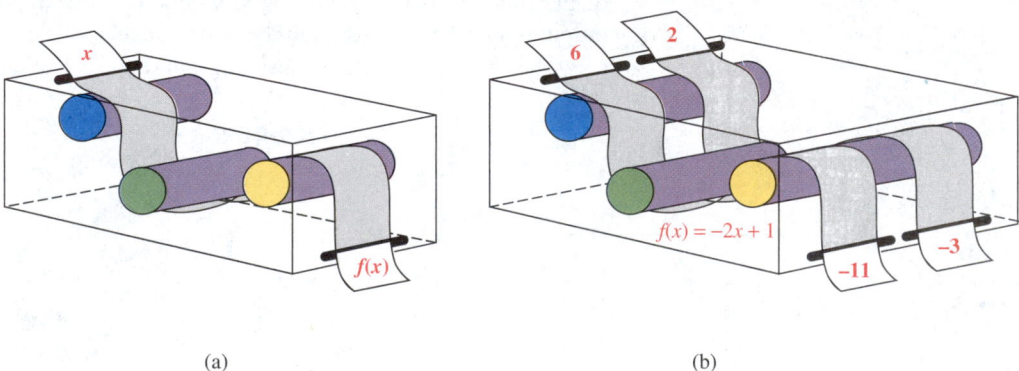

(a)   (b)

Linear Functions

Equations whose graphs are nonvertical lines define a set of basic functions called **linear functions.**

Linear functions

A **linear function** is a function defined by an equation that can be written in the form

$$f(x) = mx + b \qquad \text{or} \qquad y = mx + b$$

where m is the slope of the line graph and $(0, b)$ is the y-intercept.

EXAMPLE 10

Linear functions. Solve the equation $3x + 2y = 10$ for y to show that it defines a linear function. Then graph it to find its domain and range.

Solution   We solve the equation for y as follows:

$$3x + 2y = 10$$

$$2y = -3x + 10 \quad \text{Subtract } 3x \text{ from both sides.}$$

$$y = -\frac{3}{2}x + 5 \quad \text{Divide both sides by 2.}$$

Because the given equation can be written in the form $y = mx + b$, it defines a linear function. The slope of its line graph is $-\frac{3}{2}$, and the y-intercept is $(0, 5)$. The graph appears in Figure 8-35.

FIGURE 8-35

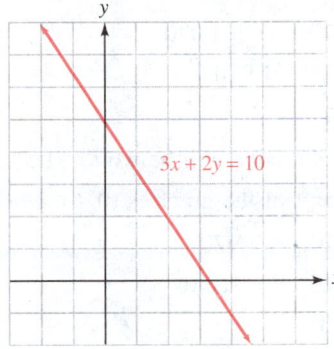

$3x + 2y = 10$

From the graph, we can see that both the domain and the range are the interval $(-\infty, \infty)$.

An Application

For tax purposes, many businesses use *straight-line depreciation* to find the declining value of aging equipment.

EXAMPLE 11

Value of a lathe. The owner of a machine shop buys a lathe for $1,970 and expects it to last for 10 years. The lathe can then be sold as scrap for an estimated *salvage value* of $270. If y represents the value of the lathe after x years of use, and y and x are related by the equation of a line,

a. find the equation of the line.

b. find the value of the lathe after $2\frac{1}{2}$ years.

c. find the economic meaning of the y-intercept of the line.

d. find the economic meaning of the slope of the line.

Solution   **a.** To find the equation of the line, we first find its slope and then use point–slope form to find its equation.

When the lathe is new, its age x is 0, and its value y is $1,970. When the lathe is 10 years old, $x = 10$, and its value is $y = \$270$. Since the line passes through the points $(0, 1,970)$ and $(10, 270)$, as shown in Figure 8-36, its slope is

$$m = \frac{y_2 - y_1}{x_2 - x_1}$$

$$= \frac{270 - 1,970}{10 - 0}$$

$$= \frac{-1,700}{10}$$

$$= -170$$

FIGURE 8-36

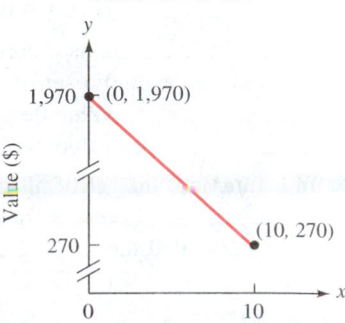

To find the equation of the line, we substitute -170 for m, 0 for x_1, and 1,970 for y_1 in the point-slope form and simplify.

$$y - y_1 = m(x - x_1)$$
$$y - 1{,}970 = -170(x - 0)$$

3. $\qquad y = -170x + 1{,}970$

The current value y of the lathe is related to its age x by the equation $y = -170x + 1{,}970$.

b. To find the value of the lathe after $2\frac{1}{2}$ years, we substitute 2.5 for x in Equation 3 and solve for y.

$$y = -170x + 1{,}970$$
$$= -170(2.5) + 1{,}970$$
$$= -425 + 1{,}970$$
$$= 1{,}545$$

In $2\frac{1}{2}$ years, the lathe will be worth \$1,545.

c. The y-intercept of the graph is $(0, b)$, where b is the value of y when $x = 0$.

$$y = -170x + 1{,}970$$
$$y = -170(0) + 1{,}970$$
$$y = 1{,}970$$

Thus, b is the value of a 0-year-old lathe, which is the lathe's original cost.

d. Each year, the value of the lathe decreases by \$170, because the slope of the line is -170. The slope of the depreciation line is the *annual depreciation rate*. ∎

STUDY SET

Section 8.6

VOCABULARY

Fill in the blanks to make the statements true.

1. The point where a graph intersects the y-axis is called the _____.

2. The point where a graph intersects the _____ is called the x-intercept.

3. A _____ is a correspondence between a set of input values and a set of output values, where each _____ value determines one _____ value.

4. In a function, the set of all inputs is called the _____ of the function. The set of output values is called the _____.

CONCEPTS

Fill in the blanks to make the statements true.

5. The graph of any equation of the form $x = a$ is a _____ line.

6. The graph of any equation of the form $y = b$ is a _____ line.

7. The formula to compute slope is $m =$

_____.

8. The slope of a _____ line is 0.

9. A _____ line has no defined slope.

10. If a line rises as x increases, its slope is _____.

11. _____ lines have the same slope.

12. The slopes of _____ lines are negative _____.

13. The point–slope form of the equation of a line is _____.

14. The slope–intercept form of the equation of a line is _____.

15. The general form of the equation of a line is _____.

16. The midpoint of a line segment joining (a, b) and (c, d) is given by the formula _____.

17. The denominator of a fraction can never be ___.

18. If a vertical line intersects a graph more than once, the graph _____ represent a function.

19. A linear function is any function that can be written in the form $f(x) =$ _____.

20. In the function $f(x) = mx + b$, m is the _____ of its graph, and b is the y-coordinate of the _____.

NOTATION

Fill in the blanks to make the statements true.

21. The symbol x_1 is read as x _____.

22. The change in x (denoted as Δx) is the _____ of the line between points P and Q.

23. The change in y (denoted as Δy) is the _____ of the line between points P and Q.

24. In the function $f(x) = \frac{1}{2}x + 7$, the notation $f(3)$ represents the value of ___ when $x = 3$.

PRACTICE

In Exercises 25–32, graph each equation. Check your work with a graphing calculator.

25. $x + y = 4$

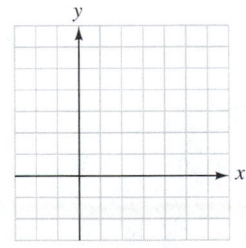

26. $2x - y = 3$

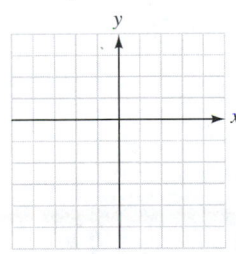

27. $3x + 4y = 12$

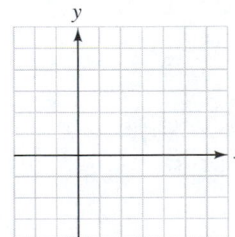

28. $4x - 3y = 12$

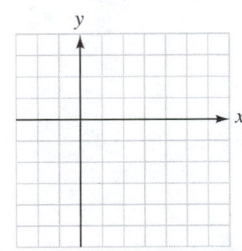

29. $x = 3$

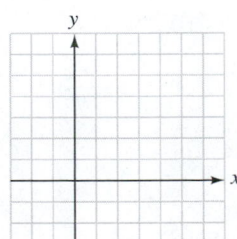

30. $y = -4$

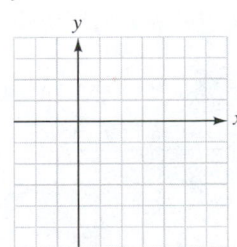

31. $-3y + 2 = 5$

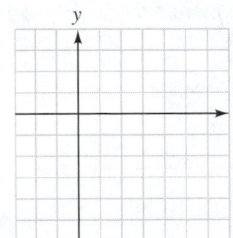

32. $-2x + 3 = 11$

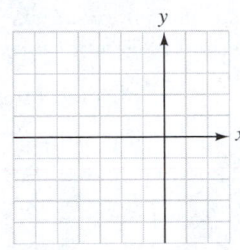

In Exercises 33–38, find the midpoint of segment PQ.

33. $P(0, 0)$, $Q(6, 8)$

34. $P(10, 12)$, $Q(0, 0)$

35. $P(6, 8)$, $Q(12, 16)$

36. $P(10, 4)$, $Q(2, -2)$

37. $P(-2, -8)$, $Q(3, 4)$

38. $P(-5, -2)$, $Q(7, 3)$

39. FINDING THE ENDPOINT OF A SEGMENT If $M(-2, 3)$ is the midpoint of segment PQ and the coordinates of P are $(-8, 5)$, find the coordinates of Q.

40. FINDING THE ENDPOINT OF A SEGMENT If $M(6, -5)$ is the midpoint of segment PQ and the coordinates of Q are $(-5, -8)$, find the coordinates of P.

In Exercises 41–48, find the slope of the line that passes through the given points, if possible.

41. $(0, 0)$, $(3, 9)$

42. $(9, 6)$, $(0, 0)$

43. $(-1, 8)$, $(6, 1)$

44. $(-5, -8)$, $(3, 8)$

45. $(3, -1)$, $(-6, 2)$

46. $(0, -8)$, $(-5, 0)$

47. $(-7, 5)$, $(-7, 2)$

48. $(3, -5)$, $(3, 8)$

In Exercises 49–52, find the slope of the line determined by each equation.

49. $3x + 2y = 12$

50. $2x - y = 6$

51. $3x = 4y - 2$

52. $x = y$

In Exercises 53–60, the slopes or equations of two lines are given. Tell whether the lines are parallel, perpendicular, or neither.

53. $3, \dfrac{6}{2}$

54. $-\dfrac{2}{3}, \dfrac{3}{2}$

55. $3x - y = 4$, $3x - y = 7$

56. $4x - y = 13$, $x - 4y = 13$

57. $x + y = 2$, $y = x + 5$

58. $x = y + 2$, $y = x + 3$

59. $3x + 6y = 1$, $y = \dfrac{1}{2}x$

60. $y = -3$, $y = -7$

In Exercises 61–66, use point–slope form to write the equation of the line with the given properties. Write each equation in general form.

61. $m = 5$, passing through $P(0, 7)$

62. $m = -8$, passing through $P(0, -2)$

63. Passing through $P(0, 0)$ and $Q(4, 4)$

64. Passing through $P(-5, -5)$ and $Q(0, 0)$

65. Passing through $P(3, 4)$ and $Q(0, -3)$

66. Passing through $P(4, 0)$ and $Q(6, -8)$

In Exercises 67–72, use slope–intercept form to write the equation of the line with the given properties. Write each equation in slope–intercept form.

67. $m = 3$, $b = 17$

68. $m = -2$, $b = 11$

69. $m = -7$, passing through $P(7, 5)$

70. $m = 3$, passing through $P(-2, -5)$

71. Passing through $P(6, 8)$ and $Q(2, 10)$

72. Passing through $P(-4, 5)$ and $Q(2, -6)$

In Exercises 73–76, write the equation of the line that passes through the given point and is parallel to the given line. Write the answer in slope–intercept form.

73. $P(0, 0)$, $y = 4x - 7$

74. $P(0, 0)$, $x = -3y - 12$

75. $P(2, 5)$, $4x - y = 7$

76. $P(-6, 3)$, $y + 3x = -12$

In Exercises 77–80, write the equation of the line that passes through the given point and is perpendicular to the given line. Write the answer in slope–intercept form.

77. $P(0, 0)$, $y = 4x - 7$

78. $P(0, 0)$, $x = -3y - 12$

79. $P(2, 5)$, $4x - y = 7$

80. $P(-6, 3)$, $y + 3x = -12$

In Exercises 81–84, use the method of Example 8 to find whether the graphs determined by each pair of equations are parallel, perpendicular, or neither.

81. $4x + 5y = 20$, $5x - 4y = 20$

82. $9x - 12y = 17$, $3x - 4y = 17$

83. $2x + 3y = 12$, $6x + 9y = 32$

84. $5x + 6y = 30$, $6x + 5y = 24$

85. Solve $Ax + By = C$ for y and thereby show that the slope of its graph is $-\frac{A}{B}$ and its y-intercept is $\left(0, \frac{C}{B}\right)$.

86. Show that the x-intercept of the graph of $Ax + By = C$ is $\left(\frac{C}{A}, 0\right)$.

In Exercises 87–90, find $f(3)$ and $f(-1)$.

87. $f(x) = 3x$

88. $f(x) = -4x$

89. $f(x) = 2x - 3$

90. $f(x) = 3x - 5$

In Exercises 91–94, find $g(w)$ and $g(w + 1)$.

91. $g(x) = 2x$

92. $g(x) = -3x$

93. $g(x) = 3x - 5$

94. $g(x) = 2x - 7$

In Exercises 95–98, each graph represents a correspondence between x and y. Tell whether the correspondence is a function. If so, give its domain and range.

95.

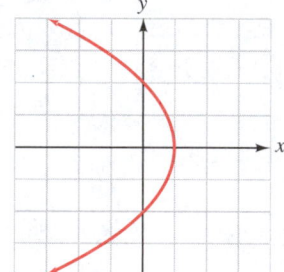

96.

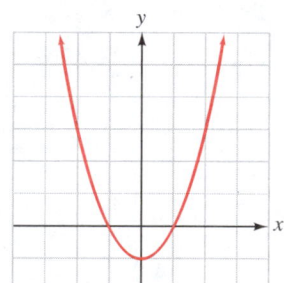

97.

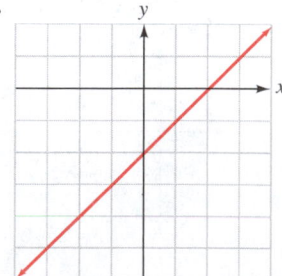

98.

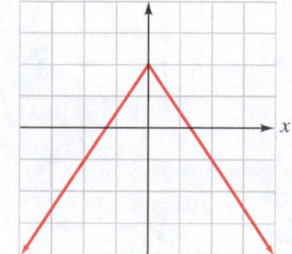

In Exercises 99–102, graph each linear function. Give the domain and range.

99. $f(x) = 2x - 1$

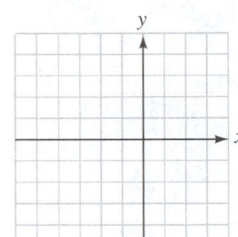

100. $f(x) = -x + 2$

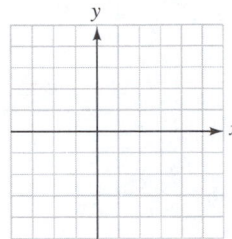

101. $2x - 3y = 6$

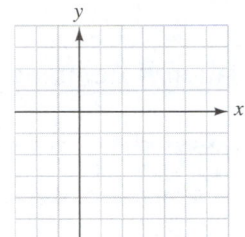

102. $3x + 2y = -6$

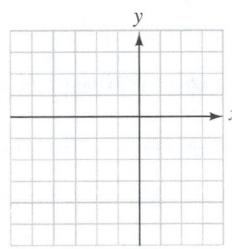

APPLICATIONS

103. GRADE OF A ROAD Find the slope of the road shown in Illustration 1. (*Hint:* 1 mi = 5,280 ft.)

ILLUSTRATION 1

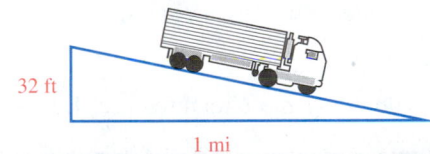

32 ft

1 mi

104. SLOPE OF A LADDER A ladder reaches 18 feet up the side of a building, with its base 5 feet from the building. Find the slope of the ladder.

105. RATE OF GROWTH When a college started an aviation program, the administration agreed to predict enrollments using a straight-line method. If the enrollment during the first year was 8 and the enrollment during the fifth year was 20, find the rate of growth per year (the slope of the line). See Illustration 2.

ILLUSTRATION 2

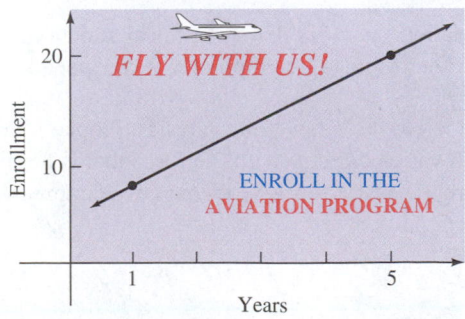

FLY WITH US!

ENROLL IN THE
AVIATION PROGRAM

Enrollment

Years

106. HOUSE APPRECIATION A house purchased for $125,000 is expected to appreciate according to the formula $y = 7{,}500x + 125{,}000$, where y is the value of the house after x years. Find the value of the house 5 years later.

107. CAR DEPRECIATION A car purchased for $17,000 is expected to depreciate according to the formula $y = -1{,}360x + 17{,}000$. When will the car be worthless?

108. CRIME PREVENTION The number n of incidents of family violence requiring police response appears to be related to d, the money spent on crisis intervention, by the equation $n = 430 - 0.005d$. What expenditure would reduce the number of incidents to 350?

109. SALVAGE VALUE A copy machine that cost $1,750 when new will be depreciated at the rate of $180 per year. If the useful life of the copier is 7 years, find its salvage value.

110. PREDICTING BURGLARIES A police department knows that city growth and the number of burglaries are related by a linear equation. City records show that 575 burglaries were reported in a year when the local population was 77,000, and 675 were reported in a year when the population was 87,000. How many burglaries can be expected when the population reaches 110,000?

111. BALLISTICS A bullet shot straight upward is s feet above the ground after t seconds, where $s = f(t) = -16t^2 + 256t$. Find the height of the bullet 3 seconds after it is shot.

112. ARTILLERY FIRE A mortar shell is s feet above the ground after t seconds, where $s = f(t) = -16t^2 + 512t + 64$. Find the height of the shell 20 seconds after it is fired.

113. SELLING TAPE RECORDERS An electronics firm manufactures tape recorders, receiving $120 for each recorder it makes. If x represents the number of recorders produced, the income received is determined by the *revenue function* $R(x) = 120x$. The manufacturer has fixed costs of $12,000 per month and variable costs of $57.50 for each recorder manufactured. Thus, the *cost function* is $C(x) = 57.50x + 12{,}000$. How many recorders must the company sell for revenue to equal cost?

114. SELLING TIRES A tire company manufactures premium tires, receiving $130 for each tire it makes. If the manufacturer has fixed costs of $15,512.50 per month and variable costs of $93.50 for each tire manufactured, how many tires must the company sell for revenue to equal cost? (*Hint:* See Exercise 113.)

WRITING

115. Explain the concept of function.

116. Explain the concepts of domain and range.

117. Explain why a vertical line has no defined slope.

118. Explain how to determine from their slopes whether two lines are parallel, perpendicular, or neither.

121. $\dfrac{5(2-x)}{3} - 1 = x + 5$

122. $\dfrac{r-1}{3} = \dfrac{r+2}{6} + 2$

REVIEW

Solve each equation.

119. $3(x+2) + x = 5x$

120. $12b + 6(3-b) = b + 3$

▶ **8.7**

Variation

In this section, you will learn about

Direct variation ■ Inverse variation ■ Joint variation ■ Combined variation

---

Introduction In this section, we will introduce some special terminology that scientists use to describe functions.

Direct Variation

To introduce direct variation, we consider the formula for the circumference of a circle

$$C = \pi D$$

where C is the circumference, D is the diameter, and $\pi \approx 3.14159$. If we double the diameter of a circle, we determine another circle with a larger circumference C_1 such that

$$C_1 = \pi(2D) = 2\pi D = 2C$$

Thus, doubling the diameter results in doubling the circumference. Likewise, if we triple the diameter, we triple the circumference.

In this formula, we say that the variables C and D *vary directly*, or that they are *directly proportional*. This is because as one variable gets larger, so does the other, in a predictable way. In this example, the constant π is called the *constant of variation* or the *constant of proportionality*.

Direct variation

The words "y varies directly with x" or "y is directly proportional to x" mean that $y = kx$ for some nonzero constant k. The constant k is called the **constant of variation** or the **constant of proportionality.**

FIGURE 8-37

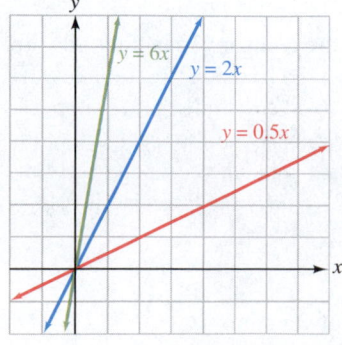

Since the formula for direct variation ($y = kx$) defines a linear function, its graph is always a line with a y-intercept at the origin. The graph of $y = kx$ appears in Figure 8-37 for three positive values of k.

One example of direct variation is Hooke's law from physics. Hooke's law states that the distance a spring will stretch varies directly with the force that is applied to it.

If d represents a distance and f represents a force, Hooke's law is expressed mathematically as

$$d = kf \qquad \text{This is a model for direct variation.}$$

where k is the constant of variation. If the spring stretches 10 inches when a weight of 6 pounds is attached, k can be found as follows:

$$d = kf$$
$$10 = k(6) \quad \text{Substitute 10 for } d \text{ and 6 for } f.$$
$$\frac{5}{3} = k \quad \text{Divide both sides by 6 and simplify.}$$

To find the force required to stretch the spring a distance of 35 inches, we can solve the equation $d = kf$ for f, with $d = 35$ and $k = \frac{5}{3}$.

$$d = kf$$
$$35 = \frac{5}{3}f \quad \text{Substitute 35 for } d \text{ and } \frac{5}{3} \text{ for } k.$$
$$105 = 5f \quad \text{Multiply both sides by 3.}$$
$$21 = f \quad \text{Divide both sides by 5.}$$

The force required to stretch the spring a distance of 35 inches is 21 pounds.

EXAMPLE 1

Direct variation. The distance traveled in a given time is directly proportional to the speed. If a car travels 70 miles at 30 mph, how far will it travel in the same time at 45 mph?

Solution　　The verbal model *distance is directly proportional to speed* can be expressed by the equation

1. $d = ks$

where d is distance, k is the constant of variation, and s is the speed. To find k, we substitute 70 for d and 30 for s, and solve for k.

$$d = ks$$
$$70 = k(30)$$
$$k = \frac{7}{3}$$

To find the distance traveled at 45 mph, we substitute $\frac{7}{3}$ for k and 45 for s in Equation 1 and simplify.

$$d = ks$$
$$d = \frac{7}{3}(45)$$
$$= 105$$

In the time it took to go 70 miles at 30 mph, the car could travel 105 miles at 45 mph.

SELF CHECK　　How far will the car travel in the same time at 60 mph?　*Answer:* 140 mi　■

Solving variation problems

To solve a variation problem:

1. Translate the verbal model into an equation.
2. Substitute the first set of values into the equation from step 1 to determine the value of k.
3. Substitute the value of k into the equation from step 1.
4. Substitute the remaining set of values into the equation from step 3 and solve for the unknown variable.

Inverse Variation

In the formula $w = \frac{12}{l}$, w gets smaller as l gets larger, and w gets larger as l gets smaller. Since these variables vary in opposite directions in a predictable way, we say that the variables *vary inversely,* or that they are *inversely proportional.* The constant 12 is the constant of variation.

Inverse variation

> The words "y varies inversely with x" or "y is inversely proportional to x" mean that $y = \frac{k}{x}$ for some nonzero constant k. The constant k is called the **constant of variation.**

Since the right-hand side of the formula for direct variation involves a fraction $\left(y = \frac{k}{x}\right)$, we say that the formula for inverse variation defines a rational function. Its graph is always a curve, with the x- and y-axes as asymptotes. The graph of $y = \frac{k}{x}$ appears in Figure 8-38 for three positive values of k.

FIGURE 8-38

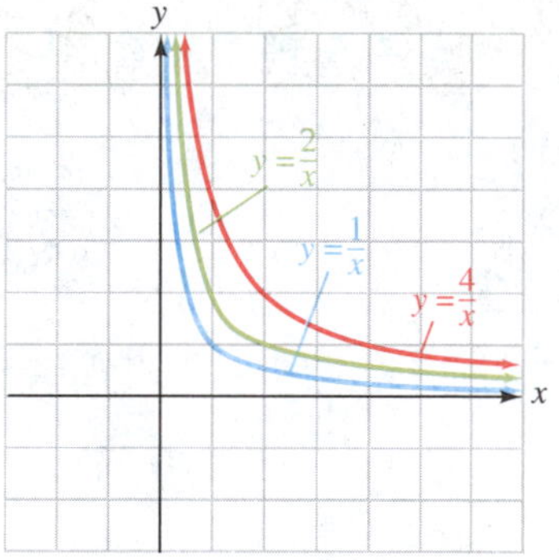

Because of gravity, an object in space is attracted to the earth. The force of this attraction varies inversely with the square of the distance from the object to the center of the earth. If f represents the force and d represents the distance, the relationship can be expressed by the equation

$$f = \frac{k}{d^2} \qquad \text{This is an inverse variation model.}$$

If we know that an object 4,000 miles from the center of the earth is attracted to the earth with a force of 90 pounds, we can find k.

$$f = \frac{k}{d^2}$$

$$90 = \frac{k}{4,000^2} \qquad \text{Substitute 90 for } f \text{ and 4,000 for } d.$$

$$k = 90(4,000^2) \qquad \text{Multiply both sides by } 4,000^2 \text{ to solve for } k.$$

$$= 1,440,000,000$$

To find the force of attraction when the object is 5,000 miles from the center of the earth, we proceed as follows:

$$f = \frac{k}{d^2}$$

$$f = \frac{1,440,000,000}{5,000^2} \qquad \text{Substitute 1,440,000,000 for } k \text{ and 5,000 for } d.$$

$$= 57.6$$

The object will be attracted to the earth with a force of 57.6 pounds when it is 5,000 miles from the earth's center.

EXAMPLE 2

Light intensity. The intensity I of light received from a light source varies inversely with the square of the distance from the light source. If a photographer, 16 feet away from her subject, has a light meter reading of 4 foot-candles of illuminance, what will the meter read when the photographer moves in for a close-up, 4 feet away from the subject?

Solution   The words *intensity varies inversely with the square of the distance d* can be expressed by the equation

$$I = \frac{k}{d^2} \quad \text{This is inverse variation.}$$

To find k, we substitute 4 for I and 16 for d and solve for k.

$$I = \frac{k}{d^2}$$

$$4 = \frac{k}{16^2}$$

$$4 = \frac{k}{256}$$

$$1{,}024 = k$$

To find the intensity when the photographer is 4 feet from the subject, we substitute 4 for d and 1,024 for k and simplify.

$$I = \frac{k}{d^2}$$

$$I = \frac{1{,}024}{4^2}$$

$$= 64$$

The intensity at 4 feet is 64 foot-candles.

SELF CHECK   See Example 2. Find the intensity when the photographer is 8 feet away from the subject.   *Answer:* 16 foot-candles ■

Joint Variation

Sometimes, one variable varies with the product of several variables. For example, the area of a triangle varies directly with the product of its base and height:

$$A = \frac{1}{2}bh$$

Such variation is called *joint variation*.

Joint variation

If one variable varies directly with the product of two or more variables, the relationship is called **joint variation.** If y varies jointly with x and z, then $y = kxz$. The nonzero constant k is called the **constant of variation.**

EXAMPLE 3

Joint variation. The volume V of a cone varies jointly with its height h and the area of its base B. If $V = 6$ cm^3 when $h = 3$ and $B = 6$ cm^2, find V when $h = 2$ cm and $B = 8$ cm^2.

Solution The words *V varies jointly with h and B* mean that *V varies directly as the product of h and B.* Thus,

$$V = khB \qquad \text{The joint variation model can also be read as "}V\text{ is directly proportional to the product of }h\text{ and }B\text{."}$$

We can find k by substituting 6 for V, 3 for h, and 6 for B.

$$V = kh B$$
$$6 = k(3)(6)$$
$$6 = k(18)$$
$$\frac{1}{3} = k \qquad \text{Divide both sides by 18.}$$

To find V when $h = 2$ and $B = 8$, we substitute these values into the formula $V = \frac{1}{3}hB$.

$$V = \frac{1}{3}h B$$
$$V = \frac{1}{3}(2)(8)$$
$$= \frac{16}{3}$$

The volume is $\frac{16}{3}$ cm^3 or $5\frac{1}{3}$ cm^3. ∎

Combined Variation

Many applied problems involve a combination of direct and inverse variation. Such variation is called **combined variation.**

EXAMPLE 4

Building highways. The time it takes to build a highway varies directly with the length of the road, but inversely with the number of workers. If it takes 100 workers 4 weeks to build 2 miles of highway, how long will it take 80 workers to build 10 miles of highway?

Solution We can let t represent the time in weeks, l represent the length in miles, and w represent the number of workers. The relationship between these variables can be expressed by the equation

$$t = \frac{kl}{w} \qquad \text{This is a combined variation model.}$$

We substitute 4 for t, 100 for w, and 2 for l to find k:

$$4 = \frac{k(2)}{100}$$
$$400 = 2k \qquad \text{Multiply both sides by 100.}$$
$$200 = k \qquad \text{Divide both sides by 2.}$$

We now substitute 80 for w, 10 for l, and 200 for k in the equation $t = \frac{kl}{w}$ and simplify:

$$t = \frac{kl}{w}$$

$$t = \frac{200(10)}{80}$$

$$= 25$$

It will take 25 weeks for 80 workers to build 10 miles of highway.

SELF CHECK   In Example 4, how long will it take 60 workers to build 6 miles of highway?   *Answer:* 20 weeks

STUDY SET

Section 8.7

VOCABULARY

Fill in the blanks to make the statements true.

1. The equation $y = kx$ defines _____ variation.

2. The equation $y = \frac{k}{x}$ defines _____ variation.

3. Inverse variation is represented by a _____ function.

4. Direct variation is represented by a _____ function.

5. The equation $y = kxz$ defines _____ variation.

6. The equation $y = \frac{kx}{z}$ defines _____ variation.

CONCEPTS

Tell whether each graph represents direct variation, inverse variation, or neither.

7.

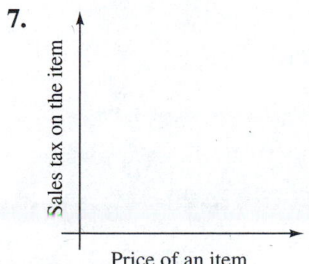

8.

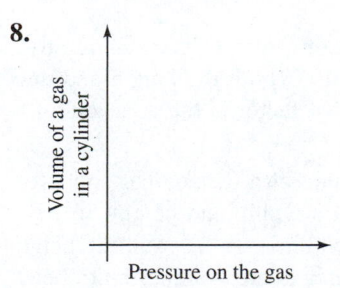

9.

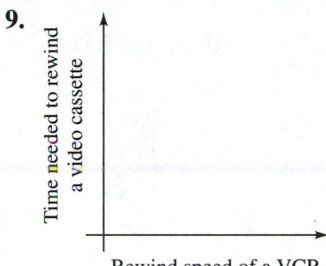

10.

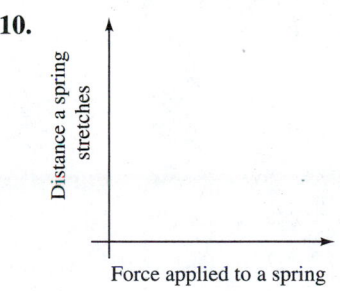

NOTATION

Complete each solution.

11. Find f if $d = 25$ and $k = \frac{7}{5}$.

$$d = kf$$

$$25 = \boxed{} f$$

$$\frac{5}{7}\left(\right) = f$$

$$\frac{125}{7} = f$$

12. Find k if $t = 8$, $w = 80$, and $l = 4$.

$$t = \frac{kl}{w}$$

$$8 = \frac{k(4)}{\boxed{}}$$

$$\boxed{}(80) = \boxed{} k$$

$$160 = k$$

PRACTICE

In Exercises 13–20, express each verbal model as a formula.

13. A varies directly with the square of p.

14. z varies inversely with the cube of t.

15. v varies inversely with the cube of r.

16. r varies directly with the square of s.

17. B varies jointly with m and n.

18. C varies jointly with x, y, and z.

19. P varies directly with the square of a, and inversely with the cube of j.

20. M varies inversely with the cube of n, and jointly with x and the square of z.

In Exercises 21–28, express each variation model in words. In each formula, k is the constant of variation.

21. $L = kmn$

22. $P = \dfrac{km}{n}$

23. $E = kab^2$

24. $U = krs^2t$

25. $X = \dfrac{kx^2}{y^2}$

26. $Z = \dfrac{kw}{xy}$

27. $R = \dfrac{kL}{d^2}$

28. $e = \dfrac{kPL}{A}$

APPLICATIONS

29. AREA OF A CIRCLE The area of a circle varies directly with the square of its radius. The constant of variation is π. Find the area of a circle with a radius of 6 inches.

30. FALLING OBJECT An object in free fall travels a distance s that is directly proportional to the square

of the time t. If an object falls 1,024 feet in 8 seconds, how far will it fall in 10 seconds?

31. FINDING DISTANCE The distance that a car can go is directly proportional to the number of gallons of gasoline it consumes. If a car can go 288 miles on 12 gallons of gasoline, how far can it go on a full tank of 18 gallons?

32. FARMING A farmer's harvest in bushels varies directly with the number of acres planted. If 8 acres can produce 144 bushels, how many acres are required to produce 1,152 bushels?

33. FARMING The length of time that a given number of bushels of corn will last when feeding cattle varies inversely with the number of animals. If x bushels will feed 25 cows for 10 days, how long will the feed last for 10 cows?

34. GEOMETRY For a fixed area, the length of a rectangle is inversely proportional to its width. A rectangle has a width of 18 feet and a length of 12 feet. If the length is increased to 16 feet, find the width.

35. GAS PRESSURE Under constant temperature, the volume occupied by a gas is inversely proportional to the pressure applied. If the gas occupies a volume of 20 cubic inches under a pressure of 6 pounds per square inch, find the volume when the gas is subjected to a pressure of 10 pounds per square inch.

36. VALUE OF A CAR The value of a car usually varies inversely with its age. If a car is worth $7,000 when it is 3 years old, how much will it be worth when it is 7 years old?

37. ORGAN PIPES The frequency of vibration of air in an organ pipe is inversely proportional to the length of the pipe. (See Illustration 1.) If a pipe 2 feet long vibrates 256 times per second, how many times per second will a 6-foot pipe vibrate?

ILLUSTRATION 1

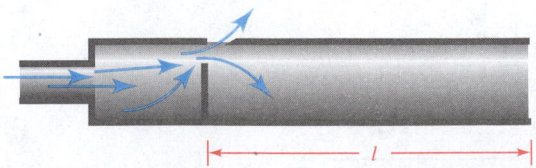

38. GEOMETRY The area of a rectangle varies jointly with its length and width. If both the length and the width are tripled, by what factor is the area multiplied?

39. GEOMETRY The volume of a rectangular solid varies jointly with its length, width, and height. If the length is doubled, the width is tripled, and the height is doubled, by what factor is the volume multiplied?

40. COSTS OF A TRUCKING COMPANY The costs incurred by a trucking company vary jointly with the number of trucks in service and the number of hours they are used. When 4 trucks are used for 6 hours each, the costs are $1,800. Find the costs of using 10 trucks, each for 12 hours.

41. STORING OIL The number of gallons of oil that can be stored in a cylindrical tank varies jointly with the height of the tank and the square of the radius of its base. The constant of proportionality is 23.5. Find the number of gallons that can be stored in the cylindrical tank in Illustration 2.

ILLUSTRATION 2

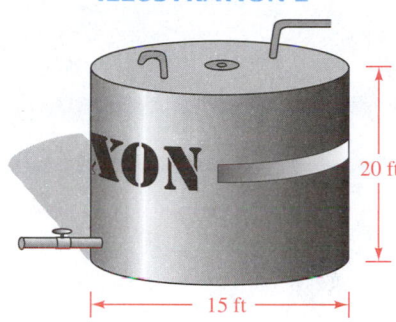

42. FINDING THE CONSTANT OF VARIATION A quantity l varies jointly with x and y and inversely with z. If the value of l is 30 when $x = 15$, $y = 5$, and $z = 10$, find k.

43. ELECTRONICS The voltage (in volts) measured across a resistor is directly proportional to the current (in amperes) flowing through the resistor. The constant of variation is the **resistance** (in ohms). If 6 volts is measured across a resistor carrying a current of 2 amperes, find the resistance.

44. ELECTRONICS The power (in watts) lost in a resistor (in the form of heat) is directly proportional to the square of the current (in amperes) passing through it. The constant of proportionality is the resistance (in ohms). What power is lost in a 5-ohm resistor carrying a 3-ampere current?

45. BUILDING CONSTRUCTION The deflection of a beam is inversely proportional to its width and the cube of its depth. If the deflection of a 4-inch-by-4-inch beam is 1.1 inches, find the deflection of a 2-inch-by-8-inch beam positioned as in Illustration 3.

ILLUSTRATION 3

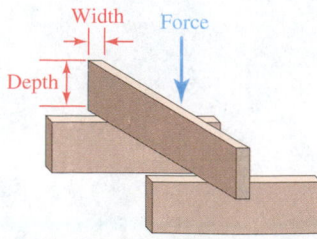

46. BUILDING CONSTRUCTION Find the deflection of the beam in Exercise 45 when the beam is positioned as in Illustration 4.

ILLUSTRATION 4

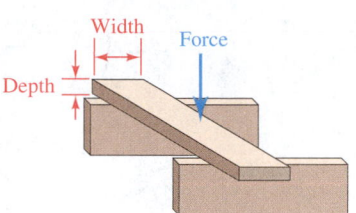

47. GAS PRESSURE The pressure of a certain amount of gas is directly proportional to the temperature (measured in degrees Kelvin) and inversely proportional to the volume. A sample of gas at a pressure of 1 atmosphere occupies a volume of 1 cubic meter at a temperature of 273 Kelvin. When heated, the gas expands to twice its volume, but the pressure remains constant. To what temperature is it heated?

48. TENSION IN A STRING When playing with a Skip It toy, a child swings a weighted ball on the end of a string in a circular motion around one leg while jumping over the revolving string with the other leg. (See Illustration 5.) The tension T in the string is directly proportional to the square of the speed s of the ball and inversely proportional to the radius r of the circle. If the tension in the string is 6 pounds when the speed of the ball is 6 feet per second and the radius is 3 feet, find the tension when the speed is 8 feet per second and the radius is 2.5 feet.

ILLUSTRATION 5

WRITING

49. Explain the term *direct variation*.

50. Explain the term *inverse variation*.

51. Explain the term *joint variation*.

52. Explain why the equation $\frac{y}{x} = k$ indicates that y varies directly with x.

REVIEW

Simplify each expression.

53. $(x^2 x^3)^2$

54. $\left(\dfrac{a^3 a^5}{a^{-2}}\right)^3$

55. $\dfrac{b^0 - 2b^0}{b^0}$

56. $\left(\dfrac{2r^{-2}r^{-3}}{4r^{-5}}\right)^3$

57. Write 35,000 in scientific notation.

58. Write 0.00035 in scientific notation.

59. Write 2.5×10^{-3} in standard notation.

60. Write 2.5×10^4 in standard notation.

Factoring

Finding the individual factors of a known product is called **factoring.** We have seen that factoring can be used to solve quadratic equations and to simplify rational expressions. A common question asked by students is, "What factoring technique should I use on a particular problem?" To identify the technique, we can follow these steps.

Factoring a random polynomial

1. Factor out all common factors.

2. If an expression has two terms, check to see whether the problem type is
   a. the **difference of two squares:** $x^2 - y^2 = (x + y)(x - y)$.
   b. the **sum of two cubes:** $x^3 + y^3 = (x + y)(x^2 - xy + y^2)$.
   c. the **difference of two cubes:** $x^3 - y^3 = (x - y)(x^2 + xy + y^2)$.

3. If an expression has three terms, check to see whether it is a **trinomial square:**
   $$x^2 + 2xy + y^2 = (x + y)(x + y) = (x + y)^2$$
   $$x^2 - 2xy + y^2 = (x - y)(x - y) = (x - y)^2$$
   If the trinomial is not a trinomial square, attempt to factor it as a **general trinomial.**

4. If an expression has four or more terms, try to factor the expression by **grouping.**

5. Continue factoring until each individual factor is prime.

6. Check the results by multiplication.

For example, to factor $x^5y^2 - xy^6$, we begin by factoring out the common factor of xy^2.

$$x^5y^2 = xy^6 = xy^2(x^4 - y^4)$$

Since the expression $x^4 - y^4$ has two terms, we check to see whether it is the difference of two squares, which it is. As the difference of two squares, it factors as

$$x^5y^2 - xy^6 = xy^2(x^4 - y^4)$$
$$= xy^2(x^2 + y^2)(x^2 - y^2)$$

The binomial $x^2 + y^2$ is the sum of two squares and cannot be factored. However, $x^2 - y^2$ is the difference of two squares and factors as $(x + y)(x - y)$.

$$x^5y^2 - xy^6 = xy^2(x^4 - y^4)$$
$$= xy^2(x^2 + y^2)(x^2 - y^2)$$
$$= xy^2(x^2 + y^2)(x + y)(x - y)$$

Since each individual factor is prime, the given expression is in completely factored form.

Factor:

1. $6t^2 + 7t - 3$ **2.** $ac + ad + bc + bd$

3. $8x^2 - 50$ **4.** $t^2 - 2t + 1$

5. $2ab^2 + 8ab - 24a$ **6.** $a^2(x - a) - b^2(x - a)$

7. $54x^3 + 250y^6$

8. $x^5 - x^3y^2 + x^2y^3 - y^5$

Accent on Teamwork

Section 8.1

Triangle inequality The **triangle inequality** states an important relationship between the sides of any triangle.

The sum of the lengths of any two sides of a triangle	$>$	the length of the third side

To illustrate this inequality, cut two pipe cleaners to lengths of 2 inches and 4 inches, to serve as two sides of a triangle. (See Illustration 1.) Cut a third pipe cleaner to a length of 5 inches. Show that there

ILLUSTRATION 1

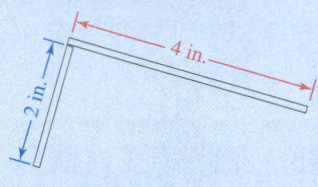

is no triangle with sides of 2, 4, and 5 inches. Cut another pipe cleaner to a length of 6 inches. Show that there is no triangle with sides of 2, 4, and 6 inches. How long must the third pipe cleaner be for you to be able to form a triangle?

Section 8.2

Tolerances Visit a machine shop or automobile repair shop and ask the employees to show you some examples of how they work with tolerances in their profession. Videotape their explanations and then play the video for your class. Show how the tolerances can be described using an absolute value inequality and interval notation.

Section 8.3

Factoring The following expressions contain variables with variable exponents. Factor each of them.

$$x^{4n} - x^{2n} - 6 \qquad 2x^{6n} - 3x^{3n} - 2$$
$$x^{2mn} - y^{2mn} \qquad ax^{4a} - by^{4b}$$

Section 8.4

Factoring The following expressions contain variables with variable exponents. Factor each of them.

$$x^{3m} - y^{3n} \qquad x^{3m} + y^{3n}$$

Section 8.5

Multiplying rational expressions Explain what is **wrong** with the student's work shown here. Then write a correct solution.

$$\frac{3}{4x} + \frac{2}{3x} = 12x\left(\frac{3}{4x} + \frac{2}{3x}\right)$$

$$= 12x \cdot \frac{3}{4x} + 12x \cdot \frac{2}{3x}$$

$$= 9 + 8$$

$$= 17$$

Complex fractions Each complex fraction in the list

$$1 + \frac{1}{2}, \, 1 + \cfrac{1}{1 + \frac{1}{2}}, \, 1 + \cfrac{1}{1 + \cfrac{1}{1 + \frac{1}{2}}},$$

$$1 + \cfrac{1}{1 + \cfrac{1}{1 + \cfrac{1}{1 + \frac{1}{2}}}}, \, \ldots$$

can be simplified by using the value of the expression preceding it. For example, to simplify the second expression in the list, replace $1 + \frac{1}{2}$ with $\frac{3}{2}$. Show that the expressions in the list simplify to the fractions $\frac{3}{2}, \frac{5}{3}, \frac{8}{5}, \frac{13}{8}, \frac{21}{13}, \frac{34}{21}, \ldots$. Do you see a pattern? Can you predict the next fraction?

Section 8.6

Warnings In this book, warnings are given to help students avoid common mistakes. For example, the following warning addresses an error that is often made by students when solving inequalities.

 WARNING! We must remember to reverse the inequality symbol every time we multiply or divide by a negative number.

Assign each member of your group an example from Section 8.6. Tell them to study the solution and identify a common mistake that might be made when solving that type of problem. Then have them write a warning for the example. When completed, each person should present the solution to his or her example, including the warning, to the other members of the group.

Section 8.7

Variation The graph of direct variation is a line with a y-intercept at the origin. The graph of inverse variation is that of a rational function with the x- and y-axes as asymptotes. See Illustration 2.

ILLUSTRATION 2

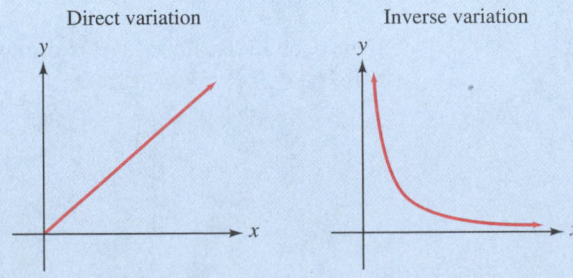

Direct variation

Inverse variation

As a group, brainstorm to come up with real-life examples of direct variation and inverse variation. Draw informal graphs of each situation and label each axis.

Section 8.1

Review of Equations and Inequalities

CONCEPTS

If $a = b$, then

$a + c = b + c$

$a - c = b - c$

$ac = bc$

$\dfrac{a}{c} = \dfrac{b}{c}$ $(c \neq 0)$

REVIEW EXERCISES

1. Solve and check each equation.

    a. $4(y - 1) = 28$ **b.** $3(x + 7) = 42$

    c. $13(x - 9) - 2 = 7x - 5$ **d.** $\dfrac{8(x - 5)}{3} = 2(x - 4)$

    e. $2x + 4 = 2(x + 3) - 2$ **f.** $\dfrac{x}{3} + \dfrac{2x}{4} = 10$

    g. $(3x - 2) + x = 2(x - 4)$

2. Solve for the indicated variable.

    a. $V = \dfrac{1}{3}\pi r^2 h$ for h

    b. $v = \dfrac{1}{6}ab(x + y)$ for x

If $a < b$, then

$a + c < b + c$

$a - c < b - c$

$ac < bc$ $(c > 0)$

$\dfrac{a}{c} < \dfrac{b}{c}$ $(c > 0)$

$ac > bc$ $(c < 0)$

$\dfrac{a}{c} > \dfrac{b}{c}$ $(c < 0)$

$a < x < b$ means
$a < x$ and $x < b$

3. Solve each inequality. Give each solution set in interval notation and graph it.

    a. $\dfrac{1}{3}y - 2 \geq \dfrac{1}{2}y + 2$

    b. $\dfrac{7}{4}(x + 3) < \dfrac{3}{8}(x - 3)$

    c. $3 < 3x + 4 < 10$

    d. $4x > 3x + 2 > x - 3$

4. CARPENTRY A carpenter wants to cut a 20-foot rafter so that one piece is 3 times as long as the other. Where should he cut the board?

5. GEOMETRY A rectangle is 4 meters longer than it is wide. If the perimeter of the rectangle is 28 meters, find its area.

Section 8.2

Solving Absolute Value Equations and Inequalities

If $x \geq 0$, then $|x| = x$.
If $x < 0$, then $|x| = -x$.

$|x| = k$ means $x = k$ or
$x = -k$.

$|a| = |b|$ means $a = b$ or
$a = -b$.

6. Solve and check each equation.

    a. $|3x + 1| = 10$ **b.** $\left|\dfrac{3}{2}x - 4\right| = 9$

    c. $|3x + 2| = |2x - 3|$ **d.** $|5x - 4| = |4x - 5|$

If $k > 0$, then

$|x| < k$ means
$-k < x < k$.

$|x| \le k$ means
$-k \le x \le k$.

If k is a nonnegative constant, then

$|x| > k$ means $x < -k$ or
$x > k$.

$|x| \ge k$ means $x \le -k$ or
$x \ge k$.

7. Solve each inequality. Give each solution in interval notation and graph it.

a. $|2x + 7| < 3$

b. $|3x - 8| \ge 4$

c. $\left| \dfrac{3}{2}x - 14 \right| \ge 0$

d. $\left| \dfrac{2}{3}x + 14 \right| < 0$

Section 8.3

$x^2 - y^2 = (x + y)(x - y)$

Review of Factoring

8. Factor each polynomial.
   a. $4x + 8$
   b. $5x^2y^3 - 10xy^2$
   c. $-8x^2y^3z^4 - 12x^4y^3z^2$
   d. $12a^6b^4c^2 + 15a^2b^4c^6$
   e. $xy + 2y + 4x + 8$
   f. $ac + bc + 3a + 3b$

9. Factor x^n from $x^{2n} + x^n$.

10. Factor y^{2n} from $y^{2n} - y^{4n}$.

11. Factor each polynomial.
   a. $x^4 + 4y + 4x^2 + x^2y$
   b. $a^5 + b^2c + a^2c + a^3b^2$
   c. $z^2 - 16$
   d. $y^2 - 121$
   e. $2x^4 - 98$
   f. $3x^6 - 300x^2$
   g. $y^2 + 21y + 20$
   h. $z^2 - 11z + 30$
   i. $-x^2 - 3x + 28$
   j. $-y^2 + 5y + 24$
   k. $y^3 + y^2 - 2y$
   l. $2a^4 + 4a^3 - 6a^2$
   m. $15x^2 - 57xy - 12y^2$
   n. $30x^2 + 65xy + 10y^2$
   o. $x^2 + 4x + 4 - 4p^4$
   p. $y^2 + 3y + 2 + 2x + xy$

Section 8.4

$x^3 + y^3 = (x + y)(x^2 - xy + y^2)$
$x^3 - y^3 = (x - y)(x^2 + xy + y^2)$

Factoring the Sum and Difference of Two Cubes

12. Factor each polynomial.
   a. $x^3 + 343$
   b. $a^3 - 125$
   c. $8y^3 - 512$
   d. $4x^3y + 108yz^3$

Section 8.5

Use the same rules to manipulate rational expressions as you would use to manipulate arithmetic fractions.

Review of Rational Expressions

13. Simplify each fraction.

a. $\dfrac{248x^2y}{576xy^2}$

b. $\dfrac{x^2 - 49}{x^2 + 14x + 49}$

14. Do the operations and simplify.

a. $\dfrac{x^2 + 4x + 4}{x^2 - x - 6} \cdot \dfrac{x^2 - 9}{x^2 + 5x + 6}$

b. $\dfrac{x^3 - 64}{x^2 + 4x + 16} \div \dfrac{x^2 - 16}{x + 4}$

c. $\dfrac{5y}{x - y} - \dfrac{3}{x - y}$

d. $\dfrac{3x - 1}{x^2 + 2} + \dfrac{3(x - 2)}{x^2 + 2}$

e. $\dfrac{3}{x + 2} + \dfrac{2}{x + 3}$

f. $\dfrac{4x}{x - 4} - \dfrac{3}{x + 3}$

g. $\dfrac{x^2 + 3x + 2}{x^2 - x - 6} \cdot \dfrac{3x^2 - 3x}{x^2 - 3x - 4} \div \dfrac{x^2 + 3x + 2}{x^2 - 2x - 8}$

h. $\dfrac{2x}{x + 1} + \dfrac{3x}{x + 2} + \dfrac{4x}{x^2 + 3x + 2}$

i. $\dfrac{3(x + 2)}{x^2 - 1} - \dfrac{2}{x + 1} + \dfrac{4(x + 3)}{x^2 - 2x + 1}$

15. Simplify each complex fraction.

a. $\dfrac{\dfrac{3}{x} - \dfrac{2}{y}}{xy}$

b. $\dfrac{\dfrac{1}{x} + \dfrac{2}{y}}{\dfrac{2}{x} - \dfrac{1}{y}}$

c. $\dfrac{2x + 3 + \dfrac{1}{x}}{x + 2 + \dfrac{1}{x}}$

d. $\dfrac{x^{-1} - y^{-1}}{x^{-1} + y^{-1}}$

Section 8.6

Graph of a vertical line:

$x = a$

x-intercept $(a, 0)$.

Graph of a horizontal line:

$y = b$

y-intercept at $(0, b)$.

Review of Linear Functions

16. Graph each equation.

a. $2x - y = 8$

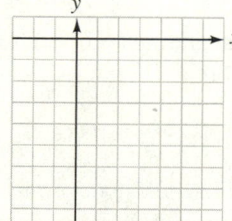

b. $x = -2$

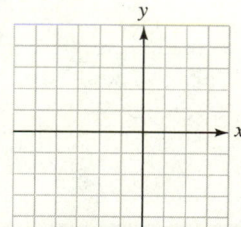

c. $y = 4$

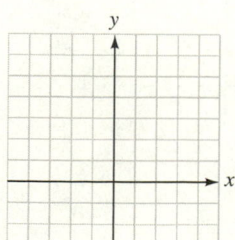

Midpoint formula:

For $P(x_1, y_1)$ and $Q(x_2, y_2)$, the midpoint of PQ is

$$M\left(\frac{x_1 + x_2}{2}, \frac{y_1 + y_2}{2}\right)$$

Slope of a nonvertical line:

If $x_2 \neq x_1$,

$$m = \frac{\Delta y}{\Delta x} = \frac{y_2 - y_1}{x_2 - x_1}$$

Horizontal lines have a slope of 0. Vertical lines have no defined slope.

Parallel lines have the same slope. The slopes of two nonvertical perpendicular lines are negative reciprocals.

Equations of a line:

Point–slope form:

$$y - y_1 = m(x - x_1)$$

Slope–intercept form:

$$y = mx + b$$

General form:

$$Ax + By = C$$

A *function* is a correspondence between a set of input values x and a set of output values y, where exactly one value of y in the range is assigned to each number x in the domain.

The *domain* of a function is the set of input values. The *range* is the set of output values.

The *vertical line test* can be used to determine whether a graph represents a function.

17. Find the midpoint of the line segment joining $P(-3, 5)$ and $Q(6, 11)$.

18. Find the slope of the line passing through points P and Q.
 a. $P(2, 5)$ and $Q(5, 8)$
 b. $P(-3, -2)$ and $Q(6, 12)$
 c. $P(-3, 4)$ and $Q(-5, -6)$
 d. $P(5, -4)$ and $Q(-6, -9)$
 e. $P(-2, 4)$ and $Q(8, 4)$
 f. $P(-5, -4)$ and $Q(-5, 8)$

19. Find the slope of the graph of each equation.
 a. $2x - 3y = 18$
 b. $2x + y = 8$

20. Tell whether the lines with the given slopes are parallel, perpendicular, or neither.

 a. $m_1 = 4, m_2 = -\dfrac{1}{4}$
 b. $m_1 = 0.5, m_2 = \dfrac{1}{2}$

21. Write the equation of the line with the given properties. Write the result in general form.
 a. Slope of 3; passing through $P(-8, 5)$
 b. Passing through $(-2, 4)$ and $(6, -9)$
 c. Passing through $(-3, -5)$; parallel to the graph of $3x - 2y = 7$
 d. Passing through $(-3, -5)$; perpendicular to the graph of $3x - 2y = 7$

22. A business purchases a copy machine for \$8,700 and will depreciate it on a straight-line basis over the next 5 years. At the end of its useful life, it will be sold as scrap for \$100. Find its depreciation equation.

23. Tell whether each equation determines y to be a function of x.
 a. $y = 6x - 4$
 b. $y = 4 - x$
 c. $y^2 = x$
 d. $|y| = x^2$

24. Assume that $f(x) = 3x + 2$ and $g(x) = x^2 - 4$ and find each value.
 a. $f(-3)$
 b. $g(8)$
 c. $g(-2)$
 d. $f(x + 1)$

25. Find the domain and range of each function.
 a. $f(x) = 4x - 1$
 b. $f(x) = 3x - 10$

26. Use the vertical line test to determine whether each graph represents a function.
 a.

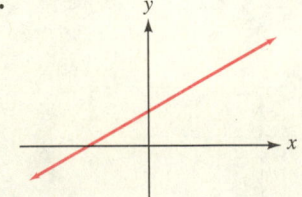

 b.

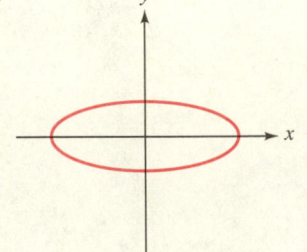

Section 8.7 Variation

Direct variation:

$y = kx$ (k is a constant)

Inverse variation:

$y = \dfrac{k}{x}$ (k is a constant)

Joint variation:

$y = kxz$ (k is a constant)

Combined variation:

$y = \dfrac{kx}{z}$ (k is a constant)

27. Assume that y varies directly with x. If $x = 12$ when $y = 2$, find x when $y = 12$.

28. Assume that y varies inversely with x. If $x = 24$ when $y = 3$, find y when $x = 12$.

29. Assume that y varies jointly with x and z. Find the constant of variation if $x = 24$ when $y = 3$ and $z = 4$.

30. Assume that y varies directly with t and inversely with x. Find the constant of variation if $x = 2$ when $t = 8$ and $y = 64$.

CHAPTER 8

Test

In Problems 1–2, solve each equation.

1. $\dfrac{y - 1}{5} + 2 = \dfrac{2y - 3}{3}$

2. Solve $n = \dfrac{360}{180 - a}$ for a.

3. A 20-foot pipe is to be cut into three pieces. One piece is to be twice as long as another, and the third piece is to be six times as long as the shortest. Find the length of the longest piece.

4. A rectangle with a perimeter of 26 centimeters is 5 centimeters longer than it is wide. Find its area.

In Problems 5–10, solve each equation or inequality.

5. $-2(2x + 3) \geq 14$

6. $-2 < \dfrac{x - 4}{3} < 4$

7. $|2x + 3| = 11$

8. $|3x + 4| = |x + 12|$

9. $|x + 3| \leq 4$

10. $|2x - 4| > 22$

In Problems 11–20, factor each expression.

11. $3xy^2 + 6x^2y$

12. $ax - xy + ay - y^2$

13. $x^2 - 49$

14. $4y^4 - 64$

15. $b^3 + 125$

16. $b^3 - 27$

17. $3u^3 - 24$

18. $6u^2 + 9u - 6$

19. $6b^{2n} + b^n - 2$

20. $x^2 + 6x + 9 - y^2$

In Problems 21–22, simplify each fraction.

21. $\dfrac{-12x^2y^3z^2}{18x^3y^4z^2}$

22. $\dfrac{2x^2 + 7x + 3}{4x + 12}$

In Problems 23–26, do the operations and simplify, if necessary. Write all answers without negative exponents.

23. $\dfrac{x^2y^{-2}}{x^3z^2} \cdot \dfrac{x^2z^4}{y^2z}$

24. $\dfrac{u^2 + 5u + 6}{u^2 - 4} \cdot \dfrac{u^2 - 5u + 6}{u^2 - 9}$

25. $\dfrac{x^3 + y^3}{4} \div \dfrac{x^2 - xy + y^2}{2x + 2y}$

26. $\dfrac{x + 2}{x + 1} - \dfrac{x + 1}{x + 2}$

In Problems 27–28, simplify each complex fraction.

27. $\dfrac{\dfrac{2u^2w^3}{v^2}}{\dfrac{4uw^4}{uv}}$

28. $\dfrac{\dfrac{x}{y} + \dfrac{1}{2}}{\dfrac{x}{2} - \dfrac{1}{y}}$

29. Find the midpoint of the line segment with endpoints at $P(-3, 5)$ and $Q(4, -6)$.

30. Find the slope of the line passing through $P(-3, 5)$ and $Q(4, -6)$.

31. Write the equation of the line passing through $P(-3, 5)$ and $Q(4, -6)$. Give the result in general form.

32. If $f(x) = 3x - 5$, find $f(-3)$.

33. Assume that y varies directly with x. If $x = 30$ when $y = 4$, find y when $x = 9$.

34. Assume that V varies inversely with t. If $V = 55$ when $t = 20$, find t when $V = 75$.

CHAPTERS 1–8

Cumulative Review Exercises

1. The diagram in Illustration 1 shows the sets that compose the set of real numbers. Which of the indicated sets make up the *rational numbers* and the *irrational numbers?*

ILLUSTRATION 1

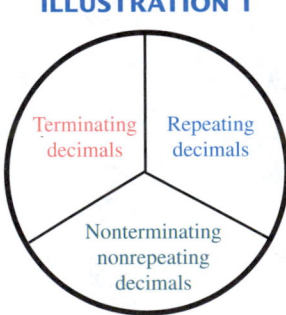

In Exercises 3–4, evaluate each expression when $x = 2$ and $y = -4$.

3. $|x| - xy$

4. $\dfrac{x^2 - y^2}{3x + y}$

In Exercises 5–6, simplify each expression.

5. $3p^2 - 6(5p^2 + p) + p^2$

6. $-(a + 2) - (a - b)$

2. SCREWS The thread profile of a screw is determined by the distance between threads. This distance, indicated by the letter p, is known as the *pitch*. If $p = 0.125$, find each of the dimensions labeled in Illustration 2.

7. PLASTIC WRAP Estimate the number of *square feet* of plastic wrap on a roll if the dimensions printed on the box describe the roll as 205 feet long by $11\frac{3}{4}$ inches wide.

8. INVESTMENTS Find the amount of money that was invested at $8\frac{7}{8}\%$ if it earned \$1,775 in simple interest in one year.

ILLUSTRATION 2

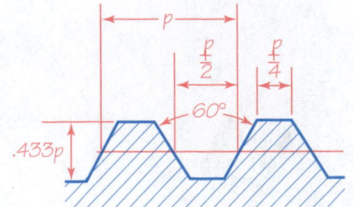

In Exercises 9–12, solve each equation, if possible.

9. $3x - 6 = 20$

10. $6(x - 1) = 2(x + 3)$

11. $\dfrac{5b}{2} - 10 = \dfrac{b}{3} + 3$

12. $2a - 5 = -2a + 4(a - 2) + 1$

In Exercises 13–14, tell whether the lines represented by the equations are parallel, perpendicular, or neither.

13. $3x + 2y = 12$, $2x - 3y = 5$

14. $3x = y + 4$, $y = 3(x - 4) - 1$

15. Write the equation of the line that passes through $P(-2, 3)$ and is perpendicular to the graph of $3x + y = 8$. Answer in slope–intercept form.

16. Find the slope of the line that passes through $(0, -8)$ and $(-5, 0)$.

17. PRISONS The graph in Illustration 3 shows the growth of the U.S. prison population from 1970 to 1995. Find the rate of change in the prison population from 1970 to 1975.

ILLUSTRATION 3

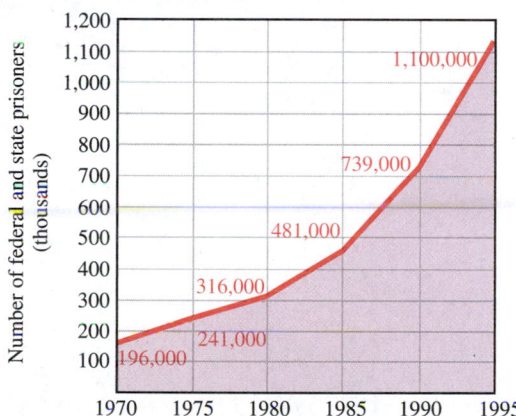

Based on data from *U.S. Statistical Abstract*

18. PRISONS Refer to the graph in Illustration 3. During what five-year period was the rate of change in the U.S. prison population the greatest? Find the rate of change.

In Exercises 19–20, $f(x) = 3x^2 - x$. Find each value.

19. $f(2)$ **20.** $f(-2)$

21. U.S. WORKERS The graph in Illustration 4 shows how the makeup of the U.S. workforce changed over the years 1900–1990. Estimate the coordinates of the points of intersection in the graph. Explain their significance.

22. WHITE-COLLAR JOBS Refer to the graph in Illustration 4. To model the growth in the percent of

white-collar workers, draw a line through $(0, 20)$ and $(90, 70)$ and write an equation for this line. Use the equation to predict the percent of white-collar workers in the year 2010.

ILLUSTRATION 4

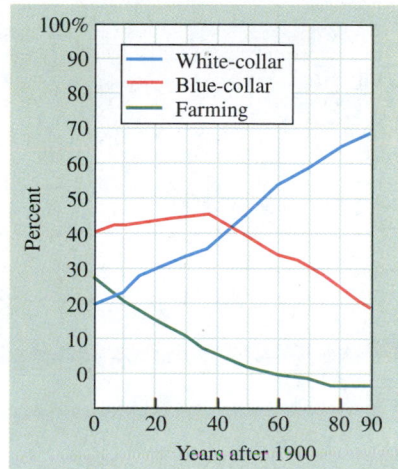

Based on data from *U.S. Statistical Abstract*

23. ENTREPRENEURS A person invests $18,375 to set up a small business producing a piece of computer software that will sell for $29.95. If each piece can be produced for $5.45, how many pieces must be sold to break even?

24. CONCERT TICKETS Tickets for a concert cost $5, $3, and $2. Twice as many $5 tickets were sold as $2 tickets. The receipts for 750 tickets were $2,625. How many tickets were sold at each price?

In Exercises 25–26, solve each equation.

25. $|4x - 3| = 9$ **26.** $|2x - 1| = |3x + 4|$

In Exercises 27–30, solve each inequality. Give the solution in interval notation and graph it.

27. $-3(x - 4) \geq x - 32$

28. $-8 < -3x + 1 < 10$

29. $|3x - 2| \leq 4$

30. $|2x + 3| - 1 > 4$

In Exercises 31–38, factor each expression.

31. $3r^2s^3 - 6rs^4$

32. $5(x - y) - a(x - y)$

33. $xu + yv + xv + yu$

34. $81x^4 - 16y^4$

35. $8x^3 - 27y^6$

36. $6x^2 + 5x - 6$

37. $9x^2 - 30x + 25$

38. $15x^2 - x - 6$

39. Solve $6x^2 + 7 = -23x$.

40. Solve $x^3 - 4x = 0$.

41. Solve $b^2x^2 + a^2y^2 = a^2b^2$ for b^2.

42. CAMPING The rectangular-shaped cooking surface of a small camping stove is 108 in.^{2}. If its length is 3 inches longer than its width, what are its dimensions?

9

Radicals and Rational Exponents

CAMPUS CONNECTION

The Physics Department

In a physics class, students study relativity theory. This theory, primarily developed by Albert Einstein in the early 20th century, explains the relationship between matter and energy as well as the properties of light. One major conclusion of relativity theory is mathematically illustrated by an equation that contains a *rational exponent.* In this chapter, we discuss the meaning of rational (fractional) exponents and consider the relationship between rational exponents and radicals. Rational exponents and radicals occur in formulas and functions used in such diverse disciplines as carpentry, law enforcement, business, theater production, and physics.

Radical expressions have the form $\sqrt[n]{a}$. In this chapter, we will see how they are used to model many real-world situations.

▶9.1

Radical Expressions and Radical Functions

In this section, you will learn about

> Square roots ■ Square roots of expressions containing variables
> ■ The square root function ■ Cube roots ■ nth roots

Introduction   In this section, we will review how to find *square roots* of numbers. Then we will generalize the concept of root and consider cube roots, fourth roots, fifth roots, and so on. We will also discuss a new family of functions, called *radical functions,* and we will see how they have application in many disciplines.

Square Roots

When solving problems, we must often find what number must be squared to obtain a second number a. If such a number can be found, it is called a **square root of a.** For example,

■   0 is a square root of 0, because $0^2 = 0$.

■   4 is a square root of 16, because $4^2 = 16$.

■ -4 is a square root of 16, because $(-4)^2 = 16$.

■ $7xy$ is a square root of $49x^2y^2$, because $(7xy)^2 = 49x^2y^2$.

■ $-7xy$ is a square root of $49x^2y^2$, because $(-7xy)^2 = 49x^2y^2$.

The preceding examples illustrate the following definition.

Square root of a

> The number b is a **square root of a** if $b^2 = a$.

All positive numbers have two real-number square roots, one that is positive and one that is negative.

EXAMPLE 1   **Square roots.** Find the two square roots of 121.

Solution   The two square roots of 121 are 11 and -11, because

$$11^2 = 121 \quad \text{and} \quad (-11)^2 = 121$$

SELF CHECK   Find the square roots of 144.   *Answer:* 12, -12 ■

In the following definition, the symbol $\sqrt{}$ is called a **radical sign,** and the number x within the radical sign is called a **radicand.**

Principal square root

If $x > 0$, the **principal square root of x** is the positive square root of x, denoted as $\sqrt{x}$.

The principal square root of 0 is 0: $\sqrt{0} = 0$.

By definition, the principal square root of a positive number is always positive. Although 5 and -5 are both square roots of 25, only 5 is the principal square root. The radical expression $\sqrt{25}$ represents 5. The radical expression $-\sqrt{25}$ represents -5. When we write $\sqrt{25} = 5$ or $-\sqrt{25} = -5$, we say that we have *simplified the radical*.

EXAMPLE 2

Finding square roots. Simplify each radical.

a. $\sqrt{1} = 1$ b. $\sqrt{81} = 9$

c. $-\sqrt{81} = -9$ d. $-\sqrt{225} = -15$

e. $\sqrt{\dfrac{1}{4}} = \dfrac{1}{2}$ f. $-\sqrt{\dfrac{16}{121}} = -\dfrac{4}{11}$

g. $\sqrt{0.04} = 0.2$ h. $-\sqrt{0.0009} = -0.03$

SELF CHECK   Simplify   a. $-\sqrt{49}$ and   b. $\sqrt{\frac{25}{49}}$.   *Answers:*   a. -7,   b. $\frac{5}{7}$ ∎

Numbers such as 4, 9, 16, 49, and 1,600 are called **integer squares,** because each one is the square of an integer. The square root of every integer square is a rational number.

$$\sqrt{4} = 2, \qquad \sqrt{9} = 3, \qquad \sqrt{16} = 4, \qquad \sqrt{49} = 7, \qquad \sqrt{1,600} = 40$$

The square roots of many positive integers are not rational numbers. For example, $\sqrt{11}$ is an *irrational number.* To find an approximate value of $\sqrt{11}$, we enter 11 into a calculator and press the square root key $\boxed{\sqrt{}}$. (On some calculators, the $\boxed{\sqrt{}}$ key must be pressed first.)

$$\sqrt{11} \approx 3.31662479$$

 WARNING! Square roots of negative numbers are not real numbers. For example, $\sqrt{-9}$ is not a real number, because no real number squared equals -9. Square roots of negative numbers come from a set called the **imaginary numbers,** which we will discuss in the next chapter.

Square Roots of Expressions Containing Variables

If $x \neq 0$, the positive number x^2 has x and $-x$ for its two square roots. To denote the positive square root of $\sqrt{x^2}$, we must know whether x is positive or negative.

If $x > 0$, we can write

$$\sqrt{x^2} = x \qquad \sqrt{x^2} \text{ represents the positive square root of } x^2, \text{ which is } x.$$

If x is negative, then $-x > 0$, and we can write

$$\sqrt{x^2} = -x \qquad \sqrt{x^2} \text{ represents the positive square root of } x^2, \text{ which is } -x.$$

If we don't know whether x is positive or negative, we can use absolute value symbols to guarantee that $\sqrt{x^2}$ is positive.

Definition of $\sqrt{x^2}$

For any real number x,

$$\sqrt{x^2} = |x|$$

EXAMPLE 3

Simplifying radical expressions containing variables. Simplify **a.** $\sqrt{16x^2}$, **b.** $\sqrt{x^2 + 2x + 1}$, and **c.** $\sqrt{m^4}$.

Solution   If x can be any real number, we have

a. $\sqrt{16x^2} = \sqrt{(4x)^2}$ Write the radicand $16x^2$ as $(4x)^2$.

$\qquad\quad = |4x|$ Because $(|4x|)^2 = 16x^2$. Since x could be negative, absolute value symbols are needed.

$\qquad\quad = 4|x|$ Since 4 is a positive constant in the product $4x$, we can write it outside the absolute value symbols.

b. $\sqrt{x^2 + 2x + 1}$

$\qquad = \sqrt{(x + 1)^2}$ Factor the radicand: $x^2 + 2x + 1 = (x + 1)^2$.

$\qquad = |x + 1|$ Since $x + 1$ can be negative (for example, when $x = -5$, $x + 1$ is -4), absolute value symbols are needed.

c. $\sqrt{m^4} = m^2$ Because $(m^2)^2 = m^4$. Since $m^2 \geq 0$, no absolute value symbols are needed.

SELF CHECK   Simplify **a.** $\sqrt{25a^2}$ and **b.** $\sqrt{16a^4}$.   *Answers:* **a.** $5|a|$, **b.** $4a^2$ ∎

If we are told that x represents a positive real number in parts a and b of Example 3, we do not need to use absolute value symbols to guarantee that the answers are positive.

$\sqrt{16x^2} = 4x$ If x is positive, $4x$ is positive.

$\sqrt{x^2 + 2x + 1} = x + 1$ If x is positive, $x + 1$ is positive.

The Square Root Function

Since there is one principal square root for every nonnegative real number x, the equation $f(x) = \sqrt{x}$ determines a function, called a **square root function.** Square root functions belong to a larger family of functions known as **radical functions.**

EXAMPLE 4

Graphing a square root function. Graph $f(x) = \sqrt{x}$ and find its domain and range.

Solution   To graph this square root function, we will evaluate it for several values of x. We begin with $x = 0$, since 0 is the smallest input for which $\sqrt{x}$ is defined.

$f(x) = \sqrt{x}$

$f(0) = \sqrt{0} = 0$

We enter the ordered pair $(0, 0)$ in the table of values in Figure 9-1. Then we continue the evaluation process for $x = 1, 4, 9,$ and 16 and list the results in the table. After plotting all the ordered pairs, we draw a smooth curve through the points. This is the graph of function f (see Figure 9-1(a)). Since the equation defines a function, its graph passes the vertical line test.

We can use a graphing calculator with window settings of $[-1, 9]$ for x and $[-1, 9]$ for y to get the graph shown in Figure 9-1(b). From either graph, we can see that the domain and the range are the set of nonnegative real numbers. Expressed in interval notation, the domain is $[0, \infty)$, and the range is $[0, \infty)$.

FIGURE 9-1

$f(x) = \sqrt{x}$

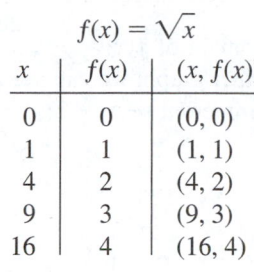

x	$f(x)$	$(x, f(x))$
0	0	$(0, 0)$
1	1	$(1, 1)$
4	2	$(4, 2)$
9	3	$(9, 3)$
16	4	$(16, 4)$

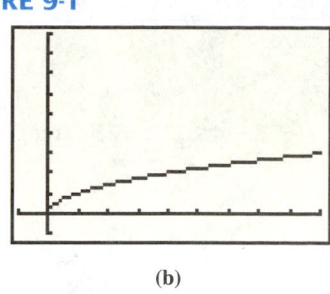

(b)

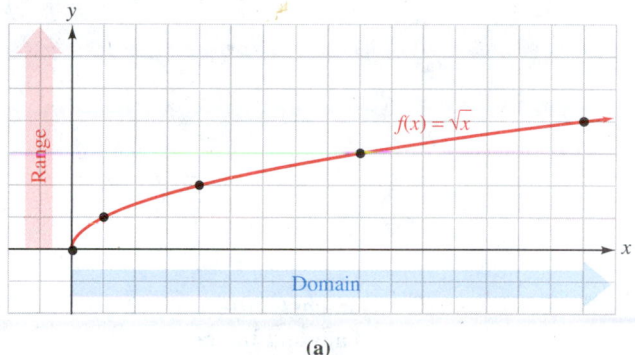

(a)

SELF CHECK

Graph $f(x) = \sqrt{x} + 2$. Then give its domain and range and compare the graph of $f(x) = \sqrt{x}$.

Answers:

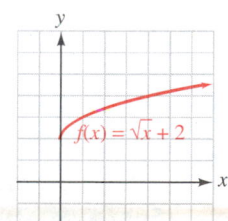

D: $[0, \infty)$, R: $[2, \infty)$; the graph is 2 units higher ■

The graphs of many radical functions are translations or reflections of the square root function, $f(x) = \sqrt{x}$. For example, if $k > 0$,

■ The graph of $f(x) = \sqrt{x} + k$ is the graph of $f(x) = \sqrt{x}$ translated k units up.
■ The graph of $f(x) = \sqrt{x} - k$ is the graph of $f(x) = \sqrt{x}$ translated k units down.
■ The graph of $f(x) = \sqrt{x + k}$ is the graph of $f(x) = \sqrt{x}$ translated k units to the left.
■ The graph of $f(x) = \sqrt{x - k}$ is the graph of $f(x) = \sqrt{x}$ translated k units to the right.
■ The graph of $f(x) = -\sqrt{x}$ is the graph of $f(x) = \sqrt{x}$ reflected about the x-axis.

EXAMPLE 5

Graphing square root functions. Graph $f(x) = -\sqrt{x + 4} - 2$ and find its domain and range.

Solution

This graph will be the reflection of $f(x) = \sqrt{x}$ about the x-axis, translated 4 units to the left and 2 units down. See Figure 9-2(a). We can confirm this graph by using a graphing calculator with window settings of $[-5, 6]$ for x and $[-6, 2]$ for y to get the graph shown in Figure 9-2(b).

We can determine the domain of the function algebraically. Since the expression $\sqrt{x+4}$ is not a real number when $x + 4$ is negative, we must require that

$$x + 4 \geq 0$$

Solving for x, we have

$x \geq -4$ The x-inputs must be real numbers greater than or equal to -4.

Thus, the domain of $f(x) = -\sqrt{x+4} - 2$ is the interval $[-4, \infty)$.

From either graph, we can see that the domain is the interval $[-4, \infty)$ and that the range is the interval $(-\infty, -2]$.

FIGURE 9-2

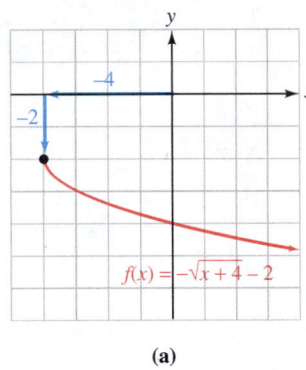
(a)

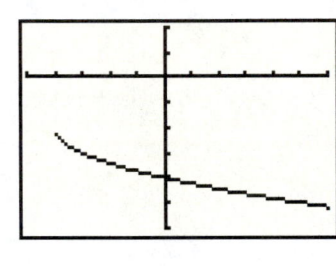
(b)

SELF CHECK   Graph $f(x) = \sqrt{x-2} - 4$. Then give the domain and the range.

Answers:

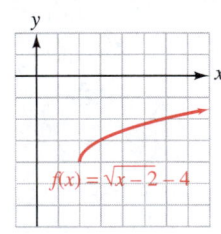

D: $[2, \infty)$, R: $[-4, \infty)$ ■

E X A M P L E   6

Square root functions. The **period of a pendulum** is the time required for the pendulum to swing back and forth to complete one cycle. The period (in seconds) is a function of the pendulum's length l (in feet) and is given by

$$f(l) = 2\pi\sqrt{\frac{l}{32}}$$

Find the period of the 5-foot-long pendulum of the clock shown in Figure 9-3.

FIGURE 9-3

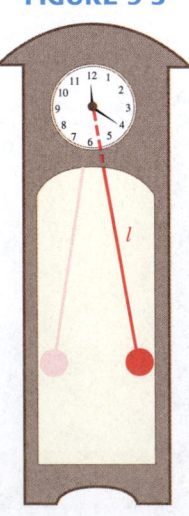

Solution   To determine the period, we substitute 5 for l.

$$f(l) = 2\pi\sqrt{\frac{l}{32}}$$

$$f(5) = 2\pi\sqrt{\frac{5}{32}}$$

≈ 2.483647066 Use a calculator to find an approximation.

The period is approximately 2.5 seconds.

SELF CHECK   To the nearest hundredth, find the period of a pendulum that is 3 feet long.   *Answer:* 1.92 sec   ■

ACCENT ON TECHNOLOGY *Evaluating a Square Root Function Using Its Graph*

To solve Example 6 with a graphing calculator with window settings of $[-2, 10]$ for x and $[-2, 10]$ for y, we graph the function $f(x) = 2\pi\sqrt{\frac{x}{32}}$, as in Figure 9-4(a). We then trace and move the cursor toward an x-value of 5 until we see the coordinates shown in Figure 9-4(b). The period is given by the y-value shown on the screen. By zooming in, we can get better results.

After entering $Y_1 = 2\pi\sqrt{\frac{x}{32}}$, we can also use the TABLE mode to evaluate the function. See Figure 9-4(c).

FIGURE 9-4

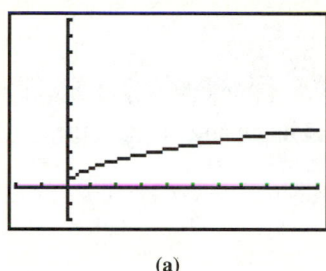

(a)

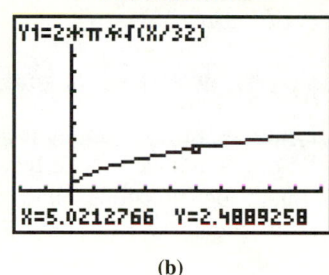

(b)

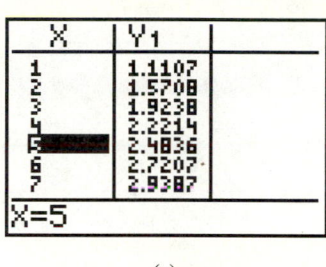

(c)

Cube Roots

The **cube root of x** is any number whose cube is x. For example,

- 4 is a cube root of 64, because $4^3 = 64$.
- $3x^2y$ is a cube root of $27x^6y^3$, because $(3x^2y)^3 = 27x^6y^3$.
- $-2y$ is a cube root of $-8y^3$, because $(-2y)^3 = -8y^3$.

Cube roots

The **cube root of x** is denoted as $\sqrt[3]{x}$ and is defined by

$$\sqrt[3]{x} = y \quad \text{if } y^3 = x$$

We note that 64 has two real-number square roots, 8 and -8. However, 64 has only one real-number cube root 4, because 4 is the only real number whose cube is 64. Since every real number has exactly one real cube root, it is unnecessary to use absolute value symbols when simplifying cube roots.

Definition of $\sqrt[3]{x^3}$

For any real number x,

$$\sqrt[3]{x^3} = x$$

EXAMPLE 7

Finding cube roots. Simplify each radical expression.

a. $\sqrt[3]{125} = 5$ Because $5^3 = 5 \cdot 5 \cdot 5 = 125$.

b. $\sqrt[3]{\dfrac{1}{8}} = \dfrac{1}{2}$ Because $\left(\dfrac{1}{2}\right)^3 = \dfrac{1}{2} \cdot \dfrac{1}{2} \cdot \dfrac{1}{2} = \dfrac{1}{8}$.

c. $\sqrt[3]{-27x^3} = -3x$ Because $(-3x)^3 = (-3x)(-3x)(-3x) = -27x^3$.

d. $\sqrt[3]{-\dfrac{8a^3}{27b^3}} = -\dfrac{2a}{3b}$ Because $\left(-\dfrac{2a}{3b}\right)^3 = \left(-\dfrac{2a}{3b}\right)\left(-\dfrac{2a}{3b}\right)\left(-\dfrac{2a}{3b}\right) = -\dfrac{8a^3}{27b^3}$.

e. $\sqrt[3]{0.216x^3y^6} = 0.6xy^2$ Because $(0.6xy^2)^3 = (0.6xy^2)(0.6xy^2)(0.6xy^2) = 0.216x^3y^6$.

SELF CHECK

Simplify **a.** $\sqrt[3]{1,000}$, **b.** $\sqrt[3]{\frac{1}{27}}$, and **c.** $\sqrt[3]{125a^3}$.

Answers: **a.** 10, **b.** $\frac{1}{3}$, **c.** $5a$

The equation $f(x) = \sqrt[3]{x}$ defines a **cube root function.** From the graph shown in Figure 9-5(a), we can see that the domain and range of the function $f(x) = \sqrt[3]{x}$ are the set of real numbers. Note that the graph of $f(x) = \sqrt[3]{x}$ passes the vertical line test. Like square root functions, cube root functions are members of the family of radical functions. Figures 9-5(b) and 9-5(c) show several translations of the cube root function.

FIGURE 9-5

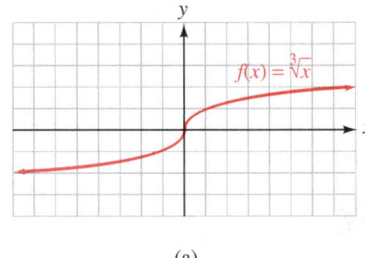

(a)

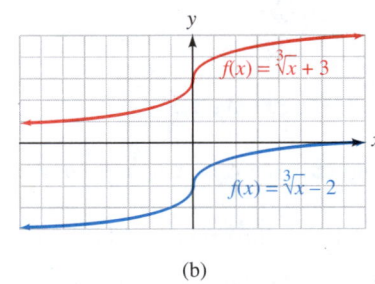

(b)

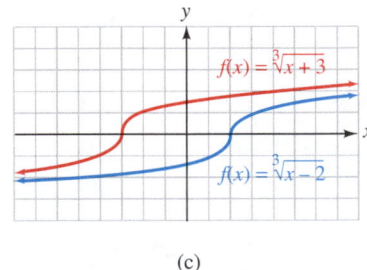

(c)

*n*th Roots

Just as there are square roots and cube roots, there are fourth roots, fifth roots, sixth roots, and so on.

When n is an odd natural number, the expression $\sqrt[n]{x}$ ($n > 1$) represents an **odd root.** Since every real number has just one real nth root when n is odd, we don't need to worry about absolute value symbols when finding odd roots. For example,

$$\sqrt[5]{243} = \sqrt[5]{3^5} = 3 \qquad \text{Because } 3^5 = 243.$$
$$\sqrt[7]{-128x^7} = \sqrt[7]{(-2x)^7} = -2x \qquad \text{Because } (-2x)^7 = -128x^7.$$

When n is an even natural number, the expression $\sqrt[n]{x}$ ($n > 1$, $x > 0$) represents an **even root.** In this case, there will be one positive and one negative real nth root. For example, the real sixth roots of 729 are 3 and -3, because $3^6 = 729$ and $(-3)^6 = 729$. When finding even roots, we can use absolute value symbols to guarantee that the nth root is positive.

$$\sqrt[4]{(-3)^4} = |-3| = 3 \qquad \begin{array}{l}\text{We could also simplify this as follows:}\\ \sqrt[4]{(-3)^4} = \sqrt[4]{81} = 3.\end{array}$$

$$\sqrt[6]{729x^6} = \sqrt[6]{(3x)^6} = |3x| = 3|x| \qquad \begin{array}{l}\text{The absolute value symbols guarantee that the}\\ \text{sixth root is positive.}\end{array}$$

In general, we have the following rules.

Rules for $\sqrt[n]{x^n}$

If x is a real number and $n > 1$, then

If n is an odd natural number, $\sqrt[n]{x^n} = x$.

If n is an even natural number, $\sqrt[n]{x^n} = |x|$.

In the radical expression $\sqrt[n]{x}$, n is called the **index** (or **order**) of the radical. When the index is 2, the radical is a square root, and we usually do not write the index.

$$\sqrt{x} = \sqrt[2]{x}$$

WARNING! When n is even ($n > 1$) and $x < 0$, $\sqrt[n]{x}$ is not a real number. For example, $\sqrt[4]{-81}$ is not a real number, because no real number raised to the fourth power is -81.

EXAMPLE 8   **Finding even and odd roots.** Simplify each radical, if possible.

a. $\sqrt[4]{625} = 5$,   because $5^4 = 625$ Read $\sqrt[4]{625}$ as "the fourth root of 625."

b. $\sqrt[4]{-1}$ is not a real number.   This is an even root of a negative number.

c. $\sqrt[5]{-32} = -2$,   because $(-2)^5 = -32$ Read $\sqrt[5]{-32}$ as "the fifth root of -32."

d. $\sqrt[6]{\dfrac{1}{64}} = \dfrac{1}{2}$,   because $\left(\dfrac{1}{2}\right)^6 = \dfrac{1}{64}$ Read $\sqrt[6]{\frac{1}{64}}$ as "the sixth root of $\frac{1}{64}$."

e. $\sqrt[7]{10^7} = 10$,   because $10^7 = 10^7$ Read $\sqrt[7]{10^7}$ as "the seventh root of 10^7."

SELF CHECK   Simplify   a. $\sqrt[4]{\frac{1}{81}}$,   b. $\sqrt[5]{10^5}$,   and   c. $\sqrt[6]{-64}$.   *Answers:*   a. $\frac{1}{3}$,   b. 10,   c. not a real number

ACCENT ON TECHNOLOGY   *Finding Roots*

The square root key $\boxed{\sqrt{}}$ on a scientific calculator can be used to evaluate square roots. To evaluate roots with an index greater than 2, we can use the root key $\boxed{\sqrt[x]{y}}$. For example, the function

$$r(V) = \sqrt[3]{\dfrac{3V}{4\pi}}$$

gives the radius of a sphere with volume V. To find the radius of the spherical propane tank shown in Figure 9-6, we substitute 113 for V to get

$$r(V) = \sqrt[3]{\dfrac{3V}{4\pi}}$$

$$r(113) = \sqrt[3]{\dfrac{3(113)}{4\pi}}$$

FIGURE 9-6

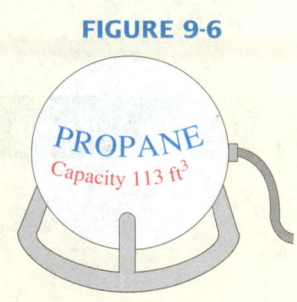

PROPANE
Capacity 113 ft³

To evaluate a root, we enter the radicand and press the root key $\boxed{\sqrt[x]{y}}$ followed by the index of the radical, which in this case is 3.

(continued)

Keystrokes 3 $\times$ 113 $\div$ $($ 4 $\times$ π $)$ $=$ $2nd$ $\sqrt[x]{y}$ 3 $=$

$$\boxed{2.999139118}$$

To evaluate the cube root of $\frac{3(113)}{4\pi}$ using a graphing calculator, we enter these numbers and press these keys.

Keystrokes $\boxed{MATH}$ 4 $($ 3 $\times$ 113 $)$ $\div$ $($ 4 $\times$ $2nd$ π $)$ $)$ $\boxed{ENTER}$

$$\sqrt[3]{((3*113)/(4*\pi)}$$
$$)$$
$$2.999139118$$

The radius of the propane tank is about 3 feet.

EXAMPLE 9

Simplifying radical expressions containing variables. Simplify each radical expression. Assume that x can be any real number.

a. $\sqrt[5]{x^5} = x$ Since n is odd, absolute value symbols aren't needed.

b. $\sqrt[4]{16x^4} = |2x| = 2|x|$ Since n is even and x can be negative, absolute value symbols are needed to guarantee that the result is positive.

c. $\sqrt[6]{(x+4)^6} = |x+4|$ Absolute value symbols are needed to guarantee that the result is positive.

d. $\sqrt[3]{(x+1)^3} = x+1$ Since n is odd, absolute value symbols aren't needed.

SELF CHECK Simplify **a.** $\sqrt[4]{16a^8}$, **b.** $\sqrt[5]{(a+5)^5}$, and **c.** $\sqrt{(x^2+4x+4)^2}$.   *Answers:* **a.** $2a^2$, **b.** $a+5$, **c.** $(x+2)^2$ ■

If we are told that x represents a positive real number in parts b and c of Example 9, we do not need to use absolute value symbols to guarantee that the results are positive.

$$\sqrt[4]{16x^4} = 2x \quad \text{If } x \text{ is positive, } 2x \text{ is positive.}$$
$$\sqrt[6]{(x+4)^6} = x+4 \quad \text{If } x \text{ is positive, } x+4 \text{ is positive.}$$

We summarize the definitions concerning $\sqrt[n]{x}$ as follows.

Summary of the definitions of $\sqrt[n]{x}$

If n is a natural number greater than 1 and x is a real number, then

If $x > 0$, then $\sqrt[n]{x}$ is the positive number such that $\left(\sqrt[n]{x}\right)^n = x$.

If $x = 0$, then $\sqrt[n]{x} = 0$.

If $x < 0$ $\begin{cases} \text{and } n \text{ is odd, then } \sqrt[n]{x} \text{ is the real number such that } \left(\sqrt[n]{x}\right)^n = x. \\ \text{and } n \text{ is even, then } \sqrt[n]{x} \text{ is not a real number.} \end{cases}$

STUDY SET

Section 9.1

VOCABULARY

In Exercises 1–8, fill in the blanks to make the statements true.

1. $5x^2$ is the _____ of $25x^4$, because $(5x^2)^2 = 25x^4$.
2. $f(x) = \sqrt{x}$ and $g(t) = \sqrt[3]{t}$ are _____ functions.
3. The symbol $\sqrt{}$ is called a _____ sign.
4. In the expression $\sqrt[3]{27x^6}$, 3 is the _____ and $27x^6$ is the _____.
5. When n is an odd number, $\sqrt[n]{x}$ represents an _____ root.
6. When n is an _____ number, $\sqrt[n]{x}$ represents an even root.
7. When we write $\sqrt{b^2 + 6b + 9} = |b + 3|$, we say that we have _____ the radical.
8. 6 is the _____ of 216 because $6^3 = 216$.

CONCEPTS

In Exercises 9–16, fill in the blanks to make the statements true.

9. b is a square root of a if _____.
10. $\sqrt{0} = $ ___ and $\sqrt[3]{0} = $ ___
11. The number 25 has _____ square roots. The principal square root of 25 is the _____ square root of 25.
12. $\sqrt{-4}$ is not a real number, because no real number _____ equals -4.
13. $\sqrt[3]{x} = y$ if $y^3 = $ ___.
14. $\sqrt{x^2} = $ ___ and $\sqrt[3]{x^3} = $ ___
15. The graph of $f(x) = \sqrt{x} + 3$ is the graph of $f(x) = \sqrt{x}$ translated ___ units ___.
16. The graph of $f(x) = \sqrt{x + 5}$ is the graph of $f(x) = \sqrt{x}$ translated ___ units to the _____.

In Exercises 17–18, complete the table of values and then graph the radical function.

17. $f(x) = -\sqrt{x}$

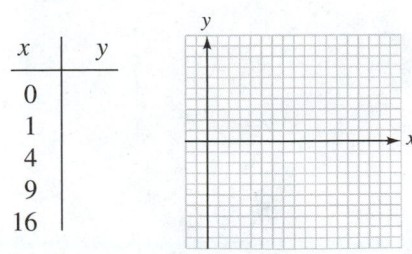

x	y
0	
1	
4	
9	
16	

18. $f(x) = -\sqrt[3]{x}$

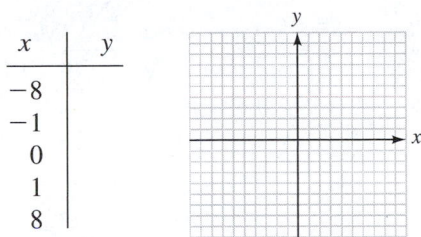

x	y
-8	
-1	
0	
1	
8	

NOTATION

In Exercises 19–22, translate each sentence into mathematical symbols.

19. The square root of x squared is the absolute value of x.
20. The cube root of x cubed is x.
21. f of x equals the square root of the quantity x minus five.
22. The fifth root of negative thirty-two is negative two.

PRACTICE

In Exercises 23–34, find each square root, if possible.

23. $\sqrt{121}$
24. $\sqrt{144}$
25. $-\sqrt{64}$
26. $-\sqrt{1}$
27. $\sqrt{\dfrac{1}{9}}$
28. $-\sqrt{\dfrac{4}{25}}$
29. $\sqrt{0.25}$
30. $\sqrt{0.16}$
31. $\sqrt{-25}$
32. $-\sqrt{-49}$
33. $\sqrt{(-4)^2}$
34. $\sqrt{(-9)^2}$

In Exercises 35–38, use a calculator to find each square root. Give the answer to four decimal places.

35. $\sqrt{12}$
36. $\sqrt{340}$
37. $\sqrt{679.25}$
38. $\sqrt{0.0063}$

In Exercises 39–46, find each square root. Assume that all variables are unrestricted, and use absolute value symbols when necessary.

39. $\sqrt{4x^2}$
40. $\sqrt{16y^4}$
41. $\sqrt{(t + 5)^2}$
42. $\sqrt{(a + 6)^2}$
43. $\sqrt{(-5b)^2}$
44. $\sqrt{(-8c)^2}$
45. $\sqrt{a^2 + 6a + 9}$
46. $\sqrt{x^2 + 10x + 25}$

In Exercises 47–54, find each value given that
$f(x) = \sqrt{x} - 4$ *and* $g(x) = \sqrt[3]{x} - 4.$

47. $f(4)$ **48.** $f(8)$

49. $f(20)$ **50.** $f(29)$

51. $g(12)$ **52.** $g(-4)$

53. $g(-996)$ **54.** $g(1,004)$

In Exercises 55–58, find each value given that
$f(x) = \sqrt{x^2 + 1}$ *and* $g(x) = \sqrt[3]{x^2 + 1}.$ *Give each answer to four decimal places.*

55. $f(4)$ **56.** $f(2.35)$

57. $g(6)$ **58.** $g(21.57)$

In Exercises 59–62, graph each radical function and find its domain and range. Check your work with a graphing calculator.

59. $f(x) = \sqrt{x + 4}$ **60.** $f(x) = -\sqrt{x - 1}$

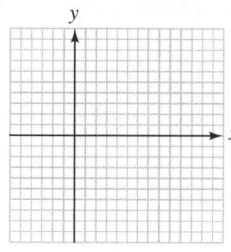 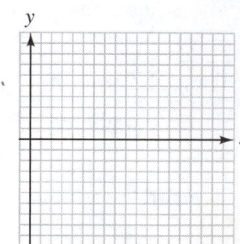

61. $f(x) = -\sqrt[3]{x} - 3$ **62.** $f(x) = \sqrt[3]{x} - 1$

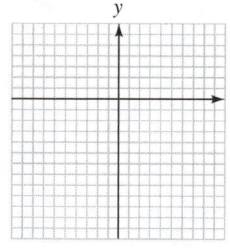 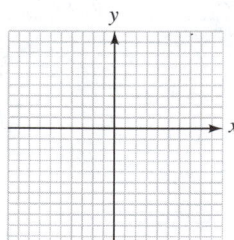

In Exercises 63–78, simplify each cube root.

63. $\sqrt[3]{1}$ **64.** $\sqrt[3]{-8}$

65. $\sqrt[3]{-125}$ **66.** $\sqrt[3]{512}$

67. $\sqrt[3]{-\dfrac{8}{27}}$ **68.** $\sqrt[3]{\dfrac{125}{216}}$

69. $\sqrt[3]{0.064}$ **70.** $\sqrt[3]{0.001}$

71. $\sqrt[3]{8a^3}$ **72.** $\sqrt[3]{-27x^6}$

73. $\sqrt[3]{-1,000p^3q^3}$ **74.** $\sqrt[3]{343a^6b^3}$

75. $\sqrt[3]{-\dfrac{1}{8}m^6n^3}$ **76.** $\sqrt[3]{0.008z^9}$

77. $\sqrt[3]{-0.064s^9t^6}$ **78.** $\sqrt[3]{\dfrac{27}{1,000}a^6b^6}$

In Exercises 79–98, simplify each radical, if possible. Assume that all variables represent positive real numbers.

79. $\sqrt[4]{81}$ **80.** $\sqrt[6]{64}$

81. $-\sqrt[5]{243}$ **82.** $-\sqrt[4]{625}$

83. $\sqrt[4]{-256}$ **84.** $\sqrt[6]{-729}$

85. $\sqrt[4]{\dfrac{16}{625}}$ **86.** $\sqrt[5]{-\dfrac{243}{32}}$

87. $-\sqrt[5]{-\dfrac{1}{32}}$ **88.** $-\sqrt[4]{\dfrac{81}{256}}$

89. $\sqrt[5]{32a^5}$ **90.** $\sqrt[5]{-32x^5}$

91. $\sqrt[4]{16a^4}$ **92.** $\sqrt[8]{x^{24}}$

93. $\sqrt[4]{k^{12}}$ **94.** $\sqrt[6]{64b^6}$

95. $\sqrt[4]{\dfrac{1}{16}m^4}$ **96.** $\sqrt[4]{\dfrac{1}{81}x^8}$

97. $\sqrt[25]{(x + 2)^{25}}$ **98.** $\sqrt[44]{(x + 4)^{44}}$

APPLICATIONS

Use a calculator to solve each problem. In each case, round to the nearest tenth.

99. EMBROIDERY The radius r of a circle is given by the formula

$$r = \sqrt{\dfrac{A}{\pi}}$$

where A is its area. Find the diameter of the embroidery hoop shown in Illustration 1 if there are 38.5 in.^{2} of stretched fabric on which to embroider.

ILLUSTRATION 1

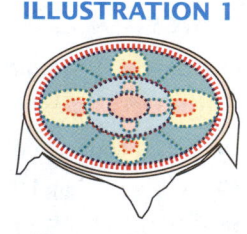

100. SOFTBALL AND BASEBALL The length of a diagonal of a square is given by the function $d(s) = \sqrt{2s^2}$, where s is the length of a side of the square. Find the distance from home plate to second base on a softball diamond and on a baseball diamond. Illustration 2 gives the dimensions of each type of infield.

ILLUSTRATION 2

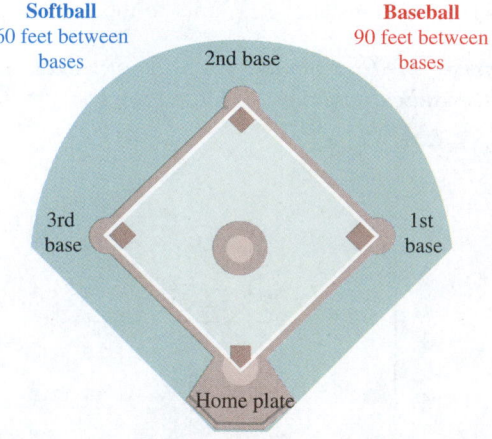

Softball
60 feet between bases

Baseball
90 feet between bases

2nd base

3rd base

1st base

Home plate

101. PULSE RATE The approximate pulse rate (in beats per minute) of an adult who is t inches tall is given by the function

$$p(t) = \frac{590}{\sqrt{t}}$$

The Guinness Book of World Records 1998 lists Ri Myong-hun of North Korea as the tallest living man, at 7 ft $8\frac{1}{2}$ in. Find his approximate pulse rate as predicted by the function.

102. THE GRAND CANYON The time t (in seconds) that it takes for an object to fall a distance of s feet is given by the formula

$$t = \frac{\sqrt{s}}{4}$$

In some places, the Grand Canyon is one mile (5,280 feet) deep. How long would it take a stone dropped over the edge of the canyon to hit bottom?

103. BIOLOGY Scientists will place five rats inside the controlled environment of a sealed hemisphere to study the rats' behavior. The function

$$d(V) = \sqrt[3]{12\left(\frac{V}{\pi}\right)}$$

gives the diameter of a hemisphere with volume V. Use the function to determine the diameter of the base of the hemisphere, if each rat requires 125 cubic feet of living space.

104. AQUARIUM The function

$$s(g) = \sqrt[3]{\frac{g}{7.5}}$$

determines how long (in feet) an edge of a cube-shaped tank must be if it is to hold g gallons of water. What dimensions should a cube-shaped aquarium have if it is to hold 1,250 gallons of water?

105. COLLECTIBLES The *effective rate of interest r* earned by an investment is given by the formula

$$r = \sqrt[n]{\frac{A}{P}} - 1$$

where P is the initial investment that grows to value A after n years. Determine the effective rate of interest earned by a collector on a Lladró porcelain figurine purchased for $800 and sold for $950 five years later.

106. LAW ENFORCEMENT The graphs of the two radical functions shown in Illustration 3 can be used to estimate the speed (in mph) of a car involved in an accident. Suppose a police accident report listed skid marks to be 220 feet long but failed to give the road conditions. Estimate the possible speeds the car was traveling prior to the brakes being applied.

ILLUSTRATION 3

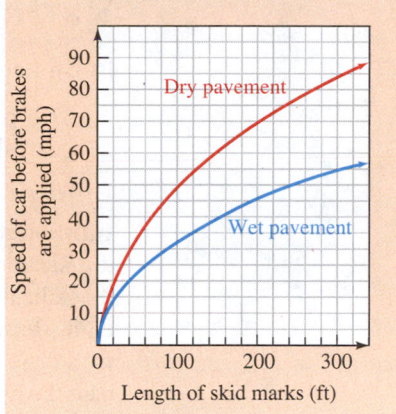

WRITING

107. Explain why 36 has two square roots, but $\sqrt{36}$ is just 6, not -6.

108. If x is any real number, then $\sqrt{x^2} = x$ is not correct. Explain.

109. Explain what is **wrong** with the graph in Illustration 4 if it is supposed to be the graph of $f(x) = \sqrt{x}$.

ILLUSTRATION 4

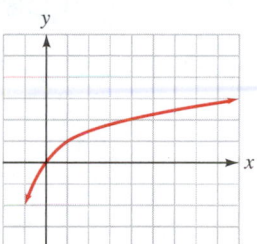

110. Explain how to estimate the domain and range of the radical function that is graphed in Illustration 5.

ILLUSTRATION 5

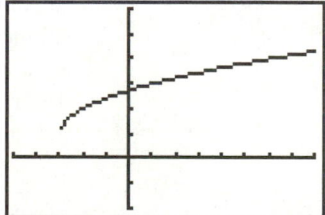

REVIEW

Do the operations.

111. $\dfrac{x^2 - x - 6}{x^2 - 2x - 3} \cdot \dfrac{x^2 - 1}{x^2 + x - 2}$

112. $\dfrac{x^2 - 3x - 4}{x^2 - 5x + 6} \div \dfrac{x^2 - 2x - 3}{x^2 - x - 2}$

113. $\dfrac{3}{m + 1} + \dfrac{3m}{m - 1}$

114. $\dfrac{2x + 3}{3x - 1} - \dfrac{x - 4}{2x + 1}$

▶ 9.2   Radical Equations

In this section, you will learn about

> The power rule ■ Equations containing one radical ■ Equations containing two radicals ■ Solving formulas containing radicals

Introduction   Many situations can be modeled by equations that contain radicals. In this section, we will develop techniques to solve such equations. For example, to solve the radical equation $\sqrt{x} = 6$, we need to isolate x by undoing the operation performed on it. Recall that $\sqrt{x}$ represents the number that, when squared, gives x. Therefore, if we square $\sqrt{x}$, we will obtain x. From this observation, it is apparent that we can eliminate the radical on the left-hand side of the equation $\sqrt{x} = 6$ by squaring that side. Intuition tells us that we should also square the right-hand side. Squaring both sides of an equation is an application of the *power rule*, which we now consider in more detail.

The Power Rule

To solve equations containing radicals, we will use the **power rule.**

The power rule

> If x, y, and n are real numbers and $x = y$, then
>
> $$x^n = y^n$$

If we raise both sides of an equation to the same power, the resulting equation might not be equivalent to the original equation. For example, if we square both sides of the equation

1. $x = 3$ With a solution set of {3}

we obtain the equation

2. $x^2 = 9$ With a solution set of {3, −3}

Equations 1 and 2 are not equivalent, because they have different solution sets, and the solution −3 of Equation 2 does not satisfy Equation 1. Since raising both sides of an equation to the same power can produce an equation with roots that don't satisfy the original equation, we must always check each apparent solution in the original equation and discard any *extraneous solutions.*

Equations Containing One Radical

Radical equations contain a radical expression with a variable radicand. To solve radical equations, we apply the power rule.

EXAMPLE 1

Squaring both sides of an equation.   Solve $\sqrt{x + 3} = 4$.

Solution   To eliminate the radical, we apply the power rule by squaring both sides of the equation and proceed as follows:

$$\sqrt{x + 3} = 4$$
$$\left(\sqrt{x + 3}\right)^2 = (4)^2 \quad \text{Square both sides.}$$
$$x + 3 = 16$$
$$x = 13 \quad \text{Subtract 3 from both sides.}$$

We must check the apparent solution of 13 to see whether it satisfies the original equation.

Check: $\sqrt{x+3} = 4$

$\sqrt{13+3} \overset{?}{=} 4$ Substitute 13 for x.

$\sqrt{16} \overset{?}{=} 4$

$4 = 4$

Since 13 satisfies the original equation, it is a solution.

SELF CHECK   Solve $\sqrt{a-2} = 3$.   *Answer:* 11   ■

To solve an equation with radicals, we follow these steps.

Solving an equation containing radicals

1. Isolate one radical expression on one side of the equation.
2. Raise both sides of the equation to the power that is the same as the index of the radical.
3. Solve the resulting equation. If it still contains a radical, go back to step 1.
4. Check the results to eliminate extraneous solutions.

EXAMPLE 2

Amusement park ride. The distance d in feet that an object will fall in t seconds is given by the formula

$$t = \sqrt{\frac{d}{16}}$$

FIGURE 9-7

If the designers of the amusement park attraction shown in Figure 9-7 want the riders to experience 3 seconds of vertical "free fall," what length of vertical drop is needed?

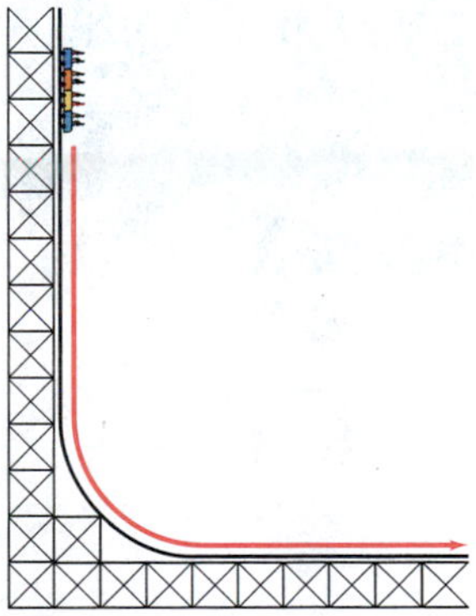

Solution   We substitute 3 for t in the formula and solve for d.

$$t = \sqrt{\frac{d}{16}}$$

$$3 = \sqrt{\frac{d}{16}}$$ Here the radical is isolated on the right-hand side.

$$(3)^2 = \left(\sqrt{\frac{d}{16}}\right)^2$$ Raise both sides to the second power.

$$9 = \frac{d}{16}$$ Simplify.

$$144 = d$$ Solve the resulting equation by multiplying both sides by 16.

The amount of vertical drop needs to be 144 feet.

SELF CHECK   In Example 2, how long a vertical drop is needed if the riders are to "free fall" for 3.5 seconds?   *Answer:* 196 ft   ■

EXAMPLE 3

Isolating the radical. Solve $\sqrt{3x + 1} + 1 = x$.

Solution   We first subtract 1 from both sides to isolate the radical. Then, to eliminate the radical, we square both sides of the equation and proceed as follows:

$$\sqrt{3x + 1} + 1 = x$$

$$\sqrt{3x + 1} = x - 1 \qquad \text{Subtract 1 from both sides.}$$

$$\left(\sqrt{3x + 1}\right)^2 = (x - 1)^2 \qquad \text{Square both sides to eliminate the square root.}$$

$$3x + 1 = x^2 - 2x + 1 \qquad \begin{array}{l}\text{On the right-hand side, use the FOIL method:}\\ (x - 1)^2 = (x - 1)(x - 1) = x^2 - x - x + 1 =\\ x^2 - 2x + 1.\end{array}$$

$$0 = x^2 - 5x \qquad \begin{array}{l}\text{Subtract } 3x \text{ and 1 from both sides. This is a quadratic}\\ \text{equation. Use factoring to solve it.}\end{array}$$

$$0 = x(x - 5) \qquad \text{Factor } x^2 - 5x.$$

$$x = 0 \quad \text{or} \quad x - 5 = 0 \qquad \text{Set each factor each to 0.}$$

$$x = 0 \qquad\qquad x = 5$$

We must check each apparent solution to see whether it satisfies the original equation.

$$\begin{array}{ll}\textbf{\textit{Check:}}\ \sqrt{3x + 1} + 1 = x & \sqrt{3x + 1} + 1 = x\\[4pt] \sqrt{3(0) + 1} + 1 \stackrel{?}{=} 0 & \sqrt{3(5) + 1} + 1 \stackrel{?}{=} 5\\[4pt] \sqrt{1} + 1 \stackrel{?}{=} 0 & \sqrt{16} + 1 \stackrel{?}{=} 5\\[4pt] 2 \neq 0 & 5 = 5 \end{array}$$

Since 0 does not check, it must be discarded. The only solution of the original equation is 5.

SELF CHECK   Solve $\sqrt{4x + 1} + 1 = x$.   *Answer:* 6, 0 is extraneous   ■

ACCENT ON TECHNOLOGY *Solving Equations Containing Radicals*

To find approximate solutions for $\sqrt{3x + 1} + 1 = x$ with a graphing calculator, we use window settings of $[-5, 10]$ for x and $[-2, 8]$ for y and graph the functions $f(x) = \sqrt{3x + 1} + 1$ and $g(x) = x$, as in Figure 9-8(a). We then trace to find the approximate x-coordinate of their intersection point, as in Figure 9-8(b). After repeated zooms, we will see that $x = 5$.

We can also use the INTERSECT feature to approximate the point of intersection of the graphs. See Figure 9-8(c). The intersection point of $(5, 5)$ implies that $x = 5$ is a solution of the radical equation. Check this result.

FIGURE 9-8

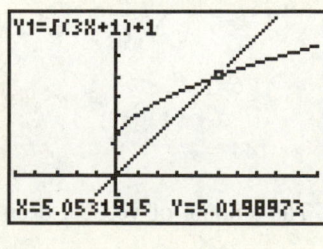

(a)   (b)

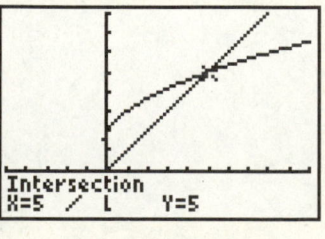

(c)

EXAMPLE 4

Cubing both sides. Solve $\sqrt[3]{x^3 + 7} = x + 1$.

Solution To eliminate the radical, we cube both sides of the equation and proceed as follows:

$$\sqrt[3]{x^3 + 7} = x + 1$$

$$\left(\sqrt[3]{x^3 + 7}\right)^3 = (x + 1)^3 \qquad \text{Cube both sides to eliminate the cube root.}$$

$$x^3 + 7 = x^3 + 3x^2 + 3x + 1 \qquad (x + 1)^3 = (x + 1)(x + 1)(x + 1).$$

$$0 = 3x^2 + 3x - 6 \qquad \text{Subtract } x^3 \text{ and 7 from both sides.}$$

$$0 = x^2 + x - 2 \qquad \text{Divide both sides by 3. To solve this quadratic equation, use factoring.}$$

$$0 = (x + 2)(x - 1) \qquad \text{Factor the trinomial.}$$

$$x + 2 = 0 \quad \text{or} \quad x - 1 = 0$$

$$x = -2 \qquad\qquad x = 1$$

We check each apparent solution to see whether it satisfies the original equation.

Check:
$$\sqrt[3]{x^3 + 7} = x + 1 \qquad\qquad \sqrt[3]{x^3 + 7} = x + 1$$
$$\sqrt[3]{(-2)^3 + 7} \stackrel{?}{=} -2 + 1 \qquad\qquad \sqrt[3]{1^3 + 7} \stackrel{?}{=} 1 + 1$$
$$\sqrt[3]{-8 + 7} \stackrel{?}{=} -1 \qquad\qquad \sqrt[3]{1 + 7} \stackrel{?}{=} 2$$
$$\sqrt[3]{-1} \stackrel{?}{=} -1 \qquad\qquad \sqrt[3]{8} \stackrel{?}{=} 2$$
$$-1 = -1 \qquad\qquad 2 = 2$$

Both solutions satisfy the original equation.

SELF CHECK Solve $\sqrt[3]{x^3 + 8} = x + 2$.   *Answer:* 0, −2 ∎

Equations Containing Two Radicals

EXAMPLE 5

Solving an equation containing two radicals. Solve $\sqrt{5x + 9} = 2\sqrt{3x + 4}$.

Solution Each radical is isolated on one side of the equation, so we square both sides to eliminate them.

$$\sqrt{5x + 9} = 2\sqrt{3x + 4}$$

$$\left(\sqrt{5x + 9}\right)^2 = \left(2\sqrt{3x + 4}\right)^2 \qquad \text{Square both sides.}$$

$$5x + 9 = 4(3x + 4) \qquad \text{On the right-hand side:}$$
$$\left(2\sqrt{3x + 4}\right)^2 = 2^2\left(\sqrt{3x + 4}\right)^2 = 4(3x + 4).$$

$$5x + 9 = 12x + 16 \qquad \text{Remove parentheses.}$$

$$-7 = 7x \qquad \text{Subtract } 5x \text{ and 16 from both sides.}$$

$$-1 = x \qquad \text{Divide both sides by 7.}$$

We check the solution by substituting -1 for x in the original equation.

$$\sqrt{5x + 9} = 2\sqrt{3x + 4}$$
$$\sqrt{5(-1) + 9} \stackrel{?}{=} 2\sqrt{3(-1) + 4} \qquad \text{Substitute } -1 \text{ for } x.$$
$$\sqrt{4} \stackrel{?}{=} 2\sqrt{1}$$
$$2 = 2$$

The solution checks.

SELF CHECK Solve $\sqrt{x-4} = 2\sqrt{x-16}$. *Answer:* 20 ∎

When more than one radical appears in an equation, it is often necessary to apply the power rule more than once.

EXAMPLE 6 **Solving equations containing two radicals.** Solve $\sqrt{x} + \sqrt{x+2} = 2$.

Solution To remove the radicals, we must square both sides of the equation. This is easier to do if one radical is on each side of the equation. So we subtract $\sqrt{x}$ from both sides to isolate $\sqrt{x+2}$ on the left-hand side of the equation.

$$\sqrt{x} + \sqrt{x+2} = 2$$

$$\sqrt{x+2} = 2 - \sqrt{x} \qquad \text{Subtract } \sqrt{x} \text{ from both sides.}$$

$$\left(\sqrt{x+2}\right)^2 = \left(2 - \sqrt{x}\right)^2 \qquad \text{Square both sides to eliminate the square root.}$$

$$x + 2 = 4 - 4\sqrt{x} + x \qquad \text{Use FOIL: } \left(2 - \sqrt{x}\right)^2 = \left(2 - \sqrt{x}\right)\left(2 - \sqrt{x}\right) = 4 - 2\sqrt{x} - 2\sqrt{x} + x = 4 - 4\sqrt{x} + x.$$

$$2 = 4 - 4\sqrt{x} \qquad \text{Subtract } x \text{ from both sides.}$$

$$-2 = -4\sqrt{x} \qquad \text{Subtract 4 from both sides.}$$

$$\frac{1}{2} = \sqrt{x} \qquad \text{Divide both sides by } -4 \text{ and simplify.}$$

$$\frac{1}{4} = x \qquad \text{Square both sides.}$$

Check: $\sqrt{x} + \sqrt{x+2} = 2$

$$\sqrt{\frac{1}{4}} + \sqrt{\frac{1}{4} + 2} \stackrel{?}{=} 2$$

$$\frac{1}{2} + \sqrt{\frac{9}{4}} \stackrel{?}{=} 2$$

$$\frac{1}{2} + \frac{3}{2} \stackrel{?}{=} 2$$

$$2 = 2$$

The solution checks.

SELF CHECK Solve $\sqrt{a} + \sqrt{a+3} = 3$. *Answer:* 1 ∎

ACCENT ON TECHNOLOGY *Solving Equations Containing Radicals*

To find approximate solutions for $\sqrt{x} + \sqrt{x+2} = 5$ (an equation similar to that in Example 6) with a graphing calculator, we use window settings of $[-2, 10]$ for x and $[-2, 8]$ for y and graph the functions $f(x) = \sqrt{x} + \sqrt{x+2}$ and $g(x) = 5$. We then trace to find an approximation of the x-coordinate of their intersection point, as in Figure 9-9(a). From the figure, we can see that $x \approx 5.15$. We can zoom to get better results.

Figure 9-9(b) shows that the INTERSECT feature gives the approximate coordinates of the point of intersection of the two graphs as $(5.29, 5)$. Therefore, an approximate solution of the radical equation is $x \approx 5.29$. Check its reasonableness.

FIGURE 9-9

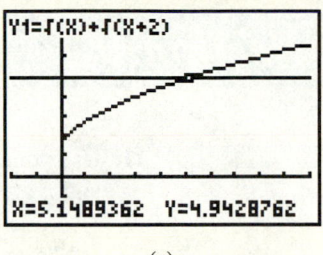

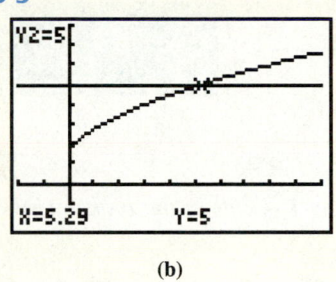

(a)   (b)

Solving Formulas Containing Radicals

To *solve a formula for a variable* means to isolate that variable on one side of the equation, with all other quantities on the other side.

EXAMPLE 7

Depreciation rate. A piece of office equipment that is now worth V dollars originally cost C dollars 3 years ago. The rate r at which it has depreciated (lost value) is given by

$$r = 1 - \sqrt[3]{\frac{V}{C}}$$

Solve the formula for C.

Solution   We begin by isolating the cube root on the right-hand side of the equation.

$$r = 1 - \sqrt[3]{\frac{V}{C}}$$

$$r - 1 = -\sqrt[3]{\frac{V}{C}} \qquad \text{Subtract 1 from both sides.}$$

$$(r - 1)^3 = \left[-\sqrt[3]{\frac{V}{C}} \right]^3 \qquad \text{Cube both sides.}$$

$$(r - 1)^3 = -\frac{V}{C} \qquad \text{Simplify the right-hand side.}$$

$$C(r - 1)^3 = -V \qquad \text{Multiply both sides by } C.$$

$$C = -\frac{V}{(r - 1)^3} \qquad \text{Divide both sides by } (r - 1)^3.$$

SELF CHECK   A formula used in statistics to determine the necessary size of a sample to obtain the desired degree of accuracy is

$$e = z_0 \sqrt{\frac{pq}{n}}$$

Solve the formula for n.

Answer: $n = \dfrac{z_0^2 pq}{e^2}$ ∎

STUDY SET

Section 9.2

VOCABULARY

In Exercises 1–4, fill in the blanks to make each statement true.

1. Equations such as $\sqrt{x+4} - 4 = 5$ and $\sqrt[3]{x+1} = 12$ are called _____ equations.

2. When solving equations containing radicals, try to _____ one radical expression on one side of the equation.

3. Squaring both sides of an equation can introduce _____ solutions.

4. To _____ an apparent solution means to substitute it into the original equation and see whether a true statement results.

CONCEPTS

5. What is the first step in solving each equation?
   a. $\sqrt{x+4} = 5$
   b. $\sqrt[3]{x+4} = 2$

6. Fill in the blank to make a true statement. $\sqrt{x}$ represents the number that, when squared, gives ____.

7. Simplify each expression.
   a. $\left(\sqrt{x}\right)^2$
   b. $\left(\sqrt{x-5}\right)^2$
   c. $\left(4\sqrt{2x}\right)^2$
   d. $\left(-\sqrt{x+3}\right)^2$

8. Simplify each expression.
   a. $\left(\sqrt[3]{x}\right)^3$
   b. $\left(\sqrt[4]{x}\right)^4$
   c. $\left(-\sqrt[3]{2x}\right)^3$
   d. $\left(2\sqrt[3]{x+3}\right)^3$

9. What is **wrong** with the student's work shown below?
   Solve $\sqrt{x+1} - 3 = 8$.
   $$\sqrt{x+1} = 11$$
   $$\left(\sqrt{x+1}\right)^2 = 11$$
   $$x + 1 = 11$$
   $$x = 10$$

10. Solve $\sqrt{x-2} + 2 = 4$ graphically, using the graphs in Illustration 1.

ILLUSTRATION 1

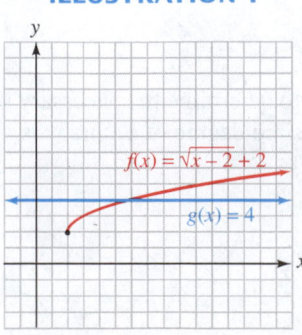

$f(x) = \sqrt{x-2} + 2$

$g(x) = 4$

11. Use your own words to restate the power rule: If x, y, and n are real numbers and $x = y$, then $x^n = y^n$.

12. The first step of a student's solution is shown below. What is a better way to begin the solution?
   Solve $\sqrt{x} + \sqrt{x+22} = 12$.
   $$\left(\sqrt{x} + \sqrt{x+22}\right)^2 = 12^2$$

NOTATION

In Exercises 13–14, complete each solution.

13. Solve $2\sqrt{x-2} = 4$.
   $$\left(\right)^2 = 4^2$$
   $$(x-2) = $$
   $$4x - = 16$$
   $$4x = $$
   $$x = 6$$

14. Solve $\sqrt{1-2x} = \sqrt{x+10}$.
   $$\left(\right)^2 = \left(\sqrt{x+10}\right)^2$$
   $$ = x + 10$$
   $$ = 9$$
   $$x = -3$$

PRACTICE

In Exercises 15–54, solve each equation. Write all apparent solutions. Cross out those that are extraneous.

15. $\sqrt{5x-6} = 2$

16. $\sqrt{7x-10} = 12$

17. $\sqrt{6x+1} + 2 = 7$

18. $\sqrt{6x+13} - 2 = 5$

19. $2\sqrt{4x+1} = \sqrt{x+4}$

20. $\sqrt{3(x+4)} = \sqrt{5x-12}$

21. $\sqrt[3]{7n-1} = 3$

22. $\sqrt[3]{12m+4} = 4$

23. $\sqrt[4]{10p+1} = \sqrt[4]{11p-7}$

24. $\sqrt[4]{10y+6} = 2\sqrt[4]{y}$

25. $x = \dfrac{\sqrt{12x-5}}{2}$

26. $x = \dfrac{\sqrt{16x-12}}{2}$

27. $\sqrt{x+2} - \sqrt{4-x} = 0$

28. $\sqrt{6-x} - \sqrt{2x+3} = 0$

29. $2\sqrt{x} = \sqrt{5x-16}$

30. $3\sqrt{x} = \sqrt{3x+54}$

31. $r - 9 = \sqrt{2r-3}$

32. $-s - 3 = 2\sqrt{5-s}$

33. $\sqrt{-5x + 24} = 6 - x$ **34.** $\sqrt{-x + 2} = x - 2$

35. $\sqrt{y + 2} = 4 - y$ **36.** $\sqrt{22y + 86} = y + 9$

37. $\sqrt[3]{x^3 - 7} = x - 1$ **38.** $\sqrt[3]{x^3 + 56} - 2 = x$

39. $\sqrt[4]{x^4 + 4x^2 - 4} = -x$ **40.** $\sqrt[4]{8x - 8} + 2 = 0$

41. $\sqrt[4]{12t + 4} + 2 = 0$ **42.** $u = \sqrt[4]{u^4 - 6u^2 + 24}$

43. $\sqrt{2y + 1} = 1 - 2\sqrt{y}$ **44.** $\sqrt{u + 3} = \sqrt{u - 3}$

45. $\sqrt{y + 7} + 3 = \sqrt{y + 4}$

46. $1 + \sqrt{z} = \sqrt{z + 3}$

47. $2 + \sqrt{u} = \sqrt{2u + 7}$

48. $5r + 4 = \sqrt{5r + 20} + 4r$

49. $\sqrt{6t + 1} - 3\sqrt{t} = -1$

50. $\sqrt{4s + 1} - \sqrt{6s} = -1$

51. $\sqrt{2x + 5} + \sqrt{x + 2} = 5$

52. $\sqrt{2x + 5} + \sqrt{2x + 1} + 4 = 0$

53. $\sqrt{x - 5} - \sqrt{x + 3} = 4$

54. $\sqrt{x + 8} - \sqrt{x - 4} = -2$

In Exercises 55–62, solve each equation for the indicated variable.

55. $v = \sqrt{2gh}$ for h

56. $d = 1.4\sqrt{h}$ for h

57. $T = 2\pi\sqrt{\dfrac{l}{32}}$ for l

58. $d = \sqrt[3]{\dfrac{12V}{\pi}}$ for V

59. $r = \sqrt[3]{\dfrac{A}{P}} - 1$ for A

60. $r = \sqrt[3]{\dfrac{A}{P}} - 1$ for P

61. $L_A = L_B\sqrt{1 - \dfrac{v^2}{c^2}}$ for v^2

62. $R_1 = \sqrt{\dfrac{A}{\pi} - R_2^2}$ for A

APPLICATIONS

63. HIGHWAY DESIGN A curved concrete road will accommodate traffic traveling s mph if the radius of the curve is r feet, according to the formula $s = 3\sqrt{r}$. If engineers expect 40-mph traffic, what radius should they specify? Give the result to the nearest foot. (See Illustration 2.)

ILLUSTRATION 2

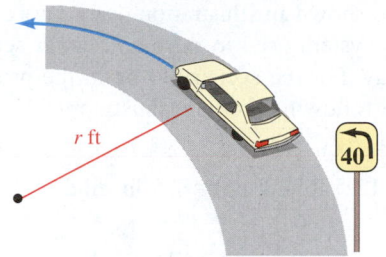

r ft

40

64. FORESTRY The higher a lookout tower is built, the farther an observer can see. See Illustration 3. That distance d (called the *horizon distance,* measured in miles) is related to the height h of the observer (measured in feet) by the formula $d = 1.4\sqrt{h}$. How tall must a lookout tower be to see the edge of the forest, 25 miles away? (Round to the nearest foot.)

ILLUSTRATION 3

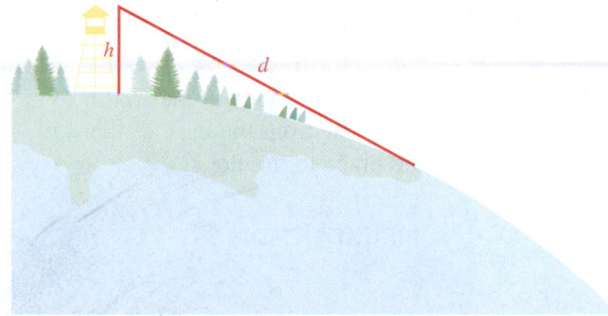

h d

65. WIND POWER The power generated by a certain windmill is related to the velocity of the wind by the formula

$$v = \sqrt[3]{\dfrac{P}{0.02}}$$

where P is the power (in watts) and v is the velocity of the wind (in mph). Find how much power the windmill is generating when the wind is 29 mph.

66. DIAMONDS The *effective rate of interest r* earned by an investment is given by the formula

$$r = \sqrt[n]{\dfrac{A}{P}} - 1$$

where P is the initial investment that grows to value A after n years. If a diamond buyer got \$4,000 for a 1.73-carat diamond that he had purchased 4 years earlier, and earned an annual rate of return of 6.5% on the investment, what did he originally pay for the diamond?

67. THEATER PRODUCTION The ropes, pulleys, and sandbags shown in Illustration 4 are part of a mechanical system used to raise and lower scenery for a stage play. For the scenery to be in the proper position, the following formula must apply:

$$w_2 = \sqrt{w_1{}^2 + w_3{}^2}$$

If $w_2 = 12.5$ lb and $w_3 = 7.5$ lb, find w_1.

ILLUSTRATION 4

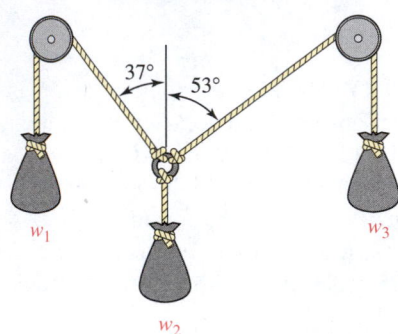

68. CARPENTRY During construction, carpenters often brace walls as shown in Illustration 5, where the length of the brace is given by the formula

$$l = \sqrt{f^2 + h^2}$$

If a carpenter nails a 10-ft brace to the wall 6 feet above the floor, how far from the base of the wall should he nail the brace to the floor?

ILLUSTRATION 5

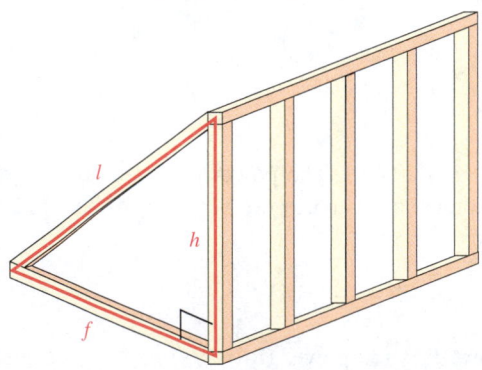

69. SUPPLY AND DEMAND The number of wrenches that will be produced at a given price can be predicted by the formula $s = \sqrt{5x}$, where s is the supply (in thousands) and x is the price (in dollars). The demand d for wrenches can be predicted by the formula $d = \sqrt{100 - 3x^2}$. Find the equilibrium price—that is, find the price at which supply will equal demand.

70. SUPPLY AND DEMAND The number of mirrors that will be produced at a given price can be predicted by the formula $s = \sqrt{23x}$, where s is the supply (in thousands) and x is the price (in dollars). The demand d for mirrors can be predicted by the formula

$d = \sqrt{312 - 2x^2}$. Find the equilibrium price—that is, find the price at which supply will equal demand.

WRITING

71. If both sides of an equation are raised to the same power, the resulting equation might not be equivalent to the original equation. Explain.

72. Explain how the radical equation $\sqrt{2x - 1} = x$ can be solved graphically.

73. Explain how the table of values in Illustration 6 can be used to solve $\sqrt{4x - 3} - 2 = \sqrt{2x - 5}$ if $Y_1 = \sqrt{4x - 3} - 2$ and $Y_2 = \sqrt{2x - 5}$.

ILLUSTRATION 6

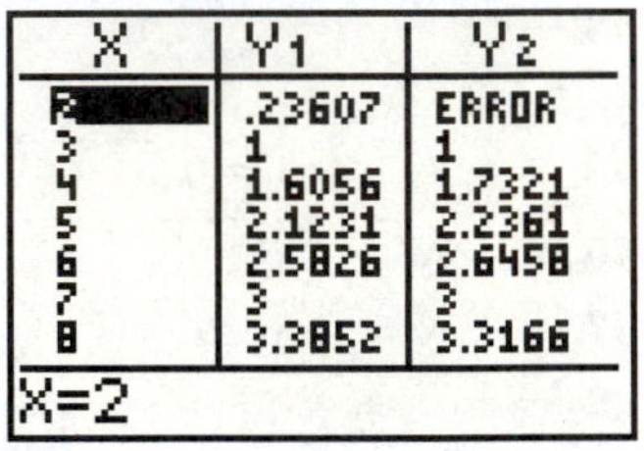

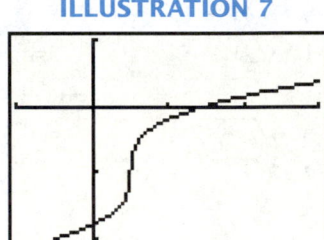

74. Explain how to use the graph of $f(x) = \sqrt[3]{x - 0.5} - 1$, shown in Illustration 7, to approximate the solution of $\sqrt[3]{x - 0.5} = 1$.

ILLUSTRATION 7

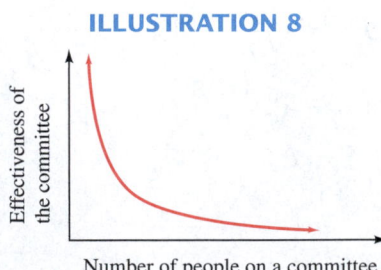

REVIEW

75. LIGHTING The intensity of the light reaching you from a light bulb varies inversely as the square of your distance from the bulb. If you are 5 feet away from a light bulb and the intensity is 40 foot-candles, what will the intensity be if you move 20 feet away from the bulb?

76. COMMITTEES What type of variation is shown in Illustration 8? As the number of people on this committee increased, what happened to its effectiveness?

ILLUSTRATION 8

Effectiveness of the committee

Number of people on a committee

77. TYPESETTING If 12-point type is 0.166044 inch tall, how tall is 30-point type?

78. GUITAR STRINGS The frequency of vibration of a string varies directly as the square root of the tension and inversely as the length of the string. Suppose a string 2.5 feet long, under a tension of 16 pounds, vibrates 25 times per second. Find k, the constant of proportionality.

▶ 9.3 Rational Exponents

In this section, you will learn about

> Rational exponents ■ Exponential expressions with variables in their bases ■ Rational exponents with numerators other than 1 ■ Negative rational exponents ■ Applying the rules for exponents ■ Simplifying radical expressions

Introduction In Chapter 4, we worked with exponential expressions containing natural-number exponents, such as 5^3 and x^2. Then the definition of exponent was extended to include zero and negative integers, which gave meaning to expressions such as 8^{-3} and $(-9xy)^0$. In this section, we will again extend the definition of exponent—this time to include rational (fractional) exponents. We will see how expressions such as $9^{1/2}$, $\left(\frac{1}{16}\right)^{3/4}$, and $(-32x^5)^{-2/5}$ can be simplified by writing them in an equivalent radical form or by using the rules for exponents.

Rational Exponents

We have seen that positive-integer exponents indicate the number of times that a base is to be used as a factor in a product. For example, x^5 means that x is to be used as a factor five times.

$$\overbrace{x^5 = x \cdot x \cdot x \cdot x \cdot x}^{\text{5 factors of } x}$$

Furthermore, we recall the following rules for exponents.

Rules for exponents

If there are no divisions by 0, then for all integers m and n,

1. $x^m x^n = x^{m+n}$ **2.** $(x^m)^n = x^{mn}$ **3.** $(xy)^n = x^n y^n$ **4.** $\left(\dfrac{x}{y}\right)^n = \dfrac{x^n}{y^n}$

5. $x^0 = 1 \ (x \neq 0)$ **6.** $x^{-n} = \dfrac{1}{x^n}$ **7.** $\dfrac{x^m}{x^n} = x^{m-n}$ **8.** $\left(\dfrac{x}{y}\right)^{-n} = \left(\dfrac{y}{x}\right)^n$

It is possible to raise many bases to fractional powers. Since we want fractional exponents to obey the same rules as integer exponents, the square of $10^{1/2}$ must be 10, because

$(10^{1/2})^2 = 10^{(1/2)2}$ Keep the base and multiply the exponents.

$\qquad\qquad = 10^1 \qquad \frac{1}{2} \cdot 2 = 1.$

$\qquad\qquad = 10 \qquad 10^1 = 10.$

However, we have seen that

$$\left(\sqrt{10}\right)^2 = 10$$

Since $(10^{1/2})^2$ and $\left(\sqrt{10}\right)^2$ both equal 10, we define $10^{1/2}$ to be $\sqrt{10}$. Likewise, we define

$10^{1/3}$ to be $\sqrt[3]{10}$ and $10^{1/4}$ to be $\sqrt[4]{10}$

Rational exponents

If n is a natural number greater than 1 and $\sqrt[n]{x}$ is a real number, then

$$x^{1/n} = \sqrt[n]{x}$$

In words, *a rational exponent of $\frac{1}{n}$ indicates that the nth root of the base should be found.*

Using this definition, we can simplify the exponential expression $8^{1/3}$. The first step is to write it as an equivalent expression in radical form and proceed as follows:

$8^{1/3} = \sqrt[3]{8}$ The base of the exponential expression, 8, is the radicand. The denominator of the fractional exponent, 3, is the index of the radical.

$\quad\quad = 2$

EXAMPLE 1

Simplifying expressions containing rational exponents. Write each expression in radical form and simplify, if possible.

a. $9^{1/2} = \sqrt{9}$
$\quad\quad = 3$

b. $-\left(\dfrac{16}{9}\right)^{1/2} = -\sqrt{\dfrac{16}{9}}$
$\quad\quad\quad\quad\quad = -\dfrac{4}{3}$

c. $(-64)^{1/3} = \sqrt[3]{-64}$
$\quad\quad\quad = -4$

d. $16^{1/4} = \sqrt[4]{16}$
$\quad\quad\quad = 2$

e. $\left(\dfrac{1}{32}\right)^{1/5} = \sqrt[5]{\dfrac{1}{32}}$
$\quad\quad\quad = \dfrac{1}{2}$

f. $0^{1/8} = \sqrt[8]{0}$
$\quad\quad = 0$

g. $y^{1/4} = \sqrt[4]{y}$

h. $-(2x^2)^{1/5} = -\sqrt[5]{2x^2}$

SELF CHECK   Write each expression in radical form and simplify, if possible: **a.** $16^{1/2}$, **b.** $\left(-\frac{27}{8}\right)^{1/3}$, and **c.** $-(6x^3)^{1/4}$.

Answers: **a.** 4, **b.** $-\frac{3}{2}$, **c.** $-\sqrt[4]{6x^3}$ ∎

EXAMPLE 2

Writing roots using rational exponents. Write $\sqrt{5xyz}$ as an exponential expression with a rational exponent.

Solution   The radicand is $5xyz$, so the base of the exponential expression is $5xyz$. The index of the radical is an understood 2, so the denominator of the fractional exponent is 2.

$$\sqrt{5xyz} = (5xyz)^{1/2}$$

SELF CHECK   Write the radical with a fractional exponent: $\sqrt[6]{7ab}$.   *Answer:* $(7ab)^{1/6}$ ∎

Rational exponents appear in formulas used in many disciplines, such as science and engineering.

EXAMPLE 3

Satellites. See Figure 9-10. The formula

$$r = \left(\frac{GMP^2}{4\pi^2}\right)^{1/3}$$

gives the orbital radius (in meters) of a satellite circling the earth, where G and M are constants and P is the time in seconds for the satellite to make one complete revolution. Write the formula using a radical.

FIGURE 9-10

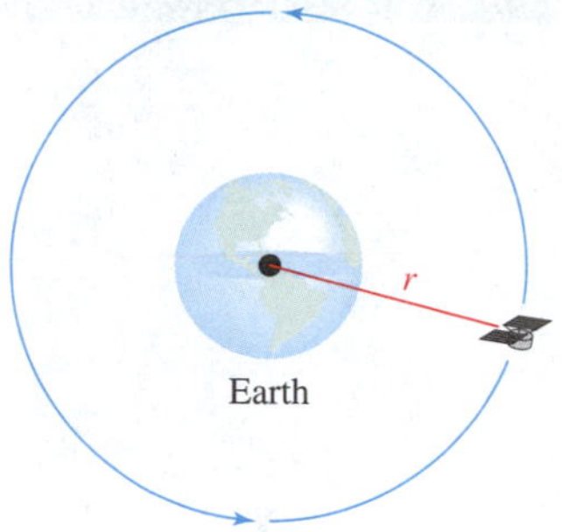

Earth

Solution The fractional exponent $\frac{1}{3}$ has a denominator of 3, which indicates that we are to find the cube root of the base of the exponential expression. So we have

$$r = \sqrt[3]{\frac{GMP^2}{4\pi^2}}$$

Exponential Expressions with Variables in Their Bases

As with radicals, when n is an *odd natural number* in the expression $x^{1/n}$ ($n > 1$), there is exactly one real nth root, and we don't have to worry about absolute value symbols.

When n is an *even natural number,* there are two nth roots. Since we want the expression $x^{1/n}$ to represent the positive nth root, we must often use absolute value symbols to guarantee that the simplified result is positive. Thus, if n is even,

$$(x^n)^{1/n} = |x|$$

When n is even and x is negative, the expression $x^{1/n}$ is not a real number.

EXAMPLE 4

A variable in the base. Simplify each exponential expression. Assume that the variables can be any real number.

a. $(-27x^3)^{1/3} = -3x$ Because $(-3x)^3 = -27x^3$. Since n is odd, no absolute value symbols are needed.

b. $(256a^8)^{1/8} = 2|a|$ Because $(2|a|)^8 = 256a^8$. Since n is even and a can be any real number, $2a$ can be negative. Thus, absolute value symbols are needed.

c. $[(y + 4)^2]^{1/2} = |y + 4|$ Because $|y + 4|^2 = (y + 4)^2$. Since n is even and y can be any real number, $y + 4$ can be negative. Thus, absolute value symbols are needed.

d. $(25b^4)^{1/2} = 5b^2$ Because $(5b^2)^2 = 25b^4$. Since $b^2 \geq 0$, no absolute value symbols are needed.

e. $(-256x^4)^{1/4}$ is not a real number. Because no real number raised to the 4th power is $-256x^4$.

SELF CHECK Simplify each expression: **a.** $(625a^4)^{1/4}$ and
b. $(b^4)^{1/2}$.   *Answers:* **a.** $5|a|$, **b.** b^2

If we are told that the variables represent positive real numbers in parts b and c of Example 4, the absolute value symbols in the answers are not needed.

$(256a^8)^{1/8} = 2a$ If a represents a positive number, then $2a$ is positive.

$[(y + 4)^2]^{1/2} = y + 4$ If y represents a positive number, then $y + 4$ is positive.

We summarize the cases as follows.

If n is a natural number greater than 1 and x is a real number,

If $x > 0$, then $x^{1/n}$ is the positive number such that $(x^{1/n})^n = x$.

If $x = 0$, then $x^{1/n} = 0$.

If $x < 0$ $\begin{cases} \text{and } n \text{ is odd, then } x^{1/n} \text{ is the real number such that } (x^{1/n})^n = x. \\ \text{and } n \text{ is even, then } x^{1/n} \text{ is not a real number.} \end{cases}$

Rational Exponents with Numerators Other Than 1

We can extend the definition of $x^{1/n}$ to include fractional exponents with numerators other than 1. For example, since $8^{2/3}$ can be written as $(8^{1/3})^2$, we have

$$8^{2/3} = (\mathbf{8^{1/3}})^2$$
$$= \left(\sqrt[3]{8}\right)^2 \quad \text{Write } 8^{1/3} \text{ in radical form.}$$
$$= 2^2 \quad \text{Find the cube root first: } \sqrt[3]{8} = 2.$$
$$= 4 \quad \text{Then find the power.}$$

Thus, we can simplify $8^{2/3}$ by finding the second power of the cube root of 8.

The numerator of the rational
exponent is the power.

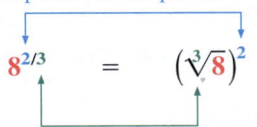

$\mathbf{8^{2/3}} \quad = \quad \left(\sqrt[3]{\mathbf{8}}\right)^2$ The base of the exponential expression is the radicand.

The denominator of the exponent
is the index of the radical.

We can also simplify $8^{2/3}$ by taking the cube root of 8 squared.

$$8^{2/3} = (\mathbf{8^2})^{1/3}$$
$$= \mathbf{64^{1/3}} \quad \text{Find the power first: } 8^2 = 64.$$
$$= \sqrt[3]{64} \quad \text{Write } 64^{1/3} \text{ in radical form.}$$
$$= 4 \quad \text{Now find the cube root.}$$

In general, we have the following rule.

If m and n are positive integers ($n \neq 1$) and $\sqrt[n]{x}$ is a real number, then

$$x^{m/n} = \sqrt[n]{x^m} = \left(\sqrt[n]{x}\right)^m$$

Because of the previous definition, we can interpret $x^{m/n}$ in two ways:

1. $x^{m/n}$ means the nth root of the mth power of x.
2. $x^{m/n}$ means the mth power of the nth root of x.

EXAMPLE 5

Rational exponents with numerators other than 1. Simplify each expression in two ways.

a. $9^{3/2} = \left(\sqrt{9}\right)^3$ or $9^{3/2} = \sqrt{9^3}$
$\phantom{9^{3/2}} = 3^3$ $\phantom{or\ 9^{3/2}} = \sqrt{729}$
$\phantom{9^{3/2}} = 27$ $\phantom{or\ 9^{3/2}} = 27$

b. $\left(\dfrac{1}{16}\right)^{3/4} = \left(\sqrt[4]{\dfrac{1}{16}}\right)^3$ or $\left(\dfrac{1}{16}\right)^{3/4} = \left[\left(\dfrac{1}{16}\right)^3\right]^{1/4}$
$\phantom{\left(\dfrac{1}{16}\right)^{3/4}} = \left(\dfrac{1}{2}\right)^3$ $\phantom{or\ \left(\dfrac{1}{16}\right)^{3/4}} = \left(\dfrac{1}{4{,}096}\right)^{1/4}$
$\phantom{\left(\dfrac{1}{16}\right)^{3/4}} = \dfrac{1}{8}$ $\phantom{or\ \left(\dfrac{1}{16}\right)^{3/4}} = \dfrac{1}{8}$

c. $(-8x^3)^{4/3} = \left(\sqrt[3]{-8x^3}\right)^4$ or $(-8x^3)^{4/3} = \sqrt[3]{(-8x^3)^4}$
$\phantom{(-8x^3)^{4/3}} = (-2x)^4$ $\phantom{or\ (-8x^3)^{4/3}} = \sqrt[3]{4{,}096x^{12}}$
$\phantom{(-8x^3)^{4/3}} = 16x^4$ $\phantom{or\ (-8x^3)^{4/3}} = 16x^4$

SELF CHECK   Simplify   **a.** $16^{3/2}$ and   **b.** $(-27x^6)^{2/3}$.   *Answers:*   **a.** 64,   **b.** $9x^4$ ∎

To avoid large numbers, it is usually better to find the root of the base first, as shown with the first solution of each part in Example 5.

ACCENT ON TECHNOLOGY *Rational Exponents*

We can evaluate exponential expressions containing rational exponents using the exponential key $\boxed{y^x}$ or $\boxed{x^y}$ on a scientific calculator. For example, to evaluate $10^{2/3}$, we enter these numbers and press these keys:

Keystrokes   10 $\boxed{y^x}$ $\boxed{(}$ 2 $\boxed{\div}$ 3 $\boxed{)}$ $\boxed{=}$ $\boxed{\text{4.641588834}}$

Note that parentheses were used when entering the power. Without them, the calculator would interpret the entry as $10^2 \div 3$.

To evaluate the exponential expression using a graphing calculator, we use the $\boxed{\wedge}$ key, which raises a base to a power. Again, we use parentheses when entering the power.

Keystrokes   10 $\boxed{\wedge}$ $\boxed{(}$ 2 $\boxed{\div}$ 3 $\boxed{)}$ $\boxed{\text{ENTER}}$ $\boxed{\begin{array}{l}\text{10^(2/3)}\\ \text{4.641588834}\end{array}}$

To the nearest hundredth, $10^{2/3} \approx 4.64$.

Negative Rational Exponents

To be consistent with the definition of negative-integer exponents, we define $x^{-m/n}$ as follows.

Definition of $x^{-m/n}$

If m and n are positive integers, $\frac{m}{n}$ is in simplified form, and $x^{1/n}$ is a real number, then

$$x^{-m/n} = \frac{1}{x^{m/n}} \quad \text{and} \quad \frac{1}{x^{-m/n}} = x^{m/n} \quad (x \neq 0)$$

EXAMPLE 6

Negative rational exponents. Write each expression without using negative exponents and simplify, if possible.

a. $64^{-1/2} = \dfrac{1}{64^{1/2}}$

$\qquad \qquad = \dfrac{1}{\sqrt{64}}$

$\qquad \qquad = \dfrac{1}{8}$

b. $(-16)^{-3/4}$ is not a real number, because $(-16)^{1/4}$ is not a real number.

c. $(-32x^5)^{-2/5} = \dfrac{1}{(-32x^5)^{2/5}}$

$\qquad \qquad = \dfrac{1}{[(-32x^5)^{1/5}]^2}$

$\qquad \qquad = \dfrac{1}{\left(\sqrt[5]{-32x^5}\right)^2}$

$\qquad \qquad = \dfrac{1}{(-2x)^2}$

$\qquad \qquad = \dfrac{1}{4x^2}$

d. $\dfrac{1}{16^{-3/2}} = 16^{3/2}$

$\qquad \qquad = (16^{1/2})^3$

$\qquad \qquad = \left(\sqrt{16}\right)^3$

$\qquad \qquad = 4^3$

$\qquad \qquad = 64$

SELF CHECK Write without using negative exponents and simplify: **a.** $25^{-3/2}$ and **b.** $(-27a^3)^{-2/3}$. *Answers:* **a.** $\dfrac{1}{125}$, **b.** $\dfrac{1}{9a^2}$ ∎

 WARNING! By definition, 0^0 is undefined. A base of 0 raised to a negative power is also undefined. For example, 0^{-2} would equal $\frac{1}{0^2}$, which is undefined because we cannot divide by 0.

Applying the Rules for Exponents

We can use the rules for exponents to simplify many expressions with fractional exponents. If all variables represent positive numbers, no absolute value symbols are necessary.

EXAMPLE 7

Simplifying expressions by using the rules for exponents. Assume that all variables represent positive numbers. Write all answers without using negative exponents.

a. $5^{2/7}5^{3/7} = 5^{2/7+3/7}$ Use the rule $x^m x^n = x^{m+n}$.

$\qquad \quad = 5^{5/7}$ Add: $\frac{2}{7} + \frac{3}{7} = \frac{5}{7}$.

b. $(5^{2/7})^3 = 5^{(2/7)(3)}$ Use the rule $(x^m)^n = x^{mn}$.

                 $= 5^{6/7}$ Multiply: $\frac{2}{7}(3) = \frac{6}{7}$.

c. $(a^{2/3}b^{1/2})^6 = (a^{2/3})^6(b^{1/2})^6$ Use the rule $(xy)^n = x^n y^n$.

                   $= a^{12/3}b^{6/2}$ Use the rule $(x^m)^n = x^{mn}$ twice.

                   $= a^4 b^3$ Simplify the exponents.

d. $\dfrac{a^{8/3}a^{1/3}}{a^2} = a^{8/3 + 1/3 - 2}$ Use the rules $x^m x^n = x^{m+n}$ and $\frac{x^m}{x^n} = x^{m-n}$.

                 $= a^{8/3 + 1/3 - 6/3}$ $2 = \frac{6}{3}$.

                 $= a^{3/3}$ $\frac{8}{3} + \frac{1}{3} - \frac{6}{3} = \frac{3}{3}$.

                 $= a$ $\frac{3}{3} = 1$.

SELF CHECK   Simplify   **a.** $(x^{1/3}y^{3/2})^6$ and   **b.** $\dfrac{x^{5/3}x^{2/3}}{x^{1/3}}$.

                           Answers: **a.** $x^2 y^9$,   **b.** x^2 ■

EXAMPLE 8   **Using the distributive property.** Assume all variables represent positive numbers and do the operations. Write all answers without negative exponents.

a. $a^{4/5}(a^{1/5} + a^{3/5}) = a^{4/5}a^{1/5} + a^{4/5}a^{3/5}$ Use the distributive property.

                         $= a^{4/5 + 1/5} + a^{4/5 + 3/5}$ Use the rule $x^m x^n = x^{m+n}$.

                         $= a^{5/5} + a^{7/5}$ Simplify the exponents.

                         $= a + a^{7/5}$ We cannot add these terms because they are not like terms.

b. $x^{1/2}(x^{-1/2} + x^{1/2}) = x^{1/2}x^{-1/2} + x^{1/2}x^{1/2}$ Use the distributive property.

                         $= x^{1/2 + (-1/2)} + x^{1/2 + 1/2}$ Use the rule $x^m x^n = x^{m+n}$.

                         $= x^0 + x^1$ Simplify each exponent.

                         $= 1 + x$ $x^0 = 1$.   ■

Simplifying Radical Expressions

We can simplify many radical expressions by using the following steps.

Using rational exponents to simplify radicals

1. Change the radical expression into an exponential expression.
2. Simplify the rational exponents.
3. Change the exponential expression back into a radical.

EXAMPLE 9   **Simplifying radical expressions.** Simplify   **a.** $\sqrt[4]{3^2}$,   **b.** $\sqrt[8]{x^6}$,   and   **c.** $\sqrt[9]{27x^6y^3}$.

Solution   **a.** $\sqrt[4]{3^2} = (3^2)^{1/4}$ Change the radical to an exponential expression.

                 $= 3^{2/4}$ Use the rule $(x^m)^n = x^{mn}$.

                 $= 3^{1/2}$ $\frac{2}{4} = \frac{1}{2}$.

                 $= \sqrt{3}$ Change back to radical form.

b. $\sqrt[8]{x^6} = (x^6)^{1/8}$ Change the radical to an exponential expression.

$= x^{6/8}$ Use the rule $(x^m)^n = x^{mn}$.

$= x^{3/4}$ $\frac{6}{8} = \frac{3}{4}$.

$= (x^3)^{1/4}$ $\frac{3}{4} = 3(\frac{1}{4})$.

$= \sqrt[4]{x^3}$ Change back to radical form.

c. $\sqrt[9]{27x^6y^3} = (3^3x^6y^3)^{1/9}$ Write 27 as 3^3 and change the radical to an exponential expression.

$= 3^{3/9}x^{6/9}y^{3/9}$ Raise each factor to the $\frac{1}{9}$ power by multiplying the fractional exponents.

$= 3^{1/3}x^{2/3}y^{1/3}$ Simplify each fractional exponent.

$= (3x^2y)^{1/3}$ Use the rule $(xy)^n = x^ny^n$.

$= \sqrt[3]{3x^2y}$ Change back to radical form.

SELF CHECK   Simplify   **a.** $\sqrt[6]{3^3}$ and   **b.** $\sqrt[4]{64x^2y^2}$.   *Answers:*   **a.** $\sqrt{3}$,   **b.** $\sqrt{8xy}$ ∎

STUDY SET

Section 9.3

VOCABULARY

In Exercises 1–4, fill in the blanks to make each statement true.

1. The expressions $4^{1/2}$ and $(-8)^{-2/3}$ have _____ exponents.

2. In the exponential expression $27^{4/3}$, 27 is the _____, and 4/3 is the _____.

3. In the radical expression $\sqrt[3]{4{,}096x^{12}}$, 3 is the _____, and $4{,}096x^{12}$ is the _____.

4. $32^{4/5}$ means the fourth _____ of the fifth _____ of 32.

CONCEPTS

5. Complete the table by writing the given expression in the alternate form.

Radical form	Exponential form
$\sqrt[5]{25}$	
	$(-27)^{2/3}$
$\left(\sqrt[4]{16}\right)^{-3}$	
	$81^{3/2}$
$-\sqrt{\frac{9}{64}}$	

6. In your own words, explain the two rules for rational exponents illustrated in the diagrams below.

a. 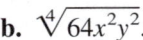 $(-32)^{1/5} = \sqrt[5]{-32}$

b. 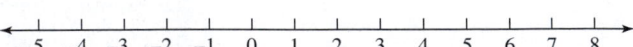 $125^{4/3} = \left(\sqrt[3]{125}\right)^4$

7. Graph each number on the number line.

$$\left\{ 8^{2/3},\ (-125)^{1/3},\ -16^{-1/4},\ 4^{3/2},\ -\left(\frac{9}{100}\right)^{-1/2} \right\}$$

$$\xleftarrow{\;\;\;\;\;}\underset{-5\;\;-4\;\;-3\;\;-2\;\;-1\;\;\;0\;\;\;1\;\;\;2\;\;\;3\;\;\;4\;\;\;5\;\;\;6\;\;\;7\;\;\;8}{\rule{10cm}{0.4pt}}\xrightarrow{\;\;\;\;\;}$$

8. Evaluate $25^{3/2}$ in two ways. Which way is easier?

In Exercises 9–14, complete each rule for exponents.

9. $x^m x^n = $ ⬜   **10.** $(x^m)^n = $ ⬜

11. $\dfrac{x^m}{x^n} = $ ⬜   **12.** $x^{-n} = $ ⬜

13. $x^{1/n} = $ ⬜   **14.** $x^{m/n} = $ ⬜ $= \sqrt[n]{x^m}$

In Exercises 15–16, complete each table of values. Then graph the function.

15. $f(x) = x^{1/2}$

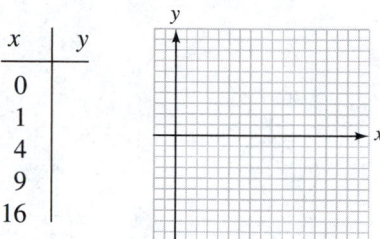

x	y
0	
1	
4	
9	
16	

16. $f(x) = x^{1/3}$

x	y
-8	
-1	
0	
1	
8	

NOTATION

In Exercises 17–18, complete each solution.

17. Simplify $(100a^4)^{3/2}$.

$$(100a^4)^{3/2} = \left(\right)^3$$
$$= \left(\right)^3$$
$$= 1{,}000a^6$$

18. Simplify $(m^{1/3}n^{1/2})^6$.

$$(m^{1/3}n^{1/2})^6 = \left(\right)^6 (n^{1/2})^6$$
$$= m^{} n^{6/2}$$
$$= m^2 n^3$$

PRACTICE

In Exercises 19–26, write each expression in radical form.

19. $x^{1/3}$

20. $b^{1/2}$

21. $(3x)^{1/4}$

22. $(4ab)^{1/6}$

23. $\left(\dfrac{1}{2}x^3 y \right)^{1/4}$

24. $\left(\dfrac{3}{4}a^2 b^2 \right)^{1/5}$

25. $(x^2 + y^2)^{1/2}$

26. $(x^3 + y^3)^{1/3}$

In Exercises 27–34, change each radical to an exponential expression.

27. $\sqrt{m}$

28. $\sqrt[3]{r}$

29. $\sqrt[4]{3a}$

30. $3\sqrt[5]{a}$

31. $\sqrt[6]{\dfrac{1}{7}abc}$

32. $\sqrt[7]{\dfrac{3}{8}p^2 q}$

33. $\sqrt[3]{a^2 - b^2}$

34. $\sqrt{x^2 + y^2}$

In Exercises 35–50, simplify each expression, if possible.

35. $4^{1/2}$

36. $25^{1/2}$

37. $8^{1/3}$

38. $125^{1/3}$

39. $16^{1/4}$

40. $625^{1/4}$

41. $32^{1/5}$

42. $0^{1/5}$

43. $\left(\dfrac{1}{4} \right)^{1/2}$

44. $\left(\dfrac{1}{16} \right)^{1/2}$

45. $-16^{1/4}$

46. $-125^{1/3}$

47. $(-64)^{1/2}$

48. $(-216)^{1/2}$

49. $(-27)^{1/3}$

50. $(-125)^{1/3}$

In Exercises 51–58, simplify each expression, if possible. Assume that all variables are unrestricted and use absolute value symbols when necessary.

51. $(25y^2)^{1/2}$

52. $(-27x^3)^{1/3}$

53. $(16x^4)^{1/4}$

54. $(-16x^4)^{1/2}$

55. $(243x^5)^{1/5}$

56. $[(x + 1)^4]^{1/4}$

57. $(-64x^8)^{1/4}$

58. $[(x + 5)^3]^{1/3}$

In Exercises 59–70, simplify each expression. Assume that all variables represent positive numbers.

59. $36^{3/2}$

60. $27^{2/3}$

61. $81^{3/4}$

62. $100^{3/2}$

63. $144^{3/2}$

64. $1{,}000^{2/3}$

65. $\left(\dfrac{1}{8} \right)^{2/3}$

66. $\left(\dfrac{4}{9} \right)^{3/2}$

67. $(25x^4)^{3/2}$

68. $(27a^3 b^3)^{2/3}$

69. $\left(\dfrac{8x^3}{27} \right)^{2/3}$

70. $\left(\dfrac{27}{64y^6} \right)^{2/3}$

In Exercises 71–82, write each expression without using negative exponents. Assume that all variables represent positive numbers.

71. $4^{-1/2}$

72. $8^{-1/3}$

73. $4^{-3/2}$

74. $25^{-5/2}$

75. $(16x^2)^{-3/2}$

76. $(81c^4)^{-3/2}$

77. $(-27y^3)^{-2/3}$

78. $(-8z^9)^{-2/3}$

79. $\left(\dfrac{27}{8} \right)^{-4/3}$

80. $\left(\dfrac{25}{49} \right)^{-3/2}$

81. $\left(-\dfrac{8x^3}{27} \right)^{-1/3}$

82. $\left(\dfrac{16}{81y^4} \right)^{-3/4}$

In Exercises 83–86, use a calculator to evaluate each expression. Round to the nearest hundredth.

83. $\sqrt[3]{15}$ **84.** $\sqrt[4]{50.5}$

85. $\sqrt[5]{1.045}$ **86.** $\sqrt[5]{-1,000}$

In Exercises 87–106, do the operations. Write the answers without negative exponents. Assume that all variables represent positive numbers.

87. $5^{3/7}5^{2/7}$ **88.** $4^{2/5}4^{2/5}$

89. $(4^{1/5})^3$ **90.** $(3^{1/3})^5$

91. $\dfrac{9^{4/5}}{9^{3/5}}$ **92.** $\dfrac{7^{2/3}}{7^{1/2}}$

93. $6^{-2/3}6^{-4/3}$ **94.** $5^{1/3}5^{-5/3}$

95. $\dfrac{3^{4/3}3^{1/3}}{3^{2/3}}$ **96.** $\dfrac{2^{5/6}2^{1/3}}{2^{1/2}}$

97. $a^{2/3}a^{1/3}$ **98.** $b^{3/5}b^{1/5}$

99. $(a^{2/3})^{1/3}$ **100.** $(t^{4/5})^{10}$

101. $(a^{1/2}b^{1/3})^{3/2}$ **102.** $(mn^{-2/3})^{-3/5}$

103. $\dfrac{(4x^3y)^{1/2}}{(9xy)^{1/2}}$ **104.** $\dfrac{(27x^3y)^{1/3}}{(8xy^2)^{2/3}}$

105. $(27x^{-3})^{-1/3}$ **106.** $(16a^{-2})^{-1/2}$

In Exercises 107–110, do the multiplications. Assume that all variables are positive.

107. $y^{1/3}(y^{2/3} + y^{5/3})$ **108.** $y^{2/5}(y^{-2/5} + y^{3/5})$

109. $x^{3/5}(x^{7/5} - x^{2/5} + 1)$ **110.** $x^{4/3}(x^{2/3} + 3x^{5/3} - 4)$

In Exercises 111–114, use rational exponents to simplify each radical. Assume that all variables represent positive numbers.

111. $\sqrt[6]{p^3}$ **112.** $\sqrt[8]{q^2}$

113. $\sqrt[4]{25b^2}$ **114.** $\sqrt[9]{-8x^6}$

APPLICATIONS

115. A BALLISTIC PENDULUM See Illustration 1. The formula

$$v = \frac{m + M}{m}(2gh)^{1/2}$$

gives the velocity (in ft/sec) of a bullet with weight m fired into a block with weight M, that raises the height of the block h feet after the collision. The letter g represents the constant, 32. Find the velocity of the bullet to the nearest ft/sec.

ILLUSTRATION 1

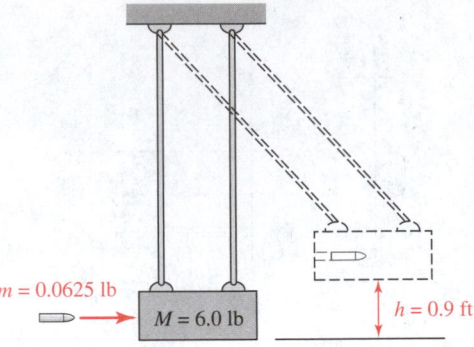

$m = 0.0625$ lb $M = 6.0$ lb $h = 0.9$ ft

116. GEOGRAPHY The formula

$$A = [s(s - a)(s - b)(s - c)]^{1/2}$$

gives the area of a triangle with sides of length a, b, and c, where s is one-half of the perimeter. Estimate the area of Virginia (to the nearest square mile) using the data given in Illustration 2.

ILLUSTRATION 2

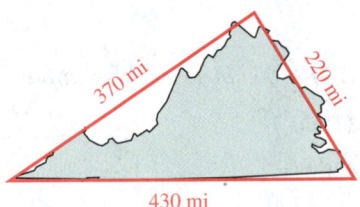

370 mi   220 mi   430 mi

117. RELATIVITY One of the concepts of relativity theory is that an object moving past an observer at a speed near the speed of light appears to have a larger mass because of its motion. If the mass of the object is m_0 when the object is at rest relative to the observer, its mass m will be given by the formula

$$m = m_0(1 - v^2/c^2)^{-1/2}$$

when it is moving with speed v (in miles per second) past the observer. The letter c represents the speed of light, 186,000 mi/sec. If a proton with a rest mass of 1 unit is accelerated by a nuclear accelerator to a speed of 160,000 mi/sec, what mass will the technicians observe it to have? Round to the nearest hundredth.

118. LOGGING See Illustration 3. The width w and height h of the strongest rectangular beam that can be cut from a cylindrical log of radius a are given by

$$w = \frac{2a}{3}(3^{1/2}) \qquad h = a\left(\frac{8}{3}\right)^{1/2}$$

Find the width, height, and cross-sectional area of the strongest beam that can be cut from a log with *diameter* 4 feet. Round to the nearest hundredth.

ILLUSTRATION 3

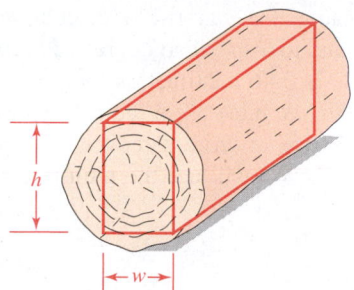

where a and b represent the widths of the hallways. Find the longest shelf that a carpenter can carry around the corner if $a = 40$ in. and $b = 64$ in. Give your result in inches and in feet. In each case, round to the nearest tenth.

ILLUSTRATION 5

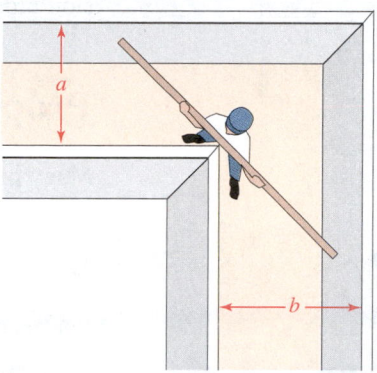

119. CUBICLES The area of the base of a cube is given by the function $A(x) = V^{2/3}$, where V is the volume of the cube. In a preschool room, 18 children's cubicles like that shown in Illustration 4 are placed on the floor around the room. Estimate how much floor space is lost to the cubicles. Give your answer in square inches and in square feet.

ILLUSTRATION 4

Mary S.

Storage capacity 4,096 in.³

WRITING

121. What is a rational exponent? Give some examples.

122. Explain how the root key $\sqrt[x]{y}$ on a scientific calculator can be used in combination with other keys to evaluate the expression $16^{3/4}$.

REVIEW

In Exercises 123–126, solve each inequality. Write each answer as an inequality.

123. $5x - 4 < 11$

124. $-2(3t - 5) \geq 8$

125. $\dfrac{4}{5}(r - 3) > \dfrac{2}{3}(r + 2)$

126. $-4 < 2x - 4 \leq 8$

120. CARPENTRY See Illustration 5. The length L of the longest board that can be carried horizontally around the right-angle corner of two intersecting hallways is given by the formula

$$L = (a^{2/3} + b^{2/3})^{3/2}$$

► **9.4**

Simplifying and Combining Radical Expressions

FIGURE 9-11

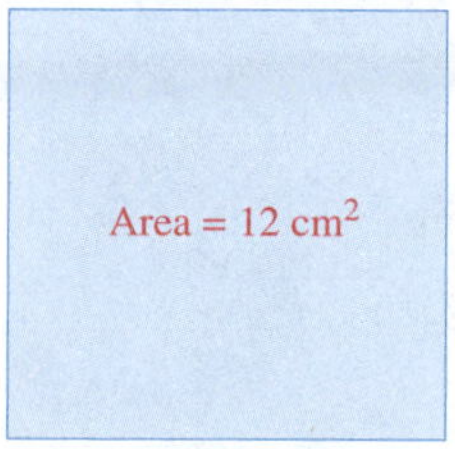

In this section, you will learn about

> Properties of radicals ▪ Simplifying radical expressions ▪ Adding and subtracting radical expressions

Area = 12 cm²

Introduction   The square shown in Figure 9-11 has an area of 12 square centimeters. We can determine the length of one side of the square in the following way:

$$A = s^2$$
$$12 = s^2 \qquad \text{Substitute 12 for } A.$$
$$\sqrt{12} = \sqrt{s^2} \qquad \text{Take the positive square root of both sides.}$$
$$\sqrt{12} = s \qquad \text{Each side of the square is } \sqrt{12} \text{ centimeters long.}$$

The form in which we express the length of a side of the square depends upon the situation. If an approximation is acceptable, we can use a calculator to find that $\sqrt{12} \approx 3.464101615$, and then we can round to a specified degree of accuracy. For example, to the nearest hundredth, each side is 3.46 centimeters long.

If the situation calls for the exact length, we must use a radical expression. As you will see in this section, it is common practice to write a radical expression such as $\sqrt{12}$ in *simplified* form. To simplify radicals, we will use the multiplication and division properties of radicals.

Properties of Radicals

Many properties of exponents have counterparts in radical notation. For example, because $a^{1/n}b^{1/n} = (ab)^{1/n}$, we have

1. $\sqrt[n]{a}\sqrt[n]{b} = \sqrt[n]{ab}$

For example, if x represents a positive real number,

$$\sqrt{5}\sqrt{5} = \sqrt{5 \cdot 5} = \sqrt{25} = 5$$

$$\sqrt[3]{7x}\sqrt[3]{49x^2} = \sqrt[3]{7x \cdot 49x^2} = \sqrt[3]{343x^3} = 7x$$

$$\sqrt[4]{2x^3}\sqrt[4]{8x} = \sqrt[4]{2x^3 \cdot 8x} = \sqrt[4]{16x^4} = 2x$$

These observations suggest the following rule.

Multiplication property of radicals

If $\sqrt[n]{a}$ and $\sqrt[n]{b}$ are real numbers, then

$$\sqrt[n]{ab} = \sqrt[n]{a}\sqrt[n]{b}$$

As long as all radical expressions represent real numbers, *the nth root of the product of two numbers is equal to the product of their nth roots.*

 WARNING! The multiplication property of radicals applies to the nth root of the product of two numbers. There is no such property for sums or differences. For example,

$$\sqrt{9 + 4} \neq \sqrt{9} + \sqrt{4} \qquad\qquad \sqrt{9 - 4} \neq \sqrt{9} - \sqrt{4}$$

$$\sqrt{13} \neq 3 + 2 \qquad\qquad\qquad \sqrt{5} \neq 3 - 2$$

$$\sqrt{13} \neq 5 \qquad\qquad\qquad\qquad \sqrt{5} \neq 1$$

Thus, $\sqrt{a + b} \neq \sqrt{a} + \sqrt{b}$ and $\sqrt{a - b} \neq \sqrt{a} - \sqrt{b}$.

A second property of radicals involves quotients. Because

$$\frac{a^{1/n}}{b^{1/n}} = \left(\frac{a}{b}\right)^{1/n}$$

it follows that

2. $\dfrac{\sqrt[n]{a}}{\sqrt[n]{b}} = \sqrt[n]{\dfrac{a}{b}} \quad (b \neq 0)$

For example, if x represents a positive real number,

$$\frac{\sqrt{8x^3}}{\sqrt{2x}} = \sqrt{\frac{8x^3}{2x}} = \sqrt{4x^2} = 2x$$

$$\frac{\sqrt[3]{54x^5}}{\sqrt[3]{2x^2}} = \sqrt[3]{\frac{54x^5}{2x^2}} = \sqrt[3]{27x^3} = 3x$$

Equation 2 suggests the following property.

Division property of radicals

If $\sqrt[n]{a}$ and $\sqrt[n]{b}$ are real numbers, then

$$\sqrt[n]{\frac{a}{b}} = \frac{\sqrt[n]{a}}{\sqrt[n]{b}} \quad (b \neq 0)$$

As long as all radical expressions represent real numbers, *the nth root of the quotient of two numbers is equal to the quotient of their nth roots.*

Simplifying Radical Expressions

A radical expression is said to be in simplest form when each of the following statements is true.

Simplified form of a radical expression

A radical expression is in simplest form when

1. No radicals appear in the denominator of a fraction.
2. The radicand contains no fractions or negative numbers.
3. Each factor in the radicand is to a power that is less than the index of the radical.

EXAMPLE 1

Simplifying radical expressions by using the multiplication property of radicals. Simplify   **a.** $\sqrt{12}$,   **b.** $\sqrt{98}$,   **c.** $\sqrt[3]{54}$,   and   **d.** $-\sqrt[4]{48}$.

Solution   **a.** Recall that numbers that are squares of integers, such as 1, 4, 9, 16, 25, and 36, are *perfect squares*. To simplify $\sqrt{12}$, we first factor 12 so that one factor is the largest perfect square that divides 12. Since 4 is the largest perfect square factor of 12, we write 12 as $4 \cdot 3$, use the multiplication property of radicals, and simplify.

$$\sqrt{12} = \sqrt{4 \cdot 3} \quad \text{Write 12 in factored form as } 4 \cdot 3.$$

$$= \sqrt{4}\sqrt{3} \quad \text{By the multiplication property of radicals, the square root of a product is equal to the product of the square roots: } \sqrt{4 \cdot 3} = \sqrt{4}\sqrt{3}.$$

$$= 2\sqrt{3} \quad \text{Simplify: } \sqrt{4} = 2.$$

We can show that $2\sqrt{3}$ is a square root of 12 as follows:

$$\left(2\sqrt{3}\right)^2 = (2)^2\left(\sqrt{3}\right)^2 \quad \text{Use a power rule of exponents: } (xy)^n = x^n y^n.$$
$$= 4(3) \qquad\qquad \sqrt{3} \text{ is the number that, when squared, gives 3.}$$
$$= 12$$

Thus, $\sqrt{12} = 2\sqrt{3}$.

b. The largest perfect square factor of 98 is 49. Thus,

$$\sqrt{98} = \sqrt{49 \cdot 2} \quad \text{Write 98 in factored form: } 98 = 49 \cdot 2.$$
$$= \sqrt{49}\sqrt{2} \quad \text{Apply the multiplication property of radicals:}$$
$$\sqrt{49 \cdot 2} = \sqrt{49}\sqrt{2}.$$
$$= 7\sqrt{2} \quad\quad \text{Simplify: } \sqrt{49} = 7.$$

Check the result.

c. Numbers that are cubes of integers, such as 1, 8, 27, 64, 125, and 216, are called *perfect cubes*. Since the largest perfect cube factor of 54 is 27, we have

$$\sqrt[3]{54} = \sqrt[3]{27 \cdot 2} \quad \text{Write 54 as } 27 \cdot 2.$$
$$= \sqrt[3]{27}\sqrt[3]{2} \quad \text{By the multiplication property of radicals, the cube root of a}$$
$$\text{product is equal to the product of the cube roots:}$$
$$\sqrt[3]{27 \cdot 2} = \sqrt[3]{27}\sqrt[3]{2}.$$
$$= 3\sqrt[3]{2} \quad\quad \text{Simplify: } \sqrt[3]{27} = 3.$$

We can show that $3\sqrt[3]{2}$ is a cube root of 54 as follows:

$$\left(3\sqrt[3]{2}\right)^3 = (3)^3\left(\sqrt[3]{2}\right)^3 = 27(2) = 54$$

d. The largest perfect fourth-power factor of 48 is 16. Thus,

$$-\sqrt[4]{48} = -\sqrt[4]{16 \cdot 3} \quad \text{Write 48 as } 16 \cdot 3.$$
$$= -\sqrt[4]{16}\sqrt[4]{3} \quad \text{Apply the multiplication property of radicals.}$$
$$= -2\sqrt[4]{3} \quad\quad \text{Simplify: } \sqrt[4]{16} = 2.$$

Check the result.

SELF CHECK Simplify **a.** $\sqrt{20}$, **b.** $\sqrt[3]{24}$, and **c.** $\sqrt[5]{-128}$.

Answers: **a.** $2\sqrt{5}$, **b.** $2\sqrt[3]{3}$, **c.** $-2\sqrt[5]{4}$

EXAMPLE 2 **Simplifying radical expressions.** Simplify **a.** $\sqrt{m^9}$, **b.** $\sqrt{128a^5}$, **c.** $\sqrt[3]{24x^5}$, and **d.** $\sqrt[5]{a^9 b^5}$. Assume that all variables represent positive real numbers.

Solution **a.** The largest perfect square factor of m^9 is m^8.

$$\sqrt{m^9} = \sqrt{m^8 \cdot m} \quad \text{Write } m^9 \text{ in factored form as } m^8 \cdot m.$$
$$= \sqrt{m^8}\sqrt{m} \quad \text{Apply the multiplication property of radicals.}$$
$$= m^4\sqrt{m} \quad\quad \text{Simplify: } \sqrt{m^8} = m^4.$$

b. Since the largest perfect square factor of 128 is 64 and the largest perfect square factor of a^5 is a^4, the largest perfect square factor of $128a^5$ is $64a^4$. We write $128a^5$ as $64a^4 \cdot 2a$ and proceed as follows:

$$\sqrt{128a^5} = \sqrt{64a^4 \cdot 2a}$$
$$= \sqrt{64a^4}\sqrt{2a} \quad \text{Use the multiplication property of radicals.}$$
$$= 8a^2\sqrt{2a} \quad\quad \text{Simplify: } \sqrt{64a^4} = 8a^2.$$

c. We write $24x^5$ as $8x^3 \cdot 3x^2$ and proceed as follows:

$$\sqrt[3]{24x^5} = \sqrt[3]{8x^3 \cdot 3x^2} \qquad \text{$8x^3$ is the largest perfect cube factor of $24x^5$.}$$
$$= \sqrt[3]{8x^3}\sqrt[3]{3x^2} \qquad \text{Use the multiplication property of radicals.}$$
$$= 2x\sqrt[3]{3x^2} \qquad \text{Simplify: $\sqrt[3]{8x^3} = 2x$.}$$

d. The largest perfect fifth-power factor of a^9 is a^5, and b^5 is a fifth power.

$$\sqrt[5]{a^9b^5} = \sqrt[5]{a^5b^5 \cdot a^4} \qquad \text{a^5b^5 is the largest perfect fifth-power factor of a^9b^5.}$$
$$= \sqrt[5]{a^5b^5}\sqrt[5]{a^4} \qquad \text{Use the multiplication property of radicals.}$$
$$= ab\sqrt[5]{a^4} \qquad \text{Simplify: $\sqrt[5]{a^5b^5} = ab$.}$$

SELF CHECK　Simplify **a.** $\sqrt{98b^3}$, **b.** $\sqrt[3]{54y^5}$, and 　　　*Answers:* **a.** $7b\sqrt{2b}$,
c. $\sqrt[4]{t^8u^{15}}$. 　　　　　　　　　　　　　　　　　　**b.** $3y\sqrt[3]{2y^2}$, **c.** $t^2u^3\sqrt[4]{u^3}$ ■

EXAMPLE 3　**Simplifying radical expressions using the division property of radicals.** Simplify **a.** $\sqrt{\dfrac{7}{64}}$, **b.** $\sqrt{\dfrac{15}{49x^2}}$, and **c.** $\sqrt[3]{\dfrac{10x^2}{27y^6}}$. Assume that the variables represent positive real numbers.

Solution　**a.** We can write the square root of the quotient as the quotient of two square roots.

$$\sqrt{\frac{7}{64}} = \frac{\sqrt{7}}{\sqrt{64}} \qquad \text{Apply the division property of radicals.}$$
$$= \frac{\sqrt{7}}{8} \qquad \text{Simplify the denominator: $\sqrt{64} = 8$.}$$

b.
$$\sqrt{\frac{15}{49x^2}} = \frac{\sqrt{15}}{\sqrt{49x^2}} \qquad \text{Apply the division property of radicals.}$$
$$= \frac{\sqrt{15}}{7x} \qquad \text{Simplify the denominator: $\sqrt{49x^2} = 7x$.}$$

c. We can write the cube root of the quotient as the quotient of two cube roots. Since $y \neq 0$, we have

$$\sqrt[3]{\frac{10x^2}{27y^6}} = \frac{\sqrt[3]{10x^2}}{\sqrt[3]{27y^6}} \qquad \text{Use the division property of radicals.}$$
$$= \frac{\sqrt[3]{10x^2}}{3y^2} \qquad \text{Simplify the denominator.}$$

SELF CHECK　Simplify **a.** $\sqrt{\dfrac{11}{36a^2}}$ $(a > 0)$　and

b. $\sqrt[4]{\dfrac{a^3}{625y^{12}}}$ $(a > 0, y > 0)$.　　　*Answers:* **a.** $\dfrac{\sqrt{11}}{6a}$, **b.** $\dfrac{\sqrt[4]{a^3}}{5y^3}$ ■

EXAMPLE 4　**Simplifying radical expressions.** Simplify each expression. Assume that all variables represent positive numbers. **a.** $\dfrac{\sqrt{45xy^2}}{\sqrt{5x}}$ and **b.** $\dfrac{\sqrt[3]{-432x^5}}{\sqrt[3]{8x}}$.

Solution **a.** We can write the quotient of the square roots as the square root of a quotient.

$$\frac{\sqrt{45xy^2}}{\sqrt{5x}} = \sqrt{\frac{45xy^2}{5x}}$$ Use the division property of radicals.

$$= \sqrt{9y^2}$$ Simplify the fraction $\frac{45xy^2}{5x}$.

$$= 3y$$ Simplify the radical.

b. We can write the quotient of the cube roots as the cube root of a quotient.

$$\frac{\sqrt[3]{-432x^5}}{\sqrt[3]{8x}} = \sqrt[3]{\frac{-432x^5}{8x}}$$ Use the division property of radicals.

$$= \sqrt[3]{-54x^4}$$ Simplify the fraction $\frac{-432x^5}{8x}$.

$$= \sqrt[3]{-27x^3 \cdot 2x}$$ $-27x^3$ is the largest perfect cube that divides $-54x^4$.

$$= \sqrt[3]{-27x^3}\sqrt[3]{2x}$$ Use the multiplication property of radicals.

$$= -3x\sqrt[3]{2x}$$ Simplify: $\sqrt[3]{-27x^3} = -3x$.

SELF CHECK   Simplify each expression (assume that all variables represent positive numbers):

a. $\dfrac{\sqrt{50ab^2}}{\sqrt{2a}}$ and **b.** $\dfrac{\sqrt[3]{-2{,}000x^5v^3}}{\sqrt[3]{2x}}$.

Answers: **a.** $5b$, **b.** $-10xv\sqrt[3]{x}$ ■

Adding and Subtracting Radical Expressions

Radical expressions with the same index and the same radicand are called **like** or **similar radicals.** For example, $3\sqrt{2}$ and $2\sqrt{2}$ are like radicals. However,

$3\sqrt{5}$ and $4\sqrt{2}$ are not like radicals, because the radicands are different.

$3\sqrt[4]{5}$ and $2\sqrt[3]{5}$ are not like radicals, because the indexes are different.

For a given expression containing two or more radical terms, we should attempt to combine like radicals, if possible. For example, to simplify the expression $3\sqrt{2} + 2\sqrt{2}$, we use the distributive property to factor out $\sqrt{2}$ and simplify.

$$3\sqrt{2} + 2\sqrt{2} = (3 + 2)\sqrt{2}$$
$$= 5\sqrt{2}$$

Radicals with the same index but different radicands can often be written as like radicals. For example, to simplify the expression $\sqrt{27} - \sqrt{12}$, we simplify both radicals first, and then we combine the like radicals.

$$\sqrt{27} - \sqrt{12} = \sqrt{9 \cdot 3} - \sqrt{4 \cdot 3}$$ Write 27 and 12 in factored form.

$$= \sqrt{9}\sqrt{3} - \sqrt{4}\sqrt{3}$$ Use the multiplication property of radicals.

$$= 3\sqrt{3} - 2\sqrt{3}$$ Simplify $\sqrt{9}$ and $\sqrt{4}$.

$$= (3 - 2)\sqrt{3}$$ Factor out $\sqrt{3}$.

$$= \sqrt{3}$$ $1\sqrt{3} = \sqrt{3}$.

As the previous examples suggest, we can use the following rule to add or subtract radicals.

Adding and subtracting radicals

To add or subtract radicals, simplify each radical expression and combine all like radicals. To add or subtract like radicals, combine the coefficients and keep the common radical.

EXAMPLE 5

Combining like radicals. Simplify $2\sqrt{12} - 3\sqrt{48} + 3\sqrt{3}$.

Solution We simplify $2\sqrt{12}$ and $3\sqrt{48}$ separately and then combine like radicals.

$$2\sqrt{12} - 3\sqrt{48} + 3\sqrt{3} = 2\sqrt{4 \cdot 3} - 3\sqrt{16 \cdot 3} + 3\sqrt{3}$$
$$= 2\sqrt{4}\sqrt{3} - 3\sqrt{16}\sqrt{3} + 3\sqrt{3}$$
$$= 2(2)\sqrt{3} - 3(4)\sqrt{3} + 3\sqrt{3}$$
$$= \mathbf{4}\sqrt{3} - \mathbf{12}\sqrt{3} + \mathbf{3}\sqrt{3}$$ All three expressions have the same index and radicand.

$$= (\mathbf{4 - 12 + 3})\sqrt{3}$$ Combine the coefficients of these like radicals and keep $\sqrt{3}$.

$$= -5\sqrt{3}$$

SELF CHECK Simplify $3\sqrt{75} - 2\sqrt{12} + 2\sqrt{48}$.   *Answer:* $19\sqrt{3}$ ∎

EXAMPLE 6

Combining like radicals. Simplify $\sqrt[3]{16} - \sqrt[3]{54} + \sqrt[3]{24}$.

Solution We begin by simplifying each radical expression separately:

$$\sqrt[3]{16} - \sqrt[3]{54} + \sqrt[3]{24} = \sqrt[3]{8 \cdot 2} - \sqrt[3]{27 \cdot 2} + \sqrt[3]{8 \cdot 3}$$
$$= \sqrt[3]{8}\sqrt[3]{2} - \sqrt[3]{27}\sqrt[3]{2} + \sqrt[3]{8}\sqrt[3]{3}$$
$$= 2\sqrt[3]{2} - 3\sqrt[3]{2} + 2\sqrt[3]{3}$$

Now we combine the two radical expressions that have the same index and radicand.

$$\sqrt[3]{16} - \sqrt[3]{54} + \sqrt[3]{24} = -\sqrt[3]{2} + 2\sqrt[3]{3}$$

SELF CHECK Simplify $\sqrt[3]{24} - \sqrt[3]{16} + \sqrt[3]{54}$.   *Answer:* $2\sqrt[3]{3} + \sqrt[3]{2}$ ∎

 WARNING! Even though the radical expressions $-\sqrt[3]{2}$ and $2\sqrt[3]{3}$ in the last line of Example 6 have the same index, we cannot combine them, because their radicands are different. Neither can we combine radical expressions having the same radicand but a different index. For example, the expression $\sqrt[3]{2} + \sqrt[4]{2}$ cannot be simplified.

EXAMPLE 7

Combining like radicals. Simplify $\sqrt[3]{16x^4} + \sqrt[3]{54x^4} - \sqrt[3]{-128x^4}$.

Solution We simplify each radical expression separately, factor out $\sqrt[3]{2x}$, and simplify.

$$\sqrt[3]{16x^4} + \sqrt[3]{54x^4} - \sqrt[3]{-128x^4}$$
$$= \sqrt[3]{8x^3 \cdot 2x} + \sqrt[3]{27x^3 \cdot 2x} - \sqrt[3]{-64x^3 \cdot 2x}$$
$$= \sqrt[3]{8x^3}\sqrt[3]{2x} + \sqrt[3]{27x^3}\sqrt[3]{2x} - \sqrt[3]{-64x^3}\sqrt[3]{2x}$$
$$= \mathbf{2x}\sqrt[3]{2x} + \mathbf{3x}\sqrt[3]{2x} + \mathbf{4x}\sqrt[3]{2x}$$ All three radicals have the same index and radicand.

$$= (\mathbf{2x + 3x + 4x})\sqrt[3]{2x}$$
$$= 9x\sqrt[3]{2x}$$ In the parentheses, combine like terms.

SELF CHECK Simplify $\sqrt{32x^3} + \sqrt{50x^3} - \sqrt{18x^3}$.   *Answer:* $6x\sqrt{2x}$ ∎

VOCABULARY

In Exercises 1–4, fill in the blanks to make each statement true.

1. Radical expressions such as $\sqrt[3]{4}$ and $6\sqrt[3]{4}$ with the same index and the same radicand are called _____ radicals.

2. Numbers such as 1, 4, 9, 16, 25, and 36 are called perfect _____. Numbers such as 1, 8, 27, 64, and 125 are called perfect _____.

3. The largest perfect square _____ of 27 is 9.

4. "To _____ $\sqrt{24}$" means to write it as $2\sqrt{6}$.

CONCEPTS

In Exercises 5–6, fill in the blanks to make each statement true.

5. $\sqrt[n]{ab} = $ [ ]

   In words, the *n*th root of the _____ of two numbers is equal to the product of their *n*th _____.

6. $\sqrt[n]{\dfrac{a}{b}} = $ [ ]

   In words, the *n*th root of the _____ of two numbers is equal to the quotient of their *n*th _____.

7. Consider the expressions
   $$\sqrt{4 \cdot 5} \text{ and } \sqrt{4}\sqrt{5}$$
   Which expression is
   a. the square root of a product?
   b. the product of square roots?
   c. How are these two expressions related?

8. Consider the expressions
   $$\dfrac{\sqrt[3]{a}}{\sqrt[3]{x^2}} \text{ and } \sqrt[3]{\dfrac{a}{x^2}}$$
   Which expression is
   a. the cube root of a quotient?
   b. the quotient of cube roots?
   c. How are these two expressions related?

9. a. Write two radical expressions that have the same radicand but a different index. Can the expressions be added?
   b. Write two radical expressions that have the same index but a different radicand. Can the expressions be added?

10. Explain the **mistake** in the student's solution shown below.
    Simplify $\sqrt[3]{54}$.
    $$\sqrt[3]{54} = \sqrt[3]{27 + 27}$$
    $$= \sqrt[3]{27} + \sqrt[3]{27}$$
    $$= 3 + 3$$
    $$= 6$$

NOTATION

In Exercises 11–12, complete each solution.

11. Simplify the radical expression.
    $$\sqrt[3]{32k^4} = \sqrt[3]{\boxed{} \cdot 4k}$$
    $$= \sqrt[3]{\boxed{}}\,\sqrt[3]{4k}$$
    $$= 2k\sqrt[3]{4k}$$

12. Simplify the radical expression.
    $$\dfrac{\sqrt{80s^2t^4}}{\sqrt{5s^2}} = \sqrt{\dfrac{80s^2t^4}{\boxed{}}}$$
    $$= \sqrt{\boxed{}}$$
    $$= 4t^2$$

PRACTICE

In Exercises 13–28, simplify each expression. Assume that all variables represent positive real numbers.

13. $\sqrt{6}\sqrt{6}$

14. $\sqrt{11}\sqrt{11}$

15. $\sqrt{t}\sqrt{t}$

16. $-\sqrt{z}\sqrt{z}$

17. $\sqrt[3]{5x^2}\sqrt[3]{25x}$

18. $\sqrt[4]{25a}\sqrt[4]{25a^3}$

19. $\dfrac{\sqrt{500}}{\sqrt{5}}$

20. $\dfrac{\sqrt{128}}{\sqrt{2}}$

21. $\dfrac{\sqrt{98x^3}}{\sqrt{2x}}$

22. $\dfrac{\sqrt{75y^5}}{\sqrt{3y}}$

23. $\dfrac{\sqrt{180ab^4}}{\sqrt{5ab^2}}$

24. $\dfrac{\sqrt{112ab^3}}{\sqrt{7ab}}$

25. $\dfrac{\sqrt[3]{48}}{\sqrt[3]{6}}$

26. $\dfrac{\sqrt[3]{64}}{\sqrt[3]{8}}$

27. $\dfrac{\sqrt[3]{189a^4}}{\sqrt[3]{7a}}$

28. $\dfrac{\sqrt[3]{243x^7}}{\sqrt[3]{9x}}$

In Exercises 29–48, simplify each radical expression.

29. $\sqrt{20}$

30. $\sqrt{8}$

31. $-\sqrt{200}$

32. $-\sqrt{250}$

33. $\sqrt[3]{80}$

34. $\sqrt[3]{270}$

35. $\sqrt[3]{-81}$

36. $\sqrt[3]{-72}$

37. $\sqrt[4]{32}$

38. $\sqrt[4]{48}$

39. $\sqrt[5]{96}$

40. $\sqrt[7]{256}$

41. $\sqrt{\dfrac{7}{9}}$

42. $\sqrt{\dfrac{3}{4}}$

43. $\sqrt[3]{\dfrac{7}{64}}$

44. $\sqrt[3]{\dfrac{4}{125}}$

45. $\sqrt[4]{\dfrac{3}{10{,}000}}$

46. $\sqrt[5]{\dfrac{4}{243}}$

47. $\sqrt[5]{\dfrac{3}{32}}$

48. $\sqrt[6]{\dfrac{5}{64}}$

In Exercises 49–68, simplify each radical expression. Assume that all variables represent positive numbers.

49. $\sqrt{50x^2}$

50. $\sqrt{75a^2}$

51. $\sqrt{32b}$

52. $\sqrt{80c}$

53. $-\sqrt{112a^3}$

54. $\sqrt{147a^5}$

55. $\sqrt{175a^2b^3}$

56. $\sqrt{128a^3b^5}$

57. $-\sqrt{300xy}$

58. $\sqrt{200x^2y}$

59. $\sqrt[3]{-54x^6}$

60. $-\sqrt[3]{-81a^3}$

61. $\sqrt[3]{16x^{12}y^3}$

62. $\sqrt[3]{40a^3b^6}$

63. $\sqrt[4]{32x^{12}y^4}$

64. $\sqrt[5]{64x^{10}y^5}$

65. $\sqrt{\dfrac{z^2}{16x^2}}$

66. $\sqrt{\dfrac{b^4}{64a^8}}$

67. $\sqrt[4]{\dfrac{5x}{16z^4}}$

68. $\sqrt[3]{\dfrac{11a^2}{125b^6}}$

In Exercises 69–104, simplify and combine like radicals. All variables represent positive numbers.

69. $4\sqrt{2x} + 6\sqrt{2x}$

70. $6\sqrt[3]{5y} + 3\sqrt[3]{5y}$

71. $8\sqrt[5]{7a^2} - 7\sqrt[5]{7a^2}$

72. $10\sqrt[6]{12xyz} - \sqrt[6]{12xyz}$

73. $\sqrt{2} - \sqrt{8}$

74. $\sqrt{20} - \sqrt{125}$

75. $\sqrt{98} - \sqrt{50}$

76. $\sqrt{72} - \sqrt{200}$

77. $3\sqrt{24} + \sqrt{54}$

78. $\sqrt{18} + 2\sqrt{50}$

79. $\sqrt[3]{24x} + \sqrt[3]{3x}$

80. $\sqrt[3]{16y} + \sqrt[3]{128y}$

81. $\sqrt[3]{32} - \sqrt[3]{108}$

82. $\sqrt[3]{80} - \sqrt[3]{10{,}000}$

83. $2\sqrt[3]{125} - 5\sqrt[3]{64}$

84. $3\sqrt[3]{27} + 12\sqrt[3]{216}$

85. $14\sqrt[4]{32} - 15\sqrt[4]{162}$

86. $23\sqrt[4]{768} + \sqrt[4]{48}$

87. $3\sqrt[4]{512} + 2\sqrt[4]{32}$

88. $4\sqrt[4]{243} - \sqrt[4]{48}$

89. $\sqrt{98} - \sqrt{50} - \sqrt{72}$

90. $\sqrt{20} + \sqrt{125} - \sqrt{80}$

91. $\sqrt{18t} + \sqrt{300t} - \sqrt{243t}$

92. $\sqrt{80m} - \sqrt{128m} + \sqrt{288m}$

93. $2\sqrt[3]{16} - \sqrt[3]{54} - 3\sqrt[3]{128}$

94. $\sqrt[4]{48} - \sqrt[4]{243} - \sqrt[4]{768}$

95. $\sqrt{25y^2z} - \sqrt{16y^2z}$

96. $\sqrt{25yz^2} + \sqrt{9yz^2}$

97. $\sqrt{36xy^2} + \sqrt{49xy^2}$

98. $3\sqrt{2x} - \sqrt{8x}$

99. $2\sqrt[3]{64a} + 2\sqrt[3]{8a}$

100. $3\sqrt[4]{x^4y} - 2\sqrt[4]{x^4y}$

101. $\sqrt{y^5} - \sqrt{9y^5} - \sqrt{25y^5}$

102. $\sqrt{8y^7} + \sqrt{32y^7} - \sqrt{2y^7}$

103. $\sqrt[5]{x^6y^2} + \sqrt[5]{32x^6y^2} + \sqrt[5]{x^6y^2}$

104. $\sqrt[3]{xy^4} + \sqrt[3]{8xy^4} - \sqrt[3]{27xy^4}$

APPLICATIONS

In Exercises 105–110, first give the exact answer, expressed as a simplified radical expression. Then give an approximation, rounded to the nearest tenth.

105. **UMBRELLA** The surface area of a cone is given by the formula $S = \pi r \sqrt{r^2 + h^2}$, where r is the radius of the base and h is its height. Use this formula to find the number of square feet of waterproof cloth used to make the umbrella shown in Illustration 1.

ILLUSTRATION 1

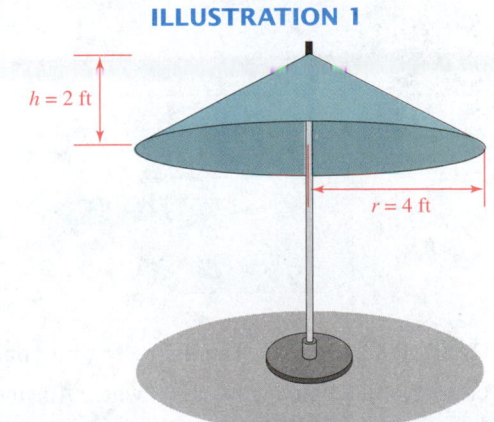

$h = 2$ ft

$r = 4$ ft

106. STRUCTURAL ENGINEERING Engineers have determined that two additional supports need to be added to strengthen a truss. See Illustration 2. Find the length L of each new support using the formula

$$L = \sqrt{\frac{b^2}{2} + \frac{c^2}{2} - \frac{a^2}{4}}$$

ILLUSTRATION 2

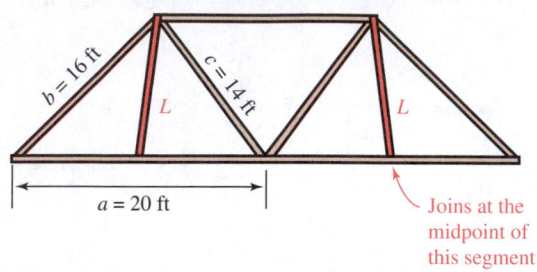

Joins at the midpoint of this segment

107. BLOW DRYER The current I (in amps), the power P (in watts), and the resistance R (in ohms) are related by the formula $I = \sqrt{\frac{P}{R}}$. What current is needed for a 1,200-watt hair dryer if the resistance is 16 ohms?

108. COMMUNICATIONS SATELLITE Engineers have determined that a spherical communications satellite needs to have a capacity of 565.2 cubic feet to house all of its operating systems. The volume V of a sphere is related to its radius r by the formula $r = \sqrt[3]{\frac{3V}{4\pi}}$. What radius must the satellite have to meet the engineer's specification? Use 3.14 for π.

109. DUCTWORK The pattern shown in Illustration 3 is laid out on a sheet of galvanized tin. Then it is cut out with snips and bent to make an air conditioning duct connection. Find the total length of the cut that must be made with the tin snips. (All measurements are in inches.)

ILLUSTRATION 3

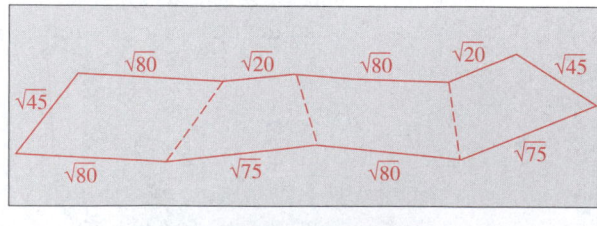

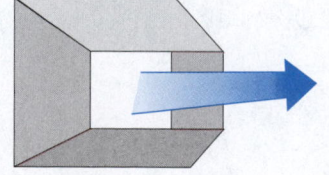

110. OUTDOOR COOKING The diameter of a circle is given by the function $d(A) = 2\sqrt{\frac{A}{\pi}}$, where A is the area

of the circle. Find the difference between the diameters of the barbecue grills shown in Illustration 4.

ILLUSTRATION 4

Cooking area 147π in.^{3}

Cooking area 48π in.^{3}

WRITING

111. Explain why $\sqrt[3]{9x^4}$ is not in simplified form.

112. How are the procedures used to simplify $3x + 4x$ and $3\sqrt{x} + 4\sqrt{x}$ similar?

113. See Illustration 5. Explain how the graphs of $Y_1 = 3\sqrt{24x} + \sqrt{54x}$ (on the left) and $Y_2 = 9\sqrt{6x}$ (on the right) can be used to verify the simplification $3\sqrt{24x} + \sqrt{54x} = 9\sqrt{6x}$. In each graph, settings of $[-5, 20]$ for x and $[-5, 100]$ for y were used.

ILLUSTRATION 5

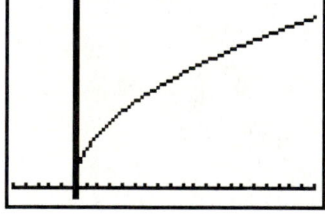 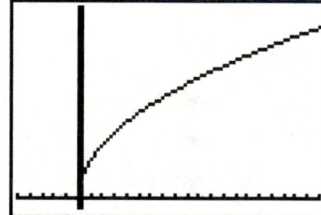

114. Explain how to verify algebraically that $\sqrt{200x^3y^5} = 10xy^2\sqrt{2xy}$

REVIEW

Do each operation.

115. $3x^2y^3(-5x^3y^{-4})$

116. $(2x^2 - 9x - 5) \cdot \frac{x}{2x^2 + x}$

117. $2p - 5)\overline{6p^2 - 7p - 25}$

118. $\dfrac{xy}{\dfrac{1}{x} - \dfrac{1}{y}}$

▶ 9.5 Multiplying and Dividing Radical Expressions

In this section, you will learn about

Multiplying a monomial by a monomial ■ Multiplying a polynomial by a monomial ■ Multiplying a polynomial by a polynomial ■ Rationalizing denominators ■ Rationalizing binomial denominators ■ Rationalizing numerators

Introduction In this section, we will discuss the methods used to multiply and divide radical expressions. Solving these problems often requires the use of procedures and properties studied earlier, such as simplifying radical expressions, combining like radicals, the FOIL method, and the distributive property.

Multiplying a Monomial by a Monomial

Radical expressions with the same index can be multiplied. For example, to find the product of $\sqrt{5}$ and $\sqrt{10}$, we proceed as follows:

$$\sqrt{5}\sqrt{10} = \sqrt{5 \cdot 10} \qquad \text{By the multiplication property of radicals: } \sqrt[n]{a}\sqrt[n]{b} = \sqrt[n]{ab}.$$
$$= \sqrt{50} \qquad \text{Multiply under the radical. Note that } \sqrt{50} \text{ can be simplified.}$$
$$= \sqrt{25 \cdot 2} \qquad \text{Begin the process of simplifying } \sqrt{50} \text{ by factoring 50.}$$
$$= 5\sqrt{2} \qquad \sqrt{25 \cdot 2} = \sqrt{25}\sqrt{2} = 5\sqrt{2}.$$

EXAMPLE 1

Multiplying radical expressions. Multiply $3\sqrt{6}$ by $2\sqrt{3}$.

Solution We use the commutative and associative properties of multiplication to multiply the coefficients and the radicals separately. Then we simplify any radicals in the product, if possible.

$$3\sqrt{6} \cdot 2\sqrt{3} = 3(2)\sqrt{6}\sqrt{3} \qquad \text{Multiply the coefficients and multiply the radicals.}$$
$$= 6\sqrt{18} \qquad 3(2) = 6 \text{ and } \sqrt{6}\sqrt{3} = \sqrt{18}.$$
$$= 6\sqrt{9}\sqrt{2} \qquad \text{Simplify: } \sqrt{18} = \sqrt{9 \cdot 2} = \sqrt{9}\sqrt{2}.$$
$$= 6(3)\sqrt{2} \qquad \sqrt{9} = 3.$$
$$= 18\sqrt{2} \qquad \text{Multiply.}$$

SELF CHECK Multiply $-2\sqrt{7}$ by $5\sqrt{2}$. *Answer:* $-10\sqrt{14}$ ■

Multiplying a Polynomial by a Monomial

To multiply a polynomial by a monomial, we use the distributive property to remove parentheses and then simplify each resulting term, if possible.

EXAMPLE 2

Using the distributive property with radical expressions. Simplify $3\sqrt{3}(4\sqrt{8} - 5\sqrt{10})$.

Solution $3\sqrt{3}\left(4\sqrt{8} - 5\sqrt{10}\right)$

$= 3\sqrt{3} \cdot 4\sqrt{8} - 3\sqrt{3} \cdot 5\sqrt{10}$ Use the distributive property.

$= 12\sqrt{24} - 15\sqrt{30}$ Multiply the coefficients and multiply the radicals.

$= 12\sqrt{4}\sqrt{6} - 15\sqrt{30}$ Simplify: $\sqrt{24} = \sqrt{4 \cdot 6} = \sqrt{4}\sqrt{6}$.

$= 12(2)\sqrt{6} - 15\sqrt{30}$ $\sqrt{4} = 2$.

$= 24\sqrt{6} - 15\sqrt{30}$

SELF CHECK   Simplify $4\sqrt{2}\left(3\sqrt{5} - 2\sqrt{8}\right)$.   *Answer:* $12\sqrt{10} - 32$ ■

Multiplying a Polynomial by a Polynomial

To multiply a binomial by a binomial, we use the FOIL method.

EXAMPLE 3   **Using the FOIL method with radical expressions.**
Multiply $\left(\sqrt{7} + \sqrt{2}\right)\left(\sqrt{7} - 3\sqrt{2}\right)$.

Solution $\left(\sqrt{7} + \sqrt{2}\right)\left(\sqrt{7} - 3\sqrt{2}\right)$

$= \sqrt{7}\sqrt{7} - 3\sqrt{7}\sqrt{2} + \sqrt{2}\sqrt{7} - 3\sqrt{2}\sqrt{2}$ Use the FOIL method.

$= 7 - 3\sqrt{14} + \sqrt{14} - 3(2)$ Do each multiplication:
$\sqrt{7}\sqrt{7} = \sqrt{49} = 7$ and
$\sqrt{2}\sqrt{2} = \sqrt{4} = 2$.

$= 7 - 2\sqrt{14} - 6$ Combine like radicals:
$-3\sqrt{14} + \sqrt{14} = -2\sqrt{14}$.

$= 1 - 2\sqrt{14}$ Combine like terms: $7 - 6 = 1$.

SELF CHECK   Multiply $\left(\sqrt{5} + 2\sqrt{3}\right)\left(\sqrt{5} - \sqrt{3}\right)$.   *Answer:* $-1 + \sqrt{15}$ ■

Technically, the expression $\sqrt{3x} - \sqrt{5}$ is not a polynomial, because the variable does not have a whole-number exponent $\left(\sqrt{3x} = 3^{1/2}x^{1/2}\right)$. However, we will multiply such expressions as if they were polynomials.

EXAMPLE 4   **Using the FOIL method with radical expressions.**
Multiply $\left(\sqrt{3x} - \sqrt{5}\right)\left(\sqrt{2x} + \sqrt{10}\right)$. Assume that $x > 0$.

Solution $\left(\sqrt{3x} - \sqrt{5}\right)\left(\sqrt{2x} + \sqrt{10}\right)$

$= \sqrt{3x}\sqrt{2x} + \sqrt{3x}\sqrt{10} - \sqrt{5}\sqrt{2x} - \sqrt{5}\sqrt{10}$ Use FOIL.

$= \sqrt{6x^2} + \sqrt{30x} - \sqrt{10x} - \sqrt{50}$ Do each multiplication.

$= \sqrt{6}\sqrt{x^2} + \sqrt{30x} - \sqrt{10x} - \sqrt{25}\sqrt{2}$ Simplify $\sqrt{6x^2}$ and $\sqrt{50}$.

$= \sqrt{6}x + \sqrt{30x} - \sqrt{10x} - 5\sqrt{2}$

SELF CHECK   Multiply $\left(\sqrt{x} + 1\right)\left(\sqrt{x} - 3\right)$. Assume that $x > 0$.   *Answer:* $x - 2\sqrt{x} - 3$ ■

WARNING! It is important to draw the radical sign carefully so that it completely covers the radicand, but no more than the radicand. To avoid confusion, we often write an expression such as $\sqrt{6}x$ in the form $x\sqrt{6}$.

Rationalizing Denominators

In Section 9.4, we saw that a radical expression is in simplified form when each of the following statements is true.

1. No radicals appear in the denominator of a fraction.
2. The radicand contains no fractions or negative numbers.
3. Each factor in the radicand appears to a power that is less than the index of the radical.

We now consider radical expressions that do not satisfy requirement 1 and radical expressions that do not satisfy requirement 2 of this list. We will introduce an algebraic technique, called *rationalizing the denominator,* that is used to write such expressions in an equivalent simplified form.

To divide radical expressions, we **rationalize the denominator** of a fraction to replace the denominator with a rational number. For example, to divide $\sqrt{70}$ by $\sqrt{3}$, we write the division as the fraction

$$\frac{\sqrt{70}}{\sqrt{3}}$$ The denominator is the irrational number $\sqrt{3}$. This radical expression is not in simplified form, because a radical appears in the denominator.

To eliminate the radical in the denominator, we multiply the numerator and the denominator by a number that will give a perfect square *under the radical in the denominator.* Because $3 \cdot 3 = 9$ and 9 is a perfect square, $\sqrt{3}$ is such a number.

$$\frac{\sqrt{70}}{\sqrt{3}} = \frac{\sqrt{70} \cdot \sqrt{3}}{\sqrt{3} \cdot \sqrt{3}}$$ Multiply numerator and denominator of the fraction by $\sqrt{3}$.

$$= \frac{\sqrt{210}}{\sqrt{9}}$$ Multiply the radicals in the numerator and in the denominator.

$$= \frac{\sqrt{210}}{3}$$ Simplify: $\sqrt{9} = 3$. The denominator is now the rational number 3.

Since there is no radical in the denominator and $\sqrt{210}$ cannot be simplified, the expression $\frac{\sqrt{210}}{3}$ is in simplest form, and the division is complete.

EXAMPLE 5

Division of radical expressions. Simplify by rationalizing the denominator:

a. $\sqrt{\dfrac{20}{7}}$ and **b.** $\dfrac{4}{\sqrt[3]{2}}$.

Solution **a.** This radical expression is not in simplified form, because the radicand contains a fraction. We begin by writing the square root of the quotient as the quotient of two square roots:

$$\sqrt{\frac{20}{7}} = \frac{\sqrt{20}}{\sqrt{7}}$$ Apply the division property of radicals: $\sqrt[n]{\dfrac{a}{b}} = \dfrac{\sqrt[n]{a}}{\sqrt[n]{b}}$

To rationalize the denominator, we proceed as follows:

$$\frac{\sqrt{20}}{\sqrt{7}} = \frac{\sqrt{20} \cdot \sqrt{7}}{\sqrt{7} \cdot \sqrt{7}}$$ Multiply the numerator and denominator by $\sqrt{7}$.

$$= \frac{\sqrt{140}}{\sqrt{49}}$$ Multiply the radicals.

$$= \frac{2\sqrt{35}}{7}$$ Simplify: $\sqrt{140} = \sqrt{4 \cdot 35} = \sqrt{4}\sqrt{35} = 2\sqrt{35}$ and $\sqrt{49} = 7$.

b. Here, we must rationalize a denominator that is a cube root. We multiply the numerator and the denominator by a number that will give a perfect cube under the radical sign. Since $2 \cdot 4 = 8$ is a perfect cube, $\sqrt[3]{4}$ is such a number.

$$\frac{4}{\sqrt[3]{2}} = \frac{4 \cdot \sqrt[3]{4}}{\sqrt[3]{2} \cdot \sqrt[3]{4}} \qquad \text{Multiply numerator and denominator by } \sqrt[3]{4}.$$

$$= \frac{4\sqrt[3]{4}}{\sqrt[3]{8}} \qquad \text{Multiply the radicals in the denominator.}$$

$$= \frac{4\sqrt[3]{4}}{2} \qquad \text{Simplify: } \sqrt[3]{8} = 2.$$

$$= 2\sqrt[3]{4} \qquad \text{Simplify: } \frac{4\sqrt[3]{4}}{2} = \frac{\overset{1}{\cancel{2}} \cdot 2\sqrt[3]{4}}{\underset{1}{\cancel{2}}} = 2\sqrt[3]{4}.$$

SELF CHECK   Rationalize the denominator:

a. $\sqrt{\dfrac{24}{5}}$ and **b.** $\dfrac{5}{\sqrt[4]{3}}$.   *Answers:* **a.** $\dfrac{2\sqrt{30}}{5}$, **b.** $\dfrac{5\sqrt[4]{27}}{3}$ ∎

E X A M P L E   6

Photography Many camera lenses (see Figure 9-12) have an adjustable opening called the **aperture,** which controls the amount of light passing through the lens. The **ƒ-number** of a lens is its **focal length** divided by the diameter of its circular aperture:

$$f\text{-number} = \frac{f}{d} \quad f \text{ is the focal length, and } d \text{ is the diameter of the aperture.}$$

A lens with a focal length of 12 centimeters and an aperture with a diameter of 6 centimeters has an ƒ-number of $\frac{12}{6}$ and is an ƒ/2 lens. If the area of the aperture is reduced to admit half as much light, the ƒ-number of the lens will change. Find the new ƒ-number.

Solution   We first find the area of the aperture when its diameter is 6 centimeters.

FIGURE 9-12

$A = \pi r^2$ The formula for the area of a circle.

$A = \pi(\mathbf{3})^2$ Since a radius is half the diameter, substitute 3 for r.

$A = 9\pi$

When the size of the aperture is reduced to admit half as much light, the area of the aperture will be $\frac{9\pi}{2}$ square centimeters. To find the diameter d of a circle with this area, we proceed as follows:

$A = \pi r^2$ The formula for the area of a circle.

$\dfrac{9\pi}{2} = \pi\left(\dfrac{d}{2}\right)^2$ Substitute $\frac{9\pi}{2}$ for A and $\frac{d}{2}$ for r.

$\dfrac{9\pi}{2} = \dfrac{\pi d^2}{4}$ $\left(\dfrac{d}{2}\right)^2 = \dfrac{d^2}{4}.$

$18 = d^2$ Multiply both sides by 4 and divide both sides by π.

$d = 3\sqrt{2}$ Take the positive square root of both sides. Then simplify: $\sqrt{18} = \sqrt{9}\sqrt{2} = 3\sqrt{2}.$

Since the focal length of the lens is still 12 centimeters and the diameter is now $3\sqrt{2}$ centimeters, the new f-number of the lens is

$$f\text{-number} = \frac{f}{d} = \frac{12}{3\sqrt{2}} \qquad \text{Substitute 12 for } f \text{ and } 3\sqrt{2} \text{ for } d.$$

$$= \frac{12\sqrt{2}}{3\sqrt{2}\sqrt{2}} \qquad \text{Rationalize the denominator.}$$

$$= \frac{12\sqrt{2}}{3(2)} \qquad \text{Simplify the denominator: } \sqrt{2}\sqrt{2} = 2.$$

$$= 2\sqrt{2} \qquad \text{Simplify: } \frac{12\sqrt{2}}{3(2)} = \frac{12\sqrt{2}}{6} = \frac{\overset{1}{\cancel{6}} \cdot 2\sqrt{2}}{\underset{1}{\cancel{6}}}.$$

$$\approx 2.828427125 \qquad \text{Use a calculator.}$$

The lens is now an $f/2.8$ lens. ∎

EXAMPLE 7

Rationalizing a variable denominator. Rationalize the denominator of $\dfrac{\sqrt{5xy^2}}{\sqrt{xy^3}}$ (x and y are positive numbers).

Solution In each case, we write the expression as a quotient of radicals and then simplify the radicand.

Method 1

$$\frac{\sqrt{5xy^2}}{\sqrt{xy^3}} = \sqrt{\frac{5xy^2}{xy^3}}$$

$$= \sqrt{\frac{5}{y}}$$

$$= \frac{\sqrt{5}}{\sqrt{y}}$$

$$= \frac{\sqrt{5}\sqrt{y}}{\sqrt{y}\sqrt{y}}$$

$$= \frac{\sqrt{5y}}{y}$$

Method 2

$$\frac{\sqrt{5xy^2}}{\sqrt{xy^3}} = \sqrt{\frac{5xy^2}{xy^3}}$$

$$= \sqrt{\frac{5}{y}}$$

$$= \sqrt{\frac{5 \cdot y}{y \cdot y}}$$

$$= \frac{\sqrt{5y}}{\sqrt{y^2}}$$

$$= \frac{\sqrt{5y}}{y}$$

SELF CHECK Rationalize the denominator of $\dfrac{\sqrt{4ab^3}}{\sqrt{2a^2b^2}}$. *Answer:* $\dfrac{\sqrt{2ab}}{a}$ ∎

EXAMPLE 8

Simplifying before rationalizing. Rationalize the denominator of $\sqrt{\dfrac{11}{20q^5}}$ ($q > 0$).

Solution   We write the expression as a quotient of two radicals. Then we simplify the radical in the denominator before rationalizing.

$$\sqrt{\frac{11}{20q^5}} = \frac{\sqrt{11}}{\sqrt{20q^5}}$$ The square root of a quotient is the quotient of the square roots.

$$= \frac{\sqrt{11}}{\sqrt{4q^4 \cdot 5q}}$$ To simplify $\sqrt{20q^5}$, write it as $\sqrt{4q^4 \cdot 5q}$.

$$= \frac{\sqrt{11}}{2q^2\sqrt{5q}}$$ $\sqrt{4q^4 \cdot 5q} = \sqrt{4q^4}\sqrt{5q} = 2q^2\sqrt{5q}$.

$$= \frac{\sqrt{11}\sqrt{5q}}{2q^2\sqrt{5q}\sqrt{5q}}$$ Rationalize the denominator.

$$= \frac{\sqrt{55q}}{2q^2(5q)}$$ Multiply the radicals: $\sqrt{5q}\sqrt{5q} = 5q$.

$$= \frac{\sqrt{55q}}{10q^3}$$ Multiply in the denominator.

SELF CHECK   Rationalize the denominator of $\sqrt[3]{\dfrac{1}{16h^4}}$.   *Answer:* $\dfrac{\sqrt[3]{4h^2}}{4h^2}$

EXAMPLE 9   **Rationalizing a variable denominator.** Rationalize the denominator of $\dfrac{\sqrt[3]{5}}{\sqrt[3]{9m}}$.

Solution   We multiply the numerator and the denominator by $\sqrt[3]{3m^2}$, which will produce a perfect cube, $27m^3$, under the radical sign in the denominator.

$$\frac{\sqrt[3]{5}}{\sqrt[3]{9m}} = \frac{\sqrt[3]{5}\sqrt[3]{3m^2}}{\sqrt[3]{9m}\sqrt[3]{3m^2}}$$ Multiply numerator and denominator by $\sqrt[3]{3m^2}$.

$$= \frac{\sqrt[3]{15m^2}}{\sqrt[3]{27m^3}}$$ Multiply the radicals.

$$= \frac{\sqrt[3]{15m^2}}{3m}$$ Simplify: $\sqrt[3]{27m^3} = 3m$.

SELF CHECK   Rationalize the denominator of $\dfrac{\sqrt{5}}{\sqrt{17b}}$.   *Answer:* $\dfrac{\sqrt{85b}}{17b}$

Rationalizing Binomial Denominators

To rationalize a denominator that contains a binomial expression involving square roots, we multiply its numerator and denominator by the *conjugate* of its denominator. **Conjugate binomials** are binomials with the same terms but with opposite signs between their terms.

Conjugate binomials

The **conjugate** of the binomial $a + b$ is $a - b$, and the conjugate of $a - b$ is $a + b$.

When we multiply a binomial expression involving square roots and its conjugate, the product does not contain a radical. For example, if we use the FOIL method to multiply $\sqrt{x} + 2$ and its conjugate, $\sqrt{x} - 2$, we have

$$\left(\sqrt{x} + 2\right)\left(\sqrt{x} - 2\right) = \sqrt{x}\sqrt{x} - 2\sqrt{x} + 2\sqrt{x} - 4$$

$$= x - 4 \qquad \text{Combine like radicals:}$$
$$-2\sqrt{x} + 2\sqrt{x} = 0.$$

EXAMPLE 10

Rationalizing binomial denominators. Rationalize the denominator of $\dfrac{1}{\sqrt{2} + 1}$.

Solution   We multiply the numerator and denominator of the fraction by $\sqrt{2} - 1$, which is the conjugate of the denominator.

$$\frac{1}{\sqrt{2} + 1} = \frac{1\left(\sqrt{2} - 1\right)}{\left(\sqrt{2} + 1\right)\left(\sqrt{2} - 1\right)}$$

$$= \frac{\sqrt{2} - 1}{2 - 1} \qquad \begin{array}{l}\text{In the denominator, multiply the binomials:} \\ \left(\sqrt{2} + 1\right)\left(\sqrt{2} - 1\right) = \sqrt{2}\sqrt{2} - \sqrt{2} + \sqrt{2} - 1 = \\ 2 - 1.\end{array}$$

$$= \sqrt{2} - 1 \qquad \text{Simplify: } \frac{\sqrt{2} - 1}{2 - 1} = \frac{\sqrt{2} - 1}{1} = \sqrt{2} - 1.$$

SELF CHECK   Rationalize the denominator of $\dfrac{2}{\sqrt{3} + 1}$.   *Answer:* $\sqrt{3} - 1$ ■

EXAMPLE 11

Rationalizing binomial denominators. Rationalize the denominator of $\dfrac{\sqrt{x} + \sqrt{2}}{\sqrt{x} - \sqrt{2}} \ (x > 0)$.

Solution   We multiply the numerator and denominator by $\sqrt{x} + \sqrt{2}$, which is the conjugate of $\sqrt{x} - \sqrt{2}$, and simplify.

$$\frac{\sqrt{x} + \sqrt{2}}{\sqrt{x} - \sqrt{2}} = \frac{\left(\sqrt{x} + \sqrt{2}\right)\left(\sqrt{x} + \sqrt{2}\right)}{\left(\sqrt{x} - \sqrt{2}\right)\left(\sqrt{x} + \sqrt{2}\right)}$$

$$= \frac{x + \sqrt{2x} + \sqrt{2x} + 2}{x - 2} \qquad \begin{array}{l}\text{In the numerator and denominator, use the} \\ \text{FOIL method.}\end{array}$$

$$= \frac{x + 2\sqrt{2x} + 2}{x - 2} \qquad \text{In the numerator, combine like radicals.}$$

SELF CHECK   Rationalize the denominator of $\dfrac{\sqrt{x} - \sqrt{2}}{\sqrt{x} + \sqrt{2}}$.   *Answer:* $\dfrac{x - 2\sqrt{2x} + 2}{x - 2}$ ■

Rationalizing Numerators

In calculus, we sometimes have to rationalize a numerator by multiplying the numerator and denominator of the fraction by the conjugate of the numerator.

EXAMPLE 12   **Rationalizing numerators.** Rationalize the numerator of $\dfrac{\sqrt{x}-3}{\sqrt{x}}$ ($x > 0$).

Solution   We multiply the numerator and denominator by $\sqrt{x}+3$, which is the conjugate of the numerator.

$$\frac{\sqrt{x}-3}{\sqrt{x}} = \frac{(\sqrt{x}-3)(\sqrt{x}+3)}{\sqrt{x}(\sqrt{x}+3)}$$

$$= \frac{x + 3\sqrt{x} - 3\sqrt{x} - 9}{x + 3\sqrt{x}}$$

$$= \frac{x - 9}{x + 3\sqrt{x}}$$

The final expression is not in simplified form. However, this nonsimplified form is often desirable in calculus.

SELF CHECK   Rationalize the numerator of $\dfrac{\sqrt{x}+3}{\sqrt{x}}$.   *Answer:* $\dfrac{x-9}{x-3\sqrt{x}}$ ∎

STUDY SET

Section 9.5

VOCABULARY

In Exercises 1–6, fill in the blanks to make each statement true.

1. To multiply $(\sqrt{3}+\sqrt{2})(\sqrt{3}-2\sqrt{2})$, we can use the _____ method.

2. To multiply $2\sqrt{5}(3\sqrt{8}+\sqrt{3})$, use the _____ property to remove parentheses.

3. The denominator of the fraction $\dfrac{4}{\sqrt{5}}$ is an _____ number.

4. The _____ of $\sqrt{x}+1$ is $\sqrt{x}-1$.

5. To obtain a _____ cube under the radical in the denominator of $\dfrac{\sqrt[3]{7}}{\sqrt[3]{5n}}$, we multiply the numerator and denominator by $\sqrt[3]{25n^2}$.

6. To _____ the denominator of $\dfrac{4}{\sqrt{5}}$, we multiply the numerator and denominator by $\sqrt{5}$.

CONCEPTS

7. Do each operation, if possible.
   a. $4\sqrt{6}+2\sqrt{6}$
   b. $4\sqrt{6}(2\sqrt{6})$
   c. $3\sqrt{2}-2\sqrt{3}$
   d. $3\sqrt{2}(-2\sqrt{3})$

8. Do each operation, if possible.
   a. $5 + 6\sqrt[3]{6}$
   b. $5(6\sqrt[3]{6})$
   c. $\dfrac{30\sqrt[3]{15}}{5}$
   d. $\dfrac{\sqrt[3]{15}}{5}$

9. Consider $\dfrac{\sqrt{3}}{\sqrt{7}} = \dfrac{\sqrt{3}\sqrt{7}}{\sqrt{7}\sqrt{7}}$. Explain why the expressions on the left-hand side and the right-hand side of the equation are equal.

10. To rationalize the denominator of $\dfrac{\sqrt[4]{2}}{\sqrt[4]{3}}$, why wouldn't we multiply the numerator and denominator by $\dfrac{\sqrt[4]{3}}{\sqrt[4]{3}}$?

11. Explain why $\dfrac{\sqrt[3]{12}}{\sqrt[3]{5}}$ is not in simplified form.

12. Explain why $\sqrt{\dfrac{3a}{11k}}$ is not in simplified form.

NOTATION

In Exercises 13–14, fill in the blanks to make each statement true.

13. Multiply $5\sqrt{8} \cdot 7\sqrt{6}$.

$$5\sqrt{8} \cdot 7\sqrt{6} = 5(7)\sqrt{8\;\boxed{}}$$
$$= 35\sqrt{\boxed{}}$$
$$= 35\sqrt{\boxed{} \cdot 3}$$
$$= 35(\boxed{})\sqrt{3}$$
$$= 140\sqrt{3}$$

14. Rationalize the denominator of $\dfrac{9}{\sqrt[3]{4a^2}}$.

$$\frac{9}{\sqrt[3]{4a^2}} = \frac{9 \cdot \sqrt[3]{2a}}{\sqrt[3]{4a^2 \cdot \boxed{}}}$$
$$= \frac{9\sqrt[3]{2a}}{\sqrt[3]{\boxed{}}}$$
$$= \frac{9\sqrt[3]{2a}}{2a}$$

PRACTICE

In Exercises 15–42, do each multiplication and simplify, if possible. All variables represent positive real numbers.

15. $\sqrt{11}\sqrt{11}$ **16.** $\sqrt{35}\sqrt{35}$

17. $\left(\sqrt{7}\right)^2$ **18.** $\left(\sqrt{23}\right)^2$

19. $\sqrt{2}\sqrt{8}$ **20.** $\sqrt{3}\sqrt{27}$

21. $\sqrt{5}\sqrt{10}$ **22.** $\sqrt{7}\sqrt{35}$

23. $2\sqrt{3}\sqrt{6}$ **24.** $-3\sqrt{11}\sqrt{33}$

25. $\sqrt[3]{5}\sqrt[3]{25}$ **26.** $-\sqrt[3]{7}\sqrt[3]{49}$

27. $\left(3\sqrt{2}\right)^2$ **28.** $\left(2\sqrt{5}\right)^2$

29. $\left(-2\sqrt{2}\right)^2$ **30.** $\left(-3\sqrt{10}\right)^2$

31. $\left(3\sqrt[3]{9}\right)\left(2\sqrt[3]{3}\right)$ **32.** $\left(2\sqrt[3]{16}\right)\left(-\sqrt[3]{4}\right)$

33. $\sqrt[3]{2}\sqrt[3]{12}$ **34.** $\sqrt[3]{3}\sqrt[3]{18}$

35. $\sqrt{ab^3}\sqrt{ab}$ **36.** $\sqrt{8x}\sqrt{2x^3y}$

37. $\sqrt{5ab}\sqrt{5a}$ **38.** $\sqrt{15rs^2}\sqrt{10r}$

39. $-4\sqrt[3]{5r^2s}\left(5\sqrt[3]{2r}\right)$ **40.** $-\sqrt[3]{3xy^2}\left(-\sqrt[3]{9x^3}\right)$

41. $\sqrt{x(x+3)}\sqrt{x^3(x+3)}$ **42.** $\sqrt{y^2(x+y)}\sqrt{(x+y)^3}$

In Exercises 43–60, do each multiplication and simplify. All variables represent positive real numbers.

43. $3\sqrt{5}\left(4 - \sqrt{5}\right)$ **44.** $2\sqrt{7}\left(3\sqrt{7} - 1\right)$

45. $3\sqrt{2}\left(4\sqrt{6} + 2\sqrt{7}\right)$ **46.** $-\sqrt{3}\left(\sqrt{7} - \sqrt{15}\right)$

47. $-2\sqrt{5x}\left(4\sqrt{2x} - 3\sqrt{3}\right)$

48. $3\sqrt{7t}\left(2\sqrt{7t} + 3\sqrt{3t^2}\right)$

49. $\left(\sqrt{2} + 1\right)\left(\sqrt{2} - 3\right)$

50. $\left(2\sqrt{3} + 1\right)\left(\sqrt{3} - 1\right)$

51. $\left(\sqrt{5z} + \sqrt{3}\right)\left(\sqrt{5z} + \sqrt{3}\right)$

52. $\left(\sqrt{3p} - \sqrt{2}\right)\left(\sqrt{3p} + \sqrt{2}\right)$

53. $\left(\sqrt{3x} - \sqrt{2y}\right)\left(\sqrt{3x} + \sqrt{2y}\right)$

54. $\left(\sqrt{3m} + \sqrt{2n}\right)\left(\sqrt{3m} + \sqrt{2n}\right)$

55. $\left(2\sqrt{3a} - \sqrt{b}\right)\left(\sqrt{3a} + 3\sqrt{b}\right)$

56. $\left(5\sqrt{p} - \sqrt{3q}\right)\left(\sqrt{p} + 2\sqrt{3q}\right)$

57. $\left(3\sqrt{2r} - 2\right)^2$ **58.** $\left(2\sqrt{3t} + 5\right)^2$

59. $-2\left(\sqrt{3x} + \sqrt{3}\right)^2$ **60.** $3\left(\sqrt{5x} - \sqrt{3}\right)^2$

In Exercises 61–84, simplify each radical expression by rationalizing the denominator. All variables represent positive real numbers.

61. $\sqrt{\dfrac{1}{7}}$ **62.** $\sqrt{\dfrac{5}{3}}$

63. $\dfrac{6}{\sqrt{30}}$ **64.** $\dfrac{8}{\sqrt{10}}$

65. $\dfrac{\sqrt{5}}{\sqrt{8}}$ **66.** $\dfrac{\sqrt{3}}{\sqrt{50}}$

67. $\dfrac{\sqrt{8}}{\sqrt{2}}$ **68.** $\dfrac{\sqrt{27}}{\sqrt{3}}$

69. $\dfrac{1}{\sqrt[3]{2}}$ **70.** $\dfrac{2}{\sqrt[3]{6}}$

71. $\dfrac{3}{\sqrt[3]{9}}$ **72.** $\dfrac{2}{\sqrt[3]{a}}$

73. $\dfrac{\sqrt[3]{2}}{\sqrt[3]{9}}$ **74.** $\dfrac{\sqrt[3]{9}}{\sqrt[3]{54}}$

75. $\dfrac{\sqrt{8}}{\sqrt{xy}}$ **76.** $\dfrac{\sqrt{9xy}}{\sqrt{3x^2y}}$

77. $\dfrac{\sqrt{10xy^2}}{\sqrt{2xy^3}}$ **78.** $\dfrac{\sqrt{5ab^2c}}{\sqrt{10abc}}$

79. $\dfrac{\sqrt[3]{4a^2}}{\sqrt[3]{2ab}}$ **80.** $\dfrac{\sqrt[3]{9x}}{\sqrt[3]{3xy}}$

81. $\dfrac{1}{\sqrt[4]{4}}$ **82.** $\dfrac{1}{\sqrt[5]{2}}$

83. $\dfrac{1}{\sqrt[5]{16}}$ **84.** $\dfrac{4}{\sqrt[4]{32}}$

In Exercises 85–98, rationalize each denominator. All variables represent positive real numbers.

85. $\dfrac{1}{\sqrt{2} - 1}$

86. $\dfrac{3}{\sqrt{3} - 1}$

87. $\dfrac{\sqrt{2}}{\sqrt{5} + 3}$

88. $\dfrac{\sqrt{3}}{\sqrt{3} - 2}$

89. $\dfrac{\sqrt{3} + 1}{\sqrt{3} - 1}$

90. $\dfrac{\sqrt{2} - 1}{\sqrt{2} + 1}$

91. $\dfrac{\sqrt{7} - \sqrt{2}}{\sqrt{2} + \sqrt{7}}$

92. $\dfrac{\sqrt{3} + \sqrt{2}}{\sqrt{3} - \sqrt{2}}$

93. $\dfrac{2}{\sqrt{x} + 1}$

94. $\dfrac{3}{\sqrt{x} - 2}$

95. $\dfrac{2z - 1}{\sqrt{2z} - 1}$

96. $\dfrac{3t - 1}{\sqrt{3t} + 1}$

97. $\dfrac{\sqrt{x} - \sqrt{y}}{\sqrt{x} + \sqrt{y}}$

98. $\dfrac{\sqrt{x} + \sqrt{y}}{\sqrt{x} - \sqrt{y}}$

In Exercises 99–104, rationalize each numerator. All variables represent positive numbers.

99. $\dfrac{\sqrt{3} + 1}{2}$

100. $\dfrac{\sqrt{5} - 1}{2}$

101. $\dfrac{\sqrt{x} + 3}{x}$

102. $\dfrac{2 + \sqrt{x}}{5x}$

103. $\dfrac{\sqrt{x} + \sqrt{y}}{\sqrt{x}}$

104. $\dfrac{\sqrt{x} - \sqrt{y}}{\sqrt{x} + \sqrt{y}}$

APPLICATIONS

105. STATISTICS An example of a normal distribution curve, or *bell-shaped* curve, is shown in Illustration 1. A fraction that is part of the equation that models this curve is

$$\dfrac{1}{\sigma\sqrt{2\pi}}$$

where σ is a letter from the Greek alphabet. Rationalize the denominator of the fraction.

106. ANALYTIC GEOMETRY See Illustration 2. The length of the perpendicular segment drawn from $(-2, 2)$ to the line with equation $2x - 4y = 4$ is given by

$$L = \dfrac{|2(-2) + (-4)(2) + (-4)|}{\sqrt{(2)^2 + (-4)^2}}$$

Find L. Express the result in simplified radical form. Then give an approximation to the nearest tenth.

ILLUSTRATION 2

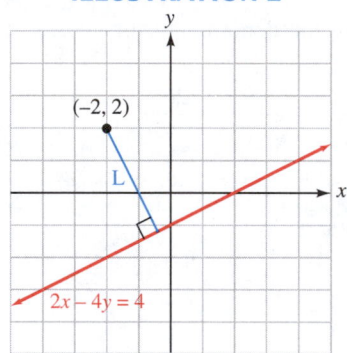

107. TRIGONOMETRY In trigonometry, we must often find the ratio of the lengths of two sides of right triangles. Use the information in Illustration 3 to find the ratio

$$\dfrac{\text{length of side } AC}{\text{length of side } AB}$$

Write the result in simplified radical form.

ILLUSTRATION 3

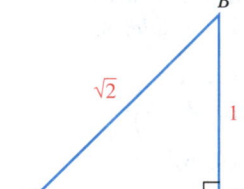

108. MECHANICAL ENGINEERING A measure of how fast the block shown in Illustration 4 will oscillate when the system is set in motion is given by the formula

$$\omega = \sqrt{\dfrac{k_1 + k_2}{m}}$$

where k_1 and k_2 indicate the stiffness of the springs and m is the mass of the block. Rationalize the right-hand side and restate the formula.

ILLUSTRATION 1

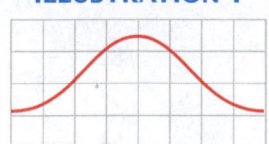

ILLUSTRATION 4

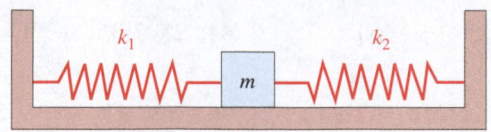

109. PHOTOGRAPHY In Example 6, we saw that a lens with a focal length of 12 centimeters and an aperture $3\sqrt{2}$ centimeters in diameter is an $f/2.8$ lens. Find the f-number if the area of the aperture is again cut in half.

110. ELECTRONICS A formula that is used when designing AC (alternating current) circuits is

$$f_0 = \frac{1}{2\pi}\sqrt{\frac{1}{LC}}$$

Rationalize the right-hand side and restate the formula.

WRITING

111. Explain why $\sqrt{m} \cdot \sqrt{m} = m$ but $\sqrt[3]{m} \cdot \sqrt[3]{m} \neq m$. (Assume that $m > 0$.)

112. Explain why the product of $\sqrt{m} + 3$ and $\sqrt{m} - 3$ does not contain a radical.

REVIEW

Solve the equation.

113. $\dfrac{2}{3-a} = 1$

114. $5(s-4) = -5(s-4)$

115. $\dfrac{8}{b-2} + \dfrac{3}{2-b} = -\dfrac{1}{b}$

116. $\dfrac{2}{x-2} + \dfrac{1}{x+1} = \dfrac{1}{(x+1)(x-2)}$

▶ 9.6 Geometric Applications of Radicals

In this section, you will learn about

The Pythagorean theorem ■ 45°–45°–90° triangles ■ 30°–60°–90° triangles ■ The distance formula

Introduction In this section, we will consider several applications of square roots in geometry. Then we will find the distance between two points on a rectangular coordinate system, using a formula that contains a square root. We begin by considering an important theorem (mathematical statement) about right triangles.

The Pythagorean Theorem

If we know the lengths of two legs of a right triangle, we can always find the length of the **hypotenuse** (the side opposite the 90° angle) by using the **Pythagorean theorem.**

Pythagorean theorem

If a and b are the lengths of two legs of a right triangle and c is the length of the hypotenuse, then

$$a^2 + b^2 = c^2$$

In words, the Pythagorean theorem is expressed as follows:

In any right triangle, the square of the hypotenuse is equal to the sum of the squares of the two legs.

FIGURE 9-13

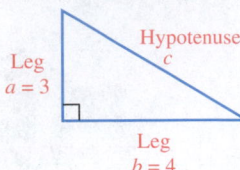

Leg
$a = 3$

Hypotenuse
c

Leg
$b = 4$

Suppose the right triangle shown in Figure 9-13 has legs of length 3 and 4 units. To find the length of the hypotenuse, we use the Pythagorean theorem.

$$a^2 + b^2 = c^2$$

$3^2 + 4^2 = c^2$ Substitute 3 for a and 4 for b.

$9 + 16 = c^2$

$25 = c^2$

To solve for c, we need to "undo" the operation performed on it. Since c is squared, we take the positive square root of both sides of the equation.

$\sqrt{25} = \sqrt{c^2}$ Since c represents the length of a side of a triangle, $c > 0$. So $\sqrt{c^2} = c$, because $c \cdot c = c^2$.

$5 = c$

The length of the hypotenuse is 5 units.

EXAMPLE 1

Firefighting. To fight a forest fire, the forestry department plans to clear a rectangular fire break around the fire, as shown in Figure 9-14. Crews are equipped with mobile communications that have a 3,000-yard range. Can crews at points A and B remain in radio contact?

Solution   Points A, B, and C form a right triangle. To find the distance c from point A to point B, we can use the Pythagorean theorem, substituting 2,400 for a and 1,000 for b and solving for c.

FIGURE 9-14

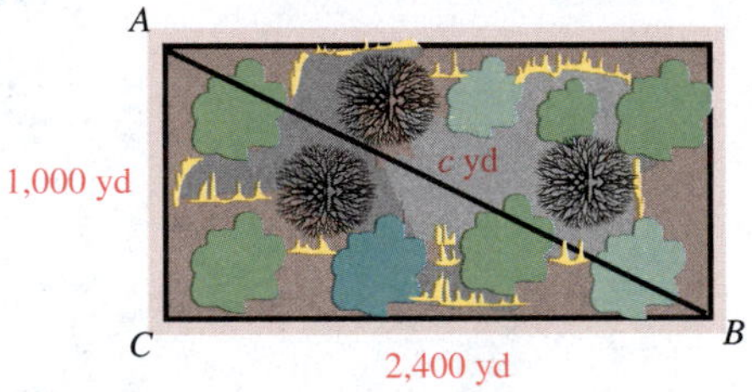

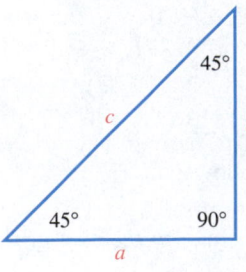

1,000 yd

c yd

A C B 2,400 yd

$$a^2 + b^2 = c^2$$

$2,400^2 + 1,000^2 = c^2$

$5,760,000 + 1,000,000 = c^2$

$6,760,000 = c^2$

$\sqrt{6,760,000} = \sqrt{c^2}$ Take the positive square root of both sides.

$2,600 = c$ Use a calculator to find the square root.

The two crews are 2,600 yards apart. Because this distance is less than the range of the radios, they can communicate by radio.

SELF CHECK   In Example 1, can the crews communicate if $b = 1,500$ yards?   *Answer:* yes ∎

45°–45°–90° Triangles

An **isosceles right triangle** is a right triangle with two legs of equal length. Isosceles right triangles have angle measures of 45°, 45°, and 90°. If we know the length of one leg of an isosceles right triangle, we can use the Pythagorean theorem to find the length of the hypotenuse. Since the triangle shown in Figure 9-15 is a right triangle, we have

FIGURE 9-15

45°

c

a

45°   90°

a

$$c^2 = a^2 + b^2$$

$c^2 = a^2 + a^2$ Both legs are a units long, so replace b with a.

$c^2 = 2a^2$ Combine like terms.

$c = \sqrt{2a^2}$ Take the positive square root of both sides.

$c = a\sqrt{2}$ Simplify the radical: $\sqrt{2a^2} = \sqrt{2}\sqrt{a^2} = \sqrt{2}a = a\sqrt{2}$.

Thus, *in an isosceles right triangle, the length of the hypotenuse is the length of one leg times* $\sqrt{2}$.

EXAMPLE 2   **45°–45°–90° triangles.** If one leg of the isosceles right triangle shown in Figure 9-15 is 10 feet long, find the length of the hypotenuse.

Solution   Since the length of the hypotenuse is the length of a leg times $\sqrt{2}$, we have

$$c = 10\sqrt{2}$$

The length of the hypotenuse is $10\sqrt{2}$ units. To two decimal places, the length is 14.14 units.

SELF CHECK   Find the length of the hypotenuse of an isosceles right triangle if one leg is 12 meters long.   *Answer:* $12\sqrt{2}$ m   ∎

If the length of the hypotenuse of an isosceles right triangle is known, we can use the Pythagorean theorem to find the length of each leg.

EXAMPLE 3   Find the exact length of each leg of the isosceles right triangle shown in Figure 9-16.

Solution   We use the Pythagorean theorem.

FIGURE 9-16

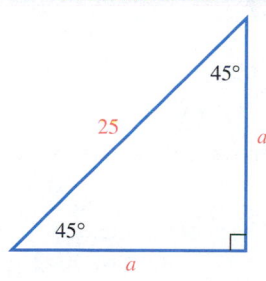

$$c^2 = a^2 + b^2$$

$$25^2 = a^2 + a^2 \qquad \text{Since both legs are } a \text{ units long, substitute } a \text{ for } b. \text{ The hypotenuse is 25 units long. Substitute 25 for } c.$$

$$25^2 = 2a^2 \qquad \text{Combine like terms.}$$

$$\frac{625}{2} = a^2 \qquad \text{Square 25 and divide both sides by 2.}$$

$$\sqrt{\frac{625}{2}} = a \qquad \text{To solve for } a, \text{ take the positive square root of both sides: } \sqrt{a^2} = a.$$

$$\frac{\sqrt{625} \cdot \sqrt{2}}{\sqrt{2} \cdot \sqrt{2}} = a \qquad \text{Write } \sqrt{\frac{625}{2}} \text{ as } \frac{\sqrt{625}}{\sqrt{2}}. \text{ Then rationalize the denominator.}$$

$$\frac{25\sqrt{2}}{2} = a \qquad \text{In the numerator, simplify the radical: } \sqrt{625} = 25. \text{ In the denominator, do the multiplication: } \sqrt{2} \cdot \sqrt{2} = 2.$$

The exact length of each leg is $\dfrac{25\sqrt{2}}{2}$ units. To two decimal places, the length is 17.68 units.   ∎

30°–60°–90° Triangles

From geometry, we know that an **equilateral triangle** is a triangle with three sides of equal length and three 60° angles. Each side of the equilateral triangle in Figure 9-17 is $2a$ units long. If an **altitude** is drawn to its base, the altitude bisects the base and divides the equilateral triangle into two 30°–60°–90° triangles. We can see that the shorter leg of each 30°–60°–90° triangle (the side *opposite* the 30° angle) is a units long. Thus,

The length of the shorter leg of a 30°–60°–90° right triangle is half as long as the hypotenuse.

We can discover another important relationship between the legs of a 30°–60°–90° triangle if we find the length of the altitude h in Figure 9-17. We begin by applying the Pythagorean theorem to one of the 30°–60°–90° triangles.

FIGURE 9-17

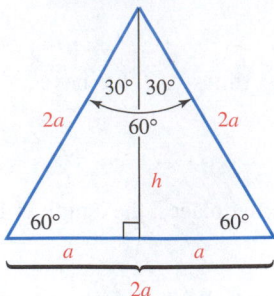

$$a^2 + \boldsymbol{b}^2 = \boldsymbol{c}^2$$

$a^2 + \boldsymbol{h}^2 = (\boldsymbol{2a})^2$ One leg is h units long, so replace b with h. The hypotenuse is $2a$ units long, so replace c with $2a$.

$a^2 + h^2 = 4a^2$ $(2a)^2 = (2a)(2a) = 4a^2.$

$h^2 = 3a^2$ Subtract a^2 from both sides.

$h = \sqrt{3a^2}$ Take the positive square root of both sides.

$h = a\sqrt{3}$ Simplify the radical: $\sqrt{3a^2} = \sqrt{3}\sqrt{a^2} = a\sqrt{3}.$

We see that the altitude—the longer side of the 30°–60°–90° triangle—is $\sqrt{3}$ times as long as the shorter leg. Thus,

The length of the longer leg of a 30°–60°–90° triangle is the length of the shorter leg times $\sqrt{3}$.

EXAMPLE 4

30°–60°–90° triangles. Find the length of the hypotenuse and the longer leg of the right triangle shown in Figure 9-18.

Solution   Since the shorter leg of a 30°–60°–90° triangle is half as long as the hypotenuse, the hypotenuse is 12 centimeters long.

Since the length of the longer leg is the length of the shorter leg times $\sqrt{3}$, the longer leg is $6\sqrt{3}$ (about 10.39) centimeters long.

FIGURE 9-18

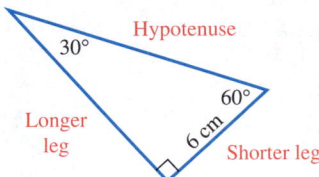

SELF CHECK   Find the length of the hypotenuse and the longer leg of a 30°–60°–90° triangle if the shorter leg is 8 centimeters long.

Answers: 16 cm, $8\sqrt{3}$ cm   ∎

EXAMPLE 5

Stretching exercises. A doctor prescribed the back-strengthening exercise shown in Figure 9-19(a) for a patient. The patient was instructed to raise his leg to an angle of 60° and hold the position for 10 seconds. If the patient's leg is 36 inches long, how high off the floor will his foot be when his leg is held at the proper angle?

Solution   In Figure 9-19(b), we see that a 30°–60°–90° triangle, which we will call triangle ABC, models the situation. Since the side opposite the 30° angle of a 30°–60°–90° triangle is half as long as the hypotenuse, side AC is 18 inches long.

Since the length of the side opposite the 60° angle is the length of the side opposite the 30° angle times $\sqrt{3}$, side BC is $18\sqrt{3}$, or about 31 inches long. So the patient's foot will be about 31 inches from the floor when his leg is in the proper stretching position.

FIGURE 9-19

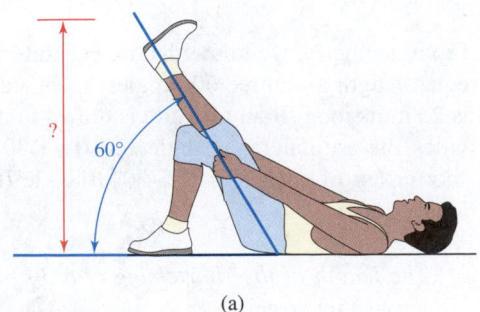

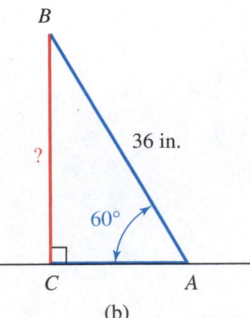

(a)   (b)   ∎

The Distance Formula

With the *distance formula*, we can find the distance between any two points that are graphed on a rectangular coordinate system.

To find the distance d between points $P(x_1, y_1)$ and $Q(x_2, y_2)$ shown in Figure 9-20, we construct the right triangle PRQ. The distance between P and R is $|x_2 - x_1|$, and the distance between R and Q is $|y_2 - y_1|$. We apply the Pythagorean theorem to the right triangle PRQ to get

$$[d(PQ)]^2 = |x_2 - x_1|^2 + |y_2 - y_1|^2$$ Read $d(PQ)$ as "the distance between P and Q."

$$= (x_2 - x_1)^2 + (y_2 - y_1)^2$$ Because $|x_2 - x_1|^2 = (x_2 - x_1)^2$ and $|y_2 - y_1|^2 = (y_2 - y_1)^2$.

We take the positive square root of both sides:

1. $d(PQ) = \sqrt{(x_2 - x_1)^2 + (y_2 - y_1)^2}$

Equation 1 is the called the **distance formula.**

FIGURE 9-20

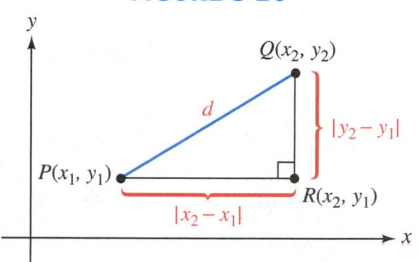

Distance formula

The distance between two points $P(x_1, y_1)$ and $Q(x_2, y_2)$ is given by the formula

$$d(PQ) = \sqrt{(x_2 - x_1)^2 + (y_2 - y_1)^2}$$

EXAMPLE 6

Using the distance formula. Find the distance between points $P(-2, 3)$ and $Q(4, -5)$.

Solution

To find the distance, we can use the distance formula by substituting 4 for x_2, -2 for x_1, -5 for y_2, and 3 for y_1.

$$d(PQ) = \sqrt{(x_2 - x_1)^2 + (y_2 - y_1)^2}$$
$$= \sqrt{[4 - (-2)]^2 + (-5 - 3)^2}$$
$$= \sqrt{(4 + 2)^2 + (-5 - 3)^2}$$
$$= \sqrt{6^2 + (-8)^2}$$
$$= \sqrt{36 + 64}$$
$$= \sqrt{100}$$
$$= 10$$

The distance between P and Q is 10 units.

SELF CHECK

Find the distance between $P(-2, -2)$ and $Q(3, 10)$. *Answer:* 13 ■

EXAMPLE 7

Robotics. Computerized robots are used to weld the parts of an automobile chassis together on an automated production line. To do this, an imaginary coordinate system is superimposed on the side of the vehicle, and the robot is programmed to move to specific positions to make each weld. See Figure 9-21, which is scaled in inches. If the welder unit moves from point to point at an average rate of speed of 48 in./sec, how long will it take it to move from position 1 to position 2?

FIGURE 9-21

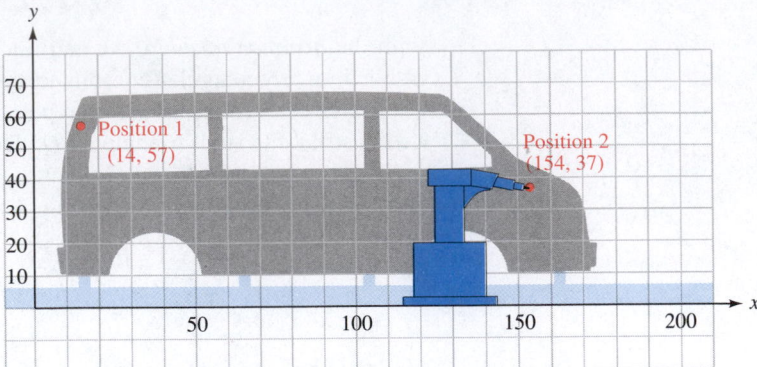

Solution   This is a uniform motion problem. We can use the formula $t = \frac{d}{r}$ to find the time it takes for the welder to move from position 1 (14, 57) to position 2 (154, 37).

We can use the distance formula to find the distance d that the welder unit moves.

$$d = \sqrt{(x_2 - x_1)^2 + (y_2 - y_1)^2}$$

$$d = \sqrt{(154 - 14)^2 + (37 - 57)^2} \quad \text{Substitute 154 for } x_2, \text{ 14 for } x_1, \text{ 37 for } y_2, \text{ and 57 for } y_1.$$

$$= \sqrt{140^2 + (-20)^2}$$

$$= \sqrt{20{,}000} \qquad\qquad 140^2 + (-20)^2 = 19{,}600 + 400 = 20{,}000.$$

$$= 100\sqrt{2} \qquad\qquad \text{Simplify: } \sqrt{20{,}000} = \sqrt{100 \cdot 100 \cdot 2} = 100\sqrt{2}.$$

The welder travels $100\sqrt{2}$ inches as it moves from position 1 to position 2. To find the time this will take, we divide the distance by the average rate of speed, 48 in./sec.

$$t = \frac{d}{r}$$

$$t = \frac{100\sqrt{2}}{48} \qquad \text{Substitute } 100\sqrt{2} \text{ for } d \text{ and 48 for } r.$$

$$t \approx 2.9 \qquad\quad \text{Use a calculator to find an approximation to the nearest tenth.}$$

It will take the welder about 2.9 seconds to travel from position 1 to position 2.   ∎

STUDY SET

Section 9.6

VOCABULARY

In Exercises 1–4, fill in the blanks to make the statements true.

1. In a right triangle, the side opposite the 90° angle is called the _____.

2. An _____ triangle is a right triangle with two legs of equal length.

3. The _____ theorem states that in any right triangle, the square of the hypotenuse is equal to the sum of the squares of the lengths of the two legs.

4. An _____ triangle has three sides of equal length and three 60° angles.

CONCEPTS

In Exercises 5–12, fill in the blanks to make the statements true.

5. If a and b are the lengths of two legs of a right triangle and c is the length of the hypotenuse, then

_____.

6. In any right triangle, the square of the hypotenuse is equal to the _____ of the squares of the two _____.

7. In an isosceles right triangle, the length of the hypotenuse is the length of one leg times _____.

8. The shorter leg of a 30°–60°–90° triangle is _____ as long as the hypotenuse.

9. The length of the longer leg of a 30°–60°–90° triangle is the length of the shorter leg times _____.

10. The formula to find the distance between two points $P(x_1, y_1)$ and $Q(x_2, y_2)$ is _____.

11. In a right triangle, the shorter leg is opposite the ___ angle, and the longer leg is opposite the ___ angle.

12. An isosceles triangle has ___ sides of equal length.

13. **a.** To solve the equation $c^2 = 20$, where c represents the length of the hypotenuse of a right triangle, how do we "undo" the operation performed on c?

   b. What is the first step when solving the equation $25 + b^2 = 81$?

14. When the lengths of the sides of the triangle in Illustration 1 are substituted into the equation $a^2 + b^2 = c^2$, the result is a false statement. Explain why.

$$a^2 + b^2 = c^2$$
$$2^2 + 4^2 = 5^2$$
$$4 + 16 = 25$$
$$20 = 25$$

ILLUSTRATION 1

NOTATION

In Exercises 15–16, complete each solution.

15. Evaluate $\sqrt{(-1-3)^2 + [2-(-4)]^2}$.

$$\sqrt{(-1-3)^2 + [2-(-4)]^2} = \sqrt{(-4)^2 + []^2}$$
$$= \sqrt{}$$
$$= \sqrt{ \cdot 13}$$
$$= \sqrt{13}$$
$$\approx 7.21$$

16. Solve $8^2 + 4^2 = c^2$.

$$ + 16 = c^2$$
$$ = c^2$$
$$\sqrt{} = \sqrt{c^2}$$
$$\sqrt{ \cdot 5} = c$$
$$\sqrt{5} = c$$
$$c \approx 8.94$$

PRACTICE

In Exercises 17–20, the lengths of two sides of the right triangle ABC shown in Illustration 2 are given. Find the length of the missing side.

17. $a = 6$ ft and $b = 8$ ft

18. $a = 10$ cm and $c = 26$ cm

19. $b = 18$ m and $c = 82$ m

20. $a = 14$ in. and $c = 50$ in.

ILLUSTRATION 2

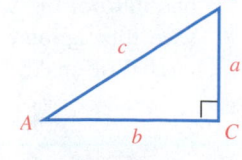

In Exercises 21–24, find the missing lengths in each triangle. Give the exact answer and then an approximation to two decimal places, when applicable.

21.

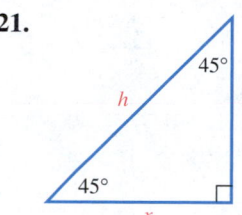

22.

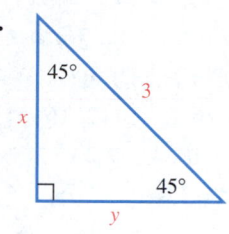

23.

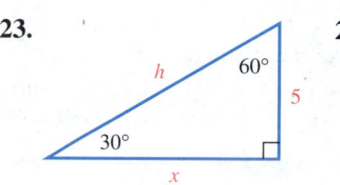

24.

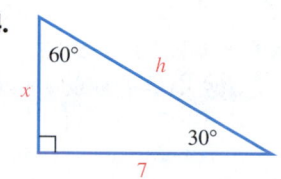

In Exercises 25–28, find the missing lengths in each triangle. Give the answer to two decimal places.

25.

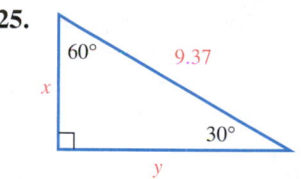

26.

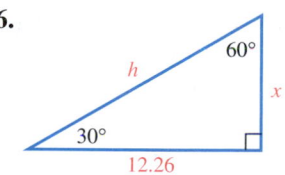

27.

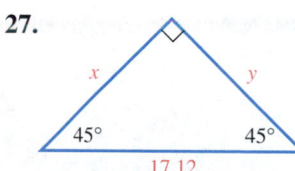

28.

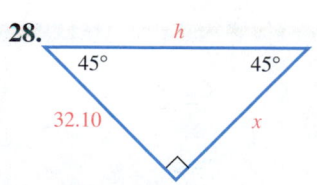

29. **GEOMETRY** Find the exact length of the diagonal (in blue) of one of the *faces* of the cube shown in Illustration 3.

30. **GEOMETRY** Find the exact length of the diagonal (in green) of the cube shown in Illustration 3.

ILLUSTRATION 3

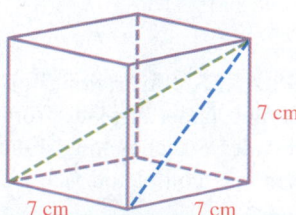

In Exercises 31–38, find the distance between P and Q.

31. $Q(0, 0)$, $P(3, -4)$

32. $Q(0, 0)$, $P(-12, 16)$

33. $P(-2, -8)$, $Q(3, 4)$

34. $P(-5, -2)$, $Q(7, 3)$

35. $P(6, 8)$, $Q(12, 16)$

36. $P(10, 4)$, $Q(2, -2)$

37. $Q(-3, 5)$, $P(-5, -5)$

38. $Q(2, -3)$, $P(4, -8)$.

39. ISOSCELES TRIANGLE Use the distance formula to show that a triangle with vertices $(-2, 4)$, $(2, 8)$, and $(6, 4)$ is isosceles.

40. RIGHT TRIANGLE Use the distance formula and the Pythagorean theorem to show that a triangle with vertices $(2, 3)$, $(-3, 4)$, and $(1, -2)$ is a right triangle.

APPLICATIONS

In Exercises 41–46, give the exact answer. Then give an approximation to two decimal places.

41. WASHINGTON, DC The square in Illustration 4 shows the 100-square-mile site selected by George Washington in 1790 to serve as a permanent capital for the United States. In 1847, the part of the district lying on the west bank of the Potomac was returned to Virginia. Find the coordinates of each corner of the original square that outlined the District of Columbia.

ILLUSTRATION 4

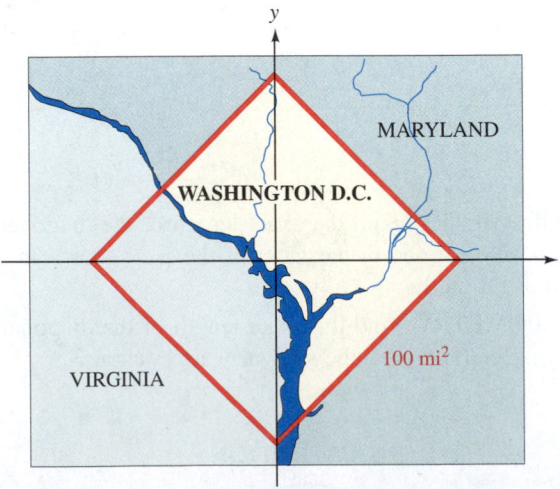

42. PAPER AIRPLANES Illustration 5 gives the directions for making a paper airplane from a square piece of paper with sides 8 inches long. Find the length l of the plane when it is completed.

ILLUSTRATION 5

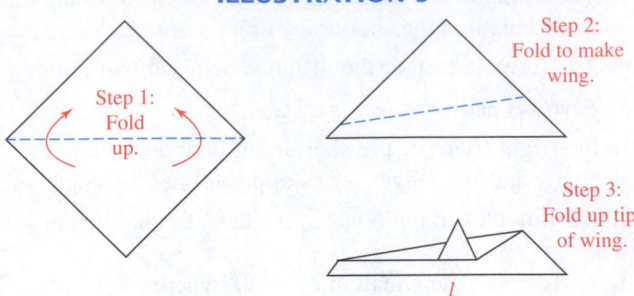

43. HARDWARE The sides of the regular hexagonal nut shown in Illustration 6 are 10 millimeters long. Find the height h of the nut.

ILLUSTRATION 6

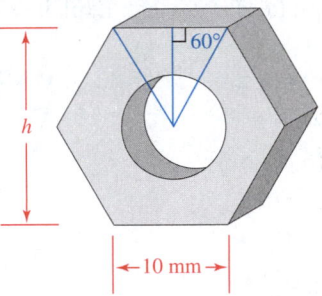

44. IRONING BOARD Find the height h of the ironing board shown in Illustration 7.

ILLUSTRATION 7

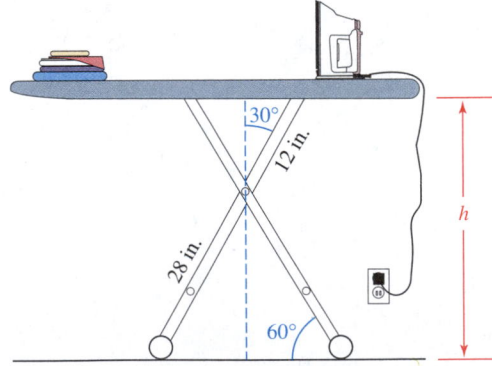

45. BASEBALL The baseball diamond shown in Illustration 8 is a square, 90 feet on a side. If the third baseman fields a ground ball 10 feet directly behind third base, how far must he throw the ball to throw a runner out at first base?

46. BASEBALL A shortstop fields a grounder at a point one-third of the way from second base to third base. (See Illustration 8.) How far will he have to throw the ball to make an out at first base?

ILLUSTRATION 8

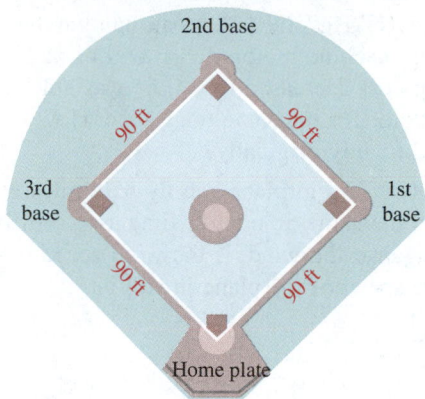

47. CLOTHESLINE A pair of damp jeans are hung on a clothesline to dry, as shown in Illustration 9. They pull the center down 1 foot. By how much is the line stretched?

ILLUSTRATION 9

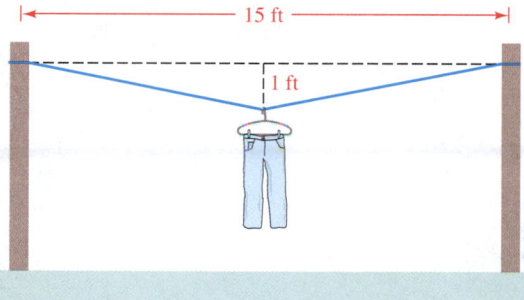

48. FIREFIGHTING The base of the 37-foot ladder in Illustration 10 is 9 feet from the wall. Will the top reach a window ledge that is 35 feet above the ground? Verify your result.

ILLUSTRATION 10

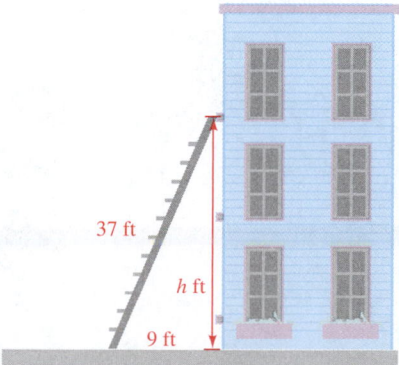

49. ART HISTORY A figure displaying some of the basic characteristics of Egyptian art is shown in Illustration 11. Use the distance formula to find the following dimensions of the drawing. Round your answers to two decimal places.

a. From the foot to the eye

b. From the belt to the hand holding the staff

c. From the shoulder to the symbol held in the hand

ILLUSTRATION 11

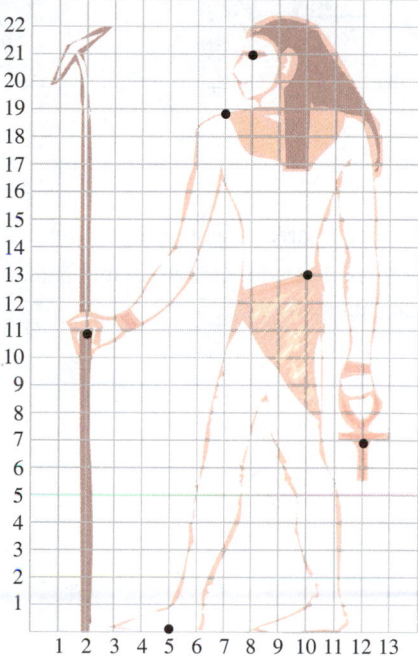

50. PACKAGING The diagonal d of a rectangular box with dimensions $a \times b \times c$ is given by

$$d = \sqrt{a^2 + b^2 + c^2}$$

Will the umbrella fit in the shipping carton in Illustration 12? Verify your result.

ILLUSTRATION 12

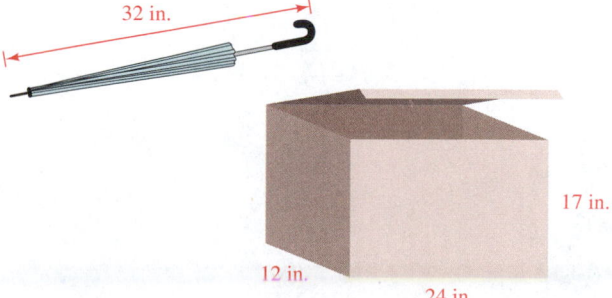

51. PACKAGING An archaeologist wants to ship a 34-inch femur bone. Will it fit in a 4-inch-tall box that has a 24-inch-square base? (See Exercise 50.) Verify your result.

52. TELEPHONE SERVICE The telephone cable in Illustration 13 (on the next page) runs from A to B to C to D. How much cable is required to run from A to D directly?

ILLUSTRATION 13

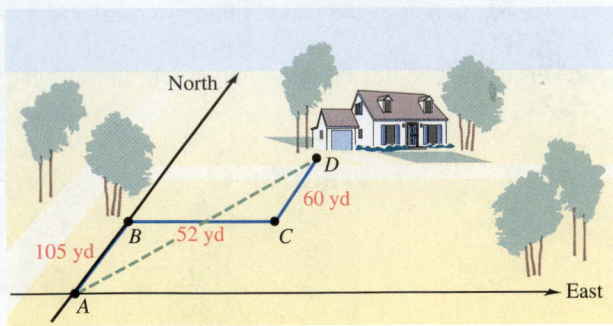

REVIEW

55. DISCOUNT BUYING A repairman purchased some washing-machine motors for a total of $224. When the unit cost decreased by $4, he was able to buy one extra motor for the same total price. How many motors did he buy originally?

56. AVIATION An airplane can fly 650 miles with the wind in the same amount of time as it can fly 475 miles against the wind. If the wind speed is 40 mph, find the speed of the plane in still air.

WRITING

53. State the Pythagorean theorem in words.

54. List the facts that you learned about special right triangles in this section.

Radicals

The expression $\sqrt[n]{a}$ is called a **radical expression.** In this chapter, we have discussed the properties and procedures used when simplifying radical expressions, solving radical equations, and writing radical expressions using rational exponents.

Expressions Containing Radicals

When working with expressions containing radicals, we must often apply the multiplication property and/or the division property of radicals to simplify the expression. Recall that

$$\sqrt[n]{ab} = \sqrt[n]{a}\sqrt[n]{b} \qquad \sqrt[n]{\frac{a}{b}} = \frac{\sqrt[n]{a}}{\sqrt[n]{b}} \quad (b \ne 0)$$

In Exercises 1–8, do each operation and simplify the expression.

1. Simplify: $\sqrt[3]{-54h^6}$.

2. Add: $2\sqrt[3]{64e} + 3\sqrt[3]{8e}$.

3. Subtract: $\sqrt{72} - \sqrt{200}$.

4. Multiply: $-4\sqrt[3]{5r^2s}\left(5\sqrt[3]{2r}\right)$.

5. Multiply: $\left(\sqrt{3s} - \sqrt{2t}\right)\left(\sqrt{3s} + \sqrt{2t}\right)$.

6. Multiply: $-\sqrt{3}\left(\sqrt{7} - \sqrt{5}\right)$.

7. Find the power: $\left(3\sqrt{2n} - 2\right)^2$.

8. Rationalize the denominator: $\dfrac{\sqrt[3]{9j}}{\sqrt[3]{3jk}}$.

Equations Containing Radicals

When solving radical equations, our objective is to rid the equation of the radical. This is achieved by using the *power rule:*

If x, y, and n are real numbers and $x = y$, then $x^n = y^n$.

If we raise both sides of an equation to the same power, the resulting equation might not be equivalent to the original equation. We must always check for extraneous solutions.

In Exercises 9–12, solve each radical equation, if possible.

9. $\sqrt{1 - 2g} = \sqrt{g + 10}$

10. $4 - \sqrt[3]{4 + 12x} = 0$

11. $\sqrt{y + 2} - 4 = -y$

12. $\sqrt[4]{12t + 4} + 2 = 0$

Radicals and Rational Exponents

Radicals can be written using rational (fractional) exponents, and exponential expressions having fractional exponents can be written in radical form. To do this, we use two rules for exponents introduced in this chapter.

$$x^{1/n} = \sqrt[n]{x} \qquad x^{m/n} = \sqrt[n]{x^m} = \left(\sqrt[n]{x}\right)^m$$

13. Express using a rational exponent: $\sqrt[3]{3}$.

14. Express in radical form: $5a^{2/5}$.

Accent on Teamwork

Section 9.1

A spiral of roots To do this project, you will need a piece of poster board, a protractor, a yardstick, and a pencil. Begin by drawing an isosceles right triangle near the right margin of the poster board. Label the length of each leg as 1 unit. (See Illustration 1.) Use the Pythagorean theorem to determine the length of the hypotenuse. Draw a second right triangle using the hypotenuse of the first triangle as one leg. Draw its second leg with a length of 1 unit. Find the length of the hypotenuse of triangle 2. Continue this process of creating right triangles, using the previous hypotenuse as one leg and drawing a new second leg of length 1 unit each time. Calculate the length of the resulting hypotenuse. What patterns, if any, do you see?

ILLUSTRATION 1

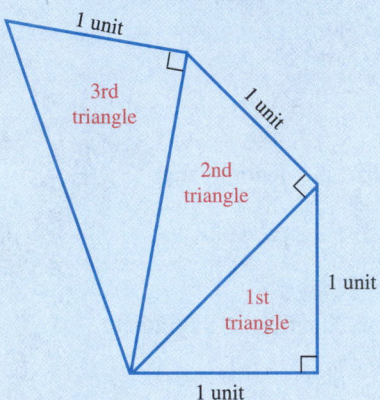

1 unit

3rd triangle

1 unit

2nd triangle

1 unit

1st triangle

1 unit

1 unit

Section 9.2

Solving radical equations In this chapter, we solved equations that contained two radicals. The radicals in those equations always had the same index. That is not the case for the following two equations:

$$\sqrt[3]{2x} = \sqrt{x} \qquad \sqrt[4]{x} = \sqrt{\frac{x}{4}}$$

Brainstorm in your group to see if you can come up with a procedure that can be used to solve these equations. What are their solutions?

Section 9.3

Exponents Use the $\boxed{y^x}$ key on your calculator to approximate each of the following exponential expressions. Then write them in order from least to greatest. Which of the exponents are not rational exponents? Could you have written the expressions in increasing order without having to approximate them? Explain.

1. $3^{0.999}$	2. $3^{3.9}$
3. $3^{1\frac{3}{4}}$	4. $3^{\sqrt{2}}$
5. $3^{1.7}$	6. 3^{π}
7. $3^{\frac{2}{3}}$	8. $3^{\frac{4}{3}}$
9. $3^{2\sqrt{3}}$	10. $3^{0.1}$

Section 9.4

Common errors In each addition or subtraction problem below, tell what **mistake** was made. Compare each problem to a similar one involving variables to help clarify your explanation. For example, compare Problem 1 to $2a + 3a$ to help explain the correct procedure that should be used to simplify the expression.

1. $2\sqrt{5x} + 3\sqrt{5x} = 5\sqrt{10x}$
2. $30 + 30\sqrt[4]{2} = 60\sqrt[4]{2}$
3. $7\sqrt[3]{y^2} - 5\sqrt[3]{y^2} = 2$
4. $6\sqrt{11ab} - 3\sqrt{5ab} = 3\sqrt{6ab}$

Section 9.5

Multiplying radicals In this chapter, when we were asked to find the product of two radical expressions, the radicals always had the same index. Brainstorm in your group to see if you can come up with a procedure that can be used to find

$$\sqrt[3]{3} \cdot \sqrt{3}$$

Keep in mind two things: The indices must be the same to apply the multiplication property of radicals, and radical expressions can be written using rational exponents.

Section 9.6

Graphing in three dimensions A point is located in three-space by plotting ordered triples of numbers (x, y, z). The point $(2, 3, 4)$ is plotted in Illustration 2. In three-space, the distance formula is $d = \sqrt{(x_2 - x_1)^2 + (y_2 - y_1)^2 + (z_2 - z_1)^2}$. Use this formula to find the distance between **a.** $(3, 2, 1)$ and $(13, 12, 11)$ and **b.** $(-1, -2, -4)$ and $(6, -2, 5)$.

ILLUSTRATION 2

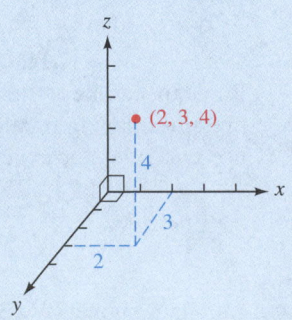

z

• $(2, 3, 4)$

4

x

3

2

y

Section 9.1

Radical Expressions and Radical Functions

CONCEPTS

The number b is a *square root of a* if $b^2 = a$.

If $x > 0$, the *principal square root of x* is the positive square root of x, denoted $\sqrt{x}$. If x can be any real number, then
$\sqrt{x^2} = |x|$.

The *cube root of x* is denoted as $\sqrt[3]{x}$ and is defined by

$$\sqrt[3]{x} = y \text{ if } y^3 = x$$

If n is an even natural number,

$$\sqrt[n]{a^n} = |a|$$

If n is an odd natural number,

$$\sqrt[n]{a^n} = a$$

If n is a natural number greater than 1 and x is a real number, then

- If $x > 0$, then $\sqrt[n]{x}$ is the positive number such that $\left(\sqrt[n]{x}\right)^n = x$.
- If $x = 0$, then $\sqrt[n]{x} = 0$.
- If $x < 0$, and n is odd, $\sqrt[n]{x}$ is the real number such that $\left(\sqrt[n]{x}\right)^n = x$.
- If $x < 0$, and n is even, $\sqrt[n]{x}$ is not a real number.

REVIEW EXERCISES

1. Simplify each radical expression, if possible. Assume that x can be any real number.
   a. $\sqrt{49}$
   b. $-\sqrt{121}$
   c. $\sqrt{\dfrac{225}{49}}$
   d. $\sqrt{-4}$
   e. $\sqrt{0.01}$
   f. $\sqrt{25x^2}$
   g. $\sqrt{x^8}$
   h. $\sqrt{x^2 + 4x + 4}$

2. Simplify each radical expression.
   a. $\sqrt[3]{-27}$
   b. $-\sqrt[3]{216}$
   c. $\sqrt[3]{64a^6b^3}$
   d. $\sqrt[3]{\dfrac{s^9}{125}}$

3. Simplify each radical expression, if possible. Assume that x and y can be any real number.
   a. $\sqrt[4]{625}$
   b. $\sqrt[5]{-32}$
   c. $\sqrt[4]{256x^8y^4}$
   d. $\sqrt{(-22y)^2}$
   e. $-\sqrt[4]{\dfrac{1}{16}}$
   f. $\sqrt[6]{-1}$
   g. $\sqrt{0}$
   h. $\sqrt[3]{0}$

4. GEOMETRY The side of a square with area A square feet is given by the function $s(A) = \sqrt{A}$. Find the *perimeter* of a square with an area of 144 square feet.

5. VOLUME OF A CUBE The total surface area of a cube is related to its volume V by the function $A(V) = 6\sqrt[3]{V^2}$. Find the surface area of a cube with a volume of 8 cm^3.

6. Graph each radical function. Find the domain and range.
   a. $f(x) = \sqrt{x + 2}$
   b. $f(x) = -\sqrt[3]{x} + 3$

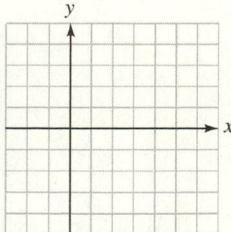

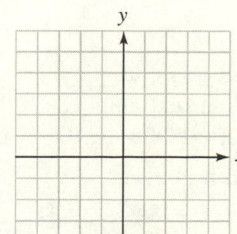

Section 9.2

The power rule:

   If $x = y$, then $x^n = y^n$.

Solving equations containing radicals:

1. Isolate one radical expression on one side of the equation.
2. Raise both sides of the equation to the power that is the same as the index.
3. Solve the resulting equation. If it still contains a radical, go back to step 1.
4. Check the solutions to eliminate *extraneous* solutions.

Radical Equations

7. Solve each equation. Write all solutions. Cross out those that are extraneous.

   a. $\sqrt{7x - 10} - 1 = 11$ **b.** $u = \sqrt{25u - 144}$

   c. $2\sqrt{y - 3} = \sqrt{2y + 1}$ **d.** $\sqrt{z + 1} + \sqrt{z} = 2$

   e. $\sqrt[3]{x^3 + 56} - 2 = x$ **f.** $\sqrt[4]{8x - 8} + 2 = 0$

8. Using the graphs of $f(x) = \sqrt{2x - 3}$ and $g(x) = -2x + 5$ in Illustration 1, estimate the solution of

   $$\sqrt{2x - 3} = -2x + 5$$

   Check the result.

ILLUSTRATION 1

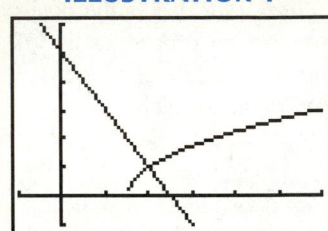

9. Solve each equation for the indicated variable.

   a. $r = \sqrt{\dfrac{A}{P} - 1}$ for P **b.** $h = \sqrt[3]{\dfrac{12I}{b}}$ for I

10. ELECTRONICS The current I (measured in amperes) and the power P (measured in watts) are related by the formula

    $$I = \sqrt{\dfrac{P}{R}}$$

    Find the resistance R in a circuit if the current used by an electrical appliance that is rated at 980 watts is 7.4 amperes.

Section 9.3

If n ($n > 1$) is a natural number and $\sqrt[n]{x}$ is a real number, then

$$x^{1/n} = \sqrt[n]{x}$$

If n is a natural number greater than 1 and x is a real number,

■ If $x > 0$, then $x^{1/n}$ is the positive number such that $(x^{1/n})^n = x$.
■ If $x = 0$, then $x^{1/n} = 0$.
■ If $x < 0$, and n is odd, then $x^{1/n}$ is the real number such that $(x^{1/n})^n = x$.
■ If $x < 0$ and n is even, then $x^{1/n}$ is not a real number.

Rational Exponents

11. Write each expression in radical form.

    a. $t^{1/2}$ **b.** $(5xy^3)^{1/4}$

12. Simplify each expression, if possible. Assume that all variables represent positive real numbers.

    a. $25^{1/2}$ **b.** $-36^{1/2}$

    c. $(-36)^{1/2}$ **d.** $1^{1/2}$

    e. $\left(\dfrac{9}{x^2}\right)^{1/2}$ **f.** $\left(\dfrac{1}{27}\right)^{1/3}$

    g. $(-8)^{1/3}$ **h.** $625^{1/4}$

    i. $(27a^3b)^{1/3}$ **j.** $(81c^4d^4)^{1/4}$

If m and n are positive integers, $x > 0$, and $\frac{m}{n}$ is in simplest form,

$$x^{m/n} = \sqrt[n]{x^m} = \left(\sqrt[n]{x}\right)^m$$

$$x^{-m/n} = \frac{1}{x^{m/n}}$$

$$\frac{1}{x^{-m/n}} = x^{m/n} \quad (x \neq 0)$$

The *rules for exponents* can be used to simplify expressions with fractional exponents.

13. Simplify each expression, if possible. Assume that all variables represent positive real numbers.
  a. $9^{3/2}$
  b. $8^{-2/3}$
  c. $-49^{5/2}$
  d. $\dfrac{1}{100^{-1/2}}$
  e. $\left(\dfrac{4}{9}\right)^{-3/2}$
  f. $\dfrac{1}{25^{5/2}}$
  g. $(25x^2y^4)^{3/2}$
  h. $(8u^6v^3)^{-2/3}$

14. Do the operations. Write answers without negative exponents. Assume that all variables represent positive real numbers.
  a. $5^{1/4}5^{1/2}$
  b. $a^{3/7}a^{-2/7}$
  c. $(k^{4/5})^{10}$
  d. $\dfrac{(4g^3h)^{1/2}}{(9gh^{-1})^{1/2}}$

15. Do the multiplications. Assume all variables represent positive real numbers.
  a. $u^{1/2}(u^{1/2} - u^{-1/2})$
  b. $v^{2/3}(v^{1/3} + v^{4/3})$

16. Simplify $\sqrt[4]{\dfrac{a^2}{25b^2}}$. (The variables represent positive real numbers.)

17. Substitute the x- and y-coordinates of each point labeled in the graph in Illustration 2 into the equation

$$x^{2/3} + y^{2/3} = 32$$

Show that each one satisfies the equation.

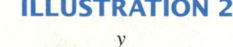
ILLUSTRATION 2

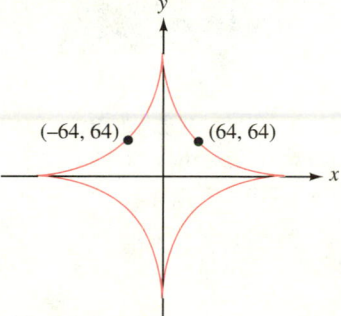

$(-64, 64)$ $(64, 64)$

Section 9.4

Simplifying and Combining Radical Expressions

A radical is in *simplest form* when:
1. No radicals appear in a denominator.
2. The radicand contains no fractions or negative numbers.
3. Each factor in the radicand appears to a power less than the index.

Properties of radicals:
1. Multiplication:

$$\sqrt[n]{ab} = \sqrt[n]{a}\sqrt[n]{b}$$

2. Division:

$$\sqrt[n]{\dfrac{a}{b}} = \dfrac{\sqrt[n]{a}}{\sqrt[n]{b}} \quad (b \neq 0)$$

18. Simplify each expression. Assume that all variables represent positive real numbers.
  a. $\sqrt{240}$
  b. $\sqrt[3]{54}$
  c. $\sqrt[4]{32}$
  d. $-2\sqrt[5]{-96}$
  e. $\sqrt{8x^5}$
  f. $\sqrt[3]{r^{17}}$
  g. $\sqrt[3]{16x^5y^4}$
  h. $3\sqrt[3]{27j^7k}$
  i. $\dfrac{\sqrt{32x^3}}{\sqrt{2x}}$
  j. $\sqrt{\dfrac{17xy}{64a^4}}$

Like radicals can be combined by addition and subtraction.

Radicals that are not similar can often be converted to radicals that are similar and then combined.

19. Simplify and combine like radicals. Assume that all variables represent positive real numbers.

a. $\sqrt{2} + 2\sqrt{2}$

b. $6\sqrt{20} - \sqrt{5}$

c. $2\sqrt[3]{3} - \sqrt[3]{24}$

d. $-\sqrt[4]{32} - 2\sqrt[4]{162}$

e. $2x\sqrt{8} + 2\sqrt{200x^2} + \sqrt{50x^2}$

f. $\sqrt[3]{54} - 3\sqrt[3]{16} + 4\sqrt[3]{128}$

20. SEWING A corner of fabric is folded over to form a collar and stitched down as shown in Illustration 3. From the dimensions given in the figure, determine the exact number of inches of stitching that must be made. Then give an approximation to one decimal place. (All measurements are in inches.)

ILLUSTRATION 3

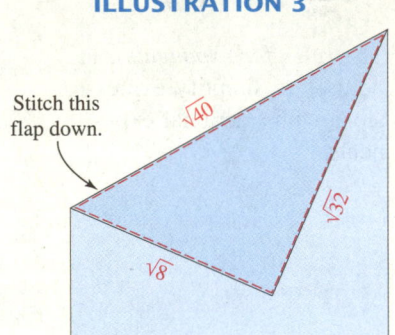

Stitch this flap down. $\sqrt{40}$ $\sqrt{32}$ $\sqrt{8}$

Section 9.5

Multiplying and Dividing Radical Expressions

If two radicals have the same index, they can be multiplied:

$$\sqrt[n]{a}\,\sqrt[n]{b} = \sqrt[n]{ab}$$

21. Simplify each expression. Assume that all variables represent positive real numbers.

a. $\sqrt{7}\sqrt{7}$

b. $(2\sqrt{5})(3\sqrt{2})$

c. $(-2\sqrt{8})^2$

d. $2\sqrt{6}\sqrt{216}$

e. $\sqrt{9x}\sqrt{x}$

f. $(\sqrt{33})^2$

g. $-\sqrt[3]{2x^2}\sqrt[3]{4x}$

h. $4\sqrt[3]{9}\sqrt[3]{9}$

i. $\sqrt{2}(\sqrt{8} - 3)$

j. $-\sqrt[4]{256x^5y^{11}}\sqrt[4]{625x^9y^3}$

k. $(\sqrt{3b} + \sqrt{3})^2$

l. $(2\sqrt{u} + 3)(3\sqrt{u} - 4)$

If a radical appears in a denominator of a fraction, or if a radicand contains a fraction, we can write the radical in simplest form by *rationalizing the denominator.*

To *rationalize the binomial denominator* of a fraction, multiply the numerator and the denominator by the conjugate of the binomial in the denominator.

22. Rationalize each denominator.

a. $\dfrac{10}{\sqrt{3}}$

b. $\sqrt{\dfrac{3}{5}}$

c. $\dfrac{x}{\sqrt{xy}}$

d. $\dfrac{\sqrt[3]{uv}}{\sqrt[3]{u^5v^7}}$

e. $\dfrac{2}{\sqrt{2} - 1}$

f. $\dfrac{\sqrt{a} + 1}{\sqrt{a} - 1}$

23. Rationalize each numerator.

a. $\dfrac{3 - \sqrt{x}}{2}$

b. $\dfrac{\sqrt{a} - \sqrt{b}}{\sqrt{a}}$

24. VOLUME The formula relating the radius r of a sphere and its volume V is $r = \sqrt[3]{\dfrac{3V}{4\pi}}$. Write the radical in simplest form.

Section 9.6

Geometric Applications of Radicals

The Pythagorean theorem:

If *a* and *b* are the lengths of the *legs* of a right triangle and *c* is the length of the *hypotenuse,* then
$a^2 + b^2 = c^2$.

25. CARPENTRY The gable end of the roof shown in Illustration 4 is divided in half by a vertical brace, 8 feet in height. Find the length of the roof line.

26. SAILING A technique called *tacking* allows a sailboat to make progress into the wind. A sailboat follows the course in Illustration 5. Find *d*, the distance the boat advances into the wind after tacking.

ILLUSTRATION 4

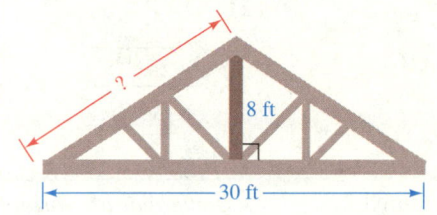

ILLUSTRATION 5

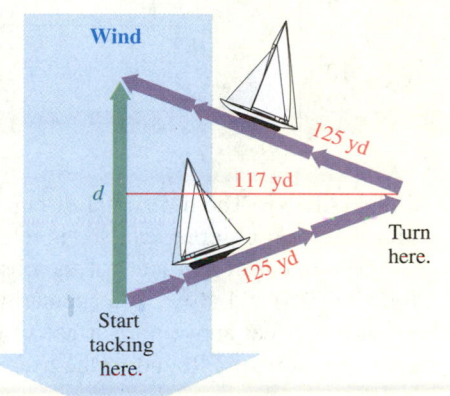

In an *isosceles right triangle,* the length of the hypotenuse is the length of one leg times $\sqrt{2}$.

27. Find the length of the hypotenuse of an isosceles right triangle whose legs measure 7 meters.

The shorter leg of a *30°–60°–90° triangle* (the side opposite the 30° angle) is half as long as the hypotenuse. The longer leg (the side opposite the 60° angle) is the length of the shorter leg times $\sqrt{3}$.

28. The hypotenuse of a 30°–60°–90° triangle measures $12\sqrt{3}$ centimeters. Find the length of each leg.

29. Find *x* to two decimal places.
  a.

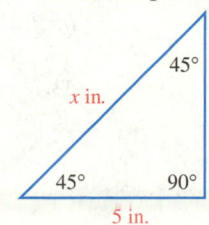

  b.

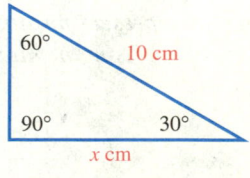

The distance formula:

$$d(PQ) = \sqrt{(x_2 - x_1)^2 + (y_2 - y_1)^2}$$

30. Find the distance between points *P* and *Q*.
  a. $P(0, 0)$ and $Q(5, -12)$
  b. $P(-4, 6)$ and $Q(-2, 8)$

1. Complete the table of values for $f(x) = \sqrt{x-1}$. Then graph the function. Round to the nearest hundredth when necessary.

x	$f(x)$
1	
2	
3	
5	
10	
12	
17	

2. SUBMARINE The horizontal distance (measured in miles) that an observer can see is related to the height h (measured in feet) of the observer by the function $d(h) = 1.4\sqrt{h}$. If a submarine's periscope extends 4.7 feet above the surface of the ocean, how far is the horizon? Round to the nearest tenth.

In Problems 3–5, solve and check each solution.

3. $2\sqrt{x} = \sqrt{x+1}$

4. $\sqrt[3]{6n+4} - 4 = 0$

5. $1 - \sqrt{u} = \sqrt{u-3}$

6. Solve $r = \sqrt[3]{\dfrac{GMt^2}{4\pi^2}}$ for G.

In Problems 7–12, simplify each expression. Assume that all variables represent positive real numbers, and write answers without using negative exponents.

7. $16^{1/4}$

8. $27^{2/3}$

9. $36^{-3/2}$

10. $\left(-\dfrac{8}{27}\right)^{-2/3}$

11. $\dfrac{2^{5/3}2^{1/6}}{2^{1/2}}$

12. $\dfrac{(8x^3y)^{1/2}(8xy^5)^{1/2}}{(x^3y^6)^{1/3}}$

In Problems 13–16, simplify each expression. Assume that the variables are unrestricted.

13. $\sqrt{x^2}$

14. $\sqrt{8x^2}$

15. $\sqrt[3]{54x^5}$

16. $\sqrt{18x^4y^8}$

In Problems 17–24, simplify each expression. Assume that all variables represent positive real numbers.

17. $\sqrt[3]{-64x^3y^6}$

18. $\sqrt{\dfrac{4a^2}{9}}$

19. $\sqrt[4]{-16}$

20. $\sqrt[5]{(t+8)^5}$

21. $\sqrt{48}$

22. $\sqrt{250x^3y^5}$

23. $\dfrac{\sqrt[3]{24x^{15}y^4}}{\sqrt[3]{y}}$

24. $\sqrt{\dfrac{3a^5}{48a^7}}$

In Problems 25–28, simplify and combine like radicals. Assume that all variables represent positive real numbers.

25. $\sqrt{12} - \sqrt{27}$

26. $2\sqrt[3]{40} - \sqrt[3]{5{,}000} + 4\sqrt[3]{625}$

27. $2\sqrt{48y^5} - 3y\sqrt{12y^3}$

28. $\sqrt[4]{768z^5} + z\sqrt[4]{48z}$

In Problems 29–30, do each operation and simplify, if possible. All variables represent positive real numbers.

29. $-2\sqrt{xy}\left(3\sqrt{x} + \sqrt{xy^3}\right)$

30. $\left(3\sqrt{2} + \sqrt{3}\right)\left(2\sqrt{2} - 3\sqrt{3}\right)$

In Problems 31–32, rationalize each denominator.

31. $\dfrac{1}{\sqrt{5}}$

32. $\dfrac{3t-1}{\sqrt{3t}-1}$

In Problems 33–34, rationalize each numerator.

33. $\dfrac{\sqrt[3]{9}}{6}$

34. $\dfrac{\sqrt{5}+3}{-4\sqrt{2}}$

In Problems 35–36, find x to two decimal places.

35.

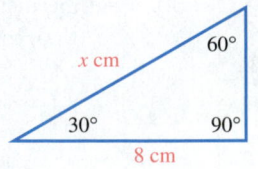

36.

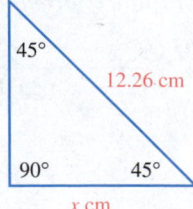

37. Find the distance between $(-2, 5)$ and $(22, 12)$.

38. PENDULUM The time t, in seconds, it takes for a pendulum to swing back and forth to complete one period is given by the formula $t = 2\pi\sqrt{\frac{l}{32}}$, where l is the length of the pendulum in feet. Find the exact period of a 4-foot-long pendulum. Then approximate the period to the nearest tenth.

39. SHIPPING CRATE The diagonal brace on the shipping crate in Illustration 1 is 53 inches. Find the height h of the crate.

40. Explain why, without having to perform any algebraic steps, it is obvious that the equation $\sqrt{x - 8} = -10$ has no solutions.

ILLUSTRATION 1

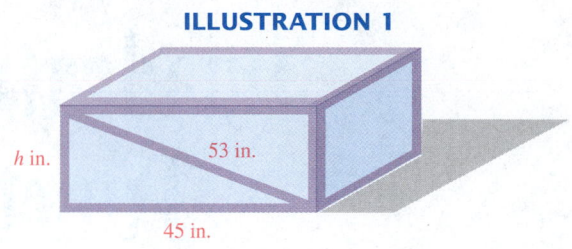

h in.   53 in.   45 in.

10

Quadratic Equations, Functions, and Inequalities

CAMPUS CONNECTION

The Art Department

In a watercolor class, students learn that light, shadow, color, and perspective are fundamental components of an attractive painting. They also learn that the appropriate matting and frame can enhance their work. In this section, we will use mathematics to determine the dimensions of a uniform matting that is to have the same area as the picture it frames. To do this, we will write a quadratic equation and then solve it by *completing the square*. The technique of completing the square can be used to derive *the quadratic formula*. This formula is a valuable algebraic tool for solving any quadratic equation.

We have previously seen how to solve quadratic equations by factoring. In this chapter, we will discuss more general methods for solving quadratic equations, and we will consider the graphs of quadratic functions.

▶ 10.1   Completing the Square

In this section, you will learn about

> A review of solving quadratic equations by factoring ■ The square root property ■ Completing the square ■ Solving equations by completing the square ■ Problem solving

Introduction   We have seen that equations that involve first-degree polynomials, such as $12x - 4 = 0$, are called *linear equations*. We have also seen that equations that involve second-degree polynomials, such as $12x^2 - 4x = 0$, are called *quadratic equations*. In Chapter 5, we learned how to solve quadratic equations by factoring and by the quadratic formula. In this section, we will introduce completing the square, a method of solving quadratic equations that can be used to develop the quadratic formula.

A Review of Solving Quadratic Equations by Factoring

A **quadratic equation** is an equation of the form $ax^2 + bx + c = 0$ $(a \neq 0)$, where a, b, and c are real numbers. We have discussed how to solve quadratic equations by factoring. For example, to solve $6x^2 - 7x - 3 = 0$, we proceed as follows:

$$6x^2 - 7x - 3 = 0$$

$$(2x - 3)(3x + 1) = 0 \qquad \text{Factor.}$$

$$2x - 3 = 0 \quad \text{or} \quad 3x + 1 = 0 \qquad \text{Set each factor equal to 0.}$$

$$x = \frac{3}{2} \qquad\qquad x = -\frac{1}{3} \qquad \text{Solve each linear equation.}$$

Many expressions do not factor as easily as $6x^2 - 7x - 3$. For example, it would be difficult to solve $2x^2 + 4x + 1 = 0$ by factoring, because $2x^2 + 4x + 1$ cannot be factored by using only integers. With this in mind, we will now develop another method of solving quadratic equations. It is based on the **square root property.**

The Square Root Property

To develop general methods for solving all quadratic equations, we first consider the equation $x^2 = c$. If $c > 0$, we can find the real solutions of $x^2 = c$ as follows:

$$x^2 = c$$

$$x^2 - c = 0 \qquad\qquad \text{Subtract } c \text{ from both sides.}$$

$$x^2 - \left(\sqrt{c}\right)^2 = 0 \qquad\qquad \text{Replace } c \text{ with } \left(\sqrt{c}\right)^2\text{, since } c = \left(\sqrt{c}\right)^2.$$

$$\left(x + \sqrt{c}\right)\left(x - \sqrt{c}\right) = 0 \qquad\qquad \text{Factor the difference of two squares.}$$

$$x + \sqrt{c} = 0 \qquad \text{or} \quad x - \sqrt{c} = 0 \qquad \text{Set each factor equal to 0.}$$

$$x = -\sqrt{c} \qquad\qquad x = \sqrt{c} \qquad \text{Solve each linear equation.}$$

The two solutions of $x^2 = c$ are $x = \sqrt{c}$ and $x = -\sqrt{c}$.

Square root property

If $c > 0$, the equation $x^2 = c$ has two real solutions. They are

$$x = \sqrt{c} \quad \text{and} \quad x = -\sqrt{c}$$

EXAMPLE 1

Solving a quadratic equation using the square root property.
Solve $x^2 - 12 = 0$.

Solution   We can write the equation as $x^2 = 12$ and use the square root property to solve it.

$$x^2 - 12 = 0$$
$$x^2 = 12 \qquad \text{Add 12 to both sides.}$$
$$x = \sqrt{12} \quad \text{or} \quad x = -\sqrt{12} \qquad \text{Use the square root property.}$$
$$x = 2\sqrt{3} \qquad\qquad x = -2\sqrt{3} \qquad \text{Simplify: } \sqrt{12} = \sqrt{4}\sqrt{3} = 2\sqrt{3}.$$

Verify that each solution satisfies the original equation.
   We can write the two solutions of the equation in a more concise form as $x = \pm 2\sqrt{3}$, where $\pm$ is read as "plus or minus."

SELF CHECK   Solve $x^2 - 18 = 0$.   *Answer:* $\pm 3\sqrt{2}$ ∎

EXAMPLE 2

Phonograph records. Before compact disc (CD) technology, one way of recording music was by engraving grooves on thin vinyl discs called records. (See Figure 10-1.) The vinyl discs used for long-playing records had a surface area of about 111 square inches per side and were played at $33\frac{1}{3}$ revolutions per minute on a turntable. What is the radius of a long-playing record?

FIGURE 10-1

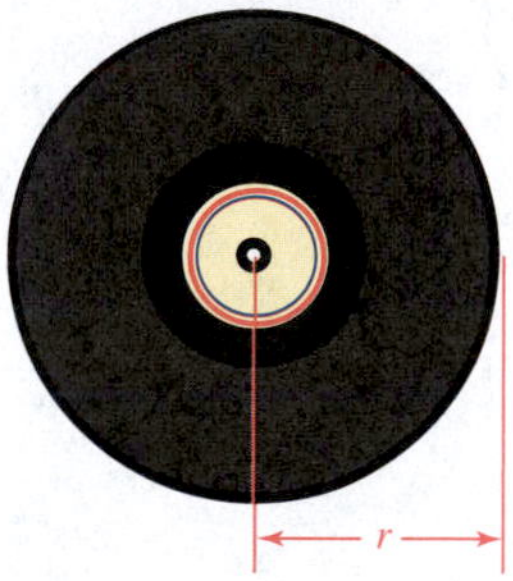

Solution   The relationship between the area of a circle and its radius is given by the formula $A = \pi r^2$. We can find the radius of a record by substituting 111 for A and solving for r.

$$A = \pi r^2 \qquad \text{The formula for the area of a circle.}$$
$$111 = \pi r^2 \qquad \text{Substitute 111 for } A.$$
$$\frac{111}{\pi} = r^2 \qquad \text{To undo the multiplication by } \pi, \text{ divide both sides by } \pi.$$
$$r = \sqrt{\frac{111}{\pi}} \quad \text{or} \quad r = -\sqrt{\frac{111}{\pi}} \qquad \text{Use the square root property. Since the radius of the record cannot be negative, discard the second solution.}$$

The radius of a record is $\sqrt{\frac{111}{\pi}}$ inches—to the nearest tenth, 5.9 inches.   ∎

EXAMPLE 3

Using the square root property to solve an equation. Solve $(x - 3)^2 = 16$.

Solution
$$(x - 3)^2 = 16$$
$$x - 3 = \sqrt{16} \quad \text{or} \quad x - 3 = -\sqrt{16} \qquad \text{Use the square root property.}$$
$$x - 3 = 4 \qquad\qquad x - 3 = -4 \qquad \text{Simplify: } \sqrt{16} = 4.$$
$$x = 3 + 4 \qquad\qquad x = 3 - 4 \qquad \text{Add 3 to both sides.}$$
$$x = 7 \qquad\qquad\quad x = -1 \qquad \text{Simplify.}$$

Verify that each solution satisfies the equation.

SELF CHECK Solve $(x + 2)^2 = 9$. *Answer:* 1, −5 ∎

Completing the Square

All quadratic equations can be solved by **completing the square.** This method involves the special products

$$x^2 + 2ax + a^2 = (x + a)^2 \quad \text{and} \quad x^2 - 2ax + a^2 = (x - a)^2$$

The trinomials $x^2 + 2ax + a^2$ and $x^2 - 2ax + a^2$ are both perfect square trinomials, because both factor as the square of a binomial. In each case, the coefficient of the first term is 1, and if we take one-half of the coefficient of x in the middle term and square it, we obtain the third term.

$$\left[\frac{1}{2}(2a)\right]^2 = a^2 \qquad \left[\frac{1}{2}(-2a)\right]^2 = (-a)^2 = a^2$$

EXAMPLE 4 **Completing the square.** Add a number to make each binomial a perfect square trinomial: **a.** $x^2 + 10x$, **b.** $x^2 - 6x$, and **c.** $x^2 - 11x$.

Solution **a.** To make $x^2 + 10x$ a perfect square trinomial, we find one-half of 10, square it, and add that result to $x^2 + 10x$.

$$x^2 + 10x + \left[\frac{1}{2}(10)\right]^2 = x^2 + 10x + (5)^2 \quad \text{Simplify: } \tfrac{1}{2}(10) = 5.$$

$$= x^2 + 10x + 25 \quad \text{Note that } x^2 + 10x + 25 = (x + 5)^2.$$

b. To make $x^2 - 6x$ a perfect square trinomial, we find one-half of −6, square it, and add that result to $x^2 - 6x$.

$$x^2 - 6x + \left[\frac{1}{2}(-6)\right]^2 = x^2 - 6x + (-3)^2 \quad \text{Simplify: } \tfrac{1}{2}(-6) = -3.$$

$$= x^2 - 6x + 9 \quad \text{Note that } x^2 - 6x + 9 = (x - 3)^2.$$

c. To make $x^2 - 11x$ a perfect square trinomial, we find one-half of −11, square it, and add that result to $x^2 - 11x$.

$$x^2 - 11x + \left[\frac{1}{2}(-11)\right]^2$$

$$= x^2 - 11x + \left(-\frac{11}{2}\right)^2 \quad \text{Simplify: } \tfrac{1}{2}(-11) = -\tfrac{11}{2}.$$

$$= x^2 - 11x + \frac{121}{4} \quad \text{Note that } x^2 - 11x + \tfrac{121}{4} = \left(x - \tfrac{11}{2}\right)^2.$$

SELF CHECK Add a number to $a^2 - 5a$ to make it a perfect square trinomial. *Answer:* $a^2 - 5a + \frac{25}{4}$ ∎

Solving Equations by Completing the Square

To solve an equation of the form $ax^2 + bx + c = 0$ by completing the square, we use the following steps.

Completing the square

1. Make sure that the coefficient of x^2 (the **leading coefficient**) is 1. If it is not, make it 1 by dividing both sides of the equation by the coefficient of x^2.
2. If necessary, add a number to both sides of the equation so that the constant term is on the right-hand side of the equals sign.
3. Complete the square:
   a. Find one-half of the coefficient of x and square it.
   b. Add that square to both sides of the equation.
4. Factor the trinomial square.
5. Solve the resulting equation using the square root property.

EXAMPLE 5

Completing the square. Use completing the square to solve $x^2 + 8x + 7 = 0$.

Solution **Step 1:** In this example, the coefficient of x^2 is an understood 1.

Step 2: We add -7 to both sides so that the constant is on the right-hand side of the equals sign:

$$x^2 + 8x + 7 = 0$$
$$x^2 + 8x = -7$$

Step 3: The coefficient of x is 8, one-half of 8 is 4, and $4^2 = 16$. To complete the square, we add 16 to both sides.

$$x^2 + 8x + \textcolor{red}{16} = \textcolor{red}{16} - 7$$
1. $x^2 + 8x + 16 = 9$ $\qquad$ Simplify: $16 - 7 = 9$.

Step 4: Since the left-hand side of Equation 1 is a perfect square trinomial, we can factor it to get $(x + 4)^2$.

$$x^2 + 8x + 16 = 9$$
2. $\qquad (x + 4)^2 = 9$

Step 5: We then solve Equation 2 by using the square root property.

$$x + 4 = \pm\sqrt{9}$$
$$x + 4 = 3 \quad \text{or} \quad x + 4 = -3$$
$$x = -1 \qquad\qquad x = -7$$

Verify that both solutions satisfy the equation. ∎

EXAMPLE 6

The leading coefficient is not 1. Solve $6x^2 + 5x - 6 = 0$.

Solution **Step 1:** To make the coefficient of x^2 equal to 1, we divide both sides of the equation by 6.

$$6x^2 + 5x - 6 = 0$$
$$\frac{6x^2}{6} + \frac{5}{6}x - \frac{6}{6} = \frac{0}{6} \qquad \text{Divide both sides by 6.}$$
$$x^2 + \frac{5}{6}x - 1 = 0 \quad \text{Simplify.}$$

Step 2: We add 1 to both sides so that the constant is on the right-hand side of the equals sign:

$$x^2 + \frac{5}{6}x = 1$$

Step 3: The coefficient of x is $\frac{5}{6}$, one-half of $\frac{5}{6}$ is $\frac{5}{12}$, and $\left(\frac{5}{12}\right)^2 = \frac{25}{144}$. To complete the square, we add $\frac{25}{144}$ to both sides.

$$x^2 + \frac{5}{6}x + \frac{25}{144} = 1 + \frac{25}{144}$$

3. $x^2 + \frac{5}{6}x + \frac{25}{144} = \frac{169}{144}$ Simplify: $1 + \frac{25}{144} = \frac{144}{144} + \frac{25}{144} = \frac{169}{144}$.

Step 4: Since the left-hand side of Equation 3 is a perfect square trinomial, we can factor it to get $\left(x + \frac{5}{12}\right)^2$.

4. $\left(x + \frac{5}{12}\right)^2 = \frac{169}{144}$

Step 5: We can solve Equation 4 by using the square root property.

$$x + \frac{5}{12} = \pm\sqrt{\frac{169}{144}}$$

$x + \frac{5}{12} = \frac{13}{12}$ or $x + \frac{5}{12} = -\frac{13}{12}$	Simplify: $\sqrt{\frac{169}{144}} = \frac{13}{12}$.
$x = -\frac{5}{12} + \frac{13}{12}$ $x = -\frac{5}{12} - \frac{13}{12}$	Subtract $\frac{5}{12}$ from both sides.
$x = \frac{8}{12}$ $x = -\frac{18}{12}$	Simplify.
$x = \frac{2}{3}$ $x = -\frac{3}{2}$	Simplify each fraction.

Verify that both solutions satisfy the original equation.   ■

E X A M P L E   7

Solving quadratic equations by completing the square. Solve $2x^2 + 4x + 1 = 0$.

Solution

$2x^2 + 4x + 1 = 0$

$x^2 + 2x + \frac{1}{2} = 0$	Divide both sides by 2 to make the coefficient of x^2 equal to 1.
$x^2 + 2x = -\frac{1}{2}$	Subtract $\frac{1}{2}$ from both sides.
$x^2 + 2x + 1 = 1 - \frac{1}{2}$	Square one-half of the coefficient of x and add it to both sides.
$(x + 1)^2 = \frac{1}{2}$	Factor and combine like terms.
$x + 1 = \pm\sqrt{\frac{1}{2}}$	Apply the square root property.

To write $\sqrt{\frac{1}{2}}$ in simplified radical form, we write it as a quotient of square roots and then rationalize the denominator.

$$x + 1 = \frac{\sqrt{2}}{2} \qquad\qquad x + 1 = -\frac{\sqrt{2}}{2} \qquad\qquad \sqrt{\frac{1}{2}} = \frac{\sqrt{1}}{\sqrt{2}} = \frac{1 \cdot \sqrt{2}}{\sqrt{2}\sqrt{2}} = \frac{\sqrt{2}}{2}.$$

$$x = -1 + \frac{\sqrt{2}}{2} \qquad\qquad x = -1 - \frac{\sqrt{2}}{2} \qquad \text{Subtract 1 from both sides.}$$

We can express each solution in an alternate form if we write -1 as a fraction with a denominator of 2.

$$x = -\frac{2}{2} + \frac{\sqrt{2}}{2} \qquad x = -\frac{2}{2} - \frac{\sqrt{2}}{2} \qquad \text{Write } -1 \text{ as } -\frac{2}{2}.$$

$$x = \frac{-2 + \sqrt{2}}{2} \qquad x = \frac{-2 - \sqrt{2}}{2} \qquad \begin{array}{l}\text{Add (subtract) the numerators and keep the}\\ \text{common denominator of 2.}\end{array}$$

The exact solutions are $x = \frac{-2 + \sqrt{2}}{2}$ or $x = \frac{-2 - \sqrt{2}}{2}$, or more concisely, $x = \frac{-2 \pm \sqrt{2}}{2}$. We can use a calculator to approximate them. To the nearest hundredth, $x \approx -0.29$ or $x \approx -1.71$.

SELF CHECK   Solve $3x^2 + 6x + 1 = 0$.   *Answer:* $\dfrac{-3 \pm \sqrt{6}}{3}$ ∎

 WARNING! Recall that to simplify a fraction, we divide out common *factors* of the numerator and denominator. In Example 7, since -2 is a *term* of the numerator of $\frac{-2 + \sqrt{2}}{2}$, no further simplification of this expression can be made. In the Self Check, 2 is a common factor and can be divided out: $\dfrac{-6 \pm 2\sqrt{6}}{6} = \dfrac{\cancel{2}(-3 \pm \sqrt{6})}{\cancel{2}(3)} = \dfrac{-3 \pm \sqrt{6}}{3}$.

ACCENT ON TECHNOLOGY   *Checking Solutions of Quadratic Equations*

We can use a graphing calculator to check the solutions of the quadratic equation $2x^2 + 4x + 1 = 0$ found in Example 7. After entering $Y_1 = 2x^2 + 4x + 1$, we call up the home screen by pressing $\boxed{2nd}$ QUIT. Then we press the $\boxed{VARS}$ key, arrow $\boxed{\blacktriangleright}$ to Y-VARS, and enter 1 to get the display shown in Figure 10-2(a). We evaluate $2x^2 + 4x + 1$ for $x = \frac{-2 + \sqrt{2}}{2}$ by inputting the solution using function notation, as shown in Figure 10-2(b). When $\boxed{ENTER}$ is pressed, the result of 0 is confirmation that $x = \frac{-2 + \sqrt{2}}{2}$ is a solution of the equation.

FIGURE 10-2

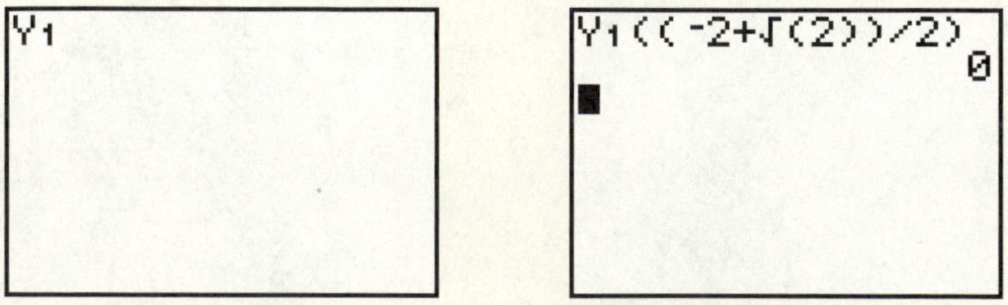

```
Y1
```

```
Y1((-2+√(2))/2)
                      0
■
```

(a)   (b)

Problem Solving

EXAMPLE 8

Graduation announcement. In creating the announcement shown in Figure 10-3, the graphic artist wants to follow two design elements.

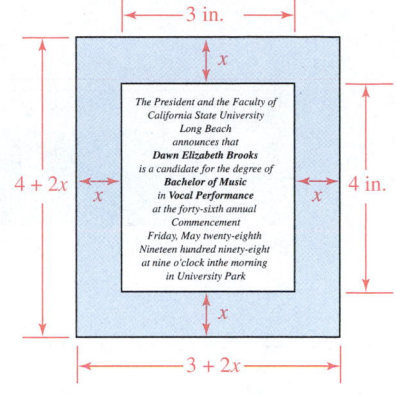

FIGURE 10-3

- A border of uniform width should surround the text.

- Equal areas should be devoted to the text and to the border.

To meet these requirements, how wide should the border be?

ANALYZE THE PROBLEM The text occupies $4 \cdot 3 = 12$ in.^{2} of space. The border must also have an area of 12 in.^{2}.

FORM AN EQUATION If we let x represent the width of the border, the length of the announcement is $(4 + 2x)$ inches and the width is $(3 + 2x)$ inches. We can now form the equation.

The area of the announcement	minus	the area of the text	equals	the area of the border.
$(4 + 2x)(3 + 2x)$	$-$	12	$=$	12

SOLVE THE EQUATION

$$(4 + 2x)(3 + 2x) - 12 = 12$$
$$12 + 8x + 6x + 4x^2 - 12 = 12 \quad \text{On the left-hand side, use the FOIL method.}$$
$$4x^2 + 14x = 12 \quad \text{Combine like terms.}$$
$$2x^2 + 7x - 6 = 0 \quad \text{Subtract 12 from both sides. Then divide both sides by 2.}$$

We note that the trinomial on the left-hand side does not factor. We will solve the equation by completing the square.

$$x^2 + \frac{7}{2}x - 3 = 0 \qquad \text{Divide both sides by 2 so that the coefficient of } x^2 \text{ is 1.}$$

$$x^2 + \frac{7}{2}x = 3 \qquad \text{Add 3 to both sides.}$$

$$x^2 + \frac{7}{2}x + \frac{49}{16} = 3 + \frac{49}{16} \qquad \text{One-half of } \frac{7}{2} \text{ is } \frac{7}{4}. \text{ Square } \frac{7}{4}, \text{ which is } \frac{49}{16}, \text{ and add it to both sides.}$$

$$\left(x + \frac{7}{4}\right)^2 = \frac{97}{16} \qquad \text{On the left-hand side, factor the trinomial. On the right-hand side, } 3 = \frac{3 \cdot 16}{1 \cdot 16} = \frac{48}{16} \text{ and } \frac{48}{16} + \frac{49}{16} = \frac{97}{16}.$$

$$x + \frac{7}{4} = \pm \frac{\sqrt{97}}{4} \qquad \text{Apply the square root property. On the right-hand side, } \sqrt{\frac{97}{16}} = \frac{\sqrt{97}}{\sqrt{16}} = \frac{\sqrt{97}}{4}.$$

$$x = -\frac{7}{4} + \frac{\sqrt{97}}{4} \quad \text{or} \quad x = -\frac{7}{4} - \frac{\sqrt{97}}{4} \qquad \text{Subtract } \frac{7}{4} \text{ from both sides.}$$

$$x = \frac{-7 + \sqrt{97}}{4} \qquad x = \frac{-7 - \sqrt{97}}{4} \qquad \text{Write each expression as a single fraction.}$$

STATE THE CONCLUSION The width of the border should be $\dfrac{-7 + \sqrt{97}}{4} \approx 0.71$ inch. (We discard the solution $\dfrac{-7 - \sqrt{97}}{4}$, since it is negative.)

CHECK THE RESULT If the border is 0.71 inch wide, the announcement has an area of about $5.42 \cdot 4.42 \approx 23.96$ in.^{2}. If we subtract the area of the text from the area of the announcement, we get $23.96 - 12 = 11.96$ in.^{2}. This represents the area of the border, which was to be 12 in.^{2}. The answer seems reasonable. ∎

STUDY SET

Section 10.1

VOCABULARY

In Exercises 1–4, fill in the blanks to make the statements true.

1. An equation of the form $ax^2 + bx + c = 0$ ($a \neq 0$) is called a _____ equation.

2. The symbol $\pm$ is read as "_____."

3. $x^2 + 6x + 9$ is called a _____ square trinomial because it factors as $(x + 3)^2$.

4. The _____ of x^2 in $x^2 - 12x + 36 = 0$ is 1, and the _____ term is 36.

CONCEPTS

In Exercises 5–8, fill in the blanks to make the statements true.

5. The solutions of $x^2 = c$, where $c > 0$, are _____ and _____ .

6. To complete the square on x in $x^2 + 6x$, find one-half of ___, square it to get ___, and add ___ to get _____.

7. In the quadratic equation $3x^2 - 2x + 5 = 0$, $a =$ ___, $b =$ ___, and $c =$ ___.

8. In the quadratic equation $ax^2 + bx + c = 0$, a cannot be ___.

9. Check to see if $-2 + \sqrt{2}$ is a solution of $x^2 + 4x + 2 = 0$.

10. Check to see if $-3\sqrt{2}$ is a solution of $x^2 - 18 = 0$.

11. Find one-half of the coefficient of x and then square it.
    a. $x^2 + 12x$
    b. $x^2 - 5x$
    c. $x^2 - \dfrac{x}{2}$

12. Add a number to make each binomial a perfect square trinomial. Then factor the result.
    a. $x^2 + 8x$
    b. $x^2 - 8x$
    c. $x^2 - x$

13. What is the first step in solving the equation $x^2 + 12x = 35$
    a. by the factoring method?
    b. by completing the square?

14. Solve the equation $x^2 = 16$
    a. by the factoring method.
    b. by the square root method.

15. Explain the **error** in the work shown below.

$$\frac{4 \pm \sqrt{3}}{8} = \frac{\overset{1}{\cancel{4}} \pm \sqrt{3}}{\underset{1}{\cancel{4} \cdot 2}} = \frac{1 \pm \sqrt{3}}{2}$$

16. Explain the **error** in the work shown below.

$$\frac{1 \pm \sqrt{5}}{5} = \frac{1 \pm \sqrt{\overset{1}{\cancel{5}}}}{\underset{1}{\cancel{5}}} = \frac{1 \pm 1}{1}$$

In Exercises 17–18, note that a and b are solutions of the equation $(x - a)(x - b) = 0$.

17. Find a quadratic equation with solutions of 3 and 5.

18. Find a quadratic equation with solutions of -4 and 6.

NOTATION

19. In solving a quadratic equation, a student obtains $x = \pm 2\sqrt{5}$.
    a. How many solutions are represented by this notation? List them.
    b. Approximate the solutions to the nearest hundredth.

20. In solving a quadratic equation, a student obtains
$$x = \frac{-5 \pm \sqrt{7}}{3}.$$

   a. How many solutions are represented by this notation? List them.

   b. Approximate the solutions to the nearest hundredth.

PRACTICE

In Exercises 21–28, use factoring to solve each equation.

21. $6x^2 + 12x = 0$ **22.** $5x^2 + 11x = 0$

23. $2y^2 - 50 = 0$ **24.** $4y^2 - 64 = 0$

25. $r^2 + 6r + 8 = 0$ **26.** $x^2 + 9x + 20 = 0$

27. $2z^2 = -2 + 5z$ **28.** $3x^2 = 8 - 10x$

In Exercises 29–40, use the square root property to solve each equation.

29. $x^2 = 36$ **30.** $x^2 = 144$

31. $z^2 = 5$ **32.** $u^2 = 24$

33. $3x^2 - 16 = 0$ **34.** $5x^2 - 49 = 0$

35. $(x + 1)^2 = 1$ **36.** $(x - 1)^2 = 4$

37. $(s - 7)^2 - 9 = 0$ **38.** $(t + 4)^2 = 16$

39. $(x + 5)^2 - 3 = 0$ **40.** $(x + 3)^2 - 7 = 0$

In Exercises 41–44, use the square root property to solve for the indicated variable. Assume that all variables represent positive numbers. Express all radicals in simplified form.

41. $2d^2 = 3h$ for d **42.** $2x^2 = d^2$ for d

43. $E = mc^2$ for c **44.** $A = \pi r^2$ for r

In Exercises 45–62, use completing the square to solve each equation.

45. $x^2 + 2x - 8 = 0$ **46.** $x^2 + 6x + 5 = 0$

47. $x + 1 = 2x^2$ **48.** $-2 = 2x^2 - 5x$

49. $6x^2 + x - 2 = 0$ **50.** $9 - 6r = 8r^2$

51. $x^2 + 8x + 6 = 0$ **52.** $x^2 + 6x + 4 = 0$

53. $x^2 - 2x - 17 = 0$ **54.** $x^2 + 10x - 7 = 0$

55. $3x^2 - 6x = 1$ **56.** $2x^2 - 6x = -3$

57. $4x^2 - 4x - 7 = 0$ **58.** $2x^2 - 8x + 5 = 0$

59. $2x^2 + 5x - 2 = 0$ **60.** $4x^2 - 4x - 1 = 0$

61. $\dfrac{7x + 1}{5} = -x^2$ **62.** $\dfrac{3x^2}{8} = \dfrac{1}{8} - x$

APPLICATIONS

63. FLAG In 1912, an executive order by President Taft fixed the overall width and length of the U.S. flag in the ratio 1 to 1.9. (See Illustration 1.) If 100 square feet of cloth are to be used to make a U.S. flag, estimate its dimensions to the nearest $\frac{1}{4}$ foot.

ILLUSTRATION 1

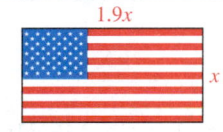
1.9x
x

64. MOVIE STUNTS According to the *Guinness Book of World Records, 1998,* stuntman Dan Koko fell a distance of 312 feet into an airbag after jumping from the Vegas World Hotel and Casino. The distance d in feet traveled by a free-falling object in t seconds is given by the formula $d = 16t^2$. To the nearest tenth of a second, how long did the stuntman's free fall last?

65. ACCIDENTS The height h (in feet) of an object that is dropped from a height of s feet is given by the formula $h = s - 16t^2$, where t is the time the object has been falling. A 5-foot-tall woman on a sidewalk looks directly overhead and sees a window washer drop a bottle from 4 stories up. How long does she have to get out of the way? Round to the nearest tenth. (A story is 12 feet.)

66. GEOGRAPHY The surface area S of a sphere is given by the formula $S = 4\pi r^2$, where r is the radius of the sphere. An almanac lists the surface area of the earth as 196,938,800 square miles. Assuming the earth to be spherical, what is its radius to the nearest mile?

67. AUTOMOBILE ENGINE As the piston shown in Illustration 2 moves upward, it pushes a "cylinder" of a gasoline/air mixture that is ignited by the spark plug. The formula that gives the volume of a cylinder is $V = \pi r^2 h$, where r is the radius and h the height. Find the radius of the piston (to the nearest hundredth of an inch) if it displaces 47.75 cubic inches of gasoline/air mixture as it moves from its lowest to its highest point.

ILLUSTRATION 2

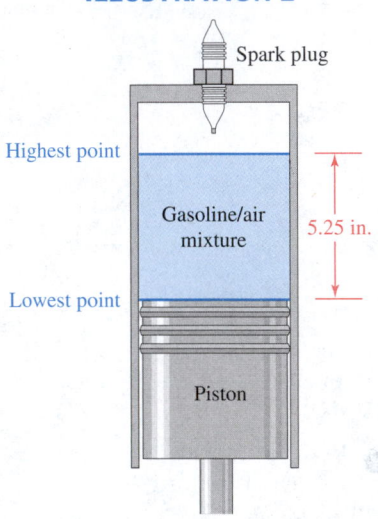

68. INVESTMENTS If P dollars are deposited in an account that pays an annual rate of interest r, then in n years, the amount of money A in the account is given by the formula $A = P(1 + r)^n$. A savings account was opened on January 3, 1996, with a deposit of $10,000 and closed on January 2, 1998, with an ending balance of $11,772.25. Find r, the rate of interest.

69. PICTURE FRAMING The matting around the picture in Illustration 3 has a uniform width. How wide is the matting if its area equals the area of the picture? Round to the nearest hundredth of an inch.

ILLUSTRATION 3

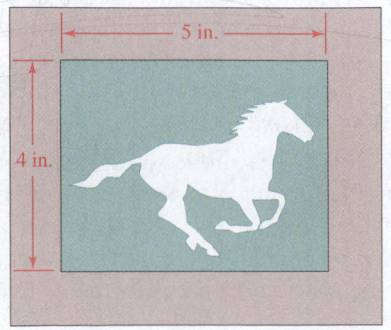

70. SWIMMING POOL See the advertisement in Illustration 4. How wide will the "free" concrete decking be if a uniform width is constructed around the perimeter of the pool? Round to the nearest hundredth of a yard. (*Hint:* Note the difference in units.)

ILLUSTRATION 4

71. DIMENSIONS OF A RECTANGLE A rectangle is 4 feet longer than it is wide, and its area is 20 square feet. Find its dimensions to the nearest tenth of a foot.

72. DIMENSIONS OF A TRIANGLE The height of a triangle is 4 meters longer than twice its base. Find the base and height if the area of the triangle is 10 square meters. Round to the nearest hundredth of a meter.

WRITING

73. Explain how to complete the square.

74. Tell why a cannot be 0 in the quadratic equation $ax^2 + bx + c = 0$.

REVIEW

Simplify each expression. All variables represent positive real numbers.

75. $\sqrt[3]{40a^3b^6}$

76. $\sqrt[3]{-27x^6}$

77. $\sqrt[8]{x^{24}}$

78. $\sqrt[4]{\dfrac{16}{625}}$

79. $\sqrt{175a^2b^3}$

80. $\sqrt{\dfrac{z^2}{16x^2}}$

▶ 10.2   The Quadratic Formula

In this section, you will learn about

> The quadratic formula ■ Solving quadratic equations using the quadratic formula ■ Problem solving

Introduction   We can solve any quadratic equation by the method of completing the square, but the work is often tedious. In this section, we will develop a formula, called the *quadratic formula,* that lets us solve quadratic equations with much less effort.

The Quadratic Formula

To develop a formula we can use to solve quadratic equations, we solve the general quadratic equation $ax^2 + bx + c = 0$ for $a > 0$.

$$ax^2 + bx + c = 0$$

$$\frac{ax^2}{a} + \frac{bx}{a} + \frac{c}{a} = \frac{0}{a} \qquad \text{Since } a > 0, \text{ we can divide both sides by } a.$$

$$x^2 + \frac{bx}{a} = -\frac{c}{a} \qquad \frac{0}{a} = 0; \text{ subtract } \frac{c}{a} \text{ from both sides.}$$

$$x^2 + \frac{b}{a}x + \left(\frac{b}{2a}\right)^2 = \left(\frac{b}{2a}\right)^2 - \frac{c}{a} \qquad \text{Complete the square on } x. \text{ Half of } \frac{b}{a} \text{ is } \frac{b}{2a}. \text{ Add } \left(\frac{b}{2a}\right)^2 \text{ to both sides.}$$

$$x^2 + \frac{b}{a}x + \frac{b^2}{4a^2} = \frac{b^2}{4a^2} - \frac{4ac}{4aa} \qquad \text{Remove parentheses and get a common denominator of } 4a^2 \text{ on the right-hand side.}$$

1.
$$\left(x + \frac{b}{2a}\right)^2 = \frac{b^2 - 4ac}{4a^2} \qquad \text{Factor the left-hand side and add the fractions on the right-hand side.}$$

We can solve Equation 1 using the square root property.

$$x + \frac{b}{2a} = \sqrt{\frac{b^2 - 4ac}{4a^2}} \qquad \text{or} \qquad x + \frac{b}{2a} = -\sqrt{\frac{b^2 - 4ac}{4a^2}}$$

$$x + \frac{b}{2a} = \frac{\sqrt{b^2 - 4ac}}{\sqrt{4a^2}} \qquad\qquad x + \frac{b}{2a} = -\frac{\sqrt{b^2 - 4ac}}{\sqrt{4a^2}}$$

$$x + \frac{b}{2a} = \frac{\sqrt{b^2 - 4ac}}{2a} \qquad\qquad x + \frac{b}{2a} = -\frac{\sqrt{b^2 - 4ac}}{2a}$$

$$x = -\frac{b}{2a} + \frac{\sqrt{b^2 - 4ac}}{2a} \qquad\qquad x = -\frac{b}{2a} - \frac{\sqrt{b^2 - 4ac}}{2a}$$

$$x = \frac{-b + \sqrt{b^2 - 4ac}}{2a} \qquad\qquad x = \frac{-b - \sqrt{b^2 - 4ac}}{2a}$$

If $a < 0$, it can be shown that the same two solutions are obtained. The result is called the **quadratic formula.**

The quadratic formula

The solutions of $ax^2 + bx + c = 0$ $(a \neq 0)$ are given by the formula

$$x = \frac{-b \pm \sqrt{b^2 - 4ac}}{2a} \qquad \text{Read the symbol } \pm \text{ as "plus or minus."}$$

WARNING! Be sure to draw the fraction bar under both parts of the numerator, and be sure to draw the radical sign exactly over $b^2 - 4ac$. Do not write the quadratic formula as

$$x = -b \pm \frac{\sqrt{b^2 - 4ac}}{2a} \quad \text{or as} \quad x = -b \pm \sqrt{\frac{b^2 - 4ac}{2a}}$$

Solving Quadratic Equations Using the Quadratic Formula

EXAMPLE 1

Solving quadratic equations. Solve $2x^2 - 3x - 5 = 0$ using the quadratic formula.

Solution   In this equation $a = 2$, $b = -3$, and $c = -5$.

$$x = \frac{-b \pm \sqrt{b^2 - 4ac}}{2a}$$

$$= \frac{-(-3) \pm \sqrt{(-3)^2 - 4(2)(-5)}}{2(2)} \qquad \text{Substitute 2 for } a, -3 \text{ for } b, \text{ and } -5 \text{ for } c.$$

$$= \frac{3 \pm \sqrt{9 + 40}}{4} \qquad \text{Simplify within the radical.}$$

$$= \frac{3 \pm \sqrt{49}}{4} \qquad \text{Simplify: } \sqrt{49} = 7.$$

$$= \frac{3 \pm 7}{4}$$

This result represents two solutions. To find the first solution, we add in the numerator. To find the second, we subtract in the numerator.

$$x = \frac{3 + 7}{4} \qquad \text{or} \qquad x = \frac{3 - 7}{4}$$

$$x = \frac{10}{4} \qquad\qquad x = \frac{-4}{4}$$

$$x = \frac{5}{2} \qquad\qquad x = -1$$

Verify that both solutions satisfy the original equation.

SELF CHECK   Solve $3x^2 - 5x - 2 = 0$.   *Answer:* $2, -\frac{1}{3}$ ■

When using the quadratic formula, we should write the equation in $ax^2 + bx + c = 0$ form (called **quadratic form**) so that a, b, and c can be determined.

EXAMPLE 2

Writing an equation in quadratic form first. Solve $\frac{1}{3}x^2 = -\frac{2}{3}x - \frac{1}{6}$.

Solution   We begin by writing the equation in quadratic form:

$$\frac{1}{3}x^2 = -\frac{2}{3}x - \frac{1}{6}$$

$$2x^2 = -4x - 1 \qquad \text{Multiply both sides by 6.}$$

$$2x^2 + 4x + 1 = 0 \qquad \text{Add } 4x \text{ and 1 to both sides.}$$

In this equation, $a = 2$, $b = 4$, and $c = 1$.

$$x = \frac{-b \pm \sqrt{b^2 - 4ac}}{2a}$$

$$= \frac{-4 \pm \sqrt{4^2 - 4(2)(1)}}{2(2)} \qquad \text{Substitute 2 for } a, \text{ 4 for } b, \text{ and 1 for } c.$$

$$= \frac{-4 \pm \sqrt{16 - 8}}{4} \qquad \text{Simplify within the radical.}$$

$$= \frac{-4 \pm \sqrt{8}}{4}$$

$$= \frac{-4 \pm 2\sqrt{2}}{4} \qquad \text{Simplify: } \sqrt{8} = \sqrt{4 \cdot 2} = 2\sqrt{2}.$$

$$= \frac{-2 \pm \sqrt{2}}{2} \qquad \frac{-4 \pm 2\sqrt{2}}{4} = \frac{2(-2 \pm \sqrt{2})}{4} = \frac{\overset{1}{2}(-2 \pm \sqrt{2})}{\underset{1}{2 \cdot 2}} = \frac{-2 \pm \sqrt{2}}{2}.$$

The solutions are $x = \dfrac{-2 + \sqrt{2}}{2}$ or $x = \dfrac{-2 - \sqrt{2}}{2}$. We can approximate the solutions using a calculator. To two decimal places, $x \approx -0.29$ or $x \approx -1.71$.

SELF CHECK   Solve $\frac{1}{2}x^2 = \frac{1}{3}x + \frac{1}{2}$.   *Answer:* $\dfrac{1 \pm \sqrt{10}}{3}$ ∎

Problem Solving

EXAMPLE 3

Taking a shortcut. Instead of using the existing hallways, students are wearing a path through a planted quad area to walk 195 feet directly from the classrooms to the cafeteria. See Figure 10-4. If the length of the hallway from the office to the cafeteria is 105 feet longer than the hallway from the office to the classrooms, how much walking are the students saving by taking the shortcut?

FIGURE 10-4

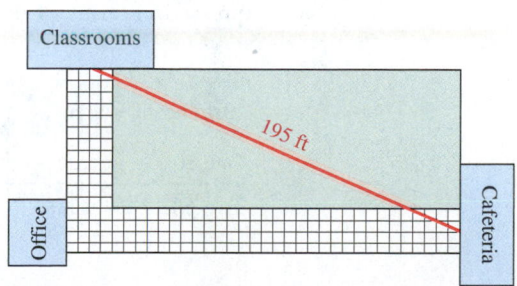

ANALYZE THE PROBLEM   The two hallways and the shortcut form a right triangle with a hypotenuse 195 feet long. We will use the Pythagorean theorem to solve this problem.

FORM AN EQUATION   If we let x represent the length (in feet) of the hallway from the classrooms to the office, then the length of the hallway from the office to the cafeteria is $(x + 105)$ feet. Substituting these lengths into the Pythagorean theorem, we have

$$a^2 + b^2 = c^2 \qquad \text{The Pythagorean theorem.}$$

$$x^2 + (x + 105)^2 = 195^2 \qquad \text{Substitute } x \text{ for } a, (x + 105) \text{ for } b, \text{ and } 195 \text{ for } c.$$

$$x^2 + x^2 + 105x + 105x + 11{,}025 = 38{,}025 \qquad \text{Use the FOIL method to find } (x + 105)^2.$$

$$2x^2 + 210x + 11{,}025 = 38{,}025 \qquad \text{Combine like terms.}$$

$$2x^2 + 210x - 27{,}000 = 0 \qquad \text{Subtract 38,025 from both sides.}$$

$$x^2 + 105x - 13{,}500 = 0 \qquad \text{Divide both sides by 2.}$$

SOLVE THE EQUATION   To solve $x^2 + 105x - 13,500 = 0$, we will use the quadratic formula with $a = 1$, $b = 105$, and $c = -13,500$.

$$x = \frac{-b \pm \sqrt{b^2 - 4ac}}{2a}$$

$$x = \frac{-105 \pm \sqrt{(105)^2 - 4(1)(-13,500)}}{2(1)}$$

$$x = \frac{-105 \pm \sqrt{65,025}}{2}$$ Simplify: $(105)^2 - 4(1)(-13,500) = 11,025 + 54,000 = 65,025$.

$$x = \frac{-105 \pm 255}{2}$$ Using a calculator, $\sqrt{65,025} = 255$.

$$x = \frac{150}{2} \quad \text{or} \quad x = \frac{-360}{2}$$

$$x = 75 \qquad\qquad x = -180$$ Since the length of the hallway can't be negative, discard the solution $x = -180$.

STATE THE CONCLUSION   The length of the hallway from the classrooms to the office is 75 feet. The length of the hallway from the office to the cafeteria is $75 + 105 = 180$ feet. Instead of using the hallways, a distance of $75 + 180 = 255$ feet, the students are taking the 195-foot short-cut to the cafeteria, a savings of $(255 - 195)$, or 60 feet.

CHECK THE RESULT   The length of the 180-foot hallway is 105 feet longer than the length of the 75-foot hallway. The sum of the squares of the lengths of the hallways is $75^2 + 180^2 = 38,025$. This equals the square of the length of the 195-foot shortcut. The answer checks. ■

E X A M P L E   4

Mass transit. A bus company has 4,000 passengers daily, each currently paying a 75¢ fare. For each 15¢ fare increase, the company estimates that it will lose 50 passengers. If the company needs to bring in $6,570 per day to stay in business, what fare must be charged to produce this amount of revenue?

ANALYZE THE PROBLEM   To understand how a fare increase affects the number of passengers, let's consider what happens if there are two fare increases. We organize the data in a table. The fares are expressed in terms of dollars.

Number of increases	New fare	Number of passengers
One $0.15 increase	$0.75 + $0.15(1) = $0.90	4,000 − 50(1) = 3,950
Two $0.15 increases	$0.75 + $0.15(2) = $1.05	4,000 − 50(2) = 3,900

In general, the new fare will be the old fare ($0.75) plus the number of fare increases times $0.15. The number of passengers who will pay the new fare is 4,000 minus 50 times the number of $0.15 fare increases.

FORM AN EQUATION   If we let x represent the number of $0.15 fare increases necessary to bring in $6,570 daily, then $(0.75 + 0.15x)$ is the fare that must be charged. The number of passengers who will pay this fare is $(4,000 - 50x)$. We can now form the equation.

The bus fare	times	the number of passengers who will pay that fare	equals	$6,570.
$(0.75 + 0.15x)$	·	$(4,000 - 50x)$	=	6,570

SOLVE THE EQUATION

$$(0.75 + 0.15x)(4{,}000 - 50x) = 6{,}570$$

$$3{,}000 - 37.5x + 600x - 7.5x^2 = 6{,}570 \quad \text{Use the FOIL method.}$$

$$-7.5x^2 + 562.5x + 3{,}000 = 6{,}570 \quad \text{Combine like terms: } -37.5x + 600x = 562.5x.$$

$$-7.5x^2 + 562.5x - 3{,}570 = 0 \quad \text{Subtract 6,570 from both sides.}$$

$$7.5x^2 - 562.5x + 3{,}570 = 0 \quad \text{Multiply both sides by } -1 \text{ so that } a, 7.5, \text{ is positive.}$$

To solve this equation, we will use the quadratic formula.

$$x = \frac{-b \pm \sqrt{b^2 - 4ac}}{2a}$$

$$x = \frac{-(-562.5) \pm \sqrt{(-562.5)^2 - 4(7.5)(3{,}570)}}{2(7.5)} \quad \text{Substitute 7.5 for } a, -562.5 \text{ for } b, \text{ and 3,570 for } c.$$

$$x = \frac{562.5 \pm \sqrt{209{,}306.25}}{15} \quad \text{Simplify: } (-562.5)^2 - 4(7.5)(3{,}570) = 316{,}406.25 - 107{,}100 = 209{,}306.25.$$

$$x = \frac{562.5 \pm 457.5}{15} \quad \text{Using a calculator, } \sqrt{209{,}306.25} = 457.5.$$

$$x = \frac{1{,}020}{15} \quad \text{or} \quad x = \frac{105}{15}$$

$$x = 68 \qquad\qquad x = 7$$

STATE THE CONCLUSION

If there are 7 fifteen-cent increases in the fare, the new fare will be $0.75 + $0.15(7) = $1.80. If there are 68 fifteen-cent increases in the fare, the new fare will be $0.75 + $0.15(68) = $10.95. Although this fare would bring in the necessary revenue, a $10.95 bus fare is unreasonable, so we discard it.

CHECK THE RESULT

A fare of $1.80 will be paid by [4,000 − 50(7)] = 3,650 bus riders. The amount of revenue brought in would be $1.80(3,650) = $6,570. The answer checks. ■

EXAMPLE 5

Lawyers. The number of lawyers N in the United States each year from 1980 to 1996 is approximated by $N = 660x^2 + 11{,}500x + 580{,}000$, where $x = 0$ corresponds to the year 1980, $x = 1$ corresponds to 1981, $x = 2$ corresponds to 1982, and so on. (Thus, $0 \le x \le 16$.) In what year does this model indicate that the United States had three-quarters of a million (750,000) lawyers?

Solution

We will substitute 750,000 for N in the equation. Then we can solve for x, which will give the number of years after 1980 that the United States had approximately 750,000 lawyers.

$$N = 660x^2 + 11{,}500x + 580{,}000$$

$$750{,}000 = 660x^2 + 11{,}500x + 580{,}000 \quad \text{Replace } N \text{ with 750,000.}$$

$$0 = 660x^2 + 11{,}500x - 170{,}000 \quad \text{Subtract 750,000 from both sides so that the equation is in quadratic form.}$$

We can simplify the computations by dividing both sides of the equation by 20, the common factor of 660, 11,500, and 170,000.

$$0 = 33x^2 + 575x - 8{,}500 \quad \text{Divide both sides by 20.}$$

We solve this equation using the quadratic formula.

$$x = \frac{-b \pm \sqrt{b^2 - 4ac}}{2a}$$

$$x = \frac{-575 \pm \sqrt{(575)^2 - 4(33)(-8,500)}}{2(33)}$$ Substitute 33 for a, 575 for b, and $-8,500$ for c.

$$x = \frac{-575 \pm \sqrt{330,625 + 1,122,000}}{66}$$ Simplify: $(575)^2 - 4(33)(-8,500) = 330,625 + 1,122,000$.

$$x = \frac{-575 \pm \sqrt{1,452,625}}{66}$$

$$x \approx \frac{630}{66} \quad \text{or} \quad x \approx \frac{-1,780}{66}$$ Use a calculator.

$$x \approx 9.5 \qquad x \approx -27.0$$ Since the model is defined only for $0 \le x \le 16$, we discard the second solution.

In 9.5 years after 1980, or midway through 1989, the United States had approximately 750,000 lawyers. ∎

STUDY SET

Section 10.2

VOCABULARY

In Exercises 1–2, fill in the blanks to make the statements true.

1. An equation of the form $ax^2 + bx + c = 0$ $(a \ne 0)$ is a _____ equation.

2. $x = \dfrac{-b \pm \sqrt{b^2 - 4ac}}{2a}$ is called the _____ formula.

CONCEPTS

3. Consider the quadratic equation $4x^2 - 2x = 0$.
  a. Solve it by factoring.
  b. What are a, b, and c? Solve it using the quadratic formula.

4. Consider the quadratic equation $x^2 = 8$.
  a. Solve it by the square root method.
  b. What are a, b, and c? Solve it using the quadratic formula.

5. Tell whether each statement is true or false.
  a. Any quadratic equation can be solved by using the quadratic formula.
  b. Any quadratic equation can be solved by completing the square.

6. What is **wrong** with the beginning of the solution shown below?
  Solve $x^2 - 3x = 2$.
  $a = 1 \qquad b = -3 \qquad c = 2$

7. What form would the quadratic formula take if, in developing it, we began with the general quadratic equation expressed as $rc^2 + sc + t = 0$ $(r \ne 0)$?

8. A student used the quadratic formula to solve a quadratic equation and obtained $x = \dfrac{-2 \pm \sqrt{3}}{2}$.
  a. How many solutions does the equation have? What are they exactly?
  b. Graph the solutions on a number line.

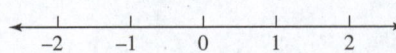

9. Solve $x^2 - 4x + 1 = 0$ using the quadratic formula. Simplify the result.

10. Solve $x^2 + 4x + 2 = 0$ using the quadratic formula. Simplify the result.

NOTATION

11. On a quiz, students were asked to write the quadratic formula. What is **wrong** with each answer shown below?

a. $x = -b \pm \dfrac{\sqrt{b^2 - 4ac}}{2a}$

b. $x = \dfrac{-b\sqrt{b^2 - 4ac}}{2a}$

12. In reading $\dfrac{-b \pm \sqrt{b^2 - 4ac}}{2a}$, we say, "The

_____ of b, plus or _____ the

square _____ of b _____

minus ____ times a times c, all _____ 2a."

PRACTICE

In Exercises 13–36, use the quadratic formula to solve each equation.

13. $x^2 + 3x + 2 = 0$ **14.** $x^2 - 3x + 2 = 0$

15. $x^2 + 12x = -36$ **16.** $y^2 - 18y = -81$

17. $2x^2 + 5x - 3 = 0$ **18.** $6x^2 - x - 1 = 0$

19. $5x^2 + 5x + 1 = 0$ **20.** $4w^2 + 6w + 1 = 0$

21. $8u = -4u^2 - 3$ **22.** $4t + 3 = 4t^2$

23. $16y^2 + 8y - 3 = 0$ **24.** $16x^2 + 16x + 3 = 0$

25. $x^2 - \dfrac{14}{15}x = \dfrac{8}{15}$ **26.** $x^2 = -\dfrac{5}{4}x + \dfrac{3}{2}$

27. $\dfrac{x^2}{2} + \dfrac{5}{2}x = -1$ **28.** $\dfrac{x^2}{8} - \dfrac{x}{4} = \dfrac{1}{2}$

29. $2x^2 - 1 = 3x$ **30.** $-9x = 2 - 3x^2$

31. $-x^2 + 10x = 18$ **32.** $-3x = \dfrac{x^2}{2} + 2$

33. $x^2 - 6x = 391$ **34.** $x^2 - 27x = 280$

35. $x^2 - \dfrac{5}{3} = -\dfrac{11}{6}x$ **36.** $x^2 - \dfrac{1}{2} = \dfrac{2}{3}x$

In Exercises 37–38, use the quadratic formula and a scientific calculator to solve each equation. Give all answers to the nearest hundredth.

37. $0.7x^2 - 3.5x - 25 = 0$

38. $-4.5x^2 + 0.2x + 3.75 = 0$

APPLICATIONS

39. MOVIE THEATER SCREEN The largest permanent movie screen is in the Panasonic Imax theater at Darling Harbor, Sydney, Australia. The rectangular screen has an area of 11,349 square feet. Find the dimensions of the screen if it is 20 feet longer than it is wide.

40. ROCK CONCERT During a 1997 tour, the rock group U2 used the world's largest LED (Light Emitting Diode) electronic screen as part of the stage backdrop. The screen, with an area of 9,520 square feet, had a length that was 2 feet more than three times its width. Find the dimensions of the LED screen.

41. CENTRAL PARK Central Park is one of New York's best-known landmarks. Rectangular in shape, its length is 5 times its width. When measured in miles, its perimeter numerically exceeds its area by 4.75. Find the dimensions of Central Park if we know that its width is less than 1 mile.

42. ANCIENT HISTORY One of the most important cities of the ancient world was Babylon. Greek historians wrote that the city was square-shaped. Measured in miles, its area numerically exceeded its perimeter by about 124. Find its dimensions. (Round to the nearest tenth.)

43. BADMINTON The person who wrote the instructions for setting up the badminton net shown in Illustration 1 forgot to give the specific dimensions for securing the pole. How long is the support string?

ILLUSTRATION 1

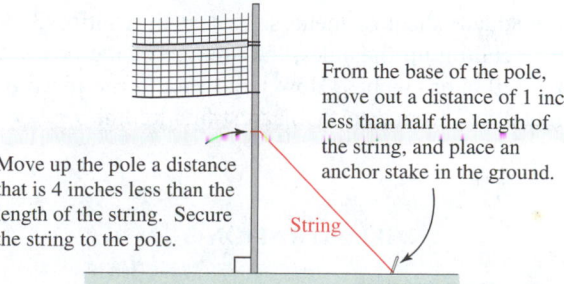

Move up the pole a distance that is 4 inches less than the length of the string. Secure the string to the pole.

From the base of the pole, move out a distance of 1 inch less than half the length of the string, and place an anchor stake in the ground.

String

44. RIGHT TRIANGLE The hypotenuse of a right triangle is 2.5 units long. The longer leg is 1.7 units longer than the shorter leg. Find the lengths of the sides of the triangle.

45. SCHOOL DANCE Tickets to a school dance cost $4, and the projected attendance is 300 persons. It is further projected that for every 10¢ increase in ticket price, the average attendance will decrease by 5. At what ticket price will the receipts from the dance be $1,248?

46. TICKET SALES A carnival at a county fair normally sells three thousand 25¢ ride tickets on a Saturday. For each 5¢ increase in price, management estimates that 80 fewer tickets will be sold. What increase in ticket price will produce $994 of revenue on Saturday?

47. MAGAZINE SALES The *Gazette's* profit is $20 per year for each of its 3,000 subscribers. Management estimates that the profit per subscriber will increase by 1¢ for each additional subscriber over the current 3,000. How many subscribers will bring a total profit of $120,000?

48. POLYGONS The five-sided polygon called a *pentagon,* shown in Illustration 2, has 5 diagonals. The number of diagonals d of a polygon of n sides is given by the formula

$$d = \frac{n(n-3)}{2}$$

ILLUSTRATION 2

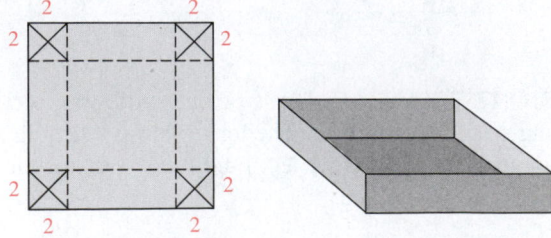

Find the number of sides of a polygon if it has 275 diagonals.

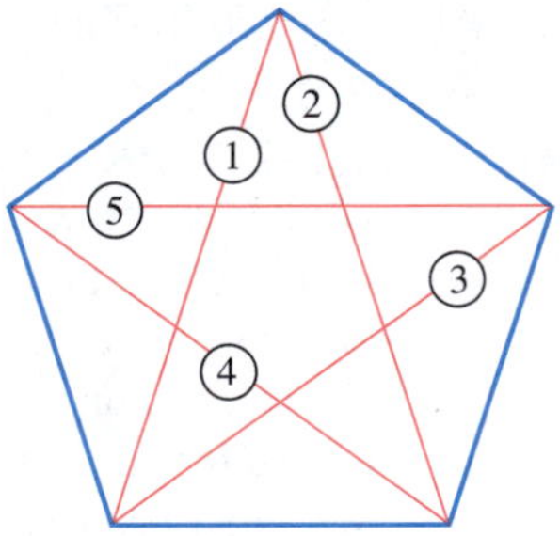

49. INVESTMENT RATES A woman invests $1,000 in a mutual fund for which interest is compounded annually at a rate r. After one year, she deposits an additional $2,000. After two years, the balance in the account is

$$\$1,000(1 + r)^2 + \$2,000(1 + r)$$

If this amount is $3,368.10, find r.

50. METAL FABRICATION A box with no top is to be made by cutting a 2-inch square from each corner of the square sheet of metal shown in Illustration 3. After bending up the sides, the volume of the box is to be 220 cubic inches. How large should the piece of metal be? Round to the nearest hundredth.

ILLUSTRATION 3

51. RETIREMENT The labor force participation rate P (in percent) for men ages 55–64 from 1970 to 1996 is approximated by the quadratic equation $P = 0.027x^2 - 1.363x + 82.5$, where $x = 0$ corresponds to the year 1970, $x = 1$ corresponds to 1971, $x = 2$ corresponds to 1972, and so on. (Thus, $0 \le x \le 26$.)

   a. To the nearest percent, what percent of men ages 55–64 were in the workforce in 1970?

   b. When does the model indicate that 75% of the men ages 55–64 were part of the workforce?

52. SPACE PROGRAM The yearly budget B (in billions of dollars) for the National Aeronautics and Space Administration (NASA) is approximated by the quadratic equation $B = -0.1492x^2 + 1.8058x + 9$, where x is the number of years since 1988 and $0 \le x \le 8$.

   a. Use the model to find the approximate budget for NASA in 1996. Round to the nearest tenth.

   b. In what year does the model indicate that NASA's budget was about $12 billion?

   c. For the years 1988–1996, was NASA's budget ever $15 billion? Explain how you know.

WRITING

53. Explain why the quadratic formula, in most cases, is less tedious to use in solving a quadratic equation than is the method of completing the square.

54. On an exam, a student was asked to solve the equation $-4w^2 - 6w - 1 = 0$. Her first step was to multiply both sides of the equation by -1. She then used the quadratic formula to solve $4w^2 + 6w + 1 = 0$ instead. Is this a valid approach? Explain.

REVIEW

In Exercises 55–58, change each radical to an exponential expression.

55. $\sqrt{n}$

56. $\sqrt[7]{\dfrac{3}{8}r^2 s}$

57. $\sqrt[4]{3b}$

58. $3\sqrt[3]{c^2 - d^2}$

In Exercises 59–62, write each expression in radical form.

59. $t^{1/3}$

60. $\left(\dfrac{3}{4}m^2 n^2\right)^{1/5}$

61. $(3t)^{1/4}$

62. $(c^2 + d^2)^{1/2}$

▶ 10.3   Quadratic Functions and Their Graphs

In this section, you will learn about

> Quadratic functions ■ Graphs of $f(x) = ax^2$ ■ Graphs of $f(x) = ax^2 + k$ ■ Graphs of $f(x) = a(x - h)^2$ ■ Graphs of $f(x) = a(x - h)^2 + k$ ■ Graphs of $f(x) = ax^2 + bx + c$ ■ Determining minimum and maximum values ■ Solving quadratic equations graphically

Introduction   The graph in Figure 10-5 shows a trampolinist's distance from the ground (in relation to time) as she bounds into the air and then falls back down to the trampoline.

FIGURE 10-5

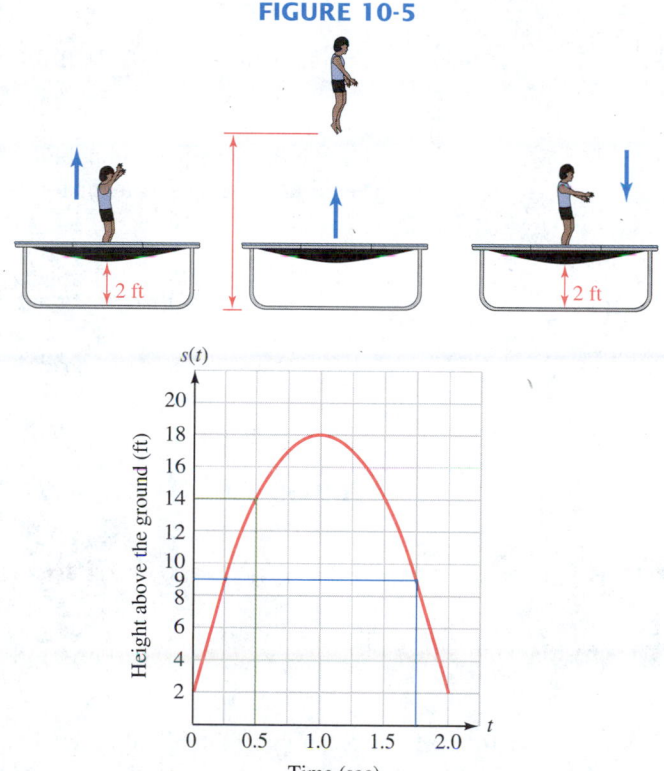

From the graph, we can see that the trampolinist is 14 feet above the ground 0.5 second after bounding upward and that her height above the ground after 1.75 seconds is 9 feet.

The parabola shown in Figure 10-5 is the graph of the *quadratic function* $s(t) = -16t^2 + 32t + 2$. In this section, we will discuss two forms in which quadratic functions are written and how to graph them.

Quadratic Functions

Quadratic functions

A **quadratic function** is a second-degree polynomial function of the form

$$f(x) = ax^2 + bx + c \quad \text{or} \quad y = ax^2 + bx + c \quad (a \neq 0)$$

where a, b, and c are real numbers.

EXAMPLE 1

Quadratic functions as models. The quadratic function $s(t) = -16t^2 + 32t + 2$ gives the distance (in feet) that the trampolinist shown in Figure 10-5 is from the ground, t seconds after bounding upward. How far is she from the ground after being in the air for $\frac{3}{4}$ second?

Solution   To find her distance from the ground, we find the value of the function for $t = \frac{3}{4} = 0.75$.

$$s(t) = -16t^2 + 32t + 2$$
$$s(0.75) = -16(0.75)^2 + 32(0.75) + 2 \quad \text{Replace } t \text{ with 0.75.}$$
$$= -9 + 24 + 2$$
$$= 17$$

The trampolinist is 17 feet off the ground $\frac{3}{4}$ second after bounding upward.

SELF CHECK   Find the distance the trampolinist is from the ground after being in the air for $1\frac{1}{2}$ seconds.   *Answer:* 14 feet   ■

We have seen that important information can be gained from the graph of a quadratic function. We now begin a discussion of how to graph such a function by considering the simplest case, quadratic functions of the form $f(x) = ax^2$.

Graphs of $f(x) = ax^2$

EXAMPLE 2

Graphing quadratic functions. Graph   **a.** $f(x) = x^2$,   **b.** $g(x) = 3x^2$, and   **c.** $h(x) = \dfrac{1}{3}x^2$.

Solution   We can make a table of ordered pairs that satisfy each equation, plot each point, and join them with a smooth curve, as in Figure 10-6. We note that the graph of $h(x) = \frac{1}{3}x^2$ is wider than the graph of $f(x) = x^2$, and that the graph of $g(x) = 3x^2$ is narrower than the graph of $f(x) = x^2$. In the function $f(x) = ax^2$, the smaller the value of $|a|$, the wider the graph.

FIGURE 10-6

$f(x) = x^2$

x	$f(x)$	$(x, f(x))$
-2	4	$(-2, 4)$
-1	1	$(-1, 1)$
0	0	$(0, 0)$
1	1	$(1, 1)$
2	4	$(2, 4)$

$g(x) = 3x^2$

x	$g(x)$	$(x, g(x))$
-2	12	$(-2, 12)$
-1	3	$(-1, 3)$
0	0	$(0, 0)$
1	3	$(1, 3)$
2	12	$(2, 12)$

$h(x) = \frac{1}{3}x^2$

x	$h(x)$	$(x, h(x))$
-2	$\frac{4}{3}$	$\left(-2, \frac{4}{3}\right)$
-1	$\frac{1}{3}$	$\left(-1, \frac{1}{3}\right)$
0	0	$(0, 0)$
1	$\frac{1}{3}$	$\left(1, \frac{1}{3}\right)$
2	$\frac{4}{3}$	$\left(2, \frac{4}{3}\right)$

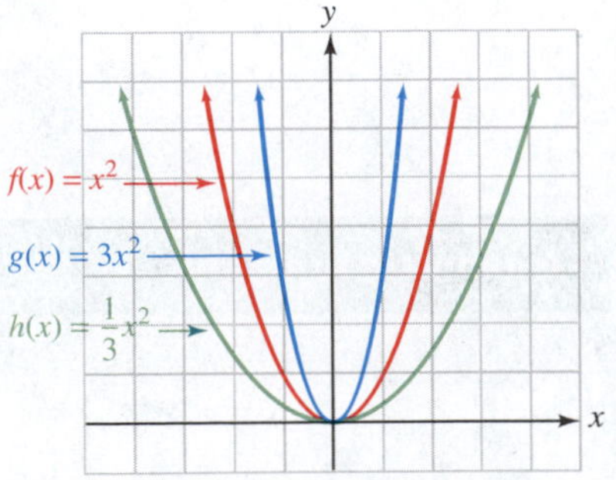

$f(x) = x^2$
$g(x) = 3x^2$
$h(x) = \frac{1}{3}x^2$

■

EXAMPLE 3

Graphing quadratic functions. Graph $f(x) = -3x^2$.

Solution   We make a table of ordered pairs that satisfy the equation, plot each point, and join them with a smooth curve, as in Figure 10-7. We see that the parabola opens downward and has the same shape as the graph of $f(x) = 3x^2$.

FIGURE 10-7

$f(x) = -3x^2$

x	$f(x)$	$(x, f(x))$
-2	-12	$(-2, -12)$
-1	-3	$(-1, -3)$
0	0	$(0, 0)$
1	-3	$(1, -3)$
2	-12	$(2, -12)$

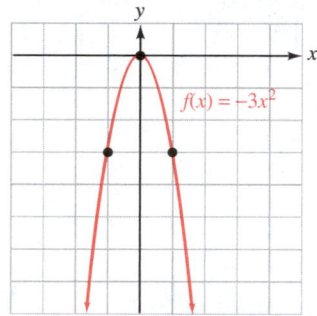

SELF CHECK   Graph $f(x) = -\dfrac{1}{3}x^2$.

Answer:

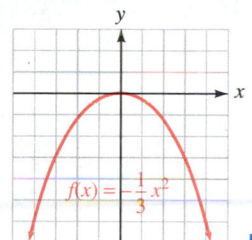

The graphs of quadratic functions of the form $f(x) = ax^2$ are parabolas. The lowest point of a parabola that opens upward, or the highest point of a parabola that opens downward, is called the **vertex** of the parabola. For example, the vertex of the parabola shown in Figure 10-7 is the point $(0, 0)$. The vertical line, called an **axis of symmetry,** that passes through the vertex divides the parabola into two congruent halves. The axis of symmetry of the parabola shown in Figure 10-7 is the line $x = 0$ (the y-axis).

The results of Examples 2 and 3 confirm the following facts.

The graph of $f(x) = ax^2$

The graph of $f(x) = ax^2$ is a parabola opening upward when $a > 0$ and downward when $a < 0$, with vertex at the point $(0, 0)$ and axis of symmetry the line $x = 0$.

Graphs of $f(x) = ax^2 + k$

EXAMPLE 4

Graphing quadratic functions. Graph   **a.** $f(x) = 2x^2$,   **b.** $g(x) = 2x^2 + 3$, and   **c.** $h(x) = 2x^2 - 3$.

Solution   We make a table of ordered pairs that satisfy each equation, plot each point, and join them with a smooth curve, as in Figure 10-8. We note that the graph of $g(x) = 2x^2 + 3$ is identical to the graph of $f(x) = 2x^2$, except that it has been translated 3 units upward. The graph of $h(x) = 2x^2 - 3$ is identical to the graph of $f(x) = 2x^2$, except that it has been translated 3 units downward. In each case, the axis of symmetry is the line $x = 0$.

FIGURE 10-8

$f(x) = 2x^2$			$g(x) = 2x^2 + 3$			$h(x) = 2x^2 - 3$		
x	$f(x)$	$(x, f(x))$	x	$g(x)$	$(x, g(x))$	x	$h(x)$	$(x, h(x))$
-2	8	$(-2, 8)$	-2	11	$(-2, 11)$	-2	5	$(-2, 5)$
-1	2	$(-1, 2)$	-1	5	$(-1, 5)$	-1	-1	$(-1, -1)$
0	0	$(0, 0)$	0	3	$(0, 3)$	0	-3	$(0, -3)$
1	2	$(1, 2)$	1	5	$(1, 5)$	1	-1	$(1, -1)$
2	8	$(2, 8)$	2	11	$(2, 11)$	2	5	$(2, 5)$

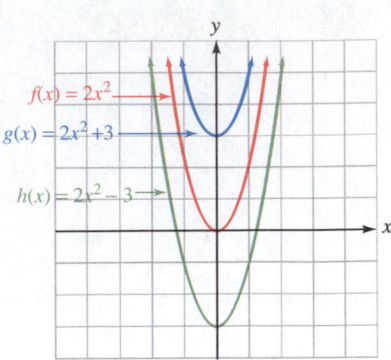

The results of Example 4 confirm the following facts.

The graph of $f(x) = ax^2 + k$

The graph of $f(x) = ax^2 + k$ is a parabola having the same shape as $f(x) = ax^2$ but translated upward k units if k is positive and downward $|k|$ units if k is negative. The vertex is at the point $(0, k)$, and the axis of symmetry is the line $x = 0$.

Graphs of $f(x) = a(x - h)^2$

EXAMPLE 5

Graphing quadratic functions. Graph **a.** $f(x) = 2x^2$, **b.** $g(x) = 2(x - 3)^2$, and **c.** $h(x) = 2(x + 3)^2$.

Solution We make a table of ordered pairs that satisfy each equation, plot each point, and join them with a smooth curve, as in Figure 10-9. We note that the graph of $g(x) = 2(x - 3)^2$ is identical to the graph of $f(x) = 2x^2$, except that it has been translated 3 units to the right. The graph of $h(x) = 2(x + 3)^2$ is identical to the graph of $f(x) = 2x^2$, except that it has been translated 3 units to the left.

FIGURE 10-9

$f(x) = 2x^2$			$g(x) = 2(x - 3)^2$			$h(x) = 2(x + 3)^2$		
x	$f(x)$	$(x, f(x))$	x	$g(x)$	$(x, g(x))$	x	$h(x)$	$(x, h(x))$
-2	8	$(-2, 8)$	1	8	$(1, 8)$	-5	8	$(-5, 8)$
-1	2	$(-1, 2)$	2	2	$(2, 2)$	-4	2	$(-4, 2)$
0	0	$(0, 0)$	3	0	$(3, 0)$	-3	0	$(-3, 0)$
1	2	$(1, 2)$	4	2	$(4, 2)$	-2	2	$(-2, 2)$
2	8	$(2, 8)$	5	8	$(5, 8)$	-1	8	$(-1, 8)$

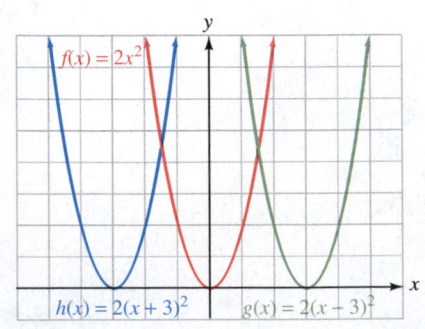

The results of Example 5 confirm the following facts.

The graph of $f(x) = a(x - h)^2$

The graph of $f(x) = a(x - h)^2$ is a parabola having the same shape as $f(x) = ax^2$ but translated h units to the right if h is positive and $|h|$ units to the left if h is negative. The vertex is at the point $(0, h)$, and the axis of symmetry is the line $x = h$.

Graphs of $f(x) = a(x - h)^2 + k$

EXAMPLE 6

Graphing quadratic functions. Graph $f(x) = 2(x - 3)^2 - 4$.

Solution The graph of $f(x) = 2(x - 3)^2 - 4$ is identical to the graph of $g(x) = 2(x - 3)^2$, except that it has been translated 4 units downward. The graph of $g(x) = 2(x - 3)^2$ is identical to the graph of $h(x) = 2x^2$, except that it has been translated 3 units to the right. Thus, to graph $f(x) = 2(x - 3)^2 - 4$, we can graph $h(x) = 2x^2$ and shift it 3 units to the right and then 4 units downward, as shown in Figure 10-10.

The vertex of the graph is the point $(3, -4)$, and the axis of symmetry is the line $x = 3$.

FIGURE 10-10

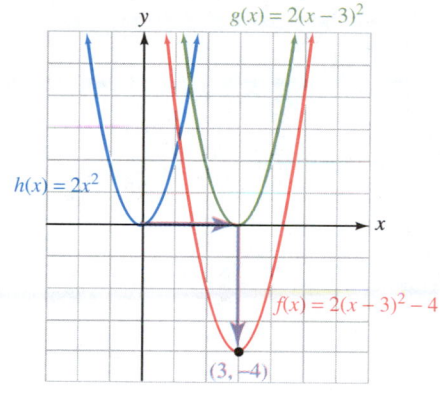

SELF CHECK

Graph $f(x) = 2(x + 3)^2 + 1$. Label the vertex and draw the axis of symmetry.

Answer:

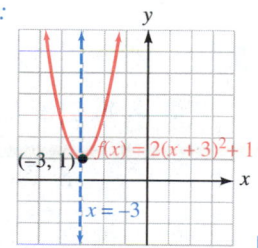

The results of Example 6 confirm the following facts.

The graph of $f(x) = a(x - h)^2 + k$

The graph of the quadratic function

$$f(x) = a(x - h)^2 + k$$

or

$$y = a(x - h)^2 + k$$

(where $a \neq 0$) is a parabola with vertex at (h, k). The parabola opens upward when $a > 0$ and downward when $a < 0$. The axis of symmetry is the line $x = h$.

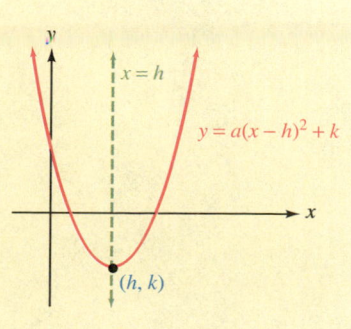

EXAMPLE 7

Determining the vertex and the axis of symmetry. Consider the graph of $f(x) = -3(x + 1)^2 - 4$.

a. Does the graph open upward or downward?

b. What are the coordinates of the vertex?

c. What is the axis of symmetry?

Solution   Rewriting the given function in $f(x) = a(x - h)^2 + k$ form, we have

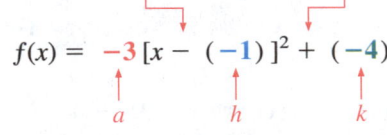

Standard form requires a minus sign here.   Standard form requires a plus sign here.

$$f(x) = \underset{a}{-3}\,[x - \underset{h}{(-1)}\,]^2 + \underset{k}{(-4)}$$

a. Since $a = -3 < 0$, the parabola opens downward.

b. The vertex is $(h, k) = (-1, -4)$.

c. The axis of symmetry is the line $x = h$. In this case, $x = -1$.

SELF CHECK   Answer parts a, b, and c of Example 7 for the graph of $y = 6(x - 5)^2 + 1$.   *Answers:* **a.** upward, **b.** $(5, 1)$, **c.** $x = 5$ ∎

Graphs of $f(x) = ax^2 + bx + c$

To graph functions of the form $f(x) = ax^2 + bx + c$, we complete the square to write the function in the form $f(x) = a(x - h)^2 + k$.

EXAMPLE 8

Completing the square to determine the vertex. Graph $f(x) = 2x^2 - 4x - 1$.

Solution   **Step 1:** We complete the square on x to write the given function in the form $f(x) = a(x - h)^2 + k$.

$$f(x) = 2x^2 - 4x - 1$$
$$f(x) = 2(x^2 - 2x) - 1 \quad \text{Factor 2 from } 2x^2 - 4x.$$

Now we complete the square on x by adding 1 inside the parentheses. Since this adds 2 to the right-hand side, we also subtract 2 from the right-hand side.

$$f(x) = 2(x^2 - 2x + 1) - 1 - 2 \quad \text{To complete the square, one-half of } -2 = -1 \text{ and } (-1)^2 = 1.$$

1. $f(x) = 2(x - 1)^2 - 3$ \qquad Factor $x^2 - 2x + 1$ and combine like terms.

Step 2: From Equation 1, we can see that $h = 1$ and $k = -3$, so the vertex will be at the point $(1, -3)$, and the axis of symmetry is $x = 1$. We plot the vertex and axis of symmetry on the coordinate system in Figure 10-11.

Step 3: Finally, we construct a table of values, plot the points, and draw the graph, which appears in the figure.

FIGURE 10-11

$f(x) = 2x^2 - 4x - 1$

x	$f(x)$	$(x, f(x))$
-1	5	$(-1, 5)$
0	-1	$(0, -1)$
2	-1	$(2, -1)$
3	5	$(3, 5)$

The x-coordinate of the vertex is 1. Choose values for x close to 1.

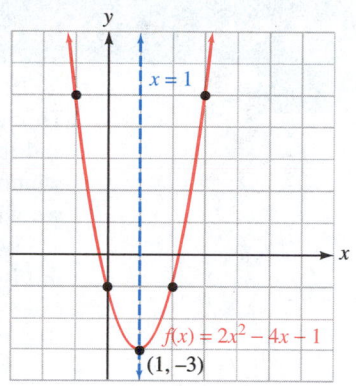

SELF CHECK   Graph $f(x) = 2x^2 - 4x + 1$.

Answer:

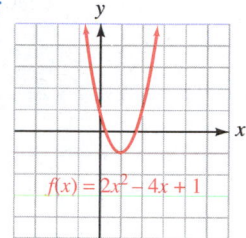

We can derive a formula for the vertex of the graph of $f(x) = ax^2 + bx + c$ by completing the square in the same manner as we did in Example 8. After using similar steps, the result is

$$f(x) = a\left[x - \left(-\frac{b}{2a}\right)\right]^2 + \frac{4ac - b^2}{4a}$$

$$\underset{h}{\uparrow} \qquad\qquad \underset{k}{\uparrow}$$

The x-coordinate of the vertex is $-\frac{b}{2a}$. The y-coordinate of the vertex is $\frac{4ac - b^2}{4a}$. However, we can also find the y-coordinate of the vertex by substituting the x-coordinate, $-\frac{b}{2a}$, for x in the quadratic function.

Formula for the vertex of a parabola

The vertex of the graph of the quadratic function $f(x) = ax^2 + bx + c$ is

$$\left(-\frac{b}{2a}, f\left(-\frac{b}{2a}\right)\right)$$

and the axis of symmetry of the parabola is the line $x = -\frac{b}{2a}$.

In Example 8, for the function $f(x) = 2x^2 - 4x - 1$, $a = 2$ and $b = -4$. To find the vertex of its graph, we compute

$$-\frac{b}{2a} = -\frac{-4}{2(2)} \qquad\qquad f\left(-\frac{b}{2a}\right) = f(\mathbf{1})$$

$$= -\frac{-4}{4} \qquad\qquad\qquad\quad = 2(\mathbf{1})^2 - 4(\mathbf{1}) - 1$$

$$= 1 \qquad\qquad\qquad\qquad\quad = -3$$

The vertex is the point $(1, -3)$. This agrees with the result we obtained in Example 8 by completing the square.

ACCENT ON TECHNOLOGY *Finding the Vertex*

To graph the function $f(x) = 2x^2 + 6x - 3$ and find the coordinates of the vertex and the axis of symmetry of the parabola, we can use a graphing calculator with window settings of $[-10, 10]$ for x and $[-10, 10]$ for y. If we enter the function, we will obtain the graph shown in Figure 10-12(a).

We then trace to move the cursor to the lowest point on the graph, as shown in Figure 10-12(b). By zooming in, we can determine that the vertex is the point $\left(-\frac{3}{2}, -\frac{15}{2}\right)$ and that the line $x = -\frac{3}{2}$ is the axis of symmetry.

FIGURE 10-12

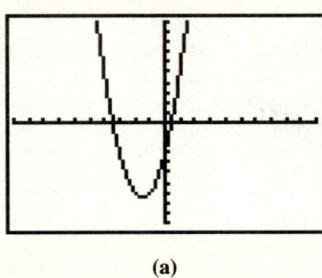

(a)

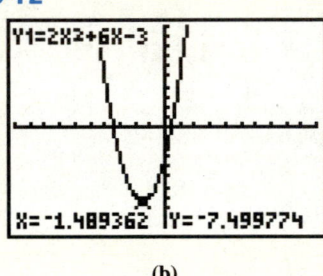

(b)

Determining Minimum and Maximum Values

It is often useful to know the smallest or largest possible value a quantity can assume. For example, companies try to minimize their costs and maximize their profits. If the quantity is expressed by a quadratic function, the vertex of the graph of the function gives its minimum or maximum value.

EXAMPLE 9

Minimizing costs. A glassworks that makes lead crystal vases has daily production costs given by the function $C(x) = 0.2x^2 - 10x + 650$, where x is the number of vases made each day. How many vases should be produced to minimize the per-day costs? What will the costs be?

Solution   The graph of $C(x) = 0.2x^2 - 10x + 650$ is a parabola opening upward. The vertex is the lowest point on the graph. To find the vertex, we compute

$$-\frac{b}{2a} = -\frac{-10}{2(0.2)} \qquad b = -10 \text{ and } a = 0.2. \qquad f\left(-\frac{b}{2a}\right) = f(25)$$

$$= -\frac{-10}{0.4} \qquad\qquad\qquad\qquad\qquad = 0.2(25)^2 - 10(25) + 650$$

$$= 25 \qquad\qquad\qquad\qquad\qquad\qquad = 525$$

The vertex is the point $(25, 525)$, and it indicates that the costs are a minimum of $525 when 25 vases are made daily.

To solve this problem with a graphing calculator with window settings of $[0, 50]$ for x and $[0, 1,000]$ for y, we graph the function $C(x) = 0.2x^2 - 10x + 650$. By using TRACE and ZOOM, we can locate the vertex of the graph. See Figure 10-13. The coordinates of the vertex indicate that the minimum cost is $525 when the number of vases produced is 25.

FIGURE 10-13

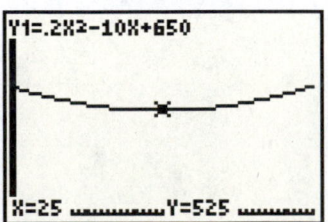

EXAMPLE 10

Maximizing area. A kennel owner wants to build the rectangular pen shown in Figure 10-14(a) to house his dog. If he uses one side of his barn, find the maximum area that he can enclose with 80 feet of fencing.

Solution   We can let the width of the area be represented by w. Then the length is represented by $80 - 2w$. The function that gives the area the pen encloses is

$$A = (80 - 2w)w \quad A = lw.$$

We can find the maximum value of A by determining the vertex of the graph of the function. This time, we will find the vertex by completing the square.

$A = 80w - 2w^2$ Use the distributive property to remove parentheses.

$\quad = -2(w^2 - 40w)$ Factor out -2.

$\quad = \mathbf{-2}(w^2 - 40w \mathbf{+ 400}) \mathbf{+ 800}$ Complete the square on w. Inside the parentheses, add $\left(\frac{-40}{2}\right)^2 = 400$. Since this subtracts 800 from the right-hand side, add 800 to the right-hand side.

$\quad = -2(w - 20)^2 + 800$ Factor $w^2 - 40w + 400$.

Thus, the coordinates of the vertex of the graph of the quadratic function are (20, 800), and the maximum area is 800 square feet.

To solve this problem with a graphing calculator with window settings of [0, 50] for x and [0, 1,000] for y, we graph the function $A = -2w^2 + 80w$ to get the graph in Figure 10-14(b). By using TRACE and ZOOM, we can determine the vertex of the graph, which indicates that the maximum area is 800 square feet when the width is 20 feet.

FIGURE 10-14

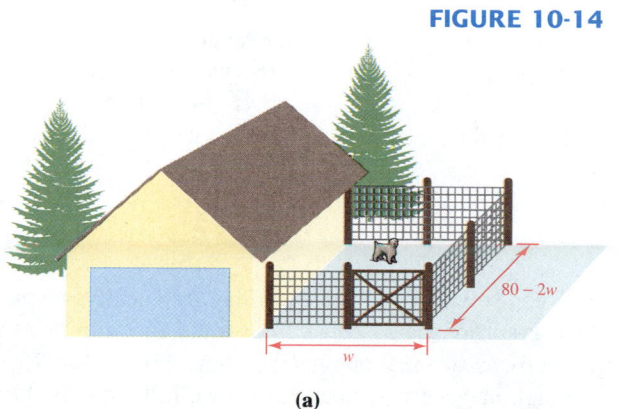

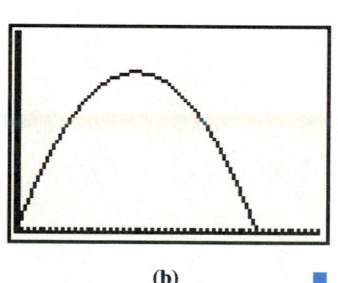

(a)   (b)   ∎

Solving Quadratic Equations Graphically

We can solve quadratic equations graphically. For example, the solutions of $x^2 - 6x + 8 = 0$ are the numbers x that will make y equal to 0 in the quadratic function $y = x^2 - 6x + 8$. To find these numbers, we inspect the graph of $y = x^2 - 6x + 8$ and locate the points on the graph with a y-coordinate of 0. In Figure 10-15, these points are (2, 0) and (4, 0), the x-intercepts of the graph. We can conclude that the x-coordinates of the x-intercepts, $x = 2$ and $x = 4$, are the solutions of $x^2 - 6x + 8 = 0$.

FIGURE 10-15

$y = x^2 - 6x + 8$

x-intercept (2,0) x-intercept (4,0)

ACCENT ON TECHNOLOGY *Solving Quadratic Equations Graphically*

We can use a graphing calculator to find approximate solutions of quadratic equations. For example, the solutions of $0.7x^2 + 2x - 3.5 = \mathbf{0}$ are the numbers x that will make $y = 0$ in the quadratic function $y = 0.7x^2 + 2x - 3.5$. To approximate these numbers, we graph the quadratic function and read the x-intercepts from the graph.

We can use the standard window settings of $[-10, 10]$ for x and $[-10, 10]$ for y and graph the function. We then trace to move the cursor to each x-intercept, as in Figures 10-16(a) and 10-16(b). From the graph, we can read the approximate value of the x-coordinate of each x-intercept. For better results, we can zoom in.

FIGURE 10-16

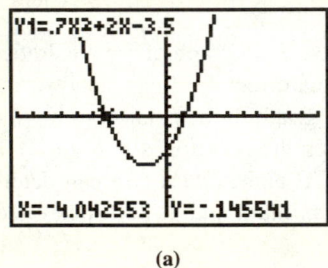 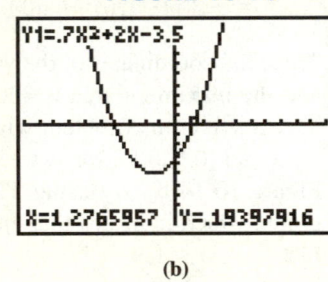 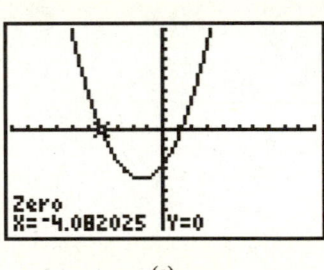

(a)　　　　　　　　　　(b)　　　　　　　　　　(c)

We can also approximate the solutions of $0.7x^2 + 2x - 3.5 = 0$ using the ZERO feature. In Figure 10-16(c), we see that x-coordinate of the leftmost x-intercept of the graph is given as -4.082025. This means that an approximate solution of the equation is -4.08. To find the positive x-intercept, we use similar steps.

When solving quadratic equations graphically, we must consider three possibilities. If the graph of the associated quadratic function has two x-intercepts, the quadratic equation has two real-number solutions. Figure 10-17(a) shows an example of this. If the graph has one x-intercept, as shown in Figure 10-17(b), the equation has one real-number solution. Finally, if the graph does not have an x-intercept, as shown in Figure 10-17(c), the equation does not have any real-number solutions.

FIGURE 10-17

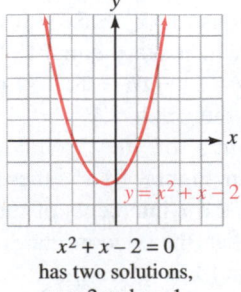 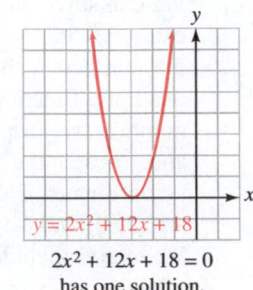 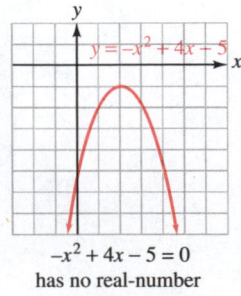

$x^2 + x - 2 = 0$ has two solutions, $x = -2$ and $x = 1$.	$2x^2 + 12x + 18 = 0$ has one solution, $x = -3$.	$-x^2 + 4x - 5 = 0$ has no real-number solutions.
(a)	(b)	(c)

STUDY SET

Section 10.3

VOCABULARY

In Exercises 1–4, refer to Illustration 1. Fill in the blanks to make the statements true.

ILLUSTRATION 1

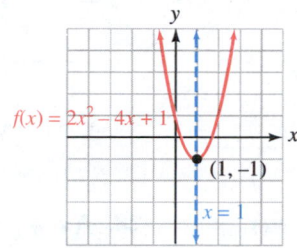

$f(x) = 2x^2 - 4x + 1$

$(1, -1)$

$x = 1$

1. The function $f(x) = 2x^2 - 4x + 1$ is graphed in Illustration 1. It is called a _____ function.

2. The graph in Illustration 1 is called a _____.

3. The lowest point on the graph is $(1, -1)$. This is called the _____.

4. The vertical line $x = 1$ divides the parabola into two halves. This line is called the _____.

CONCEPTS

In Exercises 5–6, make a table of values for each function. Then graph them on the same coordinate system.

5. $f(x) = x^2$, $g(x) = 2x^2$, $h(x) = \frac{1}{2}x^2$

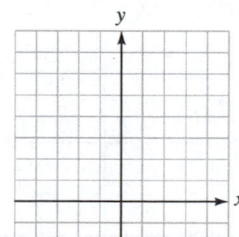

6. $f(x) = -x^2$, $g(x) = -\frac{1}{4}x^2$, $h(x) = -4x^2$

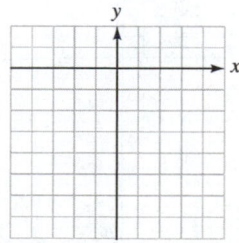

In Exercises 7–8, make a table of values to graph function f. Then use a translation to graph the other two functions on the same coordinate system.

7. $f(x) = 4x^2$, $g(x) = 4x^2 + 3$, $h(x) = 4x^2 - 2$

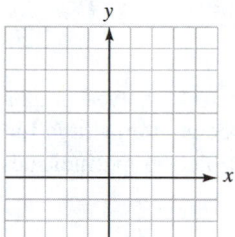

8. $f(x) = 3x^2$, $g(x) = 3(x + 2)^2$, $h(x) = 3(x - 3)^2$

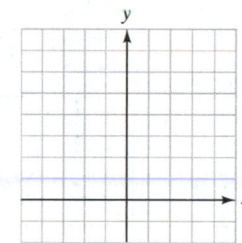

In Exercises 9–10, make a table of values to graph function f. Then use a series of translations to graph function g on the same coordinate system.

9. $f(x) = -3x^2$, $g(x) = -3(x - 2)^2 - 1$

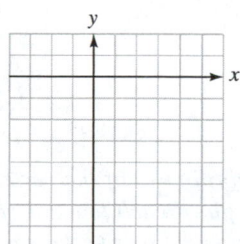

10. $f(x) = -\frac{1}{2}x^2$, $g(x) = -\frac{1}{2}(x + 1)^2 + 2$

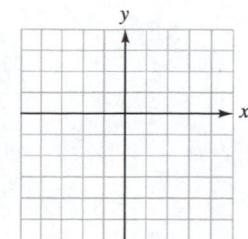

11. Use the graph of $y = \frac{x^2}{10} - \frac{x}{5} - \frac{3}{2}$, shown below, to estimate the solutions of the quadratic equation $\frac{x^2}{10} - \frac{x}{5} - \frac{3}{2} = 0$. Check your answers.

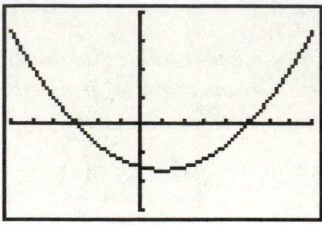

12. Three quadratic equations are to be solved graphically. The graphs of their associated quadratic functions are shown below. Tell which graph indicates that the equation has
a. two real solutions.
b. one real solution.
c. no real solutions.

i   ii   iii

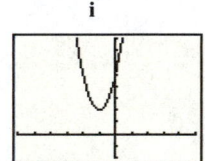 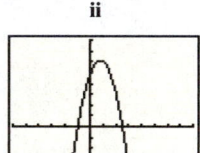 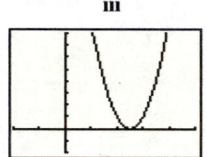

NOTATION

13. The function $f(x) = 2(x + 1)^2 + 6$ is written in the form $f(x) = a(x - h)^2 + k$. Is $h = -1$ or is $h = 1$? Explain.

14. The vertex of a quadratic function $f(x) = ax^2 + bx + c$ is given by the formula $\left(-\frac{b}{2a}, f\left(-\frac{b}{2a}\right)\right)$. Explain what is meant by the notation $f\left(-\frac{b}{2a}\right)$.

PRACTICE

*In Exercises 15–24, find the coordinates of the vertex and the axis of symmetry of the graph of each function. If necessary, complete the square on x to write the equation in the form $y = a(x - h)^2 + k$. **Do not graph the equation,** but tell whether the graph will open upward or downward.*

15. $y = (x - 1)^2 + 2$

16. $y = 2(x - 2)^2 - 1$

17. $f(x) = 2(x + 3)^2 - 4$

18. $f(x) = -3(x + 1)^2 + 3$

19. $y = 2x^2 - 4x$

20. $y = 3x^2 - 3$

21. $y = -4x^2 + 16x + 5$

22. $y = 5x^2 + 20x + 25$

23. $y = 3x^2 + 4x + 2$

24. $y = -6x^2 + 5x - 7$

In Exercises 25–40, first determine the vertex and the axis of symmetry of the graph of the function. Then plot several points and complete the graph. (See Example 8.)

25. $f(x) = (x - 3)^2 + 2$
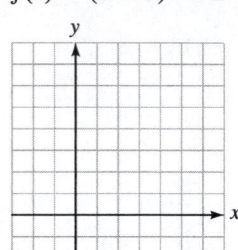

26. $f(x) = (x + 1)^2 - 2$
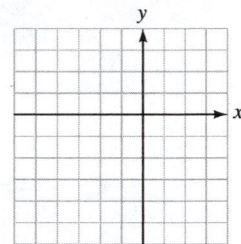

27. $f(x) = -(x - 2)^2$
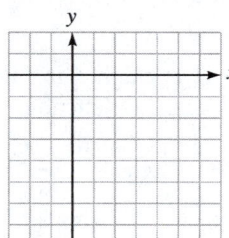

28. $f(x) = -(x + 2)^2$
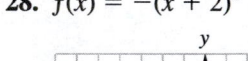
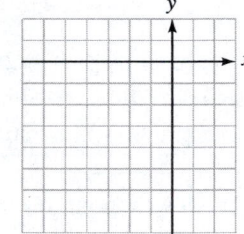

29. $f(x) = -2(x + 3)^2 + 4$
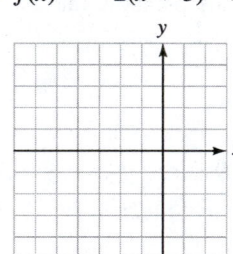

30. $f(x) = 2(x - 2)^2 - 4$
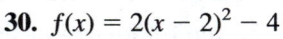
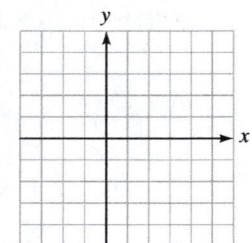

31. $f(x) = (x - 3)^2 + 2$
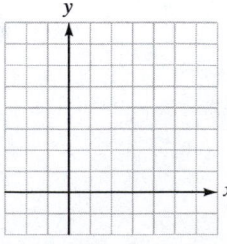

32. $f(x) = (x + 1)^2 - 2$
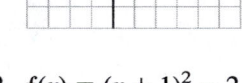
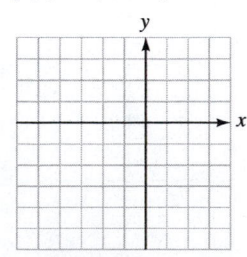

33. $f(x) = x^2 + x - 6$
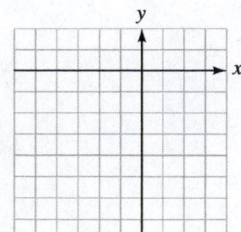

34. $y = x^2 - x - 6$
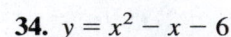
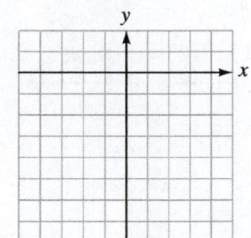

35. $y = -x^2 + 2x + 3$

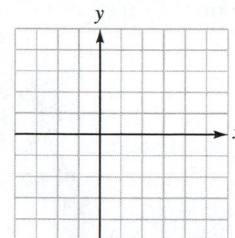

36. $f(x) = -x^2 + x + 2$

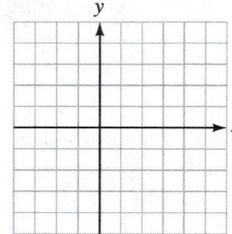

37. $f(x) = -3x^2 + 2x$

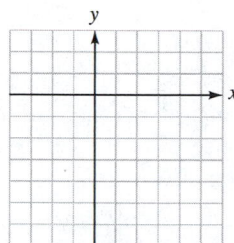

38. $f(x) = 5x + x^2$

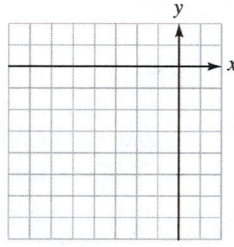

39. $y = -12x^2 - 6x + 6$

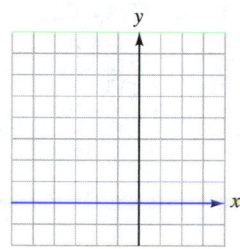

40. $f(x) = -2x^2 + 4x + 3$

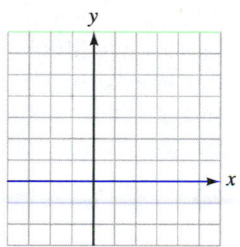

In Exercises 41–44, use a graphing calculator to find the coordinates of the vertex of the graph of each quadratic function. Round to the nearest hundredth.

41. $y = 2x^2 - x + 1$

42. $y = x^2 + 5x - 6$

43. $y = 7 + x - x^2$

44. $y = 2x^2 - 3x + 2$

In Exercises 45–48, use a graphing calculator to solve each equation. If an answer is not exact, round to the nearest hundredth.

45. $x^2 + x - 6 = 0$

46. $2x^2 - 5x - 3 = 0$

47. $0.5x^2 - 0.7x - 3 = 0$

48. $2x^2 - 0.5x - 2 = 0$

APPLICATIONS

49. FIREWORKS A fireworks shell is shot straight up with an initial velocity of 120 feet per second. Its height s after t seconds is given by the equation $s = 120t - 16t^2$. If the shell is designed to explode when it reaches its maximum height, how long after being fired, and at what height, will the fireworks appear in the sky?

50. BALLISTICS From the top of the building in Illustration 2, a ball is thrown straight up with an initial velocity of 32 feet per second. The equation

$$s = -16t^2 + 32t + 48$$

gives the height s of the ball t seconds after it is thrown. Find the maximum height reached by the ball and the time it takes for the ball to hit the ground.

ILLUSTRATION 2

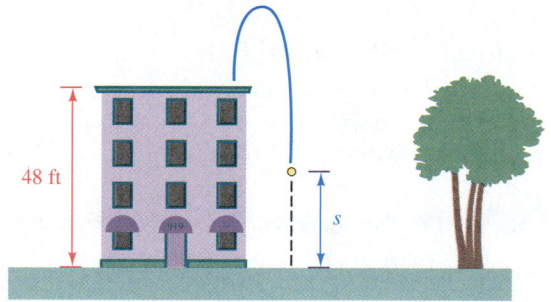

51. FENCING A FIELD See Illustration 3. A farmer wants to fence in three sides of a rectangular field with 1,000 feet of fencing. The other side of the rectangle will be a river. If the enclosed area is to be maximum, find the dimensions of the field.

ILLUSTRATION 3

52. POLICE INVESTIGATION A police officer seals off the scene of a car collision using a roll of yellow police tape that is 300 feet long, as shown in Illustration 4. What dimensions should be used to seal off the maximum rectangular area around the collision? What is the maximum area?

ILLUSTRATION 4

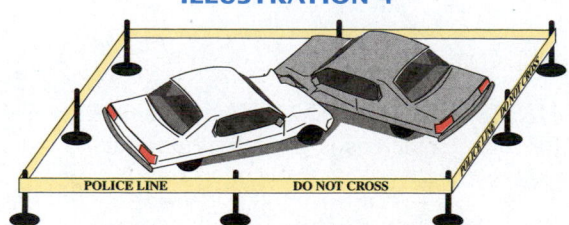

53. OPERATING COSTS The cost C in dollars of operating a certain concrete-cutting machine is related to the number of minutes n the machine is run by the function

$$C(n) = 2.2n^2 - 66n + 655$$

For what number of minutes is the cost of running the machine a minimum? What is the minimum cost?

54. WATER USAGE The height (in feet) of the water level in a reservoir over a one-year period is modeled by the function

$$H(t) = 3.3(t - 9)^2 + 14$$

where $t = 1$ represents January, $t = 2$ represents February, and so on. How low did the water level get that year, and when did it reach the low mark?

55. U.S. ARMY The function

$$N(x) = -0.0534x^2 + 0.337x + 0.969$$

gives the number of active-duty military personnel in the United States Army (in millions) for the years 1965–1972, where $x = 0$ corresponds to 1965, $x = 1$ corresponds to 1966, and so on. For this period, when was the army's personnel strength level at its highest, and what was it? Historically, can you explain why?

56. SCHOOL ENROLLMENT The total annual enrollment (in millions) in U.S. elementary and secondary schools for the years 1975–1996 is given by the model

$$E = 0.058x^2 - 1.162x + 50.604$$

where $x = 0$ corresponds to 1975, $x = 1$ corresponds to 1976, and so on.
a. For this period, when was enrollment the lowest? What was it?
b. Use the model to complete the bar graph in Illustration 5.

ILLUSTRATION 5

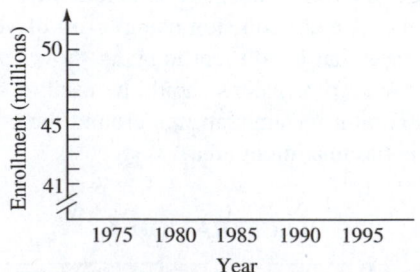

57. MAXIMIZING REVENUE The revenue R received for selling x stereos is given by the formula

$$R = -\frac{x^2}{5} + 80x - 1,000$$

How many stereos must be sold to obtain the maximum revenue? Find the maximum revenue.

58. MAXIMIZING REVENUE When priced at \$30 each, a toy has annual sales of 4,000 units. The manufacturer estimates that each \$1 increase in cost will decrease sales by 100 units. Find the unit price that will maximize total revenue. (*Hint:* Total revenue = price · the number of units sold.)

WRITING

59. Use the example of a stream of water from a drinking fountain to explain the concepts of the vertex and the axis of symmetry of a parabola.

60. What are some quantities that are good to maximize? What are some quantities that are good to minimize?

61. A table of values for $f(x) = 2x^2 - 4x + 3$ is shown in Illustration 6. Explain why it appears that the vertex of the graph of f is the point $(1, 1)$

ILLUSTRATION 6

X	Y1
.25	2.125
.5	1.5
.75	1.125
1	1
1.25	1.125
1.5	1.5
1.75	2.125

X=.25

62. Illustration 7 shows the graph of the quadratic function $y = -4x^2 + 12x$ with domain [0, 3]. Explain how the value of y changes as the value of x increases from 0 to 3.

ILLUSTRATION 7

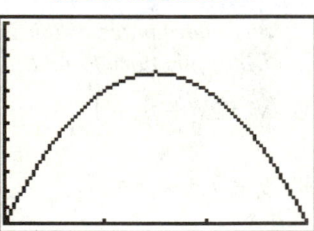

REVIEW

Simplify each expression.

63. $\sqrt{8a}\sqrt{2a^3b}$

64. $\left(\sqrt{23}\right)^2$

65. $\dfrac{\sqrt{3}}{\sqrt{50}}$

66. $\dfrac{3}{\sqrt[3]{9}}$

67. $3\left(\sqrt{5b} - \sqrt{3}\right)^2$

68. $-2\sqrt{5b}\left(4\sqrt{2b} - 3\sqrt{3}\right)$

▶ 10.4 Complex Numbers

In this section, you will learn about

> Imaginary numbers ■ Simplifying imaginary numbers ■ Complex numbers ■ Arithmetic of complex numbers ■ Complex conjugates ■ Rationalizing the denominator ■ Powers of i

Introduction We have seen that negative numbers do not have real-number square roots. For years, people believed that numbers like

$$\sqrt{-1}, \qquad \sqrt{-3}, \qquad \sqrt{-4}, \qquad \text{and} \qquad \sqrt{-9}$$

were nonsense. In the 17th century, René Descartes (1596–1650) called them *imaginary numbers*. Today, imaginary numbers have many important uses, such as describing the behavior of alternating current in electronics.

In this section, we will introduce imaginary numbers and discuss a broader set of numbers called *complex numbers*.

Imaginary Numbers

So far, all of our work with quadratic equations has involved only real numbers. However, the solutions of many quadratic equations are not real numbers.

EXAMPLE 1 **A quadratic equation whose solution is not a real number.** Solve $x^2 + x + 1 = 0$.

Solution Because $x^2 + x + 1$ is prime and cannot be factored, we will use the quadratic formula, with $a = 1$, $b = 1$, and $c = 1$:

$$x = \frac{-b \pm \sqrt{b^2 - 4ac}}{2a}$$

$$= \frac{-1 \pm \sqrt{1^2 - 4(1)(1)}}{2(1)} \qquad \text{Substitute 1 for } a, \text{ 1 for } b, \text{ and 1 for } c.$$

$$= \frac{-1 \pm \sqrt{1 - 4}}{2} \qquad \text{Simplify within the radical.}$$

$$= \frac{-1 \pm \sqrt{-3}}{2}$$

$$x = \frac{-1 + \sqrt{-3}}{2} \quad \text{or} \quad x = \frac{-1 - \sqrt{-3}}{2}$$

Each solution contains the number $\sqrt{-3}$. Since no real number squared is -3, $\sqrt{-3}$ is not a real number, and these two solutions are not real numbers.

To solve equations such as $x^2 + x + 1 = 0$, we must define the square root of a negative number.

SELF CHECK Solve $a^2 + 3a + 5 = 0$. *Answer:* $\dfrac{-3 \pm \sqrt{-11}}{2}$ ■

Square roots of negative numbers are called **imaginary numbers.** The imaginary number $\sqrt{-1}$ is often denoted by the letter i:

$$i = \sqrt{-1}$$

Because i represents the square root of -1, it follows that

$$i^2 = -1$$

Simplifying Imaginary Numbers

We can use extensions of the multiplication and division properties of radicals discussed in Chapter 9 to simplify imaginary numbers. For example,

$$\sqrt{-25} = \sqrt{25(-1)} = \sqrt{25}\sqrt{-1} = 5i \qquad \text{Replace } \sqrt{-1} \text{ with } i.$$

$$\sqrt{\frac{-100}{49}} = \sqrt{\frac{100}{49}(-1)} = \frac{\sqrt{100}}{\sqrt{49}}\sqrt{-1} = \frac{10}{7}i$$

$$\sqrt{-3} = \sqrt{3(-1)} = \sqrt{3}\sqrt{-1} = \sqrt{3}i \qquad \text{The expression } \sqrt{3}i \text{ can be written as } i\sqrt{3} \text{ to make it clear that } i \text{ is not part of the radicand. Do not confuse } \sqrt{3}i \text{ with } \sqrt{3i}.$$

These examples illustrate the following rules.

Properties of radicals

If at least one of a and b is a nonnegative real number, then

$$\sqrt{ab} = \sqrt{a}\sqrt{b} \qquad \text{and} \qquad \sqrt{\frac{a}{b}} = \frac{\sqrt{a}}{\sqrt{b}} \quad (b \neq 0)$$

EXAMPLE 2

Square roots of negative numbers. Write each expression in terms of i:
a. $\sqrt{-81}$, **b.** $-\sqrt{-8}$, and **c.** $3\sqrt{-\dfrac{98}{16}}$.

Solution **a.** $\sqrt{-81} = \sqrt{81(-1)} = \sqrt{81}\sqrt{-1} = 9i$

b. $-\sqrt{-8} = -\sqrt{8(-1)} = -\sqrt{8}\sqrt{-1} = -\sqrt{8}i = -2\sqrt{2}i$

c. $3\sqrt{-\dfrac{98}{16}} = 3\sqrt{\dfrac{98}{16}(-1)} = \dfrac{3\sqrt{49}\sqrt{2}}{\sqrt{16}}\sqrt{-1} = \dfrac{21\sqrt{2}}{4}i$

SELF CHECK

Write each expression in terms of i: **a.** $-\sqrt{-4}$, **b.** $2\sqrt{-28}$, and **c.** $\sqrt{-\dfrac{27}{100}}$.

Answers: **a.** $-2i$, **b.** $4\sqrt{7}i$, **c.** $\dfrac{3\sqrt{3}}{10}i$

EXAMPLE 3

Solving quadratic equations. Solve $x^2 + 21 = 0$.

Solution We can write the equation as $x^2 = -21$ and use the square root property to solve it.

$$x^2 + 21 = 0$$
$$x^2 = -21 \qquad \text{Subtract 21 from both sides.}$$
$$x = \pm\sqrt{-21} \qquad \text{Apply the square root property. Note the square root of a negative number.}$$

This equation has imaginary solutions, which we can express in terms of i.

$$x = \pm\sqrt{21(-1)} \qquad \text{Use the multiplication property of radicals.}$$
$$x = \pm\sqrt{21}i \qquad \sqrt{21(-1)} = \sqrt{21}\sqrt{-1} = \sqrt{21}i.$$

Verify that $x = \sqrt{21}i$ and $x = -\sqrt{21}i$ satisfy the original equation.

SELF CHECK

Solve $x^2 + 75 = 0$.

Answer: $\pm 5\sqrt{3}i$

Complex Numbers

The imaginary numbers are a subset of a set of numbers called the **complex numbers.**

Complex numbers

> A **complex number** is any number that can be written in the form $a + bi$, where a and b are real numbers and $i = \sqrt{-1}$.
>
> In the complex number $a + bi$, a is called the **real part,** and b is called the **imaginary part.**

Some examples of complex numbers are

$3 + 4i$ $a = 3$ and $b = 4$.

$-6 - 5i$ Think of this as $-6 + (-5i)$, where $a = -6$ and $b = -5$. It is common to use $a - bi$ form as a substitute for $a + (-b)i$.

$\dfrac{1}{2} + 0i$ $a = \frac{1}{2}$ and $b = 0$.

$0 + \sqrt{2}i$ $a = 0$ and $b = \sqrt{2}$.

If $b = 0$, the complex number $a + bi$ is a real number. If $b \neq 0$ and $a = 0$, the complex number $0 + bi$ (or just bi) is an imaginary number. The relationship between the real numbers, the imaginary numbers, and the complex numbers is shown in Figure 10-18.

FIGURE 10-18
COMPLEX NUMBERS

Real numbers $a + 0i$	Imaginary numbers $0 + bi$ ($b \neq 0$)
$3, \dfrac{7}{3}, \pi, 125.345$	$4i, -12i, \sqrt{-4}$

$4 + 7i,\ 5 - 16i,\ \dfrac{1}{32 - 12i},\ 15 + \sqrt{-25}$

EXAMPLE 4   **A quadratic equation with complex-number solutions.** Solve $4t^2 - 6t + 3 = 0$. Express the solutions in $a + bi$ form.

Solution   We use the quadratic formula to solve the equation.

$$t = \frac{-b \pm \sqrt{b^2 - 4ac}}{2a}$$

$$= \frac{-(-6) \pm \sqrt{(-6)^2 - 4(4)(3)}}{2(4)} \qquad \text{Substitute 4 for } a, -6 \text{ for } b, \text{ and 3 for } c.$$

$$= \frac{6 \pm \sqrt{36 - 48}}{8} \qquad \text{Simplify within the radical.}$$

$$= \frac{6 \pm \sqrt{-12}}{8}$$

$$= \frac{6 \pm 2\sqrt{3}i}{8} \qquad \sqrt{-12} = \sqrt{12(-1)} = \sqrt{12}\sqrt{-1} = 2\sqrt{3}i.$$

$$= \frac{3 \pm \sqrt{3}i}{4} \qquad \frac{6 \pm 2\sqrt{3}i}{8} = \frac{\overset{1}{\cancel{2}}(3 \pm \sqrt{3}i)}{\underset{1}{\cancel{2}} \cdot 4} = \frac{3 \pm \sqrt{3}i}{4}.$$

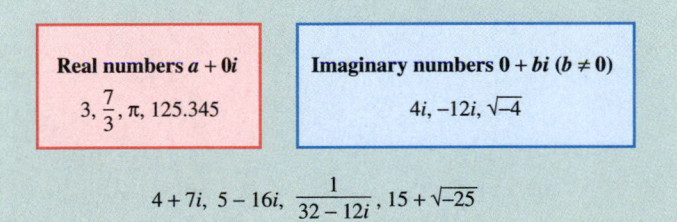

Writing each solution as a complex number in $a + bi$ form,

$$t = \frac{3}{4} + \frac{\sqrt{3}}{4}i \quad \text{or} \quad t = \frac{3}{4} - \frac{\sqrt{3}}{4}i$$

SELF CHECK   Solve $a^2 + 2a + 3 = 0$.   *Answer:* $-1 \pm \sqrt{2}i$ ■

Arithmetic of Complex Numbers

We now consider the rules used to add, subtract, multiply, and divide complex numbers.

Addition and subtraction of complex numbers

To add (or subtract) two complex numbers, add (or subtract) their real parts and add (or subtract) their imaginary parts.

$$(a + bi) + (c + di) = (a + c) + (b + d)i$$
$$(a + bi) - (c + di) = (a - c) + (b - d)i$$

EXAMPLE 5   **Adding and subtracting complex numbers.** Do the operations.

a. $(8 + 4i) + (12 + 8i) = (8 + 12) + (4 + 8)i$ Add the real parts. Add the imaginary parts.

$$= 20 + 12i$$

b. $\left(7 - \sqrt{-16}\right) + \left(9 + \sqrt{-4}\right)$

$= (7 - 4i) + (9 + 2i)$ Write $\sqrt{-16}$ and $\sqrt{-4}$ in terms of i.

$= (7 + 9) + (-4 + 2)i$ Add the real parts. Add the imaginary parts.

$= 16 - 2i$ Write $16 + (-2i)$ in the form $16 - 2i$.

c. $(-6 + i) - (3 + 2i) = (-6 - 3) + (1 - 2)i$:   Subtract the real parts. Subtract the imaginary parts.

$$= -9 - i$$

SELF CHECK   Do the operations:
a. $(3 - 5i) + (-2 + 7i)$ and   *Answers:* **a.** $1 + 2i$,
b. $\left(3 - \sqrt{-25}\right) - \left(-2 + \sqrt{-49}\right)$.   **b.** $5 - 12i$ ■

To multiply two imaginary numbers, they should first be expressed in terms of i. For example,

$$\sqrt{-2}\sqrt{-20} = \left(\sqrt{2}i\right)\left(2\sqrt{5}i\right) \quad \sqrt{-20} = \sqrt{20(-1)} = \sqrt{20}i = 2\sqrt{5}i.$$
$$= 2\sqrt{10}i^2$$
$$= -2\sqrt{10} \qquad i^2 = -1.$$

 WARNING! If a and b are both negative, then $\sqrt{ab} \neq \sqrt{a}\sqrt{b}$. For example, if $a = -16$ and $b = -4$,

$$\sqrt{(-16)(-4)} = \sqrt{64} = 8 \quad \text{but} \quad \sqrt{-16}\sqrt{-4} = (4i)(2i) = 8i^2 = 8(-1) = -8$$

To multiply a complex number by a real number, we use the distributive property to remove parentheses and then simplify. For example,

$$6(2 + 9i) = 6(2) + 6(9i)$$ Use the distributive property.
$$= 12 + 54i$$ Simplify.

To multiply a complex number by an imaginary number, we use the distributive property to remove parentheses and then simplify. For example,

$$-5i(4 - 8i) = -5i(4) - (-5i)8i$$ Use the distributive property.
$$= -20i + 40i^2$$ Simplify.
$$= -40 - 20i$$ Since $i^2 = -1$, $40i^2 = 40(-1) = -40$.

To multiply two complex numbers, we use the FOIL method to develop the following definition.

Multiplying complex numbers

Complex numbers are multiplied as if they were binomials, with $i^2 = -1$:

$$(a + bi)(c + di) = ac + adi + bci + bdi^2$$
$$= (ac - bd) + (ad + bc)i$$

EXAMPLE 6   **Using the FOIL method to multiply complex numbers.** Multiply the complex numbers.

a. $(2 + 3i)(3 - 2i) = 6 - 4i + 9i - 6i^2$ Use the FOIL method.
$$= 6 + 5i - (-6)$$ Combine the imaginary terms: $-4i + 9i = 5i$.
   Simplify: $i^2 = -1$, so $6i^2 = -6$.
$$= 6 + 5i + 6$$
$$= 12 + 5i$$ Combine like terms.

b. $(-4 + 2i)(2 + i) = -8 - 4i + 4i + 2i^2$ Use the FOIL method.
$$= -8 + 0i - 2$$ $-4i + 4i = 0i$. Since $i^2 = -1$, $2i^2 = -2$.
$$= -10 + 0i$$

SELF CHECK   Multiply $(-2 + 3i)(3 - 2i)$.   *Answer:* $0 + 13i$ ■

Complex Conjugates

Complex conjugates

The complex numbers $a + bi$ and $a - bi$ are called **complex conjugates.**

For example,

$3 + 4i$ and $3 - 4i$ are complex conjugates.

$5 - 7i$ and $5 + 7i$ are complex conjugates.

$8 + 17i$ and $8 - 17i$ are complex conjugates.

<table>
<tr><td>E X A M P L E 7</td><td>**Multiplying complex conjugates.** Find the product of $3 + i$ and its complex conjugate.</td></tr>
</table>

Solution   The complex conjugate of $3 + i$ is $3 - i$. We can find the product as follows:

$$(3 + i)(3 - i) = 9 - 3i + 3i - i^2 \quad \text{Use the FOIL method.}$$
$$= 9 - i^2 \qquad\qquad \text{Combine like terms.}$$
$$= 9 - (-1) \qquad\qquad i^2 = -1.$$
$$= 10$$

SELF CHECK   Multiply $(2 + 3i)(2 - 3i)$.   *Answer:* 13   ■

The product of the complex number $a + bi$ and its complex conjugate $a - bi$ is the real number $a^2 + b^2$, as the following work shows:

$$(a + bi)(a - bi) = a^2 - abi + abi - b^2i^2 \quad \text{Use the FOIL method.}$$
$$= a^2 - b^2(-1) \qquad\qquad \text{Combine like terms. } i^2 = -1.$$
$$= a^2 + b^2$$

Rationalizing the Denominator

To divide complex numbers, we often have to rationalize a denominator.

<table>
<tr><td>E X A M P L E 8</td><td>**Rationalizing the denominator.** Divide and write the result in $a + bi$ form: $\dfrac{1}{3 + i}$.</td></tr>
</table>

Solution   The denominator of the fraction $\frac{1}{3+i}$ is not a rational number. We can rationalize the denominator by multiplying both the numerator and the denominator of the fraction by the complex conjugate of the denominator, which is $3 - i$.

$$\frac{1}{3 + i} = \frac{1}{3 + i} \cdot \frac{3 - i}{3 - i} \qquad \tfrac{3-i}{3-i} = 1.$$

$$= \frac{3 - i}{9 - 3i + 3i - i^2} \qquad \text{Multiply the numerators and multiply the denominators.}$$

$$= \frac{3 - i}{9 - (-1)} \qquad\qquad i^2 = -1.$$

$$= \frac{3 - i}{10} \qquad\qquad \text{Simplify in the denominator.}$$

$$= \frac{3}{10} - \frac{1}{10}i \qquad\qquad \text{Write the complex number in } a + bi \text{ form.}$$

SELF CHECK   Divide 1 by $5 - i$.   *Answer:* $\frac{5}{26} + \frac{1}{26}i$ ■

EXAMPLE 9

Dividing complex numbers. Divide and write in $a + bi$ form: $\dfrac{3 - i}{2 + i}$.

Solution We multiply both the numerator and the denominator of the fraction by the complex conjugate of the denominator.

$$\frac{3 - i}{2 + i} = \frac{3 - i}{2 + i} \cdot \frac{2 - i}{2 - i} \qquad \frac{2-i}{2-i} = 1.$$

$$= \frac{6 - 3i - 2i + i^2}{4 - 2i + 2i - i^2} \qquad \text{Multiply the numerators and multiply the denominators.}$$

$$= \frac{5 - 5i}{4 - (-1)} \qquad i^2 = -1.$$

$$= \frac{5(1 - i)}{5} \qquad \text{Factor out 5 in the numerator and divide out the common factor of 5.}$$

$$= 1 - i \qquad \text{Simplify.}$$

SELF CHECK Divide $\dfrac{5 + 4i}{3 + 2i}$.

Answer: $\frac{23}{13} + \frac{2}{13}i$ ∎

EXAMPLE 10

Dividing complex numbers. Write $\dfrac{4 + \sqrt{-16}}{2 + \sqrt{-4}}$ in $a + bi$ form.

Solution $\dfrac{4 + \sqrt{-16}}{2 + \sqrt{-4}} = \dfrac{4 + 4i}{2 + 2i}$ Write the numerator and denominator in $a + bi$ form.

$$= \frac{\overset{1}{2(2 + 2i)}}{\underset{1}{2 + 2i}} \qquad \text{Factor out 2 in the numerator and divide out the common factor of } 2 + 2i.$$

$$= 2 + 0i$$

SELF CHECK Write $\dfrac{3 + \sqrt{-9}}{4 + \sqrt{-16}}$ in $a + bi$ form.

Answer: $\frac{3}{4} + 0i$ ∎

EXAMPLE 11

Division involving i. Find the quotient: $\dfrac{7}{2i}$. Express the result in $a + bi$ form.

Solution The denominator can be expressed as $0 + 2i$. Its conjugate is $0 - 2i$, or just $-2i$.

$$\frac{7}{2i} = \frac{7}{2i} \cdot \frac{-2i}{-2i} \qquad \frac{-2i}{-2i} = 1.$$

$$= \frac{-14i}{-4i^2}$$

$$= \frac{-14i}{4} \qquad i^2 = -1.$$

$$= \frac{-7i}{2} \qquad \text{Simplify.}$$

$$= 0 - \frac{7}{2}i \qquad \text{Write in } a + bi \text{ form.}$$

SELF CHECK Divide $\dfrac{5}{-i}$.

Answer: $0 + 5i$ ∎

Powers of i

The powers of i produce an interesting pattern:

$$i = \sqrt{-1} = i \qquad\qquad i^5 = i^4i = 1i = i$$
$$i^2 = \left(\sqrt{-1}\right)^2 = -1 \qquad i^6 = i^4i^2 = 1(-1) = -1$$
$$i^3 = i^2i = -1i = -i \qquad i^7 = i^4i^3 = 1(-i) = -i$$
$$i^4 = i^2i^2 = (-1)(-1) = 1 \qquad i^8 = i^4i^4 = (1)(1) = 1$$

The pattern continues: $i,\ -1,\ -i,\ 1,\ \ldots$.

Larger powers of i can be simplified by using the fact that $i^4 = 1$. For example, to simplify i^{29}, we note that 29 divided by 4 gives a quotient of 7 and a remainder of 1. Thus, $29 = 4 \cdot 7 + 1$ and

$$i^{29} = i^{4 \cdot 7 + 1} \qquad 4 \cdot 7 = 28.$$
$$= (i^4)^7 \cdot i^1$$
$$= 1^7 \cdot i \qquad i^4 = 1.$$
$$= i \qquad\quad 1 \cdot i = i.$$

The result of this example illustrates the following fact.

Powers of i

If n is a natural number that has a remainder of r when divided by 4, then

$$i^n = i^r$$

EXAMPLE 12 **Simplifying powers of i.** Simplify i^{55}.

Solution We divide 55 by 4 and get a remainder of 3. Therefore,

$$i^{55} = i^3 = -i$$

SELF CHECK Simplify i^{62}. *Answer:* -1 ■

STUDY SET

Section 10.4

VOCABULARY

In Exercises 1–6, fill in the blanks to make the statements true.

1. $\sqrt{-1}$, $\sqrt{-3}$ and $\sqrt{-4}$ are examples of _____ numbers.

2. $3 + 5i$, $2 - 7i$, and $5 - \frac{1}{2}i$ are examples of _____ numbers.

3. The _____ part of $5 + 7i$ is 5. The _____ part is $7i$.

4. $6 + 3i$ and $6 - 3i$ are called complex _____.

5. _____ is a number whose square is -1.

6. i^{25} is called a _____ of i.

CONCEPTS

In Exercises 7–12, fill in the blanks to make the statements true.

7. $i = $ ▨

8. $i^2 = $ ▨

9. $i^3 = $ ▨

10. $i^4 = $ ▨

11. We multiply two complex numbers by using the _____ method.

12. To divide two complex numbers, we _____ the denominator.

13. Use a check to see if $x = -5\sqrt{2}i$ is a solution of $x^2 + 50 = 0$.

14. Give the complex conjugate of each number.
   a. $2 - 3i$ b. 2 c. $-3i$

15. Complete Illustration 1 to show the relationship between the real numbers, the imaginary numbers, the complex numbers, the rational numbers, and the irrational numbers.

ILLUSTRATION 1

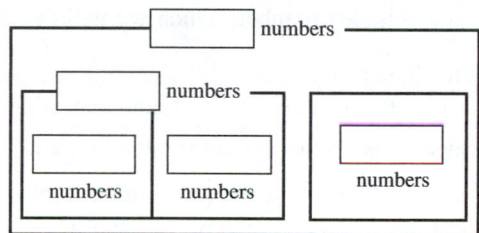

16. Tell whether each statement is true or false.
   a. Every complex number is a real number.
   b. Every real number is a complex number.
   c. Imaginary numbers can always be denoted using i.

   d. The square root of a negative number is an imaginary number.

17. Solve each equation.
   a. $x^2 - 1 = 0$ b. $x^2 + 1 = 0$

18. Is $\sqrt[3]{-64}$ an imaginary number? Explain.

NOTATION

In Exercises 19–20, complete each solution.

19. $(3 + 2i)(3 - i) = \boxed{} - 3i + \boxed{} - 2i^2$

   $= 9 + 3i + \boxed{}$

   $= 11 + 3i$

20. $\dfrac{3}{2 - i} = \dfrac{3}{2 - i} \cdot \dfrac{\boxed{}}{\boxed{}}$

   $= \dfrac{6 + \boxed{}}{4 + \boxed{} - 2i - i^2}$

   $= \dfrac{6 + 3i}{\boxed{}}$

   $= \dfrac{6}{5} + \dfrac{3}{5}i$

21. Tell whether each statement is true or false.
   a. $\sqrt{6}i = i\sqrt{6}$ b. $\sqrt{8}i = \sqrt{8i}$
   c. $\sqrt{-25} = -\sqrt{25}$ d. $-i = i$

22. Tell whether each statement is true or false.
   a. $\sqrt{-3}\sqrt{-2} = \sqrt{6}$
   b. $\sqrt{-36} + \sqrt{-25} = \sqrt{-61}$

PRACTICE

In Exercises 23–34, express each number in terms of i.

23. $\sqrt{-9}$ 24. $\sqrt{-4}$

25. $\sqrt{-7}$ 26. $\sqrt{-11}$

27. $\sqrt{-24}$ 28. $\sqrt{-28}$

29. $-\sqrt{-24}$ 30. $-\sqrt{-72}$

31. $5\sqrt{-81}$ 32. $6\sqrt{-49}$

33. $\sqrt{-\dfrac{25}{9}}$ 34. $-\sqrt{-\dfrac{121}{144}}$

In Exercises 35–42, simplify each expression.

35. $\sqrt{-1}\sqrt{-36}$ 36. $\sqrt{-9}\sqrt{-100}$

37. $\sqrt{-2}\sqrt{-6}$ 38. $\sqrt{-3}\sqrt{-6}$

39. $\dfrac{\sqrt{-25}}{\sqrt{-64}}$ 40. $\dfrac{\sqrt{-4}}{\sqrt{-1}}$

41. $-\dfrac{\sqrt{-400}}{\sqrt{-1}}$ 42. $-\dfrac{\sqrt{-225}}{\sqrt{-16}}$

In Exercises 43–54, solve each equation. Write all solutions in bi or a + bi form.

43. $x^2 + 9 = 0$ 44. $x^2 + 100 = 0$

45. $3x^2 = -16$ 46. $2x^2 = -25$

47. $x^2 + 2x + 2 = 0$ 48. $x^2 - 2x + 6 = 0$

49. $2x^2 + x + 1 = 0$ 50. $3x^2 + 2x + 1 = 0$

51. $3x^2 - 4x = -2$ 52. $2x^2 + 3x = -3$

53. $3x^2 - 2x = -3$ 54. $5x^2 = 2x - 1$

In Exercises 55–78, do the operations. Write all answers in a + bi form.

55. $(3 + 4i) + (5 - 6i)$ 56. $(5 + 3i) - (6 - 9i)$

57. $(7 - 3i) - (4 + 2i)$ 58. $(8 + 3i) + (-7 - 2i)$

59. $(6 - i) + (9 + 3i)$ **60.** $(5 - 4i) - (3 + 2i)$

61. $\left(8 + \sqrt{-25}\right) + \left(7 + \sqrt{-4}\right)$

62. $\left(-7 + \sqrt{-81}\right) - \left(-2 - \sqrt{-64}\right)$

63. $3(2 - i)$ **64.** $-4(3 + 4i)$

65. $-5i(5 - 5i)$ **66.** $2i(7 + 2i)$

67. $(2 + i)(3 - i)$ **68.** $(4 - i)(2 + i)$

69. $(3 - 2i)(2 + 3i)$ **70.** $(3 - i)(2 + 3i)$

71. $(4 + i)(3 - i)$ **72.** $(1 - 5i)(1 - 4i)$

73. $\left(2 - \sqrt{-16}\right)\left(3 + \sqrt{-4}\right)$

74. $\left(3 - \sqrt{-4}\right)\left(4 - \sqrt{-9}\right)$

75. $\left(2 + \sqrt{2}i\right)\left(3 - \sqrt{2}i\right)$ **76.** $\left(5 + \sqrt{3}i\right)\left(2 - \sqrt{3}i\right)$

77. $(2 + i)^2$ **78.** $(3 - 2i)^2$

In Exercises 79–102, write each expression in a + bi form.

79. $\dfrac{1}{i}$ **80.** $\dfrac{1}{i^3}$

81. $\dfrac{4}{5i^3}$ **82.** $\dfrac{3}{2i}$

83. $\dfrac{3i}{8\sqrt{-9}}$ **84.** $\dfrac{5i^3}{2\sqrt{-4}}$

85. $\dfrac{-3}{5i^5}$ **86.** $\dfrac{-4}{6i^7}$

87. $\dfrac{5}{2 - i}$ **88.** $\dfrac{26}{3 - 2i}$

89. $\dfrac{3}{5 + i}$ **90.** $\dfrac{-4}{7 - 2i}$

91. $\dfrac{-12}{7 - \sqrt{-1}}$ **92.** $\dfrac{4}{3 + \sqrt{-1}}$

93. $\dfrac{5i}{6 + 2i}$ **94.** $\dfrac{-4i}{2 - 6i}$

95. $\dfrac{-2i}{3 + 2i}$ **96.** $\dfrac{3i}{6 - i}$

97. $\dfrac{3 - 2i}{3 + 2i}$ **98.** $\dfrac{2 + 3i}{2 - 3i}$

99. $\dfrac{3 + 2i}{3 + i}$ **100.** $\dfrac{2 - 5i}{2 + 5i}$

101. $\dfrac{\sqrt{5} - \sqrt{3}i}{\sqrt{5} + \sqrt{3}i}$ **102.** $\dfrac{\sqrt{3} + \sqrt{2}i}{\sqrt{3} - \sqrt{2}i}$

In Exercises 103–110, simplify each expression.

103. i^{21} **104.** i^{19}

105. i^{27} **106.** i^{22}

107. i^{100} **108.** i^{42}

109. i^{97} **110.** i^{200}

APPLICATIONS

111. FRACTAL GEOMETRY Complex numbers are fundamental in the creation of the intricate geometric shape shown in Illustration 2, called a *fractal*. Fractal geometry is a rapidly expanding field with applications in science, medicine, and computer graphics. The process of creating this image is based on the following sequence of steps, which begins by picking any complex number, which we will call z.

1. Square z, and then add that result to z.
2. Square the result from step 1, and then add it to z.
3. Square the result from step 2, and then add it to z.

If we begin with the complex number i, what is the result after performing steps 1, 2, and 3?

ILLUSTRATION 2

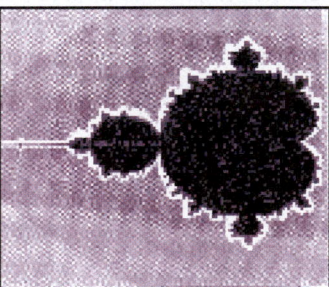

112. ELECTRONICS The impedance Z in an AC (alternating current) circuit is a measure of how much the circuit impedes (hinders) the flow of current through it. The impedance is related to the voltage V and the current I by the formula

$$V = IZ$$

If a circuit has a current of $(0.5 + 2.0i)$ amps and an impedance of $(0.4 - 3.0i)$ ohms, find the voltage.

WRITING

113. What is an imaginary number?

114. In Example 1 of this section, what unusual situation illustrated the need to define the square root of a negative number?

REVIEW

115. What are the lengths of the two legs of a 30°–60°–90° triangle if the hypotenuse is 30 units long?

116. What are the lengths of the other two sides of a 45°–45°–90° triangle if one leg is 30 units long?

117. WIND SPEED A plane that can fly 200 mph in still air makes a 330-mile flight with a tail wind and re-

turns, flying into the same wind. Find the speed of the wind if the total flying time is $3\frac{1}{3}$ hours.

118. FINDING RATES A student drove a distance of 135 miles at an average speed of 50 mph. How much faster would she have to drive on the return trip to save 30 minutes of driving time?

▶ 10.5   The Discriminant and Equations That Can Be Written in Quadratic Form

In this section, you will learn about

   The discriminant ■ Equations that can be written in quadratic form
   ■ Problem solving

Introduction   In this section, we will discuss how to predict what type of solutions a quadratic equation will have without actually solving the equation. We will then solve some special equations that can be written in quadratic form. Finally, we will use the equation-solving methods of this chapter to solve a shared-work problem.

The Discriminant

We can predict what type of solutions a particular quadratic equation will have without solving it. To see how, we suppose that the coefficients a, b, and c in the equation $ax^2 + bx + c = 0$ $(a \neq 0)$ are real numbers. Then the solutions of the equation are given by the quadratic formula

$$x = \frac{-b \pm \sqrt{b^2 - 4ac}}{2a} \quad (a \neq 0)$$

If $b^2 - 4ac \geq 0$, the solutions are real numbers. If $b^2 - 4ac < 0$, the solutions are nonreal complex numbers. Thus, the value of $b^2 - 4ac$, called the **discriminant**, determines the type of solutions for a particular quadratic equation.

The discriminant

If a, b, and c are real numbers and

if $b^2 - 4ac$ is . . .	the solutions are . . .
positive,	two different real numbers.
0,	two real numbers that are equal.
negative,	two different nonreal complex numbers that are complex conjugates.

If a, b, and c are rational numbers and

if $b^2 - 4ac$ is . . .	the solutions are . . .
a perfect square,	two different rational numbers.
positive and not a perfect square,	two different irrational numbers.

EXAMPLE 1

Using the discriminant. Determine the type of solutions for each equation:
a. $x^2 + x + 1 = 0$ and **b.** $3x^2 + 5x + 2 = 0$.

Solution   **a.** We calculate the discriminant for $x^2 + x + 1 = 0$:

$$b^2 - 4ac = 1^2 - 4(1)(1) \quad a = 1, b = 1, \text{ and } c = 1.$$
$$= -3 \qquad \text{The result is a negative number.}$$

Since $b^2 - 4ac < 0$, the solutions of $x^2 + x + 1 = 0$ are two nonreal complex numbers that are complex conjugates.

b. For $3x^2 + 5x + 2 = 0$,

$$b^2 - 4ac = 5^2 - 4(3)(2) \quad a = 3, b = 5, \text{ and } c = 2.$$
$$= 25 - 24$$
$$= 1 \qquad \text{The result is a positive number.}$$

Since $b^2 - 4ac > 0$ and $b^2 - 4ac$ is a perfect square, the two solutions of $3x^2 + 5x + 2 = 0$ are rational and unequal.

SELF CHECK

Determine the type of solutions for
a. $x^2 + x - 1 = 0$ and
b. $4x^2 - 10x + 25 = 0$.

Answers: **a.** real numbers that are irrational and unequal, **b.** nonreal numbers that are complex conjugates ■

Equations That Can Be Written in Quadratic Form

Many equations that are not quadratic can be written in quadratic form $(ax^2 + bx + c = 0)$ and then solved using the techniques discussed in previous sections. For example, a careful inspection of the equation $x^4 - 5x^2 + 4 = 0$ leads to the following observations:

In the leading term, this power of the variable is the square of . . .

$$x^4 - 5x^2 + 4 = 0$$

. . . the power of the variable in the middle term.

The last term is a constant.

Equations having these characteristics are said to be *quadratic in form*. One method used to solve them is to make a substitution.

$$x^4 - 5x^2 + 4 = 0 \quad \text{The given equation.}$$
$$(x^2)^2 - 5(x^2) + 4 = 0 \quad \text{Write } x^4 \text{ as } (x^2)^2.$$
$$y^2 - 5y + 4 = 0 \quad \text{Let } y = x^2. \text{ Replace each } x^2 \text{ with } y.$$

We can solve this quadratic equation by factoring.

$$(y - 4)(y - 1) = 0 \qquad \text{Factor } y^2 - 5y + 4.$$
$$y - 4 = 0 \quad \text{or} \quad y - 1 = 0 \quad \text{Set each factor equal to 0.}$$
$$y = 4 \qquad\qquad y = 1$$

These are *not* the solutions for x. To find x, we now "undo" the earlier substitutions by replacing each y with x^2. Then we solve for x.

$$x^2 = 4 \qquad \text{or} \quad x^2 = 1$$
$$x = \pm\sqrt{4} \qquad\quad x = \pm\sqrt{1} \quad \text{Use the square root property.}$$
$$x = \pm 2 \qquad\qquad x = \pm 1$$

This equation has four solutions: 1, −1, 2, and −2. Verify that each one satisfies the original equation.

EXAMPLE 2

Solving equations that are quadratic in form. Solve $x - 7\sqrt{x} + 12 = 0$.

Solution

This equation is quadratic in form, because the power of the variable of the leading term is the square of the variable factor of the middle term: $x = \left(\sqrt{x}\right)^2$. If we let $y = \sqrt{x}$, then $y^2 = x$. With this substitution, the equation

$$x - 7\sqrt{x} + 12 = 0$$

becomes a quadratic equation that can be solved by factoring.

$$y^2 - 7y + 12 = 0 \qquad \text{Substitute } y^2 \text{ for } x \text{ and } y \text{ for } \sqrt{x}.$$
$$(y - 3)(y - 4) = 0 \qquad \text{Factor } y^2 - 7y + 12 = 0.$$
$$y - 3 = 0 \quad \text{or} \quad y - 4 = 0 \qquad \text{Set each factor equal to 0.}$$
$$y = 3 \qquad\qquad y = 4$$

Replace each y with $\sqrt{x}$ and solve the radical equations by squaring both sides.

$$\sqrt{x} = 3 \quad \text{or} \quad \sqrt{x} = 4$$
$$x = 9 \qquad\qquad x = 16 \qquad \text{Square both sides.}$$

Verify that both solutions satisfy the original equation.

SELF CHECK   Solve $x + x^{1/2} - 6 = 0$.   *Answer:* 4   ▪

EXAMPLE 3

Solving equations that are quadratic in form. Solve $2m^{2/3} - 2 = 3m^{1/3}$.

Solution

After writing the equation in descending powers of m, we see that

$$2m^{2/3} - 3m^{1/3} - 2 = 0$$

is quadratic in form, because $m^{2/3} = (m^{1/3})^2$. We will use the substitution $y = m^{1/3}$ to write this equation in quadratic form.

$$2m^{2/3} - 3m^{1/3} - 2 = 0$$
$$2(m^{1/3})^2 - 3m^{1/3} - 2 = 0$$
$$2y^2 - 3y - 2 = 0 \qquad \text{Replace } m^{1/3} \text{ with } y.$$
$$(2y + 1)(y - 2) = 0 \qquad \text{Factor } 2y^2 - 3y - 2 = 0.$$
$$2y + 1 = 0 \quad \text{or} \quad y - 2 = 0 \qquad \text{Set each factor equal to 0.}$$
$$y = -\frac{1}{2} \qquad\qquad y = 2$$

Replace each y with $m^{1/3}$ and solve for m.

$$m^{1/3} = -\frac{1}{2} \qquad \text{or} \qquad m^{1/3} = 2$$
$$(m^{1/3})^3 = \left(-\frac{1}{2}\right)^3 \qquad (m^{1/3})^3 = (2)^3 \qquad \text{Recall that } m^{1/3} = \sqrt[3]{m}. \text{ To solve for } m, \text{ cube both sides.}$$
$$m = -\frac{1}{8} \qquad\qquad m = 8$$

Verify that both solutions satisfy the original equation.

SELF CHECK   Solve $a^{2/3} = -3a^{1/3} + 10$.   *Answer:* −125, 8   ▪

EXAMPLE 4

Solving equations that are quadratic in form.
Solve $(4t + 2)^2 - 30(4t + 2) + 224 = 0$.

Solution　If we make the substitution $y = 4t + 2$, the given equation becomes

$$y^2 - 30y + 224 = 0$$

which can be solved by using the quadratic formula.

$$y = \frac{-b \pm \sqrt{b^2 - 4ac}}{2a}$$

$$= \frac{-(-30) \pm \sqrt{(-30)^2 - 4(1)(224)}}{2(1)}$$ Substitute 1 for a, -30 for b, and 224 for c.

$$= \frac{30 \pm \sqrt{900 - 896}}{2}$$ Simplify within the radical.

$$= \frac{30 \pm 2}{2}$$ $\sqrt{900 - 896} = \sqrt{4} = 2$.

$$y = 16 \quad \text{or} \quad y = 14$$

To find t, we replace y with $4t + 2$ and solve for t.

$$4t + 2 = 16 \quad \text{or} \quad 4t + 2 = 14$$
$$4t = 14 \qquad\qquad 4t = 12$$
$$t = 3.5 \qquad\qquad t = 3$$

SELF CHECK　Solve $(n + 3)^2 - 6(n + 3) = -8$.　　　　*Answer:* $-1, 1$　　■

Problem Solving

EXAMPLE 5

Household appliances. A water tempera—ture control on a washing machine is shown in Figure 10-19. When the "warm" setting is selected, both the hot and cold water inlets open to fill the tub in 2 minutes 15 seconds. When the "cold" temperature setting is chosen, the cold water inlet fills the tub 45 seconds faster than when the "hot" setting is used. How long does it take to fill the washing machine with hot water?

FIGURE 10-19

Electronic Temperature Control

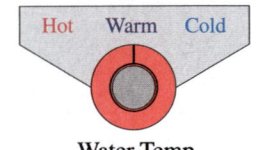

Water Temp

ANALYZE THE PROBLEM　The key to solving this problem is to determine how much of the tub is filled by each water temperature setting in 1 second. On the "warm" setting, when the hot and cold inlets are working together, the tub is filled in 2 minutes 15 seconds, or 135 seconds. So in 1 second, they fill $\frac{1}{135}$ of the tub.

FORM AN EQUATION　Let x represent the number of seconds it takes to fill the tub when the "hot" temperature is chosen. In 1 second, the hot water inlet fills $\frac{1}{x}$ of the tub. Since the cold water inlet fills the tub in 45 seconds less time, the washing machine can be filled with cold water in $(x - 45)$ seconds. In 1 second, $\frac{1}{x - 45}$ of the tub is filled by the cold water inlet. We can now form an equation.

What the hot water inlet pipe can do in 1 second	plus	what the cold water inlet pipe can do in 1 second	equals	what they can do together in 1 second.
$\dfrac{1}{x}$	$+$	$\dfrac{1}{x - 45}$	$=$	$\dfrac{1}{135}$

SOLVE THE EQUATION

$$\frac{1}{x} + \frac{1}{x - 45} = \frac{1}{135}$$

$$\mathbf{135x(x - 45)}\left(\frac{1}{x} + \frac{1}{x - 45}\right) = \mathbf{135x(x - 45)}\left(\frac{1}{135}\right)$$

Multiply both sides by $135x(x - 45)$ to clear the equation of fractions.

$$135(x - 45) + 135x = x(x - 45)$$

Simplify.

$$135x - 6{,}075 + 135x = x^2 - 45x$$

Use the distributive property to remove parentheses.

$$270x - 6{,}075 = x^2 - 45x$$

Combine like terms.

$$0 = x^2 - 315x + 6{,}075$$

Subtract $270x$ from both sides. Add 6,075 to both sides.

To solve this equation, we will use the quadratic formula, with $a = 1$, $b = -315$, and $c = 6{,}075$.

$$x = \frac{-b \pm \sqrt{b^2 - 4ac}}{2a}$$

$$= \frac{-(\mathbf{-315}) \pm \sqrt{(\mathbf{-315})^2 - 4(\mathbf{1})(\mathbf{6{,}075})}}{2(\mathbf{1})}$$

Substitute 1 for a, -315 for b, and 6,075 for c.

$$= \frac{315 \pm \sqrt{99{,}225 - 24{,}300}}{2}$$

Simplify within the radical.

$$= \frac{315 \pm \sqrt{74{,}925}}{2}$$

$\sqrt{99{,}225 - 24{,}300} = \sqrt{74{,}925}$.

$$x \approx \frac{589}{2} \quad \text{or} \quad x \approx \frac{41}{2}$$

$$x \approx 294 \qquad x \approx 21$$

STATE THE CONCLUSION

We can disregard the solution of 21 seconds, because this would imply that the cold water inlet fills the tub in a negative number of seconds ($21 - 45 = -24$). Therefore, the hot water inlet fills the washing machine tub in about 294 seconds, which is 4 minutes 54 seconds.

CHECK THE RESULT

Use estimation to check the result. ∎

STUDY SET

Section 10.5

VOCABULARY

In Exercises 1–2, fill in the blanks to make the statements true.

1. For the quadratic equation $ax^2 + bx + c = 0$, the discriminant is _____.

2. When an equation is written in the form $ax^2 + bx + c = 0$, we say that it is written in _____ form.

CONCEPTS

In Exercises 3–6, consider the equation $ax^2 + bx + c = 0$, where a, b, and c are rational numbers, and fill in the blanks to make the statements true.

3. If $b^2 - 4ac < 0$, the solutions of the equation are nonreal complex _____.

4. If $b^2 - 4ac =$ ____, the solutions of the equation are equal real numbers.

5. If $b^2 - 4ac$ is a perfect square, the solutions are _____ numbers and _____.

6. If $b^2 - 4ac$ is positive and not a perfect square, the solutions are _____ numbers and _____.

7. Consider $x^4 - 3x^2 + 2 = 0$.
   a. What is the relationship between the powers of x in the first two terms on the left-hand side?
   b. Is this equation quadratic in form?

8. Consider $x^{2/3} + 4x^{1/3} - 5 = 0$.
   a. What is the relationship between the powers of x in the first two terms on the left-hand side?

   b. Is this equation quadratic in form?

NOTATION

In Exercises 9–10, complete each solution.

9. To find the type of solutions for the equation $x^2 + 5x + 6 = 0$, we compute the discriminant.
$$b^2 - 4ac = \boxed{}^2 - 4(1)\left(\boxed{}\right)$$
$$= 25 - \boxed{}$$
$$= 1$$

Since a, b, and c are rational numbers and the value of the discriminant is a perfect square, the solutions are _____ numbers and unequal.

10. Change $\dfrac{3}{4} + x = \dfrac{3x - 50}{4(x - 6)}$ to quadratic form.

$$\boxed{}\left(\dfrac{3}{4} + x\right) = \boxed{} \dfrac{3x - 50}{4(x - 6)}$$

$$3(x - 6) + 4x\left(\boxed{}\right) = 3x - 50$$

$$3x - \boxed{} + 4x^2 - \boxed{} = 3x - 50$$

$$4x^2 - 24x + \boxed{} = 0$$

$$\boxed{} - 6x + 8 = 0$$

PRACTICE

In Exercises 11–18, use the discriminant to determine what type of solutions exist for each quadratic equation. **Do not solve the equation.**

11. $4x^2 - 4x + 1 = 0$

12. $6x^2 - 5x - 6 = 0$

13. $5x^2 + x + 2 = 0$

14. $3x^2 + 10x - 2 = 0$

15. $2x^2 = 4x - 1$

16. $9x^2 = 12x - 4$

17. $x(2x - 3) = 20$

18. $x(x - 3) = -10$

19. Use the discriminant to determine whether the solutions of $1,492x^2 + 1,776x - 2,000 = 0$ are real numbers.

20. Use the discriminant to determine whether the solutions of $1,776x^2 - 1,492x + 2,000 = 0$ are real numbers.

In Exercises 21–50, solve each equation.

21. $x^4 - 17x^2 + 16 = 0$

22. $x^4 - 10x^2 + 9 = 0$

23. $x^4 = 6x^2 - 5$

24. $2x^4 + 24 = 26x^2$

25. $t^4 + 3t^2 = 28$

26. $3h^4 + h^2 - 2 = 0$

27. $2x + \sqrt{x} - 3 = 0$

28. $2x - \sqrt{x} - 1 = 0$

29. $3x + 5\sqrt{x} + 2 = 0$

30. $3x - 4\sqrt{x} + 1 = 0$

31. $x - 6\sqrt{x} = -8$

32. $x - 5x^{1/2} + 4 = 0$

33. $x^{2/3} + 5x^{1/3} + 6 = 0$

34. $x^{2/3} - 7x^{1/3} + 12 = 0$

35. $a^{2/3} - 2a^{1/3} - 3 = 0$

36. $r^{2/3} + 4r^{1/3} - 5 = 0$

37. $2(2x + 1)^2 - 7(2x + 1) + 6 = 0$

38. $3(2 - x)^2 + 10(2 - x) - 8 = 0$

39. $(c + 1)^2 - 4(c + 1) - 8 = 0$

40. $(k - 7)^2 + 6(k - 7) + 10 = 0$

41. $x + 5 + \dfrac{4}{x} = 0$

42. $x - 4 + \dfrac{3}{x} = 0$

43. $\dfrac{1}{x + 2} + \dfrac{24}{x + 3} = 13$

44. $\dfrac{3}{x} + \dfrac{4}{x + 1} = 2$

45. $\dfrac{2}{x - 1} + \dfrac{1}{x + 1} = 3$

46. $\dfrac{3}{x - 2} - \dfrac{1}{x + 2} = 5$

47. $x^{-4} - 2x^{-2} + 1 = 0$

48. $4x^{-4} + 1 = 5x^{-2}$

49. $x + \dfrac{2}{x - 2} = 0$

50. $x + \dfrac{x + 5}{x - 3} = 0$

APPLICATIONS

51. FLOWER ARRANGING A florist needs to determine the height h of the flowers shown in Illustration 1. The radius r, the width w, and the height h of the circular-shaped arrangement are related by the formula

$$r = \frac{4h^2 + w^2}{8h}$$

If w is to be 34 inches and r is to be 18 inches, find h to the nearest tenth of an inch.

ILLUSTRATION 1

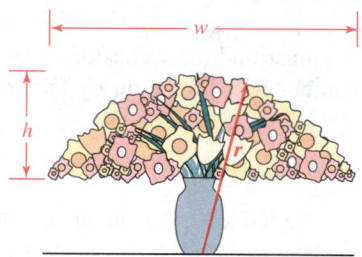

52. **ARCHITECTURE** A **golden rectangle** is said to be one of the most visually appealing of all geometric forms. The Parthenon, built by the Greeks in the 5th century B.C. and shown in Illustration 2, fits into a golden rectangle once its ruined triangular pediment is drawn in.

In a golden rectangle, the length l and width w must satisfy the equation

$$\frac{l}{w} = \frac{w}{l - w}$$

If a rectangular billboard is to have a width of 20 feet, what should its length be so that it is a golden rectangle? Round to the nearest tenth.

ILLUSTRATION 2

53. **SNOWMOBILE** A woman drives her snowmobile 150 miles at a rate of r mph. She could have gone the same distance in 2 hours less time if she had increased her speed by 20 mph. Find r.

54. **BICYCLING** Tina bicycles 160 miles at the rate of r mph. The same trip would have taken 2 hours longer if she had decreased her speed by 4 mph. Find r.

55. **CROWD CONTROL** After a sold-out performance at a county fair, security guards have found that the grandstand area can be emptied in 6 minutes if both the east and west exits are opened. If just the east exit is used, it takes 4 minutes longer to clear the grandstand than it does if just the west exit is opened. How long does it take to clear the grandstand if everyone must file through the east exit?

56. **PAPER ROUTE** When a father, in a car, and his son, on a bicycle, work together to distribute the morning edition, it takes them 35 minutes to complete a paper route. Working alone, it takes the son 25 minutes longer than the father. To the nearest minute, how long does it take the son to cover the paper route on his bicycle?

WRITING

57. Describe how to predict what type of solutions the equation $3x^2 - 4x + 5 = 0$ will have.

58. Explain how the method of substitution is used in this section to solve equations.

REVIEW

In Exercises 59–60, solve the equation.

59. $\dfrac{1}{4} + \dfrac{1}{t} = \dfrac{1}{2t}$

60. $\dfrac{p - 3}{3p} + \dfrac{1}{2p} = \dfrac{1}{4}$

61. Find the slope of the line passing through $(-2, -4)$ and $(3, 5)$.

62. Write the equation of the line passing through $(-2, -4)$ and $(3, 5)$ in general form.

▶ **10.6**

Quadratic and Other Nonlinear Inequalities

In this section, you will learn about

- Solving quadratic inequalities ■ Solving rational inequalities
- ■ Graphs of nonlinear inequalities in two variables

Introduction If $a \neq 0$, inequalities of the form $ax^2 + bx + c < 0$ and $ax^2 + bx + c > 0$ are called *quadratic inequalities*. We will begin this section by showing how to solve these inequalities by making a sign chart. We will then show how to solve other nonlinear inequalities using the same technique. To conclude, we will show how to find graphical solutions of nonlinear inequalities containing two variables.

Solving Quadratic Inequalities

To solve the inequality $x^2 + x - 6 < 0$, we must find the values of x that make the inequality true. This can be done using a number line. We begin by factoring the trinomial to obtain

$$(x + 3)(x - 2) < 0$$

Since the product of $x + 3$ and $x - 2$ must be less than 0, the values of $x + 3$ and $x - 2$ must be opposite in sign. To find the intervals where this is true, we keep track of their signs by constructing the chart in Figure 10-20. The chart shows that

- $x - 2$ is 0 when $x = 2$, is positive when $x > 2$ (indicated with $+$ signs in the figure), and is negative when $x < 2$ (indicated with $-$ signs in the figure).
- $x + 3$ is 0 when $x = -3$, is positive when $x > -3$, and is negative when $x < -3$.

The only place where the values of the binomials are opposite in sign is in the interval $(-3, 2)$. Therefore, the solution set of the inequality can be denoted as

$$-3 < x < 2$$

FIGURE 10-20

The graph of the solution set is shown on the number line in Figure 10-20.

EXAMPLE 1

Constructing a sign chart. Solve $x^2 + 2x - 3 \geq 0$.

Solution We factor the trinomial to get $(x - 1)(x + 3)$ and construct a sign chart, as in Figure 10-21.

- $x - 1$ is 0 when $x = 1$, is positive when $x > 1$, and is negative when $x < 1$.
- $x + 3$ is 0 when $x = -3$, is positive when $x > -3$, and is negative when $x < -3$.

The product of $x - 1$ and $x + 3$ will be greater than 0 when the signs of the binomial factors are the same. This occurs in the intervals $(-\infty, -3)$ and $(1, \infty)$. The numbers -3 and 1 are also included, because they make the product equal to 0. Thus, the solution set is

$$(-\infty, -3] \cup [1, \infty) \qquad \text{or} \qquad x \leq -3 \text{ or } x \geq 1$$

FIGURE 10-21

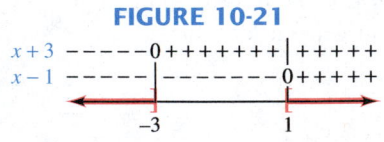

The graph of the solution set is shown on the number line in Figure 10-21.

SELF CHECK Solve $x^2 + 2x - 15 > 0$ and graph the solution set.

Answer: $(-\infty, -5) \cup (3, \infty)$

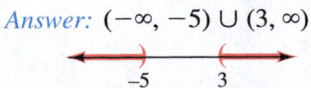

■

Solving Rational Inequalities

Making a sign chart is useful for solving many inequalities that are neither linear nor quadratic. In the next three examples, we will use a sign chart to solve *rational inequalities*.

EXAMPLE 2

Solving rational inequalities. Solve $\frac{1}{x} < 6$.

Solution

We subtract 6 from both sides to make the right-hand side equal to 0. We then find a common denominator and add:

$$\frac{1}{x} < 6$$

$$\frac{1}{x} - 6 < 0 \quad \text{\textcolor{red}{Subtract 6 from both sides.}}$$

$$\frac{1}{x} - \frac{6x}{x} < 0 \quad \text{\textcolor{red}{Get a common denominator.}}$$

$$\frac{1 - 6x}{x} < 0 \quad \text{\textcolor{red}{Subtract the numerators and keep the common denominator.}}$$

We now make a sign chart, as shown in Figure 10-22.

- The denominator x is 0 when $x = 0$, is positive when $x > 0$, and is negative when $x < 0$.
- The numerator $1 - 6x$ is 0 when $x = \frac{1}{6}$, is positive when $x < \frac{1}{6}$, and is negative when $x > \frac{1}{6}$.

The fraction $\frac{1-6x}{x}$ will be less than 0 when the numerator and denominator are opposite in sign. This occurs in the interval

FIGURE 10-22

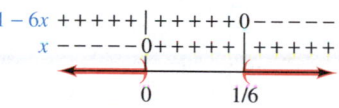

$$(-\infty, 0) \cup \left(\frac{1}{6}, \infty\right) \quad \text{or} \quad x < 0 \text{ or } x > \frac{1}{6}$$

The graph of this interval is shown in Figure 10-22.

SELF CHECK

Solve $\frac{3}{x} > 5$.

Answer: $\left(0, \frac{3}{5}\right)$

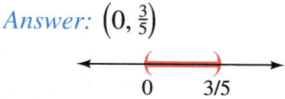

WARNING! Since we don't know whether x is positive, 0, or negative, multiplying both sides of the inequality $\frac{1}{x} < 6$ by x is a three-case situation:

- If $x > 0$, then $1 < 6x$.
- If $x = 0$, then $\frac{1}{x}$ is undefined.
- If $x < 0$, then $1 > 6x$.

If we multiply both sides by x and solve the linear inequality $1 < 6x$, we are considering only one case and will get only part of the answer.

EXAMPLE 3

Solving rational inequalities. Solve $\frac{x^2 - 3x + 2}{x - 3} \geq 0$.

Solution

We write the fraction with the numerator in factored form.

$$\frac{(x - 2)(x - 1)}{x - 3} \geq 0$$

FIGURE 10-23

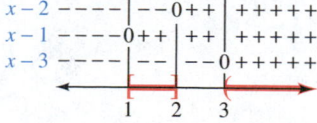

To keep track of the signs of the three binomials, we construct the sign chart shown in Figure 10-23. The fraction will be positive in the intervals where all factors are positive, or where exactly two factors are negative. The numbers 1 and 2 are included, because they make the numerator (and thus the fraction) equal to 0. The number 3 is not included, because it gives a 0 in the denominator.

The solution is the interval $[1, 2] \cup (3, \infty)$. The graph appears in Figure 10-23.

SELF CHECK   Solve $\dfrac{x + 2}{x^2 - 2x - 3} > 0$ and graph the solution set.   *Answer:* $(-2, -1) \cup (3, \infty)$

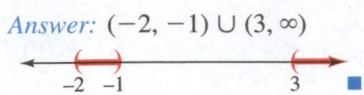

E X A M P L E   4   **Solving rational inequalities.** Solve $\dfrac{3}{x - 1} < \dfrac{2}{x}$.

Solution   We subtract $\frac{2}{x}$ from both sides to get 0 on the right-hand side and proceed as follows:

$$\frac{3}{x - 1} < \frac{2}{x}$$

$$\frac{3}{x - 1} - \frac{2}{x} < 0 \qquad \text{Subtract } \tfrac{2}{x} \text{ from both sides.}$$

$$\frac{3x}{(x - 1)x} - \frac{2(x - 1)}{x(x - 1)} < 0 \qquad \text{Get a common denominator.}$$

$$\frac{3x - 2x + 2}{x(x - 1)} < 0 \qquad \text{Keep the denominator and subtract the numerators.}$$

$$\frac{x + 2}{x(x - 1)} < 0 \qquad \text{Combine like terms.}$$

FIGURE 10-24

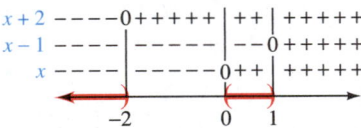

We can keep track of the signs of the three polynomials with the sign chart shown in Figure 10-24. The fraction will be negative in the intervals with either one or three negative factors. The numbers 0 and 1 are not included, because they give a 0 in the denominator, and the number -2 is not included, because it does not satisfy the inequality.

The solution is the interval $(-\infty, -2) \cup (0, 1)$, as shown in Figure 10-24.

SELF CHECK   Solve $\frac{2}{x+1} > \frac{1}{x}$ and graph the solution set.   *Answer:* $(-1, 0) \cup (1, \infty)$

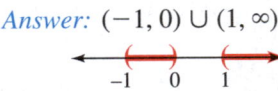

A C C E N T   O N   T E C H N O L O G Y *Solving Inequalities Graphically*

To approximate the solutions of the inequality $x^2 + 2x - 3 \geq 0$ (Example 1) by graphing, we can use the standard window settings of $[-10, 10]$ for x and $[-10, 10]$ for y and graph the quadratic function $y = x^2 + 2x - 3$, as in Figure 10-25. The solution of the inequality will be those numbers x for which the graph of $y = x^2 + 2x - 3$ lies above or on the x-axis. We can trace to find that this interval is $(-\infty, -3] \cup [1, \infty)$.

FIGURE 10-25

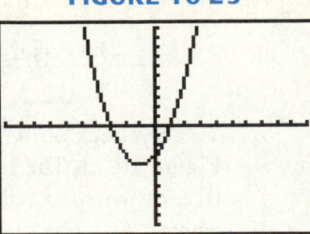

To approximate the solutions of $\frac{3}{x-1} < \frac{2}{x}$ (Example 4), we first write the inequality in the form

$$\frac{x+2}{x(x-1)} < 0$$

We use window settings of $[-5, 5]$ for x and $[-3, 3]$ for y and graph the function $y = \frac{x+2}{x(x-1)}$, as in Figure 10-26(a). The solution of the inequality will be those numbers x for which the graph lies below the x-axis.

We can trace to see that the graph is below the x-axis when x is less than -2. Since we cannot see the graph in the interval $0 < x < 1$, we redraw the graph using window settings of $[-1, 2]$ for x and $[-25, 10]$ for y. See Figure 10-26(b).

We can now see that the graph is below the x-axis in the interval $(0, 1)$. Thus, the solution of the inequality is the union of two intervals:

$$(-\infty, -2) \cup (0, 1)$$

FIGURE 10-26

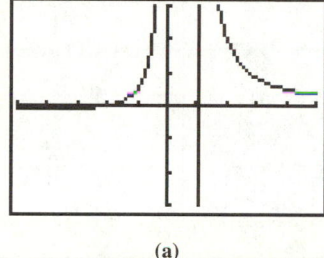 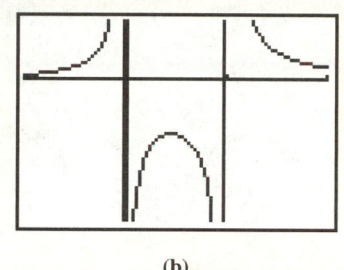

(a)   (b)

Graphs of Nonlinear Inequalities in Two Variables

We now consider the graphs of nonlinear inequalities in two variables.

EXAMPLE 5

Graphing nonlinear inequalities in two variables. Graph $y < -x^2 + 4$.

Solution   The graph of $y = -x^2 + 4$ is the parabolic boundary separating the region representing $y < -x^2 + 4$ and the region representing $y > -x^2 + 4$.

We graph the quadratic function $y = -x^2 + 4$ as a broken parabola, because equality is not permitted. Since the coordinates of the origin satisfy the inequality $y < -x^2 + 4$, the point $(0, 0)$ is in the graph. The complete graph is shown in Figure 10-27.

FIGURE 10-27

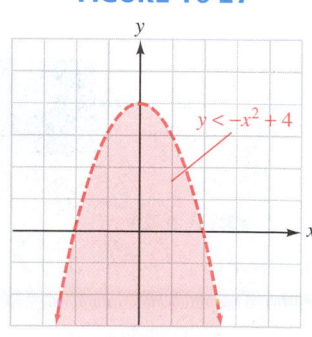

SELF CHECK   Graph $y \geq -x^2 + 4$.

Answer:

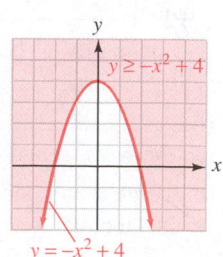

| E X A M P L E   6 | **Graphing nonlinear inequalities in two variables.** Graph $x \leq |y|$. |

Solution   We first graph $x = |y|$ as in Figure 10-28(a). We use a solid line, because equality is permitted. Because the origin is on the graph, we cannot use the origin as a test point. However, any other point, such as $(1, 0)$, will do. We substitute 1 for x and 0 for y into the inequality to get

$$x \leq |y|$$
$$1 \leq |0|$$
$$1 \leq 0$$

Since $1 \leq 0$ is a false statement, the point $(1, 0)$ does not satisfy the inequality and is not part of the graph. Thus, the graph of $x \leq |y|$ is to the left of the boundary. The complete graph is shown in Figure 10-28(b).

FIGURE 10-28

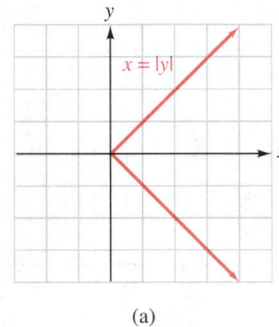
(a)

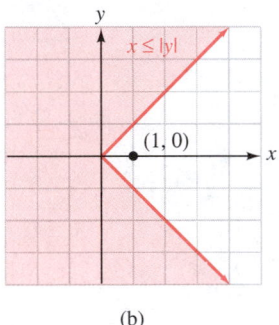
(b)

| SELF CHECK | Graph $x \geq -|y|$. | *Answer:* |

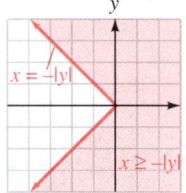

STUDY SET

Section 10.6

VOCABULARY

In Exercises 1–4, fill in the blanks to make the statements true.

1. Any inequality of the form $ax^2 + bx + c > 0$ $(a \neq 0)$ is called a _____ inequality.

2. The inequality $y < x^2 - 2x + 3$ is a nonlinear inequality in _____ variables.

3. The _____ $(3, 5)$ represents the real numbers between 3 and 5.

4. To decide which side of the boundary to shade when solving inequalities in two variables, we pick a _____ point.

CONCEPTS

In Exercises 5–8, fill in the blanks to make the statements true.

5. When $x > 3$, the binomial $x - 3$ is _____ than zero.

6. When $x < 3$, the binomial $x - 3$ is _____ than zero.

7. If $x = 0$, the fraction $\frac{1}{x}$ is _____.

8. To keep track of the signs of factors in a product or quotient, we can use a _____ chart.

9. Estimate the solution of the inequality $x^2 - x - 6 > 0$ using the graph of $y = x^2 - x - 6$, shown below.

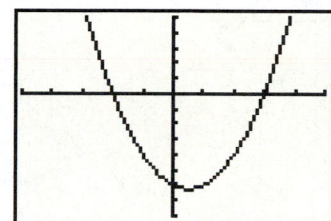

10. Estimate the solution of the inequality $\frac{x-3}{x} \le 0$ using the graph of $y = \frac{x-3}{x}$, shown below.

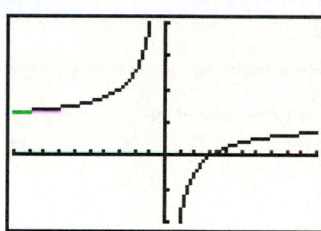

NOTATION

In Exercises 11–12, consider the inequality
$(x + 2)(x - 3) > 0$.

11. Since the product of $x + 2$ and $x - 3$ is positive, the values of $x + 2$ and $x - 3$ are both positive or both negative.
    a. $x + 2 = 0$ when $x =$ ☐
    b. $x + 2 > 0$ when $x >$ ☐
    c. $x + 2 < 0$ when $x <$ ☐
    d. $x - 3 = 0$ when $x =$ ☐
    e. $x - 3 > 0$ when $x >$ ☐
    f. $x - 3 < 0$ when $x <$ ☐

12. Use the information in Exercise 11 to make a sign chart of the data and graph the solution set.

13. Match each inequality with the corresponding interval notation.
    a. $-4 \le x < 5$ i. $(-\infty, \infty)$
    b. $x \le -4$ or $x > 5$ ii. $[-4, \infty)$
    c. $x \ge -4$ iii. $[-4, 5)$
    d. $x < 5$ or $x \ge -4$ iv. $(-\infty, -4] \cup (5, \infty)$

14. What is the meaning of the symbol $\cup$?

PRACTICE

In Exercises 15–42, solve each inequality. Give each result in interval notation and graph the solution set.

15. $x^2 - 5x + 4 < 0$

16. $x^2 - 3x - 4 > 0$

17. $x^2 - 8x + 15 > 0$

18. $x^2 + 2x - 8 < 0$

19. $x^2 + x - 12 \le 0$

20. $x^2 - 8x \le -15$

21. $x^2 + 8x < -16$
22. $x^2 + 6x \ge -9$

23. $x^2 \ge 9$

24. $x^2 \ge 16$

25. $2x^2 - 50 < 0$

26. $3x^2 - 243 < 0$

27. $\frac{1}{x} < 2$

28. $\frac{1}{x} > 3$

29. $-\frac{5}{x} < 3$

30. $\frac{4}{x} \ge 8$

31. $\frac{x^2 - x - 12}{x - 1} < 0$

32. $\frac{x^2 + x - 6}{x - 4} \ge 0$

33. $\frac{6x^2 - 5x + 1}{2x + 1} > 0$

34. $\frac{6x^2 + 11x + 3}{3x - 1} < 0$

35. $\frac{3}{x - 2} < \frac{4}{x}$

36. $\frac{-6}{x + 1} \ge \frac{1}{x}$

37. $\dfrac{7}{x-3} \geq \dfrac{2}{x+4}$

38. $\dfrac{-5}{x-4} < \dfrac{3}{x+1}$

39. $\dfrac{x}{x+4} \leq \dfrac{1}{x+1}$

40. $\dfrac{x}{x+9} \geq \dfrac{1}{x+1}$

41. $(x+2)^2 > 0$

42. $(x-3)^2 < 0$

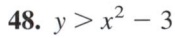 *In Exercises 43–46, use a graphing calculator to solve each inequality. Give the answer in interval notation.*

43. $x^2 - 2x - 3 < 0$

44. $x^2 + x - 6 > 0$

45. $\dfrac{x+3}{x-2} > 0$

46. $\dfrac{3}{x} < 2$

In Exercises 47–54, graph each inequality.

47. $y < x^2 + 1$

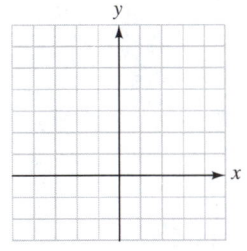

48. $y > x^2 - 3$

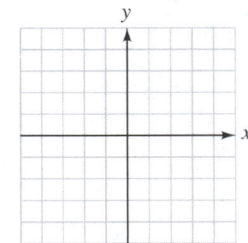

49. $y \leq x^2 + 5x + 6$

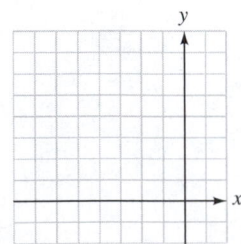

50. $y \geq x^2 + 5x + 4$

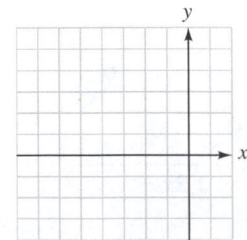

51. $y < |x+4|$

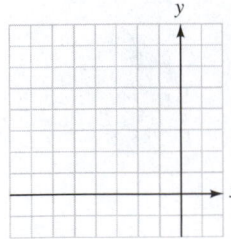

52. $y \geq |x-3|$

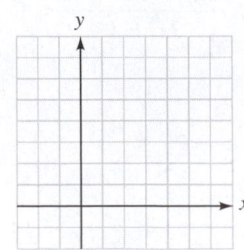

53. $y \leq -|x| + 2$

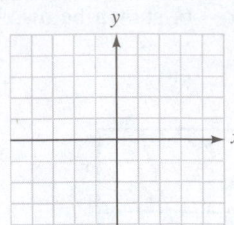

54. $y > |x| - 2$

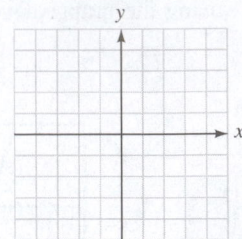

APPLICATIONS

55. SUSPENSION BRIDGE If an x-axis is superimposed over the roadway of the Golden Gate Bridge, with the origin at the center of the bridge as shown in Illustration 1, the length L in feet of a vertical support cable can be approximated by the formula

$$L = \dfrac{1}{9,000}x^2 + 5$$

For the Golden Gate Bridge, $-2,100 < x < 2,100$. For what intervals along the x-axis are the vertical cables more than 95 feet long?

ILLUSTRATION 1

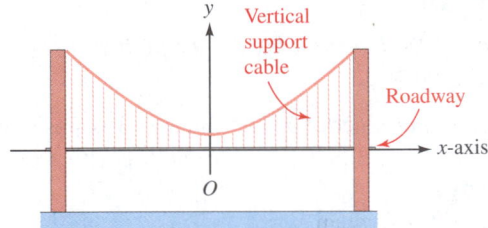

56. MALL The number of people n in a mall is modeled by the formula

$$n = -100x^2 + 1,200x$$

where x is the number of hours since the mall opened. If the mall opened at 9 A.M., when were there 2,000 or more people in it?

WRITING

57. Explain why $(x-4)(x+5)$ will be positive only when the signs of $x-4$ and $x+5$ are the same.

58. Explain how to find the graph of $y \geq x^2$.

59. The graph of $f(x) = x^2 - 3x + 4$ is shown in Illustration 2. Explain why the quadratic inequality $x^2 - 3x + 4 < 0$ has no solution.

60. What three important facts about the expression $x-1$ are indicated by the chart in Illustration 3?

ILLUSTRATION 2

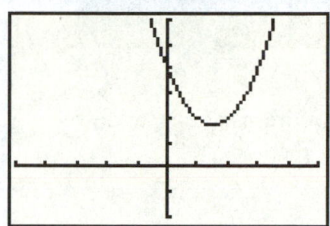

ILLUSTRATION 3

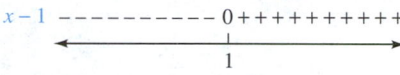

REVIEW

In Exercises 61–64, translate each statement into an equation.

61. *x* varies directly with *y*.

62. *y* varies inversely with *t*.

63. *t* varies jointly with *x* and *y*.

64. *d* varies directly with *t* and inversely with u^2.

Find the slope of the graph of each linear function.

65. $f(x) = 3x - 4$

66. $f(x) = -x$

Solving Quadratic Equations

We have discussed five methods for solving **quadratic equations.** Let's review each of them and list an advantage and a drawback of each method.

Factoring

- It can be very quick and simple if the factoring pattern is evident.
- Much of the time, $ax^2 + bx + c$ cannot be factored or is not easily factored.

In Exercises 1–3, solve the equation by factoring.

1. $4k^2 + 8k = 0$ **2.** $z^2 + 8z + 15 = 0$ **3.** $2r^2 + 5r = -3$

The Square Root Method

- If the equation can be written in the form $x^2 = a$ or $(x + d)^2 = a$, where a is a constant, the square root method is a fast method, requiring few computations.
- Most quadratic equations that we must solve are not written in either of these forms.

In Exercises 4–6, solve the equation by the square root method.

4. $u^2 = 24$ **5.** $(s - 7)^2 - 9 = 0$ **6.** $3x^2 - 16 = 0$

Completing the Square

- It can be used to solve any quadratic equation.
- Most often, it involves more steps than the other methods.

In Exercises 7–9, solve the equation by completing the square.

7. $x^2 + 10x - 7 = 0$ **8.** $4x^2 - 4x - 1 = 0$ **9.** $x^2 + 2x + 2 = 0$

The Quadratic Formula

- It simply involves an evaluation of the expression $\dfrac{-b \pm \sqrt{b^2 - 4ac}}{2a}$.
- If applicable, the factoring method and the square root method are usually faster.

In Exercises 10–12, solve the equation by using the quadratic formula.

10. $2x^2 - 1 = 3x$ **11.** $x^2 - 6x - 391 = 0$ **12.** $3x^2 + 2x + 1 = 0$

The Graphing Method

- We can solve the equation using a graphing calculator. It doesn't require any computations.
- It usually gives only approximations of the solutions.

13. Use the graph of $y = x^2 + x - 2$ to solve the equation $x^2 + x - 2 = 0$.

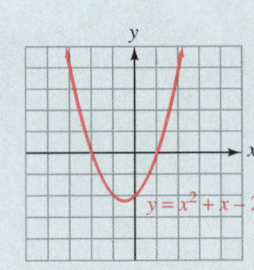

Accent on Teamwork

Section 10.1

Picture frames Each person in your group is to make an 8 in. × 10 in. collage that illustrates an important aspect of his or her life. Photographs, magazine pictures, drawings, and the like can be used to highlight family, hobbies, jobs, pets, etc. Assign each person the task of making a matting of uniform width to frame his or her collage. Some can make a matting whose area equals that of the picture. Others can make a matting that is half the area, or double the area. When completed, let each person briefly explain his or her collage to the other group members. See Example 8, Section 10.1, for some hints on how to do the mathematics.

Section 10.2

Team relay Have your group make a presentation to the class showing the derivation of the quadratic formula. (See page 731 of the text.) Begin with the general quadratic equation, $ax^2 + bx + c = 0$, and have each member of your group perform and explain several steps. Make a "baton" from a paper towel roll, like that shown in Illustration 1. As one student finishes with his or her segment of the derivation, the baton is passed to the next student to pick up where he or she left off.

ILLUSTRATION 1

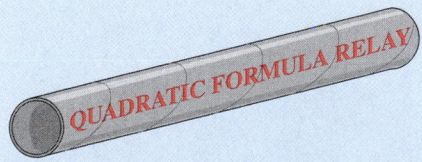

QUADRATIC FORMULA RELAY

Section 10.3

Determining the number of solutions graphically Experiment with various combinations of a's, b's, and c's to write quadratic functions of the form $f(x) = ax^2 + bx + c$ that, when graphed using a calculator, intersect the x-axis **a.** two times,   **b.** one time, and   **c.** no times. Then give the associated quadratic equation in each case and tell how many real-number solutions the equation has.

Section 10.4

Complex numbers Write a report about complex numbers that answers these questions: When were they first used and by whom? Why were they "invented"? What are some of their applications?

A book about the history of mathematics will be helpful. The school library or perhaps one of your mathematics instructors may have one you can borrow. Some colleges offer a course about complex numbers called Complex Variables. See if you can get a copy of the textbook. Germany once issued a postage stamp honoring Carl Gauss and featuring complex numbers. See if you can find a picture of it. Physics instructors or electronics instructors are two other possible resources for material.

Section 10.5

Solutions of a quadratic equation Consider the following property about the solutions of a quadratic equation: If

$$r_1 + r_2 = -\frac{b}{a} \quad \text{and} \quad r_1 r_2 = \frac{c}{a}$$

then r_1 and r_2 are the solutions of a quadratic equation $ax^2 + bx + c = 0$, with $a \neq 0$.

Use this property to show that

a. $\frac{3}{2}$ and $-\frac{1}{3}$ are solutions of $6x^2 - 7x - 3 = 0$.
b. $\sqrt{51}i$ and $-\sqrt{51}i$ are solutions of $x^2 + 51 = 0$.
c. $1 + 3\sqrt{2}$ and $1 - 3\sqrt{2}$ are solutions of $x^2 - 2x - 17 = 0$.

Section 10.6

Quadratic inequalities The fraction shown below has two factors in the numerator and two factors in the denominator.

$$\frac{(x - 1)(x + 4)}{(x + 2)(x + 1)}$$

a. With the four factors in mind, under what conditions will the fraction be positive?
b. With the four factors in mind, under what conditions will the fraction be negative?
c. Solve $\dfrac{(x - 1)(x + 4)}{(x + 2)(x + 1)} > 0$.

Section 10.1

Completing the Square

CONCEPTS

The square root property:
If $c > 0$, the equation $x^2 = c$ has two real solutions:

$$x = \sqrt{c} \text{ and } x = -\sqrt{c}$$

To *complete the square:*
1. Make sure the coefficient of x^2 is 1.
2. Make sure the constant term is on the right-hand side of the equation.
3. Add the square of one-half of the coefficient of x to both sides.
4. Factor the trinomial.
5. Use the square root property.

REVIEW EXERCISES

1. Solve each equation by factoring or using the square root property.

 a. $x^2 + 9x + 20 = 0$
 b. $6x^2 + 17x + 5 = 0$
 c. $x^2 = 28$
 d. $(t + 2)^2 = 36$
 e. $5a^2 + 11a = 0$
 f. $5x^2 - 49 = 0$

2. What number must be added to $x^2 - x$ to make it a perfect square?

3. Solve each equation by completing the square.

 a. $x^2 + 6x + 8 = 0$
 b. $2x^2 - 6x + 3 = 0$

4. HAPPY NEW YEAR As part of a New Year's Eve celebration, a huge ball is to be dropped from the top of a 605-foot-tall building at the proper moment so that it strikes the ground at exactly 12:00 midnight. The distance d in feet traveled by a free-falling object in t seconds is given by the formula $d = 16t^2$. To the nearest second, when should the ball be dropped from the building?

Section 10.2

The Quadratic Formula

The quadratic formula:
The solutions of $ax^2 + bx + c = 0$ are given by

$$x = \frac{-b \pm \sqrt{b^2 - 4ac}}{2a} \quad (a \neq 0)$$

5. Solve each equation using the quadratic formula.

 a. $-x^2 + 10x - 18 = 0$
 b. $x^2 - 10x = 0$
 c. $2x^2 + 13x = 7$
 d. $26y - 3y^2 = 2$

6. SPORTS POSTER The design specifications for a poster of tennis star Steffi Graf call for a 615-square-inch photograph to be surrounded by a blue border. (See Illustration 1.) The borders on the sides of the poster are to be half as wide as those at the top and bottom. Find the width of each border.

ILLUSTRATION 1   **ILLUSTRATION 2**

35 in.
23 in.

7. ACROBATS To begin his routine on a trapeze, an acrobat is catapulted upward as shown in Illustration 2. His distance d (in feet) from the arena floor during this maneuver is given by the formula $d = -16t^2 + 40t + 5$, where t is the time in seconds since being launched. If the trapeze bar is 25 feet in the air, at what two times will he be able to grab it? Round to the nearest tenth.

Section 10.3   Quadratic Functions and Their Graphs

A *quadratic function* is a second-degree polynomial function of the form
$f(x) = ax^2 + bx + c$.

The graph of $f(x) = ax^2$ is a *parabola* opening upward when $a > 0$ and downward when $a < 0$, with *vertex* at the point $(0, 0)$ and *axis of symmetry* the line $x = 0$.

Each of the following functions has a graph that is the same shape as $f(x) = ax^2$ but involves a vertical or horizontal translation.
1. $f(x) = ax^2 + c$: translated upward if $c > 0$, downward if $c < 0$.
2. $f(x) = a(x - h)^2$: translated right if $h > 0$, and left if $h < 0$.

If $a \neq 0$, the graph of $f(x) = a(x - h)^2 + k$ is a parabola with vertex at (h, k). It opens upward when $a > 0$ and downward when $a < 0$.

The vertex of the graph of $f(x) = ax^2 + bx + c$ is

$$\left(-\frac{b}{2a},\, f\left(-\frac{b}{2a}\right)\right)$$

and the axis of symmetry is the line $x = -\dfrac{b}{2a}$.

The vertex of the graph of a quadratic function gives the *minimum* or *maximum* value of the function.

8. AEROSPACE INDUSTRY See Illustration 3. The annual sales of the Boeing Company in billions of dollars for the years 1993–1997 can be modeled by the quadratic function

$$S(x) = 2.4375x^2 - 8.225x + 40.7$$

where x is the number of years since 1993. What were the annual sales for 1997?

ILLUSTRATION 3

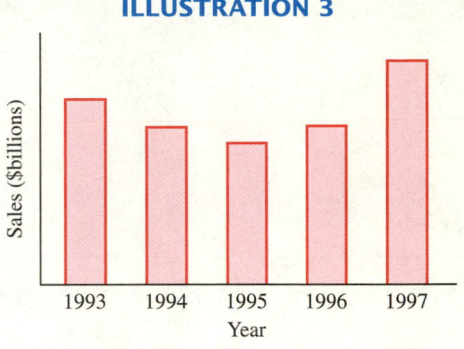

9. Make a table of values to graph function f. Then use a series of translations to graph function g on the same coordinate system.
   a. $f(x) = 2x^2$
      $g(x) = 2x^2 - 3$

   b. $f(x) = -4x^2$
      $g(x) = -4(x - 2)^2 + 1$

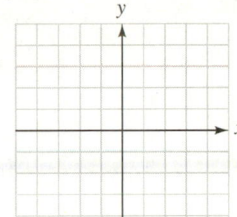
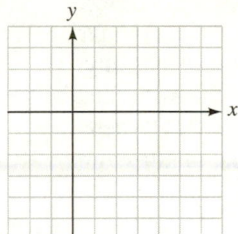

10. First determine the vertex and the axis of symmetry of the graph of each function. Then plot several points and complete the graph.
   a. $y = -\left(x + \frac{3}{2}\right)^2 + \frac{5}{2}$

   b. $f(x) = 5x^2 + 10x - 1$

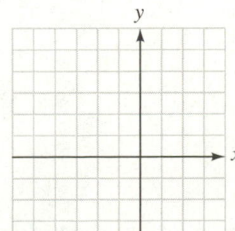
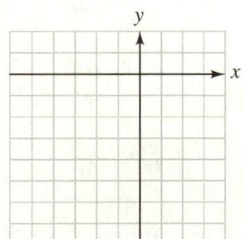

11. FARMING The number of farms in the United States for the years 1870–1970 is modeled by

$$N(x) = -1.46x^2 + 148.82x + 2{,}660$$

where $x = 0$ represents 1870, $x = 1$ represents 1871, and so on. For this period, when was the number of U.S. farms a maximum? How many farms were there?

12. Estimate the solutions of $-3x^2 - 5x + 2 = 0$ from the graph of $f(x) = -3x^2 - 5x + 2$, shown in Illustration 4.

ILLUSTRATION 4

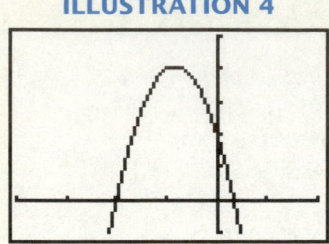

Section 10.4   Complex Numbers

Square roots of negative numbers are called *imaginary numbers*.

$i = \sqrt{-1}$ and $i^2 = -1$

13. Simplify each expression.

    a. $\sqrt{-4}$

    b. $\sqrt{-7}$

    c. $-3\sqrt{-20}$

    d. $\sqrt{-\dfrac{36}{49}}$

    e. $\sqrt{-3}\sqrt{-3}$

    f. $\dfrac{\sqrt{-64}}{\sqrt{-9}}$

A *complex number* is any number that can be written in the form $a + bi$, where a and b are real numbers and $i^2 = -1$.

14. Solve each equation.

    a. $a^2 = -25$

    b. $x^2 - 2x + 13 = 0$

Adding complex numbers:

$(a + bi) + (c + di) =$
$\qquad (a + c) + (b + d)i$

15. Do the operations and give all answers in $a + bi$ form.

    a. $(5 + 4i) + (7 - 12i)$

    b. $(-6 - 40i) - (-8 + 28i)$

    c. $\left(-8 + \sqrt{-8}\right) + \left(6 - \sqrt{-32}\right)$

    d. $2i(64 + 9i)$

Subtracting complex numbers:

$(a + bi) - (c + di) =$
$\qquad (a - c) + (b - d)i$

    e. $(2 - 7i)(-3 + 4i)$

    f. $\left(5 - \sqrt{-27}\right)\left(-6 + \sqrt{-12}\right)$

Multiplying complex numbers:

$(a + bi)(c + di) =$
$\qquad (ac - bd) + (ad + bc)i$

To divide complex numbers, we often have to *rationalize* a denominator by multiplying the numerator and the denominator by the *complex conjugate* of the denominator.

16. Write each expression in $a + bi$ form.

    a. $\dfrac{6}{2 + i}$

    b. $\dfrac{4 + i}{4 - i}$

    c. $\dfrac{\sqrt{3} + \sqrt{-4}}{\sqrt{3} - \sqrt{-4}}$

    d. $\dfrac{-2}{5i^3}$

The *powers of i* rotate through a cycle of four numbers: $i = i$, $i^2 = -1$, $i^3 = -i$, $i^4 = 1$

17. Simplify each power of i.

    a. i^{65}

    b. i^{48}

Section 10.5

The Discriminant and Equations That Can Be Written in Quadratic Form

The *discriminant* predicts the type of solutions of
$ax^2 + bx + c = 0$:

1. If $b^2 - 4ac > 0$, the solutions are unequal real numbers.
2. If $b^2 - 4ac = 0$, the solutions are equal real numbers.
3. If $b^2 - 4ac < 0$, the solutions are complex conjugates.

Many equations that are not quadratic can be written in quadratic form.

18. Use the discriminant to determine what types of solutions exist for each equation.

   a. $3x^2 + 4x - 3 = 0$

   b. $4x^2 - 5x + 7 = 0$

   c. $9x^2 - 12x + 4 = 0$

19. Solve each equation.

   a. $x - 13\sqrt{x} + 12 = 0$

   b. $a^{2/3} + a^{1/3} - 6 = 0$

   c. $6x^4 - 19x^2 + 3 = 0$

   d. $\dfrac{6}{x+2} + \dfrac{6}{x+1} = 5$

   e. $(x - 3)^2 - 8(x - 3) + 7 = 0$

20. WEEKLY CHORES Working together, two sisters can do the yard work at their house in 45 minutes. When the older girl does it all herself, she can complete the job in 20 minutes less time than it takes the younger girl working alone. How long does it take the older girl to do the yard work?

Section 10.6

Quadratic and Other Nonlinear Inequalities

To graph a *quadratic inequality in one variable,* get 0 on the right-hand side. Then factor the polynomial on the left-hand side. Use a *sign chart* to determine the solution set.

To solve rational inequalities, get 0 on the right-hand side and a single fraction on the left-hand side. Factor the numerator and denominator. Then use a sign chart to determine the solution set.

21. Solve each inequality. Give each result in interval notation and graph the solution set.

   a. $x^2 + 2x - 35 > 0$

   b. $x^2 - 81 \leq 0$

   c. $\dfrac{3}{x} \leq 5$

   d. $\dfrac{2x^2 - x - 28}{x - 1} > 0$

22. Use a graphing calculator to solve each inequality. Compare the results with Exercise 21.

   a. $x^2 + 2x - 35 > 0$

   b. $\dfrac{2x^2 - x - 28}{x - 1} > 0$

To graph a *nonlinear inequality in two variables,* first graph the boundary. Then use a test point to determine which half-plane to shade.

23. Graph each inequality.

a. $y < \dfrac{1}{2}x^2 - 1$

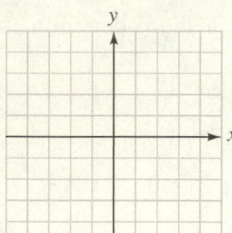

b. $y \geq -|x|$

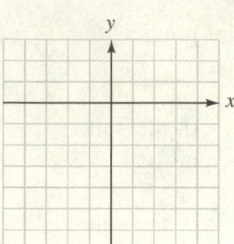

CHAPTER 10

Test

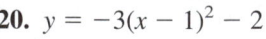

In Problems 1–2, solve each equation by factoring.

1. $3x^2 + 18x = 0$

2. $x(6x + 19) = -15$

3. Determine what number must be added to $x^2 + 24x$ to make it a perfect square.

4. Solve $x^2 - x - 1 = 0$ by completing the square.

5. Solve the equation $2x^2 - 8x = -5$ using the quadratic formula.

6. TABLECLOTH According to the *Guinness Book of World Records 1998*, the world's longest tablecloth was made in Illinois in 1990 and covered an area of 6,759 square feet. Its length was 3.5 feet more than 333 times its width. Find the dimensions of the tablecloth.

7. Simplify $\sqrt{-48}$.

8. Simplify i^{54}.

In Problems 9–14, do the operations. Give all answers in $a + bi$ form.

9. $(2 + 4i) + (-3 + 7i)$

10. $\left(3 - \sqrt{-9}\right) - \left(-1 + \sqrt{-16}\right)$

11. $2i(3 - 4i)$,

12. $(3 + 2i)(-4 - i)$

13. $\dfrac{1}{i^3}$

14. $\dfrac{2 + i}{3 - i}$

15. Determine whether the solutions of $3x^2 + 5x + 17 = 0$ are real or nonreal.

16. Solve $x^2 = -12$.

17. Solve the equation $13 = 4t - t^2$.

18. Solve the equation $2y - 3\sqrt{y} + 1 = 0$.

In Problems 19–20, determine the vertex and the axis of symmetry of the graph of the function. Then graph it.

19. $f(x) = 2x^2 + x - 1$

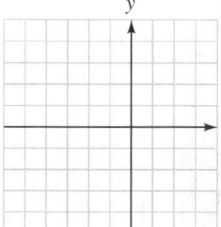

20. $y = -3(x - 1)^2 - 2$

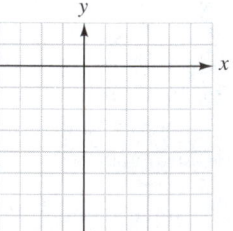

In Problems 21–22, solve the inequality and graph the solution set.

21. $x^2 - 2x - 8 > 0$

22. $\dfrac{x - 2}{x + 3} \leq 0$

23. Graph the inequality $y \le -x^2 + 3$.

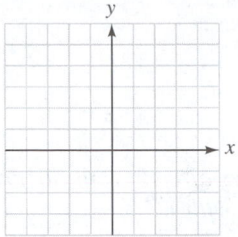

24. DRAWING An artist uses four equal-sized right triangles to block out a perspective drawing of an old hotel. See Illustration 1. For each triangle, one leg is 14 inches longer than the other, and the hypotenuse is 26 inches. On the centerline of the drawing, what is the length of the segment extending from the ground to the top of the building?

ILLUSTRATION 1

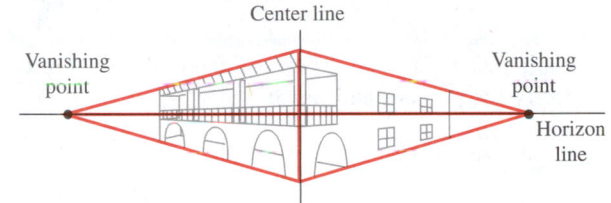

25. DISTRESS SIGNAL A flare is fired directly upward into the air from a boat that is experiencing engine problems. The height of the flare (in feet) above the water, t seconds after being fired, is given by the formula $h = -16t^2 + 112t + 15$. If the flare is designed to explode when it reaches its highest point, at what height will this occur?

26. What is an imaginary number? Give some examples.

27. The graph of a quadratic function of the form $f(x) = ax^2 + bx + c$ is shown in Illustration 2. Estimate the solutions of the corresponding quadratic equation $ax^2 + bx + c = 0$.

ILLUSTRATION 2

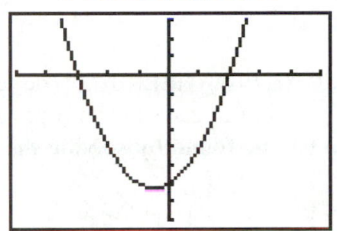

28. See Problem 27. Estimate the solution of the quadratic inequality $ax^2 + bx + c \le 0$.

CHAPTERS 1–10

Cumulative Review Exercises

1. Solve $-3 = -\dfrac{9}{8}t$.

2. Solve $\dfrac{3x - 4}{6} - \dfrac{x - 2}{2} = \dfrac{-2x - 3}{3}$.

3. AUTO SALES See the graph in Illustration 1. An automobile dealership is going to order 80 new Ford Escorts. According to the survey, exactly how many green Escorts should be purchased to meet the expected customer demand?

ILLUSTRATION 1

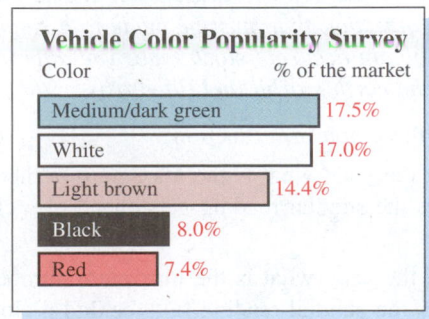

Based on data from DuPont Automotive

4. LIFE EXPECTANCY Determine the predicted rate of change in the life expectancy of females during the years 2000–2050, as shown in the graph in Illustration 2.

ILLUSTRATION 2

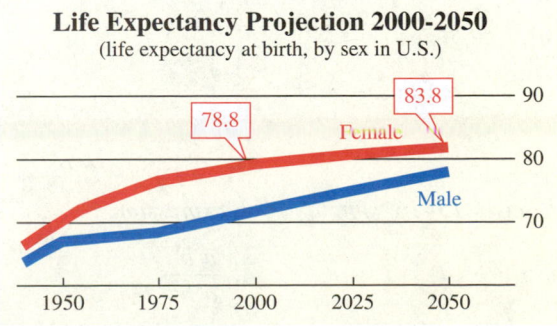

Based on data from the Social Security Administration, Office of Chief Actuary

In Exercises 5–6, write the equation of the line with the given properties. Express your answer in slope–intercept form.

5. Slope of -7, passing through $(7, 5)$

6. Passing through $(-4, 5)$ and $(2, -6)$

7. Draw the graph of the linear function $f(x) = \frac{2}{3}x - 2$. Then use interval notation to specify the domain and range.

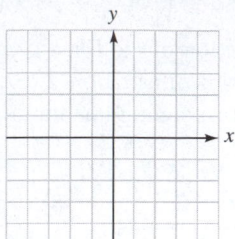

8. MECHANICAL ENGINEERING The tensions T_1 and T_2 (in pounds) in each of the ropes shown in Illustration 3 can be found by solving the system

$$\begin{cases} 0.6T_1 - 0.8T_2 = 0 \\ 0.8T_1 + 0.6T_2 = 100 \end{cases}$$

Find T_1 and T_2.

ILLUSTRATION 3

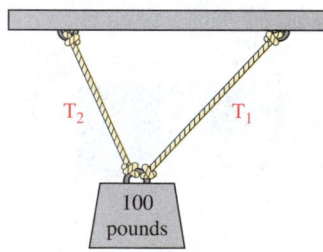

In Exercises 9–12, solve each inequality and show the solution set as a graph on the number line.

9. $-2x \le -5$

10. $\left| \dfrac{3a}{5} - 2 \right| + 1 \ge \dfrac{6}{5}$

11. $5(x + 2) \le 4(x + 1)$ and $11 + x < 0$

12. $x + 1 < -4$ or $x - 4 > 0$

In Exercises 13–16, simplify each expression.

13. $a^3b^2a^5b^2$

14. $\dfrac{a^3b^6}{a^7b^2}$

15. $\dfrac{1}{3^{-4}}$

16. $\left(\dfrac{2x^{-2}y^3}{x^2x^3y^4} \right)^{-3}$

In Exercises 17–18, write each number in standard notation.

17. 4.25×10^4

18. 7.12×10^{-4}

19. Express as a formula: *y varies directly with the product of x and z, and inversely with r.*

20. Express as a formula: *d varies jointly with x and y.*

21. Graph $y < 4 - x$.

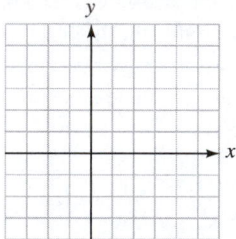

22. Graph $f(x) = 2x^2 - 3$. Then use interval notation to specify the domain and range.

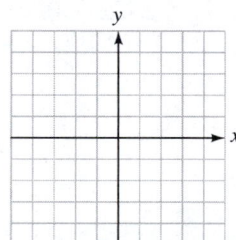

23. If $g(x) = -3x^3 + x - 4$, find $g(-2)$.

24. Find the degree of $3 + x^2y + 17x^3y^4$.

In Exercises 25–28, do the operations and simplify.

25. $(x^3 + 3x^2 - 2x + 7) + (x^3 - 2x^2 + 2x + 5)$

26. $(-5x^2 + 3x + 4) - (-2x^2 + 3x + 7)$

27. $(3x + 4)(2x - 5)$

28. $(2x^3 - 1)^2$

In Exercises 29–32, refer to Illustration 4. The graph shows the correction that must be made to a sundial reading to obtain accurate clock time. The difference is caused by the earth's orbit and tilted axis.

29. Is this the graph of a function?

30. During the year, what is the maximum number of minutes the sundial reading gets ahead of a clock?

31. During the year, what is the maximum number of minutes the sundial reading falls behind a clock?

ILLUSTRATION 4

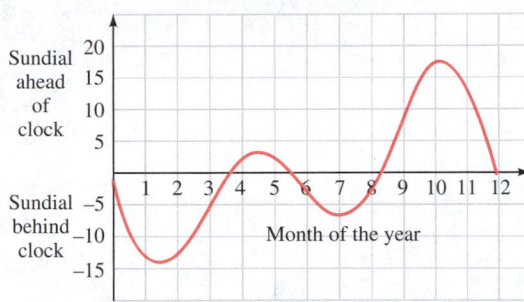

Month of the year

32. How many times during a year is the sundial reading exactly the same as a clock?

In Exercises 33–36, simplify each expression.

33. $\dfrac{2x^2y + xy - 6y}{3x^2y + 5xy - 2y}$

34. $\dfrac{p^3 - q^3}{q^2 - p^2} \cdot \dfrac{q^2 + pq}{p^3 + p^2q + pq^2}$

35. $\dfrac{2}{x + y} + \dfrac{3}{x - y} - \dfrac{x - 3y}{x^2 - y^2}$

36. $\dfrac{\dfrac{a}{b} + b}{a - \dfrac{b}{a}}$

37. Solve $\dfrac{5x - 3}{x + 2} = \dfrac{5x + 3}{x - 2}$.

38. Solve $\dfrac{3}{x - 2} + \dfrac{x^2}{(x + 3)(x - 2)} = \dfrac{x + 4}{x + 3}$.

In Exercises 39–40, do the division.

39. $(x^2 + 9x + 20) \div (x + 5)$

40. $(2x^2 + 4x - x^3 + 3) \div (x - 1)$

41. ECONOMICS See Illustration 5. The controversial Phillips curve depicts the tradeoff between unemployment and inflation as seen by one school of economists. If unemployment drops to very low levels, what does the theoretical model predict will happen to the inflation rate?

ILLUSTRATION 5

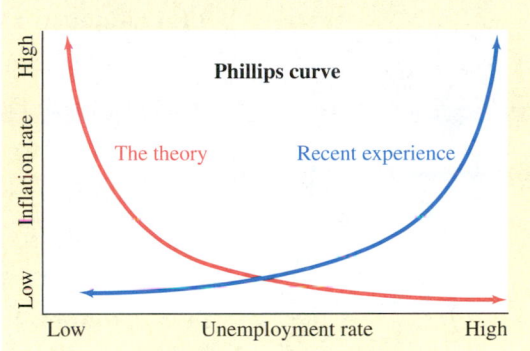

42. ECONOMICS See Exercise 41. The graph in Illustration 5 shows that economic factors have not followed the Phillips model in recent years. As unemployment has dropped to very low levels, what has happened to the inflation rate?

11

Exponential and Logarithmic Functions

CAMPUS CONNECTION

The Sociology Department

In a sociology course, students analyze surveys and statistical data to make observations and predictions about human social relations. One area of great interest to sociologists is how population growth affects social structures and institutions, such as the family, the workplace, and the schools. Population growth can often be mathematically modeled by an *exponential function*. In this chapter, we will examine exponential functions and their graphs. We will also introduce an important irrational number that is often used when writing an exponential function to describe growth or decay. Such functions have application in fields as diverse as banking, medicine, and nuclear energy.

In this chapter, we discuss the concept of function in more depth. We also introduce two new families of functions—exponential and logarithmic functions—which have applications in many areas.

▶ 11.1   Algebra and Composition of Functions

In this section, you will learn about

> Algebra of functions ■ Composition of functions ■ The identity function ■ Writing composite functions

Introduction   So far, we have defined functions with real-number domains and ranges and have graphed them on a rectangular coordinate system. Just as it is possible to perform operations on real numbers, it is also possible to perform operations on functions. We will begin this section by showing how to add, subtract, multiply, and divide functions. Then we will consider another method of combining functions, called *composition of functions.*

Algebra of Functions

We now consider how functions can be added, subtracted, multiplied, and divided.

Operations on functions

If the domains and ranges of functions f and g are subsets of the real numbers, then

The **sum** of f and g, denoted as $f + g$, is defined by

$$(f + g)(x) = f(x) + g(x)$$

The **difference** of f and g, denoted as $f - g$, is defined by

$$(f - g)(x) = f(x) - g(x)$$

The **product** of f and g, denoted as $f \cdot g$, is defined by

$$(f \cdot g)(x) = f(x)g(x)$$

The **quotient** of f and g, denoted as f/g, is defined by

$$(f/g)(x) = \frac{f(x)}{g(x)} \quad (g(x) \neq 0)$$

The domain of each of these functions is the set of real numbers x that are in the domain of both f and g. In the case of the quotient, there is the further restriction that $g(x) \neq 0$.

EXAMPLE 1   **Adding and subtracting functions.** Let $f(x) = 2x^2 + 1$ and $g(x) = 5x - 3$. Find each function and its domain: **a.** $f + g$ and   **b.** $f - g$.

Solution   **a.** $(f + g)(x) = f(x) + g(x)$

$$= (2x^2 + 1) + (5x - 3)$$

$$= 2x^2 + 5x - 2 \qquad \text{Combine like terms.}$$

The domain of $f + g$ is the set of real numbers that are in the domain of both f and g. Since the domain of both f and g is the interval $(-\infty, \infty)$, the domain of $f + g$ is also the interval $(-\infty, \infty)$.

b. $(f - g)(x) = f(x) - g(x)$

$$= (2x^2 + 1) - (5x - 3)$$

$$= 2x^2 + 1 - 5x + 3 \qquad \text{Remove parentheses.}$$

$$= 2x^2 - 5x + 4 \qquad \text{Combine like terms.}$$

Since the domain of both f and g is $(-\infty, \infty)$, the domain of $f - g$ is also the interval $(-\infty, \infty)$.

SELF CHECK   Let $f(x) = 3x - 2$ and $g(x) = 2x^2 + 3x$. Find   *Answers:*   **a.** $2x^2 + 6x - 2$,
a. $f + g$ and   **b.** $f - g$.   **b.** $-2x^2 - 2$ ■

E X A M P L E   2   **Multiplying and dividing functions.** Let $f(x) = 2x^2 + 1$ and $g(x) = 5x - 3$. Find each function and its domain: **a.** $f \cdot g$ and   **b.** f/g.

Solution   **a.** $(f \cdot g)(x) = f(x)g(x)$

$$= (2x^2 + 1)(5x - 3)$$

$$= 10x^3 - 6x^2 + 5x - 3 \quad \text{Use the FOIL method.}$$

The domain of $f \cdot g$ is the set of real numbers that are in the domain of both f and g. Since the domain of both f and g is the interval $(-\infty, \infty)$, the domain of $f \cdot g$ is also the interval $(-\infty, \infty)$.

b. $(f/g)(x) = \dfrac{f(x)}{g(x)}$

$$= \dfrac{2x^2 + 1}{5x - 3}$$

Since the denominator of the fraction cannot be 0, $x \neq \frac{3}{5}$. Thus, the domain of f/g is the interval $\left(-\infty, \frac{3}{5}\right) \cup \left(\frac{3}{5}, \infty\right)$.

SELF CHECK   Let $f(x) = 2x^2 - 3$ and $g(x) = x^2 + 1$. Find   *Answers:*   **a.** $2x^4 - x^2 - 3$,
a. $f \cdot g$ and   **b.** f/g.   **b.** $\dfrac{2x^2 - 3}{x^2 + 1}$ ■

Composition of Functions

We have seen that a function can be represented by a machine: We put in a number from the domain, and a number from the range comes out. For example, if we put the number 2 into the machine shown in Figure 11-1(a), the number $f(2) = 8$ comes out. In general, if we put x into the machine shown in Figure 11-1(b), the value $f(x)$ comes out.

Often one quantity is a function of a second quantity that depends, in turn, on a third quantity. For example, the cost of a car trip is a function of the gasoline consumed. The amount of gasoline consumed, in turn, is a function of the number of miles driven. Such chains of dependence can be analyzed mathematically as **compositions of functions.**

FIGURE 11-1

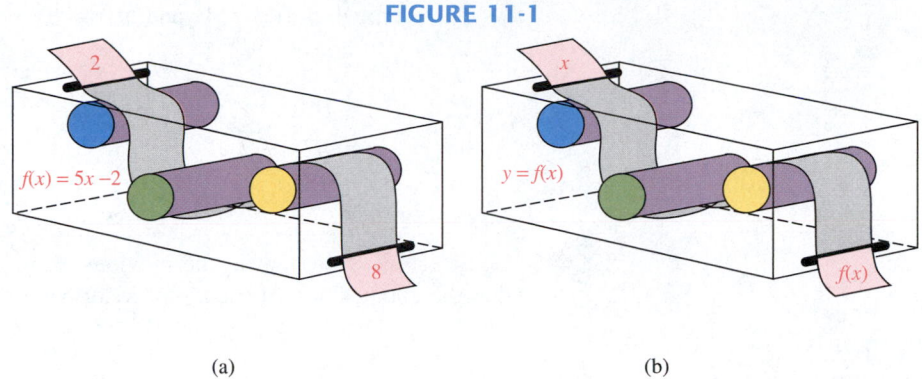

(a)   (b)

Suppose that $y = f(x)$ and $y = g(x)$ define two functions. Any number x in the domain of g will produce the corresponding value $g(x)$ in the range of g. If $g(x)$ is in the domain of function f, then $g(x)$ can be substituted into f, and a corresponding value $f(g(x))$ will be determined. This two-step process defines a new function, called a **composite function**, denoted by $f \circ g$. (This is read as "f composed with g.")

The function machines shown in Figure 11-2 illustrate the composition $f \circ g$. When we put a number x into the function g, $g(x)$ comes out. The value $g(x)$ goes into function f, which transforms $g(x)$ into $f(g(x))$. (This is read as "f of g of x.") If the function machines for g and f were connected to make a single machine, that machine would be named $f \circ g$.

FIGURE 11-2

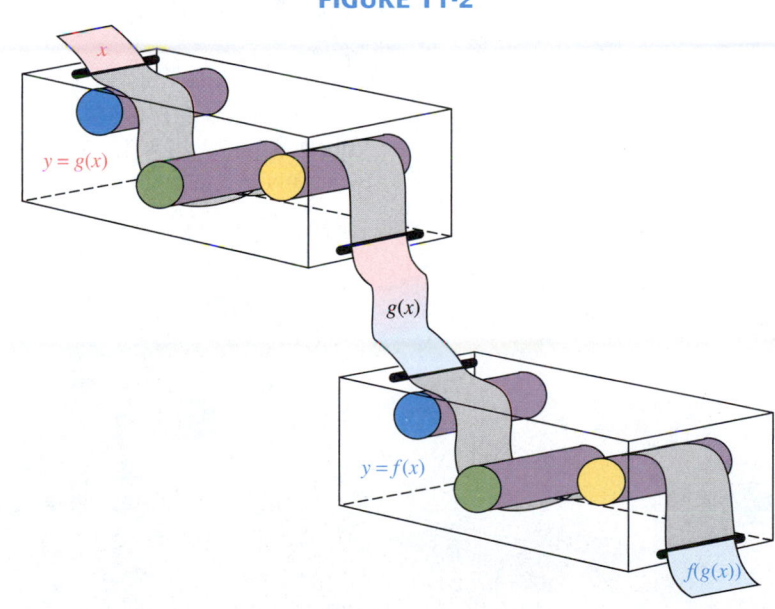

To be in the domain of the composite function $f \circ g$, a number x has to be in the domain of g. Also, the output of g must be in the domain of f. Thus, the domain of $f \circ g$ consists of those numbers x that are in the domain of g, and for which $g(x)$ is in the domain of f.

Composite functions

The **composite function $f \circ g$** is defined by

$$(f \circ g)(x) = f(g(x))$$

For example, if $f(x) = 4x$ and $g(x) = 3x + 2$, then

$$(f \circ g)(x) = f(g(x)) \qquad \text{or} \qquad (g \circ f)(x) = g(f(x))$$
$$= f(3x + 2) \qquad\qquad\qquad\qquad = g(4x)$$
$$= 4(3x + 2) \qquad\qquad\qquad\qquad = 3(4x) + 2$$
$$= 12x + 8 \qquad\qquad\qquad\qquad = 12x + 2$$

WARNING! Note that in the previous example, $(f \circ g)(x) \neq (g \circ f)(x)$. This shows that the composition of functions is not commutative.

EXAMPLE 3

Evaluating composite functions. Let $f(x) = 2x + 1$ and $g(x) = x - 4$. Find
a. $(f \circ g)(9)$, **b.** $(f \circ g)(x)$, and **c.** $(g \circ f)(-2)$.

Solution
a. $(f \circ g)(9)$ means $f(g(9))$. In Figure 11-3(a), function g receives the number 9, subtracts 4, and releases the number $g(9) = 5$. Then 5 goes into the f function, which doubles 5 and adds 1. The final result, 11, is the output of the composite function $f \circ g$:

$$(f \circ g)(9) = f(g(9)) = f(5) = 2(5) + 1 = 11$$

b. $(f \circ g)(x)$ means $f(g(x))$. In Figure 11-3(a), function g receives the number x, subtracts 4, and releases the number $x - 4$. Then $x - 4$ goes into the f function, which doubles $x - 4$ and adds 1. The final result, $2x - 7$, is the output of the composite function $f \circ g$.

$$(f \circ g)(x) = f(g(x)) = f(x - 4) = 2(x - 4) + 1 = 2x - 7$$

c. $(g \circ f)(-2)$ means $g(f(-2))$. In Figure 11-3(b), function f receives the number -2, doubles it and adds 1, and releases -3 into the g function. Function g subtracts 4 from -3 and outputs a final result of -7. Thus,

$$(g \circ f)(-2) = g(f(-2)) = g(-3) = -3 - 4 = -7$$

FIGURE 11-3

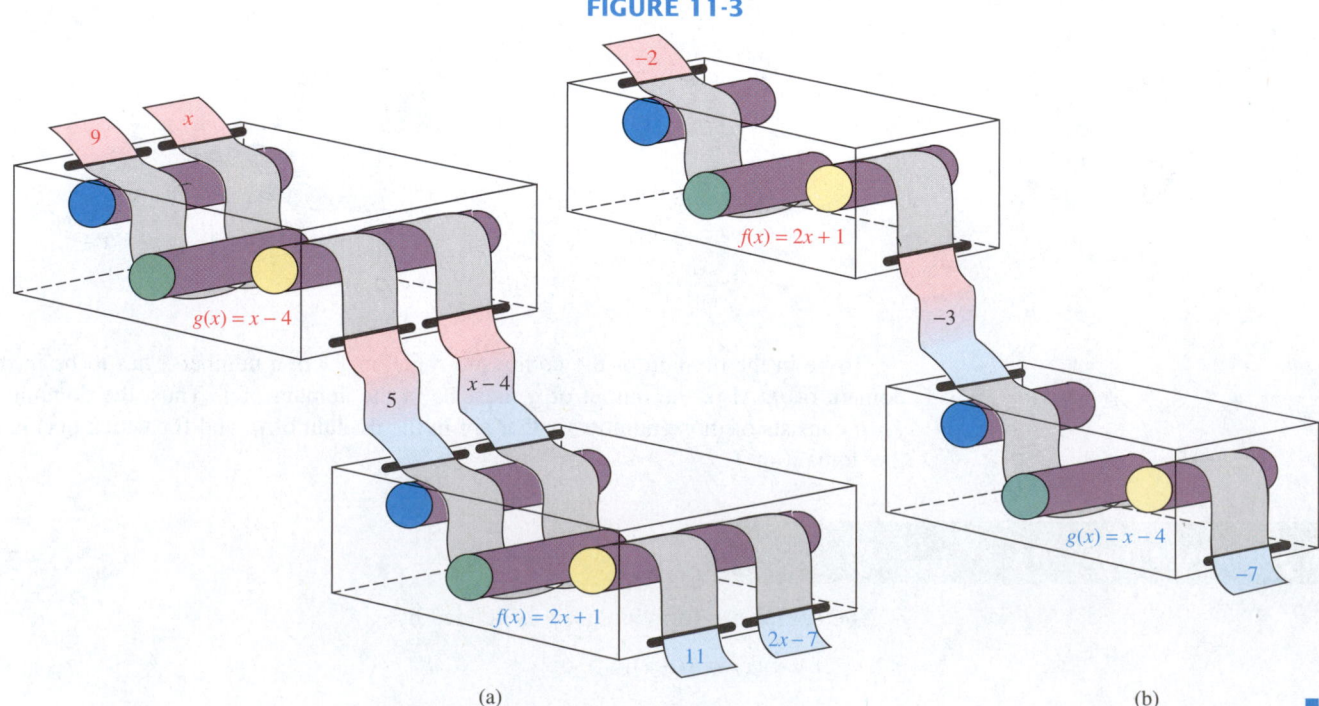

(a) (b)

The Identity Function

The **identity function** is defined by the equation $I(x) = x$. Under this function, the value that is assigned to any real number x is x itself. For example $I(2) = 2$, $I(-3) = -3$, and $I(7.5) = 7.5$. If f is any function, the composition of f with the identity function is just the function f:

$$(f \circ I)(x) = (I \circ f)(x) = f(x)$$

EXAMPLE 4

The identity function. Let f be any function and let I be the identity function, $I(x) = x$. Show that **a.** $(f \circ I)(x) = f(x)$ and **b.** $(I \circ f)(x) = f(x)$.

Solution **a.** $(f \circ I)(x)$ means $f(I(x))$. Because $I(x) = x$, we have

$$(f \circ I)(x) = f(I(x)) = f(x)$$

b. $(I \circ f)(x)$ means $I(f(x))$. Because I passes any number through unchanged, we have $I(f(x)) = f(x)$ and

$$(I \circ f)(x) = I(f(x)) = f(x) \qquad \blacksquare$$

Writing Composite Functions

EXAMPLE 5

Biological research. A laboratory specimen is stored in a refrigeration unit at a temperature of 15° Fahrenheit. Biologists remove the specimen and warm it at a controlled rate of 3° F per hour. Express the sample's Celsius temperature as a function of the time t since it was removed from refrigeration.

Solution The temperature of the specimen is 15° F when the time $t = 0$. Because it warms at a rate of 3° F per hour, its initial temperature of 15° increases by $3t°$ F in t hours. The Fahrenheit temperature of the specimen is given by the function

$$F(t) = 3t + 15$$

The Celsius temperature C is a function of this Fahrenheit temperature F, given by the function

$$C(F) = \frac{5}{9}(F - 32)$$

To express the specimen's Celsius temperature as a function of *time*, we find the composite function

$$
\begin{aligned}
(C \circ F)(t) &= C(F(t)) \\
&= C(3t + 15) \qquad &&\text{Substitute } 3t + 15 \text{ for } F(t). \\
&= \frac{5}{9}[(3t + 15) - 32] \qquad &&\text{Substitute } 3t + 15 \text{ for } F \text{ in } \tfrac{5}{9}(F - 32). \\
&= \frac{5}{9}(3t - 17) \qquad &&\text{Simplify.}
\end{aligned}
$$

The composite function, $C(t) = \frac{5}{9}(3t - 17)$, finds the temperature of the specimen in degrees Celsius t hours after it is removed from refrigeration. $\blacksquare$

Section 11.1

VOCABULARY

In Exercises 1–8, fill in the blanks to make the statements true.

1. The _____ of f and g, denoted as $f + g$, is defined by $(f + g)(x) =$ _____ .

2. The _____ of f and g, denoted as $f - g$, is defined by $(f - g)(x) =$ _____ .

3. The _____ of f and g, denoted as $f \cdot g$, is defined by $(f \cdot g)(x) =$ _____ .

4. The _____ of f and g, denoted as f/g, is defined by $(f/g)(x) =$ _____ .

5. In Exercises 1–3, the _____ of each function is the set of real numbers x that are in the domain of both f and g.

6. The _____ function $f \circ g$ is defined by $(f \circ g)(x) = f(g(x))$.

7. Under the _____ function, the value that is assigned to any real number x is x itself.

8. When reading the notation $f(g(x))$, we say "f ___ g ___ x."

CONCEPTS

9. Fill in the blanks to make the statements true.
   a. $(f \circ g)(x) = f\big(\ \ \ \ \big)$
   b. To find $f(g(x))$, we first find _____ and then substitute that value for x in $f(x)$.

10. a. If $f(x) = 3x + 1$ and $g(x) = 1 - 2x$, find $f(g(3))$ and $g(f(3))$.
    b. Is the composition of functions commutative? Explain.

11. Complete the table of values for the identity function, $I(x) = x$. Then graph it.

x	$I(x)$
-3	
-2	
-1	
0	
1	
2	
3	

12. Fill in the blanks in the drawing of the function machines in Illustration 1 that show how to compute $g(f(-2))$.

ILLUSTRATION 1

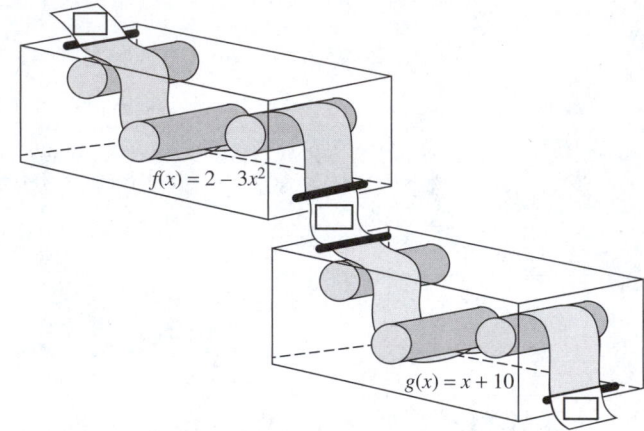

$f(x) = 2 - 3x^2$

$g(x) = x + 10$

NOTATION

In Exercises 13–14, complete each solution.

13. Let $f(x) = 3x - 1$ and $g(x) = 2x + 3$. Find $f \cdot g$.
$$(f \cdot g)(x) = f(x) \cdot g(x)$$
$$= \boxed{}(2x + 3)$$
$$= 6x^2 + \boxed{} - \boxed{} - 3$$
$$= 6x^2 + 7x - 3$$

14. Let $f(x) = 3x - 1$ and $g(x) = 2x + 3$. Find $f \circ g$.
$$(f \circ g)(x) = f(g(x))$$
$$= f\big(\boxed{}\big)$$
$$= 3\big(\boxed{}\big) - 1$$
$$= \boxed{} + \boxed{} - 1$$
$$= 6x + 8$$

PRACTICE

In Exercises 15–22, $f(x) = 3x$ and $g(x) = 4x$. Find each function and its domain.

15. $f + g$ 16. $f - g$

17. $f \cdot g$ 18. f/g

19. $g - f$ 20. $g + f$

21. g/f 22. $g \cdot f$

In Exercises 23–30, $f(x) = 2x + 1$ and $g(x) = x - 3$.
Find each function and its domain.

23. $f + g$ **24.** $f - g$

25. $f \cdot g$ **26.** f/g

27. $g - f$ **28.** $g + f$

29. g/f

30. $g \cdot f$

In Exercises 31–34, $f(x) = 3x - 2$ and $g(x) = 2x^2 + 1$.
Find each function and its domain.

31. $f - g$ **32.** $f + g$

33. f/g **34.** $f \cdot g$

In Exercises 35–38, $f(x) = x^2 - 1$ and $g(x) = x^2 - 4$.
Find each function and its domain.

35. $f - g$ **36.** $f + g$

37. g/f

38. $g \cdot f$

In Exercises 39–50, $f(x) = 2x + 1$ and $g(x) = x^2 - 1$.
Find each value.

39. $(f \circ g)(2)$ **40.** $(g \circ f)(2)$

41. $(g \circ f)(-3)$ **42.** $(f \circ g)(-3)$

43. $(f \circ g)(0)$ **44.** $(g \circ f)(0)$

45. $(f \circ g)\left(\dfrac{1}{2}\right)$ **46.** $(g \circ f)\left(\dfrac{1}{3}\right)$

47. $(f \circ g)(x)$ **48.** $(g \circ f)(x)$

49. $(g \circ f)(2x)$ **50.** $(f \circ g)(2x)$

In Exercises 51–58, $f(x) = 3x - 2$ and $g(x) = x^2 + x$.
Find each value.

51. $(f \circ g)(4)$ **52.** $(g \circ f)(4)$

53. $(g \circ f)(-3)$ **54.** $(f \circ g)(-3)$

55. $(g \circ f)(0)$ **56.** $(f \circ g)(0)$

57. $(g \circ f)(x)$ **58.** $(f \circ g)(x)$

59. If $f(x) = x + 1$ and $g(x) = 2x - 5$, show that
$(f \circ g)(x) \neq (g \circ f)(x)$.

60. If $f(x) = x^2 + 1$ and $g(x) = 3x^2 - 2$, show that
$(f \circ g)(x) \neq (g \circ f)(x)$.

APPLICATIONS

61. METALLURGY A molten alloy must be cooled slowly to control crystallization. When removed from the furnace, its temperature is 2,700° F, and it will be cooled at 200° per hour. Express the Celsius temperature as a function of the number of hours t since cooling began.

62. WEATHER FORECASTING A high-pressure area promises increasingly warmer weather for the next 48 hours. The temperature is now 34° Celsius and is expected to rise 1° every 6 hours. Express the Fahrenheit temperature as a function of the number of hours from now. $\left(Hint:\ F = \tfrac{9}{5}C + 32.\right)$

63. VACATION MILEAGE COSTS
  a. Use the graphs in Illustration 2 to determine the cost of the gasoline consumed if a family drove 500 miles on a summer vacation.
  b. Write a composition function that expresses the cost of the gasoline consumed on the vacation as a function of the miles driven.

ILLUSTRATION 2

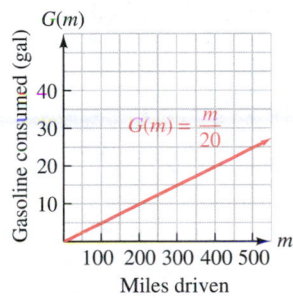 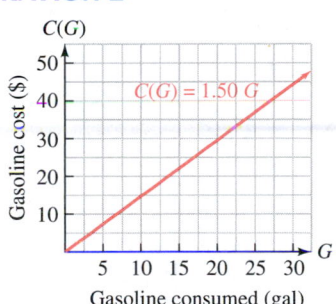

64. HALLOWEEN COSTUMES The tables on the back of the pattern package shown in Illustration 3 can be used to determine the number of yards of material needed to make a rabbit costume for a child.
  a. How many yards of material are needed if the child's chest measures 29 inches?
  b. In this exercise, one quantity is a function of a second quantity that depends, in turn, on a third quantity. Explain this dependence.

ILLUSTRATION 3

PATTERN 9810   **Simplicity**

Costumes have front zipper, long raglan sleeves, elastic sleeve and leg casings, hood. Fabrics: Fleece or suede

BODY MEASUREMENTS

Chest (in.)	21	22	23	25	26	27	28½	29	30
Pattern Size	2	3	4	6	7	8	10	11	12

YARDAGE NEEDED

Pattern Size	2-4	6-8	10-12
Yards	$2\frac{5}{8}$	$3\frac{3}{8}$	$3\frac{3}{4}$

WRITING

65. Exercise 63 illustrates a chain of dependence between the cost of the gasoline, the gasoline consumed, and the miles driven. Describe another chain of dependence that could be represented by a composition function.

66. Explain how to add, subtract, multiply, and divide two functions.

67. Write out in words how to say each of the following:

$(f \circ g)(2)$ $g(f(-8))$

68. If $Y_1 = f(x)$ and $Y_2 = g(x)$, explain how to use the tables of values shown in Illustration 4 to find $g(f(2))$.

ILLUSTRATION 4

X	Y₁			X	Y₂	
-2	-2			-2	-7	
0	2			0	-6	
2	6			2	-5	
4	10			4	-4	
6	14			6	-3	
8	18			8	-2	
10	22			10	-1	

X=-2   X=-2

REVIEW

Simplify each expression.

69. $\dfrac{3x^2 + x - 14}{4 - x^2}$

70. $\dfrac{2x^3 + 14x^2}{3 + 2x - x^2} \cdot \dfrac{x^2 - 3x}{x}$

71. $\dfrac{x^2 - 2x - 8}{3x^2 - x - 12} \div \dfrac{3x^2 + 5x - 2}{3x - 1}$

72. $\dfrac{x - 1}{1 + \dfrac{x}{x - 2}}$

▶ 11.2   Inverses of Functions

In this section, you will learn about

One-to-one functions ■ The horizontal line test ■ Finding inverses of functions

---

Introduction   The function defined by $C = \frac{5}{9}(F - 32)$ is the formula that we use to convert degrees Fahrenheit to degrees Celsius. If we input a Fahrenheit reading into the formula, the output is a Celsius reading. For example, if we substitute 41° for F, we obtain a Celsius reading of 5°:

$$C = \frac{5}{9}(F - 32)$$

$$= \frac{5}{9}(41 - 32) \quad \text{Substitute 41 for } F.$$

$$= \frac{5}{9}(9)$$

$$= 5$$

If we want to find a Fahrenheit reading from a Celsius reading, we need a formula into which we can substitute a Celsius reading and have a Fahrenheit reading come out. Such a formula is $F = \frac{9}{5}C + 32$, which takes the Celsius reading of 5° and turns it back into a Fahrenheit reading of 41°.

$$F = \frac{9}{5}C + 32$$

$$= \frac{9}{5}(5) + 32 \quad \text{Substitute 5 for } C.$$

$$= 41$$

The functions defined by these two formulas do opposite things. The first turns 41° F into 5° C, and the second turns 5° C back into 41° F. For this reason, we say that the functions are *inverses* of each other.

In this section, we will show how to find inverses of one-to-one functions.

One-to-One Functions

Recall that for each input into a function, there is a single output. For some functions, different inputs have the same output, as shown in Figure 11-4(a). For other functions, different inputs have different outputs, as shown in Figure 11-4(b).

FIGURE 11-4

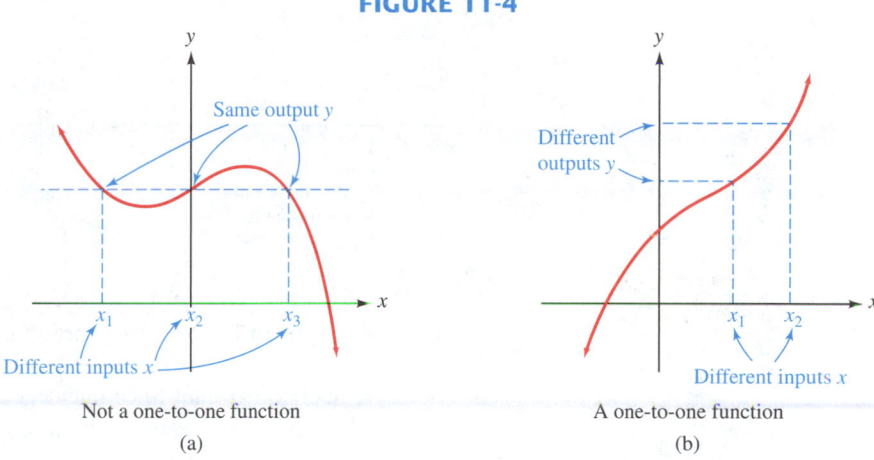

Not a one-to-one function
(a)

A one-to-one function
(b)

When every output of a function corresponds to exactly one input, we say that the function is *one-to-one*.

One-to-one functions

A function is called **one-to-one** if each input value of x in the domain determines a different output value of y in the range.

EXAMPLE 1   **One-to-one functions.** Determine whether the functions **a.** $f(x) = x^2$ and **b.** $f(x) = x^3$ are one-to-one.

Solution **a.** The function $f(x) = x^2$ is not one-to-one, because different input values x can determine the same output value y. For example, inputs of 3 and -3 produce the same output value of 9.

$$f(3) = 3^2 = 9 \qquad \text{and} \qquad f(-3) = (-3)^2 = 9$$

b. The function $f(x) = x^3$ is one-to-one, because different input values x determine different output values y. This is because different numbers have different cubes.

SELF CHECK   Determine whether $f(x) = 2x + 3$ is one-to-one and explain why or why not.

Answer: yes, because different input values determine different output values ∎

The Horizontal Line Test

The **horizontal line test** can be used to decide whether the graph of a function represents a one-to-one function. If every horizontal line that intersects the graph of a function does so only once, the function is one-to-one. Otherwise, the function is not one-to-one. See Figure 11-5.

FIGURE 11-5

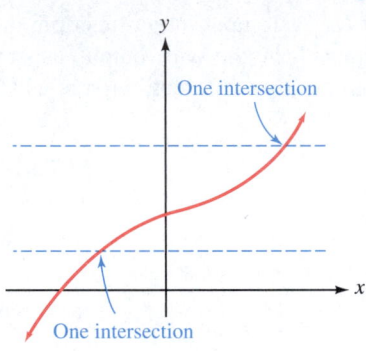
A one-to-one function

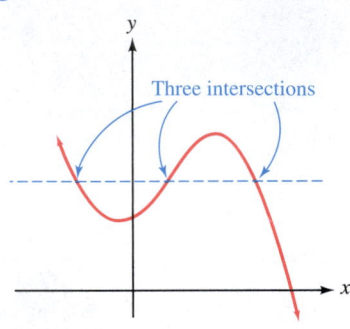
Not a one-to-one function

EXAMPLE 2 **Determining one-to-one functions.** Use the horizontal line test to decide whether the graphs in Figure 11-6 represent one-to-one functions.

Solution **a.** Because many horizontal lines intersect the graph shown in Figure 11-6(a) twice, the graph does not represent a one-to-one function.

b. Because each horizontal line that intersects the graph in Figure 11-6(b) does so exactly once, the graph represents a one-to-one function.

FIGURE 11-6

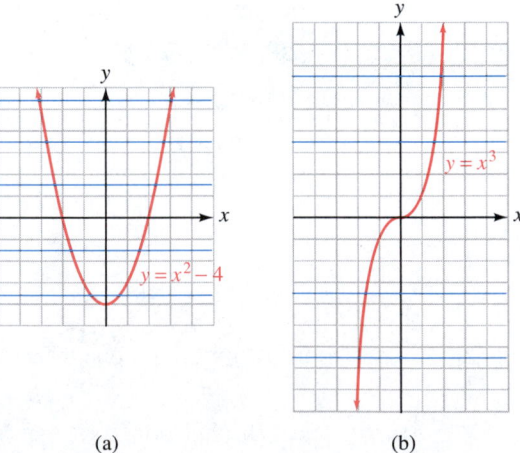

(a)   (b)

SELF CHECK Determine whether the following graphs represent one-to-one functions.

a. 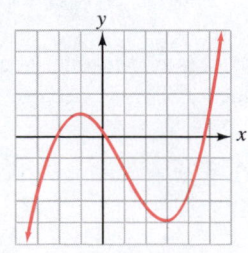   **b.**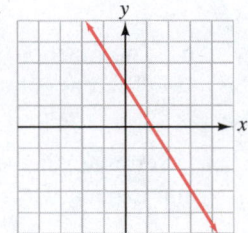

Answers: **a.** no, **b.** yes

Finding Inverses of Functions

If f is the function determined by the ordered pairs in the table shown in Figure 11-7(a), it turns the number 1 into 10, 2 into 20, and 3 into 30. Since the inverse of f must turn 10 back into 1, 20 back into 2, and 30 back into 3, it consists of the ordered pairs shown in Figure 11-7(b).

FIGURE 11-7

Function f

x	y
1	10
2	20
3	30

↑   ↑

Domain   Range

(a)

Inverse of f

x	y
10	1
20	2
30	3

↑   ↑

Domain   Range

(b)

Note that the inverse of f is also a function.

We note that the domain of f and the range of its inverse is $\{1, 2, 3\}$. The range of f and the domain of its inverse is $\{10, 20, 30\}$.

This example suggests that to form the inverse of a function f, we simply interchange the coordinates of each ordered pair that determines f. When the inverse of a function is also a function, we call it **f inverse** and denote it with the symbol f^{-1}.

WARNING! The symbol $f^{-1}(x)$ is read as "the inverse of $f(x)$" or just "f inverse." The -1 in the notation $f^{-1}(x)$ is not an exponent. Remember that $f^{-1}(x) \neq \frac{1}{f(x)}$.

Finding the inverse of a one-to-one function

If a function is one-to-one, we find its inverse as follows:

1. If the function is written using function notation, replace $f(x)$ with y.
2. Interchange the variables x and y.
3. Solve the resulting equation for y.
4. We can substitute $f^{-1}(x)$ for y.

EXAMPLE 3

Finding the inverse of a function. If $f(x) = 4x + 2$, find the inverse of f and tell whether it is a function.

Solution We proceed as follows:

$f(x) = 4x + 2$

$y = 4x + 2$ Step 1: Replace $f(x)$ with y.

$x = 4y + 2$ Step 2: Interchange the variables x and y.

To decide whether the inverse $x = 4y + 2$ is a function, we solve for y (Step 3).

$x = 4y + 2$

$x - 2 = 4y$ Subtract 2 from both sides.

1. $y = \dfrac{x - 2}{4}$ Divide both sides by 4 and write y on the left-hand side.

Because each input x that is substituted into Equation 1 gives one output y, the inverse of f is a function, so we can express it in the form

$$f^{-1}(x) = \frac{x-2}{4} \qquad \text{Step 4: Substitute } f^{-1}(x) \text{ for } y.$$

SELF CHECK   If $f(x) = -5x - 3$, find the inverse of f and tell whether it is a function.   *Answers:* $f^{-1}(x) = \frac{-x-3}{5}$, yes ∎

To emphasize an important relationship between a function and its inverse, we substitute some number x, such as $x = 3$, into the function $f(x) = 4x + 2$ of Example 3. The corresponding value of y produced is

$$f(3) = 4(3) + 2 = 14$$

If we substitute 14 into the inverse function, f^{-1}, the corresponding value of y that is produced is

$$f^{-1}(14) = \frac{14-2}{4} = 3$$

Thus, the function f turns 3 into 14, and the inverse function f^{-1} turns 14 back into 3. In general, the composition of a function and its inverse is the identity function.

To prove that $f(x) = 4x + 2$ and $f^{-1}(x) = \frac{x-2}{4}$ are inverse functions, we must show that their composition (in both directions) is the identity function:

$$(f \circ f^{-1})(x) = f(f^{-1}(x)) \qquad\qquad (f^{-1} \circ f)(x) = f^{-1}(f(x))$$

$$= f\left(\frac{x-2}{4}\right) \qquad\qquad\qquad\qquad = f^{-1}(4x+2)$$

$$= 4\left(\frac{x-2}{4}\right) + 2 \qquad\qquad\qquad = \frac{4x+2-2}{4}$$

$$= x - 2 + 2 \qquad\qquad\qquad\qquad\quad = \frac{4x}{4}$$

$$= x \qquad\qquad\qquad\qquad\qquad\qquad\quad = x$$

Thus, $(f \circ f^{-1})(x) = (f^{-1} \circ f)(x) = x$, which is the identity function $I(x)$.

EXAMPLE 4   **The graph of an inverse function.** The set of all pairs (x, y) determined by $3x + 2y = 6$ is a function. Find its inverse function and graph the function and its inverse on one coordinate system.

Solution   To find the inverse function of $3x + 2y = 6$, we interchange x and y to obtain

$$3y + 2x = 6$$

and then solve the equation for y.

$$3y + 2x = 6$$
$$3y = -2x + 6$$
$$y = -\frac{2}{3}x + 2$$

FIGURE 11-8

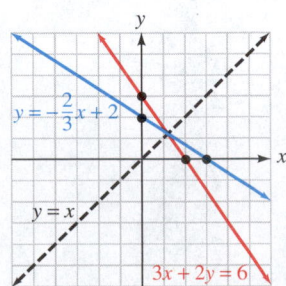

The inverse function of $3x + 2y = 6$ is $y = -\frac{2}{3}x + 2$, which can also be written $f^{-1}(x) = -\frac{2}{3}x + 2$. The graphs of $3x + 2y = 6$ and its inverse, along with the graph of $y = x$, appear in Figure 11-8.

SELF CHECK   Find the inverse of the function defined by $2x - 3y = 6$. Graph the function and its inverse on one coordinate system. Also graph $y = x$.

Answer: $f^{-1}(x) = \frac{3}{2}x + 3$

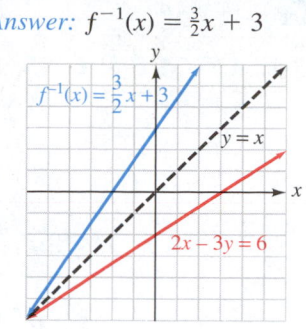

In Figure 11-8, we see that the graphs of $3x + 2y = 6$ and $y = -\frac{2}{3}x + 2$ (its inverse) are symmetric about the line $y = x$. That is, the graphs are mirror images (reflections) of each other with reference to $y = x$. This is always the case with a function and its inverse, because when the coordinates (a, b) satisfy an equation, the coordinates (b, a) will satisfy its inverse.

ACCENT ON TECHNOLOGY *Graphing the Inverse of a Function*

We can use a graphing calculator for an informal check of the result found in Example 4. First, we solve $3x + 2y = 6$ for y and enter it in the form $y = -\frac{3}{2}x + 3$. Then we enter what we believe to be the inverse function, $y = -\frac{2}{3}x + 2$, as well as the equation $y = x$. See Figure 11-9(a). Before graphing, we adjust the display so that the graphing grid will be composed of squares. (The standard window display is rectangular, having a grid with a unit width different from the unit height.) Check your owner's manual for directions on how to program the calculator to graph in the square mode.

In Figure 11-9(b), it appears that the two graphs are symmetric about the line $y = x$. Although it is not definitive, this visual check does help to validate that $y = -\frac{2}{3}x + 2$ is the inverse of the function determined by $3x + 2y = 6$.

FIGURE 11-9

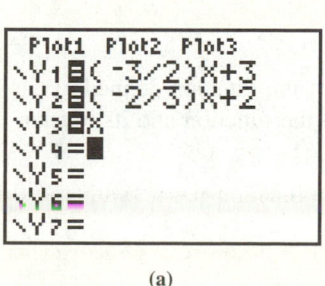

(a)

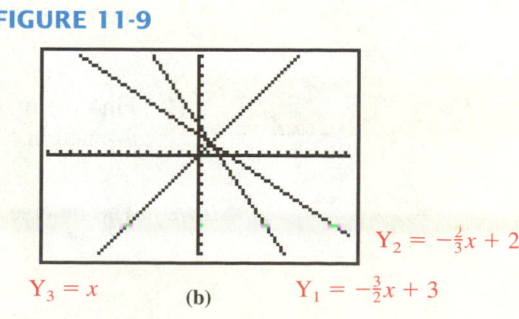

$Y_3 = x$ (b) $Y_1 = -\frac{3}{2}x + 3$

$Y_2 = -\frac{2}{3}x + 2$

In each example so far, the inverse of a function has been another function. This is not always true, as the following example will show.

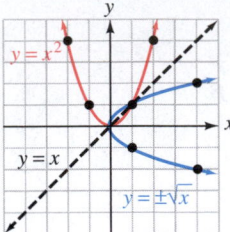

EXAMPLE 5

Finding the inverse of a function. Find the inverse of the function determined by $f(x) = x^2$.

Solution

$y = x^2$ Replace $f(x)$ with y.

$x = y^2$ Interchange x and y.

$y = \pm\sqrt{x}$ Use the square root property and write y on the left-hand side.

When the inverse $y = \pm\sqrt{x}$ is graphed, as in Figure 11-10, we see that the graph does not pass the vertical line test. Thus, it is not a function.

The graph of $y = x^2$ is also shown in the figure. Note that it is not one-to-one. The graphs of $y = x^2$ and $y = \pm\sqrt{x}$ are symmetric about the line $y = x$.

FIGURE 11-10

SELF CHECK

Find the inverse of the function determined by $f(x) = 4x^2$.

Answer: $y = \pm\dfrac{\sqrt{x}}{2}$

EXAMPLE 6

Finding the inverse of a function. Find the inverse of $f(x) = x^3$.

Solution

To find the inverse, we proceed as follows:

$y = x^3$ Replace $f(x)$ with y.

$x = y^3$ Interchange the variables x and y.

$\sqrt[3]{x} = y$ Take the cube root of both sides.

We note that to each number x there corresponds one real cube root. Thus, $y = \sqrt[3]{x}$ represents a function. Using $f^{-1}(x)$ notation, the inverse of $f(x) = x^3$ is

$$f^{-1}(x) = \sqrt[3]{x}$$

SELF CHECK

Find the inverse of $f(x) = x^5$.

Answer: $f^{-1}(x) = \sqrt[5]{x}$

If a function is not one-to-one, it is often possible to make it one-to-one by restricting its domain.

EXAMPLE 7

Find the inverse of the function defined by $y = x^2$ with $x \geq 0$. Then tell whether it is a function. Graph the function and its inverse on one set of coordinate axes.

Solution

The inverse of the function $y = x^2$ with $x \geq 0$ is

$x = y^2$ with $y \geq 0$ Interchange the variables x and y.

This equation can be written in the form

$y = \pm\sqrt{x}$ with $y \geq 0$

Since $y \geq 0$, each number x gives only one value of y: $y = \sqrt{x}$. Thus, the inverse is a function.

The graphs of the two functions appear in Figure 11-11. The line $y = x$ is included so that we can see that the graphs are symmetric about the line $y = x$.

FIGURE 11-11

$y = x^2$ and $x \geq 0$

x	y	(x, y)
0	0	$(0, 0)$
1	1	$(1, 1)$
2	4	$(2, 4)$
3	9	$(3, 9)$

$x = y^2$ and $y \geq 0$

x	y	(x, y)
0	0	$(0, 0)$
1	1	$(1, 1)$
4	2	$(4, 2)$
9	3	$(9, 3)$

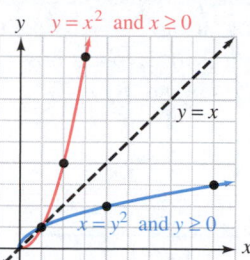

STUDY SET

Section 11.2

VOCABULARY

In Exercises 1–4, fill in the blanks to make the statements true.

1. A function is called _____ if each input determines a different output.

2. The _____ line test can be used to decide whether the graph of a function represents a one-to-one function.

3. The functions f and f^{-1} are _____.

4. An input value is an element of a function's _____. An output value is an element of a function's _____.

CONCEPTS

In Exercises 5–10, fill in the blanks to make the statements true.

5. If every horizontal line that intersects the graph of a function does so only _____, the function is one-to-one.

6. If any horizontal line that intersects the graph of a function does so more than once, the function is not _____.

7. If a function turns an input of 2 into an output of 5, the inverse function will turn an input of 5 into an output of ___.

8. The graphs of a function and its inverse are symmetrical about the line _____.

9. $(f \circ f^{-1})(x) = $ _____

10. To find the inverse of the function defined by $y = 2x - 3$, we begin by _____ x and y.

11. Use the table of values of the one-to-one function f to complete a table of values for f^{-1}.

x	$f(x)$		x	$f^{-1}(x)$
-6	-3		-3	
-4	-2		-2	
0	0		0	
2	1		1	
8	4		4	

12. How can we tell that function f is not one-to-one from the table of values?

x	$f(x)$
-2	4
-1	1
0	0
2	4
3	9

13. Is the inverse of a function always a function?

14. Name four points that the line $y = x$ passes through.

15. If f is a one-to-one function, and if $f(2) = 6$, then what is $f^{-1}(6)$?

16. If the point $(2, -4)$ is on the graph of the one-to-one function f, then what point is on the graph of f^{-1}?

17. Two functions are graphed on the square grid in Illustration 1 along with the line $y = x$. Explain why we know that the functions are not inverses of each other.

ILLUSTRATION 1

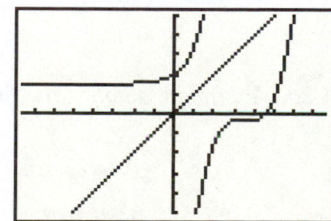

18. A table of values for a function f is shown in Illustration 2(a). A table of values for f^{-1} is shown Illustration 2(b). Explain how to use the tables to find $f^{-1}(f(4))$ and $f(f^{-1}(2))$.

ILLUSTRATION 2

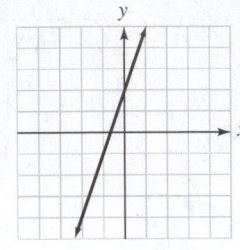

(a)

(b)

NOTATION

In Exercises 19–20, complete each solution.

19. Find the inverse of $f(x) = 2x - 3$.

$$y = \boxed{} - 3 \quad \text{Replace } f(x) \text{ with } y.$$

$$x = \boxed{} - 3 \quad \text{Interchange the variables } x \text{ and } y.$$

$$x + \boxed{} = 2y \quad \text{Add 3 to both sides.}$$

$$\frac{x+3}{2} = \boxed{} \quad \text{Divide both sides by 2.}$$

The inverse of $f(x) = 2x - 3$ is $\boxed{} = \dfrac{x+3}{2}$.

20. Find the inverse of $f(x) = \sqrt[3]{x} + 2$.

$$\boxed{} = \sqrt[3]{x} + 2 \quad \text{Replace } f(x) \text{ with } y.$$

$$x = \sqrt[3]{\boxed{}} + 2 \quad \text{Interchange the variables } x \text{ and } y.$$

$$x - \boxed{} = \sqrt[3]{y} \quad \text{Subtract 2 from both sides.}$$

$$(x-2)^3 = \boxed{} \quad \text{Cube both sides.}$$

The inverse of $f(x) = \sqrt[3]{x} + 2$ is $\boxed{} = (x-2)^3$.

21. The symbol $f^{-1}(x)$ is read as "_____ f" or "f _____."

22. Explain the difference in the meaning of the -1 in the notation $f^{-1}(x)$ as compared to x^{-1}.

PRACTICE

In Exercises 23–26, determine whether the function is one-to-one.

23. $f(x) = 2x$

24. $f(x) = |x|$

25. $f(x) = x^4$

26. $f(x) = x^3 + 1$

In Exercises 27–30, each graph represents a function. Use the horizontal line test to decide whether the function is one-to-one.

27.

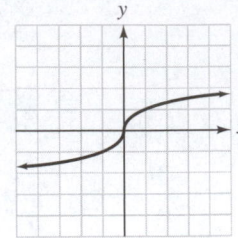

28.

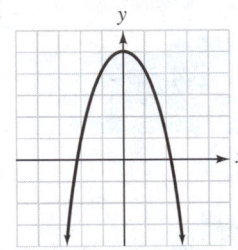

29.

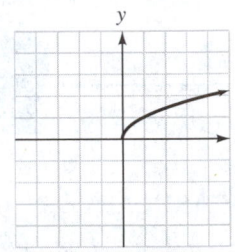

30.

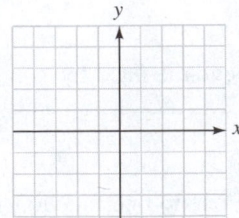

In Exercises 31–34, tell whether the set of ordered pairs (x, y) is a function. Then find the inverse of the set of ordered pairs (x, y) and tell whether the inverse is a function.

31. $\{(3, 2), (2, 1), (1, 0)\}$

32. $\{(4, 1), (5, 1), (6, 1), (7, 1)\}$

33. $\{(1, 1), (2, 1), (3, 1), (4, 1)\}$

34. $\{(1, 2), (2, 3), (1, 3), (1, 5)\}$

In Exercises 35–40, find the inverse of the function and express it using $f^{-1}(x)$ notation.

35. $4x - 5y = 20$

36. $y + 1 = 5x$

37. $x + 4 = 5y$

38. $x = 3y + 1$

39. $f(x) = \dfrac{x-4}{5}$

40. $f(x) = \dfrac{2x+6}{3}$

In Exercises 41–44, find the inverse of each function and express it in terms of x and y. Then graph the function and its inverse on one coordinate system. Show the line of symmetry on the graph.

41. $y = 4x + 3$

42. $x = 3y - 1$

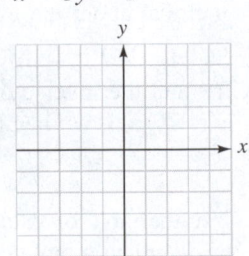

43. $2x + 3y = 9$

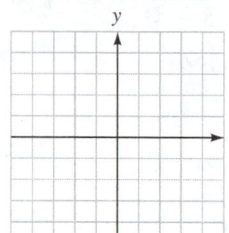

44. $3(x + y) = 2x + 4$

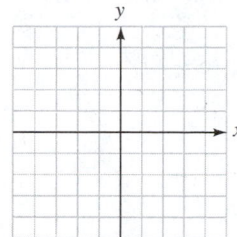

In Exercises 45–52, find the inverse of the function and tell whether it is a function. If it is a function, express it using $f^{-1}(x)$ notation.

45. $y = x^2 + 4$

46. $y = x^2 + 5$

47. $y = x^3$

48. $xy = 4$

49. $y = |x|$

50. $y = \sqrt[3]{x}$

51. $f(x) = 2x^3 - 3$

52. $f(x) = \dfrac{3}{x^3} - 1$

In Exercises 53–56, graph each equation and its inverse on one set of coordinate axes. Show the axis of symmetry.

53. $y = x^2 + 1$

54. $y = \dfrac{1}{4}x^2 - 3$

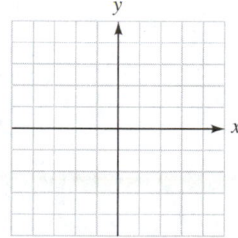

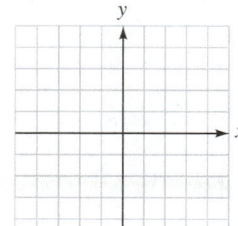

55. $y = \sqrt{x}$ $(x \geq 0)$

56. $y = |x|$

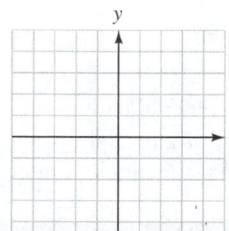

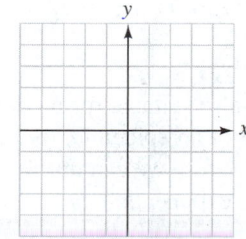

APPLICATIONS

57. INTERPERSONAL RELATIONSHIPS Feelings of anxiety in a relationship can increase or decrease, depending on what is going on in the relationship. The graph in Illustration 3 shows how a person's anxiety might vary as a relationship develops over time.

a. Is this the graph of a function? Is its inverse a function?

b. Does each anxiety level correspond to exactly one point in time? Use the dashed lined labeled *Maximum threshold* to explain.

ILLUSTRATION 3

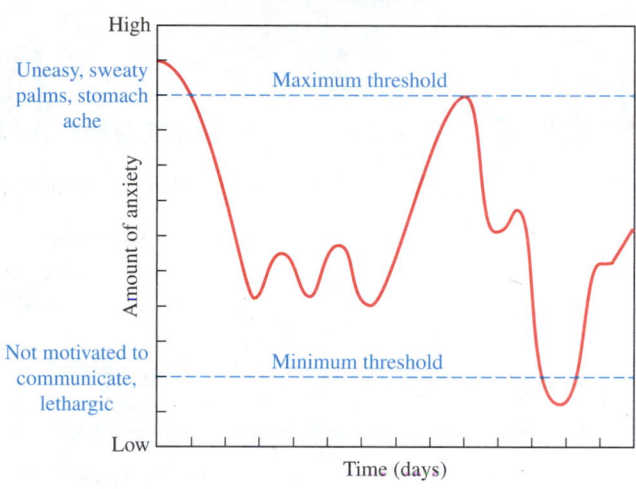

Based on information from Gudykunst, *Building Bridges: Interpersonal Skills for a Changing World* (Houghton Mifflin, 1994)

58. LIGHTING LEVELS The ability of the eye to see detail increases as the level of illumination increases. This relationship can be modeled by a function E, whose graph is shown in Illustration 4.

a. From the graph, determine $E(240)$.

b. Is function E one-to-one? Does E have an inverse?

c. If the effectiveness of seeing in an office is 7, what is the illumination in the office? How can this question be asked using inverse function notation?

ILLUSTRATION 4

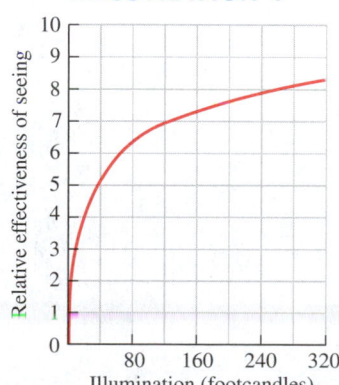

Based on information from *World Book Encyclopedia*

WRITING

59. What does it mean when we say that one graph is symmetric to the other about the line $y = x$?

60. Explain the purpose of the horizontal line test.

61. $3 - \sqrt{-64}$

62. $(2 - 3i) + (4 + 5i)$

63. $(3 + 4i)(2 - 3i)$

64. $\dfrac{6 + 7i}{3 - 4i}$

65. $(6 - 8i)^2$

66. i^{100}

▶11.3   Exponential Functions

In this section, you will learn about

■ Irrational exponents ■ Exponential functions ■ Graphing exponential functions ■ Vertical and horizontal translations ■ Compound interest ■ Exponential functions as models

Introduction The graph in Figure 11-12 shows the balance in a bank account in which \$10,000 was invested in 1998 at 9%, compounded monthly. The graph shows that in the year 2008, the value of the account will be approximately \$25,000, and in the year 2028, the value will be approximately \$147,000. The curve shown in Figure 11-12 is the graph of a function called an *exponential function.*

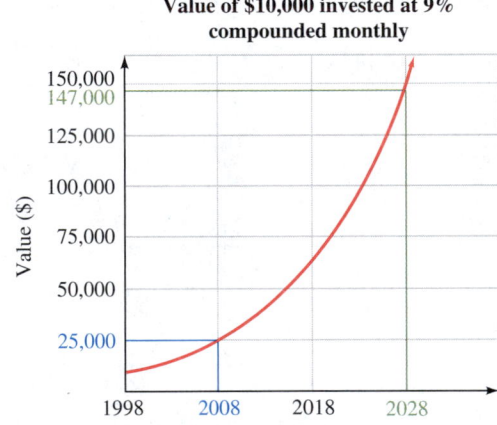

FIGURE 11-12

Value of \$10,000 invested at 9% compounded monthly

Exponential functions are also suitable for modeling many other situations, such as population growth, the spread of an epidemic, the temperature of a heated object as it cools, and radioactive decay. Before we can discuss exponential functions in more detail, we must define irrational exponents.

Irrational Exponents

We have discussed expressions of the form b^x, where x is a rational number.

$8^{1/2}$ means "the square root of 8."

$5^{1/3}$ means "the cube root of 5."

$3^{-2/5} = \dfrac{1}{3^{2/5}}$ means "the reciprocal of the fifth root of 3^2."

To give meaning to b^x when x is an irrational number, we consider the expression

$5^{\sqrt{2}}$ where $\sqrt{2}$ is the irrational number 1.414213562 . . .

Each number in the following list is defined, because each exponent is a rational number.

$5^{1.4}, \quad 5^{1.41}, \quad 5^{1.414}, \quad 5^{1.4142}, \quad 5^{1.41421}, \quad . . .$

Since the exponents are getting closer to $\sqrt{2}$, the numbers in this list are successively better approximations of $5^{\sqrt{2}}$. We can use a calculator to obtain a very good approximation.

ACCENT ON TECHNOLOGY *Evaluating Exponential Expressions*

To find the value of $5^{\sqrt{2}}$ with a scientific calculator, we enter these numbers and press these keys:

Keystrokes 5 $\boxed{y^x}$ 2 $\boxed{\sqrt{}}$ $\boxed{=}$ $\boxed{\text{9.738517742}}$

With a graphing calculator, we enter these numbers and press these keys:

Keystrokes 5 $\boxed{\wedge}$ $\boxed{\sqrt{}}$ 2 $\boxed{)}$ $\boxed{\text{Enter}}$ $\boxed{\begin{array}{l} 5 \wedge \sqrt{}(2) \\ \qquad\quad 9.738517742 \end{array}}$

In general, if $b > 0$ and x is a real number, b^x represents a positive number. It can be shown that all of the familiar rules of exponents are also true for irrational exponents.

EXAMPLE 1

Exponential expressions having irrational exponents. Use the rules of exponents to simplify **a.** $\left(5^{\sqrt{2}}\right)^{\sqrt{2}}$ and **b.** $b^{\sqrt{3}} \cdot b^{\sqrt{12}}$.

Solution **a.** $\left(5^{\sqrt{2}}\right)^{\sqrt{2}} = 5^{\sqrt{2}\sqrt{2}}$ Keep the base and multiply the exponents.

$\qquad\qquad\quad = 5^2$ $\sqrt{2}\sqrt{2} = \sqrt{4} = 2$.

$\qquad\qquad\quad = 25$

b. $b^{\sqrt{3}} \cdot b^{\sqrt{12}} = b^{\sqrt{3}+\sqrt{12}}$ Keep the base and add the exponents.

$\qquad\qquad\quad = b^{\sqrt{3}+2\sqrt{3}}$ $\sqrt{12} = \sqrt{4}\sqrt{3} = 2\sqrt{3}$.

$\qquad\qquad\quad = b^{3\sqrt{3}}$ $\sqrt{3} + 2\sqrt{3} = 3\sqrt{3}$.

SELF CHECK Simplify: **a.** $\left(3^{\sqrt{2}}\right)^{\sqrt{8}}$ and **b.** $b^{\sqrt{2}} \cdot b^{\sqrt{18}}$.   *Answers:* **a.** 81, **b.** $b^{4\sqrt{2}}$ ■

Exponential Functions

If $b > 0$ and $b \neq 1$, the function $f(x) = b^x$ is an **exponential function**. Since x can be any real number, its domain is the set of real numbers. This is the interval $(-\infty, \infty)$. Since b is positive, the value of $f(x)$ is positive, and the range is the set of positive numbers. This is the interval $(0, \infty)$.

Since $b \neq 1$, an exponential function cannot be the constant function $f(x) = 1^x$, in which $f(x) = 1$ for every real number x.

An **exponential function with base b** is defined by the equation

$$f(x) = b^x \quad \text{or} \quad y = b^x \quad (b > 0, b \neq 1, \text{ and } x \text{ is a real number})$$

The **domain of any exponential function** is the interval $(-\infty, \infty)$. The **range** is the interval $(0, \infty)$.

Graphing Exponential Functions

Since the domain and range of $f(x) = b^x$ are sets of real numbers, we can graph exponential functions on a rectangular coordinate system.

EXAMPLE 2

Graphing exponential functions. Graph $f(x) = 2^x$.

Solution To graph $f(x) = 2^x$, we find several points (x, y) whose coordinates satisfy the equation, plot the points, and join them with a smooth curve, as shown in Figure 11-13. For example, if $x = -1$, we have

$$f(x) = 2^x$$
$$f(-1) = 2^{-1}$$
$$= \frac{1}{2}$$

The point $\left(-1, \frac{1}{2}\right)$ is on the graph of $f(x) = 2^x$.

FIGURE 11-13

$f(x) = 2^x$

x	$f(x)$	$(x, f(x))$
-1	$\frac{1}{2}$	$\left(-1, \frac{1}{2}\right)$
0	1	$(0, 1)$
1	2	$(1, 2)$
2	4	$(2, 4)$
3	8	$(3, 8)$
4	16	$(4, 16)$

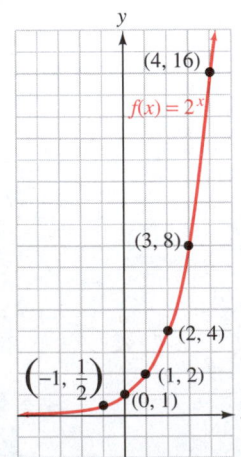

By looking at the graph, we can verify that the domain is the interval $(-\infty, \infty)$ and that the range is the interval $(0, \infty)$.

Note that as x decreases, the values of $f(x)$ decrease and approach 0. Thus, the x-axis is a horizontal asymptote of the graph.

Also note that the graph of $f(x) = 2^x$ passes through the points $(0, 1)$ and $(1, 2)$.

SELF CHECK Graph $f(x) = 4^x$.

Answer:

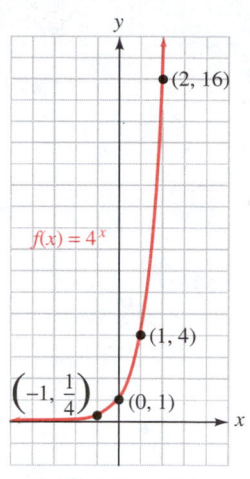

EXAMPLE 3

Graphing exponential functions. Graph $f(x) = \left(\dfrac{1}{2}\right)^x$.

Solution We find and plot pairs (x, y) that satisfy the equation. The graph of $f(x) = \left(\frac{1}{2}\right)^x$ appears in Figure 11-14. For example, if $x = -4$, we have

$$f(x) = \left(\frac{1}{2}\right)^x$$

$$f(-4) = \left(\frac{1}{2}\right)^{-4}$$

$$= \left(\frac{2}{1}\right)^4$$

$$= 16$$

The point $(-4, 16)$ is on the graph of $f(x) = \left(\frac{1}{2}\right)^x$.

FIGURE 11-14

$$f(x) = \left(\frac{1}{2}\right)^x$$

x	$f(x)$	$(x, f(x))$
-4	16	$(-4, 16)$
-3	8	$(-3, 8)$
-2	4	$(-2, 4)$
-1	2	$(-1, 2)$
0	1	$(0, 1)$
1	$\frac{1}{2}$	$\left(1, \frac{1}{2}\right)$

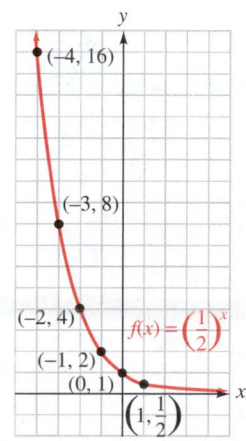

By looking at the graph, we can verify that the domain is the interval $(-\infty, \infty)$ and that the range is the interval $(0, \infty)$.

In this case, as x increases, the values of $f(x)$ decrease and approach 0. The x-axis is a horizontal asymptote. Note that the graph of $f(x) = \left(\frac{1}{2}\right)^x$ passes through the points $(0, 1)$ and $\left(1, \frac{1}{2}\right)$.

SELF CHECK   Graph $g(x) = \left(\frac{1}{3}\right)^x$.

Answer:

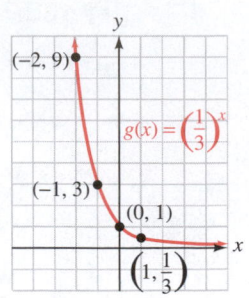

Examples 2 and 3 illustrate the following properties of exponential functions.

Properties of exponential functions

The domain of the exponential function $f(x) = b^x$ is the interval $(-\infty, \infty)$.

The range is the interval $(0, \infty)$.

The graph has a y-intercept of $(0, 1)$.

The x-axis is an asymptote of the graph.

The graph of $f(x) = b^x$ passes through the point $(1, b)$.

In Example 2 (where $b = 2$), the values of y increase as the values of x increase. Since the graph rises as we move to the right, we call the function an *increasing function*. When $b > 1$, the larger the value of b, the steeper the curve.

In Example 3 $\left(\text{where } b = \frac{1}{2}\right)$, the values of y decrease as the values of x increase. Since the graph drops as we move to the right, we call the function a *decreasing function*. When $0 < b < 1$, the smaller the value of b, the steeper the curve.

In general, the following is true.

Increasing and decreasing functions

If $b > 1$, then $f(x) = b^x$ is an **increasing function.**

If $0 < b < 1$, then $f(x) = b^x$ is a **decreasing function.**

Increasing function   Decreasing function

An exponential function with base b is either increasing (for $b > 1$) or decreasing ($0 < b < 1$). Since different real numbers x determine different values of b^x, exponential functions are one-to-one.

ACCENT ON TECHNOLOGY *Graphing Exponential Functions*

To use a graphing calculator to graph $f(x) = \left(\frac{2}{3}\right)^x$ and $f(x) = \left(\frac{3}{2}\right)^x$, we enter the right-hand sides of the equations after the symbols $Y_1 =$ and $Y_2 =$. The screen will show the following equations.

$Y_1 = (2/3)$ ^ X

$Y_2 = (3/2)$ ^ X

FIGURE 11-15

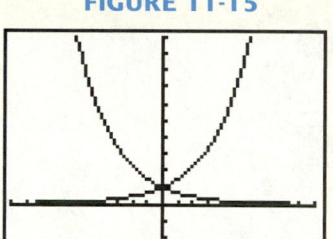

If we use window settings of $[-10, 10]$ for x and $[-2, 10]$ for y and press the $\boxed{\text{GRAPH}}$ key, we will obtain the graph shown in Figure 11-15.

We note that the graph of $f(x) = \left(\frac{2}{3}\right)^x$ passes through the points $(0, 1)$ and $\left(1, \frac{2}{3}\right)$. Since $\frac{2}{3} < 1$, the function is decreasing.

The graph of $f(x) = \left(\frac{3}{2}\right)^x$ passes through the points $(0, 1)$ and $\left(1, \frac{3}{2}\right)$. Since $\frac{3}{2} > 1$, the function is increasing.

Since both graphs pass the horizontal line test, each function is one-to-one.

Vertical and Horizontal Translations

EXAMPLE 4

Translations. On one set of axes, graph $y = 2^x$ and $y = 2^x + 3$, and describe the translation.

Solution The graph of $y = 2^x + 3$ is identical to the graph of $y = 2^x$, except that it is translated 3 units upward. (See Figure 11-16.)

FIGURE 11-16

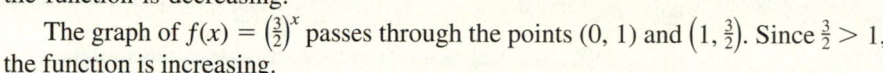

$y = 2^x$			$y = 2^x + 3$		
x	y	(x, y)	x	y	(x, y)
-4	$\frac{1}{16}$	$\left(-4, \frac{1}{16}\right)$	-4	$3\frac{1}{16}$	$\left(-4, 3\frac{1}{16}\right)$
0	1	$(0, 1)$	0	4	$(0, 4)$
2	4	$(2, 4)$	2	7	$(2, 7)$

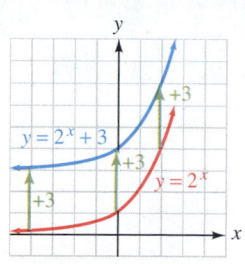

SELF CHECK Graph $f(x) = 4^x$ and $g(x) = 4^x - 3$, and describe the translation.

Answer: The graph of $f(x) = 4^x$ is translated 3 units downward.

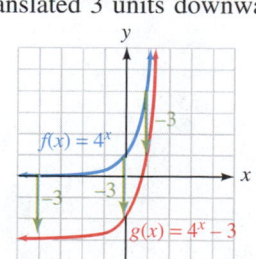

EXAMPLE 5

Translations. On one set of axes, graph $f(x) = 2^x$ and $g(x) = 2^{x+3}$, and describe the translation.

Solution The graph of $g(x) = 2^{x+3}$ is identical to the graph of $f(x) = 2^x$, except that it is translated 3 units to the left. (See Figure 11-17.)

FIGURE 11-17

	$f(x) = 2^x$			$g(x) = 2^{x+3}$	
x	$f(x)$	$(x, f(x))$	x	$g(x)$	$(x, g(x))$
-1	$\frac{1}{2}$	$\left(-1, \frac{1}{2}\right)$	-1	4	$(-1, 4)$
0	1	$(0, 1)$	0	8	$(0, 8)$
1	2	$(1, 2)$	1	16	$(1, 16)$

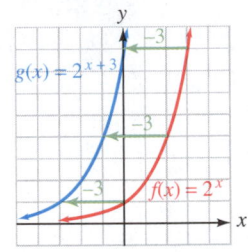

SELF CHECK On one set of axes, graph $f(x) = 4^x$ and $g(x) = 4^{x-3}$, and describe the translation.

Answer: The graph of $f(x) = 4^x$ is translated 3 units to the right.

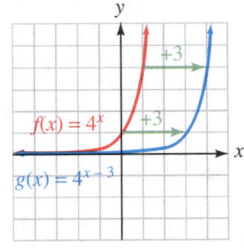

The graphs of $f(x) = kb^x$ and $f(x) = b^{kx}$ are vertical and horizontal stretchings of the graph of $f(x) = b^x$. To graph these functions, we can plot several points and join them with a smooth curve, or we can use a graphing calculator.

ACCENT ON TECHNOLOGY *Graphing Exponential Functions*

To use a graphing calculator to graph the exponential function $f(x) = 3(2^{x/3})$, we enter the right-hand side of the equation after the symbol $Y_1 =$. The display will show the equation

$$Y_1 = 3(2 \wedge (X/3))$$

FIGURE 11-18

If we use window settings of $[-10, 10]$ for x and $[-2, 18]$ for y and press the graph key, we will obtain the graph shown in Figure 11-18.

Compound Interest

If we deposit $\$P$ in an account paying an annual interest rate r, we can find the amount A in the account at the end of t years by using the formula $A = P + Prt$, or $A = P(1 + rt)$.

Suppose that we deposit $500 in such an account that pays interest every six months. Then $P = 500$, and after six months $\left(\frac{1}{2} \text{ year}\right)$, the amount in the account will be

$$A = 500(1 + rt)$$

$$= 500\left(1 + r \cdot \frac{1}{2}\right) \quad \text{Substitute } \tfrac{1}{2} \text{ for } t.$$

$$= 500\left(1 + \frac{r}{2}\right)$$

The account will begin the second six-month period with a value of $\$500\left(1 + \frac{r}{2}\right)$. After the second six-month period, the amount in the account will be

$$A = P(1 + rt)$$

$$A = \left[500\left(1 + \frac{r}{2}\right)\right]\left(1 + r \cdot \frac{1}{2}\right) \quad \text{Substitute } 500\left(1 + \tfrac{r}{2}\right) \text{ for } P \text{ and } \tfrac{1}{2} \text{ for } t.$$

$$= 500\left(1 + \frac{r}{2}\right)\left(1 + \frac{r}{2}\right)$$

$$= 500\left(1 + \frac{r}{2}\right)^2$$

At the end of a third six-month period, the amount in the account will be

$$A = 500\left(1 + \frac{r}{2}\right)^3$$

In this discussion, the earned interest is deposited back in the account and also earns interest. When this is the case, we say that the account is earning **compound interest.** The preceding discussion suggests the following formula for compound interest.

Formula for compound interest

If $\$P$ is deposited in an account and interest is paid k times a year at an annual rate r, the amount A in the account after t years is given by

$$A = P\left(1 + \frac{r}{k}\right)^{kt}$$

EXAMPLE 6

Saving for college. To save for college, parents invest $12,000 for their newborn child in a mutual fund that should average a 10% annual return. If the quarterly dividends are reinvested, how much will be available in 18 years?

Solution We substitute 12,000 for P, 0.10 for r, and 18 for t in the formula for compound interest and find A. Since interest is paid quarterly, $k = 4$.

$$A = P\left(1 + \frac{r}{k}\right)^{kt}$$

$$A = 12{,}000\left(1 + \frac{0.10}{4}\right)^{4(18)} \quad \text{Express } r = 10\% \text{ as a decimal.}$$

$$= 12{,}000(1 + 0.025)^{72}$$

$$= 12{,}000(1.025)^{72}$$

$$= 71{,}006.74 \quad \text{Use a scientific calculator and press these keys:}$$

$$1.025 \;\boxed{y^x}\; 72 \;\boxed{=}\; \boxed{\times}\; 12{,}000 \;\boxed{=}\;.$$

In 18 years, the account will be worth $71,006.74.

SELF CHECK

In Example 6, how much would be available if the parents invested $20,000?   *Answer:* $118,344.56   ■

In business applications, the initial amount of money deposited is often called the **present value** (*PV*). The amount to which the money will grow is called the **future value** (*FV*). The interest rate used for each compounding period is the **periodic interest rate** (*i*), and the number of times interest is compounded is the number of **compounding periods** (*n*). Using these definitions, we have an alternate formula for compound interest.

Formula for compound interest

$$FV = PV(1 + i)^n$$

This alternate formula appears on business calculators. To use this formula to solve Example 6, we proceed as follows:

$$FV = PV(1 + i)^n$$
$$FV = 12{,}000(1 + 0.025)^{72} \quad i = \tfrac{0.10}{4} = 0.025 \text{ and } n = 4(18) = 72.$$
$$= 71{,}006.74 \quad \text{Use a calculator to evaluate the expression.}$$

ACCENT ON TECHNOLOGY *Solving Investment Problems*

Suppose $1 is deposited in an account earning 6% annual interest, compounded monthly. To use a graphing calculator to estimate how much will be in the account in 100 years, we can substitute 1 for *P*, 0.06 for *r*, and 12 for *k* in the formula

$$A = P\left(1 + \frac{r}{k}\right)^{kt}$$

$$A = 1\left(1 + \frac{0.06}{12}\right)^{12t}$$

FIGURE 11-19

and simplify to get

$$A = (1.005)^{12t}$$

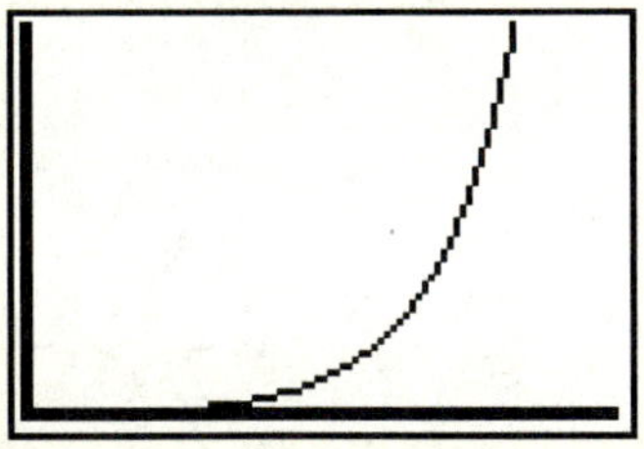

We now graph $A = (1.005)^{12t}$ using window settings of [0, 120] for *t* and [0, 400] for *A* to obtain the graph shown in Figure 11-19. We can then trace and zoom to estimate that $1 grows to be approximately $397 in 100 years. From the graph, we can see that the money grows slowly in the early years and rapidly in the later years.

Exponential Functions as Models

EXAMPLE 7

Cellular telephones. For the years 1990–1996, the U.S. cellular telephone industry experienced "exponential growth." (See Figure 11-20.) The exponential function $S(n) = 5.28(1.43)^n$ approximates the number of cellular telephone subscribers in millions, where *n* is the number of years since 1990 and $0 \le n \le 6$.

a. How many subscribers were there in 1990?

b. How many subscribers were there in 1996?

FIGURE 11-20

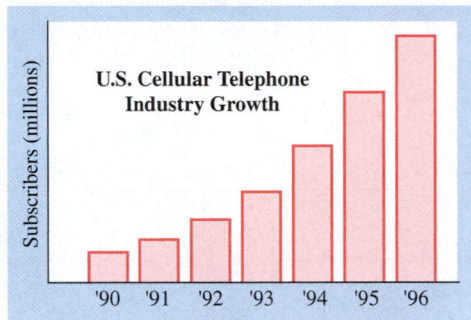

Based on information from *New York Times Almanac*, 1998

Solution **a.** The year 1990 is 0 years after 1990. To find the number of subscribers in 1990, we substitute 0 for n in the function and find $S(0)$.

$$S(n) = 5.28(1.43)^n$$
$$S(0) = 5.28(1.43)^0$$
$$= 5.28 \cdot 1 \qquad (1.43)^0 = 1.$$
$$= 5.28$$

In 1990, there were approximately 5.28 million cellular telephone subscribers.

b. The year 1996 is 6 years after 1990. We need to find $S(6)$.

$$S(n) = 5.28(1.43)^n$$
$$S(6) = 5.28(1.43)^6 \qquad \text{Substitute 6 for } n.$$
$$\approx 45.14920914 \qquad \text{Use a calculator to find an approximation.}$$

In 1996, there were approximately 45.15 million cellular telephone subscribers. ■

STUDY SET

Section 11.3

VOCABULARY

In Exercises 1–8, refer to the graph of $f(x) = 3^x$ in Illustration 1.

ILLUSTRATION 1

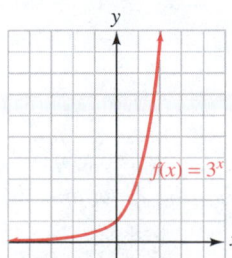

1. What type of function is $f(x) = 3^x$?

2. What is the domain of the function?

3. What is the range of the function?

4. What is the y-intercept of the graph?

5. What is the x-intercept of the graph?

6. What is an asymptote of the graph?

7. Is f an increasing or a decreasing function?

8. The graph passes through the point $(1, y)$. What is y?

CONCEPTS

9. Graph the functions $f(x) = x^2$ and $g(x) = 2^x$ on the same set of coordinate axes.

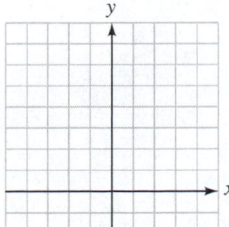

10. Graph the functions $y = x^{\frac{1}{2}}$ and $y = \left(\frac{1}{2}\right)^x$ on the same set of coordinate axes.

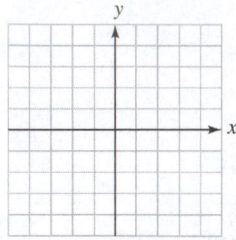

11. What are the two formulas that are used to determine the amount of money in a savings account that is earning compound interest?

12. Explain the order in which the expression $20{,}000(1.036)^{72}$ should be evaluated.

13. Illustration 2 shows the graph of $f(x) = 2^x$, as well as two vertical translations of that graph. Using the notation $g(x)$ for one translation and $h(x)$ for the other, write the defining equation for each function.

ILLUSTRATION 2

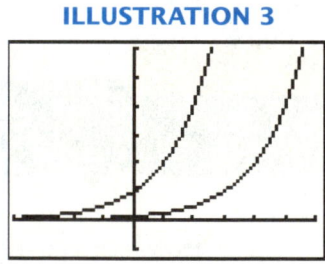

14. Illustration 3 shows the graph of $f(x) = 2^x$, as well as a horizontal translation of that graph. Using the notation $g(x)$ for the translation, write its defining equation.

ILLUSTRATION 3

NOTATION

15. In $A = P\left(1 + \frac{r}{k}\right)^{kt}$, what is the base, and what is the exponent?

16. For an exponential function of the form $f(x) = b^x$, what are the restrictions on b?

⊞ PRACTICE

In Exercises 17–20, find each value to four decimal places.

17. $2^{\sqrt{2}}$

18. $7^{\sqrt{2}}$

19. $5^{\sqrt{5}}$

20. $6^{\sqrt{3}}$

In Exercises 21–24, simplify each expression.

21. $\left(2^{\sqrt{3}}\right)^{\sqrt{3}}$

22. $3^{\sqrt{2}}3^{\sqrt{18}}$

23. $7^{\sqrt{3}}7^{\sqrt{12}}$

24. $\left(3^{\sqrt{5}}\right)^{\sqrt{5}}$

In Exercises 25–32, graph each exponential function. Check your work with a graphing calculator.

25. $f(x) = 5^x$

26. $f(x) = 6^x$

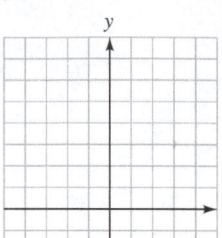

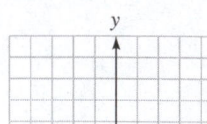

27. $y = \left(\frac{1}{4}\right)^x$

28. $y = \left(\frac{1}{5}\right)^x$

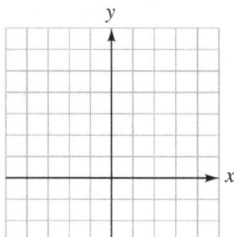

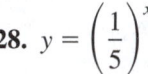

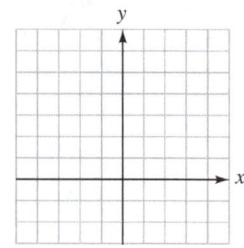

29. $f(x) = 3^x - 2$

30. $y = 2^x + 1$

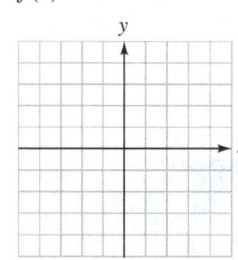

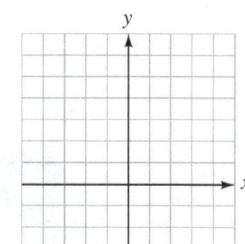

31. $f(x) = 3^{x-1}$

32. $f(x) = 2^{x+1}$

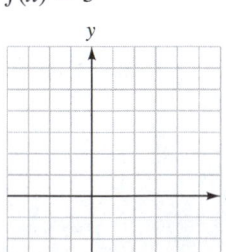

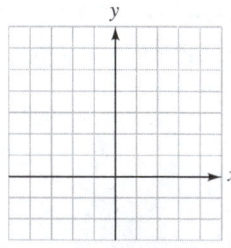

⊞ *In Exercises 33–36, use a graphing calculator to graph each function. Tell whether the function is an increasing or a decreasing function.*

33. $f(x) = \frac{1}{2}(3^{x/2})$

34. $f(x) = -3(2^{x/3})$

35. $y = 2(3^{-x/2})$ **36.** $y = -\dfrac{1}{4}(2^{-x/2})$

APPLICATIONS

In Exercises 37–42, assume that there are no deposits or withdrawals.

37. COMPOUND INTEREST An initial deposit of $10,000 earns 8% interest, compounded quarterly. How much will be in the account after 10 years?

38. COMPOUND INTEREST An initial deposit of $10,000 earns 8% interest, compounded monthly. How much will be in the account after 10 years?

39. COMPARING INTEREST RATES How much more interest could $1,000 earn in 5 years, compounded quarterly, if the annual interest rate were $5\frac{1}{2}$% instead of 5%?

40. COMPARING SAVINGS PLANS Which institution in Illustration 4 provides the better investment?

ILLUSTRATION 4

> ### *Fidelity Savings & Loan*
> Earn 5.25%
> compounded monthly

> ### Union Trust
> Money Market Account
> paying 5.35%
> compounded annually

41. COMPOUND INTEREST If $1 had been invested on July 4, 1776, at 5% interest, compounded annually, what would it be worth on July 4, 2076?

42. FREQUENCY OF COMPOUNDING $10,000 is invested in each of two accounts, both paying 6% annual interest. In the first account, interest compounds quarterly, and in the second account, interest compounds daily. Find the difference between the accounts after 20 years.

43. WORLD POPULATION See the graph in Illustration 5.

  a. Estimate when the world's population reached $\frac{1}{2}$ billion and when it reached 1 billion.

  b. Estimate the world's population in the year 2000.

  c. What type of function does it appear could be used to model the population growth?

ILLUSTRATION 5

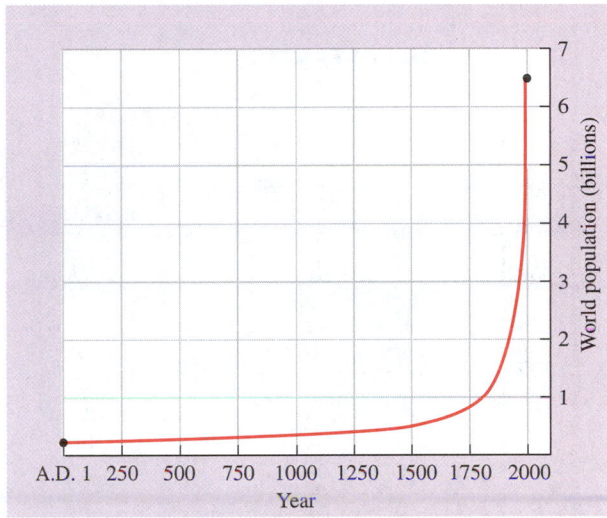

Based on data from *The Blue Planet* (Wiley, 1995)

44. THE STOCK MARKET The Dow Jones Industrial Average is a measure of how well the stock market is doing. Plot the following Dow milestones as ordered pairs of the form (year, average) on the graph in Illustration 6. What type of function does it appear could be used to model the growth of the stock market?

Year	Average	Year	Average
Jan. 1906	100	Feb. 1995	4,000
Mar. 1956	500	Nov. 1995	5,000
Nov. 1972	1,000	Oct. 1996	6,000
Jan. 1987	2,000	Feb. 1997	7,000
Apr. 1991	3,000	July 1997	8,000

ILLUSTRATION 6

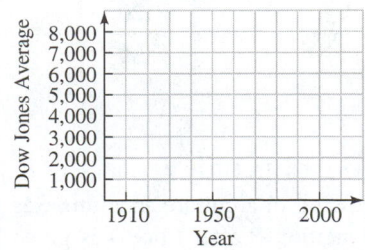

45. MARKET VALUE OF A CAR The graph in Illustration 7 shows how the value of the average car depreciates as a percent of its original value over a 10-year period. It also shows the yearly maintenance costs as a percent of the car's value.

a. When is the car worth half of its purchase price?

b. When is the car worth a quarter of its purchase price?

c. When do the average yearly maintenance costs surpass the value of the car?

ILLUSTRATION 7

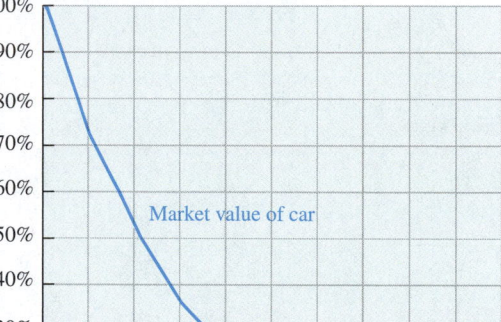

Based on data from the U.S. Department of Transportation

46. DIVING *Bottom time* is the time a scuba diver spends descending plus the actual time spent at a certain depth. On Illustration 8, graph the bottom time limits given in the table.

Bottom time limits			
Depth (ft)	Bottom time (min)	Depth (ft)	Bottom time (min)
30	no limit	80	40
35	310	90	30
40	200	100	25
50	100	110	20
60	60	120	15
70	50	130	10

47. BACTERIAL CULTURE A colony of 6 million bacteria is growing in a culture medium. See Illustration 9. The population P after t hours is given by the for-

ILLUSTRATION 8

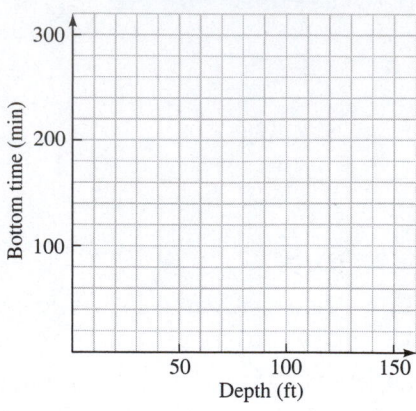

mula $P = (6 \times 10^6)(2.3)^t$. Find the population after 4 hours.

ILLUSTRATION 9

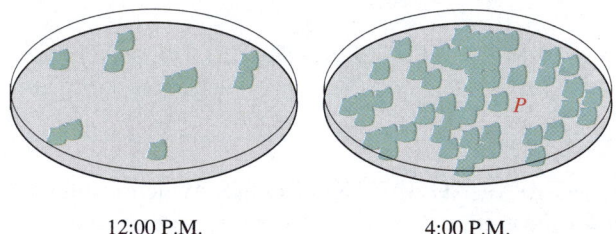

12:00 P.M.   4:00 P.M.

48. RADIOACTIVE DECAY A radioactive material decays according to the formula $A = A_0\left(\frac{2}{3}\right)^t$, where A_0 is the initial amount present and t is measured in years. Find the amount present in 5 years.

49. DISCHARGING A BATTERY The charge remaining in a battery decreases as the battery discharges. The charge C (in coulombs) after t days is given by the formula $C = (3 \times 10^{-4})(0.7)^t$. Find the charge after 5 days.

50. POPULATION GROWTH The population of North Rivers is decreasing exponentially according to the formula $P = 3,745(0.93)^t$, where t is measured in years from the present date. Find the population in 6 years, 9 months.

51. SALVAGE VALUE A small business purchased a computer for $4,700. It is expected that its value each year will be 75% of its value in the preceding year. If the business disposes of the computer after 5 years, find its salvage value (the value after 5 years).

52. LOUISIANA PURCHASE In 1803, the United States negotiated the Louisiana Purchase with France. The country doubled its territory by adding 827,000 square miles of land for $15 million. If the land appreciated at the rate of 6% each year, what would one square mile of land be worth in 1996?

WRITING

53. If world population is increasing exponentially, why is there cause for concern?

54. How do the graphs of $f(x) = 3^x$ and $g(x) = \left(\frac{1}{3}\right)^x$ differ? How are they similar?

REVIEW

In Exercises 55–58, refer to Illustration 10. Lines r and s are parallel.

55. Find x.

56. Find the measure of $\angle 1$.

57. Find the measure of $\angle 2$.

58. Find the measure of $\angle 3$.

ILLUSTRATION 10

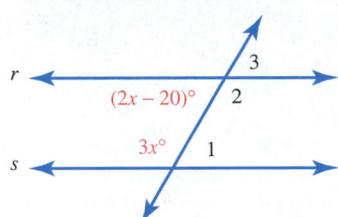

▶ 11.4   Base-*e* Exponential Functions

In this section, you will learn about

> Continuous compound interest ■ The natural exponential function ■ Graphing the natural exponential function ■ Vertical and horizontal translations ■ Malthusian population growth ■ Base-*e* exponential function models

---

Introduction   A special exponential function that appears in real-life applications is the base-*e* exponential function. In this section, we will show how to evaluate *e*, graph the base-*e* exponential function, and discuss one of its important applications in analyzing population growth.

Continuous Compound Interest

If a bank pays interest twice a year, we say that interest is compounded semiannually. If it pays interest four times a year, we say that interest is compounded quarterly. If it pays interest continuously (infinitely many times in a year), we say that interest is compounded continuously.

To develop the formula for continuous compound interest, we start with the formula

$$A = P\left(1 + \frac{r}{k}\right)^{kt}$$ The formula for compound interest: r is the annual rate and k is the number of times per year interest is paid.

and let $rn = k$. Since r and k are positive numbers, so is n.

$$A = P\left(1 + \frac{r}{rn}\right)^{rnt}$$

We can then simplify the fraction $\frac{r}{rn}$ and use the commutative property of multiplication to change the order of the exponents.

$$A = P\left(1 + \frac{1}{n}\right)^{nrt}$$

Finally, we can use a property of exponents to write the formula as

1. $A = P\left[\left(1 + \frac{1}{n}\right)^{n}\right]^{rt}$ Use the property $a^{mn} = (a^m)^n$.

To find the value of $\left(1 + \frac{1}{n}\right)^n$, we evaluate it for several values of n, as shown in Table 11-1.

TABLE 11-1

n	$\left(1 + \frac{1}{n}\right)^n$
1	2
2	2.25
4	2.44140625. . .
12	2.61303529. . .
365	2.71456748. . .
1,000	2.71692393. . .
100,000	2.71826830. . .
1,000,000	2.71828137. . .

The results suggest that as n gets larger, the value of $\left(1 + \frac{1}{n}\right)^n$ approaches the number 2.71828. . . . This number is called e, which has the following value.

$$e = 2.718281828459. . .$$

In continuous compound interest, k (the number of compoundings) is infinitely large. Since k, r, and n are all positive, r is a fixed rate, and $k = rn$, as k gets very large (approaches infinity), so does n. Therefore, we can replace $\left(1 + \frac{1}{n}\right)^n$ in Equation 1 with e to get

$$A = Pe^{rt}$$

Formula for exponential growth

If a quantity P increases or decreases at an annual rate r, compounded continuously, the amount A after t years is given by

$$A = Pe^{rt}$$

If time is measured in years, then r is called the **annual growth rate.** If r is negative, the "growth" represents a decrease.

The Natural Exponential Function

The number e is often used as the base in applications of exponential functions.

The natural exponential function

The function defined by $f(x) = e^x$ is the **natural exponential function** (or the **base-e exponential function**) where $e = 2.71828.$ . . .

The e^x key on a calculator is used to evaluate the natural exponential function.

ACCENT ON TECHNOLOGY *The Natural Exponential Function Key*

To compute the amount to which $12,000 will grow if invested for 18 years at 10% annual interest, compounded continuously, we substitute 12,000 for P, 0.10 for r, and 18 for t in the formula for exponential growth:

$$A = Pe^{rt}$$
$$A = 12{,}000e^{0.10(18)}$$
$$= 12{,}000e^{1.8}$$

To evaluate this expression, we enter these numbers and press these keys on a scientific calculator:

Keystrokes 1.8 $\boxed{e^x}$ $\boxed{\times}$ 12000 $\boxed{=}$ $\boxed{72595.76957}$

Using a graphing calculator, we enter these numbers and press these keys:

Keystrokes 12000 $\boxed{\times}$ $\boxed{2nd}$ $\boxed{e^x}$ 1.8 $\boxed{)}$ $\boxed{ENTER}$ $\boxed{\begin{array}{l}\texttt{12000*e\^{}(1.8)} \\ \texttt{72595.76957}\end{array}}$

After 18 years, the account will contain $72,595.77. This is $1,589.03 more than the result in Example 6 in the previous section, where interest was compounded quarterly.

EXAMPLE 1

Continuous compound interest. If $25,000 accumulates interest at an annual rate of 8%, compounded continuously, find the balance in the account in 50 years.

Solution

$$A = Pe^{rt}$$
$$A = 25{,}000e^{0.08(50)}$$
$$= 25{,}000e^{4}$$
$$\approx 1{,}364{,}953.75 \quad \text{Use a calculator.}$$

In 50 years, the balance will be $1,364,953.75—more than a million dollars.

SELF CHECK In Example 1, find the balance in 60 years.   *Answer:* $3,037,760.44 ■

Graphing the Natural Exponential Function

To graph the natural exponential function, we plot several points and join them with a smooth curve, as shown in Figure 11-21. By looking at the graph, we can verify that $f(x) = e^x$ is an increasing function. The domain is the interval $(-\infty, \infty)$, and the range is the interval $(0, \infty)$. Note that as x decreases, the values of $f(x)$ decrease and approach 0. Thus, the x-axis is a horizontal asymptote of the graph.

FIGURE 11-21

$$f(x) = e^x$$

x	$f(x)$	$(x, f(x))$
-2	0.1	$(-2, 0.1)$
-1	0.4	$(-1, 0.4)$
0	1	$(0, 1)$
1	2.7	$(1, 2.7)$
2	7.4	$(2, 7.4)$

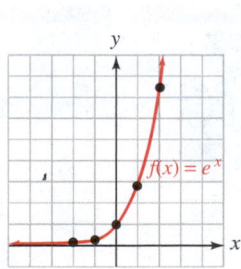

The outputs can be found using the $\boxed{e^x}$ key on a calculator. They are rounded to the nearest tenth.

Vertical and Horizontal Translations

We can illustrate the effects of vertical and horizontal translations of the natural exponential function by using a graphing calculator.

ACCENT ON TECHNOLOGY *Translations of the Natural Exponential Function*

Figure 11-22(a) shows the calculator graphs of $f(x) = e^x$, $g(x) = e^x + 5$, and $h(x) = e^x - 3$. To graph these functions with window settings of $[-3, 6]$ for x and $[-5, 15]$ for y, we enter the right-hand sides of the equations after the symbols $Y_1 =$, $Y_2 =$, and $Y_3 =$. The display will show

$$Y_1 = e \,\hat{}\, (x)$$
$$Y_2 = e \,\hat{}\, (x) + 5$$
$$Y_3 = e \,\hat{}\, (x) - 3$$

After graphing these functions, we can see that the graph of $g(x) = e^x + 5$ is 5 units above the graph of $f(x) = e^x$, and that the graph of $h(x) = e^x - 3$ is 3 units below the graph of $f(x) = e^x$.

Figure 11-22(b) shows the calculator graphs of $f(x) = e^x$, $g(x) = e^{x+5}$, and $h(x) = e^{x-3}$. To graph these functions with window settings of $[-7, 10]$ for x and $[-5, 15]$ for y, we enter the right-hand sides of the equations after the symbols $Y_1 =$, $Y_2 =$, and $Y_3 =$. The display will show

$$Y_1 = e \,\hat{}\, (x)$$
$$Y_2 = e \,\hat{}\, (x + 5)$$
$$Y_3 = e \,\hat{}\, (x - 3)$$

After graphing these functions, we can see that the graph of $g(x) = e^{x+5}$ is 5 units to the left of the graph of $f(x) = e^x$, and that the graph of $h(x) = e^{x-3}$ is 3 units to the right of the graph of $f(x) = e^x$.

FIGURE 11-22

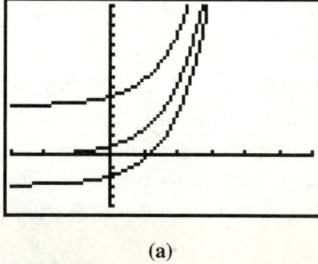 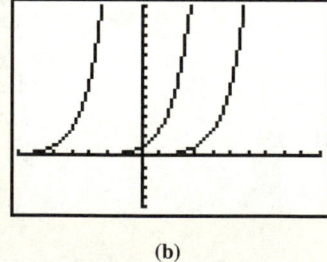

(a)   (b)

ACCENT ON TECHNOLOGY *Graphing Exponential Functions*

Figure 11-23 shows the calculator graph of $f(x) = 3e^{-x/2}$. To graph this function with window settings of $[-7, 10]$ for x and $[-5, 15]$ for y, we enter the right-hand side of the equation after the symbol $Y_1 =$. The display will show the equation

$$Y_1 = 3(e \,\hat{}\, (-x/2))$$

The calculator graph appears in Figure 11-23. Explain why the graph has a
y-intercept of $(0, 3)$.

FIGURE 11-23

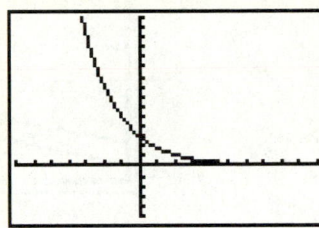

Malthusian Population Growth

An equation based on the natural exponential function provides a model for **population growth.** In the **Malthusian model for population growth,** the future population
of a colony is related to the present population by the formula $A = Pe^{rt}$.

EXAMPLE 2

City planning. The population of a city is currently 15,000, but changing economic conditions are causing the population to decrease 2% each year. If this trend
continues, find the population in 30 years.

Solution

Since the population is decreasing 2% each year, the annual growth rate is -2%, or
-0.02. We can substitute -0.02 for r, 30 for t, and 15,000 for P in the formula for exponential growth and find A.

$$A = Pe^{rt}$$
$$A = 15{,}000e^{-0.02(30)}$$
$$= 15{,}000e^{-0.6}$$
$$\approx 8{,}232.17$$

In 30 years, city planners expect a population of 8,232.

SELF CHECK

In Example 2, find the population in 50 years.   *Answer:* 5,518   ∎

The English economist Thomas Robert Malthus (1766–1834) pioneered in population study. He believed that poverty and starvation were unavoidable, because the human population tends to grow exponentially, but the food supply tends to grow linearly.

EXAMPLE 3

The Malthusian model. Suppose that a country with a population of 1,000
people is growing exponentially according to the formula

$$P = 1{,}000e^{0.02t}$$

where t is in years. Furthermore, assume that the food supply F, measured in adequate
food per day per person, is growing linearly according to the formula

$$F = 30.625t + 2{,}000 \quad (t \text{ is time in years})$$

In how many years will the population outstrip the food supply?

Solution

We can use a graphing calculator with window settings of $[0, 100]$ for x and $[0, 10{,}000]$
for y. After graphing the functions, we obtain Figure 11-24(a). If we trace, as in Figure
11-24(b), we can find the point where the two graphs intersect. From the graph, we can

see that the food supply will be adequate for about 71 years. At that time, the population of approximately 4,200 people will begin to have problems.

FIGURE 11-24

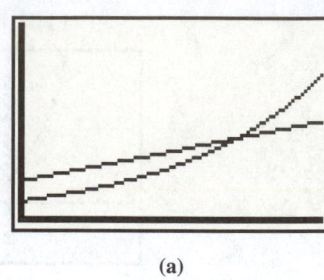 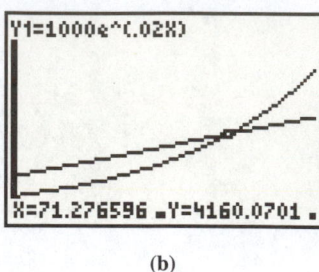

(a)   (b)

SELF CHECK

In Example 3, suppose that the population grows at a 3% rate. Use a graphing calculator to determine for how many years the food supply will be adequate.   *Answer:* about 38 years ∎

Base-*e* Exponential Function Models

EXAMPLE 4

Baking. A mother has just taken a cake out of the oven and set it on a rack to cool. The function $T(t) = 68 + 220e^{-0.2t}$ gives the cake's temperature in degrees Fahrenheit after it has cooled for t minutes. If her children will be home from school in 20 minutes, will the cake have cooled enough for the children to eat it?

Solution

When the children arrive home, the cake will have cooled for 20 minutes. To find the temperature of the cake at that time, we need to find $T(20)$.

$$T(t) = 68 + 220e^{-0.2t}$$
$$T(20) = 68 + 220e^{-0.2(20)} \quad \text{Substitute 20 for } t.$$
$$= 68 + 220e^{-4}$$
$$\approx 72.0 \qquad\qquad \text{Use a calculator.}$$

When the children return home, the temperature of the cake will be about 72°, and it can be eaten.   ∎

STUDY SET

Section 11.4

VOCABULARY

In Exercises 1–8, refer to the graph of $f(x) = e^x$ in Illustration 1.

ILLUSTRATION 1

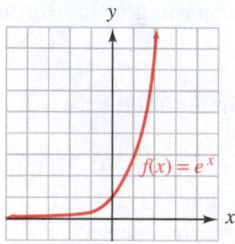

1. What is the name of the function $f(x) = e^x$?

2. What is the domain of the function?

3. What is the range of the function?

4. What is the y-intercept of the graph?

5. What is the x-intercept of the graph?

6. What is an asymptote of the graph?

7. Is f an increasing or a decreasing function?

8. The graph passes through the point $(1, y)$. What is y?

CONCEPTS

In Exercises 9–12, fill in the blanks to make the statements true.

9. In _____ compound interest, the number of compoundings is infinitely large.

10. The formula for continuous compound interest is
$A =$ _____.

11. To two decimal places, the value of *e* is
_____.

12. If *n* gets larger and larger, the value of $\left(1 + \frac{1}{n}\right)^n$ approaches the value of _____.

13. Graph each irrational number on the number line.
$$\left\{\pi, e, \sqrt{2}, \frac{\sqrt{3}}{2}\right\}$$

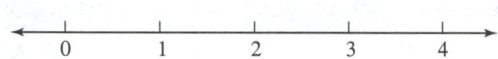

14. Complete the table of values. Round to the nearest hundredth.

x	-2	-1	0	1	2
e^x					

15. POPULATION OF THE UNITED STATES In Illustration 2, graph the U.S. census population figures shown in the table (in millions). What type of function does it appear could be used to model the population?

Year	Population	Year	Population
1790	3.9	1900	76.0
1800	5.3	1910	92.2
1810	7.2	1920	106.0
1820	9.6	1930	123.2
1830	12.9	1940	132.1
1840	17.0	1950	151.3
1850	23.1	1960	179.3
1860	31.4	1970	203.3
1870	38.5	1980	226.5
1880	50.1	1990	248.7
1890	62.9		

ILLUSTRATION 2

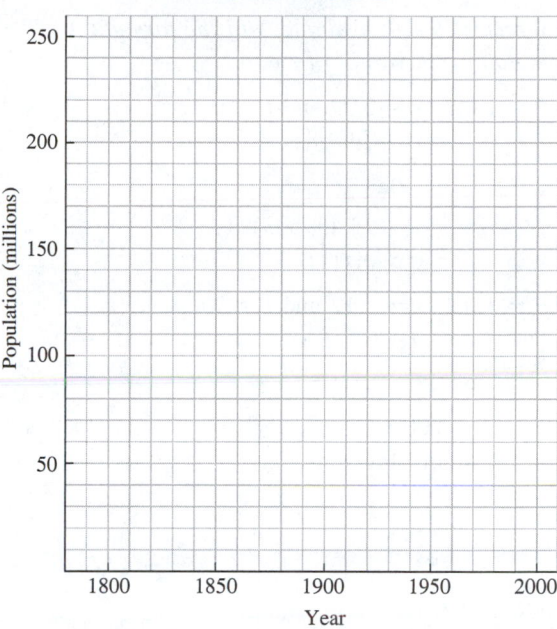

16. What is the Malthusian population growth formula?

17. The function $f(x) = e^x$ is graphed in Illustration 3, and the TRACE feature is used. What is the *y*-coordinate of the point on the graph having an *x*-coordinate of 1? What is the name given this number?

ILLUSTRATION 3

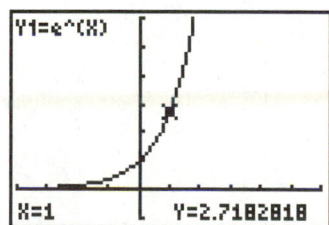

18. Illustration 4 shows a table of values for $f(x) = e^x$. As *x* decreases, what happens to the values of $f(x)$ listed in the Y_1 column? Will the value of $f(x)$ ever be 0 or negative?

ILLUSTRATION 4

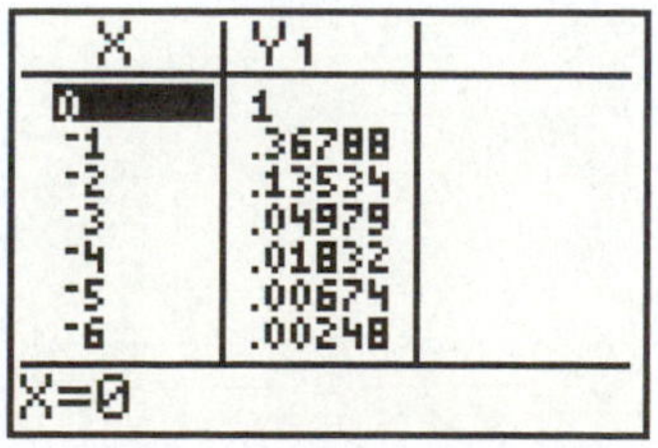

X	Y₁	
0	1	
-1	.36788	
-2	.13534	
-3	.04979	
-4	.01832	
-5	.00674	
-6	.00248	

X=0

NOTATION

In Exercises 19–20, evaluate A in the formula $A = Pe^{rt}$ for the following values of r and t.

19. $P = 1,000$, $r = 0.09$ and $t = 10$

$A = \boxed{} \, e^{(0.09)(\boxed{})}$

$ = 1,000e^{\boxed{}}$

$ \approx \boxed{} \, (2.459603111)$

$ \approx 2,459.603111$

20. $P = 1,000$, $r = 0.12$ and $t = 50$

$A = 1,000e^{(\boxed{})(50)}$

$ = \boxed{} \, e^6$

$ \approx 1,000 \left(\boxed{} \right)$

$ \approx 403,428.7935$

PRACTICE

In Exercises 21–28, graph each function. Check your work with a graphing calculator. Compare each graph to the graph of $f(x) = e^x$.

21. $f(x) = e^x + 1$

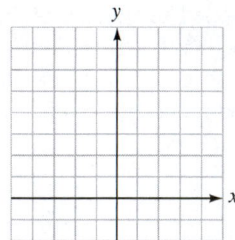

22. $f(x) = e^x - 2$

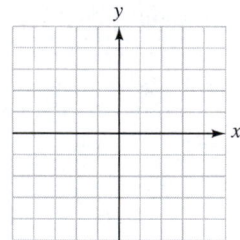

23. $y = e^{x+3}$

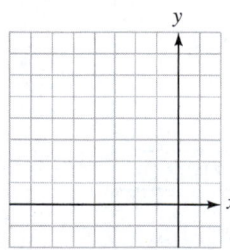

24. $y = e^{x-5}$

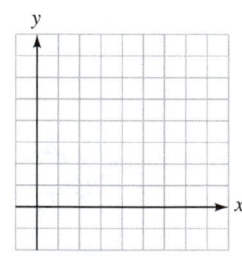

25. $f(x) = -e^x$

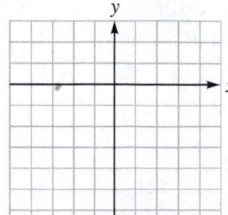

26. $f(x) = -e^x + 1$

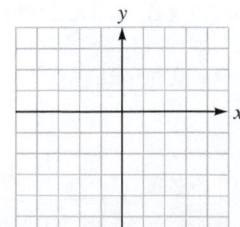

27. $f(x) = 2e^x$

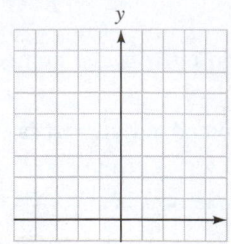

28. $f(x) = \dfrac{1}{2}e^x$

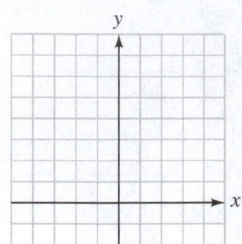

APPLICATIONS

In Exercises 29–34, assume that there are no deposits or withdrawals.

29. CONTINUOUS COMPOUND INTEREST An initial investment of $5,000 earns 8.2% interest, compounded continuously. What will the investment be worth in 12 years?

30. CONTINUOUS COMPOUND INTEREST An initial investment of $2,000 earns 8% interest, compounded continuously. What will the investment be worth in 15 years?

31. COMPARISON OF COMPOUNDING METHODS An initial deposit of $5,000 grows at an annual rate of 8.5% for 5 years. Compare the final balances resulting from annual compounding and continuous compounding.

32. COMPARISON OF COMPOUNDING METHODS An initial deposit of $30,000 grows at an annual rate of 8% for 20 years. Compare the final balances resulting from annual compounding and continuous compounding.

33. DETERMINING THE INITIAL DEPOSIT An account now contains $11,180 and has been accumulating interest at 7% annual interest, compounded continuously, for 7 years. Find the initial deposit.

34. DETERMINING THE PREVIOUS BALANCE An account now contains $3,610 and has been accumulating interest at 8% annual interest, compounded continuously. How much was in the account 4 years ago?

35. WORLD POPULATION GROWTH The population of the earth is approximately 6 billion people and is growing at an annual rate of 1.9%. Assuming a Malthusian growth model, find the world population in 30 years.

36. WORLD POPULATION GROWTH The population of the earth is approximately 6 billion people and is growing at an annual rate of 1.9%. Assuming a Malthusian growth model, find the world population in 40 years.

37. POPULATION GROWTH The growth of a population is modeled by

$$P = 173e^{0.03t}$$

How large will the population be when $t = 20$?

38. POPULATION DECLINE The decline of a population is modeled by

$$P = (1.2 \times 10^6)e^{-0.008t}$$

How large will the population be when $t = 30$?

39. EPIDEMIC The spread of ungulate fever through a herd of cattle can be modeled by the formula

$$P = P_0e^{0.27t} \quad (t \text{ is in days})$$

If a rancher does not act quickly to treat two cases, how many cattle will have the disease in 12 days?

40. OCEANOGRAPHY The width w (in millimeters) of successive growth spirals of the sea shell *Catapulus voluto*, shown in Illustration 5, is given by the exponential function

$$w = 1.54e^{0.503n}$$

where n is the spiral number. Find the width, to the nearest tenth of a millimeter, of the sixth spiral.

ILLUSTRATION 5

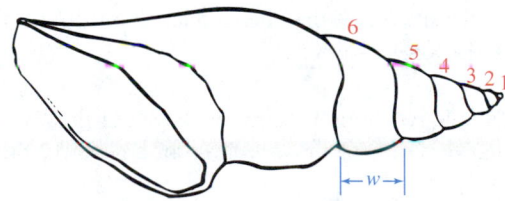

41. HALF-LIFE OF A DRUG The quantity of a prescription drug in the bloodstream of a patient t hours after it is administered can be modeled by an exponential function. (See the graph in Illustration 6.) Determine the time it takes to eliminate half of the initial dose from the body.

42. MEDICINE The concentration x of a certain prescription drug in an organ after t minutes is given by

$$x = 0.08(1 - e^{-0.1t})$$

Find the concentration of the drug at 30 minutes.

43. SKYDIVING Before the parachute opens, a skydiver's velocity v in meters per second is given by

$$v = 50(1 - e^{-0.2t})$$

Find the velocity after 20 seconds of free fall.

ILLUSTRATION 6

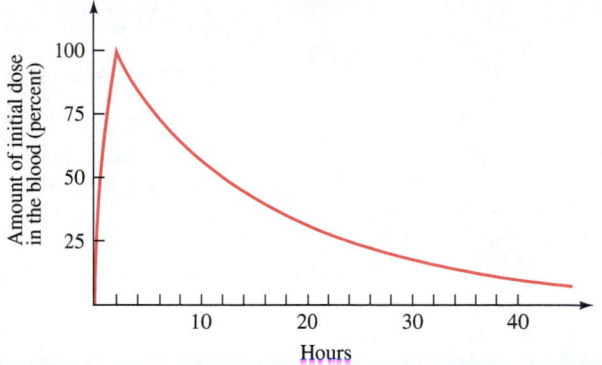

Hours

44. FREE FALL After t seconds a certain falling object has a velocity v given by

$$v = 50(1 - e^{-0.3t})$$

Which is falling faster after 2 seconds—the object or the skydiver in Exercise 43?

In Exercises 45–46, use a graphing calculator to solve each problem.

45. THE MALTHUSIAN MODEL In Example 3, suppose that better farming methods changed the formula for food growth to $y = 31x + 2,000$. How long would the food supply be adequate?

46. THE MALTHUSIAN MODEL In Example 3, suppose that a birth-control program changed the formula for population growth to $P = 1,000e^{0.01t}$. How long would the food supply be adequate?

WRITING

47. Explain why the graph of $y = e^x - 5$ is five units below the graph of $y = e^x$.

48. A feature article in a newspaper stated that the sport of snowboarding was growing *exponentially*. Explain what the author of the article meant by that.

REVIEW

Simplify each expression. Assume that all variables represent positive numbers.

49. $\sqrt{240x^5}$

50. $\sqrt[3]{-125x^5y^4}$

51. $4\sqrt{48y^3} - 3y\sqrt{12y}$

52. $\sqrt[4]{48z^5} + \sqrt[4]{768z^5}$

▶ 11.5   Logarithmic Functions

In this section, you will learn about

Logarithms ■ Graphs of logarithmic functions ■ Vertical and horizontal translations ■ Base-10 logarithms ■ Applications of logarithms

Introduction   A function that is closely related to the exponential function is the *logarithmic function*. It can be used to solve many application problems from fields such as electronics, seismology, business, and population growth.

Logarithms

Because an exponential function defined by $y = b^x$ (or $f(x) = b^x$) is one-to-one, it has an inverse function that is defined by the equation $x = b^y$. To express this inverse function in the form $y = f^{-1}(x)$, we must solve the equation $x = b^y$ for y. For this, we need the following definition.

Logarithmic functions

If $b > 0$ and $b \neq 1$, the **logarithmic function with base b** is defined by

$$y = \log_b x \quad \text{if and only if} \quad x = b^y$$

The **domain of the logarithmic function** is the interval $(0, \infty)$. The **range** is the interval $(-\infty, \infty)$.

Since the function $y = \log_b x$ is the inverse of the one-to-one exponential function $y = b^x$, the logarithmic function is also one-to-one.

 WARNING! Since the domain of the logarithmic function is the set of positive numbers, it is impossible to find the logarithm of 0 or the logarithm of a negative number.

The previous definition guarantees that any pair (x, y) that satisfies the equation $y = \log_b x$ also satisfies the exponential equation $x = b^y$. Because of this relationship, a statement written in logarithmic form can be written in an equivalent exponential form.

Logarithmic form		*Exponential form*
$\log_7 1 = \mathbf{0}$	because	$1 = 7^{\mathbf{0}}$
$\log_5 25 = \mathbf{2}$	because	$25 = 5^{\mathbf{2}}$
$\log_5 \dfrac{1}{25} = \mathbf{-2}$	because	$\dfrac{1}{25} = 5^{\mathbf{-2}}$
$\log_{16} 4 = \dfrac{\mathbf{1}}{\mathbf{2}}$	because	$4 = 16^{\mathbf{1/2}}$
$\log_2 \dfrac{1}{8} = \mathbf{-3}$	because	$\dfrac{1}{8} = 2^{\mathbf{-3}}$
$\log_b x = \mathbf{y}$	because	$x = b^{\mathbf{y}}$

In each of these examples, the logarithm of a number is an exponent. In fact,

$\log_b x$ is the exponent to which b is raised to get x.

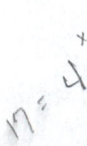

Translating this statement into symbols, we have

$$b^{\log_b x} = x$$

EXAMPLE 1

Writing an equivalent exponential statement. Find y in each equation:
a. $\log_6 1 = y$, **b.** $\log_3 27 = y$, and **c.** $\log_5 \frac{1}{5} = y$.

Solution

a. We can change the equation $\log_6 1 = y$ into the equivalent exponential equation $1 = 6^y$. Since $1 = 6^0$, it follows that $y = 0$. Thus,

$$\log_6 1 = 0$$

b. $\log_3 27 = y$ is equivalent to $27 = 3^y$. Since $27 = 3^3$, it follows that $3^y = 3^3$, and $y = 3$. Thus,

$$\log_3 27 = 3$$

c. $\log_5 \frac{1}{5} = y$ is equivalent to $\frac{1}{5} = 5^y$. Since $\frac{1}{5} = 5^{-1}$, it follows that $5^y = 5^{-1}$, and $y = -1$. Thus,

$$\log_5 \frac{1}{5} = -1$$

SELF CHECK   Find y in each equation: **a.** $\log_3 9 = y$,   *Answers:* **a.** 2, **b.** 4,
b. $\log_2 16 = y$, and **c.** $\log_5 \frac{1}{125} = y$.   **c.** -3 ■

EXAMPLE 2

Writing an equivalent exponential statement. Find the value of x in each equation: **a.** $\log_3 81 = x$, **b.** $\log_x 125 = 3$, and **c.** $\log_4 x = 3$.

Solution

a. $\log_3 81 = x$ is equivalent to $3^x = 81$. Because $3^4 = 81$, it follows that $3^x = 3^4$. Thus, $x = 4$.

b. $\log_x 125 = 3$ is equivalent to $x^3 = 125$. Because $5^3 = 125$, it follows that $x^3 = 5^3$. Thus, $x = 5$.

c. $\log_4 x = 3$ is equivalent to $4^3 = x$. Because $4^3 = 64$, it follows that $x = 64$.

SELF CHECK   Find x in each equation: **a.** $\log_2 32 = x$,
b. $\log_x 8 = 3$, and **c.** $\log_5 x = 2$.   *Answers:* **a.** 5, **b.** 2, **c.** 25 ■

EXAMPLE 3

Writing an equivalent exponential statement. Find the value of x in each equation: **a.** $\log_{1/3} x = 2$, **b.** $\log_{1/3} x = -2$, and **c.** $\log_{1/3} \frac{1}{27} = x$.

Solution

a. $\log_{1/3} x = 2$ is equivalent to $\left(\frac{1}{3}\right)^2 = x$. Thus, $x = \frac{1}{9}$.

b. $\log_{1/3} x = -2$ is equivalent to $\left(\frac{1}{3}\right)^{-2} = x$. Thus,

$$x = \left(\frac{1}{3}\right)^{-2} = 3^2 = 9$$

c. $\log_{1/3} \frac{1}{27} = x$ is equivalent to $\left(\frac{1}{3}\right)^x = \frac{1}{27}$. Because $\left(\frac{1}{3}\right)^3 = \frac{1}{27}$, it follows that $x = 3$.

SELF CHECK   Find x in each equation: **a.** $\log_{1/4} x = 3$ and
b. $\log_{1/4} x = -2$.   *Answers:* **a.** $\frac{1}{64}$, **b.** 16   ■

Graphs of Logarithmic Functions

To graph the logarithmic function $y = \log_2 x$, we calculate and plot several points with coordinates (x, y) that satisfy the equivalent equation $x = 2^y$. After joining these points with a smooth curve, we have the graph shown in Figure 11-25(a).

To graph $y = \log_{1/2} x$, we calculate and plot several points with coordinates (x, y) that satisfy the equation $x = \left(\frac{1}{2}\right)^y$. After joining these points with a smooth curve, we have the graph shown in Figure 11-25(b).

FIGURE 11-25

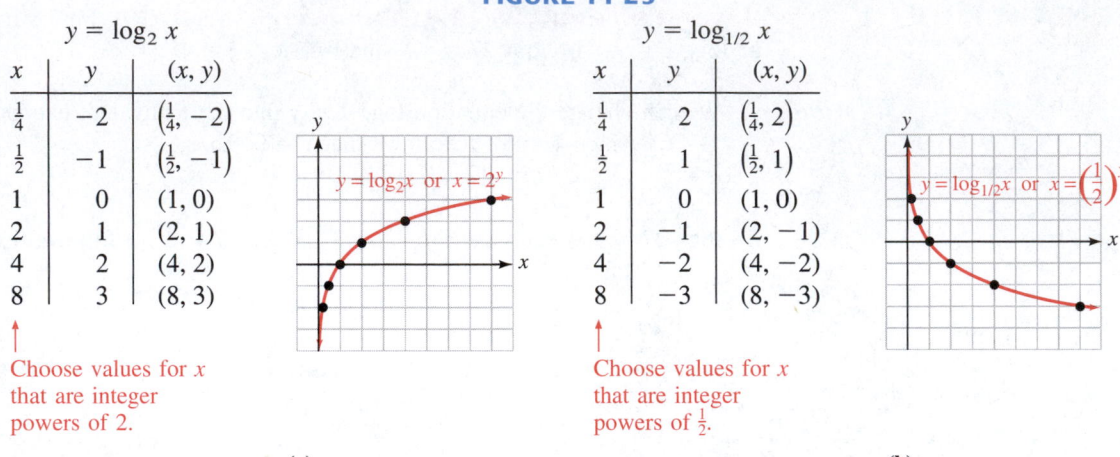

$y = \log_2 x$

x	y	(x, y)
$\frac{1}{4}$	-2	$\left(\frac{1}{4}, -2\right)$
$\frac{1}{2}$	-1	$\left(\frac{1}{2}, -1\right)$
1	0	$(1, 0)$
2	1	$(2, 1)$
4	2	$(4, 2)$
8	3	$(8, 3)$

Choose values for x that are integer powers of 2.

$y = \log_{1/2} x$

x	y	(x, y)
$\frac{1}{4}$	2	$\left(\frac{1}{4}, 2\right)$
$\frac{1}{2}$	1	$\left(\frac{1}{2}, 1\right)$
1	0	$(1, 0)$
2	-1	$(2, -1)$
4	-2	$(4, -2)$
8	-3	$(8, -3)$

Choose values for x that are integer powers of $\frac{1}{2}$.

(a)   (b)

The graphs of all logarithmic functions are similar to those in Figure 11-25. If $b > 1$, the logarithmic function is increasing, as in Figure 11-26(a). If $0 < b < 1$, the logarithmic function is decreasing, as in Figure 11-26(b).

FIGURE 11-26

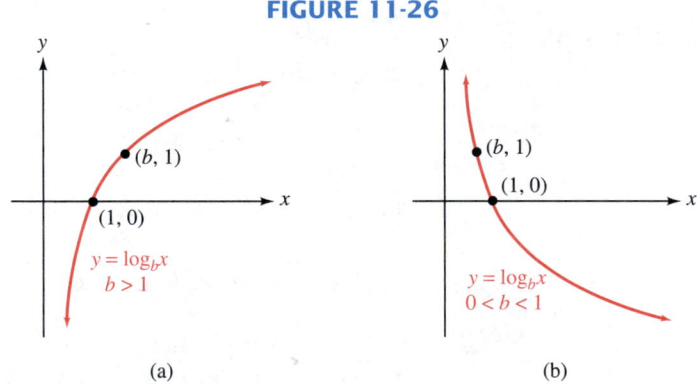

(a)   (b)

The graph of $y = \log_b x$ (or $f(x) = \log_b x$) has the following properties.

1. It passes through the point $(1, 0)$.
2. It passes through the point $(b, 1)$.
3. The y-axis (the line $x = 0$) is an asymptote.
4. The domain is $(0, \infty)$ and the range is $(-\infty, \infty)$.

The exponential and logarithmic functions are inverses of each other, so their graphs have symmetry about the line $y = x$. The graphs of $f(x) = \log_b x$ and $g(x) = b^x$ are shown in Figure 11-27(a) when $b > 1$ and in Figure 11-27(b) when $0 < b < 1$.

FIGURE 11-27

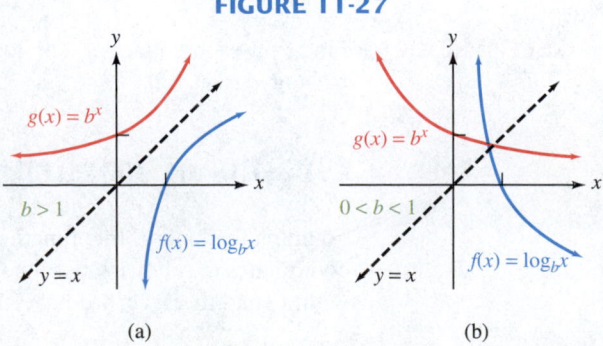

(a)   (b)

Vertical and Horizontal Translations

The graphs of many functions involving logarithms are translations of the basic logarithmic graphs.

EXAMPLE 4

Graphing logarithmic functions. Graph the function defined by $f(x) = 3 + \log_2 x$ and describe the translation.

Solution   The graph of $f(x) = 3 + \log_2 x$ is identical to the graph of $y = \log_2 x$, except that it is translated 3 units upward. (See Figure 11-28.)

FIGURE 11-28

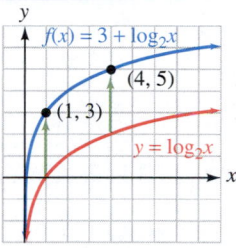

SELF CHECK   Graph $y = \log_3 x - 2$ and describe the translation.

Answer: The graph of $y = \log_3 x$ is translated 2 units downward.

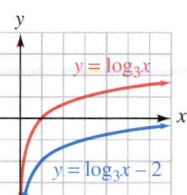

EXAMPLE 5

Graphing logarithmic functions. Graph $y = \log_{1/2} (x - 1)$ and describe the translation.

Solution   The graph of $y = \log_{1/2} (x - 1)$ is identical to the graph of $y = \log_{1/2} x$, except that it is translated 1 unit to the right. (See Figure 11-29.)

FIGURE 11-29

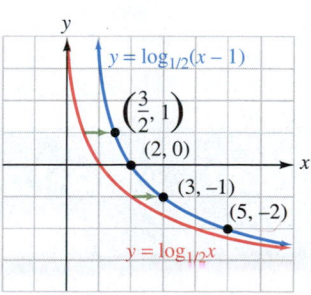

SELF CHECK   Graph $f(x) = \log_{1/3} (x + 2)$ and describe the translation.

Answer: The graph of $g(x) = \log_{1/3} x$ is translated 2 units to the left.

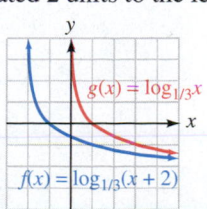

Graphing calculators can draw graphs of logarithmic functions. To use a calculator to graph the logarithmic function $y = -2 + \log_{10}\left(\frac{1}{2}x\right)$, we enter the right-hand side of the equation after the symbol $Y_1 =$. The display will show the equation

FIGURE 11-30

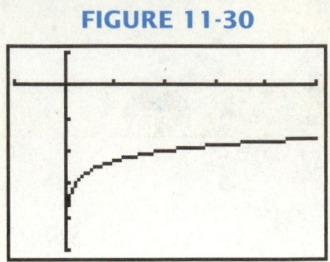

$$Y_1 = -2 + \log(1/2x)$$

If we use window settings of $[-1, 5]$ for x and $[-4, 1]$ for y and press the $\boxed{\text{GRAPH}}$ key, we will obtain the graph shown in Figure 11-30.

Base-10 Logarithms

For computational purposes and in many applications, we will use base-10 logarithms (also called **common logarithms**). When the base b is not indicated in the notation $\log x$, we assume that $b = 10$:

$\log x$ means $\log_{10} x$

Because base-10 logarithms appear so often, it is a good idea to become familiar with the following base-10 logarithms:

Logarithmic form		*Exponential form*
$\log_{10} \dfrac{1}{100} = -2$	because	$10^{-2} = \dfrac{1}{100}$
$\log_{10} \dfrac{1}{10} = -1$	because	$10^{-1} = \dfrac{1}{10}$
$\log_{10} 1 = 0$	because	$10^{0} = 1$
$\log_{10} 10 = 1$	because	$10^{1} = 10$
$\log_{10} 100 = 2$	because	$10^{2} = 100$
$\log_{10} 1{,}000 = 3$	because	$10^{3} = 1{,}000$

In general, we have

$\log_{10} 10^x = x$

Before calculators, extensive tables provided logarithms of numbers. Today, logarithms are easy to find with a calculator. For example, to find $\log 2.34$ with a scientific calculator, we enter these numbers and press these keys:

Keystrokes   2.34 $\boxed{\text{LOG}}$ $\boxed{.369215857}$

To use a graphing calculator, we enter these numbers and press these keys:

Keystrokes $\boxed{\text{LOG}}$ 2.34 $\boxed{)}$ $\boxed{\text{ENTER}}$
```
log(2.34)
        .3692158574
```

To four decimal places, $\log 2.34 = 0.3692$.

EXAMPLE 6	**Using a calculator.** Find x in the equation $\log x = 0.3568$. Round to four decimal places.

Solution The equation $\log x = 0.3568$ is equivalent to $10^{0.3568} = x$. To find x with a scientific calculator, we enter these numbers and press these keys:

10 y^x .3568 $=$

The display will read **2.274049951** . To four decimal places,

$$x = 2.2740$$

If your calculator has a 10^x key, enter .3568 and press it to get the same result.

SELF CHECK Solve $\log x = 1.87737$. Round to four decimal places.

Answer: 75.3998 ■

Applications of Logarithms

Common logarithms are used in electrical engineering to express the voltage gain (or loss) of an electronic device such as an amplifier. The unit of gain (or loss), called the **decibel,** is defined by a logarithmic relation.

Decibel voltage gain

If E_O is the output voltage of a device and E_I is the input voltage, the decibel voltage gain of the device (db gain) is given by

$$\text{db gain} = 20 \log \frac{E_O}{E_I}$$

EXAMPLE 7	**Finding db gain.** If the input to an amplifier is 0.5 volt and the output is 40 volts, find the decibel voltage gain of the amplifier.

Solution We can find the decibel voltage gain by substituting 0.5 for E_I and 40 for E_O into the formula for db gain:

$$\text{db gain} = 20 \log \frac{E_O}{E_I}$$

$$\text{db gain} = 20 \log \frac{40}{0.5}$$

$$= 20 \log 80$$

$$\approx 38 \qquad \text{Use a calculator.}$$

The amplifier provides a 38-decibel voltage gain. ■

In seismology, common logarithms are used to measure the intensity of earthquakes on the **Richter scale.** The intensity of an earthquake is given by the following logarithmic function.

Richter scale

If R is the intensity of an earthquake, A is the amplitude (measured in micrometers), and P is the period (the time of one oscillation of the earth's surface measured in seconds), then

$$R = \log \frac{A}{P}$$

EXAMPLE 8

Measuring earthquakes. Find the measure on the Richter scale of an earthquake with an amplitude of 5,000 micrometers (0.5 centimeter) and a period of 0.1 second.

Solution   We substitute 5,000 for A and 0.1 for P in the Richter scale formula and simplify:

$$R = \log \frac{A}{P}$$

$$R = \log \frac{5{,}000}{0.1}$$

$$= \log 50{,}000$$

$$\approx 4.698970004 \quad \text{Use a calculator.}$$

The earthquake measures about 4.7 on the Richter scale.   ∎

STUDY SET

Section 11.5

VOCABULARY

In Exercises 1–8, refer to the graph of $f(x) = \log_4 x$ in Illustration 1.

ILLUSTRATION 1

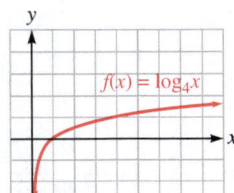

1. What type of function is $f(x) = \log_4 x$?
2. What is the domain of the function?
3. What is the range of the function?
4. What is the y-intercept of the graph?
5. What is the x-intercept of the graph?
6. What is an asymptote of the graph?
7. Is f an increasing or a decreasing function?
8. The graph passes through the point $(4, y)$. What is y?

CONCEPTS

In Exercises 9–12, fill in the blanks to make the statements true.

9. The equation $y = \log_b x$ is equivalent to the exponential equation _____.
10. $\log_b x$ is the _____ to which b is raised to get x.
11. The functions $f(x) = \log_b x$ and $f(x) = b^x$ are _____ functions.
12. The inverse of an exponential function is called a _____ function.

In Exercises 13–16, complete the table of values. If an evaluation is not possible, write "none."

13. $y = \log x$

x	y
$\frac{1}{100}$	
$\frac{1}{10}$	
1	
10	
100	

14. $f(x) = \log_5 x$

x	$f(x)$
$\frac{1}{25}$	
$\frac{1}{5}$	
1	
5	
25	

15. $f(x) = \log_6 x$

Input	Output
-6	
0	
$\frac{1}{216}$	
$\sqrt{6}$	
6^8	

16. $f(x) = \log_8 x$

x	$f(x)$
-8	
0	
$\frac{1}{8}$	
$\sqrt{8}$	
64	

17. Use a calculator to complete the table of values for $y = \log x$. Round to the nearest hundredth.

x	y
0.5	
1	
2	
3	
4	
5	
6	
7	
8	
9	
10	

18. Graph $y = \log x$ in Illustration 2. (See Exercise 17.) Note that the units on the x- and y-axes are different.

ILLUSTRATION 2

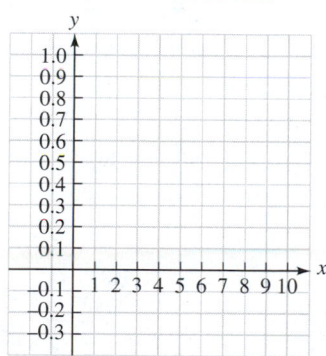

19. A table of values for $f(x) = \log x$ is shown in Illustration 3. As x decreases and gets close to 0, what happens to the values of $f(x)$?

ILLUSTRATION 3

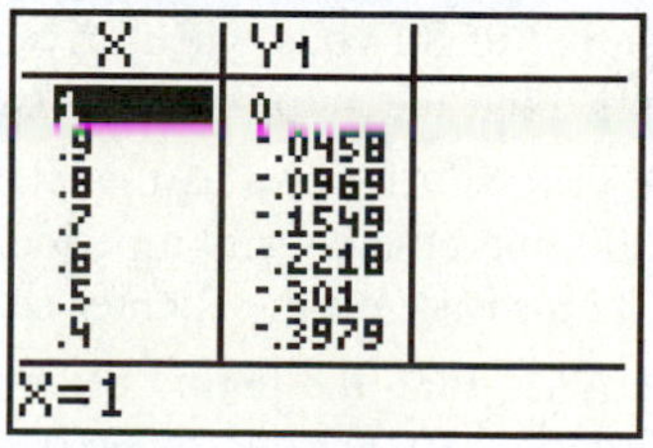

X	Y1
1	0
.9	-.0458
.8	-.0969
.7	-.1549
.6	-.2218
.5	-.301
.4	-.3979

X=1

20. For each function, determine $f^{-1}(x)$.
   a. $f(x) = 10^x$ **b.** $f(x) = 3^x$

   c. $f(x) = \log x$ **d.** $f(x) = \log_2 x$

NOTATION

In Exercises 21–22, fill in the blanks to make the statements true.

21. The notation $\log x$ means $\log$ ___ x.

22. $\log_{10} 10^x = $ ___ .

PRACTICE

In Exercises 23–30, write the equation in exponential form.

23. $\log_3 81 = 4$ **24.** $\log_7 7 = 1$

25. $\log_{12} 12 = 1$ **26.** $\log_6 36 = 2$

27. $\log_4 \frac{1}{64} = -3$ **28.** $\log_6 \frac{1}{36} = -2$

29. $\log 0.001 = -3$ **30.** $\log_3 243 = 5$

In Exercises 31–38, write the equation in logarithmic form.

31. $8^2 = 64$ **32.** $10^3 = 1{,}000$

33. $4^{-2} = \frac{1}{16}$ **34.** $3^{-4} = \frac{1}{81}$

35. $\left(\frac{1}{2}\right)^{-5} = 32$ **36.** $\left(\frac{1}{3}\right)^{-3} = 27$

37. $x^y = z$ **38.** $m^n = p$

In Exercises 39–66, find each value of x.

39. $\log_2 8 = x$ **40.** $\log_3 9 = x$

41. $\log_4 64 = x$ **42.** $\log_6 216 = x$

43. $\log_{1/2} \frac{1}{8} = x$ **44.** $\log_{1/3} \frac{1}{81} = x$

45. $\log_9 3 = x$ **46.** $\log_{125} 5 = x$

47. $\log_8 x = 2$ **48.** $\log_7 x = 0$

49. $\log_{25} x = \frac{1}{2}$ **50.** $\log_4 x = \frac{1}{2}$

51. $\log_5 x = -2$ **52.** $\log_3 x = -4$

53. $\log_{36} x = -\frac{1}{2}$ **54.** $\log_{27} x = -\frac{1}{3}$

55. $\log_{100} \frac{1}{1{,}000} = x$ **56.** $\log_{5/2} \frac{4}{25} = x$

57. $\log_{27} 9 = x$ **58.** $\log_{12} x = 0$

59. $\log_x 5^3 = 3$ **60.** $\log_x 5 = 1$

61. $\log_x \frac{9}{4} = 2$ **62.** $\log_x \frac{\sqrt{3}}{3} = \frac{1}{2}$

63. $\log_x \dfrac{1}{64} = -3$ **64.** $\log_x \dfrac{1}{100} = -2$

65. $\log_8 x = 0$ **66.** $\log_4 8 = x$

In Exercises 67–70, use a calculator to find each value. Give answers to four decimal places.

67. $\log 3.25$ **68.** $\log 0.57$

69. $\log 0.00467$ **70.** $\log 375.876$

In Exercises 71–74, use a calculator to find each value of y. Give answers to two decimal places.

71. $\log y = 1.4023$ **72.** $\log y = 0.926$

73. $\log y = -1.71$ **74.** $\log y = -0.5$

In Exercises 75–78, graph each function. Tell whether each function is an increasing or a decreasing function.

75. $f(x) = \log_3 x$ **76.** $f(x) = \log_{1/3} x$

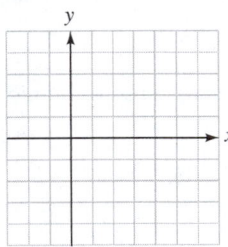 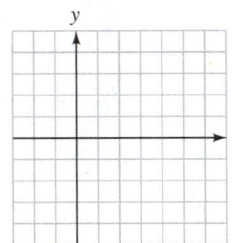

77. $y = \log_{1/2} x$ **78.** $y = \log_4 x$

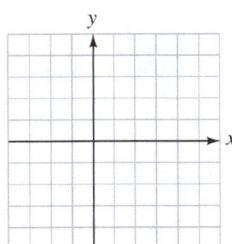 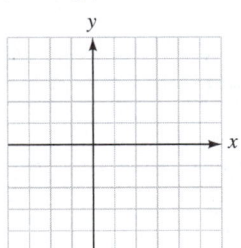

In Exercises 79–82, graph each function.

79. $f(x) = 3 + \log_3 x$ **80.** $f(x) = \log_{1/3} x - 1$

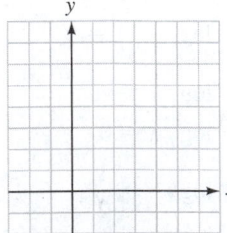 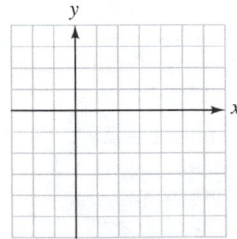

81. $y = \log_{1/2} (x - 2)$ **82.** $y = \log_4 (x + 2)$

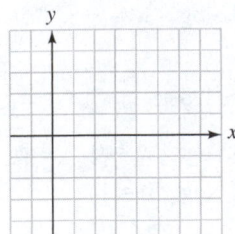 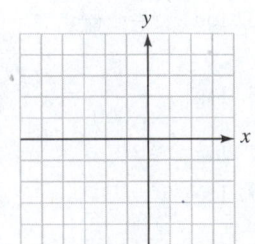

In Exercises 83–86, graph each pair of inverse functions on a single coordinate system. Draw the axis of symmetry.

83. $y = 6^x$ **84.** $y = 3^x$
    $y = \log_6 x$ $y = \log_3 x$

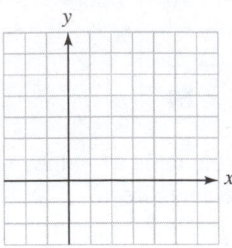 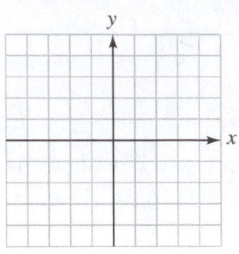

85. $f(x) = 5^x$ **86.** $f(x) = 8^x$
    $g(x) = \log_5 x$ $g(x) = \log_8 x$

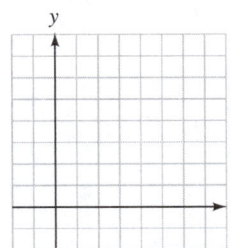 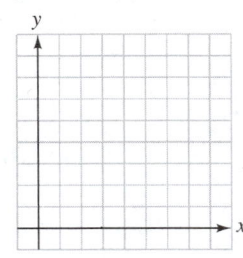

APPLICATIONS

87. FINDING INPUT VOLTAGE Find the db gain of an amplifier if the input voltage is 0.71 volts when the output voltage is 20 volts.

88. FINDING OUTPUT VOLTAGE Find the db gain of an amplifier if the output voltage is 2.8 volts when the input voltage is 0.05 volt.

89. db GAIN OF AN AMPLIFIER Find the db gain of the amplifier shown in Illustration 4.

ILLUSTRATION 4

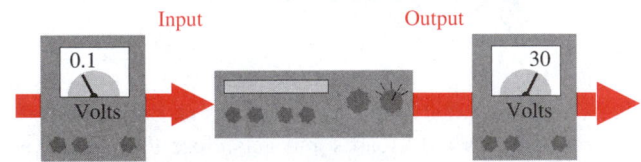

90. db GAIN OF AN AMPLIFIER An amplifier produces an output of 80 volts when driven by an input of 0.12 volts. Find the amplifier's db gain.

91. THE RICHTER SCALE An earthquake has amplitude of 5,000 micrometers and a period of 0.2 second. Find its measure on the Richter scale.

92. EARTHQUAKE Find the period of an earthquake with amplitude of 80,000 micrometers that measures 6 on the Richter scale.

93. EARTHQUAKE An earthquake with a period of $\frac{1}{4}$ second measures 4 on the Richter scale. Find its amplitude.

94. MAJOR EARTHQUAKE In 1985, Mexico City experienced an earthquake of magnitude 8.1 on the Richter scale. In 1989, the San Francisco Bay area was rocked by an earthquake measuring 7.1. By what factor must the amplitude of an earthquake change to increase its severity by 1 point on the Richter scale? (Assume that the period remains constant.)

95. DEPRECIATION In business, equipment is often depreciated using the double declining-balance method. In this method, a piece of equipment with a life expectancy of N years, costing $\$C$, will depreciate to a value of $\$V$ in n years, where n is given by the formula

$$n = \frac{\log V - \log C}{\log\left(1 - \frac{2}{N}\right)}$$

A computer that cost $\$37,000$ has a life expectancy of 5 years. If it has depreciated to a value of $\$8,000$, how old is it?

96. DEPRECIATION A typewriter worth $\$470$ when new had a life expectancy of 12 years. If it is now worth $\$189$, how old is it? (See Exercise 95.)

97. TIME FOR MONEY TO GROW If $\$P$ is invested at the end of each year in an annuity earning annual interest at a rate r, the amount in the account will be $\$A$ after n years, where

$$n = \frac{\log\left[\dfrac{Ar}{P} + 1\right]}{\log(1 + r)}$$

If $\$1,000$ is invested each year in an annuity earning 12% annual interest, how long will it take for the account to be worth $\$20,000$?

98. TIME FOR MONEY TO GROW If $\$5,000$ is invested each year in an annuity earning 8% annual interest, how long will it take for the account to be worth $\$50,000$? (See Exercise 97.)

WRITING

99. Explain the mathematical relationship between $y = \log x$ and $y = 10^x$.

100. Explain why it is impossible to find the logarithm of a negative number.

REVIEW

Solve each equation.

101. $\sqrt[3]{6x + 4} = 4$

102. $\sqrt{3x - 4} = \sqrt{-7x + 2}$

103. $\sqrt{a + 1} - 1 = 3a$

104. $3 - \sqrt{t - 3} = \sqrt{t}$

▶ # 11.6   Base-*e* Logarithms

In this section, you will learn about

> Base-*e* logarithms ■ Graphing the natural logarithmic function
> ■ An application of base-*e* logarithms

Introduction A special logarithmic function is the base-*e* logarithmic function. In this section, we will show how to evaluate base-*e* logarithms, graph the base-*e* logarithmic function, and solve some problems that involve the base-*e* logarithmic function.

Base-*e* Logarithms

We have seen the importance of the number e in mathematical models of events in nature. Base-*e* logarithms are just as important. They are called **natural logarithms** or **Napierian logarithms** after John Napier (1550–1617). They are usually written as $\ln x$ rather than $\log_e x$:

> **ln *x*   means   log*e* *x***

Like all logarithmic functions, the domain of $f(x) = \ln x$ is the interval $(0, \infty)$, and the range is the interval $(-\infty, \infty)$.

To find the base-*e* logarithms of numbers, we can use a calculator.

ACCENT ON TECHNOLOGY *Finding Base-e (Natural) Logarithms*

To use a scientific calculator to find the value of ln 2.34, we enter these numbers and press these keys:

Keystrokes   2.34 │LN│   `.850150929`

To use a graphing calculator, we enter these numbers and press these keys:

Keystrokes   │LN│ 2.34 │)│ │ENTER│   `ln(2.34)`
                                           `    .8501509294`

To four decimal places, ln 2.34 = 0.8502.

EXAMPLE 1   **Evaluating base-*e* logarithms.** Use a calculator to find each value:
a. ln 17.32   and   **b.** ln (−0.05).

Solution   **a.** Enter these numbers and press these keys:

Scientific calculator   *Graphing calculator*
17.32 │LN│   │LN│ 17.32 │)│ │ENTER│

Either way, the result is 2.851861903.

b. Enter these numbers and press these keys:

Scientific calculator   *Graphing calculator*
0.05 │+/−│ │LN│   │LN│ │(−)│ 0.05 │)│ │ENTER│

Either way, we obtain an error, because we cannot take the logarithm of a negative number.

SELF CHECK   Find each value to four decimal places: *Answers:* **a.** 1.1447, **b.** no
a. ln π and   **b.** ln 0.   value   ■

EXAMPLE 2   **Writing an equivalent exponential statement.** Solve each equation:
a. ln x = 1.335   and   **b.** ln x = −5.5. Give each result to four decimal places.

Solution   **a.** Since the base of the natural logarithmic function is e, the equation ln x = 1.335 is equivalent to $e^{1.335} = x$. To use a scientific calculator to find x, press these keys:

1.335 │e^x│

The display will read 3.799995946. To four decimal places,

$x = 3.8000$

b. The equation ln x = −5.5 is equivalent to $e^{-5.5} = x$. To use a scientific calculator to find x, press these keys:

5.5 │+/−│ │e^x│

The display will read 0.004086771. To four decimal places,

$x = 0.0041$

SELF CHECK Solve **a.** ln x = 1.9344 and **b.** −3 = ln x.   *Answers:* **a.** 6.9199,
Give each result to four decimal places.   **b.** 0.0498 ∎

Graphing the Natural Logarithmic Function

The equation $y = \ln x$ is equivalent to the equation $x = e^y$. To get the graph of ln x, we can plot points that satisfy the equation $x = e^y$ and join them with a smooth curve, as shown in Figure 11-31(a). Figure 11-31(b) shows the calculator graph of $y = \ln x$.

FIGURE 11-31

$y = \ln x$

x	y	(x, y)
$\frac{1}{e}$	-1	$\left(\frac{1}{e}, -1\right)$
1	0	$(1, 0)$
e	1	$(e, 1)$
e^2	2	$(e^2, 2)$

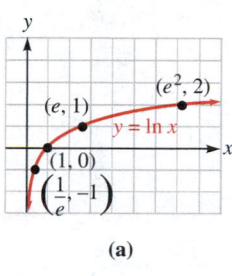

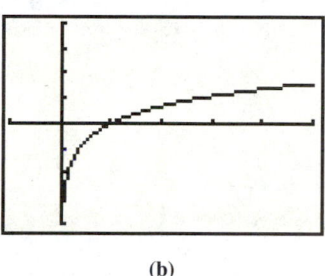

(a) (b)

FIGURE 11-32

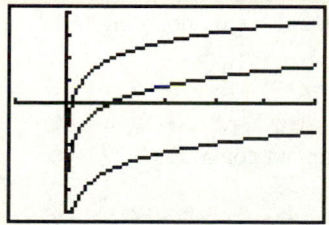

The exponential function $f(x) = e^x$ and the natural logarithmic function $f(x) = \ln x$ are inverse functions. Figure 11-32 shows that their graphs are symmetric to the line $y = x$.

ACCENT ON TECHNOLOGY *Graphing Base-e Logarithmic Functions*

Many graphs of logarithmic functions involve translations of the graph of $y = \ln x$. For example, Figure 11-33 shows calculator graphs of the functions $y = \ln x$, $y = \ln x + 2$, and $y = \ln x - 3$.

FIGURE 11-33

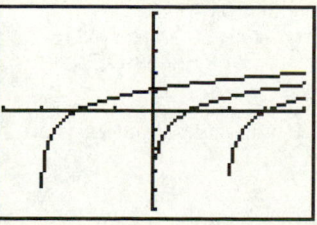

The graph of $y = \ln x + 2$ is 2 units above the graph of $y = \ln x$.

The graph of $y = \ln x - 3$ is 3 units below the graph of $y = \ln x$.

Figure 11-34 shows the calculator graph of the functions $y = \ln x$, $y = \ln (x - 2)$, and $y = \ln (x + 3)$.

FIGURE 11-34

The graph of $y = \ln (x + 3)$ is 3 units to the left of the graph of $y = \ln x$.

The graph of $y = \ln (x - 2)$ is 2 units to the right of the graph of $y = \ln x$.

An Application of Base-*e* Logarithms

Base-*e* logarithms have many applications. If a population grows exponentially at a certain annual rate, the time required for the population to double is called the **doubling time.** It is given by the following formula.

Formula for doubling time

If *r* is the annual rate, compounded continuously, and *t* is time required for a population to double, then

$$t = \frac{\ln 2}{r}$$

EXAMPLE 3

Doubling time. The population of the earth is growing at the approximate rate of 2% per year. If this rate continues, how long will it take for the population to double?

Solution　Because the population is growing at the rate of 2% per year, we substitute 0.02 for *r* in the formula for doubling time and simplify.

$$t = \frac{\ln 2}{r}$$

$$t = \frac{\ln 2}{0.02}$$

$$\approx 34.65735903 \quad \text{Use a calculator}$$

The population will double in about 35 years.

SELF CHECK　See Example 3. If the population's annual growth rate could be reduced to 1.5% per year, what would be the doubling time?　*Answer:* about 46 years ■

EXAMPLE 4

Doubling time. How long will it take $1,000 to double at an annual rate of 8%, compounded continuously?

Solution　We substitute 0.08 for *r* and simplify:

$$t = \frac{\ln 2}{r}$$

$$t = \frac{\ln 2}{0.08}$$

$$\approx 8.664339757 \quad \text{Use a calculator.}$$

It will take about $8\frac{2}{3}$ years for the money to double.

SELF CHECK　In Example 4, how long will it take at 9%, compounded continuously?　*Answer:* about 7.7 years ■

STUDY SET

Section 11.6

VOCABULARY

In Exercises 1–2, fill in the blanks to make the statements true.

1. Base-*e* logarithms are often called _____ logarithms.

2. $f(x) = \ln x$ and $f(x) = e^x$ are _____ functions.

CONCEPTS

3. Use a calculator to complete the table of values for $f(x) = \ln x$. Round to the nearest hundredth.

x	$f(x)$
0.5	
1	
2	
3	
4	
5	
6	
7	
8	
9	
10	

4. Graph $f(x) = \ln x$ in Illustration 1. (See Exercise 3.) Note that the units on the *x*- and *y*-axes are different.

ILLUSTRATION 1

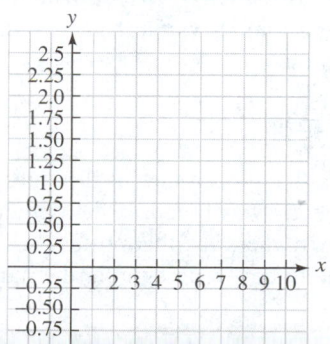

In Exercises 5–10, fill in the blanks to make the statements true.

5. The graph of $f(x) = \ln x$ has the _____ as an asymptote.

6. The domain of the function $f(x) = \ln x$ is the interval _____.

7. The range of the function $f(x) = \ln x$ is the interval _____.

8. The graph of $f(x) = \ln x$ passes through the point $(\boxed{}, 0)$.

9. The statement $y = \ln x$ is equivalent to the exponential statement _____.

10. The logarithm of a negative number is _____.

11. A table of values for $f(x) = \ln x$ is shown in Illustration 2. Explain why ERROR appears in the Y_1 column for the first three entries.

ILLUSTRATION 2

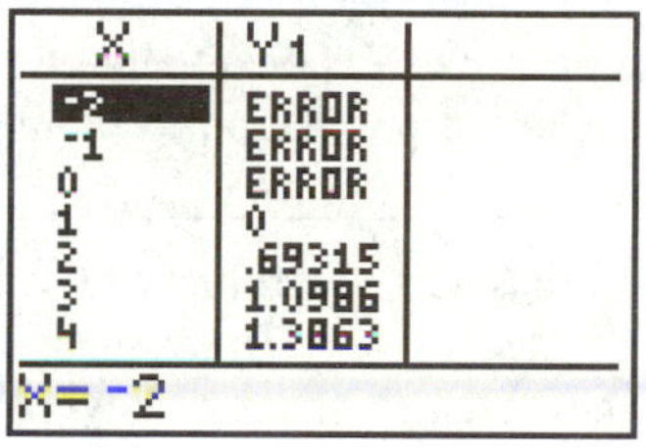

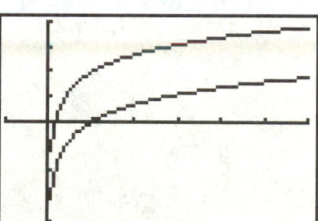

12. Illustration 3 shows the graph of $f(x) = \ln x$, as well as a vertical translation of that graph. Using the notation $g(x)$ for the translation, write the defining equation for the function.

ILLUSTRATION 3

13. In Illustration 4, $f(x) = \ln x$ was graphed, and the TRACE feature was used. What is the *x*-coordinate of the point on the graph having a *y*-coordinate of 1? What is the name given this number?

ILLUSTRATION 4

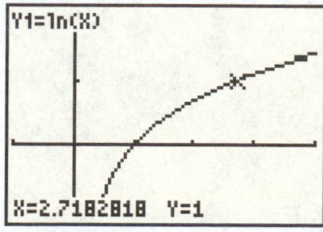

14. The graphs of $f(x) = \ln x$, $f(x) = e^x$, and $y = x$ are shown in Illustration 5. What phrase is used to describe the relationship between the graphs?

ILLUSTRATION 5

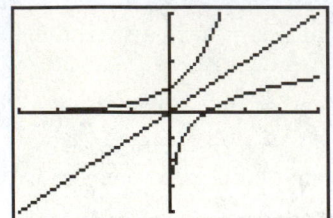

NOTATION

In Exercises 15–18, fill in the blanks to make the statements true.

15. $\ln 2$ means $\log \ \ 2$.

16. $\log 2$ means $\log \ \ 2$.

17. If a population grows exponentially at a rate r, the time it will take the population to double is given by the formula $t = $ _____.

18. To evaluate a base-10 logarithm with a calculator, use the _____ key. To evaluate a base-e logarithm, use the _____ key.

PRACTICE

In Exercises 19–26, use a calculator to find each value, if possible. Express all answers to four decimal places.

19. $\ln 35.15$ **20.** $\ln 0.675$

21. $\ln 0.00465$ **22.** $\ln 378.96$

23. $\ln 1.72$ **24.** $\ln 2.7$

25. $\ln (-0.1)$ **26.** $\ln (-10)$

 In Exercises 27–34, use a calculator to find x. Express all answers to four decimal places.

27. $\ln x = 1.4023$ **28.** $\ln x = 2.6490$

29. $\ln x = 4.24$ **30.** $\ln x = 0.926$

31. $\ln x = -3.71$ **32.** $\ln x = -0.28$

33. $1.001 = \ln x$ **34.** $\ln x = -0.001$

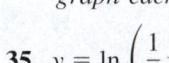 *In Exercises 35–38, use a graphing calculator to graph each function.*

35. $y = \ln \left(\dfrac{1}{2} x \right)$ **36.** $y = \ln x^2$

37. $y = \ln (-x)$ **38.** $y = \ln (3x)$

APPLICATIONS

Use a calculator to solve each problem.

39. POPULATION GROWTH How long will it take the population of River City to double? See Illustration 6.

ILLUSTRATION 6

River City
A growing community

• 6 parks   • 12% annual growth
• 10 churches   • Low crime rate

40. DOUBLING MONEY How long will it take $1,000 to double if it is invested at an annual rate of 5% compounded continuously?

41. POPULATION GROWTH A population growing continuously at an annual rate r will triple in a time t given by the formula

$$t = \frac{\ln 3}{r}$$

How long will it take the population of a town to triple if it is growing at the rate of 12% per year?

42. TRIPLING MONEY Find the length of time for $25,000 to triple when it is invested at 6% annual interest, compounded continuously. See Exercise 41.

43. FORENSIC MEDICINE To estimate the number of hours t that a murder victim had been dead, a coroner used the formula

$$t = \frac{1}{0.25} \ln \frac{98.6 - T_s}{82 - T_s}$$

where T_s is the temperature of the surroundings where the body was found. If the crime took place in an apartment where the thermostat was set at 70° F, approximately how long ago did the murder occur?

44. MAKING JELL-O After the contents of a package of JELL-O are combined with boiling water, the mixture is placed in a refrigerator whose temperature remains a constant 42° F. Estimate the number of hours t that it will take for the JELL-O to cool to 50° F using the formula

$$t = -\frac{1}{0.9} \ln \frac{50 - T_r}{200 - T_r}$$

where T_r is the temperature of the refrigerator.

45. Explain the difference between the functions $f(x) = \log x$ and $f(x) = \ln x$.

46. How are the functions $f(x) = \ln x$ and $f(x) = e^x$ related?

In Exercises 47–52, write the equation of the required line.

47. Parallel to $y = 5x - 8$ and passing through the origin

48. Having a slope of 7 and a y-intercept of 3

49. Passing through the point $(3, 2)$ and perpendicular to the line $y = \frac{2}{3}x - 12$

50. Parallel to the line $3x + 2y = 9$ and passing through the point $(-3, 5)$

51. Vertical line through the point $(2, 3)$

52. Horizontal line through the point $(2, 3)$

▶ 11.7 Properties of Logarithms

In this section, you will learn about

Properties of logarithms ■ The change-of-base formula
■ An application from chemistry

Introduction In this section, we will discuss eight properties of logarithms and use them to simplify logarithmic expressions. We will then show how to change a logarithm from one base to another. We conclude the section by solving some problems from the field of chemistry.

Properties of Logarithms

Since logarithms are exponents, the properties of exponents have counterparts in the theory of logarithms. We begin with four basic properties.

Properties of logarithms

If b is a positive number and $b \neq 1$, then

1. $\log_b 1 = 0$ **2.** $\log_b b = 1$

3. $\log_b b^x = x$ **4.** $b^{\log_b x} = x \ (x > 0)$

Properties 1 through 4 follow directly from the definition of a logarithm.

1. $\log_b 1 = 0$, because $b^0 = 1$.

2. $\log_b b = 1$, because $b^1 = b$.

3. $\log_b b^x = x$, because $b^x = b^x$.

4. $b^{\log_b x} = x$, because $\log_b x$ is the exponent to which b is raised to get x.

Properties 3 and 4 also indicate that the composition of the exponential and logarithmic functions (in both directions) is the identity function. This is expected, because the exponential and logarithmic functions are inverse functions.

EXAMPLE 1 **Applying properties of logarithms.** Simplify each expression: **a.** $\log_5 1$, **b.** $\log_3 3$, **c.** $\ln e^3$, and **d.** $b^{\log_b 7}$.

Solution **a.** By property 1, $\log_5 1 = 0$, because $5^0 = 1$.
b. By property 2, $\log_3 3 = 1$, because $3^1 = 3$.
c. By property 3, $\ln e^3 = 3$, because $e^3 = e^3$.
d. By property 4, $b^{\log_b 7} = 7$, because $\log_b 7$ is the power to which b is raised to get 7.

SELF CHECK Simplify **a.** $\log_4 1$, **b.** $\log_4 4$, *Answers:* **a.** 0, **b.** 1, **c.** 4,
c. $\log_2 2^4$, and **d.** $5^{\log_5 2}$. **d.** 2 ■

The next two properties state that

The logarithm of a product is the sum of the logarithms.
The logarithm of a quotient is the difference of the logarithms.

Properties of logarithms

If M, N, and b are positive numbers and $b \neq 1$, then

5. $\log_b MN = \log_b M + \log_b N$ **6.** $\log_b \dfrac{M}{N} = \log_b M - \log_b N$

PROOF To prove property 5, we let $x = \log_b M$ and $y = \log_b N$. We use the definition of logarithm to write each equation in exponential form.

$$M = b^x \qquad \text{and} \qquad N = b^y$$

Then $MN = b^x b^y$, and a property of exponents gives

$$MN = b^{x+y} \quad \text{Keep the base and add the exponents: } b^x b^y = b^{x+y}.$$

We write this exponential equation in logarithmic form as

$$\log_b MN = x + y$$

Substituting the values of x and y completes the proof.

$$\log_b MN = \log_b M + \log_b N \qquad \qquad \square$$

The proof of property 6 is similar.
In the following examples, we will use these properties to write a logarithm of a product as the sum of logarithms and a logarithm of a quotient as the difference of logarithms. Assume that all variables are positive and that $b \neq 1$.

EXAMPLE 2

Applying the product rule. Use the product rule for logarithms to rewrite each of the following: **a.** $\log_2 (2 \cdot 7)$ and   **b.** $\log (100x)$.

Solution   **a.** $\log_2 (2 \cdot 7) = \log_2 2 + \log_2 7$ The log of a product is the sum of the logs.

$\qquad\qquad\qquad = 1 + \log_2 7$ $\log_2 2 = 1.$

b. $\log (100x) = \log 100 + \log x$ The log of a product is the sum of the logs.

$\qquad\qquad\quad = 2 + \log x$ $\log 100 = 2.$

SELF CHECK   Rewrite   **a.** $\log_3 (4 \cdot 3)$ and   **b.** $\log (1{,}000y)$.   *Answers:*   **a.** $\log_3 4 + 1$,
$\qquad\qquad\qquad\qquad\qquad\qquad\qquad\qquad\qquad\qquad\qquad$ **b.** $3 + \log y$ ■

EXAMPLE 3

Applying the quotient rule. Use the quotient rule for logarithms to rewrite each of the following: **a.** $\ln \frac{10}{7}$ and   **b.** $\log_4 \frac{x}{64}$.

Solution   **a.** $\ln \dfrac{10}{7} = \ln 10 - \ln 7$ $\qquad$ The log of a quotient is the difference of the logs.

b. $\log_4 \dfrac{x}{64} = \log_4 x - \log_4 64$ The log of a quotient is the difference of the logs.

$\qquad\qquad = \log_4 x - 3$ $\qquad$ $\log_4 64 = 3.$

SELF CHECK   Rewrite   **a.** $\log_6 \frac{6}{5}$ and   **b.** $\ln \frac{y}{100}$. $\qquad$ *Answers:*   **a.** $1 - \log_6 5$,
$\qquad\qquad\qquad\qquad\qquad\qquad\qquad\qquad\qquad\qquad\qquad\qquad$ **b.** $\ln y - \ln 100$ ■

EXAMPLE 4

Applying the product and quotient properties. Use logarithm properties to rewrite each expression: **a.** $\log_b xyz$ and   **b.** $\log \frac{xy}{z}$.

Solution   **a.**   We observe that $\log_b xyz$ is the logarithm of a product.

$\qquad \log_b xyz = \log_b (xy)z$ $\qquad\qquad\qquad$ Group the first two factors together.

$\qquad\qquad\quad = \log_b (xy) + \log_b z$ $\qquad$ The log of a product is the sum of the logs.

$\qquad\qquad\quad = \log_b x + \log_b y + \log_b z$ The log of a product is the sum of the logs.

b.   In the expression $\log \dfrac{xy}{z}$, we have the logarithm of a quotient.

$\qquad \log \dfrac{xy}{z} = \log (xy) - \log z$ $\qquad\qquad$ The log of a quotient is the difference of the logs.

$\qquad\qquad\quad = (\log x + \log y) - \log z$ $\qquad$ The log of a product is the sum of the logs.

$\qquad\qquad\quad = \log x + \log y - \log z$ $\qquad$ Remove parentheses.

SELF CHECK   Rewrite $\log_b \dfrac{x}{yz}$.
$\qquad\qquad\qquad\qquad\qquad\qquad$ *Answer:* $\log_b x - \log_b y - \log_b z$ ■

WARNING! By property 5 of logarithms, the logarithm of a *product* is equal to the *sum* of the logarithms. The logarithm of a sum or a difference usually does not simplify. In general,

$$\log_b (M + N) \neq \log_b M + \log_b N \quad \text{and} \quad \log_b (M - N) \neq \log_b M - \log_b N$$

By property 6, the logarithm of a *quotient* is equal to the *difference* of the logarithms. The logarithm of a quotient is not the quotient of the logarithms:

$$\log_b \frac{M}{N} \neq \frac{\log_b M}{\log_b N}$$

ACCENT ON TECHNOLOGY *Verifying Properties of Logarithms*

We can use a calculator to illustrate property 5 of logarithms by showing that

$$\ln[(3.7)(15.9)] = \ln 3.7 + \ln 15.9$$

We calculate the left- and right-hand sides of the equation separately and compare the results. To use a scientific calculator to find $\ln[(3.7)(15.9)]$, we enter these numbers and press these keys:

Keystrokes   3.7 $\boxed{\times}$ 15.9 $\boxed{=}$ $\boxed{\text{LN}}$ $\boxed{4.074651929}$

To find $\ln 3.7 + \ln 15.9$, we enter these numbers and press these keys:

Keystrokes   3.7 $\boxed{\text{LN}}$ $\boxed{+}$ 15.9 $\boxed{\text{LN}}$ $\boxed{=}$ $\boxed{4.074651929}$

Since the left- and right-hand sides are equal, the equation $\ln[(3.7)(15.9)] = \ln 3.7 + \ln 15.9$ is true.

Two more properties of logarithms state that

The logarithm of a power is the power times the logarithm.

If the logarithms of two numbers are equal, the numbers are equal.

Properties of logarithms

If M, p, and b are positive numbers and $b \neq 1$, then

7. $\log_b M^p = p \log_b M$ **8.** If $\log_b x = \log_b y$, then $x = y$.

PROOF

To prove property 7, we let $x = \log_b M$, write the expression in exponential form, and raise both sides to the pth power:

$$M = b^x$$
$$(M)^p = (b^x)^p \quad \text{Raise both sides to the } p\text{th power.}$$
$$M^p = b^{px} \quad \text{Keep the base and multiply the exponents.}$$

Using the definition of logarithms gives

$$\log_b M^p = px$$

Substituting the value for x completes the proof.

$$\log_b M^p = p \log_b M \qquad \square$$

Property 8 follows from the fact that the logarithmic function is a one-to-one function. Property 8 will be important in the next section, when we solve logarithmic equations.

EXAMPLE 5   **Applying the power rule.** Use the power rule for logarithms to rewrite each of the following: **a.** $\log_5 6^2$ and   **b.** $\log\sqrt{10}$.

Solution **a.** $\log_5 6^2 = 2 \log_5 6$ The log of a power is the power times the log.

b. $\log \sqrt{10} = \log (10)^{1/2}$ Write $\sqrt{10}$ using a fractional exponent: $\sqrt{10} = (10)^{1/2}$.

$$= \frac{1}{2} \log 10$$ The log of a power is the power times the log.

$$= \frac{1}{2}$$ Simplify: $\log 10 = 1$.

SELF CHECK Rewrite **a.** $\ln x^4$ and **b.** $\log_2 \sqrt[3]{3}$.   *Answers:* **a.** $4 \ln x$, **b.** $\frac{1}{3} \log_2 3$ ∎

EXAMPLE 6   **Applying properties of logarithms.** Use logarithm properties to rewrite each expression: **a.** $\log_b (x^2 y^3 z)$ and **b.** $\ln \dfrac{y^3 \sqrt{x}}{z}$.

Solution **a.** We begin by recognizing that $\log_b (x^2 y^3 z)$ is the logarithm of a product.

$$\log_b (x^2 y^3 z) = \log_b x^2 + \log_b y^3 + \log_b z$$ The log of a product is the sum of the logs.

$$= 2 \log_b x + 3 \log_b y + \log_b z$$ The log of a power is the power times the log.

b. The expression $\ln \dfrac{y^3 \sqrt{x}}{z}$ is the logarithm of a quotient.

$$\ln \frac{y^3 \sqrt{x}}{z} = \ln \left(y^3 \sqrt{x} \right) - \ln z$$ The log of a quotient is the difference of the logs.

$$= \ln y^3 + \ln \sqrt{x} - \ln z$$ The log of a product is the sum of the logs.

$$= \ln y^3 + \ln x^{1/2} - \ln z$$ Write $\sqrt{x}$ as $x^{1/2}$.

$$= 3 \ln y + \frac{1}{2} \ln x - \ln z$$ The log of a power is the power times the log.

SELF CHECK Rewrite $\log \sqrt[4]{\dfrac{x^3 y}{z}}$.   *Answer:* $\frac{1}{4}(3 \log x + \log y - \log z)$ ∎

We can use the properties of logarithms to combine several logarithms into one logarithm.

EXAMPLE 7   **Condensing logarithmic expressions.** Write each of the given expressions as one logarithm: **a.** $3 \log_a x + \frac{1}{2} \log_a y$ and **b.** $\frac{1}{2} \log_b (x - 2) - \log_b y + 3 \log_b z$.

Solution **a.** We begin by applying the power rule to each term of the expression.

$$3 \log_a x + \frac{1}{2} \log_a y$$

$$= \log_a x^3 + \log_a y^{1/2}$$ A power times a log is the log of the power.

$$= \log_a (x^3 y^{1/2})$$ The sum of two logs is the log of the product.

b. The first and third terms of this expression can be rewritten using the power rule of logarithms.

$$\frac{1}{2} \log_b (x - 2) - \log_b y + 3 \log_b z$$

$$= \log_b (x - 2)^{1/2} - \log_b y + \log_b z^3 \qquad \text{A power times a log is the log of the power.}$$

$$= \log_b \frac{(x - 2)^{1/2}}{y} + \log_b z^3 \qquad \text{The difference of two logs is the log of the quotient.}$$

$$= \log_b \frac{z^3 \sqrt{x - 2}}{y} \qquad \text{The sum of two logs is the log of the product. Write } (x - 2)^{1/2} \text{ as } \sqrt{x - 2}.$$

SELF CHECK   Write the expression as one logarithm: $2 \log_a x + \frac{1}{2} \log_a y - 2 \log_a (x - y)$.

Answer: $\log_a \dfrac{x^2 \sqrt{y}}{(x - y)^2}$ ■

We summarize the properties of logarithms as follows.

Properties of logarithms

If b, M, and N are positive numbers and $b \neq 1$, then

1. $\log_b 1 = 0$ **2.** $\log_b b = 1$

3. $\log_b b^x = x$ **4.** $b^{\log_b x} = x$

5. $\log_b MN = \log_b M + \log_b N$ **6.** $\log_b \dfrac{M}{N} = \log_b M - \log_b N$

7. $\log_b M^p = p \log_b M$ **8.** If $\log_b x = \log_b y$, then $x = y$.

EXAMPLE 8   **Using properties of logarithms.** Given that $\log 2 \approx 0.3010$ and $\log 3 \approx 0.4771$, find approximations for **a.** $\log 6$ and **b.** $\log 18$.

Solution **a.** $\log \mathbf{6} = \log (\mathbf{2 \cdot 3})$ Write 6 using the factors 2 and 3.

$\qquad = \log 2 + \log 3$ The log of a product is the sum of the logs.

$\qquad \approx 0.3010 + 0.4771$ Substitute the value of each logarithm.

$\qquad \approx 0.7781$

b. $\log \mathbf{18} = \log (\mathbf{2 \cdot 3^2})$ Write 18 using the factors 2 and 3.

$\qquad = \log 2 + \log 3^2$ The log of a product is the sum of the logs.

$\qquad = \log 2 + 2 \log 3$ The log of a power is the power times the log.

$\qquad \approx 0.3010 + 2(0.4771)$ Substitute the value of each logarithm.

$\qquad \approx 1.2552$

SELF CHECK   Find **a.** $\log 1.5$ and **b.** $\log 0.75$.   *Answers:* **a.** 0.1761, **b.** -0.1249 ■

The Change-of-Base Formula

Most calculators can find common logarithms and natural logarithms. If we need to find a logarithm with some other base, we use a conversion formula.

If we know the base-a logarithm of a number, we can find its logarithm to some other base b by using a formula called the **change-of-base formula.**

Change-of-base formula

If a, b, and x are positive, $a \neq 1$, and $b \neq 1$, then

$$\log_b x = \frac{\log_a x}{\log_a b}$$

PROOF

To prove this formula, we begin with the equation $\log_b x = y$.

$y = \log_b x$

$x = b^y$ Change the equation from logarithmic to exponential form.

$\log_a x = \log_a b^y$ Take the base-a logarithm of both sides.

$\log_a x = y \log_a b$ The log of a power is the power times the log.

$y = \dfrac{\log_a x}{\log_a b}$ Divide both sides by $\log_a b$.

$\log_b x = \dfrac{\log_a x}{\log_a b}$ Refer to the first equation and substitute $\log_b x$ for y. $\square$

If we know logarithms to base a (for example, $a = 10$), we can find the logarithm of x to a new base b. We simply divide the base-a logarithm of x by the base-a logarithm of b.

 WARNING! $\dfrac{\log_a x}{\log_a b}$ means that one logarithm is to be divided by the other. They are not to be subtracted.

EXAMPLE 9

Using the change-of-base formula. Find $\log_3 5$.

Solution We can use base-10 logarithms to find a base-3 logarithm. To do this, we substitute 3 for b, 10 for a, and 5 for x in the change-of-base formula:

$\log_b x = \dfrac{\log_a x}{\log_a b}$

$\log_3 5 = \dfrac{\log_{10} 5}{\log_{10} 3}$ $b = 3$, $x = 5$, and $a = 10$.

≈ 1.464973521 Use a scientific calculator and press these keys: 5 log ÷ 3 log = .

To four decimal places, $\log_3 5 = 1.4650$.

We can also use the natural logarithm function (base e) in the change-of-base formula to find a base-3 logarithm.

$\log_b x = \dfrac{\log_a x}{\log_a b}$

$\log_3 5 = \dfrac{\ln 5}{\ln 3}$ $b = 3$, $x = 5$, and $a = e$. $\log_e 5 = \ln 5$ and $\log_e 3 = \ln 3$.

≈ 1.464973521 Use a calculator.

We obtain the same result.

SELF CHECK   Find $\log_5 3$ to four decimal places.   *Answer:* 0.6826   ■

An Application from Chemistry

In chemistry, common logarithms are used to express the acidity of solutions. The more acidic a solution, the greater the concentration of hydrogen ions. This concentration is indicated indirectly by the *pH scale,* or *hydrogen ion index.* The pH of a solution is defined as follows.

pH of a solution

If $[H^+]$ is the hydrogen ion concentration in gram-ions per liter, then

$$pH = -\log[H^+]$$

EXAMPLE 10

Finding the pH of a solution. Find the pH of pure water, which has a hydrogen ion concentration of 10^{-7} gram-ions per liter.

Solution   Since pure water has approximately 10^{-7} gram-ions per liter, its pH is

$$pH = -\log[\mathbf{H^+}]$$
$$pH = -\log \mathbf{10^{-7}}$$
$$= -(-7)\log 10 \quad \text{The log of a power is the power times the log.}$$
$$= -(-7)\cdot 1 \quad\quad \text{Simplify: } \log 10 = 1.$$
$$= 7$$

■

EXAMPLE 11

Finding hydrogen ion concentration. Find the hydrogen ion concentration of seawater if its pH is 8.5.

Solution   To find its hydrogen ion concentration, we solve the following equation for $[H^+]$.

$$\mathbf{pH} = -\log[H^+]$$
$$\mathbf{8.5} = -\log[H^+]$$
$$-8.5 = \log[H^+] \quad \text{Multiply both sides by } -1.$$
$$[H^+] = 10^{-8.5} \quad \text{Change the equation to exponential form.}$$

We can use a calculator to find that

$$[H^+] \approx 3.2 \times 10^{-9} \text{ gram-ions per liter}$$

■

STUDY SET

Section 11.7

VOCABULARY

In Exercises 1–4, fill in the blanks to make the statements true.

1. The expression $\log_3 (4x)$ is the logarithm of a _____.

2. The expression $\log_2 \frac{5}{x}$ is the logarithm of a _____.

3. The expression $\log 4^x$ is the logarithm of a _____.

4. In the expression $\log_5 4$, the number 5 is the _____ of the logarithm.

CONCEPTS

In Exercises 5–16, fill in the blanks to make the statements true.

5. $\log_b 1 = $ ▢

6. $\log_b b = $ ▢

7. $\log_b MN = \log_b$ ▢ $+ \log_b$ ▢

8. $b^{\log_b x} = $ ▢

9. If $\log_b x = \log_b y$, then ▢ $=$ ▢ .

10. $\log_b \dfrac{M}{N} = \log_b M$ ▢ $\log_b N$

11. $\log_b x^p = p \cdot \log_b$ ▢

12. $\log_b b^x = $ ▢

13. $\log_b (A + B)$ ▢ $\log_b A + \log_b B$

14. $\log_b A + \log_b B$ ▢ $\log_b AB$

15. $\log_b x = \dfrac{\log_a x}{\text{▢}}$

16. pH $= $ ▢

17. Three logarithmic expressions have been evaluated, and the results are shown on the calculator display in Illustration 1. Show that each result is correct by writing the equivalent base-10 exponential statement.

ILLUSTRATION 1

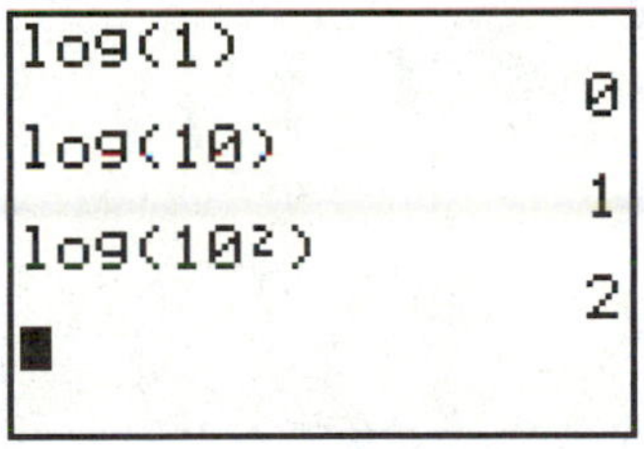

```
log(1)
                        0
log(10)
                        1
log(10²)
                        2
■
```

18. Three logarithmic expressions have been evaluated, and the results are shown on the calculator display in Illustration 2. Show that each result is correct by writing the equivalent base-e exponential statement. (The notation $\ln(e\text{^}(2))$ means $\ln e^2$.)

ILLUSTRATION 2

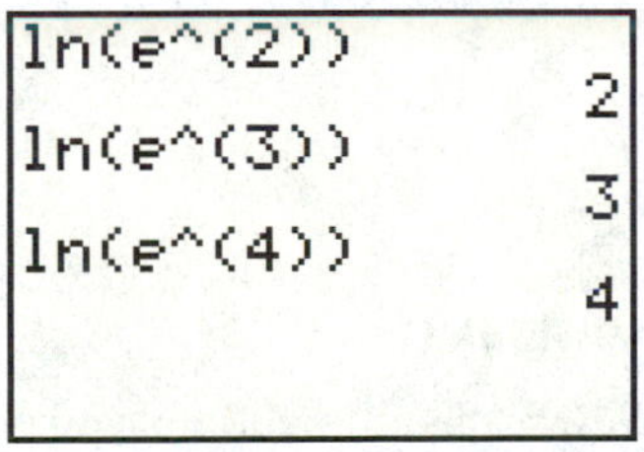

```
ln(e^(2))
                        2
ln(e^(3))
                        3
ln(e^(4))
                        4
```

In Exercises 19–30, evaluate each expression.

19. $\log_4 1$

20. $\log_4 4$

21. $\log_4 4^7$

22. $\ln e^8$

23. $5^{\log_5 10}$

24. $8^{\log_8 10}$

25. $\log_5 5^2$

26. $\log_4 4^2$

27. $\ln e$

28. $\log_7 1$

29. $\log_3 3^7$

30. $5^{\log_5 8}$

NOTATION

In Exercises 31–32, complete each solution.

31. $\log_b rst = \log_b \left(\text{▢}\right) t$

$\qquad = \log_b (rs) + \log_b$ ▢

$\qquad = \log_b$ ▢ $+ \log_b$ ▢ $+ \log_b t$

32. $\log \dfrac{r}{st} = \log r - \log \left(\text{▢}\right)$

$\qquad = \log r - \left(\log \text{▢} + \log t\right)$

$\qquad = \log r - \log s \ \text{▢} \ \log t$

PRACTICE

In Exercises 33–38, use a calculator to verify each equation.

33. $\log [(2.5)(3.7)] = \log 2.5 + \log 3.7$

34. $\log 45.37 = \dfrac{\ln 45.37}{\ln 10}$

35. $\ln (2.25)^4 = 4 \ln 2.25$

36. $\ln \dfrac{11.3}{6.1} = \ln 11.3 - \ln 6.1$

37. $\log \sqrt{24.3} = \dfrac{1}{2} \log 24.3$

38. $\ln 8.75 = \dfrac{\log 8.75}{\log e}$

In Exercises 39–46, use the properties of logarithms to rewrite each expression. Assume that x, y, and z are positive numbers.

39. $\log_2 (4 \cdot 5)$

40. $\log_3 (27 \cdot 5)$

41. $\log_6 \dfrac{x}{36}$

42. $\log_8 \dfrac{y}{8}$

43. $\ln y^4$

44. $\ln z^9$

45. $\log \sqrt{5}$

46. $\log \sqrt[3]{7}$

In Exercises 47–58, assume that x, y, z, and b are positive numbers and $b \neq 1$. Use the properties of logarithms to write each expression in terms of the logarithms of x, y, and z.

47. $\log xyz$

48. $\log 4xz$

49. $\log_2 \dfrac{2x}{y}$

50. $\log_3 \dfrac{x}{yz}$

51. $\log x^3 y^2$

52. $\log xy^2 z^3$

53. $\log_b (xy)^{1/2}$

54. $\log_b x^3 y^{1/2}$

55. $\log_a \dfrac{\sqrt[3]{x}}{\sqrt[4]{yz}}$

56. $\log_b \sqrt[4]{\dfrac{x^3 y^2}{z^4}}$

57. $\ln x\sqrt{z}$ **58.** $\ln \sqrt{xy}$

In Exercises 59–66, assume that x, y, z, and b are positive numbers and $b \neq 1$. Use the properties of logarithms to write each expression as the logarithm of a single quantity.

59. $\log_2 (x + 1) - \log_2 x$

60. $\log_3 x + \log_3 (x + 2) - \log_3 8$

61. $2 \log x + \dfrac{1}{2} \log y$

62. $-2 \log x - 3 \log y + \log z$

63. $-3 \log_b x - 2 \log_b y + \dfrac{1}{2} \log_b z$

64. $3 \log_b (x + 1) - 2 \log_b (x + 2) + \log_b x$

65. $\ln \left(\dfrac{x}{z} + x \right) - \ln \left(\dfrac{y}{z} + y \right)$

66. $\ln (xy + y^2) - \ln (xz + yz) + \ln z$

In Exercises 67–72, tell whether the given statement is true. If a statement is false, explain why.

67. $\log xy = (\log x)(\log y)$

68. $\log ab = \log a + 1$

69. $\log_b (A - B) = \dfrac{\log_b A}{\log_b B}$

70. $\dfrac{\log_b A}{\log_b B} = \log_b A - \log_b B$

71. $\log_b \dfrac{A}{B} = \log_b A - \log_b B$

72. $\log_b 2 = \log_2 b$

In Exercises 73–80, assume that $\log_b 4 = 0.6021$, $\log_b 7 = 0.8451$, and $\log_b 9 = 0.9542$. Use these values and the properties of logarithms to find each value.

73. $\log_b 28$ **74.** $\log_b \dfrac{7}{4}$

75. $\log_b \dfrac{4}{63}$ **76.** $\log_b 36$

77. $\log_b \dfrac{63}{4}$ **78.** $\log_b 2.25$

79. $\log_b 64$ **80.** $\log_b 49$

In Exercises 81–88, use the change-of-base formula to find each logarithm to four decimal places.

81. $\log_3 7$ **82.** $\log_7 3$

83. $\log_{1/3} 3$ **84.** $\log_{1/2} 6$

85. $\log_3 8$ **86.** $\log_5 10$

87. $\log_{\sqrt{2}} \sqrt{5}$ **88.** $\log_\pi e$

APPLICATIONS

89. pH OF A SOLUTION Find the pH of a solution with a hydrogen ion concentration of 1.7×10^{-5} gram-ions per liter.

90. HYDROGEN ION CONCENTRATION Find the hydrogen ion concentration of a saturated solution of calcium hydroxide whose pH is 13.2.

91. AQUARIUM To test for safe pH levels in a freshwater aquarium, a test strip is compared with the scale shown in Illustration 3. Find the corresponding range in the hydrogen ion concentration.

ILLUSTRATION 3

AquaTest pH Kit

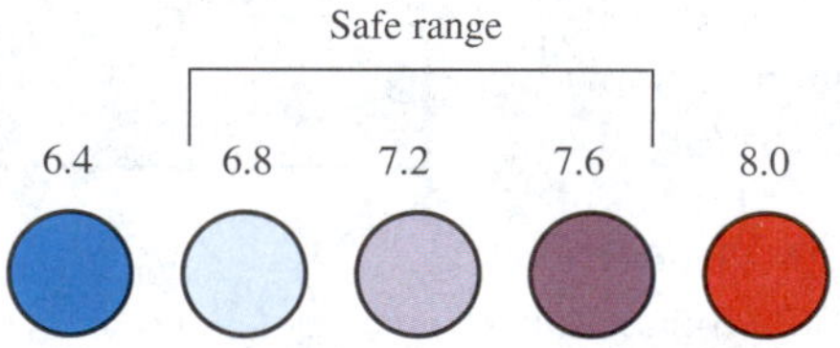

Safe range

6.4   6.8   7.2   7.6   8.0

92. pH OF SOUR PICKLES The hydrogen ion concentration of sour pickles is 6.31×10^{-4}. Find the pH.

WRITING

93. Explain the difference between a logarithm of a product and the product of logarithms.

94. How can the $\boxed{\text{LOG}}$ key on a calculator be used to find $\log_2 7$?

REVIEW

Consider the line that passes through $P(-2, 3)$ and $Q(4, -4)$.

95. Find the slope of line PQ.

96. Find the distance PQ.

97. Find the midpoint of segment PQ.

98. Write the equation of line PQ.

▶ 11.8   Exponential and Logarithmic Equations

In this section, you will learn about

Solving exponential equations ■ Solving logarithmic equations
■ Radioactive decay ■ Population growth

Introduction   An **exponential equation** is an equation that contains a variable in one of its exponents. Some examples of exponential equations are

$$3^x = 5, \qquad 6^{x-3} = 2^x, \qquad \text{and} \qquad 2^{x^2+2x} = \frac{1}{2}$$

A **logarithmic equation** is an equation with a logarithmic expression that contains a variable. Some examples of logarithmic equations are

$$\log 5x = 1, \qquad \log(3x + 2) - \log(2x - 3) = 0, \qquad \text{and} \qquad \frac{\log_2 (5x - 6)}{\log_2 x} = 2$$

In this section, we will learn how to solve many of these equations.

Solving Exponential Equations

EXAMPLE 1

Taking the logarithm of each side of an equation. Solve $3^x = 5$.

Solution   Since the logarithms of equal numbers are equal, we can take the common logarithm of each side of the equation. The power rule of logarithms then provides a way of moving the variable x from its position as an exponent to a position as a coefficient.

$$3^x = 5$$

$\log 3^x = \log 5$	Take the common logarithm of each side.
$x \log 3 = \log 5$	The log of a power is the power times the log: $\log 3^x = x \log 3$.

1. $\quad x = \dfrac{\log 5}{\log 3}$ Divide both sides by log 3.

$\qquad x \approx 1.464973521$ Use a calculator.

The solution is $\dfrac{\log 5}{\log 3}$. To four decimal places, $x = 1.4650$.

We can also take the natural logarithm of each side of the equation to solve for x.

$$3^x = 5$$

$\ln 3^x = \ln 5$	Take the natural logarithm of each side.
$x \ln 3 = \ln 5$	Use the power rule of logarithms: $\ln 3^x = x \ln 3$.
$x = \dfrac{\ln 5}{\ln 3}$	Divide both sides by ln 3.
$x \approx 1.464973521$	Use a calculator.

To check the solution, we substitute 1.4650 for x in 3^x and see if the result is close to 5. We can compute $3^{1.4650}$ by entering 3 $\boxed{y^x}$ 1.4650 $\boxed{=}$ on a scientific calculator. The result of 5.000145454 verifies that $x \approx 1.4650$ is an approximate solution of $3^x = 5$.

SELF CHECK   Solve $5^x = 4$. Give the answer to four decimal places.   *Answer:* 0.8614   ■

WARNING! A careless reading of Equation 1 leads to a common error. The right-hand side of Equation 1 calls for a division, not a subtraction.

$$\frac{\log 5}{\log 3} \quad \text{means} \quad (\log 5) \div (\log 3)$$

It is the expression $\log \frac{5}{3}$ that means $\log 5 - \log 3$.

EXAMPLE 2

Taking the logarithm of each side. Solve $6^{x-3} = 2^x$.

Solution

$$6^{x-3} = 2^x$$

$\log 6^{x-3} = \log 2^x$ Take the common logarithm of each side.

$(x - 3) \log 6 = x \log 2$ The log of a power is the power times the log.

$x \log 6 - 3 \log 6 = x \log 2$ Use the distributive property.

$x \log 6 - x \log 2 = 3 \log 6$ On both sides, add 3 log 6 and subtract x log 2.

$x (\log 6 - \log 2) = 3 \log 6$ Factor out x on the left-hand side.

$$x = \frac{3 \log 6}{\log 6 - \log 2} \quad \text{Divide both sides by log 6} - \log 2.$$

$x \approx 4.892789261$ Use a calculator.

SELF CHECK   Solve $5^{x-2} = 3^x$.   *Answer:* $\frac{2 \log 5}{\log 5 - \log 3} \approx 6.3013$ ■

EXAMPLE 3

Taking the natural logarithm of each side. Solve $e^{0.9t} = 8$.

Solution The exponential expression on the left-hand side has base e. In such cases, the computations are somewhat simpler if we take the natural logarithm of each side.

$$e^{0.9t} = 8$$

$\ln e^{0.9t} = \ln 8$ Take the natural logarithm of each side.

$0.9t \ln e = \ln 8$ Use the power rule of logarithms: $e^{0.9t} = 0.9t \ln e$.

$0.9t \cdot 1 = \ln 8$ Simplify: $\ln e = 1$.

$0.9t = \ln 8$

$$t = \frac{\ln 8}{0.9} \quad \text{Divide both sides by 0.9.}$$

$t \approx 2.310490602$ Use a calculator.

To four decimal places, $t \approx 2.3105$.

SELF CHECK   Solve $e^{2.1t} = 35$.   *Answer:* $\dfrac{\ln 35}{2.1} \approx 1.6930$ ■

EXAMPLE 4

Exponential expressions with the same base. Solve $2^{x^2+2x} = \frac{1}{2}$.

Solution Since $\frac{1}{2} = 2^{-1}$, we can write the equation in the form

$2^{x^2+2x} = 2^{-1}$ Each side of the equation can be written as an exponential expression with base 2.

Since equal quantities with equal bases have equal exponents, we have

$x^2 + 2x = -1$ Equate the exponents.

$x^2 + 2x + 1 = 0$ Add 1 to both sides.

$(x + 1)(x + 1) = 0$ Factor the trinomial.

$x + 1 = 0 \quad \text{or} \quad x + 1 = 0$ Set each factor equal to 0.

$x = -1 \qquad\qquad x = -1$

Verify that -1 satisfies the equation.

SELF CHECK Solve $3^{x^2-2x} = \frac{1}{3}$.   *Answer:* 1, 1   ■

ACCENT ON TECHNOLOGY *Solving Exponential Equations Graphically*

To use a graphing calculator to approximate the solutions of $2^{x^2+2x} = \frac{1}{2}$ (see Example 4), we can subtract $\frac{1}{2}$ from both sides of the equation to get

$$2^{x^2+2x} - \frac{1}{2} = 0$$

and graph the corresponding function

$$y = 2^{x^2+2x} - \frac{1}{2}$$

If we use window settings of $[-4, 4]$ for x and $[-2, 6]$ for y, we obtain the graph shown in Figure 11-35(a).

Since the solutions of the equation are its x-intercepts, we can approximate the solutions by zooming in on the values of the x-intercepts, as in Figure 11-35(b). Since $x = -1$ is the only x-intercept, -1 is the only solution. In this case, we have found an exact solution.

FIGURE 11-35

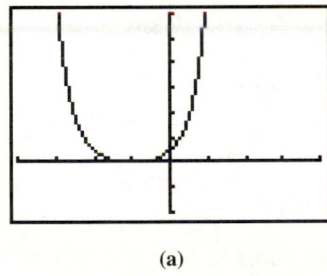   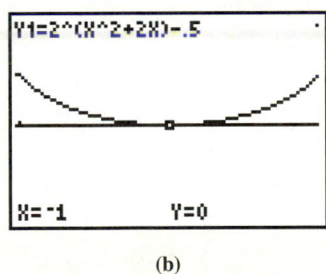   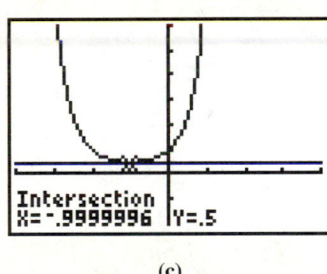

(a)   (b)   (c)

We can also solve $2^{x^2+2x} = \frac{1}{2}$ using the INTERSECT feature found on most graphing calculators. After graphing $Y_1 = 2^{x^2+2x}$ and $Y_2 = \frac{1}{2}$, we select INTERSECT, which approximates the coordinates of the point of intersection of the two graphs. From the display shown in Figure 11-35(c), we can conclude that the solution is $x = -1$. Verify this by checking.

Solving Logarithmic Equations

EXAMPLE 5   **Solving logarithmic equations.** Solve $\log 5x = 3$.

Solution   Recall that $\log 5x = \log_{10} 5x$. We can change the equation $\log 5x = 3$ into the equivalent base-10 exponential equation $10^3 = 5x$ and solve for x.

$$\log 5x = 3$$
$$10^3 = 5x$$
$$1,000 = 5x \quad \text{Simplify: } 10^3 = 1,000.$$
$$200 = x \quad \text{Divide both sides by 5.}$$

Check the result.

SELF CHECK   Solve $\log_2(x - 3) = -1$.   *Answer:* $\frac{7}{2}$ ■

EXAMPLE 6 **Solving logarithmic equations.** Solve $\log(3x + 2) - \log(2x - 3) = 0$.

Solution We isolate each logarithmic expression on one side of the equation.

$$\log(3x + 2) - \log(2x - 3) = 0$$
$$\log(3x + 2) = \log(2x - 3) \quad \text{Add } \log(2x - 3) \text{ to both sides.}$$

In the previous section, we saw that if the logarithms of two numbers are equal, the numbers are equal. So we have

$$(3x + 2) = (2x - 3) \quad \text{If } \log r = \log s, \text{ then } r = s.$$
$$x = -5 \quad \text{Subtract } 2x \text{ and } 2 \text{ from both sides.}$$

Check: $\log(3x + 2) - \log(2x - 3) = 0$
$$\log[3(-5) + 2] - \log[2(-5) - 3] \overset{?}{=} 0$$
$$\log(-13) - \log(-13) \overset{?}{=} 0$$

Since the logarithm of a negative number does not exist, the apparent solution of -5 must be discarded. This equation has no solutions.

SELF CHECK Solve $\log(5x + 2) - \log(7x - 2) = 0$. *Answer:* 2 ∎

EXAMPLE 7 **Using a property of logarithms to solve a logarithmic equation.** Solve $\log x + \log(x - 3) = 1$.

Solution $\log x + \log(x - 3) = 1$

$$\log x(x - 3) = 1 \quad \text{Use the product rule of logarithms.}$$
$$x(x - 3) = 10^1 \quad \text{Use the definition of logarithms to change the equation to exponential form:}$$
$$\log x(x - 3) = \log_{10} x(x - 3).$$
$$x^2 - 3x - 10 = 0 \quad \text{Remove parentheses and subtract 10 from both sides.}$$
$$(x + 2)(x - 5) = 0 \quad \text{Factor the trinomial.}$$
$$x + 2 = 0 \quad \text{or} \quad x - 5 = 0 \quad \text{Set each factor equal to 0.}$$
$$x = -2 \qquad \qquad x = 5$$

Check: The number -2 is not a solution, because it does not satisfy the equation (a negative number does not have a logarithm). We will check the remaining number, 5.

$$\log x + \log(x - 3) = 1$$
$$\log 5 + \log(5 - 3) \overset{?}{=} 1 \quad \text{Substitute 5 for } x.$$
$$\log 5 + \log 2 \overset{?}{=} 1$$
$$\log 10 \overset{?}{=} 1 \quad \text{Use the product rule of logarithms:}$$
$$\log 5 + \log 2 = \log(5 \cdot 2) = \log 10.$$
$$1 = 1 \quad \text{Simplify: } \log 10 = 1.$$

Since 5 satisfies the equation, it is a solution.

SELF CHECK Solve $\log x + \log(x + 3) = 1$. *Answer:* 2 ∎

WARNING! Examples 6 and 7 illustrate that we must check the solutions of a logarithmic equation.

ACCENT ON TECHNOLOGY *Solving Logarithmic Equations Graphically*

To use a graphing calculator to approximate the solutions of $\log x + \log (x - 3) = 1$ (see Example 7), we can subtract 1 from both sides of the equation to get

$$\log x + \log (x - 3) - 1 = 0$$

and graph the corresponding function

$$y = \log x + \log (x - 3) - 1$$

If we use window settings of $[0, 20]$ for x and $[-2, 2]$ for y, we obtain the graph shown in Figure 11-36(a). Since the solution of the equation is the x-intercept, we can find the solution by zooming in on the value of the x-intercept. The solution is $x = 5$.

FIGURE 11-36

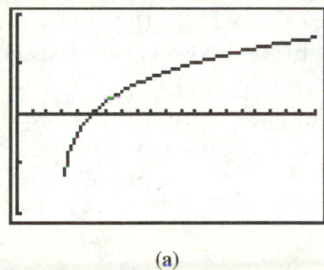
(a)

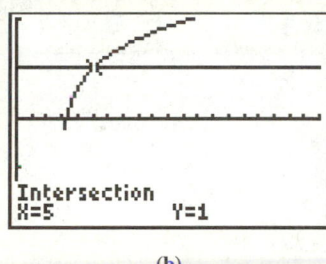
(b)

We can also solve $\log x + \log (x - 3) = 1$ using the INTERSECT feature. After graphing $Y_1 = \log x + \log (x - 3)$ and $Y_2 = 1$, we select INTERSECT, which approximates the coordinates of the point of intersection of the two graphs. From the display shown in Figure 11-36(b), we can conclude that the solution is $x = 5$. Verify this by checking.

EXAMPLE 8   **Solving logarithmic equations.** Solve $\dfrac{\log_2 (5x - 6)}{\log_2 x} = 2$.

Solution   We can multiply both sides of the equation by $\log_2 x$ to get

$$\log_2 (5x - 6) = 2 \log_2 x$$

and apply the power rule of logarithms to get

$$\log_2 (5x - 6) = \log_2 x^2$$

By property 8 of logarithms, $5x - 6 = x^2$, because they have equal logarithms. Thus,

$$5x - 6 = x^2$$
$$0 = x^2 - 5x + 6$$
$$0 = (x - 3)(x - 2)$$
$$x - 3 = 0 \quad \text{or} \quad x - 2 = 0$$
$$x = 3 \qquad\qquad x = 2$$

Verify that both 2 and 3 satisfy the equation.

SELF CHECK   Solve $\dfrac{\log_3 (5x + 6)}{\log_3 x} = 2$.

Answer: 6

Radioactive Decay

Experiments have determined the time it takes for half of a sample of a given radioactive material to decompose. This time is a constant, called the material's **half-life.**

　　When living organisms die, the oxygen/carbon dioxide cycle common to all living things ceases, and carbon-14, a radioactive isotope with a half-life of 5,700 years, is no longer absorbed. By measuring the amount of carbon-14 present in an ancient object, archaeologists can estimate the object's age by using the radioactive decay formula.

Radioactive decay formula

If A is the amount of radioactive material present at time t, A_0 was the amount present at $t = 0$, and h is the material's half-life, then

$$A = A_0 2^{-t/h}$$

E X A M P L E　9

Carbon-14 dating. How old is a wooden statue that retains only one-third of its original carbon-14 content?

Solution　To find the time t when $A = \frac{1}{3}A_0$, we substitute $\frac{A_0}{3}$ for A and 5,700 for h in the radioactive decay formula and solve for t:

$$A = A_0 2^{-t/h}$$

$$\frac{A_0}{3} = A_0 2^{-t/5,700} \qquad \text{The half-life of carbon-14 is 5,700 years.}$$

$$1 = 3(2^{-t/5,700}) \qquad \begin{array}{l}\text{Divide both sides by } A_0 \text{ and multiply both} \\ \text{sides by 3.}\end{array}$$

$$\log 1 = \log 3(2^{-t/5,700}) \qquad \text{Take the common logarithm of each side.}$$

$$0 = \log 3 + \log 2^{-t/5,700} \qquad \begin{array}{l}\log 1 = 0, \text{ and use the product rule of} \\ \text{logarithms.}\end{array}$$

$$-\log 3 = -\frac{t}{5,700}\log 2 \qquad \begin{array}{l}\text{Subtract } \log 3 \text{ from both sides and use the} \\ \text{power rule of logarithms.}\end{array}$$

$$5,700\left(\frac{\log 3}{\log 2}\right) = t \qquad \text{Multiply both sides by } -\frac{5,700}{\log 2}.$$

$$t \approx 9,034.286254 \qquad \text{Use a calculator.}$$

The statue is approximately 9,000 years old.

SELF CHECK　In Example 9, how old is a statue that retains 25% of its original carbon-14 content?　　　　*Answer:* about 11,400 years ■

Population Growth

Recall that when there is sufficient food and space, populations of living organisms tend to increase exponentially according to the Malthusian growth model.

Malthusian growth model

If P is the population at some time t, P_0 is the initial population at $t = 0$, and k depends on the rate of growth, then

$$P = P_0 e^{kt}$$

EXAMPLE 10

Population growth. The bacteria in a laboratory culture increased from an initial population of 500 to 1,500 in 3 hours. How long will it take for the population to reach 10,000?

Solution   We substitute 500 for P_0, 1,500 for P, and 3 for t and simplify to find k:

$$P = P_0 e^{kt}$$

$$1,500 = 500(e^{k3})$$ Substitute 1,500 for P, 500 for P_0, and 3 for t.

$$3 = e^{3k}$$ Divide both sides by 500.

$$3k = \ln 3$$ Change the equation from exponential to logarithmic form.

$$k = \frac{\ln 3}{3}$$ Divide both sides by 3.

To find when the population will reach 10,000, we substitute 10,000 for P, 500 for P_0, and $\frac{\ln 3}{3}$ for k in the equation $P = P_0 e^{kt}$ and solve for t:

$$P = P_0 e^{kt}$$

$$10,000 = 500 e^{[(\ln 3)/3]t}$$

$$20 = e^{[(\ln 3)/3]t}$$ Divide both sides by 500.

$$\left(\frac{\ln 3}{3}\right)t = \ln 20$$ Change the equation to logarithmic form.

$$t = \frac{3 \ln 20}{\ln 3}$$ Multiply both sides by $\frac{3}{\ln 3}$.

$$\approx 8.180499084$$ Use a calculator.

The culture will reach 10,000 bacteria in about 8 hours.

SELF CHECK   In Example 10, how long will it take the population to reach 20,000?   *Answer:* about 10 hours   ■

EXAMPLE 11

Generation time. If a medium is inoculated with a bacterial culture that contains 1,000 cells per milliliter, how many generations will pass by the time the culture has grown to a population of 1 million cells per milliliter?

Solution   During bacterial reproduction, the time required for a population to double is called the *generation time*. If b bacteria are introduced into a medium, then after the generation time of the organism has elapsed, there are $2b$ cells. After another generation, there are $2(2b)$ or $4b$ cells, and so on. After n generations, the number of cells present will be

1. $B = b \cdot 2^n$

To find the number of generations that have passed while the population grows from b bacteria to B bacteria, we solve Equation 1 for n. To do this, we can take the common logarithm or the natural logarithm of each side.

$$\ln B = \ln (b \cdot 2^n)$$ Take the natural logarithm of each side.

$$\ln B = \ln b + n \ln 2$$ Apply the product and power rules of logarithms.

$$\ln B - \ln b = n \ln 2$$ Subtract $\ln b$ from both sides.

$$n = \frac{1}{\ln 2}(\ln B - \ln b)$$ Multiply both sides by $\frac{1}{\ln 2}$.

2. $$n = \frac{1}{\ln 2}\left(\ln \frac{B}{b}\right)$$ Use the quotient rule of logarithms.

Equation 2 is a formula that gives the number of generations that will pass as the population grows from b bacteria to B bacteria.

To find the number of generations that have passed while a population of 1,000 cells per milliliter has grown to a population of 1 million cells per milliliter, we substitute 1,000 for b and 1,000,000 for B in Equation 2 and solve for n.

$$n = \frac{1}{\ln 2} \ln \frac{1,000,000}{1,000}$$

$$= \frac{1}{\ln 2} \ln 1,000 \qquad \text{Simplify.}$$

$$n \approx 9.965784285 \qquad \text{Use a calculator:}$$

2 [ln] [1/x] [×] 1000 [ln] [=].

Approximately 10 generations will have passed. ∎

STUDY SET

Section 11.8

VOCABULARY

In Exercises 1–2, fill in the blanks to make the statements true.

1. An equation with a variable in its exponent, such as $3^{2x} = 8$, is called a(n) _____ equation.

2. An equation with a logarithmic expression that contains a variable, such as $\log_5 (2x - 3) = \log_5 (x + 4)$, is a(n) _____ equation.

CONCEPTS

In Exercises 3–4, fill in the blanks to make the statements true.

3. The formula for radioactive decay is $A =$ [   ].

4. The formula for population growth is $P =$ [   ].

5. Use the graphs in Illustration 1 to estimate the solution of $2^x = 3^{-x+3}$.

ILLUSTRATION 1

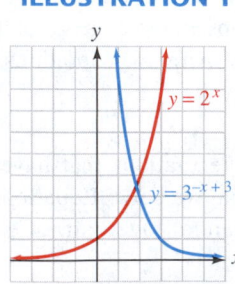

$y = 2^x$

$y = 3^{-x+3}$

6. Use the graphs in Illustration 2 to estimate the solution of $3 \log (x - 1) = 2 \log x$.

ILLUSTRATION 2

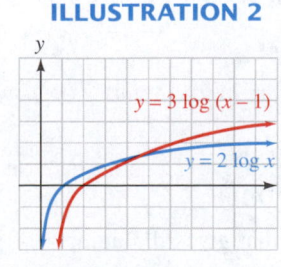

$y = 3 \log (x - 1)$

$y = 2 \log x$

7. Fill in the blanks to make the statements true. To solve $5^x = 21$, we can take the _____ of each side of the equation to get

$$\log 5^x = \log 21$$

The power rule for logarithms then provides a way of moving the variable x from its position as an _____ to a position as a coefficient.

8. Use a calculator to determine whether $x \approx 2.5646$ is a solution of $2^{2x + 1} = 70$.

9. Find $\dfrac{\log 8}{\log 5}$. Round to four decimal places.

10. Find $\dfrac{2 \ln 12}{\ln 9}$. Round to four decimal places.

11. Simplify $\ln e$.

12. Does $\dfrac{\log 7}{\log 3} = \log 7 - \log 3$?

13. Write the corresponding base-10 exponential equation for $\log (x + 1) = 2$.

14. Write the corresponding base-e exponential equation for $\ln (x + 1) = 2$.

15. Solve each equation. Round to the nearest hundredth.
  a. $x^2 = 12$ **b.** $2^x = 12$

16. Solve each equation.
  a. $\log (x - 1) = 3$ **b.** $\log (x - 1) = \log 3$

17. Perform a check to see whether $x = -4$ is a solution of $\log_5 (x + 3) = \frac{1}{5}$.

18. Perform a check to see whether $x = -2$ is a solution of $5^{2x+3} = \frac{1}{5}$.

NOTATION

In Exercises 19–20, complete each solution.

19. Solve $2^x = 7$.

$$2^x = 7$$
$$\boxed{} 2^x = \log 7$$
$$x \boxed{} = \log 7$$
$$x = \frac{\log 7}{\log 2}$$

20. Solve $\log_2 (2x - 3) = \log_2 (x + 4)$.

$$\log_2 (2x - 3) = \log_2 (x + 4)$$
$$\boxed{} = x + 4$$
$$x = 7$$

PRACTICE

In Exercises 21–44, solve each exponential equation. Give answers to four decimal places when necessary.

21. $4^x = 5$ **22.** $7^x = 12$

23. $13^{x-1} = 2$ **24.** $5^{x+1} = 3$

25. $2^{x+1} = 3^x$ **26.** $5^{x-3} = 3^{2x}$

27. $2^x = 3^x$ **28.** $3^{2x} = 4^x$

29. $7^{x^2} = 10$ **30.** $8^{x^2} = 11$

31. $8^{x^2} = 9^x$ **32.** $5^{x^2} = 2^{5x}$

33. $e^{3x} = 9$ **34.** $e^{4x} = 60$

35. $e^{-0.2t} = 14.2$ **36.** $e^{0.3t} = 9.1$

37. $2^{x-2} = 64$ **38.** $3^{-3x+1} = 243$

39. $5^{4x} = \dfrac{1}{125}$ **40.** $8^{-x+1} = \dfrac{1}{64}$

41. $2^{x^2-2x} = 8$ **42.** $3^{x^2-3x} = 81$

43. $3^{x^2+4x} = \dfrac{1}{81}$ **44.** $7^{x^2+3x} = \dfrac{1}{49}$

In Exercises 45–48, use a graphing calculator to solve each equation. Give all answers to the nearest tenth.

45. $2^{x+1} = 7$ **46.** $3^{x-1} = 2^x$

47. $4(2^{x^2}) = 8^{3x}$ **48.** $3^x - 10 = 3^{-x}$

In Exercises 49–84, solve each logarithmic equation.

49. $\log (x + 2) = 4$ **50.** $\log 5x = 4$

51. $\log (7 - x) = 2$ **52.** $\log (2 - x) = 3$

53. $\ln x = 1$ **54.** $\ln x = 5$

55. $\ln (x + 1) = 3$ **56.** $\ln 2x = 5$

57. $\log 2x = \log 4$ **58.** $\log 3x = \log 9$

59. $\ln (3x + 1) = \ln (x + 7)$

60. $\ln (x^2 + 4x) = \ln (x^2 + 16)$

61. $\log (3 - 2x) - \log (x + 24) = 0$

62. $\log (3x + 5) - \log (2x + 6) = 0$

63. $\log \dfrac{4x + 1}{2x + 9} = 0$ **64.** $\log \dfrac{2 - 5x}{2(x + 8)} = 0$

65. $\log x^2 = 2$ **66.** $\log x^3 = 3$

67. $\log x + \log (x - 48) = 2$

68. $\log x + \log (x + 9) = 1$

69. $\log x + \log (x - 15) = 2$

70. $\log x + \log (x + 21) = 2$

71. $\log (x + 90) = 3 - \log x$

72. $\log (x - 90) = 3 - \log x$

73. $\log (x - 6) - \log (x - 2) = \log \dfrac{5}{x}$

74. $\log (3 - 2x) - \log (x + 9) = 0$

75. $\dfrac{\log (3x - 4)}{\log x} = 2$ **76.** $\dfrac{\log (8x - 7)}{\log x} = 2$

77. $\dfrac{\log (5x + 6)}{2} = \log x$ **78.** $\dfrac{1}{2} \log (4x + 5) = \log x$

79. $\log_3 x = \log_3 \left(\dfrac{1}{x}\right) + 4$

80. $\log_5 (7 + x) + \log_5 (8 - x) - \log_5 2 = 2$

81. $2 \log_2 x = 3 + \log_2 (x - 2)$

82. $2 \log_3 x - \log_3 (x - 4) = 2 + \log_3 2$

83. $\log (7y + 1) = 2 \log (y + 3) - \log 2$

84. $2 \log (y + 2) = \log (y + 2) - \log 12$

In Exercises 85–88, use a graphing calculator to solve each equation. If an answer is not exact, round to the nearest tenth.

85. $\log x + \log (x - 15) = 2$

86. $\log x + \log (x + 3) = 1$

87. $\ln (2x + 5) - \ln 3 = \ln (x - 1)$

88. $2 \log (x^2 + 4x) = 1$

APPLICATIONS

89. TRITIUM DECAY The half-life of tritium is 12.4 years. How long will it take for 25% of a sample of tritium to decompose?

90. RADIOACTIVE DECAY In two years, 20% of a radioactive element decays. Find its half-life.

91. THORIUM DECAY An isotope of thorium, ^{227}Th, has a half-life of 18.4 days. How long will it take for 80% of the sample to decompose?

92. LEAD DECAY An isotope of lead, ^{201}Pb, has a half-life of 8.4 hours. How many hours ago was there 30% more of the substance?

93. CARBON-14 DATING A bone fragment analyzed by archaeologists contains 60% of the carbon-14 that it is assumed to have had initially. How old is it?

94. CARBON-14 DATING Only 10% of the carbon-14 in a small wooden bowl remains. How old is the bowl?

95. COMPOUND INTEREST If $500 is deposited in an account paying 8.5% annual interest, compounded semiannually, how long will it take for the account to increase to $800?

96. CONTINUOUS COMPOUND INTEREST In Exercise 95, how long will it take if the interest is compounded continuously?

97. COMPOUND INTEREST If $1,300 is deposited in a savings account paying 9% interest, compounded quarterly, how long will it take the account to increase to $2,100?

98. COMPOUND INTEREST A sum of $5,000 deposited in an account grows to $7,000 in 5 years. Assuming annual compounding, what interest rate is being paid?

99. RULE OF SEVENTY A rule of thumb for finding how long it takes an investment to double is called the **rule of seventy.** To apply the rule, divide 70 by the interest rate written as a percent. At 5%, doubling requires $\frac{70}{5} = 14$ years to double an investment. At 7%, it takes $\frac{70}{7} = 10$ years. Explain why this formula works.

100. BACTERIAL GROWTH A bacterial culture grows according to the formula

$$P = P_0 a^t$$

If it takes 5 days for the culture to triple in size, how long will it take to double in size?

101. RODENT CONTROL The rodent population in a city is currently estimated at 30,000. If it is expected to double every 5 years, when will the population reach 1 million?

102. POPULATION GROWTH The population of a city is expected to triple every 15 years. When can the city planners expect the present population of 140 persons to double?

103. BACTERIAL CULTURE A bacterial culture doubles in size every 24 hours. By how much will it have increased in 36 hours?

104. OCEANOGRAPHY The intensity I of a light a distance x meters beneath the surface of a lake decreases exponentially. From Illustration 3, find the depth at which the intensity will be 20%.

105. MEDICINE If a medium is inoculated with a bacterial culture containing 500 cells per milliliter, how many generations will have passed by the time the culture contains 5×10^6 cells per milliliter?

ILLUSTRATION 3

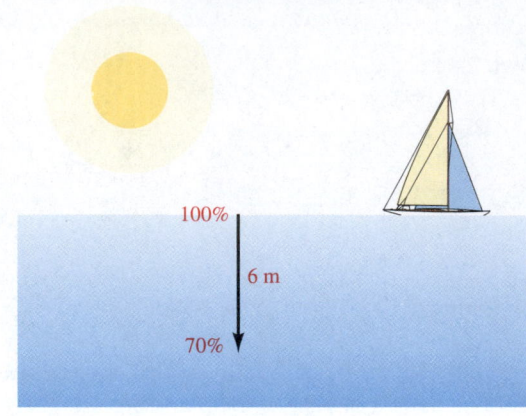

100%

6 m

70%

106. MEDICINE If a medium is inoculated with a bacterial culture containing 800 cells per milliliter, how many generations will have passed by the time the culture contains 6×10^7 cells per milliliter?

107. NEWTON'S LAW OF COOLING Water initially at 100°C is left to cool in a room at temperature 60°C. After 3 minutes, the water temperature is 90°. The water temperature T is a function of time t given by

$$T = 60 + 40e^{kt}$$

Find k.

108. NEWTON'S LAW OF COOLING Refer to Exercise 107 and find the time for the water temperature to reach 70°C.

WRITING

109. Explain how to solve the equation $2^{x+1} = 31$.

110. Explain how to solve the equation $2^{x+1} = 32$.

REVIEW

Solve each equation.

111. $5x^2 - 25x = 0$　　　**112.** $4y^2 - 25 = 0$

113. $3p^2 + 10p = 8$　　　**114.** $4t^2 + 1 = -6t$

115. In Illustration 4, find the length of leg AC.

ILLUSTRATION 4

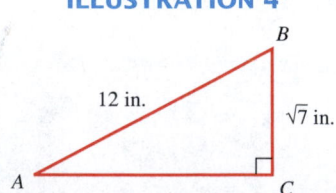

B

12 in.

$\sqrt{7}$ in.

A　　　C

116. The amount of medicine a patient should take is often proportional to his or her weight. If a patient weighing 83 kilograms needs 150 milligrams of medicine, how much will be needed by a person weighing 99.6 kilograms?

Inverse Functions

One-to-One Functions

A function is *one-to-one* if each input value x in the domain determines a different output value y in the range. In Exercises 1–3, determine whether the function is one-to-one.

1. $f(x) = x^2$

2. $f(x) = |x|$

3.

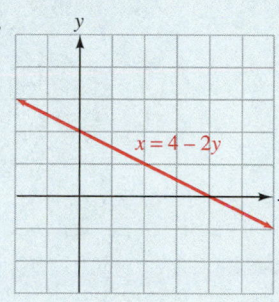

The Inverse of a Function

If a function is one-to-one, its inverse is a function. To find the inverse of a function, interchange x and y and solve for y.

4. Find $f^{-1}(x)$ if $f(x) = -2x - 1$.

5. Given the table of values for a one-to-one function f, complete the table of values for f^{-1}.

x	-2	1	3
$f(x)$	4	-2	-6

x	4	-2	-6
$f^{-1}(x)$			

Exponential and Logarithmic Functions

The exponential function $y = b^x$ and $y = \log_b x$ (where $b > 0$ and $b \neq 1$) are inverse functions.

6. Write the exponential statement $10^3 = 1,000$ in logarithmic form.

7. Write the logarithmic statement $\log_2 \frac{1}{8} = -3$ in exponential form.

8. If $\log_4 y = \frac{1}{2}$, what is y?

9. If $\log_y \frac{9}{4} = 2$, what is y?

The Natural Exponential and Natural Logarithmic Functions

A special exponential function that is used in many real-life applications involving growth and decay is the base-e exponential function, $f(x) = e^x$. Its inverse is the natural logarithm function $f(x) = \ln x$.

10. What is an approximate value of e?

11. What is the base of the logarithmic function $f(x) = \ln x$?

12. Use a calculator to find x: $\ln x = -0.28$.

13. Graph $f(x) = e^x$ and $f(x) = \ln x$ on the coordinate system in Illustration 1. Label the axis of symmetry.

ILLUSTRATION 1

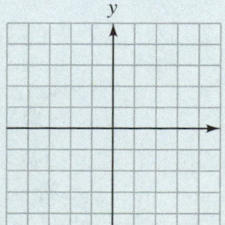

Accent on Teamwork

Section 11.1

Composition of functions Consider the functions $f(x) = x^2$, $g(x) = 2x + 1$, and $h(x) = 1 - x$. Determine whether the composition of functions f, g, and h is associative. That is, is the following true?

$$[f \circ (g \circ h)](x) \overset{?}{=} [(f \circ g) \circ h](x)$$

Section 11.2

One-to-one functions In newspapers, magazines, or books, find line graphs that are graphs of one-to-one functions and some that are not one-to-one. In each case, explain to the other members of your group what relationship the graph illustrates. Then tell whether the graph passes or fails the horizontal line test.

Section 11.3

Exponential functions On a piece of poster board, draw a rectangular coordinate system made up of a grid of 1-inch squares. Then graph each of the following exponential functions. How are the graphs similar? How are they different? For positive values of x, as the base increases, what happens to the steepness of the graph?

$$f(x) = 2^x \qquad f(x) = 3^x \qquad f(x) = 4^x$$
$$f(x) = 5^x \qquad f(x) = 6^x \qquad f(x) = 7^x$$

Section 11.4

Growth Make two copies of each of the graph shapes shown in Illustration 1.

ILLUSTRATION 1

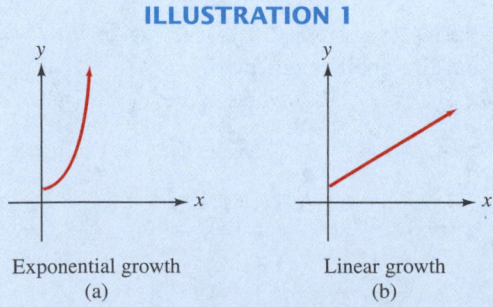

Exponential growth	Linear growth
(a)	(b)

Think of two examples of quantities that, in general, have grown exponentially over the years. Label the horizontal and vertical axes of the graphs with the proper titles.

(You do not have to scale the axes in terms of units.) Do the same for two quantities that, in general, have grown linearly over the years. Share your observations with the other members of your group. See if they agree with your models.

Section 11.5

Logarithmic functions On a piece of poster board, draw a rectangular coordinate system made up of a grid of 1-inch squares. Then graph each of the following logarithmic functions. How are the graphs similar? How are they different? For positive values of x, as the base increases, what happens to the steepness of the graph?

$$f(x) = \log_2 x \qquad f(x) = \log_3 x \qquad f(x) = \log_4 x$$
$$f(x) = \log_5 x \qquad f(x) = \log_6 x \qquad f(x) = \log_7 x$$

Section 11.6

The number e The value of e can be calculated to any degree of accuracy by adding the terms of the following pattern:

$$1, 1, \frac{1}{2}, \frac{1}{2 \cdot 3}, \frac{1}{2 \cdot 3 \cdot 4}, \frac{1}{2 \cdot 3 \cdot 4 \cdot 5}, \dots$$

The more terms that are added, the closer the sum will be to e. Write each of the first six terms in simplified form and then add them. To how many decimal places is the sum accurate?

Section 11.7

The properties of logarithms discussed in Section 11.7 hold for logarithms of any base. Use these properties to simplify each of the following natural logarithmic expressions.

a. $\ln x^y + \ln x^z$ **b.** $\ln x^y - \ln x^z$

c. $\ln e^{10}$ **d.** $\ln \dfrac{x}{y} + \ln \dfrac{y}{x}$

Section 11.8

To solve exponential equations involving e, it is convenient to take the natural logarithm (instead of the common logarithm) of both sides of the equation. Use this technique to solve each of the following equations. Give all answers to four decimal places.

a. $e^x = 10$ **b.** $e^{2x} = 15$

c. $e^{x+1} = 50.5$ **d.** $3e^{5x-2} = 3$

Section 11.1 — Algebra and Composition of Functions

CONCEPTS

Functions can be added, subtracted, multiplied, and divided:

$$(f + g)(x) = f(x) + g(x)$$
$$(f - g)(x) = f(x) - g(x)$$
$$(f \cdot g)(x) = f(x)g(x)$$
$$(f/g)(x) = \frac{f(x)}{g(x)} \quad (g(x) \neq 0)$$

Composition of functions:

$$(f \circ g)(x) = f(g(x))$$

REVIEW EXERCISES

1. Let $f(x) = 2x$ and $g(x) = x + 1$. Find each function or value.
   a. $f + g$
   b. $f - g$
   c. $f \cdot g$
   d. f/g
   e. $(f \circ g)(1)$
   f. $g(f(1))$
   g. $(f \circ g)(x)$
   h. $(g \circ f)(x)$

Section 11.2 — Inverses of Functions

A function is *one-to-one* if each input value x in the domain determines a different output value y in the range.

Horizontal line test: If every horizontal line that intersects the graph of a function does so only once, the function is one-to-one.

2. Determine whether the function is one-to-one.
   a. $f(x) = x^2 + 3$
   b. $f(x) = x + 3$

3. Use the horizontal line test to decide whether the function is one-to-one.
   a.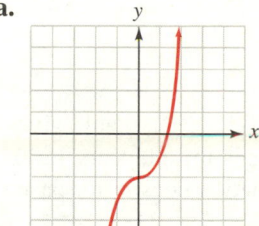
   b.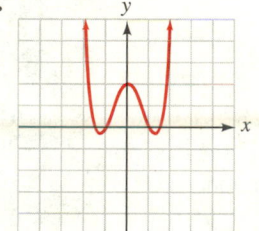

The graph of a function and its inverse are symmetric about the line $y = x$.

4. Given the graph of function f shown in Illustration 1, graph f^{-1} on the same coordinate axes. Label the axis of symmetry.

ILLUSTRATION 1

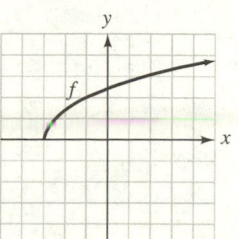

To *find the inverse of a function,* interchange the variables x and y and solve for y.

5. Find the inverse of each function.
   a. $f(x) = 6x - 3$
   b. $f(x) = 4x + 5$
   c. $f(x) = x^3$
   d. $y = 2x^2 - 1 \ (x \geq 0)$

Section 11.3

An *exponential function* with base b is defined by the equation

$$f(x) = b^x \quad (b > 0, \ b \neq 1)$$

If $b > 1$, then $f(x) = b^x$ is an *increasing function*.

If $0 < b < 1$, then $f(x) = b^x$ is a *decreasing function*.

Exponential Functions

6. Use properties of exponents to simplify each expression.

a. $5^{\sqrt{2}} \cdot 5^{\sqrt{2}}$

b. $\left(2^{\sqrt{5}}\right)^{\sqrt{2}}$

7. Graph each function. Then give the domain and the range.

a. $y = 3^x$

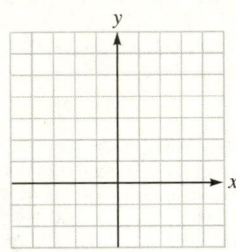

b. $f(x) = \left(\dfrac{1}{3}\right)^x$

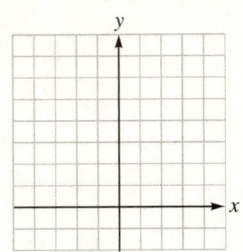

8. Graph each function by using a translation.

a. $f(x) = \left(\dfrac{1}{2}\right)^x - 2$

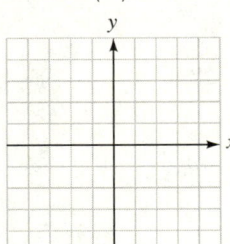

b. $g(x) = \left(\dfrac{1}{2}\right)^{x+2}$

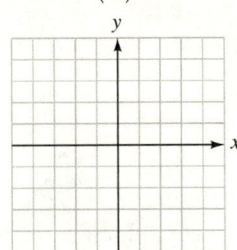

Exponential functions are suitable models for describing many situations involving the *growth* or *decay* of a quantity.

9. COAL PRODUCTION The table gives the number of tons of coal produced in the United States for the years 1800–1920. Graph the data in Illustration 2. What type of function does it appear could be used to model coal production over this period?

ILLUSTRATION 2

Year	Tons	Year	Tons
1800	108,000	1870	40,429,000
1810	178,000	1880	79,407,000
1820	881,000	1890	157,771,000
1830	1,334,000	1900	269,684,000
1840	2,474,000	1910	501,596,000
1850	8,356,000	1920	658,265,000
1860	20,041,000		

Based on information from *World Book Encyclopedia*

(continued)

ILLUSTRATION 2 *(continued)*

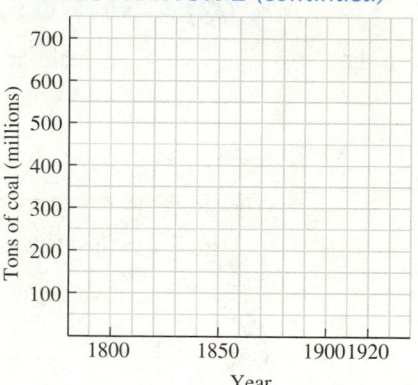

Compound interest: If $P is the deposit, and interest is paid k times a year at an annual rate r, the amount A in the account after t years is given by

$$A = P\left(1 + \frac{r}{k}\right)^{kt}$$

10. COMPOUND INTEREST How much will $10,500 become if it earns 9% annual interest, compounded quarterly, for 60 years?

Section 11.4   Base-*e* Exponential Functions

The function defined by $f(x) = e^x$ is the *natural exponential function* where $e = 2.718281828459 \ldots$.

If a quantity increases or decreases at an annual rate r, *compounded continuously,* then the amount A after t years is given by

$$A = Pe^{rt}$$

Malthusian population growth is modeled by the formula

$$P = P_0 e^{kt}$$

11. MORTGAGE RATES The average annual interest rate on a 30-year fixed-rate home mortgage for the years 1980–1996 can be approximated by the function $r(t) = 13.9e^{-0.035t}$, where t is the number of years since 1980. To the nearest hundredth of a percent, what does this model predict was the 30-year fixed rate in 1995?

12. INTEREST COMPOUNDED CONTINUOUSLY If $10,500 accumulates interest at an annual rate of 9%, compounded continuously, how much will be in the account in 60 years?

13. Graph each function. Give the domain and the range.
   a. $f(x) = e^x + 1$ **b.** $f(x) = e^{x-3}$

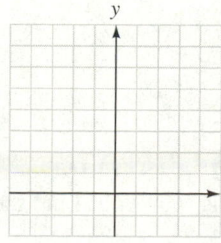 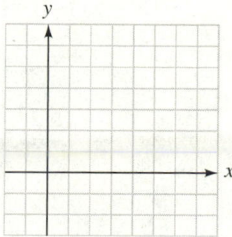

14. POPULATION GROWTH The population of the United States is approximately 275,000,000 people. Find the population in 50 years if $k = 0.015$.

Section 11.5

Logarithmic Functions

If $b > 0$ and $b \neq 1$, then the *logarithmic function with base b* is defined by
$$y = \log_b x.$$

$y = \log_b x$ means $x = b^y$.

$\log_b x$ is the exponent to which b is raised to get x.

$$b^{\log_b x} = x$$

15. Give the domain and range of the logarithmic function $f(x) = \log x$.

16. a. Write the statement $\log_4 64 = 3$ in exponential form.

    b. Write the statement $7^{-1} = \frac{1}{7}$ in logarithmic form.

17. Find each value, if possible.

    a. $\log_3 9$ **b.** $\log_9 \dfrac{1}{81}$ **c.** $\log_8 1$

    d. $\log_5 (-25)$ **e.** $\log_6 \sqrt{6}$ **f.** $\log 1{,}000$

18. Find the value of x in each equation.

    a. $\log_2 x = 5$ **b.** $\log_3 x = -4$ **c.** $\log_x 16 = 2$

    d. $\log_x \dfrac{1}{100} = -2$ **e.** $\log_9 3 = x$ **f.** $\log_{27} x = \dfrac{2}{3}$

19. Use a calculator to find the value of x to four decimal places.

    a. $\log 4.51 = x$ **b.** $\log x = 1.43$

If $b > 1$, then $f(x) = \log_b x$ is an *increasing function*. If $0 < b < 1$, then $f(x) = \log_b x$ is a *decreasing function*.

The exponential function $f(x) = b^x$ and the logarithmic function $f(x) = \log_b x$ are inverses of each other.

20. Graph each pair of equations on one set of coordinate axes.

    a. $y = 4^x$ and $y = \log_4 x$ **b.** $f(x) = \left(\dfrac{1}{3}\right)^x$ and $g(x) = \log_{1/3} x$

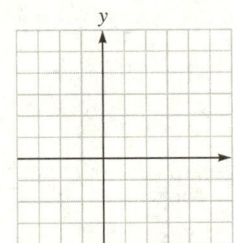 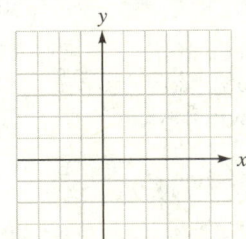

21. Graph each function.

    a. $f(x) = \log (x - 2)$ **b.** $f(x) = 3 + \log x$

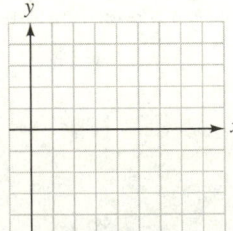 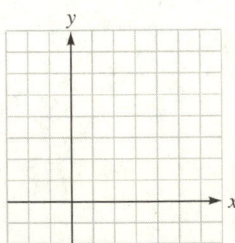

Decibel voltage gain:

$$\text{db gain} = 20 \log \frac{E_o}{E_I}$$

22. ELECTRICAL ENGINEERING An amplifier has an output of 18 volts when the input is 0.04 volt. Find the db gain.

The Richter scale:

$$R = \log \frac{A}{P}$$

23. EARTHQUAKE An earthquake had a period of 0.3 second and an amplitude of 7,500 micrometers. Find its measure on the Richter scale.

Section 11.6

Natural logarithms:

$\ln x$ means $\log_e x$

Population doubling time:

$$t = \frac{\ln 2}{r}$$

Base-*e* Logarithms

24. Use a calculator to find each value to four decimal places.
  a. $\ln 452$ **b.** $\ln 0.85$

25. Use a calculator to find the value of x to four decimal places.
  a. $\ln x = 2.336$ **b.** $\ln x = -8.8$

26. What function is the inverse of $f(x) = \ln x$?

27. Graph each function.
  a. $f(x) = 1 + \ln x$ **b.** $y = \ln(x + 1)$

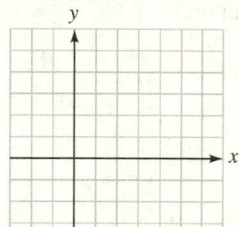   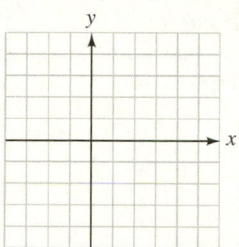

28. POPULATION GROWTH How long will it take the population of the United States to double if the growth rate is 3% per year?

Section 11.7

Properties of logarithms: If b is a positive number and $b \neq 1$,

1. $\log_b 1 = 0$ **2.** $\log_b b = 1$

3. $\log_b b^x = x$ **4.** $b^{\log_b x} = x$

5. $\log_b MN = \log_b M + \log_b N$

6. $\log_b \dfrac{M}{N} = \log_b M - \log_b N$

7. $\log_b M^p = p \log_b M$

8. If $\log_b x = \log_b y$, then $x = y$.

Change-of-base formula:

$$\log_b x = \frac{\log_a x}{\log_a b}$$

pH scale:

$$pH = -\log [H^+]$$

Properties of Logarithms

29. Simplify each expression.
  a. $\log_7 1$ **b.** $\log_7 7$
  c. $\log_7 7^3$ **d.** $7^{\log_7 4}$
  e. $\ln e^4$ **f.** $\ln e$

30. Use the properties of logarithms to rewrite each expression.
  a. $\log_3 27x$ **b.** $\log \dfrac{100}{x}$
  c. $\log_5 \sqrt{27}$ **d.** $\log 10ab$

31. Write each expression in terms of the logarithms of x, y, and z.
  a. $\log_b \dfrac{x^2 y^3}{z}$ **b.** $\ln \sqrt{\dfrac{x}{yz^2}}$

32. Write each expression as the logarithm of one quantity.
  a. $3 \log_2 x - 5 \log_2 y + 7 \log_2 z$
  b. $-3 \log_b y - 7 \log_b z + \dfrac{1}{2} \log_b (x + 2)$

33. Assume that $\log_b 5 = 1.1609$ and $\log_b 8 = 1.5000$. Use these values and the properties of logarithms to find each value to four decimal places.
  a. $\log_b 40$ **b.** $\log_b 64$

34. Find $\log_5 17$ to four decimal places.

35. pH OF GRAPEFRUIT The pH of grapefruit juice is about 3.1 Find its hydrogen ion concentration.

Section 11.8

Exponential and Logarithmic Equations

An *exponential equation* is an equation that contains a variable in one of its exponents.

36. Solve each equation for x. Give answers to four decimal places when necessary.

   a. $3^x = 7$

   b. $5^{x+2} = 625$

   c. $2^x = 3^{x-1}$

   d. $2^{x^2+4x} = \dfrac{1}{8}$

   e. $e^x = 7$

   f. $e^{-0.4t} = 25$

A *logarithmic equation* is an equation with a logarithmic expression that contains a variable.

37. Solve each equation for x.

   a. $\log (x - 4) = 2$

   b. $\ln (2x - 3) = \ln 15$

   c. $\log x + \log (29 - x) = 2$

   d. $\log_2 x + \log_2 (x - 2) = 3$

   e. $\dfrac{\log (7x - 12)}{\log x} = 2$

   f. $\log_2 (x + 2) + \log_2 (x - 1) = 2$

   g. $\log x + \log (x - 5) = \log 6$

   h. $\log 3 - \log (x - 1) = -1$

Carbon dating:

$$A = A_0 2^{-t/h}$$

38. CARBON-14 DATING A wooden statue found in Egypt has a carbon-14 content that is two-thirds of that found in living wood. If the half-life of carbon-14 is 5,700 years, how old is the statue?

39. The approximate coordinates of the points of intersection of the graphs of $f(x) = \log x$ and $g(x) = 1 - \log (7 - x)$ are shown in parts a and b of Illustration 3. Estimate the solutions of the logarithmic equation $\log x = 1 - \log (7 - x)$. Then check your answers.

ILLUSTRATION 3

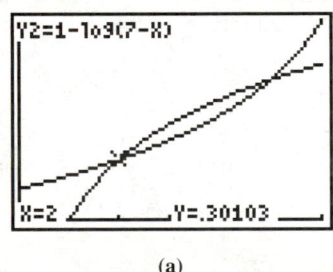
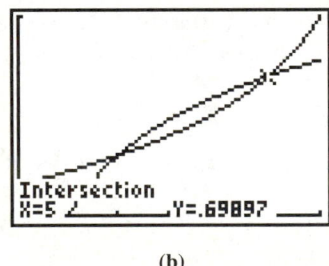

(a)   (b)

CHAPTER 11

Test

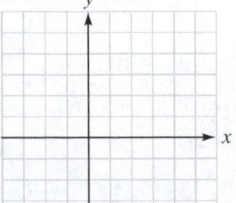

In Problems 1–4, $f(x) = 4x$ and $g(x) = x - 1$. Find each function or value.

1. $g + f$

2. $g \cdot f$

3. Find $(g \circ f)(1)$.

4. Find $f(g(x))$.

In Problems 5–6, find the inverse of each function.

5. $3x + 2y = 12$

6. $f(x) = 3x^2 + 4 \quad (x \geq 0)$

In Problems 7–8, graph each function.

7. $f(x) = 2^x + 1$

8. $f(x) = 2^{-x}$

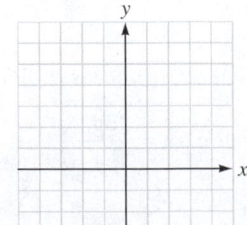

9. RADIOACTIVE DECAY A radioactive material decays according to the formula $A = A_0(2)^{-t}$. How much of a 3-gram sample will be left in 6 years?

10. COMPOUND INTEREST An initial deposit of $1,000 earns 6% interest, compounded twice a year. How much will be in the account in one year?

11. Graph the function $f(x) = e^x$.

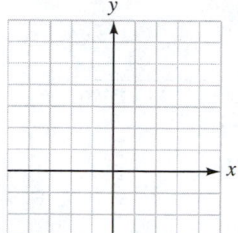

12. CONTINUOUS COMPOUNDING An account contains $2,000 and has been earning 8% interest, compounded continuously. How much will be in the account in 10 years?

In Problems 13–16, find x.

13. $\log_4 16 = x$

14. $\log_x 81 = 4$

15. $\log_3 x = -3$

16. $\ln x = 1$

17. Write the statement $\log_6 \frac{1}{36} = -2$ in exponential form.

18. Give the domain and range of the function $f(x) = \log x$.

In Problems 19–20, graph each function.

19. $f(x) = -\log_3 x$

20. $f(x) = \ln x$

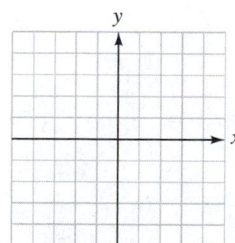 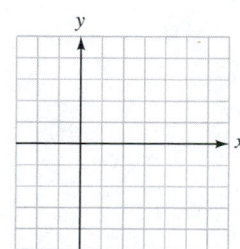

21. Write the expression $\log a^2bc^3$ in terms of the logarithms of a, b, and c.

22. Write the expression $\frac{1}{2} \ln (a + 2) + \ln b - 3 \ln c$ as a logarithm of a single quantity.

23. Use the change-of-base formula to find $\log_7 3$ to four decimal places.

24. What function is the inverse of $y = 10^x$?

25. pH Find the pH of a solution with a hydrogen ion concentration of 3.7×10^{-7}. (*Hint*: pH $= -\log [H^+]$.)

26. ELECTRONICS Find the db gain of an amplifier when $E_O = 60$ volts and $E_I = 0.3$ volt. (*Hint*: db gain $= 20 \log \left(\dfrac{E_O}{E_I} \right)$.)

In Problems 27–30, solve each equation. Round to four decimal places when necessary.

27. $5^x = 3$

28. $3^{x-1} = 27$

29. $\ln (5x + 2) = \ln (2x + 5)$

30. $\log x + \log (x - 9) = 1$

31. Illustration 1 shows the graphs of $y = \frac{1}{2} \ln (x - 1)$ and $y = \ln 2$ and the approximate coordinates of their point of intersection. Estimate the solution of the logarithmic equation $\frac{1}{2} \ln (x - 1) = \ln 2$.

ILLUSTRATION 1

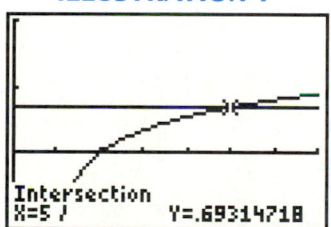

32. Show a check of your answer to Problem 31.

33. Give an example of a situation studied in this chapter that is modeled by a function with a graph that has the shape shown in Illustration 2. Label the axes. You do not have to scale the axes.

ILLUSTRATION 2

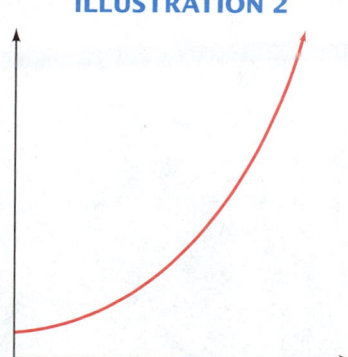

34. Consider the graph shown in Illustration 3.
   a. Is it the graph of a function?
   b. Is its inverse a function?
   c. What is $f^{-1}(260)$? What information does it give?

ILLUSTRATION 3

Relationship between car speed and tire temperature

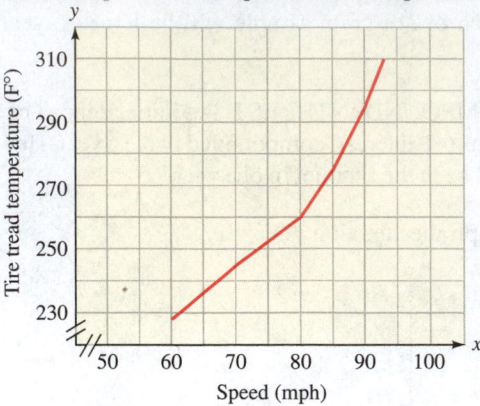

12

More on Systems of Equations

CAMPUS CONNECTION

The Cosmetology Department

Cosmetology students learn the techniques of cutting hair, facials, manicuring, and permanents. They also study business practices used in the operation of a beauty shop. In this chapter, we will discuss a mathematical procedure that a shop owner can use to determine how much business he or she must have to make a profit. To answer such a question, we must write two equations, each containing two variables. We call the pair of equations a *system of linear equations*. After learning how to solve systems of linear equations, you will be able to solve many problems that could not be solved using just one variable.

> *To solve many problems, we must use two and sometimes three variables. This requires that we solve a system of equations.*

▶ 12.1   Solving Systems with Two Variables

In this section, you will review

The graphing method ■ The substitution method ■ The addition method ■ Problem solving ■ Systems of inequalities

Introduction   In Chapter 7, we solved systems of two linear equations in two variables by graphing, by substitution, and by addition. We also solved systems of two linear inequalities in two variables by graphing. In this chapter, we will review these methods and then discuss how to solve systems of three equations in three variables. We will also discuss how to solve linear systems using matrices and determinants.

The Graphing Method

Recall that we follow these steps to solve a system of two equations in two variables by graphing.

Solving systems of equations by graphing

1. On a single set of coordinate axes, graph each equation.
2. Find the coordinates of the point (or points) where the graphs intersect. These coordinates give the solution of the system.
3. If the graphs have no point in common, the system has no solution.
4. Check the solution in both of the original equations.

EXAMPLE 1   **The graphing method.** Solve the system $\begin{cases} x + 2y = 4 \\ 2x - y = 3 \end{cases}$.

Solution   We graph both equations on one set of coordinate axes, as shown in Figure 12-1 on the next page. Although infinitely many ordered pairs (x, y) satisfy $x + 2y = 4$, and infinitely many ordered pairs (x, y) satisfy $2x - y = 3$, only the coordinates of the point where the graphs intersect satisfy both equations. Since the intersection point has coordinates of $(2, 1)$, the solution is the ordered pair $(2, 1)$, or $x = 2$ and $y = 1$.

To check the solution, we substitute 2 for x and 1 for y in both equations and verify that $(2, 1)$ satisfies each one.

SELF CHECK   Solve $\begin{cases} 2x + y = 0 \\ x - 2y = 5 \end{cases}$.

Answer:

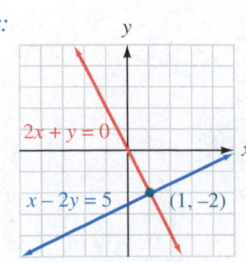

■

FIGURE 12-1

x + 2y = 4   **2x − y = 3**

x	y	(x, y)
4	0	(4, 0)
0	2	(0, 2)
−2	3	(−2, 3)

x	y	(x, y)
$\frac{3}{2}$	0	$\left(\frac{3}{2}, 0\right)$
0	−3	(0, −3)
−1	−5	(−1, −5)

Use the intercept method to graph each line.

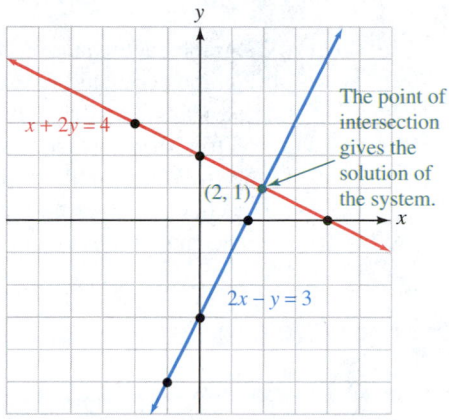

The point of intersection gives the solution of the system.

(2, 1)

x + 2y = 4

2x − y = 3

When a system of equations (as in Example 1) has a solution, it is a **consistent system of equations.** A system with no solutions is an **inconsistent system**.

EXAMPLE 2

The graphing method. Solve the system $\begin{cases} 2x + 3y = 6 \\ 4x + 6y = 24 \end{cases}$, if possible.

Solution   We graph both equations on one set of coordinate axes, as shown in Figure 12-2. Since the graphs are parallel lines, the lines do not intersect, and the system does not have a solution. It is an inconsistent system.

FIGURE 12-2

2x + 3y = 6   **4x + 6y = 24**

x	y	(x, y)
3	0	(3, 0)
0	2	(0, 2)
−3	4	(−3, 4)

x	y	(x, y)
6	0	(6, 0)
0	4	(0, 4)
−3	6	(−3, 6)

Use the intercept method to graph each line.

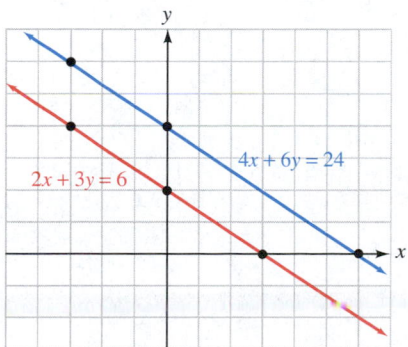

4x + 6y = 24

2x + 3y = 6

SELF CHECK   Solve $\begin{cases} 2x - 5y = 15 \\ y = \frac{2}{5}x - 2 \end{cases}$.

Answer: no solutions

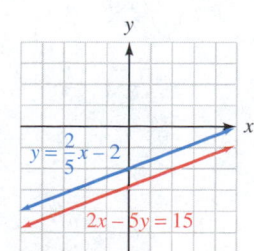

$y = \frac{2}{5}x - 2$

2x − 5y = 15

When the equations of a system have different graphs (as in Examples 1 and 2), the equations are **independent equations**. Two equations with the same graph are **dependent equations**.

| EXAMPLE 3 | **The graphing method.** Solve the system $\begin{cases} y = \frac{1}{2}x + 2 \\ 2x + 8 = 4y \end{cases}$ |

Solution   We graph each equation on one set of coordinate axes, as shown in Figure 12-3. Since the graphs coincide (are the same), the system has infinitely many solutions. Any ordered pair (x, y) that satisfies one equation also satisfies the other.

From the graph shown in Figure 12-3, we see that $(-4, 0)$, $(0, 2)$, and $(2, 3)$ are solutions. We can find infinitely many more solutions by finding additional ordered pairs (x, y) that satisfy both equations.

Because the two equations have the same graph, they are dependent equations.

FIGURE 12-3

Graph by using the slope   Graph by using the
and y-intercept.   intercept method.

$$y = \tfrac{1}{2}x + 2$$ $$2x + 8 = 4y$$

$m = \frac{1}{2}$ $b = 2$

Slope $= \frac{1}{2}$ y-intercept: $(0, 2)$

x	y	(x, y)
-4	0	$(-4, 0)$
0	2	$(0, 2)$
2	3	$(2, 3)$

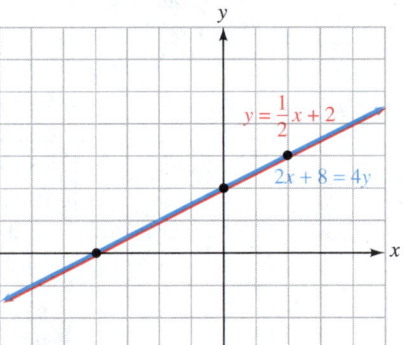

SELF CHECK   Solve $\begin{cases} x - 3y = 12 \\ y = \frac{1}{3}x - 4 \end{cases}$.   *Answer:* infinitely many solutions

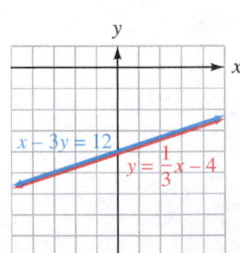

We now summarize the possibilities that can occur when two linear equations, each with two variables, are graphed.

Solving a system of equations by the graphing method

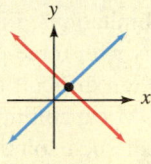

If the lines are different and intersect, the equations are independent, and the system is consistent. **One solution exists.**

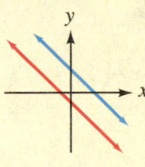

If the lines are different and parallel, the equations are independent, and the system is inconsistent. **No solution exists.**

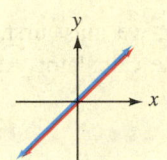

If the lines coincide, the equations are dependent, and the system is consistent. **Infinitely many solutions exist.**

Recall that two systems are **equivalent** when they have the same solution set. To solve a more difficult system such as

$$\begin{cases} \dfrac{3}{2}x - y = \dfrac{5}{2} \\[2mm] x + \dfrac{1}{2}y = 4 \end{cases}$$

we multiply both sides of $\frac{3}{2}x - y = \frac{5}{2}$ by 2 to eliminate the fractions and obtain the equation $3x - 2y = 5$. We multiply both sides of $x + \frac{1}{2}y = 4$ by 2 to eliminate the fractions and obtain the equation $2x + y = 8$.

The new system

$$\begin{cases} 3x - 2y = 5 \\ 2x + y = 8 \end{cases}$$

is equivalent to the original system and is easier to solve, since it has no fractions. If we graph each equation in the new system, as in Figure 12-4, we see that the coordinates of the point where the two lines intersect are $(3, 2)$.

FIGURE 12-4

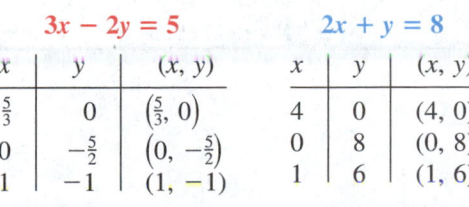

$3x - 2y = 5$

x	y	(x, y)
$\frac{5}{3}$	0	$\left(\frac{5}{3}, 0\right)$
0	$-\frac{5}{2}$	$\left(0, -\frac{5}{2}\right)$
1	-1	$(1, -1)$

$2x + y = 8$

x	y	(x, y)
4	0	$(4, 0)$
0	8	$(0, 8)$
1	6	$(1, 6)$

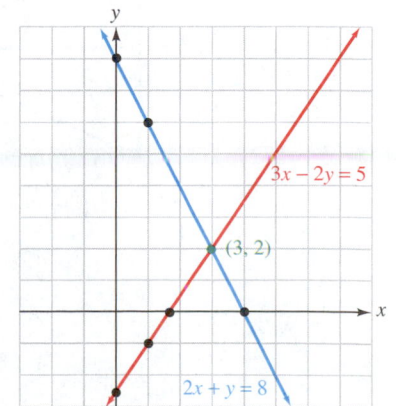

ACCENT ON TECHNOLOGY *Solving Systems by Graphing*

The graphing method has limitations. First, the method is limited to equations with two variables. Systems with three or more variables cannot be solved graphically. Second, it is often difficult to find exact solutions graphically. However, the TRACE and ZOOM capabilities of graphing calculators enable us to get very good approximations of such solutions.

To solve the system $\begin{cases} 3x + 2y = 12 \\ 2x - 3y = 12 \end{cases}$

with a graphing calculator, we must first solve each equation for y so that we can enter the equations into the calculator. After solving for y, we obtain the following equivalent system:

$$\begin{cases} y = -\dfrac{3}{2}x + 6 \\ y = \dfrac{2}{3}x - 4 \end{cases}$$

If we use window settings of $[-10, 10]$ for x and $[-10, 10]$ for y, the graphs of the equations will look like those in Figure 12-5(a). If we zoom in on the intersection point of the two lines and trace, we will get an approximate solution like the one shown in Figure 12-5(b). To get better results, we can do more zooms. We would then find that, to the nearest hundredth, the solution is $(4.63, -0.94)$. Verify that the exact solution is $x = \frac{60}{13}$ and $y = -\frac{12}{13}$.

We can also find the intersection of two lines by using the INTERSECT feature found on most graphing calculators. To learn how to use this feature, please consult your owner's manual. After graphing the lines and using INTERSECT, we obtain the graph shown in Figure 12-5(c). The display shows the approximate coordinates of the point of intersection.

FIGURE 12-5

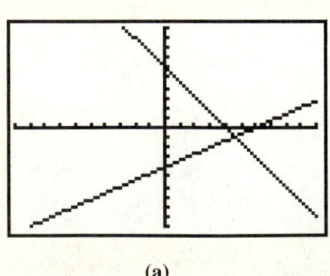

(a)

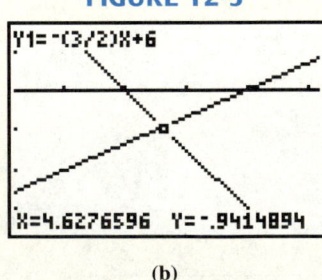

(b)

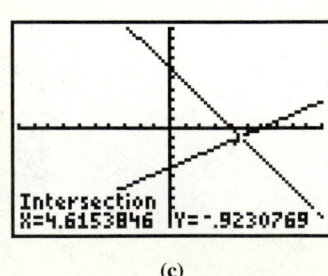

(c)

The Substitution Method

Recall that we use the following steps to solve a system of two equations in two variables by substitution.

Solving systems of equations by substitution

1. If necessary, solve one equation for one of its variables—preferably a variable with a coefficient of 1.

2. Substitute the resulting expression for the variable obtained in step 1 into the other equation and solve that equation.

3. Find the value of the other variable by substituting the value of the variable found in step 2 in any equation containing both variables.

4. Check the solution in both of the original equations.

EXAMPLE 4

The substitution method. Solve the system $\begin{cases} \frac{4}{3}x + \frac{1}{2}y = -\frac{2}{3} \\ \frac{1}{2}x + \frac{2}{3}y = \frac{5}{3} \end{cases}$.

Solution   We first find an equivalent system without fractions by multiplying each side of each equation by 6.

1. $\begin{cases} 8x + 3y = -4 \\ 3x + 4y = 10 \end{cases}$
2.

Since no variable has a coefficient of 1, it is impossible to avoid fractions when solving for a variable. If we choose Equation 2 and solve it for x, we get

$$3x + 4y = 10$$
$$3x = -4y + 10 \quad \text{Subtract } 4y \text{ from both sides.}$$

3. $\qquad x = -\frac{4}{3}y + \frac{10}{3} \quad \text{Divide both sides by 3.}$

We can then substitute $-\frac{4}{3}y + \frac{10}{3}$ for x in Equation 1 and solve for y.

$$8x + 3y = -4$$

$$8\left(-\frac{4}{3}y + \frac{10}{3}\right) + 3y = -4 \quad \text{Substitute } -\frac{4}{3}y + \frac{10}{3} \text{ for } x.$$

$$8(10 - 4y) + 9y = -12 \quad \text{Multiply both sides by 3.}$$
$$80 - 32y + 9y = -12 \quad \text{Use the distributive property to remove parentheses.}$$
$$-23y = -92 \quad \text{Combine like terms and subtract 80 from both sides.}$$
$$y = 4 \quad \text{Divide both sides by } -23.$$

We can find x by substituting 4 for y in Equation 3 and simplifying:

$$x = -\frac{4}{3}y + \frac{10}{3}$$

$$= -\frac{4}{3}(4) + \frac{10}{3} \quad \text{Substitute 4 for } y.$$

$$= -2 \qquad -\frac{16}{3} + \frac{10}{3} = -\frac{6}{3} = -2.$$

The solution is the pair $(-2, 4)$. Verify that this solution satisfies each equation in the original system.

SELF CHECK   Solve $\begin{cases} \frac{3}{2}x + \frac{1}{3}y = -5 \\ \frac{1}{2}x - \frac{2}{3}y = -4 \end{cases}$.   *Answer:* $(-4, 3)$ ∎

The Addition Method

Recall that in the addition method, we combine the equations of the system in a way that will eliminate terms involving one of the variables.

Solving systems of equations by addition

1. Write both equations of the system in general form: $Ax + By = C$.
2. Multiply the terms of one or both of the equations by constants chosen to make the coefficients of x (or y) differ only in sign.
3. Add the equations and solve the equation that results, if possible.
4. Substitute the value obtained in step 3 into either of the original equations and solve for the remaining variable.
5. The results obtained in steps 3 and 4 give the solution of the system.
6. Check the solution in both of the original equations.

EXAMPLE 5

The addition method. Solve the system $\begin{cases} \frac{4}{3}x + \frac{1}{2}y = -\frac{2}{3} \\ \frac{1}{2}x + \frac{2}{3}y = \frac{5}{3} \end{cases}$.

Solution   This system is the system discussed in Example 4. To solve it by addition, we find an equivalent system with no fractions by multiplying both sides of each equation by 6 to obtain

4. $\begin{cases} 8x + 3y = -4 \\ 3x + 4y = 10 \end{cases}$
5.

To make the y-terms drop out when we add the equations, we multiply both sides of Equation 4 by 4 and both sides of Equation 5 by -3 to get

$$\begin{cases} 32x + 12y = -16 \\ -9x - 12y = -30 \end{cases}$$

When these equations are added, the y-terms drop out, and we get

$23x = -46$

$x = -2$ Divide both sides by 23.

To find y, we substitute -2 for x in either Equation 4 or Equation 5. If we substitute -2 for x in Equation 5, we get

$3x + 4y = 10$

$3(-2) + 4y = 10$ Substitute -2 for x.

$-6 + 4y = 10$ Simplify.

$4y = 16$ Add 6 to both sides.

$y = 4$ Divide both sides by 4.

The solution is $(-2, 4)$.

SELF CHECK   Solve $\begin{cases} \frac{2}{3}x - \frac{2}{5}y = 10 \\ \frac{1}{2}x + \frac{2}{3}y = -7 \end{cases}$.   *Answer:* $(6, -15)$ ■

EXAMPLE 6

A system with no solution. Solve the system $\begin{cases} y = 2x + 4 \\ 8x - 4y = 7 \end{cases}$, if possible.

Solution   Because the first equation is already solved for y, we use the substitution method.

$8x - 4y = 7$ The second equation.

$8x - 4(2x + 4) = 7$ Substitute $2x + 4$ for y.

We then solve this equation for x:

$8x - 8x - 16 = 7$ Use the distributive property to remove parentheses.

$-16 \neq 7$ Combine like terms.

This result indicates that the equations in the system are independent and that the system is inconsistent. Since the system has no solution, the graphs of the equations in the system will be parallel.

SELF CHECK   Solve $\begin{cases} 4x - 8y = 10 \\ y = \frac{1}{2}x - \frac{9}{8} \end{cases}$.

Answer: no solution ∎

EXAMPLE 7

Dependent equations. Solve the system $\begin{cases} 4x + 6y = 12 \\ -2x - 3y = -6 \end{cases}$.

Solution   Since the equations are written in general form, we use the addition method. We copy the first equation and multiply both sides of the second equation by 2 to get

$$\begin{array}{r} 4x + 6y = 12 \\ -4x - 6y = -12 \\ \hline \end{array}$$

After adding the left-hand sides and the right-hand sides, we get

$0x + 0y = 0$

$0 = 0$

Here, both the x- and y-terms drop out. The true statement $0 = 0$ indicates that the equations are dependent and that the system is consistent.

Note that the equations of the system are equivalent, because when the second equation is multiplied by -2, it becomes the first equation. The graphs of these equations would coincide. Any ordered pair that satisfies one of the equations also satisfies the other. Some solutions are $(0, 2)$, $(3, 0)$, and $(-3, 4)$.

SELF CHECK   Solve $\begin{cases} 2(x + y) - y = 12 \\ y = -2x + 12 \end{cases}$.

Answer: There are infinitely many solutions. Any ordered pair that satisfies one equation also satisfies the other. ∎

Problem Solving

EXAMPLE 8

Retail sales. Hi-Fi Electronics advertises two types of car radios, one selling for $67 and the other for $100. If the receipts from the sale of 36 radios totaled $2,940, how many of each type were sold?

ANALYZE THE PROBLEM   We can let x represent the number of radios sold for $67 and let y represent the number of radios sold for $100. Then the receipts for the sale of the less expensive radios are $67x$, and the receipts for the sale of the more expensive radios are $100y$.

FORM TWO EQUATIONS   The information in the problem gives the following two equations:

The number of less expensive radios sold	plus	the number of more expensive radios sold	is	the total number of radios sold.
x	$+$	y	$=$	36

The value of the less expensive radios sold	plus	the value of the more expensive radios sold	is	the total receipts.
$67x$	$+$	$100y$	$=$	2,940

SOLVE THE RESULTING SYSTEM OF EQUATIONS — We can solve the following system for x and y to find out how many of each type were sold:

6. $\quad\begin{cases} x + y = 36 \\ 67x + 100y = 2{,}940 \end{cases}$
7.

We multiply both sides of Equation 6 by -100, add the resulting equation to Equation 7, and solve for x:

$$\begin{array}{rcl} -100x - 100y &=& -3{,}600 \\ 67x + 100y &=& 2{,}940 \\ \hline -33x &=& -660 \\ x &=& 20 \end{array} \quad \text{Divide both sides by } -33.$$

To find y, we substitute 20 for x in Equation 6 and solve for y:

$$x + y = 36$$
$$\mathbf{20} + y = 36 \quad \text{Substitute 20 for } x.$$
$$y = 16 \quad \text{Subtract 20 from both sides.}$$

STATE THE CONCLUSION — The store sold 20 of the less expensive radios and 16 of the more expensive radios.

CHECK THE RESULT — If 20 of one type were sold and 16 of the other type were sold, a total of 36 radios were sold. Since the value of the less expensive radios is $20(\$67) = \$1{,}340$, and the value of the more expensive radios is $16(\$100) = \$1{,}600$, the total receipts are $\$2{,}940$. ∎

Systems of Inequalities

We now review the graphing method of solving systems of two linear inequalities in two variables. Recall that the solutions are usually the intersection of half-planes.

EXAMPLE 9

Graph the solution set of the system $\begin{cases} x + y \leq 1 \\ 2x - y > 2 \end{cases}$.

Solution On the same set of coordinate axes, we graph each inequality as in Figure 12-6.

The graph of the inequality $x + y \leq 1$ includes the line graph of $x + y = 1$ and all points below it. Since the boundary line is included, it is drawn as a solid line.

The graph of the inequality $2x - y > 2$ contains only those points below the graph of $2x - y = 2$. Since the boundary line is not included, it is drawn as a broken line.

The area where the half-planes intersect represents the solutions of the given system of inequalities, because any point in that region has coordinates that will satisfy both inequalities.

FIGURE 12-6

$x + y = 1$		
x	y	(x, y)
0	1	$(0, 1)$
1	0	$(1, 0)$

$2x - y = 2$		
x	y	(x, y)
0	-2	$(0, -2)$
1	0	$(1, 0)$

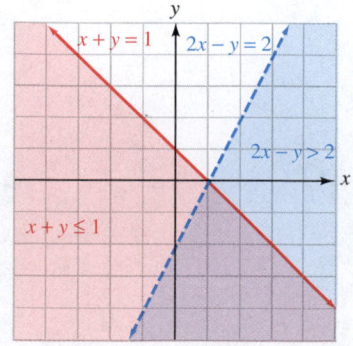

∎

STUDY SET

Section 12.1

VOCABULARY

In Exercises 1–4, fill in the blanks to make the statements true.

1. When a system of equations has one or more solutions, it is called a _____ system.

2. If a system has no solutions, it is called an _____ system.

3. If two equations have different graphs, they are called _____ equations.

4. Two equations with the same graph are called _____ equations.

CONCEPTS

5. Refer to Illustration 1. Tell whether a true or a false statement would be obtained when the coordinates of
   a. point A are substituted into the equation for line l_1.
   b. point B are substituted into the equation for line l_1.
   c. point C are substituted into the equation for line l_1.
   d. point C are substituted into the equation for line l_2.

ILLUSTRATION 1

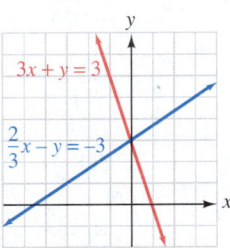

6. Refer to Illustration 2.
   a. How many ordered pairs satisfy the equation $3x + y = 3$? Name three.

   b. How many ordered pairs satisfy the equation $\frac{2}{3}x - y = -3$? Name three.

ILLUSTRATION 2

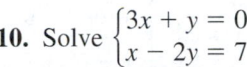

   c. How many ordered pairs satisfy both equations? Name it or them.

7. If the system $\begin{cases} 4x - 3y = 7 \\ 3x - 2y = 6 \end{cases}$ is to be solved using the addition method, by what constant should each equation be multiplied if
   a. the x-terms are to drop out?
   b. the y-terms are to drop out?

8. Consider the system $\begin{cases} \frac{2}{3}x - \frac{y}{6} = \frac{16}{9} \\ 0.03x + 0.02y = 0.03 \end{cases}$.
   a. What step should be performed to clear the first equation of fractions?
   b. What step should be performed to clear the second equation of decimals?

NOTATION

Complete each solution.

9. Solve $\begin{cases} y = 3x - 7 \\ x + y = 5 \end{cases}$.

$x + \left(\boxed{} \right) = 5$

$x + 3x - 7 = \boxed{}$

$\boxed{}x - 7 = 5$

$4x = \boxed{}$

$x = 3$

$y = 3x - 7$

$y = 3(\boxed{}) - 7$

$y = \boxed{}$

The solution is $(3, 2)$.

10. Solve $\begin{cases} 3x + y = 0 \\ x - 2y = 7 \end{cases}$.

$6x + 2y = 0$

$\underline{x - 2y = 7}$

$7x \qquad = \boxed{}$

$x = \boxed{}$

$x - 2y = 7$

$\boxed{} - 2y = 7$

$-2y = \boxed{}$

$y = \boxed{}$

The solution is $(1, -3)$.

PRACTICE

In Exercises 11–22, solve each system by graphing, if possible. Check your graphs with a graphing calculator. If a system is inconsistent or if the equations are dependent, so indicate.

11. $\begin{cases} x - y = 4 \\ 2x + y = 5 \end{cases}$

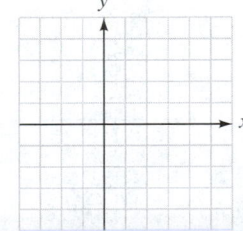

12. $\begin{cases} 2x + y = 1 \\ x - 2y = -7 \end{cases}$

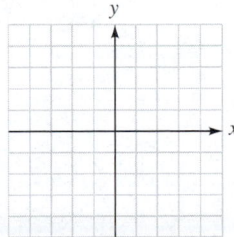

13. $\begin{cases} x = 13 - 4y \\ 3x = 4 + 2y \end{cases}$

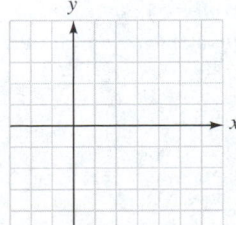

14. $\begin{cases} 3x = 7 - 2y \\ 2x = 2 + 4y \end{cases}$

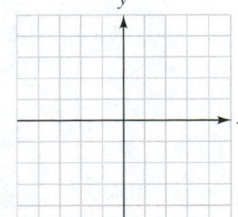

15. $\begin{cases} x = 3 - 2y \\ 2x + 4y = 6 \end{cases}$

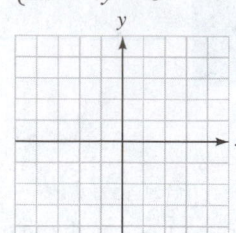

16. $\begin{cases} 3x = 5 - 2y \\ 3x + 2y = 7 \end{cases}$

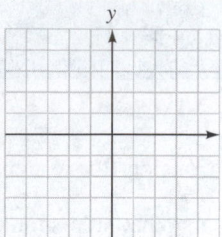

17. $\begin{cases} y = 3 \\ x = 2 \end{cases}$

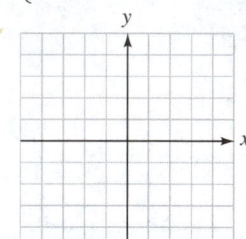

18. $\begin{cases} 2x + 3y = -15 \\ 2x + y = -9 \end{cases}$

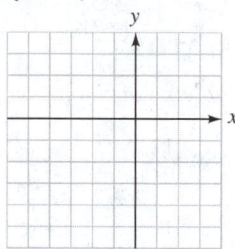

19. $\begin{cases} x = \dfrac{11 - 2y}{3} \\ y = \dfrac{11 - 6x}{4} \end{cases}$

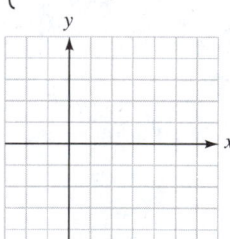

20. $\begin{cases} x = \dfrac{1 - 3y}{4} \\ y = \dfrac{12 + 3x}{2} \end{cases}$

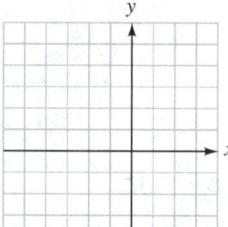

21. $\begin{cases} y = -\dfrac{5}{2}x + \dfrac{1}{2} \\ 2x - \dfrac{3}{2}y = 5 \end{cases}$

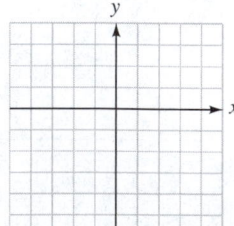

22. $\begin{cases} \dfrac{5}{2}x + 3y = 6 \\ y = -\dfrac{5}{6}x + 2 \end{cases}$

In Exercises 23–26, use a graphing calculator to solve each system. Give all answers to the nearest hundredth.

23. $\begin{cases} y = 3.2x - 1.5 \\ y = -2.7x - 3.7 \end{cases}$

24. $\begin{cases} y = -0.45x + 5 \\ y = 5.55x - 13.7 \end{cases}$

25. $\begin{cases} 1.7x + 2.3y = 3.2 \\ y = 0.25x + 8.95 \end{cases}$

26. $\begin{cases} 2.75x = 12.9y - 3.79 \\ 7.1x - y = 35.76 \end{cases}$

In Exercises 27–38, solve each system by substitution, if possible. If a system is inconsistent or if the equations are dependent, so indicate.

27. $\begin{cases} y = x \\ x + y = 4 \end{cases}$

28. $\begin{cases} y = x + 2 \\ x + 2y = 16 \end{cases}$

29. $\begin{cases} x = 2 + y \\ 2x + y = 13 \end{cases}$

30. $\begin{cases} x = -4 + y \\ 3x - 2y = -5 \end{cases}$

31. $\begin{cases} x + 2y = 6 \\ 3x - y = -10 \end{cases}$

32. $\begin{cases} 2x - y = -21 \\ 4x + 5y = 7 \end{cases}$

33. $\begin{cases} 3x = 2y - 4 \\ 6x - 4y = -4 \end{cases}$

34. $\begin{cases} 8x = 4y + 10 \\ 4x - 2y = 5 \end{cases}$

35. $\begin{cases} 3x - 4y = 9 \\ x + 2y = 8 \end{cases}$

36. $\begin{cases} 3x - 2y = -10 \\ 6x + 5y = 25 \end{cases}$

37. $\begin{cases} 2x + 2y = -1 \\ 3x + 4y = 0 \end{cases}$

38. $\begin{cases} 5x + 3y = -7 \\ 3x - 3y = 7 \end{cases}$

In Exercises 39–54, solve each system by addition, if possible. If a system is inconsistent or if the equations are dependent, so indicate.

39. $\begin{cases} x - y = 3 \\ x + y = 7 \end{cases}$

40. $\begin{cases} x + y = 1 \\ x - y = 7 \end{cases}$

41. $\begin{cases} 2x + y = -10 \\ 2x - y = -6 \end{cases}$

42. $\begin{cases} x + 2y = -9 \\ x - 2y = -1 \end{cases}$

43. $\begin{cases} 2x + 3y = 8 \\ 3x - 2y = -1 \end{cases}$

44. $\begin{cases} 5x - 2y = 19 \\ 3x + 4y = 1 \end{cases}$

45. $\begin{cases} 4(x - 2) = -9y \\ 2(x - 3y) = -3 \end{cases}$

46. $\begin{cases} 2(2x + 3y) = 5 \\ 8x = 3(1 + 3y) \end{cases}$

47. $\begin{cases} 8x - 4y = 16 \\ 2x - 4 = y \end{cases}$

48. $\begin{cases} 2y - 3x = -13 \\ 3x - 17 = 4y \end{cases}$

49. $\begin{cases} x = \frac{3}{2}y + 5 \\ 2x - 3y = 8 \end{cases}$

50. $\begin{cases} x = \frac{2}{3}y \\ y = 4x + 5 \end{cases}$

51. $\begin{cases} \frac{x}{2} + \frac{y}{2} = 6 \\ \frac{x}{2} - \frac{y}{2} = -2 \end{cases}$

52. $\begin{cases} \frac{x}{2} - \frac{y}{3} = -4 \\ \frac{x}{2} + \frac{y}{9} = 0 \end{cases}$

53. $\begin{cases} \frac{3}{4}x + \frac{2}{3}y = 7 \\ \frac{3}{5}x - \frac{1}{2}y = 18 \end{cases}$

54. $\begin{cases} \frac{2}{3}x - \frac{1}{4}y = -8 \\ \frac{1}{2}x - \frac{3}{8}y = -9 \end{cases}$

In Exercises 55–58, graph the solution set of each system of inequalities.

55. $\begin{cases} y < 3x + 2 \\ y < -2x + 3 \end{cases}$

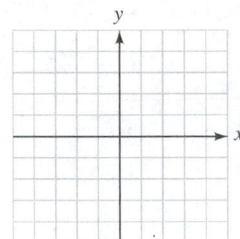

56. $\begin{cases} y \le x - 2 \\ y \ge 2x + 1 \end{cases}$

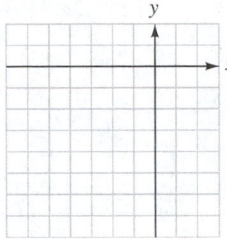

57. $\begin{cases} 3x + 2y > 6 \\ x + 3y \le 2 \end{cases}$

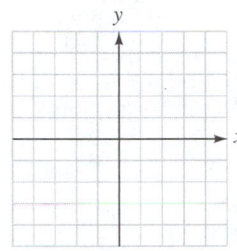

58. $\begin{cases} 3x + y \le 1 \\ -x + 2y \ge 6 \end{cases}$

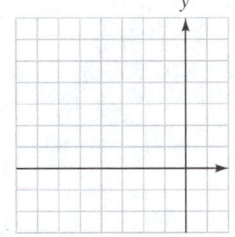

APPLICATIONS

Use two variables to solve each problem.

59. TAKE-OUT FOOD

   a. Estimate the point of intersection of the two graphs shown in Illustration 3. Express your answer in the form (year, number of meals).

   b. What information about dining out does the point of intersection give?

ILLUSTRATION 3

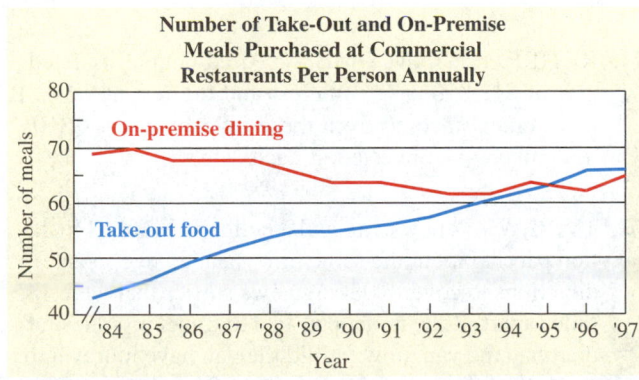

Based on data from *The Wall Street Journal Almanac* (1999)

60. ROAD ATLAS See Illustration 4. Name the cities that lie along Interstate 40. Name the cities that lie along Interstate 25. What city lies on both interstate highways?

ILLUSTRATION 4

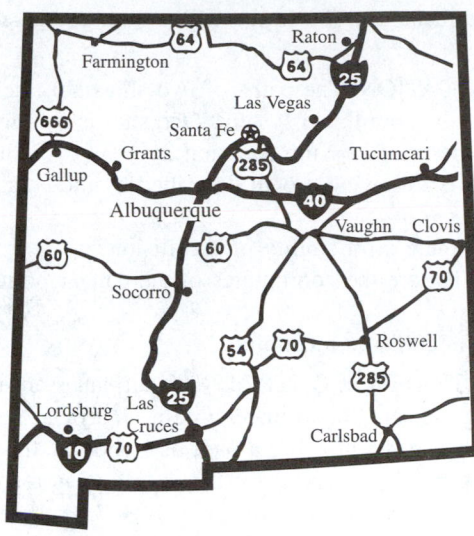

61. LAW OF SUPPLY AND DEMAND The demand function, graphed in Illustration 5, describes the relationship between the price x of a certain camera and the demand for the camera.

   a. The supply function, $S(x) = \frac{25}{4}x - 525$, describes the relationship between the price x of the camera and the number of cameras the manufacturer is willing to supply. Graph this function in Illustration 5.

   b. For what price will the supply of cameras equal the demand?

   c. As the price of the camera is increased, what happens to supply and what happens to demand?

ILLUSTRATION 5

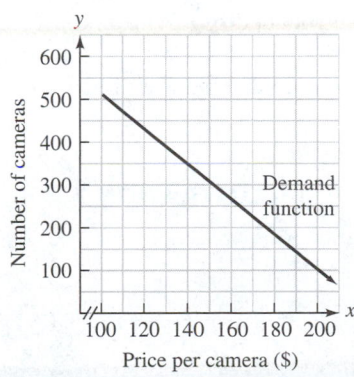

62. COST AND REVENUE The function $C(x) = 200x + 400$ gives the cost for a college to offer x sections of an introductory class in CPR (cardiopulmonary resuscitation). The function $R(x) = 280x$ gives the amount of revenue the college brings in when offering x sections of CPR.

   a. Find the *break-even point* (where cost = revenue) by graphing each function on the same coordinate system.

b. How many sections does the college need to offer to make a profit on the CPR training course?

63. NAVIGATION The paths of two ships are tracked on the same coordinate system. One ship is following a path described by the equation $2x + 3y = 6$, and the other is following a path described by the equation $y = \frac{2}{3}x - 3$.
 a. Is there a possibility of a collision?
 b. What are the coordinates of the danger point?

 c. Is a collision a certainty?

64. AIR TRAFFIC CONTROL Two airplanes are tracked using the same coordinate system on a radar screen. One plane is following a path described by the equation $y = \frac{2}{5}x - 2$, and the other is following a path described by the equation $2x = 5y + 7$. Is there a possibility of a collision?

65. ELECTRONICS In Illustration 6, two resistors in the voltage divider circuit have a total resistance of 1,375 ohms. To provide the required voltage, R_1 must be 125 ohms greater than R_2. Find both resistances.

ILLUSTRATION 6

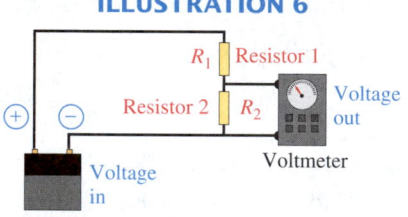

66. FENCING A FIELD The rectangular field in Illustration 7 is surrounded by 72 meters of fencing. If the field is partitioned as shown, a total of 88 meters of fencing is required. Find the dimensions of the field.

ILLUSTRATION 7

67. GEOMETRY In a right triangle, one acute angle is 15° greater than two times the other acute angle. Find the difference between the measures of the angles.

68. BRACING The bracing of a basketball backboard shown in Illustration 8 forms a parallelogram. Find the values of x and y.

ILLUSTRATION 8

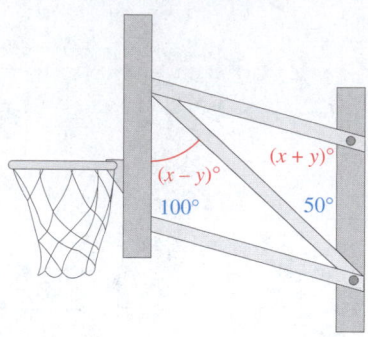

69. TRAFFIC SIGNAL In Illustration 9, brace A and brace B are perpendicular. Find the values of x and y.

ILLUSTRATION 9

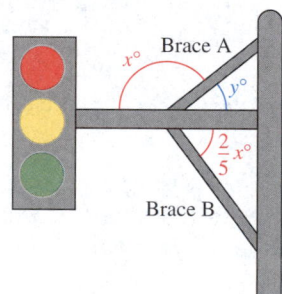

70. INVESTMENT CLUB Part of $8,000 was invested by an investment club at 10% interest and the rest at 12%. If the annual income from these investments is $900, how much was invested at each rate?

71. RETIREMENT INCOME A retired couple invested part of $12,000 at 6% interest and the rest at 7.5%. If their annual income from these investments is $810, how much was invested at each rate?

72. TV NEWS A news van and a helicopter left a TV station parking lot at the same time, headed in opposite directions to cover breaking news stories that were 145 miles apart. If the helicopter had to travel 55 miles farther than the van, how far did the van have to travel to reach the location of the news story?

73. DELIVERY SERVICE A delivery truck travels 50 miles in the same time that a cargo plane travels 180 miles. The speed of the plane is 143 mph faster than the speed of the truck. Find the speed of the delivery truck.

74. PRODUCTION PLANNING A bicycle manufacturer builds racing bikes and mountain bikes, with the per-unit manufacturing costs shown in Illustration 10. The company has budgeted $15,900 for labor and $13,075 for materials. How many bicycles of each type can be built?

ILLUSTRATION 10

Model	Cost of materials	Cost of labor
Racing	$55	$60
Mountain	$70	$90

75. FARMING A farmer keeps some animals on a strict diet. Each animal is to recieve 15 grams of protein and 7.5 grams of carbohydrates. The farmer uses two food mixes, with nutrients as shown in Illustration 11. How many grams of each mix should be used to provide the correct nutrients for each animal?

ILLUSTRATION 11

Mix	Protein	Carbohydrates
Mix A	12%	9%
Mix B	15%	5%

76. RECORDING COMPANY Three people invest a total of $105,000 to start a recording company that will produce reissues of classic jazz. Each release will be a set of 3 CDs that will retail for $45 per set. If each set can be produced for $18.95, how many sets must be sold for the investors to make a profit?

77. MACHINE SHOP Two machines can mill a brass plate. One machine has a setup cost of $300 and a cost per plate of $2. The other machine has a setup cost of $500 and a cost per plate of $1. Find the break point.

78. PUBLISHING A printer has two presses. One has a setup cost of $210 and can print the pages of a certain book for $5.98. The other press has a setup cost of $350 and can print the pages of the same book for $5.95. Find the break point.

79. MIXING CANDY How many pounds of each candy shown in Illustration 12 must be mixed to obtain 60 pounds of candy that would be worth $3 per pound?

80. COSMETOLOGY A beauty shop specializing in permanents has fixed costs of $2,101.20 per month. The owner estimates that the cost for each permanent is $23.60, which covers labor, chemicals, and electricity. If her shop can give as many permanents as she

ILLUSTRATION 12

Hard Candy $2/lb   Soft Candy $4/lb

wants at a price of $44 each, how many must be given each month to break even?

81. PRODUCTION PLANNING A paint manufacturer can choose between two processes for manufacturing house paint, with monthly costs as shown in Illustration 13. Assume that the paint sells for $18 per gallon.

ILLUSTRATION 13

Process	Fixed costs	Unit cost (per gallon)
A	$32,500	$13
B	$80,600	$5

a. Find the break point for process A.

b. Find the break point for process B.

c. If expected sales are 7,000 gallons per month, which process should the company use?

82. DERMATOLOGY Tests of an antibacterial face-wash cream showed that a mixture containing 0.3% Triclosan (active ingredient) gave the best results. How many grams of cream from each tube shown in Illustration 14 should be used to make an equal-size tube of the 0.3% cream?

ILLUSTRATION 14

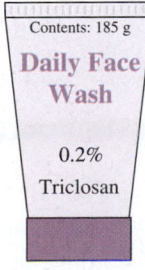   

Contents: 185 g — Daily Face Wash — 0.2% Triclosan   Contents: 185 g — Daily Face Wash — 0.7% Triclosan

83. MIXING SOLUTIONS How many ounces of the two alcohol solutions in Illustration 15 must be mixed to obtain 100 ounces of a 12.2% solution?

ILLUSTRATION 15

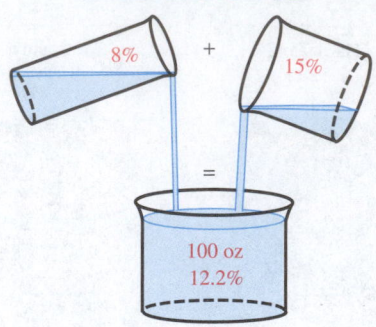

84. RETOOLING A manufacturer of automobile water pumps is considering retooling for one of two manufacturing processes, with monthly fixed costs and unit costs as indicated in Illustration 16. Each water pump can be sold for $50.

ILLUSTRATION 16

Process	Fixed costs	Unit cost
A	$12,390	$29
B	$20,460	$17

a. Find the break point for process A.

b. Find the break point for process B.

c. If expected sales are 550 per month, which process should be used?

WRITING

85. Which method would you use to solve the system
$$\begin{cases} y = 3x + 1 \\ 3x + 2y = 12 \end{cases}?$$ Explain why.

86. Which method would you use to solve the system
$$\begin{cases} 2x + 4y = 9 \\ 3x - 5y = 20 \end{cases}?$$ Explain why.

87. When graphing a system of linear inequalities, explain how to decide which region to shade.

88. Explain how a system of two linear inequalities might have no solution.

REVIEW

Simplify each expression. Write all answers without using negative exponents.

89. $(a^2a^3)^2(a^4a^2)^2$

90. $\left(\dfrac{a^2b^3c^4d}{ab^2c^3d^4}\right)^{-3}$

91. $\left(\dfrac{-3x^3y^4}{x^{-5}y^3}\right)^{-4}$

92. $\dfrac{3t^0 - 4t^0 + 5}{5t^0 + 2t^0}$

Solve each formula for the given variable.

93. $A = p + prt$ for r

94. $A = p + prt$ for p

95. $\dfrac{1}{r} = \dfrac{1}{r_1} + \dfrac{1}{r_2}$ for r

96. $\dfrac{1}{r} = \dfrac{1}{r_1} + \dfrac{1}{r_2}$ for r_1

▶ 12.2 Solving Systems with Three Variables

In this section, you will learn about

Solving three equations with three variables ■ Consistent systems ■ An inconsistent system ■ Systems with dependent equations ■ Problem solving ■ Curve fitting

Introduction In the preceding section, we solved systems of two linear equations with two variables. In this section, we will solve systems of linear equations with three variables by using a combination of the addition method and the substitution method. We will then use that procedure to solve problems involving three variables.

Solving Three Equations with Three Variables

We now extend the definition of a linear equation to include equations of the form $Ax + By + Cz = D$. The solution of a system of three linear equations with three variables is an **ordered triple** of numbers. For example, the solution of the system

$$\begin{cases} 2x + 3y + 4z = 20 \\ 3x + 4y + 2z = 17 \\ 3x + 2y + 3z = 16 \end{cases}$$

is the triple $(1, 2, 3)$, since each equation is satisfied if $x = 1$, $y = 2$, and $z = 3$.

$2x + 3y + 4z = 20$	$3x + 4y + 2z = 17$	$3x + 2y + 3z = 16$
$2(1) + 3(2) + 4(3) = 20$	$3(1) + 4(2) + 2(3) = 17$	$3(1) + 2(2) + 3(3) = 16$
$2 + 6 + 12 = 20$	$3 + 8 + 6 = 17$	$3 + 4 + 9 = 16$
$20 = 20$	$17 = 17$	$16 = 16$

The graph of an equation of the form $Ax + By + Cz = D$ is a flat surface called a *plane*. A system of three linear equations with three variables is consistent or inconsistent, depending on how the three planes corresponding to the three equations intersect. Figure 12-7 illustrates some of the possibilities.

FIGURE 12-7

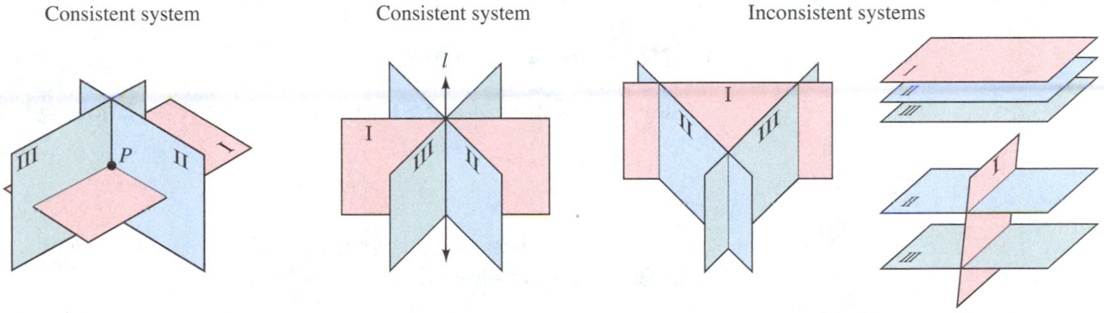

Consistent system

The three planes intersect at a single point *P*: One solution

(a)

Consistent system

The three planes have a line *l* in common: Infinitely many solutions

(b)

Inconsistent systems

The three planes have no point in common: No solutions

(c)

To solve a system of three linear equations with three variables, we follow these steps.

Solving three equations with three variables

1. Pick any two equations and eliminate a variable.
2. Pick a different pair of equations and eliminate the same variable as in step 1.
3. Solve the resulting pair of two equations in two variables.
4. To find the value of the third variable, substitute the values of the two variables found in step 3 into any equation containing all three variables and solve the equation.
5. Check the solution in all three of the original equations.

Consistent Systems

E X A M P L E   1

Solving a system of three equations. Solve the system
$$\begin{cases} 2x + y + 4z = 12 \\ x + 2y + 2z = 9 \\ 3x - 3y - 2z = 1 \end{cases}.$$

Solution   **Step 1:** We are given the system

1. $\begin{cases} 2x + y + 4z = 12 \\ \end{cases}$
2. $\begin{cases} x + 2y + 2z = 9 \\ \end{cases}$
3. $\begin{cases} 3x - 3y - 2z = 1 \end{cases}$

If we pick Equations 2 and 3 and add them, the variable z is eliminated.

2. $x + 2y + 2z = 9$

3. $\underline{3x - 3y - 2z = 1}$

4. $4x - y = 10$ This equation does not contain z.

Step 2: We now pick a different pair of equations (Equations 1 and 3) and eliminate z again. If each side of Equation 3 is multiplied by 2 and the resulting equation is added to Equation 1, z is eliminated.

1. $2x + y + 4z = 12$

   $\underline{6x - 6y - 4z = 2}$ Multiply both sides of Equation 3 by 2.

5. $8x - 5y = 14$ This equation does not contain z.

Step 3: Equations 4 and 5 form a system of two equations with two variables, x and y.

4. $\begin{cases} 4x - y = 10 \\ \end{cases}$
5. $\begin{cases} 8x - 5y = 14 \end{cases}$

To solve this system, we multiply Equation 4 by -5 and add the resulting equation to Equation 5 to eliminate y:

   $-20x + 5y = -50$ Multiply both sides of Equation 4 by -5.

5. $\underline{8x - 5y = 14}$

   $-12x = -36$

   $x = 3$ To find x, divide both sides by -12.

To find y, we substitute 3 for x in any equation containing x and y (such as Equation 5) and solve for y:

5. $8x - 5y = 14$

   $8(3) - 5y = 14$ Substitute 3 for x.

   $24 - 5y = 14$ Simplify.

   $-5y = -10$ Subtract 24 from both sides.

   $y = 2$ Divide both sides by -5.

Step 4: To find z, we substitute 3 for x and 2 for y in an equation containing x, y, and z (such as Equation 1) and solve for z:

1. $2x + y + 4z = 12$

   $2(3) + 2 + 4z = 12$ Substitute 3 for x and 2 for y.

   $8 + 4z = 12$ Simplify.

   $4z = 4$ Subtract 8 from both sides

   $z = 1$ Divide both sides by 4.

The solution of the system is $(x, y, z) = (3, 2, 1)$. Because this system has a solution, it is a consistent system.

Step 5: Verify that these values satisfy each equation in the original system.

SELF CHECK

Solve $\begin{cases} 2x + y + 4z = 16 \\ x + 2y + 2z = 11 \\ 3x - 3y - 2z = -9 \end{cases}$.

Answer: $(1, 2, 3)$ ∎

When one or more of the equations of a system is missing a term, the elimination of a variable that is normally performed in step 1 of the solution process can be skipped.

EXAMPLE 2

Missing terms. Solve the system $\begin{cases} 3x = 6 - 2y + z \\ -y - 2z = -8 - x \\ x = 1 - 2z \end{cases}$.

Solution

Step 1: First, we write each equation in $Ax + By + Cz = D$ form.

1. $\begin{cases} 3x + 2y - z = 6 \\ x - y - 2z = -8 \\ x + 2z = 1 \end{cases}$
2.
3.

Since Equation 3 does not have a y-term, we can proceed to step 2, where we will find another equation that does not contain a y-term.

Step 2: If each side of Equation 2 is multiplied by 2 and the resulting equation is added to Equation 1, y is eliminated.

1. $3x + 2y - z = 6$

   $\underline{2x - 2y - 4z = -16}$ Multiply both sides of Equation 2 by 2.
4. $5x \qquad - 5z = -10$

Step 3: Equations 3 and 4 form a system of two equations with two variables, x and z:

3. $\begin{cases} x + 2z = 1 \\ 5x - 5z = -10 \end{cases}$
4.

To solve this system, we multiply Equation 3 by -5 and add the resulting equation to Equation 4 to eliminate x:

$-5x - 10z = -5$ Multiply both sides of Equation 3 by -5.
4. $\underline{5x - 5z = -10}$

$\qquad\qquad -15z = -15$

$\qquad\qquad\qquad z = 1$ To find z, divide both sides by -15.

To find x, we substitute 1 for z in Equation 3.

3. $x + 2z = 1$

   $x + 2(1) = 1$ Substitute 1 for z.

   $x + 2 = 1$ Multiply.

   $x = -1$ Subtract 2 from both sides.

Step 4: To find y, we substitute -1 for x and 1 for z in Equation 1:

1. $3x + 2y - z = 6$

   $3(-1) + 2y - 1 = 6$ Substitute -1 for x and 1 for z.

   $-3 + 2y - 1 = 6$ Multiply.

   $2y = 10$ Add 4 to both sides.

   $y = 5$ Divide both sides by 2.

The solution of the system is $(-1, 5, 1)$.

Step 5: Check the solution in all three of the original equations.

Solve $\begin{cases} x + 2y - z = 1 \\ 2x - y + z = 3. \\ x + z = 3 \end{cases}$

Answer: $(1, 1, 2)$ ▪

An Inconsistent System

A system with no solution. Solve the system $\begin{cases} 2a + b - 3c = -3 \\ 3a - 2b + 4c = 2 \\ 4a + 2b - 6c = -7 \end{cases}$.

Solution   We can multiply the first equation of the system by 2 and add the resulting equation to the second equation to eliminate b:

$$4a + 2b - 6c = -6 \quad \text{Multiply both sides of the first equation by 2.}$$
$$\underline{3a - 2b + 4c = 2}$$
1. $7a - 2c = -4$

Now add the second and third equations of the system to eliminate b again:

$$3a - 2b + 4c = 2$$
$$\underline{4a + 2b - 6c = -7}$$
2. $7a - 2c = -5$

Equations 1 and 2 form the system

1. $\begin{cases} 7a - 2c = -4 \\ 7a - 2c = -5 \end{cases}$
2.

Since $7a - 2c$ cannot equal both -4 and -5, the system is inconsistent and has no solution.

Solve $\begin{cases} 2a + b - 3c = 8 \\ 3a - 2b + 4c = 10. \\ 4a + 2b - 6c = -5 \end{cases}$

Answer: no solution   ▪

Systems with Dependent Equations

When the equations in a system of two equations with two variables are dependent, the system has infinitely many solutions. This is not always true for systems of three equations with three variables. In fact, a system can have dependent equations and still be inconsistent. Figure 12-8 illustrates the different possibilities.

FIGURE 12-8

Consistent system

Consistent system

Inconsistent system

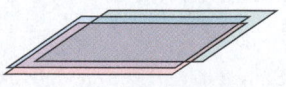

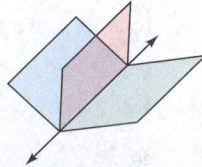

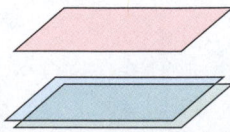

When three planes coincide, the equations are dependent, and there are infinitely many solutions.

When three planes intersect in a common line, the equations are dependent, and there are infinitely many solutions.

When two planes coincide and are parallel to a third plane, the system is inconsistent, and there are no solutions.

(a)

(b)

(c)

EXAMPLE 4

A system with infinitely many solutions. Solve the system
$$\begin{cases} 3x - 2y + z = -1 \\ 2x + y - z = 5 \\ 5x - y = 4 \end{cases}.$$

Solution We can add the first two equations to get

$$\begin{array}{r} 3x - 2y + z = -1 \\ 2x + y - z = 5 \\ \hline \mathbf{1.} \quad 5x - y = 4 \end{array}$$

Since Equation 1 is the same as the third equation of the system, the equations of the system are dependent, and there will be infinitely many solutions. From a graphical perspective, the equations represent three planes that intersect in a common line, as shown in Figure 12-8(b).

To write the general solution of this system, we can solve Equation 1 for y to get

$$5x - y = 4$$
$$-y = -5x + 4 \quad \text{\textcolor{red}{Subtract } } 5x \text{ \textcolor{red}{from both sides.}}$$
$$y = 5x - 4 \quad \text{\textcolor{red}{Multiply both sides by } } -1.$$

We can then substitute $5x - 4$ for y in the first equation of the system and solve for z to get

$$3x - 2y + z = -1$$
$$3x - 2(5x - 4) + z = -1 \qquad \text{\textcolor{red}{Substitute } } 5x - 4 \text{ \textcolor{red}{for } } y.$$
$$3x - 10x + 8 + z = -1 \qquad \text{\textcolor{red}{Use the distributive property to remove parentheses.}}$$
$$-7x + 8 + z = -1 \qquad \text{\textcolor{red}{Combine like terms.}}$$
$$z = 7x - 9 \qquad \text{\textcolor{red}{Add } } 7x \text{ \textcolor{red}{and } } -8 \text{ \textcolor{red}{to both sides.}}$$

Since we have found the values of y and z in terms of x, every solution of the system has the form $(x, 5x - 4, 7x - 9)$, where x can be any real number. For example,

If $x = 1$, a solution is $(1, 1, -2)$. $5(1) - 4 = 1$, and $7(1) - 9 = -2$.
If $x = 2$, a solution is $(2, 6, 5)$. $5(2) - 4 = 6$, and $7(2) - 9 = 5$.
If $x = 3$, a solution is $(3, 11, 12)$. $5(3) - 4 = 11$, and $7(3) - 9 = 12$.

SELF CHECK Solve $\begin{cases} 3x + 2y + z = -1 \\ 2x - y - z = 5 \\ 5x + y = 4 \end{cases}$.   *Answer:* There are infinitely many solutions. A general solution is $(x, 4 - 5x, -9 + 7x)$. Three solutions are $(1, -1, -2)$, $(2, -6, 5)$, and $(3, -11, 12)$. ∎

Problem Solving

EXAMPLE 5

Tool manufacturing. A tool company makes three types of hammers, which are marketed as "good," "better," and "best." The cost of manufacturing each type of hammer is $4, $6, and $7, respectively, and the hammers sell for $6, $9, and $12. Each day, the cost of manufacturing 100 hammers is $520, and the daily revenue from their sale is $810. How many hammers of each type are manufactured?

ANALYZE THE PROBLEM We need to find how many of each type of hammer are manufactured daily. We must write three equations to find three unknowns.

FORM THREE EQUATIONS If we let x represent the number of good hammers, y represent the number of better hammers, and z represent the number of best hammers, we know that

The total number of hammers is $x + y + z$.

The cost of manufacturing the good hammers is $4x$ ($4 times x hammers).

The cost of manufacturing the better hammers is $6y ($6 times y hammers).

The cost of manufacturing the best hammers is $7z ($7 times z hammers).

The revenue received by selling the good hammers is $6x ($6 times x hammers).

The revenue received by selling the better hammers is $9y ($9 times y hammers).

The revenue received by selling the best hammers is $12z ($12 times z hammers).

We can assemble the facts of the problem to write three equations.

The number of good hammers	plus	the number of better hammers	plus	the number of best hammers	is	the total number of hammers.
x	$+$	y	$+$	z	$=$	100

The cost of good hammers	plus	the cost of better hammers	plus	the cost of best hammers	is	the total cost.
$4x$	$+$	$6y$	$+$	$7z$	$=$	520

The revenue from good hammers	plus	the revenue from better hammers	plus	the revenue from best hammers	is	the total revenue.
$6x$	$+$	$9y$	$+$	$12z$	$=$	810

SOLVE THE SYSTEM   We must now solve the system

1. $\begin{cases} x + y + z = 100 \\ 4x + 6y + 7z = 520 \\ 6x + 9y + 12z = 810 \end{cases}$
2.
3.

If we multiply Equation 1 by -7 and add the result to Equation 2, we get

$$
\begin{array}{r}
-7x - 7y - 7z = -700 \\
4x + 6y + 7z = 520 \\
\hline
\end{array}
$$

4. $-3x - y = -180$

If we multiply Equation 1 by -12 and add the result to Equation 3, we get

$$
\begin{array}{r}
-12x - 12y - 12z = -1{,}200 \\
6x + 9y + 12z = 810 \\
\hline
\end{array}
$$

5. $-6x - 3y = -390$

If we multiply Equation 4 by -3 and add it to Equation 5, we get

$$
\begin{array}{r}
9x + 3y = 540 \\
-6x - 3y = -390 \\
\hline
3x = 150
\end{array}
$$

$x = 50$ To find x, divide both sides by 3.

To find y, we substitute 50 for x in Equation 4:

$-3\boldsymbol{x} - y = -180$

$-3(\boldsymbol{50}) - y = -180$ Substitute 50 for x.

$-150 - y = -180$ $-3(50) = -150$.

$-y = -30$ Add 150 to both sides.

$y = 30$ Divide both sides by -1.

To find z, we substitute 50 for x and 30 for y in Equation 1:

$$x + y + z = 100$$
$$50 + 30 + z = 100$$
$$z = 20 \quad \text{Subtract 80 from both sides.}$$

STATE THE CONCLUSION   The company manufactures 50 good hammers, 30 better hammers, and 20 best hammers each day.

CHECK THE RESULT   Check the solution in each equation of the original system.   ■

Curve Fitting

EXAMPLE 6

Finding the equation of a parabola. The equation of a parabola opening upward or downward is of the form $y = ax^2 + bx + c$. Find the equation of the parabola shown in Figure 12-9 by determining the values of a, b, and c.

FIGURE 12-9

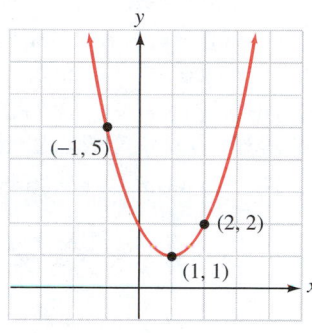

Solution   Since the parabola passes through the points $(-1, 5)$, $(1, 1)$, and $(2, 2)$, each pair of coordinates must satisfy the equation $y = ax^2 + bx + c$. If we substitute the x- and y-coordinates of each point into the equation and simplify, we obtain the following system of three equations wih three variables.

1. $\begin{cases} a - b + c = 5 \end{cases}$ Substitute the coordinates of $(-1, 5)$ into $y = ax^2 + bx + c$ and simplify.
2. $\begin{cases} a + b + c = 1 \end{cases}$ Substitute the coordinates of $(1, 1)$ into $y = ax^2 + bx + c$ and simplify.
3. $\begin{cases} 4a + 2b + c = 2 \end{cases}$ Substitute the coordinates of $(2, 2)$ into $y = ax^2 + bx + c$ and simplify.

If we add Equations 1 and 2, we obtain

$$a - b + c = 5$$
$$\underline{a + b + c = 1}$$
4. $\quad 2a \quad\quad + 2c = 6$

If we multiply Equation 1 by 2 and add the result to Equation 3, we get

$$2a - 2b + 2c = 10$$
$$\underline{4a + 2b + c = 2}$$
5. $\quad 6a \quad\quad + 3c = 12$

We can then divide both sides of Equation 4 by 2 to get Equation 6 and divide both sides of Equation 5 by 3 to get Equation 7. We now have the system

6. $\begin{cases} a + c = 3 \end{cases}$
7. $\begin{cases} 2a + c = 4 \end{cases}$

To eliminate c, we multiply Equation 6 by -1 and add the result to Equation 7. We get

$$-a - c = -3$$
$$\underline{2a + c = 4}$$
$$a \quad\quad = 1$$

To find c, we can substitute 1 for a in Equation 6 and find that $c = 2$. To find b, we can substitute 1 for a and 2 for c in Equation 2 and find that $b = -2$.

After we substitute these values of a, b, and c into the equation $y = ax^2 + bx + c$, we have the equation of the parabola.

$$y = ax^2 + bx + c$$
$$y = 1x^2 - 2x + 2$$
$$y = x^2 - 2x + 2$$

■

STUDY SET

Section 12.2

VOCABULARY

In Exercises 1–6, fill in the blanks to make the statements true.

1. $\begin{cases} 2x + y - 3z = 0 \\ 3x - y + 4z = 5 \\ 4x + 2y - 6z = 0 \end{cases}$ is called a _____ of three linear equations.

2. If the first two equations of the system in Exercise 1 are added, the variable y is _____.

3. The equation $2x + 3y + 4z = 5$ is a linear equation with _____ variables.

4. The graph of the equation $2x + 3y + 4z = 5$ is a flat surface called a _____.

5. When three planes coincide, the equations of the system are _____, and there are infinitely many solutions.

6. When three planes intersect in a line, the system will have _____ many solutions.

CONCEPTS

7. For each graph of a system of three equations, tell whether the solution set contains one solution, infinitely many solutions, or no solution.

   a. 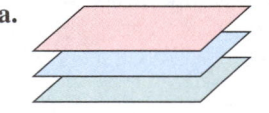 b.

8. Consider the system $\begin{cases} -2x + y + 4z = 3 \\ x - y + 2z = 1 \\ x + y - 3z = 2 \end{cases}$.

   a. What is the result if Equation 1 and Equation 2 are added?

   b. What is the result if Equation 2 and Equation 3 are added?

   c. What variable was eliminated in the steps performed in parts a and b?

NOTATION

9. Write the equation $3z - 2y = x + 6$ in $Ax + By + Cz = D$ form.

10. Fill in the blank to make a true statement: Solutions of a system of three equations in three variables, x, y, and z, are written in the form (x, y, z) and are called ordered _____.

PRACTICE

In Exercises 11–12, tell whether the given ordered triple is a solution of given system.

11. $(2, 1, 1)$, $\begin{cases} x - y + z = 2 \\ 2x + y - z = 4 \\ 2x - 3y + z = 2 \end{cases}$

12. $(-3, 2, -1)$, $\begin{cases} 2x + 2y + 3z = -1 \\ 3x + y - z = -6 \\ x + y + 2z = 1 \end{cases}$

In Exercises 13–28, solve each system. If a system is inconsistent or if the equations are dependent, so indicate.

13. $\begin{cases} x + y + z = 4 \\ 2x + y - z = 1 \\ 2x - 3y + z = 1 \end{cases}$

14. $\begin{cases} x + y + z = 4 \\ x - y + z = 2 \\ x - y - z = 0 \end{cases}$

15. $\begin{cases} 2x + 2y + 3z = 10 \\ 3x + y - z = 0 \\ x + y + 2z = 6 \end{cases}$

16. $\begin{cases} x - y + z = 4 \\ x + 2y - z = -1 \\ x + y - 3z = -2 \end{cases}$

17. $\begin{cases} b + 2c = 7 - a \\ a + c = 8 - 2b \\ 2a + b + c = 9 \end{cases}$

18. $\begin{cases} 2a = 2 - 3b - c \\ 4a + 6b + 2c - 5 = 0 \\ a + c = 3 + 2b \end{cases}$

19. $\begin{cases} 2x + y - z = 1 \\ x + 2y + 2z = 2 \\ 4x + 5y + 3z = 3 \end{cases}$

20. $\begin{cases} 4x + 3z = 4 \\ 2y - 6z = -1 \\ 8x + 4y + 3z = 9 \end{cases}$

21. $\begin{cases} a + b + c = 180 \\ \frac{a}{4} + \frac{b}{2} + \frac{c}{3} = 60 \\ 2b + 3c - 330 = 0 \end{cases}$

22. $\begin{cases} 2a + 3b - 2c = 18 \\ 5a - 6b + c = 21 \\ 4b - 2c - 6 = 0 \end{cases}$

23. $\begin{cases} 0.5a + 0.3b = 2.2 \\ 1.2c - 8.5b = -24.4 \\ 3.3c + 1.3a = 29 \end{cases}$

24. $\begin{cases} 4a - 3b = 1 \\ 6a - 8c = 1 \\ 2b - 4c = 0 \end{cases}$

25. $\begin{cases} 2x + 3y + 4z = 6 \\ 2x - 3y - 4z = -4 \\ 4x + 6y + 8z = 12 \end{cases}$

26. $\begin{cases} x - 3y + 4z = 2 \\ 2x + y + 2z = 3 \\ 4x - 5y + 10z = 7 \end{cases}$

27. $\begin{cases} x + \frac{1}{3}y + z = 13 \\ \frac{1}{2}x - y + \frac{1}{3}z = -2 \\ x + \frac{1}{2}y - \frac{1}{3}z = 2 \end{cases}$ 28. $\begin{cases} x - \frac{1}{5}y - z = 9 \\ \frac{1}{4}x + \frac{1}{5}y - \frac{1}{2}z = 5 \\ 2x + y + \frac{1}{6}z = 12 \end{cases}$

APPLICATIONS

29. MAKING STATUES An artist makes three types of ceramic statues at a monthly cost of $650 for 180 statues. The manufacturing costs for the three types are $5, $4, and $3. If the statues sell for $20, $12, and $9, respectively, how many of each type should be made to produce $2,100 in monthly revenue?

30. POTPOURRI The owner of a home decorating shop wants to mix dried rose petals selling for $6 per pound, dried lavender selling for $5 per pound, and buckwheat hulls selling for $4 per pound to get 10 pounds of a mixture that would sell for $5.50 per pound. She wants to use twice as many pounds of rose petals as lavender. How many pounds of each should she use?

31. DIETITIAN A hospital dietitian is to design a meal that will provide a patient with exactly 14 grams (g) of fat, 9 g of carbohydrates, and 9 g of protein. She is to use a combination of the three foods listed in the table. If one ounce of each of the foods has the nutrient content shown in Illustration 1, how many ounces of each should be used?

ILLUSTRATION 1

Food	Fat	Carbohydrates	Protein
A	2 g	1 g	2 g
B	3 g	2 g	1 g
C	1 g	1 g	2 g

32. NUTRITIONAL PLANNING One ounce of each of three foods has the vitamin and mineral content shown in Illustration 2. How many ounces of each must be used to provide exactly 22 milligrams (mg) of niacin, 12 mg of zinc, and 20 mg of vitamin C?

ILLUSTRATION 2

Food	Niacin	Zinc	Vitamin C
A	1 mg	1 mg	2 mg
B	2 mg	1 mg	1 mg
C	2 mg	1 mg	2 mg

33. CHAINSAW SCULPTING A north woods sculptor carves three types of statues with a chainsaw. The number of hours required for carving, sanding, and painting a totem pole, a bear, and a deer are shown in Illustration 3. How many of each should be produced to use all available labor hours?

ILLUSTRATION 3

	Totem pole	Bear	Deer	Time available
Carving	2 hr	2 hr	1 hr	14 hr
Sanding	1 hr	2 hr	2 hr	15 hr
Painting	3 hr	2 hr	2 hr	21 hr

34. CLOTHING MAKER A clothing manufacturer makes coats, shirts, and slacks. The time required for cutting, sewing, and packaging each item is shown in Illustration 4. How many of each should be made to use all available labor hours?

ILLUSTRATION 4

	Coats	Shirts	Slacks	Time available
Cutting	20 min	15 min	10 min	115 hr
Sewing	60 min	30 min	24 min	280 hr
Packaging	5 min	12 min	6 min	65 hr

35. MAIL The circle graph in Illustration 5 shows the types of mail the average American household receives each week. Use the information in the graph to determine what percent of the week's mail is advertising, is bills and statements, and is personal.

ILLUSTRATION 5

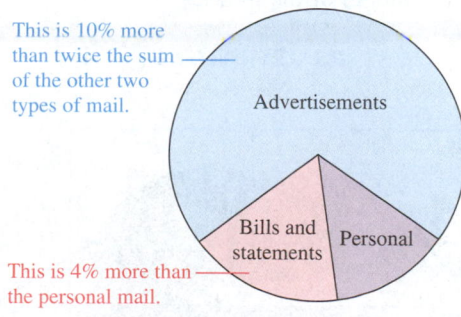

This is 10% more than twice the sum of the other two types of mail.

This is 4% more than the personal mail.

Advertisements

Bills and statements

Personal

Based on information from *Time* (July 14, 1997)

36. NFL RECORD As of the end of the 1998 National Football League season, Jerry Rice of the San Francisco 49ers had scored an all-time record 175 touchdowns. Here are some interestings facts about this feat.

- He had caught 29 more TD passes from Steve Young than from Joe Montana.
- He had caught 25 TD passes from quarterbacks other than Montana or Young.
- The number of TD passes he had caught from Joe Montana was five times the number of touchdowns he had scored on running plays.

Determine the number touchdown passes Rice had caught from Joe Montana and from Steve Young as of 1998.

37. GRAPHS OF SYSTEMS Explain how each picture in Illustration 6 could be thought of as an example of the graph of a system of three equations. Then describe the solution, if there is any.

ILLUSTRATION 6

a.

b.
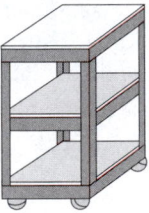

c.

d.

38. ZOOLOGY An X-ray of a laboratory mouse revealed a cancerous tumor located at the intersection of the coronal, sagittal, and transverse planes, as shown in Illustration 7. From this description, would you expect the tumor to be at the base of the tail, on the back, in the stomach, on the tip of the right ear, or in the mouth of the mouse?

ILLUSTRATION 7

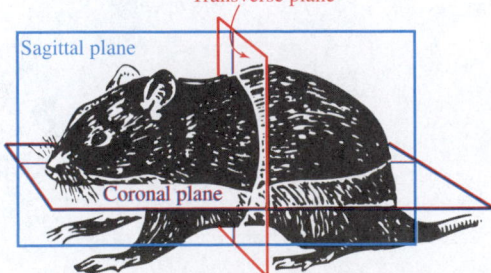

39. ASTRONOMY Comets have elliptical orbits, but the orbits of some comets are so vast that they are indistinguishable from parabolas. Find the equation of the parabola that describes the orbit of the comet shown in Illustration 8.

ILLUSTRATION 8

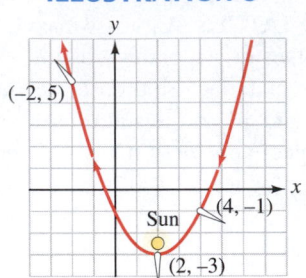

40. CURVE FITTING Find the equation of the parabola shown in Illustration 9.

ILLUSTRATION 9

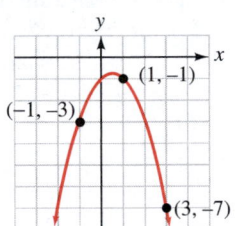

41. PARK WALKWAY A circular sidewalk is to be constructed in a city park. The walk is to pass by three particular areas of the park, as shown in the graph in Illustration 10. If an equation of a circle is of the form $x^2 + y^2 + Cx + Dy + E = 0$, find the equation that describes the path of the sidewalk by determining C, D, and E.

ILLUSTRATION 10

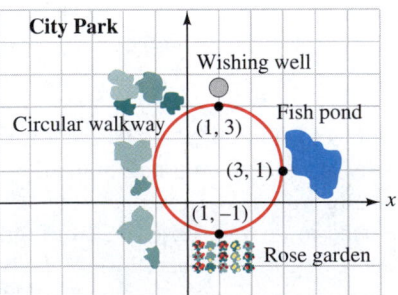

42. CURVE FITTING The equation of a circle is of the form $x^2 + y^2 + Cx + Dy + E = 0$. Find the equation of the circle shown in Illustration 11 by determining C, D, and E.

ILLUSTRATION 11

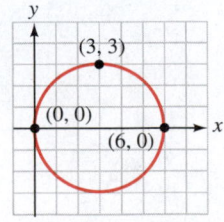

43. TRIANGLES The sum of the angles in any triangle is 180°. In triangle *ABC*, angle *A* is 100° less than the sum of angles *B* and *C*, and angle *C* is 40° less than twice angle *B*. Find each angle.

44. QUADRILATERALS The sum of the angles of any four-sided figure is 360°. In the quadrilateral shown in Illustration 12, the measures of angle *A* and angle *B* are the same, angle *C* is 20° greater than angle *A*, and angle *D* measures 40°. Find the angles.

ILLUSTRATION 12

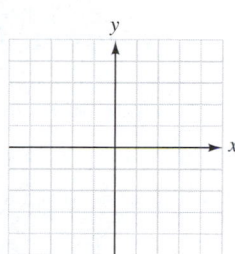

45. INTEGER PROBLEM The sum of three integers is 48. If the first integer is doubled, the sum is 60. If the second integer is doubled, the sum is 63. Find the integers.

46. INTEGER PROBLEM The sum of three integers is 18. The third integer is four times the second, and the second integer is 6 more than the first. Find the integers.

WRITING

47. Explain how a system of three equations with three variables can be reduced to a system of two equations with two variables.

48. What makes a system of three equations with three variables inconsistent?

REVIEW

In Exercises 49–52, graph each function.

49. $f(x) = |x|$

50. $g(x) = x^2$

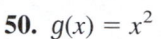

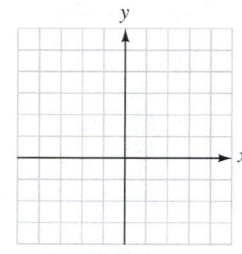

51. $h(x) = x^3$

52. $S(x) = x$

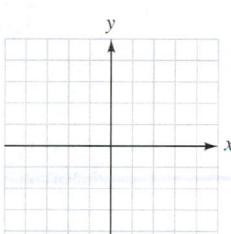

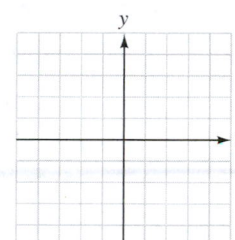

▶ 12.3   Solving Systems Using Matrices

In this section, you will learn about

> Matrices ■ Augmented matrices ■ Gaussian elimination ■ Solving a system of three equations ■ Inconsistent systems and dependent equations

Introduction In this section, we will discuss another method for solving systems of linear equations. This technique uses a mathematical tool called a *matrix* in a series of steps that are based on the addition method.

Matrices

Another method of solving systems of equations involves rectangular arrays of numbers called *matrices*.

Matrices

A **matrix** is any rectangular array of numbers arranged in rows and columns, written within brackets.

Some examples of matrices are

$$A = \begin{bmatrix} 1 & -3 & 8 \\ 2 & 5 & -1 \end{bmatrix} \begin{matrix} \leftarrow \text{Row 1} \\ \leftarrow \text{Row 2} \end{matrix} \qquad B = \begin{bmatrix} 1 & 4 & -2 & -4 \\ 6 & -2 & 6 & 1 \\ 3 & 8 & -3 & 12 \end{bmatrix} \begin{matrix} \leftarrow \text{Row 1} \\ \leftarrow \text{Row 2} \\ \leftarrow \text{Row 3} \end{matrix}$$

Column Column Column   Column Column Column Column
  1   2   3   1   2   3   4

The numbers in each matrix are called **elements.** Because matrix A has two rows and three columns, it is called a 2×3 matrix (read "2 by 3" matrix). Matrix B is a 3×4 matrix (three rows and four columns).

Augmented Matrices

To show how to use matrices to solve systems of linear equations, we consider the system

$$\begin{cases} x - y = 4 \\ 2x + y = 5 \end{cases}$$

which can be represented by the following matrix, called an **augmented matrix:**

$$\begin{bmatrix} 1 & -1 & \vdots & 4 \\ 2 & 1 & \vdots & 5 \end{bmatrix}$$

Each row of the augmented matrix represents one equation of the system. The first two columns of the augmented matrix are determined by the coefficients of x and y in the equations of the system. The last column is determined by the constants in the equations.

$$\begin{bmatrix} 1 & -1 & \vdots & 4 \\ 2 & 1 & \vdots & 5 \end{bmatrix} \begin{matrix} \leftarrow \text{This row represents the equation } x - y = 4. \\ \leftarrow \text{This row represents the equation } 2x + y = 5. \end{matrix}$$

Coefficients Coefficients Constants
   of x of y

EXAMPLE 1   **Augmented matrices.** Represent each system of equations using an augmented matrix:

a. $\begin{cases} 3x + y = 11 \\ x - 8y = 0 \end{cases}$ and **b.** $\begin{cases} 2a + b - 3c = -3 \\ 9a + 4c = 2 \\ a - b - 6c = -7 \end{cases}$.

Solution   **a.** $\begin{cases} 3x + y = 11 \\ x - 8y = 0 \end{cases} \begin{matrix} \longleftrightarrow \\ \longleftrightarrow \end{matrix} \begin{bmatrix} 3 & 1 & \vdots & 11 \\ 1 & -8 & \vdots & 0 \end{bmatrix}$

b. $\begin{cases} 2a + b - 3c = -3 \\ 9a + 4c = 2 \\ a - b - 6c = -7 \end{cases} \begin{matrix} \longleftrightarrow \\ \longleftrightarrow \\ \longleftrightarrow \end{matrix} \begin{bmatrix} 2 & 1 & -3 & \vdots & -3 \\ 9 & 0 & 4 & \vdots & 2 \\ 1 & -1 & -6 & \vdots & -7 \end{bmatrix}$

SELF CHECK   Represent each system using an augmented matrix: **a.** $\begin{cases} 2x - 4y = 9 \\ 5x - y = -2 \end{cases}$ and

b. $\begin{cases} a + b - c = -4 \\ -2b + 7c = 0 \\ 10a + 8b - 4c = 5 \end{cases}$.

Answers:

a. $\begin{bmatrix} 2 & -4 & \vdots & 9 \\ 5 & -1 & \vdots & -2 \end{bmatrix}$,

b. $\begin{bmatrix} 1 & 1 & -1 & \vdots & -4 \\ 0 & -2 & 7 & \vdots & 0 \\ 10 & 8 & -4 & \vdots & 5 \end{bmatrix}$

Gaussian Elimination

To solve a 2×2 system of equations by **Gaussian elimination,** we transform the augmented matrix into the following matrix, which has 1's down its main diagonal and a 0 below the 1 in the first column.

$$\begin{bmatrix} 1 & a & \vdots & b \\ 0 & 1 & \vdots & c \end{bmatrix} \qquad (a, b, \text{ and } c \text{ are real numbers})$$

Main diagonal

To write the augmented matrix in this form, we use three operations called **elementary row operations.**

Elementary row operations

Type 1: Any two rows of a matrix can be interchanged.
Type 2: Any row of a matrix can be multiplied by a nonzero constant.
Type 3: Any row of a matrix can be changed by adding a nonzero constant multiple of another row to it.

- A type 1 row operation corresponds to interchanging two equations of the system.
- A type 2 row operation corresponds to multiplying both sides of an equation by a nonzero constant.
- A type 3 row operation corresponds to adding a nonzero multiple of one equation to another.

None of these row operations will change the solution of the given system of equations.

EXAMPLE 2

Row operations. Consider the matrices

$$A = \begin{bmatrix} 2 & 4 & \vdots & -3 \\ 1 & -8 & \vdots & 0 \end{bmatrix} \quad B = \begin{bmatrix} 1 & -1 & \vdots & 2 \\ 4 & -8 & \vdots & 0 \end{bmatrix} \quad C = \begin{bmatrix} 2 & 1 & -8 & \vdots & 4 \\ 0 & 1 & 4 & \vdots & -2 \\ 0 & 0 & -6 & \vdots & 24 \end{bmatrix}$$

a. Interchange rows 1 and 2 of matrix A.

b. Multiply row 3 of matrix C by $-\frac{1}{6}$.

c. To the numbers in row 2 of matrix B, add the results of multiplying each number in row 1 by -4.

Solution **a.** Interchanging the rows of matrix A, we obtain $\begin{bmatrix} 1 & -8 & \vdots & 0 \\ 2 & 4 & \vdots & -3 \end{bmatrix}$.

b. We multiply each number in row 3 by $-\frac{1}{6}$. Rows 1 and 2 remain unchanged.

$$\begin{bmatrix} 2 & 1 & -8 & \vdots & 4 \\ 0 & 1 & 4 & \vdots & -2 \\ 0 & 0 & 1 & \vdots & -4 \end{bmatrix}$$ We can represent the instruction to multiply the third row by $-\frac{1}{6}$ with the symbolism $-\frac{1}{6}R_3$.

c. If we multiply each number in row 1 of matrix B by -4, we get

$$-4 \quad 4 \quad -8$$

We then add these numbers to row 2. (Note that row 1 remains unchanged.)

$$\begin{bmatrix} 1 & -1 & \vdots & 2 \\ 4 + (-4) & -8 + 4 & \vdots & 0 + (-8) \end{bmatrix}$$ We can abbreviate this procedure using the notation $-4R_1 + R_2$, which means "Multiply row 1 by -4 and add the result to row 2."

After simplifying, we have the matrix

$$\begin{bmatrix} 1 & -1 & \vdots & 2 \\ 0 & -4 & \vdots & -8 \end{bmatrix}$$

SELF CHECK

Refer to Example 2.
a. Interchange the rows of matrix B.
b. To the numbers in row 1 of matrix A, add the results of multiplying each number in row 2 by -2.
c. Interchange rows 2 and 3 of matrix C.

Answers: **a.** $\begin{bmatrix} 4 & -8 & \vdots & 0 \\ 1 & -1 & \vdots & 2 \end{bmatrix}$,

b. $\begin{bmatrix} 0 & 20 & \vdots & -3 \\ 1 & -8 & \vdots & 0 \end{bmatrix}$,

c. $\begin{bmatrix} 2 & 1 & -8 & \vdots & 4 \\ 0 & 0 & -6 & \vdots & 24 \\ 0 & 1 & 4 & \vdots & -2 \end{bmatrix}$ ∎

We now solve a system of two linear equations using the **Gaussian elimination** process, which involves a series of elementary row operations.

EXAMPLE 3

Solving a system using Gaussian elimination. Solve the system
$$\begin{cases} 2x + y = 5 \\ x - y = 4 \end{cases}.$$

Solution

We can represent the system with the following augmented matrix:

$$\begin{bmatrix} \mathbf{2} & 1 & \vdots & 5 \\ 1 & -1 & \vdots & 4 \end{bmatrix}$$

First, we want to get a 1 in the top row of the first column where the red 2 is. This can be achieved by applying a type 1 row operation: Interchange rows 1 and 2.

$$\begin{bmatrix} \mathbf{1} & -1 & \vdots & 4 \\ 2 & 1 & \vdots & 5 \end{bmatrix}$$ Interchanging row 1 and row 2 can be abbreviated as $R_1 \leftrightarrow R_2$.

To get a 0 under the 1 in the first column, we use a type 3 row operation. To row 2, we add the results of multiplying each number in row 1 by -2.

$$\begin{bmatrix} 1 & -1 & \vdots & 4 \\ \mathbf{0} & 3 & \vdots & -3 \end{bmatrix}$$ $-2R_1 + R_2$.

To get a 1 in the bottom row of the second column, we use a type 2 row operation: Multiply row 2 by $\frac{1}{3}$.

$$\begin{bmatrix} 1 & -1 & \vdots & 4 \\ 0 & \mathbf{1} & \vdots & -1 \end{bmatrix}$$ $\frac{1}{3}R_2$.

This augmented matrix represents the equations

$$1x - 1y = 4$$
$$0x + 1y = -1$$

Writing the equations without the coefficients of 1 and -1, we have

1. $x - y = 4$

2. $y = -1$

From Equation 2, we see that $y = -1$. We can **back substitute** -1 for y in Equation 1 to find x.

$$x - y = 4$$
$$x - (\mathbf{-1}) = 4 \quad \text{Substitute } -1 \text{ for } y.$$
$$x + 1 = 4 \quad -(-1) = 1.$$
$$x = 3 \quad \text{Subtract 1 from both sides.}$$

The solution of the system is $(3, -1)$. Verify that this ordered pair satisfies the original system.

Solve $\begin{cases} 3x - 2y = -5 \\ x - y = -4 \end{cases}$.

Answer: $(3, 7)$ ∎

In general, if a system of linear equations has a single solution, we can use the following steps to solve the system using matrices.

Solving systems of linear equations using matrices

1. Write an augmented matrix for the system.

2. Use elementary row operations to transform the augmented matrix into a matrix with 1's down its main diagonal and 0's under the 1's.

3. When step 2 is complete, write the resulting system. Then use back substitution to find the solution.

Solving a System of Three Equations

To show how to use matrices to solve systems of three linear equations containing three variables, we consider the system

$$\begin{cases} x - 2y - z = 6 \\ 2x + 2y - z = 1 \\ -x - y + 2z = 1 \end{cases}$$

which can be represented by the augmented matrix

$$\left[\begin{array}{ccc|c} 1 & -2 & -1 & 6 \\ 2 & 2 & -1 & 1 \\ -1 & -1 & 2 & 1 \end{array} \right]$$

To solve a 3×3 system of equations by Gaussian elimination, we transform the augmented matrix into a matrix with 1's down its main diagonal and 0's below its main diagonal.

$$\left[\begin{array}{ccc|c} 1 & a & b & c \\ 0 & 1 & d & e \\ 0 & 0 & 1 & f \end{array} \right] \qquad (a, b, c, \ldots, f \text{ are real numbers})$$

Main diagonal

E X A M P L E   4

Solving a system using Gaussian elimination.

Solve the system $\begin{cases} x - 2y - z = 6 \\ 2x + 2y - z = 1 \\ -x - y + 2z = 1 \end{cases}$.

Solution This system can be represented by the augmented matrix

$$\left[\begin{array}{ccc|c} 1 & -2 & -1 & 6 \\ \mathbf{2} & 2 & -1 & 1 \\ -1 & -1 & 2 & 1 \end{array} \right] \qquad \text{We need to get a 0 where the red 2 is.}$$

To get a 0 under the 1 in the first column where the red 2 is, we perform a type 3 row operation: Multiply row 1 by -2 and add the results to row 2.

$$\begin{bmatrix} 1 & -2 & -1 & \vdots & 6 \\ 0 & 6 & 1 & \vdots & -11 \\ \mathbf{-1} & -1 & 2 & \vdots & 1 \end{bmatrix} \quad -2R_1 + R_2.$$

To get a 0 under the 0 in the first column where the red -1 is, we perform another type 3 row operation: To row 3 we add row 1.

$$\begin{bmatrix} 1 & -2 & -1 & \vdots & 6 \\ 0 & 6 & 1 & \vdots & -11 \\ 0 & \mathbf{-3} & 1 & \vdots & 7 \end{bmatrix} \quad R_1 + R_3.$$

To get a 0 under the 6 where the red -3 is, we use another type 3 row operation: Multiply row 2 by $\frac{1}{2}$ and add it to row 3.

$$\begin{bmatrix} 1 & -2 & -1 & \vdots & 6 \\ 0 & 6 & 1 & \vdots & -11 \\ 0 & 0 & \frac{3}{2} & \vdots & \frac{3}{2} \end{bmatrix} \quad \frac{1}{2}R_2 + R_3.$$

We can use a type 2 row operation to simplify further: Multiply row 3 by $\frac{2}{3}$.

$$\begin{bmatrix} 1 & -2 & -1 & \vdots & 6 \\ 0 & 6 & 1 & \vdots & -11 \\ 0 & 0 & 1 & \vdots & 1 \end{bmatrix} \quad \frac{2}{3}R_3.$$

The final matrix represents the system

1. $\begin{cases} x - 2y - z = 6 \\ 0x + 6y + z = -11 \\ 0x + 0y + z = 1 \end{cases}$
2.
3.

From Equation 3, we can read that $z = 1$. To find y, we back substitute 1 for z in Equation 2 and solve for y:

$\quad 6y + z = -11$ Equation 2.
$\quad 6y + \mathbf{1} = -11$ Substitute 1 for z.
$\quad\quad\quad 6y = -12$ Subtract 1 from both sides.
$\quad\quad\quad\quad y = -2$ Divide both sides by 6.

Thus, $y = -2$. To find x, we back substitute 1 for z and -2 for y in Equation 1 and solve for x:

$\quad\quad x - 2y - z = 6$ Equation 1.
$\quad x - 2(\mathbf{-2}) - \mathbf{1} = 6$ Substitute 1 for z and -2 for y.
$\quad\quad\quad\quad x + 3 = 6$ Simplify.
$\quad\quad\quad\quad\quad x = 3$ Subtract 3 from both sides.

Thus, $x = 3$. The solution of the given system is $(3, -2, 1)$. Verify that this triple satisfies each equation of the original system.

SELF CHECK   Solve $\begin{cases} x - 2y - z = 2 \\ 2x + 2y - z = -5. \\ -x - y + 2z = 7 \end{cases}$ *Answer:* $(1, -2, 3)$ ∎

Inconsistent Systems and Dependent Equations

In the next example, we consider a system with no solution.

EXAMPLE 5

An inconsistent system. Using matrices, solve the system $\begin{cases} x + y = -1 \\ -3x - 3y = -5 \end{cases}$.

Solution

This system can be represented by the augmented matrix

$$\begin{bmatrix} 1 & 1 & \vdots & -1 \\ -3 & -3 & \vdots & -5 \end{bmatrix}$$

Since the matrix has a 1 in the top row of the first column, we proceed to get a 0 under it by multiplying row 1 by 3 and adding the results to row 2.

$$\begin{bmatrix} 1 & 1 & \vdots & -1 \\ 0 & 0 & \vdots & -8 \end{bmatrix} \quad 3R_1 + R_2.$$

This matrix represents the system

$$\begin{cases} x + y = -1 \\ 0 + 0 = -8 \end{cases}$$

This system has no solution, because the second equation is never true. Therefore, the system is inconsistent. It has no solutions.

SELF CHECK

Solve $\begin{cases} 4x - 8y = 9 \\ x - 2y = -5 \end{cases}$.

Answer: no solution ∎

In the next example, we consider a system with infinitely many solutions.

EXAMPLE 6

Dependent equations. Using matrices, solve the system
$$\begin{cases} 2x + 3y - 4z = 6 \\ 4x + 6y - 8z = 12 \\ -6x - 9y + 12z = -18 \end{cases}.$$

Solution

This system can be represented by the augmented matrix

$$\begin{bmatrix} 2 & 3 & -4 & \vdots & 6 \\ 4 & 6 & -8 & \vdots & 12 \\ -6 & -9 & 12 & \vdots & -18 \end{bmatrix}$$

To get a 1 in the top row of the first column, we multiply row 1 by $\frac{1}{2}$.

$$\begin{bmatrix} 1 & \frac{3}{2} & -2 & \vdots & 3 \\ 4 & 6 & -8 & \vdots & 12 \\ -6 & -9 & 12 & \vdots & -18 \end{bmatrix} \quad \frac{1}{2}R_1.$$

Next, we want to get 0's under the 1 in the first column. This can be achieved by multiplying row 1 by -4 and adding the results to row 2, and multiplying row 1 by 6 and adding the results to row 3.

$$\begin{bmatrix} 1 & \frac{3}{2} & -2 & \vdots & 3 \\ 0 & 0 & 0 & \vdots & 0 \\ 0 & 0 & 0 & \vdots & 0 \end{bmatrix} \quad \begin{matrix} -4R_1 + R_2. \\ 6R_1 + R_3. \end{matrix}$$

The last matrix represents the system

$$\begin{cases} x + \frac{3}{2}y - 2z = 3 \\ 0x + 0y + 0z = 0 \\ 0x + 0y + 0z = 0 \end{cases}$$

If we clear the first equation of fractions, we have the system

$$\begin{cases} 2x + 3y - 4z = 6 \\ 0 = 0 \\ 0 = 0 \end{cases}$$

This system has dependent equations and infinitely many solutions. Solutions of this system would be any triple (x, y, z) that satisfies the equation $2x + 3y - 4z = 6$. Two such solutions would be $(0, 2, 0)$ and $(1, 0, -1)$.

SELF CHECK   Solve $\begin{cases} 5x - 10y + 15z = 35 \\ -3x + 6y - 9z = -21 \\ 2x - 4y + 6z = 14 \end{cases}$.

Answer: There are infinitely many solutions—any triple satisfying the equation $x - 2y + 3z = 7$. ∎

STUDY SET

Section 12.3

VOCABULARY

In Exercises 1–6, fill in the blanks to make the statements true.

1. A _____ is a rectangular array of numbers.

2. The numbers in a matrix are called its _____.

3. A 3×4 matrix has 3 _____ and 4 _____.

4. Elementary _____ operations are used to produce new matrices that lead to the solution of a system.

5. A matrix that represents the equations of a system is called an _____ matrix.

6. The augmented matrix $\begin{bmatrix} 1 & 3 & \vdots & -2 \\ 0 & 1 & \vdots & 4 \end{bmatrix}$ has 1's down its main _____.

CONCEPTS

7. For each matrix, tell the number of rows and the number of columns.

a. $\begin{bmatrix} 4 & 6 & \vdots & -1 \\ \frac{1}{2} & 9 & \vdots & -3 \end{bmatrix}$

b. $\begin{bmatrix} 1 & -2 & 3 & \vdots & 1 \\ 0 & 1 & 6 & \vdots & 4 \\ 0 & 0 & 1 & \vdots & \frac{1}{3} \end{bmatrix}$

8. For each augmented matrix, give the system of equations it represents.

a. $\begin{bmatrix} 1 & 6 & \vdots & 7 \\ 0 & 1 & \vdots & 4 \end{bmatrix}$

b. $\begin{bmatrix} 2 & -2 & 9 & \vdots & 1 \\ 3 & 1 & 1 & \vdots & 0 \\ 2 & -6 & 8 & \vdots & -7 \end{bmatrix}$

9. Write the system of equations represented by the augmented matrix. Then use back substitution to find the solution.

$$\begin{bmatrix} 1 & -1 & \vdots & -10 \\ 0 & 1 & \vdots & 6 \end{bmatrix}$$

10. Write the system of equations represented by the augmented matrix. Then use back substitution to find the solution.

$$\begin{bmatrix} 1 & -2 & 1 & \vdots & -16 \\ 0 & 1 & 2 & \vdots & 8 \\ 0 & 0 & 1 & \vdots & 4 \end{bmatrix}$$

11. Matrices were used to solve a system of two linear equations. The final matrix is shown here. Explain what the result tells about the system.

$$\begin{bmatrix} 1 & 2 & \vdots & -4 \\ 0 & 0 & \vdots & 2 \end{bmatrix}$$

12. Matrices were used to solve a system of two linear equations. The final matrix is shown here. What does the result tell about the equations?

$$\begin{bmatrix} 1 & 2 & \vdots & -4 \\ 0 & 0 & \vdots & 0 \end{bmatrix}$$

NOTATION

13. Consider the matrix $A = \begin{bmatrix} 3 & 6 & -9 & \vdots & 0 \\ 1 & 5 & -2 & \vdots & 1 \\ -2 & 2 & -2 & \vdots & 5 \end{bmatrix}$.

a. Explain what is meant by $\frac{1}{3}R_1$. Then perform the operation on matrix A.

b. Explain what is meant by $-R_1 + R_2$. Then perform the operation on the answer to part a.

14. Consider the matrix $B = \begin{bmatrix} -3 & 1 & \vdots & -6 \\ 1 & -4 & \vdots & 4 \end{bmatrix}$.

a. Explain what is meant by $R_1 \longleftrightarrow R_2$. Then perform the operation on matrix B.

b. Explain what is meant by $3R_1 + R_2$. Then perform the operation on the answer to part a.

In Exercises 15–16, complete each solution.

15. Solve $\begin{cases} 4x - y = 14 \\ x + y = 6 \end{cases}$.

$$\begin{bmatrix} 4 & & \vdots & 14 \\ 1 & 1 & \vdots & 6 \end{bmatrix}$$

$$\begin{bmatrix} & 1 & \vdots & 6 \\ 4 & -1 & \vdots & 14 \end{bmatrix} \quad R_1 \longleftrightarrow R_2,$$

$$\begin{bmatrix} 1 & 1 & \vdots & 6 \\ 0 & & \vdots & -10 \end{bmatrix} \quad -4R_1 + R_2.$$

$$\begin{bmatrix} 1 & 1 & \vdots & 6 \\ 0 & 1 & \vdots & \end{bmatrix} \quad -\tfrac{1}{5}R_2.$$

This matrix represents the system

$$\begin{cases} x + y = 6 \\ = 2 \end{cases}$$

The solution is $(, 2)$.

16. Solve $\begin{cases} 2x + 2y = 18 \\ x - y = 5 \end{cases}$.

$$\begin{bmatrix} 2 & 2 & \vdots & 18 \\ & -1 & \vdots & 5 \end{bmatrix}$$

$$\begin{bmatrix} 1 & 1 & \vdots & 9 \\ & -1 & \vdots & 5 \end{bmatrix} \quad \tfrac{1}{2}R_1.$$

$$\begin{bmatrix} 1 & 1 & \vdots & 9 \\ 0 & -2 & \vdots & -4 \end{bmatrix} \quad -R_1 + R_2.$$

$$\begin{bmatrix} 1 & 1 & \vdots & 9 \\ 0 & 1 & \vdots & \end{bmatrix} \quad -\tfrac{1}{2}R_2.$$

This matrix represents the system

$$\begin{cases} x + y = \\ y = 2 \end{cases}$$

The solution is $(, 2)$.

PRACTICE

In Exercises 17–32, use matrices to solve each system of equations.

17. $\begin{cases} x + y = 2 \\ x - y = 0 \end{cases}$ **18.** $\begin{cases} x + y = 3 \\ x - y = -1 \end{cases}$

19. $\begin{cases} 2x + y = 1 \\ x + 2y = -4 \end{cases}$ **20.** $\begin{cases} 5x - 4y = 10 \\ x - 7y = 2 \end{cases}$

21. $\begin{cases} 2x - y = -1 \\ x - 2y = 1 \end{cases}$ **22.** $\begin{cases} 2x - y = 0 \\ x + y = 3 \end{cases}$

23. $\begin{cases} 3x + 4y = -12 \\ 9x - 2y = 6 \end{cases}$ **24.** $\begin{cases} 2x - 3y = 16 \\ -4x + y = -22 \end{cases}$

25. $\begin{cases} x + y + z = 6 \\ x + 2y + z = 8 \\ x + y + 2z = 9 \end{cases}$ **26.** $\begin{cases} x - y + z = 2 \\ x + 2y - z = 6 \\ 2x - y - z = 3 \end{cases}$

27. $\begin{cases} 3x + y - 3z = 5 \\ x - 2y + 4z = 10 \\ x + y + z = 13 \end{cases}$ **28.** $\begin{cases} 2x + y - 3z = -1 \\ 3x - 2y - z = -5 \\ x - 3y - 2z = -12 \end{cases}$

29. $\begin{cases} 3x - 2y + 4z = 4 \\ x + y + z = 3 \\ 6x - 2y - 3z = 10 \end{cases}$ **30.** $\begin{cases} 2x + 3y - z = -8 \\ x - y - z = -2 \\ -4x + 3y + z = 6 \end{cases}$

31. $\begin{cases} 2a + b + 3c = 3 \\ -2a - b + c = 5 \\ 4a - 2b + 2c = 2 \end{cases}$ **32.** $\begin{cases} 3a + 2b + c = 8 \\ 6a - b + 2c = 16 \\ -9a + b - c = -20 \end{cases}$

In Exercises 33–44, use matrices to solve each system of equations. If the equations of a system are dependent or if a system is inconsistent, so indicate.

33. $\begin{cases} x - 3y = 9 \\ -2x + 6y = 18 \end{cases}$ **34.** $\begin{cases} -6x + 12y = 10 \\ 2x - 4y = 8 \end{cases}$

35. $\begin{cases} 4x + 4y = 12 \\ -x - y = -3 \end{cases}$ **36.** $\begin{cases} 5x - 15y = 10 \\ 2x - 6y = 4 \end{cases}$

37. $\begin{cases} 6x + y - z = -2 \\ x + 2y + z = 5 \\ 5y - z = 2 \end{cases}$ **38.** $\begin{cases} 2x + 3y - 2z = 18 \\ 5x - 6y + z = 21 \\ 4y - 2z = 6 \end{cases}$

39. $\begin{cases} 2x + y - z = 1 \\ x + 2y + 2z = 2 \\ 4x + 5y + 3z = 3 \end{cases}$ **40.** $\begin{cases} x - 3y + 4z = 2 \\ 2x + y + 2z = 3 \\ 4x - 5y + 10z = 7 \end{cases}$

41. $\begin{cases} 5x + 3y = 4 \\ 3y - 4z = 4 \\ x + z = 1 \end{cases}$ **42.** $\begin{cases} y + 2z = -2 \\ x + y = 1 \\ 2x - z = 0 \end{cases}$

43. $\begin{cases} x - y = 1 \\ 2x - z = 0 \\ 2y - z = -2 \end{cases}$ **44.** $\begin{cases} x + y - 3z = 4 \\ 2x + 2y - 6z = 5 \\ -3x + y - z = 2 \end{cases}$

In Exercises 45–48, remember these facts from geometry. Two angles whose measures add up to 90° are complementary. Two angles whose measures add up to 180° are supplementary. The sum of the measures of the interior angles in a triangle is 180°.

45. One angle is 46° larger than its complement. Find the angles.

46. One angle is 28° larger than its supplement. Find the angles.

47. In Illustration 1, angle *B* is 25° more than angle *A*, and angle *C* is 5° less than twice angle *A*. Find each angle in the triangle.

ILLUSTRATION 1

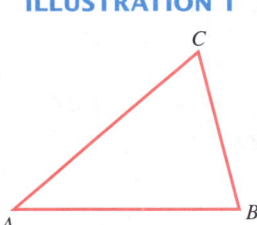

48. In Illustration 2, angle *A* is 10° less than angle *B*, and angle *B* is 10° less than angle *C*. Find each angle in the triangle.

ILLUSTRATION 2

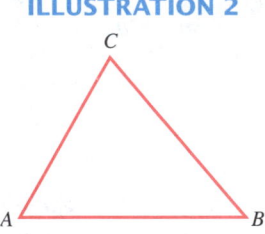

APPLICATIONS

Write a system of equations to solve each problem. Then use matrices to solve the system.

49. PHYSICAL THERAPY After an elbow injury, a volleyball player has restricted movement of her arm. (See Illustration 3.) Her range of motion (angle 1) is 28° less than angle 2. Find the measure of each angle.

ILLUSTRATION 3

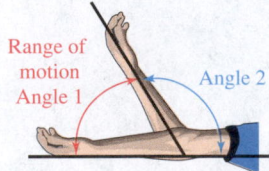

50. TRIANGLES In Illustration 4, ∠*B* (read "angle *B*") is 25° more than ∠*A*, and ∠*C* is 5° less than twice ∠*A*. Find each angle in the triangle. (*Hint:* The sum of the angles in a triangle is 180°.)

ILLUSTRATION 4

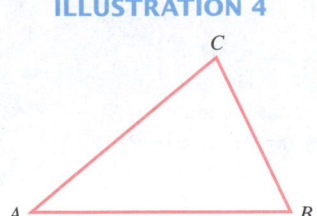

51. THEATER SEATING Illustration 5 shows the cash receipts from two sold-out performances of a play and the ticket prices. Find the number of seats in each of the three sections of the 800-seat theater.

ILLUSTRATION 5

Sunday Ticket Receipts

Matinee	$13,000
Evening	$23,000

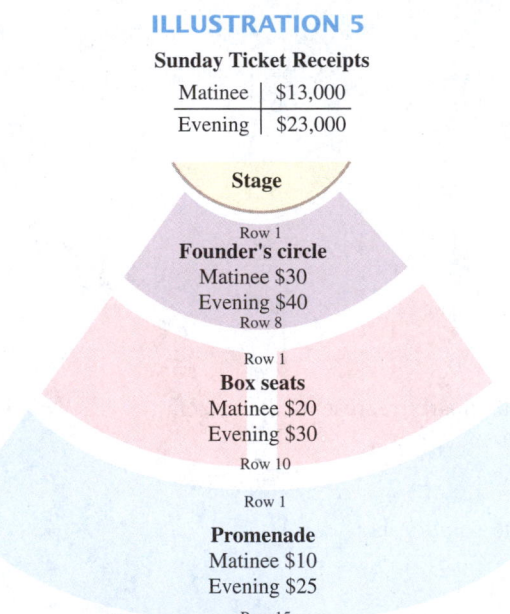

52. ICE SKATING Illustration 6 shows three circles traced out by a figure skater during her performance. If the centers of the circles are the given distances apart, find the radius of each circle.

ILLUSTRATION 6

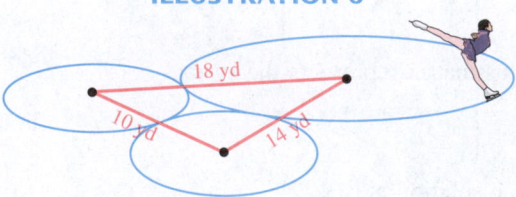

WRITING

53. Explain what is meant by the phrase *back substitution.*

54. Explain how a type 3 row operation is similar to the addition method of solving a system of equations.

REVIEW

55. What is the formula used to find the slope of a line, given two points on the line?

56. What is the form of the equation of a horizontal line? Of a vertical line?

57. What is the point–slope form of the equation of a line?

58. What is the slope–intercept form of the equation of a line?

▶ 12.4   Solving Systems Using Determinants

In this section, you will learn about

> Determinants ■ Evaluating a determinant ■ Using Cramer's rule to solve a system of two equations ■ Using Cramer's rule to solve a system of three equations

Introduction   In this section, we will discuss another method for solving systems of linear equations. With this method, called *Cramer's rule,* we work with combinations of the coefficients and the constants of the equations written as *determinants.*

Determinants

An idea closely related to the concept of matrix is the **determinant.** A determinant is a number that is associated with a **square matrix,** a matrix that has the same number of rows and columns. For any square matrix A, the symbol $|A|$ represents the determinant of A. To write a determinant, we put the elements of a square matrix between two vertical lines.

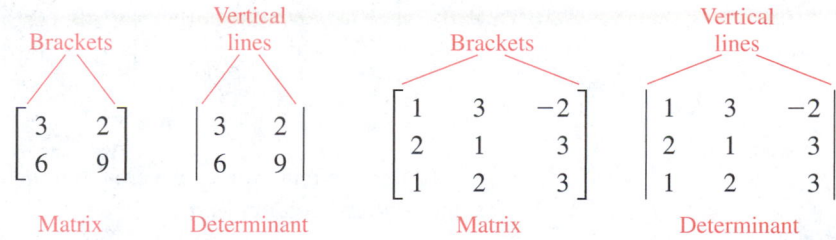

Like matrices, determinants are classified according to the number of rows and columns they contain. The determinant on the left is a 2×2 determinant. The other is a 3×3 determinant.

Evaluating a Determinant

The determinant of a 2×2 matrix is the number that is equal to the product of the numbers on the main diagonal minus the product of the numbers on the other diagonal.

$$\begin{vmatrix} a & b \\ c & d \end{vmatrix} \qquad\qquad \begin{vmatrix} a & b \\ c & d \end{vmatrix}$$

Main diagonal   Other diagonal

Value of a 2 × 2 determinant

If a, b, c, and d are numbers, the **determinant** of the matrix $\begin{bmatrix} a & b \\ c & d \end{bmatrix}$ is

$$\begin{vmatrix} a & b \\ c & d \end{vmatrix} = ad - bc$$

EXAMPLE 1

Evaluating 2 × 2 determinants. Find each value: **a.** $\begin{vmatrix} 3 & 2 \\ 6 & 9 \end{vmatrix}$ and

b. $\begin{vmatrix} -5 & \frac{1}{2} \\ -1 & 0 \end{vmatrix}$.

Solution   From the product of the numbers along the main diagonal, we subtract the product of the numbers along the other diagonal.

a. $\begin{vmatrix} 3 & 2 \\ 6 & 9 \end{vmatrix} = 3(9) - 2(6)$ **b.** $\begin{vmatrix} -5 & \frac{1}{2} \\ -1 & 0 \end{vmatrix} = -5(0) - \frac{1}{2}(-1)$

$$= 27 - 12 \qquad\qquad\qquad\qquad = 0 + \frac{1}{2}$$

$$= 15 \qquad\qquad\qquad\qquad\qquad = \frac{1}{2}$$

SELF CHECK   Evaluate $\begin{vmatrix} 4 & -3 \\ 2 & 1 \end{vmatrix}$.

Answer: 10   ■

A 3 × 3 determinant is evaluated by **expanding by minors.**

Value of a 3 × 3 determinant

| Minor of a_1 | Minor of b_1 | Minor of c_1 |

$$\begin{vmatrix} a_1 & b_1 & c_1 \\ a_2 & b_2 & c_2 \\ a_3 & b_3 & c_3 \end{vmatrix} = a_1 \begin{vmatrix} b_2 & c_2 \\ b_3 & c_3 \end{vmatrix} - b_1 \begin{vmatrix} a_2 & c_2 \\ a_3 & c_3 \end{vmatrix} + c_1 \begin{vmatrix} a_2 & b_2 \\ a_3 & b_3 \end{vmatrix}$$

To find the minor of a_1, we cross out the elements of the determinant that are in the same row and column as a_1:

$$\begin{vmatrix} a_1 & b_1 & c_1 \\ a_2 & b_2 & c_2 \\ a_3 & b_3 & c_3 \end{vmatrix}$$ The minor of a_1 is $\begin{vmatrix} b_2 & c_2 \\ b_3 & c_3 \end{vmatrix}$.

To find the minor of b_1, we cross out the elements of the determinant that are in the same row and column as b_1:

$$\begin{vmatrix} a_1 & b_1 & c_1 \\ a_2 & b_2 & c_2 \\ a_3 & b_3 & c_3 \end{vmatrix}$$ The minor of b_1 is $\begin{vmatrix} a_2 & c_2 \\ a_3 & c_3 \end{vmatrix}$.

To find the minor of c_1, we cross out the elements of the determinant that are in the same row and column as c_1:

$$\begin{vmatrix} a_1 & b_1 & c_1 \\ a_2 & b_2 & c_2 \\ a_3 & b_3 & c_3 \end{vmatrix}$$ The minor of c_1 is $\begin{vmatrix} a_2 & b_2 \\ a_3 & b_3 \end{vmatrix}$.

EXAMPLE 2

Using minors to evaluate a 3 × 3 determinant. Find the value of

$$\begin{vmatrix} 1 & 3 & -2 \\ 2 & 1 & 3 \\ 1 & 2 & 3 \end{vmatrix}.$$

Solution   We evaluate this determinant by expanding by minors along the first row of the determinant.

Minor of 1   Minor of 3   Minor of −2

$$\begin{vmatrix} 1 & 3 & -2 \\ 2 & 1 & 3 \\ 1 & 2 & 3 \end{vmatrix} = 1\begin{vmatrix} 1 & 3 \\ 2 & 3 \end{vmatrix} - 3\begin{vmatrix} 2 & 3 \\ 1 & 3 \end{vmatrix} + (-2)\begin{vmatrix} 2 & 1 \\ 1 & 2 \end{vmatrix}$$

$$= 1(3 - 6) - 3(6 - 3) - 2(4 - 1) \quad \text{Evaluate each } 2 \times 2 \text{ determinant.}$$

$$= 1(-3) - 3(3) - 2(3)$$

$$= -3 - 9 - 6$$

$$= -18$$

SELF CHECK   Evaluate $\begin{vmatrix} 2 & -1 & 3 \\ 1 & 2 & -2 \\ 3 & 1 & 1 \end{vmatrix}.$ *Answer:* 0   ■

We can evaluate a 3 × 3 determinant by expanding it along any row or column. To determine the signs between the terms of the expansion of a 3 × 3 determinant, we use the following array of signs.

Array of signs for a 3 × 3 determinant

$$\begin{matrix} + & - & + \\ - & + & - \\ + & - & + \end{matrix}$$

EXAMPLE 3

Expanding along a column. Evaluate the determinant $\begin{vmatrix} 1 & 3 & -2 \\ 2 & 1 & 3 \\ 1 & 2 & 3 \end{vmatrix}$ by expanding on the middle column.

Solution   This is the determinant of Example 2. To expand it along the middle column, we use the signs of the middle column of the array of signs:

Minor of 3   Minor of 1   Minor of 2

$$\begin{vmatrix} 1 & 3 & -2 \\ 2 & 1 & 3 \\ 1 & 2 & 3 \end{vmatrix} = -3\begin{vmatrix} 2 & 3 \\ 1 & 3 \end{vmatrix} + 1\begin{vmatrix} 1 & -2 \\ 1 & 3 \end{vmatrix} - 2\begin{vmatrix} 1 & -2 \\ 2 & 3 \end{vmatrix} \quad \text{Use the sign pattern } - + -.$$

$$= -3(6 - 3) + 1[3 - (-2)] - 2[3 - (-4)] \quad \text{Evaluate each } 2 \times 2 \text{ determinant.}$$

$$= -3(3) + 1(5) - 2(7)$$

$$= -9 + 5 - 14$$

$$= -18$$

As expected, we get the same value as in Example 2.

SELF CHECK   Evaluate $\begin{vmatrix} 1 & 3 & -2 \\ 2 & 1 & 3 \\ 1 & 2 & 3 \end{vmatrix}$ by

expanding along the last column.   *Answer:* -18 ∎

ACCENT ON TECHNOLOGY *Evaluating Determinants*

It is possible to use a graphing calculator to evaluate determinants. For example, to evaluate the determinant in Example 3, we first enter the matrix by pressing the ⃞MATRIX key, selecting EDIT, and pressing the ⃞ENTER key. We then enter the dimensions and the elements of the matrix to get Figure 12-10(a). We then press ⃞2nd ⃞QUIT to clear the screen. We then press ⃞MATRIX, select MATH, and press 1 to get Figure 12-10(b). We then press ⃞MATRIX, select NAMES, and press 1 to get Figure 12-10(c). To get the value of the determinant, we now press ⃞ENTER to get Figure 12-10(d), which shows that the value of the determinant is -18.

Figure 12-10

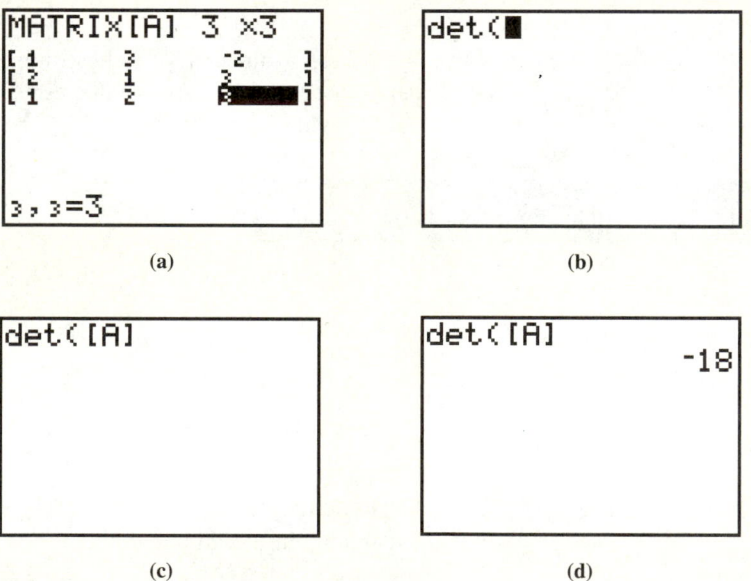

(a)   (b)

(c)   (d)

Using Cramer's Rule to Solve a System of Two Equations

The method of using determinants to solve systems of linear equations is called **Cramer's rule,** named after the 18th-century mathematician Gabriel Cramer. To develop Cramer's rule, we consider the system

$$\begin{cases} ax + by = e \\ cx + dy = f \end{cases}$$

where x and y are variables and a, b, c, d, e, and f are constants.

If we multiply both sides of the first equation by d and multiply both sides of the second equation by $-b$, we can add the equations and eliminate y:

$$\begin{array}{r} adx + bdy = ed \\ -bcx - bdy = -bf \\ \hline adx - bcx = ed - bf \end{array}$$

To solve for x, we use the distributive property to write $adx - bcx$ as $(ad - bc)x$ on the left-hand side and divide each side by $ad - bc$:

$$(ad - bc)x = ed - bf$$

$$x = \frac{ed - bf}{ad - bc} \qquad (ad - bc \neq 0)$$

We can find y in a similar manner. After eliminating the variable x, we get

$$y = \frac{af - ec}{ad - bc} \qquad (ad - bc \neq 0)$$

Determinants provide an easy way of remembering these formulas. Note that the denominator for both x and y is

$$\begin{vmatrix} a & b \\ c & d \end{vmatrix} = ad - bc$$

The numerators can be expressed as determinants also:

$$x = \frac{ed - bf}{ad - bc} = \frac{\begin{vmatrix} e & b \\ f & d \end{vmatrix}}{\begin{vmatrix} a & b \\ c & d \end{vmatrix}} \quad \text{and} \quad y = \frac{af - ec}{ad - bc} = \frac{\begin{vmatrix} a & e \\ c & f \end{vmatrix}}{\begin{vmatrix} a & b \\ c & d \end{vmatrix}}$$

If we compare these formulas with the original system

$$\begin{cases} ax + by = e \\ cx + dy = f \end{cases}$$

we note that in the expressions for x and y above, the denominator determinant is formed by using the coefficients a, b, c, and d of the variables in the equations. The numerator determinants are the same as the denominator determinant, except that the column of coefficients of the variable for which we are solving is replaced with the column of constants e and f.

Cramer's rule for two equations in two variables

The solution of the system $\begin{cases} ax + by = e \\ cx + dy = f \end{cases}$ is given by

$$x = \frac{D_x}{D} = \frac{\begin{vmatrix} e & b \\ f & d \end{vmatrix}}{\begin{vmatrix} a & b \\ c & d \end{vmatrix}} \quad \text{and} \quad y = \frac{D_y}{D} = \frac{\begin{vmatrix} a & e \\ c & f \end{vmatrix}}{\begin{vmatrix} a & b \\ c & d \end{vmatrix}}$$

If every determinant is 0, the system is consistent, but the equations are dependent.

If $D = 0$ and D_x or D_y is nonzero, the system is inconsistent. If $D \neq 0$, the system is consistent, and the equations are independent.

EXAMPLE 4

Cramer's rule. Use Cramer's rule to solve $\begin{cases} 4x - 3y = 6 \\ -2x + 5y = 4 \end{cases}$.

Solution The value of x is the quotient of two determinants, D and D_x. The denominator determinant D is made up of the coefficients of x and y:

$$D = \begin{vmatrix} 4 & -3 \\ -2 & 5 \end{vmatrix}$$

To solve for x, we form the numerator determinant D_x from D by replacing its first column (the coefficients of x) with the column of constants (6 and 4).

To solve for y, we form the numerator determinant D_y from D by replacing the second column (the coefficients of y) with the column of constants (6 and 4).

To find the values of x and y, we evaluate each determinant:

$$x = \frac{D_x}{D} = \frac{\begin{vmatrix} 6 & -3 \\ 4 & 5 \end{vmatrix}}{\begin{vmatrix} 4 & -3 \\ -2 & 5 \end{vmatrix}} = \frac{6(5) - (-3)(4)}{4(5) - (-3)(-2)} = \frac{30 + 12}{20 - 6} = \frac{42}{14} = 3$$

$$y = \frac{D_y}{D} = \frac{\begin{vmatrix} 4 & 6 \\ -2 & 4 \end{vmatrix}}{\begin{vmatrix} 4 & -3 \\ -2 & 5 \end{vmatrix}} = \frac{4(4) - 6(-2)}{14} = \frac{16 + 12}{14} = \frac{28}{14} = 2$$

The solution of this system is (3, 2). Verify that $x = 3$ and $y = 2$ satisfy both equations.

SELF CHECK Solve $\begin{cases} 2x - 3y = -16 \\ 3x + 5y = 14 \end{cases}$ using Cramer's rule.   *Answer:* $(-2, 4)$ ■

EXAMPLE 5

An inconsistent system. Use Cramer's rule to solve $\begin{cases} 7x = 8 - 4y \\ 2y = 3 - \frac{7}{2}x \end{cases}$.

Solution We multiply both sides of the second equation by 2 to eliminate the fraction and write the system in the form

$$\begin{cases} 7x + 4y = 8 \\ 7x + 4y = 6 \end{cases}$$

When we attempt to use Cramer's rule to solve this system for x, we obtain

$$x = \frac{D_x}{D} = \frac{\begin{vmatrix} 8 & 4 \\ 6 & 4 \end{vmatrix}}{\begin{vmatrix} 7 & 4 \\ 7 & 4 \end{vmatrix}} = \frac{8}{0} \quad \text{which is undefined}$$

Since the denominator determinant D is 0 and the numerator determinant D_x is not 0, the system is inconsistent. It has no solutions.

We can see directly from the system that it is inconsistent. For any values of x and y, it is impossible that 7 times x plus 4 times y could be both 8 and 6.

SELF CHECK Solve $\begin{cases} 3x = 8 - 4y \\ y = \frac{5}{2} - \frac{3}{4}x \end{cases}$ using Cramer's rule.   *Answer:* no solutions   ■

Using Cramer's Rule to Solve
a System of Three Equations

Cramer's rule can be extended to solve systems of three linear equations with three variables.

Cramer's rule for three equations with three variables

The solution of the system $\begin{cases} ax + by + cz = j \\ dx + ey + fz = k \\ gx + hy + iz = l \end{cases}$ is given by

$$x = \frac{D_x}{D}, \qquad y = \frac{D_y}{D}, \qquad \text{and} \qquad z = \frac{D_z}{D}$$

where

$$D = \begin{vmatrix} a & b & c \\ d & e & f \\ g & h & i \end{vmatrix} \qquad D_x = \begin{vmatrix} j & b & c \\ k & e & f \\ l & h & i \end{vmatrix}$$

$$D_y = \begin{vmatrix} a & j & c \\ d & k & f \\ g & l & i \end{vmatrix} \qquad D_z = \begin{vmatrix} a & b & j \\ d & e & k \\ g & h & l \end{vmatrix}$$

If every determinant is 0, the system is consistent, but the equations are dependent.

If $D = 0$ and D_x or D_y or D_z is nonzero, the system is inconsistent. If $D \neq 0$, the system is consistent, and the equations are independent.

EXAMPLE 6

A system of three equations. Use Cramer's rule to solve
$$\begin{cases} 2x + y + 4z = 12 \\ x + 2y + 2z = 9 \\ 3x - 3y - 2z = 1 \end{cases}.$$

Solution The denominator determinant D is the determinant formed by the coefficients of the variables. The numerator determinants, D_x, D_y, and D_z, are formed by replacing the coefficients of the variable being solved for by the column of constants. We form the quotients for x, y, and z and evaluate each determinant by expanding by minors about the first row:

$$x = \frac{D_x}{D} = \frac{\begin{vmatrix} 12 & 1 & 4 \\ 9 & 2 & 2 \\ 1 & -3 & -2 \end{vmatrix}}{\begin{vmatrix} 2 & 1 & 4 \\ 1 & 2 & 2 \\ 3 & -3 & -2 \end{vmatrix}}$$

$$= \frac{12 \begin{vmatrix} 2 & 2 \\ -3 & -2 \end{vmatrix} - 1 \begin{vmatrix} 9 & 2 \\ 1 & -2 \end{vmatrix} + 4 \begin{vmatrix} 9 & 2 \\ 1 & -3 \end{vmatrix}}{2 \begin{vmatrix} 2 & 2 \\ -3 & -2 \end{vmatrix} - 1 \begin{vmatrix} 1 & 2 \\ 3 & -2 \end{vmatrix} + 4 \begin{vmatrix} 1 & 2 \\ 3 & -3 \end{vmatrix}}$$

$$= \frac{12(2) - 1(-20) + 4(-29)}{2(2) - 1(-8) + 4(-9)}$$

$$= \frac{-72}{-24}$$

$$= 3$$

$$y = \frac{D_y}{D} = \frac{\begin{vmatrix} 2 & 12 & 4 \\ 1 & 9 & 2 \\ 3 & 1 & -2 \end{vmatrix}}{\begin{vmatrix} 2 & 1 & 4 \\ 1 & 2 & 2 \\ 3 & -3 & -2 \end{vmatrix}}$$

$$= \frac{2\begin{vmatrix} 9 & 2 \\ 1 & -2 \end{vmatrix} - 12\begin{vmatrix} 1 & 2 \\ 3 & -2 \end{vmatrix} + 4\begin{vmatrix} 1 & 9 \\ 3 & 1 \end{vmatrix}}{-24}$$

$$= \frac{2(-20) - 12(-8) + 4(-26)}{-24}$$

$$= \frac{-48}{-24}$$

$$= 2$$

$$z = \frac{D_z}{D} = \frac{\begin{vmatrix} 2 & 1 & 12 \\ 1 & 2 & 9 \\ 3 & -3 & 1 \end{vmatrix}}{\begin{vmatrix} 2 & 1 & 4 \\ 1 & 2 & 2 \\ 3 & -3 & -2 \end{vmatrix}}$$

$$= \frac{2\begin{vmatrix} 2 & 9 \\ -3 & 1 \end{vmatrix} - 1\begin{vmatrix} 1 & 9 \\ 3 & 1 \end{vmatrix} + 12\begin{vmatrix} 1 & 2 \\ 3 & -3 \end{vmatrix}}{-24}$$

$$= \frac{2(29) - 1(-26) + 12(-9)}{-24}$$

$$= \frac{-24}{-24}$$

$$= 1$$

The solution of this system is $(3, 2, 1)$.

SELF CHECK   Solve $\begin{cases} x + y + 2z = 6 \\ 2x - y + z = 9 \\ x + y - 2z = -6 \end{cases}$ using Cramer's rule.

Answer: $(2, -2, 3)$

STUDY SET

Section 12.4

VOCABULARY

In Exercises 1–6, fill in the blanks to make the statements true.

1. $\begin{vmatrix} 2 & 1 \\ -6 & 1 \end{vmatrix}$ is a 2×2 _____.

2. A _____ matrix has the same number of rows and columns.

3. The _____ of b_1 in $\begin{vmatrix} a_1 & b_1 & c_1 \\ a_2 & b_2 & c_2 \\ a_3 & b_3 & c_3 \end{vmatrix}$ is $\begin{vmatrix} a_2 & c_2 \\ a_3 & c_3 \end{vmatrix}$.

4. In $\begin{vmatrix} 7 & -3 \\ 1 & 2 \end{vmatrix}$, 7 and 2 lie along the main _____.

5. A 3×3 determinant has 3 _____ and 3 _____.

6. _____ rule uses determinants to solve systems of linear equations.

CONCEPTS

In Exercises 7–8, fill in the blanks to make the statements true.

7. If the denominator determinant D for a system of equations is zero, the equations of the system are _____ or the system is _____.

8. To find the minor of 5, we _____ the elements of the determinant that are in the same row and column as 5.

$$\begin{vmatrix} 3 & 5 & 1 \\ 6 & -2 & 2 \\ 8 & -1 & 4 \end{vmatrix}$$

9. What is the value of $\begin{vmatrix} a & b \\ c & d \end{vmatrix}$?

10. $\begin{vmatrix} 5 & 1 & -1 \\ 8 & 7 & 4 \\ 9 & 7 & 6 \end{vmatrix} = -1\begin{vmatrix} 8 & 7 \\ 9 & 7 \end{vmatrix} - 4\begin{vmatrix} 5 & 1 \\ 9 & 7 \end{vmatrix} + 6\begin{vmatrix} 5 & 1 \\ 8 & 7 \end{vmatrix}$

In evaluating this determinant, about what row or column was it expanded?

11. What is the denominator determinant D for the system $\begin{cases} 3x + 4y = 7 \\ 2x - 3y = 5 \end{cases}$?

12. What is the denominator determinant D for the system $\begin{cases} x + 2y = -8 \\ 3x + y - z = -2 \\ 8x + 4y - z = 6 \end{cases}$?

13. For the system $\begin{cases} 3x + 2y = 1 \\ 4x - y = 3 \end{cases}$, $D_x = -7$, $D_y = 5$, and $D = -11$. What is the solution of the system?

14. For the system $\begin{cases} 2x + 3y - z = -8 \\ x - y - z = -2 \\ -4x + 3y + z = 6 \end{cases}$, $D_x = -28$, $D_y = -14$, $D_z = 14$, and $D = 14$. What is the solution?

NOTATION

In Exercises 15–16, complete the evaluation of each determinant.

15. $\begin{vmatrix} 5 & -2 \\ -2 & 6 \end{vmatrix}$

$= 5(\ \ ) - (-2)(-2)$

$= \ \ \ \ - 4$

$= 26$

16. $\begin{vmatrix} 2 & 1 & 3 \\ 3 & 4 & 2 \\ 1 & 5 & 3 \end{vmatrix}$

$= 2\begin{vmatrix} 4 & \ \ \\ 5 & 3 \end{vmatrix} - 1\begin{vmatrix} 3 & 2 \\ \ \ & 3 \end{vmatrix} + 3\begin{vmatrix} 3 & 4 \\ 1 & \ \ \end{vmatrix}$

$= 2(\ \ - 10) - 1(9 - \ \ ) + 3(15 - \ \ )$

$= 2(2) - 1(\ \ ) + \ \ (11)$

$= 4 - 7 + \ \$

$= 30$

PRACTICE

In Exercises 17–32, evaluate each determinant.

17. $\begin{vmatrix} 2 & 3 \\ -2 & 1 \end{vmatrix}$

18. $\begin{vmatrix} 3 & -2 \\ -2 & 4 \end{vmatrix}$

19. $\begin{vmatrix} -1 & 2 \\ 3 & -4 \end{vmatrix}$

20. $\begin{vmatrix} -1 & -2 \\ -3 & -4 \end{vmatrix}$

21. $\begin{vmatrix} 10 & 0 \\ 1 & 20 \end{vmatrix}$

22. $\begin{vmatrix} 1 & 15 \\ 15 & 0 \end{vmatrix}$

23. $\begin{vmatrix} -6 & -2 \\ 15 & 4 \end{vmatrix}$

24. $\begin{vmatrix} 3 & -2 \\ 12 & -8 \end{vmatrix}$

25. $\begin{vmatrix} 1 & 2 & 0 \\ 0 & 1 & 2 \\ 0 & 0 & 1 \end{vmatrix}$

26. $\begin{vmatrix} -1 & 2 & 1 \\ 2 & 1 & -3 \\ 1 & 1 & 1 \end{vmatrix}$

27. $\begin{vmatrix} 1 & -2 & 3 \\ -2 & 1 & 1 \\ -3 & -2 & 1 \end{vmatrix}$

28. $\begin{vmatrix} 1 & 1 & 2 \\ 2 & 1 & -2 \\ 3 & 1 & 3 \end{vmatrix}$

29. $\begin{vmatrix} 1 & 0 & 1 \\ 0 & 1 & 0 \\ 1 & 1 & 1 \end{vmatrix}$

30. $\begin{vmatrix} 3 & 5 & 1 \\ 6 & -2 & 2 \\ 8 & -1 & 4 \end{vmatrix}$

31. $\begin{vmatrix} 1 & 2 & 1 \\ -3 & 7 & 3 \\ -4 & 3 & -5 \end{vmatrix}$

32. $\begin{vmatrix} 1 & 4 & 7 \\ 2 & 5 & 8 \\ 3 & 6 & 9 \end{vmatrix}$

In Exercises 33–54, use Cramer's rule to solve each system of equations, if possible. If a system is inconsistent or if the equations are dependent, so indicate.

33. $\begin{cases} x + y = 6 \\ x - y = 2 \end{cases}$

34. $\begin{cases} x - y = 4 \\ 2x + y = 5 \end{cases}$

35. $\begin{cases} 2x + 3y = 0 \\ 4x - 6y = -4 \end{cases}$

36. $\begin{cases} 4x - 3y = -1 \\ 8x + 3y = 4 \end{cases}$

37. $\begin{cases} 3x + 2y = 11 \\ 6x + 4y = 11 \end{cases}$

38. $\begin{cases} 5x + 6y = 12 \\ 10x + 12y = 24 \end{cases}$

39. $\begin{cases} y = \dfrac{-2x+1}{3} \\ 3x - 2y = 8 \end{cases}$ **40.** $\begin{cases} 2x + 3y = -1 \\ x = \dfrac{y-9}{4} \end{cases}$

41. $\begin{cases} x + y + z = 4 \\ x + y - z = 0 \\ x - y + z = 2 \end{cases}$

42. $\begin{cases} x + y + z = 4 \\ x - y + z = 2 \\ x - y - z = 0 \end{cases}$

43. $\begin{cases} x + y + 2z = 7 \\ x + 2y + z = 8 \\ 2x + y + z = 9 \end{cases}$

44. $\begin{cases} x + 2y + 2z = 10 \\ 2x + y + 2z = 9 \\ 2x + 2y + z = 1 \end{cases}$

45. $\begin{cases} 2x + y + z = 5 \\ x - 2y + 3z = 10 \\ x + y - 4z = -3 \end{cases}$

46. $\begin{cases} 3x + 2y - z = -8 \\ 2x - y + 7z = 10 \\ 2x + 2y - 3z = -10 \end{cases}$

47. $\begin{cases} 4x - 3y = 1 \\ 6x - 8z = 1 \\ 2y - 4z = 0 \end{cases}$

48. $\begin{cases} 4x + 3z = 4 \\ 2y - 6z = -1 \\ 8x + 4y + 3z = 9 \end{cases}$

49. $\begin{cases} 2x + 3y + 4z = 6 \\ 2x - 3y - 4z = -4 \\ 4x + 6y + 8z = 12 \end{cases}$

50. $\begin{cases} x - 3y + 4z - 2 = 0 \\ 2x + y + 2z - 3 = 0 \\ 4x - 5y + 10z - 7 = 0 \end{cases}$

51. $\begin{cases} 2x + y - z - 1 = 0 \\ x + 2y + 2z - 2 = 0 \\ 4x + 5y + 3z - 3 = 0 \end{cases}$

52. $\begin{cases} 2x - y + 4z + 2 = 0 \\ 5x + 8y + 7z = -8 \\ x + 3y + z + 3 = 0 \end{cases}$

53. $\begin{cases} x + y = 1 \\ \frac{1}{2}y + z = \frac{5}{2} \\ x - z = -3 \end{cases}$

54. $\begin{cases} \frac{1}{2}x + y + z + \frac{3}{2} = 0 \\ x + \frac{1}{2}y + z - \frac{1}{2} = 0 \\ x + y + \frac{1}{2}z + \frac{1}{2} = 0 \end{cases}$

APPLICATIONS

Write a system of equations to solve each problem. Then use Cramer's rule to solve the system.

55. INVENTORY Illustration 1 shows an end-of-the-year inventory report for a warehouse that supplies electronics stores. If the warehouse stocks two models of cordless telephones, one valued at $67 and the other at $100, how many of each model of phone did the warehouse have at the time of the inventory?

ILLUSTRATION 1

Item	Number	Merchandise value
Televisions	800	$1,005,450
Radios	200	$15,785
Cordless phones	360	$29,400

56. SIGNALING A system of sending signals uses two flags held in various positions to represent letters of the alphabet. Illustration 2 shows how the letter U is signaled. Find x and y, if y is to be 30° more than x.

ILLUSTRATION 2

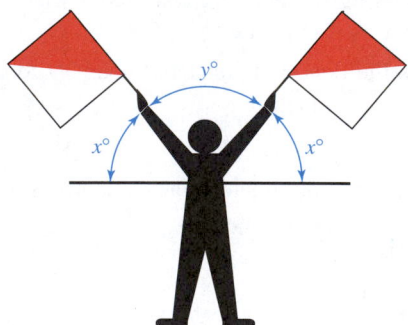

57. INVESTING A student wants to average a 6.6% return by investing $20,000 in the three stocks listed in Illustration 3. Because HiTech is a high-risk investment, he wants to invest three times as much in SaveTel and OilCo combined as he invests in HiTech. How much should he invest in each stock?

ILLUSTRATION 3

Stock	Rate of return
HiTech	10%
SaveTel	5%
OilCo	6%

58. INVESTING A woman wants to average a $7\frac{1}{3}\%$ return by investing \$30,000 in three certificates of deposit. (See Illustration 4.) She wants to invest five times as much in the 8% CD as in the 6% CD. How much should she invest in each CD?

ILLUSTRATION 4

Type of CD	Rate of return
12-month	6%
24-month	7%
36-month	8%

In Exercises 59–62, use a calculator with matrix capabilities to evaluate each determinant.

59. $\begin{vmatrix} 2 & -3 & 4 \\ -1 & 2 & 4 \\ 3 & -3 & 1 \end{vmatrix}$

60. $\begin{vmatrix} -3 & 2 & -5 \\ 3 & -2 & 6 \\ 1 & -3 & 4 \end{vmatrix}$

61. $\begin{vmatrix} 2 & 1 & -3 \\ -2 & 2 & 4 \\ 1 & -2 & 2 \end{vmatrix}$

62. $\begin{vmatrix} 4 & 2 & -3 \\ 2 & -5 & 6 \\ 2 & 5 & -2 \end{vmatrix}$

WRITING

63. Tell how to find the minor of an element of a determinant.

64. Tell how to find x when solving a system of three linear equations by Cramer's rule. Use the words *coefficients* and *constants* in your explanation.

REVIEW

65. Are the lines $y = 2x - 7$ and $x - 2y = 7$ perpendicular?

66. Are the lines $y = 2x - 7$ and $2x - y = 10$ parallel?

67. Are the equations $y = 2x - 7$ and $f(x) = 2x - 7$ the same?

68. How are the graphs of $f(x) = x^2$ and $g(x) = x^2 - 2$ related?

69. For the linear function $y = 2x - 7$, what variable is associated with the domain?

70. Is the graph of a circle the graph of a function?

71. The graph of a line passes through $(0, -3)$. Is this the x-intercept or the y-intercept of the line?

72. What is the name of the function $f(x) = |x|$?

73. For the function $y = 2x^2 + 6x + 1$, what is the independent variable and what is the dependent variable?

74. If $f(x) = x^3 - x$, what is $f(-1)$?

Systems of Equations

In this chapter, we have solved problems involving two and three variables by writing and solving a **system of equations.**

Solutions of a System of Equations

A solution of a system of equations involving two or three variables is an ordered pair or an ordered triple whose coordinates satisfy each equation of the system. In Exercises 1 and 2, decide whether the given ordered pair or ordered triple is a solution of the system.

1. $\begin{cases} 2x - y = 1 \\ 4x + 2y = 0 \end{cases} \left(\dfrac{1}{4}, -\dfrac{1}{2} \right)$

2. $\begin{cases} 2x - y + z = 9 \\ 3x + y - 4z = 8 \\ 2x - 7z = -1 \end{cases}$ $(4, 0, 1)$

Methods of Solving Systems of Linear Equations

We have studied several methods for solving systems of two and three linear equations.

3. Solve $\begin{cases} 2x + 5y = 8 \\ y = 3x + 5 \end{cases}$ using the *graphing method.*

4. Solve $\begin{cases} 9x - 8y = 1 \\ 6x + 12y = 5 \end{cases}$ using the *addition method.*

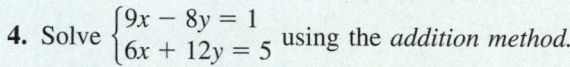

5. Solve $\begin{cases} 4x - y - 10 = 0 \\ 3x + 5y = 19 \end{cases}$ using the *substitution method.*

6. Solve $\begin{cases} -x + 3y + 2z = 5 \\ 3x + 2y + z = -1 \\ 2x - y + 3z = 4 \end{cases}$ using the *addition method.*

7. Solve $\begin{cases} x - 6y = 3 \\ x + 3y = 21 \end{cases}$ using matrices.

8. Solve $\begin{cases} x + 2z = 7 \\ 2x - y + 3z = 9 \\ y - z = 1 \end{cases}$ using Cramer's rule.

Dependent Equations and Inconsistent Systems

If the equations in a system of two linear equations are dependent, the system has infinitely many solutions. An inconsistent system has no solutions.

9. Suppose you are solving a system of two equations by the addition method, and you obtain the following.

$$2x - 3y = 4$$
$$\underline{-2x + 3y = -4}$$
$$0 = 0$$

What can you conclude?

10. Suppose you are solving a system of two equations by the substitution method, and you obtain

$$-2(x - 3) + 2x = 7$$
$$-2x + 6 + 2x = 7$$
$$6 = 7$$

What can you conclude?

Accent on Teamwork

Section 12.1

Line graphs Find five examples of line graphs where two lines intersect. (See Illustration 1.) Your school library is a good resource. Ask to look through the library's collection of recent magazines and newspapers. For each example, explain the information given by the point of intersection of the graphs.

ILLUSTRATION 1

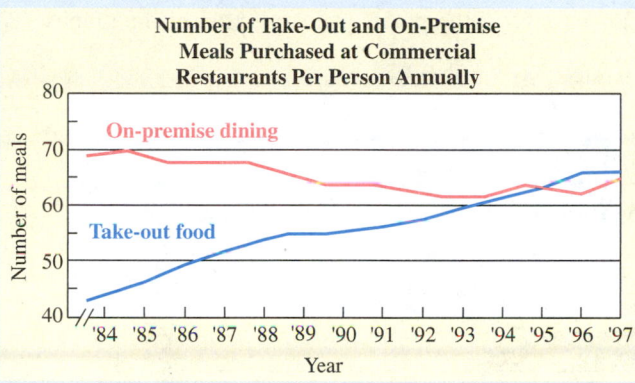

Number of Take-Out and On-Premise Meals Purchased at Commercial Restaurants Per Person Annually

Section 12.2

Graphs of systems of equations From your textbook, find and then graph examples of each of the following types of systems of two linear equations:

- Consistent system
- Inconsistent system
- Dependent equations

Then make cardboard models of each of the graphs of the systems of three linear equations illustrated in Figure 12-7 and Figure 12-8 of Section 12.2. Make a presentation to your class using the graphs and cardboard models as visual aids to help you explain whether each system has a solution, and if so, what form the solution takes.

Section 12.3

Gaussian elimination Use matrices and elementary row operations to show that the solution of this 4×4 system of linear equations is $(1, 1, 0, 1)$.

$$\begin{cases} x + y + z + w = 3 \\ x - y - z - w = -1 \\ x + y - z - w = 1 \\ x + y - z + w = 3 \end{cases}$$

Section 12.4

Methods of solution Have each person in your group solve the system

$$\begin{cases} x - y = 4 \\ 2x + y = 5 \end{cases}$$

in a different way. The methods to use are graphing, addition, substitution, matrices, and Cramer's rule. Have each person briefly explain his or her method of solution. After everyone has presented a solution, discuss the advantages and drawbacks of each method. Can your group come to a consensus? Is there a favorite method?

Section 12.1

Solving Systems with Two Variables

CONCEPTS

The graph of a linear equation is the graph of all points (x, y) on the rectangular coordinate system whose coordinates satisfy the equation.

To solve a system of two linear equations by the *graphing method*, find the coordinates of the point where the two graphs intersect.

REVIEW EXERCISES

1. See Illustration 1.
   a. Give three points that satisfy the equation $2x + y = 5$.

   b. Give three points that satisfy the equation $x - y = 4$.

   c. What is the solution of $\begin{cases} 2x + y = 5 \\ x - y = 4 \end{cases}$?

ILLUSTRATION 1

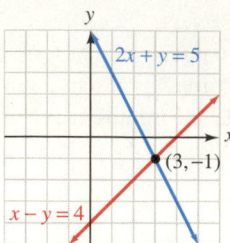

2. POLITICS Explain the importance of the points of intersection of the graphs shown in Illustration 2.

ILLUSTRATION 2

President Clinton's Job Approval Rating*

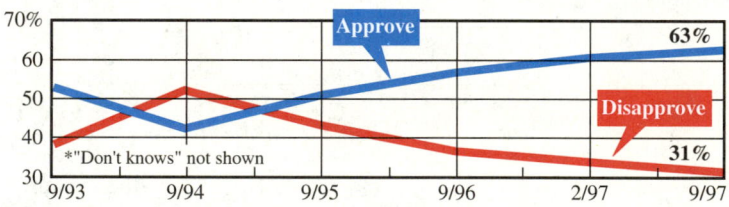

Based on data from *Los Angeles Times* (September 12, 1997)

If a system of equations has at least one solution, the system is a *consistent system*. When a system has no solution, it is called an *inconsistent system*.

If the graphs of the equations of a system are distinct, the equations are *independent equations*. Otherwise, the equations are *dependent equations*.

3. Solve each system by the graphing method, if possible. If a system is inconsistent or if the equations are dependent, so indicate.

   a. $\begin{cases} 2x + y = 11 \\ -x + 2y = 7 \end{cases}$

   b. $\begin{cases} y = -\frac{3}{2}x \\ 2x - 3y + 13 = 0 \end{cases}$

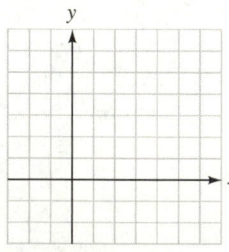

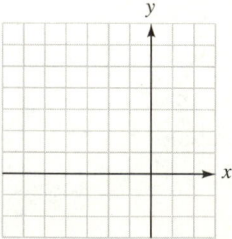

   c. $\begin{cases} \frac{1}{2}x + \frac{1}{3}y = 2 \\ y = 6 - \frac{3}{2}x \end{cases}$

   d. $\begin{cases} \frac{x}{3} - \frac{y}{2} = 1 \\ 6x - 9y = 3 \end{cases}$

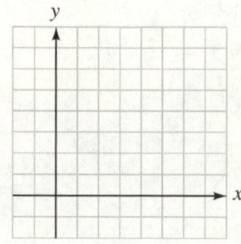

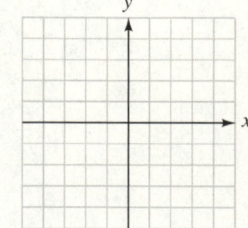

To solve a system by the *substitution method:*

1. Solve one equation for one of its variables.
2. Substitute the resulting expression for that variable into the other equation and solve that equation.
3. Find the value of the other variable by substituting the value of the variable found in step 2 into the equation from step 1.

To solve a system by the *addition method:*

1. Write both equations in general form.
2. Multiply the terms of one or both equations by constants so that the coefficients of one variable differ only in sign.
3. Add the equations from step 2 and solve the resulting equation.
4. Substitute the value obtained in step 3 into either original equation and solve for the remaining variable.

4. Solve each system using the substitution method, if possible. If a system is inconsistent or if the equations are dependent, so indicate.

a. $\begin{cases} x = y - 4 \\ 2x + 3y = 7 \end{cases}$

b. $\begin{cases} y = 2x + 5 \\ 3x - 5y = -4 \end{cases}$

c. $\begin{cases} 0.1x + 0.2y = 1.1 \\ 2x - y = 2 \end{cases}$

d. $\begin{cases} x = -2 - 3y \\ -2x - 6y = 4 \end{cases}$

5. Solve each system using the addition method, if possible.

a. $\begin{cases} x + y = -2 \\ 2x + 3y = -3 \end{cases}$

b. $\begin{cases} 2x - 3y = 5 \\ 2x - 3y = 8 \end{cases}$

c. $\begin{cases} x + \dfrac{1}{2}y = 7 \\ -2x = 3y - 6 \end{cases}$

d. $\begin{cases} y = \dfrac{x - 3}{2} \\ x = \dfrac{2y + 7}{2} \end{cases}$

6. To solve $\begin{cases} 5x - 2y = 19 \\ 3x + 4y = 1 \end{cases}$, which method, addition or substitution, would you use? Explain why.

7. Estimate the solution of the system

$$\begin{cases} y = -\frac{2}{3}x \\ 2x - 3y = -4 \end{cases}$$

from the graphs in Illustration 3. Then solve the system algebraically.

ILLUSTRATION 3

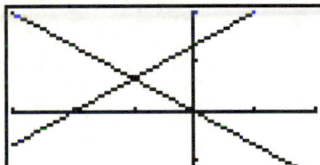

In Exercises 8–9, use two equations to solve each problem.

8. MILEAGE MAP See Illustration 4. The distance between Austin and Houston is 4 miles less than twice the distance between Austin and San Antonio. The round trip from Houston to Austin to San Antonio and back to Houston is 442 miles. Determine the mileages between Austin and Houston and between Austin and San Antonio.

ILLUSTRATION 4

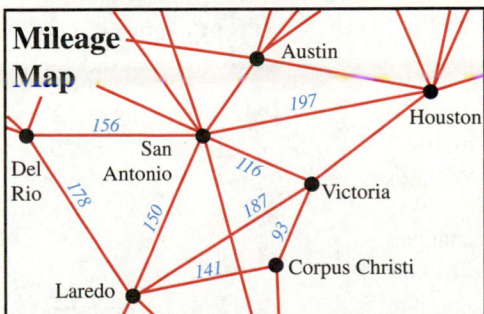

9. RIVERBOAT RIDE A Mississippi riverboat travels 30 miles downstream in three hours and then makes the return trip upstream in five hours. Find the speed of the riverboat in still water and the speed of the current.

Section 12.2

The solution of a system of three linear equations is an *ordered triple.*

To solve a system of linear equations with three variables:
1. Pick any two equations and eliminate a variable.
2. Pick a different pair of equations and eliminate the same variable.
3. Solve the resulting pair of equations.
4. Use substitution to find the value of the third variable.

Solving Systems with Three Variables

10. Tell whether $(2, -1, 1)$ is a solution of the system $\begin{cases} x - y + z = 4 \\ x + 2y - z = -1. \\ x + y - 3z = -1 \end{cases}$

11. Solve each system, if possible.

a. $\begin{cases} x + y + z = 6 \\ x - y - z = -4 \\ -x + y - z = -2 \end{cases}$

b. $\begin{cases} 2x + 3y + z = -5 \\ -x + 2y - z = -6 \\ 3x + y + 2z = 4 \end{cases}$

c. $\begin{cases} x + y - z = -3 \\ x + z = 2 \\ 2x - y + 2z = 3 \end{cases}$

d. $\begin{cases} 3x + 3y + 6z = -6 \\ -x - y - 2z = 2 \\ 2x + 2y + 4z = -4 \end{cases}$

12. MIXING NUTS The owner of a produce store wanted to mix peanuts selling for $3 per pound, cashews selling for $9 per pound, and Brazil nuts selling for $9 per pound to get 50 pounds of a mixture that would sell for $6 per pound. She used 15 fewer pounds of cashews than peanuts. How many pounds of each did she use?

Section 12.3

A *matrix* is a rectangular array of numbers.

A system of linear equations can be represented by an *augmented matrix.*

Systems of linear equations can be solved using *Gaussian elimination* and *elementary row operations:*
1. Any two rows can be interchanged.
2. Any row can be multiplied by a nonzero constant.
3. Any row can be changed by adding a nonzero constant multiple of another row to it.

Solving Systems Using Matrices

13. Represent each system of equations using an augmented matrix.

a. $\begin{cases} 5x + 4y = 3 \\ x - y = -3 \end{cases}$

b. $\begin{cases} x + 2y + 3z = 6 \\ x - 3y - z = 4 \\ 6x + y - 2z = -1 \end{cases}$

14. Solve each system using matrices, if possible.

a. $\begin{cases} x - y = 4 \\ 3x + 7y = -18 \end{cases}$

b. $\begin{cases} x + 2y - 3z = 5 \\ x + y + z = 0 \\ 3x + 4y + 2z = -1 \end{cases}$

c. $\begin{cases} 16x - 8y = 32 \\ -2x + y = -4 \end{cases}$

d. $\begin{cases} x + 2y + 2z = 2 \\ 4x + 5y + 3z = 3 \\ 2x + y - z = 1 \end{cases}$

15. INVESTING One year, a couple invested a total of $10,000 in two projects. The first investment, a mini-mall, made a 6% profit. The other investment, a skateboard park, made a 12% profit. If their investments made $960, how much was invested at each rate? To answer this question, write a system of two equations and solve it using matrices.

Section 12.4

A *determinant* of a *square matrix* is a number.

To evaluate a 2×2 determinant:

$$\begin{vmatrix} a & b \\ c & d \end{vmatrix} = ad - bc$$

To evaluate a 3×3 determinant, we expand it by *minors* along any row or column using the *array of signs*.

Cramer's rule can be used to solve systems of linear equations.

Solving Systems Using Determinants

16. Evaluate each determinant.

a. $\begin{vmatrix} 2 & 3 \\ -4 & 3 \end{vmatrix}$ **b.** $\begin{vmatrix} -3 & -4 \\ 5 & -6 \end{vmatrix}$

c. $\begin{vmatrix} -1 & 2 & -1 \\ 2 & -1 & 3 \\ 1 & -2 & 2 \end{vmatrix}$ **d.** $\begin{vmatrix} 3 & -2 & 2 \\ 1 & -2 & -2 \\ 2 & 1 & -1 \end{vmatrix}$

17. Use Cramer's rule to solve each system, if possible.

a. $\begin{cases} 3x + 4y = 10 \\ 2x - 3y = 1 \end{cases}$ **b.** $\begin{cases} -6x - 4y = -6 \\ 3x + 2y = 5 \end{cases}$

c. $\begin{cases} x + 2y + z = 0 \\ 2x + y + z = 3 \\ x + y + 2z = 5 \end{cases}$ **d.** $\begin{cases} 2x + 3y + z = 2 \\ x + 3y + 2z = 7 \\ x - y - z = -7 \end{cases}$

18. VETERINARY MEDICINE The daily requirements of a balanced diet for an animal are shown in the nutritional pyramid in Illustration 5. The number of grams per cup of nutrients in three food mixes are shown in the table. How many cups of each mix should be used to meet the daily requirements for protein, carbohydrates, and essential fatty acids in the animal's diet? To answer this problem, write a system of three equations and solve it using Cramer's rule.

	Grams per cup		
	Protein	**Carbohydrates**	**Fatty Acids**
Mix A	5	2	1
Mix B	6	3	2
Mix C	8	3	1

ILLUSTRATION 5

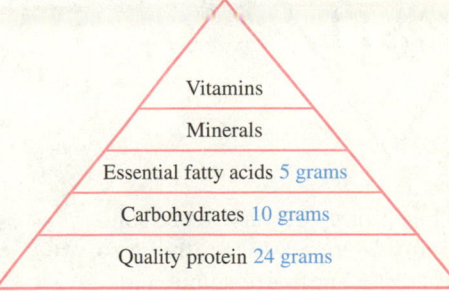

Vitamins

Minerals

Essential fatty acids 5 grams

Carbohydrates 10 grams

Quality protein 24 grams

CHAPTER 12

Test

1. Solve $\begin{cases} 2x + y = 5 \\ y = 2x - 3 \end{cases}$ by graphing.

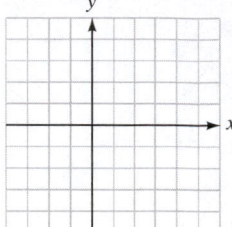

2. Use substitution to solve $\begin{cases} 2x - 4y = 14 \\ x + 2y = 7 \end{cases}$.

3. Use addition to solve $\begin{cases} 2x + 3y = -5 \\ 3x - 2y = 12 \end{cases}$.

4. Are the equations of the system
$$\begin{cases} 3(x + y) = x - 3 \\ -y = \dfrac{2x + 3}{3} \end{cases}$$
dependent or independent?

5. Is $\left(-1, -\frac{1}{2}, 5\right)$ a solution of $\begin{cases} x - 2y + z = 5 \\ 2x + 4y = -4 \\ -6y + 4z = 22 \end{cases}$?

6. Solve the system $\begin{cases} x + y + z = 4 \\ x + y - z = 6 \\ 2x - 3y + z = -1 \end{cases}$
using the addition method.

In Problems 7–8, write a system of equations to solve each problem.

7. In Illustration 1, find x and y, if y is 15 more than x.

ILLUSTRATION 1

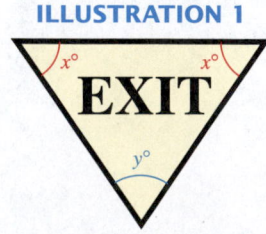

8. **ANTIFREEZE** How much of a 40% antifreeze solution must a mechanic mix with an 80% antifreeze solution if 20 gallons of a 50% antifreeze solution are needed?

In Problems 9–10, use matrices to solve each system.

9. $\begin{cases} x + y = 4 \\ 2x - y = 2 \end{cases}$

10. $\begin{cases} x + y + 2z = -1 \\ x + 3y - 6z = 7 \\ 2x - y + 2z = 0 \end{cases}$

In Problems 11–12, evaluate each determinant.

11. $\begin{vmatrix} 2 & -3 \\ 4 & 5 \end{vmatrix}$

12. $\begin{vmatrix} 1 & 2 & 0 \\ 2 & 0 & 3 \\ 1 & -2 & 2 \end{vmatrix}$

In Problems 13–16, consider the system $\begin{cases} x - y = -6 \\ 3x + y = -6 \end{cases}$, which is to be solved using Cramer's rule.

13. When solving for x, what is the numerator determinant D_x? (**Don't evaluate it.**)

14. When solving for y, what is the denominator determinant D? (**Don't evaluate it.**)

15. Solve the system for x.

16. Solve the system for y.

17. Solve the following system for z only, using Cramer's rule.
$$\begin{cases} x + y + z = 4 \\ x + y - z = 6 \\ 2x - 3y + z = -1 \end{cases}$$

18. **MOVIE TICKETS** The receipts for one showing of a movie were $410 for an audience of 100 people. The ticket prices are given in the table. If twice as many children's tickets as general admission tickets were purchased, how many of each type of ticket were sold?

Ticket prices	
Children	$3.00
General Admission	$6.00
Seniors	$5.00

19. CANDY SALES Summarize the information that can be learned from the graph in Illustration 2. What do the points of intersection of the graphs tell us?

ILLUSTRATION 2

Holiday Candy Sales

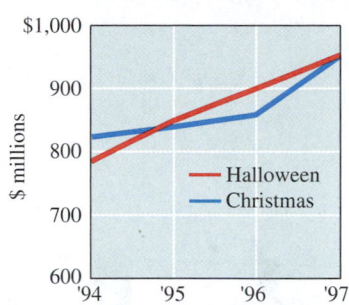

Based on data from *Los Angeles Times* (October 30, 1997)

20. Which method, substitution or addition, would you use to solve the following system?

$$\begin{cases} \dfrac{x}{2} - \dfrac{y}{4} = -4 \\ y = -2 - x \end{cases}$$

Explain your reasoning.

21. What does it mean to say that a system of two linear equations is an *inconsistent* system?

22. Suppose that two variables are used to solve an application problem. Why must two equations be written to solve the problem?

CHAPTERS 1–12

Cumulative Review Exercises

In Exercises 1–2, write the equation of the line with the given properties.

1. $m = 3$, passing through $(-2, -4)$

2. Parallel to the graph of $2x + 3y = 6$ and passing through $(0, -2)$

3. AIRPORT TRAFFIC From the graph in Illustration 1, determine the projected average rate of change in the number of takeoffs and landings at Los Angeles International Airport for the years 2000–2015.

4. Solve the system by graphing.

$$\begin{cases} y = -\dfrac{5}{2}x + \dfrac{1}{2} \\ 2x - \dfrac{3}{2}y = 5 \end{cases}$$

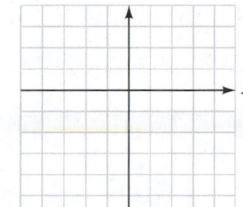

5. Solve the system using Cramer's rule.

$$\begin{cases} x - y = 3 \\ x + 2y = 0 \end{cases}$$

ILLUSTRATION 1

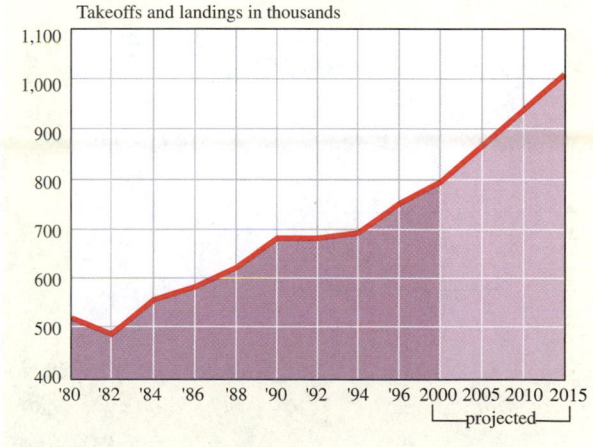

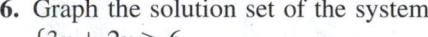

Based on data from *Los Angeles Times* (July 6, 1998) p. B3

6. Graph the solution set of the system

$$\begin{cases} 3x + 2y > 6 \\ x + 3y \le 2 \end{cases}$$

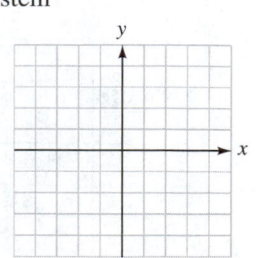

7. Solve the inequality $-9x + 6 > 16$. Give the result in interval notation and graph the solution set.

8. Solve $|2x - 5| \geq 25$. Give the result in interval notation and graph the solution set.

In Exercises 9–10, find the domain and range of each function.

9. $f(x) = 2x^2 - 3$ **10.** $y = -|x - 4|$

In Exercises 11–12, do each operation.

11. $(2a^2 + 4a - 7) - 2(3a^2 - 4a)$

12. $(3x + 2)(2x - 3)$

In Exercises 13–16, factor each expression.

13. $x^4 - 16y^4$

14. $15x^2 - 2x - 8$

15. $x^2 + 4y - xy - 4x$

16. $8x^6 + 125y^3$

In Exercises 17–20, solve each equation.

17. $x^2 - 5x - 6 = 0$ **18.** $6a^3 - 2a = a^2$

19. $\dfrac{x - 4}{x - 3} + \dfrac{x - 2}{x - 3} = x - 3$

20. $P + \dfrac{a}{V^2} = \dfrac{RT}{V - b}$ solve for b

In Exercises 21–22, simplify each expression.

21. $\dfrac{x^3 + y^3}{x^3 - y^3} \div \dfrac{x^2 - xy + y^2}{x^2 + xy + y^2}$

22. $\dfrac{1}{x + y} - \dfrac{1}{x - y} - \dfrac{2y}{y^2 - x^2}$

In Exercises 23–24, graph each function and give its domain and range.

23. $f(x) = x^3 + x^2 - 6x$ **24.** $f(x) = \dfrac{4}{x}$ for $x > 0$

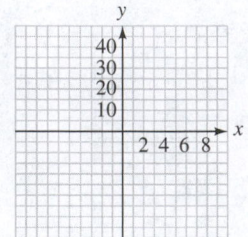

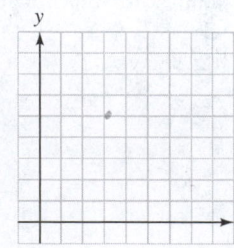

25. LIGHT As light energy radiates away from its source, its intensity varies inversely as the square of the distance from the source. Illustration 2 shows that the light energy passing through an area 1 foot from the source spreads out over 4 units of area 2 feet from the source. That energy is therefore less intense 2 feet from the source than it was 1 foot from the source. Over how many units of area will the light energy spread out 3 feet from the source?

ILLUSTRATION 2

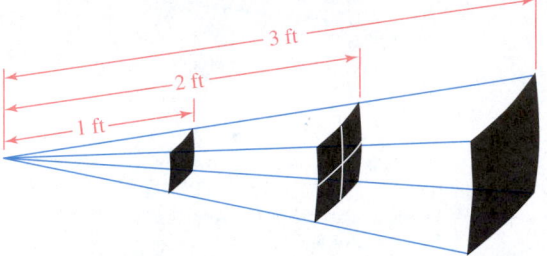

26. Graph the function $f(x) = \sqrt{x - 2}$ and give its domain and range.

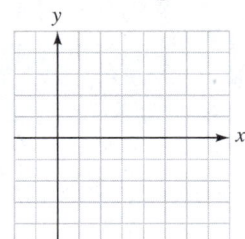

In Exercises 27–30, simplify each expression.

27. $\sqrt[3]{-27x^3}$ **28.** $\sqrt{48t^3}$

29. $64^{-2/3}$ **30.** $\dfrac{x^{5/3}x^{1/2}}{x^{3/4}}$

In Exercises 31–34, simplify each expression.

31. $-3\sqrt[4]{32} - 2\sqrt[4]{162} + 5\sqrt[4]{48}$

32. $3\sqrt{2}\left(2\sqrt{3} - 4\sqrt{12}\right)$

33. $\dfrac{\sqrt{x} + 2}{\sqrt{x} - 1}$ **34.** $\dfrac{5}{\sqrt[3]{x}}$

In Exercises 35–36, solve each equation.

35. $5\sqrt{x + 2} = x + 8$ **36.** $\sqrt{x} + \sqrt{x + 2} = 2$

37. Find the length of the hypotenuse of the right triangle shown in Illustration 3.

ILLUSTRATION 3

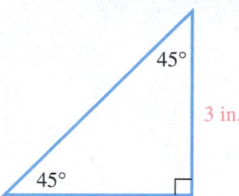

38. Find the length of the hypotenuse of the right triangle shown in Illustration 4.

ILLUSTRATION 4

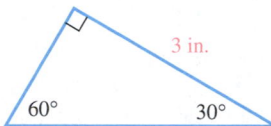

39. Find the distance between $(-2, 6)$ and $(4, 14)$.

40. What number must be added to $x^2 + 6x$ to make a perfect square trinomial ?

41. Use the method of completing the square to solve the equation $2x^2 + x - 3 = 0$.

42. Use the quadratic formula to solve the equation $3x^2 + 4x - 1 = 0$.

43. Graph $y = \frac{1}{2}x^2 - x + 1$ and find the coordinates of its vertex.

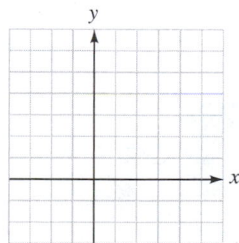

44. Graph $f(x) = -x^2 - 4x$ and find the coordinates of its vertex.

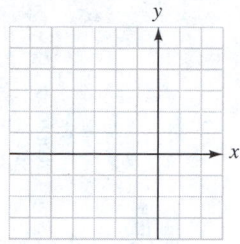

In Exercises 45–46, write each expression in $a + bi$ form.

45. $(3 + 5i) + (4 - 3i)$ **46.** $\dfrac{5}{3 - i}$

47. Simplify $\sqrt{-64}$.

48. Solve $a - 7a^{1/2} + 12 = 0$

49. The graph of $f(x) = 16x^2 + 24x + 9$ is shown in Illustration 5. Estimate the solution(s) of $16x^2 + 24x + 9 = 0$.

ILLUSTRATION 5

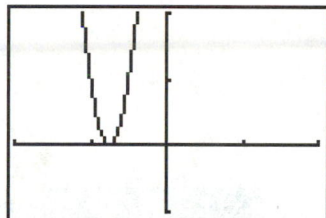

50. Use the graph in Illustration 5 to determine the solution of $16x^2 + 24x + 9 < 0$.

13

Conic Sections; More Graphing

CAMPUS CONNECTION

The Art Department

An art class wants to paint a mural on a wall at the local mall. After measuring the wall, they decide to paint the mural on an elliptical background that is 10 feet wide and 6 feet high. Fortunately, one of the students was studying conic sections in algebra class. Using her knowledge of conics, she was able to make this construction, which is discussed on page 943.

In this chapter, we will discuss some graphs that do not represent functions. Since these curves are cross sections of a cone, they are called conic sections.

▶ 13.1 The Circle and the Parabola

In this section, you will learn about

- Conic sections ■ The circle ■ Problem solving ■ The parabola
- Problem solving

Introduction   We have seen that the graphs of linear functions are straight lines and that the graphs of quadratic functions are parabolas. In this section, we will discuss some special curves, called **conic sections**.

Conic Sections

The graphs of second-degree equations in x and y represent figures that were fully investigated in the 17th century by René Descartes (1596–1650) and Blaise Pascal (1623–1662). Descartes discovered that graphs of second-degree equations fall into one of several categories: a pair of lines, a point, a circle, a parabola, an ellipse, a hyperbola, or no graph at all. Because all of these graphs can be formed by the intersection of a plane and a right-circular cone, they are called **conic sections.** See Figure 13-1.

FIGURE 13-1

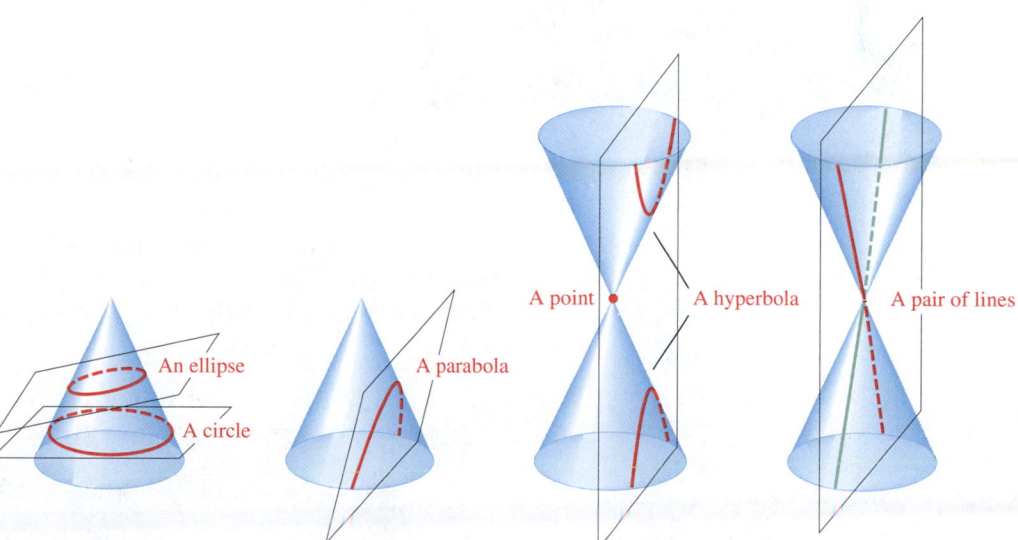

Conic sections have many applications. For example, everyone knows the importance of circular wheels and gears, pizza cutters, and Ferris wheels.

Parabolas can be rotated to generate dish-shaped surfaces called **paraboloids**. Any light or sound placed at the **focus** of a paraboloid is reflected outward in parallel paths, as shown in Figure 13-2(a) on the next page. This property makes parabolic surfaces ideal for flashlight and headlight reflectors. It also makes parabolic surfaces good antennas, because signals captured by such antennas are concentrated at the focus.

Parabolic mirrors are capable of concentrating the rays of the sun at a single point and thereby generating tremendous heat. This property is used in the design of certain solar furnaces.

Any object thrown upward and outward travels in a parabolic path, as shown in Figure 13-2(b). In architecture, many arches are parabolic in shape, because this gives strength. Cables that support suspension bridges hang in the form of a parabola. (See Figure 13-2(c).)

FIGURE 13-2

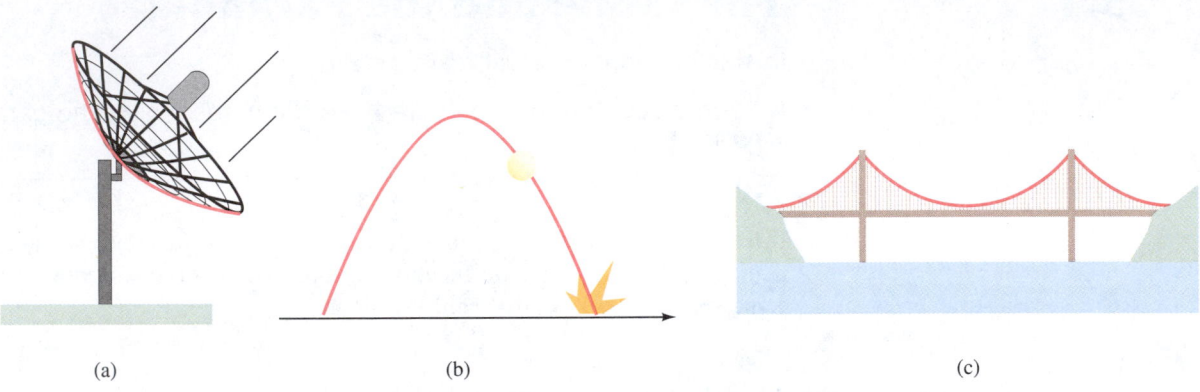

(a)   (b)   (c)

Ellipses have optical and acoustical properties that are useful in architecture and engineering. For example, many arches are portions of an ellipse, because the shape is pleasing to the eye. (See Figure 13-3(a).) The planets and many comets have elliptical orbits, as shown in Figure 13-3(b). Gears are often cut into elliptical shapes to provide nonuniform motion. (See Figure 13-3(c).)

FIGURE 13-3

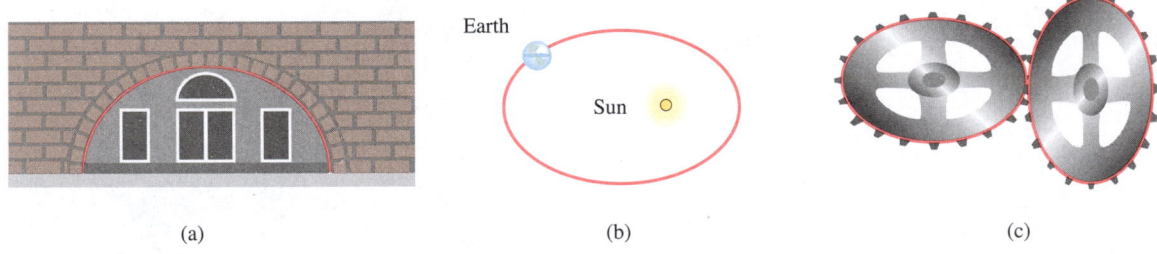

(a)   (b)   (c)

Hyperbolas serve as the basis of a navigational system known as LORAN (LOng RAnge Navigation). (See Figure 13-4.) They are also used to find the source of a distress signal, are the basis for the design of hypoid gears, and describe the orbits of some comets.

FIGURE 13-4

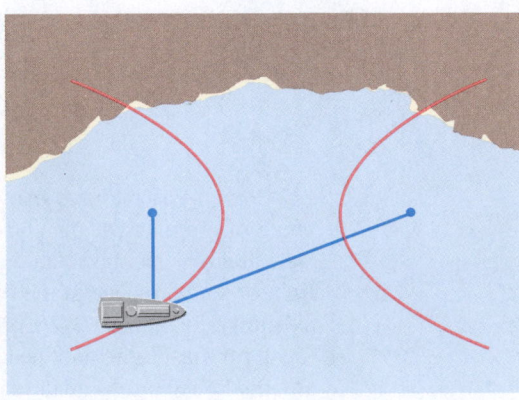

The Circle

The most common conic section is the circle.

The circle

A **circle** is the set of all points in a plane that are a fixed distance from a point called its **center.** The fixed distance is called the **radius** of the circle.

To develop the general equation of a circle, we will write the equation of a circle with a radius of r and with a center at some point $C(h, k)$, as in Figure 13-5. This task is equivalent to finding all points $P(x, y)$ such that the length of line segment CP is r. We can use the distance formula to find r.

$$r = \sqrt{(x - h)^2 + (y - k)^2}$$

We then square both sides to obtain

1. $r^2 = (x - h)^2 + (y - k)^2$

FIGURE 13-5

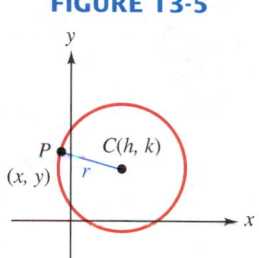

Equation 1 is called the **standard form of the equation of a circle** with a radius of r and center at the point with coordinates (h, k).

Standard equation of a circle with center at (h, k)

Any equation that can be written in the form

$$(x - h)^2 + (y - k)^2 = r^2$$

has a graph that is a circle with radius r and center at point (h, k).

If $r^2 = 0$, the graph reduces to a point called a **point circle.** If $r^2 < 0$, a circle does not exist. If both coordinates of the center are 0, the center of the circle is the origin.

Standard equation of a circle with center at (0, 0)

Any equation that can be written in the form

$$x^2 + y^2 = r^2$$

has a graph that is a circle with radius r and center at the origin.

EXAMPLE 1

Graphing circles. Graph $x^2 + y^2 = 25$.

Solution   Because this equation can be written in the form $x^2 + y^2 = r^2$, its graph is a circle with center at the origin. Since $r^2 = 25 = 5^2$, the circle has a radius of 5. The graph appears in Figure 13-6.

FIGURE 13-6

$$x^2 + y^2 = 25$$

x	y	(x, y)
-5	0	$(-5, 0)$
-4	± 3	$(-4, \pm 3)$
-3	± 4	$(-3, \pm 4)$
0	± 5	$(0, \pm 5)$
3	± 4	$(3, \pm 4)$
4	± 3	$(4, \pm 3)$
5	0	$(5, 0)$

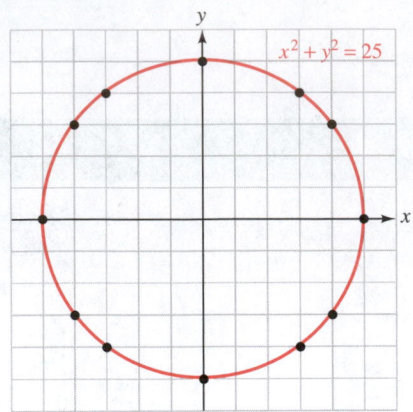

SELF CHECK   Graph $x^2 + y^2 = 4$.

Answer:

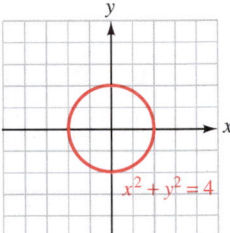

EXAMPLE 2

Finding equations of circles. Find the equation of the circle with radius 5 and center at (3, 2).

Solution   We substitute 5 for r, 3 for h, and 2 for k in standard form and simplify.

$$(x - \boldsymbol{h})^2 + (y - \boldsymbol{k})^2 = \boldsymbol{r}^2$$
$$(x - \boldsymbol{3})^2 + (y - \boldsymbol{2})^2 = \boldsymbol{5}^2$$
$$x^2 - 6x + 9 + y^2 - 4y + 4 = 25$$
$$x^2 + y^2 - 6x - 4y - 12 = 0$$

The equation is $x^2 + y^2 - 6x - 4y - 12 = 0$.

SELF CHECK   Find the equation of the circle with radius 6 and center at (2, 3).

Answer: $x^2 + y^2 - 4x - 6y - 23 = 0$ ∎

EXAMPLE 3

Graphing circles. Graph the equation $x^2 + y^2 - 4x + 2y = 20$.

Solution   To identify the curve, we complete the square on x and y and write the equation in standard form.

FIGURE 13-7

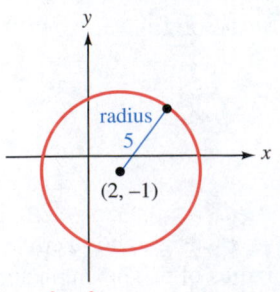

$x^2 + y^2 - 4x + 2y = 20$

$$x^2 + y^2 - 4x + 2y = 20$$
$$x^2 - 4x + y^2 + 2y = 20$$

To complete the square on x and y, add 4 and 1 to both sides.

$$x^2 - 4x \boldsymbol{+ 4} + y^2 + 2y \boldsymbol{+ 1} = 20 \boldsymbol{+ 4 + 1}$$
$$(x - 2)^2 + (y + 1)^2 = 25 \qquad \text{Factor } x^2 - 4x + 4 \text{ and } y^2 + 2y + 1.$$
$$(x - 2)^2 + [y - (-1)]^2 = 5^2$$

We can now see that this result is the equation of a circle with a radius of 5 and center at $h = 2$ and $k = -1$. If we plot the center and draw a circle with a radius of 5 units, we will obtain the circle shown in Figure 13-7.

SELF CHECK   Write $x^2 + y^2 + 2x - 4y - 11 = 0$ in standard form and graph it.

Answer: $(x + 1)^2 + (y - 2)^2 = 16$

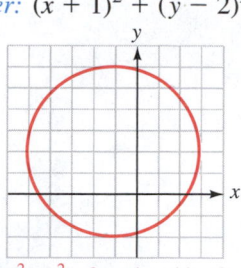

$x^2 + y^2 + 2x - 4y - 11 = 0$

ACCENT ON TECHNOLOGY *Graphing Circles*

Since the graphs of circles fail the vertical line test, their equations do not represent functions. It is somewhat more difficult to use a graphing calculator to graph equations that are not functions. For example, to graph the circle described by $(x - 1)^2 + (y - 2)^2 = 4$, we must split the equation into two functions and graph each one separately. We begin by solving the equation for y.

$$(x - 1)^2 + (y - 2)^2 = 4$$
$$(y - 2)^2 = 4 - (x - 1)^2 \qquad \text{Subtract } (x - 1)^2 \text{ from both sides.}$$
$$y - 2 = \pm \sqrt{4 - (x - 1)^2} \qquad \text{Take the square root of both sides.}$$
$$y = 2 \pm \sqrt{4 - (x - 1)^2} \qquad \text{Add 2 to both sides.}$$

This equation defines two functions. If we use window settings of $[-3, 5]$ for x and $[-3, 5]$ for y and graph the functions

$$y = 2 + \sqrt{4 - (x - 1)^2} \qquad \text{and} \qquad y = 2 - \sqrt{4 - (x - 1)^2}$$

we get the distorted circle shown in Figure 13-8(a). To get a better circle, we can use the graphing calculator's squaring feature, which gives an equal unit distance on both the x- and y-axes. Using this feature, we get the circle shown in Figure 13-8(b). Sometimes the two arcs will not join because of approximations made by the calculator at each endpoint.

FIGURE 13-8

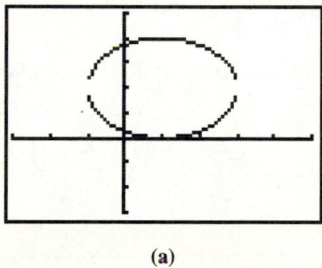

(a)

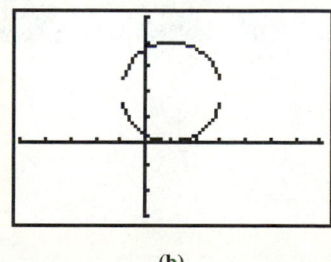

(b)

Problem Solving

EXAMPLE 4

Radio translators. The broadcast area of a television station is bounded by the circle $x^2 + y^2 = 3{,}600$, where x and y are measured in miles. A translator station picks up the signal and retransmits it from the center of a circular area bounded by $(x + 30)^2 + (y - 40)^2 = 1{,}600$. Find the location of the translator and the greatest distance from the main transmitter that the signal can be received.

Solution   The coverage of the television station is bounded by $x^2 + y^2 = 60^2$, a circle centered at the origin with a radius of 60 miles, as shown in yellow in Figure 13-9. Because the translator is at the center of the circle $(x + 30)^2 + (y - 40)^2 = 1{,}600$, it is located at $(-30, 40)$, a point 30 miles west and 40 miles north of the television station. The radius of the translator's coverage is $\sqrt{1{,}600}$, or 40 miles.

As shown in Figure 13-9, the greatest distance of reception is the sum of A, the distance from the translator to the television station, and 40 miles, the radius of the translator's coverage.

To find A, we use the distance formula to find the distance between $(x_1, y_1) = (-30, 40)$ and the origin, $(x_2, y_2) = (0, 0)$.

FIGURE 13-9

$$A = \sqrt{(x_1 - x_2)^2 + (y_1 - y_2)^2}$$
$$A = \sqrt{(-30 - 0)^2 + (40 - 0)^2}$$
$$= \sqrt{(-30)^2 + 40^2}$$
$$= \sqrt{900 + 1{,}600}$$
$$= \sqrt{2{,}500}$$
$$= 50$$

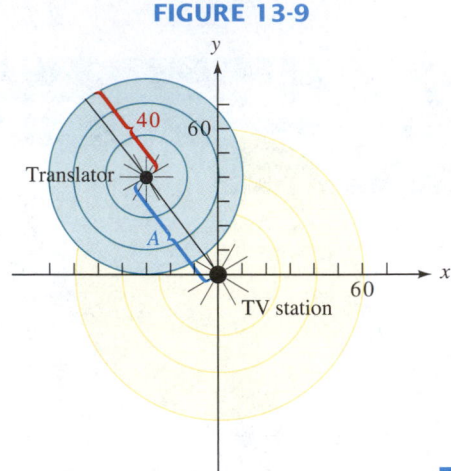

The translator is located 50 miles from the television station, and it broadcasts the signal an additional 40 miles. The greatest reception distance is $50 + 40$, or 90 miles.   ■

The Parabola

We have seen that equations of the form $y = a(x - h)^2 + k$ $(a \neq 0)$ represent parabolas with vertex at point (h, k). They open upward when $a > 0$ and downward when $a < 0$.

Equations of the form $x = a(y - k)^2 + h$ also represent parabolas with vertex at point (h, k). However, they open to the right when $a > 0$ and to the left when $a < 0$. Parabolas that open to the right or left do not represent functions, because their graphs fail the vertical line test.

Several types of parabolas are summarized in the following chart.

Equations of parabolas ($a > 0$)

Parabola opening	Vertex at origin	Vertex at (h, k)
Up	$y = ax^2$	$y = a(x - h)^2 + k$
Down	$y = -ax^2$	$y = -a(x - h)^2 + k$
Right	$x = ay^2$	$x = a(y - k)^2 + h$
Left	$x = -ay^2$	$x = -a(y - k)^2 + h$

EXAMPLE 5   **Graphing parabolas.** Graph   **a.** $x = \frac{1}{2}y^2$ and   **b.** $x = -2(y - 2)^2 + 3$.

Solution   **a.** We make a table of ordered pairs that satisfy the equation, plot each pair, and draw the parabola, as in Figure 13-10(a). Because the equation is of the form $x = ay^2$ with $a = \frac{1}{2} > 0$, the parabola opens to the right and has its vertex at the origin.

b. We make a table of ordered pairs that satisfy the equation, plot each pair, and draw the parabola, as in Figure 13-10(b). Because the equation is of the form $x = -a(y - k)^2 + h$, the parabola opens to the left and has its vertex at the point with coordinates $(3, 2)$.

FIGURE 13-10

$x = \frac{1}{2}y^2$

x	y	(x, y)
0	0	$(0, 0)$
2	2	$(2, 2)$
2	-2	$(2, -2)$
8	4	$(8, 4)$
8	-4	$(8, -4)$

$x = -2(y - 2)^2 + 3$

x	y	(x, y)
-5	0	$(-5, 0)$
1	1	$(1, 1)$
3	2	$(3, 2)$
1	3	$(1, 3)$
-5	4	$(-5, 4)$

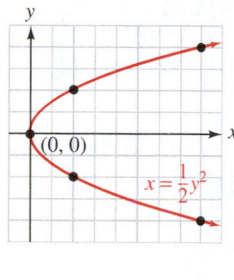

(a)

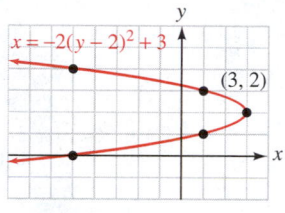

(b)

SELF CHECK Graph $x = \frac{1}{2}(y - 1)^2 - 2$.

Answer:

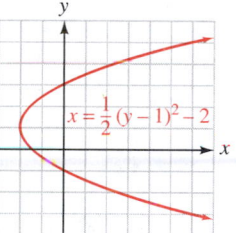

EXAMPLE 6 **Graphing parabolas.** Graph $y = -2x^2 + 12x - 15$.

Solution Because the equation is not in standard form, the coordinates of the vertex are not obvious. To write the equation in standard form, we complete the square on x.

$$y = -2x^2 + 12x - 15$$
$$y = -2(x^2 - 6x) - 15 \qquad \text{Factor out } -2 \text{ from } -2x^2 + 12x.$$
$$y = \mathbf{-2}(x^2 - 6x \mathbf{+ 9}) - 15 \mathbf{+ 18} \quad \text{Subtract and add 18; } -2(9) = -18.$$
$$y = -2(x - 3)^2 + 3$$

Since the equation is written in the form $y = -a(x - h)^2 + k$, we can see that the parabola opens downward and has its vertex at $(3, 3)$. The graph of the function is shown in Figure 13-11.

FIGURE 13-11

$y = -2x^2 + 12x - 15$

x	y	(x, y)
1	-5	$(1, -5)$
2	1	$(2, 1)$
3	3	$(3, 3)$
4	1	$(4, 1)$
5	-5	$(5, -5)$

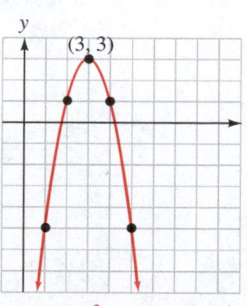

$y = -2x^2 + 12x - 15$

SELF CHECK   Graph $y = 0.5x^2 - x - 1$.   *Answer:*

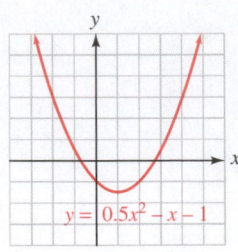

$y = 0.5x^2 - x - 1$

Problem Solving

E X A M P L E   7   **Gateway Arch.**   The shape of the Gateway Arch in St. Louis is approximately a parabola, as shown in Figure 13-12(a). How high is the arch 100 feet from its foundation?

FIGURE 13-12

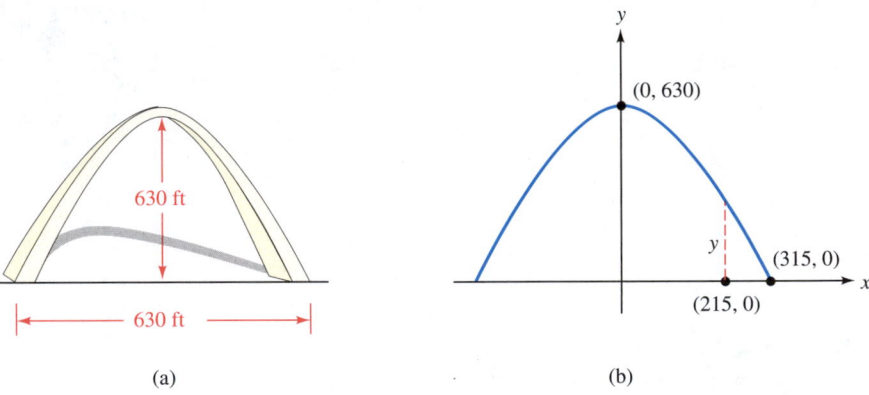

(a)   (b)

Solution   We place the parabola on a coordinate system as in Figure 13-12(b), with ground level on the x-axis and the vertex of the parabola at the point $(h, k) = (0, 630)$. The equation of this downward-opening parabola has the form

$y = -a(x - h)^2 + k$

$y = -a(x - 0)^2 + 630$ Substitute 0 for h and 630 for k.

$y = -ax^2 + 630$ Simplify.

Because the Gateway Arch is 630 feet wide at its base, the parabola passes through the point $\left(\frac{630}{2}, 0\right)$, or $(315, 0)$. To find a in the equation of the parabola, we proceed as follows:

$y = -ax^2 + 630$

$0 = -a \cdot 315^2 + 630$ Substitute 315 for x and 0 for y.

$\dfrac{-630}{315^2} = -a$ Subtract 630 from both sides and divide both sides by 315^2.

$\dfrac{2}{315} = a$ Multiply both sides by -1; $\dfrac{630}{315^2} = \dfrac{2}{315}$.

The equation of the parabola that approximates the shape of the Gateway Arch is

$$y = -\frac{2}{315}x^2 + 630$$

To find the height of the arch at a point 100 feet from its foundation, we substitute $315 - 100$, or 215, for x in the equation of the parabola and solve for y.

$$y = -\frac{2}{315}x^2 + 630$$

$$y = -\frac{2}{315}(215)^2 + 630$$

$$= 336.5079365 \qquad \text{Use a calculator.}$$

At a point 100 feet from the foundation, the height of the arch is about 337 feet. ■

STUDY SET
Section 13.1

VOCABULARY

Fill in the blanks to make the statements true.

1. Intersections of a plane and a right-circular cone are called _____ sections.

2. A _____ is the set of all points in a plane that are a fixed distance from a point.

3. The fixed distance in Exercise 2 is called the _____ of the circle, and the point is called its _____.

4. Parabolas can be rotated to form dish-shaped surfaces called _____.

CONCEPTS

Fill in the blanks to make the statements true.

5. If _____ in the equation $x^2 + y^2 = r^2$, no circle exists.

6. The graph of $y = ax^2$ is a _____ with vertex at the _____ that opens _____.

7. The graph of $x = a(y - 2)^2 + 3$ is a _____ that has its vertex at _____ and opens to the _____.

8. The graph of $x = -a(y - 1)^2 - 3$ is a _____ that has its vertex at _____ and opens to the _____.

9. The standard equation of a circle with center at (h, k) is _____.

10. The standard equation of a circle with center at $(0, 0)$ is _____.

11. The standard equation of a parabola $(a > 0)$ that opens downward and has its vertex at the origin is _____.

12. The standard equation of a parabola $(a > 0)$ that opens to the left and has its vertex at (h, k) is _____.

NOTATION

Complete each solution.

13. Find the equation of a circle with radius 6 and center at $(2, 3)$.

$$(x - h)^2 + (y - k)^2 = r^2$$
$$\left(x - \boxed{}\right)^2 + (y - 3)^2 = \boxed{}^2$$
$$x^2 - \boxed{}x + 4 + y^2 - 6y + 9 = \boxed{}$$
$$x^2 + y^2 - 4x - 6y - 23 = 0$$

14. Write the equation of a parabola with vertex at $(2, 3)$, opening to the left, and with $a = 2$.

$$x = -a(y - k)^2 + h$$
$$x = -\boxed{}\left(y - \boxed{}\right)^2 + 2$$
$$x = -2(y^2 - \boxed{} + 9) + 2$$
$$x = -2y^2 + 12y - 16$$

PRACTICE

In Exercises 15–24, graph each equation.

15. $x^2 + y^2 = 9$ **16.** $x^2 + y^2 = 16$

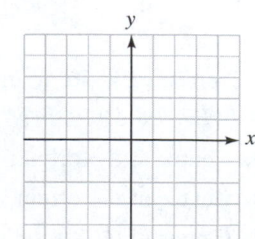

17. $(x - 2)^2 + y^2 = 9$ **18.** $x^2 + (y - 3)^2 = 4$

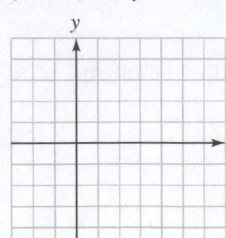   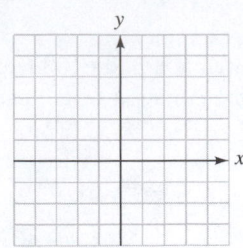

19. $(x - 2)^2 + (y - 4)^2 = 4$ **20.** $(x - 3)^2 + (y - 2)^2 = 4$

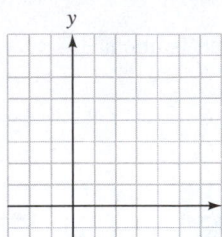   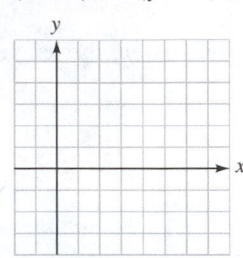

21. $(x + 3)^2 + (y - 1)^2 = 16$ **22.** $(x - 1)^2 + (y + 4)^2 = 9$

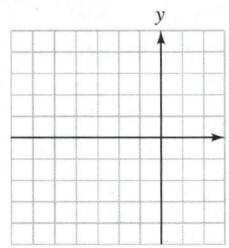   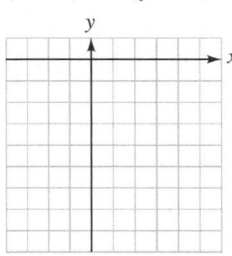

23. $x^2 + (y + 3)^2 = 1$ **24.** $(x + 4)^2 + y^2 = 1$

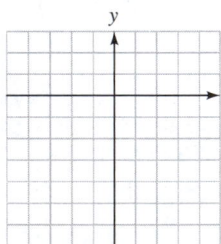   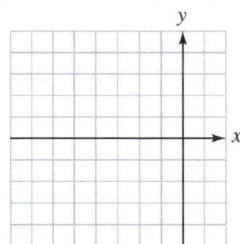

In Exercises 25–28, use a graphing calculator to graph each equation.

25. $3x^2 + 3y^2 = 16$ **26.** $2x^2 + 2y^2 = 9$

27. $(x + 1)^2 + y^2 = 16$ **28.** $x^2 + (y - 2)^2 = 4$

In Exercises 29–36, write the equation of the circle with the following properties.

29. Center at the origin; radius 1

30. Center at the origin; radius 4

31. Center at (6, 8); radius 5

32. Center at (5, 3); radius 2

33. Center at (−2, 6); radius 12

34. Center at (5, −4); radius 6

35. Center at the origin; diameter $4\sqrt{2}$

36. Center at the origin; diameter $8\sqrt{3}$

In Exercises 37–44, graph each circle. Give the coordinates of the center.

37. $x^2 + y^2 + 2x - 8 = 0$ **38.** $x^2 + y^2 - 4y = 12$

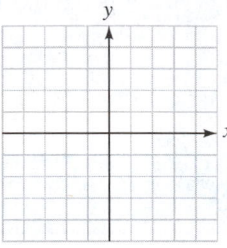   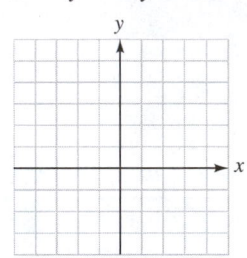

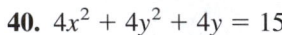

39. $9x^2 + 9y^2 - 12y = 5$ **40.** $4x^2 + 4y^2 + 4y = 15$

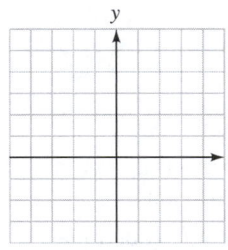   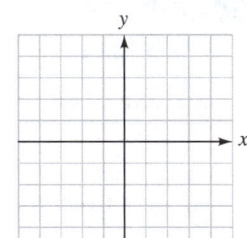

41. $x^2 + y^2 - 2x + 4y = -1$

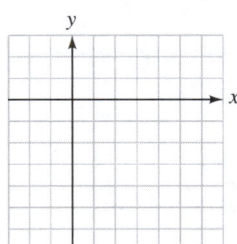

42. $x^2 + y^2 + 4x + 2y = 4$

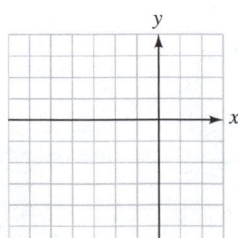

43. $x^2 + y^2 + 6x - 4y = -12$

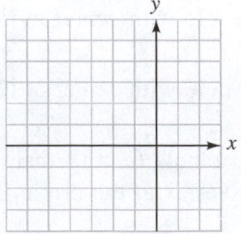

44. $x^2 + y^2 + 8x + 2y = -13$

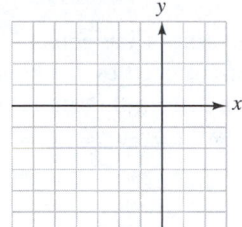

In Exercises 45–56, find the vertex of each parabola and graph it.

45. $x = y^2$

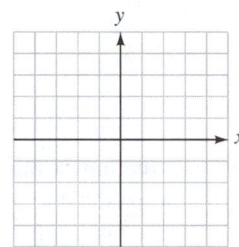

46. $x = -y^2 + 1$

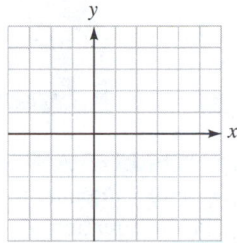

47. $x = -\dfrac{1}{4}y^2$

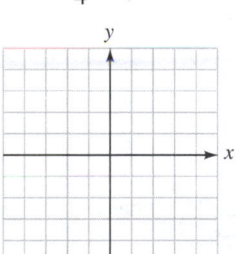

48. $x = 4y^2$

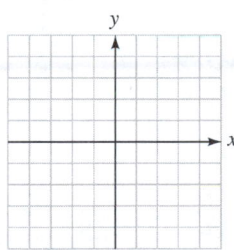

49. $y = x^2 + 4x + 5$

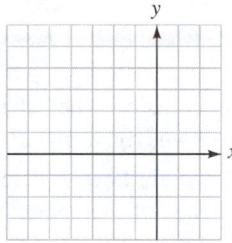

50. $y = -x^2 - 2x + 3$

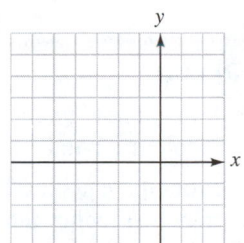

51. $y = -x^2 - x + 1$

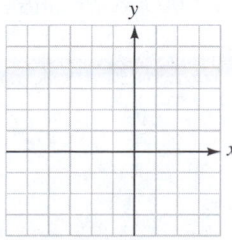

52. $x = \dfrac{1}{2}y^2 + 2y$

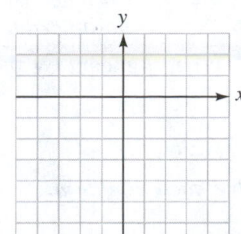

53. $y^2 + 4x - 6y = -1$

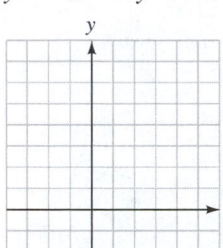

54. $x^2 - 2y - 2x = -7$

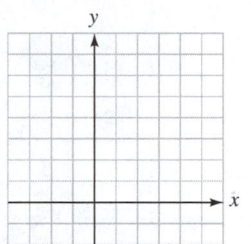

55. $y = 2(x - 1)^2 + 3$

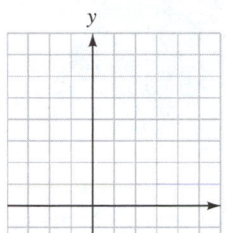

56. $y = -2(x + 1)^2 + 2$

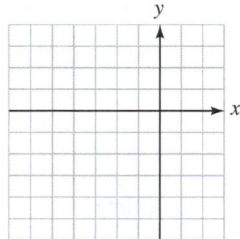

In Exercises 57–60, use a graphing calculator to graph each equation.

57. $x = 2y^2$

58. $x = y^2 - 4$

59. $x^2 - 2x + y = 6$

60. $x = -2(y - 1)^2 + 2$

APPLICATIONS

61. MESHING GEARS For design purposes, the large gear in Illustration 1 is the circle $x^2 + y^2 = 16$. The smaller gear is a circle centered at $(7, 0)$ and tangent to the larger circle. Find the equation of the smaller gear.

ILLUSTRATION 1

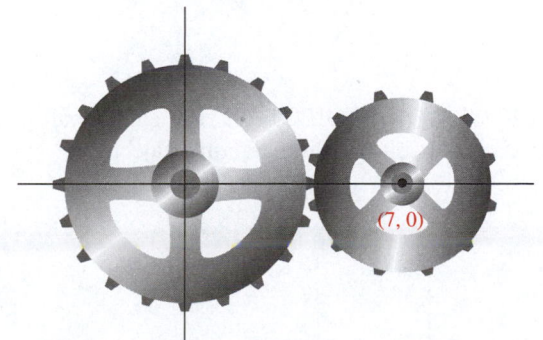

$(7, 0)$

62. WIDTH OF A WALKWAY The walkway in Illustration 2 (on the next page) is bounded by the two circles $x^2 + y^2 = 2{,}500$ and $(x - 10)^2 + y^2 = 900$, measured in feet. Find the largest and the smallest width of the walkway.

ILLUSTRATION 2

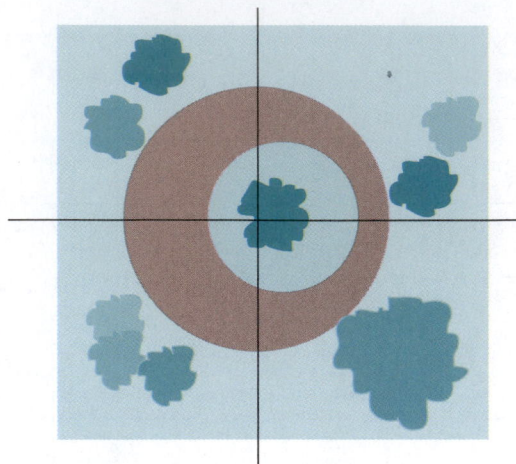

63. BROADCAST RANGES Radio stations applying for licensing may not use the same frequency if their broadcast areas overlap. One station's coverage is bounded by $x^2 + y^2 - 8x - 20y + 16 = 0$, and the other's by $x^2 + y^2 + 2x + 4y - 11 = 0$. May they be licensed for the same frequency?

64. HIGHWAY DESIGN Engineers want to join two sections of highway with a curve that is one-quarter of a circle, as in Illustration 3. The equation of the circle is $x^2 + y^2 - 16x - 20y + 155 = 0$, where distances are measured in kilometers. Find the locations (relative to the center of town) of the intersections of the highway with State and with Main.

ILLUSTRATION 3

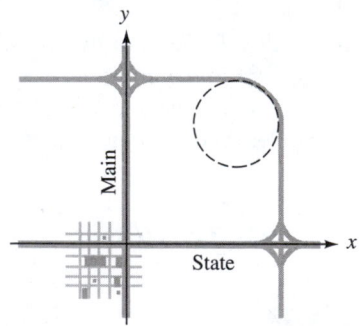

65. FLIGHT OF A PROJECTILE The cannonball in Illustration 4 follows the parabolic trajectory $y = 30x - x^2$. Where does it land?

ILLUSTRATION 4

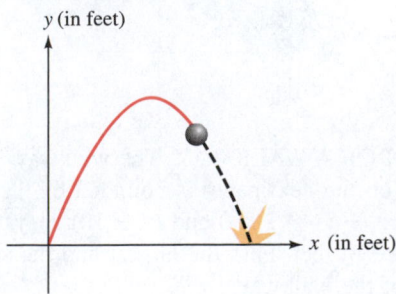

66. FLIGHT OF A PROJECTILE In Exercise 65, how high does the cannonball get?

67. ORBIT OF A COMET If the orbit of the comet shown in Illustration 5 is approximated by the equation $2y^2 - 9x = 18$, how far is it from the sun at the vertex of the orbit? Distances are measured in astronomical units (AU).

ILLUSTRATION 5

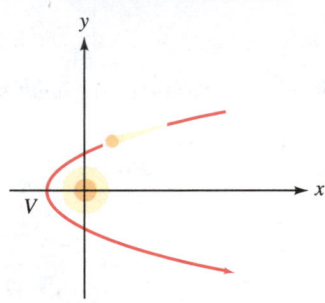

68. SATELLITE ANTENNA The cross section of the satellite antenna in Illustration 6 is a parabola given by the equation $y = \frac{1}{16}x^2$, with distances measured in feet. If the dish is 8 feet wide, how deep is it?

ILLUSTRATION 6

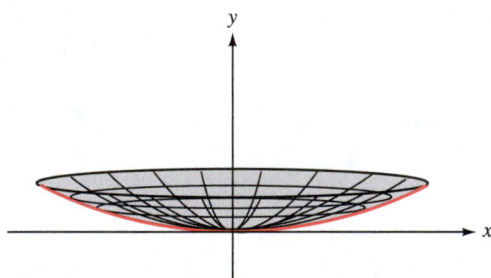

WRITING

69. Explain how to decide from its equation whether the graph of a parabola opens up, down, right, or left.

70. From the equation of a circle, explain how to determine the radius and the coordinates of the center.

71. On the day of an election, the following warning was posted in front of a school. Explain what it means.

> *No electioneering within a 1,000-foot radius of this polling place.*

72. What is meant by the *turning radius* of a truck?

REVIEW

Solve each equation.

73. $|3x - 4| = 11$

74. $\left| \dfrac{4 - 3x}{5} \right| = 12$

75. $|3x + 4| = |5x - 2|$

76. $|6 - 4x| = |x + 2|$

▶ 13.2   The Ellipse

In this section, you will learn about

> The ellipse ■ Constructing an ellipse ■ Graphing ellipses
> ■ Problem solving

Introduction   A third conic section is an oval-shaped curve called an *ellipse*. Ellipses can be nearly "round" and look almost like a circle, or they can be long and narrow. In this section, we will learn how to construct ellipses and how to graph equations that represent ellipses.

The Ellipse

The ellipse

An **ellipse** is the set of all points P in a plane the sum of whose distances from two fixed points is a constant. See Figure 13-13, in which $d_1 + d_2$ is a constant.

Each of the two points is called a **focus**. Midway between the foci is the **center** of the ellipse.

FIGURE 13-13

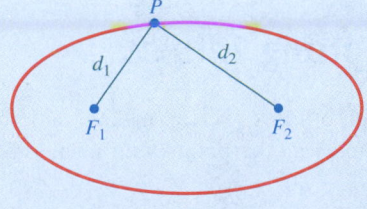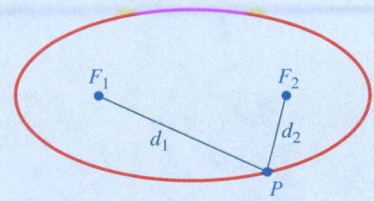

Constructing an Ellipse

We can construct an ellipse by placing two thumbtacks fairly close together, as in Figure 13-14. We then tie each end of a piece of string to a thumbtack, catch the loop with the point of a pencil, and (keeping the string taut) draw the ellipse.

FIGURE 13-14

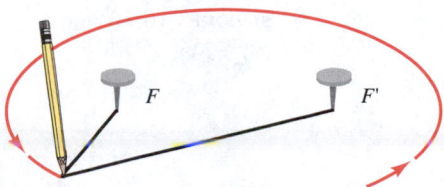

Using this method, we can construct an ellipse of any specific size. For example, to construct an ellipse that is 10 feet wide and 6 feet high, we must find the length of string to use and the distance between the thumbtacks.

To do this, we will let a represent the distance between the center and vertex V, as shown in Figure 13-15(a) on the next page. We will also let c represent the distance between the center of the ellipse and either focus. When the pencil is at vertex V, the length of the string is $c + a + (a - c)$, or just $2a$. Because $2a$ is the 10-foot width of the ellipse, the string needs to be 10 feet long. The distance $2a$ is constant for any point on the ellipse, including point B shown in Figure 13-15(b).

FIGURE 13-15

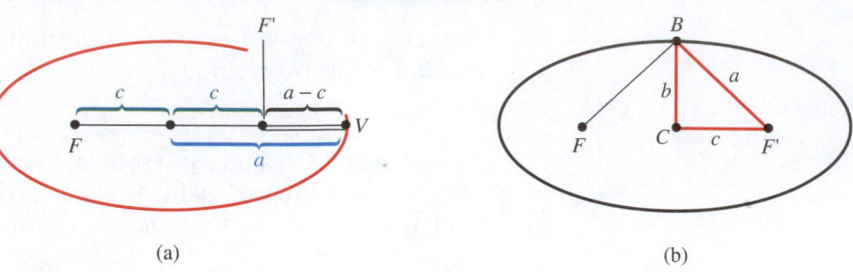

(a)   (b)

From right triangle CBF' and the Pythagorean theorem, we can find c as follows:

$$a^2 = b^2 + c^2 \quad \text{or} \quad c = \sqrt{a^2 - b^2}$$

Since distance b is one-half of the height of the ellipse, $b = 3$. Since $2a = 10$, $a = 5$. We can now substitute $a = 5$ and $b = 3$ into the formula to find c:

$$c = \sqrt{5^2 - 3^2}$$
$$= \sqrt{25 - 9}$$
$$= \sqrt{16}$$
$$= 4$$

Since $c = 4$, the distance between the thumbtacks must be 8 feet. We can construct the ellipse by tying a 10-foot string to thumbtacks that are 8 feet apart.

Graphing Ellipses

To graph ellipses, we can make a table of ordered pairs that satisfy the equation, plot them, and join the points with a smooth curve.

EXAMPLE 1

Graphing ellipses. Graph $\dfrac{x^2}{36} + \dfrac{y^2}{9} = 1$.

Solution First we note that the equation can be written in the form

$$\frac{x^2}{6^2} + \frac{y^2}{3^2} = 1 \quad 36 = 6^2 \text{ and } 9 = 3^2.$$

After making a table of ordered pairs that satisfy the equation, plotting each of them, and joining the points with a curve, we obtain the ellipse shown in Figure 13-16.

We note that the center of the ellipse is the origin, the ellipse intersects the x-axis at points $(6, 0)$ and $(-6, 0)$, and the ellipse intersects the y-axis at points $(0, 3)$ and $(0, -3)$.

FIGURE 13-16

$$\frac{x^2}{36} + \frac{y^2}{9} = 25$$

x	y	(x, y)
-6	0	$(-6, 0)$
-4	± 2.2	$(-4, \pm 2.2)$
-2	± 2.8	$(-2, \pm 2.8)$
0	± 3	$(0, \pm 3)$
2	± 2.8	$(2, \pm 2.8)$
4	± 2.2	$(4, \pm 2.2)$
6	0	$(6, 0)$

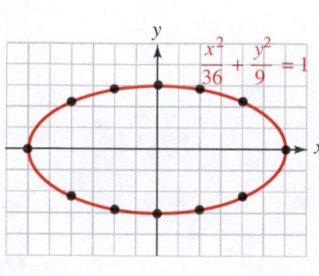

SELF CHECK Graph $\dfrac{x^2}{4} + \dfrac{y^2}{16} = 1$. *Answer:*

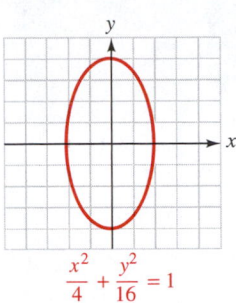

$\dfrac{x^2}{4} + \dfrac{y^2}{16} = 1$

Example 1 illustrates that the graph of

$$\frac{x^2}{a^2} + \frac{y^2}{b^2} = 1$$

is an ellipse centered at the origin. To find the *x*-intercepts of the graph, we can let $y = 0$ and solve for *x*.

$$\frac{x^2}{a^2} + \frac{\mathbf{0}^2}{b^2} = 1 \qquad \text{Substitute 0 for } y.$$

$$\frac{x^2}{a^2} + 0 = 1 \qquad \frac{0^2}{b^2} = 0.$$

$$x^2 = a^2 \qquad \text{Simplify and multiply both sides by } a^2.$$

$$x = a \quad \text{or} \quad x = -a$$

The *x*-intercepts are $(a, 0)$ and $(-a, 0)$.

To find the *y*-intercepts of the graph, we can let $x = 0$ and solve for *y*.

$$\frac{\mathbf{0}^2}{a^2} + \frac{y^2}{b^2} = 1$$

$$0 + \frac{y^2}{b^2} = 1$$

$$y^2 = b^2$$

$$y = b \quad \text{or} \quad y = -b$$

The *y*-intercepts are $(0, b)$ and $(0, -b)$.

In general, we have the following results.

Equations of an ellipse centered at the origin

The equation of an ellipse centered at the origin with *x*-intercepts at $V_1(a, 0)$ and $V_2(-a, 0)$ and with *y*-intercepts at $(0, b)$ and $(0, -b)$ is

$$\frac{x^2}{a^2} + \frac{y^2}{b^2} = 1 \quad (a > b > 0) \quad \text{See Figure 13-17(a) on the next page.}$$

The equation of an ellipse centered at the origin with *y*-intercepts at $V_1(0, a)$ and $V_2(0, -a)$ and with *x*-intercepts at $(b, 0)$ and $(-b, 0)$ is

$$\frac{x^2}{b^2} + \frac{y^2}{a^2} = 1 \quad (a > b > 0) \quad \text{See Figure 13-17(b).}$$

In Figure 13-17, the points V_1 and V_2 are the **vertices** of the ellipse, the midpoint of V_1V_2 is the **center** of the ellipse, and the distance between the center of the ellipse and either vertex is *a*. The segment V_1V_2 is called the **major axis**, and the segment joining either $(0, b)$ and $(0, -b)$ or $(b, 0)$ and $(-b, 0)$ is called the **minor axis**.

FIGURE 13-17

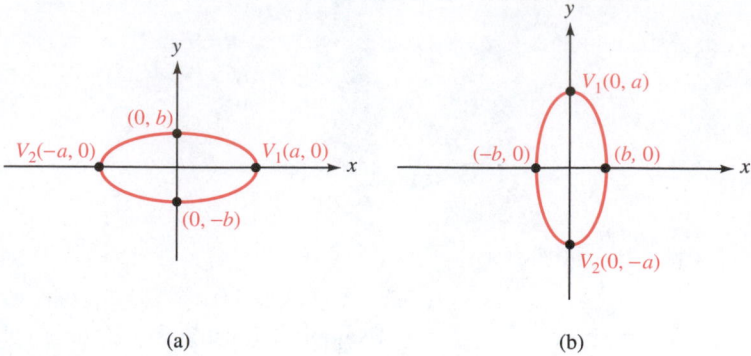

(a)   (b)

The equations for ellipses centered at (h, k) are as follows.

Equations of an ellipse centered at (h, k)

The equation of an ellipse centered at (h, k), with major axis parallel to the x-axis is

1. $\dfrac{(x - h)^2}{a^2} + \dfrac{(y - k)^2}{b^2} = 1 \quad (a > b > 0)$

The equation of an ellipse centered at (h, k), with major axis parallel to the y-axis is

2. $\dfrac{(x - h)^2}{b^2} + \dfrac{(y - k)^2}{a^2} = 1 \quad (a > b > 0)$

EXAMPLE 2

Graphing ellipses. Graph the ellipse $\dfrac{(x - 2)^2}{16} + \dfrac{(y + 3)^2}{25} = 1$.

Solution We write the equation in the form

$$\frac{(x - 2)^2}{4^2} + \frac{[y - (-3)]^2}{5^2} = 1 \quad (5 > 4)$$

This is the equation of an ellipse centered at $(h, k) = (2, -3)$, with major axis parallel to the y-axis and with $b = 4$ and $a = 5$. We first plot the center, as shown in Figure 13-18. Since a is the distance from the center to the vertex, we can locate the vertices by counting 5 units above and 5 units below the center. The vertices are at points $(2, 2)$ and $(2, -8)$.

Since $b = 4$, we can locate two more points on the ellipse by counting 4 units to the left and 4 units to the right of the center. The points $(-2, -3)$ and $(6, -3)$ are also on the graph.

Using these four points as guides, we can draw the ellipse.

FIGURE 13-18

$$\frac{(x - 2)^2}{16} + \frac{(y + 3)^2}{25} = 1$$

x	y	(x, y)
2	2	$(2, 2)$
2	-8	$(2, -8)$
6	-3	$(6, -3)$
-2	-3	$(-2, -3)$

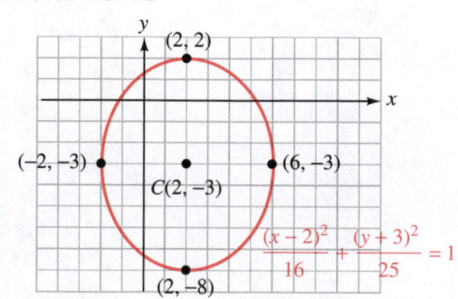

SELF CHECK Graph $\dfrac{(x-1)^2}{9} + \dfrac{(y+2)^2}{16} = 1$.

Answer:

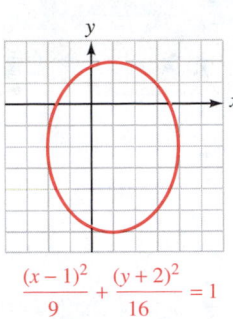

$$\dfrac{(x-1)^2}{9} + \dfrac{(y+2)^2}{16} = 1$$ ∎

ACCENT ON TECHNOLOGY *Graphing Ellipses*

To use a graphing calculator to graph

$$\frac{(x+2)^2}{4} + \frac{(y-1)^2}{25} = 1$$

we first clear the equation of fractions by multiplying both sides by 100 and solving for y.

$$25(x+2)^2 + 4(y-1)^2 = 100 \qquad \text{Multiply both sides by 100.}$$

$$4(y-1)^2 = 100 - 25(x+2)^2 \qquad \text{Subtract } 25(x+2)^2 \text{ from both sides.}$$

$$(y-1)^2 = \frac{100 - 25(x+2)^2}{4} \qquad \text{Divide both sides by 4.}$$

$$y - 1 = \pm\frac{\sqrt{100 - 25(x+2)^2}}{2} \qquad \text{Take the square root of both sides.}$$

$$y = 1 \pm \frac{\sqrt{100 - 25(x+2)^2}}{2} \qquad \text{Add 1 to both sides.}$$

If we use window settings $[-6, 6]$ for x and $[-6, 6]$ for y and graph the functions

$$y = 1 + \frac{\sqrt{100 - 25(x+2)^2}}{2} \qquad \text{and} \qquad y = 1 - \frac{\sqrt{100 - 25(x+2)^2}}{2}$$

we will obtain the ellipse shown in Figure 13-19.

FIGURE 13-19

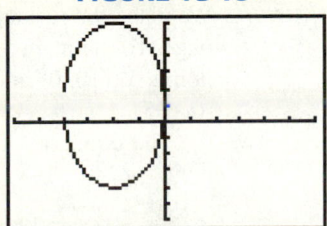

We can complete the square to write the equations of many ellipses in standard form.

EXAMPLE 3

Writing an equation in standard form. Write $4x^2 + 9y^2 - 16x - 18y = 11$ in standard form to show that the equation represents an ellipse.

Solution   We write the equation in standard form by completing the square on x and y:

$$4x^2 + 9y^2 - 16x - 18y = 11$$

$$4x^2 - 16x + 9y^2 - 18y = 11$$ Use the commutative property to rearrange terms.

$$4(x^2 - 4x) + 9(y^2 - 2y) = 11$$ Factor 4 from $4x^2 - 16x$ and factor 9 from $9y^2 - 18y$ to get coefficients of 1 for the squared terms.

$$4(x^2 - 4x + 4) + 9(y^2 - 2y + 1) = 11 + 16 + 9$$ Complete the square to make $x^2 - 4x$ and $y^2 - 2y$ perfect square trinomials. Since $16 + 9$ is added to the left-hand side, add $16 + 9$ to the right-hand side.

FIGURE 13-20

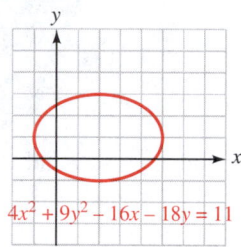

$$4(x - 2)^2 + 9(y - 1)^2 = 36$$ Factor $x^2 - 4x + 4$ and $y^2 - 2y + 1$.

$$\frac{(x - 2)^2}{9} + \frac{(y - 1)^2}{4} = 1$$ Divide both sides by 36 and simplify.

Since this equation matches Equation 1 on page 946, it represents an ellipse. Its graph is shown in Figure 13-20.

Graph $4x^2 - 8x + 9y^2 - 36y = -4$.   *Answer:*

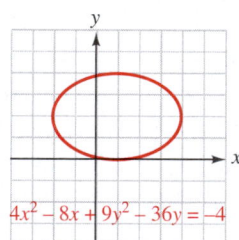

Problem Solving

EXAMPLE 4

Landscape design. A landscape architect is designing an elliptical pool that will fit in the center of a 20-by-30-foot rectangular garden, leaving at least 5 feet of space on all sides. Find the equation of the ellipse.

Solution   We place the rectangular garden in a coordinate system, as in Figure 13-21 on the next page. To maintain 5 feet of clearance at the ends of the ellipse, the vertices must be the points $V_1(10, 0)$ and $V_2(-10, 0)$. Similarly, the y-intercepts are the points $(0, 5)$ and $(0, -5)$.

The equation of the ellipse has the form

$$\frac{x^2}{a^2} + \frac{y^2}{b^2} = 1$$

with $a = 10$ and $b = 5$. Thus, the equation of the boundary of the pool is

$$\frac{x^2}{100} + \frac{y^2}{25} = 1$$

FIGURE 13-21

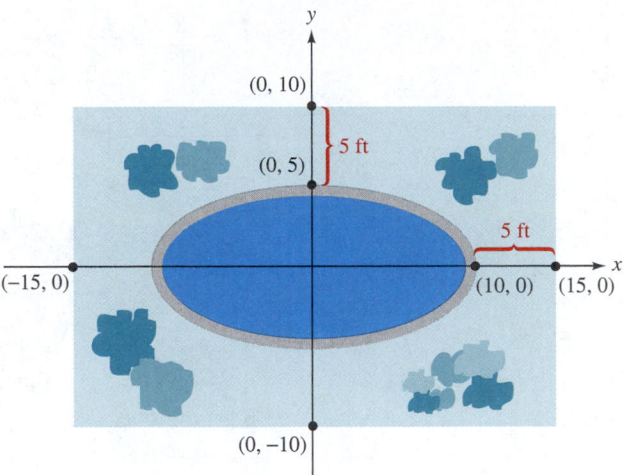

STUDY SET

Section 13.2

VOCABULARY

Fill in the blanks to make the statements true.

1. An _____ is the set of all points in a plane the _____ of whose distances from two fixed points is a constant.
2. The fixed points in Exercise 1 are the _____ of the ellipse.
3. The midpoint of the line segment joining the foci of an ellipse is called the _____ of the ellipse.
4. The segment joining the vertices of an ellipse is called the _____ axis.

CONCEPTS

Fill in the blanks to make the statements true.

5. The elliptical graph of $\frac{x^2}{a^2} + \frac{y^2}{b^2} = 1$ has x-intercepts of _____.

6. The elliptical graph of $\frac{x^2}{a^2} + \frac{y^2}{b^2} = 1$ has y-intercepts of _____.

7. The center of the ellipse with an equation of
$$\frac{x^2}{a^2} + \frac{y^2}{b^2} = 1$$
is the point _____.

8. The center of the ellipse with an equation of
$$\frac{(x-h)^2}{a^2} + \frac{(y-k)^2}{b^2} = 1$$
is the point _____.

NOTATION

Complete each solution.

9. Write $16x^2 + 9y^2 - 144 = 0$ in standard form.
$$16x^2 + 9y^2 - 144 = 0$$
$$16x^2 + 9y^2 = \boxed{}$$
$$\frac{x^2}{9} + \frac{y^2}{\boxed{}} = 1$$

10. Write $9x^2 + 4y^2 - 18x + 8y = 23$ in standard form.
$$9x^2 + 4y^2 - 18x + 8y = 23$$
$$9x^2 - 18x + \boxed{}y^2 + 8y = 23$$
$$9\left(x^2 - \boxed{}x\right) + 4\left(y^2 + \boxed{}y\right) = 23$$
$$9\left(x^2 - 2x + \boxed{}\right) + 4(y^2 + 2y + 1) = 23 + 9 + 4$$
$$9(x-1)^2 + 4(y+1)^2 = \boxed{}$$
$$\frac{(x-1)^2}{4} + \frac{(y+1)^2}{9} = 1$$

PRACTICE

In Exercises 11–20, graph each equation.

11. $\dfrac{x^2}{4} + \dfrac{y^2}{9} = 1$

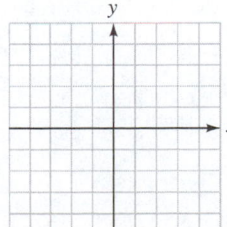

12. $x^2 + \dfrac{y^2}{9} = 1$

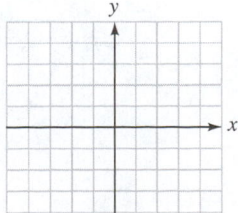

13. $x^2 + 9y^2 = 9$

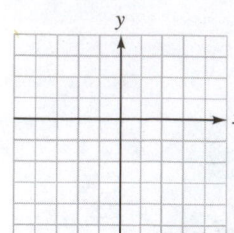

14. $25x^2 + 9y^2 = 225$

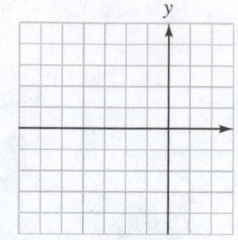

In Exercises 21–24, use a graphing calculator to graph each equation.

21. $\dfrac{x^2}{9} + \dfrac{y^2}{4} = 1$

22. $x^2 + 16y^2 = 16$

15. $16x^2 + 4y^2 = 64$

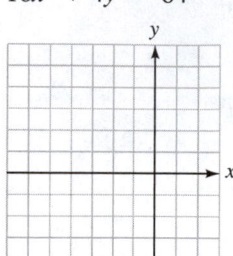

16. $4x^2 + 9y^2 = 36$

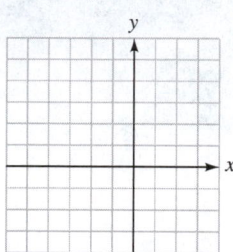

23. $\dfrac{x^2}{4} + \dfrac{(y - 1)^2}{9} = 1$

24. $\dfrac{(x + 1)^2}{9} + \dfrac{(y - 2)^2}{4} = 1$

17. $\dfrac{(x - 2)^2}{9} + \dfrac{(y - 1)^2}{4} = 1$

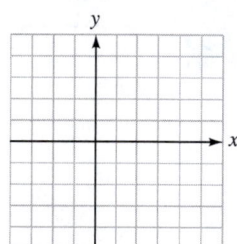

18. $\dfrac{(x - 1)^2}{9} + \dfrac{(y - 3)^2}{4} = 1$

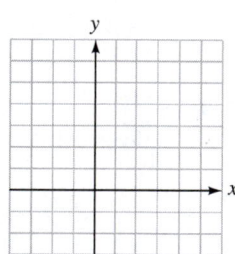

In Exercises 25–28, write each equation in standard form and graph it.

25. $x^2 + 4y^2 - 4x + 8y + 4 = 0$

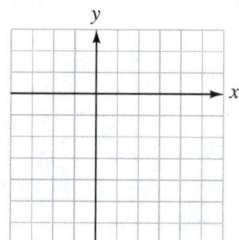

26. $x^2 + 4y^2 - 2x - 16y = -13$

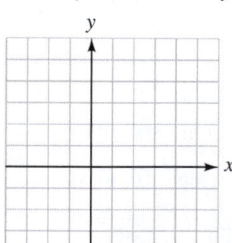

19. $(x + 1)^2 + 4(y + 2)^2 = 4$

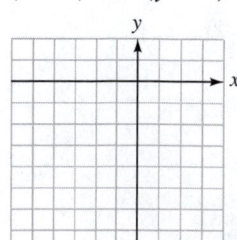

27. $9x^2 + 4y^2 - 18x + 16y = 11$

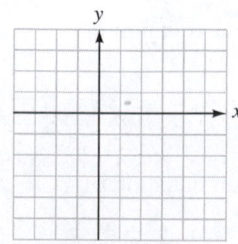

20. $25(x + 1)^2 + 9y^2 = 225$

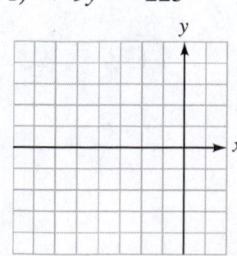

28. $16x^2 + 25y^2 - 160x - 200y + 400 = 0$

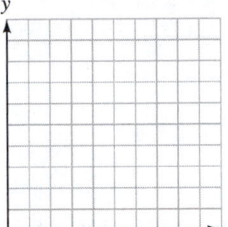

APPLICATIONS

29. DESIGNING AN UNDERPASS The arch of the underpass in Illustration 1 is a part of an ellipse. Find the equation of the arch.

ILLUSTRATION 1

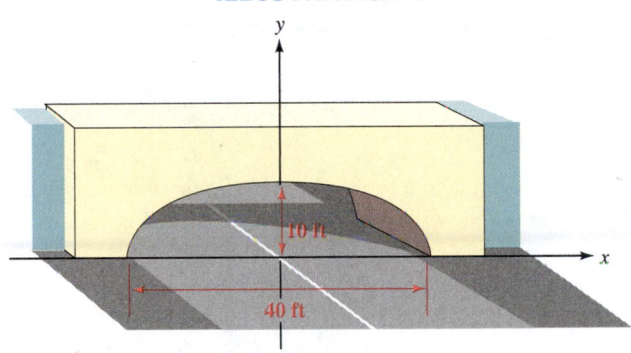

30. CALCULATING CLEARANCE Find the height of the elliptical arch in Exercise 29 at a point 10 feet from the center of the roadway.

31. AREA OF AN ELLIPSE The area A of the ellipse

$$\frac{x^2}{a^2} + \frac{y^2}{b^2} = 1$$

is given by $A = \pi ab$. Find the area of the ellipse $9x^2 + 16y^2 = 144$.

32. AREA OF A TRACK The elliptical track in Illustration 2 is bounded by the ellipses $4x^2 + 9y^2 = 576$ and $9x^2 + 25y^2 = 900$. Find the area of the track. (See Exercise 31.)

ILLUSTRATION 2

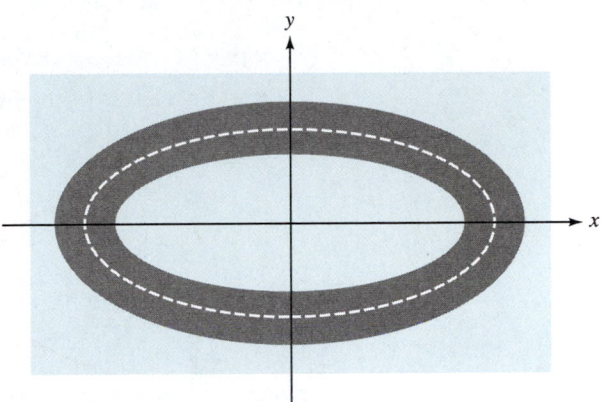

WRITING

33. Explain how to find the x- and y-intercepts of the graph of the ellipse $\frac{x^2}{a^2} + \frac{y^2}{b^2} = 1$.

34. Explain the relationship between the center, focus, and vertex of an ellipse.

REVIEW

Find each product.

35. $3x^{-2}y^2(4x^2 + 3y^{-2})$

36. $(2a^{-2} - b^{-2})(2a^{-2} + b^{-2})$

Write each expression without using negative exponents.

37. $\dfrac{x^{-2} + y^{-2}}{x^{-2} - y^{-2}}$

38. $\dfrac{2x^{-3} - 2y^{-3}}{4x^{-3} + 4y^{-3}}$

▶ **13.3**

The Hyperbola

In this section, you will learn about

The hyperbola ■ Problem solving

Introduction The final conic section, the *hyperbola*, is a curve with two branches. In this section, we will learn how to graph equations that represent hyperbolas.

The Hyperbola

The hyperbola

A **hyperbola** is the set of all points P in a plane for which the difference of the distances of each point from two fixed points is a constant. See Figure 13-22, in which $d_1 - d_2$ is a constant.

Each of the two points is called a **focus**. Midway between the foci is the **center** of the hyperbola.

FIGURE 13-22

The graph of the equation

$$\frac{x^2}{25} - \frac{y^2}{9} = 1$$

is a hyperbola. To graph the equation, we make a table of ordered pairs that satisfy the equation, plot each pair, and join the points with a smooth curve, as in Figure 13-23.

FIGURE 13-23

$$\frac{x^2}{25} - \frac{y^2}{9} = 1$$

x	y	(x, y)
-7	± 2.9	$(-7, \pm 2.9)$
-6	± 2.0	$(-6, \pm 2.0)$
-5	0	$(-5, 0)$
5	0	$(5, 0)$
6	± 2.0	$(6, \pm 2.0)$
7	± 2.9	$(7, \pm 2.9)$

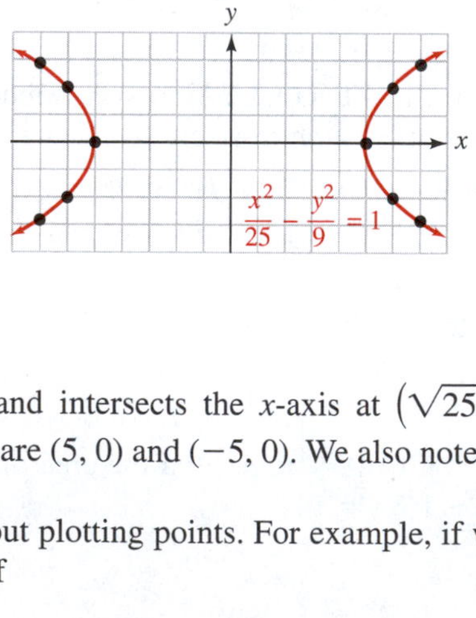

This graph is centered at the origin and intersects the x-axis at $\left(\sqrt{25}, 0\right)$ and $\left(-\sqrt{25}, 0\right)$. After simplifying, these points are $(5, 0)$ and $(-5, 0)$. We also note that the graph does not intersect the y-axis.

It is possible to draw a hyperbola without plotting points. For example, if we want to graph the hyperbola with an equation of

$$\frac{x^2}{a^2} - \frac{y^2}{b^2} = 1$$

we first look at the x- and y-intercepts. To find the x-intercepts, we let $y = 0$ and solve for x:

$$\frac{x^2}{a^2} - \frac{\mathbf{0}^2}{b^2} = 1$$

$$x^2 = a^2$$

$$x = \pm a$$

The hyperbola crosses the x-axis at the points $V_1(a, 0)$ and $V_2(-a, 0)$, called the **vertices** of the hyperbola. See Figure 13-24.

To attempt to find the y-intercepts, we let $x = 0$ and solve for y:

$$\frac{\mathbf{0}^2}{a^2} - \frac{y^2}{b^2} = 1$$

$$y^2 = -b^2$$

$$y = \pm\sqrt{-b^2}$$

FIGURE 13-24

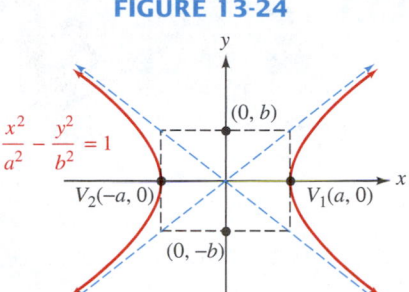

Since b^2 is always positive, $\sqrt{-b^2}$ is an imaginary number. This means that the hyperbola does not cross the y-axis.

If we construct a rectangle, called the **fundamental rectangle**, whose sides pass horizontally through $\pm b$ on the y-axis and vertically through $\pm a$ on the x-axis, the extended diagonals of the rectangle will be asymptotes of the hyperbola.

Equation of a hyperbola centered at the origin

Any equation that can be written in the form

$$\frac{x^2}{a^2} - \frac{y^2}{b^2} = 1$$

has a graph that is a hyperbola centered at the origin, as in Figure 13-25. The x-intercepts are the vertices $V_1(a, 0)$ and $V_2(-a, 0)$. There are no y-intercepts.

The asymptotes of the hyperbola are the extended diagonals of the rectangle in the figure.

FIGURE 13-25

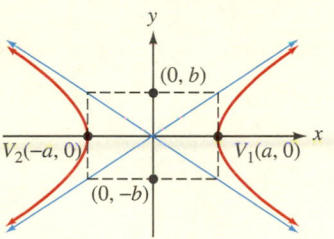

The branches of the hyperbola in previous discussions open to the left and to the right. It is possible for hyperbolas to have different orientations with respect to the x- and y-axes. For example, the branches of a hyperbola can open upward and downward. In that case, the following equation applies.

Equation of a hyperbola centered at the origin

Any equation that can be written in the form

$$\frac{y^2}{a^2} - \frac{x^2}{b^2} = 1$$

has a graph that is a hyperbola centered at the origin, as in Figure 13-26. The y-intercepts are the vertices $V_1(0, a)$ and $V_2(0, -a)$. There are no x-intercepts.

The asymptotes of the hyperbola are the extended diagonals of the rectangle shown in the figure.

FIGURE 13-26

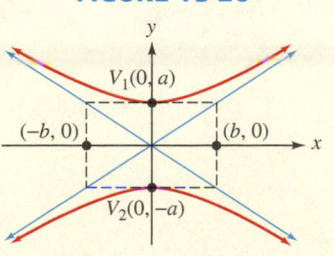

| **E X A M P L E   1** | **Graphing hyperbolas.** Graph $9y^2 - 4x^2 = 36$. |

Solution To write the equation in standard form, we divide both sides by 36 to obtain

$$\frac{9y^2}{36} - \frac{4x^2}{36} = 1$$

$$\frac{y^2}{4} - \frac{x^2}{9} = 1 \quad \text{Simplify each fraction.}$$

FIGURE 13-27

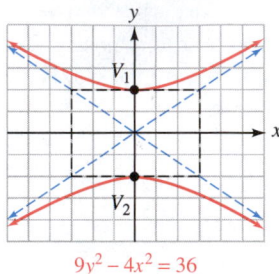

$$9y^2 - 4x^2 = 36$$

We then find the y-intercepts by letting $x = 0$ and solving for y:

$$\frac{y^2}{4} - \frac{\mathbf{0}^2}{9} = 1$$

$$y^2 = 4$$

Thus, $y = \pm 2$, and the vertices of the hyperbola are $V_1(0, 2)$ and $V_2(0, -2)$. (See Figure 13-27.)

Since $\pm\sqrt{9} = \pm 3$, we use the points $(3, 0)$ and $(-3, 0)$ on the x-axis to help draw the fundamental rectangle. We then draw its extended diagonals and sketch the hyperbola.

SELF CHECK Graph $9x^2 - 4y^2 = 36$. *Answer:*

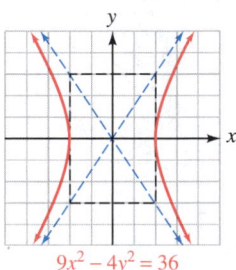

$$9x^2 - 4y^2 = 36$$

If a hyperbola is centered at a point with coordinates (h, k), the following equations apply.

Equations of hyperbolas centered at (h, k)

Any equation that can be written in the form

$$\frac{(x - h)^2}{a^2} - \frac{(y - k)^2}{b^2} = 1$$

is a hyperbola that has its center at (h, k) and opens left and right.
    Any equation of the form

$$\frac{(y - k)^2}{a^2} - \frac{(x - h)^2}{b^2} = 1$$

is a hyperbola that has its center at (h, k) and opens up and down.

| **E X A M P L E   2** | **Graphing hyperbolas.** Graph $\dfrac{(x - 3)^2}{16} - \dfrac{(y + 1)^2}{4} = 1$. |

Solution   We write the equation in the form

$$\frac{(x-3)^2}{4^2} - \frac{[y-(-1)]^2}{2^2} = 1$$

to see that its graph will be a hyperbola centered at the point $(h, k) = (3, -1)$. Its vertices are located at $a = 4$ units to the right and left of the center, at $(7, -1)$ and $(-1, -1)$. Since $b = 2$, we can count 2 units above and below the center to locate points $(3, 1)$ and $(3, -3)$. With these points, we can draw the fundamental rectangle along with its extended diagonals. We can then sketch the hyperbola, as shown in Figure 13-28.

FIGURE 13-28

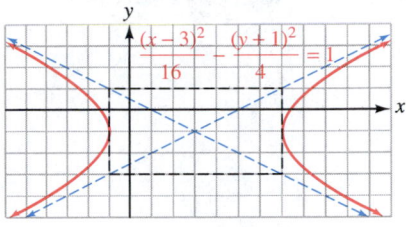

SELF CHECK   Graph $\dfrac{(x+2)^2}{9} - \dfrac{(y-1)^2}{4} = 1$.

Answer:

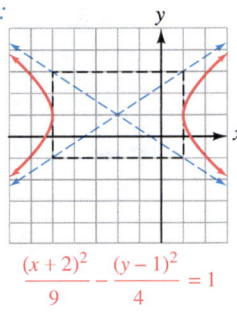

E X A M P L E   3   **Writing equations in standard form.** Write the equation $x^2 - y^2 - 2x + 4y = 12$ in standard form to show that the equation represents a hyperbola. Then graph it.

Solution   We proceed as follows.

$$x^2 - y^2 - 2x + 4y = 12$$

$$x^2 - 2x - y^2 + 4y = 12 \quad \text{Use the commutative property to rearrange terms.}$$

$$x^2 - 2x - (y^2 - 4y) = 12 \quad \text{Factor } -1 \text{ from } -y^2 + 4y.$$

FIGURE 13-29

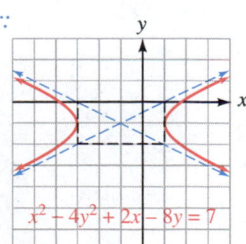

$x^2 - y^2 - 2x + 4y = 12$

We then complete the square on x and y to make $x^2 - 2x$ and $y^2 - 4y$ perfect square trinomials.

$$x^2 - 2x + 1 - (y^2 - 4y + 4) = 12 + 1 - 4$$

We then factor $x^2 - 2x + 1$ and $y^2 - 4y + 4$ to get

$$(x-1)^2 - (y-2)^2 = 9$$

$$\frac{(x-1)^2}{9} - \frac{(y-2)^2}{9} = 1 \quad \text{Divide both sides by 9.}$$

This is the equation of a hyperbola with center at $(1, 2)$. Its graph is shown in Figure 13-29.

SELF CHECK   Graph $x^2 - 4y^2 + 2x - 8y = 7$.

Answer:

There is a special type of hyperbola (also centered at the origin) that does not intersect either the x- or the y-axis. These hyperbolas have equations of the form $xy = k$, where $k \neq 0$.

EXAMPLE 4

Graphing hyperbolas. Graph $xy = -8$.

Solution   We make a table of ordered pairs, plot each pair, and join the points with a smooth curve to obtain the hyperbola in Figure 13-30.

FIGURE 13-30

$xy = -8$

x	y	(x, y)
1	-8	$(1, -8)$
2	-4	$(2, -4)$
4	-2	$(4, -2)$
8	-1	$(8, -1)$
-1	8	$(-1, 8)$
-2	4	$(-2, 4)$
-4	2	$(-4, 2)$
-8	1	$(-8, 1)$

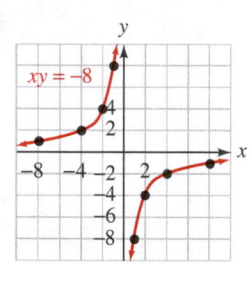

SELF CHECK   Graph $xy = 6$.   *Answer:*

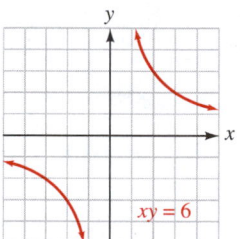

The result in Example 4 illustrates the following general equation.

Equations of hyperbolas of the form $xy = k$

Any equation of the form $xy = k$, where $k \neq 0$, has a graph that is a **hyperbola**, which does not intersect either the x- or the y-axis.

Problem Solving

EXAMPLE 5

Atomic structure. In an experiment that led to the discovery of the atomic structure of matter, Lord Rutherford (1871–1937) shot high–energy alpha particles toward a thin sheet of gold. Many of them were reflected, and Rutherford showed the existence of the nucleus of a gold atom. The alpha particle in Figure 13-31 is repelled by the nucleus at the origin; it travels along the hyperbolic path given by $4x^2 - y^2 = 16$. How close does the particle come to the nucleus?

FIGURE 13-31

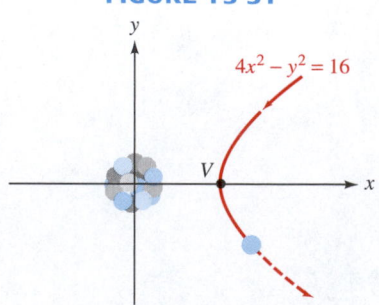

Solution To find the distance from the nucleus at the origin, we must find the coordinates of the vertex V. To do so, we write the equation of the particle's path in standard form:

$$4x^2 - y^2 = 16$$

$$\frac{4x^2}{16} - \frac{y^2}{16} = \frac{16}{16} \qquad \text{Divide both sides by 16.}$$

$$\frac{x^2}{4} - \frac{y^2}{16} = 1 \qquad \text{Simplify.}$$

$$\frac{x^2}{2^2} - \frac{y^2}{4^2} = 1 \qquad \text{Write 4 as } 2^2 \text{ and 16 as } 4^2.$$

This equation is in the form

$$\frac{x^2}{a^2} - \frac{y^2}{b^2} = 1$$

with $a = 2$. Thus, the vertex of the path is $(2, 0)$. The particle is never closer than 2 units from the nucleus. ∎

STUDY SET

Section 13.3

VOCABULARY

Fill in each blank to make the statements true.

1. A _____ is the set of all points in a plane for which the _____ of the distances from two fixed points is a constant.

2. The fixed points in Exercise 1 are the _____ of the hyperbola.

3. The midpoint of the line segment joining the foci of a hyperbola is called the _____ of the hyperbola.

4. The rectangle whose sides pass horizontally through $\pm b$ on the y-axis and vertically through $\pm a$ on the x-axis is called the _____ rectangle.

CONCEPTS

Fill in the blanks to make the statements true.

5. The hyperbolic graph of
$$\frac{x^2}{a^2} - \frac{y^2}{b^2} = 1$$
has x-intercepts of _____.

6. The hyperbolic graph of
$$\frac{x^2}{a^2} - \frac{y^2}{b^2} = 1$$
has ___ y-intercepts.

7. The center of the hyperbola with an equation of
$$\frac{x^2}{a^2} - \frac{y^2}{b^2} = 1$$
is the point _____.

8. The center of the hyperbola with an equation of
$$\frac{(x - h)^2}{a^2} - \frac{(y - k)^2}{b^2} = 1$$
is the point _____.

NOTATION

Complete each solution.

9. Write the equation $4x^2 - 9y^2 - 36 = 0$ in standard form.

$$4x^2 - 9y^2 - 36 = 0$$

$$4x^2 - 9y^2 = \boxed{}$$

$$\frac{4x^2}{\boxed{}} - \frac{9y^2}{\boxed{}} = 1$$

$$\frac{x^2}{9} - \frac{y^2}{4} = 1$$

10. Write the equation $x^2 - 4y^2 + 2x - 63 = 0$ in standard form.

$$x^2 - 4y^2 + 2x - 63 = 0$$

$$x^2 + 2x - 4y^2 = \boxed{}$$

$$x^2 + 2x + \boxed{} - 4y^2 = 63 + \boxed{}$$

$$\left(\boxed{}\right)^2 - 4y^2 = 64$$

$$\frac{(x + 1)^2}{64} - \frac{y^2}{16} = 1$$

PRACTICE

In Exercises 11–22, graph each hyperbola.

11. $\dfrac{x^2}{9} - \dfrac{y^2}{4} = 1$

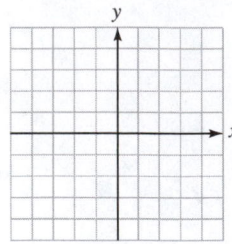

12. $\dfrac{x^2}{4} - \dfrac{y^2}{4} = 1$

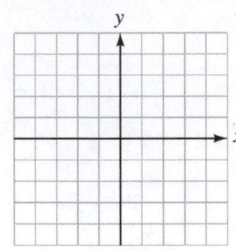

13. $\dfrac{y^2}{4} - \dfrac{x^2}{9} = 1$

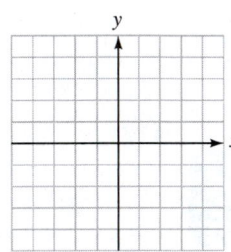

14. $\dfrac{y^2}{4} - \dfrac{x^2}{64} = 1$

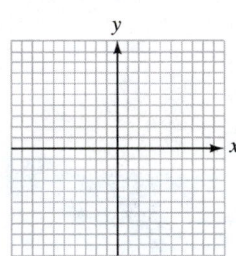

15. $25x^2 - y^2 = 25$

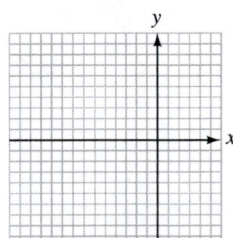

16. $9x^2 - 4y^2 = 36$

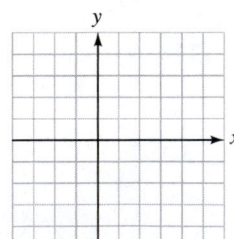

17. $\dfrac{(x-2)^2}{9} - \dfrac{y^2}{16} = 1$

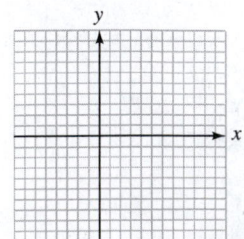

18. $\dfrac{(x+2)^2}{16} - \dfrac{(y-3)^2}{25} = 1$

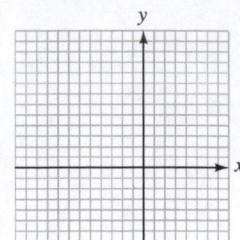

19. $4(x+3)^2 - (y-1)^2 = 4$

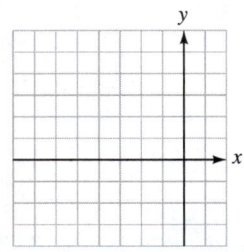

20. $(x+5)^2 - 16y^2 = 16$

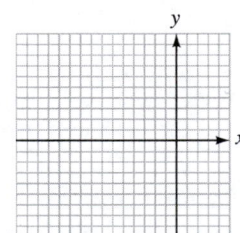

21. $xy = 8$

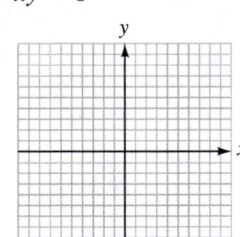

22. $xy = -10$

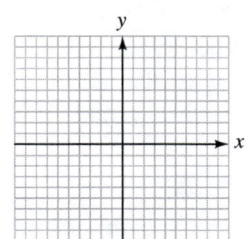

In Exercises 23–26, use a graphing calculator to graph each equation.

23. $\dfrac{x^2}{9} - \dfrac{y^2}{4} = 1$

24. $y^2 - 16x^2 = 16$

25. $\dfrac{x^2}{4} - \dfrac{(y-1)^2}{9} = 1$

26. $\dfrac{(y+1)^2}{9} - \dfrac{(x-2)^2}{4} = 1$

In Exercises 27–30, write each equation in standard form and graph it.

27. $4x^2 - y^2 + 8x - 4y = 4$

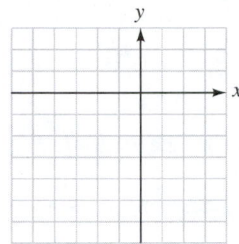

28. $x^2 - 9y^2 - 4x - 54y = 86$

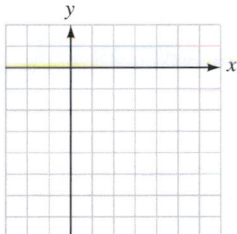

29. $4y^2 - x^2 + 8y + 4x = 4$

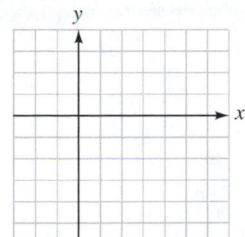

30. $y^2 - 4x^2 - 4y - 8x = 4$

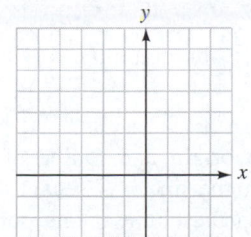

APPLICATIONS

31. ALPHA PARTICLES The particle in Illustration 1 approaches the nucleus at the origin along the path $9y^2 - x^2 = 81$. How close does the particle come to the nucleus?

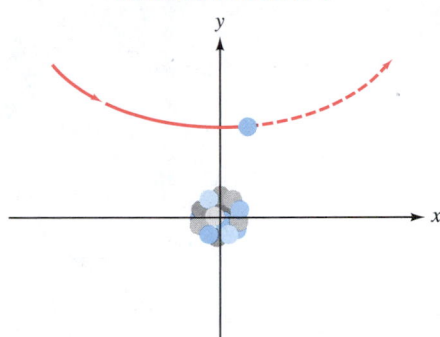

32. LORAN By determining the difference of the distances between the ship in Illustration 2 and two radio transmitters, the LORAN system places the ship on the hyperbola $x^2 - 4y^2 = 576$. If the ship is also 5 miles out to sea, find its coordinates.

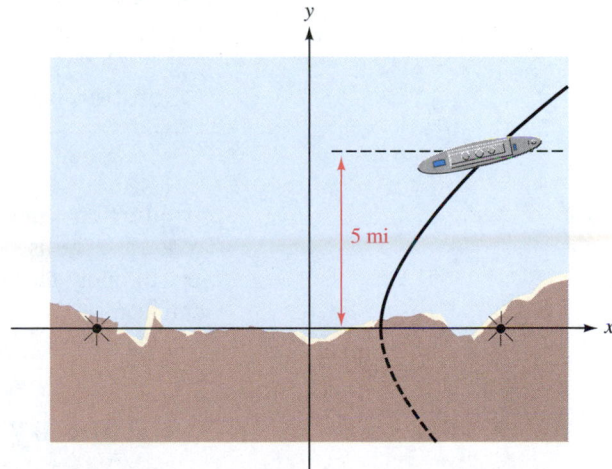

5 mi

33. SONIC BOOM The position of the sonic boom caused by the faster-than-sound aircraft in Illustration 3 (on the next page) is the hyperbola $y^2 - x^2 = 25$ in the coordinate system shown. How wide is the hyperbola 5 units from its vertex?

34. ELECTROSTATIC REPULSION Two similarly charged particles are shot together for an almost head-on collision, as in Illustration 4 on the next page. They repel each other and travel the two branches of the hyperbola given by $x^2 - 4y^2 = 4$. How close do they get?

ILLUSTRATION 3

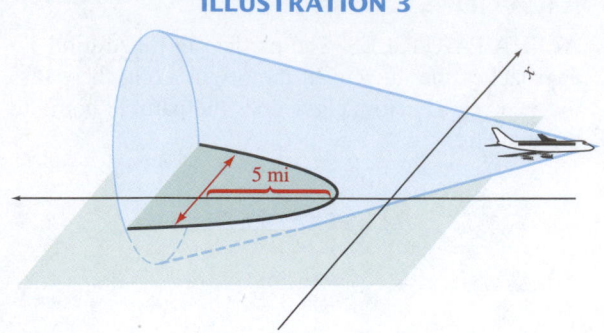

ILLUSTRATION 4

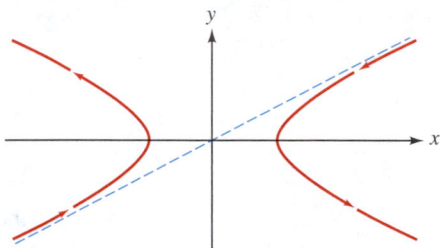

WRITING

35. Explain how to find the x- and y-intercepts of the graph of the ellipse

$$\frac{x^2}{a^2} + \frac{y^2}{b^2} = 1$$

36. Explain why the graph of the hyperbola

$$\frac{x^2}{a^2} - \frac{y^2}{b^2} = 1$$

has no y-intercept.

REVIEW

Factor each expression.

37. $-6x^4 + 9x^3 - 6x^2$

38. $4a^2 - b^2$

39. $15a^2 - 4ab - 4b^2$

40. $8p^3 - 27q^3$

▶ 13.4   Solving Simultaneous Second-Degree Equations

In this section, you will learn about

　　Solution by graphing ■ Solution by elimination

Introduction   In Chapter 7, we discussed how to solve systems of linear equations. To do this, we used the graphing, substitution, and addition methods. In this section, we will apply these methods to solve systems where at least one of the equations is nonlinear.

Solution by Graphing

We now discuss ways to solve systems of two equations in two variables where at least one of the equations is of second degree.

EXAMPLE 1

Solving systems of equations. Solve $\begin{cases} x^2 + y^2 = 25 \\ 2x + y = 10 \end{cases}$ by graphing.

Solution   The graph of $x^2 + y^2 = 25$ is a circle with center at the origin and radius of 5. The graph of $2x + y = 10$ is a line. Depending on whether the line is a secant (intersecting the circle at two points) or a tangent (intersecting the circle at one point) or does not intersect the circle at all, there are two, one, or no solutions to the system, respectively.

After graphing the circle and the line, as shown in Figure 13-32, we see that there are two intersection points, $P(3, 4)$ and $P'(5, 0)$, which is read as "P prime." The solutions of the system are

$$\begin{cases} x = 3 \\ y = 4 \end{cases} \quad \text{and} \quad \begin{cases} x = 5 \\ y = 0 \end{cases}$$

Verify the results.

FIGURE 13-32

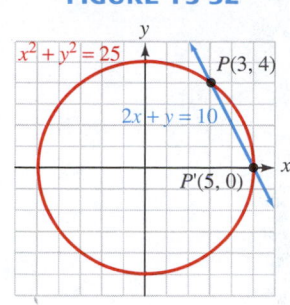

SELF CHECK Solve $\begin{cases} x^2 + y^2 = 13 \\ y = -\frac{1}{5}x + \frac{13}{5} \end{cases}$.

Answer: $(3, 2)$, $(-2, 3)$

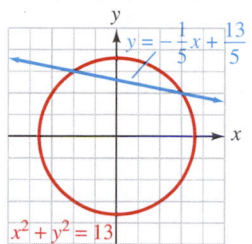

ACCENT ON TECHNOLOGY *Solving Systems of Equations*

To solve Example 1 with a graphing calculator, we graph the circle and the line on one set of coordinate axes. (See Figure 13-33(a).) We then trace to find the coordinates of the intersection points of the graphs. (See Figure 13-33(b) and Figure 13-33(c).)

We can zoom for better results.

FIGURE 13-33

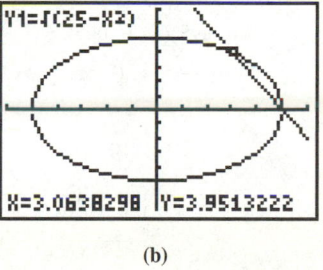

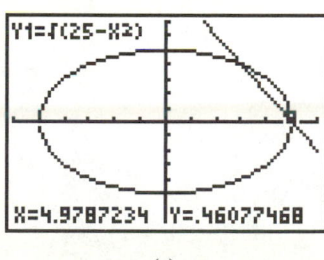

(a)   (b)   (c)

Solution by Elimination

Algebraic methods can also be used to solve systems of equations.

EXAMPLE 2   **Solving systems of equations.** Solve $\begin{cases} x^2 + y^2 = 25 \\ 2x + y = 10 \end{cases}$.

Solution   This system has one second-degree equation and one first-degree equation. We can solve this type of system by substitution. Solving the linear equation for y gives

$$2x + y = 10$$

1. $y = -2x + 10$

We can substitute $-2x + 10$ for y in the second-degree equation and solve the resulting quadratic equation for x:

$$x^2 + y^2 = 25$$
$$x^2 + (\mathbf{-2x + 10})^2 = 25$$
$$x^2 + 4x^2 - 40x + 100 = 25 \qquad (-2x + 10)(-2x + 10) = 4x^2 - 40x + 100.$$
$$5x^2 - 40x + 75 = 0 \qquad \text{Combine like terms and subtract 25 from both sides.}$$
$$x^2 - 8x + 15 = 0 \qquad \text{Divide both sides by 5.}$$
$$(x - 5)(x - 3) = 0 \qquad \text{Factor } x^2 - 8x + 15.$$
$$x - 5 = 0 \quad \text{or} \quad x - 3 = 0 \qquad \text{Set each factor equal to 0.}$$
$$x = 5 \qquad\qquad x = 3$$

If we substitute 5 for x in Equation 1, we get $y = 0$. If we substitute 3 for x in Equation 1, we get $y = 4$. The two solutions are

$$\begin{cases} x = 5 \\ y = 0 \end{cases} \quad \text{or} \quad \begin{cases} x = 3 \\ y = 4 \end{cases}$$

which can be written as $(5, 0)$ and $(3, 4)$.

SELF CHECK Solve: $\begin{cases} x^2 + y^2 = 13 \\ y = -\frac{1}{5}x + \frac{13}{5}. \end{cases}$ *Answer:* $(3, 2), (-2, 3)$ ■

E X A M P L E 3   **Solving systems of equations.** Solve $\begin{cases} 4x^2 + 9y^2 = 5 \\ y = x^2 \end{cases}.$

Solution   We can solve this system by substitution.

$$4\mathbf{x}^2 + 9y^2 = 5$$
$$4y + 9y^2 = 5 \qquad \text{Substitute } y \text{ for } x^2.$$
$$9y^2 + 4y - 5 = 0 \qquad \text{Subtract 5 from both sides.}$$
$$(9y - 5)(y + 1) = 0 \qquad \text{Factor } 9y^2 + 4y - 5.$$
$$9y - 5 = 0 \quad \text{or} \quad y + 1 = 0 \qquad \text{Set each factor equal to 0.}$$
$$y = \frac{5}{9} \qquad\qquad y = -1$$

Since $y = x^2$, the values of x are found by solving the equations

$$x^2 = \frac{5}{9} \qquad \text{and} \qquad x^2 = -1$$

Because $x^2 = -1$ has no real solutions, this possibility is discarded. The solutions of $x^2 = \frac{5}{9}$ are

$$x = \frac{\sqrt{5}}{3} \qquad \text{or} \qquad x = -\frac{\sqrt{5}}{3}$$

The solutions of the system are

$$\left(\frac{\sqrt{5}}{3}, \frac{5}{9} \right) \qquad \text{and} \qquad \left(-\frac{\sqrt{5}}{3}, \frac{5}{9} \right)$$

SELF CHECK   Solve $\begin{cases} x^2 + y^2 = 20 \\ y = x^2 \end{cases}.$ *Answer:* $(2, 4), (-2, 4)$ ■

EXAMPLE 4

Solving systems of equations. Solve $\begin{cases} 3x^2 + 2y^2 = 36 \\ 4x^2 - y^2 = 4 \end{cases}$.

Solution Since both equations are in the form $ax^2 + by^2 = c$, we can solve the system by addition.

We can copy the first equation and multiply the second equation by 2 to obtain the equivalent system

$$\begin{cases} 3x^2 + 2y^2 = 36 \\ 8x^2 - 2y^2 = 8 \end{cases}$$

We add the equations to eliminate y and solve the resulting equation for x:

$$11x^2 = 44$$
$$x^2 = 4$$
$$x = 2 \quad \text{or} \quad x = -2$$

To find y, we substitute 2 for x and then -2 for x in the first equation and proceed as follows:

For x = 2	*For x = -2*
$3x^2 + 2y^2 = 36$	$3x^2 + 2y^2 = 36$
$3(2)^2 + 2y^2 = 36$	$3(-2)^2 + 2y^2 = 36$
$12 + 2y^2 = 36$	$12 + 2y^2 = 36$
$2y^2 = 24$	$2y^2 = 24$
$y^2 = 12$	$y^2 = 12$
$y = \sqrt{12}$ or $y = -\sqrt{12}$	$y = \sqrt{12}$ or $y = -\sqrt{12}$
$y = 2\sqrt{3} \qquad y = -2\sqrt{3}$	$y = 2\sqrt{3} \qquad y = -2\sqrt{3}$

The four solutions of this system are

$$\left(2, 2\sqrt{3}\right), \quad \left(2, -2\sqrt{3}\right), \quad \left(-2, 2\sqrt{3}\right), \quad \text{and} \quad \left(-2, -2\sqrt{3}\right)$$

SELF CHECK Solve $\begin{cases} x^2 + 4y^2 = 16 \\ x^2 - y^2 = 1 \end{cases}$.

Answer: $\left(2, \sqrt{3}\right), \left(2, -\sqrt{3}\right),$ $\left(-2, \sqrt{3}\right), \left(-2, -\sqrt{3}\right)$ ∎

STUDY SET

Section 13.4

VOCABULARY

Fill in the blanks to make the statements true.

1. We can solve systems of equations by _____, addition, or _____.

2. A _____ is a line that intersects a circle at two points.

3. A _____ is a line that intersects a circle at one point.

4. The equation $y = 3x + 4$ is a linear equation. The equation $y = 3x^2 + 4$ is a _____ equation.

CONCEPTS

Fill in the blanks to make the statements true.

5. At most, a line can intersect an ellipse at _____ points.

6. At most, an ellipse can intersect a parabola at _____ points.

7. At most, an ellipse can intersect a circle at _____ points.

8. At most, a hyperbola can intersect a circle at _____ points.

NOTATION

Complete each solution.

9. Solve $\begin{cases} x^2 + y^2 = 5 \\ y = 2x \end{cases}$.

$$x^2 + y^2 = 5$$
$$x^2 + (\boxed{})^2 = 5$$
$$x^2 + 4x^2 = \boxed{}$$
$$\boxed{}x^2 = 5$$
$$x^2 = \boxed{}$$
$$x = 1 \quad \text{or} \quad x = -1$$

If $x = 1$, then
$$y = 2(\boxed{}) = 2$$
If $x = -1$, then
$$y = 2(\boxed{}) = -2$$
The solutions are $(1, 2)$ and $(-1, -2)$.

10. Solve $\begin{cases} x^2 + y^2 = 13 \\ x^2 - y^2 = 5 \end{cases}$

$$2x^2 = \boxed{} \qquad \text{Add the equations.}$$
$$x^2 = \boxed{}$$
$$x = \boxed{} \quad \text{or} \quad x = \boxed{}$$
$$2y^2 = \boxed{} \qquad \text{Subtract the equations.}$$
$$y^2 = \boxed{}$$
$$y = \boxed{} \quad \text{or} \quad y = \boxed{}$$

The solutions are
$$(3, 2), (3, -2), (-3, 2), (-3, -2)$$

PRACTICE

In Exercises 11–18, solve each system of equations by graphing.

11. $\begin{cases} 8x^2 + 32y^2 = 256 \\ x = 2y \end{cases}$

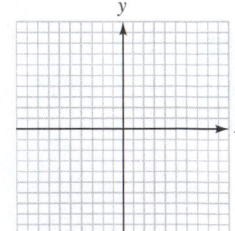

12. $\begin{cases} x^2 + y^2 = 2 \\ x + y = 2 \end{cases}$

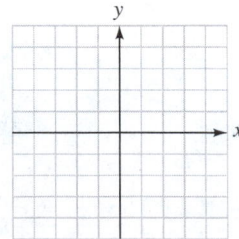

13. $\begin{cases} x^2 + y^2 = 10 \\ y = 3x^2 \end{cases}$

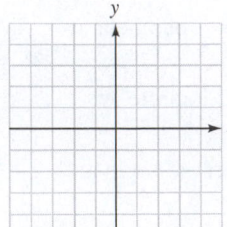

14. $\begin{cases} x^2 + y^2 = 5 \\ x + y = 3 \end{cases}$

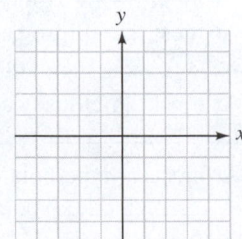

15. $\begin{cases} x^2 + y^2 = 25 \\ 12x^2 + 64y^2 = 768 \end{cases}$

16. $\begin{cases} x^2 + y^2 = 13 \\ y = x^2 - 1 \end{cases}$

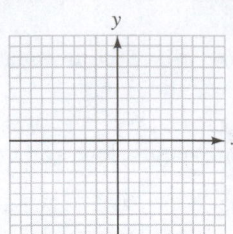
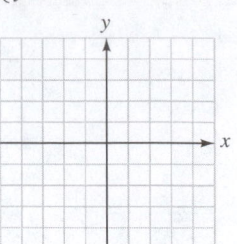

17. $\begin{cases} x^2 - 13 = -y^2 \\ y = 2x - 4 \end{cases}$

18. $\begin{cases} x^2 + y^2 = 20 \\ y = x^2 \end{cases}$

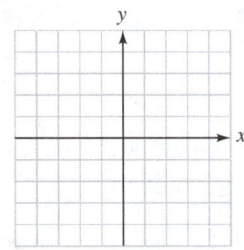
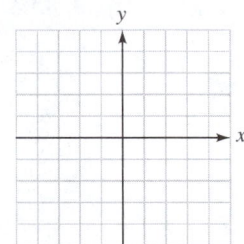

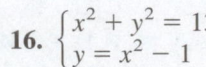 *In Exercises 19–20, use a graphing calculator to solve each system.*

19. $\begin{cases} x^2 - 6x - y = -5 \\ x^2 - 6x + y = -5 \end{cases}$

20. $\begin{cases} x^2 - y^2 = -5 \\ 3x^2 + 2y^2 = 30 \end{cases}$

In Exercises 21–46, solve each system of equations algebraically for real values of x and y.

21. $\begin{cases} 25x^2 + 9y^2 = 225 \\ 5x + 3y = 15 \end{cases}$

22. $\begin{cases} x^2 + y^2 = 20 \\ y = x^2 \end{cases}$

23. $\begin{cases} x^2 + y^2 = 2 \\ x + y = 2 \end{cases}$

24. $\begin{cases} x^2 + y^2 = 36 \\ 49x^2 + 36y^2 = 1{,}764 \end{cases}$

25. $\begin{cases} x^2 + y^2 = 5 \\ x + y = 3 \end{cases}$

26. $\begin{cases} x^2 - x - y = 2 \\ 4x - 3y = 0 \end{cases}$

27. $\begin{cases} x^2 + y^2 = 13 \\ y = x^2 - 1 \end{cases}$

28. $\begin{cases} x^2 + y^2 = 25 \\ 2x^2 - 3y^2 = 5 \end{cases}$

29. $\begin{cases} x^2 + y^2 = 30 \\ y = x^2 \end{cases}$

30. $\begin{cases} 9x^2 - 7y^2 = 81 \\ x^2 + y^2 = 9 \end{cases}$

31. $\begin{cases} x^2 + y^2 = 13 \\ x^2 - y^2 = 5 \end{cases}$

32. $\begin{cases} 2x^2 + y^2 = 6 \\ x^2 - y^2 = 3 \end{cases}$

33. $\begin{cases} x^2 + y^2 = 20 \\ x^2 - y^2 = -12 \end{cases}$

34. $\begin{cases} xy = -\dfrac{9}{2} \\ 3x + 2y = 6 \end{cases}$

35. $\begin{cases} y^2 = 40 - x^2 \\ y = x^2 - 10 \end{cases}$

36. $\begin{cases} x^2 - 6x - y = -5 \\ x^2 - 6x + y = -5 \end{cases}$

37. $\begin{cases} y = x^2 - 4 \\ x^2 - y^2 = -16 \end{cases}$

38. $\begin{cases} 6x^2 + 8y^2 = 182 \\ 8x^2 - 3y^2 = 24 \end{cases}$

39. $\begin{cases} x^2 - y^2 = -5 \\ 3x^2 + 2y^2 = 30 \end{cases}$

40. $\begin{cases} \dfrac{1}{x} + \dfrac{1}{y} = 5 \\ \dfrac{1}{x} - \dfrac{1}{y} = -3 \end{cases}$

41. $\begin{cases} \dfrac{1}{x} + \dfrac{2}{y} = 1 \\ \dfrac{2}{x} - \dfrac{1}{y} = \dfrac{1}{3} \end{cases}$

42. $\begin{cases} \dfrac{1}{x} + \dfrac{3}{y} = 4 \\ \dfrac{2}{x} - \dfrac{1}{y} = 7 \end{cases}$

43. $\begin{cases} 3y^2 = xy \\ 2x^2 + xy - 84 = 0 \end{cases}$

44. $\begin{cases} x^2 + y^2 = 10 \\ 2x^2 - 3y^2 = 5 \end{cases}$

45. $\begin{cases} xy = \dfrac{1}{6} \\ y + x = 5xy \end{cases}$

46. $\begin{cases} xy = \dfrac{1}{12} \\ y + x = 7xy \end{cases}$

47. **INTEGER PROBLEM** The product of two integers is 32, and their sum is 12. Find the integers.

48. **NUMBER PROBLEM** The sum of the squares of two numbers is 221, and the sum of the numbers is 9. Find the numbers.

APPLICATIONS

49. **GEOMETRY** The area of a rectangle is 63 square centimeters, and its perimeter is 32 centimeters. Find the dimensions of the rectangle.

50. **INVESTING** Grant receives $225 annual income from one investment. Jeff invested $500 more than Grant, but at an annual rate of 1% less. Jeff's annual income is $240. What is the amount and rate of Grant's investment?

51. **INVESTING** Carol receives $67.50 annual income from one investment. John invested $150 more than Carol at an annual rate of $1\frac{1}{2}\%$ more. John's annual income is $94.50. What is the amount and rate of Carol's investment? (*Hint:* There are two answers.)

52. **ARTILLERY** The shell fired from the base of the hill in Illustration 1 follows the parabolic path $y = -\frac{1}{6}x^2 + 2x$, with distances measured in miles. The hill has a slope of $\frac{1}{3}$. How far from the gun is the

point of impact? (*Hint*: Find the coordinates of the point and then the distance.)

ILLUSTRATION 1

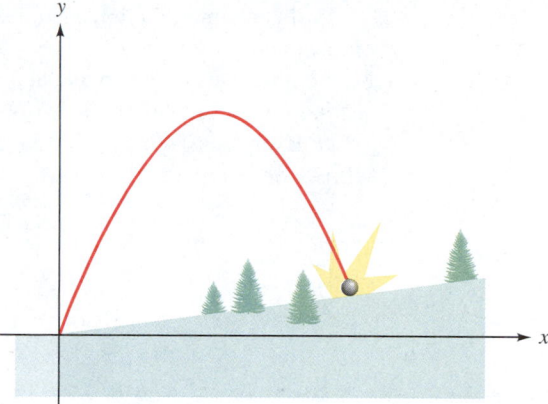

53. **DRIVING RATES** Jim drove 306 miles. Jim's brother made the same trip at a speed 17 mph slower than Jim did and required an extra $1\frac{1}{2}$ hours. What was Jim's rate and time?

54. **FENCING PASTURES** A rectangular pasture is to be fenced in along a riverbank, as shown in Illustration 2. If 260 feet of fencing is to enclose an area of 8,000 square feet, find the dimensions of the pasture.

ILLUSTRATION 2

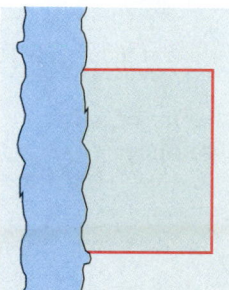

WRITING

55. Describe the benefits of the graphical method for solving a system of equations.

56. Describe the drawbacks of the graphical method.

REVIEW

Simplify each radical expression. Assume that all variables represent positive numbers.

57. $\sqrt{200x^2} - 3\sqrt{98x^2}$

58. $a\sqrt{112a} - 5\sqrt{175a^3}$

59. $\dfrac{3t\sqrt{2t} - 2\sqrt{2t^3}}{\sqrt{18t} - \sqrt{2t}}$

60. $\sqrt[3]{\dfrac{x}{4}} + \sqrt[3]{\dfrac{x}{32}} - \sqrt[3]{\dfrac{x}{500}}$

Conic Sections

In this chapter, we have studied four conic sections: the circle, the parabola, the ellipse, and the hyperbola. They are called conic sections because they are formed by the intersection of a plane and a right-circular cone. These curves are graphs of second-degree equations in two variables.

In Exercises 1–9, classify the graph of each equation as a circle, a parabola, an ellipse, or a hyperbola.

1. $\dfrac{(x-2)^2}{9} + \dfrac{(y-1)^2}{4} = 1$

2. $(x-3)^2 + (y-2)^2 = 4$

3. $y = 4x^2 - 2x + 3$

4. $\dfrac{x^2}{4} - \dfrac{(y-1)^2}{9} = 1$

5. $4x^2 - y^2 + 8x - 4y = 4$

6. $\dfrac{x^2}{4} + \dfrac{y^2}{9} = 1$

7. $9x^2 + 4y^2 - 18x + 16y = 11$

8. $x^2 - 2y - 2x = -7$

9. $x^2 + y^2 + 8x + 2y = -13$

Equations of Conic Sections

When the equation of a conic section is written in standard form, important features of its graph are apparent.

10. Consider $(x+1)^2 + (y-2)^2 = 16$.
 a. What are the coordinates of the center of the circle?
 b. What is the radius of the circle?

11. Consider $x = \frac{1}{2}(y-1)^2 - 2$.
 a. What are the coordinates of the vertex of the parabola?
 b. In which direction does the parabola open?

12. Consider $\dfrac{x^2}{4} + \dfrac{y^2}{16} = 1$.
 a. What are the coordinates of the center of the ellipse?
 b. To which axis is the major axis of the ellipse parallel?
 c. How long is the major axis?
 d. How long is the minor axis?

13. Consider $\dfrac{(x+2)^2}{9} - \dfrac{(y-1)^2}{4} = 1$.
 a. What are the coordinates of the center of the hyperbola?
 b. In which direction do the branches of the hyperbola open?
 c. What are the dimensions of the fundamental rectangle?

Graphing Conic Sections

Use the results from Exercises 10–13 to graph each conic section.

14. $(x+1)^2 + (y-2)^2 = 16$

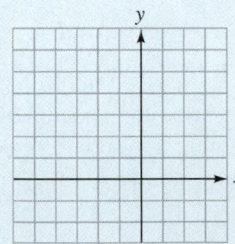

15. $x = \frac{1}{2}(y-1)^2 - 2$

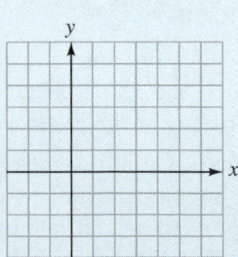

16. $\dfrac{x^2}{4} + \dfrac{y^2}{16} = 1$

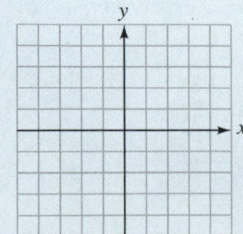

17. $\dfrac{(x+2)^2}{9} - \dfrac{(y-1)^2}{4} = 1$

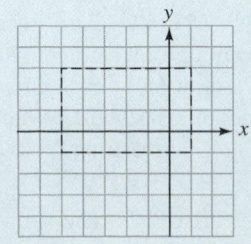

Accent on Teamwork

Section 13.1

Constructing a parabola To construct a parabola, secure one end of a piece of string that is as long as a T-square to a large piece of paper. See Illustration 1. Attach the other end of the string to the upper end of the T-square. Then, hold the string taut against the T-square with a pencil and slide the T-square along the edge of the table. As the T-square moves, the pencil will trace a parabola.

Each point on a parabola is the same distance away from a given point as it is from a given line. With this method of construction, what is the given point? What is the given line?

ILLUSTRATION 1

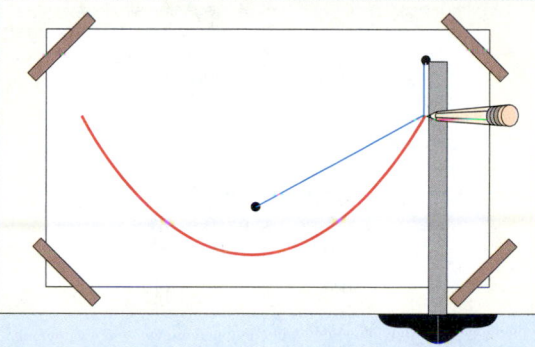

Section 13.2

Constructing ellipses We can construct an ellipse with two thumbtacks, a piece of string, and a pencil. Place the two thumbtacks fairly close together, as shown in Illustration 2. Catch the loop of the string with the point of a pencil and (keeping the string taut) draw the ellipse. Experiment by moving one of the thumbtacks farther away and then closer to the other thumbtack. How does the shape of the ellipse change?

For each point on an ellipse, the sum of the distances of the point from two given points is a constant. With this method of construction, what are the two given points? What is the constant?

ILLUSTRATION 2

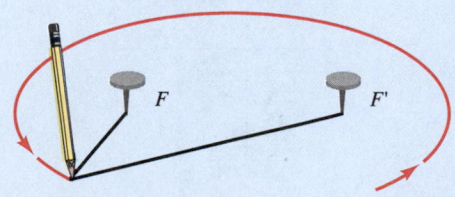

Section 13.3

Conic section models Mold some clay into the shape of a right-circular cone, as shown in Illustration 3. With both hands, pull a thin wire through the clay to slice it in such a way that a circular shape results. Experiment with ways to slice the clay to obtain an elliptical shape, a parabolic shape, and one branch of a hyperbolic shape.

ILLUSTRATION 3

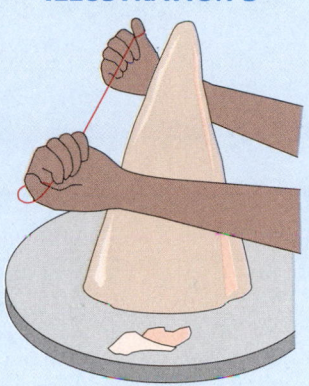

Section 13.4

Simultaneous systems Sketch the graphs of a system of second-degree equations that meet the following conditions, if possible.

The graph of one equation is a circle, the graph of the other equation is a parabola, and the system has
a. no solution.   **b.** one solution.
c. two solutions.   **d.** three solutions.
e. four solutions.   **f.** five solutions.

Simultaneous systems Sketch the graphs of a system of second-degree equations that meet the following conditions, if possible.

The graph of one equation is an ellipse, the graph of the other equation is a hyperbola, and the system has
a. no solution.   **b.** one solution.
c. two solutions.   **d.** three solutions.
e. four solutions.   **f.** five solutions.

Simultaneous systems Sketch the graphs of a system of second-degree equations that meet the following conditions, if possible.

The graph of one equation is a parabola, the graph of the other equation is a hyperbola, and the system has
a. no solution.   **b.** one solution.
c. two solutions.   **d.** three solutions.
e. four solutions.   **f.** five solutions.

Section 13.1 — The Circle and the Parabola

CONCEPTS

Equations of a circle:

$$(x - h)^2 + (y - k)^2 = r^2$$
center (h, k), radius r

$$x^2 + y^2 = r^2$$
center $(0, 0)$, radius r

REVIEW EXERCISES

1. Graph each equation.

a. $(x - 1)^2 + (y + 2)^2 = 9$

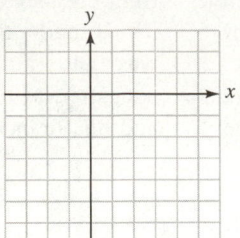

b. $x^2 + y^2 = 16$

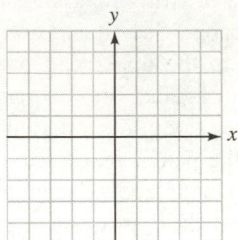

2. Write the equation in standard form and graph it.

$$x^2 + y^2 + 4x - 2y = 4$$

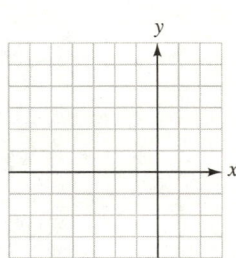

Equations of parabolas ($a > 0$):

Parabola opening	Vertex at origin
Up	$y = ax^2$
Down	$y = -ax^2$
Right	$x = ay^2$
Left	$x = -ay^2$

Parabola opening	Vertex at (h, k)
Up	$y = a(x - h)^2 + k$
Down	$y = -a(x - h)^2 + k$
Right	$x = a(y - k)^2 + h$
Left	$x = -a(y - k)^2 + h$

3. Graph each equation.

a. $x = -3(y - 2)^2 + 5$

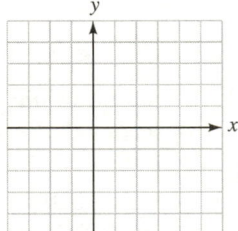

b. $x = 2(y + 1)^2 - 2$

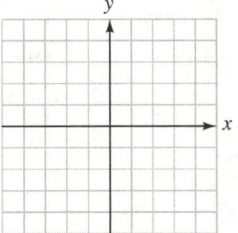

Section 13.2 — The Ellipse

Equations of an ellipse:

Center at $(0, 0)$

$$\frac{x^2}{a^2} + \frac{y^2}{b^2} = 1 \quad (a > b > 0)$$

$$\frac{x^2}{b^2} + \frac{y^2}{a^2} = 1 \quad (a > b > 0)$$

4. Graph each ellipse.

a. $9x^2 + 16y^2 = 144$

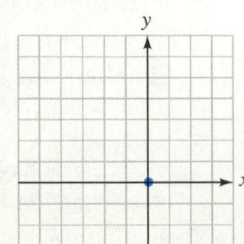

b. $\dfrac{(x - 2)^2}{4} + \dfrac{(y - 1)^2}{9} = 1$

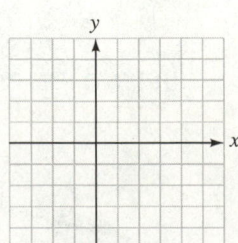

Center at (h, k)

$$\frac{(x-h)^2}{a^2} + \frac{(y-k)^2}{b^2} = 1$$

$$\frac{(x-h)^2}{b^2} + \frac{(y-k)^2}{a^2} = 1$$

5. Write the equation in standard form and graph it.

$$4x^2 + 9y^2 + 8x - 18y = 23$$

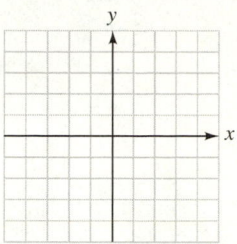

Section 13.3   The Hyperbola

Equations of a hyperbola:

Center at (0, 0)

$$\frac{x^2}{a^2} - \frac{y^2}{b^2} = 1$$

$$\frac{y^2}{b^2} - \frac{x^2}{a^2} = 1$$

Center at (h, k)

$$\frac{(x-h)^2}{a^2} - \frac{(y-k)^2}{b^2} = 1$$

$$\frac{(y-k)^2}{a^2} - \frac{(x-h)^2}{b^2} = 1$$

6. Graph each hyperbola.

    a. $9x^2 - y^2 = -9$ **b.** $xy = 9$

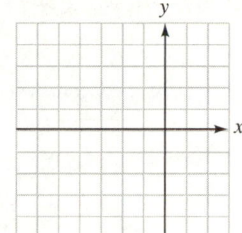 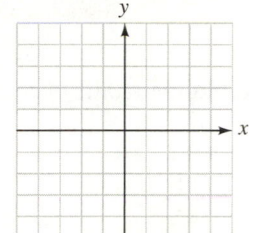

7. Write the equation $4x^2 - 2y^2 + 8x - 8y = 8$ in standard form and tell whether its graph will be an ellipse or a hyperbola.

8. Write the equation in standard form and graph it.

$$9x^2 - 4y^2 - 18x - 8y = 31$$

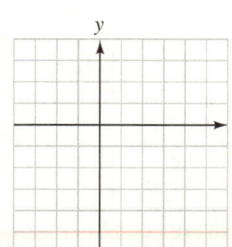

Section 13.4   Solving Simultaneous Second-Degree Equations

9. The graphs of $x^2 - y^2 = 9$ and $x^2 + y^2 = 9$ are shown in Illustration 1. Estimate the solutions of the system

$$\begin{cases} y^2 - x^2 = 9 \\ x^2 + y^2 = 9 \end{cases}$$

ILLUSTRATION 1

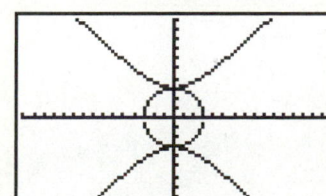

10. Solve each system.

a. $\begin{cases} 3x^2 + y^2 = 52 \\ x^2 - y^2 = 12 \end{cases}$

b. $\begin{cases} \dfrac{x^2}{16} + \dfrac{y^2}{12} = 1 \\ x^2 - \dfrac{y^2}{3} = 1 \end{cases}$

CHAPTER 13

Test

1. Find the center and the radius of the circle $(x - 2)^2 + (y + 3)^2 = 4$.

2. Find the center and the radius of the circle $x^2 + y^2 + 4x - 6y = 3$.

In Problems 3–6, graph each equation.

3. $(x + 1)^2 + (y - 2)^2 = 9$ **4.** $x = (y - 2)^2 - 1$

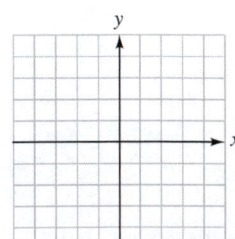

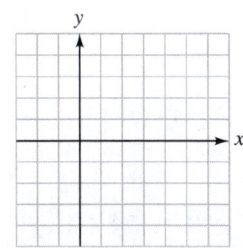

5. $9x^2 + 4y^2 = 36$ **6.** $\dfrac{(x - 2)^2}{9} - y^2 = 1$

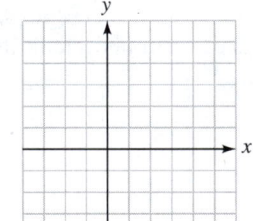

In Problems 7–8, write each equation in standard form and graph the equation.

7. $4x^2 + y^2 - 24x + 2y = -33$

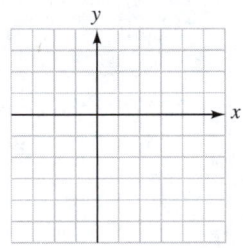

8. $x^2 - 9y^2 + 2x + 36y = 44$

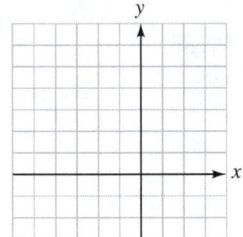

In Problems 9–10, solve each system.

9. $\begin{cases} 2x - y = -2 \\ x^2 + y^2 = 16 + 4y \end{cases}$

10. $\begin{cases} x^2 + y^2 = 25 \\ 4x^2 - 9y = 0 \end{cases}$

14

Miscellaneous Topics

The Political Science Department

In a political science class, students study the design of ballots used in city, county, state, and national elections. On a ballot, the names of the candidates running for a given office can be listed in many different orders. For example, they can be presented alphabetically, by party affiliation, or simply in random order. In this chapter, we will learn how to determine the number of ways in which a set of names can be arranged. This counting concept, known as a *permutation,* has many other important applications in areas such as designing license plates, assigning telephone numbers, and listing combinations for locks.

In this chapter, we introduce several topics that have applications in advanced mathematics and in certain occupations. The binomial theorem, permutations, and combinations are used in statistics. Arithmetic and geometric sequences are used in the mathematics of finance.

▶ 14.1

The Binomial Theorem

In this section, you will learn about

> Raising binomials to powers ▪ Pascal's triangle ▪ Factorial notation
> ▪ The binomial theorem ▪ Finding a particular term of a binomial
> expansion

Introduction   In Chapter 4, we discussed how to raise binomials to positive-integer powers. For example, we learned that

$$(a + b)^2 = a^2 + 2ab + b^2$$

and that

$$
\begin{aligned}
(a + b)^3 &= (a + b)(a + b)^2 \\
&= (a + b)(a^2 + 2ab + b^2) \\
&= a^3 + 2a^2b + ab^2 + a^2b + 2ab^2 + b^3 \\
&= a^3 + 3a^2b + 3ab^2 + b^3
\end{aligned}
$$

In this section, we will learn how to raise binomials to positive-integer powers without doing the actual multiplications.

Raising Binomials to Powers

To see how to raise binomials to positive-integer powers, we consider the following binomial expansions:

$$(a + b)^0 = 1$$
$$(a + b)^1 = a + b$$
$$(a + b)^2 = a^2 + 2ab + b^2$$
$$(a + b)^3 = a^3 + 3a^2b + 3ab^2 + b^3$$
$$(a + b)^4 = a^4 + 4a^3b + 6a^2b^2 + 4ab^3 + b^4$$
$$(a + b)^5 = a^5 + 5a^4b + 10a^3b^2 + 10a^2b^3 + 5ab^4 + b^5$$
$$(a + b)^6 = a^6 + 6a^5b + 15a^4b^2 + 20a^3b^3 + 15a^2b^4 + 6ab^5 + b^6$$

Several patterns appear in these expansions:

1. Each expansion has one more term than the power of the binomial.
2. The degree of each term in each expansion is equal to the exponent of the binomial that is being expanded.
3. The first term in each expansion is a, raised to the power of the binomial.

4. The exponents on a decrease by one in each successive term. The exponents on b, beginning with $b^0 = 1$ in the first term, increase by one in each successive term. Thus, the variables have the pattern

$$a^n, a^{n-1}b, a^{n-2}b^2, \ldots, ab^{n-1}, b^n$$

Pascal's Triangle

To see another pattern, we write the coefficients of each expansion in the following triangular array:

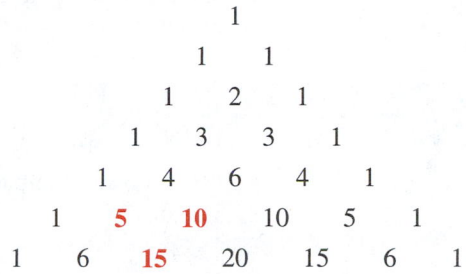

In this array, called **Pascal's triangle,** each entry between the 1's is the sum of the closest pair of numbers in the line immediately above it. For example, the first 15 in the bottom row is the sum of the 5 and 10 immediately above it. Pascal's triangle continues with the same pattern forever. The next two lines are

$$\begin{array}{ccccccccc} 1 & 7 & 21 & 35 & 35 & 21 & 7 & 1 \\ 1 & 8 & 28 & 56 & 70 & 56 & 28 & 8 & 1 \end{array}$$

EXAMPLE 1

Raising a binomial to the 5th power. Expand $(x + y)^5$.

Solution

The first term in the expansion is x^5, and the exponents on x decrease by one in each successive term. A y first appears in the second term, and the exponents on y increase by one in each successive term, concluding when the term y^5 is reached. Thus, the variables in the expansion are

$$x^5, x^4y, x^3y^2, x^2y^3, xy^4, y^5$$

The coefficients of these variables are given in Pascal's triangle in the row whose second entry is 5, the exponent of the binomial that is being expanded.

$$\begin{array}{cccccc} 1 & 5 & 10 & 10 & 5 & 1 \end{array}$$

Combining this information gives the following expansion:

$$(x + y)^5 = x^5 + 5x^4y + 10x^3y^2 + 10x^2y^3 + 5xy^4 + y^5$$

SELF CHECK   Expand $(x + y)^4$.   *Answer:* $x^4 + 4x^3y + 6x^2y^2 + 4xy^3 + y^4$ ■

EXAMPLE 2

Raising a binomial to the 4th power. Expand $(u - v)^4$.

Solution

We note that $(u - v)^4$ can be written in the form $[u + (-v)]^4$. The variables in this expansion are

$$u^4, u^3(-v), u^2(-v)^2, u(-v)^3, (-v)^4$$

and the coefficients are given in Pascal's triangle in the row whose second entry is 4:

$$\begin{array}{ccccc} 1 & 4 & 6 & 4 & 1 \end{array}$$

Thus, the required expansion is

$$(u - v)^4 = u^4 + 4u^3(-v) + 6u^2(-v)^2 + 4u(-v)^3 + (-v)^4$$
$$= u^4 - 4u^3v + 6u^2v^2 - 4uv^3 + v^4$$

SELF CHECK   Expand $(x - y)^5$.   *Answer:* $x^5 - 5x^4y + 10x^3y^2 - 10x^2y^3 + 5xy^4 - y^5$ ∎

Factorial Notation

Although Pascal's triangle gives the coefficients of the terms in a binomial expansion, it is not the best way to expand a binomial. To develop another way, we introduce **factorial notation.**

Factorial notation

> If n is a natural number, the symbol $n!$ (read as *n* **factorial** or as **factorial** *n*) is defined as
>
> $$n! = n(n - 1)(n - 2)(n - 3) \cdot \cdot \cdot (3)(2)(1)$$

EXAMPLE 3   **Evaluating factorials.** Find   **a.** $2!$,   **b.** $5!$,   **c.** $-9!$,   and   **d.** $(n - 2)!$.

Solution   **a.** $2! = 2 \cdot 1 = 2$

b. $5! = 5 \cdot 4 \cdot 3 \cdot 2 \cdot 1 = 120$

c. $-9! = -(9 \cdot 8 \cdot 7 \cdot 6 \cdot 5 \cdot 4 \cdot 3 \cdot 2 \cdot 1) = -362,880$

d. $(n - 2)! = (n - 2)(n - 3)(n - 4) \cdot \cdot \cdot \cdot 3 \cdot 2 \cdot 1$

 WARNING! According to the previous definition, part d is meaningful only if $n - 2$ is a natural number.

SELF CHECK   Find   **a.** $4!$ and   **b.** $x!$.   *Answers:*   **a.** 24,

b. $x(x - 1)(x - 2) \cdot \cdot \cdot \cdot 3 \cdot 2 \cdot 1$ ∎

We define zero factorial as follows.

Zero factorial

> $0! = 1$

We note that

$$5 \cdot 4! = 5 \cdot 4 \cdot 3 \cdot 2 \cdot 1 = 5!$$
$$7 \cdot 6! = 7 \cdot 6 \cdot 5 \cdot 4 \cdot 3 \cdot 2 \cdot 1 = 7!$$
$$10 \cdot 9! = 10 \cdot 9 \cdot 8 \cdot 7 \cdot 6 \cdot 5 \cdot 4 \cdot 3 \cdot 2 \cdot 1 = 10!$$

These examples suggest the following property of factorials.

$n(n - 1)!$

> If n is a positive integer, then $n(n - 1)! = n!$.

The Binomial Theorem

We now state the binomial theorem.

The binomial theorem

If n is any positive integer, then

$$(a + b)^n = a^n + \frac{n!}{1!(n-1)!}a^{n-1}b + \frac{n!}{2!(n-2)!}a^{n-2}b^2 + \frac{n!}{3!(n-3)!}a^{n-3}b^3$$

$$+ \cdots + \frac{n!}{r!(n-r)!}a^{n-r}b^r + \cdots + b^n$$

In the binomial theorem, the exponents on the variables follow the familiar pattern:

- The sum of the exponents on a and b in each term is n,
- the exponents on a decrease by 1 in each successive term, and
- the exponents on b increase by 1 in each successive term.

Only the method of finding the coefficients is different. Except for the first and last terms, the numerator of each coefficient is $n!$. If the exponent on b in a particular term is r, the denominator of the coefficient of that term is $r!(n-r)!$.

EXAMPLE 4   **Expanding a binomial.** Use the binomial theorem to expand $(a + b)^3$.

Solution   We can substitute directly into the binomial theorem and simplify:

$$(a + b)^3 = a^3 + \frac{3!}{1!(3-1)!}a^2b + \frac{3!}{2!(3-2)!}ab^2 + b^3$$

$$= a^3 + \frac{3!}{1! \cdot 2!}a^2b + \frac{3!}{2! \cdot 1!}ab^2 + b^3$$

$$= a^3 + \frac{3 \cdot 2 \cdot 1}{1 \cdot 2 \cdot 1}a^2b + \frac{3 \cdot 2 \cdot 1}{2 \cdot 1 \cdot 1}ab^2 + b^3$$

$$= a^3 + 3a^2b + 3ab^2 + b^3$$

SELF CHECK   Use the binomial theorem to expand $(a + b)^4$.

Answer: $a^4 + 4a^3b + 6a^2b^2 + 4ab^3 + b^4$ ■

EXAMPLE 5   **Expanding a binomial.** Use the binomial theorem to expand $(x - y)^4$.

Solution   We can write $(x - y)^4$ in the form $[x + (-y)]^4$, substitute directly into the binomial theorem, and simplify:

$$(x - y)^4 = [x + (-y)]^4$$

$$= x^4 + \frac{4!}{1!(4-1)!}x^3(-y) + \frac{4!}{2!(4-2)!}x^2(-y)^2 + \frac{4!}{3!(4-3)!}x(-y)^3 + (-y)^4$$

$$= x^4 - \frac{4 \cdot 3!}{1! \cdot 3!}x^3y + \frac{4 \cdot 3 \cdot 2!}{2! \cdot 2!}x^2y^2 - \frac{4 \cdot 3!}{3! \cdot 1!}xy^3 + y^4$$

$$= x^4 - 4x^3y + 6x^2y^2 - 4xy^3 + y^4$$

SELF CHECK   Use the binomial theorem to expand $(x - y)^3$.   *Answer:* $x^3 - 3x^2y + 3xy^2 - y^3$ ■

EXAMPLE 6

Expanding a binomial. Use the binomial theorem to expand $(3u - 2v)^4$.

Solution

We write $(3u - 2v)^4$ in the form $[3u + (-2v)]^4$ and let $a = 3u$ and $b = -2v$. Then we can use the binomial theorem to expand $(a + b)^4$.

$$(a + b)^4 = a^4 + \frac{4!}{1!(4 - 1)!}a^3b + \frac{4!}{2!(4 - 2)!}a^2b^2 + \frac{4!}{3!(4 - 3)!}ab^3 + b^4$$
$$= a^4 + 4a^3b + 6a^2b^2 + 4ab^3 + b^4$$

Now we can substitute $3u$ for a and $-2v$ for b and simplify:

$$(3u - 2v)^4 = (3u)^4 + 4(3u)^3(-2v) + 6(3u)^2(-2v)^2 + 4(3u)(-2v)^3 + (-2v)^4$$
$$= 81u^4 - 216u^3v + 216u^2v^2 - 96uv^3 + 16v^4$$

SELF CHECK

Use the binomial theorem to expand $(2a - 3b)^3$.

Answer: $8a^3 - 36a^2b + 54ab^2 - 27b^3$ ∎

Finding a Particular Term of a Binomial Expansion

To find the fourth term of the expansion of $(a + b)^9$, we could raise the binomial $a + b$ to the 9th power and look at the fourth term. However, this task would be tedious. By using the binomial theorem, we can construct the fourth term without finding the complete expansion of $(a + b)^9$.

EXAMPLE 7

Finding the 4th term of a binomial expansion. Find the fourth term of the expansion of $(a + b)^9$.

Solution

Since b^1 appears in the second term, b^2 appears in the third term, and so on, the exponent on b in the fourth term is 3. Since the exponent on b added to the exponent on a must equal 9, the exponent on a must be 6. Thus, the variables of the fourth term are

a^6b^3 The sum of the exponents must be 9.

The coefficient of these variables is

$$\frac{n!}{r!(n - r)!} = \frac{9!}{3!(9 - 3)!}$$ The numerator of each coefficient is $n!$.
   r is the exponent on b.

The complete fourth term is

$$\frac{9!}{3!(9 - 3)!}a^6b^3 = \frac{9 \cdot 8 \cdot 7 \cdot 6!}{3 \cdot 2 \cdot 1 \cdot 6!}a^6b^3$$
$$= 84a^6b^3$$

SELF CHECK

Find the third term of the expansion of $(a + b)^9$.   *Answer:* $36a^7b^2$ ∎

EXAMPLE 8

Finding the 6th term of a binomial expansion. Find the sixth term in the expansion of $(x - y)^7$.

Solution

We first find the sixth term of $[x + (-y)]^7$. In the sixth term, the exponent on $(-y)$ is 5. Thus, the variables in the sixth term are

$x^2(-y)^5$ The sum of the exponents must be 7.

The coefficient of these variables is

$$\frac{n!}{r!(n-r)!} = \frac{7!}{5!(7-5)!}$$
The numerator of each coefficient is $n!$.
r is the exponent on b.

The complete sixth term is

$$\frac{7!}{5!(7-5)!}x^2(-y)^5 = -\frac{7 \cdot 6 \cdot 5!}{5! \cdot 2 \cdot 1}x^2y^5$$
$$= -21x^2y^5$$

SELF CHECK   Find the fifth term of the expansion of $(a-b)^7$.   *Answer:* $35a^3b^4$ ■

E X A M P L E 9   **Finding the 4th term of a binomial expansion.** Find the fourth term of the expansion of $(2x-3y)^6$.

Solution   We can let $a = 2x$ and $b = -3y$ and find the fourth term of the expansion of $(a+b)^6$:

$$\frac{6!}{3!(6-3)!}a^3b^3 = \frac{6 \cdot 5 \cdot 4 \cdot 3!}{3! \cdot 3 \cdot 2 \cdot 1}a^3b^3$$
$$= 20a^3b^3$$

We can now substitute $2x$ for a and $-3y$ for b and simplify:

$$20a^3b^3 = 20(2x)^3(-3y)^3$$
$$= -4{,}320x^3y^3$$

The fourth term is $-4{,}320x^3y^3$.

SELF CHECK   Find the third term of the expansion of $(2a-3b)^6$.   *Answer:* $2{,}160a^4b^2$ ■

STUDY SET

Section 14.1

VOCABULARY

Fill in the blanks to make the statements true.

1. The expression $a + b$ is called a _____.

2. The triangular array that can be used to find the coefficients of a binomial expansion is called _____ triangle.

CONCEPTS

Fill in the blanks to make the statements true.

3. Every binomial expansion has ___ more term than the power of the binomial.

4. The first term in the expansion of $(a+b)^{20}$ is _____.

5. The degree of each term of a binomial expansion is equal to the _____ of the binomial that is being expanded.

6. In the expansion of $(a+b)^n$, the powers on a _____, and the powers on b _____.

7. In the expansion of $(a+b)^7$, the sum of the exponents on a and b is ___.

8. The coefficient of the fourth term of the expansion of $(a+b)^9$ is 9! divided by _____.

9. The symbol 5! is read as "_____."

10. $6 \cdot 5 \cdot 4 \cdot 3 \cdot 2 \cdot 1 = $ ___

11. $8! = 8 \cdot $ ___

12. $0! = $ ___

13. n_____ $= n!$

14. According to the binomial theorem, the third term of the expansion of $(a+b)^n$ is _____.

15. The exponent on b in the fourth term of the expansion of $(a+b)^6$ is ___.

16. The exponent on b in the fifth term of the expansion of $(a+b)^6$ is ___.

NOTATION

Complete the solution.

17. $(x + y)^3 = x^{\boxed{}} + \dfrac{\boxed{}}{1!(3-1)!}x^2y + \dfrac{\boxed{}}{2!(3-2)!}xy^2 + y^3$

$= x^3 + \dfrac{3 \cdot 2 \cdot 1}{1 \cdot \boxed{} \cdot 1}x^2y + \dfrac{3 \cdot 2 \cdot 1}{2 \cdot \boxed{} \cdot 1}xy^2 + y^3$

$= x^3 + 3x^2y + 3xy^2 + y^3$

18. Find the fifth term of the expansion of $(a - b)^7$.
  a. In the fifth term, the exponent on $(-b)$ is $\boxed{}$.
  b. Thus, the variables of the fifth term are $\boxed{}$.
  c. The coefficient of these variables is

  $\dfrac{\boxed{}}{} = 35.$

  d. The fifth term is $\boxed{}$.

PRACTICE

In Exercises 19–38, evaluate each expression.

19. $3!$ **20.** $7!$

21. $-5!$ **22.** $-6!$

23. $3! + 4!$ **24.** $2!(3!)$

25. $3!(4!)$ **26.** $4! + 4!$

27. $8(7!)$ **28.** $4!(5)$

29. $\dfrac{9!}{11!}$ **30.** $\dfrac{13!}{10!}$

31. $\dfrac{49!}{47!}$ **32.** $\dfrac{101!}{100!}$

33. $\dfrac{5!}{3!(5-3)!}$ **34.** $\dfrac{6!}{4!(6-4)!}$

35. $\dfrac{7!}{5!(7-5)!}$ **36.** $\dfrac{8!}{6!(8-6)!}$

37. $\dfrac{5!(8-5)!}{4! \cdot 7!}$ **38.** $\dfrac{6! \cdot 7!}{(8-3)!(7-4)!}$

In Exercises 39–54, use the binomial theorem to expand each expression.

39. $(x + y)^2$

40. $(x + y)^4$

41. $(x - y)^4$

42. $(x - y)^3$

43. $(2x + y)^3$

44. $(x + 2y)^3$

45. $(x - 2y)^3$

46. $(2x - y)^3$

47. $(2x + 3y)^3$

48. $(3x - 2y)^3$

49. $\left(\dfrac{x}{2} - \dfrac{y}{3}\right)^3$

50. $\left(\dfrac{x}{3} + \dfrac{y}{2}\right)^3$

51. $(3 + 2y)^4$

52. $(2x + 3)^4$

53. $\left(\dfrac{x}{3} - \dfrac{y}{2}\right)^4$

54. $\left(\dfrac{x}{2} + \dfrac{y}{3}\right)^4$

In Exercises 55–80, find the required term of each binomial expansion.

55. $(a + b)^3$; second term

56. $(a + b)^3$; third term

57. $(x - y)^4$; fourth term

58. $(x - y)^5$; second term

59. $(x + y)^6$; fifth term

60. $(x + y)^7$; fifth term

61. $(x - y)^8$; third term

62. $(x - y)^9$; seventh term

63. $(x + 3)^5$; third term

64. $(x - 2)^4$; second term

65. $(4x + y)^5$; third term

66. $(x + 4y)^5$; fourth term

67. $(x - 3y)^4$; second term

68. $(3x - y)^5$; third term

69. $(2x - 5)^7$; fourth term

70. $(2x + 3)^6$; sixth term

71. $(2x - 3y)^5$; fifth term

72. $(3x - 2y)^4$; second term

73. $\left(\dfrac{x}{2} - \dfrac{y}{3}\right)^4$; second term

74. $\left(\dfrac{x}{3} + \dfrac{y}{2}\right)^5$; fourth term

75. $(a + b)^n$; fourth term

76. $(a + b)^n$; third term

77. $(a - b)^n$; fifth term

78. $(a - b)^n$; sixth term

79. $(2a - 3b)^n$; fifth term

80. $(3a - 2b)^n$; fourth term

WRITING

81. Tell how to construct Pascal's triangle.

82. Tell how to find the variable part of each term in the expansion of $(r + s)^4$.

83. Tell how to find the coefficients of each term in the expansion of $(x + y)^5$.

84. Explain why the signs alternate in the expansion of $(x - y)^9$.

REVIEW

Find each value of x.

85. $\log_4 16 = x$

86. $\log_x 49 = 2$

87. $\log_{25} x = \dfrac{1}{2}$

88. $\log_{1/2} \dfrac{1}{8} = x$

Solve each system of equations.

89. $\begin{cases} 3x + 2y = 12 \\ 2x - y = 1 \end{cases}$

90. $\begin{cases} a + b + c = 6 \\ 2a + b + 3c = 11 \\ 3a - b - c = 6 \end{cases}$

Evaluate each determinant.

91. $\begin{vmatrix} 2 & -3 \\ 4 & -2 \end{vmatrix}$

92. $\begin{vmatrix} 1 & 2 & 3 \\ 4 & 5 & 0 \\ -1 & -2 & 1 \end{vmatrix}$

▶ 14.2 Arithmetic Sequences

In this section, you will learn about

- Sequences and series ■ Arithmetic sequences ■ Arithmetic means
- ■ The sum of the first *n* terms of an arithmetic sequence
- ■ Summation notation

Introduction In this section, we will discuss ordered lists called *sequences*. For example, a batting order in baseball is a sequence, because it determines who will bat first, bat second, and so on. When a sequence is a list of numbers, we can find their sum. When we put + signs between the numbers in a sequence, we form a sum that is called a *series*.

Sequences and Series

A mathematical **sequence** is a function whose domain is the set of natural numbers. For example, the function $f(n) = 3n + 2$ (where n is a natural number) is a sequence. If the natural numbers are substituted in order for n, the function $f(n) = 3n + 2$ generates the list

    5, 8, 11, 14, 17, . . .   The ellipsis . . . indicates that the list goes on forever.

Each number in the list is called a **term** of the sequence. Since this sequence has an unlimited number of terms, it is called an **infinite sequence**. A **finite sequence** has a specific number of terms.

    Examples of infinite sequences:

    $1^3, 2^3, 3^3, 4^3, 5^3, 6^3, . . .$ The ordered list of the cubes of the natural numbers.

    1, 1, 2, 3, 5, 8, 13, 21, . . .   The Fibonacci sequence.

The **Fibonacci sequence** is named after the 12th-century mathematician Leonardo of Pisa—also known as Fibonacci. Beginning with the 2, each term of the Fibonacci sequence is the sum of the two preceding terms.

    Examples of finite sequences:

    4, 8, 12, 16, 20, 24   The ordered list of the first six positive multiples of 4.

    2, 3, 5, 7, 11, 13, 17, 19, 23, 29   The ordered list of the first ten prime numbers.

When the commas between the terms of a sequence are replaced with + signs, we call the sum a **series.**

$$4 + 8 + 12 + 16 + 20 + 24$$ Since this series has a limited number of terms, it is a finite series.

$$5 + 8 + 11 + 14 + 17 + \cdots$$ Since this series has an unlimited number of terms, it is an infinite series.

Arithmetic Sequences

A sequence where each term is found by adding the same number to the preceding term is called an **arithmetic sequence.**

Arithmetic sequence

An **arithmetic sequence** is a sequence of the form

$$a, a + d, a + 2d, a + 3d, \ldots, a + (n - 1)d, \ldots$$

where a is the **first term,** $a + (n - 1)d$ is the **nth term,** and d is the **common difference.**

We note that the second term of an arithmetic sequence has an addend of $1d$, the third term has an addend of $2d$, the fourth term has an addend of $3d$, and the nth term has an addend of $(n - 1)d$. We also note that the difference between any two consecutive terms in an arithmetic sequence is d.

EXAMPLE 1

Finding terms of an arithmetic sequence. An arithmetic sequence has a first term of 5 and a common difference of 4.   **a.** Write the first six terms of the sequence.   **b.** Write the 25th term of the sequence.

Solution   **a.** Since the first term is $a = 5$ and the common difference is $d = 4$, the first six terms are

$$5, 5 + 4, 5 + 2(4), 5 + 3(4), 5 + 4(4), 5 + 5(4)$$

or

$$5, 9, 13, 17, 21, 25$$

b. The nth term is $a + (n - 1)d$. Since we want the 25th term, we let $n = 25$:

nth term $= a + (n - 1)d$

25th term $= 5 + (25 - 1)4$ Remember that $a = 5$ and $d = 4$.

$$= 5 + 24(4)$$
$$= 5 + 96$$
$$= 101$$

SELF CHECK   **a.** Write the seventh term of the sequence of Example 1.   **b.** Write the 30th term.   *Answers:*   **a.** 29,   **b.** 121   ■

EXAMPLE 2

Finding terms of an arithmetic sequence. The first three terms of an arithmetic sequence are 3, 8, and 13. Find   **a.** the 67th term and   **b.** the 100th term.

Solution   The common difference d is the difference between any two successive terms:

$$d = 8 - 3 = 13 - 8 = 5$$

a. We substitute 3 for a, 67 for n, and 5 for d in the formula for the nth term and simplify:

$$\textbf{\textit{n}th term} = a + (n - 1)d$$
$$\textbf{67th term} = \textbf{3} + (\textbf{67} - 1)\textbf{5}$$
$$= 3 + 66(5)$$
$$= 333$$

b. We substitute 3 for a, 100 for n, and 5 for d in the formula for the nth term and simplify:

$$\textbf{\textit{n}th term} = a + (n - 1)d$$
$$\textbf{100th term} = \textbf{3} + (\textbf{100} - 1)\textbf{5}$$
$$= 3 + 99(5)$$
$$= 498$$

SELF CHECK Find the 50th term of the sequence in Example 2. *Answer:* 248 ∎

EXAMPLE 3

Finding terms of an arithmetic sequence. The first term of an arithmetic sequence is 12, and the 50th term is 3,099. Write the first six terms of the sequence.

Solution The key is to find the common difference. Because the 50th term of this sequence is 3,099, we can let $n = 50$ and solve the following equation for d:

$$\textbf{50th term} = a + (n - 1)d$$
$$\textbf{3,099} = \textbf{12} + (\textbf{50} - 1)d$$
$$3,099 = 12 + 49d \qquad \text{Simplify.}$$
$$3,087 = 49d \qquad \text{Subtract 12 from both sides.}$$
$$63 = d \qquad \text{Divide both sides by 49.}$$

The first term of the sequence is 12, and the common difference is 63. Its first six terms are

12, 75, 138, 201, 264, 327

SELF CHECK The first term of an arithmetic sequence is 15, and the 12th term is 92. Write the first four terms of the sequence. *Answer:* 15, 22, 29, 36 ∎

Arithmetic Means

If numbers are inserted between two numbers a and b to form an arithmetic sequence, the inserted numbers are called **arithmetic means** between a and b. If a single number is inserted between a and b, that number is called **the arithmetic mean** between a and b.

EXAMPLE 4

Inserting arithmetic means. Insert two arithmetic means between 6 and 27.

Solution Here, the first term is $a = 6$, and the fourth term (or the last term) is $l = 27$. We must find the common difference so that the terms

$6, \mathbf{6 + \textit{d}}, \mathbf{6 + 2\textit{d}}, 27$

form an arithmetic sequence. To find d, we substitute 6 for a and 4 for n in the formula for the nth term:

$$n\text{th term} = a + (n - 1)d$$
$$4\text{th term} = 6 + (4 - 1)d$$

$$
\begin{array}{ll}
27 = 6 + 3d & \text{Simplify.} \\
21 = 3d & \text{Subtract 6 from both sides.} \\
7 = d & \text{Divide both sides by 3.}
\end{array}
$$

The two arithmetic means between 6 and 27 are

$$
\begin{array}{lll}
6 + d = 6 + 7 & \text{or} & 6 + 2d = 6 + 2(7) \\
 = 13 & & = 6 + 14 \\
& & = 20
\end{array}
$$

The numbers 6, 13, 20, and 27 are the first four terms of an arithmetic sequence.

SELF CHECK   Insert two arithmetic means between 8 and 44.   *Answer:* 20, 32   ∎

The Sum of the First n Terms of an Arithmetic Sequence

To develop a formula for evaluating the sum of the first n terms of an arithmetic sequence, we let S_n represent the sum of the first n terms of an arithmetic sequence:

$$S_n = \quad a \quad + \quad [a + d] \quad + \quad [a + 2d] \quad + \cdots + [a + (n - 1)d]$$

We write the same sum again, but in reverse order:

$$S_n = [a + (n - 1)d] + [a + (n - 2)d] + [a + (n - 3)d] + \cdots + a$$

Adding these equations together, term by term, we get

$$2S_n = [2a + (n - 1)d] + [2a + (n - 1)d] + [2a + (n - 1)d] + \cdots + [2a + (n - 1)d]$$

Because there are n equal terms on the right-hand side of the preceding equation, we can write

$$
\begin{array}{ll}
2S_n = n[2a + (n - 1)d] & \\
2S_n = n[a + a + (n - 1)d] & \text{Write } 2a \text{ as } a + a. \\
2S_n = n(a + l) & \text{Substitute } l \text{ for } a + (n - 1)d, \text{ because } a + (n - 1)d \text{ is} \\
& \text{the last term of the sequence.} \\
S_n = \dfrac{n(a + l)}{2} &
\end{array}
$$

This reasoning establishes the following formula.

Sum of the first n terms of an arithmetic sequence

The sum of the first n terms of an arithmetic sequence is given by the formula

$$S_n = \frac{n(a + l)}{2} \quad \text{with } l = a + (n - 1)d$$

where a is the first term, l is the last (or nth) term, and n is the number of terms in the sequence.

EXAMPLE 5

Find the sum of the first 40 terms of the arithmetic sequence 4, 10, 16, ...

Solution

In this example, we let $a = 4$, $n = 40$, $d = 6$, and $l = 4 + (40 - 1)6 = 238$ and substitute these values into the formula for S_n:

$$S_n = \frac{n(a + l)}{2}$$

$$S_{40} = \frac{40(4 + 238)}{2}$$

$$= 20(242)$$

$$= 4{,}840$$

The sum of the first 40 terms is 4,840.

SELF CHECK

Find the sum of the first 50 terms of the arithmetic sequence 3, 8, 13, ...   *Answer:* 6,275   ■

Summation Notation

When the general term of a sequence is known, we can use a shorthand notation to write a series. This notation, called **summation notation,** involves the Greek letter Σ (sigma). The expression

$$\sum_{k=2}^{5} 3k \quad \text{Read as "the summation of } 3k \text{ as } k \text{ runs from 2 to 5."}$$

designates the sum of all terms obtained if we successively substitute the numbers 2, 3, 4, and 5 for k, called the **index of the summation.** Thus, we have

$$\sum_{k=2}^{5} 3k = 3(2) + 3(3) + 3(4) + 3(5)$$

$$= 6 + 9 + 12 + 15$$

$$= 42$$

EXAMPLE 6

Find each sum: **a.** $\displaystyle\sum_{k=3}^{5}(2k + 1)$,   **b.** $\displaystyle\sum_{k=2}^{5} k^2$,   and   **c.** $\displaystyle\sum_{k=1}^{3}(3k^2 + 3)$.

Solution

a. $\displaystyle\sum_{k=3}^{5}(2k + 1) = [2(3) + 1] + [2(4) + 1] + [2(5) + 1]$

$$= 7 + 9 + 11$$

$$= 27$$

b. $\displaystyle\sum_{k=2}^{5} k^2 = 2^2 + 3^2 + 4^2 + 5^2$

$$= 4 + 9 + 16 + 25$$

$$= 54$$

c. $\displaystyle\sum_{k=1}^{3}(3k^2 + 3) = [3(1^2) + 3] + [3(2^2) + 3] + [3(3^2) + 3]$

$$= 6 + 15 + 30$$

$$= 51$$

SELF CHECK

Evaluate $\displaystyle\sum_{k=1}^{4}(2k^2 - 2)$.   *Answer:* 52   ■

Section 14.2

VOCABULARY

Fill in the blanks to make the statements true.

1. A _____ is a function whose domain is the set of natural numbers.
2. The sequence 1, 1, 2, 3, 5, 8, 13, 21, . . . is called the _____ sequence.
3. A sequence with an unlimited number of terms is called a(n) _____ sequence.
4. A sequence with a specific number of terms is called a(n) _____ sequence.
5. The sum of the terms of a sequence is called a _____.
6. If a number is inserted between a and b to form an arithmetic sequence, the number is called the arithmetic _____ between a and b.

CONCEPTS

Fill in the blanks to make the statements true.

7. The sequence 3, 9, 15, 21, . . . is an example of an infinite _____ sequence with a common _____ of 6.
8. The last term of an arithmetic sequence is given by the formula _____.
9. $3 + 7 + 11 + 14$ is an example of a _____ series.
10. The sum of the first n terms of an arithmetic sequence is given by the formula $S_n = $ _____.

NOTATION

In Exercises 11–12, fill in the blanks to make the statements true.

11. $\displaystyle\sum_{k=1}^{5} k$ means _____.
12. In the symbol $\displaystyle\sum_{k=1}^{5} (2k - 5)$, k is called the _____ of summation.

In Exercises 13–16, write the series associated with each summation.

13. $\displaystyle\sum_{k=1}^{3} k$
14. $\displaystyle\sum_{k=1}^{4} (3k)$
15. $\displaystyle\sum_{k=2}^{4} k^2$
16. $\displaystyle\sum_{k=3}^{5} (-2k)$

In Exercises 17–20, write the summation notation for each sum.

17. $3 + 4 + 5 + 6$
18. $1 + 4 + 9 + 16 + 25$
19. $2 + 4 + 6 + 8$
20. $-1 - 4 - 9 - 16 - 25 - 36$

PRACTICE

In Exercises 21–34, write the first five terms of each arithmetic sequence with the given properties.

21. $a = 3$, $d = 2$
22. $a = -2$, $d = 3$
23. $a = -5$, $d = -3$
24. $a = 8$, $d = -5$
25. $a = 5$, fifth term is 29
26. $a = 4$, sixth term is 39
27. $a = -4$, sixth term is -39
28. $a = -5$, fifth term is -37
29. $d = 7$, sixth term is -83
30. $d = 3$, seventh term is 12
31. $d = -3$, seventh term is 16
32. $d = -5$, seventh term is -12
33. The 19th term is 131 and the 20th term is 138.
34. The 16th term is 70 and the 18th term is 78.

35. Find the 30th term of the arithmetic sequence with $a = 7$ and $d = 12$.
36. Find the 55th term of the arithmetic sequence with $a = -5$ and $d = 4$.
37. Find the 37th term of the arithmetic sequence with a second term of -4 and a third term of -9.
38. Find the 40th term of the arithmetic sequence with a second term of 6 and a fourth term of 16.
39. Find the first term of the arithmetic sequence with a common difference of 11 if its 27th term is 263.
40. Find the common difference of the arithmetic sequence with a first term of -164 if its 36th term is -24.

41. Find the common difference of the arithmetic sequence with a first term of 40 if its 44th term is 556.

42. Find the first term of the arithmetic sequence with a common difference of -5 if its 23rd term is -625.

43. Insert three arithmetic means between 2 and 11.

44. Insert four arithmetic means between 5 and 25.

45. Insert four arithmetic means between 10 and 20.

46. Insert three arithmetic means between 20 and 30.

47. Find the arithmetic mean between 10 and 19.
48. Find the arithmetic mean between 5 and 23.
49. Find the arithmetic mean between -4.5 and 7.
50. Find the arithmetic mean between -6.3 and -5.2.

In Exercises 51–58, find the sum of the first n terms of each arithmetic sequence.

51. 1, 4, 7, . . .; $n = 30$
52. 2, 6, 10, . . .; $n = 28$
53. $-5, -1, 3, . . .; n = 17$
54. $-7, -1, 5, . . .; n = 15$
55. Second term is 7, third term is 12; $n = 12$
56. Second term is 5, fourth term is 9; $n = 16$
57. $f(n) = 2n + 1$, nth term is 31; n is a natural number

58. $f(n) = 4n + 3$, nth term is 23; n is a natural number

59. Find the sum of the first 50 natural numbers.
60. Find the sum of the first 100 natural numbers.
61. Find the sum of the first 50 odd natural numbers.

62. Find the sum of the first 50 even natural numbers.

In Exercises 63–70, find each sum.

63. $\sum_{k=1}^{4} (6k)$

64. $\sum_{k=2}^{5} (3k)$

65. $\sum_{k=3}^{4} k^3$

66. $\sum_{k=2}^{4} (-k^2)$

67. $\sum_{k=3}^{4} (k^2 + 3)$

68. $\sum_{k=2}^{6} (k^2 + 1)$

69. $\sum_{k=4}^{4} (2k + 4)$

70. $\sum_{k=3}^{5} (3k^2 - 7)$

APPLICATIONS

71. SAVING MONEY Yasmeen puts $60 into a safety deposit box. After each succeeding month, she puts $50 more in the box. Write the first six terms of an arithmetic sequence that gives the monthly amounts in her savings, and find her savings after 10 years.

72. INSTALLMENT LOANS Maria borrowed $10,000, interest-free, from her mother. She agreed to pay back the loan in monthly installments of $275. Write the first six terms of an arithmetic sequence that shows the balance due after each month, and find the balance due after 17 months.

73. DESIGNING PATIOS Each row of bricks in the triangular patio in Illustration 1 is to have one more brick than the previous row, ending with the longest row of 150 bricks. How many bricks will be needed?

ILLUSTRATION 1

74. FALLING OBJECTS The equation $s = 16t^2$ represents the distance s in feet that an object will fall in t seconds. After 1 second, the object has fallen 16 feet. After 2 seconds, it has fallen 64 feet, and so on. Find the distance that the object will fall during the second and third seconds.

75. FALLING OBJECTS Refer to Exercise 74. How far will the object fall during the 12th second?

76. INTERIOR ANGLES The sums of the angles of several polygons are given in the table shown in Illustration 2. Assuming that the pattern continues, complete the table.

ILLUSTRATION 2

Figure	Number of sides	Sum of angles
Triangle	3	180°
Quadrilateral	4	360°
Pentagon	5	540°
Hexagon	6	720°
Octagon	8	
Dodecagon	12	

WRITING

77. Define an arithmetic sequence.

78. Develop the formula for finding the sum of the first n terms of an arithmetic sequence.

81. $\dfrac{3a + 4}{a - 2} + \dfrac{3a - 4}{a + 2}$

82. $2t - 3 \overline{)8t^4 - 12t^3 + 8t^2 - 16t + 6}$

REVIEW

Do the operations and simplify, if possible.

79. $3(2x^2 - 4x + 7) + 4(3x^2 + 5x - 6)$,

80. $(2p + q)(3p^2 + 4pq - 3q^2)$

▶ 14.3   Geometric Sequences

In this section, you will learn about

> Geometric sequences ■ Geometric means ■ The sum of the first n terms of a geometric sequence ■ Population growth ■ Infinite geometric series

---

Introduction   In the previous section, we saw that the same number is added to each term of an arithmetic sequence to get the next term. In this section, we will consider another type of sequence where we multiply each term by the same number to get the next term. This type of sequence is called a *geometric sequence*. Two examples of geometric sequences are

2, 6, 18, 54, . . .   This is an infinite geometric sequence where each term is found by multiplying the previous term by 3.

$27, 9, 3, 1, \dfrac{1}{3}, \dfrac{1}{9}$ This is a finite geometric sequence where each term is found by multiplying the previous term by $\frac{1}{3}$.

Geometric Sequences

Every term of a geometric sequence is found by multiplying the previous term by the same number.

Geometric sequence

> A **geometric sequence** is a sequence of the form
>
> $$a, ar, ar^2, ar^3, \ldots, ar^{n-1}, \ldots$$
>
> where a is the **first term,** ar^{n-1} is the **nth term,** and r is the **common ratio.**

We note that the second term of a geometric sequence has a factor of r^1, the third term has a factor of r^2, the fourth term has a factor of r^3, and the nth term has a factor of r^{n-1}. We also note that r is the quotient obtained when any term is divided by the previous term.

EXAMPLE 1

Finding terms of a geometric sequence. A geometric sequence has a first term of 5 and a common ration of 3.

a. Write the first five terms of the sequence.

b. Find the ninth term.

Solution

a. Because the first term is $a = 5$ and the common ratio is $r = 3$, the first five terms are

$$5, 5(3), 5(3^2), 5(3^3), 5(3^4)$$

or

$$5, 15, 45, 135, 405$$

b. The nth term is ar^{n-1} with $a = 5$ and $r = 3$. Because we want the ninth term, we let $n = 9$:

$$n\text{th term} = ar^{n-1}$$
$$9\text{th term} = 5(3)^{9-1}$$
$$= 5(3)^8$$
$$= 5(6,561)$$
$$= 32,805$$

SELF CHECK

A geometric sequence has a first term of 3 and a common ratio of 4. **a.** Write the first four terms. **b.** Find the eighth term.

Answers: **a.** 3, 12, 48, 192, **b.** 49,152 ∎

EXAMPLE 2

Finding terms of a geometric sequence. The first three terms of a geometric sequence are 16, 4, and 1. Find the seventh term.

Solution

To determine r, we divide the second term by the first term to obtain $\frac{4}{16} = \frac{1}{4}$. We substitute 16 for a, $\frac{1}{4}$ for r, and 7 for n in the formula for the nth term and simplify:

$$n\text{th term} = ar^{n-1}$$
$$7\text{th term} = 16\left(\frac{1}{4}\right)^{7-1}$$
$$= 16\left(\frac{1}{4}\right)^6$$
$$= 16\left(\frac{1}{4,096}\right)$$
$$= \frac{1}{256}$$

SELF CHECK

Find the tenth term of the geometric sequence of Example 2.

Answer: $\frac{1}{16,384}$ ∎

Geometric Means

If numbers are inserted between two numbers a and b to form a geometric sequence, the inserted numbers are called **geometric means** between a and b. If a single number is inserted between a and b, that number is called a **geometric mean** between a and b.

EXAMPLE 3	**Inserting geometric means.** Insert two geometric means between 7 and 1,512.

Solution   In this example, the first term is $a = 7$, and the fourth term (or last term) is $l = 1,512$. To find the common ratio r so that the terms

$$7, \quad \mathbf{7r}, \quad \mathbf{7r^2}, \quad 1,512$$

form a geometric sequence, we substitute 4 for n and 7 for a in the formula for the nth term of a geometric sequence and solve for r.

$$\begin{aligned} \textbf{\textit{n}th term} &= ar^{n-1} \\ \textbf{4th term} &= 7r^{4-1} \\ 1,512 &= 7r^3 \\ 216 &= r^3 \qquad \text{Divide both sides by 7.} \\ 6 &= r \qquad \text{Take the cube root of both sides.} \end{aligned}$$

The two geometric means between 7 and 1,512 are

$$\mathbf{7r} = 7(6) = \mathbf{42}$$

and

$$\mathbf{7r^2} = 7(6)^2 = 7(36) = \mathbf{252}$$

The numbers 7, 42, 252, and 1,512 are the first four terms of a geometric sequence.

SELF CHECK   Insert three geometric means between 1 and 16.   *Answer:* 2, 4, 8   ■

EXAMPLE 4	**Finding a geometric mean.** Find a geometric mean between 2 and 20.

Solution   We want to find the middle term of the three-termed geometric sequence

$$2, \quad \mathbf{2r}, \quad 20$$

with $a = 2$, $l = 20$, and $n = 3$. To find r, we substitute these values into the formula for the nth term of a geometric sequence:

$$\begin{aligned} \textbf{\textit{n}th term} &= ar^{n-1} \\ \textbf{3rd term} &= 2r^{3-1} \\ 20 &= 2r^2 \\ 10 &= r^2 \qquad \text{Divide both sides by 2.} \\ \pm\sqrt{10} &= r \qquad \text{Take the square root of both sides.} \end{aligned}$$

Because r can be either $\sqrt{10}$ or $-\sqrt{10}$, there are two values for a geometric mean. They are

$$\mathbf{2r} = \mathbf{2\sqrt{10}}$$

and

$$\mathbf{2r} = \mathbf{-2\sqrt{10}}$$

The numbers 2, $2\sqrt{10}$, 20 and 2, $-2\sqrt{10}$, 20 both form geometric sequences. The common ratio of the first sequence is $\sqrt{10}$, and the common ratio of the second sequence is $-\sqrt{10}$.

SELF CHECK   Find the positive geometric mean between 2 and 200.   *Answer:* 20   ■

The Sum of the First n Terms of a Geometric Sequence

When we add the terms of a geometric sequence, we form a **geometric series.** There is a formula that gives the sum of the first n terms of a geometric sequence. To develop this formula, we let S_n represent the sum of the first n terms of a geometric sequence.

1. $S_n = a + ar + ar^2 + ar^3 + \cdots + ar^{n-1}$

We multiply both sides of Equation 1 by r to get

2. $S_n r = \qquad ar + ar^2 + ar^3 + \cdots + ar^{n-1} + ar^n$

We now subtract Equation 2 from Equation 1 and solve for S_n:

$$S_n - S_n r = a - ar^n$$
$$S_n(1 - r) = a - ar^n \qquad \text{Factor out } S_n \text{ from the left-hand side.}$$
$$S_n = \frac{a - ar^n}{1 - r} \qquad \text{Divide both sides by } 1 - r.$$

This reasoning establishes the following formula.

Sum of the first n terms of a geometric sequence

> The sum of the first n terms of a geometric sequence is given by the formula
>
> $$S_n = \frac{a - ar^n}{1 - r} \quad (r \neq 1)$$
>
> where S_n is the sum, a is the first term, r is the common ratio, and n is the number of terms.

EXAMPLE 5

Evaluating a geometric series. Find the sum of the first six terms of the geometric sequence 250, 50, 10, ...

Solution   In this geometric sequence, $a = 250$, $r = \frac{1}{5}$, and $n = 6$. We substitute these values into the formula for the sum of the first n terms of a geometric sequence and simplify:

$$S_n = \frac{a - ar^n}{1 - r}$$

$$S_6 = \frac{250 - 250\left(\dfrac{1}{5}\right)^6}{1 - \dfrac{1}{5}}$$

$$= \frac{250 - 250\left(\dfrac{1}{15{,}625}\right)}{\dfrac{4}{5}}$$

$$= \frac{5}{4}\left(250 - \frac{250}{15{,}625}\right)$$

$$= \frac{5}{4}\left(\frac{3{,}906{,}000}{15{,}625}\right)$$

$$= 312.48$$

The sum of the first six terms is 312.48.

SELF CHECK   Find the sum of the first five terms of the geometric sequence 100, 20, 4, ...   *Answer:* 124.96   ∎

Population Growth

E X A M P L E   6   The mayor of Leaf River (population 1,500) predicts a growth rate of 4% each year for the next ten years. Find the expected population of Leaf River ten years from now.

Solution   Let P_0 be the initial population of Leaf River. After 1 year, there will be a different population, P_1. The initial population (P_0) plus the growth (the product of P_0 and the rate of growth, r) will equal this new population, P_1:

$$P_1 = P_0 + P_0 r = P_0(1 + r)$$

The population after 2 years will be P_2, and

$$
\begin{aligned}
P_2 &= P_1 + P_1 r \\
&= P_1(1 + r) && \text{Factor out } P_1. \\
&= P_0(1 + r)(1 + r) && \text{Remember that } P_1 = P_0(1 + r). \\
&= P_0(1 + r)^2
\end{aligned}
$$

The population after 3 years will be P_3, and

$$
\begin{aligned}
P_3 &= P_2 + P_2 r \\
&= P_2(1 + r) && \text{Factor out } P_2. \\
&= P_0(1 + r)^2(1 + r) && \text{Remember that } P_2 = P_0(1 + r)^2. \\
&= P_0(1 + r)^3
\end{aligned}
$$

The yearly population figures

$$P_0, \quad P_1, \quad P_2, \quad P_3, \ldots$$

or

$$P_0, \quad P_0(1 + r), \quad P_0(1 + r)^2, \quad P_0(1 + r)^3, \ldots$$

form a geometric sequence with a first term of P_0 and a common ratio of $1 + r$. The population of Leaf River after 10 years is P_{10}, which is the 11th term of this sequence:

$$
\begin{aligned}
n\text{th term} &= ar^{n-1} \\
P_{10} = 11\text{th term} &= P_0(1 + r)^{10} \\
&= 1{,}500(1 + 0.04)^{10} \\
&= 1{,}500(1.04)^{10} \\
&\approx 1{,}500(1.480244285) && \text{Use a calculator.} \\
&\approx 2{,}220
\end{aligned}
$$

The expected population ten years from now is 2,220 people.   ∎

Infinite Geometric Series

If we form the sum of the terms of an infinite geometric sequence, we get a series called an **infinite geometric series**. For example, if the common ratio r is 3, we have

Infinite geometric sequence	*Infinite geometric series*
2, 6, 18, 54, 162, . . .	$2 + 6 + 18 + 54 + 162 + \cdots$

As the number of terms of this series gets larger, the value of the series gets larger. We can see that this is true by forming some **partial sums.**

The first partial sum, S_1, of the series is $S_1 = 2$.
The second partial sum, S_2, of the series is $S_2 = 2 + 6 = 8$.
The third partial sum, S_3, of the series is $S_3 = 2 + 6 + 18 = 26$.
The fourth partial sum, S_4, of the series is $S_4 = 2 + 6 + 18 + 54 = 80$.

We can now see that as the number of terms gets infinitely large, the value of the series will get infinitely large. The values of some infinite geometric series get closer and closer to a specific number as the number of terms approaches infinity. One such series is

$$\frac{3}{2} + \frac{3}{4} + \frac{3}{8} + \frac{3}{16} + \frac{3}{32} \ldots \qquad \text{Here, } r = \tfrac{1}{2}.$$

To see that this is true, we form some partial sums.

The first partial sum is $S_1 = \frac{3}{2} = 1.5$.
The second partial sum is $S_2 = \frac{3}{2} + \frac{3}{4} = \frac{9}{4} = 2.25$.
The third partial sum is $S_3 = \frac{3}{2} + \frac{3}{4} + \frac{3}{8} = \frac{21}{8} = 2.625$.
The fourth partial sum is $S_4 = \frac{3}{2} + \frac{3}{4} + \frac{3}{8} + \frac{3}{16} = \frac{45}{16} = 2.8125$.
The fifth partial sum is $S_5 = \frac{3}{2} + \frac{3}{4} + \frac{3}{8} + \frac{3}{16} + \frac{3}{32} = \frac{93}{32} = 2.90625$.

As the number of terms in this series gets larger, the values of the partial sums will approach the number 3. We say that 3 is the **limit** of S_n as n approaches infinity, and we say that 3 is the **sum of the infinite geometric series.**

To develop a formula for finding the sum of an infinite geometric series, we consider the formula that gives the sum of the first n terms.

$$S_n = \frac{a - ar^n}{1 - r} \quad (r \neq 1)$$

If $|r| < 1$ and a is constant, the term ar^n in the above formula approaches 0 as n becomes very large. For example,

$$a\left(\frac{1}{2}\right)^1 = \frac{1}{2}a, \qquad a\left(\frac{1}{2}\right)^2 = \frac{1}{4}a, \qquad a\left(\frac{1}{2}\right)^3 = \frac{1}{8}a$$

and so on. Thus, when n is very large, the value of ar^n is negligible, and the term ar^n in the above formula can be ignored. This reasoning justifies the following formula.

Sum of an infinite geometric sequence

If a is the first term and r is the common ratio of an infinite geometric sequence, and if $|r| < 1$, the sum of the terms of the sequence is given by the formula

$$S = \frac{a}{1 - r}$$

EXAMPLE 7

Finding the sum of an infinite geometric sequence. Find the sum of the terms of the infinite geometric sequence 125, 25, 5, ...

Solution   In this geometric sequence, $a = 125$ and $r = \frac{1}{5}$. Since $|r| = \left|\frac{1}{5}\right| = \frac{1}{5} < 1$, we can find the sum of all the terms of the sequence. We do this by substituting 125 for a and $\frac{1}{5}$ for r in the formula $S = \frac{a}{1-r}$ and simplifying:

$$S = \frac{a}{1-r} = \frac{125}{1 - \dfrac{1}{5}} = \frac{125}{\dfrac{4}{5}} = \frac{5}{4}(125) = \frac{625}{4}$$

The sum of the terms of the sequence 125, 25, 5, . . . is 156.25.

SELF CHECK   Find the sum of the terms of the infinite geometric sequence 100, 20, 4, ...   *Answer:* 125   ■

EXAMPLE 8   **Finding the sum of an infinite geometric sequence.** Find the sum of the infinite geometric sequence $64, -4, \frac{1}{4}, \ldots$.

Solution   In this geometric sequence, $a = 64$ and $r = -\frac{1}{16}$. Since $|r| = \left|-\frac{1}{16}\right| = \frac{1}{16} < 1$, we can find the sum of all the terms of the sequence. We substitute 64 for a and $-\frac{1}{16}$ for r in the formula $S = \frac{a}{1-r}$ and simplify:

$$S = \frac{a}{1-r} = \frac{64}{1 - \left(-\dfrac{1}{16}\right)} = \frac{64}{\dfrac{17}{16}} = \frac{16}{17}(64) = \frac{1{,}024}{17}$$

The sum of the terms of the geometric sequence $64, -4, \frac{1}{4}, \ldots$ is $\frac{1{,}024}{17}$.

SELF CHECK   Find the sum of the infinite geometric sequence 81, 27, 9, ...   *Answer:* $\frac{243}{2}$ ■

EXAMPLE 9   **Changing a repeating decimal to a fraction.** Change $0.\overline{8}$ to a common fraction.

Solution   The decimal $0.\overline{8}$ can be written as the infinite series

$$0.\overline{8} = 0.888 \ldots = \frac{8}{10} + \frac{8}{100} + \frac{8}{1{,}000} + \cdots$$

where $a = \frac{8}{10}$ and $r = \frac{1}{10}$. Because $|r| = \left|\frac{1}{10}\right| = \frac{1}{10} < 1$, we can find the sum as follows:

$$S = \frac{a}{1-r} = \frac{\dfrac{8}{10}}{1 - \dfrac{1}{10}} = \frac{\dfrac{8}{10}}{\dfrac{9}{10}} = \frac{8}{9}$$

Thus, $0.\overline{8} = \frac{8}{9}$. Long division will verify that $\frac{8}{9} = 0.888 \ldots$.

SELF CHECK   Change $0.\overline{6}$ to a common fraction.   *Answer:* $\frac{2}{3}$ ■

EXAMPLE 10   **Changing a repeating decimal to a fraction.** Change $0.\overline{25}$ to a common fraction.

Solution   The decimal $0.\overline{25}$ can be written as the infinite series

$$0.\overline{25} = 0.252525 \ldots = \frac{25}{100} + \frac{25}{10{,}000} + \frac{25}{1{,}000{,}000} + \cdots$$

where $a = \frac{25}{100}$ and $r = \frac{1}{100}$. Since $|r| = \left|\frac{1}{100}\right| = \frac{1}{100} < 1$, we can find the sum as follows:

$$S = \frac{a}{1-r} = \frac{\dfrac{25}{100}}{1 - \dfrac{1}{100}} = \frac{\dfrac{25}{100}}{\dfrac{99}{100}} = \frac{25}{99}$$

Thus, $0.\overline{25} = \frac{25}{99}$. Long division will verify that this is true.

SELF CHECK Change $0.\overline{15}$ to a common fraction. *Answer:* $\frac{5}{33}$ ■

STUDY SET
Section 14.3

VOCABULARY

Fill in the blanks to make the statements true.

1. A sequence of the form a, ar, ar^2, . . . is called a(n) _____ sequence.

2. If a geometric sequence has infinitely many terms, it is called a(n) _____ geometric sequence.

3. A number inserted between two numbers a and b is called a geometric _____ between a and b.

4. In a geometric sequence, r is called the _____.

CONCEPTS

Fill in the blanks to make the statements true.

5. The formula for the nth term of a geometric sequence is _____.

6. The sum of the first n terms of a geometric sequence is given by the formula _____.

7. The third partial sum of the sequence 2, 6, 18, 54, . . . is _____ = 26.

8. If we write the sum of the terms of an infinite geometric sequence, we form an infinite geometric _____.

9. The formula for the sum of the terms of an infinite geometric sequence is _____.

10. Write $0.\overline{75}$ as an infinite geometric series.

   _____.

NOTATION

Fill in the blanks to make the statements true.

11. In the formula for Exercise 6, a is the _____ term of the sequence, r is the common _____, and n is the number of _____.

12. The symbol S_3 represents the third _____ sum of an infinite geometric series.

PRACTICE

In Exercises 13–26, write the first five terms of each geometric sequence with the given properties.

13. $a = 3$, $r = 2$

14. $a = -2$, $r = 2$

15. $a = -5$, $r = \frac{1}{5}$

16. $a = 8$, $r = \frac{1}{2}$

17. $a = 2$, $r > 0$, third term is 32

18. $a = 3$, fourth term is 24

19. $a = -3$, fourth term is -192

20. $a = 2$, $r < 0$, third term is 50

21. $a = -64$, $r < 0$, fifth term is -4

22. $a = -64$, $r > 0$, fifth term is -4

23. $a = -64$, sixth term is -2

24. $a = -81$, sixth term is $\frac{1}{3}$

25. The second term is 10, and the third term is 50.

26. The third term is -27, and the fourth term is 81.

27. Find the tenth term of the geometric sequence with $a = 7$ and $r = 2$.

28. Find the 12th term of the geometric sequence with $a = 64$ and $r = \frac{1}{2}$.

29. Find the first term of the geometric sequence with a common ratio of -3 and an eighth term of -81.

30. Find the first term of the geometric sequence with a common ratio of 2 and a tenth term of 384.

31. Find the common ratio of the geometric sequence with a first term of -8 and a sixth term of $-1,944$.

32. Find the common ratio of the geometric sequence with a first term of 12 and a sixth term of $\frac{3}{8}$.

33. Insert three positive geometric means between 2 and 162.

34. Insert four geometric means between 3 and 96.

35. Insert four geometric means between -4 and $-12,500$.

36. Insert three geometric means (two positive and one negative) between -64 and $-1,024$.

37. Find the negative geometric mean between 2 and 128.

38. Find the positive geometric mean between 3 and 243.

39. Find the positive geometric mean between 10 and 20.

40. Find the negative geometric mean between 5 and 15.

41. Find a geometric mean, if possible, between -50 and 10.

42. Find a negative geometric mean, if possible, between -25 and -5.

In Exercises 43–54, find the sum of the first n terms of each geometric sequence.

43. 2, 6, 18, . . .; $n = 6$

44. 2, -6, 18, . . .; $n = 6$

45. 2, -6, 18, . . .; $n = 5$

46. 2, 6, 18, . . .; $n = 5$

47. 3, -6, 12, . . .; $n = 8$

48. 3, 6, 12, . . .; $n = 8$

49. 3, 6, 12, . . .; $n = 7$

50. 3, -6, 12, . . .; $n = 7$

51. The second term is 1 and the third term is $\frac{1}{5}$; $n = 4$.

52. The second term is 1 and the third term is 4; $n = 5$.

53. The third term is -2 and the fourth term is 1; $n = 6$.

54. The third term is -3 and the fourth term is 1; $n = 5$.

In Exercises 55–66, find the sum of each infinite geometric series, if possible.

55. $8 + 4 + 2 + \cdots$

56. $12 + 6 + 3 + \cdots$

57. $54 + 8 + 6 + \cdots$

58. $45 + 15 + 5 + \cdots$

59. $12 - 6 + 3 - \cdots$

60. $8 - 4 + 2 - \cdots$

61. $-45 + 15 - 5 + \cdots$

62. $-54 + 18 - 6 + \cdots$

63. $\frac{9}{2} + 6 + 8 + \cdots$

64. $-112 - 28 - 7 - \cdots$

65. $-\frac{27}{2} - 9 - 6 - \cdots$

66. $\frac{18}{25} + \frac{6}{5} + 2 + \cdots$

In Exercises 67–74, write each decimal in fraction form. Then check the answer by doing a long division.

67. $0.\overline{1}$ **68.** $0.\overline{2}$

69. $-0.\overline{3}$ **70.** $-0.\overline{4}$

71. $0.\overline{12}$ **72.** $0.\overline{21}$

73. $0.\overline{75}$ **74.** $0.\overline{57}$

APPLICATIONS

Use a calculator to solve each problem.

75. POPULATION GROWTH The population of Union is predicted to increase by 6% each year. What will the population of Union be 5 years from now if its current population is 500?

76. POPULATION DECLINE The population of Bensville is decreasing by 10% each year. If its current population is 98, what will the population be 8 years from now?

77. DECLINING SAVINGS John has $10,000 in a safety deposit box. Each year, he spends 12% of what is left in the box. How much will be in the box after 15 years?

78. SAVINGS GROWTH Sally has $5,000 in a savings account earning 12% annual interest. How much will be in her account 10 years from now? (Assume that Sally makes no deposits or withdrawals.)

79. HOUSE APPRECIATION A house appreciates by 6% each year. If the house is worth $70,000 today, how much will it be worth 12 years from now?

80. BOAT DEPRECIATION A motorboat that cost $5,000 when new depreciates at a rate of 9% per year. How much will the boat be worth in 5 years?

81. INSCRIBED SQUARES Each inscribed square in Illustration 1 joins the midpoints of the next larger square. The area of the first square, the largest, is 1. Find the area of the 12th square.

82. GENEALOGY The family tree in Illustration 2 spans 3 generations and lists 7 people. How many names would be listed in a family tree that spans 10 generations?

ILLUSTRATION 1

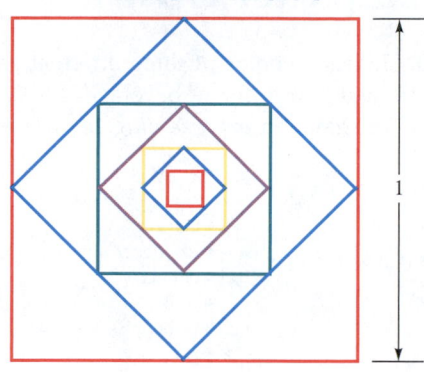

ILLUSTRATION 2

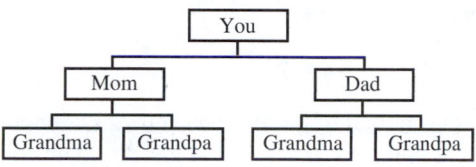

83. BOUNCING BALL On each bounce, the rubber ball in Illustration 3 rebounds to a height one-half of that from which it fell. Find the total vertical distance the ball travels.

ILLUSTRATION 3

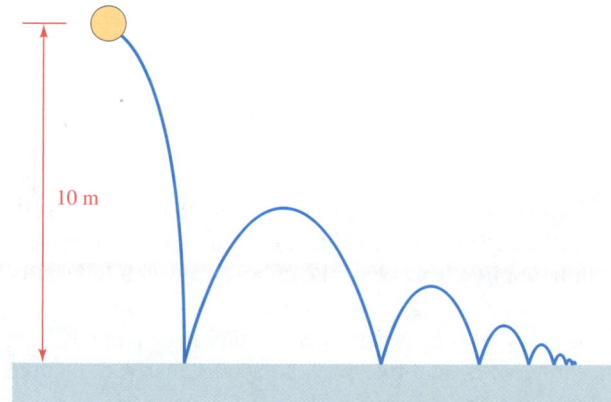

10 m

84. BOUNCING BALL A golf ball is dropped from a height of 12 feet. On each bounce, it returns to a height two-thirds of that from which it fell. Find the total vertical distance the ball travels.

85. PEST CONTROL To reduce the population of a destructive moth, biologists release 1,000 sterilized male moths each day into the environment. If 80% of these moths alive one day survive until the next, then after a long time the population of sterile males is the sum of the infinite geometric series

$$1,000 + 1,000(0.8) + 1,000(0.8)^2 + 1,000(0.8)^3 + \cdots$$

Find the long-term population.

86. PEST CONTROL If mild weather increases the day-to-day survival rate of the sterile male moths in Exercise 85 to 90%, find the long-term population.

WRITING

87. Define a geometric sequence.

88. Develop the formula for finding the sum of the first n terms of a geometric sequence.

89. Why must the common ratio be less than 1 before an infinite geometric sequence can have a sum?

90. If its common difference is not 0, can an infinite arithmetic sequence have a sum?

REVIEW

Solve each inequality.

91. $x^2 - 5x - 6 \le 0$

92. $a^2 - 7a + 12 \ge 0$

93. $\dfrac{x-4}{x+3} > 0$

94. $\dfrac{t^2 + t - 20}{t+2} < 0$

Determine whether each equation determines y to be a function of x.

95. $y = 3x^3 - 4$

96. $xy = 12$

97. $3x = y^2 + 4$

98. $x = |y|$

▶ 14.4 Permutations and Combinations

In this section, you will learn about

■ The multiplication principle for events ■ Permutations
■ Combinations ■ Alternative form of the binomial theorem

Introduction In this section, we will discuss methods of counting the different ways we can do something like lining up in a row or arranging books on a shelf. These kinds of problems are important in the fields of statistics, insurance, and telecommunications. Although one might think that counting problems are easy to solve, they can be deceptively difficult.

The Multiplication Principle for Events

Steven goes to the cafeteria for lunch. He has a choice of three different sandwiches (hamburger, hot dog, or ham and cheese) and four different beverages (cola, root beer, orange, or milk). His different options are shown in the *tree diagram* in Figure 14-1.

FIGURE 14-1

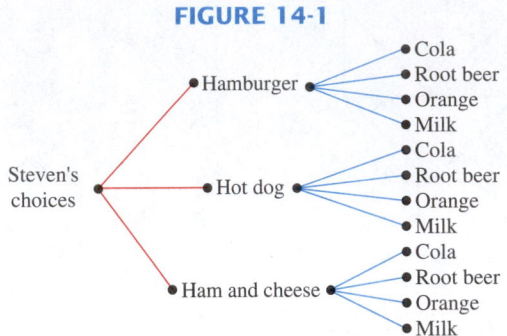

The tree diagram shows that there is a total of 12 different lunches to choose from. One possibility is a hamburger with a cola, and another is a hot dog with milk.

A situation that can have several different outcomes—such as choosing a sandwich—is called an **event.** Choosing a sandwich and choosing a beverage can be thought of as two events. The preceding example illustrates the **multiplication principle for events.**

Multiplication principle for events

Let E_1 and E_2 be two events. If E_1 can be done in a_1 ways, and if (after E_1 has occurred) E_2 can be done in a_2 ways, then the event "E_1 followed by E_2" can be done in $a_1 \cdot a_2$ ways.

EXAMPLE 1

Watching television. Before studying for an exam, Taylor plans to watch the evening news and then a situation comedy on television. If she has a choice of four news broadcasts and two comedies, in how many ways can she choose to watch television?

Solution Let E_1 be the event "watching the news" and E_2 be the event "watching a comedy." Because there are four ways to accomplish E_1 and two ways to accomplish E_2, the number of choices that Taylor has is $4 \cdot 2 = 8$.

SELF CHECK If Alex has 7 shirts and 5 pairs of pants, how many outfits could he wear?   *Answer:* 35   ■

The multiplication principle can be extended to any number of events. In Example 2, we use it to compute the number of ways in which we can arrange objects in a row.

EXAMPLE 2

Arranging books. In how many ways can we arrange five books on a shelf?

Solution We can fill the first space with any of the 5 books, the second space with any of the remaining 4 books, the third space with any of the remaining 3 books, the fourth space with any of the remaining 2 books, and the fifth space with the remaining 1 (or last) book. By the multiplication principle for events, the number of ways that the books can be arranged is

$$5 \cdot 4 \cdot 3 \cdot 2 \cdot 1 = 120$$

SELF CHECK   In how many ways can 4 people line up in a row?   *Answer:* 24   ■

E X A M P L E   3   **Signal flags.** If a sailor has six flags (each of a different color) to hang on a flag-pole, how many different signals can the sailor send by using four flags?

Solution   The sailor must find the number of arrangements of 4 flags when there are 6 flags to choose from. The sailor can hang any one of the 6 flags in the top position, any one of the remaining 5 flags in the second position, any one of the remaining 4 flags in the third position, and any one of the remaining 3 flags in the lowest position. By the multiplication principle for events, the total number of signals that can be sent is

$$6 \cdot 5 \cdot 4 \cdot 3 = 360$$

SELF CHECK   In Example 3, how many different signals can the sailor send if each signal uses three flags?   *Answer:* 120   ■

Permutations

When counting a number of possible arrangements such as books on a shelf or flags on a pole, we are finding the number of **permutations** of those objects. In Example 2, we found that the number of permutations of five books, using all five of them, is 120. In Example 3, we found that the number of permutations of six flags, using four of them, is 360.

The symbol $P(n, r)$, read as "the number of permutations of n things r at a time," is often used to express permutation problems. In Example 2, we found that $P(5, 5) = 120$. In Example 3, we found that $P(6, 4) = 360$.

E X A M P L E   4   **Signal flags.** If Sarah has seven flags (each of a different color) to hang on a flag-pole, how many different signals can she send by using three flags?

Solution   We must find $P(7, 3)$ (the number of permutations of 7 things 3 at a time). In the top position Sarah can hang any of the 7 flags, in the middle position any one of the remaining 6 flags, and in the bottom position any one of the remaining 5 flags. By the multiplication principle for events,

$$P(7, 3) = 7 \cdot 6 \cdot 5 = 210$$

Sarah can send 210 signals using only three of the seven flags.

SELF CHECK   See Example 4. How many different signals can Sarah send using four flags?   *Answer:* 840   ■

Although it is correct to write $P(7, 3) = 7 \cdot 6 \cdot 5$, there is an advantage in changing the form of this answer to obtain a formula for computing $P(7, 3)$:

$$P(7, 3) = 7 \cdot 6 \cdot 5$$
$$= \frac{7 \cdot 6 \cdot 5 \cdot \mathbf{4 \cdot 3 \cdot 2 \cdot 1}}{\mathbf{4 \cdot 3 \cdot 2 \cdot 1}} \quad \text{Multiply both the numerator and denominator by } 4 \cdot 3 \cdot 2 \cdot 1.$$
$$= \frac{7!}{4!}$$
$$= \frac{7!}{(7 - 3)!}$$

The generalization of this idea gives the following formula.

Finding $P(n, r)$

The number of permutations of n things r at a time is given by the formula

$$P(n, r) = \frac{n!}{(n - r)!}$$

EXAMPLE 5

Counting permutations. Compute **a.** $P(8, 2)$, **b.** $P(7, 5)$, **c.** $P(n, n)$, and **d.** $P(n, 0)$.

Solution

a. $P(8, 2) = \dfrac{8!}{(8 - 2)!}$

$= \dfrac{8 \cdot 7 \cdot 6!}{6!}$

$= 8 \cdot 7$

$= 56$

b. $P(7, 5) = \dfrac{7!}{(7 - 5)!}$

$= \dfrac{7 \cdot 6 \cdot 5 \cdot 4 \cdot 3 \cdot 2!}{2!}$

$= 7 \cdot 6 \cdot 5 \cdot 4 \cdot 3$

$= 2,520$

c. $P(n, n) = \dfrac{n!}{(n - n)!}$

$= \dfrac{n!}{0!}$

$= \dfrac{n!}{1}$

$= n!$

d. $P(n, 0) = \dfrac{n!}{(n - 0)!}$

$= \dfrac{n!}{n!}$

$= 1$

SELF CHECK   Compute **a.** $P(10, 6)$ and **b.** $P(10, 10)$.   *Answers:* **a.** 151,200, **b.** 1 ∎

Parts c and d of Example 5 establish the following formulas.

Finding $P(n, n)$ and $P(n, 0)$

The number of permutations of n things n at a time and n things 0 at a time are given by the formulas

$$P(n, n) = n! \quad \text{and} \quad P(n, 0) = 1$$

EXAMPLE 6

Television schedules. **a.** In how many ways can a television executive arrange the Saturday night lineup of 6 programs if there are 15 programs to choose from? **b.** If there are only 6 programs to choose from?

Solution

a. To find the number of permutations of 15 programs 6 at a time, we use the formula $P(n, r) = \frac{n!}{(n - r)!}$ with $n = 15$ and $r = 6$.

$$P(15, 6) = \frac{15!}{(15 - 6)!}$$

$$= \frac{15 \cdot 14 \cdot 13 \cdot 12 \cdot 11 \cdot 10 \cdot 9!}{9!}$$

$$= 15 \cdot 14 \cdot 13 \cdot 12 \cdot 11 \cdot 10$$

$$= 3,603,600$$

b. To find the number of permutations of 6 programs 6 at a time, we use the formula $P(n, n) = n!$ with $n = 6$.

$$P(6, 6) = 6! = 720$$

SELF CHECK

See Example 6. How many ways are there if the executive has 20 programs to choose from? *Answer:* 27,907,200 ■

Combinations

Suppose that Raul must read 4 books from a reading list of 10 books. The order in which he reads them is not important. For the moment, however, let's assume that order is important and find the number of permutations of 10 things 4 at a time:

$$
\begin{aligned}
P(10, 4) &= \frac{10!}{(10 - 4)!} \\
&= \frac{10 \cdot 9 \cdot 8 \cdot 7 \cdot 6!}{6!} \\
&= 10 \cdot 9 \cdot 8 \cdot 7 \\
&= 5{,}040
\end{aligned}
$$

If order is important, there are 5,040 ways of choosing 4 books when there are 10 books to choose from. However, because the order in which Raul reads the books does not matter, the previous result of 5,040 is too big. Since there are 24 (or 4!) ways of ordering the 4 books that are chosen, the result of 5,040 is exactly 24 (or 4!) times too big. Therefore, the number of choices that Raul has is the number of permutations of 10 things 4 at a time, divided by 24:

$$\frac{P(10, 4)}{24} = \frac{5{,}040}{24} = 210$$

Raul has 210 ways of choosing four books to read from the list of ten books.

In situations where order is *not* important, we are interested in **combinations,** not permutations. The symbols $C(n, r)$ and $\binom{n}{r}$ both mean the number of combinations of n things r at a time.

If a selection of r books is chosen from a total of n books, the number of possible selections is $C(n, r)$, and there are $r!$ arrangements of the r books in each selection. If we consider the selected books as an ordered grouping, the number of orderings is $P(n, r)$. Therefore, we have

1. $r! \cdot C(n, r) = P(n, r)$

We can divide both sides of Equation 1 by $r!$ to get the formula for finding $C(n, r)$:

$$C(n, r) = \binom{n}{r} = \frac{P(n, r)}{r!} = \frac{n!}{r!(n - r)!}$$

Finding $C(n, r)$

The number of combinations of n things r at a time is given by

$$C(n, r) = \frac{n!}{r!(n - r)!}$$

EXAMPLE 7

Counting combinations. Compute **a.** $C(8, 5)$, **b.** $\binom{7}{2}$, **c.** $C(n, n)$, and **d.** $C(n, 0)$.

Solution

a. $C(8, 5) = \dfrac{8!}{5!(8 - 5)!}$

$= \dfrac{8 \cdot 7 \cdot 6 \cdot 5!}{5! \cdot 3!}$

$= 8 \cdot 7$

$= 56$

b. $\binom{7}{2} = \dfrac{7!}{2!(7 - 2)!}$

$= \dfrac{7 \cdot 6 \cdot 5!}{2 \cdot 1 \cdot 5!}$

$= 21$

c. $C(n, n) = \dfrac{n!}{n!(n - n)!}$

$= \dfrac{n!}{n! \cdot 0!}$

$= \dfrac{1}{0!}$

$= \dfrac{1}{1}$

$= 1$

d. $C(n, 0) = \dfrac{n!}{0!(n - 0)!}$

$= \dfrac{n!}{0! \cdot n!}$

$= \dfrac{1}{0!}$

$= \dfrac{1}{1}$

$= 1$

The symbol $C(n, 0)$ indicates that we choose 0 things from the available n things.

SELF CHECK

Compute **a.** $C(9, 6)$ and **b.** $C(10, 10)$. *Answers:* **a.** 84, **b.** 1 ■

Parts c and d of Example 7 establish the following formulas.

Finding $C(n, n)$ and $C(n, 0)$

The number of combinations of n things n at a time is 1. The number of combinations of n things 0 at a time is 1.

$$C(n, n) = 1 \quad \text{and} \quad C(n, 0) = 1$$

EXAMPLE 8

Choosing committees. If 15 students want to pick a committee of 4 students to plan a party, how many different committees are possible?

Solution

Since the ordering of people on each possible committee is not important, we find the number of combinations of 15 people 4 at a time:

$$C(15, 4) = \frac{15!}{4!(15 - 4)!}$$

$$= \frac{15 \cdot 14 \cdot 13 \cdot 12 \cdot 11!}{4 \cdot 3 \cdot 2 \cdot 1 \cdot 11!}$$

$$= \frac{15 \cdot 14 \cdot 13 \cdot 12}{4 \cdot 3 \cdot 2 \cdot 1}$$

$$= 1,365$$

There are 1,365 possible committees.

SELF CHECK

In how many ways can 20 students pick a committee of 5 students to plan a party? *Answer:* 15,504 ■

EXAMPLE 9

Choosing subcommittees. A committee in Congress consists of ten Democrats and eight Republicans. In how many ways can a subcommittee be chosen if it is to contain five Democrats and four Republicans?

Solution   There are $C(10, 5)$ ways of choosing the 5 Democrats and $C(8, 4)$ ways of choosing the 4 Republicans. By the multiplication principle for events, there are $C(10, 5) \cdot C(8, 4)$ ways of choosing the subcommittee:

$$C(10, 5) \cdot C(8, 4) = \frac{10!}{5!(10 - 5)!} \cdot \frac{8!}{4!(8 - 4)!}$$

$$= \frac{10 \cdot 9 \cdot 8 \cdot 7 \cdot 6 \cdot 5!}{120 \cdot 5!} \cdot \frac{8 \cdot 7 \cdot 6 \cdot 5 \cdot 4!}{24 \cdot 4!}$$

$$= \frac{10 \cdot 9 \cdot 8 \cdot 7 \cdot 6}{120} \cdot \frac{8 \cdot 7 \cdot 6 \cdot 5}{24}$$

$$= 17,640$$

There are 17,640 possible subcommittees.

SELF CHECK   See Example 9. In how many ways can a subcommittee be chosen if it is to contain four members from each party?   *Answer:* 14,700   ∎

Alternative Form of the Binomial Theorem

We have seen that the expansion of $(x + y)^3$ is

$$(x + y)^3 = 1x^3 + 3x^2y + 3xy^2 + 1y^3$$

and that

$$\binom{3}{0} = 1, \quad \binom{3}{1} = 3, \quad \binom{3}{2} = 3, \quad \text{and} \quad \binom{3}{3} = 1$$

Putting these facts together gives the following way of writing the expansion of $(x + y)^3$:

$$(x + y)^3 = \binom{3}{0}x^3 + \binom{3}{1}x^2y + \binom{3}{2}xy^2 + \binom{3}{3}y^3$$

Likewise, we have

$$(x + y)^4 = \binom{4}{0}x^4 + \binom{4}{1}x^3y + \binom{4}{2}x^2y^2 + \binom{4}{3}xy^3 + \binom{4}{4}y^4$$

The generalization of this idea allows us to state the binomial theorem in an alternative form using combinatorial notation.

The binomial theorem

If n is any positive integer, then

$$(a + b)^n = \binom{n}{0}a^n + \binom{n}{1}a^{n-1}b + \binom{n}{2}a^{n-2}b^2 + \cdots$$

$$+ \binom{n}{r}a^{n-r}b^r + \cdots + \binom{n}{n}b^n$$

EXAMPLE 10   **Expanding binomials.** Use the alternative form of the binomial theorem to expand $(x + y)^6$.

Solution

$$(x + y)^6 = \binom{6}{0}x^6 + \binom{6}{1}x^5y + \binom{6}{2}x^4y^2 + \binom{6}{3}x^3y^3 + \binom{6}{4}x^2y^4 + \binom{6}{5}xy^5 + \binom{6}{6}y^6$$
$$= x^6 + 6x^5y + 15x^4y^2 + 20x^3y^3 + 15x^2y^4 + 6xy^5 + y^6$$

SELF CHECK   Use the alternative form of the binomial theorem to expand $(a + b)^2$.

Answer: $a^2 + 2ab + b^2$ ■

EXAMPLE 11   **Expanding binomials.** Use the alternative form of the binomial theorem to expand $(2x - y)^3$.

Solution

$$(2x - y)^3 = [2x + (-y)]^3$$

$$= \binom{3}{0}(2x)^3 + \binom{3}{1}(2x)^2(-y) + \binom{3}{2}(2x)(-y)^2 + \binom{3}{3}(-y)^3$$

$$= 1(2x)^3 + 3(4x^2)(-y) + 3(2x)(y^2) + (-y)^3$$

$$= 8x^3 - 12x^2y + 6xy^2 - y^3$$

SELF CHECK   Use the alternative form of the binomial theorem to expand $(3a + b)^3$.

Answer: $27a^3 + 27a^2b + 9ab^2 + b^3$ ■

STUDY SET

Section 14.4

VOCABULARY

Fill in the blanks to make the statements true.

1. A _____ is an arrangement of objects.

2. When selecting objects when order is not important, we count _____.

CONCEPTS

Fill in the blanks to make the statements true.

3. If an event E_1 can be done in p ways and (after it occurs) a second event E_2 can be done in q ways, the event E_1 followed by E_2 can be done in _____ ways.

4. The symbol _____ means the number of permutations of n things taken r at a time.

5. The formula for the number of permutations of n things taken r at a time is _____.

6. $P(n, n) = $ ___

7. $P(n, 0) = $ ___

8. $8! = $ _____.

9. The symbol $C(n, r)$ or _____ means the number of _____ of n things taken r at a time.

10. The formula for the number of combinations of n things taken r at a time is _____.

11. $C(n, n) = $ ___

12. $C(n, 0) = $ ___

NOTATION

Complete each solution.

13. $P(6, 2) = \dfrac{\blacksquare}{(6 - 2)!}$

$= \dfrac{6 \cdot 5 \cdot 4!}{\blacksquare}$

$= 6 \cdot \blacksquare$

$= 30$

14. $C(6, 2) = \dfrac{\blacksquare}{\blacksquare(6 - 2)!}$

$= \dfrac{6 \cdot 5 \cdot 4!}{2 \cdot 1 \cdot \blacksquare}$

$= \blacksquare \cdot 5$

$= 15$

PRACTICE

In Exercises 15–36, evaluate each permutation or combination.

15. $P(3, 3)$

16. $P(4, 4)$

17. $P(5, 3)$

18. $P(3, 2)$

19. $P(2, 2) \cdot P(3, 3)$

20. $P(3, 2) \cdot P(3, 3)$

21. $\dfrac{P(5, 3)}{P(4, 2)}$

22. $\dfrac{P(6, 2)}{P(5, 4)}$

23. $\dfrac{P(6, 2) \cdot P(7, 3)}{P(5, 1)}$

24. $\dfrac{P(8, 3)}{P(5, 3) \cdot P(4, 3)}$

25. $C(5, 3)$

26. $C(5, 4)$

27. $\dbinom{6}{3}$

28. $\dbinom{6}{4}$

29. $\dbinom{5}{4}\dbinom{5}{3}$

30. $\dbinom{6}{5}\dbinom{6}{4}$

31. $\dfrac{C(38, 37)}{C(19, 18)}$

32. $\dfrac{C(25, 23)}{C(40, 39)}$

33. $C(12, 0) \cdot C(12, 12)$

34. $\dfrac{C(8, 0)}{C(8, 1)}$

35. $C(n, 2)$

36. $C(n, 3)$

In Exercises 37–42, use the alternative form of the binomial theorem to expand each expression.

37. $(x + y)^4$

38. $(x - y)^2$

39. $(2x + y)^3$

40. $(2x + 1)^4$

41. $(3x - 2)^4$

42. $(3 - x^2)^3$

In Exercises 43–46, find the indicated term of the binomial expansion.

43. $(x - 5y)^5$; fourth term

44. $(2x - y)^5$; third term

45. $(x^2 - y^3)^4$; second term

46. $(x^3 - y^2)^4$; fourth term

APPLICATIONS

47. PLANNING AN EVENING Kristy plans to go to dinner and see a movie. In how many ways can she arrange her evening if she has a choice of five movies and seven restaurants?

48. TRAVEL CHOICES Paula has five ways to travel from New York to Chicago, three ways to travel from Chicago to Denver, and four ways to travel from Denver to Los Angeles. How many choices are available if she travels from New York to Los Angeles?

49. MAKING LICENSE PLATES How many six-digit license plates can be manufactured? Note that there are ten choices—0, 1, 2, 3, 4, 5, 6, 7, 8, 9—for each digit.

50. MAKING LICENSE PLATES How many six-digit license plates can be manufactured if no digit can be repeated?

51. MAKING LICENSE PLATES How many six-digit license plates can be manufactured if no license can begin with 0 and if no digit can be repeated?

52. MAKING LICENSE PLATES How many license plates can be manufactured with two letters followed by four digits?

53. PHONE NUMBERS How many seven-digit phone numbers are available in area code 815 if no phone number can begin with 0 or 1?

54. PHONE NUMBERS How many ten-digit phone numbers are available if area codes of 000 and 911 cannot be used and if no local number can begin with 0 or 1?

55. LINING UP In how many ways can six people be placed in a line?

56. ARRANGING BOOKS In how many ways can seven books be placed on a shelf?

57. ARRANGING BOOKS In how many ways can four novels and five biographies be arranged on a shelf if the novels are placed first?

58. MAKING A BALLOT In how many ways can six candidates for mayor and four candidates for the county board be arranged on a ballot if all of the candidates for mayor must be placed first?

59. COMBINATION LOCKS How many permutations does a combination lock have if each combination has three numbers, no two numbers of any combination are equal, and the lock has 25 numbers?

60. COMBINATION LOCKS How many permutations does a combination lock have if each combination has three numbers, no two numbers of any combination are equal, and the lock has 50 numbers?

61. ARRANGING APPOINTMENTS The receptionist at a dental office has only three appointment times available before next Tuesday, and ten patients have toothaches. In how many ways can the receptionist fill those appointments?

62. COMPUTERS In many computers, a *word* consists of 32 *bits*—a string of thirty-two 1's and 0's. How many different words are possible?

63. PALINDROMES A palindrome is any word, such as *madam* or *radar,* that reads the same backward and forward. How many five-digit numerical palindromes (like 13531) are there? (*Hint:* A leading 0 would be dropped.)

64. CALL LETTERS The call letters of a U.S. commercial radio station have 3 or 4 letters, and the first is either a W or a K. How many radio stations could this system support?

65. PLANNING A PICNIC A class of 14 students wants to pick a committee of 3 students to plan a picnic. How many committees are possible?

66. CHOOSING BOOKS Jeffrey must read 3 books from a reading list of 15 books. How many choices does he have?

67. FORMING COMMITTEES The number of three-person committees that can be formed from a group of persons is ten. How many persons are in the group?

68. FORMING COMMITTEES The number of three-person committees that can be formed from a group of persons is 20. How many persons are in the group?

69. WINNING A LOTTERY In one state lottery, anyone who picks the correct six numbers (in any order) wins. With the numbers 0 through 99 available, how many choices are possible?

70. TAKING TESTS The instructions on a test read, "Answer any ten of the following fifteen questions. Then choose one of the remaining questions for homework, and turn in its solution tomorrow." In how many ways can the questions be chosen?

71. FORMING COMMITTEES In how many ways can we select a committee of two men and two women from a group containing three men and four women?

72. FORMING COMMITTEES In how many ways can we select a committee of three men and two women from a group containing five men and three women?

73. CHOOSING CLOTHES In how many ways can we select 2 shirts and 3 neckties from a group of 12 shirts and 10 neckties?

74. CHOOSING CLOTHES In how many ways can we select five dresses and two coats from a wardrobe containing nine dresses and three coats?

WRITING

75. State the multiplication principle for events.

76. Explain why *permutation lock* would be a better name for a combination lock.

REVIEW

Find each value of x.

77. $|2x - 3| = 9$

78. $2x^2 - x = 15$

79. $\dfrac{3}{x - 5} = \dfrac{8}{x}$

80. $\dfrac{3}{x} = \dfrac{x - 2}{8}$

▶ 14.5

Probability

In this section, you will learn about

   Probability

Introduction   In this section, we will discuss probabilities. The probability that an event will occur is a measure of the likelihood of that event. A tossed coin, for example, can land in two ways, either heads or tails. Because one of these two equally likely outcomes is heads, we expect that out of several tosses, about half will be heads. We say that the probability of obtaining heads in a single toss of the coin is $\frac{1}{2}$.

If records show that out of 100 days with weather conditions like today's, 30 have received rain, the weather service will report, "There is a $\frac{30}{100}$ or 30% probability of rain today."

Probability

Activities such as tossing a coin, rolling a die, drawing a card, and predicting rain are called **experiments.** For any experiment, a list of all possible outcomes is called a **sample space.** For example, the sample space *S* for the experiment of tossing two coins is the set

$S = \{(H, H), (H, T), (T, H), (T, T)\}$ There are four possible outcomes.

where the ordered pair (H, T) represents the outcome "heads on the first coin and tails on the second coin."

An **event** is a subset of the sample space of an experiment. For example, if *E* is the event "getting at least one heads" in the experiment of tossing two coins, then

$E = \{(H, H), (H, T), (T, H)\}$ There are 3 ways of getting at least one heads.

Because the outcome of getting at least one heads can occur in 3 out of 4 possible ways, we say that the **probability** of E is $\frac{3}{4}$, and we write

$$P(E) = P(\text{at least one heads}) = \frac{3}{4}$$

Probability of an event

If a sample space of an experiment has n distinct and equally likely outcomes and E is an event that occurs in s of those ways, the **probability of E** is

$$P(E) = \frac{s}{n}$$

Since $0 \le s \le n$, it follows that $0 \le \frac{s}{n} \le 1$. This implies that all probabilities have values from 0 to 1. If an event cannot happen, its probability is 0. If an event is certain to happen, its probability is 1.

E X A M P L E   1

Sample spaces. List the sample space of the experiment "rolling two dice a single time."

Solution   We can list ordered pairs and let the first number be the result on the first die and the second number the result on the second die. The sample space S is the following set of ordered pairs:

(1, 1) (1, 2) (1, 3) (1, 4) (1, 5) (1, 6)
(2, 1) (2, 2) (2, 3) (2, 4) (2, 5) (2, 6)
(3, 1) (3, 2) (3, 3) (3, 4) (3, 5) (3, 6)
(4, 1) (4, 2) (4, 3) (4, 4) (4, 5) (4, 6)
(5, 1) (5, 2) (5, 3) (5, 4) (5, 5) (5, 6)
(6, 1) (6, 2) (6, 3) (6, 4) (6, 5) (6, 6)

By counting, we see that the experiment has 36 equally likely possible outcomes.

SELF CHECK   How many pairs in the sample space have a sum of 4?   *Answer:* 3   ■

E X A M P L E   2

Finding probabilities. Find the probability of the event "rolling a sum of 7 on one roll of two dice."

Solution   The sample space is listed in Example 1. In that sample space, the following six ordered pairs give a sum of 7:

(1, 6) (2, 5) (3, 4) (4, 3) (5, 2) (6, 1)

Since there are 6 ordered pairs whose numbers give a sum of 7 out of a total of 36 equally likely outcomes, we have

$$P(E) = P(\text{rolling a 7}) = \frac{s}{n} = \frac{6}{36} = \frac{1}{6}$$

SELF CHECK   See Example 2. Find the probability of rolling a sum of 4.   *Answer:* $\frac{1}{12}$ ■

A standard playing deck of 52 cards has two red suits, hearts and diamonds, and two black suits, clubs and spades. Each suit has 13 cards, including the ace, king, queen, jack, and cards numbered from 2 to 10. We will refer to a standard deck of cards in many examples and exercises.

EXAMPLE 3

Finding probabilities. Find the probability of drawing an ace on one draw from a standard card deck.

Solution   Since there are four aces in the deck, the number of favorable outcomes is $s = 4$. Since there are 52 cards in the deck, the total number of possible outcomes is $n = 52$. The probability of drawing an ace is the ratio of the number of favorable outcomes to the number of possible outcomes.

$$P(\text{an ace}) = \frac{s}{n} = \frac{4}{52} = \frac{1}{13}$$

The probability of drawing and ace is $\frac{1}{13}$.

SELF CHECK   Find the probability of drawing a red ace on one draw from a standard card deck.   *Answer:* $\frac{1}{26}$ ■

EXAMPLE 4

Finding probabilities. Find the probability of drawing 5 cards, all hearts, from a standard card deck.

Solution   The number of ways we can draw 5 hearts from the 13 hearts is $C(13, 5)$, the number of combinations of 13 things taken 5 at a time. The number of ways to draw 5 cards from the deck is $C(52, 5)$, the number of combinations of 52 things taken 5 at a time. The probability of drawing 5 hearts is the ratio of the number of favorable outcomes to the number of possible outcomes.

$$P(5 \text{ hearts}) = \frac{s}{n} = \frac{C(13, 5)}{C(52, 5)}$$

$$P(5 \text{ hearts}) = \frac{\dfrac{13!}{5!8!}}{\dfrac{52!}{5!47!}}$$

$$= \frac{13!}{5!8!} \cdot \frac{5!47!}{52!}$$

$$= \frac{13 \cdot 12 \cdot 11 \cdot 10 \cdot 9 \cdot 8!}{8!} \cdot \frac{47!}{52 \cdot 51 \cdot 50 \cdot 49 \cdot 48 \cdot 47!}$$

$$= \frac{13 \cdot 12 \cdot 11 \cdot 10 \cdot 9}{52 \cdot 51 \cdot 50 \cdot 49 \cdot 48}$$

$$= \frac{33}{66,640}$$

The probability of drawing 5 hearts is $\frac{33}{66,640}$.

SELF CHECK   Find the probability of drawing 6 cards, all diamonds, from a standard card deck.   *Answer:* $\frac{33}{391,510}$ ■

STUDY SET

Section 14.5

VOCABULARY

Fill in the blanks to make the statements true.

1. An _____ is any activity for which the outcome is uncertain.

2. A list of all possible outcomes for an experiment is called a _____.

CONCEPTS

Fill in the blanks to make the statements true.

3. The probability of an event E is defined as $P(E) =$ _____.

4. If an event is certain to happen, its probability is ___.

5. If an event cannot happen, its probability is ___.

6. All probability values are between ___ and ___, inclusive.

NOTATION

Complete each solution.

7. Find the probability of drawing a black face card from a standard deck.
   a. The number of black face cards is ___.
   b. The number of cards in the deck is ___.
   c. The probability is ___ or ___.

8. Find the probability of drawing 4 aces from a standard card deck.
   a. The number of ways to draw 4 aces from 4 aces is $C(4, 4) =$ ___.
   b. The number of ways to draw 4 cards from 52 cards is $C(52, 4) =$ _____.
   c. The probability is _____.

PRACTICE

In Exercises 9–12, list the sample space of each experiment.

9. Rolling one die and tossing one coin

10. Tossing three coins

11. Selecting a letter of the alphabet

12. Picking a one-digit number

In Exercises 13–16, an ordinary die is rolled once. Find the probability of each event.

13. Rolling a 2

14. Rolling a number greater than 4

15. Rolling a number larger than 1 but less than 6

16. Rolling an odd number

In Exercises 17–20, balls numbered from 1 to 42 are placed in a container and stirred. If one is drawn at random, find the probability of each result.

17. The number is less than 20.

18. The number is less than 50.

19. The number is a prime number.

20. The number is less than 10 or greater than 40.

In Exercises 21–24, refer to the spinner in Illustration 1. If the spinner is spun, find the probability of each event. Assume that the spinner never stops on a line.

ILLUSTRATION 1

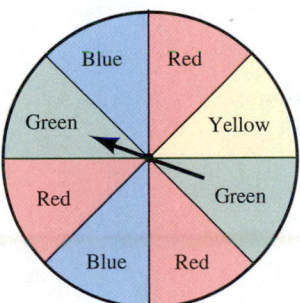

21. The spinner stops on red.

22. The spinner stops on green.

23. The spinner stops on orange.

24. The spinner stops on yellow.

In Exercises 25–32, find the probability of each event.

25. Rolling a sum of 4 on one roll of two dice

26. Drawing a diamond on one draw from a card deck

27. Drawing a red egg from a basket containing 5 red eggs and 7 blue eggs

28. Drawing a yellow egg from a basket containing 5 red eggs and 7 yellow eggs

29. Drawing 6 diamonds from a standard card deck without replacing the cards after each draw

30. Drawing 5 aces from a standard card deck without replacing the cards after each draw

31. Drawing 5 clubs from the black cards in a standard card deck

32. Drawing a face card from a standard card deck

In Exercises 33–38, assume that the probability that an airplane engine will fail during a torture test is $\frac{1}{2}$ and that the aircraft in question has 4 engines. In Exercises 34–38, find each probability.

33. Construct a sample space for the torture test.

34. All engines will survive the test.

35. Exactly 1 engine will survive.

36. Exactly 2 engines will survive.

37. Exactly 3 engines will survive.

38. No engines will survive.

39. Find the sum of the probabilities in Exercises 34 through 38.

In Exercises 40–42, assume that a survey of 282 people is taken to determine the opinions of doctors, teachers, and lawyers on a proposed piece of legislation, with the results shown in Illustration 2. A person is chosen at random from those surveyed. Refer to Illustration 2 to find each probability.

ILLUSTRATION 2

	Number that favor	Number that oppose	Number with no opinion	Total
Doctors	70	32	17	119
Teachers	83	24	10	117
Lawyers	23	15	8	46
Total	176	71	35	282

40. The person favors the legislation.

41. A doctor opposes the legislation.

42. A person who opposes the legislation is a lawyer.

APPLICATIONS

43. QUALITY CONTROL In a batch of 10 tires, 2 are known to be defective. If 4 tires are chosen at random, find the probability that all 4 tires are good.

44. MEDICINE Out of a group of 9 patients treated with a new drug, 4 suffered a relapse. Find the probability that 3 patients of this group, chosen at random, will remain disease-free.

WRITING

45. Explain why all probability values range from 0 to 1.

46. Explain the concept of probability.

REVIEW

Solve each equation.

47. $|x + 3| = 7$

48. $|3x - 2| = |2x - 3|$

In Exercises 49–50, solve each inequality. Give the answer in interval notation.

49. $|x - 3| < 7$

50. $|3x - 2| \geq 5$

The Language of Algebra

Algebra is a language in its own right. One of the keys to becoming a good algebra student is to know the vocabulary of algebra. In Exercises 1–26, match each instruction in column I with the most appropriate problem in column II. Each letter in column II is used only once.

Column I	***Column II***
1. Use the FOIL method.	**a.** 2,300,000,000
2. Apply a rule for exponents to simplify.	**b.** e^3
3. Add the rational expressions.	**c.** $f(x) = x^2 + 1$ and $g(x) = 5 - 3x$
4. Rationalize the denominator.	**d.** $-2x(3x^2 - 4x + 8)$
5. Factor completely.	**e.** $4x - 7 > -3x - 7$
6. Evaluate the expression for $a = -1$ and $b = -6$.	**f.** $\begin{cases} 2x = y - 5 \\ x + y = -1 \end{cases}$
7. Express in lowest terms.	**g.** $(x^2 - 5)(x^2 + 3)$
8. Solve for t.	**h.** $(x + 2)(x - 10) = 0$
9. Combine like terms.	**i.** $\dfrac{x - 1}{2x^2} + \dfrac{x + 1}{8x}$
10. Remove parentheses.	**j.** $\sqrt{4x^2}$
11. Solve the system by graphing.	**k.** $\ln 6 + \ln x$
12. Find $f(g(x))$.	**l.** $\dfrac{10}{\sqrt{6} - \sqrt{2}}$
13. Solve using the quadratic formula.	**m.** $2x - 8 + 6y - 14$
14. Identify the base and the exponent.	**n.** $(3, -2)$ and $(0, -5)$
15. Write without a radical sign.	**o.** $x^n \cdot x^{3n}$
16. Write the equation of the line having the given slope and y-intercept.	**p.** $\dfrac{4x^2 y}{16xy}$
17. Solve the inequality.	**q.** $h(x) = 10^x$
18. Complete the square to make a perfect square trinomial.	**r.** $\log_2 8 = 3$
19. Find the slope of the line passing through the given points.	**s.** $m = \frac{2}{3}$ and passes through $(0, 2)$
20. Use a property of logarithms to simplify.	**t.** 2, 6, 18, . . .
21. Set each factor equal to zero and solve for x.	**u.** $3y^3 - 243b^6$
22. State the solution of the compound inequality using interval notation.	**v.** $x + 7 \geq 0$ and $-x < -1$
23. Find the inverse function, $h^{-1}(x)$.	**w.** $x^2 - 3x - 4 = 0$
24. Write using scientific notation.	**x.** $Rt = cd + 2t$
25. Write the logarithmic statement in exponential form.	**y.** $-2\pi a^2 b - 3b^3$
26. Find the sum of the first 6 terms of the sequence.	**z.** $x^2 + 4x$

Accent on Teamwork

Section 14.1

Pascal's triangle Find the sum of the numbers in each of the first ten rows of Pascal's triangle. What is the pattern? Then find the sum of the numbers in the designated diagonal rows of Pascal's triangle shown in Illustration 1. What is the pattern?

ILLUSTRATION 1

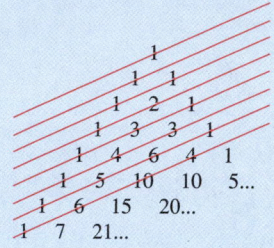

Factorial Tell whether each statement is true or false.
a. $8! - 6! = 2!$ **b.** $8! + 6! = 14!$
c. $\dfrac{8!}{6!} = 2!$ **d.** $8! \cdot 6! = 48!$

Sections 14.2 and 14.3

Sequences The terms of a sequence can be generated by substituting 1,000 for n, 2,000 for n, 3,000 for n, and so on, in the expression

$$\left(1 + \frac{1}{n}\right)^n$$

Use a calculator to make each computation for values of n from 1,000 to 20,000. The terms of the sequence get closer and closer to an important number that we studied in this course. What is that number?

Folding paper Suppose a piece of paper is 0.01 inch thick. If the paper is folded over itself 10 successive times, and if the thickness doubles after each fold, how thick will the result be?

Height of the rebound Drop a hard rubber ball from a height of 72 inches from the floor and measure how high it rebounds in inches. Plot the point (1, 72) on graph paper to represent the first time the ball is dropped. Then plot (2, rebound height in inches) to represent the second time the ball is dropped. Continue this pattern, measuring how high the ball rebounds and plotting the data. Do the points on the graph form a pattern?

Section 14.4

Permutations and combinations Explain why $C(n, r) = C(n, n - r)$. Then determine if $P(n, r) = P(n, n - r)$.

Circular permutations In how many ways may four people be seated in a row?

Suppose the same four people are seated around a circular table. Some of the possible seating arrangements are really the same, because the relative position of each person to the others is not different. For an example of this, see Illustration 2. If we agree not to count any of the duplicate arrangements, in how many different ways can four people be seated around a circular table?

ILLUSTRATION 2

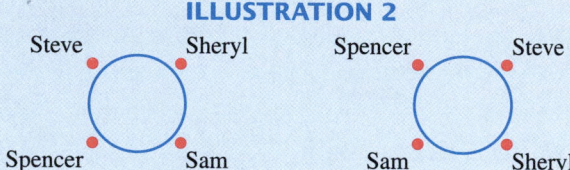

Section 14.5

Throwing dice Two standard dice are rolled, and the sum of the number of dots on the top faces of the dice is recorded.

a. Complete the table below, which gives the number of ways each sum can occur. For example, there are three ways to obtain a sum of 4: a ⏹ on die 1 and a ⏹ on die 2, a ⏹ on die 1 and a ⏹ on die 2, and a ⏹ on die 1 and a ⏹ on die 2.

Sum	2	3	4	5	6	7	8	9	10	11	12
Number of ways			3								

b. Use the table to determine the probability of obtaining a sum of 7 or 11 on one roll of the dice.

Odds For an experiment involving equally likely outcomes, the odds in favor of an event are defined to be the total number of favorable outcomes compared to the total number of unfavorable outcomes. Odds are usually expressed as a ratio of the form 1:3 or 5:2, for example. If a standard die is rolled, what are the odds in favor of the following events?
a. Getting a 1
b. Getting an even number
c. Getting a number less than 0
d. Getting a number greater than 6

Section 14.1

The Binomial Theorem

CONCEPTS

The symbol $n!$ (*n factorial*) is defined as

$$n! = n(n-1)(n-2) \cdots 2 \cdot 1$$

where n is a natural number.

$0! = 1$

$n(n-1)! = n!$ (n is a natural number)

The binomial theorem:

$$(a+b)^n =$$

$$a^n + \frac{n!}{1!(n-1)!} a^{n-1}b$$

$$+ \frac{n!}{2!(n-2)!} a^{n-2}b^2$$

$$+ \cdots + b^n$$

The binomial theorem can be used to find the nth term of a binomial expansion.

REVIEW EXERCISES

1. Evaluate each expression.

    a. $(4!)(3!)$ **b.** $\dfrac{5!}{3!}$

    c. $\dfrac{6!}{2!(6-2)!}$ **d.** $\dfrac{12!}{3!(12-3)!}$

    e. $(n-n)!$ **f.** $\dfrac{8!}{7!}$

2. Use the binomial theorem to find each expansion.
    a. $(x+y)^5$

    b. $(x-y)^4$

    c. $(4x-y)^3$

    d. $(x+4y)^3$

3. Find the specified term in each expansion.
    a. $(x+y)^4$; third term

    b. $(x-y)^5$; fourth term

    c. $(3x-4y)^3$; second term

    d. $(4x+3y)^4$; third term

Section 14.2

Arithmetic Sequences

An *arithmetic sequence* is a sequence of the form

$$a, a+d, a+2d, \ldots ,$$
$$a+(n-1)d, \ldots$$

where a is the first term, $a+(n-1)d$ is the nth term, and d is the common difference.

If numbers are inserted between two given numbers a and b to form an arithmetic sequence, the inserted numbers are *arithmetic means* between a and b.

4. Find the eighth term of an arithmetic sequence whose first term is 7 and whose common difference is 5.

5. Write the first five terms of the arithmetic sequence whose ninth term is 242 and whose seventh term is 212.

6. Find two arithmetic means between 8 and 25.

The sum of the first n terms of an arithmetic sequence is given by

$$S_n = \frac{n(a + l)}{2} \quad \text{with}$$

$$l = a + (n - 1)d$$

where a is the first term, l is the last (or nth) term, and n is the number of terms in the sequence.

$$\sum_{k=1}^{n} f(k) = f(1) + f(2) + \cdots + f(n)$$

7. Find the sum of the first 20 terms of the sequence 11, 18, 25, ...

8. Find the sum of the first ten terms of the sequence 9, $6\frac{1}{2}$, 4, ...

9. Find each sum.

a. $\displaystyle\sum_{k=4}^{6} \frac{1}{2}k$

b. $\displaystyle\sum_{k=2}^{5} 7k^2$

c. $\displaystyle\sum_{k=1}^{4} (3k - 4)$

d. $\displaystyle\sum_{k=10}^{10} 36k$

Section 14.3

Geometric Sequences

A *geometric sequence* is a sequence of the form

$$a, \ ar, \ ar^2, \ ar^3, \ . \ . \ . \ , $$
$$ar^{n-1}, \ . \ . \ . $$

where a is the first term, ar^{n-1} is the nth term, and r is the common ratio.

If numbers are inserted between a and b to form a geometric sequence, the inserted numbers are *geometric means* between a and b.

The sum of the first n terms of a geometric sequence is given by

$$S_n = \frac{a - ar^n}{1 - r} \quad (r \neq 1)$$

where S_n is the sum, a is the first term, r is the common ratio, and n is the number of terms in the sequence.

If r is the common ratio of an infinite geometric sequence, and if $|r| < 1$, the sum of the terms of the infinite geometric sequence is given by

$$S = \frac{a}{1 - r}$$

where a is the first term and r is the common ratio.

10. Write the first five terms of the geometric sequence whose fourth term is 3 and whose fifth term is $\frac{3}{2}$.

11. Find the sixth term of a geometric sequence with a first term of $\frac{1}{8}$ and a common ratio of 2.

12. Find two geometric means between -6 and 384.

13. Find the sum of the first seven terms of the sequence 162, 54, 18, ...

14. Find the sum of the first eight terms of the sequence $\dfrac{1}{8}, -\dfrac{1}{4}, \dfrac{1}{2}, ...$

15. Find the sum of the infinite geometric sequence 25, 20, 16, ...

16. Change the decimal $0.\overline{05}$ to a common fraction.

Section 14.4

Permutations and Combinations

The multiplication principle for events: If E_1 and E_2 are two events, and if E_1 can be done in a_1 ways and E_2 can be done in a_2 ways, then the event "E_1 followed by E_2" can be done in $a_1 \cdot a_2$ ways.

Formula for permutations:

$$P(n, r) = \frac{n!}{(n - r)!}$$

$P(n, n) = n!$ and $P(n, 0) = 1$

Formula for combinations:

$$C(n, r) = \binom{n}{r} = \frac{n!}{r!(n - r)!}$$

$$C(n, n) = \binom{n}{n} = 1$$

$$C(n, 0) = \binom{n}{0} = 1$$

17. If there are 17 flights from New York to Chicago and 8 flights from Chicago to San Francisco, in how many different ways could a passenger plan her trip?

18. Evaluate each expression.
  a. $P(7, 7)$
  b. $P(7, 0)$
  c. $P(8, 6)$
  d. $\dfrac{P(9, 6)}{P(10, 7)}$

19. Evaluate each expression.
  a. $C(7, 7)$
  b. $C(7, 0)$
  c. $\dbinom{8}{6}$
  d. $\dbinom{9}{6}$
  e. $C(6, 3) \cdot C(7, 3)$
  f. $\dfrac{C(7, 3)}{C(6, 3)}$

20. CAR DEPRECIATION A $5,000 car depreciates at the rate of 20% of the previous year's value. How much is the car worth after 5 years?

21. STOCK APPRECIATION The value of Mia's stock portfolio is expected to appreciate at the rate of 18% per year. How much will the portfolio be worth in 10 years if its current value is $25,700?

22. PLANTING CORN A farmer planted 300 acres in corn this year. He intends to plant an additional 75 acres in corn in each successive year until he has 1,200 acres in corn. In how many years will that be?

23. FALLING OBJECT If an object is in free fall, the sequence 16, 48, 80, . . . represents the distance in feet that object falls during the first second, during the second second, during the third second, and so on. How far will the object fall during the first ten seconds?

24. LINING UP In how many ways can five persons be arranged in a line?

25. LINING UP In how many ways can three men and five women be arranged in a line if the women are placed ahead of the men?

26. CHOOSING PEOPLE In how many ways can we pick three persons from a group of ten persons?

27. FORMING COMMITTEES In how many ways can we pick a committee of two Democrats and two Republicans from a group containing five Democrats and six Republicans?

Section 14.5 Probability

An event that cannot happen has a *probability* of 0. An event that is certain to happen has a probability of 1. All other events have probabilities between 0 and 1.

If S is the *sample space* of an experiment with n distinct and equally likely outcomes, and E is an event that occurs in s of those ways, then the probability of E is

$$P(E) = \frac{s}{n}$$

28. Find the probability of rolling an 11 on one roll of two dice.

29. Find the probability of living forever.

30. Find the probability of drawing a 10 from a standard deck of cards.

31. Find the probability of drawing a five-card poker hand that has 3 aces.

32. Find the probability of drawing 5 cards, all spades, from a standard card deck.

CHAPTER 14

Test

1. Evaluate $\dfrac{7!}{4!}$.

2. Evaluate $0!$.

3. Find the second term in the expansion of $(x - y)^5$.

4. Find the third term in the expansion of $(x + 2y)^4$.

5. Find the tenth term of an arithmetic sequence whose first three terms are 3, 10, and 17.

6. Find the sum of the first 12 terms of the sequence $-2, 3, 8, \ldots$.

7. Find two arithmetic means between 2 and 98.

8. Evaluate $\displaystyle\sum_{k=1}^{3}(2k - 3)$.

9. Find the seventh term of the geometric sequence whose first three terms are $-\frac{1}{9}$, $-\frac{1}{3}$, and -1.

10. Find the sum of the first six terms of the sequence $\frac{1}{27}$, $\frac{1}{9}$, $\frac{1}{3}, \ldots$.

11. Find two geometric means between 3 and 648.

12. Find the sum of all of the terms of the infinite geometric sequence $9, 3, 1, \ldots$.

In Problems 13–20, find the value of each expression.

13. $P(5, 4)$

14. $P(8, 8)$

15. $C(6, 4)$

16. $C(8, 3)$

17. $C(6, 0) \cdot P(6, 5)$

18. $P(8, 7) \cdot C(8, 7)$

19. $\dfrac{P(6, 4)}{C(6, 4)}$

20. $\dfrac{C(9, 6)}{P(6, 4)}$

21. CHOOSING PEOPLE In how many ways can we pick 3 persons from a group of 7 persons?

22. CHOOSING COMMITTEES From a group of 5 men and 4 women, how many 3-person committees can be chosen that will include two women?

In Problems 23–26, find each probability.

23. Rolling a 5 on one roll of a die

24. Drawing a jack or a queen from a standard card deck

25. Receiving 5 hearts for a 5-card poker hand

26. Tossing 2 heads in 5 tosses of a fair coin

27. If the probability that a person is sick is 1, find the probability that the person is well.

28. What is a sample space of an experiment? Give an example.

Cumulative Review Exercises

In Exercises 1–4, consider the set $\left\{-\frac{4}{3}, \pi, 5.6, \sqrt{2}, 0, -23, e, 7i\right\}$. *List the elements in the set that are*

1. whole numbers

2. rational numbers

3. irrational numbers

4. real numbers

5. FINANCIAL PLANNING Ana has some money to invest. Her financial planner tells her that if she can come up with $3,000 more, she will qualify for an 11% annual interest rate. Otherwise, she will have to invest the money at 7.5% annual interest. The financial planner urges her to invest the larger amount, because the 11% investment would yield twice as much annual income as the 7.5% investment. How much does she originally have on hand to invest?

6. BOATING Use the graph in Illustration 1 to determine the average rate of change in the sound level of the engine of a boat in relation to rpm of the engine.

ILLUSTRATION 1

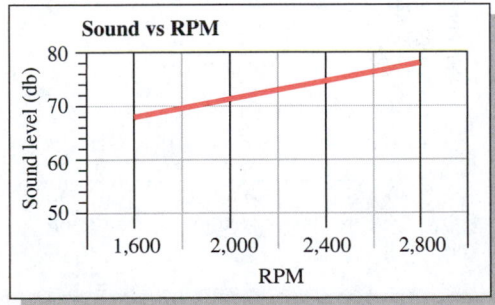

In Exercises 7–8, tell whether the graphs of the equations are parallel or perpendicular.

7. $3x - 4y = 12$, $y = \frac{3}{4}x - 5$

8. $y = 3x + 4$, $x = -3y + 4$

In Exercises 9–10, write the equation of the line with the given properties.

9. $m = -2$, passing through $(0, 5)$

10. Passing through $(8, -5)$ and $(-5, 4)$

11. Use substitution to solve $\begin{cases} 3x + y = 4 \\ 2x - 3y = -1 \end{cases}$.

12. Use addition to solve $\begin{cases} x + 2y = -2 \\ 2x - y = 6 \end{cases}$.

13. Solve using Cramer's rule: $\begin{cases} 4x - 3y = -1 \\ 3x + 4y = -7 \end{cases}$.

14. Solve $\begin{cases} x + y + z = 1 \\ 2x - y - z = -4 \\ x - 2y + z = 4 \end{cases}$.

15. The graphs of $y = 4(x - 5) - x - 2$ and $y = -(2x + 6) - 1$ are shown in Illustration 2. Use the information in the display to solve $4(x - 5) - x - 2 = -(2x + 6) - 1$ graphically.

ILLUSTRATION 2

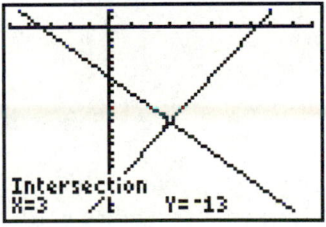

16. Evaluate $\begin{vmatrix} 3 & -2 \\ 1 & -1 \end{vmatrix}$.

17. MARTIAL ARTS Find the measure of each angle of the triangle shown in Illustration 3.

ILLUSTRATION 3

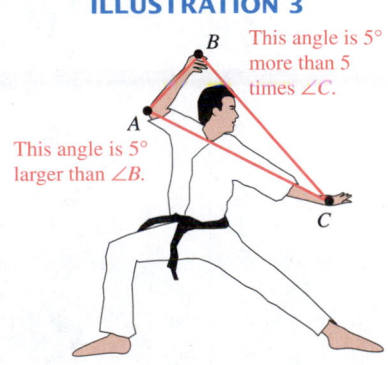

This angle is 5° more than 5 times ∠C.

This angle is 5° larger than ∠B.

18. Solve $\begin{cases} 3x - 2y \le 6 \\ y < -x + 2 \end{cases}$.

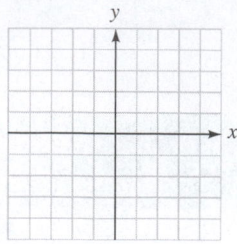

In Exercises 19–20, give the solution in interval notation and graph the solution set.

19. Solve $|5 - 3x| \le 14$.

20. Solve $4.5x - 1 < -10$ or $6 - 2x \ge 12$.

In Exercises 21–24, do the operations.

21. $(4x - 3y)(3x + y)$

22. $(-2x^2y^3 + 6xy + 5y^2) - (-4x^2y^3 - 7xy + 2y^2)$

23. $(a - 2b)^2$

24. $(a + 2)(3a^2 + 4a - 2)$

In Exercises 25–26, factor the expression completely.

25. $3x^3y - 4x^2y^2 - 6x^2y + 8xy^2$

26. $256x^4y^4 - z^8$

27. Solve for λ: $\dfrac{A\lambda}{2} + 1 = 2d + 3\lambda$.

28. Complete the table of values for
$f(x) = -x^3 - x^2 + 6x$ and then graph the function.
What are the x- and y-intercepts of the graph?

x	$f(x)$
-4	
-3	
-2	
-1	
0	
1	
2	
3	

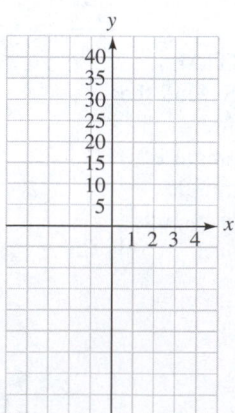

In Exercises 29–32, simplify.

29. $\left(\dfrac{4a^{-2}b}{3ab^{-3}}\right)^3$

30. $\dfrac{6x^2 + 13x + 6}{6 - 5x - 6x^2}$

31. $\dfrac{p^3 - q^3}{q^2 - p^2} \cdot \dfrac{q^2 + pq}{p^3 + p^2q + pq^2}$

32. $\dfrac{2}{a - 2} + \dfrac{3}{a + 2} - \dfrac{a - 1}{a^2 - 4}$

33. Solve $\dfrac{x - 4}{x - 3} + \dfrac{x - 2}{x - 3} = x - 3$.

34. Solve $\dfrac{1}{R} = \dfrac{1}{R_1} + \dfrac{1}{R_2} + \dfrac{1}{R_3}$ for R.

35. TIRE WEAR See Illustration 4.
  a. What type of function does it appear would model the relationship between the inflation of a tire and the percent of service it gives?
  b. At what percent(s) of inflation will a tire offer only 90% of its possible service?

ILLUSTRATION 4

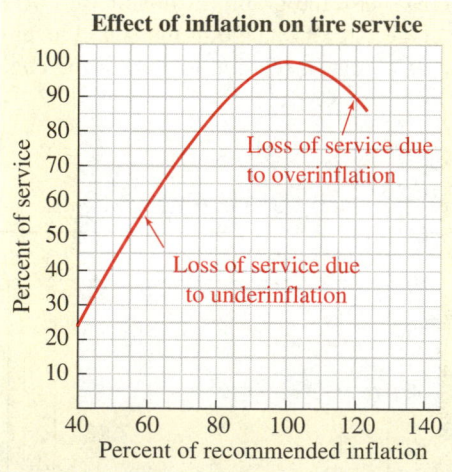

36. CHANGING DIAPERS Illustration 5 shows how to put a diaper on a baby. If the diaper is a square with sides 16 inches long, what is the largest waist size that this diaper can wrap around, assuming an overlap of 1 inch to pin the diaper?

37. Use the long division method to find
$$(2x^2 + 4x - x^3 + 3) \div (x - 1)$$

ILLUSTRATION 5

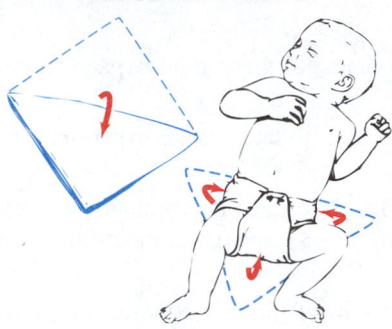

38. The graph of $f(x) = \sqrt{2x + 5} + \sqrt{x + 2} - 5$ is shown in Illustration 6. Use the information in the display to determine the solution of $\sqrt{2x + 5} = -\sqrt{x + 2} + 5$.

ILLUSTRATION 6

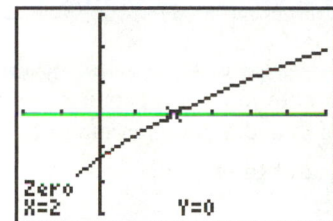

In Exercises 39–40, simplify each expression.

39. $\sqrt{98} + \sqrt{8} - \sqrt{32}$

40. $12\sqrt[3]{648x^4} + 3\sqrt[3]{81x^4}$

41. Evaluate $\left(\dfrac{25}{49}\right)^{-3/2}$.

42. Rationalize the denominator: $\dfrac{3t - 1}{\sqrt{3t} + 1}$.

In Exercises 43–44, write the expression in a + bi form.

43. $\left(-7 + \sqrt{-81}\right) - \left(-2 - \sqrt{-64}\right)$

44. $\dfrac{2 - 5i}{2 + 5i}$

In Exercises 45–50, solve each equation.

45. $\sqrt{3a + 1} = a - 1$

46. $\sqrt{x + 3} - \sqrt{3} = \sqrt{x}$

47. $6a^2 + 5a - 6 = 0$

48. $4w^2 + 6w + 1 = 0$

49. $2(2x + 1)^2 - 7(2x + 1) + 6 = 0$

50. $3x^2 - 4x = -2$

51. If $f(x) = x^2 - 2$ and $g(x) = 2x + 1$, find $(f \circ g)(x)$.

52. Find the inverse function of $f(x) = 2x^3 - 1$.

53. Graph $f(x) = \left(\dfrac{1}{2}\right)^x$.

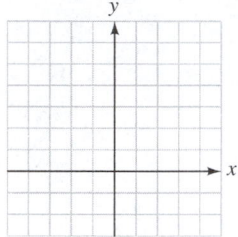

54. Graph $y = e_c^x$ and its inverse on the same coordinate system. Label the axis of symmetry.

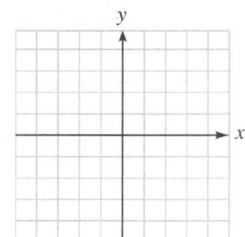

55. Write $y = \log_2 x$ as an exponential equation.

56. Apply properties of logarithms to simplify $\log_6 \dfrac{x}{36}$.

In Exercises 57–60, find x.

57. $\log_x 25 = 2$ **58.** $\log_5 125 = x$

59. $\log_3 x = -3$ **60.** $\ln e = x$

61. Find the inverse of $y = \log_2 x$.

62. If $\log_{10} 10^x = y$, then y equals what quantity?

In Exercises 63–64, log 7 = 0.8451 and log 14 = 1.1461. Evaluate each expression without using a calculator or tables.

63. $\log 98$ **64.** $\log 2$

In Exercises 65–68, solve each equation. Round to four decimal places when necessary.

65. $2^{x+2} = 3^x$

66. $\log x + \log (x + 9) = 1$

67. $5^{4x} = \dfrac{1}{125}$

68. $\log_3 x = \log_3 \left(\dfrac{1}{x}\right) + 4$

69. Write the equation of the circle that has its center at $(1, 3)$ and passes through $(-2, -1)$.

70. Write the equation in standard form and then graph it.
$$x^2 - 9y^2 + 2x + 36y = 44$$

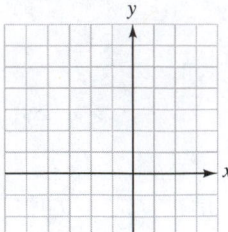

71. Graph $\dfrac{(x - 1)^2}{9} + \dfrac{(y - 3)^2}{4} = 1$.

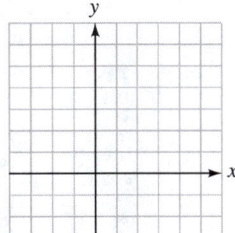

72. Graph $y^2 + 4x - 6y = -1$.

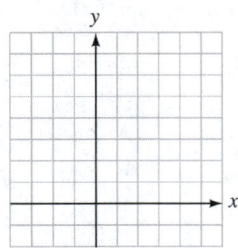

In Exercises 73–74, use a calculator.

73. BOAT DEPRECIATION How much will a $9,000 boat be worth after 9 years if it depreciates 12% per year?

74. Find $\log_6 8$.

75. Evaluate $\dfrac{6!7!}{5!}$.

76. Use the binomial theorem to expand $(3a - b)^4$.

77. Find the seventh term of the expansion of $(2x - y)^8$.

78. Find the 20th term of an arithmetic sequence with a first term of -11 and a common difference of 6.

79. Find the sum of the first 20 terms of an arithmetic sequence with a first term of 6 and a common difference of 3.

80. Insert two arithmetic means between -3 and 30.

81. Evaluate $\displaystyle\sum_{k=1}^{3} 3k^2$.

82. Evaluate $\displaystyle\sum_{k=3}^{5} (2k + 1)$.

83. Find the seventh term of a geometric sequence with a first term of $\frac{1}{27}$ and a common ratio of 3.

84. Find the sum of the first ten terms of the sequence $\frac{1}{64}$, $\frac{1}{32}$, $\frac{1}{16}$, ...

85. Insert two geometric means between -3 and 192.

86. Find the sum of all the terms of the sequence 9, 3, 1, ...

87. Evaluate $P(9, 3)$.

88. Evaluate $C(7, 4)$.

89. LINING UP In how many ways can seven people stand in a line?

90. FORMING A COMMITTEE In how many ways can a committee of three people be chosen from a group containing nine people?

Appendix I

Arithmetic Fractions

In this appendix, you will learn about

Simplifying fractions ■ Multiplying fractions ■ Dividing fractions
■ Adding fractions ■ Subtracting fractions ■ Mixed numbers
■ Decimal fractions ■ Rounding decimals ■ Applications

Introduction   In this appendix, we will review the properties of arithmetic fractions. In the **fractions**

$$\frac{1}{2}, \quad \frac{3}{5}, \quad \frac{2}{17}, \quad \text{and} \quad \frac{37}{7}$$

the number above the fraction bar is called the **numerator,** and the number below is called the **denominator.**

Fractions are used to indicate parts of a whole. In Figure I-1(a), a rectangle has been divided into 5 equal parts with 3 of the parts shaded. The fraction $\frac{3}{5}$ indicates how much of the figure is shaded. In Figure I-1(b), $\frac{5}{7}$ of the rectangle is shaded. In either example, the denominator of the fraction shows the total number of equal parts into which the whole is divided, and the numerator shows the number of these equal parts that are being considered.

FIGURE I-1

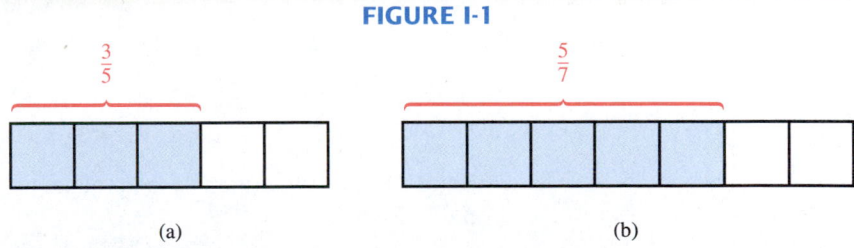

        (a)   (b)

Fractions are also used to indicate division. For example, the fraction $\frac{8}{2}$ indicates that 8 is to be divided by 2:

$$\frac{8}{2} = 8 \div 2 = 4$$

We note that $\frac{8}{2} = 4$, because $4 \cdot 2 = 8$, and that $\frac{0}{7} = 0$, because $0 \cdot 7 = 0$. However, the fraction $\frac{6}{0}$ is undefined, because no number multiplied by 0 gives 6. The fraction $\frac{0}{0}$ is indeterminate, because every number multiplied by 0 gives 0.

 WARNING! Remember that the denominator of a fraction cannot be zero.

Simplifying Fractions

A fraction is in **lowest terms** when no natural number greater than 1 will divide both its numerator and denominator exactly. The fraction $\frac{6}{11}$ is in lowest terms, because only 1 divides both 6 and 11 exactly. The fraction $\frac{6}{8}$ is not in lowest terms, because 2 divides both 6 and 8 exactly.

We can **simplify** (or **reduce**) a fraction that is not in lowest terms by dividing both its numerator and denominator by the same number. For example, to simplify the fraction $\frac{6}{8}$, we divide both numerator and denominator by 2:

$$\frac{6}{8} = \frac{6 \div 2}{8 \div 2} = \frac{3}{4}$$

From Figure I-2, we see that $\frac{6}{8}$ and $\frac{3}{4}$ are equal fractions, because each one represents the same part of the rectangle.

FIGURE I-2

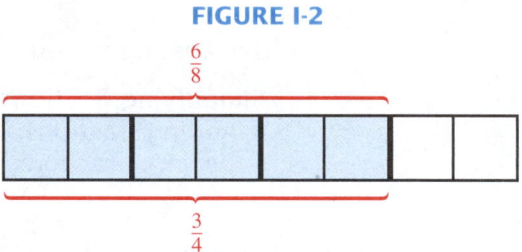

We can use the fact that nonprime numbers can be written as the product of other natural numbers to simplify fractions. For example, 12 can be written as the product of 4 and 3:

$$12 = 4 \cdot 3$$

When a number has been written as the product of other natural numbers, we say that it has been **factored.** The numbers 3 and 4 are called **factors** of 12. When a number is written as the product of prime numbers, we say that the number is written in **prime-factored form.**

EXAMPLE 1

Write 210 in prime-factored form.

Solution   We can write 210 as the product of 21 and 10 and proceed as follows:

$$210 = \mathbf{21 \cdot 10}$$
$$210 = \mathbf{3 \cdot 7 \cdot 2 \cdot 5} \quad \text{Factor 21 as } 3 \cdot 7 \text{ and factor 10 as } 2 \cdot 5.$$

Since 210 is now written as the product of prime numbers, its prime-factored form is $3 \cdot 7 \cdot 2 \cdot 5$. ∎

To simplify a fraction, we factor its numerator and denominator, and then divide out all common factors that appear in both the numerator and denominator. To simplify the fractions $\frac{6}{8}$ and $\frac{15}{18}$, for example, we proceed as follows:

$$\frac{6}{8} = \frac{3 \cdot 2}{4 \cdot 2} = \frac{3 \cdot \overset{1}{\cancel{2}}}{4 \cdot \cancel{2}} = \frac{3}{4} \qquad \text{and} \qquad \frac{15}{18} = \frac{5 \cdot 3}{6 \cdot 3} = \frac{5 \cdot \overset{1}{\cancel{3}}}{6 \cdot \cancel{3}} = \frac{5}{6}$$

 WARNING! Remember that a fraction is in lowest terms only when its numerator and denominator have no common factors.

EXAMPLE 2

a. To simplify $\frac{6}{30}$, we factor the numerator and denominator and divide out the common factor of 6.

$$\frac{6}{30} = \frac{6 \cdot 1}{6 \cdot 5} = \frac{\overset{1}{\cancel{6}} \cdot 1}{\underset{1}{\cancel{6}} \cdot 5} = \frac{1}{5}$$

b. To show that $\frac{33}{40}$ is in lowest terms, we must show that the numerator and the denominator share no common factors, by writing both the numerator and denominator in prime-factored form.

$$\frac{33}{40} = \frac{3 \cdot 11}{2 \cdot 2 \cdot 2 \cdot 5}$$

Since the numerator and denominator have no common factors, $\frac{33}{40}$ is in lowest terms.   ∎

The previous examples illustrate the **fundamental property of fractions.**

The fundamental property of fractions

If a, b, and c are real numbers, then

$$\frac{a \cdot c}{b \cdot c} = \frac{a}{b} \quad (b \neq 0, c \neq 0)$$

Multiplying Fractions

Multiplying fractions

To multiply two fractions, we multiply the numerators and multiply the denominators. In symbols, if a, b, c, and d are real numbers, then

$$\frac{a}{b} \cdot \frac{c}{d} = \frac{a \cdot c}{b \cdot d} \quad (b \neq 0, d \neq 0)$$

For example,

$$\frac{4}{7} \cdot \frac{2}{3} = \frac{4 \cdot 2}{7 \cdot 3} \qquad \text{and} \qquad \frac{4}{5} \cdot \frac{13}{9} = \frac{4 \cdot 13}{5 \cdot 9}$$

$$= \frac{8}{21} \qquad\qquad\qquad\qquad = \frac{52}{45}$$

FIGURE I-3

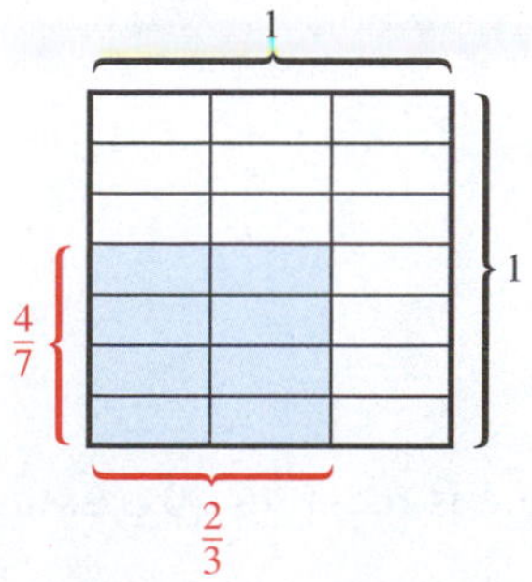

To justify the rule for multiplying fractions, we consider the square in Figure I-3. Since the length of each side of the square is 1 unit and the area is the product of the lengths of two sides, the area is 1 square unit.

If this square is divided into 3 equal parts vertically and 7 equal parts horizontally, it is divided into 21 equal parts, and each represents $\frac{1}{21}$ of the total area. The area of the shaded rectangle in the square is $\frac{8}{21}$, because it contains 8 of the 21 parts. The width w of the shaded rectangle is $\frac{4}{7}$, its length l is $\frac{2}{3}$, and its area A is the product of l and w:

$$A = l \cdot w$$

$$\frac{8}{21} = \frac{4}{7} \cdot \frac{2}{3}$$

This suggests that we can find the product of

$$\frac{4}{7} \quad \text{and} \quad \frac{2}{3}$$

by multiplying their numerators and multiplying their denominators.

Fractions with a numerator that is smaller than the denominator, such as $\frac{8}{21}$, are called **proper fractions**. Fractions with a numerator that is larger than the denominator, such as $\frac{52}{45}$, are called **improper fractions**.

EXAMPLE 3

a. $\dfrac{3}{7} \cdot \dfrac{13}{5} = \dfrac{3 \cdot 13}{7 \cdot 5}$ Multiply the numerators and multiply the denominators.

$= \dfrac{39}{35}$ Do the multiplication in the numerator.
Do the multiplication in the denominator.

b. $5 \cdot \dfrac{3}{15} = \dfrac{5}{1} \cdot \dfrac{3}{15}$ Write 5 as the improper fraction $\frac{5}{1}$.

$= \dfrac{5 \cdot 3}{1 \cdot 15}$ Multiply the numerators and multiply the denominators.

$= \dfrac{5 \cdot 3}{1 \cdot 5 \cdot 3}$ To attempt to simplify the fraction, factor the denominator.

$= \dfrac{\overset{1}{\cancel{5}} \cdot \overset{1}{\cancel{3}}}{1 \cdot \underset{1}{\cancel{5}} \cdot \underset{1}{\cancel{3}}}$ Divide out the common factors of 3 and 5.

$= \dfrac{1}{1}$

$= 1$ ∎

EXAMPLE 4

European travel. Out of 36 students in a history class, three-fourths have signed up for a trip to Europe. If there are 30 places available on the chartered flight, will there be room for one more student?

Solution We first find three-fourths of 36 by multiplying 36 by $\frac{3}{4}$.

$$\frac{3}{4} \cdot 36 = \frac{3}{4} \cdot \frac{36}{1}$$ Write 36 as $\frac{36}{1}$.

$$= \frac{3 \cdot 36}{4 \cdot 1}$$ Multiply the numerators and multiply the denominators.

$$= \frac{3 \cdot 4 \cdot 9}{4 \cdot 1}$$ To simplify the fractions, factor the numerator.

$$= \frac{3 \cdot \overset{1}{\cancel{4}} \cdot 9}{\underset{1}{\cancel{4}} \cdot 1}$$ Divide out the common factor of 4.

$$= \frac{27}{1}$$

$$= 27$$

Twenty-seven students plan to go on the trip. Since there is room for 30 passengers, there is plenty of room for one more. ∎

Dividing Fractions

One number is called the **reciprocal** of another if their product is 1. For example, $\frac{3}{5}$ is the reciprocal of $\frac{5}{3}$, because

$$\frac{3}{5} \cdot \frac{5}{3} = \frac{15}{15} = 1$$

Dividing fractions

To divide two fractions, we multiply the first fraction by the reciprocal of the second fraction. In symbols, if a, b, c, and d are real numbers, then

$$\frac{a}{b} \div \frac{c}{d} = \frac{a}{b} \cdot \frac{d}{c} \quad (b \neq 0,\ c \neq 0,\ \text{and}\ d \neq 0)$$

E X A M P L E   5

a. $\dfrac{3}{5} \div \dfrac{6}{5} = \dfrac{3}{5} \cdot \dfrac{5}{6}$ Multiply $\frac{3}{5}$ by the reciprocal of $\frac{6}{5}$.

$\qquad = \dfrac{3 \cdot 5}{5 \cdot 6}$ Multiply the numerators and multiply the denominators.

$\qquad = \dfrac{3 \cdot 5}{5 \cdot 2 \cdot 3}$ Factor the denominator.

$\qquad = \dfrac{\overset{1}{\cancel{3}} \cdot \overset{1}{\cancel{5}}}{\underset{1}{\cancel{5}} \cdot 2 \cdot \underset{1}{\cancel{3}}}$ Divide out the common factors of 3 and 5.

$\qquad = \dfrac{1}{2}$

b. $\dfrac{15}{7} \div 10 = \dfrac{15}{7} \div \dfrac{10}{1}$ Write 10 as the improper fraction $\frac{10}{1}$.

$\qquad = \dfrac{15}{7} \cdot \dfrac{1}{10}$ Multiply $\frac{15}{7}$ by the reciprocal of $\frac{10}{1}$.

$\qquad = \dfrac{15 \cdot 1}{7 \cdot 10}$ Multiply the numerators and multiply the denominators.

$\qquad = \dfrac{3 \cdot \overset{1}{\cancel{5}} \cdot 1}{7 \cdot 2 \cdot \underset{1}{\cancel{5}}}$ Factor the numerator and denominator.

$\qquad = \dfrac{3}{14}$ Divide out the common factor of 5 and simplify.   ∎

Adding Fractions

Adding fractions with like denominators

To add two fractions with the same denominator, we add the numerators and keep the common denominator. In symbols, if a, b, and d are real numbers, then

$$\frac{a}{d} + \frac{b}{d} = \frac{a+b}{d} \quad (d \neq 0)$$

For example,

$$\frac{3}{7} + \frac{2}{7} = \frac{3+2}{7}$$

$$= \frac{5}{7}$$

Figure I-4 shows why $\frac{3}{7} + \frac{2}{7} = \frac{5}{7}$.

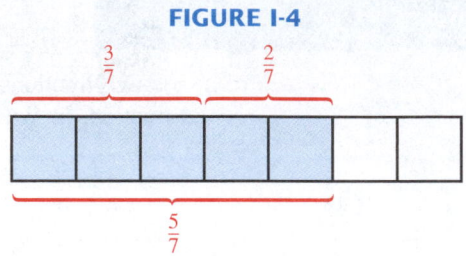

FIGURE I-4

To add fractions with unlike denominators, we rewrite the fractions so they have the same denominator. For example, we can multiply both the numerator and denominator of the fraction $\frac{1}{3}$ by 5, to obtain an equivalent fraction with a denominator of 15:

$$\frac{1}{3} = \frac{1 \cdot 5}{3 \cdot 5}$$

$$= \frac{5}{15}$$

Similarly, we can rewrite the fraction $\frac{1}{5}$ as an equal fraction with a denominator of 15. We multiply both the numerator and the denominator by 3:

$$\frac{1}{5} = \frac{1 \cdot 3}{5 \cdot 3}$$

$$= \frac{3}{15}$$

Because 15 is the smallest number that can be used as a denominator for $\frac{1}{3}$ and $\frac{1}{5}$, it is called the **least** or **lowest common denominator (LCD).**

To add the fractions $\frac{1}{3}$ and $\frac{1}{5}$, we rewrite each fraction as an equivalent fraction having a denominator of 15 and add the results:

$$\frac{1}{3} + \frac{1}{5} = \frac{1 \cdot 5}{3 \cdot 5} + \frac{1 \cdot 3}{5 \cdot 3}$$

$$= \frac{5}{15} + \frac{3}{15}$$

$$= \frac{5 + 3}{15} \qquad \text{\textcolor{red}{Add the numerators and keep the common denominator.}}$$

$$= \frac{8}{15}$$

EXAMPLE 6   Add $\dfrac{3}{10} + \dfrac{5}{28}$.

Solution   To find the least common denominator (LCD), we find the prime factorization of both denominators and use each prime factor the greatest number of times it appears in either factorization:

$$\left. \begin{array}{l} 10 = 2 \cdot 5 \\ 28 = 2 \cdot 2 \cdot 7 \end{array} \right\} \qquad \text{LCD} = 2 \cdot 2 \cdot 5 \cdot 7 = 140$$

Since 140 is the smallest number that 10 and 28 divide exactly, we write both fractions as fractions with the LCD of 140.

$$\frac{3}{10} + \frac{5}{28} = \frac{3 \cdot 14}{10 \cdot 14} + \frac{5 \cdot 5}{28 \cdot 5} \qquad \text{Write each fraction as a fraction with a denominator of 140.}$$

$$= \frac{42}{140} + \frac{25}{140} \qquad \text{Do the multiplications.}$$

$$= \frac{42 + 25}{140} \qquad \text{Add the numerators and keep the common denominator.}$$

$$= \frac{67}{140} \qquad \text{Do the addition.}$$

Since 67 is a prime number, it has no common factor with 140. Thus, $\frac{67}{140}$ is in lowest terms and cannot be simplified. ■

Subtracting Fractions

Subtracting fractions with like denominators

To subtract two fractions with the same denominator, we subtract their numerators and keep their common denominator. In symbols, if a, b, and d are real numbers, then

$$\frac{a}{d} - \frac{b}{d} = \frac{a - b}{d} \qquad (d \neq 0)$$

For example,

$$\frac{7}{9} - \frac{2}{9} = \frac{7 - 2}{9} \qquad \text{Subtract the numerators and keep the common denominator.}$$

$$= \frac{5}{9}$$

To subtract fractions with unlike denominators, we write them as equivalent fractions with a common denominator. For example, to subtract $\frac{2}{5}$ from $\frac{3}{4}$, we write $\frac{3}{4} - \frac{2}{5}$, find the LCD of 20, and proceed as follows:

$$\frac{3}{4} - \frac{2}{5} = \frac{3 \cdot 5}{4 \cdot 5} - \frac{2 \cdot 4}{5 \cdot 4}$$

$$= \frac{15}{20} - \frac{8}{20}$$

$$= \frac{15 - 8}{20} \qquad \text{Subtract the numerators and keep the common denominator.}$$

$$= \frac{7}{20}$$

EXAMPLE 7   Subtract 5 from $\dfrac{23}{3}$.

Solution

$$\dfrac{23}{3} - 5 = \dfrac{23}{3} - \dfrac{5}{1} \qquad \text{Write 5 as the improper fraction } \tfrac{5}{1}.$$

$$= \dfrac{23}{3} - \dfrac{5 \cdot 3}{1 \cdot 3} \qquad \text{Write } \tfrac{5}{1} \text{ as a fraction with a denominator of 3.}$$

$$= \dfrac{23}{3} - \dfrac{15}{3} \qquad \text{Do the multiplication.}$$

$$= \dfrac{23 - 15}{3} \qquad \text{Subtract the numerators and keep the common denominator.}$$

$$= \dfrac{8}{3}$$

Mixed Numbers

The **mixed number** $3\tfrac{1}{2}$ represents the sum of 3 and $\tfrac{1}{2}$. We can write $3\tfrac{1}{2}$ as an improper fraction as follows:

$$3\dfrac{1}{2} = 3 + \dfrac{1}{2}$$

$$= \dfrac{6}{2} + \dfrac{1}{2} \qquad 3 = \tfrac{6}{2}.$$

$$= \dfrac{6 + 1}{2} \qquad \text{Add the numerators and keep the common denominator.}$$

$$= \dfrac{7}{2}$$

To write the fraction $\tfrac{19}{5}$ as a mixed number, we divide 19 by 5 to get 3, with a remainder of 4. Thus,

$$\dfrac{19}{5} = 3 + \dfrac{4}{5}$$

$$= 3\dfrac{4}{5}$$

EXAMPLE 8   Add $2\dfrac{1}{4}$ and $1\dfrac{1}{3}$.

Solution   We first change each mixed number to an improper fraction:

$$2\dfrac{1}{4} = 2 + \dfrac{1}{4} \qquad\qquad 1\dfrac{1}{3} = 1 + \dfrac{1}{3}$$

$$= \dfrac{8}{4} + \dfrac{1}{4} \qquad\qquad = \dfrac{3}{3} + \dfrac{1}{3}$$

$$= \dfrac{9}{4} \qquad\qquad\qquad = \dfrac{4}{3}$$

and then add the fractions.

$$2\frac{1}{4} + 1\frac{1}{3} = \frac{9}{4} + \frac{4}{3}$$

$$= \frac{9 \cdot 3}{4 \cdot 3} + \frac{4 \cdot 4}{3 \cdot 4} \qquad \text{The LCD is 12.}$$

$$= \frac{27}{12} + \frac{16}{12}$$

$$= \frac{43}{12}$$

Finally, we change $\frac{43}{12}$ to a mixed number.

$$\frac{43}{12} = 3 + \frac{7}{12} = 3\frac{7}{12}$$

■

EXAMPLE 9

Fencing land. The three sides of a triangular piece of land measure $33\frac{1}{4}$, $57\frac{3}{4}$, and $72\frac{1}{2}$ meters. How much fencing will be needed to enclose the area?

Solution   We can find the sum of the lengths by adding the whole-number parts and the fractional parts of the dimensions separately:

$$33\frac{1}{4} + 57\frac{3}{4} + 72\frac{1}{2} = 33 + 57 + 72 + \frac{1}{4} + \frac{3}{4} + \frac{1}{2}$$

$$= 162 + \frac{1}{4} + \frac{3}{4} + \frac{2}{4} \qquad \text{Change } \tfrac{1}{2} \text{ to } \tfrac{2}{4} \text{ to obtain a common denominator}$$

$$= 162 + \frac{6}{4} \qquad \text{Add the fractions by adding the numerators and keeping the common denominator.}$$

$$= 162 + \frac{3}{2} \qquad \frac{6}{4} = \frac{2 \cdot 3}{2 \cdot 2} = \frac{\overset{1}{\cancel{2}} \cdot 3}{\underset{1}{\cancel{2}} \cdot 2} = \frac{3}{2}.$$

$$= 162 + 1\frac{1}{2} \qquad \text{Change } \tfrac{3}{2} \text{ to a mixed number.}$$

$$= 163\frac{1}{2}$$

It will require $163\frac{1}{2}$ meters of fencing to enclose the triangular area.

■

Decimal Fractions

Rational numbers can always be changed into decimal fractions. For example, to change $\frac{1}{4}$ and $\frac{5}{22}$ to decimal fractions, we use long division:

```
     0.25                0.22727. . .
  4)1.00             22)5.00000
     8                  44
     ──                 ──
     20                 60
     20                 44
     ──                 ──
     0                  160
                        154
                        ───
                        60
                        44
                        ──
                        160
```

The decimal fraction 0.25 is called a **terminating decimal,** and the decimal fraction 0.2272727. . . (often written as $0.2\overline{27}$) is called a **repeating decimal,** because it repeats the block of digits 27. Every rational number can be changed into either a **terminating** or a **repeating decimal.**

Terminating decimals	*Repeating decimals*
$\dfrac{1}{2} = 0.5$	$\dfrac{1}{3} = 0.3333. . .$ or $0.\overline{3}$
$\dfrac{3}{4} = 0.75$	$\dfrac{1}{6} = 0.16666. . .$ or $0.1\overline{6}$
$\dfrac{5}{8} = 0.625$	$\dfrac{5}{22} = 0.2272727. . .$ or $0.2\overline{27}$

The decimal 0.5 has one **decimal place,** because it has one digit to the right of the decimal point. The decimal 0.75 has two decimal places, and 0.625 has three.

To *add* or *subtract* decimal fractions, we first align their decimal points and then add or subtract.

$$
\begin{array}{r}
25.568 \\
2.74 \\
\hline
28.308
\end{array}
\qquad
\begin{array}{r}
25.568 \\
-\ 2.74 \\
\hline
22.828
\end{array}
$$

To do the previous operations with a calculator, we would press these keys:

25.568 + 2.74 =   and   25.568 − 2.74 =

To *multiply* decimal fractions, we multiply the numbers and then place the decimal point so that the number of decimal places in the answer is equal to the sum of the decimal places in the factors.

$$
\begin{array}{r}
3.453 \\
9.25 \\
\hline
17265 \\
6906 \\
31077 \\
\hline
31.94025
\end{array}
$$

3.453 ← Here there are three decimal places.
9.25 ← Here there are two decimal places.
31.94025 ← The product has 3 + 2 = 5 decimal places.

To do the previous multiplication with a calculator, we would press these keys:

3.453 × 9.25 =

To *divide* decimal fractions, we move the decimal point in the divisor to the right to make the divisor a whole number. We then move the decimal point in the dividend the same number of places to the right.

$1.23\overline{)30.258}$ Move the decimal point in both the divisor and the dividend two places to the right.

We align the decimal point in the quotient with the repositioned decimal point in the dividend and then use long division.

$$
\begin{array}{r}
24.6 \\
123\overline{)3025.8} \\
\underline{246} \\
565 \\
\underline{492} \\
73\ 8 \\
\underline{73\ 8} \\
0
\end{array}
$$

To do the previous division with a calculator, we would press these keys:

30.258 ÷ 1.23 =

Rounding Decimals

When decimal fractions are long, we often **round** them to a specific number of decimal places. For example, the decimal fraction 25.36124 rounded to one place (or to the nearest tenth) is 25.4. Rounded to two places (or to the nearest hundredth), the decimal is 25.36. To round decimals, we use the following rules.

Rounding decimals

1. Determine to how many decimal places you wish to round.
2. Look at the first digit to the right of that decimal place.
3. If that digit is 4 or less, drop it and all digits that follow. If it is 5 or greater, add 1 to the digit in the position to which you wish to round, and drop all digits that follow.

Applications

A **percent** is the numerator of a fraction with a denominator of 100. For example, $6\frac{1}{4}$ percent, written $6\frac{1}{4}\%$, is the fraction $\frac{6.25}{100}$, or the decimal 0.0625. In problems involving percent, the word *of* often indicates multiplication. For example, $6\frac{1}{4}\%$ of 8,500 is the product 0.0625(8,500).

EXAMPLE 10

Auto loans. Juan signs a one-year note to borrow $8,500 to buy a car. If the interest rate is $6\frac{1}{4}\%$, how much interest will he pay?

Solution For the privilege of using the bank's money for one year, Juan must pay $6\frac{1}{4}\%$ of $8,500. We calculate the interest I as follows:

$$I = 6\frac{1}{4}\% \text{ of } 8{,}500$$
$$= 0.0625 \cdot 8{,}500 \quad \text{Here, } of \text{ means } times.$$
$$= 531.25$$

He will pay $531.25 interest.

STUDY SET

Appendix I

PRACTICE

In Exercises 1–8, write each fraction in lowest terms. If the fraction is already in lowest terms, so indicate.

1. $\dfrac{6}{12}$

2. $\dfrac{3}{9}$

3. $\dfrac{15}{20}$

4. $\dfrac{22}{77}$

5. $\dfrac{24}{18}$

6. $\dfrac{35}{14}$

7. $\dfrac{72}{64}$

8. $\dfrac{26}{21}$

In Exercises 9–20, do each multiplication. Simplify each result when possible.

9. $\dfrac{1}{2} \cdot \dfrac{3}{5}$

10. $\dfrac{3}{4} \cdot \dfrac{5}{7}$

11. $\dfrac{4}{3} \cdot \dfrac{6}{5}$

12. $\dfrac{7}{8} \cdot \dfrac{6}{15}$

13. $\dfrac{5}{12} \cdot \dfrac{18}{5}$

14. $\dfrac{5}{4} \cdot \dfrac{12}{10}$

15. $\dfrac{17}{34} \cdot \dfrac{3}{6}$

16. $\dfrac{21}{14} \cdot \dfrac{3}{6}$

17. $12 \cdot \dfrac{5}{6}$ **18.** $9 \cdot \dfrac{7}{12}$

19. $\dfrac{10}{21} \cdot 14$ **20.** $\dfrac{5}{24} \cdot 16$

57. $3\dfrac{3}{4} - 2\dfrac{1}{2}$ **58.** $15\dfrac{5}{6} + 11\dfrac{5}{8}$

59. $8\dfrac{2}{9} - 7\dfrac{2}{3}$ **60.** $3\dfrac{4}{5} - 3\dfrac{1}{10}$

In Exercises 21–32, do each division. Simplify each result when possible.

21. $\dfrac{3}{5} \div \dfrac{2}{3}$ **22.** $\dfrac{4}{5} \div \dfrac{3}{7}$

23. $\dfrac{3}{4} \div \dfrac{6}{5}$ **24.** $\dfrac{3}{8} \div \dfrac{15}{28}$

25. $\dfrac{2}{13} \div \dfrac{8}{13}$ **26.** $\dfrac{4}{7} \div \dfrac{20}{21}$

27. $\dfrac{21}{35} \div \dfrac{3}{14}$ **28.** $\dfrac{23}{25} \div \dfrac{46}{5}$

29. $6 \div \dfrac{3}{14}$ **30.** $23 \div \dfrac{46}{5}$

31. $\dfrac{42}{30} \div 7$ **32.** $\dfrac{34}{8} \div 17$

In Exercises 33–60, do each addition or subtraction. Simplify each result when possible.

33. $\dfrac{3}{5} + \dfrac{3}{5}$ **34.** $\dfrac{4}{7} - \dfrac{2}{7}$

35. $\dfrac{4}{13} - \dfrac{3}{13}$ **36.** $\dfrac{2}{11} + \dfrac{9}{11}$

37. $\dfrac{1}{6} + \dfrac{1}{24}$ **38.** $\dfrac{17}{25} - \dfrac{2}{5}$

39. $\dfrac{3}{5} + \dfrac{2}{3}$ **40.** $\dfrac{4}{3} + \dfrac{7}{2}$

41. $\dfrac{9}{4} - \dfrac{5}{6}$ **42.** $\dfrac{2}{15} + \dfrac{7}{9}$

43. $\dfrac{7}{10} - \dfrac{1}{14}$ **44.** $\dfrac{7}{25} + \dfrac{3}{10}$

45. $\dfrac{5}{14} - \dfrac{4}{21}$ **46.** $\dfrac{2}{33} + \dfrac{3}{22}$

47. $3 - \dfrac{3}{4}$ **48.** $5 + \dfrac{21}{5}$

49. $\dfrac{17}{3} + 4$ **50.** $\dfrac{13}{9} - 1$

51. $\dfrac{3}{15} + \dfrac{6}{10}$ **52.** $\dfrac{7}{5} - \dfrac{2}{15}$

53. $4\dfrac{3}{5} + \dfrac{3}{5}$ **54.** $2\dfrac{1}{8} + \dfrac{3}{8}$

55. $3\dfrac{1}{3} - 1\dfrac{2}{3}$ **56.** $5\dfrac{1}{7} - 3\dfrac{2}{7}$

APPLICATIONS

61. PERIMETER OF A TRIANGLE Each side of a triangle measures $2\dfrac{3}{7}$ centimeters. Find its perimeter (the sum of the lengths of its three sides).

62. BUYING FENCING Each side of a square field measures $30\dfrac{2}{5}$ meters. How many meters of fencing is needed to enclose the field?

63. RUNNING A RACE Jim has run $6\dfrac{3}{10}$ kilometers of a 10-kilometer race. How far is the finish line?

64. SPRING PLOWING A farmer has plowed $12\dfrac{1}{3}$ acres of a $43\dfrac{1}{2}$-acre field. How much more needs to be plowed?

65. PERIMETER OF A GARDEN The four sides of a garden measure $7\dfrac{2}{3}$, $15\dfrac{1}{4}$, $19\dfrac{1}{2}$, and $10\dfrac{3}{4}$ feet. Find the length of the fence needed to enclose the garden.

66. MAKING CLOTHES A clothing designer requires $3\dfrac{1}{4}$ yards of material for each dress he makes. How much material will be used to make 14 dresses?

67. MINORITY POPULATION In Illinois, 22% of the 11,431,000 citizens are nonwhite. How many are nonwhite?

68. QUALITY CONTROL Reject rates are high in the manufacture of active-matrix color liquid crystal computer displays. If 23% of a production run of 17,500 units are defective, how many units are acceptable?

69. FREEZE DRYING Almost all of the water must be removed when food is preserved by freeze drying. Find the weight of the water removed from 750 pounds of a food that is 36% water.

70. PLANNING FOR GROWTH This year, sales at Positronics Corporation totaled $18.7 million. If the company's prediction of 12% annual growth is true, what will be next year's sales?

PRACTICE

In Exercises 71–78, do each operation.

71. $23.45 + 135.2$ **72.** $345.213 - 27.35$

73. $67.235 - 22.45$ **74.** $12.17 + 3.457$

75. $3.4 \cdot 13.2$ **76.** $4.21 \cdot 2.73$

77. $0.23 \overline{)1.0465}$ **78.** $4.7 \overline{)10.857}$

In Exercises 79–86, use a calculator to do each operation, and round the answer to two decimal places.

79. 323.24 + 27.2543

80. 843.45213 − 712.765

81. 55.77443 − 0.568245

82. 0.62317 + 1.3316

83. 25.25 · 132.179

84. 234.874 · 242.46473

85. 0.456)‾4.5694323‾

86. 43.225)‾32.465758‾

 APPLICATIONS

In Exercises 87–100, use a calculator to solve each problem. Round numeric answers to two decimal places.

87. FINDING AREAS Find the area of a square with a side that is 62.17 feet long. (*Hint: A = s · s.*)

88. SPEED SKATING In tryouts for the Olympics, a speed skater had times of 44.47, 43.24, 42.77, and 42.05 seconds. Find the average time. (*Hint:* Add the numbers and divide by 4.)

89. COST OF GASOLINE Juan drove his car 15,675.2 miles last year, averaging 25.5 miles per gallon of gasoline. The average cost of gasoline was $1.27 per gallon. Find the fuel cost to drive the car.

90. PAYING TAXES A woman earns $48,712.32 in taxable income. She must pay 15% tax on the first $23,000 and 28% on the rest. In addition, she must pay a Social Security tax of 15.4% on the total amount. How much tax will she need to pay?

91. SEALING ASPHALT A rectangular parking lot is 253.5 feet long and 178.5 feet wide. A 55-gallon drum of asphalt sealer covers 4,000 square feet and costs $97.50. Find the cost to seal the parking lot. Sealer can be purchased only in full drums.

92. INSTALLING CARPET What will it cost to carpet a 23-by-17.5-foot living room and a 17.5-by-14-foot dining room with carpet that costs $29.79 per square yard? One square yard is 9 square feet.

93. INVENTORY COSTS Each television costs $3.25 per day for warehouse storage. What are the warehousing costs to store 37 television sets for three weeks?

94. MANUFACTURING PROFITS A manufacturer of computer memory boards has a profit of $37.50 on each standard-capacity memory board and $57.35 for each high-capacity board. The sales department has orders for 2,530 standard boards and 1,670 high-capacity boards. To receive the greater profit, which order should be filled first?

95. DAIRY PRODUCTION A Holstein cow will produce 7,600 pounds of milk each year, with a $3\frac{1}{2}$% butterfat content. Each year, a Guernsey cow will produce about 6,500 pounds of milk that is 5% butterfat. Which cow produces more butterfat?

96. FEEDING COWS Each year, a typical dairy cow will eat 12,000 pounds of food that is 57% silage. To feed a herd of 30 cows, how many pounds of silage will a farmer use in a year?

97. COMPARING BIDS Two contractors bid on a home remodeling project. The first bids $9,350 for the entire job. The second contractor will work for $27.50 per hour, plus $4,500 for materials. He estimates that the job will take 150 hours. Which contractor has the lower bid?

98. CHOOSING A FURNACE A high-efficiency home heating system can be installed for $4,170, with an average monthly heating bill of $57.50. A regular furnace can be installed for $1,730, but monthly heating bills will average $107.75. After three years, which system is more expensive?

99. CHOOSING A FURNACE Refer to Exercise 98. Which furnace system is the more expensive after five years?

100. MORTGAGE INTEREST A mortgage carries an annual interest rate of 9.75%. Each month, the bank's interest charge is one-twelfth of 9.75% of the outstanding balance, currently $72,363. Find the amount of interest that will be paid this month.

WRITING

101. Describe how you would find the common denominator of two fractions.

102. Explain how to convert an improper fraction into a mixed number.

103. Explain how to convert a mixed number into an improper fraction.

104. Explain how you would decide which of two decimal fractions is the larger.

Synthetic Division

In this appendix, you will learn about

Synthetic division ■ The remainder theorem ■ The factor theorem

Introduction   In Section 4.8, we discussed how to divide polynomials by polynomials. In this appendix, we will discuss a shortcut method, called **synthetic division,** that we can use to divide a polynomial by a binomial of the form $x - r$.

Synthetic Division

To see how synthetic division works, we consider the division of $4x^3 - 5x^2 - 11x + 20$ by $x - 2$.

$$\begin{array}{r} 4x^2 + 3x\ - 5 \\ x - 2 \overline{)4x^3 - 5x^2 - 11x + 20} \\ \underline{4x^3 - 8x^2} \\ 3x^2 - 11x \\ \underline{3x^2 - 6x} \\ -\ 5x + 20 \\ \underline{-\ 5x + 10} \\ 10 \ \text{(remainder)} \end{array}$$

$$\begin{array}{r} 4\ \ 3 - 5 \\ 1 - 2 \overline{)4 - 5 - 11\ \ \ 20} \\ \underline{4 - 8} \\ 3 - 11 \\ \underline{3 - 6} \\ -\ 5\ \ 20 \\ \underline{-\ 5\ \ 10} \\ 10 \ \text{(remainder)} \end{array}$$

On the left is the long division, and on the right is the same division with the variables and their exponents removed. The various powers of x can be remembered without actually writing them, because the exponents of the terms in the divisor, dividend, and quotient were written in descending order.

We can further shorten the version on the right. The numbers printed in color need not be written, because they are duplicates of the numbers above them. Thus, we can write the division in the following form:

$$\begin{array}{r} 4\ \ 3 - 5 \\ 1 - 2 \overline{)4 - 5 - 11\ \ \ 20} \\ \underline{-\ 8} \\ 3 \\ \underline{-6} \\ -5 \\ \underline{10} \\ 10 \end{array}$$

We can shorten the process further by compressing the work vertically and eliminating the 1 (the coefficient of x in the divisor):

$$
\begin{array}{r}
4 \quad\ \ 3 \quad -5 \\
-2)\overline{4 \quad -5 \quad -11 \quad 20} \\
-8 \quad -6 \quad 10 \\
\hline
3 \quad -5 \quad 10
\end{array}
$$

If we write the 4 in the quotient on the bottom line, the bottom line gives the coefficients of the quotient and the remainder. If we eliminate the top line, the division appears as follows:

$$
\begin{array}{r}
\underline{-2|} \quad 4 \quad -5 \quad -11 \quad 20 \\
-8 \quad -\ 6 \quad 10 \\
\hline
4 \quad\ \ 3 \quad -\ 5 \quad 10
\end{array}
$$

The bottom line was obtained by subtracting the middle line from the top line. If we replace the -2 in the divisor by $+2$, the division process will reverse the signs of every entry in the middle line, and then the bottom line can be obtained by addition. This gives the final form of the synthetic division.

$$
\begin{array}{r}
\underline{+2|} \quad 4 \quad -5 \quad -11 \quad\ \ 20 \\
8 \quad\ \ \ 6 \quad -10 \\
\hline
4 \quad\ \ 3 \quad -5 \quad\ \ 10
\end{array}
$$

The coefficients of the dividend.

The coefficients of the quotient and the remainder.

Thus,

$$
\frac{4x^3 - 5x^2 - 11x + 20}{x - 2} = 4x^2 + 3x - 5 + \frac{10}{x - 2}
$$

EXAMPLE 1

Divide $6x^2 + 5x - 2$ by $x - 5$.

Solution   We write the coefficients in the dividend and the 5 in the divisor in the following form:

$$
\begin{array}{r}
\underline{5|} \quad 6 \quad 5 \quad -2 \\
\hline

\end{array}
$$

Then we follow these steps:

$$
\begin{array}{r}
\underline{5|} \quad 6 \quad 5 \quad -2 \\
\hline
6
\end{array}
$$
Begin by bringing down the 6.

$$
\begin{array}{r}
\underline{5|} \quad 6 \quad\ \ 5 \quad -2 \\
30 \\
\hline
6
\end{array}
$$
Multiply 5 by 6 to get 30.

$$
\begin{array}{r}
\underline{5|} \quad 6 \quad\ \ 5 \quad -2 \\
30 \\
\hline
6 \quad 35
\end{array}
$$
Add 5 and 30 to get 35.

$$
\begin{array}{r}
\underline{5|} \quad 6 \quad\ \ 5 \quad -2 \\
30 \quad 175 \\
\hline
6 \quad 35
\end{array}
$$
Multiply 35 by 5 to get 175.

$$
\begin{array}{r}
\underline{5|} \quad 6 \quad\ \ 5 \quad -2 \\
30 \quad 175 \\
\hline
6 \quad 35 \quad 173
\end{array}
$$
Add -2 and 175 to get 173.

The numbers 6 and 35 represent the quotient $6x + 35$, and 173 is the remainder. Thus,

$$\frac{6x^2 + 5x - 2}{x - 5} = 6x + 35 + \frac{173}{x - 5}$$

■

EXAMPLE 2 Divide $5x^3 + x^2 - 3$ by $x - 2$.

Solution We begin by writing

$$\underline{2}\begin{array}{cccc} 5 & 1 & \mathbf{0} & -3 \end{array}$$ Write 0 for the coefficient of x, the missing term.

and complete the division as follows:

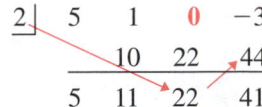

$$\underline{2}\begin{array}{cccc} 5 & 1 & \mathbf{0} & -3 \\ & 10 & & \\ \hline 5 & 11 & & \end{array} \qquad \underline{2}\begin{array}{cccc} 5 & 1 & \mathbf{0} & -3 \\ & 10 & 22 & \\ \hline 5 & 11 & 22 & \end{array} \qquad \underline{2}\begin{array}{cccc} 5 & 1 & \mathbf{0} & -3 \\ & 10 & 22 & 44 \\ \hline 5 & 11 & 22 & 41 \end{array}$$

Thus,

$$\frac{5x^3 + x^2 - 3}{x - 2} = 5x^2 + 11x + 22 + \frac{41}{x - 2}$$

■

EXAMPLE 3 Divide $5x^2 + 6x^3 + 2 - 4x$ by $x + 2$.

Solution First, we write the dividend with the exponents in descending order.

$$6x^3 + 5x^2 - 4x + 2$$

Then we write the divisor in $x - r$ form: $x - (-2)$. Using synthetic division, we begin by writing

$$\underline{-2}\begin{array}{cccc} 6 & 5 & -4 & 2 \\ \hline & & & \end{array}$$

and complete the division.

$$\underline{-2}\begin{array}{cccc} 6 & 5 & -4 & 2 \\ & -12 & 14 & -20 \\ \hline 6 & -7 & 10 & -18 \end{array}$$

Thus,

$$\frac{5x^2 + 6x^3 + 2 - 4x}{x + 2} = 6x^2 - 7x + 10 + \frac{-18}{x + 2}$$

■

The Remainder Theorem

Synthetic division is important in mathematics because of the **remainder theorem.**

Remainder theorem

If a polynomial $P(x)$ is divided by $x - r$, the remainder is $P(r)$.

We illustrate the remainder theorem in the next example.

EXAMPLE 4   Let $P(x) = 2x^3 - 3x^2 - 2x + 1$. Find   **a.** $P(3)$ and   **b.** the remainder when $P(x)$ is divided by $x - 3$.

Solution   **a.** $P(3) = 2(3)^3 - 3(3)^2 - 2(3) + 1$ Substitute 3 for x.

$\qquad\qquad = 2(27) - 3(9) - 6 + 1$

$\qquad\qquad = 54 - 27 - 6 + 1$

$\qquad\qquad = 22$

b.   We use synthetic division to find the remainder when $P(x) = 2x^3 - 3x^2 - 2x + 1$ is divided by $x - 3$.

$$
\begin{array}{r|rrrr}
3 & 2 & -3 & -2 & 1 \\
  &   & 6 & 9 & 21 \\
\hline
  & 2 & 3 & 7 & 22
\end{array}
$$

The remainder is 22.
The results of parts a and b show that when $P(x)$ is divided by $x - 3$, the remainder is $P(3)$.   ∎

It is often easier to find $P(r)$ by using synthetic division than by substituting r for x in $P(x)$. This is especially true if r is a decimal.

The Factor Theorem

Recall that if two quantities are multiplied, each is called a **factor** of the product. Thus, $x - 2$ is one factor of $6x - 12$, because $6(x - 2) = 6x - 12$. A theorem, called the **factor theorem,** tells us how to find one factor of a polynomial if the remainder of a certain division is 0.

Factor theorem

If $P(x)$ is a polynomial in x, then

$P(r) = 0$ if and only if $x - r$ is a factor of $P(x)$

If $P(x)$ is a polynomial in x and if $P(r) = 0$, r is called a **zero of the polynomial.**

EXAMPLE 5   Let $P(x) = 3x^3 - 5x^2 + 3x - 10$. Show that   **a.** $P(2) = 0$ and   **b.** $x - 2$ is a factor of $P(x)$.

Solution   **a.** Use the remainder theorem to evaluate $P(2)$ by dividing $P(x) = 3x^3 - 5x^2 + 3x - 10$ by $x - 2$.

$$
\begin{array}{r|rrrr}
2 & 3 & -5 & 3 & -10 \\
  &   & 6 & 2 & 10 \\
\hline
  & 3 & 1 & 5 & 0
\end{array}
$$

The remainder in this division is 0. By the remainder theorem, the remainder is $P(2)$. Thus, $P(2) = 0$, and 2 is a zero of the polynomial.

b. Because the remainder is 0, the numbers 3, 1, and 5 in the synthetic division in part a represent the quotient $3x^2 + x + 5$. Thus,

$$
\underbrace{(x - 2)}_{\text{Divisor} \,\cdot} \cdot \underbrace{(3x^2 + x + 5)}_{\text{quotient}} + \underbrace{0}_{+ \text{ remainder } =} = \underbrace{3x^3 - 5x^2 + 3x - 10}_{\text{the dividend, } P(x)}
$$

or

$$
(x - 2)(3x^2 + x + 5) = 3x^3 - 5x^2 + 3x - 10
$$

Thus, $x - 2$ is a factor of $P(x)$. ∎

The result in Example 5 is true, because the remainder, $P(2)$, is 0. If the remainder had not been 0, then $x - 2$ would not have been a factor of $P(x)$.

ACCENT ON TECHNOLOGY *Approximating Zeros of Polynomials*

We can use a graphing calculator to approximate the real zeros of a polynomial function $f(x)$. For example, to find the real zeros of $f(x) = 2x^3 - 6x^2 + 7x - 21$, we graph the function as in Figure II-1.

It is clear from the figure that the function f has a zero at $x = 3$.

$$
f(3) = 2(3)^3 - 6(3)^2 + 7(3) - 21 \quad \text{Substitute 3 for } x.
$$
$$
= 2(27) - 6(9) + 21 - 21
$$
$$
= 0
$$

From the factor theorem, we know that $x - 3$ is a factor of the polynomial. To find the other factor, we can synthetically divide by 3.

$$
\begin{array}{r|rrrr}
3 & 2 & -6 & 7 & -21 \\
  &   & 6 & 0 & 21 \\
\hline
  & 2 & 0 & 7 & 0
\end{array}
$$

Thus, $f(x) = (x - 3)(2x^2 + 7)$. Since $2x^2 + 7$ cannot be factored over the real numbers, we can conclude that 3 is the only real zero of the polynomial function.

FIGURE II-1

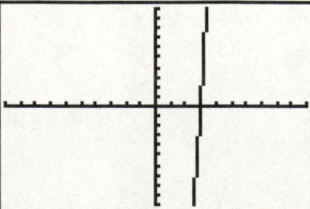

STUDY SET

Appendix II

VOCABULARY

Fill in the blanks to make the statements true.

1. The expression $3x^3 - 2x^2 + 7x + 5$ is a
   _____ in x.

2. If $P(x)$ is a polynomial in x and if $P(x) = 0$, then r is
   called a _____ of the polynomial.

CONCEPTS

Fill in the blanks to make the statements true.

3. If a polynomial $P(x)$ is divided by $x - r$, the remainder is _____.

4. If $P(x)$ is a polynomial in x, then $P(r) = 0$ if and
   only if _____ is a factor of $P(x)$.

NOTATION

Complete each synthetic division.

5.
$$2 \overline{)\begin{array}{rrrr} 6 & 1 & -23 & 2 \end{array}}$$
$$\begin{array}{rrrr} & & & 6 \\ \hline 6 & 13 & 3 & \end{array}$$

6.
$$-3 \overline{)\begin{array}{rrrr} 2 & -4 & -25 & 15 \end{array}}$$
$$\begin{array}{rrrr} & -6 & 30 & \\ \hline & & & 0 \end{array}$$

PRACTICE

*In Exercises 7–20, use synthetic division to do each
division.*

7. $(x^2 + x - 2) \div (x - 1)$

8. $(x^2 + x - 6) \div (x - 2)$

9. $(x^2 - 7x + 12) \div (x - 4)$

10. $(x^2 - 6x + 5) \div (x - 5)$

11. $(x^2 + 8 + 6x) \div (x + 4)$

12. $(x^2 - 15 - 2x) \div (x + 3)$

13. $(x^2 - 5x + 14) \div (x + 2)$

14. $(x^2 + 13x + 42) \div (x + 6)$

15. $(3x^3 - 10x^2 + 5x - 6) \div (x - 3)$

16. $(2x^3 - 9x^2 + 10x - 3) \div (x - 3)$

17. $(2x^3 - 5x - 6) \div (x - 2)$

18. $(4x^3 + 5x^2 - 1) \div (x + 2)$

19. $(5x^2 + 6x^3 + 4) \div (x + 1)$

20. $(4 - 3x^2 + x) \div (x - 4)$

*In Exercises 21–26, use a calculator and synthetic
division to do each division.*

21. $(7.2x^2 - 2.1x + 0.5) \div (x - 0.2)$

22. $(8.1x^2 + 3.2x - 5.7) \div (x - 0.4)$

23. $(2.7x^2 + x - 5.2) \div (x + 1.7)$

24. $(1.3x^2 - 0.5x - 2.3) \div (x + 2.5)$

25. $(9x^3 - 25) \div (x + 57)$

26. $(0.5x^3 + x) \div (x - 2.3)$

*In Exercises 27–34, let $P(x) = 2x^3 - 4x^2 + 2x - 1$.
Evaluate $P(x)$ by substituting the given value of x into the
polynomial and simplifying. Then evaluate the polynomial
by using the remainder theorem and synthetic division.*

27. $P(1)$ 28. $P(2)$

29. $P(-2)$ 30. $P(-1)$

31. $P(3)$ 32. $P(-4)$

33. $P(0)$ 34. $P(4)$

*In Exercises 35–42, let $Q(x) = x^4 - 3x^3 + 2x^2 + x - 3$.
Evaluate $Q(x)$ by substituting the given value of x into the
polynomial and simplifying. Then evaluate the polynomial
by using the remainder theorem and synthetic division.*

35. $Q(-1)$ 36. $Q(1)$

37. $Q(2)$ 38. $Q(-2)$

39. $Q(3)$ 40. $Q(0)$

41. $Q(-3)$ 42. $Q(-4)$

*In Exercises 43–50, use the remainder theorem and syn-
thetic division to find $P(r)$.*

43. $P(x) = x^3 - 4x^2 + x - 2; r = 2$

44. $P(x) = x^3 - 3x^2 + x + 1; r = 1$

45. $P(x) = 2x^3 + x + 2; r = 3$

46. $P(x) = x^3 + x^2 + 1; r = -2$

47. $P(x) = x^4 - 2x^3 + x^2 - 3x + 2; r = -2$

48. $P(x) = x^5 + 3x^4 - x^2 + 1; r = -1$

49. $P(x) = 3x^5 + 1; r = -\frac{1}{2}$

50. $P(x) = 5x^7 - 7x^4 + x^2 + 1; r = 2$

*In Exercises 51–54, use the factor theorem and tell
whether the first expression is a factor of $P(x)$.*

51. $x - 3; P(x) = x^3 - 3x^2 + 5x - 15$

52. $x + 1; P(x) = x^3 + 2x^2 - 2x - 3$ *(Hint: Write $x + 1$
    as $x - (-1)$.)*

53. $x + 2$; $P(x) = 3x^2 - 7x + 4$ (*Hint:* Write $x + 2$ as $x - (-2)$.)

54. x; $P(x) = 7x^3 - 5x^2 - 8x$ (*Hint:* $x = x - 0$.)

🔳 *In Exercises 55–56, use a calculator to work each problem.*

55. Find 2^6 by using synthetic division to evaluate the polynomial $P(x) = x^6$ at $x = 2$. Then check the answer by evaluating 2^6 with a calculator.

56. Find $(-3)^5$ by using synthetic division to evaluate the polynomial $P(x) = x^5$ at $x = -3$. Then check the answer by evaluating $(-3)^5$ with a calculator.

WRITING

57. If you are given $P(x)$, explain how to use synthetic division to calculate $P(a)$.

58. Explain the factor theorem.

Appendix III

Tables

TABLE A Powers and Roots

n	n^2	$\sqrt{n}$	n^3	$\sqrt[3]{n}$	n	n^2	$\sqrt{n}$	n^3	$\sqrt[3]{n}$
1	1	1.000	1	1.000	51	2,601	7.141	132,651	3.708
2	4	1.414	8	1.260	52	2,704	7.211	140,608	3.733
3	9	1.732	27	1.442	53	2,809	7.280	148,877	3.756
4	16	2.000	64	1.587	54	2,916	7.348	157,464	3.780
5	25	2.236	125	1.710	55	3,025	7.416	166,375	3.803
6	36	2.449	216	1.817	56	3,136	7.483	175,616	3.826
7	49	2.646	343	1.913	57	3,249	7.550	185,193	3.849
8	64	2.828	512	2.000	58	3,364	7.616	195,112	3.871
9	81	3.000	729	2.080	59	3,481	7.681	205,379	3.893
10	100	3.162	1,000	2.154	60	3,600	7.746	216,000	3.915
11	121	3.317	1,331	2.224	61	3,721	7.810	226,981	3.936
12	144	3.464	1,728	2.289	62	3,844	7.874	238,328	3.958
13	169	3.606	2,197	2.351	63	3,969	7.937	250,047	3.979
14	196	3.742	2,744	2.410	64	4,096	8.000	262,144	4.000
15	225	3.873	3,375	2.466	65	4,225	8.062	274,625	4.021
16	256	4.000	4,096	2.520	66	4,356	8.124	287,496	4.041
17	289	4.123	4,913	2.571	67	4,489	8.185	300,763	4.062
18	324	4.243	5,832	2.621	68	4,624	8.246	314,432	4.082
19	361	4.359	6,859	2.668	69	4,761	8.307	328,509	4.102
20	400	4.472	8,000	2.714	70	4,900	8.367	343,000	4.121
21	441	4.583	9,261	2.759	71	5,041	8.426	357,911	4.141
22	484	4.690	10,648	2.802	72	5,184	8.485	373,248	4.160
23	529	4.796	12,167	2.844	73	5,329	8.544	389,017	4.179
24	576	4.899	13,824	2.884	74	5,476	8.602	405,224	4.198
25	625	5.000	15,625	2.924	75	5,625	8.660	421,875	4.217
26	676	5.099	17,576	2.962	76	5,776	8.718	438,976	4.236
27	729	5.196	19,683	3.000	77	5,929	8.775	456,533	4.254
28	784	5.292	21,952	3.037	78	6,084	8.832	474,552	4.273
29	841	5.385	24,389	3.072	79	6,241	8.888	493,039	4.291
30	900	5.477	27,000	3.107	80	6,400	8.944	512,000	4.309
31	961	5.568	29,791	3.141	81	6,561	9.000	531,441	4.327
32	1,024	5.657	32,768	3.175	82	6,724	9.055	551,368	4.344
33	1,089	5.745	35,937	3.208	83	6,889	9.110	571,787	4.362
34	1,156	5.831	39,304	3.240	84	7,056	9.165	592,704	4.380
35	1,225	5.916	42,875	3.271	85	7,225	9.220	614,125	4.397
36	1,296	6.000	46,656	3.302	86	7,396	9.274	636,056	4.414
37	1,369	6.083	50,653	3.332	87	7,569	9.327	658,503	4.431
38	1,444	6.164	54,872	3.362	88	7,744	9.381	681,472	4.448
39	1,521	6.245	59,319	3.391	89	7,921	9.434	704,969	4.465
40	1,600	6.325	64,000	3.420	90	8,100	9.487	729,000	4.481
41	1,681	6.403	68,921	3.448	91	8,281	9.539	753,571	4.498
42	1,764	6.481	74,088	3.476	92	8,464	9.592	778,688	4.514
43	1,849	6.557	79,507	3.503	93	8,649	9.644	804,357	4.531
44	1,936	6.633	85,184	3.530	94	8,836	9.695	830,584	4.547
45	2,025	6.708	91,125	3.557	95	9,025	9.747	857,375	4.563
46	2,116	6.782	97,336	3.583	96	9,216	9.798	884,736	4.579
47	2,209	6.856	103,823	3.609	97	9,409	9.849	912,673	4.595
48	2,304	6.928	110,592	3.634	98	9,604	9.899	941,192	4.610
49	2,401	7.000	117,649	3.659	99	9,801	9.950	970,299	4.626
50	2,500	7.071	125,000	3.684	100	10,000	10.000	1,000,000	4.642

TABLE B   Base-10 Logarithms

N	0	1	2	3	4	5	6	7	8	9
1.0	.0000	.0043	.0086	.0128	.0170	.0212	.0253	.0294	.0334	.0374
1.1	.0414	.0453	.0492	.0531	.0569	.0607	.0645	.0682	.0719	.0755
1.2	.0792	.0828	.0864	.0899	.0934	.0969	.1004	.1038	.1072	.1106
1.3	.1139	.1173	.1206	.1239	.1271	.1303	.1335	.1367	.1399	.1430
1.4	.1461	.1492	.1523	.1553	.1584	.1614	.1644	.1673	.1703	.1732
1.5	.1761	.1790	.1818	.1847	.1875	.1903	.1931	.1959	.1987	.2014
1.6	.2041	.2068	.2095	.2122	.2148	.2175	.2201	.2227	.2253	.2279
1.7	.2304	.2330	.2355	.2380	.2405	.2430	.2455	.2480	.2504	.2529
1.8	.2553	.2577	.2601	.2625	.2648	.2672	.2695	.2718	.2742	.2765
1.9	.2788	.2810	.2833	.2856	.2878	.2900	.2923	.2945	.2967	.2989
2.0	.3010	.3032	.3054	.3075	.3096	.3118	.3139	.3160	.3181	.3201
2.1	.3222	.3243	.3263	.3284	.3304	.3324	.3345	.3365	.3385	.3404
2.2	.3424	.3444	.3464	.3483	.3502	.3522	.3541	.3560	.3579	.3598
2.3	.3617	.3636	.3655	.3674	.3692	.3711	.3729	.3747	.3766	.3784
2.4	.3802	.3820	.3838	.3856	.3874	.3892	.3909	.3927	.3945	.3962
2.5	.3979	.3997	.4014	.4031	.4048	.4065	.4082	.4099	.4116	.4133
2.6	.4150	.4166	.4183	.4200	.4216	.4232	.4249	.4265	.4281	.4298
2.7	.4314	.4330	.4346	.4362	.4378	.4393	.4409	.4425	.4440	.4456
2.8	.4472	.4487	.4502	.4518	.4533	.4548	.4564	.4579	.4594	.4609
2.9	.4624	.4639	.4654	.4669	.4683	.4698	.4713	.4728	.4742	.4757
3.0	.4771	.4786	.4800	.4814	.4829	.4843	.4857	.4871	.4886	.4900
3.1	.4914	.4928	.4942	.4955	.4969	.4983	.4997	.5011	.5024	.5038
3.2	.5051	.5065	.5079	.5092	.5105	.5119	.5132	.5145	.5159	.5172
3.3	.5185	.5198	.5211	.5224	.5237	.5250	.5263	.5276	.5289	.5302
3.4	.5315	.5328	.5340	.5353	.5366	.5378	.5391	.5403	.5416	.5428
3.5	.5441	.5453	.5465	.5478	.5490	.5502	.5514	.5527	.5539	.5551
3.6	.5563	.5575	.5587	.5599	.5611	.5623	.5635	.5647	.5658	.5670
3.7	.5682	.5694	.5705	.5717	.5729	.5740	.5752	.5763	.5775	.5786
3.8	.5798	.5809	.5821	.5832	.5843	.5855	.5866	.5877	.5888	.5899
3.9	.5911	.5922	.5933	.5944	.5955	.5966	.5977	.5988	.5999	.6010
4.0	.6021	.6031	.6042	.6053	.6064	.6075	.6085	.6096	.6107	.6117
4.1	.6128	.6138	.6149	.6160	.6170	.6180	.6191	.6201	.6212	.6222
4.2	.6232	.6243	.6253	.6263	.6274	.6284	.6294	.6304	.6314	.6325
4.3	.6335	.6345	.6355	.6365	.6375	.6385	.6395	.6405	.6415	.6425
4.4	.6435	.6444	.6454	.6464	.6474	.6484	.6493	.6503	.6513	.6522
4.5	.6532	.6542	.6551	.6561	.6571	.6580	.6590	.6599	.6609	.6618
4.6	.6628	.6637	.6646	.6656	.6665	.6675	.6684	.6693	.6702	.6712
4.7	.6721	.6730	.6739	.6749	.6758	.6767	.6776	.6785	.6794	.6803
4.8	.6812	.6821	.6830	.6839	.6848	.6857	.6866	.6875	.6884	.6893
4.9	.6902	.6911	.6920	.6928	.6937	.6946	.6955	.6964	.6972	.6981
5.0	.6990	.6998	.7007	.7016	.7024	.7033	.7042	.7050	.7059	.7067
5.1	.7076	.7084	.7093	.7101	.7110	.7118	.7126	.7135	.7143	.7152
5.2	.7160	.7168	.7177	.7185	.7193	.7202	.7210	.7218	.7226	.7235
5.3	.7243	.7251	.7259	.7267	.7275	.7284	.7292	.7300	.7308	.7316
5.4	.7324	.7332	.7340	.7348	.7356	.7364	.7372	.7380	.7388	.7396

TABLE B   (continued)

N	0	1	2	3	4	5	6	7	8	9
5.5	.7404	.7412	.7419	.7427	.7435	.7443	.7451	.7459	.7466	.7474
5.6	.7482	.7490	.7497	.7505	.7513	.7520	.7528	.7536	.7543	.7551
5.7	.7559	.7566	.7574	.7582	.7589	.7597	.7604	.7612	.7619	.7627
5.8	.7634	.7642	.7649	.7657	.7664	.7672	.7679	.7686	.7694	.7701
5.9	.7709	.7716	.7723	.7731	.7738	.7745	.7752	.7760	.7767	.7774
6.0	.7782	.7789	.7796	.7803	.7810	.7818	.7825	.7832	.7839	.7846
6.1	.7853	.7860	.7868	.7875	.7882	.7889	.7896	.7903	.7910	.7917
6.2	.7924	.7931	.7938	.7945	.7952	.7959	.7966	.7973	.7980	.7987
6.3	.7993	.8000	.8007	.8014	.8021	.8028	.8035	.8041	.8048	.8055
6.4	.8062	.8069	.8075	.8082	.8089	.8096	.8102	.8109	.8116	.8122
6.5	.8129	.8136	.8142	.8149	.8156	.8162	.8169	.8176	.8182	.8189
6.6	.8195	.8202	.8209	.8215	.8222	.8228	.8235	.8241	.8248	.8254
6.7	.8261	.8267	.8274	.8280	.8287	.8293	.8299	.8306	.8312	.8319
6.8	.8325	.8331	.8338	.8344	.8351	.8357	.8363	.8370	.8376	.8382
6.9	.8388	.8395	.8401	.8407	.8414	.8420	.8426	.8432	.8439	.8445
7.0	.8451	.8457	.8463	.8470	.8476	.8482	.8488	.8494	.8500	.8506
7.1	.8513	.8519	.8525	.8531	.8537	.8543	.8549	.8555	.8561	.8567
7.2	.8573	.8579	.8585	.8591	.8597	.8603	.8609	.8615	.8621	.8627
7.3	.8633	.8639	.8645	.8651	.8657	.8663	.8669	.8675	.8681	.8686
7.4	.8692	.8698	.8704	.8710	.8716	.8722	.8727	.8733	.8739	.8745
7.5	.8751	.8756	.8762	.8768	.8774	.8779	.8785	.8791	.8797	.8802
7.6	.8808	.8814	.8820	.8825	.8831	.8837	.8842	.8848	.8854	.8859
7.7	.8865	.8871	.8876	.8882	.8887	.8893	.8899	.8904	.8910	.8915
7.8	.8921	.8927	.8932	.8938	.8943	.8949	.8954	.8960	.8965	.8971
7.9	.8976	.8982	.8987	.8993	.8998	.9004	.9009	.9015	.9020	.9025
8.0	.9031	.9036	.9042	.9047	.9053	.9058	.9063	.9069	.9074	.9079
8.1	.9085	.9090	.9096	.9101	.9106	.9112	.9117	.9122	.9128	.9133
8.2	.9138	.9143	.9149	.9154	.9159	.9165	.9170	.9175	.9180	.9186
8.3	.9191	.9196	.9201	.9206	.9212	.9217	.9222	.9227	.9232	.9238
8.4	.9243	.9248	.9253	.9258	.9263	.9269	.9274	.9279	.9284	.9289
8.5	.9294	.9299	.9304	.9309	.9315	.9320	.9325	.9330	.9335	.9340
8.6	.9345	.9350	.9355	.9360	.9365	.9370	.9375	.9380	.9385	.9390
8.7	.9395	.9400	.9405	.9410	.9415	.9420	.9425	.9430	.9435	.9440
8.8	.9445	.9450	.9455	.9460	.9465	.9469	.9474	.9479	.9484	.9489
8.9	.9494	.9499	.9504	.9509	.9513	.9518	.9523	.9528	.9533	.9538
9.0	.9542	.9547	.9552	.9557	.9562	.9566	.9571	.9576	.9581	.9586
9.1	.9590	.9595	.9600	.9605	.9609	.9614	.9619	.9624	.9628	.9633
9.2	.9638	.9643	.9647	.9652	.9657	.9661	.9666	.9671	.9675	.9680
9.3	.9685	.9689	.9694	.9699	.9703	.9708	.9713	.9717	.9722	.9727
9.4	.9731	.9736	.9741	.9745	.9750	.9754	.9759	.9763	.9768	.9773
9.5	.9777	.9782	.9786	.9791	.9795	.9800	.9805	.9809	.9814	.9818
9.6	.9823	.9827	.9832	.9836	.9841	.9845	.9850	.9854	.9859	.9863
9.7	.9868	.9872	.9877	.9881	.9886	.9890	.9894	.9899	.9903	.9908
9.8	.9912	.9917	.9921	.9926	.9930	.9934	.9939	.9943	.9948	.9952
9.9	.9956	.9961	.9965	.9969	.9974	.9978	.9983	.9987	.9991	.9996

TABLE C   Base-*e* Logarithms

N	0	1	2	3	4	5	6	7	8	9
1.0	.0000	.0100	.0198	.0296	.0392	.0488	.0583	.0677	.0770	.0862
1.1	.0953	.1044	.1133	.1222	.1310	.1398	.1484	.1570	.1655	.1740
1.2	.1823	.1906	.1989	.2070	.2151	.2231	.2311	.2390	.2469	.2546
1.3	.2624	.2700	.2776	.2852	.2927	.3001	.3075	.3148	.3221	.3293
1.4	.3365	.3436	.3507	.3577	.3646	.3716	.3784	.3853	.3920	.3988
1.5	.4055	.4121	.4187	.4253	.4318	.4383	.4447	.4511	.4574	.4637
1.6	.4700	.4762	.4824	.4886	.4947	.5008	.5068	.5128	.5188	.5247
1.7	.5306	.5365	.5423	.5481	.5539	.5596	.5653	.5710	.5766	.5822
1.8	.5878	.5933	.5988	.6043	.6098	.6152	.6206	.6259	.6313	.6366
1.9	.6419	.6471	.6523	.6575	.6627	.6678	.6729	.6780	.6831	.6881
2.0	.6931	.6981	.7031	.7080	.7129	.7178	.7227	.7275	.7324	.7372
2.1	.7419	.7467	.7514	.7561	.7608	.7655	.7701	.7747	.7793	.7839
2.2	.7885	.7930	.7975	.8020	.8065	.8109	.8154	.8198	.8242	.8286
2.3	.8329	.8372	.8416	.8459	.8502	.8544	.8587	.8629	.8671	.8713
2.4	.8755	.8796	.8838	.8879	.8920	.8961	.9002	.9042	.9083	.9123
2.5	.9163	.9203	.9243	.9282	.9322	.9361	.9400	.9439	.9478	.9517
2.6	.9555	.9594	.9632	.9670	.9708	.9746	.9783	.9821	.9858	.9895
2.7	.9933	.9969	1.0006	.0043	.0080	.0116	.0152	.0188	.0225	.0260
2.8	1.0296	.0332	.0367	.0403	.0438	.0473	.0508	.0543	.0578	.0613
2.9	.0647	.0682	.0716	.0750	.0784	.0818	.0852	.0886	.0919	.0953
3.0	1.0986	.1019	.1053	.1086	.1119	.1151	.1184	.1217	.1249	.1282
3.1	.1314	.1346	.1378	.1410	.1442	.1474	.1506	.1537	.1569	.1600
3.2	.1632	.1663	.1694	.1725	.1756	.1787	.1817	.1848	.1878	.1909
3.3	.1939	.1969	.2000	.2030	.2060	.2090	.2119	.2149	.2179	.2208
3.4	.2238	.2267	.2296	.2326	.2355	.2384	.2413	.2442	.2470	.2499
3.5	1.2528	.2556	.2585	.2613	.2641	.2669	.2698	.2726	.2754	.2782
3.6	.2809	.2837	.2865	.2892	.2920	.2947	.2975	.3002	.3029	.3056
3.7	.3083	.3110	.3137	.3164	.3191	.3218	.3244	.3271	.3297	.3324
3.8	.3350	.3376	.3403	.3429	.3455	.3481	.3507	.3533	.3558	.3584
3.9	.3610	.3635	.3661	.3686	.3712	.3737	.3762	.3788	.3813	.3838
4.0	1.3863	.3888	.3913	.3938	.3962	.3987	.4012	.4036	.4061	.4085
4.1	.4110	.4134	.4159	.4183	.4207	.4231	.4255	.4279	.4303	.4327
4.2	.4351	.4375	.4398	.4422	.4446	.4469	.4493	.4516	.4540	.4563
4.3	.4586	.4609	.4633	.4656	.4679	.4702	.4725	.4748	.4770	.4793
4.4	.4816	.4839	.4861	.4884	.4907	.4929	.4951	.4974	.4996	.5019
4.5	1.5041	.5063	.5085	.5107	.5129	.5151	.5173	.5195	.5217	.5239
4.6	.5261	.5282	.5304	.5326	.5347	.5369	.5390	.5412	.5433	.5454
4.7	.5476	.5497	.5518	.5539	.5560	.5581	.5602	.5623	.5644	.5665
4.8	.5686	.5707	.5728	.5748	.5769	.5790	.5810	.5831	.5851	.5872
4.9	.5892	.5913	.5933	.5953	.5974	.5994	.6014	.6034	.6054	.6074
5.0	1.6094	.6114	.6134	.6154	.6174	.6194	.6214	.6233	.6253	.6273
5.1	.6292	.6312	.6332	.6351	.6371	.6390	.6409	.6429	.6448	.6467
5.2	.6487	.6506	.6525	.6544	.6563	.6582	.6601	.6620	.6639	.6658
5.3	.6677	.6696	.6715	.6734	.6752	.6771	.6790	.6808	.6827	.6845
5.4	.6864	.6882	.6901	.6919	.6938	.6956	.6974	.6993	.7011	.7029

TABLE C   (*continued*)

N	0	1	2	3	4	5	6	7	8	9
5.5	1.7047	.7066	.7084	.7102	.7120	.7138	.7156	.7174	.7192	.7210
5.6	.7228	.7246	.7263	.7281	.7299	.7317	.7334	.7352	.7370	.7387
5.7	.7405	.7422	.7440	.7457	.7475	.7492	.7509	.7527	.7544	.7561
5.8	.7579	.7596	.7613	.7630	.7647	.7664	.7681	.7699	.7716	.7733
5.9	.7750	.7766	.7783	.7800	.7817	.7834	.7851	.7867	.7884	.7901
6.0	1.7918	.7934	.7951	.7967	.7984	.8001	.8017	.8034	.8050	.8066
6.1	.8083	.8099	.8116	.8132	.8148	.8165	.8181	.8197	.8213	.8229
6.2	.8245	.8262	.8278	.8294	.8310	.8326	.8342	.8358	.8374	.8390
6.3	.8405	.8421	.8437	.8453	.8469	.8485	.8500	.8516	.8532	.8547
6.4	.8563	.8579	.8594	.8610	.8625	.8641	.8656	.8672	.8687	.8703
6.5	1.8718	.8733	.8749	.8764	.8779	.8795	.8810	.8825	.8840	.8856
6.6	.8871	.8886	.8901	.8916	.8931	.8946	.8961	.8976	.8991	.9006
6.7	.9021	.9036	.9051	.9066	.9081	.9095	.9110	.9125	.9140	.9155
6.8	.9169	.9184	.9199	.9213	.9228	.9242	.9257	.9272	.9286	.9301
6.9	.9315	.9330	.9344	.9359	.9373	.9387	.9402	.9416	.9430	.9445
7.0	1.9459	.9473	.9488	.9502	.9516	.9530	.9544	.9559	.9573	.9587
7.1	.9601	.9615	.9629	.9643	.9657	.9671	.9685	.9699	.9713	.9727
7.2	.9741	.9755	.9769	.9782	.9796	.9810	.9824	.9838	.9851	.9865
7.3	.9879	.9892	.9906	.9920	.9933	.9947	.9961	.9974	.9988	2.0001
7.4	2.0015	.0028	.0042	.0055	.0069	.0082	.0096	.0109	.0122	.0136
7.5	2.0149	.0162	.0176	.0189	.0202	.0215	.0229	.0242	.0255	.0268
7.6	.0281	.0295	.0308	.0321	.0334	.0347	.0360	.0373	.0386	.0399
7.7	.0412	.0425	.0438	.0451	.0464	.0477	.0490	.0503	.0516	.0528
7.8	.0541	.0554	.0567	.0580	.0592	.0605	.0618	.0631	.0643	.0656
7.9	.0669	.0681	.0694	.0707	.0719	.0732	.0744	.0757	.0769	.0782
8.0	2.0794	.0807	.0819	.0832	.0844	.0857	.0869	.0882	.0894	.0906
8.1	.0919	.0931	.0943	.0956	.0968	.0980	.0992	.1005	.1017	.1029
8.2	.1041	.1054	.1066	.1078	.1090	.1102	.1114	.1126	.1138	.1150
8.3	.1163	.1175	.1187	.1199	.1211	.1223	.1235	.1247	.1258	.1270
8.4	.1282	.1294	.1306	.1318	.1330	.1342	.1353	.1365	.1377	.1389
8.5	2.1401	.1412	.1424	.1436	.1448	.1459	.1471	.1483	.1494	.1506
8.6	.1518	.1529	.1541	.1552	.1564	.1576	.1587	.1599	.1610	.1622
8.7	.1633	.1645	.1656	.1668	.1679	.1691	.1702	.1713	.1725	.1736
8.8	.1748	.1759	.1770	.1782	.1793	.1804	.1815	.1827	.1838	.1849
8.9	.1861	.1872	.1883	.1894	.1905	.1917	.1928	.1939	.1950	.1961
9.0	2.1972	.1983	.1994	.2006	.2017	.2028	.2039	.2050	.2061	.2072
9.1	.2083	.2094	.2105	.2116	.2127	.2138	.2148	.2159	.2170	.2181
9.2	.2192	.2203	.2214	.2225	.2235	.2246	.2257	.2268	.2279	.2289
9.3	.2300	.2311	.2322	.2332	.2343	.2354	.2364	.2375	.2386	.2396
9.4	.2407	.2418	.2428	.2439	.2450	.2460	.2471	.2481	.2492	.2502
9.5	2.2513	.2523	.2534	.2544	.2555	.2565	.2576	.2586	.2597	.2607
9.6	.2618	.2628	.2638	.2649	.2659	.2670	.2680	.2690	.2701	.2711
9.7	.2721	.2732	.2742	.2752	.2762	.2773	.2783	.2793	.2803	.2814
9.8	.2824	.2834	.2844	.2854	.2865	.2875	.2885	.2895	.2905	.2915
9.9	.2925	.2935	.2946	.2956	.2966	.2976	.2986	.2996	.3006	.3016

Use the properties of logarithms and $\ln 10 \approx 2.3026$ to find logarithms of numbers less than 1 or greater than 10.

Appendix IV
Answers to Selected Exercises

1. sum **3.** product **5.** Variables **7.** expressions **9.** formula **11.** horizontal, 1 **13.** equation **15.** algebraic expression
17. algebraic expression **19.** equation **21. a.** multiplication, subtraction **b.** x **23. a.** addition, subtraction **b.** m
25. a. 90°, 60°, 45°, 30°, 0° **b.**

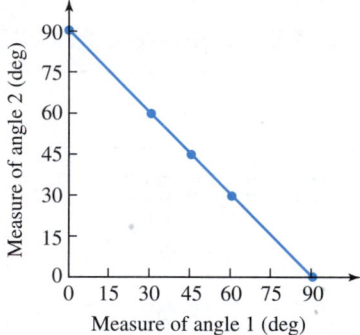

27. a. They help us determine that 15-year-old machinery is worth $35,000. **b.** It decreases. **29.** $5 \cdot 6$, $5(6)$ **31.** $34 \cdot 75$, $34(75)$ **33.** $4x$ **35.** $3rt$ **37.** lw **39.** Prt **41.** $\frac{32}{x}$ **43.** $\frac{90}{30}$
45. the product of 18 and 24 **47.** the difference of 11 and 9
49. the product of 2 and x **51.** the quotient of 66 and 11
53. $s = 100 - d$ **55.** $7d = h$ **57.** $s = 3c$ **59.** $w = e + 1{,}200$
61. $p = r - 600$ **63.** $\frac{l}{4} = m$ **65.** 390, 400, 405 **67.** 1,300,
1,200, 1,100 **69.** $d = \frac{e}{12}$; the number of dozen eggs is the
quotient of the number of eggs and 12. **71.** $i = 2c$; the
number of individuals is the product of 2 and the number of
couples. **73.** $l = 4c$, $a = 2c$, $S = c$, $b = c$, $p = 2c$, $s = 20c$
75. $d = x + 500$

77. Sales have steadily increased.

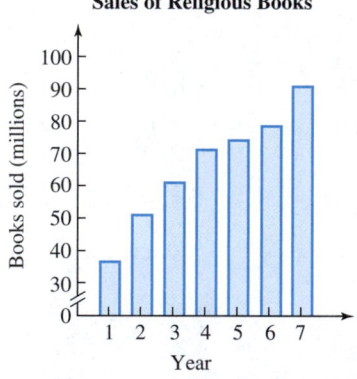

1. whole **3.** negative, positive **5.** rational, integer **7.** integers **9.** irrational **11.** absolute value **13.** $\frac{6}{1}$, $\frac{-9}{1}$, $\frac{-7}{8}$, $\frac{7}{2}$, $\frac{-3}{10}$, $\frac{283}{100}$ **15.** 13 and
-3 **17. a.** $<$ **b.** $>$ **c.** $>$, $<$ **19.**   **21.** square root

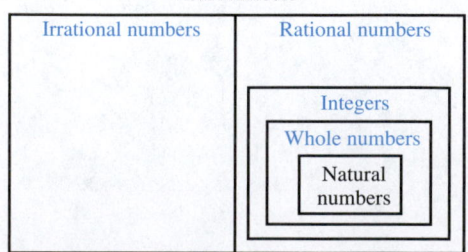

23. is approximately equal to **25.** numerator, denominator **27.** 4π means $4 \cdot \pi$; $4\pi \approx 12.6$ **29.** $>$ **31.** $>$ **33.** $<$ **35.** $>$ **37.** $=$
39. $=$ **41.** $>$ **43.** $>$ **45.** natural: none; whole: 0; integers: 0, -50; rational: $-\frac{5}{6}$, 35.99, 0, $4\frac{3}{8}$, -50, $\frac{17}{5}$; irrational: $\sqrt{2}$; real: all

47. a. true **b.** false **c.** false **d.** true **49. a.** $-5 > -6$ **b.** $-25 < 16$ **51.**

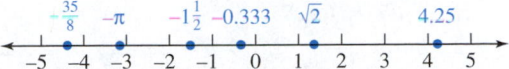

53. 2.236 **55.** 9.950 **57.** 4.472 **59.** 15.708 **61.** -5 **63.** $\frac{7}{8}$ **65.** 10 **67.** 2.3 **69.** natural, whole, integers: 750, 5,000; rational: all; irrational: none; real: all **71. a.** '90: $0.3; '91: $-$0.8; '92: $0.9; '93: $-$2.3; '94: $3.8; '95: $2.0; '96: $3.8; '97: $2.8 **b.** The company lost money. **73.** 81.7 in. **75.**

77. 55 mph, 30 mph **79.** 500 years
81. pos: A.D.; neg: B.C. **89.** The difference
of 7 and 5 is 2. **91.** The quotient of 30 and
15 is 2. **93.** 150, 180, 225

STUDY SET   SECTION 1.3 (page 33)

1. factors **3.** squared, cubed **5.** mean, average **7. a.** addition and multiplication **b.** 54, 34 **c.** 34; multiplication is to be done before addition **9. a.** Divide the sum of the scores by the number of scores. **b.** 64 **11. a.** addition, power, multiplication **b.** power, multiplication, addition **13.** multiplication; subtraction **15. a.** power **b.** addition **c.** subtraction **17.** 12^6 **19.** 9, 54 **21.** 3, 9, 18 **23.** $d = 16t^2$ **25.** 1, 8, 27, 64 **27.** b^3 **29.** 10^2k^3 **31.** $8\pi r^3$ **33.** $6x^2y^3$ **35.** 6 **37.** 41 **39.** 80 **41.** 10 **43.** 8 **45.** 194 **47.** 192 **49.** 38 **51.** 20 **53.** 26 **55.** 49 **57.** 1 **59.** 3 **61.** 50 **63.** 360 **65.** 10 **67.** 72 **69.** 615 **71.** 10,000 **73.** 384 **75.** 27 **77.** 20 **79.** 2 **81.** 12 **83.** 6 **85.** 56,120 **87.** 91,985 **89. a.** $11,875 **b.** $95 **91.** $206,250 **93.** 2, 7, 5 **95. a.** $5 = 5^1$, $25 = 5^2$, $125 = 5^3$, $625 = 5^4$ **b.** 3,125 **101.** 5 **103.** true **105.** -9 and 3

STUDY SET   SECTION 1.4 (page 44)

1. evaluate **3.** subtraction **5.** $6 + 20x$; $\frac{6-x}{20}$ (answers may vary) **7.** We would obtain $34 - 6$; it looks like 34, not 3(4). **9. a.** $x =$ weight of the car; $2x - 500 =$ weight of the van **b.** 3,500 lb **11. a.** $8x - x^2$ **b.** 3, 4, 5 **c.** 16 **13.** 5, 5, 25, 45 **15.** $l + 15$ **17.** $50x$ **19.** $\frac{w}{7}$ **21.** $P + p$ **23.** $k^2 - 2,005$ **25.** $J - 500$ **27.** $\frac{1,000}{n}$ **29.** $p + 90$ **31.** $35 + h + 300$ **33.** $p - 680$ **35.** $4d - 15$ **37.** $2(200 + t)$ **39.** $|a - 2|$ **41.** 7 less than a number **43.** the product of 7 and a number, increased by 4 **45. a.** 300 **b.** $60h$ **47. a.** $3y$ **b.** $\frac{f}{3}$ **49.** $x + 2$ **51.** $36 - x$ **53. a.** $8x$ **b.** $40x$ **55.** $5(x + 2)$ **57.** 42 **59.** 6 **61.** 1, 1, 31 **63.** 41, 11, 8 **65.** 150, 450 **67.** 49, 193 **69.** 169 **71.** 22 **73.** 40 **75.** 381.7 in.^{3} **77.** Let $n =$ original number entering program; $\frac{n-6}{8} =$ number in each squad. **79.** Let $a =$ number of apts occupied before rent was lowered; $2a - 3 =$ number now occupied. **81.** 0, 28, 48, 60, 64, 60, 48, 28, 0 **83. a.** 360, 1,152, 2,376 **b.** in.^{2} **85.** 69 ft^2 **89.** 0 **91.** $\frac{2}{3}$ **93.** c^4 **95.** 83

STUDY SET   SECTION 1.5 (page 57)

1. equation **3.** check, true **5.** equivalent **7.** isolate **9. a.** subtraction of 8, addition of 8 **b.** addition of 8, subtraction of 8 **c.** division by 8, multiplication by 8 **d.** multiplication by 8, division by 8 **11. a.** $x + 6$ **b.** neither **c.** no **d.** yes **13.** 15, 15 **15.** $12 = 0.40 \cdot x$ **17. a.** 0.35 **b.** 0.035 **c.** 3.5 **d.** 0.005 **19.** yes **21.** no **23.** no **25.** yes **27.** no **29.** no **31.** yes **33.** yes **35.** yes **37.** 3 **39.** 71 **41.** 9 **43.** 81 **45.** 0 **47.** 239 **49.** 4 **51.** 41 **53.** 0 **55.** 1 **57.** 45 **59.** 48 **61.** 140 **63.** 0 **65.** 312 **67.** 26% **69.** 300 **71.** 46.2 **73.** 2.5% **75.** 1,464 **77. a.** 6 ft^2 **b.** 1.2 ft^2 **c.** 20% **79.** 0.0018% **81.** $240 billion **83.**

STEAK STAMPEDE
Bloomington, MN
Server #12\ AT

| VISA | 67463777288 |
| NAME | DALTON/ LIZ |

AMOUNT	$75.18
GRATUITY $	12.00
TOTAL $	87.18

85. 12% **87.** $143 billion **89.** $56.50 **91. a.** 1994–1995; 10% **b.** 1996–1997; 25% **93.** 30,114% **95.** .878, 1.000, .763 **101.** 0 **103.** $45 - x$ **105.** 9.4

STUDY SET   SECTION 1.6 (page 66)

1. variable **3.** equation **5.** analyze, form, solve, state, check **7.** $x =$ the length of the running portion of the triathlon; $1 + 10 + x = 16$ **9.** $x =$ the number of pages in the preface; $4 + x + 400 + 12 = 430$ **11.** $x =$ the amount of interest earned on the savings account; $x + 73 = 678$ **13.** disbursed equally; division **15.** melted; subtraction **17.** reclaimed; addition **19.** eroded; subtraction **21.** sectioned off uniformly; division **23.** cut; subtraction **25.** dilated to twice their size; multiplication **27.** $1,000 + x = 1,525$ **29.** $x - 4 = 8$ **31.** $\frac{x}{12} = 5$ **33.** $15x = 495$ **35.** the old unit requirement, 6, 28; the old unit requirement, subtract, x, 28; x, 6, 6; 34; 34 **37.** 27 pints **39.** 54 **41.** $188 million **43.** $322.00 **45.** 27 min **47.** 16 **49.** 60 **51.** 975 mi **53.** 54 ft **55.** 135° **61.** 29% **63.** 2,730 **65.** no **67.** 20

KEY CONCEPT (page 71)

1. $b - c$ **2.** cb **3.** $C = p + t$ **4.** $b = 2t$ **5.** $x + 4 =$ amount of business (\$ millions) the year with the celebrity **6.** 1, 41, 97
7. Let $x =$ the percent of the U.S. total that Maine's area represents. **8.** Let $x =$ the number of city blocks canvassed.

CHAPTER REVIEW (page 73)

1. The production of wide-screen TVs is increasing. **2. a.** 1 hr; 100 cars **b.** 100 **c.** 7 P.M. **3. a.** The difference of 15 and 3 is 12.
b. The sum of 15 and 3 is 18. **c.** The quotient of 15 and 3 is 5. **d.** The product of 15 and 3 is 45. **4. a.** $4 \cdot 9$; $4(9)$ **b.** $\frac{9}{3}$ **5. a.** $8b$
b. xy **c.** $2lw$ **d.** Prt **6. a.** equation **b.** algebraic expression **c.** algebraic expression **d.** equation **7.** $T = \frac{d}{350}$ **8.** 10, 15, 25
9. $f = 50c$; the total fees are the product of 50 and the number of children. **10.** 0
11. 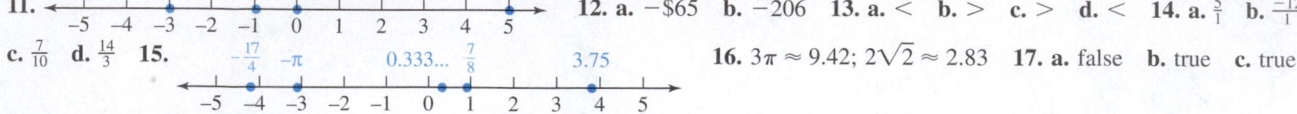 **12. a.** $-\$65$ **b.** -206 **13. a.** $<$ **b.** $>$ **c.** $>$ **d.** $<$ **14. a.** $\frac{5}{1}$ **b.** $\frac{-12}{1}$
c. $\frac{7}{10}$ **d.** $\frac{14}{3}$ **15.** **16.** $3\pi \approx 9.42$; $2\sqrt{2} \approx 2.83$ **17. a.** false **b.** true **c.** true
d. true **18.** natural: 8; whole: 0, 8; integers: 0, -12, 8; rational: $-\frac{4}{5}$, 99.99, 0, -12, $4\frac{1}{2}$, 0.666. . . , 8; irrat: $\sqrt{2}$; real: all **19. a.** -10
b. 3 **c.** $\frac{9}{16}$ **d.** 0 **20. a.** $>$ **b.** $=$ **c.** $>$ **d.** $<$ **21. a.** 8^5 **b.** 2^3 **c.** $5^3 \cdot 9^2$ **d.** a^4 **e.** $9\pi r^2$ **f.** $x^3 y^4$ **g.** 100^2 **h.** 1^6 **22. a.** 81
b. 8 **c.** 32 **d.** 15 **23.** 225 in.^{2} **24.** 5.4 ft^3 **25.** 4; power, multiplication, subtraction, addition **26. a.** 6 **b.** 1,009 **c.** 28 **d.** 120
e. 32 **f.** 29 **g.** 300 **h.** 13 **27.** parentheses, brackets, fraction bar **28. a.** 2 **b.** 5 **c.** 110 **d.** 2 **e.** 9 **f.** 27 **29.** \$21,134
30. \$20 **31. a.** $h + 25$ **b.** $s - 15$ **c.** $\frac{1}{2}t$ **d.** $6x$ **32.** $p - 500$ **33.** $x =$ time to complete test; $3x =$ time spent standing in line
34. a. $n + 4$ **b.** $b - 4$ **35. a.** $10d$ **b.** $\frac{x}{12}$ **c.** $x - 5$ **36.** 5, 30; 10, 10d **37.** 0, 19, 16 **38.** 0, 2, 8 **39. a.** 36 **b.** 110 **c.** 40
d. 432 **e.** 736 **f.** 18 **40.** 17.7 in.^{3} **41. a.** yes **b.** no **c.** yes **d.** no **e.** no **f.** yes **42.** variable, true **43. a.** 21 **b.** 17 **c.** 5
d. 93 **e.** 20 **f.** 8 **g.** 11 **h.** 0 **i.** 96 **j.** 78 **k.** 0 **l.** 0 **44.** $16 = 0.05x$ **45.** \$26.74 **46.** 192.4 **47.** no **48.** 2,688,424
49. 1,567% **50.** $x - 28 = 165$ **51.** 1,500 **52.** 429 mi **53.** 36 **54.** 60°

CHAPTER 1 TEST (page 79)

1. \$24 **2.** 5 hr **3.** 3, 20, 70 **4.** **5. a.** true **b.** false **c.** true **d.** true **6. a.** $>$
b. $<$ **c.** $<$ **d.** $>$ **7.** 1,000 in.^{3} **8. a.** 9^5 **b.** $3x^2z^3$ **9.** 15,625 **10.** 170 **11.** 36 **12.** 2 **13.** A real number is any number that can
be written as a decimal. **14.** An equation is a mathematical sentence that contains an $=$ sign. An expression does not contain an $=$
sign. **15.** 4 **16.** 4, 17, 59 **17.** 128 **18.** $x - 2 =$ number of songs on the CD **19.** $25q$ **20.** no **21.** 13 **22.** 5 **23.** 30 **24.** 550
25. \$76,000 **26.** 37%. There is an increase in the number of videos sold during the holiday season (4th quarter). Then sales fall back
to normal levels in the 1st quarter of the next year. **27.** 1,125 **28.** 32,000 mi^2

STUDY SET   SECTION 2.1 (page 90)

1. positive **3.** zero **5.** commutative **7.** 5 **9.** 1 **11.** same, absolute **13.** opposite **15. a.** 0 **b.** 0 **c.** 0 **d.** 0 **17.** -13, 10
19. 23, 23 **21.** -2 **23.** 2 **25.** -77 **27.** -8.2 **29.** $-\frac{1}{8}$ **31.** $\frac{5}{12}$ **33.** 16 **35.** -15 **37.** -21 **39.** $-1,519$ **41.** -579.84 **43.** 11
45. -21 **47.** -2 **49.** -2.3 **51.** 0 **53.** -4 **55.** $-\frac{1}{2}$ **57.** $-\frac{5}{16}$ **59.** 12,466 **61.** $-44,571.251$ **63.** -4, 0, 24 **65.** 0, 10, -5
67. yes **69.** no **71.** -19 **73.** 51 **75.** -14 **77.** 4 **79.** 2,150 m **81. a.** -18, -6, -5, -4 **b.** 12 strokes **83.** 2,772 **85.** 160°F
87. 1,030 ft **89.** a gain of \$2.2 million **91.** southward, 132 km **93.** $+29$, $+35$, -23 **99.** c^4 **101.** a^2b^3

STUDY SET   SECTION 2.2 (page 102)

1. product **3.** like **5.** commutative **7.** undefined **9.** $-5(-5) = 25$ **11.** unlike **13.** 0 **15.** 3 **17. a.** NEG **b.** not possible to tell
c. POS **d.** NEG **19.** yes **21.** 5, 10 **23.** -3, -3, -12 **25.** 54 **27.** -60 **29.** -24 **31.** -800 **33.** 2.4 **35.** -0.48 **37.** $-\frac{3}{8}$
39. $\frac{15}{16}$ **41.** 0 **43.** 0 **45.** 60 **47.** 84 **49.** 120 **51.** 3 **53.** -2 **55.** -4 **57.** 1 **59.** -4 **61.** -20 **63.** 0 **65.** undefined **67.** $-\frac{5}{12}$
69. $\frac{9}{32}$ **71.** $-1,109.2$ **73.** -417.3624 **75.** -5.5 **77.** -3, -15, -30 **79.** -1, -3, -4 **81.** -1 **83.** 3 **85.** -15 **87.** 42
89. $-72°$ **91.** $-\$160$ **93.** $-193°$ F **95.** $-\$55.8$ million **97.** $-\$614,516$ **99. a.** 5, -10 **b.** 2.5, -5 **c.** 7.5, -15 **d.** 10, -20
103. 64 in.^{3} **105.** any number that can be written as a decimal **107.** 208.08

STUDY SET   SECTION 2.3 (page 112)

1. order **3.** evaluate **5.** base, 3 **7.** parentheses, brackets, absolute value **9.** exponential **11. a.** 9, 16, 25. When a negative number
is raised to an even power, the result is a positive number. **b.** -27, -64, -125. When a negative number is raised to an odd power,
the result is a negative number. **13. a.** 5 **b.** subtraction, power, multiplication, subtraction, multiplication **15.** 8, -3, 36, -6
17. 36 **19.** -256 **21.** -125 **23.** $-1,296$ **25.** 0.16 **27.** $-\frac{8}{125}$ **29.** -17 **31.** -3 **33.** 28 **35.** 64 **37.** -9 **39.** 24 **41.** -2
43. $\frac{1}{8}$ **45.** -8 **47.** 31 **49.** -39 **51.** -27 **53.** -8 **55.** 11 **57.** $-\frac{8}{9}$ **59.** 8 **61.** 10 **63.** -12 **65.** 5 **67.** 156 **69.** $\frac{1}{5}$ **71.** 17
73. 36 **75.** 230 **77.** 983.4496 **79.** -1.762341683 **81.** 1,443.54 **83.** $-37°$C, $-64°$C **85.** $80 - 5(21) = -25$ **87.** 77.8 in.^{2}
93. **95.** a^2b^3 **97.** $x = 5$

STUDY SET   SECTION 2.4 (page 123)

1. simplify **3.** coefficient, variable **5.** remove **7.** the distributive property **9.** $2(3 + 4) = 2 \cdot 3 + 2 \cdot 4$ **11.** $x + 20 - x = 20$; 20 ft
13. $(90 - a)°$ **15.** -7, -7, -7, -35, $7a$ **17. a.** no **b.** yes **19.** $63m$ **21.** $-35q$ **23.** $20bp$ **25.** $40r^2$ **27.** $5x + 15$ **29.** $-2b + 2$

31. $24t - 16$ **33.** $12y - 6$ **35.** $0.4x - 1.6$ **37.** $-2w + 4$ **39.** $r^2 - 10r$ **41.** $-x + 7$ **43.** $34x - 17y + 34$ **45.** $14 - 3p + t$
47. $-5r, 4s; -5, 4$ **49.** $-15r^2s; -15$ **51.** $50a, 2; 50, 2$ **53.** $x^3, -125; 1, -125$ **55.** -1 **57.** $\frac{1}{4}$ **59.** term   **61.** factor   **63.** $20x$
65. $3x^2$ **67.** 0 **69.** 0 **71.** x **73.** $10x^2$ **75.** $1.1h$ **77.** $\frac{4}{3}t$ **79.** $-2y + 36$ **81.** $7z - 15$ **83.** $6c + 62$ **85.** $7X - 2x$ **87.** $2 + b$
89. $-2x^2 + 3x$ **91.** $12x$ **93.** $(4x + 8)$ ft   **99.** 0 **101.** 2

STUDY SET   SECTION 2.5 (page 134)

1. equation   **3.** numbers   **5.** distributive   **7.** subtraction, multiplication   **9.** subtracting   **11. a.** $2x + 5$ **b.** 2   **c.** 23   **13. a.** balcony
b. 200   **c.** $3x + 150$ **15.** 30   **17.** 7, 7, 28, 2, 2   **19. a.** -1 **b.** $\frac{3}{5}$ **21.** 6   **23.** $-\frac{3}{5}$ **25.** 3.5   **27.** 4   **29.** $-\frac{8}{3}$ **31.** -12 **33.** -12
35. 9   **37.** 28   **39.** -19 **41.** 5   **43.** 1   **45.** 3   **47.** $\frac{1}{7}$ **49.** -2.5 **51.** 1   **53.** $\frac{5}{4}$ **55.** $\frac{1}{8}$ **57.** -20 **59.** -41 **61.** 9   **63.** -1 **65.** 3
67. 0   **69.** identity   **71.** impossible equation   **73.** impossible equation   **75.** identity   **77.** 4 ft and 8 ft   **79.** 5 ft, 9 ft, 4 ft
81. Australia: 12 wk; Japan: 16 wk; Sweden: 10 wk   **83.** 250   **85.** \$26,000   **87.** 7.3 ft and 10.7 ft   **89.** \$1.50   **91.** 9 mo   **95.** 137.76
97. 15%

STUDY SET   SECTION 2.6 (page 146)

1. formula   **3.** perimeter   **5.** radius   **7.** circumference   **9. a.** volume   **b.** circumference   **c.** area   **d.** perimeter   **11.** $(12x - 8)$ mm^2
13. a. Isolate w on one side of the equation.   **b.** a is not isolated on one side of the equation.   **15. a.** Multiply r by 2.   **b.** Divide D
by 2.   **17.** 11,176,920 mi, 65,280 mi   **19.** 3, 3, Bh, h, h, B **21.** \$7.50   **23.** \$24.55   **25.** \$65 million   **27.** 3.5%   **29.** 2.5 mph
31. 4,014°F   **33.** $R = \frac{E}{I}$ **35.** $w = \frac{V}{lh}$ **37.** $x = \frac{y-b}{m}$ **39.** $t = \frac{A-P}{Pr}$ **41.** $h = \frac{3V}{\pi r^2}$ **43.** $F = \frac{9C + 160}{5}$ **45.** 36 ft, 48 ft^2 **47.** 50.3 in.,

201.1 in.^{2} **49.** 56 in., 144 in.^{2} **51.** 2,450 ft^2 **53.** 27.75 in., 47.8125 in.^{2} **55.** 32 ft^2, 128 ft^3 **57.** 348 ft^3 **59.** 254 in.^{2} **61.** $I = \dfrac{E}{R}$;

$I = 4$ A   **63.** $R = \dfrac{P}{I^2}$; $R = 13.78\ \Omega$ **65.** $p = \dfrac{G - U + TS}{V}$ **69.** 1, -3, 6   **71.** equation   **73.** expression

STUDY SET   SECTION 2.7 (page 156)

1. perimeter   **3.** vertex   **5.** 180°   **7. a.** \$2x, \$5x, \$7(x + 10)   **b.** 3 in.   **c.** 1 in.   **d.** \$(14x + 70)   **9. a.** 0.04x, (0.06)2x   **b.** twice as
much   **c.** $0.04x + 0.12x = 0.16x$ **11. a.** $(x + 42)$ gal   **b.** 32%   **13. a.** \$0.38x, \$1.12, \$0.21(14 + x)   **b.** \$0.21   **15.** 6,000   **17.** 22°,
68°   **19.** 15   **21.** 19 ft   **23.** 7 ft, 7 ft, and 11 ft   **25.** 75 m by 480 m   **27.** 20°   **29.** 12   **31.** 90   **33.** \$5,500   **35.** \$4,900 in each
account   **37.** \$7,500   **39.** 2 hr   **41.** 65 mph and 45 mph   **43.** 4 hr into the flights   **45.** 50   **47.** 7.5 oz   **49.** 20   **51.** 40 lb lemon
drops and 60 lb jelly beans   **53.** 80   **59.** $-50x + 125$ **61.** $19p + 11q$

STUDY SET   SECTION 2.8 (page 170)

1. inequality   **3.** solution   **5.** same   **7.** opposite   **9. a.** a true statement   **b.** a false statement   **11. a.** all real numbers greater than 8
b. ⟵———(——————→   **c.** $(8, \infty)$ **13.** is less than   **15.** is greater than   **17.** is not equal to   **19.** $x > -2$ **21.** $-2 \le 17$ **23.** 5, 5, 12,

4, 4   **25.** ⟵———)————→ $(-\infty, 5)$ **27.** ⟵—(——]→ $(-3, 1]$ **29.** $x < -1, (-\infty, -1)$ **31.** $-7 < x \le 2, (-7, 2]$

33. $x > 3, (3, \infty)$ **35.** $x \ge -10, [-10, \infty)$ **37.** $x < -1, (-\infty, -1)$

39. $x \le 0.4, (-\infty, 0.4]$ **41.** $x < -2, (-\infty, -2)$

43. $x < -\frac{11}{4}, \left(-\infty, -\frac{11}{4}\right)$ **45.** $x \ge 3, [3, \infty)$ **47.** $x \ge -24, [-24, \infty)$

49. $x \le -1, (-\infty, -1]$ **51.** $x \ge -13, [-13, \infty)$ **53.** $x > 0, (0, \infty)$

55. $x < -2, (-\infty, -2)$ **57.** $x \ge \frac{9}{4}, \left[\frac{9}{4}, \infty\right)$ **59.** $x > -15, (-15, \infty)$

61. $x \le 20, (-\infty, 20]$ **63.** $7 < x < 10, (7, 10)$

65. $-9 < x \le 3, (-9, 3]$ **67.** $-10 \le x \le 0, [-10, 0]$

69. $-5 < x < -2, (-5, -2)$ **71.** $-6 \le x \le 10, [-6, 10]$

73. $2 \le x < 3, [2, 3)$ **75.** $-1 \le x < 2, [-1, 2)$ **77.** $s \ge 98\%$ **79.** $r \ge 27$ mpg

81. $t \ge 420$ min   **83. a.** $0° < a \le 18°$ **b.** $18° \le a \le 50°$ **c.** $30° \le a \le 37°$ **d.** $75° \le a < 90°$ **85. a.** $470\text{ ft} \le x \le 13{,}143$ ft
b. $0.1\text{ mi} \le x \le 2.5$ mi   **87.** $1.496\text{ in.} \le w \le 1.498$ in.; $1.5000\text{ in.} \le w \le 1.5010$ in.   **91.** -125 **93.** -256

KEY CONCEPT (page 175)

1. a. $2x - 8$ **b.** $x = 6$ **2. a.** $y + 5$ **b.** $y = -5$ **3. a.** $\frac{2}{3}a$ **b.** $a = \frac{1}{2}$ **4. a.** $-2x - 10$ **b.** $x \leq 5$ **5.** The mistake is on the third line. The student made an equation out of the answer, which is $x - 6$, by writing "0 =" on the left and then solved that equation.

CHAPTER REVIEW (page 177)

1. a. 45 **b.** -82 **c.** 22 **d.** 12 **e.** -11 **f.** -4 **g.** -12.3 **h.** $-\frac{3}{16}$ **i.** 11 **j.** -45 **k.** -7 **l.** 0 **2. a.** commutative property of addition **b.** associative property of addition **3. a.** -19 **b.** -49 **c.** -15 **d.** 5.7 **4. a.** 3, 5, -4 **b.** -1, 7, 9 **5. a.** -29 **b.** -0.4 **c.** -25 **d.** 3 **6. a.** 14 **b.** 13.78 **7.** 65,233 ft **8. a.** -56 **b.** 54 **c.** 12 **d.** -24 **e.** -24 **f.** 36 **g.** 6.36 **h.** -2 **i.** $-\frac{2}{15}$ **j.** $-\frac{3}{4}$ **k.** 0 **l.** -3 **9. a.** associative property of multiplication **b.** commutative property of multiplication **10. a.** 2 **b.** -4 **c.** 3 **d.** 0 **e.** $-\frac{6}{5}$ **f.** undefined **g.** -4.5 **h.** 1 **11. a.** -2 **b.** -3 **c.** 20 **d.** -51 **12. a.** 32 **b.** -32 **c.** 81 **d.** -125 **e.** 0.64 **f.** $-\frac{8}{27}$ **13. a.** 44 **b.** -430 **c.** $-\frac{14}{19}$ **d.** 113 **e.** -7 **f.** 0 **14. a.** 36 **b.** -18 **15.** 33.5 in.^{3} **16. a.** 34, -164, 6 **b.** -8, -996, -1 **17. a.** $-28w$ **b.** $15r^2$ **c.** $24xy$ **d.** $2.08f$ **18. a.** $5x + 15$ **b.** $-4x - 6 + 2y$ **c.** $-a + 4$ **d.** $3c - 6$ **19. a.** 3 **b.** 1 **20. a.** 2, -5 **b.** 16, -5, 25 **c.** $\frac{1}{2}$, 1 **d.** 9.6, -1 **21. a.** $9p$ **b.** $-7m$ **c.** $-2a - 10b$ **d.** $-p - 18q$ **e.** x **f.** $-8a^3$ **22.** $(4x + 4)$ ft **23. a.** 2 **b.** -1 **c.** 30 **d.** -19 **e.** 4 **f.** 1 **g.** $\frac{5}{4}$ **h.** -6 **i.** identity, all values of a **j.** impossible equation, no solution **24.** 8 ft **25.** \$16,098 **26.** \$176 **27.** \$11,800 **28.** 3.38 hr **29.** 1,949°F **30.** 168 in. **31.** 1,440 in.^{2} **32.** 76.5 m^2 **33.** 144 in.^{2} **34.** 50.27 cm **35.** 201.06 cm^2 **36.** 4,320 in.^{3} **37.** 9.4 ft^3 **38.** 120 ft^3 **39.** 381.70 in.^{3} **40. a.** $h = \frac{A}{2\pi r}$ **b.** $l = \frac{P - 2w}{2}$ **41.** 147 **42.** 24.875 in. × 29.875 in. $\left(24\frac{7}{8}\text{ in.} \times 29\frac{7}{8}\text{ in.}\right)$ **43.** 76.5°, 76.5° **44.** \$45x **45.** \$16,000 at 7%, \$11,000 at 9% **46.** 20 **47.** 10 lb of each **48.** 0.12x gal **49. a.** $x < 1$, $(-\infty, 1)$ **b.** $x < -3$, $(-\infty, -3)$

c. $x \geq 4$, $[4, \infty)$ **d.** $x \geq 6$, $[6, \infty)$ **e.** $x \geq 3$, $[3, \infty)$

f. $x \leq 12$, $(-\infty, 12]$ **g.** $6 < x < 11$, $(6, 11)$

h. $-2 < x \leq 1$, $(-2, 1]$ **50.** **51.** 2.40 g $< w <$ 2.53 g

CHAPTER 2 TEST (page 183)

1. -2 **2.** 2.2 **3.** -30 **4.** 2 **5.** -483 **6.** -3 **7.** -2 **8.** -4 **9.** 6 **10.** 3 **11.** $-20x$ **12.** $224t^2$ **13.** $6x - 6$ **14.** $-4.9d^2$ **15.** -12 **16.** -5 **17.** -7 **18.** 1 **19.** $-\frac{1}{2}$ **20.** $\frac{7}{6}$ **21.** -3 **22.** 1.1 **23.** $t = \frac{d}{r}$ **24.** $r = \frac{A - P}{Pt}$ **25.** 0.0475 million ft^2 **26.** 393 in.^{3} **27.** \$5,250 **28.** $\frac{3}{5}$ hr **29.** 10 **30.** 68° **31.** $x \geq -3$, $[-3, \infty)$ **32.** $-3 \leq x < 4$, $[-3, 4)$

CHAPTERS 1–2 CUMULATIVE REVIEW EXERCISES (page 184)

1. no **2.** yes **3.** no **4.** yes **5.** natural number, whole number, integer, rational, real **6.** rational, real **7.** **8.** **9.** 12 **10.** 15 **11.** -15 **12.** -12 **13.** 8.77 **14.** 0.51 **15.** 1.77 **16.** 1.18 **17.** 3^3 **18.** $8\pi r^2$ **19.** $4x^2y^2$ **20.** m^3n **21.** 1 **22.** -98 **23.** 13 **24.** 21 **25.** $w + 12$ **26.** $n - 4$ **27.** -4, -3, 5 **28.** 16, 9, 1 **29.** 2, 3, 4 **30.** 0, 7, 14 **31.** 9 **32.** 77 **33.** 1 **34.** 18 **35.** 4 **36.** 5 **37.** 20 **38.** 22 **39.** 288 **40.** 15% **41.** 300 **42.** 1,200 **43.** 0 **44.** -2 **45.** 16 **46.** 0 **47.** -216 **48.** -36 **49.** -25 **50.** $-\frac{8}{27}$ **51.** 8 **52.** 9 **53.** 11 **54.** $\frac{8}{9}$ **55.** $-32d$ **56.** $10x - 15y + 5$ **57.** $5x$ **58.** $-8a$ **59.** $8q^2 - 5q$ **60.** $8t - 20$ **61.** $(x + 3)$ ft **62.** $3x$ ft **63.** 9 **64.** 20 **65.** -0.6 **66.** 19 **67.** -20 **68.** -2 **69.** 65 m^2 **70.** 78.54 cm^2 **71.** 9.42 ft^3 **72.** 376.99 cm^3 **73.** $w = \frac{P - 2l}{2}$ **74.** $t = \frac{A - P}{Pr}$ **75.** 37.5 ft-lb **76.** 9.45 lb **77.** 55°, 55° **78.** 65° **79.** $x > -2$, $(-2, \infty)$ **80.** $x \leq 2$, $(-\infty, 2]$ **81.** $x \geq -1$, $[-1, \infty)$ **82.** $-1 \leq x < 2$, $[-1, 2)$

STUDY SET SECTION 3.1 (page 193)

1. ordered **3.** origin **5.** rectangular **7.** origin, left, up **9.** no **11.** quadrant II **13.** 3, -3, 5, -5, 4, -4, 3, -3, 0 **15.** 10 min before the workout, her heart rate was 60 beats/min. **17.** 150 beats/min **19.** approximately 5 min and 50 min after starting **21.** 10 beats/min faster after cool-down **23.** (3, 5) is an ordered pair, 3(5) indicates multiplication, and 5(3 + 5) is an expression containing grouping symbols. **25.** yes **27.**

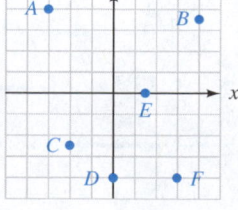

29. rivets: $(-6, 0)$, $(-2, 0)$, $(2, 0)$, $(6, 0)$; welds: $(-4, 3)$, $(0, 3)$, $(4, 3)$; anchors: $(-6, -3)$, $(6, -3)$ **31.** $(-3, 10)$, $(-2, 7)$, $(-1, 4.8)$, $(0, 3)$, $(1, 1.8)$, $(2.5, 0.5)$, $(4, 0)$ **33. a.** \$2 **b.** \$4 **c.** \$7 **d.** \$9

35.

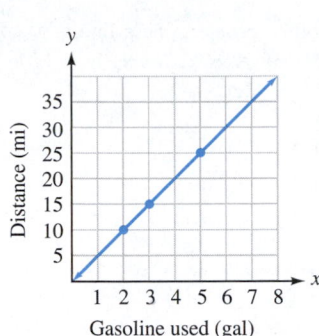

a. 35 mi   **b.** 4   **c.** 32.5 mi   **37.** Rockford (5, B), Mount Carroll (1, C), Harvard (7, A), Intersection (5, E)   **43.** 12   **45.** 8   **47.** 7   **49.** −49

STUDY SET   SECTION 3.2 (page 205)

1. two   **3.** independent, dependent   **5. a.** 2   **b.** yes   **c.** yes   **d.** infinitely many   **7. a.** yes   **b.** no   **9.** Not enough ordered pairs were found—the correct graph is not a line.   **11.** one, one

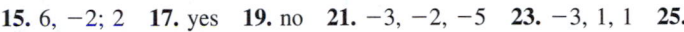

13. two, infinitely many   **15.** 6, −2; 2   **17.** yes   **19.** no   **21.** −3, −2, −5   **23.** −3, 1, 1   **25.**

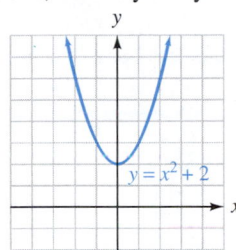

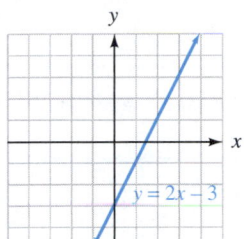

27.

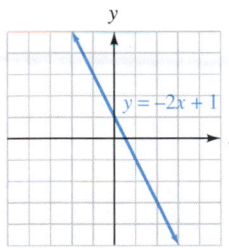

29. 1 unit higher

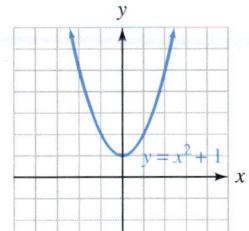

31. 2 units to the right

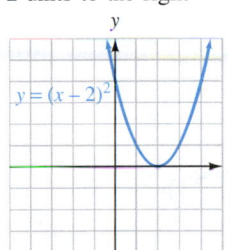

33. It is turned upside down.

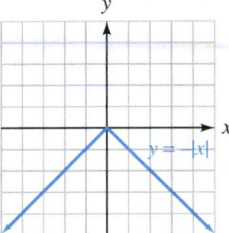

35. 2 units to the left

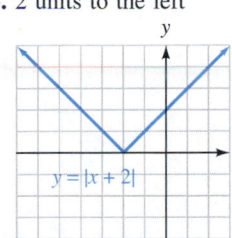

37. It is turned upside down.   **39.** 2 units lower

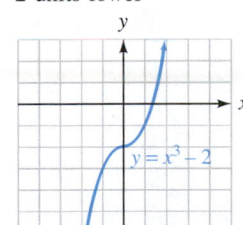

41. a. 2   **b.**

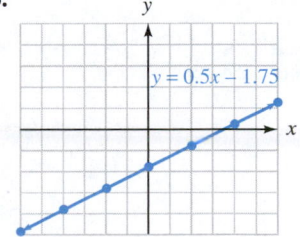

43. 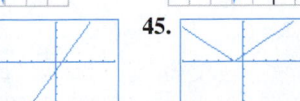   **45.** 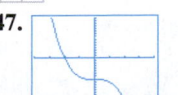   **47.** 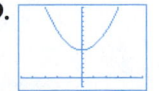   **49.**

51. 0, 1, 4, 1, 4   **53. a.** It costs 8¢ to make a 2-in. bolt.   **b.** 12¢   **c.** a 4-in. bolt   **d.** It decreases as the length approaches 4 in., then increases as the length increases to 7 in.   **55. a.** $90,000   **b.** the 3rd yr after being bought   **c.** after the 6th yr   **d.** It decreased in value for 3 yr, then increased in value for 5 yr.   **61.** −96   **63.** an expression   **65.** 1.25   **67.** 0.1

STUDY SET   SECTION 3.3 (page 219)

1. linear   **3.** y-intercept   **5.** parallel   **7. a.** nonlinear   **b.** linear   **c.** nonlinear   **d.** linear   **e.** nonlinear   **9.** 6, −5, 4   **11.** −2, 4, $-\frac{3}{2}$ **13.** because A is on the line   **15.** 1st power   **17.** x-intercept: $(-3, 0)$; y-intercept: $(0, -1)$ **19.** $y = b$ **21.** The student made a

mistake; the points should lie on a line.   **23.** x, y **25. a.** $(0, 0)$ **b.** $(0, 0)$ **c.** It takes two distinct points to determine a line.
27. The x-coordinate of the x-intercept is the solution: -1.   **29. a.** $4x - y = 6$ **b.** $x - 2y = 0$ **c.** $x + 3y = 9$ **d.** $x + 0y = 12$

31. 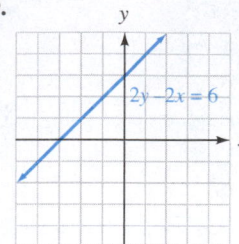   **33.** 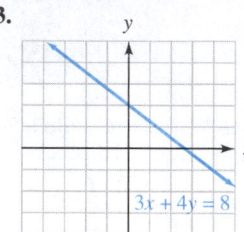   **35.** 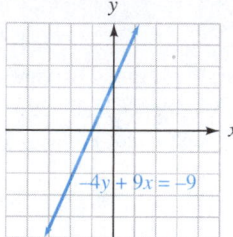   **37.**

39. 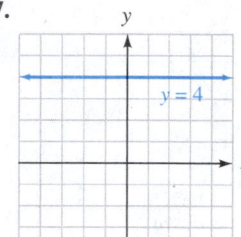   **41.** 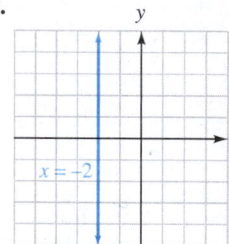   **43.** 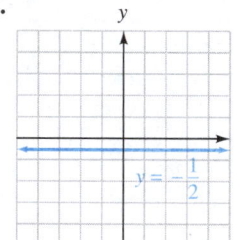   **45.**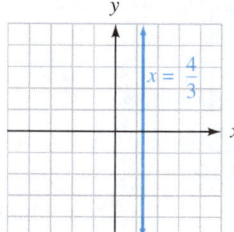

47. 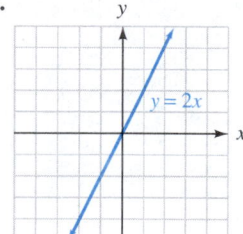   **49.**   **51.**   **53.**

55. 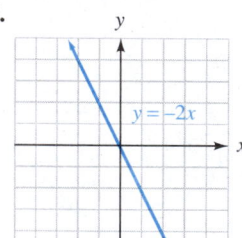   **57.**   **59.** 3   **61.** -1

63. a. $c = 50 + 25u$ **b.** 150, 250, 400   **c.** The service fee is \$50.   **d.** \$850

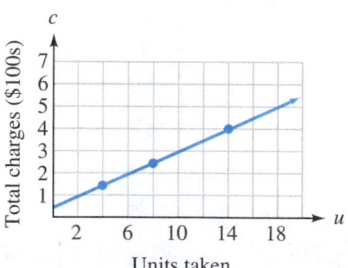

65. a. 56.2, 62.1, 64.0   **b.** taller the woman is.   **c.** 58 in.   **71.** $5 + 4c$ **73.** -4
75. profit = revenue − costs   **77.** 491

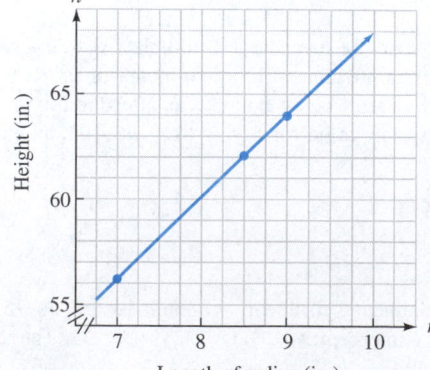

STUDY SET   SECTION 3.4 (page 231)

1. ratio **3.** slope **5.** change **7. a.** l_2 **b.** l_1 **c.** l_4 **d.** l_3 **9.** -1 **11.** 3 in./yr **13. a.** "The Lost World"; it reached $100 million in the fewest days, 6. **b.** LW: $\frac{100}{6} \approx \$16.7$ million/day; ID: $\frac{100}{7} \approx \$14.3$ million/day; JP: $\frac{100}{9} \approx \$11.1$ million/day **15.** 0; sales of 7-Up are not changing—each year the same number cases have been sold. **17.** $m = \dfrac{y_2 - y_1}{x_2 - x_1}$ or $m = \dfrac{y_1 - y_2}{x_1 - x_2}$ **19.** 1 **21.** -3 **23.** $\frac{5}{4}$ **25.** $-\frac{1}{2}$ **27.** $\frac{3}{5}$ **29.** 0 **31.** undefined **33.** $-\frac{2}{3}$ **35.** -4.75 **37.** $m = \frac{2}{3}$ **39.** $m = -\frac{7}{8}$ **41.**

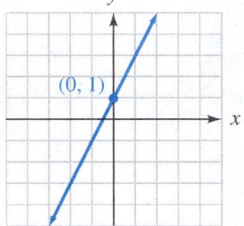
(0, 1)

43.

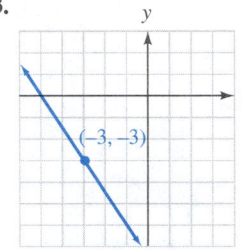
(−3, −3)

45.

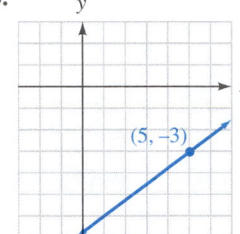
(5, −3)

47.

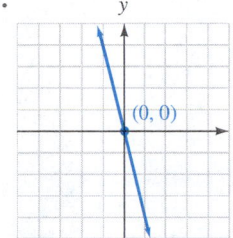
(0, 0)

49.

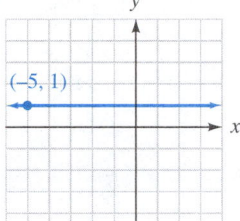
(−5, 1)

51.

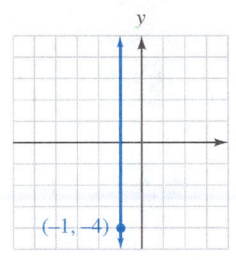
(−1, −4)

53. $\frac{2}{5}$ **55.** $\frac{1}{20}$; 5% **57. a.** $\frac{1}{8}$ **b.** $\frac{1}{12}$ **c.** 1: less expensive, steeper; 2: not as steep, more expensive **59.** 3 hp/40 rpm **65.** quadrant II **67.** no **69.** linear

STUDY SET   SECTION 3.5 (page 244)

1. slope–intercept **3.** parallel **5.** reciprocals **7.** no, because the graph is not a straight line **9. a.** When there are no head waves, the ship could travel at 18 knots **b.** $-\frac{1}{2}$ knot/ft **c.** $y = -\frac{1}{2}x + 18$ **11. a.** (0, 0) **b.** same slope; different y-intercepts **13. a.** $-\frac{2}{3}$ **b.** $\frac{1}{4}$ **c.** -8 **d.** 3 **e.** 1 **f.** -1 **15.** It shows that the y-intercept is (0, −1.25). **17.** $6x$, $-6x$, $3x$, 3, (0, −5) **19. a.** 4, (0, 2) **b.** −4, (0, −2) **c.** $\frac{1}{4}$, $\left(0, -\frac{1}{2}\right)$ **d.** 4, (0, −2)

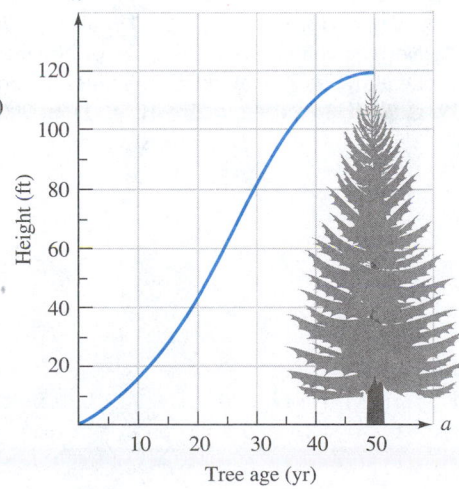

Height (ft)

Tree age (yr)

21. $y = 5x - 3$

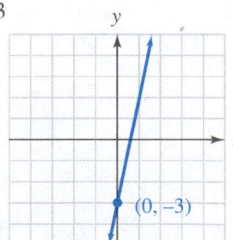
(0, −3)

23. $y = \frac{1}{4}x - 2$

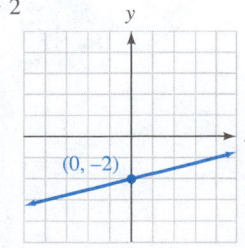
(0, −2)

25. $y = -\frac{8}{3}x + 5$

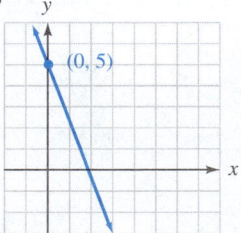
(0, 5)

27. $m = -\frac{3}{4}$, $(0, 4)$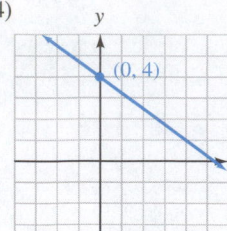

29. $m = 2$, $(0, -1)$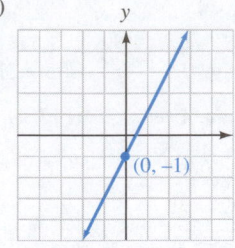

31.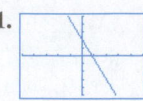

33. $(-1.63, 0)$ **35.** same slope **37. a.** $y = 2{,}000x + 5{,}000$ **b.** \$21,000 **39.** $y = 5x - 10$
41. a. $y = 0.20x + 1.00$ **b.**

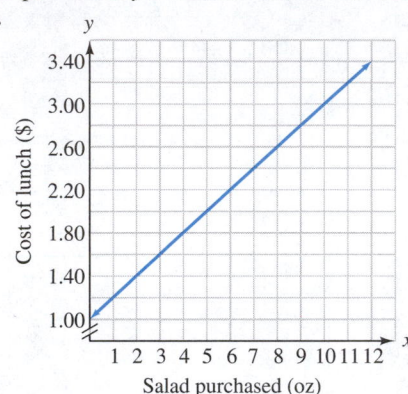

Cost of lunch (\$) vs. Salad purchased (oz)

c. same slope, different y-intercept **d.** same y-intercept, steeper slope **43.** $y = -20x + 500$ **45.** not quite: $(0.128)(-7.615) = -0.97472 \neq -1$ **51.** $-\frac{1}{4}$ **53.** 0
55. subtraction **57.** 25%

STUDY SET SECTION 3.6 (page 253)

1. point–slope **3.** x-coordinate, y-coordinate **5. a.** The slope is 2; the y-intercept is $(0, -3)$. **b.** The graph passes through $(5, 4)$; the slope is 6. **7. a.** In a group of 8,000 women, 108 would have given birth in 1997. **b.** $\frac{108}{8{,}000} = \frac{27}{2{,}000}$; the birth rate is 27 per 2,000 or 13.5 per 1,000. **9. a.** no **b.** no **c.** yes **11.** sub **13.** $2, -3x$ **15.** $5, -1, -2x$ **17.** $y - 1 = 3(x - 2)$ **19.** $y + 1 = -\frac{4}{5}(x + 5)$
21. $y = \frac{1}{5}x - 1$ **23.** $y = -5x - 37$ **25.** $y = -\frac{4}{3}x + 4$ **27.** $y = -\frac{2}{3}x + 2$ **29.** $y = 8x + 4$ **31.** $y = -3x$ **33.** $y = 2x + 5$
35. $y = -\frac{1}{2}x + 1$ **37.** $y = 5$ **39.** $x = 4$ **41.** $y = 5$ **43. a.** position 1: $(0, 0)$, $(-5, 2)$; position 2: $(0, 0)$, $(-3, 6)$; position 3: $(0, 0)$, $(-1, 7)$; position 4: $(0, 0)$, $(0, 10)$ **b.** $y = -\frac{2}{5}x$, $y = -7x$, $x = 0$ **c.** The pole is not in the shape of a straight line.
45. a. $y = -40x + 920$ **b.** 440 yd³ **47.** $c = 30t + 45$ **49. a.** $(0, 32)$; $(100, 212)$ **b.** $F = \frac{9}{5}C + 32$ **51.** $y = -\frac{4}{15}x + 83$ **57.** $-\frac{1}{2}$
59. 113.1 ft² **61.** -1 **63.** 6

STUDY SET SECTION 3.7 (page 263)

1. function **3.** independent, dependent **5. a.** positive numbers **b.** positive numbers **c.** 0 **d.** D: all reals; R: real numbers greater than or equal to 0 **7. a.** $(-2, 4)$, $(-2, -4)$ **b.** No; the x-value -2 is assigned to more than one y-value (4 and -4). **9. a.** 0 **b.** 1
c. 3 **11.** x **13.** $x, -5, 4, -5$ **15.** of, is **17.** yes **19.** yes **21.** no; $(4, 2)$, $(4, -2)$ **23.** yes **25.** yes **27.** no; $(3, 4)$, $(3, -1)$
29. No; you could have a shoe size of 10 when you are 25 and 26 years old: $(10, 25)$, $(10, 26)$. **31.** D: all reals; R: all reals
33. D: all reals; R: real numbers greater than or equal to 0 **35. a.** 3 **b.** -9 **c.** 0 **d.** 199 **37. a.** 0.32 **b.** 18 **c.** 2,000,000 **d.** $\frac{1}{32}$
39. a. 7 **b.** 14 **c.** 0 **d.** 1 **41. a.** 0 **b.** 990 **c.** -24 **d.** 210
43. $-2, -5, 1, 4$; D: all reals, R: all reals

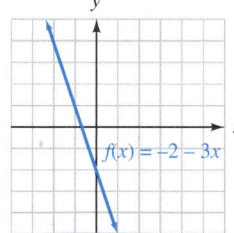

45. $2, 1, -2, 1, -2$; D: all reals, R: all real numbers less than or equal to 2

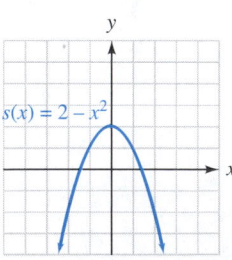

47. $f(x) = |x|$ **49. a.** $0 \leq x \leq 24$ **b.** 0.5 **c.** 1.5 **d.** -1.5 **e.** The low tide mark was -2.5 m. **f.** 1.6 **51.** 78.5 ft², 314.2 ft²,
1,256.6 ft² **57.** $y = 6$ **59.** profit = revenue − costs **61.** $-6x + 12$ **63.** $12d$

KEY CONCEPT (page 267)

1. $y = -3x - 4$ **2.** $y = \frac{1}{5}x + 1$ **3.** $-2, 4, -3$

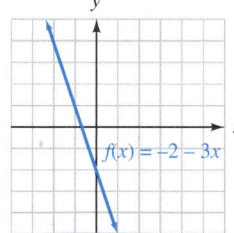
$2x - 4y = 8$

4. a. When new, the press cost \$40,000. **b.** $-5{,}000$; the value of the press decreased \$5,000/yr **5.** 130; the cost to rent the mixer for 3 days **6.** 80°F

CHAPTER REVIEW (page 269)

1. a. 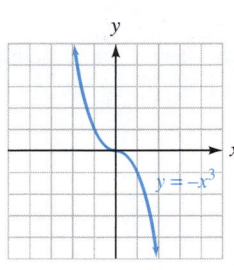 **b.** $-1, 0, 1$ **2.** quadrant III **3. a.** 2 ft **b.** 2 ft **c.** 6 ft **4. a.** 2,500; week 2 **b.** 1,000 **c.** 1st week and 5th week **5.** not a solution **6. a.** 8, $(-2, 8)$; 1, $(-1, 1)$; 0, $(0, 0)$; -1, $(1, -1)$; -8, $(2, -8)$

b. It would be 2 units higher.

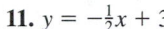

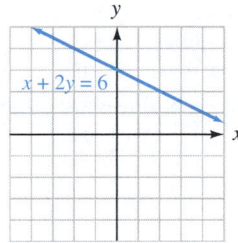

7. a. 9,000 **b.** It tells us that 40 trees on an acre give the highest yield, 18,000 oranges. **8. a.** nonlinear **b.** linear **c.** linear **d.** nonlinear **9.** $A = 5, B = 2, C = 10$ **10.** $-6, -6; -8, -8$

11. $y = -\frac{1}{2}x + 3$ **12.** x-intercept: $(-2, 0)$; y-intercept: $(0, 4)$

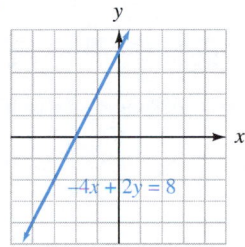

13. a. 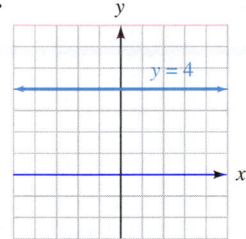 **b.** 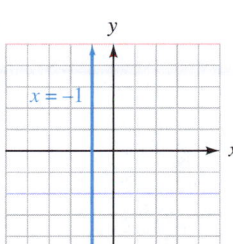 **14.** -2 **15. a.** $\frac{1}{4}$ **b.** -7 **c.** 0 **d.** $-\frac{3}{2}$ **16.**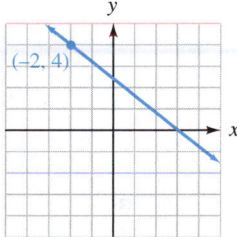

17. a. 1985–1987; -8.5 million salmon/yr **b.** 1987–1989; 6 million salmon/yr **18. a.** $m = \frac{3}{4}$; y-intercept: $(0, -2)$ **b.** $m = -4$; y-intercept: $(0, 0)$ **19.** $m = 3$; y-intercept: $(0, -5)$ **20. a.** $c = 300w + 75,000$ **b.** 90,600 **21. a.** parallel

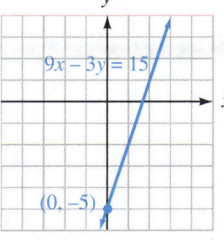

b. perpendicular **22. a.** $y = 3x + 2$ **b.** $y = -\frac{1}{2}x - 3$ **23. a.** $y = \frac{2}{3}x + 5$

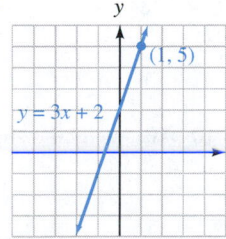

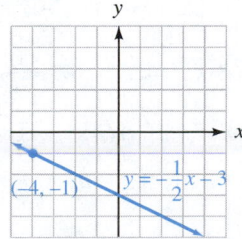

b. $y = -8$ **24.** $f = -35x + 450$ **25. a.** yes **b.** no **26. a.** D: all reals; R: real numbers less than or equal to 0 **b.** D: all reals;

R: real numbers greater than or equal to 0   **27. a.** -5 **b.** 37   **c.** -2 **d.** -8 **28.** 1, 0, -1, 0, -1, -2
29. a. no   **b.** yes   **30.** 1,004.8 in.^{3}

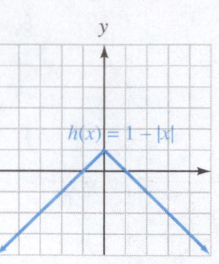

CHAPTER 3 TEST (page 274)

1. 10   **2.** 60   **3.** 1 day before and the 3rd day of the holiday   **4.** 50 dogs were in the kennel when the holiday began.
5.   **6.**   **7.** yes   **8.** no   **9.** x-intercept: (3, 0); y-intercept: (0, -2)   **10.** $m = -\frac{1}{2}$; (0, 4)

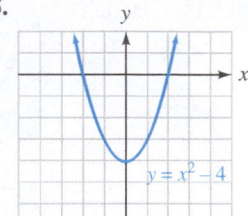   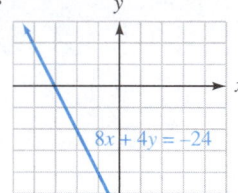

11.   **12.**   **13.** -1 **14.** undefined   **15.** $\frac{8}{7}$ **16.** parallel   **17.** the 15–20 mi segment:
30 ft/mi   **18.** the 22–25 mi segment: $-\frac{100}{3}$ ft/mi $= -33\frac{1}{3}$ ft/mi
19. $v = 15,000 - 1,500x$ **20.** $y = 7x + 19$ **21.** no   **22.** D: all reals;
R: real numbers less than or equal to 0   **23.** yes   **24.** -13 **25.** 756

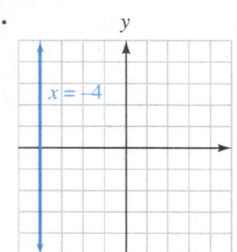   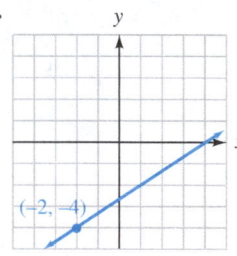

STUDY SET   SECTION 4.1 (page 282)

1. base, exponent   **3.** power   **5.** $3x, 3x, 3x, 3x$ **7.** x^{m+n} **9.** $\dfrac{a^n}{b^n}$ **11.** x^{m-n} **13.** 1   **15. a.** $2x^2$ **b.** 0   **c.** x^4 **d.** 1

17. a. doesn't simplify   **b.** doesn't simplify   **c.** $12x^5$ **d.** $3x$ **19.** a^{10} mi^2 **21.** $36x^7$ m^3 **23.** 3, 9, 27, 81   **25.** -4, 16, -64, 256
27. 2, 4, 8, 16   **29.** x^6 **31.** base 4, exponent 3   **33.** base x, exponent 5   **35.** base $-3x$, exponent 2   **37.** base y, exponent 6
39. $5 \cdot 5 \cdot 5$ **41.** $x \cdot x \cdot x \cdot x \cdot x \cdot x$ **43.** $-\frac{3}{4} \cdot x \cdot x \cdot x \cdot x \cdot x$ **45.** $\left(\frac{1}{3}t\right)\left(\frac{1}{3}t\right)\left(\frac{1}{3}t\right)$ **47.** $(4t)^4$ **49.** $-4t^3$ **51.** 1,122   **53.** $-3,625$ **55.** x^7
57. a^9 **59.** y^9 **61.** $12x^7$ **63.** $-4y^5$ **65.** 3^8 **67.** y^{15} **69.** x^{25} **71.** $243x^{30}$ **73.** x^{31} **75.** x^3y^3 **77.** r^6s^4 **79.** $16a^2b^4$ **81.** $-8r^6s^9$
83. $\dfrac{a^3}{b^3}$ **85.** $\dfrac{x^{10}}{y^{15}}$ **87.** $\dfrac{-32a^5}{b^5}$ **89.** $\dfrac{b^6}{27a^3}$ **91.** x^2 **93.** y^4 **95.** $3a$ **97.** ab^4 **99.** $\dfrac{10r^{13}s^3}{3}$ **101.** $\dfrac{y^3}{8}$ **103. a.** $25x^2$ ft^2 **b.** $9\pi x^2$ ft^2
105. b. 16 ft, 8 ft, 4 ft, 2 ft   **107.** $12^1, 12^2, 12^3, 12^4$ **109.** \$16,000   **115.** c   **117.** d

STUDY SET   SECTION 4.2 (page 290)

1. base, exponent   **3.** negative   **5. a.** $4 - 4$, 0   **b.** 6, 6, 6, 6, 1   **c.** 1, 1   **7.** 9, 3, 1, $\frac{1}{3}$, $\frac{1}{9}$ **9.** 81, -9, 1, $-\frac{1}{9}$, $\frac{1}{81}$ **11.** 4, 2^2; 2, 2^1; 1, 2^0;
$\frac{1}{2}$, 2^{-1}; $\frac{1}{4}$, 2^{-2} **13.** y^8, -40 **15.** base x, exponent -2 **17. a.** 4; 2; -16 **b.** 4; -2; $\frac{1}{16}$ **c.** 4; -2; $-\frac{1}{16}$ **19.** 1   **21.** 1   **23.** 2   **25.** 1
27. 1   **29.** $\dfrac{5}{2}$ **31.** $\dfrac{1}{144}$ **33.** $-\dfrac{1}{4}$ **35.** 125   **37.** $\dfrac{3}{16}$ **39.** $-\dfrac{1}{64}$ **41.** $\dfrac{1}{64}$ **43.** $\dfrac{1}{x^2}$ **45.** $-\dfrac{1}{b^5}$ **47.** $\dfrac{1}{16y^4}$ **49.** $\dfrac{1}{a^3b^6}$ **51.** 8   **53.** 1
55. $\dfrac{8}{7}$ **57.** 1   **59.** $\dfrac{1}{y}$ **61.** $\dfrac{1}{r^6}$ **63.** y^5 **65.** 2   **67.** $\dfrac{1}{a^2b^4}$ **69.** $\dfrac{1}{x^6y^3}$ **71.** $\dfrac{1}{x^3}$ **73.** $\dfrac{a^8}{b^{12}}$ **75.** $-\dfrac{y^{10}}{32x^{15}}$ **77.** a^{14} **79.** $\dfrac{1}{b^{14}}$ **81.** $\dfrac{256x^{28}}{81}$
83. $\dfrac{16y^{14}}{z^{10}}$ **85.** x^{3m} **87.** $\dfrac{1}{u^m}$ **89.** y^{2m+2} **91.** y^m **93.** $\dfrac{1}{x^{3n}}$ **95.** x^{2m+2} **97.** $10^2, 10^1, 10^0, 10^{-1}, 10^{-2}, 10^{-3}, 10^{-4}$
99. approximately \$4,603.09   **103.** 13.5 yr   **105.** $y = \frac{3}{4}x - 5$

STUDY SET   SECTION 4.3 (page 297)

1. scientific   **3.** 250   **5.** 0.000025   **7.** 10^5 **9.** 10^{-3} **11.** 5, right   **13.** 10^{-4} **15.** positive   **17.** 6.37×10^1, 1   **19.** 2.3×10^4
21. 1.7×10^6 **23.** 6.2×10^{-2} **25.** 5.1×10^{-6} **27.** 4.25×10^3 **29.** 2.5×10^{-3} **31.** 230   **33.** 812,000   **35.** 0.00115
37. 0.000976   **39.** 25,000,000   **41.** 0.00051   **43.** 2.57×10^{13} mi   **45.** 63,800,000 mi^2 **47.** 6.22×10^{-3} mi   **49.** 714,000
51. 30,000   **53.** 200,000   **55.** $9.038030748 \times 10^{15}$ **57.** $1.881676423 \times 10^{12}$ **59.** 5.85×10^{15} **61.** g, x, u, v, i, m, r
63. 1.7×10^{-18} g   **65.** 3.099363×10^{16} ft   **67.** 1.368×10^{11} dollars   **69. a.** 1.7×10^6; 1,700,000   **b.** 1.5×10^6, 1,500,000;
2.05×10^6; 2,050,000   **75.** 5   **77.** commutative property of addition   **79.** 6

STUDY SET   SECTION 4.4 (page 307)

1. polynomial **3.** term **5.** degree **7.** in, decreasing **9.** of **11.** yes **13.** no **15.** binomial **17.** trinomial **19.** monomial
21. binomial **23.** trinomial **25.** none of these **27.** 4th **29.** 2nd **31.** 1st **33.** 4th **35.** 12th **37.** 0th **39.** $(-2, -5)$
41. 2, 2, 4, 6, 1, -2 **43.** because $x^3 + 2x^2 - 3$ is a polynomial **45.** 7 **47.** -8 **49.** -2 **51.** -7.5 **53.** -4 **55.** -5 **57.** -5.69
59. -188.96 **61.** 3 **63.** -1 **65.** 1.929 **67.** -512.241 **69.** 3, 0, -1, 0, 3

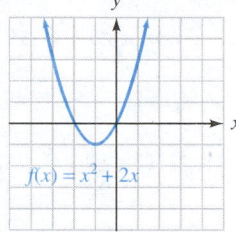

71. -16, 0, 4, 2, 0, 4 **73.** 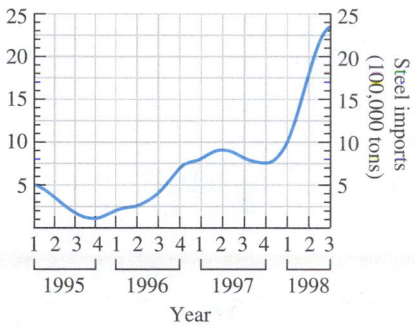 **75.** about 5.4 m **77.** 22 ft, 2 ft **79.** about 404 ft

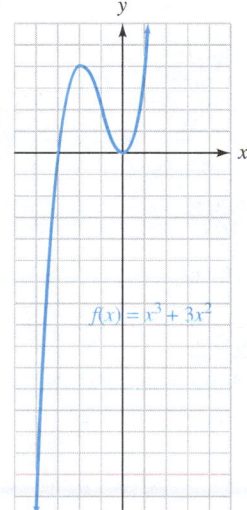

81. 18.75 ft, 20 ft, 15 ft **83.** the 4th quarter of 1995; the 3rd quarter of 1998
87. $y \geq -3$ **89.** x^{18} **91.** y^9

STUDY SET   SECTION 4.5 (page 315)

1. polynomials **3.** coefficients **5.** terms **7.** like **9.** changing **11.** $2x^2 + 3x - 4$ **13.** $(11x - 12)$ ft **15.** $3x$, $2x$, $7x^2$ **17.** $9y$
19. $12t^2$ **21.** $-48u^3$ **23.** $-0.1x$ **25.** $2s$ **27.** $6r$ **29.** $-ab$ **31.** $15x^2$ **33.** $7x + 4$ **35.** $2a + 7$ **37.** $7x - 7y$ **39.** $3x - 4y$
41. $6x^2 + x - 5$ **43.** $7b + 4$ **45.** $3x + 1$ **47.** $5x^2 + x + 11$ **49.** $-7x^3 - 7x^2 - x - 1$ **51.** $2x^2 + 4x + 13$ **53.** $5x^2 + 6x - 8$
55. $-x^3 + 6x^2 + x + 14$ **57.** $-12x^2y^2 - 9xy + 36y^2$ **59.** $t^3 + 3t^2 + 6t - 5$ **61.** $-3x^2 + 5x - 7$ **63.** $6x - 2$ **65.** $-5x^2 - 8x - 4$
67. $4y^3 - 12y^2 + 8y + 8$ **69.** $3a^2b^2 - 6ab + b^2 - 6ab^2$ **71.** $(3x^2 + 6x - 2)$ yd **73.** $(x^2 - 8x + 12)$ ft **75.** $(3x^2 + 11x + 4.5\pi)$ in.
77. \$114,000 **79.** $y = 1,900x + 225,000$ **81.** $y = -1,100x + 6,600$ **83.** $y = -2,800x + 15,800$ **89.** $180°$
91. $x \leq 2$

STUDY SET   SECTION 4.6 (page 325)

1. binomials **3.** distributive **5.** $6x^2$ **7.** $15x$ **9.** $(6x^2 + x - 1)$ cm^2 **11.** $7x$, $7x$, $7x$ **13.** $12x^5$ **15.** $-24b^6$ **17.** $6x^5y^5$ **19.** $-3x^4y^5z^8$
21. $3x + 12$ **23.** $-4t - 28$ **25.** $3x^2 - 6x$ **27.** $-6x^4 + 2x^3$ **29.** $3x^2y + 3xy^2$ **31.** $6x^4 + 8x^3 - 14x^2$ **33.** $-6x^4 - 24x^3$
35. $a^2 + 9a + 20$ **37.** $3x^2 + 10x - 8$ **39.** $6a^2 + 2a - 20$ **41.** $6x^2 - 7x - 5$ **43.** $2x^2 + 3x - 9$ **45.** $6t^2 + 7st - 3s^2$
47. $x^2 + xz + xy + yz$ **49.** $-12t^2 + 7tu - u^2$ **51.** $8x^2 - 12x - 8$ **53.** $3a^3 - 3ab^2$ **55.** $5t^2 - 11t$ **57.** $2x^2 + xy - y^2$
59. $x^2 + 8x + 16$ **61.** $t^2 - 6t + 9$ **63.** $r^2 - 16$ **65.** $16x^2 - 25$ **67.** $4s^2 + 4s + 1$ **69.** $x^2 + 10x + 25$ **71.** $x^2 - 4xy + 4y^2$
73. $4a^2 - 12ab + 9b^2$ **75.** $16x^2 + 40xy + 25y^2$ **77.** $(4x^2 - 6x + 2)$ cm^2 **79.** $(x^2 + 6x + 9)\pi$ in.^{2} **81.** $x^3 - x + 6$
83. $4t^3 + 11t^2 + 18t + 9$ **85.** $-3x^3 + 25x^2y - 56xy^2 + 16y^3$ **87.** $x^3 - 3x + 2$ **89.** $12x^3 + 17x^2 - 6x - 8$ **91.** -3 **93.** -8

95. -1 **97.** 0 **99.** $(24x + 14)$ cm, $(35x^2 + 43x + 12)$ cm^2 **101. a.** x^2 ft^2, $6x$ ft^2, $5x$ ft^2, 30 ft^2; $(x^2 + 11x + 30)$ ft^2 **b.** $(x + 6)$ ft, $(x + 5)$ ft; $(x^2 + 11x + 30)$ ft^2 **c.** They are the same. **103.** 5 and 6 **105.** 4 m **107.** 90 ft **111.** 1 **113.** $-\frac{2}{3}$ **115.** $(0, 2)$

STUDY SET   SECTION 4.7 (page 332)

1. polynomial **3.** two **5.** reciprocal **7.** $\frac{a}{b}$ **9.** In the numerator and denominator, a common factor of 2 was divided out. **11. a.** $t = \frac{d}{r}$
b. $3x^2$ **13.** $2x + 7$ **15.** b, b, b, a, a, a, b, a **17.** $\frac{1}{3}$ **19.** $-\frac{5}{3}$ **21.** $\frac{3}{4}$ **23.** 1 **25.** x^3 **27.** $\frac{r^2}{s}$ **29.** $\frac{2x^2}{y}$ **31.** $-\frac{3u^3}{v^2}$ **33.** $\frac{4r}{y^2}$
35. $-\frac{13}{3rs}$ **37.** $\frac{x^4}{y^6}$ **39.** $a^8 b^8$ **41.** $-\frac{3r}{s^9}$ **43.** $-\frac{x^3}{4y^3}$ **45.** $-\frac{16}{y^6}$ **47.** a^8 **49.** $2x + 3$ **51.** $\frac{1}{5y} - \frac{2}{5x}$ **53.** $\frac{1}{y^2} + \frac{2y}{x^2}$
55. $3a - 2b$ **57.** $\frac{1}{y} - \frac{1}{2x} + \frac{2z}{xy}$ **59.** $3x^2 y - 2x - \frac{1}{y}$ **61.** $5x - 6y + 1$ **63.** $\frac{10x^2}{y} - 5x$ **65.** $-\frac{4x}{3} + \frac{3x^2}{2}$ **67.** $xy - 1$
69. $\frac{x}{y} - \frac{11}{6} + \frac{y}{2x}$ **71.** 2 **73.** $(2x^2 - x + 3)$ in. **75.** $(3x - 2)$ ft **77.** yes **79.** no **83.** binomial **85.** none of the above **87.** 2

STUDY SET   SECTION 4.8 (page 338)

1. divisor, dividend **3.** remainder **5.** $4x^3 - 2x^2 + 7x + 6$ **7.** $6x^4 - x^3 + 2x^2 + 9x$ **9.** $0x^3$ and $0x$ **11.** $x^3 + 3x^2 + 9x + 27$
13. a. $r = \frac{d}{t}$ **b.** $x - 3$ **15.** It is correct. **17.** $x, 2x, 2x$ **19.** $x + 6$ **21.** $y + 12$ **23.** $3a - 2$ **25.** $b + 3$ **27.** $2x + 1$ **29.** $x - 7$
31. $3x + 2$ **33.** $2x - 1$ **35.** $x^2 + 2x - 1$ **37.** $2x^2 + 2x + 1$ **39.** $x^2 + x + 1$ **41.** $x + 1 + \frac{-1}{2x+3}$ **43.** $2x + 2 + \frac{-3}{2x+1}$
45. $x^2 + 2x + 1$ **47.** $x^2 + 2x - 1 + \frac{6}{2x+3}$ **49.** $2x^2 + 8x + 14 + \frac{31}{x-2}$ **51.** $x + 1$ **53.** $2x - 3$ **55.** $x^2 - x + 1$
57. $a^2 - 3a + 10 + \frac{-30}{a+3}$ **59.** $5x^2 - x + 4 + \frac{16}{3x-4}$ **61. a.** $(x - 6)$ in. **b.** $(4x - 4)$ in. **63.** $4x^2 + 3x + 7$ **67.** x^{22}
69. $8x^2 - 6x + 1$ **71.** They are the same.

KEY CONCEPT (page 342)

1. combine **2.** change **3.** each, each **4.** long **5.** $3x - 5$ **6.** $x + 11$ **7.** $2x^2 - 13x - 24$ **8.** $4x^4 + 12x^2 + 9$ **9.** $y^2 + 2y - 3$
10. $y^2 - 9$ **11.** $y^3 + 4y^2 - 3y - 18$ **12.** $y - 2$ **13.** 7, $(-2, 7)$; 0, $(-1, 0)$; -1, $(0, -1)$; -2, $(1, -2)$;
-9, $(2, -9)$ **14.** $(2, 5.5)$; for 1990–1997, youth drug use was the lowest in 1992, at 5.5%

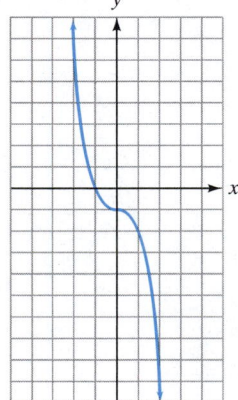

CHAPTER REVIEW (page 344)

1. a. $-3 \cdot x \cdot x \cdot x \cdot x$ **b.** $\left(\frac{1}{2}pq\right)\left(\frac{1}{2}pq\right)\left(\frac{1}{2}pq\right)$ **2. a.** 125 **b.** 64 **c.** -64 **d.** 4 **3. a.** x^5 **b.** $-3y^6$ **c.** y^{21} **d.** $81x^4$ **e.** b^{12} **f.** $-y^2 z^5$
g. $256s^3$ **h.** $4x^4 y^2$ **i.** x^{15} **j.** $\frac{x^2}{y^2}$ **k.** x^4 **l.** $5yz^4$ **4. a.** $64x^{12}$ in.^{3} **b.** y^4 m^2 **5. a.** 1 **b.** 1 **c.** 9 **d.** $\frac{1}{1,000}$ **e.** $\frac{4}{3}$ **f.** $-\frac{1}{25}$ **g.** $\frac{1}{x^5}$
h. $-\frac{6}{y}$ **i.** $\frac{1}{x^{10}}$ **j.** x^{14} **k.** $\frac{1}{x^5}$ **l.** $\frac{1}{9z^2}$ **6. a.** y^{7n} **b.** $\frac{1}{z^{2c}}$ **7. a.** 7.28×10^2 **b.** 9.37×10^6 **c.** 1.36×10^{-2} **d.** 9.42×10^{-3}
e. 1.8×10^{-4} **f.** 7.53×10^5 **8. a.** 726,000 **b.** 0.000391 **c.** 2.68 **d.** 57.6 **9. a.** 0.03 **b.** 160 **10.** 5,927,000,000; 5.927×10^9
11. $1.0 \times 10^5 = 100,000$ **12. a.** yes **b.** no **c.** no **d.** yes **13. a.** 7th, monomial **b.** 3rd, monomial **c.** 2nd, binomial **d.** 5th,
trinomial **e.** 6th, binomial **f.** 4th, none of these **14. a.** 34 **b.** 1 **c.** 9 **d.** 0.72 **15.** -16, $(-3, -16)$; 0, $(-2, 0)$; 4, $(-1, 4)$; 2,
$(0, 2)$; 0, $(1, 0)$; 4, $(2, 4)$; 20, $(3, 20)$ **16.** 8 in. **17. a.** $2x^6 + 5x^5$ **b.** $-2x^2 y^2$ **c.** $8x^2 - 6x$ **d.** $5x^2 + 19x + 3$
18. a. $4x^2 + 2x + 8$ **b.** $8x^3 - 7x^2 + 19x$ **19. a.** $10x^3$ **b.** $-6x^{10} z^5$
c. $-6r^3 s^4 t^5$ **d.** $120b^{11}$ **20. a.** $5x + 15$ **b.** $3x^4 - 5x^2$ **c.** $x^2 y^3 - x^3 y^2$
d. $-2y^4 + 10y^3$ **e.** $6x^6 + 12x^5$ **f.** $-3x^3 + 3x^2 - 6x$ **21. a.** $x^2 + 5x + 6$
b. $2x^2 - x - 1$ **c.** $6a^2 - 6$ **d.** $6a^2 - 6$ **e.** $2a^2 - ab - b^2$
f. $-6x^2 - 5xy - y^2$ **22. a.** $x^2 + 6x + 9$ **b.** $x^2 - 25$ **c.** $a^2 - 6a + 9$
d. $x^2 + 8x + 16$ **e.** $4y^2 - 4y + 1$ **f.** $y^4 - 1$
23. a. $3x^3 + 7x^2 + 5x + 1$ **b.** $8a^3 - 27$ **24. a.** 1 **b.** -1 **c.** 7 **d.** 0
25. $(6x + 10)$ in.; $(2x^2 + 11x - 6)$ in.^{2}; $(6x^3 + 33x^2 - 18x)$ in.^{3}
26. a. $-\frac{2x}{3y^2}$ **b.** $\frac{1}{x}$ **27. a.** $4x + 3$ **b.** $2 - \frac{3}{y}$ **c.** $3a + 4b - 5$

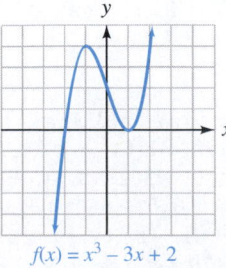

$f(x) = x^3 - 3x + 2$

d. $-\dfrac{x}{y} - \dfrac{y}{x}$ **28.** $x + 5$ **29. a.** $x + 1 + \dfrac{3}{x+2}$ **b.** $x - 5$ **c.** $2x + 1$ **d.** $x + 5 + \dfrac{3}{3x-1}$ **e.** $3x^2 + 2x + 1 + \dfrac{2}{2x-1}$ **f.** $3x^2 - x - 4$
31. $(4x + 3)$ in./min

CHAPTER 4 TEST (page 348)

1. $2x^3y^4$ **2.** 64 **3.** y^6 **4.** $32x^{21}$ **5.** 3 **6.** $\dfrac{2}{y^3}$ **7.** y^3 **8.** $\dfrac{64a^3}{b^3}$ **9.** $1{,}000y^{12}$ in.^{3} **10.** $\dfrac{1}{4^2}, \dfrac{1}{16}$ **11.** 6.25×10^{18} **12.** 0.000093
13. binomial   **14.** 5th degree   **15.** 0 **16.** $-3x^2y^2$ **17.** $-7x + 2y$ **18.** $-x^2 - 5x + 4$ **19.** $-4x^5y$ **20.** $3y^4 - 6y^3 + 9y^2$
21. $x^2 - 81$ **22.** $9y^2 - 24y + 16$ **23.** $6x^2 - 7x - 20$ **24.** $2x^3 - 7x^2 + 14x - 12$ **25.** $\frac{1}{2}$ **26.** $\frac{y}{2x}$ **27.** $\frac{a}{4b} - \frac{b}{2a}$ **28.** $x - 2$
30. $(x - 5)$ ft   **31.** $15, (-2, 15); -1, (-1, -1); -5, (0, -5); -3, (1, -3); -1, (2, -1); -5, (3, -5)$
32. $(1, 9), (6, 21)$; the smallest number of tons, about 9 million, was sold in 1991. The largest number of tons, about 21 million, was sold in 1996.

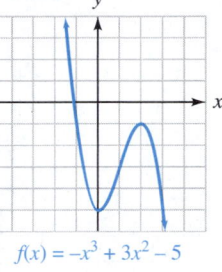

CHAPTERS 1–4 CUMULATIVE REVIEW EXERCISES (page 349)

1. The negative savings rate means that Americans spent more than they earned that month.   **2.** $19, 16$ **3.** 5 **4.** $37y$ **5.** $-2x + 2y$
6. $x - 5$ **7.** x^2y^3 **8.** $4x^2$ **9.** 13 **10.** 41 **11.** $h = \frac{2A}{b+B}$ **12.** $x = \frac{y-b}{m}$ **13.** -9 **14.** 1 **15.** 4 **16.** -2
17. $x < -14$ 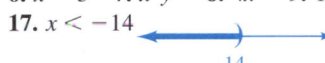   **18.** $-5 < x \le -1$ 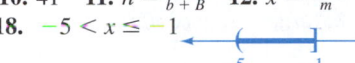   **19.** 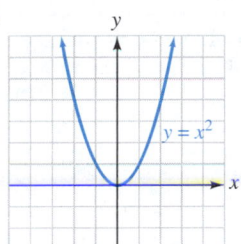   **20.**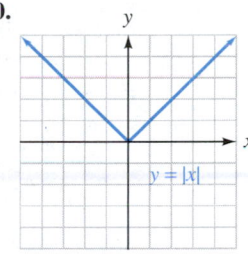

21. 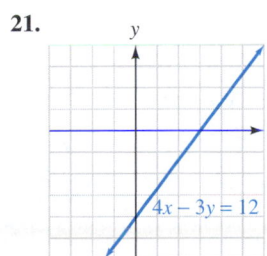   **22.**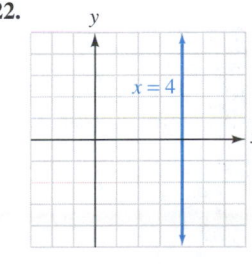

23. $\frac{1}{2}$ **24.** 0 **25.** -4 **26.** $\frac{2}{3}$ **27.** $y = \frac{2}{3}x + 5$ **28.** $3x - 4y = -22$
29. $y = 4$ **30.** $x = 2$ **31.** perpendicular   **32.** parallel   **33.** yes   **34.** no
35. -3 **36.** 15 **37.** 5 **38.** -2.5 **39.** D: all real numbers, R: all real numbers greater than or equal to 0   **40.** no   **41.** The temperature is rising at a rate of $3°$/hr; the temperature is falling at a rate of $-3°$/hr   **42.** of
43. y^{14} **44.** x^{14} **45.** x^2 **46.** a^7 **47.** $\dfrac{1}{x^5}$ **48.** $\dfrac{1}{16y^4}$ **49.** $\dfrac{1}{x^8}$ **50.** $-x^{15}$
51. 6.15×10^5 **52.** 1.3×10^{-6} **53.** 0.000525 **54.** $2{,}770$ **55.** 2 **56.** 5
57. 1.5 in.

58. $11, (-1, 11); -1, (0, -1); -5, (1, -5); -1, (2, -1); 11, (3, 11)$

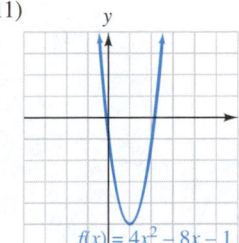

59. $x^2 + 4x - 14$ **60.** $4x^2 - x - 1$
61. $-35x^5 + 10x^4 + 10x^2$ **62.** $-12x^5y^5$
63. $6x^2 + 10x - 56$
64. $15x^2 - 2xy - 8y^2$
65. $9x^2 + 6x + 1$ **66.** $x^3 - 8$
67. $3x - 4$ **68.** $2x + 1$

STUDY SET   SECTION 5.1 (page 360)

1. prime   **3.** largest   **5.** factoring   **7.** The 0 in the first line should be a 1.   **9.** The GCF is $6a^2$, not $6a$.   **11. a.** the distributive property   **13.** $3x$ **15.** Use the FOIL method to multiply $(j + 2)(3j^2 + 2)$. The result should be $3j^3 + 6j^2 + 2j + 4$.   **17.** $b^2, (b - 6),$
$(b^2 + 2)$ **19.** 1 **21.** $2^2 \cdot 3$ **23.** $3 \cdot 5$ **25.** $2^3 \cdot 5$ **27.** $2 \cdot 7^2$ **29.** $3^2 \cdot 5^2$ **31.** $2^5 \cdot 3^2$ **33.** 4 **35.** $4, x$ **37.** $3(x + 2)$
39. $2\pi(R - r)$ **41.** $t^2(t + 2)$ **43.** $a^2(a - 1)$ **45.** $8xy^2(3xy + 1)$ **47.** $6uvw^2(2w - 3v)$ **49.** $3(x + y - 2z)$ **51.** $a(b + c - d)$
53. $3r(4r - s + 3rs^2)$ **55.** $\pi(R^2 - ab)$ **57.** $(x + 2)(3 - x)$ **59.** $(14 + r)(h^2 + 1)$ **61.** $-(a + b)$ **63.** $-(2x - 5y)$
65. $-(3m + 4n - 1)$ **67.** $-(3ab + 5ac - 9bc)$ **69.** $-3x(x + 2)$ **71.** $-4a^2b^2(b - 3a)$ **73.** $-2ab^2c(2ac - 7a + 5c)$
75. $(x + y)(2 + a)$ **77.** $(r + s)(7 - k)$ **79.** $(r + s)(x + y)$ **81.** $(2x + 3)(a + b)$ **83.** $(b + c)(2a + 3)$ **85.** $(3x - 1)(2x - 5)$
87. $(3p + q)(3m - n)$ **89.** $(2x + y)(y - 1)$ **91.** $(2z^3 + 3)(4z^2 - 5)$ **93.** $x^2(a + b)(x + 2y)$ **95.** $4a(b + 3)(a - 2)$

97. $y(x^2 - y)(x - 1)$ **99. a.** $12x^3$ in.^{2} **b.** $20x^2$ in.^{2} **c.** $4x^2(3x - 5)$ in.^{2} **101. a.** $4r$ in.; $16r$ in.^{2} **b.** $4\pi r^2$ in.^{2}
c. $16r^2 - 4\pi r^2 = 4r^2(4 - \pi)$ in.^{2} **107.** 13   **109.** yes

STUDY SET   SECTION 5.2 (page 365)

1. difference   **3.** $F - L$ **5. a.** $6x$ **b.** $2y^2$ **c.** $7ab^2$ **7.** x^2 is the square of x, and 25 is the square of 5.   **9.** $4y^2$, 5, $2y$, $2y$ **11.** 7, 7
13. t, w **15.** $6u, 5t$ **17.** $(x + 4)(x - 4)$ **19.** $(2y + 1)(2y - 1)$ **21.** $(3x + y)(3x - y)$ **23.** $(4a + 5b)(4a - 5b)$ **25.** prime
27. $(a^2 + 2b)(a^2 - 2b)$ **29.** $8(x + 2y)(x - 2y)$ **31.** $2(a + 1)(a - 1)$ **33.** $3(r + 2s)(r - 2s)$ **35.** $x(x + y)(x - y)$
37. $x(2a + 3b)(2a - 3b)$ **39.** $(x^2 + 9)(x + 3)(x - 3)$ **41.** $(a^2 + 4)(a + 2)(a - 2)$ **43.** $(9r^2 + 16s^2)(3r + 4s)(3r - 4s)$
45. $2(x^2 + y^2)(x + y)(x - y)$ **47.** $(a + 3)^2(a - 3)$ **49.** $(y + 4)(y - 4)(y - 3)$ **51.** $3(x + 2)(x - 2)(x + 1)$
53. $3(m + n)(m - n)(m + a)$ **55.** $2(m + 4)(m - 4)(mn^2 + 4)$ **57.** $16(4 + t)(4 - t)$ **59.** $\pi(R + r)(R - r)$ **63.** 12   **65.** -12 **67.** $\frac{3}{2}$
69. $(-15, \infty)$ **71.** $F = \frac{9}{5}C + 32$

STUDY SET   SECTION 5.3 (page 377)

1. trinomial   **3.** factors   **5.** prime   **7.** $4 \cdot 1, 2 \cdot 2$ **9.** $(x + y)^2$ **11.** descending   **13.** $1 + 8 = 9, 2 + 4 = 6, -1 + (-8) = -9,$
$-2 + (-4) = -6$ **15. a.** 1   **b.** -15; -5 and 3   **c.** -2; -5 and 3   **17.** 3, 3, 2   **19.** 2, 1   **21.** $t, 2$ **23.** $+, -$ **25.** $(z + 11)(z + 1)$
27. $(m - 3)(m - 2)$ **29.** $(a - 5)(a + 1)$ **31.** $(x + 8)(x - 3)$ **33.** $(a - 13)(a + 3)$ **35.** prime   **37.** $(s + 13)(s - 2)$ **39.** prime
41. $(m - 4)(m + 3)$ **43.** $(x + 3)^2$ **45.** $(y - 4)^2$ **47.** $(t + 10)^2$ **49.** $(u - 9)^2$ **51.** $-(x + 5)(x + 2)$ **53.** $-(t + 17)(t - 2)$
55. $-(r - 10)(r - 4)$ **57.** 1   **59.** 3, $4z$ **61.** $-, +$ **63.** $(2x - 1)(x - 1)$ **65.** $(3y + 2)(2y + 1)$ **67.** $(3x - 2)(2x - 1)$
69. $(2x + 1)(x - 2)$ **71.** $(5y + 1)(2y - 1)$ **73.** $(3y - 2)(4y + 1)$ **75.** $(5t + 3)(t + 2)$ **77.** $(8m - 3)(2m - 1)$ **79.** $(x - 4)(x - 1)$
81. $(y + 9)(y + 1)$ **83.** $-(r - 2)(r + 1)$ **85.** $2(x + 3)(x + 2)$ **87.** $3y(y + 1)(y + 1)$ **89.** $-5(a - 3)(a - 2)$ **91.** $3(z - 4)(z - 1)$
97. 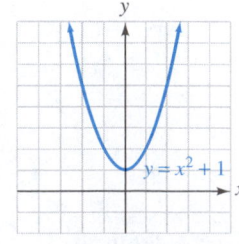   **99.**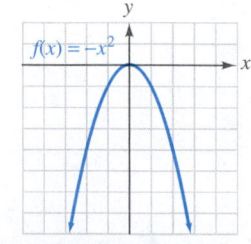

STUDY SET   SECTION 5.4 (page 384)

1. quadratic   **3.** zero, 0, 0   **5.** zero   **7. a.** quadratic   **b.** linear   **c.** linear   **d.** quadratic   **9. a.** -16 **b.** $(x - 2)(x + 8)$ **c.** $2, -8$
11. a. Add 6 to both sides.   **b.** Multiply both sides by 2.   **13.** $7y, y + 2$ **15.** $2, -3$ **17.** $\frac{5}{2}, -6$ **19.** $1, -2, 3$ **21.** $0, 3$ **23.** $0, \frac{5}{2}$
25. $0, 7$ **27.** $0, -\frac{8}{3}$ **29.** $0, 2$ **31.** $-5, 5$ **33.** $-\frac{1}{2}, \frac{1}{2}$ **35.** $-\frac{2}{3}, \frac{2}{3}$ **37.** $-10, 10$ **39.** $-\frac{9}{2}, \frac{9}{2}$ **41.** $12, 1$ **43.** $-3, 7$ **45.** $8, 1$
47. $-3, -5$ **49.** $-4, 2$ **51.** $0, -1, -2$ **53.** $0, 9, -3$ **55.** $1, -2, -3$ **57.** $\frac{1}{2}, 2$ **59.** $\frac{1}{5}, 1$ **61.** $-\frac{1}{2}, -\frac{1}{2}$ **63.** $\frac{2}{3}, -\frac{1}{5}$ **65.** $-\frac{5}{2}, 4$
67. $\frac{1}{8}, 1$ **69.** $0, -3, -\frac{1}{3}$ **71.** $0, -3, -3$ **73.** 9 sec   **75.** $\frac{3}{2} = 1.5$ sec   **77.** 2 sec   **79.** 8   **81.** 4 m by 9 m   **83.** 10 in., 16 in.
85. 50 m   **87.** $h = 5$ ft, $b = 12$ ft   **89.** 5 in.   **91.** 4 m   **95.** $15 \min \le E < 30 \min$ **97.** 675 cm^2

STUDY SET   SECTION 5.5 (page 397)

1. square root   **3.** quadratic   **5.** quadratic   **7.** two   **9. a.** ± 9 **b.** ± 9 **11.** because $x^2 - 2x - 1 = 0$ doesn't factor   **13.** quadratic
15. $-4, 8, 0$ **17.** 1.47, 0.53   **19.** $\sqrt{9}, 1; y - 1, -3$ **21.** two; 3.2, -3.2 **23.** The student didn't extend the fraction bar so that it
underlines the complete numerator.   **25.** 8.06   **27.** 0.95   **29.** ± 1 **31.** ± 5.2 **33.** ± 4.5 **35.** ± 3 **37.** ± 6.8 **39.** 7.5, -11.5
41. $-6, 4$ **43.** 7, -11 **45.** 4.83, -0.83 **47.** 0, 4   **49.** $2, -\frac{2}{3}$ **51.** $a = 1, b = 4, c = 3$ **53.** $a = 3, b = -2, c = 7$ **55.** $a = 4$,
$b = -2, c = 1$ **57.** $a = 3, b = -5, c = 2$ **59.** 2, 3   **61.** $-3, -4$ **63.** $1, -\frac{1}{2}$ **65.** $-1, -\frac{2}{3}$ **67.** $\frac{1}{2}, -\frac{3}{2}$ **69.** 7.646, 2.354   **71.** 0.414,
-2.414 **73.** $-0.382, -2.618$ **75.** $-2, 3$ **77.** ± 4.24 **79.** 2.41, -0.41 **81.** 1.35, -1.85 **83.** $-5, 7$ **85.** 0.82, 3.13

87. 20 ft by 40 ft   **89.** 4 ft by 8 ft   **91.** 6 in.   **93.** 10 ft by 18 ft   **95.** about 9.5 sec   **97.** 7%   **99.** 600   **101.** 2 in.   **109.** $M = \dfrac{Fd^2}{Gm}$

111. $4x - 3y = 0$

STUDY SET   SECTION 5.6 (page 409)

1. quadratic   **3.** y-intercept   **5.** symmetry   **7.** $a > 0$ **9.** $(0, c)$ **11. a.** a parabola   **b.** $(1, 0), (3, 0)$ **c.** $(0, -3)$ **d.** $(2, 1)$
15. The most cases of flu (25) were reported the fifth week.   **17.** 1, 3   **19.** true
21. moved up 1   **23.** opens the opposite direction   **25.** $(3, 1)$ **27.** $(1, -1)$ **29.** $(1, 0); (0, 1)$
                                                                      31. $(-3, 0), (-7, 0); (0, -21)$

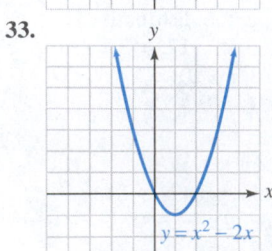

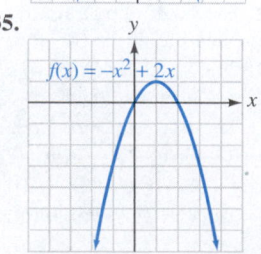

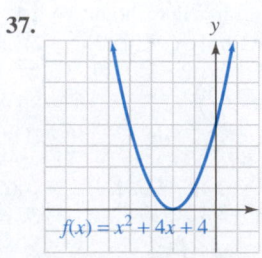

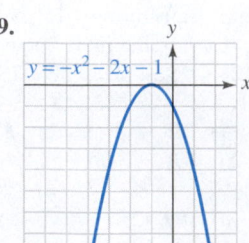

41.

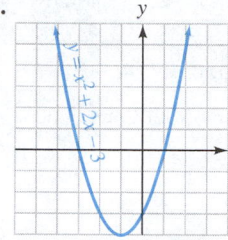

43.

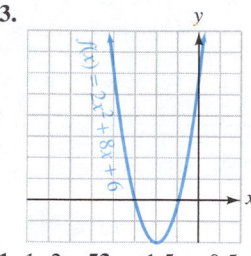

45.

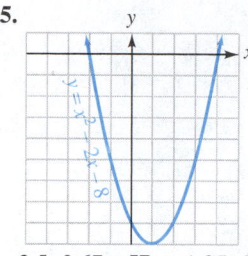

47.

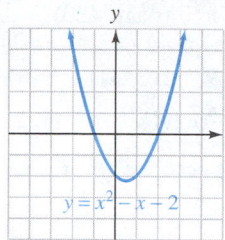

49.

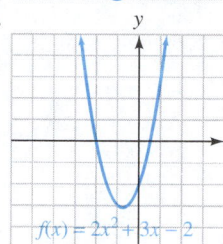

51. 1, 2   **53.** −1.5, −0.5   **55.** −2.5, 0.67   **57.** −1.85, 3.25   **59.** 0.5, 0.5

61.

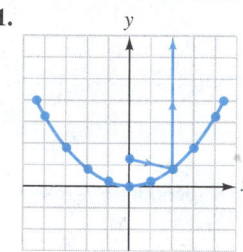

63. 32, 18, 8, 2, 0, 2, 8, 18, 32   **65. a.** 1,350   **b.** $303,750   **71.** $3x^2 − 6x$ **73.** $6x^2 − x − 2$

KEY CONCEPT (page 414)

1. no   **2.** no   **3.** yes   **4.** yes   **5.** no   **6.** no   **7.** yes   **8.** no   **9.** no   **10.** yes   **11.** no   **12.** no   **13.** The student divided both sides by 2 and incorrectly thought that $\frac{x^2}{2}$ equals x.   **14.** The student factored the left-hand side instead of first subtracting 10 from both sides and then factoring.   **15.** $0, \frac{1}{4}$ **16.** −0.38, −2.62   **17.** ±6   **18.** 8, −7   **19.** −0.27, −3.73   **20.** 1, −7   **21.**

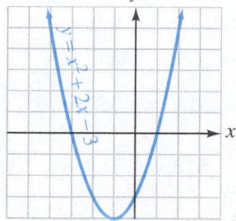

22. vertex (2, 4); the company should manufacture 200 chairs to get a maximum revenue of $40,000.

CHAPTER REVIEW (page 416)

1. a. $5 \cdot 7$ **b.** $3^2 \cdot 5$ **c.** $2^5 \cdot 3$ **d.** $3^2 \cdot 11$ **e.** $2 \cdot 5^2 \cdot 41$ **f.** 2^{12} **2. a.** $3(x + 3y)$ **b.** $5a(x^2 + 3)$ **c.** $7s(s + 2)$ **d.** $\pi a(b − c)$ **e.** $2x(x^2 + 2x − 4)$ **f.** $xyz(x + y + 1)$ **g.** $−5ab(b − 2a + 3)$ **h.** $(x − 2)(4 − x)$ **3. a.** $−(a + 7)$ **b.** $−(4t^2 − 3t + 1)$ **4. a.** $(c + d)(2 + a)$ **b.** $(y + 3)(3x − 2)$ **c.** $(2a^2 − 1)(a + 1)$ **d.** $4m(n + 3)(m − 2)$ **5. a.** $(x + 3)(x − 3)$ **b.** $(7t + 5y)(7t − 5y)$ **c.** $(xy + 20)(xy − 20)$ **d.** $8a(t + 2)(t − 2)$ **e.** $(c^2 + 16)(c + 4)(c − 4)$ **f.** prime   **6. a.** $6y(x + 2y)(x − 2y)$ **b.** $2(x^2 + 9)(x + 3)(x − 3)$ **c.** $−(m + 10)(m − 10)$ **7.** 7; 3, 5; −1; −7; −5   **8. a.** $(x + 6)(x − 4)$ **b.** $(x − 6)(x + 2)$ **c.** $(n − 5)(n − 2)$ **d.** prime   **e.** $−(y − 5)(y − 4)$ **f.** $(y + 9)(y + 1)$ **g.** $(c + 5d)(c − 2d)$ **h.** $(m − 2n)(m − n)$ **9. a.** Use the FOIL method to see if $(x − 4)(x + 5) = x^2 + x − 20$.   **10. a.** $5(a + 10)(a − 1)$ **b.** $−4x(x + 3y)(x − 2y)$ **11. a.** $(2x + 1)(x − 3)$ **b.** $(2y + 5)(5y − 2)$ **c.** $−(3x + 1)(x − 5)$ **d.** $2(2a + 3)(2a + 1)$ **e.** $3p(2p + 1)(p − 2)$ **f.** $(4b − c)(b + 4c)$ **g.** prime   **h.** $(7r^2 + 3)(r^2 + 4)$ **12.** $(4x + 1)$ in., $(3x − 1)$ in.   **13. a.** 0, −2   **b.** 0, 3   **c.** −3, 3   **d.** $−\frac{5}{6}, \frac{5}{6}$ **e.** 3, 4   **f.** −2, −2   **g.** 6, −4   **h.** $\frac{1}{8}$, 1   **i.** −1, 12   **j.** 0, −1, 2   **14.** 15 m   **15.** 3 ft by 9 ft   **16. a.** 5, −3   **b.** 7, −1   **c.** $\frac{3}{2}, −\frac{1}{3}$ **d.** $3 \pm \sqrt{2}$ **17.** $\dfrac{−1 \pm \sqrt{7}}{3}$, −1.22, 0.55

18. no real solutions   **19.** 10 ft, 24 ft   **20.** 15 sec   **21. a.** $\dfrac{−5 \pm \sqrt{17}}{2}$ **b.** 1, −7   **c.** 0, −5   **d.** $\dfrac{−1 \pm \sqrt{41}}{4}$ **e.** $\pm\sqrt{21}$ **f.** 2, 2

22. a. (−3, 0), (1, 0)   **b.** (0, −3)   **c.** (−1, −4)   **23.** The maximum profit of $16,000 is obtained from the sale of 400 units.   **24. a.** (1, 5); upward   **b.** (3, 16); downward   **25.** (−5, 0), (−1, 0); (0, 5)   **26. a.**

b.

27. −3, 1

CHAPTER 5 TEST (page 420)

1. $2^2 \cdot 7^2$ **2.** $3 \cdot 37$ **3.** $4(x + 4)$ **4.** $5ab(6ab^2 - 4a^2b + c)$ **5.** $(q + 9)(q - 9)$ **6.** prime   **7.** $(4x^2 + 9)(2x + 3)(2x - 3)$
8. $(x + 3)(x + 1)$ **9.** $-(x - 11)(x + 2)$ **10.** $(3x + 1)(x + 4)$ **11.** $(2a - 3)(a + 4)$ **12.** $6(2p - 3)(p + 2)$ **13.** $0, \frac{1}{6}$ **14.** $-3, -3$
15. $\frac{1}{3}, -\frac{1}{2}$ **16.** $\frac{1}{5}, -\frac{9}{2}$ **17.** $4, -4$ **18.** $2 \pm \sqrt{3}$ **19.** $2, -5$ **20.** $-\frac{3}{2}, 4$ **21.** $\dfrac{5 \pm \sqrt{33}}{2}$ **22.** 10 m   **23.** The most air conditioners
sold in a week (18) occurred when 3 ads were run.   **24.**   **25.** $(-5, -4)$

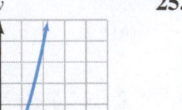

$y = x^2 + x - 2$

STUDY SET   SECTION 6.1 (page 428)

1. comparison, quotient   **3.** equal   **7. a.** $\frac{3}{11}$ **b.** $\frac{27}{32}$ **9.** $\frac{5}{7}$ **11.** $\frac{1}{2}$ **13.** $\frac{2}{3}$ **15.** $\frac{2}{7}$ **17.** $\frac{1}{3}$ **19.** $\frac{1}{5}$ **21.** $\frac{3}{7}$ **23.** $\frac{3}{4}$ **25.** \$1,825   **27.** $\frac{22}{365}$
29. \$8,725   **31.** $\frac{336}{1,745}$ **33.** $\frac{1}{16}$ **35.** $\frac{80}{45}, \frac{120}{80}, \frac{140}{85}$ **37.** $\frac{\$21.59}{17 \text{ gal}}$; \$1.27 per gallon   **39.** 57¢ per ride or attraction   **41.** the 6-ounce can   **43.** the
first student   **45.** $\frac{11,880 \text{ gallons}}{27 \text{ minutes}}$; 440 gallons per minute   **47.** 5%   **49.** 65 mph   **51.** the second car   **53. a.** $\frac{7.5 \text{ marriages}}{1 \text{ divorce}}$; for every 7.5
marriages, there was 1 divorce   **b.** 2, 1   **c.** The number of divorces for a given number of marriages is increasing.   **55.** One inch on
the map represents 1,000,000 inches on the earth.   **59.** -1 **61.** 6   **63.** 4

STUDY SET   SECTION 6.2 (page 437)

1. proportion   **3.** means   **5.** shape   **7.** ad, bc **9. a.** U.S.: $\frac{278}{7,316} = \frac{139}{3,658}$; China: $\frac{32}{561}$ **b.** No; their ratios of defense spending to gross
domestic product are different. China's is larger.   **11.** x, 288, 18, 18   **13.** triangle   **15.** no   **17.** yes   **19.** no   **21.** yes   **23.** 4   **25.** 6
27. 3   **29.** 9   **31.** 0   **33.** -17 **35.** $-\frac{3}{2}$ **37.** $\frac{83}{2}$ **39.** \$17   **41.** \$1.8 million   **43.** 24   **45.** $7\frac{1}{2}, 1\frac{2}{3}, \frac{5}{8}, 1\frac{1}{4}, 2\frac{1}{2}, 5, \frac{5}{16}$ **47.** 47   **49.** $7\frac{1}{2}$ gal
51. \$1,456.80   **53.** 65 ft, 3 in.   **55.** 568, 13, 14   **57.** not exactly, but close   **59.** 39 ft   **61.** $46\frac{7}{8}$ ft   **63.** 6,750 ft   **67.** 90%   **69.** $\frac{1}{3}$
71. 480   **73.** \$73.50

STUDY SET   SECTION 6.3 (page 448)

1. numerator, denominator   **3.** 0   **5.** lowest   **7.** $\frac{a}{b}$ **9.** factor, common   **11.** 6, 3, 1, 0.5   **13.** 6, 1   **15.** $\frac{4}{5}$ **17.** $\frac{4}{5}$ **19.** $\frac{2}{13}$ **21.** $\frac{2}{9}$ **23.** $-\frac{1}{3}$
25. $2x$ **27.** $-\frac{x}{3}$ **29.** $\frac{5}{a}$ **31.** $\frac{2}{z}$ **33.** $\frac{a}{9}$ **35.** $\frac{2}{3}$ **37.** $\frac{3}{2}$ **39.** in lowest terms   **41.** $\frac{3x}{y}$ **43.** $\frac{7x}{8y}$ **45.** $\frac{1}{3}$ **47.** 5   **49.** $\frac{x}{2}$ **51.** $\frac{3x}{5y}$ **53.** $\frac{2}{3}$ **55.** -1
57. -1 **59.** -1 **61.** $\frac{x+1}{x-1}$ **63.** $\frac{x-5}{x+2}$ **65.** $\frac{2x}{x-2}$ **67.** $\frac{x}{y}$ **69.** $\frac{x+2}{x^2}$ **71.** $\frac{x-4}{x+4}$ **73.** $\frac{2(x+2)}{x-1}$ **75.** in lowest terms   **77.** $\frac{3-x}{3+x}$ or $-\frac{x-3}{x+3}$ **79.** $\frac{4}{3}$
81. $x + 3$ **83.** 4, 2, 1.3, 1, 0.8, 0.7, 0.5   **85.** $\frac{x+2}{x-2}$ **87. a.** As the x-values (distance from the light bulb) get
larger, the y-values (intensity of the light) get smaller.   **b.** As the
x-values (distance from the light bulb) get very large, the y-values
(intensity of the light) are very small, almost zero.
93. $(a + b) + c = a + (b + c)$ **95.** one of them is zero.   **97.** $\frac{5}{3}$
99. the number

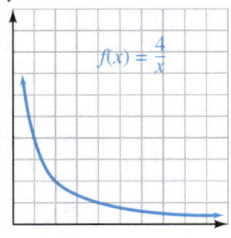
$f(x) = \dfrac{4}{x}$

STUDY SET   SECTION 6.4 (page 456)

1. numerator   **3.** numerators, denominators   **5.** 1   **7.** divisor, multiply   **9.** $(x - 2), (3x - 6), x(x + 1), 3(x - 2)$ **11.** $\dfrac{45}{91}$ **13.** $-\dfrac{3}{11}$
15. $\dfrac{5}{7}$ **17.** $\dfrac{3x}{2}$ **19.** $\dfrac{xy}{z}$ **21.** $\dfrac{14}{9}$ **23.** x^2y^2 **25.** $2xy^2$ **27.** $-3y^2$ **29.** $\dfrac{b^3c}{a^4}$ **31.** $\dfrac{r^3t^4}{s}$ **33.** $\dfrac{(z+7)(z+2)}{7z}$ **35.** x **37.** $\dfrac{x}{5}$ **39.** $x + 2$
41. $\dfrac{3}{2x}$ **43.** $x - 2$ **45.** x **47.** $\dfrac{(x-2)^2}{x}$ **49.** $\dfrac{(m-2)(m-3)}{2(m+2)}$ **51.** $\dfrac{c^2}{ab}$ **53.** $\dfrac{x+1}{2(x-2)}$ **55.** $\dfrac{2}{3}$ **57.** $\dfrac{3}{5}$ **59.** $\dfrac{3}{2y}$ **61.** 3   **63.** $\dfrac{6}{y}$ **65.** 6
67. $\dfrac{2x}{3}$ **69.** $\dfrac{2}{y}$ **71.** $\dfrac{2}{3x}$ **73.** $\dfrac{2(z-2)}{z}$ **75.** $\dfrac{5z(z-7)}{z+2}$ **77.** $\dfrac{x+2}{3}$ **79.** 1   **81.** $\dfrac{2(x-7)}{x+9}$ **83.** $d + 5$ **85.** $\dfrac{9}{2x}$ **87.** $\dfrac{x}{36}$ **89.** 2
91. $\dfrac{2x(1-x)}{5(x-2)}$ **93.** $\dfrac{y^2}{3}$ **95.** $\dfrac{x+2}{x-2}$ **97.** $\dfrac{12x^2 + 12x + 3}{2}$ in.²   **103.** $-6x^5y^6$ **105.** $\dfrac{1}{81y^4}$ **107.** $4y^3 + 16y^2 - 8y + 8$

STUDY SET   SECTION 6.5 (page 467)

1. LCD   **3.** numerators, common denominator   **5.** $2a + 3, 2, (4a + 1)$ **7.** $\dfrac{x}{3}$ **9.** $\dfrac{4x}{y}$ **11.** $\dfrac{2}{y}$ **13.** $\dfrac{y+3}{5z}$ **15.** 9   **17.** $\dfrac{1}{a+2}$ **19.** $-\dfrac{y}{8}$ **21.** $\dfrac{x}{y}$
23. $\dfrac{y}{x}$ **25.** $\dfrac{1}{y}$ **27.** $\dfrac{1}{2}$ **29.** $\dfrac{2}{c+d}$ **31.** $\dfrac{4x}{3}$ **33.** 0   **35.** $\dfrac{4x-2y}{y+2}$ **37.** $\dfrac{2x+10}{x-2}$ **39.** $\dfrac{125x}{20x}$ **41.** $\dfrac{8xy}{x^2y}$ **43.** $\dfrac{3x(x+1)}{(x+1)^2}$ **45.** $\dfrac{2y(x+1)}{x^2+x}$
47. $\dfrac{z(z+1)}{z^2-1}$ **49.** $\dfrac{2(x+2)}{x^2+3x+2}$ **51.** $6x$ **53.** $18xy^2$ **55.** $x^2 - 1$ **57.** $x^2 + 6x$ **59.** $(x+1)(x+5)(x-5)$ **61.** $\dfrac{5y}{9}$ **63.** $\dfrac{53x}{42}$

65. $\dfrac{4xy + 6x}{3y}$ **67.** $\dfrac{2 - 3x^2}{x}$ **69.** $\dfrac{y^2 + 7y + 6}{15y^2}$ **71.** $\dfrac{x^2 + 4x + 1}{x^2y}$ **73.** $\dfrac{2x^2 - 1}{x(x + 1)}$ **75.** $\dfrac{2xy + x - y}{xy}$ **77.** $\dfrac{x + 2}{x - 2}$ **79.** $\dfrac{2x^2 + 2}{(x - 1)(x + 1)}$
81. $-\dfrac{2}{a - 4}$ **83.** $\dfrac{2(t + 1)}{t - 7}$ **85.** $\dfrac{2(2x + 1)}{x - 2}$ **87.** $\dfrac{b - 1}{2(b + 1)}$ **89.** $\dfrac{1}{a + 1}$ **91.** $-\dfrac{1}{2(x - 2)}$ **93.** $\dfrac{x}{x - 2}$ **95.** $\dfrac{5x + 3}{x + 1}$ **97.** $\dfrac{20x + 9}{6x^2}$ cm
103. 7^2 **105.** $2^3 \cdot 17$

STUDY SET   SECTION 6.6 (page 473)

1. complex fraction   **3.** single, divide   **5.** $2b - a,\ b + 2a,\ \div,\ ab,\ ab,\ (b + 2a)$ **7.** $\frac{8}{9}$ **9.** $\frac{3}{8}$ **11.** $\frac{5}{4}$ **13.** $\frac{5}{7}$ **15.** $\dfrac{x^2}{y}$ **17.** $\dfrac{5t^2}{27}$
19. $\dfrac{1 - 3x}{5 + 2x}$ **21.** $\dfrac{1 + x}{2 + x}$ **23.** $\dfrac{3 - x}{x - 1}$ **25.** $\dfrac{1}{x + 2}$ **27.** $\dfrac{1}{x + 3}$ **29.** $\dfrac{xy}{y + x}$ **31.** $\dfrac{y}{x - 2y}$ **33.** $\dfrac{x^2}{(x - 1)^2}$ **35.** $\dfrac{7x + 3}{-x - 3}$ **37.** $\dfrac{x - 2}{x + 3}$
39. -1 **41.** $\dfrac{y}{x^2}$ **43.** $\dfrac{x + 1}{1 - x}$ **45.** $\dfrac{1 + a^3}{a^3}$ **47.** 2 **49.** $\dfrac{y - 5}{y + 5}$ **51.** $\frac{3}{14}$ **53.** $\dfrac{R_1R_2}{R_2 + R_1}$ **57.** t^9 **59.** $-2r^7$ **61.** $\dfrac{81}{256r^8}$ **63.** $\dfrac{r^{10}}{9}$

STUDY SET   SECTION 6.7 (page 479)

1. extraneous   **3.** LCD   **5.** x **7. a.** $\dfrac{11r - 18}{6r}$ **b.** $\dfrac{18}{7}$ **9.** $2a, 2a, 2a, 4, 4, 4$ **11.** 4 **13.** -20 **15.** -12 **17.** -1 **19.** 0 **21.** -3
23. 3 **25.** No solution; 0 is extraneous.   **27.** $\frac{11}{4}$ **29.** $-4, 4$ **31.** 1 **33.** 5 **35.** No solution; -2 is extraneous.   **37.** No solution; 5 is extraneous.   **39.** -1 **41.** 6 **43.** 2 **45.** -3 **47.** 1 **49.** No solution; -2 is extraneous.   **51.** $1, 2$ **53.** $3; -3$ is extraneous.
55. $3, -4$ **57.** 1 **59.** 0 **61.** $-2, 1$ **63.** $a = \dfrac{b}{b - 1}$ **65.** $f = \dfrac{d_1 d_2}{d_1 + d_2}$ **67.** $R = \dfrac{HB}{B - H}$ **71.** $x(x + 4)$ **73.** $(2x + 3)(x - 1)$
75. $(x^2 + 4)(x + 2)(x - 2)$

STUDY SET   SECTION 6.8 (page 485)

1. interest, principal, rate   **3.** LCD   **5.** analyze, form, solve, state, check   **7. a.** $r = \frac{d}{t}$ **b.** $t = \frac{d}{r}$ **9. a.** $\frac{1}{6}, \frac{1}{5}$ **b.** $\frac{11}{30}$ **11.** $\frac{1}{x}$ **13.** 2 **15.** 5
17. $\frac{2}{3}, \frac{3}{2}$ **19.** $2\frac{2}{9}$ hr   **21.** $2\frac{6}{11}$ days   **23.** $7\frac{1}{2}$ hr   **25.** 4 mph   **27.** 7% and 8%   **29.** 5 **31.** 30 **33.** 25 mph   **37.** $-1, 6$ **39.** $-2, -3, -4$
41. $0, 0, 1$ **43.** $1, -1, 2, -2$

KEY CONCEPT (page 488)

1. g   **2.** n   **3.** i   **4.** s   **5.** q   **6.** o   **7.** a   **8.** l   **9.** d   **10.** f   **11.** c   **12.** p   **13.** b   **14.** j   **15.** r   **16.** e   **17.** m   **18.** k   **19.** h

CHAPTER REVIEW (page 490)

1. a. $\frac{1}{2}$ **b.** $\frac{4}{5}$ **c.** $\frac{2}{3}$ **d.** $\frac{5}{6}$ **2.** $20{:}7;\ 40{:}17;\ 2{:}1$ **3.** \$2.93   **4.** 568.75 kwh per week   **5.** \$3,156   **6. a.** no   **b.** yes   **7. a.** $\frac{9}{2}$ **b.** 0   **c.** 7
d. 1   **8.** 20 ft   **9.** 255   **10. a.** $\frac{2}{5}$ **b.** $-\frac{2}{3}$ **c.** $\frac{1}{2x}$ **d.** $\frac{5}{2x}$ **e.** $\frac{x}{x + 1}$ **f.** $\frac{1}{x - 2}$ **g.** -1 **h.** $-\frac{1}{x + 3}$ **i.** $\frac{x}{x - 1}$ **j.** in lowest terms   **11.** x is not a
common factor of the numerator and the denominator.   **12.** $8, 4, 2.7, 2, 1.6, 1.3, 1.1, 1$ **13. a.** $\frac{3x}{y}$ **b.** $\frac{6}{x^2}$

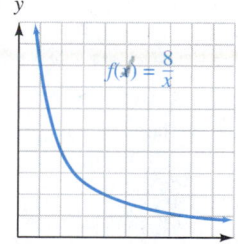

c. $\frac{x - 1}{x + 2}$ **d.** $\frac{2x}{x + 1}$ **14. a.** $\frac{3y}{2}$ **b.** $\frac{1}{x}$ **c.** $x + 2$ **d.** $\frac{2 - x}{x - 1}$ **e.** $b + 2$ **15. a.** 1   **b.** $\dfrac{2x + 2}{x - 7}$ **c.** $\dfrac{1}{a - 4}$ **16. a.** $4x^2$ **b.** $18xy^2$
c. $(x + 1)(x + 2)$ **d.** $y^2 - 25$ **17. a.** $\dfrac{x^2 + x - 1}{x(x - 1)}$ **b.** $\dfrac{c - 7}{7c}$ **c.** $\dfrac{x^2 + 4x - 4}{2x^2}$ **d.** $\dfrac{1}{t + 1}$ **e.** $\dfrac{x + 1}{x}$ **f.** $\dfrac{b + 6}{b - 1}$ **18. a.** $\frac{9}{4}$ **b.** $\frac{3}{2}$
c. $\dfrac{1 + y}{1 - y}$ **d.** $\dfrac{x(x + 3)}{2x^2 - 1}$ **e.** $x^2 + 3$ **f.** $\dfrac{1 + x^2}{1 - x^2}$ **19. a.** 3   **b.** 1   **c.** 3   **d.** $4, -\frac{3}{2}$ **e.** $2, 4$ **f.** 0   **20.** $T_1 = \dfrac{T_2}{1 - E}$ **21.** $r_1 = \dfrac{rr_2}{r_2 - r}$
22. 3   **23.** $\frac{1}{4}$ **24.** $5\frac{5}{6}$ days   **25.** 5%   **26.** 5 mph   **27.** 40 mph

CHAPTER 6 TEST (page 494)

1. $\frac{2}{3}$ **2.** yes   **3.** $\frac{2}{3}$ **4.** yes   **5.** $\frac{8x}{9y}$ **6.** $\frac{x + 1}{2x + 3}$ **7.** 3   **8.** $-\frac{5y^2}{4}$ **9.** $\frac{x + 1}{3(x - 2)}$ **10.** $\frac{3}{5}$ **11.** $-\frac{x^2}{3}$ **12.** $x + 2$ **13.** $\dfrac{10x - 1}{x - 1}$ **14.** $\dfrac{13}{2y + 3}$
15. $\dfrac{2x^2 + x + 1}{x(x + 1)}$ **16.** $\dfrac{2a + 7}{a - 1}$ **17.** $\dfrac{4n - 5mn}{10m}$ **18.** $\dfrac{x + y}{y - x}$ **19.** $1, 2$ **20.** 6   **21.** 4   **22.** $B = \dfrac{HR}{R - H}$ **23.** $3\frac{15}{16}$ hr   **24.** 5 mph

25. 8,050 ft   **26.** We can only divide out common factors, as in the first expression. We can't divide out common terms, as in the second expression.   **27.** 4, 2, 1, 0.7, 0.5, 0.4, 0.3

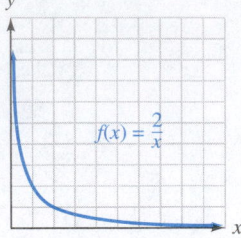

CHAPTERS 1–6 CUMULATIVE REVIEW EXERCISES (page 495)

1. 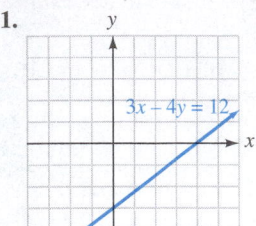   **2.** 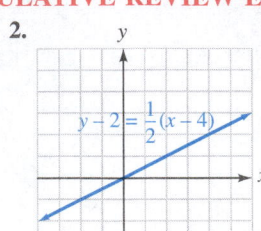   **3.** 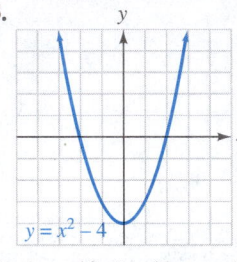   **4.** 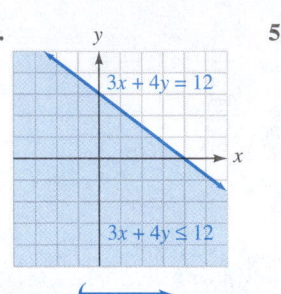   **5.** $\frac{1}{2}$

6. $x - 2y = -12$ **7.** 4.8×10^{18} m   **8.** 4 in.   **9.** \$50 billion/yr   **10.** 7   **11.** $x > 2$

12. $x < 2$    **13.** $-2 \le x < 5$ 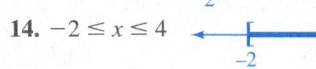   **14.** $-2 \le x \le 4$

15. $3xy(x - 2y)$ **16.** $(x + 8)(x - 8)$ **17.** $(5p^2 + 4q)(5p^2 - 4q)$ **18.** $3x(x + 9)(x - 9)$ **19.** $(x - 12)(x + 1)$ **20.** $(x - 3)(x + 2)$
21. $(3a + 4)(2a - 5)$ **22.** $2(4m + 1)(2m - 3)$ **23.** 15   **24.** 4   **25.** $\frac{2}{3}, -\frac{1}{2}$ **26.** 0, 2   **27.** $-1, -2$ **28.** $\frac{3}{2}, -4$ **29.** 2, -4
30. $-3 \pm \sqrt{5}$ (-0.8 and -5.2)   **31.** 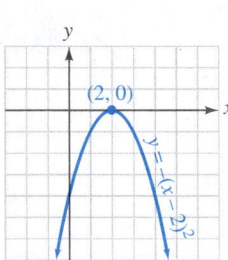   **32.** 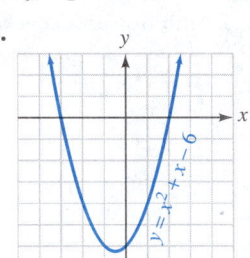   **33.** $-\frac{3}{2}$ **34.** $\frac{8}{11}$ **35.** $\frac{x + 1}{x - 1}$ **36.** $-\frac{3}{5a}$

37. $\frac{x + 1}{x + 3}$ **38.** $-\frac{3y}{7}$ **39.** $\frac{(x - 2)^2}{x - 1}$ **40.** $\frac{(p + 2)(p - 3)}{3(p + 3)}$ **41.** $\frac{xy^2 d}{c^3}$ **42.** $x + 2$ **43.** $\frac{10x + 4y}{21}$ **44.** $\frac{7x + 29}{(x + 5)(x + 7)}$ **45.** 1
46. $\frac{9a^2 - 4b^2}{6ab}$ **47.** $-\frac{1}{2(a - 2)}$ **48.** $\frac{y + x}{y - x}$ **49.** 2   **50.** 2

STUDY SET   SECTION 7.1 (page 506)

1. system   **3.** independent   **5.** inconsistent   **7.** true   **9.** false   **11.** true   **13. a.** \$2,000   **b.** \$1,200   **c.** lose \$800   **15.** 1 solution; consistent   **17.** The method is not accurate enough to find a solution such as $\left(\frac{2}{5}, -\frac{1}{3}\right)$.   **19.** 6, 6, 6, $\frac{1}{3}y$ **21.** yes   **23.** yes   **25.** no
27. no   **29.** no   **31.** yes   **33.** 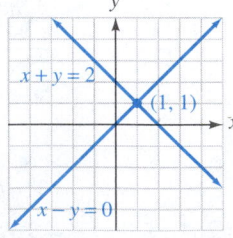   **35.** 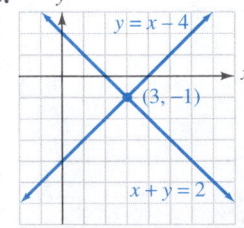   **37.**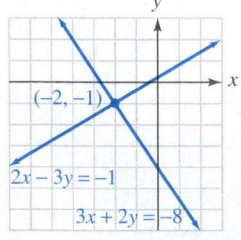

39. 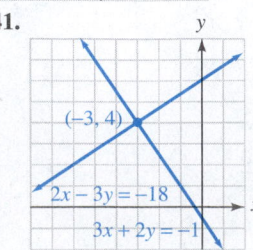   **41.** 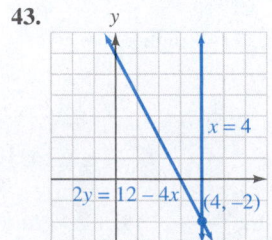   **43.** 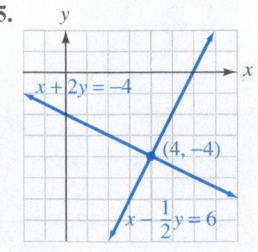   **45.**

47.
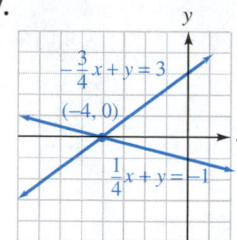

49.
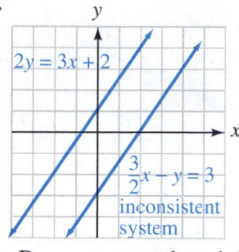

51.
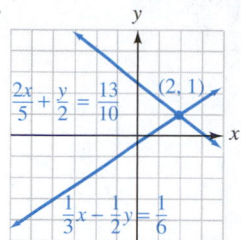

53. $(1, 3)$ **55.** no solution   **57. a.** Donors outnumbered those needing a transplant.   **b.** 1994; 4,100   **c.** People needing a transplant outnumber the donors.   **59. a.** Houston, New Orleans, St. Augustine   **b.** St. Louis, Memphis, New Orleans   **c.** New Orleans
61. yes, $(2, 2)$

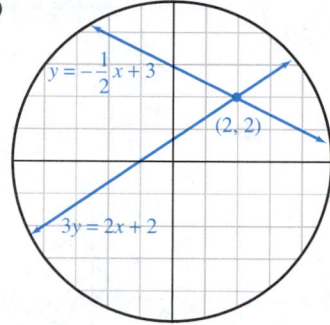

65. -3; $(0, 4)$ **67.** -21 **69.** $x = 0$ **71.** $(5, 2)$

STUDY SET   SECTION 7.2 (page 515)

1. y, terms   **3.** remove   **5.** infinitely   **7. a.** 2   **b.** 1   **9. a.** $x = 2y - 10$ **b.** $y = \frac{x}{2} + 5$ **c.** x; it involved only one step.
11. a. Parentheses must be written around $x - 4$ in line 2.   **b.** $\frac{13}{3}$ **13. a.** The coordinates of the intersection point are not integers.

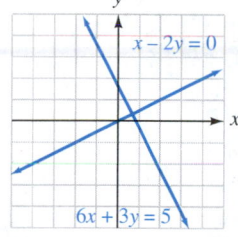

b. $\left(\frac{2}{3}, \frac{1}{3}\right)$ **15.** $3x$, 4, -2, $(-2, -6)$ **17.** $(2, 4)$ **19.** $(3, 0)$ **21.** $(-3, -1)$ **23.** inconsistent system

25. $(-2, 3)$ **27.** $(3, 2)$ **29.** $(3, -2)$ **31.** $\left(\frac{2}{3}, -\frac{1}{3}\right)$ **33.** $(-1, -1)$ **35.** dependent equations   **37.** $(4, -2)$ **39.** $\left(\frac{1}{2}, \frac{1}{3}\right)$ **41.** $(1, 4)$
43. inconsistent system   **45.** $(4, 2)$ **47.** $(-6, 4)$ **49.** $\left(\frac{1}{5}, 4\right)$ **51.** $(5, 5)$ **53.** melon, because it's the same price as hash browns
59. $-\frac{5}{8}$ **61.** $(0, -6)$ **63.** no

STUDY SET   SECTION 7.3 (page 522)

1. coefficient   **3.** general   **5.** The second equation should be written in general form: $3x - 2y = 10$.   **7.** Multiply both sides by 15.
9. a. $(2, -3)$

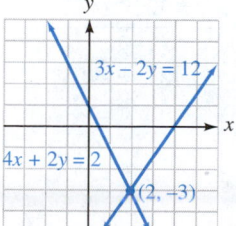

b. $(2, -3)$ **c.** $(2, -3)$ **11.** $2x$, 1, 1, $(1, 4)$ **13.** $(-2, 3)$ **15.** $(-1, 1)$ **17.** $(-3, 4)$ **19.** $(0, 8)$

21. $(2, 3)$ **23.** $(3, -2)$ **25.** $(2, 7)$ **27.** inconsistent system   **29.** $\left(1, -\frac{5}{2}\right)$ **31.** $\left(\frac{10}{3}, \frac{10}{3}\right)$ **33.** $(5, -6)$ **35.** $(-1, 2)$ **37.** dependent
equations   **39.** $(4, 0)$ **41.** $(-1, 2)$ **43.** $(0, 1)$ **45.** $(-2, 3)$ **47.** $(2, 2)$ **53.** 4   **55.** 0   **57.** 7.5 ft^2 **59.** $x - 10$

STUDY SET   SECTION 7.4 (page 531)

1. variable   **3.** system   **5.** $x - c$, $x + c$ **7. a.** $\$y$ **b.** 5%   **c.** $0.05x$, $0.11y$ **9. a.** two   **b.** graphing, substitution, addition   **11.** $A = lw$
13. $d = rt$ **15.** $2l + 2w = 90$ **17.** 32, 64   **19.** 8, 5   **21.** 22 ft, 29 ft   **23.** president: $\$200,000$; vice president: $\$171,500$ **25.** $\$15$, $\$5$

27. 40 **29.** 65°, 115° **31.** 6,720 ft² **33. a.** 400 tires **b.**

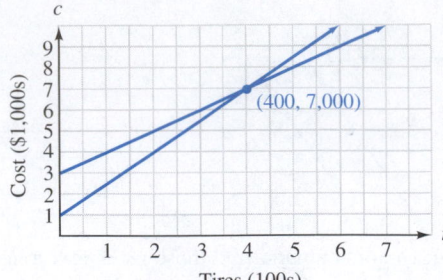

c. the second mold

(400, 7,000)

35. nursing: $2,000; business: $3,000 **37.** 5 mph **39.** 50 mph **41.** 4 L 6% salt water, 12 L 2% salt water **43.** 32 lb peanuts, 16 lb cashews **45.** $640 **49.** ⟵————) 4 **51.** ⟵—(————]—→ −1 2 **53.** −2, 2 **55.** 2, 2

STUDY SET   SECTION 7.5 (page 540)

1. inequality **3.** boundary **5. a.** yes **b.** no **c.** yes **d.** no **7. a.** no **b.** yes **9. a.** no **b.** yes **c.** yes **d.** no **11.** The test point must be on one side of the boundary.

13.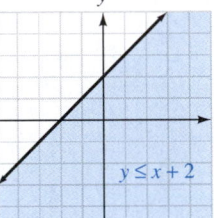

15.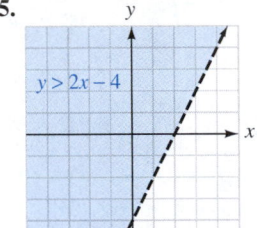

17.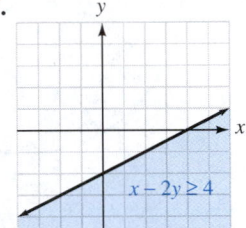

19.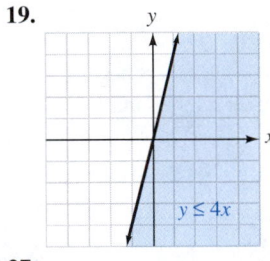

21.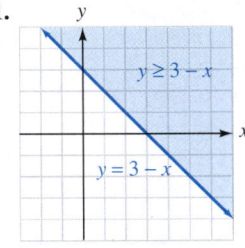

23.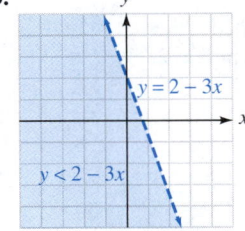

25.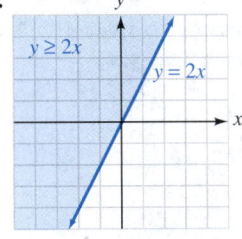

27.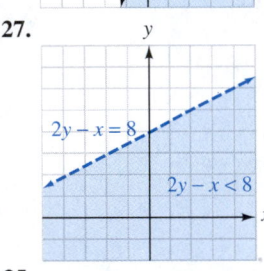

29.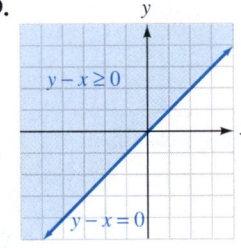

31.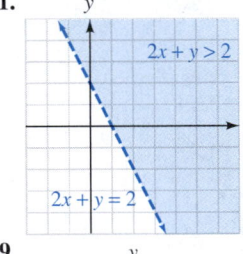

33.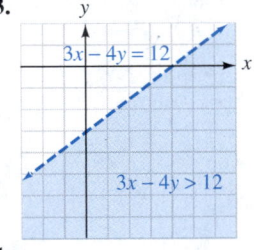

35.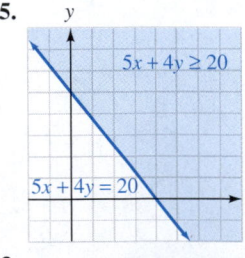

37.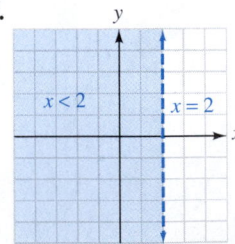

39.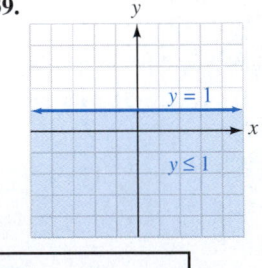

41.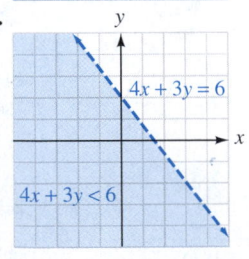

43.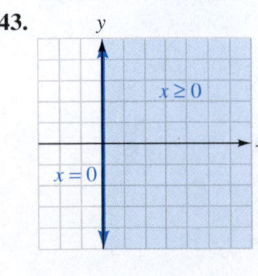

45.

47. (10, 10), (20, 10), (10, 20)

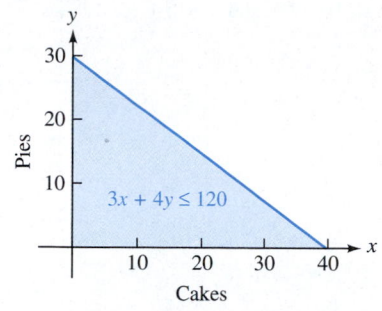

49. (50, 50), (30, 40), (40, 40)

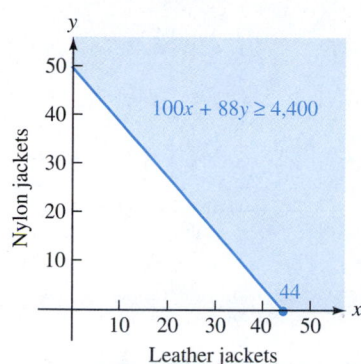

51. (80, 40), (80, 80), (120, 40)

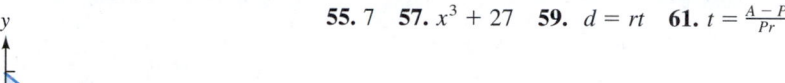 **55.** 7   **57.** $x^3 + 27$ **59.** $d = rt$ **61.** $t = \frac{A - P}{Pr}$

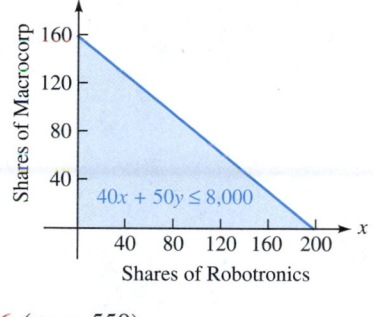

STUDY SET   SECTION 7.6 (page 550)

1. inequalities   **3.** doubly shaded   **5. a.** true   **b.** false   **c.** false   **d.** true   **e.** true   **f.** true   **7. a.** yes   **b.** no   **c.** no   **9.** *ABC*

11. 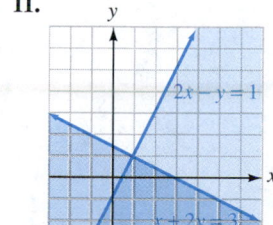 **13.** 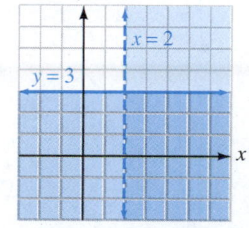 **15.** 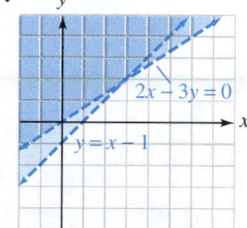 **17.**

19. 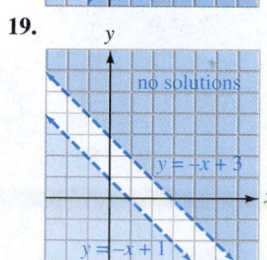 **21.** 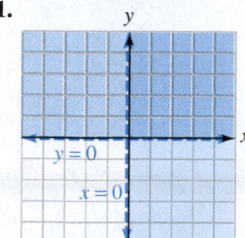 **23.** 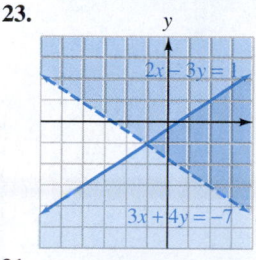 **25.**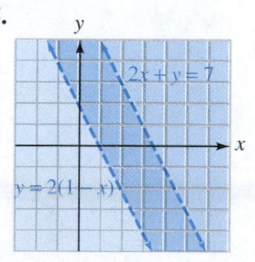

27. 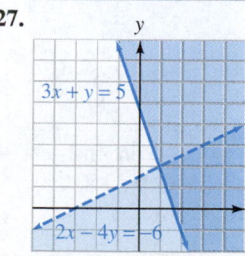 **29.** 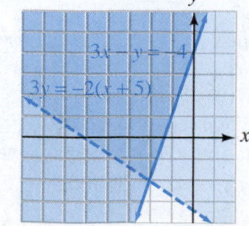 **31.** 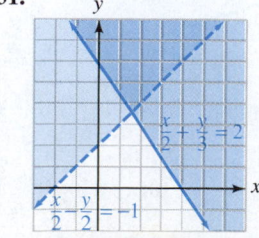 **33.**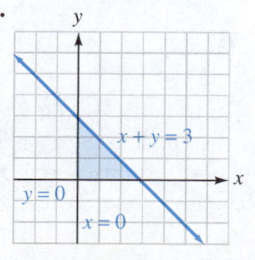

35.

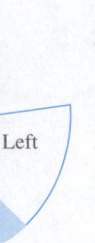

Right   Left

(c)

37. 1 $10 CD and 2 $15 CDs; 4 $10 CDs and 1 $15 CD

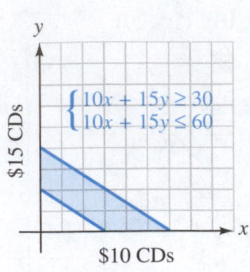

$$\begin{cases} 10x + 15y \geq 30 \\ 10x + 15y \leq 60 \end{cases}$$

$15 CDs

$10 CDs

39. 2 desk chairs and 4 side chairs; 1 desk chair and 5 side chairs

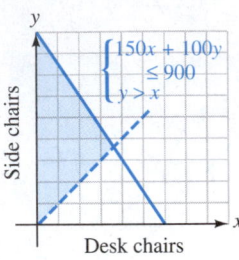

$$\begin{cases} 150x + 100y \leq 900 \\ y > x \end{cases}$$

Side chairs

Desk chairs

41.

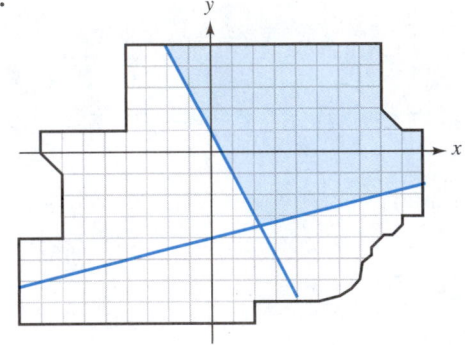

47. 128, 8   **49.** 4, −23

KEY CONCEPT (page 554)

1. (500, 3,000)

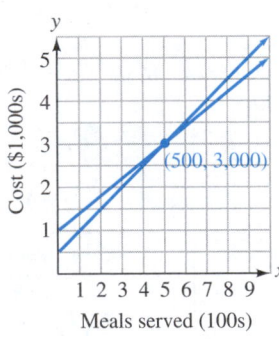

Cost ($1,000s)

(500, 3,000)

Meals served (100s)

2. (5, 1)   **3.** (−1, −1)   **4.** (3, −2)   **5.** (−3, 4)
6. (1, 2), (2, 2), (3, 1) (answers may vary)

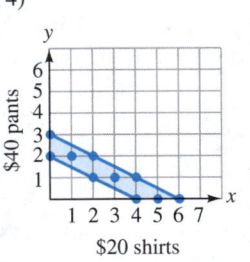

$40 pants

$20 shirts

CHAPTER REVIEW (page 556)

1. a. yes   **b.** yes   **2.** 2000; 60,000 per year
3. a.

$2x - y = 5$
(4, 3)
$x + y = 7$

b.

$2x + y = 5$
$y = -\dfrac{x}{3}$
(3, −1)

c.

$3x + 6y = 6$
$x + 2y = 2$
dependent equations

d.

$6x + 3y = 12$
$2x + y = 2$
inconsistent system

4. a. (3, 3)   **b.** (5, 0)   **c.** $\left(-\frac{1}{2}, \frac{7}{2}\right)$ **d.** (−2, 1)   **e.** dependent equations   **f.** (12, 10)   **5. a.** no solutions   **b.** two parallel lines
c. inconsistent system   **6. a.** (3, −5)   **b.** $\left(3, \frac{1}{2}\right)$ **c.** (−1, 7)   **d.** (0, 9)   **e.** dependent equations   **f.** (0, 0)   **g.** (−5, 2)   **h.** inconsistent
system   **7. a.** Addition, no variables have a coefficient of 1 or −1.   **b.** Substitution; equation 1 is solved for x.   **8.** stroke: 150,000;
heart disease: 750,000   **9.** base: 21 ft; extension: 14 ft   **10.** 10,800 yd^2 **11. a.** $y = 5x + 30,000$, $y = 10x + 20,000$ **b.** 2,000

c. the athlete   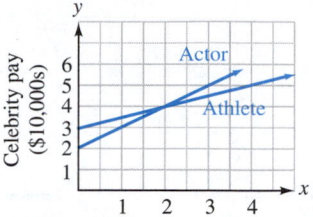   **12.** 12 lb worms, 18 lb bears   **13.** 3 mph   **14.** $5.40, $6.20   **15.** $750

16. $13\frac{1}{3}$ gal 40%, $6\frac{2}{3}$ gal 70%   **17. a.** yes   **b.** yes   **c.** yes   **d.** no

18. a. 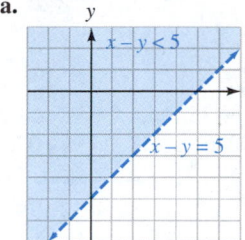   **b.** 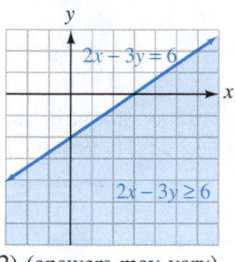   **c.** 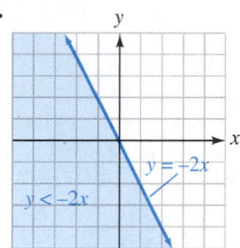   **d.** 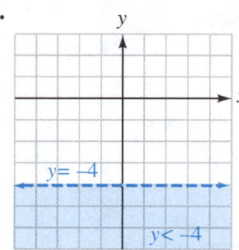   **19. a.** true   **b.** false   **c.** false

20. $3x + 5y \le 30$; (2, 4), (5, 3), (6, 2) (answers may vary)   **21. a.**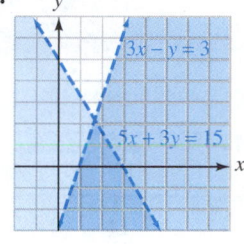

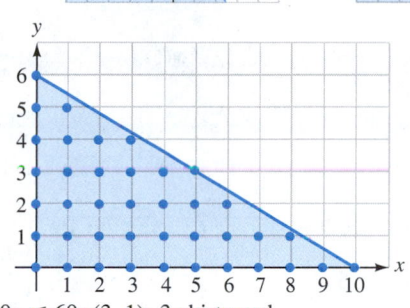

b. 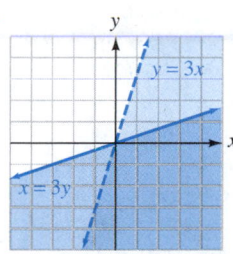   **22.** $10x + 20y \ge 40$, $10x + 20y \le 60$; (3, 1): 3 shirts and 1 pair of pants; (1, 2): 1 shirt and 2 pairs of pants (answers may vary)

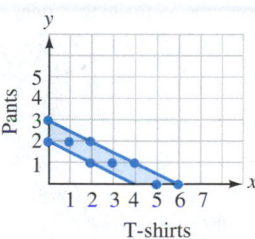

CHAPTER 7 TEST (page 560)

1. yes   **2.** no   **3.** (2, 1)   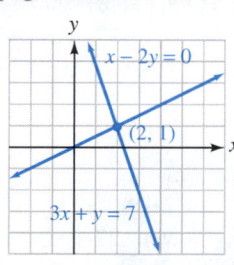   **4.** The lines appear to be parallel. Since the lines do not intersect, the system does not have a solution.   **5.** (−2, −3)   **6.** (−1, −1)   **7.** (2, 4)   **8.** (−3, 3)   **9.** inconsistent   **10.** consistent   **11.** Addition method; the terms involving y can be eliminated easily.   **12.** $4,000   **13.** yes   **14.** no

15. 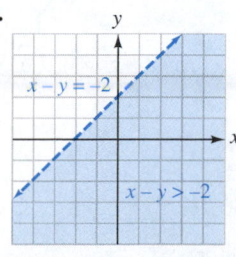   **16.** 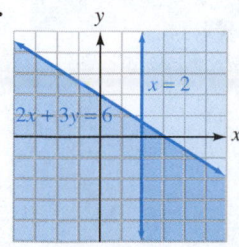   **17.** (30, 3); if 30 items are sold, the salesperson gets paid the same by both plans, $3,000.   **18.** If sales of less than 30 items are anticipated, Plan 1 is better. Otherwise, Plan 2 is more profitable.

STUDY SET   SECTION 8.1 (page 573)

1. equation   **3.** impossible   **5.** interval   **7.** added, subtracted   **9.** positive   **11.** endpoints   **13.** $a < c$ **15.** is less than   **17.** infinity   **19.** $137 < b < 165$ **21.** 6   **23.** 4   **25.** 2   **27.** 6   **29.** −4   **31.** −6   **33.** −11   **35.** 24   **37.** 6   **39.** 0   **41.** identity   **43.** −6

45. impossible   **47.** identity   **49.** $B = \dfrac{3V}{h}$ **51.** $w = \dfrac{p - 2l}{2}$ **53.** $x = z\sigma + \mu$ **55.** $x = \dfrac{y - b}{m}$ **57.** $s = \dfrac{f(P - L)}{i}$

59. $(2, \infty)$ **61.** $[-2, \infty)$ **63.** $(-\infty, 1]$ **65.** $\left(-\infty, -\frac{8}{5}\right)$

67. $[-36, \infty)$ **69.** $(-2, 5)$ **71.** $(8, 11)$ **73.** $[-4, 6)$

75. no solution   **77.** $[-2, 4]$ **79.** $(-\infty, 2) \cup (7, \infty)$ **81.** $(-\infty, 1)$

83. no solution   **85.** 7 ft, 15 ft   **87.** 12 m by 24 m   **89.** 20 ft by 45 ft   **91.** $p < \$15$ **93.** 18   **95. a.** 128, 192   **b.** $32 \le s \le 48$
101. $\dfrac{1}{t^{12}}$ **103.** 471 or more

STUDY SET   SECTION 8.2 (page 583)

1. equation   **3.** isolate   **5.** 0   **7.** more than   **9.** 5   **11. a.** yes   **b.** no   **c.** yes   **d.** no   **13. a.** ii   **b.** iii   **c.** i   **15.** $|x| < 4$
17. $|x + 3| > 6$ **19.** 8   **21.** -0.02 **23.** $-\frac{31}{16}$ **25.** π **27.** 23, -23 **29.** 9.1, -2.9 **31.** $\frac{14}{3}$, -6 **33.** no solution   **35.** 2, $-\frac{1}{2}$ **37.** -8
39. -4, -28 **41.** 0, -6 **43.** -2, $-\frac{4}{5}$ **45.** 0, -2 **47.** 0   **49.** $\frac{4}{3}$ **51.** $(-4, 4)$ **53.** $[-21, 3]$

55. $\left[-\frac{8}{3}, 4\right]$ **57.** no solution   **59.** $(-\infty, -3) \cup (3, \infty)$

61. $(-\infty, -12) \cup (36, \infty)$ **63.** $\left(-\infty, -\frac{16}{3}\right) \cup (4, \infty)$ **65.** $(-\infty, \infty)$

67. $(-\infty, -2] \cup \left[\frac{10}{3}, \infty\right)$ **69.** $(-\infty, -2) \cup (5, \infty)$ **71.** $[-10, 14]$

73. $\left(-\frac{5}{3}, 1\right)$ **75.** $(-\infty, -24) \cup (-18, \infty)$ **77.** $70° \le t \le 86°$ **79. a.** $|c - 0.6°| \le 0.5°$
b. $[0.1°, 1.1°]$ **81. a.** 26.45%, 24.76%   **b.** It is less than or equal to 1%.   **87.** $50°, 130°$

STUDY SET   SECTION 8.3 (page 594)

1. factored   **3.** squares   **5.** trinomial   **7.** $ab + ac$ **9.** perfect   **11. a.** The terms inside the parentheses have a common factor of 2.
b. The terms inside the parentheses have a common factor of t.   **13. a.** positive   **b.** negative   **c.** positive   **15.** 4   **17.** $p - q$
19. $2(x + 4)$ **21.** $2x(x - 3)$ **23.** $5x^2y(3 - 2y)$ **25.** $3z(z^2 + 4z + 1)$ **27.** $6s(4s^2 - 2st + t^2)$ **29.** $-3(a + 2)$ **31.** $-3x(2x + y)$
33. $-7u^2v^3(9uv^3 - 4v^4 + 3u)$ **35.** $x^2(x^n + x^{n+1})$ **37.** $y^n(2y^2 - 3y^3)$ **39.** $(x + y)(a + b)$ **41.** $(x + 2)(x + y)$ **43.** $(3 - c)(c + d)$
45. $r_1 = \dfrac{rr_2}{r_2 - r}$ **47.** $r = \dfrac{S - a}{S - l}$ **49.** $(x + 2)(x - 2)$ **51.** $(3y + 8)(3y - 8)$ **53.** $(9a^2 + 7b)(9a^2 - 7b)$ **55.** $(x + y + z)(x + y - z)$
57. $(x^2 + y^2)(x + y)(x - y)$ **59.** $2(x + 12)(x - 12)$ **61.** $2x(x + 4)(x - 4)$ **63.** $(x^m + y^{2n})(x^m - y^{2n})$ **65.** $(a + b)(a - b + 1)$
67. $(2x + y)(1 + 2x - y)$ **69.** $(x + 3)(x + 2)$ **71.** $(x - 2)(x - 5)$ **73.** prime   **75.** $3(x + 7)(x - 3)$ **77.** $b^2(a - 11)(a - 2)$
79. $-(a - 8)(a + 4)$ **81.** $-3(x - 3)(x - 2)$ **83.** $(3y + 2)(2y + 1)$ **85.** $(4a - 3)(2a + 3)$ **87.** prime   **89.** $(4x - 3)(2x - 1)$
91. $(a + b)(a - 4b)$ **93.** $(2y - 3t)(y + 2t)$ **95.** $-(3a + 2b)(a - b)$ **97.** $x(3x - 1)(x - 3)$ **99.** $-x(2x - 3)^2$ **101.** $(x^2 + 5)(x^2 + 3)$
103. $(y^2 - 10)(y^2 - 3)$ **105.** $(a + 3)(a - 3)(a + 2)(a - 2)$ **107.** $(x^n + 1)^2$ **109.** $(2a^{3n} + 1)(a^{3n} - 2)$ **111.** $(x^{2n} + y^{2n})^2$
113. $(3x^n - 1)(2x^n + 3)$ **115.** $(x + 2 + y)(x + 2 - y)$ **117.** $(x + 1 + 3z)(z + 1 - 3z)$ **119.** $(c + 2a - b)(c - 2a + b)$
121. $(a - 16)(a - 1)$ **123.** $(2u + 3)(u + 1)$ **125.** $(5r + 2s)(4r - 3s)$ **127.** $(5u + v)(4u + 3v)$ **129. a.** $\frac{1}{2}b_1h$ **b.** $\frac{1}{2}b_2h$
c. $\frac{1}{2}h(b_1 + b_2)$; the formula for the area of a trapezoid   **131.** $r^2(4 - \pi)$ **133.** $\frac{4}{3}\pi(r_1 - r_2)(r_1^2 + r_1r_2 + r_2^2)$ **135.** $x + 3$ **141.** $x^3 + 1$
143. $r^3 - 8$ **145.** 12

STUDY SET   SECTION 8.4 (page 600)

1. difference   **3.** $(x^2 - xy + y^2)$ **5.** $a + b$ **7.** $(y + 1)(y^2 - y + 1)$ **9.** $(a - 3)(a^2 + 3a + 9)$ **11.** $(2 + x)(4 - 2x + x^2)$
13. $(s - t)(s^2 + st + t^2)$ **15.** $(3x + y)(9x^2 - 3xy + y^2)$ **17.** $(a + 2b)(a^2 - 2ab + 4b^2)$ **19.** $(4x - 3)(16x^2 + 12x + 9)$
21. $(3x - 5y)(9x^2 + 15xy + 25y^2)$ **23.** $(a^2 - b)(a^4 + a^2b + b^2)$ **25.** $(x^3 + y^2)(x^6 - x^3y^2 + y^4)$ **27.** $2(x + 3)(x^2 - 3x + 9)$
29. $-(x - 6)(x^2 + 6x + 36)$ **31.** $8x(2m - n)(4m^2 + 2mn + n^2)$ **33.** $xy(x + 6y)(x^2 - 6xy + 36y^2)$
35. $3rs^2(3r - 2s)(9r^2 + 6rs + 4s^2)$ **37.** $a^3b^2(5a + 4b)(25a^2 - 20ab + 16b^2)$ **39.** $yz(y^2 - z)(y^4 + y^2z + z^2)$
41. $2mp(p + 2q)(p^2 - 2pq + 4q^2)$ **43.** $(x + 1)(x^2 - x + 1)(x - 1)(x^2 + x + 1)$ **45.** $(x^2 + y)(x^4 - x^2y + y^2)(x^2 - y)(x^4 + x^2y + y^2)$
47. $(x + y)(x^2 - xy + y^2)(3 - z)$ **49.** $(m + 2n)(m^2 - 2mn + 4n^2)(1 + x)$ **51.** $(a + 3)(a^2 - 3a + 9)(a - b)$
53. $(y + 1)(y - 1)(y - 3)(y^2 + 3y + 9)$ **55.** $2a + 3b$ **61.** 0.0000000000001 cm

STUDY SET   SECTION 8.5 (page 610)

1. numerator   **3.** complex   **5.** 0   **7.** $\frac{a}{b}$ **9.** $\frac{ad}{bc}$ **11.** $(x - 5)$, $(5x - 25)$, $x(x + 3)$, $5(x - 5)$ **13.** $4x^2$ **15.** $-\dfrac{4y}{3x}$ **17.** $\dfrac{3y}{7(y - z)}$ **19.** 1

21. $\dfrac{1}{x - y}$ **23.** $\dfrac{-3(x + 2)}{x + 1}$ **25.** $x + 2$ **27.** $\dfrac{x + 1}{x + 3}$ **29.** $\dfrac{m - 2n}{n - 2m}$ **31.** $\dfrac{x + 4}{2(2x - 3)}$ **33.** $\dfrac{3(x - y)}{x + 2}$ **35.** $\dfrac{1}{x^2 + xy + y^2 - 1}$ **37.** $\dfrac{xy^2d}{c^3}$

39. $-\dfrac{x^{10}}{y^2}$ **41.** $x + 1$ **43.** 1   **45.** $\dfrac{x - 4}{x + 5}$ **47.** $\dfrac{(a + 7)^2(a - 5)}{12x^2}$ **49.** $\dfrac{t - 1}{t + 1}$ **51.** $\dfrac{n + 2}{n + 1}$ **53.** $x - 5$ **55.** $\dfrac{x + y}{x - y}$ **57.** $-\dfrac{x + 3}{x + 2}$

59. $\dfrac{a+b}{(x-3)(c+d)}$ **61.** $-\dfrac{x+1}{x+3}$ **63.** $-\dfrac{2x+y}{3x+y}$ **65.** $-\dfrac{x^7}{18y^4}$ **67.** $x^2(x+3)$ **69.** $\dfrac{3x}{2}$ **71.** $\dfrac{x-7}{x+7}$ **73.** 1   **75.** $\dfrac{3-a}{a+b}$ **77.** 2   **79.** 3

81. $\dfrac{6x}{(x-3)(x-2)}$ **83.** $\dfrac{9a}{10}$ **85.** $\dfrac{17}{12x}$ **87.** $\dfrac{10a+4b}{21}$ **89.** $\dfrac{8x-2}{(x+2)(x-4)}$ **91.** $\dfrac{7x+29}{(x+5)(x+7)}$ **93.** $\dfrac{x^2+1}{x}$ **95.** 2

97. $\dfrac{2x^2+x}{(x+3)(x+2)(x-2)}$ **99.** $\dfrac{-4x^2+14x+54}{x(x+3)(x-3)}$ **101.** $\dfrac{x^2-5x-5}{x-5}$ **103.** $\dfrac{2}{x+1}$ **105.** $\dfrac{3x+1}{x(x+3)}$ **107.** $\dfrac{2y}{3z}$ **109.** $-\dfrac{1}{y}$ **111.** $\dfrac{b+a}{b}$

113. $y-x$ **115.** $-\dfrac{1}{a+b}$ **117.** $\dfrac{x+2}{x-3}$ **119.** $\dfrac{a-1}{a+1}$ **121.** $\dfrac{y+x}{y-x}$ **123.** xy **125.** $\dfrac{x^2(xy^2-1)}{y^2(x^2y-1)}$ **127.** $\dfrac{(b+a)(b-a)}{b(b-a-ab)}$ **129.** $\dfrac{3a^2+2a}{2a+1}$

131. $k_1(k_1+2)$, k_2+6 **133.** $\dfrac{10r+20}{r}$, $\dfrac{3t+9}{t}$ **135.** $\dfrac{k_1k_2}{k_2+k_1}$ **141.** ⟵)-4 [5 ⟶   **143.** $w=\dfrac{P-2l}{2}$ **145.** 2, -2, 3, -3

STUDY SET   SECTION 8.6 (page 626)

1. y-intercept   **3.** function, input, output   **5.** vertical   **7.** $\dfrac{y_2-y_1}{x_2-x_1}$ **9.** vertical   **11.** parallel   **13.** $y-y_1=m(x-x_1)$

15. $Ax+By=C$ **17.** 0   **19.** $mx+b$ **21.** sub 1   **23.** rise

25. 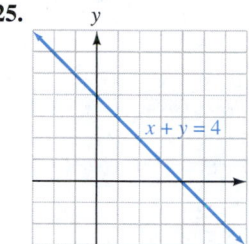   **27.** 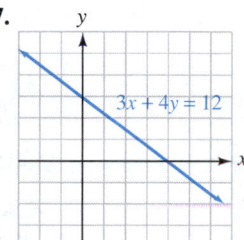   **29.** 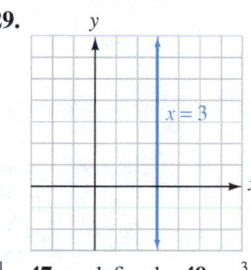   **31.** 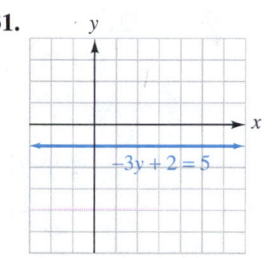   **33.** $(3,4)$

35. $(9,12)$ **37.** $\left(\frac{1}{2},-2\right)$ **39.** $(4,1)$ **41.** 3   **43.** -1 **45.** $-\frac{1}{3}$ **47.** undefined   **49.** $-\frac{3}{2}$ **51.** $\frac{3}{4}$ **53.** parallel   **55.** parallel
57. perpendicular   **59.** neither   **61.** $5x-y=-7$ **63.** $x-y=0$ **65.** $7x-3y=9$ **67.** $y=3x+17$ **69.** $y=-7x+54$
71. $y=-\frac{1}{2}x+11$ **73.** $y=4x$ **75.** $y=4x-3$ **77.** $y=-\frac{1}{4}x$ **79.** $y=-\frac{1}{4}x+\frac{11}{2}$ **81.** perpendicular   **83.** parallel   **85.** $y=-\frac{A}{B}x+\frac{C}{B}$
87. $9,-3$ **89.** $3,-5$ **91.** $2w, 2w+2$ **93.** $3w-5, 3w-2$ **95.** not a function   **97.** a function; $D=(-\infty,\infty)$, $R=(-\infty,\infty)$
99. 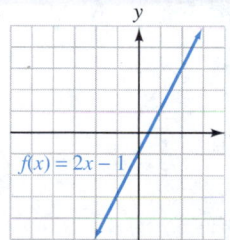 $D=(-\infty,\infty)$, $R=(-\infty,\infty)$ **101.** 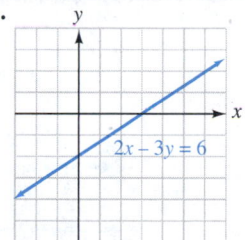 $D=(-\infty,\infty)$, $R=(-\infty,\infty)$ **103.** $\frac{1}{165}$

105. 3 students per yr   **107.** 12.5 yr   **109.** \$490   **111.** 624 ft   **113.** 192   **119.** 6   **121.** -1

STUDY SET   SECTION 8.7 (page 635)

1. direct   **3.** rational   **5.** joint   **7.** 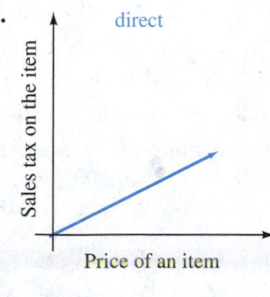   **9.** 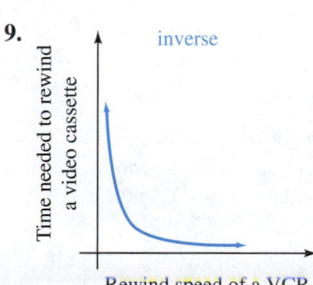   **11.** $\frac{7}{5}$, 25   **13.** $A=kp^2$ **15.** $v=\dfrac{k}{r^3}$

17. $B=kmn$ **19.** $P=\dfrac{ka^2}{j^3}$ **21.** L varies jointly with m and n.   **23.** E varies jointly with a and the square of b.   **25.** X varies directly
with x^2 and inversely with y^2.   **27.** R varies directly with L and inversely with d^2.   **29.** 36π in.^{2} **31.** 432 mi   **33.** 25 days   **35.** 12 in.^{3}
37. 85.3   **39.** 12   **41.** 26,437.5 gal   **43.** 3 ohms   **45.** 0.275 in.   **47.** 546 Kelvin   **53.** x^{10} **55.** -1 **57.** 3.5×10^4 **59.** 0.0025

KEY CONCEPT (page 639)

1. $(3t-1)(2t+3)$ **2.** $(a+b)(c+d)$ **3.** $2(2x+5)(2x-5)$ **4.** $(t-1)^2$ **5.** $2a(b+6)(b-2)$ **6.** $(x-a)(a+b)(a-b)$
7. $2(3x+5y^2)(9x^2-15xy^2+25y^4)$ **8.** $(x+y)(x-y)(x+y)(x^2-xy+y^2)$

CHAPTER REVIEW (page 641)

1. a. 8   **b.** 7   **c.** 19   **d.** 8   **e.** identity   **f.** 12   **g.** impossible equation   **2. a.** $h = \dfrac{3V}{\pi r^2}$ **b.** $x = \dfrac{6v}{ab} - y$

3. a. $(-\infty, -24]$ 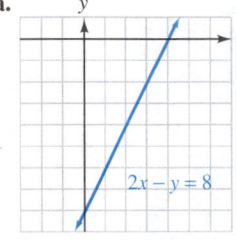 -24 **b.** $\left(-\infty, -\frac{51}{11}\right)$ $-51/11$ **c.** $\left(-\frac{1}{3}, 2\right)$ $-1/3$ 2 **d.** $(2, \infty)$ 2 **4.** 5 ft from

one end   **5.** 45 m^2 **6. a.** 3, $-\frac{11}{3}$ **b.** $\frac{26}{3}$, $-\frac{10}{3}$ **c.** $\frac{1}{5}$, -5 **d.** -1, 1   **7. a.** $(-5, -2)$ -5 -2

b. $\left(-\infty, \frac{4}{3}\right] \cup [4, \infty)$ $4/3$ 4 **c.** $(-\infty, \infty)$ 0 **d.** no solutions   **8. a.** $4(x + 2)$ **b.** $5xy^2(xy - 2)$

c. $-4x^2y^3z^2(2z^2 + 3x^2)$ **d.** $3a^2b^4c^2(4a^4 + 5c^4)$ **e.** $(x + 2)(y + 4)$ **f.** $(a + b)(c + 3)$ **9.** $x^n(x^n + 1)$ **10.** $y^{2n}(1 - y^{2n})$
11. a. $(x^2 + 4)(x^2 + y)$ **b.** $(a^3 + c)(a^2 + b^2)$ **c.** $(z + 4)(z - 4)$ **d.** $(y + 11)(y - 11)$ **e.** $2(x^2 + 7)(x^2 - 7)$ **f.** $3x^2(x^2 + 10)(x^2 - 10)$
g. $(y + 20)(y + 1)$ **h.** $(z - 5)(z - 6)$ **i.** $-(x + 7)(x - 4)$ **j.** $-(y - 8)(y + 3)$ **k.** $y(y + 2)(y - 1)$ **l.** $2a^2(a + 3)(a - 1)$
m. $3(5x + y)(x - 4y)$ **n.** $5(6x + y)(x + 2y)$ **o.** $(x + 2 + 2p^2)(x + 2 - 2p^2)$ **p.** $(y + 2)(y + 1 + x)$ **12. a.** $(x + 7)(x^2 - 7x + 49)$

b. $(a - 5)(a^2 + 5a + 25)$ **c.** $8(y - 4)(y^2 + 4y + 16)$ **d.** $4y(x + 3z)(x^2 - 3xz + 9z^2)$ **13. a.** $\dfrac{31x}{72y}$ **b.** $\dfrac{x - 7}{x + 7}$ **14. a.** 1   **b.** 1

c. $\dfrac{5y - 3}{x - y}$ **d.** $\dfrac{6x - 7}{x^2 + 2}$ **e.** $\dfrac{5x + 13}{(x + 2)(x + 3)}$ **f.** $\dfrac{4x^2 + 9x + 12}{(x - 4)(x + 3)}$ **g.** $\dfrac{3x(x - 1)}{(x - 3)(x + 1)}$ **h.** $\dfrac{5x^2 + 11x}{(x + 1)(x + 2)}$ **i.** $\dfrac{5x^2 + 23x + 4}{(x + 1)(x - 1)(x - 1)}$

15. a. $\dfrac{3y - 2x}{x^2y^2}$ **b.** $\dfrac{y + 2x}{2y - x}$ **c.** $\dfrac{2x + 1}{x + 1}$ **d.** $\dfrac{y - x}{y + x}$

16. a.   **b.**   **c.**   **17.** $\left(\frac{3}{2}, 8\right)$ **18. a.** 1   **b.** $\frac{14}{9}$ **c.** 5   **d.** $\frac{5}{11}$ **e.** 0

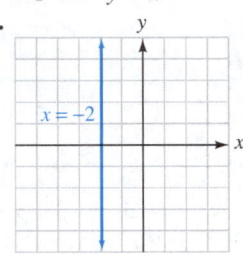
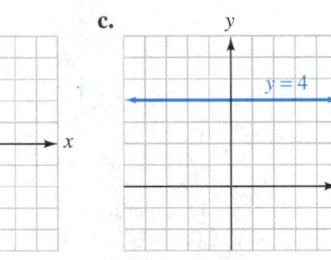

f. no defined slope   **19. a.** $\frac{2}{3}$ **b.** -2 **20. a.** perpendicular   **b.** parallel   **21. a.** $3x - y = -29$ **b.** $13x + 8y = 6$ **c.** $3x - 2y = 1$
d. $2x + 3y = -21$ **22.** $y = -1{,}720x + 8{,}700$ **23. a.** yes   **b.** yes   **c.** no   **d.** no   **24. a.** -7 **b.** 60   **c.** 0   **d.** $3x + 5$
25. a. $D = (-\infty, \infty)$, $R = (-\infty, \infty)$ **b.** $D = (-\infty, \infty)$, $R = (-\infty, \infty)$ **26. a.** a function   **b.** not a function   **27.** 2   **28.** 6   **29.** $\frac{1}{32}$
30. 16

CHAPTER 8 TEST (page 645)

1. 6   **2.** $a = \dfrac{180n - 360}{n}$ **3.** $13\frac{1}{3}$ ft   **4.** 36 m^2 **5.** $(-\infty, -5]$ -5 **6.** $(-2, 16)$ -2 16 **7.** 4, -7 **8.** 4, -4

9. $[-7, 1]$ -7 1 **10.** $(-\infty, -9) \cup (13, \infty)$ -9 13 **11.** $3xy(y + 2x)$ **12.** $(a - y)(x + y)$

13. $(x + 7)(x - 7)$ **14.** $4(y^2 + 4)(y + 2)(y - 2)$ **15.** $(b + 5)(b^2 - 5b + 25)$ **16.** $(b - 3)(b^2 + 3b + 9)$ **17.** $3(u - 2)(u^2 + 2u + 4)$
18. $3(u + 2)(2u - 1)$ **19.** $(3b^n + 2)(2b^n - 1)$ **20.** $(x + 3 + y)(x + 3 - y)$ **21.** $\dfrac{-2}{3xy}$ **22.** $-\dfrac{2}{3xy}$ **23.** $\dfrac{xz}{y^4}$ **24.** 1   **25.** $\dfrac{(x + y)^2}{2}$

26. $\dfrac{2x + 3}{(x + 1)(x + 2)}$ **27.** $\dfrac{u^2}{2vw}$ **28.** $\dfrac{2x + y}{xy - 2}$ **29.** $\left(\frac{1}{2}, -\frac{1}{2}\right)$ **30.** $-\frac{11}{7}$ **31.** $11x + 7y = 2$ **32.** -14 **33.** $\frac{6}{5}$ **34.** $\frac{44}{3}$

CHAPTERS 1–8 CUMULATIVE REVIEW EXERCISES (page 646)

1. rational numbers: terminating and repeating decimals; irrational numbers: nonterminating, nonrepeating decimals   **2.** 0.125, 0.0625,
0.03125, 0.054125   **3.** 10   **4.** -6 **5.** $-26p^2 - 6p$ **6.** $-2a + b - 2$ **7.** 201 ft^2 **8.** \$20,000   **9.** $\frac{26}{3}$ **10.** 3   **11.** 6
12. impossible   **13.** perpendicular   **14.** parallel   **15.** $y = \frac{1}{3}x + \frac{11}{3}$ **16.** $-\frac{8}{5}$ **17.** 9,000 prisoners/yr   **18.** 1990–1995, 72,200 prisoners/yr
19. 10   **20.** 14   **21.** (7, 23), in 1907 the percent of U.S. workers in white-collar and farming jobs was the same (23%); (45, 42), in
1945 the percent of U.S. workers in white-collar and blue-collar jobs was the same (42%)   **22.** $y = \frac{5}{9}x + 20$, about 81%   **23.** 750
24. 250 \$5 tickets, 375 \$3 tickets, 125 \$2 tickets   **25.** 3, $-\frac{3}{2}$ **26.** -5, $-\frac{3}{5}$ **27.** $(-\infty, 11]$ 11

28. $(-3, 3)$ -3 3 **29.** $\left[-\frac{2}{3}, 2\right]$ $-2/3$ 2 **30.** $(-\infty, -4) \cup (1, \infty)$ -4 1 **31.** $3rs^3(r - 2s)$

32. $(x - y)(5 - a)$ **33.** $(x + y)(u + v)$ **34.** $(9x^2 + 4y^2)(3x + 2y)(3x - 2y)$ **35.** $(2x - 3y^2)(4x^2 + 6xy^2 + 9y^4)$ **36.** $(2x + 3)(3x - 2)$

37. $(3x - 5)^2$ **38.** $(5x + 3)(3x - 2)$ **39.** $-\frac{1}{3}$, $-\frac{7}{2}$ **40.** 0, 2, -2 **41.** $b^2 = \dfrac{a^2y^2}{a^2 - x^2}$ **42.** 9 in. by 12 in.

STUDY SET   SECTION 9.1 (page 659)
1. square root   **3.** radical   **5.** odd   **7.** simplified   **9.** $b^2 = a$ **11.** two, positive   **13.** x **15.** 3, up
17. 0, −1, −2, −3, −4   **19.** $\sqrt{x^2} = |x|$ **21.** $f(x) = \sqrt{x-5}$ **23.** 11   **25.** −8   **27.** $\frac{1}{3}$ **29.** 0.5

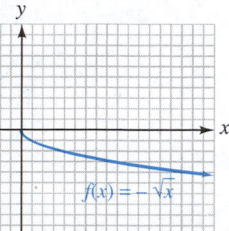

31. not real   **33.** 4   **35.** 3.4641   **37.** 26.0624   **39.** $2|x|$ **41.** $|t+5|$ **43.** $5|b|$ **45.** $|a+3|$ **47.** 0   **49.** 4   **51.** 2   **53.** −10
55. 4.1231   **57.** 3.3322   **59.** D: $[-4, \infty)$, R: $[0, \infty)$ **61.** D: $(-\infty, \infty)$, R: $(-\infty, \infty)$

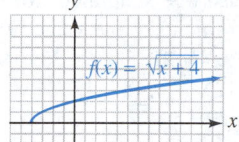

63. 1   **65.** −5   **67.** $-\frac{2}{3}$ **69.** 0.4   **71.** $2a$ **73.** $-10pq$ **75.** $-\frac{1}{2}m^2n$ **77.** $-0.4s^3t^2$ **79.** 3   **81.** −3   **83.** not real   **85.** $\frac{2}{5}$ **87.** $\frac{1}{2}$
89. $2a$ **91.** $2a$ **93.** k^3 **95.** $\frac{1}{2}m$ **97.** $x+2$ **99.** 7.0 in.   **101.** about 61.3 beats/min   **103.** 13.4 ft   **105.** 3.5%   **111.** 1
113. $\dfrac{3(m^2 + 2m - 1)}{(m+1)(m-1)}$

STUDY SET   SECTION 9.2 (page 668)
1. radical   **3.** extraneous   **5. a.** Square both sides.   **b.** Cube both sides.   **7. a.** x **b.** $x-5$ **c.** $32x$ **d.** $x+3$ **9.** Only one side of
the equation was squared.   **11.** If we raise two equal quantities to the same power, the results are equal quantities.   **13.** $2\sqrt{x-2}$, 4,
16, 8, 24   **15.** 2   **17.** 4   **19.** 0   **21.** 4   **23.** 8   **25.** $\frac{5}{2}, \frac{1}{2}$ **27.** 1   **29.** 16   **31.** 14, 6̶   **33.** 4, 3   **35.** 2, 7̶   **37.** 2, −1   **39.** −1, 1̶
41. 1̶, no solutions   **43.** 0, 4̶   **45.** −3̶, no solutions   **47.** 1, 9   **49.** 4, 0̶   **51.** 2, 12̶   **53.** 6̶, no solutions   **55.** $h = \dfrac{v^2}{2g}$ **57.** $l = \dfrac{8T^2}{\pi^2}$
59. $A = P(r+1)^3$ **61.** $v^2 = c^2\left(1 - \dfrac{L_A^2}{L_B^2}\right)$ **63.** 178 ft   **65.** about 488 watts   **67.** 10 lb   **69.** \$5   **75.** 2.5 foot-candles
77. 0.41511 in.

STUDY SET   SECTION 9.3 (page 678)
1. rational (or fractional)   **3.** index, radicand   **5.** $25^{1/5}$, $\left(\sqrt[3]{-27}\right)^2$, $16^{-3/4}$, $\left(\sqrt{81}\right)^3$, $-\left(\frac{9}{64}\right)^{1/2}$
7.

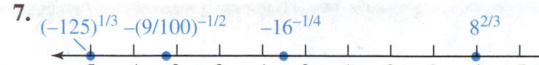

9. x^{m+n} **11.** x^{m-n} **13.** $\sqrt[n]{x}$
15. 0, 1, 2, 3, 4   **17.** $\sqrt{100a^4}$, $10a^2$ **19.** $\sqrt[3]{x}$ **21.** $\sqrt[4]{3x}$ **23.** $\sqrt[4]{\frac{1}{2}x^3y}$ **25.** $\sqrt{x^2 + y^2}$ **27.** $m^{1/2}$

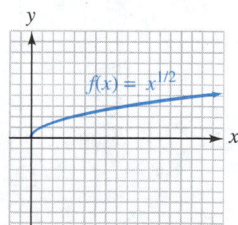

29. $(3a)^{1/4}$ **31.** $\left(\frac{1}{7}abc\right)^{1/6}$ **33.** $(a^2-b^2)^{1/3}$ **35.** 2   **37.** 2   **39.** 2   **41.** 2   **43.** $\frac{1}{2}$ **45.** −2   **47.** not real   **49.** −3   **51.** $5|y|$ **53.** $2|x|$
55. $3x$ **57.** not real   **59.** 216   **61.** 27   **63.** 1,728   **65.** $\frac{1}{4}$ **67.** $125x^6$ **69.** $\dfrac{4x^2}{9}$ **71.** $\frac{1}{2}$ **73.** $\frac{1}{8}$ **75.** $\dfrac{1}{64x^3}$ **77.** $\dfrac{1}{9y^2}$ **79.** $\dfrac{16}{81}$
81. $-\dfrac{3}{2x}$ **83.** 2.47   **85.** 1.01   **87.** $5^{5/7}$ **89.** $4^{3/5}$ **91.** $9^{1/5}$ **93.** $\dfrac{1}{36}$ **95.** 3   **97.** a **99.** $a^{2/9}$ **101.** $a^{3/4}b^{1/2}$ **103.** $\dfrac{2x}{3}$ **105.** $\dfrac{1}{3}x$
107. $y + y^2$ **109.** $x^2 - x + x^{3/5}$ **111.** $\sqrt{p}$ **113.** $\sqrt{5b}$ **115.** 736 ft/sec   **117.** 1.96 units   **119.** 4,608 in.², 32 ft²   **123.** $x < 3$
125. $r > 28$

STUDY SET SECTION 9.4 (page 688)

1. like **3.** factor **5.** $\sqrt[n]{a}\sqrt[n]{b}$, product, roots **7. a.** $\sqrt{4 \cdot 5}$ **b.** $\sqrt{4}\sqrt{5}$ **c.** $\sqrt{4 \cdot 5} = \sqrt{4}\sqrt{5}$ **9. a.** $\sqrt{5}, \sqrt[3]{5}$ (answers may vary); no
b. $\sqrt{5}, \sqrt{6}$ (answers may vary); no **11.** $8k^3, 8k^3$ **13.** 6 **15.** t **17.** $5x$ **19.** 10 **21.** $7x$ **23.** $6b$ **25.** 2 **27.** $3a$ **29.** $2\sqrt{5}$
31. $-10\sqrt{2}$ **33.** $2\sqrt[3]{10}$ **35.** $-3\sqrt[3]{3}$ **37.** $2\sqrt[4]{2}$ **39.** $2\sqrt[5]{3}$ **41.** $\dfrac{\sqrt{7}}{3}$ **43.** $\dfrac{\sqrt[3]{7}}{4}$ **45.** $\dfrac{\sqrt[4]{3}}{10}$ **47.** $\dfrac{\sqrt[5]{3}}{2}$ **49.** $5x\sqrt{2}$ **51.** $4\sqrt{2b}$
53. $-4a\sqrt{7a}$ **55.** $5ab\sqrt{7b}$ **57.** $-10\sqrt{3xy}$ **59.** $-3x^2\sqrt[3]{2}$ **61.** $2x^4y\sqrt[3]{2}$ **63.** $2x^3y\sqrt[4]{2}$ **65.** $\dfrac{z}{4x}$ **67.** $\dfrac{\sqrt[4]{5x}}{2z}$ **69.** $10\sqrt{2x}$ **71.** $\sqrt[5]{7a^2}$
73. $-\sqrt{2}$ **75.** $2\sqrt{2}$ **77.** $9\sqrt{6}$ **79.** $3\sqrt[3]{3x}$ **81.** $-\sqrt[3]{4}$ **83.** -10 **85.** $-17\sqrt[4]{2}$ **87.** $16\sqrt[4]{2}$ **89.** $-4\sqrt{2}$ **91.** $3\sqrt{2t} + \sqrt{3t}$
93. $-11\sqrt[3]{2}$ **95.** $y\sqrt{z}$ **97.** $13y\sqrt{x}$ **99.** $12\sqrt[3]{a}$ **101.** $-7y^2\sqrt{y}$ **103.** $4x\sqrt[5]{xy^2}$ **105.** $8\pi\sqrt{5}$ ft²; 56.2 ft² **107.** $5\sqrt{3}$ amps; 8.7 amps
109. $\left(26\sqrt{5} + 10\sqrt{3}\right)$ in.; 75.5 in. **115.** $\dfrac{-15x^5}{y}$ **117.** $3p + 4 - \dfrac{5}{2p - 5}$

STUDY SET SECTION 9.5 (page 698)

1. FOIL **3.** irrational **5.** perfect **7. a.** $6\sqrt{6}$ **b.** 48 **c.** can't be simplified **d.** $-6\sqrt{6}$ **9.** When the numerator and the
denominator of a fraction are multiplied by the same nonzero number, the value of the fraction is not changed. **11.** A radical appears
in the denominator. **13.** $\sqrt{6}, 48, 16, 4$ **15.** 11 **17.** 7 **19.** 4 **21.** $5\sqrt{2}$ **23.** $6\sqrt{2}$ **25.** 5 **27.** 18 **29.** 8 **31.** 18 **33.** $2\sqrt[3]{3}$
35. ab^2 **37.** $5a\sqrt{b}$ **39.** $-20r\sqrt[3]{10s}$ **41.** $x^2(x + 3)$ **43.** $12\sqrt{5} - 15$ **45.** $24\sqrt{3} + 6\sqrt{14}$ **47.** $-8x\sqrt{10} + 6\sqrt{15x}$
49. $-1 - 2\sqrt{2}$ **51.** $5z + 2\sqrt{15z} + 3$ **53.** $3x - 2y$ **55.** $6a + 5\sqrt{3ab} - 3b$ **57.** $18r - 12\sqrt{2r} + 4$ **59.** $-6x - 12\sqrt{x} - 6$
61. $\dfrac{\sqrt{7}}{7}$ **63.** $\dfrac{\sqrt{30}}{5}$ **65.** $\dfrac{\sqrt{10}}{4}$ **67.** 2 **69.** $\dfrac{\sqrt[3]{4}}{2}$ **71.** $\sqrt[3]{3}$ **73.** $\dfrac{\sqrt[3]{6}}{3}$ **75.** $\dfrac{2\sqrt{2xy}}{xy}$ **77.** $\dfrac{\sqrt{5y}}{y}$ **79.** $\dfrac{\sqrt[3]{2ab^2}}{b}$ **81.** $\dfrac{\sqrt[4]{4}}{2}$ **83.** $\dfrac{\sqrt[5]{2}}{2}$
85. $\sqrt{2} + 1$ **87.** $\dfrac{3\sqrt{2} - \sqrt{10}}{4}$ **89.** $2 + \sqrt{3}$ **91.** $\dfrac{9 - 2\sqrt{14}}{5}$ **93.** $\dfrac{2\left(\sqrt{x} - 1\right)}{x - 1}$ **95.** $\sqrt{2z} + 1$ **97.** $\dfrac{x - 2\sqrt{xy} + y}{x - y}$ **99.** $\dfrac{1}{\sqrt{3} - 1}$
101. $\dfrac{x - 9}{x\left(\sqrt{x} - 3\right)}$ **103.** $\dfrac{x - y}{\sqrt{x}\left(\sqrt{x} - \sqrt{y}\right)}$ **105.** $\dfrac{\sqrt{2\pi}}{2\pi\sigma}$ **107.** $\dfrac{\sqrt{2}}{2}$ **109.** $f/4$ **113.** 1 **115.** $\tfrac{1}{3}$

STUDY SET SECTION 9.6 (page 706)

1. hypotenuse **3.** Pythagorean **5.** $a^2 + b^2 = c^2$ **7.** $\sqrt{2}$ **9.** $\sqrt{3}$ **11.** 30°, 60° **13. a.** Take the positive square root of both sides.
b. Subtract 25 from both sides. **15.** 6, 52, 4, 2 **17.** 10 ft **19.** 80 m **21.** $h = 2\sqrt{2} \approx 2.83$, $x = 2$ **23.** $x = 5\sqrt{3} \approx 8.66$, $h = 10$
25. $x = 4.69$, $y = 8.11$ **27.** $x = 12.11$, $y = 12.11$ **29.** $7\sqrt{2}$ cm **31.** 5 **33.** 13 **35.** 10 **37.** $2\sqrt{26}$ **41.** $\left(5\sqrt{2}, 0\right)$, $\left(0, 5\sqrt{2}\right)$,
$\left(-5\sqrt{2}, 0\right)$, $\left(0, -5\sqrt{2}\right)$; (7.07, 0), (0, 7.07), (−7.07, 0), (0, −7.07) **43.** $10\sqrt{3}$ mm, 17.32 mm **45.** $10\sqrt{181}$ ft, 134.54 ft **47.** about
0.13 ft **49. a.** 21.21 units **b.** 8.25 units **c.** 13.00 units **51.** yes **55.** 7

KEY CONCEPT (page 711)

1. $-3h^2\sqrt[3]{2}$ **2.** $14\sqrt[3]{e}$ **3.** $-4\sqrt{2}$ **4.** $-20r\sqrt[3]{10s}$ **5.** $3s - 2t$ **6.** $-\sqrt{21} + \sqrt{15}$ **7.** $18n - 12\sqrt{2n} + 4$ **8.** $\dfrac{\sqrt[3]{3k^2}}{k}$ **9.** -3 **10.** 5
11. 2; 7 is extraneous **12.** no solutions **13.** $3^{1/3}$ **14.** $5\sqrt[5]{a^2}$

CHAPTER REVIEW (page 713)

1. a. 7 **b.** -11 **c.** $\tfrac{15}{7}$ **d.** not real **e.** 0.1 **f.** $5|x|$ **g.** x^4 **h.** $|x + 2|$ **2. a.** -3 **b.** -6 **c.** $4a^2b$ **d.** $\dfrac{s^3}{5}$ **3. a.** 5 **b.** -2
c. $4x^2|y|$ **d.** $22|y|$ **e.** $-\tfrac{1}{2}$ **f.** not real **g.** 0 **h.** 0 **4.** 48 ft **5.** 24 cm² **6. a.** D $= [-2, \infty)$, R $= [0, \infty)$

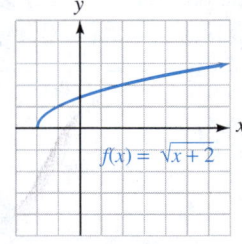

b. D $= (-\infty, \infty)$, R $= (-\infty, \infty)$

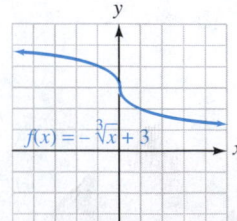

7. a. 22 **b.** 16, 9 **c.** $\tfrac{13}{2}$ **d.** $\tfrac{9}{16}$ **e.** 2, -4 **f.** 3, no solutions **8.** 2

9. a. $P = \dfrac{A}{(r + 1)^2}$ **b.** $I = \dfrac{h^3b}{12}$ **10.** about 17.9 ohms **11. a.** $\sqrt{t}$ **b.** $\sqrt[4]{5xy^3}$ **12. a.** 5 **b.** -6 **c.** not real **d.** 1 **e.** $\dfrac{3}{x}$ **f.** $\dfrac{1}{3}$
g. -2 **h.** 5 **i.** $3ab^{1/3}$ **j.** $3cd$ **13. a.** 27 **b.** $\dfrac{1}{4}$ **c.** $-16{,}807$ **d.** 10 **e.** $\dfrac{27}{8}$ **f.** $\dfrac{1}{3{,}125}$ **g.** $125x^3y^6$ **h.** $\dfrac{1}{4u^4v^2}$ **14. a.** $5^{3/4}$ **b.** $a^{1/7}$

c. k^8 **d.** $\dfrac{2gh}{3}$ **15. a.** $u - 1$ **b.** $v + v^2$ **16.** $\sqrt{\dfrac{a}{5b}} = \dfrac{\sqrt{5ab}}{5b}$ **17.** Two true statements result: $16 = 16$.   **18. a.** $4\sqrt{15}$ **b.** $3\sqrt[3]{2}$

c. $2\sqrt[4]{2}$ **d.** $4\sqrt[5]{3}$ **e.** $2x^2\sqrt{2x}$ **f.** $r^5\sqrt[3]{r^2}$ **g.** $2xy\sqrt[3]{2x^2y}$ **h.** $9j^2\sqrt[3]{jk}$ **i.** $4x$ **j.** $\dfrac{\sqrt{17xy}}{8a^2}$ **19. a.** $3\sqrt{2}$ **b.** $11\sqrt{5}$ **c.** 0 **d.** $-8\sqrt[4]{2}$

e. $29x\sqrt{2}$ **f.** $13\sqrt[3]{2}$ **20.** $\left(6\sqrt{2} + 2\sqrt{10}\right)$ in., 14.8 in.   **21. a.** 7 **b.** $6\sqrt{10}$ **c.** 32 **d.** 72 **e.** $3x$ **f.** 33 **g.** $-2x$ **h.** $12\sqrt[3]{3}$

i. $4 - 3\sqrt{2}$ **j.** $-20x^3y^3\sqrt{xy}$ **k.** $3b + 6\sqrt{b} + 3$ **l.** $6u - 12 + \sqrt{u}$ **22. a.** $\dfrac{10\sqrt{3}}{3}$ **b.** $\dfrac{\sqrt{15}}{5}$ **c.** $\dfrac{\sqrt{xy}}{y}$ **d.** $\dfrac{\sqrt[3]{u^2}}{u^2v^2}$ **e.** $2\left(\sqrt{2} + 1\right)$

f. $\dfrac{a + 2\sqrt{a} + 1}{a - 1}$ **23. a.** $\dfrac{9 - x}{2\left(3 + \sqrt{x}\right)}$ **b.** $\dfrac{a - b}{a + \sqrt{ab}}$ **24.** $r = \dfrac{\sqrt[3]{6\pi^2V}}{2\pi}$ **25.** 17 ft   **26.** 88 yd   **27.** $7\sqrt{2}$ m   **28.** $6\sqrt{3}$ cm, 18 cm

29. a. 7.07 in.   **b.** 8.66 cm   **30. a.** 13   **b.** $2\sqrt{2}$

CHAPTER 9 TEST (page 718)

1. 0, 1, 1.41, 2, 3, 3.32, 4   **2.** 3.0 mi   **3.** $\dfrac{1}{3}$ **4.** 10   **5.** 4 is extraneous.   **6.** $G = \dfrac{4\pi^2 r^3}{Mt^2}$ **7.** 2   **8.** 9   **9.** $\dfrac{1}{216}$

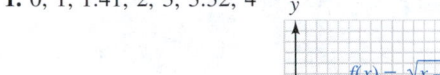

10. $\dfrac{9}{4}$ **11.** $2^{4/3}$ **12.** $8xy$ **13.** $|x|$ **14.** $2|x|\sqrt{2}$ **15.** $3x\sqrt[3]{2x^2}$ **16.** $3x^2y^4\sqrt{2}$ **17.** $-4xy^2$ **18.** $\dfrac{2}{3}a$ **19.** not real   **20.** $t + 8$

21. $4\sqrt{3}$ **22.** $5xy^2\sqrt{10xy}$ **23.** $2x^5y^3\sqrt[3]{3}$ **24.** $\dfrac{1}{4a}$ **25.** $-\sqrt{3}$ **26.** $14\sqrt[3]{5}$ **27.** $2y^2\sqrt{3y}$ **28.** $6z\sqrt[4]{3z}$ **29.** $-6x\sqrt{y} - 2xy^2$

30. $3 - 7\sqrt{6}$ **31.** $\dfrac{\sqrt{5}}{5}$ **32.** $\sqrt{3t} + 1$ **33.** $\dfrac{1}{2\sqrt[3]{3}}$ **34.** $\dfrac{1}{\sqrt{2}\left(\sqrt{5} - 3\right)}$ **35.** 9.24 cm   **36.** 8.67 cm   **37.** 25   **38.** $\dfrac{\pi\sqrt{2}}{2}$ sec, 2.22 sec

39. 28 in.

STUDY SET   SECTION 10.1 (page 728)

1. quadratic   **3.** perfect   **5.** $x = \sqrt{c}, x = -\sqrt{c}$ **7.** 3, -2, 5   **9.** It is a solution.   **11. a.** 36   **b.** $\dfrac{25}{4}$ **c.** $\dfrac{1}{16}$ **13. a.** Subtract 35 from both sides.   **b.** Add 36 to both sides.   **15.** 4 is not a factor of the numerator. It cannot be divided out.   **17.** $x^2 - 8x + 15 = 0$

19. a. 2; $2\sqrt{5}, -2\sqrt{5}$ **b.** ± 4.47 **21.** 0, -2 **23.** 5, -5 **25.** -2, -4 **27.** 2, $\dfrac{1}{2}$ **29.** ± 6 **31.** $\pm\sqrt{5}$ **33.** $\pm\dfrac{4\sqrt{3}}{3}$ **35.** 0, -2

37. 4, 10   **39.** $-5 \pm \sqrt{3}$ **41.** $d = \dfrac{\sqrt{6h}}{2}$ **43.** $c = \dfrac{\sqrt{Em}}{m}$ **45.** 2, -4 **47.** 1, $-\dfrac{1}{2}$ **49.** $\dfrac{1}{2}, -\dfrac{2}{3}$ **51.** $-4 \pm \sqrt{10}$ **53.** $1 \pm 3\sqrt{2}$

55. $\dfrac{3 \pm 2\sqrt{3}}{3}$ **57.** $\dfrac{1 \pm 2\sqrt{2}}{2}$ **59.** $\dfrac{-5 \pm \sqrt{41}}{4}$ **61.** $\dfrac{-7 \pm \sqrt{29}}{10}$ **63.** width: $7\dfrac{1}{4}$ ft; length: $13\dfrac{3}{4}$ ft   **65.** 1.6 sec   **67.** 1.70 in.

69. 0.92 in.   **71.** 2.9 ft, 6.9 ft   **75.** $2ab^2\sqrt[3]{5}$ **77.** x^3 **79.** $5ab\sqrt{7b}$

STUDY SET   SECTION 10.2 (page 736)

1. quadratic   **3. a.** 0, $\dfrac{1}{2}$ **b.** 4, -2, 0; 0, $\dfrac{1}{2}$ **5. a.** true   **b.** true   **7.** $c = \dfrac{-s \pm \sqrt{s^2 - 4rt}}{2r}$ **9.** $2 \pm \sqrt{3}$ **11. a.** The fraction bar wasn't

drawn under both parts of the numerator.   **b.** A $\pm$ sign wasn't written between b and the radical.   **13.** -1, -2 **15.** -6, -6

17. $\dfrac{1}{2}$, -3 **19.** $\dfrac{-5 \pm \sqrt{5}}{10}$ **21.** $-\dfrac{3}{2}, -\dfrac{1}{2}$ **23.** $\dfrac{1}{4}, -\dfrac{3}{4}$ **25.** $\dfrac{4}{3}, -\dfrac{2}{5}$ **27.** $\dfrac{-5 \pm \sqrt{17}}{2}$ **29.** $\dfrac{3 \pm \sqrt{17}}{4}$ **31.** $5 \pm \sqrt{7}$ **33.** 23, -17

35. $\dfrac{2}{3}, -\dfrac{5}{2}$ **37.** 8.98, -3.98 **39.** 97 ft by 117 ft   **41.** 0.5 mi by 2.5 mi   **43.** 34 in.   **45.** $4.80 or $5.20   **47.** 4,000   **49.** 9%

51. a. 83%   **b.** early 1976   **55.** $n^{1/2}$ **57.** $(3b)^{1/4}$ **59.** $\sqrt[3]{t}$ **61.** $\sqrt[4]{3t}$

STUDY SET   SECTION 10.3 (page 749)

1. quadratic   **3.** vertex   **5.**   **7.**   **9.**

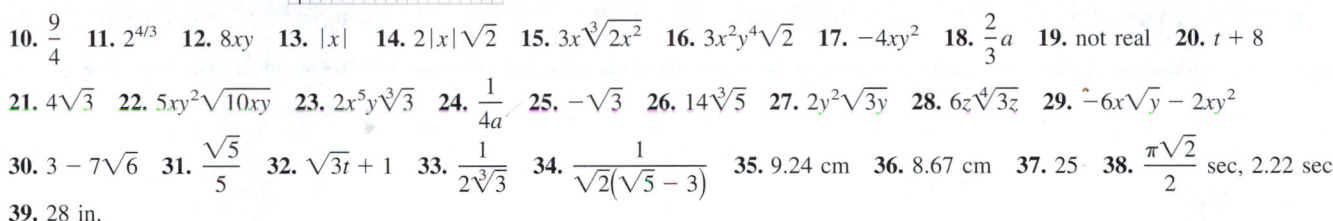

11. $-3, 5$ **13.** $h = -1$; $f(x) = 2[x - (-1)]^2 + 6$ **15.** $(1, 2)$, $x = 1$, upward   **17.** $(-3, -4)$, $x = -3$, upward
19. $(1, -2)$, $x = 1$, upward   **21.** $(2, 21)$, $x = 2$, downward   **23.** $\left(-\frac{2}{3}, \frac{2}{3}\right)$, $x = -\frac{2}{3}$, upward

25. 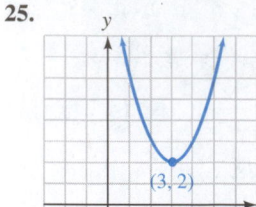   **27.** 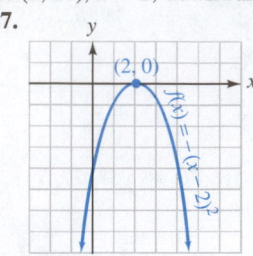   **29.** 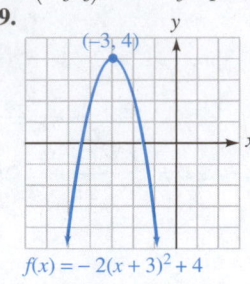   **31.**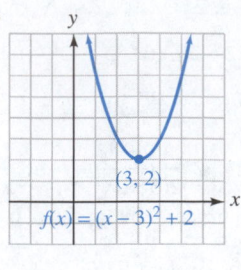

33. 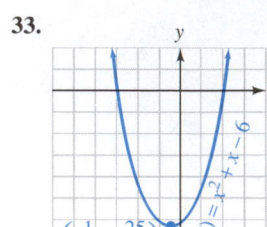   **35.** 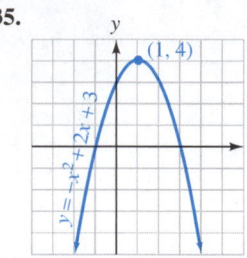   **37.** 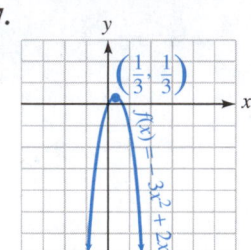   **39.**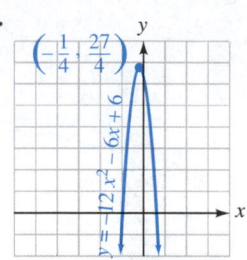

41. $(0.25, 0.88)$ **43.** $(0.50, 7.25)$ **45.** $2, -3$ **47.** $-1.85, 3.25$ **49.** 3.75 sec, 225 ft   **51.** 250 ft by 500 ft   **53.** 15 min, \$160

55. 1968, 1.5 million; the U.S. involvement in the war in Vietnam was at its peak   **57.** 200, \$7,000   **63.** $4a^2\sqrt{b}$ **65.** $\dfrac{\sqrt{6}}{10}$

67. $15b - 6\sqrt{15b} + 9$

STUDY SET   SECTION 10.4 (page 760)

1. imaginary   **3.** real, imaginary   **5.** i or $\sqrt{-1}$ **7.** $\sqrt{-1}$ **9.** $-i$ **11.** FOIL   **13.** It is a solution.

15.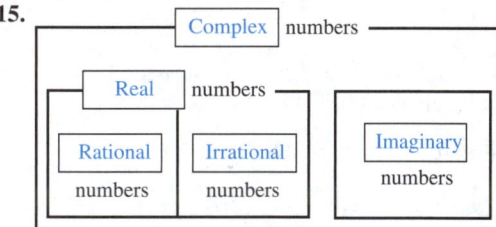

17. a. ± 1 **b.** $\pm i$ **19.** $9, 6i, 2$ **21. a.** true   **b.** false   **c.** false   **d.** false   **23.** $3i$
25. $\sqrt{7}i$ **27.** $2\sqrt{6}i$ **29.** $-2\sqrt{6}i$ **31.** $45i$ **33.** $\frac{5}{3}i$ **35.** -6 **37.** $-2\sqrt{3}$
39. $\frac{5}{8}$ **41.** -20 **43.** $\pm 3i$ **45.** $\pm\dfrac{4\sqrt{3}}{3}i$ **47.** $-1 \pm i$ **49.** $-\dfrac{1}{4} \pm \dfrac{\sqrt{7}}{4}i$
51. $\dfrac{2}{3} \pm \dfrac{\sqrt{2}}{3}i$ **53.** $\dfrac{1}{3} \pm \dfrac{2\sqrt{2}}{3}i$ **55.** $8 - 2i$ **57.** $3 - 5i$ **59.** $15 + 2i$
61. $15 + 7i$ **63.** $6 - 3i$ **65.** $-25 - 25i$ **67.** $7 + i$ **69.** $12 + 5i$ **71.** $13 - i$
73. $14 - 8i$ **75.** $8 + \sqrt{2}i$ **77.** $3 + 4i$ **79.** $0 - i$ **81.** $0 + \dfrac{4}{5}i$ **83.** $\dfrac{1}{8} + 0i$

85. $0 + \dfrac{3}{5}i$ **87.** $2 + i$ **89.** $\dfrac{15}{26} - \dfrac{3}{26}i$ **91.** $-\dfrac{42}{25} - \dfrac{6}{25}i$ **93.** $\dfrac{1}{4} + \dfrac{3}{4}i$ **95.** $-\dfrac{4}{13} - \dfrac{6}{13}i$ **97.** $\dfrac{5}{13} - \dfrac{12}{13}i$ **99.** $\dfrac{11}{10} + \dfrac{3}{10}i$

101. $\dfrac{1}{4} - \dfrac{\sqrt{15}}{4}i$ **103.** i **105.** $-i$ **107.** 1 **109.** i **111.** $-1 + i$ **115.** 15 units, $15\sqrt{3}$ units   **117.** 20 mph

STUDY SET   SECTION 10.5 (page 767)

1. $b^2 - 4ac$ **3.** conjugates   **5.** rational, unequal   **7. a.** $x^4 = (x^2)^2$ **b.** yes   **9.** 5, 6, 24, rational   **11.** rational, equal   **13.** complex
conjugates   **15.** irrational, unequal   **17.** rational, unequal   **19.** yes   **21.** $1, -1, 4, -4$ **23.** $1, -1, \sqrt{5}, -\sqrt{5}$

25. $2, -2, \sqrt{7}i, -\sqrt{7}i$ **27.** 1 **29.** no solution   **31.** $16, 4$ **33.** $-8, -27$ **35.** $-1, 27$ **37.** $\dfrac{1}{4}, \dfrac{1}{2}$ **39.** $1 - 2\sqrt{3}, 1 + 2\sqrt{3}$

41. $-1, -4$ **43.** $-1, -\dfrac{27}{13}$ **45.** $\dfrac{3 + \sqrt{57}}{6}, \dfrac{3 - \sqrt{57}}{6}$ **47.** $1, 1, -1, -1$ **49.** $1 \pm i$ **51.** 12.1 in.   **53.** 30 mph   **55.** 14.3 min

59. -2 **61.** $\frac{9}{5}$

STUDY SET   SECTION 10.6 (page 774)

1. quadratic   **3.** interval   **5.** greater   **7.** undefined   **9.** $(-\infty, -2) \cup (3, \infty)$ **11. a.** -2 **b.** -2 **c.** -2 **d.** 3   **e.** 3   **f.** 3   **13. a.** iii
b. iv   **c.** ii   **d.** i   **15.** $(1, 4)$,   **17.** $(-\infty, 3) \cup (5, \infty)$,   **19.** $[-4, 3]$,

21. no solutions   **23.** $(-\infty, -3] \cup [3, \infty)$,   **25.** $(-5, 5)$,

27. $(-\infty, 0) \cup \left(\frac{1}{2}, \infty\right)$,   **29.** $\left(-\infty, -\frac{5}{3}\right) \cup (0, \infty)$,   **31.** $(-\infty, -3) \cup (1, 4)$,

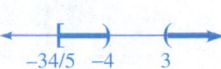

33. $\left(-\frac{1}{2}, \frac{1}{3}\right) \cup \left(\frac{1}{2}, \infty\right)$,   **35.** $(0, 2) \cup (8, \infty)$,   **37.** $\left[-\frac{34}{5}, -4\right) \cup (3, \infty)$,

39. $(-4, -2] \cup (-1, 2]$,   **41.** $(-\infty, -2) \cup (-2, \infty)$,   **43.** $(-1, 3)$ **45.** $(-\infty, -3) \cup (2, \infty)$

47. 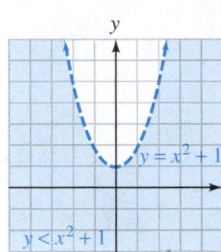   **49.** 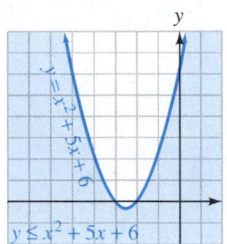   **51.** 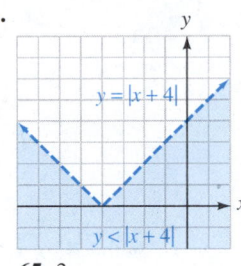   **53.**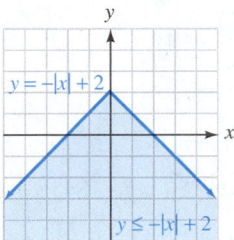

55. $(-2,100, -900) \cup (900, 2,100)$ **61.** $x = ky$ **63.** $t = kxy$ **65.** 3

KEY CONCEPT (page 778)

1. $0, -2$ **2.** $-3, -5$ **3.** $-\frac{3}{2}, -1$ **4.** $\pm 2\sqrt{6}$ **5.** $4, 10$ **6.** $\pm\frac{4\sqrt{3}}{3}$ **7.** $-5 \pm 4\sqrt{2}$ **8.** $\frac{1 \pm \sqrt{2}}{2}$ **9.** $-1 \pm i$ **10.** $\frac{3 \pm \sqrt{17}}{4}$

11. $23, -17$ **12.** $-\frac{1}{3} \pm \frac{\sqrt{2}}{3}i$ **13.** $1, -2$

CHAPTER REVIEW (page 780)

1. a. $-5, -4$ **b.** $-\frac{1}{3}, -\frac{5}{2}$ **c.** $\pm 2\sqrt{7}$ **d.** $4, -8$ **e.** $0, -\frac{11}{5}$ **f.** $\pm\frac{7\sqrt{5}}{5}$ **2.** $\frac{1}{4}$ **3. a.** $-4, -2$ **b.** $\frac{3 \pm \sqrt{3}}{2}$ **4.** 6 seconds before

midnight   **5. a.** $5 \pm \sqrt{7}$ **b.** $0, 10$ **c.** $\frac{1}{2}, -7$ **d.** $\frac{13 \pm \sqrt{163}}{3}$ **6.** sides: 1.25 in. wide; top/bottom: 2.5 in. wide   **7.** 0.7 sec, 1.8 sec

8. \$46.8 billion   **9. a.** 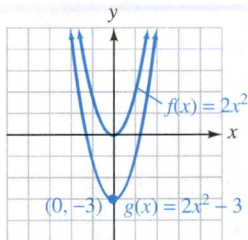   **b.** 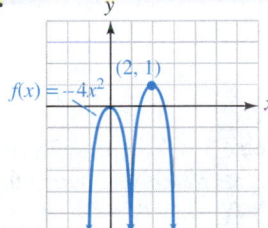   **10. a.**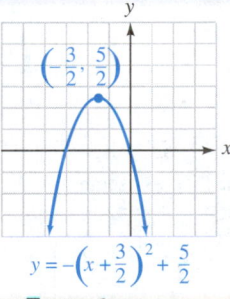

b. 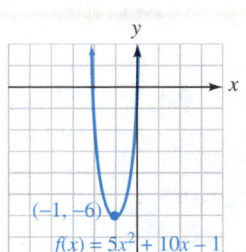   **11.** 1921, about 6,452   **12.** $-2, \frac{1}{3}$ **13. a.** $2i$ **b.** $\sqrt{7}i$ **c.** $-6\sqrt{5}i$ **d.** $\frac{6}{7}i$ **e.** -3 **f.** $\frac{8}{3}$ **14. a.** $\pm 5i$

b. $1 \pm 2\sqrt{3}i$ **15. a.** $12 - 8i$ **b.** $2 - 68i$ **c.** $-2 - 2\sqrt{2}i$ **d.** $-18 + 128i$ **e.** $22 + 29i$ **f.** $-12 + 28\sqrt{3}i$ **16. a.** $\frac{12}{5} - \frac{6}{5}i$

b. $\frac{15}{17} + \frac{8}{17}i$ **c.** $-\frac{1}{7} + \frac{4\sqrt{3}}{7}i$ **d.** $0 - \frac{2}{5}i$ **17. a.** i **b.** 1   **18. a.** irrational, unequal   **b.** complex conjugates   **c.** equal rational

numbers   **19. a.** $1, 144$ **b.** $8, -27$ **c.** $\sqrt{3}, -\sqrt{3}, \frac{\sqrt{6}}{6}, -\frac{\sqrt{6}}{6}$ **d.** $1, -\frac{8}{5}$ **e.** $10, 4$ **20.** about 81 min

21. a. $(-\infty, -7) \cup (5, \infty)$,   **b.** $[-9, 9]$,   **c.** $(-\infty, 0) \cup [3/5, \infty)$,

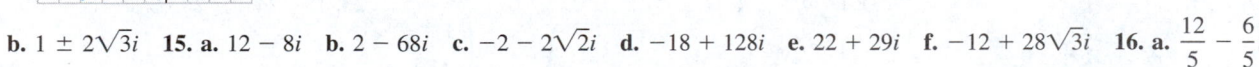

d. $(-7/2, 1) \cup (4, \infty)$,   **22. a.**    **b.**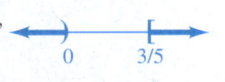

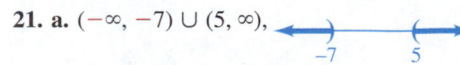

23. a.

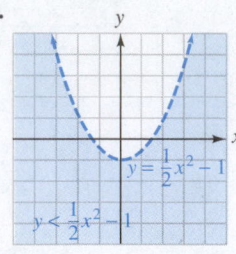

b.

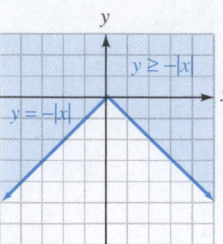

CHAPTER 10 TEST (page 784)

1. $0, -6$ **2.** $-\frac{3}{2}, -\frac{5}{3}$ **3.** 144 **4.** $\dfrac{1 \pm \sqrt{5}}{2}$ **5.** $\dfrac{4 \pm \sqrt{6}}{2}$ **6.** 4.5 ft by 1,502 ft **7.** $4\sqrt{3}i$ **8.** -1 **9.** $-1 + 11i$ **10.** $4 - 7i$

11. $8 + 6i$ **12.** $-10 - 11i$ **13.** $0 + i$ **14.** $\frac{1}{2} + \frac{1}{2}i$ **15.** nonreal **16.** $\pm 2\sqrt{3}i$ **17.** $2 \pm 3i$ **18.** $1, \frac{1}{4}$

19.

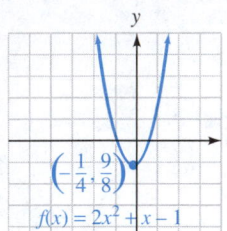

20.

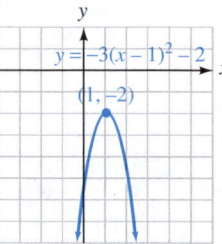

21. $(-\infty, -2) \cup (4, \infty)$,

22. $(-3, 2]$,

23.

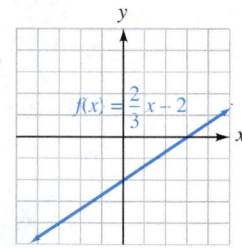

24. 20 in. **25.** 211 ft **27.** $-3, 2$ **28.** $[-3, 2]$

CHAPTERS 1–10 CUMULATIVE REVIEW EXERCISES (page 785)

1. $\frac{8}{3}$ **2.** -2 **3.** 14 **4.** Life expectancy will increase 0.1 year each year during this period. **5.** $y = -7x + 54$ **6.** $y = -\frac{11}{6}x - \frac{7}{3}$

7. D: $(-\infty, \infty)$; R: $(-\infty, \infty)$

8. $T_1 = 80, T_2 = 60$ **9.**

10.

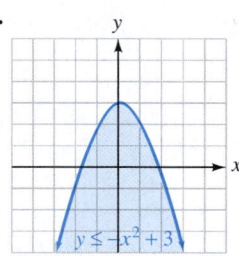

11.

12.

13. $a^8 b^4$ **14.** $\dfrac{b^4}{a^4}$ **15.** 81 **16.** $\dfrac{x^{21} y^3}{8}$ **17.** $42,500$ **18.** 0.000712 **19.** $y = \dfrac{kxz}{r}$

20. $d = kxy$ **21.**

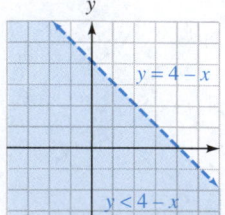

22. D: $(-\infty, \infty)$; R: $[-3, \infty)$

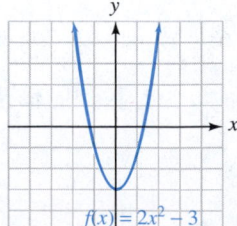

23. 18 **24.** 7 **25.** $2x^3 + x^2 + 12$

26. $-3x^2 - 3$ **27.** $6x^2 - 7x - 20$ **28.** $4x^6 - 4x^3 + 1$ **29.** yes **30.** about 17 **31.** about 14 **32.** 4 **33.** $\dfrac{2x - 3}{3x - 1}$ **34.** $-\dfrac{q}{p}$

35. $\dfrac{4}{x - y}$ **36.** $\dfrac{a^2 + ab^2}{a^2 b - b^2}$ **37.** 0 **38.** -17 **39.** $x + 4$ **40.** $-x^2 + x + 5 + \dfrac{8}{x - 1}$ **41.** It will rise sharply. **42.** It has dropped.

STUDY SET   SECTION 11.1 (page 794)

1. sum, $f(x) + g(x)$ **3.** product, $f(x)g(x)$ **5.** domain   **7.** identity   **9. a.** $g(x)$ **b.** $g(x)$
11. $-3, -2, -1, 0, 1, 2, 3$ **13.** $(3x - 1)$, $9x$, $2x$ **15.** $7x$, $(-\infty, \infty)$ **17.** $12x^2$, $(-\infty, \infty)$ **19.** x, $(-\infty, \infty)$

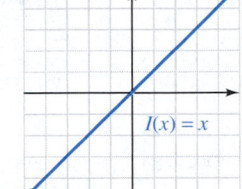

21. $\frac{4}{3}$, $(-\infty, 0) \cup (0, \infty)$ **23.** $3x - 2$, $(-\infty, \infty)$ **25.** $2x^2 - 5x - 3$, $(-\infty, \infty)$ **27.** $-x - 4$, $(-\infty, \infty)$
29. $\frac{x - 3}{2x + 1}$, $(-\infty, -1/2) \cup (-1/2, \infty)$ **31.** $-2x^2 + 3x - 3$, $(-\infty, \infty)$ **33.** $\frac{3x - 2}{2x^2 + 1}$, $(-\infty, \infty)$ **35.** 3, $(-\infty, \infty)$
37. $\frac{x^2 - 4}{x^2 - 1}$, $(-\infty, -1) \cup (-1, 1) \cup (1, \infty)$ **39.** 7 **41.** 24 **43.** -1 **45.** $-\frac{1}{2}$ **47.** $2x^2 - 1$ **49.** $16x^2 + 8x$ **51.** 58 **53.** 110 **55.** 2
57. $9x^2 - 9x + 2$ **61.** $C(t) = \frac{5}{9}(2{,}668 - 200t)$ **63. a.** about \$37.50   **b.** $C(m) = \frac{1.50m}{20}$ **69.** $-\frac{3x + 7}{x + 2}$ **71.** $\frac{x - 4}{3x^2 - x - 12}$

STUDY SET   SECTION 11.2 (page 803)

1. one-to-one   **3.** inverses   **5.** once   **7.** 2 **9.** x **11.** $-6, -4, 0, 2, 8$ **13.** no   **15.** 2 **17.** The graphs are not symmetric about the line
$y = x$.   **19.** $2x, 2y, 3, y, f^{-1}(x)$ **21.** the inverse of, inverse   **23.** yes   **25.** no   **27.**   **29.**

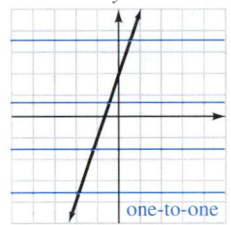
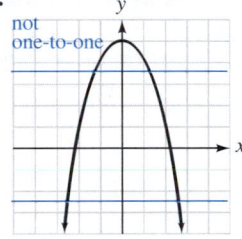

31. yes, $\{(2, 3), (1, 2), (0, 1)\}$, yes   **33.** yes, $\{(1, 1), (1, 2), (1, 3), (1, 4)\}$, no   **35.** $f^{-1}(x) = \frac{5}{4}x + 5$ **37.** $f^{-1}(x) = 5x - 4$
39. $f^{-1}(x) = 5x + 4$ **41.**   **43.**   **45.** $y = \pm\sqrt{x - 4}$, no   **47.** $f^{-1}(x) = \sqrt[3]{x}$, yes

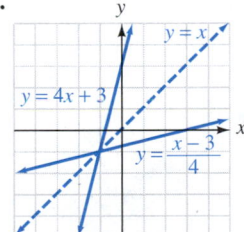
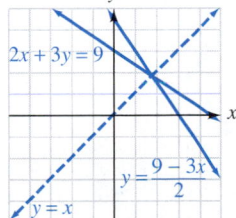

49. $x = |y|$, no   **51.** $f^{-1}(x) = \sqrt[3]{\dfrac{x + 3}{2}}$, yes   **53.**   **55.**   **57. a.** yes, no

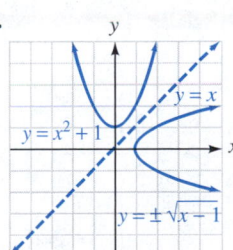
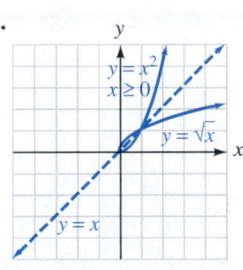

b. No. Twice during this period, the person's anxiety level was at the maximum threshold value.   **61.** $3 - 8i$ **63.** $18 - i$ **65.** $-28 - 96i$

STUDY SET   SECTION 11.3 (page 815)

1. exponential   **3.** $(0, \infty)$ **5.** none   **7.** increasing   **9.**   **11.** $A = P\left(1 + \frac{r}{k}\right)^{kt}$, $FV = PV(1 + i)^n$

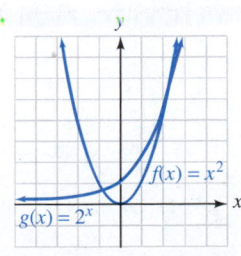

13. $g(x) = 2^x + 3$; $h(x) = 2^x - 2$ **15.** $\left(1 + \frac{r}{k}\right)$, kt **17.** 2.6651 **19.** 36.5548 **21.** 8 **23.** $7^{3\sqrt{3}}$

25.   **27.**   **29.**   **31.**

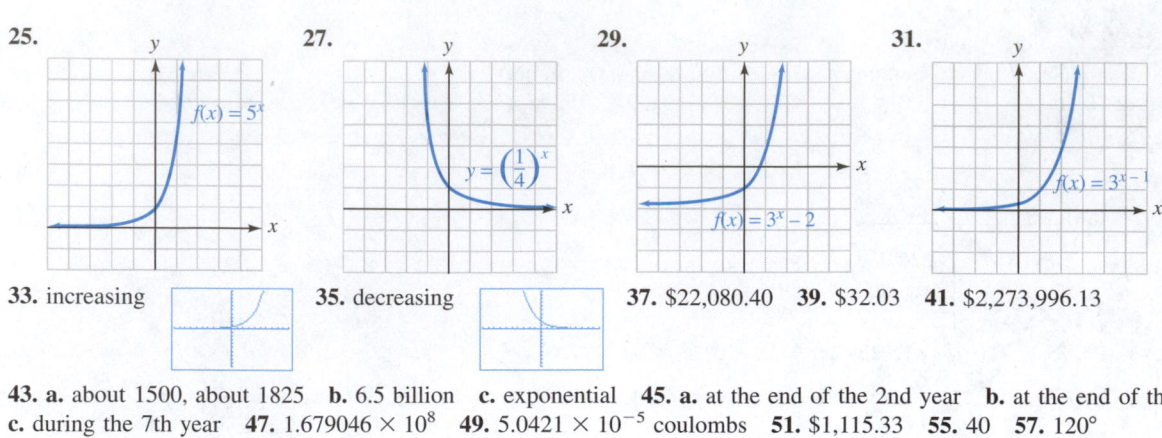

33. increasing   **35.** decreasing   **37.** $22,080.40   **39.** $32.03   **41.** $2,273,996.13

43. a. about 1500, about 1825   **b.** 6.5 billion   **c.** exponential   **45. a.** at the end of the 2nd year   **b.** at the end of the 4th year
c. during the 7th year   **47.** 1.679046×10^8 **49.** 5.0421×10^{-5} coulombs   **51.** $1,115.33   **55.** 40   **57.** 120°

STUDY SET   SECTION 11.4 (page 824)

1. the natural exponential function   **3.** $(0, \infty)$ **5.** none   **7.** increasing   **9.** continuous   **11.** 2.72
13.

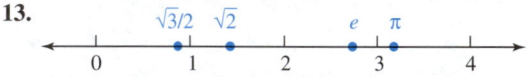

15. an exponential function

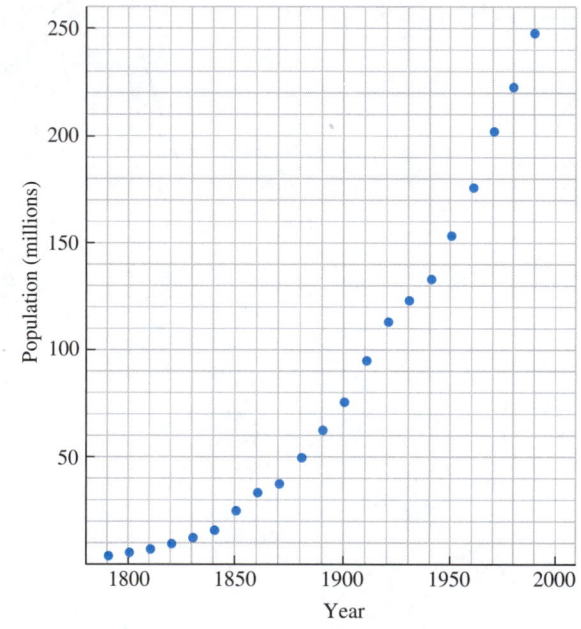

17. 2.7182818. . .; e **19.** 1,000, 10, 0.9, 1,000

21.   **23.**   **25.**   **27.**

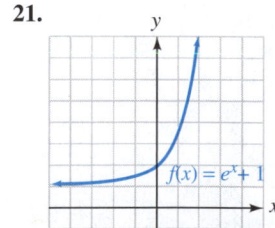 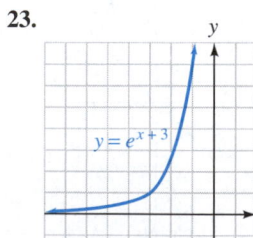 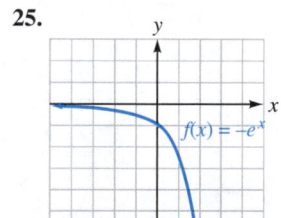 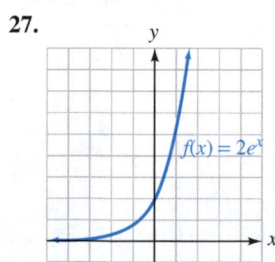

29. $13,375.68   **31.** $7,518.28 from annual compounding, $7,647.95 from continuous compounding   **33.** $6,849.16   **35.** 10.6 billion
37. 315   **39.** 51   **41.** 12 hr   **43.** 49 mps   **45.** about 72 yr   **49.** $4x^2\sqrt{15x}$ **51.** $10y\sqrt{3y}$

STUDY SET   SECTION 11.5 (page 834)

1. logarithmic   **3.** $(-\infty, \infty)$ **5.** $(1, 0)$ **7.** increasing   **9.** $x = b^y$ **11.** inverse   **13.** $-2, -1, 0, 1, 2$ **15.** none, none, $-3, \frac{1}{2}, 8$
17. $-0.30, 0, 0.30, 0.48, 0.60, 0.70, 0.78, 0.85, 0.90, 0.95, 1$ **19.** They decrease.   **21.** 10   **23.** $3^4 = 81$ **25.** $12^1 = 12$ **27.** $4^{-3} = \frac{1}{64}$
29. $10^{-3} = 0.001$ **31.** $\log_8 64 = 2$ **33.** $\log_4 \frac{1}{16} = -2$ **35.** $\log_{1/2} 32 = -5$ **37.** $\log_x z = y$ **39.** 3   **41.** 3   **43.** 3   **45.** $\frac{1}{2}$ **47.** 64
49. 5   **51.** $\frac{1}{25}$ **53.** $\frac{1}{6}$ **55.** $-\frac{3}{2}$ **57.** $\frac{2}{3}$ **59.** 5   **61.** $\frac{3}{2}$ **63.** 4   **65.** 1   **67.** 0.5119   **69.** -2.3307 **71.** 25.25   **73.** 0.02

75. increasing

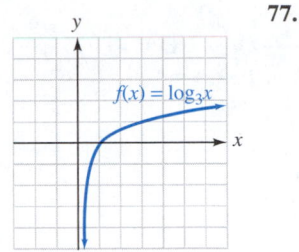

77. decreasing

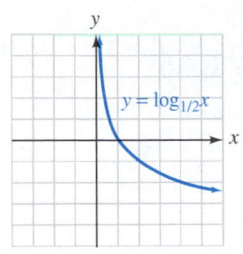

79.

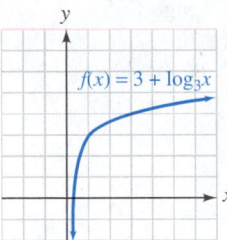

81.

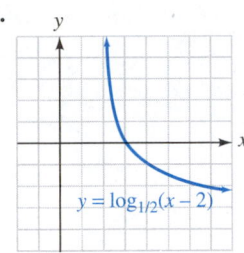

83.

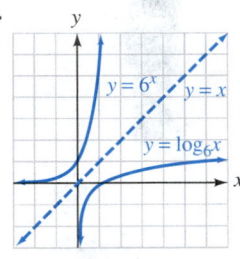

85.

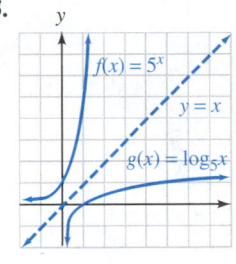

87. 29 db   **89.** 49.5 db   **91.** 4.4
93. 2,500 micrometers   **95.** 3 yr old
97. 10.8 yr   **101.** 10   **103.** 0; $-\frac{5}{9}$ does not
check

STUDY SET   SECTION 11.6 (page 841)

1. natural   **3.** $-0.69, 0, 0.69, 1.10, 1.39, 1.61, 1.79, 1.95, 2.08, 2.20, 2.30$ **5.** y-axis   **7.** $(-\infty, \infty)$ **9.** $e^y = x$ **11.** The logarithm of a
negative number or zero is not defined.   **13.** 2.7182818. . .; e **15.** e **17.** $\frac{\ln 2}{r}$ **19.** 3.5596   **21.** -5.3709 **23.** 0.5423   **25.** undefined
27. 4.0645   **29.** 69.4079   **31.** 0.0245   **33.** 2.7210   **35.**   **37.**   **39.** 5.8 yr   **41.** 9.2 yr   **43.** about 3.5 hr

47. $y = 5x$ **49.** $y = -\frac{3}{2}x + \frac{13}{2}$ **51.** $x = 2$

STUDY SET   SECTION 11.7 (page 850)

1. product   **3.** power   **5.** 0   **7.** M, N **9.** x, y **11.** x **13.** $\neq$ **15.** $\log_a b$ **17.** $10^0 = 1, 10^1 = 10, 10^2 = 10^2$ **19.** 0   **21.** 7   **23.** 10
25. 2   **27.** 1   **29.** 7   **31.** rs, t, r, s **39.** $2 + \log_2 5$ **41.** $\log_6 x - 2$ **43.** $4 \ln y$ **45.** $\frac{1}{2} \log 5$ **47.** $\log x + \log y + \log z$
49. $1 + \log_2 x - \log_2 y$ **51.** $3 \log x + 2 \log y$ **53.** $\frac{1}{2}(\log_b x + \log_b y)$ **55.** $\frac{1}{3} \log_a x - \frac{1}{4} \log_a y - \frac{1}{4} \log_a z$ **57.** $\ln x + \frac{1}{2} \ln z$

59. $\log_2 \dfrac{x + 1}{x}$ **61.** $\log x^2 y^{1/2}$ **63.** $\log_b \dfrac{z^{1/2}}{x^3 y^2}$ **65.** $\ln \dfrac{\frac{x}{z} + x}{\frac{y}{z} + y} = \ln \dfrac{x}{y}$ **67.** false   **69.** false   **71.** true   **73.** 1.4472   **75.** -1.1972
77. 1.1972   **79.** 1.8063   **81.** 1.7712   **83.** -1.0000 **85.** 1.8928   **87.** 2.3219   **89.** 4.8   **91.** from 2.5×10^{-8} to 1.6×10^{-7} **95.** $-\frac{7}{6}$
97. $\left(1, -\frac{1}{2}\right)$

STUDY SET   SECTION 11.8 (page 860)

1. exponential   **3.** $A_0 2^{-t/h}$ **5.** about 1.8   **7.** logarithm, exponent   **9.** 1.2920   **11.** 1   **13.** $10^2 = x + 1$ **15. a.** ± 3.46 **b.** 3.58
17. not a solution   **19.** log, log 2   **21.** 1.1610   **23.** 1.2702   **25.** 1.7095   **27.** 0   **29.** ± 1.0878 **31.** 0, 1.0566   **33.** 0.7324
35. -13.2662 **37.** 8   **39.** $-\frac{3}{4}$ **41.** 3, -1 **43.** $-2, -2$ **45.** 1.8   **47.** 8.8, 0.2   **49.** 9,998   **51.** -93 **53.** $e \approx 2.7183$ **55.** 19.0855
57. 2   **59.** 3   **61.** -7 **63.** 4   **65.** 10, -10 **67.** 50   **69.** 20   **71.** 10   **73.** 10   **75.** no solution   **77.** 6   **79.** 9   **81.** 4   **83.** 1, 7
85. 20   **87.** 8   **89.** 5.1 yr   **91.** 42.7 days   **93.** about 4,200 yr   **95.** 5.6 yr   **97.** 5.4 yr   **99.** because $\ln 2 \approx 0.7$ **101.** 25.3 yr
103. 2.828 times larger   **105.** 13.3   **107.** $\frac{1}{3} \ln 0.75$ **111.** 0, 5   **113.** $\frac{2}{3}, -4$ **115.** $\sqrt{137}$ in.

KEY CONCEPT (page 863)

1. no   **2.** no   **3.** yes   **4.** $f^{-1}(x) = \dfrac{-x-1}{2}$ **5.** $-2, 1, 3$ **6.** log 1,000 = 3   **7.** $2^{-3} = \frac{1}{8}$ **8.** 2   **9.** $\frac{3}{2}$ **10.** 2.71828   **11.** e **12.** 0.7558
13.

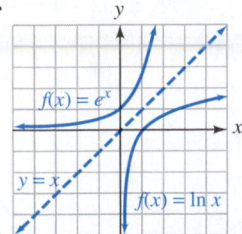

CHAPTER REVIEW (page 865)

1. a. $(f + g)(x) = 3x + 1$ **b.** $(f - g)(x) = x - 1$ **c.** $(f \cdot g)(x) = 2x^2 + 2x$ **d.** $(f/g)(x) = \frac{2x}{x+1}$ **e.** 4   **f.** 3   **g.** $(f \circ g)(x) = 2(x + 1)$
h. $(g \circ f)(x) = 2x + 1$ **2. a.** no   **b.** yes   **3. a.** yes   **b.** no   **4.**

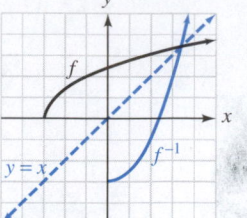

5. a. $f^{-1}(x) = \frac{x+3}{6}$ **b.** $f^{-1}(x) = \frac{x-5}{4}$

c. $f^{-1}(x) = \sqrt[3]{x}$ **d.** $y = \sqrt{\frac{x+1}{2}}$ **6. a.** $5^{2\sqrt{2}}$ **b.** $2^{\sqrt{10}}$ **7. a.** D: $(-\infty, \infty)$, R: $(0, \infty)$

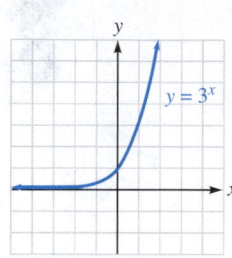

b. D: $(-\infty, \infty)$, R: $(0, \infty)$ **8. a.**   **b.**

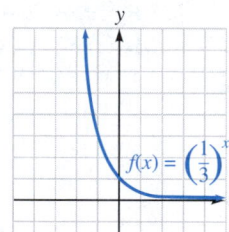   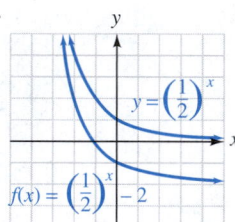   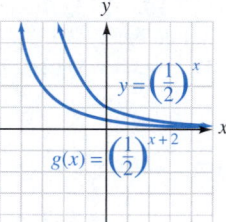

9. an exponential function   **10.** $2,189,703.45   **11.** 8.22%   **12.** $2,324,767.37

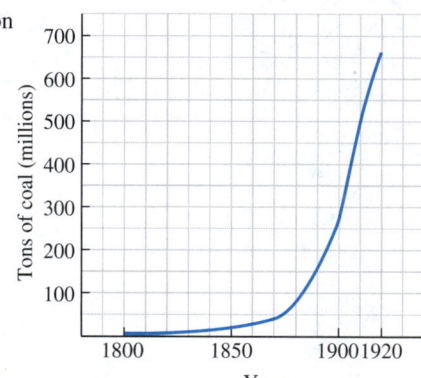
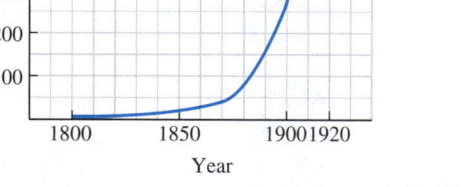

13. a. D: $(-\infty, \infty)$, R: $(1, \infty)$ **b.** D: $(-\infty, \infty)$, R: $(0, \infty)$ **14.** 582,175,004

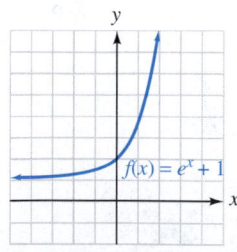   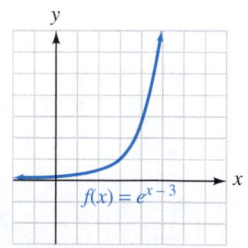

15. D: $(0, \infty)$, R: $(-\infty, \infty)$ **16. a.** $4^3 = 64$ **b.** $\log_7 \frac{1}{7} = -1$ **17. a.** 2   **b.** -2 **c.** 0   **d.** not possible   **e.** $\frac{1}{2}$ **f.** 3   **18. a.** 32   **b.** $\frac{1}{81}$
c. 4   **d.** 10   **e.** $\frac{1}{2}$ **f.** 9   **19. a.** 0.6542   **b.** 26.9153   **20. a.**   **b.**

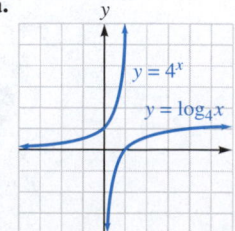

21. a. 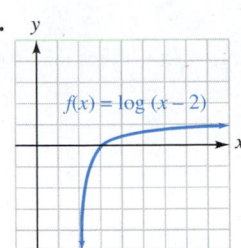   **b.**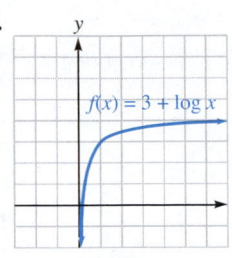

22. about 53   **23.** about 4.4   **24. a.** 6.1137   **b.** −0.1625

25. a. 10.3398   **b.** 0.0002   **26.** $f(x) = e^x$ **27. a.** 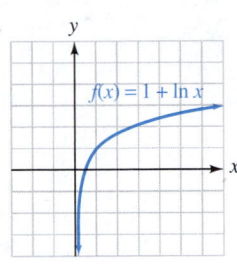   **b.** 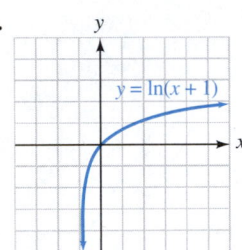   **28.** about 23 yr

29. a. 0   **b.** 1   **c.** 3   **d.** 4   **e.** 4   **f.** 1   **30. a.** $3 + \log_3 x$ **b.** $2 - \log x$ **c.** $\frac{1}{2} \log_5 27$ **d.** $1 + \log a + \log b$

31. a. $2 \log_b x + 3 \log_b y - \log_b z$ **b.** $\frac{1}{2}(\ln x - \ln y - 2 \ln z)$ **32. a.** $\log_2 \dfrac{x^3 z^7}{y^5}$ **b.** $\log_b \dfrac{\sqrt{x+2}}{y^3 z^7}$ **33. a.** 2.6609   **b.** 3.0000

34. 1.7604   **35.** about 7.9×10^{-4} gram-ions/liter   **36. a.** 1.7712   **b.** 2   **c.** 2.7095   **d.** −3, −1   **e.** 1.9459   **f.** −8.0472
37. a. 104   **b.** 9   **c.** 25, 4   **d.** 4   **e.** 4, 3   **f.** 2   **g.** 6   **h.** 31   **38.** about 3,300 yr   **39.** 2, 5

CHAPTER 11 TEST (page 870)

1. $(g + f)(x) = 5x - 1$ **2.** $(g \cdot f)(x) = 4x^2 - 4x$ **3.** 3   **4.** $4(x - 1)$ **5.** $y = \frac{12 - 2x}{3}$ **6.** $f^{-1}(x) = \sqrt{\dfrac{x-4}{3}}$

7. 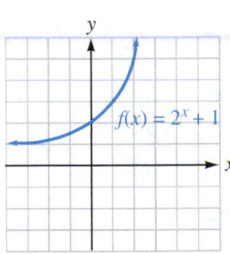   **8.** 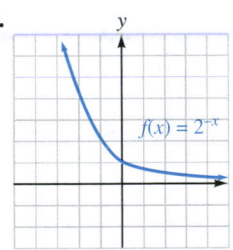   **9.** $\frac{3}{64}$ g = 0.046875 g   **10.** \$1,060.90   **11.**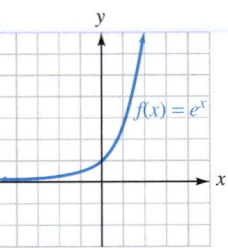

12. \$4,451.08   **13.** 2   **14.** 3   **15.** $\frac{1}{27}$ **16.** e **17.** $6^{-2} = \frac{1}{36}$ **18.** D: $(0, \infty)$, R: $(-\infty, \infty)$ **19.**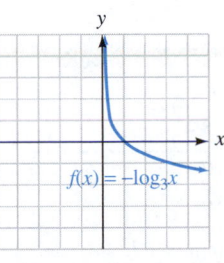

20. 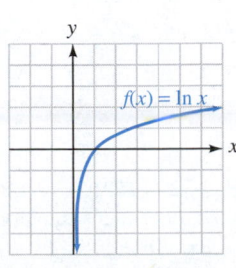   **21.** $2 \log a + \log b + 3 \log c$ **22.** $\ln \dfrac{b\sqrt{a+2}}{c^3}$ **23.** 0.5646   **24.** $y = \log x$ **25.** 6.4   **26.** 46

27. 0.6826   **28.** 4   **29.** 1   **30.** 10   **31.** 5   **32.** $\frac{1}{2} \ln (5 - 1) = \frac{1}{2} \ln 4 = 0.69314718. . .$, which is $\ln 2$.   **34. a.** yes   **b.** yes   **c.** 80; when the temperature of the tire tread is 260°, the vehicle is traveling 80 mph

STUDY SET   SECTION 12.1 (page 883)

1. consistent   **3.** independent   **5. a.** true   **b.** false   **c.** true   **d.** true   **7. a.** 3; -4 (answers may vary)   **b.** 2; -3 (answers may vary)

9. $3x - 7$, 5, 4, 12, 3, 2   **11.**   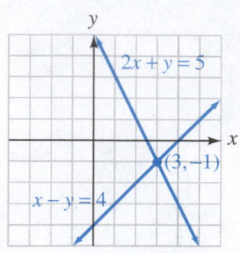   **13.**   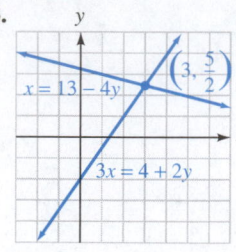   **15.**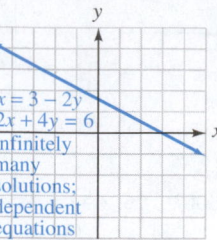

17.   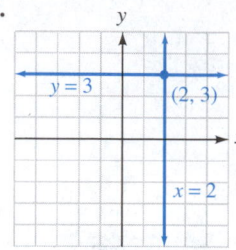   **19.**   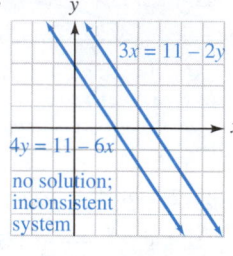   **21.**   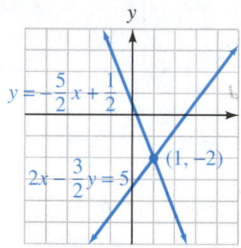   **23.** $(-0.37, -2.69)$ **25.** $(-7.64, 7.04)$

27. $(2, 2)$ **29.** $(5, 3)$ **31.** $(-2, 4)$ **33.** no solution, inconsistent system   **35.** $\left(5, \frac{3}{2}\right)$ **37.** $\left(-2, \frac{3}{2}\right)$ **39.** $(5, 2)$ **41.** $(-4, -2)$ **43.** $(1, 2)$
45. $\left(\frac{1}{2}, \frac{2}{3}\right)$ **47.** infinitely many solutions, dependent equations   **49.** no solution, inconsistent system   **51.** $(4, 8)$ **53.** $(20, -12)$

55.   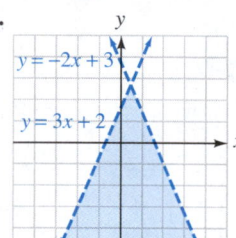   **57.**   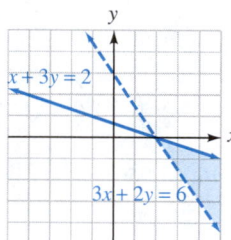   **59. a.** $(95, 64)$ **b.** In 1995, the number of take-out meals and on-premise meals were the same: 64

61. a.   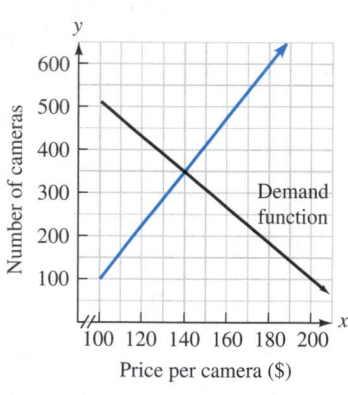   **b.** \$140   **c.** Supply increases, and demand decreases.   **63. a.** yes   **b.** $(3.75, -0.5)$ **c.** no

65. 750 ohms, 625 ohms   **67.** $40°$ **69.** $150°, 30°$ **71.** \$6,000 at 6%, \$6,000 at 7.5%   **73.** 55 mph   **75.** 50 g of A, 60 g of B
77. 200 plates   **79.** 30 lb of each   **81. a.** 6,500 gal per month   **b.** 6,200 gal per month   **c.** B   **83.** 40 oz of 8% solution, 60 oz of
15% solution   **89.** a^{22} **91.** $\dfrac{1}{81x^{32}y^4}$ **93.** $r = \dfrac{A - P}{pt}$ **95.** $r = \dfrac{r_1 r_2}{r_2 + r_1}$

STUDY SET   SECTION 12.2 (page 896)

1. system   **3.** three   **5.** dependent   **7. a.** no solution   **b.** no solution   **9.** $x + 2y - 3z = -6$ **11.** yes   **13.** $(1, 1, 2)$ **15.** $(0, 2, 2)$
17. $(3, 2, 1)$ **19.** no solution, inconsistent system   **21.** $(60, 30, 90)$ **23.** $(2, 4, 8)$ **25.** infinitely many solutions, dependent equations
27. $(2, 6, 9)$ **29.** 30 expensive, 50 middle-priced, 100 inexpensive   **31.** 2, 3, 1   **33.** 3 poles, 2 bears, 4 deer   **35.** 70%, 17%, 13%
37. a. infinitely many solutions, all lying on the line running down the binding   **b.** 3 parallel planes (shelves); no solution   **c.** each
pair of planes (cards) intersect; no solution   **d.** 3 planes (faces of die) intersect at a corner; 1 solution   **39.** $y = \frac{1}{2}x^2 - 2x - 1$

41. $x^2 + y^2 - 2x - 2y - 2 = 0$ **43.** $A = 40°$, $B = 60°$, $C = 80°$ **45.** 12, 15, 21

49.

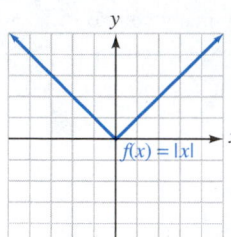

51.

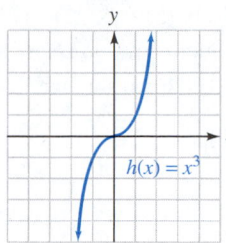

STUDY SET   SECTION 12.3 (page 906)

1. matrix **3.** rows, columns **5.** augmented **7. a.** 2×3 **b.** 3×4 **9.** $\begin{cases} x - y = -10 \\ y = 6 \end{cases}$; $(-4, 6)$ **11.** It has no solution. The system is

inconsistent. **13. a.** multiply row 1 by $\frac{1}{3}$; $\begin{bmatrix} 1 & 2 & -3 & \vdots & 0 \\ 1 & 5 & -2 & \vdots & 1 \\ -2 & 2 & -2 & \vdots & 5 \end{bmatrix}$ **b.** to row 2, add -1 times row 1; $\begin{bmatrix} 1 & 2 & -3 & \vdots & 0 \\ 0 & 3 & 1 & \vdots & 1 \\ -2 & 2 & -2 & \vdots & 5 \end{bmatrix}$

15. -1, 1, -5, 2, y, 4 **17.** $(1, 1)$ **19.** $(2, -3)$ **21.** $(-1, -1)$ **23.** $(0, -3)$ **25.** $(1, 2, 3)$ **27.** $(4, 5, 4)$ **29.** $(2, 1, 0)$
31. $(-1, -1, 2)$ **33.** no solution, inconsistent system **35.** infinitely many solutions, dependent equations **37.** $(0, 1, 3)$
39. no solution, inconsistent system **41.** $(-4, 8, 5)$ **43.** infinitely many solutions, dependent equations **45.** $22°$, $68°$ **47.** $40°$, $65°$, $75°$

49. $76°$, $104°$ **51.** founder's circle: 100; box seats: 300; promenade: 400 **55.** $m = \dfrac{y_2 - y_1}{x_2 - x_1}$ $(x_2 \neq x_1)$ **57.** $y - y_1 = m(x - x_1)$

STUDY SET   SECTION 12.4 (page 916)

1. determinant **3.** minor **5.** rows, columns **7.** dependent, inconsistent **9.** $ad - bc$ **11.** $\begin{vmatrix} 3 & 4 \\ 2 & -3 \end{vmatrix}$ **13.** $\left(\frac{7}{11}, -\frac{5}{11}\right)$ **15.** 6, 30

17. 8 **19.** -2 **21.** 200 **23.** 6 **25.** 1 **27.** 26 **29.** 0 **31.** -79 **33.** $(4, 2)$ **35.** $\left(-\frac{1}{2}, \frac{1}{3}\right)$ **37.** no solution, inconsistent system
39. $(2, -1)$ **41.** $(1, 1, 2)$ **43.** $(3, 2, 1)$ **45.** $(3, -2, 1)$ **47.** $\left(-\frac{1}{2}, -1, -\frac{1}{2}\right)$ **49.** infinitely many solutions, dependent equations
51. no solution, inconsistent system **53.** $(-2, 3, 1)$ **55.** 200 of the \$67 phones, 160 of the \$100 phones **57.** \$5,000 in HiTech,
\$8,000 in SaveTel, \$7,000 in OilCo **59.** -23 **61.** 26 **65.** no **67.** yes **69.** x **71.** y-intercept **73.** x; y

KEY CONCEPT (page 920)

1. yes **2.** no **3.**

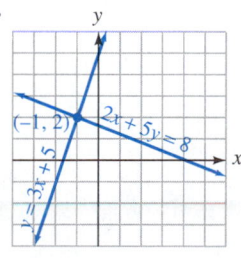

4. $\left(\frac{1}{3}, \frac{1}{4}\right)$ **5.** $(3, 2)$ **6.** $(-1, 0, 2)$ **7.** $(15, 2)$ **8.** $(3, 3, 2)$ **9.** The equations of the
system are dependent. There are infinitely many solutions.
10. The system is inconsistent. There are no solutions.

CHAPTER REVIEW (page 922)

1. a. $(1, 3)$, $(2, 1)$, $(4, -3)$ (answers may vary) **b.** $(0, -4)$, $(2, -2)$, $(4, 0)$ (answers may vary) **c.** $(3, -1)$ **2.** President Clinton's job
approval and disapproval ratings were the same: approximately 47% in 5/94 and approximately 48% in 5/95.

3. a.

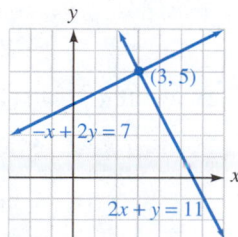

b.

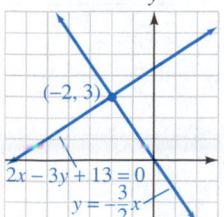

c.

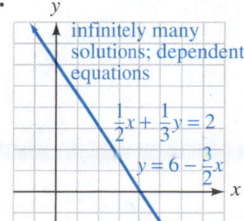

d.

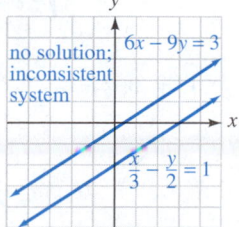

4. a. $(-1, 3)$ **b.** $(-3, -1)$ **c.** $(3, 4)$ **d.** infinitely many solutions, dependent equations **5. a.** $(-3, 1)$ **b.** no solution,
inconsistent system **c.** $(9, -4)$ **d.** $\left(4, \frac{1}{2}\right)$ **6.** Using the addition method, the computations are easier. **7.** $(-1, 0.7)$ (answers may
vary); $\left(-1, \frac{2}{3}\right)$ **8.** A-H: 162 mi, A-SA: 83 mi **9.** 8 mph, 2 mph **10.** no **11. a.** $(1, 2, 3)$ **b.** no solution, inconsistent system
c. $(-1, 1, 3)$ **d.** infinitely many solutions, dependent equations **12.** 25 lb peanuts, 10 lb cashews, 15 lb Brazil nuts

13. a. $\begin{bmatrix} 5 & 4 & \vdots & 3 \\ 1 & -1 & \vdots & -3 \end{bmatrix}$ **b.** $\begin{bmatrix} 1 & 2 & 3 & \vdots & 6 \\ 1 & -3 & -1 & \vdots & 4 \\ 6 & 1 & -2 & \vdots & -1 \end{bmatrix}$ **14. a.** $(1, -3)$ **b.** $(5, -3, -2)$ **c.** infinitely many solutions, dependent equations **d.** no solution, inconsistent system

15. \$4,000 at 6%, \$6,000 at 12% **16. a.** 18 **b.** 38 **c.** -3 **d.** 28 **17. a.** $(2, 1)$ **b.** no solution, inconsistent system **c.** $(1, -2, 3)$ **d.** $(-3, 2, 2)$ **18.** 2 cups mix A, 1 cup mix B, 1 cup mix C

CHAPTER 12 TEST (page 926)

1.

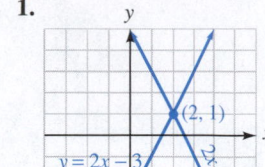

2. $(7, 0)$ **3.** $(2, -3)$ **4.** dependent **5.** no **6.** $(3, 2, -1)$ **7.** 55, 70 **8.** 15 gal 40%, 5 gal 80%

9. $(2, 2)$ **10.** $(1, 0, -1)$ **11.** 22 **12.** 4 **13.** $\begin{vmatrix} -6 & -1 \\ -6 & 1 \end{vmatrix}$ **14.** $\begin{vmatrix} 1 & -1 \\ 3 & 1 \end{vmatrix}$ **15.** -3 **16.** 3

17. -1 **18.** C: 60, GA: 30, S: 10 **19.** Sales of Halloween and Christmas candy were the same in 1995 (about \$840 million) and in 1997 (about \$950 million). **21.** The system has no solution.

CHAPTERS 1–12 CUMULATIVE REVIEW EXERCISES (page 927)

1. $y = 3x + 2$ **2.** $y = -\frac{2}{3}x - 2$ **3.** an increase of about 13,333 a year **4.**

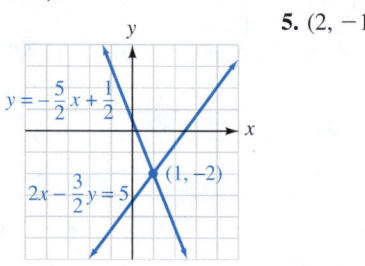

5. $(2, -1)$

6.

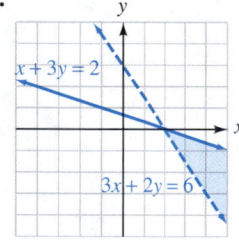

7. $\left(-\infty, -\frac{10}{9}\right)$,

8. $(-\infty, -10] \cup [15, \infty)$,

9. $D = (-\infty, \infty), R = [-3, \infty)$

10. $D = (-\infty, \infty), R = (-\infty, 0]$ **11.** $-4a^2 + 12a - 7$ **12.** $6x^2 - 5x - 6$ **13.** $(x^2 + 4y^2)(x + 2y)(x - 2y)$

14. $(3x + 2)(5x - 4)$ **15.** $(x - y)(x - 4)$ **16.** $(2x^2 + 5y)(4x^4 - 10x^2y + 25y^2)$ **17.** $6, -1$ **18.** $0, \frac{2}{3}, -\frac{1}{2}$

19. 5; 3 is extraneous **20.** $b = \dfrac{-RTV^2 + aV + PV^3}{PV^2 + a}$ **21.** $\dfrac{x + y}{x - y}$ **22.** 0

23. $D = (-\infty, \infty), R = (-\infty, \infty)$

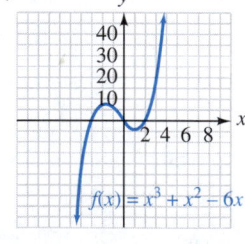

24. $D = (0, \infty), R = (0, \infty)$

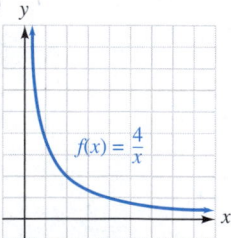

25. 9 **26.** $D = [2, \infty), R = [0, \infty)$

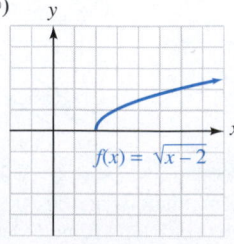

27. $-3x$ **28.** $4t\sqrt{3t}$ **29.** $\frac{1}{16}$ **30.** $x^{17/12}$ **31.** $-12\sqrt[4]{2} + 10\sqrt[4]{3}$

32. $-18\sqrt{6}$ **33.** $\dfrac{x + 3\sqrt{x} + 2}{x - 1}$ **34.** $\dfrac{5\sqrt[3]{x^2}}{x}$ **35.** 2, 7 **36.** $\frac{1}{4}$ **37.** $3\sqrt{2}$ in. **38.** $2\sqrt{3}$ in. **39.** 10 **40.** 9 **41.** $1, -\frac{3}{2}$ **42.** $\dfrac{-2 \pm \sqrt{7}}{3}$

43.
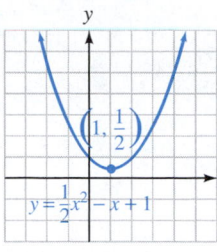

44.
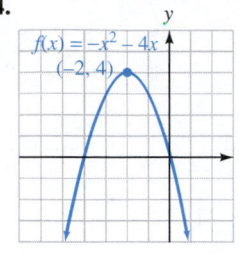

45. $7 + 2i$ **46.** $\frac{3}{2} + \frac{1}{2}i$ **47.** $8i$ **48.** 9, 16   **49.** $-\frac{3}{4}$ **50.** no solution

STUDY SET   SECTION 13.1 (page 939)

1. conic   **3.** radius, center   **5.** $r^2 < 0$ **7.** parabola, (3, 2), right   **9.** $(x - h)^2 + (y - k)^2 = r^2$ **11.** $y = -ax^2$ **13.** 2, 6, 4, 36

15.
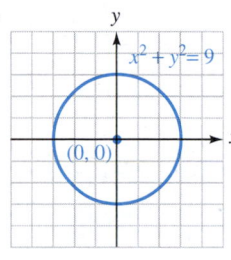

17.
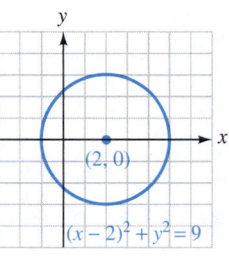

19.
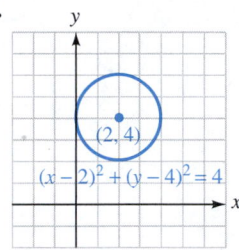

21.
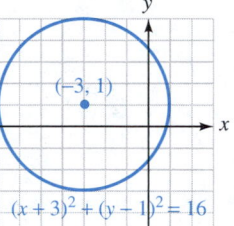

23.
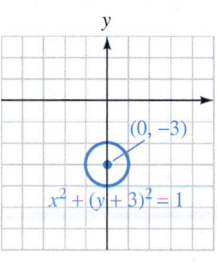

25.

27.
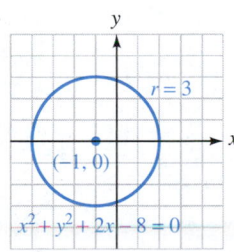

29. $x^2 + y^2 = 1$ **31.** $(x - 6)^2 + (y - 8)^2 = 25$

33. $(x + 2)^2 + (y - 6)^2 = 144$ **35.** $x^2 + y^2 = 8$ **37.**
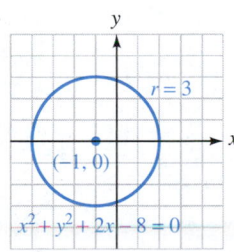

39.
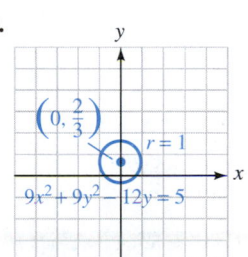

41.
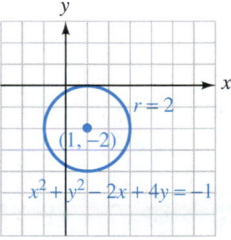

43.
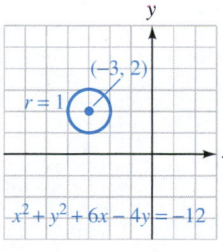

45.
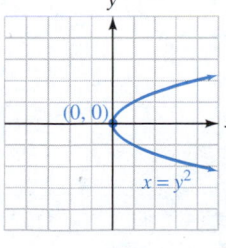

47.
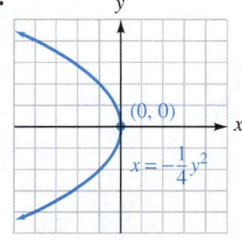

49.
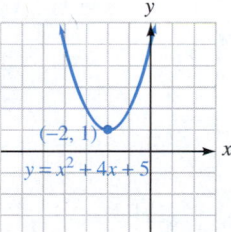

51.
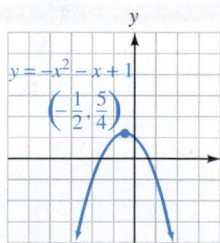

53.
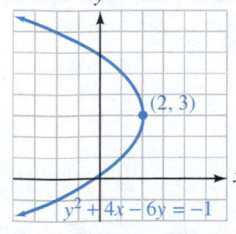

55.
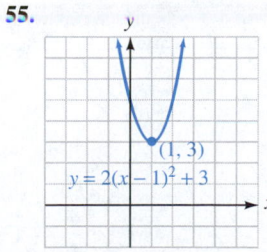

57.

59.

61. $(x - 7)^2 + y^2 = 9$ **63.** no   **65.** 30 ft away   **67.** 2 AU   **73.** 5, $-\frac{7}{3}$ **75.** 3, $-\frac{1}{4}$

STUDY SET SECTION 13.2 (page 949)

1. ellipse, sum **3.** center **5.** $(\pm a, 0)$ **7.** $(0, 0)$ **9.** 144, 16 **11.**

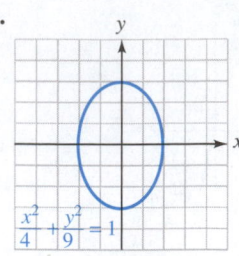

13.

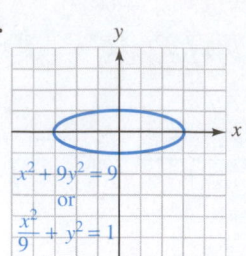

$x^2 + 9y^2 = 9$
or
$\dfrac{x^2}{9} + y^2 = 1$

15.

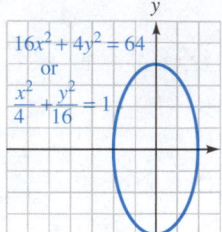

$16x^2 + 4y^2 = 64$
or
$\dfrac{x^2}{4} + \dfrac{y^2}{16} = 1$

17.

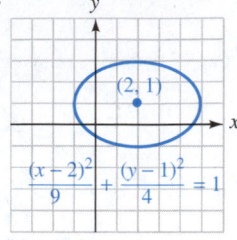

$(2, 1)$

$\dfrac{(x-2)^2}{9} + \dfrac{(y-1)^2}{4} = 1$

19.

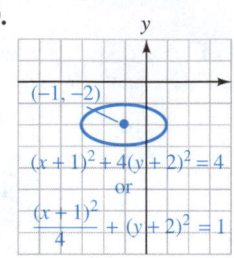

$(-1, -2)$

$(x+1)^2 + 4(y+2)^2 = 4$
or
$\dfrac{(x+1)^2}{4} + (y+2)^2 = 1$

21. 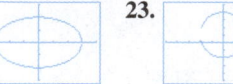   **23.**

25.

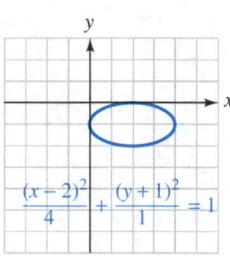

$\dfrac{(x-2)^2}{4} + \dfrac{(y+1)^2}{1} = 1$

27.

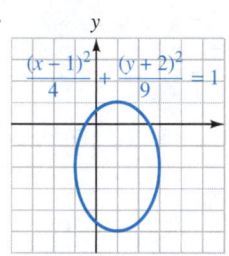

$\dfrac{(x-1)^2}{4} + \dfrac{(y+2)^2}{9} = 1$

29. $y = \frac{1}{2}\sqrt{400 - x^2}$ **31.** 12π sq. units   **35.** $12y^2 + \dfrac{9}{x^2}$ **37.** $\dfrac{y^2 + x^2}{y^2 - x^2}$

STUDY SET SECTION 13.3 (page 957)

1. hyperbola, difference **3.** center **5.** $(\pm a, 0)$ **7.** $(0, 0)$ **9.** 36, 36, 36 **11.**

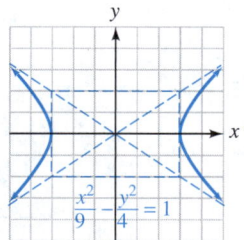

$\dfrac{x^2}{9} - \dfrac{y^2}{4} = 1$

13.

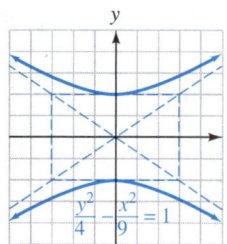

$\dfrac{y^2}{4} - \dfrac{x^2}{9} = 1$

15.

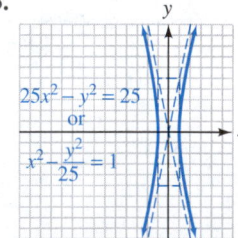

$25x^2 - y^2 = 25$
or
$x^2 - \dfrac{y^2}{25} = 1$

17.

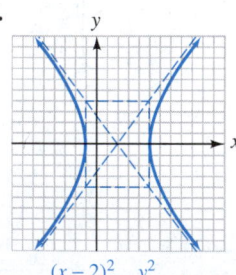

$\dfrac{(x-2)^2}{9} - \dfrac{y^2}{16} = 1$

19.

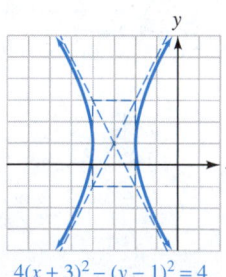

$4(x+3)^2 - (y-1)^2 = 4$
or $(x+3)^2 - \dfrac{(y-1)^2}{4} = 1$

21.

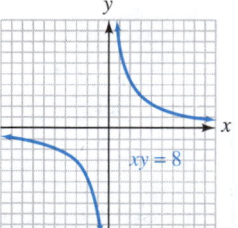

$xy = 8$

23.    **25.**    **27.**

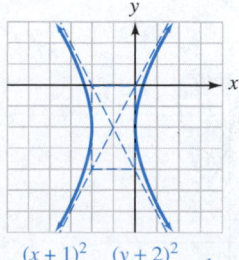

$\dfrac{(x+1)^2}{1} - \dfrac{(y+2)^2}{4} = 1$

29.

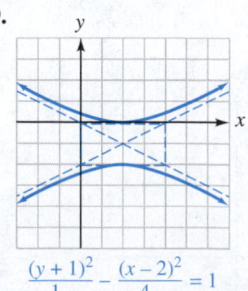

$\dfrac{(y+1)^2}{1} - \dfrac{(x-2)^2}{4} = 1$

31. 3 units   **33.** $10\sqrt{3}$ units

37. $-3x^2(2x^2 - 3x + 2)$ **39.** $(5a + 2b)(3a - 2b)$

STUDY SET   SECTION 13.4 (page 963)

1. graphing, substitution   **3.** tangent   **5.** two   **7.** four   **9.** $2x$, 5, 5, 1, 1, -1 **11.** 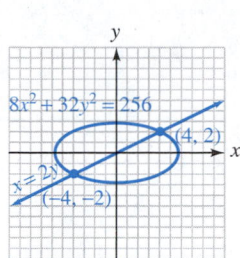   **13.**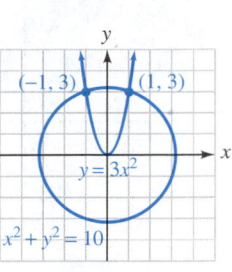

15. 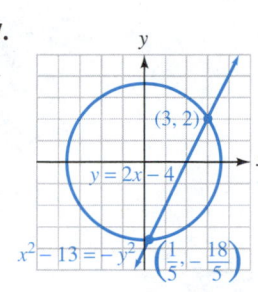   **17.** 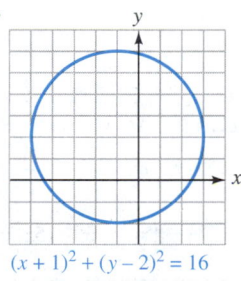   **19.** (1, 0), (5, 0) 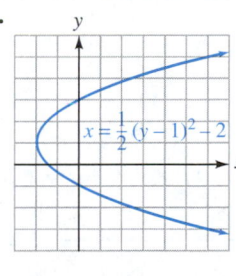   **21.** (3, 0), (0, 5)   **23.** (1, 1)

25. (1, 2), (2, 1)   **27.** (-2, 3), (2, 3)   **29.** $\left(\sqrt{5}, 5\right), \left(-\sqrt{5}, 5\right)$ **31.** (3, 2), (3, -2), (-3, 2), (-3, -2)   **33.** (2, 4), (2, -4), (-2, 4), (-2, -4)   **35.** $\left(-\sqrt{15}, 5\right), \left(\sqrt{15}, 5\right)$, ($-2$, -6), (2, -6)   **37.** (0, -4), (-3, 5), (3, 5)   **39.** (-2, 3), (2, 3), (-2, -3), (2, -3)   **41.** (3, 3)   **43.** (6, 2), (-6, -2), $\left(\sqrt{42}, 0\right), \left(-\sqrt{42}, 0\right)$ **45.** $\left(\frac{1}{2}, \frac{1}{3}\right), \left(\frac{1}{3}, \frac{1}{2}\right)$ **47.** 4, 8   **49.** 7 cm by 9 cm   **51.** either \$750 at 9% or \$900 at 7.5%   **53.** 68 mph, 4.5 hr   **57.** $-11x\sqrt{2}$ **59.** $\frac{1}{2}$

KEY CONCEPT (page 966)

1. ellipse   **2.** circle   **3.** parabola   **4.** hyperbola   **5.** hyperbola   **6.** ellipse   **7.** ellipse   **8.** parabola   **9.** circle   **10. a.** (-1, 2)   **b.** 4
11. a. (-2, 1)   **b.** right   **12. a.** (0, 0)   **b.** the y-axis   **c.** 8 units   **d.** 4 units   **13. a.** (-2, 1)   **b.** left and right   **c.** 6 units horizontally,
4 units vertically   **14.** 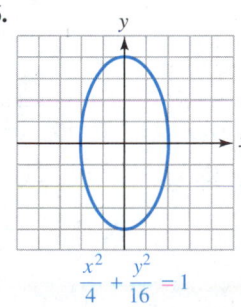   **15.** 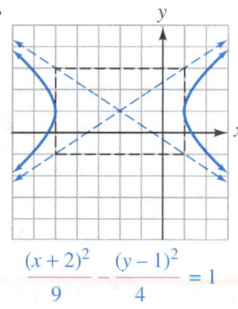   **16.** 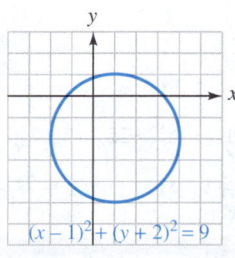   **17.**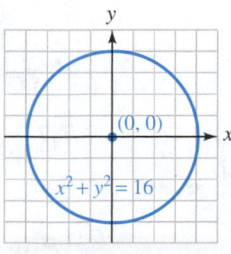

CHAPTER REVIEW (page 968)

1. a. 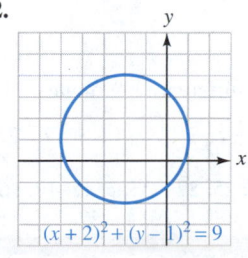   **b.** 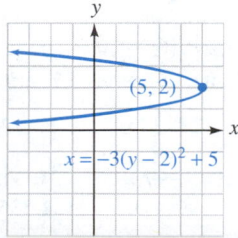   **2.** 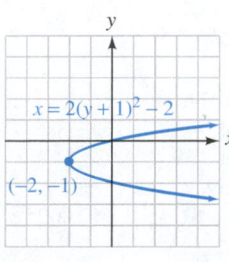   **3. a.**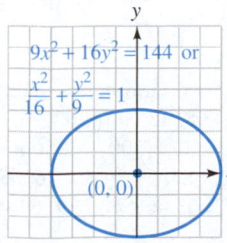

b. 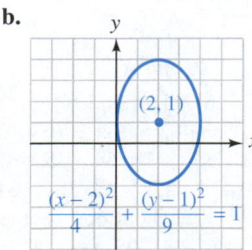   **4. a.** 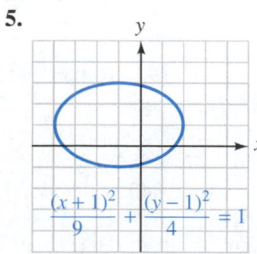   **b.**   **5.**

6. a. 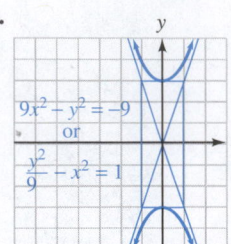 **b.** 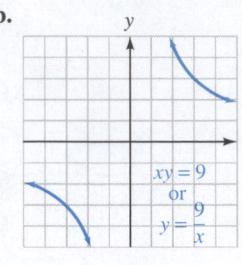 **7.** hyperbola **8.** 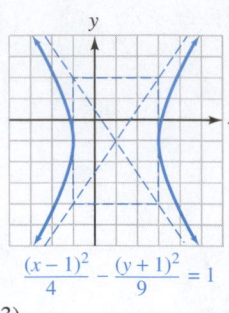 **9.** $(0, 3)$, $(0, -3)$

10. a. $(4, 2)$, $(4, -2)$, $(-4, 2)$, $(-4, -2)$ **b.** $(2, 3)$, $(2, -3)$, $(-2, 3)$, $(-2, -3)$

CHAPTER 13 TEST (page 970)

1. $(2, -3)$, 2 **2.** $(-2, 3)$, 4 **3.** 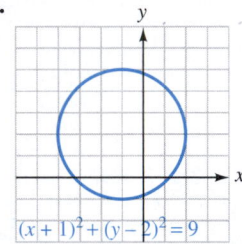 **4.** 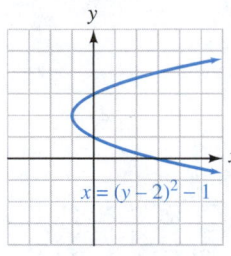 **5.**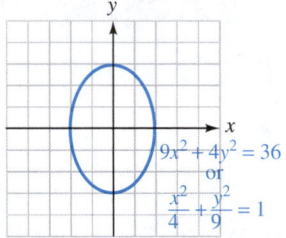

6. 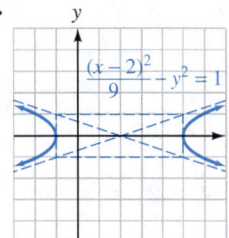 **7.** 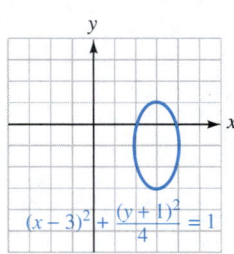 **8.** 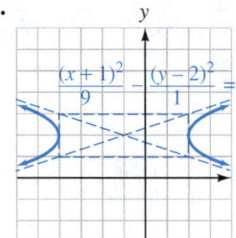 **9.** $(2, 6)$, $(-2, -2)$ **10.** $(3, 4)$, $(-3, 4)$

STUDY SET SECTION 14.1 (page 977)

1. binomial **3.** one **5.** exponent **7.** 7 **9.** five factorial **11.** 7! **13.** $(n - 1)!$ **15.** 3 **17.** 3, 3!, 3!, 2, 1 **19.** 6 **21.** -120 **23.** 30
25. 144 **27.** 40,320 **29.** $\frac{1}{110}$ **31.** 2,352 **33.** 10 **35.** 21 **37.** $\frac{1}{168}$ **39.** $x^2 + 2xy + y^2$ **41.** $x^4 - 4x^3y + 6x^2y^2 - 4xy^3 + y^4$

43. $8x^3 + 12x^2y + 6xy^2 + y^3$ **45.** $x^3 - 6x^2y + 12xy^2 - 8y^3$ **47.** $8x^3 + 36x^2y + 54xy^2 + 27y^3$ **49.** $\frac{x^3}{8} - \frac{x^2y}{4} + \frac{xy^2}{6} - \frac{y^3}{27}$

51. $81 + 216y + 216y^2 + 96y^3 + 16y^4$ **53.** $\frac{x^4}{81} - \frac{2x^3y}{27} + \frac{x^2y^2}{6} - \frac{xy^3}{6} + \frac{y^4}{16}$ **55.** $3a^2b$ **57.** $-4xy^3$ **59.** $15x^2y^4$ **61.** $28x^6y^2$

63. $90x^3$ **65.** $640x^3y^2$ **67.** $-12x^3y$ **69.** $-70,000x^4$ **71.** $810xy^4$ **73.** $-\frac{1}{6}x^3y$ **75.** $\frac{n!}{3!(n - 3)!}a^{n-3}b^3$ **77.** $\frac{n!}{4!(n - 4)!}a^{n-4}b^4$

79. $\frac{81(2^{n-4})n!}{4!(n - 4)!}a^{n-4}b^4$ **85.** 2 **87.** 5 **89.** $(2, 3)$ **91.** 8

STUDY SET SECTION 14.2 (page 984)

1. sequence **3.** infinite **5.** series **7.** arithmetic, difference **9.** finite **11.** $1 + 2 + 3 + 4 + 5$ **13.** $1 + 2 + 3$ **15.** $4 + 9 + 16$
17. $\sum_{k=3}^{6} k$ **19.** $\sum_{k=1}^{4} 2k$ **21.** 3, 5, 7, 9, 11 **23.** $-5, -8, -11, -14, -17$ **25.** 5, 11, 17, 23, 29 **27.** $-4, -11, -18, -25, -32$
29. $-118, -111, -104, -97, -90$ **31.** 34, 31, 28, 25, 22 **33.** 5, 12, 19, 26, 33 **35.** 355 **37.** -179 **39.** -23 **41.** 12 **43.** $\frac{17}{4}, \frac{13}{2}, \frac{35}{4}$
45. 12, 14, 16, 18 **47.** $\frac{29}{2}$ **49.** $\frac{5}{4}$ **51.** 1,335 **53.** 459 **55.** 354 **57.** 255 **59.** 1,275 **61.** 2,500 **63.** 60 **65.** 91 **67.** 31 **69.** 12
71. 60, 110, 160, 210, 260, 310; $6,060 **73.** 11,325 **75.** 368 ft **79.** $18x^2 + 8x - 3$ **81.** $\frac{6a^2 + 16}{(a + 2)(a - 2)}$

STUDY SET SECTION 14.3 (page 993)

1. geometric **3.** mean **5.** nth term $= ar^{n-1}$ **7.** $2 + 6 + 18$ **9.** $S = \frac{a}{1 - r}$ **11.** first, ratio, terms **13.** 3, 6, 12, 24, 48 **15.** $-5, -1$,
$-\frac{1}{5}, -\frac{1}{25}, -\frac{1}{125}$ **17.** 2, 8, 32, 128, 512 **19.** $-3, -12, -48, -192, -768$ **21.** $-64, 32, -16, 8, -4$ **23.** $-64, -32, -16, -8, -4$
25. 2, 10, 50, 250, 1,250 **27.** 3,584 **29.** $\frac{1}{27}$ **31.** 3 **33.** 6, 18, 54 **35.** $-20, -100, -500, -2,500$ **37.** -16 **39.** $10\sqrt{2}$
41. No geometric mean exists. **43.** 728 **45.** 122 **47.** -255 **49.** 381 **51.** $\frac{156}{25}$ **53.** $-\frac{21}{4}$ **55.** 16 **57.** 81 **59.** 8 **61.** $-\frac{135}{4}$
63. no sum **65.** $-\frac{81}{2}$ **67.** $\frac{1}{9}$ **69.** $-\frac{1}{3}$ **71.** $\frac{4}{33}$ **73.** $\frac{25}{33}$ **75.** about 669 people **77.** $1,469.74 **79.** $140,853.75 **81.** $\left(\frac{1}{2}\right)^{11} \approx 0.0005$
83. 30 m **85.** 5,000 **91.** $[-1, 6]$ **93.** $(-\infty, -3) \cup (4, \infty)$ **95.** yes **97.** no

STUDY SET   SECTION 14.4 (page 1002)

1. permutation   **3.** $p \cdot q$ **5.** $P(n, r) = \frac{n!}{(n-r)!}$ **7.** 1   **9.** $\binom{n}{r}$, combinations   **11.** 1   **13.** 6!, 4!, 5   **15.** 6   **17.** 60   **19.** 12   **21.** 5

23. 1,260   **25.** 10   **27.** 20   **29.** 50   **31.** 2   **33.** 1   **35.** $\frac{n!}{2!(n-2)!}$ **37.** $x^4 + 4x^3y + 6x^2y^2 + 4xy^3 + y^4$ **39.** $8x^3 + 12x^2y + 6xy^2 + y^3$
41. $81x^4 - 216x^3 + 216x^2 - 96x + 16$ **43.** $-1,250x^2y^3$ **45.** $-4x^6y^3$ **47.** 35   **49.** 1,000,000   **51.** 136,080   **53.** 8,000,000
55. 720   **57.** 2,880   **59.** 13,800   **61.** 720   **63.** 900   **65.** 364   **67.** 5   **69.** 1,192,052,400   **71.** 18   **73.** 7,920   **77.** 6, -3 **79.** 8

STUDY SET   SECTION 14.5 (page 1007)

1. experiment   **3.** $\frac{s}{n}$ **5.** 0   **7.** 6, 52, $\frac{6}{52}, \frac{3}{26}$ **9.** {(1, H), (2, H), (3, H), (4, H), (5, H), (6, H), (1, T), (2, T), (3, T), (4, T), (5, T), (6, T)}
11. (a, b, c, d, e, f, g, h, i, j, k, l, m, n, o, p, q, r, s, t, u, v, w, x, y, z}   **13.** $\frac{1}{6}$ **15.** $\frac{2}{3}$ **17.** $\frac{19}{42}$ **19.** $\frac{13}{42}$ **21.** $\frac{3}{8}$ **23.** 0   **25.** $\frac{1}{12}$ **27.** $\frac{5}{12}$
29. $\frac{33}{391,510}$ **31.** $\frac{9}{460}$ **35.** $\frac{1}{4}$ **37.** $\frac{1}{4}$ **39.** 1   **41.** $\frac{32}{119}$ **43.** $\frac{1}{3}$ **47.** $-10, 4$ **49.** $(-4, 10)$

KEY CONCEPT (page 1009)

1. g   **2.** o   **3.** i   **4.** l   **5.** u   **6.** y   **7.** p   **8.** x   **9.** m   **10.** d   **11.** f   **12.** c   **13.** w   **14.** b   **15.** j   **16.** s   **17.** e   **18.** z   **19.** n   **20.** k
21. h   **22.** v   **23.** q   **24.** a   **25.** r   **26.** t

CHAPTER REVIEW (page 1011)

1. a. 144   **b.** 20   **c.** 15   **d.** 220   **e.** 1   **f.** 8   **2. a.** $x^5 + 5x^4y + 10x^3y^2 + 10x^2y^3 + 5xy^4 + y^5$ **b.** $x^4 - 4x^3y + 6x^2y^2 - 4xy^3 + y^4$
c. $64x^3 - 48x^2y + 12xy^2 - y^3$ **d.** $x^3 + 12x^2y + 48xy^2 + 64y^3$ **3. a.** $6x^2y^2$ **b.** $-10x^2y^3$ **c.** $-108x^2y$ **d.** $864x^2y^2$ **4.** 42   **5.** 122,
137, 152, 167, 182   **6.** $\frac{41}{3}, \frac{58}{3}$ **7.** 1,550   **8.** $-\frac{45}{2}$ **9. a.** $\frac{15}{2}$ **b.** 378   **c.** 14   **d.** 360   **10.** 24, 12, 6, 3, $\frac{3}{2}$ **11.** 4   **12.** 24, -96 **13.** $\frac{2,186}{9}$
14. $-\frac{85}{8}$ **15.** 125   **16.** $\frac{5}{99}$ **17.** 136   **18. a.** 5,040   **b.** 1   **c.** 20,160   **d.** $\frac{1}{10}$ **19. a.** 1   **b.** 1   **c.** 28   **d.** 84   **e.** 700   **f.** $\frac{7}{4}$ **20.** $1,638.40
21. $134,509.57   **22.** 12 yr   **23.** 1,600 ft   **24.** 120   **25.** 720   **26.** 120   **27.** 150   **28.** $\frac{1}{18}$ **29.** 0   **30.** $\frac{1}{13}$ **31.** $\frac{1}{649,640}$ **32.** $\frac{63}{66,640}$

CHAPTER 14 TEST (page 1014)

1. 210   **2.** 1   **3.** $-5x^4y$ **4.** $24x^2y^2$ **5.** 66   **6.** 306   **7.** 34, 66   **8.** 3   **9.** -81 **10.** $\frac{364}{27}$ **11.** 18, 108   **12.** $\frac{27}{2}$ **13.** 120   **14.** 40,320
15. 15   **16.** 56   **17.** 720   **18.** 322,560   **19.** 24   **20.** $\frac{7}{30}$ **21.** 35   **22.** 30   **23.** $\frac{1}{5}$ **24.** $\frac{2}{13}$ **25.** $\frac{33}{66,640}$ **26.** $\frac{5}{16}$ **27.** 0   **28.** a list of all
possible outcomes

CHAPTERS 1–14 CUMULATIVE REVIEW EXERCISES (page 1015)

1. 0   **2.** $-\frac{4}{3}$, 5.6, 0, -23 **3.** π, $\sqrt{2}$, e **4.** $-\frac{4}{3}$, π, 5.6, $\sqrt{2}$, 0, -23, e **5.** $8,250   **6.** $\frac{1}{120}$ db/rpm   **7.** parallel   **8.** perpendicular
9. $y = -2x + 5$ **10.** $y = -\frac{9}{13}x + \frac{7}{13}$ **11.** $(1, 1)$ **12.** $(2, -2)$ **13.** $(-1, -1)$ **14.** $(-1, -1, 3)$ **15.** 3   **16.** -1 **17.** 85°, 80°, 15°
18.   **19.** $\left[-3, \frac{19}{3}\right]$ **20.** $(-\infty, -2)$ **21.** $12x^2 - 5xy - 3y^2$

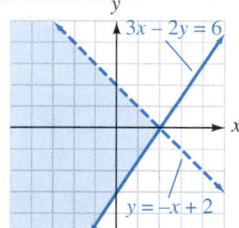

22. $2x^2y^3 + 13xy + 3y^2$ **23.** $a^2 - 4ab + 4b^2$ **24.** $3a^3 + 10a^2 + 6a - 4$ **25.** $xy(3x - 4y)(x - 2)$
26. $(16x^2y^2 + z^4)(4xy + z^2)(4xy - z^2)$ **27.** $\lambda = \frac{4d-2}{A-6}$

28. $(-3, 0), (0, 0), (2, 0); (0, 0); 24, 0, -8, -6, 0, 4, 0, -18$

29. $\frac{64b^{12}}{27a^9}$ **30.** $-\frac{3x+2}{3x-2}$ **31.** $-\frac{q}{p}$
32. $\frac{4a-1}{(a+2)(a-2)}$ **33.** 5; 3 is extraneous

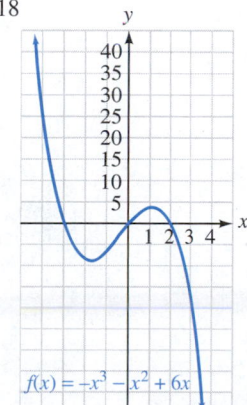

34. $R = \frac{R_1R_2R_3}{R_2R_3 + R_1R_3 + R_1R_2}$ **35. a.** a quadratic function   **b.** at about 85% and 120% of the suggested inflation   **36.** about $21\frac{1}{2}$ in.
37. $-x^2 + x + 5 + \frac{8}{x-1}$ **38.** 2   **39.** $5\sqrt{2}$ **40.** $81x\sqrt[3]{3x}$ **41.** $\frac{343}{125}$ **42.** $\sqrt{3t} - 1$ **43.** $-5 + 17i$ **44.** $-\frac{21}{29} - \frac{20}{29}i$
45. 5, 0 does not check   **46.** 0   **47.** $\frac{2}{3}, -\frac{3}{2}$ **48.** $\frac{-3 \pm \sqrt{5}}{4}$ **49.** $\frac{1}{4}, \frac{1}{2}$ **50.** $\frac{2 \pm \sqrt{2}i}{3}$ **51.** $4x^2 + 4x - 1$ **52.** $f^{-1}(x) = \sqrt[3]{\frac{x+1}{2}}$

53.

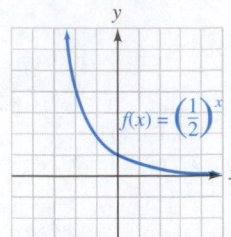

54.

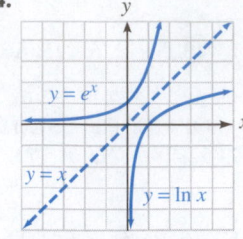

55. $2^y = x$ **56.** $\log_6 x - 2$ **57.** 5   **58.** 3   **59.** $\frac{1}{27}$ **60.** 1   **61.** $y = 2^x$

62. x **63.** 1.9912   **64.** 0.301   **65.** 3.4190   **66.** 1, -10 does not check   **67.** $-\frac{3}{4}$ **68.** 9   **69.** $x^2 + y^2 - 2x - 6y - 15 = 0$

70.

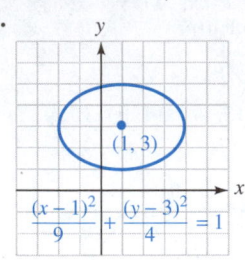

71.

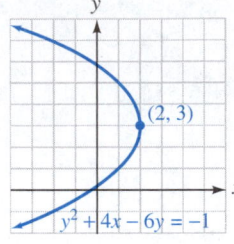

72.

73. \$2,848.31   **74.** 1.16056   **75.** 30,240

76. $81a^4 - 108a^3b + 54a^2b^2 - 12ab^3 + b^4$ **77.** $112x^2y^6$ **78.** 103   **79.** 690   **80.** 8, 19   **81.** 42   **82.** 27   **83.** 27   **84.** $\frac{1,023}{64}$
85. 12, -48 **86.** $\frac{27}{2}$ **87.** 504   **88.** 35   **89.** 5,040   **90.** 84

STUDY SET   APPENDIX I   (page A-11)

1. $\frac{1}{2}$ **3.** $\frac{3}{4}$ **5.** $\frac{4}{3}$ **7.** $\frac{9}{8}$ **9.** $\frac{3}{10}$ **11.** $\frac{8}{5}$ **13.** $\frac{3}{2}$ **15.** $\frac{1}{4}$ **17.** 10   **19.** $\frac{20}{3}$ **21.** $\frac{9}{10}$ **23.** $\frac{5}{8}$ **25.** $\frac{1}{4}$ **27.** $\frac{14}{5}$ **29.** 28   **31.** $\frac{1}{5}$ **33.** $\frac{6}{5}$ **35.** $\frac{1}{13}$ **37.** $\frac{5}{24}$
39. $\frac{19}{15}$ **41.** $\frac{17}{12}$ **43.** $\frac{22}{35}$ **45.** $\frac{1}{6}$ **47.** $\frac{9}{4}$ **49.** $\frac{29}{3}$ **51.** $\frac{4}{5}$ **53.** $5\frac{1}{5}$ **55.** $1\frac{2}{3}$ **57.** $1\frac{1}{4}$ **59.** $\frac{5}{9}$ **61.** $7\frac{2}{7}$ cm   **63.** $3\frac{7}{10}$ km   **65.** $53\frac{1}{6}$ feet
67. 2,514,820   **69.** 270 lb   **71.** 158.65   **73.** 44.785   **75.** 44.88   **77.** 4.55   **79.** 350.49   **81.** 55.21   **83.** 3,337.52   **85.** 10.02
87. 3,865.11 ft^2 **89.** \$780.69   **91.** \$1,170   **93.** \$2,525.25   **95.** the Guernsey   **97.** the second contractor   **99.** the regular furnace

STUDY SET   APPENDIX II   (page A-19)

1. polynomial   **3.** $P(r)$ **5.** 12, 26, 8   **7.** $x + 2$ **9.** $x - 3$ **11.** $x + 2$ **13.** $x - 7 + \frac{28}{x+2}$ **15.** $3x^2 - x + 2$ **17.** $2x^2 + 4x + 3$
19. $6x^2 - x + 1 + \frac{3}{x+1}$ **21.** $7.2x - 0.66 + \frac{0.368}{x-0.2}$ **23.** $2.7x - 3.59 + \frac{0.903}{x+1.7}$ **25.** $9x^2 - 513x + 29,241 + \frac{-1,666,762}{x+57}$ **27.** -1 **29.** -37
31. 23   **33.** -1 **35.** 2   **37.** -1 **39.** 18   **41.** 174   **43.** -8 **45.** 59   **47.** 44   **49.** $\frac{29}{32}$ **51.** yes   **53.** no   **55.** 64

Index